# PRINCIPLES OF HEATING VENTILATING AND AIR CONDITIONING

## ABOUT THE AUTHORS

**Ronald H. Howell, PhD, PE,** *Fellow ASHRAE*, retired as professor and chair of mechanical engineering at the University of South Florida and is also professor emeritus of the University of Missouri-Rolla. For 40 years he taught courses in refrigeration, heating and air conditioning, thermal analysis, and related areas. He has been the principal or co-principal investigator of 12 ASHRAE-funded research projects. His industrial and consulting engineering experience ranges from ventilation and condensation problems to the development and implementation of a complete air curtain test program.

**Harry J. Sauer, Jr., PhD, PE,** *Fellow ASHRAE*, was a professor of mechanical and aerospace engineering at the University of Missouri-Rolla. He taught courses in air conditioning, refrigeration, environmental quality analysis and control, and related areas. His research ranged from experimental boiling/condensing heat transfer and energy recovery equipment for HVAC systems to computer simulations of building energy use and actual monitoring of residential energy use. He served as an advisor to the Missouri state government and conducted energy auditor training programs for the US Department of Energy.

**William J. Coad, PE,** *Fellow ASHRAE*, was ASHRAE president in 2001-2002. He has been with McClure Engineering Associates, St. Louis, Mo., for 40 years and is currently a consulting principal. He is also president of Coad Engineering Enterprises. He has served as a consultant to the Missouri state government and was a lecturer in mechanical engineering for 12 years and an affiliate professor in the graduate program for 17 years at Washington University, St. Louis. He is the author of *Energy Engineering and Management for Building Systems* (Van Nostrand Reinhold).

---

**Updates/errata for this publication will be posted on the ASHRAE Web site at www.ashrae.org/publicationupdates.**

# PRINCIPLES OF HEATING VENTILATING AND AIR CONDITIONING

7th Edition

---

A Textbook with Design Data Based on the
*2013 ASHRAE Handbook—Fundamentals*

---

Ronald H. Howell   William J. Coad   Harry J. Sauer, Jr.

ISBN 978-1-936504-57-2

© 2013 ASHRAE
1791 Tullie Circle, N.E.
Atlanta, GA 30329
www.ashrae.org

All rights reserved.

Printed in the United States of America

ASHRAE is a registered trademark in the U.S. Patent and Trademark Office, owned by the American Society of Heating, Refrigerating and Air-Conditioning Engineers, Inc.

ASHRAE has compiled this publication with care, but ASHRAE has not investigated, and ASHRAE expressly disclaims any duty to investigate, any product, service, process, procedure, design, or the like that may be described herein. The appearance of any technical data or editorial material in this publication does not constitute endorsement, warranty, or guaranty by ASHRAE of any product, service, process, procedure, design, or the like. ASHRAE does not warrant that the information in the publication is free of errors, and ASHRAE does not necessarily agree with any statement or opinion in this publication. The entire risk of the use of any information in this publication is assumed by the user.

No part of this book may be reproduced without permission in writing from ASHRAE, except by a reviewer who may quote brief passages or reproduce illustrations in a review with appropriate credit; nor may any part of this book be reproduced, stored in a retrieval system, or transmitted in any way or by any means—electronic, photocopying, recording, or other—without permission in writing from ASHRAE.

Howell, Ronald H. (Ronald Hunter), 1935-
  Principles of heating ventilating and air conditioning : a textbook with design data based on the 2013 ASHRAE handbook fundamentals / Ronald H. Howell, William J. Coad, Harry J. Sauer, Jr. -- 7th edition.
      pages cm
  Some editions have Sauer's name first.
  Includes bibliographical references and index.
  Summary: "A textbook with design data based on the 2013 ASHRAE handbook of fundamentals"-- Provided by publisher.
  ISBN 978-1-936504-57-2 (hardcover : alk. paper)
  1. Heating--Textbooks. 2. Ventilation--Textbooks. 3. Air conditioning--Textbooks. I. Coad, William J. II. Sauer, Harry J., Jr., 1935- III. Title.
TH7012.H73 2013
697--dc23
                        2013036759

| | | |
|---|---|---|
| **ASHRAE Staff** | Special Publications | Mark S. Owen, *Editor/Group Manager of Handbook and Special Publications* |
| | | Cindy Sheffield Michaels, *Managing Editor* |
| | | James Madison Walker, *Associate Editor* |
| | | Roberta Hirschbuehler, *Assistant Editor* |
| | | Sarah Boyle, *Editorial Assistant* |
| | | Michshell Phillips, *Editorial Coordinator* |
| | Publishing Services | David Soltis, *Group Manager of Publishing Services and Electronic Communications* |
| | | Jayne Jackson, *Publication Traffic Administrator* |
| | | Tracy Becker, *Graphics Specialist* |
| | Publisher | W. Stephen Comstock |

# Contents

**Part I**      **General Concepts**

**Chapter 1**      **Background**
- Introduction ................................................................................................. 1
- Historical Notes ........................................................................................... 2
- Building Energy Use ................................................................................... 5
- Conceptualizing an HVAC System .............................................................. 7
- Sustainability and Green Buildings ............................................................. 7
- Problems ....................................................................................................... 8
- Bibliography ................................................................................................. 9

**Chapter 2**      **Thermodynamics and Psychrometrics**
- Fundamental Concepts and Principles ...................................................... 11
- Properties of a Substance ........................................................................... 13
- Forms of Energy ......................................................................................... 36
- First Law of Thermodynamics ................................................................... 40
- Second Law of Thermodynamics ............................................................... 42
- Third Law of Thermodynamics .................................................................. 44
- Basic Equations of Thermodynamics ........................................................ 44
- Thermodynamics Applied to Refrigeration ............................................... 44
- Applying Thermodynamics to Heat Pumps ............................................... 49
- Absorption Refrigeration Cycle ................................................................. 49
- Problems ..................................................................................................... 50
- Bibliography ............................................................................................... 55
- SI Tables and Figures ................................................................................. 55

**Chapter 3**      **Basic HVAC System Calculations**
- Applying Thermodynamics to HVAC Processes ....................................... 67
- Single-Path Systems ................................................................................... 72
- Air-Volume Equations for Single-Path Systems ....................................... 72
- Psychrometric Representation of Single-Path Systems ............................ 74
- Sensible Heat Factor (Sensible Heat Ratio) .............................................. 74
- Problems ..................................................................................................... 76
- Bibliography ............................................................................................... 80

**Chapter 4**      **Design Conditions**
- Indoor Design Conditions .......................................................................... 81
- Outdoor Design Conditions: Weather Data ............................................... 89
- Other Factors Affecting Design ............................................................... 136
- Temperatures in Adjacent Unconditioned Spaces ................................... 140
- Problems ................................................................................................... 141
- Bibliography ............................................................................................. 142
- SI Tables and Figures ............................................................................... 143

**Chapter 5**      **Load Estimating Fundamentals**
- General Considerations ............................................................................ 145
- Outdoor Air Load Components ................................................................ 146
- Heat-Transfer Coefficients ....................................................................... 158
- Calculating Surface Temperatures ........................................................... 169
- Problems ................................................................................................... 173

| | | |
|---|---|---|
| | Bibliography | 176 |
| | SI Figures and Tables | 177 |
| **Chapter 6** | **Residential Cooling and Heating Load Calculations** | |
| | Background | 189 |
| | General Guidelines | 190 |
| | Cooling Load Methodology | 195 |
| | Heating Load Methodology | 198 |
| | Nomenclature | 204 |
| | Load Calculation Example | 205 |
| | Problems | 207 |
| | Bibliography | 210 |
| | SI Figures and Tables | 211 |
| **Chapter 7** | **Nonresidential Cooling and Heating Load Calculations** | |
| | Principles | 217 |
| | Initial Design Considerations | 221 |
| | Heat Gain Calculation Concepts | 221 |
| | Description of Radiant Time Series (RTS) | 247 |
| | Cooling Load Calculation Using RTS | 249 |
| | Heating Load Calculations | 250 |
| | Design Loads Calculation Example | 252 |
| | Problems | 266 |
| | Bibliography | 269 |
| | SI Figures and Tables | 274 |
| **Chapter 8** | **Energy Estimating Methods** | |
| | General Considerations | 289 |
| | Component Modeling and Loads | 290 |
| | Overall Modeling Strategies | 291 |
| | Integration of System Models | 292 |
| | Degree-Day Methods | 293 |
| | Bin Method (Heating and Cooling) | 302 |
| | Problems | 304 |
| | Bibliography | 308 |
| **Chapter 9** | **Duct and Pipe Sizing** | |
| | Duct Systems | 309 |
| | Fans | 346 |
| | Air-Diffusing Equipment | 354 |
| | Pipe, Tube, and Fittings | 356 |
| | Pumps | 361 |
| | Problems | 363 |
| | References | 367 |
| | SI Figures and Tables | 369 |
| **Chapter 10** | **Economic Analyses and Life-Cycle Costs** | |
| | Introduction | 377 |
| | Owning Costs | 377 |
| | Service Life | 377 |
| | Depreciation | 380 |
| | Interest or Discount Rate | 380 |
| | Periodic Costs | 380 |
| | Operating Costs | 381 |

|  |  |
|---|---|
| Economic Analysis Techniques | 385 |
| Reference Equations | 388 |
| Problems | 388 |
| Symbols | 389 |
| References | 390 |
| Bibliography | 390 |

# Part II  HVAC Systems

## Chapter 11  Air-Conditioning System Concepts

|  |  |
|---|---|
| System Objectives and Categories | 391 |
| System Selection and Design | 392 |
| Design Parameters | 392 |
| Performance Requirements | 393 |
| Focusing on System Options | 393 |
| Narrowing the Choice | 394 |
| Energy Considerations of Air Systems | 395 |
| Basic Central Air-Conditioning and Distribution System | 396 |
| Smoke Management | 398 |
| Components | 398 |
| Air Distribution | 401 |
| Space Heating | 403 |
| Primary Systems | 403 |
| Space Requirements | 405 |
| Problems | 408 |
| Bibliography | 410 |

## Chapter 12  System Configurations

|  |  |
|---|---|
| Introduction | 411 |
| Selecting the System | 412 |
| Multiple-Zone Control Systems | 412 |
| Ventilation and Dedicated Outdoor Air Systems (DOAS) | 415 |
| All-Air System with DOAS Unit | 416 |
| Air-and-Water Systems with DOAS Unit | 416 |
| In-Space Temperature Control Systems | 416 |
| Problems | 422 |
| Bibliography | 425 |

## Chapter 13  Hydronic Heating and Cooling System Design

|  |  |
|---|---|
| Introduction | 427 |
| Closed Water Systems | 428 |
| Design Considerations | 436 |
| Design Procedures | 444 |
| Problems | 447 |
| Bibliography | 448 |

## Chapter 14  Unitary and Room Air Conditioners

|  |  |
|---|---|
| Unitary Air Conditioners | 449 |
| Combined Unitary and Dedicated Outdoor Air Systems | 451 |
| Window Air Conditioners | 451 |
| Through-the-Wall Conditioner System | 452 |
| Typical Performance | 453 |
| Minisplits, Multisplits, and Variable-Refrigerant-Flow (VRF) Systems | 453 |

|  |  | Water-Source Heat Pumps | 454 |
|---|---|---|---|
|  |  | Problems | 455 |
|  |  | Bibliography | 455 |
| **Chapter 15** | **Panel Heating and Cooling Systems** |  |  |
|  |  | General | 457 |
|  |  | Types | 458 |
|  |  | Design Steps | 460 |
|  |  | Problems | 461 |
|  |  | Bibliography | 461 |
| **Chapter 16** | **Heat Pump, Cogeneration, and Heat Recovery Systems** |  |  |
|  |  | General | 463 |
|  |  | Types of Heat Pumps | 463 |
|  |  | Heat Sources and Sinks | 465 |
|  |  | Cogeneration | 468 |
|  |  | Heat Recovery Terminology and Concepts | 469 |
|  |  | Heat Recovery Systems | 471 |
|  |  | Problems | 474 |
|  |  | Bibliography | 474 |
|  |  | SI Figures | 474 |
| **Part III** | **HVAC Equipment** |  |  |
| **Chapter 17** | **Air-Processing Equipment** |  |  |
|  |  | Air-Handling Equipment | 477 |
|  |  | Cooling Coils | 477 |
|  |  | Heating Coils | 482 |
|  |  | Evaporative Air-Cooling Equipment | 483 |
|  |  | Air Washers | 484 |
|  |  | Dehumidification | 484 |
|  |  | Humidification | 486 |
|  |  | Sprayed Coil Humidifiers/Dehumidifiers | 488 |
|  |  | Air Cleaners | 488 |
|  |  | Air-to-Air Energy Recovery Equipment | 493 |
|  |  | Economizers | 500 |
|  |  | Problems | 500 |
|  |  | Bibliography | 502 |
|  |  | SI Table | 503 |
| **Chapter 18** | **Refrigeration Equipment** |  |  |
|  |  | Mechanical Vapor Compression | 505 |
|  |  | Absorption Air-Conditioning and Refrigeration Equipment | 523 |
|  |  | Cooling Towers | 529 |
|  |  | Problems | 532 |
|  |  | Bibliography | 533 |
|  |  | SI Tables | 534 |
| **Chapter 19** | **Heating Equipment** |  |  |
|  |  | Fuels and Combustion | 537 |
|  |  | Burners | 540 |
|  |  | Residential Furnaces | 541 |
|  |  | Commercial Furnaces | 543 |
|  |  | Boilers | 546 |

|  |  |  |
|---|---|---|
|  | Terminal Units | 548 |
|  | Electric Heating | 549 |
|  | Problems | 551 |
|  | Bibliography | 552 |
| **Chapter 20** | **Heat Exchange Equipment** |  |
|  | Modes of Heat Transfer | 555 |
|  | Heat Exchangers | 561 |
|  | Basic Heat Exchanger Design Equation | 563 |
|  | Estimation of Heat Load | 563 |
|  | Mean Temperature Difference | 563 |
|  | Estimation of the Overall Heat Transfer Coefficient $U$ | 564 |
|  | Extended Surfaces, Fin Efficiency, and Fin-Tube Contact Resistance | 565 |
|  | Fouling Factors | 566 |
|  | Convective Heat Transfer Coefficients $h_i$ and $h_o$ | 567 |
|  | Calculation of Heat Exchanger Surface Area and Overall Size | 570 |
|  | Fluids and Their Thermophysical Properties | 570 |
|  | Example Finned-Tube Heat Exchanger Design | 570 |
|  | Problems | 570 |
|  | Bibliography | 572 |

# Appendices

| | | |
|---|---|---|
| **Appendix A** | **SI for HVAC&R** | |
| | General | 573 |
| | Units | 573 |
| | Symbols | 574 |
| | Prefixes | 575 |
| | Numbers | 575 |
| | Words | 576 |
| **Appendix B** | **Systems Design Problems** | |
| | Combination Water Chillers | 579 |
| | Absorption Chiller Selection | 579 |
| | Owning and Operating Costs | 580 |
| | Animal Rooms | 580 |
| | Greenhouse | 582 |
| | Drying Room | 583 |
| | Air Washer | 583 |
| | Two-Story Building | 583 |
| | Motel | 584 |
| | Building Renovation | 584 |
| | Building with Neutral Deck Multizone | 585 |

**This book includes access to a web site containing the Radiant Time Series (RTS) Method Load Calculation Spreadsheets. See www.ashrae.org/PrinciplesofHVAC7.**

# PREFACE

*Principles of Heating, Ventilating, and Air Conditioning*, a textbook based on the 2013 *ASHRAE Handbook—Fundamentals*, should provide an attractive text for air-conditioning courses at engineering colleges and technical institutes. The text has been developed to give broad and current coverage of the heating, ventilation, and air-conditioning field when combined with the 2013 *ASHRAE Handbook—Fundamentals*.

The book should prove most suitable as a textbook and subsequent reference book for (a) undergraduate engineering courses in the general field of HVAC, (b) similar courses at technical institutes, (c) continuing education and refresher short courses for engineers, and (d) adult education courses for nonengineers. It contains more material than can normally be covered in a one-semester course. However, several different single-semester or shorter courses can be easily planned by merely eliminating the chapters and/or parts that are least applicable to the objectives of the particular course. This text will also readily aid in self-instruction of the 2013 *ASHRAE Handbook—Fundamentals* by engineers wishing to develop their competence in the HVAC&R field.

Although numerous references are made to the other ASHRAE Handbook volumes, sufficient material has been included from these to make this text complete enough for various courses in the HVAC&R field. The material covered for various audiences in regular university courses, technical institute courses, and short courses can and will vary greatly. This textbook needed to be complete to satisfy all of these anticipated uses and needs. Toward this end, the following major sections are included:

- Part I  General Concepts, Chapters 1–10
- Part II  Air-Conditioning Systems, Chapters 11–16
- Part III  HVAC&R Equipment, Chapters 17–20

Although the 2013 *ASHRAE Handbook—Fundamentals* is published in an SI edition, which uses international units, and an inch-pound (I-P) edition, this single version of *Principles of Heating, Ventilating, and Air Conditioning* is designed to serve the I-P edition with some SI interspersed throughout.

There are several significant changes in this edition. Chapter 4 has a new format as well as new values for climatic design information. Chapter 7 has been extensively revised with new design data. These changes make *Principles* compatible with the 2013 *ASHRAE Handbook—Fundamentals*. In addition, the chapters on system design and equipment have been significantly revised to reflect recent changes and concepts in contemporary heating and air-conditioning system practices, and Chapter 10 has been significantly rewritten to reflect methodologies, techniques, and analyses as presented in the current Chapter 37 of the 2011 *ASHRAE Handbook—Applications*. Also, the *Solutions Manual* has been extensively edited.

A particular point of confusion must be pointed out. Because this book was developed to be used with the ASHRAE Handbook's *Fundamentals* volume, a number of tables and figures have been reproduced in the original form, complete with references to material elsewhere in *Fundamentals* (not in this book). Thus, if the subheading in the table or figure indicates that it is a *Fundamentals* table or figure, then all references to other locations, equations, tables, etc., refer to those in *Fundamentals*, not in *Principles*.

Dr. Harry Sauer, Jr., one of the co-authors of this textbook, passed away in June 2008. Dr. Sauer's name still appears as a co-author, because he made significant contributions to *Principles*.

September 2013

Ronald H. Howell
William J. Coad

# Chapter 1

# BACKGROUND

**This chapter provides a brief background on the heating, ventilating, air-conditioning, and refrigeration (HVAC&R) field and industry, including the early history and some significant developments. An introduction to a few basic concepts is included along with suggestions for further reading.**

## 1.1 Introduction

On the National Academy of Engineering's list of engineering achievements "that had the greatest impact on the quality of life in the 20th century," *air conditioning and refrigeration* came in tenth, indicating the great significance of this field in the world. With many people in the United States spending nearly 90% of their time indoors, it is hardly surprising that providing a comfortable and healthy indoor environment is a major factor in life today. In fact, over $33 billion of air-conditioning equipment was sold in the US during the year 2010 alone.

Air-conditioning systems usually provide year-round control of several air *conditions*, namely, temperature, humidity, cleanliness, and air motion. These systems may also be referred to as *environmental control systems*, although today they are usually called heating, ventilating, and air-conditioning (HVAC) systems.

The primary function of an HVAC system is either (1) the generation and maintenance of comfort for occupants in a conditioned space; or (2) the supplying of a set of environmental conditions (high temperature and high humidity, low temperature and high humidity, etc.) for a process or product within a space. Human comfort design conditions are quite different from conditions required in textile mills or for grain storage and vary with factors such as time of year and the activity and clothing levels of the occupants.

If improperly sized equipment or the wrong type of equipment is used, the desired environmental conditions usually will not be met. Furthermore, improperly selected and/or sized equipment normally requires excess power and/or energy and may have a higher initial cost. The design of an HVAC system includes calculation of the maximum heating and cooling loads for the spaces to be served, selection of the type of system to be used, calculation of piping and/or duct sizes, selection of the type and size of equipment (heat exchangers, boilers, chillers, fans, etc.), and a layout of the system, with cost, indoor air quality, and energy conservation being considered along the way. Some criteria to be considered are

- Temperature, humidity, and space pressure requirements
- Capacity requirements
- Equipment redundancy
- Spatial requirements
- First cost
- Operating cost
- Maintenance cost
- Reliability
- Flexibility
- Life-cycle cost analysis

The following details should be considered to properly design an air-conditioning system:

1. The location, elevation, and orientation of the structure so that the effects of the weather (wind, sun, and precipitation) on the building heating and cooling loads can be anticipated.
2. The building size (wall area, roof area, glass area, floor area, and so forth).
3. The building shape (L-shaped, A-shaped, rectangular, etc.), which influences equipment location, type of heating and cooling system used, and duct or piping locations.
4. The space use characteristics. Will there be different users (office, bank, school, dance studios, etc.) of the space from year to year? Will there be different concurrent requirements from the tenants? Will there be night setback of the temperature controller or intermittent use of the building's facilities?
5. The type of material (wood, masonry, metal, and so forth) used in the construction of the building. What is the expected quality of the construction?
6. The type of fenestration (light transmitting partition) used, its location in the building, and how it might be shaded. Is glass heat absorbing, reflective, colored, etc.?
7. The types of doors (sliding, swinging, revolving) and windows (sealed, wood or metal frames, etc.) used. What is their expected use? This will affect the amount of infiltration air.
8. The expected occupancy for the space and the time schedule of this occupancy.
9. Type and location of lighting. Types of appliances and electrical machinery in the space and their expected use.
10. Location of electric, gas, and water services. These services should be integrated with the locations of the heating and air-conditioning duct, piping, and equipment.

11. Ventilation requirements for the structure. Does it require 100% outdoor air, a given number of CFM per person, or a given number of CFM per square foot of floor area?
12. Local and/or national codes relating to ventilation, gas, and/or electric piping.
13. Outside design temperatures and wind velocities for the location.
14. The environmental conditions that are maintained. Will fluctuations of these conditions with load be detrimental to the purpose served by the structure?
15. The heating and cooling loads (also consider the moisture load, air contaminants, and noise).
16. The type of heating and cooling system to be used in the structure. Is it forced air, circulated water, or direct expansion? Will it be a multizone, single zone, reheat, variable air volume, or another type of system? What method of control will be used? Will a dedicated outdoor air system be considered?
17. The heating and cooling equipment size that will maintain the inside design conditions for the selected outside design condition. Electric heat or fossil fuel? Mechanical vapor compression or absorption chiller?
18. The advantages and disadvantages of oversizing and undersizing the equipment as applied to the structure. Survey any economic tradeoffs to be made. Should a different type of unit be installed in order to reduce operating costs? Should a more sophisticated control system be used to give more exact control of humidity and temperature or should an on-off cycle be used? Fuel economy as related to design will become an even more important factor in system selection and operation.
19. What is the estimated annual energy usage?

In general, no absolute rules dictate correct selections or specifications for each of the above items, so only engineering estimates or educated guesses can be made. However, estimates must be based on sound fundamental principles and concepts. This book presents a basic philosophy of environmental control as well as the basic concepts of design. These ideas relate directly to the *ASHRAE Handbook* series: 2010 *Refrigeration*, 2011 *HVAC Applications*, 2012 *HVAC Systems and Equipment*, and most directly to 2013 *Fundamentals*.

## 1.2 Historical Notes

Knowing something of the past helps in understanding current design criteria and trends. As in other fields of technology, the accomplishments and failures of the past affect current and future design concepts. The following paragraphs consist mainly of edited excerpts from *ASHRAE Journal* articles: "A History of Heating" by John W. James, "The History of Refrigeration" by Willis R. Woolrich, and "Milestones in Air Conditioning" by Walter A. Grant, with additional information obtained from ASHRAE's Historical Committee. These excerpts provide a synopsis of the history of environmental control.

Obviously, the earliest form of heating was the open fire. The addition of a chimney to carry away combustion byproducts was the first important step in the evolution of heating systems. By the time of the Romans, there was sufficient knowledge of ventilation to allow for the installation of ventilating and panel heating in baths. Leonardo da Vinci had invented a ventilating fan by the end of the 15th century. Robert Boyle's law was established in 1659; John Dalton's in 1800. In 1775, Dr. William Cullen made ice by pumping a vacuum in a vessel of water. A few years later, Benjamin Franklin wrote his treatise on Pennsylvania fireplaces, detailing their construction, installation, and operation.

Although warming and ventilating techniques had greatly improved by the 19th century, manufacturers were unable to exploit these techniques because

- Data available on such subjects as transmission coefficients, air and water friction in pipes, and brine and ammonia properties were sparse and unreliable.
- Neither set design conditions nor reliable psychrometric charts existed.
- A definitive rational theory that would permit performance calculation and prediction of results had not yet been developed.
- Little was known about physical, thermodynamic, and fluid dynamic properties of air, water, brines, and refrigerants.
- No authoritative information existed on heat transmission involving combustion, conduction, convection, radiation, evaporation, and condensation.
- No credible performance information for manufactured equipment was available.

Thanks to Thomas Edison, the first electric power plant opened in New York in 1882, making it possible for the first time to have an inexpensive source of energy for residential and commercial buildings.

### 1.2.1 Furnaces

By 1894, the year the American Society of Heating and Ventilating Engineers (ASH&VE) was born, central heating was fairly well developed. The basic heat sources were warm air furnaces and boilers. The combustion chambers of the first warm air furnaces were made of cast iron. Circulation in a gravity warm air furnace system is caused by the difference in air density in the many parts of the system. As the force of combustion is small, the system was designed to allow air to circulate freely. The addition of fans (circa 1899) to furnace systems provided a mechanical means of air circulation. Other additions to the modern furnace include cooling systems, humidification apparatuses, air distributors, and air filters. Another important step for the modern heating industry was the conversion of furnaces from coal to oil and gas, and from manual to automatic firing.

# Background

## 1.2.2 Steam Systems

James Watt developed the first steam heating system in 1770. However, the first real breakthrough in design did not occur until the early 1900s when circulation problems in these systems were improved with the introduction of a fluid-operated thermostatic trap.

From 1900 to 1925, two-pipe steam systems with thermostatic traps at the outlet of each radiator and at drip points in the piping gained wide acceptance. In smaller buildings, gravity systems were commonly installed to remove condensate. For larger systems, boiler return traps and condensate pumps with receivers were used. By 1926, the vacuum return line system was perfected for installation in large and moderate-sized buildings.

Hot water heating systems were developed in parallel with steam systems. As mentioned before, the first hot water heating system was the gravity system. In 1927, the circulator, which forced water through the system, was added to two-pipe heating systems. A few years later, a diverting tee was added to the one-pipe system, allowing for forced circulation.

During the 1930s, radiators and convectors were commonly concealed by enclosures, shields, and cabinets. In 1944, the baseboard radiator was developed. Baseboard heating improved comfort conditions as it reduced floor-to-ceiling temperature stratification.

Unit heaters and unit ventilators are two other forms of convection heating developed in the 1920s. Unit heaters were available in suspended and floor types and were classified according to the heating medium used (e.g., steam, hot water, electricity, gas, oil, or coal combustion). In addition to the heating element and fan, unit ventilators were often equipped with an air filter. Many designs provided air recirculation and were equipped with a separate outdoor air connection.

Panel heating, another form of heat distribution, was developed in the 1920s. In panel heating, a fluid such as hot water, steam, air, or electricity, circulates through distribution units embedded in the building components.

## 1.2.3 Early Refrigeration

Early forms of refrigeration included the use of snow, pond and lake ice, chemical mixture cooling to form freezing baths, and the manufacture of ice by evaporative and radiation cooling of water on clear nights.

By the 18th century, certain mixtures were known to lower temperatures. One such mixture, calcium chloride and snow, was introduced for commercial use. This particular mixture made possible a temperature down to $-27°F$ ($-33°C$). In Great Britain, machines using chemical mixtures to produce low temperatures were introduced. However, by the time these machines were ready for commercial exploitation, mechanical ice-making processes had been perfected to such an extent that chemical mixture freezing was rendered obsolete except for such batch processes as ice cream making.

## 1.2.4 Mechanical and Chemical Refrigeration

In 1748, in Scotland, Dr. William Cullen and Joseph Black lectured on the latent heat of fusion and evaporation and "fixed air" (later identified as carbon dioxide). These discoveries served as the foundation on which modern refrigeration is based.

In 1851, Dr. John Gorrie, was granted US Patent No. 8080 for a refrigeration machine that produced ice and refrigerated air with compressed air in an open cycle. Also in 1851, Ferdinand Carre designed the first ammonia absorption unit.

In 1853, Professor Alexander Twining of New Haven, Connecticut, produced 1600 lb of ice a day with a doubleacting vacuum and compression pump that used sulfuric ether as the refrigerant.

Daniel L. Holden improved the Carre machine by designing and building reciprocating compressors. These compressors were applied to ice making, brewing, and meat packing. In 1872, David Boyle developed an ammonia compression machine that produced ice.

Until 1880, mechanical refrigeration was primarily used to make ice and preserve meat and fish. Notable exceptions were the use of these machines in the United States, Europe, and Australia for beer making, oil dewaxing, and wine cooling. At this time, comfort air cooling was obtained by ice or by chilling machines that used either lake or manufactured ice.

## 1.2.5 History of ASHRAE

The American Society of Heating and Ventilating Engineers (ASHVE) was formed in New York City in 1894 to conduct research, develop standards, hold technical meetings, and publish technical articles in journals and handbooks. Its scope was limited to the fields of heating and ventilating for commercial and industrial applications, with secondary emphasis on residential heating. Years later the Society's name was changed to the American Society of Heating and Air-Conditioning Engineers (ASHAE) to recognize the increasing importance of air conditioning.

In 1904, the American Society of Refrigerating Engineers (ASRE) was organized and headquartered at the American Society of Mechanical Engineers (ASME). The new Society had 70 charter members and was the only engineering group in the world that confined its activities to refrigeration, which at that time consisted mainly of ammonia systems.

In 1905, ASME established 288,000 Btu in 24 hrs as the commercial ton of refrigeration (within the United States). In the same year, the New York Stock Exchange was cooled by refrigeration. In 1906, Stuart W. Cramer coined the term "air conditioning."

The First International Congress on Refrigeration was organized in Paris in 1908 and a delegation of 26 was sent from the United States. Most of the participants were members of ASRE.

ASHAE and ASRE merged in 1959, creating the American Society of Heating, Refrigerating and Air-Conditioning Engineers.

Figure 1-1 depicts ASHRAE's history. ASHRAE celebrated its Centennial Year during society year 1994-1995. In commemoration of the centennial, two books on the history of ASHRAE and of the HVAC industry were published, *Proclaiming the Truth* and *Heat and Cold: Mastering the Great Indoors*.

## 1.2.6 Willis H. Carrier

Willis H. Carrier (1876-1950) has often been referred to as the "Father of Air Conditioning." His analytical and practical accomplishments contributed greatly to the development of the refrigeration industry.

Carrier graduated from Cornell University in 1901 and was employed by the Buffalo Forge Company. He realized that satisfactory refrigeration could not be installed due to the inaccurate data that were available. By 1902, he developed formulas to optimize forced-draft boiler fans, conducted tests and developed multirating performance tables on indirect pipe coil heaters, and set up the first research laboratory in the heating and ventilating industry.

In 1902, Carrier was asked to solve the problem faced by the lithographic industry of poor color register caused by weather changes. Carrier's solution was to design, test, and install at the Sackett-Wilhelms Lithographing Company of Brooklyn a scientifically engineered, year-round air-conditioning system that provided heating, cooling, humidifying, and dehumidifying.

By 1904, Carrier had adapted atomizing nozzles and developed eliminators for air washers to control dew-point temperature by heating or cooling a system's recirculated water. Soon after this development, over 200 industries were using year-round air conditioning.

At the 1911 ASME meeting, Carrier presented his paper, "Rational Psychrometric Formulae," which related dry-bulb, wet-bulb, and dew-point temperatures of air, as well as its sensible, latent, and total heat load, and set forth the theory of adiabatic saturation. The formulas and psychrometric chart presented in this paper became the basis for all fundamental calculations used by the air-conditioning industry.

By 1922, Carrier's centrifugal refrigeration machine, together with the development of nonhazardous, low-pressure refrigerants, made water chilling for large and medium-size commercial and industrial applications both economical and practical. A conduit induction system for multiroom buildings, was invented in 1937 by Carrier and his associate, Carlyle Ashley.

## 1.2.7 Comfort Cooling

Although comfort air-cooling systems had been built as of the 1890s, no real progress was made in mechanical air cooling until after the turn of the century. At that time, several scientifically designed air-conditioning plants were installed in buildings. One such installation included a theater in Cologne, Germany. In 1902, Alfred Wolff designed a 400-ton system for the New York Stock Exchange. Installed in 1902, this system was in operation for 20 years. The Boston Floating Hospital, in 1908, was the first hospital to be equipped with modern air conditioning. Mechanical air cooling was installed in a Texas church in 1914. In 1922, Grauman's Metropolitan Theater, the first air-conditioned movie theater, opened in Los Angeles. The first office building designed with and built for comfort air-conditioning specifications was the Milam Building, in San Antonio, Texas, which was completed in 1928. Also in 1928, the Chamber of the House of Representatives became air conditioned. The Senate became air conditioned the following year and in 1930, the White House and the Executive Office Building were air-conditioned.

The system of air bypass control, invented in 1924 by L. Logan Lewis, solved the difficult problem of humidity control under varying load. By the end of the 1920s, the first room air conditioner was introduced by Frigidaire. Other important inventions of the 1920s include lightweight extended surface coils and the first unit heater and cold diffuser.

Thomas Midgley, Jr. developed the halocarbon refrigerants in 1930. These refrigerants were found to be safe and economical for the small reciprocating compressors used in commercial and residential markets. Manufacturers were soon producing mass market room air conditioners that used Refrigerant 12.

Fluorinated refrigerants were also applied to centrifugal compression, which required only half the number of impellers for the same head as chlorinated hydrocarbons. Space and materials were saved when pressure-formed extended-surface tubes in shell-and-tube exchangers were introduced by Walter Jones. This invention was an important advance for centrifugal and reciprocating equipment.

Other achievements of the 1930s included

- The first residential lithium bromide absorption machine was introduced in 1931 by Servel.

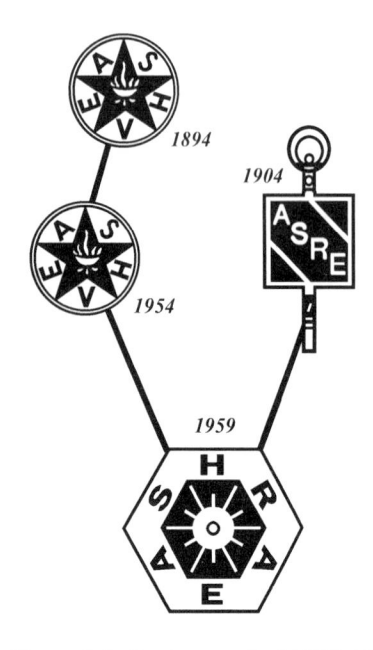

*Figure 1-1 Background of ASHRAE*

## Background

- In 1931, Carrier marketed steam ejector cooling units for railroad passenger cars.
- As of the mid-1930s, General Electric introduced the heat pump; the electrostatic air cleaner was put out by Westinghouse; Charles Neeson of Airtemp invented the high-speed radial compressor; and W.B. Connor discovered that odors could be removed by using activated carbon.

With the end of World War II, air-conditioning technology advanced rapidly. Among the advances were air-source heat pumps, large lithium-bromide water chillers, automobile air conditioners, rooftop heating and cooling units, small, outdoor-installed ammonia absorption chillers, air purifiers, a vapor cycle aircraft cabin cooling unit, and a large-capacity Lysholm rotary compressor.

Improvements on and expansions of products that already existed include

- Dual-duct central systems for office buildings
- Change from open to hermetic compressors from the smallest reciprocating units to large-capacity centrifugals
- Resurgence of electric heating in all kinds of applications
- Use of heat pumps to reclaim heat in large buildings
- Application of electrostatic cleaners to residences
- Self-contained variable volume air terminals for multiple interior rooms
- Increasing use of total energy systems for large buildings and clusters of buildings
- Larger sizes of centrifugals, now over 5000 tons in a single unit
- Central heating and cooling plants for shopping centers, colleges, and apartment and office building complexes

In the late 1940s and into the early 1950s, development work continued on unitary heat pumps for residential and small commercial installations. These factory-engineered and assembled units, like conventional domestic boilers, could be easily and cheaply installed in the home or small commercial businesses by engineers. In 1952, heat pumps were placed on the market for mass consumption. Early heat pumps lacked the durability needed to withstand winter temperatures. Low winter temperatures placed severe stress on the components of these heat pumps (compressors, outdoor fans, reversing valves, and control hardware). Improvements in the design of heat pumps has continued, resulting in more-reliable compressors and lubricating systems, improved reversing valves, and refined control systems.

In the 1950s came the rooftop unit for commercial buildings. Multizone packaged rooftop units were popular during the 1960s; however, most were very energy inefficient and lost favor during the 1970s. Beginning with the oil embargo of 1973, the air-conditioning field could no longer conduct "business as usual," with concern mainly for the initial cost of the building and its conditioning equipment. The use of crude rules of thumb, which significantly oversized equipment and wasted energy, was largely replaced with reliance upon more scientifically sound, and often computer-assisted, design, sizing, and selection procedures. Variable air volume (VAV) designs rapidly became the most popular type of HVAC system for offices, hospitals, and some school buildings. Although energy-efficient, VAV systems proved to have their own set of problems related to indoor air quality (IAQ), sick building syndrome (SBS), and building related illness (BRI). Solutions to these problems are only now being realized.

In 1987, the United Nations Montreal Protocol for protecting the earth's ozone layer was signed, establishing the phase-out schedule for the production of chlorofluorocarbon (CFC) and hydrochlorofluorocarbon (HCFC) refrigerants. Contemporary buildings and their air-conditioning equipment must now provide improved indoor air quality as well as comfort, while consuming less energy and using alternative refrigerants.

## 1.3 Building Energy Use

Energy is generally used in buildings to perform functions of heating, lighting, mechanical drives, cooling, and special applications. A typical breakdown of the relative energy use in a commercial building is given as Figure 1-2.

Energy is available in limited forms, such as electricity, fossil fuels, and solar energy, and these energy forms must be converted within a building to serve the end use of the various functions. A degradation of energy is associated with any conversion process. In energy conservation efforts, two avenues of approach were taken: (1) reducing the amount of use and/or (2) reducing conversion losses. For example, the furnace that heats a building produces unusable and toxic flue gas that must be vented to the outside and in this process some of the energy is lost. Table 1-1 presents typical values for building heat losses and gains at design conditions for a mid-America climate. Actual values will vary significantly with climate and building construction.

The projected total U.S. energy consumption by end-user sector: transportation, industrial, commercial, and residential is shown in Figure 1-3. The per capita energy consumption for the U.S. and the world is shown in Figure 1-4, showing that in

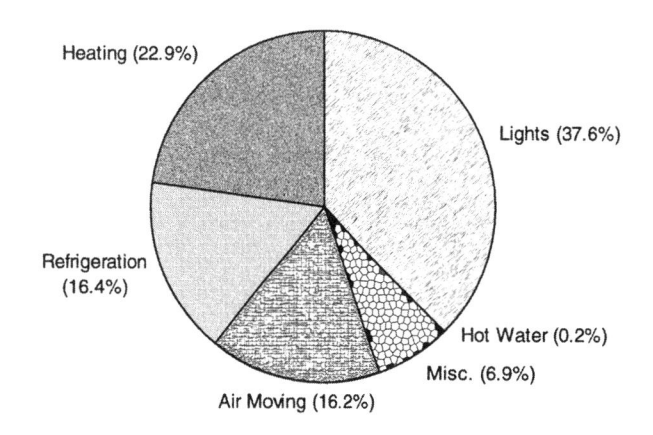

*Figure 1-2 Energy Use in a Commercial Building*

2007 the U.S. consumption was the same as in 1965. This has been achieved through application of energy conservation principles as well as increased energy costs and changes in the economy.

The efficient use of energy in buildings can be achieved by implementing (1) optimum energy designs, (2) well-developed energy use policies, and (3) dedicated management backed up by a properly trained and motivated operating staff. Optimum energy conservation is attained when the least amount of energy is used to achieve a desired result. If this is not fully realizable, the next best method is to move excess energy from where it is not wanted to where it can be used or stored for future use, which generally results in a minimum expenditure of new energy. A system should be designed so that it cannot heat and cool the same locations simultaneously.

ASHRAE Standard 90.1, "Energy Standard for Buildings Except Low-Rise Residential Buildings," and the 100 series standards, "Energy Conservation in Existing Buildings," provide minimum guidelines for energy conservation design and operation. They incorporate these types of energy standards:

Table 1.1  Typical Building Design Heat Losses or Gains

| Building Type | Air Conditioning ft$^2$/ton | Air Conditioning m$^2$/kW | Heating Btu/h·ft$^3$ | Heating W/m$^3$ |
|---|---|---|---|---|
| Apartment | 450 | 12 | 4.5 | 45 |
| Bank | 250 | 7 | 3.0 | 30 |
| Department Store | 250 | 7 | 1.0 | 10 |
| Dormitories | 450 | 12 | 4.5 | 45 |
| House | 700 | 18 | 3.0 | 30 |
| Medical Center | 300 | 8 | 4.5 | 45 |
| Night Club | 250 | 7 | 3.0 | 30 |
| Office Interior | 350 | 9 | 3.0 | 30 |
| Exterior | 275 | 7 | 3.0 | 30 |
| Post Office | 250 | 7 | 3.0 | 30 |
| Restaurant | 250 | 7 | 3.0 | 30 |
| Schools | 275 | 7 | 3.0 | 30 |
| Shopping Center | 250 | 7 | 3.0 | 30 |

(1) prescriptive, which specifies the materials and methods for design and construction of buildings; (2) system performance, which sets requirements for each component, system, or subsystem within a building; and (3) building energy, which considers the performance of the building as a whole. In this last type, a design goal is set for the annual energy requirements of the entire building on basis such as Btu/ft$^2$ per year (GJ/m$^2$ per year). Any combination of materials, systems, and operating procedures can be applied, as long as design energy usage does not exceed the building's annual energy budget goal. "Standard 90.1-2010 User's Manual" is extremely helpful in understanding and applying the requirements of ASHRAE Standard 90.1

This approach allows greater flexibility while promoting the goals of energy efficiency. It also allows and encourages the use of innovative techniques and the development of new methods for saving energy. Means for its implementation are still being developed. They are different for new and for existing buildings; in both cases, an accurate data base is required as well as an accurate, verifiable means of measuring consumption.

As energy prices have risen, more sophisticated schemes for reducing energy consumption have been conceived. Included in such schemes are cogeneration, energy management systems (EMS), direct digital control (DDS), daylighting, closed water-loop heat pumps, variable air volume (VAV) systems, variable frequency drives, thermal storage, dessicant dehumidication, and heat recovery in commercial and institutional buildings and industrial plants.

As detailed in a 1992 Department of Energy Report, "Commercial Buildings Consumption and Expenditures, 1989," more than seventy percent of the commercial-industrial-institutional (C-I-I) buildings recently built in the United States made use of energy conservation measures for heating and cooling.

The type of building and its use strongly affects the energy use as shown in Table 1-2.

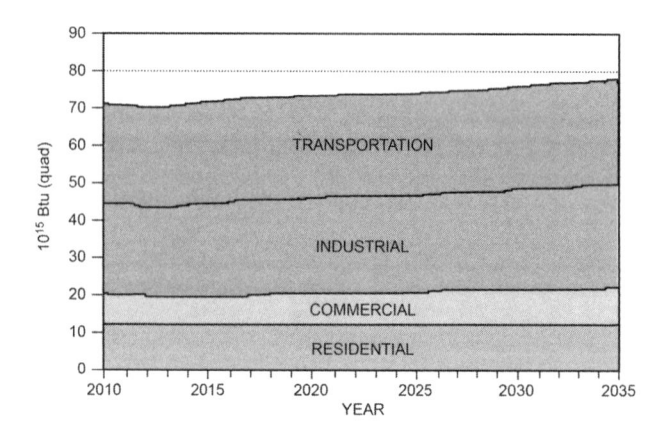

*Figure 1-3  Projected total U.S. Energy Consumption by End-User Sector*
*(EIA 2011)*

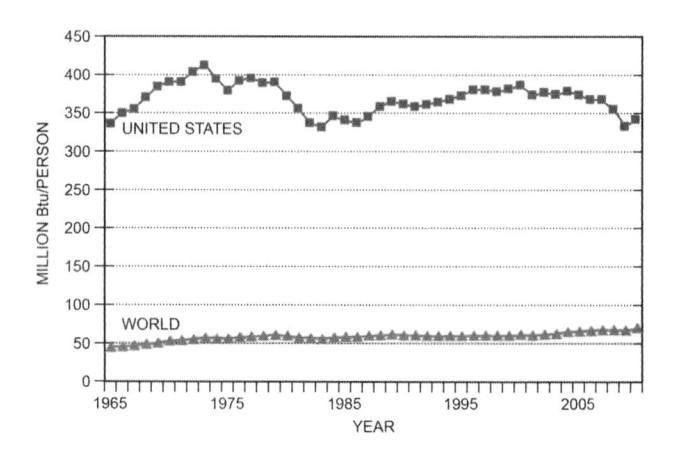

*Figure 1-4  U.S Per Capita Energy Consumption*
*(BP 2012)*

# Background

Heating and air-conditioning systems that are simple in design and of the proper size for a given building generally have relatively low maintenance and operating costs. For optimum results, as much inherent thermal control as is economically possible should be built into the basic structure. Such control might include materials with high thermal properties, insulation, and multiple or special glazing and shading devices. The relationship between the shape, orientation, and air-conditioning requirement of a building should also be considered. Since the exterior load may vary from 30 to 60% of the total air-conditioning load when the fenestration (light transmitting) area ranges from 25 to 75% of the floor area, it may be desirable to minimize the perimeter area. For example, a rectangular building with a 4-to-1 aspect ratio requires substantially more refrigeration than a square building with the same floor area.

When a structure is characterized by several exposures and multipurpose use, especially with wide load swings and non-coincident energy use in certain areas, multiunit or unitary systems may be considered for such areas, but not necessarily for the entire building. The benefits of transferring heat absorbed by cooling from one area to other areas, processes, or services that require heat may enhance the selection of such systems.

Buildings in the US consume significant quantities of energy each year. According to the US Department of Energy (DOE), buildings account for 36% of all the energy used in the US, and 66% of all the electricity used. Beyond economics, energy use in the buildings sector has significant implications for our environment. Emissions related to building energy use account for 35% of carbon dioxide emissions, 47% of sulfur dioxide emissions, and 22% of nitrogen oxide emissions.

## 1.4 Conceptualizing an HVAC System

An important tool for the HVAC design engineer is the ability to develop a quick overview or "concept" of the magnitude of the project at hand. Toward this goal, the industry has developed a number of "rules of thumb," some more accurate than others. As handy as they might be, these approximations must be treated as just that—approximations. Don't use them as "rules of dumb."

Table 1.2 Annual Energy Use Per Unit Floor Area

| Building Type | Annual Energy Use kWh/ft² |
|---|---|
| Assembly | 18.7 |
| Education | 25.5 |
| Food Sales | 51.5 |
| Health Care | 64.0 |
| Lodging | 38.8 |
| Mercantile | 24.8 |
| Office | 30.5 |
| Warehouse | 16.9 |
| Vacant | 6.9 |
| All Buildings | 26.7 |

Tables 1-1 and 1-2 are examples of such rules-of-thumb, providing data for a quick estimate of heating and cooling equipment sizes and of building energy use, requiring knowledge only of the size and intended use of the building. Other rules-of-thumb include using a face velocity of 500 fpm in determining the face area for a cooling coil, the use of 400 cfm/ton for estimating the required cooling airflow rate, the use of 2.5 gpm/ton for determining the water flow rate through the cooling coil and chiller unit, using 1.2 cfm/sq ft of gross floor area for estimating the required conditioned airflow rate for comfort cooling, and the estimation of 0.6 kW/ton as the power requirement for air conditioning. Table 1-3 provides very approximate data related to the cost of HVAC equipment and systems.

Table 1.4 provides approximate energy costs for commercial consumers in the United States for 2013. Keep in mind that these energy costs are very volatile at this time.

Table 1.5 gives approximate total building costs for offices and medical offices averaged for twenty U.S. locations in 2007.

The material presented in this book will enable the reader to validate appropriate rules as well as to improve upon these approximations for the final design.

## 1.5 Sustainability and Green Buildings

The following discussion concerning sustainable design and green buildings has been extracted from Chapters 34 and 35 of the 2013 *ASHRAE Handbook—Fundamentals*.

Pollution, toxic waste creation, waste disposal, global climate change, ozone depletion, deforestation, and resource depletion are recognized as results of uncontrolled technological and population growth. Without mitigation, current

Table 1.3 Capital Cost Estimating Factors

**Cooling Systems**
- $1,600/installed ton of cooling

**Heating Systems**
- $2.65/cfm of installed heating

**Fans/Ducting/Coils/Dampers/Filters**
- $7.60/cfm all-system

Table 1.4 Approximate Energy Costs to Commercial Consumers (2009)

| | |
|---|---|
| Electricity ($/kWh) | 0.088 |
| Natural Gas ($/therm) | 0.85 |
| LPG ($/gal) | 2.90 |
| No. 2 Fuel Oil ($/gas) | 3.40 |

Table 1.5 Approximate Total Building Costs ($/sq. ft.)
(Adapted from RSMeans Costs Comparisons 2007)

| | 2–4-Story Office Building | 5–10-Story Office Building | 11–20-Story Office Building | Medical Office Building |
|---|---|---|---|---|
| **High** | 194 | 181 | 167 | 219 |
| **Average** | 149 | 130 | 121 | 169 |
| **Low** | 117 | 110 | 98 | 132 |

trends will adversely affect the ability of the earth's ecosystem to regenerate and remain viable for future generations.

The built environment contributes significantly to these effects, accounting for one-sixth of the world's fresh water use, one-quarter of its wood harvest, and two-fifths of its material and energy flows. Air quality, transportation patterns, and watersheds are also affected. The resources required to serve this sector are considerable and many of them are diminishing.

Recognition of how the building industry affects the environment is changing the approach to design, construction, operation, maintenance, reuse, and demolition of buildings and focusing on environmental and long-term economic consequences. Although this sustainable design ethic—*sustainability*—covers things beyond the HVAC industry alone, efficient use of energy resources is certainly a key element of any sustainable design and is very much under the control of the HVAC designer.

Research over the years has shown that new commercial construction can reduce annual energy consumption by about 50% using integrated design procedures and energy conservation techniques. In the past few years several programs promoting energy efficiency in building design and operation have been developed. One of these is Energy Star Label (www.energystar.gov) and another one, which is becoming well known, is Leadership in Energy and Environmental Design (LEED) (www.usbc.org/leed).

In 1999 the Environmental Protection Agency of the US government introduced the Energy Star Label for buildings. This is a set of performance standards that compare a building's adjusted energy use to that of similar buildings nationwide. The buildings that perform in the top 25%, while conforming to standards for temperature, humidity, illumination, outdoor air requirements, and air cleanliness, earn the Energy Star Label.

LEED is a voluntary points-based national standard for developing a high-performance building using an integrated design process. LEED evaluates "greenness" in five categories: sustainable sites, water efficiency, energy and atmosphere, materials and resources, and the indoor air environmental quality.

In the energy and atmosphere category, building systems commissioning and minimum energy usages are necessary requirements. The latter requires meeting the requirements *ANSI/ASHRAE/IESNA Standard 90.1-2010, Energy Standard for Buildings Except Low-Rise Residential Buildings*, or the local energy code, whichever is more stringent.

Basically LEED defines what makes a building "green" while the Energy Star Label is concerned only with energy performance. Both of these programs require adherence to ASHRAE standards. Chapter 35 of the 2009 *ASHRAE Handbook—Fundamentals* provides guidance in achieving sustainable designs.

The basic approach to energy-efficient design is reducing loads (power), improving transport systems, and providing efficient components and "intelligent" controls. Important design concepts include understanding the relationship between energy and power, maintaining simplicity, using self-imposed budgets, and applying energy-smart design practices.

Just as an engineer must work to a cost budget with most designs, self-imposed power budgets can be similarly helpful in achieving energy-efficient design. For example, the following are possible goals for mid-rise to high-rise office buildings in a typical midwestern or northeastern temperature climate:

- Installed lighting (overall)         0.8 W/ft$^2$
- Space sensible cooling               15 Btu/h·ft$^2$
- Space heating load                   10 Btu/h·ft$^2$
- Electric power (overall)             3 W/ft$^2$
- Thermal power (overall)              20 Btu/h·ft$^2$
- Hydronic system head                 70 ft of water
- Water chiller (water-cooled)         0.5 kW/ton
- Chilled-water system auxiliaries     0.15 kW/ton
- Unitary air-conditioning systems     1.0 kW/ton
- Annual electric energy               15 kWh/ft$^2$·yr
- Annual thermal energy                5 Btu/ft$^2$·yr·°F·day

These goals, however, may not be realistic for all projects.

As the building and systems are designed, all decisions become interactive as the result of each subsystem's power or energy performance being continually compared to the "budget."

Energy efficiency should be considered at the beginning of building design because energy-efficient features are most easily and effectively incorporated at that time. Active participation of all members of the design team (including owner, architect, engineer, and often the contractor) should be sought early. Consider building attributes such as building function, form, orientation, window/wall ratio, and HVAC system types early because each has major energy implications.

## 1.6 Problems

**1.1** Estimate whether ice will form on a clear night when ambient air temperature is 45°F (7.2°C), if the water is placed in a shallow pan in a sheltered location where the convective heat transfer coefficient is 0.5 Btu/h·ft$^2$·°F [2.8 W/(m$^2$·K)].

**1.2** Obtain a sketch or drawing of Gorrie's refrigeration machine and describe its operation.

**1.3** Plot the history of the annual energy use per square foot of floor space for nonresidential buildings and predict the values for the years 2014 and 2015.

**1.4** Estimate the size of cooling and heating equipment that is needed for a new bank building in middle America that is 140 ft by 220 ft by 12 ft high (42.7 m by 67 m by 3.7 m high). [Answer: 123 tons cooling, 11,109,000 Btu/h heating]

**1.5** Estimate the size of heating and cooling equipment that will be needed for a residence in middle America that is 28 ft by 78 ft by 8 ft high (8.5 m by 23.8 m by 2.4 m high).

**1.6** Estimate the initial cost of the complete HVAC system (heating, cooling, and air moving) for an office building, 40 ft by 150 ft by 10 ft high (12.2 m by 45.7 m by 3.1 m high).

**1.7** Estimate the annual operating cost for the building in Problem 1.6 if it is all-electric. [Answer: $14,640]

**1.8** Conceptualize, as completely as possible, using information only from Sections 1.3, 1.4, and 1.5, the building of Project 8 Two-Story Building, Appendix B, SYSTEMS DESIGN PROBLEMS.

## 1.7 Bibliography

ASHRAE. 2010. *2010 ASHRAE Handbook—Refrigeration.*

ASHRAE. 2011. *2011 ASHRAE Handbook—HVAC Applications.*

ASHRAE. 2012. *2012 ASHRAE Handbook—HVAC Systems and Equipment.*

ASHRAE. 2013. *2013 ASHRAE Handbook—Fundamentals.*

ASHRAE. 1995. *Proclaiming the Truth.*

BP. 2012. *Statistical review of world energy 2012.* http://www.bp.com/sectionbodycopy.do?categoryId=7500&contentId=7068481.

EIA. 2001. *Annual energy review 2000.* DOE/EIA-0384(2000). Energy Information Administration, U.S. Department of Energy, Washington, D.C.

EIA. 2011. *International energy statistics.* U.S. Energy Information Administration, Washington, D.C. http://www.eia.gov/cfapps/ipdbproject/IED Index3.cfm.

EIA. 2012. *Annual energy outlook 2012 with projects to 2035.* http://www.eia.gov/oiaf/aeo/tablebrowser/#release =AEO2012&subject=0-AEO2012&table=1 -AEO2012&region+0-0&cases=ref2012=d020112c.

Coad, W.J. 1997. Designing for Tomorrow, *Heating/Piping/Air Conditioning*, February.

Donaldson, B. and B. Nagengast. 1995. *Heat and Cold: Mastering the Great Indoors.* ASHRAE.

Downing, R. 1984. Development of Chlorofluorocarbon Refrigerants. *ASHRAE Transactions* 90(2).

Faust, F.H. 1992. The Merger of ASHAE and ASRE: The Author Presents An Overview on Events Leading up to ASHRAE's Founding. *ASHRAE Insights* 7(5).

Ivanovich, M.G. 1997. HVAC&R and the Internet: Where to Go, *Heating/Piping/Air Conditioning*, May.

Nagengast, B.A. 1988. A historical look at CFC refrigerants. *ASHRAE Journal* 30(11).

Nagengast, B.A. 1993. The 1920s: The first realization of public air conditioning. *ASHRAE Journal* 35(1).

Nelson, L.W. 1989. Residential comfort: A historical look at early residential HVAC systems. *ASHRAE Journal* 31(1).

Woolrich, W.R. 1969. The History of Refrigeration; 220 Years of Mechanical and Chemical Cold: 1748-1968. *ASHRAE Journal* 33(7).

# Chapter 2

# THERMODYNAMICS AND PSYCHROMETRICS

This chapter reviews the principles of thermodynamics, evaluates thermodynamic properties, and applies thermodynamics and psychrometrics to air-conditioning and refrigeration processes and systems. Greater detail on thermodynamics, particularly relating to the Second Law and irreversibility, is found in Chapter 2, 2013 *ASHRAE Handbook—Fundamentals*. Details on psychrometric properties can be found in Chapter 1 of the 2013 *ASHRAE Handbook—Fundamentals*.

## 2.1 Fundamental Concepts and Principles

### 2.1.1 Thermodynamics

Thermodynamics is the science devoted to the study of energy, its transformations, and its relation to status of matter. Since every engineering operation involves an interaction between energy and materials, the principles of thermodynamics can be found in all engineering activities.

Thermodynamics may be considered the description of the behavior of matter in equilibrium and its changes from one equilibrium state to another. The important concepts of thermodynamics are energy and entropy; the two major principles of thermodynamics are called the first and second laws of thermodynamics. The first law of thermodynamics deals with energy. The idea of energy represents the attempt to find an invariant in the physical universe, something that remains constant in the midst of change. The second law of thermodynamics explains the concept of entropy; e.g., every naturally occurring transformation of energy is accompanied somewhere by a loss in the availability of energy for future performance of work.

The German physicist, Rudolf Clausius (1822–1888), devised the concept of entropy to quantitatively describe the loss of available energy in all naturally occurring transformations. Although the natural tendency is for heat to flow from a hot to a colder body with which it is in contact, corresponding to an increase in entropy, it is possible to make heat flow from a colder body to a hot body, as is done every day in a refrigerator. However, this does not take place naturally or without effort exerted somewhere.

According to the fundamental principles of thermodynamics, the energy of the world stays constant and the entropy of the world increases without limit. If the essence of the first principle in everyday life is that one cannot get something for nothing, the second principle emphasizes that every time one does get something, the opportunity to get that something in the future is reduced by a measurable amount, until ultimately, there will be no more "getting." This "heat death," envisioned by Clausius, will be a time when the universe reaches a level temperature; and though the total amount of energy will be the same as ever, there will be no means of making it available, as entropy will have reached its maximum value.

Like all sciences, the basis of thermodynamics is experimental observation. Findings from these experimental observations have been formalized into basic laws. In the sections that follow, these laws and their related thermodynamic properties will be presented and applied to various examples. These examples should give the student an understanding of the basic concepts and an ability to apply these fundamentals to thermodynamic problems. It is not necessary to memorize numerous equations, for problems are best solved by applying the definitions and laws of thermodynamics.

Thermodynamic reasoning is always from the general law to the specific case; that is, the reasoning is deductive rather than inductive. To illustrate the elements of thermodynamic reasoning, the analytical processes may be divided into two steps:

1. The idealization or substitution of an analytical model for a real system. This step is taken in all engineering sciences. Therefore, skill in making idealizations is an essential part of the engineering art.
2. The second step, unique to thermodynamics, is the deductive reasoning from the first and second laws of thermodynamics.

These steps involve (a) an energy balance, (b) a suitable properties relation, and (c) accounting for entropy changes.

### 2.1.2 System and Surroundings

Most applications of thermodynamics require the definition of a system and its surroundings. A system can be an object, any quantity of matter, or any region of space selected for study and set apart (mentally) from everything else, which then becomes the surroundings. The systems of interest in thermodynamics are finite, and the point of view taken is macroscopic rather than microscopic. No account is taken of the detailed structure of matter, and only the coarse characteristics of the system, such as its temperature and pressure, are regarded as thermodynamic coordinates.

Everything external to the system is the surroundings, and the system is separated from the surroundings by the system boundaries. These boundaries may be either movable or fixed; either real or imaginary.

### 2.1.3 Properties and State

A property of a system is any observable characteristic of the system. The more common thermodynamic properties are temperature, pressure, specific volume or density, internal energy, enthalpy, and entropy.

The state of a system is its condition or configuration described in sufficient detail so that one state may be distinguished from all other states. A listing of a sufficient number of independent properties constitutes a complete definition of the state of a system.

The state may be identified or described by observable, macroscopic properties such as temperature, pressure, and density. Each property of a substance in a given state has only one value; this property always has the same value for a given state, regardless of how the substance arrived at that state. In fact, a property can be defined as any quantity that depends on the state of the system and is independent of the path (i.e., the prior history) by which the system arrived at that given state. Conversely, the state is specified or described by its properties.

The state of a macroscopic system is the condition of the system as characterized by the values of its properties. This chapter directs attention to **equilibrium states**, with equilibrium used in its generally accepted context—the equality of forces, or the state of balance. In future discussion, the term *state* refers to an equilibrium state unless otherwise noted. The concept of equilibrium is important, as it is only in an equilibrium state that thermodynamic properties have meaning. A system is in thermodynamic equilibrium if it is incapable of finite, spontaneous change to another state without a finite change in the state of the surroundings.

Included in the many types of equilibria are thermal, mechanical, and chemical. A system in thermal equilibrium is at the same temperature as the surroundings and the temperature is the same throughout. A system in mechanical equilibrium has no part accelerating ($\Sigma F = 0$) and the pressure within the system is the same as in the surroundings. A system in chemical equilibrium does not tend to undergo a chemical reaction; the matter in the system is said to be inert.

Any property of a thermodynamic system has a fixed value in a given equilibrium state, regardless of how the system arrives at that state. Therefore, the change that occurs in the value of a property when a system is altered from one equilibrium state to another is always the same. This is true regardless of the method used to bring about a change between the two end states. The converse of this statement is equally true. If a measured quantity always has the same value between two given states, that quantity is a measure of the change in a property. This latter assertion is useful in connection with the conservation of energy principle introduced in the next section.

The uniqueness of a property value for a given state can be described mathematically in the following manner. The integral of an exact differential $dY$ is given by

$$\int_1^2 dY = Y_2 - Y_1 = \Delta Y$$

Thus the value of the integral depends solely on the initial and final states. Likewise, the change in the value of a property depends only on the end states. Hence the differential change $dY$ in a property $Y$ is an exact differential. Throughout this text, the infinitesimal variation of a property will be identified by the differential symbol $d$ preceding the property symbol. For example, the infinitesimal change in the pressure $p$ of a system is given by $dp$. The finite change in a property is denoted by the symbol $\Delta$ (capital delta), for example, $\Delta p$. The change in a property value $\Delta Y$ always represents the final value minus the initial value. This convention must be kept in mind.

The symbol $\delta$ is used instead of the usual differential operator $d$ as a reminder that some quantities depend on the process and are not a property of the system. $\delta Q$ represents only a small quantity of heat, not a differential. $\delta m$ represents only a small quantity of matter.

The same qualifications for $\delta$ hold in the case of thermodynamic work. As there is no exact differential $dW$, small quantities of $W$ similar in magnitude to differentials are expressed as $\delta W$.

### 2.1.4 Processes and Cycles

A process is a change in state which can be defined as *any change in the properties of a system*. A process is described in part by the series of states passed through by the system. Often, but not always, some interaction between the system and surroundings occurs during a process; the specification of this interaction completes the description of the process.

Describing a process typically involves specifying the initial and final equilibrium states, the path (if identifiable), and the interactions which take place across the boundaries of the system during the process. The following terms define special processes:

**isobaric** or **constant pressure**—process wherein the pressure does not change;

**isothermal**—process that occurs at constant temperature;

**isometric**—process with constant volume;

**adiabatic**—process in which no heat is transferred to or from the system;

**isentropic**—process with no change in entropy.

A **cycle** is a process, or more frequently, a series of processes wherein the initial and final states of the system are identical. Therefore, at the conclusion of a cycle, all the properties have the same value they had at the beginning.

### 2.1.5 Reversibility

All naturally occurring changes or processes are irreversible. Like a clock, they tend to run down and cannot rewind themselves without other changes in the surroundings occurring. Familiar examples are the transfer of heat with a finite temperature difference, the mixing of two gases, a waterfall, and a chemical reaction. *All of the above changes can be reversed, however.* Heat can be transferred from a region of low temperature to one of higher temperature; gas can be separated into its components; water can be forced to flow uphill. The important point is that these things can be done *only at the expense of some other system*, which itself becomes run down.

A process is reversible if its direction can be reversed at any stage by an infinitesimal change in external conditions. If a connected series of equilibrium states is considered, each representing only an infinitesimal displacement from the adjacent one, but with the overall result a finite change, then a reversible process exists.

All actual processes can be made to approach a reversible process by a suitable choice of conditions; but like the absolute zero of temperature, the strictly reversible process is only a concept that aids in the analysis of problems. The approach of actual processes to this ideal limit can be made almost as close as is desired. However, the closeness of approach is generally limited by economic factors rather than physical ones. The truly reversible process would require an infinite time for completion. The sole reason for the concept of the reversible process is to establish a standard for the comparison of actual processes. The reversible process is one that gives the maximum accomplishment, i.e., yields the greatest amount of work or requires the least amount of work to bring about a given change. It gives the maximum efficiency toward which to strive, but which will never be equalled. The reversible process is the standard for judging the efficiency of an actual process.

Since the reversible process represents a succession of equilibrium states, each only a differential step from its neighbor, the reversible process can be represented as a continuous line on a state diagram ($p$-$v$, $T$-$s$, etc.). The irreversible process cannot be so represented. The terminal states and general direction of change can be noted, but the complete path of change is an indeterminate, irreversible process and cannot be drawn as a line on a thermodynamic diagram.

Irreversibilities always lower the efficiencies of processes. Their effect is identical to that of friction, which is one cause of irreversibility. Conversely, no process more efficient than a reversible process can be imagined. The reversible process represents a standard of perfection that cannot be exceeded because

1. It places an upper limit on the work that may be obtained for a given work-producing process;
2. It places a lower limit on the work input for a given work-requiring process.

### 2.1.6 Conservation of Mass

From relativistic considerations, mass $m$ and energy $E$ are related by the well-known equation:

$$E = mc^2$$

where $c$ = velocity of light.

This equation shows that the mass of a system does change when its energy changes. However, for other than nuclear reactions, the change is quite small and even the most accurate chemical balance cannot detect the change in mass. Thus, conservation of mass and conservation of energy are treated as separate laws in basic thermodynamics.

The mass rate of flow of a fluid passing through a cross-sectional area $A$ is

$$m = AV/v \qquad (2\text{-}1)$$

where $V$ is the average velocity of the fluid in a direction normal to the plane of the area $A$, and $v$ is the specific volume of the fluid. For steady flow with fluid entering a system at a section 1 and leaving at a section 2,

$$m_1 = m_2 = A_1 V_1 / v_1 = A_2 V_2 / v_2 \qquad (2\text{-}1a)$$

This is the continuity equation of steady flow. It can readily be extended to any number of system inlets and outlets and is used in nearly all energy analyses.

## 2.2 Properties of a Substance

### 2.2.1 Specific Volume and Density

The specific volume of a substance $v$ is the volume per unit mass. The density of a substance $\rho$ is the mass per unit volume, and is therefore the reciprocal of the specific volume. Specific volume and density are intensive properties in that they are independent of the size of the system.

### 2.2.2 Pressure

When dealing with liquids and gases, we ordinarily speak of *pressure*; in solids we speak of *stresses*. The pressure in a fluid at rest at a given point is the same in all directions. Pressure is defined as the normal component of force per unit area.

Absolute pressure is the quantity of interest in most thermodynamic investigations. Most pressure and vacuum gages, however, read the difference between absolute pressure and the atmospheric pressure existing at the gage (Figure 2-1).

### 2.2.3 Temperature

Because temperature is difficult to define, equality of temperature is defined instead. Two bodies have **equality of temperature** if no change in any observable property occurs when they are in thermal communication.

The **zeroth law of thermodynamics** states that when two bodies have equality of temperature with a third body, they in turn have equality of temperature with each other. Since this fact is not derivable from other laws, and since in the logical presentation of thermodynamics it precedes the first and second laws of thermodynamics, it has been called the zeroth law of thermodynamics. This law is the basis of temperature measurement. Every time a body has equality of temperature with a thermometer, it is said that the body has the temperature read on the thermometer. The problem remains, however, of relating temperatures that might be read on different thermometers, or that are obtained when different temperature-measuring devices are used, such as thermocouples and resistance thermometers. The need for a standard scale for temperature measurements is apparent.

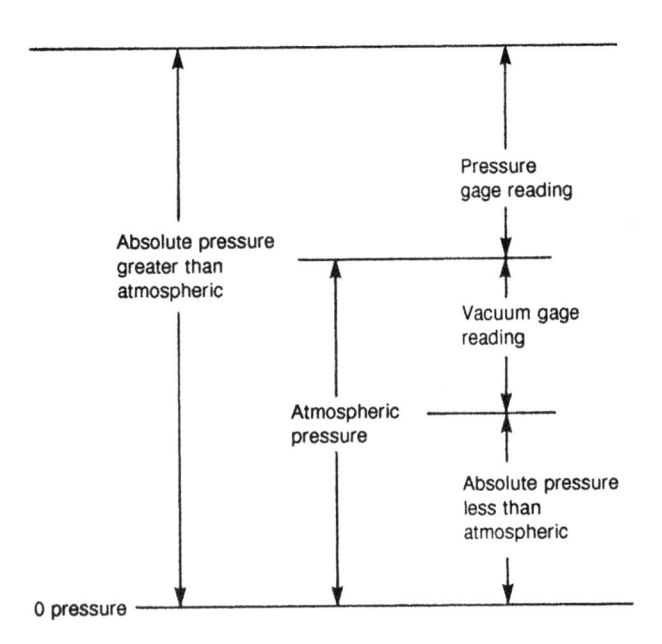

*Fig. 2-1 Terms Used in Pressure Measurement*

**Fahrenheit** and **Celsius** are two commonly used temperature measuring scales. The Celsius scale was formerly called the Centigrade scale.

In this text, the abbreviations °F and °C denote the Fahrenheit and Celsius scales, respectively. The symbols $t$ and $T$ are both used in the literature for temperature on all temperature scales. Unfortunately, little uniformity exists with nomenclature in engineering.

The absolute scale related to the Celsius scale is referred to as the Kelvin scale and is designated K. For SI units, the degree sign is not used with the Kelvin scale. The relation between the SI temperature scales is

$$K = °C + 273.15$$

The absolute scale related to the Fahrenheit scale is referred to as the Rankine scale and is designated °R. The relation between these scales is

$$°R = °F + 459.67$$

### 2.2.4 Internal Energy

Internal energy refers to the energy possessed by a material due to the motion and/or position of the molecules. This form of energy may be divided into two parts: (1) kinetic internal energy, which is due to the velocity of the molecules; and (2) potential internal energy, which is due to the attractive forces existing between molecules. Changes in the average velocity of molecules are indicated by temperature changes of the system; variations in relative distance between molecules are denoted by changes in phase of the system.

The symbol $U$ designates the internal energy of a given mass of a substance. Following the convention used with other extensive properties, the symbol $u$ designates the internal energy per unit mass. As in the case of specific volume, $u$ can represent specific internal energy.

### 2.2.5 Enthalpy

In analyzing specific types of processes, certain combinations of thermodynamic properties, which are therefore also properties of the substance undergoing the change of state, are frequently encountered. One such combination is $U + pV$. It is convenient to define a new extensive property, called *enthalpy*:

$$H = U + pV$$

or, per unit mass

$$h = u + pv \qquad (2\text{-}2)$$

As in the case of internal energy, specific enthalpy can be referred to as $h$, and total enthalpy $H$. However, both may be called enthalpy, since the context makes it clear which is meant.

### 2.2.6 Entropy

**Entropy** $S$ is a measure of the molecular disorder or of the probability of a given state. The more disordered a system, the greater is its entropy; conversely, an orderly or unmixed configuration is one of low entropy.

By applying the theory of probability to molecular systems, Boltzmann showed a simple relationship between the entropy of a given system of molecules and the probability of its occurrence. This relationship is given as

$$S = k \ln W$$

where $k$ is the Boltzmann constant and $W$ is the thermodynamic probability.

Since entropy is the property used in quantifying the Second Law of Thermodynamics, additional discussion from a classical thermodynamic viewpoint will be presented when the Second Law is discussed.

### 2.2.7 Specific Heats

The constant-volume specific heat and the constant-pressure specific heat are useful functions for thermodynamic calculations—particularly for gases.

The constant-volume specific heat $c_v$ is defined by the relation

$$c_v = (\partial u / \partial T)_v \qquad (2\text{-}3)$$

The constant-pressure specific heat $c_p$ is defined by the relation

$$c_p = (\partial h / \partial T)_p \qquad (2\text{-}4)$$

Note that each of these quantities is defined in terms of properties. Thus, the constant-volume and constant-pressure specific heats are thermodynamic properties of a substance.

### 2.2.8 Dimensions and Units

The fundamental and primitive concepts which underlie all physical measurements and all properties are time, length, mass, absolute temperature, electric current, and amount of substance. Arbitrary scales of measurement must be established for these primary dimensions, with each scale divided into specific units of size. The internationally accepted base units for the six quantities are as follows:

| | |
|---|---|
| length | metre (m) |
| mass | kilogram (kg) |
| time | second (s) |
| electric current | ampere (A) |
| thermodynamic temperature | kelvin (K) |
| amount of substance | mole (mol) |

Each of these has a precise definition according to international agreement. They form the basis for the SI from the French document, *Le Système International d'Unités* (*SI*), or International System of Units.

The mass of a system is often given by stating the number of moles it contains. A mole is the mass of a chemical species equal numerically to its molecular mass. Thus, a kilogram mole of oxygen ($O_2$) contains 32 kilograms. In addition, the number of molecules in a kilogram mole is the same for all substances. This is also true for a gram mole, and in this case the number of molecules is Avogadro's number, equal to $6.0225 \times 10^{23}$ molecules.

Many derived units are important in thermodynamics. Examples are force, pressure, and density. Force is determined through Newton's second law of motion, $F = ma$, and has the basic unit $(kg \cdot m)/s^2$. The SI unit for this composite set is the newton (N). Pressure is defined as force per unit area ($N/m^2$), called the pascal (Pa); and density is mass per unit volume ($kg/m^3$).

The US customary engineering system of units also recognizes the second as the basic unit of time, and the ampere as the unit of current. However, absolute temperature is measured in degrees Rankine (°R). The foot (ft) is the usual unit of length and the pound mass ($lb_m$) is the unit of mass. The molar unit is the pound mole. ASHRAE calls this system the inch-pound (I-P) unit system.

The unit of force, the pound force ($lb_f$), is defined without reference to Newton's second law, so this law must be written to include a dimensional proportionality constant:

$$F = ma/g_c$$

where $g_c$ is the proportionality constant. In the I-P system, the proportionality constant is

$$g_c = 32.174 \ (lb_m/lb_f)(ft/s^2)$$

The unit of density is $lb_m/ft^3$, and the unit of pressure is $lb_f/ft^2$ or $lb_f/in^2$, often written psi. Pressure gages usually measure pressure relative to atmospheric pressure. The term **absolute pressure** is often used to distinguish thermodynamic (actual) pressure (psia) from gage (relative) pressure (psig).

In SI units, the proportionality constant $g_c$ in Newton's law is unity or

$$g_c = 1 \ (kg/N)(m/s^2)$$

In this book, all equations that derive from Newton's law carry the constant $g_c$.

### 2.2.9 Pure Substance

A pure substance is one that has a homogeneous and invariable chemical composition. It may exist in more than one phase, but the chemical composition is the same in all phases. Thus, liquid water, a mixture of liquid water and water vapor (steam), or a mixture of ice and liquid water are all pure substances, for every phase has the same chemical composition. On the other hand, a mixture of liquid air and gaseous air is not a pure substance, since the composition of the liquid phase is different from that of the vapor phase.

Sometimes a mixture of gases is considered a pure substance as long as there is no change of phase. Strictly speaking, this is not true. A mixture of gases, such as air, exhibits some of the characteristics of a pure substance as long as there is no change of phase.

Consider as a system that water is contained in the piston-cylinder arrangement of Figure 2-2. Suppose that the piston maintains a pressure of 14.7 $lb_f$/in. (101.3 kPa) in the cylinder containing $H_2O$, and that the initial temperature is 59°F (15°C). As heat is transferred to the water, the temperature increases appreciably, the specific volume increases slightly, and the pressure remains constant. When the temperature reaches 212°F (100°C), additional heat transfer results in a change of phase. That is, some of the liquid becomes vapor, and during this process both the temperature and pressure remain constant, while the specific volume increases considerably. When the last drop of liquid has vaporized, further heat transfer results in an increase in both temperature and specific volume of the vapor.

*Saturation temperature* designates the temperature at which vaporization takes place at a given pressure; this pressure is called the *saturation pressure* for the given temperature. Thus for water at 212°F (100°C), the saturation pressure is 14.7 $lb_f$/in.² (101.3 kPa), and for water at 14.7 $lb_f$/in.² (101.3 kPa), the saturation temperature is 212°F (100°C).

If a substance exists as liquid at the saturation temperature and pressure, it is called saturated liquid. If the temperature of the liquid is lower than the saturation temperature for the existing pressure, it is called a subcooled liquid (implying that the temperature is lower than the saturation temperature for the given pressure) or a compressed liquid (implying that the pressure is greater than the saturation pressure for the given temperature).

When a substance exists as part liquid and part vapor at the saturation temperature, its quality is defined as the ratio of the mass of vapor to the total mass. The quality may be considered as an intensive property, and it has the symbol $x$. Quality has meaning only when the substance is in a saturated state, i.e., at saturation pressure and temperature.

If a substance exists as vapor at the saturation temperature, it is called **saturated vapor**. (Sometimes the term dry saturated vapor is used to emphasize that the quality is 100%.) When the vapor is at a temperature greater than the saturation temperature,

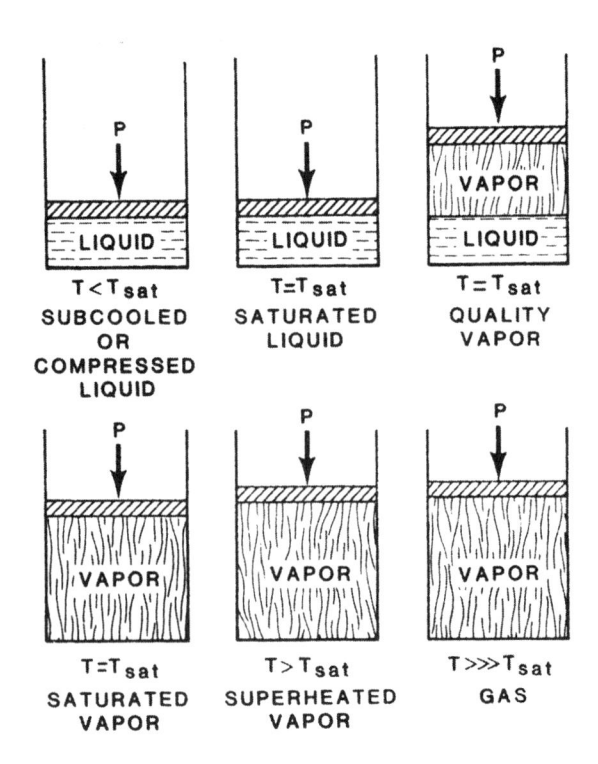

*Fig. 2-2 Thermodynamic Fluid States*

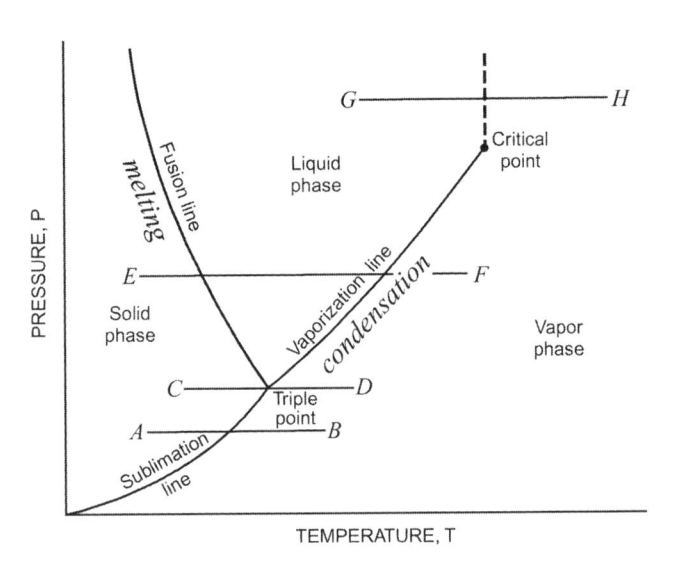

*Fig. 2-3 The Pure Substance*

it is said to exist as superheated vapor. The pressure and temperature of superheated vapor are independent properties because the temperature may increase while the pressure remains constant. Actually, gases are highly superheated vapors.

The entire range of phases is summarized by Figure 2-3, which shows how the solid, liquid, and vapor phases may exist together in equilibrium. Along the sublimation line, the solid and vapor phases are in equilibrium, along the fusion line, the solid and liquid phases are in equilibrium, and along the vaporization line, the liquid and vapor phases are in equilibrium. The only point at which all three phases may exist in equilibrium is the triple point. The vaporization line ends at the critical point because there is no distinct change from the liquid phase to the vapor phase above the critical point.

Consider a solid in state A, Figure 2-3. When the temperature is increased while the pressure (which is less than the triple point pressure) is constant, the substance passes directly from the solid to the vapor phase. Along the constant pressure line EF, the substance first passes from the solid to the liquid phase at one temperature, and then from the liquid to the vapor phase at a higher temperature. Constant-pressure line CD passes through the triple point, and it is only at the triple point that the three phases may exist together in equilibrium. At a pressure above the critical pressure, such as GH, there is no sharp distinction between the liquid and vapor phases.

One important reason for introducing the concept of a pure substance is that the state of a simple compressible pure substance is defined by two independent properties. This means, for example, if the specific volume and temperature of superheated steam are specified, the state of the steam is determined.

To understand the significance of the term **independent property**, consider the saturated-liquid and saturated-vapor states of a pure substance. These two states have the same pressure and the same temperature, but are definitely not the same state. Therefore, in a saturation state, pressure and temperature are not independent properties. Two independent properties such as pressure and specific volume, or pressure and quality, are required to specify a saturation state of a pure substance.

Thus, a mixture of gases, such as air, has the same characteristics as a pure substance, as long as only one phase is present. The state of air, which is a mixture of gases of definite composition, is determined by specifying two properties as long as it remains in the gaseous phase, and in this regard, air can be treated as a pure substance.

### 2.2.10 Tables and Graphs of Thermodynamic Properties

Tables of thermodynamic properties of many substances are available, and they all generally have the same form. This section refers to the tables for $H_2O$ and R-134a, as well as their respective Mollier diagrams, the *h-s* chart for steam, and the *p-h* diagram for R-134a.

Table 3 in Chapter 1 of the 2013 *ASHRAE Handbook—Fundamentals* gives thermodynamic properties of water at saturation and is reproduced in part as Table 2-1 on the following pages. In Table 2-1, the first two columns after the temperature give the corresponding saturation pressure in pounds force per square inch and in inches of mercury. The next three columns give specific volume in cubic feet per pound mass. The first of these gives the specific volume of the saturated solid $(v_i)$ or liquid $(v_f)$; the third column gives the specific volume of saturated vapor $v_g$. The difference between these two values, $v_g - v_i$ or $v_g - v_f$, represents the increase in specific volume when the state changes from saturated solid or liquid to saturated vapor, and is designated $v_{ig}$ or $v_{fg}$.

# Thermodynamics and Psychrometrics

### Table 2-1 Thermodynamic Properties of Water
*(Table 3, Chapter 1, 2013 ASHRAE Handbook—Fundamentals)*

| Temp., °F $t$ | Absolute Pressure $p$, psi | $p$, in. Hg | Sat. Solid $v_i$ | Evap. $v_{ig}$ | Sat. Vapor $v_g$ | Sat. Solid $h_i$ | Evap. $h_{ig}$ | Sat. Vapor $h_g$ | Sat. Solid $s_i$ | Evap. $s_{ig}$ | Sat. Vapor $s_g$ | Temp., °F $t$ |
|---|---|---|---|---|---|---|---|---|---|---|---|---|
| −80 | 0.000116 | 0.000236 | 0.01732 | 1953234 | 1953234 | −193.50 | 1219.19 | 1025.69 | −0.4067 | 3.2112 | 2.8045 | −80 |
| −79 | 0.000125 | 0.000254 | 0.01732 | 1814052 | 1814052 | −193.11 | 1219.24 | 1026.13 | −0.4056 | 3.2029 | 2.7972 | −79 |
| −78 | 0.000135 | 0.000275 | 0.01732 | 1685445 | 1685445 | −192.71 | 1219.28 | 1026.57 | −0.4046 | 3.1946 | 2.7900 | −78 |
| −77 | 0.000145 | 0.000296 | 0.01732 | 1566663 | 1566663 | −192.31 | 1219.33 | 1027.02 | −0.4036 | 3.1964 | 2.7828 | −77 |
| −76 | 0.000157 | 0.000319 | 0.01732 | 1456752 | 1456752 | −191.92 | 1219.38 | 1027.46 | −0.4025 | 3.1782 | 2.7757 | −76 |
| −75 | 0.000169 | 0.000344 | 0.01733 | 1355059 | 1355059 | −191.52 | 1219.42 | 1027.90 | −0.4015 | 3.1701 | 2.7685 | −75 |
| −74 | 0.000182 | 0.000371 | 0.01733 | 1260977 | 1260977 | −191.12 | 1219.47 | 1028.34 | −0.4005 | 3.1619 | 2.7615 | −74 |
| −73 | 0.000196 | 0.000399 | 0.01733 | 1173848 | 1173848 | −190.72 | 1219.51 | 1028.79 | −0.3994 | 3.1539 | 2.7544 | −73 |
| −72 | 0.000211 | 0.000430 | 0.01733 | 1093149 | 1093149 | −190.32 | 1219.55 | 1029.23 | −0.3984 | 3.1459 | 2.7475 | −72 |
| −71 | 0.000227 | 0.000463 | 0.01733 | 1018381 | 1018381 | −189.92 | 1219.59 | 1029.67 | −0.3974 | 3.1379 | 2.7405 | −71 |
| −70 | 0.000245 | 0.000498 | 0.01733 | 949067 | 949067 | −189.52 | 1219.63 | 1030.11 | −0.3963 | 3.1299 | 2.7336 | −70 |
| −69 | 0.000263 | 0.000536 | 0.01733 | 884803 | 884803 | −189.11 | 1219.67 | 1030.55 | −0.3953 | 3.1220 | 2.7267 | −69 |
| −68 | 0.000283 | 0.000576 | 0.01733 | 825187 | 825187 | −188.71 | 1219.71 | 1031.00 | −0.3943 | 3.1141 | 2.7199 | −68 |
| −67 | 0.000304 | 0.000619 | 0.01734 | 769864 | 769864 | −188.30 | 1219.74 | 1031.44 | −0.3932 | 3.1063 | 2.7131 | −67 |
| −66 | 0.000326 | 0.000664 | 0.01734 | 718508 | 718508 | −187.90 | 1219.78 | 1031.88 | −0.3922 | 3.0985 | 2.7063 | −66 |
| −65 | 0.000350 | 0.000714 | 0.01734 | 670800 | 670800 | −187.49 | 1219.82 | 1032.32 | −0.3912 | 3.0907 | 2.6996 | −65 |
| −64 | 0.000376 | 0.000766 | 0.01734 | 626503 | 626503 | −187.08 | 1219.85 | 1032.77 | −0.3901 | 3.0830 | 2.6929 | −64 |
| −63 | 0.000404 | 0.000822 | 0.01734 | 585316 | 585316 | −186.67 | 1219.88 | 1033.21 | −0.3891 | 3.0753 | 2.6862 | −63 |
| −62 | 0.000433 | 0.000882 | 0.01734 | 548041 | 547041 | −186.26 | 1219.91 | 1033.65 | −0.3881 | 3.0677 | 2.6730 | −62 |
| −61 | 0.000464 | 0.000945 | 0.01734 | 511446 | 511446 | −185.85 | 1219.95 | 1034.09 | −0.3870 | 3.0601 | 2.6730 | −61 |
| −60 | 0.000498 | 0.001013 | 0.01734 | 478317 | 478317 | −185.44 | 1219.98 | 1034.54 | −0.3860 | 3.0525 | 2.6665 | −60 |
| −59 | 0.000533 | 0.001086 | 0.01735 | 447495 | 447495 | −185.03 | 1220.01 | 1034.98 | −0.3850 | 3.0449 | 2.6600 | −59 |
| −58 | 0.000571 | 0.001163 | 0.01735 | 418803 | 418803 | −184.61 | 1220.03 | 1035.42 | −0.3839 | 3.0374 | 2.6535 | −58 |
| −57 | 0.000612 | 0.001246 | 0.01735 | 392068 | 392068 | −184.20 | 1220.06 | 1035.86 | −0.3829 | 3.0299 | 2.6470 | −57 |
| −56 | 0.000655 | 0.001333 | 0.01735 | 367172 | 367172 | −183.78 | 1220.09 | 1036.30 | −0.3819 | 3.0225 | 2.6406 | −56 |
| −55 | 0.000701 | 0.001427 | 0.01735 | 343970 | 343970 | −183.37 | 1220.11 | 1036.75 | −0.3808 | 3.0151 | 2.6342 | −55 |
| −54 | 0.000750 | 0.001526 | 0.01735 | 322336 | 322336 | −182.95 | 1220.14 | 1037.19 | −0.3798 | 3.0077 | 2.6279 | −54 |
| −53 | 0.000802 | 0.001632 | 0.01735 | 302157 | 302157 | −182.53 | 1220.16 | 1037.63 | −0.3788 | 3.0004 | 2.6216 | −53 |
| −52 | 0.000857 | 0.001745 | 0.01735 | 283335 | 283335 | −182.11 | 1220.18 | 1038.07 | −0.3778 | 2.9931 | 2.6153 | −52 |
| −51 | 0.000916 | 0.001865 | 0.01736 | 265773 | 265773 | −181.69 | 1220.21 | 1038.52 | −0.3767 | 2.9858 | 2.6091 | −51 |
| −50 | 0.000979 | 0.001992 | 0.01736 | 249381 | 249381 | −181.27 | 1220.23 | 1038.96 | −0.3757 | 2.9786 | 2.6029 | −50 |
| −49 | 0.001045 | 0.002128 | 0.01736 | 234067 | 234067 | −180.85 | 1220.25 | 1039.40 | −0.3747 | 2.9714 | 2.5967 | −49 |
| −48 | 0.001116 | 0.002272 | 0.01736 | 219766 | 219766 | −180.42 | 1220.26 | 1039.84 | −0.3736 | 2.9642 | 2.5906 | −48 |
| −47 | 0.001191 | 0.002425 | 0.01736 | 206398 | 206398 | −180.00 | 1220.28 | 1040.28 | −0.3726 | 2.9570 | 2.5844 | −47 |
| −46 | 0.001271 | 0.002587 | 0.01736 | 193909 | 193909 | −179.57 | 1220.30 | 1040.73 | −0.3716 | 2.9499 | 2.5784 | −46 |
| −45 | 0.001355 | 0.002760 | 0.01736 | 182231 | 182231 | −179.14 | 1220.31 | 1041.17 | −0.3705 | 2.9429 | 2.5723 | −45 |
| −44 | 0.001445 | 0.002943 | 0.01736 | 171304 | 171304 | −178.72 | 1220.33 | 1041.61 | −0.3695 | 2.9358 | 2.5663 | −44 |
| −43 | 0.001541 | 0.003137 | 0.01737 | 161084 | 161084 | −178.29 | 1220.34 | 1042.05 | −0.3685 | 2.9288 | 2.5603 | −43 |
| −42 | 0.001642 | 0.003343 | 0.01737 | 151518 | 151518 | −177.86 | 1220.36 | 1042.50 | −0.3675 | 2.9218 | 2.5544 | −42 |
| −41 | 0.001749 | 0.003562 | 0.01737 | 142566 | 142566 | −177.43 | 1220.37 | 1042.94 | −0.3664 | 2.9149 | 2.5485 | −41 |
| −40 | 0.001863 | 0.003793 | 0.01737 | 134176 | 134176 | −177.00 | 1220.38 | 1043.38 | −0.3654 | 2.9080 | 2.5426 | −40 |
| −39 | 0.001984 | 0.004039 | 0.01737 | 126322 | 126322 | −176.57 | 1220.39 | 1043.82 | −0.3644 | 2.9011 | 2.5367 | −39 |
| −38 | 0.002111 | 0.004299 | 0.01737 | 118959 | 118959 | −176.13 | 1220.40 | 1044.27 | −0.3633 | 2.8942 | 2.5309 | −38 |
| −37 | 0.002247 | 0.004574 | 0.01737 | 112058 | 112058 | −175.70 | 1220.40 | 1044.71 | −0.3623 | 2.8874 | 2.5251 | −37 |
| −36 | 0.002390 | 0.004866 | 0.01738 | 105592 | 105592 | −175.26 | 1220.41 | 1045.15 | −0.3613 | 2.8806 | 2.5193 | −36 |
| −35 | 0.002542 | 0.005175 | 0.01738 | 99522 | 99522 | −174.83 | 1220.42 | 1045.59 | −0.3603 | 2.8738 | 2.5136 | −35 |
| −34 | 0.002702 | 0.005502 | 0.01738 | 93828 | 93828 | −174.39 | 1220.42 | 1046.03 | −0.3592 | 2.8671 | 2.5078 | −34 |
| −33 | 0.002872 | 0.005848 | 0.01738 | 88489 | 88489 | −173.95 | 1220.43 | 1046.48 | −0.3582 | 2.8604 | 2.5022 | −33 |
| −32 | 0.003052 | 0.006213 | 0.01738 | 83474 | 83474 | −173.51 | 1220.43 | 1046.92 | −0.3572 | 2.8537 | 2.4965 | −32 |
| −31 | 0.003242 | 0.006600 | 0.01738 | 78763 | 78763 | −173.07 | 1220.43 | 1047.36 | −0.3561 | 2.8470 | 2.4909 | −31 |
| −30 | 0.003443 | 0.007009 | 0.01738 | 74341 | 74341 | −172.63 | 1220.43 | 1047.80 | −0.3551 | 2.8404 | 2.4853 | −30 |
| −29 | 0.003655 | 0.007441 | 0.01738 | 70187 | 70187 | −172.19 | 1220.43 | 1048.25 | −0.3541 | 2.8338 | 2.4797 | −29 |
| −28 | 0.003879 | 0.007898 | 0.01739 | 66282 | 66282 | −171.74 | 1220.43 | 1048.69 | −0.3531 | 2.8272 | 2.4742 | −28 |
| −27 | 0.004116 | 0.008380 | 0.01739 | 62613 | 62613 | −171.30 | 1220.43 | 1049.13 | −0.3520 | 2.8207 | 2.4687 | −27 |
| −26 | 0.004366 | 0.008890 | 0.01739 | 59161 | 59161 | −170.86 | 1220.43 | 1049.57 | −0.3510 | 2.8142 | 2.4632 | −26 |
| −25 | 0.004630 | 0.009428 | 0.01739 | 55915 | 55915 | −170.41 | 1220.42 | 1050.01 | −0.3500 | 2.8077 | 2.4577 | −25 |
| −24 | 0.004909 | 0.009995 | 0.01739 | 52861 | 52861 | −169.96 | 1220.42 | 1050.46 | −0.3489 | 2.8013 | 2.4523 | −24 |
| −23 | 0.005203 | 0.010594 | 0.01739 | 49986 | 49986 | −169.51 | 1220.41 | 1050.90 | −0.3479 | 2.7948 | 2.4469 | −23 |
| −22 | 0.005514 | 0.011226 | 0.01739 | 47281 | 47281 | −169.07 | 1220.41 | 1051.34 | −0.3469 | 2.7884 | 2.4415 | −22 |
| −21 | 0.005841 | 0.011892 | 0.01740 | 44733 | 44733 | −168.62 | 1220.40 | 1051.78 | −0.3459 | 2.7820 | 2.4362 | −21 |
| −20 | 0.006186 | 0.012595 | 0.01740 | 42333 | 42333 | −168.16 | 1220.39 | 1052.22 | −0.3448 | 2.7757 | 2.4309 | −20 |
| −19 | 0.006550 | 0.013336 | 0.01740 | 40073 | 40073 | −167.71 | 1220.38 | 1052.67 | −0.3438 | 2.7694 | 2.4256 | −19 |
| −18 | 0.006933 | 0.014117 | 0.01740 | 37943 | 37943 | −167.26 | 1220.37 | 1053.11 | −0.3428 | 2.7631 | 2.4203 | −18 |
| −17 | 0.007337 | 0.014939 | 0.01740 | 35934 | 35934 | −166.81 | 1220.36 | 1053.55 | −0.3418 | 2.7568 | 2.4151 | −17 |
| −16 | 0.007763 | 0.015806 | 0.01740 | 34041 | 34041 | −166.35 | 1220.34 | 1053.99 | −0.3407 | 2.7506 | 2.4098 | −16 |
| −15 | 0.008211 | 0.016718 | 0.01740 | 32256 | 32256 | −165.90 | 1220.33 | 1054.43 | −0.3397 | 2.7444 | 2.4046 | −15 |
| −14 | 0.008683 | 0.017678 | 0.01741 | 30572 | 30572 | −165.44 | 1220.31 | 1054.87 | −0.3387 | 2.7382 | 2.3995 | −14 |

**Table 2-1 Thermodynamic Properties of Water (*Continued*)**
*(Table 3, Chapter 1, 2013 ASHRAE Handbook—Fundamentals)*

| Temp., °F  t | Absolute Pressure p, psi | Absolute Pressure p, in. Hg | Specific Volume, ft³/lb_w Sat. Solid/Liq. $v_i/v_f$ | Specific Volume Evap. $v_{ig}$ | Specific Volume Sat. Vapor $v_g$ | Specific Enthalpy, Btu/lb_w Sat. Solid/Liq. $h_i/h_f$ | Specific Enthalpy Evap. $h_{ig}$ | Specific Enthalpy Sat. Vapor $h_g$ | Specific Entropy, Btu/lb_w·°F Sat. Solid/Liq. $s_i/s_f$ | Specific Entropy Evap. $s_{ig}$ | Specific Entropy Sat. Vapor $s_g$ | Temp., °F  t |
|---|---|---|---|---|---|---|---|---|---|---|---|---|
| −13 | 0.009179 | 0.018689 | 0.01741 | 28983 | 28983 | −164.98 | 1220.30 | 1055.32 | −0.3377 | 2.7320 | 2.3943 | −13 |
| −12 | 0.009702 | 0.019753 | 0.01741 | 27483 | 27483 | −164.52 | 1220.28 | 1055.76 | −0.3366 | 2.7259 | 2.3892 | −12 |
| −11 | 0.010252 | 0.020873 | 0.01741 | 26067 | 26067 | −164.26 | 1220.26 | 1056.20 | −0.3356 | 2.7197 | 2.3841 | −11 |
| −10 | 0.010830 | 0.022050 | 0.01741 | 24730 | 24730 | −163.60 | 1220.24 | 1056.64 | −0.3346 | 2.7136 | 2.3791 | −10 |
| −9 | 0.011438 | 0.023288 | 0.01741 | 23467 | 23467 | −163.14 | 1220.22 | 1057.08 | −0.3335 | 2.7076 | 2.3740 | −9 |
| −8 | 0.012077 | 0.024590 | 0.01741 | 22274 | 22274 | −162.68 | 1220.20 | 1057.53 | −0.3325 | 2.7015 | 2.3690 | −8 |
| −7 | 0.012749 | 0.025958 | 0.01742 | 21147 | 21147 | −162.21 | 1220.18 | 1057.97 | −0.3315 | 2.6955 | 2.3640 | −7 |
| −6 | 0.013456 | 0.027396 | 0.01742 | 20081 | 20081 | −161.75 | 1220.16 | 1058.41 | −0.3305 | 2.6895 | 2.3591 | −6 |
| −5 | 0.014197 | 0.028906 | 0.01742 | 19074 | 19074 | −161.28 | 1220.13 | 1058.85 | −0.3294 | 2.6836 | 2.3541 | −5 |
| −4 | 0.014977 | 0.030493 | 0.01742 | 18121 | 18121 | −160.82 | 1220.11 | 1059.29 | −0.3284 | 2.6776 | 2.3492 | −4 |
| −3 | 0.015795 | 0.032159 | 0.01742 | 17220 | 17220 | −160.35 | 1220.08 | 1059.73 | −0.3274 | 2.6717 | 2.3443 | −3 |
| −2 | 0.016654 | 0.033908 | 0.01742 | 16367 | 16367 | −159.88 | 1220.05 | 1060.17 | −0.3264 | 2.6658 | 2.3394 | −2 |
| −1 | 0.017556 | 0.035744 | 0.01742 | 15561 | 15561 | −159.41 | 1220.02 | 1060.62 | −0.3253 | 2.6599 | 2.3346 | −1 |
| 0 | 0.018502 | 0.037671 | 0.01743 | 14797 | 14797 | −158.94 | 1220.00 | 1061.06 | −0.3243 | 2.6541 | 2.3298 | 0 |
| 1 | 0.019495 | 0.039693 | 0.01743 | 14073 | 14073 | −158.47 | 1219.96 | 1061.50 | −0.3233 | 2.6482 | 2.3249 | 1 |
| 2 | 0.020537 | 0.041813 | 0.01743 | 13388 | 13388 | −157.99 | 1219.93 | 1061.94 | −0.3223 | 2.6424 | 2.3202 | 2 |
| 3 | 0.021629 | 0.044037 | 0.01743 | 12740 | 12740 | −157.52 | 1219.90 | 1062.38 | −0.3212 | 2.6367 | 2.3154 | 3 |
| 4 | 0.022774 | 0.046369 | 0.01743 | 12125 | 12125 | −157.05 | 1219.87 | 1062.82 | −0.3202 | 2.6309 | 2.3107 | 4 |
| 5 | 0.023975 | 0.048813 | 0.01743 | 11543 | 11543 | −156.57 | 1219.83 | 1063.26 | −0.3192 | 2.6252 | 2.3060 | 5 |
| 6 | 0.025233 | 0.051375 | 0.01743 | 10991 | 10991 | −156.09 | 1219.80 | 1063.70 | −0.3182 | 2.6194 | 2.3013 | 6 |
| 7 | 0.026552 | 0.054059 | 0.01744 | 10468 | 10468 | −155.62 | 1219.76 | 1064.14 | −0.3171 | 2.6138 | 2.2966 | 7 |
| 8 | 0.027933 | 0.056872 | 0.01744 | 9971 | 9971 | −155.14 | 1219.72 | 1064.58 | −0.3161 | 2.6081 | 2.2920 | 8 |
| 9 | 0.029379 | 0.059817 | 0.01744 | 9500 | 9500 | −154.66 | 1219.68 | 1065.03 | −0.3151 | 2.6024 | 2.2873 | 9 |
| 10 | 0.030894 | 0.062901 | 0.01744 | 9054 | 9054 | −154.18 | 1219.64 | 1065.47 | −0.3141 | 2.5968 | 2.2827 | 10 |
| 11 | 0.032480 | 0.066131 | 0.01744 | 8630 | 8630 | −153.70 | 1219.60 | 1065.91 | −0.3130 | 2.5912 | 2.2782 | 11 |
| 12 | 0.034140 | 0.069511 | 0.01744 | 8228 | 8228 | −153.21 | 1219.56 | 1066.35 | −0.3120 | 2.5856 | 2.2736 | 12 |
| 13 | 0.035878 | 0.073047 | 0.01745 | 7846 | 7846 | −152.73 | 1219.52 | 1066.79 | −0.3110 | 2.5801 | 2.2691 | 13 |
| 14 | 0.037696 | 0.076748 | 0.01745 | 7483 | 7483 | −152.24 | 1219.47 | 1067.23 | −0.3100 | 2.5745 | 2.2645 | 14 |
| 15 | 0.039597 | 0.080621 | 0.01745 | 7139 | 7139 | −151.76 | 1219.43 | 1067.67 | −0.3089 | 2.5690 | 2.2600 | 15 |
| 16 | 0.041586 | 0.084671 | 0.01745 | 6811 | 6811 | −151.27 | 1219.38 | 1068.11 | −0.3079 | 2.5635 | 2.2556 | 16 |
| 17 | 0.043666 | 0.088905 | 0.01745 | 6501 | 6501 | −150.78 | 1219.33 | 1068.55 | −0.3069 | 2.5580 | 2.2511 | 17 |
| 18 | 0.045841 | 0.093332 | 0.01745 | 6205 | 6205 | −150.30 | 1219.28 | 1068.99 | −0.3059 | 2.5526 | 2.2467 | 18 |
| 19 | 0.048113 | 0.097960 | 0.01745 | 5924 | 5924 | −149.81 | 1219.23 | 1069.43 | −0.3049 | 2.5471 | 2.2423 | 19 |
| 20 | 0.050489 | 0.102796 | 0.01746 | 5657 | 5657 | −149.32 | 1219.18 | 1069.87 | −0.3038 | 2.5417 | 2.2379 | 20 |
| 21 | 0.052970 | 0.107849 | 0.01746 | 5404 | 5404 | −148.82 | 1219.13 | 1070.31 | −0.3028 | 2.5363 | 2.2335 | 21 |
| 22 | 0.055563 | 0.113128 | 0.01746 | 5162 | 5162 | −148.33 | 1219.08 | 1070.75 | −0.3018 | 2.5309 | 2.2292 | 22 |
| 23 | 0.058271 | 0.118641 | 0.01746 | 4932 | 4932 | −147.84 | 1219.02 | 1071.19 | −0.3008 | 2.5256 | 2.2248 | 23 |
| 24 | 0.061099 | 0.124398 | 0.01746 | 4714 | 4714 | −147.34 | 1218.97 | 1071.63 | −0.2997 | 2.5203 | 2.2205 | 24 |
| 25 | 0.064051 | 0.130408 | 0.01746 | 4506 | 4506 | −146.85 | 1218.91 | 1072.07 | −0.2987 | 2.5149 | 2.2162 | 25 |
| 26 | 0.067133 | 0.136684 | 0.01747 | 4308 | 4308 | −146.35 | 1218.85 | 1072.50 | −0.2977 | 2.5096 | 2.2119 | 26 |
| 27 | 0.070349 | 0.143233 | 0.01747 | 4119 | 4119 | −145.85 | 1218.80 | 1072.94 | −0.2967 | 2.5044 | 2.2077 | 27 |
| 28 | 0.073706 | 0.150066 | 0.01747 | 3940 | 3940 | −145.35 | 1218.74 | 1073.38 | −0.2956 | 2.4991 | 2.2035 | 28 |
| 29 | 0.077207 | 0.157195 | 0.01747 | 3769 | 3769 | −144.85 | 1218.68 | 1073.82 | −0.2946 | 2.4939 | 2.1992 | 29 |
| 30 | 0.080860 | 0.164632 | 0.01747 | 3606 | 3606 | −144.35 | 1218.61 | 1074.26 | −0.2936 | 2.4886 | 2.1951 | 30 |
| 31 | 0.084669 | 0.172387 | 0.01747 | 3450 | 3450 | −143.85 | 1218.55 | 1074.70 | −0.2926 | 2.4834 | 2.1909 | 31 |
| 32 | 0.088640 | 0.180474 | 0.01747 | 3302 | 3302 | −143.35 | 1218.49 | 1075.14 | −0.2915 | 2.4783 | 2.1867 | 32 |
| 32* | 0.08865 | 0.18049 | 0.01602 | 3302.07 | 3302.09 | −0.02 | 1075.15 | 1075.14 | 0.0000 | 2.1867 | 2.1867 | 32 |
| 33 | 0.09229 | 0.18791 | 0.01602 | 3178.15 | 3178.16 | 0.99 | 1074.59 | 1075.58 | 0.0020 | 2.1811 | 2.1832 | 33 |
| 34 | 0.09607 | 0.19559 | 0.01602 | 3059.47 | 3059.49 | 2.00 | 1074.02 | 1076.01 | 0.0041 | 2.1756 | 2.1796 | 34 |
| 35 | 0.09998 | 0.20355 | 0.01602 | 2945.66 | 2945.68 | 3.00 | 1073.45 | 1076.45 | 0.0061 | 2.1700 | 2.1761 | 35 |
| 36 | 0.10403 | 0.21180 | 0.01602 | 2836.60 | 2836.61 | 4.01 | 1072.88 | 1076.89 | 0.0081 | 2.1645 | 2.1726 | 36 |
| 37 | 0.10822 | 0.22035 | 0.01602 | 2732.13 | 2732.15 | 5.02 | 1072.32 | 1077.33 | 0.0102 | 2.1590 | 2.1692 | 37 |
| 38 | 0.11257 | 0.22919 | 0.01602 | 2631.88 | 2631.89 | 6.02 | 1071.75 | 1077.77 | 0.0122 | 2.1535 | 2.1657 | 38 |
| 39 | 0.11707 | 0.23835 | 0.01602 | 2535.86 | 2535.88 | 7.03 | 1071.18 | 1078.21 | 0.0142 | 2.1481 | 2.1623 | 39 |
| 40 | 0.12172 | 0.24783 | 0.01602 | 2443.67 | 2443.69 | 8.03 | 1070.62 | 1078.65 | 0.0162 | 2.1426 | 2.1589 | 40 |
| 41 | 0.12654 | 0.25765 | 0.01602 | 2355.22 | 2355.24 | 9.04 | 1070.05 | 1079.09 | 0.0182 | 2.1372 | 2.1554 | 41 |
| 42 | 0.13153 | 0.26780 | 0.01602 | 2270.42 | 2270.43 | 10.04 | 1069.48 | 1079.52 | 0.0202 | 2.1318 | 2.1521 | 42 |
| 43 | 0.13669 | 0.27831 | 0.01602 | 2189.02 | 2189.04 | 11.04 | 1068.92 | 1079.96 | 0.0222 | 2.1265 | 2.1487 | 43 |
| 44 | 0.14203 | 0.28918 | 0.01602 | 2110.92 | 2110.94 | 12.05 | 1068.35 | 1080.40 | 0.0242 | 2.1211 | 2.1454 | 44 |
| 45 | 0.14755 | 0.30042 | 0.01602 | 2035.91 | 2035.92 | 13.05 | 1067.79 | 1080.84 | 0.0262 | 2.1158 | 2.1420 | 45 |
| 46 | 0.15326 | 0.31205 | 0.01602 | 1963.85 | 1963.87 | 14.05 | 1067.22 | 1081.28 | 0.0282 | 2.1105 | 2.1387 | 46 |
| 47 | 0.15917 | 0.32407 | 0.01602 | 1894.71 | 1894.73 | 15.06 | 1066.66 | 1081.71 | 0.0302 | 2.1052 | 2.1354 | 47 |
| 48 | 0.16527 | 0.33650 | 0.01602 | 1828.28 | 1828.30 | 16.06 | 1066.09 | 1082.15 | 0.0321 | 2.1000 | 2.1321 | 48 |
| 49 | 0.17158 | 0.34935 | 0.01602 | 1764.44 | 1764.46 | 17.06 | 1065.53 | 1082.59 | 0.0341 | 2.0947 | 2.1288 | 49 |
| 50 | 0.17811 | 0.36263 | 0.01602 | 1703.18 | 1703.20 | 18.06 | 1064.96 | 1083.03 | 0.0361 | 2.0895 | 2.1256 | 50 |
| 51 | 0.18484 | 0.37635 | 0.01602 | 1644.25 | 1644.26 | 19.06 | 1064.40 | 1083.46 | 0.0381 | 2.0843 | 2.1224 | 51 |
| 52 | 0.19181 | 0.39053 | 0.01603 | 1587.64 | 1587.65 | 20.07 | 1063.83 | 1083.90 | 0.0400 | 2.0791 | 2.1191 | 52 |

*Extrapolated to represent metastable equilibrium with undercooled liquid.

## Table 2-1  Thermodynamic Properties of Water (*Continued*)

*(Table 3, Chapter 1, 2013 ASHRAE Handbook—Fundamentals)*

| Temp., °F t | Absolute Pressure p, psi | Absolute Pressure p, in. Hg | Specific Volume, ft³/lb_w Sat. Liquid v_f | Specific Volume, ft³/lb_w Evap. v_fg | Specific Volume, ft³/lb_w Sat. Vapor v_g | Specific Enthalpy, Btu/lb_w Sat. Liquid h_f | Specific Enthalpy, Btu/lb_w Evap. h_fg | Specific Enthalpy, Btu/lb_w Sat. Vapor h_g | Specific Entropy, Btu/lb_w·°F Sat. Liquid s_f | Specific Entropy, Btu/lb_w·°F Evap. s_fg | Specific Entropy, Btu/lb_w·°F Sat. Vapor s_g | Temp., °F t |
|---|---|---|---|---|---|---|---|---|---|---|---|---|
| 53 | 0.19900 | 0.40516 | 0.01603 | 1533.22 | 1533.24 | 21.07 | 1063.27 | 1084.34 | 0.0420 | 2.0740 | 2.1159 | 53 |
| 54 | 0.20643 | 0.42029 | 0.01603 | 1480.89 | 1480.91 | 22.07 | 1062.71 | 1084.77 | 0.0439 | 2.0689 | 2.1128 | 54 |
| 55 | 0.21410 | 0.43591 | 0.01603 | 1430.61 | 1430.62 | 23.07 | 1062.14 | 1085.21 | 0.0459 | 2.0637 | 2.1096 | 55 |
| 56 | 0.22202 | 0.45204 | 0.01603 | 1382.19 | 1382.21 | 24.07 | 1061.58 | 1085.65 | 0.0478 | 2.0586 | 2.1064 | 56 |
| 57 | 0.23020 | 0.46869 | 0.01603 | 1335.65 | 1335.67 | 25.07 | 1061.01 | 1086.08 | 0.0497 | 2.0536 | 2.1033 | 57 |
| 58 | 0.23864 | 0.48588 | 0.01603 | 1290.85 | 1290.87 | 26.07 | 1060.45 | 1086.52 | 0.0517 | 2.0485 | 2.1002 | 58 |
| 59 | 0.24735 | 0.50362 | 0.01603 | 1247.76 | 1247.78 | 27.07 | 1059.89 | 1086.96 | 0.0536 | 2.0435 | 2.0971 | 59 |
| 60 | 0.25635 | 0.52192 | 0.01604 | 1206.30 | 1206.32 | 28.07 | 1059.32 | 1087.39 | 0.0555 | 2.0385 | 2.0940 | 60 |
| 61 | 0.26562 | 0.54081 | 0.01604 | 1166.38 | 1166.40 | 29.07 | 1058.76 | 1087.83 | 0.0575 | 2.0334 | 2.0909 | 61 |
| 62 | 0.27519 | 0.56029 | 0.01604 | 1127.93 | 1127.95 | 30.07 | 1058.19 | 1088.27 | 0.0594 | 2.0285 | 2.0878 | 62 |
| 63 | 0.28506 | 0.58039 | 0.01604 | 1090.94 | 1090.96 | 31.07 | 1057.63 | 1088.70 | 0.0613 | 2.0235 | 2.0848 | 63 |
| 64 | 0.29524 | 0.60112 | 0.01604 | 1055.32 | 1055.33 | 32.07 | 1057.07 | 1089.14 | 0.0632 | 2.0186 | 2.0818 | 64 |
| 65 | 0.30574 | 0.62249 | 0.01604 | 1020.98 | 1021.00 | 33.07 | 1056.50 | 1089.57 | 0.0651 | 2.0136 | 2.0787 | 65 |
| 66 | 0.31656 | 0.64452 | 0.01604 | 987.95 | 987.97 | 34.07 | 1055.94 | 1090.01 | 0.0670 | 2.0087 | 2.0758 | 66 |
| 67 | 0.32772 | 0.66724 | 0.01605 | 956.11 | 956.12 | 35.07 | 1055.37 | 1090.44 | 0.0689 | 2.0039 | 2.0728 | 67 |
| 68 | 0.33921 | 0.69065 | 0.01605 | 925.44 | 925.45 | 36.07 | 1054.81 | 1090.88 | 0.0708 | 1.9990 | 2.0698 | 68 |
| 69 | 0.35107 | 0.71478 | 0.01605 | 895.86 | 895.87 | 37.07 | 1054.24 | 1091.31 | 0.0727 | 1.9941 | 2.0668 | 69 |
| 70 | 0.36328 | 0.73964 | 0.01605 | 867.34 | 867.36 | 38.07 | 1053.68 | 1091.75 | 0.0746 | 1.9893 | 2.0639 | 70 |
| 71 | 0.37586 | 0.76526 | 0.01605 | 839.87 | 839.88 | 39.07 | 1053.11 | 1092.18 | 0.0765 | 1.9845 | 2.0610 | 71 |
| 72 | 0.38882 | 0.79164 | 0.01606 | 813.37 | 813.39 | 40.07 | 1052.55 | 1092.61 | 0.0783 | 1.9797 | 2.0580 | 72 |
| 73 | 0.40217 | 0.81883 | 0.01606 | 787.85 | 787.87 | 41.07 | 1051.98 | 1093.05 | 0.0802 | 1.9749 | 2.0552 | 73 |
| 74 | 0.41592 | 0.84682 | 0.01606 | 763.19 | 763.21 | 42.06 | 1051.42 | 1093.48 | 0.0821 | 1.9702 | 2.0523 | 74 |
| 75 | 0.43008 | 0.87564 | 0.01606 | 739.42 | 739.44 | 43.06 | 1050.85 | 1093.92 | 0.0840 | 1.9654 | 2.0494 | 75 |
| 76 | 0.44465 | 0.90532 | 0.01606 | 716.51 | 726.53 | 44.06 | 1050.29 | 1094.35 | 0.0858 | 1.9607 | 2.0465 | 76 |
| 77 | 0.45966 | 0.93587 | 0.01607 | 694.38 | 699.80 | 45.06 | 1049.72 | 1094.78 | 0.0877 | 1.9560 | 2.0437 | 77 |
| 78 | 0.47510 | 0.96732 | 0.01607 | 673.05 | 673.06 | 46.06 | 1049.16 | 1095.22 | 0.0896 | 1.9513 | 2.0409 | 78 |
| 79 | 0.49100 | 0.99968 | 0.01607 | 652.44 | 652.46 | 47.06 | 1048.59 | 1095.65 | 0.0914 | 1.9466 | 2.0380 | 79 |
| 80 | 0.50736 | 1.03298 | 0.01607 | 632.54 | 632.56 | 48.06 | 1048.03 | 1096.08 | 0.0933 | 1.9420 | 2.0352 | 80 |
| 81 | 0.52419 | 1.06725 | 0.01608 | 613.35 | 613.37 | 49.06 | 1047.46 | 1096.51 | 0.0951 | 1.9373 | 2.0324 | 81 |
| 82 | 0.54150 | 1.10250 | 0.01608 | 594.82 | 594.84 | 50.05 | 1046.89 | 1096.95 | 0.0970 | 1.9327 | 2.0297 | 82 |
| 83 | 0.55931 | 1.13877 | 0.01608 | 576.90 | 576.92 | 51.05 | 1046.33 | 1097.38 | 0.0988 | 1.9281 | 2.0269 | 83 |
| 84 | 0.57763 | 1.17606 | 0.01608 | 559.63 | 559.65 | 52.05 | 1045.76 | 1097.81 | 0.1006 | 1.9235 | 2.0242 | 84 |
| 85 | 0.59647 | 1.21442 | 0.01609 | 542.93 | 542.94 | 53.05 | 1045.19 | 1098.24 | 0.1025 | 1.9189 | 2.0214 | 85 |
| 86 | 0.61584 | 1.25385 | 0.01609 | 526.80 | 526.81 | 54.05 | 1044.63 | 1098.67 | 0.1043 | 1.9144 | 2.0187 | 86 |
| 87 | 0.63575 | 1.29440 | 0.01609 | 511.21 | 511.22 | 55.05 | 1044.06 | 1099.11 | 0.1061 | 1.9098 | 2.0160 | 87 |
| 88 | 0.65622 | 1.33608 | 0.01609 | 496.14 | 496.15 | 56.05 | 1043.49 | 1099.54 | 0.1080 | 1.9053 | 2.0133 | 88 |
| 89 | 0.67726 | 1.37892 | 0.01610 | 481.60 | 481.61 | 57.04 | 1042.92 | 1099.97 | 0.1098 | 1.9008 | 2.0106 | 89 |
| 90 | 0.69889 | 1.42295 | 0.01610 | 467.52 | 467.53 | 58.04 | 1042.36 | 1100.40 | 0.1116 | 1.8963 | 2.0079 | 90 |
| 91 | 0.72111 | 1.46820 | 0.01610 | 453.91 | 453.93 | 59.04 | 1041.79 | 1100.83 | 0.1134 | 1.8918 | 2.0053 | 91 |
| 92 | 0.74394 | 1.51468 | 0.01611 | 440.76 | 440.78 | 60.04 | 1041.22 | 1101.26 | 0.1152 | 1.8874 | 2.0026 | 92 |
| 93 | 0.76740 | 1.56244 | 0.01611 | 428.04 | 428.06 | 61.04 | 1040.65 | 1101.69 | 0.1170 | 1.8829 | 2.0000 | 93 |
| 94 | 0.79150 | 1.61151 | 0.01611 | 415.74 | 415.76 | 62.04 | 1040.08 | 1102.12 | 0.1188 | 1.8785 | 1.9973 | 94 |
| 95 | 0.81625 | 1.66189 | 0.01612 | 403.84 | 403.86 | 63.03 | 1039.51 | 1102.55 | 0.1206 | 1.8741 | 1.9947 | 95 |
| 96 | 0.84166 | 1.71364 | 0.01612 | 392.33 | 392.34 | 64.03 | 1038.95 | 1102.98 | 0.1224 | 1.8697 | 1.9921 | 96 |
| 97 | 0.86776 | 1.76678 | 0.01612 | 381.20 | 381.21 | 65.03 | 1038.38 | 1103.41 | 0.1242 | 1.8653 | 1.9895 | 97 |
| 98 | 0.89456 | 1.82134 | 0.01612 | 370.42 | 370.44 | 66.03 | 1037.81 | 1103.84 | 0.1260 | 1.8610 | 1.9870 | 98 |
| 99 | 0.92207 | 1.87736 | 0.01613 | 359.99 | 360.01 | 67.03 | 1037.24 | 1104.26 | 0.1278 | 1.8566 | 1.9844 | 99 |
| 100 | 0.95031 | 1.93485 | 0.01613 | 349.91 | 349.92 | 68.03 | 1036.67 | 1104.69 | 0.1296 | 1.8523 | 1.9819 | 100 |
| 101 | 0.97930 | 1.99387 | 0.01613 | 340.14 | 340.15 | 69.03 | 1036.10 | 1105.12 | 0.1314 | 1.8479 | 1.9793 | 101 |
| 102 | 1.00904 | 2.05443 | 0.01614 | 330.69 | 330.71 | 70.02 | 1035.53 | 1105.55 | 0.1332 | 1.8436 | 1.9768 | 102 |
| 103 | 1.03956 | 2.11667 | 0.01614 | 321.53 | 321.55 | 71.02 | 1034.95 | 1105.98 | 0.1349 | 1.8393 | 1.9743 | 103 |
| 104 | 1.07088 | 2.18034 | 0.01614 | 312.67 | 312.69 | 72.02 | 1034.38 | 1106.40 | 0.1367 | 1.8351 | 1.9718 | 104 |
| 105 | 1.10301 | 2.24575 | 0.01615 | 304.08 | 304.10 | 73.02 | 1033.81 | 1106.83 | 0.1385 | 1.8308 | 1.9693 | 105 |
| 106 | 1.13597 | 2.31285 | 0.01615 | 295.76 | 295.77 | 74.02 | 1033.24 | 1107.26 | 0.1402 | 1.8266 | 1.9668 | 106 |
| 107 | 1.16977 | 2.38168 | 0.01616 | 287.71 | 287.73 | 75.01 | 1032.67 | 1107.68 | 0.1420 | 1.8223 | 1.9643 | 107 |
| 108 | 1.20444 | 2.45226 | 0.01616 | 279.91 | 279.92 | 76.01 | 1032.10 | 1108.11 | 0.1438 | 1.8181 | 1.9619 | 108 |
| 109 | 1.23999 | 2.52464 | 0.01616 | 272.34 | 272.36 | 77.01 | 1031.52 | 1108.54 | 0.1455 | 1.8139 | 1.9594 | 109 |
| 110 | 1.27644 | 2.59885 | 0.01617 | 265.02 | 265.03 | 78.01 | 1030.95 | 1108.96 | 0.1473 | 1.8097 | 1.9570 | 110 |
| 111 | 1.31381 | 2.67494 | 0.01617 | 257.91 | 257.93 | 79.01 | 1030.38 | 1109.39 | 0.1490 | 1.8055 | 1.9546 | 111 |
| 112 | 1.35212 | 2.75293 | 0.01617 | 251.02 | 251.04 | 80.01 | 1029.80 | 1109.81 | 0.1508 | 1.8014 | 1.9521 | 112 |
| 113 | 1.39138 | 2.83288 | 0.01618 | 244.36 | 244.38 | 81.01 | 1029.23 | 1110.24 | 0.1525 | 1.7972 | 1.9497 | 113 |
| 114 | 1.43162 | 2.91481 | 0.01618 | 237.89 | 237.90 | 82.00 | 1028.66 | 1110.66 | 0.1543 | 1.7931 | 1.9474 | 114 |
| 115 | 1.47286 | 2.99878 | 0.01619 | 231.62 | 231.63 | 83.00 | 1028.08 | 1111.09 | 0.1560 | 1.7890 | 1.9450 | 115 |
| 116 | 1.51512 | 3.08481 | 0.01619 | 225.53 | 225.55 | 84.00 | 1027.51 | 1111.51 | 0.1577 | 1.7849 | 1.9426 | 116 |
| 117 | 1.55842 | 3.17296 | 0.01619 | 219.63 | 219.65 | 85.00 | 1026.93 | 1111.93 | 0.1595 | 1.7808 | 1.9402 | 117 |
| 118 | 1.60277 | 3.26327 | 0.01620 | 213.91 | 213.93 | 86.00 | 1026.36 | 1112.36 | 0.1612 | 1.7767 | 1.9379 | 118 |
| 119 | 1.64820 | 3.35577 | 0.01620 | 208.36 | 208.37 | 87.00 | 1025.78 | 1112.78 | 0.1629 | 1.7726 | 1.9356 | 119 |
| 120 | 1.69474 | 3.45052 | 0.01620 | 202.98 | 202.99 | 88.00 | 1025.20 | 1113.20 | 0.1647 | 1.7686 | 1.9332 | 120 |

## Table 2-1 Thermodynamic Properties of Water (*Continued*)
*(Table 3, Chapter 1, 2013 ASHRAE Handbook—Fundamentals)*

| Temp., °F t | Absolute Pressure p, psi | p, in. Hg | Sat. Liquid $v_f$ | Evap. $v_{fg}$ | Sat. Vapor $v_g$ | Sat. Liquid $h_f$ | Evap. $h_{fg}$ | Sat. Vapor $h_g$ | Sat. Liquid $s_f$ | Evap. $s_{fg}$ | Sat. Vapor $s_g$ | Temp., °F t |
|---|---|---|---|---|---|---|---|---|---|---|---|---|
| 121 | 1.74240 | 3.54755 | 0.01621 | 197.76 | 197.76 | 89.00 | 1024.63 | 1113.62 | 0.1664 | 1.7645 | 1.9309 | 121 |
| 122 | 1.79117 | 3.64691 | 0.01621 | 192.69 | 192.69 | 90.00 | 1024.05 | 1114.05 | 0.1681 | 1.7605 | 1.9286 | 122 |
| 123 | 1.84117 | 3.74863 | 0.01622 | 187.78 | 187.78 | 90.99 | 1023.47 | 1114.47 | 0.1698 | 1.7565 | 1.9263 | 123 |
| 124 | 1.89233 | 3.85282 | 0.01622 | 182.98 | 182.99 | 91.99 | 1022.90 | 1114.89 | 0.1715 | 1.7525 | 1.9240 | 124 |
| 125 | 1.94470 | 3.95945 | 0.01623 | 178.34 | 178.36 | 92.99 | 1022.32 | 1115.31 | 0.1732 | 1.7485 | 1.9217 | 125 |
| 126 | 1.99831 | 4.06860 | 0.01623 | 173.85 | 173.86 | 93.99 | 1021.74 | 1115.73 | 0.1749 | 1.7445 | 1.9195 | 126 |
| 127 | 2.05318 | 4.18032 | 0.01623 | 169.47 | 169.49 | 94.99 | 1021.16 | 1116.15 | 0.1766 | 1.7406 | 1.9172 | 127 |
| 128 | 2.10934 | 4.29465 | 0.01624 | 165.23 | 165.25 | 95.99 | 1020.58 | 1116.57 | 0.1783 | 1.7366 | 1.9150 | 128 |
| 129 | 2.16680 | 4.41165 | 0.01624 | 161.11 | 161.12 | 96.99 | 1020.00 | 1116.99 | 0.1800 | 1.7327 | 1.9127 | 129 |
| 130 | 2.22560 | 4.53136 | 0.01625 | 157.11 | 157.12 | 97.99 | 1019.42 | 1117.41 | 0.1817 | 1.7288 | 1.9105 | 130 |
| 131 | 2.28576 | 4.65384 | 0.01625 | 153.22 | 153.23 | 98.99 | 1018.84 | 1117.83 | 0.1834 | 1.7249 | 1.9083 | 131 |
| 132 | 2.34730 | 4.77914 | 0.01626 | 149.44 | 149.46 | 99.99 | 1018.26 | 1118.25 | 0.1851 | 1.7210 | 1.9061 | 132 |
| 133 | 2.41025 | 4.90730 | 0.01626 | 145.77 | 145.78 | 100.99 | 1017.68 | 1118.67 | 0.1868 | 1.7171 | 1.9039 | 133 |
| 134 | 2.47463 | 5.03839 | 0.01627 | 142.21 | 142.23 | 101.99 | 1017.10 | 1119.08 | 0.1885 | 1.7132 | 1.9017 | 134 |
| 135 | 2.54048 | 5.17246 | 0.01627 | 138.74 | 138.76 | 102.99 | 1016.52 | 1119.50 | 0.1902 | 1.7093 | 1.8995 | 135 |
| 136 | 2.60782 | 5.30956 | 0.01627 | 135.37 | 135.39 | 103.98 | 1015.93 | 1119.92 | 0.1919 | 1.7055 | 1.8974 | 136 |
| 137 | 2.67667 | 5.44975 | 0.01628 | 132.10 | 132.12 | 104.98 | 1015.35 | 1120.34 | 0.1935 | 1.7017 | 1.8952 | 137 |
| 138 | 2.74707 | 5.59308 | 0.01628 | 128.92 | 128.94 | 105.98 | 1014.77 | 1120.75 | 0.1952 | 1.6978 | 1.8930 | 138 |
| 139 | 2.81903 | 5.73961 | 0.01629 | 125.83 | 125.85 | 106.98 | 1014.18 | 1121.17 | 0.1969 | 1.6940 | 1.8909 | 139 |
| 140 | 2.89260 | 5.88939 | 0.01629 | 122.82 | 122.84 | 107.98 | 1013.60 | 1121.58 | 0.1985 | 1.6902 | 1.8888 | 140 |
| 141 | 2.96780 | 6.04250 | 0.01630 | 119.90 | 119.92 | 108.98 | 1013.01 | 1122.00 | 0.2002 | 1.6864 | 1.8867 | 141 |
| 142 | 3.04465 | 6.19897 | 0.01630 | 117.05 | 117.07 | 109.98 | 1012.43 | 1122.41 | 0.2019 | 1.6827 | 1.8845 | 142 |
| 143 | 3.12320 | 6.35888 | 0.01631 | 114.29 | 114.31 | 110.98 | 1011.84 | 1122.83 | 0.2035 | 1.6789 | 1.8824 | 143 |
| 144 | 3.20345 | 6.52229 | 0.01631 | 111.60 | 111.62 | 111.98 | 1011.26 | 1123.24 | 0.2052 | 1.6752 | 1.8803 | 144 |
| 145 | 3.28546 | 6.68926 | 0.01632 | 108.99 | 109.00 | 112.98 | 1010.67 | 1123.66 | 0.2068 | 1.6714 | 1.8783 | 145 |
| 146 | 3.36924 | 6.85984 | 0.01632 | 106.44 | 106.45 | 113.98 | 1010.09 | 1124.07 | 0.2085 | 1.6677 | 1.8762 | 146 |
| 147 | 3.45483 | 7.03410 | 0.01633 | 103.96 | 103.98 | 114.98 | 1009.50 | 1124.48 | 0.2101 | 1.6640 | 1.8741 | 147 |
| 148 | 3.54226 | 7.21211 | 0.01633 | 101.55 | 101.57 | 115.98 | 1008.91 | 1124.89 | 0.2118 | 1.6603 | 1.8721 | 148 |
| 149 | 3.63156 | 7.39393 | 0.01634 | 99.21 | 99.22 | 116.98 | 1008.32 | 1125.31 | 0.2134 | 1.6566 | 1.8700 | 149 |
| 150 | 3.72277 | 7.57962 | 0.01634 | 96.93 | 96.94 | 117.98 | 1007.73 | 1125.72 | 0.2151 | 1.6529 | 1.8680 | 150 |
| 151 | 3.81591 | 7.76925 | 0.01635 | 94.70 | 94.72 | 118.99 | 1007.14 | 1126.13 | 0.2167 | 1.6492 | 1.8659 | 151 |
| 152 | 3.91101 | 7.96289 | 0.01635 | 92.54 | 92.56 | 119.99 | 1006.55 | 1126.54 | 0.2184 | 1.6455 | 1.8639 | 152 |
| 153 | 4.00812 | 8.16061 | 0.01636 | 90.44 | 90.46 | 120.99 | 1005.96 | 1126.95 | 0.2200 | 1.6419 | 1.8619 | 153 |
| 154 | 4.10727 | 8.36247 | 0.01636 | 88.39 | 88.41 | 121.99 | 1005.37 | 1127.36 | 0.2216 | 1.6383 | 1.8599 | 154 |
| 155 | 4.20848 | 8.56854 | 0.01637 | 86.40 | 86.41 | 122.99 | 1004.78 | 1127.77 | 0.2233 | 1.6346 | 1.8579 | 155 |
| 156 | 4.31180 | 8.77890 | 0.01637 | 84.45 | 84.47 | 123.99 | 1004.19 | 1128.18 | 0.2249 | 1.6310 | 1.8559 | 156 |
| 157 | 4.41725 | 8.99360 | 0.01638 | 82.56 | 82.58 | 124.99 | 1003.60 | 1128.59 | 0.2265 | 1.6274 | 1.8539 | 157 |
| 158 | 4.52488 | 9.21274 | 0.01638 | 80.72 | 80.73 | 125.99 | 1003.00 | 1128.99 | 0.2281 | 1.6238 | 1.8519 | 158 |
| 159 | 4.63472 | 9.43637 | 0.01639 | 78.92 | 78.94 | 126.99 | 1002.41 | 1129.40 | 0.2297 | 1.6202 | 1.8500 | 159 |
| 160 | 4.7468 | 9.6646 | 0.01639 | 77.175 | 77.192 | 127.99 | 1001.82 | 1129.81 | 0.2314 | 1.6167 | 1.8480 | 160 |
| 161 | 4.8612 | 9.8974 | 0.01640 | 75.471 | 75.488 | 128.99 | 1001.22 | 1130.22 | 0.2330 | 1.6131 | 1.8461 | 161 |
| 162 | 4.9778 | 10.1350 | 0.01640 | 73.812 | 73.829 | 130.00 | 1000.63 | 1130.62 | 0.2346 | 1.6095 | 1.8441 | 162 |
| 163 | 5.0969 | 10.3774 | 0.01641 | 72.196 | 72.213 | 131.00 | 1000.03 | 1131.03 | 0.2362 | 1.6060 | 1.8422 | 163 |
| 164 | 5.2183 | 10.6246 | 0.01642 | 70.619 | 70.636 | 132.00 | 999.43 | 1131.43 | 0.2378 | 1.6025 | 1.8403 | 164 |
| 165 | 5.3422 | 10.8768 | 0.01642 | 69.084 | 69.101 | 133.00 | 998.84 | 1131.84 | 0.2394 | 1.5989 | 1.8383 | 165 |
| 166 | 5.4685 | 11.1340 | 0.01643 | 67.587 | 67.604 | 134.00 | 998.24 | 1132.24 | 0.2410 | 1.5954 | 1.8364 | 166 |
| 167 | 5.5974 | 11.3963 | 0.01643 | 66.130 | 66.146 | 135.00 | 997.64 | 1132.64 | 0.2426 | 1.5919 | 1.8345 | 167 |
| 168 | 5.7287 | 11.6638 | 0.01644 | 64.707 | 64.723 | 136.01 | 997.04 | 1133.05 | 0.2442 | 1.5884 | 1.8326 | 168 |
| 169 | 5.8627 | 11.9366 | 0.01644 | 63.320 | 63.336 | 137.01 | 996.44 | 1133.45 | 0.2458 | 1.5850 | 1.8308 | 169 |
| 170 | 5.9993 | 12.2148 | 0.01645 | 61.969 | 61.986 | 138.01 | 995.84 | 1133.85 | 0.2474 | 1.5815 | 1.8289 | 170 |
| 171 | 6.1386 | 12.4983 | 0.01646 | 60.649 | 60.666 | 139.01 | 995.24 | 1134.25 | 0.2490 | 1.5780 | 1.8270 | 171 |
| 172 | 6.2806 | 12.7874 | 0.01646 | 59.363 | 59.380 | 140.01 | 994.64 | 1134.66 | 0.2506 | 1.5746 | 1.8251 | 172 |
| 173 | 6.4253 | 13.0821 | 0.01647 | 58.112 | 58.128 | 141.02 | 994.04 | 1135.06 | 0.2521 | 1.5711 | 1.8233 | 173 |
| 174 | 6.5729 | 13.3825 | 0.01647 | 56.887 | 56.904 | 142.02 | 993.44 | 1135.46 | 0.2537 | 1.5677 | 1.8214 | 174 |
| 175 | 6.7232 | 13.6886 | 0.01648 | 55.694 | 55.711 | 143.02 | 992.83 | 1135.86 | 0.2553 | 1.5643 | 1.8196 | 175 |
| 176 | 6.8765 | 14.0006 | 0.01648 | 54.532 | 54.549 | 144.02 | 992.23 | 1136.26 | 0.2569 | 1.5609 | 1.8178 | 176 |
| 177 | 7.0327 | 14.3186 | 0.01649 | 53.397 | 53.414 | 145.03 | 991.63 | 1136.65 | 0.2585 | 1.5575 | 1.8159 | 177 |
| 178 | 7.1918 | 14.6426 | 0.01650 | 52.290 | 52.307 | 146.03 | 991.02 | 1137.05 | 0.2600 | 1.5541 | 1.8141 | 178 |
| 179 | 7.3539 | 14.9727 | 0.01650 | 51.210 | 51.226 | 147.03 | 990.42 | 1137.45 | 0.2616 | 1.5507 | 1.8123 | 179 |
| 180 | 7.5191 | 15.3091 | 0.01651 | 50.155 | 50.171 | 148.04 | 989.81 | 1137.85 | 0.2632 | 1.5473 | 1.8105 | 180 |
| 181 | 7.6874 | 15.6518 | 0.01651 | 49.126 | 49.143 | 149.04 | 989.20 | 1138.24 | 0.2647 | 1.5440 | 1.8087 | 181 |
| 182 | 7.8589 | 16.0008 | 0.01652 | 48.122 | 48.138 | 150.04 | 988.60 | 1138.64 | 0.2663 | 1.5406 | 1.8069 | 182 |
| 183 | 8.0335 | 16.3564 | 0.01653 | 47.142 | 47.158 | 151.05 | 987.99 | 1139.03 | 0.2679 | 1.5373 | 1.8051 | 183 |
| 184 | 8.2114 | 16.7185 | 0.01653 | 46.185 | 46.202 | 152.05 | 987.38 | 1139.43 | 0.2694 | 1.5339 | 1.8034 | 184 |
| 185 | 8.3926 | 17.0874 | 0.01654 | 45.251 | 45.267 | 153.05 | 986.77 | 1139.82 | 0.2710 | 1.5306 | 1.8016 | 185 |
| 186 | 8.5770 | 17.4630 | 0.01654 | 44.339 | 44.356 | 154.06 | 986.16 | 1140.22 | 0.2725 | 1.5273 | 1.7998 | 186 |
| 187 | 8.7649 | 17.8455 | 0.01655 | 43.448 | 43.465 | 155.06 | 985.55 | 1140.61 | 0.2741 | 1.5240 | 1.7981 | 187 |
| 188 | 8.9562 | 18.2350 | 0.01656 | 42.579 | 42.595 | 156.07 | 984.94 | 1141.00 | 0.2756 | 1.5207 | 1.7963 | 188 |
| 189 | 9.1510 | 18.6316 | 0.01656 | 41.730 | 41.746 | 157.07 | 984.32 | 1141.39 | 0.2772 | 1.5174 | 1.7946 | 189 |

# Thermodynamics and Psychrometrics

### Table 2-1 Thermodynamic Properties of Water (*Continued*)
(Table 3, Chapter 1, 2013 ASHRAE Handbook—Fundamentals)

| Temp., °F $t$ | Absolute Pressure $p$, psi | $p$, in. Hg | Specific Volume, ft³/lb$_w$ Sat. Liquid $v_f$ | Evap. $v_{fg}$ | Sat. Vapor $v_g$ | Specific Enthalpy, Btu/lb$_w$ Sat. Liquid $h_f$ | Evap. $h_{fg}$ | Sat. Vapor $h_g$ | Specific Entropy, Btu/lb$_w$·°F Sat. Liquid $s_f$ | Evap. $s_{fg}$ | Sat. Vapor $s_g$ | Temp., °F $t$ |
|---|---|---|---|---|---|---|---|---|---|---|---|---|
| 190 | 9.3493 | 19.0353 | 0.01657 | 40.901 | 40.918 | 158.07 | 983.71 | 1141.78 | 0.2787 | 1.5141 | 1.7929 | 190 |
| 191 | 9.5512 | 19.4464 | 0.01658 | 40.092 | 40.108 | 159.08 | 983.10 | 1142.18 | 0.2803 | 1.5109 | 1.7911 | 191 |
| 192 | 9.7567 | 19.8648 | 0.01658 | 39.301 | 39.317 | 160.08 | 982.48 | 1142.57 | 0.2818 | 1.5076 | 1.7894 | 192 |
| 193 | 9.9659 | 20.2907 | 0.01659 | 38.528 | 38.544 | 161.09 | 981.87 | 1142.95 | 0.2834 | 1.5043 | 1.7877 | 193 |
| 194 | 10.1788 | 20.7242 | 0.01659 | 37.774 | 37.790 | 162.09 | 981.25 | 1143.34 | 0.2849 | 1.5011 | 1.7860 | 194 |
| 195 | 10.3955 | 21.1653 | 0.01660 | 37.035 | 37.052 | 163.10 | 980.63 | 1143.73 | 0.2864 | 1.4979 | 1.7843 | 195 |
| 196 | 10.6160 | 21.6143 | 0.01661 | 36.314 | 36.331 | 164.10 | 980.02 | 1144.12 | 0.2880 | 1.4946 | 1.7826 | 196 |
| 197 | 10.8404 | 22.0712 | 0.01661 | 35.611 | 35.628 | 165.11 | 979.40 | 1144.51 | 0.2895 | 1.4914 | 1.7809 | 197 |
| 198 | 11.0687 | 22.5361 | 0.01662 | 34.923 | 34.940 | 166.11 | 978.78 | 1144.89 | 0.2910 | 1.4882 | 1.7792 | 198 |
| 199 | 11.3010 | 23.0091 | 0.01663 | 34.251 | 34.268 | 167.12 | 978.16 | 1145.28 | 0.2926 | 1.4850 | 1.7776 | 199 |
| 200 | 11.5374 | 23.4904 | 0.01663 | 33.594 | 33.610 | 168.13 | 977.54 | 1145.66 | 0.2941 | 1.4818 | 1.7759 | 200 |
| 201 | 11.7779 | 23.9800 | 0.01664 | 32.951 | 32.968 | 169.13 | 976.92 | 1146.05 | 0.2956 | 1.4786 | 1.7742 | 201 |
| 202 | 12.0225 | 24.4780 | 0.01665 | 32.324 | 32.340 | 170.14 | 976.29 | 1146.43 | 0.2971 | 1.4755 | 1.7726 | 202 |
| 203 | 12.2713 | 24.9847 | 0.01665 | 31.710 | 31.726 | 171.14 | 975.67 | 1146.81 | 0.2986 | 1.4723 | 1.7709 | 203 |
| 204 | 12.5244 | 25.5000 | 0.01666 | 31.110 | 31.127 | 172.15 | 975.05 | 1147.20 | 0.3002 | 1.4691 | 1.7693 | 204 |
| 205 | 12.7819 | 26.0241 | 0.01667 | 30.523 | 30.540 | 173.16 | 974.42 | 1147.58 | 0.3017 | 1.4660 | 1.7677 | 205 |
| 206 | 13.0436 | 26.5571 | 0.01667 | 29.949 | 29.965 | 174.16 | 973.80 | 1147.96 | 0.3032 | 1.4628 | 1.7660 | 206 |
| 207 | 13.3099 | 27.0991 | 0.01668 | 29.388 | 29.404 | 175.17 | 973.17 | 1148.34 | 0.3047 | 1.4597 | 1.7644 | 207 |
| 208 | 13.5806 | 27.6503 | 0.01669 | 28.839 | 28.856 | 176.18 | 972.54 | 1148.72 | 0.3062 | 1.4566 | 1.7628 | 208 |
| 209 | 13.8558 | 28.2108 | 0.01669 | 28.303 | 28.319 | 177.18 | 971.92 | 1149.10 | 0.3077 | 1.4535 | 1.7612 | 209 |
| 210 | 14.1357 | 28.7806 | 0.01670 | 27.778 | 27.795 | 178.19 | 971.29 | 1149.48 | 0.3092 | 1.4503 | 1.7596 | 210 |
| 212 | 14.7096 | 29.9489 | 0.01671 | 26.763 | 26.780 | 180.20 | 970.03 | 1150.23 | 0.3122 | 1.4442 | 1.7564 | 212 |
| 214 | 15.3025 | 31.1563 | 0.01673 | 25.790 | 25.807 | 182.22 | 968.76 | 1150.98 | 0.3152 | 1.4380 | 1.7532 | 214 |
| 216 | 15.9152 | 32.4036 | 0.01674 | 24.861 | 24.878 | 184.24 | 967.50 | 1151.73 | 0.3182 | 1.4319 | 1.7501 | 216 |
| 218 | 16.5479 | 33.6919 | 0.01676 | 23.970 | 23.987 | 186.25 | 966.23 | 1152.48 | 0.3212 | 1.4258 | 1.7469 | 218 |
| 220 | 17.2013 | 35.0218 | 0.01677 | 23.118 | 23.134 | 188.27 | 964.95 | 1153.22 | 0.3241 | 1.4197 | 1.7438 | 220 |
| 222 | 17.8759 | 36.3956 | 0.01679 | 22.299 | 22.316 | 190.29 | 963.67 | 1153.96 | 0.3271 | 1.4136 | 1.7407 | 222 |
| 224 | 18.5721 | 37.8131 | 0.01680 | 21.516 | 21.533 | 192.31 | 962.39 | 1154.70 | 0.3301 | 1.4076 | 1.7377 | 224 |
| 226 | 19.2905 | 39.2758 | 0.01682 | 20.765 | 20.782 | 194.33 | 961.11 | 1155.43 | 0.3330 | 1.4016 | 1.7347 | 226 |
| 228 | 20.0316 | 40.7848 | 0.01683 | 20.045 | 20.062 | 196.35 | 959.82 | 1156.16 | 0.3359 | 1.3957 | 1.7316 | 228 |
| 230 | 20.7961 | 42.3412 | 0.01684 | 19.355 | 19.372 | 198.37 | 958.52 | 1156.89 | 0.3389 | 1.3898 | 1.7287 | 230 |
| 232 | 21.5843 | 43.9461 | 0.01686 | 18.692 | 18.709 | 200.39 | 957.22 | 1157.62 | 0.3418 | 1.3839 | 1.7257 | 232 |
| 234 | 22.3970 | 45.6006 | 0.01688 | 18.056 | 18.073 | 202.41 | 955.92 | 1158.34 | 0.3447 | 1.3780 | 1.7227 | 234 |
| 236 | 23.2345 | 47.3060 | 0.01689 | 17.446 | 17.463 | 204.44 | 954.62 | 1159.06 | 0.3476 | 1.3722 | 1.7198 | 236 |
| 238 | 24.0977 | 49.0633 | 0.01691 | 16.860 | 16.877 | 206.46 | 953.31 | 1159.77 | 0.3505 | 1.3664 | 1.7169 | 238 |
| 240 | 24.9869 | 50.8738 | 0.01692 | 16.298 | 16.314 | 208.49 | 952.00 | 1160.48 | 0.3534 | 1.3606 | 1.7140 | 240 |
| 242 | 25.9028 | 52.7386 | 0.01694 | 15.757 | 15.774 | 210.51 | 950.68 | 1161.19 | 0.3563 | 1.3548 | 1.7111 | 242 |
| 244 | 26.8461 | 54.6591 | 0.01695 | 15.238 | 15.255 | 212.54 | 949.35 | 1161.90 | 0.3592 | 1.3491 | 1.7083 | 244 |
| 246 | 27.8172 | 56.6364 | 0.01697 | 14.739 | 14.756 | 214.57 | 948.03 | 1162.60 | 0.3621 | 1.3434 | 1.7055 | 246 |
| 248 | 28.8169 | 58.6717 | 0.01698 | 14.259 | 14.276 | 216.60 | 946.70 | 1163.29 | 0.3649 | 1.3377 | 1.7026 | 248 |
| 250 | 29.8457 | 60.7664 | 0.01700 | 13.798 | 13.815 | 218.63 | 945.36 | 1163.99 | 0.3678 | 1.3321 | 1.6998 | 250 |
| 252 | 30.9043 | 62.9218 | 0.01702 | 13.355 | 13.372 | 220.66 | 944.02 | 1164.68 | 0.3706 | 1.3264 | 1.6971 | 252 |
| 254 | 31.9934 | 65.1391 | 0.01703 | 12.928 | 12.945 | 222.69 | 942.68 | 1165.37 | 0.3735 | 1.3208 | 1.6943 | 254 |
| 256 | 33.1135 | 67.4197 | 0.01705 | 12.526 | 12.147 | 226.73 | 939.99 | 1166.72 | 0.3764 | 1.3153 | 1.6916 | 256 |
| 258 | 34.2653 | 69.7649 | 0.01707 | 12.123 | 12.140 | 226.76 | 939.97 | 1166.73 | 0.3792 | 1.3097 | 1.6889 | 258 |
| 260 | 35.4496 | 72.1760 | 0.01708 | 11.742 | 11.759 | 228.79 | 938.61 | 1167.40 | 0.3820 | 1.3042 | 1.6862 | 260 |
| 262 | 36.6669 | 74.6545 | 0.01710 | 11.376 | 11.393 | 230.83 | 937.25 | 1168.08 | 0.3848 | 1.2987 | 1.6835 | 262 |
| 264 | 37.9180 | 77.2017 | 0.01712 | 11.024 | 11.041 | 232.87 | 935.88 | 1168.74 | 0.3876 | 1.2932 | 1.6808 | 264 |
| 266 | 39.2035 | 79.8190 | 0.01714 | 10.684 | 10.701 | 234.90 | 934.50 | 1169.41 | 0.3904 | 1.2877 | 1.6781 | 266 |
| 268 | 40.5241 | 82.5078 | 0.01715 | 10.357 | 10.374 | 236.94 | 933.12 | 1170.07 | 0.3932 | 1.2823 | 1.6755 | 268 |
| 270 | 41.8806 | 85.2697 | 0.01717 | 10.042 | 10.059 | 238.98 | 931.74 | 1170.72 | 0.3960 | 1.2769 | 1.6729 | 270 |
| 272 | 43.2736 | 88.1059 | 0.01719 | 9.737 | 9.755 | 241.03 | 930.35 | 1171.38 | 0.3988 | 1.2715 | 1.6703 | 272 |
| 274 | 44.7040 | 91.0181 | 0.01721 | 9.445 | 9.462 | 243.07 | 928.95 | 1172.02 | 0.4016 | 1.2661 | 1.6677 | 274 |
| 276 | 46.1723 | 94.0076 | 0.01722 | 9.162 | 9.179 | 245.11 | 927.55 | 1172.67 | 0.4044 | 1.2608 | 1.6651 | 276 |
| 278 | 47.6794 | 97.0761 | 0.01724 | 8.890 | 8.907 | 247.16 | 926.15 | 1173.31 | 0.4071 | 1.2554 | 1.6626 | 278 |
| 280 | 49.2260 | 100.2250 | 0.01726 | 8.627 | 8.644 | 249.20 | 924.74 | 1173.94 | 0.4099 | 1.2501 | 1.6600 | 280 |
| 282 | 50.8128 | 103.4558 | 0.01728 | 8.373 | 8.390 | 251.25 | 923.32 | 1174.57 | 0.4127 | 1.2448 | 1.6575 | 282 |
| 284 | 52.4406 | 106.7701 | 0.01730 | 8.128 | 8.146 | 253.30 | 921.90 | 1175.20 | 0.4154 | 1.2396 | 1.6550 | 284 |
| 286 | 54.1103 | 110.1695 | 0.01731 | 7.892 | 7.910 | 255.35 | 920.47 | 1175.82 | 0.4182 | 1.2343 | 1.6525 | 286 |
| 288 | 55.8225 | 113.6556 | 0.01733 | 7.664 | 7.681 | 257.40 | 919.03 | 1176.44 | 0.4209 | 1.2291 | 1.6500 | 288 |
| 290 | 57.5780 | 117.2299 | 0.01735 | 7.444 | 7.461 | 259.45 | 917.59 | 1177.05 | 0.4236 | 1.2239 | 1.6476 | 290 |
| 292 | 59.3777 | 120.8941 | 0.01737 | 7.231 | 7.248 | 261.51 | 916.15 | 1177.66 | 0.4264 | 1.2187 | 1.6451 | 292 |
| 294 | 61.2224 | 124.6498 | 0.01739 | 7.026 | 7.043 | 263.56 | 914.69 | 1178.26 | 0.4291 | 1.2136 | 1.6427 | 294 |
| 296 | 63.1128 | 128.4987 | 0.01741 | 6.827 | 6.844 | 265.62 | 913.24 | 1178.86 | 0.4318 | 1.2084 | 1.6402 | 296 |
| 298 | 65.0498 | 132.4425 | 0.01743 | 6.635 | 6.652 | 267.68 | 911.77 | 1179.45 | 0.4345 | 1.2033 | 1.6378 | 298 |
| 300 | 67.0341 | 136.4827 | 0.01745 | 6.450 | 6.467 | 269.74 | 910.30 | 1180.04 | 0.4372 | 1.1982 | 1.6354 | 300 |

The specific volume of a substance having a given quality can be found by using the definition of quality. **Quality** is defined as the ratio of the mass of vapor to total mass of liquid plus vapor when a substance is in a saturation state. Consider a mass of 1 kg having a quality $x$. The specific volume is the sum of the volume of the liquid and the volume of the vapor. The volume of the liquid is $(1-x)v_f$, and the volume of the vapor is $xv_g$. Therefore, the specific volume $v$ is

$$v = xv_g + (1-x)v_f \qquad (2\text{-}5)$$

Since $v_f + v_{fg} = v_g$, Equation 2-5 can also be written in the following form:

$$v = v_f + xv_{fg} \qquad (2\text{-}6)$$

The same procedure is followed for determining the enthalpy and the entropy for quality conditions:

$$h = xh_g + (1-x)h_f \qquad (2\text{-}7)$$

$$s = xs_g + (1-x)s_f \qquad (2\text{-}8)$$

Internal energy can then be obtained from the definition of enthalpy as $u = h - pv$.

If the substance is a compressed or subcooled liquid, the thermodynamic properties of specific volume, enthalpy, internal energy, and entropy are strongly temperature dependent (rather than pressure dependent). If compressed liquid tables are unavailable, they may be approximated by the corresponding values for saturated liquid ($v_f$, $h_f$, $u_f$, and $s_f$) at the existing temperature.

In the superheat region, thermodynamic properties must be obtained from superheat tables or a plot of the thermodynamic properties, called a Mollier diagram (Figure 2-4).

The thermodynamic and transport properties of the refrigerants used in vapor compression systems are found in similar tables typified by Table 2-2, which is a section of the R-134a property tables from Chapter 30 of the 2013 *ASHRAE Handbook—Fundamentals*. However, for these refrigerants the common Mollier plot is the *p-h* diagram as illustrated in Figure 2-5.

For fluids used in absorption refrigeration systems, the thermodynamic properties are commonly found on a different type of plot—the enthalpy-concentration diagram, as illustrated in Figure 2-6 for aqua-ammonia and in Figure 2-7 for lithium-bromide/water.

### 2.2.11 Property Equations for Ideal Gases

An ideal gas is defined as a gas at sufficiently low density so that intermolecular forces are negligible. As a result, an ideal gas has the equation of state

$$pv = RT \qquad (2\text{-}9)$$

For an ideal gas, the internal energy is a function of temperature only, which means that regardless of the pressure, an ideal gas at a given temperature has a certain definite specific internal energy $u$.

The relation between the internal energy $u$ and the temperature can be established by using the definition of constant-volume specific heat given by

$$c_v = (\partial u/\partial T)_v$$

Since the internal energy of an ideal gas is not a function of volume, an ideal gas can be written as

$$c_v = du/dT$$

$$du = c_v\, dt \qquad (2\text{-}10)$$

This equation is always valid for an ideal gas regardless of the kind of process considered.

From the definition of enthalpy and the equation of state of an ideal gas, it follows that

$$h = u + pv = u + RT$$

Since $R$ is a constant and $u$ is a function of temperature only, the enthalpy $h$ of an ideal gas is also a function of temperature only.

The relation between enthalpy and temperature is found from the constant pressure specific heat as defined by

$$c_p = (\partial h/\partial T)_p$$

Since the enthalpy of an ideal gas is a function of the temperature only, and is independent of the pressure, it follows that

$$c_p = dh/dT$$

$$dh = c_p\, dT \qquad (2\text{-}11)$$

This equation is always valid for an ideal gas regardless of the kind of process considered.

Entropy, however, remains a function of both temperature and pressure, and is given by the equation

$$ds = c_p(dT/T) - R(dp/p) \qquad (2\text{-}12)$$

where $c_p$ is frequently treated as being constant.

The ratio of heat capacities is often denoted by

$$k = c_p/c_v \qquad (2\text{-}13)$$

and is a useful quantity in calculations for ideal gases. Ideal gas values for some common gases are listed in Table 2-3.

No real gas exactly satisfies these equations over any finite range of temperature and pressure. However, all real gases approach ideal behavior at low pressures, and in the limit as $p \to 0$ do in fact meet the above requirements.

Thus, in solving problems, ideal behavior is assumed in two cases. First, at very low pressures, ideal gas behavior can be assumed with good accuracy, regardless of the temperature. Second, at temperatures that are double the critical temperature or above (the critical temperature of nitrogen is 126 K), ideal gas behavior can be assumed with good accuracy to pressures of at least 1000 lb$_f$/in$^2$ (7000 kPa). In the superheated vapor region, when the temperature is less than twice the critical temperature and the pressure is above a very

# Thermodynamics and Psychrometrics

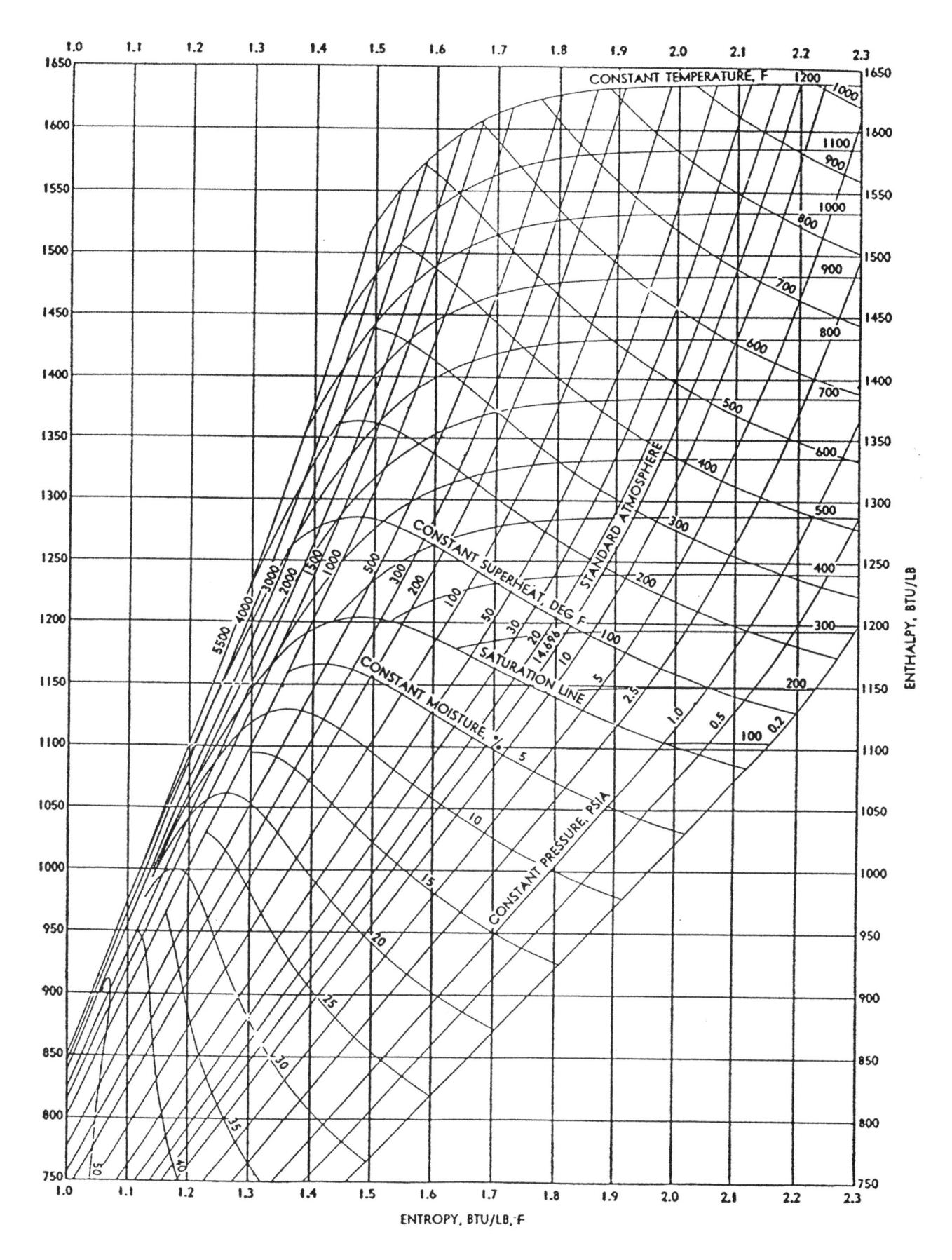

Fig. 2-4  Mollier (h,s) Diagram for Steam

## Table 2-2 Refrigerant 134a Properties of Saturated Liquid and Saturated Vapor

*(Table Refrigerant 134a, Chapter 30, 2013 ASHRAE Handbook—Fundamentals)*

| Temp.,* °F | Pressure, psia | Density, lb/ft³ Liquid | Volume, ft³/lb Vapor | Enthalpy, Btu/lb Liquid | Enthalpy, Btu/lb Vapor | Entropy, Btu/lb·°F Liquid | Entropy, Btu/lb·°F Vapor | Specific Heat $c_p$, Btu/lb·°F Liquid | Specific Heat $c_p$, Btu/lb·°F Vapor | $c_p/c_v$ Vapor | Vel. of Sound, ft/s Liquid | Vel. of Sound, ft/s Vapor | Viscosity, lb$_m$/ft·h Liquid | Viscosity, lb$_m$/ft·h Vapor | Thermal Cond., Btu/h·ft·°F Liquid | Thermal Cond., Btu/h·ft·°F Vapor | Surface Tension, dyne/cm | Temp.,* °F |
|---|---|---|---|---|---|---|---|---|---|---|---|---|---|---|---|---|---|---|
| −153.94a | 0.057 | 99.33 | 568.59 | −32.992 | 80.362 | −0.09154 | 0.27923 | 0.2829 | 0.1399 | 1.1637 | 3674. | 416.0 | 5.262 | 0.0156 | 0.0840 | 0.00178 | 28.07 | −153.94 |
| −150.00 | 0.072 | 98.97 | 452.12 | −31.878 | 80.907 | −0.08791 | 0.27629 | 0.2830 | 0.1411 | 1.1623 | 3638. | 418.3 | 4.790 | 0.0159 | 0.0832 | 0.00188 | 27.69 | −150.00 |
| −140.00 | 0.129 | 98.05 | 260.63 | −29.046 | 82.304 | −0.07891 | 0.26941 | 0.2834 | 0.1443 | 1.1589 | 3545. | 424.2 | 3.880 | 0.0164 | 0.0813 | 0.00214 | 26.74 | −140.00 |
| −130.00 | 0.221 | 97.13 | 156.50 | −26.208 | 83.725 | −0.07017 | 0.26329 | 0.2842 | 0.1475 | 1.1559 | 3452. | 429.9 | 3.238 | 0.0170 | 0.0794 | 0.00240 | 25.79 | −130.00 |
| −120.00 | 0.365 | 96.20 | 97.481 | −23.360 | 85.168 | −0.06166 | 0.25784 | 0.2853 | 0.1508 | 1.1532 | 3360. | 435.5 | 2.762 | 0.0176 | 0.0775 | 0.00265 | 24.85 | −120.00 |
| −110.00 | 0.583 | 95.27 | 62.763 | −20.500 | 86.629 | −0.05337 | 0.25300 | 0.2866 | 0.1540 | 1.1509 | 3269. | 440.8 | 2.396 | 0.0182 | 0.0757 | 0.00291 | 23.92 | −110.00 |
| −100.00 | 0.903 | 94.33 | 41.637 | −17.626 | 88.107 | −0.04527 | 0.24871 | 0.2881 | 0.1573 | 1.1490 | 3178. | 446.0 | 2.105 | 0.0187 | 0.0739 | 0.00317 | 22.99 | −100.00 |
| −90.00 | 1.359 | 93.38 | 28.381 | −14.736 | 89.599 | −0.03734 | 0.24490 | 0.2898 | 0.1607 | 1.1475 | 3087. | 450.9 | 1.869 | 0.0193 | 0.0722 | 0.00343 | 22.07 | −90.00 |
| −80.00 | 1.993 | 92.42 | 19.825 | −11.829 | 91.103 | −0.02959 | 0.24152 | 0.2916 | 0.1641 | 1.1465 | 2998. | 455.6 | 1.673 | 0.0199 | 0.0705 | 0.00369 | 21.16 | −80.00 |
| −75.00 | 2.392 | 91.94 | 16.711 | −10.368 | 91.858 | −0.02577 | 0.23998 | 0.2925 | 0.1658 | 1.1462 | 2954. | 457.8 | 1.587 | 0.0201 | 0.0696 | 0.00382 | 20.71 | −75.00 |
| −70.00 | 2.854 | 91.46 | 14.161 | −8.903 | 92.614 | −0.02198 | 0.23854 | 0.2935 | 0.1676 | 1.1460 | 2909. | 460.0 | 1.509 | 0.0204 | 0.0688 | 0.00395 | 20.26 | −70.00 |
| −65.00 | 3.389 | 90.97 | 12.060 | −7.432 | 93.372 | −0.01824 | 0.23718 | 0.2945 | 0.1694 | 1.1459 | 2866. | 462.1 | 1.436 | 0.0207 | 0.0680 | 0.00408 | 19.81 | −65.00 |
| −60.00 | 4.002 | 90.49 | 10.321 | −5.957 | 94.131 | −0.01452 | 0.23590 | 0.2955 | 0.1713 | 1.1460 | 2822. | 464.1 | 1.369 | 0.0210 | 0.0671 | 0.00420 | 19.36 | −60.00 |
| −55.00 | 4.703 | 90.00 | 8.8733 | −4.476 | 94.890 | −0.01085 | 0.23470 | 0.2965 | 0.1731 | 1.1462 | 2778. | 466.0 | 1.306 | 0.0212 | 0.0663 | 0.00433 | 18.92 | −55.00 |
| −50.00 | 5.501 | 89.50 | 7.6621 | −2.989 | 95.650 | −0.00720 | 0.23358 | 0.2976 | 0.1751 | 1.1466 | 2735. | 467.8 | 1.248 | 0.0215 | 0.0655 | 0.00446 | 18.47 | −50.00 |
| −45.00 | 6.406 | 89.00 | 6.6438 | −1.498 | 96.409 | −0.00358 | 0.23252 | 0.2987 | 0.1770 | 1.1471 | 2691. | 469.6 | 1.193 | 0.0218 | 0.0647 | 0.00460 | 18.03 | −45.00 |
| −40.00 | 7.427 | 88.50 | 5.7839 | 0.000 | 97.167 | 0.00000 | 0.23153 | 0.2999 | 0.1790 | 1.1478 | 2648. | 471.2 | 1.142 | 0.0221 | 0.0639 | 0.00473 | 17.60 | −40.00 |
| −35.00 | 8.576 | 88.00 | 5.0544 | 1.503 | 97.924 | 0.00356 | 0.23060 | 0.3010 | 0.1811 | 1.1486 | 2605. | 472.8 | 1.095 | 0.0223 | 0.0632 | 0.00486 | 17.16 | −35.00 |
| −30.00 | 9.862 | 87.49 | 4.4330 | 3.013 | 98.679 | 0.00708 | 0.22973 | 0.3022 | 0.1832 | 1.1496 | 2563. | 474.2 | 1.050 | 0.0226 | 0.0624 | 0.00499 | 16.73 | −30.00 |
| −25.00 | 11.299 | 86.98 | 3.9014 | 4.529 | 99.433 | 0.01058 | 0.22892 | 0.3035 | 0.1853 | 1.1508 | 2520. | 475.6 | 1.007 | 0.0229 | 0.0616 | 0.00512 | 16.30 | −25.00 |
| −20.00 | 12.898 | 86.47 | 3.4449 | 6.051 | 100.184 | 0.01406 | 0.22816 | 0.3047 | 0.1875 | 1.1521 | 2477. | 476.8 | 0.968 | 0.0231 | 0.0608 | 0.00525 | 15.87 | −20.00 |
| −15.00 | 14.671 | 85.95 | 3.0514 | 7.580 | 100.932 | 0.01751 | 0.22744 | 0.3060 | 0.1898 | 1.1537 | 2435. | 477.9 | 0.930 | 0.0234 | 0.0601 | 0.00538 | 15.44 | −15.00 |
| −14.93b | 14.696 | 85.94 | 3.0465 | 7.600 | 100.942 | 0.01755 | 0.22743 | 0.3061 | 0.1898 | 1.1537 | 2434. | 477.9 | 0.929 | 0.0234 | 0.0601 | 0.00538 | 15.44 | −14.93 |
| −10.00 | 16.632 | 85.43 | 2.7109 | 9.115 | 101.677 | 0.02093 | 0.22678 | 0.3074 | 0.1921 | 1.1554 | 2393. | 478.9 | 0.894 | 0.0237 | 0.0593 | 0.00552 | 15.02 | −10.00 |
| −5.00 | 18.794 | 84.90 | 2.4154 | 10.657 | 102.419 | 0.02433 | 0.22615 | 0.3088 | 0.1945 | 1.1573 | 2350. | 479.8 | 0.860 | 0.0240 | 0.0586 | 0.00565 | 14.60 | −5.00 |
| 0.00 | 21.171 | 84.37 | 2.1579 | 12.207 | 103.156 | 0.02771 | 0.22557 | 0.3102 | 0.1969 | 1.1595 | 2308. | 480.5 | 0.828 | 0.0242 | 0.0578 | 0.00578 | 14.18 | 0.00 |
| 5.00 | 23.777 | 83.83 | 1.9330 | 13.764 | 103.889 | 0.03107 | 0.22502 | 0.3117 | 0.1995 | 1.1619 | 2266. | 481.1 | 0.798 | 0.0245 | 0.0571 | 0.00592 | 13.76 | 5.00 |
| 10.00 | 26.628 | 83.29 | 1.7357 | 15.328 | 104.617 | 0.03440 | 0.22451 | 0.3132 | 0.2021 | 1.1645 | 2224. | 481.6 | 0.769 | 0.0248 | 0.0564 | 0.00605 | 13.35 | 10.00 |
| 15.00 | 29.739 | 82.74 | 1.5623 | 16.901 | 105.339 | 0.03772 | 0.22403 | 0.3147 | 0.2047 | 1.1674 | 2182. | 482.0 | 0.741 | 0.0250 | 0.0556 | 0.00619 | 12.94 | 15.00 |
| 20.00 | 33.124 | 82.19 | 1.4094 | 18.481 | 106.056 | 0.04101 | 0.22359 | 0.3164 | 0.2075 | 1.1705 | 2140. | 482.2 | 0.715 | 0.0253 | 0.0549 | 0.00632 | 12.53 | 20.00 |
| 25.00 | 36.800 | 81.63 | 1.2742 | 20.070 | 106.767 | 0.04429 | 0.22317 | 0.3181 | 0.2103 | 1.1740 | 2098. | 482.2 | 0.689 | 0.0256 | 0.0542 | 0.00646 | 12.12 | 25.00 |
| 30.00 | 40.784 | 81.06 | 1.1543 | 21.667 | 107.471 | 0.04755 | 0.22278 | 0.3198 | 0.2132 | 1.1777 | 2056. | 482.2 | 0.665 | 0.0258 | 0.0535 | 0.00660 | 11.72 | 30.00 |
| 35.00 | 45.092 | 80.49 | 1.0478 | 23.274 | 108.167 | 0.05079 | 0.22241 | 0.3216 | 0.2163 | 1.1818 | 2014. | 481.9 | 0.642 | 0.0261 | 0.0528 | 0.00674 | 11.32 | 35.00 |
| 40.00 | 49.741 | 79.90 | 0.9528 | 24.890 | 108.856 | 0.05402 | 0.22207 | 0.3235 | 0.2194 | 1.1862 | 1973. | 481.5 | 0.620 | 0.0264 | 0.0521 | 0.00688 | 10.92 | 40.00 |
| 45.00 | 54.749 | 79.32 | 0.8680 | 26.515 | 109.537 | 0.05724 | 0.22174 | 0.3255 | 0.2226 | 1.1910 | 1931. | 481.0 | 0.598 | 0.0267 | 0.0514 | 0.00703 | 10.53 | 45.00 |
| 50.00 | 60.134 | 78.72 | 0.7920 | 28.150 | 110.209 | 0.06044 | 0.22144 | 0.3275 | 0.2260 | 1.1961 | 1889. | 480.3 | 0.578 | 0.0270 | 0.0507 | 0.00717 | 10.14 | 50.00 |
| 55.00 | 65.913 | 78.11 | 0.7238 | 29.796 | 110.871 | 0.06362 | 0.22115 | 0.3297 | 0.2294 | 1.2018 | 1847. | 479.4 | 0.558 | 0.0273 | 0.0500 | 0.00732 | 9.75 | 55.00 |
| 60.00 | 72.105 | 77.50 | 0.6625 | 31.452 | 111.524 | 0.06680 | 0.22088 | 0.3319 | 0.2331 | 1.2079 | 1805. | 478.3 | 0.539 | 0.0275 | 0.0493 | 0.00747 | 9.36 | 60.00 |
| 65.00 | 78.729 | 76.87 | 0.6072 | 33.120 | 112.165 | 0.06996 | 0.22062 | 0.3343 | 0.2368 | 1.2145 | 1763. | 477.0 | 0.520 | 0.0278 | 0.0486 | 0.00762 | 8.98 | 65.00 |
| 70.00 | 85.805 | 76.24 | 0.5572 | 34.799 | 112.796 | 0.07311 | 0.22037 | 0.3368 | 0.2408 | 1.2217 | 1721. | 475.6 | 0.503 | 0.0281 | 0.0479 | 0.00777 | 8.60 | 70.00 |
| 75.00 | 93.351 | 75.59 | 0.5120 | 36.491 | 113.414 | 0.07626 | 0.22013 | 0.3394 | 0.2449 | 1.2296 | 1679. | 474.0 | 0.485 | 0.0284 | 0.0472 | 0.00793 | 8.23 | 75.00 |
| 80.00 | 101.39 | 74.94 | 0.4710 | 38.195 | 114.019 | 0.07939 | 0.21989 | 0.3422 | 0.2492 | 1.2382 | 1636. | 472.2 | 0.469 | 0.0287 | 0.0465 | 0.00809 | 7.86 | 80.00 |
| 85.00 | 109.93 | 74.27 | 0.4338 | 39.913 | 114.610 | 0.08252 | 0.21966 | 0.3451 | 0.2537 | 1.2475 | 1594. | 470.1 | 0.453 | 0.0291 | 0.0458 | 0.00825 | 7.49 | 85.00 |
| 90.00 | 119.01 | 73.58 | 0.3999 | 41.645 | 115.186 | 0.08565 | 0.21944 | 0.3482 | 0.2585 | 1.2578 | 1551. | 467.9 | 0.437 | 0.0294 | 0.0451 | 0.00842 | 7.13 | 90.00 |
| 95.00 | 128.65 | 72.88 | 0.3690 | 43.392 | 115.746 | 0.08877 | 0.21921 | 0.3515 | 0.2636 | 1.2690 | 1509. | 465.4 | 0.422 | 0.0297 | 0.0444 | 0.00860 | 6.77 | 95.00 |
| 100.00 | 138.85 | 72.17 | 0.3407 | 45.155 | 116.289 | 0.09188 | 0.21898 | 0.3551 | 0.2690 | 1.2813 | 1466. | 462.7 | 0.407 | 0.0301 | 0.0437 | 0.00878 | 6.41 | 100.00 |
| 105.00 | 149.65 | 71.44 | 0.3148 | 46.934 | 116.813 | 0.09500 | 0.21875 | 0.3589 | 0.2747 | 1.2950 | 1423. | 459.8 | 0.393 | 0.0304 | 0.0431 | 0.00897 | 6.06 | 105.00 |
| 110.00 | 161.07 | 70.69 | 0.2911 | 48.731 | 117.317 | 0.09811 | 0.21851 | 0.3630 | 0.2809 | 1.3101 | 1380. | 456.7 | 0.378 | 0.0308 | 0.0424 | 0.00916 | 5.71 | 110.00 |
| 115.00 | 173.14 | 69.93 | 0.2693 | 50.546 | 117.799 | 0.10123 | 0.21826 | 0.3675 | 0.2875 | 1.3268 | 1337. | 453.2 | 0.365 | 0.0312 | 0.0417 | 0.00936 | 5.36 | 115.00 |
| 120.00 | 185.86 | 69.14 | 0.2493 | 52.382 | 118.258 | 0.10435 | 0.21800 | 0.3723 | 0.2948 | 1.3456 | 1294. | 449.6 | 0.351 | 0.0316 | 0.0410 | 0.00958 | 5.03 | 120.00 |
| 125.00 | 199.28 | 68.32 | 0.2308 | 54.239 | 118.690 | 0.10748 | 0.21772 | 0.3775 | 0.3026 | 1.3666 | 1250. | 445.6 | 0.338 | 0.0320 | 0.0403 | 0.00981 | 4.69 | 125.00 |
| 130.00 | 213.41 | 67.49 | 0.2137 | 56.119 | 119.095 | 0.11062 | 0.21742 | 0.3833 | 0.3112 | 1.3903 | 1206. | 441.4 | 0.325 | 0.0324 | 0.0396 | 0.01005 | 4.36 | 130.00 |
| 135.00 | 228.28 | 66.62 | 0.1980 | 58.023 | 119.468 | 0.11376 | 0.21709 | 0.3897 | 0.3208 | 1.4173 | 1162. | 436.8 | 0.313 | 0.0329 | 0.0389 | 0.01031 | 4.04 | 135.00 |
| 140.00 | 243.92 | 65.73 | 0.1833 | 59.954 | 119.807 | 0.11692 | 0.21673 | 0.3968 | 0.3315 | 1.4481 | 1117. | 432.0 | 0.301 | 0.0334 | 0.0382 | 0.01058 | 3.72 | 140.00 |
| 145.00 | 260.36 | 64.80 | 0.1697 | 61.915 | 120.108 | 0.12010 | 0.21634 | 0.4048 | 0.3435 | 1.4837 | 1072. | 426.8 | 0.288 | 0.0339 | 0.0375 | 0.01089 | 3.40 | 145.00 |
| 150.00 | 277.61 | 63.83 | 0.1571 | 63.908 | 120.366 | 0.12330 | 0.21591 | 0.4138 | 0.3571 | 1.5250 | 1027. | 421.2 | 0.276 | 0.0344 | 0.0368 | 0.01122 | 3.09 | 150.00 |
| 155.00 | 295.73 | 62.82 | 0.1453 | 65.936 | 120.576 | 0.12653 | 0.21542 | 0.4242 | 0.3729 | 1.5738 | 980. | 415.3 | 0.264 | 0.0350 | 0.0361 | 0.01158 | 2.79 | 155.00 |
| 160.00 | 314.73 | 61.76 | 0.1343 | 68.005 | 120.731 | 0.12979 | 0.21488 | 0.4362 | 0.3914 | 1.6318 | 934. | 409.1 | 0.253 | 0.0357 | 0.0354 | 0.01199 | 2.50 | 160.00 |
| 165.00 | 334.65 | 60.65 | 0.1239 | 70.118 | 120.823 | 0.13309 | 0.21426 | 0.4504 | 0.4133 | 1.7022 | 886. | 402.4 | 0.241 | 0.0364 | 0.0346 | 0.01245 | 2.21 | 165.00 |
| 170.00 | 355.53 | 59.47 | 0.1142 | 72.283 | 120.842 | 0.13644 | 0.21356 | 0.4675 | 0.4400 | 1.7889 | 837. | 395.3 | 0.229 | 0.0372 | 0.0339 | 0.01297 | 1.93 | 170.00 |
| 175.00 | 377.41 | 58.21 | 0.1051 | 74.509 | 120.773 | 0.13985 | 0.21274 | 0.4887 | 0.4733 | 1.8984 | 786. | 387.7 | 0.218 | 0.0381 | 0.0332 | 0.01358 | 1.66 | 175.00 |
| 180.00 | 400.34 | 56.86 | 0.0964 | 76.807 | 120.598 | 0.14334 | 0.21180 | 0.5156 | 0.5159 | 2.0405 | 734. | 379.6 | 0.206 | 0.0391 | 0.0325 | 0.01430 | 1.39 | 180.00 |
| 185.00 | 424.36 | 55.38 | 0.0881 | 79.193 | 120.294 | 0.14693 | 0.21069 | 0.5512 | 0.5729 | 2.2321 | 680. | 371.0 | 0.194 | 0.0403 | 0.0318 | 0.01516 | 1.14 | 185.00 |
| 190.00 | 449.52 | 53.76 | 0.0801 | 81.692 | 119.822 | 0.15066 | 0.20935 | 0.6012 | 0.6532 | 2.5041 | 624. | 361.8 | 0.182 | 0.0417 | 0.0311 | 0.01623 | 0.90 | 190.00 |
| 195.00 | 475.91 | 51.91 | 0.0724 | 84.343 | 119.123 | 0.15459 | 0.20771 | 0.6768 | 0.7751 | 2.9192 | 565. | 352.0 | 0.169 | 0.0435 | 0.0304 | 0.01760 | 0.67 | 195.00 |
| 200.00 | 503.59 | 49.76 | 0.0647 | 87.214 | 118.097 | 0.15880 | 0.20562 | 0.8062 | 0.9835 | 3.6309 | 502. | 341.3 | 0.155 | 0.0457 | 0.0300 | 0.01949 | 0.45 | 200.00 |
| 205.00 | 532.68 | 47.08 | 0.0567 | 90.454 | 116.526 | 0.16353 | 0.20275 | 1.083 | 1.425 | 5.136 | 436. | 329.4 | 0.140 | 0.0489 | 0.0300 | 0.02240 | 0.26 | 205.00 |
| 210.00 | 563.35 | 43.20 | 0.0477 | 94.530 | 113.746 | 0.16945 | 0.19814 | 2.113 | 3.008 | 10.512 | 363. | 315.5 | 0.120 | 0.0543 | 0.0316 | 0.02848 | 0.09 | 210.00 |
| 213.91c | 588.75 | 31.96 | 0.0313 | 103.894 | 103.894 | 0.18320 | 0.18320 | • | • | • | 0. | 0.0 | — | — | • | • | 0.00 | 213.91 |

*Temperatures are on the ITS-90 scale     a = triple point     b = normal boiling point     c = critical point

# Thermodynamics and Psychrometrics

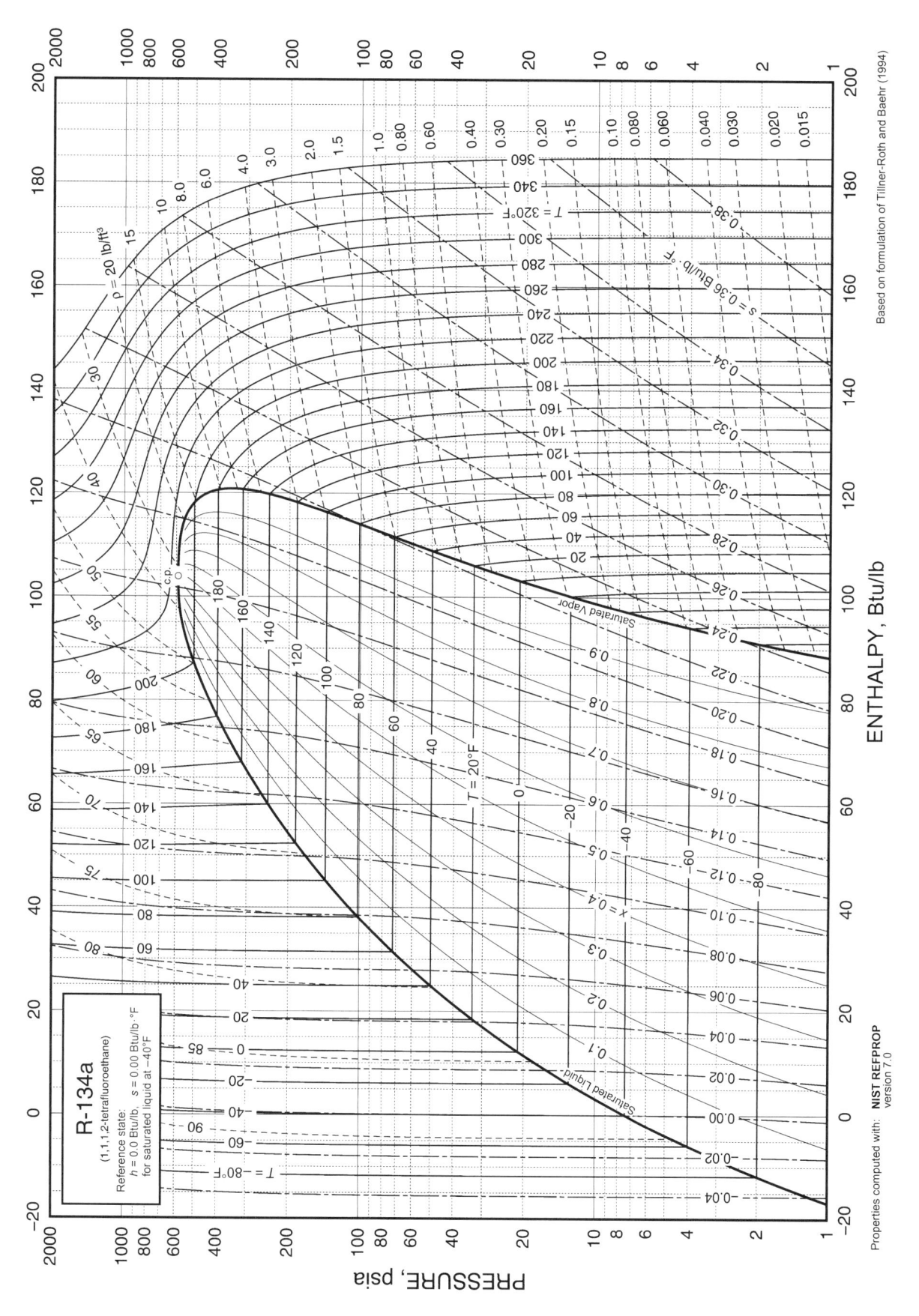

*Fig. 2-5 Pressure-Enthalpy Diagram for Refrigerant 134a*
*(Figure 8, Chapter 30, 2013 ASHRAE Handbook—Fundamentals)*

26　Principles of HVAC

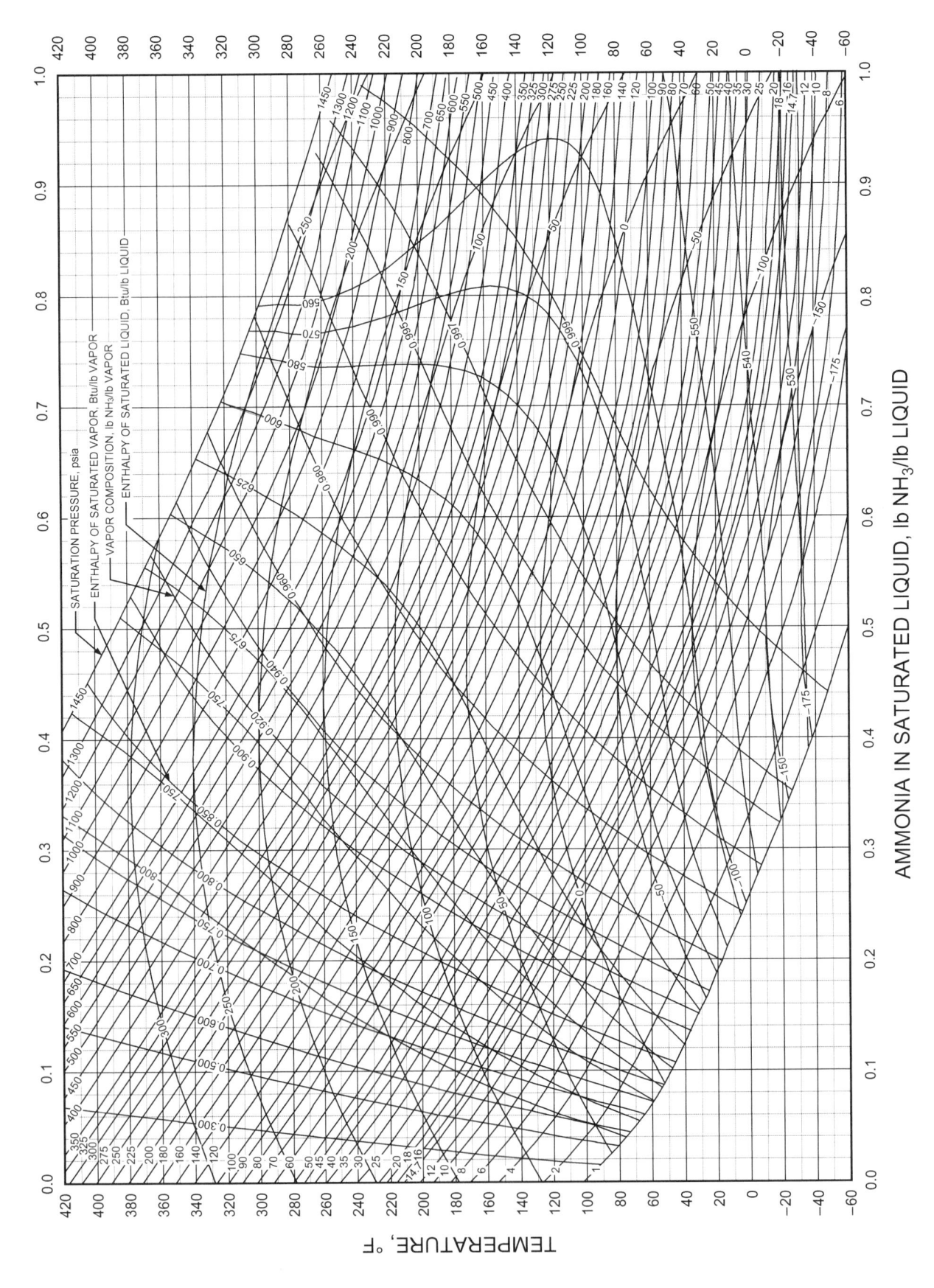

Fig. 2-6 Enthalpy-Concentration Diagram for Aqua-Ammonia
(Figure 33, Chapter 30, 2013 ASHRAE Handbook—Fundamentals)

# Thermodynamics and Psychrometrics

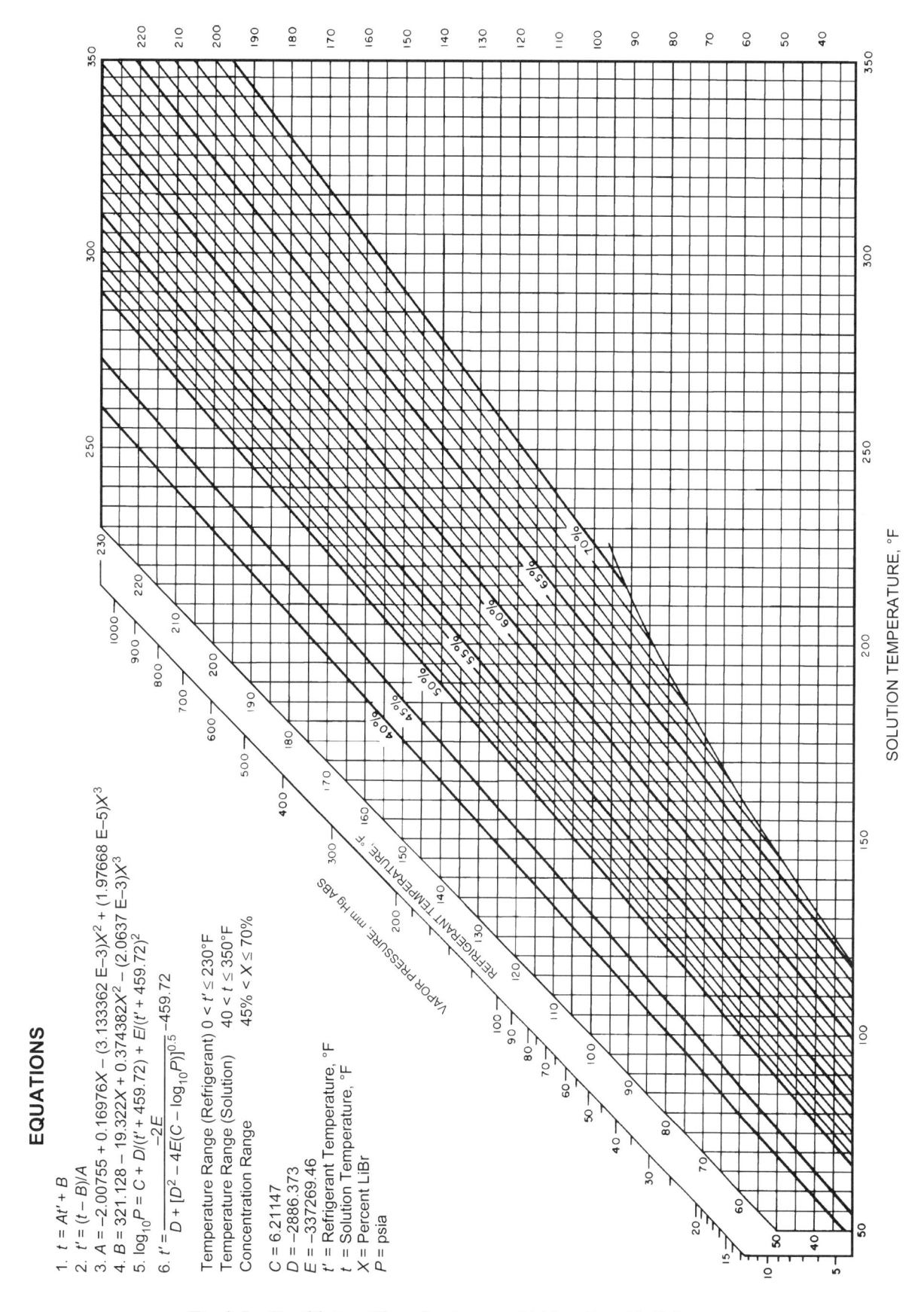

## EQUATIONS

1. $t = At' + B$
2. $t' = (t - B)/A$
3. $A = -2.00755 + 0.16976X - (3.133362\,E{-3})X^2 + (1.97668\,E{-5})X^3$
4. $B = 321.128 - 19.322X + 0.374382X^2 - (2.0637\,E{-3})X^3$
5. $\log_{10} P = C + D/(t' + 459.72) + E/(t' + 459.72)^2$
6. $t' = \dfrac{-2E}{D + [D^2 - 4E(C - \log_{10}P)]^{0.5}} - 459.72$

Temperature Range (Refrigerant) $0 < t' \leq 230°F$
Temperature Range (Solution) $40 < t \leq 350°F$
Concentration Range $45\% < X \leq 70\%$

$C = 6.21147$
$D = -2886.373$
$E = -337269.46$
$t'$ = Refrigerant Temperature, °F
$t$ = Solution Temperature, °F
$X$ = Percent LiBr
$P$ = psia

*Fig. 2-7a  Equilibrium Chart for Aqueous Lithium Bromide Solutions*
(Figure 35, Chapter 30, 2013 ASHRAE Handbook—Fundamentals)

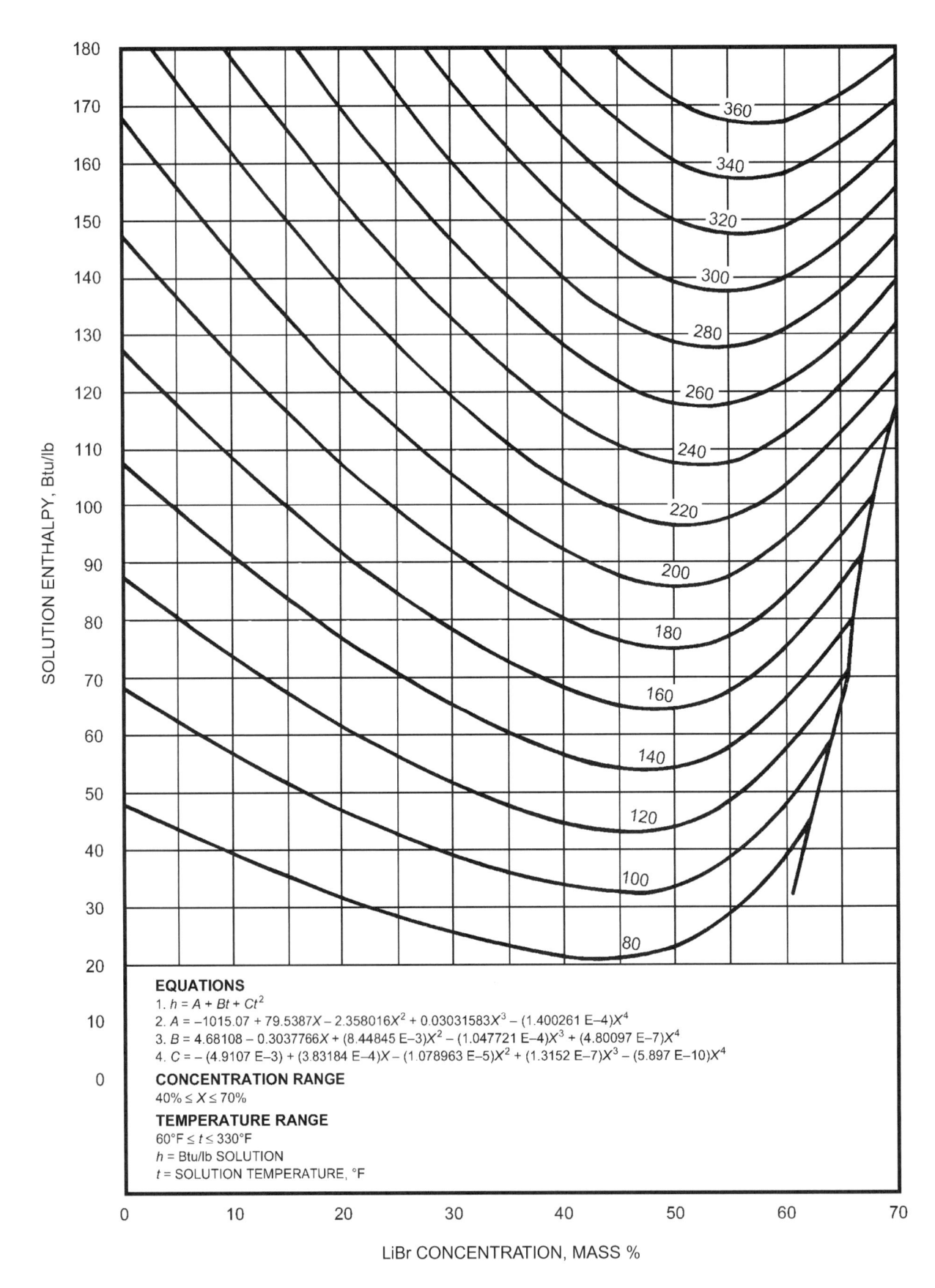

*Fig. 2-7b Enthalpy-Concentration Diagram for Water/Lithium Bromide Solutions*
*(Figure 34, Chapter 30, 2013 ASHRAE Handbook—Fundamentals)*

**EQUATIONS**
1. $h = A + Bt + Ct^2$
2. $A = -1015.07 + 79.5387X - 2.358016X^2 + 0.03031583X^3 - (1.400261\text{ E}{-4})X^4$
3. $B = 4.68108 - 0.3037766X + (8.44845\text{ E}{-3})X^2 - (1.047721\text{ E}{-4})X^3 + (4.80097\text{ E}{-7})X^4$
4. $C = -(4.9107\text{ E}{-3}) + (3.83184\text{ E}{-4})X - (1.078963\text{ E}{-5})X^2 + (1.3152\text{ E}{-7})X^3 - (5.897\text{ E}{-10})X^4$

**CONCENTRATION RANGE**
$40\% \le X \le 70\%$

**TEMPERATURE RANGE**
$60°F \le t \le 330°F$
$h$ = Btu/lb SOLUTION
$t$ = SOLUTION TEMPERATURE, °F

# Thermodynamics and Psychrometrics

**Table 2-3  Properties of Gases**

| Gas | Chemical Formula | Relative Molecular Mass | $R$, ft·lb$_f$/lb$_m$°R | $c_p$, Btu/lb$_m$°R | $c_p$, kJ/kg·K | $c_v$, Btu/lb$_m$°R | $c_v$, kJ/kg·K | $k$ |
|---|---|---|---|---|---|---|---|---|
| Air | — | 28.97 | 53.34 | 0.240 | 1.0 | 0.171 | 0.716 | 1.400 |
| Argon | Ar | 39.94 | 38.66 | 0.125 | 0.523 | 0.075 | 0.316 | 1.667 |
| Carbon Dioxide | CO$_2$ | 44.01 | 35.10 | 0.203 | 0.85 | 0.158 | 0.661 | 1.285 |
| Carbon Monoxide | CO | 28.01 | 55.16 | 0.249 | 1.04 | 0.178 | 0.715 | 1.399 |
| Helium | He | 4.003 | 386.0 | 1.25 | 5.23 | 0.753 | 3.153 | 1.667 |
| Hydrogen | H$_2$ | 2.016 | 766.4 | 3.43 | 14.36 | 2.44 | 10.22 | 1.404 |
| Methane | CH$_4$ | 16.04 | 96.35 | 0.532 | 2.23 | 0.403 | 1.69 | 1.32 |
| Nitrogen | N$_2$ | 28.016 | 55.15 | 0.248 | 1.04 | 0.177 | 0.741 | 1.400 |
| Oxygen | O$_2$ | 32.000 | 48.28 | 0.219 | 0.917 | 0.157 | 0.657 | 1.395 |
| Steam | H$_2$O | 18.016 | 85.76 | 0.445 | 1.863 | 0.335 | 1.402 | 1.329 |

low value (e.g., atmospheric pressure), the deviation from ideal gas behavior may be considerable. In this region, tables of thermodynamic properties or charts for a particular substance should be used.

## 2.2.12 Mixtures

A large number of thermodynamic problems involve mixtures of different pure substances. A pure substance is a substance which is homogeneous and unchanging in chemical composition. Homogeneous mixtures of gases that do not react with each other are therefore pure substances, and the properties of such mixtures can be determined, correlated, and either tabulated or fitted by equations just like the properties of any other pure substance. This work has been done for common mixtures such as air and certain combustion products, but, as an unlimited number of mixtures is possible, properties of all of them cannot be determined experimentally and tabulated. Thus, it is important to be able to calculate the properties of any mixture from the properties of its constituents. Such calculations are discussed in this section, first for gas mixtures and then for gas-vapor mixtures.

Since individual gases can often be approximated as ideal gases, the study of mixtures of ideal gases and their properties is of considerable importance. Each constituent gas in a mixture has its own pressure called the **partial pressure** of the particular gas. The Gibbs-Dalton law states that in a mixture of ideal gases, the pressure of the mixture is equal to the sum of the partial pressures of the individual constituent gases. In equation form

$$p_m = p_a + p_b + p_c \qquad (2\text{-}14)$$

$$p_a = p_m(n_a/n_m), \; p_b = p_m(n_b/n_m), \; p_c = (n_c/n_m)$$

where $p_m$ is the total pressure of the mixture of gases $a$, $b$, and $c$, and $p_a$, $p_b$, and $p_c$ are the partial pressures. In a mixture of ideal gases, the partial pressure of each constituent equals the pressure that constituent would exert if it existed alone at the temperature and volume of the mixture.

Generally, in gas mixtures, each constituent gas behaves as though the other gases were not present; each gas occupies the total volume of the mixture at the temperature of the mixture and the partial pressure of the gas. Thus, if $V_m$ is the volume of the mixture, then the volume of each component is also $V_m$, or

$$V_m = V_a = V_b = V_c \qquad (2\text{-}15)$$

However, the volume of a mixture of ideal gases equals the sum of the volumes of its constituents if each existed alone at the temperature and pressure of the mixture. This property is known as **Amagat's law**, **Leduc's law**, or the **law of additive volumes**. Like Dalton's law, it is strictly true only for ideal gases, but holds approximately for real-gas mixtures, even those in some ranges of pressure and temperature where $pv = RT$ is inaccurate. When the temperature of a real-gas mixture is well above the critical temperatures of all its constituents, the additive volume law is usually more accurate than the additive pressure law.

For ideal-gas mixtures, volumetric analyses are frequently used. The volume fraction is defined as

$$\text{Volume fraction of A} = \frac{V_a(p_m, T_m)}{V_m}$$

$$= \frac{\text{Volume of A existing alone at } p_m, T_m}{\text{Volume of mixture at } p_m, T_m}$$

Note that in a gas mixture, each constituent occupies the total volume; thus volume fraction is not defined as the ratio of a constituent volume to the mixture volume because this ratio is always unity.

Avogadro's law goes on to state, *equal volumes of ideal gases held under exactly the same temperature and pressure have equal numbers of molecules.* If $T_m$ is the temperature of the mixture,

$$T_m = T_a = T_b = T_c \qquad (2\text{-}16)$$

for the temperature relationship.

The analysis of a gas mixture based on mass is called a **gravimetric analysis**. It is based on the fact that the mass of a mixture equals the sum of the masses of its constituents:

$$m_m = m_a + m_b + m_c \qquad (2\text{-}17)$$

where the subscript m refers to the mixture and the subscripts $a$, $b$, and $c$ refer to individual constituents of the mixture. The ratio $m_a/m_m$ is called the mass fraction of constituent $a$.

The total number of moles in a mixture is defined as the sum of the number of moles of its constituents:

$$n_m = n_a + n_b + n_c \quad (2\text{-}18)$$

The mole fraction $x$ is defined as $n/n_m$, and

$$M_m = x_a M_a + x_b M_b + x_c M_c \quad (2\text{-}19)$$

where $M_m$ is called the apparent (or average) molecular weight of the mixture. The second part of the Gibbs-Dalton law can be taken as a basic definition:

$$U_m = U_a + U_b + U_c \quad (2\text{-}20)$$

$$H_m = H_a + H_b + H_c \quad (2\text{-}21)$$

$$S_m = S_a + S_b + S_c \quad (2\text{-}22)$$

Remember that the constituent entropies here must be evaluated at the temperature and volume of the mixture or at the mixture temperature and the constituent partial pressures. The entropy of any constituent at the volume and temperature of the mixture (and hence at its partial pressure) is greater than its entropy when existing at the pressure and temperature of the mixture (and hence at its partial volume).

Consider the constituents as perfect gases:

$$R_m = (m_a R_a + m_b R_b + m_c R_c)/m_m \quad (2\text{-}23)$$

$$c_{vm} = (m_a c_{va} + m_b c_{vb} + m_c c_{vc})/m_m \quad (2\text{-}24)$$

$$c_{pm} = (m_a c_{pa} + m_b c_{pb} + m_c c_{pc})/m_m \quad (2\text{-}25)$$

## 2.2.13 Psychrometrics: Moist Air Properties

Consider a simplification of the problem involving a mixture of ideal gases that is in contact with a solid or liquid phase of one of the components. The most familiar example is a mixture of air and water vapor in contact with liquid water or ice, such as the problems encountered in air conditioning or drying. This, and a number of similar problems can be analyzed simply and with considerable accuracy if the following assumptions are made:

1. The solid or liquid phase contains no dissolved gases.
2. The gaseous phase can be treated as a mixture of ideal gases.
3. When the mixture and the condensed phase are at a given pressure and temperature, the equilibrium between the condensed phase and its vapor is not influenced by the presence of the other component. This means that when equilibrium is achieved, the partial pressure of the vapor equals the saturation pressure corresponding to the temperature of the mixture.

If the vapor is at the saturation pressure and temperature, the mixture is referred to as a **saturated mixture**. For an air-water vapor mixture, the term **saturated air** is used.

Psychrometrics is the science involving thermodynamic properties of moist air and the effect of atmospheric moisture on materials and human comfort. As it applies in this text, the definition is broadened to include the method of controlling the thermal properties of moist air.

When moist air is considered to be a mixture of independent, perfect gases, dry air, and water vapor, each is assumed to obey the perfect gas equation of state:

Dry air: $p_a V = n_a RT$

Water vapor: $p_w V = n_w RT$

where

$p_a$ = partial pressure of dry air
$p_w$ = partial pressure of water vapor
$V$ = total mixture volume
$n_a$ = number of moles of dry air
$n_w$ = number of moles of water vapor
$R$ = universal gas constant
 (8.31441 J/g-mol·K or 1545.32 ft·lb$_f$/ lb·mol·°R)
$T$ = absolute temperature

The mixture also obeys the perfect gas equation:

$$pV = nRT \text{ or } (p_a + p_w)V = (n_a + n_w)RT$$

**Dry-bulb temperature** $t$ is the temperature of air as registered by an ordinary thermometer.

**Thermodynamic wet-bulb temperature** $t^*$ is the temperature at which water (liquid or solid), by evaporating into moist air at a given dry-bulb temperature $t$ and humidity ratio $W$, can bring the air to saturation adiabatically at the same temperature $t^*$, while the pressure $p$ is maintained constant. Figure 2-8 may be used as a schematic representation of the adiabatic saturation process, where the leaving air is saturated and at a temperature equal to that of the injected water. A device used in place of the adiabatic saturator is the psychrometer.

The psychrometer consists of two thermometers or other temperature-sensing elements, one of which has a wetted cotton wick covering the bulb (Figure 2-9). When the wet bulb is placed in an airstream, water may evaporate from the wick. The equilibrium temperature the water eventually reaches is called the **wet-bulb temperature**. This process is not one of adiabatic saturation which defines the thermodynamic wet-bulb temperature, but is one of simultaneous heat and mass transfer from the wet-bulb thermometer. Fortunately, the corrections applied to

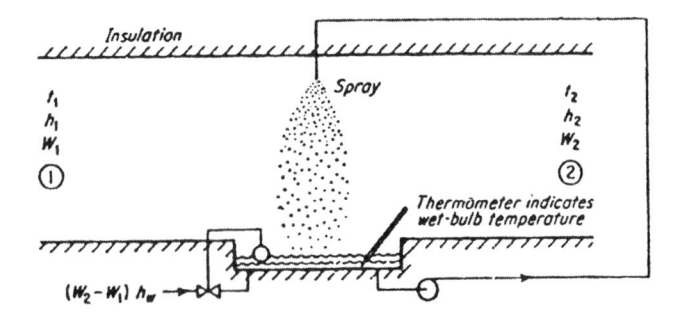

*Fig. 2-8 Adiabatic Saturator*

wet-bulb thermometer readings to obtain the thermodynamic wet-bulb temperature are usually small.

**Humidity ratio** $W$ of a given moist air sample is defined as the ratio of the mass of water vapor to the mass of dry air contained in the sample:

$$W = m_w/m_a$$
$$W = 0.62198 p_w/(p - p_w) \quad (2\text{-}26)$$

$$W = \frac{(2501 - 2.381 t^*)W_s^* - (t - t^*)}{2501 + 1.805 t - 4.186 t^*} \quad (2\text{-}27\text{a})$$

where $t$ and $t^*$ are in °C.

In inch-pound units

$$W = \frac{(1093 - 0.556 t^*)W_s^* - 0.240(t - t^*)}{1093 + 0.444 t - t^*} \quad (2\text{-}27\text{b})$$

where $t$ and $t^*$ are in °F.

The term $W_s^*$ indicates the humidity ratio if saturated at the wet bulb temperature.

**Degree of saturation** $\mu$ is the ratio of the actual humidity ratio $W$ to the humidity ratio $W_s$ of saturated air at the same temperature and pressure.

$$\mu = \frac{W}{W_s}\bigg|_{t,p} \quad (2\text{-}28)$$

Relative humidity $\phi$ is the ratio of the mole fraction of water vapor $x_w$ in a given moist air sample to the mole fraction $x_{ws}$ in an air sample which is saturated at the same temperature and pressure:

$$\phi = \frac{x_w}{x_{ws}}\bigg|_{t,p}$$
$$\phi = \frac{p_w}{p_{ws}}\bigg|_{t,p} \quad (2\text{-}29)$$

The term $p_{ws}$ is the saturation pressure of water vapor at the given temperature $t$.

**Dew-point temperature** $t_d$ is the temperature of moist air which is saturated at the same pressure $p$ and has the same humidity ratio $W$ as that of the given sample of moist air. It corresponds to the saturation temperature (Column 1) of Table 2-1 for the vapor pressure found in Column 2. As an alternate to using the table, equations have been formulated for the relationship.

For 0°C to 70°C

$$t_d = -35.957 - 1.8726 a + 1.1689 a^2 \quad (2\text{-}30\text{a})$$

and for –60°C to 0°C

$$t_d = -60.45 + 7.0322 a + 0.3700 a^2 \quad (2\text{-}30\text{b})$$

with $t_d$ in °C and $a = \ln p_w$, with $p_w$ in pascals.

For the temperature range of 32°F to 150°F

$$t_d = 79.047 + 30.5790 a + 1.8893 a^2 \quad (2\text{-}31\text{a})$$

and for temperatures below 32°F

$$t_d = 71.98 + 24.873 a + 0.8927 a^2 \quad (2\text{-}31\text{b})$$

where $t_d$ is the dew-point temperature in °F and $a = \ln p_w$, with $p_w$ the water vapor partial pressure (in. Hg).

The **volume** $v$ of a moist air mixture is expressed in terms of a unit mass of dry air, with the relation $p = p_a + p_w$, or

$$v = R_a T/(p - p_w) \quad (2\text{-}32)$$

The **enthalpy** of a mixture of perfect gases is equal to the sum of the individual partial enthalpies of the components. The enthalpy of moist air is then

$$h = h_a + W h_g$$

where $h_a$ is the specific enthalpy for dry air and $h_g$ is the specific enthalpy for water vapor at the temperature of the mixture. Approximately

$$h_a = t \text{ (kJ/kg)} \quad (2\text{-}33)$$

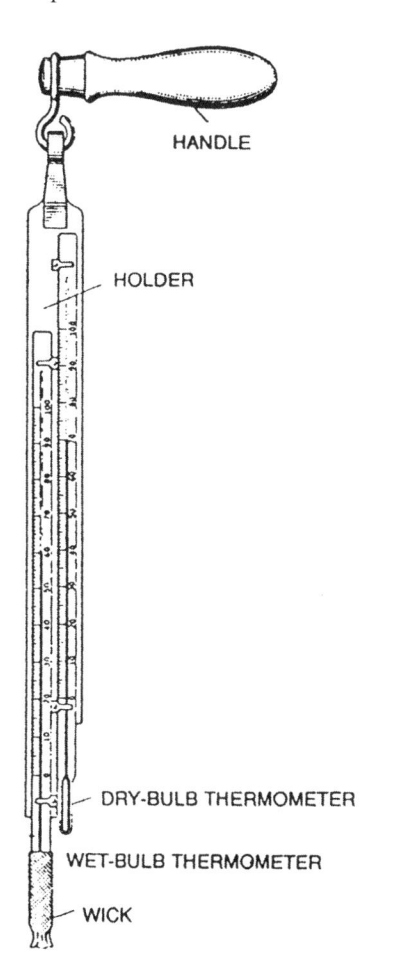

*Fig. 2-9 Sling Psychrometer*

Table 2-4  Thermodynamic Properties of Moist Air at Standard Atmospheric Pressure, 14.696 psia

(Table 2, Chapter 1, 2013 ASHRAE Handbook—Fundamentals)

| Temp., °F $t$ | Humidity Ratio $W_s$, lb$_w$/lb$_{da}$ | Specific Volume, ft³/lb$_{da}$ $v_{da}$ | $v_{as}$ | $v_s$ | Specific Enthalpy, Btu/lb$_{da}$ $h_{da}$ | $h_{as}$ | $h_s$ | Specific Entropy, Btu/lb$_{da}$·°F $s_{da}$ | $s_s$ | Temp., °F $t$ |
|---|---|---|---|---|---|---|---|---|---|---|
| −80 | 0.0000049 | 9.553 | 0.000 | 9.553 | −19.221 | 0.005 | −19.215 | −0.04594 | −0.04592 | −80 |
| −79 | 0.0000053 | 9.579 | 0.000 | 9.579 | −18.980 | 0.005 | −18.975 | −0.04531 | −0.04529 | −79 |
| −78 | 0.0000057 | 9.604 | 0.000 | 9.604 | −18.740 | 0.006 | −18.734 | −0.04468 | −0.04466 | −78 |
| −77 | 0.0000062 | 9.629 | 0.000 | 9.629 | −18.500 | 0.007 | −18.493 | −0.04405 | −0.04403 | −77 |
| −76 | 0.0000067 | 9.655 | 0.000 | 9.655 | −18.259 | 0.007 | −18.252 | −0.04342 | −0.04340 | −76 |
| −75 | 0.0000072 | 9.680 | 0.000 | 9.680 | −18.019 | 0.007 | −18.011 | −0.04279 | −0.04277 | −75 |
| −74 | 0.0000078 | 9.705 | 0.000 | 9.705 | −17.778 | 0.008 | −17.770 | −0.04217 | −0.04215 | −74 |
| −73 | 0.0000084 | 9.731 | 0.000 | 9.731 | −17.538 | 0.009 | −17.529 | −0.04155 | −0.04152 | −73 |
| −72 | 0.0000090 | 9.756 | 0.000 | 9.756 | −17.298 | 0.010 | −17.288 | −0.04093 | −0.04090 | −72 |
| −71 | 0.0000097 | 9.781 | 0.000 | 9.782 | −17.057 | 0.010 | −17.047 | −0.04031 | −0.04028 | −71 |
| −70 | 0.0000104 | 9.807 | 0.000 | 9.807 | −16.806 | 0.011 | −16.817 | −0.03969 | −0.03966 | −70 |
| −69 | 0.0000112 | 9.832 | 0.000 | 9.832 | −16.577 | 0.012 | −16.565 | −0.03907 | −0.03904 | −69 |
| −68 | 0.0000120 | 9.857 | 0.000 | 9.858 | −16.336 | 0.013 | −16.324 | −0.03846 | −0.03843 | −68 |
| −67 | 0.0000129 | 9.883 | 0.000 | 9.883 | −16.096 | 0.013 | −16.083 | −0.03785 | −0.03781 | −67 |
| −66 | 0.0000139 | 9.908 | 0.000 | 9.908 | −15.856 | 0.015 | −15.841 | −0.03724 | −0.03720 | −66 |
| −65 | 0.0000149 | 9.933 | 0.000 | 9.934 | −15.616 | 0.015 | −15.600 | −0.03663 | −0.03659 | −65 |
| −64 | 0.0000160 | 9.959 | 0.000 | 9.959 | −15.375 | 0.017 | −15.359 | −0.03602 | −0.03597 | −64 |
| −63 | 0.0000172 | 9.984 | 0.000 | 9.984 | −15.117 | 0.018 | −15.135 | −0.03541 | −0.03536 | −63 |
| −62 | 0.0000184 | 10.009 | 0.000 | 10.010 | −14.895 | 0.019 | −14.876 | −0.03481 | −0.03476 | −62 |
| −61 | 0.0000198 | 10.035 | 0.000 | 10.035 | −14.654 | 0.021 | −14.634 | −0.03420 | −0.03415 | −61 |
| −60 | 0.0000212 | 10.060 | 0.000 | 10.060 | −14.414 | 0.022 | −14.392 | −0.03360 | −0.03354 | −60 |
| −59 | 0.0000227 | 10.085 | 0.000 | 10.086 | −14.174 | 0.024 | −14.150 | −0.03300 | −0.03294 | −59 |
| −58 | 0.0000243 | 10.111 | 0.000 | 10.111 | −13.933 | 0.025 | −13.908 | −0.03240 | −0.03233 | −58 |
| −57 | 0.0000260 | 10.136 | 0.000 | 10.137 | −13.693 | 0.027 | −13.666 | −0.03180 | −0.03173 | −57 |
| −56 | 0.0000279 | 10.161 | 0.000 | 10.162 | −13.453 | 0.029 | −13.424 | −0.03121 | −0.03113 | −56 |
| −55 | 0.0000298 | 10.187 | 0.000 | 10.187 | −13.213 | 0.031 | −13.182 | −0.03061 | −0.03053 | −55 |
| −54 | 0.0000319 | 10.212 | 0.001 | 10.213 | −12.972 | 0.033 | −12.939 | −0.03002 | −0.02993 | −54 |
| −53 | 0.0000341 | 10.237 | 0.001 | 10.238 | −12.732 | 0.035 | −12.697 | −0.02943 | −0.02934 | −53 |
| −52 | 0.0000365 | 10.263 | 0.001 | 10.263 | −12.492 | 0.038 | −12.454 | −0.02884 | −0.02874 | −52 |
| −51 | 0.0000390 | 10.288 | 0.001 | 10.289 | −12.251 | 0.041 | −12.211 | −0.02825 | −0.02814 | −51 |
| −50 | 0.0000416 | 10.313 | 0.001 | 10.314 | −12.011 | 0.043 | −11.968 | −0.02766 | −0.02755 | −50 |
| −49 | 0.0000445 | 10.339 | 0.001 | 10.340 | −11.771 | 0.046 | −11.725 | −0.02708 | −0.02696 | −49 |
| −48 | 0.0000475 | 10.364 | 0.001 | 10.365 | −11.531 | 0.050 | −11.481 | −0.02649 | −0.02636 | −48 |
| −47 | 0.0000507 | 10.389 | 0.001 | 10.390 | −11.290 | 0.053 | −11.237 | −0.02591 | −0.02577 | −47 |
| −46 | 0.0000541 | 10.415 | 0.001 | 10.416 | −11.050 | 0.056 | −10.994 | −0.02533 | −0.02518 | −46 |
| −45 | 0.0000577 | 10.440 | 0.001 | 10.441 | −10.810 | 0.060 | −10.750 | −0.02475 | −0.02459 | −45 |
| −44 | 0.0000615 | 10.465 | 0.001 | 10.466 | −10.570 | 0.064 | −10.505 | −0.02417 | −0.02400 | −44 |
| −43 | 0.0000656 | 10.491 | 0.001 | 10.492 | −10.329 | 0.068 | −10.261 | −0.02359 | −0.02342 | −43 |
| −42 | 0.0000699 | 10.516 | 0.001 | 10.517 | −10.089 | 0.073 | −10.016 | −0.02302 | −0.02283 | −42 |
| −41 | 0.0000744 | 10.541 | 0.001 | 10.543 | −9.849 | 0.078 | −9.771 | −0.02244 | −0.02224 | −41 |
| −40 | 0.0000793 | 10.567 | 0.001 | 10.568 | −9.609 | 0.083 | −9.526 | −0.02187 | −0.02166 | −40 |
| −39 | 0.0000844 | 10.592 | 0.001 | 10.593 | −9.368 | 0.088 | −9.280 | −0.02130 | −0.02107 | −39 |
| −38 | 0.0000898 | 10.617 | 0.002 | 10.619 | −9.128 | 0.094 | −9.034 | −0.02073 | −0.02049 | −38 |
| −37 | 0.0000956 | 10.643 | 0.002 | 10.644 | −8.888 | 0.100 | −8.788 | −0.02016 | −0.01991 | −37 |
| −36 | 0.0001017 | 10.668 | 0.002 | 10.670 | −8.648 | 0.106 | −8.541 | −0.01959 | −0.01932 | −36 |
| −35 | 0.0001081 | 10.693 | 0.002 | 10.695 | −8.407 | 0.113 | −8.294 | −0.01902 | −0.01874 | −35 |
| −34 | 0.0001150 | 10.719 | 0.002 | 10.721 | −8.167 | 0.120 | −8.047 | −0.01846 | −0.01816 | −34 |
| −33 | 0.0001222 | 10.744 | 0.002 | 10.746 | −7.927 | 0.128 | −7.799 | −0.01790 | −0.01758 | −33 |
| −32 | 0.0001298 | 10.769 | 0.002 | 10.772 | −7.687 | 0.136 | −7.551 | −0.01733 | −0.01699 | −32 |
| −31 | 0.0001379 | 10.795 | 0.002 | 10.797 | −7.447 | 0.145 | −7.302 | −0.01677 | −0.01641 | −31 |
| −30 | 0.0001465 | 10.820 | 0.003 | 10.822 | −7.206 | 0.154 | −7.053 | −0.01621 | −0.01583 | −30 |
| −29 | 0.0001555 | 10.845 | 0.003 | 10.848 | −6.966 | 0.163 | −6.803 | −0.01565 | −0.01525 | −29 |
| −28 | 0.0001650 | 10.871 | 0.003 | 10.873 | −6.726 | 0.173 | −6.553 | −0.01510 | −0.01467 | −28 |
| −27 | 0.0001751 | 10.896 | 0.003 | 10.899 | −6.486 | 0.184 | −6.302 | −0.01454 | −0.01409 | −27 |
| −26 | 0.0001858 | 10.921 | 0.003 | 10.924 | −6.245 | 0.195 | −6.051 | −0.01399 | −0.01351 | −26 |
| −25 | 0.0001970 | 10.947 | 0.003 | 10.950 | −6.005 | 0.207 | −5.798 | −0.01343 | −0.01293 | −25 |
| −24 | 0.0002088 | 10.972 | 0.004 | 10.976 | −5.765 | 0.220 | −5.545 | −0.01288 | −0.01235 | −24 |
| −23 | 0.0002214 | 10.997 | 0.004 | 11.001 | −5.525 | 0.233 | −5.292 | −0.01233 | −0.01176 | −23 |
| −22 | 0.0002346 | 11.022 | 0.004 | 11.027 | −5.284 | 0.247 | −5.038 | −0.01178 | −0.01118 | −22 |
| −21 | 0.0002485 | 11.048 | 0.004 | 11.052 | −5.044 | 0.261 | −4.783 | −0.01123 | −0.01060 | −21 |
| −20 | 0.0002632 | 11.073 | 0.005 | 11.078 | −4.804 | 0.277 | −4.527 | −0.01069 | −0.01002 | −20 |
| −19 | 0.0002786 | 11.098 | 0.005 | 11.103 | −4.564 | 0.293 | −4.271 | −0.01014 | −0.00943 | −19 |
| −18 | 0.0002950 | 11.124 | 0.005 | 11.129 | −4.324 | 0.311 | −4.013 | −0.00960 | −0.00885 | −18 |
| −17 | 0.0003121 | 11.149 | 0.006 | 11.155 | −4.084 | 0.329 | −3.754 | −0.00905 | −0.00826 | −17 |
| −16 | 0.0003303 | 11.174 | 0.006 | 11.180 | −3.843 | 0.348 | −3.495 | −0.00851 | −0.00768 | −16 |
| −15 | 0.0003493 | 11.200 | 0.006 | 11.206 | −3.603 | 0.368 | −3.235 | −0.00797 | −0.00709 | −15 |
| −14 | 0.0003694 | 11.225 | 0.007 | 11.232 | −3.363 | 0.390 | −2.973 | −0.00743 | −0.00650 | −14 |
| −13 | 0.0003905 | 11.250 | 0.007 | 11.257 | −3.123 | 0.412 | −2.710 | −0.00689 | −0.00591 | −13 |
| −12 | 0.0004128 | 11.276 | 0.007 | 11.283 | −2.882 | 0.436 | −2.447 | −0.00635 | −0.00532 | −12 |
| −11 | 0.0004362 | 11.301 | 0.008 | 11.309 | −2.642 | 0.460 | −2.182 | −0.00582 | −0.00473 | −11 |

**Table 2-4  Thermodynamic Properties of Moist Air at Standard Sea Level Pressure, 14.696 psi (29.921 in. Hg)** (*Continued*)

*(Table 2, Chapter 1, 2013 ASHRAE Handbook—Fundamentals)*

| Temp., °F t | Humidity Ratio, lb_w/lb_da W_s | Specific Volume, ft³/lb_da v_da | v_as | v_s | Specific Enthalpy, Btu/lb_da h_da | h_as | h_s | Specific Entropy, Btu/lb_da·°F s_da | s_as | s_s | Condensed Water Specific Enthalpy, Btu/lb_w h_w | Specific Entropy, Btu/lb_w·°F s_w | Vapor Pressure, in. Hg p_s | Temp., °F t |
|---|---|---|---|---|---|---|---|---|---|---|---|---|---|---|
| −10 | 0.0004608 | 11.326 | 0.008 | 11.335 | −2.402 | 0.487 | −1.915 | −0.00528 | 0.00115 | −0.00414 | −163.55 | −0.3346 | 0.022050 | −10 |
| −9 | 0.0004867 | 11.351 | 0.009 | 11.360 | −2.162 | 0.514 | −1.647 | −0.00475 | 0.00121 | −0.00354 | −163.09 | −0.3335 | 0.023289 | −9 |
| −8 | 0.0005139 | 11.377 | 0.009 | 11.386 | −1.922 | 0.543 | −1.378 | −0.00422 | 0.00127 | −0.00294 | −162.63 | −0.3325 | 0.024591 | −8 |
| −7 | 0.0005425 | 11.402 | 0.010 | 11.412 | −1.681 | 0.574 | −1.108 | −0.00369 | 0.00134 | −0.00234 | −162.17 | −0.3315 | 0.025959 | −7 |
| −6 | 0.0005726 | 11.427 | 0.010 | 11.438 | −1.441 | 0.606 | −0.835 | −0.00316 | 0.00141 | −0.00174 | −161.70 | −0.3305 | 0.027397 | −6 |
| −5 | 0.0006041 | 11.453 | 0.011 | 11.464 | −1.201 | 0.640 | −0.561 | −0.00263 | 0.00149 | −0.00114 | −161.23 | −0.3294 | 0.028907 | −5 |
| −4 | 0.0006373 | 11.478 | 0.012 | 11.490 | −0.961 | 0.675 | −0.286 | −0.00210 | 0.00157 | −0.00053 | −160.77 | −0.3284 | 0.030494 | −4 |
| −3 | 0.0006722 | 11.503 | 0.012 | 11.516 | −0.721 | 0.712 | −0.008 | −0.00157 | 0.00165 | 0.00008 | −160.30 | −0.3274 | 0.032160 | −3 |
| −2 | 0.0007088 | 11.529 | 0.013 | 11.542 | −0.480 | 0.751 | 0.271 | −0.00105 | 0.00174 | 0.00069 | −159.83 | −0.3264 | 0.033909 | −2 |
| −1 | 0.0007472 | 11.554 | 0.014 | 11.568 | −0.240 | 0.792 | 0.552 | −0.00052 | 0.00183 | 0.00130 | −159.36 | −0.3253 | 0.035744 | −1 |
| 0 | 0.0007875 | 11.579 | 0.015 | 11.594 | 0.0 | 0.835 | 0.835 | 0.00000 | 0.00192 | 0.00192 | −158.89 | −0.3243 | 0.037671 | 0 |
| 1 | 0.0008298 | 11.604 | 0.015 | 11.620 | 0.240 | 0.880 | 1.121 | 0.00052 | 0.00202 | 0.00254 | −158.42 | −0.3233 | 0.039694 | 1 |
| 2 | 0.0008742 | 11.630 | 0.016 | 11.646 | 0.480 | 0.928 | 1.408 | 0.00104 | 0.00212 | 0.00317 | −157.95 | −0.3223 | 0.041814 | 2 |
| 3 | 0.0009207 | 11.655 | 0.017 | 11.672 | 0.721 | 0.978 | 1.699 | 0.00156 | 0.00223 | 0.00380 | −157.47 | −0.3212 | 0.044037 | 3 |
| 4 | 0.0009695 | 11.680 | 0.018 | 11.699 | 0.961 | 1.030 | 1.991 | 0.00208 | 0.00235 | 0.00443 | −157.00 | −0.3202 | 0.046370 | 4 |
| 5 | 0.0010207 | 11.706 | 0.019 | 11.725 | 1.201 | 1.085 | 2.286 | 0.00260 | 0.00247 | 0.00506 | −156.52 | −0.3192 | 0.048814 | 5 |
| 6 | 0.0010743 | 11.731 | 0.020 | 11.751 | 1.441 | 1.143 | 2.584 | 0.00311 | 0.00259 | 0.00570 | −156.05 | −0.3182 | 0.051375 | 6 |
| 7 | 0.0011306 | 11.756 | 0.021 | 11.778 | 1.681 | 1.203 | 2.884 | 0.00363 | 0.00635 | 0.00272 | −155.57 | −0.3171 | 0.054060 | 7 |
| 8 | 0.0011895 | 11.782 | 0.022 | 11.804 | 1.922 | 1.266 | 3.188 | 0.00414 | 0.00286 | 0.00700 | −155.09 | −0.3161 | 0.056872 | 8 |
| 9 | 0.0012512 | 11.807 | 0.024 | 11.831 | 2.162 | 1.332 | 3.494 | 0.00466 | 0.00300 | 0.00766 | −154.61 | −0.3151 | 0.059819 | 9 |
| 10 | 0.0013158 | 11.832 | 0.025 | 11.857 | 2.402 | 1.402 | 3.804 | 0.00517 | 0.00315 | 0.00832 | −154.13 | −0.3141 | 0.062901 | 10 |
| 11 | 0.0013835 | 11.857 | 0.026 | 11.884 | 2.642 | 1.474 | 4.117 | 0.00568 | 0.00330 | 0.00898 | −153.65 | −0.3130 | 0.066131 | 11 |
| 12 | 0.0014544 | 11.883 | 0.028 | 11.910 | 2.882 | 1.550 | 4.433 | 0.00619 | 0.00347 | 0.00966 | −153.17 | −0.3120 | 0.069511 | 12 |
| 13 | 0.0015286 | 11.908 | 0.029 | 11.937 | 3.123 | 1.630 | 4.753 | 0.00670 | 0.00364 | 0.01033 | −152.68 | −0.3110 | 0.073049 | 13 |
| 14 | 0.0016062 | 11.933 | 0.031 | 11.964 | 3.363 | 1.714 | 5.077 | 0.00721 | 0.00381 | 0.01102 | −152.20 | −0.3100 | 0.076751 | 14 |
| 15 | 0.0016874 | 11.959 | 0.032 | 11.991 | 3.603 | 1.801 | 5.404 | 0.00771 | 0.00400 | 0.01171 | −151.71 | −0.3089 | 0.080623 | 15 |
| 16 | 0.0017724 | 11.984 | 0.034 | 12.018 | 3.843 | 1.892 | 5.736 | 0.00822 | 0.00419 | 0.01241 | −151.22 | −0.3079 | 0.084673 | 16 |
| 17 | 0.0018613 | 12.009 | 0.036 | 12.045 | 4.084 | 1.988 | 6.072 | 0.00872 | 0.00439 | 0.01312 | −150.74 | −0.3069 | 0.088907 | 17 |
| 18 | 0.0019543 | 12.035 | 0.038 | 12.072 | 4.324 | 2.088 | 6.412 | 0.00923 | 0.00460 | 0.01383 | −150.25 | −0.3059 | 0.093334 | 18 |
| 19 | 0.0020515 | 12.060 | 0.040 | 12.099 | 4.564 | 2.193 | 6.757 | 0.00973 | 0.00482 | 0.01455 | −149.76 | −0.3049 | 0.097962 | 19 |
| 20 | 0.0021531 | 12.085 | 0.042 | 12.127 | 4.804 | 2.303 | 7.107 | 0.01023 | 0.00505 | 0.01528 | −149.27 | −0.3038 | 0.102798 | 20 |
| 21 | 0.0022592 | 12.110 | 0.044 | 12.154 | 5.044 | 2.417 | 7.462 | 0.01073 | 0.00529 | 0.01602 | −148.78 | −0.3028 | 0.107849 | 21 |
| 22 | 0.0023703 | 12.136 | 0.046 | 12.182 | 5.285 | 2.537 | 7.822 | 0.01123 | 0.00554 | 0.01677 | −148.28 | −0.3018 | 0.113130 | 22 |
| 23 | 0.0024863 | 12.161 | 0.048 | 12.209 | 5.525 | 2.662 | 8.187 | 0.01173 | 0.00580 | 0.01753 | −147.79 | −0.3008 | 0.118645 | 23 |
| 24 | 0.0026073 | 12.186 | 0.051 | 12.237 | 5.765 | 2.793 | 8.558 | 0.01223 | 0.00607 | 0.01830 | −147.30 | −0.2997 | 0.124396 | 24 |
| 25 | 0.0027339 | 12.212 | 0.054 | 12.265 | 6.005 | 2.930 | 8.935 | 0.01272 | 0.00636 | 0.01908 | −146.80 | −0.2987 | 0.130413 | 25 |
| 26 | 0.0028660 | 12.237 | 0.056 | 12.293 | 6.246 | 3.073 | 9.318 | 0.01322 | 0.00665 | 0.01987 | −146.30 | −0.2977 | 0.136684 | 26 |
| 27 | 0.0030039 | 12.262 | 0.059 | 12.321 | 6.486 | 3.222 | 9.708 | 0.01371 | 0.00696 | 0.02067 | −145.81 | −0.2967 | 0.143233 | 27 |
| 28 | 0.0031480 | 12.287 | 0.062 | 12.349 | 6.726 | 3.378 | 10.104 | 0.01420 | 0.00728 | 0.02148 | −145.31 | −0.2956 | 0.150066 | 28 |
| 29 | 0.0032984 | 12.313 | 0.065 | 12.378 | 6.966 | 3.541 | 10.507 | 0.01470 | 0.00761 | 0.02231 | −144.81 | −0.2946 | 0.157198 | 29 |
| 30 | 0.0034552 | 12.338 | 0.068 | 12.406 | 7.206 | 3.711 | 10.917 | 0.01519 | 0.00796 | 0.02315 | −144.31 | −0.2936 | 0.164631 | 30 |
| 31 | 0.0036190 | 12.363 | 0.072 | 12.435 | 7.447 | 3.888 | 11.335 | 0.01568 | 0.00832 | 0.02400 | −143.80 | −0.2926 | 0.172390 | 31 |
| 32 | 0.0037895 | 12.389 | 0.075 | 12.464 | 7.687 | 4.073 | 11.760 | 0.01617 | 0.00870 | 0.02487 | −143.30 | −0.2915 | 0.180479 | 32 |
| 32* | 0.003790 | 12.389 | 0.075 | 12.464 | 7.687 | 4.073 | 11.760 | 0.01617 | 0.00870 | 0.02487 | 0.02 | 0.0000 | 0.18050 | 32 |
| 33 | 0.003947 | 12.414 | 0.079 | 12.492 | 7.927 | 4.243 | 12.170 | 0.01665 | 0.00905 | 0.02570 | 1.03 | 0.0020 | 0.18791 | 33 |
| 34 | 0.004109 | 12.439 | 0.082 | 12.521 | 8.167 | 4.420 | 12.587 | 0.01714 | 0.00940 | 0.02655 | 2.04 | 0.0041 | 0.19559 | 34 |
| 35 | 0.004277 | 12.464 | 0.085 | 12.550 | 8.408 | 4.603 | 13.010 | 0.01763 | 0.00977 | 0.02740 | 3.05 | 0.0061 | 0.20356 | 35 |
| 36 | 0.004452 | 12.490 | 0.089 | 12.579 | 8.648 | 4.793 | 13.441 | 0.01811 | 0.01016 | 0.02827 | 4.05 | 0.0081 | 0.21181 | 36 |
| 37 | 0.004633 | 12.515 | 0.093 | 12.608 | 8.888 | 4.990 | 13.878 | 0.01860 | 0.01055 | 0.02915 | 5.06 | 0.0102 | 0.22035 | 37 |
| 38 | 0.004820 | 12.540 | 0.097 | 12.637 | 9.128 | 5.194 | 14.322 | 0.01908 | 0.01096 | 0.03004 | 6.06 | 0.0122 | 0.22920 | 38 |
| 39 | 0.005014 | 12.566 | 0.101 | 12.667 | 9.369 | 5.405 | 14.773 | 0.01956 | 0.01139 | 0.03095 | 7.07 | 0.0142 | 0.23835 | 39 |
| 40 | 0.005216 | 12.591 | 0.105 | 12.696 | 9.609 | 5.624 | 15.233 | 0.02004 | 0.01183 | 0.03187 | 8.07 | 0.0162 | 0.24784 | 40 |
| 41 | 0.005424 | 12.616 | 0.110 | 12.726 | 9.849 | 5.851 | 15.700 | 0.02052 | 0.01228 | 0.03281 | 9.08 | 0.0182 | 0.25765 | 41 |
| 42 | 0.005640 | 12.641 | 0.114 | 12.756 | 10.089 | 6.086 | 16.175 | 0.02100 | 0.01275 | 0.03375 | 10.08 | 0.0202 | 0.26781 | 42 |
| 43 | 0.005863 | 12.667 | 0.119 | 12.786 | 10.330 | 6.330 | 16.660 | 0.02148 | 0.01324 | 0.03472 | 11.09 | 0.0222 | 0.27831 | 43 |
| 44 | 0.006094 | 12.692 | 0.124 | 12.816 | 10.570 | 6.582 | 17.152 | 0.02196 | 0.01374 | 0.03570 | 12.09 | 0.0242 | 0.28918 | 44 |
| 45 | 0.006334 | 12.717 | 0.129 | 12.846 | 10.810 | 6.843 | 17.653 | 0.02244 | 0.01426 | 0.03669 | 13.09 | 0.0262 | 0.30042 | 45 |
| 46 | 0.006581 | 12.743 | 0.134 | 12.877 | 11.050 | 7.114 | 18.164 | 0.02291 | 0.01479 | 0.03770 | 14.10 | 0.0282 | 0.31206 | 46 |
| 47 | 0.006838 | 12.768 | 0.140 | 12.908 | 11.291 | 7.394 | 18.685 | 0.02339 | 0.01534 | 0.03873 | 15.10 | 0.0302 | 0.32408 | 47 |
| 48 | 0.007103 | 12.793 | 0.146 | 12.939 | 11.531 | 7.684 | 19.215 | 0.02386 | 0.01592 | 0.03978 | 16.10 | 0.0321 | 0.33651 | 48 |
| 49 | 0.007378 | 12.818 | 0.152 | 12.970 | 11.771 | 7.984 | 19.756 | 0.02433 | 0.01651 | 0.04084 | 17.10 | 0.0341 | 0.34937 | 49 |
| 50 | 0.007661 | 12.844 | 0.158 | 13.001 | 12.012 | 8.295 | 20.306 | 0.02480 | 0.01712 | 0.04192 | 18.11 | 0.0361 | 0.36264 | 50 |
| 51 | 0.007955 | 12.869 | 0.164 | 13.033 | 12.252 | 8.616 | 20.868 | 0.02528 | 0.01775 | 0.04302 | 19.11 | 0.0381 | 0.37636 | 51 |
| 52 | 0.008259 | 12.894 | 0.171 | 13.065 | 12.492 | 8.949 | 21.441 | 0.02575 | 0.01840 | 0.04415 | 20.11 | 0.0400 | 0.39054 | 52 |
| 53 | 0.008573 | 12.920 | 0.178 | 13.097 | 12.732 | 9.293 | 22.025 | 0.02622 | 0.01907 | 0.04529 | 21.11 | 0.0420 | 0.40518 | 53 |
| 54 | 0.008897 | 12.945 | 0.185 | 13.129 | 12.973 | 9.648 | 22.621 | 0.02668 | 0.01976 | 0.04645 | 22.11 | 0.0439 | 0.42030 | 54 |
| 55 | 0.009233 | 12.970 | 0.192 | 13.162 | 13.213 | 10.016 | 23.229 | 0.02715 | 0.02048 | 0.04763 | 23.11 | 0.0459 | 0.43592 | 55 |
| 56 | 0.009580 | 12.995 | 0.200 | 13.195 | 13.453 | 10.397 | 23.850 | 0.02762 | 0.02122 | 0.04884 | 24.11 | 0.0478 | 0.45205 | 56 |
| 57 | 0.009938 | 13.021 | 0.207 | 13.228 | 13.694 | 10.790 | 24.484 | 0.02808 | 0.02198 | 0.05006 | 25.11 | 0.0497 | 0.46870 | 57 |
| 58 | 0.010309 | 13.046 | 0.216 | 13.262 | 13.934 | 11.197 | 25.131 | 0.02855 | 0.02277 | 0.05132 | 26.11 | 0.0517 | 0.48589 | 58 |
| 59 | 0.010692 | 13.071 | 0.224 | 13.295 | 14.174 | 11.618 | 25.792 | 0.02901 | 0.02358 | 0.05259 | 27.11 | 0.0536 | 0.50363 | 59 |

*Extrapolated to represent metastable equilibrium with undercooled liquid.

**Table 2-4  Thermodynamic Properties of Moist Air at Standard Sea Level Pressure, 14.696 psi (29.921 in. Hg) (*Continued*)**

*(Table 2, Chapter 1, 2013 ASHRAE Handbook—Fundamentals)*

| Temp., °F t | Humidity Ratio, lb_w/lb_da W_s | Specific Volume, ft³/lb_da v_da | v_as | v_s | Specific Enthalpy, Btu/lb_da h_da | h_as | h_s | Specific Entropy, Btu/lb_da·°F s_da | s_as | s_s | Condensed Water Specific Enthalpy, Btu/lb_w h_w | Specific Entropy, Btu/lb_w·°F s_w | Vapor Pressure, in. Hg p_s | Temp., °F t |
|---|---|---|---|---|---|---|---|---|---|---|---|---|---|---|
| 60 | 0.011087 | 13.096 | 0.233 | 13.329 | 14.415 | 12.052 | 26.467 | 0.02947 | 0.02442 | 0.05389 | 28.11 | 0.0555 | 0.52193 | 60 |
| 61 | 0.011496 | 13.122 | 0.242 | 13.364 | 14.655 | 12.502 | 27.157 | 0.02994 | 0.02528 | 0.05522 | 29.12 | 0.0575 | 0.54082 | 61 |
| 62 | 0.011919 | 13.147 | 0.251 | 13.398 | 14.895 | 12.966 | 27.862 | 0.03040 | 0.02617 | 0.05657 | 30.11 | 0.0594 | 0.56032 | 62 |
| 63 | 0.012355 | 13.172 | 0.261 | 13.433 | 15.135 | 13.446 | 28.582 | 0.03086 | 0.02709 | 0.05795 | 31.11 | 0.0613 | 0.58041 | 63 |
| 64 | 0.012805 | 13.198 | 0.271 | 13.468 | 15.376 | 13.942 | 29.318 | 0.03132 | 0.02804 | 0.05936 | 32.11 | 0.0632 | 0.60113 | 64 |
| 65 | 0.013270 | 13.223 | 0.281 | 13.504 | 15.616 | 14.454 | 30.071 | 0.03178 | 0.02902 | 0.06080 | 33.11 | 0.0651 | 0.62252 | 65 |
| 66 | 0.013750 | 13.248 | 0.292 | 13.540 | 15.856 | 14.983 | 30.840 | 0.03223 | 0.03003 | 0.06226 | 34.11 | 0.0670 | 0.64454 | 66 |
| 67 | 0.014246 | 13.273 | 0.303 | 13.577 | 16.097 | 15.530 | 31.626 | 0.03269 | 0.03107 | 0.06376 | 35.11 | 0.0689 | 0.66725 | 67 |
| 68 | 0.014758 | 13.299 | 0.315 | 13.613 | 16.337 | 16.094 | 32.431 | 0.03315 | 0.03214 | 0.06529 | 36.11 | 0.0708 | 0.69065 | 68 |
| 69 | 0.015286 | 13.324 | 0.326 | 13.650 | 16.577 | 16.677 | 33.254 | 0.03360 | 0.03325 | 0.06685 | 37.11 | 0.0727 | 0.71479 | 69 |
| 70 | 0.015832 | 13.349 | 0.339 | 13.688 | 16.818 | 17.279 | 34.097 | 0.03406 | 0.03438 | 0.06844 | 38.11 | 0.0746 | 0.73966 | 70 |
| 71 | 0.016395 | 13.375 | 0.351 | 13.726 | 17.058 | 17.901 | 34.959 | 0.03451 | 0.03556 | 0.07007 | 39.11 | 0.0765 | 0.76567 | 71 |
| 72 | 0.016976 | 13.400 | 0.365 | 13.764 | 17.299 | 18.543 | 35.841 | 0.03496 | 0.03677 | 0.07173 | 40.11 | 0.0783 | 0.79167 | 72 |
| 73 | 0.017575 | 13.425 | 0.378 | 13.803 | 17.539 | 19.204 | 36.743 | 0.03541 | 0.03801 | 0.07343 | 41.11 | 0.0802 | 0.81882 | 73 |
| 74 | 0.018194 | 13.450 | 0.392 | 13.843 | 17.779 | 19.889 | 37.668 | 0.03586 | 0.03930 | 0.07516 | 42.11 | 0.0821 | 0.84684 | 74 |
| 75 | 0.018833 | 13.476 | 0.407 | 13.882 | 18.020 | 20.595 | 38.615 | 0.03631 | 0.04062 | 0.07694 | 43.11 | 0.0840 | 0.87567 | 75 |
| 76 | 0.019491 | 13.501 | 0.422 | 13.923 | 18.260 | 21.323 | 39.583 | 0.03676 | 0.04199 | 0.07875 | 44.10 | 0.0858 | 0.90533 | 76 |
| 77 | 0.020170 | 13.526 | 0.437 | 13.963 | 18.500 | 22.075 | 40.576 | 0.03721 | 0.04339 | 0.08060 | 45.10 | 0.0877 | 0.93589 | 77 |
| 78 | 0.020871 | 13.551 | 0.453 | 14.005 | 18.741 | 22.851 | 41.592 | 0.03766 | 0.04484 | 0.08250 | 46.10 | 0.0896 | 0.96733 | 78 |
| 79 | 0.021594 | 13.577 | 0.470 | 14.046 | 18.981 | 23.652 | 42.633 | 0.03811 | 0.04633 | 0.08444 | 47.10 | 0.0914 | 0.99970 | 79 |
| 80 | 0.022340 | 13.602 | 0.487 | 14.089 | 19.222 | 24.479 | 43.701 | 0.03855 | 0.04787 | 0.08642 | 48.10 | 0.0933 | 1.03302 | 80 |
| 81 | 0.023109 | 13.627 | 0.505 | 14.132 | 19.462 | 25.332 | 44.794 | 0.03900 | 0.04945 | 0.08844 | 49.10 | 0.0951 | 1.06728 | 81 |
| 82 | 0.023902 | 13.653 | 0.523 | 14.175 | 19.702 | 26.211 | 45.913 | 0.03944 | 0.05108 | 0.09052 | 50.10 | 0.0970 | 1.10252 | 82 |
| 83 | 0.024720 | 13.678 | 0.542 | 14.220 | 19.943 | 27.120 | 47.062 | 0.03988 | 0.05276 | 0.09264 | 51.09 | 0.0988 | 1.13882 | 83 |
| 84 | 0.025563 | 13.703 | 0.561 | 14.264 | 20.183 | 28.055 | 48.238 | 0.04033 | 0.05448 | 0.09481 | 52.09 | 0.1006 | 1.17608 | 84 |
| 85 | 0.026433 | 13.728 | 0.581 | 14.310 | 20.424 | 29.021 | 49.445 | 0.04077 | 0.05626 | 0.09703 | 53.09 | 0.1025 | 1.21445 | 85 |
| 86 | 0.027329 | 13.754 | 0.602 | 14.356 | 20.664 | 30.017 | 50.681 | 0.04121 | 0.05809 | 0.09930 | 54.09 | 0.1043 | 1.25388 | 86 |
| 87 | 0.028254 | 13.779 | 0.624 | 14.403 | 20.905 | 31.045 | 51.949 | 0.04165 | 0.05998 | 0.10163 | 55.09 | 0.1061 | 1.29443 | 87 |
| 88 | 0.029208 | 13.804 | 0.646 | 14.450 | 21.145 | 32.105 | 53.250 | 0.04209 | 0.06192 | 0.10401 | 56.09 | 0.1080 | 1.33613 | 88 |
| 89 | 0.030189 | 13.829 | 0.669 | 14.498 | 21.385 | 33.197 | 54.582 | 0.04253 | 0.06392 | 0.10645 | 57.09 | 0.1098 | 1.37893 | 89 |
| 90 | 0.031203 | 13.855 | 0.692 | 14.547 | 21.626 | 34.325 | 55.951 | 0.04297 | 0.06598 | 0.10895 | 58.08 | 0.1116 | 1.42298 | 90 |
| 91 | 0.032247 | 13.880 | 0.717 | 14.597 | 21.866 | 35.489 | 57.355 | 0.06810 | 0.04340 | 0.11150 | 59.08 | 0.1134 | 1.46824 | 91 |
| 92 | 0.033323 | 13.905 | 0.742 | 14.647 | 22.107 | 36.687 | 58.794 | 0.04384 | 0.07028 | 0.11412 | 60.08 | 0.1152 | 1.51471 | 92 |
| 93 | 0.034433 | 13.930 | 0.768 | 14.699 | 22.347 | 37.924 | 60.271 | 0.04427 | 0.07253 | 0.11680 | 61.08 | 0.1170 | 1.56248 | 93 |
| 94 | 0.035577 | 13.956 | 0.795 | 14.751 | 22.588 | 39.199 | 61.787 | 0.04471 | 0.07484 | 0.11955 | 62.08 | 0.1188 | 1.61154 | 94 |
| 95 | 0.036757 | 13.981 | 0.823 | 14.804 | 22.828 | 40.515 | 63.343 | 0.04514 | 0.07722 | 0.12237 | 63.08 | 0.1206 | 1.66196 | 95 |
| 96 | 0.037972 | 14.006 | 0.852 | 14.858 | 23.069 | 41.871 | 64.940 | 0.04558 | 0.07968 | 0.12525 | 64.07 | 0.1224 | 1.71372 | 96 |
| 97 | 0.039225 | 14.032 | 0.881 | 14.913 | 23.309 | 43.269 | 66.578 | 0.04601 | 0.08220 | 0.12821 | 65.07 | 0.1242 | 1.76685 | 97 |
| 98 | 0.040516 | 14.057 | 0.912 | 14.969 | 23.550 | 44.711 | 68.260 | 0.04644 | 0.08480 | 0.13124 | 66.07 | 0.1260 | 1.82141 | 98 |
| 99 | 0.041848 | 14.082 | 0.944 | 15.026 | 23.790 | 46.198 | 69.988 | 0.04687 | 0.08747 | 0.13434 | 67.07 | 0.1278 | 1.87745 | 99 |
| 100 | 0.043219 | 14.107 | 0.976 | 15.084 | 24.031 | 47.730 | 71.761 | 0.04730 | 0.09022 | 0.13752 | 68.07 | 0.1296 | 1.93492 | 100 |
| 101 | 0.044634 | 14.133 | 1.010 | 15.143 | 24.271 | 49.312 | 73.583 | 0.04773 | 0.09306 | 0.14079 | 69.07 | 0.1314 | 1.99396 | 101 |
| 102 | 0.046090 | 14.158 | 1.045 | 15.203 | 24.512 | 50.940 | 75.452 | 0.04816 | 0.09597 | 0.14413 | 70.06 | 0.1332 | 2.05447 | 102 |
| 103 | 0.047592 | 14.183 | 1.081 | 15.264 | 24.752 | 52.621 | 77.373 | 0.04859 | 0.09897 | 0.14756 | 71.06 | 0.1349 | 2.11661 | 103 |
| 104 | 0.049140 | 14.208 | 1.118 | 15.326 | 24.993 | 54.354 | 79.346 | 0.04901 | 0.10206 | 0.15108 | 72.06 | 0.1367 | 2.18037 | 104 |
| 105 | 0.050737 | 14.234 | 1.156 | 15.390 | 25.233 | 56.142 | 81.375 | 0.04944 | 0.10525 | 0.15469 | 73.06 | 0.1385 | 2.24581 | 105 |
| 106 | 0.052383 | 14.259 | 1.196 | 15.455 | 25.474 | 57.986 | 83.460 | 0.04987 | 0.10852 | 0.15839 | 74.06 | 0.1402 | 2.31297 | 106 |
| 107 | 0.054077 | 14.284 | 1.236 | 15.521 | 25.714 | 59.884 | 85.599 | 0.05029 | 0.11189 | 0.16218 | 75.06 | 0.1420 | 2.38173 | 107 |
| 108 | 0.055826 | 14.309 | 1.279 | 15.588 | 25.955 | 61.844 | 87.799 | 0.05071 | 0.11537 | 0.16608 | 76.05 | 0.1438 | 2.45232 | 108 |
| 109 | 0.057628 | 14.335 | 1.322 | 15.657 | 26.195 | 63.866 | 90.061 | 0.05114 | 0.11894 | 0.17008 | 77.05 | 0.1455 | 2.52473 | 109 |
| 110 | 0.059486 | 14.360 | 1.367 | 15.727 | 26.436 | 65.950 | 92.386 | 0.05156 | 0.12262 | 0.17418 | 78.05 | 0.1473 | 2.59891 | 110 |
| 111 | 0.061401 | 14.385 | 1.414 | 15.799 | 26.677 | 68.099 | 94.776 | 0.05198 | 0.12641 | 0.17839 | 79.05 | 0.1490 | 2.67500 | 111 |
| 112 | 0.063378 | 14.411 | 1.462 | 15.872 | 26.917 | 70.319 | 97.237 | 0.05240 | 0.13032 | 0.18272 | 80.05 | 0.1508 | 2.75310 | 112 |
| 113 | 0.065411 | 14.436 | 1.511 | 15.947 | 27.158 | 72.603 | 99.760 | 0.05282 | 0.13434 | 0.18716 | 81.05 | 0.1525 | 2.83291 | 113 |
| 114 | 0.067512 | 14.461 | 1.562 | 16.023 | 27.398 | 74.964 | 102.362 | 0.05324 | 0.13847 | 0.19172 | 82.04 | 0.1543 | 2.91491 | 114 |
| 115 | 0.069676 | 14.486 | 1.615 | 16.101 | 27.639 | 77.396 | 105.035 | 0.05366 | 0.14274 | 0.19640 | 83.04 | 0.1560 | 2.99883 | 115 |
| 116 | 0.071908 | 14.512 | 1.670 | 16.181 | 27.879 | 79.906 | 107.786 | 0.05408 | 0.14713 | 0.20121 | 84.04 | 0.1577 | 3.08488 | 116 |
| 117 | 0.074211 | 14.537 | 1.726 | 16.263 | 28.120 | 82.497 | 110.617 | 0.05450 | 0.15165 | 0.20615 | 85.04 | 0.1595 | 3.17305 | 117 |
| 118 | 0.076586 | 14.562 | 1.784 | 16.346 | 28.361 | 85.169 | 113.530 | 0.05492 | 0.15631 | 0.21122 | 86.04 | 0.1612 | 3.26335 | 118 |
| 119 | 0.079036 | 14.587 | 1.844 | 16.432 | 28.601 | 87.927 | 116.528 | 0.05533 | 0.16111 | 0.21644 | 87.04 | 0.1629 | 3.35586 | 119 |
| 120 | 0.081560 | 14.613 | 1.906 | 16.519 | 28.842 | 90.770 | 119.612 | 0.05575 | 0.16605 | 0.22180 | 88.04 | 0.1647 | 3.45052 | 120 |
| 121 | 0.084169 | 14.638 | 1.971 | 16.609 | 29.083 | 93.709 | 122.792 | 0.05616 | 0.17115 | 0.22731 | 89.04 | 0.1664 | 3.54764 | 121 |
| 122 | 0.086860 | 14.663 | 2.037 | 16.700 | 29.323 | 96.742 | 126.065 | 0.05658 | 0.17640 | 0.23298 | 90.03 | 0.1681 | 3.64704 | 122 |
| 123 | 0.089633 | 14.688 | 2.106 | 16.794 | 29.564 | 99.868 | 129.432 | 0.05699 | 0.18181 | 0.23880 | 91.03 | 0.1698 | 3.74871 | 123 |
| 124 | 0.092500 | 14.714 | 2.176 | 16.890 | 29.805 | 103.102 | 132.907 | 0.05740 | 0.18739 | 0.24480 | 92.03 | 0.1715 | 3.85298 | 124 |
| 125 | 0.095456 | 14.739 | 2.250 | 16.989 | 30.045 | 106.437 | 136.482 | 0.05781 | 0.19314 | 0.25096 | 93.03 | 0.1732 | 3.95961 | 125 |
| 126 | 0.098504 | 14.764 | 2.325 | 17.090 | 30.286 | 109.877 | 140.163 | 0.05823 | 0.19907 | 0.25729 | 94.03 | 0.1749 | 4.06863 | 126 |
| 127 | 0.101657 | 14.789 | 2.404 | 17.193 | 30.527 | 113.438 | 143.965 | 0.05864 | 0.20519 | 0.26382 | 95.03 | 0.1766 | 4.18046 | 127 |
| 128 | 0.104910 | 14.815 | 2.485 | 17.299 | 30.767 | 117.111 | 147.878 | 0.05905 | 0.21149 | 0.27054 | 96.03 | 0.1783 | 4.29477 | 128 |
| 129 | 0.108270 | 14.840 | 2.569 | 17.409 | 31.008 | 120.908 | 151.916 | 0.05946 | 0.21810 | 0.27745 | 97.03 | 0.1800 | 4.41181 | 129 |

## Thermodynamics and Psychrometrics

**Table 2-4  Thermodynamic Properties of Moist Air at Standard Sea Level Pressure, 14.696 psi (29.921 in. Hg)** (*Continued*)

*(Table 2, Chapter 1, 2013 ASHRAE Handbook—Fundamentals)*

| Temp., °F  t | Humidity Ratio, lb$_w$/lb$_{da}$  $W_s$ | Specific Volume, ft³/lb$_{da}$  $v_{da}$ | $v_{as}$ | $v_s$ | Specific Enthalpy, Btu/lb$_{da}$  $h_{da}$ | $h_{as}$ | $h_s$ | Specific Entropy, Btu/lb$_{da}$·°F  $s_{da}$ | $s_{as}$ | $s_s$ | Condensed Water Specific Enthalpy, Btu/lb$_w$  $h_w$ | Specific Entropy, Btu/lb$_w$·°F  $s_w$ | Vapor Pressure, in. Hg  $p_s$ | Temp., °F  t |
|---|---|---|---|---|---|---|---|---|---|---|---|---|---|---|
| 130 | 0.111738 | 14.865 | 2.655 | 17.520 | 31.249 | 124.828 | 156.076 | 0.05986 | 0.22470 | 0.28457 | 98.03 | 0.1817 | 4.53148 | 130 |
| 131 | 0.115322 | 14.891 | 2.745 | 17.635 | 31.489 | 128.880 | 160.370 | 0.06027 | 0.23162 | 0.29190 | 99.02 | 0.1834 | 4.65397 | 131 |
| 132 | 0.119023 | 14.916 | 2.837 | 17.753 | 31.730 | 133.066 | 164.796 | 0.06068 | 0.23876 | 0.29944 | 100.02 | 0.1851 | 4.77919 | 132 |
| 133 | 0.122855 | 14.941 | 2.934 | 17.875 | 31.971 | 137.403 | 169.374 | 0.06109 | 0.24615 | 0.30723 | 101.02 | 0.1868 | 4.90755 | 133 |
| 134 | 0.126804 | 14.966 | 3.033 | 17.999 | 32.212 | 141.873 | 174.084 | 0.06149 | 0.25375 | 0.31524 | 102.02 | 0.1885 | 5.03844 | 134 |
| 135 | 0.130895 | 14.992 | 3.136 | 18.127 | 32.452 | 146.504 | 178.957 | 0.06190 | 0.26161 | 0.32351 | 103.02 | 0.1902 | 5.17258 | 135 |
| 136 | 0.135124 | 15.017 | 3.242 | 18.259 | 32.693 | 151.294 | 183.987 | 0.06230 | 0.26973 | 0.33203 | 104.02 | 0.1919 | 5.30973 | 136 |
| 137 | 0.139494 | 15.042 | 3.352 | 18.394 | 32.934 | 156.245 | 189.179 | 0.06271 | 0.27811 | 0.34082 | 105.02 | 0.1935 | 5.44985 | 137 |
| 138 | 0.144019 | 15.067 | 3.467 | 18.534 | 33.175 | 161.374 | 194.548 | 0.06311 | 0.28707 | 0.35018 | 106.02 | 0.1952 | 5.59324 | 138 |
| 139 | 0.148696 | 15.093 | 3.585 | 18.678 | 33.415 | 166.677 | 200.092 | 0.06351 | 0.29602 | 0.35954 | 107.02 | 0.1969 | 5.73970 | 139 |
| 140 | 0.153538 | 15.118 | 3.708 | 18.825 | 33.656 | 172.168 | 205.824 | 0.06391 | 0.30498 | 0.36890 | 108.02 | 0.1985 | 5.88945 | 140 |
| 141 | 0.158643 | 15.143 | 3.835 | 18.978 | 33.897 | 177.857 | 211.754 | 0.06431 | 0.31456 | 0.37887 | 109.02 | 0.2002 | 6.04256 | 141 |
| 142 | 0.163748 | 15.168 | 3.967 | 19.135 | 34.138 | 183.754 | 217.892 | 0.06471 | 0.32446 | 0.38918 | 110.02 | 0.2019 | 6.19918 | 142 |
| 143 | 0.169122 | 15.194 | 4.103 | 19.297 | 34.379 | 189.855 | 224.233 | 0.06511 | 0.33470 | 0.39981 | 111.02 | 0.2035 | 6.35898 | 143 |
| 144 | 0.174694 | 15.219 | 4.245 | 19.464 | 34.620 | 196.183 | 230.802 | 0.06551 | 0.34530 | 0.41081 | 112.02 | 0.2052 | 6.52241 | 144 |
| 145 | 0.180467 | 15.244 | 4.392 | 19.637 | 34.860 | 202.740 | 237.600 | 0.06591 | 0.35626 | 0.42218 | 113.02 | 0.2068 | 6.68932 | 145 |
| 146 | 0.186460 | 15.269 | 4.545 | 19.815 | 35.101 | 209.550 | 244.651 | 0.06631 | 0.36764 | 0.43395 | 114.02 | 0.2085 | 6.86009 | 146 |
| 147 | 0.192668 | 15.295 | 4.704 | 19.999 | 35.342 | 216.607 | 251.949 | 0.06671 | 0.37941 | 0.44611 | 115.02 | 0.2101 | 7.03435 | 147 |
| 148 | 0.199110 | 15.320 | 4.869 | 20.189 | 35.583 | 223.932 | 259.514 | 0.06710 | 0.39160 | 0.45871 | 116.02 | 0.2118 | 7.21239 | 148 |
| 149 | 0.205792 | 15.345 | 5.040 | 20.385 | 35.824 | 231.533 | 267.356 | 0.06750 | 0.40424 | 0.47174 | 117.02 | 0.2134 | 7.39413 | 149 |
| 150 | 0.212730 | 15.370 | 5.218 | 20.589 | 36.064 | 239.426 | 275.490 | 0.06790 | 0.41735 | 0.48524 | 118.02 | 0.2151 | 7.57977 | 150 |
| 151 | 0.219945 | 15.396 | 5.404 | 20.799 | 36.305 | 247.638 | 283.943 | 0.06829 | 0.43096 | 0.49925 | 119.02 | 0.2167 | 7.76958 | 151 |
| 152 | 0.227429 | 15.421 | 5.596 | 21.017 | 36.546 | 256.158 | 292.705 | 0.06868 | 0.44507 | 0.51375 | 120.02 | 0.2184 | 7.96366 | 152 |
| 153 | 0.235218 | 15.446 | 5.797 | 21.243 | 36.787 | 265.028 | 301.816 | 0.06908 | 0.45973 | 0.52881 | 121.02 | 0.2200 | 8.16087 | 153 |
| 154 | 0.243309 | 15.471 | 6.005 | 21.477 | 37.028 | 274.245 | 311.273 | 0.06947 | 0.47494 | 0.54441 | 122.02 | 0.2216 | 8.36256 | 154 |
| 155 | 0.251738 | 15.497 | 6.223 | 21.720 | 37.269 | 283.849 | 321.118 | 0.06986 | 0.49077 | 0.56064 | 123.02 | 0.2233 | 8.56871 | 155 |
| 156 | 0.260512 | 15.522 | 6.450 | 21.972 | 37.510 | 293.849 | 331.359 | 0.07025 | 0.50723 | 0.57749 | 124.02 | 0.2249 | 8.77915 | 156 |
| 157 | 0.269644 | 15.547 | 6.686 | 22.233 | 37.751 | 304.261 | 342.012 | 0.07065 | 0.52434 | 0.59499 | 125.02 | 0.2265 | 8.99378 | 157 |
| 158 | 0.279166 | 15.572 | 6.933 | 22.505 | 37.992 | 315.120 | 353.112 | 0.07104 | 0.54217 | 0.61320 | 126.02 | 0.2281 | 9.21297 | 158 |
| 159 | 0.289101 | 15.598 | 7.190 | 22.788 | 38.233 | 326.452 | 364.685 | 0.07143 | 0.56074 | 0.63216 | 127.02 | 0.2297 | 9.43677 | 159 |
| 160 | 0.29945 | 15.623 | 7.459 | 23.082 | 38.474 | 338.263 | 376.737 | 0.07181 | 0.58007 | 0.65188 | 128.02 | 0.2314 | 9.6648 | 160 |
| 161 | 0.31027 | 15.648 | 7.740 | 23.388 | 38.715 | 350.610 | 389.325 | 0.07220 | 0.60025 | 0.67245 | 129.02 | 0.2330 | 9.8978 | 161 |
| 162 | 0.32156 | 15.673 | 8.034 | 23.707 | 38.956 | 363.501 | 402.457 | 0.07259 | 0.62128 | 0.69388 | 130.03 | 0.2346 | 10.1353 | 162 |
| 163 | 0.33336 | 15.699 | 8.341 | 24.040 | 39.197 | 376.979 | 416.175 | 0.07298 | 0.64325 | 0.71623 | 131.03 | 0.2362 | 10.3776 | 163 |
| 164 | 0.34572 | 15.724 | 8.664 | 24.388 | 39.438 | 391.095 | 430.533 | 0.07337 | 0.66622 | 0.73959 | 132.03 | 0.2378 | 10.6250 | 164 |
| 165 | 0.35865 | 15.749 | 9.001 | 24.750 | 39.679 | 405.865 | 445.544 | 0.07375 | 0.69022 | 0.76397 | 133.03 | 0.2394 | 10.8771 | 165 |
| 166 | 0.37220 | 15.774 | 9.355 | 25.129 | 39.920 | 421.352 | 461.271 | 0.07414 | 0.71535 | 0.78949 | 134.03 | 0.2410 | 11.1343 | 166 |
| 167 | 0.38639 | 15.800 | 9.726 | 25.526 | 40.161 | 437.578 | 477.739 | 0.07452 | 0.74165 | 0.81617 | 135.03 | 0.2426 | 11.3965 | 167 |
| 168 | 0.40131 | 15.825 | 10.117 | 25.942 | 40.402 | 454.630 | 495.032 | 0.07491 | 0.76925 | 0.84415 | 136.03 | 0.2442 | 11.6641 | 168 |
| 169 | 0.41698 | 15.850 | 10.527 | 26.377 | 40.643 | 472.554 | 513.197 | 0.07529 | 0.79821 | 0.87350 | 137.04 | 0.2458 | 11.9370 | 169 |
| 170 | 0.43343 | 15.875 | 10.959 | 26.834 | 40.884 | 491.372 | 532.256 | 0.07567 | 0.82858 | 0.90425 | 138.04 | 0.2474 | 12.2149 | 170 |
| 171 | 0.45079 | 15.901 | 11.414 | 27.315 | 41.125 | 511.231 | 552.356 | 0.07606 | 0.86058 | 0.93664 | 139.04 | 0.2490 | 12.4988 | 171 |
| 172 | 0.46905 | 15.926 | 11.894 | 27.820 | 41.366 | 532.138 | 573.504 | 0.07644 | 0.89423 | 0.97067 | 140.04 | 0.2506 | 12.7880 | 172 |
| 173 | 0.48829 | 15.951 | 12.400 | 28.352 | 41.607 | 554.160 | 595.767 | 0.07682 | 0.92962 | 1.00644 | 141.04 | 0.2521 | 13.0823 | 173 |
| 174 | 0.50867 | 15.976 | 12.937 | 28.913 | 41.848 | 577.489 | 619.337 | 0.07720 | 0.96707 | 1.04427 | 142.04 | 0.2537 | 13.3831 | 174 |
| 175 | 0.53019 | 16.002 | 13.504 | 29.505 | 42.089 | 602.139 | 644.229 | 0.07758 | 1.00657 | 1.08416 | 143.05 | 0.2553 | 13.6894 | 175 |
| 176 | 0.55294 | 16.027 | 14.103 | 30.130 | 42.331 | 628.197 | 670.528 | 0.07796 | 1.04828 | 1.12624 | 144.05 | 0.2569 | 14.0010 | 176 |
| 177 | 0.57710 | 16.052 | 14.741 | 30.793 | 42.572 | 655.876 | 698.448 | 0.07834 | 1.09253 | 1.17087 | 145.05 | 0.2585 | 14.3191 | 177 |
| 178 | 0.60274 | 16.078 | 15.418 | 31.496 | 42.813 | 685.260 | 728.073 | 0.07872 | 1.13943 | 1.21815 | 146.05 | 0.2600 | 14.6430 | 178 |
| 179 | 0.63002 | 16.103 | 16.139 | 32.242 | 43.054 | 716.524 | 759.579 | 0.07910 | 1.18927 | 1.26837 | 147.06 | 0.2616 | 14.9731 | 179 |
| 180 | 0.65911 | 16.128 | 16.909 | 33.037 | 43.295 | 749.871 | 793.166 | 0.07947 | 1.24236 | 1.32183 | 148.06 | 0.2632 | 15.3097 | 180 |
| 181 | 0.69012 | 16.153 | 17.730 | 33.883 | 43.536 | 785.426 | 828.962 | 0.07985 | 1.29888 | 1.37873 | 149.06 | 0.2647 | 15.6522 | 181 |
| 182 | 0.72331 | 16.178 | 18.609 | 34.787 | 43.778 | 823.487 | 867.265 | 0.08023 | 1.35932 | 1.43954 | 150.06 | 0.2663 | 16.0014 | 182 |
| 183 | 0.75885 | 16.204 | 19.551 | 35.755 | 44.019 | 864.259 | 908.278 | 0.08060 | 1.42396 | 1.50457 | 151.07 | 0.2679 | 16.3569 | 183 |
| 184 | 0.79703 | 16.229 | 20.564 | 36.793 | 44.260 | 908.061 | 952.321 | 0.08098 | 1.49332 | 1.57430 | 152.07 | 0.2694 | 16.7190 | 184 |
| 185 | 0.83817 | 16.254 | 21.656 | 37.910 | 44.501 | 955.261 | 999.763 | 0.08135 | 1.56797 | 1.64932 | 153.07 | 0.2710 | 17.0880 | 185 |
| 186 | 0.88251 | 16.280 | 22.834 | 39.113 | 44.742 | 1006.149 | 1050.892 | 0.08172 | 1.64834 | 1.73006 | 154.08 | 0.2725 | 17.4634 | 186 |
| 187 | 0.93057 | 16.305 | 24.111 | 40.416 | 44.984 | 1061.314 | 1106.298 | 0.08210 | 1.73534 | 1.81744 | 155.08 | 0.2741 | 17.8462 | 187 |
| 188 | 0.98272 | 16.330 | 25.498 | 41.828 | 45.225 | 1121.174 | 1166.399 | 0.08247 | 1.82963 | 1.91210 | 156.08 | 0.2756 | 18.2357 | 188 |
| 189 | 1.03951 | 16.355 | 27.010 | 43.365 | 45.466 | 1186.382 | 1231.848 | 0.08284 | 1.93221 | 2.01505 | 157.09 | 0.2772 | 18.6323 | 189 |
| 190 | 1.10154 | 16.381 | 28.661 | 45.042 | 45.707 | 1257.614 | 1303.321 | 0.08321 | 2.04412 | 2.12733 | 158.09 | 0.2787 | 19.0358 | 190 |
| 191 | 1.16965 | 16.406 | 30.476 | 46.882 | 45.949 | 1335.834 | 1381.783 | 0.08359 | 2.16684 | 2.25043 | 159.09 | 0.2803 | 19.4468 | 191 |
| 192 | 1.24471 | 16.431 | 32.477 | 48.908 | 46.190 | 1422.047 | 1468.238 | 0.08396 | 2.30193 | 2.38589 | 160.10 | 0.2818 | 19.8652 | 192 |
| 193 | 1.32788 | 16.456 | 34.695 | 51.151 | 46.431 | 1517.581 | 1564.013 | 0.08433 | 2.45144 | 2.53576 | 161.10 | 0.2834 | 20.2913 | 193 |
| 194 | 1.42029 | 16.481 | 37.161 | 53.642 | 46.673 | 1623.758 | 1670.430 | 0.08470 | 2.61738 | 2.70208 | 162.11 | 0.2849 | 20.7244 | 194 |
| 195 | 1.52396 | 16.507 | 39.928 | 56.435 | 46.914 | 1742.879 | 1789.793 | 0.08506 | 2.80332 | 2.88838 | 163.11 | 0.2864 | 21.1661 | 195 |
| 196 | 1.64070 | 16.532 | 43.046 | 59.578 | 47.155 | 1877.032 | 1924.188 | 0.08543 | 3.01244 | 3.09787 | 164.12 | 0.2880 | 21.6152 | 196 |
| 197 | 1.77299 | 16.557 | 46.580 | 63.137 | 47.397 | 2029.069 | 2076.466 | 0.08580 | 3.24914 | 3.33494 | 165.12 | 0.2895 | 22.0714 | 197 |
| 198 | 1.92472 | 16.583 | 50.636 | 67.218 | 47.638 | 2203.464 | 2251.102 | 0.08617 | 3.52030 | 3.60647 | 166.13 | 0.2910 | 22.5367 | 198 |
| 199 | 2.09975 | 16.608 | 55.316 | 71.923 | 47.879 | 2404.668 | 2452.547 | 0.08653 | 3.83275 | 3.91929 | 167.13 | 0.2926 | 23.0092 | 199 |
| 200 | 2.30454 | 16.633 | 60.793 | 77.426 | 48.121 | 2640.084 | 2688.205 | 0.08690 | 4.19787 | 4.28477 | 168.13 | 0.2941 | 23.4906 | 200 |

$$h_g = 2501 + 1.805t \text{ (kJ/kg)} \tag{2-34}$$

where $t$ is the dry-bulb temperature, °C. The moist air enthalpy then becomes

$$h = t + W(2501 + 1.805t) \text{ (kJ/kg dry air)} \tag{2-35}$$

In conventional (I-P) units

$$h_a = 0.240t \text{ (Btu/lb)} \tag{2-36}$$

$$h_g = 1061 + 0.444t \text{ (Btu/lb)} \tag{2-37}$$

$$h = 0.240t + W(1061 + 0.444t) \text{ (Btu/lb)} \tag{2-38}$$

where $t$ is the dry-bulb temperature, °F.

Table 2-4 is a tabulation of the thermodynamic properties of moist air at sea level standard atmospheric pressure. Table 2-5 shows variation of atmospheric pressure with altitude.

### 2.2.14 Psychrometric Chart

The ASHRAE psychrometric chart may be used to solve numerous process problems with moist air. Processes performed with air can be plotted on the chart for quick visualization, as well as for determining changes in significant properties such as temperature, humidity ratio, and enthalpy for the process. ASHRAE Psychrometric Chart No. 1 in I-P units is shown in Figure 2-10. Some basic air-conditioning processes are shown in Figure 2-11.

*Sensible heating* only (C) or *sensible cooling* (G) shows a change in dry-bulb temperature with no change in humidity ratio. For either sensible heat change process, the temperature changes but not the moisture content of the air.

*Humidifying* only (A) or *dehumidifying* only (E) shows a change in humidity ratio with no change in dry-bulb temperature. For these latent heat processes, the moisture content of the air is changed but not the temperature.

**Table 2-5 Standard Atmospheric Data with Altitude**
*(Table 1, Chapter 1, 2013 ASHRAE Handbook—Fundamentals)*

| Altitude, ft | Temperature, °F | Pressure in. Hg | Pressure psia |
|---|---|---|---|
| −1000 | 62.6 | 31.02 | 15.236 |
| −500 | 60.8 | 30.47 | 14.966 |
| 0 | 59.0 | 29.921 | 14.696 |
| 500 | 57.2 | 29.38 | 14.430 |
| 1000 | 55.4 | 28.86 | 14.175 |
| 2000 | 51.9 | 27.82 | 13.664 |
| 3000 | 48.3 | 26.82 | 13.173 |
| 4000 | 44.7 | 25.82 | 12.682 |
| 5000 | 41.2 | 24.90 | 12.230 |
| 6000 | 37.6 | 23.98 | 11.778 |
| 7000 | 34.0 | 23.09 | 11.341 |
| 8000 | 30.5 | 22.22 | 10.914 |
| 9000 | 26.9 | 21.39 | 10.506 |
| 10,000 | 23.4 | 20.58 | 10.108 |
| 15,000 | 5.5 | 16.89 | 8.296 |
| 20,000 | −12.3 | 13.76 | 6.758 |
| 30,000 | −47.8 | 8.90 | 4.371 |
| 40,000 | −69.7 | 5.56 | 2.731 |
| 50,000 | −69.7 | 3.44 | 1.690 |
| 60,000 | −69.7 | 2.14 | 1.051 |

*Source:* Adapted from NASA (1976).

*Cooling and dehumidifying* (F) result in a reduction of both the dry-bulb temperature and the humidity ratio. Cooling coils generally perform this type of process.

*Heating and humidifying* (B) result in an increase of both the dry-bulb temperature and the humidity ratio.

*Chemical dehumidifying* (D) is a process in which moisture from the air is absorbed or adsorbed by a hygroscopic material. Generally, the process essentially occurs at constant enthalpy.

*Evaporative cooling* only (H) is an adiabatic heat transfer process in which the wet-bulb temperature of the air remains constant but the dry-bulb temperature drops as the humidity rises.

*Adiabatic mixing* of air at one condition with air at some other condition is represented on the psychrometric chart by a straight line drawn between the points representing the two air conditions (Figure 2-12).

## 2.3 Forms of Energy

### 2.3.1 Energy

Energy is the capacity for producing an effect. Thermodynamics is founded on the *law of conservation of energy*, which says that energy can neither be created nor destroyed. Heat and work are transitory forms of energy, losing their identity as soon as they are absorbed by the body or region to which they are delivered. Work and heat are not possessed by a system and, therefore, are not properties. Thus, if there is a net transfer of energy across the boundary from a system (as heat and/or work), from where did this energy come? The only answer is that it must have come from a store of energy in the given system. These stored forms of energy may be assumed to reside within the bodies or regions with which they are associated. In thermodynamics, accent is placed on the changes of stored energy rather than on absolute quantities.

### 2.3.2 Stored Forms of Energy

Energy may be stored in such forms as thermal (internal), mechanical, electrical, chemical, and atomic (nuclear).

**Internal Energy, $U$.** Internal (thermal) energy is the energy possessed by matter due to the motion and/or position of its molecules. This energy is comprised of two components: (1) kinetic internal energy—due to the velocity of the molecules and manifested by temperature; and (2) potential internal energy—due to the attractive forces existing between molecules and manifested by the phase of the system.

**Potential Energy, P.E.** Potential energy is the energy possessed by the system due to the elevation or position of the system. This potential energy is equivalent to the work required to lift the system from the arbitrary datum (0 elevation) to its elevation $z$ in the absence of friction.

# Thermodynamics and Psychrometrics

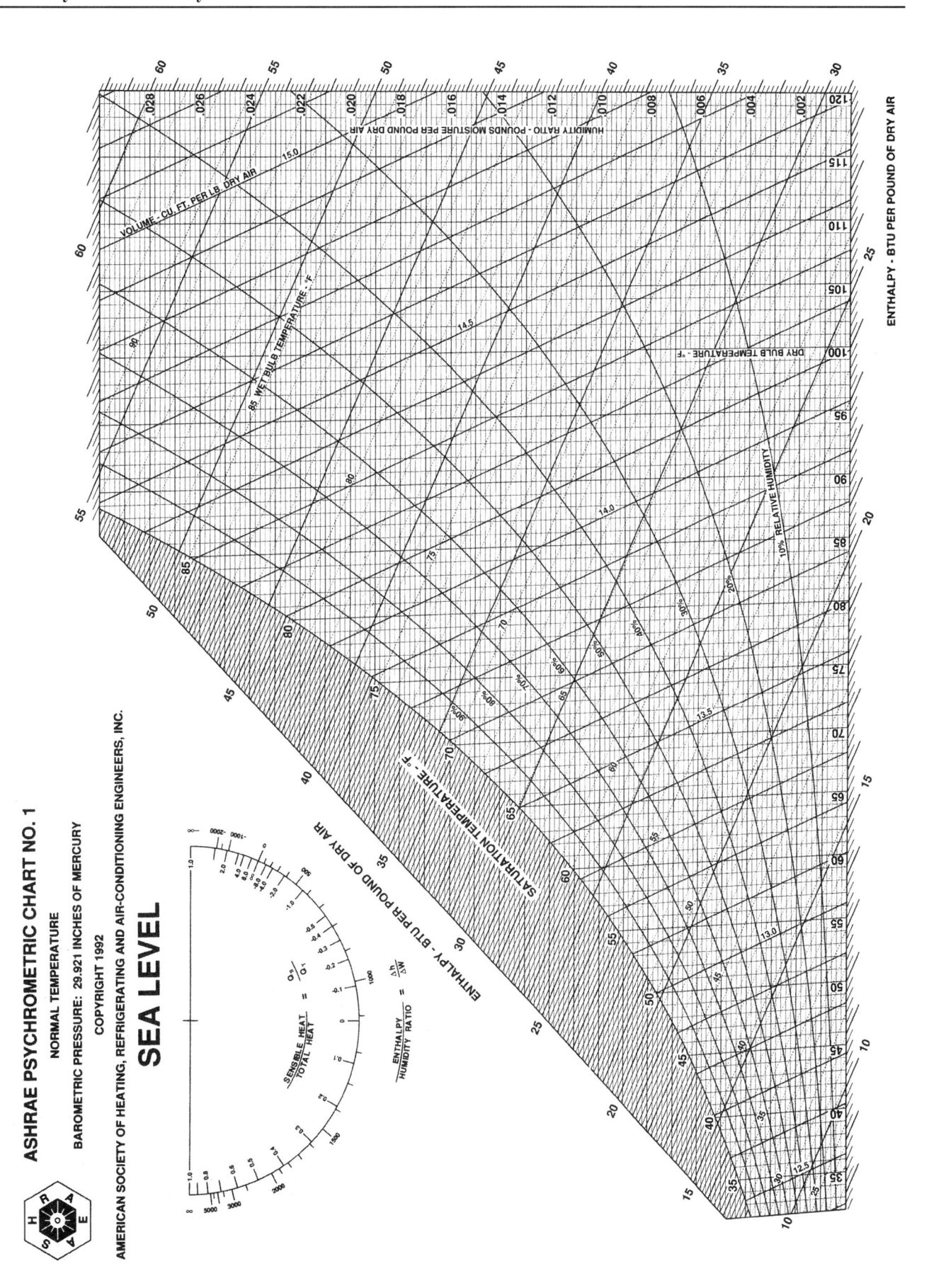

*Fig. 2-10 ASHRAE Psychrometric Chart No. 1*

$$F = \frac{ma}{g_c} = \frac{mg}{g_c}$$

$$\text{P.E.} = W = \int_0^x F\,dx = \int_0^z m\frac{g}{g_c}dx = m\frac{g}{g_c}z \quad (2\text{-}39)$$

**Kinetic Energy, K.E.** Kinetic energy is the energy possessed by the system as a result of the velocity of the system. It is equal to the work that could be done in bringing to rest a system that is in motion, with a velocity $V$, in the absence of gravity.

$$F = \frac{ma}{g_c} = \frac{m}{g_c}\frac{dV}{dt} \quad (2\text{-}40)$$

$$\text{K.E.} = W = \int_0^x F\,dx = -\int_V^0 \frac{m}{g_c}\frac{dV}{dt}dx$$

$$= -\int_V^0 \frac{m}{g_c}V\,dV = \frac{mV^2}{2g_c}$$

**Chemical Energy, $E_c$.** Chemical energy is possessed by the system because of the arrangement of the atoms composing the molecules. Reactions that liberate energy are termed *exothermic* and those that absorb energy are termed *endothermic*.

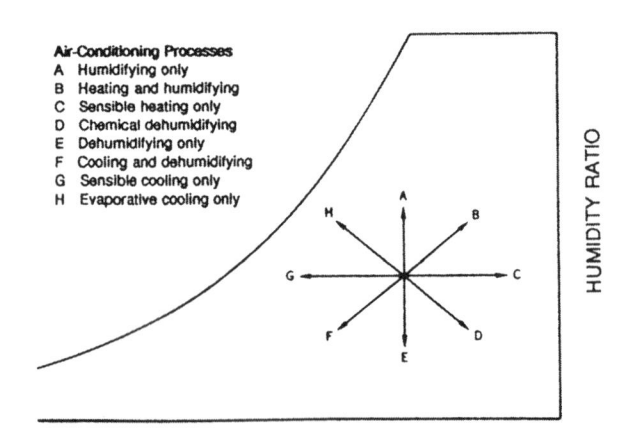

*Fig. 2-11 Psychrometric Representations of Basic Air-Conditioning Process*

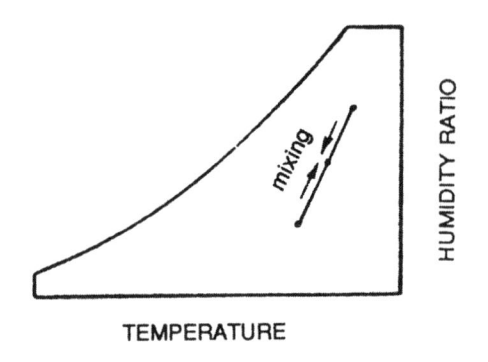

*Fig. 2-12 Adiabatic Mixing*

**Nuclear (Atomic) Energy, $E_a$.** Nuclear energy is possessed by the system due to the cohesive forces holding the protons and neutrons together as the nucleus of the atom.

*Stored* energy is concerned with

- Molecules of the system (internal energy)
- The system as a unit (kinetic and potential energy)
- Arrangement of the atoms (chemical)
- Nucleus of the atom (nuclear)

Molecular stored energy is associated with the relative position and velocity of the molecules; the total effect is called internal energy. Kinetic energy and potential energy are both forms of mechanical energy, and they can be converted readily and completely into work. Chemical, electrical, and atomic energy would be included in any accounting of stored energy; however, engineering thermodynamics frequently confines itself to systems not undergoing changes in these three forms of energy.

### 2.3.3 Transient Forms of Energy

**Heat, $Q$.** Heat is the mechanism by which energy is transferred across the boundary between systems by reason of the difference in temperature of the two systems, and always in the direction of the lower temperature. Being transitory, heat is not a property. It is redundant to speak of heat as being transferred, for the term *heat* signifies energy in transit. Nevertheless, in keeping with common usage, this text refers to heat as being transferred.

Although a body or system cannot contain heat, it is useful, when discussing many processes, to speak of heat received or heat rejected, so that the direction of heat transfer relative to the system is obvious. This should not be construed as treating heat as a substance.

Heat *transferred* to a system is considered to be positive, and *heat transferred* from a system, negative. Thus, positive heat represents energy transferred to a system, and negative heat represents energy transferred from a system. A process in which there is no heat transfer ($Q = 0$) is called an **adiabatic process**.

**Work.** Work is the mechanism by which energy is transferred across the boundary between systems by reason of the difference in pressure (or force of any kind) of the two systems, and is in the direction of the lower pressure.

If the total effect produced in the system can be reduced to the raising of a weight, then nothing but work has crossed the boundary. Work, like heat, is not possessed by the system, but occurs only as energy being transferred.

Work is, by definition, the energy resulting from a force having moved through a distance. If the force varies with distance $x$, work may be expressed as $\delta W = F\,dx$.

In thermodynamics, work is often done by a force distributed over an area, i.e., by pressure $p$ acting through volume $V$, as in the case of fluid pressure exerted on a piston. In this event,

$$\delta W = p\,dV$$

where $p$ is an external pressure exerted on the system.

**Mechanical** or **shaft work** $W$ is the energy delivered or absorbed by a mechanism, such as a turbine, air compressor, or internal combustion engine. Shaft work can always be evaluated from the basic relation for work.

**Power** is the rate of doing work.

Work done *by* a system is considered positive; work done *on* a system is considered negative. The symbol $W$ designates the work done by a system.

Work may be done on or by a system in a variety of ways. In addition to mechanical work and flow work (the types most frequently encountered in thermodynamics), work may be done due to surface tension, the flow of electricity, magnetic fields, and in other ways.

For nonflow processes, the form of mechanical work most frequently encountered is that done at the moving boundary of a system, such as the work done in moving the piston in a cylinder. It may be expressed in equation form for reversible processes as $W = \int p\,dV$. Generally, for the nonflow process, work can be expressed as

$$W = \int p\,dv\ldots \quad (2\text{-}41)$$

where the dots indicate other ways in which work can be done by or on the system.

The following section shows the derivation of a useful expression for the work of a frictionless steady-flow process. The derivation procedure is

1. Make a free-body diagram of an element of fluid.
2. Evaluate the external forces on the free body.
3. Relate the sum of the external forces to the mass and acceleration of the free body.
4. Solve the resulting relation for the force by which work is done on the fluid.
5. Apply the definition of work as $\int F\,ds$. A free-body diagram is illustrated in Figure 2-13.

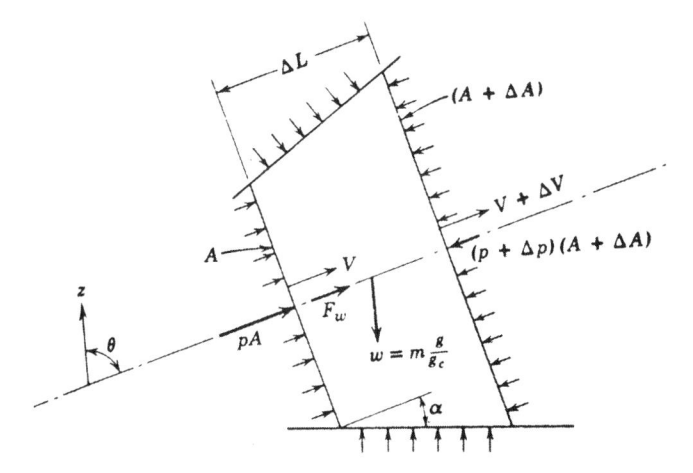

*Fig. 2-13 Element of Fluid in Frictionless Steady Flow*

Applying Newton's second law of motion, the sum of the external forces on the fluid element must equal $ma/g_c$. The mass of the element is $\rho(A + \Delta A/2)\Delta L$, and the acceleration is approximately $\Delta V/\Delta \tau$. Thus

$$\Sigma F = \frac{ma}{g_c} = \frac{\rho}{g_c}\left(A + \frac{\Delta A}{2}\right)\Delta L \frac{\Delta V}{\Delta \tau}$$

The sum or resultant of the forces is

$$\Sigma F = pA - (p + \Delta p)(A + \Delta A) - m\frac{g}{g_c}\cos\theta$$

$$+ \left(p + \frac{\Delta p}{2}\right)\Delta A + F_W$$

$$= -A\Delta p - \frac{\Delta p \Delta A}{3} - m\frac{g}{g_c}\cos\theta + F_w$$

$$\text{Work}_{in} = F_W \Delta L = (\text{Volume})\Delta p + \frac{V\Delta V}{g_C} + \frac{g}{g_c}\Delta z$$

and per unit mass

$$\text{Work}_{in} = v\Delta p + \frac{V\Delta V}{g_C} + \frac{g}{g_c}\Delta Z$$

Now, if $\Delta L$ is made to approach $dL$, then the other differences also approach differentials, and the work (per unit mass) done on the fluid in the distance $dL$ is

$$\delta\text{Work}_{in} = v\,dp + \frac{V\,dV}{g_c} + \frac{g}{g_c}dz \quad (2\text{-}42)$$

or, for flow between sections a finite distance apart

$$\text{Work}_{in} = \int v\,dp + \Delta\left(\frac{V^2}{2g_c}\right) + \frac{g}{g_c}\Delta z$$

This important relation shows the mechanical work done on a unit mass of fluid in a frictionless steady-flow process.

In addition to commonly encountering work done at a moving boundary in an **open system**, flow work must be considered. **Flow work** consists of the energy carried into or transmitted across the system boundary because of the work done by the fluid just outside the system on the adjacent fluid entering the system to force or push it into the system. Flow work also occurs as fluid leaves the system, this time by fluid in the system on the fluid just leaving the system. As an analogy, consider two people as particles of fluid, one in the doorway and one just outside. Flow work would be done by the person outside if he or she shoved the person in the doorway into the room (system).

$$\text{Flow work} = \int F\,dx = \int p\,dAx, \quad p = c$$
$$= p\int_0^v dV = pv \quad (2\text{-}43)$$

where flow work is per unit mass and $v$ is the specific volume, or the volume displaced per unit mass.

This analysis shows that work must be done in causing fluid to flow into or out of a system. This work is called flow work. Such terms as flow energy and displacement energy are sometimes used.

## 2.4 First Law of Thermodynamics

The first law of thermodynamics is often called the *law of conservation of energy*. From the first law, it can be concluded that for any system, open or closed, there is an energy balance as

$$\begin{array}{c}\text{Net amount of}\\\text{energy added}\\\text{to system}\end{array} = \begin{array}{c}\text{Net increase in}\\\text{stored energy}\\\text{of system}\end{array}$$

or

Energy in − Energy out =

Increase in energy in system     (2-44)

With both open and closed systems, energy can be added to or taken from the system by means of heat and work. In an open system, there is an additional mechanism for increasing or decreasing the stored energy of the system. When mass enters a system, the stored energy of the system is increased by the stored energy of the entering mass. The stored energy of a system is decreased whenever mass leaves the system because the mass leaving the system takes some stored energy with it. To distinguish this transfer of stored energy of mass crossing the system boundary from heat and work, consider

$$\begin{array}{c}\text{Stored energy}\\\text{of mass}\\\text{entering system}\end{array} - \begin{array}{c}\text{Stored energy}\\\text{of mass}\\\text{leaving system}\end{array}$$
$$+ \begin{array}{c}\text{Net energy added}\\\text{to system}\\\text{heat and work}\end{array} = \begin{array}{c}\text{Net increase in}\\\text{stored energy}\\\text{of system}\end{array}$$

The net exchange of energy between the system and its surroundings must be balanced by the change in energy of the system. Exchange of energy in transition includes either work or heat. However, what is meant by the energy of the system and the energy associated with any matter entering or leaving the system must be described further.

The energy $E$ of the system is a property of the system and consists of all of the various forms in which energy is characteristic. These forms include potential energy (due to position), kinetic energy (due to any motion), electrical energy (due to charge), and so forth. Because work and heat are energy in transition and are not characteristic of the system, they are not included here.

All the energy of a system—exclusive of kinetic and potential energy—is called *internal energy*. The symbol for internal energy per unit mass is $u$. The symbol for internal energy contained in a mass of $m$ pounds or kilograms is $U$. Each unit of mass flowing into or out of the system carries with it the energy characteristic of that unit of mass. This energy includes the internal energy $u$ plus the kinetic and potential energy.

Work is always done on or by a system where fluid flows across the system boundary. Therefore, work in an energy balance for an open system is usually separated into two parts: (1) the work required to push a fluid into or out of the system; and (2) all other forms of work.

The work flow per unit mass crossing the boundary of a system is $pv$. If the pressure, specific volume, or both vary as a fluid flows across a system boundary, the flow work is calculated by integrating $\int pv\delta m$, where $\delta m$ is an infinitesimal mass crossing the boundary. The symbol $\delta m$ is used instead of $dm$ because the amount of mass crossing the boundary is not a property. The mass within the system is a property, so the infinitesimal change in mass within the system is properly represented by $dm$.

Since the work term in an energy balance for an open system is usually separated into two parts, (1) flow work and (2) all other forms of work, the term work $W$ without modifiers stands for all other forms of work except flow work, and the complete, two-word name is always used when referring to flow work.

An equation representing the first law can be written using the symbols defined for the general system of Figure 2-14. Referring to Figure 2-14, let $\delta m_1$ be the mass entering the system and $\delta m_2$ be the mass leaving. The first law in differential or incremental form becomes

$$[\delta m(e + pv)]_{\text{in}} - [\delta m(e + pv)]_{\text{out}} + \delta Q - \delta W = dE \quad (2\text{-}45)$$

or

$$\delta m_1 \left( u_1 + p_1 v_1 + \frac{V_1^2}{2g_c} + z_1 \frac{g}{g_c} \right)$$
$$- \delta m_2 \left( u_2 + p_2 v_2 + \frac{V_2^2}{2g_c} + z_2 \frac{g}{g_c} \right) + \delta Q - \delta W = dE$$

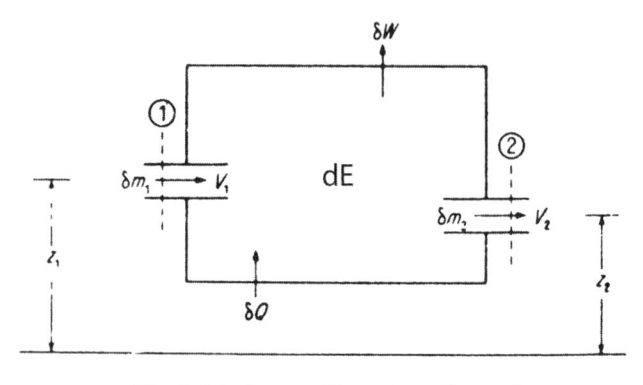

*Fig. 2-14 Energy Flows in a General Thermodynamic System*

where $\delta Q$ and $\delta W$ are the increments of work and heat, and $dE$ is the differential change in the energy of the system. Because $E$ or $U$ (or $e$ or $u$) are properties of the system, they are treated like any other property such as temperature, pressure, density, or viscosity.

The combination of properties $u + pv$ is also a property, which has been defined as enthalpy. The symbol $H$ stands for the total enthalpy associated with a mass $m$; $h$ stands for specific enthalpy, or enthalpy per unit mass. In equation form

$$h = u + pv$$

In terms of enthalpy, the first law equation becomes

$$\delta m_1 \left( h_1 + \frac{V_1^2}{2g_c} + \frac{g}{g_c} z_1 \right)$$
$$- \delta m_2 \left( h_2 + \frac{V_2^2}{2g_c} + \frac{g}{g_c} z_2 \right) + \delta Q - \delta W = dE \quad (2\text{-}46)$$

or in integrated form,

$$\int_0^{m_1} \delta m_1 \left( h_1 + \frac{V_1^2}{2g_c} + \frac{g}{g_c} z_1 \right)$$
$$- \int_0^{m_2} \delta m_2 \left( h_2 + \frac{V_2^2}{2g_c} + \frac{g}{g_c} z_2 \right) \quad (2\text{-}47)$$
$$+ \delta Q - \delta W = E_{\text{final}} - E_{\text{initial}}$$

or, if divided by the time interval $\Delta \tau$,

$$\frac{\delta m_1}{\Delta \tau} \left( h_1 + \frac{V_1^2}{2g_c} + \frac{g}{g_c} z_1 \right) - \frac{\delta m_2}{\Delta \tau} \left( h_2 + \frac{V_2^2}{2g_c} + \frac{g}{g_c} z_2 \right)$$
$$+ \frac{\delta Q}{\Delta \tau} - \frac{\delta W}{\Delta \tau} = \frac{dE}{\Delta \tau}$$

$\Delta \tau \to 0$, $\dfrac{\delta Q}{\Delta \tau} \to \dot{Q}$, $\dfrac{\delta W}{\Delta \tau} \to \dot{W}$

$\dfrac{\delta m_1}{\Delta \tau} \to \dot{m}_1$, $\dfrac{\delta m_2}{\Delta \tau} \to \dot{m}_2$, $\dfrac{dE}{\Delta \tau} \to \dfrac{dE}{d\tau}$

$$\dot{m}_1 \left( h_1 + \frac{V_1^2}{2g_c} + \frac{g}{g_c} z_1 \right) - \dot{m}_2 \left( h_2 + \frac{V_2^2}{2g_c} + \frac{g}{g_c} z_2 \right) \quad (2\text{-}48)$$
$$+ \dot{Q} - \dot{W} = \frac{dE}{d\tau}$$

where $\dot{Q}$ is heat flow and $\dot{W}$ is the work rate or power.

The most general case in which the prior integration of the first law is possible has the following conditions:

1. The properties of the fluids crossing the boundary remain constant at each point on the boundary.

2. The flow rate at each section where mass crosses the boundary is constant. (The flow rate cannot change as long as all properties, including velocity, remain constant at each point.)

3. All interactions with the surroundings occur at a steady rate.

Integration then yields

$$\sum m_{\text{in}} \left( u + pv + \frac{V_1^2}{2g_c} + \frac{g}{g_c} z \right)_{\text{in}}$$
$$- \sum m_{\text{out}} \left( u + pv + \frac{V_2^2}{2g_c} + \frac{g}{g_c} z \right)_{\text{out}} + Q - W \quad (2\text{-}49)$$
$$= \left[ m_f \left( u + \frac{V^2}{2g_c} + \frac{gz}{g_c} \right) - m_i \left( u + \frac{V^2}{2g_c} + \frac{gz}{g_c} \right)_i \right]_{\text{system}}$$

A special case in engineering applications is the **steady-flow process**. In steady flow, all quantities associated with the system do not vary with time. Consequently,

$$\sum \dot{m} \left( h + \frac{V^2}{2g_c} + \frac{g}{g_c} z \right)_{\text{in}}$$
$$- \sum \dot{m} \left( h + \frac{V^2}{2g_c} + \frac{g}{g_c} z \right)_{\text{out}} + \dot{Q} - \dot{W} = 0 \quad (2\text{-}50)$$

A second common application is the **closed-stationary system**, for which the first law equation reduces to

$$Q - W = [m(u_f - u_i)]_{\text{system}} \quad (2\text{-}51)$$

**Example 2-1** Nitrogen having a mass of 0.85 kg expands in an irreversible manner doing 572 J of work. The temperature of the nitrogen drops from 81°C to 34°C. Determine the heat transfer.

**Solution:**

First law for stationary closed system

$$Q - W = m(u_f - u_i)$$

For gases

$$U_f - U_i = mc_v(t_f - t_i)$$

$$Q - 572/1000 = 0.85(0.741)(34 - 81)$$

$$Q = -29.03 \text{ kJ} \; (-27.5 \text{ Btu})$$

**Example 2-2** A tank having a volume of 40 ft³ is initially evacuated. This tank is connected to an air line. The air in this line has an internal energy of 80 Btu/lb, a pressure of 100 psia, and a specific volume of 32 ft³/lb. The valve is opened, the tank fills with air until the pressure is 100 psia, and then the valve is closed. The process takes place adiabatically, and kinetic and potential energies are negligible. Determine the final internal energy of the air in the tank in Btu/lb.

**Solution:**

First law

$$m_{in}(u + pv)_{in} = m_f u_f$$

Conservation of mass

$$m_f = m_{in}$$

$$(u + pv)_{in} = u_f = 80 + (100)(144)(32)/778$$

$$= 673 \text{ Btu/lb } (1565 \text{ kJ/kg})$$

## 2.5 Second Law of Thermodynamics

### 2.5.1 Second Law from Classical Thermodynamics

As a generality, the second law deals with the fact that many processes proceed only in one direction, and not in the opposite direction. Everyday examples include the fact that a cup of hot coffee cools to its surroundings, but the cooler surroundings never heat up the warmer cup of coffee; or the fact that the chemical energy in gasoline is used as a car is driven up a hill, but coasting down the hill does not restore the gasoline. The second law of thermodynamics is a formalized statement of such observations.

A system that undergoes a series of processes and returns to its initial state is said to go through a cycle. For the closed system undergoing a cycle, from the first law of thermodynamics,

$$\oint \delta Q = \oint \delta W$$

The symbol $\oint$ stands for the cyclical integral of the increment of heat or work. Any heat supplied to a cycling system must be balanced by an equivalent amount of work done by the system. Conversely, any work done on the cycling system gives off an equivalent amount of heat.

Many examples exist of work being completely converted into heat. However, a cycling system that completely converts heat into work has never been observed, although such complete conversion would not be a violation of the first law. The fact that heat cannot be completely converted into work is the basis for the second law of thermodynamics. The justification for the second law is based on empirical observations.

The second law has been stated in different ways; one is the *Kelvin-Planck statement* of the second law which states, *It is impossible for any cycling device to exchange heat with only a single reservoir and produce work.*

In this context, reservoir refers to a body whose temperature remains constant regardless of how much heat is added to or taken from it. In other words, the Kelvin-Planck statement says that heat cannot be continuously and completely converted into work; a fraction of the heat must be rejected to another reservoir at a lower temperature. The second law thus restricts the first law in relation to the way energy is transferred. Work can be continuously and completely converted into heat, but not vice versa.

If the Kelvin-Planck statement were not true and heat could be completely converted to work, heat might be obtained from a low-temperature source, converted into work, and the work converted back into heat in a region of higher temperature. The net result of this series of events would be the flow of heat from a low-temperature region to a high-temperature region with no other effect. This phenomenon has never been observed and is contrary to all experience.

The Clausius statement of the second law is, *No process is possible whose sole result is the removal of heat from a reservoir at one temperature and the absorption of an equal quantity of heat by a reservoir at a higher temperature.*

Two major consequences related to the Kelvin-Planck and Clausius statements of the second law are the limiting values for thermal efficiency of power systems operating on heat and for the coefficients of performance for heat pumps. In accordance with the Kelvin-Planck concept, the maximum possible thermal efficiency (work output/heat input) of a heat engine operating between temperature reservoirs at $T_H$ and $T_L$ is

$$\eta_{\text{thermal, max}} = [1 - T_L/T_H]\,100$$

The maximum possible coefficient of performance for cooling (cooling effect/work input) is

$$\text{COP}_{c,\text{max}} = T_L/(T_H - T_L)$$

and the maximum possible coefficient of performance for heating (heating effect/work input) is

$$\text{COP}_{h,\text{max}} = T_H/(T_H - T_L)$$

### 2.5.2 Second Law from Statistical Thermodynamics

To help understand the significance of the second law of thermodynamics, consider the molecular nature of matter. Although a sample of a gas may be at rest, its molecules are not. Rather, they are in a state of continuous, random motion, with an average speed of the same order of magnitude as the speed of sound waves in the gas. For air, this is about 1100 ft/s (335 m/s) at room temperature. Some of the molecules move more rapidly than this and some more slowly.

Due to collisions with one another and with the walls of the containing vessel, the velocity of any one molecule is continually being changed in magnitude and direction. The number of molecules traveling in a given direction with a given speed, however, remains constant. If the gas as a whole is at rest, the molecular velocities are distributed randomly in any direction. From the standpoint of conservation of energy (the first law), molecules of a gas, flying about in all directions and with a wide range of speeds, could get together in a cooperative effort and simultaneously acquire a common velocity component in the same direction, although it is unexpected. Nevertheless, these phenomena have been observed and are described as fluctuations.

Hence, processes whose sole result is the flow of heat from a heat reservoir and the performance of an equivalent amount of work, do not occur with sufficient frequency or with objects of sufficient size to make them useful.

Thus, an accurate statement of the second law should replace the term *impossible* with *improbable*. Therefore, the second law is a statement of the improbability of the spontaneous passage of the system from a highly probable state (random or disordered) to one of lower probability.

### 2.5.3 Physical Meaning of Entropy

Entropy is a measure of the random mix or of the probability of a given state. The more completely shuffled any system is, the greater is its entropy; conversely, an orderly or unmixed configuration is one of low entropy. Thus, when a substance reaches a state in which all randomness disappears, it then has zero entropy.

In 1851, Rankine analytically demonstrated that the ratio of the heat exchanged in a reversible process to the temperature of the interaction defined a thermodynamic function that was not consumed in a reversible cycle. The following year, Clausius independently derived the same result, but identified the function as a property of a system, and designated it as the entropy.

The concept of energy serves as a measure of the quantity of heat, but entropy serves as a measure of its quality. Clausius also concluded that although the energy of the world is constant, the entropy increases to a maximum, due to the irreversible nature of real processes. This extreme represents a condition when there are no potential differences in the universe.

In an irreversible process, the entropy of the universe is irretrievably increased, but no energy is lost. From the engineer's point of view, however, the opportunity to convert internal energy to mechanical energy is lost. An example is the mixing of hot and cold water from two reservoirs into a single reservoir. The internal energy of the system is the same before as after mixing, but at the end of the process, no heat can be withdrawn from the single reservoir to operate a cyclic machine. An engine could have been operated between the original hot and cold reservoirs, but once the reservoirs have come to the same temperature, this opportunity is lost.

Entropy is not conserved, except in reversible processes. When a beaker of hot water is mixed with a beaker of cold water, the heat lost by the hot water equals the heat gained by the cold water and energy is conserved. On the other hand, while the entropy of the hot water decreases, the entropy of the cold water increases a greater amount, and the total entropy of the system is greater at the end of the process than it is at the beginning. Where did this additional entropy come from? It was created in the process. Furthermore, once entropy has been created, it can never be destroyed. *Energy can neither be created nor destroyed*, states the first law of thermodynamics. *Entropy cannot be destroyed, but it can be created*, states the second law.

### 2.5.4 Entropy Equation of the Second Law of Thermodynamics

For the general case of an open system, the second law can be written

$$dS_{system} = (\delta Q/T)_{rev} + \delta m_i s_i - \delta m_e s_e + dS_{irr} \qquad (2\text{-}52)$$

where

$\delta m_i s_i$ = entropy increase due to the mass entering
$\delta m_e s_e$ = entropy decrease due to the mass leaving
$\delta Q/T$ = entropy change due to reversible heat transfer between the system and surroundings
$dS_{irr}$ = entropy created due to irreversibilities

The equation accounts for all the entropy changes in the system. Rearranging,

$$\delta Q = T[\delta m_e s_e - \delta m_i s_i) + dS_{sys} - dS_{irr}] \qquad (2\text{-}53)$$

In integrated form—subject to the restrictions that inlet and outlet properties, mass flow rates, and interactions with the surroundings do not vary with time—the general equation for the second law is

$$(S_f - S_i)_{out} = \int_{rev} \delta Q/T + \Sigma(ms)_{in} - \Sigma(ms)_{out} + \Delta S_{produced} \qquad (2\text{-}54)$$

**Example 2-3** A contact feedwater heater operates on the principle of mixing steam and water. Steam enters the heater at 100 psia and 98% quality. Water enters the heater at 100 psia, 80°F. As a result, 25,000 lb$_m$/h of water at 95 psia and 290°F leave the heater. No heat transfers between the heater and the surroundings. Evaluate each term in the general entropy equation for the second law.

**Solution:**
First law

$$m_1 h_1 + m_2 h_2 - m_3 h_3 = 0$$

From Table 2-1

$$h_1 = 298.61 + 0.98(889.2) = 1170$$
$$h_2 = 48.05$$
$$h_3 = 259.4 \text{ Btu/lb}$$
$$s_1 = 0.47439 + 0.98(1.1290) = 1.581$$
$$s_2 = 0.09325$$
$$s_3 = 0.4236 \text{ Btu/lb·°R}$$
$$m_1(1170) + (25000 - m_1)(48.05) = 25000(259.4)$$
$$m_1 = 4710 \text{ lb}_m/h; \quad m_2 = 20{,}290 \text{ lb}_m/h$$

Second law

$$m_f s_f - m_i s_i = \int \delta Q/T + m_1 s_1 + m_2 s_2 - m_3 s_3 + \Delta S_{irr}$$
$$0 = 0 + 4710(1.581) + 20290(0.9325) - 25000(0.4236) + \Delta S_{irr}$$
$$\Delta S_{irr} = 1150$$
$$m_f s_f - m_i s_i = \int \delta Q/T + m_1 s_1 + m_2 s_2 - m_3 s_3 + \Delta S_{irr}$$
$$0 = 0 + 7450 + 2000 - 10600 + 1150$$

**First Law of Thermodynamics** (Energy Balance)

$$\sum m_{in} \left( u + \frac{Pv}{J} + \frac{V^2}{2g_c J} + \frac{g}{g_c} \frac{z}{J} \right)_{in}$$

$$- \sum m_{out} \left( u + \frac{Pv}{J} + \frac{V^2}{2g_c J} + \frac{g}{g_c} \frac{z}{J} \right)_{out} + Q - W$$

$$= \left[ m_f \left( u + \frac{V^2}{2g_c J} + \frac{g}{g_c} \frac{z}{J} \right)_f - m_i \left( u + \frac{V^2}{2g_c J} + \frac{g}{g_c} \frac{z}{J} \right)_i \right]_{system}$$

**Second Law of Thermodynamics** (Availability)
General Equation:

$$(S_f - S_i)_{system} = \int_{rev} \frac{\delta Q}{T} + \sum (ms)_{in}$$
$$- \sum (ms)_{out} + \Delta S_{produced}$$

**Mass Flow** (Continuity)

$$m = \frac{AV}{v} = \rho A V$$

**Perfect Gas Relations**

$$Pv = RT; \; R = \frac{1544}{M}; \; c_p - c_v = \frac{R}{J}; \; K = \frac{c_p}{c_v}$$

$$\Delta h = h_2 - h_1 = \int_{T_1}^{T_2} c_p dT = c_p(T_2 - T_1)$$

$$\Delta u = u_2 - u_1 = \int_{T_1}^{T_2} c_v dT = c_v(T_2 - T_1)$$

$$\Delta s = s_2 - s_1 = \int_{T_1}^{T_2} c_p \frac{dT}{T} - \frac{R}{J} \ln \frac{P_2}{P_1}$$

$$= c_p \ln \frac{T_2}{T_1} - \frac{R}{J} \ln \frac{P_2}{P_1} \text{ or } c_v \ln \frac{T_2}{T_1} + \frac{R}{J} \ln \frac{v_2}{v_1}$$

**Reversible Polytropic Processes**

$$pv^n = C = p_1 v_1^n = p_2 v_2^n = \ldots$$

$$\frac{T_2}{T_1} = \left( \frac{P_2}{P_1} \right)^{\frac{n-1}{n}} = \left( \frac{v_1}{v_2} \right)^{n-1}$$

where, if
isentropic, $n = k$    constant volume, $n = \infty$
constant pressure, $n = 0$
constant temperature, $n = 1$ or for general case,
   $n = n$ as specified by problem

**Definitions**

Enthalpy, $h = u + Pv/J$
Work, $W = \int F dx$
$\quad = \int P dv$ (reversible, closed system)
$\quad = -\int v dP - \Delta ke - \Delta pe$
$\quad$ (reversible, steady flow system)

*Fig. 2-15 Basic Equations of Thermodynamics*

## 2.6 Third Law of Thermodynamics

In the late 1800s, Amontons found that a given volume of air, when heated or cooled, expands or contracts by the same amount for each degree of temperature change. He determined this change as about 1/240 of the gas volume at 0°C, which suggests that at –240°C, the volume becomes zero, and at still lower temperature, it has to become negative. Since this observation makes no sense, he concluded that an ultimate low temperature of absolute zero exists. Today, it is known that absolute zero is –273.15°C.

In 1889, Dewar approached absolute zero within 20°C. In 1906, Nernst formulated the third law of thermodynamics that states, *While absolute zero can be approached to an arbitrary degree, it can never be reached*. Postulates to this law conclude that it is not the energy, but the entropy that tends to zero, and that a residual amount of energy is left in any substance, even at absolute zero.

## 2.7 Basic Equations of Thermodynamics

Figure 2-15 contains the basic set of equations required for most applications of thermodynamics.

## 2.8 Thermodynamics Applied to Refrigeration

Continuous refrigeration can be accomplished by several processes. In most applications, and almost exclusively in the smaller power range, the vapor compression system (commonly termed the mechanical vapor compression cycle) is used for the refrigeration process. However, absorption systems and steam-jet vacuum systems are being successfully used in many applications. In larger equipment, centrifugal systems are an adaptation of the compression cycle.

A larger number of working fluids (refrigerants) are used in vapor-compression refrigeration systems than in vapor power cycles. Ammonia and sulfur dioxide were first used as vapor-compression refrigerants. Today, the main refrigerants are the halogenated hydrocarbons. Two important considerations in selecting a refrigerant are the temperature at which refrigeration is desired and the type of equipment to be used.

Refrigerants used in most mechanical refrigeration systems are R-22, which boils at –41.4°F (–40.8°C), R-134a, which boils at –15.1°F (–26.2°C) at atmospheric pressure, and R-123 with a boiling point of +82.2°F (+27.9°C) at atmospheric pressure. Past favorites R-12 and R-11 are phased out due to adverse effects on the ozone in the stratosphere. R-22 will be phased out in 2010.

The basic vapor compression cycle is illustrated in Figure 2-16. Cool, low-pressure liquid refrigerant enters the evaporator and evaporates. As it does so, it absorbs heat from another substance, such as air or water, thereby accomplishing refrigeration. The refrigerant then leaves the evaporator as a cool, low-pressure gas and proceeds to the compressor. Here, its pressure and temperature are increased, and this hot, high-pressure gas is discharged to the condenser. In the condenser, the hot gas is condensed into a liquid. The condensing agent, air or water, is at a temperature lower than the refrigerant gas. This hot, high-pressure liquid flows from the condenser through the expansion valve to the evaporator. The expansion valve reduces the pressure and meters the liquid flow, reducing the hot, high-pressure liquid to a cool, low-pressure liquid as it enters the evaporator.

The basic refrigeration cycle is plotted on the Pressure-Enthalpy diagram as Figures 2-17. Subcooled liquid, at Point A, begins losing pressure as it goes through the metering valve, located at the point where the vertical liquid line meets the saturation curve. As it leaves the metering point, some of the liquid flashes into vapor and cools the liquid entering the evaporator at Point B. Notice that there is additional reduction in pressure from the metering point to Point B, but no change in enthalpy.

As it passes from Point B to C, the remaining liquid picks up heat and changes from a liquid to a gas but does not increase in pressure. Enthalpy, however, does increase. Superheat is added between Point C, where the vapor passes the saturation curve, and Point D.

As it passes through the compressor, Point D to E, the temperature and the pressure are markedly increased, as is the enthalpy, due to the heat of compression. Line E-F indicates that the vapor must be desuperheated within the condenser before it attains a saturated condition and begins condensing. Line F-G represents the change from vapor to liquid within the condenser. Line G-A represents subcooling within the liquid line prior to flow through the metering device.

Note that the pressure remains essentially constant as the refrigerant passes through the evaporator, but due to superheat, its temperature is increased beyond the saturation point before it enters the compressor.

Likewise, the pressure remains constant as the refrigerant enters the condenser as a vapor and leaves as a liquid. While the temperature is constant through the condenser, it is reduced as the liquid is subcooled before entering the metering valve.

The change in enthalpy as the refrigerant passes through the evaporator is almost all latent heat since the temperature does not change appreciably.

Figure 2-18 provides a somewhat more completely labeled *p-h* diagram of the refrigeration cycle.

Applying the first law of thermodynamics to the vapor compression refrigeration system as a whole requires that the sum of all energy in must equal the sum of all energy out when the unit is operating at a steady state rate; hence

$$Q_L + W = Q_H$$

The rate of heat being rejected at the condenser $Q_H$ is numerically greater than the rate at which work is delivered to drive the compressor $W$ and is also greater than the rate of heat absorption into the evaporator $Q_L$.

This relation shows that every refrigeration cycle operates as a heat pump. The household refrigerator absorbs a quantity of heat $Q_L$ at a low temperature in the vicinity of the ice-making section (the evaporator) and rejects heat $Q_H$ at a higher temperature to the air in the room. The rate of heat rejection $Q_H$ is greater than the rate of absorption $Q_L$ by the power input W to drive the compressor.

In air-conditioning applications the desired effect for cooling is the heat absorbed at the evaporator located inside the conditioned space. Heat is rejected through the condenser outside the conditioned space.

A heat pump uses this same basic cycle for heating. In this application, the condenser is located inside the building and the evaporator is located outside the building where it absorbs heat. Both modes of the cycle are in Figure 2-19.

**Example 2-4** An R-134a air-conditioning unit contains a 1 hp (0.746 kW) motor and operates between pressures of 0.2 MPa (in evaporator) and 1.02 MPa (in condenser). Estimate the maximum cooling effect, in kW, that can be expected from this unit.

**Solution:**

$$\text{COP} = Q_i/W = Q_i/(Q_o - Q_i) = 1/[(Q_o/Q_i) - 1]$$

$$\text{COP}_{max} = 1/(T_o/T_i - 1) = 1/(313/263 - 1) = 1/0.19 = 5.26$$

$$Q_i = (\text{COP})W = 5.26(0.746)$$
$$= 3.92 \text{ kW } (13,400 \text{ Btu/h})$$

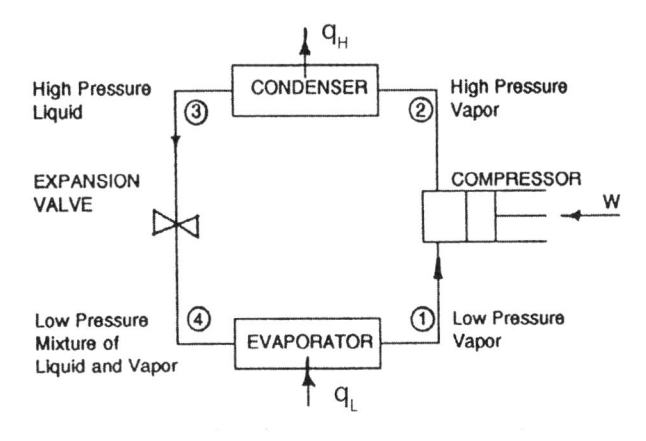

*Fig. 2-16 Basic Vapor Compression Refrigeration Cycle*

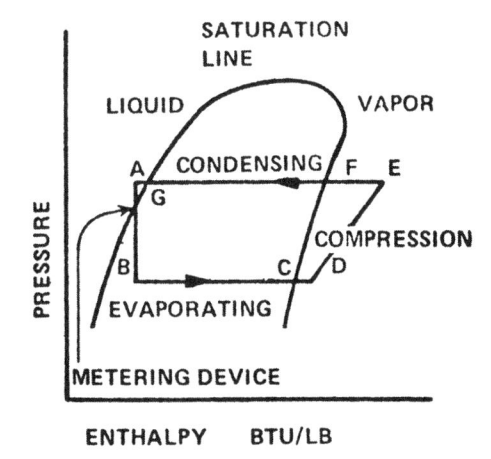

*Fig. 2-17 Simplified Pressure-Enthalpy Plot*

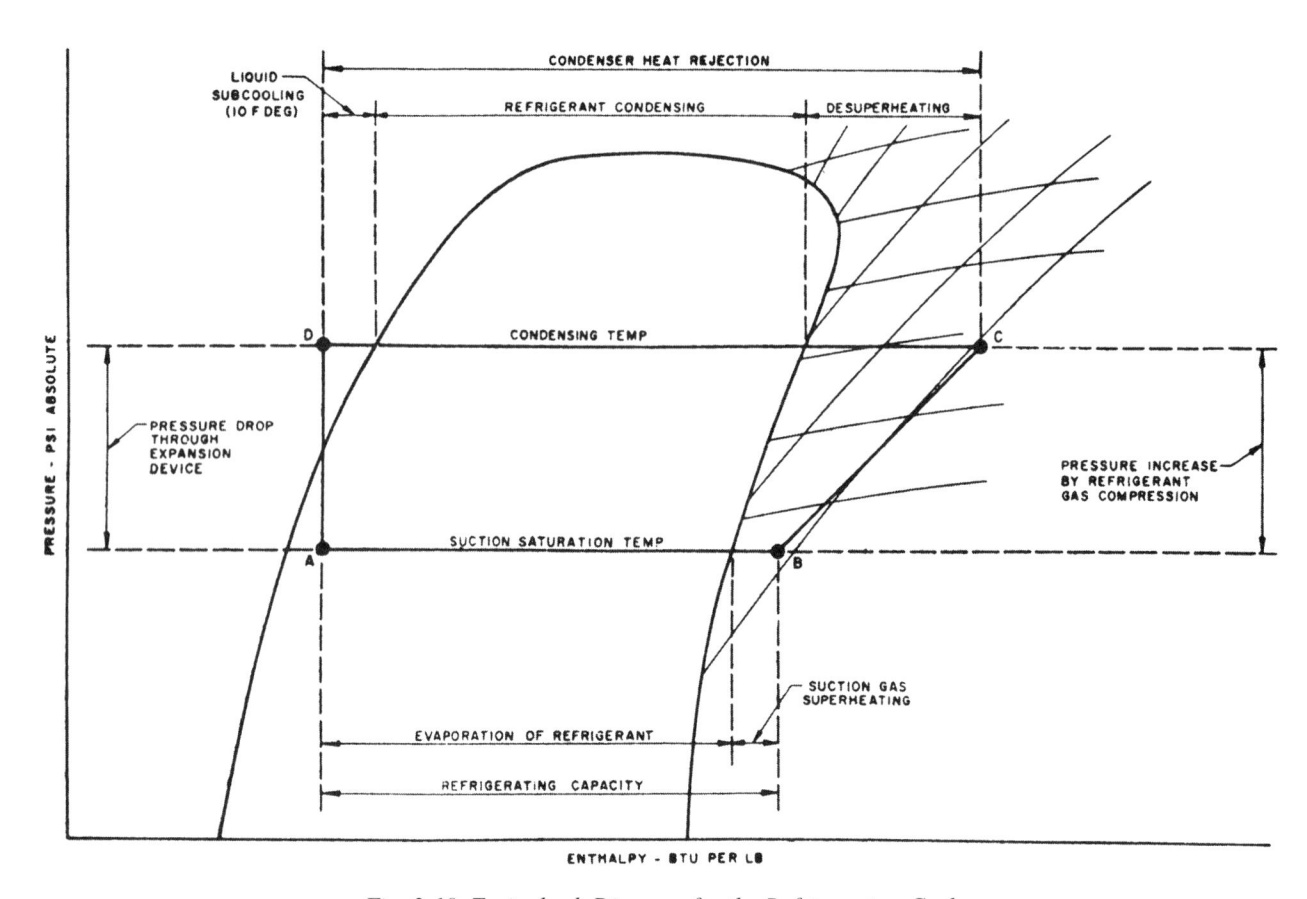

*Fig. 2-18 Typical p-h Diagram for the Refrigeration Cycle*

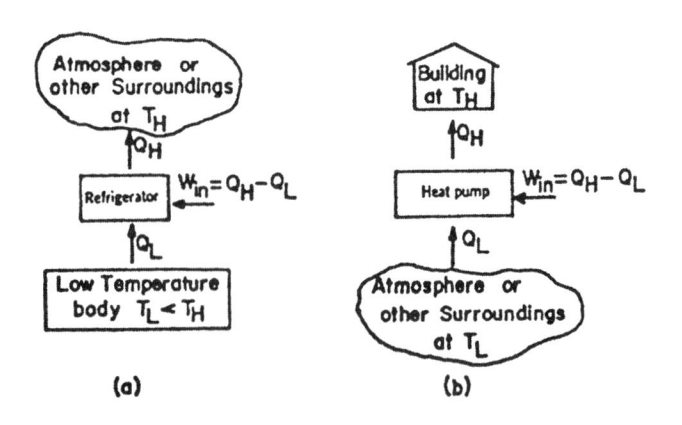

*Fig. 2-19 Refrigeration Cycle (a) as Refrigerator, (b) as Heat Pump*

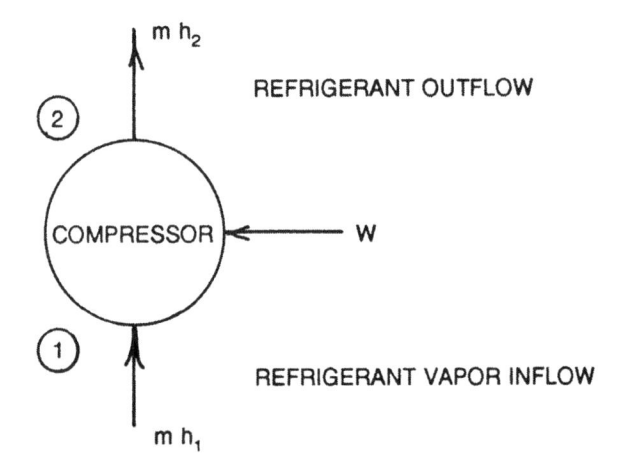

*Fig. 2-20 Compressor Flows*

## 2.8.1 Energy Relations for the Basic Refrigeration Cycle

The first law of thermodynamics can be applied to each component of the system individually, since energy must be conserved at each of these, as well as for the entire system.

**Compressor.** The mass and major energy flows for a compressor are shown in Figure 2-20.

The rate of energy inflow must equal the rate of energy outflow during steady state operation; hence

$$mh_1 + W = mh_2$$

$$W = m(h_2 - h_1) \qquad (2\text{-}55)$$

where
  $m$ = rate of refrigerant flow
  $h_1, h_2$ = enthalpies of refrigerant at compressor inlet and outlet

**Condenser.** The mass and energy flow for a condenser are illustrated in Figure 2-21.

$$mh_2 = mh_3 + Q_R \quad \text{or} \quad Q_R = m(h_2 - h_3) \qquad (2\text{-}56)$$

# Thermodynamics and Psychrometrics

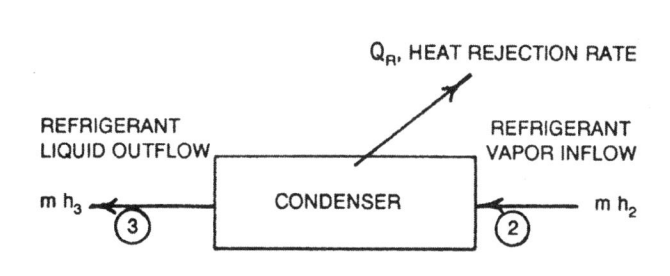

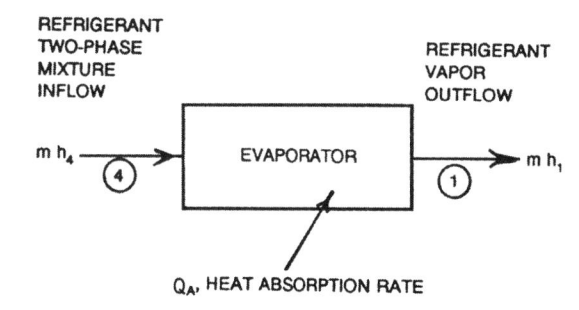

*Fig. 2-21  Condenser Flows*

*Fig. 2-22  Evaporator Flows*

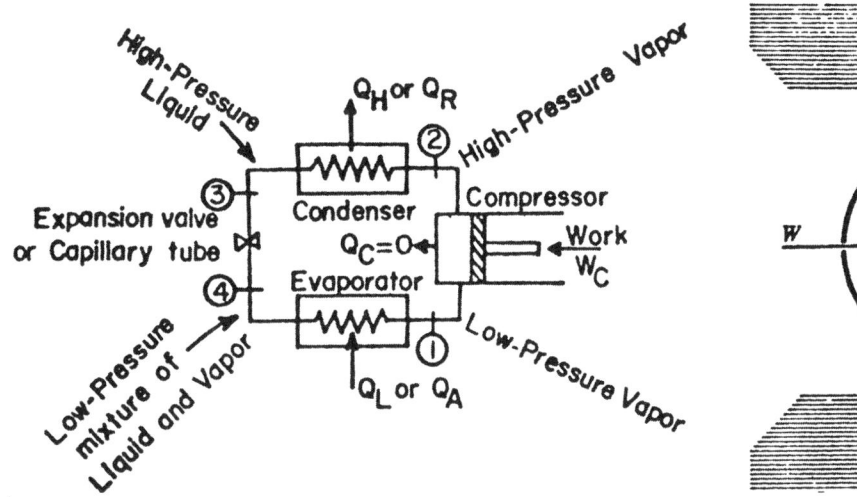

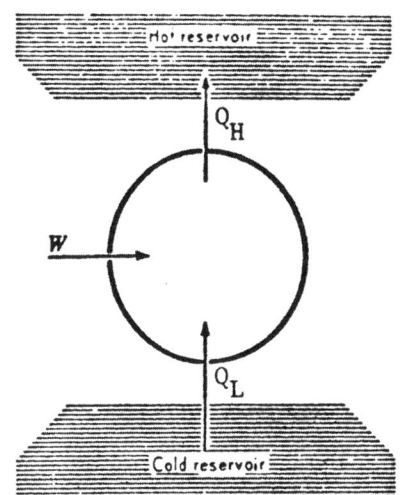

Compressor:   $m(h_1 - h_2) + {}_1Q_2 - {}_1W_2 = 0$    $W_C = -{}_1W_2$
Condenser:    $m(h_2 - h_3) + {}_2Q_3 = 0$    $Q_R = -{}_2Q_3 = Q_H$
Expansion Device: $h_3 - h_4 = 0$
Evaporator:   $m(h_4 - h_1) + {}_4Q_1 = 0$    $Q_A = {}_4Q_1 = Q_L$
Overall:      $Q_A + W_C = Q_C + Q_R$
              or $({}_1Q_2 + {}_2Q_3 + {}_4Q_1) - ({}_1W_2) = 0$

Coefficient of Performance for Cooling:

$$\text{COP}_c = \frac{\text{Useful effect}}{\text{Input that costs}} = \frac{Q_L}{W} = \frac{1}{Q_H/Q_L - 1} \qquad \text{COP}_{c,\max} = \frac{1}{T_H/T_L - 1}$$

Coefficient of Performance for Heating:

$$\text{COP}_h = \frac{Q_h}{W} \qquad \text{COP}_{h,\max} = \frac{1}{1 - T_L/T_H}$$

*Fig. 2-22  The Vapor Compression System*

Typically, heat from the condensing refrigerant $Q_R$ is rejected to another fluid. The $Q_R$ leaving the condensing refrigerant must equal the heat absorbed by the fluid receiving it. Thus, from the viewpoint of the condenser cooling fluid,

$$Q_R = m_{\text{fluid}}(c_p)_{\text{fluid}}(t_{\text{out}} - t_{\text{in}}) \qquad (2\text{-}57)$$

**Expansion Device.** An expansion device is a throttling device or a flow restrictor—a small valve seat opening or a long length of small bore tubing—so neither work nor any significant amount of heat transfer occurs. Hence

$$mh_3 = mh_4$$

or, dividing both sides by $m$ gives

$$h_3 = h_4 \qquad (2\text{-}58)$$

**Evaporator.** Major mass and energy flows for an evaporator are shown in Figure 2-22.

$$mh_4 + Q_A = mh_1 \text{ or}$$

$$Q_A = m(h_1 - h_4) \tag{2-59}$$

Typically, the evaporator receives the heat flow quantity $Q_A$ by heat transfer to it from another fluid—usually water or air. From the viewpoint of that other fluid,

$$Q_A = m_{\text{fluid}} (c_p)_{\text{fluid}} (t_{\text{in}} - t_{\text{out}})_{\text{fluid}}$$

The results of applying the laws of thermodynamics to the basic vapor-compression refrigeration system are summarized in Figure 2-23.

**Example 2-5** A window air conditioner using R-134A is rated at 24,000 Btu/h when operating between an evaporating temperature of 40°F and a condensing temperature of 125°F. A thermostatic expansion valve is used so that the refrigerant vapor leaving the evaporator is superheated by 20°F to safeguard against any liquid entering the compressor. If the compressor efficiency is 62%, determine the adiabatic discharge temperature from the compressor (°F) and the power input (kW). Show the cycle on the p-h diagram.

**Solution:**
Fig. 2-16 provides a sketch with notation for the cycle. Using the p-h diagram for R-134A shown below, the following thermodynamic properties are found:

$p_H$ = 200 psia (saturation pressure at 125°F)
$p_L$ = 50 psia (saturation pressure at 40°F)

$h_3 = h_4 = h_{f \text{ at } 125°F}$ = 54.2 Btu/lb

$h_1$ (50 psia, 60°F) = 113.0 Btu/lb

$s_1$ = 0.230

$h_{2, \text{ideal}}$ (200 psia, s = 0.230) = 126 Btu/lb

From first law, $m(h_1 - h_2) - W = 0$ for the compressor and, $m(h_4 - h_1) + q_L = 0$ for the evaporator.
Thus, $m(54.2 - 113.0) + 24000 = 0$, or

$$m = 408 \text{ lb/h}$$

The ideal compressor work is

$$W_i = 408(113 - 126) = -5304 \text{ Btu/h,}$$

and the actual work using the compressor efficiency is

$$W_a = W_i/\eta = -5304/.62 = -8555 \text{ Btu/h or 2.5 kW}$$

The actual enthalpy leaving the compressor can be determined using the efficiency as

$$h_{2a} = h_1 + (h_{2i} - h_i)/\eta = 113 + (126 - 113)/0.62$$
$$= 134 \text{ Btu/lb}$$

$t_2$ {200 psia, h = 134} = 180°F (adiabatic discharge)

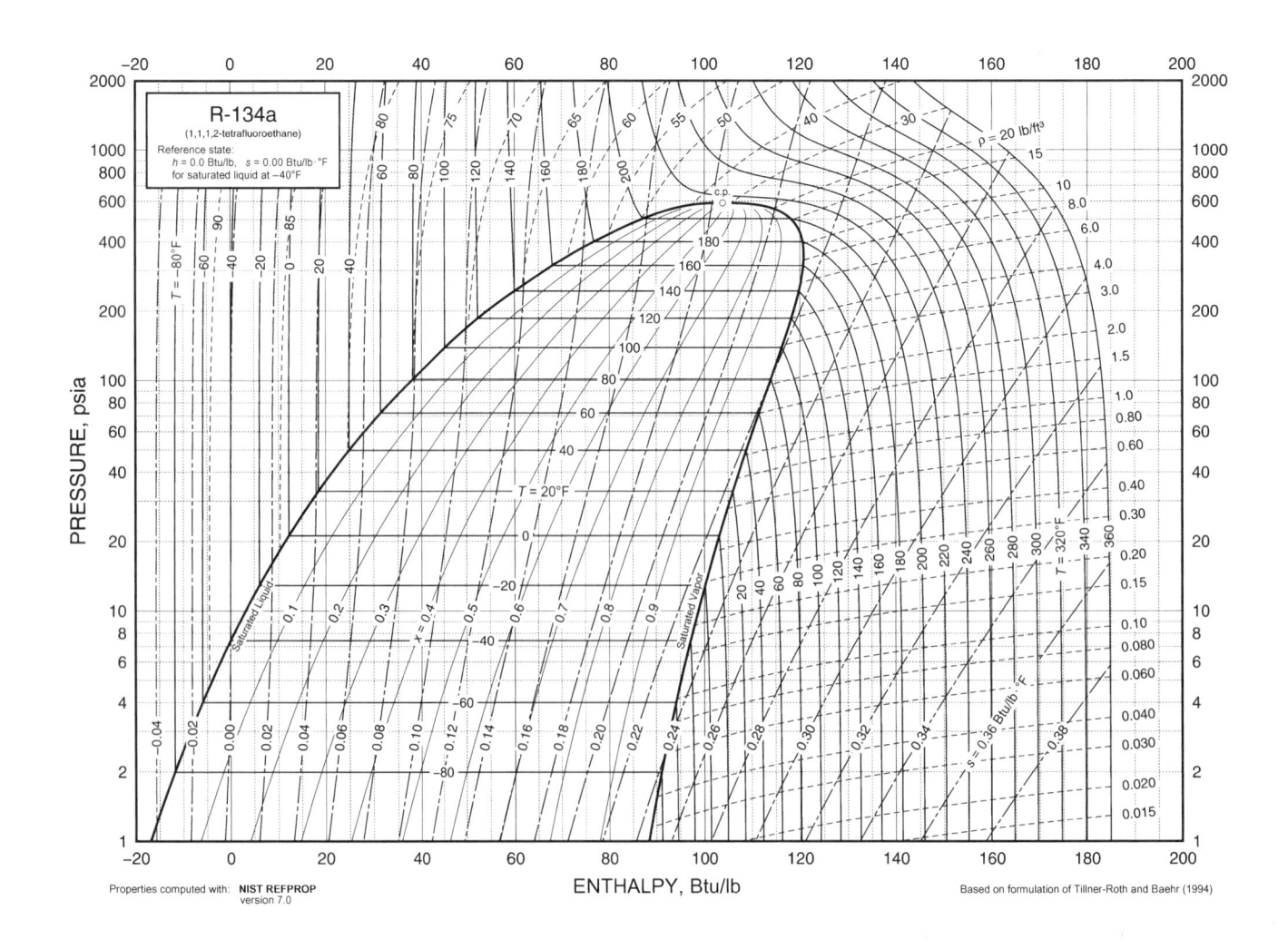

# Thermodynamics and Psychrometrics

## 2.9 Applying Thermodynamics to Heat Pumps

A heat pump is a thermodynamic device that operates in a cycle requiring work; it transfers heat from a low-temperature body to a high-temperature body. The heat pump cycle is identical to a refrigeration cycle in principle, but differs in that the primary purpose of the heat pump is to *supply* heat rather than *remove* it from an enclosed space.

The heat pump cycle can be reversed to provide space cooling (Figure 2-24). A four-way valve switches the heat exchangers so that the indoor exchanger becomes the evaporator and the outdoor heat exchanger becomes the condenser.

The four basic components of a heat pump are the compressor, condenser, expansion device, and evaporator (Figure 2-16). The thermodynamic cycle for a heat pump is identical to the conventional vapor-compression refrigeration cycle (Figure 2-25).

Superheated refrigerant vapor with low pressure and temperature at state 1 is compressed to a much higher pressure and temperature at state 2. The high- pressure, high-temperature gas then passes through the condenser (indoor coil of a heat pump), where it transfers heat to the high-temperature environment and changes from vapor to liquid at high pressure. At state 3, the refrigerant exits the condenser (usually as a subcooled liquid). Next, the refrigerant passes through an expansion device where its pressure drops. This drop in pressure is accompanied by a drop in temperature such that the refrigerant leaves the expansion device and enters the evaporator (outdoor coil of a heat pump) as a low-pressure, low-temperature mixture of liquid and vapor at state 4. Finally, the refrigerant passes through the evaporator, where it picks up heat from the low-temperature environment, changes to all vapor, and exits at state 1.

An energy balance on the system shown in Figure 2-25 gives

$$Q_H = Q_L + W$$

where

$Q_H$ = heat energy rejected to the high-temperature environment

$Q_L$ = heat energy taken from the low-temperature environment

$W$ = input work required to move $Q_L$ from the low-temperature environment to the high-temperature environment

The coefficient of performance (COP) for heating equals the heat output divided by the work input:

$$COP = Q_H/W = (Q_L + W)/W$$

$$COP = 1 + Q_L/W \qquad (2\text{-}60)$$

The COP of a heat pump is always greater than one. That is, a heat pump always produces more heat energy than work energy consumed because a net gain of energy $Q_L$ is transferred from the low-temperature to the high-temperature environment.

The heat pump is a reverse heat engine and is therefore limited by the Carnot cycle COP:

$$COP_{Carnot} = T_H/(T_H - T_L) \qquad (2\text{-}61)$$

where

$T_L$ = low temperature in cycle
$T_H$ = high temperature in cycle

The maximum possible COP for a heat pump maintaining a fixed temperature in the heated space is therefore a function of source temperature (Figure 2-26). However, any real heat transfer system must have finite temperature differences across the heat exchangers. The Carnot COP for a typical air-to-air heat pump, as well as the actual COP for the same heat pump, are illustrated in Figure 2-26. The difference between Carnot and actual COPs is due to the nature of real working fluids, flow losses, and compressor efficiency.

## 2.10 Absorption Refrigeration Cycle

Absorption refrigeration cycles are heat-operated cycles in which a secondary fluid, the absorbent, is used to absorb the primary fluid, gaseous refrigerant, that has been vaporized in the evaporator. The basic absorption cycle is shown in Figure 2-27.

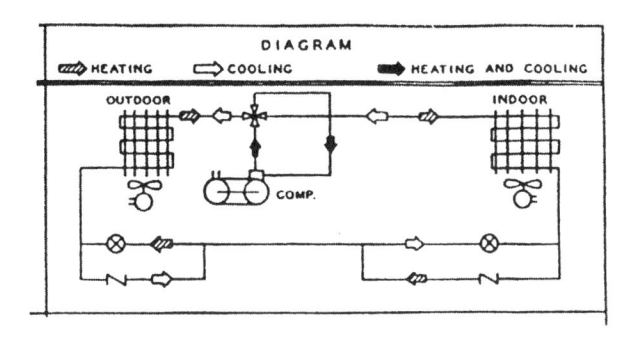

*Fig. 2-23 Basic Heat Pump Cycle*

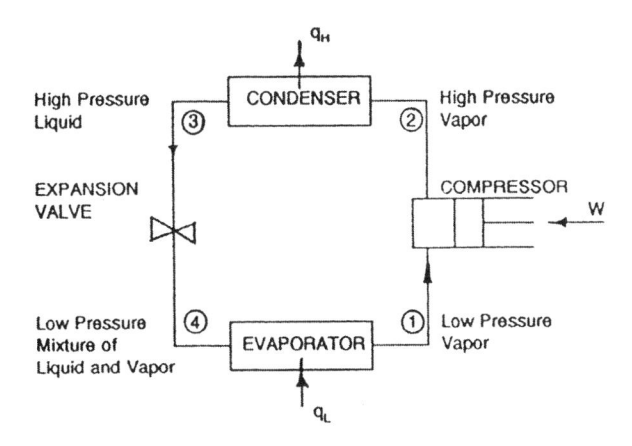

*Fig. 2-24 Basic Heat Pump Components*

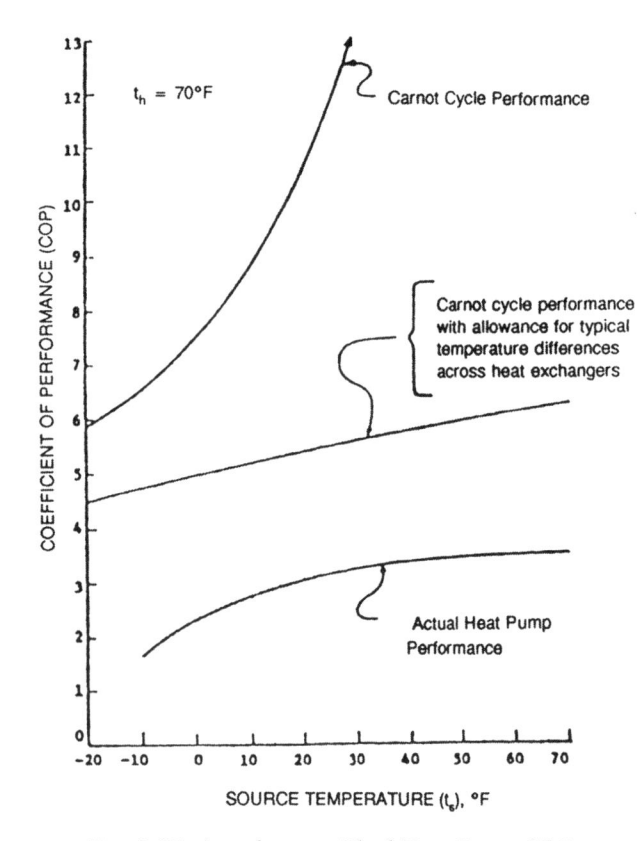

*Fig. 2-25 Actual versus Ideal Heat Pump COPs*

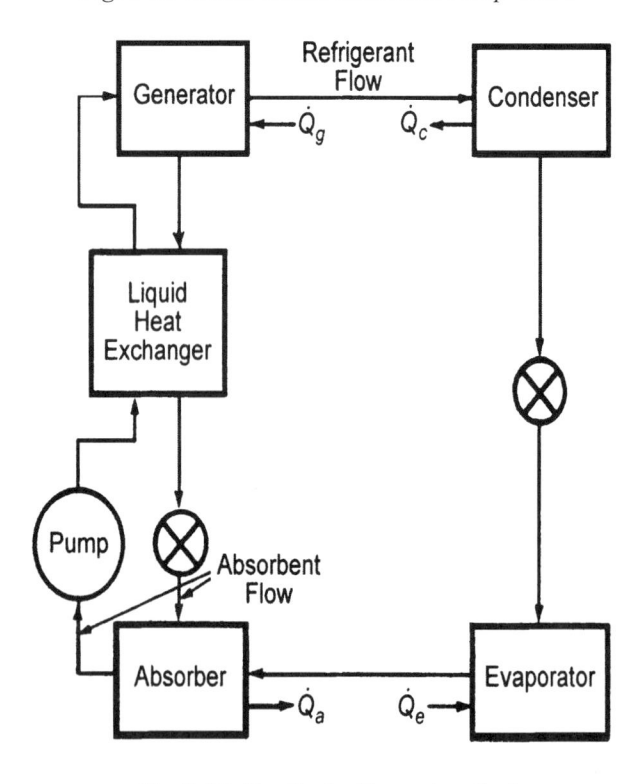

*Fig. 2-26 The Basic Absorption Cycle*

The low-pressure vapor leaving the evaporator enters the absorber where it is absorbed in the weak solution. Since the temperature is slightly above that of the surroundings, heat is transferred to the surroundings during this process. The strong solution is then pumped through a heat exchanger to the generator at a higher pressure and temperature. As a result of heat transfer, the refrigerant evaporates from the solution. The refrigerant vapor goes to the condenser (as in a vapor-compression system) and then to the expansion valve and evaporator. The weak solution is returned to the absorber through the heat exchanger.

The distinctive feature of the absorption system is that little work input is required because liquids (rather than vapors) are pumped. However, more equipment is involved in an absorption system than in the vapor-compression cycle. Thus, an absorption system is economically feasible only where a low-cost source of heat is available. Additional material on absorption systems is given in Chapter 18 with much greater details available in the 2013 *ASHRAE Handbook—Fundamentals*, Chapter 2, and in the 2010 *ASHRAE Handbook—Refrigeration*, Chapter 41.

## 2.11 Problems

**2.1** Write the first law of thermodynamics in general integrated form.

**2.2** Write the second law of thermodynamics in general integrated form.

**2.3** Write the following perfect gas relations:

(a) the equation of state
(b) the equation for entropy change
(c) the equation for enthalpy change
(d) the equation for internal energy change

**2.4** Write the continuity (mass flow) equation.

**2.5** Write the equations for work for

(a) a reversible, closed system
(b) a reversible, steady-flow system

**2.6** Two pounds of air contained in a cylinder expand without friction against a piston. The pressure on the back side of the piston is constant at 200 psia. The air initially occupies a volume of 0.50 ft$^3$. What is the work done by the air in ft·lb$_f$ if the expansion continues until the temperature of the air reaches 100°F? [Ans: 45300 ft·lb$_f$]

**2.7** Determine the specific volume, enthalpy, and entropy of 1 kg of R-134a at a saturation temperature of –5°C and a quality of 14%.

**2.8** Saturated R-134a vapor at 42°C is superheated at constant pressure to a final temperature of 72°C. What is the pressure? What are the changes in specific volume, enthalpy, entropy, and internal energy?

**2.9** A tank having a volume of 200 ft$^3$ contains saturated vapor (steam) at a pressure of 20 psia. Attached to this tank is a line in which vapor at 100 psia, 400°F flows. Steam from this line enters the vessel until the pressure is 100 psia. If there

# Thermodynamics and Psychrometrics

is no heat transfer from the tank and the heat capacity of the tank is neglected, calculate the mass of steam that enters the tank. [Ans: 24 lb$_m$]

**2.10** Determine the heat required to vaporize 50 kg of water at a saturation temperature of 100°C.

**2.11** The temperature of 150 kg of water is raised from 15°C to 85°C by the addition of heat. How much heat is supplied?

**2.12** Three cubic meters per second of water are cooled from 30°C to 2°C. Compute the rate of heat transfer in kilojoules per second (kilowatts). [Ans: 351 000 kW]

**2.13** Consider 10 lb$_m$ of air that is initially at 14.7 psia, 100°F. Heat is transferred to the air until the temperature reaches 500°F. Determine the change of internal energy, the change in enthalpy, the heat transfer, and the work done for

(a) a constant-volume process
(b) a constant-pressure process.

**2.14** The discharge of a pump is 10 ft above the inlet. Water enters at a pressure of 20 psia and leaves at a pressure of 200 psia. The specific volume of the water is 0.016 ft$^3$/lb. If there is no heat transfer and no change in kinetic or internal energy, what is the work per pound? [Ans: –0.546 Btu]

**2.15** The discharge of a pump is 3 m above the inlet. Water enters at a pressure of 138 kPa and leaves at a pressure of 1380 kPa. The specific volume of the water is 0.001 m$^3$/kg. If there is no heat transfer and no change in kinetic or internal energy, what is the work per unit mass? [Ans: –30.7 J/kg]

**2.16** Air is compressed in a reversible, isothermal, steady-flow process from 15 psia, 100°F to 100 psia. Calculate the work of compression per pound, the change of entropy, and the heat transfer per pound of air compressed.

**2.17** Liquid nitrogen at a temperature of –240°F exists in a container, and both the liquid and vapor phases are present. The volume of the container is 3 ft$^3$ and the mass of nitrogen in the container has been determined as 44.5 lb$_m$. What is the mass of liquid and the mass of vapor present in the container?

**2.18** A fan in an air-conditioning system is drawing 1.25 hp at 1760 rpm. The capacity through the fan is 0.85 m$^3$/s of 24°C air and the inlet and outlet ducts are 0.31 m in diameter. What is the temperature rise of the air due to this fan? [Ans: 0.9°C]

**2.19** Air is contained in a cylinder. Initially, the cylinder contains 1.5 m$^3$ of air at 150 kPa, 20°C. The air is then compressed reversibly according to the relationship $pv^n$ = constant until the final pressure is 600 kPa, at which point the temperature is 120°C. For this process, determine

(a) the polytropic exponent $n$
(b) the final volume of the air
(c) the work done on the air and the heat transfer

**2.20** Water at 20°C is pumped from ground level to an elevated storage tank above ground level; the volume of the tank is 50 m$^3$. Initially, the tank contains air at 100 kPa, 20°C, and the tank is closed so that the air is compressed as the water enters the bottom of the tank. The pump is operated until the tank is three-quarters full. The temperature of the air and water remain constant at 20°C. Determine the work input to the pump.

**2.21** A centrifugal pump delivers liquid oxygen to a rocket engine at the rate of 100 lb$_m$/s. The oxygen enters the pump as liquid at 15 psia and the discharge pressure is 500 psia. The density of liquid oxygen is 66.7 lb$_m$/ft$^3$. Determine the minimum size motor (in horsepower) to drive this pump. [Ans: 190.4 hp]

**2.22** Air undergoes a steady-flow, reversible adiabatic process. The initial state is 200 psia, 1500°F, and the final pressure is 20 psia. Changes in kinetic and potential energy are negligible. Determine

(a) final temperature
(b) final specific volume
(c) change in internal energy per lb$_m$
(d) change in enthalpy per lb$_m$
(e) work per lb$_m$

**2.23** Air undergoes a steady-flow, reversible adiabatic process. The initial state is 1400 kPa, 815°C, and the final pressure is 140 kPa. Changes in kinetic and potential energy are negligible. Determine

(a) final temperature
(b) final specific volume
(c) change in specific internal energy
(d) change in specific enthalpy
(e) specific work

**2.24** A fan provides fresh air to the welding area in an industrial plant. The fan takes in outdoor air at 80°F and 14.7 psia at the rate of 1200 cfm with negligible inlet velocity. In the 10 ft$^2$ duct leaving the fan, air pressure is 1 psig. If the process is assumed to be reversible and adiabatic, determine the size motor needed to drive the fan. [Ans: $W$ = 5.1 hp]

**2.25** If the fan in the previous problem has an efficiency of 64% and is driven by a motor having an efficiency of 78%, determine the required power, kW.

**2.26** A fan provides fresh air to the welding area in an industrial plant. The fan takes in outdoor air at 32.2°C and 101.4 kPa at the rate of 566 L/s with negligible inlet velocity. In the 0.93 m$^2$ duct leaving the fan, air pressure is 102 kPa. Determine the minimum size motor needed to drive the fan.

**2.27** In an insulated feedwater heater, steam condenses at a constant temperature of 220°F. The feedwater is heated from 60°F to 150°F at constant pressure.

(a) Assuming the specific heat at constant pressure of the feedwater is unity, how many Btu are absorbed by each pound in its passage through the heater? [Ans: 90 Btu/lb]

(b) What is the change in entropy of the condensing steam per pound of feedwater heated? [Ans: –0.1324 Btu/lb·R]

(c) What is the change in entropy of 1 lb of feedwater as it passes through the heater? [Ans: +0.1595 Btu/lb·R]

(d) What is the change in entropy of the combined system? Does this violate the second law? Explain. [Ans: +0.0271 Btu/lb·R, No]

**2.28** Steam at 124 kPa and 96% quality enters a radiator. The steam is condensed as it flows through the radiator and leaves as condensate at 88°C. If the radiator is to have a heating capacity of 1.85 kW, how many kilograms per hour of steam must be supplied to the radiator?

**2.29** Solve the following:

(a) Air at 50 psia and 90°F flows through a restriction in a 2 in. ID pipe. The velocity of the air upstream from the restriction is 450 fpm. If 58°F air is desired, what must the velocity downstream of the restriction be? Comment on this as a method of cooling.

(b) Air at 50 psia and 90°F flows through an insulated turbine at the rate of 1.6 $lb_m$/s. If the air delivers 11.5 hp to the turbine blades, at what temperature does the air leave the turbine?

(c) Air at 50 psia and 90°F flows through an insulated turbine at the rate of 1.6 $lb_m$/s to an exit pressure of 14.7 psia. What is the lowest temperature attainable at exit?

**2.30** Liquid water at a pressure of 10 psia and a temperature of 80°F enters a 1 in. diameter tube at the rate of 0.8 ft³/min. Heat is transferred to the water so that it leaves as saturated vapor at 9 psia. Determine the heat transfer per minute. [Ans: 95,800 Btu/min]

**2.31** A refrigerator uses R-134a as the refrigerant and handles 200 $lb_m$/h. Condensing temperature is 110°F and evaporating temperature is 5°F. For a cooling effect of 11,000 Btu/h, determine the minimum size motor (hp) required to drive the compressor.

**2.32** A heat pump is used in place of a furnace for heating a house. In winter, when the outdoor air temperature is 10°F, the heat loss from the house is 60,000 Btu/h if the inside is maintained at 70°F. Determine the minimum electric power required to operate the heat pump (in kW).

**2.33** A heat pump is used in place of a furnace for heating a house. In winter, when the outdoor air temperature is –10°C, the heat loss from the house is 200 kW if the inside is maintained at 21°C. Determine the minimum electric power required to operate the heat pump. [Ans: 21.1 kW]

**2.34** Refrigerant-134a vapor enters a compressor at 25 psia, 40°F, and the mass rate of flow is 5 $lb_m$/min. What is the smallest diameter tubing that can be used if the velocity of refrigerant must not exceed 20 ft/s?

**2.35** An R-134a refrigerating system is operating with a condensing temperature of 86°F and evaporating temperature of 25°F. If the liquid line from the condenser is soldered to the suction line from the evaporator to form a simple heat exchanger, and if as a result of this, saturated liquid leaving the condenser is subcooled 6°F, how many degrees will the saturated vapor leaving the evaporator be superheated? (Use tables.)

**2.36** Ammonia is heated in the evaporator of a refrigeration system from inlet conditions of 10°F, 10% quality, to saturated vapor. The pressure remains constant during the process. For each pound, determine the changes in enthalpy and volume. [Ans: 505 Btu/lb; 6.55 ft³/lb]

**2.37** For a compressor using R-134a with an evaporator temperature of 20°F and a condensing temperature of 80°F, calculate per ton of refrigeration

(a) displacement

(b) mass flow

(c) horsepower required

**2.38** For a compressor using an R-22 system operating between 100°F condensing temperature and –10°F evaporator temperature, calculate per ton

(a) displacement

(b) mass flow

(c) horsepower required

**2.39** An industrial plant has available a 4 cylinder, 3 in. bore by 4 in. stroke, 800 rpm, single-acting compressor for use with R-134a. Proposed operating conditions for the compressor are 100°F condensing temperature and 40°F evaporating temperature. It is estimated that the refrigerant will enter the expansion valve as a saturated liquid, that vapor will leave the evaporator at a temperature of 45°F, and that vapor will enter the compressor at a temperature of 55°F. Assume a compressor-volumetric efficiency of 70% and frictionless flow. Calculate the refrigerating capacity in tons for a system equipped with this compressor. Plot the cycle on the p-h diagram. [Ans: 12 tons]

**2.40** A mechanical refrigeration system with R-134a is operating under such conditions that the evaporator pressure is 160 kPa and the liquid approaching the refrigerant control valve is at a temperature of 41°C. If the system has a capacity of 15 kW, determine

(a) the refrigerating effect per kilogram of refrigerant circulated

(b) the mass flow rate in kilograms per second per kilowatt

(c) the volume flow rate in liters per second per kilowatt at the compressor inlet

(d) the total mass flow rate in kilograms per second

(e) the total volume flow rate in liters per second at the compressor inlet

**2.41** A vapor-compression R-22 refrigeration system is being designed to provide 50 kW of cooling when operating between evaporating and condensing temperatures of 0°C and 34°C, respectively. The refrigerant leaving the condenser is subcooled 3 degrees and the vapor leaving the evaporator is superheated 5 degrees. Determine

(a) ideal compressor discharge temperature, °C
(b) refrigerant flow rate, kg/s
(c) compressor motor size, kW
(d) COP for cooling
(e) compressor discharge temperature if compression efficiency is 60%

**2.42** For a line of ammonia compressors, the actual volumetric efficiency is given by

$$\eta_{va} = 94 - 6.1(p_d/p_s), \%$$

The compression efficiency is fairly constant at 82%. A compressor in this line has two cylinders, each having a 92 mm bore and a 74 mm stroke. The compressor has 4.5% clearance and operates at 28 r/s. The system is being selected for an air-conditioning unit and will therefore operate between an evaporating temperature of 0°C and a condensing temperature of 35°C. There is 5°C of subcooling in the condenser and 10°C of superheating in the evaporator. Sketch and label the system, including appropriate values for the thermodynamic properties, starting with state 1 at the compressor inlet. Determine

(a) refrigerant flow rate [Ans: 0.0676 kg/s]
(b) refrigerating capacity [Ans: 77.4 kW]
(c) compressor motor size [Ans: 14.4 kW]
(d) compressor discharge temperature [Ans: 110°C]
(e) $COP_c$ [Ans: 5.4]

**2.43** An ammonia refrigerating system is operating with a condensing temperature of 30°C and an evaporating temperature of –4°C. For the ideal standard vapor compression cycle, determine

(a) refrigerating effect
(b) COP

Sketch and label a *p-h* diagram showing values.

**2.44** A single-cylinder R-22 compressor has a 50 mm bore, a 40 mm stroke, and operates at 1725 rpm. Clearance volume is 4%. Determine as close as possible the actual refrigerating capacity, kW, and the required motor size, in hp, if the compressor is used in a system operating between 10°C and 40°C, evaporating and condensing temperatures, respectively.

**2.45** For the lithium-bromide/water absorption refrigeration system shown below, determine

(a) heat required at the generator per ton of cooling [Ans: 516 Btu/min]

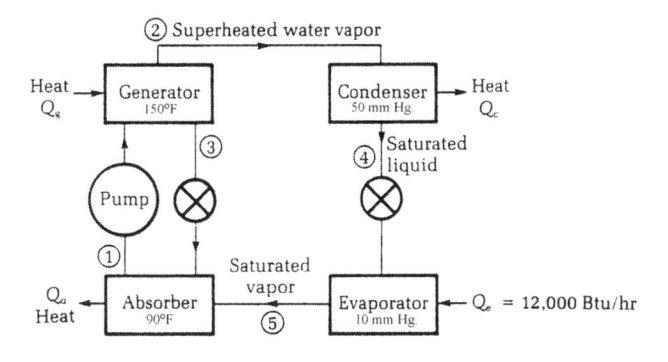

(b) COP [Ans: 0.39]
(c) heat rejection ratio $(Q_{absorber} + Q_{condenser})/Q_{evaporator}$ [Ans: 3.58]

**2.46** In the basic lithium-bromide water absorption system, the generator operates at 170°F while the evaporator is at 47°F. The absorbing temperature is 75°F and the condensing temperature is 88°F. Calculate the heat rejection ratio for these conditions.

**2.47** For the aqua-ammonia absorption refrigeration system shown in the sketch below, complete the table of properties.

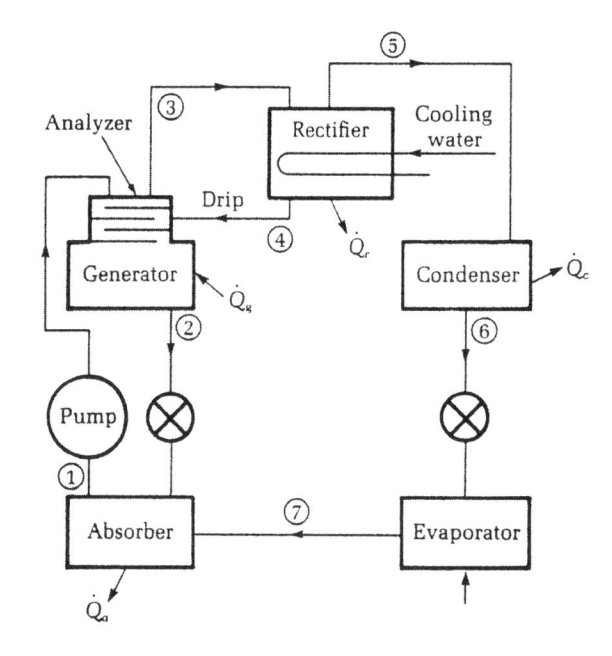

| Point | p, psia | t, °F | x, lb NH$_3$/lb mix | h, Btu/lb |
|---|---|---|---|---|
| 1 | | 80 | | |
| 2 | | | | |
| 3 | 200 | 260 | | |
| 4 | | | | |
| 5 | | 160 | | |
| 6 | | | | |
| 7 | 25 | 20 | | |

**2.48** Solar energy is to be used to warm a large collector plate. This energy will, in turn, be transferred as heat to a fluid within a heat engine, and the engine will reject energy as heat to the atmosphere. Experiments indicate that about 200 Btu/h·ft² of energy can be collected when the plate is operating at 190°F. Estimate the minimum collector area that will be required for a plant producing 1 kW of useful shaft power, when the atmospheric temperature is 70°F. [Ans: 92.7 ft²]

**2.49** What are the Seebeck, Peltier, Thomson, Joule, and Fourier effects? Which are reversible and which are irreversible?

**2.50** A 20 ft by 12 ft by 8 ft (6.1 m by 3.6 m by 2.4 m) room contains an air-water vapor mixture at 80°F (26.7°C). The barometric pressure is standard and the partial pressure of the water vapor is measured to be 0.2 psia (1.38 kPa). Calculate

(a) relative humidity

(b) humidity ratio

(c) dew-point temperature

(d) mass of water vapor contained in the room

**2.51** Given room conditions of 75°F (23.9°C) dry bulb and 60% rh, determine for the air vapor mixture *without using* the psychrometric charts

(a) humidity ratio

(b) enthalpy

(c) dew-point temperature

(d) specific volume

(e) degree of saturation

**2.52** For the conditions of *Problem 2.51* (above), *using* the ASHRAE Psychrometric Chart, find

(a) wet-bulb temperature [Ans: 65.2°F (18.4°C)]

(b) enthalpy [Ans: 30.2 Btu/lb$_m$ (70.2 J/g)]

(c) humidity ratio [Ans: 0.0112 lb/lb (0.0112 kg/kg)]

**2.53** Using the ASHRAE Psychrometric Chart, complete the following table.

| Dry Bulb, °F | Wet Bulb, °F | Dew Point, °F | Humidity W, lb/lb$_{air}$ | Enthalpy h, Btu/lb$_{air}$ | Relative Humidity φ, % | Specific Volume v, ft³/lb$_{air}$ |
|---|---|---|---|---|---|---|
| 85 | 60 | | | | | |
| 75 | | 50 | | | | |
| | | | | 30 | 60 | |
| | 70 | | 0.01143 | | | |
| | | 82 | | 50 | | |

**2.54** Using the ASHRAE Psychrometric Chart complete the following table:

| Dry Bulb, °F | Wet Bulb, °F | Dew Point, °F | Humidity Ratio, lb$_v$/lb$_a$ | Relative Humidity, % | Enthalpy, Btu/lb$_{air}$ | Specific Volume, ft³/lb$_{air}$ |
|---|---|---|---|---|---|---|
| 80 | | | | | | 13.8 |
| 70 | 55 | | | | | |
| 100 | 70 | | | | | |
| | | | | 40 | 40 | |
| | | | 0.01 | | | 13.8 |
| | 60 | 40 | | | | |
| 40 | | | | 20 | | |
| | | 60 | | | 30 | |
| 85 | | | 0.012 | | | |
| 80 | 80 | | | | | |

**2.55** Complete the following table using the Psychrometric Chart.

| Dry Bulb, °C | Wet Bulb, °C | Dew Point, °C | Humidity Ratio, kg/kg | Relative Humidity, % | Enthalpy, kJ/kg | Specific Volume, m³/kg |
|---|---|---|---|---|---|---|
| 26.5 | | | | | | 0.86 |
| 21 | 13 | | | | | |
| 38 | 21 | | | | | |
| | | | | 40 | 95 | |
| | | | 0.01 | | | 0.85 |
| | 16 | 4 | | | | |
| 4 | | | | 20 | | |
| | | 16 | | | 70 | |
| 30 | | | 0.012 | | | |
| 27 | 27 | | | | | |

**2.56** Complete the following table.

| Dry Bulb, °C | Wet Bulb, °C | Dew Point, °C | Humidity Ratio, kg/kg | Relative Humidity, % | Enthalpy, kJ/kg | Specific Volume, m³/kg |
|---|---|---|---|---|---|---|
| 32 | 24 | | | | | |
| 40 | | | | | 81 | |
| | | 18 | | 30 | | |
| | | | 0.022 | | | 0.9 |
| 7 | 7 | | | | | |

**2.57** Without using the psychrometric chart, determine the humidity ratio and relative humidity of an air-water vapor mixture with a dry-bulb temperature of 90°F and thermodynamic wet-bulb temperature of 78°F. The barometric pressure is 14.7 psia. Check your result using the psychrometric chart. [Ans: $W$ = 0.018 lb/lb, relative humidity $\phi$ = 59%]

## 2.12 Bibliography

ASHRAE. 2010. *2010 ASHRAE Handbook—Refrigeration*.
ASHRAE. 2013. *2013 ASHRAE Handbook—Fundamentals*.
Look, D.C., Jr. and H.J. Sauer, Jr. 1986. *Engineering Thermodynamics*. PWS Engineering, Boston.
Sauer, H.J., Jr. and R.H. Howell. 1985. *Heat Pump Systems*. Wiley Interscience, New York.

## SI Tables and Figures

**Table 2-5 SI   Standard Atmospheric Data with Altitude**
*(Table 1, Chapter 1, 2013 ASHRAE Handbook—Fundamentals)*

| Altitude, m | Temperature, °C | Pressure, kPa |
|---|---|---|
| –500 | 18.2 | 107.478 |
| 0 | 15.0 | 101.325 |
| 500 | 11.8 | 95.461 |
| 1000 | 8.5 | 89.875 |
| 1500 | 5.2 | 84.556 |
| 2000 | 2.0 | 79.495 |
| 2500 | –1.2 | 74.682 |
| 3000 | –4.5 | 70.108 |
| 4000 | –11.0 | 61.640 |
| 5000 | –17.5 | 54.020 |
| 6000 | –24.0 | 47.181 |
| 7000 | –30.5 | 41.061 |
| 8000 | –37.0 | 35.600 |
| 9000 | –43.5 | 30.742 |
| 10000 | –50 | 26.436 |
| 12000 | –63 | 19.284 |
| 14000 | –76 | 13.786 |
| 16000 | –89 | 9.632 |
| 18000 | –102 | 6.556 |
| 20000 | –115 | 4.328 |

Data adapted from NASA (1976).

**Table 2-1 SI  Thermodynamic Properties of Water**
*(Table 3, Chapter 1, 2013 ASHRAE Handbook—Fundamentals)*

| Temp., °C  t | Absolute Pressure, kPa  p | Specific Volume, m³/kg (water) Sat. Solid $v_i$ | Evap. $v_{ig}$ | Sat. Vapor $v_g$ | Specific Enthalpy, kJ/kg (water) Sat. Solid $h_i$ | Evap. $h_{ig}$ | Sat. Vapor $h_g$ | Specific Entropy, kJ/(kg·K) (water) Sat. Solid $s_i$ | Evap. $s_{ig}$ | Sat. Vapor $s_g$ | Temp., °C  t |
|---|---|---|---|---|---|---|---|---|---|---|---|
| −60 | 0.00108 | 0.001082 | 90942.00 | 90942.00 | −446.40 | 2836.27 | 2389.87 | −1.6854 | 13.3065 | 11.6211 | −60 |
| −59 | 0.00124 | 0.001082 | 79858.69 | 79858.69 | −444.74 | 2836.46 | 2391.72 | −1.7667 | 13.2452 | 11.5677 | −59 |
| −58 | 0.00141 | 0.001082 | 70212.37 | 70212.37 | −443.06 | 2836.64 | 2393.57 | −1.6698 | 13.8145 | 11.5147 | −58 |
| −57 | 0.00161 | 0.001082 | 61805.35 | 61805.35 | −441.38 | 2836.81 | 2395.43 | −1.6620 | 13.1243 | 11.4623 | −57 |
| −56 | 0.00184 | 0.001082 | 54469.39 | 54469.39 | −439.69 | 2836.97 | 2397.28 | −1.6542 | 13.0646 | 11.4104 | −56 |
| −55 | 0.00209 | 0.001082 | 48061.05 | 48061.05 | −438.00 | 2837.13 | 2399.12 | −1.6464 | 13.0054 | 11.3590 | −55 |
| −54 | 0.00238 | 0.001082 | 42455.57 | 42455.57 | −436.29 | 2837.27 | 2400.98 | −1.6386 | 12.9468 | 11.3082 | −54 |
| −53 | 0.00271 | 0.001083 | 37546.09 | 37546.09 | −434.59 | 2837.42 | 2402.83 | −1.6308 | 12.8886 | 11.2578 | −53 |
| −52 | 0.00307 | 0.001083 | 33242.14 | 33242.14 | −432.87 | 2837.55 | 2404.68 | −1.6230 | 12.8309 | 11.2079 | −52 |
| −51 | 0.00348 | 0.001083 | 29464.67 | 29464.67 | −431.14 | 2837.68 | 2406.53 | −1.6153 | 12.7738 | 11.1585 | −51 |
| −50 | 0.00394 | 0.001083 | 26145.01 | 26145.01 | −429.41 | 2837.80 | 2408.39 | −1.6075 | 12.7170 | 11.1096 | −50 |
| −49 | 0.00445 | 0.001083 | 23223.69 | 23223.70 | −427.67 | 2837.91 | 2410.24 | −1.5997 | 12.6608 | 11.0611 | −49 |
| −48 | 0.00503 | 0.001083 | 20651.68 | 20651.69 | −425.93 | 2838.02 | 2412.09 | −1.5919 | 12.6051 | 11.0131 | −48 |
| −47 | 0.00568 | 0.001083 | 18383.50 | 18383.51 | −424.27 | 2838.12 | 2413.94 | −1.5842 | 12.5498 | 10.9656 | −47 |
| −46 | 0.00640 | 0.001083 | 16381.35 | 16381.36 | −422.41 | 2838.21 | 2415.79 | −1.5764 | 12.4949 | 10.9185 | −46 |
| −45 | 0.00721 | 0.001984 | 14612.35 | 14512.36 | −420.65 | 2838.29 | 2417.65 | −1.5686 | 12.4405 | 10.8719 | −45 |
| −44 | 0.00811 | 0.001084 | 13047.65 | 13047.66 | −418.87 | 2838.37 | 2419.50 | −1.5609 | 12.3866 | 10.8257 | −44 |
| −43 | 0.00911 | 0.001084 | 11661.85 | 11661.85 | −417.09 | 2838.44 | 2421.35 | −1.5531 | 12.3330 | 10.7799 | −43 |
| −42 | 0.01022 | 0.001084 | 10433.85 | 10433.85 | −415.30 | 2838.50 | 2423.20 | −1.5453 | 12.2799 | 10.7346 | −42 |
| −41 | 0.01147 | 0.001084 | 9344.25 | 9344.25 | −413.50 | 2838.55 | 2425.05 | −1.5376 | 12.2273 | 10.6897 | −41 |
| −40 | 0.01285 | 0.001084 | 8376.33 | 8376.33 | −411.70 | 2838.60 | 2426.90 | −1.5298 | 12.1750 | 10.6452 | −40 |
| −39 | 0.01438 | 0.001085 | 7515.86 | 7515.87 | −409.88 | 2838.64 | 2428.76 | −1.5221 | 12.1232 | 10.6011 | −39 |
| −38 | 0.01608 | 0.001085 | 6750.36 | 6750.36 | −508.07 | 2838.67 | 1430.61 | −1.5143 | 12.0718 | 10.5575 | −38 |
| −37 | 0.01796 | 0.001085 | 6068.16 | 6068.17 | −406.24 | 2838.70 | 2432.46 | −1.5066 | 12.0208 | 10.5142 | −37 |
| −36 | 0.02004 | 0.001085 | 5459.82 | 5459.82 | −404.40 | 2838.71 | 2434.31 | −1.4988 | 11.9702 | 10.4713 | −36 |
| −35 | 0.02235 | 0.001085 | 4917.09 | 4917.10 | −402.56 | 2838.73 | 2436.16 | −1.4911 | 11.9199 | 10.4289 | −35 |
| −34 | 0.02490 | 0.001085 | 4432.36 | 4432.37 | −400.72 | 2838.73 | 2438.01 | −1.4833 | 11.8701 | 10.3868 | −34 |
| −33 | 0.02771 | 0.001085 | 3998.71 | 3998.71 | −398.86 | 2838.72 | 2439.86 | −1.4756 | 11.8207 | 10.3451 | −33 |
| −32 | 0.03082 | 0.001086 | 3610.71 | 3610.71 | −397.00 | 2838.71 | 2441.72 | −1.4678 | 11.7716 | 10.3037 | −32 |
| −31 | 0.03424 | 0.001086 | 3263.20 | 3263.20 | −395.12 | 2838.69 | 2443.57 | −1.4601 | 11.7229 | 10.2628 | −31 |
| −30 | 0.03802 | 0.001086 | 2951.64 | 2951.64 | −393.25 | 2838.66 | 2445.42 | −1.4524 | 11.6746 | 10.2222 | −30 |
| −29 | 0.04217 | 0.001086 | 2672.03 | 2672.03 | −391.36 | 2838.63 | 2447.27 | −1.4446 | 11.6266 | 10.1820 | −29 |
| −28 | 0.04673 | 0.001086 | 2420.89 | 2420.89 | −389.47 | 2838.59 | 2449.12 | −1.4369 | 11.4790 | 10.1421 | −28 |
| −27 | 0.05174 | 0.001086 | 2195.23 | 2195.23 | −387.57 | 2838.53 | 2450.97 | −1.4291 | 11.5318 | 10.1026 | −27 |
| −26 | 0.05725 | 0.001087 | 1992.15 | 1992.15 | −385.66 | 2838.48 | 2452.82 | −1.4214 | 11.4849 | 10.0634 | −26 |
| −25 | 0.06329 | 0.001087 | 1809.35 | 1809.35 | −383.74 | 2838.41 | 2454.67 | −1.4137 | 11.4383 | 10.0246 | −25 |
| −24 | 0.06991 | 0.001087 | 1644.59 | 1644.59 | −381.34 | 2838.34 | 2456.52 | −1.4059 | 11.3921 | 9.9862 | −24 |
| −23 | 0.07716 | 0.001087 | 1495.98 | 1495.98 | −379.89 | 2838.26 | 2458.37 | −1.3982 | 11.3462 | 9.9480 | −23 |
| −22 | 0.08510 | 0.001087 | 1361.94 | 1361.94 | −377.95 | 2838.17 | 2460.22 | −1.3905 | 11.3007 | 9.9102 | −22 |
| −21 | 0.09378 | 0.001087 | 1240.77 | 1240.77 | −376.01 | 2838.07 | 2462.06 | −1.3828 | 11.2555 | 9.8728 | −21 |
| −20 | 0.10326 | 0.001087 | 1131.27 | 1131.27 | −374.06 | 2837.97 | 2463.91 | −1.3750 | 11.2106 | 9.8356 | −20 |
| −19 | 0.11362 | 0.001088 | 1032.18 | 1032.18 | −372.10 | 2837.86 | 2465.76 | −1.3673 | 11.1661 | 9.7988 | −19 |
| −18 | 0.12492 | 0.001088 | 942.46 | 942.47 | −370.13 | 2837.74 | 2467.61 | −1.3596 | 11.1218 | 9.7623 | −18 |
| −17 | 0.13725 | 0.001088 | 861.17 | 861.18 | −368.15 | 2837.61 | 2469.46 | −1.3518 | 11.0779 | 9.7261 | −17 |
| −16 | 0.15068 | 0.001088 | 787.48 | 787.49 | −366.17 | 2837.47 | 2471.30 | −1.3441 | 11.0343 | 9.6902 | −16 |
| −15 | 0.16530 | 0.001088 | 720.59 | 720.59 | −364.18 | 2837.33 | 2473.15 | −1.3364 | 10.9910 | 9.6546 | −15 |
| −14 | 0.18122 | 0.001088 | 659.86 | 659.86 | −362.18 | 2837.18 | 2474.99 | −1.3287 | 10.9480 | 9.6193 | −14 |
| −13 | 0.19852 | 0.001089 | 604.65 | 604.65 | −360.18 | 2837.02 | 2476.84 | −1.3210 | 10.9053 | 9.5844 | −13 |
| −12 | 0.21732 | 0.001089 | 554.45 | 554.45 | −358.17 | 2836.85 | 2478.68 | −1.3232 | 10.8629 | 9.5497 | −12 |
| −11 | 0.23774 | 0.001089 | 508.75 | 508.75 | −356.15 | 2836.68 | 2480.53 | −1.3055 | 10.8208 | 9.5153 | −11 |
| −10 | 0.25990 | 0.001089 | 467.14 | 467.14 | −354.12 | 2836.49 | 2482.37 | −1.2978 | 10.7790 | 9.4812 | −10 |
| −9 | 0.28393 | 0.001089 | 429.21 | 429.21 | −352.08 | 2836.30 | 2484.22 | −1.2901 | 10.7375 | 9.4474 | −9 |
| −8 | 0.30998 | 0.001090 | 394.64 | 394.64 | −350.04 | 2836.10 | 2486.06 | −1.2824 | 10.6962 | 9.4139 | −8 |
| −7 | 0.33819 | 0.001090 | 363.07 | 363.07 | −347.99 | 2835.89 | 2487.90 | −1.2746 | 10.6552 | 9.3806 | −7 |
| −6 | 0.36874 | 0.001090 | 334.25 | 334.25 | −345.93 | 2835.68 | 2489.74 | −1.2669 | 10.6145 | 9.3476 | −6 |
| −5 | 0.40176 | 0.001090 | 307.91 | 307.91 | −343.87 | 2835.45 | 2491.58 | −2.2592 | 10.4741 | 9.3149 | −5 |
| −4 | 0.43747 | 0.001090 | 283.83 | 283.83 | −341.80 | 2835.22 | 2493.42 | −1.2515 | 10.5340 | 9.2825 | −4 |
| −3 | 0.47606 | 0.001090 | 261.79 | 261.79 | −339.72 | 2834.98 | 2495.26 | −1.2438 | 10.4941 | 9.2503 | −3 |
| −2 | 0.51772 | 0.001091 | 241.60 | 241.60 | −337.63 | 2834.72 | 2497.10 | −1.2361 | 10.4544 | 9.2184 | −2 |
| −1 | 0.56267 | 0.001091 | 223.11 | 223.11 | −335.53 | 2834.47 | 2498.93 | −1.2284 | 10.4151 | 9.1867 | −1 |
| 0 | 0.61115 | 0.001091 | 206.16 | 206.16 | −333.43 | 2834.20 | 2500.77 | −1.2206 | 10.3760 | 9.1553 | 0 |

# Thermodynamics and Psychrometrics

**Table 2-1 SI   Thermodynamic Properties of Water (*Continued*)**
*(Table 3, Chapter 1, 2013 ASHRAE Handbook—Fundamentals)*

| Temp., °C $t$ | Absolute Pressure, kPa $p$ | Sat. Liquid $v_f$ | Evap. $v_{fg}$ | Sat. Vapor $v_g$ | Sat. Liquid $h_f$ | Evap. $h_{fg}$ | Sat. Vapor $h_g$ | Sat. Liquid $s_f$ | Evap. $s_{fg}$ | Sat. Vapor $s_g$ | Temp., °C $t$ |
|---|---|---|---|---|---|---|---|---|---|---|---|
| 0  | 0.6112 | 0.001000 | 206.141 | 206.143 | −0.04 | 2500.81 | 2500.77 | −0.0002 | 9.1555 | 9.1553 | 0 |
| 1  | 0.6571 | 0.001000 | 192.455 | 192.456 | 4.18  | 2498.43 | 2502.61 | 0.0153 | 9.1134 | 9.1286 | 1 |
| 2  | 0.7060 | 0.001000 | 179.769 | 179.770 | 8.39  | 2496.05 | 2504.45 | 0.0306 | 9.0716 | 9.1022 | 2 |
| 3  | 0.7580 | 0.001000 | 168.026 | 168.027 | 12.60 | 2493.68 | 2506.28 | 0.0459 | 9.0302 | 9.0761 | 3 |
| 4  | 0.8135 | 0.001000 | 157.137 | 157.138 | 16.81 | 2491.31 | 2508.12 | 0.0611 | 8.9890 | 9.0501 | 4 |
| 5  | 0.8725 | 0.001000 | 147.032 | 147.033 | 21.02 | 2488.94 | 2509.96 | 0.0763 | 8.9482 | 9.0244 | 5 |
| 6  | 0.9353 | 0.001000 | 137.653 | 137.654 | 25.22 | 2486.57 | 2511.79 | 0.0913 | 8.9077 | 8.9990 | 6 |
| 7  | 1.0020 | 0.001000 | 128.947 | 128.948 | 29.42 | 2484.20 | 2513.62 | 0.1064 | 8.8674 | 8.9738 | 7 |
| 8  | 1.0728 | 0.001000 | 120.850 | 120.851 | 33.62 | 2481.84 | 2515.46 | 0.1213 | 8.8273 | 8.9488 | 8 |
| 9  | 1.1481 | 0.001000 | 113.326 | 113.327 | 37.82 | 2479.47 | 2517.29 | 0.1362 | 8.7878 | 8.9245 | 9 |
| 10 | 1.2280 | 0.001000 | 106.328 | 106.329 | 42.01 | 2477.11 | 2519.12 | 0.1511 | 8.7484 | 8.8995 | 10 |
| 11 | 1.3127 | 0.001000 | 99.812  | 99.813  | 46.21 | 2474.74 | 2520.95 | 0.1659 | 8.7093 | 8.8752 | 11 |
| 12 | 1.4026 | 0.001001 | 93.743  | 93.744  | 50.40 | 2472.38 | 2522.78 | 0.1806 | 8.6705 | 8.8511 | 12 |
| 13 | 1.4978 | 0.001001 | 88.088  | 88.089  | 54.59 | 2470.02 | 2524.61 | 0.1953 | 8.6319 | 8.8272 | 13 |
| 14 | 1.5987 | 0.001001 | 82.815  | 82.816  | 58.78 | 2467.66 | 2526.44 | 0.2099 | 8.5936 | 3.8035 | 14 |
| 15 | 1.7055 | 0.001001 | 77.897  | 77.898  | 62.97 | 2465.30 | 2528.26 | 0.2244 | 8.5556 | 8.7801 | 15 |
| 16 | 1.8184 | 0.001001 | 73.307  | 73.308  | 67.16 | 2462.93 | 2530.09 | 0.2389 | 8.5178 | 8.7568 | 16 |
| 17 | 1.9380 | 0.001001 | 69.021  | 69.022  | 71.34 | 2460.57 | 2531.92 | 0.2534 | 8.4804 | 8.7338 | 17 |
| 18 | 2.0643 | 0.001002 | 65.017  | 65.018  | 75.53 | 2458.21 | 2533.74 | 0.2678 | 8.4431 | 8.7109 | 18 |
| 19 | 2.1978 | 0.001002 | 65.274  | 61.273  | 79.72 | 2455.85 | 2535.56 | 0.2821 | 8.4061 | 8.6883 | 19 |
| 20 | 2.3388 | 0.001002 | 57.774  | 57.773  | 83.90 | 2453.48 | 2537.38 | 0.2964 | 8.3694 | 8.6658 | 20 |
| 21 | 2.4877 | 0.001002 | 54.450  | 54.500  | 88.08 | 2451.12 | 2539.20 | 0.3107 | 8.3329 | 8.6436 | 21 |
| 22 | 2.6448 | 0.001002 | 51.433  | 51.434  | 92.27 | 2448.75 | 2541.02 | 0.3249 | 8.2967 | 8.6215 | 22 |
| 23 | 2.8104 | 0.001003 | 48.562  | 48.563  | 96.45 | 2446.39 | 2542.84 | 0.3390 | 8.2607 | 8.5996 | 23 |
| 24 | 2.9851 | 0.001003 | 45.872  | 45.873  | 100.63| 2444.02 | 2544.65 | 0.3531 | 8.2249 | 8.5780 | 24 |
| 25 | 3.1692 | 0.001003 | 43.350  | 43.351  | 104.81| 2441.66 | 2546.47 | 0.3672 | 8.1894 | 8.5565 | 25 |
| 26 | 3.3631 | 0.001003 | 40.985  | 40.986  | 108.99| 2439.29 | 2548.28 | 0.3812 | 8.1541 | 8.5352 | 26 |
| 27 | 3.5673 | 0.001004 | 38.766  | 38.767  | 113.18| 2436.92 | 2550.09 | 0.3951 | 8.1190 | 8.5141 | 27 |
| 28 | 3.7822 | 0.001004 | 36.682  | 36.683  | 117.36| 2434.55 | 2551.90 | 0.4090 | 8.0842 | 8.4932 | 28 |
| 29 | 4.0083 | 0.001004 | 34.726  | 34.727  | 121.54| 2432.17 | 2553.71 | 0.4229 | 8.0496 | 8.4724 | 29 |
| 30 | 4.2460 | 0.001004 | 32.889  | 32.889  | 125.72| 2429.80 | 2555.52 | 0.4367 | 8.0152 | 8.4519 | 30 |
| 31 | 4.4959 | 0.001005 | 31.160  | 31.161  | 129.90| 2427.43 | 2557.32 | 0.4505 | 7.9810 | 8.4315 | 31 |
| 32 | 4.7585 | 0.001005 | 29.535  | 29.536  | 134.08| 2425.05 | 2559.13 | 0.4642 | 7.9471 | 8.4112 | 32 |
| 33 | 5.0343 | 0.001005 | 28.006  | 28.007  | 138.26| 2422.67 | 2560.93 | 0.4779 | 7.9133 | 8.3912 | 33 |
| 34 | 5.3239 | 0.001006 | 26.567  | 26.568  | 142.44| 2410.29 | 2562.73 | 0.4915 | 7.8790 | 8.3713 | 34 |
| 35 | 5.6278 | 0.001006 | 25.212  | 25.213  | 146.62| 2417.91 | 2564.53 | 0.5051 | 7.8465 | 8.3516 | 35 |
| 36 | 5.9466 | 0.001006 | 23.935  | 23.936  | 150.80| 2415.53 | 2566.33 | 0.5186 | 7.8134 | 8.3320 | 36 |
| 37 | 6.2810 | 0.001007 | 22.733  | 22.734  | 154.98| 2413.14 | 2568.12 | 0.5321 | 7.7805 | 8.3127 | 37 |
| 38 | 6.6315 | 0.001007 | 21.599  | 21.600  | 159.16| 2410.76 | 2569.91 | 0.5456 | 7.7479 | 8.2934 | 38 |
| 39 | 6.9987 | 0.001008 | 20.529  | 20.530  | 163.34| 2408.37 | 2571.71 | 0.5590 | 7.7154 | 8.2744 | 39 |
| 40 | 7.3835 | 0.001008 | 19.520  | 19.521  | 167.52| 2405.98 | 2573.50 | 0.5724 | 7.6831 | 8.2555 | 40 |
| 41 | 7.7863 | 0.001008 | 18.567  | 18.568  | 171.70| 2403.58 | 2575.28 | 0.5857 | 7.6510 | 8.2367 | 41 |
| 42 | 8.2080 | 0.001009 | 17.667  | 17.668  | 175.88| 2401.19 | 2577.07 | 0.5990 | 7.6191 | 8.2181 | 42 |
| 43 | 8.6492 | 0.001009 | 16.818  | 16.819  | 180.06| 2398.79 | 2578.85 | 0.6122 | 7.5875 | 8.1997 | 43 |
| 44 | 9.1107 | 0.001010 | 16.014  | 16.015  | 184.24| 2396.39 | 2580.63 | 0.6254 | 7.3560 | 8.1814 | 44 |
| 45 | 9.5932 | 0.001010 | 15.255  | 15.256  | 188.42| 2393.99 | 2582.41 | 0.6386 | 7.5247 | 8.1632 | 45 |
| 46 | 10.0976| 0.001010 | 14.537  | 14.538  | 192.60| 2391.59 | 2584.19 | 0.6517 | 7.4936 | 8.1452 | 46 |
| 47 | 10.6246| 0.001011 | 13.858  | 13.859  | 196.78| 2389.18 | 2585.96 | 0.6648 | 7.4626 | 8.1274 | 47 |
| 48 | 11.1751| 0.001011 | 13.214  | 13.215  | 200.97| 2386.77 | 2587.74 | 0.6778 | 7.4319 | 8.1097 | 48 |
| 49 | 11.7500| 0.001012 | 12.606  | 12.607  | 205.15| 2384.36 | 2589.51 | 0.6908 | 7.4013 | 8.0921 | 49 |
| 50 | 12.3499| 0.001012 | 12.029  | 12.029  | 209.33| 2381.94 | 2591.27 | 0.7038 | 7.3709 | 8.0747 | 50 |
| 51 | 12.9759| 0.001013 | 11.482  | 11.483  | 213.51| 2379.53 | 2593.04 | 0.7167 | 7.3407 | 8.0574 | 51 |
| 52 | 13.6290| 0.001013 | 10.964  | 10.965  | 217.70| 2377.10 | 2594.80 | 0.7296 | 7.3107 | 8.0403 | 52 |
| 53 | 14.3100| 0.001014 | 10.473  | 10.474  | 221.88| 2374.68 | 2596.56 | 0.7424 | 7.2809 | 8.0233 | 53 |
| 54 | 15.0200| 0.001014 | 10.001  | 10.008  | 226.06| 2372.26 | 2598.32 | 0.7552 | 7.2512 | 8.0064 | 54 |
| 55 | 15.7597| 0.001015 | 9.563   | 9.5663  | 230.25| 2369.83 | 2600.07 | 0.7680 | 7.2217 | 7.9897 | 55 |
| 56 | 16.5304| 0.001015 | 9.147   | 9.1468  | 234.43| 2367.39 | 2601.82 | 0.7807 | 7.1924 | 7.9731 | 56 |
| 57 | 17.3331| 0.001016 | 8.744   | 8.7489  | 238.61| 2364.96 | 2603.57 | 0.7934 | 7.1632 | 7.9566 | 57 |
| 58 | 18.1690| 0.001016 | 8.3690  | 8.3700  | 242.80| 2362.52 | 2605.32 | 0.8061 | 7.1342 | 7.9403 | 58 |
| 59 | 19.0387| 0.001017 | 8.0094  | 8.0114  | 246.99| 2360.08 | 2607.06 | 0.8187 | 7.1054 | 7.9240 | 59 |
| 60 | 19.944 | 0.001017 | 7.6677  | 7.6697  | 251.17| 2357.63 | 2608.80 | 0.8313 | 7.0767 | 7.9079 | 60 |
| 61 | 20.885 | 0.001018 | 7.3428  | 7.3438  | 255.36| 2355.19 | 2610.54 | 0.8438 | 7.0482 | 7.8920 | 61 |
| 62 | 21.864 | 0.001018 | 7.0337  | 7.0347  | 259.54| 2352.73 | 2612.28 | 0.8563 | 7.0198 | 7.8761 | 62 |
| 63 | 22.882 | 0.001019 | 6.7397  | 6.7407  | 263.73| 2350.28 | 2614.01 | 0.8688 | 6.9916 | 7.8604 | 63 |
| 64 | 23.940 | 0.001019 | 6.4599  | 6.4609  | 267.92| 2347.82 | 2615.74 | 0.8812 | 6.9636 | 7.8448 | 64 |
| 65 | 25.040 | 0.001020 | 6.1935  | 6.1946  | 272.11| 2345.36 | 2617.46 | 0.8936 | 6.9357 | 7.8293 | 65 |
| 66 | 26.180 | 0.001020 | 5.9397  | 5.9409  | 276.30| 2342.89 | 2619.19 | 0.9060 | 6.9080 | 7.8140 | 66 |
| 67 | 27.366 | 0.001021 | 5.6982  | 5.6992  | 280.49| 2340.42 | 2620.90 | 0.9183 | 6.8804 | 7.7987 | 67 |
| 68 | 28.596 | 0.001022 | 5.4680  | 5.4690  | 284.68| 2337.95 | 2622.62 | 0.9306 | 2.8530 | 7.7836 | 68 |
| 69 | 29.873 | 0.001022 | 5.2485  | 5.2495  | 288.87| 2335.47 | 2624.33 | 0.9429 | 6.8257 | 7.7686 | 69 |

**Table 2-1 SI Thermodynamic Properties of Water (*Continued*)**

*(Table 3, Chapter 1, 2013 ASHRAE Handbook—Fundamentals)*

| Temp., °C t | Absolute Pressure, kPa p | Specific Volume, m³/kg (water) Sat. Liquid $v_f$ | Evap. $v_{fg}$ | Sat. Vapor $v_g$ | Specific Enthalpy, kJ/kg (water) Sat. Liquid $h_f$ | Evap. $h_{fg}$ | Sat. Vapor $h_g$ | Specific Entropy, kJ/(kg·K) (water) Sat. Liquid $s_f$ | Evap. $s_{fg}$ | Sat. Vapor $s_g$ | Temp., °C t |
|---|---|---|---|---|---|---|---|---|---|---|---|
| 70 | 31.198 | 0.001023 | 5.0392 | 5.0402 | 293.06 | 2332.99 | 2626.04 | 0.9551 | 6.7986 | 7.7537 | 70 |
| 71 | 32.572 | 0.001023 | 4.8396 | 4.8407 | 297.25 | 2330.50 | 2627.75 | 0.9673 | 6.7716 | 7.7389 | 71 |
| 72 | 33.997 | 0.001024 | 4.6492 | 4.6502 | 301.44 | 2328.01 | 2629.45 | 0.9795 | 6.7448 | 7.7242 | 72 |
| 73 | 35.475 | 0.001025 | 4.4675 | 4.4685 | 305.63 | 2325.51 | 2631.15 | 0.9916 | 6.7181 | 7.7097 | 73 |
| 74 | 37.006 | 0.001025 | 4.2940 | 4.2951 | 309.83 | 2323.02 | 2632.84 | 1.0037 | 6.6915 | 7.6952 | 74 |
| 75 | 38.592 | 0.001026 | 4.1284 | 4.1294 | 314.02 | 2320.51 | 2634.53 | 1.0157 | 6.6651 | 7.6809 | 75 |
| 76 | 40.236 | 0.001026 | 3.9702 | 3.9712 | 318.22 | 2318.01 | 2636.22 | 1.0278 | 6.6389 | 7.6666 | 76 |
| 77 | 41.938 | 0.001027 | 3.8190 | 3.8201 | 322.41 | 2315.49 | 2637.90 | 1.0398 | 6.6127 | 7.6525 | 77 |
| 78 | 43.700 | 0.001028 | 3.6746 | 3.6756 | 326.61 | 2312.98 | 2639.58 | 1.0517 | 6.5867 | 7.6384 | 78 |
| 79 | 45.524 | 0.001028 | 3.5365 | 3.5375 | 330.81 | 2310.46 | 2641.26 | 1.0636 | 6.5609 | 7.6245 | 79 |
| 80 | 47.412 | 0.001029 | 3.4044 | 3.4055 | 335.00 | 2307.93 | 2642.93 | 1.0755 | 6.5351 | 7.6107 | 80 |
| 81 | 49.364 | 0.001030 | 3.2781 | 3.2792 | 339.20 | 2305.40 | 2644.60 | 1.0874 | 6.5095 | 7.5969 | 81 |
| 82 | 51.384 | 0.001030 | 3.1573 | 3.1583 | 343.40 | 2902.86 | 2646.26 | 1.0993 | 6.4841 | 7.5833 | 82 |
| 83 | 53.473 | 0.001031 | 3.0417 | 3.0427 | 347.60 | 2300.32 | 2647.92 | 1.1111 | 6.4587 | 7.5698 | 83 |
| 84 | 55.633 | 0.001032 | 2.9310 | 2.9320 | 351.80 | 2297.78 | 2649.58 | 1.1228 | 6.4335 | 7.5563 | 84 |
| 85 | 57.865 | 0.001032 | 2.8250 | 2.8260 | 356.01 | 2295.22 | 2651.23 | 1.1346 | 6.4084 | 7.5430 | 85 |
| 86 | 60.171 | 0.001033 | 2.7235 | 2.7245 | 350.21 | 2292.67 | 2652.88 | 1.1463 | 6.3834 | 7.5297 | 86 |
| 87 | 62.554 | 0.001034 | 2.6263 | 2.6273 | 364.41 | 2290.11 | 2654.52 | 1.1580 | 6.3586 | 7.5166 | 87 |
| 88 | 65.015 | 0.001035 | 2.5331 | 2.5341 | 368.62 | 2287.54 | 2656.16 | 1.1696 | 6.3339 | 7.5035 | 88 |
| 89 | 67.556 | 0.001035 | 2.4438 | 2.4448 | 372.82 | 2284.97 | 2657.79 | 1.1812 | 6.3093 | 7.4905 | 89 |
| 90 | 70.180 | 0.001036 | 2.3582 | 2.3592 | 377.03 | 2282.39 | 2659.42 | 1.1928 | 6.2848 | 7.4776 | 90 |
| 91 | 72.888 | 0.001037 | 2.2760 | 2.2771 | 381.24 | 2279.81 | 2661.04 | 1.2044 | 6.2605 | 7.4648 | 91 |
| 92 | 75.683 | 0.001037 | 2.1973 | 2.1983 | 385.45 | 2277.22 | 2662.66 | 1.2159 | 6.2362 | 7.4521 | 92 |
| 93 | 78.566 | 0.001038 | 2.1217 | 2.1228 | 389.66 | 2274.62 | 2664.28 | 1.2274 | 6.2121 | 7.4395 | 93 |
| 94 | 81.541 | 0.001039 | 2.0492 | 2.0502 | 393.87 | 2272.02 | 2665.89 | 1.2389 | 6.1881 | 7.4270 | 94 |
| 95 | 84.608 | 0.001040 | 1.9796 | 1.9806 | 398.08 | 2269.41 | 2667.49 | 1.2504 | 6.1642 | 7.4146 | 95 |
| 96 | 87.770 | 0.001040 | 1.9128 | 1.9138 | 402.29 | 2266.80 | 2669.09 | 1.2618 | 6.1404 | 7.4022 | 96 |
| 97 | 91.030 | 0.001041 | 1.8486 | 1.8496 | 406.51 | 2264.18 | 2670.69 | 1.2732 | 6.1168 | 7.3899 | 97 |
| 98 | 94.390 | 0.001042 | 1.7869 | 1.7880 | 410.72 | 2261.55 | 2672.28 | 1.2845 | 6.0932 | 7.3777 | 98 |
| 99 | 97.852 | 0.001044 | 1.7277 | 1.7287 | 414.94 | 2258.92 | 2673.86 | 1.2959 | 6.0697 | 7.3656 | 99 |
| 100 | 101.419 | 0.001044 | 1.6708 | 1.6718 | 419.16 | 2256.28 | 2675.44 | 1.3072 | 6.0464 | 7.3536 | 100 |
| 101 | 105.092 | 0.001044 | 1.6161 | 1.6171 | 423.38 | 2253.64 | 2677.02 | 1.3185 | 6.0232 | 7.3416 | 101 |
| 102 | 108.875 | 0.001045 | 1.5635 | 1.5645 | 427.60 | 2250.99 | 2678.58 | 1.3297 | 6.0000 | 7.3298 | 102 |
| 103 | 112.770 | 0.001046 | 1.5129 | 1.5139 | 431.82 | 2248.33 | 2680.15 | 1.3410 | 5.9770 | 7.3180 | 103 |
| 104 | 116.779 | 0.001047 | 1.4642 | 1.4652 | 436.04 | 2245.66 | 2681.71 | 1.3522 | 5.9541 | 7.3062 | 104 |
| 105 | 120.906 | 0.001047 | 1.4174 | 1.4184 | 440.27 | 2242.99 | 2683.26 | 1.3634 | 5.9313 | 7.2946 | 105 |
| 106 | 125.152 | 0.001048 | 1.3723 | 1.3734 | 444.49 | 2240.31 | 2684.80 | 1.3745 | 5.9086 | 7.2830 | 106 |
| 107 | 129.520 | 0.001049 | 1.3290 | 1.3300 | 448.72 | 2237.63 | 2686.35 | 1.3856 | 5.8860 | 7.2716 | 107 |
| 108 | 134.012 | 0.001050 | 1.2872 | 1.2883 | 452.95 | 2234.93 | 2687.88 | 1.3967 | 5.8635 | 7.2601 | 108 |
| 109 | 138.633 | 0.001051 | 1.2470 | 1.2481 | 457.18 | 2232.23 | 2689.41 | 1.4078 | 5.8410 | 7.2488 | 109 |
| 110 | 143.384 | 0.001052 | 1.2083 | 1.2093 | 461.41 | 2229.52 | 2690.93 | 1.4188 | 5.8187 | 7.2375 | 110 |
| 111 | 148.267 | 0.001052 | 1.1710 | 1.1720 | 465.64 | 2226.81 | 2692.45 | 1.4298 | 5.7965 | 7.2263 | 111 |
| 112 | 153.287 | 0.001053 | 1.1350 | 1.1361 | 469.88 | 2224.09 | 2693.96 | 1.4408 | 5.7744 | 7.2152 | 112 |
| 113 | 158.445 | 0.001054 | 1.1004 | 1.1015 | 474.11 | 2221.35 | 2695.47 | 1.4518 | 5.7524 | 7.2402 | 113 |
| 114 | 163.745 | 0.001055 | 1.0670 | 1.0681 | 478.35 | 2218.62 | 2696.97 | 1.4627 | 5.7304 | 7.1931 | 114 |
| 115 | 169.190 | 0.001056 | 1.0348 | 1.0359 | 482.59 | 2215.87 | 2698.46 | 1.4737 | 5.7086 | 7.1822 | 115 |
| 116 | 174.782 | 0.001057 | 1.0038 | 1.0048 | 486.83 | 2213.12 | 2699.95 | 1.4846 | 5.6868 | 7.1714 | 116 |
| 117 | 180.525 | 0.001058 | 0.9739 | 0.9749 | 491.07 | 2210.35 | 2701.43 | 1.4954 | 5.6652 | 7.1606 | 117 |
| 118 | 186.420 | 0.001059 | 0.9450 | 0.9460 | 495.32 | 2207.58 | 2702.90 | 1.5063 | 5.6436 | 7.1499 | 118 |
| 119 | 192.473 | 0.001059 | 0.9171 | 0.9182 | 499.56 | 2204.80 | 2704.37 | 1.5171 | 5.6221 | 7.1392 | 119 |
| 120 | 198.685 | 0.001060 | 0.8902 | 0.8913 | 503.81 | 2202.02 | 2705.83 | 1.5279 | 5.6007 | 7.1286 | 120 |
| 122 | 211.601 | 0.001062 | 0.8391 | 0.8402 | 512.31 | 2196.42 | 2706.73 | 1.5494 | 5.5582 | 7.1076 | 122 |
| 124 | 225.194 | 0.001064 | 0.7916 | 0.7927 | 520.82 | 2190.78 | 2711.60 | 1.5709 | 5.5160 | 7.0869 | 124 |
| 126 | 239.490 | 0.001066 | 0.7472 | 0.7483 | 529.33 | 2185.11 | 2714.44 | 1.5922 | 5.4742 | 7.0664 | 126 |
| 128 | 254.515 | 0.001068 | 0.7057 | 0.7068 | 537.86 | 2179.40 | 2717.26 | 1.6135 | 5.4326 | 7.0461 | 128 |
| 130 | 270.298 | 0.001070 | 0.6670 | 0.6681 | 546.39 | 2173.66 | 2720.05 | 1.6347 | 5.3914 | 7.0261 | 130 |
| 132 | 286.866 | 0.001072 | 0.6308 | 0.6318 | 554.93 | 2167.87 | 2722.80 | 1.6557 | 5.3505 | 7.0063 | 132 |
| 134 | 304.247 | 0.001074 | 0.5969 | 0.5979 | 563.48 | 2162.05 | 2725.53 | 1.6767 | 5.3099 | 6.9867 | 134 |
| 136 | 322.470 | 0.001076 | 0.5651 | 0.5662 | 572.04 | 2156.18 | 2728.22 | 1.6977 | 5.2697 | 6.9673 | 136 |
| 138 | 341.566 | 0.001078 | 0.5354 | 0.5364 | 580.60 | 2150.28 | 2730.88 | 1.7185 | 5.2296 | 6.9481 | 138 |
| 140 | 361.565 | 0.001080 | 0.5075 | 0.5085 | 589.18 | 2144.33 | 2733.51 | 1.7393 | 5.1899 | 6.9292 | 140 |
| 142 | 382.497 | 0.001082 | 0.4813 | 0.4824 | 597.76 | 2138.34 | 2736.11 | 1.7599 | 5.1505 | 6.9104 | 142 |
| 144 | 404.394 | 0.001084 | 0.4567 | 0.4578 | 606.36 | 2132.31 | 2738.67 | 1.7805 | 5.1113 | 6.8918 | 144 |
| 146 | 427.288 | 0.001086 | 0.4336 | 0.4347 | 614.97 | 2126.23 | 2741.19 | 1.8011 | 5.0724 | 6.8735 | 146 |
| 148 | 451.211 | 0.001088 | 0.4119 | 0.4130 | 623.58 | 2120.10 | 2743.68 | 1.8215 | 5.0338 | 6.8553 | 148 |
| 150 | 476.198 | 0.001091 | 0.3914 | 0.3925 | 632.21 | 2113.92 | 2746.13 | 1.8419 | 4.9954 | 6.8373 | 150 |
| 152 | 502.281 | 0.001093 | 0.3722 | 0.3733 | 640.85 | 2107.70 | 2748.55 | 1.8622 | 4.9573 | 6.8194 | 152 |
| 154 | 529.495 | 0.001095 | 0.3541 | 0.3552 | 649.50 | 2101.43 | 2750.93 | 1.8824 | 4.9194 | 6.8017 | 154 |
| 156 | 557.875 | 0.001097 | 0.3370 | 0.3381 | 658.16 | 2095.11 | 2753.27 | 1.9026 | 4.8817 | 6.7842 | 156 |
| 158 | 587.456 | 0.001100 | 0.3209 | 0.3220 | 666.83 | 2088.73 | 2755.57 | 1.9226 | 4.8443 | 6.7669 | 158 |
| 160 | 618.275 | 0.001102 | 0.3058 | 0.3069 | 675.52 | 2082.31 | 2757.82 | 1.9427 | 4.8070 | 6.7497 | 160 |

Thermodynamics and Psychrometrics 59

### Table 2-2 SI Refrigerant 134a Properties of Saturated Liquid and Saturated Vapor

*(Table Refrigerant 134a, Chapter 30, 2013 ASHRAE Handbook—Fundamentals)*

| Temp.,* °C | Pressure, MPa | Density, kg/m³ Liquid | Volume, m³/kg Vapor | Enthalpy, kJ/kg Liquid | Enthalpy, kJ/kg Vapor | Entropy, kJ/(kg·K) Liquid | Entropy, kJ/(kg·K) Vapor | Specific Heat $c_p$, kJ/(kg·K) Liquid | Specific Heat $c_p$, kJ/(kg·K) Vapor | $c_p/c_v$ Vapor | Velocity of Sound, m/s Liquid | Velocity of Sound, m/s Vapor | Viscosity, μPa·s Liquid | Viscosity, μPa·s Vapor | Thermal Cond., mW/(m·K) Liquid | Thermal Cond., mW/(m·K) Vapor | Surface Tension, mN/m | Temp., °C |
|---|---|---|---|---|---|---|---|---|---|---|---|---|---|---|---|---|---|---|
| –103.30a | 0.00039 | 1591.1 | 35.496 | 71.46 | 334.94 | 0.4126 | 1.9639 | 1.184 | 0.585 | 1.164 | 1120. | 126.8 | 2175. | 6.46 | 145.2 | 3.08 | 28.07 | –103.30 |
| –100.00 | 0.00056 | 1582.4 | 25.193 | 75.36 | 336.85 | 0.4354 | 1.9456 | 1.184 | 0.593 | 1.162 | 1103. | 127.9 | 1893. | 6.60 | 143.2 | 3.34 | 27.50 | –100.00 |
| –90.00 | 0.00152 | 1555.8 | 9.7698 | 87.23 | 342.76 | 0.5020 | 1.8972 | 1.189 | 0.617 | 1.156 | 1052. | 131.0 | 1339. | 7.03 | 137.3 | 4.15 | 25.79 | –90.00 |
| –80.00 | 0.00367 | 1529.0 | 4.2682 | 99.16 | 348.83 | 0.5654 | 1.8580 | 1.198 | 0.642 | 1.151 | 1002. | 134.0 | 1018. | 7.46 | 131.5 | 4.95 | 24.10 | –80.00 |
| –70.00 | 0.00798 | 1501.9 | 2.0590 | 111.20 | 355.02 | 0.6262 | 1.8264 | 1.210 | 0.667 | 1.148 | 952. | 136.8 | 809.2 | 7.89 | 126.0 | 5.75 | 22.44 | –70.00 |
| –60.00 | 0.01591 | 1474.3 | 1.0790 | 123.36 | 361.31 | 0.6846 | 1.8010 | 1.223 | 0.692 | 1.146 | 903. | 139.4 | 663.1 | 8.30 | 120.7 | 6.56 | 20.80 | –60.00 |
| –50.00 | 0.02945 | 1446.3 | 0.60620 | 135.67 | 367.65 | 0.7410 | 1.7806 | 1.238 | 0.720 | 1.146 | 855. | 141.7 | 555.1 | 8.72 | 115.6 | 7.36 | 19.18 | –50.00 |
| –40.00 | 0.05121 | 1417.7 | 0.36108 | 148.14 | 374.00 | 0.7956 | 1.7643 | 1.255 | 0.749 | 1.148 | 807. | 143.6 | 472.2 | 9.12 | 110.6 | 8.17 | 17.60 | –40.00 |
| –30.00 | 0.08438 | 1388.4 | 0.22594 | 160.79 | 380.32 | 0.8486 | 1.7515 | 1.273 | 0.781 | 1.152 | 760. | 145.2 | 406.4 | 9.52 | 105.8 | 8.99 | 16.04 | –30.00 |
| –28.00 | 0.09270 | 1382.4 | 0.20680 | 163.34 | 381.57 | 0.8591 | 1.7492 | 1.277 | 0.788 | 1.153 | 751. | 145.4 | 394.9 | 9.60 | 104.8 | 9.15 | 15.73 | –28.00 |
| –26.07b | 0.10133 | 1376.7 | 0.19018 | 165.81 | 382.78 | 0.8690 | 1.7472 | 1.281 | 0.794 | 1.154 | 742. | 145.7 | 384.2 | 9.68 | 103.9 | 9.31 | 15.44 | –26.07 |
| –26.00 | 0.10167 | 1376.5 | 0.18958 | 165.90 | 382.82 | 0.8694 | 1.7471 | 1.281 | 0.794 | 1.154 | 742. | 145.7 | 383.8 | 9.68 | 103.9 | 9.32 | 15.43 | –26.00 |
| –24.00 | 0.11130 | 1370.4 | 0.17407 | 168.47 | 384.07 | 0.8798 | 1.7451 | 1.285 | 0.801 | 1.155 | 732. | 145.9 | 373.1 | 9.77 | 102.9 | 9.48 | 15.12 | –24.00 |
| –22.00 | 0.12165 | 1364.4 | 0.16006 | 171.05 | 385.32 | 0.8900 | 1.7432 | 1.289 | 0.809 | 1.156 | 723. | 146.1 | 362.9 | 9.85 | 102.0 | 9.65 | 14.82 | –22.00 |
| –20.00 | 0.13273 | 1358.3 | 0.14739 | 173.64 | 386.55 | 0.9002 | 1.7413 | 1.293 | 0.816 | 1.158 | 714. | 146.3 | 353.0 | 9.92 | 101.1 | 9.82 | 14.51 | –20.00 |
| –18.00 | 0.14460 | 1352.1 | 0.13592 | 176.23 | 387.79 | 0.9104 | 1.7396 | 1.297 | 0.823 | 1.159 | 705. | 146.4 | 343.5 | 10.01 | 100.1 | 9.98 | 14.21 | –18.00 |
| –16.00 | 0.15728 | 1345.9 | 0.12551 | 178.83 | 389.02 | 0.9205 | 1.7379 | 1.302 | 0.831 | 1.161 | 695. | 146.6 | 334.3 | 10.09 | 99.2 | 10.15 | 13.91 | –16.00 |
| –14.00 | 0.17082 | 1339.7 | 0.11605 | 181.44 | 390.24 | 0.9306 | 1.7363 | 1.306 | 0.838 | 1.163 | 686. | 146.7 | 325.4 | 10.17 | 98.3 | 10.32 | 13.61 | –14.00 |
| –12.00 | 0.18524 | 1333.4 | 0.10744 | 184.07 | 391.46 | 0.9407 | 1.7348 | 1.311 | 0.846 | 1.165 | 677. | 146.8 | 316.9 | 10.25 | 97.4 | 10.49 | 13.32 | –12.00 |
| –10.00 | 0.20060 | 1327.1 | 0.09959 | 186.70 | 392.66 | 0.9506 | 1.7334 | 1.316 | 0.854 | 1.167 | 668. | 146.9 | 308.6 | 10.33 | 96.5 | 10.66 | 13.02 | –10.00 |
| –8.00 | 0.21693 | 1320.8 | 0.09242 | 189.34 | 393.87 | 0.9606 | 1.7320 | 1.320 | 0.863 | 1.169 | 658. | 146.9 | 300.6 | 10.41 | 95.6 | 10.83 | 12.72 | –8.00 |
| –6.00 | 0.23428 | 1314.3 | 0.08587 | 191.99 | 395.06 | 0.9705 | 1.7307 | 1.325 | 0.871 | 1.171 | 649. | 147.0 | 292.9 | 10.49 | 94.7 | 11.00 | 12.43 | –6.00 |
| –4.00 | 0.25268 | 1307.9 | 0.07987 | 194.65 | 396.25 | 0.9804 | 1.7294 | 1.330 | 0.880 | 1.174 | 640. | 147.0 | 285.4 | 10.57 | 93.8 | 11.17 | 12.14 | –4.00 |
| –2.00 | 0.27217 | 1301.4 | 0.07436 | 197.32 | 397.43 | 0.9902 | 1.7282 | 1.336 | 0.888 | 1.176 | 631. | 147.0 | 278.1 | 10.65 | 92.9 | 11.34 | 11.85 | –2.00 |
| 0.00 | 0.29280 | 1294.8 | 0.06931 | 200.00 | 398.60 | 1.0000 | 1.7271 | 1.341 | 0.897 | 1.179 | 622. | 146.9 | 271.1 | 10.73 | 92.0 | 11.51 | 11.56 | 0.00 |
| 2.00 | 0.31462 | 1288.1 | 0.06466 | 202.69 | 399.77 | 1.0098 | 1.7260 | 1.347 | 0.906 | 1.182 | 612. | 146.9 | 264.3 | 10.81 | 91.1 | 11.69 | 11.27 | 2.00 |
| 4.00 | 0.33766 | 1281.4 | 0.06039 | 205.40 | 400.92 | 1.0195 | 1.7250 | 1.352 | 0.916 | 1.185 | 603. | 146.8 | 257.6 | 10.90 | 90.2 | 11.86 | 10.99 | 4.00 |
| 6.00 | 0.36198 | 1274.7 | 0.05644 | 208.11 | 402.06 | 1.0292 | 1.7240 | 1.358 | 0.925 | 1.189 | 594. | 146.7 | 251.2 | 10.98 | 89.4 | 12.04 | 10.70 | 6.00 |
| 8.00 | 0.38761 | 1267.9 | 0.05280 | 210.84 | 403.20 | 1.0388 | 1.7230 | 1.364 | 0.935 | 1.192 | 585. | 146.5 | 244.9 | 11.06 | 88.5 | 12.22 | 10.42 | 8.00 |
| 10.00 | 0.41461 | 1261.0 | 0.04944 | 213.58 | 404.32 | 1.0485 | 1.7221 | 1.370 | 0.945 | 1.196 | 576. | 146.4 | 238.8 | 11.15 | 87.6 | 12.40 | 10.14 | 10.00 |
| 12.00 | 0.44301 | 1254.0 | 0.04633 | 216.33 | 405.43 | 1.0581 | 1.7212 | 1.377 | 0.956 | 1.200 | 566. | 146.2 | 232.9 | 11.23 | 86.7 | 12.58 | 9.86 | 12.00 |
| 14.00 | 0.47288 | 1246.9 | 0.04345 | 219.09 | 406.53 | 1.0677 | 1.7204 | 1.383 | 0.967 | 1.204 | 557. | 146.0 | 227.1 | 11.32 | 85.9 | 12.77 | 9.58 | 14.00 |
| 16.00 | 0.50425 | 1239.8 | 0.04078 | 221.87 | 407.61 | 1.0772 | 1.7196 | 1.390 | 0.978 | 1.209 | 548. | 145.7 | 221.5 | 11.40 | 85.0 | 12.95 | 9.30 | 16.00 |
| 18.00 | 0.53718 | 1232.6 | 0.03830 | 224.66 | 408.69 | 1.0867 | 1.7188 | 1.397 | 0.989 | 1.214 | 539. | 145.5 | 216.0 | 11.49 | 84.1 | 13.14 | 9.03 | 18.00 |
| 20.00 | 0.57171 | 1225.3 | 0.03600 | 227.47 | 409.75 | 1.0962 | 1.7180 | 1.405 | 1.001 | 1.219 | 530. | 145.1 | 210.7 | 11.58 | 83.3 | 13.33 | 8.76 | 20.00 |
| 22.00 | 0.60789 | 1218.0 | 0.03385 | 230.29 | 410.79 | 1.1057 | 1.7173 | 1.413 | 1.013 | 1.224 | 520. | 144.8 | 205.5 | 11.67 | 82.4 | 13.53 | 8.48 | 22.00 |
| 24.00 | 0.64578 | 1210.5 | 0.03186 | 233.12 | 411.82 | 1.1152 | 1.7166 | 1.421 | 1.025 | 1.230 | 511. | 144.5 | 200.4 | 11.76 | 81.6 | 13.72 | 8.21 | 24.00 |
| 26.00 | 0.68543 | 1202.9 | 0.03000 | 235.97 | 412.84 | 1.1246 | 1.7159 | 1.429 | 1.038 | 1.236 | 502. | 144.1 | 195.4 | 11.85 | 80.7 | 13.92 | 7.95 | 26.00 |
| 28.00 | 0.72688 | 1195.2 | 0.02826 | 238.84 | 413.84 | 1.1341 | 1.7152 | 1.437 | 1.052 | 1.243 | 493. | 143.6 | 190.5 | 11.95 | 79.8 | 14.13 | 7.68 | 28.00 |
| 30.00 | 0.77020 | 1187.5 | 0.02664 | 241.72 | 414.82 | 1.1435 | 1.7145 | 1.446 | 1.065 | 1.249 | 483. | 143.2 | 185.8 | 12.04 | 79.0 | 14.33 | 7.42 | 30.00 |
| 32.00 | 0.81543 | 1179.6 | 0.02513 | 244.62 | 415.78 | 1.1529 | 1.7138 | 1.456 | 1.080 | 1.257 | 474. | 142.7 | 181.1 | 12.14 | 78.1 | 14.54 | 7.15 | 32.00 |
| 34.00 | 0.86263 | 1171.6 | 0.02371 | 247.54 | 416.72 | 1.1623 | 1.7131 | 1.466 | 1.095 | 1.265 | 465. | 142.1 | 176.6 | 12.24 | 77.3 | 14.76 | 6.89 | 34.00 |
| 36.00 | 0.91185 | 1163.4 | 0.02238 | 250.48 | 417.65 | 1.1717 | 1.7124 | 1.476 | 1.111 | 1.273 | 455. | 141.6 | 172.1 | 12.34 | 76.4 | 14.98 | 6.64 | 36.00 |
| 38.00 | 0.96315 | 1155.1 | 0.02113 | 253.43 | 418.55 | 1.1811 | 1.7118 | 1.487 | 1.127 | 1.282 | 446. | 141.0 | 167.7 | 12.44 | 75.6 | 15.21 | 6.38 | 38.00 |
| 40.00 | 1.0166 | 1146.7 | 0.01997 | 256.41 | 419.43 | 1.1905 | 1.7111 | 1.498 | 1.145 | 1.292 | 436. | 140.3 | 163.4 | 12.55 | 74.7 | 15.44 | 6.13 | 40.00 |
| 42.00 | 1.0722 | 1138.2 | 0.01887 | 259.41 | 420.28 | 1.1999 | 1.7103 | 1.510 | 1.163 | 1.303 | 427. | 139.7 | 159.2 | 12.65 | 73.9 | 15.68 | 5.88 | 42.00 |
| 44.00 | 1.1301 | 1129.5 | 0.01784 | 262.43 | 421.11 | 1.2092 | 1.7096 | 1.523 | 1.182 | 1.314 | 418. | 138.9 | 155.1 | 12.76 | 73.0 | 15.93 | 5.63 | 44.00 |
| 46.00 | 1.1903 | 1120.6 | 0.01687 | 265.47 | 421.92 | 1.2186 | 1.7089 | 1.537 | 1.202 | 1.326 | 408. | 138.2 | 151.0 | 12.88 | 72.1 | 16.18 | 5.38 | 46.00 |
| 48.00 | 1.2529 | 1111.5 | 0.01595 | 268.53 | 422.69 | 1.2280 | 1.7081 | 1.551 | 1.223 | 1.339 | 399. | 137.4 | 147.0 | 13.00 | 71.3 | 16.45 | 5.13 | 48.00 |
| 50.00 | 1.3179 | 1102.3 | 0.01509 | 271.62 | 423.44 | 1.2375 | 1.7072 | 1.566 | 1.246 | 1.354 | 389. | 136.6 | 143.1 | 13.12 | 70.4 | 16.72 | 4.89 | 50.00 |
| 52.00 | 1.3854 | 1092.9 | 0.01428 | 274.74 | 424.15 | 1.2469 | 1.7064 | 1.582 | 1.270 | 1.369 | 379. | 135.7 | 139.2 | 13.24 | 69.6 | 17.01 | 4.65 | 52.00 |
| 54.00 | 1.4555 | 1083.2 | 0.01351 | 277.89 | 424.83 | 1.2563 | 1.7055 | 1.600 | 1.296 | 1.386 | 370. | 134.7 | 135.4 | 13.37 | 68.7 | 17.31 | 4.41 | 54.00 |
| 56.00 | 1.5282 | 1073.4 | 0.01278 | 281.06 | 425.47 | 1.2658 | 1.7045 | 1.618 | 1.324 | 1.405 | 360. | 133.8 | 131.6 | 13.51 | 67.8 | 17.63 | 4.18 | 56.00 |
| 58.00 | 1.6036 | 1063.2 | 0.01209 | 284.27 | 426.07 | 1.2753 | 1.7035 | 1.638 | 1.354 | 1.425 | 350. | 132.7 | 127.9 | 13.65 | 67.0 | 17.96 | 3.95 | 58.00 |
| 60.00 | 1.6818 | 1052.9 | 0.01144 | 287.50 | 426.63 | 1.2848 | 1.7024 | 1.660 | 1.387 | 1.448 | 340. | 131.7 | 124.2 | 13.79 | 66.1 | 18.31 | 3.72 | 60.00 |
| 62.00 | 1.7628 | 1042.2 | 0.01083 | 290.78 | 427.14 | 1.2944 | 1.7013 | 1.684 | 1.422 | 1.473 | 331. | 130.5 | 120.6 | 13.95 | 65.2 | 18.68 | 3.49 | 62.00 |
| 64.00 | 1.8467 | 1031.2 | 0.01024 | 294.09 | 427.61 | 1.3040 | 1.7000 | 1.710 | 1.461 | 1.501 | 321. | 129.4 | 117.0 | 14.11 | 64.3 | 19.07 | 3.27 | 64.00 |
| 66.00 | 1.9337 | 1020.0 | 0.00969 | 297.44 | 428.02 | 1.3137 | 1.6987 | 1.738 | 1.504 | 1.532 | 311. | 128.1 | 113.5 | 14.28 | 63.4 | 19.50 | 3.05 | 66.00 |
| 68.00 | 2.0237 | 1008.3 | 0.00916 | 300.84 | 428.36 | 1.3234 | 1.6972 | 1.769 | 1.552 | 1.567 | 301. | 126.8 | 109.9 | 14.46 | 62.6 | 19.95 | 2.83 | 68.00 |
| 70.00 | 2.1168 | 996.2 | 0.00865 | 304.28 | 428.65 | 1.3332 | 1.6956 | 1.804 | 1.605 | 1.607 | 290. | 125.5 | 106.4 | 14.65 | 61.7 | 20.45 | 2.61 | 70.00 |
| 72.00 | 2.2132 | 983.8 | 0.00817 | 307.78 | 428.86 | 1.3430 | 1.6939 | 1.843 | 1.665 | 1.653 | 280. | 124.0 | 102.9 | 14.85 | 60.8 | 20.98 | 2.40 | 72.00 |
| 74.00 | 2.3130 | 970.8 | 0.00771 | 311.33 | 429.00 | 1.3530 | 1.6920 | 1.887 | 1.734 | 1.705 | 269. | 122.6 | 99.5 | 15.07 | 59.9 | 21.56 | 2.20 | 74.00 |
| 76.00 | 2.4161 | 957.3 | 0.00727 | 314.94 | 429.04 | 1.3631 | 1.6899 | 1.938 | 1.812 | 1.766 | 259. | 121.0 | 96.0 | 15.30 | 59.0 | 22.21 | 1.99 | 76.00 |
| 78.00 | 2.5228 | 943.1 | 0.00685 | 318.63 | 428.98 | 1.3733 | 1.6876 | 1.996 | 1.904 | 1.838 | 248. | 119.4 | 92.5 | 15.56 | 58.1 | 22.92 | 1.80 | 78.00 |
| 80.00 | 2.6332 | 928.2 | 0.00645 | 322.39 | 428.81 | 1.3836 | 1.6850 | 2.065 | 2.012 | 1.924 | 237. | 117.7 | 89.0 | 15.84 | 57.2 | 23.72 | 1.60 | 80.00 |
| 85.00 | 2.9258 | 887.2 | 0.00550 | 332.22 | 427.76 | 1.4104 | 1.6771 | 2.306 | 2.397 | 2.232 | 207. | 113.1 | 80.2 | 16.67 | 54.9 | 26.22 | 1.14 | 85.00 |
| 90.00 | 3.2442 | 837.8 | 0.00461 | 342.93 | 425.42 | 1.4390 | 1.6662 | 2.756 | 3.121 | 2.820 | 176. | 107.9 | 70.9 | 17.81 | 52.8 | 29.91 | 0.71 | 90.00 |
| 95.00 | 3.5912 | 772.7 | 0.00374 | 355.25 | 420.67 | 1.4715 | 1.6492 | 3.938 | 5.020 | 4.369 | 141. | 101.9 | 60.4 | 19.61 | 51.7 | 36.40 | 0.33 | 95.00 |
| 100.00 | 3.9724 | 651.2 | 0.00268 | 373.30 | 407.68 | 1.5188 | 1.6109 | 17.59 | 25.35 | 20.81 | 101. | 94.0 | 45.1 | 24.21 | 59.9 | 60.58 | 0.04 | 100.00 |
| 101.06c | 4.0593 | 511.9 | 0.00195 | 389.64 | 389.64 | 1.5621 | 1.5621 | • | • | • | 0. | 0.0 | — | — | • | • | 0.00 | 101.06 |

*Temperatures are on the ITS-90 scale a = triple point b = normal boiling point c = critical point

## Table 2-4 SI   Thermodynamic Properties of Moist Air at Standard Atmospheric Pressure, 101.325 kPa

*(Table 2, Chapter 1, 2013 ASHRAE Handbook—Fundamentals)*

| Temp., °C  t | Humidity Ratio, kg(w)/kg(da) $W_s$ | Specific Volume, m³/kg (dry air) $v_{da}$ | $v_{as}$ | $v_s$ | Specific Enthalpy, kJ/kg (dry air) $h_{da}$ | $h_{as}$ | $h_s$ | Specific Entropy, kJ/(kg·K) (dry air) $s_{da}$ | $s_{as}$ | $s_s$ | Condensed Water Specific Enthalpy, kJ/kg $h_w$ | Specific Entropy, kJ/(kg·K) $s_w$ | Vapor Pressure, kPa $p_s$ | Temp., °C  t |
|---|---|---|---|---|---|---|---|---|---|---|---|---|---|---|
| −60 | 0.0000067 | 0.6027 | 0.0000 | 0.6027 | −60.351 | 0.017 | −60.334 | −0.2495 | 0.0001 | −0.2494 | −446.29 | −1.6854 | 0.00108 | −60 |
| −59 | 0.0000076 | 0.6056 | 0.0000 | 0.6056 | −59.344 | 0.018 | −59.326 | −0.2448 | 0.0001 | −0.2447 | −444.63 | −1.6776 | 0.00124 | −59 |
| −58 | 0.0000087 | 0.6084 | 0.0000 | 0.6084 | −58.338 | 0.021 | −58.317 | −0.2401 | 0.0001 | −0.2400 | −442.95 | −1.6698 | 0.00141 | −58 |
| −57 | 0.0000100 | 0.6113 | 0.0000 | 0.6113 | −57.332 | 0.024 | −57.308 | −0.2354 | 0.0001 | −0.2353 | −441.27 | −1.6620 | 0.00161 | −57 |
| −56 | 0.0000114 | 0.6141 | 0.0000 | 0.6141 | −56.326 | 0.028 | −56.298 | −0.2308 | 0.0001 | −0.2306 | −439.58 | −1.6542 | 0.00184 | −56 |
| −55 | 0.0000129 | 0.6170 | 0.0000 | 0.6170 | −55.319 | 0.031 | −55.288 | −0.2261 | 0.0002 | −0.2260 | −437.89 | −1.6464 | 0.00209 | −55 |
| −54 | 0.0000147 | 0.6198 | 0.0000 | 0.6198 | −54.313 | 0.036 | −54.278 | −0.2215 | 0.0002 | −0.2214 | −436.19 | −1.6386 | 0.00238 | −54 |
| −53 | 0.0000167 | 0.6226 | 0.0000 | 0.6227 | −53.307 | 0.041 | −53.267 | −0.2170 | 0.0002 | −0.2168 | −434.48 | −1.6308 | 0.00271 | −53 |
| −52 | 0.0000190 | 0.6255 | 0.0000 | 0.6255 | −52.301 | 0.046 | −52.255 | −0.2124 | 0.0002 | −0.2122 | −432.76 | −1.6230 | 0.00307 | −52 |
| −51 | 0.0000215 | 0.6283 | 0.0000 | 0.6284 | −51.295 | 0.052 | −51.243 | −0.2079 | 0.0002 | −0.2076 | −431.03 | −1.6153 | 0.00348 | −51 |
| −50 | 0.0000243 | 0.6312 | 0.0000 | 0.6312 | −50.289 | 0.059 | −50.230 | −0.2033 | 0.0003 | −0.2031 | −429.30 | −1.6075 | 0.00394 | −50 |
| −49 | 0.0000275 | 0.6340 | 0.0000 | 0.6341 | −49.283 | 0.067 | −49.216 | −0.1988 | 0.0003 | −0.1985 | −427.56 | −1.5997 | 0.00445 | −49 |
| −48 | 0.0000311 | 0.6369 | 0.0000 | 0.6369 | −48.277 | 0.075 | −48.202 | −0.1944 | 0.0004 | −0.1940 | −425.82 | −1.5919 | 0.00503 | −48 |
| −47 | 0.0000350 | 0.6397 | 0.0000 | 0.6398 | −47.271 | 0.085 | −47.186 | −0.1899 | 0.0004 | −0.1895 | −424.06 | −1.5842 | 0.00568 | −47 |
| −46 | 0.0000395 | 0.6426 | 0.0000 | 0.6426 | −46.265 | 0.095 | −46.170 | −0.1855 | 0.0004 | −0.1850 | −422.30 | −1.5764 | 0.00640 | −46 |
| −45 | 0.0000445 | 0.6454 | 0.0000 | 0.6455 | −45.259 | 0.108 | −45.151 | −0.1811 | 0.0005 | −0.1805 | −420.54 | −1.5686 | 0.00721 | −45 |
| −44 | 0.0000500 | 0.6483 | 0.0001 | 0.6483 | −44.253 | 0.121 | −44.132 | −0.1767 | 0.0006 | −0.1761 | −418.76 | −1.5609 | 0.00811 | −44 |
| −43 | 0.0000562 | 0.6511 | 0.0001 | 0.6512 | −43.247 | 0.137 | −43.111 | −0.1723 | 0.0006 | −0.1716 | −416.98 | −1.5531 | 0.00911 | −43 |
| −42 | 0.0000631 | 0.6540 | 0.0001 | 0.6540 | −42.241 | 0.153 | −42.088 | −0.1679 | 0.0007 | −0.1672 | −415.19 | −1.5453 | 0.01022 | −42 |
| −41 | 0.0000708 | 0.6568 | 0.0001 | 0.6569 | −41.235 | 0.172 | −41.063 | −0.1636 | 0.0008 | −0.1628 | −413.39 | −1.5376 | 0.01147 | −41 |
| −40 | 0.0000793 | 0.6597 | 0.0001 | 0.6597 | −40.229 | 0.192 | −40.037 | −0.1592 | 0.0009 | −0.1584 | −411.59 | −1.5298 | 0.01285 | −40 |
| −39 | 0.0000887 | 0.6625 | 0.0001 | 0.6626 | −39.224 | 0.216 | −39.007 | −0.1549 | 0.0010 | −0.1540 | −409.77 | −1.5221 | 0.01438 | −39 |
| −38 | 0.0000992 | 0.6653 | 0.0001 | 0.6654 | −38.218 | 0.241 | −37.976 | −0.1507 | 0.0011 | −0.1496 | −407.96 | −1.5143 | 0.01608 | −38 |
| −37 | 0.0001108 | 0.6682 | 0.0001 | 0.6683 | −37.212 | 0.270 | −36.942 | −0.1464 | 0.0012 | −0.1452 | −406.13 | −1.5066 | 0.01796 | −37 |
| −36 | 0.0001237 | 0.6710 | 0.0001 | 0.6712 | −36.206 | 0.302 | −35.905 | −0.1421 | 0.0014 | −0.1408 | −404.29 | −1.4988 | 0.02005 | −36 |
| −35 | 0.0001379 | 0.6739 | 0.0001 | 0.6740 | −35.200 | 0.336 | −34.864 | −0.1379 | 0.0015 | −0.1364 | −402.45 | −1.4911 | 0.02235 | −35 |
| −34 | 0.0001536 | 0.6767 | 0.0002 | 0.6769 | −34.195 | 0.375 | −33.820 | −0.1337 | 0.0017 | −0.1320 | −400.60 | −1.4833 | 0.02490 | −34 |
| −33 | 0.0001710 | 0.6796 | 0.0002 | 0.6798 | −33.189 | 0.417 | −32.772 | −0.1295 | 0.0018 | −0.1276 | −398.75 | −1.4756 | 0.02772 | −33 |
| −32 | 0.0001902 | 0.6824 | 0.0002 | 0.6826 | −32.183 | 0.464 | −31.718 | −0.1253 | 0.0020 | −0.1233 | −396.89 | −1.4678 | 0.03082 | −32 |
| −31 | 0.0002113 | 0.6853 | 0.0002 | 0.6855 | −31.178 | 0.517 | −30.661 | −0.1212 | 0.0023 | −0.1189 | −395.01 | −1.4601 | 0.03425 | −31 |
| −30 | 0.0002346 | 0.6881 | 0.0003 | 0.6884 | −30.171 | 0.574 | −29.597 | −0.1170 | 0.0025 | −0.1145 | −393.14 | −1.4524 | 0.03802 | −30 |
| −29 | 0.0002602 | 0.6909 | 0.0003 | 0.6912 | −29.166 | 0.636 | −28.529 | −0.1129 | 0.0028 | −0.1101 | −391.25 | −1.4446 | 0.04217 | −29 |
| −28 | 0.0002883 | 0.6938 | 0.0003 | 0.6941 | −28.160 | 0.707 | −27.454 | −0.1088 | 0.0031 | −0.1057 | −389.36 | −1.4369 | 0.04673 | −28 |
| −27 | 0.0003193 | 0.6966 | 0.0004 | 0.6970 | −27.154 | 0.782 | −26.372 | −0.1047 | 0.0034 | −0.1013 | −387.46 | −1.4291 | 0.05175 | −27 |
| −26 | 0.0003533 | 0.6995 | 0.0004 | 0.6999 | −26.149 | 0.867 | −25.282 | −0.1006 | 0.0037 | −0.0969 | −385.55 | −1.4214 | 0.05725 | −26 |
| −25 | 0.0003905 | 0.7023 | 0.0004 | 0.7028 | −25.143 | 0.959 | −24.184 | −0.0965 | 0.0041 | −0.0924 | −383.63 | −1.4137 | 0.06329 | −25 |
| −24 | 0.0004314 | 0.7052 | 0.0005 | 0.7057 | −24.137 | 1.059 | −23.078 | −0.0925 | 0.0045 | −0.0880 | −381.71 | −1.4059 | 0.06991 | −24 |
| −23 | 0.0004762 | 0.7080 | 0.0005 | 0.7086 | −23.132 | 1.171 | −21.961 | −0.0885 | 0.0050 | −0.0835 | −379.78 | −1.3982 | 0.07716 | −23 |
| −22 | 0.0005251 | 0.7109 | 0.0006 | 0.7115 | −22.126 | 1.292 | −20.834 | −0.0845 | 0.0054 | −0.0790 | −377.84 | −1.3905 | 0.08510 | −22 |
| −21 | 0.0005787 | 0.7137 | 0.0007 | 0.7144 | −21.120 | 1.425 | −19.695 | −0.0805 | 0.0060 | −0.0745 | −375.90 | −1.3828 | 0.09378 | −21 |
| −20 | 0.0006373 | 0.7165 | 0.0007 | 0.7173 | −20.115 | 1.570 | −18.545 | −0.0765 | 0.0066 | −0.0699 | −373.95 | −1.3750 | 0.10326 | −20 |
| −19 | 0.0007013 | 0.7194 | 0.0008 | 0.7202 | −19.109 | 1.729 | −17.380 | −0.0725 | 0.0072 | −0.0653 | −371.99 | −1.3673 | 0.11362 | −19 |
| −18 | 0.0007711 | 0.7222 | 0.0009 | 0.7231 | −18.103 | 1.902 | −16.201 | −0.0686 | 0.0079 | −0.0607 | −370.02 | −1.3596 | 0.12492 | −18 |
| −17 | 0.0008473 | 0.7251 | 0.0010 | 0.7261 | −17.098 | 2.092 | −15.006 | −0.0646 | 0.0086 | −0.0560 | −368.04 | −1.3518 | 0.13725 | −17 |
| −16 | 0.0009303 | 0.7279 | 0.0011 | 0.7290 | −16.092 | 2.299 | −13.793 | −0.0607 | 0.0094 | −0.0513 | −366.06 | −1.3441 | 0.15068 | −16 |
| −15 | 0.0010207 | 0.7308 | 0.0012 | 0.7320 | −15.086 | 2.524 | −12.562 | −0.0568 | 0.0103 | −0.0465 | −364.07 | −1.3364 | 0.16530 | −15 |
| −14 | 0.0011191 | 0.7336 | 0.0013 | 0.7349 | −14.080 | 2.769 | −11.311 | −0.0529 | 0.0113 | −0.0416 | −362.07 | −1.3287 | 0.18122 | −14 |
| −13 | 0.0012262 | 0.7364 | 0.0014 | 0.7379 | −13.075 | 3.036 | −10.039 | −0.0490 | 0.0123 | −0.0367 | −360.07 | −1.3210 | 0.19852 | −13 |
| −12 | 0.0013425 | 0.7393 | 0.0016 | 0.7409 | −12.069 | 3.327 | −8.742 | −0.0452 | 0.0134 | −0.0318 | −358.06 | −1.3132 | 0.21732 | −12 |
| −11 | 0.0014690 | 0.7421 | 0.0017 | 0.7439 | −11.063 | 3.642 | −7.421 | −0.0413 | 0.0146 | −0.0267 | −356.04 | −1.3055 | 0.23775 | −11 |
| −10 | 0.0016062 | 0.7450 | 0.0019 | 0.7469 | −10.057 | 3.986 | −6.072 | −0.0375 | 0.0160 | −0.0215 | −354.01 | −1.2978 | 0.25991 | −10 |
| −9 | 0.0017551 | 0.7478 | 0.0021 | 0.7499 | −9.052 | 4.358 | −4.693 | −0.0337 | 0.0174 | −0.0163 | −351.97 | −1.2901 | 0.28395 | −9 |
| −8 | 0.0019166 | 0.7507 | 0.0023 | 0.7530 | −8.046 | 4.764 | −3.283 | −0.0299 | 0.0189 | −0.0110 | −349.93 | −1.2824 | 0.30999 | −8 |
| −7 | 0.0020916 | 0.7535 | 0.0025 | 0.7560 | −7.040 | 5.202 | −1.838 | −0.0261 | 0.0206 | −0.0055 | −347.88 | −1.2746 | 0.33821 | −7 |
| −6 | 0.0022811 | 0.7563 | 0.0028 | 0.7591 | −6.035 | 5.677 | −0.357 | −0.0223 | 0.0224 | −0.0000 | −345.82 | −1.2669 | 0.36874 | −6 |
| −5 | 0.0024862 | 0.7592 | 0.0030 | 0.7622 | −5.029 | 6.192 | 1.164 | −0.0186 | 0.0243 | −0.0057 | −343.76 | −1.2592 | 0.40178 | −5 |
| −4 | 0.0027081 | 0.7620 | 0.0033 | 0.7653 | −4.023 | 6.751 | 2.728 | −0.0148 | 0.0264 | −0.0115 | −341.69 | −1.2515 | 0.43748 | −4 |
| −3 | 0.0029480 | 0.7649 | 0.0036 | 0.7685 | −3.017 | 7.353 | 4.336 | −0.0111 | 0.0286 | −0.0175 | −339.61 | −1.2438 | 0.47606 | −3 |
| −2 | 0.0032074 | 0.7677 | 0.0039 | 0.7717 | −2.011 | 8.007 | 5.995 | −0.0074 | 0.0310 | −0.0236 | −337.52 | −1.2361 | 0.51773 | −2 |
| −1 | 0.0034874 | 0.7705 | 0.0043 | 0.7749 | −1.006 | 8.712 | 7.706 | −0.0037 | 0.0336 | −0.0299 | −335.42 | −1.2284 | 0.56268 | −1 |
| 0 | 0.0037895 | 0.7734 | 0.0047 | 0.7781 | 0.000 | 9.473 | 9.473 | 0.0000 | 0.0364 | 0.0364 | −333.32 | −1.2206 | 0.61117 | 0 |
| 0* | 0.003789 | 0.7734 | 0.0047 | 0.7781 | 0.000 | 9.473 | 9.473 | 0.0000 | 0.0364 | 0.0364 | 0.06 | −0.0001 | 0.6112 | 0 |
| 1 | 0.004076 | 0.7762 | 0.0051 | 0.7813 | 1.006 | 10.197 | 11.203 | 0.0037 | 0.0391 | 0.0427 | 4.28 | 0.0153 | 0.6571 | 1 |
| 2 | 0.004381 | 0.7791 | 0.0055 | 0.7845 | 2.012 | 10.970 | 12.982 | 0.0073 | 0.0419 | 0.0492 | 8.49 | 0.0306 | 0.7060 | 2 |
| 3 | 0.004707 | 0.7819 | 0.0059 | 0.7878 | 3.018 | 11.793 | 14.811 | 0.0110 | 0.0449 | 0.0559 | 12.70 | 0.0459 | 0.7581 | 3 |
| 4 | 0.005054 | 0.7848 | 0.0064 | 0.7911 | 4.024 | 12.672 | 16.696 | 0.0146 | 0.0480 | 0.0627 | 16.91 | 0.0611 | 0.8135 | 4 |
| 5 | 0.005424 | 0.7876 | 0.0068 | 0.7944 | 5.029 | 13.610 | 18.639 | 0.0182 | 0.0514 | 0.0697 | 21.12 | 0.0762 | 0.8725 | 5 |
| 6 | 0.005818 | 0.7904 | 0.0074 | 0.7978 | 6.036 | 14.608 | 20.644 | 0.0219 | 0.0550 | 0.0769 | 25.32 | 0.0913 | 0.9353 | 6 |
| 7 | 0.006237 | 0.7933 | 0.0079 | 0.8012 | 7.041 | 15.671 | 22.713 | 0.0255 | 0.0588 | 0.0843 | 29.52 | 0.1064 | 1.0020 | 7 |
| 8 | 0.006683 | 0.7961 | 0.0085 | 0.8046 | 8.047 | 16.805 | 24.852 | 0.0290 | 0.0628 | 0.0919 | 33.72 | 0.1213 | 1.0729 | 8 |
| 9 | 0.007157 | 0.7990 | 0.0092 | 0.8081 | 9.053 | 18.010 | 27.064 | 0.0326 | 0.0671 | 0.0997 | 37.92 | 0.1362 | 1.1481 | 9 |
| 10 | 0.007661 | 0.8018 | 0.0098 | 0.8116 | 10.059 | 19.293 | 29.352 | 0.0362 | 0.0717 | 0.1078 | 42.11 | 0.1511 | 1.2280 | 10 |
| 11 | 0.008197 | 0.8046 | 0.0106 | 0.8152 | 11.065 | 20.658 | 31.724 | 0.0397 | 0.0765 | 0.1162 | 46.31 | 0.1659 | 1.3128 | 11 |
| 12 | 0.008766 | 0.8075 | 0.0113 | 0.8188 | 12.071 | 22.108 | 34.179 | 0.0433 | 0.0816 | 0.1248 | 50.50 | 0.1806 | 1.4026 | 12 |
| 13 | 0.009370 | 0.8103 | 0.0122 | 0.8225 | 13.077 | 23.649 | 36.726 | 0.0468 | 0.0870 | 0.1337 | 54.69 | 0.1953 | 1.4979 | 13 |

*Extrapolated to represent metastable equilibrium with undercooled liquid.

## Thermodynamics and Psychrometrics

### Table 2-4 SI  Thermodynamic Properties of Moist Air at Standard Atmospheric Pressure, 101.325 kPa (*Continued*)
*(Table 2, Chapter 1, 2013 ASHRAE Handbook—Fundamentals)*

| Temp., °C  t | Humidity Ratio, $W_s$ kg(w)/kg(da) | Specific Volume, m³/kg (dry air) $v_{da}$ | $v_{as}$ | $v_s$ | Specific Enthalpy, kJ/kg (dry air) $h_{da}$ | $h_{as}$ | $h_s$ | Specific Entropy, kJ/(kg·K) (dry air) $s_{da}$ | $s_{as}$ | $s_s$ | Condensed Water Specific Enthalpy, kJ/kg $h_w$ | Specific Entropy, kJ/(kg·K) $s_w$ | Vapor Pressure, kPa $p_s$ | Temp., °C  t |
|---|---|---|---|---|---|---|---|---|---|---|---|---|---|---|
| 14 | 0.010012 | 0.8132 | 0.0131 | 0.8262 | 14.084 | 25.286 | 39.370 | 0.0503 | 0.0927 | 0.1430 | 58.88 | 0.2099 | 1.5987 | 14 |
| 15 | 0.010692 | 0.8160 | 0.0140 | 0.8300 | 15.090 | 27.023 | 42.113 | 0.0538 | 0.0987 | 0.1525 | 63.07 | 0.2244 | 1.7055 | 15 |
| 16 | 0.011413 | 0.8188 | 0.0150 | 0.8338 | 16.096 | 28.867 | 44.963 | 0.0573 | 0.1051 | 0.1624 | 67.26 | 0.2389 | 1.8185 | 16 |
| 17 | 0.012178 | 0.8217 | 0.0160 | 0.8377 | 17.102 | 30.824 | 47.926 | 0.0607 | 0.1119 | 0.1726 | 71.44 | 0.2534 | 1.9380 | 17 |
| 18 | 0.012989 | 0.8245 | 0.0172 | 0.8417 | 18.108 | 32.900 | 51.008 | 0.0642 | 0.1190 | 0.1832 | 75.63 | 0.2678 | 2.0643 | 18 |
| 19 | 0.013848 | 0.8274 | 0.0184 | 0.8457 | 19.114 | 35.101 | 54.216 | 0.0677 | 0.1266 | 0.1942 | 79.81 | 0.2821 | 2.1979 | 19 |
| 20 | 0.014758 | 0.8302 | 0.0196 | 0.8498 | 20.121 | 37.434 | 57.555 | 0.0711 | 0.1346 | 0.2057 | 84.00 | 0.2965 | 2.3389 | 20 |
| 21 | 0.015721 | 0.8330 | 0.0210 | 0.8540 | 21.127 | 39.908 | 61.035 | 0.0745 | 0.1430 | 0.2175 | 88.18 | 0.3107 | 2.4878 | 21 |
| 22 | 0.016741 | 0.8359 | 0.0224 | 0.8583 | 22.133 | 42.527 | 64.660 | 0.0779 | 0.1519 | 0.2298 | 92.36 | 0.3249 | 2.6448 | 22 |
| 23 | 0.017821 | 0.8387 | 0.0240 | 0.8627 | 23.140 | 45.301 | 68.440 | 0.0813 | 0.1613 | 0.2426 | 96.55 | 0.3390 | 2.8105 | 23 |
| 24 | 0.018963 | 0.8416 | 0.0256 | 0.8671 | 24.146 | 48.239 | 72.385 | 0.0847 | 0.1712 | 0.2559 | 100.73 | 0.3531 | 2.9852 | 24 |
| 25 | 0.020170 | 0.8444 | 0.0273 | 0.8717 | 25.153 | 51.347 | 76.500 | 0.0881 | 0.1817 | 0.2698 | 104.91 | 0.3672 | 3.1693 | 25 |
| 26 | 0.021448 | 0.8472 | 0.0291 | 0.8764 | 26.159 | 54.638 | 80.798 | 0.0915 | 0.1927 | 0.2842 | 109.09 | 0.3812 | 3.3633 | 26 |
| 27 | 0.022798 | 0.8501 | 0.0311 | 0.8811 | 27.165 | 58.120 | 85.285 | 0.0948 | 0.2044 | 0.2992 | 113.27 | 0.3951 | 3.5674 | 27 |
| 28 | 0.024226 | 0.8529 | 0.0331 | 0.8860 | 28.172 | 61.804 | 89.976 | 0.0982 | 0.2166 | 0.3148 | 117.45 | 0.4090 | 3.7823 | 28 |
| 29 | 0.025735 | 0.8558 | 0.0353 | 0.8910 | 29.179 | 65.699 | 94.878 | 0.1015 | 0.2296 | 0.3311 | 121.63 | 0.4229 | 4.0084 | 29 |
| 30 | 0.027329 | 0.8586 | 0.0376 | 0.8962 | 30.185 | 69.820 | 100.006 | 0.1048 | 0.2432 | 0.3481 | 125.81 | 0.4367 | 4.2462 | 30 |
| 31 | 0.029014 | 0.8614 | 0.0400 | 0.9015 | 31.192 | 74.177 | 105.369 | 0.1082 | 0.2576 | 0.3658 | 129.99 | 0.4505 | 4.4961 | 31 |
| 32 | 0.030793 | 0.8643 | 0.0426 | 0.9069 | 32.198 | 78.780 | 110.979 | 0.1115 | 0.2728 | 0.3842 | 134.17 | 0.4642 | 4.7586 | 32 |
| 33 | 0.032674 | 0.8671 | 0.0454 | 0.9125 | 33.205 | 83.652 | 116.857 | 0.1148 | 0.2887 | 0.4035 | 138.35 | 0.4779 | 5.0345 | 33 |
| 34 | 0.034660 | 0.8700 | 0.0483 | 0.9183 | 34.212 | 88.799 | 123.011 | 0.1180 | 0.3056 | 0.4236 | 142.53 | 0.4915 | 5.3242 | 34 |
| 35 | 0.036756 | 0.8728 | 0.0514 | 0.9242 | 35.219 | 94.236 | 129.455 | 0.1213 | 0.3233 | 0.4446 | 146.71 | 0.5051 | 5.6280 | 35 |
| 36 | 0.038971 | 0.8756 | 0.0546 | 0.9303 | 36.226 | 99.983 | 136.209 | 0.1246 | 0.3420 | 0.4666 | 150.89 | 0.5186 | 5.9468 | 36 |
| 37 | 0.041309 | 0.8785 | 0.0581 | 0.9366 | 37.233 | 106.058 | 143.290 | 0.1278 | 0.3617 | 0.4895 | 155.07 | 0.5321 | 6.2812 | 37 |
| 38 | 0.043778 | 0.8813 | 0.0618 | 0.9431 | 38.239 | 112.474 | 150.713 | 0.1311 | 0.3824 | 0.5135 | 159.25 | 0.5456 | 6.6315 | 38 |
| 39 | 0.046386 | 0.8842 | 0.0657 | 0.9498 | 39.246 | 119.258 | 158.504 | 0.1343 | 0.4043 | 0.5386 | 163.43 | 0.5590 | 6.9988 | 39 |
| 40 | 0.049141 | 0.8870 | 0.0698 | 0.9568 | 40.253 | 126.430 | 166.683 | 0.1375 | 0.4273 | 0.5649 | 167.61 | 0.5724 | 7.3838 | 40 |
| 41 | 0.052049 | 0.8898 | 0.0741 | 0.9640 | 41.261 | 134.005 | 175.265 | 0.1407 | 0.4516 | 0.5923 | 171.79 | 0.5857 | 7.7866 | 41 |
| 42 | 0.055119 | 0.8927 | 0.0788 | 0.9714 | 42.268 | 142.007 | 184.275 | 0.1439 | 0.4771 | 0.6211 | 175.97 | 0.5990 | 8.2081 | 42 |
| 43 | 0.058365 | 0.8955 | 0.0837 | 0.9792 | 43.275 | 150.475 | 193.749 | 0.1471 | 0.5041 | 0.6512 | 180.15 | 0.6122 | 8.6495 | 43 |
| 44 | 0.061791 | 0.8983 | 0.0888 | 0.9872 | 44.282 | 159.417 | 203.699 | 0.1503 | 0.5325 | 0.6828 | 184.33 | 0.6254 | 9.1110 | 44 |
| 45 | 0.065411 | 0.9012 | 0.0943 | 0.9955 | 45.289 | 168.874 | 214.164 | 0.1535 | 0.5624 | 0.7159 | 188.51 | 0.6386 | 9.5935 | 45 |
| 46 | 0.069239 | 0.9040 | 0.1002 | 1.0042 | 46.296 | 178.882 | 225.179 | 0.1566 | 0.5940 | 0.7507 | 192.69 | 0.6517 | 10.0982 | 46 |
| 47 | 0.073282 | 0.9069 | 0.1063 | 1.0132 | 47.304 | 189.455 | 236.759 | 0.1598 | 0.6273 | 0.7871 | 196.88 | 0.6648 | 10.6250 | 47 |
| 48 | 0.077556 | 0.9097 | 0.1129 | 1.0226 | 48.311 | 200.644 | 248.955 | 0.1629 | 0.6624 | 0.8253 | 201.06 | 0.6778 | 11.1754 | 48 |
| 49 | 0.082077 | 0.9125 | 0.1198 | 1.0323 | 49.319 | 212.485 | 261.803 | 0.1661 | 0.6994 | 0.8655 | 205.24 | 0.6908 | 11.7502 | 49 |
| 50 | 0.086858 | 0.9154 | 0.1272 | 1.0425 | 50.326 | 225.019 | 275.345 | 0.1692 | 0.7385 | 0.9077 | 209.42 | 0.7038 | 12.3503 | 50 |
| 51 | 0.091918 | 0.9182 | 0.1350 | 1.0532 | 51.334 | 238.290 | 289.624 | 0.1723 | 0.7798 | 0.9521 | 213.60 | 0.7167 | 12.9764 | 51 |
| 52 | 0.097272 | 0.9211 | 0.1433 | 1.0643 | 52.341 | 252.340 | 304.682 | 0.1754 | 0.8234 | 0.9988 | 217.78 | 0.7296 | 13.6293 | 52 |
| 53 | 0.102948 | 0.9239 | 0.1521 | 1.0760 | 53.349 | 267.247 | 320.596 | 0.1785 | 0.8695 | 1.0480 | 221.97 | 0.7424 | 14.3108 | 53 |
| 54 | 0.108954 | 0.9267 | 0.1614 | 1.0882 | 54.357 | 283.031 | 337.388 | 0.1816 | 0.9182 | 1.0998 | 226.15 | 0.7552 | 15.0205 | 54 |
| 55 | 0.115321 | 0.9296 | 0.1713 | 1.1009 | 55.365 | 299.772 | 355.137 | 0.1847 | 0.9698 | 1.1544 | 230.33 | 0.7680 | 15.7601 | 55 |
| 56 | 0.122077 | 0.9324 | 0.1819 | 1.1143 | 56.373 | 317.549 | 373.922 | 0.1877 | 1.0243 | 1.2120 | 234.52 | 0.7807 | 16.5311 | 56 |
| 57 | 0.129243 | 0.9353 | 0.1932 | 1.1284 | 57.381 | 336.417 | 393.798 | 0.1908 | 1.0820 | 1.2728 | 238.70 | 0.7934 | 17.3337 | 57 |
| 58 | 0.136851 | 0.9381 | 0.2051 | 1.1432 | 58.389 | 356.461 | 414.850 | 0.1938 | 1.1432 | 1.3370 | 242.88 | 0.8061 | 18.1691 | 58 |
| 59 | 0.144942 | 0.9409 | 0.2179 | 1.1588 | 59.397 | 377.788 | 437.185 | 0.1969 | 1.2081 | 1.4050 | 247.07 | 0.8187 | 19.0393 | 59 |
| 60 | 0.15354 | 0.9438 | 0.2315 | 1.1752 | 60.405 | 400.458 | 460.863 | 0.1999 | 1.2769 | 1.4768 | 251.25 | 0.8313 | 19.9439 | 60 |
| 61 | 0.16269 | 0.9466 | 0.2460 | 1.1926 | 61.413 | 424.624 | 486.036 | 0.2029 | 1.3500 | 1.5530 | 255.44 | 0.8438 | 20.8858 | 61 |
| 62 | 0.17244 | 0.9494 | 0.2614 | 1.2109 | 62.421 | 450.377 | 512.798 | 0.2059 | 1.4278 | 1.6337 | 259.62 | 0.8563 | 21.8651 | 62 |
| 63 | 0.18284 | 0.9523 | 0.2780 | 1.2303 | 63.429 | 477.837 | 541.266 | 0.2089 | 1.5104 | 1.7194 | 263.81 | 0.8688 | 22.8826 | 63 |
| 64 | 0.19393 | 0.9551 | 0.2957 | 1.2508 | 64.438 | 507.177 | 571.615 | 0.2119 | 1.5985 | 1.8105 | 268.00 | 0.8812 | 23.9405 | 64 |
| 65 | 0.20579 | 0.9580 | 0.3147 | 1.2726 | 65.446 | 538.548 | 603.995 | 0.2149 | 1.6925 | 1.9074 | 272.18 | 0.8936 | 25.0397 | 65 |
| 66 | 0.21848 | 0.9608 | 0.3350 | 1.2958 | 66.455 | 572.116 | 638.571 | 0.2179 | 1.7927 | 2.0106 | 276.37 | 0.9060 | 26.1810 | 66 |
| 67 | 0.23207 | 0.9636 | 0.3568 | 1.3204 | 67.463 | 608.103 | 675.566 | 0.2209 | 1.8999 | 2.1208 | 280.56 | 0.9183 | 27.3664 | 67 |
| 68 | 0.24664 | 0.9665 | 0.3803 | 1.3467 | 68.472 | 646.724 | 715.196 | 0.2238 | 2.0147 | 2.2385 | 284.75 | 0.9306 | 28.5967 | 68 |
| 69 | 0.26231 | 0.9693 | 0.4055 | 1.3749 | 69.481 | 688.261 | 757.742 | 0.2268 | 2.1378 | 2.3646 | 288.94 | 0.9429 | 29.8741 | 69 |
| 70 | 0.27916 | 0.9721 | 0.4328 | 1.4049 | 70.489 | 732.959 | 803.448 | 0.2297 | 2.2699 | 2.4996 | 293.13 | 0.9551 | 31.1986 | 70 |
| 71 | 0.29734 | 0.9750 | 0.4622 | 1.4372 | 71.498 | 781.208 | 852.706 | 0.2327 | 2.4122 | 2.6448 | 297.32 | 0.9673 | 32.5734 | 71 |
| 72 | 0.31698 | 0.9778 | 0.4941 | 1.4719 | 72.507 | 833.335 | 905.842 | 0.2356 | 2.5655 | 2.8010 | 301.51 | 0.9794 | 33.9983 | 72 |
| 73 | 0.33824 | 0.9807 | 0.5287 | 1.5093 | 73.516 | 889.807 | 963.323 | 0.2385 | 2.7311 | 2.9696 | 305.70 | 0.9916 | 35.4759 | 73 |
| 74 | 0.36130 | 0.9835 | 0.5662 | 1.5497 | 74.525 | 951.077 | 1025.603 | 0.2414 | 2.9104 | 3.1518 | 309.89 | 1.0037 | 37.0063 | 74 |
| 75 | 0.38641 | 0.9863 | 0.6072 | 1.5935 | 75.535 | 1017.841 | 1093.375 | 0.2443 | 3.1052 | 3.3496 | 314.08 | 1.0157 | 38.5940 | 75 |
| 76 | 0.41377 | 0.9892 | 0.6519 | 1.6411 | 76.543 | 1090.628 | 1167.172 | 0.2472 | 3.3171 | 3.5644 | 318.28 | 1.0278 | 40.2369 | 76 |
| 77 | 0.44372 | 0.9920 | 0.7010 | 1.6930 | 77.553 | 1170.328 | 1247.881 | 0.2501 | 3.5486 | 3.7987 | 322.47 | 1.0398 | 41.9388 | 77 |
| 78 | 0.47663 | 0.9948 | 0.7550 | 1.7498 | 78.562 | 1257.921 | 1336.483 | 0.2530 | 3.8023 | 4.0553 | 326.67 | 1.0517 | 43.7020 | 78 |
| 79 | 0.51284 | 0.9977 | 0.8145 | 1.8121 | 79.572 | 1354.347 | 1433.918 | 0.2559 | 4.0810 | 4.3368 | 330.86 | 1.0636 | 45.5248 | 79 |
| 80 | 0.55295 | 1.0005 | 0.8805 | 1.8810 | 80.581 | 1461.200 | 1541.781 | 0.2587 | 4.3890 | 4.6477 | 335.06 | 1.0755 | 47.4135 | 80 |
| 81 | 0.59751 | 1.0034 | 0.9539 | 1.9572 | 81.591 | 1579.961 | 1661.552 | 0.2616 | 4.7305 | 4.9921 | 339.25 | 1.0874 | 49.3670 | 81 |
| 82 | 0.64724 | 1.0062 | 1.0360 | 2.0422 | 82.600 | 1712.547 | 1795.148 | 0.2644 | 5.1108 | 5.3753 | 343.45 | 1.0993 | 51.3860 | 82 |
| 83 | 0.70311 | 1.0090 | 1.1283 | 2.1373 | 83.610 | 1861.548 | 1945.158 | 0.2673 | 5.5372 | 5.8045 | 347.65 | 1.1111 | 53.4746 | 83 |
| 84 | 0.76624 | 1.0119 | 1.2328 | 2.2446 | 84.620 | 2029.983 | 2114.603 | 0.2701 | 6.0181 | 6.2882 | 351.85 | 1.1228 | 55.6337 | 84 |
| 85 | 0.83812 | 1.0147 | 1.3518 | 2.3666 | 85.630 | 2221.806 | 2307.436 | 0.2729 | 6.5644 | 6.8373 | 356.05 | 1.1346 | 57.8658 | 85 |
| 86 | 0.92062 | 1.0175 | 1.4887 | 2.5062 | 86.640 | 2442.036 | 2528.677 | 0.2757 | 7.1901 | 7.4658 | 360.25 | 1.1463 | 60.1727 | 86 |
| 87 | 1.01611 | 1.0204 | 1.6473 | 2.6676 | 87.650 | 2697.016 | 2784.666 | 0.2785 | 7.9128 | 8.1914 | 364.45 | 1.1580 | 62.5544 | 87 |
| 88 | 1.12800 | 1.0232 | 1.8333 | 2.8565 | 88.661 | 2995.890 | 3084.551 | 0.2813 | 8.7580 | 9.0393 | 368.65 | 1.1696 | 65.0166 | 88 |
| 89 | 1.26064 | 1.0261 | 2.0540 | 3.0800 | 89.671 | 3350.254 | 3439.925 | 0.2841 | 9.7577 | 10.0419 | 372.86 | 1.1812 | 67.5581 | 89 |
| 90 | 1.42031 | 1.0289 | 2.3199 | 3.3488 | 90.681 | 3776.918 | 3867.599 | 0.2869 | 10.9586 | 11.2455 | 377.06 | 1.1928 | 70.1817 | 90 |

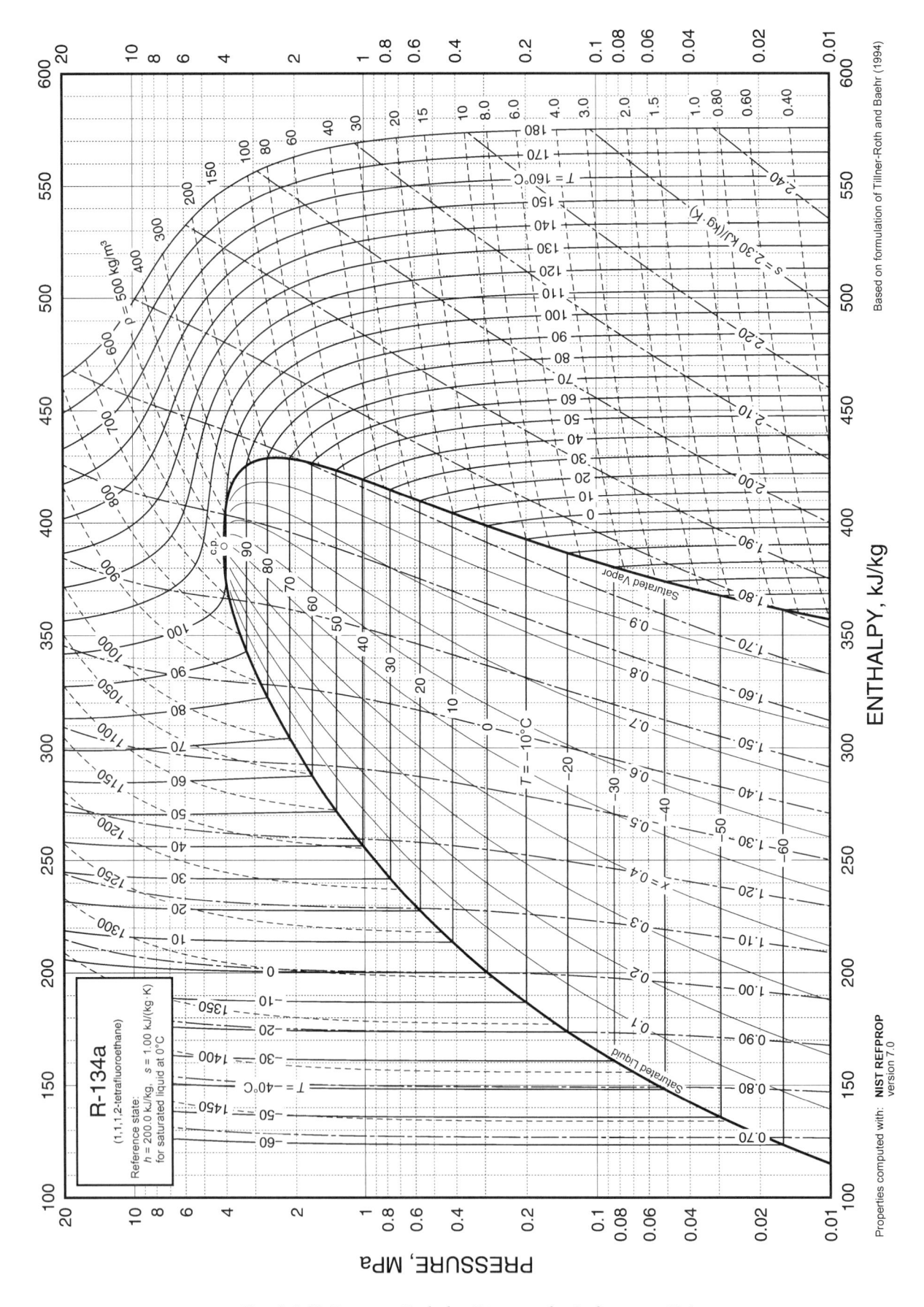

*Fig. 2-5 SI Pressure-Enthalpy Diagram for Refrigerant 134a*

# Thermodynamics and Psychrometrics

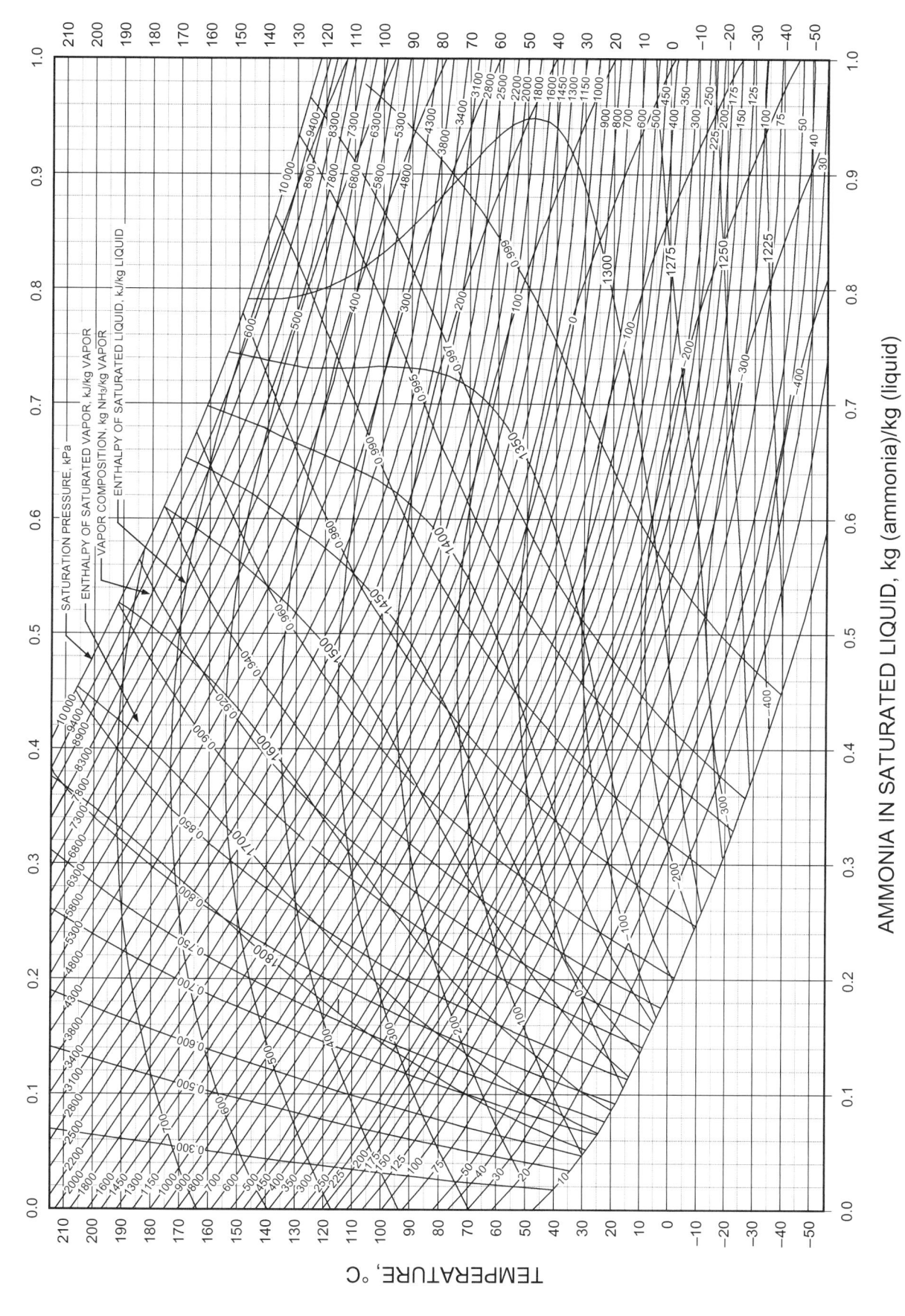

*Fig. 2-6 SI Enthalpy-Concentration Diagram for Aqua-Ammonia*

# Principles of HVAC

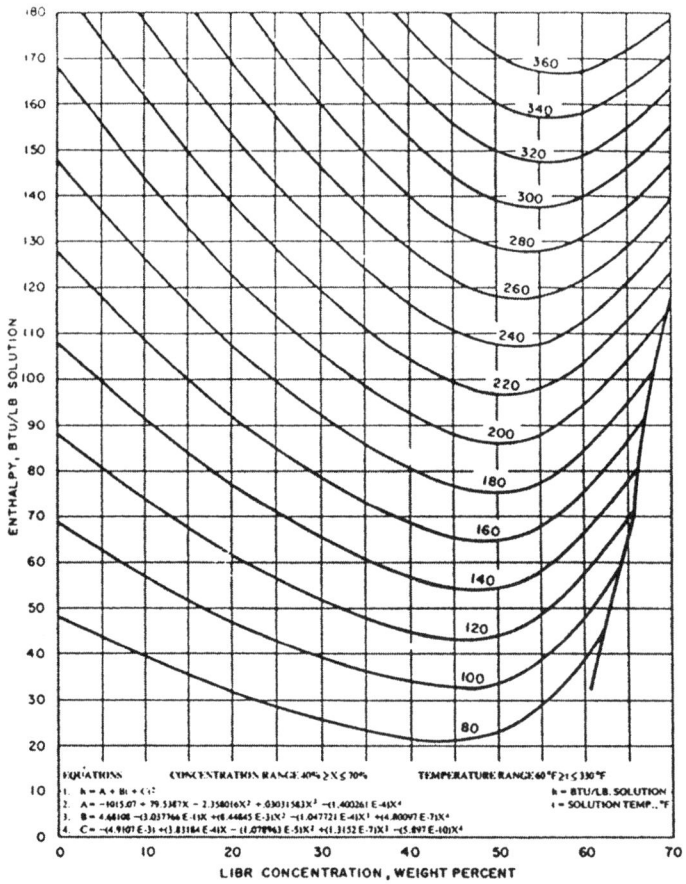

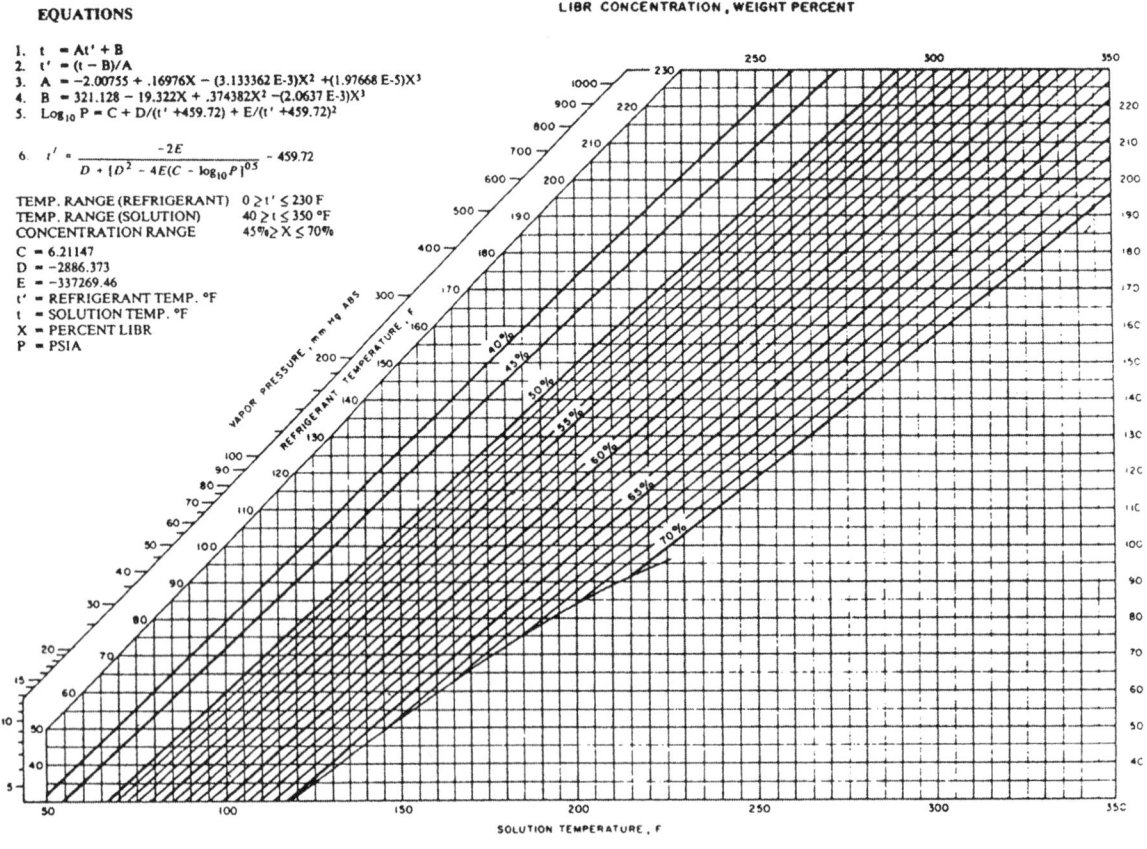

**EQUATIONS**

1. $t = At' + B$
2. $t' = (t - B)/A$
3. $A = -2.00755 + .16976X - (3.133362 \text{ E-}3)X^2 + (1.97668 \text{ E-}5)X^3$
4. $B = 321.128 - 19.322X + .374382X^2 - (2.0637 \text{ E-}3)X^3$
5. $\log_{10} P = C + D/(t' + 459.72) + E/(t' + 459.72)^2$
6. $t' = \dfrac{-2E}{D + [D^2 - 4E(C - \log_{10} P)]^{0.5}} - 459.72$

TEMP. RANGE (REFRIGERANT)  $0 \geq t' \leq 230\ F$
TEMP. RANGE (SOLUTION)  $40 \geq t \leq 350\ °F$
CONCENTRATION RANGE  $45\% \geq X \leq 70\%$
C = 6.21147
D = −2886.373
E = −337269.46
t' = REFRIGERANT TEMP. °F
t = SOLUTION TEMP. °F
X = PERCENT LIBR
P = PSIA

*Fig. 2-7 SI Enthalpy-Concentration and Equilibrium Charts for Li-Br/Water*

# Thermodynamics and Psychrometrics

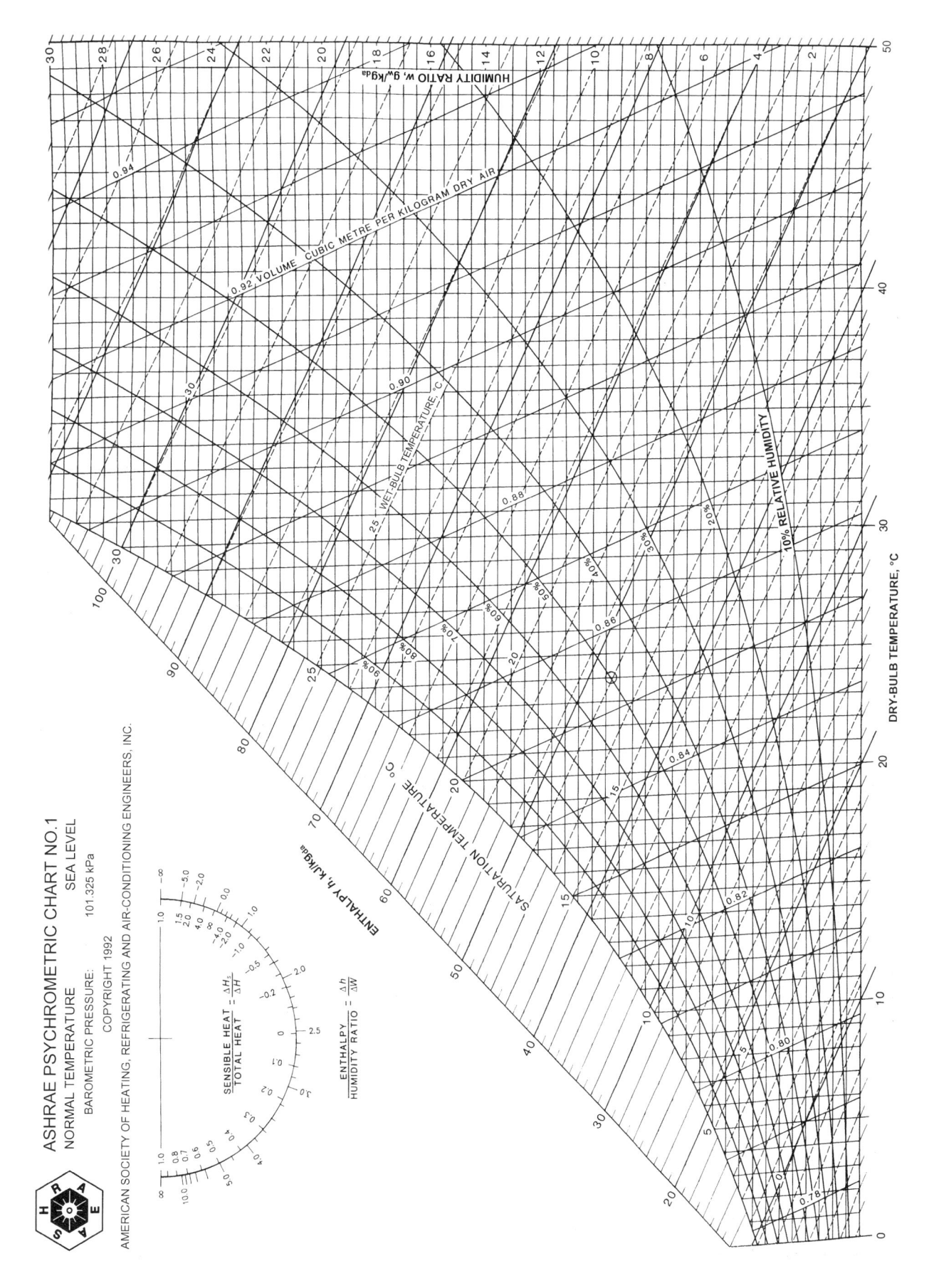

*Fig. 2-10 SI ASHRAE Psychrometric Chart*

# Chapter 3

# BASIC HVAC SYSTEM CALCULATIONS

This chapter illustrates the application of the principles of thermodynamics and psychrometrics to the various processes found in air-conditioning systems. It relates to material in Chapter 1 of the 2013 *ASHRAE Handbook—Fundamentals*.

## 3.1 Applying Thermodynamics to HVAC Processes

A simple but complete air-conditioning system is given schematically in Figure 3-1, which shows various space heat and moisture transfers. The symbol $q_S$ represents a sensible heat transfer rate; the symbol $m_w$ represents a moisture transfer rate. The symbol $q_L$ designates the transfer of energy that accompanies moisture transfer and is given by $\Sigma m_w h_w$, where $h_w$ is the specific enthalpy of the added (or removed) moisture. Solar radiation and internal loads are always gains for the space. Heat transmission through solid construction components due to a temperature difference, as well as energy transfer because of infiltration, may represent a gain or a loss.

Note that the energy $q_c$ and moisture $m_c$ transfers at the conditioner cannot be determined from the space heat and moisture transfers alone. The effect of the outdoor ventilation air, as well as other system load components, must be included. The designer must recognize that items such as fan energy, duct transmission, roof and ceiling transmissions, heat from lights, bypass and leakage, type of return air system, location of main fans, and actual versus design room conditions are all related to one another, to component sizing, and to system arrangement.

The first law of thermodynamics (energy balance) and the conservation of mass (mass balance) are the basis for the analysis of moist air processes. The following sections demonstrate the application of these laws to specific HVAC processes.

In many air-conditioning systems, air is removed from the room, returned to the air-conditioning apparatus where it is reconditioned, and then supplied again to the room. In most systems, the return air from the room is mixed with outdoor air required for ventilation. A typical air-conditioning system and the corresponding psychrometric representation of the processes for cooling are illustrated in Figure 3-2. Outdoor air $o$ is mixed with return air $r$ from the room and enters the apparatus at condition $m$. Air flows through the conditioner and is supplied to the space $s$. The air supplied to the space absorbs heat $q_s$ and moisture $m_w$, and the cycle is repeated.

A typical psychrometric representation of the previous system operating under conditions of heating followed by humidification is given as Figure 3-3.

### 3.1.1 Absorption of Space Heat and Moisture Gains

The problem of air conditioning a space usually reduces to determining the quantity of moist air that must be supplied and the condition it must have to remove given amounts of energy and water from the space and be withdrawn at a specified condition. A space with incident rates of energy and moisture gains is shown in Figure 3-4. The quantity $q_s$ denotes

*Fig. 3-1 Schematic of Air-Conditioning System*

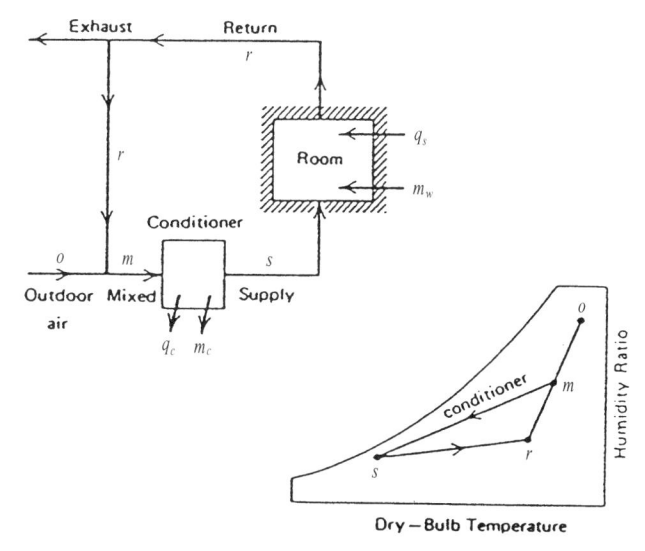

*Fig. 3-2 Typical Air-Conditioning Cooling Processes*

the net sum of all rates of heat gain upon the space arising from transfers through boundaries and from sources within the space. This **sensible heat gain** involves the addition of energy alone, and does not include energy contributions due to the addition of water (or water vapor). The quantity $m_w$ denotes the net sum of all rates of moisture gain upon the space arising from transfers through boundaries and from sources within the space. Each pound (kilogram) of moisture injected into the space adds an amount of energy equal to its specific enthalpy. A typical value of $h_w$ is 1100 Btu/lb.

Assuming steady-state conditions, the governing equations are

$$m_a h_1 + m_w h_w - m_a h_2 + q_s = 0 \qquad (3\text{-}1)$$

$$m_a W_1 + m_w = m_a W_2 \qquad (3\text{-}2)$$

### 3.1.2 Heating or Cooling of Air

When air is heated or cooled without moisture loss or gain, the process yields a straight horizontal line on the psychrometric chart because the humidity ratio is constant. Such processes can occur when moist air flows through a heat exchanger (Figure 3-5).

For steady-flow conditions, the governing equations are

$$m_a h_1 - m_a h_2 + q = 0 \qquad (3\text{-}3)$$

$$W_2 = W_1 \qquad (3\text{-}4)$$

### 3.1.3 Cooling and Dehumidifying Air

When moist air is cooled to a temperature below its dew point, some of the water vapor condenses and leaves the air stream. A schematic cooling and dehumidifying device is shown in Figure 3-6.

Although the actual process path varies depending on the type of surface, surface temperature, and flow conditions, the heat and mass transfer can be expressed in terms of the initial and final states.

Although water may condense out at various temperatures ranging from the initial dew point to the final saturation temperature, it is assumed that condensed water is cooled to the final air temperature $t_2$ before it drains from the system. For the system shown in Figure 3-6, the steady-flow energy and material balance equations are

$$m_a h_1 = m_a h_2 + q + m_w h_{w_2} \qquad (3\text{-}5)$$

$$m_a W_1 = m_a W_2 + m_w \qquad (3\text{-}6)$$

Thus

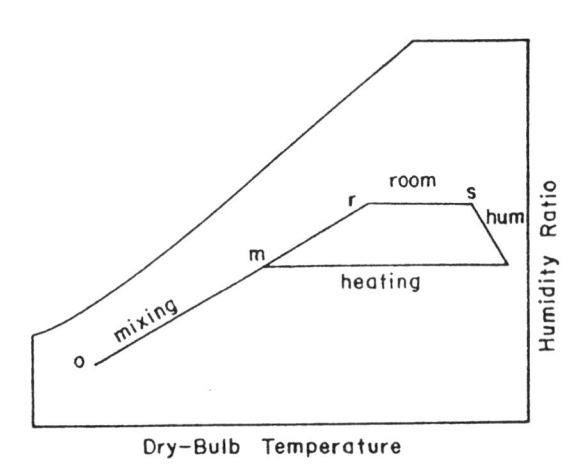

$$m_w = m_a (W_1 - W_2) \qquad (3\text{-}7)$$

*Fig. 3-3 Psychrometric Representation of Heating/Humidifying Process*

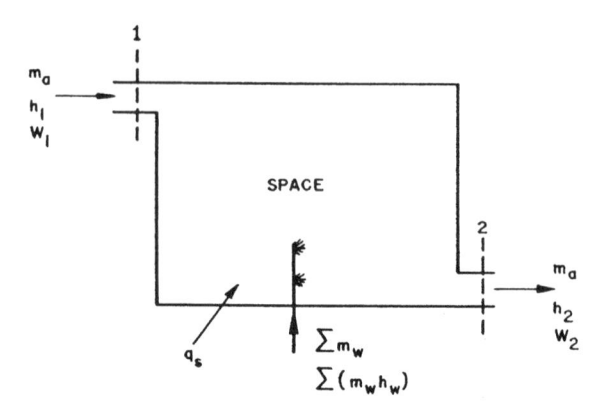

*Fig. 3-4 Space HVAC Process*

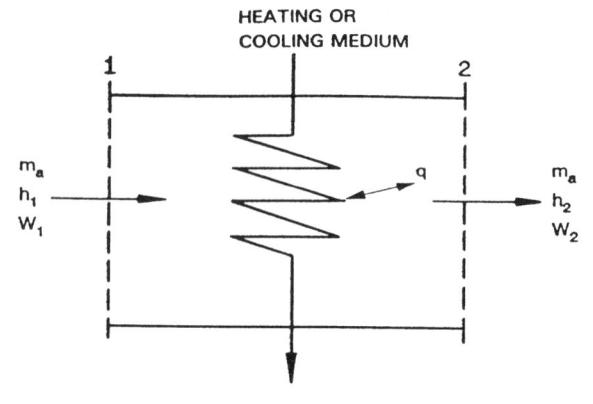

*Fig. 3-5 Schematic Heating or Cooling Device*

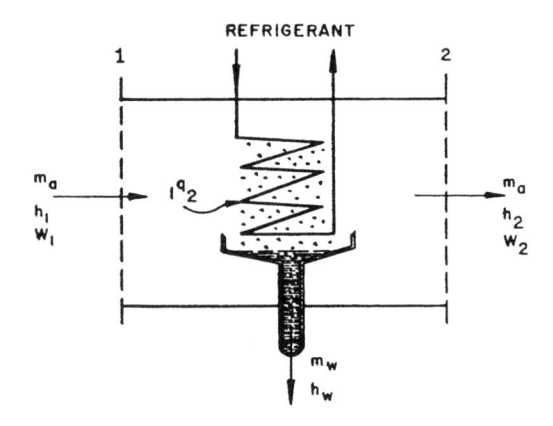

*Fig. 3-6 Schematic Cooling and Dehumidifying Device*

# Basic HVAC System Calculations

$$q = m_a[(h_1 - h_2) - (W_1 - W_2)h_{w_2}] \quad (3\text{-}8)$$

The cooling and dehumidifying process involves both sensible and latent heat transfer where sensible heat transfer is associated with the decrease in dry-bulb temperature and the latent heat transfer is associated with the decrease in humidity ratio. These quantities may be expressed as

$$q_S = m_a c_p (t_1 - t_2) \quad (3\text{-}9)$$

$$q_L = m_a (W_1 - W_2) h_{fg} \quad (3\text{-}10)$$

**Example 3-1.** In a steady-flow process, a cooling and dehumidifying coil receives an air-water vapor mixture at 16 psia, 95°F, and 83% rh and discharges the air-water vapor mixture at 14.7 psia, 50°F, and 96% rh. The condensate leaves the unit at 50°F. Calculate the heat transfer per pound of dry air flowing through the unit.

**Solution:**

The first law of thermodynamics for a steady-state, steady-flow system is given by

$$_1\dot{q}_2 + \left(\frac{V_1^2}{2g_c} + \frac{g}{g_c}z_1 + h_1\right)m_{a_1} = {_1W_2} + \left(\frac{V_2^2}{2g_c} + \frac{g}{g_c}z_2 + h_2\right)m_{a_2}$$

The continuity equation is

$$m = A\rho V$$

For the air flowing through the apparatus this becomes

$$m_{a_1} = m_{a_2} = m_a$$

For the water vapor this becomes

$$m_{w_1} - m_{w_2} = m_{\text{condensate}}$$

Neglecting any kinetic or potential energy changes of the flowing fluid and noting that there is no mechanical work being done on or by the system, the First Law equation reduces to

$$_1\dot{q}_2 = m_a(h_2 - h_1) + m_{\text{condensate}} h_{\text{condensate}}$$

and the enthalpy terms may be expanded, so that

$$\frac{_1\dot{q}_2}{m_a} = h_{a_2} - h_{a_1} + W_2 h_{w_2} - W_1 h_{w_1} + \frac{m_{\text{condensate}}}{m_a} h_{\text{condensate}}$$

With a mass balance on the water this becomes

$$\frac{_1\dot{q}_2}{m_a} = h_{a_2} - h_{a_1} + W_2 h_{w_2} - W_1 h_{w_1} + (W_1 - W_2) h_{\text{condensate}}$$

From Table 3, Chapter 1 of the 2013 *ASHRAE Handbook—Fundamentals*,

$$h_{w_2} = 1083.07 \text{ Btu/lb}$$

$$h_{w_1} = 1102.57 \text{ Btu/lb}$$

$$h_{\text{condensate}} = 18.07 \text{ Btu/lb}$$

$W_1$ and $W_2$ can be calculated as

$$p_{w_1} = \phi_1 p_{ws_1} = 0.83(0.8156) = 0.678 \text{ psia}$$

$$W_1 = 0.622 p_{w_1}/(p_1 - p_{w_1})$$
$$= 0.622 \times 0.678/(16 - 0.678) = 0.0275$$

$$p_{w_2} = 0.96(0.178) = 0.171 \text{ psia}$$

$$W_2 = 0.622 p_{w_2}/(p_2 - p_{w_2})$$
$$= 0.622 \times 0.171/(14.7 - 0.171) = 0.0073$$

Substituting into the energy equation,

$$\frac{_1\dot{q}_2}{m_a} = 0.24(50 - 95) + 0.0073(1083.07) - 0.0275(1102.57)$$
$$+ (0.0275 - 0.0073)18.07$$
$$= -10.8 + 7.91 - 30.3 \div 0.36 = 32.8 \text{ Btu/lb}_{\text{air}}$$

### 3.1.4 Heating and Humidifying Air

A device to heat and humidify moist air is shown in Figure 3-7. This process is generally required during the cold months of the year. An energy balance on the device yields

$$m_a h_1 + q + m_w h_w = m_a h_2 \quad (3\text{-}11)$$

and a mass balance on the water gives

$$m_a W_1 + m_w = m_a W_2 \quad (3\text{-}12)$$

### 3.1.5 Adiabatic Mixing of Two Streams of Air

A common process involved in air conditioning is the adiabatic mixing of two streams of moist air (Figure 3-8).

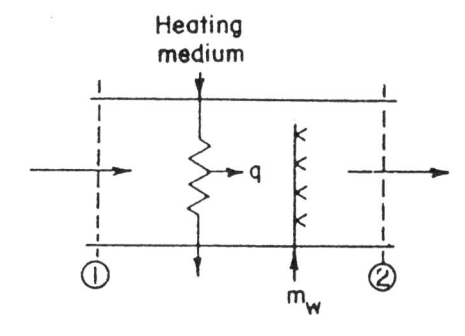

*Fig. 3-7 Schematic Heating and Humidifying Device*

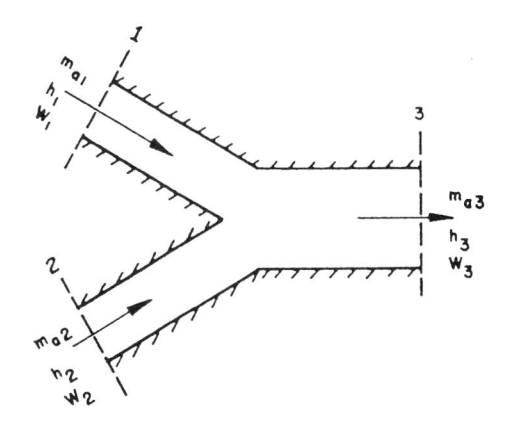

*Fig. 3-8 Adiabatic Mixing of Two Streams of Moist Air*

If the mixing is adiabatic, it must be governed by these three equations:

$$m_{a_1}h_1 + m_{a_2}h_2 = m_{a_3}h_3 \quad (3\text{-}13)$$

$$m_{a_1} + m_{a_2} = m_{a_3} \quad (3\text{-}14)$$

$$m_{a_1}W_1 + m_{a_2}W_2 = m_{a_3}W_3 \quad (3\text{-}15)$$

### 3.1.6 Adiabatic Mixing of Moist Air with Injected Water

Injecting steam or liquid water into a moist air stream to raise the humidity ratio of the moist air is a frequent air-conditioning process (Figure 3-9). If the mixing is adiabatic, the following equations apply:

$$m_a h_1 + m_w h_w = m_a h_2 \quad (3\text{-}16)$$

$$m_a W_1 + m_w = m_a W_2 \quad (3\text{-}17)$$

### 3.1.7 Moving Air

In all HVAC systems, a fan or blower is required to move the air. Under steady-flow conditions for the fan shown in Figure 3-10, the conservation equations are

$$m_a h_1 - m_a h_2 - W_k = 0 \quad (3\text{-}18)$$

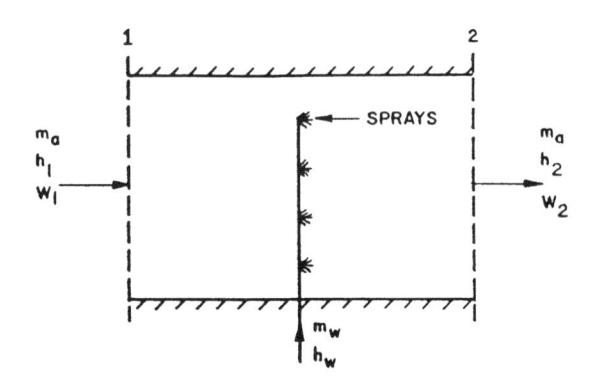

$$W_1 = W_2 \quad (3\text{-}19)$$

*Fig. 3-9 Schematic Injection of Water into Moist Air*

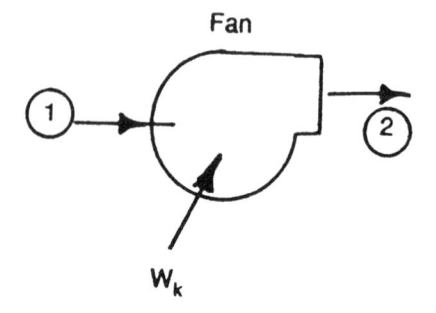

*Fig. 3-10 Air Moving*

### 3.1.8 Approximate Equations Using Volume Flow Rates

Since the specific volume of air varies appreciably with temperature, all calculations should be made with the mass of air instead of the volume. Nevertheless, volume values are required when selecting coils, fans, ducts, and other components.

In Chapter 1 of the 2013 *ASHRAE Handbook—Fundamentals*, basic calculations for air system loads, flow rates, and psychrometric representations are based on mass flow. Chapters 17 and 18 on load estimating give equations based on volumetric flow. These volumetric equations are commonly used and generally apply to most air systems.

One method of using volume while still including mass is to use volume values based on measurement at standard air conditions. ASHRAE defines one standard condition as dry air at 20°C and 101.325 kPa (68°F and 14.7 psia). Under that condition the density of dry air is about 1.204 kg/m³ (0.075 lb/ft³) and the specific volume is 0.83 m³/kg (13.3 ft³/lb). Saturated air at 15°C (59.5°F) has about the same density or specific volume. Thus, in the range at which air usually passes through the coils, fans, ducts, and other equipment, its density is close to standard and is not likely to require correction.

When the actual volumetric airflow is desired at any particular condition or point, the corresponding specific volume is obtained from the psychrometric chart and the volume at standard conditions is multiplied by the ratio of the actual specific volume to the standard value of 0.83 (13.3). To illustrate, assume the outdoor airflow rate at ASHRAE standard conditions is 470 L/s (1000 cfm). The actual outdoor air condition is 35°C (95°F) dry bulb and 23.8°C (75°F) wet bulb [$v$ = 0.89 m³/kg (14.3 ft³/lb)]. The actual volume flow rate at this condition would be 470(0.89/0.83) = 500 L/s [1000(14.3/13.3) = 1080 cfm].

Air-conditioning design often requires the calculation of sensible, latent, and total energy gains as follow:

1. *Sensible heat gain corresponding to the change of dry-bulb temperature for a given airflow (standard conditions).* The sensible heat gain in watts (Btu/h) as a result of a difference in temperature $\Delta t$ between the incoming air and leaving air flowing at ASHRAE standard conditions is

$$q_s = Q(1.204)(1.00 + 1.872W)\Delta t \quad (3\text{-}20)$$

where
- $q_s$ = sensible heat gain, W
- $Q$ = airflow, L/s
- 1.204 = density of standard dry air, kg/m³
- 1.00 = specific heat of dry air, kJ/(kg·K)
- 1.872 = specific heat of water vapor, kJ/(kg·K)
- $W$ = humidity ratio, mass of water per mass of dry air, kg/kg
- $\Delta t$ = temperature difference, °C

In I-P units

$$q_s = Q(60)(0.075)(0.24 + 0.45W)\Delta t \quad (3\text{-}21)$$

# Basic HVAC System Calculations

where
- $q_s$ = sensible heat gain, Btu/h
- $Q$ = airflow, ft³/min (cfm)
- 60 = minutes per hour
- 0.075 = density of dry air, lb/ft³
- 0.24 = specific heat of dry air, Btu/lb·°F
- 0.45 = specific heat of water vapor, Btu/lb·°F
- $W$ = humidity ratio, pounds of water per pound of dry air
- $\Delta t$ = temperature difference, °F

Since $W \cong 0.01$ in many air-conditioning problems, the sensible heat gain may be approximated by

$$q_s \approx 1.23 Q \, \Delta t \text{ in which } Q \text{ is in L/s} \quad (3\text{-}22)$$

and in I-P units

$$q_s \approx 1.10 Q \, \Delta t \text{ in which } Q \text{ is in CFM} \quad (3\text{-}23)$$

2. *Latent heat gain corresponding to the change of humidity ratio W for given airflow (standard conditions).* The latent heat gain in watts (Btu/h) as a result of a difference in humidity ratio $\Delta W$ between the incoming and leaving air flowing at ASHRAE standard conditions is

$$q_l = (\text{L/s})(1000)(0.001204)(2500) \, \Delta W \quad (3\text{-}24)$$

and in I-P units

$$q_l = (\text{cfm})(60)(0.075)(1076) \, \Delta W \quad (3\text{-}25)$$

In Equations 3-24 and 3-25, respectively, 2500 kJ/kg (1076 Btu/lb) is the approximate energy content of the superheated water vapor at 23.8°C (75°F) (1093.95 Btu/lb), less the energy content of water at 10°C (50°F) (18.07 Btu/lb). This difference is rounded up to 1076 Btu/lb (2500 kJ/kg) in Equation 3-25. A temperature of 24°C (75°F) is a common design condition for the space and 10°C (50°F) is normal condensate temperature from cooling and dehumidifying coils. Combining the constants, the latent heat gain is

$$q_l = 3010 Q \, \Delta W \quad (3\text{-}26)$$

In I-P units

$$q_l = 4840 Q \, \Delta W \quad (3\text{-}27)$$

3. *Total heat gain corresponding to the change of dry-bulb temperature and humidity ratio W for given airflow (standard conditions).* Total heat gain in watts (Btu/h) as a result of a difference in enthalpy $\Delta h$ between the incoming and leaving air flowing at ASHRAE standard conditions is

$$q = Q(1000)(0.001204) \, \Delta h \quad (3\text{-}28)$$

In I-P units

$$q = Q(60)(0.075) \, \Delta h \quad (3\text{-}29)$$

If the product of the two constants is used as a single number, the total energy exchange is

$$q = 1.204 Q \, \Delta h \quad (3\text{-}30)$$

In I-P units

$$q = 4.5 Q \, \Delta h \quad (3\text{-}31)$$

The values 1.23 (1.10), 3010 (4840), and 1.204 (4.5) are useful in air-conditioning calculations for an atmospheric pressure of approximately 101.3 kPa (14.7 psia) and normal temperatures and moisture ratios. For other conditions, calculations should use more precise values. For frequent computations at other altitudes, it may be desirable to calculate appropriate values in the same manner. For example, at an altitude of 1500 m (5000 ft) with a pressure of 84.15 kPa (12.2 psia), appropriate values are 1.03 (0.92), 2500 (4020), and 0.998 (3.73), respectively.

**Example 3-2.** A hospital operating room is being designed to use the type of HVAC system shown in the following sketch. To avoid recirculating bacteria, 100% outdoor air is used. For summer operation, the air leaving the cooling coil and supplied to the space is at 55°F, 100% relative humidity. The summer design loads are: 37,500 Btu/h (sensible) and 8,800 Btu/h (latent). The indoor design temperature is 75°F. Outdoor design conditions are: 94°F (dry bulb) and 75°F (wet bulb). Atmospheric pressure is close to sea level standard of 14.7 psia. Neglecting the effect of the fan, determine the size of cooling unit required, Btu/h, and determine the relative humidity of the air leaving the operating room, %.

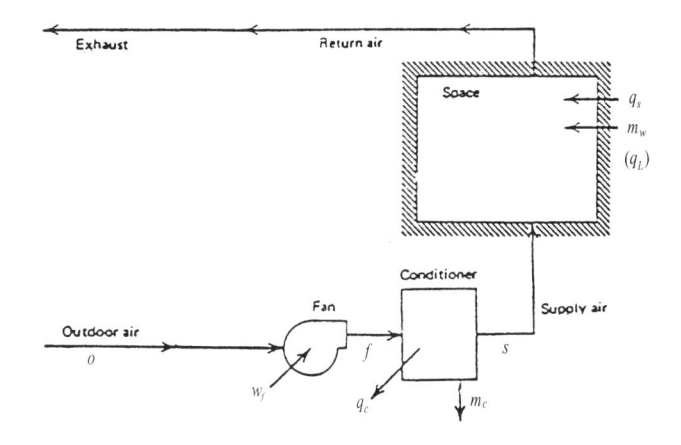

**Solution:**

Using either the psychrometric chart or corresponding psychrometric equations, the following properties are determined:

$$h_o = 38.4 \text{ Btu/lb}; \quad W_o = 0.0144 \text{ lb/lb}$$

$$h_s = 23.2 \text{ Btu/lb}; \quad W_s = 0.0092 \text{ lb/lb}$$

Using $\quad q_s = 0.244 m_a (t_r - t_s)$

$$m_a = 37500/[0.244(75 - 55)] = 7680 \text{ lb/h}$$

A water vapor balance across the space may be written

$$W_r = W_s + m_w / m_a$$

where

$$m_w = q_l / 1100 = 8800/1100 = 8 \text{ lb/h}$$

where 1100 Btu/lb approximates the enthalpy of the moisture added, causing the latent heat load. It is an approximation for 1076 Btu/lb in Equation 3-25.

Thus, $W_r = 0.0092 + 8/7680 = 0.01024 \text{ lb/lb}$

At 75°F and 0.01024 lb/lb humidity ratio, the relative humidity is found to be 55%.

For the conditioner (cooling coil), the first law of thermodynamics may be written

$$m_a[h_o - h_s - (W_o - W_s)h_c] + q_c = 0$$

$$7684\,[38.4 - 23.2 - (0.0144 - 0.0092)23] + Q_c = 0$$

$$q_c = -115,900 \text{ Btu/h}$$

Alternatively,

$$q_c = Q[1.10\,(t_o - t_s) + 4840\,(W_o - W_s)]$$

where

$$Q = m_a 13.33/60 = 7680(13.33/60) = 1706 \text{ cfm}$$

and

$$q_c = 1706\,[1.10(94 - 55) + 4840(0.0144 - 0.0092)]$$

$$= 116,100 \text{ Btu/h}$$

## 3.2 Single-Path Systems

In the following discussions various pieces of HVAC equipment and systems will be mentioned. These are discussed later in the text when the various HVAC systems are described and analyzed. The reader can jump ahead to Chapter 11 or read Chapters 1 and 4 in the 2012 *ASHRAE Handbook—Systems and Equipment* or review items in the 1991 *ASHRAE Terminology of Heating, Ventilation, Air Conditioning, and Refrigeration*.

The simplest form of an all-air HVAC system is a single conditioner serving a single temperature control zone. The unit may be installed within, or remote from, the space it serves, and it may operate either with or without distributing ductwork. Well-designed systems can maintain temperature and humidity closely and efficiently and can be shut down when desired without affecting the operation of adjacent areas.

A single-zone system responds to only one set of space conditions. Its use is limited to situations where variations occur approximately uniformly throughout the zone served or where the load is stable; but when installed in multiple, it can handle a variety of conditions efficiently. Single-zone systems are used in such applications as small department stores, small individual shops in a shopping center, individual classrooms for a small school, and computer rooms. For example, a rooftop unit complete with a refrigeration system serving an individual space is a single-zone system. The refrigeration system, however, may be remote and serving several single-zone units in a larger installation.

A schematic of the single-zone central unit is shown in Figure 3-11. The return fan is necessary if 100% outdoor air is used for cooling purposes, but may be eliminated if air may be relieved from the space with very little pressure loss through the relief system. In general, a return air fan is needed if the resistance of the return air system (grilles and ductwork) exceeds 60 Pa (0.25 in. water gage).

Control of the single-zone system can be affected by on-off operation, varying the quantity of cooling medium providing reheat, face and bypass dampers, or a combination of these. The single-duct system with reheat satisfies variations in load by providing independent sources of heating and cooling. When a humidifier is included in the system, humidity control completely responsive to space needs is available. Since control is directly from space temperature and humidity, close regulation of the system conditions may be achieved. Single-duct systems without reheat offer cooling flexibility but cannot control summer humidity independent of temperature requirements.

## 3.3 Air-Volume Equations for Single-Path Systems

Basic equations for individual rooms (zones) are the same for all single-path systems. Air supplied to each room must be adequate to take care of each room's peak load conditions whether or not it occurs simultaneously in all rooms. The peak may be governed by sensible or latent room cooling loads, heating loads, outdoor air requirements, air motion, or exhaust.

**Supply Air for Cooling:**

$$(Q_{sRS1})^S = \frac{(q_{SR1})^S}{C_1(t_R - t_s)} \qquad (3\text{-}32)$$

**Supply Air for Dehumidification:**

$$(Q_{sRL1})^S = \frac{(q_{LR1})^S}{C_2(W_R - W_s)} \qquad (3\text{-}33)$$

**Supply Air for Heating:**

$$(Q_{sRS1})^W = \frac{(q_{SR1})^W}{C_1(t_s - t_R)} \qquad (3\text{-}34)$$

where

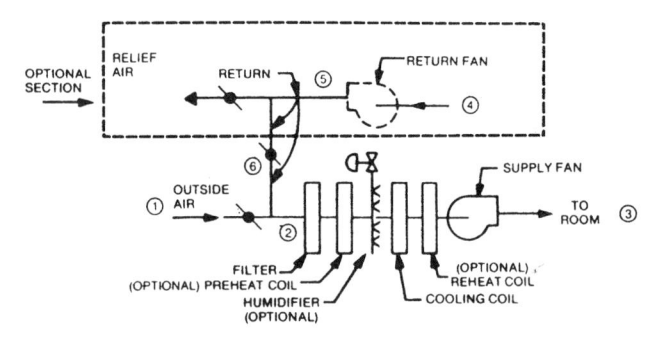

*Fig. 3-11 Single-Duct System*

# Basic HVAC System Calculations

$(Q_{sRS1})^S$ = summer room supply air volume required to satisfy peak sensible load of each room
$(Q_{sRL1})^S$ = summer room supply air volume required to satisfy peak latent load of each room
$(q_{SR1})^S$ = peak summer sensible room load for each room
$(q_{LR1})^S$ = peak summer latent room load for each room
$(Q_{sRS1})^W$ = winter room supply air volume required to satisfy peak heating load of each room
$(q_{SR1})^W$ = peak winter sensible room load (less any auxiliary heat) for each room
$W_R$ = room humidity ratio
$W_S$ = humidity ratio of dehumidified supply air
$t_R$ = room air temperature
$t_s$ = supply air temperature required to satisfy the summer or winter peak loads
$C_1$ = 1.23 (SI); 1.10 (I-P)
$C_2$ = 3010 (SI); 4840 (I-P)

**Supply Air for Ventilation.** Ventilation requirements, rather than sensible or latent loads, may govern when the supply air is deficient in any of the following ways:

1. If it does not contain adequate outdoor air, as determined by the outdoor ratio $X_o = Q_o/Q_s$, then, for such a room, supply air for outdoor air ventilation must be determined from the required room outdoor air $Q_{oR}$ and

$$Q_{sRv} = Q_{oR}/X_o \quad (3\text{-}35)$$

where
$Q_{sRv}$ = room supply air required to satisfy ventilation requirements
$Q_{oR}$ = minimum outdoor air required in a particular room
$X_o$ = ratio of the system's total outdoor air to its total supply air that satisfies outdoor air requirements in most rooms

2. The supply air may not be adequate to serve as make-up for exhaust requirements in the room. In such cases, no return air comes from the room, and only conditioned supply air is assumed to be used as make-up (no supplementary ventilation supply system). This entire volume of make-up ventilation air would become an outdoor air burden to the system, in the form of a larger $X_o$ distributed to all rooms, even though all the air supplied to this particular room is not outdoor air and

$$Q_{sRv} = Q_{sR} \quad (3\text{-}36)$$

where
$Q_{sR}$ = air exhausted or relieved from a room and not returned to the conditioned air system

3. If the desired rate of air change in the room is not satisfied, then supply air is determined as follows (in SI units):

$Q_{sRv}$ (L/s) = (Room volume in m³)
  × (No. of air changes/hour)/3.6 (3-37 SI)

and in I-P units

$Q_{sRv}$ (cfm) = (Room volume in ft³)
  × (No. of air changes/hour)/60 (3-37 I-P)

4. If air movement, as measured by an area index instead of an air change index, is not satisfied, or

$$Q_{sRv} = K \times [(\text{L/s})/\text{m}^2] \quad (3\text{-}38 \text{ SI})$$

$$Q_{sRv} = K \times (\text{cfm/ft}^2) \quad (3\text{-}38 \text{ I-P})$$

where $K$ = a constant, greater or less than one.

Both the rate of air change and $K$ are empirical values that vary according to designers' experiences and local building code requirements. For example, 5 air changes per hour in a room with a 12 ft (3.7 m) ceiling corresponds to 1 cfm/ft² (5 L/s·m²), while the same air change rate with an 8 ft (2.4 m) ceiling is only 0.66 cfm/ft² (3.3 L/s·m²). Physiologically, one may have no advantage over the other.

Case 1 is to be used when outdoor air $Q_o$ governs, Cases 3 and 4 when air movement governs, and Case 2 when exhaust governs.

**Example 3-3.** A space is designed to have a summer inside temperature of 75°F and relative humidity of 50% and a winter inside temperature of 72°F and relative humidity of 25%. The summer supply air conditions are 55°F, 90% rh, while the winter supply air temperature is 110°F with a humidity ratio of 0.0065 lb/lb. The summer design loads are:

$$q_{\text{sensible},s} = 10{,}000 \text{ Btu/h}$$

$$q_{\text{latent},s} = 1500 \text{ Btu/h}$$

The winter design loads are:

$$q_{\text{sensible},w} = 12{,}000 \text{ Btu/h}$$

$$q_{\text{latent},w} = 1{,}000 \text{ Btu/h}$$

The outdoor air requirement is 80 cfm and the ratio of outdoor air to total supply air is 0.33. Determine the required supply air in CFM to satisfy summer, winter, and ventilation conditions.

**Solution:**
Summer: Sensible load

$$q_{\text{sensible},s} = 1.1 \text{ CFM}_{s,s} \Delta t$$
$$10{,}000 = 1.1 \times \text{CFM}_{s,s} \times (75 - 55)$$
$$\text{CFM}_{s,s} = 455 \text{ cfm}$$

Latent load

$$q_{\text{latent},s} = 4840 \times \text{CFM}_{s,L} \times \Delta W$$
$$W_i = 0.0093 \text{ lb/lb at } 75°F, 50\% \text{ rh}$$
$$W_s = 0.0083 \text{ lb/lb at } 55°F, 90\% \text{ rh}$$
$$1500 = 4840 \times \text{CFM}_{s,L} \times (0.0093 - 0.0083)$$
$$\text{CFM}_{s,l} = 310 \text{ cfm}$$

Winter: Sensible load

$$q_{\text{sensible},w} = 1.1 \times \text{CFM}_{w,s} \times \Delta t$$
$$12{,}000 = 1.1 \times \text{CFM}_{w,s} \times (110 - 72)$$
$$\text{CFM}_{w,s} = 287 \text{ cfm}$$

Latent load

$$q_{latent,w} = 4840 \times CFM_{L,w} \times (\Delta W)$$

$$W_s = 0.0065 \text{ lb/lb}$$

$$W_r = 0.0042 \text{ lb/lb at } 72°F, 25\% \text{ rh}$$

$$1000 = 4840 \times CFM_{L,w} \times (0.0065 - 0.0042)$$

$$CFM_{L,w} = 90 \text{ cfm}$$

Ventilation:

$$Q_{sRV} = Q_{o,r} / X_o$$

$$Q_{sRV} = 80 / 0.33$$

$$Q_{sRV} = 242 \text{ cfm}$$

For satisfaction of all design parameters, the design volume flow rate should be selected as the maximum flow requirement of 455 cfm. This is what is required for the summer sensible design load. All of the other design parameters will be satisfied with this volume flow but would require some form of control to maintain temperature, relative humidity, and outside ventilation air quality.

## 3.4 Psychrometric Representation of Single-Path Systems

The operation of a single path system is illustrated in Figure 3-12. Each state point is shown with corresponding nomenclature in the cycle diagram and in the summer and winter representations. Each change in temperature $\Delta t$ or humidity ratio $\Delta W$ is a result of sensible or latent heat loss or gain $q_S$ or $q_L$. The cycle diagram in Figure 3-12 shows the full symbol and subscript for each state, while psychrometric charts show only the corresponding state points. This section describes the flow paths in a typical single-duct, single-zone, draw-through system. In this illustration all return air is assumed to pass from the room through a hung-ceiling return air plenum.

In Figure 3-12, supply air $cfm_S$ at the fan discharge temperature $t_{sf}$ in the summer mode absorbs transmitted supply duct heat $q_{sd}$ and supply air fan velocity pressure energy $q_{sfvp}$, thereby raising the temperature to $t_s$. Room supply air absorbs room sensible and latent heat $q_{SR}$ and $q_{LR}$ along the room sensible heat factor (SHF) line **s-R**, reaching desired room state, $t_R$ and $W_R$. Room (internal) sensible loads which determine $cfm_S$ consist of

1. Ceiling transmission $c_{clg}$ (shown) from the hung ceiling above the room, and floor transmission $q_D$ (not shown) from the floor deck below (the $q_D$ shown is the heat gain to the hung ceiling from the roof deck above, not the floor deck below);
2. Direct light heat emissions to the room $q_{lR}$ (without the direct light heat emission to the plenum $q_{lp}$);
3. Transmissions $q_{tran}$ from other surfaces such as walls, windows, etc.;
4. Appliance heat $(q_S)_{aux}$ and occupancy heat $(q_S)_{occ}$; and
5. Infiltration load $(q_S)_{inf}$ taken here as zero, with exfiltration air $cfm_{eR}$ shown instead.

The remaining air returned from the room ($CFM_S - CFM_{er} = CFM_R$) passes through the ceiling plenum along line **R-rp**, absorbing some ceiling heat $q_{rp}$ while simultaneously retransmitting some of it back to the room $q_{clg}$. Air is picked up by the return duct system at $t_{rp}$, after passing through the ceiling at average temperature $t_p$.

Return volume $cfm_R$ picks up return duct transmissions $q_{rd}$ and return fan heat gains $q_{rf}$ (including both static and velocity energy) by the time it reaches the intake plenum entrance at $t_r$. $CFM_r$ may be less than $cfm_R$ if leakage occurs from the ducts or through the exhaust damper from the return air system. Mixing of outdoor air, $cfm_o$ at state $o$, with final return air occurs along process line **r-o** to mixture state $m$. Total system air $cfm_s$ passes through the cooling coil releasing the total sensible heat factor, $SHR_{cc}$ line **m-cc**, terminating in state $cc$. The temperature rise through the supply fan from $(q_{sf})_{sp}$ results in $t_{sf}$, which completes the cycle.

The winter cycle (Figure 3-12) is similar, except for the interposition of the heating coil energy $q_{hc}$ added in process line **m-hc**; the deletion of the cooling coil action; and the temperature drop in process line **sf-s**, resulting from duct transmission losses $q_{sd}$ instead of gains.

## 3.5 Sensible Heat Factor (Sensible Heat Ratio)

The **sensible heat factor** (SHF), also called the **sensible heat ratio** (SHR), is the ratio of the sensible heat for a process to the summation of the sensible and latent heat for the process. The sum of sensible and latent heat is also called the **total heat**. On the ASHRAE Psychrometric Chart, values of sensible heat factors are given on the protractor as $\Delta H_s/\Delta H_T$, and they may be used to establish the process line for changes in the condition of the air across either the room or the conditioner on the psychrometric chart.

The supply air to a conditioned space must have the capacity to offset simultaneously both the room sensible and room latent heat loads. The room and the supply air conditions to the space may be plotted on the psychrometric chart and connected with a straight line called the room sensible heat factor line.

As air passes through the cooling coil of an air conditioner, the sensible heat factor line represents the simultaneous cooling and dehumidifying that occurs. If the cooling process involves removing only sensible and no latent heat, i.e., no moisture removal, the sensible heat factor line is horizontal and the sensible heat factor is 1.0. If 50% is sensible and 50% latent, the SHF is 0.5.

**Example 3-4.** An air-conditioned space has a summer sensible design heat load of 100,000 Btu/h, a summer design latent load of 20,000 Btu/h, and is maintained at 75°F and 55% rh. Conditioned air leaves the apparatus and enters the room at 58°F. The outdoor air is at 96°F, 77°F wet bulb, and is 20% of total flow to the conditioning apparatus.
(a) Draw and label the schematic flow diagram for the system.
(b) Complete a table of properties and flow rates at various locations in the system.

# Basic HVAC System Calculations

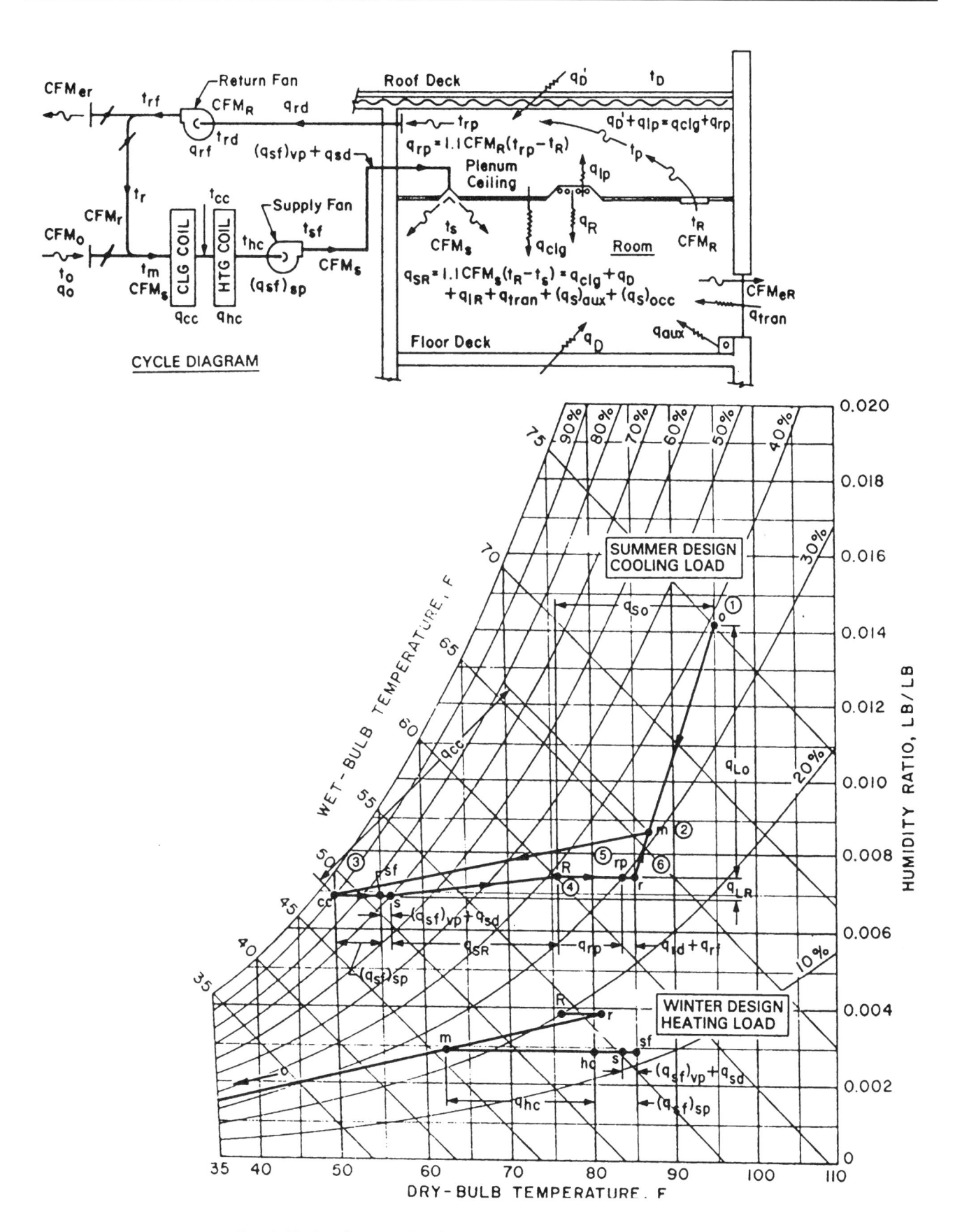

*Fig. 3-12 Single-Duct, Single-Zone Cycle and Psychrometric Chart*

(c) Show all the processes on the psychrometric coordinates.
(d) Determine the size of cooling unit needed in Btu/h and tons.
(e) What percent of the required cooling is for sensible cooling and what percent is for dehumidification?
(f) What percent of the required cooling is due to the outdoor air load?

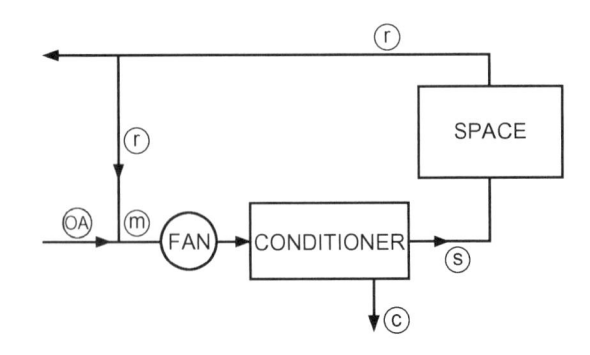

**Solution:**
(a)

(b)

| Point | $t$, °F | $\phi$, % | WB, °F | $h$, Btu/lb | $W$, lb/lb | $v_a$, ft³/lb | $m$, lb/h | CFM | SCFM |
|---|---|---|---|---|---|---|---|---|---|
| OA | **96** | 43 | 77 | 40.2 | 0.0157 | 14.36 | 4822 | 1151 | 1071 |
| r | **75** | **55** | 64 | 29.2 | 0.0102 | 13.7 | 19,286 | 4403 | 4285 |
| m | 79.1 | 52 | 66.8 | 31.4 | 0.0113 | 13.83 | 24,108 | 5557 | 5356 |
| s | **58** | 89 | 56.1 | 24 | 0.00944 | 13.24 | 24,108 | 5320 | 5356 |

Note: 1. Assume that the thermal conditions leaving the fan are the same as those entering the fan.
2. The properties in bold are given values.

The psychrometric properties for OA and r are read directly from the psychrometric chart.

$$q_{sensible} = m \times c_{pa} \times (t_r - t_s)$$

$$100{,}000 = m \times 0.244 \text{ Btu/lb°F} \times (75 - 58)$$

$$m = 24{,}108 \text{ lb/h}$$

Mass flow rate for outdoor air is

$$24{,}108 \times 0.2 = 4822 \text{ lb/h}$$

Mass flow rate of return air mixed with outdoor air is

$$24{,}108 - 4822 = 19{,}286 \text{ lb/h}$$

For the mixed air state, $m$:

Energy balance at $m$  $0.8(29.2) + 0.2(40.2) = 31.4$ Btu/lb

Moisture balance at $m$  $0.8(0.0102) + 0.2(0.015) = 0.0113$ lb/lb

The other conditions at $m$ come from the psychrometric chart.

A moisture balance on the space provides $W_s$.

$$W_s = W_r - \frac{m_w}{m_a} = 0.0102 - \frac{20{,}000/1100 \text{ Btu/lb}}{24{,}108}$$

$$W_s = 0.0102 - 0.000\,7542$$

$$W_s = 0.00944 \text{ lb/lb}$$

With $W_s$ and $t_s$ the other properties are found from the psychrometric chart from

$$m = \text{cfm}/v \Rightarrow \text{cfm} = m \times v/60$$

The scfm values at the four states are found from the same equation; however, using

$$v = v_s = 13.33 \text{ ft}^3/\text{lb}.$$

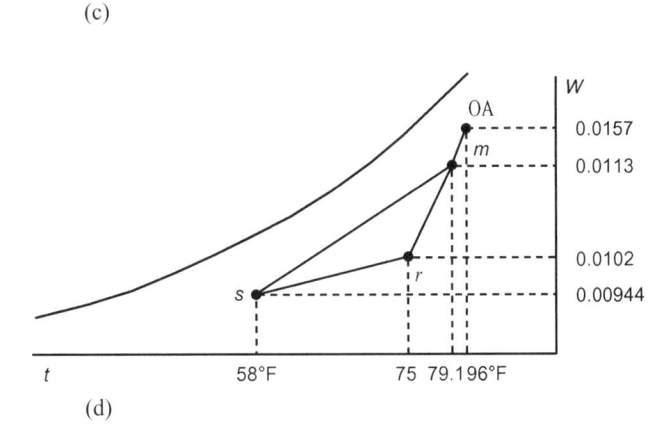

(c)

(d)

$$q = m\,[h_m - h_s - (w_m - w_s)\,h_c]$$

$$q = 24{,}108\,[31.4 - 24 - (0.0113 - 0.00944)\,26]$$

$$q = 177{,}240 \text{ Btu/h or } 14.8 \text{ tons}$$

(e)

$$\%\text{ sensible} = \frac{24{,}108(0.244)(79.1 - 58)}{177{,}240} = 0.70$$

$$\%\text{ latent} = 0.30, 30\%$$

(f)

$$\%\text{ Due to OA} = \frac{1071[1.1(96 - 75) + 4840(0.0157 - 0.0102)]}{177{,}240}$$

$$= 0.30, 30\%$$

## 3.6 Problems

**3.1** One of the many methods used for drying air is to cool the air below the dew point so that condensation or freezing of the moisture takes place. To what temperature must atmospheric air be cooled in order to have a humidity ratio of 0.000017 lb/lb (17 mg/kg)? To what temperature must this air be cooled if its pressure is 10 atm?

**3.2** One method of removing moisture from atmospheric air is to cool the air so that the moisture condenses or freezes out. Suppose an experiment requires a humidity ratio of 0.0001. To what temperature must the air be cooled at a pressure of 0.1 kPa in order to achieve this humidity?

**3.3** A room of dimensions 4 m by 6 m by 2.4 m contains an air-water vapor mixture at a total pressure of 100 kPa and a temperature of 25°C. The partial pressure of the water vapor is 1.4 kPa. Calculate

(a) humidity ratio [Ans: 0.0088 kg/kg$_{air}$]
(b) dew point [Ans: 11.8°C]
(c) total mass of water vapor in the room [Ans: 0.584 kg]

**3.4** The air conditions at the intake of an air compressor are 70°F (21.1°C), 50% rh, and 14.7 psia (101.3 kPa). The air is compressed to 50 psia (344.7 kPa), then sent to an intercooler. If condensation of water vapor from the air is to be prevented, what is the lowest temperature to which the air can be cooled in the intercooler?

**3.5** Humid air enters a dehumidifier with an enthalpy of 21.6 Btu/lb$_m$ of dry air and 1100 Btu/lb$_m$ of water vapor. There are 0.02 lb$_m$ of vapor per pound of dry air at entrance and 0.009 lb$_m$ of vapor per pound of dry air at exit. The dry air at exit has an enthalpy of 13.2 Btu/lb$_m$, and the vapor at exit has an enthalpy of 1085 Btu/lb$_m$. Condensate leaves with an enthalpy of 22 Btu/lb$_m$. The rate of flow of dry air is 287 lb$_m$/min. Determine

(a) the amount of moisture removed from the air (lb$_m$/min)
(b) the rate of heat removal required
   [Ans: 3.16 lb/min, −5860 Btu/min]

**3.6** Air is supplied to a room from the outside, where the temperature is 20°F (−6.7°C) and the relative humidity is 60%. The room is to be maintained at 70°F (21.1°C) and 50% rh. How many pounds of water must be supplied per pound of air supplied to the room?

**3.7** Air is heated to 80°F (26.7°C) without adding water, from 60°F (15.6°C) dry-bulb and 50°F (10°C) wet-bulb temperature. Use the psychrometric chart to find

(a) relative humidity of the original mixture
(b) original dew-point temperature
(c) original humidity ratio
(d) initial enthalpy
(e) final enthalpy
(f) the heat added
(g) final relative humidity

**3.8** Saturated air at 40°F (4.4°C) is first preheated and then saturated adiabatically. This saturated air is then heated to a final condition of 105°F (40.6°C) and 28% rh. To what temperature must the air initially be heated in the preheat coil?
   [Ans: 101°F (37.8°C)]

**3.9** Atmospheric air at 100°F (37.8°C) dry-bulb and 65°F (18.3°C) wet-bulb temperature is humidified adiabatically with steam. The supply steam contains 10% moisture and is at 16 psia (110.3 kPa). What is the dry-bulb temperature of the humidified air if enough steam is added to bring the air to 70% rh?

**3.10** The conditions on a day in New Orleans, Louisiana, are 95°F (35°C) dry-bulb and 80°F (26.7°C) wet-bulb temperature. In Tucson, Arizona, the air conditions are 105°F (40.6°C) dry-bulb and 72°F (22.2°C) wet-bulb temperature. What is the lowest air temperature that could theoretically be attained in an evaporative cooler at these summer conditions in these two cities?

**3.11** Air at 29.92 in. Hg enters an adiabatic saturator at 80°F dry-bulb and 66°F wet-bulb temperature. Water is supplied at 66°F. Find (without using the psychrometric chart) the humidity ratio, degree of saturation, enthalpy, and specific volume of entering air.
   [Ans: 0.0104 lb/lb air, $\mu$ = 0.47, h = 30.7 Btu/lb air, $v$ = 13.83 ft$^3$/lb air]

**3.12** An air-water vapor mixture enters an air-conditioning unit at a pressure of 150 kPa, a temperature of 30°C, and a relative humidity of 80%. The mass flow of dry air entering is 1 kg/s. The air-vapor mixture leaves the air-conditioning unit at 125 kPa, 10°C, 100% rh. The moisture condensed leaves at 10°C. Determine the heat transfer rate for the process.

**3.13** Air at 40°C, 300 kPa, with a relative humidity of 35% is to be expanded in a reversible adiabatic nozzle. How low a pressure can the gas be expanded to if no condensation is to take place? What is the exit velocity at this condition?

**3.14** By using basic definitions and Dalton's Law of partial pressure, show that

$$v = R_a T/(p - p_w)$$

**3.15** In an air-conditioning unit, 71,000 cfm at 80°F dry bulb, 60% rh, and standard atmospheric pressure, enter the unit. The air leaves the unit at 57°F dry bulb and 90% relative humidity. Calculate the following:

(a) cooling capacity of the air-conditioning unit, in Btu/h
(b) rate of water removal from the unit
(c) sensible heat load on the conditioner, in Btu/h
(d) latent heat load on the conditioner, in Btu/h
(e) the dew point of the air leaving the conditioner

**3.16** Four pounds of air at 80°F (26.7°C) dry bulb and 50% rh are mixed with one pound of air at 50°F (15.6°C) and 50% rh. Determine

(a) relative humidity of the mixture
(b) dew-point temperature of the mixture
   [Ans: 52%, 55.5°F (13°C)]

**3.17** Air is compressed from 85°F, 60% rh, 14.7 psia to 60 psia and then cooled in an intercooler before entering a second stage of compression. What is the minimum temperature to which the air can be cooled so that condensation does not take place?

**3.18** An air-water vapor mixture flowing at a rate of 4000 cfm (1890 L/s) enters a perfect refrigeration coil at 84°F (28.9°C) and 70°F (21.1°C) wet-bulb temperature. The air leaves the coil at 53°F (11.7°C). How many Btu/h of refrigeration are required?

**3.19** Air at 40°F dry bulb and 35°F wet bulb is mixed with air at 100°F dry bulb and 77°F wet bulb in the ratio of 2 lb of cool air to 1 lb of warm air. Compute the resultant humidity ratio and enthalpy of the mixed air. [Ans: 0.007 lb/lb, 22.2 Btu/lb]

**3.20** Outdoor air at 90°F (32.2°C) and 78°F (25.6°C) wet bulb is mixed with return air at 75°F (23.9°C) and 52% rh. There are 1000 lb (454 kg) of outdoor air for every 5000 lb (2265 kg) of return air. What are the dry- and wet-bulb temperatures for the mixed airstream?

**3.21** In a mixing process of two streams of air, 10,000 cfm of air at 75°F and 50% rh mix with 4000 cfm of air at 98°F dry-bulb and 78°F wet-bulb temperature. Calculate the following conditions after mixing at atmospheric pressure:

(a) dry-bulb temperature
(b) humidity ratio
(c) relative humidity
(d) enthalpy
(e) dew-point temperature

**3.22** Solve the following:

(a) Determine the humidity ratio and relative humidity of an air-water vapor mixture that has a dry-bulb temperature of 30°C, an adiabatic saturation temperature of 25°C, and a pressure of 100 kPa.
(b) Use the psychrometric chart to determine the humidity ratio and relative humidity of an air-water vapor mixture that has a dry-bulb temperature of 30°C, a wet-bulb temperature of 25°C, and a pressure of 100 kPa.
[Ans: 0.0183 kg/kg, 67%, 0.018 kg/kg, 67%]

**3.23** An air-water vapor mixture at 100 kPa, 35°C, and 70% rh is contained in a 0.5 m³ closed tank. The tank is cooled until the water just begins to condense. Determine the temperature at which condensation begins and the heat transfer for the process.

**3.24** A room is to be maintained at 76°F and 40% rh. Supply air at 39°F is to absorb 100,000 Btu sensible heat and 35 lb of moisture per hour. Assume the moisture has an enthalpy of 1100 Btu/lb. How many pounds of dry air per hour are required? What should the dew-point temperature and relative humidity of the supply air be? Assume $h_w = 1100$ Btu/lb$_m$.
[Ans: 11,260 lb/h, 36°F, 90%]

**3.25** Moist air enters a chamber at 40°F dry-bulb and 36°F wet-bulb temperature at a rate of 3000 cfm. In passing through the chamber, the air absorbs sensible heat at a rate of 116,000 Btu/h and picks up 83 lb/h of saturated steam at 230°F. Determine the dry-bulb and wet-bulb temperatures of the leaving air. [Ans: 74°F db, 64°F wb]

**3.26** In an auditorium maintained at a temperature not to exceed 77°F, and a relative humidity not to exceed 55%, a sensible-heat load of 350,000 Btu and 1,000,000 grains of moisture per hour must be removed. Air is supplied to the auditorium at 67°F.

(a) How much air must be supplied, in lb/h?
(b) What is the dew-point temperature of the entering air, and what is its relative humidity?
(c) How much latent heat is picked up in the auditorium?
(d) What is the sensible heat ratio?

**3.27** A meeting hall is maintained at 75°F dry bulb and 65°F wet bulb. The barometric pressure is 29.92 in. Hg. The space has a load of 200,000 Btu/h sensible, and 200,000 Btu/h latent. The temperature of the supply air to the space cannot be lower than 65°F dry bulb.

(a) How much air must be supplied, in lb/h?
(b) What is the required wet-bulb temperature of the supply air?
(c) What is the sensible heat ratio?
[Ans: 81,970 lb/h, 58°F, 0.5]

**3.28** A structure to be air conditioned has a sensible heat load of 20,000 Btu/h at a time when the total load is 100,000 Btu/h. If the inside state is to be at 80°F, 50% rh, is it possible to meet the load conditions by supplying air to the room at 100°F and 60% rh? If not, discuss the direction in which the inside state would be expected to move if such air were supplied.

**3.29** A flow rate of 30,000 lb/h of conditioned air at 60°F and 85% rh is added to a space that has a sensible load of 120,000 Btu/h and a latent load of 30,000 Btu/h.

(a) What are the dry- and wet-bulb temperatures in the space?
(b) If a mixture of 50% return air and 50% outdoor air at 98°F dry bulb and 77°F wet bulb enters the air conditioner, what is the refrigeration load?

**3.30** An air-water vapor mixture enters a heater-humidifier unit at 5°C, 100 kPa, 50% rh. The flow rate of dry air is 0.1 kg/s. Liquid water at 10°C is sprayed into the mixture at the rate of 0.0022 kg/s. The mixture leaves the unit at 30°C, 100 kPa. Calculate

(a) the relative humidity at the outlet
(b) the rate of heat transfer to the unit
[Ans: 91%, 7.94 kW]

**3.31** A room is being maintained at 75°F and 50% rh. The outdoor air conditions are 40°F and 50% rh at this time. Return air from the room is cooled and dehumidified by mixing it with fresh ventilation air from the outside. The total airflow to the room is 60% outdoor and 40% return air by mass. Determine the temperature, relative humidity, and humidity content of the mixed air going to the room.
[Ans: 54.5°F, 58%, 0.00524 lb/lb]

**3.32** A room with a sensible load of 20,000 Btu/h is maintained at 75°F and 50% rh. Outdoor air at 95°F and 80°F wet bulb was mixed with the room return air. The outdoor air, which is mixed, is 25% by mass of the total flow going to the conditioner. This air is then cooled and dehumidified by a coil and leaves the coil saturated at 50°F, which is on the condition line for the room. The air is then mixed with some room return air so that the temperature of the air entering the room is at 60°F. Find the following:

(a) the air-conditioning processes on the psychrometric chart
(b) ratio of latent to sensible load
(c) airflow rate
(d) the percent by mass of room return air mixed with air leaving the cooling coil

**3.33** An air-water vapor mixture at 14.7 psia (101.5 kPa), 85°F (29.4°C), and 50% rh is contained within a 15 ft³ (0.425 m³) tank. At what temperature will condensation begin? If the tank and mixture are cooled an additional 15°F (8.3°C), how much water will condense from the mixture?

**3.34** Air flowing at 1000 cfm and at 14.7 psia, 90°F, and 60% rh passes over a coil with a mean surface temperature of 40°F. A spray on the coil assures that the leaving air is saturated at the coil temperature. What is the required cooling capacity of the coil? [Ans: 9.3 tons]

**3.35** An air-vapor mixture at 100°F (37.8°C) dry bulb contains 0.02 lb water vapor per pound of dry air (20 g/kg). The barometric pressure is 28.561 in. Hg (96.7 kPa). Calculate the relative humidity, dew-point temperature, and degree of saturation.

**3.36** Air enters a space at 20°F and 80% rh. Within the space, sensible heat is added at the rate of 45,000 Btu/h and latent heat is added at the rate of 20,000 Btu/h. The conditions to be maintained inside the space are 50°F and 75% rh. What must the air exhaust rate (lb/h) from the space be to maintain a 50°F temperature? What must be the air exhaust rate (lb/h) from the space to maintain a 75% rh? Discuss the difference.

**3.37** Moist air at a low pressure of 11 psia is flowing through a duct at a low velocity of 200 fpm. The duct is 1 ft in diameter and has negligible heat transfer to the surroundings. The dry-bulb temperature is 85°F and the wet-bulb temperature is 70°F. Calculate the following:

(a) humidity ratio, lb/lb
(b) dew-point temperature, °F
(c) relative humidity, %
[Ans: 0.0177 lb/lb, 52°F, 32%]

**3.38** If an air compressor takes in moist air (at about 90% rh) at room temperature and pressure and compresses this to 120 psig (827 kPa) (and slightly higher temperature), would you expect some condensation to occur? Why? If yes, where would the condensation form? How would you remove it?

**3.39** Does a sling psychrometer give an accurate reading of the adiabatic saturation temperature? Explain.

**3.40** An air processor handles 2000 cfm of air with initial conditions of 50°F and 50% rh. The air is heated with a finned heat exchanger with 78 ft² of heat transfer surface area and a $UA$ value of 210 Btu/h·°F.

Also, a steam spray system adds moisture to the air from saturated steam at 16 psia. The outlet air is at 100°F and 50% rh. Do the following:

(a) Show the processes on the psychrometric chart.
(b) Calculate the mass flow rate, lb/min.
(c) Calculate the pounds per minute of steam required.
(d) Calculate the heat added by the coil, Btu/min.
[Ans: 155 lb/min, 2.65 lb/min, 1900 Btu/min]

**3.41** At an altitude of 5000 ft (1500 m), a sling psychrometer reads 80°F (26.7°C) and 67°F (19.4°C) wet bulb. Determine correct values of relative humidity and enthalpy from the chart. Compare these to the corresponding values for the same readings at sea level.

**3.42** The average person gives off sensible heat at the rate of 250 Btu/h and perspires and respires about 0.27 lb/h of moisture. Estimate the sensible and latent load for a room with 25 people in it (the lights give off 9000 Btu/h). If the room conditions are to be 78°F and 50% rh, what flow rate of air would be required if the supply air came in at 63°F? What would be the supply air relative humidity?

**3.43** A space in an industrial building has a winter sensible heat loss of 200,000 Btu/h and a negligible latent heat load (latent losses to outside are made up by latent gains within the space). The space is to be maintained precisely at 75°F and 50% rh. Due to the nature of the process, 100% outdoor air is required for ventilation. The outdoor air conditions can be taken as saturated air at 20°F. The amount of ventilation air is 7000 scfm and the air is to be preheated, humidified with an adiabatic saturator to the desired humidity, and then reheated. The temperature out of the adiabatic saturator is to be maintained at 60°F dry bulb. Determine the following:

(a) temperature of air entering the space to be heated, °F
(b) heat supplied to preheat coil, Btu/h
(c) heat supplied to reheat coil, Btu/h
(d) amount of water required for humidification, gpm

**3.44** Using the SI psychrometric chart at standard atmospheric pressure, find

(a) dew point and humidity ratio for air at 28°C dry bulb and 22°C wet bulb [Ans: 19.5°C, 0.014 kg/kg]
(b) enthalpy and specific volume [Ans: 64.7 kJ/kg, 0.87 m³/kg]

**3.45** Using the SI chart, find

(a) moisture that must be removed in cooling air from 24°C dry bulb, 21°C wet bulb to 13°C dry bulb, saturated
(b) total, sensible, and latent heat removal for the process

**3.46** An air-conditioned space has a sensible heat load of 200,000 Btu/h, a latent load of 50,000 Btu/h, and is maintained at 78°F dry bulb and 60% rh. On a mass basis, 25% outdoor air is mixed with return air. Outdoor air is at 95°F dry bulb and 76°F wet bulb. Conditioned air leaves the apparatus and enters the room at 60°F dry bulb. The fan must produce a pressure increase of 3.5 in. water to overcome the system pressure loss. Fan efficiency is estimated as 55%.

(a) Draw and label the schematic flow diagram for the complete system. (Hint: See Fig. 3-1)
(b) Complete the table below.
(c) Plot and draw all processes on a psychrometric chart.
(d) Specify the fan size, scfm, and fan motor rating, HP.
(e) Determine the size refrigeration unit needed, in Btu/h and tons.
(f) What percent of the required refrigeration is for (1) sensible cooling and (2) for dehumidification?
(g) What percent of the required refrigeration is due to the outdoor air load?

| Point | Dry Bulb $t$, °F | $\phi$, % | Enthalpy $h$, Btu/lb | $W$, lb/lb | $m_a$, lb/h | scfm | CFM |
|---|---|---|---|---|---|---|---|
| OA | | | | | | | |
| r | | | | | | | |
| m | | | | | | | |
| f | | | | | | | |
| s | | | | | | | |

## 3.7 Bibliography

ASHRAE. 2013. *2013 ASHRAE Handbook—Fundamentals.*

ASHRAE. 1991. *ASHRAE Terminology of Heating, Ventilation, Air Conditioning, and Refrigeration.*

Goff, J.A. and S. Gratch. 1945. Thermodynamic properties of moist air. *ASHVE Transactions* 51:125.

Kusuda, T. 1970. Algorithms for psychrometric calculations. NBS Publication BSS21 (January), Superintendent of Documents, US Government Printing Office, Washington, D.C.

Olivieri, J., T. Singh, and S. Lorodocky. 1996. *Psychrometrics: Theory and Practice.* ASHRAE.

# Chapter 4

# DESIGN CONDITIONS

This chapter covers the selection, specification, and determination of the indoor and outdoor environmental conditions that are to be expected at "design time" or the conditions that will govern the sizing of the heating, cooling, and ventilating equipment. Additional details related to design conditions are provided in Chapters 9, 10, and 14 of the 2013 *ASHRAE Handbook—Fundamentals*.

## 4.1 Indoor Design Conditions

### 4.1.1 Physiological Principles

A principal purpose of HVAC is to provide conditions for human comfort, "that condition of mind that expresses satisfaction with the thermal environment" (ASHRAE Standard 55). This definition leaves open what is meant by "condition of mind" or "satisfaction," but it correctly emphasizes that judgment of comfort is a cognitive process involving many inputs influenced by physical, physiological, psychological, and other processes. The conscious mind appears to reach conclusions about thermal comfort and discomfort from direct temperature and moisture sensations from the skin, deep body temperatures, and the effects necessary to regulate body temperatures. Surprisingly, although climates, living conditions, and cultures differ widely throughout the world, the temperatures that people choose for comfort under similar conditions of clothing, activity, humidity, and air movement have been found to be very similar. Definitions of comfort do vary. Comfort encompasses perception of the environment (e.g., hot, cold, noisy) and a value rating of affective implications (e.g., too high, too cold). Acceptability is the foundation of a number of standards covering thermal comfort and acoustics.

Operational definitions of health and discomfort are controversial. However, the most generally accepted definition is that in the constitution of the World Health Organization (WHO): "Health is a state of complete physical, mental, and social well-being and not merely the absence of disease or infirmity." Concern about the health effects associated with indoor air dates back several hundred years, and it has increased dramatically in recent decades, particularly since the energy crisis in the early 1970s. This attention was partially the result of increased reporting by building occupants about poor health associated with exposure to indoor air. Since then, two types of diseases associated with exposure to indoor air have been identified: **sick building syndrome (SBS)** and **building-related illness (BRI)**.

SBS describes a number of adverse health symptoms related to occupancy in a "sick" building, including mucosal irritation, fatigue, headache, and, occasionally, lower respiratory symptoms and nausea. Sick building syndrome is characterized by an absence of routine physical signs and clinical laboratory abnormalities. The term *nonspecific* is sometimes used to imply that the pattern of symptoms reported by afflicted building occupants is not consistent with the pattern of symptoms for a particular disease. Additional symptoms can include nosebleeds, chest tightness, and fever.

Building-related illnesses, in contrast, have a known origin, may have a different set of symptoms, and are often accompanied by physical signs and abnormalities that can be clinically identified with laboratory measurements. For example, hypersensitivity illnesses, including humidifier fever, asthma, and allergic rhinitis, are caused by individual sensitization to bioaerosols.

The thermal environment affects human health in that it affects body temperature regulation and heat exchange with the environment. In the normal, healthy, resting adult, internal or core body temperatures are very stable, with variations seldom exceeding 1°F. The internal temperature of a resting adult, measured orally, averages about 98.6°F; measured rectally, it is about 1°F higher. In contrast, skin temperature is basically unregulated and can (depending on environmental temperature) vary from about 88°F to 96.8°F in normal environments and activities.

To design an environmental control system that is effective for comfort and health, the engineer must understand physiological principles. In its broadest sense, the term air conditioning implies control of any or all of the physical and chemical qualities of air. Herein, the definition of air conditioning pertains only to those conditions of air relating to health and comfort requirements of the occupants of the conditioned space.

Significant variations in the percentage composition of the normal constituents of air may make it unfit for human use. The presence of foreign materials classified as contaminants may also make air unfit. Air conditioning can control most climatological environmental factors for the service and comfort of people.

The objective of a comfort air-conditioning system is to provide a comfortable environment for the occupants of residential or commercial buildings. A comfortable environment is created by simultaneously controlling temperature, humidity, air cleanliness, and air distribution within the occupant's vicinity. These factors include mean radiant temperature, as well as air temperature, odor control, and control of the proper acoustic level within the occupant's vicinity.

Both the air and surfaces of the enclosure surrounding the occupant are sinks for the metabolic heat emitted by the occupant (Figure 4-1). Air circulates around the occupant and the surfaces. The occupant also exchanges radiant heat with the surrounding surfaces (e.g., glass and outside walls). Air is brought into motion within a given space thermally or by mechanical forces.

Whether nude or clothed, humans feel comfortable at a mean skin temperature of 91.5°F (33°C). The range of skin temperature within which no discomfort is experienced is about ±2.5°F (±1.4°C). The necessary criteria, indices, and standards for use where human occupancy is concerned are given in Chapter 9, "Thermal Comfort" and in Chapter 10, "Indoor Environmental Health," of the 2013 *ASHRAE Handbook—Fundamentals*.

The environmental indices used to evaluate the sensation of comfort for the human body are classified as direct, rationally derived, and empirical indices, which include the following factors:

**Direct Indices**
  Dry-bulb temperature
  Dew-point temperature
  Wet-bulb temperature
  Relative humidity
  Air movement

**Rationally Derived Indices**
  Mean radiant temperature
  Operative temperature
  Humid operative temperature
  Heat stress index
  Index of skin wettedness

**Empirical Indices**
  Effective temperature
  Black globe temperature
  Corrected effective temperature
  Wet-bulb-globe temperature index
  Wind chill index

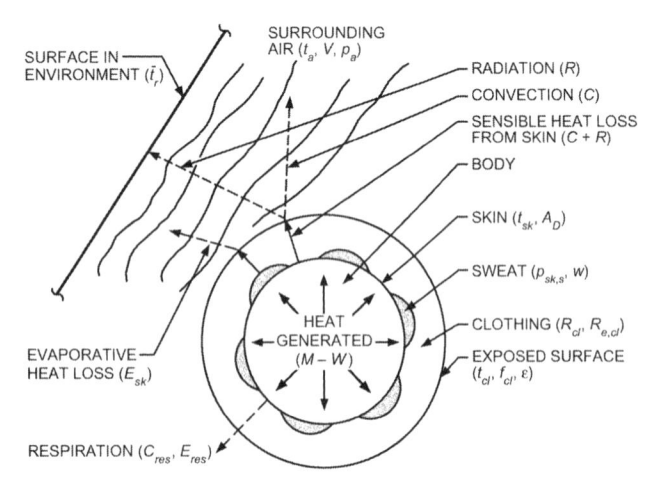

*Fig. 4-1 Cylindrical Model of Body's Interaction with Environment.*
(Figure 1, Chapter 9, 2013 ASHRAE Handbook—Fundamentals)

The **mean radiant temperature** $\bar{T}_r$ is a key variable in thermal calculations for the human body. It is the uniform temperature of an imaginary enclosure in which radiant heat transfer from the human body equals the radiant heat transfer in the actual uniform enclosure. It is calculated from the measured temperature of surrounding walls and surfaces and their positions with respect to the person. For most building surfaces, the emittance is high and the following equation can be used:

$$\bar{T}_r^4 = T_1^4 F_{p-1} + T_2^4 F_{p-2} + \ldots T_N^4 F_{p-N}$$

where
$\bar{T}_r$ = mean radiant temperature, °R
$T_N$ = surface temperature of surface N, °R
$F_{p-N}$ = angle factor between a person and surface N

*(See Chapter 8 in 2013 ASHRAE Handbook—Fundamentals)*

A simplification is often used for many HVAC applications:

$$\text{MRT} = \bar{T}_r = \frac{A_1 t_1 + A_2 t_2 + \ldots A_N t_N}{A_1 + A_2 + \ldots A_N}$$

where
MRT = approximate mean radiant temperature for surface temperatures not significantly different from each other, °F
$t_N$ = surface temperature of surface N, °F
$A_N$ = area of surface N

The **operative temperature** is the uniform temperature of a radiantly black enclosure in which an occupant exchanges the same amount of heat by radiation plus convection as in the actual nonuniform environment. Numerically, operative temperature is the average, weighted by respective heat-transfer coefficients, of the air and mean radiant temperatures. At air speeds of 0.4 m/s (80 fpm) or less and mean radiant temperature less than 50°C (120°F), operative temperature is approximately the simple average of the air and mean radiant temperatures:

$$t_o = \frac{\text{MRT} + t_a}{2}$$

Physiologists recognize that sensations of comfort and temperature may have different physiological and physical bases, thus each type should be considered separately. This dichotomy was recognized in ANSI/ASHRAE Standard 55, where thermal comfort is defined as *"that state of mind which expresses satisfaction with the thermal environment."* In contrast, most current predictive charts are based on comfort defined as a sensation *"that is neither slightly warm nor slightly cool."*

Most research on comfort has been limited to lightly clothed, sedentary people. This research has proven sound since about 90% of people's indoor occupation and leisure time is spent at or near the sedentary activity level. The predictive methods to be described are all believed to be accurate depending upon the limitations stated. During physical activity, a change occurs in a person's physiology. Physiological

# Design Conditions

thermal neutrality in the sedentary sense does not exist. Some form of thermal regulation is always occurring during the sedentary condition. The same skin and body temperatures, if used as indices of comfort, prove false during moderate to heavy activity.

The main purpose of the HVAC system is to maintain comfortable indoor conditions for occupants. However, the purpose of load calculations is to obtain data for sizing the system components. In most cases, the system will rarely be set to operate at design conditions. Therefore, the use and occupancy of the space is a general consideration from the design temperature point of view. Later, when the energy requirements of the building are computed, the actual conditions in the space and outdoor environment must be considered.

The indoor design temperature should be selected at the lower end of the acceptable temperature range so that the heating equipment will not be oversized. Even properly sized equipment usually operates under partial load, at reduced efficiency, most of the time; therefore, any oversizing aggravates this condition and lowers the overall system efficiency. A maximum design dry-bulb temperature of 72°F is recommended for most occupancies. The indoor design value of relative humidity should be compatible with a healthful environment and the thermal and moisture integrity of the building envelope. A maximum relative humidity of 30% is recommended for most situations.

The conscious mind appears to reach conclusions about thermal comfort and discomfort from direct temperature and moisture sensations from the skin, deep body temperatures, and the efforts necessary to regulate body temperatures. In general, comfort occurs when body temperatures are held within narrow ranges, skin moisture is low, and the physiological effort of regulation is minimized.

The ASHRAE thermal sensation scale ($Y$), with its numerical representation, is

| | |
|---|---|
| +3 | hot |
| +2 | warm |
| +1 | slightly warm |
| 0 | neutral |
| −1 | slightly cool |
| −2 | cool |
| −3 | cold |

The equations in Table 4-1 indicate that women are more sensitive to temperature and less sensitive to humidity than men. However, in general, about a 5.4°F change in temperature or a 0.44 psi change in water vapor pressure is necessary to change a thermal sensation vote by one unit or temperature category.

## 4.1.2 Metabolic Rate

In choosing optimal conditions for comfort and health, the energy expended during the course of routine physical activities must be known, since body heat production increases in proportion to exercise intensity. Table 4-2 provides metabolic rates for various activities, on a per unit body surface area basis. The most useful measure of body surface area, proposed by DuBois, is described by

$$A_D = 0.108 m^{0.425} l^{0.725} \qquad (4\text{-}1)$$

where

$A_D$ = body surface area, ft²
$m$ = mass of body, lb
$l$ = height of body, in.

An average-sized male has a mass of 70 kg (154 lb) and a height of 1.73 m (5 ft, 8 in.), so his body surface area is 1.83 m² (19.7 ft²).

In choosing optimal conditions for comfort and health, the rate of work done during routine physical activities must be known, because metabolic power increases in proportion to exercise intensity. Metabolic rate varies over a wide range, depending on the activity, the person, and the conditions under which the activity is performed. Table 4-2 lists typical metabolic rates for an average adult male ($A_D$ = 19.7 ft²) for activities performed continuously. The highest power a person can maintain for any continuous period is approximately 50% of the maximal capacity to use oxygen (maximum energy capacity).

A unit used to express the metabolic rate per unit DuBois area is the met, defined as the metabolic rate of a sedentary person (seated, quiet): 1 met = 18.4 Btu/h·ft²) = 50 kcal/(h·m²). A normal, healthy man has a maximum capacity of approximately $M_{act}$ = 12 met at age 20, which drops to 7 met at age 70. Maximum rates for women are about 30% lower. Long-distance runners and trained athletes have maximum rates as high as 20 met. An average 35 year-old who does not exercise has a maximum rate of about 10 met and activities with $M_{act}$ > 5 met are likely to prove exhausting.

The metabolic activities of the body result almost completely in heat that must be continuously dissipated and regulated to prevent abnormal body temperatures. Insufficient heat loss leads to over-heating, called hyperthermia, and excessive heat loss results in body cooling, called hypothermia. Skin temperatures associated with comfort at sedentary activities are 91.5 to 93°F and decrease with increasing activity.

The heat produced by a resting adult is about 340 Btu/h. Because most of this heat is transferred to the environment

**Table 4-1  Equations for Predicting Thermal Sensation ($Y$)[a,b]**
*(Table 9, Chapter 9, 2013 Handbook—Fundamentals)*

| Exposure Period, h | Sex | Regression Equations<br>$t$ = dry-bulb temperature, °F<br>$p$ = vapor pressure, psi |
|---|---|---|
| 1.0 | Male | $Y = 0.122\,t + 1.61\,p - 9.584$ |
|  | Female | $Y = 0.151\,t + 1.71\,p - 12.080$ |
|  | Combined | $Y = 0.136\,t + 1.71\,p - 10.880$ |
| 2.0 | Male | $Y = 0.123\,t + 1.86\,p - 9.953$ |
|  | Female | $Y = 0.157\,t + 1.45\,p - 12.725$ |
|  | Combined | $Y = 0.140\,t + 1.65\,p - 11.339$ |
| 3.0 | Male | $Y = 0.118\,t + 2.02\,p - 9.718$ |
|  | Female | $Y = 0.153\,t + 1.76\,p - 13.511$ |
|  | Combined | $Y = 0.135\,t + 1.92\,p - 11.122$ |

[a] $Y$ values refer to the ASHRAE thermal sensation scale.
[b] For young adult subjects with sedentary activity and wearing clothing with a thermal resistance of approximately 0.5 clo, $\bar{t}_r \approx \bar{t}_a$ and air velocities are <40 fpm.

**Table 4-2  Typical Heat Generation Rates**
*(Table 4, Chapter 9, 2013 ASHRAE Handbook—Fundamentals)*

|  | Btu/h·ft² | met[a] |
|---|---|---|
| Resting | | |
|   Sleeping | 13 | 0.7 |
|   Reclining | 15 | 0.8 |
|   Seated, quiet | 18 | 1.0 |
|   Standing, relaxed | 22 | 1.2 |
| Walking (on level surface) | | |
|   2.9 ft/s (2 mph) | 37 | 2.0 |
|   4.4 ft/s (3 mph) | 48 | 2.6 |
|   5.9 ft/s (4 mph) | 70 | 3.8 |
| Office Activities | | |
|   Reading, seated | 18 | 1.0 |
|   Writing | 18 | 1.0 |
|   Typing | 20 | 1.1 |
|   Filing, seated | 22 | 1.2 |
|   Filing, standing | 26 | 1.4 |
|   Walking about | 31 | 1.7 |
|   Lifting/packing | 39 | 2.1 |
| Driving/Flying | | |
|   Car | 18 to 37 | 1.0 to 2.0 |
|   Aircraft, routine | 22 | 1.2 |
|   Aircraft, instrument landing | 33 | 1.8 |
|   Aircraft, combat | 44 | 2.4 |
|   Heavy vehicle | 59 | 3.2 |
| Miscellaneous Occupational Activities | | |
|   Cooking | 29 to 37 | 1.6 to 2.0 |
|   Housecleaning | 37 to 63 | 2.0 to 3.4 |
|   Seated, heavy limb movement | 41 | 2.2 |
|   Machine work | | |
|     sawing (table saw) | 33 | 1.8 |
|     light (electrical industry) | 37 to 44 | 2.0 to 2.4 |
|     heavy | 74 | 4.0 |
|   Handling 110 lb bags | 74 | 4.0 |
|   Pick and shovel work | 74 to 88 | 4.0 to 4.8 |
| Miscellaneous Leisure Activities | | |
|   Dancing, social | 44 to 81 | 2.4 to 4.4 |
|   Calisthenics/exercise | 55 to 74 | 3.0 to 4.0 |
|   Tennis, singles | 66 to 74 | 3.6 to 4.0 |
|   Basketball | 90 to 140 | 5.0 to 7.6 |
|   Wrestling, competitive | 130 to 160 | 7.0 to 8.7 |

[a] 1 met = 18.4 Btu/h·ft²

through the skin, it is often convenient to characterize metabolic activity in terms of heat production per unit area of skin. For the resting person, this is about 18.4 Btu/h·ft² (58 W/m²) (the average male has a skin surface area of about 19.7 ft²) and is called 1 met. Higher metabolic rates are often described in terms of the resting rate. Thus, a person working at a metabolic rate five times the resting rate would have a metabolic rate of 5 met.

### 4.1.3 Clothing Level

Clothing insulation value may be expressed in clo units. In order to avoid confusion, the symbol $I$ is used with the clo unit instead of the symbol $R$, normally used for thermal resistance per unit area. The relationship between the two is

$$R = 0.88I \qquad (4\text{-}2)$$

or 1.0 clo is equivalent to 0.88 ft²·h·°F/Btu. (0.155 (m²·K)/W).

Often it is not possible to find an already measured clothing ensemble that matches the one in question. In this case, the ensemble insulation can be estimated from the insulation of individual garments. Table 4-3 gives a list of individual garments commonly worn. The insulation of an ensemble is estimated from the individual values using a summation formula:

$$I_{cl} = \sum_i I_{clu,i} \qquad (4\text{-}3)$$

The main source of inaccuracy is in determining the appropriate values for individual garments. Overall accuracies are on the order of ±25% if the tables are used carefully. Where $I_{clu,i}$ is the effective insulation of garment $i$, and $I_{cl}$, as before is the insulation for the entire ensemble.

### 4.1.4 Conditions for Thermal Comfort

Environmental conditions for good thermal comfort minimize effort of the physiological control system. For a resting person wearing trousers and a long-sleeved shirt, thermal comfort in a steady state is experienced in a still-air environment at 75°F. A zone of comfort extends about 3°F above and below this optimum level.

ANSI/ASHRAE Standard 55-2010, *Thermal Environmental Conditions for Human Occupancy* specifies conditions of the indoor thermal environment that a majority of the occupants will find acceptable. The body of the standard clearly defines "majority" such that the requirements are based on 80% overall acceptability, while specific dissatisfaction limits vary for different sources of local discomfort. A space that meets the criteria of the standard likely will have individual occupants who are not satisfied due to large individual differences in preference and sensitivity. The standard is intended for use in designing, commissioning, and testing of buildings and other occupied spaces and their HVAC systems and for the evaluation of existing thermal environments. Standard 55 deals exclusively with thermal comfort in the indoor environment. The scope is not limited to any specific building type, so it may be used for residential or commercial buildings and for new or existing buildings. It also can apply to other occupied spaces such as cars, trains, planes, and ships.

The standard does not cover hot or cold stress in thermally extreme environments or comfort in outdoor spaces. It also does not address nonthermal environmental conditions (e.g., air quality or acoustics) or the effect of any environmental factors on nonthermal human responses (e.g., the effect of humidity on health). The scope clearly states that its criteria are based only on thermal comfort. Thus, a minimum humidity level is not specified since no lower humidity limits relate exclusively to thermal comfort. The form of the upper limit of humidity has changed throughout the standard's history. The

# Design Conditions

**Table 4-3 Garment Insulation Values**
*(Table 8, Chapter 9, 2013 Handbook—Fundamentals)*

| Garment Description[a] | $I_{clu,i}$, clo[b] | Garment Description[a] | $I_{clu,i}$, clo[b] | Garment Description[a] | $I_{clu,i}$, clo[b] |
|---|---|---|---|---|---|
| **Underwear** | | Long-sleeve, flannel shirt | 0.34 | Long-sleeve (thin) | 0.25 |
| Men's briefs | 0.04 | Short-sleeve, knit sport shirt | 0.17 | Long-sleeve (thick) | 0.36 |
| Panties | 0.03 | Long-sleeve, sweat shirt | 0.34 | **Dresses and skirts[c]** | |
| Bra | 0.01 | **Trousers and Coveralls** | 0.06 | Skirt (thin) | 0.14 |
| T-shirt | 0.08 | Short shorts | 0.08 | Skirt (thick) | 0.23 |
| Full slip | 0.16 | Walking shorts | 0.15 | Long-sleeve shirtdress (thin) | 0.33 |
| Half slip | 0.14 | Straight trousers (thin) | 0.24 | Long-sleeve shirtdress (thick) | 0.47 |
| Long underwear top | 0.20 | Straight trousers (thick) | 0.28 | Short-sleeve shirtdress (thin) | 0.29 |
| Long underwear bottoms | 0.15 | Sweatpants | 0.30 | Sleeveless, scoop neck (thin) | 0.23 |
| **Footwear** | | Overalls | 0.49 | Sleeveless, scoop neck (thick), i.e., jumper | 0.27 |
| Ankle-length athletic socks | 0.02 | Coveralls | | | |
| Calf-length socks | 0.03 | **Suit jackets and vests (lined)** | | **Sleepwear and Robes** | |
| Knee socks (thick) | 0.06 | Single-breasted (thin) | 0.36 | Sleeveless, short gown (thin) | 0.18 |
| Panty hose | 0.02 | Single-breasted (thick) | 0.44 | Sleeveless, long gown (thin) | 0.20 |
| Sandals/thongs | 0.02 | Double-breasted (thin) | 0.42 | Short-sleeve hospital gown | 0.31 |
| Slippers (quilted, pile-lined) | 0.03 | Double-breasted (thick) | 0.48 | Long-sleeve, long gown (thick) | 0.46 |
| Boots | 0.10 | Sleeveless vest (thin) | 0.10 | Long-sleeve pajamas (thick) | 0.57 |
| **Shirts and Blouses** | | Sleeveless vest (thick) | 0.17 | Short-sleeve pajamas (thin) | 0.42 |
| Sleeveless, scoop-neck blouse | 0.12 | **Sweaters** | | Long-sleeve, long wrap robe (thick) | 0.69 |
| Short-sleeve, dress shirt | 0.19 | Sleeveless vest (thin) | 0.13 | Long-sleeve, short wrap robe (thick) | 0.48 |
| Long-sleeve, dress shirt | 0.25 | Sleeveless vest (thick) | 0.22 | Short-sleeve, short robe (thin) | 0.34 |

[a] "Thin" garments are made of light, thin fabrics worn in summer; "thick" garments are made of heavy, thick fabrics worn in winter.
[b] 1 clo = 0.880°F·ft²·h/Btu
[c] Knee-length

current upper limit is specified in terms of absolute humidity as the limiting parameter at a humidity ratio of 0.012.

The winter and summer comfort zones specified in ANSI/ASHRAE Standard 55 are given in Figure 4-2. The temperature ranges are appropriate for current seasonal clothing habits in the United States. Summer clothing is considered to be light slacks and a short-sleeved shirt or a comparable ensemble with an insulation value of 0.5 clo. Winter clothing is considered heavy slacks, long-sleeved shirt, and sweater or jacket with an insulation value of 0.9 clo. The temperature ranges are for sedentary and slightly active people.

The winter zone is for air speeds less than 0.15 m/s; the summer zone is for air movements less than 0.25 m/s. The standard allows the summer comfort zone to extend above 26°C if the average air movement is increased 0.275 m/s for each °C of temperature increase to a maximum temperature of 28°C and air movement of 0.8 m/s.

The temperature boundaries of the comfort zones in Figure 4-2 can be shifted −1°F (−0.6°C) per 0.1 clo for clothing levels other than 0.5 and 0.9. The zones can also be shifted lower for increased activity levels.

Thermal comfort conditions must be fairly uniform over the body to prevent local discomfort. The radiant temperature

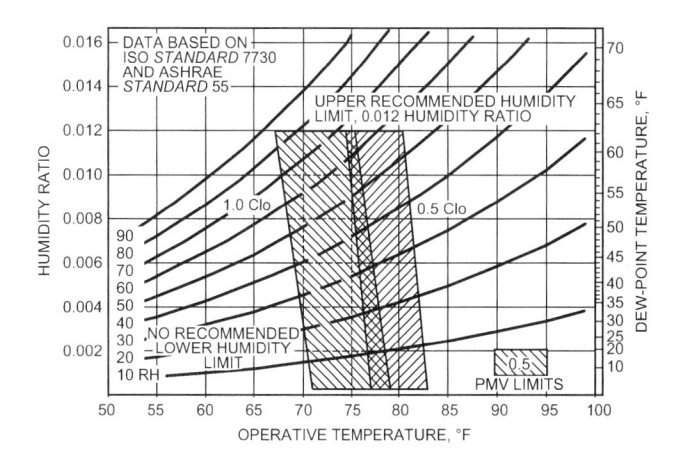

*Fig. 4-2 ASHRAE Summer and Winter Comfort Zones*
*(Acceptable ranges of operative temperature and humidity with air speed ≤ 40 fpm for people wearing 1.0 and 0.5 clothing during primarily sedentary activity.)*
*(Figure 5, Chapter 9, 2013 ASHRAE Handbook—Fundamentals)*

asymmetry should be less than 5°C in the vertical direction and 10°C in the horizontal direction. The vertical temperature difference between head and foot should not exceed 3°C.

Figure 4-2 applies generally to altitudes from sea level to 3000 m (10,000 ft), and to the most common indoor thermal environments in which mean radiant temperature is nearly equal to the dry-bulb air temperature and where the air velocity is less than 30 fpm in winter and 50 fpm in summer. For these cases, the thermal environment can be specified by operative temperature and humidity variables.

Comfort zones for other clothing levels can be approximated by decreasing the temperature borders of the zone by 1°F for each 0.1 clo increase in clothing insulation and vise-versa. Similarly, a zone's temperatures can be decreased by 2.5°F per met increase in activity above 1.2 met. The upper and lower humidity levels of the comfort zones are less precise. Low humidity can lead to drying of the skin and mucous surfaces. Comfort complaints about dry nose, throat, eyes, and skin occur in low humidity conditions, typically when the dew point is less than 32°F. In compliance with these and other discomfort observations, Standard 55 recommends that the dew point temperature of occupied spaces not be less than 35°F. In contrast, at high humidity levels too much skin moisture tends to increase discomfort, particularly skin moisture that is physiological in origin (water diffusion and perspiration). On the warm side of the comfort zone, the relative humidity should not exceed 60%.

Table 4-4 provides design criteria covering factors that apply to many different building types.

### 4.1.5 Adjustments for Clothing and/or Activity Levels

**Activity.** The indoor designated temperatures of Figure 4-2 should be decreased when the average steady-state activity level of the occupants is higher than light, primarily sedentary (>1.2 met).

This temperature can be calculated from the operative temperature at sedentary conditions with the following equation:

$$t_{i,\text{active}} = t_{i,\text{sedentary}} - 3(1 + \text{clo})(\text{met} - 1.2)(°C) \quad (4\text{-}4)$$

$$t_{i,\text{active}} = t_{i,\text{sedentary}} - 5.4(1 + \text{clo})(\text{met} - 1.2)(°F) \quad (4\text{-}5)$$

The equation is only appropriate between 1.2 and 3 met. The minimum allowable operative temperature for these activities is 15°C (59°F). The acceptable range (based on a 10% dissatisfaction criterion) will increase with activity and clothing. The ranges are approximately ±1.4°C (2.7°F) for 0.1 clo, ±2°C (3.5°F) for 0.5 clo, and ±3°C (5.4°F) for 0.9 clo.

**Clothing.** The temperatures of Figure 4.2, after being corrected for activity level, can be corrected for clothing level using a decrease of 1°F for each 0.1 clo increase.

A wide range of environmental applications are covered by the ANSI/ASHRAE Standard 55. The comfort envelope defined by the standard applies only for sedentary and slightly active, normally clothed people at low air velocities, when the mean radiant temperature (MRT) equals the air temperature. For other clothing, activities, air temperatures, etc., the standard recommends use of Fanger's General Comfort Charts. Examples of these are shown in Figure 4-3.

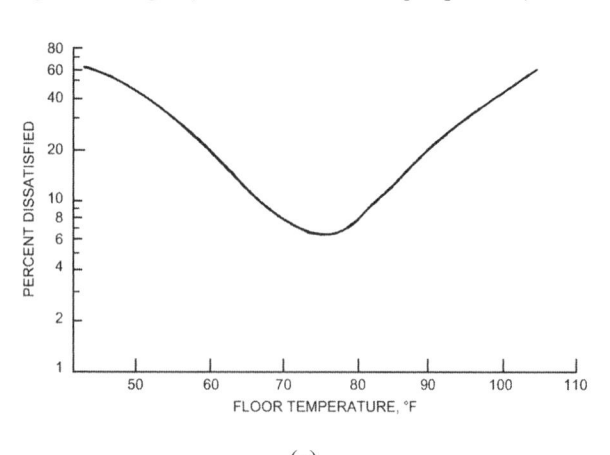

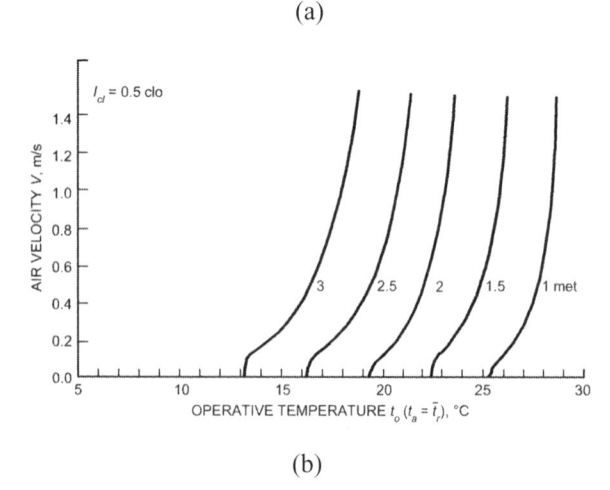

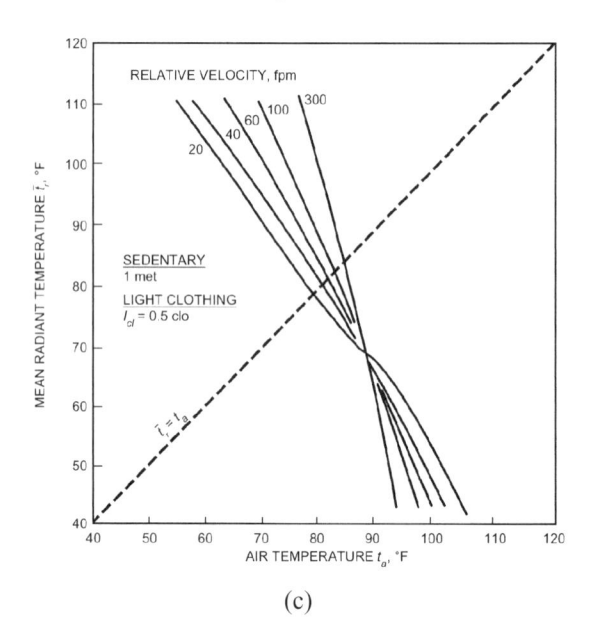

*Fig. 4-3 Examples of Fanger's Charts*
(Figures 13, 14, and 15, Chapter 9, 2013 ASHRAE Handbook—Fundamentals)

# Design Conditions

**Table 4-4 General Design Criteria**
*(Adapted from Table 1, Chapter 3, 2007 Handbook—HVAC Applications)*

| General Category | Specific Category | Inside Design Conditions — Winter | Inside Design Conditions — Summer | Air Movement | Circulation, Air Changes per Hour | Load Profile |
|---|---|---|---|---|---|---|
| Dining and Entertainment Centers | Cafeterias and Luncheonettes | 70°F to 74°F  20% to 30% rh | 78°F[e]  50% rh | 50 fpm at 6 ft above floor | 12 to 15 | Peak at 1 to 2 p.m. |
| | Restaurants | 70°F to 74°F  20% to 30% rh | 74°F to 78°F  55% to 60% rh | 25 to 30 fpm | 8 to 12 | Peak at 1 to 2 p.m. |
| | Bars | 70°F to 74°F  20% to 30% rh | 74°F to 78°F  50% to 60% rh | 30 fpm at 6 ft above floor | 15 to 20 | Peak at 5 to 7 p.m. |
| | Nightclubs | 70°F to 74°F  20% to 30% rh | 74°F to 78°F  50% to 60% rh | below 25 fpm at 5 ft above floor | 20 to 30 | Peak after 8 p.m., off from 2 a.m. to 4 p.m. |
| | Kitchens | 70°F to 74°F | 85°F to 88°F | 30 to 50 fpm | 12 to 15[h] | |
| Office Buildings | | 70°F to 74°F  20% to 30% rh | 74°F to 78°F  50% to 60% rh | 25 to 45 fpm  0.75 to 2 cfm/ft² | 4 to 10 | Peak at 4 p.m. |
| Libraries and Museums | Average | 68°F to 72°F  40% to 55% rh | | below 25 fpm | 8 to 12 | Peak at 3 p.m. |
| | Archival | See Chapter 20, 1999 *Handbook—HVAC Applications* | | below 25 fpm | 8 to 12 | Peak at 3 p.m. |
| Bowling Centers | | 70°F to 74°F  20% to 30% rh | 75°F to 78°F  50% to 55% rh | 50 fpm at 6 ft above floor | 10 to 15 | Peak at 6 to 8 p.m. |
| Communication Centers | Telephone Terminal Rooms | 72°F to 78°F  40% to 50% rh | 72°F to 78°F  40% to 50% rh | 25 to 30 fpm | 8 to 20 | Varies with location and use |
| | Teletype Centers | 70°F to 74°F  40% to 50% rh | 74°F to 78°F  45% to 55% rh | 25 to 30 fpm | 8 to 20 | Varies with location and use |
| | Radio and Television Studios | 74°F to 78°F  30% to 40% rh | 74°F to 78°F  40% to 55% rh | below 25 fpm at 12 ft above floor | 15 to 40 | Varies widely due to changes in lighting and people |
| Transportation Centers | Airport Terminals | 70°F to 74°F  20% to 30% rh | 74°F to 78°F  50% to 60% rh | 25 to 30 fpm at 6 ft above floor | 8 to 12 | Peak at 10 a.m. to 9 p.m. |
| | Ship Docks | 70°F to 74°F  20% to 30% rh | 74°F to 78°F  50% to 60% rh | 25 to 30 fpm at 6 ft above floor | 8 to 12 | Peak at 10 a.m. to 5 p.m. |
| | Bus Terminals | 70°F to 74°F  20% to 30% rh | 74°F to 78°F  50% to 60% rh | 25 to 30 fpm at 6 ft above floor | 8 to 12 | Peak at 10 a.m. to 5 p.m. |
| | Garages[l] | 40°F to 55°F | 80°F to 100°F | 30 to 75 fpm | 4 to 6  Refer to NFPA | Peak at 10 a.m. to 5 p.m. |
| Warehouses | | Inside design temperatures for warehouses often depend on the materials stored. | | | 1 to 4 | Peak at 10 a.m. to 3 p.m. |

**Example 4-1:** Determine the optimal comfort conditions for a conference room under summer conditions. Occupants wear light clothing (0.5 clo) during summer. Air movement is 0.1 m/s (20 fpm), MRT = air temperature, and summer rh = 70%.

**Solution:**

From Figure 4-3(b):
Air Temperature = MRT = 78°F (26°C)

**Example 4-2:** If, in winter, neither the air temperature nor the level of clothing for Example 4.1 change, what activity level must the occupant move up to in order to be comfortable?

**Solution:**

From Figure 4-3(a): Given conditions of 0.33 ft/s and 78°F fall on comfort line at approximately 1 met (only slightly more active, e.g., reading or writing).

The indoor conditions to be maintained within a building are the dry-bulb temperature and relative humidity of the air at the breathing line, 1 to 1.5 m (3 to 5 ft) above the floor, in an area that would indicate average conditions at that level and which would not be affected by abnormal or unusual heat gains or losses from the interior or exterior.

### 4.1.6 Moisture and Humidity

Too often, the behavior of moisture is given insufficient attention in building design and construction. Moisture is present as a vapor in all air and as absorbed moisture in most building materials. Problems involving moisture may arise from changes in moisture content, from the presence of excessive moisture, or from effects associated with its changes in state.

Water vapor originates from such activities as cooking, laundering, bathing, and people breathing and perspiring. Some typical values of moisture production are given in Table 4-5.

Exterior and interior building materials should allow vapor to pass five times more rapidly than materials inside the wall. Provided this condition is met, any moisture that may get into a wall will move on through it.

Selecting and applying humidification or dehumidification equipment involves considering both the environmental requirements of the occupancy or process and the limitations imposed by the thermal and permeable characteristics of the building enclosure. As these may not always be compatible, a compromise solution may be necessary, particularly in the case of existing buildings.

The environmental requirements for a particular occupancy or process may dictate a specific relative humidity, a required range of relative humidity, or certain limiting maximum or minimum values. The following classifications give guidance for most applications.

**Human Comfort.** The effect of relative humidity on human comfort has not been completely established. Nevertheless, humidity extremes are assumed to be undesirable and, for human comfort, relative humidities should be kept within a broad range of 30 to 60%.

**Static Electricity.** Electrostatic charges are generated when materials of high electrical resistance move against each other. Such charges may cause unpleasant sparks for people walking over carpets; difficulties in handling sheets of paper, fibers, and fabrics; objectionable clinging of dust to oppositely charged objects; or dangerous situations when explosive gases are present. Increasing the relative humidity of an environment tends to prevent the accumulation of such charges, but the optimum level of humidity depends to some extent on the materials involved. With many materials, relative humidities of 45% or more are usually required to reduce or eliminate electrostatic effects. Hospital operating rooms, where explosive mixtures of anesthetics are used, constitute a special and critical case in regard to electrostatic charges. A relative humidity of 50% or more is usually required and other special grounding arrangements and restrictions are imposed as to types of clothing the occupants wear. From a consideration of both comfort and safety, conditions of 72°F (22°C) and 55% rh are usually recommended in operating rooms.

**Prevention and Treatment of Disease.** Relative humidity has a significant effect on the control of airborne infection. At 50% rh, the mortality rate of certain organisms is highest (e.g., influenza virus loses much of its virulence). The mortality rate decreases both above and below this value. A relative humidity of 65% is regarded as optimum for nurseries for premature infants, while a value of 50% is suitable for full-terms and observational nurseries. In the treatment of allergic disorders, humidities well below 50% have proven satisfactory.

**Visible Condensation.** Condensation occurs on any interior surface when the dew point of the air in contact with it exceeds the surface temperature. The maximum permissible relative humidity that may be maintained without condensation is thus influenced by the thermal properties of the enclosure and the interior and exterior environment. In general, windows present the lowest surface temperature in most buildings and provide the best guide to permissible indoor humidity levels for no condensation (Table 4-6).

**Concealed Condensation.** The humidity level a building can tolerate without serious difficulties from concealed condensation may be much lower than indicated by visible

**Table 4-5 Moisture Production in Residences**

| Operation | Moisture | |
|---|---|---|
| | lb | (kg) |
| Floor mopping—80 ft² (7.4 m²) kitchen | 2.40 | (1.09) |
| Clothes drying* (not vented) | 26.40 | (11.97) |
| Clothes washing* | 4.33 | (1.96) |
| Cooking (not vented)*   *From Food*   *From Gas* | | |
| Breakfast    0.34   (0.16)   0.56   (0.25) | 0.90 | (0.41) |
| Lunch    0.51   (0.23)   0.66   (0.33) | 1.17 | (0.53) |
| Dinner    1.17   (0.53)   1.52   (0.69) | 2.69 | (1.22) |
| Bathing—shower | 0.50 | (0.23) |
| Bathing—tub | 0.12 | (0.05) |
| Dishwashing* | | |
| Breakfast | 0.20 | (0.09) |
| Lunch | 0.15 | (0.07) |
| Dinner | 0.65 | (0.29) |
| Human Contribution—Adults | | |
| When resting ............................... Per hour | 0.2 | (0.09) |
| Working hard............................... Per hour | 0.6 | (0.27) |
| Average......................................... Per hour | 0.4 | (0.18) |
| House plants............................... Per hour | 0.04 | (0.02) |

*Based on family of four

condensation criteria. The migration of water vapor through the inner envelope by diffusion or air leakage brings it into contact with surfaces at temperatures approaching the outside temperature. Unless the building has been designed to eliminate or effectively reduce this possibility, the permissible humidity may be limited by the ability of the building enclosure to handle internal moisture rather than prevent the occurrence of moisture.

## 4.2 Outdoor Design Conditions: Weather Data

The 2013 *ASHRAE Handbook—Fundamentals*, Chapter 14, "Climatic Design Information," and its accompanying CD-ROM provide design weather information for 6443 locations in the United States, Canada, and around the world. The large number of stations made printing all the tables impractical. However, 31 of the locations required for the solution of the problems in this textbook have been included. These 31 locations make up Figure 4-4. The complete tables are contained on a CD-ROM distributed with the 2013 *ASHRAE Handbook—Fundamentals*.

This climatic design information is commonly used for design, sizing, distribution, installation, and marketing of heating, ventilating, air-conditioning, and dehumidification equipment, as well as for other energy-related processes in residential, agricultural, commercial, and industrial applications. These summaries include values of dry-bulb, wet-bulb, and dew-point temperature, and wind speed with direction at various frequencies of occurrence. Also included in this edition are monthly degree-days to various bases, and parameters to calculate clear-sky irradiance.

Design information in this chapter includes design values of dry-bulb with mean coincident wet-bulb temperature, design wet-bulb with mean coincident dry-bulb temperature, and design dew-point with mean coincident dry-bulb temperature and corresponding humidity ratio. These data allow the designer to consider various operational peak conditions. Design values of wind speed facilitate the design of smoke management systems in buildings.

Warm-season temperature and humidity conditions are based on annual percentiles of 0.4, 1.0, and 2.0. Cold-season conditions are based on annual percentiles of 99.6 and 99.0. The use of annual percentiles to define design conditions ensures that they represent the same probability of occurrence in any climate, regardless of the seasonal distribution of extreme temperature and humidity.

Monthly information including percentiles is compiled in addition to annual percentiles, to provide seasonally representative combinations of temperature, humidity, and solar conditions. The tables also list heating and cooling degree-days for bases 65°F and 50°F, as well as cooling degree-hours for bases 74°F and 80°F. The calculation of daily dry-bulb and wet-bulb temperature profiles, which are useful for generating 24 h weather data sequences suitable as input to many HVAC

**Table 4-6 Maximum Relative Humidity without Window Condensation**

| Natural Convection, Indoor Air at 23.3°C (74°F) |||
|---|---|---|
| Outdoor Temp., °C (°F) | Single Glazing | Double Glazing |
| 4.4 (40) | 39 | 59 |
| −1.1 (30) | 29 | 50 |
| −6.7 (20) | 21 | 43 |
| −12.2 (10) | 15 | 36 |
| −17.8 (0) | 10 | 30 |
| −23.3 (−10) | 7 | 26 |
| −28.9 (−20) | 5 | 21 |
| −34.4 (−30) | 3 | 17 |

**Table 4-7 Nomenclature for Tables of Climatic Design Conditions**
*(Table 1A, Chapter 14, 2013 ASHRAE Handbook)*

| | |
|---|---|
| CDD$n$ | Cooling degree-days base $n$°F, °F-day |
| CDH$n$ | Cooling degree-hours base $n$°F, °F-hour |
| DB | Dry-bulb temperature, °F |
| DP | Dew-point temperature, °F |
| Ebn,noon | Clear sky beam normal irradiances at solar noon, Btu/h·ft$^2$ |
| Edh,noon | Clear sky diffuse horizontal irradiance at solar noon, Btu/h·ft$^2$ |
| Elev | Elevation, ft |
| Enth | Enthalpy, Btu/lb |
| HDD$n$ | Heating degree-days base $n$°F, °F-day |
| Hours 8/4 & 55/69 | Number of hours between 8 a.m. and 4 p.m. with DB between 55°F and 69°F |
| HR | Humidity ratio, gr$_{moisture}$/lb$_{dry\ air}$ |
| Lat | Latitude, ° |
| Long | Longitude, ° |
| MCDB | Mean coincident dry bulb temperature, °F |
| MCDBR | Mean coincident dry bulb temp. range, °F |
| MCDP | Mean coincident dew point temperature, °F |
| MCWB | Mean coincident wet bulb temperature, °F |
| MCWBR | Mean coincident wet bulb temp. range, °F |
| MCWS | Mean coincident wind speed, mph |
| MDBR | Mean dry bulb temp. range, °F |
| PCWD | Prevailing coincident wind direction, ° (0 = North; 90 = East) |
| Period | Years used to calculate the design conditions |
| PrecAvg | Average precipitation, in. |
| PrecSD | Standard deviation of precipitation, in. |
| PrecMin | Minimum precipitation, in. |
| PrecMax | Maximum precipitation, in. |
| Sd | Standard deviation of daily average temperature, °F |
| StdP | Standard pressure at station elevation, psi |
| taub | Clear sky optical depth for beam irradiance |
| taud | Clear sky optical depth for diffuse irradiance |
| Tavg | Average temperature, °F |
| Time Zone | Hours ahead or behind UTC, and time zone code |
| WB | Wet bulb temperature, °F |
| WBAN | Weather Bureau Army Navy number |
| WMO# | Station identifier from the World Meteorological Organization |
| WS | Wind speed, mph |

Note: Numbers *(1)* to *(41)* and letters *(a)* to *(p)* are row and column references to quickly point to an element in the table. For example, the 5% design wet-bulb temperature for July can be found in row *(29)*, column *(k)*.

**90**                                                                                   **Principles of HVAC**

2013 ASHRAE Handbook - Fundamentals (IP)            © 2013 ASHRAE, Inc.

**ATLANTA MUNICIPAL, GA, USA**      WMO#: 722190

Lat: **33.64N**    Long: **84.43W**    Elev: **1027**    StdP: **14.16**    Time Zone: **-5 (NAE)**    Period: **86-10**    WBAN: **13874**

### Annual Heating and Humidification Design Conditions

| Coldest Month | Heating DB | | Humidification DP/MCDB and HR | | | | | | Coldest month WS/MCDB | | | | MCWS/PCWD to 99.6% DB | |
|---|---|---|---|---|---|---|---|---|---|---|---|---|---|---|
| | 99.6% | 99% | 99.6% | | | 99% | | | 0.4% | | 1% | | | |
| | | | DP | HR | MCDB | DP | HR | MCDB | WS | MCDB | WS | MCDB | MCWS | PCWD |
| (a) | (b) | (c) | (d) | (e) | (f) | (g) | (h) | (i) | (j) | (k) | (l) | (m) | (n) | (o) |
| 1 | 21.5 | 26.4 | 4.2 | 7.1 | 28.6 | 9.1 | 9.1 | 32.2 | 24.9 | 39.9 | 23.5 | 40.0 | 11.9 | 320 |

### Annual Cooling, Dehumidification, and Enthalpy Design Conditions

| Hottest Month | Hottest Month DB Range | Cooling DB/MCWB | | | | | | Evaporation WB/MCDB | | | | | | MCWS/PCWD to 0.4% DB | |
|---|---|---|---|---|---|---|---|---|---|---|---|---|---|---|---|
| | | 0.4% | | 1% | | 2% | | 0.4% | | 1% | | 2% | | | |
| | | DB | MCWB | DB | MCWB | DB | MCWB | WB | MCDB | WB | MCDB | WB | MCDB | MCWS | PCWD |
| (a) | (b) | (c) | (d) | (e) | (f) | (g) | (h) | (i) | (j) | (k) | (l) | (m) | (n) | (o) | (p) |
| 7 | 17.0 | 93.9 | 74.2 | 91.7 | 73.9 | 89.8 | 73.5 | 77.3 | 88.5 | 76.4 | 86.7 | 75.4 | 85.0 | 8.7 | 300 |

| Dehumidification DP/MCDB and HR | | | | | | | | | Enthalpy/MCDB | | | | | | Hours 8 to 4 & 55/69 |
|---|---|---|---|---|---|---|---|---|---|---|---|---|---|---|---|
| 0.4% | | | 1% | | | 2% | | | 0.4% | | 1% | | 2% | | |
| DP | HR | MCDB | DP | HR | MCDB | DP | HR | MCDB | Enth | MCDB | Enth | MCDB | Enth | MCDB | |
| (a) | (b) | (c) | (d) | (e) | (f) | (g) | (h) | (i) | (j) | (k) | (l) | (m) | (n) | (o) | (p) |
| 74.3 | 133.1 | 81.3 | 73.3 | 128.7 | 80.2 | 72.6 | 125.5 | 79.6 | 41.4 | 88.5 | 40.4 | 86.7 | 39.5 | 85.6 | 800 |

### Extreme Annual Design Conditions

| Extreme Annual WS | | | Extreme Max WB | Extreme Annual DB | | | | n-Year Return Period Values of Extreme DB | | | | | | | |
|---|---|---|---|---|---|---|---|---|---|---|---|---|---|---|---|
| | | | | Mean | | Standard deviation | | n=5 years | | n=10 years | | n=20 years | | n=50 years | |
| 1% | 2.5% | 5% | | Min | Max | Min | Max | Min | Max | Min | Max | Min | Max | Min | Max |
| (a) | (b) | (c) | (d) | (e) | (f) | (g) | (h) | (i) | (j) | (k) | (l) | (m) | (n) | (o) | (p) |
| 21.5 | 19.0 | 17.1 | 82.4 | 14.1 | 96.7 | 4.4 | 3.3 | 10.9 | 99.1 | 8.3 | 101.0 | 5.8 | 102.9 | 2.6 | 105.3 |

### Monthly Climatic Design Conditions

| | | | Annual | Jan | Feb | Mar | Apr | May | Jun | Jul | Aug | Sep | Oct | Nov | Dec |
|---|---|---|---|---|---|---|---|---|---|---|---|---|---|---|---|
| | | | (d) | (e) | (f) | (g) | (h) | (i) | (j) | (k) | (l) | (m) | (n) | (o) | (p) |
| (5) | Temperatures, Degree-Days and Degree-Hours | Tavg | 62.9 | 44.5 | 47.7 | 54.8 | 62.3 | 70.4 | 77.2 | 80.3 | 79.5 | 73.8 | 63.2 | 54.1 | 45.6 |
| (6) | | Sd | | 9.42 | 9.11 | 8.92 | 7.69 | 5.89 | 4.52 | 3.46 | 3.93 | 5.54 | 7.10 | 8.13 | 8.92 |
| (7) | | HDD50 | 672 | 222 | 138 | 55 | 8 | 0 | 0 | 0 | 0 | 0 | 4 | 52 | 193 |
| (8) | | HDD65 | 2671 | 637 | 484 | 329 | 135 | 22 | 1 | 0 | 0 | 8 | 118 | 335 | 602 |
| (9) | | CDD50 | 5370 | 50 | 75 | 204 | 378 | 634 | 817 | 938 | 916 | 713 | 414 | 174 | 57 |
| (10) | | CDD65 | 1893 | 0 | 1 | 12 | 55 | 190 | 368 | 473 | 451 | 271 | 63 | 8 | 1 |
| (11) | | CDH74 | 16504 | 0 | 5 | 98 | 467 | 1453 | 3331 | 4587 | 4215 | 1985 | 335 | 27 | 1 |
| (12) | | CDH80 | 6259 | 0 | 0 | 9 | 85 | 390 | 1340 | 2009 | 1760 | 627 | 38 | 1 | 0 |
| (13) | Precipitation | PrecAvg | 50.8 | 4.8 | 4.8 | 5.8 | 4.3 | 4.3 | 3.5 | 5.0 | 3.7 | 3.4 | 3.0 | 3.9 | 4.3 |
| (14) | | PrecMax | 64.9 | 10.2 | 12.8 | 11.7 | 11.9 | 8.4 | 7.4 | 8.5 | 8.7 | 6.1 | 7.5 | 7.2 | 9.9 |
| (15) | | PrecMin | 37.7 | 1.7 | 0.8 | 2.4 | 1.5 | 0.4 | 1.0 | 0.8 | 0.5 | 0.7 | 0.1 | 0.9 | 0.7 |
| (16) | | PrecSD | 7.2 | 2.1 | 2.8 | 2.7 | 2.4 | 2.3 | 1.8 | 2.2 | 2.2 | 1.6 | 2.1 | 1.6 | 2.4 |
| (17) | Monthly Design Dry Bulb and Mean Coincident Wet Bulb Temperatures | 0.4% DB | | 70.5 | 73.5 | 80.8 | 85.8 | 90.0 | 94.5 | 97.8 | 97.4 | 92.6 | 83.7 | 77.7 | 72.1 |
| (18) | | 0.4% MCWB | | 59.7 | 61.7 | 62.5 | 66.1 | 71.6 | 73.0 | 74.7 | 75.0 | 72.5 | 69.4 | 63.9 | 63.0 |
| (19) | | 2% DB | | 66.1 | 69.2 | 76.7 | 82.4 | 86.9 | 91.8 | 94.4 | 93.3 | 88.8 | 80.7 | 73.6 | 67.5 |
| (20) | | 2% MCWB | | 58.0 | 58.9 | 60.1 | 64.3 | 69.8 | 72.9 | 74.6 | 74.7 | 71.4 | 66.5 | 62.1 | 61.0 |
| (21) | | 5% DB | | 62.8 | 65.9 | 73.3 | 79.5 | 84.3 | 90.0 | 91.9 | 90.9 | 86.3 | 78.0 | 70.8 | 63.8 |
| (22) | | 5% MCWB | | 56.5 | 57.1 | 58.7 | 62.8 | 68.7 | 72.4 | 74.4 | 74.6 | 70.9 | 64.8 | 61.0 | 58.5 |
| (23) | | 10% DB | | 59.3 | 62.8 | 69.8 | 76.2 | 81.8 | 87.7 | 89.5 | 88.5 | 83.7 | 74.9 | 67.7 | 60.2 |
| (24) | | 10% MCWB | | 53.1 | 54.8 | 57.4 | 61.5 | 67.5 | 71.9 | 74.3 | 73.8 | 70.5 | 64.0 | 59.4 | 54.0 |
| (25) | Monthly Design Wet Bulb and Mean Coincident Dry Bulb Temperatures | 0.4% WB | | 64.0 | 65.4 | 66.5 | 70.4 | 75.0 | 77.3 | 78.8 | 78.4 | 76.3 | 72.7 | 69.3 | 66.1 |
| (26) | | 0.4% MCDB | | 67.3 | 67.6 | 73.1 | 79.1 | 83.5 | 88.6 | 89.7 | 90.1 | 85.9 | 80.0 | 72.3 | 69.2 |
| (27) | | 2% WB | | 61.0 | 62.6 | 64.1 | 68.1 | 72.8 | 75.8 | 77.4 | 77.3 | 74.6 | 70.4 | 66.5 | 63.1 |
| (28) | | 2% MCDB | | 64.0 | 66.6 | 71.5 | 76.2 | 82.3 | 86.5 | 88.5 | 88.7 | 83.0 | 76.1 | 70.4 | 65.9 |
| (29) | | 5% WB | | 58.0 | 59.9 | 62.1 | 66.2 | 71.3 | 74.9 | 76.5 | 76.3 | 73.5 | 68.8 | 64.0 | 59.9 |
| (30) | | 5% MCDB | | 61.5 | 64.1 | 69.4 | 74.1 | 80.4 | 84.9 | 86.9 | 86.5 | 81.1 | 73.5 | 68.0 | 63.5 |
| (31) | | 10% WB | | 54.6 | 56.4 | 59.9 | 64.2 | 69.9 | 73.8 | 75.4 | 75.3 | 72.6 | 67.0 | 61.3 | 55.3 |
| (32) | | 10% MCDB | | 58.0 | 60.7 | 66.4 | 72.2 | 78.2 | 83.0 | 85.1 | 84.5 | 79.7 | 72.0 | 65.9 | 58.4 |
| (33) | Mean Daily Temperature Range | MDBR | | 17.3 | 18.1 | 19.5 | 20.2 | 18.5 | 17.4 | 17.0 | 16.5 | 16.4 | 18.2 | 18.3 | 16.8 |
| (34) | | 5% DB MCDBR | | 20.4 | 21.0 | 23.0 | 22.8 | 20.3 | 20.2 | 20.5 | 19.4 | 19.1 | 20.3 | 20.8 | 19.8 |
| (35) | | 5% DB MCWBR | | 14.2 | 13.0 | 11.4 | 9.7 | 7.7 | 6.5 | 6.2 | 6.0 | 6.8 | 9.1 | 11.6 | 13.9 |
| (36) | | 5% WB MCDBR | | 16.7 | 17.2 | 18.0 | 18.3 | 17.3 | 17.3 | 17.7 | 17.1 | 15.6 | 14.7 | 16.4 | 17.1 |
| (37) | | 5% WB MCWBR | | 14.0 | 13.3 | 11.6 | 10.0 | 7.7 | 6.8 | 6.5 | 6.1 | 7.0 | 8.9 | 12.4 | 14.3 |
| (38) | Clear Sky Solar Irradiance | taub | | 0.334 | 0.324 | 0.355 | 0.383 | 0.379 | 0.406 | 0.440 | 0.427 | 0.388 | 0.358 | 0.354 | 0.335 |
| (39) | | taud | | 2.614 | 2.580 | 2.474 | 2.328 | 2.324 | 2.270 | 2.202 | 2.269 | 2.428 | 2.514 | 2.523 | 2.618 |
| (40) | | Ebn,noon | | 281 | 296 | 292 | 287 | 288 | 278 | 268 | 270 | 277 | 279 | 269 | 273 |
| (41) | | Edh,noon | | 24 | 28 | 33 | 40 | 41 | 43 | 46 | 42 | 34 | 29 | 26 | 23 |

Nomenclature: See separate page

*Fig. 4-4  Example tables of climatic design information included on CD-ROM of 2013 ASHRAE Handbook—Fundamentals*

# Design Conditions

2013 ASHRAE Handbook - Fundamentals (IP) © 2013 ASHRAE, Inc.

**BALTIMORE-WASHINGTO, MD, USA** WMO#: 724060

Lat: 39.17N  Long: 76.68W  Elev: 154  StdP: 14.61  Time Zone: -5 (NAE)  Period: 86-10  WBAN: 93721

### Annual Heating and Humidification Design Conditions

| Coldest Month | Heating DB | | Humidification DP/MCDB and HR | | | | | | Coldest month WS/MCDB | | | | MCWS/PCWD to 99.6% DB | |
|---|---|---|---|---|---|---|---|---|---|---|---|---|---|---|
| | 99.6% | 99% | 99.6% | | | 99% | | | 0.4% | | 1% | | | |
| | | | DP | HR | MCDB | DP | HR | MCDB | WS | MCDB | WS | MCDB | MCWS | PCWD |
| (a) | (b) | (c) | (d) | (e) | (f) | (g) | (h) | (i) | (j) | (k) | (l) | (m) | (n) | (o) |
| 1 | 14.0 | 17.9 | -2.5 | 4.8 | 19.0 | 1.4 | 5.9 | 22.3 | 26.1 | 35.6 | 24.0 | 34.4 | 8.6 | 290 |

### Annual Cooling, Dehumidification, and Enthalpy Design Conditions

| Hottest Month | Hottest Month DB Range | Cooling DB/MCWB | | | | | | Evaporation WB/MCDB | | | | | | MCWS/PCWD to 0.4% DB | |
|---|---|---|---|---|---|---|---|---|---|---|---|---|---|---|---|
| | | 0.4% | | 1% | | 2% | | 0.4% | | 1% | | 2% | | | |
| | | DB | MCWB | DB | MCWB | DB | MCWB | WB | MCDB | WB | MCDB | WB | MCDB | MCWS | PCWD |
| (a) | (b) | (c) | (d) | (e) | (f) | (g) | (h) | (i) | (j) | (k) | (l) | (m) | (n) | (o) | (p) |
| 7 | 18.6 | 94.0 | 74.9 | 91.3 | 74.1 | 88.7 | 73.1 | 78.1 | 88.6 | 76.8 | 86.6 | 75.7 | 84.3 | 9.9 | 280 |

| Dehumidification DP/MCDB and HR | | | | | | | | Enthalpy/MCDB | | | | | | Hours 8 to 4 & 55/69 |
|---|---|---|---|---|---|---|---|---|---|---|---|---|---|---|
| 0.4% | | | 1% | | | 2% | | 0.4% | | 1% | | 2% | | |
| DP | HR | MCDB | DP | HR | MCDB | DP | HR | MCDB | Enth | MCDB | Enth | MCDB | Enth | MCDB | |
| (a) | (b) | (c) | (d) | (e) | (f) | (g) | (h) | (i) | (j) | (k) | (l) | (m) | (n) | (o) | (p) |
| 75.3 | 133.2 | 82.1 | 74.1 | 127.8 | 80.7 | 73.0 | 123.1 | 79.7 | 41.5 | 89.1 | 40.2 | 86.6 | 39.1 | 84.6 | 708 |

### Extreme Annual Design Conditions

| Extreme Annual WS | | | Extreme Max WB | Extreme Annual DB | | | | n-Year Return Period Values of Extreme DB | | | | | | | |
|---|---|---|---|---|---|---|---|---|---|---|---|---|---|---|---|
| 1% | 2.5% | 5% | | Mean | | Standard deviation | | n=5 years | | n=10 years | | n=20 years | | n=50 years | |
| | | | | Min | Max | Min | Max | Min | Max | Min | Max | Min | Max | Min | Max |
| (a) | (b) | (c) | (d) | (e) | (f) | (g) | (h) | (i) | (j) | (k) | (l) | (m) | (n) | (o) | (p) |
| 22.3 | 19.1 | 17.1 | 84.6 | 6.9 | 98.2 | 4.6 | 3.5 | 3.6 | 100.8 | 0.9 | 102.8 | -1.7 | 104.8 | -5.0 | 107.4 |

### Monthly Climatic Design Conditions

| | | | Annual | Jan | Feb | Mar | Apr | May | Jun | Jul | Aug | Sep | Oct | Nov | Dec |
|---|---|---|---|---|---|---|---|---|---|---|---|---|---|---|---|
| | | | (d) | (e) | (f) | (g) | (h) | (i) | (j) | (k) | (l) | (m) | (n) | (o) | (p) |
| | | Tavg | 56.0 | 34.6 | 36.3 | 44.6 | 54.5 | 63.8 | 73.2 | 77.8 | 75.9 | 68.6 | 56.7 | 47.2 | 37.4 |
| | | Sd | | 9.89 | 8.63 | 9.45 | 8.30 | 7.61 | 6.35 | 5.05 | 5.20 | 6.58 | 7.82 | 8.33 | 8.68 |
| Temperatures, | | HDD50 | 1717 | 487 | 391 | 223 | 42 | 1 | 0 | 0 | 0 | 0 | 22 | 151 | 400 |
| Degree-Days | | HDD65 | 4552 | 942 | 804 | 636 | 331 | 118 | 11 | 0 | 1 | 39 | 280 | 536 | 854 |
| and | | CDD50 | 3902 | 10 | 7 | 55 | 178 | 428 | 697 | 861 | 802 | 558 | 228 | 67 | 11 |
| Degree-Hours | | CDD65 | 1261 | 0 | 0 | 3 | 16 | 80 | 258 | 396 | 339 | 146 | 22 | 1 | 0 |
| | | CDH74 | 11735 | 0 | 0 | 43 | 218 | 826 | 2422 | 3941 | 3053 | 1055 | 168 | 8 | 1 |
| | | CDH80 | 4507 | 0 | 0 | 8 | 65 | 282 | 943 | 1711 | 1172 | 295 | 31 | 0 | 0 |
| | | PrecAvg | 41.4 | 3.1 | 3.0 | 3.7 | 3.2 | 3.8 | 3.6 | 3.9 | 3.7 | 3.8 | 3.2 | 3.2 | 3.5 |
| Precipitation | | PrecMax | 54.1 | 7.8 | 7.2 | 5.5 | 6.4 | 7.1 | 10.0 | 8.2 | 10.9 | 8.6 | 8.1 | 7.1 | 7.4 |
| | | PrecMin | 29.2 | 0.9 | 0.6 | 0.9 | 1.3 | 0.4 | 1.2 | 0.7 | 1.3 | 0.2 | 0.7 | 0.6 | 0.6 |
| | | PrecSD | 7.3 | 1.8 | 2.0 | 1.3 | 1.5 | 1.8 | 2.3 | 2.0 | 2.5 | 2.7 | 2.0 | 2.0 | 1.9 |
| | 0.4% | DB | | 66.0 | 68.6 | 80.1 | 87.4 | 91.0 | 95.2 | 98.7 | 96.8 | 91.2 | 84.0 | 74.6 | 67.0 |
| Monthly Design | | MCWB | | 58.1 | 57.2 | 62.4 | 66.7 | 71.7 | 74.8 | 76.5 | 75.7 | 71.6 | 69.7 | 62.3 | 59.7 |
| Dry Bulb | 2% | DB | | 60.8 | 61.3 | 71.7 | 80.6 | 87.2 | 91.7 | 95.0 | 92.7 | 86.7 | 78.8 | 69.4 | 60.7 |
| and | | MCWB | | 55.1 | 53.3 | 58.2 | 62.7 | 69.6 | 74.0 | 75.7 | 74.9 | 71.5 | 66.5 | 59.7 | 54.5 |
| Mean Coincident | 5% | DB | | 54.4 | 54.9 | 65.5 | 74.6 | 82.8 | 88.8 | 91.8 | 89.7 | 83.2 | 74.5 | 64.9 | 55.4 |
| Wet Bulb | | MCWB | | 48.4 | 47.0 | 54.5 | 60.4 | 68.2 | 73.0 | 74.8 | 74.0 | 70.1 | 64.3 | 58.1 | 50.3 |
| Temperatures | 10% | DB | | 48.4 | 49.9 | 60.4 | 70.0 | 78.5 | 85.6 | 88.8 | 86.3 | 80.2 | 70.6 | 61.2 | 50.5 |
| | | MCWB | | 43.1 | 43.8 | 51.1 | 57.7 | 65.8 | 72.0 | 73.9 | 72.7 | 68.7 | 62.9 | 55.0 | 45.1 |
| | 0.4% | WB | | 60.4 | 60.0 | 64.7 | 68.9 | 74.9 | 78.8 | 80.2 | 79.5 | 76.9 | 72.2 | 66.3 | 61.3 |
| Monthly Design | | MCDB | | 63.6 | 65.3 | 76.9 | 80.6 | 86.1 | 88.4 | 91.5 | 90.0 | 84.2 | 78.3 | 69.9 | 66.3 |
| Wet Bulb | 2% | WB | | 56.0 | 53.6 | 60.2 | 65.7 | 72.3 | 76.6 | 78.3 | 77.6 | 74.9 | 70.0 | 62.9 | 56.5 |
| and | | MCDB | | 59.9 | 58.9 | 69.0 | 76.3 | 83.2 | 86.4 | 89.5 | 88.0 | 81.1 | 76.1 | 67.0 | 59.6 |
| Mean Coincident | 5% | WB | | 49.6 | 48.4 | 56.3 | 62.5 | 69.9 | 75.3 | 77.1 | 76.3 | 73.2 | 67.0 | 59.6 | 50.9 |
| Dry Bulb | | MCDB | | 54.0 | 53.8 | 63.0 | 71.8 | 79.8 | 84.7 | 87.5 | 85.1 | 78.8 | 72.0 | 63.8 | 54.1 |
| Temperatures | 10% | WB | | 44.0 | 44.2 | 52.3 | 59.7 | 67.6 | 73.8 | 75.9 | 75.0 | 71.6 | 63.8 | 56.1 | 45.9 |
| | | MCDB | | 47.7 | 49.4 | 59.2 | 67.9 | 76.0 | 82.1 | 85.1 | 82.6 | 77.2 | 69.0 | 60.4 | 49.7 |
| | | MDBR | | 15.7 | 16.8 | 18.9 | 20.3 | 20.2 | 19.7 | 18.6 | 18.1 | 18.4 | 20.1 | 18.4 | 15.7 |
| Mean Daily | 5% DB | MCDBR | | 22.9 | 25.3 | 27.2 | 28.0 | 26.1 | 23.1 | 22.2 | 21.6 | 21.9 | 24.2 | 23.4 | 22.5 |
| Temperature | | MCWBR | | 17.3 | 17.8 | 16.6 | 14.4 | 12.1 | 9.3 | 7.8 | 8.1 | 9.7 | 13.6 | 16.1 | 17.8 |
| Range | 5% WB | MCDBR | | 21.4 | 22.3 | 24.2 | 23.8 | 22.7 | 19.5 | 18.9 | 18.5 | 17.1 | 19.5 | 19.6 | 19.6 |
| | | MCWBR | | 18.1 | 17.8 | 16.9 | 13.8 | 11.7 | 9.0 | 8.0 | 8.0 | 9.1 | 12.6 | 16.3 | 17.7 |
| | | taub | 0.320 | 0.333 | 0.364 | 0.406 | 0.419 | 0.477 | 0.455 | 0.440 | 0.405 | 0.353 | 0.331 | 0.325 |
| Clear Sky | | taud | 2.486 | 2.458 | 2.396 | 2.232 | 2.170 | 2.009 | 2.129 | 2.186 | 2.323 | 2.473 | 2.547 | 2.534 |
| Solar Irradiance | | Ebn,noon | 270 | 282 | 283 | 276 | 273 | 257 | 262 | 263 | 265 | 271 | 265 | 260 |
| | | Edh,noon | 25 | 29 | 34 | 43 | 47 | 55 | 49 | 45 | 37 | 28 | 23 | 22 |

Nomenclature: See separate page

*Fig. 4-4 Example tables of climatic design information included on CD-ROM of 2013 ASHRAE Handbook—Fundamentals*

2013 ASHRAE Handbook - Fundamentals (IP)  © 2013 ASHRAE, Inc.

**BOISE MUNICIPAL, ID, USA**  WMO#: 726810

Lat: 43.57N  Long: 116.22W  Elev: 2867  StdP: 13.24  Time Zone: -7 (NAM)  Period: 86-10  WBAN: 24131

### Annual Heating and Humidification Design Conditions

| Coldest Month | Heating DB | | Humidification DP/MCDB and HR | | | | | | Coldest month WS/MCDB | | | | MCWS/PCWD to 99.6% DB | |
|---|---|---|---|---|---|---|---|---|---|---|---|---|---|---|
| | 99.6% | 99% | 99.6% | | | 99% | | | 0.4% | | 1% | | | |
| | | | DP | HR | MCDB | DP | HR | MCDB | WS | MCDB | WS | MCDB | MCWS | PCWD |
| (a) | (b) | (c) | (d) | (e) | (f) | (g) | (h) | (i) | (j) | (k) | (l) | (m) | (n) | (o) |
| 12 | 8.7 | 15.5 | 1.8 | 6.7 | 12.2 | 8.1 | 9.3 | 20.9 | 24.6 | 41.7 | 22.2 | 41.2 | 5.1 | 120 |

### Annual Cooling, Dehumidification, and Enthalpy Design Conditions

| Hottest Month | Hottest Month DB Range | Cooling DB/MCWB | | | | | | Evaporation WB/MCDB | | | | | | MCWS/PCWD to 0.4% DB | |
|---|---|---|---|---|---|---|---|---|---|---|---|---|---|---|---|
| | | 0.4% | | 1% | | 2% | | 0.4% | | 1% | | 2% | | | |
| | | DB | MCWB | DB | MCWB | DB | MCWB | WB | MCDB | WB | MCDB | WB | MCDB | MCWS | PCWD |
| (a) | (b) | (c) | (d) | (e) | (f) | (g) | (h) | (i) | (j) | (k) | (l) | (m) | (n) | (o) | (p) |
| 7 | 30.0 | 98.6 | 63.9 | 95.4 | 62.9 | 92.5 | 61.9 | 66.2 | 92.3 | 64.7 | 90.5 | 63.4 | 88.5 | 9.1 | 330 |

| Dehumidification DP/MCDB and HR | | | | | | | | | Enthalpy/MCDB | | | | | | Hours 8 to 4 & 55/69 |
|---|---|---|---|---|---|---|---|---|---|---|---|---|---|---|---|
| 0.4% | | | 1% | | | 2% | | | 0.4% | | 1% | | 2% | | |
| DP | HR | MCDB | DP | HR | MCDB | DP | HR | MCDB | Enth | MCDB | Enth | MCDB | Enth | MCDB | |
| (a) | (b) | (c) | (d) | (e) | (f) | (g) | (h) | (i) | (j) | (k) | (l) | (m) | (n) | (o) | (p) |
| 57.2 | 77.5 | 71.6 | 54.9 | 71.3 | 71.4 | 53.0 | 66.4 | 70.8 | 32.2 | 92.1 | 31.1 | 90.2 | 30.0 | 88.2 | 686 |

### Extreme Annual Design Conditions

| Extreme Annual WS | | | Extreme Max WB | Extreme Annual DB | | | | n-Year Return Period Values of Extreme DB | | | | | | | |
|---|---|---|---|---|---|---|---|---|---|---|---|---|---|---|---|
| 1% | 2.5% | 5% | | Mean | | Standard deviation | | n=5 years | | n=10 years | | n=20 years | | n=50 years | |
| | | | | Min | Max | Min | Max | Min | Max | Min | Max | Min | Max | Min | Max |
| (a) | (b) | (c) | (d) | (e) | (f) | (g) | (h) | (i) | (j) | (k) | (l) | (m) | (n) | (o) | (p) |
| 21.9 | 19.0 | 17.1 | 73.0 | 3.5 | 104.2 | 9.7 | 2.3 | -3.5 | 105.8 | -9.1 | 107.2 | -14.6 | 108.4 | -21.6 | 110.1 |

### Monthly Climatic Design Conditions

| | | Annual | Jan | Feb | Mar | Apr | May | Jun | Jul | Aug | Sep | Oct | Nov | Dec |
|---|---|---|---|---|---|---|---|---|---|---|---|---|---|---|
| | Tavg | 52.7 | 31.6 | 36.7 | 44.6 | 50.8 | 59.1 | 67.6 | 76.0 | 74.5 | 65.4 | 53.1 | 40.1 | 31.4 |
| | Sd | | 7.92 | 8.05 | 6.78 | 7.42 | 8.53 | 8.06 | 6.90 | 6.80 | 7.94 | 8.07 | 7.87 | 8.74 |
| Temperatures, | HDD50 | 2175 | 571 | 373 | 195 | 78 | 16 | 0 | 0 | 0 | 2 | 57 | 307 | 576 |
| Degree-Days | HDD65 | 5453 | 1035 | 792 | 633 | 428 | 225 | 64 | 8 | 11 | 92 | 378 | 746 | 1041 |
| and | CDD50 | 3152 | 0 | 1 | 26 | 103 | 299 | 528 | 806 | 761 | 464 | 152 | 11 | 1 |
| Degree-Hours | CDD65 | 957 | 0 | 0 | 0 | 4 | 44 | 142 | 348 | 306 | 104 | 9 | 0 | 0 |
| | CDH74 | 13211 | 0 | 0 | 2 | 90 | 680 | 1990 | 4879 | 4053 | 1369 | 148 | 0 | 0 |
| | CDH80 | 6612 | 0 | 0 | 0 | 16 | 247 | 922 | 2718 | 2130 | 545 | 34 | 0 | 0 |
| | PrecAvg | 12.1 | 1.5 | 1.1 | 1.3 | 1.2 | 1.1 | 0.8 | 0.4 | 0.4 | 0.8 | 0.8 | 1.5 | 1.4 |
| Precipitation | PrecMax | 16.1 | 3.9 | 2.6 | 2.1 | 2.8 | 3.8 | 2.0 | 1.1 | 2.4 | 2.1 | 2.0 | 2.4 | 3.2 |
| | PrecMin | 9.7 | 0.4 | 0.2 | 0.2 | 0.2 | 0.1 | 0.1 | 0.0 | 0.0 | 0.0 | 0.2 | 0.2 | 0.1 |
| | PrecSD | 1.9 | 1.0 | 0.6 | 0.6 | 0.7 | 0.9 | 0.6 | 0.3 | 0.6 | 0.5 | 0.5 | 0.7 | 0.8 |
| | 0.4% DB | 53.2 | 61.2 | 71.4 | 82.0 | 91.9 | 98.4 | 103.4 | 101.3 | 94.6 | 84.3 | 65.6 | 55.8 | |
| Monthly Design | 0.4% MCWB | 44.2 | 48.5 | 50.8 | 56.3 | 61.0 | 63.5 | 65.4 | 64.6 | 61.9 | 57.0 | 49.0 | 45.0 | |
| Dry Bulb | 2% DB | 48.3 | 55.0 | 66.1 | 76.0 | 86.8 | 93.8 | 99.3 | 97.5 | 90.1 | 78.1 | 59.4 | 50.8 | |
| and | 2% MCWB | 42.1 | 44.8 | 48.9 | 54.1 | 59.7 | 61.8 | 64.1 | 63.4 | 60.9 | 55.1 | 47.2 | 43.6 | |
| Mean Coincident | 5% DB | 45.1 | 51.3 | 61.5 | 71.0 | 81.8 | 89.9 | 96.6 | 94.5 | 86.3 | 72.8 | 55.0 | 46.3 | |
| Wet Bulb | 5% MCWB | 39.5 | 42.6 | 46.5 | 51.8 | 57.2 | 61.0 | 63.6 | 62.5 | 59.3 | 52.8 | 45.2 | 40.7 | |
| Temperatures | 10% DB | 42.4 | 47.9 | 57.2 | 65.6 | 76.6 | 85.4 | 93.1 | 91.2 | 82.3 | 67.6 | 51.7 | 42.7 | |
| | 10% MCWB | 37.8 | 40.9 | 44.8 | 49.9 | 55.7 | 59.6 | 62.4 | 61.4 | 57.4 | 50.6 | 43.7 | 38.2 | |
| | 0.4% WB | 46.2 | 49.9 | 52.2 | 58.2 | 63.7 | 66.4 | 67.3 | 66.0 | 58.8 | 52.3 | 46.7 | | |
| Monthly Design | 0.4% MCDB | 50.6 | 59.1 | 66.8 | 78.3 | 86.1 | 90.8 | 94.6 | 95.2 | 86.5 | 79.3 | 60.5 | 53.2 | |
| Wet Bulb | 2% WB | 42.4 | 45.6 | 50.0 | 55.2 | 60.7 | 64.1 | 66.4 | 65.6 | 62.6 | 56.2 | 49.2 | 44.0 | |
| and | 2% MCDB | 47.4 | 53.4 | 63.3 | 72.8 | 81.8 | 87.6 | 93.0 | 92.7 | 84.7 | 74.9 | 56.6 | 50.0 | |
| Mean Coincident | 5% WB | 40.2 | 43.6 | 47.9 | 53.0 | 58.7 | 62.3 | 65.1 | 64.1 | 60.4 | 53.7 | 46.8 | 41.4 | |
| Dry Bulb | 5% MCDB | 44.5 | 49.9 | 59.6 | 68.1 | 78.3 | 85.3 | 91.6 | 90.2 | 82.7 | 70.7 | 53.8 | 46.1 | |
| Temperatures | 10% WB | 38.1 | 41.5 | 45.9 | 50.8 | 56.7 | 60.4 | 63.7 | 62.5 | 58.7 | 51.5 | 44.3 | 38.7 | |
| | 10% MCDB | 41.8 | 47.2 | 56.1 | 63.5 | 74.1 | 81.9 | 89.8 | 87.6 | 80.2 | 65.6 | 50.7 | 42.4 | |
| | MDBR | 12.8 | 16.3 | 20.0 | 22.4 | 24.4 | 26.6 | 30.0 | 29.1 | 27.1 | 23.4 | 16.0 | 12.9 | |
| Mean Daily | 5% DB MCDBR | 14.5 | 20.4 | 26.4 | 30.3 | 31.6 | 32.3 | 32.9 | 32.0 | 31.1 | 29.5 | 20.2 | 15.0 | |
| Temperature | 5% DB MCWBR | 9.5 | 12.2 | 13.7 | 13.7 | 12.4 | 11.4 | 12.0 | 12.4 | 13.1 | 13.6 | 11.4 | 9.8 | |
| Range | 5% WB MCDBR | 13.8 | 19.3 | 23.3 | 27.1 | 28.9 | 29.6 | 30.3 | 29.3 | 28.9 | 27.1 | 17.8 | 13.4 | |
| | 5% WB MCWBR | 9.3 | 11.9 | 12.6 | 12.8 | 11.9 | 10.7 | 11.2 | 11.9 | 12.3 | 13.0 | 10.7 | 9.2 | |
| | taub | 0.271 | 0.276 | 0.291 | 0.302 | 0.302 | 0.307 | 0.307 | 0.328 | 0.312 | 0.298 | 0.287 | 0.278 | |
| Clear Sky | taud | 2.520 | 2.549 | 2.534 | 2.479 | 2.461 | 2.451 | 2.465 | 2.423 | 2.530 | 2.589 | 2.635 | 2.569 | |
| Solar Irradiance | Ebn,noon | 279 | 297 | 304 | 306 | 307 | 304 | 304 | 294 | 292 | 284 | 272 | 266 | |
| | Edh,noon | 21 | 25 | 29 | 33 | 34 | 35 | 34 | 34 | 28 | 24 | 19 | 19 | |

Nomenclature: See separate page

*Fig. 4-4 Example tables of climatic design information included on CD-ROM of 2013 ASHRAE Handbook—Fundamentals*

# Design Conditions

2013 ASHRAE Handbook - Fundamentals (IP) © 2013 ASHRAE, Inc.

**CAMP PENDLETON MCAS, CA, USA** WMO#: 722926

Lat: 33.30N  Long: 117.35W  Elev: 79  StdP: 14.65  Time Zone: -8 (NAP)  Period: 87-10  WBAN: 3154

### Annual Heating and Humidification Design Conditions

| Coldest Month | Heating DB |  | Humidification DP/MCDB and HR |  |  |  |  |  | Coldest month WS/MCDB |  |  |  | MCWS/PCWD to 99.6% DB |  |
|---|---|---|---|---|---|---|---|---|---|---|---|---|---|---|
|  | 99.6% | 99% | 99.6% |  |  | 99% |  |  | 0.4% |  | 1% |  | MCWS | PCWD |
|  |  |  | DP | HR | MCDB | DP | HR | MCDB | WS | MCDB | WS | MCDB |  |  |
| (a) | (b) | (c) | (d) | (e) | (f) | (g) | (h) | (i) | (j) | (k) | (l) | (m) | (n) | (o) |
| 12 | 32.3 | 35.5 | 11.6 | 10.0 | 56.7 | 17.8 | 13.5 | 57.2 | 20.5 | 60.9 | 17.8 | 60.9 | 0.7 | 30 |

### Annual Cooling, Dehumidification, and Enthalpy Design Conditions

| Hottest Month | Hottest Month DB Range | Cooling DB/MCWB |  |  |  |  |  | Evaporation WB/MCDB |  |  |  |  |  | MCWS/PCWD to 0.4% DB |  |
|---|---|---|---|---|---|---|---|---|---|---|---|---|---|---|---|
|  |  | 0.4% |  | 1% |  | 2% |  | 0.4% |  | 1% |  | 2% |  | MCWS | PCWD |
|  |  | DB | MCWB | DB | MCWB | DB | MCWB | WB | MCDB | WB | MCDB | WB | MCDB |  |  |
| (a) | (b) | (c) | (d) | (e) | (f) | (g) | (h) | (i) | (j) | (k) | (l) | (m) | (n) | (o) | (p) |
| 8 | 20.1 | 92.1 | 66.2 | 88.0 | 65.7 | 84.1 | 65.5 | 71.6 | 83.6 | 70.3 | 82.0 | 69.1 | 80.4 | 9.1 | 210 |

| Dehumidification DP/MCDB and HR |  |  |  |  |  |  |  |  | Enthalpy/MCDB |  |  |  |  |  | Hours 8 to 4 & 55/69 |
|---|---|---|---|---|---|---|---|---|---|---|---|---|---|---|---|
| 0.4% |  |  | 1% |  |  | 2% |  |  | 0.4% |  | 1% |  | 2% |  |  |
| DP | HR | MCDB | DP | HR | MCDB | DP | HR | MCDB | Enth | MCDB | Enth | MCDB | Enth | MCDB |  |
| (a) | (b) | (c) | (d) | (e) | (f) | (g) | (h) | (i) | (j) | (k) | (l) | (m) | (n) | (o) | (p) |
| 67.7 | 102.2 | 76.7 | 66.0 | 96.0 | 76.1 | 64.4 | 90.8 | 74.1 | 35.4 | 83.7 | 34.2 | 82.3 | 33.2 | 80.6 | 1349 |

### Extreme Annual Design Conditions

| Extreme Annual WS |  |  | Extreme Max WB | Extreme Annual DB |  |  |  | n-Year Return Period Values of Extreme DB |  |  |  |  |  |  |  |
|---|---|---|---|---|---|---|---|---|---|---|---|---|---|---|---|
| 1% | 2.5% | 5% |  | Mean |  | Standard deviation |  | n=5 years |  | n=10 years |  | n=20 years |  | n=50 years |  |
|  |  |  |  | Min | Max | Min | Max | Min | Max | Min | Max | Min | Max | Min | Max |
| (a) | (b) | (c) | (d) | (e) | (f) | (g) | (h) | (i) | (j) | (k) | (l) | (m) | (n) | (o) | (p) |
| 16.8 | 14.4 | 12.5 | 87.1 | 27.2 | 101.9 | 3.7 | 3.2 | 24.5 | 104.2 | 22.4 | 106.1 | 20.3 | 107.9 | 17.7 | 110.2 |

### Monthly Climatic Design Conditions

|  |  |  | Annual | Jan | Feb | Mar | Apr | May | Jun | Jul | Aug | Sep | Oct | Nov | Dec |
|---|---|---|---|---|---|---|---|---|---|---|---|---|---|---|---|
|  |  |  | (a) | (b) | (c) | (d) | (e) | (f) | (g) | (h) | (i) | (j) | (k) | (l) | (m) |
| Temperatures, Degree-Days and Degree-Hours | | Tavg | 62.1 | 54.6 | 54.8 | 57.0 | 59.4 | 63.2 | 66.5 | 70.1 | 71.5 | 69.8 | 64.8 | 58.9 | 53.8 |
| | | Sd |  | 5.40 | 5.01 | 4.45 | 5.09 | 3.81 | 3.63 | 3.76 | 3.53 | 4.73 | 4.86 | 5.16 | 4.95 |
| | | HDD50 | 52 | 15 | 11 | 3 | 1 | 0 | 0 | 0 | 0 | 0 | 0 | 3 | 19 |
| | | HDD65 | 1764 | 326 | 288 | 250 | 179 | 80 | 23 | 3 | 1 | 9 | 63 | 195 | 347 |
| | | CDD50 | 4462 | 157 | 145 | 220 | 284 | 410 | 494 | 623 | 668 | 595 | 460 | 268 | 138 |
| | | CDD65 | 695 | 3 | 2 | 2 | 11 | 25 | 66 | 161 | 203 | 154 | 57 | 10 | 1 |
| | | CDH74 | 6525 | 179 | 118 | 157 | 342 | 230 | 400 | 975 | 1441 | 1343 | 801 | 421 | 118 |
| | | CDH80 | 2082 | 36 | 30 | 46 | 140 | 68 | 105 | 220 | 406 | 522 | 335 | 151 | 23 |
| Precipitation | | PrecAvg | 12.4 | 2.5 | 2.4 | 2.2 | 0.9 | 0.2 | 0.1 | 0.1 | 0.1 | 0.3 | 0.3 | 1.5 | 1.7 |
| | | PrecMax | 87.6 | 6.4 | 6.4 | 6.3 | 6.9 | 7.6 | 8.1 | 8.9 | 9.2 | 9.3 | 8.0 | 7.2 | 6.3 |
| | | PrecMin | 73.0 | 4.2 | 0.4 | 0.0 | 0.1 | 0.0 | 6.5 | 0.1 | 6.3 | 6.3 | 6.1 | 5.4 | 3.9 |
| | | PrecSD | 3.7 | 0.5 | 1.2 | 1.4 | 1.8 | 2.0 | 0.5 | 1.8 | 0.7 | 0.7 | 0.6 | 0.5 | 0.7 |
| Monthly Design Dry Bulb and Mean Coincident Wet Bulb Temperatures | 0.4% | DB |  | 84.1 | 84.2 | 86.0 | 92.6 | 89.7 | 90.4 | 89.8 | 93.4 | 99.3 | 96.6 | 91.1 | 83.0 |
| | | MCWB |  | 57.4 | 59.1 | 59.9 | 60.6 | 62.0 | 67.8 | 70.5 | 68.9 | 68.4 | 64.7 | 60.2 | 56.1 |
| | 2% | DB |  | 79.3 | 77.9 | 78.6 | 82.9 | 79.4 | 81.6 | 84.4 | 87.9 | 90.9 | 88.4 | 84.1 | 77.1 |
| | | MCWB |  | 55.9 | 56.6 | 58.1 | 60.9 | 64.7 | 67.2 | 69.2 | 69.5 | 69.4 | 62.4 | 58.3 | 54.3 |
| | 5% | DB |  | 74.0 | 71.6 | 72.5 | 75.8 | 75.1 | 77.9 | 81.9 | 83.9 | 85.5 | 82.2 | 78.8 | 72.4 |
| | | MCWB |  | 54.5 | 55.6 | 58.2 | 60.3 | 63.5 | 65.3 | 68.8 | 68.8 | 68.1 | 62.0 | 56.7 | 54.1 |
| | 10% | DB |  | 68.0 | 66.0 | 67.9 | 70.9 | 72.4 | 75.1 | 79.4 | 81.2 | 81.3 | 77.1 | 73.0 | 67.7 |
| | | MCWB |  | 54.1 | 54.9 | 58.0 | 59.6 | 62.3 | 64.3 | 67.4 | 68.3 | 67.4 | 63.0 | 56.6 | 53.1 |
| Monthly Design Wet Bulb and Mean Coincident Dry Bulb Temperatures | 0.4% | WB |  | 61.9 | 62.0 | 65.5 | 65.6 | 68.3 | 70.4 | 73.1 | 73.2 | 74.0 | 69.9 | 64.1 | 61.7 |
| | | MCDB |  | 72.7 | 76.5 | 73.1 | 80.7 | 78.2 | 81.5 | 82.9 | 88.2 | 87.4 | 84.1 | 75.9 | 71.7 |
| | 2% | WB |  | 59.6 | 59.5 | 62.0 | 63.4 | 66.0 | 68.1 | 71.1 | 71.3 | 71.3 | 67.5 | 62.2 | 59.1 |
| | | MCDB |  | 70.2 | 68.8 | 72.7 | 76.7 | 76.2 | 79.7 | 81.9 | 83.5 | 84.2 | 80.6 | 74.2 | 67.8 |
| | 5% | WB |  | 58.0 | 58.0 | 59.8 | 61.6 | 64.3 | 66.5 | 69.4 | 70.0 | 69.7 | 65.8 | 60.6 | 57.3 |
| | | MCDB |  | 67.0 | 66.5 | 68.3 | 73.5 | 73.9 | 77.0 | 80.4 | 81.6 | 82.0 | 76.5 | 72.0 | 65.9 |
| | 10% | WB |  | 56.3 | 56.5 | 58.4 | 59.8 | 62.7 | 65.0 | 68.2 | 68.9 | 68.3 | 64.3 | 59.2 | 55.5 |
| | | MCDB |  | 65.2 | 64.3 | 66.1 | 69.8 | 71.1 | 74.0 | 78.4 | 79.7 | 79.8 | 73.6 | 69.6 | 64.5 |
| Mean Daily Temperature Range | | MDBR |  | 27.4 | 22.9 | 21.8 | 21.9 | 18.1 | 17.5 | 18.5 | 20.1 | 23.0 | 25.5 | 29.2 | 28.7 |
| | 5% DB | MCDBR |  | 38.4 | 35.4 | 33.7 | 35.2 | 24.7 | 23.9 | 23.0 | 25.9 | 33.3 | 39.1 | 40.3 | 38.1 |
| | | MCWBR |  | 19.9 | 19.3 | 18.9 | 18.1 | 13.7 | 12.6 | 11.6 | 12.8 | 16.0 | 19.7 | 20.8 | 21.3 |
| | 5% WB | MCDBR |  | 26.9 | 23.8 | 23.9 | 26.6 | 21.5 | 21.8 | 20.9 | 21.9 | 26.6 | 26.4 | 29.1 | 27.4 |
| | | MCWBR |  | 16.5 | 14.9 | 15.2 | 15.3 | 12.6 | 11.8 | 10.8 | 11.3 | 13.9 | 15.3 | 17.5 | 17.7 |
| Clear Sky Solar Irradiance | | taub |  | 0.334 | 0.329 | 0.336 | 0.338 | 0.338 | 0.335 | 0.351 | 0.363 | 0.365 | 0.360 | 0.332 | 0.331 |
| | | taud |  | 2.607 | 2.601 | 2.527 | 2.469 | 2.438 | 2.430 | 2.420 | 2.437 | 2.474 | 2.479 | 2.609 | 2.617 |
| | | Ebn,noon |  | 281 | 295 | 299 | 301 | 300 | 299 | 294 | 290 | 285 | 278 | 279 | 276 |
| | | Edh,noon |  | 25 | 27 | 32 | 35 | 36 | 37 | 37 | 36 | 33 | 30 | 24 | 23 |

Nomenclature:  See separate page

*Fig. 4-4  Example tables of climatic design information included on CD-ROM of 2013 ASHRAE Handbook—Fundamentals*

**94**                                                      **Principles of HVAC**

2013 ASHRAE Handbook - Fundamentals (IP)            © 2013 ASHRAE, Inc.

**CEDAR RAPIDS MUNI, IA, USA**       WMO#: 725450

Lat: **41.88N**    Long: **91.71W**    Elev: **873**    StdP: **14.24**    Time Zone: **-6 (NAC)**    Period: **86-10**    WBAN: **14990**

### Annual Heating and Humidification Design Conditions

| Coldest Month | Heating DB | | Humidification DP/MCDB and HR | | | | | | Coldest month WS/MCDB | | | | MCWS/PCWD to 99.6% DB | |
|---|---|---|---|---|---|---|---|---|---|---|---|---|---|---|
| | 99.6% | 99% | 99.6% | | | 99% | | | 0.4% | | 1% | | | |
| | | | DP | HR | MCDB | DP | HR | MCDB | WS | MCDB | WS | MCDB | MCWS | PCWD |
| (a) | (b) | (c) | (d) | (e) | (f) | (g) | (h) | (i) | (j) | (k) | (l) | (m) | (n) | (o) |
| 1 | -8.4 | -2.8 | -14.9 | 2.5 | -7.8 | -9.6 | 3.4 | -2.6 | 32.3 | 18.5 | 28.4 | 18.4 | 10.1 | 300 |

### Annual Cooling, Dehumidification, and Enthalpy Design Conditions

| Hottest Month | Hottest Month DB Range | Cooling DB/MCWB | | | | | | Evaporation WB/MCDB | | | | | | MCWS/PCWD to 0.4% DB | |
|---|---|---|---|---|---|---|---|---|---|---|---|---|---|---|---|
| | | 0.4% | | 1% | | 2% | | 0.4% | | 1% | | 2% | | | |
| | | DB | MCWB | DB | MCWB | DB | MCWB | WB | MCDB | WB | MCDB | WB | MCDB | MCWS | PCWD |
| (a) | (b) | (c) | (d) | (e) | (f) | (g) | (h) | (i) | (j) | (k) | (l) | (m) | (n) | (o) | (p) |
| 7 | 19.2 | 91.0 | 76.0 | 88.0 | 74.4 | 85.1 | 72.6 | 78.5 | 86.9 | 76.9 | 85.0 | 75.1 | 82.5 | 10.9 | 190 |

| Dehumidification DP/MCDB and HR | | | | | | | | | Enthalpy/MCDB | | | | | | Hours 8 to 4 & 55/69 |
|---|---|---|---|---|---|---|---|---|---|---|---|---|---|---|---|
| 0.4% | | | 1% | | | 2% | | | 0.4% | | 1% | | 2% | | |
| DP | HR | MCDB | DP | HR | MCDB | DP | HR | MCDB | Enth | MCDB | Enth | MCDB | Enth | MCDB | |
| (a) | (b) | (c) | (d) | (e) | (f) | (g) | (h) | (i) | (j) | (k) | (l) | (m) | (n) | (o) | (p) |
| 76.2 | 141.3 | 83.9 | 74.3 | 132.4 | 81.9 | 72.7 | 125.1 | 80.0 | 42.7 | 87.1 | 40.8 | 84.7 | 39.1 | 82.7 | 631 |

### Extreme Annual Design Conditions

| Extreme Annual WS | | | Extreme Max WB | Extreme Annual DB | | | | n-Year Return Period Values of Extreme DB | | | | | | | |
|---|---|---|---|---|---|---|---|---|---|---|---|---|---|---|---|
| 1% | 2.5% | 5% | | Mean | | Standard deviation | | n=5 years | | n=10 years | | n=20 years | | n=50 years | |
| | | | | Min | Max | Min | Max | Min | Max | Min | Max | Min | Max | Min | Max |
| (a) | (b) | (c) | (d) | (e) | (f) | (g) | (h) | (i) | (j) | (k) | (l) | (m) | (n) | (o) | (p) |
| 26.5 | 23.6 | 20.4 | 88.3 | -15.3 | 94.4 | 5.6 | 3.4 | -19.3 | 96.8 | -22.6 | 98.8 | -25.7 | 100.7 | -29.7 | 103.2 |

### Monthly Climatic Design Conditions

| | | Annual | Jan | Feb | Mar | Apr | May | Jun | Jul | Aug | Sep | Oct | Nov | Dec |
|---|---|---|---|---|---|---|---|---|---|---|---|---|---|---|
| | Tavg | 48.8 | 20.8 | 24.9 | 37.0 | 49.7 | 60.5 | 70.0 | 73.3 | 71.2 | 63.2 | 51.0 | 37.8 | 24.4 |
| | Sd | | 12.52 | 12.10 | 11.23 | 9.75 | 8.11 | 6.54 | 5.57 | 6.01 | 7.83 | 9.48 | 10.30 | 11.79 |
| Temperatures, Degree-Days and Degree-Hours | HDD50 | 3454 | 905 | 703 | 427 | 120 | 11 | 0 | 0 | 0 | 4 | 104 | 386 | 794 |
| | HDD65 | 6705 | 1370 | 1123 | 868 | 468 | 188 | 27 | 4 | 15 | 124 | 444 | 817 | 1257 |
| | CDD50 | 3007 | 0 | 0 | 25 | 111 | 335 | 601 | 722 | 656 | 401 | 136 | 19 | 1 |
| | CDD65 | 785 | 0 | 0 | 1 | 9 | 47 | 178 | 261 | 206 | 72 | 11 | 0 | 0 |
| | CDH74 | 6984 | 0 | 0 | 11 | 119 | 446 | 1593 | 2392 | 1685 | 625 | 112 | 1 | 0 |
| | CDH80 | 2250 | 0 | 0 | 1 | 27 | 102 | 524 | 865 | 540 | 166 | 25 | 0 | 0 |
| Precipitation | PrecAvg | 34.4 | 1.0 | 1.0 | 2.2 | 3.3 | 3.9 | 4.6 | 4.5 | 4.1 | 3.7 | 2.3 | 2.1 | 1.6 |
| | PrecMax | 48.2 | 2.6 | 2.1 | 6.1 | 5.7 | 8.7 | 11.5 | 10.7 | 9.8 | 10.2 | 5.0 | 3.6 | 3.6 |
| | PrecMin | 24.2 | 0.2 | 0.1 | 0.4 | 1.2 | 1.5 | 0.2 | 0.6 | 0.6 | 0.4 | 0.0 | 0.2 | 0.2 |
| | PrecSD | 8.0 | 0.8 | 0.5 | 1.5 | 1.4 | 1.9 | 2.6 | 2.4 | 2.9 | 2.9 | 1.3 | 1.0 | 1.0 |
| Monthly Design Dry Bulb and Mean Coincident Wet Bulb Temperatures | 0.4% DB | 52.3 | 56.9 | 74.7 | 84.1 | 87.5 | 92.5 | 95.7 | 94.8 | 88.9 | 83.3 | 70.3 | 57.3 |
| | 0.4% MCWB | 45.9 | 49.1 | 60.0 | 65.1 | 69.5 | 74.5 | 78.1 | 76.9 | 73.5 | 67.3 | 58.4 | 52.4 |
| | 2% DB | 44.4 | 50.5 | 66.4 | 77.3 | 82.7 | 89.0 | 91.3 | 89.6 | 84.3 | 76.7 | 63.2 | 49.6 |
| | 2% MCWB | 39.9 | 44.5 | 56.0 | 61.7 | 66.8 | 73.2 | 77.0 | 76.6 | 70.6 | 63.1 | 55.3 | 44.6 |
| | 5% DB | 39.3 | 44.5 | 60.7 | 71.7 | 79.2 | 85.9 | 88.1 | 85.6 | 81.3 | 71.8 | 57.5 | 43.2 |
| | 5% MCWB | 36.0 | 39.6 | 53.4 | 58.1 | 65.0 | 71.8 | 76.0 | 74.9 | 68.5 | 60.8 | 51.3 | 39.3 |
| | 10% DB | 35.9 | 39.5 | 54.5 | 66.1 | 75.2 | 82.5 | 84.5 | 82.2 | 77.2 | 66.3 | 53.5 | 38.6 |
| | 10% MCWB | 33.5 | 36.3 | 47.0 | 54.8 | 63.4 | 70.0 | 73.7 | 72.5 | 66.2 | 58.0 | 48.5 | 35.7 |
| Monthly Design Wet Bulb and Mean Coincident Dry Bulb Temperatures | 0.4% WB | 46.3 | 50.2 | 62.4 | 67.3 | 73.9 | 78.3 | 82.0 | 80.9 | 76.4 | 70.2 | 61.1 | 54.4 |
| | 0.4% MCDB | 49.3 | 54.9 | 70.6 | 80.2 | 83.1 | 87.6 | 90.3 | 89.0 | 85.7 | 78.9 | 66.7 | 56.5 |
| | 2% WB | 39.7 | 44.9 | 58.5 | 63.3 | 71.1 | 76.3 | 79.4 | 78.1 | 73.3 | 65.9 | 57.1 | 45.6 |
| | 2% MCDB | 43.9 | 50.1 | 64.5 | 72.5 | 78.6 | 85.2 | 87.9 | 85.8 | 80.8 | 73.6 | 61.9 | 47.9 |
| | 5% WB | 36.5 | 40.1 | 53.7 | 60.3 | 68.1 | 74.4 | 77.5 | 76.3 | 71.0 | 62.7 | 52.3 | 39.7 |
| | 5% MCDB | 39.3 | 44.0 | 60.6 | 69.4 | 75.4 | 82.6 | 86.1 | 83.6 | 77.5 | 69.1 | 56.9 | 43.0 |
| | 10% WB | 33.6 | 36.5 | 47.6 | 56.6 | 65.4 | 72.3 | 75.4 | 74.5 | 68.8 | 59.2 | 47.9 | 35.8 |
| | 10% MCDB | 35.3 | 39.4 | 53.4 | 64.6 | 72.3 | 80.1 | 82.5 | 81.0 | 75.0 | 66.0 | 52.4 | 38.3 |
| Mean Daily Temperature Range | MDBR | 15.4 | 16.1 | 18.4 | 21.2 | 20.3 | 19.5 | 19.2 | 19.1 | 21.6 | 20.7 | 17.3 | 15.1 |
| | 5% DB MCDBR | 19.2 | 21.2 | 26.9 | 28.9 | 24.6 | 22.5 | 22.0 | 21.6 | 25.0 | 26.9 | 23.3 | 20.0 |
| | 5% DB MCWBR | 15.1 | 15.3 | 17.2 | 16.1 | 12.9 | 10.3 | 10.8 | 10.7 | 12.5 | 15.0 | 16.4 | 15.8 |
| | 5% WB MCDBR | 17.5 | 20.3 | 24.4 | 25.1 | 20.4 | 19.2 | 19.7 | 19.0 | 20.4 | 22.0 | 21.1 | 19.2 |
| | 5% WB MCWBR | 14.3 | 15.3 | 16.8 | 15.4 | 12.4 | 10.5 | 10.9 | 10.9 | 11.8 | 14.5 | 16.8 | 15.6 |
| Clear Sky Solar Irradiance | taub | 0.282 | 0.294 | 0.322 | 0.340 | 0.363 | 0.377 | 0.408 | 0.376 | 0.370 | 0.324 | 0.302 | 0.289 |
| | taud | 2.467 | 2.387 | 2.447 | 2.376 | 2.320 | 2.299 | 2.218 | 2.347 | 2.389 | 2.537 | 2.620 | 2.559 |
| | Ebn,noon | 278 | 290 | 294 | 295 | 289 | 284 | 273 | 280 | 273 | 277 | 270 | 267 |
| | Edh,noon | 23 | 30 | 32 | 37 | 40 | 41 | 44 | 37 | 33 | 26 | 21 | 20 |

Nomenclature     See separate page

*Fig. 4-4  Example tables of climatic design information included on CD-ROM of 2013 ASHRAE Handbook—Fundamentals*

# Design Conditions

2013 ASHRAE Handbook - Fundamentals (IP) © 2013 ASHRAE, Inc.

**CHARLOTTE/DOUGLAS, NC, USA** WMO#: 723140

Lat: 35.21N  Long: 80.94W  Elev: 768  StdP: 14.29  Time Zone: -5 (NAE)  Period: 86-10  WBAN: 13881

### Annual Heating and Humidification Design Conditions

| Coldest Month | Heating DB | | Humidification DP/MCDB and HR | | | | | | Coldest month WS/MCDB | | | | MCWS/PCWD to 99.6% DB | |
|---|---|---|---|---|---|---|---|---|---|---|---|---|---|---|
| | 99.6% | 99% | 99.6% | | | 99% | | | 0.4% | | 1% | | | |
| | | | DP | HR | MCDB | DP | HR | MCDB | WS | MCDB | WS | MCDB | MCWS | PCWD |
| (a) | (b) | (c) | (d) | (e) | (f) | (g) | (h) | (i) | (j) | (k) | (l) | (m) | (n) | (o) |
| 1 | 21.0 | 25.0 | 2.2 | 6.3 | 30.6 | 6.6 | 7.9 | 32.6 | 22.4 | 51.2 | 19.2 | 46.9 | 5.4 | 20 |

### Annual Cooling, Dehumidification, and Enthalpy Design Conditions

| Hottest Month | Hottest Month DB Range | Cooling DB/MCWB | | | | | | Evaporation WB/MCDB | | | | | | MCWS/PCWD to 0.4% DB | |
|---|---|---|---|---|---|---|---|---|---|---|---|---|---|---|---|
| | | 0.4% | | 1% | | 2% | | 0.4% | | 1% | | 2% | | | |
| | | DB | MCWB | DB | MCWB | DB | MCWB | WB | MCDB | WB | MCDB | WB | MCDB | MCWS | PCWD |
| (a) | (b) | (c) | (d) | (e) | (f) | (g) | (h) | (i) | (j) | (k) | (l) | (m) | (n) | (o) | (p) |
| 7 | 18.6 | 94.3 | 74.5 | 92.0 | 74.0 | 89.9 | 73.4 | 77.2 | 88.4 | 76.2 | 86.8 | 75.3 | 85.4 | 7.2 | 240 |

| Dehumidification DP/MCDB and HR | | | | | | | | | Enthalpy/MCDB | | | | | | Hours 8 to 4 & 55/69 |
|---|---|---|---|---|---|---|---|---|---|---|---|---|---|---|---|
| 0.4% | | | 1% | | | 2% | | | 0.4% | | 1% | | 2% | | |
| DP | HR | MCDB | DP | HR | MCDB | DP | HR | MCDB | Enth | MCDB | Enth | MCDB | Enth | MCDB | |
| (a) | (b) | (c) | (d) | (e) | (f) | (g) | (h) | (i) | (j) | (k) | (l) | (m) | (n) | (o) | (p) |
| 74.1 | 130.8 | 81.0 | 73.2 | 126.7 | 80.0 | 72.3 | 123.2 | 79.4 | 40.9 | 88.6 | 40.0 | 87.0 | 39.1 | 85.7 | 790 |

### Extreme Annual Design Conditions

| Extreme Annual WS | | | Extreme Max WB | Extreme Annual DB | | | | n-Year Return Period Values of Extreme DB | | | | | | | |
|---|---|---|---|---|---|---|---|---|---|---|---|---|---|---|---|
| 1% | 2.5% | 5% | | Mean | | Standard deviation | | n=5 years | | n=10 years | | n=20 years | | n=50 years | |
| | | | | Min | Max | Min | Max | Min | Max | Min | Max | Min | Max | Min | Max |
| (a) | (b) | (c) | (d) | (e) | (f) | (g) | (h) | (i) | (j) | (k) | (l) | (m) | (n) | (o) | (p) |
| 18.6 | 16.5 | 14.3 | 81.9 | 13.8 | 97.7 | 4.5 | 2.9 | 10.6 | 99.8 | 7.9 | 101.5 | 5.4 | 103.2 | 2.1 | 105.3 |

### Monthly Climatic Design Conditions

| | | | Annual | Jan | Feb | Mar | Apr | May | Jun | Jul | Aug | Sep | Oct | Nov | Dec |
|---|---|---|---|---|---|---|---|---|---|---|---|---|---|---|---|
| | | | (d) | (e) | (f) | (g) | (h) | (i) | (j) | (k) | (l) | (m) | (n) | (o) | (p) |
| (5) | | Tavg | 61.3 | 42.4 | 45.5 | 52.8 | 61.0 | 68.7 | 76.5 | 79.8 | 78.6 | 72.2 | 61.3 | 52.2 | 43.8 |
| (6) | | Sd | | 9.37 | 8.76 | 8.80 | 7.80 | 6.36 | 4.92 | 4.02 | 4.01 | 5.70 | 7.42 | 8.14 | 8.67 |
| (7) | Temperatures, | HDD50 | 846 | 272 | 177 | 74 | 10 | 0 | 0 | 0 | 0 | 0 | 7 | 72 | 234 |
| (8) | Degree-Days | HDD65 | 3065 | 702 | 548 | 388 | 165 | 38 | 2 | 0 | 0 | 13 | 160 | 390 | 659 |
| (9) | and | CDD50 | 4968 | 35 | 50 | 159 | 340 | 580 | 794 | 923 | 885 | 667 | 358 | 136 | 41 |
| (10) | Degree-Hours | CDD65 | 1713 | 0 | 1 | 8 | 45 | 153 | 346 | 458 | 421 | 230 | 46 | 4 | 1 |
| (11) | | CDH74 | 15524 | 1 | 5 | 93 | 468 | 1287 | 3206 | 4552 | 3847 | 1698 | 335 | 29 | 3 |
| (12) | | CDH80 | 6107 | 0 | 0 | 11 | 99 | 367 | 1320 | 2075 | 1634 | 541 | 59 | 1 | 0 |
| (13) | | PrecAvg | 43.1 | 3.7 | 3.8 | 4.4 | 2.7 | 3.8 | 3.4 | 3.9 | 3.7 | 3.5 | 3.4 | 3.2 | 3.5 |
| (14) | Precipitation | PrecMax | 57.8 | 7.4 | 7.6 | 8.7 | 6.5 | 12.5 | 8.3 | 7.6 | 9.2 | 9.7 | 8.4 | 5.4 | 6.0 |
| (15) | | PrecMin | 34.1 | 1.7 | 0.8 | 2.1 | 0.3 | 0.9 | 1.9 | 0.8 | 0.6 | 0.0 | 0.0 | 0.5 | 0.4 |
| (16) | | PrecSD | 6.2 | 1.7 | 2.0 | 1.9 | 1.3 | 2.3 | 1.8 | 1.9 | 2.3 | 2.5 | 2.3 | 1.5 | 1.6 |
| (17) | | 0.4% DB | | 70.0 | 73.3 | 81.5 | 86.7 | 90.1 | 95.8 | 98.3 | 97.0 | 92.2 | 85.2 | 78.1 | 72.8 |
| (18) | Monthly Design | 0.4% MCWB | | 60.7 | 60.7 | 62.4 | 66.3 | 71.4 | 74.5 | 74.3 | 74.4 | 71.4 | 69.8 | 63.6 | 62.6 |
| (19) | Dry Bulb | 2% DB | | 65.7 | 68.5 | 76.4 | 82.9 | 87.1 | 92.6 | 94.9 | 93.4 | 88.9 | 81.2 | 73.4 | 66.7 |
| (20) | and | 2% MCWB | | 58.0 | 56.3 | 60.1 | 64.2 | 70.0 | 73.5 | 74.9 | 74.3 | 71.3 | 66.5 | 60.9 | 59.8 |
| (21) | Mean Coincident | 5% DB | | 62.0 | 64.2 | 72.1 | 79.6 | 84.2 | 89.9 | 92.4 | 91.0 | 85.8 | 77.9 | 69.9 | 62.7 |
| (22) | Wet Bulb | 5% MCWB | | 54.7 | 53.8 | 57.8 | 62.1 | 69.0 | 72.7 | 74.6 | 74.2 | 71.0 | 65.0 | 60.3 | 56.3 |
| (23) | Temperatures | 10% DB | | 57.6 | 60.5 | 67.9 | 75.8 | 81.4 | 87.6 | 89.9 | 88.3 | 82.9 | 74.3 | 66.4 | 58.6 |
| (24) | | 10% MCWB | | 50.4 | 51.8 | 55.7 | 61.0 | 67.2 | 71.9 | 74.1 | 73.6 | 69.8 | 63.5 | 58.4 | 51.5 |
| (25) | | 0.4% WB | | 64.1 | 63.6 | 66.2 | 70.2 | 75.0 | 77.3 | 78.5 | 78.4 | 76.1 | 72.9 | 68.5 | 65.8 |
| (26) | Monthly Design | 0.4% MCDB | | 67.1 | 70.2 | 75.2 | 80.5 | 85.3 | 89.7 | 90.5 | 89.2 | 85.0 | 81.0 | 71.5 | 69.4 |
| (27) | Wet Bulb | 2% WB | | 60.1 | 60.5 | 63.4 | 67.5 | 72.7 | 75.9 | 77.2 | 77.2 | 74.5 | 70.3 | 65.5 | 61.9 |
| (28) | and | 2% MCDB | | 63.9 | 65.5 | 71.6 | 77.2 | 82.9 | 87.3 | 88.7 | 88.0 | 82.6 | 76.2 | 69.6 | 65.5 |
| (29) | Mean Coincident | 5% WB | | 56.3 | 57.3 | 60.9 | 65.3 | 70.9 | 74.8 | 76.2 | 76.0 | 73.3 | 68.2 | 63.0 | 58.2 |
| (30) | Dry Bulb | 5% MCDB | | 60.8 | 61.6 | 68.4 | 74.5 | 80.5 | 85.6 | 87.3 | 85.9 | 80.8 | 73.9 | 67.4 | 62.0 |
| (31) | Temperatures | 10% WB | | 51.7 | 53.2 | 58.4 | 63.3 | 69.2 | 73.6 | 75.3 | 75.0 | 72.1 | 66.1 | 60.1 | 53.0 |
| (32) | | 10% MCDB | | 55.9 | 58.1 | 64.9 | 72.3 | 78.0 | 83.7 | 85.9 | 84.2 | 79.0 | 71.8 | 65.1 | 56.6 |
| (33) | | MDBR | | 18.3 | 19.5 | 20.9 | 22.1 | 20.2 | 18.9 | 18.6 | 17.9 | 18.0 | 20.5 | 20.5 | 18.2 |
| (34) | Mean Daily | 5% DB MCDBR | | 22.9 | 24.5 | 26.0 | 26.0 | 22.9 | 22.0 | 21.6 | 21.5 | 21.6 | 23.5 | 24.9 | 22.6 |
| (35) | Temperature | 5% DB MCWBR | | 15.7 | 15.8 | 13.3 | 11.3 | 9.1 | 7.5 | 6.6 | 6.5 | 7.9 | 11.1 | 14.5 | 16.3 |
| (36) | Range | 5% WB MCDBR | | 19.6 | 19.9 | 21.2 | 21.4 | 20.1 | 19.4 | 19.1 | 18.5 | 17.0 | 17.2 | 18.7 | 19.0 |
| (37) | | 5% WB MCWBR | | 15.9 | 15.4 | 13.4 | 11.2 | 9.0 | 7.5 | 6.9 | 6.6 | 7.6 | 10.2 | 14.2 | 16.3 |
| (38) | Clear Sky | taub | | 0.331 | 0.342 | 0.350 | 0.392 | 0.392 | 0.425 | 0.448 | 0.441 | 0.385 | 0.353 | 0.334 | 0.333 |
| (39) | Solar | taud | | 2.580 | 2.520 | 2.455 | 2.290 | 2.253 | 2.179 | 2.163 | 2.195 | 2.412 | 2.499 | 2.581 | 2.584 |
| (40) | Irradiance | Ebn,noon | | 278 | 286 | 292 | 283 | 283 | 272 | 265 | 265 | 277 | 278 | 274 | 269 |
| (41) | | Edh,noon | | 25 | 29 | 33 | 41 | 44 | 47 | 47 | 45 | 35 | 29 | 24 | 23 |

Nomenclature: See separate page

*Fig. 4-4 Example tables of climatic design information included on CD-ROM of 2013 ASHRAE Handbook—Fundamentals*

# 96

**Principles of HVAC**

2013 ASHRAE Handbook - Fundamentals (IP) © 2013 ASHRAE, Inc.

**CHICAGO/O'HARE ARPT, IL, USA** WMO#: 725300

Lat: 41.99N  Long: 87.91W  Elev: 673  StdP: 14.34  Time Zone: -6 (NAC)  Period: 86-10  WBAN: 94846

### Annual Heating and Humidification Design Conditions

| Coldest Month | Heating DB |  | Humidification DP/MCDB and HR |  |  |  |  |  | Coldest month WS/MCDB |  |  |  | MCWS/PCWD to 99.6% DB |  |
|---|---|---|---|---|---|---|---|---|---|---|---|---|---|---|
|  | 99.6% | 99% | 99.6% |  |  | 99% |  |  | 0.4% |  | 1% |  |  |  |
|  |  |  | DP | HR | MCDB | DP | HR | MCDB | WS | MCDB | WS | MCDB | MCWS | PCWD |
| (a) | (b) | (c) | (d) | (e) | (f) | (g) | (h) | (i) | (j) | (k) | (l) | (m) | (n) | (o) |
| (1) 1 | -1.5 | 3.7 | -11.2 | 3.1 | -0.1 | -6.0 | 4.1 | 5.3 | 27.0 | 31.2 | 24.9 | 31.0 | 10.9 | 270 | (1) |

### Annual Cooling, Dehumidification, and Enthalpy Design Conditions

| Hottest Month | Hottest Month DB Range | Cooling DB/MCWB |  |  |  |  |  | Evaporation WB/MCDB |  |  |  |  |  | MCWS/PCWD to 0.4% DB |  |
|---|---|---|---|---|---|---|---|---|---|---|---|---|---|---|---|
|  |  | 0.4% |  | 1% |  | 2% |  | 0.4% |  | 1% |  | 2% |  |  |  |
|  |  | DB | MCWB | DB | MCWB | DB | MCWB | WB | MCDB | WB | MCDB | WB | MCDB | MCWS | PCWD |
| (a) | (b) | (c) | (d) | (e) | (f) | (g) | (h) | (i) | (j) | (k) | (l) | (m) | (n) | (o) | (p) |
| (2) 7 | 18.3 | 91.4 | 74.3 | 88.7 | 73.2 | 86.0 | 71.8 | 77.8 | 87.8 | 76.0 | 84.8 | 74.3 | 82.5 | 11.3 | 220 | (2) |

| Dehumidification DP/MCDB and HR |  |  |  |  |  |  |  |  | Enthalpy/MCDB |  |  |  |  |  | Hours 8 to 4 & 55/69 |
|---|---|---|---|---|---|---|---|---|---|---|---|---|---|---|---|
| 0.4% |  |  | 1% |  |  | 2% |  |  | 0.4% |  | 1% |  | 2% |  |  |
| DP | HR | MCDB | DP | HR | MCDB | DP | HR | MCDB | Enth | MCDB | Enth | MCDB | Enth | MCDB |  |
| (a) | (b) | (c) | (d) | (e) | (f) | (g) | (h) | (i) | (j) | (k) | (l) | (m) | (n) | (o) | (p) |
| (3) 74.7 | 133.3 | 83.7 | 73.0 | 125.8 | 81.7 | 71.5 | 119.3 | 79.8 | 41.6 | 88.0 | 39.8 | 84.9 | 38.1 | 82.5 | 619 | (3) |

### Extreme Annual Design Conditions

| Extreme Annual WS |  |  | Extreme Max WB | Extreme Annual DB |  |  |  | n-Year Return Period Values of Extreme DB |  |  |  |  |  |  |  |
|---|---|---|---|---|---|---|---|---|---|---|---|---|---|---|---|
| 1% | 2.5% | 5% |  | Mean |  | Standard deviation |  | n=5 years |  | n=10 years |  | n=20 years |  | n=50 years |  |
|  |  |  |  | Min | Max | Min | Max | Min | Max | Min | Max | Min | Max | Min | Max |
| (a) | (b) | (c) | (d) | (e) | (f) | (g) | (h) | (i) | (j) | (k) | (l) | (m) | (n) | (o) | (p) |
| (4) 24.6 | 21.0 | 19.1 | 83.3 | -8.0 | 96.0 | 5.7 | 3.6 | -12.2 | 98.6 | -15.5 | 100.6 | -18.7 | 102.6 | -22.9 | 105.2 | (4) |

### Monthly Climatic Design Conditions

|  |  |  | Annual | Jan | Feb | Mar | Apr | May | Jun | Jul | Aug | Sep | Oct | Nov | Dec |
|---|---|---|---|---|---|---|---|---|---|---|---|---|---|---|---|
| (5) | Temperatures, Degree-Days and Degree-Hours | Tavg | 50.4 | 24.9 | 28.2 | 38.4 | 49.3 | 59.4 | 69.3 | 74.2 | 72.5 | 64.9 | 52.7 | 40.7 | 28.5 |
| (6) |  | Sd |  | 11.44 | 10.55 | 10.61 | 9.43 | 8.66 | 7.63 | 5.81 | 5.93 | 7.54 | 8.85 | 9.45 | 10.60 |
| (7) |  | HDD50 | 2967 | 778 | 613 | 389 | 122 | 13 | 0 | 0 | 0 | 2 | 75 | 306 | 669 |
| (8) |  | HDD65 | 6209 | 1243 | 1032 | 827 | 480 | 222 | 44 | 3 | 9 | 94 | 394 | 730 | 1131 |
| (9) |  | CDD50 | 3101 | 1 | 2 | 29 | 102 | 305 | 579 | 750 | 698 | 450 | 157 | 26 | 2 |
| (10) |  | CDD65 | 864 | 0 | 0 | 1 | 9 | 49 | 174 | 288 | 241 | 91 | 11 | 0 | 0 |
| (11) |  | CDH74 | 7836 | 0 | 0 | 10 | 107 | 505 | 1729 | 2722 | 1953 | 713 | 97 | 0 | 0 |
| (12) |  | CDH80 | 2656 | 0 | 0 | 1 | 23 | 143 | 617 | 1030 | 637 | 188 | 17 | 0 | 0 |
| (13) | Precipitation | PrecAvg | 35.8 | 1.5 | 1.4 | 2.7 | 3.6 | 3.3 | 3.8 | 3.7 | 4.2 | 3.8 | 2.4 | 2.9 | 2.5 |
| (14) |  | PrecMax | 45.6 | 4.1 | 3.2 | 5.9 | 5.5 | 7.1 | 8.7 | 7.6 | 8.5 | 14.2 | 4.7 | 5.0 | 5.4 |
| (15) |  | PrecMin | 20.8 | 0.3 | 0.2 | 1.0 | 1.0 | 1.4 | 1.6 | 1.2 | 0.7 | 0.0 | 0.2 | 0.6 | 0.3 |
| (16) |  | PrecSD | 7.0 | 1.2 | 0.8 | 1.2 | 1.3 | 1.5 | 1.9 | 1.8 | 2.7 | 3.5 | 1.2 | 1.1 | 1.4 |
| (17) | Monthly Design Dry Bulb and Mean Coincident Wet Bulb Temperatures | 0.4% DB | | 56.5 | 59.5 | 74.3 | 83.2 | 88.3 | 93.4 | 96.2 | 94.5 | 89.6 | 82.4 | 69.4 | 59.4 |
| (18) |  | 0.4% MCWB | | 53.0 | 51.2 | 61.3 | 64.5 | 70.1 | 72.6 | 75.8 | 77.2 | 71.8 | 67.2 | 57.7 | 56.6 |
| (19) |  | 2% DB | | 48.3 | 52.2 | 67.6 | 76.5 | 84.1 | 90.2 | 92.1 | 90.0 | 84.9 | 76.1 | 63.9 | 51.9 |
| (20) |  | 2% MCWB | | 44.3 | 46.0 | 57.4 | 61.6 | 68.5 | 72.2 | 75.8 | 75.2 | 69.4 | 63.3 | 56.4 | 47.5 |
| (21) |  | 5% DB | | 43.2 | 45.7 | 61.2 | 71.5 | 80.1 | 86.8 | 88.8 | 86.4 | 81.5 | 71.9 | 60.2 | 46.4 |
| (22) |  | 5% MCWB | | 39.5 | 40.6 | 52.8 | 58.2 | 65.9 | 71.1 | 74.0 | 73.6 | 67.4 | 60.8 | 54.4 | 42.4 |
| (23) |  | 10% DB | | 39.1 | 41.5 | 55.1 | 65.9 | 75.3 | 83.3 | 85.8 | 83.2 | 78.2 | 67.3 | 55.4 | 41.8 |
| (24) |  | 10% MCWB | | 36.4 | 37.8 | 48.0 | 54.8 | 63.4 | 69.8 | 72.5 | 71.4 | 66.0 | 58.7 | 49.5 | 38.5 |
| (25) | Monthly Design Wet Bulb and Mean Coincident Dry Bulb Temperatures | 0.4% WB | | 53.5 | 53.2 | 63.8 | 66.8 | 74.3 | 77.4 | 80.4 | 79.6 | 75.1 | 70.0 | 61.6 | 57.1 |
| (26) |  | 0.4% MCDB | | 55.4 | 55.3 | 72.4 | 78.9 | 84.4 | 87.9 | 90.9 | 90.2 | 83.6 | 78.3 | 65.2 | 59.0 |
| (27) |  | 2% WB | | 45.6 | 46.3 | 59.0 | 63.6 | 71.2 | 75.2 | 78.1 | 77.6 | 72.3 | 66.0 | 58.5 | 48.7 |
| (28) |  | 2% MCDB | | 47.7 | 50.8 | 65.4 | 72.3 | 79.8 | 85.7 | 88.2 | 87.4 | 80.1 | 73.0 | 63.0 | 50.5 |
| (29) |  | 5% WB | | 40.1 | 41.3 | 53.7 | 59.9 | 68.4 | 73.3 | 76.5 | 75.5 | 70.5 | 62.7 | 54.6 | 42.7 |
| (30) |  | 5% MCDB | | 42.6 | 45.3 | 60.6 | 70.2 | 76.5 | 82.8 | 85.7 | 83.1 | 77.8 | 69.8 | 59.4 | 45.9 |
| (31) |  | 10% WB | | 36.7 | 37.8 | 48.2 | 55.9 | 65.2 | 71.4 | 74.5 | 73.8 | 68.6 | 59.5 | 50.5 | 38.6 |
| (32) |  | 10% MCDB | | 39.1 | 41.1 | 54.8 | 64.8 | 73.4 | 80.7 | 82.9 | 80.5 | 75.4 | 65.9 | 55.0 | 41.5 |
| (33) | Mean Daily Temperature Range | MDBR | | 13.3 | 13.9 | 16.4 | 18.8 | 20.0 | 19.9 | 18.3 | 17.1 | 18.8 | 17.8 | 14.5 | 12.8 |
| (34) |  | 5% DB MCDBR | | 16.6 | 19.6 | 24.8 | 26.8 | 25.1 | 23.4 | 21.3 | 20.0 | 22.3 | 24.0 | 19.8 | 16.9 |
| (35) |  | 5% DB MCWBR | | 14.3 | 14.9 | 16.9 | 15.5 | 13.0 | 10.7 | 9.5 | 8.9 | 10.1 | 13.3 | 15.1 | 13.9 |
| (36) |  | 5% WB MCDBR | | 16.0 | 18.1 | 23.5 | 23.8 | 22.0 | 20.7 | 19.3 | 17.2 | 18.3 | 20.3 | 18.0 | 16.0 |
| (37) |  | 5% WB MCWBR | | 14.6 | 14.7 | 17.5 | 15.4 | 13.3 | 11.3 | 10.1 | 8.9 | 10.0 | 13.4 | 15.4 | 14.2 |
| (38) | Clear Sky Solar Irradiance | taub | | 0.288 | 0.305 | 0.325 | 0.359 | 0.369 | 0.383 | 0.415 | 0.416 | 0.392 | 0.344 | 0.306 | 0.294 |
| (39) |  | taud | | 2.524 | 2.474 | 2.473 | 2.343 | 2.310 | 2.285 | 2.209 | 2.242 | 2.343 | 2.487 | 2.629 | 2.584 |
| (40) |  | Ebn,noon | | 276 | 288 | 293 | 288 | 287 | 282 | 271 | 267 | 266 | 269 | 269 | 265 |
| (41) |  | Edh,noon | | 22 | 27 | 31 | 38 | 40 | 41 | 44 | 42 | 35 | 27 | 20 | 20 |

Nomenclature:    See separate page

*Fig. 4-4 Example tables of climatic design information included on CD-ROM of 2013 ASHRAE Handbook—Fundamentals*

# Design Conditions

2013 ASHRAE Handbook - Fundamentals (IP)    © 2013 ASHRAE, Inc.

**CINCINNATI MUNI LUN, OH, USA**    WMO#: 724297

Lat: 39.10N    Long: 84.42W    Elev: 499    StdP: 14.43    Time Zone: -5 (NAE)    Period: 86-10    WBAN: 93812

### Annual Heating and Humidification Design Conditions

| Coldest Month | Heating DB | | Humidification DP/MCDB and HR | | | | | | Coldest month WS/MCDB | | | | MCWS/PCWD to 99.6% DB | |
|---|---|---|---|---|---|---|---|---|---|---|---|---|---|---|
| | 99.6% | 99% | 99.6% | | | 99% | | | 0.4% | | 1% | | MCWS | PCWD |
| | | | DP | HR | MCDB | DP | HR | MCDB | WS | MCDB | WS | MCDB | | |
| (a) | (b) | (c) | (d) | (e) | (f) | (g) | (h) | (i) | (j) | (k) | (l) | (m) | (n) | (o) |
| 1 | 8.1 | 13.4 | -2.4 | 4.9 | 9.6 | 2.9 | 6.5 | 15.4 | 24.3 | 38.9 | 21.0 | 38.3 | 7.0 | 300 |

### Annual Cooling, Dehumidification, and Enthalpy Design Conditions

| Hottest Month | Hottest Month DB Range | Cooling DB/MCWB | | | | | | Evaporation WB/MCDB | | | | | | MCWS/PCWD to 0.4% DB | |
|---|---|---|---|---|---|---|---|---|---|---|---|---|---|---|---|
| | | 0.4% | | 1% | | 2% | | 0.4% | | 1% | | 2% | | MCWS | PCWD |
| | | DB | MCWB | DB | MCWB | DB | MCWB | WB | MCDB | WB | MCDB | WB | MCDB | | |
| (a) | (b) | (c) | (d) | (e) | (f) | (g) | (h) | (i) | (j) | (k) | (l) | (m) | (n) | (o) | (p) |
| 7 | 19.9 | 92.8 | 74.5 | 90.3 | 74.2 | 88.1 | 73.3 | 78.0 | 87.9 | 76.7 | 86.2 | 75.4 | 84.1 | 9.7 | 240 |

| Dehumidification DP/MCDB and HR | | | | | | | | | Enthalpy/MCDB | | | | | | Hours 8 to 4 & 55/69 |
|---|---|---|---|---|---|---|---|---|---|---|---|---|---|---|---|
| 0.4% | | | 1% | | | 2% | | | 0.4% | | 1% | | 2% | | |
| DP | HR | MCDB | DP | HR | MCDB | DP | HR | MCDB | Enth | MCDB | Enth | MCDB | Enth | MCDB | |
| (a) | (b) | (c) | (d) | (e) | (f) | (g) | (h) | (i) | (j) | (k) | (l) | (m) | (n) | (o) | (p) |
| 75.1 | 134.4 | 82.6 | 73.8 | 128.4 | 81.1 | 72.8 | 123.7 | 80.1 | 41.6 | 88.1 | 40.3 | 86.3 | 39.1 | 84.2 | 685 |

### Extreme Annual Design Conditions

| Extreme Annual WS | | | Extreme Max WB | Extreme Annual DB | | | | n-Year Return Period Values of Extreme DB | | | | | | | |
|---|---|---|---|---|---|---|---|---|---|---|---|---|---|---|---|
| 1% | 2.5% | 5% | | Mean | | Standard deviation | | n=5 years | | n=10 years | | n=20 years | | n=50 years | |
| | | | | Min | Max | Min | Max | Min | Max | Min | Max | Min | Max | Min | Max |
| (a) | (b) | (c) | (d) | (e) | (f) | (g) | (h) | (i) | (j) | (k) | (l) | (m) | (n) | (o) | (p) |
| 20.4 | 18.3 | 16.6 | 84.2 | 0.5 | 96.3 | 7.4 | 3.2 | -4.9 | 98.6 | -9.2 | 100.5 | -13.4 | 102.3 | -18.8 | 104.6 |

### Monthly Climatic Design Conditions

| | | | Annual | Jan | Feb | Mar | Apr | May | Jun | Jul | Aug | Sep | Oct | Nov | Dec |
|---|---|---|---|---|---|---|---|---|---|---|---|---|---|---|---|
| | | Tavg | 55.2 | 33.1 | 35.8 | 44.6 | 54.8 | 64.0 | 72.6 | 76.3 | 75.1 | 68.0 | 55.7 | 45.5 | 35.4 |
| | | Sd | | 11.25 | 10.12 | 10.08 | 8.71 | 7.41 | 5.79 | 4.78 | 5.18 | 6.81 | 8.01 | 8.93 | 9.95 |
| Temperatures, Degree-Days and Degree-Hours | | HDD50 | 1914 | 536 | 408 | 229 | 49 | 3 | 0 | 0 | 0 | 0 | 34 | 192 | 463 |
| | | HDD65 | 4744 | 989 | 818 | 635 | 320 | 111 | 12 | 0 | 2 | 47 | 305 | 586 | 919 |
| | | CDD50 | 3803 | 12 | 10 | 62 | 193 | 437 | 677 | 816 | 780 | 539 | 211 | 57 | 9 |
| | | CDD65 | 1155 | 0 | 0 | 2 | 14 | 80 | 239 | 351 | 317 | 136 | 16 | 0 | 0 |
| | | CDH74 | 10719 | 0 | 0 | 32 | 202 | 777 | 2172 | 3310 | 2851 | 1181 | 186 | 8 | 0 |
| | | CDH80 | 3990 | 0 | 0 | 2 | 34 | 213 | 790 | 1359 | 1166 | 391 | 35 | 0 | 0 |
| Precipitation | | PrecAvg | 41.1 | 2.6 | 2.4 | 3.8 | 3.7 | 4.5 | 3.6 | 4.0 | 3.9 | 3.4 | 2.8 | 3.3 | 3.1 |
| | | PrecMax | 52.6 | 4.8 | 6.5 | 11.1 | 6.3 | 10.1 | 6.8 | 7.8 | 6.6 | 8.5 | 5.0 | 5.8 | 5.9 |
| | | PrecMin | 29.7 | 0.8 | 0.3 | 1.3 | 0.7 | 0.6 | 1.5 | 1.8 | 0.7 | 0.5 | 0.6 | 0.6 | 0.4 |
| | | PrecSD | 6.7 | 1.1 | 1.8 | 2.4 | 1.6 | 2.3 | 1.6 | 2.1 | 1.6 | 2.6 | 1.3 | 1.5 | 1.5 |
| Monthly Design Dry Bulb and Mean Coincident Wet Bulb Temperatures | 0.4% | DB | 65.5 | 68.1 | 78.7 | 84.1 | 88.4 | 93.0 | 96.8 | 96.8 | 93.0 | 84.0 | 74.5 | 65.8 | |
| | | MCWB | 58.9 | 56.5 | 61.7 | 65.7 | 70.9 | 73.0 | 76.1 | 75.7 | 70.9 | 68.0 | 60.7 | 60.2 | |
| | 2% | DB | 60.7 | 61.9 | 72.7 | 79.7 | 85.0 | 90.1 | 92.8 | 92.6 | 88.3 | 79.2 | 69.6 | 59.6 | |
| | | MCWB | 55.8 | 52.7 | 58.4 | 63.7 | 68.9 | 73.9 | 75.7 | 74.9 | 70.9 | 65.0 | 58.8 | 53.9 | |
| | 5% | DB | 55.1 | 56.8 | 67.1 | 75.3 | 82.2 | 87.8 | 90.1 | 89.8 | 84.3 | 74.8 | 65.3 | 55.2 | |
| | | MCWB | 49.9 | 50.0 | 55.6 | 61.5 | 68.4 | 73.0 | 75.0 | 74.4 | 69.6 | 63.3 | 58.0 | 50.5 | |
| | 10% | DB | 49.7 | 51.9 | 62.3 | 71.6 | 78.7 | 84.8 | 87.7 | 86.8 | 81.1 | 70.5 | 61.2 | 50.4 | |
| | | MCWB | 45.1 | 46.3 | 53.9 | 59.4 | 66.7 | 71.8 | 74.0 | 73.1 | 68.2 | 61.0 | 54.9 | 46.0 | |
| Monthly Design Wet Bulb and Mean Coincident Dry Bulb Temperatures | 0.4% | WB | 60.2 | 59.2 | 64.1 | 68.8 | 74.7 | 78.5 | 80.0 | 79.4 | 76.1 | 71.5 | 64.7 | 60.2 | |
| | | MCDB | 63.8 | 64.7 | 74.5 | 79.3 | 83.4 | 88.0 | 90.7 | 88.9 | 85.0 | 79.9 | 70.2 | 64.7 | |
| | 2% | WB | 56.7 | 55.3 | 60.8 | 65.8 | 72.3 | 76.5 | 78.1 | 77.8 | 74.3 | 68.3 | 61.5 | 55.4 | |
| | | MCDB | 60.4 | 59.7 | 69.1 | 75.6 | 81.2 | 85.9 | 88.3 | 87.7 | 82.6 | 75.5 | 66.6 | 58.1 | |
| | 5% | WB | 50.8 | 51.0 | 57.9 | 63.6 | 70.4 | 75.0 | 76.9 | 76.4 | 72.5 | 65.5 | 58.8 | 51.2 | |
| | | MCDB | 54.4 | 56.0 | 65.5 | 72.3 | 78.7 | 84.3 | 86.5 | 85.7 | 79.9 | 72.0 | 64.3 | 54.5 | |
| | 10% | WB | 45.5 | 46.0 | 54.2 | 61.3 | 68.3 | 73.5 | 75.6 | 75.1 | 70.8 | 62.7 | 55.4 | 46.2 | |
| | | MCDB | 49.2 | 51.0 | 61.4 | 69.6 | 75.9 | 82.0 | 84.2 | 83.2 | 77.4 | 68.5 | 60.9 | 49.9 | |
| Mean Daily Temperature Range | | MDBR | 15.7 | 17.5 | 20.2 | 22.4 | 21.3 | 20.4 | 19.9 | 20.7 | 21.9 | 22.7 | 19.3 | 15.3 | |
| | 5% DB | MCDBR | 21.0 | 26.6 | 28.9 | 28.2 | 26.0 | 23.8 | 22.5 | 23.9 | 26.4 | 28.2 | 26.0 | 22.4 | |
| | | MCWBR | 16.5 | 19.0 | 17.2 | 15.2 | 12.4 | 10.3 | 8.5 | 9.0 | 11.1 | 15.3 | 17.7 | 18.0 | |
| | 5% WB | MCDBR | 17.6 | 23.0 | 24.5 | 24.0 | 21.4 | 20.4 | 19.6 | 20.5 | 19.7 | 22.9 | 21.7 | 20.0 | |
| | | MCWBR | 15.6 | 18.4 | 16.2 | 14.0 | 11.2 | 10.0 | 8.5 | 8.9 | 10.0 | 14.1 | 16.8 | 17.8 | |
| Clear Sky Solar Irradiance | | taub | 0.314 | 0.332 | 0.348 | 0.396 | 0.391 | 0.443 | 0.453 | 0.440 | 0.401 | 0.367 | 0.327 | 0.319 | |
| | | taud | 2.484 | 2.440 | 2.419 | 2.255 | 2.246 | 2.112 | 2.118 | 2.182 | 2.309 | 2.423 | 2.570 | 2.533 | |
| | | Ebn,noon | 273 | 282 | 289 | 279 | 282 | 266 | 262 | 263 | 267 | 266 | 267 | 263 | |
| | | Edh,noon | 25 | 30 | 34 | 42 | 43 | 50 | 49 | 45 | 37 | 30 | 23 | 22 | |

Nomenclature: See separate page

*Fig. 4-4 Example tables of climatic design information included on CD-ROM of 2013 ASHRAE Handbook—Fundamentals*

## CLEVELAND, OH, USA

2013 ASHRAE Handbook - Fundamentals (IP) © 2013 ASHRAE, Inc.

WMO#: 725240
Lat: 41.41N  Long: 81.85W  Elev: 804  StdP: 14.27  Time Zone: -5 (NAE)  Period: 86-10  WBAN: 14820

### Annual Heating and Humidification Design Conditions

| Coldest Month | Heating DB | | Humidification DP/MCDB and HR | | | | | | Coldest month WS/MCDB | | | | MCWS/PCWD to 99.6% DB | |
|---|---|---|---|---|---|---|---|---|---|---|---|---|---|---|
| | 99.6% | 99% | 99.6% | | | 99% | | | 0.4% | | 1% | | | |
| | | | DP | HR | MCDB | DP | HR | MCDB | WS | MCDB | WS | MCDB | MCWS | PCWD |
| (a) | (b) | (c) | (d) | (e) | (f) | (g) | (h) | (i) | (j) | (k) | (l) | (m) | (n) | (o) |
| 1 | 4.1 | 9.7 | -3.5 | 4.7 | 5.8 | 1.3 | 6.0 | 11.2 | 28.0 | 32.8 | 25.7 | 32.6 | 11.0 | 230 |

### Annual Cooling, Dehumidification, and Enthalpy Design Conditions

| Hottest Month | Hottest Month DB Range | Cooling DB/MCWB | | | | | | Evaporation WB/MCDB | | | | | | MCWS/PCWD to 0.4% DB | |
|---|---|---|---|---|---|---|---|---|---|---|---|---|---|---|---|
| | | 0.4% | | 1% | | 2% | | 0.4% | | 1% | | 2% | | | |
| | | DB | MCWB | DB | MCWB | DB | MCWB | WB | MCDB | WB | MCDB | WB | MCDB | MCWS | PCWD |
| (a) | (b) | (c) | (d) | (e) | (f) | (g) | (h) | (i) | (j) | (k) | (l) | (m) | (n) | (o) | (p) |
| 7 | 17.3 | 89.7 | 73.7 | 87.0 | 72.4 | 84.2 | 71.0 | 76.2 | 85.4 | 74.7 | 83.0 | 73.2 | 81.1 | 11.3 | 230 |

| Dehumidification DP/MCDB and HR | | | | | | | | Enthalpy/MCDB | | | | | | Hours 8 to 4 & 55/69 |
|---|---|---|---|---|---|---|---|---|---|---|---|---|---|---|
| 0.4% | | | 1% | | | 2% | | 0.4% | | 1% | | 2% | | |
| DP | HR | MCDB | DP | HR | MCDB | DP | HR | MCDB | Enth | MCDB | Enth | MCDB | Enth | MCDB |
| (a) | (b) | (c) | (d) | (e) | (f) | (g) | (h) | (i) | (j) | (k) | (l) | (m) | (n) | (o) | (p) |
| 73.2 | 127.3 | 81.3 | 72.0 | 121.9 | 79.6 | 70.5 | 115.6 | 78.0 | 40.1 | 85.4 | 38.6 | 83.0 | 37.2 | 81.2 | 673 |

### Extreme Annual Design Conditions

| Extreme Annual WS | | | Extreme Max WB | Extreme Annual DB | | | | n-Year Return Period Values of Extreme DB | | | | | | | |
|---|---|---|---|---|---|---|---|---|---|---|---|---|---|---|---|
| | | | | Mean | | Standard deviation | | n=5 years | | n=10 years | | n=20 years | | n=50 years | |
| 1% | 2.5% | 5% | | Min | Max | Min | Max | Min | Max | Min | Max | Min | Max | Min | Max |
| (a) | (b) | (c) | (d) | (e) | (f) | (g) | (h) | (i) | (j) | (k) | (l) | (m) | (n) | (o) | (p) |
| 24.5 | 20.9 | 18.8 | 84.0 | -2.0 | 93.6 | 6.7 | 3.2 | -6.8 | 95.8 | -10.7 | 97.7 | -14.5 | 99.5 | -19.3 | 101.8 |

### Monthly Climatic Design Conditions

| | | Annual | Jan | Feb | Mar | Apr | May | Jun | Jul | Aug | Sep | Oct | Nov | Dec |
|---|---|---|---|---|---|---|---|---|---|---|---|---|---|---|
| Temperatures, Degree-Days and Degree-Hours | Tavg | 51.1 | 28.5 | 30.0 | 38.1 | 49.3 | 59.4 | 69.0 | 73.0 | 71.6 | 64.6 | 53.2 | 43.2 | 32.0 |
| | Sd | | 11.30 | 10.16 | 11.03 | 9.90 | 8.49 | 7.19 | 5.40 | 5.58 | 6.90 | 8.26 | 9.14 | 9.93 |
| | HDD50 | 2652 | 671 | 564 | 402 | 132 | 12 | 0 | 0 | 0 | 1 | 60 | 245 | 565 |
| | HDD65 | 5850 | 1132 | 980 | 837 | 480 | 219 | 43 | 4 | 9 | 90 | 378 | 654 | 1024 |
| | CDD50 | 3049 | 4 | 4 | 32 | 111 | 304 | 570 | 713 | 668 | 439 | 158 | 41 | 5 |
| | CDD65 | 774 | 0 | 0 | 2 | 10 | 47 | 162 | 252 | 213 | 77 | 11 | 0 | 0 |
| | CDH74 | 6181 | 0 | 0 | 11 | 103 | 444 | 1344 | 2104 | 1591 | 517 | 66 | 1 | 0 |
| | CDH80 | 1814 | 0 | 0 | 0 | 17 | 106 | 421 | 685 | 464 | 115 | 6 | 0 | 0 |
| Precipitation | PrecAvg | 36.9 | 2.2 | 2.2 | 2.9 | 3.2 | 3.6 | 3.5 | 3.5 | 3.4 | 3.5 | 2.6 | 3.2 | 3.0 |
| | PrecMax | 48.0 | 4.5 | 4.0 | 5.2 | 6.6 | 5.7 | 9.1 | 6.5 | 9.0 | 4.9 | 4.0 | 5.2 | 4.8 |
| | PrecMin | 19.1 | 0.4 | 0.5 | 1.7 | 1.2 | 1.0 | 1.2 | 1.9 | 0.5 | 0.8 | 0.7 | 0.8 | 1.1 |
| | PrecSD | 6.2 | 1.1 | 1.0 | 1.0 | 1.3 | 1.3 | 1.8 | 1.1 | 2.1 | 1.1 | 0.9 | 1.3 | 1.0 |
| Monthly Design Dry Bulb and Mean Coincident Wet Bulb Temperatures | 0.4% DB | | 61.5 | 62.6 | 75.3 | 82.4 | 86.6 | 91.4 | 93.6 | 91.6 | 88.2 | 80.5 | 70.6 | 61.6 |
| | 0.4% MCWB | | 55.8 | 54.1 | 60.7 | 64.1 | 70.0 | 71.9 | 75.8 | 76.1 | 71.0 | 66.9 | 59.6 | 57.1 |
| | 2% DB | | 55.3 | 55.0 | 68.0 | 76.9 | 83.1 | 88.1 | 89.9 | 88.4 | 83.2 | 75.0 | 65.5 | 55.5 |
| | 2% MCWB | | 51.6 | 49.5 | 56.1 | 61.9 | 68.2 | 72.3 | 74.4 | 74.0 | 69.8 | 63.0 | 57.6 | 51.0 |
| | 5% DB | | 49.8 | 49.1 | 62.6 | 71.8 | 79.5 | 84.9 | 86.8 | 84.9 | 79.8 | 70.8 | 61.9 | 50.1 |
| | 5% MCWB | | 45.4 | 43.6 | 53.4 | 59.0 | 66.2 | 71.0 | 72.9 | 72.3 | 67.9 | 61.3 | 55.3 | 45.7 |
| | 10% DB | | 43.7 | 44.0 | 56.4 | 66.3 | 74.8 | 81.7 | 83.7 | 82.0 | 76.4 | 66.7 | 57.5 | 45.4 |
| | 10% MCWB | | 39.7 | 39.6 | 49.7 | 56.1 | 64.1 | 69.4 | 71.3 | 70.9 | 66.6 | 58.9 | 51.9 | 41.6 |
| Monthly Design Wet Bulb and Mean Coincident Dry Bulb Temperatures | 0.4% WB | | 56.3 | 55.9 | 62.1 | 67.9 | 73.9 | 76.4 | 78.5 | 78.3 | 74.5 | 69.3 | 62.7 | 57.4 |
| | 0.4% MCDB | | 60.2 | 60.7 | 72.0 | 76.9 | 82.4 | 86.1 | 88.8 | 88.7 | 81.9 | 77.6 | 67.3 | 60.9 |
| | 2% WB | | 52.1 | 50.1 | 57.9 | 63.9 | 70.9 | 74.5 | 76.6 | 76.0 | 72.1 | 65.7 | 59.0 | 51.7 |
| | 2% MCDB | | 55.3 | 54.2 | 65.7 | 74.0 | 79.6 | 84.0 | 86.4 | 84.5 | 79.2 | 72.4 | 64.3 | 54.6 |
| | 5% WB | | 45.8 | 44.7 | 54.1 | 60.3 | 68.3 | 72.8 | 74.7 | 74.4 | 70.3 | 62.9 | 55.9 | 46.2 |
| | 5% MCDB | | 48.8 | 48.5 | 61.2 | 69.7 | 76.5 | 81.9 | 83.2 | 81.8 | 76.6 | 68.8 | 61.0 | 49.3 |
| | 10% WB | | 40.0 | 39.6 | 49.8 | 57.0 | 65.4 | 71.1 | 73.1 | 72.7 | 68.5 | 60.1 | 52.6 | 41.5 |
| | 10% MCDB | | 43.7 | 43.9 | 56.9 | 65.4 | 73.4 | 79.2 | 81.0 | 79.7 | 74.6 | 65.7 | 57.2 | 45.2 |
| Mean Daily Temperature Range | MDBR | | 12.8 | 13.5 | 15.9 | 18.0 | 18.6 | 17.9 | 17.3 | 16.9 | 17.3 | 16.5 | 13.7 | 11.7 |
| | 5% DB MCDBR | | 17.8 | 20.5 | 25.7 | 25.2 | 23.9 | 21.2 | 20.7 | 19.9 | 21.5 | 21.8 | 19.1 | 17.8 |
| | 5% DB MCWBR | | 15.2 | 16.5 | 16.7 | 15.4 | 12.7 | 9.8 | 9.1 | 9.4 | 11.0 | 12.8 | 14.0 | 15.4 |
| | 5% WB MCDBR | | 17.0 | 19.2 | 23.0 | 23.0 | 20.8 | 18.6 | 17.8 | 17.3 | 16.9 | 18.2 | 18.0 | 16.9 |
| | 5% WB MCWBR | | 15.7 | 16.7 | 16.4 | 15.1 | 12.4 | 10.0 | 8.9 | 9.2 | 10.1 | 12.2 | 14.6 | 16.1 |
| Clear Sky Solar Irradiance | taub | 0.307 | 0.318 | 0.339 | 0.380 | 0.397 | 0.436 | 0.427 | 0.434 | 0.392 | 0.363 | 0.321 | 0.299 |
| | taud | 2.476 | 2.442 | 2.430 | 2.302 | 2.231 | 2.118 | 2.182 | 2.192 | 2.344 | 2.437 | 2.573 | 2.513 |
| | Ebn,noon | | 269 | 283 | 289 | 282 | 279 | 267 | 268 | 262 | 267 | 262 | 263 | 263 |
| | Edh,noon | | 24 | 29 | 33 | 40 | 44 | 49 | 46 | 44 | 35 | 29 | 22 | 21 |

Nomenclature: See separate page

*Fig. 4-4 Example tables of climatic design information included on CD-ROM of 2013 ASHRAE Handbook—Fundamentals*

# Design Conditions

2013 ASHRAE Handbook - Fundamentals (IP)  © 2013 ASHRAE, Inc.

**DENVER/STAPLETON, CO, USA**  WMO#: 724690

Lat: 39.75N  Long: 104.87W  Elev: 5289  StdP: 12.1  Time Zone: -7 (NAM)  Period: 86-95  WBAN: 23062

### Annual Heating and Humidification Design Conditions

| Coldest Month | Heating DB | | Humidification DP/MCDB and HR | | | | | | Coldest month WS/MCDB | | | | MCWS/PCWD to 99.6% DB | |
|---|---|---|---|---|---|---|---|---|---|---|---|---|---|---|
| | 99.6% | 99% | 99.6% | | | 99% | | | 0.4% | | 1% | | | |
| | | | DP | HR | MCDB | DP | HR | MCDB | WS | MCDB | WS | MCDB | MCWS | PCWD |
| (a) | (b) | (c) | (d) | (e) | (f) | (g) | (h) | (i) | (j) | (k) | (l) | (m) | (n) | (o) |
| 12 | -1.4 | 5.1 | -7.5 | 4.5 | 5.4 | -2.4 | 5.9 | 14.0 | 27.2 | 40.8 | 23.8 | 40.8 | 5.0 | 180 |

### Annual Cooling, Dehumidification, and Enthalpy Design Conditions

| Hottest Month | Hottest Month DB Range | Cooling DB/MCWB | | | | | | Evaporation WB/MCDB | | | | | | MCWS/PCWD to 0.4% DB | |
|---|---|---|---|---|---|---|---|---|---|---|---|---|---|---|---|
| | | 0.4% | | 1% | | 2% | | 0.4% | | 1% | | 2% | | | |
| | | DB | MCWB | DB | MCWB | DB | MCWB | WB | MCDB | WB | MCDB | WB | MCDB | MCWS | PCWD |
| (a) | (b) | (c) | (d) | (e) | (f) | (g) | (h) | (i) | (j) | (k) | (l) | (m) | (n) | (o) | (p) |
| 7 | 27.9 | 93.9 | 60.7 | 91.2 | 60.0 | 88.5 | 59.6 | 64.5 | 81.8 | 63.4 | 80.7 | 62.3 | 79.7 | 9.1 | 50 |

| Dehumidification DP/MCDB and HR | | | | | | | | Enthalpy/MCDB | | | | | | Hours 8 to 4 & 55/69 |
|---|---|---|---|---|---|---|---|---|---|---|---|---|---|---|
| 0.4% | | | 1% | | | 2% | | 0.4% | | 1% | | 2% | | |
| DP | HR | MCDB | DP | HR | MCDB | DP | HR | MCDB | Enth | MCDB | Enth | MCDB | Enth | MCDB |
| (a) | (b) | (c) | (d) | (e) | (f) | (g) | (h) | (i) | (j) | (k) | (l) | (m) | (n) | (o) | (p) |
| 60.1 | 94.7 | 67.2 | 58.5 | 89.1 | 67.0 | 56.8 | 83.9 | 67.4 | 32.3 | 82.0 | 31.3 | 80.3 | 30.5 | 80.1 | 754 |

### Extreme Annual Design Conditions

| Extreme Annual WS | | | Extreme Max WB | Extreme Annual DB | | | | n-Year Return Period Values of Extreme DB | | | | | | | |
|---|---|---|---|---|---|---|---|---|---|---|---|---|---|---|---|
| 1% | 2.5% | 5% | | Mean | | Standard deviation | | n=5 years | | n=10 years | | n=20 years | | n=50 years | |
| | | | | Min | Max | Min | Max | Min | Max | Min | Max | Min | Max | Min | Max |
| (a) | (b) | (c) | (d) | (e) | (f) | (g) | (h) | (i) | (j) | (k) | (l) | (m) | (n) | (o) | (p) |
| 24.3 | 19.7 | 17.2 | 71.4 | -10.4 | 99.7 | 8.2 | 2.5 | -16.3 | 101.5 | -21.1 | 103.0 | -25.7 | 104.4 | -31.7 | 106.3 |

### Monthly Climatic Design Conditions

| | | Annual | Jan | Feb | Mar | Apr | May | Jun | Jul | Aug | Sep | Oct | Nov | Dec |
|---|---|---|---|---|---|---|---|---|---|---|---|---|---|---|
| | | (d) | (e) | (f) | (g) | (h) | (i) | (j) | (k) | (l) | (m) | (n) | (o) | (p) |
| Temperatures, Degree-Days and Degree-Hours | Tavg | 51.4 | 32.4 | 34.4 | 42.3 | 50.1 | 59.1 | 69.2 | 73.0 | 71.5 | 63.4 | 51.6 | 38.8 | 30.4 |
| | Sd | | 10.51 | 12.19 | 9.56 | 9.38 | 7.41 | 6.60 | 5.07 | 4.90 | 7.37 | 9.29 | 10.57 | 11.59 |
| | HDD50 | 2446 | 546 | 441 | 268 | 113 | 16 | 0 | 0 | 0 | 6 | 90 | 357 | 609 |
| | HDD65 | 5667 | 1010 | 857 | 704 | 451 | 205 | 34 | 7 | 11 | 113 | 416 | 787 | 1072 |
| | CDD50 | 2971 | 1 | 4 | 28 | 116 | 299 | 576 | 713 | 666 | 407 | 140 | 20 | 1 |
| | CDD65 | 721 | 0 | 0 | 0 | 3 | 24 | 161 | 255 | 212 | 64 | 2 | 0 | 0 |
| | CDH74 | 9191 | 0 | 0 | 8 | 139 | 484 | 2066 | 3023 | 2291 | 988 | 187 | 5 | 0 |
| | CDH80 | 3800 | 0 | 0 | 0 | 19 | 110 | 921 | 1448 | 957 | 315 | 30 | 0 | 0 |
| Precipitation | PrecAvg | 15.4 | 0.5 | 0.6 | 1.3 | 1.7 | 2.4 | 1.8 | 1.9 | 1.5 | 1.2 | 1.0 | 0.9 | 0.6 |
| | PrecMax | 24.1 | 1.3 | 2.5 | 2.3 | 3.9 | 6.1 | 5.2 | 5.9 | 5.8 | 4.9 | 4.1 | 1.9 | 2.8 |
| | PrecMin | 7.9 | 0.0 | 0.0 | 0.4 | 0.6 | 0.0 | 0.1 | 0.4 | 0.2 | 0.0 | 0.0 | 0.2 | 0.1 |
| | PrecSD | 5.2 | 0.4 | 0.6 | 0.5 | 0.9 | 1.9 | 1.5 | 1.3 | 1.3 | 1.3 | 0.9 | 0.5 | 0.7 |
| Monthly Design Dry Bulb and Mean Coincident Wet Bulb Temperatures | 0.4% DB | | 63.7 | 67.9 | 75.4 | 82.4 | 87.3 | 97.7 | 98.3 | 95.0 | 90.8 | 83.3 | 73.7 | 64.2 |
| | MCWB | | 42.7 | 45.6 | 48.1 | 52.9 | 55.3 | 60.0 | 61.3 | 61.1 | 60.0 | 52.7 | 47.3 | 43.4 |
| | 2% DB | | 58.5 | 62.1 | 70.5 | 78.3 | 83.0 | 93.0 | 94.4 | 91.8 | 87.2 | 79.3 | 67.1 | 58.4 |
| | MCWB | | 40.6 | 43.0 | 46.2 | 51.4 | 54.6 | 59.3 | 61.0 | 60.4 | 57.7 | 51.7 | 46.2 | 41.1 |
| | 5% DB | | 53.9 | 57.5 | 65.6 | 73.4 | 79.6 | 89.4 | 91.7 | 89.1 | 83.7 | 75.1 | 61.5 | 52.5 |
| | MCWB | | 38.1 | 40.7 | 44.1 | 50.2 | 54.5 | 59.0 | 60.7 | 60.2 | 56.7 | 50.4 | 43.5 | 38.2 |
| | 10% DB | | 49.4 | 52.8 | 60.3 | 68.7 | 75.8 | 85.5 | 88.6 | 85.8 | 80.3 | 70.0 | 55.7 | 47.1 |
| | MCWB | | 36.2 | 38.6 | 42.3 | 47.9 | 53.6 | 58.8 | 60.5 | 60.0 | 55.9 | 48.6 | 40.9 | 35.1 |
| Monthly Design Wet Bulb and Mean Coincident Dry Bulb Temperatures | 0.4% WB | | 43.5 | 46.5 | 49.3 | 54.7 | 59.1 | 64.9 | 66.5 | 66.4 | 63.3 | 55.3 | 50.1 | 44.3 |
| | MCDB | | 61.6 | 66.0 | 73.0 | 76.7 | 75.4 | 82.7 | 86.1 | 80.8 | 84.9 | 73.2 | 67.1 | 61.4 |
| | 2% WB | | 40.8 | 43.3 | 46.6 | 52.5 | 57.4 | 63.3 | 64.5 | 64.4 | 60.4 | 53.0 | 47.1 | 41.4 |
| | MCDB | | 57.3 | 60.4 | 67.7 | 74.3 | 74.0 | 81.9 | 82.7 | 80.0 | 78.1 | 73.8 | 64.3 | 57.0 |
| | 5% WB | | 38.6 | 41.2 | 44.7 | 50.8 | 56.2 | 61.8 | 63.5 | 63.2 | 59.0 | 51.4 | 44.5 | 38.7 |
| | MCDB | | 53.2 | 56.7 | 63.6 | 70.9 | 73.6 | 80.0 | 81.2 | 79.1 | 76.6 | 71.3 | 60.3 | 52.0 |
| | 10% WB | | 36.3 | 38.9 | 42.8 | 48.9 | 54.9 | 60.5 | 62.3 | 62.1 | 57.4 | 49.8 | 41.6 | 35.3 |
| | MCDB | | 48.6 | 52.2 | 59.1 | 67.2 | 72.1 | 79.0 | 80.6 | 78.2 | 75.3 | 68.7 | 54.3 | 46.4 |
| Mean Daily Temperature Range | MDBR | | 24.4 | 23.1 | 24.6 | 25.1 | 25.3 | 27.1 | 27.9 | 26.4 | 27.8 | 27.5 | 23.9 | 24.4 |
| | 5% DB MCDBR | | 29.7 | 30.2 | 33.1 | 31.8 | 31.5 | 32.8 | 32.8 | 31.2 | 32.6 | 35.4 | 31.8 | 30.9 |
| | MCWBR | | 16.6 | 15.8 | 15.4 | 12.9 | 11.5 | 10.0 | 9.5 | 9.3 | 11.4 | 14.8 | 16.2 | 17.0 |
| | 5% WB MCDBR | | 28.9 | 29.4 | 31.3 | 29.8 | 27.4 | 27.4 | 28.1 | 25.9 | 27.8 | 32.2 | 30.5 | 30.4 |
| | MCWBR | | 16.4 | 15.7 | 14.9 | 12.5 | 10.9 | 10.3 | 9.2 | 8.9 | 10.8 | 14.1 | 15.7 | 17.1 |
| Clear Sky Solar Irradiance | taub | | 0.250 | 0.259 | 0.276 | 0.273 | 0.277 | 0.301 | 0.312 | 0.323 | 0.300 | 0.277 | 0.268 | 0.263 |
| | taud | | 2.659 | 2.539 | 2.484 | 2.542 | 2.536 | 2.472 | 2.480 | 2.474 | 2.570 | 2.650 | 2.574 | 2.567 |
| | Ebn,noon | | 302 | 311 | 313 | 319 | 317 | 308 | 304 | 298 | 301 | 301 | 290 | 286 |
| | Edh,noon | | 20 | 26 | 31 | 31 | 32 | 35 | 34 | 33 | 28 | 23 | 22 | 21 |

Nomenclature: See separate page

*Fig. 4-4 Example tables of climatic design information included on CD-ROM of 2013 ASHRAE Handbook—Fundamentals*

**100**                                                                     **Principles of HVAC**

2013 ASHRAE Handbook - Fundamentals (IP)                  © 2013 ASHRAE, Inc.

**DES MOINES INTL, IA, USA**      WMO#: 725460

Lat: 41.54N    Long: 93.67W    Elev: 965    StdP: 14.19    Time Zone: -6 (NAC)    Period: 86-10    WBAN: 14933

**Annual Heating and Humidification Design Conditions**

| Coldest Month | Heating DB 99.6% | 99% | Humidification DP/MCDB and HR 99.6% DP | HR | MCDB | 99% DP | HR | MCDB | Coldest month WS/MCDB 0.4% WS | MCDB | 1% WS | MCDB | MCWS/PCWD to 99.6% DB MCWS | PCWD |
|---|---|---|---|---|---|---|---|---|---|---|---|---|---|---|
| (a) | (b) | (c) | (d) | (e) | (f) | (g) | (h) | (i) | (j) | (k) | (l) | (m) | (n) | (o) |
| 1 | -5.3 | -0.2 | -13.1 | 2.8 | -4.2 | -8.7 | 3.6 | 0.6 | 30.2 | 13.3 | 26.9 | 20.9 | 9.6 | 310 |

**Annual Cooling, Dehumidification, and Enthalpy Design Conditions**

| Hottest Month | Hottest Month DB Range | Cooling DB/MCWB 0.4% DB | MCWB | 1% DB | MCWB | 2% DB | MCWB | Evaporation WB/MCDB 0.4% WB | MCDB | 1% WB | MCDB | 2% WB | MCDB | MCWS/PCWD to 0.4% DB MCWS | PCWD |
|---|---|---|---|---|---|---|---|---|---|---|---|---|---|---|---|
| (a) | (b) | (c) | (d) | (e) | (f) | (g) | (h) | (i) | (j) | (k) | (l) | (m) | (n) | (o) | (p) |
| 7 | 17.9 | 92.5 | 76.4 | 89.6 | 75.1 | 86.9 | 73.3 | 78.5 | 88.5 | 77.1 | 86.8 | 75.5 | 84.1 | 12.2 | 180 |

| Dehumidification DP/MCDB and HR 0.4% DP | HR | MCDB | 1% DP | HR | MCDB | 2% DP | HR | MCDB | Enthalpy/MCDB 0.4% Enth | MCDB | 1% Enth | MCDB | 2% Enth | MCDB | Hours 8 to 4 & 55/69 |
|---|---|---|---|---|---|---|---|---|---|---|---|---|---|---|---|
| (a) | (b) | (c) | (d) | (e) | (f) | (g) | (h) | (i) | (j) | (k) | (l) | (m) | (n) | (o) | (p) |
| 75.6 | 138.7 | 84.7 | 74.1 | 131.9 | 83.4 | 72.6 | 125.0 | 81.6 | 42.8 | 88.8 | 41.2 | 86.8 | 39.7 | 84.6 | 621 |

**Extreme Annual Design Conditions**

| Extreme Annual WS 1% | 2.5% | 5% | Extreme Max WB | Extreme Annual DB Mean Min | Max | Standard deviation Min | Max | n=5 years Min | Max | n=10 years Min | Max | n=20 years Min | Max | n=50 years Min | Max |
|---|---|---|---|---|---|---|---|---|---|---|---|---|---|---|---|
| (a) | (b) | (c) | (d) | (e) | (f) | (g) | (h) | (i) | (j) | (k) | (l) | (m) | (n) | (o) | (p) |
| 25.4 | 22.3 | 19.5 | 85.1 | -11.4 | 96.8 | 5.2 | 3.1 | -15.1 | 99.0 | -18.2 | 100.8 | -21.1 | 102.6 | -24.8 | 104.9 |

**Monthly Climatic Design Conditions**

| | | Annual (d) | Jan (e) | Feb (f) | Mar (g) | Apr (h) | May (i) | Jun (j) | Jul (k) | Aug (l) | Sep (m) | Oct (n) | Nov (o) | Dec (p) |
|---|---|---|---|---|---|---|---|---|---|---|---|---|---|---|
| Temperatures, Degree-Days and Degree-Hours | Tavg | 50.9 | 23.2 | 27.3 | 39.4 | 51.6 | 62.2 | 71.7 | 75.8 | 73.8 | 65.5 | 53.0 | 39.3 | 26.8 |
| | Sd | | 12.23 | 12.36 | 11.79 | 9.99 | 7.93 | 6.39 | 5.44 | 5.98 | 7.87 | 9.51 | 10.51 | 11.50 |
| | HDD50 | 3085 | 832 | 638 | 367 | 95 | 5 | 0 | 0 | 0 | 2 | 78 | 346 | 722 |
| | HDD65 | 6172 | 1297 | 1057 | 795 | 415 | 150 | 17 | 1 | 7 | 90 | 388 | 770 | 1185 |
| | CDD50 | 3421 | 0 | 1 | 39 | 144 | 383 | 652 | 801 | 737 | 467 | 170 | 26 | 1 |
| | CDD65 | 1034 | 0 | 0 | 2 | 14 | 63 | 219 | 337 | 279 | 105 | 15 | 0 | 0 |
| | CDH74 | 9260 | 0 | 0 | 22 | 165 | 550 | 1956 | 3178 | 2388 | 857 | 141 | 3 | 0 |
| | CDH80 | 3231 | 0 | 0 | 4 | 38 | 124 | 679 | 1250 | 861 | 249 | 26 | 0 | 0 |
| Precipitation | PrecAvg | 33.1 | 1.0 | 1.1 | 2.3 | 3.4 | 3.7 | 4.5 | 3.8 | 4.2 | 3.5 | 2.6 | 1.8 | 1.3 |
| | PrecMax | 44.6 | 4.3 | 2.9 | 5.0 | 7.8 | 7.2 | 11.4 | 9.2 | 13.7 | 9.5 | 5.2 | 3.3 | 2.6 |
| | PrecMin | 22.7 | 0.2 | 0.1 | 0.4 | 0.9 | 1.6 | 1.1 | 0.0 | 1.1 | 0.8 | 0.3 | 0.0 | 0.1 |
| | PrecSD | 5.3 | 0.9 | 0.9 | 1.3 | 1.7 | 1.5 | 2.4 | 2.2 | 3.2 | 2.5 | 1.5 | 1.0 | 0.7 |
| Monthly Design Dry Bulb and Mean Coincident Wet Bulb Temperatures | 0.4% DB | | 56.0 | 61.6 | 76.6 | 85.0 | 87.5 | 93.9 | 96.9 | 96.6 | 90.3 | 83.6 | 72.7 | 58.8 |
| | 0.4% MCWB | | 46.7 | 51.1 | 59.2 | 64.9 | 70.5 | 75.4 | 77.7 | 76.6 | 71.9 | 67.2 | 59.3 | 52.3 |
| | 2% DB | | 48.6 | 54.9 | 69.2 | 79.0 | 83.5 | 89.7 | 93.2 | 91.3 | 86.2 | 78.1 | 65.4 | 51.7 |
| | 2% MCWB | | 41.8 | 46.2 | 55.9 | 62.3 | 67.3 | 73.9 | 76.8 | 76.7 | 71.1 | 64.0 | 57.0 | 45.1 |
| | 5% DB | | 43.3 | 48.6 | 63.2 | 73.3 | 80.3 | 87.0 | 90.1 | 88.0 | 82.5 | 73.0 | 60.0 | 46.1 |
| | 5% MCWB | | 38.1 | 41.7 | 54.0 | 58.9 | 65.7 | 72.2 | 76.3 | 75.3 | 69.2 | 61.1 | 52.2 | 40.7 |
| | 10% DB | | 38.4 | 43.4 | 57.5 | 68.4 | 76.6 | 83.9 | 86.8 | 84.5 | 79.1 | 68.4 | 55.1 | 41.4 |
| | 10% MCWB | | 34.5 | 38.3 | 48.4 | 55.8 | 63.5 | 70.6 | 74.6 | 73.0 | 67.4 | 58.9 | 48.3 | 36.9 |
| Monthly Design Wet Bulb and Mean Coincident Dry Bulb Temperatures | 0.4% WB | | 48.2 | 52.0 | 62.3 | 68.1 | 74.4 | 78.4 | 81.2 | 80.2 | 75.9 | 71.1 | 62.1 | 54.5 |
| | 0.4% MCDB | | 53.3 | 57.1 | 71.3 | 81.1 | 83.8 | 88.8 | 92.4 | 90.4 | 85.6 | 79.5 | 67.8 | 57.2 |
| | 2% WB | | 42.0 | 46.4 | 58.8 | 64.3 | 71.3 | 76.5 | 79.0 | 78.2 | 73.5 | 66.6 | 58.2 | 46.1 |
| | 2% MCDB | | 48.1 | 55.0 | 65.4 | 75.2 | 79.6 | 86.5 | 89.3 | 88.0 | 82.3 | 74.2 | 64.4 | 49.8 |
| | 5% WB | | 38.2 | 43.4 | 54.5 | 61.1 | 68.4 | 74.6 | 77.5 | 76.7 | 71.6 | 63.4 | 53.2 | 40.8 |
| | 5% MCDB | | 43.2 | 47.9 | 63.6 | 70.8 | 76.6 | 84.0 | 87.5 | 85.8 | 79.5 | 70.7 | 58.5 | 45.7 |
| | 10% WB | | 34.8 | 38.3 | 49.4 | 57.6 | 65.7 | 72.6 | 76.0 | 75.0 | 69.4 | 59.9 | 48.6 | 37.1 |
| | 10% MCDB | | 37.9 | 42.9 | 57.3 | 66.6 | 73.7 | 81.4 | 84.7 | 83.1 | 76.4 | 67.1 | 54.5 | 40.7 |
| Mean Daily Temperature Range | MDBR | | 15.8 | 16.5 | 18.8 | 20.4 | 19.2 | 18.4 | 17.9 | 17.8 | 20.2 | 19.5 | 16.9 | 15.2 |
| | 5% DB MCDBR | | 22.1 | 23.6 | 26.7 | 27.4 | 23.1 | 21.3 | 21.0 | 20.8 | 22.9 | 25.1 | 23.9 | 21.7 |
| | 5% DB MCWBR | | 16.3 | 16.0 | 16.0 | 15.0 | 12.1 | 9.9 | 9.2 | 9.5 | 11.2 | 13.7 | 15.4 | 16.0 |
| | 5% WB MCDBR | | 21.0 | 22.4 | 24.2 | 23.7 | 19.7 | 18.9 | 18.9 | 18.6 | 19.6 | 21.3 | 20.6 | 20.3 |
| | 5% WB MCWBR | | 15.7 | 15.8 | 15.1 | 14.4 | 11.7 | 10.4 | 9.5 | 9.8 | 11.2 | 13.9 | 15.3 | 15.8 |
| Clear Sky Solar Irradiance | taub | | 0.282 | 0.294 | 0.324 | 0.341 | 0.345 | 0.377 | 0.389 | 0.372 | 0.350 | 0.320 | 0.302 | 0.289 |
| | taud | | 2.480 | 2.389 | 2.429 | 2.384 | 2.380 | 2.297 | 2.280 | 2.375 | 2.440 | 2.550 | 2.620 | 2.557 |
| | Ebn,noon | | 279 | 291 | 294 | 295 | 294 | 284 | 279 | 281 | 281 | 280 | 271 | 268 |
| | Edh,noon | | 23 | 30 | 32 | 37 | 38 | 41 | 41 | 36 | 32 | 25 | 21 | 20 |

Nomenclature:    See separate page

*Fig. 4-4 Example tables of climatic design information included on CD-ROM of 2013 ASHRAE Handbook—Fundamentals*

# Design Conditions

2013 ASHRAE Handbook - Fundamentals (IP) © 2013 ASHRAE, Inc.

**HARTFORD/BRADLEY IN, CT, USA**     WMO#: 725080

Lat: 41.94N   Long: 72.68W   Elev: 180   StdP: 14.6   Time Zone: -5 (NAE)   Period: 86-10   WBAN: 14740

### Annual Heating and Humidification Design Conditions

| Coldest Month | Heating DB | | Humidification DP/MCDB and HR | | | | | | Coldest month WS/MCDB | | | | MCWS/PCWD to 99.6% DB | |
|---|---|---|---|---|---|---|---|---|---|---|---|---|---|---|
| | 99.6% | 99% | 99.6% | | | 99% | | | 0.4% | | 1% | | | |
| | | | DP | HR | MCDB | DP | HR | MCDB | WS | MCDB | WS | MCDB | MCWS | PCWD |
| (a) | (b) | (c) | (d) | (e) | (f) | (g) | (h) | (i) | (j) | (k) | (l) | (m) | (n) | (o) |
| 1 | 4.1 | 9.2 | -10.7 | 3.1 | 8.1 | -6.4 | 3.9 | 12.3 | 26.5 | 27.5 | 24.3 | 26.1 | 7.7 | 0 |

### Annual Cooling, Dehumidification, and Enthalpy Design Conditions

| Hottest Month | Hottest Month DB Range | Cooling DB/MCWB | | | | | | Evaporation WB/MCDB | | | | | | MCWS/PCWD to 0.4% DB | |
|---|---|---|---|---|---|---|---|---|---|---|---|---|---|---|---|
| | | 0.4% | | 1% | | 2% | | 0.4% | | 1% | | 2% | | | |
| | | DB | MCWB | DB | MCWB | DB | MCWB | WB | MCDB | WB | MCDB | WB | MCDB | MCWS | PCWD |
| (a) | (b) | (c) | (d) | (e) | (f) | (g) | (h) | (i) | (j) | (k) | (l) | (m) | (n) | (o) | (p) |
| 7 | 20.1 | 91.4 | 73.3 | 88.5 | 72.0 | 85.5 | 70.5 | 76.3 | 86.7 | 74.8 | 83.9 | 73.3 | 81.2 | 10.2 | 220 |

| Dehumidification DP/MCDB and HR | | | | | | | | | Enthalpy/MCDB | | | | | | Hours 8 to 4 & 55/69 |
|---|---|---|---|---|---|---|---|---|---|---|---|---|---|---|---|
| 0.4% | | | 1% | | | 2% | | | 0.4% | | 1% | | 2% | | |
| DP | HR | MCDB | DP | HR | MCDB | DP | HR | MCDB | Enth | MCDB | Enth | MCDB | Enth | MCDB | |
| (a) | (b) | (c) | (d) | (e) | (f) | (g) | (h) | (i) | (j) | (k) | (l) | (m) | (n) | (o) | (p) |
| 73.2 | 124.3 | 80.3 | 72.1 | 119.4 | 79.2 | 70.6 | 113.6 | 77.9 | 39.8 | 87.0 | 38.3 | 84.1 | 36.9 | 81.4 | 702 |

### Extreme Annual Design Conditions

| Extreme Annual WS | | | Extreme Max WB | Extreme Annual DB | | | | n-Year Return Period Values of Extreme DB | | | | | | | |
|---|---|---|---|---|---|---|---|---|---|---|---|---|---|---|---|
| | | | | Mean | | Standard deviation | | n=5 years | | n=10 years | | n=20 years | | n=50 years | |
| 1% | 2.5% | 5% | | Min | Max | Min | Max | Min | Max | Min | Max | Min | Max | Min | Max |
| (a) | (b) | (c) | (d) | (e) | (f) | (g) | (h) | (i) | (j) | (k) | (l) | (m) | (n) | (o) | (p) |
| 22.8 | 19.3 | 17.5 | 84.6 | -2.3 | 96.5 | 5.2 | 2.7 | -6.1 | 98.4 | -9.1 | 100.0 | -12.1 | 101.6 | -15.9 | 103.6 |

### Monthly Climatic Design Conditions

| | | | Annual | Jan | Feb | Mar | Apr | May | Jun | Jul | Aug | Sep | Oct | Nov | Dec |
|---|---|---|---|---|---|---|---|---|---|---|---|---|---|---|---|
| | | | (d) | (e) | (f) | (g) | (h) | (i) | (j) | (k) | (l) | (m) | (n) | (o) | (p) |
| (5) | Temperatures, Degree-Days and Degree-Hours | Tavg | 50.8 | 27.3 | 29.4 | 38.1 | 49.6 | 59.5 | 68.6 | 73.5 | 72.0 | 64.0 | 52.3 | 42.4 | 31.8 |
| (6) | | Sd | | 10.24 | 8.89 | 9.22 | 8.11 | 7.42 | 6.64 | 5.48 | 5.95 | 6.99 | 7.79 | 8.34 | 9.00 |
| (7) | | HDD50 | 2664 | 704 | 577 | 384 | 103 | 7 | 0 | 0 | 0 | 1 | 64 | 257 | 567 |
| (8) | | HDD65 | 5935 | 1168 | 996 | 834 | 469 | 206 | 38 | 4 | 11 | 100 | 401 | 679 | 1029 |
| (9) | | CDD50 | 2966 | 1 | 1 | 16 | 89 | 302 | 558 | 729 | 682 | 422 | 136 | 27 | 3 |
| (10) | | CDD65 | 765 | 0 | 0 | 1 | 6 | 36 | 147 | 268 | 228 | 71 | 8 | 0 | 0 |
| (11) | | CDH74 | 6936 | 0 | 0 | 12 | 112 | 487 | 1363 | 2419 | 1960 | 525 | 56 | 2 | 0 |
| (12) | | CDH80 | 2389 | 0 | 0 | 2 | 40 | 159 | 468 | 913 | 678 | 121 | 8 | 0 | 0 |
| (13) | Precipitation | PrecAvg | 45.4 | 3.4 | 3.1 | 3.7 | 3.8 | 4.1 | 4.0 | 3.8 | 3.7 | 4.2 | 3.8 | 3.9 | 3.8 |
| (14) | | PrecMax | 65.8 | 9.6 | 5.3 | 6.7 | 6.6 | 7.5 | 9.7 | 8.2 | 6.3 | 9.0 | 5.4 | 8.5 | 8.4 |
| (15) | | PrecMin | 29.7 | 0.4 | 1.0 | 1.5 | 1.4 | 0.9 | 0.7 | 1.5 | 1.1 | 1.4 | 0.4 | 0.5 | 0.8 |
| (16) | | PrecSD | 9.0 | 2.4 | 1.3 | 1.3 | 1.5 | 1.6 | 2.0 | 1.7 | 1.4 | 2.4 | 1.4 | 1.8 | 2.0 |
| (17) | Monthly Design Dry Bulb and Mean Coincident Wet Bulb Temperatures | 0.4% DB | | 58.1 | 57.7 | 73.3 | 86.1 | 90.1 | 93.0 | 96.1 | 94.5 | 88.7 | 80.3 | 70.2 | 61.4 |
| (18) | | MCWB | | 53.8 | 49.4 | 56.4 | 64.6 | 67.9 | 73.7 | 76.2 | 75.2 | 71.8 | 66.3 | 59.2 | 56.0 |
| (19) | | 2% DB | | 50.8 | 51.3 | 63.9 | 75.6 | 84.5 | 89.2 | 91.6 | 90.3 | 83.1 | 74.2 | 64.0 | 54.6 |
| (20) | | MCWB | | 46.5 | 44.5 | 51.0 | 57.2 | 66.2 | 72.1 | 74.4 | 73.7 | 69.2 | 63.3 | 57.8 | 49.6 |
| (21) | | 5% DB | | 45.1 | 46.2 | 57.7 | 69.8 | 79.5 | 85.4 | 88.3 | 87.0 | 79.8 | 70.3 | 60.5 | 49.4 |
| (22) | | MCWB | | 40.0 | 40.0 | 47.8 | 54.6 | 63.5 | 70.3 | 72.6 | 72.3 | 67.5 | 61.5 | 55.0 | 45.0 |
| (23) | | 10% DB | | 40.4 | 41.9 | 52.3 | 64.2 | 74.5 | 81.8 | 85.0 | 83.5 | 76.5 | 65.9 | 56.4 | 44.6 |
| (24) | | MCWB | | 36.4 | 36.8 | 45.0 | 50.6 | 60.4 | 67.9 | 70.7 | 70.2 | 66.1 | 57.9 | 50.5 | 39.6 |
| (25) | Monthly Design Wet Bulb and Mean Coincident Dry Bulb Temperatures | 0.4% WB | | 55.6 | 52.5 | 60.7 | 65.5 | 72.2 | 76.8 | 78.4 | 78.2 | 74.9 | 70.3 | 63.8 | 56.5 |
| (26) | | MCDB | | 57.4 | 55.7 | 70.6 | 82.6 | 85.9 | 88.6 | 90.3 | 90.0 | 82.4 | 75.2 | 66.2 | 59.7 |
| (27) | | 2% WB | | 47.0 | 45.7 | 54.1 | 60.8 | 68.9 | 74.6 | 76.6 | 76.0 | 72.5 | 66.6 | 59.4 | 50.5 |
| (28) | | MCDB | | 49.7 | 50.6 | 60.2 | 72.0 | 79.4 | 84.7 | 87.4 | 85.5 | 78.8 | 72.0 | 63.1 | 52.9 |
| (29) | | 5% WB | | 40.8 | 40.6 | 49.7 | 57.2 | 65.8 | 72.7 | 75.0 | 74.4 | 70.7 | 63.1 | 56.0 | 45.3 |
| (30) | | MCDB | | 44.3 | 45.4 | 56.3 | 64.5 | 75.9 | 81.3 | 84.3 | 82.6 | 76.3 | 67.9 | 59.7 | 49.0 |
| (31) | | 10% WB | | 37.1 | 37.2 | 45.6 | 53.5 | 62.8 | 70.6 | 73.3 | 72.8 | 68.6 | 59.7 | 51.3 | 40.0 |
| (32) | | MCDB | | 39.9 | 40.9 | 52.1 | 61.2 | 71.2 | 78.5 | 81.4 | 80.2 | 74.2 | 64.6 | 55.2 | 43.6 |
| (33) | Mean Daily Temperature Range | MDBR | | 15.2 | 16.1 | 18.4 | 20.5 | 21.9 | 20.7 | 20.1 | 20.1 | 20.4 | 20.3 | 17.0 | 14.8 |
| (34) | | 5% DB MCDBR | | 19.3 | 20.9 | 27.1 | 30.5 | 29.7 | 25.5 | 23.6 | 23.1 | 23.1 | 25.3 | 22.2 | 20.7 |
| (35) | | MCWBR | | 15.8 | 15.6 | 17.0 | 15.4 | 14.2 | 11.9 | 9.5 | 9.4 | 11.4 | 14.9 | 16.8 | 16.7 |
| (36) | | 5% WB MCDBR | | 18.1 | 19.3 | 24.2 | 24.7 | 26.3 | 21.5 | 20.6 | 19.5 | 18.4 | 20.6 | 19.8 | 19.7 |
| (37) | | MCWBR | | 16.0 | 15.3 | 17.2 | 14.9 | 13.8 | 11.4 | 9.4 | 9.0 | 10.9 | 14.6 | 17.4 | 17.7 |
| (38) | Clear Sky Solar Irradiance | taub | | 0.315 | 0.325 | 0.341 | 0.383 | 0.409 | 0.465 | 0.431 | 0.418 | 0.399 | 0.331 | 0.327 | 0.305 |
| (39) | | taud | | 2.484 | 2.443 | 2.419 | 2.299 | 2.198 | 2.036 | 2.175 | 2.247 | 2.319 | 2.554 | 2.560 | 2.627 |
| (40) | | Ebn,noon | | 264 | 279 | 288 | 281 | 275 | 259 | 267 | 267 | 263 | 275 | 259 | 262 |
| (41) | | Edh,noon | | 24 | 28 | 33 | 40 | 45 | 53 | 46 | 41 | 36 | 25 | 22 | 19 |

Nomenclature: See separate page

*Fig. 4-4 Example tables of climatic design information included on CD-ROM of 2013 ASHRAE Handbook—Fundamentals*

2013 ASHRAE Handbook - Fundamentals (IP)  © 2013 ASHRAE, Inc.

**HOUSTON/INTERCONTIN, TX, USA**  WMO#: 722430

Lat: 29.99N  Long: 95.36W  Elev: 105  StdP: 14.64  Time Zone: -6 (NAC)  Period: 86-10  WBAN: 12960

### Annual Heating and Humidification Design Conditions

| Coldest Month | Heating DB | | Humidification DP/MCDB and HR | | | | | | Coldest month WS/MCDB | | | | MCWS/PCWD to 99.6% DB | |
|---|---|---|---|---|---|---|---|---|---|---|---|---|---|---|
| | 99.6% | 99% | 99.6% | | | 99% | | | 0.4% | | 1% | | | |
| | | | DP | HR | MCDB | DP | HR | MCDB | WS | MCDB | WS | MCDB | MCWS | PCWD |
| (a) | (b) | (c) | (d) | (e) | (f) | (g) | (h) | (i) | (j) | (k) | (l) | (m) | (n) | (o) |
| 1 | 30.3 | 33.8 | 16.1 | 12.5 | 42.1 | 20.8 | 15.6 | 44.6 | 22.3 | 57.5 | 20.0 | 58.2 | 5.5 | 0 |

### Annual Cooling, Dehumidification, and Enthalpy Design Conditions

| Hottest Month | Hottest Month DB Range | Cooling DB/MCWB | | | | | | Evaporation WB/MCDB | | | | | | MCWS/PCWD to 0.4% DB | |
|---|---|---|---|---|---|---|---|---|---|---|---|---|---|---|---|
| | | 0.4% | | 1% | | 2% | | 0.4% | | 1% | | 2% | | | |
| | | DB | MCWB | DB | MCWB | DB | MCWB | WB | MCDB | WB | MCDB | WB | MCDB | MCWS | PCWD |
| (a) | (b) | (c) | (d) | (e) | (f) | (g) | (h) | (i) | (j) | (k) | (l) | (m) | (n) | (o) | (p) |
| 7 | 18.0 | 97.2 | 76.6 | 95.2 | 76.7 | 93.4 | 76.6 | 80.2 | 88.9 | 79.4 | 88.2 | 78.8 | 87.4 | 7.7 | 170 |

| Dehumidification DP/MCDB and HR | | | | | | | | | Enthalpy/MCDB | | | | | | Hours 8 to 4 & 55/69 |
|---|---|---|---|---|---|---|---|---|---|---|---|---|---|---|---|
| 0.4% | | | 1% | | | 2% | | | 0.4% | | 1% | | 2% | | |
| DP | HR | MCDB | DP | HR | MCDB | DP | HR | MCDB | Enth | MCDB | Enth | MCDB | Enth | MCDB | |
| (a) | (b) | (c) | (d) | (e) | (f) | (g) | (h) | (i) | (j) | (k) | (l) | (m) | (n) | (o) | (p) |
| 78.2 | 147.1 | 82.9 | 77.3 | 142.7 | 82.5 | 76.7 | 139.8 | 82.2 | 43.6 | 88.2 | 42.9 | 88.2 | 42.3 | 87.5 | 648 |

### Extreme Annual Design Conditions

| Extreme Annual WS | | | Extreme Max WB | Extreme Annual DB | | | | n-Year Return Period Values of Extreme DB | | | | | | | |
|---|---|---|---|---|---|---|---|---|---|---|---|---|---|---|---|
| 1% | 2.5% | 5% | | Mean | | Standard deviation | | n=5 years | | n=10 years | | n=20 years | | n=50 years | |
| | | | | Min | Max | Min | Max | Min | Max | Min | Max | Min | Max | Min | Max |
| (a) | (b) | (c) | (d) | (e) | (f) | (g) | (h) | (i) | (j) | (k) | (l) | (m) | (n) | (o) | (p) |
| 19.6 | 17.8 | 16.2 | 83.8 | 25.6 | 100.6 | 4.9 | 2.7 | 22.1 | 102.6 | 19.2 | 104.2 | 16.5 | 105.7 | 12.9 | 107.7 |

### Monthly Climatic Design Conditions

| | | Annual | Jan | Feb | Mar | Apr | May | Jun | Jul | Aug | Sep | Oct | Nov | Dec |
|---|---|---|---|---|---|---|---|---|---|---|---|---|---|---|
| | | (d) | (e) | (f) | (g) | (h) | (i) | (j) | (k) | (l) | (m) | (n) | (o) | (p) |
| Temperatures, Degree-Days and Degree-Hours | Tavg | 69.6 | 53.4 | 56.8 | 62.5 | 69.2 | 76.8 | 81.9 | 84.0 | 84.0 | 79.4 | 70.9 | 61.7 | 54.1 |
| | Sd | | 9.32 | 9.30 | 8.25 | 6.73 | 4.91 | 3.37 | 2.73 | 3.25 | 4.94 | 7.30 | 8.87 | 9.80 |
| | HDD50 | 194 | 70 | 36 | 10 | 0 | 0 | 0 | 0 | 0 | 0 | 1 | 12 | 65 |
| | HDD65 | 1371 | 374 | 257 | 145 | 37 | 1 | 0 | 0 | 0 | 0 | 30 | 166 | 361 |
| | CDD50 | 7360 | 177 | 226 | 397 | 576 | 832 | 958 | 1053 | 1055 | 883 | 648 | 362 | 193 |
| | CDD65 | 3059 | 15 | 27 | 67 | 163 | 368 | 508 | 588 | 590 | 433 | 211 | 65 | 24 |
| | CDH74 | 30921 | 48 | 116 | 331 | 1137 | 3321 | 5548 | 6925 | 6939 | 4268 | 1810 | 402 | 76 |
| | CDH80 | 13353 | 1 | 15 | 36 | 256 | 1190 | 2511 | 3376 | 3485 | 1854 | 570 | 56 | 3 |
| Precipitation | PrecAvg | 49.7 | 3.6 | 2.9 | 3.5 | 3.6 | 5.3 | 5.4 | 3.5 | 3.9 | 4.7 | 5.4 | 4.1 | 3.7 |
| | PrecMax | 71.2 | 9.8 | 6.1 | 8.5 | 10.9 | 14.4 | 19.2 | 9.8 | 9.4 | 12.3 | 16.1 | 11.7 | 9.3 |
| | PrecMin | 22.9 | 0.4 | 0.4 | 0.1 | 0.4 | 0.0 | 0.1 | 0.3 | 0.3 | 0.8 | 0.1 | 0.4 | 0.6 |
| | PrecSD | 10.5 | 2.3 | 1.8 | 2.1 | 2.8 | 3.5 | 4.8 | 2.3 | 2.3 | 3.1 | 4.1 | 2.7 | 1.9 |
| Monthly Design Dry Bulb and Mean Coincident Wet Bulb Temperatures | 0.4% DB | | 78.6 | 81.7 | 83.6 | 89.1 | 93.5 | 97.3 | 99.3 | 99.7 | 97.4 | 91.3 | 84.6 | 79.5 |
| | 0.4% MCWB | | 66.6 | 66.2 | 68.8 | 69.8 | 74.0 | 76.6 | 75.9 | 76.8 | 75.9 | 74.7 | 71.9 | 69.2 |
| | 2% DB | | 75.0 | 78.0 | 80.6 | 85.3 | 90.7 | 94.5 | 96.8 | 97.4 | 93.6 | 88.8 | 81.5 | 76.0 |
| | 2% MCWB | | 65.1 | 66.6 | 67.3 | 70.7 | 74.8 | 76.5 | 76.6 | 76.7 | 75.4 | 73.6 | 70.1 | 68.1 |
| | 5% DB | | 72.3 | 74.2 | 77.9 | 82.9 | 88.7 | 92.6 | 94.7 | 95.5 | 91.4 | 86.1 | 78.7 | 73.2 |
| | 5% MCWB | | 65.2 | 64.8 | 66.5 | 69.9 | 74.7 | 76.5 | 76.9 | 77.0 | 75.5 | 72.3 | 68.9 | 67.6 |
| | 10% DB | | 69.6 | 71.5 | 75.0 | 80.6 | 86.6 | 90.8 | 92.9 | 93.3 | 89.6 | 83.1 | 75.7 | 70.6 |
| | 10% MCWB | | 64.2 | 64.8 | 65.4 | 68.9 | 74.1 | 76.5 | 77.0 | 77.1 | 75.0 | 71.2 | 68.4 | 65.7 |
| Monthly Design Wet Bulb and Mean Coincident Dry Bulb Temperatures | 0.4% WB | | 70.4 | 70.8 | 72.5 | 76.2 | 79.3 | 81.0 | 80.7 | 80.8 | 80.2 | 78.9 | 75.3 | 71.9 |
| | 0.4% MCDB | | 73.3 | 75.8 | 78.3 | 82.1 | 88.2 | 89.3 | 89.2 | 90.3 | 87.8 | 83.6 | 80.1 | 75.5 |
| | 2% WB | | 68.8 | 69.3 | 70.9 | 74.7 | 77.9 | 79.8 | 79.9 | 80.0 | 79.1 | 77.4 | 73.4 | 70.6 |
| | 2% MCDB | | 72.4 | 73.9 | 76.0 | 80.3 | 86.0 | 87.8 | 88.7 | 89.7 | 86.8 | 83.5 | 77.7 | 74.2 |
| | 5% WB | | 67.2 | 68.1 | 69.7 | 73.2 | 76.8 | 78.8 | 79.2 | 79.2 | 78.3 | 76.1 | 71.9 | 69.0 |
| | 5% MCDB | | 71.2 | 72.6 | 74.5 | 78.8 | 84.4 | 86.8 | 88.1 | 88.9 | 86.0 | 82.1 | 76.4 | 72.7 |
| | 10% WB | | 64.9 | 66.2 | 68.4 | 71.8 | 75.8 | 78.1 | 78.5 | 78.4 | 77.5 | 74.5 | 70.1 | 66.7 |
| | 10% MCDB | | 68.8 | 70.2 | 73.2 | 77.4 | 83.0 | 86.1 | 87.2 | 87.8 | 85.2 | 80.2 | 74.8 | 70.2 |
| Mean Daily Temperature Range | MDBR | | 18.8 | 19.9 | 19.8 | 19.4 | 17.9 | 17.3 | 18.0 | 18.9 | 19.1 | 20.7 | 19.9 | 19.1 |
| | 5% DB MCDBR | | 21.4 | 20.9 | 20.2 | 20.2 | 18.9 | 19.4 | 20.4 | 21.4 | 20.8 | 21.2 | 21.1 | 20.1 |
| | 5% DB MCWBR | | 14.0 | 11.8 | 10.2 | 8.9 | 6.7 | 5.6 | 5.3 | 5.4 | 6.5 | 8.4 | 11.7 | 13.3 |
| | 5% WB MCDBR | | 18.0 | 16.8 | 16.8 | 15.5 | 16.4 | 16.6 | 17.9 | 18.5 | 17.0 | 16.5 | 18.1 | 17.5 |
| | 5% WB MCWBR | | 13.7 | 12.0 | 10.7 | 8.4 | 6.8 | 6.0 | 5.7 | 5.6 | 6.3 | 8.3 | 12.7 | 13.6 |
| Clear Sky Solar Irradiance | taub | | 0.361 | 0.374 | 0.370 | 0.393 | 0.379 | 0.383 | 0.399 | 0.400 | 0.397 | 0.370 | 0.386 | 0.365 |
| | taud | | 2.562 | 2.487 | 2.447 | 2.344 | 2.384 | 2.404 | 2.381 | 2.404 | 2.442 | 2.529 | 2.489 | 2.561 |
| | Ebn,noon | | 277 | 282 | 291 | 286 | 288 | 286 | 281 | 280 | 278 | 280 | 265 | 269 |
| | Edh,noon | | 27 | 32 | 35 | 40 | 39 | 38 | 39 | 37 | 35 | 30 | 29 | 26 |

Nomenclature:   See separate page

*Fig. 4-4  Example tables of climatic design information included on CD-ROM of 2013 ASHRAE Handbook—Fundamentals*

# Design Conditions

2013 ASHRAE Handbook - Fundamentals (IP)  © 2013 ASHRAE, Inc.

**INDIANAPOLIS/I.-MUN, IN, USA**  WMO#: 724380

Lat: 39.71N  Long: 86.27W  Elev: 807  StdP: 14.27  Time Zone: -5 (NAE)  Period: 86-10  WBAN: 93819

## Annual Heating and Humidification Design Conditions

| Coldest Month | Heating DB | | Humidification DP/MCDB and HR | | | | | | Coldest month WS/MCDB | | | | MCWS/PCWD to 99.6% DB | |
|---|---|---|---|---|---|---|---|---|---|---|---|---|---|---|
| | 99.6% | 99% | 99.6% | | | 99% | | | 0.4% | | 1% | | MCWS | PCWD |
| | | | DP | HR | MCDB | DP | HR | MCDB | WS | MCDB | WS | MCDB | | |
| (a) | (b) | (c) | (d) | (e) | (f) | (g) | (h) | (i) | (j) | (k) | (l) | (m) | (n) | (o) |
| 1 | 2.0 | 8.1 | -6.0 | 4.1 | 3.2 | -0.3 | 5.6 | 9.5 | 27.4 | 32.5 | 25.4 | 31.1 | 9.8 | 280 |

## Annual Cooling, Dehumidification, and Enthalpy Design Conditions

| Hottest Month | Hottest Month DB Range | Cooling DB/MCWB | | | | | | Evaporation WB/MCDB | | | | | | MCWS/PCWD to 0.4% DB | |
|---|---|---|---|---|---|---|---|---|---|---|---|---|---|---|---|
| | | 0.4% | | 1% | | 2% | | 0.4% | | 1% | | 2% | | MCWS | PCWD |
| | | DB | MCWB | DB | MCWB | DB | MCWB | WB | MCDB | WB | MCDB | WB | MCDB | | |
| (a) | (b) | (c) | (d) | (e) | (f) | (g) | (h) | (i) | (j) | (k) | (l) | (m) | (n) | (o) | (p) |
| 7 | 18.0 | 91.0 | 75.1 | 88.7 | 74.0 | 86.4 | 72.9 | 78.2 | 87.5 | 76.8 | 85.4 | 75.3 | 82.9 | 10.7 | 220 |

| Dehumidification DP/MCDB and HR | | | | | | | | Enthalpy/MCDB | | | | | | Hours 8 to 4 & 55/69 |
|---|---|---|---|---|---|---|---|---|---|---|---|---|---|---|
| 0.4% | | | 1% | | | 2% | | 0.4% | | 1% | | 2% | | |
| DP | HR | MCDB | DP | HR | MCDB | DP | HR | MCDB | Enth | MCDB | Enth | MCDB | Enth | MCDB | |
| (a) | (b) | (c) | (d) | (e) | (f) | (g) | (h) | (i) | (j) | (k) | (l) | (m) | (n) | (o) | (p) |
| 75.3 | 136.8 | 83.6 | 74.0 | 130.6 | 82.0 | 72.7 | 124.9 | 80.4 | 42.2 | 88.0 | 40.6 | 85.2 | 39.2 | 83.1 | 665 |

## Extreme Annual Design Conditions

| Extreme Annual WS | | | Extreme Max WB | Extreme Annual DB | | | | n-Year Return Period Values of Extreme DB | | | | | | | |
|---|---|---|---|---|---|---|---|---|---|---|---|---|---|---|---|
| 1% | 2.5% | 5% | | Mean | | Standard deviation | | n=5 years | | n=10 years | | n=20 years | | n=50 years | |
| | | | | Min | Max | Min | Max | Min | Max | Min | Max | Min | Max | Min | Max |
| (a) | (b) | (c) | (d) | (e) | (f) | (g) | (h) | (i) | (j) | (k) | (l) | (m) | (n) | (o) | (p) |
| 24.7 | 20.9 | 18.8 | 84.4 | -5.3 | 94.3 | 7.9 | 2.9 | -10.9 | 96.4 | -15.5 | 98.1 | -20.0 | 99.7 | -25.7 | 101.8 |

## Monthly Climatic Design Conditions

| | | | Annual | Jan | Feb | Mar | Apr | May | Jun | Jul | Aug | Sep | Oct | Nov | Dec |
|---|---|---|---|---|---|---|---|---|---|---|---|---|---|---|---|
| | | | (d) | (e) | (f) | (g) | (h) | (i) | (j) | (k) | (l) | (m) | (n) | (o) | (p) |
| Temperatures, Degree-Days and Degree-Hours | | Tavg | 53.5 | 29.2 | 32.5 | 42.7 | 53.5 | 63.0 | 72.2 | 75.4 | 74.3 | 67.3 | 54.9 | 43.8 | 32.3 |
| | | Sd | | 11.93 | 10.90 | 10.84 | 9.42 | 7.92 | 6.20 | 4.74 | 5.40 | 7.29 | 8.72 | 9.66 | 10.87 |
| | | HDD50 | 2341 | 651 | 494 | 282 | 69 | 5 | 0 | 0 | 0 | 1 | 48 | 235 | 556 |
| | | HDD65 | 5272 | 1109 | 909 | 693 | 362 | 135 | 16 | 1 | 4 | 62 | 330 | 636 | 1015 |
| | | CDD50 | 3630 | 6 | 5 | 56 | 172 | 408 | 667 | 788 | 753 | 519 | 201 | 49 | 6 |
| | | CDD65 | 1087 | 0 | 0 | 2 | 16 | 73 | 232 | 324 | 292 | 130 | 18 | 0 | 0 |
| | | CDH74 | 9191 | 0 | 0 | 16 | 129 | 579 | 1961 | 2858 | 2484 | 1022 | 140 | 2 | 0 |
| | | CDH80 | 3009 | 0 | 0 | 1 | 17 | 131 | 655 | 1023 | 859 | 298 | 25 | 0 | 0 |
| Precipitation | | PrecAvg | 39.9 | 2.3 | 2.5 | 3.8 | 3.7 | 4.0 | 3.5 | 4.5 | 3.6 | 2.9 | 2.6 | 3.2 | 3.3 |
| | | PrecMax | 44.6 | 6.2 | 5.4 | 9.0 | 8.1 | 9.3 | 7.1 | 11.1 | 8.4 | 5.7 | 5.7 | 5.6 | 6.0 |
| | | PrecMin | 31.7 | 1.0 | 0.4 | 1.3 | 1.0 | 1.4 | 0.8 | 1.2 | 0.7 | 0.2 | 0.0 | 0.8 | 0.4 |
| | | PrecSD | 4.2 | 1.3 | 1.4 | 2.0 | 2.0 | 2.2 | 1.7 | 2.6 | 2.1 | 1.6 | 1.5 | 1.3 | 1.5 |
| Monthly Design Dry Bulb and Mean Coincident Wet Bulb Temperatures | 0.4% | DB | | 62.2 | 64.1 | 76.1 | 82.4 | 87.1 | 92.1 | 95.1 | 93.6 | 91.2 | 83.3 | 71.9 | 62.9 |
| | | MCWB | | 57.6 | 54.6 | 61.0 | 65.2 | 70.0 | 72.6 | 76.3 | 76.3 | 72.3 | 68.2 | 59.2 | 57.2 |
| | 2% | DB | | 56.7 | 58.1 | 70.7 | 78.3 | 83.7 | 89.2 | 90.9 | 90.5 | 87.1 | 78.2 | 67.1 | 57.1 |
| | | MCWB | | 53.5 | 51.4 | 58.1 | 63.4 | 69.5 | 72.9 | 76.6 | 75.9 | 69.8 | 64.9 | 58.8 | 52.7 |
| | 5% | DB | | 51.6 | 53.4 | 65.6 | 73.4 | 80.7 | 86.6 | 88.3 | 87.9 | 83.3 | 73.3 | 63.2 | 52.4 |
| | | MCWB | | 48.6 | 47.4 | 55.9 | 60.6 | 67.5 | 72.5 | 75.1 | 74.5 | 69.3 | 62.3 | 56.6 | 48.7 |
| | 10% | DB | | 45.3 | 47.8 | 60.3 | 69.5 | 76.9 | 84.0 | 85.7 | 84.8 | 80.1 | 69.5 | 59.1 | 47.0 |
| | | MCWB | | 41.7 | 42.9 | 52.5 | 58.7 | 66.2 | 71.3 | 73.6 | 72.7 | 67.1 | 60.4 | 53.9 | 43.2 |
| Monthly Design Wet Bulb and Mean Coincident Dry Bulb Temperatures | 0.4% | WB | | 58.4 | 58.6 | 63.5 | 69.1 | 74.8 | 78.2 | 80.3 | 79.8 | 75.7 | 71.3 | 63.6 | 58.8 |
| | | MCDB | | 61.2 | 62.0 | 71.0 | 78.6 | 82.4 | 87.3 | 90.6 | 89.3 | 85.6 | 78.6 | 67.2 | 62.5 |
| | 2% | WB | | 54.2 | 52.9 | 60.2 | 65.3 | 72.1 | 76.2 | 78.5 | 78.0 | 73.5 | 67.6 | 60.3 | 53.9 |
| | | MCDB | | 56.4 | 56.1 | 68.2 | 73.2 | 80.0 | 85.1 | 87.8 | 87.6 | 82.1 | 74.4 | 64.7 | 56.2 |
| | 5% | WB | | 48.8 | 47.9 | 57.3 | 62.8 | 69.8 | 74.7 | 77.1 | 76.5 | 71.6 | 64.9 | 57.8 | 48.8 |
| | | MCDB | | 51.4 | 52.7 | 63.7 | 70.6 | 77.4 | 83.2 | 85.5 | 85.0 | 78.7 | 71.2 | 62.5 | 51.8 |
| | 10% | WB | | 42.0 | 43.4 | 53.1 | 60.2 | 67.8 | 73.1 | 75.5 | 74.9 | 69.8 | 61.9 | 54.3 | 43.4 |
| | | MCDB | | 44.9 | 47.6 | 59.5 | 67.9 | 74.4 | 80.9 | 82.6 | 82.0 | 75.9 | 67.6 | 58.9 | 47.1 |
| Mean Daily Temperature Range | | MDBR | 14.2 | 15.4 | 18.1 | 19.7 | 19.0 | 18.5 | 18.0 | 18.3 | 20.1 | 19.6 | 16.1 | 13.7 |
| | 5% DB | MCDBR | 18.2 | 22.1 | 24.7 | 23.7 | 22.2 | 20.9 | 20.5 | 20.1 | 23.1 | 23.9 | 21.2 | 18.5 |
| | | MCWBR | 15.4 | 16.7 | 15.0 | 13.6 | 11.3 | 9.2 | 8.8 | 8.5 | 10.0 | 12.8 | 15.3 | 15.6 |
| | 5% WB | MCDBR | 17.0 | 20.0 | 21.8 | 20.0 | 19.2 | 17.8 | 18.3 | 18.1 | 18.3 | 19.5 | 18.8 | 17.4 |
| | | MCWBR | 15.6 | 16.7 | 14.4 | 13.1 | 10.8 | 9.2 | 9.2 | 8.7 | 9.6 | 11.6 | 15.6 | 16.4 |
| Clear Sky Solar Irradiance | | taub | 0.311 | 0.327 | 0.345 | 0.375 | 0.388 | 0.403 | 0.437 | 0.442 | 0.393 | 0.347 | 0.325 | 0.316 |
| | | taud | 2.483 | 2.422 | 2.414 | 2.304 | 2.252 | 2.220 | 2.143 | 2.154 | 2.333 | 2.485 | 2.555 | 2.526 |
| | | Ebn,noon | 272 | 283 | 289 | 285 | 282 | 277 | 266 | 261 | 269 | 272 | 266 | 262 |
| | | Edh,noon | 25 | 30 | 34 | 40 | 43 | 45 | 48 | 46 | 36 | 28 | 23 | 22 |

Nomenclature:  See separate page

*Fig. 4-4  Example tables of climatic design information included on CD-ROM of 2013 ASHRAE Handbook—Fundamentals*

**104**                                                                  **Principles of HVAC**

2013 ASHRAE Handbook - Fundamentals (IP)                  © 2013 ASHRAE, Inc.

**JACKSONVILLE/INTNL., FL, USA**         WMO#: 722060

Lat: 30.49N     Long: 81.69W     Elev: 33     StdP: 14.68     Time Zone: -5 (NAE)     Period: 86-10     WBAN: 13889

### Annual Heating and Humidification Design Conditions

| Coldest Month | Heating DB | | Humidification DP/MCDB and HR | | | | | | Coldest month WS/MCDB | | | | MCWS/PCWD to 99.6% DB | |
|---|---|---|---|---|---|---|---|---|---|---|---|---|---|---|
| | 99.6% | 99% | 99.6% | | | 99% | | | 0.4% | | 1% | | | |
| | | | DP | HR | MCDB | DP | HR | MCDB | WS | MCDB | WS | MCDB | MCWS | PCWD |
| (a) | (b) | (c) | (d) | (e) | (f) | (g) | (h) | (i) | (j) | (k) | (l) | (m) | (n) | (o) |
| 1 | 29.4 | 32.5 | 16.2 | 12.5 | 38.9 | 20.5 | 15.4 | 41.0 | 23.7 | 55.8 | 20.7 | 56.7 | 4.0 | 320 |

### Annual Cooling, Dehumidification, and Enthalpy Design Conditions

| Hottest Month | Hottest Month DB Range | Cooling DB/MCWB | | | | | | Evaporation WB/MCDB | | | | | | MCWS/PCWD to 0.4% DB | |
|---|---|---|---|---|---|---|---|---|---|---|---|---|---|---|---|
| | | 0.4% | | 1% | | 2% | | 0.4% | | 1% | | 2% | | | |
| | | DB | MCWB | DB | MCWB | DB | MCWB | WB | MCDB | WB | MCDB | WB | MCDB | MCWS | PCWD |
| (a) | (b) | (c) | (d) | (e) | (f) | (g) | (h) | (i) | (j) | (k) | (l) | (m) | (n) | (o) | (p) |
| 7 | 18.2 | 94.6 | 77.3 | 92.8 | 77.0 | 91.0 | 76.6 | 80.0 | 89.6 | 79.1 | 88.4 | 78.4 | 87.1 | 7.9 | 260 |

| Dehumidification DP/MCDB and HR | | | | | | | | | Enthalpy/MCDB | | | | | | Hours 8 to 4 & 55/69 |
|---|---|---|---|---|---|---|---|---|---|---|---|---|---|---|---|
| 0.4% | | | 1% | | | 2% | | | 0.4% | | 1% | | 2% | | |
| DP | HR | MCDB | DP | HR | MCDB | DP | HR | MCDB | Enth | MCDB | Enth | MCDB | Enth | MCDB | |
| (a) | (b) | (c) | (d) | (e) | (f) | (g) | (h) | (i) | (j) | (k) | (l) | (m) | (n) | (o) | (p) |
| 77.4 | 142.9 | 83.4 | 76.7 | 139.4 | 82.8 | 76.0 | 136.2 | 82.2 | 43.2 | 89.3 | 42.5 | 88.4 | 41.7 | 87.5 | 690 |

### Extreme Annual Design Conditions

| Extreme Annual WS | | | Extreme Max WB | Extreme Annual DB | | | | n-Year Return Period Values of Extreme DB | | | | | | | |
|---|---|---|---|---|---|---|---|---|---|---|---|---|---|---|---|
| | | | | Mean | | Standard deviation | | n=5 years | | n=10 years | | n=20 years | | n=50 years | |
| 1% | 2.5% | 5% | | Min | Max | Min | Max | Min | Max | Min | Max | Min | Max | Min | Max |
| (a) | (b) | (c) | (d) | (e) | (f) | (g) | (h) | (i) | (j) | (k) | (l) | (m) | (n) | (o) | (p) |
| 20.0 | 17.9 | 16.2 | 83.5 | 23.8 | 98.0 | 3.1 | 2.2 | 21.6 | 99.6 | 19.8 | 100.9 | 18.1 | 102.2 | 15.8 | 103.8 |

### Monthly Climatic Design Conditions

| | | Annual | Jan | Feb | Mar | Apr | May | Jun | Jul | Aug | Sep | Oct | Nov | Dec |
|---|---|---|---|---|---|---|---|---|---|---|---|---|---|---|
| | | (d) | (e) | (f) | (g) | (h) | (i) | (j) | (k) | (l) | (m) | (n) | (o) | (p) |
| | Tavg | 68.6 | 53.9 | 56.5 | 61.7 | 66.9 | 74.1 | 79.7 | 82.1 | 81.5 | 78.2 | 70.1 | 62.1 | 55.2 |
| | Sd | | 9.20 | 8.60 | 7.97 | 6.64 | 5.09 | 3.45 | 2.57 | 2.48 | 3.52 | 6.65 | 7.92 | 9.24 |
| Temperatures, Degree-Days and Degree-Hours | HDD50 | 168 | 67 | 32 | 10 | 0 | 0 | 0 | 0 | 0 | 0 | 0 | 7 | 52 |
| | HDD65 | 1327 | 355 | 257 | 155 | 57 | 4 | 0 | 0 | 0 | 0 | 30 | 144 | 325 |
| | CDD50 | 6951 | 189 | 215 | 374 | 507 | 747 | 892 | 995 | 977 | 847 | 623 | 372 | 213 |
| | CDD65 | 2632 | 12 | 20 | 53 | 114 | 286 | 442 | 530 | 512 | 397 | 188 | 58 | 20 |
| | CDH74 | 23105 | 72 | 143 | 408 | 992 | 2593 | 3982 | 5224 | 4818 | 3185 | 1264 | 328 | 96 |
| | CDH80 | 8850 | 2 | 13 | 61 | 254 | 944 | 1708 | 2375 | 2037 | 1119 | 297 | 36 | 4 |
| Precipitation | PrecAvg | 51.3 | 3.3 | 3.9 | 3.7 | 2.8 | 3.6 | 5.7 | 5.6 | 7.9 | 7.1 | 2.9 | 2.2 | 2.7 |
| | PrecMax | 70.6 | 7.3 | 8.9 | 10.2 | 11.6 | 10.4 | 12.9 | 16.2 | 16.2 | 17.8 | 9.8 | 4.6 | 5.9 |
| | PrecMin | 37.5 | 0.3 | 0.5 | 0.7 | 0.4 | 0.9 | 2.6 | 2.0 | 2.4 | 1.0 | 0.2 | 0.1 | 0.2 |
| | PrecSD | 8.1 | 2.0 | 2.1 | 2.8 | 2.5 | 3.1 | 3.2 | 3.1 | 3.8 | 4.0 | 2.4 | 1.4 | 1.6 |
| Monthly Design Dry Bulb and Mean Coincident Wet Bulb Temperatures | 0.4% DB | 79.7 | 82.1 | 84.9 | 89.5 | 93.4 | 97.3 | 97.3 | 95.9 | 92.8 | 89.2 | 83.7 | 80.0 |
| | 0.4% MCWB | 68.3 | 67.8 | 68.3 | 69.7 | 73.4 | 77.1 | 77.7 | 78.2 | 76.4 | 75.5 | 72.7 | 68.9 |
| | 2% DB | 75.9 | 78.7 | 81.6 | 85.9 | 90.6 | 93.9 | 94.6 | 93.4 | 90.3 | 85.5 | 80.6 | 76.9 |
| | 2% MCWB | 66.2 | 66.4 | 66.9 | 68.3 | 72.5 | 76.5 | 77.9 | 78.2 | 76.0 | 74.0 | 70.2 | 67.8 |
| | 5% DB | 72.3 | 74.7 | 78.9 | 82.9 | 88.3 | 91.3 | 92.8 | 91.5 | 88.2 | 83.3 | 78.1 | 73.2 |
| | 5% MCWB | 64.3 | 63.9 | 65.3 | 67.6 | 71.9 | 76.0 | 77.7 | 77.8 | 76.0 | 72.7 | 68.6 | 66.3 |
| | 10% DB | 68.7 | 70.8 | 75.4 | 80.0 | 85.5 | 88.8 | 90.7 | 89.6 | 86.3 | 81.1 | 75.0 | 68.9 |
| | 10% MCWB | 62.4 | 61.8 | 63.8 | 66.4 | 71.3 | 75.8 | 77.5 | 77.7 | 75.7 | 71.5 | 67.1 | 64.2 |
| Monthly Design Wet Bulb and Mean Coincident Dry Bulb Temperatures | 0.4% WB | 70.4 | 69.9 | 72.1 | 74.4 | 77.2 | 80.1 | 81.2 | 81.1 | 79.6 | 78.0 | 74.7 | 71.8 |
| | 0.4% MCDB | 76.0 | 78.0 | 80.2 | 82.4 | 86.8 | 91.0 | 91.5 | 90.9 | 86.4 | 84.1 | 80.2 | 76.9 |
| | 2% WB | 68.3 | 68.1 | 70.0 | 72.1 | 75.6 | 78.6 | 80.0 | 80.0 | 78.4 | 76.6 | 72.7 | 69.4 |
| | 2% MCDB | 73.4 | 75.5 | 77.5 | 79.7 | 84.9 | 88.6 | 90.0 | 89.5 | 85.2 | 82.3 | 77.6 | 74.2 |
| | 5% WB | 66.0 | 66.3 | 68.3 | 70.5 | 74.6 | 77.7 | 79.1 | 79.1 | 77.8 | 75.4 | 70.9 | 67.5 |
| | 5% MCDB | 70.7 | 72.4 | 74.9 | 78.0 | 83.5 | 87.1 | 88.8 | 88.2 | 84.6 | 80.6 | 75.3 | 71.9 |
| | 10% WB | 63.8 | 64.3 | 66.4 | 69.0 | 73.3 | 76.9 | 78.3 | 78.3 | 77.1 | 73.9 | 69.2 | 65.3 |
| | 10% MCDB | 68.0 | 68.6 | 72.4 | 76.3 | 81.7 | 85.5 | 87.6 | 86.7 | 83.5 | 78.5 | 73.4 | 69.5 |
| Mean Daily Temperature Range | MDBR | 21.5 | 21.9 | 22.2 | 23.1 | 21.4 | 18.6 | 18.2 | 16.9 | 16.0 | 18.8 | 21.3 | 21.1 |
| | 5% DB MCDBR | 23.5 | 24.1 | 24.7 | 24.9 | 23.7 | 21.8 | 20.4 | 19.6 | 18.5 | 19.1 | 21.2 | 21.8 |
| | 5% DB MCWBR | 14.5 | 13.0 | 11.9 | 10.5 | 8.7 | 7.7 | 7.0 | 7.0 | 7.1 | 8.8 | 11.5 | 13.2 |
| | 5% WB MCDBR | 20.3 | 20.7 | 20.3 | 20.1 | 20.2 | 19.0 | 18.6 | 17.6 | 15.6 | 15.3 | 17.7 | 19.1 |
| | 5% WB MCWBR | 13.5 | 12.4 | 11.4 | 9.8 | 8.2 | 7.5 | 7.2 | 7.1 | 6.8 | 7.8 | 10.9 | 12.5 |
| Clear Sky Solar Irradiance | taub | 0.351 | 0.361 | 0.370 | 0.382 | 0.371 | 0.393 | 0.444 | 0.411 | 0.398 | 0.391 | 0.371 | 0.351 |
| | taud | 2.642 | 2.559 | 2.486 | 2.392 | 2.414 | 2.384 | 2.236 | 2.382 | 2.483 | 2.514 | 2.585 | 2.643 |
| | Ebn,noon | 281 | 287 | 290 | 289 | 291 | 283 | 268 | 277 | 277 | 272 | 270 | 275 |
| | Edh,noon | 25 | 29 | 34 | 38 | 38 | 39 | 45 | 38 | 33 | 30 | 26 | 24 |

Nomenclature:     See separate page

*Fig. 4-4  Example tables of climatic design information included on CD-ROM of 2013 ASHRAE Handbook—Fundamentals*

# Design Conditions

2013 ASHRAE Handbook - Fundamentals (IP) © 2013 ASHRAE, Inc.

**KANSAS CITY INTL, MO, USA** WMO#: 724460

Lat: 39.30N  Long: 94.72W  Elev: 1024  StdP: 14.16  Time Zone: -6 (NAC)  Period: 86-10  WBAN: 3947

### Annual Heating and Humidification Design Conditions

| Coldest Month | Heating DB | | Humidification DP/MCDB and HR | | | | | | Coldest month WS/MCDB | | | | MCWS/PCWD to 99.6% DB | |
|---|---|---|---|---|---|---|---|---|---|---|---|---|---|---|
| | 99.6% | 99% | 99.6% | | | 99% | | | 0.4% | | 1% | | | |
| | | | DP | HR | MCDB | DP | HR | MCDB | WS | MCDB | WS | MCDB | MCWS | PCWD |
| (a) | (b) | (c) | (d) | (e) | (f) | (g) | (h) | (i) | (j) | (k) | (l) | (m) | (n) | (o) |
| 1 | 2.0 | 7.2 | -6.4 | 4.1 | 4.9 | -1.8 | 5.2 | 9.3 | 28.2 | 42.2 | 26.1 | 40.2 | 9.6 | 310 |

### Annual Cooling, Dehumidification, and Enthalpy Design Conditions

| Hottest Month | Hottest Month DB Range | Cooling DB/MCWB | | | | | | Evaporation WB/MCDB | | | | | | MCWS/PCWD to 0.4% DB | |
|---|---|---|---|---|---|---|---|---|---|---|---|---|---|---|---|
| | | 0.4% | | 1% | | 2% | | 0.4% | | 1% | | 2% | | | |
| | | DB | MCWB | DB | MCWB | DB | MCWB | WB | MCDB | WB | MCDB | WB | MCDB | MCWS | PCWD |
| (a) | (b) | (c) | (d) | (e) | (f) | (g) | (h) | (i) | (j) | (k) | (l) | (m) | (n) | (o) | (p) |
| 7 | 18.5 | 95.8 | 76.8 | 92.5 | 76.2 | 89.7 | 75.4 | 79.8 | 90.5 | 78.3 | 88.9 | 77.0 | 87.2 | 12.1 | 190 |

| Dehumidification DP/MCDB and HR | | | | | | | | | Enthalpy/MCDB | | | | | | Hours 8 to 4 & 55/69 |
|---|---|---|---|---|---|---|---|---|---|---|---|---|---|---|---|
| 0.4% | | | 1% | | | 2% | | | 0.4% | | 1% | | 2% | | |
| DP | HR | MCDB | DP | HR | MCDB | DP | HR | MCDB | Enth | MCDB | Enth | MCDB | Enth | MCDB | |
| (a) | (b) | (c) | (d) | (e) | (f) | (g) | (h) | (i) | (j) | (k) | (l) | (m) | (n) | (o) | (p) |
| 76.9 | 145.3 | 86.6 | 75.2 | 137.2 | 85.2 | 73.7 | 130.3 | 83.2 | 44.1 | 90.6 | 42.6 | 89.4 | 41.1 | 87.1 | 657 |

### Extreme Annual Design Conditions

| Extreme Annual WS | | | Extreme Max WB | Extreme Annual DB | | | | n-Year Return Period Values of Extreme DB | | | | | | | |
|---|---|---|---|---|---|---|---|---|---|---|---|---|---|---|---|
| | | | | Mean | | Standard deviation | | n=5 years | | n=10 years | | n=20 years | | n=50 years | |
| 1% | 2.5% | 5% | | Min | Max | Min | Max | Min | Max | Min | Max | Min | Max | Min | Max |
| (a) | (b) | (c) | (d) | (e) | (f) | (g) | (h) | (i) | (j) | (k) | (l) | (m) | (n) | (o) | (p) |
| 25.6 | 23.0 | 20.0 | 84.4 | -4.5 | 99.7 | 5.7 | 3.4 | -8.6 | 102.1 | -11.9 | 104.1 | -15.2 | 106.0 | -19.3 | 108.5 |

### Monthly Climatic Design Conditions

| | | Annual | Jan | Feb | Mar | Apr | May | Jun | Jul | Aug | Sep | Oct | Nov | Dec |
|---|---|---|---|---|---|---|---|---|---|---|---|---|---|---|
| | | (d) | (e) | (f) | (g) | (h) | (i) | (j) | (k) | (l) | (m) | (n) | (o) | (p) |
| Temperatures, Degree-Days and Degree-Hours | Tavg | 55.0 | 29.9 | 33.9 | 44.6 | 55.2 | 65.0 | 73.9 | 78.2 | 77.1 | 68.2 | 56.5 | 44.0 | 32.5 |
| | Sd | | 11.78 | 11.77 | 11.37 | 9.92 | 7.62 | 6.04 | 5.49 | 6.24 | 8.01 | 9.24 | 10.40 | 11.27 |
| | HDD50 | 2213 | 628 | 459 | 241 | 56 | 2 | 0 | 0 | 0 | 1 | 41 | 236 | 549 |
| | HDD65 | 5012 | 1089 | 870 | 636 | 322 | 97 | 7 | 0 | 2 | 58 | 291 | 632 | 1008 |
| | CDD50 | 4048 | 3 | 9 | 74 | 211 | 466 | 718 | 876 | 839 | 547 | 243 | 56 | 6 |
| | CDD65 | 1372 | 0 | 0 | 4 | 27 | 96 | 275 | 411 | 376 | 154 | 27 | 2 | 0 |
| | CDH74 | 13301 | 0 | 1 | 31 | 239 | 800 | 2588 | 4276 | 3813 | 1317 | 229 | 7 | 0 |
| | CDH80 | 5322 | 0 | 0 | 3 | 55 | 206 | 969 | 1890 | 1705 | 442 | 52 | 0 | 0 |
| Precipitation | PrecAvg | 37.8 | 1.1 | 1.2 | 2.5 | 3.4 | 4.9 | 4.6 | 4.4 | 3.8 | 4.6 | 3.2 | 2.0 | 1.7 |
| | PrecMax | 63.2 | 3.0 | 2.6 | 9.1 | 6.9 | 10.1 | 10.6 | 10.3 | 8.7 | 11.6 | 8.6 | 5.6 | 5.4 |
| | PrecMin | 24.3 | 0.0 | 0.4 | 0.7 | 0.8 | 2.2 | 2.2 | 0.2 | 0.3 | 0.4 | 0.3 | 0.1 | 0.0 |
| | PrecSD | 10.7 | 0.9 | 0.7 | 2.0 | 1.7 | 1.8 | 2.6 | 3.2 | 2.1 | 3.8 | 2.6 | 1.5 | 1.3 |
| Monthly Design Dry Bulb and Mean Coincident Wet Bulb Temperatures | 0.4% DB | | 62.6 | 67.7 | 77.9 | 86.4 | 88.7 | 95.1 | 99.0 | 100.6 | 94.2 | 85.4 | 74.2 | 64.1 |
| | 0.4% MCWB | | 52.7 | 53.7 | 59.8 | 66.8 | 72.4 | 74.6 | 77.4 | 76.6 | 74.5 | 69.1 | 61.7 | 56.5 |
| | 2% DB | | 56.8 | 61.6 | 72.6 | 80.4 | 85.1 | 91.0 | 95.6 | 96.3 | 88.6 | 80.2 | 69.2 | 57.9 |
| | 2% MCWB | | 47.7 | 50.2 | 58.5 | 64.1 | 70.4 | 75.4 | 77.2 | 76.9 | 72.6 | 65.4 | 59.0 | 50.5 |
| | 5% DB | | 51.9 | 56.4 | 67.8 | 75.9 | 82.0 | 88.6 | 92.4 | 92.5 | 84.9 | 75.4 | 64.3 | 52.8 |
| | 5% MCWB | | 43.9 | 46.9 | 56.5 | 61.9 | 68.8 | 74.5 | 77.1 | 76.3 | 71.7 | 63.7 | 55.4 | 45.8 |
| | 10% DB | | 46.1 | 51.4 | 63.2 | 72.0 | 78.7 | 85.8 | 89.4 | 89.0 | 81.5 | 71.6 | 60.4 | 47.9 |
| | 10% MCWB | | 40.0 | 43.7 | 53.8 | 60.2 | 66.7 | 73.1 | 76.2 | 75.6 | 69.3 | 61.8 | 53.4 | 41.8 |
| Monthly Design Wet Bulb and Mean Coincident Dry Bulb Temperatures | 0.4% WB | | 56.0 | 57.9 | 64.4 | 69.5 | 76.1 | 79.6 | 82.2 | 81.3 | 77.2 | 71.8 | 64.0 | 58.8 |
| | 0.4% MCDB | | 60.0 | 65.1 | 71.9 | 81.3 | 85.2 | 89.7 | 91.7 | 91.9 | 88.8 | 81.1 | 70.8 | 62.6 |
| | 2% WB | | 49.3 | 51.6 | 61.5 | 67.1 | 73.3 | 77.8 | 80.3 | 79.5 | 75.2 | 68.5 | 60.7 | 52.2 |
| | 2% MCDB | | 55.8 | 58.8 | 69.4 | 76.6 | 81.7 | 87.5 | 90.7 | 90.8 | 85.3 | 76.1 | 67.1 | 55.2 |
| | 5% WB | | 44.1 | 47.6 | 58.4 | 64.4 | 70.9 | 76.1 | 78.8 | 77.9 | 73.4 | 65.7 | 57.7 | 46.1 |
| | 5% MCDB | | 51.2 | 55.7 | 66.5 | 73.5 | 79.2 | 85.6 | 89.3 | 89.2 | 82.1 | 73.9 | 62.8 | 51.5 |
| | 10% WB | | 40.4 | 43.8 | 54.0 | 61.5 | 68.5 | 74.6 | 77.4 | 76.4 | 71.5 | 62.8 | 53.8 | 42.1 |
| | 10% MCDB | | 46.0 | 50.9 | 62.6 | 70.4 | 76.3 | 83.6 | 87.5 | 87.0 | 79.2 | 69.6 | 59.6 | 47.5 |
| Mean Daily Temperature Range | MDBR | | 17.1 | 18.1 | 20.6 | 20.2 | 19.2 | 18.6 | 18.5 | 19.3 | 20.4 | 19.9 | 18.0 | 16.5 |
| | 5% DB MCDBR | | 24.6 | 25.9 | 26.3 | 24.9 | 22.1 | 20.8 | 21.5 | 22.9 | 22.6 | 24.1 | 23.3 | 23.2 |
| | 5% DB MCWBR | | 17.1 | 17.0 | 15.8 | 14.0 | 11.5 | 9.4 | 8.8 | 9.1 | 10.2 | 13.1 | 15.1 | 16.7 |
| | 5% WB MCDBR | | 22.9 | 24.1 | 22.9 | 21.7 | 19.5 | 18.8 | 19.2 | 20.3 | 19.7 | 19.9 | 20.2 | 20.7 |
| | 5% WB MCWBR | | 16.8 | 16.9 | 14.9 | 13.5 | 11.6 | 9.8 | 9.5 | 9.6 | 10.4 | 12.7 | 14.9 | 16.4 |
| Clear Sky Solar Irradiance | taub | | 0.296 | 0.309 | 0.330 | 0.345 | 0.352 | 0.382 | 0.395 | 0.392 | 0.358 | 0.326 | 0.307 | 0.295 |
| | taud | | 2.559 | 2.487 | 2.459 | 2.386 | 2.379 | 2.303 | 2.285 | 2.340 | 2.456 | 2.564 | 2.632 | 2.584 |
| | $E_{bn,noon}$ | | 281 | 291 | 295 | 295 | 293 | 283 | 278 | 277 | 281 | 281 | 276 | 273 |
| | $E_{dh,noon}$ | | 23 | 28 | 32 | 37 | 38 | 41 | 42 | 38 | 32 | 26 | 21 | 21 |

Nomenclature: See separate page

*Fig. 4-4 Example tables of climatic design information included on CD-ROM of 2013 ASHRAE Handbook—Fundamentals*

## 106

## Principles of HVAC

2013 ASHRAE Handbook - Fundamentals (IP) © 2013 ASHRAE, Inc.

### KIRKSVILLE RGNL, MO, USA

WMO#: 724455

Lat: 40.10N  Long: 92.54W  Elev: 965  StdP: 14.19  Time Zone: -6 (NAC)  Period: 86-10  WBAN: 14938

**Annual Heating and Humidification Design Conditions**

| | Coldest Month | Heating DB | | Humidification DP/MCDB and HR | | | | | | Coldest month WS/MCDB | | | | MCWS/PCWD to 99.6% DB | |
|---|---|---|---|---|---|---|---|---|---|---|---|---|---|---|---|
| | | 99.6% | 99% | 99.6% | | | 99% | | | 0.4% | | 1% | | MCWS | PCWD |
| | | | | DP | HR | MCDB | DP | HR | MCDB | WS | MCDB | WS | MCDB | | |
| | (a) | (b) | (c) | (d) | (e) | (f) | (g) | (h) | (i) | (j) | (k) | (l) | (m) | (n) | (o) |
| (1) | 1 | -1.5 | 3.1 | -9.2 | 3.5 | 0.7 | -5.6 | 4.2 | 4.4 | 26.2 | 36.1 | 24.6 | 34.5 | 8.1 | 300 |

**Annual Cooling, Dehumidification, and Enthalpy Design Conditions**

| | Hottest Month | Hottest Month DB Range | Cooling DB/MCWB | | | | | | Evaporation WB/MCDB | | | | | | MCWS/PCWD to 0.4% DB | |
|---|---|---|---|---|---|---|---|---|---|---|---|---|---|---|---|---|
| | | | 0.4% | | 1% | | 2% | | 0.4% | | 1% | | 2% | | MCWS | PCWD |
| | | | DB | MCWB | DB | MCWB | DB | MCWB | WB | MCDB | WB | MCDB | WB | MCDB | | |
| | (a) | (b) | (c) | (d) | (e) | (f) | (g) | (h) | (i) | (j) | (k) | (l) | (m) | (n) | (o) | (p) |
| (2) | 7 | 18.9 | 92.7 | 76.0 | 90.1 | 75.4 | 87.6 | 74.0 | 78.3 | 88.7 | 77.0 | 87.3 | 75.5 | 84.4 | 11.0 | 210 |

| | Dehumidification DP/MCDB and HR | | | | | | | | Enthalpy/MCDB | | | | | | Hours 8 to 4 & |
|---|---|---|---|---|---|---|---|---|---|---|---|---|---|---|---|
| | 0.4% | | | 1% | | | 2% | | 0.4% | | 1% | | 2% | | 55/69 |
| | DP | HR | MCDB | DP | HR | MCDB | DP | HR | MCDB | Enth | MCDB | Enth | MCDB | Enth | MCDB | |
| | (a) | (b) | (c) | (d) | (e) | (f) | (g) | (h) | (i) | (j) | (k) | (l) | (m) | (n) | (o) | (p) |
| (3) | 75.2 | 136.8 | 84.1 | 73.4 | 128.7 | 82.4 | 72.5 | 124.8 | 81.4 | 42.4 | 89.1 | 41.0 | 87.4 | 39.6 | 84.8 | 655 |

**Extreme Annual Design Conditions**

| | Extreme Annual WS | | | Extreme Max WB | Extreme Annual DB | | | | n-Year Return Period Values of Extreme DB | | | | | | |
|---|---|---|---|---|---|---|---|---|---|---|---|---|---|---|---|
| | 1% | 2.5% | 5% | | Mean | | Standard deviation | | n=5 years | | n=10 years | | n=20 years | | n=50 years | |
| | | | | | Min | Max | Min | Max | Min | Max | Min | Max | Min | Max | Min | Max |
| | (a) | (b) | (c) | (d) | (e) | (f) | (g) | (h) | (i) | (j) | (k) | (l) | (m) | (n) | (o) | (p) |
| (4) | 23.8 | 20.4 | 18.7 | 83.1 | -6.1 | 96.3 | 4.2 | 3.5 | -9.1 | 98.8 | -11.6 | 100.8 | -13.9 | 102.8 | -17.0 | 105.3 |

**Monthly Climatic Design Conditions**

| | | | Annual | Jan | Feb | Mar | Apr | May | Jun | Jul | Aug | Sep | Oct | Nov | Dec |
|---|---|---|---|---|---|---|---|---|---|---|---|---|---|---|---|
| | | | (d) | (e) | (f) | (g) | (h) | (i) | (j) | (k) | (l) | (m) | (n) | (o) | (p) |
| (5) | | Tavg | 52.5 | 27.4 | 29.6 | 43.0 | 53.2 | 63.0 | 72.1 | 75.5 | 73.3 | 66.0 | 53.7 | 42.6 | 28.9 |
| (6) | | Sd | | 11.28 | 11.43 | 10.77 | 9.93 | 7.79 | 6.01 | 5.77 | 6.01 | 7.53 | 8.92 | 10.27 | 11.57 |
| (7) | Temperatures, | HDD50 | 2624 | 703 | 573 | 277 | 78 | 5 | 0 | 0 | 0 | 1 | 63 | 265 | 659 |
| (8) | Degree-Days | HDD65 | 5597 | 1166 | 991 | 685 | 368 | 132 | 13 | 1 | 7 | 78 | 363 | 673 | 1120 |
| (9) | and | CDD50 | 3531 | 2 | 2 | 60 | 176 | 408 | 663 | 791 | 723 | 482 | 178 | 43 | 3 |
| (10) | Degree-Hours | CDD65 | 1028 | 0 | 0 | 3 | 16 | 69 | 226 | 327 | 265 | 109 | 13 | 0 | 0 |
| (11) | | CDH74 | 9146 | 1 | 0 | 26 | 186 | 543 | 1891 | 3116 | 2357 | 893 | 131 | 2 | 0 |
| (12) | | CDH80 | 3218 | 0 | 0 | 3 | 42 | 113 | 617 | 1257 | 901 | 258 | 27 | 0 | 0 |
| (13) | | PrecAvg | 36.3 | 1.3 | 0.9 | 2.6 | 3.4 | 4.3 | 4.1 | 3.8 | 3.9 | 4.6 | 3.3 | 2.2 | 1.9 |
| (14) | Precipitation | PrecMax | 63.2 | 6.4 | 1.8 | 6.4 | 7.6 | 8.9 | 8.9 | 9.9 | 13.6 | 11.7 | 7.2 | 4.7 | 6.1 |
| (15) | | PrecMin | 25.2 | 0.2 | 0.2 | 0.8 | 1.0 | 1.5 | 0.8 | 0.2 | 0.3 | 0.2 | 0.2 | 0.1 | 0.1 |
| (16) | | PrecSD | 9.9 | 1.4 | 0.5 | 1.5 | 1.9 | 2.2 | 2.1 | 2.5 | 3.1 | 3.5 | 1.9 | 1.2 | 1.4 |
| (17) | | 0.4% DB | 61.1 | 63.3 | 77.8 | 85.8 | 87.7 | 93.2 | 97.8 | 95.4 | 91.3 | 83.9 | 72.9 | 62.8 |
| (18) | Monthly Design | MCWB | 51.2 | 53.2 | 61.9 | 67.8 | 72.9 | 75.9 | 76.5 | 76.5 | 73.9 | 68.5 | 60.8 | 57.7 |
| (19) | Dry Bulb | 2% DB | 53.6 | 55.3 | 71.5 | 79.3 | 83.7 | 89.6 | 93.4 | 91.3 | 86.5 | 77.8 | 67.9 | 55.3 |
| (20) | and | MCWB | 46.3 | 47.5 | 59.5 | 63.2 | 69.0 | 74.4 | 76.6 | 75.7 | 71.8 | 64.0 | 58.2 | 49.3 |
| (21) | Mean Coincident | 5% DB | 47.8 | 50.0 | 65.9 | 74.5 | 80.6 | 86.5 | 90.4 | 89.5 | 82.4 | 73.2 | 63.2 | 48.4 |
| (22) | Wet Bulb | MCWB | 42.4 | 43.6 | 57.2 | 60.9 | 67.5 | 72.9 | 76.6 | 75.3 | 69.9 | 62.7 | 54.4 | 43.9 |
| (23) | Temperatures | 10% DB | 43.1 | 45.3 | 61.0 | 70.1 | 77.0 | 83.7 | 87.5 | 84.5 | 79.3 | 69.7 | 57.5 | 44.9 |
| (24) | | MCWB | 38.5 | 40.1 | 53.0 | 59.0 | 64.6 | 71.7 | 74.7 | 72.9 | 67.8 | 60.7 | 50.7 | 40.6 |
| (25) | | 0.4% WB | 53.8 | 53.5 | 64.2 | 68.6 | 74.9 | 79.0 | 80.2 | 80.7 | 76.3 | 70.7 | 62.5 | 58.5 |
| (26) | Monthly Design | MCDB | 58.1 | 60.9 | 72.9 | 82.5 | 83.8 | 88.1 | 90.5 | 89.6 | 86.8 | 79.5 | 69.7 | 62.2 |
| (27) | Wet Bulb | 2% WB | 46.5 | 48.0 | 61.2 | 65.4 | 72.2 | 76.6 | 78.3 | 78.2 | 74.3 | 67.3 | 59.7 | 50.3 |
| (28) | and | MCDB | 50.8 | 55.1 | 67.9 | 76.3 | 80.7 | 85.4 | 89.5 | 88.0 | 83.2 | 74.1 | 67.2 | 52.8 |
| (29) | Mean Coincident | 5% WB | 42.2 | 44.0 | 58.5 | 63.0 | 69.4 | 74.9 | 77.3 | 76.5 | 72.4 | 64.6 | 55.9 | 45.0 |
| (30) | Dry Bulb | MCDB | 47.8 | 49.5 | 65.5 | 71.7 | 77.6 | 83.8 | 88.4 | 86.0 | 79.9 | 71.3 | 61.4 | 48.4 |
| (31) | Temperatures | 10% WB | 38.6 | 40.2 | 53.5 | 60.3 | 67.1 | 73.2 | 75.8 | 74.6 | 70.2 | 61.1 | 52.3 | 40.2 |
| (32) | | MCDB | 42.7 | 45.0 | 61.1 | 68.3 | 74.6 | 81.8 | 85.3 | 82.9 | 77.2 | 68.5 | 57.6 | 43.8 |
| (33) | | MDBR | 16.4 | 17.6 | 19.7 | 20.7 | 19.8 | 18.7 | 18.9 | 19.9 | 21.1 | 20.6 | 18.6 | 16.3 |
| (34) | Mean Daily | 5% DB MCDBR | 23.4 | 24.0 | 25.8 | 26.7 | 22.6 | 21.0 | 21.0 | 22.8 | 23.4 | 25.6 | 24.5 | 23.0 |
| (35) | Temperature Range | MCWBR | 17.8 | 17.3 | 16.4 | 15.2 | 12.1 | 10.0 | 8.7 | 10.2 | 11.6 | 15.1 | 16.9 | 18.5 |
| (36) | | 5% WB MCDBR | 21.8 | 22.5 | 22.8 | 23.6 | 19.4 | 18.8 | 19.4 | 19.9 | 19.5 | 22.0 | 22.1 | 22.0 |
| (37) | | MCWBR | 17.4 | 17.2 | 16.0 | 15.3 | 11.3 | 10.3 | 9.3 | 10.2 | 11.0 | 15.4 | 16.7 | 19.0 |
| (38) | | taub | 0.293 | 0.306 | 0.329 | 0.343 | 0.366 | 0.397 | 0.392 | 0.394 | 0.355 | 0.326 | 0.307 | 0.297 |
| (39) | Clear Sky Solar | taud | 2.541 | 2.471 | 2.451 | 2.386 | 2.323 | 2.245 | 2.282 | 2.315 | 2.454 | 2.562 | 2.632 | 2.597 |
| (40) | Irradiance | Ebn,noon | 280 | 291 | 294 | 295 | 289 | 279 | 279 | 276 | 281 | 280 | 274 | 271 |
| (41) | | Edh,noon | 23 | 28 | 32 | 37 | 40 | 43 | 42 | 39 | 32 | 26 | 21 | 20 |

Nomenclature: See separate page

*Fig. 4-4 Example tables of climatic design information included on CD-ROM of 2013 ASHRAE Handbook—Fundamentals*

# Design Conditions

**2013 ASHRAE Handbook - Fundamentals (IP)** © 2013 ASHRAE, Inc.

## KNOXVILLE MUNICIPAL, TN, USA

Lat: 35.82N  Long: 83.99W  Elev: 981  StdP: 14.18  Time Zone: -5 (NAE)  Period: 86-10  WMO#: 723260  WBAN: 13891

### Annual Heating and Humidification Design Conditions

| Coldest Month | Heating DB | | Humidification DP/MCDB and HR | | | | | | Coldest month WS/MCDB | | | | MCWS/PCWD to 99.6% DB | |
|---|---|---|---|---|---|---|---|---|---|---|---|---|---|---|
| | 99.6% | 99% | 99.6% | | | 99% | | | 0.4% | | 1% | | MCWS | PCWD |
| | | | DP | HR | MCDB | DP | HR | MCDB | WS | MCDB | WS | MCDB | | |
| (a) | (b) | (c) | (d) | (e) | (f) | (g) | (h) | (i) | (j) | (k) | (l) | (m) | (n) | (o) |
| 1 | 16.5 | 20.8 | 4.0 | 7.0 | 21.5 | 8.6 | 8.9 | 25.6 | 25.0 | 55.7 | 21.5 | 53.7 | 5.7 | 50 |

### Annual Cooling, Dehumidification, and Enthalpy Design Conditions

| Hottest Month | Hottest Month DB Range | Cooling DB/MCWB | | | | | | Evaporation WB/MCDB | | | | | | MCWS/PCWD to 0.4% DB | |
|---|---|---|---|---|---|---|---|---|---|---|---|---|---|---|---|
| | | 0.4% | | 1% | | 2% | | 0.4% | | 1% | | 2% | | MCWS | PCWD |
| | | DB | MCWB | DB | MCWB | DB | MCWB | WB | MCDB | WB | MCDB | WB | MCDB | | |
| (a) | (b) | (c) | (d) | (e) | (f) | (g) | (h) | (i) | (j) | (k) | (l) | (m) | (n) | (o) | (p) |
| 7 | 18.5 | 93.0 | 73.9 | 90.6 | 73.7 | 88.5 | 73.0 | 77.1 | 87.8 | 76.1 | 86.1 | 75.1 | 84.5 | 7.1 | 260 |

| Dehumidification DP/MCDB and HR | | | | | | | | | Enthalpy/MCDB | | | | | | Hours 8 to 4 & 55/69 |
|---|---|---|---|---|---|---|---|---|---|---|---|---|---|---|---|
| 0.4% | | | 1% | | | 2% | | | 0.4% | | 1% | | 2% | | |
| DP | HR | MCDB | DP | HR | MCDB | DP | HR | MCDB | Enth | MCDB | Enth | MCDB | Enth | MCDB | |
| (a) | (b) | (c) | (d) | (e) | (f) | (g) | (h) | (i) | (j) | (k) | (l) | (m) | (n) | (o) | (p) |
| 74.0 | 131.5 | 81.4 | 73.0 | 127.2 | 80.4 | 72.2 | 123.7 | 79.6 | 41.0 | 87.8 | 40.1 | 86.7 | 39.1 | 84.8 | 722 |

### Extreme Annual Design Conditions

| Extreme Annual WS | | | Extreme Max WB | Extreme Annual DB | | | | n-Year Return Period Values of Extreme DB | | | | | | | |
|---|---|---|---|---|---|---|---|---|---|---|---|---|---|---|---|
| 1% | 2.5% | 5% | | Mean | | Standard deviation | | n=5 years | | n=10 years | | n=20 years | | n=50 years | |
| | | | | Min | Max | Min | Max | Min | Max | Min | Max | Min | Max | Min | Max |
| (a) | (b) | (c) | (d) | (e) | (f) | (g) | (h) | (i) | (j) | (k) | (l) | (m) | (n) | (o) | (p) |
| 20.4 | 17.7 | 15.3 | 83.3 | 8.7 | 95.3 | 5.9 | 3.2 | 4.5 | 97.6 | 1.1 | 99.5 | -2.2 | 101.3 | -6.5 | 103.6 |

### Monthly Climatic Design Conditions

| | | Annual | Jan | Feb | Mar | Apr | May | Jun | Jul | Aug | Sep | Oct | Nov | Dec |
|---|---|---|---|---|---|---|---|---|---|---|---|---|---|---|
| | Tavg | 59.3 | 39.2 | 42.7 | 50.4 | 59.0 | 67.3 | 75.0 | 78.3 | 77.7 | 71.3 | 59.3 | 49.6 | 40.7 |
| | Sd | | 10.02 | 9.46 | 9.35 | 8.52 | 6.66 | 4.64 | 3.54 | 3.92 | 5.95 | 7.72 | 8.54 | 9.02 |
| Temperatures, Degree-Days and Degree-Hours | HDD50 | 1166 | 358 | 237 | 112 | 22 | 1 | 0 | 0 | 0 | 0 | 14 | 111 | 311 |
| | HDD65 | 3594 | 801 | 624 | 456 | 212 | 57 | 2 | 0 | 0 | 19 | 207 | 463 | 753 |
| | CDD50 | 4562 | 22 | 33 | 126 | 293 | 536 | 749 | 877 | 859 | 640 | 304 | 100 | 23 |
| | CDD65 | 1514 | 0 | 0 | 5 | 33 | 127 | 301 | 412 | 394 | 209 | 31 | 2 | 0 |
| | CDH74 | 12949 | 0 | 1 | 47 | 314 | 1001 | 2562 | 3706 | 3494 | 1579 | 229 | 16 | 0 |
| | CDH80 | 4780 | 0 | 0 | 3 | 48 | 232 | 944 | 1541 | 1455 | 527 | 30 | 0 | 0 |
| Precipitation | PrecAvg | 48.1 | 4.3 | 4.1 | 5.0 | 4.0 | 4.2 | 4.0 | 4.9 | 3.3 | 3.2 | 2.7 | 3.8 | 4.6 |
| | PrecMax | 61.9 | 7.4 | 8.0 | 10.4 | 7.2 | 11.0 | 7.6 | 10.1 | 6.7 | 6.9 | 6.0 | 5.7 | 11.6 |
| | PrecMin | 35.7 | 2.6 | 0.8 | 2.7 | 0.4 | 0.8 | 1.0 | 1.1 | 1.3 | 0.5 | 0.8 | 1.4 | 0.4 |
| | PrecSD | 7.3 | 1.4 | 1.9 | 2.5 | 1.8 | 2.3 | 1.6 | 2.5 | 1.5 | 1.4 | 1.5 | 1.3 | 2.5 |
| Monthly Design Dry Bulb and Mean Coincident Wet Bulb Temperatures | 0.4% DB | 69.0 | 71.5 | 79.0 | 84.4 | 88.0 | 93.4 | 96.3 | 96.2 | 92.2 | 83.3 | 76.4 | 69.0 | |
| | 0.4% MCWB | 59.5 | 60.1 | 62.0 | 66.2 | 70.2 | 73.2 | 74.7 | 74.8 | 71.0 | 68.9 | 63.6 | 62.3 | |
| | 2% DB | 62.9 | 66.2 | 74.5 | 81.1 | 85.1 | 90.4 | 93.2 | 92.6 | 88.7 | 79.8 | 72.0 | 63.1 | |
| | 2% MCWB | 56.1 | 56.7 | 59.9 | 64.2 | 69.7 | 72.9 | 74.6 | 74.2 | 70.6 | 66.0 | 60.8 | 57.3 | |
| | 5% DB | 58.4 | 62.1 | 70.7 | 77.9 | 82.5 | 88.1 | 90.4 | 90.3 | 85.9 | 76.3 | 68.3 | 59.2 | |
| | 5% MCWB | 52.3 | 53.6 | 57.5 | 62.8 | 68.7 | 72.2 | 74.4 | 74.2 | 70.6 | 64.1 | 59.0 | 54.4 | |
| | 10% DB | 54.1 | 57.6 | 66.4 | 74.1 | 80.1 | 85.7 | 88.0 | 87.8 | 82.7 | 72.9 | 64.2 | 55.1 | |
| | 10% MCWB | 48.9 | 50.3 | 55.5 | 61.1 | 67.4 | 71.7 | 73.8 | 73.5 | 69.5 | 62.4 | 56.7 | 50.0 | |
| Monthly Design Wet Bulb and Mean Coincident Dry Bulb Temperatures | 0.4% WB | 60.8 | 62.2 | 65.0 | 69.9 | 74.3 | 77.3 | 78.6 | 78.2 | 75.3 | 71.4 | 66.9 | 63.3 | |
| | 0.4% MCDB | 66.4 | 68.3 | 75.2 | 80.0 | 83.1 | 88.3 | 89.3 | 89.0 | 85.3 | 79.4 | 72.9 | 67.7 | |
| | 2% WB | 57.6 | 59.1 | 62.2 | 67.1 | 72.4 | 75.7 | 77.3 | 77.0 | 73.8 | 68.8 | 63.4 | 59.1 | |
| | 2% MCDB | 62.3 | 64.5 | 70.7 | 76.9 | 81.5 | 85.7 | 87.7 | 88.0 | 83.1 | 75.0 | 68.7 | 61.9 | |
| | 5% WB | 53.8 | 55.6 | 59.7 | 64.8 | 70.8 | 74.6 | 76.3 | 75.8 | 72.5 | 66.8 | 60.7 | 55.5 | |
| | 5% MCDB | 57.2 | 60.0 | 67.2 | 73.9 | 79.7 | 84.0 | 86.1 | 86.1 | 81.1 | 73.1 | 66.4 | 58.5 | |
| | 10% WB | 50.0 | 51.7 | 57.2 | 62.7 | 69.1 | 73.4 | 75.3 | 74.9 | 71.3 | 64.7 | 57.8 | 50.9 | |
| | 10% MCDB | 53.3 | 55.8 | 64.5 | 71.2 | 76.9 | 82.1 | 84.5 | 84.4 | 79.0 | 70.7 | 63.4 | 54.1 | |
| Mean Daily Temperature Range | MDBR | 17.0 | 18.4 | 20.8 | 21.6 | 20.6 | 19.4 | 18.5 | 19.1 | 19.6 | 21.7 | 20.1 | 16.9 | |
| | 5% DB MCDBR | 22.3 | 24.5 | 27.2 | 26.1 | 22.9 | 22.7 | 22.1 | 22.2 | 23.5 | 24.6 | 25.4 | 22.3 | |
| | 5% DB MCWBR | 15.8 | 16.4 | 15.2 | 12.8 | 9.9 | 8.0 | 7.3 | 7.4 | 8.4 | 12.1 | 15.1 | 16.6 | |
| | 5% WB MCDBR | 18.7 | 20.2 | 21.6 | 20.9 | 19.7 | 18.9 | 18.9 | 19.3 | 18.9 | 18.7 | 20.5 | 19.6 | |
| | 5% WB MCWBR | 14.8 | 15.6 | 13.7 | 11.7 | 9.2 | 7.9 | 7.4 | 7.2 | 8.0 | 10.3 | 14.4 | 16.7 | |
| Clear Sky Solar Irradiance | taub | 0.331 | 0.342 | 0.369 | 0.393 | 0.411 | 0.429 | 0.455 | 0.443 | 0.404 | 0.372 | 0.353 | 0.336 | |
| | taud | 2.605 | 2.523 | 2.413 | 2.299 | 2.209 | 2.182 | 2.134 | 2.198 | 2.354 | 2.443 | 2.525 | 2.601 | |
| | Ebn,noon | 277 | 285 | 285 | 282 | 277 | 271 | 263 | 264 | 270 | 270 | 265 | 267 | |
| | Edh,noon | 24 | 29 | 35 | 41 | 46 | 47 | 49 | 45 | 37 | 31 | 26 | 23 | |

Nomenclature: See separate page

*Fig. 4-4 Example tables of climatic design information included on CD-ROM of 2013 ASHRAE Handbook—Fundamentals*

**108**                                **Principles of HVAC**

2013 ASHRAE Handbook - Fundamentals (IP)      © 2013 ASHRAE, Inc.

**LOUISVILLE/STANDIFO, KY, USA**     WMO#: 724230

Lat: 38.18N    Long: 85.73W    Elev: 489    StdP: 14.44    Time Zone: -5 (NAE)    Period: 86-10    WBAN: 93821

### Annual Heating and Humidification Design Conditions

| Coldest Month | Heating DB | | Humidification DP/MCDB and HR | | | | | | Coldest month WS/MCDB | | | | MCWS/PCWD to 99.6% DB | |
|---|---|---|---|---|---|---|---|---|---|---|---|---|---|---|
| | 99.6% | 99% | 99.6% | | | 99% | | | 0.4% | | 1% | | | |
| | | | DP | HR | MCDB | DP | HR | MCDB | WS | MCDB | WS | MCDB | MCWS | PCWD |
| (a) | (b) | (c) | (d) | (e) | (f) | (g) | (h) | (i) | (j) | (k) | (l) | (m) | (n) | (o) |
| 1 | 10.2 | 15.9 | -1.5 | 5.2 | 13.1 | 4.1 | 6.9 | 18.8 | 25.5 | 41.2 | 23.4 | 41.8 | 8.5 | 300 |

### Annual Cooling, Dehumidification, and Enthalpy Design Conditions

| Hottest Month | Hottest Month DB Range | Cooling DB/MCWB | | | | | | Evaporation WB/MCDB | | | | | | MCWS/PCWD to 0.4% DB | |
|---|---|---|---|---|---|---|---|---|---|---|---|---|---|---|---|
| | | 0.4% | | 1% | | 2% | | 0.4% | | 1% | | 2% | | | |
| | | DB | MCWB | DB | MCWB | DB | MCWB | WB | MCDB | WB | MCDB | WB | MCDB | MCWS | PCWD |
| (a) | (b) | (c) | (d) | (e) | (f) | (g) | (h) | (i) | (j) | (k) | (l) | (m) | (n) | (o) | (p) |
| 7 | 17.3 | 93.8 | 75.3 | 91.5 | 75.0 | 89.6 | 74.2 | 78.7 | 89.1 | 77.5 | 87.9 | 76.4 | 85.8 | 9.1 | 240 |

| Dehumidification DP/MCDB and HR | | | | | | | | Enthalpy/MCDB | | | | | | Hours 8 to 4 & 55/69 |
|---|---|---|---|---|---|---|---|---|---|---|---|---|---|---|
| 0.4% | | | 1% | | | 2% | | 0.4% | | 1% | | 2% | | |
| DP | HR | MCDB | DP | HR | MCDB | DP | HR | MCDB | Enth | MCDB | Enth | MCDB | Enth | MCDB |
| (a) | (b) | (c) | (d) | (e) | (f) | (g) | (h) | (i) | (j) | (k) | (l) | (m) | (n) | (o) | (p) |
| 75.7 | 136.8 | 84.7 | 74.4 | 131.0 | 83.2 | 73.3 | 126.0 | 81.9 | 42.5 | 89.0 | 41.2 | 88.0 | 40.0 | 86.2 | 672 |

### Extreme Annual Design Conditions

| Extreme Annual WS | | | Extreme Max WB | Extreme Annual DB | | | | n-Year Return Period Values of Extreme DB | | | | | | | |
|---|---|---|---|---|---|---|---|---|---|---|---|---|---|---|---|
| | | | | Mean | | Standard deviation | | n=5 years | | n=10 years | | n=20 years | | n=50 years | |
| 1% | 2.5% | 5% | | Min | Max | Min | Max | Min | Max | Min | Max | Min | Max | Min | Max |
| (a) | (b) | (c) | (d) | (e) | (f) | (g) | (h) | (i) | (j) | (k) | (l) | (m) | (n) | (o) | (p) |
| 21.0 | 18.7 | 16.8 | 84.2 | 3.2 | 97.1 | 7.7 | 3.7 | -2.3 | 99.8 | -6.9 | 101.9 | -11.2 | 104.0 | -16.8 | 106.7 |

### Monthly Climatic Design Conditions

| | | | Annual | Jan | Feb | Mar | Apr | May | Jun | Jul | Aug | Sep | Oct | Nov | Dec |
|---|---|---|---|---|---|---|---|---|---|---|---|---|---|---|---|
| (5) | Temperatures, Degree-Days and Degree-Hours | Tavg | 58.1 | 35.5 | 38.8 | 47.8 | 57.9 | 66.9 | 75.4 | 78.9 | 78.2 | 71.0 | 58.9 | 48.4 | 37.7 |
| (6) | | Sd | | 11.44 | 10.64 | 10.43 | 9.27 | 7.43 | 5.72 | 4.47 | 5.09 | 7.01 | 8.41 | 9.52 | 10.48 |
| (7) | | HDD50 | 1567 | 467 | 337 | 169 | 32 | 1 | 0 | 0 | 0 | 0 | 18 | 144 | 399 |
| (8) | | HDD65 | 4109 | 913 | 734 | 540 | 247 | 69 | 5 | 0 | 1 | 26 | 226 | 501 | 847 |
| (9) | | CDD50 | 4506 | 19 | 23 | 100 | 269 | 525 | 763 | 896 | 873 | 629 | 295 | 96 | 18 |
| (10) | | CDD65 | 1572 | 0 | 0 | 6 | 35 | 128 | 318 | 431 | 409 | 205 | 38 | 2 | 0 |
| (11) | | CDH74 | 14896 | 0 | 1 | 51 | 313 | 1024 | 2995 | 4412 | 4047 | 1740 | 297 | 16 | 0 |
| (12) | | CDH80 | 5700 | 0 | 0 | 3 | 61 | 266 | 1140 | 1854 | 1710 | 606 | 60 | 0 | 0 |
| (13) | Precipitation | PrecAvg | 44.4 | 2.9 | 3.3 | 4.7 | 4.2 | 4.6 | 3.5 | 4.5 | 3.5 | 3.2 | 2.7 | 3.7 | 3.6 |
| (14) | | PrecMax | 57.7 | 5.9 | 6.6 | 14.9 | 11.1 | 9.0 | 10.1 | 10.0 | 8.8 | 10.5 | 6.1 | 7.6 | 7.6 |
| (15) | | PrecMin | 32.6 | 1.1 | 0.8 | 1.0 | 0.8 | 1.4 | 0.8 | 1.9 | 0.9 | 0.9 | 0.6 | 0.7 | 0.6 |
| (16) | | PrecSD | 7.6 | 1.5 | 1.9 | 3.2 | 2.8 | 1.8 | 2.2 | 2.5 | 2.1 | 2.2 | 1.5 | 1.8 | 1.7 |
| (17) | Monthly Design Dry Bulb and Mean Coincident Wet Bulb Temperatures | 0.4% DB | 67.1 | 69.6 | 79.4 | 85.6 | 88.5 | 93.5 | 96.7 | 98.1 | 94.8 | 85.6 | 76.3 | 67.3 | |
| (18) | | 0.4% MCWB | 60.9 | 57.9 | 61.5 | 66.0 | 71.8 | 75.0 | 76.8 | 76.0 | 71.5 | 69.4 | 62.3 | 59.9 | |
| (19) | | 2% DB | 62.6 | 64.4 | 74.5 | 81.5 | 85.7 | 91.1 | 93.5 | 93.9 | 90.1 | 81.1 | 71.9 | 61.6 | |
| (20) | | 2% MCWB | 57.1 | 55.0 | 59.5 | 64.0 | 70.5 | 74.7 | 76.2 | 75.3 | 71.4 | 66.3 | 60.1 | 55.2 | |
| (21) | | 5% DB | 57.6 | 60.2 | 70.0 | 77.7 | 83.2 | 89.1 | 91.3 | 91.3 | 86.5 | 77.2 | 67.9 | 57.7 | |
| (22) | | 5% MCWB | 52.6 | 53.0 | 57.9 | 62.3 | 69.2 | 73.8 | 75.7 | 75.2 | 70.7 | 64.7 | 59.3 | 53.1 | |
| (23) | | 10% DB | 52.8 | 55.4 | 64.9 | 73.7 | 80.2 | 86.7 | 89.0 | 88.6 | 83.2 | 73.1 | 63.9 | 53.8 | |
| (24) | | 10% MCWB | 47.3 | 48.5 | 55.0 | 60.8 | 67.8 | 72.9 | 74.8 | 74.1 | 69.1 | 62.5 | 56.8 | 48.9 | |
| (25) | Monthly Design Wet Bulb and Mean Coincident Dry Bulb Temperatures | 0.4% WB | 62.0 | 61.5 | 64.7 | 69.6 | 75.3 | 78.8 | 80.8 | 80.4 | 76.3 | 73.0 | 66.0 | 61.9 | |
| (26) | | 0.4% MCDB | 65.5 | 66.3 | 73.9 | 80.3 | 84.3 | 88.8 | 91.2 | 91.0 | 86.4 | 80.1 | 72.4 | 65.8 | |
| (27) | | 2% WB | 58.6 | 57.6 | 62.1 | 66.9 | 73.2 | 77.4 | 78.9 | 78.7 | 74.6 | 69.5 | 63.4 | 57.4 | |
| (28) | | 2% MCDB | 62.1 | 62.2 | 70.9 | 76.2 | 81.8 | 87.8 | 89.8 | 88.9 | 84.2 | 76.9 | 68.0 | 60.2 | |
| (29) | | 5% WB | 53.8 | 55.4 | 59.5 | 64.6 | 71.4 | 76.0 | 77.7 | 77.3 | 73.3 | 67.0 | 60.4 | 53.6 | |
| (30) | | 5% MCDB | 57.1 | 59.0 | 67.8 | 73.8 | 80.1 | 85.6 | 88.2 | 87.4 | 81.6 | 74.2 | 66.4 | 56.6 | |
| (31) | | 10% WB | 48.1 | 49.2 | 56.4 | 62.5 | 69.6 | 74.7 | 76.5 | 75.9 | 71.8 | 64.4 | 57.3 | 49.2 | |
| (32) | | 10% MCDB | 52.3 | 55.0 | 64.1 | 71.5 | 77.5 | 83.6 | 85.9 | 85.2 | 79.3 | 70.9 | 63.3 | 53.0 | |
| (33) | Mean Daily Temperature Range | MDBR | 15.0 | 16.2 | 18.6 | 20.1 | 18.5 | 17.7 | 17.3 | 18.2 | 19.2 | 19.9 | 17.0 | 14.3 | |
| (34) | | 5% DB MCDBR | 19.5 | 22.0 | 25.1 | 24.5 | 21.0 | 20.2 | 19.7 | 21.0 | 22.5 | 23.4 | 22.6 | 19.7 | |
| (35) | | 5% DB MCWBR | 15.4 | 15.9 | 14.3 | 12.0 | 9.2 | 8.1 | 7.0 | 7.4 | 8.5 | 11.5 | 14.6 | 16.3 | |
| (36) | | 5% WB MCDBR | 17.0 | 18.6 | 21.4 | 19.9 | 17.6 | 17.6 | 17.4 | 18.0 | 17.3 | 18.3 | 19.7 | 18.0 | |
| (37) | | 5% WB MCWBR | 14.9 | 15.4 | 14.1 | 11.6 | 9.2 | 8.2 | 7.6 | 7.5 | 8.2 | 10.9 | 15.4 | 16.7 | |
| (38) | Clear Sky Solar Irradiance | taub | 0.321 | 0.338 | 0.352 | 0.397 | 0.410 | 0.427 | 0.461 | 0.448 | 0.404 | 0.352 | 0.330 | 0.325 | |
| (39) | | taud | 2.548 | 2.469 | 2.427 | 2.258 | 2.193 | 2.166 | 2.089 | 2.154 | 2.312 | 2.485 | 2.562 | 2.543 | |
| (40) | | Ebn,noon | 274 | 282 | 288 | 280 | 277 | 271 | 261 | 261 | 267 | 273 | 268 | 264 | |
| (41) | | Edh,noon | 24 | 29 | 34 | 42 | 46 | 47 | 51 | 46 | 37 | 29 | 24 | 23 | |

Nomenclature: See separate page

*Fig. 4-4 Example tables of climatic design information included on CD-ROM of 2013 ASHRAE Handbook—Fundamentals*

# Design Conditions

2013 ASHRAE Handbook - Fundamentals (IP)    © 2013 ASHRAE, Inc.

**MANHATTAN RGNL, KS, USA**    WMO#: 724555

Lat: 39.13N  Long: 96.68W  Elev: 1070  StdP: 14.14  Time Zone: -6 (NAC)  Period: 96-10  WBAN: 3936

### Annual Heating and Humidification Design Conditions

| | Coldest Month | Heating DB | | Humidification DP/MCDB and HR | | | | | | Coldest month WS/MCDB | | | | MCWS/PCWD to 99.6% DB | |
|---|---|---|---|---|---|---|---|---|---|---|---|---|---|---|---|
| | | 99.6% | 99% | 99.6% DP | HR | MCDB | 99% DP | HR | MCDB | 0.4% WS | MCDB | 1% WS | MCDB | MCWS | PCWD |
| | (a) | (b) | (c) | (d) | (e) | (f) | (g) | (h) | (i) | (j) | (k) | (l) | (m) | (n) | (o) |
| (1) | 1 | 1.4 | 8.6 | -6.1 | 4.1 | 5.7 | -0.5 | 5.6 | 10.6 | 24.1 | 51.5 | 20.8 | 43.0 | 3.2 | 350 |

### Annual Cooling, Dehumidification, and Enthalpy Design Conditions

| | Hottest Month | Hottest Month DB Range | Cooling DB/MCWB | | | | | | Evaporation WB/MCDB | | | | | | MCWS/PCWD to 0.4% DB | |
|---|---|---|---|---|---|---|---|---|---|---|---|---|---|---|---|---|
| | | | 0.4% DB | MCWB | 1% DB | MCWB | 2% DB | MCWB | 0.4% WB | MCDB | 1% WB | MCDB | 2% WB | MCDB | MCWS | PCWD |
| | (a) | (b) | (c) | (d) | (e) | (f) | (g) | (h) | (i) | (j) | (k) | (l) | (m) | (n) | (o) | (p) |
| (2) | 7 | 22.2 | 99.6 | 75.8 | 96.7 | 75.9 | 92.7 | 74.9 | 78.7 | 92.7 | 77.7 | 91.5 | 76.5 | 89.4 | 11.5 | 200 |

| | Dehumidification DP/MCDB and HR | | | | | | | | | Enthalpy/MCDB | | | | | | Hours 8 to 4 & 55/69 |
|---|---|---|---|---|---|---|---|---|---|---|---|---|---|---|---|---|
| | 0.4% DP | HR | MCDB | 1% DP | HR | MCDB | 2% DP | HR | MCDB | 0.4% Enth | MCDB | 1% Enth | MCDB | 2% Enth | MCDB | |
| | (a) | (b) | (c) | (d) | (e) | (f) | (g) | (h) | (i) | (j) | (k) | (l) | (m) | (n) | (o) | (p) |
| (3) | 75.0 | 136.4 | 85.8 | 73.3 | 128.6 | 83.7 | 72.6 | 125.6 | 83.1 | 43.0 | 92.8 | 41.8 | 90.8 | 40.6 | 89.5 | 655 |

### Extreme Annual Design Conditions

| | Extreme Annual WS | | | Extreme Max WB | Extreme Annual DB | | | | n-Year Return Period Values of Extreme DB | | | | | | | |
|---|---|---|---|---|---|---|---|---|---|---|---|---|---|---|---|---|
| | 1% | 2.5% | 5% | | Mean Min | Max | Standard deviation Min | Max | n=5 years Min | Max | n=10 years Min | Max | n=20 years Min | Max | n=50 years Min | Max |
| | (a) | (b) | (c) | (d) | (e) | (f) | (g) | (h) | (i) | (j) | (k) | (l) | (m) | (n) | (o) | (p) |
| (4) | 24.2 | 20.5 | 18.3 | 83.7 | -7.3 | 105.0 | 5.2 | 3.4 | -11.1 | 107.4 | -14.2 | 109.4 | -17.1 | 111.3 | -20.9 | 113.8 |

### Monthly Climatic Design Conditions

| | | | Annual (d) | Jan (e) | Feb (f) | Mar (g) | Apr (h) | May (i) | Jun (j) | Jul (k) | Aug (l) | Sep (m) | Oct (n) | Nov (o) | Dec (p) |
|---|---|---|---|---|---|---|---|---|---|---|---|---|---|---|---|
| (5) | | Tavg | 54.9 | 29.6 | 33.8 | 43.3 | 54.7 | 64.6 | 74.1 | 79.5 | 77.7 | 68.2 | 56.2 | 43.6 | 31.5 |
| (6) | | Sd | | 11.05 | 10.98 | 10.84 | 9.79 | 8.33 | 6.66 | 6.07 | 6.86 | 8.05 | 9.50 | 9.52 | 10.57 |
| (7) | Temperatures, | HDD50 | 2284 | 633 | 463 | 261 | 60 | 2 | 0 | 0 | 0 | 1 | 46 | 240 | 578 |
| (8) | Degree-Days | HDD65 | 5144 | 1097 | 873 | 674 | 332 | 114 | 9 | 1 | 2 | 57 | 302 | 644 | 1039 |
| (9) | and | CDD50 | 4056 | 1 | 10 | 54 | 201 | 455 | 722 | 916 | 859 | 548 | 238 | 47 | 5 |
| (10) | Degree-Hours | CDD65 | 1440 | 0 | 0 | 2 | 23 | 102 | 281 | 451 | 396 | 155 | 29 | 1 | 0 |
| (11) | | CDH74 | 16441 | 0 | 2 | 29 | 303 | 1127 | 3040 | 5326 | 4531 | 1710 | 355 | 18 | 0 |
| (12) | | CDH80 | 7728 | 0 | 0 | 3 | 74 | 405 | 1305 | 2788 | 2348 | 700 | 104 | 1 | 0 |
| (13) | | PrecAvg | 33.9 | 0.8 | 0.9 | 2.4 | 3.0 | 4.6 | 5.5 | 3.4 | 3.3 | 4.1 | 3.0 | 1.8 | 1.1 |
| (14) | Precipitation | PrecMax | 50.2 | 3.2 | 2.2 | 7.4 | 6.0 | 9.8 | 12.0 | 8.1 | 7.2 | 9.9 | 6.5 | 4.7 | 3.4 |
| (15) | | PrecMin | 15.0 | 0.0 | 0.0 | 0.0 | 1.1 | 1.8 | 1.7 | 0.7 | 0.3 | 0.6 | 0.1 | 0.0 | 0.0 |
| (16) | | PrecSD | 8.5 | 0.8 | 0.7 | 1.6 | 1.5 | 2.2 | 3.0 | 2.3 | 1.9 | 2.9 | 1.9 | 1.4 | 1.0 |
| (17) | | 0.4% DB | | 64.0 | 70.1 | 78.8 | 87.9 | 92.8 | 97.2 | 102.5 | 105.8 | 98.7 | 89.7 | 76.7 | 65.9 |
| (18) | Monthly Design | MCWB | | 49.6 | 55.2 | 59.1 | 66.3 | 71.9 | 75.4 | 76.9 | 74.0 | 72.5 | 69.0 | 58.9 | 59.2 |
| (19) | Dry Bulb | 2% DB | | 57.3 | 63.5 | 72.6 | 81.5 | 88.4 | 93.2 | 99.6 | 100.0 | 91.0 | 82.0 | 71.7 | 58.7 |
| (20) | and | MCWB | | 45.5 | 50.1 | 57.4 | 63.0 | 71.0 | 74.8 | 76.7 | 75.2 | 71.9 | 64.5 | 57.1 | 48.7 |
| (21) | Mean Coincident | 5% DB | | 52.3 | 56.9 | 67.8 | 77.3 | 84.1 | 90.8 | 96.9 | 96.8 | 88.1 | 77.4 | 65.7 | 54.1 |
| (22) | Wet Bulb | MCWB | | 43.4 | 47.5 | 56.4 | 61.1 | 69.3 | 74.5 | 76.8 | 75.6 | 71.0 | 63.2 | 55.2 | 46.1 |
| (23) | Temperatures | 10% DB | | 46.3 | 52.1 | 63.3 | 72.7 | 80.9 | 88.1 | 92.8 | 91.2 | 83.7 | 72.8 | 61.3 | 47.9 |
| (24) | | MCWB | | 39.7 | 43.9 | 52.7 | 59.5 | 67.6 | 73.4 | 75.8 | 74.7 | 69.1 | 61.4 | 52.6 | 41.4 |
| (25) | | 0.4% WB | | 52.7 | 58.1 | 63.9 | 69.8 | 75.6 | 79.7 | 80.4 | 79.8 | 76.3 | 72.1 | 62.4 | 59.0 |
| (26) | Monthly Design | MCDB | | 58.1 | 66.6 | 71.3 | 80.1 | 87.5 | 92.4 | 93.4 | 93.5 | 89.4 | 83.0 | 69.9 | 63.9 |
| (27) | Wet Bulb | 2% WB | | 47.3 | 52.4 | 61.1 | 66.3 | 73.4 | 77.7 | 79.2 | 78.1 | 74.4 | 68.8 | 59.6 | 51.4 |
| (28) | and | MCDB | | 56.5 | 60.2 | 67.1 | 76.6 | 84.5 | 90.3 | 93.5 | 92.1 | 87.5 | 78.1 | 66.7 | 55.3 |
| (29) | Mean Coincident | 5% WB | | 44.0 | 47.8 | 57.8 | 62.9 | 71.3 | 76.1 | 78.1 | 76.9 | 72.6 | 65.9 | 56.8 | 45.6 |
| (30) | Dry Bulb | MCDB | | 52.3 | 57.1 | 67.2 | 74.5 | 81.1 | 87.2 | 92.3 | 90.6 | 84.7 | 74.0 | 64.6 | 52.1 |
| (31) | Temperatures | 10% WB | | 40.2 | 43.9 | 52.8 | 60.6 | 68.9 | 74.5 | 77.0 | 75.6 | 70.8 | 63.0 | 53.0 | 41.4 |
| (32) | | MCDB | | 47.3 | 51.6 | 63.0 | 71.4 | 78.1 | 84.6 | 90.6 | 88.9 | 81.1 | 70.6 | 60.8 | 47.9 |
| (33) | | MDBR | | 22.1 | 22.6 | 24.2 | 24.2 | 23.2 | 22.0 | 22.2 | 23.4 | 25.0 | 24.5 | 23.4 | 21.9 |
| (34) | Mean Daily | 5% DB MCDBR | | 33.1 | 32.9 | 32.5 | 30.3 | 28.1 | 25.1 | 26.7 | 28.6 | 27.9 | 30.3 | 31.8 | 30.7 |
| (35) | Temperature | MCWBR | | 22.5 | 21.0 | 18.9 | 16.1 | 13.5 | 10.7 | 9.0 | 8.7 | 11.7 | 15.5 | 19.1 | 21.2 |
| (36) | Range | 5% WB MCDBR | | 31.3 | 29.5 | 26.0 | 25.6 | 23.9 | 21.9 | 23.4 | 24.3 | 24.3 | 22.4 | 25.6 | 28.3 |
| (37) | | MCWBR | | 21.9 | 19.9 | 16.6 | 15.8 | 13.1 | 10.8 | 9.4 | 10.1 | 11.7 | 13.5 | 17.1 | 20.7 |
| (38) | | taub | | 0.293 | 0.303 | 0.328 | 0.339 | 0.344 | 0.360 | 0.369 | 0.389 | 0.353 | 0.323 | 0.304 | 0.294 |
| (39) | Clear Sky Solar | taud | | 2.565 | 2.451 | 2.449 | 2.407 | 2.397 | 2.352 | 2.361 | 2.343 | 2.444 | 2.536 | 2.622 | 2.582 |
| (40) | Irradiance | Ebn,noon | | 283 | 293 | 296 | 297 | 296 | 290 | 286 | 278 | 283 | 283 | 277 | 274 |
| (41) | | Edh,noon | | 23 | 29 | 33 | 36 | 37 | 39 | 38 | 38 | 32 | 27 | 22 | 21 |

Nomenclature:  See separate page

*Fig. 4-4 Example tables of climatic design information included on CD-ROM of 2013 ASHRAE Handbook—Fundamentals*

**110**      Principles of HVAC

2013 ASHRAE Handbook - Fundamentals (IP)      © 2013 ASHRAE, Inc.

## MINNEAPOLIS/ST.PAUL, MN, USA

WMO#: **726580**

Lat: **44.88N**   Long: **93.23W**   Elev: **837**   StdP: **14.26**   Time Zone: **-6 (NAC)**   Period: **86-10**   WBAN: **14922**

### Annual Heating and Humidification Design Conditions

| Coldest Month | Heating DB | | Humidification DP/MCDB and HR | | | | | | Coldest month WS/MCDB | | | | MCWS/PCWD to 99.6% DB | |
|---|---|---|---|---|---|---|---|---|---|---|---|---|---|---|
| | 99.6% | 99% | 99.6% | | | 99% | | | 0.4% | | 1% | | | |
| | | | DP | HR | MCDB | DP | HR | MCDB | WS | MCDB | WS | MCDB | MCWS | PCWD |
| (a) | (b) | (c) | (d) | (e) | (f) | (g) | (h) | (i) | (j) | (k) | (l) | (m) | (n) | (o) |
| 1 | -11.2 | -6.2 | -19.8 | 1.9 | -9.9 | -15.2 | 2.5 | -5.3 | 26.7 | 18.6 | 24.3 | 18.5 | 8.4 | 310 |

### Annual Cooling, Dehumidification, and Enthalpy Design Conditions

| Hottest Month | Hottest Month DB Range | Cooling DB/MCWB | | | | | | Evaporation WB/MCDB | | | | | | MCWS/PCWD to 0.4% DB | |
|---|---|---|---|---|---|---|---|---|---|---|---|---|---|---|---|
| | | 0.4% | | 1% | | 2% | | 0.4% | | 1% | | 2% | | | |
| | | DB | MCWB | DB | MCWB | DB | MCWB | WB | MCDB | WB | MCDB | WB | MCDB | MCWS | PCWD |
| (a) | (b) | (c) | (d) | (e) | (f) | (g) | (h) | (i) | (j) | (k) | (l) | (m) | (n) | (o) | (p) |
| 7 | 17.6 | 90.9 | 72.9 | 88.0 | 71.9 | 84.8 | 70.3 | 76.8 | 87.0 | 74.8 | 84.0 | 72.8 | 81.7 | 13.2 | 180 |

| Dehumidification DP/MCDB and HR | | | | | | | | Enthalpy/MCDB | | | | | | Hours 8 to 4 & 55/69 |
|---|---|---|---|---|---|---|---|---|---|---|---|---|---|---|
| 0.4% | | | 1% | | | 2% | | | 0.4% | | 1% | | 2% | |
| DP | HR | MCDB | DP | HR | MCDB | DP | HR | MCDB | Enth | MCDB | Enth | MCDB | Enth | MCDB |
| (a) | (b) | (c) | (d) | (e) | (f) | (g) | (h) | (i) | (j) | (k) | (l) | (m) | (n) | (o) | (p) |
| 73.4 | 128.3 | 83.3 | 71.7 | 120.6 | 81.0 | 69.7 | 112.6 | 78.5 | 40.8 | 87.0 | 38.8 | 83.9 | 36.9 | 81.9 | 610 |

### Extreme Annual Design Conditions

| Extreme Annual WS | | | Extreme Max WB | Extreme Annual DB | | | | n-Year Return Period Values of Extreme DB | | | | | | | |
|---|---|---|---|---|---|---|---|---|---|---|---|---|---|---|---|
| | | | | Mean | | Standard deviation | | n=5 years | | n=10 years | | n=20 years | | n=50 years | |
| 1% | 2.5% | 5% | | Min | Max | Min | Max | Min | Max | Min | Max | Min | Max | Min | Max |
| (a) | (b) | (c) | (d) | (e) | (f) | (g) | (h) | (i) | (j) | (k) | (l) | (m) | (n) | (o) | (p) |
| 24.4 | 20.9 | 19.1 | 83.5 | -17.2 | 95.9 | 5.8 | 3.4 | -21.4 | 98.3 | -24.8 | 100.3 | -28.0 | 102.2 | -32.3 | 104.6 |

### Monthly Climatic Design Conditions

| | | Annual | Jan | Feb | Mar | Apr | May | Jun | Jul | Aug | Sep | Oct | Nov | Dec |
|---|---|---|---|---|---|---|---|---|---|---|---|---|---|---|
| | | (d) | (e) | (f) | (g) | (h) | (i) | (j) | (k) | (l) | (m) | (n) | (o) | (p) |
| Temperatures, Degree-Days and Degree-Hours | Tavg | 46.6 | 16.4 | 20.8 | 33.0 | 47.7 | 59.3 | 69.3 | 73.6 | 71.0 | 62.4 | 49.1 | 34.2 | 20.9 |
| | Sd | | 12.35 | 12.12 | 11.52 | 9.85 | 8.57 | 7.23 | 5.94 | 5.88 | 8.25 | 9.24 | 10.53 | 11.67 |
| | HDD50 | 4091 | 1040 | 817 | 536 | 157 | 17 | 0 | 0 | 0 | 6 | 132 | 485 | 901 |
| | HDD65 | 7472 | 1505 | 1237 | 991 | 524 | 218 | 40 | 6 | 14 | 145 | 501 | 925 | 1366 |
| | CDD50 | 2859 | 0 | 0 | 0 | 10 | 88 | 306 | 578 | 733 | 652 | 379 | 103 | 10 |
| | CDD65 | 765 | 0 | 0 | 0 | 0 | 5 | 42 | 167 | 274 | 202 | 68 | 7 | 0 |
| | CDH74 | 6630 | 0 | 0 | 4 | 65 | 435 | 1488 | 2479 | 1567 | 525 | 67 | 0 | 0 |
| | CDH80 | 2152 | 0 | 0 | 0 | 15 | 128 | 497 | 898 | 471 | 132 | 11 | 0 | 0 |
| Precipitation | PrecAvg | 28.3 | 1.0 | 0.9 | 1.9 | 2.4 | 3.4 | 4.1 | 3.5 | 3.6 | 2.7 | 2.2 | 1.6 | 1.1 |
| | PrecMax | 39.3 | 3.6 | 2.1 | 4.8 | 5.4 | 8.0 | 8.0 | 6.5 | 9.3 | 6.1 | 5.7 | 4.8 | 2.2 |
| | PrecMin | 16.2 | 0.2 | 0.1 | 0.8 | 0.7 | 0.6 | 1.1 | 0.6 | 0.8 | 0.5 | 0.2 | 0.1 | 0.2 |
| | PrecSD | 6.6 | 0.9 | 0.6 | 1.0 | 1.2 | 1.9 | 2.1 | 1.5 | 2.1 | 1.6 | 1.7 | 1.2 | 0.6 |
| Monthly Design Dry Bulb and Mean Coincident Wet Bulb Temperatures | 0.4% DB | | 44.3 | 49.9 | 69.7 | 82.0 | 89.8 | 93.5 | 96.2 | 93.2 | 88.4 | 81.4 | 66.8 | 48.9 |
| | 0.4% MCWB | | 38.1 | 42.9 | 57.9 | 62.1 | 66.7 | 73.6 | 75.7 | 74.5 | 71.1 | 65.9 | 55.6 | 43.7 |
| | 2% DB | | 38.5 | 44.3 | 60.2 | 73.5 | 83.1 | 89.3 | 91.3 | 88.4 | 83.8 | 74.4 | 59.6 | 42.4 |
| | 2% MCWB | | 34.3 | 39.4 | 51.3 | 56.8 | 64.8 | 70.5 | 74.0 | 73.7 | 69.4 | 61.6 | 52.4 | 37.7 |
| | 5% DB | | 35.5 | 40.1 | 54.1 | 68.9 | 79.0 | 85.6 | 88.2 | 84.7 | 80.3 | 68.9 | 54.0 | 38.2 |
| | 5% MCWB | | 32.7 | 35.8 | 46.6 | 54.0 | 62.5 | 69.6 | 72.9 | 71.7 | 67.1 | 57.6 | 46.8 | 34.6 |
| | 10% DB | | 33.0 | 36.8 | 48.5 | 63.9 | 74.2 | 82.1 | 84.9 | 81.9 | 76.2 | 63.8 | 48.7 | 35.3 |
| | 10% MCWB | | 30.9 | 33.3 | 42.2 | 51.1 | 60.1 | 67.1 | 71.3 | 69.4 | 65.1 | 55.1 | 42.9 | 32.6 |
| Monthly Design Wet Bulb and Mean Coincident Dry Bulb Temperatures | 0.4% WB | | 38.4 | 43.9 | 58.9 | 65.0 | 71.4 | 77.3 | 80.0 | 78.9 | 74.1 | 69.1 | 58.9 | 45.2 |
| | 0.4% MCDB | | 43.7 | 47.7 | 66.3 | 77.2 | 83.0 | 89.5 | 89.8 | 88.8 | 83.8 | 76.6 | 64.3 | 47.2 |
| | 2% WB | | 35.0 | 39.6 | 52.7 | 59.9 | 67.9 | 74.2 | 77.4 | 76.0 | 71.4 | 63.6 | 53.0 | 37.8 |
| | 2% MCDB | | 37.8 | 44.1 | 59.2 | 70.2 | 77.9 | 84.8 | 88.2 | 84.7 | 79.8 | 70.7 | 58.6 | 41.5 |
| | 5% WB | | 32.9 | 36.0 | 47.1 | 55.9 | 65.3 | 71.8 | 75.4 | 74.0 | 69.3 | 59.6 | 47.6 | 34.9 |
| | 5% MCDB | | 35.0 | 39.7 | 53.3 | 66.3 | 75.2 | 81.7 | 84.7 | 82.3 | 77.2 | 67.1 | 53.4 | 38.0 |
| | 10% WB | | 31.0 | 33.4 | 42.5 | 52.4 | 62.3 | 69.5 | 73.2 | 71.7 | 66.8 | 55.7 | 43.4 | 32.8 |
| | 10% MCDB | | 32.9 | 36.4 | 47.9 | 62.6 | 72.1 | 78.8 | 82.1 | 79.3 | 74.4 | 62.9 | 48.7 | 35.0 |
| Mean Daily Temperature Range | MDBR | | 14.7 | 15.2 | 16.0 | 19.1 | 19.0 | 18.4 | 17.6 | 17.3 | 18.1 | 17.3 | 13.8 | 13.5 |
| | 5% DB MCDBR | | 15.7 | 17.5 | 23.2 | 26.1 | 25.2 | 22.1 | 20.6 | 20.0 | 22.1 | 24.9 | 20.3 | 16.7 |
| | 5% DB MCWBR | | 12.8 | 13.1 | 15.0 | 14.1 | 11.7 | 9.9 | 9.0 | 9.3 | 10.7 | 14.2 | 13.8 | 13.1 |
| | 5% WB MCDBR | | 14.4 | 17.0 | 21.8 | 23.0 | 21.5 | 18.9 | 18.8 | 18.0 | 19.4 | 21.7 | 18.3 | 15.6 |
| | 5% WB MCWBR | | 12.3 | 13.1 | 14.8 | 14.0 | 11.8 | 10.2 | 9.9 | 10.2 | 11.4 | 14.3 | 14.1 | 12.5 |
| Clear Sky Solar Irradiance | taub | | 0.271 | 0.266 | 0.293 | 0.314 | 0.323 | 0.340 | 0.361 | 0.364 | 0.342 | 0.316 | 0.292 | 0.277 |
| | taud | | 2.483 | 2.404 | 2.429 | 2.449 | 2.431 | 2.392 | 2.354 | 2.392 | 2.464 | 2.551 | 2.598 | 2.524 |
| | Ebn,noon | | 273 | 296 | 301 | 301 | 300 | 294 | 286 | 281 | 279 | 274 | 265 | 260 |
| | Edh,noon | | 21 | 28 | 31 | 33 | 35 | 37 | 38 | 35 | 30 | 24 | 19 | 19 |

Nomenclature: See separate page

*Fig. 4-4 Example tables of climatic design information included on CD-ROM of 2013 ASHRAE Handbook—Fundamentals*

# Design Conditions

2013 ASHRAE Handbook - Fundamentals (IP) © 2013 ASHRAE, Inc.

**NEW ORLEANS/MOISANT, LA, USA** WMO#: 722310

Lat: 29.99N  Long: 90.25W  Elev: 20  StdP: 14.69  Time Zone: -6 (NAC)  Period: 86-10  WBAN: 12916

### Annual Heating and Humidification Design Conditions

| Coldest Month | Heating DB | | Humidification DP/MCDB and HR | | | | | | Coldest month WS/MCDB | | | | MCWS/PCWD to 99.6% DB | |
|---|---|---|---|---|---|---|---|---|---|---|---|---|---|---|
| | 99.6% | 99% | 99.6% DP | HR | MCDB | 99% DP | HR | MCDB | 0.4% WS | MCDB | 1% WS | MCDB | MCWS | PCWD |
| (a) | (b) | (c) | (d) | (e) | (f) | (g) | (h) | (i) | (j) | (k) | (l) | (m) | (n) | (o) |
| 1 | 33.1 | 36.3 | 17.3 | 13.2 | 37.7 | 21.5 | 16.2 | 42.2 | 23.9 | 57.3 | 20.9 | 54.1 | 8.7 | 0 |

### Annual Cooling, Dehumidification, and Enthalpy Design Conditions

| Hottest Month | Hottest Month DB Range | Cooling DB/MCWB | | | | | | Evaporation WB/MCDB | | | | | | MCWS/PCWD to 0.4% DB | |
|---|---|---|---|---|---|---|---|---|---|---|---|---|---|---|---|
| | | 0.4% DB | MCWB | 1% DB | MCWB | 2% DB | MCWB | 0.4% WB | MCDB | 1% WB | MCDB | 2% WB | MCDB | MCWS | PCWD |
| (a) | (b) | (c) | (d) | (e) | (f) | (g) | (h) | (i) | (j) | (k) | (l) | (m) | (n) | (o) | (p) |
| 7 | 15.0 | 93.8 | 78.1 | 92.2 | 77.7 | 90.8 | 77.5 | 80.9 | 88.9 | 80.2 | 88.0 | 79.5 | 87.1 | 7.4 | 0 |

| Dehumidification DP/MCDB and HR | | | | | | | | Enthalpy/MCDB | | | | | | Hours 8 to 4 & 55/69 |
|---|---|---|---|---|---|---|---|---|---|---|---|---|---|---|
| 0.4% DP | HR | MCDB | 1% DP | HR | MCDB | 2% DP | HR | MCDB | 0.4% Enth | MCDB | 1% Enth | MCDB | 2% Enth | MCDB | |
| (a) | (b) | (c) | (d) | (e) | (f) | (g) | (h) | (i) | (j) | (k) | (l) | (m) | (n) | (o) | (p) |
| 79.0 | 150.6 | 84.4 | 78.2 | 146.5 | 83.8 | 77.4 | 142.5 | 83.3 | 44.4 | 89.0 | 43.5 | 87.7 | 42.8 | 87.0 | 682 |

### Extreme Annual Design Conditions

| Extreme Annual WS | | | Extreme Max WB | Extreme Annual DB | | | | n-Year Return Period Values of Extreme DB | | | | | | | |
|---|---|---|---|---|---|---|---|---|---|---|---|---|---|---|---|
| 1% | 2.5% | 5% | | Mean Min | Max | Standard deviation Min | Max | n=5 years Min | Max | n=10 years Min | Max | n=20 years Min | Max | n=50 years Min | Max |
| (a) | (b) | (c) | (d) | (e) | (f) | (g) | (h) | (i) | (j) | (k) | (l) | (m) | (n) | (o) | (p) |
| 20.7 | 18.7 | 17.0 | 84.7 | 27.0 | 97.0 | 4.5 | 2.1 | 23.7 | 98.6 | 21.1 | 99.8 | 18.5 | 101.0 | 15.2 | 102.5 |

### Monthly Climatic Design Conditions

| | | | Annual (d) | Jan (e) | Feb (f) | Mar (g) | Apr (h) | May (i) | Jun (j) | Jul (k) | Aug (l) | Sep (m) | Oct (n) | Nov (o) | Dec (p) |
|---|---|---|---|---|---|---|---|---|---|---|---|---|---|---|---|
| (5) | | Tavg | 69.5 | 53.9 | 56.8 | 62.4 | 68.7 | 76.5 | 81.1 | 82.9 | 82.9 | 79.7 | 70.8 | 62.2 | 55.2 |
| (6) | | Sd | | 9.20 | 8.76 | 7.68 | 6.30 | 4.25 | 2.91 | 2.47 | 2.69 | 4.07 | 6.83 | 7.93 | 9.34 |
| (7) | Temperatures, | HDD50 | 160 | 65 | 31 | 8 | 0 | 0 | 0 | 0 | 0 | 0 | 0 | 6 | 50 |
| (8) | Degree-Days | HDD65 | 1286 | 361 | 252 | 140 | 37 | 0 | 0 | 0 | 0 | 0 | 26 | 144 | 326 |
| (9) | and | CDD50 | 7273 | 185 | 222 | 391 | 563 | 821 | 934 | 1021 | 1018 | 890 | 644 | 372 | 212 |
| (10) | Degree-Hours | CDD65 | 2925 | 16 | 22 | 58 | 149 | 356 | 484 | 556 | 553 | 441 | 205 | 61 | 24 |
| (11) | | CDH74 | 27414 | 33 | 57 | 230 | 918 | 2980 | 4970 | 6091 | 6101 | 4228 | 1467 | 277 | 62 |
| (12) | | CDH80 | 9968 | 0 | 1 | 9 | 159 | 932 | 1941 | 2533 | 2562 | 1478 | 326 | 25 | 2 |
| (13) | | PrecAvg | 61.9 | 5.1 | 6.0 | 4.9 | 4.5 | 4.6 | 5.8 | 6.1 | 6.2 | 5.5 | 3.1 | 4.4 | 5.8 |
| (14) | Precipitation | PrecMax | 80.4 | 11.9 | 14.5 | 10.3 | 22.8 | 13.2 | 11.2 | 9.1 | 15.0 | 16.7 | 11.3 | 9.7 | 9.1 |
| (15) | | PrecMin | 36.6 | 1.1 | 1.4 | 0.7 | 0.7 | 0.2 | 0.4 | 2.4 | 2.4 | 1.4 | 0.0 | 0.3 | 1.8 |
| (16) | | PrecSD | 12.3 | 3.0 | 3.3 | 2.6 | 5.3 | 3.4 | 3.2 | 2.0 | 2.9 | 4.1 | 2.6 | 2.7 | 2.1 |
| (17) | | 0.4% DB | | 77.4 | 79.0 | 81.3 | 86.8 | 91.2 | 94.5 | 96.3 | 96.4 | 93.6 | 89.4 | 83.0 | 79.3 |
| (18) | Monthly Design | MCWB | | 69.4 | 69.6 | 70.0 | 71.1 | 75.0 | 77.2 | 78.7 | 78.8 | 76.8 | 76.1 | 72.9 | 70.8 |
| (19) | Dry Bulb | 2% DB | | 74.2 | 75.6 | 79.1 | 83.8 | 88.9 | 92.1 | 93.7 | 93.7 | 91.0 | 86.0 | 79.9 | 75.5 |
| (20) | and | MCWB | | 67.2 | 67.6 | 68.4 | 71.6 | 74.6 | 76.8 | 78.4 | 78.4 | 76.5 | 73.9 | 71.3 | 68.9 |
| (21) | Mean Coincident | 5% DB | | 71.3 | 72.9 | 76.7 | 81.7 | 87.3 | 90.5 | 91.9 | 92.0 | 89.4 | 83.6 | 77.2 | 72.7 |
| (22) | Wet Bulb | MCWB | | 65.4 | 66.2 | 67.4 | 70.4 | 74.3 | 76.7 | 78.2 | 78.3 | 76.2 | 73.2 | 69.8 | 68.1 |
| (23) | Temperatures | 10% DB | | 68.3 | 70.3 | 74.2 | 79.5 | 85.2 | 88.9 | 90.2 | 90.3 | 87.6 | 81.3 | 74.7 | 70.0 |
| (24) | | MCWB | | 64.2 | 64.8 | 66.7 | 69.3 | 73.5 | 76.6 | 78.1 | 78.2 | 76.0 | 71.8 | 68.5 | 66.2 |
| (25) | | 0.4% WB | | 71.1 | 72.0 | 73.4 | 76.2 | 79.0 | 80.9 | 81.9 | 82.3 | 80.7 | 78.9 | 76.0 | 72.9 |
| (26) | Monthly Design | MCDB | | 74.8 | 76.0 | 77.8 | 81.8 | 86.8 | 89.0 | 90.7 | 89.9 | 87.9 | 84.3 | 79.4 | 76.2 |
| (27) | Wet Bulb | 2% WB | | 69.1 | 70.2 | 71.8 | 74.6 | 77.5 | 79.8 | 80.5 | 81.2 | 79.6 | 77.3 | 73.9 | 71.0 |
| (28) | and | MCDB | | 72.7 | 73.6 | 75.7 | 79.8 | 84.8 | 87.7 | 89.0 | 89.0 | 86.1 | 82.6 | 77.0 | 74.1 |
| (29) | Mean Coincident | 5% WB | | 67.3 | 68.4 | 70.3 | 73.2 | 76.5 | 78.9 | 79.9 | 80.3 | 78.6 | 76.0 | 72.0 | 69.2 |
| (30) | Dry Bulb | MCDB | | 70.1 | 71.6 | 74.4 | 78.1 | 83.3 | 86.5 | 88.3 | 87.9 | 84.9 | 80.8 | 75.7 | 72.0 |
| (31) | Temperatures | 10% WB | | 64.7 | 66.1 | 68.4 | 71.9 | 75.5 | 78.1 | 79.2 | 79.6 | 77.9 | 74.7 | 70.0 | 66.6 |
| (32) | | MCDB | | 67.4 | 69.5 | 72.8 | 76.9 | 81.7 | 85.5 | 87.2 | 87.0 | 84.1 | 79.4 | 74.0 | 69.4 |
| (33) | | MDBR | | 16.3 | 16.7 | 17.4 | 17.5 | 16.3 | 15.1 | 15.0 | 15.1 | 14.6 | 16.8 | 17.4 | 16.6 |
| (34) | Mean Daily | 5% DB MCDBR | | 18.3 | 17.8 | 17.7 | 17.8 | 17.1 | 17.0 | 17.1 | 17.2 | 16.6 | 16.9 | 18.3 | 18.1 |
| (35) | Temperature | MCWBR | | 12.6 | 11.4 | 9.5 | 8.6 | 6.5 | 6.1 | 6.0 | 6.1 | 6.4 | 7.6 | 10.8 | 12.7 |
| (36) | Range | 5% WB MCDBR | | 15.6 | 15.4 | 15.3 | 15.0 | 15.1 | 15.0 | 15.6 | 15.3 | 14.4 | 13.4 | 15.9 | 16.3 |
| (37) | | MCWBR | | 12.2 | 11.4 | 10.6 | 8.7 | 6.7 | 6.5 | 6.4 | 6.4 | 6.4 | 7.3 | 12.1 | 12.9 |
| (38) | Clear Sky | taub | | 0.367 | 0.361 | 0.371 | 0.381 | 0.370 | 0.394 | 0.433 | 0.423 | 0.400 | 0.373 | 0.372 | 0.368 |
| (39) | Solar | taud | | 2.603 | 2.559 | 2.489 | 2.416 | 2.437 | 2.377 | 2.281 | 2.344 | 2.463 | 2.550 | 2.566 | 2.607 |
| (40) | Irradiance | Ebn,noon | | 275 | 287 | 290 | 290 | 291 | 283 | 271 | 273 | 277 | 279 | 271 | 269 |
| (41) | | Edh,noon | | 26 | 29 | 34 | 37 | 37 | 39 | 43 | 40 | 34 | 29 | 27 | 25 |

Nomenclature: See separate page

*Fig. 4-4 Example tables of climatic design information included on CD-ROM of 2013 ASHRAE Handbook—Fundamentals*

# 112 — Principles of HVAC

2013 ASHRAE Handbook - Fundamentals (IP)  © 2013 ASHRAE, Inc.

**NEW YORK/JOHN F. KE, NY, USA**  WMO#: 744860

Lat: 40.66N  Long: 73.80W  Elev: 23  StdP: 14.68  Time Zone: -5 (NAE)  Period: 86-10  WBAN: 94789

### Annual Heating and Humidification Design Conditions

| Coldest Month | Heating DB | | Humidification DP/MCDB and HR | | | | | | Coldest month WS/MCDB | | | | MCWS/PCWD to 99.6% DB | |
|---|---|---|---|---|---|---|---|---|---|---|---|---|---|---|
| | 99.6% | 99% | 99.6% DP | HR | MCDB | 99% DP | HR | MCDB | 0.4% WS | MCDB | 1% WS | MCDB | MCWS | PCWD |
| (a) | (b) | (c) | (d) | (e) | (f) | (g) | (h) | (i) | (j) | (k) | (l) | (m) | (n) | (o) |
| 1 | 13.8 | 17.8 | -5.6 | 4.1 | 16.8 | -1.7 | 5.0 | 20.8 | 31.7 | 29.7 | 28.6 | 30.5 | 16.6 | 310 |

### Annual Cooling, Dehumidification, and Enthalpy Design Conditions

| Hottest Month | Hottest Month DB Range | Cooling DB/MCWB | | | | | | Evaporation WB/MCDB | | | | | | MCWS/PCWD to 0.4% DB | |
|---|---|---|---|---|---|---|---|---|---|---|---|---|---|---|---|
| | | 0.4% DB | MCWB | 1% DB | MCWB | 2% DB | MCWB | 0.4% WB | MCDB | 1% WB | MCDB | 2% WB | MCDB | MCWS | PCWD |
| (a) | (b) | (c) | (d) | (e) | (f) | (g) | (h) | (i) | (j) | (k) | (l) | (m) | (n) | (o) | (p) |
| 7 | 13.3 | 89.8 | 72.9 | 86.5 | 71.8 | 83.7 | 71.1 | 76.7 | 83.8 | 75.4 | 81.6 | 74.3 | 80.0 | 13.0 | 230 |

| Dehumidification DP/MCDB and HR | | | | | | | | Enthalpy/MCDB | | | | | | Hours 8 to 4 & 55/69 |
|---|---|---|---|---|---|---|---|---|---|---|---|---|---|---|
| 0.4% DP | HR | MCDB | 1% DP | HR | MCDB | 2% DP | HR | MCDB | 0.4% Enth | MCDB | 1% Enth | MCDB | 2% Enth | MCDB | |
| (a) | (b) | (c) | (d) | (e) | (f) | (g) | (h) | (i) | (j) | (k) | (l) | (m) | (n) | (o) | (p) |
| 74.6 | 129.4 | 80.1 | 73.4 | 124.2 | 78.6 | 72.3 | 119.7 | 77.6 | 40.0 | 84.3 | 38.8 | 82.1 | 37.6 | 79.8 | 753 |

### Extreme Annual Design Conditions

| Extreme Annual WS | | | Extreme Max WB | Extreme Annual DB | | | | n-Year Return Period Values of Extreme DB | | | | | | | |
|---|---|---|---|---|---|---|---|---|---|---|---|---|---|---|---|
| 1% | 2.5% | 5% | | Mean Min | Mean Max | Std dev Min | Std dev Max | n=5 yrs Min | n=5 yrs Max | n=10 yrs Min | n=10 yrs Max | n=20 yrs Min | n=20 yrs Max | n=50 yrs Min | n=50 yrs Max |
| (a) | (b) | (c) | (d) | (e) | (f) | (g) | (h) | (i) | (j) | (k) | (l) | (m) | (n) | (o) | (p) |
| 27.2 | 24.6 | 21.6 | 82.4 | 8.8 | 95.7 | 4.5 | 3.2 | 5.6 | 98.0 | 2.9 | 99.9 | 0.4 | 101.6 | -2.8 | 103.9 |

### Monthly Climatic Design Conditions

| | | Annual | Jan | Feb | Mar | Apr | May | Jun | Jul | Aug | Sep | Oct | Nov | Dec |
|---|---|---|---|---|---|---|---|---|---|---|---|---|---|---|
| Temperatures, Degree-Days and Degree-Hours | Tavg | 54.4 | 33.5 | 34.6 | 41.6 | 51.1 | 60.4 | 70.2 | 75.6 | 74.8 | 68.1 | 56.9 | 47.4 | 37.6 |
| | Sd | | 9.09 | 7.78 | 7.77 | 6.52 | 6.47 | 5.94 | 4.80 | 4.80 | 5.73 | 6.86 | 7.29 | 8.13 |
| | HDD50 | 1824 | 514 | 432 | 277 | 61 | 1 | 0 | 0 | 0 | 0 | 16 | 133 | 390 |
| | HDD65 | 4843 | 977 | 851 | 725 | 420 | 176 | 20 | 0 | 1 | 33 | 264 | 528 | 848 |
| | CDD50 | 3441 | 2 | 2 | 17 | 93 | 323 | 606 | 794 | 768 | 543 | 231 | 55 | 7 |
| | CDD65 | 984 | 0 | 0 | 0 | 3 | 33 | 175 | 329 | 304 | 126 | 14 | 0 | 0 |
| | CDH74 | 6142 | 0 | 0 | 4 | 27 | 222 | 1016 | 2358 | 1972 | 500 | 43 | 0 | 0 |
| | CDH80 | 1658 | 0 | 0 | 0 | 6 | 62 | 290 | 718 | 497 | 78 | 7 | 0 | 0 |
| Precipitation | PrecAvg | 41.6 | 3.2 | 3.0 | 3.6 | 3.9 | 3.8 | 3.7 | 3.8 | 3.4 | 3.3 | 2.9 | 3.7 | 3.4 |
| | PrecMax | 56.7 | 8.4 | 4.9 | 8.2 | 7.5 | 6.5 | 7.1 | 8.5 | 8.3 | 9.7 | 5.0 | 9.5 | 6.1 |
| | PrecMin | 25.6 | 0.6 | 1.1 | 1.3 | 1.3 | 0.6 | 0.2 | 0.6 | 0.4 | 0.9 | 0.2 | 0.3 | 0.9 |
| | PrecSD | 7.3 | 2.0 | 1.1 | 1.5 | 1.7 | 1.5 | 1.7 | 2.3 | 2.2 | 2.3 | 1.2 | 2.4 | 1.7 |
| Monthly Design Dry Bulb and Mean Coincident Wet Bulb Temperatures | 0.4% DB | | 56.4 | 59.5 | 69.1 | 77.7 | 87.0 | 91.8 | 95.6 | 92.6 | 86.7 | 79.5 | 67.4 | 61.0 |
| | 0.4% MCWB | | 51.3 | 47.9 | 54.7 | 60.0 | 69.1 | 73.1 | 76.4 | 74.5 | 70.7 | 67.4 | 59.2 | 53.9 |
| | 2% DB | | 52.3 | 51.9 | 60.6 | 70.3 | 80.0 | 87.4 | 90.2 | 88.1 | 81.9 | 73.4 | 63.3 | 55.4 |
| | 2% MCWB | | 49.4 | 45.1 | 49.6 | 56.2 | 65.6 | 71.5 | 73.4 | 72.9 | 68.9 | 65.0 | 57.5 | 51.5 |
| | 5% DB | | 48.9 | 48.4 | 55.6 | 65.2 | 74.8 | 82.9 | 86.9 | 84.9 | 79.2 | 70.3 | 60.8 | 52.3 |
| | 5% MCWB | | 45.3 | 43.3 | 47.4 | 53.1 | 62.9 | 69.3 | 72.0 | 72.5 | 68.6 | 63.3 | 56.5 | 48.9 |
| | 10% DB | | 45.7 | 45.6 | 52.1 | 61.2 | 71.0 | 79.6 | 83.7 | 82.3 | 76.6 | 67.4 | 58.4 | 49.3 |
| | 10% MCWB | | 41.7 | 41.1 | 46.1 | 51.1 | 61.0 | 68.2 | 71.5 | 71.2 | 67.9 | 61.3 | 54.5 | 45.2 |
| Monthly Design Wet Bulb and Mean Coincident Dry Bulb Temperatures | 0.4% WB | | 53.5 | 51.1 | 57.5 | 63.0 | 71.7 | 76.2 | 78.7 | 78.3 | 75.5 | 70.6 | 62.7 | 56.3 |
| | 0.4% MCDB | | 54.8 | 54.2 | 64.3 | 73.3 | 82.2 | 86.4 | 88.7 | 86.8 | 79.6 | 75.0 | 64.3 | 58.9 |
| | 2% WB | | 50.3 | 47.3 | 52.5 | 59.0 | 68.1 | 74.0 | 77.0 | 76.6 | 73.6 | 68.1 | 59.8 | 52.7 |
| | 2% MCDB | | 51.9 | 50.2 | 56.8 | 66.4 | 76.7 | 82.6 | 84.6 | 82.9 | 77.1 | 71.9 | 61.7 | 54.6 |
| | 5% WB | | 46.2 | 46.4 | 50.0 | 55.8 | 65.4 | 72.3 | 75.6 | 75.4 | 72.3 | 65.4 | 57.6 | 49.8 |
| | 5% MCDB | | 48.3 | 47.2 | 54.1 | 62.0 | 72.4 | 79.4 | 82.4 | 80.9 | 75.8 | 68.8 | 60.0 | 52.0 |
| | 10% WB | | 42.4 | 41.6 | 47.2 | 53.3 | 62.8 | 70.6 | 74.2 | 74.4 | 70.6 | 62.7 | 55.3 | 46.0 |
| | 10% MCDB | | 45.1 | 45.3 | 50.8 | 58.1 | 68.8 | 76.7 | 80.5 | 79.5 | 74.6 | 66.0 | 57.9 | 48.4 |
| Mean Daily Temperature Range | MDBR | | 12.0 | 12.8 | 13.9 | 14.4 | 14.7 | 14.2 | 13.3 | 13.0 | 13.3 | 13.7 | 12.5 | 11.6 |
| | 5% DB MCDBR | | 14.6 | 15.9 | 19.1 | 21.1 | 21.2 | 19.2 | 17.0 | 15.4 | 15.0 | 16.0 | 14.3 | 14.9 |
| | 5% DB MCWBR | | 12.6 | 12.5 | 11.9 | 10.9 | 10.5 | 8.7 | 7.4 | 7.6 | 8.0 | 10.5 | 11.9 | 13.9 |
| | 5% WB MCDBR | | 14.4 | 15.2 | 17.3 | 17.8 | 19.4 | 16.7 | 14.2 | 12.8 | 12.1 | 13.9 | 12.8 | 14.1 |
| | 5% WB MCWBR | | 14.0 | 13.7 | 13.1 | 11.2 | 11.0 | 8.9 | 7.0 | 7.0 | 8.1 | 10.9 | 12.4 | 15.1 |
| Clear Sky Solar Irradiance | taub | | 0.339 | 0.333 | 0.363 | 0.391 | 0.418 | 0.458 | 0.434 | 0.439 | 0.403 | 0.352 | 0.329 | 0.325 |
| | taud | | 2.470 | 2.481 | 2.402 | 2.297 | 2.179 | 2.079 | 2.188 | 2.197 | 2.331 | 2.498 | 2.567 | 2.568 |
| | Ebn,noon | | 258 | 280 | 282 | 280 | 273 | 261 | 267 | 262 | 264 | 269 | 262 | 256 |
| | Edh,noon | | 25 | 28 | 34 | 40 | 46 | 51 | 46 | 44 | 36 | 27 | 22 | 21 |

Nomenclature: See separate page

*Fig. 4-4 Example tables of climatic design information included on CD-ROM of 2013 ASHRAE Handbook—Fundamentals*

# Design Conditions

2013 ASHRAE Handbook - Fundamentals (IP) © 2013 ASHRAE, Inc.

**OKLAHOMA CITY/W. RO, OK, USA** WMO#: 723530

Lat: 35.39N   Long: 97.60W   Elev: 1306   StdP: 14.02   Time Zone: -6 (NAC)   Period: 86-10   WBAN: 13967

### Annual Heating and Humidification Design Conditions

| | Coldest Month | Heating DB | | Humidification DP/MCDB and HR | | | | | | Coldest month WS/MCDB | | | | MCWS/PCWD to 99.6% DB | |
|---|---|---|---|---|---|---|---|---|---|---|---|---|---|---|---|
| | | 99.6% | 99% | 99.6% | | | 99% | | | 0.4% | | 1% | | | |
| | | | | DP | HR | MCDB | DP | HR | MCDB | WS | MCDB | WS | MCDB | MCWS | PCWD |
| | (a) | (b) | (c) | (d) | (e) | (f) | (g) | (h) | (i) | (j) | (k) | (l) | (m) | (n) | (o) |
| (1) | 1 | 14.1 | 18.9 | 1.8 | 6.3 | 18.0 | 6.9 | 8.2 | 23.6 | 31.8 | 44.4 | 28.5 | 44.0 | 12.5 | 0 |

### Annual Cooling, Dehumidification, and Enthalpy Design Conditions

| | Hottest Month | Hottest Month DB Range | Cooling DB/MCWB | | | | | | Evaporation WB/MCDB | | | | | | MCWS/PCWD to 0.4% DB | |
|---|---|---|---|---|---|---|---|---|---|---|---|---|---|---|---|---|
| | | | 0.4% | | 1% | | 2% | | 0.4% | | 1% | | 2% | | | |
| | | | DB | MCWB | DB | MCWB | DB | MCWB | WB | MCDB | WB | MCDB | WB | MCDB | MCWS | PCWD |
| | (a) | (b) | (c) | (d) | (e) | (f) | (g) | (h) | (i) | (j) | (k) | (l) | (m) | (n) | (o) | (p) |
| (2) | 7 | 21.0 | 99.6 | 74.2 | 96.9 | 74.2 | 94.0 | 74.0 | 77.8 | 91.0 | 76.9 | 89.9 | 75.8 | 88.8 | 12.3 | 190 |

| | Dehumidification DP/MCDB and HR | | | | | | | | Enthalpy/MCDB | | | | | | Hours 8 to 4 & 55/69 |
|---|---|---|---|---|---|---|---|---|---|---|---|---|---|---|---|
| | 0.4% | | | 1% | | | 2% | | | 0.4% | | 1% | | 2% | | |
| | DP | HR | MCDB | DP | HR | MCDB | DP | HR | MCDB | Enth | MCDB | Enth | MCDB | Enth | MCDB | |
| | (a) | (b) | (c) | (d) | (e) | (f) | (g) | (h) | (i) | (j) | (k) | (l) | (m) | (n) | (o) | (p) |
| (3) | 74.2 | 134.3 | 83.6 | 73.2 | 129.3 | 82.3 | 72.3 | 125.4 | 81.5 | 42.2 | 90.9 | 41.1 | 89.9 | 40.1 | 88.8 | 710 |

### Extreme Annual Design Conditions

| | Extreme Annual WS | | | Extreme Max WB | Extreme Annual DB | | | | n-Year Return Period Values of Extreme DB | | | | | | |
|---|---|---|---|---|---|---|---|---|---|---|---|---|---|---|---|
| | 1% | 2.5% | 5% | | Mean | | Standard deviation | | n=5 years | | n=10 years | | n=20 years | | n=50 years | |
| | | | | | Min | Max | Min | Max | Min | Max | Min | Max | Min | Max | Min | Max |
| | (a) | (b) | (c) | (d) | (e) | (f) | (g) | (h) | (i) | (j) | (k) | (l) | (m) | (n) | (o) | (p) |
| (4) | 27.6 | 24.9 | 22.7 | 83.3 | 7.5 | 102.7 | 5.1 | 3.6 | 3.8 | 105.4 | 0.8 | 107.5 | -2.1 | 109.5 | -5.8 | 112.2 |

### Monthly Climatic Design Conditions

| | | | Annual (d) | Jan (e) | Feb (f) | Mar (g) | Apr (h) | May (i) | Jun (j) | Jul (k) | Aug (l) | Sep (m) | Oct (n) | Nov (o) | Dec (p) |
|---|---|---|---|---|---|---|---|---|---|---|---|---|---|---|---|
| (5) | Temperatures, Degree-Days and Degree-Hours | Tavg | 60.9 | 39.2 | 43.4 | 51.6 | 60.4 | 69.4 | 77.4 | 82.0 | 81.3 | 73.0 | 61.7 | 50.2 | 40.5 |
| (6) | | Sd | | 10.10 | 10.96 | 10.36 | 8.24 | 7.01 | 5.04 | 4.49 | 5.30 | 7.27 | 8.11 | 9.47 | 9.71 |
| (7) | | HDD50 | 1149 | 356 | 234 | 105 | 14 | 0 | 0 | 0 | 0 | 0 | 10 | 112 | 318 |
| (8) | | HDD65 | 3438 | 801 | 607 | 425 | 182 | 38 | 0 | 0 | 0 | 19 | 157 | 448 | 761 |
| (9) | | CDD50 | 5137 | 20 | 48 | 155 | 325 | 600 | 821 | 993 | 970 | 690 | 374 | 119 | 22 |
| (10) | | CDD65 | 1950 | 0 | 1 | 10 | 43 | 173 | 371 | 528 | 505 | 259 | 55 | 5 | 0 |
| (11) | | CDH74 | 20913 | 2 | 18 | 107 | 406 | 1486 | 3757 | 6341 | 5874 | 2372 | 501 | 49 | 0 |
| (12) | | CDH80 | 9914 | 0 | 3 | 18 | 94 | 491 | 1664 | 3390 | 3129 | 997 | 125 | 3 | 0 |
| (13) | Precipitation | PrecAvg | 34.1 | 1.1 | 1.5 | 2.8 | 2.9 | 5.1 | 4.3 | 2.7 | 3.0 | 3.8 | 3.2 | 2.0 | 1.6 |
| (14) | | PrecMax | 41.2 | 3.4 | 3.2 | 6.8 | 5.7 | 10.1 | 9.9 | 7.7 | 6.8 | 9.7 | 7.2 | 5.5 | 2.8 |
| (15) | | PrecMin | 18.4 | 0.1 | 0.2 | 0.1 | 0.6 | 0.9 | 0.8 | 0.4 | 0.2 | 0.7 | 0.4 | 0.1 | 0.2 |
| (16) | | PrecSD | 6.1 | 0.9 | 0.9 | 1.4 | 1.3 | 2.7 | 2.3 | 2.2 | 1.6 | 2.5 | 1.7 | 1.5 | 0.8 |
| (17) | Monthly Design Dry Bulb and Mean Coincident Wet Bulb Temperatures | 0.4% DB | 71.5 | 76.4 | 82.1 | 87.8 | 93.4 | 97.6 | 104.1 | 102.2 | 99.5 | 88.3 | 79.4 | 70.1 | |
| (18) | | MCWB | 53.4 | 56.5 | 61.1 | 67.0 | 71.9 | 74.9 | 73.6 | 73.1 | 72.9 | 68.1 | 61.9 | 57.4 | |
| (19) | | 2% DB | 64.2 | 70.4 | 76.6 | 82.3 | 89.0 | 94.2 | 100.0 | 99.5 | 93.0 | 83.4 | 74.3 | 64.7 | |
| (20) | | MCWB | 51.4 | 54.3 | 59.9 | 64.8 | 71.5 | 74.2 | 74.5 | 74.0 | 72.7 | 66.2 | 60.4 | 53.5 | |
| (21) | | 5% DB | 60.2 | 65.3 | 72.4 | 78.9 | 85.4 | 91.3 | 97.0 | 97.1 | 89.4 | 79.6 | 70.0 | 59.9 | |
| (22) | | MCWB | 49.0 | 53.0 | 58.9 | 63.3 | 70.3 | 74.3 | 74.6 | 74.0 | 71.9 | 64.5 | 58.8 | 50.9 | |
| (23) | | 10% DB | 55.2 | 60.7 | 68.3 | 74.9 | 82.0 | 89.0 | 93.9 | 93.9 | 85.9 | 75.8 | 65.8 | 55.4 | |
| (24) | | MCWB | 46.2 | 50.6 | 56.7 | 61.7 | 69.3 | 73.8 | 74.4 | 74.2 | 71.0 | 63.9 | 56.8 | 47.3 | |
| (25) | Monthly Design Wet Bulb and Mean Coincident Dry Bulb Temperatures | 0.4% WB | 59.4 | 61.9 | 66.4 | 71.5 | 75.7 | 79.1 | 79.4 | 78.6 | 76.8 | 73.3 | 66.9 | 62.7 | |
| (26) | | MCDB | 64.0 | 68.2 | 73.7 | 80.0 | 86.9 | 91.5 | 93.1 | 91.9 | 87.8 | 81.0 | 73.3 | 66.8 | |
| (27) | | 2% WB | 54.9 | 58.1 | 63.9 | 68.6 | 74.1 | 77.4 | 77.8 | 77.3 | 75.2 | 70.7 | 63.8 | 57.1 | |
| (28) | | MCDB | 60.9 | 65.3 | 72.2 | 77.1 | 84.6 | 89.1 | 91.8 | 90.9 | 86.8 | 77.8 | 70.9 | 61.4 | |
| (29) | | 5% WB | 50.5 | 54.9 | 61.6 | 66.8 | 72.6 | 76.2 | 76.8 | 76.3 | 74.0 | 68.1 | 61.3 | 51.9 | |
| (30) | | MCDB | 57.7 | 63.7 | 69.1 | 74.8 | 82.1 | 87.5 | 90.7 | 90.2 | 84.6 | 75.5 | 67.3 | 57.5 | |
| (31) | | 10% WB | 46.6 | 51.0 | 58.5 | 64.4 | 71.0 | 74.9 | 75.8 | 75.2 | 72.8 | 65.7 | 58.2 | 47.6 | |
| (32) | | MCDB | 54.4 | 59.8 | 66.7 | 72.2 | 79.5 | 85.7 | 89.3 | 89.2 | 82.4 | 73.4 | 64.0 | 54.2 | |
| (33) | Mean Daily Temperature Range | MDBR | 19.7 | 20.7 | 21.4 | 21.6 | 19.4 | 19.5 | 21.0 | 21.1 | 20.4 | 21.4 | 20.7 | 19.2 | |
| (34) | | 5% DB MCDBR | 27.9 | 29.0 | 27.1 | 26.2 | 22.9 | 22.0 | 25.7 | 25.7 | 24.1 | 25.9 | 26.4 | 25.8 | |
| (35) | | MCWBR | 16.9 | 16.4 | 14.8 | 13.4 | 10.2 | 8.2 | 7.7 | 7.4 | 8.6 | 12.6 | 15.1 | 16.5 | |
| (36) | | 5% WB MCDBR | 23.8 | 24.5 | 21.5 | 21.0 | 20.1 | 19.8 | 21.5 | 21.2 | 19.8 | 20.0 | 21.6 | 22.6 | |
| (37) | | MCWBR | 16.6 | 16.0 | 13.7 | 13.1 | 10.5 | 8.9 | 8.0 | 7.7 | 8.2 | 12.1 | 15.3 | 17.5 | |
| (38) | Clear Sky Solar Irradiance | taub | 0.301 | 0.317 | 0.331 | 0.342 | 0.351 | 0.380 | 0.384 | 0.383 | 0.362 | 0.330 | 0.307 | 0.302 | |
| (39) | | taud | 2.638 | 2.541 | 2.479 | 2.404 | 2.396 | 2.329 | 2.357 | 2.356 | 2.457 | 2.547 | 2.627 | 2.618 | |
| (40) | | Ebn,noon | 290 | 295 | 299 | 298 | 295 | 285 | 284 | 282 | 284 | 286 | 285 | 282 | |
| (41) | | Edh,noon | 23 | 28 | 33 | 37 | 38 | 40 | 39 | 38 | 33 | 28 | 23 | 22 | |

Nomenclature: See separate page

*Fig. 4-4 Example tables of climatic design information included on CD-ROM of 2013 ASHRAE Handbook—Fundamentals*

# 114

**Principles of HVAC**

2013 ASHRAE Handbook - Fundamentals (IP)  © 2013 ASHRAE, Inc.

## PURDUE UNIV, IN, USA

WMO#: 724386

Lat: 40.41N  Long: 86.94W  Elev: 636  StdP: 14.36  Time Zone: -5 (NAE)  Period: 86-10  WBAN: 14835

### Annual Heating and Humidification Design Conditions

| Coldest Month | Heating DB | | Humidification DP/MCDB and HR | | | | | | Coldest month WS/MCDB | | | | MCWS/PCWD to 99.6% DB | |
|---|---|---|---|---|---|---|---|---|---|---|---|---|---|---|
| | 99.6% | 99% | 99.6% | | | 99% | | | 0.4% | | 1% | | | |
| | | | DP | HR | MCDB | DP | HR | MCDB | WS | MCDB | WS | MCDB | MCWS | PCWD |
| (a) | (b) | (c) | (d) | (e) | (f) | (g) | (h) | (i) | (j) | (k) | (l) | (m) | (n) | (o) |
| 1 | 0.1 | 5.7 | -8.4 | 3.6 | 1.6 | -2.2 | 5.0 | 7.2 | 26.5 | 27.5 | 24.1 | 28.7 | 8.5 | 260 |

### Annual Cooling, Dehumidification, and Enthalpy Design Conditions

| Hottest Month | Hottest Month DB Range | Cooling DB/MCWB | | | | | | Evaporation WB/MCDB | | | | | | MCWS/PCWD to 0.4% DB | |
|---|---|---|---|---|---|---|---|---|---|---|---|---|---|---|---|
| | | 0.4% | | 1% | | 2% | | 0.4% | | 1% | | 2% | | | |
| | | DB | MCWB | DB | MCWB | DB | MCWB | WB | MCDB | WB | MCDB | WB | MCDB | MCWS | PCWD |
| (a) | (b) | (c) | (d) | (e) | (f) | (g) | (h) | (i) | (j) | (k) | (l) | (m) | (n) | (o) | (p) |
| 7 | 19.5 | 91.6 | 75.5 | 89.9 | 74.5 | 87.5 | 73.3 | 78.6 | 88.3 | 77.1 | 86.1 | 75.4 | 83.5 | 9.7 | 230 |

| Dehumidification DP/MCDB and HR | | | | | | | | Enthalpy/MCDB | | | | | | Hours 8 to 4 & 55/69 |
|---|---|---|---|---|---|---|---|---|---|---|---|---|---|---|
| 0.4% | | | 1% | | | 2% | | 0.4% | | 1% | | 2% | | |
| DP | HR | MCDB | DP | HR | MCDB | DP | HR | MCDB | Enth | MCDB | Enth | MCDB | Enth | MCDB | |
| (a) | (b) | (c) | (d) | (e) | (f) | (g) | (h) | (i) | (j) | (k) | (l) | (m) | (n) | (o) | (p) |
| 75.5 | 136.6 | 84.3 | 74.1 | 130.3 | 82.4 | 72.7 | 124.3 | 80.6 | 42.6 | 88.6 | 40.8 | 86.1 | 39.3 | 83.6 | 654 |

### Extreme Annual Design Conditions

| Extreme Annual WS | | | Extreme Max WB | Extreme Annual DB | | | | n-Year Return Period Values of Extreme DB | | | | | | | |
|---|---|---|---|---|---|---|---|---|---|---|---|---|---|---|---|
| 1% | 2.5% | 5% | | Mean | | Standard deviation | | n=5 years | | n=10 years | | n=20 years | | n=50 years | |
| | | | | Min | Max | Min | Max | Min | Max | Min | Max | Min | Max | Min | Max |
| (a) | (b) | (c) | (d) | (e) | (f) | (g) | (h) | (i) | (j) | (k) | (l) | (m) | (n) | (o) | (p) |
| 22.8 | 19.9 | 18.2 | 88.5 | -7.2 | 96.2 | 7.7 | 3.2 | -12.8 | 98.5 | -17.3 | 100.3 | -21.6 | 102.1 | -27.2 | 104.5 |

### Monthly Climatic Design Conditions

| | | Annual | Jan | Feb | Mar | Apr | May | Jun | Jul | Aug | Sep | Oct | Nov | Dec |
|---|---|---|---|---|---|---|---|---|---|---|---|---|---|---|
| | | (d) | (e) | (f) | (g) | (h) | (i) | (j) | (k) | (l) | (m) | (n) | (o) | (p) |
| Temperatures, Degree-Days and Degree-Hours | Tavg | 52.6 | 28.0 | 31.1 | 41.5 | 52.6 | 62.5 | 71.7 | 74.8 | 73.3 | 66.5 | 54.3 | 43.0 | 31.1 |
| | Sd | | 12.11 | 10.98 | 10.83 | 9.65 | 8.31 | 6.61 | 5.23 | 5.70 | 7.42 | 8.87 | 9.86 | 10.92 |
| | HDD50 | 2520 | 686 | 533 | 311 | 80 | 6 | 0 | 0 | 0 | 2 | 55 | 255 | 592 |
| | HDD65 | 5524 | 1147 | 949 | 731 | 387 | 151 | 20 | 1 | 6 | 71 | 349 | 660 | 1052 |
| | CDD50 | 3484 | 4 | 4 | 47 | 159 | 392 | 650 | 769 | 724 | 497 | 188 | 45 | 5 |
| | CDD65 | 1014 | 0 | 0 | 2 | 16 | 73 | 220 | 305 | 265 | 116 | 17 | 0 | 0 |
| | CDH74 | 9413 | 0 | 0 | 14 | 149 | 705 | 2110 | 2836 | 2372 | 1059 | 166 | 2 | 0 |
| | CDH80 | 3326 | 0 | 0 | 1 | 26 | 204 | 791 | 1085 | 850 | 336 | 33 | 0 | 0 |
| Precipitation | PrecAvg | 35.6 | 1.7 | 1.5 | 2.6 | 3.7 | 3.7 | 4.1 | 3.4 | 4.1 | 3.0 | 2.4 | 2.8 | 2.9 |
| | PrecMax | 45.3 | 4.4 | 3.2 | 5.7 | 8.7 | 6.5 | 7.9 | 7.1 | 10.5 | 8.8 | 5.0 | 5.5 | 7.8 |
| | PrecMin | 28.9 | 0.3 | 0.3 | 1.1 | 0.6 | 1.5 | 1.7 | 0.6 | 0.9 | 0.5 | 0.2 | 0.3 | 0.3 |
| | PrecSD | 4.2 | 1.4 | 0.9 | 1.4 | 2.3 | 1.3 | 1.9 | 2.0 | 2.4 | 2.4 | 1.3 | 1.3 | 1.9 |
| Monthly Design Dry Bulb and Mean Coincident Wet Bulb Temperatures | 0.4% DB | 62.5 | 63.3 | 75.8 | 83.1 | 89.1 | 94.0 | 96.4 | 94.0 | 91.2 | 84.2 | 72.4 | 63.1 |
| | 0.4% MCWB | 57.6 | 55.4 | 61.2 | 66.3 | 70.0 | 74.4 | 76.2 | 77.3 | 71.6 | 67.5 | 58.6 | 58.1 |
| | 2% DB | 55.1 | 56.8 | 70.4 | 79.0 | 85.4 | 90.7 | 91.8 | 90.8 | 87.6 | 79.2 | 66.3 | 56.1 |
| | 2% MCWB | 51.7 | 50.6 | 58.5 | 63.5 | 69.1 | 73.6 | 76.6 | 76.7 | 70.4 | 63.8 | 58.5 | 51.9 |
| | 5% DB | 49.7 | 51.7 | 64.8 | 73.3 | 81.7 | 88.1 | 89.3 | 88.1 | 83.8 | 74.1 | 63.1 | 51.6 |
| | 5% MCWB | 46.0 | 46.0 | 55.7 | 60.4 | 67.5 | 72.5 | 75.7 | 75.1 | 69.1 | 62.2 | 56.4 | 47.2 |
| | 10% DB | 43.5 | 45.6 | 59.9 | 69.8 | 77.7 | 84.4 | 86.1 | 84.4 | 80.9 | 70.0 | 58.9 | 45.4 |
| | 10% MCWB | 39.7 | 41.1 | 51.4 | 58.6 | 65.8 | 71.1 | 73.9 | 72.6 | 67.0 | 60.3 | 53.0 | 41.8 |
| Monthly Design Wet Bulb and Mean Coincident Dry Bulb Temperatures | 0.4% WB | 58.0 | 58.0 | 64.0 | 69.4 | 75.2 | 78.4 | 81.6 | 80.5 | 75.7 | 71.0 | 63.5 | 58.6 |
| | 0.4% MCDB | 61.3 | 62.1 | 71.8 | 79.5 | 83.8 | 89.4 | 90.7 | 90.0 | 86.0 | 79.5 | 67.3 | 62.5 |
| | 2% WB | 53.2 | 51.4 | 60.3 | 65.6 | 72.3 | 76.6 | 79.1 | 78.3 | 73.5 | 67.3 | 60.1 | 53.1 |
| | 2% MCDB | 55.0 | 55.6 | 68.2 | 74.4 | 81.1 | 86.7 | 88.6 | 88.2 | 81.9 | 74.3 | 64.8 | 55.6 |
| | 5% WB | 46.0 | 45.8 | 56.9 | 62.8 | 70.0 | 74.9 | 77.4 | 76.8 | 71.5 | 64.4 | 57.5 | 46.8 |
| | 5% MCDB | 48.6 | 50.0 | 64.2 | 71.4 | 78.0 | 84.2 | 86.0 | 85.4 | 78.9 | 71.5 | 62.3 | 49.4 |
| | 10% WB | 40.1 | 41.5 | 52.2 | 59.9 | 67.6 | 73.1 | 75.5 | 75.0 | 69.6 | 61.5 | 53.3 | 41.7 |
| | 10% MCDB | 43.4 | 45.8 | 59.1 | 68.3 | 74.9 | 81.6 | 83.5 | 81.9 | 76.7 | 68.9 | 58.4 | 45.6 |
| Mean Daily Temperature Range | MDBR | 14.3 | 15.6 | 18.3 | 21.0 | 21.2 | 20.6 | 19.5 | 19.9 | 22.3 | 21.0 | 16.4 | 13.9 |
| | 5% DB MCDBR | 18.3 | 22.9 | 25.2 | 25.1 | 25.1 | 24.1 | 22.3 | 21.8 | 25.7 | 26.1 | 21.1 | 18.8 |
| | 5% DB MCWBR | 15.1 | 17.4 | 15.7 | 14.4 | 12.5 | 10.7 | 9.9 | 9.9 | 11.6 | 13.8 | 15.2 | 15.7 |
| | 5% WB MCDBR | 17.3 | 20.4 | 21.7 | 21.6 | 20.6 | 20.0 | 19.3 | 18.9 | 20.4 | 21.4 | 18.6 | 17.3 |
| | 5% WB MCWBR | 15.7 | 16.9 | 14.8 | 13.7 | 12.0 | 10.2 | 10.0 | 9.7 | 11.0 | 12.9 | 15.4 | 15.7 |
| Clear Sky Solar Irradiance | taub | 0.306 | 0.324 | 0.344 | 0.362 | 0.388 | 0.403 | 0.418 | 0.419 | 0.392 | 0.347 | 0.325 | 0.298 |
| | taud | 2.440 | 2.395 | 2.406 | 2.331 | 2.250 | 2.218 | 2.192 | 2.237 | 2.331 | 2.483 | 2.536 | 2.582 |
| | Ebn,noon | 271 | 282 | 288 | 289 | 282 | 277 | 271 | 268 | 268 | 271 | 264 | 269 |
| | Edh,noon | 25 | 30 | 34 | 39 | 43 | 45 | 45 | 42 | 36 | 28 | 23 | 21 |

Nomenclature:  See separate page

*Fig. 4-4 Example tables of climatic design information included on CD-ROM of 2013 ASHRAE Handbook—Fundamentals*

# Design Conditions

2013 ASHRAE Handbook - Fundamentals (IP) © 2013 ASHRAE, Inc.

## ROANOKE MUNICIPAL, VA, USA

WMO#: 724110

Lat: 37.32N   Long: 79.97W   Elev: 1175   StdP: 14.08   Time Zone: -5 (NAE)   Period: 86-10   WBAN: 13741

### Annual Heating and Humidification Design Conditions

| Coldest Month | Heating DB | | Humidification DP/MCDB and HR | | | | | | Coldest month WS/MCDB | | | | MCWS/PCWD to 99.6% DB | |
|---|---|---|---|---|---|---|---|---|---|---|---|---|---|---|
| | 99.6% | 99% | 99.6% | | | 99% | | | 0.4% | | 1% | | | |
| | | | DP | HR | MCDB | DP | HR | MCDB | WS | MCDB | WS | MCDB | MCWS | PCWD |
| (a) | (b) | (c) | (d) | (e) | (f) | (g) | (h) | (i) | (j) | (k) | (l) | (m) | (n) | (o) |
| 1 | 15.7 | 19.6 | -2.7 | 5.0 | 22.9 | 1.3 | 6.1 | 25.4 | 27.1 | 36.4 | 24.9 | 36.5 | 9.0 | 300 |

### Annual Cooling, Dehumidification, and Enthalpy Design Conditions

| Hottest Month | Hottest Month DB Range | Cooling DB/MCWB | | | | | | Evaporation WB/MCDB | | | | | | MCWS/PCWD to 0.4% DB | |
|---|---|---|---|---|---|---|---|---|---|---|---|---|---|---|---|
| | | 0.4% | | 1% | | 2% | | 0.4% | | 1% | | 2% | | | |
| | | DB | MCWB | DB | MCWB | DB | MCWB | WB | MCDB | WB | MCDB | WB | MCDB | MCWS | PCWD |
| (a) | (b) | (c) | (d) | (e) | (f) | (g) | (h) | (i) | (j) | (k) | (l) | (m) | (n) | (o) | (p) |
| 7 | 19.0 | 92.3 | 72.8 | 90.0 | 72.2 | 87.7 | 71.5 | 75.4 | 86.5 | 74.5 | 85.1 | 73.4 | 83.4 | 8.7 | 280 |

| Dehumidification DP/MCDB and HR | | | | | | | | Enthalpy/MCDB | | | | | | Hours 8 to 4 & 55/69 |
|---|---|---|---|---|---|---|---|---|---|---|---|---|---|---|
| 0.4% | | | 1% | | | 2% | | 0.4% | | 1% | | 2% | | |
| DP | HR | MCDB | DP | HR | MCDB | DP | HR | MCDB | Enth | MCDB | Enth | MCDB | Enth | MCDB | |
| (a) | (b) | (c) | (d) | (e) | (f) | (g) | (h) | (i) | (j) | (k) | (l) | (m) | (n) | (o) | (p) |
| 72.4 | 125.3 | 79.6 | 71.5 | 121.3 | 78.5 | 70.4 | 116.9 | 77.7 | 39.7 | 87.2 | 38.7 | 85.2 | 37.7 | 83.7 | 756 |

### Extreme Annual Design Conditions

| Extreme Annual WS | | | Extreme Max WB | Extreme Annual DB | | | | n-Year Return Period Values of Extreme DB | | | | | | | |
|---|---|---|---|---|---|---|---|---|---|---|---|---|---|---|---|
| 1% | 2.5% | 5% | | Mean | | Standard deviation | | n=5 years | | n=10 years | | n=20 years | | n=50 years | |
| | | | | Min | Max | Min | Max | Min | Max | Min | Max | Min | Max | Min | Max |
| (a) | (b) | (c) | (d) | (e) | (f) | (g) | (h) | (i) | (j) | (k) | (l) | (m) | (n) | (o) | (p) |
| 22.9 | 19.0 | 16.8 | 79.7 | 8.4 | 96.1 | 5.1 | 3.0 | 4.7 | 98.2 | 1.7 | 100.0 | -1.2 | 101.6 | -4.9 | 103.8 |

### Monthly Climatic Design Conditions

| | | Annual (d) | Jan (e) | Feb (f) | Mar (g) | Apr (h) | May (i) | Jun (j) | Jul (k) | Aug (l) | Sep (m) | Oct (n) | Nov (o) | Dec (p) |
|---|---|---|---|---|---|---|---|---|---|---|---|---|---|---|
| Temperatures, Degree-Days and Degree-Hours | Tavg | 57.3 | 37.6 | 39.9 | 47.9 | 57.1 | 64.8 | 73.0 | 76.8 | 75.6 | 68.4 | 57.7 | 48.4 | 39.2 |
| | Sd | | 9.93 | 9.12 | 9.59 | 8.69 | 7.23 | 5.59 | 4.48 | 4.87 | 6.30 | 7.74 | 8.56 | 8.96 |
| | HDD50 | 1390 | 400 | 300 | 156 | 31 | 1 | 0 | 0 | 0 | 0 | 20 | 132 | 350 |
| | HDD65 | 4044 | 848 | 702 | 537 | 264 | 97 | 9 | 1 | 1 | 38 | 248 | 500 | 799 |
| | CDD50 | 4049 | 17 | 18 | 89 | 242 | 459 | 689 | 830 | 793 | 552 | 260 | 83 | 17 |
| | CDD65 | 1230 | 0 | 0 | 6 | 26 | 89 | 249 | 366 | 330 | 140 | 23 | 1 | 0 |
| | CDH74 | 10391 | 1 | 2 | 52 | 299 | 768 | 2060 | 3216 | 2779 | 1004 | 193 | 16 | 1 |
| | CDH80 | 3690 | 0 | 0 | 6 | 66 | 198 | 710 | 1305 | 1082 | 290 | 33 | 0 | 0 |
| Precipitation | PrecAvg | 40.7 | 2.9 | 3.0 | 3.6 | 3.2 | 3.9 | 3.7 | 3.9 | 3.8 | 3.5 | 3.3 | 3.1 | 3.0 |
| | PrecMax | 54.3 | 8.0 | 8.0 | 7.9 | 11.4 | 10.1 | 10.3 | 10.1 | 9.5 | 11.7 | 9.9 | 12.4 | 8.2 |
| | PrecMin | 24.9 | 0.3 | 0.6 | 0.4 | 0.5 | 1.0 | 0.6 | 0.5 | 0.8 | 0.2 | 0.0 | 0.2 | 0.2 |
| | PrecSD | 7.6 | 1.7 | 1.7 | 1.8 | 2.0 | 2.0 | 2.3 | 1.8 | 1.9 | 2.7 | 2.4 | 2.1 | 1.5 |
| Monthly Design Dry Bulb and Mean Coincident Wet Bulb Temperatures | 0.4% DB | | 67.9 | 70.7 | 80.5 | 86.3 | 88.5 | 92.9 | 96.4 | 96.7 | 90.8 | 83.8 | 76.4 | 68.9 |
| | 0.4% MCWB | | 57.6 | 56.5 | 60.7 | 63.6 | 69.8 | 73.1 | 73.4 | 73.1 | 70.2 | 67.0 | 60.3 | 56.8 |
| | 2% DB | | 62.6 | 64.1 | 73.7 | 81.4 | 85.0 | 89.8 | 92.6 | 91.9 | 87.0 | 79.3 | 71.0 | 62.3 |
| | 2% MCWB | | 54.3 | 51.8 | 57.2 | 60.9 | 68.2 | 72.0 | 73.2 | 72.8 | 69.5 | 64.5 | 57.7 | 53.6 |
| | 5% DB | | 57.8 | 59.5 | 68.8 | 77.3 | 81.9 | 87.3 | 90.0 | 89.3 | 83.2 | 75.1 | 66.6 | 57.5 |
| | 5% MCWB | | 49.3 | 49.5 | 54.7 | 59.8 | 66.5 | 71.2 | 72.9 | 72.3 | 68.4 | 62.3 | 57.0 | 50.0 |
| | 10% DB | | 53.2 | 55.0 | 64.0 | 72.8 | 78.6 | 84.2 | 87.4 | 86.2 | 80.0 | 71.6 | 62.9 | 53.6 |
| | 10% MCWB | | 45.6 | 45.8 | 52.4 | 58.0 | 64.9 | 70.0 | 72.2 | 71.4 | 67.2 | 60.2 | 54.0 | 46.3 |
| Monthly Design Wet Bulb and Mean Coincident Dry Bulb Temperatures | 0.4% WB | | 60.5 | 59.3 | 63.3 | 66.8 | 73.3 | 75.7 | 77.3 | 77.1 | 74.3 | 70.2 | 64.9 | 60.4 |
| | 0.4% MCDB | | 65.2 | 67.4 | 74.7 | 79.4 | 83.7 | 87.1 | 89.2 | 88.7 | 83.8 | 78.2 | 69.2 | 65.5 |
| | 2% WB | | 55.9 | 55.1 | 60.1 | 64.4 | 70.8 | 74.2 | 75.8 | 75.3 | 72.4 | 67.5 | 61.7 | 56.0 |
| | 2% MCDB | | 61.1 | 61.4 | 70.5 | 75.7 | 81.2 | 85.1 | 86.9 | 86.3 | 80.8 | 74.4 | 67.1 | 60.4 |
| | 5% WB | | 51.1 | 51.0 | 57.1 | 62.3 | 68.7 | 73.0 | 74.7 | 74.2 | 71.1 | 65.1 | 59.0 | 51.8 |
| | 5% MCDB | | 56.2 | 56.3 | 66.0 | 72.1 | 77.9 | 83.3 | 85.4 | 84.6 | 78.6 | 72.4 | 64.8 | 56.8 |
| | 10% WB | | 45.8 | 46.5 | 53.8 | 60.0 | 66.9 | 71.6 | 73.6 | 73.1 | 69.6 | 62.5 | 55.4 | 46.8 |
| | 10% MCDB | | 51.8 | 54.1 | 61.8 | 69.9 | 75.4 | 81.3 | 83.7 | 82.7 | 75.8 | 69.4 | 62.0 | 52.0 |
| Mean Daily Temperature Range | MDBR | | 16.7 | 17.7 | 19.9 | 21.2 | 20.3 | 19.3 | 19.0 | 18.9 | 18.9 | 21.0 | 19.2 | 16.1 |
| | 5% DB MCDBR | | 23.1 | 25.6 | 27.4 | 27.6 | 24.7 | 22.7 | 22.7 | 23.0 | 23.7 | 25.5 | 25.6 | 22.9 |
| | 5% DB MCWBR | | 15.8 | 16.6 | 15.1 | 13.1 | 11.3 | 8.9 | 8.0 | 7.9 | 9.6 | 13.2 | 15.6 | 16.4 |
| | 5% WB MCDBR | | 20.2 | 22.6 | 22.7 | 23.2 | 21.5 | 19.9 | 19.6 | 19.7 | 18.2 | 19.9 | 21.0 | 20.0 |
| | 5% WB MCWBR | | 16.3 | 16.9 | 14.6 | 12.2 | 10.9 | 8.6 | 8.0 | 7.6 | 8.7 | 11.9 | 15.7 | 17.0 |
| Clear Sky Solar Irradiance | taub | | 0.325 | 0.336 | 0.362 | 0.399 | 0.405 | 0.450 | 0.461 | 0.436 | 0.397 | 0.349 | 0.330 | 0.310 |
| | taud | | 2.565 | 2.507 | 2.410 | 2.255 | 2.201 | 2.095 | 2.109 | 2.196 | 2.344 | 2.478 | 2.562 | 2.648 |
| | Ebn,noon | | 275 | 285 | 286 | 279 | 278 | 265 | 261 | 265 | 270 | 276 | 270 | 275 |
| | Edh,noon | | 24 | 29 | 34 | 43 | 46 | 51 | 50 | 45 | 36 | 29 | 24 | 21 |

Nomenclature: See separate page

*Fig. 4-4  Example tables of climatic design information included on CD-ROM of 2013 ASHRAE Handbook—Fundamentals*

# 116

**Principles of HVAC**

2013 ASHRAE Handbook - Fundamentals (IP)  © 2013 ASHRAE, Inc.

### ROCHESTER-MONROE CO, NY, USA   WMO#: 725290

Lat: 43.12N  Long: 77.68W  Elev: 554  StdP: 14.4  Time Zone: -5 (NAE)  Period: 86-10  WBAN: 14768

**Annual Heating and Humidification Design Conditions**

| | Coldest Month | Heating DB | | Humidification DP/MCDB and HR | | | | | | Coldest month WS/MCDB | | | | MCWS/PCWD to 99.6% DB | | |
|---|---|---|---|---|---|---|---|---|---|---|---|---|---|---|---|---|
| | | 99.6% | 99% | 99.6% | | | 99% | | | 0.4% | | 1% | | MCWS | PCWD | |
| | | | | DP | HR | MCDB | DP | HR | MCDB | WS | MCDB | WS | MCDB | | | |
| | (a) | (b) | (c) | (d) | (e) | (f) | (g) | (h) | (i) | (j) | (k) | (l) | (m) | (n) | (o) | |
| (1) | 1 | 2.9 | 6.9 | -4.1 | 4.5 | 4.3 | -0.5 | 5.5 | 8.9 | 29.2 | 33.2 | 26.9 | 29.7 | 8.9 | 240 | (1) |

**Annual Cooling, Dehumidification, and Enthalpy Design Conditions**

| | Hottest Month | Hottest Month DB Range | Cooling DB/MCWB | | | | | | Evaporation WB/MCDB | | | | | | MCWS/PCWD to 0.4% DB | | |
|---|---|---|---|---|---|---|---|---|---|---|---|---|---|---|---|---|---|
| | | | 0.4% | | 1% | | 2% | | 0.4% | | 1% | | 2% | | MCWS | PCWD | |
| | | | DB | MCWB | DB | MCWB | DB | MCWB | WB | MCDB | WB | MCDB | WB | MCDB | | | |
| | (a) | (b) | (c) | (d) | (e) | (f) | (g) | (h) | (i) | (j) | (k) | (l) | (m) | (n) | (o) | | |
| (2) | 7 | 18.9 | 88.7 | 73.2 | 85.6 | 71.2 | 82.7 | 69.7 | 75.4 | 84.5 | 73.5 | 81.9 | 71.9 | 79.7 | 11.6 | 240 | (2) |

| | Dehumidification DP/MCDB and HR | | | | | | | | Enthalpy/MCDB | | | | | | Hours 8 to 4 & 55/69 | |
|---|---|---|---|---|---|---|---|---|---|---|---|---|---|---|---|---|
| | 0.4% | | | 1% | | | 2% | | 0.4% | | 1% | | 2% | | | |
| | DP | HR | MCDB | DP | HR | MCDB | DP | HR | MCDB | Enth | MCDB | Enth | MCDB | Enth | MCDB | | |
| | (a) | (b) | (c) | (d) | (e) | (f) | (g) | (h) | (i) | (j) | (k) | (l) | (m) | (n) | (o) | (p) |
| (3) | 72.5 | 122.7 | 80.6 | 70.8 | 115.8 | 78.2 | 69.3 | 110.0 | 76.5 | 39.2 | 84.7 | 37.4 | 82.2 | 36.0 | 80.0 | 696 | (3) |

**Extreme Annual Design Conditions**

| | Extreme Annual WS | | | Extreme Max WB | Extreme Annual DB | | | | n-Year Return Period Values of Extreme DB | | | | | | | |
|---|---|---|---|---|---|---|---|---|---|---|---|---|---|---|---|---|
| | 1% | 2.5% | 5% | | Mean | | Standard deviation | | n=5 years | | n=10 years | | n=20 years | | n=50 years | | |
| | | | | | Min | Max | Min | Max | Min | Max | Min | Max | Min | Max | Min | Max | |
| | (a) | (b) | (c) | (d) | (e) | (f) | (g) | (h) | (i) | (j) | (k) | (l) | (m) | (n) | (o) | (p) |
| (4) | 25.1 | 21.4 | 18.9 | 82.2 | -2.7 | 92.3 | 5.9 | 3.2 | -7.0 | 94.6 | -10.5 | 96.5 | -13.8 | 98.3 | -18.1 | 100.7 | (4) |

**Monthly Climatic Design Conditions**

| | | | Annual | Jan | Feb | Mar | Apr | May | Jun | Jul | Aug | Sep | Oct | Nov | Dec | |
|---|---|---|---|---|---|---|---|---|---|---|---|---|---|---|---|---|
| | | | (d) | (e) | (f) | (g) | (h) | (i) | (j) | (k) | (l) | (m) | (n) | (o) | (p) | |
| (5) | | Tavg | 48.6 | 25.9 | 26.3 | 34.7 | 46.5 | 57.3 | 66.7 | 70.9 | 69.4 | 61.9 | 50.6 | 40.7 | 30.3 | (5) |
| (6) | | Sd | | 11.28 | 9.71 | 10.73 | 9.42 | 8.38 | 6.90 | 5.56 | 5.77 | 7.27 | 8.20 | 8.84 | 9.34 | (6) |
| (7) | Temperatures, | HDD50 | 3119 | 749 | 664 | 490 | 176 | 23 | 0 | 0 | 0 | 5 | 95 | 305 | 612 | (7) |
| (8) | Degree-Days | HDD65 | 6558 | 1211 | 1084 | 940 | 559 | 270 | 63 | 11 | 23 | 141 | 451 | 730 | 1075 | (8) |
| (9) | and | CDD50 | 2593 | 2 | 1 | 16 | 72 | 248 | 501 | 648 | 601 | 363 | 114 | 25 | 2 | (9) |
| (10) | Degree-Hours | CDD65 | 555 | 0 | 0 | 0 | 5 | 30 | 113 | 194 | 159 | 49 | 5 | 0 | 0 | (10) |
| (11) | | CDH74 | 4551 | 0 | 0 | 10 | 75 | 306 | 972 | 1618 | 1179 | 348 | 43 | 0 | 0 | (11) |
| (12) | | CDH80 | 1289 | 0 | 0 | 1 | 16 | 68 | 286 | 507 | 333 | 75 | 3 | 0 | 0 | (12) |
| (13) | | PrecAvg | 32.0 | 2.1 | 2.1 | 2.3 | 2.6 | 2.7 | 3.0 | 2.7 | 3.4 | 3.0 | 2.4 | 2.9 | 2.7 | (13) |
| (14) | Precipitation | PrecMax | 39.4 | 5.8 | 4.5 | 3.8 | 4.1 | 6.6 | 6.8 | 5.6 | 5.9 | 6.3 | 4.7 | 4.8 | 4.7 | (14) |
| (15) | | PrecMin | 22.9 | 0.9 | 0.8 | 1.1 | 1.3 | 0.4 | 0.2 | 1.1 | 1.8 | 0.3 | 0.2 | 0.4 | 1.0 | (15) |
| (16) | | PrecSD | 4.9 | 1.2 | 1.1 | 0.8 | 0.8 | 1.3 | 1.8 | 1.3 | 1.1 | 1.6 | 1.2 | 1.3 | 1.1 | (16) |
| (17) | | 0.4% DB | | 61.0 | 58.3 | 73.4 | 82.2 | 86.3 | 90.8 | 92.7 | 91.5 | 87.7 | 79.3 | 68.3 | 58.9 | (17) |
| (18) | Monthly Design | MCWB | | 54.3 | 48.5 | 58.0 | 64.0 | 68.9 | 73.0 | 74.5 | 74.6 | 71.3 | 64.9 | 57.9 | 52.0 | (18) |
| (19) | Dry Bulb | 2% DB | | 52.8 | 50.2 | 63.4 | 74.5 | 81.5 | 86.5 | 88.8 | 87.2 | 81.5 | 73.2 | 64.1 | 53.0 | (19) |
| (20) | and | MCWB | | 48.1 | 43.5 | 53.3 | 59.1 | 66.0 | 71.2 | 74.1 | 72.9 | 68.8 | 62.0 | 55.9 | 47.7 | (20) |
| (21) | Mean Coincident | 5% DB | | 45.7 | 44.7 | 57.1 | 67.9 | 77.2 | 83.3 | 85.4 | 83.6 | 78.0 | 68.4 | 60.0 | 47.5 | (21) |
| (22) | Wet Bulb | MCWB | | 41.3 | 39.3 | 49.2 | 55.7 | 64.2 | 69.7 | 71.9 | 70.8 | 67.1 | 59.5 | 53.7 | 42.5 | (22) |
| (23) | Temperatures | 10% DB | | 40.7 | 39.7 | 50.9 | 62.7 | 72.6 | 79.9 | 82.3 | 80.6 | 74.3 | 64.0 | 54.6 | 43.1 | (23) |
| (24) | | MCWB | | 36.9 | 36.0 | 44.1 | 52.5 | 61.7 | 67.6 | 70.0 | 69.1 | 65.1 | 56.6 | 49.1 | 39.0 | (24) |
| (25) | | 0.4% WB | | 54.5 | 49.9 | 60.5 | 65.4 | 72.2 | 76.4 | 78.2 | 77.6 | 73.2 | 67.8 | 60.1 | 54.1 | (25) |
| (26) | Monthly Design | MCDB | | 59.6 | 54.9 | 70.9 | 78.9 | 81.5 | 85.9 | 87.9 | 88.4 | 82.6 | 75.6 | 65.6 | 57.7 | (26) |
| (27) | Wet Bulb | 2% WB | | 48.5 | 44.3 | 54.7 | 61.2 | 68.9 | 73.7 | 75.9 | 74.7 | 70.7 | 63.8 | 57.3 | 48.4 | (27) |
| (28) | and | MCDB | | 51.9 | 48.9 | 62.2 | 71.4 | 77.7 | 83.0 | 85.2 | 83.2 | 78.3 | 71.5 | 62.6 | 51.8 | (28) |
| (29) | Mean Coincident | 5% WB | | 42.1 | 41.0 | 49.7 | 57.5 | 66.1 | 71.7 | 73.9 | 72.8 | 68.8 | 60.5 | 53.8 | 42.8 | (29) |
| (30) | Dry Bulb | MCDB | | 45.5 | 44.3 | 56.9 | 66.5 | 75.0 | 79.9 | 82.4 | 80.5 | 75.5 | 67.1 | 59.6 | 46.8 | (30) |
| (31) | Temperatures | 10% WB | | 37.0 | 36.2 | 44.3 | 53.4 | 62.9 | 69.5 | 72.0 | 71.1 | 66.9 | 57.7 | 50.0 | 39.1 | (31) |
| (32) | | MCDB | | 40.3 | 39.5 | 50.0 | 61.2 | 71.1 | 76.5 | 79.8 | 78.1 | 73.1 | 63.4 | 54.5 | 42.8 | (32) |
| (33) | | MDBR | 13.3 | 14.5 | 16.2 | 18.2 | 19.9 | 19.3 | 18.9 | 18.4 | 18.5 | 16.8 | 14.2 | 12.2 | | (33) |
| (34) | Mean Daily | 5% DB MCDBR | 17.8 | 20.7 | 26.0 | 27.3 | 25.1 | 23.5 | 22.0 | 21.7 | 22.8 | 23.2 | 21.2 | 18.0 | | (34) |
| (35) | Temperature | MCWBR | 14.9 | 15.9 | 17.2 | 15.7 | 13.4 | 11.5 | 10.3 | 10.3 | 12.2 | 14.0 | 15.3 | 14.6 | | (35) |
| (36) | Range | 5% WB MCDBR | 17.5 | 19.0 | 25.4 | 24.6 | 22.6 | 20.0 | 19.0 | 18.5 | 18.8 | 20.8 | 19.3 | 17.1 | | (36) |
| (37) | | MCWBR | 15.5 | 15.7 | 18.0 | 15.5 | 13.5 | 11.3 | 10.1 | 10.0 | 11.2 | 13.6 | 15.2 | 15.1 | | (37) |
| (38) | | taub | | 0.301 | 0.314 | 0.332 | 0.378 | 0.406 | 0.426 | 0.406 | 0.413 | 0.370 | 0.343 | 0.319 | 0.293 | (38) |
| (39) | Clear Sky Solar | taud | | 2.459 | 2.392 | 2.418 | 2.299 | 2.202 | 2.148 | 2.235 | 2.252 | 2.398 | 2.484 | 2.546 | 2.570 | (39) |
| (40) | Irradiance | Ebn,noon | | 265 | 280 | 289 | 281 | 275 | 269 | 273 | 267 | 272 | 267 | 258 | 261 | (40) |
| (41) | | Edh,noon | | 23 | 29 | 32 | 39 | 45 | 47 | 43 | 41 | 33 | 27 | 21 | 19 | (41) |

Nomenclature:   See separate page

*Fig. 4-4  Example tables of climatic design information included on CD-ROM of 2013 ASHRAE Handbook—Fundamentals*

# Design Conditions

2013 ASHRAE Handbook - Fundamentals (IP) © 2013 ASHRAE, Inc.

**SPRINGFIELD MUNI, MO, USA** WMO#: 724400

Lat: 37.24N  Long: 93.39W  Elev: 1270  StdP: 14.03  Time Zone: -6 (NAC)  Period: 86-10  WBAN: 13995

### Annual Heating and Humidification Design Conditions

| Coldest Month | Heating DB | | Humidification DP/MCDB and HR | | | | | | Coldest month WS/MCDB | | | | MCWS/PCWD to 99.6% DB | |
|---|---|---|---|---|---|---|---|---|---|---|---|---|---|---|
| | 99.6% | 99% | 99.6% | | | 99% | | | 0.4% | | 1% | | | |
| | | | DP | HR | MCDB | DP | HR | MCDB | WS | MCDB | WS | MCDB | MCWS | PCWD |
| (a) | (b) | (c) | (d) | (e) | (f) | (g) | (h) | (i) | (j) | (k) | (l) | (m) | (n) | (o) |
| 1 | 6.6 | 12.3 | -1.9 | 5.2 | 9.4 | 2.9 | 6.7 | 15.3 | 25.1 | 44.3 | 23.6 | 42.4 | 8.5 | 340 |

### Annual Cooling, Dehumidification, and Enthalpy Design Conditions

| Hottest Month | Hottest Month DB Range | Cooling DB/MCWB | | | | | | Evaporation WB/MCDB | | | | | | MCWS/PCWD to 0.4% DB | |
|---|---|---|---|---|---|---|---|---|---|---|---|---|---|---|---|
| | | 0.4% | | 1% | | 2% | | 0.4% | | 1% | | 2% | | | |
| | | DB | MCWB | DB | MCWB | DB | MCWB | WB | MCDB | WB | MCDB | WB | MCDB | MCWS | PCWD |
| (a) | (b) | (c) | (d) | (e) | (f) | (g) | (h) | (i) | (j) | (k) | (l) | (m) | (n) | (o) | (p) |
| 7 | 19.1 | 94.8 | 74.6 | 91.7 | 74.7 | 89.3 | 74.3 | 77.8 | 88.9 | 76.8 | 87.6 | 75.5 | 85.8 | 8.4 | 180 |

| Dehumidification DP/MCDB and HR | | | | | | | | Enthalpy/MCDB | | | | | | Hours 8 to 4 & 55/69 |
|---|---|---|---|---|---|---|---|---|---|---|---|---|---|---|
| 0.4% | | | 1% | | | 2% | | 0.4% | | 1% | | 2% | | |
| DP | HR | MCDB | DP | HR | MCDB | DP | HR | MCDB | Enth | MCDB | Enth | MCDB | Enth | MCDB | |
| (a) | (b) | (c) | (d) | (e) | (f) | (g) | (h) | (i) | (j) | (k) | (l) | (m) | (n) | (o) | (p) |
| 74.6 | 135.6 | 83.6 | 73.4 | 130.1 | 82.3 | 72.5 | 126.0 | 81.1 | 42.1 | 89.0 | 40.9 | 87.7 | 39.9 | 86.0 | 720 |

### Extreme Annual Design Conditions

| Extreme Annual WS | | | Extreme Max WB | Extreme Annual DB | | | | n-Year Return Period Values of Extreme DB | | | | | | | |
|---|---|---|---|---|---|---|---|---|---|---|---|---|---|---|---|
| 1% | 2.5% | 5% | | Mean | | Standard deviation | | n=5 years | | n=10 years | | n=20 years | | n=50 years | |
| | | | | Min | Max | Min | Max | Min | Max | Min | Max | Min | Max | Min | Max |
| (a) | (b) | (c) | (d) | (e) | (f) | (g) | (h) | (i) | (j) | (k) | (l) | (m) | (n) | (o) | (p) |
| 23.3 | 20.1 | 18.3 | 81.9 | -0.4 | 98.3 | 6.4 | 3.5 | -5.0 | 100.8 | -8.7 | 102.8 | -12.3 | 104.7 | -16.9 | 107.2 |

### Monthly Climatic Design Conditions

| | | | Annual | Jan | Feb | Mar | Apr | May | Jun | Jul | Aug | Sep | Oct | Nov | Dec |
|---|---|---|---|---|---|---|---|---|---|---|---|---|---|---|---|
| | | | (d) | (e) | (f) | (g) | (h) | (i) | (j) | (k) | (l) | (m) | (n) | (o) | (p) |
| | | Tavg | 56.6 | 34.0 | 38.1 | 46.8 | 56.3 | 65.1 | 73.6 | 77.9 | 77.4 | 68.8 | 57.6 | 46.4 | 35.8 |
| | | Sd | | 11.48 | 11.22 | 10.83 | 9.31 | 7.22 | 5.49 | 4.84 | 5.77 | 7.64 | 8.67 | 10.02 | 10.67 |
| Temperatures, | | HDD50 | 1771 | 507 | 355 | 192 | 44 | 2 | 0 | 0 | 0 | 1 | 30 | 185 | 455 |
| Degree-Days | | HDD65 | 4442 | 960 | 754 | 568 | 287 | 90 | 8 | 1 | 2 | 50 | 257 | 558 | 907 |
| and | | CDD50 | 4170 | 12 | 22 | 92 | 231 | 471 | 707 | 865 | 851 | 564 | 264 | 78 | 13 |
| Degree-Hours | | CDD65 | 1366 | 0 | 0 | 4 | 25 | 94 | 265 | 400 | 388 | 163 | 26 | 1 | 0 |
| | | CDH74 | 12583 | 0 | 1 | 35 | 216 | 687 | 2299 | 3947 | 3821 | 1349 | 217 | 11 | 0 |
| | | CDH80 | 4999 | 0 | 0 | 2 | 35 | 139 | 824 | 1746 | 1762 | 453 | 38 | 0 | 0 |
| | | PrecAvg | 43.1 | 1.8 | 2.2 | 3.9 | 4.2 | 4.4 | 5.1 | 2.9 | 3.5 | 4.6 | 3.6 | 3.8 | 3.2 |
| Precipitation | | PrecMax | 57.1 | 4.4 | 5.0 | 9.0 | 7.4 | 9.4 | 11.3 | 8.0 | 6.3 | 11.3 | 8.7 | 8.1 | 6.2 |
| | | PrecMin | 27.6 | 0.1 | 0.6 | 0.9 | 1.1 | 1.5 | 1.3 | 0.7 | 0.8 | 1.4 | 0.4 | 0.2 | 0.4 |
| | | PrecSD | 8.1 | 1.2 | 1.0 | 2.2 | 1.7 | 2.1 | 2.6 | 1.9 | 1.7 | 3.0 | 2.0 | 2.3 | 1.5 |
| | 0.4% | DB | | 67.0 | 70.3 | 78.1 | 84.1 | 86.6 | 92.9 | 98.3 | 98.9 | 94.3 | 84.3 | 75.5 | 67.0 |
| Monthly Design | | MCWB | | 55.9 | 55.4 | 60.8 | 66.6 | 72.3 | 74.0 | 74.9 | 74.3 | 72.6 | 67.1 | 62.4 | 58.1 |
| Dry Bulb | 2% | DB | | 61.4 | 64.6 | 73.2 | 80.0 | 83.7 | 90.0 | 93.9 | 95.6 | 88.4 | 79.5 | 70.4 | 61.9 |
| and | | MCWB | | 52.9 | 52.3 | 58.7 | 63.7 | 70.3 | 74.0 | 75.3 | 74.4 | 71.0 | 64.8 | 59.5 | 54.7 |
| Mean Coincident | 5% | DB | | 56.7 | 60.6 | 69.0 | 75.8 | 81.2 | 87.7 | 91.2 | 92.6 | 85.1 | 75.6 | 66.1 | 56.6 |
| Wet Bulb | | MCWB | | 49.9 | 51.2 | 56.7 | 62.1 | 68.9 | 73.4 | 75.3 | 74.6 | 70.6 | 63.4 | 57.7 | 49.6 |
| Temperatures | 10% | DB | | 51.5 | 55.9 | 64.4 | 72.1 | 78.2 | 85.0 | 88.9 | 89.4 | 81.8 | 72.1 | 62.4 | 51.6 |
| | | MCWB | | 45.4 | 48.0 | 54.6 | 60.2 | 67.3 | 72.5 | 75.1 | 74.2 | 68.6 | 61.8 | 55.2 | 46.0 |
| | 0.4% | WB | | 59.3 | 59.7 | 64.4 | 68.7 | 74.4 | 77.9 | 79.4 | 78.5 | 75.4 | 71.3 | 64.8 | 61.1 |
| Monthly Design | | MCDB | | 64.4 | 64.5 | 73.1 | 79.4 | 84.0 | 87.9 | 90.0 | 90.4 | 85.3 | 79.7 | 71.0 | 65.5 |
| Wet Bulb | 2% | WB | | 55.5 | 56.1 | 61.5 | 66.4 | 72.6 | 76.3 | 78.1 | 77.4 | 74.0 | 68.4 | 61.9 | 56.1 |
| and | | MCDB | | 59.8 | 61.9 | 69.8 | 76.1 | 81.0 | 86.3 | 88.8 | 89.5 | 83.9 | 75.1 | 67.7 | 60.5 |
| Mean Coincident | 5% | WB | | 50.7 | 52.2 | 58.8 | 64.3 | 70.6 | 74.9 | 77.1 | 76.4 | 72.4 | 65.9 | 59.3 | 51.2 |
| Dry Bulb | | MCDB | | 55.5 | 58.2 | 66.6 | 72.8 | 78.5 | 84.4 | 87.6 | 88.0 | 81.5 | 72.6 | 64.6 | 55.0 |
| Temperatures | 10% | WB | | 45.0 | 48.2 | 55.6 | 62.1 | 68.7 | 73.6 | 75.9 | 75.2 | 71.0 | 63.5 | 56.2 | 46.0 |
| | | MCDB | | 51.1 | 55.8 | 63.4 | 69.9 | 75.8 | 82.4 | 86.0 | 86.0 | 78.8 | 70.0 | 61.7 | 50.4 |
| | | MDBR | | 18.4 | 19.5 | 21.0 | 21.2 | 19.2 | 19.2 | 19.1 | 20.5 | 20.7 | 21.0 | 19.3 | 17.8 |
| Mean Daily | 5% DB | MCDBR | | 25.4 | 26.7 | 25.9 | 25.0 | 21.1 | 21.6 | 22.2 | 23.7 | 23.0 | 24.2 | 23.6 | 25.1 |
| Temperature | | MCWBR | | 17.6 | 17.3 | 15.2 | 13.7 | 10.7 | 9.0 | 7.9 | 7.5 | 9.5 | 12.6 | 15.6 | 17.7 |
| Range | 5% WB | MCDBR | | 22.0 | 22.7 | 21.9 | 21.3 | 18.8 | 19.4 | 19.4 | 20.4 | 19.1 | 19.9 | 20.4 | 22.2 |
| | | MCWBR | | 17.2 | 16.8 | 14.7 | 13.1 | 10.5 | 9.3 | 8.4 | 7.9 | 9.1 | 12.1 | 15.2 | 17.9 |
| | | taub | | 0.317 | 0.309 | 0.331 | 0.346 | 0.367 | 0.381 | 0.407 | 0.395 | 0.356 | 0.328 | 0.308 | 0.300 |
| Clear Sky | | taud | | 2.532 | 2.537 | 2.454 | 2.390 | 2.326 | 2.302 | 2.255 | 2.325 | 2.477 | 2.550 | 2.637 | 2.606 |
| Solar | | Ebn,noon | | 277 | 295 | 297 | 296 | 290 | 285 | 276 | 277 | 284 | 284 | 281 | 278 |
| Irradiance | | Edh,noon | | 25 | 28 | 33 | 37 | 40 | 41 | 43 | 39 | 32 | 27 | 22 | 22 |

Nomenclature: See separate page

*Fig. 4-4 Example tables of climatic design information included on CD-ROM of 2013 ASHRAE Handbook—Fundamentals*

# Principles of HVAC

2013 ASHRAE Handbook - Fundamentals (IP) © 2013 ASHRAE, Inc.

**ST. LOUIS/LAMBERT, MO, USA**  WMO#: 724340

Lat: 38.75N  Long: 90.37W  Elev: 709  StdP: 14.32  Time Zone: -6 (NAC)  Period: 86-10  WBAN: 13994

### Annual Heating and Humidification Design Conditions

| Coldest Month | Heating DB | | Humidification DP/MCDB and HR | | | | | | Coldest month WS/MCDB | | | | MCWS/PCWD to 99.6% DB | |
|---|---|---|---|---|---|---|---|---|---|---|---|---|---|---|
| | 99.6% | 99% | 99.6% | | | 99% | | | 0.4% | | 1% | | | |
| | | | DP | HR | MCDB | DP | HR | MCDB | WS | MCDB | WS | MCDB | MCWS | PCWD |
| (a) | (b) | (c) | (d) | (e) | (f) | (g) | (h) | (i) | (j) | (k) | (l) | (m) | (n) | (o) |
| 1 | 6.6 | 11.7 | -4.8 | 4.4 | 9.4 | 0.4 | 5.8 | 14.1 | 27.7 | 30.7 | 25.0 | 33.7 | 10.3 | 300 |

### Annual Cooling, Dehumidification, and Enthalpy Design Conditions

| Hottest Month | Hottest Month DB Range | Cooling DB/MCWB | | | | | | Evaporation WB/MCDB | | | | | | MCWS/PCWD to 0.4% DB | |
|---|---|---|---|---|---|---|---|---|---|---|---|---|---|---|---|
| | | 0.4% | | 1% | | 2% | | 0.4% | | 1% | | 2% | | | |
| | | DB | MCWB | DB | MCWB | DB | MCWB | WB | MCDB | WB | MCDB | WB | MCDB | MCWS | PCWD |
| (a) | (b) | (c) | (d) | (e) | (f) | (g) | (h) | (i) | (j) | (k) | (l) | (m) | (n) | (o) | (p) |
| 7 | 16.8 | 95.5 | 76.8 | 93.0 | 76.1 | 90.7 | 75.0 | 79.4 | 90.8 | 78.1 | 89.2 | 77.0 | 87.6 | 10.1 | 240 |

| Dehumidification DP/MCDB and HR | | | | | | | | Enthalpy/MCDB | | | | | | Hours 8 to 4 & 55/69 |
|---|---|---|---|---|---|---|---|---|---|---|---|---|---|---|
| 0.4% | | | 1% | | | 2% | | 0.4% | | 1% | | 2% | | |
| DP | HR | MCDB | DP | HR | MCDB | DP | HR | MCDB | Enth | MCDB | Enth | MCDB | Enth | MCDB | |
| (a) | (b) | (c) | (d) | (e) | (f) | (g) | (h) | (i) | (j) | (k) | (l) | (m) | (n) | (o) | (p) |
| 76.2 | 140.6 | 85.8 | 74.9 | 134.2 | 84.9 | 73.5 | 127.9 | 83.3 | 43.4 | 90.6 | 42.1 | 89.3 | 40.8 | 87.4 | 639 |

### Extreme Annual Design Conditions

| Extreme Annual WS | | | Extreme Max WB | Extreme Annual DB | | | | n-Year Return Period Values of Extreme DB | | | | | | | |
|---|---|---|---|---|---|---|---|---|---|---|---|---|---|---|---|
| 1% | 2.5% | 5% | | Mean | | Standard deviation | | n=5 years | | n=10 years | | n=20 years | | n=50 years | |
| | | | | Min | Max | Min | Max | Min | Max | Min | Max | Min | Max | Min | Max |
| (a) | (b) | (c) | (d) | (e) | (f) | (g) | (h) | (i) | (j) | (k) | (l) | (m) | (n) | (o) | (p) |
| 23.7 | 20.1 | 18.1 | 85.1 | 0.7 | 99.9 | 5.9 | 2.6 | -3.6 | 101.8 | -7.0 | 103.4 | -10.4 | 104.8 | -14.6 | 106.8 |

### Monthly Climatic Design Conditions

| | | Annual | Jan | Feb | Mar | Apr | May | Jun | Jul | Aug | Sep | Oct | Nov | Dec |
|---|---|---|---|---|---|---|---|---|---|---|---|---|---|---|
| | Tavg | 57.4 | 32.8 | 36.7 | 46.8 | 57.6 | 67.1 | 76.1 | 80.0 | 78.6 | 70.5 | 58.7 | 46.9 | 35.4 |
| | Sd | | 11.59 | 11.14 | 11.08 | 9.79 | 7.60 | 6.32 | 5.37 | 5.90 | 7.51 | 8.94 | 10.15 | 10.76 |
| Temperatures, | HDD50 | 1826 | 542 | 389 | 194 | 36 | 1 | 0 | 0 | 0 | 0 | 23 | 176 | 465 |
| Degree-Days | HDD65 | 4436 | 997 | 793 | 571 | 263 | 68 | 6 | 0 | 1 | 34 | 238 | 546 | 919 |
| and | CDD50 | 4515 | 10 | 16 | 95 | 264 | 531 | 782 | 929 | 887 | 615 | 291 | 84 | 11 |
| Degree-Hours | CDD65 | 1650 | 0 | 0 | 7 | 41 | 134 | 338 | 464 | 423 | 199 | 41 | 3 | 0 |
| | CDH74 | 16623 | 0 | 2 | 64 | 370 | 1094 | 3398 | 5188 | 4452 | 1722 | 311 | 22 | 0 |
| | CDH80 | 6773 | 0 | 0 | 8 | 94 | 315 | 1370 | 2366 | 1955 | 594 | 70 | 1 | 0 |
| | PrecAvg | 37.5 | 1.8 | 2.1 | 3.6 | 3.5 | 4.0 | 3.7 | 3.9 | 2.9 | 3.1 | 2.7 | 3.3 | 3.0 |
| Precipitation | PrecMax | 44.4 | 5.4 | 4.2 | 6.7 | 9.1 | 7.2 | 8.7 | 7.1 | 6.5 | 6.2 | 5.8 | 5.8 | 6.5 |
| | PrecMin | 25.0 | 0.2 | 0.2 | 1.1 | 1.0 | 1.0 | 0.9 | 0.6 | 0.1 | 0.8 | 0.2 | 0.4 | 0.7 |
| | PrecSD | 5.3 | 1.4 | 1.1 | 1.5 | 2.1 | 1.8 | 2.0 | 1.8 | 1.7 | 1.5 | 1.3 | 1.6 | 1.6 |
| | 0.4% DB | 67.5 | 70.5 | 80.6 | 87.9 | 90.3 | 95.3 | 99.0 | 99.8 | 93.5 | 86.5 | 77.3 | 67.1 | |
| | 0.4% MCWB | 57.2 | 57.8 | 61.9 | 67.2 | 72.9 | 75.3 | 77.6 | 77.2 | 74.5 | 68.9 | 63.5 | 60.4 | |
| Monthly Design | 2% DB | 60.0 | 64.1 | 74.8 | 82.5 | 86.8 | 92.4 | 95.6 | 95.7 | 89.8 | 81.4 | 71.9 | 60.9 | |
| Dry Bulb and | 2% MCWB | 53.0 | 54.1 | 60.5 | 64.7 | 71.4 | 75.1 | 77.2 | 77.0 | 72.9 | 66.2 | 60.1 | 54.6 | |
| Mean Coincident | 5% DB | 55.0 | 58.7 | 69.9 | 78.6 | 83.6 | 90.3 | 93.2 | 92.6 | 86.4 | 77.3 | 67.1 | 55.8 | |
| Wet Bulb | 5% MCWB | 48.8 | 50.4 | 57.8 | 63.4 | 69.7 | 74.2 | 76.6 | 76.4 | 71.2 | 64.6 | 58.0 | 49.3 | |
| Temperatures | 10% DB | 49.3 | 53.4 | 64.7 | 73.7 | 80.5 | 87.8 | 90.7 | 89.6 | 83.1 | 73.1 | 63.2 | 50.7 | |
| | 10% MCWB | 43.0 | 46.2 | 55.4 | 61.0 | 67.3 | 73.0 | 75.6 | 75.1 | 69.3 | 62.5 | 55.6 | 45.0 | |
| | 0.4% WB | 60.0 | 61.1 | 65.8 | 70.2 | 76.2 | 79.2 | 81.2 | 81.3 | 77.4 | 72.3 | 65.9 | 62.2 | |
| | 0.4% MCDB | 65.2 | 66.5 | 76.1 | 82.9 | 85.7 | 89.9 | 93.0 | 92.3 | 88.7 | 81.4 | 73.5 | 67.0 | |
| Monthly Design | 2% WB | 55.3 | 56.0 | 62.7 | 67.5 | 73.7 | 77.6 | 79.6 | 79.5 | 75.3 | 69.3 | 62.4 | 56.1 | |
| Wet Bulb and | 2% MCDB | 58.4 | 62.4 | 70.7 | 77.5 | 83.2 | 89.0 | 91.3 | 90.7 | 84.9 | 77.1 | 68.0 | 59.8 | |
| Mean Coincident | 5% WB | 49.5 | 50.9 | 59.8 | 65.3 | 71.9 | 76.1 | 78.3 | 78.0 | 73.8 | 66.5 | 59.9 | 50.3 | |
| Dry Bulb | 5% MCDB | 54.6 | 57.7 | 68.6 | 75.0 | 80.7 | 86.7 | 89.7 | 88.7 | 82.7 | 73.9 | 66.0 | 54.2 | |
| Temperatures | 10% WB | 43.4 | 46.5 | 55.6 | 62.9 | 69.7 | 74.7 | 77.2 | 76.7 | 72.1 | 64.2 | 56.5 | 45.6 | |
| | 10% MCDB | 48.4 | 52.9 | 64.3 | 72.1 | 77.6 | 84.6 | 88.0 | 86.7 | 80.2 | 72.1 | 62.6 | 50.0 | |
| | MDBR | 15.0 | 16.3 | 18.5 | 19.3 | 17.7 | 17.3 | 16.8 | 17.2 | 18.4 | 18.6 | 16.3 | 14.4 | |
| Mean Daily | 5% DB MCDBR | 23.5 | 25.8 | 24.6 | 24.0 | 20.3 | 19.1 | 18.7 | 19.8 | 20.6 | 22.4 | 22.6 | 21.8 | |
| Temperature | 5% DB MCWBR | 17.2 | 17.4 | 14.0 | 12.3 | 9.4 | 7.4 | 6.7 | 6.9 | 8.4 | 11.3 | 14.3 | 16.5 | |
| Range | 5% WB MCDBR | 21.9 | 23.1 | 21.6 | 21.5 | 18.1 | 17.4 | 17.5 | 18.0 | 17.4 | 18.6 | 19.8 | 19.4 | |
| | 5% WB MCWBR | 18.0 | 17.2 | 13.8 | 12.6 | 9.5 | 8.0 | 7.6 | 7.6 | 8.2 | 11.4 | 14.8 | 16.8 | |
| | taub | | 0.317 | 0.318 | 0.348 | 0.364 | 0.369 | 0.406 | 0.419 | 0.421 | 0.378 | 0.346 | 0.311 | 0.305 |
| Clear Sky | taud | | 2.546 | 2.530 | 2.426 | 2.338 | 2.329 | 2.225 | 2.199 | 2.240 | 2.400 | 2.504 | 2.638 | 2.618 |
| Solar Irradiance | Ebn,noon | | 274 | 290 | 289 | 289 | 288 | 277 | 272 | 268 | 275 | 274 | 276 | 272 |
| | Edh,noon | | 24 | 27 | 34 | 39 | 40 | 44 | 45 | 42 | 34 | 28 | 22 | 21 |

Nomenclature: See separate page

*Fig. 4-4 Example tables of climatic design information included on CD-ROM of 2013 ASHRAE Handbook—Fundamentals*

# Design Conditions

2013 ASHRAE Handbook - Fundamentals (IP) © 2013 ASHRAE, Inc.

## TUCSON INTL, AZ, USA

WMO#: 722740
Lat: 32.13N  Long: 110.96W  Elev: 2556  StdP: 13.39  Time Zone: -7 (NAZ)  Period: 86-10  WBAN: 23160

### Annual Heating and Humidification Design Conditions

| | Coldest Month | Heating DB | | Humidification DP/MCDB and HR | | | | | | Coldest month WS/MCDB | | | | MCWS/PCWD to 99.6% DB | | |
|---|---|---|---|---|---|---|---|---|---|---|---|---|---|---|---|---|
| | | 99.6% | 99% | 99.6% | | | 99% | | | 0.4% | | 1% | | | | |
| | | | | DP | HR | MCDB | DP | HR | MCDB | WS | MCDB | WS | MCDB | MCWS | PCWD | |
| | (a) | (b) | (c) | (d) | (e) | (f) | (g) | (h) | (i) | (j) | (k) | (l) | (m) | (n) | (o) | |
| (1) | 12 | 31.6 | 34.3 | 0.5 | 6.2 | 56.6 | 4.9 | 7.8 | 58.9 | 24.9 | 58.5 | 22.2 | 58.7 | 5.4 | 140 | (1) |

### Annual Cooling, Dehumidification, and Enthalpy Design Conditions

| | Hottest Month | Hottest Month DB Range | Cooling DB/MCWB | | | | | | Evaporation WB/MCDB | | | | | | MCWS/PCWD to 0.4% DB | |
|---|---|---|---|---|---|---|---|---|---|---|---|---|---|---|---|---|
| | | | 0.4% | | 1% | | 2% | | 0.4% | | 1% | | 2% | | | |
| | | | DB | MCWB | DB | MCWB | DB | MCWB | WB | MCDB | WB | MCDB | WB | MCDB | MCWS | PCWD |
| | (a) | (b) | (c) | (d) | (e) | (f) | (g) | (h) | (i) | (j) | (k) | (l) | (m) | (n) | (o) | (p) |
| (2) | 7 | 24.3 | 106.0 | 66.2 | 103.6 | 66.0 | 101.5 | 65.7 | 72.6 | 88.6 | 71.8 | 88.0 | 71.1 | 87.4 | 11.0 | 300 | (2) |

| | Dehumidification DP/MCDB and HR | | | | | | | | Enthalpy/MCDB | | | | | | Hours 8 to 4 & 55/69 |
|---|---|---|---|---|---|---|---|---|---|---|---|---|---|---|---|
| | 0.4% | | | 1% | | | 2% | | | 0.4% | | 1% | | 2% | | |
| | DP | HR | MCDB | DP | HR | MCDB | DP | HR | MCDB | Enth | MCDB | Enth | MCDB | Enth | MCDB | |
| | (a) | (b) | (c) | (d) | (e) | (f) | (g) | (h) | (i) | (j) | (k) | (l) | (m) | (n) | (o) | (p) |
| (3) | 69.3 | 118.7 | 76.2 | 68.0 | 113.3 | 76.6 | 66.6 | 107.7 | 77.6 | 37.9 | 88.2 | 37.1 | 87.6 | 36.5 | 86.7 | 701 | (3) |

### Extreme Annual Design Conditions

| | Extreme Annual WS | | | Extreme Max WB | Extreme Annual DB | | | | n-Year Return Period Values of Extreme DB | | | | | | | |
|---|---|---|---|---|---|---|---|---|---|---|---|---|---|---|---|---|
| | 1% | 2.5% | 5% | | Mean | | Standard deviation | | n=5 years | | n=10 years | | n=20 years | | n=50 years | |
| | | | | | Min | Max | Min | Max | Min | Max | Min | Max | Min | Max | Min | Max |
| | (a) | (b) | (c) | (d) | (e) | (f) | (g) | (h) | (i) | (j) | (k) | (l) | (m) | (n) | (o) | (p) |
| (4) | 21.5 | 18.8 | 16.8 | 77.7 | 26.1 | 110.1 | 3.4 | 2.6 | 23.6 | 111.9 | 21.6 | 113.5 | 19.7 | 114.9 | 17.3 | 116.8 | (4) |

### Monthly Climatic Design Conditions

| | | | Annual | Jan | Feb | Mar | Apr | May | Jun | Jul | Aug | Sep | Oct | Nov | Dec | |
|---|---|---|---|---|---|---|---|---|---|---|---|---|---|---|---|---|
| | | | (d) | (e) | (f) | (g) | (h) | (i) | (j) | (k) | (l) | (m) | (n) | (o) | (p) | |
| (5) | | Tavg | 70.1 | 53.2 | 55.9 | 61.0 | 67.7 | 76.5 | 85.4 | 87.5 | 85.7 | 82.1 | 71.9 | 60.6 | 52.5 | (5) |
| (6) | | Sd | | 6.21 | 6.23 | 7.04 | 6.94 | 5.99 | 5.06 | 4.37 | 3.97 | 4.12 | 6.75 | 7.27 | 6.38 | (6) |
| (7) | Temperatures, | HDD50 | 121 | 39 | 17 | 6 | 1 | 0 | 0 | 0 | 0 | 0 | 0 | 9 | 49 | (7) |
| (8) | Degree-Days | HDD65 | 1416 | 366 | 258 | 161 | 51 | 3 | 0 | 0 | 0 | 0 | 22 | 166 | 389 | (8) |
| (9) | and | CDD50 | 7452 | 138 | 183 | 347 | 533 | 822 | 1063 | 1164 | 1106 | 964 | 681 | 326 | 125 | (9) |
| (10) | Degree-Hours | CDD65 | 3273 | 1 | 4 | 37 | 133 | 361 | 613 | 699 | 641 | 514 | 237 | 33 | 0 | (10) |
| (11) | | CDH74 | 46014 | 59 | 162 | 759 | 2111 | 5285 | 9651 | 9992 | 8295 | 6391 | 2731 | 541 | 37 | (11) |
| (12) | | CDH80 | 26563 | 5 | 24 | 211 | 895 | 2965 | 6342 | 6243 | 4840 | 3614 | 1283 | 140 | 1 | (12) |
| (13) | | PrecAvg | 11.5 | 0.9 | 0.8 | 0.7 | 0.3 | 0.2 | 0.2 | 2.2 | 2.2 | 1.5 | 0.9 | 0.6 | 1.0 | (13) |
| (14) | Precipitation | PrecMax | 24.1 | 5.3 | 3.5 | 2.9 | 1.6 | 1.0 | 1.9 | 6.0 | 4.9 | 3.9 | 4.4 | 2.3 | 5.8 | (14) |
| (15) | | PrecMin | 9.5 | 0.0 | 0.0 | 0.1 | 0.0 | 0.0 | 0.0 | 0.3 | 0.7 | 0.1 | 0.0 | 0.0 | 0.0 | (15) |
| (16) | | PrecSD | 3.1 | 1.0 | 0.8 | 0.8 | 0.4 | 0.2 | 0.4 | 1.2 | 1.0 | 0.9 | 1.1 | 0.6 | 1.4 | (16) |
| (17) | | 0.4% DB | | 79.8 | 83.6 | 89.4 | 96.6 | 103.1 | 110.3 | 108.9 | 106.0 | 102.2 | 97.5 | 88.2 | 78.2 | (17) |
| (18) | Monthly Design | MCWB | | 52.2 | 53.8 | 56.4 | 59.0 | 63.2 | 67.2 | 66.2 | 68.5 | 66.4 | 62.4 | 57.2 | 52.3 | (18) |
| (19) | Dry Bulb | 2% DB | | 75.0 | 78.9 | 85.2 | 91.9 | 99.4 | 106.2 | 106.1 | 102.7 | 99.7 | 94.2 | 84.0 | 74.1 | (19) |
| (20) | and | MCWB | | 50.8 | 52.0 | 55.2 | 57.3 | 61.3 | 65.1 | 66.7 | 68.4 | 65.5 | 61.6 | 55.4 | 50.7 | (20) |
| (21) | Mean Coincident | 5% DB | | 71.4 | 75.1 | 81.9 | 88.8 | 96.6 | 103.8 | 103.5 | 100.2 | 97.8 | 91.1 | 80.5 | 70.1 | (21) |
| (22) | Wet Bulb | MCWB | | 49.2 | 50.7 | 53.4 | 56.3 | 60.3 | 64.6 | 67.0 | 68.5 | 65.1 | 60.4 | 54.4 | 49.7 | (22) |
| (23) | Temperatures | 10% DB | | 67.2 | 71.0 | 78.6 | 85.5 | 93.4 | 101.6 | 100.7 | 97.7 | 95.4 | 87.8 | 76.4 | 66.0 | (23) |
| (24) | | MCWB | | 48.1 | 49.3 | 52.4 | 55.3 | 59.0 | 63.5 | 67.1 | 68.6 | 65.2 | 59.5 | 53.0 | 48.0 | (24) |
| (25) | | 0.4% WB | | 56.9 | 57.0 | 59.2 | 61.3 | 66.2 | 71.0 | 73.2 | 74.0 | 73.0 | 67.5 | 60.3 | 56.5 | (25) |
| (26) | Monthly Design | MCDB | | 65.3 | 74.1 | 81.3 | 90.3 | 93.9 | 94.7 | 88.2 | 89.8 | 88.9 | 82.8 | 73.1 | 64.8 | (26) |
| (27) | Wet Bulb | 2% WB | | 53.8 | 54.9 | 57.0 | 59.2 | 64.0 | 69.0 | 72.2 | 73.0 | 71.3 | 65.3 | 58.3 | 53.9 | (27) |
| (28) | and | MCDB | | 65.9 | 69.5 | 79.9 | 85.8 | 90.6 | 94.2 | 88.0 | 89.0 | 86.3 | 82.3 | 74.5 | 65.7 | (28) |
| (29) | Mean Coincident | 5% WB | | 51.9 | 53.2 | 55.1 | 57.6 | 62.4 | 67.6 | 71.4 | 72.2 | 70.2 | 63.4 | 56.5 | 52.0 | (29) |
| (30) | Dry Bulb | MCDB | | 65.2 | 67.8 | 76.9 | 83.9 | 89.1 | 93.4 | 87.8 | 88.4 | 85.0 | 82.2 | 74.6 | 64.4 | (30) |
| (31) | Temperatures | 10% WB | | 50.2 | 51.4 | 53.3 | 56.0 | 60.8 | 66.0 | 70.8 | 71.4 | 69.0 | 61.9 | 54.6 | 50.1 | (31) |
| (32) | | MCDB | | 64.0 | 66.0 | 74.2 | 82.2 | 87.9 | 94.3 | 87.5 | 87.7 | 84.2 | 81.9 | 73.4 | 63.4 | (32) |
| (33) | | MDBR | | 25.4 | 25.5 | 27.5 | 29.5 | 30.3 | 30.0 | 24.3 | 23.3 | 25.2 | 26.9 | 27.0 | 25.0 | (33) |
| (34) | Mean Daily | 5% DB MCDBR | | 31.3 | 32.3 | 33.1 | 33.6 | 33.2 | 32.7 | 28.6 | 26.5 | 28.2 | 30.8 | 31.3 | 30.0 | (34) |
| (35) | Temperature | MCWBR | | 15.1 | 14.5 | 14.1 | 13.7 | 13.0 | 11.9 | 8.1 | 6.2 | 8.4 | 11.8 | 13.8 | 14.7 | (35) |
| (36) | Range | 5% WB MCDBR | | 24.6 | 26.0 | 29.8 | 30.9 | 29.8 | 27.3 | 22.0 | 22.6 | 22.3 | 24.2 | 25.9 | 24.2 | (36) |
| (37) | | MCWBR | | 12.4 | 11.7 | 12.4 | 12.7 | 11.2 | 9.5 | 5.2 | 5.6 | 6.3 | 9.2 | 12.0 | 12.4 | (37) |
| (38) | | taub | | 0.302 | 0.296 | 0.316 | 0.292 | 0.290 | 0.298 | 0.343 | 0.357 | 0.334 | 0.318 | 0.299 | 0.301 | (38) |
| (39) | Clear Sky Solar | taud | | 2.661 | 2.651 | 2.513 | 2.528 | 2.509 | 2.473 | 2.431 | 2.436 | 2.532 | 2.556 | 2.643 | 2.642 | (39) |
| (40) | Irradiance | Ebn,noon | | 296 | 308 | 307 | 317 | 315 | 311 | 297 | 292 | 297 | 295 | 294 | 290 | (40) |
| (41) | | Edh,noon | | 24 | 26 | 32 | 33 | 34 | 35 | 37 | 36 | 31 | 28 | 24 | 23 | (41) |

Nomenclature: See separate page

*Fig. 4-4 Example tables of climatic design information included on CD-ROM of 2013 ASHRAE Handbook—Fundamentals*

# 120 Principles of HVAC

## TULSA INTL ARPT(AW), OK, USA

2013 ASHRAE Handbook - Fundamentals (IP)  © 2013 ASHRAE, Inc.  WMO#: 723560

Lat: 36.20N  Long: 95.89W  Elev: 676  StdP: 14.34  Time Zone: -6 (NAC)  Period: 86-10  WBAN: 13968

### Annual Heating and Humidification Design Conditions

| Coldest Month | Heating DB | | Humidification DP/MCDB and HR | | | | | | Coldest month WS/MCDB | | | | MCWS/PCWD to 99.6% DB | |
|---|---|---|---|---|---|---|---|---|---|---|---|---|---|---|
| | 99.6% | 99% | 99.6% DP | HR | MCDB | 99% DP | HR | MCDB | 0.4% WS | MCDB | 1% WS | MCDB | MCWS | PCWD |
| (a) | (b) | (c) | (d) | (e) | (f) | (g) | (h) | (i) | (j) | (k) | (l) | (m) | (n) | (o) |
| 1 | 13.2 | 18.3 | 1.1 | 5.9 | 17.5 | 6.2 | 7.8 | 22.5 | 26.4 | 53.9 | 24.5 | 50.3 | 9.0 | 350 |

### Annual Cooling, Dehumidification, and Enthalpy Design Conditions

| Hottest Month | Hottest Month DB Range | Cooling DB/MCWB | | | | | | Evaporation WB/MCDB | | | | | | MCWS/PCWD to 0.4% DB | |
|---|---|---|---|---|---|---|---|---|---|---|---|---|---|---|---|
| | | 0.4% DB | MCWB | 1% DB | MCWB | 2% DB | MCWB | 0.4% WB | MCDB | 1% WB | MCDB | 2% WB | MCDB | MCWS | PCWD |
| (a) | (b) | (c) | (d) | (e) | (f) | (g) | (h) | (i) | (j) | (k) | (l) | (m) | (n) | (o) | |
| 7 | 18.6 | 99.4 | 75.9 | 96.8 | 76.0 | 94.0 | 75.6 | 79.2 | 92.5 | 78.2 | 91.2 | 77.3 | 89.9 | 10.7 | 190 |

| Dehumidification DP/MCDB and HR | | | | | | | | Enthalpy/MCDB | | | | | | Hours 8 to 4 & 55/69 |
|---|---|---|---|---|---|---|---|---|---|---|---|---|---|---|
| 0.4% DP | HR | MCDB | 1% DP | HR | MCDB | 2% DP | HR | MCDB | 0.4% Enth | MCDB | 1% Enth | MCDB | 2% Enth | MCDB | |
| (a) | (b) | (c) | (d) | (e) | (f) | (g) | (h) | (i) | (j) | (k) | (l) | (m) | (n) | (o) | (p) |
| 75.4 | 136.6 | 85.5 | 74.4 | 131.8 | 84.7 | 73.4 | 127.3 | 83.8 | 43.1 | 92.6 | 42.1 | 91.9 | 41.0 | 90.0 | 698 |

### Extreme Annual Design Conditions

| Extreme Annual WS | | | Extreme Max WB | Extreme Annual DB | | | | n-Year Return Period Values of Extreme DB | | | | | | | |
|---|---|---|---|---|---|---|---|---|---|---|---|---|---|---|---|
| 1% | 2.5% | 5% | | Mean Min | Mean Max | SD Min | SD Max | n=5 Min | n=5 Max | n=10 Min | n=10 Max | n=20 Min | n=20 Max | n=50 Min | n=50 Max |
| (a) | (b) | (c) | (d) | (e) | (f) | (g) | (h) | (i) | (j) | (k) | (l) | (m) | (n) | (o) | (p) |
| 24.6 | 21.5 | 19.4 | 85.3 | 6.4 | 103.0 | 6.2 | 3.6 | 1.9 | 105.6 | -1.7 | 107.7 | -5.2 | 109.7 | -9.7 | 112.3 |

### Monthly Climatic Design Conditions

| | | Annual | Jan | Feb | Mar | Apr | May | Jun | Jul | Aug | Sep | Oct | Nov | Dec |
|---|---|---|---|---|---|---|---|---|---|---|---|---|---|---|
| Temperatures, Degree-Days and Degree-Hours | Tavg | 61.2 | 38.8 | 42.9 | 51.6 | 60.9 | 69.7 | 77.9 | 82.9 | 82.2 | 73.1 | 61.9 | 50.7 | 40.3 |
| | Sd | | 10.79 | 11.09 | 10.62 | 8.71 | 6.98 | 5.37 | 5.04 | 5.81 | 7.56 | 8.39 | 9.80 | 10.22 |
| | HDD50 | 1187 | 371 | 245 | 110 | 14 | 0 | 0 | 0 | 0 | 0 | 9 | 111 | 327 |
| | HDD65 | 3455 | 813 | 618 | 429 | 178 | 35 | 0 | 0 | 0 | 20 | 159 | 437 | 766 |
| | CDD50 | 5258 | 24 | 47 | 158 | 341 | 610 | 836 | 1018 | 997 | 692 | 378 | 131 | 26 |
| | CDD65 | 2051 | 0 | 1 | 12 | 55 | 180 | 387 | 553 | 532 | 262 | 62 | 7 | 0 |
| | CDH74 | 22972 | 1 | 11 | 108 | 485 | 1548 | 4123 | 6983 | 6584 | 2539 | 532 | 58 | 0 |
| | CDH80 | 10738 | 0 | 2 | 18 | 119 | 463 | 1790 | 3651 | 3506 | 1049 | 136 | 4 | 0 |
| Precipitation | PrecAvg | 41.1 | 1.5 | 1.8 | 3.3 | 3.9 | 5.7 | 4.8 | 3.5 | 3.1 | 4.5 | 3.7 | 2.8 | 2.4 |
| | PrecMax | 67.1 | 3.4 | 4.2 | 11.9 | 8.3 | 9.3 | 8.9 | 10.9 | 6.1 | 18.8 | 8.0 | 7.3 | 6.3 |
| | PrecMin | 27.4 | 0.2 | 0.4 | 0.1 | 1.3 | 1.7 | 0.5 | 0.1 | 0.6 | 0.1 | 0.2 | 0.3 | 0.4 |
| | PrecSD | 10.7 | 0.9 | 1.1 | 2.5 | 2.0 | 2.4 | 2.2 | 3.3 | 1.6 | 4.4 | 2.4 | 2.4 | 1.5 |
| Monthly Design Dry Bulb and Mean Coincident Wet Bulb Temperatures | 0.4% DB | | 71.3 | 74.9 | 82.0 | 88.5 | 91.3 | 97.2 | 102.1 | 102.3 | 99.3 | 89.2 | 79.6 | 71.3 |
| | 0.4% MCWB | | 55.5 | 58.8 | 62.3 | 66.0 | 73.7 | 75.6 | 76.4 | 75.2 | 73.8 | 70.2 | 62.5 | 61.1 |
| | 2% DB | | 64.7 | 70.2 | 76.7 | 83.2 | 88.2 | 93.9 | 99.0 | 99.9 | 92.9 | 83.7 | 74.9 | 65.6 |
| | 2% MCWB | | 54.2 | 55.7 | 60.9 | 65.9 | 72.3 | 75.7 | 76.6 | 75.4 | 73.7 | 67.6 | 61.4 | 56.1 |
| | 5% DB | | 60.8 | 65.6 | 72.7 | 79.6 | 85.1 | 91.3 | 96.5 | 97.4 | 89.6 | 79.8 | 70.9 | 60.9 |
| | 5% MCWB | | 51.3 | 53.7 | 58.9 | 64.3 | 71.0 | 75.2 | 76.7 | 75.8 | 72.8 | 65.9 | 60.2 | 52.4 |
| | 10% DB | | 55.6 | 60.8 | 69.0 | 75.7 | 82.1 | 89.1 | 93.6 | 94.6 | 86.2 | 76.2 | 66.5 | 56.1 |
| | 10% MCWB | | 47.0 | 51.1 | 57.7 | 62.5 | 70.0 | 74.6 | 76.0 | 75.6 | 71.9 | 64.5 | 57.6 | 48.3 |
| Monthly Design Wet Bulb and Mean Coincident Dry Bulb Temperatures | 0.4% WB | | 60.7 | 63.5 | 67.0 | 72.0 | 76.3 | 79.9 | 81.1 | 80.2 | 77.5 | 73.6 | 67.5 | 63.9 |
| | 0.4% MCDB | | 66.2 | 69.6 | 75.7 | 80.9 | 86.7 | 91.4 | 93.6 | 93.7 | 89.3 | 82.7 | 74.3 | 68.4 |
| | 2% WB | | 57.4 | 59.3 | 64.2 | 69.4 | 74.7 | 78.2 | 79.6 | 78.8 | 76.2 | 71.2 | 64.6 | 58.6 |
| | 2% MCDB | | 62.1 | 66.3 | 72.4 | 77.9 | 84.4 | 89.9 | 93.1 | 92.4 | 87.4 | 78.3 | 70.9 | 64.0 |
| | 5% WB | | 52.4 | 56.1 | 61.8 | 67.5 | 73.2 | 77.1 | 78.4 | 77.7 | 74.8 | 68.8 | 62.1 | 53.8 |
| | 5% MCDB | | 59.0 | 63.5 | 70.1 | 76.1 | 82.4 | 88.4 | 91.6 | 91.5 | 85.3 | 76.4 | 68.7 | 58.1 |
| | 10% WB | | 47.4 | 51.0 | 58.9 | 65.1 | 71.6 | 75.8 | 77.4 | 76.7 | 73.4 | 66.4 | 58.9 | 48.6 |
| | 10% MCDB | | 54.8 | 59.9 | 67.3 | 73.2 | 80.1 | 86.2 | 90.2 | 90.1 | 82.8 | 74.5 | 65.4 | 55.0 |
| Mean Daily Temperature Range | MDBR | | 18.6 | 20.1 | 20.7 | 20.7 | 18.4 | 18.1 | 18.6 | 19.9 | 20.1 | 21.0 | 19.7 | 18.2 |
| | 5% DB MCDBR | | 25.6 | 27.9 | 25.8 | 24.5 | 20.7 | 19.9 | 20.9 | 23.0 | 22.8 | 24.7 | 25.0 | 25.0 |
| | 5% DB MCWBR | | 16.5 | 16.9 | 14.3 | 12.6 | 9.5 | 7.4 | 6.7 | 6.5 | 8.4 | 12.0 | 14.7 | 16.8 |
| | 5% WB MCDBR | | 21.8 | 24.5 | 20.8 | 19.7 | 18.2 | 18.3 | 18.9 | 19.8 | 19.0 | 19.6 | 21.5 | 21.0 |
| | 5% WB MCWBR | | 16.0 | 17.1 | 13.7 | 12.4 | 9.7 | 8.2 | 7.5 | 6.9 | 8.1 | 11.5 | 15.5 | 17.4 |
| Clear Sky Solar Irradiance | taub | | 0.323 | 0.322 | 0.352 | 0.364 | 0.374 | 0.387 | 0.404 | 0.403 | 0.366 | 0.335 | 0.315 | 0.307 |
| | taud | | 2.567 | 2.540 | 2.421 | 2.355 | 2.324 | 2.302 | 2.302 | 2.323 | 2.460 | 2.551 | 2.618 | 2.604 |
| | Ebn,noon | | 278 | 292 | 291 | 291 | 288 | 283 | 277 | 276 | 282 | 283 | 280 | 277 |
| | Edh,noon | | 24 | 28 | 34 | 39 | 41 | 41 | 41 | 40 | 33 | 27 | 23 | 22 |

Nomenclature: See separate page

*Fig. 4-4 Example tables of climatic design information included on CD-ROM of 2013 ASHRAE Handbook—Fundamentals*

**Design Conditions**

analysis methods, has been significantly updated, with the inclusion of mean dry-bulb and wet-bulb temperature ranges coincident with the 5% monthly dry-bulb and wet-bulb design temperatures.

Design conditions are provided for locations for which long-term hourly observations were available (1986–2010 for most stations in the United States and Canada).

Figure 4-4 shows climatic design conditions for 31 locations, to illustrate the format of the data available on the CD-ROM. A subset of the United States and Canada weather stations containing 23 annual data elements is provided for conveniences in Table 4-8.

The top part of Figure 4-4 contains station information as follows:

- Name of the observing station, state (USA) or province (Canada), country.
- World Meteorological Organization (WMO) station identifier.
- Weather Bureau Army Navy (WBAN) number (–99999 denotes missing).
- Latitude of station, °N/S.
- Longitude of station, °E/W.
- Elevation of station, ft.
- Standard pressure at elevation, in psia (see Chapter 2 for equations used to calculate standard pressure).
- Time zone, h ± UTC
- Time zone code (e.g., NAE = Eastern Time, USA and Canada). The CD-ROM contains a list of all time zone codes used in the tables.
- Period analyzed (e.g., 86–10 = data from 1986 to 2010 were used).
- Table 4-7 gives the nomenclature for tables of climatic design conditions given in Figure 4-4.

**Annual Design Conditions.** Annual climatic design conditions are contained in the first three sections following the top part of the table. They contain information as follows:

**Annual Heating and Humidification Design Conditions.**

- Coldest month (i.e., month with lowest average dry-bulb temperature; 1 = January, 12 = December).
- Dry-bulb temperature corresponding to 99.6 and 99.0% annual cumulative frequency of occurrence (cold conditions), °F.
- Dew-point temperature corresponding to 99.6 and 99.0% annual cumulative frequency of occurrence, °F; corresponding humidity ratio, calculated at the standard atmospheric pressure at elevation of station, grains of moisture per lb of dry air; mean coincident dry-bulb temperature, °F.
- Wind speed corresponding to 0.4 and 1.0% cumulative frequency of occurrence for coldest month, mph; mean coincident dry-bulb temperature, °F.
- Mean wind speed coincident with 99.6% dry-bulb temperature, mph; corresponding most frequent wind direction, degrees from north (east = 90°).

**Annual Cooling, Dehumidification, and Enthalpy Design Conditions.**

- Hottest month (i.e., month with highest average dry-bulb temperature; 1 = January, 12 = December).
- Daily temperature range for hottest month, °F [defined as mean of the difference between daily maximum and daily minimum dry-bulb temperatures for hottest month].
- Dry-bulb temperature corresponding to 0.4, 1.0, and 2.0% annual cumulative frequency of occurrence (warm conditions), °F; mean coincident wet-bulb temperature, °F.
- Wet-bulb temperature corresponding to 0.4, 1.0, and 2.0% annual cumulative frequency of occurrence, °F; mean coincident dry-bulb temperature, °F.
- Mean wind speed coincident with 0.4% dry-bulb temperature, mph; corresponding most frequent wind direction, degrees true from north (east = 90°).
- Dew-point temperature corresponding to 0.4, 1.0, and 2.0% annual cumulative frequency of occurrence, °F; corresponding humidity ratio, calculated at the standard atmospheric pressure at elevation of station, grains of moisture per lb of dry air; mean coincident dry-bulb temperature, °F.
- Enthalpy corresponding to 0.4, 1.0, and 2.0% annual cumulative frequency of occurrence, Btu/lb; mean coincident dry-bulb temperature, °F.
- Number of hours between 8 AM and 4 PM (inclusive) with dry-bulb temperature between 55°F and 69°F.

**Extreme Annual Design Conditions.**

- Wind speed corresponding to 1.0, 2.5, and 5.0% annual cumulative frequency of occurrence, mph.
- Extreme maximum wet-bulb temperature, °F.
- Mean and standard deviation of extreme annual minimum and maximum dry-bulb temperature, °F.
- 5-, 10-, 20-, and 50-year return period values for minimum and maximum extreme dry-bulb temperature, °F.

**Monthly Design Conditions.** Monthly design conditions are divided into subsections as follows:

**Temperatures, Degree-Days, and Degree-Hours.**

- Average temperature, °F. This parameter is a prime indicator of climate and is also useful to calculate heating and cooling degree-days to any base.
- Standard deviation of average daily temperature, °F. This parameter is useful to calculate heating and cooling degree-days to any base. Its use is explained in the section on Estimation of Degree-Days.
- Heating and cooling degree-days (bases 50°F and 65°F). These parameters are useful in energy estimating methods. They are also used to classify locations into climate zones in ASHRAE Standard 169.
- Cooling degree-hours (bases 74°F and 80°F). These are used in various standards, such as Standard 90.2-2004.

**Precipitation.**

- Average precipitation, in. This parameter is used to calculate climate zones for Standard 169, and is of interest in some green building technologies (e.g., vegetative roofs).
- Standard deviation of precipitation, in. This parameter indicates the variability of precipitation at the site.

## Table 4-8  Design Conditions for Selected Locations

Meaning of acronyms:
DB: Dry bulb temperature, °F
WB: Wet bulb temperature, °F
MCWB: Mean coincident wet bulb temperature, °F

Lat: Latitude, °
Long: Longitude, °
DP: Dew point temperature, °F
HR: Humidity ratio, grains of moisture per lb of dry air
MCDB: Mean coincident dry bulb temperature, °F

Elev: Elevation, ft
WS: Wind speed, mph
HDD and CDD 65: Annual heating and cooling degree-days, base 65°F, °F-day

| Station | Lat | Long | Elev | Heating DB 99.6% | Heating DB 99% | Cooling DB/MCWB 0.4% DB/MCWB | 1% DB/MCWB | 2% DB/MCWB | Evaporation WB/MCDB 0.4% WB/MCDB | 1% WB/MCDB | Dehumidification DP/HR/MCDB 0.4% DP/HR/MCDB | 1% DP/HR/MCDB | Extreme Annual WS 1% | 2.5% | 5% | Heat./Cool. Degree-Days HDD / CDD 65 |
|---|---|---|---|---|---|---|---|---|---|---|---|---|---|---|---|---|
| **United States of America** | | | | | | | | | | | | | | | | |
| *Alabama* | | | | | | | | | | | | | *542 sites, 864 more on CD-ROM* | | | |
| AUBURN OPELIKA ROBE | 32.62N | 85.43W | 778 | 23.5 | 27.6 | 93.9 74.3 | 91.4 74.2 | 90.2 73.9 | 78.0 88.4 | 77.0 87.2 | 75.1 135.7 81.9 | 73.5 128.4 80.4 | 17.6 15.5 | 12.9 | 2377 1953 |
| BIRMINGHAM MUNI | 33.56N | 86.75W | 630 | 20.5 | 24.8 | 95.5 74.9 | 93.0 74.5 | 90.9 74.3 | 78.4 88.5 | 77.5 87.6 | 75.9 138.7 82.6 | 74.8 133.3 81.7 | 18.4 16.4 | 14.5 | 2653 2014 |
| CAIRNS AAF | 31.28N | 85.71W | 302 | 26.6 | 30.0 | 96.2 76.9 | 94.2 76.5 | 92.2 76.1 | 81.1 89.4 | 79.8 88.3 | 79.1 152.8 84.4 | 77.4 144.3 82.8 | 16.9 14.3 | 12.4 | 1785 2471 |
| DOTHAN RGNL | 31.32N | 85.45W | 354 | 27.4 | 30.7 | 96.6 76.1 | 94.7 75.3 | 91.5 75.2 | 79.9 89.7 | 78.7 88.3 | 77.4 144.3 83.3 | 76.3 138.9 82.5 | 19.3 17.5 | 15.5 | 1743 2512 |
| GADSDEN MUNI | 33.97N | 86.08W | 568 | 18.5 | 21.5 | 93.5 74.6 | 91.3 74.5 | 90.0 74.3 | 78.1 89.1 | 77.1 88.0 | 74.9 133.6 83.6 | 73.3 126.5 82.1 | 16.7 14.2 | 12.3 | 3216 1586 |
| HUNTSVILLE/MADISON | 34.64N | 86.79W | 643 | 18.4 | 22.5 | 95.3 75.1 | 92.8 74.6 | 90.6 74.1 | 78.4 88.4 | 77.6 87.6 | 75.9 138.7 82.6 | 74.9 133.8 81.7 | 20.8 18.6 | 16.7 | 3093 1819 |
| MAXWELL AFB | 32.38N | 86.36W | 171 | 27.0 | 32.0 | 97.3 76.7 | 95.4 76.6 | 93.5 76.3 | 80.6 91.2 | 79.7 90.2 | 78.0 146.4 84.9 | 77.0 141.7 84.2 | 18.0 15.6 | 13.1 | 1898 2615 |
| MOBILE/BATES FIELD | 30.69N | 88.25W | 220 | 27.7 | 31.0 | 93.8 76.9 | 92.0 76.5 | 90.5 76.1 | 80.1 88.5 | 79.1 87.3 | 78.0 146.6 83.4 | 77.0 141.9 82.5 | 20.2 18.2 | 16.4 | 1652 2499 |
| MONTGOMERY/DANNELLY | 32.30N | 86.39W | 203 | 24.3 | 27.6 | 96.8 76.1 | 94.5 76.0 | 92.6 75.7 | 79.7 90.7 | 78.6 89.2 | 76.8 140.7 84.3 | 75.8 135.9 83.2 | 18.6 16.5 | 14.3 | 2149 2320 |
| NORTHWEST ALABAMA R | 34.75N | 87.61W | 561 | 19.1 | 23.1 | 96.5 75.3 | 93.4 74.9 | 91.2 74.6 | 78.7 89.7 | 77.8 88.6 | 75.7 137.1 83.2 | 74.9 133.4 82.3 | 18.8 16.8 | 14.7 | 3045 1876 |
| TUSCALOOSA RGNL | 33.21N | 87.62W | 187 | 21.9 | 26.2 | 97.0 76.0 | 94.3 75.9 | 92.3 75.6 | 79.5 90.8 | 78.5 89.3 | 76.8 140.6 83.3 | 75.5 134.3 82.7 | 17.2 14.8 | 12.3 | 2477 2163 |
| *Alaska* | | | | | | | | | | | | | *7 sites, 87 more on CD-ROM* | | | |
| FAIRBANKS INTL ARPT | 64.82N | 147.86W | 453 | -43.5 | -38.3 | 81.3 61.0 | 78.3 60.0 | 74.8 58.6 | 63.2 76.9 | 61.6 74.2 | 58.4 74.1 65.4 | 56.6 69.4 64.3 | 17.2 14.9 | 12.3 | 13517 72 |
| FT. RICHARDSON/BRYA | 61.27N | 149.65W | 377 | -19.6 | -14.0 | 74.8 60.3 | 71.6 58.9 | 68.3 57.1 | 61.7 72.7 | 59.6 69.5 | 56.3 68.5 65.4 | 54.9 65.1 62.6 | 19.3 14.6 | 11.5 | 10677 5 |
| ANCHORAGE/ELMENDORF | 61.25N | 149.79W | 213 | -15.4 | -10.4 | 73.9 58.5 | 71.5 57.7 | 68.1 56.3 | 60.3 70.0 | 59.3 67.3 | 57.3 70.6 61.1 | 55.5 66.1 60.3 | 19.0 15.9 | 12.9 | 10313 11 |
| ANCHORAGE LAKE HOOD | 61.18N | 149.96W | 131 | -8.6 | -4.0 | 73.9 59.5 | 71.1 58.4 | 67.9 57.0 | 61.1 71.7 | 59.5 68.0 | 56.6 68.7 63.1 | 55.1 65.0 62.4 | 18.5 16.0 | 13.0 | 9764 16 |
| ANCHORAGE INTL ARPT | 61.18N | 149.99W | 131 | -9.3 | -4.8 | 71.5 58.9 | 68.3 57.4 | 65.8 56.2 | 60.4 68.9 | 58.9 66.2 | 56.5 68.2 62.7 | 55.2 65.2 61.5 | 20.8 18.5 | 16.7 | 10121 5 |
| MERRILL FLD | 61.22N | 149.86W | 138 | -11.3 | -7.8 | 73.0 59.4 | 70.4 58.3 | 67.9 57.1 | 61.3 70.4 | 59.8 67.1 | 57.2 70.2 63.1 | 55.6 66.1 62.1 | 15.2 12.3 | 10.6 | 10045 10 |
| JUNEAU | 58.36N | 134.58W | 23 | 4.5 | 9.0 | 73.8 59.6 | 70.1 58.2 | 66.5 56.7 | 61.1 71.3 | 59.5 | 57.2 69.9 61.8 | 55.9 66.6 60.6 | 26.7 23.7 | 19.8 | 8304 3 |
| *Arizona* | | | | | | | | | | | | | *9 sites, 12 more on CD-ROM* | | | |
| CASA GRANDE MUNI | 32.96N | 111.77W | 1463 | 32.1 | 35.3 | 108.5 69.5 | 106.8 69.1 | 104.9 68.7 | 73.7 94.4 | 73.0 93.2 | 69.8 115.9 79.3 | 66.9 104.6 79.8 | 20.5 17.7 | 15.0 | 1508 3545 |
| DAVIS MONTHAN AFB | 32.17N | 110.88W | 2703 | 32.3 | 35.7 | 105.8 64.8 | 103.7 64.8 | 100.4 64.4 | 73.3 83.2 | 72.3 84.6 | 71.8 129.9 76.7 | 69.7 120.8 76.6 | 20.4 18.0 | 16.0 | 1405 3345 |
| FLAGSTAFF AIRPORT | 35.14N | 111.67W | 7018 | 3.9 | 9.5 | 85.7 55.6 | 83.3 55.1 | 81.2 54.9 | 61.3 73.0 | 60.2 72.5 | 57.9 93.2 63.7 | 56.5 88.5 63.3 | 23.7 19.7 | 17.2 | 6830 123 |
| LUKE AFB | 33.54N | 112.38W | 1086 | 35.3 | 37.8 | 111.0 70.8 | 108.6 70.7 | 106.4 70.4 | 77.0 97.8 | 75.8 96.5 | 72.2 123.9 82.7 | 70.4 116.3 83.7 | 20.0 17.3 | 14.7 | 1193 3979 |
| PHOENIX/SKY HARBOR | 33.44N | 111.99W | 1106 | 38.7 | 41.6 | 110.3 69.6 | 108.3 69.4 | 106.4 69.3 | 77.0 97.7 | 75.8 95.8 | 71.3 120.1 82.3 | 69.4 112.8 84.5 | 18.5 16.1 | 13.0 | 923 4626 |
| ERNEST A LOVE FLD | 34.65N | 112.42W | 5052 | 17.7 | 20.7 | 94.4 60.8 | 91.5 60.2 | 90.0 59.8 | 66.5 81.4 | 65.4 80.1 | 63.1 104.4 70.4 | 61.3 97.8 70.0 | 21.0 18.6 | 16.8 | 4174 982 |
| TUCSON INTL | 32.13N | 110.96W | 2556 | 31.6 | 34.3 | 106.0 66.2 | 103.6 66.0 | 101.5 65.7 | 72.6 88.6 | 71.8 88.0 | 69.3 118.7 76.2 | 68.0 113.3 76.6 | 21.5 18.8 | 16.8 | 1416 3273 |
| YUMA INTL AIRPORT | 32.65N | 114.60W | 207 | 42.0 | 44.8 | 110.7 73.8 | 108.6 73.4 | 106.6 72.7 | 79.9 96.8 | 78.5 96.1 | 75.5 134.6 87.5 | 73.8 127.1 88.2 | 20.8 18.4 | 16.3 | 666 4728 |
| YUMA MCAS | 32.62N | 114.60W | 213 | 41.6 | 44.7 | 110.9 73.1 | 108.8 72.8 | 107.3 72.3 | 79.7 96.6 | 78.3 95.8 | 75.3 133.8 87.0 | 73.4 125.2 88.1 | 20.9 18.5 | 16.4 | 665 4717 |
| *Arkansas* | | | | | | | | | | | | | *11 sites, 15 more on CD-ROM* | | | |
| BENTONVILLE MUNI THA | 36.35N | 94.22W | 1296 | 10.3 | 16.1 | 94.6 74.7 | 91.3 75.0 | 89.9 74.4 | 77.6 89.8 | 76.5 88.2 | 73.4 130.3 84.3 | 72.7 127.5 83.5 | 19.6 17.5 | 15.7 | 4045 1372 |
| DRAKE FLD | 36.01N | 94.17W | 1260 | 10.0 | 16.2 | 95.1 74.9 | 92.5 74.6 | 90.2 74.2 | 77.9 89.3 | 76.8 88.1 | 74.6 135.9 83.3 | 73.2 129.5 82.1 | 20.6 18.7 | 17.2 | 3897 1424 |
| FORT SMITH MUNI | 35.33N | 94.37W | 463 | 17.0 | 21.7 | 99.4 76.5 | 96.8 76.3 | 93.9 76.1 | 79.2 92.6 | 78.6 91.1 | 76.4 140.0 84.7 | 75.3 135.1 83.8 | 20.5 18.1 | 16.2 | 3158 2061 |
| JONESBORO MUNI | 35.83N | 90.65W | 276 | 16.2 | 19.5 | 97.0 77.0 | 94.0 76.3 | 91.9 75.7 | 80.2 91.2 | 79.1 90.2 | 76.4 143.1 85.7 | 75.4 134.6 84.2 | 21.9 19.0 | 17.2 | 3504 1952 |
| ADAMS FLD | 34.75N | 92.23W | 256 | 19.3 | 23.9 | 98.7 77.3 | 95.6 77.0 | 93.1 76.5 | 80.2 92.0 | 79.2 91.0 | 77.1 142.6 85.3 | 76.1 137.7 84.3 | 18.8 16.9 | 15.3 | 2918 2170 |
| LITTLE ROCK AFB | 34.92N | 92.15W | 312 | 17.5 | 21.7 | 99.5 77.4 | 96.7 77.0 | 93.8 76.5 | 81.1 92.4 | 80.1 91.3 | 78.6 150.2 86.1 | 77.2 143.0 85.3 | 17.8 15.2 | 12.9 | 3108 2069 |
| LITTLE ROCK/ADAMS F | 34.83N | 92.25W | 568 | 18.5 | 23.3 | 95.4 76.6 | 93.0 76.3 | 90.9 75.6 | 79.1 90.7 | 78.1 88.8 | 76.0 138.9 84.7 | 75.0 134.1 83.9 | 18.5 16.5 | 14.7 | 3158 1938 |
| GRIDER FLD | 34.18N | 91.94W | 213 | 21.3 | 25.5 | 97.3 77.4 | 95.0 77.4 | 92.9 76.8 | 80.4 91.9 | 79.3 91.0 | 77.3 143.0 85.9 | 76.0 136.9 84.9 | 18.9 17.1 | 15.3 | 2700 2230 |
| ROGERS MUNI CARTER F | 36.37N | 94.10W | 1352 | 10.1 | 15.5 | 93.3 73.6 | 91.4 73.9 | 89.5 73.5 | 77.1 88.1 | 76.0 86.4 | 73.3 130.4 81.3 | 72.7 127.7 81.4 | 21.8 19.0 | 17.0 | 4040 1385 |
| SMITH FLD | 36.19N | 94.48W | 1194 | 10.4 | 16.2 | 95.6 74.7 | 92.6 74.5 | 90.2 74.2 | 77.5 90.2 | 76.4 88.4 | 75.5 134.6 84.6 | 72.6 126.3 83.8 | 23.3 20.2 | 18.2 | 3970 1441 |
| TEXARKANA RGNL WEBB | 33.45N | 94.01W | 400 | 23.3 | 27.1 | 98.8 76.3 | 96.3 76.3 | 93.5 75.7 | 79.5 91.4 | 78.7 90.2 | 76.6 140.8 84.3 | 75.5 135.5 83.4 | 18.8 16.9 | 15.7 | 2440 2335 |
| *California* | | | | | | | | | | | | | *55 sites, 38 more on CD-ROM* | | | |
| ALAMEDA(USN) | 37.73N | 122.32W | 13 | 39.9 | 42.1 | 83.0 64.7 | 79.0 63.6 | 75.4 62.5 | 66.4 78.7 | 65.1 76.1 | 62.3 84.0 68.6 | 61.0 80.3 67.6 | 20.6 18.5 | 16.7 | 2105 209 |
| BAKERSFIELD/MEADOWS | 35.43N | 119.06W | 492 | 32.2 | 35.0 | 102.8 70.6 | 100.4 69.5 | 98.2 68.6 | 73.6 97.5 | 72.0 96.0 | 65.2 94.8 87.4 | 62.9 87.4 85.2 | 18.4 15.9 | 13.2 | 2095 2253 |
| BEALE AFB | 39.14N | 121.44W | 112 | 32.2 | 34.8 | 100.8 70.9 | 98.3 69.9 | 95.1 68.1 | 72.7 96.4 | 71.3 93.6 | 64.4 91.1 83.7 | 63.0 86.5 81.4 | 21.7 18.5 | 15.1 | 2356 1532 |
| BURBANK/GLENDALE | 34.20N | 118.36W | 732 | 38.6 | 41.0 | 97.7 67.4 | 93.8 66.7 | 90.9 66.3 | 72.4 89.7 | 70.9 87.4 | 66.5 100.4 77.4 | 65.6 97.0 76.3 | 18.4 15.1 | 12.9 | 1353 1423 |
| CAMARILLO | 34.22N | 119.10W | 75 | 37.1 | 39.2 | 85.8 62.3 | 81.8 63.0 | 79.3 63.4 | 69.2 78.7 | 67.9 77.1 | 65.6 94.7 74.6 | 63.9 89.2 71.8 | 25.2 20.1 | 16.3 | 1872 374 |
| CAMP PENDLETON MCAS | 33.30N | 117.35W | 79 | 32.3 | 35.5 | 92.1 68.2 | 89.0 65.7 | 84.1 65.5 | 71.6 83.6 | 70.3 82.0 | 67.7 102.2 76.7 | 66.0 96.0 76.1 | 16.8 14.4 | 12.2 | 1764 695 |
| MC CLELLAN PALOMAR | 33.13N | 117.28W | 328 | 43.1 | 44.9 | 82.4 62.7 | 80.7 63.5 | 77.1 63.8 | 68.7 80.7 | 68.7 77.1 | 66.5 98.7 73.2 | 66.0 97.1 72.8 | 14.0 12.1 | 10.8 | 1701 481 |
| CASTLE AFB/MERCED | 37.37N | 120.57W | 197 | 30.6 | 32.5 | 99.5 70.1 | 97.1 69.3 | 94.4 68.3 | 72.3 94.9 | 70.9 93.3 | 63.7 89.0 84.3 | 61.9 83.3 83.9 | 18.2 14.8 | 12.6 | 2629 1474 |
| EL TORO MCAS | 33.68N | 117.73W | 384 | 40.7 | 43.4 | 91.7 67.5 | 88.3 67.2 | 85.3 66.4 | 71.9 85.6 | 70.1 83.9 | 66.2 89.0 83.9 | 64.7 93.0 77.3 | 14.9 12.0 | 10.4 | 1142 1067 |
| FRESNO AIR TERMINAL | 36.78N | 119.72W | 328 | 31.4 | 33.7 | 103.5 70.9 | 100.8 69.3 | 98.6 68.3 | 73.5 97.4 | 71.9 95.3 | 65.3 94.7 85.7 | 63.2 87.9 84.1 | 18.2 16.3 | 14.2 | 2266 2097 |
| FULLERTON MUNICIPAL | 33.87N | 117.98W | 95 | 39.2 | 42.8 | 93.4 67.0 | 90.6 66.8 | 88.0 66.7 | 72.4 86.5 | 71.0 84.3 | 67.6 101.7 79.6 | 65.9 95.8 78.1 | 12.9 11.0 | 10.2 | 1202 1240 |
| S CALIF LOGISTICS | 34.60N | 117.38W | 2884 | 27.5 | 30.3 | 100.7 65.3 | 98.4 64.7 | 96.0 63.9 | 69.8 88.4 | 68.4 88.2 | 64.5 101.2 76.9 | 61.9 92.1 78.6 | 22.4 18.8 | 16.2 | 2661 1911 |
| HAYWARD AIR TERM | 37.66N | 122.12W | 46 | 36.9 | 39.0 | 87.8 65.7 | 82.3 64.2 | 78.8 63.2 | 67.8 82.3 | 66.0 79.2 | 62.8 85.6 73.0 | 61.1 80.5 70.0 | 19.6 17.7 | 16.2 | 2572 288 |

## Design Conditions

### Table 4-8 Design Conditions for Selected Locations (*Continued*)

Meaning of acronyms:
DB: Dry bulb temperature, °F
MCWB: Mean coincident wet bulb temperature, °F
WB: Wet bulb temperature, °F
Lat: Latitude, °
DP: Dew point temperature, °F
MCDB: Mean coincident dry bulb temperature, °F
Long: Longitude, °
HR: Humidity ratio, grains of moisture per lb of dry air
HDD and CDD 65: Annual heating and cooling degree-days, base 65°F, °F-day
Elev: Elevation, ft
WS: Wind speed, mph

| Station | Lat | Long | Elev | Heating DB 99.6% | Heating DB 99% | Cooling DB/MCWB 0.4% DB/MCWB | Cooling DB/MCWB 1% DB/MCWB | Cooling DB/MCWB 2% DB/MCWB | Evaporation WB/MCDB 0.4% WB/MCDB | Evaporation WB/MCDB 1% WB/MCDB | Dehumidification DP/HR/MCDB 0.4% DP/HR/MCDB | Dehumidification DP/HR/MCDB 1% DP/HR/MCDB | Extreme Annual WS 1% | Extreme Annual WS 2.5% | Extreme Annual WS 5% | Heat./Cool. Degree-Days HDD/CDD 65 |
|---|---|---|---|---|---|---|---|---|---|---|---|---|---|---|---|---|
| IMPERIAL CO | 32.83N | 115.58W | -56 | 35.7 | 37.5 | 111.2 72.9 | 109.0 72.7 | 107.7 72.6 | 81.1 97.4 | 79.7 96.8 | 77.1 140.7 88.5 | 75.0 131.0 89.1 | 26.1 | 22.0 | 18.8 | 949 / 4149 |
| JACK NORTHROP FLD H | 33.92N | 118.33W | 62 | 44.7 | 45.7 | 88.2 63.2 | 83.7 63.4 | 81.1 63.5 | 69.9 79.5 | 68.7 77.8 | 66.2 96.7 74.3 | 64.4 90.7 71.9 | 16.3 | 14.0 | 12.5 | 1135 / 782 |
| LANCASTER/FOX FIELD | 34.74N | 118.22W | 2339 | 21.3 | 24.9 | 104.2 66.0 | 100.0 64.4 | 97.8 63.5 | 68.6 95.9 | 67.0 94.8 | 58.6 80.0 80.6 | 55.7 72.1 79.3 | 29.5 | 27.0 | 25.0 | 2954 / 1830 |
| LEMOORE NAS | 36.33N | 119.95W | 233 | 28.4 | 31.8 | 103.0 71.6 | 100.3 70.1 | 98.5 69.1 | 74.7 97.6 | 72.7 95.7 | 66.4 98.1 89.2 | 63.9 89.8 86.0 | 20.3 | 17.3 | 14.9 | 2260 / 1821 |
| LIVERMORE MUNICIPAL | 37.69N | 121.82W | 397 | 30.2 | 33.5 | 99.0 67.8 | 94.9 66.6 | 90.8 65.1 | 70.1 94.2 | 68.1 90.7 | 61.1 81.6 77.4 | 59.2 76.2 73.6 | 19.5 | 17.9 | 16.3 | 2773 / 796 |
| LOMPOC | 34.67N | 120.47W | 89 | 32.7 | 35.5 | 81.4 60.8 | 77.2 61.1 | 73.5 60.3 | 65.4 75.3 | 63.8 73.4 | 61.1 81.5 69.1 | 59.4 76.0 67.7 | 20.4 | 18.6 | 17.2 | 2838 / 53 |
| LONG BEACH/LB AIRP. | 33.83N | 118.16W | 39 | 41.3 | 43.6 | 91.1 66.7 | 87.6 66.5 | 84.1 65.8 | 72.0 83.0 | 70.5 81.0 | 68.6 105.1 76.0 | 66.9 99.1 74.9 | 17.0 | 14.6 | 12.5 | 1190 / 1062 |
| LOS ANGELES INTL | 33.94N | 118.41W | 325 | 44.5 | 46.4 | 83.7 63.3 | 80.4 63.6 | 77.5 64.2 | 69.9 77.2 | 68.7 75.5 | 67.3 101.6 73.7 | 66.0 97.0 72.5 | 19.9 | 17.4 | 16.0 | 1295 / 582 |
| RIVERSIDE/MARCH AFB | 33.90N | 117.25W | 1535 | 32.1 | 35.6 | 100.2 67.0 | 98.8 66.6 | 95.5 65.6 | 71.5 93.3 | 70.2 91.1 | 65.8 88.0 73.7 | 63.9 84.1 72.9 | 17.8 | 15.5 | 12.8 | 1861 / 1590 |
| MC CLELLAN AFLD | 38.67N | 121.40W | 75 | 31.1 | 34.0 | 102.1 70.2 | 99.2 69.2 | 96.1 68.0 | 72.3 97.5 | 70.8 95.3 | 63.2 87.1 80.2 | 61.4 81.5 79.4 | 20.4 | 17.0 | 14.4 | 2269 / 1605 |
| MODESTO CITY CO HAR | 37.63N | 120.95W | 98 | 31.1 | 33.8 | 101.6 70.3 | 98.8 68.6 | 95.8 67.5 | 72.1 96.6 | 70.4 94.5 | 63.0 84.8 84.8 | 60.9 80.2 81.3 | 19.2 | 17.0 | 15.5 | 2386 / 1621 |
| MONTEREY PENINSULA | 36.59N | 121.85W | 220 | 36.7 | 38.8 | 79.0 60.3 | 73.3 59.1 | 71.7 58.8 | 62.8 72.3 | 61.7 70.7 | 59.1 75.5 64.6 | 57.3 70.7 63.7 | 17.0 | 14.9 | 12.7 | 3281 / 49 |
| MOUNTAIN VIEW (SUNN | 37.42N | 122.05W | 33 | 36.2 | 38.6 | 88.4 65.5 | 83.7 64.6 | 80.7 64.1 | 68.3 82.4 | 66.7 79.8 | 63.1 86.6 74.4 | 61.4 81.4 71.8 | 18.9 | 17.3 | 15.5 | 2164 / 464 |
| NAPA CO | 38.21N | 122.28W | 56 | 29.6 | 32.1 | 91.3 65.8 | 87.4 65.3 | 83.2 63.7 | 68.4 86.5 | 66.6 83.3 | 61.4 81.4 74.2 | 60.5 78.9 73.2 | 21.2 | 19.0 | 17.5 | 3239 / 246 |
| SAN BERNARDINO INTL | 34.08N | 117.23W | 1158 | 33.9 | 36.5 | 102.9 69.7 | 100.2 69.5 | 97.4 68.8 | 74.5 95.1 | 73.0 93.6 | 68.1 107.7 83.1 | 66.1 100.5 83.3 | 16.7 | 12.9 | 10.9 | 1652 / 1811 |
| OAKLAND/METROP. OAK | 37.76N | 122.22W | 89 | 36.7 | 39.0 | 82.3 64.3 | 78.7 63.2 | 74.6 62.4 | 66.4 78.5 | 65.0 75.4 | 62.0 83.4 69.9 | 61.0 80.4 68.5 | 23.4 | 19.9 | 18.3 | 2637 / 155 |
| ONTARIO INTL ARPT | 34.05N | 117.57W | 942 | 37.1 | 39.7 | 100.1 69.9 | 97.4 68.9 | 94.6 68.4 | 74.0 93.6 | 72.4 91.4 | 68.1 106.8 80.8 | 66.1 99.7 78.3 | 23.1 | 18.3 | 16.3 | 1387 / 1769 |
| PALM SPRINGS INTL | 33.83N | 116.51W | 449 | 41.4 | 44.5 | 112.1 72.1 | 109.1 70.8 | 107.9 70.7 | 79.1 99.4 | 77.4 98.0 | 73.0 124.7 92.3 | 71.7 119.0 92.1 | 23.2 | 20.1 | 18.0 | 783 / 4336 |
| JACQUELINE COCHRAN | 33.63N | 116.16W | -118 | 31.2 | 34.4 | 111.2 72.5 | 108.9 72.0 | 107.3 71.7 | 79.8 97.6 | 78.3 95.5 | 74.8 129.9 89.5 | 72.8 120.9 89.4 | 19.9 | 17.5 | 15.4 | 1095 / 3874 |
| POINT ARGUELLO | 38.21N | 120.63W | 112 | 45.5 | 47.5 | 71.1 N/A | 67.7 N/A | 65.2 N/A | N/A N/A | N/A N/A | 67.3 100.0 72.3 | N/A N/A 71.3 | 41.6 | 34.8 | 31.5 | 3397 / 21 |
| PT MUGU (NAWS) | 34.12N | 119.12W | 13 | 38.8 | 41.2 | 81.7 60.1 | 78.5 61.8 | 75.2 62.9 | 69.4 75.0 | 67.8 73.5 | 67.3 100.5 72.3 | 66.1 94.1 71.3 | 23.1 | 18.9 | 16.4 | 1975 / 223 |
| PORTERVILLE MUNI | 36.03N | 119.05W | 443 | 30.2 | 33.5 | 100.4 70.1 | 99.2 69.3 | 97.0 68.1 | 72.9 96.8 | 71.1 94.2 | 63.8 90.1 86.1 | 62.7 86.5 84.9 | 12.9 | 11.4 | 10.4 | 2551 / 1671 |
| REDDING MUNICIPAL | 40.52N | 122.31W | 502 | 28.3 | 30.8 | 105.9 69.3 | 102.2 67.6 | 99.3 66.2 | 71.7 97.2 | 70.2 95.2 | 63.7 90.0 79.9 | 61.6 83.4 78.7 | 24.9 | 19.9 | 17.0 | 2724 / 1888 |
| RIVERSIDE MUNI | 33.95N | 117.44W | 830 | 36.1 | 37.3 | 100.0 69.5 | 98.5 68.9 | 94.7 67.8 | 73.2 93.8 | 71.5 91.5 | 66.0 98.9 81.5 | 64.0 92.1 79.0 | 20.0 | 16.7 | 14.0 | 1567 / 1606 |
| SACRAMENTO/EXECUTIV | 38.51N | 121.49W | 26 | 31.1 | 33.9 | 100.1 69.9 | 97.2 68.7 | 93.6 67.5 | 72.6 95.7 | 70.6 93.2 | 63.8 88.9 84.4 | 61.6 82.0 79.8 | 20.3 | 18.1 | 16.2 | 2495 / 1213 |
| SACRAMENTO MATHER FL | 38.55N | 121.29W | 95 | 29.7 | 32.0 | 101.1 69.8 | 98.7 67.3 | 95.0 66.4 | 70.9 96.9 | 69.2 93.8 | 61.3 81.4 75.3 | 59.9 77.3 75.4 | 20.6 | 17.2 | 14.1 | 2687 / 1215 |
| SACRAMENTO INTL | 38.70N | 121.59W | 33 | 30.4 | 33.7 | 100.5 70.5 | 98.2 69.6 | 94.8 68.4 | 73.2 96.6 | 71.7 94.2 | 64.2 89.8 85.1 | 62.6 85.1 82.8 | 23.3 | 19.3 | 17.3 | 2425 / 1390 |
| SALINAS MUNI | 36.66N | 121.61W | 79 | 33.9 | 36.3 | 82.8 62.1 | 78.7 61.1 | 74.9 60.6 | 64.9 77.0 | 63.5 74.7 | 60.5 79.1 67.3 | 59.1 75.1 66.2 | 20.8 | 18.7 | 17.3 | 2741 / 105 |
| SAN DIEGO/LINDBERGH | 32.74N | 117.17W | 30 | 44.8 | 46.4 | 83.1 65.0 | 80.2 65.4 | 77.8 65.5 | 71.0 77.6 | 69.7 75.6 | 68.5 104.7 67.4 | 67.4 100.7 73.5 | 17.1 | 15.5 | 13.3 | 1197 / 673 |
| MIRAMAR MCAS | 32.87N | 117.15W | 479 | 38.9 | 41.4 | 90.6 66.5 | 87.6 66.3 | 83.9 66.1 | 71.6 83.2 | 70.4 81.6 | 67.9 104.7 76.4 | 66.2 98.4 75.1 | 15.1 | 12.5 | 11.0 | 1495 / 844 |
| NORTH ISLAND NAS | 32.70N | 117.20W | 26 | 44.6 | 46.0 | 83.4 64.2 | 80.7 65.1 | 78.0 65.1 | 70.9 77.1 | 69.5 76.0 | 68.5 104.9 74.0 | 66.9 99.1 72.6 | 18.8 | 16.5 | 14.5 | 1186 / 684 |
| BROWN FLD MUNI | 32.57N | 116.98W | 525 | 38.9 | 42.5 | 89.7 64.7 | 84.3 64.7 | 81.7 64.4 | 70.1 81.7 | 69.5 79.2 | 67.3 102.3 75.3 | 65.9 97.4 74.0 | 13.1 | 12.1 | 11.1 | 1681 / 653 |
| MONTGOMERY FLD | 32.82N | 117.14W | 423 | 40.7 | 43.1 | 90.2 65.9 | 86.2 65.2 | 82.3 64.6 | 71.2 82.5 | 69.9 79.9 | 66.5 99.0 75.3 | 65.8 96.8 74.8 | 15.8 | 13.0 | 12.0 | 1529 / 822 |
| SAN FRANCISCO INTL | 37.62N | 122.40W | 20 | 39.1 | 41.4 | 82.8 62.9 | 78.1 61.9 | 74.2 61.2 | 65.5 77.6 | 64.0 74.6 | 61.2 80.8 68.2 | 59.8 76.9 66.8 | 28.6 | 25.7 | 23.6 | 2689 / 144 |
| NORMAN Y MINETA SAN | 37.36N | 121.93W | 49 | 35.8 | 37.7 | 91.6 65.6 | 88.2 65.5 | 84.2 64.7 | 69.4 86.1 | 67.7 83.5 | 62.8 85.8 75.9 | 61.3 81.1 74.3 | 19.5 | 17.9 | 16.2 | 2076 / 663 |
| SAN LUIS CO RGNL | 35.24N | 120.64W | 207 | 34.1 | 36.4 | 89.5 64.0 | 85.6 63.2 | 81.3 62.7 | 67.3 83.1 | 65.8 80.4 | 61.5 82.2 71.0 | 60.5 79.4 70.0 | 25.3 | 22.7 | 19.7 | 2213 / 295 |
| SANTA BARBARA MUNI | 34.43N | 119.84W | 20 | 34.5 | 36.7 | 82.3 63.5 | 79.4 63.5 | 76.8 63.0 | 68.4 76.9 | 67.7 75.2 | 64.7 91.7 71.3 | 63.6 88.2 69.9 | 19.1 | 16.6 | 13.5 | 2246 / 196 |
| SANTA MARIA PUBLIC | 34.92N | 120.47W | 240 | 32.6 | 35.1 | 83.8 61.9 | 79.8 61.0 | 76.2 60.5 | 65.9 77.8 | 64.3 75.4 | 61.4 81.9 68.7 | 59.9 77.8 67.5 | 20.4 | 18.3 | 17.0 | 2760 / 94 |
| C M SCHULZ SONOMA CO | 38.51N | 122.81W | 148 | 29.6 | 32.1 | 95.3 66.6 | 91.1 65.9 | 87.8 64.8 | 69.2 90.4 | 67.5 87.8 | 61.0 80.5 76.6 | 59.0 75.0 74.0 | 17.1 | 15.0 | 12.7 | 3047 / 375 |
| STOCKTON/METROPOLIT | 37.89N | 121.24W | 26 | 30.5 | 33.0 | 100.8 69.9 | 97.9 68.9 | 94.8 68.2 | 73.3 95.8 | 70.9 94.1 | 65.1 92.8 85.8 | 61.7 82.4 80.4 | 22.8 | 19.3 | 17.4 | 2448 / 1382 |
| FAIRFIELD/TRAVIS AF | 38.27N | 121.95W | 72 | 30.0 | 33.5 | 99.3 67.2 | 95.2 66.1 | 91.1 65.4 | 69.7 93.7 | 68.2 90.8 | 61.1 80.7 73.4 | 59.3 75.5 72.5 | 29.8 | 27.5 | 25.7 | 2608 / 965 |
| VISALIA MUNI | 36.32N | 119.38W | 295 | 29.9 | 32.8 | 99.9 71.8 | 98.6 71.1 | 69.5 | 74.8 95.2 | 73.0 93.1 | 68.1 104.5 85.5 | 65.7 95.9 84.3 | 15.3 | 12.5 | 10.9 | 2551 / 1591 |

*10 sites, 19 more on CD-ROM*

### Colorado

| Station | Lat | Long | Elev | 99.6% | 99% | | | | | | | | | | | |
|---|---|---|---|---|---|---|---|---|---|---|---|---|---|---|---|---|
| BUCKLEY AFB | 39.70N | 104.75W | 5663 | 2.6 | 8.8 | 93.4 58.6 | 91.1 58.6 | 88.6 58.3 | 64.4 78.5 | 63.1 78.4 | 61.2 100.0 65.6 | 59.1 92.7 65.8 | 24.0 | 19.9 | 17.2 | 5734 / 787 |
| COLORADO SPRINGS/MU | 38.81N | 104.71W | 6171 | 1.3 | 7.1 | 90.4 58.7 | 88.0 58.4 | 85.1 58.3 | 63.4 78.5 | 62.2 77.3 | 59.4 95.5 65.6 | 58.2 90.5 65.3 | 28.0 | 24.7 | 20.9 | 6160 / 459 |
| DENVER INTERNATIONA | 39.83N | 104.66W | 5430 | 0.5 | 6.6 | 94.4 60.0 | 91.7 59.8 | 89.3 59.6 | 65.0 80.9 | 63.6 80.5 | 61.0 98.1 68.1 | 58.2 92.2 67.8 | 26.9 | 23.5 | 19.8 | 5959 / 777 |
| DENVER/STAPLETON | 39.75N | 104.87W | 5289 | -1.4 | 5.1 | 93.9 60.7 | 91.2 60.0 | 88.5 59.9 | 64.5 81.8 | 63.4 80.7 | 60.1 94.7 67.2 | 58.5 89.1 67.0 | 24.3 | 19.7 | 17.2 | 5667 / 721 |
| CENTENNIAL | 39.57N | 104.85W | 5827 | 0.0 | 6.2 | 91.4 59.9 | 89.8 59.9 | 86.5 59.4 | 65.1 80.9 | 63.9 79.6 | 61.0 99.7 68.5 | 58.7 91.8 68.0 | 21.4 | 21.4 | 18.6 | 6103 / 583 |
| FORT COLLINS (AWOS) | 40.45N | 105.00W | 5016 | 0.0 | 5.6 | 93.6 60.5 | 90.8 60.5 | 88.2 60.0 | 65.1 82.1 | 63.9 81.8 | 60.7 95.9 69.6 | 57.5 85.0 69.5 | 25.9 | 21.4 | 18.0 | 6235 / 618 |
| FORT COLLINS(SAWRS) | 40.58N | 105.08W | 5003 | -2.6 | 4.8 | 90.1 60.9 | 87.2 60.3 | 84.4 60.1 | 64.6 80.7 | 63.5 80.0 | 59.7 93.2 69.7 | 58.4 88.1 69.2 | 20.0 | 16.8 | 13.6 | 6096 / 462 |
| GRAND JUNCTION/WALK | 39.13N | 108.54W | 4839 | 5.1 | 10.2 | 97.7 61.5 | 95.1 60.4 | 92.6 59.8 | 65.1 85.7 | 63.9 84.3 | 59.7 95.0 68.2 | 58.3 87.0 68.7 | 19.2 | 16.9 | 13.8 | 5430 / 1212 |
| GREELEY WELD CO | 40.44N | 104.62W | 4649 | -6.3 | 0.4 | 95.3 62.3 | 91.5 61.9 | 89.8 62.1 | 65.1 85.7 | 65.8 84.3 | 62.7 101.3 72.9 | 60.8 94.9 71.9 | 28.1 | 23.8 | 19.2 | 6579 / 611 |
| PUEBLO MEMORIAL(AW) | 38.29N | 104.50W | 4721 | -0.4 | 6.5 | 98.5 62.4 | 95.8 62.2 | 92.9 61.8 | 66.9 85.2 | 65.9 84.4 | 62.9 102.4 69.2 | 61.3 96.6 69.1 | 28.7 | 24.8 | 19.7 | 5473 / 915 |

### Connecticut

*5 sites, 3 more on CD-ROM*

| BRIDGEPORT/IGOR I. | 41.18N | 73.15W | 16 | 11.4 | 15.8 | 87.7 73.1 | 84.4 71.5 | 82.1 70.3 | 75.9 82.9 | 74.7 80.8 | 73.5 124.7 78.1 | 72.6 121.0 78.1 | 24.5 | 20.9 | 18.7 | 5274 / 830 |
| HARTFORD/BRADLEY IN | 41.94N | 72.68W | 180 | 4.1 | 9.2 | 91.4 73.3 | 88.5 72.0 | 85.5 70.5 | 76.3 86.7 | 74.8 83.9 | 73.2 124.3 80.3 | 72.1 119.4 79.2 | 22.8 | 19.3 | 17.5 | 5935 / 765 |
| HARTFORD BRAINARD | 41.74N | 72.65W | 20 | 8.5 | 12.2 | 90.7 73.2 | 88.2 72.4 | 84.6 70.9 | 76.9 85.7 | 75.3 83.2 | 74.2 127.9 81.1 | 72.7 121.4 79.8 | 19.7 | 18.1 | 16.5 | 5510 / 862 |

**Table 4-8 Design Conditions for Selected Locations** *(Continued)*

Meaning of acronyms:
DB: Dry bulb temperature, °F
MCWB: Mean coincident wet bulb temperature, °F
WB: Wet bulb temperature, °F
DP: Dew point temperature, °F
MCDB: Mean coincident dry bulb temperature, °F
Lat: Latitude, °
Long: Longitude, °
HR: Humidity ratio, grains of moisture per lb of dry air
HDD and CDD 65: Annual heating and cooling degree-days, base 65°F, °F-day
Elev: Elevation, ft
WS: Wind speed, mph

| Station | Lat | Long | Elev | Heating DB 99.6% | Heating DB 99% | Cooling DB/MCWB 0.4% DB/MCWB | Cooling DB/MCWB 1% DB/MCWB | Cooling DB/MCWB 2% DB/MCWB | Evaporation WB/MCDB 0.4% WB/MCDB | Evaporation WB/MCDB 1% WB/MCDB | Evaporation WB/MCDB 2% WB/MCDB | Dehumidification DP/HR/MCDB 0.4% DP/HR/MCDB | Dehumidification DP/HR/MCDB 1% DP/HR/MCDB | Extreme Annual WS 1% | Extreme Annual WS 2.5% | Extreme Annual WS 5% | Heat./Cool. Degree-Days HDD/CDD 65 |
|---|---|---|---|---|---|---|---|---|---|---|---|---|---|---|---|---|---|
| WATERBURY OXFORD | 41.48N | 73.13W | 725 | 4.1 | 9.1 | 87.7/72.7 | 83.8/71.2 | 81.4/69.6 | 75.4/83.4 | 73.7/80.5 | — | 72.9/125.3/79.0 | 72.0/121.6/77.8 | 19.6 | 17.1 | 15.0 | 6360/475 |
| WINDHAM AIRPORT | 41.74N | 72.18W | 246 | 3.5 | 9.4 | 89.9/73.1 | 86.4/72.0 | 83.7/70.8 | 76.0/84.9 | 74.5/82.1 | — | 73.1/124.1/79.4 | 72.4/120.9/78.4 | 19.4 | 17.2 | 15.5 | 5998/617 |

**Delaware** *2 sites, 1 more on CD-ROM*

| DOVER AFB | 39.12N | 75.47W | 30 | 15.5 | 18.4 | 91.0/75.6 | 89.6/75.5 | 86.3/73.7 | 78.4/86.5 | 77.2/84.4 | — | 76.6/138.8/81.7 | 75.0/131.5/80.6 | 25.2 | 22.1 | 19.4 | 4503/1170 |
| WILMINGTON NEW CAST | 39.67N | 75.60W | 79 | 13.3 | 17.3 | 91.9/75.9 | 89.4/73.9 | 86.9/73.1 | 78.0/86.2 | 76.7/85.0 | — | 75.4/133.3/81.7 | 74.3/128.3/80.5 | 24.5 | 20.6 | 18.4 | 4756/1142 |

**Florida** *32 sites, 28 more on CD-ROM*

| CECIL FLD | 30.22N | 81.87W | 82 | 27.6 | 31.8 | 96.0/76.4 | 94.0/76.2 | 92.3/75.9 | 80.1/89.2 | 79.0/88.5 | — | 77.4/143.1/82.9 | 76.8/140.2/82.5 | 18.7 | 16.6 | 14.6 | 1286/2711 |
| DAYTONA BEACH INTL | 29.18N | 81.06W | 43 | 35.6 | 39.2 | 92.8/76.9 | 90.9/76.8 | 89.6/76.8 | 80.1/87.1 | 79.1/87.1 | — | 77.7/144.2/83.7 | 77.0/140.6/83.1 | 20.3 | 18.0 | 16.3 | 748/2992 |
| FORT LAUDERDALE HOL | 26.07N | 80.15W | 10 | 46.7 | 51.6 | 91.7/78.2 | 90.6/78.2 | 90.0/78.1 | 81.1/87.9 | 80.4/87.2 | — | 79.3/152.0/84.7 | 78.6/148.6/84.4 | 22.2 | 19.7 | 18.2 | 133/4566 |
| FORT MYERS/PAGE FLD | 26.59N | 81.86W | 20 | 42.5 | 46.0 | 93.5/76.7 | 92.6/76.4 | 91.9/76.5 | 80.3/88.1 | 79.5/87.5 | — | 78.4/147.3/83.3 | 77.4/142.6/82.7 | 18.1 | 17.1 | 15.4 | 281/3923 |
| GAINESVILLE RGNL | 29.69N | 82.27W | 164 | 29.6 | 33.4 | 93.4/76.4 | 91.9/76.2 | 90.5/75.9 | 79.7/88.3 | 78.7/87.1 | — | 77.4/143.4/83.1 | 76.5/139.1/82.2 | 18.4 | 16.5 | 14.4 | 1176/2629 |
| HOMESTEAD ARB | 25.48N | 80.38W | 7 | 45.9 | 49.9 | 90.4/78.7 | 90.0/78.7 | 88.4/78.5 | 81.4/85.6 | 80.9/85.3 | — | 80.8/160.2/83.2 | 79.4/152.4/82.8 | 20.3 | 18.3 | 16.5 | 151/4050 |
| JACKSONVILLE/INTNL | 30.49N | 81.69W | 33 | 29.4 | 32.5 | 94.6/77.3 | 92.8/77.0 | 91.4/76.6 | 80.0/89.6 | 79.1/88.4 | — | 77.4/142.9/83.4 | 76.7/139.4/82.8 | 20.0 | 17.9 | 16.2 | 1327/2632 |
| JACKSONVILLE NAS | 30.23N | 81.68W | 23 | 33.5 | 36.8 | 95.7/77.1 | 93.4/76.3 | 91.5/76.0 | 80.5/88.4 | 79.5/87.5 | — | 78.9/150.0/83.7 | 77.3/142.2/83.1 | 20.6 | 18.2 | 16.4 | 995/3175 |
| JACKSONVILLE/CRAIG | 30.34N | 81.52W | 43 | 32.3 | 35.9 | 93.5/77.2 | 91.4/76.9 | 90.2/76.8 | 80.2/88.7 | 79.3/87.6 | — | 78.1/146.3/83.8 | 77.1/141.3/83.1 | 18.9 | 17.9 | 15.8 | 1217/2645 |
| MACDILL AFB/TAMPA | 27.85N | 82.52W | 13 | 39.1 | 43.0 | 92.6/77.6 | 91.1/77.5 | 90.4/77.6 | 81.5/86.8 | 80.9/86.5 | — | 80.7/159.6/84.2 | 79.2/151.8/83.5 | 19.5 | 17.4 | 15.8 | 536/3506 |
| MAYPORT NS | 30.40N | 81.42W | 13 | 34.2 | 38.1 | 94.4/77.4 | 91.9/77.4 | 90.3/77.1 | 80.7/88.8 | 79.8/88.1 | — | 78.8/149.4/85.1 | 77.4/142.5/84.5 | 20.4 | 18.1 | 16.2 | 1044/2990 |
| MELBOURNE REGIONAL | 28.10N | 80.65W | 26 | 38.6 | 43.1 | 92.2/77.8 | 90.6/77.8 | 89.7/77.5 | 80.8/87.2 | 80.0/87.2 | — | 79.1/151.2/84.3 | 77.9/143.5/83.9 | 21.0 | 19.1 | 18.0 | 467/3496 |
| MIAMI | 25.82N | 80.30W | 30 | 47.6 | 51.9 | 91.8/77.6 | 90.8/77.6 | 90.0/77.5 | 80.3/88.5 | 79.7/87.9 | — | 78.5/148.1/83.5 | 77.6/145.5/83.3 | 20.4 | 18.6 | 17.0 | 126/4537 |
| KENDALL TAMIAMI EXEC | 25.65N | 80.43W | 10 | 45.2 | 48.5 | 92.7/77.0 | 91.2/77.7 | 90.4/77.6 | 80.4/87.9 | 79.7/87.9 | — | 79.0/150.6/83.3 | 77.5/143.0/82.9 | 20.7 | 18.9 | 17.7 | 176/4110 |
| NAPLES MUNI | 26.15N | 81.78W | 23 | 43.3 | 46.6 | 91.8/77.6 | 90.5/77.4 | 89.9/77.3 | 80.7/87.4 | 80.0/86.6 | — | 79.2/151.4/84.0 | 78.4/147.0/83.5 | 18.9 | 17.0 | 15.1 | 290/3747 |
| NASA SHUTTLE LANDING | 28.62N | 80.69W | 10 | 37.5 | 42.5 | 91.7/78.1 | 90.4/77.7 | 89.6/77.9 | 81.0/87.5 | 80.1/87.5 | — | 79.3/152.0/84.0 | 78.7/148.8/83.6 | 18.8 | 16.7 | 14.8 | 565/3151 |
| OCALA INTL J TAYLOR | 29.17N | 82.22W | 89 | 28.8 | 33.7 | 93.3/75.5 | 91.5/75.5 | 90.7/75.3 | 79.1/87.8 | 78.2/87.3 | — | 77.0/140.8/82.7 | 75.3/133.2/82.0 | 18.8 | 15.3 | 12.6 | 1052/2772 |
| EXECUTIVE | 28.55N | 81.33W | 112 | 38.7 | 43.2 | 93.5/75.9 | 92.6/75.9 | 91.0/75.7 | 79.5/86.8 | 78.7/86.1 | — | 78.1/146.4/82.2 | 77.0/141.1/81.6 | 19.3 | 17.7 | 15.9 | 512/3560 |
| ORLANDO/JETPORT | 28.43N | 81.33W | 105 | 37.8 | 42.3 | 93.8/76.5 | 92.5/76.2 | 91.1/76.0 | 79.6/87.5 | 78.8/86.7 | — | 77.6/144.2/81.8 | 76.9/140.9/81.4 | 20.2 | 18.1 | 16.4 | 550/3386 |
| ORLANDO SANFORD | 28.78N | 81.24W | 56 | 36.7 | 40.9 | 94.8/75.6 | 93.0/75.2 | 91.2/75.2 | 79.7/88.1 | 78.5/88.1 | — | 75.5/133.5/83.7 | 75.2/132.6/81.8 | 20.4 | 18.1 | 15.0 | 646/3314 |
| PANAMA CITY BAY CO | 30.21N | 85.68W | 20 | 31.8 | 35.9 | 92.7/76.1 | 91.0/76.9 | 86.9/76.9 | 80.8/88.0 | 80.3/87.9 | — | 79.5/153.1/83.7 | 79.0/150.7/83.6 | 18.7 | 16.7 | 15.0 | 1238/2842 |
| PENSACOLA NAS | 30.35N | 87.32W | 30 | 29.5 | 33.2 | 93.2/78.6 | 91.4/77.4 | 90.2/78.0 | 81.9/88.5 | 80.8/87.7 | — | 80.2/156.7/85.4 | 79.1/150.9/84.8 | 20.9 | 18.7 | 16.8 | 1459/2647 |
| PENSACOLA RGNL | 30.47N | 87.19W | 118 | 29.7 | 33.7 | 93.9/77.7 | 91.9/77.7 | 90.4/77.1 | 81.0/88.5 | 80.1/87.4 | — | 79.2/152.3/84.3 | 78.2/147.1/83.6 | 20.1 | 18.1 | 16.5 | 1453/2687 |
| SARASOTA BRADENTON | 27.40N | 82.56W | 26 | 39.2 | 44.0 | 92.3/78.8 | 91.1/78.7 | 90.3/78.6 | 82.6/88.8 | 81.5/87.9 | — | 81.2/162.3/86.5 | 79.6/153.5/85.2 | 21.0 | 18.6 | 16.9 | 462/3445 |
| SOUTHWEST FLORIDA I | 26.54N | 81.76W | 30 | 40.5 | 45.0 | 93.5/77.8 | 92.6/76.8 | 91.1/76.8 | 81.1/88.0 | 80.4/87.8 | — | 79.0/150.5/82.9 | 77.6/143.8/82.6 | 20.6 | 18.4 | 16.5 | 323/3764 |
| ST PETERSBURG CLEAR | 27.91N | 82.69W | 10 | 42.4 | 45.4 | 92.1/77.8 | 91.0/77.7 | 90.3/77.3 | 81.7/87.3 | 80.6/86.5 | — | 80.4/157.8/84.6 | 79.2/151.4/83.8 | 20.8 | 18.7 | 17.2 | 456/3677 |
| TALLAHASSEE MUNICIP | 30.39N | 84.35W | 69 | 25.7 | 29.0 | 96.0/76.5 | 93.8/75.9 | 91.5/75.5 | 79.8/86.9 | 78.8/88.0 | — | 77.4/143.0/82.8 | 76.5/138.7/82.1 | 18.0 | 16.0 | 13.6 | 1553/2599 |
| TAMPA INTL AIRPORT | 27.96N | 82.54W | 10 | 38.8 | 42.9 | 92.6/77.2 | 91.4/77.0 | 90.4/77.1 | 80.5/88.0 | 79.9/87.7 | — | 78.4/147.7/84.8 | 77.5/142.9/84.1 | 18.0 | 16.0 | 13.6 | 527/3563 |
| TYNDALL AFB | 30.07N | 85.58W | 16 | 31.5 | 35.6 | 91.3/78.8 | 90.2/78.8 | 88.9/78.6 | 82.5/87.7 | 81.4/86.9 | — | 81.2/162.1/86.1 | 79.8/154.9/85.0 | 19.5 | 17.3 | 15.4 | 1309/2620 |
| VENICE PIER | 27.07N | 82.45W | 16 | 41.4 | 45.4 | 88.1/76.3 | 86.8/76.8 | 86.1/77.0 | 82.9/79.7 | 82.4/79.7 | — | 81.3/162.8/84.1 | 78.4/147.4/82.2 | 23.6 | 19.6 | 16.4 | 502/2966 |
| VERO BEACH MUNI | 27.66N | 80.42W | 30 | 38.7 | 43.0 | 91.5/77.8 | 90.5/77.8 | 89.6/77.8 | 81.6/88.0 | 80.4/87.9 | — | 78.6/148.6/84.3 | 77.3/142.4/83.8 | 20.3 | 18.5 | 17.1 | 420/3464 |
| WEST PALM BEACH/IN | 26.69N | 80.10W | 20 | 43.9 | 48.0 | 91.4/77.6 | 90.4/77.6 | 89.4/77.7 | 80.2/87.7 | 79.5/87.1 | — | 78.1/146.1/83.6 | 77.3/142.3/83.4 | 23.1 | 20.2 | 18.6 | 222/4085 |

**Georgia** *19 sites, 8 more on CD-ROM*

| ALBANY MUNICIPAL | 31.54N | 84.19W | 194 | 26.9 | 29.8 | 97.0/76.1 | 94.8/76.1 | 92.8/75.4 | 79.8/90.6 | 78.7/89.2 | — | 77.2/142.5/83.4 | 75.9/136.3/82.3 | 18.6 | 16.7 | 14.6 | 1764/2551 |
| ATHENS MUNICIPAL | 33.95N | 82.56W | 801 | 22.4 | 26.5 | 94.9/74.0 | 93.0/74.1 | 90.7/73.7 | 77.8/89.2 | 76.8/87.5 | — | 74.8/134.1/82.4 | 73.7/129.3/81.6 | 18.3 | 16.1 | 14.1 | 2781/1804 |
| PEACHTREE CITY FALCO | 33.36N | 84.57W | 797 | 19.1 | 23.1 | 93.1/73.5 | 91.0/73.4 | 89.9/73.3 | 77.4/87.6 | 76.4/86.1 | — | 74.7/133.8/80.9 | 73.3/127.3/79.3 | 17.4 | 15.0 | 12.4 | 3054/1540 |
| ATLANTA MUNICIPAL | 33.64N | 84.43W | 1027 | 21.5 | 26.4 | 93.9/74.2 | 91.7/73.9 | 89.8/73.5 | 77.3/88.5 | 76.4/86.7 | — | 74.3/133.1/81.3 | 73.3/128.7/80.2 | 21.5 | 19.0 | 17.1 | 2671/1893 |
| AUGUSTA/BUSH FIELD | 33.37N | 81.97W | 148 | 22.5 | 26.1 | 97.3/76.0 | 94.8/75.9 | 91.0/78.4 | 79.5/91.0 | 78.4/89.3 | — | 76.7/139.8/82.6 | 75.6/134.5/82.6 | 18.8 | 16.6 | 14.3 | 2407/2078 |
| DANIEL FIELD | 33.47N | 82.04W | 423 | 27.2 | 29.7 | 97.1/74.4 | 94.4/74.5 | 91.3/73.4 | 77.6/89.4 | 77.0/88.2 | — | 74.8/132.5/81.0 | 73.4/126.0/80.0 | 16.7 | 14.7 | 12.6 | 2135/2316 |
| COLUMBUS METROPOLIT | 32.52N | 84.94W | 394 | 25.9 | 29.2 | 96.6/74.7 | 94.2/74.5 | 92.3/74.1 | 78.1/89.4 | 77.0/88.4 | — | 75.3/134.5/81.9 | 74.4/130.5/81.1 | 18.3 | 16.3 | 14.3 | 2083/2339 |
| DEKALB PEACHTREE | 33.88N | 84.30W | 991 | 21.0 | 25.4 | 93.5/73.4 | 91.4/73.5 | 90.3/73.5 | 77.0/88.4 | 76.1/86.1 | — | 73.4/128.8/82.1 | 73.0/127.2/79.1 | 18.5 | 16.4 | 14.0 | 2871/1827 |
| MARIETTA/DOBBINS AF | 33.92N | 84.52W | 1070 | 18.9 | 24.5 | 93.6/74.2 | 91.3/74.2 | 89.5/73.8 | 77.3/88.0 | 76.4/86.6 | — | 74.4/133.6/81.7 | 73.3/128.7/80.6 | 18.7 | 16.4 | 14.1 | 2970/1758 |
| FORT BENNING | 32.33N | 85.00W | 233 | 22.9 | 26.6 | 97.1/76.0 | 94.6/76.1 | 91.5/75.8 | 81.0/89.7 | 79.4/89.2 | — | 78.9/153.8/84.8 | 77.0/142.0/83.3 | 17.1 | 14.6 | 12.4 | 2251/2131 |
| FULTON CO ARPT BROW | 33.78N | 84.52W | 840 | 20.8 | 25.2 | 94.0/74.6 | 91.5/74.3 | 90.2/73.8 | 77.7/88.0 | 76.7/87.4 | — | 74.5/133.2/82.1 | 73.4/128.0/81.1 | 17.7 | 15.5 | 13.4 | 2869/1742 |
| LEE GILMER MEM | 34.27N | 83.83W | 1276 | 21.0 | 26.6 | 92.5/73.4 | 90.5/73.2 | 88.4/72.6 | 76.6/86.8 | 75.5/85.0 | — | 73.4/130.1/79.6 | 72.8/127.5/79.1 | 19.0 | 17.0 | 15.3 | 3019/1642 |
| HUNTER AAF | 32.01N | 81.15W | 43 | 27.9 | 31.7 | 95.5/77.5 | 93.4/77.1 | 91.3/76.9 | 81.2/88.8 | 80.1/88.3 | — | 79.3/152.3/83.9 | 77.8/144.7/83.2 | 18.9 | 16.7 | 14.5 | 1632/2582 |
| MACON/LEWIS B WILSO | 32.69N | 83.65W | 361 | 23.9 | 27.4 | 96.9/75.6 | 94.7/75.6 | 92.4/74.9 | 79.0/90.2 | 78.1/88.9 | — | 76.1/138.3/83.0 | 75.2/134.0/82.1 | 18.1 | 16.0 | 13.4 | 2263/2179 |
| MOODY AFB/VALDOSTA | 30.97N | 83.19W | 236 | 29.4 | 33.1 | 96.0/76.7 | 94.0/76.7 | 92.5/76.3 | 80.2/90.7 | 79.1/89.5 | — | 77.3/143.5/83.9 | 76.6/139.8/83.3 | 17.1 | 14.4 | 12.4 | 1438/2683 |
| ROME/RUSSELL(RAMOS) | 34.35N | 85.16W | 643 | 18.8 | 22.9 | 93.4/74.0 | 91.2/74.0 | 91.2/74.0 | 77.2/89.9 | 77.2/89.9 | — | 75.0/134.4/82.6 | 73.5/127.6/81.6 | 15.8 | 12.9 | 11.4 | 3111/1762 |
| SAVANNAH MUNICIPAL | 32.12N | 81.20W | 52 | 27.4 | 30.4 | 95.5/77.2 | 93.3/76.9 | 91.4/76.3 | 80.2/89.5 | 79.3/88.3 | — | 78.1/146.1/83.5 | 77.1/141.3/82.7 | 18.9 | 16.9 | 15.5 | 1761/2455 |

124  Principles of HVAC

# Table 4-8 Design Conditions for Selected Locations *(Continued)*

*Meaning of acronyms:*
*DB: Dry bulb temperature, °F* · *WB: Wet bulb temperature, °F* · *MCWB: Mean coincident wet bulb temperature, °F*
*Lat: Latitude, °* · *DP: Dew point temperature, °F* · *MCDB: Mean coincident dry bulb temperature, °F*
*Long: Longitude, °* · *HR: Humidity ratio, grains of moisture per lb of dry air* · *HDD and CDD 65: Annual heating and cooling degree-days, base 65°F, °F-day*
*Elev: Elevation, ft* · *WS: Wind speed, mph*

| Station | Lat | Long | Elev | Heating DB | | Cooling DB/MCWB | | | | | | Evaporation WB/MCDB | | | | Dehumidification DP/HR/MCDB | | | | | | Extreme Annual WS | | | Heat./Cool. Degree-Days | |
|---|---|---|---|---|---|---|---|---|---|---|---|---|---|---|---|---|---|---|---|---|---|---|---|---|---|---|
| | | | | | | 0.4% | | 1% | | 2% | | 0.4% | | 1% | | 0.4% | | | 1% | | | | | | | |
| | | | | 99.6% | 99% | DB | MCWB | DB | MCWB | DB | MCWB | WB | MCDB | WB | MCDB | DP | HR | MCDB | DP | HR | MCDB | 1% | 2.5% | 5% | HDD | CDD 65 |
| VALDOSTA RGNL | 30.78N | 83.28W | 197 | 27.6 | 30.6 | 95.6 | 77.3 | 93.5 | 76.5 | 92.1 | 76.1 | 80.4 | 89.9 | 79.4 | 88.8 | 78.5 | 149.0 | 83.6 | 77.1 | 142.2 | 82.6 | 16.9 | 14.7 | 12.8 | 1527 | 2559 |
| ROBINS AFB | 32.63N | 83.60W | 295 | 25.0 | 27.9 | 96.9 | 75.5 | 94.6 | 75.4 | 91.4 | 74.8 | 79.4 | 90.4 | 78.4 | 88.7 | 77.0 | 142.0 | 83.1 | 75.4 | 134.3 | 81.4 | 18.4 | 16.0 | 13.0 | 2130 | 2231 |
| ***Hawaii*** | | | | | | | | | | | | | | | | | | | | | | | | | | *4 sites, 4 more on CD-ROM* |
| KALAELOA ARPT | 21.30N | 158.07W | 33 | 59.5 | 61.8 | 90.9 | 73.2 | 89.9 | 73.2 | 88.9 | 73.1 | 78.0 | 85.8 | 76.8 | 85.3 | 75.4 | 133.4 | 82.9 | 74.1 | 127.2 | 82.3 | 19.4 | 17.7 | 16.2 | 0 | 4450 |
| HILO INTL | 19.72N | 155.05W | 36 | 61.5 | 62.8 | 85.7 | 74.1 | 84.7 | 73.8 | 83.9 | 73.6 | 76.6 | 82.1 | 75.9 | 81.5 | 75.1 | 131.7 | 79.2 | 74.1 | 127.5 | 78.6 | 17.4 | 15.7 | 13.3 | 0 | 3264 |
| HONOLULU INTL | 21.33N | 157.94W | 16 | 62.0 | 63.9 | 89.8 | 74.0 | 88.9 | 73.6 | 88.1 | 73.3 | 77.2 | 84.8 | 76.3 | 84.1 | 75.0 | 131.2 | 81.2 | 73.8 | 126.0 | 80.6 | 22.2 | 20.2 | 18.8 | 0 | 4679 |
| KANEOHE BAY (MCAF) | 21.45N | 157.77W | 20 | 64.0 | 65.9 | 84.9 | 74.4 | 84.1 | 74.1 | 83.3 | 73.8 | 77.1 | 81.9 | 76.2 | 81.5 | 75.3 | 132.6 | 80.2 | 74.4 | 128.8 | 79.9 | 18.8 | 17.0 | 15.8 | 0 | 4243 |
| ***Idaho*** | | | | | | | | | | | | | | | | | | | | | | | | | | *7 sites, 10 more on CD-ROM* |
| BOISE MUNICIPAL | 43.57N | 116.22W | 2867 | 8.7 | 15.5 | 98.6 | 63.9 | 95.4 | 62.9 | 92.5 | 61.9 | 66.2 | 92.3 | 64.7 | 90.5 | 57.2 | 77.5 | 71.6 | 54.9 | 71.3 | 71.4 | 21.9 | 19.0 | 17.1 | 5453 | 957 |
| CALDWELL (AWOS) | 43.64N | 116.63W | 2431 | 11.5 | 16.3 | 97.0 | 66.4 | 93.1 | 64.7 | 90.5 | 63.8 | 68.2 | 92.3 | 66.5 | 89.9 | 59.3 | 82.6 | 77.8 | 56.9 | 75.4 | 77.4 | 22.1 | 19.1 | 16.9 | 5729 | 660 |
| COEUR D ALENE AIR TE | 47.77N | 116.82W | 2320 | 5.5 | 10.3 | 91.4 | 63.0 | 88.5 | 62.4 | 84.2 | 60.9 | 65.8 | 86.4 | 64.0 | 84.0 | 57.4 | 76.7 | 71.3 | 55.4 | 71.3 | 70.0 | 22.2 | 18.9 | 16.7 | 6908 | 300 |
| IDAHO FALLS RGNL | 43.52N | 112.07W | 4744 | -6.7 | -0.3 | 91.5 | 60.9 | 89.6 | 60.6 | 86.4 | 59.5 | 64.5 | 83.4 | 62.8 | 82.6 | 57.8 | 85.1 | 69.9 | 55.4 | 78.0 | 68.3 | 27.1 | 24.2 | 20.6 | 7701 | 272 |
| JOSLIN FLD MAGIC VA | 42.48N | 114.49W | 4190 | 9.0 | 12.2 | 94.7 | 63.2 | 91.2 | 62.3 | 89.7 | 61.9 | 66.4 | 88.8 | 64.9 | 86.4 | 58.8 | 86.5 | 75.5 | 56.5 | 79.6 | 74.6 | 27.9 | 24.6 | 20.9 | 6128 | 729 |
| LEWISTON NEZ PERCE | 46.38N | 117.01W | 1437 | 12.0 | 18.6 | 98.2 | 65.3 | 94.5 | 64.4 | 90.9 | 63.1 | 67.5 | 92.4 | 65.9 | 90.0 | 59.4 | 79.7 | 72.5 | 57.1 | 73.4 | 71.8 | 20.8 | 17.9 | 15.0 | 5020 | 839 |
| POCATELLO MUNICIPAL | 42.92N | 112.57W | 4478 | -2.0 | 3.8 | 94.6 | 61.6 | 91.4 | 60.9 | 88.6 | 60.0 | 65.1 | 86.8 | 63.4 | 84.8 | 58.2 | 85.5 | 71.0 | 55.4 | 77.2 | 70.7 | 28.3 | 25.3 | 22.3 | 6938 | 426 |
| ***Illinois*** | | | | | | | | | | | | | | | | | | | | | | | | | | *14 sites, 14 more on CD-ROM* |
| AURORA MUNICIPAL | 41.77N | 88.48W | 715 | -5.6 | 0.5 | 90.4 | 74.2 | 88.2 | 73.4 | 84.4 | 71.6 | 77.5 | 86.4 | 75.8 | 83.9 | 74.7 | 133.5 | 82.9 | 72.9 | 125.5 | 80.8 | 25.9 | 22.9 | 19.8 | 6508 | 701 |
| CAHOKIA/ST. LOUIS | 38.57N | 90.16W | 413 | 9.1 | 12.4 | 93.4 | 77.1 | 91.3 | 76.2 | 90.2 | 75.6 | 80.1 | 90.3 | 78.4 | 88.8 | 77.2 | 143.9 | 85.1 | 75.2 | 134.2 | 83.9 | 20.7 | 18.5 | 16.6 | 4545 | 1398 |
| CHICAGO/MIDWAY | 41.79N | 87.75W | 617 | 0.2 | 5.4 | 91.5 | 74.6 | 89.5 | 73.3 | 86.5 | 72.0 | 78.0 | 88.1 | 76.1 | 85.1 | 74.9 | 134.0 | 84.1 | 73.0 | 125.4 | 82.0 | 24.5 | 21.2 | 19.2 | 5872 | 1034 |
| CHICAGO/O'HARE ARPT | 41.99N | 87.91W | 673 | -1.5 | 3.7 | 91.4 | 74.3 | 88.7 | 73.2 | 86.0 | 71.8 | 77.8 | 87.8 | 76.0 | 84.8 | 74.7 | 133.3 | 83.7 | 73.0 | 125.8 | 81.7 | 24.6 | 21.0 | 19.1 | 6209 | 864 |
| DECATUR | 39.98N | 88.87W | 679 | 0.9 | 6.6 | 92.9 | 76.6 | 90.6 | 75.5 | 88.3 | 74.3 | 79.3 | 89.7 | 77.8 | 87.7 | 76.2 | 140.3 | 85.9 | 74.8 | 133.5 | 84.2 | 24.8 | 21.6 | 19.7 | 5442 | 1100 |
| GLENVIEW NAS | 42.08N | 87.82W | 653 | -0.7 | 4.8 | 93.7 | 75.0 | 90.2 | 73.3 | 87.1 | 72.1 | 77.9 | 90.2 | 76.2 | 87.0 | 74.2 | 130.7 | 85.1 | 72.4 | 123.1 | 83.6 | 20.2 | 18.0 | 16.2 | 6104 | 909 |
| MOLINE/QUAD CITY | 41.47N | 90.52W | 594 | -3.9 | 1.3 | 92.9 | 76.1 | 90.2 | 74.8 | 87.5 | 73.3 | 79.1 | 89.2 | 77.3 | 86.9 | 76.2 | 139.6 | 85.2 | 74.5 | 131.9 | 83.1 | 24.1 | 20.3 | 18.3 | 6074 | 994 |
| GREATER PEORIA MUNI | 40.67N | 89.68W | 663 | -1.5 | 3.3 | 92.2 | 76.2 | 89.8 | 75.1 | 87.2 | 73.6 | 79.2 | 88.5 | 77.5 | 86.6 | 76.4 | 141.4 | 85.0 | 74.8 | 133.6 | 83.0 | 23.4 | 19.9 | 18.0 | 5756 | 1040 |
| QUINCY RGNL BALDWIN | 39.94N | 91.19W | 768 | -0.2 | 4.8 | 92.7 | 76.5 | 90.1 | 75.3 | 87.7 | 74.1 | 78.6 | 89.1 | 77.4 | 87.4 | 75.5 | 137.2 | 84.8 | 74.2 | 131.3 | 83.3 | 24.5 | 20.8 | 18.9 | 5501 | 1101 |
| GREATER ROCKFORD | 42.20N | 89.09W | 745 | -5.8 | 0.0 | 91.1 | 74.6 | 88.2 | 73.2 | 85.5 | 71.7 | 78.0 | 87.4 | 76.0 | 84.4 | 75.1 | 135.6 | 83.5 | 73.2 | 126.9 | 81.7 | 24.4 | 20.9 | 19.0 | 6608 | 775 |
| SCOTT AFB MIDAMERIC | 38.53N | 89.83W | 459 | 9.0 | 12.4 | 94.8 | 76.5 | 91.4 | 75.5 | 90.1 | 75.2 | 80.3 | 88.5 | 78.7 | 87.2 | 78.6 | 151.0 | 84.6 | 76.6 | 141.3 | 83.1 | 23.1 | 19.8 | 17.7 | 4579 | 1401 |
| SPRINGFIELD/CAPITAL | 39.85N | 89.68W | 614 | 0.4 | 6.4 | 92.4 | 76.6 | 90.3 | 75.5 | 88.0 | 74.1 | 79.4 | 89.4 | 77.9 | 87.2 | 76.4 | 141.1 | 85.9 | 74.9 | 134.0 | 84.1 | 24.7 | 21.4 | 19.2 | 5360 | 1137 |
| UNIV OF ILLINOIS WI | 40.04N | 88.28W | 764 | -0.5 | 4.2 | 92.0 | 76.0 | 90.0 | 75.1 | 87.7 | 74.1 | 79.6 | 88.8 | 77.7 | 86.5 | 76.9 | 144.3 | 86.1 | 75.0 | 135.0 | 83.3 | 27.5 | 24.6 | 21.8 | 5681 | 1008 |
| DUPAGE | 41.91N | 88.25W | 758 | -2.5 | 1.6 | 90.3 | 74.9 | 87.9 | 74.0 | 84.4 | 72.2 | 78.2 | 87.0 | 76.4 | 84.3 | 75.3 | 136.3 | 84.1 | 73.4 | 127.6 | 81.4 | 24.6 | 21.2 | 19.1 | 6429 | 738 |
| ***Indiana*** | | | | | | | | | | | | | | | | | | | | | | | | | | *8 sites, 5 more on CD-ROM* |
| EVANSVILLE REGIONAL | 38.04N | 87.54W | 387 | 8.1 | 13.8 | 93.7 | 76.2 | 91.4 | 75.7 | 89.7 | 74.9 | 79.4 | 89.9 | 78.2 | 88.1 | 76.4 | 139.9 | 85.2 | 75.3 | 134.3 | 83.7 | 20.6 | 18.3 | 16.4 | 4424 | 1437 |
| FORT WAYNE/BAER FLD | 41.01N | 85.21W | 827 | -0.7 | 5.0 | 90.8 | 74.3 | 88.2 | 73.1 | 85.5 | 71.9 | 77.6 | 86.8 | 75.9 | 84.0 | 74.8 | 134.5 | 82.9 | 73.2 | 127.4 | 80.9 | 24.8 | 21.0 | 18.9 | 5991 | 825 |
| GRISSOM ARB | 40.65N | 86.15W | 810 | -0.2 | 5.4 | 91.6 | 75.3 | 88.8 | 74.6 | 86.2 | 73.0 | 79.4 | 87.9 | 77.3 | 85.3 | 76.9 | 144.5 | 84.7 | 74.9 | 134.8 | 82.3 | 25.0 | 20.9 | 18.7 | 5777 | 978 |
| INDIANAPOLIS/I.-MUN | 39.71N | 86.27W | 807 | 2.0 | 8.1 | 91.0 | 75.1 | 88.7 | 74.0 | 86.4 | 72.9 | 78.2 | 87.5 | 76.8 | 85.4 | 75.3 | 136.8 | 83.6 | 74.0 | 130.6 | 82.0 | 24.7 | 20.9 | 18.8 | 5272 | 1087 |
| PURDUE UNIV | 40.41N | 86.94W | 636 | 0.1 | 5.7 | 91.6 | 75.5 | 89.9 | 74.5 | 87.5 | 73.3 | 78.6 | 88.3 | 77.1 | 86.1 | 75.5 | 136.6 | 84.3 | 74.1 | 130.3 | 82.4 | 22.8 | 19.9 | 18.2 | 5524 | 1014 |
| MONROE CO | 39.14N | 86.62W | 846 | 3.5 | 9.7 | 90.7 | 75.5 | 89.6 | 75.4 | 87.5 | 74.1 | 78.6 | 86.8 | 77.4 | 85.9 | 75.5 | 137.7 | 83.2 | 74.7 | 134.0 | 82.3 | 19.4 | 17.3 | 15.7 | 5047 | 1015 |
| SOUTH BEND/ST.JOSEP | 41.71N | 86.33W | 774 | 0.2 | 5.8 | 90.6 | 74.0 | 88.0 | 72.5 | 85.1 | 71.3 | 77.3 | 86.4 | 75.4 | 83.7 | 74.5 | 133.0 | 83.0 | 72.8 | 125.2 | 80.7 | 23.8 | 20.4 | 18.5 | 6182 | 796 |
| TERRE HAUTE INTL HU | 39.45N | 87.30W | 591 | 1.1 | 8.4 | 91.8 | 76.4 | 90.0 | 75.7 | 87.8 | 74.5 | 79.5 | 88.7 | 77.9 | 86.7 | 76.8 | 142.7 | 85.3 | 75.1 | 134.5 | 83.3 | 23.0 | 19.6 | 17.9 | 5166 | 1078 |
| ***Iowa*** | | | | | | | | | | | | | | | | | | | | | | | | | | *9 sites, 38 more on CD-ROM* |
| AMES MUNI | 41.99N | 93.62W | 955 | -6.4 | -0.4 | 90.3 | 76.2 | 88.1 | 74.7 | 84.4 | 72.8 | 78.9 | 87.0 | 77.1 | 85.1 | 76.6 | 143.7 | 84.5 | 74.7 | 134.4 | 82.5 | 26.4 | 23.5 | 20.2 | 6547 | 787 |
| ANKENY REGIONAL ARP | 41.69N | 93.55W | 909 | -3.7 | 1.2 | 93.3 | 75.4 | 90.3 | 74.8 | 87.6 | 73.4 | 78.3 | 88.4 | 76.9 | 86.7 | 75.4 | 137.7 | 84.2 | 73.4 | 128.3 | 82.9 | 21.7 | 18.9 | 17.0 | 5992 | 1005 |
| BOONE MUNI | 42.05N | 93.84W | 1161 | -5.8 | 0.2 | 91.0 | 77.6 | 89.5 | 76.5 | 86.1 | 74.0 | 80.5 | 88.3 | 78.5 | 85.6 | 78.7 | 155.8 | 86.2 | 76.6 | 145.0 | 84.4 | 26.3 | 23.3 | 20.3 | 6424 | 882 |
| CEDAR RAPIDS MUNI | 41.88N | 91.71W | 873 | -8.4 | -2.8 | 91.0 | 76.0 | 88.0 | 74.4 | 85.1 | 72.6 | 78.5 | 86.9 | 76.9 | 85.0 | 76.2 | 141.3 | 83.9 | 74.3 | 132.4 | 81.9 | 26.5 | 23.6 | 20.4 | 6705 | 785 |
| DAVENPORT MUNI | 41.61N | 90.59W | 755 | -5.8 | 0.3 | 90.1 | 76.0 | 88.0 | 75.0 | 84.3 | 72.9 | 78.3 | 86.8 | 76.8 | 84.8 | 75.3 | 136.5 | 83.5 | 73.4 | 127.6 | 81.8 | 26.6 | 23.7 | 20.4 | 6311 | 794 |
| DES MOINES INTL | 41.54N | 93.67W | 965 | -5.3 | -0.2 | 92.5 | 76.4 | 89.6 | 75.1 | 86.9 | 73.3 | 78.5 | 88.5 | 77.1 | 86.8 | 75.6 | 138.7 | 84.7 | 74.1 | 131.9 | 83.4 | 25.4 | 22.3 | 19.5 | 6172 | 1034 |
| DUBUQUE MUNICIPAL | 42.40N | 90.70W | 1079 | -8.4 | -2.9 | 88.8 | 74.6 | 85.9 | 73.2 | 83.2 | 71.2 | 77.4 | 85.5 | 75.4 | 82.9 | 75.0 | 136.5 | 82.6 | 73.0 | 127.7 | 80.3 | 25.7 | 22.8 | 19.9 | 7023 | 638 |
| SIOUX CITY MUNI | 42.39N | 96.38W | 1102 | -7.8 | -2.8 | 93.1 | 75.1 | 90.2 | 74.2 | 87.7 | 73.1 | 78.7 | 88.2 | 77.1 | 86.8 | 76.1 | 141.8 | 84.8 | 74.1 | 132.4 | 83.4 | 28.6 | 25.3 | 22.3 | 6682 | 916 |
| WATERLOO MUNICIPAL | 42.55N | 92.40W | 879 | -9.9 | -4.8 | 91.2 | 75.3 | 88.4 | 73.7 | 85.7 | 72.2 | 78.6 | 87.4 | 76.7 | 85.1 | 76.0 | 140.5 | 84.3 | 74.1 | 131.4 | 82.1 | 26.0 | 23.3 | 20.1 | 6988 | 775 |
| ***Kansas*** | | | | | | | | | | | | | | | | | | | | | | | | | | *10 sites, 19 more on CD-ROM* |
| FT RILEY/MARSHALL A | 39.06N | 96.76W | 1063 | 2.9 | 9.5 | 99.9 | 75.3 | 96.7 | 75.1 | 93.5 | 74.8 | 78.7 | 92.0 | 77.4 | 91.1 | 75.1 | 137.1 | 86.3 | 73.4 | 129.2 | 84.3 | 21.0 | 18.6 | 16.6 | 4834 | 1585 |
| LAWRENCE MUNI | 39.01N | 95.21W | 833 | 3.4 | 9.3 | 98.8 | 77.2 | 94.7 | 76.5 | 91.1 | 75.5 | 80.0 | 92.4 | 78.4 | 90.7 | 76.6 | 143.1 | 87.3 | 74.9 | 135.0 | 85.7 | 25.0 | 21.8 | 19.2 | 4933 | 1440 |
| MANHATTAN RGNL | 39.13N | 96.68W | 1070 | 1.4 | 8.6 | 99.6 | 75.8 | 96.7 | 75.9 | 92.7 | 74.9 | 78.7 | 92.7 | 77.7 | 91.5 | 75.0 | 136.4 | 85.8 | 73.3 | 128.6 | 83.7 | 24.2 | 20.5 | 18.3 | 5144 | 1440 |
| MC CONNELL AFB | 37.62N | 97.27W | 1371 | 6.5 | 12.3 | 99.4 | 73.9 | 96.4 | 73.8 | 93.0 | 73.6 | 77.8 | 91.0 | 76.6 | 89.7 | 74.1 | 133.8 | 83.7 | 72.8 | 128.0 | 82.7 | 26.9 | 24.3 | 21.1 | 4312 | 1695 |
| JOHNSON CO EXECUTIVE | 38.85N | 94.74W | 1073 | 4.9 | 9.5 | 95.2 | 76.9 | 91.3 | 76.2 | 89.9 | 75.6 | 79.1 | 89.9 | 77.9 | 88.8 | 75.4 | 138.5 | 85.3 | 74.7 | 135.3 | 84.4 | 23.4 | 20.2 | 18.3 | 4779 | 1388 |
| SALINA MUNI | 38.81N | 97.66W | 1283 | 3.4 | 9.2 | 101.0 | 73.8 | 98.0 | 73.7 | 94.7 | 73.2 | 77.4 | 92.4 | 76.2 | 90.6 | 73.1 | 129.1 | 83.9 | 72.2 | 125.0 | 83.0 | 27.8 | 24.9 | 22.5 | 4773 | 1664 |
| FORBES FLD | 38.95N | 95.66W | 1079 | 2.8 | 8.9 | 95.7 | 75.7 | 92.6 | 75.2 | 90.2 | 74.6 | 78.4 | 90.2 | 77.3 | 89.0 | 75.1 | 137.2 | 85.5 | 73.4 | 129.3 | 83.4 | 26.0 | 23.6 | 20.5 | 4929 | 1314 |

126 Principles of HVAC

## Table 4-8 Design Conditions for Selected Locations *(Continued)*

*Meaning of acronyms:*
*DB: Dry bulb temperature, °F* *WB: Wet bulb temperature, °F*
*MCWB: Mean coincident wet bulb temperature, °F*
*Lat: Latitude, °*
*DP: Dew point temperature, °F*
*MCDB: Mean coincident dry bulb temperature, °F*
*Long: Longitude, °*
*HR: Humidity ratio, grains of moisture per lb of dry air*
*HDD and CDD 65: Annual heating and cooling degree-days, base 65°F, °F-day*
*Elev: Elevation, ft*
*WS: Wind speed, mph*

| Station | Lat | Long | Elev | Heating DB | | Cooling DB/MCWB | | | | | | Evaporation WB/MCDB | | | | Dehumidification DP/HR/MCDB | | | | | | Extreme Annual WS | | | Heat./Cool. Degree-Days | |
|---|---|---|---|---|---|---|---|---|---|---|---|---|---|---|---|---|---|---|---|---|---|---|---|---|---|---|
| | | | | | | 0.4% | | 1% | | 2% | | 0.4% | | 1% | | 0.4% | | | 1% | | | | | | | |
| | | | | 99.6% | 99% | DB | MCWB | DB | MCWB | DB | MCWB | WB | MCDB | WB | MCDB | DP | HR | MCDB | DP | HR | MCDB | 1% | 2.5% | 5% | HDD | CDD 65 |
| TOPEKA/BILLARD MUNI | 39.07N | 95.63W | 886 | 3.1 | 8.7 | 97.1 | 76.2 | 93.9 | 75.9 | 91.0 | 75.0 | 79.0 | 91.3 | 77.8 | 90.1 | 75.5 | 137.9 | 86.3 | 74.2 | 132.1 | 84.7 | 23.5 | 20.1 | 18.3 | 4902 | 1446 |
| WICHITA/MID-CONTINE | 37.65N | 97.43W | 1339 | 7.4 | 12.2 | 100.1 | 73.7 | 97.0 | 73.8 | 93.5 | 73.7 | 77.7 | 90.5 | 76.5 | 89.5 | 74.2 | 134.2 | 83.6 | 72.9 | 128.5 | 82.1 | 28.2 | 25.6 | 23.4 | 4464 | 1682 |
| COL JAMES JABARA | 37.75N | 97.22W | 1421 | 7.1 | 11.5 | 99.4 | 74.0 | 96.8 | 74.4 | 92.6 | 73.9 | 77.4 | 91.2 | 76.4 | 89.7 | 73.2 | 130.2 | 83.4 | 72.5 | 126.9 | 82.4 | 27.7 | 25.0 | 22.6 | 4495 | 1577 |
| ***Kentucky*** | | | | | | | | | | | | | | | | | | | | | | | | | *8 sites, 5 more on CD-ROM* | |
| BOWLING GREEN WARRE | 36.98N | 86.44W | 538 | 11.2 | 16.7 | 93.4 | 75.0 | 91.1 | 75.2 | 89.5 | 74.7 | 78.4 | 88.6 | 77.4 | 87.3 | 75.5 | 136.0 | 83.6 | 74.5 | 131.5 | 82.5 | 19.8 | 17.9 | 16.1 | 4063 | 1427 |
| CINCINNATI/GREATER | 39.04N | 84.67W | 883 | 5.4 | 11.3 | 91.4 | 74.2 | 89.2 | 73.5 | 86.7 | 72.5 | 77.4 | 87.1 | 76.1 | 85.0 | 74.5 | 133.1 | 82.3 | 73.2 | 127.6 | 80.7 | 21.8 | 19.1 | 17.2 | 4954 | 1107 |
| FORT CAMPBELL (AAF) | 36.67N | 87.50W | 571 | 12.3 | 18.1 | 93.3 | 76.0 | 91.0 | 76.0 | 89.8 | 75.8 | 79.6 | 87.6 | 78.4 | 86.5 | 77.3 | 145.3 | 83.0 | 76.6 | 141.8 | 82.4 | 20.1 | 17.7 | 15.7 | 3818 | 1548 |
| HENDERSON CITY CO | 37.81N | 87.68W | 387 | 8.8 | 14.2 | 93.2 | 76.4 | 91.1 | 76.0 | 90.0 | 75.4 | 79.4 | 91.0 | 78.1 | 89.0 | 75.5 | 135.4 | 86.9 | 74.6 | 131.3 | 85.8 | 21.0 | 18.7 | 16.7 | 4444 | 1384 |
| LEXINGTON/BLUE GRAS | 38.04N | 84.61W | 988 | 8.3 | 13.6 | 91.6 | 73.9 | 89.6 | 73.6 | 87.3 | 72.8 | 77.3 | 87.5 | 76.1 | 85.4 | 74.2 | 132.6 | 82.7 | 73.1 | 127.5 | 81.1 | 20.3 | 18.0 | 16.3 | 4567 | 1201 |
| BOWMAN FLD | 38.23N | 85.66W | 558 | 9.7 | 15.7 | 93.3 | 75.1 | 91.1 | 74.7 | 89.6 | 74.0 | 78.5 | 88.6 | 77.4 | 87.3 | 75.4 | 136.0 | 83.3 | 74.6 | 132.0 | 82.7 | 18.7 | 16.8 | 14.7 | 4201 | 1459 |
| LOUISVILLE/STANDIFO | 38.18N | 85.73W | 489 | 10.2 | 15.9 | 93.8 | 75.3 | 91.5 | 75.0 | 89.6 | 74.2 | 78.7 | 89.1 | 77.5 | 87.9 | 75.7 | 136.8 | 84.7 | 74.4 | 131.0 | 83.2 | 21.0 | 18.7 | 16.8 | 4109 | 1572 |
| SOMERSET PULASKI CO | 37.05N | 84.60W | 928 | 12.3 | 17.9 | 94.6 | 74.9 | 91.6 | 74.3 | 90.3 | 73.7 | 78.0 | 90.5 | 76.8 | 88.4 | 74.1 | 131.8 | 84.1 | 73.0 | 126.9 | 82.4 | 17.9 | 15.4 | 12.6 | 3866 | 1460 |
| ***Louisiana*** | | | | | | | | | | | | | | | | | | | | | | | | | *12 sites, 8 more on CD-ROM* | |
| ESLER RGNL | 31.40N | 92.30W | 118 | 26.6 | 28.3 | 97.8 | 76.7 | 95.3 | 77.2 | 93.0 | 76.7 | 80.3 | 89.7 | 79.5 | 89.6 | 78.3 | 147.8 | 83.8 | 77.1 | 141.8 | 83.3 | 16.5 | 13.8 | 12.0 | 2004 | 2485 |
| ALEXANDRIA INT | 31.34N | 92.56W | 79 | 27.4 | 29.9 | 97.2 | 77.1 | 94.7 | 77.3 | 92.8 | 76.8 | 80.7 | 89.5 | 79.8 | 89.4 | 78.9 | 150.3 | 84.4 | 77.3 | 142.3 | 84.0 | 18.6 | 16.5 | 14.1 | 1835 | 2621 |
| BARKSDALE AFB | 32.50N | 93.66W | 167 | 23.6 | 27.2 | 97.3 | 76.6 | 95.0 | 76.5 | 92.8 | 76.5 | 80.0 | 90.8 | 79.0 | 89.9 | 77.2 | 142.5 | 84.0 | 76.1 | 137.4 | 83.2 | 19.1 | 17.0 | 14.8 | 2291 | 2305 |
| BATON ROUGE METRO R | 30.54N | 91.15W | 75 | 28.5 | 31.8 | 94.6 | 77.6 | 93.1 | 77.3 | 91.5 | 76.9 | 80.4 | 89.0 | 79.8 | 88.2 | 78.5 | 148.3 | 83.8 | 77.5 | 143.5 | 83.2 | 18.7 | 16.7 | 15.0 | 1573 | 2709 |
| LAFAYETTE RGNL | 30.21N | 91.99W | 43 | 29.9 | 33.6 | 94.6 | 77.8 | 92.9 | 77.6 | 91.3 | 77.3 | 80.7 | 88.9 | 80.0 | 88.3 | 78.9 | 150.3 | 83.7 | 77.8 | 144.9 | 83.3 | 20.4 | 18.3 | 16.5 | 1463 | 2806 |
| LAKE CHARLES MUNI | 30.13N | 93.23W | 10 | 30.3 | 33.8 | 94.4 | 77.8 | 92.8 | 77.7 | 91.2 | 77.6 | 81.4 | 88.5 | 80.4 | 87.6 | 79.4 | 152.8 | 84.2 | 78.7 | 149.2 | 83.7 | 20.5 | 18.4 | 16.7 | 1453 | 2806 |
| MONROE RGNL | 32.51N | 92.04W | 82 | 25.2 | 28.1 | 97.5 | 77.8 | 95.2 | 77.7 | 93.1 | 77.0 | 81.0 | 91.4 | 80.1 | 90.6 | 78.5 | 148.2 | 85.4 | 77.3 | 142.6 | 84.6 | 19.0 | 17.0 | 15.0 | 2189 | 2462 |
| NEW ORLEANS NAS JRB | 29.83N | 90.03W | 0 | 30.7 | 34.2 | 92.8 | 78.1 | 91.2 | 77.8 | 90.2 | 77.6 | 81.9 | 87.5 | 80.6 | 86.6 | 80.4 | 157.8 | 84.6 | 79.2 | 151.3 | 83.6 | 18.1 | 16.0 | 13.6 | 1444 | 2626 |
| NEW ORLEANS/MOISANT | 29.99N | 90.25W | 20 | 33.1 | 36.3 | 93.8 | 78.1 | 92.2 | 77.7 | 90.8 | 77.5 | 80.9 | 88.9 | 80.2 | 88.0 | 79.0 | 150.6 | 84.4 | 78.2 | 146.5 | 83.8 | 20.7 | 18.7 | 17.0 | 1286 | 2925 |
| LAKEFRONT | 30.04N | 90.03W | 10 | 35.6 | 38.6 | 93.3 | 78.7 | 91.8 | 78.2 | 90.7 | 77.9 | 81.4 | 89.3 | 80.6 | 88.3 | 79.3 | 152.3 | 85.3 | 78.8 | 149.7 | 85.0 | 24.9 | 21.0 | 18.9 | 1138 | 3232 |
| SHREVEPORT DOWNTOWN | 32.54N | 93.74W | 180 | 26.9 | 29.6 | 99.2 | 76.6 | 96.9 | 76.5 | 93.5 | 76.2 | 79.6 | 91.3 | 78.8 | 90.2 | 76.9 | 141.0 | 83.1 | 75.5 | 134.3 | 82.7 | 18.8 | 16.7 | 14.9 | 2149 | 2628 |
| SHREVEPORT REGIONAL | 32.45N | 93.82W | 259 | 25.2 | 28.4 | 98.5 | 76.2 | 96.0 | 76.3 | 93.6 | 76.0 | 79.4 | 91.2 | 78.6 | 89.9 | 76.4 | 139.2 | 83.2 | 75.7 | 135.8 | 82.7 | 19.7 | 17.7 | 16.1 | 2117 | 2535 |
| ***Maine*** | | | | | | | | | | | | | | | | | | | | | | | | | *5 sites, 16 more on CD-ROM* | |
| AUBURN LEWISTON MUNI | 44.05N | 70.28W | 289 | -6.2 | -0.1 | 87.9 | 70.7 | 83.6 | 69.2 | 81.0 | 67.5 | 73.6 | 83.4 | 71.4 | 80.2 | 70.4 | 113.1 | 78.5 | 68.5 | 106.0 | 76.4 | 20.8 | 18.5 | 16.4 | 7632 | 308 |
| BANGOR INTL | 44.81N | 68.82W | 194 | -7.3 | -2.0 | 87.9 | 70.7 | 84.1 | 69.0 | 81.1 | 67.0 | 73.2 | 83.1 | 71.3 | 80.6 | 70.2 | 111.7 | 78.1 | 68.2 | 104.2 | 75.4 | 23.5 | 19.7 | 18.0 | 7665 | 355 |
| BRUNSWICK (NAS) | 43.90N | 69.93W | 75 | -2.2 | 2.1 | 86.3 | 70.7 | 82.8 | 68.9 | 80.5 | 67.3 | 73.5 | 82.4 | 71.5 | 79.8 | 70.4 | 112.4 | 78.0 | 69.0 | 106.7 | 76.0 | 23.2 | 19.5 | 17.4 | 7202 | 367 |
| PORTLAND/INTNL. JET | 43.64N | 70.30W | 62 | 0.1 | 4.9 | 86.8 | 71.3 | 83.4 | 69.9 | 80.4 | 68.2 | 74.1 | 83.2 | 72.2 | 80.2 | 71.0 | 114.7 | 78.8 | 69.5 | 108.8 | 76.6 | 23.2 | 19.6 | 17.6 | 7023 | 370 |
| SANFORD RGNL | 43.39N | 70.70W | 243 | -6.2 | 0.3 | 89.5 | 71.0 | 85.2 | 69.4 | 82.0 | 67.8 | 74.1 | 84.6 | 72.1 | 81.9 | 71.5 | 117.4 | 80.1 | 69.6 | 109.6 | 77.7 | 20.9 | 18.5 | 16.3 | 7470 | 350 |
| ***Maryland*** | | | | | | | | | | | | | | | | | | | | | | | | | *3 sites, 4 more on CD-ROM* | |
| ANDREWS AFB/CAMP SP | 38.82N | 76.85W | 289 | 15.6 | 18.4 | 92.6 | 74.1 | 90.2 | 73.4 | 87.9 | 72.8 | 77.6 | 86.6 | 76.3 | 84.9 | 75.1 | 133.1 | 80.4 | 73.4 | 125.3 | 79.2 | 24.7 | 20.7 | 18.3 | 4419 | 1199 |
| BALTIMORE-WASHINGTO | 39.17N | 76.68W | 154 | 14.0 | 17.9 | 94.0 | 74.9 | 91.3 | 74.1 | 88.7 | 73.1 | 78.1 | 88.6 | 76.8 | 86.6 | 75.3 | 133.2 | 82.1 | 74.1 | 127.8 | 80.7 | 22.3 | 19.1 | 17.1 | 4552 | 1261 |
| THOMAS POINT | 38.90N | 76.43W | 39 | 17.6 | 21.4 | 86.8 | 74.7 | 84.8 | 74.6 | 83.1 | 74.0 | 79.6 | 82.7 | 77.8 | 81.5 | 78.7 | 149.0 | 81.2 | 76.9 | 140.3 | 80.0 | 37.7 | 31.6 | 26.5 | 4196 | 1236 |
| ***Massachusetts*** | | | | | | | | | | | | | | | | | | | | | | | | | *11 sites, 10 more on CD-ROM* | |
| BARNSTABLE MUNI BOA | 41.67N | 70.28W | 52 | 9.9 | 15.6 | 84.0 | 73.0 | 81.5 | 71.5 | 79.3 | 70.2 | 75.5 | 81.3 | 74.2 | 78.9 | 73.3 | 124.1 | 77.6 | 72.7 | 121.5 | 77.0 | 24.7 | 21.2 | 19.2 | 5872 | 511 |
| BOSTON/LOGAN INTL | 42.36N | 71.01W | 30 | 8.1 | 13.0 | 90.6 | 72.7 | 87.6 | 71.7 | 84.2 | 70.2 | 75.9 | 85.7 | 74.3 | 83.0 | 72.8 | 122.0 | 80.6 | 71.5 | 116.3 | 78.6 | 26.9 | 24.2 | 20.9 | 5596 | 750 |
| BUZZARDS BAY | 41.38N | 71.03W | 56 | 12.4 | 16.7 | 76.0 | N/A | 74.4 | N/A | 73.1 | N/A | N/A | N/A | N/A | N/A | N/A | N/A | N/A | N/A | N/A | N/A | 43.8 | 38.3 | 34.0 | 5552 | 302 |
| CHATHAM MUNI | 41.69N | 69.99W | 69 | 11.7 | 17.1 | 82.1 | 72.3 | 80.7 | 71.6 | 78.1 | 70.5 | 75.1 | 80.2 | 73.9 | 78.2 | 73.2 | 123.5 | 77.6 | 72.5 | 120.5 | 76.7 | 21.7 | 18.7 | 16.7 | 5688 | 457 |
| LAWRENCE MUNI | 42.72N | 71.12W | 148 | 3.4 | 9.6 | 90.4 | 72.8 | 87.9 | 71.9 | 83.9 | 70.7 | 75.7 | 84.8 | 74.3 | 82.8 | 72.8 | 122.3 | 79.7 | 72.1 | 119.3 | 79.1 | 20.4 | 18.1 | 16.2 | 6091 | 652 |
| MARTHAS VINEYARD | 41.39N | 70.62W | 69 | 9.5 | 14.0 | 83.8 | 72.5 | 81.3 | 71.2 | 79.1 | 70.0 | 75.2 | 80.9 | 73.9 | 78.6 | 73.1 | 123.3 | 77.8 | 72.4 | 120.3 | 76.8 | 26.0 | 23.5 | 20.4 | 5886 | 429 |
| NEW BEDFORD RGNL | 41.68N | 70.96W | 79 | 8.7 | 12.1 | 88.2 | 73.2 | 84.1 | 71.7 | 81.7 | 70.1 | 75.8 | 83.7 | 74.4 | 80.9 | 73.1 | 123.4 | 78.7 | 72.4 | 120.4 | 77.8 | 23.1 | 19.9 | 17.9 | 5833 | 570 |
| NORWOOD MEM | 42.19N | 71.17W | 49 | 3.1 | 9.1 | 90.4 | 73.3 | 88.0 | 72.6 | 84.1 | 70.8 | 76.5 | 85.6 | 74.9 | 82.9 | 73.3 | 124.1 | 79.5 | 72.5 | 120.7 | 78.7 | 20.5 | 18.2 | 16.3 | 6233 | 581 |
| PLYMOUTH MUNICIPAL | 41.91N | 70.73W | 148 | 5.3 | 9.9 | 89.5 | 73.1 | 84.4 | 71.6 | 82.0 | 69.9 | 75.8 | 84.1 | 74.4 | 81.3 | 73.2 | 123.9 | 78.8 | 72.4 | 120.6 | 78.0 | 23.3 | 19.9 | 17.9 | 6154 | 553 |
| SOUTH WEYMOUTH NAS | 42.15N | 70.93W | 161 | 5.9 | 10.4 | 91.2 | 73.8 | 87.7 | 72.3 | 84.7 | 70.7 | 76.9 | 86.8 | 74.9 | 83.8 | 74.1 | 127.9 | 81.9 | 72.2 | 119.7 | 79.4 | 18.5 | 16.5 | 14.5 | 5832 | 646 |
| WORCESTER REGIONAL ARPT | 42.27N | 71.88W | 1017 | 1.9 | 6.7 | 85.7 | 71.2 | 83.0 | 69.7 | 80.7 | 68.1 | 74.0 | 81.7 | 72.3 | 79.4 | 71.7 | 121.5 | 77.8 | 70.1 | 115.1 | 76.1 | 25.9 | 22.9 | 19.7 | 6706 | 462 |
| ***Michigan*** | | | | | | | | | | | | | | | | | | | | | | | | | *15 sites, 44 more on CD-ROM* | |
| DETROIT CITY | 42.41N | 83.01W | 627 | 5.2 | 9.6 | 90.7 | 73.4 | 88.2 | 72.1 | 85.5 | 70.8 | 76.4 | 86.4 | 74.6 | 83.8 | 73.2 | 126.3 | 81.9 | 71.9 | 120.5 | 80.3 | 20.4 | 18.5 | 17.1 | 5989 | 884 |
| DETROIT/METROPOLITA | 42.22N | 83.35W | 663 | 2.9 | 8.0 | 90.4 | 73.8 | 87.6 | 72.6 | 84.7 | 71.1 | 76.9 | 86.4 | 75.0 | 83.4 | 73.8 | 129.2 | 82.3 | 72.3 | 122.7 | 80.2 | 25.3 | 22.2 | 19.5 | 6103 | 807 |
| WILLOW RUN | 42.24N | 83.53W | 715 | 0.8 | 6.3 | 90.3 | 74.0 | 87.8 | 72.5 | 84.3 | 70.8 | 76.7 | 86.3 | 74.8 | 83.6 | 73.3 | 127.3 | 81.3 | 72.2 | 122.2 | 80.0 | 24.7 | 21.5 | 19.0 | 6415 | 679 |
| FLINT/BISHOP INTL | 42.97N | 83.75W | 768 | -0.2 | 4.5 | 89.7 | 73.8 | 86.7 | 72.0 | 83.9 | 70.3 | 76.3 | 85.3 | 74.4 | 82.9 | 73.3 | 127.4 | 81.9 | 71.7 | 120.4 | 79.3 | 23.8 | 20.4 | 18.5 | 6741 | 594 |
| GRAND RAPIDS/KENT C | 42.88N | 85.52W | 804 | 2.2 | 6.9 | 89.4 | 73.1 | 86.5 | 71.7 | 83.8 | 70.1 | 76.3 | 85.1 | 74.4 | 82.6 | 73.4 | 128.2 | 81.4 | 71.8 | 121.0 | 79.3 | 24.7 | 21.0 | 19.0 | 6615 | 639 |
| GROSSE ILE MUNI | 42.10N | 83.17W | 591 | 7.1 | 10.2 | 89.7 | 74.2 | 85.9 | 73.5 | 82.5 | 71.8 | 77.9 | 84.5 | 76.2 | 82.7 | 75.3 | 135.8 | 81.1 | 73.4 | 127.2 | 79.6 | 20.9 | 18.6 | 16.7 | 5804 | 863 |
| TULIP CITY | 42.75N | 86.10W | 689 | 7.2 | 10.1 | 88.4 | 73.3 | 85.7 | 72.0 | 82.3 | 70.3 | 76.1 | 84.2 | 74.4 | 82.2 | 73.1 | 126.0 | 80.9 | 72.2 | 122.3 | 79.5 | 25.7 | 22.1 | 19.2 | 6234 | 617 |
| JACKSON CO REYNOLDS | 42.26N | 84.46W | 1020 | 0.5 | 5.4 | 88.3 | 73.4 | 85.6 | 71.9 | 82.7 | 70.1 | 76.0 | 84.4 | 74.3 | 82.2 | 73.2 | 127.9 | 80.9 | 71.8 | 122.2 | 79.4 | 20.2 | 18.4 | 16.9 | 6619 | 565 |
| KALAMAZOO BATTLE CR | 42.24N | 85.55W | 873 | 2.8 | 8.6 | 90.0 | 72.8 | 87.7 | 71.8 | 83.9 | 70.4 | 75.9 | 84.6 | 74.4 | 82.8 | 72.8 | 125.9 | 81.0 | 72.0 | 122.4 | 79.9 | 21.7 | 19.0 | 17.2 | 6251 | 709 |

# Design Conditions

**Table 4-8  Design Conditions for Selected Locations (Continued)**

Meaning of acronyms:
DB: Dry bulb temperature, °F
MCWB: Mean coincident wet bulb temperature, °F
WB: Wet bulb temperature, °F
Lat: Latitude, °
DP: Dew point temperature, °F
MCDB: Mean coincident dry bulb temperature, °F
Long: Longitude, °
HR: Humidity ratio, grains of moisture per lb of dry air
HDD and CDD 65: Annual heating and cooling degree-days, base 65°F, °F-day
Elev: Elevation, ft
WS: Wind speed, mph

| Station | Lat | Long | Elev | Heating DB 99.6% | Heating DB 99% | Cooling DB/MCWB 0.4% DB/MCWB | Cooling DB/MCWB 1% DB/MCWB | Cooling DB/MCWB 2% DB/MCWB | Evaporation WB/MCDB 0.4% WB/MCDB | Evaporation WB/MCDB 1% WB/MCDB | Dehumidification DP/HR/MCDB 0.4% DP/HR/MCDB | Dehumidification DP/HR/MCDB 1% DP/HR/MCDB | Extreme Annual WS 1% | Extreme Annual WS 2.5% | Extreme Annual WS 5% | Heat/Cool Degree-Days HDD/CDD 65 |
|---|---|---|---|---|---|---|---|---|---|---|---|---|---|---|---|---|
| LANSING/CAPITAL CIT | 42.78N | 84.58W | 873 | -1.0 | 4.2 | 89.5/73.3 | 86.4/71.9 | 83.6/70.2 | 76.2/85.2 | 74.3/82.6 | 73.2/127.6/81.4 | 71.6/120.6/79.2 | 24.3 | 20.6 | 18.6 | 6815/575 |
| MUSKEGON | 43.17N | 86.24W | 633 | 5.2 | 9.3 | 86.2/72.4 | 83.7/71.0 | 81.5/69.5 | 75.5/82.4 | 73.9/80.3 | 73.2/126.4/79.8 | 71.8/120.1/78.2 | 25.3 | 23.0 | 20.0 | 6619/524 |
| OAKLAND CO INTL | 42.67N | 83.42W | 981 | 1.1 | 5.7 | 89.7/73.2 | 86.2/71.3 | 83.5/69.9 | 75.3/84.8 | 73.5/82.2 | 72.4/124.6/80.6 | 70.9/117.9/78.5 | 24.3 | 20.8 | 18.8 | 6633/641 |
| MBS INTL | 43.53N | 84.08W | 669 | 0.4 | 4.6 | 89.9/73.3 | 86.6/71.6 | 83.7/70.3 | 76.2/85.6 | 74.2/82.9 | 73.1/126.1/81.3 | 71.7/120.1/79.5 | 24.2 | 20.7 | 18.8 | 6908/580 |
| SELFRIDGE ANGB | 42.60N | 82.83W | 581 | 2.9 | 7.4 | 90.1/73.2 | 86.5/71.9 | 83.9/70.7 | 75.7/85.0 | 74.1/83.0 | 72.8/124.1/80.0 | 71.9/120.6/79.1 | 21.0 | 18.8 | 17.0 | 6460/645 |
| ST CLAIR CO INTL | 42.91N | 82.52W | 650 | 0.6 | 5.5 | 90.0/73.5 | 85.9/71.1 | 82.4/69.5 | 75.7/84.6 | 73.8/81.8 | 72.9/125.0/79.9 | 71.9/120.8/78.7 | 18.6 | 16.6 | 14.7 | 6731/465 |
| **Minnesota** | | | | | | | | | | | | | | | | *11 sites, 68 more on CD-ROM* |
| SKY HARBOR | 46.72N | 92.03W | 610 | -10.7 | -6.4 | 85.9/71.6 | 82.1/69.4 | 79.3/67.5 | 75.6/82.7 | 73.1/79.5 | 73.2/126.0/78.7 | 71.8/120.1/77.4 | 28.0 | 24.8 | 21.4 | 8572/298 |
| DULUTH INTL AIRPORT | 46.84N | 92.19W | 1417 | -17.9 | -12.5 | 84.3/69.6 | 81.3/67.3 | 78.4/65.4 | 72.5/81.1 | 70.1/78.2 | 69.5/114.4/78.0 | 67.1/105.2/75.1 | 24.7 | 21.1 | 19.2 | 9325/210 |
| FLYING CLOUD | 44.83N | 93.47W | 945 | -9.4 | -6.0 | 90.5/73.9 | 88.1/72.7 | 84.3/70.7 | 77.3/86.9 | 75.2/83.9 | 73.5/129.0/82.8 | 72.4/124.1/81.3 | 21.9 | 19.2 | 17.4 | 7343/773 |
| MANKATO RGNL ARPT | 44.22N | 93.92W | 1020 | -13.0 | -8.3 | 89.6/73.5 | 85.9/71.4 | 82.4/69.5 | 76.7/84.7 | 74.4/82.6 | 73.4/129.2/81.6 | 72.0/123.0/80.3 | 26.8 | 24.1 | 20.8 | 7714/601 |
| ANOKA CO BLAINE | 45.15N | 93.20W | 912 | -9.1 | -5.8 | 90.1/74.3 | 87.6/73.0 | 83.8/71.0 | 77.8/85.7 | 75.4/83.0 | 75.0/135.8/82.9 | 72.8/125.7/80.2 | 23.0 | 19.7 | 17.8 | 7536/624 |
| CRYSTAL | 45.06N | 93.35W | 869 | -9.1 | -5.8 | 90.5/73.1 | 88.1/72.0 | 84.2/70.1 | 76.6/86.3 | 74.4/83.9 | 73.0/126.6/82.5 | 71.9/121.7/81.0 | 21.3 | 19.0 | 17.2 | 7521/692 |
| MINNEAPOLIS/ST.PAUL | 44.88N | 93.23W | 837 | -11.2 | -6.2 | 90.9/72.9 | 88.0/71.9 | 84.8/70.3 | 76.8/87.0 | 74.8/84.0 | 73.4/128.3/83.3 | 71.7/120.6/81.0 | 24.4 | 20.9 | 19.1 | 7472/765 |
| ROCHESTER MUNICIPAL | 43.90N | 92.49W | 1319 | -13.1 | -8.2 | 88.0/73.4 | 84.8/71.7 | 82.2/70.1 | 76.7/84.3 | 74.5/81.7 | 74.2/134.2/81.6 | 72.2/125.0/79.5 | 28.5 | 25.8 | 23.5 | 7868/515 |
| SOUTH ST PAUL MUNI | 44.85N | 93.03W | 820 | -9.1 | -5.5 | 90.5/73.1 | 88.0/71.7 | 84.3/70.0 | 77.1/85.5 | 74.7/83.0 | 74.6/133.5/82.3 | 72.3/123.3/79.7 | 18.6 | 16.5 | 14.3 | 7401/730 |
| ST. CLOUD MUNICIPAL | 45.55N | 94.05W | 1024 | -17.2 | -11.4 | 89.9/72.5 | 86.6/70.8 | 83.6/69.0 | 76.2/85.9 | 74.1/82.0 | 73.1/127.8/82.0 | 70.9/118.2/79.7 | 23.4 | 19.8 | 17.8 | 8424/477 |
| ST PAUL DOWNTOWN HO | 44.93N | 93.05W | 712 | -11.0 | -6.4 | 90.4/73.5 | 87.8/72.1 | 84.0/70.4 | 76.8/86.0 | 74.7/83.6 | 73.4/127.5/82.4 | 72.0/121.7/80.6 | 23.1 | 20.1 | 18.2 | 7462/722 |
| **Mississippi** | | | | | | | | | | | | | | | | *6 sites, 7 more on CD-ROM* |
| HATTIESBURG LAUREL | 31.47N | 89.34W | 299 | 25.1 | 27.8 | 96.6/75.8 | 93.2/75.0 | 91.2/74.8 | 78.5/89.7 | 77.7/88.9 | 75.3/134.1/82.8 | 74.7/131.2/82.4 | 16.1 | 13.2 | 11.7 | 2081/2292 |
| JACKSON/ALLEN C. TH | 32.32N | 90.08W | 331 | 23.2 | 26.7 | 96.4/76.2 | 94.0/76.2 | 92.2/76.0 | 79.8/90.4 | 77.8/88.5 | 77.1/142.9/83.5 | 76.2/138.4/82.7 | 18.6 | 16.5 | 14.6 | 2282/2294 |
| KEESLER AFB | 30.41N | 88.92W | 33 | 30.7 | 35.1 | 93.5/79.8 | 91.6/79.2 | 90.4/78.9 | 83.2/89.8 | 81.9/88.1 | 81.4/163.5/86.8 | 80.6/159.0/86.0 | 17.7 | 15.6 | 13.5 | 1447/2757 |
| MERIDIAN/KEY FIELD | 32.33N | 88.75W | 312 | 22.5 | 26.1 | 96.3/75.9 | 93.8/75.7 | 92.0/75.7 | 79.6/89.7 | 78.6/88.6 | 77.0/142.5/83.4 | 76.2/137.3/82.5 | 18.5 | 16.6 | 14.7 | 2344/2161 |
| MERIDIAN NAS | 32.55N | 88.57W | 318 | 22.5 | 26.7 | 97.4/76.6 | 95.3/76.6 | 93.1/75.7 | 79.3/91.6 | 79.0/90.5 | 77.3/143.6/85.9 | 75.7/136.3/84.2 | 15.7 | 12.7 | 10.9 | 2307/2290 |
| TUPELO C.D. LEMONS | 34.26N | 88.77W | 361 | 19.1 | 23.4 | 96.4/76.0 | 93.5/75.6 | 91.5/75.4 | 79.2/89.9 | 78.3/88.7 | 76.4/139.5/83.7 | 75.4/134.7/82.9 | 18.9 | 17.0 | 15.4 | 2915/2003 |
| **Missouri** | | | | | | | | | | | | | | | | *9 sites, 10 more on CD-ROM* |
| CAPE GIRARDEAU RGNL | 37.23N | 89.57W | 351 | 9.7 | 15.5 | 94.5/77.3 | 92.3/76.8 | 90.3/76.1 | 80.3/90.2 | 78.8/88.8 | 77.4/144.3/86.1 | 75.9/137.1/84.4 | 21.2 | 19.0 | 17.3 | 4182/1531 |
| COLUMBIA REGIONAL | 38.82N | 92.22W | 899 | 2.8 | 8.6 | 94.2/76.4 | 91.3/76.0 | 88.8/74.9 | 79.3/89.3 | 77.9/87.8 | 76.4/142.6/85.4 | 75.0/135.5/83.8 | 24.2 | 20.6 | 18.6 | 4937/1247 |
| JEFFERSON CITY MEM | 38.59N | 92.16W | 574 | 7.1 | 11.8 | 95.3/76.5 | 91.4/75.8 | 90.1/75.1 | 79.4/89.4 | 78.0/88.3 | 76.4/142.9/85.0 | 74.9/133.9/83.2 | 20.9 | 18.5 | 16.5 | 4560/1397 |
| JOPLIN RGNL | 37.15N | 94.50W | 984 | 8.5 | 13.9 | 90.6/75.7 | 93.4/75.4 | 91.0/75.1 | 78.6/90.3 | 77.7/89.5 | 75.3/137.5/85.4 | 74.1/132.1/84.2 | 24.9 | 21.7 | 19.3 | 4033/1638 |
| CHARLES B WHEELER D | 39.12N | 94.59W | 751 | 5.0 | 9.8 | 96.8/76.4 | 93.3/75.9 | 90.9/74.9 | 79.5/91.8 | 78.2/89.9 | 75.9/139.4/86.6 | 74.8/134.0/85.7 | 22.5 | 19.6 | 18.3 | 4542/1637 |
| KANSAS CITY INTL | 39.30N | 94.72W | 1024 | 2.0 | 7.2 | 95.8/76.8 | 92.5/76.1 | 89.7/75.3 | 79.8/90.5 | 78.3/88.9 | 75.9/145.3/86.6 | 75.2/137.2/85.2 | 25.6 | 23.0 | 20.0 | 5012/1372 |
| SPRINGFIELD MUNI | 37.24N | 93.39W | 1270 | 6.6 | 12.3 | 94.8/74.6 | 91.7/74.7 | 89.3/74.3 | 77.8/88.9 | 76.8/87.6 | 74.6/135.6/83.6 | 73.4/130.1/82.3 | 23.3 | 20.1 | 18.3 | 4442/1366 |
| ST. LOUIS/LAMBERT | 38.75N | 90.37W | 709 | 6.6 | 11.7 | 95.5/76.8 | 93.0/76.1 | 90.7/75.0 | 79.4/90.8 | 78.1/89.2 | 76.2/140.6/85.8 | 74.9/134.2/84.9 | 23.7 | 20.1 | 18.1 | 4436/1650 |
| SPIRIT OF ST LOUIS | 38.66N | 90.68W | 463 | 5.3 | 10.6 | 95.3/77.3 | 92.6/76.3 | 90.3/75.2 | 79.9/90.8 | 78.3/89.0 | 76.9/142.3/86.2 | 75.2/134.3/84.4 | 20.7 | 18.5 | 16.7 | 4679/1389 |
| **Montana** | | | | | | | | | | | | | | | | *7 sites, 14 more on CD-ROM* |
| BILLINGS/LOGAN INT. | 45.81N | 108.54W | 3570 | -9.4 | -3.2 | 94.5/62.9 | 91.2/61.9 | 88.0/61.3 | 65.3/85.3 | 63.8/83.8 | 60.3/89.3/72.1 | 58.0/82.0/70.5 | 27.1 | 24.5 | 21.1 | 6705/630 |
| GALLATIN FLD | 45.79N | 111.15W | 4449 | -15.8 | -8.0 | 92.0/60.7 | 88.6/60.7 | 85.0/59.5 | 64.5/83.3 | 62.7/81.7 | 58.5/86.5/69.7 | 56.1/79.1/68.4 | 20.8 | 18.0 | 15.2 | 8184/233 |
| BERT MOONEY | 45.95N | 112.51W | 5535 | -18.2 | -9.5 | 88.0/57.5 | 84.4/56.5 | 81.6/55.8 | 60.4/79.1 | 59.0/77.5 | 54.7/78.2/62.9 | 52.3/71.7/62.1 | 22.0 | 19.0 | 17.2 | 9104/77 |
| GREAT FALLS | 47.45N | 111.38W | 3707 | -13.5 | -6.6 | 90.4/60.3 | 86.9/59.4 | 83.7/58.8 | 63.3/82.9 | 61.7/80.5 | 57.1/80.0/66.4 | 55.1/74.1/67.2 | 31.3 | 27.1 | 24.6 | 7733/311 |
| GREAT FALLS INTL | 47.48N | 111.38W | 3658 | -16.3 | -9.1 | 92.2/61.1 | 88.7/60.3 | 85.2/59.5 | 64.1/84.6 | 62.4/82.3 | 57.6/81.1/66.7 | 55.6/75.4/67.1 | 31.4 | 27.5 | 24.8 | 7470/326 |
| MALMSTROM AFHP | 47.50N | 111.19W | 3471 | -16.3 | -9.2 | 92.2/61.7 | 88.9/61.0 | 85.5/60.3 | 64.6/84.7 | 63.1/83.3 | 57.6/80.6/69.2 | 55.8/75.5/68.7 | 29.8 | 26.5 | 23.4 | 6886/394 |
| MISSOULA/JOHNSON-BE | 46.92N | 114.09W | 3189 | -3.8 | 3.1 | 92.8/62.1 | 89.7/61.5 | 86.0/60.5 | 65.0/85.3 | 63.4/83.6 | 58.7/83.0/68.9 | 56.5/76.6/67.8 | 21.4 | 18.8 | 16.6 | 7372/314 |
| **Nebraska** | | | | | | | | | | | | | | | | *5 sites, 20 more on CD-ROM* |
| GRAND ISLAND COUNTY | 40.96N | 98.31W | 1857 | -4.3 | 1.1 | 95.7/74.1 | 92.4/73.2 | 89.5/72.0 | 77.4/89.1 | 75.8/87.7 | 74.1/136.2/84.1 | 63.2/128.5/82.4 | 28.6 | 25.5 | 22.8 | 6081/1037 |
| LINCOLN MUNICIPAL | 40.83N | 96.76W | 1188 | -3.5 | 1.5 | 96.9/75.1 | 93.2/74.5 | 90.4/73.1 | 78.3/90.6 | 76.9/89.1 | 74.9/136.6/85.5 | 73.2/129.0/83.7 | 27.1 | 24.3 | 20.9 | 5917/1185 |
| OFFUTT AFB | 41.12N | 95.90W | 1053 | -1.5 | 2.6 | 94.8/76.1 | 91.0/75.0 | 88.5/73.8 | 79.8/88.3 | 77.9/86.5 | 77.3/147.6/84.2 | 75.2/137.5/82.2 | 24.9 | 21.0 | 18.7 | 5874/1151 |
| OMAHA EPPLEY FIELD | 41.31N | 95.90W | 981 | -4.3 | 0.6 | 94.5/76.4 | 91.4/75.2 | 88.8/73.6 | 79.3/89.6 | 77.6/87.8 | 76.4/142.7/85.8 | 74.6/134.3/84.0 | 26.3 | 23.7 | 20.4 | 6025/1132 |
| OMAHA | 41.37N | 96.02W | 1332 | -6.1 | -0.1 | 94.0/75.0 | 90.9/74.0 | 88.0/73.0 | 77.7/89.0 | 76.3/87.3 | 74.4/135.3/84.2 | 72.9/128.2/83.0 | 23.3 | 19.2 | 17.8 | 5981/1093 |
| **Nevada** | | | | | | | | | | | | | | | | *3 sites, 10 more on CD-ROM* |
| LAS VEGAS/MCCARRAN | 36.08N | 115.16W | 2182 | 31.0 | 33.8 | 108.4/67.8 | 106.3/67.0 | 104.1/66.5 | 72.6/96.7 | 71.1/95.1 | 65.7/103.0/81.7 | 63.2/94.2/84.6 | 26.3 | 23.3 | 20.0 | 2015/3486 |
| NELLIS AFB | 36.24N | 115.03W | 1867 | 27.7 | 30.9 | 109.2/67.4 | 107.1/66.9 | 104.7/66.2 | 72.3/95.2 | 71.0/95.5 | 65.8/101.8/80.8 | 63.0/92.4/83.7 | 25.8 | 22.5 | 19.3 | 2130/3303 |
| RENO/CANNON INTL | 39.48N | 119.77W | 4400 | 12.1 | 17.6 | 96.3/61.6 | 93.4/60.2 | 91.1/59.3 | 64.0/89.3 | 62.3/87.7 | 55.1/76.0/71.4 | 51.9/67.5/70.9 | 25.5 | 21.1 | 18.7 | 5043/791 |
| **New Hampshire** | | | | | | | | | | | | | | | | *4 sites, 8 more on CD-ROM* |
| CONCORD MUNICIPAL | 43.20N | 71.50W | 348 | -3.6 | 1.5 | 90.1/71.4 | 87.1/69.9 | 84.1/68.7 | 74.8/85.0 | 73.0/82.2 | 71.8/118.9/78.5 | 70.2/112.4/77.0 | 20.9 | 18.6 | 16.6 | 7141/469 |
| JAFFREY ARPT SILVER | 42.81N | 72.00W | 1040 | -2.2 | 1.4 | 86.5/69.8 | 83.6/68.8 | 81.2/67.1 | 73.0/81.1 | 71.6/79.1 | 71.6/121.3/76.5 | 70.0/114.5/75.2 | 16.3 | 13.7 | 12.1 | 7323/362 |

128 Principles of HVAC

## Table 4-8 Design Conditions for Selected Locations *(Continued)*

*Meaning of acronyms:*
*DB: Dry bulb temperature, °F* | *WB: Wet bulb temperature, °F* | *Lat: Latitude, °* | *Long: Longitude, °* | *Elev: Elevation, ft*
*MCWB: Mean coincident wet bulb temperature, °F* | *DP: Dew point temperature, °F* | *HR: Humidity ratio, grains of moisture per lb of dry air* | *WS: Wind speed, mph*
*MCDB: Mean coincident dry bulb temperature, °F* | *HDD and CDD 65: Annual heating and cooling degree-days, base 65°F, °F-day*

| Station | Lat | Long | Elev | Heating DB | | Cooling DB/MCWB | | | | | | Evaporation WB/MCDB | | | | Dehumidification DP/HR/MCDB | | | | | | Extreme Annual WS | | | Heat./Cool. Degree-Days | |
|---|---|---|---|---|---|---|---|---|---|---|---|---|---|---|---|---|---|---|---|---|---|---|---|---|---|---|
| | | | | | | 0.4% | | 1% | | 2% | | 0.4% | | 1% | | 0.4% | | | 1% | | | | | | | |
| | | | | 99.6% | 99% | DB | MCWB | DB | MCWB | DB | MCWB | WB | MCDB | WB | MCDB | DP | HR | MCDB | DP | HR | MCDB | 1% | 2.5% | 5% | HDD | CDD 65 |
| MANCHESTER | 42.93N | 71.44W | 233 | 1.4 | 7.1 | 91.1 | 71.9 | 88.5 | 70.6 | 85.7 | 69.5 | 75.5 | 85.7 | 73.8 | 83.2 | 72.4 | 121.0 | 80.3 | 71.2 | 116.1 | 78.8 | 19.3 | 17.7 | 15.7 | 6214 | 730 |
| PEASE INTL TRADEPOR | 43.08N | 70.82W | 102 | 2.8 | 8.4 | 89.6 | 72.8 | 86.0 | 71.2 | 82.4 | 69.5 | 75.4 | 84.8 | 73.6 | 82.2 | 72.4 | 120.6 | 80.2 | 71.0 | 114.6 | 78.4 | 23.1 | 19.7 | 17.5 | 6418 | 545 |
| ***New Jersey*** | | | | | | | | | | | | | | | | | | | | | | | | | *7 sites, 3 more on CD-ROM* | |
| ATLANTIC CITY INTL | 39.46N | 74.58W | 66 | 11.4 | 15.9 | 92.2 | 74.9 | 89.5 | 73.8 | 86.5 | 72.8 | 77.9 | 87.4 | 76.7 | 84.9 | 75.2 | 132.5 | 81.8 | 74.2 | 128.0 | 80.6 | 24.9 | 21.3 | 18.9 | 4913 | 1014 |
| MONMOUTH EXECUTIVE | 40.19N | 74.12W | 161 | 11.6 | 16.0 | 90.9 | 73.6 | 88.3 | 72.3 | 84.4 | 70.9 | 76.3 | 86.8 | 74.9 | 84.2 | 73.0 | 123.3 | 80.9 | 72.1 | 119.6 | 80.2 | 25.2 | 22.0 | 19.3 | 5105 | 894 |
| MC GUIRE AFB | 40.02N | 74.58W | 131 | 11.9 | 15.9 | 92.6 | 75.9 | 90.2 | 74.7 | 87.8 | 73.8 | 79.2 | 87.4 | 77.7 | 85.6 | 77.1 | 141.7 | 82.5 | 75.2 | 132.6 | 81.4 | 23.2 | 19.8 | 17.8 | 4864 | 1050 |
| MILLVILLE MUNI | 39.37N | 75.08W | 75 | 11.0 | 15.8 | 92.1 | 74.8 | 89.7 | 74.1 | 87.3 | 73.1 | 78.0 | 87.2 | 76.8 | 85.1 | 75.4 | 133.4 | 81.6 | 74.4 | 128.8 | 80.6 | 20.2 | 18.3 | 16.6 | 4891 | 1059 |
| NEWARK INTL AIRPORT | 40.68N | 74.17W | 30 | 12.3 | 16.6 | 94.2 | 74.6 | 91.1 | 73.1 | 88.4 | 72.0 | 77.7 | 88.7 | 76.3 | 85.8 | 74.7 | 130.0 | 82.1 | 73.4 | 124.4 | 80.8 | 25.0 | 21.8 | 19.3 | 4687 | 1257 |
| TETERBORO | 40.85N | 74.06W | 7 | 11.5 | 15.8 | 92.4 | 74.5 | 89.9 | 73.5 | 87.4 | 72.1 | 77.7 | 87.7 | 75.9 | 85.1 | 74.8 | 130.2 | 82.5 | 73.1 | 122.9 | 80.3 | 20.7 | 18.6 | 17.0 | 4996 | 1050 |
| TRENTON MERCER | 40.28N | 74.81W | 213 | 11.9 | 16.1 | 92.8 | 74.2 | 90.2 | 73.2 | 87.8 | 72.4 | 77.2 | 88.3 | 75.8 | 85.3 | 73.4 | 125.4 | 81.4 | 72.7 | 122.4 | 80.7 | 21.3 | 19.0 | 17.5 | 4982 | 1049 |
| ***New Mexico*** | | | | | | | | | | | | | | | | | | | | | | | | | *8 sites, 11 more on CD-ROM* | |
| ALAMOGORDO WHITE SA | 32.84N | 105.98W | 4199 | 21.3 | 25.2 | 99.8 | 63.9 | 98.6 | 64.2 | 95.2 | 64.0 | 71.0 | 86.6 | 69.5 | 85.3 | 66.3 | 113.4 | 75.3 | 65.6 | 110.5 | 75.2 | 22.2 | 18.7 | 16.4 | 2856 | 1904 |
| ALBUQUERQUE INTL | 35.04N | 106.62W | 5315 | 18.2 | 21.6 | 95.3 | 60.1 | 92.9 | 59.8 | 90.6 | 59.7 | 65.3 | 81.3 | 64.4 | 80.4 | 61.6 | 100.0 | 68.0 | 60.4 | 95.6 | 68.6 | 28.2 | 24.7 | 20.6 | 3994 | 1370 |
| CANNON AFB | 34.38N | 103.32W | 4295 | 12.6 | 17.8 | 97.8 | 63.4 | 94.8 | 63.8 | 92.1 | 64.2 | 70.4 | 83.8 | 69.2 | 83.1 | 67.0 | 116.7 | 73.3 | 65.5 | 110.7 | 73.3 | 28.2 | 24.8 | 21.5 | 3776 | 1355 |
| CLOVIS MUNI | 34.43N | 103.07W | 4213 | 13.9 | 17.9 | 97.0 | 63.9 | 93.3 | 63.8 | 91.1 | 63.9 | 69.5 | 84.3 | 68.5 | 83.6 | 65.7 | 111.2 | 74.3 | 64.1 | 104.8 | 72.7 | 31.8 | 27.2 | 24.3 | 4084 | 1191 |
| FOUR CORNERS RGNL | 36.74N | 108.23W | 5502 | 7.3 | 11.8 | 95.4 | 59.8 | 92.8 | 59.1 | 90.5 | 59.0 | 65.0 | 81.7 | 64.0 | 80.9 | 61.2 | 99.4 | 67.5 | 59.3 | 92.6 | 67.9 | 24.8 | 21.3 | 18.5 | 5328 | 912 |
| HOLLOMAN AFB | 32.85N | 106.10W | 4094 | 18.9 | 22.2 | 99.4 | 63.0 | 97.0 | 62.9 | 94.5 | 62.7 | 68.8 | 85.7 | 67.9 | 85.0 | 64.4 | 105.6 | 71.9 | 63.4 | 101.7 | 72.2 | 24.0 | 19.7 | 17.0 | 3228 | 1715 |
| ROSWELL/INDUSTRIAL | 33.31N | 104.54W | 3668 | 18.0 | 21.4 | 100.1 | 64.9 | 97.8 | 65.0 | 95.5 | 64.9 | 70.9 | 86.8 | 69.8 | 86.0 | 67.0 | 114.0 | 74.1 | 65.9 | 109.4 | 73.8 | 25.9 | 21.2 | 18.5 | 3116 | 1892 |
| WHITE SANDS | 32.38N | 106.48W | 4081 | 18.4 | 22.5 | 99.0 | 63.7 | 96.5 | 63.9 | 94.2 | 63.8 | 69.8 | 87.4 | 68.9 | 86.1 | 65.9 | 111.2 | 72.1 | 64.6 | 106.4 | 72.3 | 18.7 | 16.2 | 13.3 | 2946 | 1811 |
| ***New York*** | | | | | | | | | | | | | | | | | | | | | | | | | *19 sites, 17 more on CD-ROM* | |
| ALBANY COUNTY AIRPO | 42.75N | 73.80W | 292 | -0.9 | 3.9 | 89.2 | 73.0 | 86.2 | 71.2 | 83.4 | 70.1 | 75.5 | 84.8 | 74.0 | 82.2 | 72.7 | 122.5 | 80.3 | 71.3 | 116.7 | 78.7 | 24.1 | 20.5 | 18.5 | 6562 | 619 |
| AMBROSE LIGHT | 40.45N | 73.80W | 69 | 13.8 | 17.8 | 83.9 | N/A | 80.8 | N/A | 78.5 | N/A | N/A | N/A | N/A | N/A | N/A | N/A | N/A | N/A | N/A | N/A | 42.3 | 36.9 | 33.3 | 4916 | 704 |
| BINGHAMTON/BROOME C | 42.21N | 75.98W | 1637 | -0.2 | 4.1 | 85.5 | 70.1 | 82.3 | 68.3 | 80.0 | 67.2 | 72.8 | 80.9 | 71.1 | 78.5 | 70.2 | 118.2 | 76.6 | 68.7 | 112.0 | 74.9 | 20.9 | 18.8 | 17.3 | 7097 | 399 |
| GREATER BUFFALO INT | 42.94N | 78.74W | 705 | 3.6 | 7.4 | 86.4 | 71.3 | 83.9 | 70.1 | 81.6 | 68.8 | 74.8 | 81.9 | 73.2 | 80.1 | 72.4 | 123.1 | 79.0 | 70.7 | 115.9 | 77.6 | 27.4 | 24.5 | 20.8 | 6508 | 563 |
| ELMIRA CORNING RGNL | 42.16N | 76.89W | 955 | -0.3 | 4.7 | 89.9 | 71.9 | 86.5 | 70.0 | 83.7 | 68.8 | 74.7 | 84.4 | 72.8 | 82.2 | 71.9 | 122.1 | 80.1 | 70.1 | 114.8 | 77.5 | 20.3 | 18.2 | 16.3 | 6766 | 470 |
| GRIFFISS AIRPARK | 43.23N | 75.41W | 518 | -5.5 | 0.6 | 88.5 | 72.5 | 85.5 | 70.3 | 82.4 | 68.8 | 75.0 | 85.0 | 73.0 | 81.8 | 71.9 | 120.4 | 80.8 | 70.1 | 112.9 | 78.4 | 22.8 | 19.1 | 16.9 | 7054 | 473 |
| LONG ISLAND MAC ART | 40.79N | 73.10W | 98 | 11.5 | 15.7 | 88.5 | 73.4 | 85.7 | 72.2 | 82.7 | 70.9 | 76.6 | 83.5 | 75.3 | 81.2 | 74.6 | 129.8 | 79.7 | 73.3 | 124.2 | 78.3 | 24.0 | 20.4 | 18.6 | 5294 | 809 |
| CHATAUQUA CO JAMESTO | 42.15N | 79.25W | 1722 | 1.0 | 5.2 | 82.4 | 69.5 | 81.0 | 68.5 | 78.8 | 66.8 | 72.2 | 79.8 | 70.5 | 77.7 | 70.1 | 118.0 | 77.4 | 67.9 | 109.4 | 75.1 | 21.5 | 18.9 | 17.3 | 7166 | 295 |
| NEW YORK/JOHN F. KE | 40.66N | 73.80W | 23 | 13.8 | 17.8 | 89.8 | 72.9 | 86.5 | 71.8 | 83.7 | 71.1 | 76.7 | 83.8 | 75.4 | 81.6 | 74.6 | 129.4 | 80.1 | 73.4 | 124.2 | 78.6 | 27.2 | 24.6 | 21.6 | 4843 | 984 |
| NEW YORK/LA GUARDIA | 40.78N | 73.88W | 30 | 13.9 | 18.0 | 92.4 | 74.0 | 89.7 | 72.7 | 86.9 | 71.8 | 77.0 | 87.0 | 75.8 | 84.5 | 74.2 | 127.9 | 81.0 | 73.0 | 122.8 | 80.2 | 27.2 | 24.6 | 21.5 | 4555 | 1259 |
| STEWART INTL | 41.50N | 74.10W | 492 | 4.6 | 9.5 | 90.2 | 72.9 | 86.4 | 71.9 | 83.9 | 70.7 | 76.0 | 85.1 | 74.4 | 82.9 | 73.0 | 124.9 | 79.8 | 72.2 | 121.3 | 79.1 | 24.3 | 20.3 | 18.4 | 5933 | 722 |
| NIAGARA FALLS INTL | 43.11N | 78.95W | 587 | 3.0 | 7.3 | 88.0 | 72.9 | 85.3 | 71.3 | 82.4 | 69.7 | 75.5 | 83.7 | 73.9 | 81.6 | 72.9 | 124.6 | 80.2 | 71.7 | 119.6 | 78.7 | 26.4 | 23.6 | 20.3 | 6584 | 590 |
| PLATTSBURGH INTL | 44.65N | 73.47W | 233 | -9.6 | -5.1 | 86.5 | 71.3 | 83.2 | 69.5 | 80.3 | 68.2 | 74.1 | 82.3 | 72.2 | 80.0 | 71.4 | 117.0 | 79.0 | 69.5 | 109.5 | 76.6 | 20.6 | 18.4 | 16.3 | 7823 | 360 |
| DUTCHESS CO | 41.63N | 73.88W | 161 | 1.7 | 7.5 | 91.4 | 73.8 | 88.7 | 72.6 | 85.7 | 71.3 | 76.7 | 87.3 | 75.1 | 84.5 | 73.3 | 124.6 | 82.0 | 72.2 | 119.9 | 80.7 | 18.5 | 16.8 | 14.3 | 6149 | 702 |
| REPUBLIC | 40.73N | 73.42W | 82 | 12.3 | 17.7 | 90.2 | 73.7 | 86.5 | 71.9 | 83.7 | 71.3 | 76.7 | 84.6 | 75.4 | 82.1 | 74.7 | 130.1 | 79.9 | 73.1 | 123.4 | 78.3 | 24.7 | 21.4 | 19.0 | 5041 | 912 |
| ROCHESTER-MONROE CO | 43.12N | 77.68W | 554 | 2.9 | 6.9 | 88.7 | 73.2 | 85.6 | 71.2 | 82.7 | 69.7 | 75.4 | 84.5 | 73.5 | 81.9 | 72.5 | 122.7 | 80.6 | 70.8 | 115.8 | 78.2 | 25.1 | 21.4 | 18.9 | 6558 | 555 |
| SYRACUSE/HANCOCK | 43.11N | 76.10W | 417 | -1.2 | 4.3 | 89.2 | 73.2 | 86.3 | 71.3 | 83.6 | 70.0 | 75.3 | 85.2 | 73.6 | 82.6 | 72.2 | 121.1 | 80.9 | 70.5 | 114.0 | 78.7 | 24.3 | 20.5 | 18.3 | 6577 | 594 |
| ONEIDA CO | 43.15N | 75.38W | 745 | -5.0 | 1.0 | 87.3 | 72.5 | 84.3 | 70.6 | 81.8 | 69.0 | 75.0 | 83.2 | 73.2 | 80.8 | 72.4 | 123.4 | 79.2 | 70.6 | 115.8 | 77.4 | 20.7 | 18.7 | 17.2 | 7074 | 463 |
| WESTCHESTER CO | 41.07N | 73.71W | 397 | 9.0 | 12.8 | 89.9 | 73.9 | 86.5 | 72.1 | 83.7 | 70.8 | 76.4 | 84.9 | 74.8 | 82.3 | 73.5 | 126.4 | 79.4 | 72.5 | 122.4 | 78.4 | 21.8 | 18.6 | 16.5 | 5559 | 749 |
| ***North Carolina*** | | | | | | | | | | | | | | | | | | | | | | | | | *14 sites, 22 more on CD-ROM* | |
| ASHEVILLE MUNICIPAL | 35.43N | 82.54W | 2169 | 14.7 | 18.9 | 88.3 | 71.2 | 85.9 | 70.6 | 83.8 | 69.8 | 73.9 | 83.1 | 72.8 | 81.6 | 71.4 | 125.8 | 77.4 | 70.3 | 120.8 | 76.3 | 23.1 | 19.5 | 17.5 | 4144 | 844 |
| CHARLOTTE/DOUGLAS | 35.21N | 80.94W | 768 | 21.0 | 25.0 | 94.3 | 74.5 | 92.0 | 74.0 | 89.9 | 73.4 | 77.2 | 88.4 | 76.2 | 86.8 | 74.1 | 130.8 | 81.0 | 73.2 | 126.7 | 80.0 | 18.6 | 16.5 | 14.3 | 3065 | 1713 |
| FAYETTEVILLE RGNL G | 34.99N | 78.88W | 197 | 22.2 | 26.4 | 96.5 | 76.3 | 93.2 | 75.1 | 91.1 | 74.8 | 79.2 | 89.8 | 78.2 | 88.2 | 76.7 | 140.0 | 82.5 | 75.4 | 133.9 | 81.5 | 20.2 | 17.8 | 15.8 | 2764 | 1957 |
| FORT BRAGG/SIMMONS | 35.13N | 78.94W | 243 | 21.9 | 25.8 | 97.0 | 76.3 | 94.7 | 75.9 | 92.1 | 75.1 | 79.4 | 90.9 | 78.2 | 89.4 | 76.4 | 138.8 | 84.3 | 75.2 | 133.4 | 83.4 | 17.6 | 14.8 | 12.6 | 2786 | 2071 |
| GREENSBORO/G.-HIGH | 36.10N | 79.94W | 886 | 18.4 | 22.2 | 92.6 | 74.3 | 90.4 | 73.7 | 88.3 | 72.9 | 77.0 | 87.8 | 75.8 | 85.9 | 73.7 | 129.6 | 81.1 | 72.8 | 125.7 | 80.0 | 19.9 | 17.7 | 15.8 | 3606 | 1446 |
| HICKORY RGNL | 35.74N | 81.39W | 1188 | 19.2 | 23.4 | 92.9 | 72.4 | 90.4 | 72.4 | 88.2 | 71.8 | 75.7 | 86.0 | 74.8 | 84.7 | 72.9 | 127.5 | 79.2 | 72.1 | 124.1 | 78.3 | 17.5 | 14.9 | 13.0 | 3509 | 1377 |
| JACKSONVILLE (AWOS) | 34.83N | 77.61W | 95 | 20.5 | 24.8 | 94.0 | 76.9 | 91.4 | 75.8 | 90.1 | 75.2 | 79.6 | 90.9 | 78.2 | 88.5 | 76.5 | 138.7 | 85.0 | 75.1 | 132.2 | 83.2 | 19.9 | 17.6 | 15.5 | 2966 | 1721 |
| NEW RIVER MCAS | 34.70N | 77.43W | 26 | 23.2 | 26.8 | 93.1 | 78.1 | 90.9 | 77.5 | 89.3 | 76.9 | 80.6 | 89.0 | 79.4 | 87.7 | 78.7 | 149.3 | 85.1 | 77.0 | 140.9 | 83.8 | 20.1 | 17.8 | 15.9 | 2547 | 1937 |
| PITT GREENVILLE | 35.64N | 77.38W | 26 | 20.8 | 24.8 | 95.1 | 76.3 | 93.0 | 75.1 | 91.0 | 74.5 | 79.0 | 91.2 | 77.5 | 88.6 | 75.3 | 132.8 | 83.0 | 74.6 | 129.8 | 82.2 | 18.8 | 16.4 | 14.1 | 2930 | 1923 |
| POPE AFB | 35.17N | 79.03W | 200 | 20.9 | 24.8 | 97.1 | 75.5 | 93.4 | 74.7 | 91.1 | 74.5 | 79.6 | 87.9 | 78.5 | 87.0 | 77.5 | 144.0 | 81.1 | 76.7 | 140.4 | 80.9 | 18.7 | 16.5 | 14.1 | 2880 | 1991 |
| RALEIGH/RALEIGH-DUR | 35.87N | 78.79W | 436 | 19.6 | 23.6 | 94.8 | 75.7 | 92.4 | 75.2 | 90.2 | 74.5 | 78.3 | 89.7 | 77.3 | 88.1 | 75.3 | 134.8 | 82.7 | 74.3 | 130.2 | 81.5 | 18.9 | 16.8 | 15.0 | 3275 | 1666 |
| SEYMOUR JOHNSON AFB | 35.34N | 77.96W | 112 | 22.4 | 26.3 | 96.9 | 76.7 | 94.4 | 75.9 | 91.8 | 75.5 | 80.0 | 89.9 | 78.9 | 88.4 | 77.4 | 143.2 | 83.3 | 76.6 | 139.0 | 82.7 | 18.5 | 16.2 | 14.0 | 2734 | 2053 |
| WILMINGTON | 34.27N | 77.90W | 33 | 24.6 | 27.7 | 92.6 | 77.0 | 90.2 | 76.3 | 88.2 | 75.6 | 79.2 | 87.9 | 78.3 | 86.1 | 76.9 | 140.4 | 83.1 | 76.1 | 136.5 | 82.4 | 21.3 | 18.9 | 16.8 | 2444 | 2030 |
| SMITH REYNOLDS | 36.13N | 80.22W | 971 | 18.9 | 23.3 | 92.9 | 73.6 | 90.6 | 73.0 | 88.5 | 72.2 | 76.4 | 87.0 | 75.3 | 85.5 | 73.1 | 127.4 | 80.6 | 72.3 | 124.1 | 79.8 | 18.2 | 15.9 | 13.4 | 3468 | 1481 |
| ***North Dakota*** | | | | | | | | | | | | | | | | | | | | | | | | | *6 sites, 7 more on CD-ROM* | |
| BISMARCK MUNICIPAL | 46.77N | 100.75W | 1660 | -18.5 | -13.1 | 93.9 | 69.6 | 90.2 | 68.7 | 86.7 | 67.6 | 74.6 | 86.1 | 72.2 | 84.4 | 70.9 | 121.3 | 81.9 | 68.2 | 110.4 | 79.0 | 27.2 | 24.3 | 20.8 | 8396 | 546 |
| FARGO/HECTOR FIELD | 46.93N | 96.81W | 899 | -19.3 | -14.5 | 90.7 | 72.0 | 87.6 | 70.4 | 84.5 | 68.8 | 75.4 | 85.3 | 73.4 | 83.5 | 72.3 | 123.7 | 81.9 | 70.0 | 114.0 | 80.2 | 28.2 | 25.4 | 23.0 | 8729 | 555 |

**Design Conditions** 129

## Table 4-8 Design Conditions for Selected Locations *(Continued)*

*Meaning of acronyms:*
DB: Dry bulb temperature, °F
MCWB: Mean coincident wet bulb temperature, °F
WB: Wet bulb temperature, °F
DP: Dew point temperature, °F
Lat: Latitude, °
Long: Longitude, °
HR: Humidity ratio, grains of moisture per lb of dry air
MCDB: Mean coincident dry bulb temperature, °F
HDD and CDD 65: Annual heating and cooling degree-days, base 65°F, °F-day
Elev: Elevation, ft
WS: Wind speed, mph

| Station | Lat | Long | Elev | Heating DB 99.6% | Heating DB 99% | Cooling DB/MCWB 0.4% DB/MCWB | 1% DB/MCWB | 2% DB/MCWB | Evaporation WB/MCDB 0.4% WB/MCDB | 1% WB/MCDB | Dehumidification DP/HR/MCDB 0.4% DP/HR/MCDB | 1% DP/HR/MCDB | Extreme Annual WS 1% | 2.5% | 5% | Heat./Cool. Degree-Days HDD/CDD 65 |
|---|---|---|---|---|---|---|---|---|---|---|---|---|---|---|---|---|
| GRAND FORKS AFB | 47.95N | 97.40W | 912 | -18.3 | -14.7 | 89.9 73.8 | 86.1 70.9 | 82.4 68.4 | 77.1 84.3 | 74.2 82.0 | 74.9 135.5 81.2 | 72.2 123.4 78.2 | 27.7 24.9 22.0 | | | 9197 406 |
| GRAND FORKS INTL | 47.95N | 97.18W | 833 | -22.0 | -17.1 | 89.9 71.1 | 86.4 69.4 | 83.5 68.0 | 74.8 84.4 | 72.5 82.4 | 71.7 120.8 81.2 | 69.2 110.7 78.7 | 26.9 24.3 20.9 | | | 9320 425 |
| MINOT AFB | 48.43N | 101.36W | 1667 | -23.4 | -17.7 | 93.1 68.4 | 89.3 67.5 | 85.5 66.4 | 73.0 86.0 | 70.7 83.1 | 69.5 115.4 79.4 | 66.4 103.3 76.7 | 28.9 25.9 22.8 | | | 9024 430 |
| MINOT INTL | 48.26N | 101.28W | 1713 | -19.1 | -13.9 | 91.2 68.5 | 87.8 68.0 | 84.0 66.0 | 73.4 84.1 | 70.9 81.8 | 70.2 118.4 79.9 | 67.4 107.5 77.0 | 27.9 24.9 21.9 | | | 8696 444 |
| **Ohio** | | | | | | | | | | | | | 13 sites, 15 more on CD-ROM | | | |
| AKRON/AKRON-CANTON | 40.92N | 81.44W | 1237 | 2.8 | 7.9 | 88.8 72.8 | 86.1 71.8 | 83.5 70.3 | 75.5 84.4 | 73.9 82.1 | 72.7 127.1 80.4 | 71.4 121.4 78.4 | 23.2 19.8 18.1 | | | 6054 688 |
| CINCINNATI MUNI LUN | 39.10N | 84.42W | 499 | 8.1 | 13.4 | 92.8 74.5 | 90.3 74.2 | 88.1 73.3 | 78.0 87.9 | 76.7 86.2 | 75.1 134.4 82.6 | 73.8 128.4 81.1 | 20.4 18.3 16.6 | | | 4744 1155 |
| CLEVELAND | 41.41N | 81.85W | 804 | 4.1 | 9.7 | 89.7 73.7 | 87.0 72.4 | 84.2 71.0 | 76.2 85.4 | 74.7 83.0 | 73.2 127.3 81.3 | 72.0 121.9 79.6 | 24.5 20.9 18.8 | | | 5850 774 |
| COLUMBUS/PORT COLUM | 39.99N | 82.88W | 817 | 5.0 | 10.4 | 91.1 73.6 | 89.0 72.9 | 86.5 71.7 | 76.8 87.0 | 75.3 84.5 | 73.6 129.1 81.3 | 72.4 123.8 80.2 | 22.3 18.9 16.9 | | | 5255 1015 |
| DAYTON/JAMES M COX | 39.91N | 84.22W | 1004 | 2.0 | 8.1 | 90.4 73.5 | 88.0 72.8 | 85.5 71.4 | 76.5 86.2 | 75.1 84.0 | 73.4 128.7 81.9 | 72.2 123.8 80.5 | 24.5 20.8 18.8 | | | 5512 945 |
| FINDLAY | 41.01N | 83.67W | 814 | 1.0 | 6.8 | 90.5 73.2 | 88.1 72.3 | 85.1 70.7 | 76.5 86.2 | 74.8 83.5 | 73.2 127.2 82.2 | 72.0 122.2 80.5 | 24.7 21.0 18.9 | | | 5930 808 |
| FAIRFIELD CO | 39.76N | 82.66W | 869 | 1.5 | 9.4 | 90.5 73.4 | 88.4 73.0 | 85.9 71.7 | 76.1 85.9 | 75.3 83.8 | 73.2 127.6 81.1 | 72.5 124.2 80.2 | 20.1 17.8 15.9 | | | 5459 810 |
| MANSFIELD LAHM RGNL | 40.82N | 82.52W | 1312 | 1.0 | 6.7 | 88.1 73.0 | 85.5 71.8 | 83.0 70.3 | 75.6 84.5 | 74.1 82.3 | 72.9 128.0 80.9 | 71.5 122.2 79.2 | 24.3 20.8 18.9 | | | 6152 659 |
| OHIO STATE UNIVERSI | 40.08N | 83.08W | 906 | 4.9 | 10.0 | 90.4 73.3 | 88.2 72.8 | 85.6 71.5 | 76.4 85.9 | 74.9 83.5 | 73.0 126.8 81.2 | 72.3 123.7 80.0 | 21.8 19.1 17.3 | | | 5429 911 |
| RICKENBACKER INTL | 39.80N | 82.92W | 745 | 6.6 | 11.7 | 92.7 75.2 | 90.5 74.5 | 88.3 73.5 | 80.5 86.8 | 78.5 85.6 | 79.1 155.3 84.2 | 76.5 142.2 82.5 | 23.7 20.0 17.7 | | | 4971 1156 |
| TOLEDO EXPRESS | 41.59N | 83.80W | 692 | 1.3 | 6.8 | 91.3 74.0 | 88.6 72.6 | 85.9 71.4 | 77.2 85.9 | 75.4 84.1 | 74.2 130.8 82.8 | 72.6 124.1 80.6 | 24.3 20.7 18.6 | | | 6074 798 |
| DAYTON/WRIGHT-PATTE | 39.83N | 84.05W | 823 | 3.1 | 9.4 | 90.6 73.8 | 88.3 73.3 | 86.0 72.0 | 77.0 87.0 | 75.6 85.3 | 74.8 134.3 80.7 | 73.0 126.1 79.1 | 21.4 18.8 16.9 | | | 5301 958 |
| YOUNGSTOWN MUNI | 41.25N | 80.67W | 1188 | 2.9 | 7.9 | 89.6 71.8 | 85.8 71.1 | 83.4 69.6 | 75.0 84.4 | 73.4 81.8 | 72.2 124.8 79.5 | 70.7 118.1 77.6 | 21.3 18.9 17.2 | | | 6198 583 |
| **Oklahoma** | | | | | | | | | | | | | 9 sites, 11 more on CD-ROM | | | |
| FORT SILL | 34.65N | 98.40W | 1188 | 14.3 | 20.2 | 100.8 73.1 | 98.6 73.4 | 95.5 73.4 | 77.5 91.0 | 76.5 89.8 | 74.1 132.8 82.5 | 72.9 127.6 81.6 | 24.7 21.3 19.3 | | | 3197 2117 |
| LAWTON MUNICIPAL | 34.57N | 98.42W | 1109 | 17.9 | 20.8 | 102.4 73.9 | 100.2 73.7 | 98.8 73.8 | 77.4 92.6 | 77.1 91.3 | 73.5 129.9 82.8 | 73.0 127.8 82.5 | 26.0 23.1 20.1 | | | 3168 2271 |
| OKLAHOMA CITY/W. RO | 35.39N | 97.60W | 1306 | 14.1 | 18.9 | 99.6 74.2 | 96.9 74.2 | 94.0 74.0 | 77.8 91.0 | 76.9 89.9 | 73.2 134.3 83.6 | 73.2 129.3 82.3 | 27.6 24.9 22.7 | | | 3438 1950 |
| OKLAHOMA CITY/WILEY | 35.53N | 97.65W | 1299 | 12.5 | 18.2 | 99.7 73.9 | 97.2 73.9 | 93.8 73.7 | 77.4 91.2 | 76.3 89.7 | 72.5 134.3 83.3 | 72.5 129.0 82.4 | 26.7 24.5 22.0 | | | 3487 2047 |
| STILLWATER RGNL | 36.16N | 97.09W | 984 | 13.7 | 18.2 | 101.7 75.4 | 99.1 75.6 | 95.4 75.8 | 79.1 93.9 | 78.0 92.5 | 75.1 136.5 86.1 | 73.3 128.6 84.0 | 24.7 22.0 19.7 | | | 3589 2001 |
| TINKER AFB | 35.42N | 97.38W | 1302 | 15.8 | 18.9 | 99.3 72.9 | 96.8 73.0 | 93.1 73.0 | 77.3 89.2 | 76.2 88.0 | 75.1 130.5 81.0 | 72.9 128.1 80.6 | 26.3 23.7 20.6 | | | 3383 1916 |
| TULSA INTL ARPT(AW) | 36.20N | 95.89W | 676 | 13.2 | 18.3 | 101.4 73.1 | 98.4 73.0 | 96.0 73.0 | 79.2 92.5 | 78.1 91.2 | 75.4 136.1 85.5 | 74.4 131.8 84.7 | 24.6 21.5 19.4 | | | 3455 2051 |
| RICHARD LLOYD JONES | 36.04N | 95.98W | 627 | 15.8 | 18.8 | 100.0 76.7 | 98.6 76.9 | 95.0 76.6 | 79.6 94.1 | 78.5 92.6 | 75.4 136.6 85.4 | 74.8 133.5 84.9 | 19.5 17.8 16.1 | | | 3503 2020 |
| VANCE AFB | 36.34N | 97.99W | 1306 | 10.0 | 15.7 | 100.4 73.4 | 98.5 73.5 | 95.1 73.7 | 77.4 91.6 | 76.3 90.4 | 73.4 130.3 82.0 | 72.5 126.5 81.7 | 27.5 24.7 21.7 | | | 3936 1864 |
| **Oregon** | | | | | | | | | | | | | 9 sites, 18 more on CD-ROM | | | |
| AURORA STATE | 45.25N | 122.77W | 197 | 26.6 | 28.2 | 91.3 66.9 | 88.3 66.6 | 83.9 65.5 | 70.1 86.3 | 68.2 84.1 | 63.8 89.4 75.7 | 62.9 86.6 73.8 | 18.0 15.8 12.8 | | | 4415 379 |
| CORVALLIS MUNI | 44.50N | 123.29W | 246 | 25.0 | 27.7 | 92.9 67.0 | 89.0 65.9 | 85.6 64.2 | 68.6 89.1 | 67.0 86.9 | 61.0 80.7 76.7 | 58.5 73.9 75.5 | 19.8 17.8 16.0 | | | 4255 397 |
| EUGENE/MAHLON SWEET | 44.13N | 123.21W | 374 | 27.3 | 29.3 | 92.9 66.5 | 88.9 65.6 | 84.1 64.5 | 68.8 87.4 | 67.1 84.7 | 60.4 79.5 74.6 | 60.4 79.5 72.2 | 19.6 17.5 15.9 | | | 4638 270 |
| MC MINNVILLE MUNI | 45.20N | 123.13W | 161 | 26.8 | 28.2 | 91.4 66.0 | 88.4 65.8 | 84.0 64.7 | 68.7 87.2 | 67.0 85.1 | 62.6 85.4 74.0 | 60.8 80.1 71.9 | 20.7 17.7 15.5 | | | 4673 287 |
| MEDFORD-JACKSON COU | 42.39N | 122.87W | 1329 | 23.1 | 26.1 | 99.2 66.9 | 95.6 65.8 | 92.2 64.6 | 68.8 94.3 | 67.4 91.5 | 60.3 82.2 74.1 | 58.3 76.4 73.7 | 18.3 15.4 12.4 | | | 4264 834 |
| PORTLAND INTL ARPT | 45.59N | 122.60W | 108 | 25.2 | 29.5 | 91.4 67.3 | 87.5 66.6 | 83.6 65.3 | 69.5 86.9 | 67.5 84.5 | 63.2 87.0 75.1 | 61.6 82.3 73.1 | 23.6 19.7 17.5 | | | 4214 433 |
| PORTLAND/HILLSBORO | 45.54N | 122.95W | 230 | 23.2 | 27.0 | 92.2 68.0 | 88.3 67.0 | 84.0 65.5 | 70.5 88.0 | 68.3 85.3 | 63.7 89.2 76.8 | 61.7 82.8 73.6 | 18.7 16.8 14.3 | | | 4744 283 |
| ROBERTS FLD | 44.25N | 121.15W | 3084 | 5.6 | 12.6 | 93.0 61.7 | 90.2 60.8 | 87.0 59.7 | 63.7 88.6 | 62.1 85.3 | 54.8 71.5 67.2 | 52.8 66.5 66.9 | 20.7 18.6 16.7 | | | 6470 237 |
| SALEM/MCNARY | 44.91N | 123.00W | 200 | 23.5 | 27.4 | 92.7 66.8 | 88.2 65.8 | 84.3 64.6 | 68.7 88.4 | 67.1 85.2 | 61.5 82.2 73.0 | 59.9 79.7 72.3 | 20.9 18.4 16.3 | | | 4533 313 |
| **Pennsylvania** | | | | | | | | | | | | | 14 sites, 14 more on CD-ROM | | | |
| ALLENTOWN/A-BETHLE | 40.65N | 75.45W | 384 | 8.4 | 12.6 | 91.0 72.5 | 88.3 72.5 | 85.7 71.3 | 75.2 86.4 | 73.6 83.7 | 73.8 127.8 81.2 | 72.5 122.1 79.7 | 23.4 19.7 17.7 | | | 5552 838 |
| ALTOONA BLAIR CO | 40.30N | 78.32W | 1470 | 5.9 | 10.0 | 88.3 71.8 | 85.6 70.8 | 82.8 69.4 | 74.7 83.9 | 73.1 81.8 | 72.0 125.2 79.9 | 70.3 118.0 77.8 | 23.0 19.1 17.1 | | | 5950 612 |
| BUTLER CO SCHOLTER F | 40.78N | 79.94W | 1247 | 3.2 | 8.8 | 88.1 71.9 | 84.5 70.3 | 82.2 68.9 | 74.5 83.4 | 72.9 81.6 | 72.1 124.3 79.8 | 70.3 116.9 77.3 | 18.0 15.6 13.0 | | | 6089 549 |
| ERIE INTL AIRPORT | 42.08N | 80.18W | 738 | 6.8 | 10.4 | 86.7 73.0 | 84.2 71.8 | 81.8 70.5 | 74.7 82.7 | 73.3 81.1 | 71.4 119.1 78.7 | 70.3 116.9 78.3 | 24.5 21.3 19.4 | | | 6080 659 |
| HARRISBURG/CAPITAL | 40.22N | 76.85W | 348 | 10.7 | 15.4 | 92.5 73.8 | 89.9 72.6 | 87.5 71.8 | 76.6 87.1 | 75.3 84.6 | 73.4 125.8 80.5 | 72.5 121.8 79.5 | 20.6 18.4 16.5 | | | 5109 1056 |
| HARRISBURG INTL | 40.19N | 76.76W | 312 | 11.6 | 15.6 | 92.3 73.5 | 89.7 74.0 | 86.9 72.8 | 78.0 88.0 | 76.5 85.3 | 73.4 125.8 83.0 | 73.5 126.1 81.0 | 25.6 23.1 19.0 | | | 5046 1110 |
| PHILADELPHIA INTL | 39.87N | 75.23W | 30 | 13.8 | 18.0 | 93.1 75.1 | 90.3 75.1 | 88.4 73.9 | 78.3 88.6 | 76.7 86.4 | 74.3 128.2 82.5 | 74.3 128.2 81.4 | 24.7 20.9 18.7 | | | 4512 1332 |
| NORTHEAST PHILADELPH | 40.08N | 75.01W | 118 | 12.3 | 16.8 | 93.2 75.3 | 90.6 74.2 | 88.3 73.0 | 78.4 88.7 | 76.8 86.7 | 75.4 133.4 82.8 | 73.5 125.4 80.9 | 21.8 18.9 17.3 | | | 4754 1177 |
| ALLEGHENY CO | 40.36N | 79.92W | 1273 | 5.6 | 10.0 | 90.0 72.3 | 87.5 71.3 | 84.4 69.7 | 75.0 84.8 | 73.6 82.8 | 73.5 125.2 79.9 | 70.8 118.8 78.2 | 20.3 18.4 16.8 | | | 5438 851 |
| GREATER PITTSBURGH I | 40.50N | 80.23W | 1204 | 5.2 | 9.9 | 89.7 72.4 | 87.0 71.1 | 84.4 69.8 | 75.2 84.8 | 73.6 82.6 | 72.3 125.0 80.0 | 70.9 119.2 78.3 | 23.1 19.6 17.6 | | | 5583 782 |
| READING RGNL CARL A | 40.37N | 75.96W | 354 | 9.9 | 14.3 | 92.6 74.1 | 90.0 73.0 | 87.6 72.0 | 77.1 87.6 | 75.4 84.8 | 73.4 125.9 82.0 | 72.5 121.9 80.7 | 22.8 19.3 17.8 | | | 5171 997 |
| WASHINGTON CO | 40.14N | 80.28W | 1184 | 3.0 | 8.8 | 86.7 70.7 | 85.5 69.6 | 82.9 68.4 | 73.6 83.1 | 72.2 82.0 | 70.4 117.1 78.9 | 69.7 114.0 77.7 | 19.3 17.0 14.7 | | | 5964 539 |
| WILKES-BARRE-SCRANT | 41.34N | 75.73W | 961 | 4.4 | 9.1 | 89.3 71.9 | 86.2 70.3 | 83.5 69.0 | 74.9 84.0 | 73.2 81.6 | 72.1 123.0 79.1 | 70.5 116.6 77.3 | 20.0 18.0 16.3 | | | 6086 637 |
| WILLOW GROVE NAS JR | 40.20N | 75.15W | 361 | 11.7 | 15.7 | 92.6 74.7 | 90.1 73.4 | 87.6 72.2 | 77.6 88.5 | 76.1 85.9 | 74.3 129.9 83.2 | 72.9 127.9 81.6 | 18.8 16.5 14.1 | | | 4907 1074 |
| **Rhode Island** | | | | | | | | | | | | | 1 site, 2 more on CD-ROM | | | |
| PROVIDENCE/GREEN ST | 41.72N | 71.43W | 62 | 8.5 | 12.9 | 90.1 73.3 | 86.7 71.9 | 83.8 70.6 | 76.4 85.2 | 74.9 82.0 | 73.9 126.5 80.2 | 72.7 121.3 78.6 | 24.3 20.7 18.7 | | | 5562 743 |

130	Principles of HVAC

## Table 4-8  Design Conditions for Selected Locations *(Continued)*

Meaning of acronyms:  
DB: Dry bulb temperature, °F    WB: Wet bulb temperature, °F    Lat: Latitude, °    Long: Longitude, °    Elev: Elevation, ft    WS: Wind speed, mph  
MCWB: Mean coincident wet bulb temperature, °F    DP: Dew point temperature, °F    HR: Humidity ratio, grains of moisture per lb of dry air  
MCDB: Mean coincident dry bulb temperature, °F    HDD and CDD 65: Annual heating and cooling degree-days, base 65°F, °F-day

| Station | Lat | Long | Elev | Heating DB 99.6% | Heating DB 99% | Cooling DB/MCWB 0.4% DB/MCWB | 1% DB/MCWB | 2% DB/MCWB | Evaporation WB/MCDB 0.4% WB/MCDB | 1% WB/MCDB | 2% WB/MCDB | Dehumidification DP/HR/MCDB 0.4% DP/HR/MCDB | 1% DP/HR/MCDB | Extreme Annual WS 1% | 2.5% | 5% | Heat./Cool. Degree-Days HDD/CDD65 |
|---|---|---|---|---|---|---|---|---|---|---|---|---|---|---|---|---|---|
| **South Carolina** | | | | | | | | | | | | | | | | | *6 sites, 8 more on CD-ROM* |
| CHARLESTON MUNI | 32.90N | 80.04W | 49 | 27.3 | 30.4 | 94.3 78.2 | 92.1 77.6 | 90.4 77.1 | 80.8 89.1 | 79.9 87.9 | 78.9 150.0 84.4 | 77.7 144.3 83.5 | 20.4 18.3 | 16.5 | 1880 | 2357 |
| COLUMBIA METRO | 33.94N | 81.12W | 226 | 22.8 | 26.5 | 97.2 75.2 | 94.8 75.0 | 92.6 74.5 | 78.5 89.9 | 77.7 88.7 | 75.7 135.6 82.2 | 74.8 131.7 81.4 | 19.4 17.0 | 15.2 | 2500 | 2166 |
| FLORENCE RGNL | 34.19N | 79.73W | 151 | 23.8 | 27.1 | 96.2 76.7 | 93.4 76.0 | 91.3 75.5 | 79.3 90.5 | 78.3 88.9 | 76.6 139.4 83.6 | 75.4 133.8 82.2 | 19.4 17.7 | 15.8 | 2429 | 2102 |
| FOLLY ISLAND | 32.68N | 79.88W | 16 | 31.4 | 34.6 | 87.5 77.8 | 86.2 77.7 | 85.1 77.5 | 80.0 85.0 | 79.3 84.4 | 78.9 150.0 84.0 | 77.8 144.7 82.9 | 33.1 26.2 | 23.0 | 1923 | 2126 |
| GREENVILLE/GREENVIL | 34.90N | 82.22W | 971 | 21.2 | 25.1 | 94.4 73.8 | 91.8 73.5 | 89.8 72.9 | 77.1 88.0 | 76.1 86.2 | 74.2 132.2 80.4 | 73.2 127.8 79.6 | 19.5 17.5 | 15.8 | 3080 | 1630 |
| SHAW AFB/SUMTER | 33.97N | 80.48W | 240 | 24.6 | 27.3 | 95.5 75.3 | 93.1 75.2 | 90.9 74.8 | 79.1 90.3 | 78.2 88.3 | 76.6 139.9 82.8 | 75.2 133.4 81.4 | 19.2 17.0 | 15.3 | 2422 | 2080 |
| **South Dakota** | | | | | | | | | | | | | | | | | *3 sites, 16 more on CD-ROM* |
| ELLSWORTH AFB | 44.15N | 103.10W | 3278 | −9.4 | −3.6 | 95.7 65.6 | 93.0 65.5 | 88.2 64.6 | 70.7 85.7 | 68.9 84.7 | 66.0 108.2 78.0 | 63.9 100.4 75.7 | 34.6 28.7 | 25.1 | 6882 | 668 |
| RAPID CITY/REGIONAL | 44.05N | 103.05W | 3169 | −9.2 | −3.4 | 97.2 65.8 | 93.0 65.5 | 89.4 64.8 | 69.2 84.7 | 66.6 109.5 77.7 | 64.3 101.6 75.7 | 35.2 30.6 | 26.2 | 7000 | 671 |
| SIOUX FALLS/FOSS FI | 43.58N | 96.75W | 1427 | −12.3 | −7.3 | 92.2 73.6 | 88.9 73.0 | 86.0 71.3 | 77.2 87.2 | 75.4 85.4 | 74.3 135.4 83.4 | 72.4 126.4 81.6 | 27.5 24.6 | 21.2 | 7470 | 745 |
| **Tennessee** | | | | | | | | | | | | | | | | | *7 sites, 3 more on CD-ROM* |
| TRI CITIES RGNL | 36.48N | 82.40W | 1526 | 12.9 | 17.7 | 90.5 71.8 | 88.2 71.1 | 86.0 71.1 | 75.1 85.0 | 74.0 83.7 | 72.2 126.3 79.1 | 71.2 121.8 77.9 | 18.9 16.5 | 13.9 | 4214 | 1033 |
| CHATTANOOGA/LOVELL | 35.03N | 85.20W | 689 | 19.0 | 23.1 | 95.0 74.9 | 92.6 74.4 | 90.4 73.8 | 77.9 89.1 | 76.9 87.6 | 75.0 134.7 81.5 | 73.9 129.7 80.6 | 17.9 16.0 | 13.6 | 3145 | 1763 |
| MC KELLAR SIPES RGN | 35.59N | 88.92W | 423 | 15.4 | 19.3 | 94.9 76.6 | 92.8 76.4 | 90.7 75.9 | 79.9 90.3 | 78.6 88.9 | 74.0 142.6 85.5 | 73.0 135.5 84.0 | 19.5 17.8 | 16.0 | 3427 | 1746 |
| KNOXVILLE MUNICIPAL | 35.82N | 83.99W | 981 | 16.5 | 20.8 | 93.0 73.7 | 90.6 73.7 | 88.5 73.0 | 77.1 87.8 | 76.1 86.1 | 74.0 131.5 81.4 | 73.0 127.2 80.4 | 20.4 17.7 | 15.3 | 3594 | 1514 |
| MEMPHIS INTL ARPT | 35.06N | 83.99W | 331 | 18.7 | 22.9 | 96.7 77.2 | 94.3 76.6 | 92.4 76.1 | 80.0 91.5 | 79.0 90.1 | 76.9 141.9 85.8 | 75.8 136.7 84.8 | 20.2 18.2 | 16.5 | 2898 | 2253 |
| MILLINGTON MUNI ARP | 35.35N | 89.87W | 322 | 17.6 | 21.4 | 98.2 78.7 | 95.4 77.4 | 92.9 76.3 | 81.4 93.0 | 79.6 91.3 | 77.8 151.1 87.0 | 76.2 138.2 86.2 | 18.5 16.3 | 14.1 | 3123 | 2031 |
| NASHVILLE/METROPOLI | 36.12N | 86.69W | 604 | 14.8 | 19.3 | 94.8 74.9 | 92.4 74.7 | 90.3 74.1 | 78.2 88.9 | 77.2 87.9 | 75.2 135.0 82.9 | 74.0 128.8 81.8 | 19.4 17.3 | 15.6 | 3518 | 1729 |
| **Texas** | | | | | | | | | | | | | | | | | *51 sites, 34 more on CD-ROM* |
| ABILENE DYESS AFB | 32.43N | 99.85W | 1788 | 18.9 | 23.4 | 101.7 72.0 | 99.3 71.9 | 96.9 71.9 | 77.2 91.1 | 76.0 89.9 | 73.6 133.7 81.4 | 72.4 128.5 80.7 | 25.3 22.2 | 19.4 | 2507 | 2537 |
| ABILENE MUNICIPAL | 32.41N | 99.68W | 1791 | 20.1 | 24.6 | 99.4 70.8 | 97.3 70.8 | 95.0 70.9 | 75.5 88.9 | 74.6 87.9 | 72.2 127.4 80.1 | 71.2 123.0 79.4 | 26.0 23.7 | 20.7 | 2482 | 2389 |
| AMARILLO INTL | 35.22N | 101.71W | 3606 | 9.8 | 15.6 | 97.3 66.2 | 94.7 66.3 | 92.2 66.2 | 71.3 86.1 | 70.2 85.3 | 67.3 114.9 75.3 | 66.1 110.1 74.4 | 29.2 26.4 | 24.2 | 4102 | 1366 |
| AUSTIN/MUELLER MUNI | 30.18N | 97.68W | 495 | 26.6 | 29.8 | 99.8 74.5 | 98.2 74.5 | 96.1 74.4 | 79.1 89.7 | 78.3 88.9 | 76.7 141.9 81.8 | 75.8 137.3 81.0 | 21.2 19.0 | 17.2 | 1671 | 2962 |
| BROWNSVILLE INTL | 25.91N | 97.43W | 23 | 38.1 | 42.1 | 95.4 77.8 | 94.3 77.7 | 93.1 77.7 | 80.9 89.7 | 80.3 87.9 | 79.3 152.2 83.0 | 78.7 149.1 82.8 | 26.2 23.8 | 20.7 | 538 | 3986 |
| AUSTIN CAMP MABRY | 30.32N | 97.77W | 659 | 28.4 | 32.4 | 99.7 74.2 | 97.9 74.4 | 95.8 74.5 | 78.5 89.0 | 77.9 88.4 | 76.3 140.8 81.0 | 75.4 136.6 80.4 | 19.2 17.1 | 15.5 | 1498 | 3093 |
| EASTERWOOD FLD | 30.59N | 96.36W | 328 | 27.8 | 31.6 | 99.6 75.7 | 97.4 75.7 | 95.3 75.8 | 79.8 90.8 | 78.8 89.2 | 77.2 143.5 83.2 | 76.4 139.3 82.3 | 20.2 18.3 | 16.6 | 1588 | 3030 |
| CORPUS CHRISTI/INT. | 27.77N | 97.51W | 43 | 34.3 | 38.0 | 96.3 77.8 | 94.6 77.9 | 93.0 77.8 | 81.1 89.3 | 80.4 88.3 | 79.3 152.1 83.1 | 78.7 149.4 82.9 | 27.2 24.9 | 22.8 | 861 | 3529 |
| DRAUGHON MILLER CEN | 31.15N | 97.40W | 682 | 25.0 | 28.1 | 99.7 74.2 | 98.0 74.3 | 95.9 74.3 | 78.2 90.9 | 77.4 90.0 | 75.9 159.8 84.4 | 73.7 128.6 81.0 | 25.6 22.4 | 19.9 | 1975 | 2734 |
| EL PASO INTL ARPT | 31.81N | 106.38W | 3917 | 23.9 | 27.5 | 100.7 64.5 | 98.5 64.6 | 96.2 64.0 | 70.2 86.0 | 69.3 85.2 | 66.8 114.3 72.9 | 65.5 109.2 73.0 | 26.5 22.3 | 18.7 | 2383 | 2379 |
| ROBERT GRAY AAF | 31.07N | 97.83W | 1014 | 27.1 | 29.9 | 100.1 71.9 | 97.2 72.2 | 95.7 72.5 | 76.8 87.2 | 76.0 86.9 | 74.9 135.6 78.3 | 73.3 128.6 77.7 | 22.8 19.7 | 17.0 | 1816 | 2816 |
| FORT WORTH ALLIANCE | 32.97N | 97.32W | 722 | 22.4 | 26.7 | 101.8 74.5 | 99.5 74.6 | 97.3 74.4 | 78.3 92.0 | 77.2 90.7 | 74.8 134.0 83.7 | 73.3 127.4 82.3 | 23.6 20.8 | 18.9 | 2363 | 2668 |
| FORT WORTH MEACHAM | 32.82N | 97.36W | 705 | 21.9 | 26.7 | 100.6 74.6 | 98.6 74.7 | 96.8 74.6 | 78.5 91.5 | 77.6 90.8 | 75.2 135.5 83.7 | 74.1 130.4 82.5 | 22.4 19.8 | 18.3 | 2253 | 2723 |
| FORT WORTH NAS JRB | 32.77N | 97.44W | 650 | 22.0 | 27.4 | 101.6 72.3 | 99.6 72.7 | 97.8 72.2 | 79.0 92.0 | 78.0 91.1 | 74.9 137.9 84.1 | 73.5 129.5 81.1 | 24.1 20.7 | 18.5 | 2149 | 2785 |
| GALVESTON | 29.27N | 94.86W | 10 | 36.0 | 39.2 | 91.5 79.1 | 90.5 79.0 | 89.0 78.9 | 81.6 86.6 | 80.8 86.0 | 79.2 151.6 83.9 | 78.5 148.1 84.0 | 25.1 22.1 | 19.8 | 1011 | 3242 |
| DRAUGHON MILLER CEN | 31.09N | 97.69W | 850 | 26.5 | 29.5 | 99.9 74.2 | 97.8 74.3 | 95.6 74.5 | 80.4 89.4 | 78.9 88.4 | 77.4 146.3 83.5 | 76.8 143.3 82.9 | 20.2 19.8 | 15.7 | 1392 | 3183 |
| LACKLAND AFB KELLY | 31.09N | 97.69W | 850 | 26.5 | 29.5 | 99.6 74.2 | 97.8 74.3 | 95.6 74.5 | 80.4 89.4 | 78.9 88.4 | 77.4 146.3 83.5 | 76.8 143.3 82.9 | 22.1 19.8 | 15.7 | 1392 | 3183 |
| KILLEEN MUNI (AWOS) | 31.09N | 97.69W | 850 | 26.5 | 29.5 | 99.6 74.2 | 97.8 74.3 | 95.6 74.5 | 80.4 89.4 | 78.9 88.4 | 77.4 146.3 83.5 | 76.8 143.3 82.9 | 22.1 19.8 | 15.7 | 1392 | 3183 |
| LAREDO INTL AIRPORT | 27.55N | 99.47W | 509 | 34.4 | 38.6 | 102.3 73.4 | 100.4 73.5 | 98.9 73.1 | 79.1 90.7 | 78.4 90.7 | 75.7 137.0 81.4 | 75.0 134.0 81.1 | 25.1 22.0 | 20.1 | 839 | 4149 |
| LAUGHLIN AFB | 29.36N | 100.78W | 1083 | 30.3 | 34.1 | 104.1 72.8 | 101.6 72.9 | 99.3 73.5 | 78.8 91.3 | 77.5 90.7 | 75.5 138.9 83.2 | 74.3 133.2 82.6 | 22.5 19.6 | 17.6 | 1218 | 3518 |
| LONGVIEW | 32.39N | 94.71W | 374 | 24.9 | 27.9 | 99.3 75.5 | 97.5 75.3 | 95.5 75.3 | 79.1 90.7 | 78.3 89.9 | 75.3 138.4 82.7 | 75.3 134.5 82.0 | 19.6 17.8 | 16.0 | 2109 | 2531 |
| LUBBOCK/LUBBOCK INT | 33.67N | 101.82W | 3241 | 15.9 | 19.9 | 99.0 66.7 | 96.6 67.4 | 94.0 67.6 | 73.2 87.6 | 72.1 86.5 | 69.6 122.7 77.3 | 68.3 117.2 76.2 | 28.9 25.9 | 23.5 | 3275 | 1846 |
| ANGELINA CO | 31.23N | 94.75W | 315 | 27.1 | 29.8 | 99.0 75.7 | 97.1 75.7 | 95.3 75.7 | 80.0 90.4 | 79.2 89.5 | 75.2 135.3 83.1 | 76.9 141.6 82.5 | 17.8 16.1 | 14.3 | 1847 | 2646 |
| MC GREGOR EXECUTIVE | 31.49N | 97.30W | 591 | 25.2 | 28.0 | 100.1 74.7 | 98.5 74.7 | 96.7 74.7 | 78.5 91.7 | 77.7 91.1 | 75.2 135.3 83.1 | 74.5 132.0 82.7 | 20.4 20.0 | 18.5 | 2082 | 2721 |
| MC ALLEN MILLER INT | 26.18N | 98.24W | 112 | 37.9 | 42.0 | 100.2 76.3 | 99.0 76.5 | 97.0 76.5 | 80.4 91.0 | 79.8 89.7 | 78.8 150.1 82.6 | 77.7 144.5 82.2 | 24.9 22.7 | 20.4 | 546 | 4465 |
| COLLIN CO RGNL | 33.18N | 96.59W | 584 | 21.4 | 26.5 | 100.1 75.1 | 99.0 75.3 | 96.8 75.3 | 78.5 92.0 | 77.9 91.3 | 75.2 134.9 82.9 | 74.5 132.0 82.4 | 23.1 19.8 | 17.8 | 2486 | 2492 |

# Design Conditions

**Table 4-8  Design Conditions for Selected Locations** *(Continued)*

*Meaning of acronyms:*
DB: Dry bulb temperature, °F
MCWB: Mean coincident wet bulb temperature, °F
WB: Wet bulb temperature, °F
DP: Dew point temperature, °F
MCDB: Mean coincident dry bulb temperature, °F
Lat: Latitude, °
Long: Longitude, °
HR: Humidity ratio, grains of moisture per lb of dry air
HDD and CDD 65: Annual heating and cooling degree-days, base 65°F, °F-day
Elev: Elevation, ft
WS: Wind speed, mph

| Station | Lat | Long | Elev | Heating DB 99.6% | Heating DB 99% | Cooling DB/MCWB 0.4% DB/MCWB | Cooling DB/MCWB 1% DB/MCWB | Cooling DB/MCWB 2% DB/MCWB | Evaporation WB/MCDB 0.4% WB/MCDB | Evaporation WB/MCDB 1% WB/MCDB | Dehumidification DP/HR/MCDB 0.4% DP/HR/MCDB | Dehumidification DP/HR/MCDB 1% DP/HR/MCDB | Extreme Annual WS 1% | Extreme Annual WS 2.5% | Extreme Annual WS 5% | Heat./Cool. Degree-Days HDD/CDD 65 |
|---|---|---|---|---|---|---|---|---|---|---|---|---|---|---|---|---|
| MIDLAND/MIDLAND REG | 31.93N | 102.21W | 2861 | 19.9 | 24.1 | 98.2 67.1 | 95.9 67.4 | 93.0 67.7 | 73.3 87.0 | 72.3 86.4 | 70.2 123.8 76.6 | 69.0 118.4 76.1 | 26.6 | 24.0 | 20.7 | 2617 2260 |
| A L MANGHAM JR RGNL | 31.58N | 94.70W | 354 | 25.2 | 27.8 | 98.5 75.8 | 95.3 76.0 | 93.0 75.7 | 79.1 89.6 | 78.3 88.9 | 76.8 141.3 82.3 | 75.3 134.4 81.5 | 18.3 | 16.1 | 13.7 | 2121 2426 |
| PORT ARANSAS | 27.83N | 97.07W | 20 | 36.8 | 41.1 | 86.1 78.0 | 85.4 78.1 | 84.9 78.0 | 80.4 84.0 | 80.1 83.8 | 79.4 152.4 83.1 | 78.7 149.1 82.9 | 38.4 | 32.2 | 26.7 | 829 3047 |
| PORT ARTHUR/JEFFERS | 29.95N | 94.02W | 16 | 31.4 | 34.7 | 94.5 78.0 | 92.9 78.1 | 91.3 78.0 | 81.5 88.8 | 80.6 87.9 | 79.5 153.0 84.5 | 78.9 149.9 83.9 | 21.4 | 19.1 | 17.6 | 1356 2899 |
| RANDOLPH AFB | 29.53N | 98.28W | 761 | 27.8 | 31.7 | 99.5 74.3 | 97.7 74.3 | 95.8 74.6 | 78.6 89.3 | 78.0 88.6 | 76.4 141.7 81.2 | 75.3 136.6 80.8 | 20.8 | 18.8 | 16.9 | 1482 3066 |
| REESE AFB/LUBBOCK | 33.60N | 102.05W | 3337 | 14.7 | 19.4 | 101.0 67.0 | 97.8 67.3 | 95.1 67.2 | 73.2 87.2 | 72.0 86.7 | 69.5 122.9 78.7 | 67.9 116.4 77.6 | 27.4 | 24.1 | 20.6 | 3182 1831 |
| SABINE | 29.67N | 94.05W | 20 | 32.2 | 35.9 | 88.6 77.8 | 87.3 77.5 | 86.4 77.5 | 80.2 85.4 | 80.2 85.0 | 79.7 154.1 83.9 | 78.9 148.3 83.6 | 34.9 | 27.5 | 23.8 | 1455 2605 |
| SAN ANGELO/MATHIS | 31.35N | 100.49W | 1893 | 21.9 | 25.9 | 100.4 70.3 | 98.7 70.1 | 96.5 70.1 | 75.3 88.7 | 74.3 87.9 | 72.0 126.9 79.8 | 70.8 121.9 79.0 | 24.7 | 21.5 | 19.4 | 2241 2509 |
| SAN ANTONIO INTL | 29.53N | 98.46W | 810 | 29.2 | 32.7 | 99.0 73.5 | 97.2 73.7 | 95.3 73.9 | 78.1 87.9 | 77.4 87.1 | 76.0 139.9 80.2 | 75.4 135.6 79.9 | 20.2 | 18.3 | 16.6 | 1418 3157 |
| STINSON MUNI | 29.34N | 98.47W | 577 | 30.4 | 34.1 | 100.1 74.2 | 99.0 74.0 | 96.9 74.0 | 79.0 89.2 | 78.2 88.3 | 76.7 142.2 82.1 | 75.4 135.8 81.1 | 18.9 | 17.1 | 15.8 | 1283 3298 |
| SAN MARCOS MUNI | 29.89N | 97.85W | 597 | 27.8 | 30.4 | 99.4 74.3 | 97.8 74.3 | 95.8 74.2 | 78.3 90.2 | 77.7 89.8 | 75.2 135.1 83.0 | 74.5 131.9 82.6 | 24.5 | 21.1 | 19.1 | 1617 3003 |
| VICTORIA/VICTORIA R | 28.86N | 96.93W | 118 | 31.0 | 34.5 | 97.1 77.6 | 95.2 76.7 | 93.4 76.7 | 80.3 87.8 | 79.7 87.3 | 78.1 146.5 82.0 | 78.1 146.5 82.0 | 24.3 | 20.9 | 19.1 | 1185 3193 |
| WACO RGNL | 31.61N | 97.23W | 509 | 24.6 | 28.1 | 100.5 75.0 | 99.0 75.1 | 96.9 75.2 | 78.7 91.5 | 78.1 90.8 | 75.8 137.4 82.3 | 75.0 133.8 81.8 | 24.6 | 21.4 | 19.5 | 2010 2856 |
| WICHITA FALLS/SHEPS | 33.98N | 98.49W | 1030 | 18.2 | 22.6 | 102.5 73.2 | 100.1 73.3 | 97.7 73.4 | 77.8 92.0 | 76.8 91.0 | 74.0 131.7 82.8 | 73.0 127.1 81.7 | 26.9 | 24.3 | 20.9 | 2811 2456 |

## Utah
*5 sites, 7 more on CD-ROM*

| Station | Lat | Long | Elev | 99.6% | 99% | DB/MCWB | DB/MCWB | DB/MCWB | WB/MCDB | WB/MCDB | DP/HR/MCDB | DP/HR/MCDB | 1% | 2.5% | 5% | HDD/CDD |
|---|---|---|---|---|---|---|---|---|---|---|---|---|---|---|---|---|
| HILL AFB | 41.12N | 111.97W | 4790 | 9.5 | 12.3 | 95.2 61.7 | 92.7 60.5 | 90.2 59.7 | 65.0 86.3 | 63.5 85.0 | 57.4 84.2 72.8 | 54.7 76.1 72.9 | 22.9 | 19.8 | 18.2 | 6041 1030 |
| LOGAN CACHE | 41.79N | 111.85W | 4455 | -5.9 | 0.4 | 94.5 62.1 | 91.3 60.9 | 89.9 60.4 | 65.1 86.6 | 63.8 85.0 | 58.9 87.9 69.5 | 55.4 77.1 70.2 | 19.7 | 16.6 | 13.1 | 7264 466 |
| PROVO MUNI | 40.22N | 111.72W | 4498 | 7.2 | 11.5 | 94.7 62.4 | 91.3 62.2 | 89.9 61.9 | 66.4 86.7 | 65.1 85.2 | 59.8 90.8 75.3 | 57.2 82.5 74.8 | 24.1 | 20.1 | 17.4 | 6030 791 |
| ST GEORGE MUNI | 37.09N | 113.58W | 2940 | 26.6 | 28.1 | 106.3 66.2 | 103.7 65.1 | 100.9 64.4 | 69.1 94.4 | 68.0 93.6 | 62.9 95.8 76.7 | 59.5 84.6 79.1 | 26.8 | 23.3 | 19.6 | 2971 2735 |
| SALT LAKE CITY INTL | 40.79N | 111.97W | 4226 | 9.6 | 14.2 | 97.7 62.8 | 95.1 62.2 | 92.6 61.5 | 66.3 88.1 | 65.1 86.9 | 60.1 90.7 72.6 | 57.5 82.6 73.1 | 25.0 | 20.9 | 19.6 | 5507 1218 |

## Vermont
*1 site, 5 more on CD-ROM*

| BURLINGTON INTL | 44.47N | 73.15W | 341 | -7.8 | -2.7 | 88.4 71.3 | 85.5 69.9 | 82.4 68.4 | 74.4 84.0 | 72.6 81.4 | 71.3 117.1 78.9 | 69.7 110.6 77.5 | 23.6 | 20.3 | 18.4 | 7352 505 |

## Virginia
*17 sites, 22 more on CD-ROM*

| DANVILLE RGNL | 36.57N | 79.34W | 591 | 18.2 | 21.3 | 94.6 74.6 | 91.3 74.0 | 90.0 73.5 | 77.8 87.5 | 76.7 87.5 | 74.7 132.9 82.8 | 73.2 126.3 81.0 | 18.6 | 16.5 | 14.2 | 3609 1481 |
| DINWIDDIE CO | 37.18N | 77.50W | 194 | 16.1 | 19.3 | 97.3 77.3 | 94.6 76.4 | 91.5 74.9 | 80.7 91.9 | 79.2 90.8 | 77.5 144.0 86.2 | 75.4 133.9 84.0 | 18.1 | 15.8 | 13.0 | 3732 1555 |
| DAVISON AAF | 38.72N | 77.32W | 75 | 13.5 | 18.1 | 96.9 76.0 | 93.7 75.1 | 91.1 74.3 | 79.6 89.0 | 78.1 87.9 | 76.3 137.9 85.4 | 74.9 131.4 83.9 | 17.6 | 14.2 | 13.4 | 4304 1436 |
| LANGLEY AFB/HAMPTON | 37.08N | 76.35W | 20 | 20.7 | 24.8 | 96.9 77.4 | 94.6 76.4 | 91.4 75.3 | 80.3 88.3 | 79.0 86.4 | 79.0 150.8 83.3 | 77.2 141.8 82.1 | 23.8 | 20.2 | 18.3 | 3449 1555 |
| LEESBURG EXECUTIVE | 39.08N | 77.55W | 390 | 14.5 | 18.1 | 95.1 75.8 | 92.2 75.1 | 90.4 74.0 | 79.1 90.6 | 77.7 88.4 | 75.4 135.0 83.1 | 74.7 132.0 82.5 | 22.7 | 18.8 | 16.3 | 4433 1350 |
| LYNCHBURG/MUN. P. G | 37.34N | 79.21W | 938 | 15.3 | 19.0 | 92.3 73.8 | 89.9 73.0 | 87.6 72.4 | 76.5 87.0 | 75.3 85.4 | 73.4 128.4 80.5 | 72.4 124.1 79.5 | 17.7 | 15.9 | 13.5 | 4228 1132 |
| MANASSAS RGNL DAVIS | 38.72N | 77.50W | 194 | 11.8 | 16.4 | 93.0 74.1 | 90.7 73.8 | 88.4 72.8 | 77.2 88.7 | 75.8 86.6 | 73.3 124.8 82.2 | 72.5 121.2 81.3 | 21.6 | 18.6 | 16.3 | 4774 1073 |
| NEWPORT NEWS WILLIA | 37.13N | 76.49W | 52 | 19.2 | 23.2 | 94.6 77.1 | 92.0 76.1 | 90.0 75.2 | 79.5 90.3 | 78.3 88.1 | 76.8 140.0 83.9 | 76.8 140.0 83.9 | 20.0 | 18.3 | 16.8 | 3527 1589 |
| NORFOLK INTL ARPT | 36.90N | 76.19W | 30 | 22.5 | 26.2 | 93.7 76.7 | 91.3 76.1 | 89.2 75.2 | 79.1 88.7 | 78.0 87.1 | 76.2 139.1 83.0 | 75.6 133.9 81.6 | 24.8 | 21.0 | 19.1 | 3230 1700 |
| NORFOLK NS | 36.93N | 76.28W | 16 | 23.8 | 27.3 | 94.1 77.2 | 91.4 76.2 | 89.9 75.8 | 80.0 89.4 | 78.6 87.7 | 77.4 142.5 84.1 | 76.1 136.2 83.0 | 25.4 | 22.0 | 19.2 | 3059 1843 |
| OCEANA NAS | 36.82N | 76.03W | 23 | 21.5 | 25.5 | 92.9 77.2 | 90.5 76.3 | 88.3 75.8 | 79.4 88.8 | 78.1 87.5 | 76.9 140.1 83.9 | 75.0 131.3 82.4 | 24.5 | 20.7 | 18.6 | 3308 1569 |
| QUANTICO MCAF | 38.50N | 77.30W | 13 | 16.3 | 19.6 | 92.6 76.4 | 90.3 75.7 | 88.1 74.7 | 79.4 89.0 | 77.9 87.3 | 76.6 139.1 85.0 | 75.0 131.4 83.1 | 19.5 | 17.0 | 15.1 | 4180 1363 |
| RICHMOND/BYRD FIELD | 37.51N | 77.32W | 164 | 17.8 | 21.2 | 95.1 75.9 | 92.6 75.0 | 90.2 74.1 | 78.4 89.3 | 77.4 87.7 | 75.7 135.2 82.8 | 74.6 130.1 81.5 | 20.8 | 18.6 | 16.8 | 3729 1532 |
| ROANOKE MUNICIPAL | 37.32N | 79.97W | 1175 | 15.7 | 19.6 | 92.3 73.8 | 90.0 72.2 | 87.7 71.5 | 75.4 86.5 | 74.5 85.1 | 72.4 125.3 79.6 | 71.5 121.3 78.5 | 22.9 | 19.0 | 16.8 | 4044 1230 |
| SHENANDOAH VALLEY RG | 38.26N | 78.88W | 1201 | 11.8 | 16.3 | 91.1 73.8 | 88.9 73.7 | 85.9 73.2 | 78.4 87.4 | 77.1 86.1 | 75.4 139.3 82.6 | 74.7 135.7 81.7 | 17.6 | 15.2 | 12.6 | 4422 1182 |
| VIRGINIA TECH ARPT | 37.21N | 80.40W | 2133 | 10.4 | 15.8 | 89.8 72.6 | 87.6 71.2 | 84.0 70.2 | 75.6 83.5 | 74.1 82.0 | 73.1 132.2 79.0 | 72.0 128.4 78.3 | 20.4 | 18.0 | 15.8 | 4823 789 |
| WASHINGTON/DULLES | 38.94N | 77.45W | 325 | 12.1 | 16.5 | 94.5 75.7 | 91.8 74.7 | 88.6 73.9 | 77.6 89.0 | 76.6 87.0 | 74.4 130.1 81.9 | 73.3 125.4 80.6 | 20.8 | 18.4 | 16.5 | 4675 1183 |
| WASHINGTON/NATIONAL | 38.87N | 77.03W | 66 | 17.3 | 20.7 | 94.5 75.7 | 91.8 74.8 | 89.5 73.7 | 77.8 88.9 | 77.4 87.7 | 75.9 135.6 83.1 | 74.7 130.3 82.0 | 23.4 | 20.0 | 18.1 | 3996 1555 |

## Washington
*20 sites, 18 more on CD-ROM*

| ARLINGTON MUNI | 48.16N | 122.15W | 138 | 20.6 | 24.7 | 82.2 66.0 | 79.3 64.3 | 75.3 62.9 | 67.3 80.7 | 65.4 77.7 | 62.3 84.3 73.0 | 60.7 79.8 71.1 | 20.9 | 18.1 | 15.6 | 5371 60 |
| BELLINGHAM INTL | 48.79N | 122.54W | 151 | 19.0 | 23.9 | 79.5 63.8 | 76.0 63.8 | 73.0 62.1 | 66.7 77.7 | 64.8 74.5 | 62.2 84.1 71.1 | 60.7 79.8 69.3 | 25.4 | 20.7 | 18.4 | 5338 53 |
| BREMERTON NATIONAL | 47.49N | 122.76W | 440 | 22.6 | 26.6 | 85.9 65.7 | 81.8 63.5 | 78.7 62.1 | 66.4 83.1 | 64.5 80.0 | 59.1 76.1 71.5 | 57.3 71.2 68.8 | 18.9 | 16.8 | 14.6 | 5615 101 |
| FAIRCHILD AFB | 47.62N | 117.65W | 2461 | 6.8 | 11.7 | 92.0 60.2 | 90.1 61.1 | 86.2 60.4 | 64.6 85.5 | 63.0 84.1 | 57.3 76.6 66.0 | 55.1 70.8 65.5 | 24.7 | 20.7 | 18.3 | 6776 462 |
| FELTS FLD | 47.68N | 117.32W | 1969 | 7.6 | 13.8 | 94.3 63.6 | 90.8 63.6 | 87.9 62.6 | 67.4 89.6 | 65.4 86.7 | 60.0 83.2 71.3 | 57.4 75.5 71.2 | 19.9 | 17.5 | 15.1 | 6130 439 |
| FORT LEWIS/GRAY AAF | 47.12N | 122.55W | 302 | 19.8 | 24.8 | 87.7 65.4 | 83.3 64.0 | 79.7 62.6 | 67.0 83.9 | 65.3 80.3 | 61.3 81.8 69.2 | 59.4 76.5 68.8 | 18.2 | 15.7 | 12.9 | 5111 147 |
| KELSO LONGVIEW | 46.12N | 122.89W | 20 | 21.3 | 26.2 | 88.1 67.7 | 82.5 65.7 | 77.9 64.0 | 68.0 84.6 | 66.8 81.2 | 63.0 86.2 75.5 | 61.1 80.6 72.5 | 17.5 | 15.0 | 12.8 | 4825 185 |
| TACOMA/MC CHORD AFB | 47.15N | 122.48W | 285 | 21.3 | 25.0 | 86.3 64.4 | 82.2 63.2 | 79.2 61.9 | 66.2 82.7 | 64.4 79.2 | 60.8 80.3 69.3 | 58.9 75.0 68.2 | 20.3 | 17.7 | 15.6 | 5288 123 |
| OLYMPIA | 46.97N | 122.90W | 200 | 20.1 | 24.5 | 87.6 66.0 | 83.4 64.8 | 79.9 63.3 | 67.8 84.8 | 65.8 81.2 | 61.4 81.9 71.0 | 59.9 77.6 69.5 | 18.8 | 16.3 | 14.6 | 5372 106 |
| TRI CITIES | 46.27N | 119.12W | 404 | 7.4 | 15.6 | 99.4 66.1 | 96.6 68.0 | 92.5 66.5 | 71.7 93.9 | 69.7 91.5 | 64.0 90.6 79.1 | 62.5 85.9 77.5 | 24.6 | 20.7 | 18.3 | 4936 805 |
| PEARSON FLD | 45.62N | 122.66W | 20 | 24.9 | 27.6 | 90.9 66.0 | 87.8 65.8 | 82.4 64.6 | 69.2 85.6 | 67.4 83.8 | 63.1 86.6 74.4 | 61.3 81.1 72.4 | 16.6 | 13.8 | 12.1 | 4415 374 |
| BOEING FLD KING CO | 47.53N | 122.30W | 30 | 24.8 | 28.4 | 86.0 65.7 | 81.8 65.2 | 79.2 63.7 | 66.9 83.2 | 65.0 79.5 | 61.2 80.8 69.5 | 59.5 75.9 69.3 | 18.6 | 16.8 | 14.5 | 4320 264 |
| SEATTLE-TACOMA INTL | 47.46N | 122.31W | 433 | 25.2 | 29.6 | 85.3 65.2 | 81.6 63.7 | 78.3 62.6 | 66.5 82.6 | 65.0 79.0 | 61.0 81.1 70.0 | 59.4 76.8 68.0 | 20.3 | 18.2 | 16.4 | 4705 188 |
| SANDERSON FLD | 47.24N | 123.15W | 269 | 23.1 | 26.7 | 87.9 65.1 | 82.7 64.3 | 79.2 62.8 | 67.2 83.8 | 65.3 80.5 | 61.2 81.5 69.9 | 59.4 76.3 68.1 | 20.4 | 18.2 | 16.4 | 5465 103 |
| SNOHOMISH CO | 47.91N | 122.28W | 607 | 25.5 | 29.6 | 80.8 63.8 | 75.5 62.0 | 72.9 60.9 | 65.4 77.0 | 63.5 74.0 | 61.1 82.4 68.4 | 59.1 76.4 67.2 | 24.4 | 20.1 | 17.5 | 5208 79 |

131

**Table 4-8 Design Conditions for Selected Locations (Continued)**

Meaning of acronyms:
DB: Dry bulb temperature, °F
MCWB: Mean coincident wet bulb temperature, °F

WB: Wet bulb temperature, °F
DP: Dew point temperature, °F
MCDB: Mean coincident dry bulb temperature, °F

Lat: Latitude, °
Long: Longitude, °
HR: Humidity ratio, grains of moisture per lb of dry air
HDD and CDD 65: Annual heating and cooling degree-days, base 65°F, °F-day

Elev: Elevation, ft
WS: Wind speed, mph

| Station | Lat | Long | Elev | Heating DB 99.6% | Heating DB 99% | Cooling DB/MCWB 0.4% DB/MCWB | Cooling DB/MCWB 1% DB/MCWB | Cooling DB/MCWB 2% DB/MCWB | Evaporation WB/MCDB 0.4% WB/MCDB | Evaporation WB/MCDB 1% WB/MCDB | Dehumidification DP/HR/MCDB 0.4% DP/HR/MCDB | Dehumidification DP/HR/MCDB 1% DP/HR/MCDB | Extreme Annual WS 1% | Extreme Annual WS 2.5% | Extreme Annual WS 5% | Heat./Cool. Degree-Days HDD / CDD 65 |
|---|---|---|---|---|---|---|---|---|---|---|---|---|---|---|---|---|
| SPOKANE INTL ARPT | 47.62N | 117.53W | 2365 | 4.7 | 11.0 | 92.8 / 63.0 | 89.6 / 61.7 | 86.0 / 60.5 | 65.0 / 87.2 | 63.4 / 84.9 | 57.6 / 77.3 / 67.4 | 55.5 / 71.6 / 67.5 | 25.6 | 22.0 | 19.1 | 6627 / 434 |
| TACOMA NARROWS | 47.27N | 122.58W | 315 | 27.4 | 31.2 | 83.8 / 64.5 | 80.8 / 63.0 | 76.8 / 61.6 | 66.0 / 80.9 | 64.2 / 77.3 | 61.0 / 81.2 / 68.4 | 59.1 / 75.8 / 66.9 | 19.5 | 17.4 | 15.5 | 4771 / 145 |
| WALLA WALLA RGNL | 46.10N | 118.29W | 1204 | 10.4 | 18.0 | 98.7 / 66.2 | 94.6 / 65.1 | 90.9 / 63.8 | 68.4 / 92.7 | 66.6 / 90.4 | 60.7 / 82.8 / 73.5 | 57.6 / 74.0 / 72.4 | 24.0 | 19.9 | 17.8 | 4825 / 910 |
| WEST POINT (LS) | 47.67N | 122.43W | 30 | 29.7 | 33.4 | 70.5 / 60.0 | 68.1 / 60.0 | 66.1 / 59.3 | 62.1 / 67.5 | 60.9 / 65.7 | 59.8 / 77.0 / 64.0 | 58.9 / 74.3 / 62.9 | 36.7 | 30.9 | 26.0 | 4906 / 8 |
| YAKIMA AIR TERMINAL | 46.56N | 120.53W | 1066 | 7.8 | 13.7 | 96.0 / 66.4 | 92.7 / 65.3 | 89.5 / 63.8 | 68.4 / 91.0 | 66.6 / 88.8 | 60.4 / 81.4 / 76.0 | 57.8 / 74.3 / 74.5 | 23.2 | 19.1 | 16.4 | 5898 / 509 |
| **West Virginia** | | | | | | | | | | | | | | | | *3 sites, 8 more on CD-ROM* |
| YEAGER | 38.38N | 81.59W | 981 | 10.1 | 15.5 | 91.3 / 72.9 | 89.1 / 72.7 | 86.6 / 72.1 | 76.7 / 86.0 | 75.3 / 84.1 | 74.0 / 131.5 / 80.7 | 72.8 / 126.0 / 79.3 | 17.2 | 14.8 | 12.3 | 4444 / 1076 |
| HUNTINGTON/TRI STAT | 38.38N | 82.56W | 837 | 10.1 | 15.5 | 91.9 / 73.5 | 89.6 / 73.2 | 87.2 / 72.6 | 77.3 / 86.6 | 75.9 / 84.8 | 74.5 / 133.1 / 81.4 | 73.2 / 127.4 / 80.1 | 16.8 | 14.8 | 12.7 | 4426 / 1156 |
| MID OHIO VALLEY RGN | 39.35N | 81.44W | 863 | 7.3 | 12.3 | 90.9 / 73.6 | 88.4 / 72.9 | 86.0 / 72.0 | 76.9 / 86.5 | 75.4 / 84.0 | 73.9 / 130.5 / 81.2 | 72.7 / 125.1 / 79.7 | 18.2 | 16.0 | 13.5 | 4940 / 949 |
| **Wisconsin** | | | | | | | | | | | | | | | | *14 sites, 31 more on CD-ROM* |
| OUTAGAMIE CO RGNL | 44.26N | 88.52W | 919 | -6.3 | -0.9 | 88.2 / 74.3 | 84.4 / 72.2 | 82.1 / 70.3 | 76.9 / 84.8 | 74.9 / 82.0 | 74.6 / 134.0 / 81.2 | 72.5 / 124.7 / 79.3 | 24.7 | 21.2 | 18.9 | 7273 / 587 |
| CHIPPEWA VALLEY RGN | 44.87N | 91.49W | 896 | -13.5 | -8.3 | 90.5 / 73.0 | 87.2 / 71.1 | 84.1 / 69.2 | 75.8 / 85.7 | 73.9 / 83.4 | 72.7 / 125.3 / 81.8 | 70.7 / 117.0 / 79.5 | 19.9 | 17.9 | 16.3 | 7801 / 600 |
| FOND DU LAC CO | 43.77N | 88.49W | 807 | -5.7 | -0.1 | 88.3 / 73.5 | 84.4 / 71.4 | 82.2 / 69.8 | 75.7 / 84.6 | 74.3 / 82.1 | 73.0 / 126.4 / 81.9 | 72.1 / 122.1 / 80.2 | 23.5 | 20.1 | 18.2 | 7071 / 588 |
| GREEN BAY/A-STRAUB | 44.51N | 88.12W | 702 | -8.2 | -3.0 | 88.5 / 73.5 | 85.3 / 71.9 | 82.5 / 70.1 | 76.3 / 85.0 | 74.3 / 82.2 | 73.5 / 127.8 / 81.3 | 71.7 / 120.0 / 79.5 | 23.5 | 20.0 | 18.1 | 7599 / 479 |
| KENOSHA RGNL | 42.60N | 87.94W | 745 | -1.7 | 2.8 | 89.9 / 74.7 | 87.0 / 73.3 | 83.7 / 71.8 | 76.6 / 86.3 | 75.1 / 83.5 | 73.3 / 127.2 / 81.6 | 72.4 / 123.3 / 80.5 | 24.9 | 21.7 | 19.2 | 6681 / 614 |
| LA CROSSE MUNICIPAL | 43.88N | 91.25W | 656 | -9.3 | -4.5 | 91.7 / 74.8 | 88.8 / 73.2 | 85.9 / 71.8 | 77.9 / 87.8 | 75.7 / 84.9 | 74.9 / 134.1 / 83.9 | 72.8 / 124.5 / 81.3 | 23.1 | 19.6 | 18.0 | 7010 / 818 |
| MADISON/DANE COUNTY | 43.14N | 89.35W | 866 | -7.0 | -1.6 | 89.6 / 74.2 | 86.6 / 72.6 | 83.8 / 71.0 | 77.0 / 86.1 | 75.0 / 83.3 | 73.9 / 130.4 / 82.9 | 72.2 / 123.1 / 80.5 | 22.6 | 19.6 | 17.9 | 7104 / 620 |
| MANITOWOC CO | 44.13N | 87.67W | 650 | -4.2 | 0.5 | 84.4 / 71.6 | 81.7 / 70.1 | 79.3 / 68.5 | 74.7 / 82.1 | 73.0 / 79.2 | 72.3 / 122.5 / 79.8 | 70.3 / 114.2 / 77.2 | 24.1 | 20.7 | 18.8 | 7541 / 344 |
| MILWAUKEE/GEN. MITC | 42.95N | 87.90W | 692 | -1.4 | 3.2 | 90.0 / 74.3 | 86.5 / 72.4 | 83.5 / 70.8 | 76.8 / 86.5 | 74.9 / 83.3 | 73.6 / 128.4 / 82.2 | 72.1 / 121.8 / 80.4 | 25.2 | 22.3 | 19.8 | 6684 / 690 |
| CENTRAL WISCONSIN | 44.78N | 89.67W | 1276 | -10.8 | -6.5 | 86.4 / 71.8 | 83.0 / 69.6 | 81.2 / 68.0 | 74.0 / 82.6 | 72.0 / 80.0 | 71.8 / 123.4 / 79.3 | 69.8 / 115.0 / 77.6 | 23.0 | 19.6 | 17.6 | 8223 / 363 |
| SHEBOYGAN CO MEM | 43.77N | 87.85W | 748 | -3.7 | 0.6 | 88.1 / 71.8 | 84.0 / 71.5 | 81.6 / 69.7 | 75.6 / 83.9 | 73.8 / 81.8 | 72.8 / 124.9 / 81.1 | 71.8 / 122.9 / 79.4 | 24.3 | 20.7 | 18.6 | 7375 / 423 |
| SHEBOYGAN | 43.75N | 87.68W | 620 | -2.2 | 2.8 | 83.0 / 71.2 | 79.3 / 70.2 | 76.5 / 69.7 | 76.2 / 79.0 | 74.1 / 76.8 | 75.3 / 135.9 / 77.3 | 73.3 / 126.6 / 75.7 | 40.7 | 33.3 | 28.0 | 7272 / 322 |
| WAUSAU DOWNTOWN | 44.93N | 89.63W | 1198 | -11.8 | -6.9 | 88.1 / 71.6 | 84.5 / 69.5 | 81.9 / 67.6 | 74.3 / 83.2 | 72.5 / 81.0 | 71.8 / 122.8 / 79.0 | 69.8 / 114.7 / 77.4 | 20.0 | 17.7 | 15.9 | 7973 / 462 |
| WITTMAN RGNL | 43.98N | 88.56W | 840 | -6.0 | -0.4 | 88.2 / 73.7 | 84.3 / 71.8 | 82.0 / 69.9 | 76.1 / 84.3 | 74.2 / 82.0 | 73.1 / 122.6 / 81.2 | 72.0 / 121.1 / 79.8 | 23.1 | 20.0 | 18.0 | 7286 / 548 |
| **Wyoming** | | | | | | | | | | | | | | | | *2 sites, 16 more on CD-ROM* |
| CASPER/NATRONA COUN | 42.90N | 106.47W | 5289 | -8.3 | -0.7 | 93.8 / 59.7 | 91.1 / 59.0 | 88.3 / 58.4 | 63.2 / 83.2 | 61.8 / 82.2 | 57.4 / 85.9 / 66.3 | 55.2 / 79.1 / 66.2 | 32.1 | 28.1 | 25.5 | 7285 / 461 |
| CHEYENNE/WARREN AFB | 41.16N | 104.81W | 6142 | -3.7 | 2.9 | 89.7 / 58.3 | 86.6 / 57.7 | 83.8 / 57.2 | 62.6 / 77.3 | 61.5 / 76.9 | 58.8 / 93.1 / 65.6 | 56.9 / 86.9 / 65.2 | 33.6 | 28.9 | 25.9 | 7050 / 338 |
| **Canada** | | | | | | | | | | | | | | | | *100 sites, 462 more on CD-ROM* |
| **Alberta** | | | | | | | | | | | | | | | | *13 sites, 50 more on CD-ROM* |
| CALGARY INTL A | 51.11N | 114.02W | 3556 | -19.8 | -13.1 | 83.5 / 60.7 | 79.9 / 59.7 | 76.4 / 58.4 | 63.6 / 78.0 | 61.7 / 75.6 | 58.3 / 83.1 / 69.3 | 56.2 / 76.8 / 66.8 | 27.0 | 23.1 | 20.3 | 9093 / 64 |
| COP UPPER | 51.08N | 114.22W | 4052 | -17.8 | -12.0 | 82.7 / 58.8 | 78.8 / 57.7 | 75.2 / 56.7 | 62.9 / 75.3 | 60.7 / 73.2 | 58.7 / 85.7 / 67.2 | 56.2 / 78.2 / 64.7 | 23.1 | 20.0 | 17.5 | 9048 / 73 |
| EDMONTON CITY CENTRE AWOS | 53.57N | 113.52W | 2201 | -20.5 | -14.8 | 83.0 / 64.4 | 79.6 / 62.7 | 76.5 / 60.8 | 66.5 / 79.3 | 64.5 / 76.3 | 61.9 / 89.7 / 72.3 | 59.9 / 83.4 / 69.8 | 22.0 | 18.8 | 16.5 | 9356 / 121 |
| EDMONTON INTL A | 53.32N | 113.58W | 2372 | -26.7 | -20.6 | 82.0 / 63.9 | 78.5 / 62.3 | 75.6 / 60.6 | 66.5 / 78.5 | 64.4 / 75.6 | 61.9 / 90.5 / 73.1 | 59.9 / 84.0 / 70.5 | 22.9 | 19.7 | 17.1 | 10321 / 42 |
| FORT MCMURRAY A | 53.67N | 113.47W | 2257 | -22.7 | -16.7 | 84.0 / 64.0 | 78.7 / 62.2 | 75.5 / 60.2 | 66.0 / 78.5 | 64.1 / 75.6 | 61.4 / 88.6 / 71.3 | 59.4 / 82.1 / 69.2 | 23.1 | 20.0 | 17.4 | 9893 / 68 |
| GRANDE PRAIRIE A | 56.65N | 111.21W | 1211 | -33.4 | -28.3 | 84.1 / 63.9 | 80.5 / 61.8 | 77.5 / 60.0 | 65.9 / 79.7 | 64.1 / 76.7 | 61.2 / 84.4 / 70.4 | 59.2 / 78.5 / 68.5 | 18.6 | 16.2 | 14.1 | 11405 / 82 |
| LACOMBE CDA 2 | 55.18N | 118.88W | 2195 | -32.8 | -24.5 | 81.4 / 61.9 | 78.0 / 60.2 | 75.0 / 58.7 | 64.3 / 77.3 | 62.2 / 74.5 | 59.7 / 82.8 / 68.9 | 57.5 / 76.5 / 66.2 | 24.8 | 21.6 | 18.7 | 10552 / 45 |
| LETHBRIDGE AWOS A | 52.45N | 113.76W | 2822 | -26.5 | -19.4 | 83.0 / 64.9 | 79.2 / 62.8 | 76.0 / 61.0 | 66.9 / 79.3 | 64.7 / 76.5 | 62.2 / 92.9 / 73.9 | 59.9 / 85.5 / 71.2 | 21.2 | 18.2 | 15.6 | 10304 / 43 |
| LETHBRIDGE CDA | 49.63N | 112.80W | 3048 | -21.1 | -14.5 | 83.8 / 62.2 | 85.0 / 61.2 | 81.4 / 60.4 | 65.7 / 81.4 | 63.7 / 79.5 | 60.0 / 86.6 / 72.7 | 57.9 / 80.1 / 69.6 | 35.6 | 30.5 | 27.0 | 8320 / 153 |
| MEDICINE HAT RCS | 49.70N | 110.72W | 2986 | -18.5 | -12.5 | 89.4 / 62.3 | 85.5 / 61.3 | 81.9 / 60.5 | 65.8 / 81.5 | 64.0 / 79.6 | 60.4 / 87.6 / 72.1 | 58.2 / 80.8 / 69.7 | 29.8 | 26.5 | 23.4 | 8108 / 205 |
| RED DEER A | 50.03N | 110.72W | 2346 | -23.0 | -16.4 | 90.8 / 64.0 | 87.3 / 62.8 | 83.8 / 61.5 | 66.2 / 84.7 | 64.3 / 82.6 | 60.3 / 84.3 / 72.2 | 58.0 / 78.4 / 70.4 | 25.1 | 21.6 | 18.9 | 8354 / 301 |
| SPRINGBANK A | 52.18N | 113.89W | 2969 | -26.0 | -18.1 | 82.4 / 63.1 | 78.9 / 61.3 | 75.7 / 59.8 | 64.8 / 78.6 | 63.3 / 75.6 | 60.3 / 87.4 / 71.8 | 58.3 / 81.1 / 69.2 | 20.3 | 18.2 | 16.4 | 10196 / 42 |
| | 51.10N | 114.37W | 3940 | -25.0 | -18.3 | 80.3 / 60.0 | 76.7 / 58.3 | 73.6 / 57.3 | 62.3 / 75.5 | 60.4 / 73.2 | 57.1 / 80.6 / 68.0 | 55.2 / 75.1 / 65.3 | 24.8 | 21.2 | 18.6 | 10293 / 8 |
| **British Columbia** | | | | | | | | | | | | | | | | *27 sites, 52 more on CD-ROM* |
| ABBOTSFORD A | 49.03N | 122.36W | 194 | 17.9 | 22.9 | 85.7 / 67.2 | 82.0 / 65.9 | 78.4 / 64.2 | 68.7 / 83.4 | 66.7 / 79.9 | 62.6 / 85.5 / 77.2 | 60.9 / 80.3 / 73.4 | 19.8 | 16.7 | 14.3 | 5256 / 134 |
| AGASSIZ CS | 49.24N | 121.76W | 62 | 18.8 | 23.3 | 86.6 / 68.7 | 83.1 / 67.4 | 75.2 / 66.3 | 71.0 / 82.8 | 68.9 / 80.2 | 66.5 / 97.9 / 77.8 | 64.5 / 91.2 / 75.0 | 22.9 | 17.9 | 14.2 | 5150 / 203 |
| BALLENAS ISLAND | 49.35N | 124.16W | 43 | 30.8 | 33.6 | 64.4 / 66.8 | 79.6 / 62.7 | 76.5 / 60.8 | 66.5 / 80.8 | 66.6 / 71.3 | 61.9 / 89.7 / 71.7 | 64.6 / 91.4 / 69.6 | 35.7 | 30.9 | 27.2 | 4627 / 109 |
| COMOX A | 49.72N | 124.90W | 85 | 23.4 | 27.1 | 83.0 / 63.9 | 78.5 / 62.5 | 75.6 / 61.4 | 65.2 / 76.8 | 63.8 / 73.9 | 60.8 / 96.9 / 68.9 | 59.6 / 76.3 / 67.4 | 22.9 | 19.7 | 17.1 | 5541 / 94 |
| DISCOVERY ISLAND | 48.42N | 123.23W | 49 | 30.6 | 34.8 | 73.5 / N/A | 69.8 / N/A | 66.8 / N/A | 66.0 / N/A | 64.1 / N/A | 61.4 / 88.6 / 71.3 | 59.4 / 82.1 / 69.2 | 23.1 | 20.0 | 17.4 | 4802 / 23 |
| ENTRANCE ISLAND CS | 49.22N | 123.80W | 16 | 28.9 | 32.1 | 75.1 / N/A | 72.2 / N/A | 70.0 / N/A | 65.9 / 79.7 | 64.1 / 76.7 | 61.2 / 84.5 / 70.4 | 59.2 / 78.5 / 68.5 | 18.6 | 16.2 | 14.1 | 4814 / 108 |
| ESQUIMALT HARBOUR | 48.43N | 123.44W | 10 | 27.1 | 30.9 | 72.1 / N/A | 69.0 / N/A | 66.4 / N/A | 62.3 / 69.1 | 61.0 / 66.8 | 59.4 / N/A / 64.4 | 58.4 / 72.8 / 63.1 | 21.7 | 18.8 | 16.5 | 5403 / 12 |
| HOWE SOUND - PAM ROCKS | 49.49N | 123.30W | 16 | 26.9 | 30.4 | 76.9 / 66.3 | 73.8 / 64.7 | 71.5 / 63.8 | 68.0 / 74.0 | 66.2 / 72.3 | 65.5 / 94.1 / 72.3 | 63.7 / 88.4 / 70.2 | 40.4 | 35.5 | 30.2 | 4781 / 142 |
| KAMLOOPS AUT | 50.70N | 120.44W | 1132 | -3.4 | 4.0 | 93.0 / 64.7 | 89.2 / 63.7 | 85.3 / 62.3 | 66.4 / 88.3 | 64.8 / 85.0 | 59.4 / 78.7 / 70.6 | 57.4 / 73.5 / 69.6 | 22.6 | 19.8 | 17.9 | 6329 / 482 |
| KELOWNA A | 49.96N | 119.38W | 1411 | -0.2 | 6.9 | 91.4 / 64.8 | 87.9 / 63.7 | 84.1 / 62.0 | 66.6 / 86.4 | 64.8 / 83.8 | 59.8 / 80.9 / 70.7 | 57.9 / 75.6 / 69.6 | 17.2 | 14.3 | 11.9 | 7014 / 236 |
| MALAHAT | 48.57N | 123.53W | 1201 | 22.0 | 26.5 | 81.9 / 62.7 | 78.3 / 61.9 | 75.3 / 60.7 | 65.9 / 77.4 | 64.1 / 74.6 | 61.1 / 84.0 / 73.1 | 59.4 / 79.0 / 70.6 | 14.9 | 12.8 | 11.0 | 5852 / 174 |
| PENTICTON A | 49.46N | 119.60W | 1129 | 7.4 | 12.5 | 91.0 / 65.4 | 87.4 / 64.2 | 84.3 / 62.8 | 66.7 / 86.9 | 65.1 / 83.6 | 59.6 / 79.5 / 72.4 | 57.9 / 74.7 / 72.0 | 23.1 | 20.1 | 18.0 | 6161 / 391 |
| PITT MEADOWS CS | 49.21N | 122.69W | 16 | 18.7 | 23.3 | 86.9 / 68.1 | 83.1 / N/A | 79.6 / N/A | 69.7 / 83.2 | 67.6 / 80.2 | 64.5 / 90.9 / 75.8 | 62.7 / 85.1 / 72.7 | 12.2 | 10.3 | 8.9 | 5367 / 141 |
| POINT ATKINSON | 49.33N | 123.26W | 115 | 29.1 | 32.5 | 76.7 / N/A | 74.3 / N/A | 72.4 / N/A | N/A | N/A | N/A | N/A | 30.5 | 26.0 | 22.2 | 4173 / 214 |

# Design Conditions 133

**Table 4-8 Design Conditions for Selected Locations (*Continued*)**

*Meaning of acronyms:*
*DB:* Dry bulb temperature, °F
*MCWB:* Mean coincident wet bulb temperature, °F
*WB:* Wet bulb temperature, °F
*DP:* Dew point temperature, °F
*MCDB:* Mean coincident dry bulb temperature, °F
*Lat:* Latitude, °
*Long:* Longitude, °
*HR:* Humidity ratio, grains of moisture per lb of dry air
*HDD and CDD 65:* Annual heating and cooling degree-days, base 65°F, °F-day
*Elev:* Elevation, ft
*WS:* Wind speed, mph

| Station | Lat | Long | Elev | Heating DB 99.6% | Heating DB 99% | Cooling DB/MCWB 0.4% DB/MCWB | 1% DB/MCWB | 2% DB/MCWB | Evaporation WB/MCDB 0.4% WB/MCDB | 1% WB/MCDB | Dehumidification DP/HR/MCDB 0.4% DP/HR/MCDB | 1% DP/HR/MCDB | Extreme Annual WS 1% | 2.5% | 5% | Heat./Cool. Degree-Days HDD/CDD 65 |
|---|---|---|---|---|---|---|---|---|---|---|---|---|---|---|---|---|
| PRINCE GEORGE AIRPORT AUTO | 53.89N | 122.67W | 2231 | −22.4 | −14.0 | 82.1 61.3 | 78.3 59.7 | 74.9 58.1 | 62.9 78.3 | 61.0 75.1 | 57.1 75.6 66.2 | 55.4 70.9 64.6 | 21.3 | 18.6 | 16.4 | 9174 38 |
| SANDHEADS CS | 49.11N | 123.30W | 36 | 25.5 | 29.8 | 72.3 63.9 | 70.3 N/A | 68.4 N/A | 66.8 85.3 | 65.0 82.8 | N/A N/A N/A | N/A N/A N/A | 30.6 | 27.0 | 24.1 | 4951 55 |
| SUMMERLAND CS | 49.56N | 119.64W | 1490 | 6.0 | 12.4 | 91.5 63.9 | 88.1 63.1 | 84.7 62.0 | 66.8 85.3 | 65.0 82.8 | 60.5 83.0 71.5 | 58.3 76.8 71.5 | 18.6 | 14.8 | 12.0 | 6311 466 |
| VANCOUVER HARBOUR CS | 49.30N | 123.12W | 10 | 26.5 | 30.5 | 78.6 65.1 | 75.9 64.8 | 73.4 73.4 | N/A N/A | N/A N/A | N/A N/A N/A | N/A N/A N/A | N/A | N/A | N/A | 4807 124 |
| VANCOUVER INT'L A | 49.20N | 123.18W | 13 | 26.0 | 26.0 | 77.3 65.1 | 74.5 64.0 | 72.2 62.8 | 66.3 75.2 | 64.8 73.1 | 62.4 84.4 71.6 | 61.1 80.5 69.6 | 23.6 | 20.2 | 17.5 | 5225 80 |
| VERNON AUTO | 50.22N | 119.19W | 1581 | 3.2 | 9.3 | 91.7 65.1 | 87.9 64.3 | 83.9 62.9 | 67.5 85.6 | 65.7 83.3 | 62.1 88.5 70.6 | 60.1 82.3 69.3 | 14.2 | 11.8 | 10.0 | 6790 370 |
| VICTORIA GONZALES CS | 48.41N | 123.33W | 230 | 26.9 | 31.1 | 76.4 62.6 | 72.1 61.0 | 68.9 59.7 | 63.9 73.3 | 62.2 69.8 | 60.3 78.7 65.9 | 59.1 75.3 64.8 | 27.3 | 23.3 | 20.5 | 5146 42 |
| VICTORIA HARTLAND CS | 48.53N | 123.46W | 505 | 25.5 | 29.3 | 83.4 65.8 | 80.0 64.4 | 76.9 63.2 | 68.0 79.8 | 66.3 77.0 | 63.8 90.2 72.5 | 62.2 85.2 70.3 | 21.0 | 18.2 | 15.6 | 5055 177 |
| VICTORIA INT'L A | 48.65N | 123.43W | 66 | 24.4 | 27.9 | 80.2 63.7 | 76.4 62.4 | 73.3 61.1 | 64.6 78.0 | 63.3 74.7 | 59.0 74.6 68.4 | 57.7 71.2 67.3 | 19.9 | 16.6 | 14.2 | 5417 44 |
| ESQUIMALT HARBOUR | 48.43N | 123.44W | 10 | 27.1 | 30.9 | 72.1 60.8 | 69.0 59.7 | 66.4 58.8 | 62.3 69.1 | 61.0 66.8 | 59.4 58.4 72.8 | 57.7 58.4 63.1 | 21.7 | 18.8 | 16.5 | 5403 12 |
| VICTORIA UNIVERSITY CS | 48.46N | 123.30W | 197 | 27.4 | 31.6 | 80.8 65.1 | 77.4 63.9 | 74.3 62.8 | 67.1 77.6 | 65.4 74.8 | 63.3 87.8 70.5 | 61.8 83.0 68.6 | 12.8 | 11.1 | 9.7 | 4901 70 |
| WEST VANCOUVER AUT | 49.35N | 123.19W | 551 | 21.6 | 26.3 | 81.0 65.6 | 77.6 64.8 | 74.6 63.5 | 67.1 78.0 | 66.0 75.4 | 63.6 89.8 71.3 | 62.0 84.9 71.3 | 11.1 | 9.5 | 7.8 | 5408 135 |
| WHITE ROCK CAMPBELL SCIENTIFI | 49.02N | 122.78W | 43 | 22.3 | 26.6 | 76.8 65.9 | 73.9 64.6 | 71.6 63.5 | 67.6 74.4 | 65.8 72.2 | 64.9 92.4 71.5 | 63.2 86.8 69.2 | 14.1 | 11.6 | 9.4 | 5020 55 |

**Manitoba** *1 site, 38 more on CD-ROM*
| WINNIPEG RICHARDSON INT'L A | 49.92N | 97.23W | 784 | −25.9 | −21.5 | 87.1 70.0 | 83.8 68.6 | 80.8 67.0 | 73.3 83.0 | 70.9 80.6 | 70.1 114.0 79.5 | 67.5 104.0 76.5 | 28.0 | 24.7 | 22.0 | 10309 292 |

**New Brunswick** *3 sites, 10 more on CD-ROM*
| FREDERICTON A | 45.87N | 66.53W | 69 | −10.3 | −5.6 | 85.6 69.9 | 82.2 67.6 | 79.2 66.0 | 72.1 82.0 | 70.2 77.0 | 68.7 105.7 77.0 | 67.1 99.8 74.8 | 22.1 | 19.2 | 17.3 | 8399 242 |
| MONCTON A | 46.10N | 64.69W | 233 | −8.5 | −4.2 | 83.3 69.4 | 80.3 67.4 | 77.4 64.8 | 71.8 79.8 | 69.8 77.0 | 69.1 107.9 76.0 | 67.4 101.7 74.1 | 28.1 | 24.4 | 21.4 | 8556 182 |
| SAINT JOHN A | 45.32N | 65.89W | 358 | −8.3 | −3.5 | 79.0 65.5 | 75.9 63.9 | 73.1 62.3 | 68.1 75.2 | 66.2 72.7 | 65.4 95.0 71.1 | 63.8 89.7 68.9 | 27.3 | 23.5 | 20.8 | 8554 55 |

**Newfoundland and Labrador** *1 site, 37 more on CD-ROM*
| ST JOHN'S A | 47.62N | 52.74W | 463 | 4.3 | 8.1 | 76.3 66.1 | 73.5 64.5 | 71.0 63.2 | 68.7 73.7 | 66.8 71.2 | 66.8 100.1 71.5 | 65.0 94.1 69.5 | 35.5 | 30.1 | 27.0 | 8727 54 |

**Northwest Territories** *1 site, 38 more on CD-ROM*
| YELLOWKNIFE A | 62.46N | 114.44W | 676 | −41.2 | −37.2 | 77.4 60.7 | 74.4 59.2 | 71.5 58.1 | 62.9 72.8 | 61.2 71.1 | 59.1 76.6 67.1 | 56.9 70.8 65.2 | 21.0 | 18.7 | 16.8 | 14741 62 |

**Nova Scotia** *3 sites, 16 more on CD-ROM*
| HALIFAX STANFIELD INT'L A | 44.88N | 63.52W | 476 | −1.1 | 2.6 | 82.0 68.7 | 78.8 66.7 | 76.0 65.1 | 71.1 78.2 | 69.3 75.4 | 68.9 107.9 74.4 | 67.2 101.9 72.1 | 27.6 | 23.6 | 20.9 | 7794 185 |
| SHEARWATER RCS | 44.63N | 63.51W | 79 | 2.0 | 5.9 | 79.0 67.3 | 76.0 65.4 | 73.6 64.2 | 69.9 75.2 | 68.2 72.9 | 68.1 103.5 72.0 | 66.4 97.6 70.1 | 26.8 | 23.2 | 20.7 | 7514 124 |
| SYDNEY A | 46.17N | 60.05W | 203 | −0.2 | 4.1 | 81.4 68.5 | 78.4 67.1 | 75.3 65.3 | 70.8 78.6 | 68.8 75.5 | 68.0 103.7 74.7 | 66.4 97.8 72.4 | 28.0 | 24.6 | 21.8 | 8245 145 |

**Nunavut** *1 site, 41 more on CD-ROM*
| IQALUIT CLIMATE | 63.75N | 68.54W | 112 | −39.0 | −35.6 | 62.7 52.7 | 57.6 50.0 | 54.0 48.0 | 53.6 61.0 | 50.8 56.9 | 48.9 51.4 56.1 | 46.7 47.3 53.0 | 34.4 | 28.9 | 25.2 | 17863 0 |

**Ontario** *20 sites, 49 more on CD-ROM*
| BEAUSOLEIL | 44.85N | 79.87W | 600 | −10.8 | −4.7 | 85.9 74.0 | 82.6 71.7 | 79.7 70.3 | 75.9 82.6 | 73.9 79.9 | 73.9 129.2 79.5 | 71.9 120.8 77.2 | 14.0 | 12.1 | 10.7 | 7850 382 |
| BELLE RIVER | 42.30N | 82.70W | 604 | 5.7 | 10.1 | 88.9 75.6 | 86.0 74.6 | 83.3 73.0 | 78.6 85.3 | 76.8 83.1 | 76.7 142.5 83.0 | 74.9 133.8 80.6 | 29.0 | 25.3 | 22.2 | 5983 810 |
| BURLINGTON PIERS (AUT) | 43.30N | 79.80W | 253 | 5.0 | 9.1 | 86.4 70.3 | 83.5 69.2 | 80.6 67.9 | 73.7 80.8 | 72.0 79.4 | 71.6 117.6 77.5 | 69.6 110.0 75.7 | 23.4 | 20.2 | 17.6 | 6408 557 |
| ERIEAU (AUT) | 42.25N | 81.90W | 584 | 5.7 | 9.8 | 80.2 73.0 | 78.4 71.9 | 76.9 70.5 | 76.2 77.8 | 74.6 76.7 | 75.6 137.1 77.2 | 73.9 129.3 75.9 | 28.5 | 25.0 | 22.0 | 6470 502 |
| LAGOON CITY | 44.55N | 79.40W | 725 | −10.4 | −4.5 | 81.4 73.2 | 79.1 71.2 | 77.0 70.5 | 75.6 79.2 | 73.8 77.3 | 74.6 132.8 78.0 | 72.6 124.2 76.4 | 28.3 | 24.9 | 21.9 | 7940 340 |
| LONDON CS | 43.03N | 81.15W | 912 | 0.0 | 4.2 | 86.4 72.2 | 83.6 70.9 | 81.0 69.2 | 74.5 82.9 | 72.8 80.7 | 71.9 122.0 79.2 | 70.2 115.0 77.3 | 23.5 | 21.0 | 18.7 | 7117 433 |
| NORTH BAY A | 46.36N | 79.42W | 1214 | −17.2 | −12.2 | 82.2 68.2 | 79.2 66.5 | 76.6 65.0 | 71.2 78.7 | 69.3 75.3 | 68.8 110.9 74.5 | 67.1 104.4 72.8 | 21.6 | 18.9 | 17.0 | 9345 221 |
| OTTAWA MACDONALD-CARTIER INT | 45.32N | 75.67W | 374 | −11.5 | −6.6 | 87.1 71.1 | 84.1 69.5 | 81.1 68.0 | 73.8 83.0 | 71.9 80.3 | 71.0 108.8 76.6 | 69.2 108.8 76.6 | 22.3 | 19.6 | 17.4 | 8142 428 |
| PETERBOROUGH AWOS | 44.23N | 78.37W | 627 | −10.1 | −3.8 | 86.1 72.1 | 83.1 70.2 | 80.5 68.8 | 74.5 82.7 | 72.6 80.1 | 72.0 121.0 79.1 | 70.1 113.2 76.5 | 20.2 | 17.6 | 15.4 | 7866 269 |
| PORT WELLER (AUT) | 43.25N | 79.22W | 259 | 8.6 | 12.2 | 84.4 73.5 | 81.7 72.2 | 79.2 70.6 | 76.2 80.9 | 74.4 78.9 | 73.1 124.2 77.2 | 71.3 124.2 77.2 | 32.2 | 28.4 | 25.0 | 6328 562 |
| SAULT STE MARIE A | 46.48N | 84.51W | 630 | −12.3 | −6.7 | 83.4 70.1 | 80.0 67.6 | 77.1 66.3 | 72.2 79.9 | 70.0 77.1 | 71.4 111.4 76.6 | 67.6 103.8 74.2 | 23.1 | 20.0 | 17.7 | 8910 165 |
| SUDBURY A | 46.62N | 80.80W | 1142 | −17.8 | −12.5 | 84.7 68.4 | 81.3 66.3 | 78.3 64.6 | 70.9 80.6 | 69.0 77.3 | 67.9 106.8 74.4 | 66.0 100.1 73.0 | 22.7 | 20.1 | 18.0 | 9433 238 |
| THUNDER BAY CS | 48.37N | 89.33W | 653 | −20.7 | −15.6 | 84.2 68.9 | 80.6 66.6 | 77.4 64.8 | 71.3 80.7 | 68.3 77.2 | 65.8 105.5 76.6 | 65.8 97.6 73.6 | 21.8 | 18.9 | 16.6 | 10069 123 |
| TIMMINS VICTOR POWER A | 48.57N | 81.38W | 968 | −27.5 | −21.8 | 85.2 69.7 | 81.6 65.3 | 78.4 64.0 | 70.5 81.2 | 68.3 77.7 | 66.9 110.5 75.3 | 64.9 95.4 73.3 | 18.7 | 17.4 | 14.7 | 10830 157 |
| TORONTO BUTTONVILLE A | 43.86N | 79.37W | 650 | −3.6 | 1.5 | 88.9 71.2 | 85.5 70.4 | 82.4 68.9 | 74.5 85.3 | 72.5 82.5 | 71.2 118.0 80.0 | 69.3 110.4 77.8 | 21.5 | 18.8 | 17.2 | 7352 456 |
| TORONTO CITY CENTRE | 43.63N | 79.40W | 253 | 3.0 | 8.1 | 83.3 71.2 | 80.2 70.5 | 77.5 69.5 | 75.6 79.2 | 73.8 77.5 | 72.8 122.8 77.2 | 71.2 116.0 75.6 | 29.7 | 26.5 | 23.4 | 6698 427 |
| TORONTO LESTER B. PEARSON INT | 43.68N | 79.63W | 568 | −0.5 | 4.0 | 88.5 72.3 | 85.2 70.6 | 82.2 69.1 | 74.7 84.4 | 72.9 82.1 | 71.7 119.7 80.1 | 69.9 112.1 78.1 | 27.1 | 23.4 | 20.7 | 7006 526 |
| TRENTON A | 44.12N | 77.53W | 282 | −6.5 | −1.2 | 84.7 71.9 | 82.1 70.5 | 79.6 69.1 | 74.4 81.6 | 72.7 79.4 | 70.3 119.5 78.8 | 70.3 112.8 77.0 | 23.5 | 20.6 | 18.1 | 7455 380 |
| WELCOME ISLAND (AUT) | 48.37N | 89.12W | 692 | −14.4 | −10.2 | 75.5 65.5 | 72.6 63.9 | 70.1 63.1 | 68.2 72.1 | 66.1 70.3 | 66.8 101.3 70.5 | 64.5 93.4 68.1 | 34.6 | 29.4 | 26.1 | 9664 68 |
| WINDSOR A | 42.28N | 82.96W | 623 | 4.0 | 8.4 | 89.7 73.2 | 86.8 72.0 | 84.2 70.7 | 76.0 85.8 | 74.2 83.2 | 73.0 125.4 81.7 | 71.4 118.5 79.3 | 25.4 | 22.4 | 20.0 | 6200 781 |

**Prince Edward Island** *1 site, 4 more on CD-ROM*
| CHARLOTTETOWN A | 46.29N | 63.13W | 161 | −4.7 | −0.3 | 80.1 69.2 | 77.5 67.3 | 75.0 65.8 | 71.1 77.6 | 69.3 75.1 | 68.8 106.5 75.2 | 67.1 100.3 73.3 | 26.1 | 22.5 | 19.8 | 8389 181 |

**Québec** *23 sites, 71 more on CD-ROM*
| BAGOTVILLE A | 48.33N | 71.00W | 522 | −21.7 | −17.2 | 84.7 67.0 | 81.1 65.4 | 77.8 64.0 | 70.2 79.6 | 68.2 77.1 | 67.0 101.3 74.2 | 65.1 94.7 72.3 | 26.9 | 23.5 | 21.0 | 10247 176 |
| JONQUIERE | 48.42N | 71.15W | 420 | −20.0 | −15.6 | 84.3 67.9 | 80.8 66.0 | 77.6 64.8 | 71.4 79.6 | 69.4 76.7 | 66.7 107.0 75.4 | 66.7 99.9 73.1 | 23.6 | 21.1 | 19.0 | 9893 175 |
| LA BAIE | 48.30N | 70.92W | 499 | −22.4 | −17.9 | 84.4 67.5 | 80.7 66.1 | 77.4 64.8 | 71.1 79.5 | 69.1 76.6 | 68.4 106.3 75.2 | 66.5 99.2 72.8 | 23.0 | 20.4 | 18.1 | 10285 126 |
| LAC SAINT-PIERRE | 46.18N | 72.92W | 52 | −12.5 | −7.0 | 81.9 69.8 | 79.2 68.0 | 77.0 67.0 | 72.6 78.6 | 70.9 76.6 | 70.4 112.0 76.4 | 68.7 105.7 74.8 | 29.8 | 26.5 | 23.6 | 8375 325 |

134  Principles of HVAC

**Table 4-8 Design Conditions for Selected Locations *(Continued)***

Meaning of acronyms:
DB: Dry bulb temperature, °F
MCWB: Mean coincident wet bulb temperature, °F
WB: Wet bulb temperature, °F
Lat: Latitude, °
DP: Dew point temperature, °F
MCDB: Mean coincident dry bulb temperature, °F
Long: Longitude, °
HR: Humidity ratio, grains of moisture per lb of dry air
HDD and CDD 65: Annual heating and cooling degree-days, base 65°F, °F-day
Elev: Elevation, ft
WS: Wind speed, mph

| Station | Lat | Long | Elev | Heating DB | | Cooling DB/MCWB | | | | | | Evaporation WB/MCDB | | | | Dehumidification DP/HR/MCDB | | | | | | Extreme Annual WS | | | Heat./Cool. Degree-Days | |
|---|---|---|---|---|---|---|---|---|---|---|---|---|---|---|---|---|---|---|---|---|---|---|---|---|---|---|
| | | | | 99.6% | 99% | 0.4% DB/MCWB | | 1% DB/MCWB | | 2% DB/MCWB | | 0.4% WB/MCDB | | 1% WB/MCDB | | 0.4% DP/HR/MCDB | | | 1% DP/HR/MCDB | | | 1% | 2.5% | 5% | HDD | CDD 65 |
| L'ACADIE | 45.29N | 73.35W | 144 | -10.9 | -6.4 | 86.2 | 71.1 | 83.4 | 70.0 | 80.8 | 68.7 | 74.7 | 82.3 | 72.7 | 79.4 | 72.1 | 119.6 | 79.2 | 70.5 | 112.8 | 77.0 | 22.8 | 19.6 | 16.9 | 7926 | 404 |
| L'ASSOMPTION | 45.81N | 73.43W | 69 | -14.1 | -8.6 | 86.7 | 71.4 | 83.7 | 69.6 | 80.9 | 68.2 | 74.2 | 82.7 | 72.2 | 79.8 | 71.4 | 116.3 | 78.4 | 69.7 | 109.4 | 76.6 | 18.9 | 16.5 | 14.4 | 8309 | 366 |
| LENNOXVILLE | 45.37N | 71.82W | 594 | -14.1 | -8.1 | 85.0 | 70.8 | 82.3 | 69.4 | 79.7 | 67.9 | 73.8 | 81.2 | 71.9 | 79.0 | 71.5 | 119.1 | 77.8 | 69.6 | 111.2 | 75.9 | 20.2 | 17.7 | 15.7 | 8291 | 266 |
| MCTAVISH | 45.50N | 73.58W | 240 | -7.1 | -2.4 | 86.2 | 71.6 | 83.4 | 69.7 | 81.0 | 68.2 | 74.0 | 82.9 | 72.1 | 79.8 | 71.0 | 115.3 | 79.0 | 69.4 | 109.1 | 77.3 | 11.3 | 9.8 | 8.9 | 7460 | 533 |
| MONT-JOLI A | 48.60N | 68.22W | 171 | -10.8 | -6.8 | 80.2 | 67.6 | 77.0 | 65.6 | 74.3 | 64.1 | 69.3 | 77.5 | 67.3 | 74.9 | 66.1 | 96.9 | 75.1 | 64.1 | 90.2 | 72.3 | 28.1 | 24.8 | 22.1 | 9623 | 123 |
| MONT-ORFORD | 45.31N | 72.24W | 2776 | -19.0 | -13.2 | 77.2 | 65.3 | 74.3 | 63.9 | 71.6 | 62.8 | 69.0 | 73.6 | 66.7 | 70.7 | 67.4 | 111.8 | 71.2 | 65.4 | 104.2 | 69.0 | 35.1 | 30.3 | 27.2 | 10169 | 96 |
| MONTREAL/MIRABEL INTL A | 45.67N | 74.03W | 269 | -14.9 | -9.6 | 85.2 | 71.6 | 82.3 | 69.4 | 79.5 | 67.9 | 73.4 | 82.3 | 71.4 | 79.4 | 70.4 | 113.1 | 79.0 | 68.5 | 105.9 | 76.4 | 18.9 | 16.4 | 14.2 | 8630 | 307 |
| MONTREAL/PIERRE ELLIOTT TRUDE | 45.47N | 73.74W | 105 | -9.8 | -5.3 | 86.1 | 71.9 | 83.3 | 70.0 | 80.8 | 68.5 | 73.9 | 82.9 | 72.2 | 80.1 | 71.0 | 114.5 | 79.0 | 69.3 | 108.3 | 77.5 | 25.2 | 22.0 | 19.5 | 7885 | 470 |
| MONTREAL/ST-HUBERT A | 45.52N | 73.42W | 89 | -10.9 | -6.1 | 86.2 | 71.8 | 83.4 | 70.1 | 80.8 | 68.7 | 74.4 | 82.8 | 72.4 | 80.2 | 71.6 | 117.3 | 79.1 | 69.8 | 109.8 | 77.1 | 25.1 | 22.0 | 19.6 | 8111 | 397 |
| MONTREAL-EST | 45.63N | 73.55W | 164 | -9.4 | -4.4 | 86.9 | 69.8 | 84.2 | 68.1 | 81.7 | 66.9 | 72.9 | 81.9 | 71.1 | 79.1 | 70.1 | 111.4 | 76.6 | 68.4 | 104.9 | 75.7 | 19.3 | 17.0 | 15.2 | 7765 | 511 |
| NICOLET | 46.23N | 72.66W | 26 | -13.7 | -8.4 | 83.8 | 72.5 | 80.9 | 70.4 | 78.4 | 68.9 | 74.4 | 81.1 | 72.4 | 78.5 | 72.1 | 118.9 | 78.5 | 70.2 | 111.4 | 76.2 | 21.2 | 18.3 | 15.9 | 8425 | 292 |
| POINTE-AU-PERE (INRS) | 48.51N | 68.47W | 16 | -7.5 | -2.8 | 73.3 | 65.4 | 70.5 | 63.6 | 68.1 | 62.0 | 67.4 | 71.7 | 65.0 | 69.2 | 65.5 | 94.1 | 70.6 | 63.1 | 86.4 | 68.2 | 29.0 | 25.5 | 22.6 | 9584 | 20 |
| QUEBEC/JEAN LESAGE INTL | 46.79N | 71.38W | 243 | -14.9 | -9.9 | 84.0 | 70.3 | 81.0 | 68.4 | 78.2 | 66.4 | 72.8 | 80.8 | 70.6 | 77.9 | 70.0 | 111.5 | 77.5 | 68.0 | 103.9 | 75.3 | 25.2 | 22.0 | 19.6 | 9104 | 238 |
| SHERBROOKE A | 45.43N | 71.68W | 791 | -18.1 | -12.4 | 83.8 | 70.0 | 81.0 | 68.5 | 78.5 | 66.8 | 72.6 | 80.8 | 70.5 | 78.2 | 69.8 | 112.7 | 77.5 | 67.8 | 105.2 | 75.1 | 20.2 | 17.6 | 15.4 | 9011 | 178 |
| ST-ANICET 1 | 45.12N | 74.29W | 161 | -12.3 | -7.1 | 86.6 | 72.8 | 83.8 | 71.1 | 81.3 | 69.5 | 75.5 | 83.4 | 73.6 | 81.0 | 72.9 | 123.0 | 80.4 | 71.2 | 115.6 | 77.9 | 20.8 | 18.2 | 16.1 | 8022 | 361 |
| STE-ANNE-DE-BELLEVUE 1 | 45.43N | 73.93W | 128 | -10.7 | -5.5 | 86.0 | 71.4 | 83.2 | 69.9 | 80.6 | 68.5 | 74.3 | 82.3 | 72.5 | 79.7 | 71.8 | 117.9 | 78.6 | 70.1 | 111.2 | 76.4 | 20.0 | 17.7 | 15.8 | 7963 | 405 |
| STE-FOY (U. LAVAL) | 46.78N | 71.29W | 299 | -12.2 | -7.3 | 84.4 | 69.4 | 81.5 | 67.5 | 78.6 | 65.6 | 72.6 | 80.7 | 70.6 | 77.7 | 69.9 | 111.1 | 76.8 | 68.2 | 104.6 | 74.5 | 21.0 | 18.0 | 15.3 | 8717 | 259 |
| TROIS-RIVIERES | 46.35N | 72.52W | 20 | -10.8 | -6.0 | 81.3 | 70.5 | 79.1 | 69.7 | 77.0 | 68.5 | 73.3 | 78.4 | 71.8 | 76.8 | 71.6 | 116.6 | 76.7 | 70.0 | 110.2 | 75.2 | 23.8 | 20.9 | 18.5 | 8229 | 330 |
| VARENNES | 45.72N | 73.38W | 59 | -10.3 | -5.7 | 86.6 | 71.3 | 83.5 | 69.7 | 80.8 | 68.2 | 74.3 | 82.5 | 72.4 | 80.0 | 71.7 | 117.1 | 78.8 | 69.9 | 110.2 | 76.9 | 24.5 | 21.2 | 18.8 | 8085 | 367 |
| **Saskatchewan** | | | | | | | | | | | | | | | | | | | | | | | | | 5 sites, 41 more on CD-ROM | |
| MOOSE JAW CS | 50.33N | 105.54W | 1893 | -25.2 | -19.5 | 89.9 | 65.5 | 86.1 | 64.4 | 82.3 | 63.0 | 69.6 | 82.4 | 67.3 | 80.2 | 65.7 | 101.6 | 74.3 | 63.0 | 92.4 | 72.0 | 28.3 | 25.1 | 22.4 | 9482 | 254 |
| PRINCE ALBERT A | 53.22N | 105.67W | 1404 | -32.8 | -26.7 | 84.6 | 65.6 | 81.0 | 64.0 | 77.9 | 62.1 | 68.1 | 80.3 | 66.1 | 77.5 | 63.8 | 93.2 | 73.5 | 61.7 | 86.6 | 71.2 | 21.0 | 18.6 | 16.8 | 11090 | 123 |
| REGINA RCS | 50.43N | 104.67W | 1893 | -28.5 | -22.9 | 88.2 | 65.9 | 84.5 | 65.0 | 81.1 | 63.3 | 70.0 | 82.2 | 67.5 | 79.6 | 66.1 | 103.1 | 76.0 | 63.2 | 93.1 | 73.5 | 29.8 | 26.3 | 23.4 | 10244 | 211 |
| SASKATOON RCS | 52.17N | 106.72W | 1654 | -30.3 | -24.7 | 87.2 | 65.6 | 83.5 | 64.3 | 79.9 | 62.9 | 69.1 | 81.8 | 66.8 | 78.9 | 64.8 | 97.8 | 74.6 | 62.6 | 90.2 | 72.0 | 25.0 | 22.0 | 19.5 | 10508 | 180 |
| SASKATOON KERNEN FARM | 52.15N | 106.55W | 1673 | -28.3 | -23.0 | 87.2 | 63.8 | 83.4 | 62.4 | 80.2 | 61.0 | 68.9 | 80.5 | 66.5 | 76.9 | 65.1 | 98.8 | 74.6 | 62.6 | 90.3 | 71.6 | 24.0 | 21.2 | 19.5 | 10626 | 182 |
| **Yukon Territory** | | | | | | | | | | | | | | | | | | | | | | | | | 1 site, 15 more on CD-ROM | |
| WHITEHORSE A | 60.71N | 135.07W | 2316 | -39.5 | -30.3 | 78.2 | 57.5 | 74.1 | 55.9 | 70.3 | 54.2 | 58.7 | 74.5 | 57.0 | 71.4 | 52.2 | 63.2 | 61.1 | 50.5 | 59.4 | 60.3 | 23.0 | 20.8 | 18.7 | 12155 | 12 |

- Minimum and maximum precipitation, in. These parameters give extremes of precipitation and are useful for green building technologies and stormwater management.

**Monthly Design Dry-Bulb, Wet-Bulb, and Mean Coincident Temperatures.** These values are derived from the same analysis that results in the annual design conditions. The monthly summaries are useful when seasonal variations in solar geometry and intensity, building or facility occupancy, or building use patterns require consideration. In particular, these values can be used when determining air-conditioning loads during periods of maximum solar radiation. The values listed in the tables include

- Dry-bulb temperature corresponding to 0.4, 2.0, 5.0, and 10.0% cumulative frequency of occurrence for indicated month, °F; mean coincident wet-bulb temperature, °F.
- Wet-bulb temperature corresponding to 0.4, 2.0, 5.0, and 10.0% cumulative frequency of occurrence for indicated month, °F; mean coincident dry-bulb temperature, °F.

For a 30-day month, the 0.4, 2.0, 5.0 and 10.0% values of occurrence represent the value that occurs or is exceeded for a total of 3, 14, 36, or 72 h, respectively, per month on average over the period of record. Monthly percentile values of dry- or wet-bulb temperature may be higher or lower than the annual design conditions corresponding to the same nominal percentile, depending on the month and the seasonal distribution of the parameter at that location. Generally, for the hottest or most humid months of the year, the monthly percentile value exceeds the design condition for the same element corresponding to the same nominal percentile. For example, Table 4-4 shows that the annual 0.4% design dry-bulb temperature at Atlanta, GA, is 93.9°F; the 0.4% monthly dry-bulb temperature exceeds 93.4°F for June, July, and August, with values of 94.5°F, 97.8°F, and 97.4°F, respectively. Two new percentiles were added to this chapter (5.0 and 10.0% values) to give a greater range in the frequency of occurrence, in particular providing less extreme options to select for design calculations.

A general, very approximate rule of thumb is that the $n\%$ annual cooling design condition is roughly equivalent to the $5n\%$ monthly cooling condition for the hottest month; that is, the 0.4% annual design dry-bulb temperature is roughly equivalent to the 2% monthly design dry-bulb temperature for the hottest month; the 1% annual value is roughly equivalent to the 5% monthly value for the hottest month, and the 2% annual value is roughly equivalent to the 10% monthly value for the hottest month.

**Mean Daily Temperature Range.** These values are useful in calculating daily dry- and wet-bulb temperature profiles, as explained in the section on Generating Design-Day Data. Three kinds of profiles are defined:

- Mean daily temperature range for month indicated, °F (defined as mean of difference between daily maximum and minimum dry-bulb temperatures).
- Mean daily dry- and wet-bulb temperature ranges coincident with the 5% monthly design dry-bulb temperature. This is the difference between daily maximum and minimum dry- or wet-bulb temperatures, respectively, averaged over all days where the maximum daily dry-bulb temperature exceeds the 5% monthly design dry-bulb temperature.
- Mean daily dry- and wet-bulb temperature ranges coincident with the 5% monthly design wet-bulb temperature. This is the difference between daily maximum and minimum dry- or wet-bulb temperatures, respectively, averaged over all days where the maximum daily wet-bulb temperature exceeds the 5% monthly design wet-bulb temperature.

**Clear-Sky Solar Irradiance.** Clear-sky irradiance parameters are useful in calculating solar-related air conditioning loads for any time of any day of the year. Parameters are provided for the 21st day of each month. The 21st of the month is usually a convenient day for solar calculations because June 21 and December 21 represent the solstices (longest and shortest days) and March 21 and September 21 are close to the equinox (days and nights have the same length). Parameters listed in the tables are

- Clear-sky optical depths for beam and diffuse irradiances, which are used to calculate beam and diffuse irradiance as explained in the section on Calculating Clear-Sky Solar Radiation.
- Clear-sky beam normal and diffuse horizontal irradiances at solar noon. These two values can be calculated from the clear-sky optical depths but are listed here for convenience.

### Differences from Previously Published Design Conditions

- Climatic design conditions in this chapter are generally similar to those in previous editions, because similar if not identical analysis procedures were used. There are some differences, however, owing to a more recent period of record (generally 1986–2010 versus 1982–2006). For example, compared to the 2009 edition, 99.6% heating dry-bulb temperatures have increased by 0.31°F on average, and 0.4% cooling dry-bulb temperatures have increased by 0.18°F on average. Similar trends are observed for other design temperatures. The root mean square differences are 0.74°F for the 99.6% heating dry-bulb values and 0.70°F for 0.4% cooling dry-bulb. The increases noted here are generally consistent with the discussion in the section on Impacts of Climate Change.

### Applicability and Characteristics of Design Conditions

Climatic design values in this chapter represent different psychrometric conditions. Design data based on dry-bulb temperature represent peak occurrences of the sensible component of ambient outdoor conditions. Design values based on wet-bulb temperature are related to the enthalpy of the outdoor air. Conditions based on dew point relate to the peaks of the humidity ratio. The designer, engineer, or other user must decide which set(s) of conditions and probability of occurrence apply to the design situation under consideration.

**Annual Heating and Humidification Design Conditions.** The month with the lowest mean dry-bulb temperature is used, for example, to determine the time of year where the maximum heating load occurs.

The 99.6 and 99.0% design conditions are often used in sizing heating equipment.

The humidification dew point and mean coincident dry-bulb temperatures and humidity ratio provide information for cold-season humidification applications.

Wind design data provide information for estimating peak loads accounting for infiltration: extreme wind speeds for the coldest month, with the mean coincident dry-bulb temperature; and mean wind speed and direction coincident to the 99.6% design dry-bulb temperature.

**Annual Cooling, Dehumidification, and Enthalpy Design Conditions.** The month with the highest mean dry-bulb temperature is used, for example, to determine the time of year where the maximum sensible cooling load occurs, not taking into account solar loads.

The mean daily dry-bulb temperature range for the hottest month is the mean difference between the daily maximum and minimum temperatures during the hottest month and is calculated from the extremes of the hourly temperature observations. The true maximum and minimum temperatures for any day generally occur between hourly readings. Thus, the mean maximum and minimum temperatures calculated in this way are about 1°F less extreme than the mean daily extreme temperatures observed with maximum and minimum thermometers. This results in the true daily temperature range generally about 2°F greater than that calculated from hourly data. The mean daily dry-bulb temperature range is used in cooling load calculations.

The 0.4, 1.0, and 2.0% dry-bulb temperatures and mean coincident wet-bulb temperatures often represent conditions on hot, mostly sunny days. These are often used in sizing cooling equipment such as chillers or air-conditioning units.

Design conditions based on wet-bulb temperature represent extremes of the total sensible plus latent heat of outdoor air. This information is useful for design of cooling towers, evaporative coolers, and outdoor air ventilation systems.

The mean wind speed and direction coincident with the 0.4% design dry-bulb temperature is used for estimating peak loads accounting for infiltration.

Design conditions based on dew-point temperatures are directly related to extremes of humidity ratio, which represent peak moisture loads from the weather. Extreme dew-point conditions may occur on days with moderate dry-bulb temperatures, resulting in high relative humidity. These values are especially useful for humidity control applications, such as desiccant cooling and dehumidification, cooling-based dehumidification, and fresh-air ventilation systems. The values are also used as a check point when analyzing the behavior of cooling systems at part-load conditions, particularly when such systems are used for humidity control as a secondary function. Humidity ratio values are calculated from the corresponding dew-point temperature and the standard pressure at the location's elevation.

Annual enthalpy design conditions give the annual enthalpy for the cooling season; this is used for calculating cooling loads caused by infiltration and/or ventilation into buildings. Enthalpy represents the total heat content of air (the sum of its sensible and latent energies). Cooling loads can be calculated knowing the conditions of both the outdoor ambient and the building's interior air.

**Extreme Annual Design Conditions.** Extreme annual design wind speeds are used in designing smoke management systems.

**General Design Conditions Discussion.**

Minimum temperatures usually occur between 6:00 A.M. and 8:00 A.M. suntime on clear days when the daily range is greatest. For residential or other applications where the occupancy is continuous throughout the day, the recommended design temperatures apply. With commercial applications or other applications where occupancy is only during hours near the middle of the day, design temperatures above the recommended minimum may apply.

Maximum temperatures usually occur between 2:00 P.M. and 4:00 P.M. suntime with deviations on cloudy days when the daily range is less. Typically, the design dry-bulb temperatures should be used with the coincident wet-bulb temperatures in computing building cooling loads. For residential or other applications where the occupancy is continuous throughout the day, the recommended design temperatures apply. For commercial applications or other applications where occupancy is only during hours near the middle of the day, design temperatures below the recommended maximum might apply. In some cases, the peak occupancy load occurs before the effect of the outdoor maximum temperature has reached the space by conduction through the building mass. In other cases, the peak occupancy loads may be in months other than the three or four summer months when the maximum outdoor temperature is expected; here design temperatures from other months will apply.

When determining the heat loss for below grade components of a building (e.g., basement walls and floor), the average winter outdoor air temperature needed for the current ASHRAE loads methodology. Thus, Table 4-9 is provided. The heating load for the below grade structure also uses the amplitude of the ground temperature to determine the design ground surface temperature (see Chapter 6).

Although not a direct design criteria, another environmental index of general interest is the wind chill index. The wind chill index reliably expresses combined effects of temperature and wind on subjective discomfort. However, rather than using the numerical value of the wind chill index, meteorologists use an index derived from the WCI called the equivalent wind chill temperature. Table 4-10 shows a typical wind chill chart, expressed in equivalent wind chill temperature.

## 4.3  Other Factors Affecting Design

In the interest of energy conservation, new buildings used primarily for human occupancy must meet certain minimum design requirements that enable the efficient use of energy in

# Design Conditions

Table 4-9  Average Winter Temperature and Yearly Degree Days[a,b,c]

| State | Station | | Avg. Winter Temp,[d] °F | Degree-Days Yearly Total | State | Station | | Avg. Winter Temp,[d] °F | Degree-Days Yearly Total |
|---|---|---|---|---|---|---|---|---|---|
| Ala. | Birmingham | A | 54.2 | 2551 | Fla. | Orlando | A | 65.7 | 766 |
| | Huntsville | A | 51.3 | 3070 | (Cont'd | Pensacola | A | 60.4 | 1463 |
| | Mobile | A | 59.9 | 1560 | | Tallahassee | A | 60.1 | 1485 |
| | Montgomery | A | 55.4 | 2291 | | Tampa | A | 66.4 | 683 |
| | | | | | | West Palm Beach | A | 68.4 | 253 |
| Alaska | Anchorage | A | 23.0 | 10,864 | | | | | |
| | Fairbanks | A | 6.7 | 14,276 | Ga. | Athens | A | 51.8 | 2929 |
| | Juneau | A | 32.1 | 9075 | | Atlanta | A | 51.7 | 2961 |
| | Nome | A | 13.1 | 14,171 | | Augusta | A | 54.5 | 2397 |
| | | | | | | Columbus | A | 54.8 | 2383 |
| Ariz. | Flagstaff | A | 35.6 | 7152 | | Macon | A | 56.2 | 2136 |
| | Phoenix | A | 58.5 | 1765 | | Rome | A | 49.9 | 3326 |
| | Tucson | A | 58.1 | 1800 | | Savannah | A | 57.8 | 1819 |
| | Winslow | A | 43.0 | 4782 | | Thomasville | C | 60.0 | 1529 |
| | Yuma | A | 64.2 | 974 | Hawaii | Lihue | A | 72.7 | 0 |
| Ark. | Fort Smith | A | 50.3 | 3292 | | Honolulu | A | 74.2 | 0 |
| | Little Rock | A | 50.5 | 3219 | | Hilo | A | 71.9 | 0 |
| | Texarkana | A | 54.2 | 2533 | | | | | |
| | | | | | Idaho | Boise | A | 39.7 | 5809 |
| Calif. | Bakersfield | A | 55.4 | 2122 | | Lewiston | A | 41.0 | 5542 |
| | Bishop | A | 46.0 | 4275 | | Pocatello | A | 34.8 | 7033 |
| | Blue Canyon | A | 42.2 | 5596 | | | | | |
| | Burbank | A | 58.6 | 1646 | Ill. | Cairo | C | 47.9 | 3821 |
| | Eureka | C | 49.9 | 4643 | | Chicago (O'Hare) | A | 35.8 | 6639 |
| | Fresno | A | 53.3 | 2611 | | Chicago (Midway) | A | 37.5 | 6155 |
| | Long Beach | A | 57.8 | 1803 | | Chicago | C | 38.9 | 5882 |
| | Los Angeles | A | 57.4 | 2061 | | Moline | A | 36.4 | 6408 |
| | Los Angeles | C | 60.3 | 1349 | | Peoria | A | 38.1 | 6025 |
| | Mt. Shasta | C | 41.2 | 5722 | | Rockford | A | 34.8 | 6830 |
| | Oakland | A | 53.5 | 2870 | | Springfield | A | 40.6 | 5429 |
| | Red Bluff | A | 53.8 | 2515 | | | | | |
| | Sacramento | A | 53.9 | 2502 | Ind. | Evansville | A | 45.0 | 4435 |
| | Sacramento | C | 54.4 | 2419 | | Fort Wayne | A | 37.3 | 6205 |
| | Sandberg | C | 46.8 | 4209 | | Indianapolis | A | 39.6 | 5699 |
| | San Diego | A | 59.5 | 1458 | | South Bend | A | 36.6 | 6439 |
| | San Francisco | A | 53.4 | 3015 | | | | | |
| | San Francisco | C | 55.1 | 3001 | Iowa | Burlington | A | 37.6 | 6114 |
| | Santa Maria | A | 54.3 | 2967 | | Des Moines | A | 35.5 | 6588 |
| | | | | | | Dubuque | A | 32.7 | 7376 |
| Colo. | Alamosa | A | 29.7 | 8529 | | Sioux City | A | 34.0 | 6951 |
| | Colorado Springs | A | 37.3 | 6423 | | Waterloo | A | 32.6 | 7320 |
| | Denver | A | 37.6 | 6283 | | | | | |
| | Denver | C | 40.8 | 5524 | Kans. | Concordia | A | 40.4 | 5479 |
| | Grand Junction | A | 39.3 | 5641 | | Dodge City | A | 42.5 | 4986 |
| | Pueblo | A | 40.4 | 5462 | | Goodland | A | 37.8 | 6141 |
| | | | | | | Topeka | A | 41.7 | 5182 |
| Conn. | Bridgeport | A | 39.9 | 5617 | | Wichita | A | 44.2 | 4620 |
| | Hartford | A | 37.3 | 6235 | | | | | |
| | New Haven | A | 39.0 | 5897 | Ky. | Covington | A | 41.1 | 5265 |
| | | | | | | Lexington | A | 43.8 | 4683 |
| Del. | Wilmington | A | 42.5 | 4930 | | Louisville | A | 44.0 | 4660 |
| D.C. | Washington | A | 45.7 | 4224 | | | | | |
| | | | | | La. | Alexandria | A | 57.5 | 1921 |
| Fla. | Apalachicola | C | 61.2 | 1308 | | Baton Rouge | A | 59.8 | 1560 |
| | Daytona Beach | A | 64.5 | 879 | | Lake Charles | A | 60.5 | 1459 |
| | Fort Myers | A | 68.6 | 442 | | New Orleans | A | 61.0 | 1385 |
| | Jacksonville | A | 61.9 | 1239 | | New Orleans | C | 61.8 | 1254 |
| | Key West | A | 73.1 | 108 | | Shreveport | A | 56.2 | 2184 |
| | Lakeland | C | 66.7 | 661 | | | | | |
| | Miami | A | 71.1 | 214 | Mass. | Boston | A | 40.0 | 5634 |
| | Miami Beach | C | 72.5 | 141 | | Nantucket | A | 40.2 | 5891 |

[a]Data for US cities from a publication of the US Weather Bureau, *Monthly Normals of Temperature, Precipitation and Heating Degree Days,* 1962, are for the period 1931 to 1960 inclusive. These data also include information from the 1963 revisions to this publication, where available.
[b]Data for airport station, A, and city stations, C, are both given where available.
[c]Data for Canadian cities were computed by the Climatology Division, Department of Transport, from normal monthly mean temperatures, and the monthly values of heating days data were obtained using the National Research Council computer and a method devised by H.C.S. Thom of the US Weather Bureau. The heating days are based on the period from 1931 to 1960.
[d]For period October to April, inclusive.

Table 4-9 Average Winter Temperature and Yearly Degree Days[a,b,c] *(Continued)*

| State | Station | | Avg. Winter Temp,[d] °F | Degree-Days Yearly Total | State | Station | | Avg. Winter Temp,[d] °F | Degree-Days Yearly Total |
|---|---|---|---|---|---|---|---|---|---|
| Mass. | Pittsfield | A | 32.6 | 7578 | N.M. | Raton | A | 38.1 | 6228 |
| (Cont'd | Worcester | A | 34.7 | 6969 | (Cont'd | Roswell | A | 47.5 | 3793 |
| | | | | | | Silver City | A | 48.0 | 3705 |
| Md. | Baltimore | A | 43.7 | 4654 | | | | | |
| | Baltimore | C | 46.2 | 4111 | N.Y. | Albany | A | 34.6 | 6875 |
| | Frederick | A | 42.0 | 5087 | | Albany | C | 37.2 | 6201 |
| | | | | | | Binghamton | A | 33.9 | 7286 |
| Me. | Caribou | A | 24.2 | 9767 | | Binghamton | C | 36.6 | 6451 |
| | Portland | A | 33.0 | 7511 | | Buffalo | A | 34.5 | 7062 |
| | | | | | | New York (Cent. Park) | C | 42.8 | 4871 |
| Mich. | Alpena | A | 29.7 | 8506 | | New York (La Guardia) | A | 43.1 | 4811 |
| | Detroit (City) | A | 37.2 | 6232 | | New York (Kennedy) | A | 41.4 | 5219 |
| | Detroit (Wayne) | A | 37.1 | 6293 | | Rochester | A | 35.4 | 6748 |
| | Detroit (Willow Run) | A | 37.2 | 6258 | | Schenectady | C | 35.4 | 6650 |
| | Escanaba | C | 29.6 | 8481 | | Syracuse | A | 35.2 | 6756 |
| | Flint | A | 33.1 | 7377 | | | | | |
| | Grand Rapids | A | 34.9 | 6894 | N.C. | Asheville | C | 46.7 | 4042 |
| | Lansing | A | 34.8 | 6909 | | Cape Hatteras | | 53.3 | 2612 |
| | Marquette | C | 30.2 | 8393 | | Charlotte | A | 50.4 | 3191 |
| | Muskegon | A | 36.0 | 6696 | | Greensboro | A | 47.5 | 3805 |
| | Sault Ste. Marie | A | 27.7 | 9048 | | Raleigh | A | 49.4 | 3393 |
| | | | | | | Wilmington | A | 54.6 | 2347 |
| Minn. | Duluth | A | 23.4 | 10,000 | | Winston-Salem | A | 48.4 | 3595 |
| | Minneapolis | A | 28.3 | 8382 | | | | | |
| | Rochester | A | 28.8 | 8295 | N.D. | Bismarck | A | 26.6 | 8851 |
| | | | | | | Devils Lake | C | 22.4 | 9901 |
| Miss. | Jackson | A | 55.7 | 2239 | | Fargo | A | 24.8 | 9226 |
| | Meridian | A | 55.4 | 2289 | | Williston | A | 25.2 | 9243 |
| | Vicksburg | C | 56.9 | 2041 | | | | | |
| | | | | | Ohio | Akron-Canton | A | 38.1 | 6037 |
| Mo. | Columbia | A | 42.3 | 5046 | | Cincinnati | C | 45.1 | 4410 |
| | Kansas City | A | 43.9 | 4711 | | Cleveland | A | 37.2 | 6351 |
| | St. Joseph | A | 40.3 | 5484 | | Columbus | A | 39.7 | 5660 |
| | St. Louis | A | 43.1 | 4900 | | Columbus | C | 41.5 | 5211 |
| | St. Louis | C | 44.8 | 4484 | | Dayton | A | 39.8 | 5622 |
| | Springfield | A | 44.5 | 4900 | | Mansfield | A | 36.9 | 6403 |
| | | | | | | Sandusky | C | 39.1 | 5796 |
| Mont. | Billings | A | 34.5 | 7049 | | Toledo | A | 36.4 | 6494 |
| | Glasgow | A | 26.4 | 8996 | | Youngstown | A | 36.8 | 6417 |
| | Great Falls | A | 32.8 | 7750 | | | | | |
| | Havre | A | 28.1 | 8700 | Okla. | Oklahoma City | A | 48.3 | 3725 |
| | Havre | C | 29.8 | 8182 | | Tulsa | A | 47.7 | 3860 |
| | Helena | A | 31.1 | 8129 | | | | | |
| | Kalispell | A | 31.4 | 8191 | Ore. | Astoria | A | 45.6 | 5186 |
| | Miles City | A | 31.2 | 7723 | | Burns | C | 35.9 | 6957 |
| | Missoula | A | 31.5 | 8125 | | Eugene | A | 45.6 | 4726 |
| | | | | | | Meacham | A | 34.2 | 7874 |
| Neb. | Grand Island | A | 36.0 | 6530 | | Medford | A | 43.2 | 5008 |
| | Lincoln | C | 38.8 | 5864 | | Pendleton | A | 42.6 | 5127 |
| | Norfolk | A | 34.0 | 6979 | | Portland | A | 45.6 | 4635 |
| | North Platte | A | 35.5 | 6684 | | Portland | C | 47.4 | 4109 |
| | Omaha | A | 35.6 | 6612 | | Roseburg | A | 46.3 | 4491 |
| | Scottsbluff | A | 35.9 | 6673 | | Salem | A | 45.4 | 4486 |
| | Valentine | A | 32.6 | 7425 | | | | | |
| | | | | | Pa. | Allentown | A | 38.9 | 5810 |
| Nev. | Elko | A | 34.0 | 7433 | | Erie | A | 36.8 | 6451 |
| | Ely | A | 33.1 | 7733 | | Harrisburg | A | 41.2 | 5251 |
| | Las Vegas | A | 53.3 | 2709 | | Philadelphia | A | 41.8 | 5144 |
| | Reno | A | 39.3 | 6332 | | Philadelphia | C | 44.5 | 4486 |
| | Winnemucca | A | 36.7 | 6761 | | Pittsburgh | A | 38.4 | 5987 |
| | | | | | | Pittsburgh | C | 42.2 | 5053 |
| N.H. | Concord | A | 33.0 | 7383 | | Reading | C | 42.4 | 4945 |
| | Mt. Washington Obsv. | | 15.2 | 13,817 | | Scranton | A | 37.2 | 6254 |
| | | | | | | Williamsport | A | 38.5 | 5934 |
| N.J. | Atlantic City | A | 43.2 | 4812 | | | | | |
| | Newark | A | 42.8 | 4589 | R.I. | Block Island | A | 40.1 | 5804 |
| | Trenton | C | 42.4 | 4980 | | Providence | A | 38.8 | 5954 |
| N.M. | Albuquerque | A | 45.0 | 4348 | S.C. | Charleston | A | 56.4 | 2033 |
| | Clayton | A | 42.0 | 5158 | | Charleston | C | 57.9 | 1794 |

# Design Conditions

**Table 4-9  Average Winter Temperature and Yearly Degree Days[a,b,c]** *(Continued)*

| State | Station | | Avg. Winter Temp.,[d] °F | Degree-Days Yearly Total | State | Station | | Avg. Winter Temp.,[d] °F | Degree-Days Yearly Total |
|---|---|---|---|---|---|---|---|---|---|
| S.C. (Cont'd | Columbia | A | 54.0 | 2484 | Wyo. | Casper | A | 33.4 | 7410 |
| | Florence | A | 54.5 | 2387 | | Cheyenne | A | 34.2 | 7381 |
| | Greenville-Spartenburg | A | 51.6 | 2980 | | Lander | A | 31.4 | 7870 |
| | | | | | | Sheridan | A | 32.5 | 7680 |
| S.D. | Huron | A | 28.8 | 8223 | | | | | |
| | Rapid City | A | 33.4 | 7345 | Alta. | Banff | C | — | 10,551 |
| | Sioux Falls | A | 30.6 | 7839 | | Calgary | A | — | 9703 |
| | | | | | | Edmonton | A | — | 10,268 |
| Tenn. | Bristol | A | 46.2 | 4143 | | Lethbridge | A | — | 8644 |
| | Chattanooga | A | 50.3 | 3254 | | | | | |
| | Knoxville | A | 49.2 | 3494 | B.C. | Kamloops | A | — | 6799 |
| | Memphis | A | 50.5 | 3232 | | Prince George* | A | — | 9755 |
| | Memphis | C | 51.6 | 3015 | | Prince Rupert | C | — | 7029 |
| | Nashville | A | 48.9 | 3578 | | Vancouver* | A | — | 5515 |
| | Oak Ridge | C | 47.7 | 3817 | | Victoria* | A | — | 5699 |
| Tex. | Abilene | A | 53.9 | 2624 | | Victoria | C | — | 5579 |
| | Amarillo | A | 47.0 | 3985 | | | | | |
| | Austin | A | 59.1 | 1711 | Man. | Brandon* | A | — | 11,036 |
| | Brownsville | A | 67.7 | 600 | | Churchill | A | — | 16,728 |
| | Corpus Christi | A | 64.6 | 914 | | The Pas | C | — | 12,281 |
| | Dallas | A | 55.3 | 2363 | | Winnipeg | A | — | 10,679 |
| | El Paso | A | 52.9 | 2700 | | | | | |
| | Fort Worth | A | 55.1 | 2405 | N.B. | Fredericton* | A | — | 8671 |
| | Galveston | A | 62.2 | 1274 | | Moncton | C | — | 8727 |
| | Galveston | C | 62.0 | 1235 | | St. John | C | — | 8219 |
| | Houston | A | 61.0 | 1396 | | | | | |
| | Houston | C | 62.0 | 1278 | Nfld. | Argentia | A | — | 8440 |
| | Laredo | A | 66.0 | 797 | | Corner Brook | C | — | 8978 |
| | Lubbock | A | 48.8 | 3578 | | Gander | A | — | 9254 |
| | Midland | A | 53.8 | 2591 | | Goose* | A | — | 11,887 |
| | Port Arthur | A | 60.5 | 1447 | | St. John's* | A | — | 8991 |
| | San Angelo | A | 56.0 | 2255 | | | | | |
| | San Antonio | A | 60.1 | 1546 | N.W.T. | Aklavik | C | — | 18,017 |
| | Victoria | A | 62.7 | 1173 | | Fort Norman | C | — | 16,109 |
| | Waco | A | 57.2 | 2030 | | Resolution Island | C | — | 16,021 |
| | Wichita Falls | A | 53.0 | 2832 | | | | | |
| | | | | | N.S. | Halifax | C | — | 7361 |
| Utah | Milford | A | 36.5 | 6497 | | Sydney | A | — | 8049 |
| | Salt Lake City | A | 38.4 | 6052 | | Yarmouth | A | — | 7340 |
| | Wendover | A | 39.1 | 5778 | | | | | |
| | | | | | Ont. | Cochrane | C | — | 11,412 |
| Vt. | Burlington | A | 29.4 | 8269 | | Fort William | A | — | 10,405 |
| | | | | | | Kapuskasing | C | — | 11,572 |
| Va. | Cape Henry | C | 50.0 | 3279 | | Kitchener | C | — | 7566 |
| | Lynchburg | A | 46.0 | 4166 | | London | A | — | 7349 |
| | Norfolk | A | 49.2 | 3421 | | North Bay | C | — | 9219 |
| | Richmond | A | 47.3 | 3865 | | Ottawa | C | — | 8735 |
| | Roanoke | A | 46.1 | 4150 | | Toronto | C | — | 6827 |
| Wash. | Olympia | A | 44.2 | 5236 | P.E.I. | Charlottetown | C | — | 8164 |
| | Seattle-Tacoma | A | 44.2 | 5145 | | Summerside | C | — | 8488 |
| | Seattle | C | 46.9 | 4424 | | | | | |
| | Spokane | A | 36.5 | 6655 | Que. | Arvida | C | — | 10,528 |
| | Walla Walla | C | 43.8 | 4805 | | Montreal* | A | — | 8203 |
| | Yakima | A | 39.1 | 5941 | | Montreal | C | — | 7899 |
| | | | | | | Quebec* | A | — | 9372 |
| W.Va. | Charleston | A | 44.8 | 4476 | | Quebec | C | — | 8937 |
| | Elkins | A | 40.1 | 5675 | | | | | |
| | Huntington | A | 45.0 | 4446 | Sask. | Prince Albert | A | — | 11,630 |
| | Parkersburg | C | 43.5 | 4754 | | Regina | A | — | 10,806 |
| | | | | | | Saskatoon | C | — | 10,870 |
| Wisc. | Green Bay | A | 30.3 | 8029 | | | | | |
| | La Crosse | A | 31.5 | 7589 | Y.T. | Dawson | C | — | 15,067 |
| | Madison | A | 30.9 | 7893 | | Mayo Landing | C | — | 14,454 |
| | Milwaukee | A | 32.6 | 7635 | | | | | |

*The data for these normals were from the full 10-year period 1951–1960, adjusted for the standard journal period 1931–1960.

**Table 4-10 Equivalent Wind Chill Temperatures**
*(Table 12, Chapter 9, 2013 ASHRAE Handbook—Fundamentals)*

| Wind Speed, mph | Actual Thermometer Reading, °F ||||||||||||
|---|---|---|---|---|---|---|---|---|---|---|---|---|
| | 50 | 40 | 30 | 20 | 10 | 0 | −10 | −20 | −30 | −40 | −50 | −60 |
| | Equivalent Chill Temperature, °F ||||||||||||
| 0 | 50 | 40 | 30 | 20 | 10 | 0 | −10 | −20 | −30 | −40 | −50 | −60 |
| 5 | 48 | 37 | 27 | 16 | 6 | −5 | −15 | −26 | −36 | −47 | −57 | −68 |
| 10 | 40 | 28 | 16 | 3 | −9 | −21 | −34 | −46 | −58 | −71 | −83 | −95 |
| 15 | 36 | 22 | 9 | −5 | −18 | −32 | −45 | −59 | −72 | −86 | −99 | −113 |
| 20 | 32 | 18 | 4 | −11 | −25 | −39 | −53 | −68 | −82 | −96 | −110 | −125 |
| 25 | 30 | 15 | 0 | −15 | −30 | −44 | −59 | −74 | −89 | −104 | −119 | −134 |
| 30 | 28 | 13 | −3 | −18 | −33 | −48 | −64 | −79 | −94 | −110 | −125 | −140 |
| 35 | 27 | 11 | −4 | −20 | −36 | −51 | −67 | −83 | −98 | −114 | −129 | −145 |
| 40 | 26 | 10 | −6 | −22 | −38 | −53 | −69 | −85 | −101 | −117 | −133 | −148 |

**Little danger:** In less than 5 h, with dry skin. Maximum danger from false sense of security. (WCI less than 1400)

**Increasing danger:** Danger of freezing exposed flesh within one minute. (WCI between 1400 and 2000)

**Great danger:** Flesh may freeze within 30 seconds. (WCI greater than 2000)

*Notes:* Cooling power of environment expressed as an equivalent temperature under calm conditions [Equation (79)].

Winds greater than 43 mph have little added chilling effect.
*Source:* US Army Research Institute of Environmental Medicine.

---

such new buildings. ASHRAE/IES Standard 90.1, "Energy Efficient Design of New Buildings Except Low-Rise Residential Buildings," has been widely adopted. In fact, the U. S. Energy Policy Act of 1992 requires that each state have energy policies in place that require buildings to conform, on the minimum, to Standard 90.1.

Although a heating or cooling system is sized to meet design conditions, it usually functions at only partial capacity. The proper use and application of controls should receive primary consideration at the time the heating or cooling system is being designed. Control devices that produce almost any degree of control can be used, but it is useless to provide such controls unless the air-conditioning system is capable of properly responding to the demands of the controllers. For example, it is impossible to maintain close control of temperature and humidity by starting and stopping a refrigeration compressor or by opening and closing a refrigerant valve. Instead, refrigeration equipment that permits proportional control, or a chilled-water system with a cooling coil or spray dehumidifier that permits proportional control, must be used. It is neither economical nor good practice to select equipment capable of producing far more precise control than the application requires, or to complicate the system to obtain special sequences or cycles of operation when they are not necessary. Because the system must be adjusted and maintained in operation for many years, the simplest system which produces the necessary results is usually the best.

Another factor at least as important as comfort, first cost, and owning and operating cost, is the application, which includes such factors as

- Flexibility for change
- Suitability for all spaces
- Appearance of completed building
- Special requirements of project income potential
- Durability, reliability, and serviceability
- Fire and smoke control
- Pollution control

**Example 4-3:** Select appropriate summer and winter, indoor and outdoor, design temperatures and humidities for an office building near the Philadelphia International Airport, Pennsylvania.

**Solution:**
Based upon the climatic data of Table 4.8, the *outdoor design values* are
SUMMER: 90.8°F db and 74.4°F wb (1% values)
WINTER: 13.8°F (0.4% value) and 100% rh
Based upon the comfort envelope of Standard 55, the selected *indoor design values* are
SUMMER: 78°F db, humidity ratio of 0.012
WINTER: 72°F db, 30% rh

## 4.4 Temperatures in Adjacent Unconditioned Spaces

The heat loss or gain between conditioned rooms and unconditioned rooms or spaces must be based on the estimated or assumed temperature in such unconditioned spaces. This temperature normally lies in the range between the indoor and outdoor temperatures. The temperature in the unconditioned space may be estimated by

$$t_u = [\, t_i(A_1 U_1 + A_2 U_2 + \ldots + \text{etc.}) \\ + t_o(K V_o + A_a U_a + A_b U_b + \ldots + \text{etc.})] \\ \div ([\, A_1 U_1 + A_2 U_2 + \ldots + \text{etc.}]) \\ + (K V_o + A_a U_a + A_b U_b + \ldots + \text{etc.})] \quad (4\text{-}6)$$

where

$t_u$ = temperature in unheated space, °F (°C)

# Design Conditions

$t_i$ = indoor design temperature of heated room, °F (°C)

$t_o$ = outdoor design temperature, °F (°C)

$A_1, A_2$, etc. = areas of surface of unheated space adjacent to heated space, ft² (m²)

$A_a, A_b$, etc. = areas of surface of unheated space exposed to outdoors, ft² (m²)

$U_1, U_2$, etc. = heat transfer coefficients of surfaces of $A_1, A_2$, etc., Btu/h·ft²·°F (W/[m²·°C])

$U_a, U_b$, etc. = heat transfer coefficients of surfaces of $A_a, A_b$, etc. Btu/h·ft²·°F (W/[m²·°C])

$V_o$ = rate of introduction of outdoor air into the unheated space by infiltration and/or ventilation, cfm (L/s)

$K$ = 1.10 (1200)

Reasonable accuracy for ordinary unconditioned spaces may be attained if the following approximations for adjacent rooms are used:

1. Cooling with adjacent unconditioned room. Select for computation a temperature equal to $t_i + 0.667(t_o - t_i)$ in the unconditioned space.

2. Heating with adjacent room unheated. Select for computation a temperature equal to $t_i - 0.50(t_i - t_o)$ in the unconditioned space.

Temperatures in unconditioned spaces having large glass areas and two or more surfaces exposed to the outdoors (such as sleeping porches or sun parlors) and/or large amounts of infiltration (such as garages with poor fitting doors) are generally assumed to be that of the outdoors.

**Example 4-4:** Calculate the temperature in an unheated space adjacent to a conditioned room with three common surface areas of 100, 120, and 140 ft² and overall heat transfer coefficients of 0.15, 0.20, and 0.25 Btu/h·ft²·°F, respectively. The surface areas of the unheated space exposed to the outdoors are 100 and 140 ft² with corresponding overall heat transfer coefficients are 0.10 and 0.30 Btu/h·ft²·°F. The sixth surface is on the ground and can be neglected for this example as can be the effect of any outdoor air entering the space. Inside and outside design temperatures are 70°F and –10°F, respectively.

**Solution:**

Substituting into Equation 4-6,

$$t_u = [70(100 \times 0.15 + 120 \times 0.20 + 140 \times 0.25)$$
$$+ (-10)(100 \times 0.10 + 140 \times 0.30)]/$$
$$[100 \times 0.15 + 120 \times 0.20 + 140 \times 0.25$$
$$+ 100 \times 0.10 + 140 \times 0.30]$$
$$= 4660/126 = 37°F$$

## 4.5 Problems

**4.1** Describe the various processes of heat transfer involved in maintaining heat balance between the human body and the surrounding space.

**4.2** Discuss the following:
(a) What single factor governs most specifications of outdoor air ratio?
(b) How does the fresh air rate vary with the allowable air space per person?
(c) What are typical outdoor air rates for large air spaces? for small (<100 ft³ per person) air spaces?

**4.3** Discuss the following:
(a) What factors affect the temperature and humidity required for comfort?
(b) Of what is "effective temperature" (ET*) a function?
(c) What are the criteria on which the values given on the ET* chart are based?

**4.4** How must the ASHRAE Comfort Chart Data be altered for (a) women, (b) older people, (c) people in warm climates, and (d) hot or cold radiative walls or windows?

**4.5** Discuss the disadvantages of specifying a high humidity? A low humidity? Answer with respect to both equipment and room occupancy consideration. What are recommended limits of humidity?

**4.6** For an office building in St. Louis, Missouri, the inside dry-bulb temperature is maintained at 24°C (75°F) during the summer. What is the maximum allowable humidity ratio consistent with the ASHRAE Comfort Standard? [Ans: 0.012]

**4.7** A person reclining nude would probably consider which of the following interior environments comfortable?
(a) $t = 78°F$; $\phi = 50\%$; 25 fpm; MRT = 78°F
(b) $t = 75°F$; $\phi = 10\%$; 25 fpm; MRT = 75°F
(c) $t = 75°F$; $\phi = 70\%$; 30 fpm; MRT = 82°F
(d) $t = 80°F$; $\phi = 50\%$; 20 fpm; MRT = 80°F
(e) $t = 90°F$; $\phi = 20\%$; 30 fpm; MRT = 80°F
(f) $t = 70°F$; $\phi = 70\%$; 20 fpm; MRT = 70°F

**4.8** For the person in Problem 4.7 (5 ft, 5 in., 120 lb), compute the body surface area (ft²). [Ans: 17.1 ft²]

**4.9** Compute your body surface area based on the DuBois formula and estimate your current rate of dry heat exchange with the surroundings.

**4.10** From weather data, estimate the lowest equivalent wind chill temperature which will be exceeded 22 h per year for the following locations:
(a) Fairbanks, Alaska
(b) Los Angeles, California
(c) Denver, Colorado
(d) Hartford, Connecticut
(e) Minneapolis, Minnesota
(f) Wichita, Kansas

**4.11** Specify comfortable environments for the following space usage:
(a) senior citizens retirement home in Florida
(b) gymnasium in Missouri
(c) office building in Colorado
(d) residence in Nevada

**4.12** The living room in a home is occupied by adults at rest wearing medium clothing. The mean radiant temperature is 18°C (64°F). Determine the air temperature necessary for comfort. [Ans: 30°C (86°F)]

**4.13** A room has a net outside wall area of 275 ft$^2$ with a surface temperature of 54°F, 45 ft$^2$ of glass with a surface temperature of 20°F, 540 ft$^2$ of ceiling with a surface temperature of 60°F, 670 ft$^2$ of partitions with a surface temperature of 70°F, and 540 ft$^2$ of floor with a surface temperature of 70°F. If the air movement is 20 fpm and light clothing is being worn, determine the air temperature necessary for comfort.

**4.14** Workers on an assembly line making electronic equipment dissipate 700 Btu/h, of which 310 Btu/h is latent heat. When the MRT for the area is 69°F, what air temperature must the heating system maintain for comfort of the workers if the air movement is 40 fpm?

**4.15** A room has 1000 ft$^2$ of surface, of which 120 ft$^2$ is to be heated, and the balance has an average surface temperature of 60°F. The air temperature in the room is 68°F. The room is occupied by light clothed adults at rest. Determine the surface temperature of the heated panel necessary to produce comfort if the air velocity is 20 fpm.
[Ans: 243°F ? unfeasible]

**4.16** Assume that in Problem 4.15, the maximum allowable panel temperature is 120°F. The average temperature of other surfaces in the room remains at 60°F and the air temperature is still 68°F. What panel area will be required if the room is occupied by adults at rest?

**4.17** In an auditorium, 17,150 students are watching slides. The MRT is 80°F and the average room air temperature is 72°F. Air enters the room at 57°F. Assume the lights are out and no heat gain or loss occurs through the walls, floor, and ceiling.
(a) How much air (CFM) should be supplied to remove the sensible heat?
(b) Explain what must be done to remove the latent heat.

**4.18** Two hundred people attend a theater matinee. Air is supplied at 60°F. Determine the required flow rate (lb/h) to handle the heat gain from the occupants if the return air temperature is not to exceed 75°F. [Ans: 11,475 lb/h]

**4.19** Determine the increase in humidity ratio due to 80 people in a dance hall if air is circulated at the rate of 0.64 m$^3$/s (1350 cfm).

**4.20** Specify the MRT for comfort in a space where the air temperature is 68°F and the relative velocity is 20 fpm, for sedentary activity and light clothing. [Ans: 92°F]

**4.21** Specify completely a suitable set of indoor and outdoor design conditions for each of the following cases:
(a) winter; apartment building; St. Louis, Missouri
(b) summer; apartment building; St. Louis, Missouri
(c) winter; factory (medium activity); Rochester, Minnesota

**4.22** The mean radiant temperature in a bus is 6°C lower in winter than the air temperature. For passengers seated without coats, determine the desired air temperature if the relative air velocity is 0.2 m/s. [Ans: 28°C]

**4.23** For Atlanta, Georgia, specify the normal indoor design conditions listed below for

(a) Winter: Dry bulb = _____ °C; W = _____ kg/kg

(b) Summer: Dry bulb = _____ °C; W = _____ kg/kg

**4.24** Specify completely indoor and outdoor design conditions for winter for a clean room in Kansas City, Missouri, having a 1.2 m by 1.2 m radiant panel at 49°C on each of the four walls. The room is 6 m by 4 m by 3 m high and the other surfaces are all at 22°C. Assume very little activity and light clothing.

## 4.6 Bibliography

ASHRAE. 2011. *2011 ASHRAE Handbook—HVAC Applications*.
ASHRAE. Thermal environmental conditions for human occupancy. Standard 55.
ASHRAE. Energy-efficient design of new buildings except low-rise residential buildings. *Standard* 90.1.
ASHRAE. Energy-efficient design of new low-rise residential buildings. *Standard* 90.2.

# Design Conditions

## SI Tables and Figures

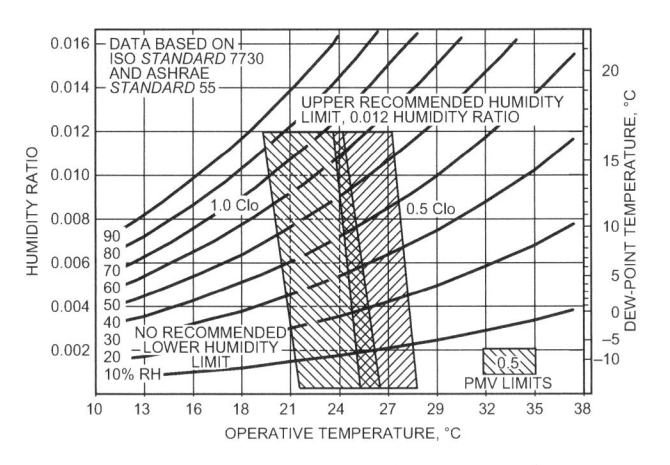

*Fig. 4-2 SI  ASHRAE Summer and Winter Comfort Zones*
*(Acceptable ranges of operative temperature and humidity for people in typical summer and winter clothing during primarily sedentary activity.)*
*(Figure 5, Chapter 9, 2013 ASHRAE Handbook—Fundamentals)*

(a)

(b)

(c)

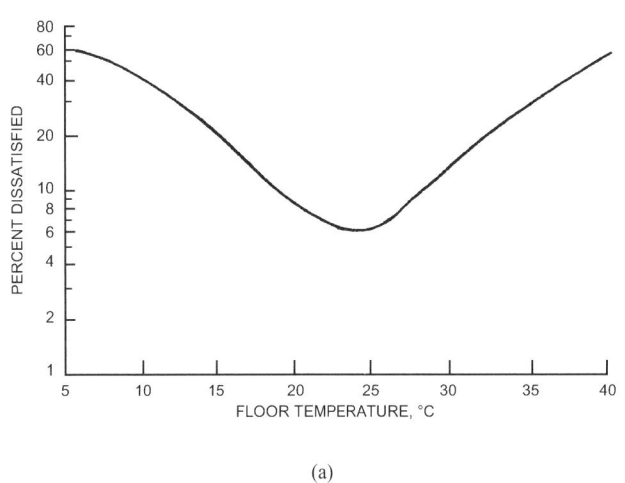

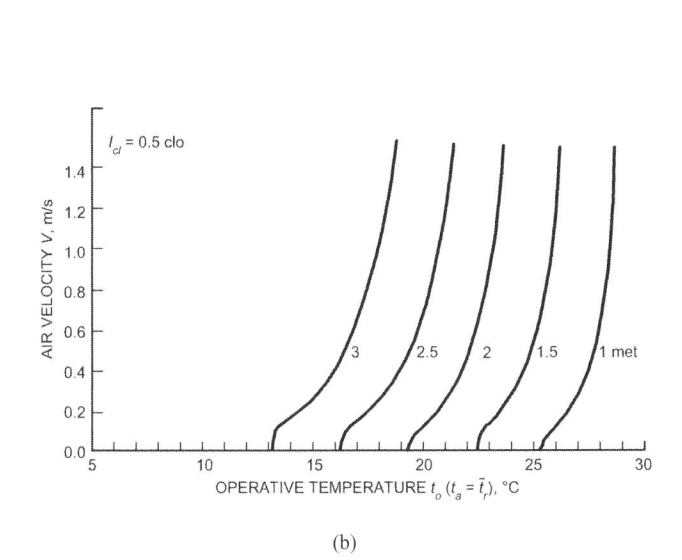

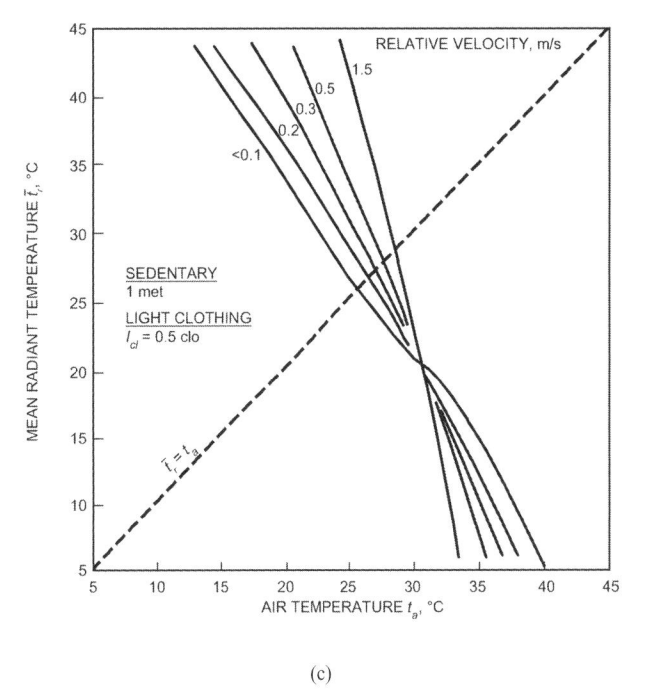

*Fig. 4-3 SI  Examples of Fanger's Charts*
*(Figures 13, 14, and 15, Chapter 9, 2013 ASHRAE Handbook—Fundamentals)*

### Table 4-1 SI  Equations for Prediction Thermal Sensation $(Y)$[a,b]
*(Table 9, Chapter 9, 2013 ASHRAE Handbook—Fundamentals)*

| Exposure Period, h | Sex | Regression Equations $t$ = dry-bulb temperature, °C; $p$ = vapor pressure, kPa |
|---|---|---|
| 1.0 | Male | $Y = 0.220\,t + 0.233\,p - 5.673$ |
|  | Female | $Y = 0.272\,t + 0.248\,p - 7.245$ |
|  | Combined | $Y = 0.245\,t + 0.248\,p - 6.475$ |
| 2.0 | Male | $Y = 0.221\,t + 0.270\,p - 6.024$ |
|  | Female | $Y = 0.283\,t + 0.210\,p - 7.694$ |
|  | Combined | $Y = 0.252\,t + 0.240\,p - 6.859$ |
| 3.0 | Male | $Y = 0.212\,t + 0.293\,p - 5.949$ |
|  | Female | $Y = 0.275\,t + 0.255\,p - 8.622$ |
|  | Combined | $Y = 0.243\,t + 0.278\,p - 6.802$ |

[a] $Y$ values refer to the ASHRAE thermal sensation scale.
[b] For young adult subjects with sedentary activity and wearing clothing with a thermal resistance of approximately 0.5 clo, $\bar{t}_r \approx \bar{t}_a$ and air velocities are < 0.2 m/s.

### Table 4-10 SI  Equivalent Wind Chill Temperatures[a]
*(Table 12, Chapter 9, 2013 ASHRAE Handbook—Fundamentals)*

| Wind Speed, km/h | Actual Thermometer Reading, °C ||||||||||||
|---|---|---|---|---|---|---|---|---|---|---|---|---|
|  | 10 | 5 | 0 | −5 | −10 | −15 | −20 | −25 | −30 | −35 | −40 | −45 | −50 |
|  | Equivalent Chill Temperature, °C |
| Calm | 10 | 5 | 0 | −5 | −10 | −15 | −20 | −25 | −30 | −35 | −40 | −45 | −50 |
| 10 | 8 | 2 | −3 | −9 | −14 | −20 | −25 | −31 | −37 | −42 | −48 | −53 | −59 |
| 20 | 3 | −3 | −10 | −16 | −23 | −29 | −35 | −42 | −48 | −55 | −61 | −68 | −74 |
| 30 | 1 | −6 | −13 | −20 | −27 | −34 | −42 | −49 | −56 | −63 | −70 | −77 | −84 |
| 40 | −1 | −8 | −16 | −23 | −31 | −38 | −46 | −53 | −60 | −68 | −75 | −83 | −90 |
| 50 | −2 | −10 | −18 | −25 | −33 | −41 | −48 | −56 | −64 | −71 | −79 | −87 | −94 |
| 60 | −3 | −11 | −19 | −27 | −35 | −42 | −50 | −58 | −66 | −74 | −82 | −90 | −97 |
| 70 | −4 | −12 | −20 | −28 | −35 | −43 | −51 | −59 | −67 | −75 | −83 | −91 | −99 |

**Little danger:** In less than 5 h, with dry skin. Maximum danger from false sense of security. (WCI less than 1400)

**Increasing danger:** Danger of freezing exposed flesh within one minute. (WCI between 1400 and 2000)

**Great danger:** Flesh may freeze within 30 seconds. (WCI greater than 2000)

*Note:* Cooling power of environment expressed as an equivalent temperature under calm conditions [Equation (79)].

Winds greater than 70 km/h have little added chilling effect.
*Source:* US Army Research Institute of Environmental Medicine.

# Chapter 5

# LOAD ESTIMATING FUNDAMENTALS

In this chapter, the fundamental elements that accompany the load calculations for sizing heating and cooling systems are presented. The material includes the estimation of outdoor air quantities and the evaluation of the overall coefficient of heat transfer for building components. Chapters 16, 23, 25, 26, and 27 of the 2013 *ASHRAE Handbook—Fundamentals* and ASHRAE's *Cooling and Heating Load Calculation Principles* are the major sources of information.

## 5.1 General Considerations

The basic components of heating and cooling loads are illustrated in Figure 5-1.

Proper design of space heating, air-conditioning, or refrigeration systems, and other industrial applications requires a knowledge of thermal insulations and the thermal behavior of building structures. Chapters 25, 26, and 27 of the 2013 *ASHRAE Handbook—Fundamentals*, dealing with thermal insulation and vapor retarders, provide the fundamentals and properties of thermal insulating materials, water vapor barriers, economic thickness of insulation, general practices for building and industrial insulation, and the insulating of mobile equipment and environmental spaces.

Flowing fluids such as air, water, and refrigerants are used in heating, ventilating, air-conditioning, and refrigeration systems to carry heat or mass. An understanding of fluid flow and the nature of its mechanisms is vital to engineers working in these fields. Chapter 3, "Fluid Flow," of the 2013 *ASHRAE Handbook—Fundamentals* introduces the principles of fluid mechanics relevant to these processes.

Heating or cooling air involves only heat transfer, resulting in a temperature change of the air. However, in a true air-conditioning process, there is a simultaneous transfer of heat and mass (water vapor). Chapter 6 of the 2013 *ASHRAE Handbook—Fundamentals* presents the elementary principles of mass transfer to provide a basic understanding of the air-conditioning processes involving mass transfer.

The humidification or dehumidification load depends primarily on the ventilation rate of the space to be conditioned, but other sources of moisture gain or loss should be considered. Chapter 1, "Psychrometrics," of the 2013 *ASHRAE Handbook—Fundamentals* contains equations and examples for determining loads and energy requirements associated with humidification or dehumidification.

## 5.2 Outdoor Air Load Components

### 5.2.1 Basic Concepts

Outdoor air that flows through a building is often used to dilute and remove indoor air contaminants. However, the energy required to condition this outdoor air can be a significant portion of the total space-conditioning load. The magnitude of the outdoor airflow into the building must be known for proper sizing of the HVAC equipment and evaluation of energy consumption. For buildings without mechanical cooling and dehumidification, proper ventilation and infiltration airflows are important for providing comfort for occupants.

A conditioned space may be ventilated by natural infiltration, alone or in combination with intentional mechanical ventilation. Natural infiltration varies with indoor-outdoor temperature difference, wind velocity, and the tightness of the construction, as discussed in Chapter 16, "Ventilation and Infiltration," of the 2013 *ASHRAE Handbook—Fundamentals*. Related information is presented in Chapter 24, "Airflow Around Buildings," of the 2013 *ASHRAE Handbook—Fundamentals*.

*Air exchange* of outdoor air with the air already in a building can be divided into two broad classifications: ventilation and infiltration. **Ventilation** is the intentional introduction of air from the outside into a building; it is further subdivided into natural ventilation and forced ventilation. **Natural ventilation** is the intentional flow of air through open windows, doors, grilles, and other planned building envelope penetrations, and it is driven by natural and/or artificially produced pressure differentials. **Forced ventilation** is the intentional movement of air into and out of a building using fans and intake and exhaust vents; it is also called **mechanical ventilation**.

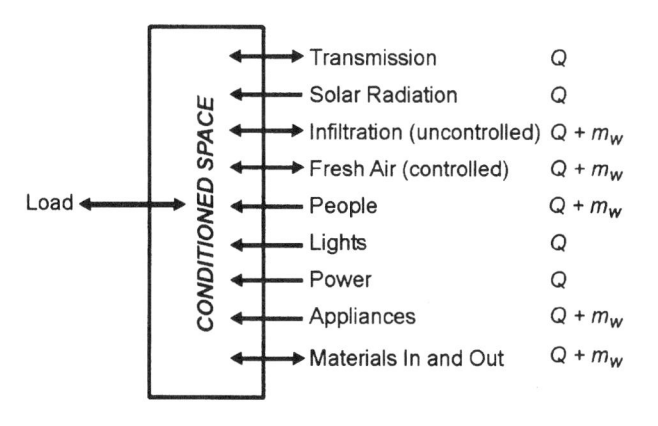

*Fig. 5-1 Components of Heating and Cooling Loads*

**Infiltration** is the uncontrolled flow of outdoor air into a building through cracks and other unintentional openings and through the normal use of exterior doors for entrance and egress. Infiltration is also known as **air leakage** into a building. **Exfiltration** is the leakage of indoor air out of a building. Like natural ventilation, infiltration and exfiltration are driven by natural and/or artificial pressure differences. **Transfer air** (ta) is air that moves from one interior space to another, either intentionally or not.

These modes of air exchange differ significantly in how they affect energy, air quality, and thermal comfort, and they can each vary with weather conditions, building operation, and use. Although one mode may be expected to dominate in a particular building, all must be considered for the proper design and operation of an HVAC system.

Modern commercial and institutional buildings are normally required to have forced ventilation and are usually pressurized somewhat (approximately 0.05 in $H_2O$) to reduce or eliminate infiltration. Forced ventilation has the greatest potential for control of air exchange when the system is properly designed, installed, and operated; it should provide acceptable indoor air quality when ASHRAE Standard 62.1 requirements are followed. Forced ventilation equipment and systems are described in Chapters 1, 2, and 9 of the 2012 *ASHRAE Handbook—HVAC Systems and Equipment*.

In commercial and institutional buildings, uncontrolled natural ventilation, such as through operable windows, may not be desirable from the point of view of energy conservation and comfort. In commercial and institutional buildings with mechanical cooling and forced ventilation, an air- or water-side economizer cycle may be preferable to operable windows for taking advantage of cool outdoor conditions when interior cooling is required. Infiltration may be significant in commercial and institutional buildings, especially in tall, leaky, or underpressurized buildings.

In most of the United States, residential buildings typically rely on infiltration and natural ventilation to meet their ventilation needs. Infiltration is not reliable for ventilation purposes because it depends on weather conditions, building construction, and maintenance. Natural ventilation, usually through operable windows, is dependent on weather and building design but allows occupants to control airborne contaminants and interior air temperature. However, natural ventilation can have a substantial energy cost if used while the residence's heating or cooling equipment is operating.

In place of operable windows, small exhaust fans may be provided for localized venting in residential spaces such as kitchens and bathrooms. Not all local buildings codes require that the exhaust be vented to the outside. Instead, the code may allow the air to be treated and returned to the space or to be discharged to an attic space. Poor maintenance of these treatment devices can make nonducted vents ineffective for ventilation purposes. Condensation in attics should be avoided. In northern Europe and in Canada, some building codes require general forced ventilation in residences, and heat recovery heat exchangers are popular for reducing the energy impact. Residential buildings with low rates of infiltration and natural ventilation require forced ventilation at rates given in ASHRAE Standard 62.2.

Regardless of these complexities and uncertainties, designers and operators need guidance on ventilation and indoor air quality.

ASHRAE Standard 62.1 provides guidance on ventilation and indoor air quality in the form of several alternative procedures. The **Ventilation Rate Procedure** (VRP), the **Indoor Air Quality Procedure** (IAQP), and/or the **Natural Ventilation Procedure** (NVP) are required to satisfy the requirements of Standard 62. In the **Ventilation Rate Procedure**, indoor air quality is assumed to be acceptable if (1) the concentrations of six pollutants in the incoming outdoor air meet the United States national ambient air quality standards, and (2) the outdoor air supply rates meet or exceed values (which vary depending on the type of space) provided in a table. The minimum outdoor air supply for most types of space is 5 cfm (2.5 L/s) per person plus 0.06 cfm/ft$^2$ (0.3 L/s·m$^2$). This minimum rate will maintain an indoor $CO_2$ concentration below 0.1% (1000 parts per million) assuming a typical $CO_2$ generation rate per occupant.

The second alternative in ASHRAE Standard 62 is the **Indoor Air Quality Procedure**. In this procedure, any outdoor air supply rate is acceptable if (1) the indoor concentrations of nine pollutants are maintained below specified values, and (2) the air is deemed acceptable via subjective evaluations of odor. If users of the IAQ Procedure control pollutant source strengths or use air cleaning of local exhaust ventilation, they may be able to reduce the outdoor air supply rates to below those specified in the ventilation rate procedure.

The **Natural Ventilation Procedure** (NVP) allows some exceptions using natural ventilation in conjunction with the VRP on the IAQP.

*As a rough guideline, the minimum infiltration outdoor air allowance may be taken as 0.5 air changes per hour (actually to produce a slight positive pressure within the structure producing exfiltration from the conditioned spaces). The minimum ventilation air allowance based on ASHRAE Standard 62.1 is 5 cfm (2.5 L/s) per person plus 0.06 cfm/ft$^2$ [0.3 L/(s·m$^2$)]. However, local ventilation ordinances must be checked as they may require greater quantities of outdoor air.*

### 5.2.2 Terminology

Figure 5-2 shows a simple **air-handling unit** (AHU) that conditions air for a building. Air brought back to the air handler from the conditioned space is *return air* (ra). The portion of the return air that is discharged to the environment is *exhaust air* (ea), and the part of the return air that is reused is *recirculated air* (ca). Air brought in intentionally from the environment is *outdoor* or *outdoor air* (oa). Because outdoor air may need treatment to be acceptable for use in a building, it should not be called "fresh air." The outdoor air and the recirculated air are combined to form *mixed air* (ma), which is then conditioned and delivered to the thermal zone as *supply air* (sa). Any portion of the mixed air that intentionally or unintentionally circumvents conditioning is *bypass air* (ba). *Ventilation air* is

# Load Estimating Fundamentals

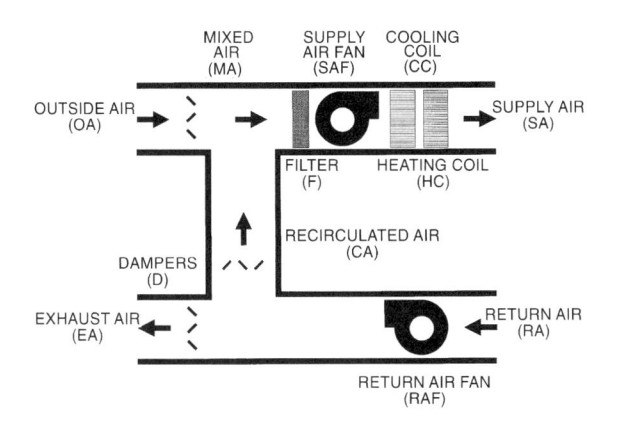

*Fig. 5-2 Simple All-Air Air-Handling Unit with Associated Airflows*
*(Figure 2, Chapter 16, 2013 ASHRAE Handbook—Fundamentals)*

used to provide acceptable indoor air quality. It may be composed of forced or natural ventilation air, infiltration air, suitable treated recirculated air, transfer air, or an appropriate combination. Due to the wide variety of air-handling systems, these airflows may not all be present in a particular system as defined here. Also, more complex systems may have additional airflows.

The outdoor airflow being introduced to a building or zone by an air-handling unit can also be described by the *outdoor air fraction* $X_{oa}$, which is the ratio of the flow rate of outdoor air brought in by the air handler to the total supply airflow rate. When expressed as a percentage, the outdoor air fraction is called the *percent outdoor air*. The design outdoor airflow rate for a building's ventilation system is found through evaluating the requirements of ASHRAE Standard 62. The supply airflow rate is that required to meet the thermal load. The outdoor air fraction and percent outdoor air then describe the degree of recirculation, where a low value indicates a high rate of recirculation, and a high value shows little recirculation. Conventional all-air-handling systems for commercial and institutional buildings have approximately 10% to 40% outdoor air. *100% outdoor air* means no recirculation of return air through the air-handling system. Instead, all the supply air is treated outdoor air, also known as *makeup air*, and all return air is discharged directly to the outside as *relief air*. An air-handling unit that provides 100% outdoor air is typically called a **makeup air unit** (MAU).

Outdoor air introduced into a building constitutes a large part of the total space-conditioning (heating, cooling, humidification, and dehumidification) load, which is one reason to limit air exchange rates in buildings to the minimum required. Air exchange typically represents 20% to 40% of a shell-dominated building's thermal load. First, the incoming air must be heated or cooled from the outdoor air temperature to the indoor air temperature. The rate of energy consumption due to this sensible heating or cooling is given by

$$q_s = Q\rho c_p \Delta t \quad (5\text{-}1)$$

where

$q_s$ = sensible heat load

$Q$ = airflow rate

$\rho$ = air density

$c_p$ = specific heat of air

$\Delta t$ = indoor-outdoor temperature difference

Second, air exchange modifies the moisture content of the air in a building. This is particularly important in some locations in the summer when the outdoor air must be dehumidified. In the winter, when the relative humidity of the indoor air is below 30%, humidification may be needed. The rate of energy consumption associated with these latent loads is given by

$$q_l = Q\rho h_{fg} \Delta W \quad (5\text{-}2a)$$

where

$q_l$ = latent heat load

$h_{fg}$ = latent heat of water

$\Delta W$ = humidity ratio of indoor air minus humidity ratio of outdoor air

### 5.2.3 Infiltration

**Infiltration** or **exfiltration** is air leakage through cracks and interstices, around windows and doors, and through floors and walls of any type of building. The magnitude of infiltration/exfiltration depends on the type of construction, the workmanship, and the condition of the building.

Outdoor air infiltration/exfiltration may account for a significant proportion of the heating or cooling requirements for buildings. Thus, it is important to make an adequate estimate of its contribution with respect to both design loads and seasonal energy requirements. Air infiltration is also an important factor in determining the relative humidity that occurs in buildings or, conversely, the amount of humidification or dehumidification required to maintain given humidities.

The rate of airflow into and out of a building due to either infiltration, exfiltration, or natural ventilation depends on the magnitude of the pressure difference between the inside and outside of the structure and on the resistances to airflow offered by openings and interstices in the building. The pressure difference exerted on the building enclosure by the air may be caused either by wind or by a difference in density of the inside and outdoor air. The effect of the difference in density is often called the **chimney** or **stack effect**, and is often the major factor contributing to air leakage. The pattern of airflow through any part of the structure depends on both the pressure difference and the area of openings.

When the pressure difference is the result of wind pressure, air enters the building through openings in the windward walls and leaves through openings in the leeward walls or, as may be the case in one-story commercial buildings, through ventilating ducts in the roof. When the pressure difference is caused by the indoor-outdoor temperature difference, the flow is along the path of least resistance from inlets at lower levels to outlets at higher levels in a heated building or in the opposite direction for an air-conditioned building, as may be the case in multistory skyscrapers.

In most instances, the pressure difference between inside and outdoor air results from temperature difference forces. Depending on the design, mechanical ventilation and exhaust systems can affect pressure differences across the building enclosure.

The principles of infiltration calculations are discussed in Chapter 16 of the 2013 *ASHRAE Handbook—Fundamentals*, with emphasis placed on the heating season. For the cooling season, infiltration calculations are usually limited to doors and windows. However, in multistory commercial buildings, a reversed chimney effect may exist.

Heat gain due to infiltration must be included whenever the outdoor air mechanically introduced by the system is unable to maintain a positive pressure within the enclosure to prevent infiltration. Most buildings require input of ventilation air for occupants or processes and by properly balancing or controlling the input air versus exhaust/relief air, a positive pressure can be maintained to minimize infiltration. This excess outdoor air introduced through the air-conditioning equipment will maintain a constant outward escape of air, thereby eliminating the infiltration portion of the gain. The positive pressure maintained must be sufficient to overcome wind pressure through cracks and door openings. When this condition prevails, it is unnecessary to include any infiltration component of heat gain. However, on a typical building, infiltration is usually not eliminated due to the pressure that a 25–30 mph wind can exert and, as buildings age, they develop cracks and leak. When the quantity of outdoor air introduced through the cooling equipment is unable to build up the pressure needed to eliminate infiltration, the entire infiltration load should be included in the space heat gain calculations.

Two methods are used to estimate air infiltration in buildings. In one case, the estimate is based on measured leakage characteristics of the building components and selected pressure differences. This is known as the **crack method**, since cracks around windows and doors are usually the major source of air leakage. The other method is known as the air change method and consists of estimating a certain number of air changes per hour for each room, the number of changes assumed being dependent upon the type, use, and location of the room.

The air change method is used often by engineers and designers both for its simplicity and because either method requires the estimation of at least one appropriate numerical value. The method requires the assumption, based on the performance of similar construction, of the number of air changes per hour (ACH) that a space will experience. The infiltration rate is then obtained as follows:

$$Q = ACH \times VOL/60 \qquad (5\text{-}2b)$$

where

$Q$ = infiltration rate, cfm
VOL = gross space volume, ft$^3$

Table 5-1 presents some values for residential type buildings.

**Table 5-1 Change Rates as a Function of Airtightness**

| Class | \multicolumn{10}{c}{Outdoor Design Temperature, °F} |
| --- | --- | --- | --- | --- | --- | --- | --- | --- | --- | --- |
|  | 50 | 40 | 30 | 20 | 10 | 0 | –10 | –20 | –30 | –40 |
| Tight | 0.41 | 0.43 | 0.45 | 0.47 | 0.49 | 0.51 | 0.53 | 0.55 | 0.57 | 0.59 |
| Medium | 0.69 | 0.73 | 0.77 | 0.81 | 0.85 | 0.89 | 0.93 | 0.97 | 1.00 | 1.05 |
| Loose | 1.11 | 1.15 | 1.20 | 1.23 | 1.27 | 1.30 | 1.35 | 1.40 | 1.43 | 1.47 |

*Note*: Values are for 15 mph wind and indoor temperature of 68 °F.

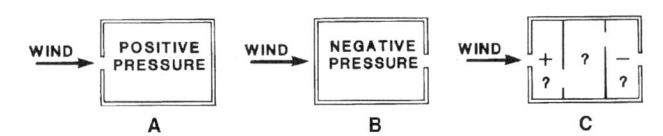

**Pressures in Building Resulting from Wind:**
A. With upstream opening only, pressure is positive.
B. With downstream opening only, pressure is negative.
C. Pressures are as shown if openings are equal in shape and area. With unequal openings, pressures can be either negative or positive in each space, depending on relative areas of openings.

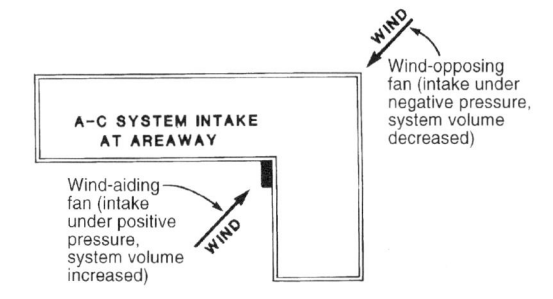

*Fig. 5-3 Building Pressure due to Wind Effect*
*(Figure 11, Chapter 24, 2013 ASHRAE Handbook—Fundamentals)*

The crack method is more firmly based on scientific principles and is generally regarded as being more accurate, *provided that leakage characteristics and pressure differences can be properly evaluated*. Otherwise, the air change method may be justified. The accuracy of estimating infiltration for design load calculations by the crack or component method is restricted by limitations in information on air leakage characteristics of components and by the difficulty of estimating pressure differences under appropriate design conditions of temperature and wind.

A building with only upwind openings is under a *positive pressure* (Figure 5-3). Building pressures are *negative* when there are only downwind openings. A building with internal partitions and openings is under various pressures depending on the relative sizes of the openings and the wind direction. With larger openings on the windward face, the building interior tends to remain under positive pressure; the reverse is also true. Airflow through a wall opening results from positive or negative external and internal pressures. Such differential pressures may exceed 0.5 in. of water during high winds. Supply and exhaust systems, openings, dampers, louvers, doors, and windows make the building flow conditions too complex for most calculations. The opening and closing of doors and windows by building occupants adds further complications.

It is impossible to accurately predict infiltration from theory alone because of the many unknowns. However, it is possible to

# Load Estimating Fundamentals

develop relationships describing the general nature of the problem based on theory and add numerical constants and exponents from experience and experimentation. These semi-empirical expressions are then useful in estimating infiltration rates.

Infiltration is caused by a greater air pressure on the outside of the building than on the inside. The amount of infiltrated air depends on the pressure difference, the nature of the flow through gaps and cracks (laminar versus turbulent), and the size and shape of the cracks. The relationship between the airflow Q through an opening in the building envelope and the pressure difference $\Delta p$ across it is called the **leakage function** of the opening. The fundamental equation for the airflow rate through an opening is

$$Q = C_D A (2\Delta p/\rho)^n \quad (5\text{-}3a)$$

where

$Q$ = airflow rate
$C_D$ = discharge coefficient for the opening
$A$ = cross-section area of the opening
$\rho$ = air density
$\Delta p$ = pressure difference across opening
$n$ = flow exponent (1.0 if laminar; 0.5 if turbulent)

The discharge coefficient $C_D$ is a dimensionless number that depends on the opening geometry and the Reynolds number of the flow.

The above equation is often simplified when evaluating infiltration by combining the discharge coefficient and the crack area into a flow coefficient $C$ yielding

$$Q = C(\Delta p)^n \quad (5\text{-}3b)$$

The pressure difference is given by

$$\Delta p = \Delta p_s + \Delta p_w + \Delta p_p$$

where

$\Delta p_s$ = pressure difference caused by stack effect
$\Delta p_w$ = pressure difference caused by wind
$\Delta p_p$ = pressure difference due to building pressurization

Stack effect occurs when air densities are different on the inside and outside of a building. The air density decreases with increasing temperature and decreases slightly with increasing humidity. Because the pressure of the air is due to the weight of a column of air, on winter days the air pressure at ground level will be less inside the building due to warmer inside air than outdoor air. As a result of this pressure difference, air will infiltrate at ground level and flow upward inside the building. Under summer conditions when the air is cooler inside, outdoor air enters the top of the building and flows downward on the inside. With the stack effect there will be a vertical location in the building where the inside pressure equals the outside pressure, called the neutral pressure level of the building. Unless there is detailed information on the vertical distribution of cracks and other openings, it is assumed that the neutral pressure will be at

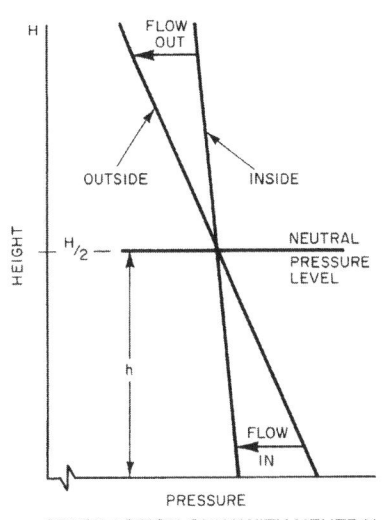

*Fig. 5-4 Winter Stack Effect Showing Theoretical Pressure Difference*
(Figure 6A, Chapter 16, 2013 ASHRAE Handbook—Fundamentals)

the building mid-height when under the influence of the stack effect alone.

The theoretical pressure difference due to the stack effect can be found from:

$$\Delta p_s = 0.52\, p_b h[(1/T_o) - (1/T_i)] \quad (5\text{-}4)$$

where

$\Delta p_s$ = theoretical pressure difference, in. water
$p_b$ = outside absolute (barometric) pressure, psi
$h$ = vertical distance from neutral pressure level, ft
$T_o$ = outside absolute temperature, °R
$T_i$ = inside absolute temperature, °R

Figure 5-4 shows the stack effect pressure variations under winter conditions.

The pressure associated with the wind velocity, called velocity pressure, is

$$p_v = \rho V_w^2 / 2$$

Even on the windward side of a building the wind velocity does not go to zero as the air comes in contact with the building. To account for this, a wind pressure coefficient $C_p$ is used when determining the static pressure obtained from the velocity pressure of the wind:

$$\Delta p_w = 0.5 C_p \rho V_w^2 \quad (5\text{-}5)$$

The pressure coefficient will always have a value less than 1.0 and can be negative when the wind causes outdoor pressures below atmospheric such as on the leeward side of buildings. Figure 5-5 provides average pressure coefficients for tall buildings.

The building pressure and the corresponding pressure difference $\Delta p_p$ depend on the design and operation of the HVAC

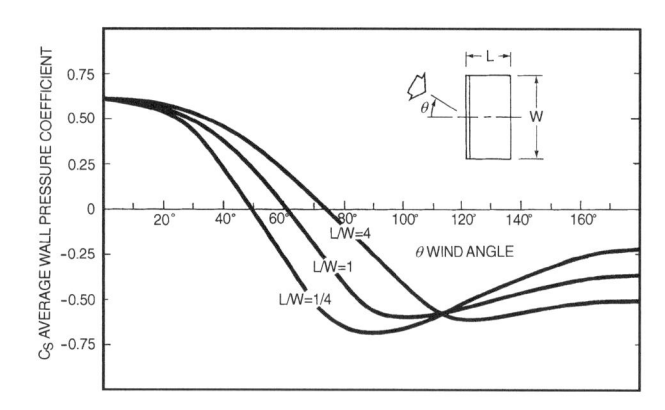

*Fig. 5-5 Wind Pressure Coefficients for Tall Buildings*

system. A building can be operated either at a positive or negative pressure depending upon the relative flow resistances of the supply and return duct systems. A positive building pressure results in a negative pressure difference $\Delta p_p$ and a reduction in infiltration from wind and stack effects. While building pressurization is usually desired and assumed to occur, the air circulation system must be carefully designed and balanced to achieve this effect. Care must be taken to estimate a realistic value that the system can actually achieve.

The flow coefficient $C$ in Equation (5-3b) has a particular value for each crack and each window and door perimeter gap. Although values of $C$ are determined experimentally for window and door gaps, this procedure does not work for cracks. Cracks occur at random in fractures of building materials and at the interface of materials. The number and size of cracks depend on the type of construction, the workmanship during construction, and the maintenance of the building after construction. To determine a value of $C$ for each crack is impractical; however, an overall leakage coefficient can used by modifying Equation (5-3b) into the following form:

$$Q = KA(\Delta p)^n \qquad (5\text{-}6)$$

where
  $A$ = wall area
  $K$ = leakage coefficient

Table 5-2 provides typical values of the leakage coefficient for various types of curtain wall construction. The associated infiltration can be determined from Figure 5-6.

Although the terms infiltration and air leakage are sometimes used synonymously, they are different, though related, quantities. Infiltration is the rate of uncontrolled air exchange through unintentional openings that occur under given conditions, while air leakage is a measure of the airtightness of the building shell. The greater the air leakage of a building, the greater its infiltration rate. The infiltration rate of an individual building depends on weather conditions, equipment operation, and occupant activities. The rate can vary by a factor of five from weather effects alone.

Typical infiltration values in housing in North America vary by about a factor of ten, from tight housing with seasonal

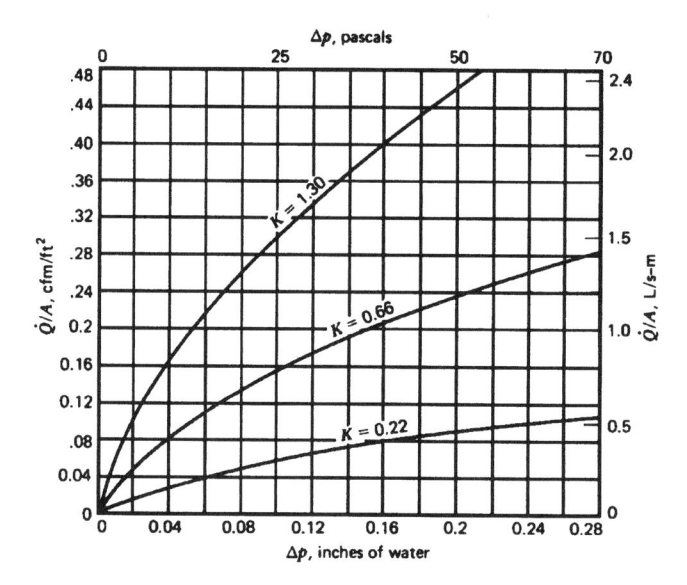

*Fig. 5-6 Curtain Wall Infiltration Rates*

**Table 5-2 Curtain Wall Leakage Coefficients**

| Leakage Coefficient | Description | Curtain Wall Construction |
|---|---|---|
| $K = 0.22$ | Tight-fitting wall | Constructed under close supervision of workmanship on wall joints. When joints seals appear inadequate they must be redone |
| $K = 0.66$ | Average-fitting wall | Conventional construction procedures are used |
| $K = 1.30$ | Loose-fitting wall | Poor construction quality control or an older building having separated joints |

air change rates of 0.2 per hour (ACH), to housing with infiltration rates as great as 2.0 per hour. Histograms of infiltration rates measured in two different samples of North American housing are shown in Figures 5-7 and 5-8. The average seasonal infiltration of 312 houses located in many different areas in North America is shown in Figure 5-7. The median infiltration value of this sample is 0.50 ach. Measurements in 266 houses located in 16 cities in the United States are represented in Figure 5-8. The median value of this sample is 0.90 ach. The group of houses in the Figure 5-7 sample is biased toward new energy-efficient houses, while the group in Figure 5-8 represents older, low-income housing in the United States. While these two samples are not a valid random sample of North American housing, they indicate the distribution expected in an appropriate building sample. Note that infiltration values listed are appropriate for unoccupied structures. *Although occupancy influences have not been measured directly, estimates add an average of 0.10 to 0.15 ach to unoccupied values.*

When estimating infiltration rates for a large group of houses to determine gross energy loads, an adequate assumption for infiltration rates is obtained by using Figures 5-7 and 5-8. A combination of the median values in these figures, with an additional 0.10 to 0.15 ach to account for occupancy

# Load Estimating Fundamentals

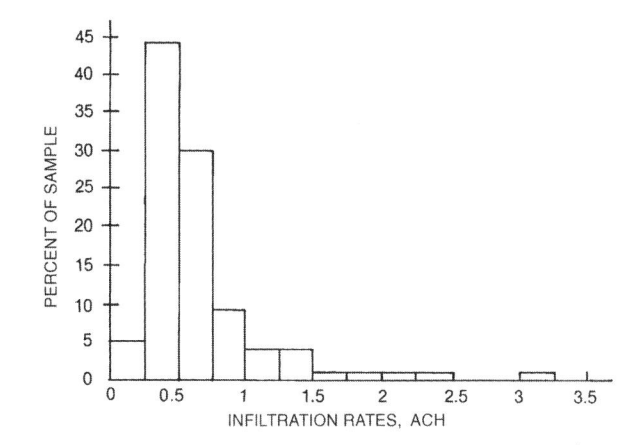

*Fig. 5-7 Histogram of Infiltration Values—New Construction*

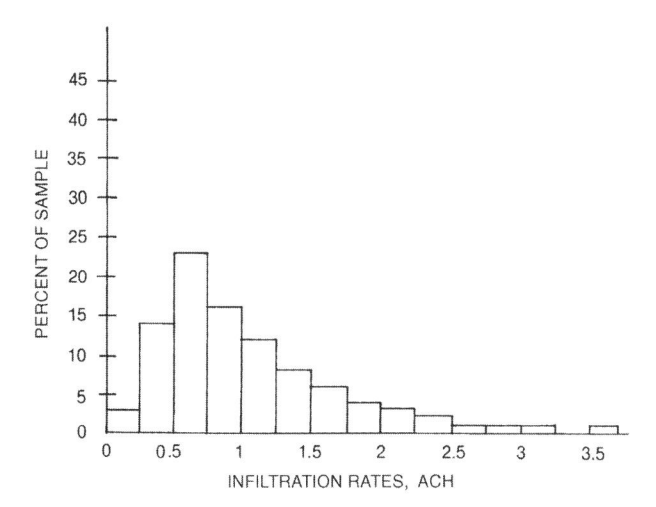

*Fig. 5-8 Histogram of Infiltration Values—Low-Income Housing*

**Table 5-3  Total Ventilation Air Requirements**

| Area Based | Occupancy Based |
|---|---|
| 1 cfm/100 ft² of floor space | 7.5 cfm per person, based on normal occupancy |

effects, represents average infiltration for a large group of houses.

The building envelopes of large commercial buildings are often thought to be nearly airtight. The National Association of Architectural Metal Manufacturers specifies a maximum leakage per unit of exterior wall area of 0.060 cfm/ft² at a pressure difference of 0.30 in. of water, exclusive of leakage through operable windows. Recent measurements on eight US office buildings ranged from 0.213 to 1.028 cfm/ft² at 0.30 in. of water (0.1 to 0.6 air changes per hour) with no outdoor air intake. Therefore, office building envelopes may leak more than expected. Typical air leakage values per unit wall area at 0.30 in. of water are 0.10, 0.30, and 0.60 for tight, average, and leaky walls, respectively.

Infiltration may also be estimated by statistical analysis of long-term data of infiltration measurements for specific sites. Weather-related pressures that drive infiltration are estimated by finding the regression constants in a function of the form:

$$I = K_1 + K_2 \Delta t + K_3 V \qquad (5\text{-}7)$$

where

$I$ = air change rate, h$^{-1}$

$\Delta t$ = indoor-outdoor temperature difference, °F

$V$ = wind speed, mph

$K_1, K_2, K_3$ = empirical regression constants derived from measurements at the site

The analysis reveals that correlation with weather variables are only moderately successful. Because the regression coefficients reflect structural characteristics, as well as shielding effects and occupants' behavior, the values of $K_1$, $K_2$, and $K_3$ have varied by 20:1 between similar residences, and the model may be inappropriate as a design tool or for inclusion in computer simulations for building energy analysis.

Several procedures have been developed that treat the building as a well-mixed zone. But a study of a large variety of single-cell models found that these models can only be used to calculate air change rates in structures that can be assumed to have a uniform internal pressure.

Multicell models treat the actual complexity of flows in a building by recognizing effects of internal flow restrictions. They require extensive information on flow characteristics and pressure distributions, and, in many cases, are too complex to justify their use to predict flow for simple structures such as single-family residences.

A simple, single-zone approach to calculating air infiltration rates in houses requires an effective leakage area at 0.016 in. of water. Although control of significant sources of pollution in a dwelling is important, whole house ventilation may still be needed. Each dwelling should be provided with outdoor air according to Table 5-3. The rate is the sum of the "Area Based" and "Occupancy Based" columns. Design occupancy can be based on the number of bedrooms as follows: first bedroom, two persons; each additional bedroom, one person. Additional ventilation should be considered when occupancy densities exceed 1/250 ft². ASHRAE Standard 62.2 provides guidance for ventilation and acceptable indoor air quality in low-rise residential buildings.

Using the effective leakage area, the airflow rate due to infiltration is calculated according to the following equation:

$$Q = A_L(C_S \Delta t + C_W V^2)^{0.5} \qquad (5\text{-}8)$$

where

$Q$ = airflow rate, cfm

$A_L$ = effective leakage area, in²

$C_S$ = stack coefficient, cfm²/(in⁴·°F)

$\Delta t$ = average indoor-outdoor temperature difference, °F

$C_W$ = wind coefficient, cfm²/(in⁴·mph²)

$V$ = average wind speed, mph

**Table 5-4 Local Shelter Classes**
*(Table 5, Chapter 16, 2013 ASHRAE Handbook—Fundamentals)*

| Shelter Class | Description |
|---|---|
| 1 | No obstructions or local shielding |
| 2 | Typical shelter for an isolated rural house |
| 3 | Typical shelter caused by other buildings across street from building under study |
| 4 | Typical shelter for urban buildings on larger lots where sheltering obstacles are more than one building height away |
| 5 | Typical shelter produced by buildings or other structures immediately adjacent (closer than one house height): e.g., neighboring houses on same side of street, trees, bushes, etc. |

**Table 5-5 Stack Coefficient $C_S$**
*(Table 4, Chapter 16, 2013 ASHRAE Handbook—Fundamentals)*

| | House Height (Stories) | | |
|---|---|---|---|
| | One | Two | Three |
| Stack coefficient | 0.0150 | 0.0299 | 0.0449 |

**Table 5-6 Wind Coefficient $C_W$**
*(Table 6, Chapter 16, 2013 ASHRAE Handbook—Fundamentals)*

| Shielding Class | House Height (Stories) | | |
|---|---|---|---|
| | One | Two | Three |
| 1 | 0.0119 | 0.0157 | 0.0184 |
| 2 | 0.0092 | 0.0121 | 0.0143 |
| 3 | 0.0065 | 0.0086 | 0.0101 |
| 4 | 0.0039 | 0.0051 | 0.0060 |
| 5 | 0.0012 | 0.0016 | 0.0018 |

The infiltration rate of the building is obtained by dividing $Q$ by the building volume. The value of $C_W$ depends on the local shielding class of the building. Five different shielding classes are listed in Table 5-4. Values for $C_S$ for one-, two-, and three-story houses are presented in Table 5-5. Values of $C_W$ for one-, two-, and three-story houses in shielding classes one through five are found in Table 5-6. The heights of the one-, two-, and three-story buildings are 8, 16, and 24 ft, respectively.

**Example 5-1** Calculate the average infiltration during a one-week period in January for a one-story house in Portland, Oregon. During this period, the average indoor-outdoor temperature difference is 30°F, and the average wind speed is 6 mph. The house has a volume of 9000 ft$^3$ and an effective air leakage area of 107 in.$^2$, and it is located in an area with buildings and trees within 30 ft in most directions (shelter class 4).

**Solution:** From Equation (5-8), the airflow rate due to infiltration is

$$Q = 107\sqrt{(0.0150)(30) + (0.0039)(6^2)} = 82.2 \text{ cfm} = 4930 \text{ ft}^3/\text{h}$$

The air exchange rate is therefore

$$I = 4930 / 9000 = 0.55 \text{ h}^{-1} = 0.55 \text{ ach}$$

**Example 5-2** Estimate the infiltration at design conditions for a two-story house in Nebraska. The house has an effective leakage area of 77 in.$^2$, and is surrounded by a thick hedge. The design temperature is –2°F.

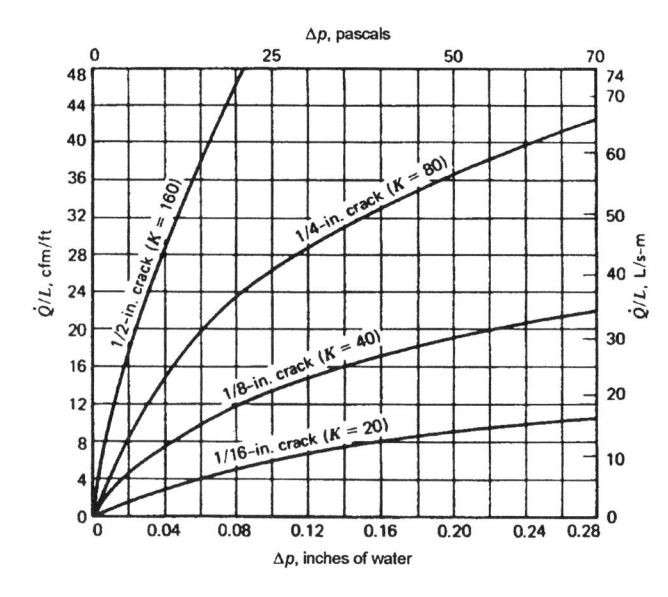

*Fig. 5-9 Infiltration through Closed Swinging Door Cracks*

**Solution:** Assume indoor temperature is 75°F. Design wind speed is 15 mph. Shielding Class 3 is due to hedge. From Table 5-5, CS = 0.0299, and from Table 5-6, $C_W$ = 0.0086. The airflow rate due to infiltration is thus:

$$Q = 77[(0.0299 \times 77) + (0.0086 \times 15^2)]^{0.5}$$
$$= 158 \text{ cfm} = 9510 \text{ ft}^3/\text{h}$$

Infiltration rate $I$ equals $Q$ divided by the building volume:

$$I = 9510/12000 = 0.79 \text{ ach}$$

Commercial-type doors have different infiltration characteristics than those given in Table 5-3 in that they normally have larger cracks and are used more often.

An example of data for the infiltration through cracks in swinging doors is given in Figure 5-9. Commercial buildings often have a large number of people entering and leaving with associated infiltration. Figures 5-10 and 5-11 provide examples of data for estimating the infiltration due to traffic in and out of the building. The figures are based on a standard sized (3 by 7 ft) door. Automatic doors stay open two to four times longer than manually operated doors. Thus, doubling of the infiltration obtained for manually operated doors would be a reasonable estimate for automatic doors.

The total infiltration is the infiltration through the cracks when the door is closed added to the infiltration due to traffic.

Figure 5-12 shows the infiltration due to a pressure difference across the seals of a standard sized revolving door. Figure 5-13 and 5-14 provide data for estimating the infiltration caused by the rotation of the standard-sized door. The amount of infiltration depends upon the inside-outside temperature difference and the rotational speed of the door. The total infiltration is the infiltration due to leakage through the door seals and the infiltration due to the mechanical interchange of air due to the rotation of the door.

# Load Estimating Fundamentals

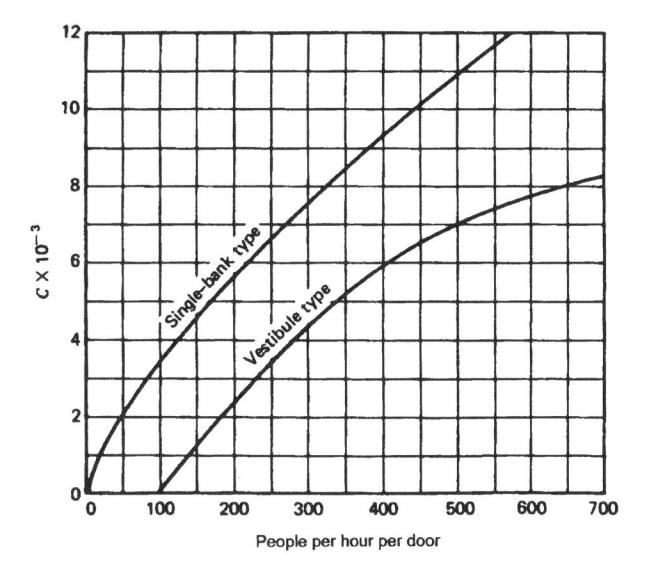

*Fig. 5-10 Flow Coefficients for Swinging Doors*

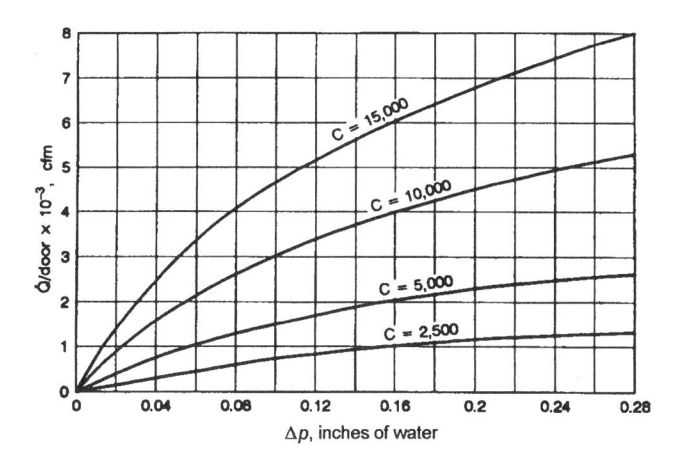

*Fig. 5-11 Infiltration for Swinging Doors with Traffic*

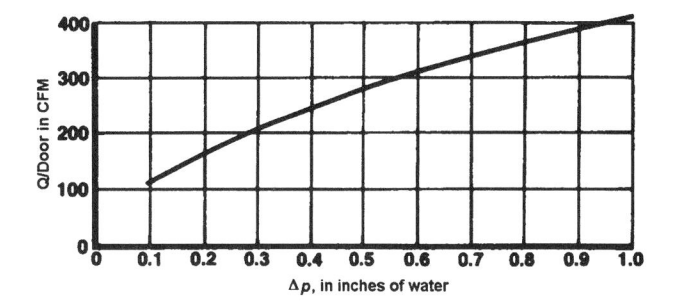

*Fig. 5-12 Infiltration Through Revolving Door Seals while Stationary*

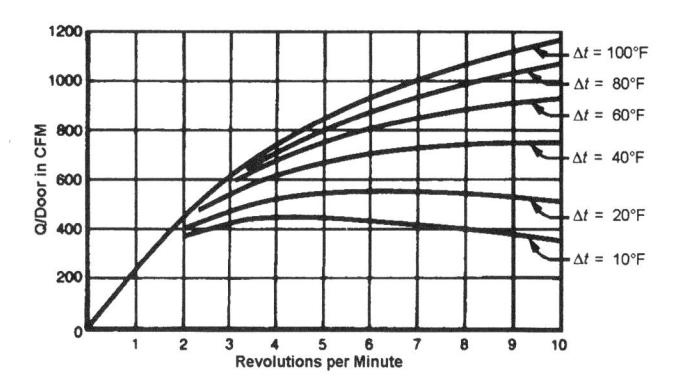

*Fig. 5-13 Infiltration for Motor-Operated Revolving Door*

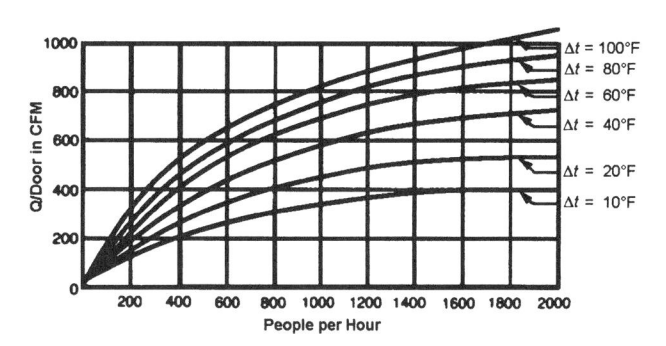

*Fig. 5-14 Infiltration for Manual Revolving Door*

## 5.2.4 Ventilation Air

Outdoor air requirements for acceptable indoor air quality (IAQ) have long been debated and different rationales have produced quite different ventilation standards. Historically, the major considerations have included the amount of outdoor air required to control moisture, carbon dioxide, odors, and tobacco smoke generated by occupants. These considerations have led to the minimum rate of outdoor air supply per occupant. More recently, the maintenance of acceptable indoor concentrations of a variety of additional pollutants that are not generated primarily by occupants has been a major concern. The most common pollutants of concern and their sources are given in Table 5-7.

Regardless of the complexities and uncertainties that exist regarding proper ventilation, building designers and operators require guidance on minimum outdoor air quantities and indoor air quality. ASHRAE Standard 62 provides guidance on ventilation and indoor air quality in the form of alternative procedures. In the **ventilation rate procedure**, indoor air quality is assumed to be acceptable if the concentrations of pollutants in the incoming outdoor air meet the US national ambient air quality standards (Table 5-8) and the outdoor air supply rates meet or exceed values provided in Table 5-9. ANSI/ASHRAE Standard 62.1-2011 is the latest edition of Standard 62, which has been given the new designation of 62.1 to distinguish it from ANSI/ASHRAE Standard 62.2-2013, *Ventilation and Acceptable Indoor Air Quality in Low-Rise Residential Buildings*. The purpose of this standard is to specify minimum ventilation rates and indoor air quality that will be acceptable to human occupants and are intended to minimize the potential for adverse health effects. It is intended for regulatory application to new buildings and additions and some changes to existing buildings. It is also intended to be used to guide the improvement of indoor air quality in existing buildings.

**Table 5-7  Indoor Air Pollutants and Sources**

| Sources | Contaminants |
|---|---|
| **OUTDOOR** | |
| Ambient air | $SO_2$, NO, $NO_2$, $O_3$, hydrocarbons, CO, particulates, bioaerosols |
| Motor vehicles | CO, Pb, hydrocarbons, particulates |
| Soil | Radon organics |
| **INDOOR** | |
| Building construction materials | |
| Concrete, stone | Radon |
| Particleboard, plywood | Formaldehyde |
| Insulation | Formaldehyde, fiberglass |
| Fire retardant | Asbestos |
| Adhesives | Organics |
| Paint | Mercury, organics |
| Building Contents | |
| Heating and cooking combustion appliances | CO, NO, $NO_2$, formaldehyde, particulates, organics |
| Furnishings | Organics |
| Water service, natural gas | Radon |
| Human Occupants | |
| Tobacco smoke | CO, $NO_2$, organics, particulates, odors |
| Aerosol sprays | Fluorocarbons, vinyl chloride, organics |
| Cleaning and cooking products | Organics, $NH_2$, odors |
| Hobbies and crafts | Organics |
| Damp organic materials, stagnant water | |
| Coil drain pans | Bioaerosols |
| Humidifiers | |

**Table 5-8  United States Ambient Air Quality Standards**

| Contaminant | Averaging Time | Primary Standard Levels | Secondary Standard Levels |
|---|---|---|---|
| Particulate matter | Annual (geometric mean) | 75 $\mu g/m^3$ | 60 $\mu g/m^3$ |
| | 24 h | 260 $\mu g/m^3$ | 150 $\mu g/m^3$ |
| Sulfur oxides | Annual (arithmetic mean) | 80 $\mu g/m^3$ (0.03 ppm) | — |
| | 24 h | 365 $\mu g/m^3$ (0.14 ppm) | — |
| | 3 h | — | 1300 $\mu g/m^3$ (0.5 ppm) |
| Carbon monoxide | 8 h | 10 $mg/m^3$ (9 ppm) | 10 $mg/m^3$ (9 ppm) |
| | 1 h | 40 $mg/m^3$ (35 ppm) | 40 $mg/m^3$ (35 ppm) |
| Nitrogen dioxide | Annual (arithmetic mean) | 100 $\mu g/m^3$ (0.05 ppm) | 100 $\mu g/m^3$ (0.05 ppm) |
| Ozone | 1 h | 240 $\mu g/m^3$ (0.12 ppm) | 240 $\mu g/m^3$ (0.12 ppm) |
| Hydrocarbons (nonmethane) | 3 h (6 A.M. to 9 A.M.) | 160 $\mu g/m^3$ (0.24 ppm) | 160 $\mu g/m^3$ (0.24 ppm) |
| Lead | 3 months | 1.5 $\mu g/m^3$ | 1.5 $\mu g/m^3$ |

Air movement within spaces affects the diffusion of ventilation air and, therefore, indoor air quality and comfort. **Ventilation effectiveness** is a description of an air distribution system's ability to remove internally generated pollutants from a building, zone, or space. **Air change effectiveness** is a description of an air distribution system's ability to deliver ventilation air to a building, zone, or space. The HVAC design engineer does not have knowledge or control of actual pollutant sources within buildings, so Table 5-9 (Table 6-1 from ASHRAE Standard 62.1-2010) defines outdoor air requirements for typical, expected building uses. For most projects, therefore, the air change effectiveness is of more relevance to HVAC system design than the ventilation effectiveness.

Currently, the HVAC design engineer must assume that a properly designed, installed, operated, and maintained air distribution system provides an air change effectiveness of about 1. Therefore, the Table 5-9 values from ASHRAE Standard 62.1 are appropriate for design of commercial and institutional buildings when the ventilation rate procedure is used. If the indoor air quality procedure of Standard 62.1 is used, then actual pollutant sources and the air change effectiveness must be known for the successful design of HVAC systems that have fixed ventilation airflow rates. Where appropriate, the table lists the estimated density of people for design purposes. The requirements for ventilation air quantities given in Table 5-9 are for 100% outdoor air when the outdoor air quality meets the national specifications for acceptable outdoor air quality.

Properly cleaned air may be recirculated. Under the ventilation rate procedure, for other than intermittent variable occupancy, outdoor air flow rates may not be reduced below the requirements in Table 5-9. If cleaned, recirculated air is used to reduce the outdoor air flow rate below the values of Table 5-9, the indoor air quality procedure must be used. The *indoor air quality procedure* provides an alternative performance method to the *ventilation rate procedure* for achieving acceptable air quality. The ventilation rate procedure is deemed to provide acceptable indoor air quality. Nevertheless that procedure, through prescription of required ventilation rates, provides only an indirect solution to the control of indoor contaminants. The indoor air quality procedure provides a direct solution by restricting contaminant concentration concern to acceptable levels.

The amount of outdoor air specified in Table 5-9 may be reduced by recirculating air from which offending contaminants have been removed or converted to less objectionable forms. The amount of outdoor air required depends on the contaminant concentrations in the indoor and outdoor air, the contaminant generation in the space, the filter location, the filter efficiency for the contaminants in question, the ventilation effectiveness, the supply air circulation rate, and the fraction recirculated.

In ASHRAE Standard 62.1 a third procedure is defined. This is the Natural Ventilation Procedure and allows some exceptions using natural ventilation in conjunction with the Ventilation Rate Procedure or the Indoor Air Quality Procedure.

Figure 5-15 shows a representative HVAC ventilation system. A filter may be located in the recirculated airstream (location A) or in the supply (mixed) airstream (location B). Variable air volume (VAV) systems reduce the circulation rate when the

# Load Estimating Fundamentals

**Table 5-9 Minimum Ventilation Rates in Breathing Zone**
(This table is not valid in isolation; it must be used in conjunction with the accompanying notes.)
*(Table 6-1, ASHRAE Standard 62.1-2010)*

| Occupancy Category | People Outdoor Air Rate $R_p$ CFM/person | L/s·person | Area Outdoor Air Rate $R_a$ CFM/ft² | L/s·m² | Notes | Default Values Occupant Density (see Note 4) #/1000 ft² or #/100 m² | Combined Outdoor Air Rate (see Note 5) CFM/person | L/s·person | Air Class |
|---|---|---|---|---|---|---|---|---|---|
| **Correctional Facilities** | | | | | | | | | |
| Cell | 5 | 2.5 | 0.12 | 0.6 | | 25 | 10 | 4.9 | 2 |
| Dayroom | 5 | 2.5 | 0.06 | 0.3 | | 30 | 7 | 3.5 | 1 |
| Guard stations | 5 | 2.5 | 0.06 | 0.3 | | 15 | 9 | 4.5 | 1 |
| Booking/waiting | 7.5 | 3.8 | 0.06 | 0.3 | | 50 | 9 | 4.4 | 2 |
| **Educational Facilities** | | | | | | | | | |
| Daycare (through age 4) | 10 | 5 | 0.18 | 0.9 | | 25 | 17 | 8.6 | 2 |
| Daycare sickroom | 10 | 5 | 0.18 | 0.9 | | 25 | 17 | 8.6 | 3 |
| Classrooms (ages 5–8) | 10 | 5 | 0.12 | 0.6 | | 25 | 15 | 7.4 | 1 |
| Classrooms (age 9 plus) | 10 | 5 | 0.12 | 0.6 | | 35 | 13 | 6.7 | 1 |
| Lecture classroom | 7.5 | 3.8 | 0.06 | 0.3 | | 65 | 8 | 4.3 | 1 |
| Lecture hall (fixed seats) | 7.5 | 3.8 | 0.06 | 0.3 | | 150 | 8 | 4.0 | 1 |
| Art classroom | 10 | 5 | 0.18 | 0.9 | | 20 | 19 | 9.5 | 2 |
| Science laboratories | 10 | 5 | 0.18 | 0.9 | | 25 | 17 | 8.6 | 2 |
| University/college laboratories | 10 | 5 | 0.18 | 0.9 | | 25 | 17 | 8.6 | 2 |
| Wood/metal shop | 10 | 5 | 0.18 | 0.9 | | 20 | 19 | 9.5 | 2 |
| Computer lab | 10 | 5 | 0.12 | 0.6 | | 25 | 15 | 7.4 | 1 |
| Media center | 10 | 5 | 0.12 | 0.6 | A | 25 | 15 | 7.4 | 1 |
| Music/theater/dance | 10 | 5 | 0.06 | 0.3 | | 35 | 12 | 5.9 | 1 |
| Multi-use assembly | 7.5 | 3.8 | 0.06 | 0.3 | | 100 | 8 | 4.1 | 1 |
| **Food and Beverage Service** | | | | | | | | | |
| Restaurant dining rooms | 7.5 | 3.8 | 0.18 | 0.9 | | 70 | 10 | 5.1 | 2 |
| Cafeteria/fast-food dining | 7.5 | 3.8 | 0.18 | 0.9 | | 100 | 9 | 4.7 | 2 |
| Bars, cocktail lounges | 7.5 | 3.8 | 0.18 | 0.9 | | 100 | 9 | 4.7 | 2 |
| Kitchen (cooking) | 7.5 | 3.8 | 0.12 | 0.6 | | 20 | 14 | 7.0 | 2 |
| **General** | | | | | | | | | |
| Break rooms | 5 | 2.5 | 0.06 | 0.3 | | 25 | 7 | 3.5 | 1 |
| Coffee stations | 5 | 2.5 | 0.06 | 0.3 | | 20 | 8 | 4 | 1 |
| Conference/meeting | 5 | 2.5 | 0.06 | 0.3 | | 50 | 6 | 3.1 | 1 |
| Corridors | — | — | 0.06 | 0.3 | | — | | | 1 |
| Occupiable storage rooms for liquids or gels | 5 | 2.5 | 0.12 | 0.6 | B | 2 | 65 | 32.5 | 2 |
| **Hotels, Motels, Resorts, Dormitories** | | | | | | | | | |
| Bedroom/living room | 5 | 2.5 | 0.06 | 0.3 | | 10 | 11 | 5.5 | 1 |
| Barracks sleeping areas | 5 | 2.5 | 0.06 | 0.3 | | 20 | 8 | 4.0 | 1 |
| Laundry rooms, central | 5 | 2.5 | 0.12 | 0.6 | | 10 | 17 | 8.5 | 2 |
| Laundry rooms within dwelling units | 5 | 2.5 | 0.12 | 0.6 | | 10 | 17 | 8.5 | 1 |
| Lobbies/prefunction | 7.5 | 3.8 | 0.06 | 0.3 | | 30 | 10 | 4.8 | 1 |
| Multipurpose assembly | 5 | 2.5 | 0.06 | 0.3 | | 120 | 6 | 2.8 | 1 |
| **Office Buildings** | | | | | | | | | |
| Breakrooms | 5 | 2.5 | 0.12 | 0.6 | | 50 | 7 | 3.5 | 1 |
| Main entry lobbies | 5 | 2.5 | 0.06 | 0.3 | | 10 | 11 | 5.5 | 1 |
| Occupiable storage rooms for dry materials | 5 | 2.5 | 0.06 | 0.3 | | 2 | 35 | 17.5 | 1 |
| Office space | 5 | 2.5 | 0.06 | 0.3 | | 5 | 17 | 8.5 | 1 |
| Reception areas | 5 | 2.5 | 0.06 | 0.3 | | 30 | 7 | 3.5 | 1 |
| Telephone/data entry | 5 | 2.5 | 0.06 | 0.3 | | 60 | 6 | 3.0 | 1 |
| **Miscellaneous Spaces** | | | | | | | | | |
| Bank vaults/safe deposit | 5 | 2.5 | 0.06 | 0.3 | | 5 | 17 | 8.5 | 2 |
| Banks or bank lobbies | 7.5 | 3.8 | 0.06 | 0.3 | | 15 | 12 | 6.0 | 1 |
| Computer (not printing) | 5 | 2.5 | 0.06 | 0.3 | | 4 | 20 | 10.0 | 1 |
| General manufacturing (excludes heavy industrial and processes using chemicals) | 10 | 5.0 | 0.18 | 0.9 | | 7 | 36 | 18 | 3 |
| Pharmacy (prep. area) | 5 | 2.5 | 0.18 | 0.9 | | 10 | 23 | 11.5 | 2 |
| Photo studios | 5 | 2.5 | 0.12 | 0.6 | | 10 | 17 | 8.5 | 1 |
| Shipping/receiving | 10 | 5 | 0.12 | 0.6 | B | 2 | 70 | 35 | 2 |
| Sorting, packing, light assembly | 7.5 | 3.8 | 0.12 | 0.6 | | 7 | 25 | 12.5 | 2 |
| Telephone closets | — | — | 0.00 | 0.0 | | — | | | 1 |
| Transportation waiting | 7.5 | 3.8 | 0.06 | 0.3 | | 100 | 8 | 4.1 | 1 |
| Warehouses | 10 | 5 | 0.06 | 0.3 | B | — | | | 2 |
| **Public Assembly Spaces** | | | | | | | | | |
| Auditorium seating area | 5 | 2.5 | 0.06 | 0.3 | | 150 | 5 | 2.7 | 1 |
| Places of religious worship | 5 | 2.5 | 0.06 | 0.3 | | 120 | 6 | 2.8 | 1 |
| Courtrooms | 5 | 2.5 | 0.06 | 0.3 | | 70 | 6 | 2.9 | 1 |
| Legislative chambers | 5 | 2.5 | 0.06 | 0.3 | | 50 | 6 | 3.1 | 1 |
| Libraries | 5 | 2.5 | 0.12 | 0.6 | | 10 | 17 | 8.5 | 1 |
| Lobbies | 5 | 2.5 | 0.06 | 0.3 | | 150 | 5 | 2.7 | 1 |
| Museums (children's) | 7.5 | 3.8 | 0.12 | 0.6 | | 40 | 11 | 5.3 | 1 |
| Museums/galleries | 7.5 | 3.8 | 0.06 | 0.3 | | 40 | 9 | 4.6 | 1 |
| **Residential** | | | | | | | | | |
| Dwelling unit | 5 | 2.5 | 0.06 | 0.3 | F,G | F | | | 1 |

**Table 5-9 Minimum Ventilation Rates in Breathing Zone** *(Continued)*
**(This table is not valid in isolation; it must be used in conjunction with the accompanying notes.)**
*(Table 6-1, ASHRAE Standard 62.1-2010)*

| Occupancy Category | People Outdoor Air Rate $R_p$ CFM/person | People Outdoor Air Rate $R_p$ L/s·person | Area Outdoor Air Rate $R_a$ CFM/ft² | Area Outdoor Air Rate $R_a$ L/s·m² | Notes | Default Values Occupant Density (see Note 4) #/1000 ft² or #/100 m² | Default Values Combined Outdoor Air Rate (see Note 5) CFM/person | Default Values Combined Outdoor Air Rate (see Note 5) L/s·person | Air Class |
|---|---|---|---|---|---|---|---|---|---|
| Common corridors | – | – | 0.06 | 0.3 | | | | | 1 |
| **Retail** | | | | | | | | | |
| Sales (except as below) | 7.5 | 3.8 | 0.12 | 0.6 | | 15 | 16 | 7.8 | 2 |
| Mall common areas | 7.5 | 3.8 | 0.06 | 0.3 | | 40 | 9 | 4.6 | 1 |
| Barbershop | 7.5 | 3.8 | 0.06 | 0.3 | | 25 | 10 | 5.0 | 2 |
| Beauty and nail salons | 20 | 10 | 0.12 | 0.6 | | 25 | 25 | 12.4 | 2 |
| Pet shops (animal areas) | 7.5 | 3.8 | 0.18 | 0.9 | | 10 | 26 | 12.8 | 2 |
| Supermarket | 7.5 | 3.8 | 0.06 | 0.3 | | 8 | 15 | 7.6 | 1 |
| Coin-operated laundries | 7.5 | 3.8 | 0.12 | 0.6 | | 20 | 14 | 7.0 | 2 |
| **Sports and Entertainment** | | | | | | | | | |
| Sports arena (play area) | – | – | 0.30 | 1.5 | E | – | | | 1 |
| Gym, stadium (play area) | – | – | 0.30 | 1.5 | | 30 | | | 2 |
| Spectator areas | 7.5 | 3.8 | 0.06 | 0.3 | | 150 | 8 | 4.0 | 1 |
| Swimming (pool & deck) | – | – | 0.48 | 2.4 | C | – | | | 2 |
| Disco/dance floors | 20 | 10 | 0.06 | 0.3 | | 100 | 21 | 10.3 | 2 |
| Health club/aerobics room | 20 | 10 | 0.06 | 0.3 | | 40 | 22 | 10.8 | 2 |
| Health club/weight rooms | 20 | 10 | 0.06 | 0.3 | | 10 | 26 | 13.0 | 2 |
| Bowling alley (seating) | 10 | 5 | 0.12 | 0.6 | | 40 | 13 | 6.5 | 1 |
| Gambling casinos | 7.5 | 3.8 | 0.18 | 0.9 | | 120 | 9 | 4.6 | 1 |
| Game arcades | 7.5 | 3.8 | 0.18 | 0.9 | | 20 | 17 | 8.3 | 1 |
| Stages, studios | 10 | 5 | 0.06 | 0.3 | D | 70 | 11 | 5.4 | 1 |

GENERAL NOTES FOR TABLE 5-9

1. **Related requirements:** The rates in this table are based on all other applicable requirements of this standard being met.
2. **Environmental Tobacco Smoke:** This table applies to ETS-free areas. Refer to Section 5.17 for requirements for buildings containing ETS areas and ETS-free areas.
3. **Air density:** Volumetric airflow rates are based on an air density of 0.075 $lb_{da}/ft^3$ (1.2 $kg_{da}/m^3$), which corresponds to dry air at a barometric pressure of 1 atm (101.3 kPa) and an air temperature of 70°F (21°C). Rates may be adjusted for actual density but such adjustment is not required for compliance with this standard.
4. **Default occupant density:** The default occupant density shall be used when actual occupant density is not known.
5. **Default combined outdoor air rate (per person):** This rate is based on the default occupant density.
6. **Unlisted occupancies:** If the occupancy category for a proposed space or zone is not listed, the requirements for the listed occupancy category that is most similar in terms of occupant density, activities and building construction shall be used.

ITEM-SPECIFIC NOTES FOR TABLE 5-9

A. For high school and college libraries, use values shown for Public Assembly Spaces—Libraries.
B. Rate may not be sufficient when stored materials include those having potentially harmful emissions.
C. Rate does not allow for humidity control. Additional ventilation or dehumidification may be required to remove moisture. "Deck area" refers to the area surrounding the pool that would be expected to be wetted during normal pool use, i.e., when the pool is occupied. Deck area that is not expected to be wetted shall be designated as a space type (for example, "spectator area").
D. Rate does not include special exhaust for stage effects, e.g., dry ice vapors, smoke.
E. When combustion equipment is intended to be used on the playing surface, additional dilution ventilation and/or source control shall be provided.
F. Default occupancy for dwelling units shall be two persons for studio and one-bedroom units, with one additional person for each additional bedroom.
G. Air from one residential dwelling shall not be recirculated or transferred to any other space outside of that dwelling.

thermal load is satisfied. This is accounted for by a flow reduction factor $F_r$. A mass balance for the contaminant may be written to determine the space contaminant concentration for each of the system arrangements. The various permutations for the air-handling and distribution systems are described in Table 5-10. The mass balance equations for computing the space contaminant concentration for each system are included in Table 5-10.

If the allowable space contamination is specified, the equations in Table 5-10 may be solved for the outdoor flow rate. When the outdoor flow rate is specified, the equations may be used to determine the resulting contaminant concentration.

Local codes and ordinances frequently specify outdoor air ventilation requirements for public places and for industrial installations and must be checked and complied with if their requirements are for greater quantities of outdoor air than provided above.

**Example 5-3** A small fast-food cafeteria building, 30 × 100 × 9 ft, located in downtown Chicago, IL, has windows and doors on the east and north sides, but none on the south and west. The HVAC system is to include a humidifier. Estimate the winter design heat losses due to ventilation and/or infiltration. Indoor and outdoor design conditions are specified as 72°F, 30% rh and −3°F, respectively.

Based on ASHRAE Standard 62, minimum ventilation is to be at the rate of 7.5 cfm/person with a density of 100 people per 1000 ft² plus 0.18 cfm/ft². Thus,

$$\text{Ventilation} = [(30 \times 100)/1000] \times 100 \times 7.5 + (30 \times 100)(0.18) = 2790 \text{ cfm}$$

With the limited number of openings, infiltration is estimated as 1/2 ach (see Table 5-1) plus that due to traffic through swinging doors, approximated as 2000 cfm from Figures 5-10 and 5-11. Thus,

$$\text{Infiltration} = (1/2)(30 \times 100 \times 9)/60 + 2000 = 2225 \text{ cfm}$$

The ventilation airflow of 2790 cfm is larger than the estimated 2225 cfm of infiltration and should be sufficient to pressurize the building, actually producing exfiltration, rather than having infiltration.

The design heat losses due to outdoor air entering the building due to ventilation are determined as

$$q_s = 1.10(2790)[72 - (-3)] = 230{,}000 \text{ Btu/h}$$

$$q_l = 4840(2790)(0.005 - 0.0007) = 58{,}000 \text{ Btu/h}$$

$$q_{\text{total}} = 230{,}000 + 58{,}000 = 288{,}000 \text{ Btu/h}$$

# Load Estimating Fundamentals

*Fig. 5-15 HVAC Ventilation System Schematic*

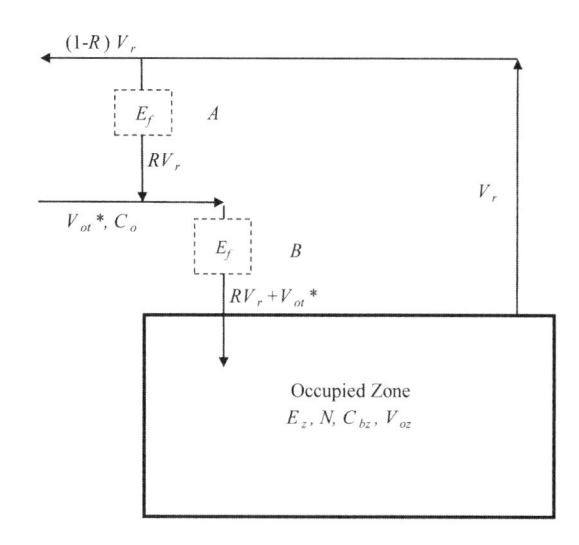

| Symbol or Subscript | Definition |
|---|---|
| A, B | filter location |
| V | volumetric flow |
| C | contaminant concentration |
| $E_z$ | zone air distribution effectiveness |
| $E_f$ | filter efficiency |
| $F_r$ | design flow reduction fraction factor |
| N | contaminant generation rate |
| R | recirculation flow factor |
| Subscript: $o$ | outdoor |
| Subscript: $r$ | return |
| Subscript: $b$ | breathing |
| Subscript: $z$ | zone |

**Table 5-10  Required Zone Outdoor Airflow or Space Breathing Zone Contaminant Concentration with Recirculation and Filtration for Single-Zone Systems**

| Filter Location | Flow | Outdoor Airflow (Required Recirculation Rate) | Required Zone Outdoor Airflow ($V_{oz}$ in Section 6) | Space Breathing Zone Contaminant Concentration |
|---|---|---|---|---|
| None | VAV | 100% | $V_{oz} = \dfrac{N}{E_z F_r (C_{bz} - C_o)}$ | $C_{bz} = C_o + \dfrac{N}{E_z F_r V_{oz}}$ |
| A | Constant | Constant | $V_{oz} = \dfrac{1 - E_z R V_r E_f C_{bz}}{E_z (C_{bz} - C_o)}$ | $C_{bz} = \dfrac{N + E_z V_{oz} C_o}{E_z (V_{oz} + R V_r E_f)}$ |
| A | VAV | Constant | $V_{oz} = \dfrac{N - E_z F_r R V_r E_f C_{bz}}{E_z (C_{bz} - C_o)}$ | $C_{bz} = \dfrac{N + E_z V_{oz} C_o}{E_z (V_{oz} + F_r R V_r E_f)}$ |
| B | Constant | Constant | $V_{oz} = \dfrac{N - E_z R V_r E_f C_{bz}}{E_z [C_{bz} - (1 - E_f)(C_o)]}$ | $C_{bz} = \dfrac{N + E_z V_{oz}(1 - E_f) C_o}{E_z (V_{oz} + R V_r E_f)}$ |
| B | VAV | 100% | $V_{oz} = \dfrac{N}{E_z F_r [C_{bz} - (1 - E_f)(C_o)]}$ | $C_{bz} = \dfrac{N + E_z F_r V_{oz}(1 - E_f) C_o}{E_z F_r V_{oz}}$ |
| B | VAV | Constant | $V_{oz} = \dfrac{N - E_z F_r R V_r E_f C_{bz}}{E_z [C_{bz} - (1 - E_f)(C_o)]}$ | $C_{bz} = \dfrac{N + E_z V_{oz}(1 - E_f) C_o}{E_z (V_{oz} + F_r R V_r E_f)}$ |

## 5.3 Heat Transfer Coefficients

### 5.3.1 Modes of Heat Transfer

The design of a heating, refrigerating, or air-conditioning system, including selection of building insulation or sizing of piping and ducts, or the evaluation of the thermal performance of system parts such as chillers, heat exchangers, and fans, is based on the principles of heat transfer given in Chapter 4, "Heat Transfer," of the 2013 *ASHRAE Handbook—Fundamentals*.

Whenever a temperature difference between two areas (indoor-outdoor) exists, heat flows from the warmer area to the cooler area. The flow or transfer of heat takes place by one or more of three modes—conduction, convection, or radiation.

**Conduction** is the transfer of heat through a solid. When a poker is left in a fire, the handle is also warmed even though it is not in direct contact with the flame. The flow of heat along the poker is by conduction. The rate of flow is influenced by the temperature difference, the area of the material, the distance through the material from warm side to cool side, and the thermal conductivity of the material. Insulating materials have low thermal conductivity, which, combined with their thickness, provide a barrier that slows conductive heat transfer.

**Convection** transfers heat by movement through liquids or gases. As a gas (e.g., air) is heated, it expands, becomes lighter, and rises. It is then displaced with cooler air which follows the same cycle, carrying heat with it. The continuous cycle of rising warm air and descending cool air is a convection current. By dividing a large space into many small spaces and providing barriers that restrict convection, the flow of heat can be slowed. The mat of random fibers in batt-type insulation, such as mineral wool or wood fiber insulation, provides such barriers.

**Radiation** is a method of heat transfer whereby a warm object can heat a cool object without the need of a solid, liquid, or gas between them. An example of radiation is heat from the sun passing through the vacuum of space and warming the earth. In another example, standing in front of a campfire results in the warming of those parts of the body that face the fire. For other body parts to be warmed, the body position must be reversed. The body can also be the radiator and radiate heat to cooler objects.

The **U-factor** and the **R-value** are used to indicate the relative insulating value of materials and sections of walls, floors, and ceilings. The U-factor indicates the rate at which heat flows through a specific material or a building section (Figure 5-16). The smaller the U-factor, the better the insulating value of the material or group of materials making up the wall, ceiling, or floor.

The R-value indicates the ability of one specific material, or a group of materials in a building section, to resist heat flow through them. Many insulating materials now have their R-value stamped on the outside of the package, batt, or blanket. The R-value and relative heat-resisting values of several of these materials are listed in Table 5-11. The R-value for batt, blanket, and loose-fill insulation as listed in this figure is for a thickness of 1 in. The R-value for greater thicknesses can be determined by multiplying the thickness desired, in inches, by the R-value listed.

The greater the R-value, the greater the insulating value of the material, and the lower the heat loss. Thus, a high R-value means lower heating and cooling costs and less energy used to maintain a comfortable temperature.

Thermal conductivity $k$ is a property of a homogeneous material. Building materials, such as lumber, brick, and stone, are usually considered homogeneous. Most thermal insulation and many other building materials are porous and consist of combinations of solid matter with small voids. For most insulating materials, conduction is not the only mode of heat transfer. Consequently, the term **apparent thermal conductivity** describes the heat flow properties of most materials. Some materials with low thermal conductivities are almost purely conductive (silica opacified aerogel, corkboard, etc.). The apparent thermal conductivity of insulation varies with form and physical structure, environment, and application conditions. Form and physical structure vary with the basic material and manufacturing process. Typical variations include density, cell size, diameter and arrangement of fibers or particles, degree and extent of bonding materials, transparency to thermal radiation, and the type and pressure of gas within the insulation.

The method of calculating an overall coefficient of heat transmission requires knowledge of (1) the apparent thermal conductivity and thickness of homogeneous components, (2) thermal conductance of nonhomogeneous components, (3) surface conductances of both sides of the construction, and (4) conductance of air spaces in the construction.

Surface conductance is the heat transfer to or from the surface by the combined effects of radiation, convection, and conduction. Each of these transport modes can vary independently. Heat transfer by radiation between two surfaces is controlled by the character of the surfaces (emittance and reflectivity), the temperature difference between them, and the solid angle through which they see each other. Heat transfer by convection and conduction is controlled by surface roughness, air movement, and temperature difference between the air and surface.

Heat transfer across an air space is affected by the nature of the boundary surfaces, as well as the intervening air space, the orientation of the air space, the distance between boundary sur-

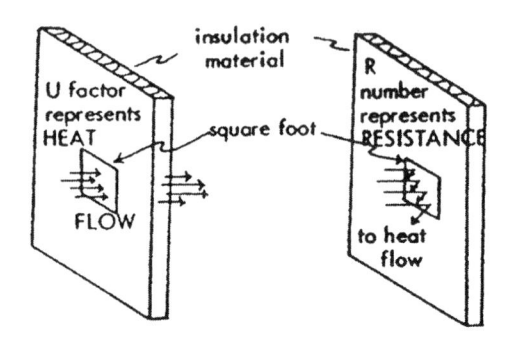

*Fig. 5-16 U-Factor and R-value.*

**Load Estimating Fundamentals**

faces, and the direction of heat flow. Air space conductance coefficients represent the total conductance from one surface bounding the air space to the other. The total conductance is the sum of a radiation component and a convection and conduction component. In all cases, the spaces are considered airtight with no through air leakage.

The combined effect of the emittances of the boundary surfaces of an air space is expressed by the effective emittance $E$ of the air space. The radiation component is affected only slightly by the thickness of the space, its orientation, the direction of heat flow, or the order of emittance (hot or cold surface). The heat transfer by convection and conduction combined is affected markedly by orientation of the air space and the direction of heat flow, by the temperature difference across the space, and, in some cases, by the thickness of the space. It is also slightly affected by the mean temperature of its surfaces.

The steady-state thermal resistances (R-values) of building components (walls, floors, windows, roof systems, etc.) can be calculated from the thermal properties of the materials in the component; or the heat flow through the assembled component can be measured directly with laboratory equipment such as the guarded hot box (ASTM Standard C236) or the calibrated hot box (ASTM Standard C976).

Tables 5-12 through 5-15 list thermal values which may be used to calculate thermal resistances of building wall, floors, and ceilings. The values shown in these tables were developed under ideal conditions. In practice, overall thermal performance can be reduced significantly by such factors as improper installation and shrinkage, settling, or compression of the insulation.

The performance of materials fabricated in the field is especially subject to the quality of workmanship during construction and installation. Good workmanship becomes increasingly important as the insulation requirement becomes greater. Therefore, some engineers include additional insulation or other safety factors based on experience in their design.

When installing insulation, irregular areas must be given careful attention. Blanket-type insulation should be sealed by stapling or taping it to the floor, ceiling plates, and studs. Insulating material should be carefully fitted around all plumbing, wiring, and other projections. The proper thickness should be maintained throughout the walls, ceilings, and floors. To get the most value from reflective materials, such as aluminum foil facing on batts or blankets, allow a 0.6 in. (15 mm) airspace between the foil and the wallboard.

In order to obtain the proper density of insulation, competent operators using special equipment blow loose fill insulation (be it for new or remodeled homes) into the walls, floors, and ceilings. Insulation should be blown into each wall cavity at both top and bottom so that all spaces are filled and variations in density are minimized. Variations in insulation density affect the resistance value and the R-value of material. Loose fill insulation tends to settle, resulting in high density insulation below and none above. Therefore, batt or blanket insulation is preferable for vertical spaces in new construction.

Insulation resists heat flow in proportion to its R-value *only* if it is installed according to these general recommendations and the manufacturer's instructions. Wall voids are frequently left open at the top of the stud space, thereby allowing outdoor air to enter and greatly affect the U-factor of the wall.

Common variations in conditions, materials, workmanship, and so forth, can introduce much greater variations in U-values than the variations resulting from assumed mean temperatures and temperature differences. *Therefore, stating a U-factor of more than two significant figures may assume more precision than can possibly exist.*

Shading devices, such as venetian blinds, draperies, and roller shades, substantially reduce the U-factor for windows and/or glass doors if they fit tightly to the window jambs, head, and sill, and are made of nonporous material. As a rough approximation, tight-fitting shading devices may be considered to reduce the U-factor of vertical exterior single glazing by 25% and of vertical exterior double glazing and glass block by 15%. These adjustments should not be considered in choosing heating equipment, but may be used for calculating design cooling loads.

### 5.3.2 Determining U-Factors

The total resistance to heat flow through building construction such as a flat ceiling, floor, or wall (or curved surface if the curvature is small) is the numerical sum of the resistances (R-values) of all parts of the construction in series:

$$R = R_1 + R_2 + R_3 + R_4 + ... + R_n$$

where

$R_1, R_2, ..., R_n$ = individual resistances of the parts
$R$ = resistance of the construction from inside surface to outside surface

However, in buildings, to obtain the overall resistance $R_T$, the air film resistances $R_i$ and $R_o$ must be added to $R$.

$$R_T = R_i + R + R_o \qquad (5\text{-}9)$$

The U-factor (thermal transmittance) is the reciprocal of $R_T$:

$$U = \frac{1}{R_T}$$

Thus, $U$ is computed by adding up all of the R-values, including those of inside- and outside-air films, the air gap, and all building materials.

With the use of higher values of $R_T$, the corresponding values of $U$ become very small. This is one reason why it is sometimes preferable to specify resistance rather than transmittance. Also, a whole number is more understandable to an insulation buyer than is a decimal or fraction.

For a wall with air space construction, consisting of two homogeneous materials of conductivities $k_1$ and $k_2$ and thickness $x_1$ and $x_2$, respectively, and separated by an air space of conductance $C$, the overall resistance would be determined from

$$R_T = \frac{1}{h_i} + \frac{x_1}{k_1} + ... + \frac{x_n}{k_n} + \frac{1}{h_0}$$

where $h_i$ and $h_0$ are the heat transfer film coefficients.

Table 5-11  Relative Thermal Resistances of Building Material

| Material Description | Material Density, lb/ft$^3$ | Material Thickness, in. | Resistance for Thickness Listed, °F·ft$^2$·h/Btu |
|---|---|---|---|
| Building paper | – | – | 0.06 |
| Gypsum plaster, sand aggregate | 105 | 1/2 | 0.09 |
| Structural glass | – | – | 0.10 |
| Air surface, 15 mph wind, outside surface | – | – | 0.17 |
| Gypsum or plaster board | 50 | 3/8 | 0.32 |
| Stone, lime, or sand | – | 4 | 0.32 |
| Concrete, sand-gravel aggregate | 140 | 4 | 0.32 |
| Built-up roofing | 70 | 3/8 | 0.33 |
| Brick, face | 130 | 4 | 0.44 |
| Still air surface, horiz., ordinary materials, heat flow up | – | – | 0.61 |
| Aluminum, steel, or vinyl over sheathing, hollow backed | – | – | 0.61 |
| Plywood | 34 | 1/2 | 0.63 |
| Still air surface, vertical, ordinary mtrls, horiz. heat flow | – | – | 0.68 |
| Wood siding, bevel, 1/2 in  8 in. lapped | – | – | 0.81 |
| Wood shingle siding, 16 in., 7 1/2 in. exposure | – | – | 0.87 |
| Oak, maple, and similar hardwoods | 45 | 1 | 0.91 |
| Air space, vertical, ordinary materials, horiz. heat flow | – | 3/4 to 4 | 0.97 |
| Clay tile, one cell deep | – | 4 | 1.11 |
| Concrete block, 3 core, sand-gravel aggregate | – | 8 | 1.11 |
| Acoustical tile, wood or cane fiber | – | 1/2 | 1.19 |
| Fir, pine, and similar softwoods | 32 | 1 | 1.25 |
| Insulation board, impregnated | 20 | 1/2 | 1.32 |
| Concrete, lightweight aggregate | 80 | 4 | 1.50 |
| Air space, vertical, bounded by reflective material | – | 3/4 to 4 | 1.70 |
| Concrete block, 3 core, cinder aggregate | – | 8 | 1.72 |
| Concrete block, 3 core, lightweight aggregate | – | 8 | 2.00 |
| Vermiculite, expanded | 7 | 1 | 2.08 |
| Carpet and fibrous pad | – | – | 2.08 |
| Cellular glass insulation board | 9 | 1 | 2.50 |
| Roof insulation, preformed for above deck | – | 1 | 2.78 |
| Mineral wool, loose fill, from slag glass or rock | 2–5 | 1 | 3.33 |
| Wood fiber, loose fill, hemlock, fir or redwood | 2–3.5 | 1 | 3.33 |
| Plastic, foamed | 1.62 | 1 | 3.45 |
| Macerated paper or pulp | 2–3.5 | 1 | 3.57 |
| Corkboard, without added binder | 6.5–8 | 1 | 3.70 |
| **Batt and Blankets Bounded by Nonreflective Materials** | | | |
| Mineral wool, fibrous form, rock, slag, or glass | 1.5–4 | 1 | 3.70 |
| Wood fiber, multilayer, stitched expanded | 1.5–2 | 1 | 3.70 |
| Cotton fiber | 0.8–2 | 1 | 3.85 |
| Wood fiber | 3.2–3.6 | 1 | 4.00 |

# Load Estimating Fundamentals

**Table 5-12  Surface Film Coefficients/Resistances**
*(Table 10, Chapter 26, 2013 ASHRAE Handbook—Fundamentals)*

| Position of Surface | Direction of Heat Flow | Non-reflective ε = 0.90 $h_i$ | R | Reflective ε = 0.20 $h_i$ | R | ε = 0.05 $h_i$ | R |
|---|---|---|---|---|---|---|---|
| STILL AIR | | | | | | | |
| Horizontal | Upward | 1.63 | 0.61 | 0.91 | 1.10 | 0.76 | 1.32 |
| Sloping—45° | Upward | 1.60 | 0.62 | 0.88 | 1.14 | 0.73 | 1.37 |
| Vertical | Horizontal | 1.46 | 0.68 | 0.74 | 1.35 | 0.59 | 1.70 |
| Sloping—45° | Downward | 1.32 | 0.76 | 0.60 | 1.67 | 0.45 | 2.22 |
| Horizontal | Downward | 1.08 | 0.92 | 0.37 | 2.70 | 0.22 | 4.55 |
| MOVING AIR (Any position) | | $h_o$ | R | | | | |
| 15 mph Wind (for winter) | Any | 6.00 | 0.17 | — | — | — | — |
| 7.5 mph Wind (for summer) | Any | 4.00 | 0.25 | — | — | — | — |

*Notes:* (References are to Chapter 26 in the 2013 *ASHRAE Handbook—Fundamentals*)
1. Surface conductance $h_i$ and $h_o$ measured in Btu/h·ft²·°F; resistance R in °F·ft²·h/Btu.
2. No surface has both an air space resistance value and a surface resistance value.
3. Conductances are for surfaces of the stated emittance facing virtual blackbody surroundings at the same temperature as the ambient air. Values are based on a surface-air temperature difference of 10°F and for surface temperatures of 70°F.
4. See Chapter 4 in the 2013 *ASHRAE Handbook—Fundamentals* for more detailed information.
5. Condensate can have a significant impact on surface emittance.

**Table 5-13  Emissivity of Various Surfaces and Effective Emittances of Facing Air Spaces[a]**
*(Table 2, Chapter 26, 2013 ASHRAE Handbook—Fundamentals)*

| Surface | Average Emissivity ε | Effective Emittance $\varepsilon_{eff}$ of Air Space — One Surface's Emittance ε; Other, 0.9 | Both Surfaces' Emittance ε |
|---|---|---|---|
| Aluminum foil, bright | 0.05 | 0.05 | 0.03 |
| Aluminum foil, with condensate just visible (>0.7 g/ft²)(>0.5 g/m²) | 0.30[b] | 0.29 | — |
| Aluminum foil, with condensate clearly visible (>2.9 g/ft²)(>2.0 g/m²) | 0.70[b] | 0.65 | — |
| Aluminum sheet | 0.12 | 0.12 | 0.06 |
| Aluminum-coated paper, polished | 0.20 | 0.20 | 0.11 |
| Brass, nonoxidized | 0.04 | 0.038 | 0.02 |
| Copper, black oxidized | 0.74 | 0.41 | 0.59 |
| Copper, polished | 0.04 | 0.038 | 0.02 |
| Iron and steel, polished | 0.2 | 0.16 | 0.11 |
| Iron and steel, oxidized | 0.58 | 0.35 | 0.41 |
| Lead, oxidized | 0.27 | 0.21 | 0.16 |
| Nickel, nonoxidized | 0.06 | 0.056 | 0.03 |
| Silver, polished | 0.03 | 0.029 | 0.015 |
| Steel, galvanized, bright | 0.25 | 0.24 | 0.15 |
| Tin, nonoxidized | 0.05 | 0.047 | 0.026 |
| Aluminum paint | 0.50 | 0.47 | 0.35 |
| Building materials: wood, paper, masonry, nonmetallic paints | 0.90 | 0.82 | 0.82 |
| Regular glass | 0.84 | 0.77 | 0.72 |

[a] Values apply in 4 to 40 μm range of electromagnetic spectrum. Also, oxidation, corrosion, and accumulation of dust and dirt can dramatically increase surface emittance. Emittance values of 0.05 should only be used where the highly reflective surface can be maintained over the service life of the assembly. Except as noted, data from VDI (1999).
[b] Values based on data in Bassett and Trethowen (1984).

**Series and Parallel Heat Flow Paths.** In many installations, components are arranged so that heat flows in parallel paths of different conductances. If no heat flows between lateral paths, the U-factor for each path is calculated. The average transmittance is then

$$U_{av} = aU_a + bU_b + \ldots + nU_n \quad (5\text{-}10)$$

where $a, b, \ldots, n$ are respective fractions of a typical basic area composed of several different paths with transmittances $U_a, U_b, \ldots, U_n$.

If heat can flow laterally with little resistance in any continuous layer so that transverse isothermal planes result, total average resistance $R_{T(av)}$ is the sum of the resistance of the layers between such planes. This is a series combination of layers, of which one (or more) provides parallel paths.

The average overall R-values and U-factors of wood frame walls can be calculated by assuming parallel heat flow paths through areas with different thermal resistances.

The following equation is recommended to correct for the effect of framing members.

$$U_{av} = (S/100)U_s + (1 - S/100)U_i \quad (5\text{-}11)$$

where

$U_{av}$ = average U-factor for building section
$U_i$ = U-factor for area between framing members
$U_s$ = U-factor for area backed by framing members
$S$ = percentage of area backed by framing members

The framing factor or fraction of the building component that is framing depends on the specific type of construction, and it may vary based on local construction practices even for the same type of construction. For stud walls 16 in. on center (OC) the fraction of insulated cavity may be as low as 0.75, where the fraction of studs, plates, and sills is 0.21 and the fraction of headers is 0.04, For studs 24 in. OC, the respective values are 0.78, 0.18, and 0.04. These fractions contain an allowance for multiple studs, plates, sills, extra framing around windows, headers, and band joists.

**Unequal Areas.** A construction may be made up of two or more layers of unequal area, separated by an airspace, and arranged so that heat flows through the layers in series. The most common such construction is a ceiling and roof combination where the attic space is unheated and unventilated. A combined coefficient based on the most convenient area, say the ceiling area, from air inside to air outside can be calculated from

$$R_{T,c} = R_c + R_r/n \text{ and } U_c = 1/R_{T,c} \quad (5\text{-}12)$$

where

$U_c$ = combined coefficient based on ceiling area
$n$ = ratio of roof to ceiling area, $A_r/A_c$

**Windows and Doors.** Table 5-16 lists U-factors for various fenestration products. Tables 5-17 through 5-20 provide U-factors for various exterior doors. All U-factors are approximate, because a significant portion of the resistance of a window or door is contained in the air film resistances, and some parameters that may have important effects are not considered. For

**Table 5-14  Effective Thermal Resistance of Plane Air Spaces,[a,b,c] h·ft²·°F/Btu**
*(Table 3, Chapter 26, 2013 ASHRAE Handbook—Fundamentals)*

| Position of Air Space | Direction of Heat Flow | Mean Temp.[d], °F | Temp. Diff.,[d] °F | \multicolumn{5}{c|}{0.5 in. Air Space[c]} | \multicolumn{5}{c|}{0.75 in. Air Space[c]} |
|---|---|---|---|---|---|---|---|---|---|---|---|---|---|
| | | | | 0.03 | 0.05 | 0.2 | 0.5 | 0.82 | 0.03 | 0.05 | 0.2 | 0.5 | 0.82 |
| Horiz. | Up ↑ | 90 | 10 | 2.13 | 2.03 | 1.51 | 0.99 | 0.73 | 2.34 | 2.22 | 1.61 | 1.04 | 0.75 |
| | | 50 | 30 | 1.62 | 1.57 | 1.29 | 0.96 | 0.75 | 1.71 | 1.66 | 1.35 | 0.99 | 0.77 |
| | | 50 | 10 | 2.13 | 2.05 | 1.60 | 1.11 | 0.84 | 2.30 | 2.21 | 1.70 | 1.16 | 0.87 |
| | | 0 | 20 | 1.73 | 1.70 | 1.45 | 1.12 | 0.91 | 1.83 | 1.79 | 1.52 | 1.16 | 0.93 |
| | | 0 | 10 | 2.10 | 2.04 | 1.70 | 1.27 | 1.00 | 2.23 | 2.16 | 1.78 | 1.31 | 1.02 |
| | | −50 | 20 | 1.69 | 1.66 | 1.49 | 1.23 | 1.04 | 1.77 | 1.74 | 1.55 | 1.27 | 1.07 |
| | | −50 | 10 | 2.04 | 2.00 | 1.75 | 1.40 | 1.16 | 2.16 | 2.11 | 1.84 | 1.46 | 1.20 |
| 45° Slope | Up ↗ | 90 | 10 | 2.44 | 2.31 | 1.65 | 1.06 | 0.76 | 2.96 | 2.78 | 1.88 | 1.15 | 0.81 |
| | | 50 | 30 | 2.06 | 1.98 | 1.56 | 1.10 | 0.83 | 1.99 | 1.92 | 1.52 | 1.08 | 0.82 |
| | | 50 | 10 | 2.55 | 2.44 | 1.83 | 1.22 | 0.90 | 2.90 | 2.75 | 2.00 | 1.29 | 0.94 |
| | | 0 | 20 | 2.20 | 2.14 | 1.76 | 1.30 | 1.02 | 2.13 | 2.07 | 1.72 | 1.28 | 1.00 |
| | | 0 | 10 | 2.63 | 2.54 | 2.03 | 1.44 | 1.10 | 2.72 | 2.62 | 2.08 | 1.47 | 1.12 |
| | | −50 | 20 | 2.08 | 2.04 | 1.78 | 1.42 | 1.17 | 2.05 | 2.01 | 1.76 | 1.41 | 1.16 |
| | | −50 | 10 | 2.62 | 2.56 | 2.17 | 1.66 | 1.33 | 2.53 | 2.47 | 2.10 | 1.62 | 1.30 |
| Vertical | Horiz. → | 90 | 10 | 2.47 | 2.34 | 1.67 | 1.06 | 0.77 | 3.50 | 3.24 | 2.08 | 1.22 | 0.84 |
| | | 50 | 30 | 2.57 | 2.46 | 1.84 | 1.23 | 0.90 | 2.91 | 2.77 | 2.01 | 1.30 | 0.94 |
| | | 50 | 10 | 2.66 | 2.54 | 1.88 | 1.24 | 0.91 | 3.70 | 3.46 | 2.35 | 1.43 | 1.01 |
| | | 0 | 20 | 2.82 | 2.72 | 2.14 | 1.50 | 1.13 | 3.14 | 3.02 | 2.32 | 1.58 | 1.18 |
| | | 0 | 10 | 2.93 | 2.82 | 2.20 | 1.53 | 1.15 | 3.77 | 3.59 | 2.64 | 1.73 | 1.26 |
| | | −50 | 20 | 2.90 | 2.82 | 2.35 | 1.76 | 1.39 | 2.90 | 2.83 | 2.36 | 1.77 | 1.39 |
| | | −50 | 10 | 3.20 | 3.10 | 2.54 | 1.87 | 1.46 | 3.72 | 3.60 | 2.87 | 2.04 | 1.56 |
| 45° Slope | Down ↘ | 90 | 10 | 2.48 | 2.34 | 1.67 | 1.06 | 0.77 | 3.53 | 3.27 | 2.10 | 1.22 | 0.84 |
| | | 50 | 30 | 2.64 | 2.52 | 1.87 | 1.24 | 0.91 | 3.43 | 3.23 | 2.24 | 1.39 | 0.99 |
| | | 50 | 10 | 2.67 | 2.55 | 1.89 | 1.25 | 0.92 | 3.81 | 3.57 | 2.40 | 1.45 | 1.02 |
| | | 0 | 20 | 2.91 | 2.80 | 2.19 | 1.52 | 1.15 | 3.75 | 3.57 | 2.63 | 1.72 | 1.26 |
| | | 0 | 10 | 2.94 | 2.83 | 2.21 | 1.53 | 1.15 | 4.12 | 3.91 | 2.81 | 1.80 | 1.30 |
| | | −50 | 20 | 3.16 | 3.07 | 2.52 | 1.86 | 1.45 | 3.78 | 3.65 | 2.90 | 2.05 | 1.57 |
| | | −50 | 10 | 3.26 | 3.16 | 2.58 | 1.89 | 1.47 | 4.35 | 4.18 | 3.22 | 2.21 | 1.66 |
| Horiz. | Down ↓ | 90 | 10 | 2.48 | 2.34 | 1.67 | 1.06 | 0.77 | 3.55 | 3.29 | 2.10 | 1.22 | 0.85 |
| | | 50 | 30 | 2.66 | 2.54 | 1.88 | 1.24 | 0.91 | 3.77 | 3.52 | 2.38 | 1.44 | 1.02 |
| | | 50 | 10 | 2.67 | 2.55 | 1.89 | 1.25 | 0.92 | 3.84 | 3.59 | 2.41 | 1.45 | 1.02 |
| | | 0 | 20 | 2.94 | 2.83 | 2.20 | 1.53 | 1.15 | 4.18 | 3.96 | 2.83 | 1.81 | 1.30 |
| | | 0 | 10 | 2.96 | 2.85 | 2.22 | 1.53 | 1.16 | 4.25 | 4.02 | 2.87 | 1.82 | 1.31 |
| | | −50 | 20 | 3.25 | 3.15 | 2.58 | 1.89 | 1.47 | 4.60 | 4.41 | 3.36 | 2.28 | 1.69 |
| | | −50 | 10 | 3.28 | 3.18 | 2.60 | 1.90 | 1.47 | 4.71 | 4.51 | 3.42 | 2.30 | 1.71 |

| Position of Air Space | Direction of Heat Flow | Mean Temp., °F | Temp. Diff., °F | \multicolumn{5}{c|}{1.5 in. Air Space[c]} | \multicolumn{5}{c|}{3.5 in. Air Space[c]} |
|---|---|---|---|---|---|---|---|---|---|---|---|---|---|
| Horiz. | Up ↑ | 90 | 10 | 2.55 | 2.41 | 1.71 | 1.08 | 0.77 | 2.84 | 2.66 | 1.83 | 1.13 | 0.80 |
| | | 50 | 30 | 1.87 | 1.81 | 1.45 | 1.04 | 0.80 | 2.09 | 2.01 | 1.58 | 1.10 | 0.84 |
| | | 50 | 10 | 2.50 | 2.40 | 1.81 | 1.21 | 0.89 | 2.80 | 2.66 | 1.95 | 1.28 | 0.93 |
| | | 0 | 20 | 2.01 | 1.95 | 1.63 | 1.23 | 0.97 | 2.25 | 2.18 | 1.79 | 1.32 | 1.03 |
| | | 0 | 10 | 2.43 | 2.35 | 1.90 | 1.38 | 1.06 | 2.71 | 2.62 | 2.07 | 1.47 | 1.12 |
| | | −50 | 20 | 1.94 | 1.91 | 1.68 | 1.36 | 1.13 | 2.19 | 2.14 | 1.86 | 1.47 | 1.20 |
| | | −50 | 10 | 2.37 | 2.31 | 1.99 | 1.55 | 1.26 | 2.65 | 2.58 | 2.18 | 1.67 | 1.33 |
| 45° Slope | Up ↗ | 90 | 10 | 2.92 | 2.73 | 1.86 | 1.14 | 0.80 | 3.18 | 2.96 | 1.97 | 1.18 | 0.82 |
| | | 50 | 30 | 2.14 | 2.06 | 1.61 | 1.12 | 0.84 | 2.26 | 2.17 | 1.67 | 1.15 | 0.86 |
| | | 50 | 10 | 2.88 | 2.74 | 1.99 | 1.29 | 0.94 | 3.12 | 2.95 | 2.10 | 1.34 | 0.96 |
| | | 0 | 20 | 2.30 | 2.23 | 1.82 | 1.34 | 1.04 | 2.42 | 2.35 | 1.90 | 1.38 | 1.06 |
| | | 0 | 10 | 2.79 | 2.69 | 2.12 | 1.49 | 1.13 | 2.98 | 2.87 | 2.23 | 1.54 | 1.16 |
| | | −50 | 20 | 2.22 | 2.17 | 1.88 | 1.49 | 1.21 | 2.34 | 2.29 | 1.97 | 1.54 | 1.25 |
| | | −50 | 10 | 2.71 | 2.64 | 2.23 | 1.69 | 1.35 | 2.87 | 2.79 | 2.33 | 1.75 | 1.39 |
| Vertical | Horiz. → | 90 | 10 | 3.99 | 3.66 | 2.25 | 1.27 | 0.87 | 3.69 | 3.40 | 2.15 | 1.24 | 0.85 |
| | | 50 | 30 | 2.58 | 2.46 | 1.84 | 1.23 | 0.90 | 2.67 | 2.55 | 1.89 | 1.25 | 0.91 |
| | | 50 | 10 | 3.79 | 3.55 | 2.39 | 1.45 | 1.02 | 3.63 | 3.40 | 2.32 | 1.42 | 1.01 |
| | | 0 | 20 | 2.76 | 2.66 | 2.10 | 1.48 | 1.12 | 2.88 | 2.78 | 2.17 | 1.51 | 1.14 |
| | | 0 | 10 | 3.51 | 3.35 | 2.51 | 1.67 | 1.23 | 3.49 | 3.33 | 2.50 | 1.67 | 1.23 |
| | | −50 | 20 | 2.64 | 2.58 | 2.18 | 1.66 | 1.33 | 2.82 | 2.75 | 2.30 | 1.73 | 1.37 |
| | | −50 | 10 | 3.31 | 3.21 | 2.62 | 1.91 | 1.48 | 3.40 | 3.30 | 2.67 | 1.94 | 1.50 |

162    Principles of HVAC

# Load Estimating Fundamentals

**Table 5-14 Effective Thermal Resistance of Plane Air Spaces,**[a,b,c] **h·ft²·°F/Btu** *(Continued)*
*(Table 3, Chapter 26, 2013 ASHRAE Handbook—Fundamentals)*

| Position of Air Space | Direction of Heat Flow | Mean Temp.[d], °F | Temp. Diff.,[d] °F | \multicolumn{5}{c}{1.5 in. Air Space[c]} | \multicolumn{5}{c}{3.5 in. Air Space[c]} |
|---|---|---|---|---|---|---|---|---|---|---|---|---|---|
| | | | | 0.03 | 0.05 | 0.2 | 0.5 | 0.82 | 0.03 | 0.05 | 0.2 | 0.5 | 0.82 |
| 45° Slope | Down | 90 | 10 | 5.07 | 4.55 | 2.56 | 1.36 | 0.91 | 4.81 | 4.33 | 2.49 | 1.34 | 0.90 |
| | | 50 | 30 | 3.58 | 3.36 | 2.31 | 1.42 | 1.00 | 3.51 | 3.30 | 2.28 | 1.40 | 1.00 |
| | | 50 | 10 | 5.10 | 4.66 | 2.85 | 1.60 | 1.09 | 4.74 | 4.36 | 2.73 | 1.57 | 1.08 |
| | | 0 | 20 | 3.85 | 3.66 | 2.68 | 1.74 | 1.27 | 3.81 | 3.63 | 2.66 | 1.74 | 1.27 |
| | | 0 | 10 | 4.92 | 4.62 | 3.16 | 1.94 | 1.37 | 4.59 | 4.32 | 3.02 | 1.88 | 1.34 |
| | | −50 | 20 | 3.62 | 3.50 | 2.80 | 2.01 | 1.54 | 3.77 | 3.64 | 2.90 | 2.05 | 1.57 |
| | | −50 | 10 | 4.67 | 4.47 | 3.40 | 2.29 | 1.70 | 4.50 | 4.32 | 3.31 | 2.25 | 1.68 |
| Horiz. | Down | 90 | 10 | 6.09 | 5.35 | 2.79 | 1.43 | 0.94 | 10.07 | 8.19 | 3.41 | 1.57 | 1.00 |
| | | 50 | 30 | 6.27 | 5.63 | 3.18 | 1.70 | 1.14 | 9.60 | 8.17 | 3.86 | 1.88 | 1.22 |
| | | 50 | 10 | 6.61 | 5.90 | 3.27 | 1.73 | 1.15 | 11.15 | 9.27 | 4.09 | 1.93 | 1.24 |
| | | 0 | 20 | 7.03 | 6.43 | 3.91 | 2.19 | 1.49 | 10.90 | 9.52 | 4.87 | 2.47 | 1.62 |
| | | 0 | 10 | 7.31 | 6.66 | 4.00 | 2.22 | 1.51 | 11.97 | 10.32 | 5.08 | 2.52 | 1.64 |
| | | −50 | 20 | 7.73 | 7.20 | 4.77 | 2.85 | 1.99 | 11.64 | 10.49 | 6.02 | 3.25 | 2.18 |
| | | −50 | 10 | 8.09 | 7.52 | 4.91 | 2.89 | 2.01 | 12.98 | 11.56 | 6.36 | 3.34 | 2.22 |

| Position of Air Space | Direction of Heat Flow | Mean Temp.[d], °F | Temp. Diff.,[d] °F | \multicolumn{5}{c}{5.5 in. Air Space[c]} |
|---|---|---|---|---|---|---|---|---|
| | | | | 0.03 | 0.05 | 0.2 | 0.5 | 0.82 |
| Horiz. | Up | 90 | 10 | 3.01 | 2.82 | 1.90 | 1.15 | 0.81 |
| | | 50 | 30 | 2.22 | 2.13 | 1.65 | 1.14 | 0.86 |
| | | 50 | 10 | 2.97 | 2.82 | 2.04 | 1.31 | 0.95 |
| | | 0 | 20 | 2.40 | 2.33 | 1.89 | 1.37 | 1.06 |
| | | 0 | 10 | 2.90 | 2.79 | 2.18 | 1.52 | 1.15 |
| | | −50 | 20 | 2.31 | 2.27 | 1.95 | 1.53 | 1.24 |
| | | −50 | 10 | 2.80 | 2.73 | 2.29 | 1.73 | 1.37 |
| 45° Slope | Up | 90 | 10 | 3.26 | 3.04 | 2.00 | 1.19 | 0.83 |
| | | 50 | 30 | 2.19 | 2.10 | 1.64 | 1.13 | 0.85 |
| | | 50 | 10 | 3.16 | 2.99 | 2.12 | 1.35 | 0.97 |
| | | 0 | 20 | 2.35 | 2.28 | 1.86 | 1.35 | 1.05 |
| | | 0 | 10 | 3.00 | 2.88 | 2.24 | 1.54 | 1.16 |
| | | −50 | 20 | 2.16 | 2.12 | 1.84 | 1.46 | 1.20 |
| | | −50 | 10 | 2.78 | 2.71 | 2.27 | 1.72 | 1.37 |
| Vertical | Horiz. → | 90 | 10 | 3.76 | 3.46 | 2.17 | 1.25 | 0.86 |
| | | 50 | 30 | 2.83 | 2.69 | 1.97 | 1.28 | 0.93 |
| | | 50 | 10 | 3.72 | 3.49 | 2.36 | 1.44 | 1.01 |
| | | 0 | 20 | 3.08 | 2.95 | 2.28 | 1.57 | 1.17 |
| | | 0 | 10 | 3.66 | 3.49 | 2.59 | 1.70 | 1.25 |
| | | −50 | 20 | 3.03 | 2.95 | 2.44 | 1.81 | 1.42 |
| | | −50 | 10 | 3.59 | 3.47 | 2.78 | 2.00 | 1.53 |
| 45° Slope | Down | 90 | 10 | 4.90 | 4.41 | 2.51 | 1.35 | 0.91 |
| | | 50 | 30 | 3.86 | 3.61 | 2.42 | 1.46 | 1.02 |
| | | 50 | 10 | 4.93 | 4.52 | 2.80 | 1.59 | 1.09 |
| | | 0 | 20 | 4.24 | 4/-1 | 2.86 | 1.82 | 1.31 |
| | | 0 | 10 | 4.93 | 4.63 | 3.16 | 1.94 | 1.37 |
| | | −50 | 20 | 4.28 | 4.12 | 3.19 | 2.19 | 1.65 |
| | | −50 | 10 | 4.93 | 4.71 | 3.53 | 2.35 | 1.74 |
| Horiz. | Down | 90 | 10 | 11.72 | 9.24 | 3.58 | 1.61 | 1.01 |
| | | 50 | 30 | 10.61 | 8.89 | 4.02 | 1.92 | 1.23 |
| | | 50 | 10 | 12.70 | 10.32 | 4.28 | 1.98 | 1.25 |
| | | 0 | 20 | 12.10 | 10.42 | 5.10 | 2.52 | 1.64 |
| | | 0 | 10 | 13.80 | 11.65 | 5.38 | 2.59 | 1.67 |
| | | −50 | 20 | 12.45 | 11.14 | 6.22 | 3.31 | 2.20 |
| | | −50 | 10 | 14.60 | 12.83 | 6.72 | 3.44 | 2.26 |

[a] See Chapter 25 in the 2013 *ASHRAE Handbook—Fundamentals*. Thermal resistance values were determined from $R = 1/C$, where $C = h_c + \varepsilon_{eff} h_r$, $h_c$ is conduction/convection coefficient, $\varepsilon_{eff} h_r$ is radiation coefficient $\approx 0.0068\varepsilon_{eff}[(t_m + 460)/100]^3$, and $t_m$ is mean temperature of air space. Values for $h_c$ were determined from data developed by Robinson et al. (1954). Equations (5) to (7) in Yarbrough (1983) show data in this table in analytic form. For extrapolation from this table to air spaces less than 0.5 in. (e.g., insulating window glass), assume $h_c = 0.159(1 + 0.0016t_m)/l$, where $l$ is air space thickness in in., and $h_c$ is heat transfer through air space only.

[b] Values based on data presented by Robinson et al. (1954). (Also see Chapter 4, Tables 5 and 6, and Chapter 33). Values apply for ideal conditions (i.e., air spaces of uniform thickness bounded by plane, smooth, parallel surfaces with no air leakage to or from the space). **This table should not be used for hollow siding or profiled cladding: see Table 1.** For greater accuracy, use overall U-factors determined through guarded hot box (ASTM Standard C1363) testing. Thermal resistance values for multiple air spaces must be based on careful estimates of mean temperature differences for each air space.

[c] A single resistance value cannot account for multiple air spaces; each air space requires a separate resistance calculation that applies only for established boundary conditions. Resistances of horizontal spaces with heat flow downward are substantially independent of temperature difference.

[d] Interpolation is permissible for other values of mean temperature, temperature difference, and effective emittance $\varepsilon_{eff}$. Interpolation and moderate extrapolation for air spaces greater than 3.5 in. are also permissible.

[e] Effective emittance $\varepsilon_{eff}$ of air space is given by $1/\varepsilon_{eff} = 1/\varepsilon_1 + 1/\varepsilon_2 − 1$, where $\varepsilon_1$ and $\varepsilon_2$ are emittances of surfaces of air space (see Table 2). **Also, oxidation, corrosion, and accumulation of dust and dirt can dramatically increase surface emittance. Emittance values of 0.05 should only be used where the highly reflective surface can be maintained over the service life of the assembly.**

**Table 5-15 Building and Insulating Materials: Design Values**[a]
*(Table 1, Chapter 26, 2013 ASHRAE Handbook—Fundamentals)*

| Description | Density, lb/ft³ | Conductivity[b] $k$, Btu·in/h·ft²·°F | Resistance $R$, h·ft²·°F/Btu | Specific Heat, Btu/lb·°F | Reference[l] |
|---|---|---|---|---|---|
| **Insulating Materials** | | | | | |
| *Blanket and batt*[c,d] | | | | | |
| Glass-fiber batts | | | | 0.2 | Kumaran (2002) |
| | 0.47 to 0.51 | 0.32 to 0.33 | — | — | Four manufacturers (2011) |
| | 0.61 to 0.75 | 0.28 to 0.30 | — | — | Four manufacturers (2011) |
| | 0.79 to 0.85 | 0.26 to 0.27 | — | — | Four manufacturers (2011) |
| | 1.4 | 0.23 | — | — | Four manufacturers (2011) |
| Rock and slag wool batts | — | — | — | 0.2 | Kumaran (1996) |
| | 2 to 2.3 | 0.25 to 0.26 | — | — | One manufacturer (2011) |
| | 2.8 | 0.23 to 0.24 | — | — | One manufacturer (2011) |
| Mineral wool, felted | 1 to 3 | 0.28 | — | — | CIBSE (2006), NIST (2000) |
| | 1 to 8 | 0.24 | — | — | NIST (2000) |
| *Board and slabs* | | | | | |
| Cellular glass | 7.5 | 0.29 | — | 0.20 | One manufacturer (2011) |
| Cement fiber slabs, shredded wood with Portland cement binder | 25 to 27 | 0.50 to 0.53 | — | — | |
| with magnesia oxysulfide binder | 22 | 0.57 | — | 0.31 | |
| Glass fiber board | — | — | — | 0.2 | Kumaran (1996) |
| | 1.5 to 6.0 | 0.23 to 0.24 | — | — | One manufacturer (2011) |
| Expanded rubber (rigid) | 4 | 0.2 | — | 0.4 | Nottage (1947) |
| Extruded polystyrene, smooth skin | — | — | — | 0.35 | Kumaran (1996) |
| aged per Can/ULC Standard S770-2003 | 1.4 to 3.6 | 0.18 to 0.20 | — | — | Four manufacturers (2011) |
| aged 180 days | 1.4 to 3.6 | 0.20 | — | — | One manufacturer (2011) |
| European product | 1.9 | 0.21 | — | — | One manufacturer (2011) |
| aged 5 years at 75°F | 2 to 2.2 | 0.21 | — | — | One manufacturer (2011) |
| blown with low global warming potential (GWP) (<5) blowing agent | | 0.24 to 0.25 | — | — | One manufacturer (2011) |
| Expanded polystyrene, molded beads | — | — | — | 0.35 | Kumaran (1996) |
| | 1.0 to 1.5 | 0.24 to 0.26 | — | — | Independent test reports (2008) |
| | 1.8 | 0.23 | — | — | Independent test reports (2008) |
| Mineral fiberboard, wet felted | 10 | 0.26 | — | 0.2 | Kumaran (1996) |
| Rock wool board | — | — | — | 0.2 | Kumaran (1996) |
| floors and walls | 4.0 to 8.0 | 0.23 to 0.25 | — | — | Five manufacturers (2011) |
| roofing | 10. to 11. | 0.27 to 0.29 | — | 0.2 | Five manufacturers (2011) |
| Acoustical tile[e] | 21 to 23 | 0.36 to 0.37 | — | 0.14 to 0.19 | |
| Perlite board | 9 | 0.36 | — | — | One manufacturer (2010) |
| Polyisocyanurate | — | — | — | 0.35 | Kumaran (1996) |
| unfaced, aged per Can/ULC Standard S770-2003 | 1.6 to 2.3 | 0.16 to 0.17 | — | — | Seven manufacturers (2011) |
| with foil facers, aged 180 days | — | 0.15 to 0.16 | — | — | Two manufacturers (2011) |
| Phenolic foam board with facers, aged | — | 0.14 to 0.16 | — | — | One manufacturer (2011) |
| *Loose fill* | | | | | |
| Cellulose fiber, loose fill | — | — | — | 0.33 | NIST (2000), Kumaran (1996) |
| attic application up to 4 in. | 1.0 to 1.2 | 0.31 to 0.32 | — | — | Four manufacturers (2011) |
| attic application > 4 in. | 1.2 to 1.6 | 0.27 to 0.28 | — | — | Four manufacturers (2011) |
| wall application, densely packed | 3.5 | 0.27 – 0.28 | — | — | One manufacturer (2011) |
| Perlite, expanded | 2 to 4 | 0.27 to 0.31 | — | 0.26 | (Manufacturer, pre-2001) |
| | 4 to 7.5 | 0.31 to 0.36 | — | — | (Manufacturer, pre-2001) |
| | 7.5 to 11 | 0.36 to 0.42 | — | — | (Manufacturer, pre-2001) |
| Glass fiber[d] | | | | | |
| attics, ~4 to 12 in. | 0.4 to 0.5 | 0.36 to 0.38 | — | — | Four manufacturers (2011) |
| attics, ~12 to 22 in. | 0.5 to 0.6 | 0.34 to 0.36 | — | — | Four manufacturers (2011) |
| closed attic or wall cavities | 1.8 to 2.3 | 0.24 to 0.25 | — | — | Four manufacturers (2011) |
| Rock and slag wool[d] | | | | | |
| attics, ~3.5 to 4.5 in. | 1.5 to 1.6 | 0.34 | — | — | Three manufacturers (2011) |
| attics, ~5 to 17 in. | 1.5 to 1.8 | 0.32 to 0.33 | — | — | Three manufacturers (2011) |
| closed attic or wall cavities | 4.0 | 0.27 to 0.29 | — | — | Three manufacturers (2011) |
| Vermiculite, exfoliated | 7.0 to 8.2 | 0.47 | — | 0.32 | Sabine et al. (1975) |
| | 4.0 to 6.0 | 0.44 | — | — | Manufacturer (pre-2001) |
| *Spray applied* | | | | | |
| Cellulose, sprayed into open wall cavities | 1.6 to 2.6 | 0.27 to 0.28 | — | — | Two manufacturers (2011) |
| Glass fiber, sprayed into open wall or attic cavities | 1.0 | 0.27 to 0.29 | — | — | Manufacturers' association (2011) |
| | 1.8 to 2.3 | 0.23 to 0.26 | — | — | Four manufacturers (2011) |
| Polyurethane foam | — | — | — | 0.35 | Kumaran (2002) |
| low density, open cell | 0.45 to 0.65 | 0.26 to 0.29 | — | — | Three manufacturers (2011) |

## Load Estimating Fundamentals

**Table 5-15  Building and Insulating Materials: Design Values[a]** *(Continued)*
*(Table 1, Chapter 26, 2013 ASHRAE Handbook—Fundamentals)*

| Description | Density, lb/ft$^3$ | Conductivity[b] $k$, Btu·in/h·ft$^2$·°F | Resistance $R$, h·ft$^2$·°F/Btu | Specific Heat, Btu/lb·°F | Reference[l] |
|---|---|---|---|---|---|
| medium density, closed cell, aged 180 days | 1.9 to 3.2 | 0.14 to 0.20 | — | — | Five manufacturers (2011) |
| **Building Board and Siding** | | | | | |
| *Board* | | | | | |
| Asbestos/cement board | 120 | 4 | — | 0.24 | Nottage (1947) |
| Cement board | 71 | 1.7 | — | 0.2 | Kumaran (2002) |
| Fiber/cement board | 88 | 1.7 | — | 0.2 | Kumaran (2002) |
|  | 61 | 1.3 | — | 0.2 | Kumaran (1996) |
|  | 26 | 0.5 | — | 0.45 | Kumaran (1996) |
|  | 20 | 0.4 | — | 0.45 | Kumaran (1996) |
| Gypsum or plaster board | 40 | 1.1 | — | 0.21 | Kumaran (2002) |
| Oriented strand board (OSB) ............ 7/16 in. | 41 | — | 0.62 | 0.45 | Kumaran (2002) |
| ............................................................ 1/2 in. | 41 | — | 0.68 | 0.45 | Kumaran (2002) |
| Plywood (douglas fir) ......................... 1/2 in. | 29 | — | 0.79 | 0.45 | Kumaran (2002) |
| ............................................................ 5/8 in. | 34 | — | 0.85 | 0.45 | Kumaran (2002) |
| Plywood/wood panels ......................... 3/4 in. | 28 | — | 1.08 | 0.45 | Kumaran (2002) |
| Vegetable fiber board | | | | | |
| sheathing, regular density ............. 1/2 in. | 18 | — | 1.32 | 0.31 | Lewis (1967) |
| intermediate density ................. 1/2 in. | 22 | — | 1.09 | 0.31 | Lewis (1967) |
| nail-based sheathing .................... 1/2 in. | 25 | — | 1.06 | 0.31 | |
| shingle backer ............................... 3/8 in. | 18 | — | 0.94 | 0.3 | |
| sound-deadening board ................ 1/2 in. | 15 | — | 1.35 | 0.3 | |
| tile and lay-in panels, plain or acoustic | 18 | 0.4 | — | 0.14 | |
| laminated paperboard | 30 | 0.5 | — | 0.33 | Lewis (1967) |
| homogeneous board from repulped paper | 30 | 0.5 | — | 0.28 | |
| Hardboard | | | | | |
| medium density | 50 | 0.73 | — | 0.31 | Lewis (1967) |
| high density, service-tempered and service grades | 55 | 0.82 | — | 0.32 | Lewis (1967) |
| high density, standard-tempered grade | 63 | 1.0 | — | 0.32 | Lewis (1967) |
| Particleboard | | | | | |
| low density | 37 | 0.71 | — | 0.31 | Lewis (1967) |
| medium density | 50 | 0.94 | — | 0.31 | Lewis (1967) |
| high density | 62 | 1.18 | 0.85 | — | Lewis (1967) |
| underlayment ................................ 5/8 in. | 44 | 0.73 | 0.82 | 0.29 | Lewis (1967) |
| Waferboard | 37 | 0.63 | 0.21 | 0.45 | Kumaran (1996) |
| *Shingles* | | | | | |
| Asbestos/cement | 120 | — | 0.21 | — | |
| Wood, 16 in., 7 1/2 in. exposure | — | — | 0.87 | 0.31 | |
| Wood, double, 16 in., 12 in. exposure | — | — | 1.19 | 0.28 | |
| Wood, plus ins. backer board ........... 5/16 in. | — | — | 1.4 | 0.31 | |
| *Siding* | | | | | |
| Asbestos/cement, lapped ..................... 1/4 in. | — | — | 0.21 | 0.24 | |
| Asphalt roll siding | — | — | 0.15 | 0.35 | |
| Asphalt insulating siding (1/2 in. bed) | — | — | 0.21 | 0.24 | |
| Hardboard siding ................................ 7/16 in. | — | — | 0.15 | 0.35 | |
| Wood, drop, 8 in. ..................................... 1 in. | — | — | 0.79 | 0.28 | |
| Wood, bevel | | | | | |
| 8 in., lapped .................................... 1/2 in. | — | — | 0.81 | 0.28 | |
| 10 in., lapped .................................. 3/4 in. | — | — | 1.05 | 0.28 | |
| Wood, plywood, 3/8 in., lapped | — | — | 0.59 | 0.29 | |
| Aluminum, steel, or vinyl,[h, i] over sheathing | | | | — | |
| hollow-backed | — | — | 0.62 | 0.29[i] | |
| insulating-board-backed .................. 3/8 in. | — | — | 1.82 | 0.32 | |
| foil-backed ........................................ 3/8 in. | — | — | 2.96 | — | |
| Architectural (soda-lime float) glass | 158 | 6.9 | — | 0.21 | |
| **Building Membrane** | | | | | |
| Vapor-permeable felt | — | — | 0.06 | — | |
| Vapor: seal, 2 layers of mopped 15 lb felt | — | — | 0.12 | — | |
| Vapor: seal, plastic film | — | — | Negligible | — | |
| **Finish Flooring Materials** | | | | | |
| Carpet and rebounded urethane pad .......... 3/4 in. | 7 | — | 2.38 | — | NIST (2000) |
| Carpet and rubber pad (one-piece) ............. 3/8 in. | 20 | — | 0.68 | — | NIST (2000) |
| Pile carpet with rubber pad ............ 3/8 to 1/2 in. | 18 | — | 1.59 | — | NIST (2000) |
| Linoleum/cork tile ........................................ 1/4 in. | 29 | — | 0.51 | — | NIST (2000) |
| PVC/rubber floor covering | — | 2.8 | — | — | CIBSE (2006) |
| rubber tile ................................................. 1.0 in. | 119 | — | 0.34 | — | NIST (2000) |
| terrazzo ..................................................... 1.0 in. | — | — | 0.08 | 0.19 | |
| **Metals** (See Chapter 33, Table 3) | | | | | |

**Table 5-15 Building and Insulating Materials: Design Values[a] (Continued)**
*(Table 1, Chapter 26, 2013 ASHRAE Handbook—Fundamentals)*

| Description | Density, lb/ft³ | Conductivity[b] k, Btu·in/h·ft²·°F | Resistance R, h·ft²·°F/Btu | Specific Heat, Btu/lb·°F | Reference[l] |
|---|---|---|---|---|---|
| **Roofing** | | | | | |
| Asbestos/cement shingles | 120 | — | 0.21 | 0.24 | |
| Asphalt (bitumen with inert fill) | 100 | 2.98 | — | — | CIBSE (2006) |
| | 119 | 4.0 | — | — | CIBSE (2006) |
| | 144 | 7.97 | — | — | CIBSE (2006) |
| Asphalt roll roofing | 70 | — | 0.15 | 0.36 | |
| Asphalt shingles | 70 | — | 0.44 | 0.3 | |
| Built-up roofing............3/8 in. | 70 | — | 0.33 | 0.35 | |
| Mastic asphalt (heavy, 20% grit) | 59 | 1.32 | — | — | CIBSE (2006) |
| Reed thatch | 17 | 0.62 | — | — | CIBSE (2006) |
| Roofing felt | 141 | 8.32 | — | — | CIBSE (2006) |
| Slate............1/2 in. | — | — | 0.05 | 0.3 | |
| Straw thatch | 15 | 0.49 | — | — | CIBSE (2006) |
| Wood shingles, plain and plastic-film-faced | — | — | 0.94 | 0.31 | |
| **Plastering Materials** | | | | | |
| Cement plaster, sand aggregate | 116 | 5.0 | — | 0.2 | |
| Sand aggregate............3/8 in. | — | — | 0.08 | 0.2 | |
| ............3/4 in. | — | — | 0.15 | 0.2 | |
| Gypsum plaster | 70 | 2.63 | — | — | CIBSE (2006) |
| | 80 | 3.19 | — | — | CIBSE (2006) |
| Lightweight aggregate............1/2 in. | 45 | — | 0.32 | — | |
| ............5/8 in. | 45 | — | 0.39 | — | |
| on metal lath............3/4 in. | — | — | 0.47 | — | |
| Perlite aggregate | 45 | 1.5 | — | 0.32 | |
| Sand aggregate | 105 | 5.6 | — | 0.2 | |
| on metal lath............3/4 in. | — | — | 0.13 | — | |
| Vermiculite aggregate | 30 | 1.0 | — | — | CIBSE (2006) |
| | 40 | 1.39 | — | — | CIBSE (2006) |
| | 45 | 1.7 | — | — | CIBSE (2006) |
| | 50 | 1.8 | — | — | CIBSE (2006) |
| | 60 | 2.08 | — | — | CIBSE (2006) |
| Perlite plaster | 25 | 0.55 | — | — | CIBSE (2006) |
| | 38 | 1.32 | — | — | CIBSE (2006) |
| Pulpboard or paper plaster | 38 | 0.48 | — | — | CIBSE (2006) |
| Sand/cement plaster, conditioned | 98 | 4.4 | — | — | CIBSE (2006) |
| Sand/cement/lime plaster, conditioned | 90 | 3.33 | — | — | CIBSE (2006) |
| Sand/gypsum (3:1) plaster, conditioned | 97 | 4.5 | — | — | CIBSE (2006) |
| **Masonry Materials** | | | | | |
| *Masonry units* | | | | | |
| Brick, fired clay | 150 | 8.4 to 10.2 | — | — | Valore (1988) |
| | 140 | 7.4 to 9.0 | — | — | Valore (1988) |
| | 130 | 6.4 to 7.8 | — | — | Valore (1988) |
| | 120 | 5.6 to 6.8 | — | 0.19 | Valore (1988) |
| | 110 | 4.9 to 5.9 | — | — | Valore (1988) |
| | 100 | 4.2 to 5.1 | — | — | Valore (1988) |
| | 90 | 3.6 to 4.3 | — | — | Valore (1988) |
| | 80 | 3.0 to 3.7 | — | — | Valore (1988) |
| | 70 | 2.5 to 3.1 | — | — | Valore (1988) |
| Clay tile, hollow | | | | | |
| 1 cell deep............3 in. | — | — | 0.80 | 0.21 | Rowley and Algren (1937) |
| ............4 in. | — | — | 1.11 | — | Rowley and Algren (1937) |
| 2 cells deep............6 in. | — | — | 1.52 | — | Rowley and Algren (1937) |
| ............8 in. | — | — | 1.85 | — | Rowley and Algren (1937) |
| ............10 in. | — | — | 2.22 | — | Rowley and Algren (1937) |
| 3 cells deep............12 in. | — | — | 2.50 | — | Rowley and Algren (1937) |
| Lightweight brick | 50 | 1.39 | — | — | Kumaran (1996) |
| | 48 | 1.51 | — | — | Kumaran (1996) |
| *Concrete blocks*[f, g] | | | | | |
| Limestone aggregate | | | | | |
| 8 in., 36 lb, 138 lb/ft³ concrete, 2 cores | — | — | — | — | |
| with perlite-filled cores | — | — | 2.1 | — | Valore (1988) |
| 12 in., 55 lb, 138 lb/ft³ concrete, 2 cores | — | — | — | — | |
| with perlite-filled cores | — | — | 3.7 | — | Valore (1988) |
| Normal-weight aggregate (sand and gravel) | | | | | |
| 8 in., 33 to 36 lb, 126 to 136 lb/ft³ concrete, 2 or 3 cores | — | — | 1.11 to 0.97 | 0.22 | Van Geem (1985) |
| with perlite-filled cores | — | — | 2.0 | — | Van Geem (1985) |
| with vermiculite-filled cores | — | — | 1.92 to 1.37 | — | Valore (1988) |

## Load Estimating Fundamentals

**Table 5-15  Building and Insulating Materials: Design Values[a] *(Continued)***

*(Table 1, Chapter 26, 2013 ASHRAE Handbook—Fundamentals)*

| Description | Density, lb/ft³ | Conductivity[b] $k$, Btu·in/h·ft²·°F | Resistance $R$, h·ft²·°F/Btu | Specific Heat, Btu/lb·°F | Reference[l] |
|---|---|---|---|---|---|
| 12 in., 50 lb, 125 lb/ft³ concrete, 2 cores | — | — | 1.23 | 0.22 | Valore (1988) |
| Medium-weight aggregate (combinations of normal and lightweight aggregate) | | | | | |
| 8 in., 26 to 29 lb, 97 to 112 lb/ft³ concrete, 2 or 3 cores | — | — | 1.71 to 1.28 | — | Van Geem (1985) |
| with perlite-filled cores | — | — | 3.7 to 2.3 | — | Van Geem (1985) |
| with vermiculite-filled cores | — | — | 3.3 | — | Van Geem (1985) |
| with molded-EPS-filled (beads) cores | — | — | 3.2 | — | Van Geem (1985) |
| with molded EPS inserts in cores | — | — | 2.7 | — | Van Geem (1985) |
| Lightweight aggregate (expanded shale, clay, slate or slag, pumice) | | | | | |
| 6 in., 16 to 17 lb, 85 to 87 lb/ft³ concrete, 2 or 3 cores | — | — | 1.93 to 1.65 | — | Van Geem (1985) |
| with perlite-filled cores | — | — | 4.2 | — | Van Geem (1985) |
| with vermiculite-filled cores | — | — | 3.0 | — | Van Geem (1985) |
| 8 in., 19 to 22 lb, 72 to 86 lb/ft³ concrete | — | — | 3.2 to 1.90 | 0.21 | Van Geem (1985) |
| with perlite-filled cores | — | — | 6.8 to 4.4 | — | Van Geem (1985) |
| with vermiculite-filled cores | — | — | 5.3 to 3.9 | — | Shu et al. (1979) |
| with molded-EPS-filled (beads) cores | — | — | 4.8 | — | Shu et al. (1979) |
| with UF foam-filled cores | — | — | 4.5 | — | Shu et al. (1979) |
| with molded EPS inserts in cores | — | — | 3.5 | — | Shu et al. (1979) |
| 12 in., 32 to 36 lb, 80 to 90 lb/ft³, concrete, 2 or 3 cores | — | — | 2.6 to 2.3 | — | Van Geem (1985) |
| with perlite-filled cores | — | — | 9.2 to 6.3 | — | Van Geem (1985) |
| with vermiculite-filled cores | — | — | 5.8 | — | Valore (1988) |
| Stone, lime, or sand | 180 | 72 | — | — | Valore (1988) |
| Quartzitic and sandstone | 160 | 43 | — | — | Valore (1988) |
|  | 140 | 24 | — | — | Valore (1988) |
|  | 120 | 13 | — | 0.19 | Valore (1988) |
| Calcitic, dolomitic, limestone, marble, and granite | 180 | 30 | — | — | Valore (1988) |
|  | 160 | 22 | — | — | Valore (1988) |
|  | 140 | 16 | — | — | Valore (1988) |
|  | 120 | 11 | — | 0.19 | Valore (1988) |
|  | 100 | 8 | — | — | Valore (1988) |
| Gypsum partition tile | | | | | |
| 3 by 12 by 30 in., solid | — | — | 1.26 | 0.19 | Rowley and Algren (1937) |
| 4 cells | — | — | 1.35 | — | Rowley and Algren (1937) |
| 4 by 12 by 30 in., 3 cells | — | — | 1.67 | — | Rowley and Algren (1937) |
| Limestone | 150 | 3.95 | — | 0.2 | Kumaran (2002) |
|  | 163 | 6.45 | — | 0.2 | Kumaran (2002) |
| *Concretes*[i] | | | | | |
| Sand and gravel or stone aggregate concretes | 150 | 10.0 to 20.0 | — | — | Valore (1988) |
| (concretes with >50% quartz or quartzite sand have | 140 | 9.0 to 18.0 | — | 0.19 to 0.24 | Valore (1988) |
| conductivities in higher end of range) | 130 | 7.0 to 13.0 | — | — | Valore (1988) |
| Lightweight aggregate or limestone concretes | 120 | 6.4 to 9.1 | — | — | Valore (1988) |
| expanded shale, clay, or slate; expanded slags; cinders; | 100 | 4.7 to 6.2 | — | 0.2 | Valore (1988) |
| pumice (with density up to 100 lb/ft³); scoria (sanded | 80 | 3.3 to 4.1 | — | 0.2 | Valore (1988) |
| concretes have conductivities in higher end of range) | 60 | 2.1 to 2.5 | — | — | Valore (1988) |
|  | 40 | 1.3 | — | — | Valore (1988) |
| Gypsum/fiber concrete (87.5% gypsum, 12.5% wood chips) | 51 | 1.66 | — | 0.2 | Rowley and Algren (1937) |
| Cement/lime, mortar, and stucco | 120 | 9.7 | — | — | Valore (1988) |
|  | 100 | 6.7 | — | — | Valore (1988) |
|  | 80 | 4.5 | — | — | Valore (1988) |
| Perlite, vermiculite, and polystyrene beads | 50 | 1.8 to 1.9 | — | — | Valore (1988) |
|  | 40 | 1.4 to 1.5 | — | 0.15 to 0.23 | Valore (1988) |
|  | 30 | 1.1 | — | — | Valore (1988) |
|  | 20 | 0.8 | — | — | Valore (1988) |
| Foam concretes | 120 | 5.4 | — | — | Valore (1988) |
|  | 100 | 4.1 | — | — | Valore (1988) |
|  | 80 | 3.0 | — | — | Valore (1988) |
|  | 70 | 2.5 | — | — | Valore (1988) |
| Foam concretes and cellular concretes | 60 | 2.1 | — | — | Valore (1988) |
|  | 40 | 1.4 | — | — | Valore (1988) |
|  | 20 | 0.8 | — | — | Valore (1988) |
| Aerated concrete (oven-dried) | 27 to 50 | 1.4 | — | 0.2 | Kumaran (1996) |
| Polystyrene concrete (oven-dried) | 16 to 50 | 2.54 | — | 0.2 | Kumaran (1996) |
| Polymer concrete | 122 | 11.4 | — | — | Kumaran (1996) |
|  | 138 | 7.14 | — | — | Kumaran (1996) |
| Polymer cement | 117 | 5.39 | — | — | Kumaran (1996) |
| Slag concrete | 60 | 1.5 | — | — | Touloukian et al (1970) |
|  | 80 | 2.25 | — | — | Touloukian et al. (1970) |
|  | 100 | 3 | — | — | Touloukian et al. (1970) |

**168**  Principles of HVAC

**Table 5-15  Building and Insulating Materials: Design Values[a] *(Continued)***
*(Table 1, Chapter 26, 2013 ASHRAE Handbook—Fundamentals)*

| Description | Density, lb/ft$^3$ | Conductivity[b] $k$, Btu·in/h·ft$^2$·°F | Resistance $R$, h·ft$^2$·°F/Btu | Specific Heat, Btu/lb·°F | Reference[l] |
|---|---|---|---|---|---|
| | 125 | 8.53 | — | — | Touloukian et al. (1970) |
| **Woods** (12% moisture content)[j] | | | | | |
| *Hardwoods* | — | — | — | 0.39[k] | Wilkes (1979) |
| Oak | 41 to 47 | 1.12 to 1.25 | — | — | Cardenas and Bible (1987) |
| Birch | 43 to 45 | 1.16 to 1.22 | — | — | Cardenas and Bible (1987) |
| Maple | 40 to 44 | 1.09 to 1.19 | — | — | Cardenas and Bible (1987) |
| Ash | 38 to 42 | 1.06 to 1.14 | — | — | Cardenas and Bible (1987) |
| *Softwoods* | — | — | — | 0.39[k] | Wilkes (1979) |
| Southern pine | 36 to 41 | 1.00 to 1.12 | — | — | Cardenas and Bible (1987) |
| Southern yellow pine | 31 | 1.06 to 1.16 | — | — | Kumaran (2002) |
| Eastern white pine | 25 | 0.85 to 0.94 | — | — | Kumaran (2002) |
| Douglas fir/larch | 34 to 36 | 0.95 to 1.01 | — | — | Cardenas and Bible (1987) |
| Southern cypress | 31 to 32 | 0.90 to 0.92 | — | — | Cardenas and Bible (1987) |
| Hem/fir, spruce/pine/fir | 24 to 31 | 0.74 to 0.90 | — | — | Cardenas and Bible (1987) |
| Spruce | 25 | 0.74 to 0.85 | — | — | Kumaran (2002) |
| Western red cedar | 22 | 0.83 to 0.86 | — | — | Kumaran (2002) |
| West coast woods, cedars | 22 to 31 | 0.68 to 0.90 | — | — | Cardenas and Bible (1987) |
| Eastern white cedar | 23 | 0.82 to 0.89 | — | — | Kumaran (2002) |
| California redwood | 24 to 28 | 0.74 to 0.82 | — | — | Cardenas and Bible (1987) |
| Pine (oven-dried) | 23 | 0.64 | — | 0.45 | Kumaran (1996) |
| Spruce (oven-dried) | 25 | 0.69 | — | 0.45 | Kumaran (1996) |

**Notes for Table 5-15**

[a] Values are for mean temperature of 75°F (24°C). Representative values for dry materials are intended as design (not specification) values for materials in normal use. Thermal values of insulating materials may differ from design values depending on in-situ properties (e.g., density and moisture content, orientation, etc.) and manufacturing variability. For properties of specific product, use values supplied by manufacturer or unbiased tests.

[b] Symbol λ also used to represent thermal conductivity.

[c] Does not include paper backing and facing, if any. Where insulation forms boundary (reflective or otherwise) of airspace, see Tables 2 and 3 in Chapter 26 of 2013 *ASHRAE Handbook—Fundamentals* for insulating value of airspace with appropriate effective emittance and temperature conditions of space.

[d] Conductivity varies with fiber diameter (see Chapter 25). Batt, blanket, and loose-fill mineral fiber insulations are manufactured to achieve specified R-values, the most common of which are listed in the table. Because of differences in manufacturing processes and materials, the product thicknesses, densities, and thermal conductivities vary over considerable ranges for a specified R-value.

[e] Insulating values of acoustical tile vary, depending on density of board and on type, size, and depth of perforations.

[f] Values for fully grouted block may be approximated using values for concrete with similar unit density.

[g] Values for concrete block and concrete are at moisture contents representative of normal use.

[h] Values for metal or vinyl siding applied over flat surfaces vary widely, depending on ventilation of the airspace beneath the siding; whether airspace is reflective or nonreflective; and on thickness, type, and application of insulating backing-board used. Values are averages for use as design guides, and were obtained from several guarded hot box tests (ASTM Standard C1363) on hollow-backed types and types made using backing of wood fiber, foamed plastic, and glass fiber. Departures of ±50% or more from these values may occur.

[i] Vinyl specific heat = 0.25 Btu/lb·°F 1.0 kJ/(kg·K)

[j] See Adams (1971), MacLean (1941), and Wilkes (1979). Conductivity values listed are for heat transfer across the grain. Thermal conductivity of wood varies linearly with density, and density ranges listed are those normally found for wood species given. If density of wood species is not known, use mean conductivity value. For extrapolation to other moisture contents, the following empirical equation developed by Wilkes (1979) may be used:

$$k = 0.1791 + \frac{(1.874 \times 10^{-2} + 5.753 \times 10^{-4} M)\rho}{1 + 0.01 M}$$

where ρ is density of moist wood in lb/ft$^3$kg/m$^3$, and $M$ is moisture content in percent.

[k] From Wilkes (1979), an empirical equation for specific heat of moist wood at 75°F (24°C) is as follows:

$$c_p = \frac{(0.299 + 0.01 M)}{(1 + 0.01 M)} + \Delta c_p$$

where $\Delta c_p$ accounts for heat of sorption and is denoted by

$$\Delta c_p = M(1.921 \times 10^{-3} - 3.168 \times 10^{-5} M)$$

where $M$ is moisture content in percent by mass.

[l] Blank space in reference column indicates historical values from previous volumes of *ASHRAE Handbook*. Source of information could not be determined.

---

example, the listed U-factors assume the surface temperatures of surrounding bodies are equal to the ambient air temperature.

Most fenestration products consist of transparent multipane glazing units and opaque elements comprising the sash and frame (hereafter called frame). The glazing unit's heat transfer paths include a one-dimensional center-of-glass contribution and a two-dimensional edge contribution. The frame contribution is primarily two-dimensional.

Consequently, the total heat transfer can be determined by calculating the separate heat transfer contributions of the center glass, edge glass, and frame. (When present, glazing dividers, such as decorative grilles and muntins, also affect heat transfer, and their contribution must be considered.) The overall U-factor may be estimated by adding the area-weighted U-factors for each contribution.

Table 5-16 lists computed U-factors for a variety of generic fenestration products, *which should only be used as an estimating tool for the early phases of design*. The table is based on ASHRAE-sponsored research involving laboratory testing and computer simulation of various fenestration products. Consequently, computer simulations (with high/low validation by testing) are now accepted as a standard method for determining accurate, product-specific U-factors.

While these U-factors have been determined for winter conditions, they can also be used to estimate heat gain during peak cooling conditions, because conductive gain is usually a small portion of the total heat gain for fenestration in direct sunlight. Glazing designs and framing materials may be compared in choosing a product that needs a specific winter design U-factor.

Table 5-16 lists 48 types of glazing. The multiple glazing categories are appropriate for sealed glass units and the

# Load Estimating Fundamentals

combinations of storm sash and other glazing units. Unless other-wise noted, all multiple-glazed units are filled with dry air.

Several frame types are listed (through not all for any one category), in order of improving thermal performance. The most conservative frame to assume is the aluminum frame **without thermal break** (although some products on the market have higher U-factors). The aluminum frame with thermal break has at least 3/8 in. (10 mm) thermal break between the inside and outside for all members including both the frame and the operable sash, if applicable. (Products are available with significantly wider thermal breaks, which reduce heat flow considerably.)

The **reinforced vinyl/aluminum clad wood** category represents vinyl-frame products, such as sliding glass doors or large windows. These units have extensive metal reinforcing in the frame and wood products with extensive metal, usually on the exterior surface of the frame. The metal, of course, degrades the thermal performance of the frame material.

The **wood/vinyl frame** represents improved thermal performance over reinforced vinyl/aluminum clad wood. **Insulated fiberglass/vinyl frames** do not have metal reinforcing and the frame cavities are filled with insulation.

Shading devices, such as venetian blinds, draperies, and roller shades, substantially reduce the U-factor for windows and/or glass doors if they fit tightly to the window jambs, head, and sill, and are made of nonporous material. As a rough approximation, tight-fitting shading devices may be considered to reduce the U-factor of vertical exterior single glazing by 25% and of vertical exterior double glazing and glass block by 15%. These adjustments should not be considered in choosing heating equipment, but may be used for calculating design cooling loads.

## 5.3.3 The Overall Thermal Transmittance

$U_o$ is the combined thermal transmittance of the respective areas of gross exterior wall, roof or ceiling or both, and floor assemblies. The overall thermal transmittance of the building envelope assembly shall be calculated from

$$U_o = \Sigma U_i A_i / A_o$$
$$= (U_1 A_1 + U_2 A_2 + \ldots + U_n A_n)/A_o \quad (5\text{-}13)$$

where

$U_o$ = the area-weighted average thermal transmittance of the gross area of an envelope assembly; i.e., the exterior wall assembly including fenestration and doors, the roof and ceiling assembly, and the floor assembly, Btu/(h·ft²·°F)

$A_o$ = The gross area of the envelope assembly, ft²

$U_i$ = the thermal transmittance of each individual path of the envelope assembly, i.e., the opaque portion or the fenestration, Btu/(h·ft²·°F)

$U_i = 1/R_i$ (where $R_i$ is the total resistance to heat flow of an individual path through an envelope assembly).

$A_i$ = the area of each individual element of the envelope assembly, ft²

## 5.4 Calculating Surface Temperatures

The temperature at any interface can be calculated, since the temperature drop through any component of the wall is proportional to its resistance. Thus, the temperature drop $\Delta t$ through $R_1$ is

$$\Delta t_1 = \frac{R_1(t_i - t_0)}{R_T} \quad (5\text{-}14)$$

where $t_i$ and $t_0$ are the indoor and outdoor temperatures, respectively. Hence, the temperature at the interface between $R_1$ and $R_2$ is

$$t_{1-2} = t_i - \Delta t_1$$

If the resistances of materials in a wall are highly dependent on temperature, the mean temperature must be known to assign the correct value. In such cases, it is perhaps most convenient to use a trial-and-error procedure for the calculation of the total resistance $R_T$. First the mean operating temperature for each layer is estimated, and R-values for the particular materials are selected. The total resistance $R_T$ is then calculated and the temperature at each interface is calculated.

The mean temperature of each component (arithmetic mean of its surface temperatures) can then be used to obtain second generation R-values. This procedure can then be repeated until the R-values have been correctly selected for the resulting mean temperatures. Generally, this can be done in two or three trial calculations.

Figure 5-17 illustrates the procedure for determining the temperatures throughout the structure.

| Item | R | $R_T$ | U |
|---|---|---|---|
| ① Outdoor air film | 0.17 | 0.17 | |
| ② 8 in. concrete block, light aggregate vermiculite filled core—specified | 4.6 | 4.77 | |
| ③ 3 1/2 in. mineral fiber insulation—specified | 11.0 | 15.77 | |
| ④ 1/2 in. mineral fiber insulation—specified | 0.45 | 16.22 | |
| ⑤ Inside air film | 0.68 | 16.9 | 0.059 |

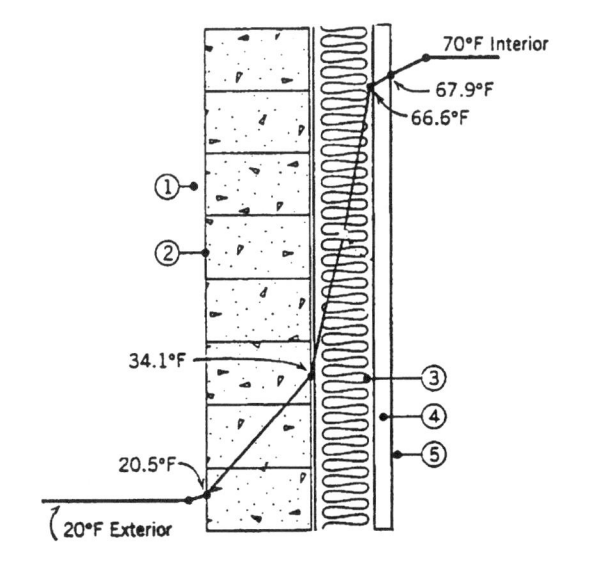

*Fig. 5-17 Temperatures Throughout Wall*

## Table 5-16 U-Factors for Various Fenestration Products in Btu/h·ft²·°F

*(Table 4, Chapter 15, 2013 ASHRAE Handbook—Fundamentals)*

**Vertical Installation**

| | | Glass Only | | Operable (including sliding and swinging glass doors) | | | | | Fixed | | | | |
|---|---|---|---|---|---|---|---|---|---|---|---|---|---|
| | Frame Type | Center of Glass | Edge of Glass | Aluminum Without Thermal Break | Aluminum with Thermal Break | Reinforced Vinyl/ Aluminum Clad Wood | Wood/ Vinyl | Insulated Fiberglass/ Vinyl | Aluminum Without Thermal Break | Aluminum with Thermal Break | Reinforced Vinyl/ Aluminum Clad Wood | Wood/ Vinyl | Insulated Fiberglass/ Vinyl |
| ID | Glazing Type | | | | | | | | | | | | |
| | **Single Glazing** | | | | | | | | | | | | |
| 1 | 1/8 in. glass | 1.04 | 1.04 | 1.23 | 1.07 | 0.93 | 0.91 | 0.85 | 1.12 | 1.07 | 0.98 | 0.98 | 1.04 |
| 2 | 1/4 in. acrylic/polycarbonate | 0.88 | 0.88 | 1.10 | 0.94 | 0.81 | 0.80 | 0.74 | 0.98 | 0.92 | 0.84 | 0.84 | 0.88 |
| 3 | 1/8 in. acrylic/polycarbonate | 0.96 | 0.96 | 1.17 | 1.01 | 0.87 | 0.86 | 0.79 | 1.05 | 0.99 | 0.91 | 0.91 | 0.96 |
| | **Double Glazing** | | | | | | | | | | | | |
| 4 | 1/4 in. air space | 0.55 | 0.64 | 0.81 | 0.64 | 0.57 | 0.55 | 0.50 | 0.68 | 0.62 | 0.56 | 0.56 | 0.55 |
| 5 | 1/2 in. air space | 0.48 | 0.59 | 0.76 | 0.58 | 0.52 | 0.50 | 0.45 | 0.62 | 0.56 | 0.50 | 0.50 | 0.48 |
| 6 | 1/4 in. argon space | 0.51 | 0.61 | 0.78 | 0.61 | 0.54 | 0.52 | 0.47 | 0.65 | 0.59 | 0.53 | 0.52 | 0.51 |
| 7 | 1/2 in. argon space | 0.45 | 0.57 | 0.73 | 0.56 | 0.50 | 0.48 | 0.43 | 0.60 | 0.53 | 0.48 | 0.47 | 0.45 |
| | **Double Glazing, $e = 0.60$ on surface 2 or 3** | | | | | | | | | | | | |
| 8 | 1/4 in. air space | 0.52 | 0.62 | 0.79 | 0.61 | 0.55 | 0.53 | 0.48 | 0.66 | 0.59 | 0.54 | 0.53 | 0.52 |
| 9 | 1/2 in. air space | 0.44 | 0.56 | 0.72 | 0.55 | 0.49 | 0.48 | 0.43 | 0.59 | 0.53 | 0.47 | 0.47 | 0.44 |
| 10 | 1/4 in. argon space | 0.47 | 0.58 | 0.75 | 0.57 | 0.51 | 0.50 | 0.45 | 0.61 | 0.55 | 0.49 | 0.49 | 0.47 |
| 11 | 1/2 in. argon space | 0.41 | 0.54 | 0.70 | 0.53 | 0.47 | 0.45 | 0.41 | 0.56 | 0.50 | 0.44 | 0.44 | 0.41 |
| | **Double Glazing, $e = 0.40$ on surface 2 or 3** | | | | | | | | | | | | |
| 12 | 1/4 in. air space | 0.49 | 0.60 | 0.76 | 0.59 | 0.53 | 0.51 | 0.46 | 0.63 | 0.57 | 0.51 | 0.51 | 0.49 |
| 13 | 1/2 in. air space | 0.40 | 0.54 | 0.69 | 0.52 | 0.47 | 0.45 | 0.40 | 0.55 | 0.49 | 0.44 | 0.43 | 0.40 |
| 14 | 1/4 in. argon space | 0.43 | 0.56 | 0.72 | 0.54 | 0.49 | 0.47 | 0.42 | 0.58 | 0.52 | 0.46 | 0.46 | 0.43 |
| 15 | 1/2 in. argon space | 0.36 | 0.51 | 0.66 | 0.49 | 0.44 | 0.42 | 0.37 | 0.52 | 0.46 | 0.40 | 0.40 | 0.36 |
| | **Double Glazing, $e = 0.20$ on surface 2 or 3** | | | | | | | | | | | | |
| 16 | 1/4 in. air space | 0.45 | 0.57 | 0.73 | 0.56 | 0.50 | 0.48 | 0.43 | 0.60 | 0.53 | 0.48 | 0.47 | 0.45 |
| 17 | 1/2 in. air space | 0.35 | 0.50 | 0.65 | 0.48 | 0.43 | 0.41 | 0.37 | 0.51 | 0.45 | 0.39 | 0.39 | 0.35 |
| 18 | 1/4 in. argon space | 0.38 | 0.52 | 0.68 | 0.51 | 0.45 | 0.43 | 0.39 | 0.54 | 0.47 | 0.42 | 0.42 | 0.38 |
| 19 | 1/2 in. argon space | 0.30 | 0.46 | 0.61 | 0.45 | 0.39 | 0.38 | 0.33 | 0.47 | 0.41 | 0.35 | 0.35 | 0.30 |
| | **Double Glazing, $e = 0.10$ on surface 2 or 3** | | | | | | | | | | | | |
| 20 | 1/4 in. air space | 0.42 | 0.55 | 0.71 | 0.54 | 0.48 | 0.46 | 0.41 | 0.57 | 0.51 | 0.45 | 0.45 | 0.42 |
| 21 | 1/2 in. air space | 0.32 | 0.48 | 0.63 | 0.46 | 0.41 | 0.39 | 0.34 | 0.49 | 0.42 | 0.37 | 0.37 | 0.32 |
| 22 | 1/4 in. argon space | 0.35 | 0.50 | 0.65 | 0.48 | 0.43 | 0.41 | 0.37 | 0.51 | 0.45 | 0.39 | 0.39 | 0.35 |
| 23 | 1/2 in. argon space | 0.27 | 0.44 | 0.59 | 0.42 | 0.37 | 0.36 | 0.31 | 0.44 | 0.38 | 0.33 | 0.32 | 0.27 |
| | **Double Glazing, $e = 0.05$ on surface 2 or 3** | | | | | | | | | | | | |
| 24 | 1/4 in. air space | 0.41 | 0.54 | 0.70 | 0.53 | 0.47 | 0.45 | 0.41 | 0.56 | 0.50 | 0.44 | 0.44 | 0.41 |
| 25 | 1/2 in. air space | 0.30 | 0.46 | 0.61 | 0.45 | 0.39 | 0.38 | 0.33 | 0.47 | 0.41 | 0.35 | 0.35 | 0.30 |
| 26 | 1/4 in. argon space | 0.33 | 0.48 | 0.64 | 0.47 | 0.42 | 0.40 | 0.35 | 0.49 | 0.43 | 0.38 | 0.37 | 0.33 |
| 27 | 1/2 in. argon space | 0.25 | 0.42 | 0.57 | 0.41 | 0.36 | 0.34 | 0.30 | 0.43 | 0.36 | 0.31 | 0.31 | 0.25 |
| | **Triple Glazing** | | | | | | | | | | | | |
| 28 | 1/4 in. air spaces | 0.38 | 0.52 | 0.67 | 0.49 | 0.43 | 0.43 | 0.38 | 0.53 | 0.47 | 0.42 | 0.42 | 0.38 |
| 29 | 1/2 in. air spaces | 0.31 | 0.47 | 0.61 | 0.44 | 0.38 | 0.38 | 0.34 | 0.47 | 0.41 | 0.36 | 0.36 | 0.31 |
| 30 | 1/4 in. argon spaces | 0.34 | 0.49 | 0.63 | 0.46 | 0.41 | 0.40 | 0.36 | 0.50 | 0.44 | 0.38 | 0.38 | 0.34 |
| 31 | 1/2 in. argon spaces | 0.29 | 0.45 | 0.59 | 0.42 | 0.37 | 0.36 | 0.32 | 0.45 | 0.40 | 0.34 | 0.34 | 0.29 |
| | **Triple Glazing, $e = 0.20$ on surface 2, 3, 4, or 5** | | | | | | | | | | | | |
| 32 | 1/4 in. air spaces | 0.33 | 0.48 | 0.62 | 0.45 | 0.40 | 0.39 | 0.35 | 0.49 | 0.43 | 0.37 | 0.37 | 0.33 |
| 33 | 1/2 in. air spaces | 0.25 | 0.42 | 0.56 | 0.39 | 0.34 | 0.33 | 0.29 | 0.42 | 0.36 | 0.31 | 0.31 | 0.25 |
| 34 | 1/4 in. argon spaces | 0.28 | 0.45 | 0.58 | 0.41 | 0.36 | 0.36 | 0.31 | 0.45 | 0.39 | 0.33 | 0.33 | 0.28 |
| 35 | 1/2 in. argon spaces | 0.22 | 0.40 | 0.54 | 0.37 | 0.32 | 0.31 | 0.27 | 0.39 | 0.33 | 0.28 | 0.28 | 0.22 |
| | **Triple Glazing, $e = 0.20$ on surfaces 2 or 3 and 4 or 5** | | | | | | | | | | | | |
| 36 | 1/4 in. air spaces | 0.29 | 0.45 | 0.59 | 0.42 | 0.37 | 0.36 | 0.32 | 0.45 | 0.40 | 0.34 | 0.34 | 0.29 |
| 37 | 1/2 in. air spaces | 0.20 | 0.39 | 0.52 | 0.35 | 0.31 | 0.30 | 0.26 | 0.38 | 0.32 | 0.26 | 0.26 | 0.20 |
| 38 | 1/4 in. argon spaces | 0.23 | 0.41 | 0.54 | 0.37 | 0.33 | 0.32 | 0.28 | 0.40 | 0.34 | 0.29 | 0.29 | 0.23 |
| 39 | 1/2 in. argon spaces | 0.17 | 0.36 | 0.49 | 0.33 | 0.28 | 0.28 | 0.24 | 0.35 | 0.29 | 0.24 | 0.24 | 0.17 |
| | **Triple Glazing, $e = 0.10$ on surfaces 2 or 3 and 4 or 5** | | | | | | | | | | | | |
| 40 | 1/4 in. air spaces | 0.27 | 0.44 | 0.58 | 0.40 | 0.36 | 0.35 | 0.31 | 0.44 | 0.38 | 0.32 | 0.32 | 0.27 |
| 41 | 1/2 in. air spaces | 0.18 | 0.37 | 0.50 | 0.34 | 0.29 | 0.28 | 0.25 | 0.36 | 0.30 | 0.25 | 0.25 | 0.18 |
| 42 | 1/4 in. argon spaces | 0.21 | 0.39 | 0.53 | 0.36 | 0.31 | 0.31 | 0.27 | 0.38 | 0.33 | 0.27 | 0.27 | 0.21 |
| 43 | 1/2 in. argon spaces | 0.14 | 0.34 | 0.47 | 0.30 | 0.26 | 0.26 | 0.22 | 0.32 | 0.27 | 0.21 | 0.21 | 0.14 |
| | **Quadruple Glazing, $e = 0.10$ on surfaces 2 or 3 and 4 or 5** | | | | | | | | | | | | |
| 44 | 1/4 in. air spaces | 0.22 | 0.40 | 0.54 | 0.37 | 0.32 | 0.31 | 0.27 | 0.39 | 0.33 | 0.28 | 0.28 | 0.22 |
| 45 | 1/2 in. air spaces | 0.15 | 0.35 | 0.48 | 0.31 | 0.27 | 0.26 | 0.23 | 0.33 | 0.27 | 0.22 | 0.22 | 0.15 |
| 46 | 1/4 in. argon spaces | 0.17 | 0.36 | 0.49 | 0.33 | 0.28 | 0.28 | 0.24 | 0.35 | 0.29 | 0.24 | 0.24 | 0.17 |
| 47 | 1/2 in. argon spaces | 0.12 | 0.32 | 0.45 | 0.29 | 0.25 | 0.24 | 0.20 | 0.31 | 0.25 | 0.20 | 0.20 | 0.12 |
| 48 | 1/4 in. krypton spaces | 0.12 | 0.32 | 0.45 | 0.29 | 0.25 | 0.24 | 0.20 | 0.31 | 0.25 | 0.20 | 0.20 | 0.12 |

*Notes*:
1. All heat transmission coefficients in this table include film resistances and are based on winter conditions of 0°F outdoor air temperature and 70°F indoor air temperature, with 15 mph outdoor air velocity and zero solar flux. With the exception of single glazing, small changes in indoor and outdoor temperatures will not significantly affect overall U-factors. Coefficients are for vertical position except skylight values, which are for 20° from horizontal with heat flow up.
2. Glazing layer surfaces are numbered from outdoor to indoor. Double, triple, and quadruple refer to the number of glazing panels. All data are based on 1/8 in. glass, unless otherwise noted. Thermal conductivities are: 0.53 Btu/h·ft·°F for glass, and 0.11 Btu/h·ft·°F for acrylic and polycarbonate.
3. Standard spacers are metal. Edge-of-glass effects are assumed to extend over the 2 1/2 in. band around perimeter of each glazing unit.

# Load Estimating Fundamentals

Table 5-16 U-Factors for Various Fenestration Products in Btu/h·ft$^2$·°F *(Continued)*

| Vertical Installation ||||| Sloped Installation |||||||||
|---|---|---|---|---|---|---|---|---|---|---|---|---|---|
| Garden Windows || Curtain Wall ||| Glass Only (Skylights) || Manufactured Skylight |||| Site-Assembled Sloped/Overhead Glazing ||||
| Aluminum Without Thermal Break | Wood/ Vinyl | Aluminum Without Thermal Break | Aluminum with Thermal Break | Structural Glazing | Center of Glass | Edge of Glass | Aluminum Without Thermal Break | Aluminum with Thermal Break | Reinforced Vinyl/ Aluminum Clad Wood | Wood/ Vinyl | Aluminum Without Thermal Break | Aluminum with Thermal Break | Structural Glazing | ID |
|---|---|---|---|---|---|---|---|---|---|---|---|---|---|---|
| 2.50 | 2.10 | 1.21 | 1.10 | 1.10 | 1.19 | 1.19 | 1.77 | 1.70 | 1.61 | 1.42 | 1.35 | 1.34 | 1.25 | 1 |
| 2.24 | 1.84 | 1.06 | 0.96 | 0.96 | 1.03 | 1.03 | 1.60 | 1.54 | 1.45 | 1.31 | 1.20 | 1.20 | 1.10 | 2 |
| 2.37 | 1.97 | 1.13 | 1.03 | 1.03 | 1.11 | 1.11 | 1.68 | 1.62 | 1.53 | 1.39 | 1.27 | 1.27 | 1.18 | 3 |
| 1.72 | 1.32 | 0.77 | 0.67 | 0.63 | 0.58 | 0.66 | 1.10 | 0.96 | 0.92 | 0.84 | 0.80 | 0.83 | 0.66 | 4 |
| 1.62 | 1.22 | 0.71 | 0.61 | 0.57 | 0.57 | 0.65 | 1.09 | 0.95 | 0.91 | 0.84 | 0.79 | 0.82 | 0.65 | 5 |
| 1.66 | 1.26 | 0.74 | 0.63 | 0.59 | 0.53 | 0.63 | 1.05 | 0.91 | 0.87 | 0.80 | 0.76 | 0.80 | 0.62 | 6 |
| 1.57 | 1.17 | 0.68 | 0.58 | 0.54 | 0.53 | 0.63 | 1.05 | 0.91 | 0.87 | 0.80 | 0.76 | 0.80 | 0.62 | 7 |
| 1.68 | 1.28 | 0.74 | 0.64 | 0.60 | 0.54 | 0.63 | 1.06 | 0.92 | 0.88 | 0.81 | 0.77 | 0.80 | 0.63 | 8 |
| 1.56 | 1.16 | 0.68 | 0.57 | 0.53 | 0.53 | 0.63 | 1.05 | 0.91 | 0.87 | 0.80 | 0.76 | 0.80 | 0.62 | 9 |
| 1.60 | 1.20 | 0.70 | 0.60 | 0.56 | 0.49 | 0.60 | 1.01 | 0.87 | 0.83 | 0.76 | 0.72 | 0.77 | 0.58 | 10 |
| 1.51 | 1.11 | 0.65 | 0.55 | 0.51 | 0.49 | 0.60 | 1.01 | 0.87 | 0.83 | 0.76 | 0.72 | 0.77 | 0.58 | 11 |
| 1.63 | 1.23 | 0.72 | 0.62 | 0.58 | 0.51 | 0.61 | 1.03 | 0.89 | 0.85 | 0.78 | 0.74 | 0.78 | 0.60 | 12 |
| 1.50 | 1.10 | 0.64 | 0.54 | 0.50 | 0.50 | 0.61 | 1.02 | 0.88 | 0.84 | 0.77 | 0.73 | 0.78 | 0.59 | 13 |
| 1.54 | 1.14 | 0.67 | 0.56 | 0.52 | 0.44 | 0.56 | 0.96 | 0.83 | 0.78 | 0.72 | 0.68 | 0.74 | 0.54 | 14 |
| 1.44 | 1.04 | 0.61 | 0.50 | 0.46 | 0.46 | 0.58 | 0.98 | 0.85 | 0.80 | 0.74 | 0.70 | 0.75 | 0.56 | 15 |
| 1.57 | 1.17 | 0.68 | 0.58 | 0.54 | 0.46 | 0.58 | 0.98 | 0.85 | 0.80 | 0.74 | 0.70 | 0.75 | 0.56 | 16 |
| 1.43 | 1.03 | 0.60 | 0.50 | 0.45 | 0.46 | 0.58 | 0.98 | 0.85 | 0.80 | 0.74 | 0.70 | 0.75 | 0.56 | 17 |
| 1.47 | 1.07 | 0.62 | 0.52 | 0.48 | 0.39 | 0.53 | 0.91 | 0.78 | 0.74 | 0.68 | 0.64 | 0.70 | 0.50 | 18 |
| 1.35 | 0.95 | 0.55 | 0.45 | 0.41 | 0.40 | 0.54 | 0.92 | 0.79 | 0.75 | 0.68 | 0.64 | 0.71 | 0.51 | 19 |
| 1.53 | 1.13 | 0.66 | 0.56 | 0.51 | 0.44 | 0.56 | 0.96 | 0.83 | 0.78 | 0.72 | 0.68 | 0.74 | 0.54 | 20 |
| 1.38 | 0.98 | 0.57 | 0.47 | 0.43 | 0.44 | 0.56 | 0.96 | 0.83 | 0.78 | 0.72 | 0.68 | 0.74 | 0.54 | 21 |
| 1.43 | 1.03 | 0.60 | 0.50 | 0.45 | 0.36 | 0.51 | 0.88 | 0.75 | 0.71 | 0.65 | 0.61 | 0.68 | 0.47 | 22 |
| 1.30 | 0.90 | 0.53 | 0.43 | 0.38 | 0.38 | 0.52 | 0.90 | 0.77 | 0.73 | 0.67 | 0.63 | 0.69 | 0.49 | 23 |
| 1.51 | 1.11 | 0.65 | 0.55 | 0.51 | 0.42 | 0.55 | 0.94 | 0.81 | 0.76 | 0.70 | 0.66 | 0.72 | 0.52 | 24 |
| 1.35 | 0.95 | 0.55 | 0.45 | 0.41 | 0.43 | 0.56 | 0.95 | 0.82 | 0.77 | 0.71 | 0.67 | 0.73 | 0.53 | 25 |
| 1.40 | 1.00 | 0.58 | 0.48 | 0.44 | 0.34 | 0.49 | 0.86 | 0.73 | 0.69 | 0.63 | 0.59 | 0.66 | 0.45 | 26 |
| 1.27 | 0.87 | 0.51 | 0.41 | 0.37 | 0.36 | 0.51 | 0.88 | 0.75 | 0.71 | 0.65 | 0.61 | 0.68 | 0.47 | 27 |
| see note 7 | see note 7 | 0.61 | 0.51 | 0.46 | 0.39 | 0.53 | 0.90 | 0.75 | 0.71 | 0.64 | 0.62 | 0.69 | 0.48 | 28 |
| | | 0.55 | 0.45 | 0.40 | 0.36 | 0.51 | 0.87 | 0.72 | 0.68 | 0.61 | 0.60 | 0.67 | 0.45 | 29 |
| | | 0.58 | 0.48 | 0.43 | 0.35 | 0.50 | 0.86 | 0.71 | 0.67 | 0.60 | 0.59 | 0.66 | 0.44 | 30 |
| | | 0.53 | 0.43 | 0.38 | 0.33 | 0.48 | 0.84 | 0.69 | 0.65 | 0.59 | 0.57 | 0.65 | 0.42 | 31 |
| see note 7 | see note 7 | 0.57 | 0.47 | 0.42 | 0.34 | 0.49 | 0.85 | 0.70 | 0.66 | 0.59 | 0.58 | 0.65 | 0.43 | 32 |
| | | 0.50 | 0.40 | 0.35 | 0.31 | 0.47 | 0.82 | 0.67 | 0.63 | 0.57 | 0.56 | 0.63 | 0.41 | 33 |
| | | 0.53 | 0.43 | 0.37 | 0.28 | 0.45 | 0.80 | 0.64 | 0.60 | 0.54 | 0.53 | 0.61 | 0.38 | 34 |
| | | 0.47 | 0.37 | 0.32 | 0.27 | 0.44 | 0.79 | 0.63 | 0.59 | 0.53 | 0.52 | 0.60 | 0.37 | 35 |
| see note 7 | see note 7 | 0.53 | 0.43 | 0.38 | 0.29 | 0.45 | 0.81 | 0.65 | 0.61 | 0.55 | 0.54 | 0.62 | 0.39 | 36 |
| | | 0.46 | 0.36 | 0.30 | 0.27 | 0.44 | 0.79 | 0.63 | 0.59 | 0.53 | 0.52 | 0.60 | 0.37 | 37 |
| | | 0.48 | 0.38 | 0.33 | 0.24 | 0.42 | 0.76 | 0.60 | 0.57 | 0.50 | 0.49 | 0.58 | 0.35 | 38 |
| | | 0.43 | 0.33 | 0.28 | 0.22 | 0.40 | 0.74 | 0.58 | 0.55 | 0.49 | 0.48 | 0.57 | 0.33 | 39 |
| see note 7 | see note 7 | 0.52 | 0.42 | 0.37 | 0.27 | 0.44 | 0.79 | 0.63 | 0.59 | 0.53 | 0.52 | 0.60 | 0.37 | 40 |
| | | 0.44 | 0.34 | 0.29 | 0.25 | 0.42 | 0.77 | 0.61 | 0.57 | 0.51 | 0.50 | 0.59 | 0.36 | 41 |
| | | 0.46 | 0.36 | 0.31 | 0.21 | 0.39 | 0.73 | 0.57 | 0.54 | 0.48 | 0.47 | 0.56 | 0.32 | 42 |
| | | 0.40 | 0.30 | 0.25 | 0.20 | 0.39 | 0.72 | 0.56 | 0.53 | 0.47 | 0.46 | 0.55 | 0.31 | 43 |
| | | 0.47 | 0.37 | 0.32 | 0.22 | 0.40 | 0.74 | 0.58 | 0.55 | 0.49 | 0.48 | 0.57 | 0.33 | 44 |
| see note 7 | see note 7 | 0.41 | 0.31 | 0.26 | 0.19 | 0.38 | 0.71 | 0.55 | 0.52 | 0.46 | 0.45 | 0.54 | 0.30 | 45 |
| | | 0.43 | 0.33 | 0.28 | 0.18 | 0.37 | 0.70 | 0.54 | 0.51 | 0.45 | 0.44 | 0.54 | 0.29 | 46 |
| | | 0.39 | 0.29 | 0.23 | 0.16 | 0.35 | 0.68 | 0.52 | 0.49 | 0.43 | 0.42 | 0.52 | 0.28 | 47 |
| | | 0.39 | 0.29 | 0.23 | 0.13 | 0.33 | 0.65 | 0.49 | 0.46 | 0.40 | 0.40 | 0.50 | 0.25 | 48 |

4. Product sizes are described in Figure 4, and frame U-factors are from Table 1.
5. Use $U = 0.6$ Btu/(h·ft$^2$·°F) for glass block with mortar but without reinforcing or framing.
6. Use of this table should be limited to that of an estimating tool for the early phases of design.
7. Values for triple- and quadruple-glazed garden windows are not listed, because these are not common products.
8. U-factors in this table were determined using NFRC 100-91. They have not been updated to the current rating methodology in NFRC 100-2004.

**Table 5-17** Design U-Factors of Swinging Doors in Btu/h·ft²·°F

*(Table 6, Chapter 15, 2013 ASHRAE Handbook—Fundamentals)*

| Door Type (Rough Opening = 38 × 82 in.) | No Glazing | Single Glazing | Double Glazing with 1/2 in. Air Space | Double Glazing with e = 0.10, 1/2 in. Argon |
|---|---|---|---|---|
| *Slab Doors* | | | | |
| Wood slab in wood frame[a] | 0.46 | | | |
|   6% glazing (22 × 8 in. lite) | — | 0.48 | 0.46 | 0.44 |
|   25% glazing (22 × 36 in. lite) | — | 0.58 | 0.46 | 0.42 |
|   45% glazing (22 × 64 in. lite) | — | 0.69 | 0.46 | 0.39 |
|   More than 50% glazing | Use Table 5-16 (operable) | | | |
| Insulated steel slab with wood edge in wood frame[b] | 0.16 | | | |
|   6% glazing (22 × 8 in. lite) | — | 0.21 | 0.19 | 0.18 |
|   25% glazing (22 × 36 in. lite) | — | 0.39 | 0.26 | 0.23 |
|   45% glazing (22 × 64 in. lite) | — | 0.58 | 0.35 | 0.26 |
|   More than 50% glazing | Use Table 5-16 (operable) | | | |
| Foam insulated steel slab with metal edge in steel frame[c] | 0.37 | | | |
|   6% glazing (22 × 8 in. lite) | — | 0.44 | 0.41 | 0.39 |
|   25% glazing (22 × 36 in. lite) | — | 0.55 | 0.48 | 0.44 |
|   45% glazing (22 × 64 in. lite) | — | 0.71 | 0.56 | 0.48 |
|   More than 50% glazing | Use Table 5-16 (operable) | | | |
| Cardboard honeycomb slab with metal edge in steel frame | 0.61 | | | |
| *Stile-and-Rail Doors* | | | | |
| Sliding glass doors/French doors | Use Table 5-16 (operable) | | | |
| *Site-Assembled Stile-and-Rail Doors* | | | | |
| Aluminum in aluminum frame | — | 1.32 | 0.93 | 0.79 |
| Aluminum in aluminum frame with thermal break | — | 1.13 | 0.74 | 0.63 |

*Notes*:
[a] Thermally broken sill [add 0.03 Btu/h·ft²·°F for non-thermally broken sill]
[b] Non-thermally broken sill
[c] Nominal U-factors are through center of insulated panel before consideration of thermal bridges around edges of door sections and because of frame.

**Table 5-18** Design U-Factors for Revolving Doors in Btu/h·ft²·°F

*(Table 7, Chapter 15, 2013 ASHRAE Handbook—Fundamentals)*

| Type | Size (Width × Height) | U-Factor |
|---|---|---|
| 3-wing | 8 × 7 ft | 0.79 |
|  | 10 × 8 ft | 0.80 |
| 4-wing | 7 × 6.5 ft | 0.63 |
|  | 7 × 7.5 ft | 0.64 |
| Open* | 82 × 84 in. | 1.32 |

*Notes*:
*U-factor of Open door determined using NFRC *Technical Document* 100-91. It has not been updated to current rating methodology in NFRC *Technical Document* 100-2004.

**Example 5-4** Determine the winter U-factor for a cavity wall consisting of face brick, 8 in. concrete block (three oval core, lightweight aggregate), a 3/4 in. airspace as the cavity, another layer of the same type of concrete block, 2 in. of rigid organic bonded glass fiber insulation, and 1/2 in. plasterboard.

**Solution:**

| Component | Resistance | |
|---|---|---|
| Outdoor air | 0.17 | (Table 5-12) |
| Brick, 4 in. face | 0.44 | (Table 5-11) |
| Concrete block, 8 in. | 2.00 | (Table 5-11) |
| Airspace, 3/4 in. (estimated) | 1.00 | (Table 5-14) |

**Table 5-19** Design U-Factors for Double-Skin Steel Emergency Exit Doors in Btu/h·ft²·°F

*(Table 8, Chapter 15, 2013 ASHRAE Handbook—Fundamentals)*

| Core Insulation | | Rough Opening Size | |
|---|---|---|---|
| Thickness, in. | Type | 3 ft × 6 ft 8 in. | 6 ft × 6 ft 8 in. |
| 1 3/8 | Honeycomb kraft paper | 0.57 | 0.52 |
|  | Mineral wool, steel ribs | 0.44 | 0.36 |
|  | Polyurethane foam | 0.34 | 0.28 |
| 1 3/4 | Honeycomb kraft paper | 0.57 | 0.54 |
|  | Mineral wool, steel ribs | 0.41 | 0.33 |
|  | Polyurethane foam | 0.31 | 0.26 |
| 1 3/8 | Honeycomb kraft paper | 0.60 | 0.55 |
|  | Mineral wool, steel ribs | 0.47 | 0.39 |
|  | Polyurethane foam | 0.37 | 0.31 |
| 1 3/4 | Honeycomb kraft paper | 0.60 | 0.57 |
|  | Mineral wool, steel ribs | 0.44 | 0.37 |
|  | Polyurethane foam | 0.34 | 0.30 |

*With thermal break

| | | |
|---|---|---|
| Concrete block, 8 in. | 2.00 | (Table 5-15) |
| Glass fiber insulation, 2 in. | 8.00 | (Table 5-14, 5-15) |
| Plasterboard, 1/2 in. | 0.45 | (Table 5-15) |
| Inside air | 0.68 | (Table 5-12) |
| Total $R$ = | 14.74 | |

$$U = 1/R_T = 1/14.74 = 0.068 \text{ Btu/h} \cdot \text{ft}^2 \cdot °\text{F}$$

**Example 5-5** Determine the overall coefficients (U-factors) for winter for (a) a solid wood door, (b) the flat ceiling/roof of an industrial building, which consists of a painted sheet metal exterior, 1/2 in. nail-base fiberboard sheathing, and wood rafters.

**Solution:**

(a) Use Table 5-17 to find the U-factor.

$$U = 0.46 \text{ Btu/h} \cdot \text{ft}^2 \cdot °\text{F}$$

(b) To calculate the U-factor of the ceiling, add the resistance values of each element.

| Component | Resistance | |
|---|---|---|
| Inside air | 0.61 | (Table 5-12) |
| Sheathing | 1.06 | (Table 5-15) |
| Outdoor air | 0.17 | (Table 5-12) |
| Total $R$ = | 1.84 | |

$$U = 1/1.84 = 0.54 \text{ Btu/h} \cdot \text{ft}^2 \cdot °\text{F}$$

**Example 5-6** Calculate the U-factor of the 2 by 4 stud wall shown in Figure 5-18. The studs are at 16 in. OC. There is 3.5 in. mineral fiber batt insulation (R-13) in the stud space. The inside finish is 0.5 in. gypsum wallboard; the outside is finished with rigid foam insulating sheathing (R-4) and 0.5 in. by 8 in. wood bevel lapped siding. The insulated cavity occupies approximately 75% of the transmission area; the studs, plates, and sills occupy 21%; and the headers occupy 4%.

**Solution.** Obtain the R-values of the various building elements from Tables 5-12 and 5-15. Assume the $R = 1.25$ per inch for the wood framing. Also, assume the headers are solid wood, in this case, and group them with the studs, plates, and sills.

Since the U-factor is the reciprocal of R-value, $U_1 = 0.052$ and $U_2 = 0.095$ Btu/h·ft²·°F.

# Load Estimating Fundamentals

| Element | R (Insulated Cavity) | R (Studs, Plates, and Headers) |
|---|---|---|
| 1. Outside surface, 15 mph wind | 0.17 | 0.17 |
| 2. Wood bevel lapped siding | 0.81 | 0.81 |
| 3. Rigid foam insulating sheathing | 4.0 | 4.0 |
| 4. Mineral fiber batt insulation, 3.5 in. | 13.0 | — |
| 5. Wood stud, nominal 2 × 4 (est.) | — | 4.38 |
| 6. Gypsum wallboard, 0.5 in. | 0.45 | 0.45 |
| 7. Inside surface, still air | 0.68 | 0.68 |
|  | $R_1 = 19.11$ | $R_2 = 10.49$ |

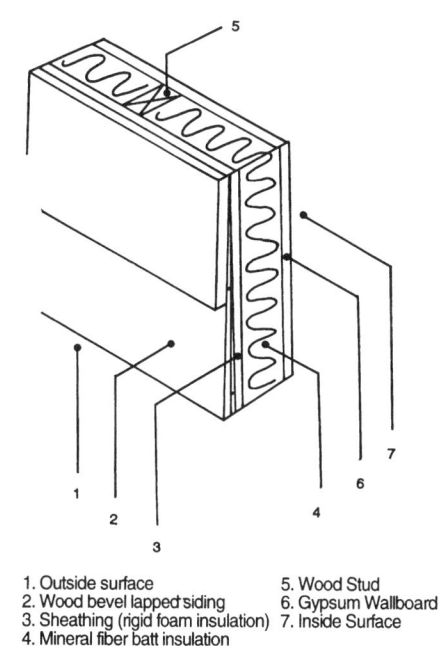

*Fig. 5-18 Insulated Wood Frame Wall (Example 5-6)*

1. Outside surface
2. Wood bevel lapped siding
3. Sheathing (rigid foam insulation)
4. Mineral fiber batt insulation
5. Wood Stud
6. Gypsum Wallboard
7. Inside Surface

If the wood framing is accounted for using the parallel-path flow method, the U-factor of the wall is determined using the following equation

$$U_{av} = (0.75 \times 0.052) + (0.25 \times 0.095) = 0.063 \text{ Btu/h·ft}^2\cdot°F$$

If the wood framing (thermal bridging) is not included, Equation (3) from Chapter 22 may be used to calculate the U-factor of the wall as follows:

$$U_{av} = U_1 = \frac{1}{R_1} = 0.052 \text{ Btu/h·ft}^2\cdot°F$$

## 5.5 Problems

**5.1** With an 11.2 m/s wind blowing uniformly against one face of a building, what pressure differential would be used to calculate the air leakage into the building?

**5.2** Can wind forces and Δt forces cancel each other when predicting infiltration? Reinforce? How can infiltration be prevented entirely?

**5.3** A double door has a 1/8 in. crack on all sides except between the two doors, which has a 1/4 in. crack. What would be the leakage rate for the building of Problem 5.1?
[Ans: 453 cfm]

**5.4** Determine the heat loss due to infiltration of 236 L/s of outdoor air at 9°C when the indoor air is 24°C.

**5.5** Give an expression for (a) the sensible load due to infiltration, and (b) the latent load due to infiltration.

**5.6** A building is 75 ft wide by 100 ft long and 10 ft high. The indoor conditions are 75°F and 24% rh, and the outside conditions are 0°F and saturated. The infiltration rate is estimated to be 0.75 ach. Calculate the sensible and latent heat loss. [Ans: 77,300 Btu/h; 17,700 Btu/h]

**5.7** How can infiltration rates be reduced? Does a reduction in the infiltration rate always result in a reduction of the air-conditioning load? In air-conditioning equipment size? Explain.

**5.8** A 3 ft by 3 ft ventilation opening is in a wall facing in the prevalent wind direction. There are adequate openings in the roof for the passage of exhaust air. Estimate the ventilation rate for a 25 mph wind.

**5.9** A building 20 ft by 40 ft by 9 ft has an anticipated infiltration rate of 0.75 air changes per hour. Indoor design conditions are 75°F, 30% rh minimum. Outdoor design temperature is 5°F.

(a) Determine sensible, latent, and total heat loads (Btu/h) due to infiltration.
(b) Specify the necessary humidifier size (lb/h).
[Ans: 8890 Btu/h, 1.8 lb/h]

**5.10** A small factory with a 10 ft high ceiling is shown. There are 22 employees normally in the shop area and 4 employees in the office area. On a winter day when the outside temperature is 0°F, the office is maintained at 75°F, 25% rh, and the shop is kept at 68°F with no humidity control. Determine for each area

(a) infiltration, CFM
(b) minimum required outdoor air, CFM
(c) sensible heat loss due to infiltration, Btu/h
(d) latent heat loss due to infiltration, Btu/h

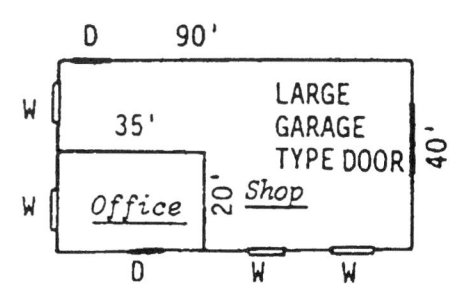

**5.11** Specify an acceptable amount of outdoor air for ventilation of the following:

(a) 12 ft by 12 ft private office with 8 ft high walls
(b) department store with 20,000 ft² of floor area

**5.12** A wall consists of 4 in. of face brick, 1/2 in. of cement mortar, 8 in. hollow clay tile, an airspace 1 5/8 in. wide, and wood

**Table 5-20 Design U-Factors for Double-Skin Steel Garage and Aircraft Hangar Doors in Btu/h·ft²·°F**
*(Table 9, Chapter 15, 2009 ASHRAE Handbook—Fundamentals)*

| Insulation Thickness, in. | Type | One-Piece Tilt-Up[a] 8 × 7 ft | One-Piece Tilt-Up[a] 16 × 7 ft | Sectional Tilt-Up[b] 9 × 7 ft | Aircraft Hangar 72 × 12 ft[c] | Aircraft Hangar 240 × 50 ft[d] |
|---|---|---|---|---|---|---|
| 1 3/8 | EPS, steel ribs[e] | 0.36 | 0.33 | 0.34 – 0.39 | | |
|  | XPS, steel ribs[f] | 0.33 | 0.31 | 0.31 – 0.36 | | |
| 2 | EPS, steel ribs[e] | 0.31 | 0.28 | 0.29 – 0.33 | | |
|  | XPS, steel ribs[f] | 0.29 | 0.26 | 0.27 – 0.31 | | |
| 3 | EPS, steel ribs[e] | 0.26 | 0.23 | 0.25 – 0.28 | | |
|  | XPS, steel ribs[f] | 0.24 | 0.21 | 0.24 – 0.27 | | |
| 4 | EPS, steel ribs[e] | 0.23 | 0.20 | 0.23 – 0.25 | | |
|  | XPS, steel ribs[f] | 0.21 | 0.19 | 0.21 – 0.24 | | |
| 6 | EPS, steel ribs[e] | 0.20 | 0.16 | 0.20 – 0.21 | | |
|  | XPS, steel ribs[f] | 0.19 | 0.15 | 0.19 – 0.21 | | |
| 4 | XPS | | | | 0.25 | 0.16 |
|  | Mineral wool, steel ribs | | | | 0.25 | 0.16 |
|  | EPS | | | | 0.23 | 0.15 |
| 6 | XPS | | | | 0.21 | 0.13 |
|  | Mineral wool, steel ribs | | | | 0.23 | 0.13 |
|  | EPS | | | | 0.20 | 0.12 |
| — | Uninsulated[g] | | | | 1.10 | 1.23 |
|  | All products[f] | 1.15[g] | | | | |

Notes:
[a] Values are for thermally broken or thermally unbroken doors.
[b] Lower values are for thermally broken doors; upper values are for doors with no thermal break.
[c] Typical size for a small private airplane (single- or twin-engine.)
[d] Typical hangar door for a midsize commercial jet airliner.
[e] EPS = extruded polystyrene; XPS = expanded polystyrene.
[f] U-factor determined using NFRC *Technical Document* 100-91. Not updated to current rating methodology in NFRC *Technical Document* 100-2004.
[g] U-factor determined for 10 × 10 ft sectional door, but is representative of similar products of different size.

lath and plaster totaling 3/4 in. Find the U-factors for both winter and summer.

**5.13** The ceiling of a house is 3/4 in. (19 mm) acoustical tile on furring strips with highly reflective aluminum foil across the top of the ceiling joists. Determine the U-factor for cooling load calculations.

**5.14** Calculate the winter U-factor for a wall consisting of 4 in. (100 mm) face brick, 4 in. (100 mm) common brick, and 1/2 in. (13 mm) of gypsum plaster (sand aggregate).

[Ans: 0.459 Btu/h· ft²·°F [2.61 W/(m²· K)]]

**5.15** Find the overall coefficient of heat transmission $U$ for a wall consisting of 4 in. of face brick, 1/2 in. of cement mortar, 8 in. of stone, and 3/4 in. of gypsum plaster. The outdoor air velocity is 15 mph and the inside air is still.

**5.16** A wall has an overall coefficient $U$ = 1.31 W/(m²·K). What is the conductance of the wall when its outside surface is exposed to a wind velocity of 6.7 m/s and the inside air is still?

**5.17** Compute the U-factor for a wall of frame construction consisting of 1/2 by 8 bevel siding, permeable felt building paper, 25/32 in. wood fiber sheathing, 2 by 4 studding on 16 in. centers, and 3/4 in. metal lath and sand plaster. Out-side wind velocity is 15 mph. [Ans: 0.206 Btu/ h·ft²·°F]

**5.18** For the wall of Problem 5.17, determine $U$ if the space between the studs is filled with fiberglass blanket insulation. Neglect the effect of the studs.

**5.19** Rework Problem 5.18 including the effect of the studs.

**5.20** A concrete wall 250 mm thick is exposed to outdoor air at –15°C with a velocity of 6.7 m/s. Inside air temperature is 15.6°C. Determine the heat flow through 14.9 m² of this wall. [Ans: 1548 W]

**5.21** Find the overall coefficient of heat transfer and the total thermal resistance for the following exterior wall exposed to a 25 mph wind: face brick veneer, 25/32 in. insulating board sheathing, 3 in. fiberglass insulation in stud space, and 1/4 in. walnut veneer plywood panels for the interior.

**5.22** What is the thermal resistance of 12.1 cm (4 3/4 in.) thick precast concrete (stone aggregate, oven dried)?

**5.23** A composite wall structure experiences a –10°F air temperature on the outside and a 75°F air temperature on the inside. The wall consists of a 4 in. thick outer facebrick, a 2 in. batt of fiberglass insulation, and a 3/8 in. sheet of gypsum board. Determine the U-factor and the heat flow rate per ft². Plot the steady-state temperature profile across the wall. [Ans: 0.118 Btu/h·ft²·°F, 10.07 Btu/h·ft²]

**5.24** Find the overall heat transmission coefficient for a floor-ceiling sandwich (heat flow up) having the following construction. [Ans: 0.22 Btu/h· ft·°F]

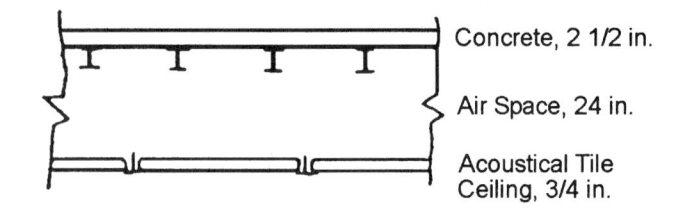

Concrete, 2 1/2 in.

Air Space, 24 in.

Acoustical Tile Ceiling, 3/4 in.

# Load Estimating Fundamentals

**5.25** The exterior windows of a house are of double insulating glass with 1/4 in. airspace and have metal sashes. Determine the design U-factor for heating.

**5.26** In designing a house, the total heat loss is calculated as 17.9 kW. The heat loss through the outside walls is 28% of this total when the overall coefficient for the outside walls is 1.4 W/(m²·K). If 50 mm organic bonded fiberglass is added to the wall in the stud space, determine the new total heat loss for the house. [Ans: 14.6 kW]

**5.27** The top floor ceiling of a building 30 ft by 36 ft is constructed of 2 in. by 4 in. joists on 18 in. centers. On the underside is metal lath with plaster, 3/4 in. thick. On top of the joists there are only scattered walking planks, but the space between the joists is filled with rock wool. The air temperature at the ceiling in the room is 78°F and the attic temperature is 25°F. Find the overall coefficient of heat transfer for the ceiling.

**5.28** Determine the U-factor and the temperature at each point of change of material for the flat roof shown below. The roof has 3/8 in. built-up roofing, 1 1/2 in. roof insulation, 2 in. thick, 80 lb/ft³ lightweight aggregate concrete on corrugated metal over steel joists, with a metal lath and 3/4 in. (sand) plaster ceiling. Omit correction for framing.

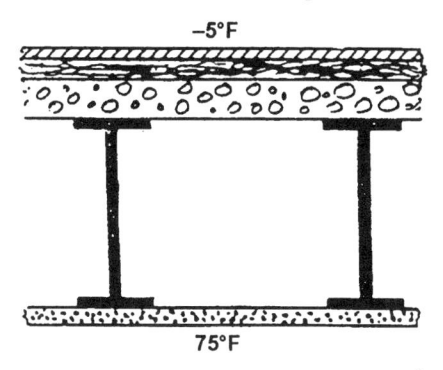

**5.29** Calculate the heat loss through a roof of 100 ft² area where the inside air temperature is to be 70°F, outdoor air 10°F, and the composition from outside to inside: 3/8 in. built-up roofing, 1 in. cellular glass insulation, 4 in. concrete slab, and 3/4 in. acoustical tile. [Ans: 1030 Btu/h (0.30 kW)]

**5.30** Calculate the heat loss through 100 ft² (9.29 m²) of 1/4 in. (6.5 mm) plate glass with inside and outdoor air temperatures of 70°F and 10°F (21.1°C and –12.2°C), respectively.

**5.31** A building has single-glass windows and an indoor temperature of 75°F. The outdoor air temperature is 40°F. With a 15 mph outside wind, still air inside, and after sundown, what can the maximum relative humidity of the inside air be without condensation forming on the glass?

**5.32** Repeat Problem 5.31 for a double-glass window with a 1/2 in. airspace. [Ans: 67% rh]

**5.33** A wall is constructed of 4 in. face brick, pressed fiber board sheathing ($k = 0.44$), 3 1/2 in. airspace, and 1/2 in. lightweight gypsum plaster on 1/2 in. plasterboard. When the inside air temperature is 70°F and the outside temperature is –15°F, how thick must the sheathing be to prevent water pipes in the stud space from freezing?

**5.34** The roof of a rapid transit car is constructed of 3/8 in. plywood ($C = 2.12$), a vapor seal having negligible thermal resistance, expanded polystyrene insulation ($k = 0.24$), 3/4 in. airspace, 1/16 in. steel with welded joints and aluminum paint. If the car is traveling at 60 mph (film coefficient is 20.0 Btu/h·ft²·°F) when the ambient temperature is –20°F, what thickness of insulation is necessary to prevent condensation when the inside conditions in the car are 72°F dry bulb and 55% rh?

**5.35** A roof is constructed of 2 in. wood decking, insulation on top of deck, and 3/8 in. built-up roofing. It has no ceiling. Assuming that the insulation forms a perfect vapor barrier, determine the required resistance of the insulation to prevent condensation from occurring at the deck-insulation interface when indoor conditions are 70°F and 40% rh, and the outside temperature is 20°F. [Ans: 2.48 ft²·°F·h/Btu]

**5.36** Determine the summer U-factor for each of the following:

(a) Building wall consisting of face brick veneer, 3/4 in. plywood sheathing, 2 by 6 studs on 24 in. centers, no insulation, and 5/32 in. plywood paneling
(b) Ceiling/roof where the ceiling is composed of 1/2 in. plasterboard nailed to 2 by 6 joists on 16 in. centers and the roof consists of asphalt shingles on 3/4 in. plywood on 2 by 4 rafters on 16 in. centers. The roof area is 2717 ft² while the ceiling area is 1980 ft².
(c) Sliding patio door with insulating glass (double) having a 0.50 in. airspace in a metal frame
(d) A 2 in. solid wood door

**5.37** A prefabricated commercial building has exterior walls constructed of 2 in. expanded polyurethane bonded between 1/8 in. aluminum sheet and 1/4 in. veneer plywood. Design conditions include 105°F outdoor air temperature, 72°F indoor air, and 7.5 mph wind. Determine

(a) overall thermal resistance
(b) value of $U$
(c) heat gain per ft²
[Ans: (a) 13.74 ft²·°F·h/Btu; (b) 0.073 Btu/h·ft²·°F; (c) 2.40 Btu/h·ft²]

**5.38** An outside wall consists of 4 in. face brick, 25/32 in. insulating board sheathing 2 in. mineral fiber batt between the 2 by 4 studs, and 1/2 in. plasterboard. Determine the winter U-factor.

**5.39** Solve the following:

(a) Compute the winter U-factor for the wall of Problem 5.38 if the wind velocity is 30 mph.
(b) Compute the summer U-factor for the wall of Problem 5.38.

(c) If full wall insulation is used, compute the summer U-factor for the wall of Problem 5.38.

**5.40** An exterior wall contains a 3 ft by 7 ft solid wood door, 1 3/8 in. thick, and a 6 ft by 7 ft sliding patio door with double glass having a 1/2 in. airspace and metal frame. Determine the summer U-factor for each door.

[Ans: 0.46 Btu/h·ft²·°F; 0.76 Btu/h·ft²·°F]

**5.41** If the doors of Problem 5.40 are between the residence and a completely enclosed swimming pool area, determine the U-factor for each door.

**5.42** Determine the winter U-factor in W/(m²·K) for the wall of a building that has the following construction: face brick, 4 in.; airspace, 3/4 in.; concrete, 9 in.; cellular glass board insulation, 1 in.; plywood paneling, 1/4 in.

**5.43** Determine the summer U-factor for the following building components

(a) Wall: wood drop siding, 1 by 8 in.; 1/2 in. nail-base insulating board sheathing; 2 by 4 studs (16 in. oc) with full wall fiberglass insulation; 1/4 in. paneling
(b) Door: wood, 1 1/2 in. thick, with 25% single-pane glass

**5.44** Determine the combined ceiling and roof winter U-factor for the following construction: The ceiling consists of 3/8 in. gypsum board on 2 by 6 in. ceiling joists. Six inches of fiberglass (mineral/glass wool) insulation fills the space between the joists. The pitched roof has asphalt shingles on 25/32 in. solid wood sheathing with no insulation between the rafters. The ratio of roof area to ceiling area is 1.3. The attic is unvented in winter.

**5.45** The west wall of a residence is 70 ft long by 8 ft high. The wall contains four 3 ft by 5 ft wood sash 80% glass single-pane windows each with a storm window; one double-glazed (1/2 in. airspace) picture window, 5 1/2 ft by 10 ft; and one 1 3/4 in. thick solid wood door, 3 ft by 7 ft. The wall itself has the construction of Problem 5.21. Specify the U-factor and corresponding area for each of the various parts of the wall with normal winter air velocities.

**5.46** A wall is 20 m by 3 m, which includes 14% double glazed glass windows with a 6 mm airspace. The wall proper consists of one layer of face brick backed by 250 mm of concrete with 12 mm of gypsum plaster on the inside. For indoor and outdoor design temperatures of 22°C and –15°C, respectively, determine the heat loss through this wall, kW.

## 5.6  Bibliography

ASHRAE Standard 55. Thermal Environmental Conditions for Human Occupancy.

ASHRAE Standard 62.1. Ventilation for Acceptable Indoor Air Quality.

ASHRAE Standard 62.2. Acceptable Indoor Air Quality in Low-Rise Residential Buildings.

ASHRAE Standard 90.1. Energy Efficient Design of New Buildings Except New Low-Rise Residential Buildings.

ASHRAE. 1998. *Cooling and Heating Load Calculation Principles*.

# SI Figures and Tables

**Table 5-1 SI Change Rates as a Function of Airtightness**

| Class | \multicolumn{10}{c}{Outdoor Design Temperature, °C} |
|---|---|---|---|---|---|---|---|---|---|---|
| | 10 | 4 | −1 | −7 | −12 | −18 | −23 | −29 | −34 | −40 |
| Tight | 0.41 | 0.43 | 0.45 | 0.47 | 0.49 | 0.51 | 0.53 | 0.55 | 0.57 | 0.59 |
| Medium | 0.69 | 0.73 | 0.77 | 0.81 | 0.85 | 0.89 | 0.93 | 0.97 | 1.00 | 1.05 |
| Loose | 1.11 | 1.15 | 1.20 | 1.23 | 1.27 | 1.30 | 1.35 | 1.40 | 1.43 | 1.47 |

*Note*: Values are for 6.7 m/s (24 km/h) wind and indoor temperature of 20°C.

**Table 5-5 SI Basic Model Stack Coefficient $C_s$**
*(Table 4, Chapter 16, 2013 ASHRAE Handbook—Fundamentals, SI Version)*

| | House Height (Stories) | | |
|---|---|---|---|
| | One | Two | Three |
| Stack coefficient | 0.000 145 | 0.000 290 | 0.000 435 |

**Table 5-6 SI Basic Model Wind Coefficient $C_w$**
*(Table 6, Chapter 16, 2013 ASHRAE Handbook—Fundamentals, SI Version)*

| Shelter Class | House Height (Stories) | | |
|---|---|---|---|
| | One | Two | Three |
| 1 | 0.000 319 | 0.000 420 | 0.000 494 |
| 2 | 0.000 246 | 0.000 325 | 0.000 382 |
| 3 | 0.000 174 | 0.000 231 | 0.000 271 |
| 4 | 0.000 104 | 0.000 137 | 0.000 161 |
| 5 | 0.000 032 | 0.000 042 | 0.000 049 |

**Table 5-12 SI Surface Film Coefficients/Resistances**
*(Table 1, Chapter 26, 2013 ASHRAE Handbook—Fundamentals, SI Version)*

| Position of Surface | Direction of Heat Flow | Nonreflective $\varepsilon = 0.90$ | | Reflective $\varepsilon = 0.20$ | | $\varepsilon = 0.05$ | |
|---|---|---|---|---|---|---|---|
| | | $h_i$ | $R_i$ | $h_i$ | $R_i$ | $h_i$ | $R_i$ |
| **Indoor** | | | | | | | |
| Horizontal | Upward | 9.26 | 0.11 | 5.17 | 0.19 | 4.32 | 0.23 |
| Sloping at 45° | Upward | 9.09 | 0.11 | 5.00 | 0.20 | 4.15 | 0.24 |
| Vertical | Horizontal | 8.29 | 0.12 | 4.20 | 0.24 | 3.35 | 0.30 |
| Sloping at 45° | Downward | 7.50 | 0.13 | 3.41 | 0.29 | 2.56 | 0.39 |
| Horizontal | Downward | 6.13 | 0.16 | 2.10 | 0.48 | 1.25 | 0.80 |
| **Outdoor** (any position) | | $h_o$ | $R_o$ | | | | |
| Wind (for winter) at 6.7 m/s | Any | 34.0 | 0.030 | — | — | — | — |
| Wind (for summer) at 3.4 m/s | Any | 22.7 | 0.044 | — | — | — | — |

*Notes*:
1. Surface conductance $h_i$ and $h_o$ measured in W/(m²·K); resistance $R_i$ and $R_o$ in (m²·K)/W.
2. No surface has both an air space resistance value and a surface resistance value.
3. Conductances are for surfaces of the stated emittance facing virtual blackbody surroundings at same temperature as ambient air. Values based on surface/air temperature difference of 5.6 K and surface temperatures of 21°C.
4. See Chapter 4 for more detailed information.
5. Condensate can have significant effect on surface emittance. Also, oxidation, corrosion, and accumulation of dust and dirt can dramatically increase surface emittance. Emittance values of 0.05 should only be used where highly reflective surface can be maintained over the service life of the assembly.

**Table 5-13 SI Emissivity of Various Surfaces and Effective Emittances of Facing Air Spaces[a]**
*(Table 2, Chapter 26, 2013 ASHRAE Handbook—Fundamentals, SI Version)*

| Surface | Average Emissivity $\varepsilon$ | Effective Emittance $\varepsilon_{eff}$ of Air Space | |
|---|---|---|---|
| | | One Surface's Emittance $\varepsilon$; Other, 0.9 | Both Surfaces' Emittance $\varepsilon$ |
| Aluminum foil, bright | 0.05 | 0.05 | 0.03 |
| Aluminum foil, with condensate just visible (>0.7 g/ft²)(>0.5 g/m²) | 0.30[b] | 0.29 | — |
| Aluminum foil, with condensate clearly visible (>2.9 g/ft²)(>2.0 g/m²) | 0.70[b] | 0.65 | — |
| Aluminum sheet | 0.12 | 0.12 | 0.06 |
| Aluminum-coated paper, polished | 0.20 | 0.20 | 0.11 |
| Brass, nonoxidized | 0.04 | 0.038 | 0.02 |
| Copper, black oxidized | 0.74 | 0.41 | 0.59 |
| Copper, polished | 0.04 | 0.038 | 0.02 |
| Iron and steel, polished | 0.2 | 0.16 | 0.11 |
| Iron and steel, oxidized | 0.58 | 0.35 | 0.41 |
| Lead, oxidized | 0.27 | 0.21 | 0.16 |
| Nickel, nonoxidized | 0.06 | 0.056 | 0.03 |
| Silver, polished | 0.03 | 0.029 | 0.015 |
| Steel, galvanized, bright | 0.25 | 0.24 | 0.15 |
| Tin, nonoxidized | 0.05 | 0.047 | 0.026 |
| Aluminum paint | 0.50 | 0.47 | 0.35 |
| Building materials: wood, paper, masonry, nonmetallic paints | 0.90 | 0.82 | 0.82 |
| Regular glass | 0.84 | 0.77 | 0.72 |

[a] Values apply in 4 to 40 μm range of electromagnetic spectrum. Also, oxidation, corrosion, and accumulation of dust and dirt can dramatically increase surface emittance. Emittance values of 0.05 should only be used where the highly reflective surface can be maintained over the service life of the assembly. Except as noted, data from VDI (1999).
[b] Values based on data in Bassett and Trethowen (1984).

**Table 5-14 SI  Effective Thermal Resistances of Plane Air Spaces,[a,b,c] ($m^2 \cdot K$)/W**

*(Table 3, Chapter 26, 2013 ASHRAE Handbook—Fundamentals, SI Version)*

| Position of Air Space | Direction of Heat Flow | Air Space Mean Temp.[d], °C | Temp. Diff.,[d] K | 13 mm Air Space[c] 0.03 | 0.05 | 0.2 | 0.5 | 0.82 | 20 mm Air Space[c] 0.03 | 0.05 | 0.2 | 0.5 | 0.82 |
|---|---|---|---|---|---|---|---|---|---|---|---|---|---|
| Horiz. | Up ↑ | 32.2 | 5.6 | 0.37 | 0.36 | 0.27 | 0.17 | 0.13 | 0.41 | 0.39 | 0.28 | 0.18 | 0.13 |
| | | 10.0 | 16.7 | 0.29 | 0.28 | 0.23 | 0.17 | 0.13 | 0.30 | 0.29 | 0.24 | 0.17 | 0.14 |
| | | 10.0 | 5.6 | 0.37 | 0.36 | 0.28 | 0.20 | 0.15 | 0.40 | 0.39 | 0.30 | 0.20 | 0.15 |
| | | −17.8 | 11.1 | 0.30 | 0.30 | 0.26 | 0.20 | 0.16 | 0.32 | 0.32 | 0.27 | 0.20 | 0.16 |
| | | −17.8 | 5.6 | 0.37 | 0.36 | 0.30 | 0.22 | 0.18 | 0.39 | 0.38 | 0.31 | 0.23 | 0.18 |
| | | −45.6 | 11.1 | 0.30 | 0.29 | 0.26 | 0.22 | 0.18 | 0.31 | 0.31 | 0.27 | 0.22 | 0.19 |
| | | −45.6 | 5.6 | 0.36 | 0.35 | 0.31 | 0.25 | 0.20 | 0.38 | 0.37 | 0.32 | 0.26 | 0.21 |
| 45° Slope | Up ↗ | 32.2 | 5.6 | 0.43 | 0.41 | 0.29 | 0.19 | 0.13 | 0.52 | 0.49 | 0.33 | 0.20 | 0.14 |
| | | 10.0 | 16.7 | 0.36 | 0.35 | 0.27 | 0.19 | 0.15 | 0.35 | 0.34 | 0.27 | 0.19 | 0.14 |
| | | 10.0 | 5.6 | 0.45 | 0.43 | 0.32 | 0.21 | 0.16 | 0.51 | 0.48 | 0.35 | 0.23 | 0.17 |
| | | −17.8 | 11.1 | 0.39 | 0.38 | 0.31 | 0.23 | 0.18 | 0.37 | 0.36 | 0.30 | 0.23 | 0.18 |
| | | −17.8 | 5.6 | 0.46 | 0.45 | 0.36 | 0.25 | 0.19 | 0.48 | 0.46 | 0.37 | 0.26 | 0.20 |
| | | −45.6 | 11.1 | 0.37 | 0.36 | 0.31 | 0.25 | 0.21 | 0.36 | 0.35 | 0.31 | 0.25 | 0.20 |
| | | −45.6 | 5.6 | 0.46 | 0.45 | 0.38 | 0.29 | 0.23 | 0.45 | 0.43 | 0.37 | 0.29 | 0.23 |
| Vertical | Horiz. → | 32.2 | 5.6 | 0.43 | 0.41 | 0.29 | 0.19 | 0.14 | 0.62 | 0.57 | 0.37 | 0.21 | 0.15 |
| | | 10.0 | 16.7 | 0.45 | 0.43 | 0.32 | 0.22 | 0.16 | 0.51 | 0.49 | 0.35 | 0.23 | 0.17 |
| | | 10.0 | 5.6 | 0.47 | 0.45 | 0.33 | 0.22 | 0.16 | 0.65 | 0.61 | 0.41 | 0.25 | 0.18 |
| | | −17.8 | 11.1 | 0.50 | 0.48 | 0.38 | 0.26 | 0.20 | 0.55 | 0.53 | 0.41 | 0.28 | 0.21 |
| | | −17.8 | 5.6 | 0.52 | 0.50 | 0.39 | 0.27 | 0.20 | 0.66 | 0.63 | 0.46 | 0.30 | 0.22 |
| | | −45.6 | 11.1 | 0.51 | 0.50 | 0.41 | 0.31 | 0.24 | 0.51 | 0.50 | 0.42 | 0.31 | 0.24 |
| | | −45.6 | 5.6 | 0.56 | 0.55 | 0.45 | 0.33 | 0.26 | 0.65 | 0.63 | 0.51 | 0.36 | 0.27 |
| 45° Slope | Down ↘ | 32.2 | 5.6 | 0.44 | 0.41 | 0.29 | 0.19 | 0.14 | 0.62 | 0.58 | 0.37 | 0.21 | 0.15 |
| | | 10.0 | 16.7 | 0.46 | 0.44 | 0.33 | 0.22 | 0.16 | 0.60 | 0.57 | 0.39 | 0.24 | 0.17 |
| | | 10.0 | 5.6 | 0.47 | 0.45 | 0.33 | 0.22 | 0.16 | 0.67 | 0.63 | 0.42 | 0.26 | 0.18 |
| | | −17.8 | 11.1 | 0.51 | 0.49 | 0.39 | 0.27 | 0.20 | 0.66 | 0.63 | 0.46 | 0.30 | 0.22 |
| | | −17.8 | 5.6 | 0.52 | 0.50 | 0.39 | 0.27 | 0.20 | 0.73 | 0.69 | 0.49 | 0.32 | 0.23 |
| | | −45.6 | 11.1 | 0.56 | 0.54 | 0.44 | 0.33 | 0.25 | 0.67 | 0.64 | 0.51 | 0.36 | 0.28 |
| | | −45.6 | 5.6 | 0.57 | 0.56 | 0.45 | 0.33 | 0.26 | 0.77 | 0.74 | 0.57 | 0.39 | 0.29 |
| Horiz. | Down ↓ | 32.2 | 5.6 | 0.44 | 0.41 | 0.29 | 0.19 | 0.14 | 0.62 | 0.58 | 0.37 | 0.21 | 0.15 |
| | | 10.0 | 16.7 | 0.47 | 0.45 | 0.33 | 0.22 | 0.16 | 0.66 | 0.62 | 0.42 | 0.25 | 0.18 |
| | | 10.0 | 5.6 | 0.47 | 0.45 | 0.33 | 0.22 | 0.16 | 0.68 | 0.63 | 0.42 | 0.26 | 0.18 |
| | | −17.8 | 11.1 | 0.52 | 0.50 | 0.39 | 0.27 | 0.20 | 0.74 | 0.70 | 0.50 | 0.32 | 0.23 |
| | | −17.8 | 5.6 | 0.52 | 0.50 | 0.39 | 0.27 | 0.20 | 0.75 | 0.71 | 0.51 | 0.32 | 0.23 |
| | | −45.6 | 11.1 | 0.57 | 0.55 | 0.45 | 0.33 | 0.26 | 0.81 | 0.78 | 0.59 | 0.40 | 0.30 |
| | | −45.6 | 5.6 | 0.58 | 0.56 | 0.46 | 0.33 | 0.26 | 0.83 | 0.79 | 0.60 | 0.40 | 0.30 |

| Position of Air Space | Direction of Heat Flow | Air Space Mean Temp., °C | Temp. Diff., K | 40 mm Air Space[c] 0.03 | 0.05 | 0.2 | 0.5 | 0.82 | 90 mm Air Space[c] 0.03 | 0.05 | 0.2 | 0.5 | 0.82 |
|---|---|---|---|---|---|---|---|---|---|---|---|---|---|
| Horiz. | Up ↑ | 32.2 | 5.6 | 0.45 | 0.42 | 0.30 | 0.19 | 0.14 | 0.50 | 0.47 | 0.32 | 0.20 | 0.14 |
| | | 10.0 | 16.7 | 0.33 | 0.32 | 0.26 | 0.18 | 0.14 | 0.27 | 0.35 | 0.28 | 0.19 | 0.15 |
| | | 10.0 | 5.6 | 0.44 | 0.42 | 0.32 | 0.21 | 0.16 | 0.49 | 0.47 | 0.34 | 0.23 | 0.16 |
| | | −17.8 | 11.1 | 0.35 | 0.34 | 0.29 | 0.22 | 0.17 | 0.40 | 0.38 | 0.32 | 0.23 | 0.18 |
| | | −17.8 | 5.6 | 0.43 | 0.41 | 0.33 | 0.24 | 0.19 | 0.48 | 0.46 | 0.36 | 0.26 | 0.20 |
| | | −45.6 | 11.1 | 0.34 | 0.34 | 0.30 | 0.24 | 0.20 | 0.39 | 0.38 | 0.33 | 0.26 | 0.21 |
| | | −45.6 | 5.6 | 0.42 | 0.41 | 0.35 | 0.27 | 0.22 | 0.47 | 0.45 | 0.38 | 0.29 | 0.23 |
| 45° Slope | Up ↗ | 32.2 | 5.6 | 0.51 | 0.48 | 0.33 | 0.20 | 0.14 | 0.56 | 0.52 | 0.35 | 0.21 | 0.14 |
| | | 10.0 | 16.7 | 0.38 | 0.36 | 0.28 | 0.20 | 0.15 | 0.40 | 0.38 | 0.29 | 0.20 | 0.15 |
| | | 10.0 | 5.6 | 0.51 | 0.48 | 0.35 | 0.23 | 0.17 | 0.55 | 0.52 | 0.37 | 0.24 | 0.17 |
| | | −17.8 | 11.1 | 0.40 | 0.39 | 0.32 | 0.24 | 0.18 | 0.43 | 0.41 | 0.33 | 0.24 | 0.19 |
| | | −17.8 | 5.6 | 0.49 | 0.47 | 0.37 | 0.26 | 0.20 | 0.52 | 0.51 | 0.39 | 0.27 | 0.20 |
| | | −45.6 | 11.1 | 0.39 | 0.38 | 0.33 | 0.26 | 0.21 | 0.41 | 0.40 | 0.35 | 0.27 | 0.22 |
| | | −45.6 | 5.6 | 0.48 | 0.46 | 0.39 | 0.30 | 0.24 | 0.51 | 0.49 | 0.41 | 0.31 | 0.24 |
| Vertical | Horiz. → | 32.2 | 5.6 | 0.70 | 0.64 | 0.40 | 0.22 | 0.15 | 0.65 | 0.60 | 0.38 | 0.22 | 0.15 |
| | | 10.0 | 16.7 | 0.45 | 0.43 | 0.32 | 0.22 | 0.16 | 0.47 | 0.45 | 0.33 | 0.22 | 0.16 |
| | | 10.0 | 5.6 | 0.67 | 0.62 | 0.42 | 0.26 | 0.18 | 0.64 | 0.60 | 0.41 | 0.25 | 0.18 |
| | | −17.8 | 11.1 | 0.49 | 0.47 | 0.37 | 0.26 | 0.20 | 0.51 | 0.49 | 0.38 | 0.27 | 0.20 |
| | | −17.8 | 5.6 | 0.62 | 0.59 | 0.44 | 0.29 | 0.22 | 0.61 | 0.59 | 0.44 | 0.29 | 0.22 |
| | | −45.6 | 11.1 | 0.46 | 0.45 | 0.38 | 0.29 | 0.23 | 0.50 | 0.48 | 0.40 | 0.30 | 0.24 |
| | | −45.6 | 5.6 | 0.58 | 0.56 | 0.46 | 0.34 | 0.26 | 0.60 | 0.58 | 0.47 | 0.34 | 0.26 |

# Load Estimating Fundamentals

Table 5-14 SI  Effective Thermal Resistances of Plane Air Spaces,[a,b,c] (m²·K)/W *(Continued)*
*(Table 3, Chapter 26, 2013 ASHRAE Handbook—Fundamentals, SI Version)*

| Position of Air Space | Direction of Heat Flow | Mean Temp.[d], °C | Temp. Diff.,[d] K | 40 mm Air Space[c] 0.03 | 0.05 | 0.2 | 0.5 | 0.82 | 90 mm Air Space[c] 0.03 | 0.05 | 0.2 | 0.5 | 0.82 |
|---|---|---|---|---|---|---|---|---|---|---|---|---|---|
| 45° Slope | Down | 32.2 | 5.6 | 0.89 | 0.80 | 0.45 | 0.24 | 0.16 | 0.85 | 0.76 | 0.44 | 0.24 | 0.16 |
|  |  | 10.0 | 16.7 | 0.63 | 0.59 | 0.41 | 0.25 | 0.18 | 0.62 | 0.58 | 0.40 | 0.25 | 0.18 |
|  |  | 10.0 | 5.6 | 0.90 | 0.82 | 0.50 | 0.28 | 0.19 | 0.83 | 0.77 | 0.48 | 0.28 | 0.19 |
|  |  | −17.8 | 11.1 | 0.68 | 0.64 | 0.47 | 0.31 | 0.22 | 0.67 | 0.64 | 0.47 | 0.31 | 0.22 |
|  |  | −17.8 | 5.6 | 0.87 | 0.81 | 0.56 | 0.34 | 0.24 | 0.81 | 0.76 | 0.53 | 0.33 | 0.24 |
|  |  | −45.6 | 11.1 | 0.64 | 0.62 | 0.49 | 0.35 | 0.27 | 0.66 | 0.64 | 0.51 | 0.36 | 0.28 |
|  |  | −45.6 | 5.6 | 0.82 | 0.79 | 0.60 | 0.40 | 0.30 | 0.79 | 0.76 | 0.58 | 0.40 | 0.30 |
| Horiz. | Down | 32.2 | 5.6 | 1.07 | 0.94 | 0.49 | 0.25 | 0.17 | 1.77 | 1.44 | 0.60 | 0.28 | 0.18 |
|  |  | 10.0 | 16.7 | 1.10 | 0.99 | 0.56 | 0.30 | 0.20 | 1.69 | 1.44 | 0.68 | 0.33 | 0.21 |
|  |  | 10.0 | 5.6 | 1.16 | 1.04 | 0.58 | 0.30 | 0.20 | 1.96 | 1.63 | 0.72 | 0.34 | 0.22 |
|  |  | −17.8 | 11.1 | 1.24 | 1.13 | 0.69 | 0.39 | 0.26 | 1.92 | 1.68 | 0.86 | 0.43 | 0.29 |
|  |  | −17.8 | 5.6 | 1.29 | 1.17 | 0.70 | 0.39 | 0.27 | 2.11 | 1.82 | 0.89 | 0.44 | 0.29 |
|  |  | −45.6 | 11.1 | 1.36 | 1.27 | 0.84 | 0.50 | 0.35 | 2.05 | 1.85 | 1.06 | 0.57 | 0.38 |
|  |  | −45.6 | 5.6 | 1.42 | 1.32 | 0.86 | 0.51 | 0.35 | 2.28 | 2.03 | 1.12 | 0.59 | 0.39 |

| Position of Air Space | Direction of Heat Flow | Mean Temp., °C | Temp. Diff., K | 143 mm Air Space[c] 0.03 | 0.05 | 0.2 | 0.5 | 0.82 |
|---|---|---|---|---|---|---|---|---|
| Horiz. | Up | 32.2 | 5.6 | 0.53 | 0.50 | 0.33 | 0.20 | 0.14 |
|  |  | 10.0 | 16.7 | 0.39 | 0.38 | 0.29 | 0.20 | 0.15 |
|  |  | 10.0 | 5.6 | 0.52 | 0.50 | 0.36 | 0.23 | 0.17 |
|  |  | 17.8 | 11.1 | 0.42 | 0.41 | 0.33 | 0.24 | 0.19 |
|  |  | 17.8 | 5.6 | 0.51 | 0.49 | 0.38 | 0.27 | 0.20 |
|  |  | 45.6 | 11.1 | 0.41 | 0.40 | 0.34 | 0.27 | 0.22 |
|  |  | 45.6 | 5.6 | 0.49 | 0.48 | 0.40 | 0.30 | 0.24 |
| 45° Slope | Up | 32.2 | 5.6 | 0.57 | 0.54 | 0.35 | 0.21 | 0.15 |
|  |  | 10.0 | 16.7 | 0.39 | 0.37 | 0.29 | 0.20 | 0.15 |
|  |  | 10.0 | 5.6 | 0.56 | 053 | 0.37 | 0.24 | 0.17 |
|  |  | 17.8 | 11.1 | 0.41 | 0.40 | 0.33 | 0.24 | 0.18 |
|  |  | 17.8 | 5.6 | 0.53 | 0.51 | 0.39 | 0.27 | 0.20 |
|  |  | 45.6 | 11.1 | 0.38 | 0.37 | 0.32 | 0.26 | 0.21 |
|  |  | 45.6 | 11.1 | 0.49 | 0.48 | 0.40 | 0.30 | 0.24 |
| Vertical | Horiz. | 32.2 | 5.6 | 0.66 | 0.61 | 0.38 | 0.22 | 0.15 |
|  |  | 10.0 | 16.7 | 0.50 | 0.47 | 0.35 | 0.23 | 0.16 |
|  |  | 10.0 | 5.6 | 0.66 | 0.61 | 0.42 | 0.25 | 0.18 |
|  |  | 17.8 | 11.1 | 0.54 | 0.52 | 0.40 | 0.28 | 0.21 |
|  |  | 17.8 | 5.6 | 0.64 | 0.61 | 0.46 | 0.30 | 0.22 |
|  |  | 45.6 | 11.1 | 0.53 | 0.52 | 0.43 | 0.32 | 0.25 |
|  |  | 45.6 | 11.1 | 0.63 | 0.61 | 0.49 | 0.35 | 0.27 |
| 45° Slope | Down | 32.2 | 5.6 | 0.865 | 0.78 | 0.44 | 0.24 | 0.16 |
|  |  | 10.0 | 16.7 | 0.68 | 0.64 | 0.43 | 0.26 | 0.18 |
|  |  | 10.0 | 5.6 | 0.87 | 0.80 | 0.49 | 0.28 | 0.19 |
|  |  | 17.8 | 11.1 | 0.75 | 0.71 | 0.50 | 0.32 | 0.23 |
|  |  | 17.8 | 5.6 | 0.87 | 0.82 | 0.56 | 0.34 | 0.24 |
|  |  | 45.6 | 11.1 | 0.75 | 0.73 | 0.56 | 0.39 | 0.29 |
|  |  | 45.6 | 11.1 | 0.87 | 0.83 | 0.62 | 0.41 | 0.31 |
| Horiz. | Down | 32.2 | 5.6 | 2.06 | 1.63 | 0.63 | 0.28 | 0.18 |
|  |  | 10.0 | 16.7 | 1.87 | 1.57 | 0.71 | 0.34 | 0.22 |
|  |  | 10.0 | 5.6 | 2.24 | 1.82 | 0.75 | 0.35 | 0.22 |
|  |  | 17.8 | 11.1 | 2.13 | 1.84 | 0.90 | 0.44 | 0.29 |
|  |  | 17.8 | 5.6 | 2.43 | 2.05 | 0.95 | 0.46 | 0.29 |
|  |  | 45.6 | 11.1 | 2.19 | 1.96 | 1.10 | 0.58 | 0.39 |
|  |  | 45.6 | 11.1 | 2.57 | 2.26 | 1.18 | 0.61 | 0.40 |

[a]See Chapter 25. Thermal resistance values were determined from $R = 1/C$, where $C = h_c + \varepsilon_{eff} h_r$, $h_c$ is conduction/convection coefficient, $\varepsilon_{eff} h_r$ is radiation coefficient $\approx 0.227\varepsilon_{eff}[(t_m + 273)/100]^3$, and $t_m$ is mean temperature of air space. Values for $h_c$ were determined from data developed by Robinson et al. (1954). Equations (5) to (7) in Yarbrough (1983) show data in this table in analytic form. For extrapolation from this table to air spaces less than 12.5 mm (e.g., insulating window glass), assume $h_c = 21.8(1 + 0.00274 t_m)/l$, where $l$ is air space thickness in mm, and $h_c$ is heat transfer in W/(m²·K) through air space only.

[b]Values based on data presented by Robinson et al. (1954). (Also see Chapter 4, Tables 5 and 6, and Chapter 33). Values apply for ideal conditions (i.e., air spaces of uniform thickness bounded by plane, smooth, parallel surfaces with no air leakage to or from the space). **This table should not be used for hollow siding or profiled cladding: see Table 1.** For greater accuracy, use overall U-factors determined through guarded hot box (ASTM Standard C1363) testing. Thermal resistance values for multiple air spaces must be based on careful estimates of mean temperature differences for each air space.

[c]A single resistance value cannot account for multiple air spaces; each air space requires a separate resistance calculation that applies only for established boundary conditions. Resistances of horizontal spaces with heat flow downward are substantially independent of temperature difference.

[d]Interpolation is permissible for other values of mean temperature, temperature difference, and effective emittance $\varepsilon_{eff}$. Interpolation and moderate extrapolation for air spaces greater than 90 mm are also permissible.

[e]Effective emittance $\varepsilon_{eff}$ of air space is given by $1/\varepsilon_{eff} = 1/\varepsilon_1 + 1/\varepsilon_2 - 1$, where $\varepsilon_1$ and $\varepsilon_2$ are emittances of surfaces of air space (see Table 2). **Also, oxidation, corrosion, and accumulation of dust and dirt can dramatically increase surface emittance. Emittance values of 0.05 should only be used where the highly reflective surface can be maintained over the service life of the assembly.**

### Table 5-15 SI  Building and Insulating Materials: Design Values[a]
*(Table 1, Chapter 26, 2013 ASHRAE Handbook—Fundamentals)*

| Description | Density, kg/m³ | Conductivity[b] k, W/(m·K) | Resistance R, (m²·K)/W | Specific Heat, kJ/(kg·K) | Reference[l] |
|---|---|---|---|---|---|
| **Insulating Materials** | | | | | |
| *Blanket and batt*[c,d] | | | | | |
| Glass-fiber batts................................................ | — | — | — | 0.8 | Kumaran (2002) |
| | 7.5 to 8.2 | 0.046 to 0.048 | — | — | Four manufacturers (2011) |
| | 9.8 to 12 | 0.040 to 0.043 | — | — | Four manufacturers (2011) |
| | 13 to 14 | 0.037 to 0.039 | — | — | Four manufacturers (2011) |
| | 22 | 0.033 | — | — | Four manufacturers (2011) |
| Rock and slag wool batts.................................. | — | — | — | 0.8 | Kumaran (1996) |
| | 32 to 37 | 0.036 to 0.037 | — | — | One manufacturer (2011) |
| | 45 | 0.033 to 0.035 | — | — | One manufacturer (2011) |
| Mineral wool, felted ........................................ | 16 to 48 | 0.040 | — | — | CIBSE (2006), NIST (2000) |
| | 16 to 130 | 0.035 | — | — | NIST (2000) |
| *Board and slabs* | | | | | |
| Cellular glass ................................................... | 120 | 0.042 | — | 0.8 | One manufacturer (2011) |
| Cement fiber slabs, shredded wood with Portland cement | | | | | |
| binder........................................................... | 400 to 430 | 0.072 to 0.076 | — | — | |
| with magnesia oxysulfide binder................... | 350 | 0.082 | — | 1.3 | |
| Glass fiber board ............................................. | — | — | — | 0.8 | Kumaran (1996) |
| | 24 to 96 | 0.033 to 0.035 | — | — | One manufacturer (2011) |
| Expanded rubber (rigid) ................................... | 64 | 0.029 | — | 1.7 | Nottage (1947) |
| Extruded polystyrene, smooth skin ................... | — | — | — | 1.5 | Kumaran (1996) |
| aged per Can/ULC Standard S770-2003 ...... | 22 to 58 | 0.026 to 0.029 | — | — | Four manufacturers (2011) |
| aged 180 days ............................................. | 22 to 58 | 0.029 | — | — | One manufacturer (2011) |
| European product ........................................ | 30 | 0.030 | — | — | One manufacturer (2011) |
| aged 5 years at 24°C................................... | 32 to 35 | 0030 | — | — | One manufacturer (2011) |
| blown with low global warming potential (GWP) (<5) | | | | | |
| blowing agent ......................................... | | 0.035 to 0.036 | — | — | One manufacturer (2011) |
| Expanded polystyrene, molded beads ............... | — | — | — | 1.5 | Kumaran (1996) |
| | 16 to 24 | 0.035 to 0.037 | — | — | Independent test reports (2008) |
| | 29 | 0.033 | — | — | Independent test reports (2008) |
| Mineral fiberboard, wet felted......................... | 160 | 0.037 | — | 0.8 | Kumaran (1996) |
| Rock wool board .............................................. | — | — | — | 0.8 | Kumaran (1996) |
| floors and walls .......................................... | 64 to 130 | 0.033 to 0.036 | — | — | Five manufacturers (2011) |
| roofing ........................................................ | 160 to 180. | 0.039 to 0.042 | — | 0.8 | Five manufacturers (2011) |
| Acoustical tile[e] ................................................ | 340 to 370 | 0.052 to 0.053 | 0.6 to 0.8 | | |
| Perlite board ..................................................... | 140 | 0.052 | — | — | One manufacturer (2010) |
| Polyisocyanurate ............................................... | — | — | — | 1.5 | Kumaran (1996) |
| unfaced, aged per Can/ULC Standard S770-2003 ........ | 26 to 37 | 0.023 to 0.025 | — | — | Seven manufacturers (2011) |
| with foil facers, aged 180 days................... | — | 0.022 to 0.023 | — | — | Two manufacturers (2011) |
| Phenolic foam board with facers, aged ............. | — | 0.020 to 0.023 | — | — | One manufacturer (2011) |
| *Loose fill* | | | | | |
| Cellulose fiber, loose fill ................................... | — | — | — | 1.4 | NIST (2000), Kumaran (1996) |
| attic application up to 100 mm .................. | 16 to 19 | 0.045 to 0.046 | — | — | Four manufacturers (2011) |
| attic application > 100 mm........................ | 19 to 26 | 0.039 to 0.040 | — | — | Four manufacturers (2011) |
| wall application, dense packed................... | 56 | 0.039 to 0.040 | — | — | One manufacturer (2011) |
| Perlite, expanded .............................................. | 32 to 64 | 0.039 to 0.045 | — | 1.1 | (Manufacturer, pre 2001) |
| | 64 to 120 | 0.045 to 0.052 | — | — | (Manufacturer, pre 2001) |
| | 120 to 180 | 0.052 to 0.061 | — | — | (Manufacturer, pre 2001) |
| Glass fiber[d] | | | | | |
| attics, ~100 to 600 mm ............................... | 6.4 to 8.0 | 0.052 to 0.055 | — | — | Four manufacturers (2011) |
| attics, ~600 to 1100 mm ............................. | 8 to 9.6 | 0.049 to 0.052 | — | — | Four manufacturers (2011) |
| closed attic or wall cavities ........................ | 29 to 37 | 0.035 to 0.036 | — | — | Four manufacturers (2011) |
| Rock and slag wool[d] | | | | | |
| attics, ~90 to 115 mm ................................. | 24 to 26 | 0.049 | — | — | Three manufacturers (2011) |
| attics, ~125 to 430 mm ............................... | 24 to 29 | 0.046 to 0.048 | — | — | Three manufacturers (2011) |
| closed attic or wall cavities ........................ | 64 | 0.039 to 0.042 | — | — | Three manufacturers (2011) |
| Vermiculite, exfoliated...................................... | 112 to 131 | 0.068 | — | 1.3 | Sabine et al. (1975) |
| | 64 to 96 | 0.063 | — | — | Manufacturer (pre 2001) |
| *Spray-applied* | | | | | |
| Cellulose, sprayed into open wall cavities ....... | 26 to 42 | 0.039 to 0.040 | — | — | Two manufacturers (2011) |
| Glass fiber, sprayed into open wall or attic cavities...... | 16 | 0.039 to 0.042 | — | — | Manufacturers' association (2011) |
| | 29 to 37 | 0.033 to 0.037 | — | — | Four manufacturers (2011) |
| Polyurethane foam ............................................ | — | — | — | 1.5 | Kumaran (2002) |
| low density, open cell................................. | 7.2 to 10 | 0.037 to 0.042 | — | — | Three manufacturers (2011) |
| medium density, closed cell, aged 180 days ................. | 30 to 51 | 0.020 to 0.029 | — | — | Five manufacturers (2011) |
| **Building Board and Siding** | | | | | |

# Load Estimating Fundamentals

**Table 5-15 SI   Building and Insulating Materials: Design Values[a]** *(Continued)*
*(Table 1, Chapter 26, 2013 ASHRAE Handbook—Fundamentals)*

| Description | Density, kg/m³ | Conductivity[b] $k$, W/(m·K) | Resistance $R$, (m²·K)/W | Specific Heat, kJ/(kg·K) | Reference[l] |
|---|---|---|---|---|---|
| *Board* | | | | | |
| Asbestos/cement board | 1900 | 0.57 | — | 1.00 | Nottage (1947) |
| Cement board | 1150 | 0.25 | — | 0.84 | Kumaran (2002) |
| Fiber/cement board | 1400 | 0.25 | — | 0.84 | Kumaran (2002) |
|  | 1000 | 0.19 | — | 0.84 | Kumaran (1996) |
|  | 400 | 0.07 | — | 1.88 | Kumaran (1996) |
|  | 300 | 0.06 | — | 1.88 | Kumaran (1996) |
| Gypsum or plaster board | 640 | 0.16 | — | 1.15 | Kumaran (2002) |
| Oriented strand board (OSB) ............ 9 to 11 mm | 650 | — | 0.11 | 1.88 | Kumaran (2002) |
| ............................................................. 12.7 mm | 650 | — | 0.12 | 1.88 | Kumaran (2002) |
| Plywood (douglas fir) ........................... 12.7 mm | 460 | — | 0.14 | 1.88 | Kumaran (2002) |
| ............................................................. 15.9 mm | 540 | — | 0.15 | 1.88 | Kumaran (2002) |
| Plywood/wood panels ........................... 19.0 mm | 450 | — | 0.19 | 1.88 | Kumaran (2002) |
| Vegetable fiber board | 650 | — | 0.11 | 1.88 | Kumaran (2002) |
| sheathing, regular density ............... 12.7 mm | 290 | — | 0.23 | 1.30 | Lewis (1967) |
| intermediate density ................ 12.7 mm | 350 | — | 0.19 | 1.30 | Lewis (1967) |
| nail-based sheathing ........................ 12.7 mm | 400 | — | 0.19 | 1.30 |  |
| shingle backer .................................... 9.5 mm | 290 | — | 0.17 | 1.30 |  |
| sound deadening board ................... 12.7 mm | 240 | — | 0.24 | 1.26 |  |
| tile and lay-in panels, plain or acoustic | 290 | 0.058 | — | 0.59 |  |
| laminated paperboard | 480 | 0.072 | — | 1.38 | Lewis (1967) |
| homogeneous board from repulped paper | 480 | 0.072 | — | 1.17 |  |
| Hardboard | | | | | |
| medium density | 800 | 0.105 | — | 1.30 | Lewis (1967) |
| high density, service-tempered grade and service grade | 880 | 0.12 | — | 1.34 | Lewis (1967) |
| high density, standard-tempered grade | 1010 | 0.144 | — | 1.34 | Lewis (1967) |
| Particleboard | | | | | |
| low density | 590 | 0.102 | — | 1.30 | Lewis (1967) |
| medium density | 800 | 0.135 | — | 1.30 | Lewis (1967) |
| high density | 1000 | 1.18 | — | — | Lewis (1967) |
| underlayment ..................................... 15.9 mm | 640 | — | 1.22 | 1.21 | Lewis (1967) |
| Waferboard | 700 | 0.072 | — | 1.88 | Kumaran (1996) |
| *Shingles* | | | | | |
| Asbestos/cement | 1900 | — | 0.037 | — |  |
| Wood, 400 mm, 190 mm exposure | — | — | 0.15 | 1.30 |  |
| Wood, double, 400 mm, 300 mm exposure | — | — | 0.21 | 1.17 |  |
| Wood, plus ins. backer board ................ 8 mm | — | — | 0.25 | 1.30 |  |
| *Siding* | | | | | |
| Asbestos/cement, lapped ..................... 6.4 mm | — | — | 0.037 | 1.01 |  |
| Asphalt roll siding | — | — | 0.026 | 1.47 |  |
| Asphalt insulating siding (12.7 mm bed) | — | — | 0.26 | 1.47 |  |
| Hardboard siding ..................................... 11 mm | — | — | — | 0.12 | 1.17 |
| Wood, drop, 200 mm ............................... 25 mm | — | — | 0.14 | 1.17 |  |
| Wood, bevel | | | | | |
| 200 mm, lapped ................................... 13 mm | — | — | 0.14 | 1.17 |  |
| 250 mm, lapped ................................... 19 mm | — | — | 0.18 | 1.17 |  |
| Wood, plywood, lapped ....................... 9.5 mm | — | — | 0.10 | 1.22 |  |
| Aluminum, steel, or vinyl,[h, i] over sheathing | — | — | — |  |  |
| hollow-backed | — | — | 0.11 | 1.22[i] |  |
| insulating-board-backed ................... 9.5 mm | — | — | 0.32 | 1.34 |  |
| foil-backed ........................................... 9.5 mm | — | — | 0.52 | — |  |
| Architectural (soda-lime float) glass | 2500 | 1.0 | — | 0.84 |  |
| **Building Membrane** | | | | | |
| Vapor-permeable felt | — | — | 0.011 | — |  |
| Vapor: seal, 2 layers of mopped 0.73 kg/m² felt | — | — | 0.21 | — |  |
| Vapor: seal, plastic film | — | — | Negligible | — |  |
| **Finish Flooring Materials** | | | | | |
| Carpet and rebounded urethane pad ........ 19 mm | 110 | — | 0.42 | — | NIST (2000) |
| Carpet and rubber pad (one-piece) ........ 9.5 mm | 320 | — | 0.12 | — | NIST (2000) |
| Pile carpet with rubber pad ........ 9.5 to 12.7 mm | 290 | — | 0.28 | — | NIST (2000) |
| Linoleum/cork tile .................................... 6.4 mm | 465 | — | 0.09 | — | NIST (2000) |
| PVC/rubber floor covering | — | 0.40 | — | — | CIBSE (2006) |
| rubber tile ............................................... 25 mm | 1900 | — | 0.06 | — | NIST (2000) |
| terrazzo .................................................... 25 mm | — | — | 0.014 | 0.80 |  |
| **Metals** (See Chapter 33, Table 3) | | | | | |
| **Roofing** | | | | | |
| Asbestos/cement shingles | 1920 | — | 0.037 | 1.00 |  |

**Table 5-15 SI  Building and Insulating Materials: Design Values[a] *(Continued)***
*(Table 1, Chapter 26, 2013 ASHRAE Handbook—Fundamentals)*

| Description | Density, kg/m³ | Conductivity[b] k, W/(m·K) | Resistance R, (m²·K)/W | Specific Heat, kJ/(kg·K) | Reference[l] |
|---|---|---|---|---|---|
| Asphalt (bitumen with inert fill) | 1600 | 0.43 | — | — | CIBSE (2006) |
|  | 1900 | 0.58 | — | — | CIBSE (2006) |
|  | 2300 | 1.15 | — | — | CIBSE (2006) |
| Asphalt roll roofing | 920 | — | 0.027 | 1.51 |  |
| Asphalt shingles | 920 | — | 0.078 | 1.26 |  |
| Built-up roofing ............ 10 mm | 920 | — | 0.059 | 1.47 |  |
| Mastic asphalt (heavy, 20% grit) | 950 | 0.19 | — | — | CIBSE (2006) |
| Reed thatch | 270 | 0.09 | — | — | CIBSE (2006) |
| Roofing felt | 2250 | 1.20 | — | — | CIBSE (2006) |
| Slate ............ 13 mm | — | — | 0.009 | 1.26 |  |
| Straw thatch | 240 | 0.07 | — | — | CIBSE (2006) |
| Wood shingles, plain and plastic-film-faced | — | — | 0.166 | 1.30 |  |
| **Plastering Materials** |  |  |  |  |  |
| Cement plaster, sand aggregate | 1860 | 0.72 | — | 0.84 |  |
| Sand aggregate ............ 10 mm | — | — | 0.013 | 0.84 |  |
| ............ 20 mm | — | — | 0.026 | 0.84 |  |
| Gypsum plaster | 1120 | 0.38 | — | — | CIBSE (2006) |
|  | 1280 | 0.46 | — | — | CIBSE (2006) |
| Lightweight aggregate ............ 13 mm |  | 720 | — | 0.056 | — |
| ............ 16 mm |  | 720 | — | 0.066 | — |
| on metal lath ............ 19 mm | — | — | 0.083 | — |  |
| Perlite aggregate | 720 | 0.22 | — | 1.34 |  |
| Sand aggregate | 1680 | 0.81 | — | 0.84 |  |
| on metal lath ............ 19 mm | — | — | 0.023 | — |  |
| Vermiculite aggregate | 480 | 0.14 | — | — | CIBSE (2006) |
|  | 600 | 0.20 | — | — | CIBSE (2006) |
|  | 720 | 0.25 | — | — | CIBSE (2006) |
|  | 840 | 0.26 | — | — | CIBSE (2006) |
|  | 960 | 0.30 | — | — | CIBSE (2006) |
| Perlite plaster | 400 | 0.08 | — | — | CIBSE (2006) |
|  | 600 | 0.19 | — | — | CIBSE (2006) |
| Pulpboard or paper plaster | 600 | 0.07 | — | — | CIBSE (2006) |
| Sand/cement plaster, conditioned | 1560 | 0.63 | — | — | CIBSE (2006) |
| Sand/cement/lime plaster, conditioned | 1440 | 0.48 | — | — | CIBSE (2006) |
| Sand/gypsum (3:1) plaster, conditioned | 1550 | 0.65 | — | — | CIBSE (2006) |
| **Masonry Materials** |  |  |  |  |  |
| *Masonry units* |  |  |  |  |  |
| Brick, fired clay | 2400 | 1.21 to 1.47 | — | — | Valore (1988) |
|  | 2240 | 1.07 to 1.30 | — | — | Valore (1988) |
|  | 2080 | 0.92 to 1.12 | — | — | Valore (1988) |
|  | 1920 | 0.81 to 0.98 | — | 0.80 | Valore (1988) |
|  | 1760 | 0.71 to 0.85 | — | — | Valore (1988) |
|  | 1600 | 0.61 to 0.74 | — | — | Valore (1988) |
|  | 1440 | 0.52 to 0.62 | — | — | Valore (1988) |
|  | 1280 | 0.43 to 0.53 | — | — | Valore (1988) |
|  | 1120 | 0.36 to 0.45 | — | — | Valore (1988) |
| Clay tile, hollow |  |  |  |  |  |
| 1 cell deep ............ 75 mm | — | — | 0.14 | 0.88 | Rowley and Algren (1937) |
| ............ 100 mm | — | — | 0.20 | — | Rowley and Algren (1937) |
| 2 cells deep ............ 150 mm | — | — | 0.27 | — | Rowley and Algren (1937) |
| ............ 200 mm | — | — | 0.33 | — | Rowley and Algren (1937) |
| ............ 250 mm | — | — | 0.39 | — | Rowley and Algren (1937) |
| 3 cells deep ............ 300 mm | — | — | 0.44 | — | Rowley and Algren (1937) |
| Lightweight brick | 800 | 0.20 | — | — | Kumaran (1996) |
|  | 770 | 0.22 | — | — | Kumaran (1996) |
| *Concrete blocks*[f, g] |  |  |  |  |  |
| Limestone aggregate |  |  |  |  |  |
| ~200 mm, 16.3 kg, 2200 kg/m³ concrete, 2 cores | — | — | — | — |  |
| with perlite-filled cores | — | — | 0.37 | — | Valore (1988) |
| ~300 mm, 25 kg, 2200 kg/m³ concrete, 2 cores | — | — | — | — |  |
| with perlite-filled cores | — | — | 0.65 | — | Valore (1988) |
| Normal-weight aggregate (sand and gravel) |  |  |  |  |  |
| ~200 mm, 16 kg, 2100 kg/m³ concrete, 2 or 3 cores | — | — | 0.20 to 0.17 | 0.92 | Van Geem (1985) |
| with perlite-filled cores | — | — | 0.35 | — | Van Geem (1985) |
| with vermiculite-filled cores | — | — | 0.34 to 0.24 | — | Valore (1988) |
| ~300 mm, 22.7 kg, 2000 kg/m³ concrete, 2 cores | — | — | 0.217 | 0.92 | Valore (1988) |
| Medium-weight aggregate (combinations of normal and lightweight aggregate) |  |  |  |  |  |

## Load Estimating Fundamentals

**Table 5-15 SI  Building and Insulating Materials: Design Values[a] *(Continued)***
*(Table 1, Chapter 26, 2013 ASHRAE Handbook—Fundamentals)*

| Description | Density, kg/m$^3$ | Conductivity[b] $k$, W/(m·K) | Resistance $R$, (m$^2$·K)/W | Specific Heat, kJ/(kg·K) | Reference[l] |
|---|---|---|---|---|---|
| ~200 mm, 13 kg, 1550 to 1800 kg/m$^3$ concrete, 2 or 3 cores | — | — | 0.30 to 0.22 | — | Van Geem (1985) |
| with perlite-filled cores | — | — | 0.65 to 0.41 | — | Van Geem (1985) |
| with vermiculite-filled cores | — | — | 0.58 | — | Van Geem (1985) |
| with molded-EPS-filled (beads) cores | — | — | 0.56 | — | Van Geem (1985) |
| with molded EPS inserts in cores | — | — | 0.47 | — | Van Geem (1985) |
| Low-mass aggregate (expanded shale, clay, slate or slag, pumice) | | | | | |
| ~150 mm, 7 1/2 kg, 1400 kg/m$^2$ concrete, 2 or 3 cores | — | — | 0.34 to 0.29 | — | Van Geem (1985) |
| with perlite-filled cores | — | — | 0.74 | — | Van Geem (1985) |
| with vermiculite-filled cores | — | — | 0.53 | — | Van Geem (1985) |
| 200 mm, 8 to 10 kg, 1150 to 1380 kg/m$^2$ concrete | — | — | 0.56 to 0.33 | 0.88 | Van Geem (1985) |
| with perlite-filled cores | — | — | 1.20 to 0.77 | — | Van Geem (1985) |
| with vermiculite-filled cores | — | — | 0.93 to 0.69 | — | Shu et al. (1979) |
| with molded-EPS-filled (beads) cores | — | — | 0.85 | — | Shu et al. (1979) |
| with UF foam-filled cores | — | — | 0.79 | — | Shu et al. (1979) |
| with molded EPS inserts in cores | — | — | 0.62 | — | Shu et al. (1979) |
| 300 mm, 16 kg, 1400 kg/m$^3$, concrete, 2 or 3 cores | — | — | 0.46 to 0.40 | — | Van Geem (1985) |
| with perlite-filled cores | — | — | 1.6 to 1.1 | — | Van Geem (1985) |
| with vermiculite-filled cores | — | — | 1.0 | — | Valore (1988) |
| Stone, lime, or sand | 2880 | 10.4 | — | — | Valore (1988) |
| Quartzitic and sandstone | 2560 | 6.2 | — | — | Valore (1988) |
| | 2240 | 3.46 | — | — | Valore (1988) |
| | 1920 | 1.88 | — | 0.88 | Valore (1988) |
| Calcitic, dolomitic, limestone, marble, and granite | 2880 | 4.33 | — | — | Valore (1988) |
| | 2560 | 3.17 | — | — | Valore (1988) |
| | 2240 | 2.31 | — | — | Valore (1988) |
| | 1920 | 1.59 | — | 0.88 | Valore (1988) |
| | 1600 | 1.15 | — | — | Valore (1988) |
| Gypsum partition tile | | | | | |
| 75 by 300 by 760 mm, solid | — | — | 0.222 | 0.79 | Rowley and Algren (1937) |
| 4 cells | — | — | 0.238 | — | Rowley and Algren (1937) |
| 100 by 300 by 760 mm, 3 cells | — | — | 0.294 | — | Rowley and Algren (1937) |
| Limestone | 2400 | 0.57 | — | 0.84 | Kumaran (2002) |
| | 2600 | 0.93 | — | 0.84 | Kumaran (2002) |
| *Concretes*[i] | | | | | |
| Sand and gravel or stone aggregate concretes | 2400 | 1.4 to 2.9 | — | — | Valore (1988) |
| (concretes with >50% quartz or quartzite sand have | 2240 | 1.3 to 2.6 | — | 0.80 to 1.00 | Valore (1988) |
| conductivities in higher end of range) | 2080 | 1.0 to 1.9 | — | — | Valore (1988) |
| Low-mass aggregate or limestone concretes | 1920 | 0.9 to 1.3 | — | — | Valore (1988) |
| expanded shale, clay, or slate; expanded slags; cinders; | 1600 | 0.68 to 0.89 | — | 0.84 | Valore (1988) |
| pumice (with density up to 1600 kg/m$^3$); scoria (sanded | 1280 | 0.48 to 0.59 | — | 0.84 | Valore (1988) |
| concretes have conductivities in higher end of range) | 960 | 0.30 to 0.36 | — | — | Valore (1988) |
| | 640 | 0.18 | — | — | Valore (1988) |
| Gypsum/fiber concrete (87.5% gypsum, 12.5% wood chips) | 800 | 0.24 | — | 0.84 | Rowley and Algren (1937) |
| Cement/lime, mortar, and stucco | 1920 | 1.40 | — | — | Valore (1988) |
| | 1600 | 0.97 | — | — | Valore (1988) |
| | 1280 | 0.65 | — | — | Valore (1988) |
| Perlite, vermiculite, and polystyrene beads | 800 | 0.26 to 0.27 | — | — | Valore (1988) |
| | 640 | 0.20 to 0.22 | — | 0.63 to 0.96 | Valore (1988) |
| | 480 | 0.16 | — | — | Valore (1988) |
| | 320 | 0.12 | — | — | Valore (1988) |
| Foam concretes | 1920 | 0.75 | — | — | Valore (1988) |
| | 1600 | 0.60 | — | — | Valore (1988) |
| | 1280 | 0.44 | — | — | Valore (1988) |
| | 1120 | 0.36 | — | — | Valore (1988) |
| Foam concretes and cellular concretes | 960 | 0.30 | — | — | Valore (1988) |
| | 640 | 0.20 | — | — | Valore (1988) |
| | 320 | 0.12 | — | — | Valore (1988) |
| Aerated concrete (oven-dried) | 430 to 800 | 0.20 | — | 0.84 | Kumaran (1996) |
| Polystyrene concrete (oven-dried) | 255 to 800 | 0.37 | — | 0.84 | Kumaran (1996) |
| Polymer concrete | 1950 | 1.64 | — | — | Kumaran (1996) |
| | 2200 | 1.03 | — | — | Kumaran (1996) |
| Polymer cement | 1870 | 0.78 | — | — | Kumaran (1996) |
| Slag concrete | 960 | 0.22 | — | — | Touloukian et al (1970) |
| | 1280 | 0.32 | — | — | Touloukian et al. (1970) |
| | 1600 | 0.43 | — | — | Touloukian et al. (1970) |
| | 2000 | 1.23 | — | — | Touloukian et al. (1970) |

**Table 5-15 SI  Building and Insulating Materials: Design Values[a] *(Continued)***
*(Table 1, Chapter 26, 2013 ASHRAE Handbook—Fundamentals)*

| Description | Density, kg/m³ | Conductivity[b] $k$, W/(m·K) | Resistance $R$, (m²·K)/W | Specific Heat, kJ/(kg·K) | Reference[l] |
|---|---|---|---|---|---|
| **Woods** (12% moisture content)[j] | | | | | |
| *Hardwoods* | — | — | — | 1.63[k] | Wilkes (1979) |
| Oak | 660 to 750 | 0.16 to 0.18 | — | — | Cardenas and Bible (1987) |
| Birch | 680 to 725 | 0.17 to 0.18 | — | — | Cardenas and Bible (1987) |
| Maple | 635 to 700 | 0.16 to 0.17 | — | — | Cardenas and Bible (1987) |
| Ash | 615 to 670 | 0.15 to 0.16 | — | — | Cardenas and Bible (1987) |
| *Softwoods* | — | — | — | 1.63[k] | Wilkes (1979) |
| Southern pine | 570 to 660 | 0.14 to 0.16 | — | — | Cardenas and Bible (1987) |
| Southern yellow pine | 500 | 0.13 | — | — | Kumaran (2002) |
| Eastern white pine | 400 | 0.10 | — | — | Kumaran (2002) |
| Douglas fir/larch | 535 to 580 | 0.14 to 0.15 | — | — | Cardenas and Bible (1987) |
| Southern cypress | 500 to 515 | 0.13 | — | — | Cardenas and Bible (1987) |
| Hem/fir, spruce/pine/fir | 390 to 500 | 0.11 to 0.13 | — | — | Cardenas and Bible (1987) |
| Spruce | 400 | 0.09 | — | — | Kumaran (2002) |
| Western red cedar | 350 | 0.09 | — | — | Kumaran (2002) |
| West coast woods, cedars | 350 to 500 | 0.10 to 0.13 | — | — | Cardenas and Bible (1987) |
| Eastern white cedar | 360 | 0.10 | — | — | Kumaran (2002) |
| California redwood | 390 to 450 | 0.11 to 0.12 | — | — | Cardenas and Bible (1987) |
| Pine (oven-dried) | 370 | 0.092 | — | 1.88 | Kumaran (1996) |
| Spruce (oven-dried) | 395 | 0.10 | — | 1.88 | Kumaran (1996) |

**Notes for Table 5-15 SI**
*(Chapter references for 2013 ASHRAE Handbook—Fundamentals)*

[a]Values are for mean temperature of 75°F (24°C). Representative values for dry materials are intended as design (not specification) values for materials in normal use. Thermal values of insulating materials may differ from design values depending on in-situ properties (e.g., density and moisture content, orientation, etc.) and manufacturing variability. For properties of specific product, use values supplied by manufacturer or unbiased tests.

[b]Symbol $\lambda$ also used to represent thermal conductivity.

[c]Does not include paper backing and facing, if any. Where insulation forms boundary (reflective or otherwise) of airspace, see Tables 2 and 3 for insulating value of airspace with appropriate effective emittance and temperature conditions of space.

[d]Conductivity varies with fiber diameter (see Chapter 25). Batt, blanket, and loose-fill mineral fiber insulations are manufactured to achieve specified R-values, the most common of which are listed in the table. Because of differences in manufacturing processes and materials, the product thicknesses, densities, and thermal conductivities vary over considerable ranges for a specified R-value.

[e]Insulating values of acoustical tile vary, depending on density of board and on type, size, and depth of perforations.

[f]Values for fully grouted block may be approximated using values for concrete with similar unit density.

[g]Values for concrete block and concrete are at moisture contents representative of normal use.

[h]Values for metal or vinyl siding applied over flat surfaces vary widely, depending on ventilation of the airspace beneath the siding; whether airspace is reflective or nonreflective; and on thickness, type, and application of insulating backing-board used. Values are averages for use as design guides, and were obtained from several guarded hot box tests (ASTM Standard C1363) on hollow-backed types and types made using backing of wood fiber, foamed plastic, and glass fiber. Departures of ±50% or more from these values may occur.

[i]Vinyl specific heat = 0.25 Btu/lb·°F 1.0 kJ/(kg·K)

[j]See Adams (1971), MacLean (1941), and Wilkes (1979). Conductivity values listed are for heat transfer across the grain. Thermal conductivity of wood varies linearly with density, and density ranges listed are those normally found for wood species given. If density of wood species is not known, use mean conductivity value. For extrapolation to other moisture contents, the following empirical equation developed by Wilkes (1979) may be used:

$$k = 0.1791 + \frac{(1.874 \times 10^{-2} + 5.753 \times 10^{-4} M)\rho}{1 + 0.01 M}$$

where $\rho$ is density of moist wood in lb/ft³ kg/m³, and $M$ is moisture content in percent.

[k]From Wilkes (1979), an empirical equation for specific heat of moist wood at 75°F (24°C) is as follows:

$$c_p = \frac{(0.299 + 0.01 M)}{(1 + 0.01 M)} + \Delta c_p$$

where $\Delta c_p$ accounts for heat of sorption and is denoted by

$$\Delta c_p = M(1.921 \times 10^{-3} - 3.168 \times 10^{-5} M)$$

where $M$ is moisture content in percent by mass.

[l]Blank space in reference column indicates historical values from previous volumes of *ASHRAE Handbook*. Source of information could not be determined.

# Load Estimating Fundamentals

### Table 5-16 SI  U-Factors for Various Fenestration Products in W/(m²·K)
*(Table 1, Chapter 26, 2013 ASHRAE Handbook—Fundamentals)*

| | | Glass Only | | Vertical Installation — Operable (including sliding and swinging glass doors) | | | | | Fixed | | | | |
|---|---|---|---|---|---|---|---|---|---|---|---|---|---|
| **Frame Type** | | Center of Glass | Edge of Glass | Aluminum Without Thermal Break | Aluminum With Thermal Break | Reinforced Vinyl/ Aluminum Clad Wood | Wood/ Vinyl | Insulated Fiberglass/ Vinyl | Aluminum Without Thermal Break | Aluminum With Thermal Break | Reinforced Vinyl/ Aluminum Clad Wood | Wood/ Vinyl | Insulated Fiberglass/ Vinyl |
| **ID** | **Glazing Type** | | | | | | | | | | | | |
| | **Single Glazing** | | | | | | | | | | | | |
| 1 | 3 mm glass | 5.91 | 5.91 | 7.01 | 6.08 | 5.27 | 5.20 | 4.83 | 6.38 | 6.06 | 5.58 | 5.58 | 5.40 |
| 2 | 6 mm acrylic/polycarb | 5.00 | 5.00 | 6.23 | 5.35 | 4.59 | 4.52 | 4.18 | 5.55 | 5.23 | 4.77 | 4.77 | 4.61 |
| 3 | 3.2 mm acrylic/polycarb | 5.45 | 5.45 | 6.62 | 5.72 | 4.93 | 4.86 | 4.51 | 5.96 | 5.64 | 5.18 | 5.18 | 5.01 |
| | **Double Glazing** | | | | | | | | | | | | |
| 4 | 6 mm airspace | 3.12 | 3.63 | 4.62 | 3.61 | 3.24 | 3.14 | 2.84 | 3.88 | 3.52 | 3.18 | 3.16 | 3.04 |
| 5 | 13 mm airspace | 2.73 | 3.36 | 4.30 | 3.31 | 2.96 | 2.86 | 2.58 | 3.54 | 3.18 | 2.85 | 2.83 | 2.72 |
| 6 | 6 mm argon space | 2.90 | 3.48 | 4.43 | 3.44 | 3.08 | 2.98 | 2.69 | 3.68 | 3.33 | 3.00 | 2.98 | 2.86 |
| 7 | 13 mm argon space | 2.56 | 3.24 | 4.16 | 3.18 | 2.84 | 2.74 | 2.46 | 3.39 | 3.04 | 2.71 | 2.69 | 2.58 |
| | **Double Glazing, $e = 0.60$ on surface 2 or 3** | | | | | | | | | | | | |
| 8 | 6 mm airspace | 2.95 | 3.52 | 4.48 | 3.48 | 3.12 | 3.02 | 2.73 | 3.73 | 3.38 | 3.04 | 3.02 | 2.90 |
| 9 | 13 mm airspace | 2.50 | 3.20 | 4.11 | 3.14 | 2.80 | 2.70 | 2.42 | 3.34 | 2.99 | 2.67 | 2.65 | 2.53 |
| 10 | 6 mm argon space | 2.67 | 3.32 | 4.25 | 3.27 | 2.92 | 2.82 | 2.54 | 3.49 | 3.13 | 2.81 | 2.79 | 2.67 |
| 11 | 13 mm argon space | 2.33 | 3.08 | 3.98 | 3.01 | 2.68 | 2.58 | 2.31 | 3.20 | 2.84 | 2.52 | 2.50 | 2.39 |
| | **Double Glazing, $e = 0.40$ on surface 2 or 3** | | | | | | | | | | | | |
| 12 | 6 mm airspace | 2.78 | 3.40 | 4.34 | 3.35 | 3.00 | 2.90 | 2.61 | 3.59 | 3.23 | 2.90 | 2.88 | 2.77 |
| 13 | 13 mm airspace | 2.27 | 3.04 | 3.93 | 2.96 | 2.64 | 2.54 | 2.27 | 3.15 | 2.79 | 2.48 | 2.46 | 2.35 |
| 14 | 6 mm argon space | 2.44 | 3.16 | 4.07 | 3.09 | 2.76 | 2.66 | 2.38 | 3.30 | 2.94 | 2.62 | 2.60 | 2.49 |
| 15 | 13 mm argon space | 2.04 | 2.88 | 3.75 | 2.79 | 2.48 | 2.38 | 2.11 | 2.95 | 2.60 | 2.29 | 2.27 | 2.16 |
| | **Double Glazing, $e = 0.20$ on surface 2 or 3** | | | | | | | | | | | | |
| 16 | 6 mm airspace | 2.56 | 3.24 | 4.16 | 3.18 | 2.84 | 2.74 | 2.46 | 3.39 | 3.04 | 2.71 | 2.69 | 2.58 |
| 17 | 13 mm airspace | 1.99 | 2.83 | 3.70 | 2.75 | 2.44 | 2.34 | 2.07 | 2.91 | 2.55 | 2.24 | 2.22 | 2.12 |
| 18 | 6 mm argon space | 2.16 | 2.96 | 3.84 | 2.88 | 2.56 | 2.46 | 2.19 | 3.05 | 2.70 | 2.38 | 2.36 | 2.26 |
| 19 | 13 mm argon space | 1.70 | 2.62 | 3.47 | 2.53 | 2.24 | 2.14 | 1.88 | 2.66 | 2.30 | 2.00 | 1.98 | 1.88 |
| | **Double Glazing, $e = 0.10$ on surface 2 or 3** | | | | | | | | | | | | |
| 20 | 6 mm airspace | 2.39 | 3.12 | 4.02 | 3.05 | 2.72 | 2.62 | 2.34 | 3.25 | 2.89 | 2.57 | 2.55 | 2.44 |
| 21 | 13 mm airspace | 1.82 | 2.71 | 3.56 | 2.62 | 2.32 | 2.22 | 1.96 | 2.76 | 2.40 | 2.10 | 2.08 | 1.98 |
| 22 | 6 mm argon space | 1.99 | 2.83 | 3.70 | 2.75 | 2.44 | 2.34 | 2.07 | 2.91 | 2.55 | 2.24 | 2.22 | 2.12 |
| 23 | 13 mm argon space | 1.53 | 2.49 | 3.33 | 2.40 | 2.12 | 2.02 | 1.76 | 2.51 | 2.16 | 1.86 | 1.84 | 1.74 |
| | **Double Glazing, $e = 0.05$ on surface 2 or 3** | | | | | | | | | | | | |
| 24 | 6 mm airspace | 2.33 | 3.08 | 3.98 | 3.01 | 2.68 | 2.58 | 2.31 | 3.20 | 2.84 | 2.52 | 2.50 | 2.39 |
| 25 | 13 mm airspace | 1.70 | 2.62 | 3.47 | 2.53 | 2.24 | 2.14 | 1.88 | 2.66 | 2.30 | 2.00 | 1.98 | 1.88 |
| 26 | 6 mm argon space | 1.87 | 2.75 | 3.61 | 2.66 | 2.36 | 2.26 | 2.00 | 2.81 | 2.45 | 2.15 | 2.12 | 2.02 |
| 27 | 13 mm argon space | 1.42 | 2.41 | 3.24 | 2.31 | 2.04 | 1.94 | 1.69 | 2.42 | 2.06 | 1.76 | 1.74 | 1.65 |
| | **Triple Glazing** | | | | | | | | | | | | |
| 28 | 6 mm airspace | 2.16 | 2.96 | 3.78 | 2.78 | 2.46 | 2.42 | 2.17 | 3.02 | 2.68 | 2.36 | 2.36 | 2.25 |
| 29 | 13 mm airspace | 1.76 | 2.67 | 3.46 | 2.47 | 2.18 | 2.14 | 1.90 | 2.68 | 2.34 | 2.03 | 2.03 | 1.92 |
| 30 | 6 mm argon space | 1.93 | 2.79 | 3.60 | 2.60 | 2.30 | 2.26 | 2.02 | 2.82 | 2.49 | 2.17 | 2.17 | 2.06 |
| 31 | 13 mm argon space | 1.65 | 2.58 | 3.36 | 2.39 | 2.10 | 2.06 | 1.83 | 2.58 | 2.24 | 1.93 | 1.93 | 1.83 |
| | **Triple Glazing, $e = 0.20$ on surface 2, 3, 4, or 5** | | | | | | | | | | | | |
| 32 | 6 mm airspace | 1.87 | 2.75 | 3.55 | 2.56 | 2.26 | 2.22 | 1.98 | 2.78 | 2.44 | 2.12 | 2.12 | 2.01 |
| 33 | 13 mm airspace | 1.42 | 2.41 | 3.18 | 2.21 | 1.94 | 1.90 | 1.67 | 2.38 | 2.05 | 1.74 | 1.74 | 1.64 |
| 34 | 6 mm argon space | 1.59 | 2.54 | 3.32 | 2.34 | 2.06 | 2.02 | 1.79 | 2.53 | 2.20 | 1.89 | 1.89 | 1.78 |
| 35 | 13 mm argon space | 1.25 | 2.28 | 3.04 | 2.08 | 1.82 | 1.78 | 1.55 | 2.24 | 1.90 | 1.60 | 1.60 | 1.50 |
| | **Triple Glazing, $e = 0.20$ on surfaces 2 or 3 and 4 or 5** | | | | | | | | | | | | |
| 36 | 6 mm airspace | 1.65 | 2.58 | 3.36 | 2.39 | 2.10 | 2.06 | 1.83 | 2.58 | 2.24 | 1.93 | 1.93 | 1.83 |
| 37 | 13 mm airspace | 1.14 | 2.19 | 2.95 | 1.99 | 1.74 | 1.69 | 1.48 | 2.14 | 1.80 | 1.50 | 1.50 | 1.40 |
| 38 | 6 mm argon space | 1.31 | 2.32 | 3.09 | 2.12 | 1.86 | 1.82 | 1.59 | 2.29 | 1.95 | 1.65 | 1.65 | 1.55 |
| 39 | 13 mm argon space | 0.97 | 2.05 | 2.81 | 1.86 | 1.62 | 1.57 | 1.36 | 1.99 | 1.65 | 1.36 | 1.36 | 1.26 |
| | **Triple Glazing, $e = 0.10$ on surfaces 2 or 3 and 4 or 5** | | | | | | | | | | | | |
| 40 | 6 mm airspace | 1.53 | 2.49 | 3.27 | 2.30 | 2.02 | 1.98 | 1.75 | 2.48 | 2.15 | 1.84 | 1.84 | 1.73 |
| 41 | 13 mm airspace | 1.02 | 2.10 | 2.85 | 1.90 | 1.66 | 1.61 | 1.40 | 2.04 | 1.70 | 1.41 | 1.41 | 1.31 |
| 42 | 6 mm argon space | 1.19 | 2.23 | 2.99 | 2.04 | 1.78 | 1.73 | 1.52 | 2.19 | 1.85 | 1.55 | 1.55 | 1.45 |
| 43 | 13 mm argon space | 0.80 | 1.92 | 2.67 | 1.73 | 1.49 | 1.45 | 1.24 | 1.84 | 1.51 | 1.22 | 1.22 | 1.12 |
| | **Quadruple Glazing, $e = 0.10$ on surfaces 2 or 3 and 4 or 5** | | | | | | | | | | | | |
| 44 | 6 mm airspaces | 1.25 | 2.28 | 3.04 | 2.08 | 1.82 | 1.78 | 1.55 | 2.24 | 1.90 | 1.60 | 1.60 | 1.50 |
| 45 | 13 mm airspaces | 0.85 | 1.96 | 2.71 | 1.77 | 1.54 | 1.49 | 1.28 | 1.89 | 1.55 | 1.26 | 1.26 | 1.17 |
| 46 | 6 mm argon spaces | 0.97 | 2.05 | 2.81 | 1.86 | 1.62 | 1.57 | 1.36 | 1.99 | 1.65 | 1.36 | 1.36 | 1.26 |
| 47 | 13 mm argon spaces | 0.68 | 1.83 | 2.57 | 1.64 | 1.41 | 1.37 | 1.16 | 1.74 | 1.41 | 1.12 | 1.12 | 1.03 |
| 48 | 6 mm krypton spaces | 0.68 | 1.83 | 2.57 | 1.64 | 1.41 | 1.37 | 1.16 | 1.74 | 1.41 | 1.12 | 1.12 | 1.03 |

*Notes:*

1. All heat transmission coefficients in this table include film resistances and are based on winter conditions of −18°C outdoor air temperature and 21°C indoor air temperature, with 6.7 m/s outdoor air velocity and zero solar flux. Except for single glazing, small changes in the indoor and outdoor temperatures do not significantly affect overall U-factors. Coefficients are for vertical position except skylight values, which are for 20° from horizontal with heat flow up.

2. Glazing layer surfaces are numbered from outdoor to indoor. Double, triple, and quadruple refer to number of glazing panels. All data are based on 3 mm glass, unless otherwise noted. Thermal conductivities are: 0.917 W/(m·K) for glass, and 0.19 W/(m·K) for acrylic and polycarbonate.

3. Standard spacers are metal. Edge-of-glass effects are assumed to extend over the 63.5 mm band around perimeter of each glazing unit.

**Table 5-16  U-Factors for Various Fenestration Products in W/(m²·K) (*Concluded*)**

|  Vertical Installation |||||||| Sloped Installation ||||||
| Garden Windows || Curtainwall ||| Glass Only (Skylights) || Manufactured Skylight ||| Site-Assembled Sloped/Overhead Glazing |||
| Aluminum Without Thermal Break | Wood/ Vinyl | Aluminum Without Thermal Break | Aluminum With Thermal Break | Structural Glazing | Center of Glass | Edge of Glass | Aluminum Without Thermal Break | Aluminum With Thermal Break | Reinforced Vinyl/ Aluminum Clad Wood | Wood/ Vinyl | Aluminum Without Thermal Break | Aluminum With Thermal Break | Structural Glazing | ID |
|---|---|---|---|---|---|---|---|---|---|---|---|---|---|---|
| 14.21 | 11.94 | 6.86 | 6.27 | 6.27 | *6.76* | *6.76* | 10.03 | 9.68 | 9.16 | 8.05 | 7.66 | 7.64 | 7.10 | 1 |
| 12.70 | 10.42 | 6.03 | 5.44 | 5.44 | *5.85* | *5.85* | 9.09 | 8.74 | 8.23 | 7.45 | 6.83 | 6.80 | 6.27 | 2 |
| 13.45 | 11.18 | 6.44 | 5.86 | 5.86 | *6.30* | *6.30* | 9.56 | 9.21 | 8.70 | 7.89 | 7.24 | 7.22 | 6.68 | 3 |
| 9.78 | 7.50 | 4.38 | 3.79 | 3.56 | *3.29* | *3.75* | 6.23 | 5.46 | 5.21 | 4.79 | 4.54 | 4.71 | 3.75 | 4 |
| 9.19 | 6.92 | 4.03 | 3.45 | 3.22 | *3.24* | *3.71* | 6.17 | 5.41 | 5.16 | 4.74 | 4.49 | 4.68 | 3.70 | 5 |
| 9.44 | 7.17 | 4.18 | 3.60 | 3.37 | *3.01* | *3.56* | 5.96 | 5.19 | 4.94 | 4.54 | 4.30 | 4.52 | 3.51 | 6 |
| 8.94 | 6.67 | 3.89 | 3.30 | 3.07 | *3.01* | *3.56* | 5.96 | 5.19 | 4.94 | 4.54 | 4.30 | 4.52 | 3.51 | 7 |
| 9.53 | 7.25 | 4.23 | 3.65 | 3.41 | *3.07* | *3.60* | 6.01 | 5.24 | 4.99 | 4.59 | 4.35 | 4.56 | 3.55 | 8 |
| 8.86 | 6.58 | 3.84 | 3.25 | 3.02 | *3.01* | *3.56* | 5.96 | 5.19 | 4.94 | 4.54 | 4.30 | 4.52 | 3.51 | 9 |
| 9.11 | 6.84 | 3.99 | 3.40 | 3.17 | *2.78* | *3.40* | 5.74 | 4.97 | 4.72 | 4.34 | 4.10 | 4.37 | 3.31 | 10 |
| 8.61 | 6.33 | 3.69 | 3.11 | 2.88 | *2.78* | *3.40* | 5.74 | 4.97 | 4.72 | 4.34 | 4.10 | 4.37 | 3.31 | 11 |
| 9.28 | 7.00 | 4.08 | 3.50 | 3.27 | *2.90* | *3.48* | 5.85 | 5.08 | 4.83 | 4.44 | 4.20 | 4.45 | 3.41 | 12 |
| 8.52 | 6.25 | 3.64 | 3.06 | 2.83 | *2.84* | *3.44* | 5.79 | 5.02 | 4.78 | 4.39 | 4.15 | 4.41 | 3.36 | 13 |
| 8.77 | 6.50 | 3.79 | 3.21 | 2.97 | *2.50* | *3.20* | 5.46 | 4.69 | 4.45 | 4.09 | 3.86 | 4.18 | 3.07 | 14 |
| 8.18 | 5.91 | 3.45 | 2.86 | 2.63 | *2.61* | *3.28* | 5.57 | 4.80 | 4.56 | 4.19 | 3.96 | 4.25 | 3.17 | 15 |
| 8.94 | 6.67 | 3.89 | 3.30 | 3.07 | *2.61* | *3.28* | 5.57 | 4.80 | 4.56 | 4.19 | 3.96 | 4.25 | 3.17 | 16 |
| 8.10 | 5.82 | 3.40 | 2.81 | 2.58 | *2.61* | *3.28* | 5.57 | 4.80 | 4.56 | 4.19 | 3.96 | 4.25 | 3.17 | 17 |
| 8.35 | 6.08 | 3.54 | 2.96 | 2.73 | *2.22* | *3.00* | 5.19 | 4.42 | 4.18 | 3.84 | 3.61 | 3.98 | 2.83 | 18 |
| 7.67 | 5.39 | 3.15 | 2.56 | 2.33 | *2.27* | *3.04* | 5.24 | 4.47 | 4.24 | 3.89 | 3.66 | 4.02 | 2.88 | 19 |
| 8.69 | 6.42 | 3.74 | 3.16 | 2.92 | *2.50* | *3.20* | 5.46 | 4.69 | 4.45 | 4.09 | 3.86 | 4.18 | 3.07 | 20 |
| 7.84 | 5.57 | 3.25 | 2.66 | 2.43 | *2.50* | *3.20* | 5.46 | 4.69 | 4.45 | 4.09 | 3.86 | 4.18 | 3.07 | 21 |
| 8.10 | 5.82 | 3.40 | 2.81 | 2.58 | *2.04* | *2.88* | 5.02 | 4.25 | 4.02 | 3.69 | 3.46 | 3.86 | 2.68 | 22 |
| 7.41 | 5.14 | 3.00 | 2.42 | 2.18 | *2.16* | *2.96* | 5.13 | 4.36 | 4.13 | 3.79 | 3.56 | 3.94 | 2.78 | 23 |
| 8.61 | 6.33 | 3.69 | 3.11 | 2.88 | *2.39* | *3.12* | 5.35 | 4.58 | 4.34 | 3.99 | 3.76 | 4.10 | 2.97 | 24 |
| 7.67 | 5.39 | 3.15 | 2.56 | 2.33 | *2.44* | *3.16* | 5.41 | 4.64 | 4.40 | 4.04 | 3.81 | 4.14 | 3.02 | 25 |
| 7.93 | 5.65 | 3.30 | 2.71 | 2.48 | *1.93* | *2.79* | 4.91 | 4.14 | 3.91 | 3.58 | 3.37 | 3.77 | 2.58 | 26 |
| 7.24 | 4.96 | 2.90 | 2.32 | 2.09 | *2.04* | *2.88* | 5.02 | 4.25 | 4.02 | 3.69 | 3.46 | 3.86 | 2.68 | 27 |
| see | see | 3.48 | 2.91 | 2.62 | *2.22* | *3.00* | 5.13 | 4.24 | 4.03 | 3.63 | 3.55 | 3.92 | 2.70 | 28 |
| note | note | 3.14 | 2.57 | 2.27 | *2.04* | *2.88* | 4.96 | 4.07 | 3.87 | 3.48 | 3.40 | 3.80 | 2.56 | 29 |
| 7 | 7 | 3.28 | 2.71 | 2.42 | *1.99* | *2.83* | 4.91 | 4.01 | 3.81 | 3.43 | 3.35 | 3.76 | 2.51 | 30 |
|  |  | 3.04 | 2.47 | 2.17 | *1.87* | *2.75* | 4.80 | 3.90 | 3.70 | 3.33 | 3.25 | 3.68 | 2.41 | 31 |
| see | see | 3.23 | 2.66 | 2.37 | *1.93* | *2.79* | 4.85 | 3.96 | 3.76 | 3.38 | 3.30 | 3.72 | 2.46 | 32 |
| note | note | 2.84 | 2.27 | 1.97 | *1.76* | *2.67* | 4.68 | 3.79 | 3.59 | 3.22 | 3.16 | 3.59 | 2.31 | 33 |
| 7 | 7 | 2.99 | 2.42 | 2.12 | *1.59* | *2.54* | 4.52 | 3.63 | 3.43 | 3.07 | 3.01 | 3.47 | 2.17 | 34 |
|  |  | 2.69 | 2.12 | 1.83 | *1.53* | *2.49* | 4.46 | 3.57 | 3.37 | 3.02 | 2.96 | 3.43 | 2.12 | 35 |
| see | see | 3.04 | 2.47 | 2.17 | *1.65* | *2.58* | 4.57 | 3.68 | 3.48 | 3.12 | 3.06 | 3.51 | 2.22 | 36 |
| note | note | 2.59 | 2.02 | 1.73 | *1.53* | *2.49* | 4.46 | 3.57 | 3.37 | 3.02 | 2.96 | 3.43 | 2.12 | 37 |
| 7 | 7 | 2.74 | 2.17 | 1.87 | *1.36* | *2.36* | 4.29 | 3.40 | 3.21 | 2.86 | 2.81 | 3.30 | 1.97 | 38 |
|  |  | 2.44 | 1.87 | 1.58 | *1.25* | *2.28* | 4.18 | 3.29 | 3.10 | 2.76 | 2.71 | 3.22 | 1.87 | 39 |
| see | see | 2.94 | 2.37 | 2.07 | *1.53* | *2.49* | 4.46 | 3.57 | 3.37 | 3.02 | 2.96 | 3.43 | 2.12 | 40 |
| note | note | 2.49 | 1.92 | 1.63 | *1.42* | *2.41* | 4.35 | 3.46 | 3.27 | 2.91 | 2.86 | 3.34 | 2.02 | 41 |
| 7 | 7 | 2.64 | 2.07 | 1.78 | *1.19* | *2.23* | 4.13 | 3.24 | 3.04 | 2.71 | 2.66 | 3.18 | 1.82 | 42 |
|  |  | 2.29 | 1.72 | 1.43 | *1.14* | *2.19* | 4.07 | 3.18 | 2.99 | 2.66 | 2.61 | 3.13 | 1.77 | 43 |
|  |  | 2.69 | 2.12 | 1.83 | *1.25* | *2.28* | 4.18 | 3.29 | 3.10 | 2.76 | 2.71 | 3.22 | 1.87 | 44 |
| see | see | 2.34 | 1.77 | 1.48 | *1.08* | *2.14* | 4.02 | 3.12 | 2.93 | 2.60 | 2.56 | 3.09 | 1.72 | 45 |
| note | note | 2.44 | 1.87 | 1.58 | *1.02* | *2.10* | 3.96 | 3.07 | 2.88 | 2.55 | 2.51 | 3.05 | 1.67 | 46 |
| 7 | 7 | 2.19 | 1.62 | 1.33 | *0.91* | *2.01* | 3.85 | 2.96 | 2.77 | 2.45 | 2.41 | 2.96 | 1.58 | 47 |
|  |  | 2.19 | 1.62 | 1.33 | *0.74* | *1.87* | 3.68 | 2.79 | 2.60 | 2.29 | 2.26 | 2.83 | 1.43 | 48 |

4. Product sizes are described in Figure 4, Chapter 15, 2013 *ASHRAE Handbook—Fundamentals* and frame U-factors are from Table 1.
5. Use $U = 3.40$ W/(m²·K) for glass block with mortar but without reinforcing or framing.
6. Use of this table should be limited to that of an estimating tool for early phases of design.
7. Values for triple- and quadruple-glazed garden windows are not listed, because these are not common products.
8. U-factors in this table were determined using NFRC 100-91. They have not been updated to the current rating methodology in NFRC 100-2010.

# Load Estimating Fundamentals

### Table 5-17 SI  Design U-Factors of Swinging Doors in W/(m²·K)

*(Table 6, Chapter 15, 2013 ASHRAE Handbook—Fundamentals)*

| Door Type (Rough Opening = 970 × 2080 mm) | No Glazing | Single Glazing | Double Glazing with 12.7 mm Air Space | Double Glazing with $e = 0.10$, 12.7 mm Argon |
|---|---|---|---|---|
| *Slab Doors* | | | | |
| Wood slab in wood frame[a] | 2.61 | | | |
| 6% glazing (560 × 200 lite) | — | 2.73 | 2.61 | 2.50 |
| 25% glazing (560 × 910 lite) | — | 3.29 | 2.61 | 2.38 |
| 45% glazing (560 × 1620 lite) | — | 3.92 | 2.61 | 2.21 |
| More than 50% glazing | | Use Table 4 (operable) | | |
| Insulated steel slab with wood edge in wood frame[b] | 0.91 | | | |
| 6% glazing (560 × 200 lite) | — | 1.19 | 1.08 | 1.02 |
| 25% glazing (560 × 910 lite) | — | 2.21 | 1.48 | 1.31 |
| 45% glazing (560 × 1630 lite) | — | 3.29 | 1.99 | 1.48 |
| More than 50% glazing | | Use Table 4 (operable) | | |
| Foam-insulated steel slab with metal edge in steel frame[c] | 2.10 | | | |
| 6% glazing (560 × 200 lite) | — | 2.50 | 2.33 | 2.21 |
| 25% glazing (560 × 910 lite) | — | 3.12 | 2.73 | 2.50 |
| 45% glazing (560 × 1630 lite) | — | 4.03 | 3.18 | 2.73 |
| More than 50% glazing | | Use Table 4 (operable) | | |
| Cardboard honeycomb slab with metal edge in steel frame | 3.46 | | | |
| *Stile-and-Rail Doors* | | | | |
| Sliding glass doors/French doors | | Use Table 4 (operable) | | |
| *Site-Assembled Stile-and-Rail Doors* | | | | |
| Aluminum in aluminum frame | — | 7.49 | 5.28 | 4.49 |
| Aluminum in aluminum frame with thermal break | — | 6.42 | 4.20 | 3.58 |

*Notes*:
[a] Thermally broken sill [add 0.17 W/(m²·K) for non-thermally broken sill]
[b] Non-thermally broken sill
[c] Nominal U-factors are through center of insulated panel before consideration of thermal bridges around edges of door sections and because of frame.

### Table 5-18 SI  Design U-factors for Revolving Doors in W/(m²·K)

*(Table 7, Chapter 15, 2013 ASHRAE Handbook—Fundamentals)*

| Type | Size (Width × Height) | U-Factor |
|---|---|---|
| 3-wing | 2.44 × 2.13 m | 4.46 |
|  | 3.28 × 2.44 m | 4.53 |
| 4-wing | 2.13 × 1.98 m | 3.56 |
|  | 2.13 × 2.29 m | 3.63 |
| Open* | 2.08 × 2.13 m | 7.49 |

*U-factor of Open door determined using NFRC *Technical Document* 100-91. It has not been updated to current rating methodology in NFRC *Technical Document* 100-2004.

**Example 5-5** Calculate the U-factor of the 38 mm by 90 mm stud wall shown in Figure 5-18. The studs are at 400 mm OC. There is 90 mm mineral fiber batt insulation ($R = 2.3$ (m²·K)/W) in the stud space. The inside finish is 13 mm gypsum wallboard; the outside is finished with rigid foam insulating sheathing ($R = 0.7$ (m²·K)/W) and 13 mm by 200 mm wood bevel lapped siding. The insulated cavity occupies approximately 75% of the transmission area; the studs, plates, and sills occupy 21%; and the headers occupy 4%.

**Solution.** Obtain the R-values of the various building elements from Tables 5-12 and 5-15. Assume the $R = 7.0$ (m²·K)/W for the wood framing. Also, assume the headers are solid wood, in this case, and group them with the studs, plates, and sills.

| Element | R (Insulated Cavity) | R (Studs, Plates, and Headers) |
|---|---|---|
| 1. Outside surface, 24 km/h wind | 0.03 | 0.03 |
| 2. Wood bevel lapped siding | 0.14 | 0.14 |
| 3. Rigid foam insulating sheathing | 0.70 | 0.70 |
| 4. Mineral fiber batt insulation | 2.30 | — |
| 5. Wood stud | — | 0.63 |
| 6. Gypsum wallboard | 0.08 | 0.08 |
| 7. Inside surface, still air | 0.12 | 0.12 |
|  | 3.37 | 1.70 |

Since the U-factor is the reciprocal of R-value, $U_1 = 0.297$ W/(m²·K) and $U_2 = 0.588$ W/(m²·K).

If the wood framing (thermal bridging) is not included, Equation (3) from Chapter 22 may be used to calculate the U-factor of the wall as follows:

$$U_{av} = U_1 = \frac{1}{R_1} = 0.30 \text{ W/(m}^2 \cdot \text{K)}$$

If the wood framing is accounted for using the parallel-path flow method, the U-factor of the wall is determined using Equation (5) from Chapter 22 as follows:

$$U_{av} = (0.75 \times 0.297) + (0.25 \times 0.588) = 0.37 \text{ W/(m}^2 \cdot \text{K)}$$

If the wood framing is included using the isothermal planes method, the U-factor of the wall is determined using Equations (2) and (3) from Chapter 22 as follows:

$$R_{T(av)} = 4.98 + 1/[(0.75/2.30) + (0.25/0.63)] + 0.22$$
$$= 2.47 \text{ K} \cdot \text{m}^2/\text{W}$$
$$U_{av} = 0.40 \text{ W/(m}^2 \cdot \text{K)}$$

For a frame wall with a 600 mm OC stud space, the average overall R-value is 0.25 m²·K/W. Similar calculation procedures may be used to evaluate other wall designs, except those with thermal bridges.

**Table 5-19 SI  Design U-factors for Double-Skin Steel Emergency Exit Doors in W/(m$^2$·K)**

*(Table 8, Chapter 15, 2013 ASHRAE Handbook—Fundamentals)*

| Core Insulation Thickness, mm | Type | Rough Opening Size 0.9 × 2 m | 1.8 × 2 m |
|---|---|---|---|
| 35* | Honeycomb kraft paper | 3.23 | 2.97 |
|  | Mineral wool, steel ribs | 2.50 | 2.05 |
|  | Polyurethane foam | 1.92 | 1.60 |
| 44* | Honeycomb kraft paper | 3.25 | 3.06 |
|  | Mineral wool, steel ribs | 2.30 | 1.90 |
|  | Polyurethane foam | 1.77 | 1.50 |
| 35 | Honeycomb kraft paper | 3.38 | 3.11 |
|  | Mineral wool, steel ribs | 2.67 | 2.21 |
|  | Polyurethane foam | 2.10 | 1.77 |
| 44 | Honeycomb kraft paper | 3.38 | 3.22 |
|  | Mineral wool, steel ribs | 2.47 | 2.08 |
|  | Polyurethane foam | 1.95 | 1.69 |

*With thermal break

**Table 5-20 SI  Design U-factors for Double-Skin Steel Garage and Aircraft Hanger Doors in W/(m$^2$·K)**

*(Table 9, Chapter 15, 2013 ASHRAE Handbook—Fundamentals)*

| Insulation Thickness, mm | Type | One-Piece Tilt-Up[a] 2.44 □ 2.1 m | 4.9 □ 2.1 m | Sectional Tilt-Up[b] 2.74 □ 2.1 m | Aircraft Hangar 22 □ 3.7 m[c] | 73 □ 15.2 m[d] |
|---|---|---|---|---|---|---|
| 35 | EPS, steel ribs[e] | 2.03 | 1.90 | 1.94 to 2.19 |  |  |
|  | XPS, steel ribs[f] | 1.90 | 1.74 | 1.76 to 2.05 |  |  |
| 50 | EPS, steel ribs[e] | 1.74 | 1.58 | 1.66 to 1.87 |  |  |
|  | XPS, steel ribs[f] | 1.62 | 1.46 | 1.53 to 1.77 |  |  |
| 76 | EPS, steel ribs[e] | 1.46 | 1.29 | 1.43 to 1.60 |  |  |
|  | XPS, steel ribs[f] | 1.37 | 1.21 | 1.34 to 1.52 |  |  |
| 102 | EPS, steel ribs[e] | 1.29 | 1.13 | 1.29 to 1.43 |  |  |
|  | XPS, steel ribs[f] | 1.22 | 1.06 | 1.22 to 1.36 |  |  |
|  | EPS, steel ribs[e] | 1.11 | 0.93 | 1.13 to 1.22 |  |  |
|  | XPS, steel ribs[f] | 1.06 | 0.88 | 1.08 to 1.17 |  |  |
| 89 | XPS |  |  |  | 1.40 | 0.91 |
|  | Mineral wool, steel ribs |  |  |  | 1.45 | 0.92 |
|  | EPS |  |  |  | 1.32 | 0.83 |
| 140 | XPS |  |  |  | 1.18 | 0.72 |
|  | Mineral wool, steel ribs |  |  |  | 1.28 | 0.73 |
|  | EPS |  |  |  | 1.14 | 0.67 |
| — | Uninsulated[g] |  |  |  | 6,27 | 7.00 |
|  | All products[f] | 6.53[g] |  |  |  |  |

*Notes:*
[a] Values are for thermally broken or thermally unbroken doors.
[b] Lower values are for thermally broken doors; upper values are for doors with no thermal break.
[c] Typical size for a small private airplane (single- or twin-engine).
[d] Typical hangar door for a mid-sized commercial jet airliner.
[e] EPS = extruded polystyrene; XPS = expanded polystyrene.
[f] U-factor determined using NFRC *Technical Document* 100-91. Not updated to current rating methodology in NFRC *Technical Document* 100-2004.
[g] U-factor determined for 3.05 × 3.05 m sectional door, but is representative of similar products of different size.

# Chapter 6

# RESIDENTIAL COOLING AND HEATING LOAD CALCULATIONS

The procedures described in this chapter are for calculating the design cooling and heating loads for residential buildings. Additional details can be found in Chapter 17 of the 2013 *ASHRAE Handbook—Fundamentals*, Chapter 1 of the 2011 *ASHRAE Handbook—HVAC Applications*, and Chapter 10 of the 2012 *ASHRAE Handbook—HVAC Systems and Equipment*.

## 6.1 Background

The primary function of heat loss and heat gain calculations is to estimate the capacity that will be required for the heating and cooling components of the air-conditioning systems in order to maintain comfort in the building. These calculations are therefore based on peak load conditions for heating and cooling and correspond to weather conditions that are near the extremes normally encountered (see Chapter 4). A number of load calculation procedures have been developed over the years. Most procedures, even though they differ in some respects, are based on a systematic evaluation of the components of heat loss and heat gain. The most recent procedures developed by ASHRAE, as included in Chapter 17 of the 2013 *ASHRAE Handbook—Fundamentals*, are used here.

Residences and small commercial buildings have heat gains and cooling loads that are dominated by the building envelope (walls, roof, windows, and doors) whereas internal gains from occupants, lights, equipment, and appliances play a significant role and often dominate in commercial buildings. For all buildings, the energy required to heat/cool and/or to humidify/dehumidify any outdoor air entering the space, whether intentional for ventilation or leaking in through cracks and openings (infiltration) is often significant and must be included. ASHRAE Standard 62.2-2010 should be consulted for recommended residential ventilation rates.

A key common element of all cooling methods is attention to temperature swing via empirical data or suitable models. Throughout the literature, it is repeatedly emphasized that direct application of nonresidential methods (based on a fixed setpoint) results in unrealistically high cooling loads for residential applications.

The procedures in this chapter are based on the same fundamentals as the nonresidential methods presented in Chapter 7. However, many characteristics distinguish residential loads, and the Chapter 7 procedures should be applied with care to residential applications.

With respect to heating and cooling load calculation and equipment sizing, the unique features distinguishing residences from other types of buildings are the following:

- **Smaller Internal Heat Gains.** Residential system loads are primarily imposed by heat gain or loss through structural components and by air leakage or ventilation. Internal heat gains, particularly those from occupants and lights, are small compared to those in commercial or industrial structures.
- **Varied Use of Spaces.** Use of spaces in residences is more flexible than in commercial buildings. Localized or temporary temperature excursions are often tolerable.
- **Fewer Zones.** Residences are generally conditioned as a single zone or, at most, a few zones. Typically, a thermostat located in one room controls unit output for multiple rooms, and capacity cannot be redistributed from one area to another as loads change over the day. This results in some hour-to-hour temperature variation or "swing" that has a significant moderating effect on peak loads because of heat storage in building components.
- **Greater Distribution Losses.** Residential ducts are frequently installed in attics or other unconditioned buffer spaces. Duct leakage and heat gain or loss can require significant increases in unit capacity. Residential distribution gains and losses cannot be neglected or estimated with simple rules of thumb.
- **Partial Loads.** Most residential cooling systems use units of relatively small capacity (about 12,000 to 60,000 Btu/h cooling, 40,000 to 120,000 Btu/h heating). Because loads are largely determined by outside conditions, and few days each season are design days, the unit operates at partial load during most of the season; thus, an oversized unit is detrimental to good system performance, especially for cooling in areas of high wet-bulb temperature.
- **Dehumidification Issues.** Dehumidification occurs during cooling unit operation only, and space condition control is usually limited to use of room thermostats (sensible heat-actuated devices). Excessive sensible capacity results in short-cycling and severely degraded dehumidification performance.

In addition to these general features, residential buildings can be categorized according to their exposure:

- **Single-Family Detached.** A house in this category usually has exposed walls in four directions, often more than one story, and a roof. The cooling system is a single-zone, unitary system with a single thermostat. Two-story houses may have a separate cooling system for each floor. Rooms are reasonably open and generally have a centralized air return. In this configuration, both air and load from rooms are mixed, and a load-leveling effect, which requires a distribution of air to each room that is different from a pure commercial system, results. Because the amount of air supplied to each room is based on the load for that room, proper load calculation procedures must be used.

- **Multifamily.** Unlike single-family detached units, multifamily units generally do not have exposed surfaces facing in all directions. Rather, each unit typically has a maximum of three exposed walls and possibly a roof. Both east and west walls might not be exposed in a given living unit. Each living unit has a single unitary cooling system or a single fan-coil unit and the rooms are relatively open to one another. This configuration does not have the same load-leveling effect as a single-family detached house.
- **Other.** Many buildings do not fall into either of the preceding categories. Critical to the designation of a single-family detached building is well-distributed exposure so there is not a short-duration peak; however, if fenestration exposure is predominantly east or west, the cooling load profile resembles that of a multifamily unit. On the other hand, multifamily units with both east and west exposures or neither east nor west exposure exhibit load profiles similar to single-family detached.

Variations in the characteristics of residences can lead to surprisingly complex load calculations. Time-varying heat flows combine to produce a time-varying load. The relative magnitude and pattern of the heat flows depends on the building characteristics and exposure, resulting in a building-specific load profile. In general, an hour-by-hour analysis is required to determine that profile and find its peak.

In theory, cooling and heating processes are identical; a common analysis procedure should apply to either. Acceptable simplifications are possible for heating; however, for cooling, different approaches are used.

Heating calculations use simple worst-case assumptions: no solar or internal gains and no heat storage (with all heat losses evaluated instantaneously). With these simplifications, the heating problem is reduced to a basic $UA\Delta t$ calculation. The heating procedures in this chapter use this long-accepted approach and thus differ only in details from prior methods put forth by ASHRAE and others.

In contrast, the cooling procedures in this edition are extensively revised, based on the results of ASHRAE research project RP-1199 and supported by the Air-Conditioning Contractors of America (ACCA). Now that computing power is routinely available, it is appropriate to promulgate 24 h, equation-based procedures.

A key common element of all cooling methods is attention to temperature swing via empirical data or suitable models. Throughout the literature, it is repeatedly emphasized that direct application of nonresidential methods (based on a fixed setpoint) results in unrealistically high cooling loads for residential applications. The procedure presented in this chapter is the Residential Load Factor (RLF) method. RLF is a simplified procedure derived from a detailed residential heat balance (RHB) analysis of prototypical buildings across a range of climates. The method is tractable by hand but is best applied using a spreadsheet. The RLF method should not be applied to situations outside the range of underlying cases, as shown in Table 6-1.

## 6.2 General Guidelines

### 6.2.1 Overview

The following guidelines, data requirements, and procedures apply to all load calculation approaches, whether heating or cooling, hand-tractable or computerized.

Table 6-1  RLF Limitations

| Item | Valid Range | Notes |
|---|---|---|
| Latitude | 20°N to 60°N | Also approximately valid for 20°S to 60°S with N and S orientations reversed for southern hemisphere. |
| Date | July 21 | Application must be summer peaking. Buildings in mild climates with significant SE/S/SW glazing may experience maximum cooling load in fall or even winter. Use RHB if local experience indicates this is a possibility. |
| Elevation | Less than 6500 ft | RLF factors assume 164 ft elevation. With elevation-corrected $C_s$, method is acceptably accurate except at very high elevations. |
| Climate | Warm/hot | Design-day average outdoor temperature assumed to be above indoor design temperature. |
| Construction | Lightweight residential construction (wood or metal framing, wood or stucco siding) | May be applied to masonry veneer over frame construction; results are conservative. Use RHB for structural masonry or unconventional construction. |
| Fenestration area | 0% to 15% of floor area on any façade, 0% to 30% of floor area total | Spaces with high fenestration fraction should be analyzed with RHB. |
| Fenestration tilt | Vertical or horizontal | Skylights with tilt less than 30° can be treated as horizontal. Buildings with significant sloped glazing areas should be analyzed with RHB. |
| Occupancy | Residential | Applications with high internal gains and/or high occupant density should be analyzed with RHB or nonresidential procedures. |
| Temperature swing | 3°F | |
| Distribution losses | Typical | Applications with extensive duct runs in unconditioned spaces should be analyzed with RHB. |

# Residential Cooling and Heating Load Calculations

**Design for Typical Building Use.** In general, residential systems should be designed to meet representative maximum-load conditions, not extreme conditions. Normal occupancy should be assumed, not the maximum that might occur during an occasional social function. Intermittently operated ventilation fans should be assumed to be off. These considerations are especially important for cooling-system sizing.

**Building Codes and Standards.** This chapter presentation is necessarily general. Codes and regulations take precedence; consult local authorities to determine applicable requirements.

**Designer Judgment.** Designer experience with local conditions, building practices, and prior projects should be considered when applying the procedures in this chapter. For equipment-replacement projects, occupant knowledge concerning performance of the existing system can often provide useful guidance in achieving a successful design.

**Verification.** Postconstruction commissioning and verification are important steps in achieving design performance. Designers should encourage pressurization testing and other procedures that allow identification and repair of construction shortcomings.

**Uncertainty and Safety Allowances.** Residential load calculations are inherently approximate. Many building characteristics are estimated during design and ultimately determined by construction quality and occupant behavior. These uncertainties apply to all calculation methods, including first-principles procedures such as RHB. It is therefore tempting to include safety allowances for each aspect of a calculation. However, this practice has a compounding effect and often produces oversized results. Typical conditions should be assumed; safety allowances, if applied at all, should be added to the final calculated loads rather than to intermediate components. In addition, temperature swing provides a built-in safety factor for sensible cooling: a 20% capacity shortfall typically results in a temperature excursion of at most about one or two degrees.

## 6.2.2 Basic Relationships

Common air-conditioning processes involve transferring heat via air transport or leakage. The sensible, latent, and total heat conveyed by air on a volumetric basis is:

$$q_s = C_s Q \Delta t \tag{6-1}$$

$$q_l = C_l Q \Delta W \tag{6-2}$$

$$q_t = C_t Q \Delta h \tag{6-3}$$

$$q_t = q_s + q_l \tag{6-4}$$

where

$q_s, q_l, q_t$ = sensible, latent, total heat transfer rates, Btu/h

$C_s$ = air sensible heat factor, Btu/h·°F·cfm (1.1 at sea level)

$C_l$ = air latent heat factor, Btu/h·cfm (4840 at sea level)

$C_t$ = air total heat factor, Btu/h·cfm per Btu/lb enthalpy, $h$ (4.5 at sea level)

$Q$ = air volumetric flow rate, cfm

$\Delta t$ = air temperature difference across process, °F

$\Delta W$ = air humidity ratio difference across process, lb/lb

$\Delta h$ = air enthalpy difference across process, Btu/lb

The heat factors $C_s$, $C_l$, and $C_t$ are elevation dependent. The sea level values given in the preceding definitions are appropriate for elevations up to about 1000 ft. Procedures are provided in Chapter 7 for calculating adjusted values for higher elevations.

## 6.2.3 Design Conditions

The initial step in the load calculation is selecting indoor and outdoor design conditions.

**Indoor Conditions.** Indoor conditions assumed for design purposes depend on building use, type of occupancy, and/or code requirements. Chapter 4 and ASHRAE Standard 55 define the relationship between indoor conditions and comfort.

Typical practice for cooling is to design for indoor conditions of 75°F db and a maximum of 50% to 65% rh. For heating, 68°F to 72°F db and 30% rh are common design values.

**Outdoor Conditions.** Outdoor design conditions for load calculations should be selected from location-specific climate data in Chapter 4 or according to local code requirements as applicable.

*Cooling.* The 1% values from Figure 4-4 or Table 4-8 climate data are generally appropriate. As previously emphasized, oversized cooling equipment results in poor system performance. Extremely hot events are necessarily of short duration (conditions always moderate each night); therefore, sacrificing comfort under typical conditions to meet occasional extremes is not recommended.

Load calculations also require the daily range of the dry-bulb and coincident wet-bulb temperatures and wind speed. These values can also be found in Figure 4-4 or Table 4-8, although wind speed is commonly assumed to be 7.5 mph.

Typical buildings in middle latitudes generally experience maximum cooling requirements in midsummer (July in the northern hemisphere and January in the southern hemisphere). For this reason, the RLF method is based on midsummer solar gains. However, this pattern does not always hold. Buildings at low latitudes or with significant south-facing glazing (north-facing in the southern hemisphere) should be analyzed at several times of the year using the RHB method. Local experience can provide guidance as to when maximum cooling is probable. For example, it is common for south-facing buildings in mild northern-hemisphere climates to have peak cooling loads in the fall because of low sun angles.

*Heating.* General practice is to use the 99% condition from Figure 4-4 or Table 4-8. Heating load calculations ignore solar and internal gains, providing a built-in safety factor. However, the designer should consider two additional factors:

- Many locations experience protracted (several-day) cold periods during which the outdoor temperature remains below the 99% value.
- Wind is a major determinant of infiltration. Residences with significant leakage (e.g., older houses) may have peak heating demand under conditions other than extreme cold, depending on site wind patterns.

Depending on the application and system type, the designer should consider using the 99.6% value or the mean minimum extreme as the heating design temperature. Alternatively, the heating load can be calculated at the 99% condition and a safety factor applied when equipment is selected. This additional capacity can also serve to meet pickup loads under non-extreme conditions.

**Adjacent Buffer Spaces.** Residential buildings often include unconditioned buffer spaces, such as garages, attics, crawlspaces, basements, or enclosed porches. Accurate load calculations require the adjacent air temperature. In many cases, a simple, conservative estimate is adequate, especially for heating calculations. For example, it is generally reasonable to assume that, under heating design conditions, adjacent uninsulated garages, porches, and attics are at outdoor temperature. Another reasonable assumption is that the temperature in an adjacent, unheated, *insulated* room is the mean of the indoor and outdoor temperatures.

### 6.2.4 Building Data

**Component Areas.** To perform load calculations efficiently and reliably, standard methods must be used for determining building surface areas. For fenestration, the definition of component area must be consistent with associated ratings.

*Gross area.* It is both efficient and conservative to derive gross surface areas from outside building dimensions, ignoring wall and floor thicknesses. Thus, floor areas should be measured to the outside of adjacent exterior walls or to the centerline of adjacent partitions. When apportioning to rooms, facade area should be divided at partition centerlines. Wall height should be taken as floor-to-floor.

Using outside dimensions avoids separate accounting of floor edge and wall corner conditions. Further, it is standard practice in residential construction to define floor area in terms of outside dimensions, so outside-dimension takeoffs yield areas that can be readily checked against building plans (e.g., the sum of room areas should equal the plan floor area). Although outside-dimension procedures are recommended as expedient for load calculations, they are not consistent with rigorous definitions used in building-related standards. However, the inconsistencies are not significant in the load calculation context.

*Fenestration area.* Fenestration includes exterior windows, skylights, and doors. Fenestration U-factor and SHGC ratings (discussed below) are based on the entire product area, including frames. Thus, for load calculations, fenestration area is the area of the rough opening in the wall or roof, less installation clearances (projected product area $A_{pf}$). Installation clearances can be neglected; it is acceptable to use the rough opening as an approximation of $A_{pf}$

*Net area.* Net surface area is the gross surface area less fenestration area (rough opening or $A_{pf}$ contained within the surface).

*Volume.* Building volume is expediently calculated by multiplying floor area by floor-to-floor height. This produces a conservative estimate of enclosed air volume because wall and floor volumes are included in the total. More precise calculations are possible but are generally not justified in this context.

**Construction Characteristics.**

*U-factors.* Except for fenestration, construction U-factors should be calculated using procedures in Chapter 5 or taken from manufacturer's data, if available. U-factors should be evaluated under heating (winter) conditions.

*Fenestration.* Fenestration is characterized by U-factor and solar heat gain coefficient (SHGC), which apply to the entire assembly (including frames). If available, rated values should be used. Ratings can be obtained from product literature, product label, or published listings. For unrated products (e.g., in existing construction), the U-factor and SHGC can be estimated using Table 6-2. Fenestration U-factors are evaluated under heating (winter) design conditions but are used for both heating and cooling calculations.

Relatively few types of glazing are encountered in residential applications. Single-glazed clear, double-glazed clear, and double-glazed low-emissivity ("low-e") glass predominate. Single glazing is now rare in new construction but common in older homes. Triple-glazing, reflective glass, and heat-absorbing glass are encountered occasionally. The properties of windows equipped with storm windows should be estimated from data for a similar configuration with an additional pane. For example, data for clear, double glazing should be used for a clear single-glazed window with a storm window. Fenestration interior and exterior shading must be included in cooling load calculations.

### 6.2.5 Load Components

**Below-Grade Surfaces.** For cooling calculations, heat flow into the ground is usually ignored because it is difficult to quantify. Surfaces adjacent to the ground are modeled as if well insulated on the outside, so there is no overall heat transfer, but diurnal heat storage effects are included. Heating calculations must include loss via slabs and basement walls and floors, as discussed in the heating load section.

**Infiltration.** Infiltration is generally a significant component of both cooling and heating loads. Refer to Chapter 5 for a discussion of residential air leakage. The simplified residential models found in that chapter can be used to calculate infiltration rates for load calculations. Infiltration should be evaluated for the entire building, not individual rooms or zones. Natural infiltration leakage rates are modified by mechanical pressurization caused by unbalanced ventilation or duct leakage.

# Residential Cooling and Heating Load Calculations

Table 6-2  Typical Fenestration Characteristics

| Glazing Type | Glazing Layers | ID[b] | Property[c,d] | Center of Glazing | Operable Aluminum | Operable Aluminum with Thermal Break | Operable Reinforced Vinyl/Aluminum Clad Wood | Operable Wood/Vinyl | Operable Insulated Fiberglass/Vinyl | Fixed Aluminum | Fixed Aluminum with Thermal Break | Fixed Reinforced Vinyl/Aluminum Clad Wood | Fixed Wood/Vinyl | Fixed Insulated Fiberglass/Vinyl |
|---|---|---|---|---|---|---|---|---|---|---|---|---|---|---|
| Clear | 1 | 1a | U | 1.04 | 1.27 | 1.08 | 0.90 | 0.89 | 0.81 | 1.13 | 1.07 | 0.98 | 0.98 | 0.94 |
|  |  |  | SHGC | 0.86 | 0.75 | 0.75 | 0.64 | 0.64 | 0.64 | 0.78 | 0.78 | 0.75 | 0.75 | 0.75 |
|  | 2 | 5a | U | 0.48 | 0.81 | 0.60 | 0.53 | 0.51 | 0.44 | 0.64 | 0.57 | 0.50 | 0.50 | 0.48 |
|  |  |  | SHGC | 0.76 | 0.67 | 0.67 | 0.57 | 0.57 | 0.57 | 0.69 | 0.69 | 0.67 | 0.67 | 0.67 |
|  | 3 | 29a | U | 0.31 | 0.67 | 0.46 | 0.40 | 0.39 | 0.34 | 0.49 | 0.42 | 0.36 | 0.35 | 0.34 |
|  |  |  | SHGC | 0.68 | 0.60 | 0.60 | 0.51 | 0.51 | 0.51 | 0.62 | 0.62 | 0.60 | 0.60 | 0.60 |
| Low-e, low-solar | 2 | 25a | U | 0.30 | 0.67 | 0.47 | 0.41 | 0.39 | 0.33 | 0.48 | 0.41 | 0.36 | 0.35 | 0.33 |
|  |  |  | SHGC | 0.41 | 0.37 | 0.37 | 0.31 | 0.31 | 0.31 | 0.38 | 0.38 | 0.36 | 0.36 | 0.36 |
|  | 3 | 40c | U | 0.27 | 0.64 | 0.43 | 0.37 | 0.36 | 0.31 | 0.45 | 0.39 | 0.33 | 0.32 | 0.31 |
|  |  |  | SHGC | 0.27 | 0.25 | 0.25 | 0.21 | 0.21 | 0.21 | 0.25 | 0.25 | 0.24 | 0.24 | 0.24 |
| Low-e, high-solar | 2 | 17c | U | 0.35 | 0.71 | 0.51 | 0.44 | 0.42 | 0.36 | 0.53 | 0.46 | 0.40 | 0.39 | 0.37 |
|  |  |  | SHGC | 0.70 | 0.62 | 0.62 | 0.52 | 0.52 | 0.52 | 0.64 | 0.64 | 0.61 | 0.61 | 0.61 |
|  | 3 | 32c | U | 0.33 | 0.69 | 0.47 | 0.41 | 0.40 | 0.35 | 0.50 | 0.44 | 0.38 | 0.37 | 0.36 |
|  |  |  | SHGC | 0.62 | 0.55 | 0.55 | 0.46 | 0.46 | 0.46 | 0.56 | 0.56 | 0.54 | 0.54 | 0.54 |
| Heat-absorbing | 1 | 1c | U | 1.04 | 1.27 | 1.08 | 0.90 | 0.89 | 0.81 | 1.13 | 1.07 | 0.98 | 0.98 | 0.94 |
|  |  |  | SHGC | 0.73 | 0.64 | 0.64 | 0.54 | 0.54 | 0.54 | 0.66 | 0.66 | 0.64 | 0.64 | 0.64 |
|  | 2 | 5c | U | 0.48 | 0.81 | 0.60 | 0.53 | 0.51 | 0.44 | 0.64 | 0.57 | 0.50 | 0.50 | 0.48 |
|  |  |  | SHGC | 0.62 | 0.55 | 0.55 | 0.46 | 0.46 | 0.46 | 0.56 | 0.56 | 0.54 | 0.54 | 0.54 |
|  | 3 | 29c | U | 0.31 | 0.67 | 0.46 | 0.40 | 0.39 | 0.34 | 0.49 | 0.42 | 0.36 | 0.35 | 0.34 |
|  |  |  | SHGC | 0.34 | 0.31 | 0.31 | 0.26 | 0.26 | 0.26 | 0.31 | 0.31 | 0.30 | 0.30 | 0.30 |
| Reflective | 1 | 11 | U | 1.04 | 1.27 | 1.08 | 0.90 | 0.89 | 0.81 | 1.13 | 1.07 | 0.98 | 0.98 | 0.94 |
|  |  |  | SHGC | 0.31 | 0.28 | 0.28 | 0.24 | 0.24 | 0.24 | 0.29 | 0.29 | 0.27 | 0.27 | 0.27 |
|  | 2 | 5p | U | 0.48 | 0.81 | 0.60 | 0.53 | 0.51 | 0.44 | 0.64 | 0.57 | 0.50 | 0.50 | 0.48 |
|  |  |  | SHGC | 0.29 | 0.27 | 0.27 | 0.22 | 0.22 | 0.22 | 0.27 | 0.27 | 0.26 | 0.26 | 0.26 |
|  | 3 | 29c | U | 0.31 | 0.67 | 0.46 | 0.40 | 0.39 | 0.34 | 0.49 | 0.42 | 0.36 | 0.35 | 0.34 |
|  |  |  | SHGC | 0.34 | 0.31 | 0.31 | 0.26 | 0.26 | 0.26 | 0.31 | 0.31 | 0.30 | 0.30 | 0.30 |

[b] ID = Chapter 15 in the 2013 *ASHRAE Handbook—Fundamentals* glazing type identifier.  [c] $U$ = U-factor, Btu/h·ft$^2$·°F  [d] SHGC = solar heat gain coefficient

***Leakage rate.*** Air leakage rates are specified either as airflow rate $Q$ or air exchanges per hour (ACH), related as follows:

$$Q_i = \text{ACH}\,(V/60) \qquad (6\text{-}5)$$

$$\text{ACH} = \frac{60 Q_i}{V} \qquad (6\text{-}6)$$

where

$Q_i$ = infiltration airflow rate, cfm

ACH = air exchange rate, ach

$V$ = building volume, ft$^3$

The infiltration airflow rate depends on two factors:

- The building effective leakage area (envelope leaks plus other air leakage paths, notably flues) and its distribution among ceilings, walls, floors, and flues.
- The driving pressure caused by buoyancy (stack effect) and wind.

These factors can be evaluated separately and combined using Equation 6-7.

$$Q_i = A_L \text{IDF} \qquad (6\text{-}7)$$

where

$A_L$ = building effective leakage area (including flue) at reference pressure difference = 0.016 in. of water, assuming discharge coefficient $C_D$ = 1, in$^2$

IDF = infiltration driving force, cfm/in$^2$

The following sections provide procedures for determining $A_L$ and IDF.

***Leakage area.*** There are several ways to characterize building leakage, depending on reference pressure differences and assumed discharge coefficient. The only accurate procedure for determining $A_L$ is by measurement using a pressurization test (commonly called a blower door test). For buildings in design, a pressurization test is not possible and leakage area must be assumed for design purposes. Leakage can be estimated using a simple approach based on an assumed average leakage per unit of building surface area:

$$A_L = A_{es} A_{ul} \qquad (6\text{-}8)$$

Table 6-3  Unit Leakage Areas

| Construction | Description | $A_{ul}$ (in.$^2$/ft$^2$) |
|---|---|---|
| Tight | Construction supervised by air-sealing specialist | 0.01 |
| Good | Carefully sealed construction by knowledgeable builder | 0.02 |
| Average | Typical current production housing | 0.04 |
| Leaky | Typical pre-1970 houses | 0.08 |
| Very leaky | Old houses in original condition | 0.15 |

where

$A_{es}$ = building exposed surface area, ft$^2$

$A_{ul}$ = unit leakage area, in$^2$/ft$^2$ (from Table 6-3)

$A_{ul}$ is the leakage area per unit surface area; suitable design values are found in Table 6-3.

In Equation 6-8, $A_{es}$ is the total building surface area at the envelope pressure boundary, defined as all above-grade surface area that separates the outdoors from conditioned or semiconditioned space. Table 6-4 provides guidance for evaluating $A_{es}$.

**IDF.** To determine IDF, Barnaby and Spitler (2005) derived the following relationship:

$$\text{IDF} = \frac{I_0 + H|\Delta t|[I_1 + I_2(A_{L,\text{flue}}/A_L)]}{1000} \quad (6\text{-}9)$$

where

| | Cooling 7.5 mph | Heating 15 mph |
|---|---|---|
| $I_0$ | 343 | 698 |
| $I_1$ | 0.88 | 0.81 |
| $I_2$ | 0.28 | 0.53 |

$H$ = building average stack height, ft (typically 8 to 10 ft per story)

$\Delta t$ = difference between indoor and outdoor temperatures, °F

$A_{L,\text{flue}}$ = flue effective leakage area at reference pressure difference = 0.016 in. of water, assuming CD = 1, in$^2$ (total for flues serving furnaces, domestic water heaters, fireplaces, or other vented equipment, evaluated assuming associated equipment is not operating and with dampers in closed position)

The building stack height $H$ is the average height difference between the ceiling and floor (or grade, if the floor is below grade). Thus, for buildings with vented crawlspaces, the crawlspace height is not included. For basement or slab-on-grade construction, $H$ is the average height of the ceiling above grade. Generally, there is significant leakage between basements and spaces above, so above-grade basement height should be included whether or not the basement is fully conditioned. With suitable adjustments for grade level, $H$ can also be estimated as $V/A_{cf}$ (conditioned floor area).

Table 6-5 shows *IDF* values, assuming $A_{L,\text{flue}} = 0$.

**Ventilation.**

*Whole-building ventilation.* Because of energy efficiency concerns, residential construction has become significantly

Table 6-4  Evaluation of Exposed Surface Area

| Situation | Include | Exclude |
|---|---|---|
| Ceiling/roof combination (e.g., cathedral ceiling without attic) | Gross surface area | |
| Ceiling or wall adjacent to attic | Ceiling or wall area | Roof area |
| Wall exposed to ambient | Gross wall area (including fenestration area) | |
| Wall adjacent to unconditioned buffer space (e.g., garage or porch) | Common wall area | Exterior wall area |
| Floor over open or vented crawlspace | Floor area | Crawlspace wall area |
| Floor over sealed crawlspace | Crawlspace wall area | Floor area |
| Floor over conditioned or semiconditioned basement | Above-grade basement wall area | Floor area |
| Slab floor | | Slab area |

Table 6-5  Typical IDF Values, cfm/in.$^2$

| | Heating Design Temperature, °F | | | | | Cooling Design Temperature, °F | | |
|---|---|---|---|---|---|---|---|---|
| H, ft | −40 | −20 | 0 | 20 | 40 | 85 | 95 | 105 |
| 8 | 1.40 | 1.27 | 1.14 | 1.01 | 0.88 | 0.41 | 0.48 | 0.55 |
| 10 | 1.57 | 1.41 | 1.25 | 1.09 | 0.92 | 0.43 | 0.52 | 0.61 |
| 12 | 1.75 | 1.55 | 1.36 | 1.16 | 0.97 | 0.45 | 0.55 | 0.66 |
| 14 | 1.92 | 1.70 | 1.47 | 1.24 | 1.02 | 0.47 | 0.59 | 0.71 |
| 16 | 2.10 | 1.84 | 1.58 | 1.32 | 1.06 | 0.48 | 0.62 | 0.76 |
| 18 | 2.27 | 1.98 | 1.69 | 1.40 | 1.11 | 0.50 | 0.66 | 0.82 |
| 20 | 2.45 | 2.12 | 1.80 | 1.48 | 1.15 | 0.52 | 0.69 | 0.87 |
| 22 | 2.62 | 2.27 | 1.91 | 1.55 | 1.20 | 0.54 | 0.73 | 0.92 |
| 24 | 2.80 | 2.41 | 2.02 | 1.63 | 1.24 | 0.55 | 0.76 | 0.98 |

tighter over the last several decades. Natural leakage rates are often insufficient to maintain acceptable indoor air quality. ASHRAE Standard 62.2-2010 specifies the required minimum whole-building ventilation rate as

$$Q_v = 0.01 A_{cf} + 7.5 (N_{br} + 1) \quad (6\text{-}10)$$

where

$Q_v$ = required ventilation flow rate, cfm

$A_{cf}$ = building conditioned floor area, ft$^2$

$N_{br}$ = number of bedrooms (not less than 1)

Certain mild climates are exempted from this standard; local building authorities ultimately dictate actual requirements. It is expected that whole-building ventilation will become more common because of a combination of regulation and consumer demand. The load effect of $Q_v$ must be included in both cooling and heating calculations.

**Distribution Losses.** Air leakage and heat losses from duct systems frequently impose substantial equipment loads in excess of building requirements. The magnitude of losses depends on the location of duct runs, their surface areas, surrounding temperatures, duct wall insulation, and duct airtightness. These values are usually difficult to accurately determine at the time of preconstruction load calculations and must be estimated using assumed values so that selected equipment capacity is sufficient.

## 6.3 Cooling Load Methodology

A cooling load calculation determines total sensible cooling load from heat gain (1) through opaque surfaces (walls, floors, ceilings, and doors), (2) through transparent fenestration surfaces (windows, skylights, and glazed doors), (3) caused by infiltration and ventilation, and (4) because of occupancy. The latent portion of the cooling load is evaluated separately. Although the entire structure may be considered a single zone, equipment selection and system design should be based on room-by-room calculations. For proper design of the distribution system, the conditioned airflow required by each room must be known.

### 6.3.1 Peak Load Computation

To select a properly sized cooling unit, the peak or maximum load (block load) for each zone must be computed. The block load for a single-family detached house with one central system is the sum of all the room loads. If the house has a separate system for each zone, each zone block load is required. When a house is zoned with one central cooling system, the system size is based on the entire house block load, whereas zone components, such as distribution ducts, are sized using zone block loads.

In multifamily structures, each living unit has a zone load that equals the sum of the room loads. For apartments with separate systems, the block load for each unit establishes the system size. Apartment buildings having a central cooling system with fan-coils in each apartment require a block load calculation for the complete structure to size the central system; each unit load establishes the size of the fan-coil and air distribution system for each apartment.

### 6.3.2 Opaque Surfaces

Heat gain through walls, floors, ceilings, and doors is caused by (1) the air temperature difference across such surfaces and (2) solar gains incident on the surfaces. The heat capacity of typical construction moderates and delays building heat gain. This effect is modeled in detail in the computerized RHB method, resulting in accurate simultaneous load estimates. The RLF method uses the following to estimate cooling load:

$$q_{opq} = A \times CF_{opq} \quad (6\text{-}11)$$

$$CF_{opq} = U(OF_t \Delta t + OF_b + OF_r DR) \quad (6\text{-}12)$$

where
- $q_{opq}$ = opaque surface cooling load, Btu/h
- $A$ = net surface area, ft²
- $CF$ = surface cooling factor, Btu/h·ft²
- $U$ = construction U-factor, Btu/h·ft²·°F
- $\Delta t$ = cooling design temperature difference, °F
- $OF_t, OF_b, OF_r$ = opaque-surface cooling factors (Table 6-6)
- $DR$ = cooling daily range, °F

OF factors, found in Table 6-6, represent construction-specific physical characteristics. OF values less than 1 capture the buffering effect of attics and crawlspaces, $OF_b$ represents incident solar gain, and $OF_r$ captures heat storage effects by

Table 6-6  Opaque Surface Cooling Factor Coefficients

| Surface Type | $OF_t$ | $OF_b$, °F | $OF_r$ |
|---|---|---|---|
| Ceiling or wall adjacent to vented attic | 0.62 | $25.7\alpha_{roof} - 8.1$ | –0.19 |
| Ceiling/roof assembly | 1 | $68.9\alpha_{roof} - 12.6$ | –0.36 |
| Wall (wood frame) or door with solar exposure | 1 | 14.8 | –0.36 |
| Wall (wood frame) or door (shaded) | 1 | 0 | –0.36 |
| Floor over ambient | 1 | 0 | –0.06 |
| Floor over crawlspace | 0.33 | 0 | –0.28 |
| Slab floor (see Slab Floor section) | | | |

$\alpha_{roof}$ = roof solar absorptance (see Table 6-7)

Table 6-7  Roof Solar Absorptance $\alpha_{roof}$

| | Color | | | |
|---|---|---|---|---|
| Material | White | Light | Medium | Dark |
| Asphalt shingles | 0.75 | 0.75 | 0.85 | 0.92 |
| Tile | 0.30 | 0.40 | 0.80 | 0.80 |
| Metal | 0.35 | 0.50 | 0.70 | 0.90 |
| Elastomeric coating | 0.30 | | | |

*Source*: Summarized from Parker et al. 2000

reducing the effective temperature difference. Note also that CF can be viewed as CF = $U \times$ CLTD, the formulation used in prior residential and nonresidential methods.

As shown in Table 6-6, roof solar absorptance has a significant effect on ceiling cooling load contribution. Table 6-7 shows typical values for solar absorptance of residential roofing materials.

### 6.3.3 Slab Floors

Slab floors produce a slight reduction in cooling load, as follows:

$$q_{apq} = A \times CF_{slab} \quad (6\text{-}13)$$

$$CF_{slab} = 0.51 - 2.5 h_{srf} \quad (6\text{-}14)$$

where
- $A$ = area of slab, ft²
- $CF_{slab}$ = slab cooling factor, Btu/h·ft²
- $h_{srf}$ = effective surface conductance, including resistance of slab covering material such as carpet = $1/(R_{cvr} + 0.68)$, Btu/h·ft²·°F. Representative $R_{cvr}$ values are found in Table 6-8.

### 6.3.4 Transparent Fenestration Surfaces

Cooling load associated with nondoor fenestration is calculated as follows:

$$q_{fen} = A \times CF_{fen} \quad (6\text{-}15)$$

$$CF_{fen} = U(\Delta t - 0.46 DR) + PXI \times SHGC \times IAC \times FF_s \quad (6\text{-}16)$$

where
- $q_{fen}$ = fenestration cooling load, Btu/h
- $A$ = fenestration area (including frame), ft²
- $CF_{fen}$ = surface cooling factor, Btu/h·ft
- $U$ = fenestration NFRC heating U-factor, Btu/h·ft²·°F (Table 6-2)

### Table 6-8 Thermal Resistance of Floor Coverings

| Description | Thermal Resistance $r_c$, ft$^2$·h·°F/Btu |
|---|---|
| Bare concrete, no covering | 0 |
| Asphalt tile | 0.05 |
| Rubber tile | 0.05 |
| Light carpet | 0.60 |
| Light carpet with rubber pad | 1.00 |
| Light carpet with light pad | 1.40 |
| Light carpet with heavy pad | 1.70 |
| Heavy carpet | 0.80 |
| Heavy carpet with rubber pad | 1.20 |
| Heavy carpet with light pad | 1.60 |
| Heavy carpet with heavy pad | 1.90 |
| 3/8 in. hardwood | 0.54 |
| 5/8 in. wood floor (oak) | 0.57 |
| 1/2 in. oak parquet and pad | 0.68 |
| Linoleum | 0.12 |
| Marble floor and mudset | 0.18 |
| Rubber pad | 0.62 |
| Prime urethane underlayment, 3/8 in. | 1.61 |
| 48 oz. waffled sponge rubber | 0.78 |
| Bonded urethane, 1/2 in. | 2.09 |

*Notes*:
1. Carpet pad thickness should not be more than 1/4 in.
2. Total thermal resistance of carpet is more a function of thickness than of fiber type.
3. Generally, thermal resistance (R-value) is approximately 2.6 times the total carpet thickness in inches.
4. Before carpet is installed, verify that the backing is resistant to long periods of continuous heat up to 120°F

### Table 6-9 Peak Irradiance, Btu/h·ft$^2$

| Exposure | | 20° | 25° | 30° | 35° | 40° | 45° | 50° | 55° | 60° |
|---|---|---|---|---|---|---|---|---|---|---|
| North | $E_D$ | 40 | 34 | 29 | 26 | 26 | 27 | 30 | 36 | 43 |
| | $E_d$ | 41 | 36 | 33 | 30 | 27 | 24 | 22 | 20 | 18 |
| | $E_t$ | 80 | 70 | 62 | 56 | 53 | 51 | 52 | 55 | 61 |
| Northeast/Northwest | $E_D$ | 146 | 142 | 139 | 135 | 131 | 127 | 122 | 118 | 114 |
| | $E_d$ | 56 | 54 | 51 | 50 | 48 | 47 | 45 | 44 | 43 |
| | $E_t$ | 202 | 196 | 190 | 184 | 179 | 173 | 168 | 163 | 158 |
| East/West | $E_D$ | 168 | 172 | 175 | 177 | 178 | 177 | 176 | 173 | 170 |
| | $E_d$ | 63 | 62 | 61 | 60 | 60 | 60 | 59 | 59 | 59 |
| | $E_t$ | 231 | 234 | 236 | 237 | 237 | 237 | 235 | 233 | 230 |
| Southeast/Southwest | $E_D$ | 89 | 104 | 117 | 128 | 138 | 147 | 154 | 160 | 164 |
| | $E_d$ | 65 | 64 | 64 | 65 | 65 | 66 | 66 | 67 | 68 |
| | $E_t$ | 154 | 168 | 181 | 193 | 203 | 213 | 220 | 227 | 232 |
| South | $E_D$ | 0 | 19 | 44 | 68 | 90 | 110 | 129 | 147 | 163 |
| | $E_d$ | 53 | 61 | 62 | 63 | 65 | 66 | 68 | 70 | 71 |
| | $E_t$ | 53 | 80 | 106 | 131 | 155 | 177 | 197 | 217 | 235 |
| Horizontal | $E_D$ | 268 | 266 | 262 | 255 | 246 | 234 | 219 | 202 | 182 |
| | $E_d$ | 54 | 54 | 54 | 54 | 54 | 54 | 54 | 54 | 54 |
| | $E_t$ | 322 | 320 | 316 | 309 | 300 | 288 | 273 | 256 | 236 |

$\Delta t$ = cooling design temperature difference, °F

PXI = peak exterior irradiance, including shading modifications, Btu/h·ft$^2$ [see Equations 6-17 and 6-18]

SHGC = fenestration rated or estimated NFRC solar heat gain coefficient (Table 6-2)

IAC = interior shading attenuation coefficient

$FF_s$ = fenestration solar load factor, Table 6-13

Although solar gain occurs throughout the day, the cooling load contribution of fenestration correlates well

### Table 6-10 Peak Irradiance Equations

Horizontal surfaces

$$E_t = 302 + 2.06L - 0.0526L^2$$
$$E_d = \min(E_t, 53.9)$$
$$E_D = E_t - E_d$$

Vertical surfaces

$$\phi = \left|\frac{\psi}{180}\right| \text{(normalized exposure, 0 – 1)}$$

$$E_t = 143.7 + 425.1\phi - 1673\phi^3 + 1033\phi^4 - 10.81\phi L + 0.0838\phi L^2 - 4.067L - 0.2671L^2 + [0.3118L^2/(\phi + 1)]$$

$$E_d = \min\left(E_t, 113.2 - 27.57\phi^2 + 0.559\phi L - \frac{34.35\sqrt[4]{L}}{\phi + 1}\right)$$

$$E_D = E_t - E_d$$

where

$E_t, E_d, E_D$ = peak hourly total, diffuse, and direct irradiance, Btu/h·ft$^2$
$L$ = site latitude, °N
$\psi$ = exposure (surface azimuth), ° from south (–180 to +180)

### Table 6-11 Exterior Attachment Transmission

| Attachment | $T_x$ |
|---|---|
| None | 1.0 |
| Exterior insect screen | 0.64 |
| Shade screen | Manufacturer shading coefficient (SC) value, typically 0.4 to 0.6 |

*Note*: See Brunger et al. (1999) regarding insect screens

### Table 6-12 Shade Line Factors (SLFs)

| Exposure | 20° | 25° | 30° | 35° | 40° | 45° | 50° | 55° | 60° |
|---|---|---|---|---|---|---|---|---|---|
| North | 2.8 | 2.1 | 1.4 | 1.5 | 1.7 | 1.0 | 0.8 | 0.9 | 0.8 |
| Northeast/Northwest | 1.4 | 1.5 | 1.6 | 1.2 | 1.3 | 1.3 | 0.9 | 0.9 | 0.8 |
| East/West | 1.2 | 1.2 | 1.1 | 1.1 | 1.1 | 1.0 | 1.0 | 0.9 | 0.8 |
| Southeast/Southwest | 2.1 | 1.8 | 2.0 | 1.7 | 1.5 | 1.6 | 1.4 | 1.2 | 1.1 |
| South | 20.0 | 14.0 | 6.9 | 4.7 | 3.3 | 2.7 | 2.1 | 1.7 | 1.4 |

*Note*: Shadow length below overhang = SLF × $D_{oh}$

with the peak hour irradiance incident on the fenestration exterior. PXI is calculated as follows:

$$\text{PXI} = T_x E_t \text{ (unshaded fenestration)} \quad (6\text{-}17)$$

$$\text{PXI} = T_x[E_d + (1 - F_{shd})E_D] \text{ (shaded fenestration)} \quad (6\text{-}18)$$

where

PXI = peak exterior irradiance, Btu/h·ft$^2$
$E_t, E_d, E_D$ = peak total, diffuse, and direct irradiance (Table 6-9 or 6-10), Btu/h·ft$^2$
$T_x$ = transmission of exterior attachment (insect screen or shade screen) (Table 6-11)
$F_{shd}$ = fraction of fenestration shaded by permanent overhangs, fins, or environmental obstacles (Equation 6-19)

For horizontal or vertical surfaces, peak irradiance values can be obtained from Table 6-9 for primary exposures or from Table 6-10 equations for any exposure. Skylights with slope less than 30° from horizontal should be treated as horizontal.

# Residential Cooling and Heating Load Calculations

Common window coverings can significantly reduce fenestration solar gain. Table 6-11 shows transmission values for typical attachments.

The shaded fraction $F_{\text{shd}}$ can be taken as 1 for any fenestration shaded by adjacent structures during peak hours. Simple overhang shading can be estimated using the following:

$$F_{\text{shd}} = \min\left[1, \max\left(0, \frac{\text{SLF} \times D_{\text{oh}} - X_{\text{oh}}}{h}\right)\right] \quad (6\text{-}19)$$

where

SLF = shade line factor from Table 6-12
$D_{\text{oh}}$ = depth of overhang (from plane of fenestration), ft
$X_{\text{oh}}$ = vertical distance from top of fenestration to overhang, ft
$h$ = height of fenestration, ft

The shade line factor (SLF) is the ratio of the vertical distance a shadow falls beneath the edge of an overhang to the depth of the overhang, so the shade line equals the SLF times the overhang depth. Table 6-12 shows SLFs for July 21 averaged over the hours of greatest solar intensity on each exposure.

Fenestration solar load factors $FF_S$ depend on fenestration exposure and are found in Table 6-13. The values represent the fraction of transmitted solar gain that contributes to peak cooling load. It is thus understandable that morning (east) values are lower than afternoon (west) values. Higher values are included for multifamily buildings with limited exposure.

Interior shading significantly reduces solar gain and is ubiquitous in residential buildings. Field studies have shown that a large fraction of windows feature some sort of shading. Therefore, in all but special circumstances, interior shading should be assumed when calculating cooling loads. In the RLF method, the interior attenuation coefficient (IAC) model is used, as described in Chapter 15, 2013 *Fundamentals*. Residential values from that chapter are consolidated in Table 6-14.

In some cases, it is reasonable to assume that a shade is partially open. For example, drapes are often partially open to admit daylight. IAC values are computed as follows:

$$\text{IAC} = 1 + F_{\text{cl}}(\text{IAC}_{\text{cl}} - 1) \quad (6\text{-}20)$$

where

IAC = interior attenuation coefficient of fenestration with partially closed shade
$F_{\text{cl}}$ = shade fraction closed (0 to 1)
$\text{IAC}_{\text{cl}}$ = interior attenuation coefficient of fully closed configuration (Table 6-14)

## 6.3.5 Internal Gain

The contributions of occupants, lighting, and appliance gains to peak sensible and latent loads can be estimated as

$$q_{\text{ig},s} = 464 + 0.7 A_{\text{cf}} + 75 N_{\text{oc}} \quad (6\text{-}21)$$

**Table 6-13  Fenestration Solar Load Factors $FF_s$**

| Exposure | Single Family Detached | Multifamily |
|---|---|---|
| North | 0.44 | 0.27 |
| Northeast | 0.21 | 0.43 |
| East | 0.31 | 0.56 |
| Southeast | 0.37 | 0.54 |
| South | 0.47 | 0.53 |
| Southwest | 0.58 | 0.61 |
| West | 0.56 | 0.65 |
| Northwest | 0.46 | 0.57 |
| Horizontal | 0.58 | 0.73 |

$$q_{\text{ig},l} = 68 + 0.07 A_{\text{cf}} + 41 N_{\text{oc}} \quad (6\text{-}22)$$

where

$q_{\text{ig},s}$ = sensible cooling load from internal gains, Btu/h
$q_{\text{ig},l}$ = latent cooling load from internal gains, Btu/h
$A_{\text{cf}}$ = conditioned floor area of building, ft$^2$
$N_{\text{oc}}$ = number of occupants (unknown, estimate as $N_{\text{br}} + 1$)

Predicted gains are typical for US homes. Further allowances should be considered when unusual lighting intensities or other equipment are in continuous use during peak cooling hours. In critical situations where intermittent high occupant density or other internal gains are expected, a parallel cooling system should be considered. For room-by-room calculations, $q_{\text{ig},s}$ should be evaluated for the entire conditioned area and allocated to kitchen and living spaces.

## 6.3.6 Total Latent Load

The latent cooling load is the result of three predominant moisture sources: outdoor air (infiltration and ventilation), occupants, and miscellaneous sources, such as cooking, laundry, and bathing.

These components, discussed in previous sections, combine to yield the total latent load:

$$q_l = q_{\text{vi},l} + q_{\text{ig},l} \quad (6\text{-}23)$$

where

$q_l$ = total latent load, Btu/h
$q_{\text{vi},l}$ = ventilation/infiltration latent gain, Btu/h
$q_{\text{ig},l}$ = internal latent gain, Btu/h

Because air conditioning systems are usually controlled by a thermostat, latent cooling is a side effect of equipment operation. During periods of significant latent gain but mild temperatures, there is little cooling operation, resulting in unacceptable indoor humidity. Multispeed equipment, combined temperature/humidity control, and dedicated dehumidification should be considered to address this condition.

## 6.3.7 Summary of RLF Cooling Load Equations

Table 6-15 contains a brief list of equations used in the cooling load calculation procedure described in this chapter.

### Table 6-14  Interior Attenuation Coefficients (IAC$_{cl}$)

| Glazing Layers | Glazing Type (ID*) | Drapes Open-Weave Light | Drapes Closed-Weave Dark | Drapes Closed-Weave Light | Roller Shades Opaque Dark | Roller Shades Opaque White | Roller Shades Translucent Light | Blinds Medium | Blinds White |
|---|---|---|---|---|---|---|---|---|---|
| 1 | Clear (1a) | 0.64 | 0.71 | 0.45 | 0.64 | 0.34 | 0.44 | 0.74 | 0.66 |
|   | Heat absorbing (1c) | 0.68 | 0.72 | 0.50 | 0.67 | 0.40 | 0.49 | 0.76 | 0.69 |
| 2 | Clear (5a) | 0.72 | 0.81 | 0.57 | 0.76 | 0.48 | 0.55 | 0.82 | 0.74 |
|   | Low-e high-solar (17c) | 0.76 | 0.86 | 0.64 | 0.82 | 0.57 | 0.62 | 0.86 | 0.79 |
|   | Low-e low-solar (25a) | 0.79 | 0.88 | 0.68 | 0.85 | 0.60 | 0.66 | 0.88 | 0.82 |
|   | Heat absorbing (5c) | 0.73 | 0.82 | 0.59 | 0.77 | 0.51 | 0.58 | 0.83 | 0.76 |

*Chapter 15 glazing identifier

### Table 6-15  Summary of RLF Cooling Load Equations

| Load Source | Equation | Tables and Notes |
|---|---|---|
| Exterior opaque surfaces | $q_{opq} = A \times CF$ <br> $CF = U(OF_t \Delta t + OF_b + OF_r DR)$ | OF factors from Table 6-6 |
| Exterior transparent surfaces | $q_{fen} = A \times CF$ <br> $CF = U(\Delta t - 0.49 DR) + PXI \times SHGC \times IAC \times FF_s$ | PXI from Table 6-9 plus adjustments <br> $FF_s$ from Table 6-13 |
| Partitions to unconditioned space | $q = AU\Delta t$ | $\Delta t$ = temperature difference across partition |
| Ventilation/infiltration | $q_s = C_s Q \Delta t$ | See Chapter 5 |
| Occupants and appliances | $q_{ig,s} = 464 + 0.7 A_{cf} + 75 N_{oc}$ | |
| Distribution | $q_d = F_{dl} \Sigma q$ | $F_{dl}$ from Table 6-16 |
| Total sensible load | $q_s = q_d + \Sigma q$ | |
| Latent load | $q_l = q_{vi,l} + q_{ig}$ | |
|    Ventilation/infiltration | $q_{vi,l} = C_l Q \Delta W$ | |
|    Internal gain | $q_{ig,l} = 68 + 0.07 A_{cf} + 41 N_{oc}$ | |

### Table 6-16  Typical Duct Loss/Gain Factor

| | | 1 Story 11%/11% R-0 | R-4 | R-8 | 1 Story 5%/5% R-0 | R-4 | R-8 | 2+ Stories 11%/11% R-0 | R-4 | R-8 | 2+ Stories 5%/5% R-0 | R-4 | R-8 |
|---|---|---|---|---|---|---|---|---|---|---|---|---|---|
| **Duct Location** Supply/Return Leakage <br> Insulation ft$^2 \cdot$h$\cdot$°F/Btu | | | | | | | | | | | | | |
| Conditioned space | | No loss ($F_{dl} = 0$) | | | | | | | | | | | |
| Attic | C | 1.26 | 0.71 | 0.63 | 0.68 | 0.33 | 0.27 | 1.02 | 0.66 | 0.60 | 0.53 | 0.29 | 0.25 |
|  | H/F | 0.49 | 0.29 | 0.25 | 0.34 | 0.16 | 0.13 | 0.41 | 0.26 | 0.24 | 0.27 | 0.14 | 0.12 |
|  | H/HP | 0.56 | 0.37 | 0.34 | 0.34 | 0.19 | 0.16 | 0.49 | 0.35 | 0.33 | 0.28 | 0.17 | 0.15 |
| Basement | C | 0.12 | 0.09 | 0.09 | 0.07 | 0.05 | 0.04 | 0.11 | 0.09 | 0.09 | 0.06 | 0.04 | 0.04 |
|  | H/F | 0.28 | 0.18 | 0.16 | 0.19 | 0.10 | 0.08 | 0.24 | 0.17 | 0.15 | 0.16 | 0.09 | 0.08 |
|  | H/HP | 0.23 | 0.17 | 0.16 | 0.14 | 0.09 | 0.08 | 0.20 | 0.16 | 0.15 | 0.12 | 0.08 | 0.07 |
| Crawlspace | C | 0.16 | 0.12 | 0.11 | 0.10 | 0.06 | 0.05 | 0.14 | 0.12 | 0.11 | 0.08 | 0.06 | 0.05 |
|  | H/F | 0.49 | 0.29 | 0.25 | 0.34 | 0.16 | 0.13 | 0.41 | 0.26 | 0.24 | 0.27 | 0.14 | 0.12 |
|  | H/HP | 0.56 | 0.37 | 0.34 | 0.34 | 0.19 | 0.16 | 0.49 | 0.35 | 0.33 | 0.28 | 0.17 | 0.15 |

Values calculated for ASHRAE Standard 152 default duct system surface area using model of Francisco and Palmiter (1999). Values are provided as guidance only; losses can differ substantially for other conditions and configurations. Assumed surrounding temperatures:

Cooling (C): $t_o$ = 95°F, $t_{attic}$ = 120°F, $t_b$ = 68°F, $t_{crawl}$ = 72°F     Heating/furnace (H/F) and heating/heating pump (H/HP): $t_o$ = 32°F, $t_{attic}$ = 32°F, $t_b$ = 64°F, $t_{crawl}$ = 32°F

## 6.4 Heating Load Methodology

During the coldest months, sustained periods of cold and cloudy weather with relatively small variation in outdoor conditions may occur. In this situation, heat loss from the space will be relatively constant and in the absence of internal gains will peak during the early morning hours. Therefore, for design purposes, the heat loss is usually estimated based on steady-state heat transfer for some reasonable design temperature. This simplified approach can be used to estimate a heating load for the "worst case" conditions that can reasonably be anticipated during a heating season. Traditionally this is considered as the load that must be met under design interior and exterior conditions, including infiltration and/or ventilation, but in the absence of solar effect (at night or on cloudy winter days) and before the periodic presence of people, lights, and appliances can begin to have an offsetting effect. The primary orientation is thus toward identification of adequately sized heating equipment to handle the normal worst-case condition.

Prior to designing a heating system, the engineer must estimate the maximum probable (design) heat loss of each room or space to be heated, based on maintaining a selected indoor air temperature during periods of design outdoor

# Residential Cooling and Heating Load Calculations

**Table 6-17 Summary of Heating Load Calculation Equations**

| Load Source | Equation | Tables and Notes |
|---|---|---|
| Exterior surfaces above grade | $q = UA\Delta t$ | $\Delta t = t_i - t_o$ |
| Partitions to unconditioned buffer space | $q = UA\Delta t$ | $\Delta t =$ temp. difference across partition |
| Walls below grade | $q = U_{avg,bw} A(t_{in} - t_{gr})$ | |
| Floors on grade | $q = F_p p \Delta t$ | |
| Floors below grade | $q = U_{avg,bf} A(t_{in} - t_{gr})$ | |
| Ventilation/infiltration | $q_{vi} = C_s Q \Delta t$ | From Chapter 5 |
| Total sensible load | $q_s = \Sigma q$ | |

weather conditions. Heat losses may be divided into two groups: (1) transmission losses or heat transmitted through the confining walls, floor, ceiling, glass, or other surfaces and (2) infiltration losses or heat required to warm outdoor air that leaks in through cracks and crevices, around doors and windows, or through open doors and windows, or heat required to warm outdoor air used for ventilation.

The ideal solution is to select a system with a capacity at maximum output just equal to the heating load that develops when the most severe weather conditions for the locality occur. However, where night setback is used, some excess capacity may be needed unless the owner understands that under some conditions of operation, it may be impossible to elevate the set-back temperature.

The heating load is normally estimated for the winter design temperature, which usually occurs at night; therefore, no credit is taken for the heat given off by such internal sources as persons, lights, and equipment.

Table 6-17 contains a brief list of equations used in the heating load calculation procedures described in this chapter.

## 6.4.1 General Procedure

To calculate a design heating load, prepare the following information about building design and weather data at design conditions.

1. Select outdoor design weather conditions: temperature, wind direction, and wind speed.
2. Select the indoor air temperature to be maintained in each space and the humidity level of the return air, if a humidifier is to be installed, during design weather conditions.
3. Estimate temperatures in adjacent unheated spaces and, if there are below-grade spaces, determine the ground surface temperature at design winter conditions.
4. Select or compute heat transfer coefficients for outside walls and glass; for inside walls, non-basement floors, and ceilings, if these are next to unheated spaces; and for the roof if it is next to heated spaces.
5. Determine the net area of outside wall, glass, and roof next to heated spaces, as well as any cold walls, floors, or ceilings next to unheated spaces. These determinations can be made from building plans or from the actual buildings, using inside dimensions.
6. Compute heat transmission losses for each kind of wall, glass, floor, ceiling, and roof in the building by multiplying the overall heat transfer coefficient in each case by the area of the surface and the temperature difference between indoor air and outdoor air, adjacent space air, or ground surface, as appropriate.
7. Compute heat losses from grade-level slab floors using the heat loss rate per unit length of exposed perimeter, using the method in this chapter.
8. Compute the energy associated with infiltration of cold air around outside doors, windows, porous building materials, and other openings. These unit values depend on the kind or width of crack, wind speed, and the temperature difference between indoor and outdoor air.
9. When positive ventilation using outdoor air is provided by the conditioning unit, the energy required to warm and humidify the ventilation outdoor air to the space conditions must be provided by the unit. Unless mechanical ventilation is sufficient to maintain the building at a slightly positive pressure at all time (producing exfiltration and preventing infiltration), the heating unit must provide for both ventilation and natural infiltration losses.
10. Sum the coincidental transmission losses or heat transmitted through the confining walls, floor, ceiling, glass, and other surfaces, and the energy associated with cold air entering by infiltration and/or the ventilation air, to obtain the total heating load.
11. Include the pickup loads that may be required in intermittently heated buildings or when using night thermostat setback. Pickup loads frequently require an increase in heating capacity to bring the temperature of structure, air, and material contents to the specified temperature.

With the exception of slab-on-grade heat transfer, the basic formula for the heat loss by conductive and convective heat transfer through any surface is

$$q = AU(t_i - t_o) \quad (6\text{-}24)$$

where

$q$ = heat transfer through wall, roof, ceiling, floor, or glass, Btu/h (W)

$A$ = area of wall, glass, roof, ceiling, floor, or other exposed surface, ft² (m²)

$U$ = air-to-air (ground surface) heat transfer coefficient, Btu/h·ft²·°F [W/(m²·K)]

$t_i$ = indoor air temperature near surface involved, °F (°C)

$t_o$ = outdoor air temperature, temperature of adjacent unheated space, or ground surface temperature, °F (°C)

**Example 6-1** Calculate the design heat loss $q_t$ for the following wall:

*Wall size*: 10 ft high by 50 ft long

*Wall construction*: 4 in. face brick; building paper, vapor seal, two layers of mopped 15 lb felt; 4 in. concrete block (normal weight); 1 in. glass fiber, organic bonded; 3/8 in. painted plasterboard.

*Design conditions*: indoor, 72°F dry bulb, relative humidity = 30%; outdoor, 32°F dry bulb

**Solution:**

| | R-Value |
|---|---|
| Outdoor air | 0.17 |
| Brick | 0.44 |
| Felt | 0.12 |
| Block | 1.11 |
| Glass fiber | 4.00 |
| Plasterboard | 0.32 |
| Inside air | 0.68 |
| ΣR = | 6.84 |

$U = 1/R = 1/6.84 = 0.146$ Btu/h·ft$^2$·°F
$A = 10 \times 50 = 500$ ft$^2$
$\Delta t = 72°F - 32°F = 40°F$
$q_t = UA\Delta t = 0.146 \times 500 \times 40$
$= 2920$ Btu/h (0.856 kW)

## 6.4.2 Selecting Heating Design Conditions

The ideal solution to a basic heating system design is a plant with a maximum output capacity equal to the heating load that develops with the most severe local weather conditions. However, this solution is usually uneconomical. Weather records show that severe weather conditions do not repeat annually. If heating systems were designed for maximum weather conditions, excess capacity would exist during most of the system's operating life.

In many cases, an occasional failure of a heating plant to maintain a preselected indoor design temperature during brief periods of severe weather is not critical. However, the successful completion of some industrial or commercial processes may depend on close regulation of indoor temperatures. The specific requirements for each building should be carefully evaluated.

Before selecting an outdoor design temperature from Chapter 4, the designer should consider the following:

- Is the type of structure heavy, medium, or light?
- Is the structure insulated?
- Is the structure exposed to high wind?
- Is the load from infiltration or ventilation high?
- Is there more glass area than normal?
- During what part of the day will the structure be used?
- What is the nature of occupancy?
- Will there be long periods of operation at reduced indoor temperature?
- What is the amplitude between local maximum and minimum daily temperatures?
- Are there local conditions that cause significant variation from temperatures reported by the Weather Bureau?
- What auxiliary heating devices will be in the building?

Before selecting an outdoor design temperature, the designer must keep in mind that, if the outdoor to indoor design temperature difference is exceeded, the indoor temperature may fall, depending on the thermal mass of the structure and its contents, whether or not the internal load was included in calculations, and the duration of the cold period.

The effect of wind on the heating requirements of any building should be considered because:

- Wind movement increases the heat transmission of walls, glass, and roof, affecting poorly insulated walls to a much greater extent than well-insulated walls.
- Wind increases the infiltration of cold air through cracks around doors and windows and even through building materials themselves.

Although 72°F to 75°F are the most commonly selected indoor temperatures for comfort heating design, local code requirements must be checked. ASHRAE Standard 55 and Chapter 9 of the 2013 *Fundamentals* provide additional details on selecting indoor design conditions, as well as Table 4-4.

## 6.4.3 Heat Loss from Above-Grade Exterior Surfaces

All above-grade surfaces exposed to outdoor conditions (walls, doors, ceilings, fenestration, and raised floors) are treated identically, as follows:

$$q = A \times \text{HF} \qquad (6\text{-}25)$$

$$\text{HF} = U \Delta t$$

where HF is the heating load factor in Btu/h·ft$^2$.

Two ceiling configurations are common:

- For ceiling/roof combinations (e.g., flat roof or cathedral ceiling), the U-factor should be evaluated for the entire assembly.
- For well-insulated ceilings (or walls) adjacent to vented attic space, the U-factor should be that of the insulated assembly only (the roof is omitted) and the attic temperature assumed to equal the heating design outdoor temperature.

## 6.4.4 Heat Loss Through Below-Grade Surfaces

Heat transfer through basement walls and floors to the ground depends on (1) the difference between the air temperature within the room and that of the ground and outdoor air, (2) the material of the walls or floor, and (3) conductivity of the surrounding earth. Conductivity of the earth is usually unknown. Because of thermal inertia, ground temperature varies with depth, and there is a substantial time lag between changes in outdoor air temperatures and corresponding changes in ground temperature. As a result, ground-coupled heat transfer is less amenable to steady-state representation than is the case for above-grade building elements.

An approximate method for estimating below-grade heat loss finds the steady-state heat loss to the ground surface, as follows:

$$\text{HF} = U_{avg}(t_{in} - t_{gr}) \quad (6\text{-}26)$$

where

$U_{avg}$ = average U-factor for below-grade surface, Btu/h·ft²·°F

$t_{in}$ = below-grade space air temperature, °F

$t_{gr}$ = design ground surface temperature, °F

The effect of soil heat capacity means that none of the usual external design air temperatures are suitable values for $t_{gr}$. Ground surface temperature fluctuates about an annual mean value by amplitude $A$, which varies with geographic location and surface cover. The minimum ground surface temperature, suitable for heat loss estimates, is therefore

$$t_{gr} = t_m - A \quad (6\text{-}27)$$

where

$t_m$ = mean winter temperature, estimated from the winter average air temperature or from well-water temperature

$A$ = ground surface temperature amplitude from Figure 6-1

The value of the soil thermal conductivity $k$ varies widely with soil type and moisture content. A typical value of 0.8 Btu/h·ft·°F was used to tabulate U-factors with an R-value of approximately 1.47 h·ft²·°F/Btu for uninsulated concrete walls. For these parameters, representative values for $U_{avg,bw}$ are shown in Table 6-18. Representative values of $U_{avg,bf}$ for uninsulated basement floors are shown in Table 6-19.

### 6.4.5 Heat Loss From On-Grade Surfaces

Concrete slab floors may be (1) unheated, relying for warmth on heat delivered above floor level by the heating system, or (2) heated, containing heated pipes or ducts that constitute a radiant slab or a portion of it for complete or partial heating of the house. Heat loss from a concrete slab floor is mostly through the perimeter rather than through the floor and into the ground. Total heat loss is more nearly proportional to the length of the perimeter than to the area of the floor. The simplified approach that treats heat loss as proportional to slab perimeter allows slab heat loss to be estimated for both unheated and heated slab floors:

$$q = P \times \text{HF}$$

$$\text{HF} = F_p \Delta t, \text{ or}$$

$$q = F_p P(t_i - t_o) \quad (6\text{-}28)$$

where

$q$ = heat loss through the perimeter, Btu/h (W)

$F_p$ = heat loss coefficient, Btu/h·°F·ft of perimeter [W/(m·K)]

$P$ = perimeter of exposed edge of floor, ft (m)

$t_i$ = indoor temperature, °F (°C)

$t_o$ = outdoor design temperature, °F (°C)

Representative heat loss coefficients for slab-on-grade floors are available from Table 6-20.

**Example 6-2** Determine the heat loss for a basement in St. Louis, Missouri, which is 60 ft by 40 ft by 8 ft high, of standard concrete construction and entirely below grade. Average winter temperature in St. Louis is 44°F.

Design $\Delta t = t_i - (t_a - A) = 70 - (44 - 22) = 48$ (Figure 6-1)

Wall Average U-Factors (Table 6-18): 0.157 Btu/h·ft°F

$$\text{HF} = U \times \Delta t = 0.157(48) = 7.54$$

**Table 6-18  Average U-Factor for Basement Walls with Uniform Insulation**

| Depth, ft | Uninsulated | R-5 | R-10 | R-15 |
|---|---|---|---|---|
| 1 | 0.4321 | 0.1351 | 0.080 | 0.057 |
| 2 | 0.331 | 0.121 | 0.075 | 0.054 |
| 3 | 0.273 | 0.110 | 0.070 | 0.052 |
| 4 | 0.235 | 0.101 | 0.066 | 0.050 |
| 5 | 0.208 | 0.094 | 0.063 | 0.048 |
| 6 | 0.187 | 0.088 | 0.060 | 0.046 |
| 7 | 0.170 | 0.083 | 0.057 | 0.044 |
| 8 | 0.157 | 0.078 | 0.055 | 0.043 |

$U_{avg,bw}$ from grade to depth, Btu/h·ft²·°F

Soil conductivity = 0.8 Btu/h·ft·°F; insulation is over entire depth.

**Table 6-19  Average U-Factor for Basement Floors**

$U_{avg,bf}$, Btu/h·ft²·F

| $z_f$ (depth of floor below grade), ft | $w_b$ (shortest width of basement), ft |||| 
|---|---|---|---|---|
| | 20 | 24 | 28 | 32 |
| 1 | 0.064 | 0.057 | 0.052 | 0.047 |
| 2 | 0.054 | 0.048 | 0.044 | 0.040 |
| 3 | 0.047 | 0.042 | 0.039 | 0.036 |
| 4 | 0.042 | 0.038 | 0.035 | 0.033 |
| 5 | 0.038 | 0.035 | 0.032 | 0.030 |
| 6 | 0.035 | 0.032 | 0.030 | 0.028 |
| 7 | 0.032 | 0.030 | 0.028 | 0.026 |

Soil conductivity is 0.8 Btu/h·ft·°F; floor is uninsulated.

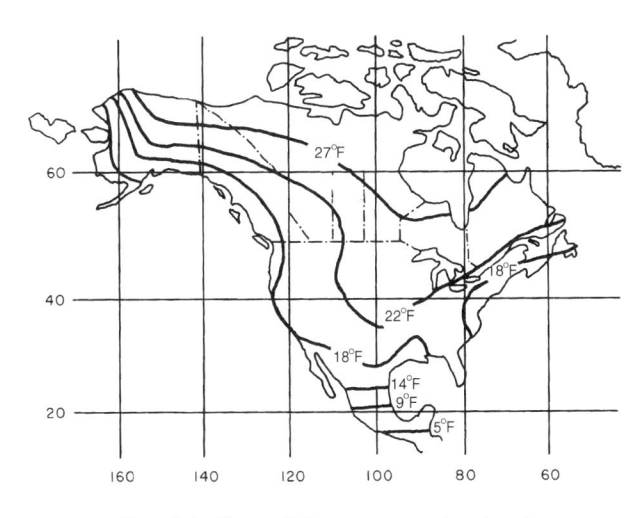

*Fig. 6-1 Ground Temperature Amplitude*

Wall loss = HF × $A$ = 7.54 (200 × 8) = 12060 Btu/h

Floor Heat Loss (Table 6-20)

$$U_{\text{floor}} = 0.026 \text{ Btu/h·ft}^2\text{·°F}$$

$$\text{Area} = 60 \times 40 = 2400 \text{ ft}^2$$

Floor heat loss = 0.026 (2400) 48 = 3000 Btu/h

$q_{\text{total}}$ = 12060 + 3000 = 15,060 Btu/h (4.41 kW)

### 6.4.6 Heat Loss To Buffer Spaces

Heat loss to adjacent unconditioned or semiconditioned spaces can be calculated using a heating load factor based on the partition temperature difference:

$$Q = \text{HF} \times A \quad (6\text{-}29)$$

where

$$\text{HF} = U(t_i - t_b).$$

Buffer space air temperature $t_b$ can be estimated using procedures discussed in Section 4.4. Generally, simple approximations are sufficient except where the partition surface is poorly insulated.

### 6.4.7 Infiltration Heat Loss

Heat loss due to infiltration of outdoor air can be divided into sensible and latent components. The energy quantity that raises the temperature of outdoor infiltrating air up to indoor air temperature is the sensible component. The energy quantity associated with net loss of moisture from the space is the latent component. These calculations are discussed in Chapter 5.

**Example 6-3** The west brick wall of a residence in Louisville, Kentucky, has a net area (excluding windows and doors) of 506 ft² and a U-factor of 0.067 Btu/h·ft²·°F. Determine the heating load for the wall, Btu/h.

**Solution:**

$$q_s = UA\Delta t = 0.067 \, (506)(72 - 8) = 2170 \text{ Btu/h}$$

**Example 6-4** A large window, essentially all glass, 10 ft by 5 ft, is located in the west wall of a residence in Kansas City, Missouri, where the indoor and outdoor winter design conditions have been selected as 72°F and 6°F, respectively. To reduce energy

**Table 6-20 Heat Loss Coefficient $F_p$ of Slab Floor Construction**

| Construction | Insulation | $F_p$, Btu/h·ft·°F |
|---|---|---|
| 8 in. block wall, brick facing | Uninsulated floor | 0.68 |
| | R-5.4 from edge to footer | 0.50 |
| 4 in. block wall, brick facing | Uninsulated floor | 0.84 |
| | R-5.4 from edge to footer | 0.49 |
| Studded wall, stucco | Uninsulated floor | 1.20 |
| | R-5.4 from edge to footer | 0.53 |
| Poured concrete wall with duct near perimeter* | Uninsulated floor | 2.12 |
| | R-5.4 from edge to footer | 0.72 |

*Weighted average temperature of the heating duct was assumed at 110°F during heating season (outdoor air temperature less than 65°F).

requirements, the window is heat-absorbing double glass ($e$ = 0.40), 1/2 in. air space, wood framing, and has inside draperies that are closed at peak conditions. Determine the heating load for the window, Btu/h.

**Solution:**
From the Chapter 5 table for fenestration U-factors,

$$U_{\text{window w/o draperies}} = 0.45 \text{ Btu/h·ft}^2\text{·°F}$$

The major effect of the draperies will be to create an air space that will add approximately 1°F·ft²·F/Btu of thermal resistance. Thus,

$$R_{\text{with drapes}} = 1/U_{w/o} = 1/0.45 + 1 = 3.22$$

$$U_{\text{with drapes}} = 1/3.22 = 0.31 \text{ Btu/h·ft}^2\text{·°F}$$

$$q_s = UA\Delta t = 0.31 \, (50)(72 - 6) = 1023 \text{ Btu/h}$$

**Example 6-5** For the light commercial/residential building described below, determine the design heating load. The outside winter design temperature is 6°F and inside winter design conditions are 72°F, 30% rh. The 8 in. concrete floor is a slab-on-grade with $R$ = 5.4 insulation, properly installed. There are 4900 (base 65°F) degree-days.

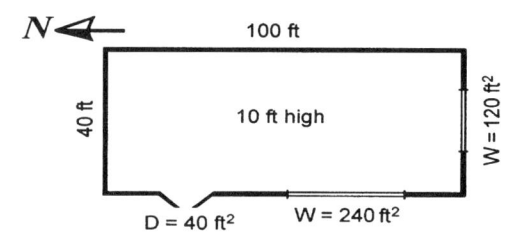

[Diagram: Rectangular building 100 ft × 40 ft, 10 ft high, with N arrow pointing left. W = 120 ft², D = 40 ft², W = 240 ft²]

*Infiltration:* 0.5 air changes per hour (ACH)
*Wall:* 1 in. dark stucco on 4 in. regular concrete, $U$ = 0.350 Btu/h·ft²·°F
*Roof:* 8 in. lightweight concrete, $U$ = 0.12 Btu/h·ft²·°F
*Door:* 2 in. solid wood, $U$ = 0.42 Btu/h·ft²·°F, $A$ = 40 ft²
*Windows:*
South: 1/8 in. regular sheet, $A$ = 120 ft², $U$ = 1.30 Btu/h·ft²·°F
West: Insulating glass, regular sheet out/regular sheet in, 1/4 in. air space, 1/4 in. thick glass; $A$ = 240 ft², $U$ = 0.69 Btu/h·ft²·°F

**Solution:**

| Construction Element | $U$ | $A$ | $\Delta t$ | $q$ |
|---|---|---|---|---|
| East wall | 0.35 | 1000 | 66 | 23,100 |
| North wall | 0.35 | 400 | 66 | 9,240 |
| West wall | 0.35 | 720 | 66 | 16,632 |
| South wall | 0.35 | 280 | 66 | 6,468 |
| Door | 0.42 | 40 | 66 | 1,135 |
| South glass | 1.30 | 120 | 66 | 10,296 |
| West glass | 0.69 | 240 | 66 | 10,930 |
| Roof | 0.12 | 4000 | 66 | 31,680 |
| Floor | $F_2$ = 0.49 | $P$ = 280 | 66 | 9,055 |
| | | | **Subtotal** = | 118,500 |

Infiltration:

$$q_s = 1.10 Q(t_i - t_o)$$

$$q_l = 4840 Q(W_i - W_o)$$

**Residential Cooling and Heating Load Calculations** 203

where, at 0.5 ach,

$$Q = 0.5(40 \times 100 \times 10)/60 = 333$$

Thus,

$$q_s = 1.10 \times 333 \times 66 = 24{,}200 \text{ Btu/h}$$

$$q_l = 4840 \times 333 \times (0.005 - 0.001) = 6447 \text{ Btu/h}$$

Total design heating load =

$$118{,}500 + 24{,}200 + 6400 = 149{,}100 \text{ Btu/h}$$

**Example 6-6** Redo the design heating load for the building of Example 6-5 if, instead of a slab floor, the building is constructed over a full, conditioned basement that has a height of 8 ft, all below grade. Basement walls are insulated with $R = 10$.

**Solution:**

All values are the same as for the previous problem except that the previously calculated floor loss must be replaced with losses for below-grade walls and floor.

$$\text{Below-grade wall loss} = U_w A[t_i - (t_a - A)]$$

$$\text{Below-grade floor loss} = U_f A[t_i - (t_a - A)]$$

The average winter air temperature $t_a$ (Chapter 4) is found to be 43.1°F, and the ground temperature amplitude from Fig. 6-1 is 22°F.

For the wall, $U_w = 0.055$ (Table 6-18).

$$q_{\text{wall}} = 0.055(280 \times 8)[72 - (43.1 - 22)] = 6270 \text{ Btu/h}$$

For the floor, $U \cong 0.026$.

## HEAT LOSS CALCULATION SHEET

| Job Name _____ | Design Conditions |
|---|---|
| Location _____ | $t_o$ = ___  $t_i$ = ___  $\Delta t$ = ___ |
| Date _____ | $rh_o$ = ___  $rh_i$ = ___ |
|  | $W_o$ = ___  $W_i$ = ___  $\Delta W$ = ___ |

| Component | No. | U or U' | Δt or ΔW | Area | Btu/h Sens. | Btu/h Lat. | Area | Btu/h Sens. | Btu/h Lat. | Area | Btu/h Sens. | Btu/h Lat. |
|---|---|---|---|---|---|---|---|---|---|---|---|---|
| Room Number or Name | | | | | | | | | | | | |
| Length of Exposed Slab Perim., ft | | | | | | | | | | | | |
| Room Dimension, Height, Length, Width, ft | | | | | | | | | | | | |
| Gross Exposed Walls and Partitions | | | | | | | | | | | | |
| Windows and Doors | | | | | | | | | | | | |
| Net Exposed Walls and Partitions | | | | | | | | | | | | |
| Ceilings and Roofs | | | | | | | | | | | | |
| Floors | | | | | | | | | | | | |
| Infiltration | | | | | | | | | | | | |
| Heat Loss, Subtotal | | | | | | | | | | | | |
| Duct Heat Loss | | | | | | | | | | | | |
| Total Heat Loss | | | | | | | | | | | | |
| Air Quantity, cfm | | | | | | | | | | | | |

*Fig. 6-2 Heat Loss Calculation Sheet*

$$q_{\text{floor}} = 0.026 \times 4000 \times 50.9 = 5294 \text{ Btu/h}$$

Thus, total load =

$$149{,}100 - 9055 + 6270 + 5294 = 151{,}609 \text{ Btu/h}$$

**Example 6-7** Redo the design heating load for the building of Example 6-5 if, instead of a slab floor on grade, the 8 in. concrete floor is above an open parking garage.

**Solution:**

All values are the same as for Example 6-5 except that the previously calculated floor loss must be replaced with one appropriately calculated for air-to-air heat transfer.

The air beneath the floor in the garage may be considered to be at outdoor air temperature and with the normal winter wind velocity. Thus, the U-factor for the floor can be determined as

$$U = \frac{1}{0.92 + 0.075 \times 8 + 0.17} = 0.59 \text{ Btu/h} \cdot \text{ft}^2 \cdot {}^\circ\text{F}$$

The design heat loss becomes

$$q = 0.59 \times 4000 \times 66 = 156{,}213 \text{ Btu/h}$$

Total design heating load =

$$149{,}100 - 9055 + 156{,}213 = 296{,}300 \text{ Btu/h}$$

*Note:* Floor needs several inches of insulation.

### 6.4.8 Heating Load Summary Sheet

When preparing a set of design heating loads for a multi-space building, either a computer spreadsheet software program or a summary sheet such as provided in Figure 6-2 might prove useful. A summary of heating load calculations is given in Table 6-17.

## 6.5 Nomenclature

**Symbols**

$A$ = area, ft$^2$; ground surface temperature amplitude, °F
$A_{cf}$ = building conditioned floor area, ft$^2$
$A_L$ = building effective leakage area (including flue) at 0.016 in. of water assuming $C_D = 1$, in.$^2$
$C_l$ = air latent heat factor, 4840 Btu/h·cfm at sea level
$C_s$ = air sensible heat factor, 1.1 Btu/h·cfm·°F at sea level
$C_t$ = air total heat factor, 4.5 Btu/h·cfm·(Btu/lb) at sea level
CF = cooling load factor, Btu/h·ft$^2$
$D_{oh}$ = depth of overhang (from plane of fenestration), ft
DR = daily range of outdoor dry-bulb temperature, °F
$E$ = peak irradiance for exposure, Btu/h·ft$^2$
$F_{dl}$ = distribution loss factor
$F_p$ = heat loss coefficient per unit length of perimeter, Btu/h·ft·°F
$F_{shd}$ = shaded fraction
FF = coefficient for CF$_{fen}$
$G$ = internal gain coefficient
$h_{srf}$ = effective surface conductance, including resistance of slab covering material such as carpet, 1/($R_{cvr}$ + 0.68) Btu/h·ft$^2$·°F
$\Delta h$ = indoor/outdoor enthalpy difference, Btu/lb

$H$ = height, ft
HF = heating (load) factor, Btu/h·ft$^2$
$I$ = infiltration coefficient
IAC = interior shading attenuation coefficient
$k$ = conductivity, Btu/h·ft·°F
LF = load factor, Btu/h·ft$^2$
OF = coefficient for CF$_{opq}$
$p$ = perimeter or exposed edge of floor, ft
PXI = peak exterior irradiance, including shading modifications, Btu/h·ft$^2$
$q$ = heating or cooling load, Btu/h
$Q$ = air volumetric flow rate, cfm
$R$ = insulation thermal resistance, h·ft$^2$·°F/Btu
SHGC = fenestration rated or estimated NFRC solar heat gain coefficient
SLF = shade line factor
$t$ = temperature, °F
$T_x$ = solar transmission of exterior attachment
$\Delta t$ = design dry-bulb temperature difference (cooling or heating), °F
$U$ = construction U-factor, Btu/h·ft$^2$·°F (for fenestration, NFRC rated *heating* U-factor)
$w$ = width, ft
$\Delta W$ = indoor-outdoor humidity ratio difference, lb$_w$/lb$_{da}$
$X_{oh}$ = vertical distance from top of fenestration to overhang, ft
$z$ = depth below grade, ft
$\alpha_{roof}$ = roof solar absorptance
$\varepsilon$ = heat/energy recovery ventilation (HRV/ERV) effectiveness

**Subscripts**

avg = average
$b$ = base (as in OF$_b$) or basement
bal = balanced
bf = basement floor
bw = basement wall
br = bedrooms
ceil = ceiling
cf = conditioned floor
cvr = floor covering
$D$ = direct
$d$ = diffuse
da = dry air
dl = distribution loss
env = envelope
es = exposed surface
exh = exhaust
fen = fenestration
floor = floor
gr = ground
hr = heat recovery
$i$ = infiltration
in = indoor
ig = internal gain
$l$ = latent
$o$ = outdoor

# Residential Cooling and Heating Load Calculations

### Table 6-21 Example House Characteristics

| Component | Description | Factors |
|---|---|---|
| Roof/ceiling | Flat wood frame ceiling (insulated with R-30 fiberglass) beneath vented attic with medium asphalt shingle roof | $U = 0.031$ Btu/h·ft²·°F<br>$\alpha_{roof} = 0.85$ (Table 6-7) |
| Exterior walls | Wood frame, exterior wood sheathing, interior gypsum board, R-13 fiberglass insulation | $U = 0.090$ Btu/h·ft²·°F |
| Doors | Wood, solid core | $U = 0.40$ Btu/h·ft²·°F |
| Floor | Slab on grade with heavy carpet over rubber pad; R-5 edge insulation to 3 ft below grade | $R_{cvr} = 1.2$ /h·ft²·°F/Btu (Table 3, Chapter 6, 2008 *ASHRAE Handbook—HVAC Systems and Equipment*)<br>$F_p = 0.5$ Btu/h·ft·°F (estimated from Table 6-20) |
| Windows | Clear double-pane glass in wood frames. Half fixed, half operable with insect screens (except living room picture window, which is fixed). 2 ft eave overhang on east and west with eave edge at same height as top of glazing for all windows. Allow for typical interior shading, half closed. | Fixed: $U = 0.50$ Btu/h·ft²·°F; SHGC = 0.67 (Table 6-2)<br>Operable: $U = 0.51$ Btu/h·ft²·°F; SHGC = 0.57 (Table 6-2)<br>$T_x = 0.6$ (Table 6-11)<br>$IAC_{cl} = 0.6$ (estimated from Table 6-14) |
| Construction | Good | $A_{ul} = 0.02$ in²/ft² (Table 6-3) |

oc = occupant
oh = overhang
opq = opaque
oth = other
r = daily range (as in $OF_r$)
rhb = calculated with RHB method
s = sensible or solar
shd = shaded
slab = slab
srf = surface
sup = supply
t = total or temperature (as in $OF_t$)
ul = unit leakage
unbal = unbalanced
v = ventilation
vi = ventilation/infiltration
w = water
wall = wall

## 6.6 Load Calculation Example

A single-family detached house with floor plan shown in Figure 6-3 is located in Atlanta, Georgia, USA. Construction characteristics are documented in Table 6-21. Using the RLF method, find the block (whole-house) design cooling and heating loads. A furnace/air-conditioner forced-air system is planned with a well-sealed and well-insulated (R-8 wrap) attic duct system.

**Solution**

**Design Conditions.** Table 6-22 summarizes design conditions. Typical indoor conditions are assumed. Outdoor conditions should be determined from Chapter 4, 2013 *ASHRAE Handbook—Fundamentals*, however slightly modified values were selected.

**Component Quantities.** Areas and lengths required for load calculations are derived from plan dimensions (Figure 6-3). Table 6-23 summarizes these quantities.

**Opaque Surface Factors.** Heating and cooling factors are derived for each component condition. Table 6-24 shows the resulting factors and their sources.

### Table 6-22 Example House Design Conditions

| Item | Heating | Cooling | Notes |
|---|---|---|---|
| Latitude | — | — | 33.64°N |
| Elevation | — | — | 1027 ft |
| Indoor temperature | 68°F | 75°F | |
| Indoor relative humidity | N/A | 50% | No humidification |
| Outdoor temperature | 26°F | 92°F | Cooling: 1% value<br>Rounded values<br>Heating: 99% |
| Daily range | N/A | 18°F | |
| Outdoor wet bulb | N/A | 74°F | MCWB* at 1% |
| Wind speed | 15 mph | 7.5 mph | Default assumption |
| Design $\Delta t$ | 42°F | 17°F | |
| Moisture difference | | 0.0052 lb/lb | Psychrometric chart |

*MCWB = mean coincident wet bulb

**Window Factors.** Deriving cooling factor values for windows requires identifying all unique glazing configurations in the house. Equation 6-16 input items indicate that the variations for this case are exposure, window height (with overhang shading), and frame type (which determines U-factor, SHGC, and the presence of insect screen). CF derivation for all configurations is summarized in Table 6-25.

For example, CF for operable 3 ft high windows facing west (the second row in Table 6-25) is derived as follows:

- U-factor and SHGC are found in Table 6-2.
- Each operable window is equipped with an insect screen. From Table 6-12, $T_x = 0.6$ for this arrangement.
- Overhang shading is evaluated with Equation 6-19. For west exposure and latitude 34°, Table 6-12 shows SLF = 1.1. Overhang depth ($D_{oh}$) is 2 ft and the window-overhang distance ($X_{oh}$) is 0 ft. With window height h of 3 ft, $F_s = 0.73$ (73% shaded).
- PXI depends on peak irradiance and shading. Approximating site latitude as 35°N, Table 6-9 shows $E_D = 177$ and $E_d = 60$ Btu/h·ft² for west exposure. Equation 6-18 combines these values with $T_x$ and $F_s$ to find PXI = 0.6[52 + (1 – 0.73)177] = 69 Btu/h·ft².

**Table 6-23  Example House Opaque Surface Factors**

| Component | $U$, Btu/h·ft²·°F or $F_p$, Btu/h·ft·°F | Heating HF | Reference | Cooling $OF_t$ | $OF_b$ | $OF_r$ | CF | Reference |
|---|---|---|---|---|---|---|---|---|
| Ceiling | 0.031 | 1.30 | Equation 34 | 0.62 | 13.75 | −0.19 | 0.65 | Table 7 Equation 21 |
| Wall | 0.090 | 3.78 | | 1 | 14.80 | −0.36 | 2.31 | |
| Garage wall | 0.090 | 3.78 | | 1 | 0.00 | −0.36 | 0.98 | |
| Door | 0.400 | 16.8 | | 1 | 14.80 | −0.36 | 10.27 | |
| Floor perimeter | 0.500 | 21.0 | Chapter 18, Equation 42 | | | | | |
| Floor area | | | | 0.59 | −2.5/(0.68 + 1.20) = −1.33 | | −0.74 | Equation 23 |

**Table 6-24  Example House Window Factors**

| Exposure | Height, ft | Frame | $U$, Btu/h·ft²·°F Table 2 | HF Eq. 34 | $T_x$ Table 12 | $F_{shd}$ Eq. 28 | PXI Eq. 27 | SHGC Table 2 | IAC Eq. 29 | $FF_s$ Table 14 | CF Eq. 25 |
|---|---|---|---|---|---|---|---|---|---|---|---|
| West | 3 | Fixed | 0.50 | 21.0 | 1 | 0.73 | 108 | 0.67 | 0.80 | 0.56 | 36.9 |
| | 3 | Operable | 0.51 | 21.4 | 0.64 | 0.73 | 69 | 0.57 | 0.80 | 0.56 | 22.3 |
| | 6 | Fixed | 0.50 | 21.0 | 1 | 0.37 | 172 | 0.67 | 0.80 | 0.56 | 56.1 |
| | 6 | Operable | 0.51 | 21.4 | 0.64 | 0.37 | 110 | 0.57 | 0.80 | 0.56 | 32.7 |
| | 8 | Fixed | 0.50 | 21.0 | 1 | 0.28 | 187 | 0.67 | 0.80 | 0.56 | 60.9 |
| South | 4 | Fixed | 0.50 | 21.0 | 1 | 0.00 | 131 | 0.67 | 0.80 | 0.47 | 37.6 |
| | 4 | Operable | 0.51 | 21.4 | 0.64 | 0.00 | 84 | 0.57 | 0.80 | 0.47 | 22.7 |
| East | 3 | Fixed | 0.50 | 21.0 | 1 | 0.73 | 108 | 0.67 | 0.80 | 0.31 | 22.5 |
| | 3 | Operable | 0.51 | 21.4 | 0.64 | 0.73 | 69 | 0.57 | 0.80 | 0.31 | 14.4 |
| | 4 | Fixed | 0.50 | 21.0 | 1 | 0.55 | 140 | 0.67 | 0.80 | 0.31 | 27.8 |
| | 4 | Operable | 0.51 | 21.4 | 0.64 | 0.55 | 89 | 0.57 | 0.80 | 0.31 | 17.3 |

**Table 6-25  Example House Envelope Loads**

| Component | HF | CF | Quantity, ft² or ft | Heating Load, Btu/h | Cooling Load, Btu/h |
|---|---|---|---|---|---|
| Ceiling | 1.30 | 0.65 | 2088 | 2714 | 1363 |
| Wall | 3.78 | 2.31 | 1180 | 4460 | 2727 |
| Garage wall | 3.78 | 0.98 | 384 | 1452 | 376 |
| Door | 16.8 | 10.27 | 42 | 706 | 431 |
| Floor perimeter | 21.0 | | 220 | 4620 | |
| Floor area | | −0.74 | 2088 | | −1545 |
| W-Fixed-3 | 21.0 | 36.9 | 4.5 | 95 | 166 |
| W-Operable-3 | 21.4 | 22.3 | 4.5 | 96 | 100 |
| W-Fixed-6 | 21.0 | 56.1 | 12 | 252 | 673 |
| W-Operable-6 | 21.4 | 32.7 | 12 | 257 | 393 |
| W-Fixed-8 | 21.0 | 60.9 | 48 | 1008 | 2921 |
| S-Fixed-4 | 21.0 | 37.6 | 8 | 168 | 301 |
| S-Operable-4 | 21.4 | 22.7 | 8 | 171 | 181 |
| E-Fixed-3 | 21.0 | 22.5 | 4.5 | 95 | 101 |
| E-Operable-3 | 21.4 | 14.4 | 4.5 | 96 | 65 |
| E-Fixed-4 | 21.0 | 27.8 | 24 | 504 | 667 |
| E-Operable-4 | 21.4 | 17.3 | 24 | 514 | 416 |
| Envelope totals | | | | 17,207 | 9336 |

**Table 6-26  Example House Total Sensible Loads**

| Item | Heating Load, Btu/h | Cooling Load, Btu/h |
|---|---|---|
| Envelope | 17,207 | 9336 |
| Infiltration/ventilation | 5914 | 1627 |
| Internal gain | | 2226 |
| Subtotal | 23,121 | 13,189 |
| Distribution loss | 3006 | 3561 |
| Total sensible load | 26,126 | 16,750 |

- All windows are assumed to have some sort of interior shading in the half-closed position. Use (Equation 6-20) with $F_{cl}$ = 0.5 and $IAC_{cl}$ = 0.6 (per Table 6-21) to derive IAC = 0.8.

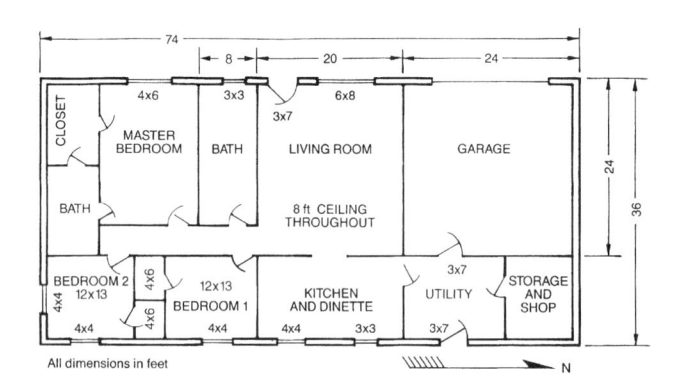

*Fig. 6-3 Example House*

- $FF_s$ is taken from Table 6-13 for west exposure.
- Finally, inserting the preceding values into Equation 6-16 gives CF = 0.51(17 − 0.46 × 17) + 69 × 0.57 × 0.80 × 0.56 = 22.3 Btu/h·ft².

**Envelope Loads.** Given the load factors and component quantities, heating and cooling loads are calculated for each envelope element, as shown in Table 6-26.

**Infiltration and Ventilation.** From Table 6-3, $A_{ul}$ for this house is 0.02 in.²/ft² of exposed surface area. Applying Equation 6-8 yields $A_L = A_{es} \times A_{ul}$ = 3848 × 0.02 = 77 in². Using Table 6-5, estimate heating and cooling IDF to be 1.0 and 0.48 cfm/in², respectively [alternatively, Equation 6-9 could be used to find IDF values]. Apply Equation 6-7 to find the infiltration leakage rates and Equation 6-6 to convert the rate to air changes per hour:

$$Q_{i,h} = 77 \times 1.0 = 77 \text{ cfm } (0.28 \text{ ach})$$

$Q_{i,c} = 77 \times 0.48 = 36$ cfm (0.13 ach)

Calculate the ventilation outdoor air requirement with Equation 6-10 using $A_{cf} = 2088$ ft$^2$ and $N_{br} = 3$, resulting in $Q_v = 51$ cfm. For design purposes, assume that this requirement is met by a mechanical system with balanced supply and exhaust flow rates ($Q_{unbal} = 0$).

Find the combined infiltration/ventilation flow rates with Equation 14, Chapter 17, 2013 *ASHRAE Handbook—Fundamentals*:

$$Q_{vi,h} = 51 + \max(0, 77 + 0.5 \times 0) = 128 \text{ cfm}$$

$$Q_{vi,c} = 51 + \max(0, 36 + 0.5 \times 0) = 87 \text{ cfm}$$

At Atlanta's elevation of 1027 ft, elevation adjustment of heat factors results in a small (4%) reduction in air heat transfer; thus, adjustment is unnecessary, resulting in $C_s = 1.10$ Btu/h·°F·cfm. Use Equation 6-1 with $Q_{bal,hr} = 0$ and $Q_{bal,oth} = 0$ to calculate the sensible infiltration/ventilation loads:

$$q_{vi,s,h} = 1.1 \times 128 \times 42 = 5914 \text{ Btu/h}$$

$$q_{vi,s,c} = 1.1 \times 87 \times 17 = 1627 \text{ Btu/h}$$

*Note:* Using an estimate of 0.4 ach for good construction would result in an infiltration rate of 142 cfm, slightly higher than the 128 calculated by this procedure.

**Internal Gain.** Apply Equation 6-21 to find the sensible cooling load from internal gain:

$$q_{ig,s} = 464 + 0.7 \times 2088 + 75(3+1) = 2226 \text{ Btu/h}$$

**Distribution Losses and Total Sensible Load.** Table 6-27 summarizes the sensible load components. Distribution loss factors $F_{dl}$ are estimated at 0.13 for heating and 0.27 for cooling.

**Latent Load.** Use Equation 6-2 with $C_l = 4840$ Btu/h·cfm, $Q_{vi,c} = 87$ cfm, $Q_{bal,oth} = 0$, and $\Delta W = 0.0052$ to calculate the infiltration/ventilation latent load = 2187 Btu/h. Use Equation 6-23 to find the latent load from internal gains = 378 Btu/h. Therefore, the total latent cooling load is 2565 Btu/h.

## 6.7 Problems

**6.1** Determine which of the following walls of 150 ft$^2$ gross area will have the greatest heat loss:
(a) Wall of 25% single glass and the remainder brick veneer ($U = 0.25$ Btu/h·ft$^2$·°F)
(b) Wall of 50% double-glazed windows with the remainder of the wall brick veneer ($U = 0.25$ Btu/h·ft$^2$·°F)
(c) Wall of 10% single-pane glass and 90% of 6 in. poured concrete with $h_o = 6.0$ and $h_i = 1.6$ Btu/h·ft$^2$

**6.2** A house has a pitched roof with an area of 159 m$^2$ and a $U$ of 1.6 W/(m$^2$·K). The ceiling beneath the roof has an area of 133 m$^2$ and a $U$ of 0.42 W/(m$^2$·K). The attic is unvented in winter for which the design conditions are –19°C outside and 22°C inside. Determine the heat loss through the ceiling. [Ans: 1.88 kW (6400 Btu/h)]

**6.3** Determine the design winter heat loss through each of the following components of a building located in Minneapolis, Minnesota:
(a) Wall having 648 ft$^2$ of area and construction of 4 in. face brick; 3/4 in. plywood sheathing; 2 1/2 in. glass fiber insulation in 2 by 4 stud space (16 in. on centers); 1/2 in. plasterboard interior wall.
(b) A 2185 ft$^2$ ceiling topped by a 2622 ft$^2$ hip roof. The ceiling consists of 1/2 in. acoustical tile with R-19 insulation between the 2 by 6 (16 in. on centers) ceiling joists. The roof has asphalt shingles on 3/4 in. plywood sheathing on the roof rafters. The attic is unvented in winter.
(c) Two 4 ft by 6 ft single-pane glass windows with storm windows.

**6.4** If the building of Problem 6.3 is a residence having a volume of 17,480 ft$^3$ that is equipped with a humidifier set for 25% rh, determine:
(a) Sensible heat load due to infiltration
(b) Latent heat load due to infiltration

**6.5** For a frame building with design conditions of 72°F indoor and 12°F outdoor, determine the heat loss through each of the following components:
(a) Slab floor, 56 ft by 28 ft, on grade without perimeter insulation [Ans: 12,100 Btu/h]
(b) Single-glass double-hung window, 3 ft by 5 ft, with storm window in common metal frame [Ans: 729 Btu/h]
(c) 1 3/8 in. thick solid wood door, 3 ft by 7 ft, with wood storm door [Ans: 731 Btu/h]
(d) Sliding patio door, 6 ft by 7 ft, metal frame with double insulating glass having 1/4 in. air space [Ans: 2041 Btu/h]

**6.6** Determine the heat loss for a basement in Chicago, Illinois, which is 8 m by 12 m by 2.1 m high, of standard concrete construction, and entirely below grade.

**6.7** A residence located in Chicago, Illinois, has a total ceiling area of 1960 ft$^2$ and consists of 3/8 in. gypsum board on 2 by 6 ceiling joists. Six inches of fiberglass (mineral/glass wool) insulation fills the space between the joists. The pitched roof has asphalt shingles on 25/32 in. solid wood sheathing with no insulation between the rafters. The ratio of roof area to ceiling area is 1:3. The attic is unvented in winter. For winter design conditions, including a 72°F inside dry bulb at the 5 ft line, determine
(a) Outside design temperature, °F
(b) Appropriate temperature difference, °F
(c) Appropriate overall coefficient $U$, Btu/h·ft$^2$·°F
(d) Ceiling heat loss $q$, Btu/h

**6.8** A residential building, 30 ft by 100 ft, located in Des Moines, Iowa, has a conditioned space that extends 9 ft below grade level. Determine the design heat loss from the uninsulated below-grade concrete walls and floor.

**6.9** Determine the heating load and specify the furnace for the following residence (located in St. Louis, Missouri) with

(a) 1 in. fiberglass wall insulation and 2 in. fiberglass ceiling insulation

(b) Full wall fiberglass insulation and 4 in. fiberglass ceiling insulation

*Basic Plan*

*Wall Construction*: Face brick, 25/32 in. insulating board sheathing, 2 by 4 studs on 16 in. centers, 3/8 in. gypsum board interior

*Ceiling*: 2 by 6 ceiling joists, 16 in. on center, no flooring above, 3/8 in. gypsum board ceiling

*Roof*: Asphalt shingles on solid wood sheathing, 2 by 6 rafters, no insulation between rafters, no ceiling applied to rafters, 1:4 pitch, 1 ft overhang on eaves, no overhang on gables

*Full basement*: Heated, 10 in. concrete walls, all below grade, 4 in. concrete floor over 4 in. gravel

*Fireplaces:* One in living room on first floor

*Garage:* Attached but unheated

*Windows*:

   W1: 3 ft by 5 ft single-glazed, double-hung wood sash, weather stripped with storm window

   W2: 10 ft by 5 1/2 ft picture window, double glazed, 1/2 in. airspace

   W3: 5 ft by 3 ft wood sash casement, double glazed, 1/2 in. airspace

   W4: 3 ft by 3 ft wood sash casement, double glazed, 1/2 in. airspace

*Doors*:

   D1: 3 ft by 6 ft 8 in., 1 3/4 in. solid with glass storm door

   D2: Sliding glass door, two section, each 3 ft by 6 ft, 8 in. double-glazed, 1/2 in. airspace, aluminum frame

[Ans: (b) 51,000 Btu/h (14.7 kW)]

**6.10** Determine the total conductance loss through the wall panel as shown below. The window has a wooden sill and the plate glass ($U = 1.06$) covers 85% of the window area.

[Ans: 9640 Btu/h (2.77 kW)]

**6.11** Calculate, for design purposes, the heat losses from a room of a building as shown in the diagram, if the outside ambient is 0°F.

[Ans: 46,800 Btu/h (13.7 kW)]

**6.12** A room has three 760 mm by 1520 mm well-fitted double-hung windows. For design conditions of –1°C and 21°C, calculate (a) heating load from air leakage and (b) heating load from transmission through the windows.

**6.13** A residence has a total ceiling area of 1960 ft$^2$ and consists of 3/8 in. gypsum board on 2 in. by 6 in. ceiling joists. Six inches of fiberglass (mineral/glass wool) insulation fills the space between the joists. The effect of the joists themselves can be neglected. The pitched roof has asphalt shingles on 5/8 in. plywood with no insulation between the rafters. The ratio of roof area to ceiling area is 1:3. The attic contains louvers that remain open all year. The residence is located in Louisville, Kentucky. For winter design conditions, determine: (a) appropriate temperature difference $\Delta t$, (b) overall coefficient $U$, and (c) ceiling heat loss.

**6.14** Estimate the heat loss from the uninsulated slab floor of a frame house having dimensions of 18 m by 38 m. The house is maintained at 22°C. Outdoor design temperature is –15°C in a region with 5400 kelvin days.

[Ans: 8.6 kW]

**6.15** Repeat Problem 6.14 for the case where insulation [$R = 0.9$ (m$^2$·K)/W] is applied to the slab edge and extended below grade to the frost line.

**6.16** To preclude attic condensation, an attic ventilation rate of 59 L/s is provided with outdoor air at –13°C. The roof area is 244 m$^2$ and $U_{roof} = 2.7$ W/(m$^2$·K). The ceiling area is 203 m$^2$ and $U_{clg} = 0.30$ W/(m$^2$·K). Inside design temperature is

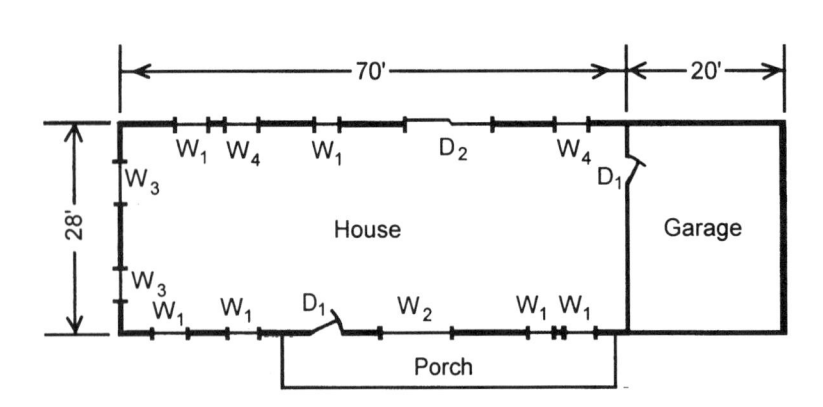

*Diagram for Problem 6.9*

22°C. Determine the ceiling heat loss $W$ with ventilation and compare to the loss if there had been no ventilation.

**6.17** For a residence in Roanoke, Virginia, the hip roof consisting of asphalt shingles on 1/2 in. plywood has an area of 2950 ft$^2$. The 2300 ft$^2$ ceiling consists of 3/8 in. plasterboard on 2 by 6 joists on 24 in. centers. The attic has forced ventilation at the rate of 325 cfm. Determine the attic air temperature at winter design conditions.

**6.18** Solve the following:

(a) A 115 ft by 10 ft high wall in Minneapolis, Minnesota, consists of face brick, a 3/4 in. air gap, 8 in. cinder aggregate concrete blocks, 1 in. organic bonded glass fiber insulation, and 4 in. clay tile interior. Determine the design heat loss through the wall in winter, Btu/h.

(b) If the wall of Part (a) is converted to 60% single-glazed glass, what is the winter design heat loss through the total wall, Btu/h?

**6.19** Determine the design heating load for a residence, 30 ft by 100 ft by 10 ft, to be located in Windsor Locks, Connecticut, which has an uninsulated slab-on-grade concrete floor. The construction consists of

*Walls:* 4 in. face brick, 3/4 in. plywood sheathing, 4 in. cellular glass insulation, and 1/2 in. plasterboard

*Ceiling/roof:* 3 in. lightweight concrete deck, built-up roofing, 2 in. of rigid, expanded rubber insulation, and a drop ceiling of 1/2 in. acoustical tiles, some 18 in. below the roof.

*Windows:* 45% of each wall is double-pane, nonoperable, metal-framed glass (1/4 in. air gap).

*Doors:* Two 3 ft by 7 ft, 1.75 in. thick, solid wood doors are located in each wall.

**6.20** As an attempt to minimize energy requirements, a new residence has been constructed in Dallas, Texas (100°F dry bulb; 20°F daily range; $W = 0.0156$). Size the air-conditioning unit (Btu/h) for this residence with the following features:

- No windows
- Inside design conditions 75°F, 60% rh ($W = 0.0112$)
- One 3 ft by 7 ft, 2 in. thick wood door with storm door on south side, $U = 0.26$ Btu/h·ft$^2$·°F, one 3 ft by 7 ft, 2 in. thick wood door with storm door on east side, $U = 0.26$ Btu/h·ft$^2$·°F
- Overall size: 70 ft by 28 ft by 8 ft high
- Walls are frame construction with 2 by 6 studs and full insulation for $U = 0.043$
- Attic has natural ventilation, 12 in. fiberglass insulation, and overall $U$ of roof and ceiling of 0.021 based on ceiling area, light-colored roof
- Infiltration is so small that outdoor air must be brought in at a rate of 9 cfm/person
- Estimated occupancy at design condition is 15 people
- Fluorescent tights rated at 320 W will be on all the time

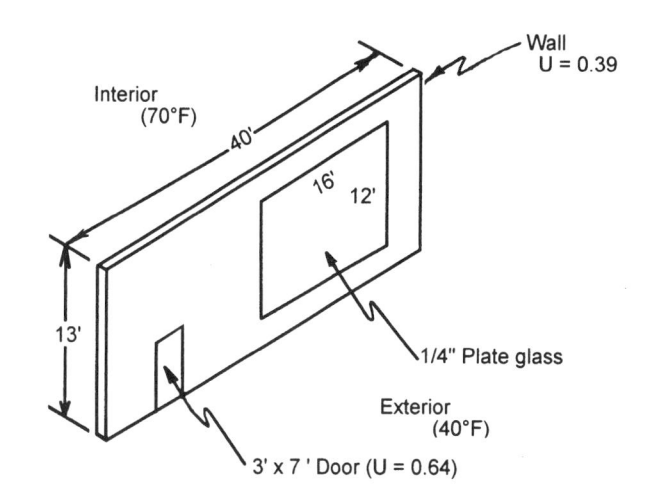

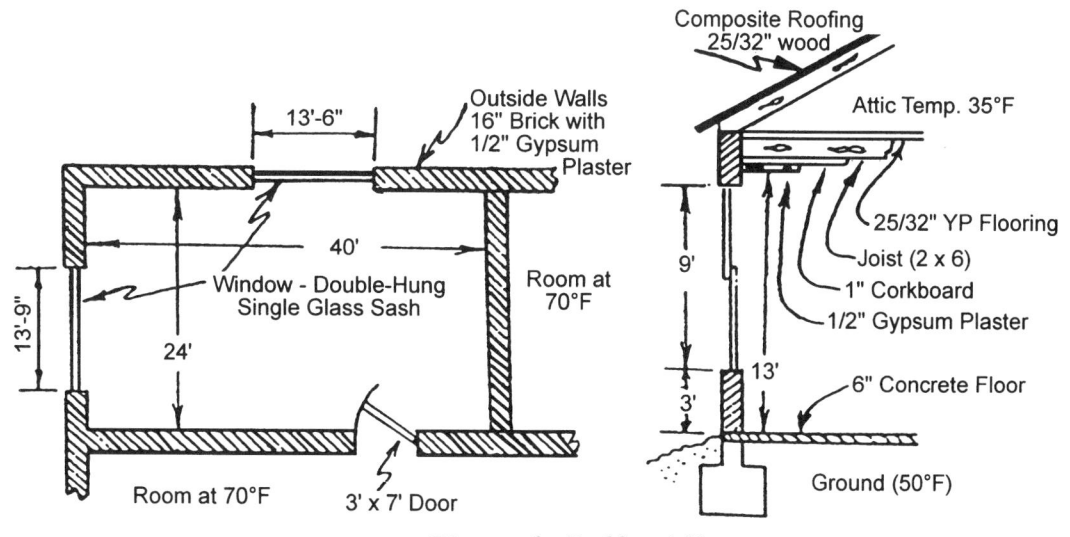

*Diagram for Problem 6.11*

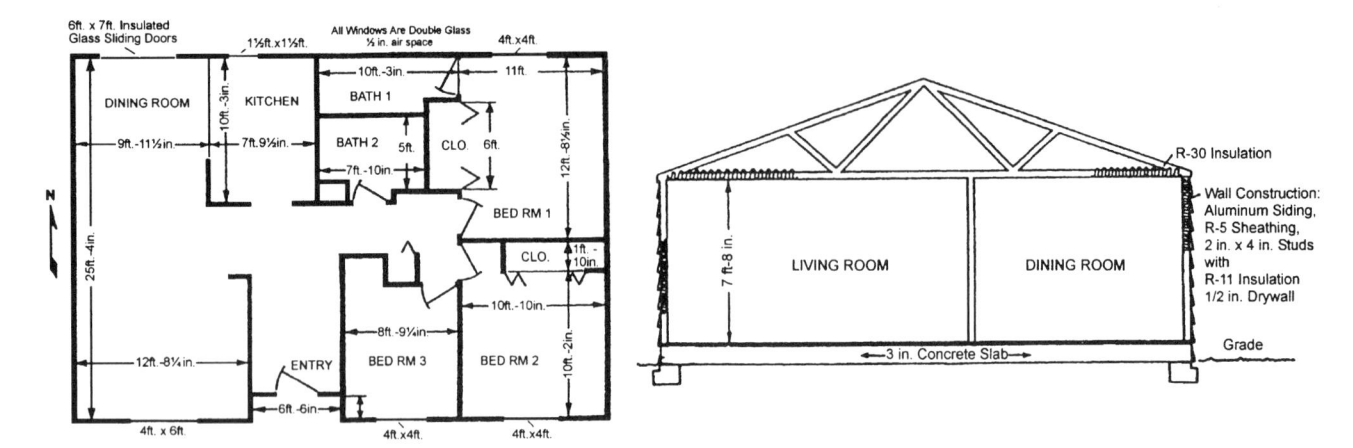

*Diagram for Problem 6.21*

- Floor is concrete slab on ground.

**6.21** Determine design heating and cooling loads for the residence shown in the figure and located in Manhattan, Kansas.

**6.22** Determine the cooling load and specify the central air-conditioning system for the following residence in St. Louis, Missouri, having:

(a) black asphalt shingles, full-wall fiberglass insulation, 4 in. fiberglass ceiling insulation, no drapes, no attic fan

(b) same as (a) except silver-white asphalt shingles

(c) same as (a) except lined drapes at all windows

(d) same as (a) except large attic vent fan

*Basic Plan*

   *Wall construction*: Face brick, 25/32 in. insulating board sheathing, 2 by 4 studs on 16 in. centers, 3/8 in. gypsum board interior

   *Ceiling*: 2 by 6 ceiling joists, 16 in. on-canter, no flooring above, 3/8 in. gypsum board ceiling

   *Roof*: Asphalt shingles on solid wood sheathing, 2 by 6 rafters, no insulation between rafters, no ceiling applied to rafters, 1:4 pitch, 1 ft overhang on eaves, no overhang on gables

   *Full basement*: Heated, 10 in. concrete watts, all below grade, 4 in. concrete floor over 4 in. gravel

   *Fireplace*: One in living room of first floor

   *Garage*: Attached but unheated

   *Windows*:

      W1: 3 ft by 5 ft single-glazed, double-hung wood sash, weather stripped with storm window

      W2: 10 ft by 5.5 ft picture window, double-glazed, 1/2 in. airspace

      W3: 5 ft by 3 ft wood-sash casement, double-glazed, 1/2 in. airspace

      W4: 3 ft by 3 ft wood-sash casement, double-glazed, 1/2 in. airspace

   *Doors*:

      D1: 3 ft by 6 ft, 8 in., 1 3/4 in. solid with glass storm door

      D2: Sliding glass door, two 3 ft by 6 ft 8 in sections, double-glazed, 1/2 in. airspace, aluminum frame

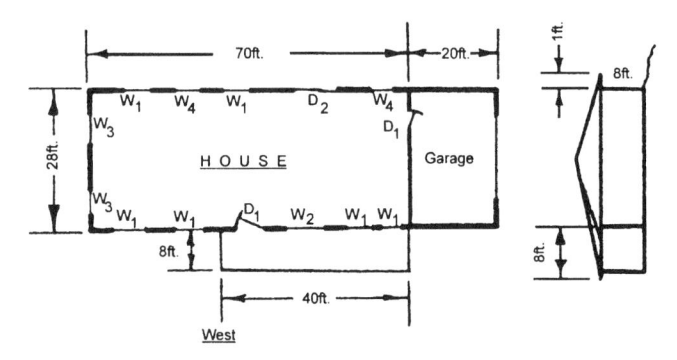

## 6.8 Bibliography

ACCA. 1986. *Load Calculation for Residential Winter and Summer Air Conditioning—Manual J, 7th edition.* Arlington, VA: Air Conditioning Contractors of America.

ACCA. 2003. *Manual J. Residential Load Calculations, 8th Edition.* Arlington, VA: Air Conditioning Contractors of America.

ASHRAE. 2013. *2013 ASHRAE Handbook—Fundamentals.*

Barnaby, C.S., and J.D. Spitler, 2005. Development of the residential load factor method for heating and cooling load calculation. *ASHRAE Transactions* 111(1): 291–307.

Kusuda, T., and J.W. Bean. 1984. Simplified methods for determining seasonal heat loss from uninsulated slab-on-grade floors. *ASHRAE Transactions* 89(2A).

Mitalas, G.P. 1983. Calculation of basement heat loss. *ASHRAE Transactions* 89(1B).

Parker, D.S., J.E.R. McIlvaine, S.F. Barkaszi, D.J. Beal, and M.T. Anello, 2000. Laboratory testing of the reflectance properties of roofing materials. FSEC-CR 670-00, Florida Solar Energy Center, Cocoa Beach, FL.

Pedersen, C.O., D.E. Fisher, J.D. Spitler, and R.J. Liesen. 1998. *Cooling and Heating Load Calculation Principles.* Atlanta: American Society of Heating, Refrigerating and Air-Conditioning Engineers, Inc.

# Residential Cooling and Heating Load Calculations

## SI Figures and Tables

### Table 6-1 SI  RLF Limitations

| Item | Valid Range | Notes |
|---|---|---|
| Latitude | 20°S to 60°N | Also approximately valid for 20°S to 60°S with N and S orientations reversed for southern hemisphere. |
| Date | July 21 | Application must be summer peaking. Buildings in mild climates with significant SE/S/SW glazing may experience maximum cooling load in fall or even winter. Use RHB if local experience indicates this is a possibility. |
| Elevation | Less than 2000 m | RLF factors assume 50 m elevation. With elevation-corrected $C_s$, method is acceptably accurate except at very high elevations. |
| Climate | Warm/hot | Design-day average outdoor temperature assumed to be above indoor design temperature. |
| Construction | Lightweight residential construction (wood or metal framing, wood or stucco siding) | May be applied to masonry veneer over frame construction; results are conservative. Use RHB for structural masonry or unconventional construction. |
| Fenestration area | 0% to 15% of floor area on any façade, 0% to 30% of floor area total | Spaces with high fenestration fraction should be analyzed with RHB. |
| Fenestration tilt | Vertical or horizontal | Skylights with tilt less than 30° can be treated as horizontal. Buildings with significant sloped glazing areas should be analyzed with RHB. |
| Occupancy | Residential | Applications with high internal gains and/or high occupant density should be analyzed with RHB or nonresidential procedures. |
| Temperature swing | 1.7 Ks | |
| Distribution losses | Typical | Applications with extensive duct runs in unconditioned spaces should be analyzed with RHB. |

Table 6-2 SI  Typical Fenestration Characteristics

| Glazing Type | Glazing Layers | ID[b] | Property[c,d] | Center of Glazing | Operable Aluminum | Operable Aluminum with Thermal Break | Operable Reinforced Vinyl/Aluminum Clad Wood | Operable Wood/Vinyl | Operable Insulated Fiberglass/Vinyl | Fixed Aluminum | Fixed Aluminum with Thermal Break | Fixed Reinforced Vinyl/Aluminum Clad Wood | Fixed Wood/Vinyl | Fixed Insulated Fiberglass/Vinyl |
|---|---|---|---|---|---|---|---|---|---|---|---|---|---|---|
| Clear | 1 | 1a | U | 5.91 | 7.24 | 6.12 | 5.14 | 5.05 | 4.61 | 6.42 | 6.07 | 5.55 | 5.55 | 5.35 |
| | | | SHGC | 0.86 | 0.75 | 0.75 | 0.64 | 0.64 | 0.64 | 0.78 | 0.78 | 0.75 | 0.75 | 0.75 |
| | 2 | 5a | U | 2.73 | 4.62 | 3.42 | 3.00 | 2.87 | 5.83 | 3.61 | 3.22 | 2.86 | 2.84 | 2.72 |
| | | | SHGC | 0.76 | 0.67 | 0.67 | 0.57 | 0.57 | 0.57 | 0.69 | 0.69 | 0.67 | 0.67 | 0.67 |
| | 3 | 29a | U | 1.76 | 3.80 | 2.60 | 2.25 | 2.19 | 1.91 | 2.76 | 2.39 | 2.05 | 2.01 | 1.93 |
| | | | SHGC | 0.68 | 0.60 | 0.60 | 0.51 | 0.51 | 0.51 | 0.62 | 0.62 | 0.60 | 0.60 | 0.60 |
| Low-e, low-solar | 2 | 25a | U | 1.70 | 3.83 | 2.68 | 2.33 | 2.21 | 1.89 | 2.75 | 2.36 | 2.03 | 2.01 | 1.90 |
| | | | SHGC | 0.41 | 0.37 | 0.37 | 0.31 | 0.31 | 0.31 | 0.38 | 0.38 | 0.36 | 0.36 | 0.36 |
| | 3 | 40c | U | 1.02 | 3.22 | 2.07 | 1.76 | 1.71 | 1.45 | 2.13 | 1.76 | 1.44 | 1.40 | 1.33 |
| | | | SHGC | 0.27 | 0.25 | 0.25 | 0.21 | 0.21 | 0.21 | 0.25 | 0.25 | 0.24 | 0.24 | 0.24 |
| Low-e, high-solar | 2 | 17c | U | 1.99 | 4.05 | 2.89 | 2.52 | 2.39 | 2.07 | 2.99 | 2.60 | 2.26 | 2.24 | 2.13 |
| | | | SHGC | 0.70 | 0.62 | 0.62 | 0.52 | 0.52 | 0.52 | 0.64 | 0.64 | 0.61 | 0.61 | 0.61 |
| | 3 | 32c | U | 1.42 | 3.54 | 2.36 | 2.02 | 1.97 | 1.70 | 2.47 | 2.10 | 1.77 | 1.73 | 1.66 |
| | | | SHGC | 0.62 | 0.55 | 0.55 | 0.46 | 0.46 | 0.46 | 0.56 | 0.56 | 0.54 | 0.54 | 0.54 |
| Heat-absorbing | 1 | 1c | U | 5.91 | 7.24 | 6.12 | 5.14 | 5.05 | 4.61 | 6.42 | 6.07 | 5.55 | 5.55 | 5.35 |
| | | | SHGC | 0.73 | 0.64 | 0.64 | 0.54 | 0.54 | 0.54 | 0.66 | 0.66 | 0.64 | 0.64 | 0.64 |
| | 2 | 5c | U | 2.73 | 4.62 | 3.42 | 3.00 | 2.87 | 2.53 | 3.61 | 3.22 | 2.86 | 2.84 | 2.72 |
| | | | SHGC | 0.62 | 0.55 | 0.55 | 0.46 | 0.46 | 0.46 | 0.56 | 0.56 | 0.54 | 0.54 | 0.54 |
| | 3 | 29c | U | 1.76 | 3.80 | 2.60 | 2.25 | 2.19 | 1.91 | 2.76 | 2.39 | 2.05 | 2.01 | 1.93 |
| | | | SHGC | 0.34 | 0.31 | 0.31 | 0.26 | 0.26 | 0.26 | 0.31 | 0.31 | 0.30 | 0.30 | 0.30 |
| Reflective | 1 | 1l | U | 5.91 | 7.24 | 6.12 | 5.14 | 5.05 | 4.61 | 6.42 | 6.07 | 5.55 | 5.55 | 5.35 |
| | | | SHGC | 0.31 | 0.28 | 0.28 | 0.24 | 0.24 | 0.24 | 0.29 | 0.29 | 0.27 | 0.27 | 0.27 |
| | 2 | 5p | U | 2.73 | 4.62 | 3.42 | 3.00 | 2.87 | 2.53 | 3.61 | 3.22 | 2.86 | 2.84 | 2.72 |
| | | | SHGC | 0.29 | 0.27 | 0.27 | 0.22 | 0.22 | 0.22 | 0.27 | 0.27 | 0.26 | 0.26 | 0.26 |
| | 3 | 29c | U | 1.76 | 3.80 | 2.60 | 2.25 | 2.19 | 1.91 | 2.76 | 2.39 | 2.05 | 2.01 | 1.93 |
| | | | SHGC | 0.34 | 0.31 | 0.31 | 0.26 | 0.26 | 0.26 | 0.31 | 0.31 | 0.30 | 0.30 | 0.30 |

[a] Data are from Chapter 15 in the 2013 *ASHRAE Handbook—Fundamentals*, Tables 4 and 14 for selected combinations.  [b] ID = Chapter 15 in the 2013 *ASHRAE Handbook—Fundamentals* glazing type identifier.  [c] $U$ = U-factor, W/(m$^2$·K)  [d] SHGC = solar heat gain coefficient

# Residential Cooling and Heating Load Calculations

### Table 6-3 SI  Unit Leakage Areas

| Construction | Description | $A_{ul}$ (cm²/m²) |
|---|---|---|
| Tight | Construction supervised by air-sealing specialist | 0.7 |
| Good | Carefully sealed construction by knowledgeable builder | 1.4 |
| Average | Typical current production housing | 2.8 |
| Leaky | Typical pre-1970 houses | 5.6 |
| Very leaky | Old houses in original condition | 10.4 |

### Table 6-4 SI  Evaluation of Exposed Surface Area

| Situation | Include | Exclude |
|---|---|---|
| Ceiling/roof combination (e.g., cathedral ceiling without attic) | Gross surface area | |
| Ceiling or wall adjacent to attic | Ceiling or wall area | Roof area |
| Wall exposed to ambient | Gross wall area (including fenestration area) | |
| Wall adjacent to unconditioned buffer space (e.g., garage or porch) | Common wall area | Exterior wall area |
| Floor over open or vented crawlspace | Floor area | Crawlspace wall area |
| Floor over sealed crawlspace | Crawlspace wall area | Floor area |
| Floor over conditioned or semiconditioned basement | Above-grade basement wall area | Floor area |
| Slab floor | | Slab area |

### Table 6-5 SI  Typical IDF Values, L/(s·cm²)

| | Heating Design Temperature, °C | | | | | Cooling Design Temperature, °C | | | |
|---|---|---|---|---|---|---|---|---|---|
| H, m | −40 | −30 | −20 | −10 | 0 | 10 | 30 | 35 | 40 |
| 2.5 | 0.10 | 0.095 | 0.086 | 0.077 | 0.069 | 0.060 | 0.031 | 0.035 | 0.040 |
| 3 | 0.11 | 0.10 | 0.093 | 0.083 | 0.072 | 0.061 | 0.032 | 0.038 | 0.043 |
| 4 | 0.14 | 0.12 | 0.11 | 0.093 | 0.079 | 0.065 | 0.034 | 0.042 | 0.049 |
| 5 | 0.16 | 0.14 | 0.12 | 0.10 | 0.086 | 0.069 | 0.036 | 0.046 | 0.055 |
| 6 | 0.18 | 0.16 | 0.14 | 0.11 | 0.093 | 0.072 | 0.039 | 0.050 | 0.061 |
| 7 | 0.20 | 0.17 | 0.15 | 0.12 | 0.10 | 0.075 | 0.041 | 0.051 | 0.068 |
| 8 | 0.22 | 0.19 | 0.16 | 0.14 | 0.11 | 0.079 | 0.043 | 0.058 | 0.074 |

### Table 6-6 SI  Opaque Surface Cooling Factor Coefficients

| Surface Type | $OF_t$ | $OF_b$, °F | $OF_r$ |
|---|---|---|---|
| Ceiling or wall adjacent to vented attic | 0.62 | 25.7 $\alpha_{roof}$ − 8.1 | −0.19 |
| Ceiling/roof assembly | 1 | 68.9 $\alpha_{roof}$ − 12.6 | −0.36 |
| Wall (wood frame) or door with solar exposure | 1 | 14.8 | −0.36 |
| Wall (wood frame) or door (shaded) | 1 | 0 | −0.36 |
| Floor over ambient | 1 | 0 | −0.06 |
| Floor over crawlspace | 0.33 | 0 | −0.28 |
| Slab floor (see Slab Floor section) | | | |

$\alpha_{roof}$ = roof solar absorptance (see Table 6-7)

### Table 6-7 SI  Roof Solar Absorptance $a_{roof}$

| | Color | | | |
|---|---|---|---|---|
| Material | White | Light | Medium | Dark |
| Asphalt shingles | 0.75 | 0.75 | 0.85 | 0.92 |
| Tile | 0.30 | 0.40 | 0.80 | 0.80 |
| Metal | 0.35 | 0.50 | 0.70 | 0.90 |
| Elastomeric coating | 0.30 | | | |

*Source*: Summarized from Parker et al. 2000

### Table 6-8 SI  Thermal Resistance of Floor Coverings

| Description | Thermal Resistance $r_c$, |
|---|---|
| Bare concrete, no covering | 0 |
| Asphalt tile | |
| Rubber tile | |
| Light carpet | |
| Light carpet with rubber pad | |
| Light carpet with light pad | |
| Light carpet with heavy pad | |
| Heavy carpet | |
| Heavy carpet with rubber pad | |
| Heavy carpet with light pad | |
| Heavy carpet with heavy pad | |
| hardwood | |
| wood floor (oak) | |
| oak parquet and pad | |
| Linoleum | |
| Marble floor and mudset | |
| Rubber pad | |
| Prime urethane underlayment, waffled sponge rubber | |
| Bonded urethane, | |

*Notes*:
1. Carpet pad thickness should not be more than.
2. Total thermal resistance of carpet is more a function of thickness than of fiber type.
3. Generally, thermal resistance (R-value) is approximately times the total carpet thickness in.
4. Before carpet is installed, verify that the backing is resistant to long periods of continuous heat up to.

### Table 6-9 SI  Peak Irradiance, W/m²

| | | Latitude | | | | | | | | |
|---|---|---|---|---|---|---|---|---|---|---|
| Exposure | | 20° | 25° | 30° | 35° | 40° | 45° | 50° | 55° | 60° |
| North | $E_D$ | 125 | 106 | 92 | 84 | 81 | 85 | 96 | 112 | 136 |
| | $E_d$ | 128 | 115 | 103 | 93 | 84 | 76 | 69 | 62 | 55 |
| | $E_t$ | 253 | 221 | 195 | 177 | 166 | 162 | 164 | 174 | 191 |
| Northeast/Northwest | $E_D$ | 460 | 449 | 437 | 425 | 412 | 399 | 386 | 374 | 361 |
| | $E_d$ | 177 | 169 | 162 | 156 | 151 | 147 | 143 | 140 | 137 |
| | $E_t$ | 637 | 618 | 599 | 581 | 563 | 546 | 529 | 513 | 498 |
| East/West | $E_D$ | 530 | 543 | 552 | 558 | 560 | 559 | 555 | 547 | 537 |
| | $E_d$ | 200 | 196 | 193 | 190 | 189 | 188 | 187 | 187 | 187 |
| | $E_t$ | 730 | 739 | 745 | 748 | 749 | 747 | 742 | 734 | 724 |
| Southeast/Southwest | $E_D$ | 282 | 328 | 369 | 405 | 436 | 463 | 485 | 503 | 517 |
| | $E_d$ | 204 | 203 | 203 | 204 | 205 | 207 | 210 | 212 | 215 |
| | $E_t$ | 485 | 531 | 572 | 609 | 641 | 670 | 695 | 715 | 732 |
| South | $E_D$ | 0 | 60 | 139 | 214 | 283 | 348 | 408 | 464 | 515 |
| | $E_d$ | 166 | 193 | 196 | 200 | 204 | 209 | 214 | 219 | 225 |
| | $E_t$ | 166 | 253 | 335 | 414 | 487 | 557 | 622 | 683 | 740 |
| Horizontal | $E_D$ | 845 | 840 | 827 | 806 | 776 | 738 | 691 | 637 | 574 |
| | $E_d$ | 170 | 170 | 170 | 170 | 170 | 170 | 170 | 170 | 170 |
| | $E_t$ | 1015 | 1010 | 997 | 976 | 946 | 908 | 861 | 807 | 744 |

### Table 6-10 SI    Peak Irradiance Equations

Horizontal surfaces

$$E_t = 302 + 2.06L - 0.0526L^2$$

$$E_d = \min(E_t, 53.9)$$

$$E_D = E_t - E_d$$

Vertical surfaces

$$\phi = \left|\frac{\psi}{180}\right| \text{ (normalized exposure, 0 – 1)}$$

$$E_t = 143.7 + 425.1\phi - 1673\phi^3 + 1033\phi^4 - 10.81\phi L + 0.0838\phi L^2 - 4.067L - 0.2671L^2 + [0.3118L^2/(\phi + 1)]$$

$$E_d = \min\left(E_t,\; 113.2 - 27.57\phi^2 + 0.559\phi L - \frac{34.35\sqrt[4]{L}}{\phi + 1}\right)$$

$$E_D = E_t - E_d$$

where

$E_t, E_d, E_D$ = peak hourly total, diffuse, and direct irradiance, Btu/h·ft²
$L$ = site latitude, °N
$\psi$ = exposure (surface azimuth), ° from south (–180 to +180)

### Table 6-11 SI    Exterior Attachment Transmission

| Attachment | $T_x$ |
|---|---|
| None | 1.0 |
| Exterior insect screen | 0.64 |
| Shade screen | Manufacturer SC value, typically 0.4 to 0.6 |

*Note*: See Brunger et al. (1999) regarding insect screens

### Table 6-12 SI    Shade Line Factors (SLFs)

| Exposure | 20° | 25° | 30° | 35° | 40° | 45° | 50° | 55° | 60° |
|---|---|---|---|---|---|---|---|---|---|
| North | 2.8 | 2.1 | 1.4 | 1.5 | 1.7 | 1.0 | 0.8 | 0.9 | 0.8 |
| Northeast/Northwest | 1.4 | 1.5 | 1.6 | 1.2 | 1.3 | 1.3 | 0.9 | 0.9 | 0.8 |
| East/West | 1.2 | 1.2 | 1.1 | 1.1 | 1.1 | 1.0 | 1.0 | 0.9 | 0.8 |
| Southeast/Southwest | 2.1 | 1.8 | 2.0 | 1.7 | 1.5 | 1.6 | 1.4 | 1.2 | 1.1 |
| South | 20.0 | 14.0 | 6.9 | 4.7 | 3.3 | 2.7 | 2.1 | 1.7 | 1.4 |

*Note*: Shadow length below overhang = SLF × $D_{oh}$

### Table 6-13 SI    Fenestration Solar Load Factors $FF_s$

| Exposure | Single Family Detached | Multifamily |
|---|---|---|
| North | 0.44 | 0.27 |
| Northeast | 0.21 | 0.43 |
| East | 0.31 | 0.56 |
| Southeast | 0.37 | 0.54 |
| South | 0.47 | 0.53 |
| Southwest | 0.58 | 0.61 |
| West | 0.56 | 0.65 |
| Northwest | 0.46 | 0.57 |
| Horizontal | 0.58 | 0.73 |

# Residential Cooling and Heating Load Calculations

### Table 6-14 SI  Interior Attenuation Coefficients (IAC$_{cl}$)

| Glazing Layers | Glazing Type (ID*) | Drapes Open-Weave Light | Drapes Closed-Weave Dark | Drapes Closed-Weave Light | Roller Shades Opaque Dark | Roller Shades Opaque White | Roller Shades Translucent Light | Blinds Medium | Blinds White |
|---|---|---|---|---|---|---|---|---|---|
| 1 | Clear (1a) | 0.64 | 0.71 | 0.45 | 0.64 | 0.34 | 0.44 | 0.74 | 0.66 |
|   | Heat absorbing (1c) | 0.68 | 0.72 | 0.50 | 0.67 | 0.40 | 0.49 | 0.76 | 0.69 |
| 2 | Clear (5a) | 0.72 | 0.81 | 0.57 | 0.76 | 0.48 | 0.55 | 0.82 | 0.74 |
|   | Low-e high-solar (17c) | 0.76 | 0.86 | 0.64 | 0.82 | 0.57 | 0.62 | 0.86 | 0.79 |
|   | Low-e low-solar (25a) | 0.79 | 0.88 | 0.68 | 0.85 | 0.60 | 0.66 | 0.88 | 0.82 |
|   | Heat absorbing (5c) | 0.73 | 0.82 | 0.59 | 0.77 | 0.51 | 0.58 | 0.83 | 0.76 |

*Chapter 15 glazing identifier

### Table 6-15 SI  Summary of RLF Cooling Load Equations

| Load Source | Equation | Tables and Notes |
|---|---|---|
| Exterior opaque surfaces | $q_{opq} = A \times CF$ <br> $CF = U(OF_t \Delta t + OF_b + OF_r DR)$ | OF factors from Table 7, Chapter 29, 2005 *ASHRAE Handbook—Fundamentals* |
| Exterior transparent surfaces | $q_{fen} = A \times CF$ <br> $CF = U(\Delta t - 0.46 DR) + PXI \times SHGC \times IAC \times FF_s$ | PXI from Table 9 plus adjustments, Chapter 29, 2005 *ASHRAE Handbook—Fundamentals* <br> FF$_s$ from Table 13, Chapter 29, 2005 *ASHRAE Handbook—Fundamentals* |
| Partitions to unconditioned space | $q = AU\Delta t$ | $\Delta t$ = temperature difference across partition |
| Ventilation/infiltration | $q_s = C_s Q \Delta t$ | See Common Data and Procedures section |
| Occupants and appliances | $q_{ig,s} = 136 + 2.2 A_{cf} + 22 N_{oc}$ | |
| Distribution | $q_d = F_{dl} \Sigma q$ | $F_{dl}$ from Table 6, Chapter 17, 2009 *ASHRAE Handbook—Fundamentals* |
| Total sensible load | $q_s = q_d + \Sigma q$ | |
| Latent load | $q_l = q_{vi,l} + q_{ig,l}$ | |
|    Ventilation/infiltration | $q_{vi,l} = C_l Q \Delta W$ | |
|    Internal gain | $q_{ig,l} = 20 + 0.22 A_{cf} + 12 N_{oc}$ | |

### Table 6-16 SI  Typical Duct Loss/Gain Factors

| | | 1 Story 11%/11% R-0 | R- | R- | 1 Story 5%/5% R-0 | R- | R- | 2+ Stories 11%/11% R-0 | R- | R- | 2+ Stories 5%/5% R-0 | R- | R- |
|---|---|---|---|---|---|---|---|---|---|---|---|---|---|
| Duct Location | Insulation | | | | | | | | | | | | |
| Conditioned space | | No loss ($F_{dl} = 0$) | | | | | | | | | | | |
| Attic | C | 1.26 | 0.71 | 0.63 | 0.68 | 0.33 | 0.27 | 1.02 | 0.66 | 0.60 | 0.53 | 0.29 | 0.25 |
|  | H/F | 0.49 | 0.29 | 0.25 | 0.34 | 0.16 | 0.13 | 0.41 | 0.26 | 0.24 | 0.27 | 0.14 | 0.12 |
|  | H/HP | 0.56 | 0.37 | 0.34 | 0.34 | 0.19 | 0.16 | 0.49 | 0.35 | 0.33 | 0.28 | 0.17 | 0.15 |
| Basement | C | 0.12 | 0.09 | 0.09 | 0.07 | 0.05 | 0.04 | 0.11 | 0.09 | 0.09 | 0.06 | 0.04 | 0.04 |
|  | H/F | 0.28 | 0.18 | 0.16 | 0.19 | 0.10 | 0.08 | 0.24 | 0.17 | 0.15 | 0.16 | 0.09 | 0.08 |
|  | H/HP | 0.23 | 0.17 | 0.16 | 0.14 | 0.09 | 0.08 | 0.20 | 0.16 | 0.15 | 0.12 | 0.08 | 0.07 |
| Crawlspace | C | 0.16 | 0.12 | 0.11 | 0.10 | 0.06 | 0.05 | 0.14 | 0.12 | 0.11 | 0.08 | 0.06 | 0.05 |
|  | H/F | 0.49 | 0.29 | 0.25 | 0.34 | 0.16 | 0.13 | 0.41 | 0.26 | 0.24 | 0.27 | 0.14 | 0.12 |
|  | H/HP | 0.56 | 0.37 | 0.34 | 0.34 | 0.19 | 0.16 | 0.49 | 0.35 | 0.33 | 0.28 | 0.17 | 0.15 |

Values calculated for ASHRAE *Standard* 152 default duct system surface area using model of Francisco and Palmiter (1999). Values are provided as guidance only; losses can differ substantially for other conditions and configurations. Assumed surrounding temperatures:
Cooling (C): $t_o =$ , $t_{attic} =$ , $t_b =$ , $t_{crawl} =$   Heating/furnace (H/F) and heating/heating pump (H/HP): $t_o =$ , $t_{attic} =$ , $t_b =$ , $t_{crawl} =$

**Table 6-17 SI  Average U-Factor for Basement Walls with Uniform Insulation**

| Depth, m | $U_{avg,bw}$ from grade to depth, W/(m²·K) | | | |
|---|---|---|---|---|
| | Uninsulated | R-0.88 | R-1.76 | R-2.64 |
| 0.3 | 2.468 | 0.769 | 0.458 | 0.326 |
| 0.6 | 1.898 | 0.689 | 0.427 | 0.310 |
| 0.9 | 1.571 | 0.628 | 0.401 | 0.296 |
| 1.2 | 1.353 | 0.579 | 0.379 | 0.283 |
| 1.5 | 1.195 | 0.539 | 0.360 | 0.272 |
| 1.8 | 1.075 | 0.505 | 0.343 | 0.262 |
| 2.1 | 0.980 | 0.476 | 0.328 | 0.252 |
| 2.4 | 0.902 | 0.450 | 0.315 | 0.244 |

Soil conductivity = 1.4 W/(m·K); insulation is over entire depth.

**Table 6-18 SI  Average U-Factor for Basement Floors**

| $z_f$ (depth of floor below grade), m | $U_{avg,bf}$, W/(m²·K) | | | |
|---|---|---|---|---|
| | $w_b$ (shortest width of basement), m | | | |
| | 6 | 7 | 8 | 9 |
| 0.3 | 0.370 | 0.335 | 0.307 | 0.283 |
| 0.6 | 0.310 | 0.283 | 0.261 | 0.242 |
| 0.9 | 0.271 | 0.249 | 0.230 | 0.215 |
| 1.2 | 0.242 | 0.224 | 0.208 | 0.195 |
| 1.5 | 0.220 | 0.204 | 0.190 | 0.179 |
| 1.8 | 0.202 | 0.188 | 0.176 | 0.166 |
| 2.1 | 0.187 | 0.175 | 0.164 | 0.155 |

Soil conductivity is 1.4 W/(m·K); floor is uninsulated.

**Table 6-19 SI  Heat Loss Coefficient $F_p$ of Slab Floor Construction**

| Construction | Insulation | $F_p$, W/(m·K) |
|---|---|---|
| 200 mm block wall, brick facing | Uninsulated floor | 1.17 |
| | R-0.95 (m²·K)/W from edge to footer | 0.86 |
| 4 in. block wall, brick facing | Uninsulated floor | 1.45 |
| | R-0.95 (m²·K)/W from edge to footer | 0.85 |
| Studded wall, stucco | Uninsulated floor | 2.07 |
| | R-0.95 (m²·K)/W from edge to footer | 0.92 |
| Poured concrete wall with duct near perimeter* | Uninsulated floor | 3.67 |
| | R-0.95 (m²·K)/W from edge to footer | 1.24 |

*Weighted average temperature of the heating duct was assumed at 43°C during heating season (outdoor air temperature less than 18°C).

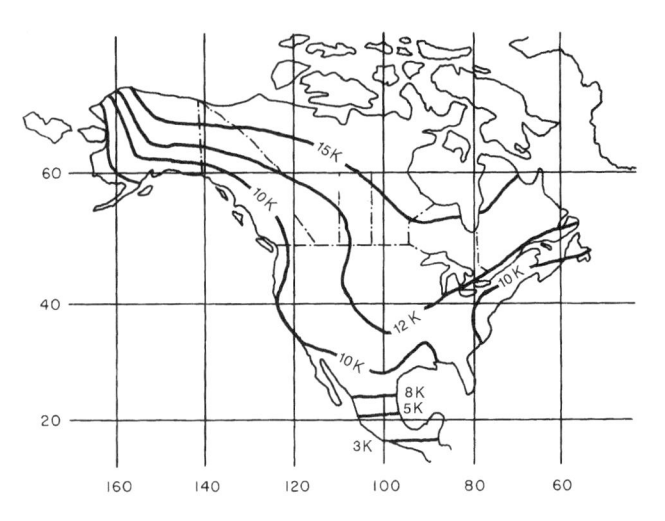

*Fig. 6-1 SI  Ground Temperature Amplitude*

# Chapter 7

# NONRESIDENTIAL COOLING AND HEATING LOAD CALCULATIONS

This chapter presents the methodology for determining the air-conditioning cooling and heating loads used for sizing cooling and heating equipment for nonresidential buildings. A more detailed discussion of the cooling and heating loads for buildings is given in Chapters 15 and 18 of the 2013 *ASHRAE Handbook—Fundamentals*. Another excellent source of information is the *Load Calculation Applications Manual* (Spitler 2009).

## 7.1 Principles

Heating and cooling load calculations are the primary basis for the design and selection of most heating and air-conditioning systems and components. These calculations are necessary to determine the size of piping, ducting, diffusers, air handlers, boilers, chillers, coils, compressors, fans, and every other component of the systems that condition indoor environments. Cooling and heating load calculations will directly or indirectly affect the first cost of building construction, the comfort and productivity of building occupants, and operation and energy consumption.

Heating and cooling loads are the rates of energy input (heating) or removal (cooling) required to maintain an indoor environment at a desired combination of temperature and humidity. Heating and cooling systems are designed, sized, and controlled to accomplish that energy transfer. The amount of heating or cooling required at any particular time varies widely, depending on external (e.g., outside temperature) and internal (e.g., number of people present) factors. Peak design heating and cooling load calculations seek to determine the maximum rate of heating and of cooling energy transfer needed at any time in the year. This chapter discusses common elements of load calculations and several methods of making load estimates, but it focuses on ASHRAE's Radiant Time Series (RTS) method.

Cooling loads result from many conductive, convective, and radiative heat transfer processes through the building envelope and from internal sources and system components. Building components or contents that may affect cooling loads include the following:

**External**: Walls, roofs, windows, partitions, ceilings, and floors
**Internal**: Lights, people, appliances, and equipment
**Infiltration**: Air leakage and moisture migration
**System**: Ventilation air, duct leakage, reheat, and fan and pump power.

The variables affecting cooling load calculations are numerous, often difficult to define precisely, and always intricately interrelated. Many cooling load components vary in magnitude over a wide range during a 24-hour period. Since these cyclic changes in load components are often not in phase with each other, each must be analyzed to establish the resultant maximum cooling load for a building or zone. A zoned system (a system of conditioning equipment serving several independent areas, each with its own temperature control) needs no greater total cooling load capacity than the largest hourly summary of simultaneous zone loads throughout a design day; however, it must handle the peak cooling load for each zone at its individual peak hour. At certain times of the day during the heating or intermediate seasons, some zones may need heating while others need cooling.

**Calculation Accuracy.** The concept of determining the cooling load for a given building must be kept in perspective. A proper cooling load calculation gives values adequate for proper performance. Variation in the heat transmission coefficient of typical building materials and composite assemblies, the differing motivations and skills of those who physically construct the building, and the manner in which the building is actually operated are some of the variables that make a numerically precise calculation impossible. While the designer uses reasonable procedures to account for these factors, the calculation can never be more than a good estimate of the actual cooling load.

**Heat Flow Rates.** In air-conditioning design, four related heat flow rates, each of which varies with time, must be differentiated: (1) space heat gain, (2) space cooling load, (3) space heat extraction rate, and (4) cooling coil load.

*Space Heat Gain.* This instantaneous rate of heat gain is the rate heat enters into and/or is generated within a space at a given instant. Heat gain is classified by (1) the mode in which it enters the space and (2) whether it is a sensible or latent gain.

*Mode of Entry.* The modes of heat gain may be (1) solar radiation through transparent surfaces; (2) heat conduction through exterior walls and roofs; (3) heat conduction through interior partitions, ceilings, and floors; (4) heat generated within the space by occupants, lights, and appliances; (5) energy transfer as a result of ventilation and infiltration of outdoor air; or (6) miscellaneous heat gains.

*Sensible or Latent Heat.* Sensible heat gain is directly added to the conditioned space by conduction, convection,

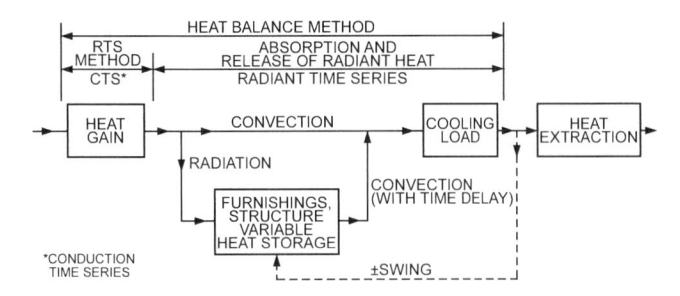

*Fig. 7-1 Origin of Difference Between Magnitude of Instantaneous Heat Gain and Instantaneous Cooling Load (Figure 1, Chapter 18, 2013 ASHRAE Handbook—Fundamentals)*

and/or radiation. Latent heat gain occurs when moisture is added to the space (e.g., from vapor emitted by occupants and equipment). To maintain a constant humidity ratio, water vapor must condense on the cooling apparatus at a rate equal to its rate of addition into the space. The amount of energy required to offset the latent heat gain essentially equals the product of the rate of condensation and the latent heat of condensation. In selecting cooling apparatus, it is necessary to distinguish between sensible and latent heat gain. Every cooling apparatus has a maximum latent heat removal capacity for particular operating conditions.

*Space Cooling Load.* This is the rate at which heat must be removed from the space to maintain a constant space air temperature. The sum of all space instantaneous heat gains at any given time does not necessarily (or even frequently) equal the cooling load for the space at that same time.

*Radiant Heat Gain.* Space heat gain by radiation is not immediately converted into cooling load. Radiant energy must first be absorbed by the surfaces that enclose the space (walls, floor, and ceiling) and the objects in the space (furniture, etc.). As soon as these surfaces and objects become warmer than the space air, some of their heat is transferred in the air space by convection. The composite heat storage capacity of these surfaces and objects determines the rate at which their respective surface temperatures increase for a given radiant input and thus governs the relationship between the radiant portion of heat gain and its corresponding part of the space cooling load (Figure 7-1). The thermal storage effect is critically important in differentiating between instantaneous heat gain for a given space and its cooling load for that moment. Predicting the nature and magnitude of this elusive phenomenon in order to estimate a realistic cooling load for a particular combination of circumstances has long been a subject of major interest to design engineers.

*Space Heat Extraction Rate.* The rate at which heat is removed from the conditioned space equals the space cooling load only to the degree that room air temperature is held constant. In conjunction with intermittent operation of the cooling equipment, the control system characteristics usually permit a minor cyclic variation or swing in room temperature. Therefore, a proper simulation of the control system gives a more realistic value of energy removal over a fixed period than using the values of the space cooling load. This concept is primarily important for estimating energy use over time; however, it is not needed to calculate design peak cooling load for equipment selection.

*Cooling Coil Load.* The rate at which energy is removed at the cooling coil that serves one or more conditioned spaces equals the sum of the instantaneous space cooling loads (or space heat extraction rate if it is assumed that the space temperature does not vary) for all the spaces served by the coil, plus any external loads. Such external loads include heat gain by the distribution system between the individual spaces and the cooling equipment, and outdoor air heat and moisture introduced into the distribution system through the cooling equipment.

**Cooling Load Estimation in Practice.** Frequently, a cooling load must be calculated before every parameter in the conditioned space can be properly or completely defined. An example is a cooling load estimate for a new building with many floors of unleased spaces where detailed partition requirements, furnishings, lighting selection, and layout cannot be predefined. Potential tenant modifications once the building is occupied also must be considered. The load estimating process requires proper engineering judgment that includes a thorough understanding of heat balance fundamentals.

### 7.1.1 Heat Balance Fundamentals

The calculation of cooling load for a space involves calculating a surface-by-surface conductive, convective, and radiative heat balance for each room surface and a convective heat balance for the room air. Sometimes called "the exact solution," these principles form the foundation for all other methods described in this chapter.

To calculate space cooling load directly by heat balance procedures requires a laborious solution of energy balance equations involving the space air, surrounding walls and windows, infiltration and ventilation air, and internal energy sources. To demonstrate the calculation principle, consider a sample room enclosed by four walls, a ceiling, and a floor, with infiltration air and normal internal energy sources. The energy exchange at each inside surface at a given time can be calculated from the following equation:

$$q_{i,\theta} = \left[ h_{ci}(t_{a,\theta} - t_{i,\theta}) + \sum_{j=1 \neq i}^{m} g_{ij}(t_{j,\theta} - t_{i,\theta}) \right] A_i$$
$$+ RS_{i,\theta} + RL_{i,\theta} + RE_{i,\theta} \qquad (7\text{-}1)$$
$$\text{for } i = 1, 2, 3, 4, 5, 6$$

where

$m$ = number of surfaces in room (six in this case)

$q_{i,\theta}$ = rate of heat conducted into surface $i$ at inside surface at time $\theta$

$A_i$ = area of surface $i$

$h_{ci}$ = convective heat transfer coefficient at interior surface $i$

$g_{ij}$ = a radiation heat transfer factor between interior surface $i$ and interior surface $j$

$t_{a,\theta}$ = inside air temperature at time $\theta$

$t_{i,\theta}$ = average temperature of interior surface $i$ at time $\theta$

$RS_{i,\theta}$ = rate of solar energy coming through windows and absorbed by surface $i$ at time $\theta$

$RL_{i,\theta}$ = rate of heat radiated from lights and absorbed by surface $i$ at time $\theta$

$RE_{i,\theta}$ = rate of heat radiated from equipment and occupants and absorbed by surface $i$ at time $\theta$

**Conduction Transfer Functions.** The equations governing conduction within the six surfaces cannot be solved independently of Equation 7-1, since the energy exchanges occurring within the room affect the inside surface conditions, in turn affecting the internal conduction. Consequently, the six formulations of Equation 7-1 must be solved simultaneously with the governing equations of conduction within the six surfaces in order to calculate the space cooling load. Typically, these equations are formulated as conduction transfer functions (CTFs) in the form

$$q_{\text{in},\theta} = \sum_{m=1}^{M} Y_{k,m} t_{o,\theta-m+1} - \sum_{m=1}^{M} Z_{k,m} t_{o,\theta-m+1} + \sum_{m=1}^{M} F_m q_{\text{in},\theta-m} \quad (7\text{-}2)$$

where

$q$ = rate of heat conducted into a specific surface at a specific hour

in = inside surface subscript

$k$ = order of CTF

$m$ = time index variable

$M$ = number of nonzero CTF values

$o$ = outside surface subscript

$t$ = temperature

$\theta$ = time

$Y$ = cross CTF values

$Z$ = interior CTF values

$F_m$ = flux history coefficients

**Space Air Energy Balance.** Note that the interior surface temperature, $t_{i,\theta}$ in Equation 7-1 and $t_{\text{in},\theta}$ in Equation 7-2, requires simultaneous solution. In addition, Equation 7-3, which represents an energy balance on the space air, must also be solved simultaneously.

$$Q_{L,\theta} = \left[\sum_{i=1}^{6} h_{ci}(t_{i,\theta} - t_{a,\theta})\right] A_i + \rho C V_{L,\theta}(t_{o,\theta} - t_{a,\theta}) + \rho C V_{v,\theta}(t_{v,\theta} - t_{a,\theta}) + RS_{a,\theta} + RL_{a,\theta} + RE_{a,\theta} \quad (7\text{-}3)$$

where

$\rho$ = air density

$C$ = air specific heat

$V_{L,\theta}$ = volume flow rate of outdoor air infiltrating into room at time $\theta$

$t_{o,\theta}$ = outdoor air temperature at time $\theta$

$V_{v,\theta}$ = volume rate of flow of ventilation air at time $\theta$

$t_{v,\theta}$ = ventilation air temperature at time $\theta$

$RS_{a,\theta}$ = rate of solar heat coming through windows and convected into room air at time $\theta$

$RL_{a,\theta}$ = rate of heat from lights convected into room air at time $\theta$

$RE_{a,\theta}$ = rate of heat from equipment and occupants and convected into room air at time $\theta$

Note that the space air temperature is allowed to float. By fixing the space air temperature, the cooling load need not be determined simultaneously.

This rigorous approach to calculating space cooling load is impractical without the speed at which computations can be done by digital computers. Computer programs that calculate instantaneous space cooling loads in this exact manner are primarily oriented to energy use calculations over extended periods (Mitalas and Stephenson 1967; Buchberg 1958).

The transfer function concept is a simplification to the strict heat balance calculation procedure. In the transfer function concept, Mitalas and Stephenson (1967) used room thermal response factors. In their procedure, room surface temperatures and cooling load were first calculated by the rigorous method just described for several typical constructions representing offices, schools, and dwellings of heavy, medium, and light construction. In these calculations, components such as solar heat gain, conductive heat gain, or heat gain from the lighting, equipment, and occupants were simulated by pulses of unit strength. The transfer functions were then calculated as numerical constants representing the cooling load corresponding to the input excitation pulses. Once these transfer functions were determined for typical constructions they were assumed independent of input pulses, thus permitting cooling loads to be determined without the more rigorous calculation. Instead, the calculation requires simple multiplication of the transfer functions by a time-series representation of heat gain and subsequent summation of these products. The same transfer function concept can be applied to calculating heat gain components themselves, as explained later.

### 7.1.2 Total Equivalent Temperature Differential Method

In the total equivalent temperature differential (TETD) method, the response factor technique was used with a number of representative wall and roof assemblies from which data were derived to calculate TETD values as functions of sol-air temperature and maintained room temperature. Various components of space heat gain are calculated using associated TETD values, and the results are added to internal heat gain elements to get an instantaneous total rate of space heat gain. This gain is converted to an instantaneous space cooling load by the time-averaging (TA) technique of averaging the radiant portions of the heat gain load components for the current hour with related values from an appropriate period of immediately preceding hours. This technique provides a rational means to deal quantitatively with the thermal storage phenomenon, but it is best solved by computer because of its complexity.

### 7.1.3 Transfer Function Method

Although similar in principle to TETD/TA, the transfer function method (TFM) (Mitalas 1972) applies a series of weighting factors, or conduction transfer function (CTF) coefficients, to the various exterior opaque surfaces and to differences between sol-air temperature and inside space temperature to determine heat gain with appropriate reflection of thermal inertia of such surfaces. Solar heat gain through glass and various forms of internal heat gain are calculated directly for the load hour of interest. The TFM next applies a second series of weighting factors, or coefficients of room transfer functions (RTF), to heat gain and cooling load values from all load elements having radiant components to account for the thermal storage effect in converting heat gain to cooling load. Both evaluation series consider data from several previous hours as well as the current hour. RTF coefficients relate specifically to the spatial geometry, configuration, mass, and other characteristics of the space so as to reflect weighted variations in thermal storage effect on a time basis rather than a straight-line average.

**Transfer Functions.** These coefficients relate an output function at a given time to the value of one or more driving functions at a given time and at a set period immediately preceding. The CTF described in this chapter is no different from the thermal response factor used for calculating wall or roof heat conduction, while the RTF is the weighting factor for obtaining cooling load components (ASHRAE 1975). While the TFM is scientifically appropriate and technically sound for a specific cooling load analysis, its computational complexity requires computer use for effective application in a commercial design environment.

### 7.1.4 Heat Balance Method (HB)

The estimation of cooling load for a space involves calculating a surface-by-surface conductive, convective, and radiative heat balance for each room surface and a convective heat balance for the room air. Sometimes called the **exact solution**, these principles form the foundation for all methods described in this chapter.

Some of the computations required by this rigorous approach to calculating space cooling load make the use of digital computers essential. The heat balance procedure is not new. Many energy calculation programs have used it in some form for many years. The first implementation that incorporated all the elements to form a complete method was NBSLD (Kusuda 1967). The heat balance procedure is also implemented in both the BLAST and TARP energy analysis programs (Walton 1983). Prior to the implementation of ASHRAE Research Project 875, the method had never been described completely or in a form applicable to cooling load calculations. The papers resulting from RP-875 describe the heat balance procedure in detail (Pedersen et al. 1997; Liesen and Pedersen 1997; McClellan and Pedersen 1997).

**Description of Heat Balance Model.** All calculation procedures involve some kind of model. All models require simplifying assumptions and therefore are approximate. The most fundamental assumption is that the air in the thermal zone can be modeled as **well mixed**, meaning it has a uniform temperature throughout the zone. ASHRAE Research Project 664 established that this assumption is valid over a wide range of conditions.

The next major assumption is that the surfaces of the room (walls, windows, floor, etc.) can be treated as having:

- Uniform surface temperatures
- Uniform longwave (LW) and shortwave (SW) irradiation
- Diffuse radiating surfaces
- One-dimensional heat conduction within

The resulting formulation is called the **heat balance model**. It is important to note that the foregoing assumptions, although common, are quite restrictive and set certain limits on the information that can be obtained from the model.

### 7.1.5 Radiant Time Series Method (RTS)

The radiant time series (RTS) method is a new simplified method for performing design cooling load calculations that is derived from the heat balance (HB) method described above. It effectively replaces all other simplified (non-heat-balance) methods, such as the transfer function method (TFM), the cooling load temperature difference/cooling load factor (CLTD/CLF) method, and the total equivalent temperature difference/time averaging (TETD/TA) method.

The casual observer might well ask why yet another load calculation method is necessary. This method was developed in response to the desire to offer a method that is rigorous, yet does not require iterative calculations, and that quantifies each component contribution to the total cooling load. In addition, it is desirable for the user to be able to inspect and compare the coefficients for different construction and zone types in a form illustrating their relative impact on the result. These characteristics of the RTS method make it easier to apply engineering judgment during the cooling load calculation process.

The RTS method is suitable for peak design load calculations, but it should not be used for annual energy simulations due to its inherent limiting assumptions. The RTS method, while simple in concept, involves too many calculations to be used practically as a manual method, although it can easily be implemented in a simple computerized spreadsheet. For a manual cooling load calculation method, refer to the CLTD/CLF method included in the 1997 *ASHRAE Handbook—Fundamentals*.

## 7.2 Initial Design Considerations

To calculate a space cooling load, detailed building design information and weather data at selected design conditions are required. Generally, the following steps should be followed:

**Building Characteristics.** Characteristics of the building, such as building materials, component size, external surface colors, and shape, can usually be obtained from building plans and specifications.

**Configuration.** Determine building location, orientation, and external shading from plans and specifications. Shading from adjacent buildings can be determined by a site plan or by visiting the proposed site. The probable permanence of shading should be evaluated before it is included in the calculations. Possible high ground-reflected solar radiation from adjacent water, sand, parking lots, or solar load from adjacent reflective buildings should not be overlooked.

**Outdoor Design Conditions.** Obtain appropriate weather data and select outdoor design conditions. Weather data can be obtained from local weather stations or (in the United States) from the National Climatic Data Center (NCDC), Asheville, North Carolina 28801. (See Chapter 4 for outdoor design conditions for a large number of weather stations.) The designer should exercise judgment to ensure that results are consistent with expectations. Prevailing wind velocity and the relationship of a project site to a selected weather station should also be considered.

**Indoor Design Conditions.** Select indoor design conditions such as indoor dry-bulb temperature, indoor wet-bulb temperature, and ventilation rate. Include permissible variations and control limits.

**Internal Heat Gains and Operating Schedules.** Obtain planned density and a proposed schedule of lighting, occupancy, internal equipment, appliances, and processes that contribute to the internal thermal load.

**Areas.** Use consistent methods for calculation of building areas. For fenestration, the definition of a component's area must be consistent with associated ratings.

*Gross surface area.* It is efficient and conservative to derive gross surface areas from outside building dimensions, ignoring wall and floor thicknesses and avoiding separate accounting for floor edge and wall corner conditions. Measure floor areas to the outside of adjacent walls or to the center line of adjacent partitions. When apportioning to rooms, façade area should be divided at partition centerlines. Wall height should be taken as floor-to-floor height. The outside dimension is recommended as expedient for load calculations, but it may not be consistent with rigorous definitions used in building-related standards. However, the resulting differences do not introduce significant errors in the estimated design cooling and heating loads.

*Fenestration area.* Fenestration ratings [U-factor and solar heat gain coefficient (SHGC)] are based on the entire product area, including frames. Thus, for load calculations, fenestration area is the area of the rough opening in the wall or roof.

*Net surface area.* Net surface area is the gross surface area minus any fenestration area.

**Additional Considerations.** The proper design and sizing of all-air or air-and-water central air-conditioning systems require more than calculation of the cooling load in the space to be conditioned. The type of air-conditioning system, fan energy, fan location, duct heat loss and gain, duct leakage, heat-extraction lighting systems, and type of return air system all affect system load and component sizing. Adequate system design and component sizing require that system performance be analyzed as a series of psychrometric processes.

## 7.3 Heat Gain Calculation Concepts

The primary weather-related variable influencing a building's cooling load is solar radiation. The effect of solar radiation is more pronounced and immediate on exposed, nonopaque surfaces. The calculation of solar heat gain and conductive heat transfer through various glazing materials and associated mounting frames is discussed in Chapter 18 of the 2013 *ASHRAE Handbook—Fundamentals*.

### 7.3.1 Heat Gain Through Exterior Walls and Roofs

Heat gain through exterior opaque surfaces is derived in the same way as for fenestration areas. It differs primarily as a function of the mass and nature of wall or roof construction, since those elements affect the rate of conductive transfer through the composite assembly to the interior surface.

**Sol-Air Temperature.** This is the temperature of the outdoor air that, in the absence of all radiation changes, gives the same rate of heat entry into the surface as would the combination of incident solar radiation, radiant energy exchange with the sky and other outdoor surroundings, and convective heat exchange with outdoor air.

**Heat Gain Through Exterior Surfaces.** The heat balance at a sunlit surface gives the heat flux into the surface $q/A$ in Btu/h·ft² (W/m²), as

$$q/A = \alpha I_t + h_o(t_o - t_s) - \varepsilon \Delta R \qquad (7\text{-}4)$$

where

$\alpha$ = absorptance of surface for solar radiation

$I_t$ = total solar radiation incident on surface, Btu/h·ft² (W/m²)

$h_o$ = coefficient of heat transfer by longwave radiation and convection at outer surface, Btu/h·ft²·°F [W/(m²·K)]

$t_o$ = outdoor air temperature, °F (°C)

$t_s$ = surface temperature, °F (°C)

$\varepsilon$ = hemispherical emittance of surface

$\Delta R$ = difference between longwave radiation incident on surface from sky and surroundings and radiation emitted by blackbody at outdoor air temperature, Btu/h·ft² (W/m²)

**Table 7-1  Sol-Air Temperature ($t_e$) for July 21, 40° N Latitude**

$$t_e = t_o + \alpha I_t/h_o - \varepsilon \Delta R/h_o$$

| Time | Air Temp. $t_o$, °F | \multicolumn{9}{c}{Light Colored Surface, $\alpha/h_o = 0.15$} | Time | Air Temp. $t_o$, °F | \multicolumn{9}{c}{Dark Colored Surface, $\alpha/h_o = 0.30$} |
|---|---|---|---|---|---|---|---|---|---|---|---|---|---|---|---|---|---|---|---|
| | | N | NE | E | SE | S | SW | W | NW | HOR | | | N | NE | E | SE | S | SW | W | NW | HOR |
| 1 | 76 | 76 | 76 | 76 | 76 | 76 | 76 | 76 | 76 | 69 | 1 | 76 | 76 | 76 | 76 | 76 | 76 | 76 | 76 | 76 | 69 |
| 2 | 76 | 76 | 76 | 76 | 76 | 76 | 76 | 76 | 76 | 69 | 2 | 76 | 76 | 76 | 76 | 76 | 76 | 76 | 76 | 76 | 69 |
| 3 | 75 | 75 | 75 | 75 | 75 | 75 | 75 | 75 | 75 | 68 | 3 | 75 | 75 | 75 | 75 | 75 | 75 | 75 | 75 | 75 | 68 |
| 4 | 74 | 74 | 74 | 74 | 74 | 74 | 74 | 74 | 74 | 67 | 4 | 74 | 74 | 74 | 74 | 74 | 74 | 74 | 74 | 74 | 67 |
| 5 | 74 | 74 | 74 | 74 | 74 | 74 | 74 | 74 | 74 | 67 | 5 | 74 | 74 | 75 | 75 | 74 | 74 | 74 | 74 | 74 | 67 |
| 6 | 74 | 80 | 93 | 95 | 84 | 76 | 76 | 76 | 76 | 72 | 6 | 74 | 85 | 112 | 115 | 94 | 77 | 77 | 77 | 77 | 77 |
| 7 | 75 | 80 | 99 | 106 | 94 | 78 | 78 | 78 | 78 | 81 | 7 | 75 | 84 | 124 | 136 | 113 | 81 | 81 | 81 | 81 | 94 |
| 8 | 77 | 81 | 99 | 109 | 101 | 82 | 81 | 81 | 81 | 92 | 8 | 77 | 85 | 121 | 142 | 125 | 86 | 85 | 85 | 85 | 114 |
| 9 | 80 | 85 | 96 | 109 | 106 | 88 | 85 | 85 | 85 | 102 | 9 | 80 | 90 | 112 | 138 | 131 | 96 | 89 | 89 | 89 | 131 |
| 10 | 83 | 88 | 91 | 105 | 107 | 95 | 88 | 88 | 88 | 111 | 10 | 83 | 94 | 100 | 127 | 131 | 107 | 94 | 94 | 94 | 145 |
| 11 | 87 | 93 | 93 | 99 | 106 | 102 | 93 | 93 | 93 | 118 | 11 | 87 | 98 | 99 | 111 | 125 | 118 | 100 | 98 | 98 | 156 |
| 12 | 90 | 96 | 96 | 96 | 102 | 106 | 102 | 96 | 96 | 122 | 12 | 90 | 101 | 101 | 102 | 114 | 123 | 114 | 102 | 101 | 162 |
| 13 | 93 | 99 | 99 | 99 | 99 | 108 | 112 | 105 | 99 | 124 | 13 | 93 | 104 | 104 | 104 | 106 | 124 | 131 | 117 | 105 | 162 |
| 14 | 94 | 99 | 99 | 99 | 99 | 106 | 118 | 116 | 102 | 122 | 14 | 94 | 105 | 105 | 105 | 105 | 118 | 142 | 138 | 111 | 156 |
| 15 | 95 | 100 | 100 | 100 | 100 | 103 | 121 | 124 | 111 | 117 | 15 | 95 | 105 | 104 | 104 | 104 | 111 | 146 | 153 | 127 | 146 |
| 16 | 94 | 98 | 98 | 98 | 98 | 99 | 118 | 126 | 116 | 109 | 16 | 94 | 102 | 102 | 102 | 102 | 103 | 142 | 159 | 138 | 131 |
| 17 | 93 | 98 | 96 | 96 | 96 | 96 | 112 | 124 | 117 | 99 | 17 | 93 | 102 | 99 | 99 | 99 | 99 | 131 | 154 | 142 | 112 |
| 18 | 91 | 97 | 93 | 93 | 93 | 93 | 101 | 112 | 110 | 89 | 18 | 91 | 102 | 94 | 94 | 94 | 94 | 111 | 132 | 129 | 94 |
| 19 | 87 | 87 | 87 | 87 | 87 | 87 | 87 | 87 | 87 | 80 | 19 | 87 | 87 | 87 | 87 | 87 | 87 | 87 | 88 | 88 | 80 |
| 20 | 85 | 85 | 85 | 85 | 85 | 85 | 85 | 85 | 85 | 78 | 20 | 85 | 85 | 85 | 85 | 85 | 85 | 85 | 85 | 85 | 78 |
| 21 | 83 | 83 | 83 | 83 | 83 | 83 | 83 | 83 | 83 | 76 | 21 | 83 | 83 | 83 | 83 | 83 | 83 | 83 | 83 | 83 | 76 |
| 22 | 81 | 81 | 81 | 81 | 81 | 81 | 81 | 81 | 81 | 74 | 22 | 81 | 81 | 81 | 81 | 81 | 81 | 81 | 81 | 81 | 74 |
| 23 | 79 | 79 | 79 | 79 | 79 | 79 | 79 | 79 | 79 | 72 | 23 | 79 | 79 | 79 | 79 | 79 | 79 | 79 | 79 | 79 | 72 |
| 24 | 77 | 77 | 77 | 77 | 77 | 77 | 77 | 77 | 77 | 70 | 24 | 77 | 77 | 77 | 77 | 77 | 77 | 77 | 77 | 77 | 70 |
| Avg. | 83 | 86 | 88 | 90 | 90 | 87 | 90 | 90 | 88 | 90 | Avg. | 83 | 89 | 94 | 99 | 97 | 93 | 97 | 99 | 94 | 104 |

*Note*: Sol-air temperatures are calculated based on $\varepsilon \Delta R/h_o = 7$°F for horizontal surfaces and 0°F for vertical surfaces.

Assuming the rate of heat transfer can be expressed in terms of the sol-air temperature $t_e$:

$$q/A = h_o(t_e - t_s) \tag{7-5}$$

From Equations (7-4) and (7-5):

$$t_e = t_o + \alpha I_t/h_o - \varepsilon \Delta R/h_o \tag{7-6}$$

For horizontal surfaces that receive longwave radiation from the sky only, an appropriate value of $\Delta R$ is about 20 Btu/h·ft², so if $\varepsilon = 1$ and $h_o = 3.0$ Btu/h·ft²·°F, the longwave correction term is about –7°F.

Because vertical surfaces receive longwave radiation from the ground and surrounding buildings, as well as from the sky, accurate $\Delta R$ values are difficult to determine. When solar radiation intensity is high, surfaces of terrestrial objects usually have a higher temperature than the outdoor air; thus, their longwave radiation compensates to some extent for the sky's low emittance. Therefore, it is assumed that $\Delta R = 0$ for vertical surfaces.

The sol-air temperatures in Table 7-1 are calculated based on $\varepsilon \Delta R/h_o = -7$°F for horizontal surfaces and 0°F for vertical surfaces; total solar intensity values for the calculation were the same as those used to evaluate the solar heat gain factors (SHGF) for July 21 at 40° N latitude. These values of $I_t$ incorporate diffuse radiation from a clear sky and ground reflection but make no allowance for reflection from adjacent walls.

**Table 7-2  Percentage of Daily Range**

| Time, h | % | Time, h | % | Time, h | % |
|---|---|---|---|---|---|
| 1 | 87 | 9 | 71 | 17 | 10 |
| 2 | 92 | 10 | 56 | 18 | 21 |
| 3 | 96 | 11 | 39 | 19 | 34 |
| 4 | 99 | 12 | 23 | 20 | 47 |
| 5 | 100 | 13 | 11 | 21 | 58 |
| 6 | 98 | 14 | 3 | 22 | 68 |
| 7 | 93 | 15 | 0 | 23 | 76 |
| 8 | 84 | 16 | 3 | 24 | 82 |

**Surface Colors.** Sol-air temperature values are given for two values of the parameter $\alpha/h_o$ (Table 7-1); 0.15 is appropriate for a light-colored surface, while 0.30 is the usual maximum value for this parameter (i.e., for a dark-colored surface or any surface for which the permanent lightness cannot be reliably anticipated).

**Air Temperature Cycle.** The air temperature cycle used to calculate sol-air temperatures is given in Column 2, Table 7-1. These values are obtained by using the daily temperature range and the percent (%) difference from Table 7-2. Sol-air temperatures can be adjusted to any other air temperature cycle by adding or subtracting the difference between the desired air temperature and the air temperature value given in Column 2.

# Nonresidential Cooling and Heating Load Calculations

**Average Sol-Air Temperature.** Average daily sol-air temperature tea can be calculated by

$$t_{ea} = t_{oa} + \frac{\alpha}{h_o}\left(\frac{I_{DT}}{24}\right) - \frac{\varepsilon\Delta R}{h_o} \quad (7\text{-}7)$$

where $I_{DT}$ is the sum of two appropriate half-day totals of solar heat gain in Btu/h·ft². For example, the average sol-air temperature for a wall facing southeast at 40° N latitude on August 21 would be

$$t_{ea} = t_{oa} + \frac{\alpha}{h_o}\left[\frac{1.15(205 + 956)}{24}\right] \quad (7\text{-}8)$$

The daily solar heat gain of double-strength sheet glass is (205 + 956) Btu/ft² in a southeast facade at this latitude and date, and $\varepsilon\Delta R/h_o$ is assumed to be zero for this vertical surface.

**Hourly Air Temperatures.** The hourly air temperatures in Column 2, Table 7-1 are for a location with a design temperature of 95°F and a range of 21°F. To compute corresponding temperatures for other locations, select a suitable design temperature and note the outdoor daily range.

For each hour, take the percentage of the daily range indicated in Table 7-2 and subtract from the design temperature.

## 7.3.2 Heat Gain Through Fenestration

Fenestration is the term used here to designate any light-transmitting opening in a building wall or roof. The opening may be glazed with single or multiple sheet, plate or float glass, pattern glass, plastic panels, or glass block. Interior or exterior shading devices are usually employed, and some glazing systems incorporate integral sun control devices.

Calculating heat transfer through fenestration is explained in detail in Chapter 15 of the 2013 *ASHRAE Handbook—Fundamentals* while Chapter 18 presents only the portion of this operation required in the calculation of space cooling load due to heat transfer through fenestration.

Heat admission or loss through fenestration areas is affected by many factors, including

- Solar radiation intensity and incident angle
- Outdoor-indoor temperature difference
- Velocity and direction of airflow across the exterior and interior fenestration surfaces
- Low-temperature radiation exchange between the surfaces of the fenestration and the surroundings
- Exterior and/or interior shading

When solar radiation strikes an unshaded window (Figure 7-2), part of the radiant energy (8% for uncoated clear glass) is reflected back outdoors, part is absorbed within the glass (from 5 to 50%, depending upon the composition and thickness of the glass), and the remainder is transmitted directly indoors to become part of the cooling load. The solar heat gain is the sum of the transmitted radiation and the portion of the absorbed radiation that flows inward.

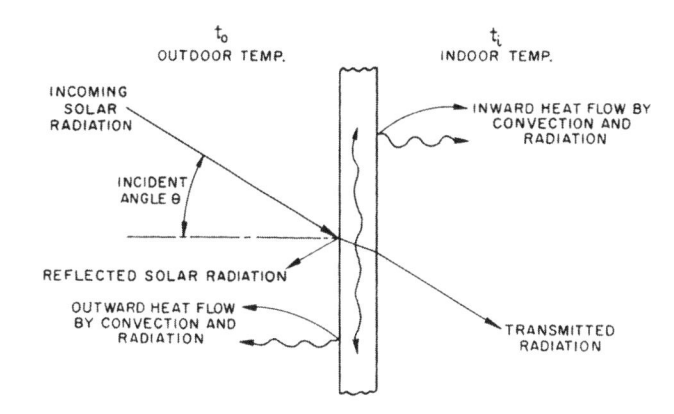

*Fig. 7-2 Instantaneous Heat Balance for Sunlit Glazing Material*
*(Figure 14, Chapter 15, 2013 ASHRAE Handbook—Fundamentals)*

The total instantaneous rate of heat gain through a glazing material can be obtained from the heat balance between a unit area of fenestration and its thermal environment:

| Total heat transmission through glass | = | Heat flow due to outdoor-indoor temp. difference | + | Inward flow of absorbed solar radiation | + | Radiation transmitted through glass |

In this equation, the last two terms on the right are present only when the fenestration is irradiated and are therefore related to the incident radiation. The first term occurs whether or not the sun is shining, since it represents the heat flow through fenestration by thermal conduction.

Combining the last two terms,

| Total heat transmission through glass | = | Conductive heat gain | + | Solar heat gain |

In this way, heat gain is divided into two components: (1) the conductive heat gain (or loss), due to differences in outdoor and indoor air temperature, and (2) the solar heat gain (SHG), due to transmitted and absorbed solar energy. The total load through fenestration is the sum of the load due to conductive heat gain and the load due to solar heat gain.

Whether or not sunlight is present, heat flows through fenestration by thermal conduction, as expressed by

| Conductive heat flow | = | Overall coefficient of heat transfer | × | Outdoor-indoor temperature difference |

or

$$q/A = U(t_o - t_i) \quad (7\text{-}9)$$

where

$q/A$ = instantaneous rate of heat transfer through fenestration

$U$ = overall coefficient of heat transfer for the glazing

$t_o$ = outdoor air temperature

$t_i$ = inside air temperature

Values of the overall coefficient of heat transfer for a number of widely used fenestrations are in Chapter 15 of the 2013

## Table 7-3 Solar Heat Gain
*(Table 14, Chapter 30, 2005 ASHRAE Handbook—Fundamentals)*

**Solar Angles**

All angles are in degrees. The solar azimuth $\phi$ and the surface azimuth $\psi$ are measured in degrees from south; angles to the east of south are negative, and angles to the west of south are positive. Calculate solar altitude, azimuth, and surface incident angles as follows:

Apparent solar time AST, in decimal hours:
$$AST = LST + ET/60 + (LSM - LON)/15$$

Hour angle $H$, degrees:
$$H = 15(\text{hours of time from local solar noon}) = 15(AST - 12)$$

Solar altitude $\beta$:
$$\sin \beta = \cos L \cos \delta \cos H + \sin L \sin \delta$$

Solar azimuth $\phi$:
$$\cos \phi = (\sin \beta \sin L - \sin \delta)/(\cos \beta \cos L)$$

Surface-solar azimuth $\gamma$:
$$\gamma = \phi - \psi$$

Incident angle $\theta$:
$$\cos \theta = \cos \beta \cos \gamma \sin \Sigma + \sin \beta \cos \Sigma$$

where
- ET = equation of time, decimal minutes
- $L$ = latitude
- LON = local longitude, decimal degrees of arc
- LSM = local standard time meridian, decimal degrees of arc
  - = 60° for Atlantic Standard Time
  - = 75° for Eastern Standard Time
  - = 90° for Central Standard Time
  - = 105° for Mountain Standard Time
  - = 120° for Pacific Standard Time
  - = 135° for Alaska Standard Time
  - = 150° for Hawaii-Aleutian Standard Time
- LST = local standard time, decimal hours
- $\delta$ = solar declination, °
- $\psi$ = surface azimuth, °
- $\Sigma$ = surface tilt from horizontal, horizontal = 0°

Values of ET and $\delta$ are given in Table 2 of Chapter 35, 2011 *ASHRAE Handbook—HVAC Applications* for the 21st day of each month.

**Direct, Diffuse, and Total Solar Irradiance**

Direct normal irradiance $E_{DN}$

If $\beta > 0$ $\quad E_{DN} = \left[\dfrac{A}{\exp(B/\sin \beta)}\right]CN$

Otherwise, $\quad E_{DN} = 0$

Surface direct irradiance $E_D$

If $\cos \theta > 0 \quad E_D = E_{DN} \cos \theta$

Otherwise, $\quad E_D = 0$

Ratio $Y$ of sky diffuse on vertical surface to sky diffuse on horizontal surface

If $\cos \theta > -0.2 \quad Y = 0.55 + 0.437 \cos \theta + 0.313 \cos^2 \theta$

Otherwise, $\quad Y = 0.45$

Diffuse irradiance $E_d$

Vertical surfaces $\quad E_d = CYE_{DN}$

Surfaces other than vertical $\quad E_d = CE_{DN}(1 + \cos \Sigma)/2$

Ground-reflected irradiance $\quad E_r = E_{DN}(C + \sin \beta)\rho_g(1 - \cos \Sigma)/2$

Total surface irradiance $\quad E_t = E_D + E_d + E_r$

where
- $A$ = apparent solar constant
- $B$ = atmospheric extinction coefficient
- $C$ = sky diffuse factor
- CN = clearness number multiplier for clear/dry or hazy/humid locations. See Figure 5 in Chapter 33 of the 2003 *ASHRAE Handbook—HVAC Applications* for CN values.
- $E_d$ = diffuse sky irradiance
- $E_r$ = diffuse ground-reflected irradiance
- $\rho_g$ = ground reflectivity

Values of $A$, $B$, and $C$ are given in Table 1 of Chapter 35, 2011 *ASHRAE Handbook—HVAC Applications* for the 21st day of each month. Values of ground reflectivity $\rho_g$ are given in Table 10 of Chapter 31.

---

*ASHRAE Handbook—Fundamentals* (also see Chapter 5). Table 7-3 includes some useful solar equations.

For fenestration heat gain, the following equations apply:

Direct beam solar heat gain $q_b$:
$$q_b = AE_D \text{SHGC}(\Theta)\text{IAC} \tag{7-10}$$

Diffuse solar heat gain $q_d$:
$$q_d = A(E_d + E_r)(\text{SHGC})_D \text{IAC} \tag{7-11}$$

Conductive heat gain $q_c$:
$$q_c = UA(T_{out} - T_{in}) \tag{7-12}$$

Total fenestration heat gain $q$:
$$q = q_b + q_d + q_c \tag{7-13}$$

where
- $A$ = window area, ft²
- $E_D, E_d,$ and $E_r$ = direct, diffuse, and ground-reflected irradiance, calculated using the equations in Table 7-3
- SHGC($\Theta$) = direct solar heat gain coefficient as a function of incident angle $q$; may be interpolated between values in Table 7-4
- (SHGC)$_D$ = diffuse solar heat gain coefficient (also referred to as hemispherical SHGC); from Table 7-4
- $T_{in}$ = inside temperature, °F
- $T_{out}$ = outside temperature, °F
- $U$ = overall U-factor, including frame and mounting orientation from Table 5-16, Btu/h·ft²·°F
- IAC = inside shading attenuation coefficient, = 1.0 if no inside shading device

If specific window manufacturer's SHGC and U-factor data are available, those should be used. For fenestration equipped with inside shading (blinds, drapes, or shades), IAC is listed in Table 7-8. The inside shading attenuation coefficients given are used to calculate both direct and diffuse solar heat gains.

Fenestration ratings (U-factor and SHGC) are based on the entire product area, including frames. Thus, for load calculations, fenestration area is the area of the entire opening in the wall or roof.

Nonuniform exterior shading, caused by roof overhangs, side fins, or building projections, requires separate hourly calculations for the externally shaded and unshaded areas of the window in question, with the inside shading SHGC still used to account for any internal shading devices. The areas, shaded

**Table 7-4** Visible Transmittance ($T_v$), Solar Heat Gain Coefficient (SHGC), Solar Transmittance ($T$), Front Reflectance ($R^f$), Back Reflectance ($R^b$), and Layer Absorptances ($A_n^f$) for Glazing and Window Systems
*(From Table 10, Chapter 15, 2013 ASHRAE Handbook—Fundamentals)*

| | | | | | | | | | | | | | Total Window SHGC at Normal Incidence | | | | Total Window $T_v$ at Normal Incidence | | | |
|---|---|---|---|---|---|---|---|---|---|---|---|---|---|---|---|---|---|---|---|---|
| | | | | | | Center-of-Glazing Properties | | | | | | | | | | | | | | |
| | Glazing System | | | | | Incidence Angles | | | | | | | Aluminum | | Other Frames | | Aluminum | | Other Frames | |
| ID | Glass Thick., in. | | Center Glazing $T_v$ | | Normal 0.00 | 40.00 | 50.00 | 60.00 | 70.00 | 80.00 | Hemis., Diffuse | | Operable | Fixed | Operable | Fixed | Operable | Fixed | Operable | Fixed |
| *Uncoated Single Glazing* | | | | | | | | | | | | | | | | | | | | |
| 1a | 1/8 | CLR | 0.90 | SHGC | 0.86 | 0.84 | 0.82 | 0.78 | 0.67 | 0.42 | 0.78 | | 0.78 | 0.79 | 0.70 | 0.76 | 0.80 | 0.81 | 0.72 | 0.79 |
| | | | | $T$ | 0.83 | 0.82 | 0.80 | 0.75 | 0.64 | 0.39 | 0.75 | | | | | | | | | |
| | | | | $R^f$ | 0.08 | 0.08 | 0.10 | 0.14 | 0.25 | 0.51 | 0.14 | | | | | | | | | |
| | | | | $R^b$ | 0.08 | 0.08 | 0.10 | 0.14 | 0.25 | 0.51 | 0.14 | | | | | | | | | |
| | | | | $A_1^f$ | 0.09 | 0.10 | 0.10 | 0.11 | 0.11 | 0.11 | 0.10 | | | | | | | | | |
| 1b | 1/4 | CLR | 0.88 | SHGC | 0.81 | 0.80 | 0.78 | 0.73 | 0.62 | 0.39 | 0.73 | | 0.74 | 0.74 | 0.66 | 0.72 | 0.78 | 0.79 | 0.70 | 0.77 |
| | | | | $T$ | 0.77 | 0.75 | 0.73 | 0.68 | 0.58 | 0.35 | 0.69 | | | | | | | | | |
| | | | | $R^f$ | 0.07 | 0.08 | 0.09 | 0.13 | 0.24 | 0.48 | 0.13 | | | | | | | | | |
| | | | | $R^b$ | 0.07 | 0.08 | 0.09 | 0.13 | 0.24 | 0.48 | 0.13 | | | | | | | | | |
| | | | | $A_1^f$ | 0.16 | 0.17 | 0.18 | 0.19 | 0.19 | 0.17 | 0.17 | | | | | | | | | |
| 1c | 1/8 | BRZ | 0.68 | SHGC | 0.73 | 0.71 | 0.68 | 0.64 | 0.55 | 0.34 | 0.65 | | 0.67 | 0.67 | 0.59 | 0.65 | 0.61 | 0.61 | 0.54 | 0.60 |
| | | | | $T$ | 0.65 | 0.62 | 0.59 | 0.55 | 0.46 | 0.27 | 0.56 | | | | | | | | | |
| | | | | $R^f$ | 0.06 | 0.07 | 0.08 | 0.12 | 0.22 | 0.45 | 0.12 | | | | | | | | | |
| | | | | $R^b$ | 0.06 | 0.07 | 0.08 | 0.12 | 0.22 | 0.45 | 0.12 | | | | | | | | | |
| | | | | $A_1^f$ | 0.29 | 0.31 | 0.32 | 0.33 | 0.33 | 0.29 | 0.31 | | | | | | | | | |
| 1d | 1/4 | BRZ | 0.54 | SHGC | 0.62 | 0.59 | 0.57 | 0.53 | 0.45 | 0.29 | 0.54 | | 0.57 | 0.57 | 0.50 | 0.55 | 0.48 | 0.49 | 0.43 | 0.48 |
| | | | | $T$ | 0.49 | 0.45 | 0.43 | 0.39 | 0.32 | 0.18 | 0.41 | | | | | | | | | |
| | | | | $R^f$ | 0.05 | 0.06 | 0.07 | 0.11 | 0.19 | 0.42 | 0.10 | | | | | | | | | |
| | | | | $R^b$ | 0.05 | 0.68 | 0.66 | 0.62 | 0.53 | 0.33 | 0.10 | | | | | | | | | |
| | | | | $A_1^f$ | 0.46 | 0.49 | 0.50 | 0.51 | 0.49 | 0.41 | 0.48 | | | | | | | | | |
| 1e | 1/8 | GRN | 0.82 | SHGC | 0.70 | 0.68 | 0.66 | 0.62 | 0.53 | 0.33 | 0.63 | | 0.64 | 0.64 | 0.57 | 0.62 | 0.73 | 0.74 | 0.66 | 0.72 |
| | | | | $T$ | 0.61 | 0.58 | 0.56 | 0.52 | 0.43 | 0.25 | 0.53 | | | | | | | | | |
| | | | | $R^f$ | 0.06 | 0.07 | 0.08 | 0.12 | 0.21 | 0.45 | 0.11 | | | | | | | | | |
| | | | | $R^b$ | 0.06 | 0.07 | 0.08 | 0.12 | 0.21 | 0.45 | 0.11 | | | | | | | | | |
| | | | | $A_1^f$ | 0.33 | 0.35 | 0.36 | 0.37 | 0.36 | 0.31 | 0.35 | | | | | | | | | |
| 1f | 1/4 | GRN | 0.76 | SHGC | 0.60 | 0.58 | 0.56 | 0.52 | 0.45 | 0.29 | 0.54 | | 0.55 | 0.55 | 0.49 | 0.53 | 0.68 | 0.68 | 0.61 | 0.67 |
| | | | | $T$ | 0.47 | 0.44 | 0.42 | 0.38 | 0.32 | 0.18 | 0.40 | | | | | | | | | |
| | | | | $R^f$ | 0.05 | 0.06 | 0.07 | 0.11 | 0.20 | 0.42 | 0.10 | | | | | | | | | |
| | | | | $R^b$ | 0.05 | 0.06 | 0.07 | 0.11 | 0.20 | 0.42 | 0.10 | | | | | | | | | |
| | | | | $A_1^f$ | 0.47 | 0.50 | 0.51 | 0.51 | 0.49 | 0.40 | 0.49 | | | | | | | | | |
| 1g | 1/8 | GRY | 0.62 | SHGC | 0.70 | 0.68 | 0.66 | 0.61 | 0.53 | 0.33 | 0.63 | | 0.64 | 0.64 | 0.57 | 0.62 | 0.55 | 0.56 | 0.50 | 0.55 |
| | | | | $T$ | 0.61 | 0.58 | 0.56 | 0.51 | 0.42 | 0.24 | 0.53 | | | | | | | | | |
| | | | | $R^f$ | 0.06 | 0.07 | 0.08 | 0.12 | 0.21 | 0.44 | 0.11 | | | | | | | | | |
| | | | | $R^b$ | 0.06 | 0.07 | 0.08 | 0.12 | 0.21 | 0.44 | 0.11 | | | | | | | | | |
| | | | | $A_1^f$ | 0.33 | 0.36 | 0.37 | 0.37 | 0.37 | 0.32 | 0.35 | | | | | | | | | |
| 1h | 1/4 | GRY | 0.46 | SHGC | 0.59 | 0.57 | 0.55 | 0.51 | 0.44 | 0.28 | 0.52 | | 0.54 | 0.54 | 0.48 | 0.52 | 0.41 | 0.41 | 0.37 | 0.40 |
| | | | | $T$ | 0.46 | 0.42 | 0.40 | 0.36 | 0.29 | 0.16 | 0.38 | | | | | | | | | |
| | | | | $R^f$ | 0.05 | 0.06 | 0.07 | 0.10 | 0.19 | 0.41 | 0.10 | | | | | | | | | |
| | | | | $R^b$ | 0.05 | 0.06 | 0.07 | 0.10 | 0.19 | 0.41 | 0.10 | | | | | | | | | |
| | | | | $A_1^f$ | 0.49 | 0.52 | 0.54 | 0.54 | 0.52 | 0.43 | 0.51 | | | | | | | | | |
| 1i | 1/4 | BLUGRN | 0.75 | SHGC | 0.62 | 0.59 | 0.57 | 0.54 | 0.46 | 0.30 | 0.55 | | 0.57 | 0.57 | 0.50 | 0.55 | 0.67 | 0.68 | 0.60 | 0.66 |
| | | | | $T$ | 0.49 | 0.46 | 0.44 | 0.40 | 0.33 | 0.19 | 0.42 | | | | | | | | | |
| | | | | $R^f$ | 0.06 | 0.06 | 0.07 | 0.11 | 0.20 | 0.43 | 0.11 | | | | | | | | | |
| | | | | $R^b$ | 0.06 | 0.06 | 0.07 | 0.11 | 0.20 | 0.43 | 0.11 | | | | | | | | | |
| | | | | $A_1^f$ | 0.45 | 0.48 | 0.49 | 0.49 | 0.47 | 0.38 | 0.48 | | | | | | | | | |
| *Reflective Single Glazing* | | | | | | | | | | | | | | | | | | | | |
| 1j | 1/4 | SS on CLR 8% | 0.08 | SHGC | 0.19 | 0.19 | 0.19 | 0.18 | 0.16 | 0.10 | 0.18 | | 0.18 | 0.18 | 0.16 | 0.17 | 0.07 | 0.07 | 0.06 | 0.07 |
| | | | | $T$ | 0.06 | 0.06 | 0.06 | 0.05 | 0.04 | 0.03 | 0.05 | | | | | | | | | |
| | | | | $R^f$ | 0.33 | 0.34 | 0.35 | 0.37 | 0.44 | 0.61 | 0.36 | | | | | | | | | |
| | | | | $R^b$ | 0.50 | 0.50 | 0.51 | 0.53 | 0.58 | 0.71 | 0.52 | | | | | | | | | |
| | | | | $A_1^f$ | 0.61 | 0.61 | 0.60 | 0.58 | 0.52 | 0.37 | 0.57 | | | | | | | | | |
| 1k | 1/4 | SS on CLR 14% | 0.14 | SHGC | 0.25 | 0.25 | 0.24 | 0.23 | 0.20 | 0.13 | 0.23 | | 0.24 | 0.24 | 0.21 | 0.22 | 0.12 | 0.13 | 0.11 | 0.12 |
| | | | | $T$ | 0.11 | 0.10 | 0.10 | 0.09 | 0.07 | 0.04 | 0.09 | | | | | | | | | |
| | | | | $R^f$ | 0.26 | 0.27 | 0.28 | 0.31 | 0.38 | 0.57 | 0.30 | | | | | | | | | |

**Table 7-4** Visible Transmittance ($T_v$), Solar Heat Gain Coefficient (SHGC), Solar Transmittance ($T$), Front Reflectance ($R^f$), Back Reflectance ($R^b$), and Layer Absorptances ($A_n^f$) for Glazing and Window Systems (*Continued*)

*(From Table 10, Chapter 15, 2013 ASHRAE Handbook—Fundamentals)*

| | | | | | | Center-of-Glazing Properties | | | | | | Total Window SHGC at Normal Incidence | | | | Total Window $T_v$ at Normal Incidence | | | |
|---|---|---|---|---|---|---|---|---|---|---|---|---|---|---|---|---|---|---|---|
| | | Glazing System | | | | Incidence Angles | | | | | | Aluminum | | Other Frames | | Aluminum | | Other Frames | |
| ID | Glass Thick., in. | | Center Glazing $T_v$ | | Normal 0.00 | 40.00 | 50.00 | 60.00 | 70.00 | 80.00 | Hemis., Diffuse | Operable | Fixed | Operable | Fixed | Operable | Fixed | Operable | Fixed |
| | | | | $R^b$ | 0.44 | 0.44 | 0.45 | 0.47 | 0.52 | 0.67 | 0.46 | | | | | | | | |
| | | | | $A_1^f$ | 0.63 | 0.63 | 0.62 | 0.60 | 0.55 | 0.39 | 0.60 | | | | | | | | |
| 1l | 1/4 | SS on CLR 20% | 0.20 | SHGC | 0.31 | 0.30 | 0.30 | 0.28 | 0.24 | 0.16 | 0.28 | 0.29 | 0.29 | 0.26 | 0.28 | 0.18 | 0.18 | 0.16 | 0.18 |
| | | | | $T$ | 0.15 | 0.15 | 0.14 | 0.13 | 0.11 | 0.06 | 0.13 | | | | | | | | |
| | | | | $R^f$ | 0.21 | 0.22 | 0.23 | 0.26 | 0.34 | 0.54 | 0.25 | | | | | | | | |
| | | | | $R^b$ | 0.38 | 0.38 | 0.39 | 0.41 | 0.48 | 0.64 | 0.41 | | | | | | | | |
| | | | | $A_1^f$ | 0.64 | 0.64 | 0.63 | 0.61 | 0.56 | 0.40 | 0.60 | | | | | | | | |
| 1m | 1/4 | SS on GRN 14% | 0.12 | SHGC | 0.25 | 0.25 | 0.24 | 0.23 | 0.21 | 0.14 | 0.23 | 0.24 | 0.24 | 0.21 | 0.22 | 0.11 | 0.11 | 0.10 | 0.11 |
| | | | | $T$ | 0.06 | 0.06 | 0.06 | 0.06 | 0.04 | 0.03 | 0.06 | | | | | | | | |
| | | | | $R^f$ | 0.14 | 0.14 | 0.16 | 0.19 | 0.27 | 0.49 | 0.18 | | | | | | | | |
| | | | | $R^b$ | 0.44 | 0.44 | 0.45 | 0.47 | 0.52 | 0.67 | 0.46 | | | | | | | | |
| | | | | $A_1^f$ | 0.80 | 0.80 | 0.78 | 0.76 | 0.68 | 0.48 | 0.75 | | | | | | | | |
| 1n | 1/4 | TI on CLR 20% | 0.20 | SHGC | 0.29 | 0.29 | 0.28 | 0.27 | 0.23 | 0.15 | 0.27 | 0.27 | 0.27 | 0.24 | 0.26 | 0.18 | 0.18 | 0.16 | 0.18 |
| | | | | $T$ | 0.14 | 0.13 | 0.13 | 0.12 | 0.09 | 0.06 | 0.12 | | | | | | | | |
| | | | | $R^f$ | 0.22 | 0.22 | 0.24 | 0.26 | 0.34 | 0.54 | 0.26 | | | | | | | | |
| | | | | $R^b$ | 0.40 | 0.40 | 0.42 | 0.44 | 0.50 | 0.65 | 0.43 | | | | | | | | |
| | | | | $A_1^f$ | 0.65 | 0.65 | 0.64 | 0.62 | 0.57 | 0.40 | 0.62 | | | | | | | | |
| 1o | 1/4 | TI on CLR 30% | 0.30 | SHGC | 0.39 | 0.38 | 0.37 | 0.35 | 0.30 | 0.20 | 0.35 | 0.36 | 0.36 | 0.32 | 0.35 | 0.27 | 0.27 | 0.24 | 0.26 |
| | | | | $T$ | 0.23 | 0.22 | 0.21 | 0.19 | 0.16 | 0.09 | 0.20 | | | | | | | | |
| | | | | $R^f$ | 0.15 | 0.15 | 0.17 | 0.20 | 0.28 | 0.50 | 0.19 | | | | | | | | |
| | | | | $R^b$ | 0.32 | 0.33 | 0.34 | 0.36 | 0.43 | 0.60 | 0.36 | | | | | | | | |
| | | | | $A_1^f$ | 0.63 | 0.65 | 0.64 | 0.62 | 0.57 | 0.40 | 0.62 | | | | | | | | |
| ***Uncoated Double Glazing*** | | | | | | | | | | | | | | | | | | | |
| 5a | 1/8 | CLR CLR | 0.81 | SHGC | 0.76 | 0.74 | 0.71 | 0.64 | 0.50 | 0.26 | 0.66 | 0.69 | 0.70 | 0.62 | 0.67 | 0.72 | 0.73 | 0.65 | 0.71 |
| | | | | $T$ | 0.70 | 0.68 | 0.65 | 0.58 | 0.44 | 0.21 | 0.60 | | | | | | | | |
| | | | | $R^f$ | 0.13 | 0.14 | 0.16 | 0.23 | 0.36 | 0.61 | 0.21 | | | | | | | | |
| | | | | $R^b$ | 0.13 | 0.14 | 0.16 | 0.23 | 0.36 | 0.61 | 0.21 | | | | | | | | |
| | | | | $A_1^f$ | 0.10 | 0.11 | 0.11 | 0.12 | 0.13 | 0.13 | 0.11 | | | | | | | | |
| | | | | $A_2^f$ | 0.07 | 0.08 | 0.08 | 0.08 | 0.07 | 0.05 | 0.07 | | | | | | | | |
| 5b | 1/4 | CLR CLR | 0.78 | SHGC | 0.70 | 0.67 | 0.64 | 0.58 | 0.45 | 0.23 | 0.60 | 0.64 | 0.64 | 0.57 | 0.62 | 0.69 | 0.70 | 0.62 | 0.69 |
| | | | | $T$ | 0.61 | 0.58 | 0.55 | 0.48 | 0.36 | 0.17 | 0.51 | | | | | | | | |
| | | | | $R^f$ | 0.11 | 0.12 | 0.15 | 0.20 | 0.33 | 0.57 | 0.18 | | | | | | | | |
| | | | | $R^b$ | 0.11 | 0.12 | 0.15 | 0.20 | 0.33 | 0.57 | 0.18 | | | | | | | | |
| | | | | $A_1^f$ | 0.17 | 0.18 | 0.19 | 0.20 | 0.21 | 0.20 | 0.19 | | | | | | | | |
| | | | | $A_2^f$ | 0.11 | 0.12 | 0.12 | 0.12 | 0.10 | 0.07 | 0.11 | | | | | | | | |
| 5c | 1/8 | BRZ CLR | 0.62 | SHGC | 0.62 | 0.60 | 0.57 | 0.51 | 0.39 | 0.20 | 0.53 | 0.57 | 0.57 | 0.50 | 0.55 | 0.55 | 0.56 | 0.50 | 0.55 |
| | | | | $T$ | 0.55 | 0.51 | 0.48 | 0.42 | 0.31 | 0.14 | 0.45 | | | | | | | | |
| | | | | $R^f$ | 0.09 | 0.10 | 0.12 | 0.16 | 0.27 | 0.49 | 0.15 | | | | | | | | |
| | | | | $R^b$ | 0.12 | 0.13 | 0.15 | 0.21 | 0.35 | 0.59 | 0.19 | | | | | | | | |
| | | | | $A_1^f$ | 0.30 | 0.33 | 0.34 | 0.36 | 0.37 | 0.34 | 0.33 | | | | | | | | |
| | | | | $A_2^f$ | 0.06 | 0.06 | 0.06 | 0.06 | 0.05 | 0.03 | 0.06 | | | | | | | | |
| 5d | 1/4 | BRZ CLR | 0.47 | SHGC | 0.49 | 0.46 | 0.44 | 0.39 | 0.31 | 0.17 | 0.41 | 0.45 | 0.45 | 0.40 | 0.43 | 0.42 | 0.42 | 0.38 | 0.41 |
| | | | | $T$ | 0.38 | 0.35 | 0.32 | 0.27 | 0.20 | 0.08 | 0.30 | | | | | | | | |
| | | | | $R^f$ | 0.07 | 0.08 | 0.09 | 0.13 | 0.22 | 0.44 | 0.12 | | | | | | | | |
| | | | | $R^b$ | 0.10 | 0.11 | 0.13 | 0.19 | 0.31 | 0.55 | 0.17 | | | | | | | | |
| | | | | $A_1^f$ | 0.48 | 0.51 | 0.52 | 0.53 | 0.53 | 0.45 | 0.50 | | | | | | | | |
| | | | | $A_2^f$ | 0.07 | 0.07 | 0.07 | 0.07 | 0.06 | 0.04 | 0.07 | | | | | | | | |
| 5e | 1/8 | GRN CLR | 0.75 | SHGC | 0.60 | 0.57 | 0.54 | 0.49 | 0.38 | 0.20 | 0.51 | 0.55 | 0.55 | 0.49 | 0.53 | 0.67 | 0.68 | 0.60 | 0.66 |
| | | | | $T$ | 0.52 | 0.49 | 0.46 | 0.40 | 0.30 | 0.13 | 0.43 | | | | | | | | |
| | | | | $R^f$ | 0.09 | 0.10 | 0.12 | 0.16 | 0.27 | 0.50 | 0.15 | | | | | | | | |
| | | | | $R^b$ | 0.12 | 0.13 | 0.15 | 0.21 | 0.35 | 0.60 | 0.19 | | | | | | | | |
| | | | | $A_1^f$ | 0.34 | 0.37 | 0.38 | 0.39 | 0.39 | 0.35 | 0.37 | | | | | | | | |
| | | | | $A_2^f$ | 0.05 | 0.05 | 0.05 | 0.04 | 0.04 | 0.03 | 0.04 | | | | | | | | |
| 5f | 1/4 | GRN CLR | 0.68 | SHGC | 0.49 | 0.46 | 0.44 | 0.39 | 0.31 | 0.17 | 0.41 | 0.45 | 0.45 | 0.40 | 0.43 | 0.61 | 0.61 | 0.54 | 0.60 |
| | | | | $T$ | 0.39 | 0.36 | 0.33 | 0.29 | 0.21 | 0.09 | 0.31 | | | | | | | | |
| | | | | $R^f$ | 0.08 | 0.08 | 0.10 | 0.14 | 0.23 | 0.45 | 0.13 | | | | | | | | |

# Nonresidential Cooling and Heating Load Calculations

**Table 7-4** Visible Transmittance ($T_v$), Solar Heat Gain Coefficient (SHGC), Solar Transmittance ($T$), Front Reflectance ($R^f$), Back Reflectance ($R^b$), and Layer Absorptances ($A_n^f$) for Glazing and Window Systems (*Continued*)

*(From Table 10, Chapter 15, 2013 ASHRAE Handbook—Fundamentals)*

| | | | | | | | Center-of-Glazing Properties | | | | | | Total Window SHGC at Normal Incidence | | | | Total Window $T_v$ at Normal Incidence | | | |
|---|---|---|---|---|---|---|---|---|---|---|---|---|---|---|---|---|---|---|---|---|
| | **Glazing System** | | | | | **Incidence Angles** | | | | | | | Aluminum | | Other Frames | | Aluminum | | Other Frames | |
| ID | Glass Thick., in. | | Center Glazing $T_v$ | | Normal 0.00 | 40.00 | 50.00 | 60.00 | 70.00 | 80.00 | Hemis., Diffuse | Operable | Fixed | Operable | Fixed | Operable | Fixed | Operable | Fixed |
| | | | | $R^b$ | 0.10 | 0.11 | 0.13 | 0.19 | 0.31 | 0.55 | 0.17 | | | | | | | | |
| | | | | $A_1^f$ | 0.49 | 0.51 | 0.05 | 0.53 | 0.52 | 0.43 | 0.50 | | | | | | | | |
| | | | | $A_2^f$ | 0.05 | 0.05 | 0.05 | 0.05 | 0.04 | 0.03 | 0.05 | | | | | | | | |
| 5g | 1/8 | GRY CLR | 0.56 | SHGC | 0.60 | 0.57 | 0.54 | 0.48 | 0.37 | 0.20 | 0.51 | 0.55 | 0.55 | 0.49 | 0.53 | 0.50 | 0.50 | 0.45 | 0.49 |
| | | | | $T$ | 0.51 | 0.48 | 0.45 | 0.39 | 0.29 | 0.12 | 0.42 | | | | | | | | |
| | | | | $R^f$ | 0.09 | 0.09 | 0.11 | 0.16 | 0.26 | 0.48 | 0.14 | | | | | | | | |
| | | | | $R^b$ | 0.12 | 0.13 | 0.15 | 0.21 | 0.34 | 0.59 | 0.19 | | | | | | | | |
| | | | | $A_1^f$ | 0.34 | 0.37 | 0.39 | 0.40 | 0.41 | 0.37 | 0.37 | | | | | | | | |
| | | | | $A_2^f$ | 0.05 | 0.06 | 0.06 | 0.05 | 0.05 | 0.03 | 0.05 | | | | | | | | |
| 5h | 1/4 | GRY CLR | 0.41 | SHGC | 0.47 | 0.44 | 0.42 | 0.37 | 0.29 | 0.16 | 0.39 | 0.43 | 0.43 | 0.38 | 0.42 | 0.36 | 0.37 | 0.33 | 0.36 |
| | | | | $T$ | 0.36 | 0.32 | 0.29 | 0.25 | 0.18 | 0.07 | 0.28 | | | | | | | | |
| | | | | $R^f$ | 0.07 | 0.07 | 0.08 | 0.12 | 0.21 | 0.43 | 0.12 | | | | | | | | |
| | | | | $R^b$ | 0.10 | 0.11 | 0.13 | 0.18 | 0.31 | 0.55 | 0.17 | | | | | | | | |
| | | | | $A_1^f$ | 0.51 | 0.54 | 0.56 | 0.57 | 0.56 | 0.47 | 0.53 | | | | | | | | |
| | | | | $A_2^f$ | 0.07 | 0.07 | 0.07 | 0.06 | 0.05 | 0.03 | 0.06 | | | | | | | | |
| 5i | 1/4 | BLUGRN CLR | 0.67 | SHGC | 0.50 | 0.47 | 0.45 | 0.40 | 0.32 | 0.17 | 0.43 | 0.46 | 0.46 | 0.41 | 0.44 | 0.60 | 0.60 | 0.54 | 0.59 |
| | | | | $T$ | 0.40 | 0.37 | 0.34 | 0.30 | 0.22 | 0.10 | 0.32 | | | | | | | | |
| | | | | $R^f$ | 0.08 | 0.08 | 0.10 | 0.14 | 0.24 | 0.46 | 0.13 | | | | | | | | |
| | | | | $R^b$ | 0.11 | 0.11 | 0.14 | 0.19 | 0.31 | 0.55 | 0.17 | | | | | | | | |
| | | | | $A_1^f$ | 0.47 | 0.49 | 0.50 | 0.51 | 0.50 | 0.42 | 0.48 | | | | | | | | |
| | | | | $A_2^f$ | 0.06 | 0.06 | 0.06 | 0.05 | 0.04 | 0.03 | 0.05 | | | | | | | | |
| 5j | 1/4 | HI-P GRN CLR | 0.59 | SHGC | 0.39 | 0.37 | 0.35 | 0.31 | 0.25 | 0.14 | 0.33 | 0.36 | 0.36 | 0.32 | 0.35 | 0.53 | 0.53 | 0.47 | 0.52 |
| | | | | $T$ | 0.28 | 0.26 | 0.24 | 0.20 | 0.15 | 0.06 | 0.22 | | | | | | | | |
| | | | | $R^f$ | 0.06 | 0.07 | 0.08 | 0.12 | 0.21 | 0.43 | 0.11 | | | | | | | | |
| | | | | $R^b$ | 0.10 | 0.11 | 0.13 | 0.19 | 0.31 | 0.55 | 0.17 | | | | | | | | |
| | | | | $A_1^f$ | 0.62 | 0.65 | 0.65 | 0.65 | 0.62 | 0.50 | 0.63 | | | | | | | | |
| | | | | $A_2^f$ | 0.03 | 0.03 | 0.03 | 0.03 | 0.02 | 0.01 | 0.03 | | | | | | | | |

***Reflective Double Glazing***

| ID | Glass Thick., in. | System | $T_v$ | | Normal | 40 | 50 | 60 | 70 | 80 | Hemis. | | | | | | | | |
|---|---|---|---|---|---|---|---|---|---|---|---|---|---|---|---|---|---|---|---|
| 5k | 1/4 | SS on CLR 8%, CLR | 0.07 | SHGC | 0.13 | 0.12 | 0.12 | 0.11 | 0.10 | 0.06 | 0.11 | 0.13 | 0.13 | 0.11 | 0.12 | 0.06 | 0.06 | 0.06 | 0.06 |
| | | | | $T$ | 0.05 | 0.05 | 0.04 | 0.04 | 0.03 | 0.01 | 0.04 | | | | | | | | |
| | | | | $R^f$ | 0.33 | 0.34 | 0.35 | 0.37 | 0.44 | 0.61 | 0.37 | | | | | | | | |
| | | | | $R^b$ | 0.38 | 0.37 | 0.38 | 0.40 | 0.46 | 0.61 | 0.40 | | | | | | | | |
| | | | | $A_1^f$ | 0.61 | 0.61 | 0.60 | 0.58 | 0.53 | 0.37 | 0.56 | | | | | | | | |
| | | | | $A_2^f$ | 0.01 | 0.01 | 0.01 | 0.01 | 0.01 | 0.01 | 0.01 | | | | | | | | |
| 5l | 1/4 | SS on CLR 14%, CLR | 0.13 | SHGC | 0.17 | 0.17 | 0.16 | 0.15 | 0.13 | 0.08 | 0.16 | 0.17 | 0.16 | 0.14 | 0.15 | 0.12 | 0.12 | 0.10 | 0.11 |
| | | | | $T$ | 0.08 | 0.08 | 0.08 | 0.07 | 0.05 | 0.02 | 0.07 | | | | | | | | |
| | | | | $R^f$ | 0.26 | 0.27 | 0.28 | 0.31 | 0.38 | 0.57 | 0.30 | | | | | | | | |
| | | | | $R^b$ | 0.34 | 0.33 | 0.34 | 0.37 | 0.44 | 0.60 | 0.36 | | | | | | | | |
| | | | | $A_1^f$ | 0.63 | 0.64 | 0.64 | 0.63 | 0.61 | 0.56 | 0.60 | | | | | | | | |
| | | | | $A_2^f$ | 0.02 | 0.02 | 0.02 | 0.02 | 0.02 | 0.02 | 0.02 | | | | | | | | |
| 5m | 1/4 | SS on CLR 20%, CLR | 0.18 | SHGC | 0.22 | 0.21 | 0.21 | 0.19 | 0.16 | 0.09 | 0.20 | 0.21 | 0.21 | 0.18 | 0.20 | 0.16 | 0.16 | 0.14 | 0.16 |
| | | | | $T$ | 0.12 | 0.11 | 0.11 | 0.09 | 0.07 | 0.03 | 0.10 | | | | | | | | |
| | | | | $R^f$ | 0.21 | 0.22 | 0.23 | 0.26 | 0.34 | 0.54 | 0.25 | | | | | | | | |
| | | | | $R^b$ | 0.30 | 0.30 | 0.31 | 0.34 | 0.41 | 0.59 | 0.33 | | | | | | | | |
| | | | | $A_1^f$ | 0.64 | 0.64 | 0.63 | 0.62 | 0.57 | 0.41 | 0.61 | | | | | | | | |
| | | | | $A_2^f$ | 0.03 | 0.03 | 0.03 | 0.03 | 0.02 | 0.02 | 0.03 | | | | | | | | |
| 5n | 1/4 | SS on GRN 14%, CLR | 0.11 | SHGC | 0.16 | 0.16 | 0.15 | 0.14 | 0.12 | 0.08 | 0.14 | 0.16 | 0.16 | 0.14 | 0.14 | 0.10 | 0.10 | 0.09 | 0.10 |
| | | | | $T$ | 0.05 | 0.05 | 0.05 | 0.04 | 0.03 | 0.01 | 0.04 | | | | | | | | |
| | | | | $R^f$ | 0.14 | 0.14 | 0.16 | 0.19 | 0.27 | 0.49 | 0.18 | | | | | | | | |
| | | | | $R^b$ | 0.34 | 0.33 | 0.34 | 0.37 | 0.44 | 0.60 | 0.36 | | | | | | | | |
| | | | | $A_1^f$ | 0.80 | 0.80 | 0.79 | 0.76 | 0.69 | 0.49 | 0.76 | | | | | | | | |
| | | | | $A_2^f$ | 0.01 | 0.01 | 0.01 | 0.01 | 0.01 | 0.01 | 0.01 | | | | | | | | |

**Table 7-4  Visible Transmittance ($T_v$), Solar Heat Gain Coefficient (SHGC), Solar Transmittance ($T$), Front Reflectance ($R^f$), Back Reflectance ($R^b$), and Layer Absorptances ($A_n^f$) for Glazing and Window Systems (*Continued*)**

*(From Table 10, Chapter 15, 2013 ASHRAE Handbook—Fundamentals)*

|  |  |  |  |  | \multicolumn{7}{c|}{Center-of-Glazing Properties} | \multicolumn{4}{c|}{Total Window SHGC at Normal Incidence} | \multicolumn{4}{c|}{Total Window $T_v$ at Normal Incidence} |
|---|---|---|---|---|---|---|---|---|---|---|---|---|---|---|---|---|---|---|---|
|  | \multicolumn{3}{c|}{Glazing System} |  |  | \multicolumn{6}{c|}{Incidence Angles} |  | \multicolumn{2}{c|}{Aluminum} | \multicolumn{2}{c|}{Other Frames} | \multicolumn{2}{c|}{Aluminum} | \multicolumn{2}{c|}{Other Frames} |
| ID | Glass Thick., in. |  | Center Glazing $T_v$ |  | Normal 0.00 | 40.00 | 50.00 | 60.00 | 70.00 | 80.00 | Hemis., Diffuse | Operable | Fixed | Operable | Fixed | Operable | Fixed | Operable | Fixed |
| 5o | 1/4 | TI on CLR 20%, CLR | 0.18 | SHGC | 0.21 | 0.20 | 0.19 | 0.18 | 0.15 | 0.09 | 0.18 | 0.20 | 0.20 | 0.18 | 0.19 | 0.16 | 0.16 | 0.14 | 0.16 |
|  |  |  |  | $T$ | 0.11 | 0.10 | 0.10 | 0.08 | 0.06 | 0.03 | 0.09 |  |  |  |  |  |  |  |  |
|  |  |  |  | $R^f$ | 0.22 | 0.22 | 0.24 | 0.27 | 0.34 | 0.54 | 0.26 |  |  |  |  |  |  |  |  |
|  |  |  |  | $R^b$ | 0.32 | 0.31 | 0.32 | 0.35 | 0.42 | 0.59 | 0.35 |  |  |  |  |  |  |  |  |
|  |  |  |  | $A_1^f$ | 0.65 | 0.66 | 0.65 | 0.63 | 0.58 | 0.41 | 0.62 |  |  |  |  |  |  |  |  |
|  |  |  |  | $A_2^f$ | 0.02 | 0.02 | 0.02 | 0.02 | 0.02 | 0.01 | 0.02 |  |  |  |  |  |  |  |  |
| 5p | 1/4 | TI on CLR 30%, CLR | 0.27 | SHGC | 0.29 | 0.28 | 0.27 | 0.25 | 0.20 | 0.12 | 0.25 | 0.27 | 0.27 | 0.24 | 0.26 | 0.24 | 0.24 | 0.22 | 0.24 |
|  |  |  |  | T | 0.18 | 0.17 | 0.16 | 0.14 | 0.10 | 0.05 | 0.15 |  |  |  |  |  |  |  |  |
|  |  |  |  | $R^f$ | 0.15 | 0.15 | 0.17 | 0.20 | 0.29 | 0.51 | 0.19 |  |  |  |  |  |  |  |  |
|  |  |  |  | $R^b$ | 0.27 | 0.27 | 0.28 | 0.31 | 0.40 | 0.58 | 0.31 |  |  |  |  |  |  |  |  |
|  |  |  |  | $A_1^f$ | 0.64 | 0.64 | 0.63 | 0.62 | 0.58 | 0.43 | 0.61 |  |  |  |  |  |  |  |  |
|  |  |  |  | $A_2^f$ | 0.04 | 0.04 | 0.04 | 0.04 | 0.03 | 0.02 | 0.04 |  |  |  |  |  |  |  |  |

**KEY:**
CLR = clear, BRZ = bronze, GRN = green, GRY = gray, BLUGRN = blue-green, SS = stainless steel reflective coating, TI = titanium reflective coating
Reflective coating descriptors include percent visible transmittance as *x*%.
HI-P GRN = high-performance green tinted glass, LE = low-emissivity coating

$T_v$ = visible transmittance, $T$ = solar transmittance, SHGC = solar heat gain coefficient, and H. = hemispherical SHGC
ID #s refer to U-factors in Table 4, except for products 49 and 50.

**Table 7-5  Solar Heat Gain Coefficients for Domed Horizontal Skylights**

*(Table 11, Chapter 15, 2013 ASHRAE Handbook—Fundamentals)*

| Dome | Light Diffuser (Translucent) | Curb Height, in. | Curb Width-to-Height Ratio | Solar Heat Gain Coefficient | Visible Transmittance |
|---|---|---|---|---|---|
| Clear | Yes | 0 | ∞ | 0.53 | 0.56 |
| τ = 0.86 | τ = 0.58 | 9 | 5 | 0.50 | 0.58 |
|  |  | 12 | 2.5 | 0.44 | 0.59 |
| Clear | None | 0 | ∞ | 0.86 | 0.91 |
| τ = 0.86 |  | 9 | 5 | 0.77 | 0.91 |
|  |  | 12 | 2.5 | 0.70 | 0.91 |
| Translucent | None | 0 | ∞ | 0.50 | 0.46 |
| τ = 0.52 |  | 12 | 2.5 | 0.40 | 0.32 |
| Translucent | None | 0 | ∞ | 0.30 | 0.25 |
| τ = 0.27 |  | 9 | 5 | 0.26 | 0.21 |
|  |  | 12 | 2.5 | 0.24 | 0.18 |

*Source*: Laouadi et al. (2003), Schutrum and Ozisik (1961).

and unshaded, depend on the location of the shadow line on a surface in the plane of the glass.

To account for the different types of fenestration and shading devices, used the inside shading attenuation coefficient IAC, which relates the solar heat gain through a glazing system under a specific set of conditions to the solar heat gain through the reference glazing material under the same conditions.

Most fenestration has some type of internal shading to provide privacy and aesthetic effects, as well as to give varying degrees of sun control. The IAC values applicable to typical internal shading devices are given in Tables 7-4 though 7-11 for various fenestrations and shading device combinations.

Table 7-8 gives values of IAC (derived from measurements) for a variety of glazing and shading combinations.

The IAC bears a certain similarity to the shading coefficient. There is, however, an important difference: we must calculate the solar heat flux through the unshaded glazing at the appropriate angle before applying the IAC. With the shading coefficient, only the angular dependence of single glazing was included (through the now-discarded SHGF). The effectiveness of any internal shading device depends on its ability to reflect incoming solar radiation back through the fenestration before it can be absorbed and converted into heat within the building. Table 7-10 lists approximate values of solar-optical properties for the typical indoor shading devices described in Tables 7-8 and 7-9.

During certain seasons of the year and for some exposures, horizontal projections can result in considerable reductions in solar heat gain by providing shade. This is particularly applicable to south, southeast, and southwest exposures during the late spring, summer, and early fall. On east and west exposures during the entire year, and on southerly exposures during the winter, the solar altitude is

## Table 7-6 Solar Heat Gain Coefficients and U-Factors for Standard Hollow Glass Block Wall Panels

*(Table 12, Chapter 15, 2013 ASHRAE Handbook—Fundamentals)*

| Type of Glass Block[a] | Description of Glass Block | SHGC In Sun | SHGC In Shade[b] | U-Factor[c] Btu/(h·ft²·°F) |
|---|---|---|---|---|
| Type I | Glass colorless or aqua<br>A, D: Smooth<br>B, C: Smooth or wide ribs, or flutes horizontal or vertical, or shallow configuration<br>E: None | 0.57 | 0.35 | 0.51 |
| Type IA | Same as Type I except ceramic enamel on A | 0.23 | 0.17 | 0.51 |
| Type II | Same as Type I except glass fiber screen partition E | 0.38 | 0.30 | 0.48 |
| Type III | Glass colorless or aqua<br>A, D: Narrow vertical ribs or flutes.<br>B, C: Horizontal light-diffusing prisms, or horizontal light-directing prisms<br>E: Glass fiber screen | 0.29 | 0.23 | 0.48 |
| Type IIIA | Same as Type III except<br>E: Glass fiber screen with green ceramic spray coating or glass fiber screen and gray glass or glass fiber screen with light-selecting prisms | 0.22 | 0.16 | 0.48 |
| Type IV | Same as Type I except reflective oxide coating on A | 0.14 | 0.10 | 0.51 |

[a]All values are for 7 3/4 by 7 3/4 by 3 7/8 in. block, set in light-colored mortar. For 11 3/4 by 11 3/4 by 3 7/8 in. block, increase coefficients by 15%, and for 5 3/4 by 5 3/4 by 3 7/8 in. block reduce coefficients by 15%.

[b]For NE, E, and SE panels in shade, add 50% to values listed for panels in shade.

[c]Values shown are identical for all size block.

## Table 7-7 Unshaded Fractions ($F_u$) and Exterior Solar Attenuation Coefficients (EAC) for Louvered Sun Screens

| Profile Angle | Group 1 Transmittance | $F_u$ | EAC | Group 2 Transmittance | $F_u$ | EAC | Group 3 Transmittance | $F_u$ | EAC | Group 4 Transmittance | $F_u$ | EAC |
|---|---|---|---|---|---|---|---|---|---|---|---|---|
| 10° | 0.23 | 0.20 | 0.15 | 0.25 | 0.13 | 0.02 | 0.4 | 0.33 | 0.18 | 0.48 | 0.29 | 0.3 |
| 20° | 0.06 | 0.02 | 0.15 | 0.14 | 0.03 | 0.02 | 0.32 | 0.24 | 0.18 | 0.39 | 0.2 | 0.3 |
| 30° | 0.04 | 0.00 | 0.15 | 0.12 | 0.01 | 0.02 | 0.21 | 0.13 | 0.18 | 0.28 | 0.08 | 0.3 |
| ≥ 40° | 0.04 | 0.00 | 0.15 | 0.11 | 0.00 | 0.02 | 0.07 | 0.00 | 0.18 | 0.2 | 0.00 | 0.3 |

Group 1: Black, width over spacing ratio 1.15/1; 23 louvers/in. Group 2: Light color; high reflectance, otherwise same as Group 1. Group 3: Black or dark color; w/s ratio 0.85/1; 17 louvers/in. Group 4: Light color or unpainted aluminum; high reflectance; otherwise same as Group 3. U-factor = 0.85 Btu/(h·ft²·°F) for all groups when used with single glazing.

generally so low that horizontal projections, in order to be effective, would have to be excessively long.

A vertical fin or projection $P_V$ required to produce a given shadow width $S_V$ on a window or wall for any given time of day and year is related to the vertical surface-solar azimuth (see Table 7-12) by

$$P_V = S_V/|\tan\gamma| \qquad (7\text{-}14)$$

The horizontal projection $P_H$ required to produce a given shadow height $S_H$ on a window or wall for any time of day or year is related to the profile angle $\Omega$ (Table 7-13) by

$$P_H = S_H/\tan\Omega \qquad (7\text{-}15)$$

The profile angle, also called the vertical shadow line angle, is the angle between a line perpendicular to the plane of the window and the projection of the earth-sun line in a vertical plane that is also normal to the window and is given by

$$\tan\Omega = \tan\beta/\cos\gamma$$

The use of an overhang to shade glass is an excellent way of reducing the solar heat gain. Use the values for north glass when reducing solar heat gain for a shaded window. North glass does not have sun shining directly on it, so the SHGFs listed for it are based on only diffuse solar radiation (as on shaded glass). If a window is partially shaded, use the north value of SHGF for the shaded portion and the regular (east, west, south, etc.) SHGF for the rest of the window.

A window with a significant depth of reveal generally has part of its glass area shaded by the mullions and the transom. The area that is shaded varies throughout the day, but it can be estimated by treating mullions and transoms as vertical and horizontal projections.

The width $S_V$ of the shadow from a mullion that projects a distance $P_V$ beyond the plane of the glass is

$$S_V = P_V|\tan\gamma| \qquad (7\text{-}16)$$

where $\gamma$ is the wall-solar azimuth. When $\gamma$ is greater than 90°, the entire window is in the shade.

Similarly, the height of the shadow cast by a transom that projects $P_H$ beyond the plane of the glass is

$$S_H = P_H\tan\beta/\cos\gamma \qquad (7\text{-}17)$$

where $\beta$ is the solar altitude. The solar angles are shown in Figure 7-3.

Table 7-8 Interior Solar Attenuation Coefficients (IAC) for Single or Double Glazings Shaded by Interior Venetian Blinds or Roller Shades

| Glazing System[a] | Nominal Thickness[b] Each Pane, in. | Glazing Solar Transmittance[b] Outer Pane | Single or Inner Pane | Glazing SHGC[b] | IAC Venetian Blinds Medium | Light | Roller Shades Opaque Dark | Opaque White | Translucent Light |
|---|---|---|---|---|---|---|---|---|---|
| **Single Glazing Systems** | | | | | | | | | |
| Clear, residential | 1/8[c] | | 0.87 to 0.80 | 0.86 | 0.75[d] | 0.68[d] | 0.82 | 0.40 | 0.45 |
| Clear, commercial | 1/4 to 1/2 | | 0.80 to 0.71 | 0.82 | | | | | |
| Clear, pattern | 1/8 to 1/2 | | 0.87 to 0.79 | | | | | | |
| Heat absorbing, pattern | 1/8 | | | 0.59 | | | | | |
| Tinted | 3/16, 7/32 | | 0.74, 0.71 | | | | | | |
| Above glazings, automated blinds[e] | | | | 0.86 | 0.64 | 0.59 | | | |
| Above glazings, tightly closed vertical blinds | | | | 0.85 | 0.30 | 0.26 | | | |
| Heat absorbing[f] | 1/4 | | 0.46 | 0.59 | 0.84 | 0.78 | 0.66 | 0.44 | 0.47 |
| Heat absorbing, pattern | 1/4 | | | | | | | | |
| Tinted | 1/8, 1/4 | | 0.59, 0.45 | | | | | | |
| Heat absorbing or pattern | | | 0.44 to 0.30 | 0.59 | 0.79 | 0.76 | 0.59 | 0.41 | 0.47 |
| Heat absorbing | 3/8 | | 0.34 | | | | | | |
| Heat absorbing or pattern | | | 0.29 to 0.15 | | | | | | |
| | | | 0.24 | 0.37 | 0.99 | 0.94 | 0.85 | 0.66 | 0.73 |
| Reflective coated glass | | | | 0.26 to 0.52 | 0.83 | 0.75 | | | |
| **Double Glazing Systems**[g] | | | | | | | | | |
| Clear double, residential | 1/8 | 0.87 | 0.87 | 0.76 | 0.71[d] | 0.66[d] | 0.81 | 0.40 | 0.46 |
| Clear double, commercial | 1/4 | 0.80 | 0.80 | 0.70 | | | | | |
| Heat absorbing double[f] | 1/4 | 0.46 | 0.8 | 0.47 | 0.72 | 0.66 | 0.74 | 0.41 | 0.55 |
| Reflective double | | | | 0.17 to 0.35 | 0.90 | 0.86 | | | |
| **Other Glazings** (Approximate) | | | | | 0.83 | 0.77 | 0.74 | 0.45 | 0.52 |
| ± Range of Variation[h] | | | | | 0.15 | 0.17 | 0.16 | 0.21 | 0.21 |

[a] Systems listed in the same table block have same IAC.
[b] Values or ranges given for identification of appropriate IAC value; where paired, solar transmittances and thicknesses correspond. SHGC is for unshaded glazing at normal incidence.
[c] Typical thickness for residential glass.
[d] From measurements by Van Dyke and Konen (1982) for 45° open venetian blinds, 35° solar incidence, and 35° profile angle.
[e] Use these values only when operation is automated for exclusion of beam solar (as opposed to daylight maximization). Also applies to tightly closed horizontal blinds.
[f] Refers to gray, bronze, and green tinted heat-absorbing glass (on exterior pane in double glazing)
[g] Applies either to factory-fabricated insulating glazing units or to prime windows plus storm windows.
[h] The listed approximate IAC value may be higher or lower by this amount, due to glazing/shading interactions and variations in the shading properties (e.g., manufacturing tolerances).

Table 7-9 Between-Glass Solar Attenuation Coefficients (BAC) for Double Glazing with Between-Glass Shading

| Type of Glass | Nominal Thickness, Each Pane | Solar Transmittance[a] Outer Pane | Inner Pane | Description of Air Space | Type of Shading Venetian Blinds Light | Medium | Louvered Sun Screen |
|---|---|---|---|---|---|---|---|
| Clear out, Clear in | 3/32, 1/8 in. | 0.87 | 0.87 | Shade in contact with glass or shade separated from glass by air space. | 0.33 | 0.36 | 0.43 |
| Clear out, Clear in | 1/4 in. | 0.80 | 0.80 | Shade in contact with glass-voids filled with plastic. | — | — | 0.49 |
| Heat-absorbing[b] out, Clear in | | | | Shade in contact with glass or shade separated from glass by air space. | 0.28 | 0.30 | 0.37 |
| | 1/4 in. | 0.46 | 0.80 | Shade in contact with glass-voids filled with plastic. | — | — | 0.41 |

[a] Refer to manufacturers' literature for exact values.
[b] Refers to gray, bronze and green tinted heat-absorbing glass.

Table 7-10 Properties of Representative Indoor Shading Devices Shown in Tables 7-8 and 7-9

| Indoor Shade | Solar-Optical Properties (Normal Incidence) Transmittance | Reflectance | Absorptance |
|---|---|---|---|
| Venetian blinds[a] (ratio of slat width to slat spacing 1.2, slat angle 45°) | | | |
| Light-colored slat | 0.05 | 0.55 | 0.40 |
| Medium-colored slat | 0.05 | 0.35 | 0.60 |
| Vertical blinds | | | |
| White louvers | 0.00 | 0.77 | 0.23 |
| Roller shades | | | |
| Light shades (translucent) | 0.25 | 0.60 | 0.15 |
| White shade (opaque) | 0.00 | 0.65 | 0.35 |
| Dark-colored shade (opaque) | 0.00 | 0.20 | 0.80 |

[a] Values in this table and Tables 19 and 20 are based on horizontal venetian blinds. However, tests show that these values can be used for vertical blinds with good accuracy.

Thus, the sunlit area of the window is

$$\text{Area in Sun} = (\text{Width} - S_V)(\text{Height} - S_H) \quad (7\text{-}18)$$

### 7.3.3 Heat Gain Through Interior Surfaces

Whenever a conditioned space is adjacent to a space with a different temperature, heat transfer through the separating physical section must be considered. The heat transfer rate $q$, in Btu/h, is given by

$$q = UA(t_b - t_i) \quad (7\text{-}19)$$

where

$U$ = coefficient of overall heat transfer between the adjacent and the conditioned space, Btu/h·ft²·°F
$A$ = area of separating section concerned, ft²
$t_b$ = average air temperature in adjacent space, °F
$t_i$ = air temperature in conditioned space, °F

Temperature $t_b$ may have any value over a considerable range according to conditions in the adjacent space. The temperature in a kitchen or boiler room may be as much as 15°F to 50°F above the outdoor air temperature. The actual temperatures in adjoining spaces should be measured wherever practicable; where nothing is known except that the adjacent space is of conventional construction and contains no heat sources, $t_b - t_i$ may be considered to be the difference between the outdoor air and conditioned-space design dry-bulb temperatures minus 5°F (2.8°C). In some cases, the air temperature in the adjacent space corresponds to the outdoor air temperature or higher.

For floors in direct contact with the ground, or over an underground basement that is neither ventilated nor conditioned, heat transfer may be neglected for cooling load estimates.

### 7.3.4 Heat Sources in Conditioned Spaces

**People.** Representative rates at which heat and moisture are given off by human beings in different states of activity are listed in Table 7-14. Often, these sensible and latent heat gains constitute a large fraction of the total load. For short occupancy, the extra heat and moisture brought in by people may be significant.

The conversion of sensible heat gain from people to space cooling load is affected by the thermal storage characteristics of that space, and it is thus subject to application of appropriate room transfer functions (RTF). Latent heat gains are instantaneous.

**Lighting.** Since lighting is often the major component of space load, an accurate estimate of the space heat gain it imposes is needed. The rate of heat gain at any moment can be different from the heat equivalent of power supplied instantaneously to those lights.

Only part of the energy from lights is in the form of convective heat, which is immediately picked up by the air-conditioning apparatus. The remaining portion is in the form of radiation that affects the conditioned space once it has been absorbed and re-released by the walls, floors, furniture, etc. This absorbed energy contributes to space cooling load only after a time lag, so part of this energy is reradiating after the lights have been turned off (Figure 7-4).

Time lag effect should always be considered when calculating cooling load, since load felt by the space can be lower than the instantaneous heat gain generated, and peak load for the space may be affected.

The primary source of heat from lighting comes from the light-emitting elements (lamps), although additional heat may be generated from associated components in the light fixtures housing such lamps.

Generally, the instantaneous rate of heat gain from electric lighting may be calculated from

$$q_{el} = 3.41 W F_{ul} F_{sa} \quad (7\text{-}20)$$

where

$q_{el}$ = heat gain, Btu/h
$W$ = total light wattage
$F_{ul}$ = lighting use factor
$F_{sa}$ = lighting special allowance factor

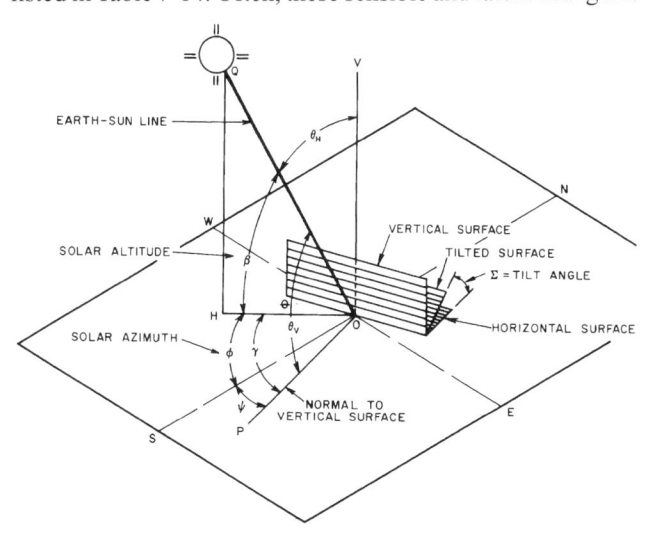

*Fig. 7-3 Solar Angles*

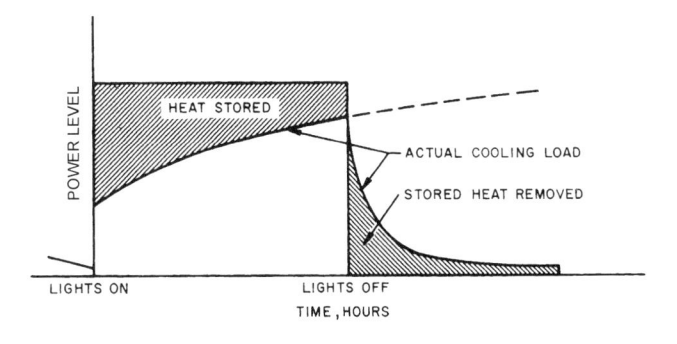

*Fig. 7-4 Thermal Storage Effect in Cooling Load from Lights*
(Figure 2, Chapter 18, 2013 ASHRAE Handbook—Fundamentals)

Table 7-11 Interior Solar Attenuation Coefficients for Single and Insulating Glass with Draperies

| Glazing | Glass Transmission | Glazing SHGC (No Drapes) | IAC A | B | C | D | E | F | G | H | I | J |
|---|---|---|---|---|---|---|---|---|---|---|---|---|
| Single glass | | | | | | | | | | | | |
| 1/8 in. clear | 0.86 | 0.87 | 0.87 | 0.82 | 0.74 | 0.69 | 0.64 | 0.59 | 0.53 | 0.48 | 0.42 | 0.37 |
| 1/4 in. clear | 0.8 | 0.83 | 0.84 | 0.79 | 0.74 | 0.68 | 0.63 | 0.58 | 0.53 | 0.47 | 0.42 | 0.37 |
| 1/2 in. clear | 0.71 | 0.77 | 0.84 | 0.80 | 0.75 | 0.69 | 0.64 | 0.59 | 0.55 | 0.49 | 0.44 | 0.40 |
| 1/4 in. heat absorbing | 0.46 | 0.58 | 0.85 | 0.81 | 0.78 | 0.73 | 0.69 | 0.66 | 0.61 | 0.57 | 0.54 | 0.49 |
| 1/2 in. heat absorbing | 0.24 | 0.44 | 0.86 | 0.84 | 0.80 | 0.78 | 0.76 | 0.72 | 0.68 | 0.66 | 0.64 | 0.60 |
| Reflective coated | — | 0.52 | 0.95 | 0.90 | 0.85 | 0.82 | 0.77 | 0.72 | 0.68 | 0.63 | 0.60 | 0.55 |
| | — | 0.44 | 0.92 | 0.88 | 0.84 | 0.82 | 0.78 | 0.76 | 0.72 | 0.68 | 0.66 | 0.62 |
| | — | 0.35 | 0.90 | 0.88 | 0.85 | 0.83 | 0.80 | 0.75 | 0.73 | 0.70 | 0.68 | 0.65 |
| | — | 0.26 | 0.83 | 0.80 | 0.80 | 0.77 | 0.77 | 0.77 | 0.73 | 0.70 | 0.70 | 0.67 |
| Insulating glass, 1/4-in. air space | | | | | | | | | | | | |
| (1/8 in. out and 1/8 in. in) | 0.76 | 0.77 | 0.84 | 0.80 | 0.73 | 0.71 | 0.64 | 0.60 | 0.54 | 0.51 | 0.43 | 0.40 |
| Insulating glass 1/2-in. air space | | | | | | | | | | | | |
| Clear out and clear in | 0.64 | 0.72 | 0.80 | 0.75 | 0.70 | 0.67 | 0.63 | 0.58 | 0.54 | 0.51 | 0.45 | 0.42 |
| Heat absorbing out and clear in | 0.37 | 0.48 | 0.89 | 0.85 | 0.82 | 0.78 | 0.75 | 0.71 | 0.67 | 0.64 | 0.60 | 0.58 |
| Reflective coated | — | 0.35 | 0.95 | 0.93 | 0.93 | 0.90 | 0.85 | 0.80 | 0.78 | 0.73 | 0.70 | 0.70 |
| | — | 0.26 | 0.97 | 0.93 | 0.90 | 0.90 | 0.87 | 0.87 | 0.83 | 0.83 | 0.80 | 0.80 |
| | — | 0.17 | 0.95 | 0.95 | 0.90 | 0.90 | 0.85 | 0.85 | 0.80 | 0.80 | 0.75 | 0.75 |

**Interior Solar Attenuation (IAC)**

*Notes*:
1. Interior attenuation coefficients are for draped fabrics.
2. Other properties are for fabrics in flat orientation.
3. Use fabric reflectance and transmittance to obtain accurate IAC values.
4. Use openness and yarn reflectance or openness and fabric reflectance to obtain the various environmental characteristics, or to obtain approximate IAC values.

**Classification of Fabrics**

I = Open weave
II = Semiopen weave
III = Closed weave

D = Dark color
M = Medium color
L = Light color

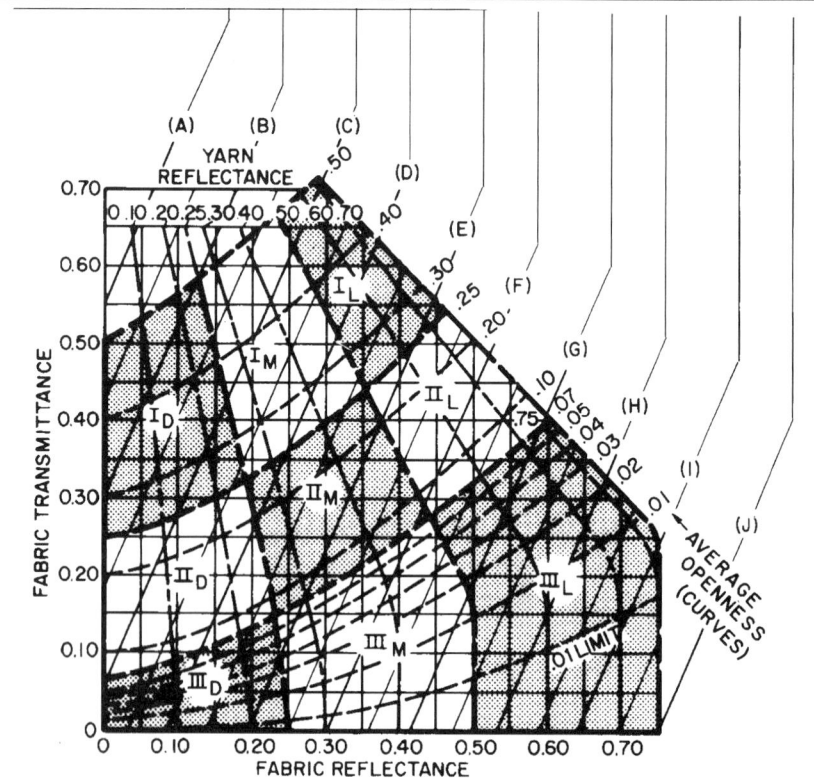

**To obtain fabric designator (III$_L$, I$_M$, etc.).** Using either (1) fabric transmittance and fabric reflectance coordinates, or (2) openness and yarn reflectance coordinates, find a point on the chart and note the designator for that area. If properties are not known, the classification may be approximated by eye as described in the note in Figure 31.

**To obtain interior attenuation (IAC).** (1) Locate drapery fabric as a point using its known properties, or approximate using its fabric classification designator. For accuracy, use fabric transmittance and fabric reflectance; (2) follow diagonal IAC lines to lettered columns in the table. Find IAC value in selected column on line corresponding to glazing used. For example, IAC is 0.4 for 1/4-in. clear single glass with III$_L$ drapery (Column H).

# Nonresidential Cooling and Heating Load Calculations

Total light wattage is obtained from the ratings of all lamps installed, both for general illumination and for display use.

The **lighting use factor** is the ratio of the wattage in use for the conditions under which the load estimate is being made to the total installed wattage. For commercial applications, such as stores, the use factor generally is unity.

The **special allowance factor** is for fluorescent fixtures and/or fixtures that are either ventilated or installed so that only part of their heat goes to the conditioned space. For fluorescent or high-intensity discharge fixtures, the special allowance factor accounts primarily for ballast losses. Table 7-15 shows that the special allowance factor for a two-lamp fluorescent fixture ranges from 0.94 for T8 lamps with an electronic ballast to 1.21 for energy-saver T12 lamps with a standard electromagnetic ballast. High-intensity discharge fixtures, such as metal halide, may have special allowance factors varying from 1.07 to 1.44, depending on the lamp wattage and quantity of lamps per fixture, and should be dealt with individually. There is a wide variety of lamp and ballast combinations is available, and ballast catalog data provide the overall fixture wattage.

An alternative procedure is to estimate the lighting heat gain on a per square foot basis. Such an approach may be required when final lighting plans are not available. Table 7-16 shows the maximum lighting power density (LPD) (lighting heat gain per square foot) allowed by ASHRAE Standard 90.1-2010 for a range of space types.

In addition to determining the lighting heat gain, the fraction of lighting heat gain that enters the conditioned space may need to be distinguished from the fraction that enters an

**Table 7-12  Surface Orientations and Azimuths, Measured from South**

|  | N | NE | E | SE | S | SW | W | NW |
|---|---|---|---|---|---|---|---|---|
| Surface azimuth, ψ | 180° | −135° | −90° | −45° | 0° | 45° | 90° | 135° |

**Table 7-13  Solar Position and Profile Angles for 40° N Latitude**

| DATE | Solar Time | ALT | AZ | N | NNE | NE | ENE | E | ESE | SE | SSE | S | SSW | SW | WSW | N | NNE | NE | ENE | E | ESE | SE | SSE | S | SSW | SW | WSW | HOR | Solar Time |
|---|---|---|---|---|---|---|---|---|---|---|---|---|---|---|---|---|---|---|---|---|---|---|---|---|---|---|---|---|---|
| DEC | 0800 | 5 | 53 |  |  | 35 | 11 | 7 | 6 | 6 | 6 | 9 | 21 |  |  |  |  | 82 | 60 | 37 | 16 | 10 | 31 | 53 | 76 |  |  | 85 | 1600 |
|  | 0900 | 14 | 42 |  |  |  | 37 | 20 | 15 | 14 | 15 | 18 | 30 | 78 |  |  |  |  |  | 71 | 50 | 29 | 14 | 24 | 44 | 65 | 87 |  | 76 | 1500 |
|  | 1000 | 21 | 29 |  |  |  | 72 | 38 | 26 | 21 | 21 | 23 | 31 | 54 |  |  |  |  |  | 84 | 63 | 43 | 26 | 22 | 35 | 55 | 75 |  | 69 | 1400 |
|  | 1100 | 25 | 15 |  |  |  |  | 61 | 37 | 28 | 25 | 26 | 31 | 43 | 75 |  |  |  |  |  | 76 | 56 | 38 | 26 | 29 | 44 | 63 | 83 | 65 | 1300 |
|  | 1200 | 27 | 0 |  |  |  |  | 90 | 53 | 35 | 28 | 27 | 28 | 35 | 53 |  |  |  |  |  | 90 | 70 | 51 | 34 | 27 | 34 | 51 | 70 | 63 | 1200 |
| JAN | 0800 | 8 | 55 |  |  | 38 | 15 | 10 | 8 | 8 | 10 | 14 | 34 |  |  |  |  |  | 80 | 58 | 36 | 15 | 13 | 34 | 56 | 78 |  |  | 82 | 1600 |
| + | 0900 | 17 | 44 |  |  |  | 40 | 24 | 18 | 17 | 18 | 23 | 37 | 87 |  |  |  |  |  | 70 | 48 | 29 | 17 | 27 | 46 | 68 | 89 |  | 73 | 1500 |
| NOV | 1000 | 24 | 31 |  |  |  | 72 | 41 | 29 | 24 | 24 | 27 | 36 | 61 |  |  |  |  |  | 82 | 62 | 43 | 27 | 25 | 38 | 57 | 77 |  | 66 | 1400 |
|  | 1100 | 21 | 16 |  |  |  |  | 63 | 41 | 32 | 29 | 29 | 35 | 48 | 78 |  |  |  |  |  | 76 | 57 | 40 | 29 | 32 | 47 | 65 | 84 | 62 | 1300 |
|  | 1200 | 30 | 0 |  |  |  |  | 90 | 56 | 39 | 32 | 30 | 32 | 39 | 56 |  |  |  |  |  | 90 | 71 | 52 | 37 | 30 | 37 | 52 | 71 | 60 | 1200 |
| FEB | 0700 | 4 | 72 |  | 43 | 9 | 6 | 4 | 4 | 5 | 7 | 14 |  |  |  |  | 85 | 63 | 41 | 18 | 6 | 27 | 50 | 72 |  |  |  | 86 | 1700 |
| + | 0800 | 15 | 62 |  |  | 43 | 23 | 17 | 15 | 15 | 19 | 29 | 69 |  |  |  |  | 74 | 52 | 32 | 16 | 22 | 41 | 63 | 84 |  |  | 75 | 1600 |
| OCT | 0900 | 24 | 50 |  | 80 | 45 | 31 | 25 | 24 | 27 | 35 | 56 |  |  |  |  | 86 | 65 | 46 | 30 | 25 | 36 | 54 | 74 |  |  | 66 | 1500 |
|  | 1000 | 32 | 35 |  |  | 70 | 47 | 36 | 32 | 33 | 38 | 50 | 75 |  |  |  |  |  | 79 | 61 | 44 | 33 | 34 | 46 | 63 | 82 |  | 58 | 1400 |
|  | 1100 | 37 | 19 |  |  |  | 67 | 49 | 40 | 37 | 39 | 45 | 60 | 85 |  |  |  |  | 75 | 58 | 45 | 37 | 41 | 53 | 69 | 87 | 53 | 1300 |
|  | 1200 | 39 | 0 |  |  |  | 90 | 65 | 49 | 41 | 39 | 41 | 49 | 65 |  |  |  |  | 90 | 73 | 57 | 44 | 39 | 44 | 57 | 73 | 51 | 1200 |
| MAR | 0700 | 11 | 80 | 43 | 13 | 13 | 12 | 12 | 14 | 21 | 50 |  |  |  |  | 78 | 56 | 34 | 15 | 17 | 37 | 58 | 80 |  |  |  | 79 | 1700 |
| + | 0800 | 23 | 70 | 85 | 45 | 29 | 24 | 23 | 25 | 31 | 50 |  |  |  |  | 88 | 67 | 47 | 30 | 23 | 33 | 51 | 71 |  |  |  | 67 | 1600 |
| SEP | 0900 | 33 | 57 |  | 72 | 48 | 37 | 33 | 33 | 38 | 50 | 75 |  |  |  |  | 80 | 61 | 45 | 34 | 35 | 46 | 63 | 81 |  |  | 57 | 1500 |
|  | 1000 | 42 | 42 |  |  | 69 | 53 | 45 | 42 | 45 | 50 | 64 | 87 |  |  |  |  | 76 | 60 | 48 | 42 | 45 | 56 | 71 | 88 | 48 | 1400 |
|  | 1100 | 48 | 23 |  |  | 90 | 71 | 57 | 50 | 48 | 50 | 57 | 71 |  |  |  |  | 90 | 75 | 62 | 52 | 48 | 52 | 62 | 75 | 42 | 1300 |
|  | 1200 | 50 | 0 |  |  |  | 90 | 72 | 59 | 52 | 50 | 52 | 59 | 72 |  |  |  |  | 90 | 76 | 63 | 54 | 50 | 54 | 63 | 76 | 40 | 1200 |
| APR | 0600 | 7 | 99 | 40 | 14 | 9 | 8 | 8 | 9 | 12 | 29 |  |  |  |  | 81 | 59 | 37 | 15 | 12 | 32 | 54 | 77 |  |  |  | 83 | 1800 |
| + | 0700 | 19 | 89 |  | 42 | 26 | 20 | 19 | 20 | 26 | 41 | 88 |  |  |  |  | 69 | 48 | 29 | 19 | 29 | 48 | 68 | 89 |  |  | 71 | 1700 |
| AUG | 0800 | 30 | 79 |  | 71 | 46 | 35 | 31 | 31 | 35 | 47 | 72 |  |  |  |  | 80 | 61 | 44 | 32 | 32 | 44 | 62 | 81 |  |  | 60 | 1600 |
|  | 0900 | 41 | 67 |  |  | 67 | 51 | 44 | 41 | 43 | 51 | 66 | 90 |  |  |  |  | 74 | 58 | 46 | 41 | 46 | 58 | 73 | 90 | 49 | 1500 |
|  | 1000 | 51 | 51 |  |  | 85 | 69 | 58 | 52 | 51 | 55 | 63 | 77 |  |  |  |  | 86 | 72 | 61 | 53 | 51 | 57 | 67 | 80 | 39 | 1400 |
|  | 1100 | 59 | 29 |  |  |  | 86 | 73 | 64 | 60 | 59 | 62 | 69 | 81 |  |  |  |  | 87 | 75 | 66 | 60 | 59 | 63 | 71 | 82 | 31 | 1300 |
|  | 1200 | 62 | 0 |  |  |  | 90 | 78 | 69 | 63 | 62 | 63 | 69 | 78 |  |  |  |  | 90 | 80 | 70 | 64 | 62 | 64 | 70 | 80 | 28 | 1200 |
| MAY | 0500 | 2 | 115 | 5 | 3 | 2 | 2 | 2 | 3 | 6 |  |  |  |  | 65 | 43 | 20 | 3 | 25 | 47 | 70 |  |  |  |  | 88 | 1900 |
| + | 0600 | 13 | 106 | 40 | 20 | 15 | 13 | 13 | 16 | 25 | 62 |  |  |  |  | 75 | 53 | 32 | 14 | 20 | 40 | 61 | 83 |  |  | 77 | 1800 |
| JUL | 0700 | 24 | 97 | 75 | 42 | 30 | 25 | 24 | 27 | 36 | 58 |  |  |  |  | 84 | 64 | 44 | 28 | 25 | 37 | 55 | 76 |  |  | 66 | 1700 |
|  | 0800 | 35 | 87 |  | 65 | 47 | 38 | 35 | 37 | 44 | 59 | 86 |  |  |  |  | 74 | 57 | 43 | 36 | 40 | 53 | 70 | 88 |  |  | 55 | 1600 |
|  | 0900 | 47 | 76 |  | 82 | 64 | 53 | 48 | 47 | 51 | 61 | 77 |  |  |  |  | 84 | 69 | 57 | 48 | 47 | 54 | 66 | 80 |  |  | 43 | 1500 |
|  | 1000 | 57 | 61 |  |  | 80 | 68 | 61 | 58 | 58 | 63 | 73 | 86 |  |  |  |  | 82 | 70 | 62 | 58 | 59 | 65 | 75 | 86 | 33 | 1400 |
|  | 1100 | 66 | 37 |  |  |  | 84 | 75 | 69 | 66 | 67 | 71 | 77 | 87 |  |  |  |  | 84 | 76 | 70 | 66 | 67 | 71 | 78 | 87 | 24 | 1300 |
|  | 1200 | 70 | 0 |  |  |  | 90 | 82 | 76 | 71 | 70 | 71 | 76 | 82 |  |  |  |  | 90 | 82 | 76 | 72 | 70 | 72 | 76 | 82 | 20 | 1200 |
| JUN | 0500 | 4 | 117 | 9 | 6 | 4 | 4 | 5 | 7 | 14 |  |  |  |  | 63 | 40 | 18 | 6 | 28 | 50 | 72 |  |  |  |  | 86 | 1900 |
|  | 0600 | 15 | 108 | 40 | 22 | 16 | 15 | 16 | 19 | 31 | 75 |  |  |  |  | 72 | 51 | 30 | 15 | 23 | 43 | 64 | 86 |  |  | 75 | 1800 |
|  | 0700 | 26 | 100 | 71 | 42 | 31 | 27 | 26 | 30 | 40 | 66 |  |  |  |  | 81 | 61 | 43 | 29 | 28 | 40 | 59 | 79 |  |  | 64 | 1700 |
|  | 0800 | 37 | 91 | 89 | 63 | 47 | 39 | 37 | 40 | 48 | 64 |  |  |  |  | 89 | 72 | 55 | 42 | 37 | 43 | 56 | 73 |  |  | 53 | 1600 |
|  | 0900 | 49 | 80 |  | 79 | 63 | 54 | 49 | 50 | 54 | 65 | 82 |  |  |  |  | 82 | 68 | 56 | 50 | 50 | 57 | 69 | 84 |  |  | 41 | 1500 |
|  | 1000 | 60 | 66 |  |  | 78 | 68 | 62 | 60 | 61 | 67 | 77 | 89 |  |  |  |  | 80 | 70 | 63 | 60 | 62 | 69 | 78 | 89 | 30 | 1400 |
|  | 1100 | 69 | 42 |  |  |  | 83 | 76 | 71 | 69 | 70 | 74 | 81 | 89 |  |  |  |  | 83 | 76 | 71 | 69 | 70 | 75 | 81 | 89 | 21 | 1300 |
|  | 1200 | 73 | 0 |  |  |  | 90 | 84 | 78 | 75 | 73 | 75 | 78 | 84 |  |  |  |  | 90 | 84 | 78 | 75 | 73 | 75 | 78 | 84 | 17 | 1200 |
|  |  | N | NNW | NW | WNW | W | WSW | SW | SSW | S | SSE | SE | ESE | N | NNW | NW | WNW | W | WSW | SW | SSW | S | SSE | SE | ESE | HOR |  |

Dates vary year to year within plus or minus two days of the twenty-first day of the month.

## Table 7-14  Rates of Heat Gain from Occupants
*(Table 1, Chapter 18, 2013 ASHRAE Handbook—Fundamentals)*

| Degree of Activity | | Total Heat, Btu/h Adult Male | Adjusted, M/F[a] | Sensible Heat, Btu/h | Latent Heat, Btu/h | % Sensible Heat that is Radiant[b] Low $V$ | High $V$ |
|---|---|---|---|---|---|---|---|
| Seated at theater | Theater, matinee | 390 | 330 | 225 | 105 | | |
| Seated at theater, night | Theater, night | 390 | 350 | 245 | 105 | 60 | 27 |
| Seated, very light work | Offices, hotels, apartments | 450 | 400 | 245 | 155 | | |
| Moderately active office work | Offices, hotels, apartments | 475 | 450 | 250 | 200 | | |
| Standing, light work; walking | Department store; retail store | 550 | 450 | 250 | 200 | 58 | 38 |
| Walking, standing | Drug store, bank | 550 | 500 | 250 | 250 | | |
| Sedentary work | Restaurant[c] | 490 | 550 | 275 | 275 | | |
| Light bench work | Factory | 800 | 750 | 275 | 475 | | |
| Moderate dancing | Dance hall | 900 | 850 | 305 | 545 | 49 | 35 |
| Walking 3 mph; light machine work | Factory | 1000 | 1000 | 375 | 625 | | |
| Bowling[d] | Bowling alley | 1500 | 1450 | 580 | 870 | | |
| Heavy work | Factory | 1500 | 1450 | 580 | 870 | 54 | 19 |
| Heavy machine work; lifting | Factory | 1600 | 1600 | 635 | 965 | | |
| Athletics | Gymnasium | 2000 | 1800 | 710 | 1090 | | |

*Notes:*
1. Tabulated values are based on 75°F room dry-bulb temperature. For 80°F room dry bulb, the total heat remains the same, but the sensible heat values should be decreased by approximately 20%, and the latent heat values increased accordingly.
2. Also refer to Table 4, Chapter 8, for additional rates of metabolic heat generation.
3. All values are rounded to nearest 5 Btu/h.

[a] Adjusted heat gain is based on normal percentage of men, women, and children for the application listed, with the postulate that the gain from an adult female is 85% of that for an adult male, and that the gain from a child is 75% of that for an adult male.
[b] Values approximated from data in Table 6, Chapter 8, where is air velocity with limits shown in that table.
[c] Adjusted heat gain includes 60 Btu/h for food per individual (30 Btu/h sensible and 30 Btu/h latent).
[d] Figure one person per alley actually bowling, and all others as sitting (400 Btu/h) or standing or walking slowly (550 Btu/h).

unconditioned space; of the former category, the distribution between radiative and convective heat gain must be established.

Fisher and Chantrasrisalai (2006) experimentally studied 12 luminaire types and recommended five different categories of luminaires, as shown in Table 7-17. The table provides a range of design data for the conditioned space fraction, short-wave radiative fraction, and long-wave radiative fraction under typical operating conditions: airflow rate of 1 cfm/ft², supply air temperature between 59°F and 62°F, and room air temperature between 72°F and 75°F. The recommended fractions in Table 7-17 are based on lighting heat input rates range of 0.9 to 2.6 W/ft². For higher design power input, the lower bounds of the space and short-wave fractions should be used; for design power input below this range, the upper bounds of the space and short-wave fractions should be used. The **space fraction** in the table is the fraction of lighting heat gain that goes to the room; the fraction going to the plenum can be computed as 1 − the space fraction. The **radiative fraction** is the radiative part of the lighting heat gain that goes to the room. The convective fraction of the lighting heat gain that goes to the room is 1 − the radiative fraction. Using values in the middle of the range yields sufficiently accurate results. However, values that better suit a specific situation may be determined according to the notes for Table 7-17.

Table 7-17's data apply to both ducted and nonducted returns. However, application of the data, particularly the ceiling plenum fraction, may vary for different return configurations. For instance, for a room with a ducted return, although a portion of the lighting energy initially dissipated to the ceiling plenum is quantitatively equal to the plenum fraction, a large portion of this energy would likely end up as the conditioned space cooling load and a small portion would end up as the cooling load to the return air.

If the space airflow rate is different from the typical condition (i.e., about 1 cfm/ft²), Figure 7-5 can be used to estimate the lighting heat gain parameters. Design data shown in Figure 7-5 are only applicable for the recessed fluorescent luminaire without lens.

Although design data presented in Table 7-17 and Figure 7-5 can be used for a vented luminaire with side-slot returns, they are likely not applicable for a vented luminaire with lamp compartment returns, because in the latter case, all heat convected in the vented luminaire is likely to go directly to the ceiling plenum, resulting in zero convective fraction and a much lower space fraction. Therefore, the design data should

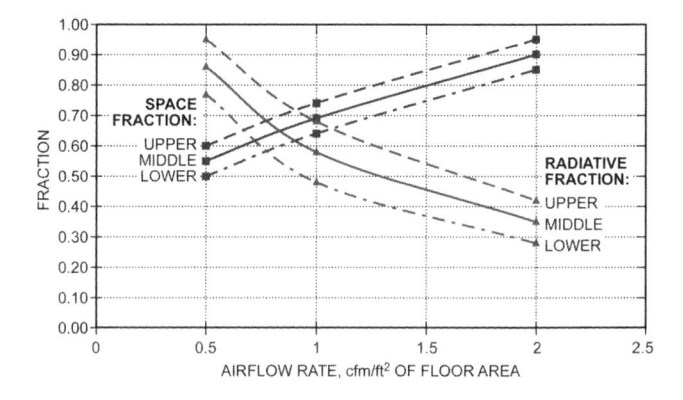

*Fig. 7-5 Lighting Heat Gain Parameters for Recessed Fluorescent Luminaire Without Lens*
*(Figure 3, Chapter 18, 2013 ASHRAE Handbook—Fundamentals)*

# Nonresidential Cooling and Heating Load Calculations

only be used for a configuration where conditioned air is returned through the ceiling grille or luminaire side slots.

For other luminaire types, it may be necessary to estimate the heat gain for each component as a fraction of the total lighting heat gain by using judgment to estimate heat-to-space and heat-to-return percentages.

Because of the directional nature of downlight luminaires, a large portion of the short-wave radiation typically falls on the floor. When converting heat gains to cooling loads in the RTS method, the solar **radiant time factors (RTFs)** may be more appropriate than nonsolar RTFs. (Solar RTFs are calculated assuming most solar radiation is intercepted by the floor; nonsolar RTFs assume uniform distribution by area over all interior surfaces.) This effect may be significant for rooms where lighting heat gain is high and for which solar RTFs are significantly different from nonsolar RTFs.

For ventilated or recessed fixtures, manufacturers' or other data must be sought to establish the fraction of the total wattage that may be expected to enter the conditioned space directly (and subject to time lag effect) versus that which must be picked up by return air or in some other appropriate manner.

**Light Heat Components.** Cooling load caused by lights recessed into ceiling cavities is made up of two components: one part (known as the **heat-to-space** load) comes from the light heat directly contributing to the space heat gain, and the other is the light heat released into the above-ceiling cavity, which (if used as a return air plenum) is mostly picked up by the return air that passes over or through the light fixtures. In such a ceiling return air plenum, this second part of the load (sometimes referred to as **heat-to-return**) never enters the conditioned space. It does, however, add to the overall load and significantly influences the load calculation.

Even though the total cooling load imposed on the cooling coil from these two components remains the same, the larger the fraction of heat output picked up by the return air, the more the space cooling load is reduced. The minimum required airflow rate for the conditioned space decreases as the space cooling load decreases. Supply fan power decreases accordingly, which results in reduced energy consumption for the system and possibly reduced equipment size as well.

For ordinary design load estimation, the heat gain for each component may be calculated simply as a fraction of the total lighting load by using judgment to estimate heat-to-space and heat-to-return percentages (Mitalas and Kimura 1971).

**Return Air Light Fixtures.** Two generic types of return air light fixture are available—those that allow and those that do not allow return air to flow through the lamp chamber. The first type is sometimes called a heat-of-light fixture. The percentage of light heat released through the plenum side of various ventilated fixtures can be obtained from lighting fixture manufacturers. For representative data, see Nevins et al. (1971). Even unventilated fixtures lose some heat to plenum spaces; however, most of the heat ultimately enters the conditioned space from a dead-air plenum or is picked up by return air via ceiling return air openings. The percentage of heat to

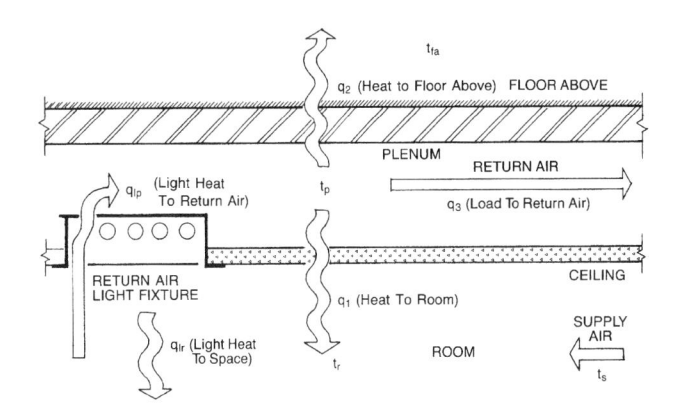

*Fig. 7-6 Heat Balance of Typical Ceiling Return Plenum*
*(Figure 15, Chapter 18, 2013 ASHRAE Handbook—Fundamentals)*

return air ranges from 40% to 60% for heat-to-return ventilated fixtures or 15% to 25% for unventilated fixtures.

**Plenum Temperatures.** As heat from lighting is picked up by the return air, the temperature differential between the ceiling space and the conditioned space causes part of that heat to flow from the ceiling back to the conditioned space. Return air from the conditioned space can be ducted to capture light heat without passing through a ceiling plenum as such, or the ceiling space can be used as a return air plenum, causing the distribution of light heat to be handled in distinctly different ways. Most plenum temperatures do not rise more than 1 to 3°F above space temperature, thus generating only a relatively small thermal gradient for heat transfer through plenum surfaces but a relatively large percentage reduction in space cooling load.

**Energy Balance.** Where the ceiling space is used as a return air plenum, an energy balance requires that the heat picked up from the lights into the return air do one or more of the following:

1. Become a part of the cooling load to the return air (represented by a temperature rise of the return air as it passes through the ceiling space)
2. Be partially transferred back into the conditioned space through the ceiling material below
3. Be partially lost (from the space) through the floor surfaces above the plenum

In a multistory building, the conditioned space frequently gains heat through its floor from a similar plenum below, offsetting the loss just mentioned. The radiant component of heat leaving the ceiling or floor surface of a plenum is normally so small that all such heat transfer is considered convective in calculations.

Figure 7-6 shows a schematic diagram of a typical return air plenum. Equations (7-21) through (7-25), using the sign convention as shown in Figure 7-5, represent the heat balance of a return air plenum design for a typical interior room in a multifloor building:

$$q_1 = U_c A_c (t_p - t_r) \qquad (7\text{-}21)$$

Table 7-15  Typical Nonincandescent Light Fixtures

| Description | Ballast | Watts/Lamp | Lamps/Fixture | Lamp Watts | Fixture Watts | Special Allowance Factor | Description | Ballast | Watts/Lamp | Lamps/Fixture | Lamp Watts | Fixture Watts | Special Allowance Factor |
|---|---|---|---|---|---|---|---|---|---|---|---|---|---|
| **Compact Fluorescent Fixtures** | | | | | | | | | | | | | |
| Twin, (1) 5 W lamp | Mag-Std | 5 | 1 | 5 | 9 | 1.80 | Twin, (2) 40 W lamp | Mag-Std | 40 | 2 | 80 | 85 | 1.06 |
| Twin, (1) 7 W lamp | Mag-Std | 7 | 1 | 7 | 10 | 1.43 | Quad, (1) 13 W lamp | Electronic | 13 | 1 | 13 | 15 | 1.15 |
| Twin, (1) 9 W lamp | Mag-Std | 9 | 1 | 9 | 11 | 1.22 | Quad, (1) 26 W lamp | Electronic | 26 | 1 | 26 | 27 | 1.04 |
| Quad, (1) 13 W lamp | Mag-Std | 13 | 1 | 13 | 17 | 1.31 | Quad, (2) 18 W lamp | Electronic | 18 | 2 | 36 | 38 | 1.06 |
| Quad, (2) 18 W lamp | Mag-Std | 18 | 2 | 36 | 45 | 1.25 | Quad, (2) 26 W lamp | Electronic | 26 | 2 | 52 | 50 | 0.96 |
| Quad, (2) 22 W lamp | Mag-Std | 22 | 2 | 44 | 48 | 1.09 | Twin or multi, (2) 32 W lamp | Electronic | 32 | 2 | 64 | 62 | 0.97 |
| Quad, (2) 26 W lamp | Mag-Std | 26 | 2 | 52 | 66 | 1.27 | | | | | | | |
| **Fluorescent Fixtures** | | | | | | | | | | | | | |
| (1) 18 in., T8 lamp | Mag-Std | 15 | 1 | 15 | 19 | 1.27 | (4) 48 in., T8 lamp | Electronic | 32 | 4 | 128 | 120 | 0.94 |
| (1) 18 in., T12 lamp | Mag-Std | 15 | 1 | 15 | 19 | 1.27 | (1) 60 in., T12 lamp | Mag-Std | 50 | 1 | 50 | 63 | 1.26 |
| (2) 18 in., T8 lamp | Mag-Std | 15 | 2 | 30 | 36 | 1.20 | (2) 60 in., T12 lamp | Mag-Std | 50 | 2 | 100 | 128 | 1.28 |
| (2) 18 in., T12 lamp | Mag-Std | 15 | 2 | 30 | 36 | 1.20 | (1) 60 in., T12 HO lamp | Mag-Std | 75 | 1 | 75 | 92 | 1.23 |
| (1) 24 in., T8 lamp | Mag-Std | 17 | 1 | 17 | 24 | 1.41 | (2) 60 in., T12 HO lamp | Mag-Std | 75 | 2 | 150 | 168 | 1.12 |
| (1) 24 in., T12 lamp | Mag-Std | 20 | 1 | 20 | 28 | 1.40 | (1) 60 in., T12 ES VHO lamp | Mag-Std | 135 | 1 | 135 | 165 | 1.22 |
| (2) 24 in., T12 lamp | Mag-Std | 20 | 2 | 40 | 56 | 1.40 | (2) 60 in., T12 ES VHO lamp | Mag-Std | 135 | 2 | 270 | 310 | 1.15 |
| (1) 24 in., T12 HO lamp | Mag-Std | 35 | 1 | 35 | 62 | 1.77 | (1) 60 in., T12 HO lamp | Mag-ES | 75 | 1 | 75 | 88 | 1.17 |
| (2) 24 in., T12 HO lamp | Mag-Std | 35 | 2 | 70 | 90 | 1.29 | (2) 60 in., T12 HO lamp | Mag-ES | 75 | 2 | 150 | 176 | 1.17 |
| (1) 24 in., T8 lamp | Electronic | 17 | 1 | 17 | 16 | 0.94 | (1) 60 in., T12 lamp | Electronic | 50 | 1 | 50 | 44 | 0.88 |
| (2) 24 in., T8 lamp | Electronic | 17 | 2 | 34 | 31 | 0.91 | (2) 60 in., T12 lamp | Electronic | 50 | 2 | 100 | 88 | 0.88 |
| (1) 36 in., T12 lamp | Mag-Std | 30 | 1 | 30 | 46 | 1.53 | (1) 60 in., T12 HO lamp | Electronic | 75 | 1 | 75 | 69 | 0.92 |
| (2) 36 in., T12 lamp | Mag-Std | 30 | 2 | 60 | 81 | 1.35 | (2) 60 in., T12 HO lamp | Electronic | 75 | 2 | 150 | 138 | 0.92 |
| (1) 36 in., T12 ES lamp | Mag-Std | 25 | 1 | 25 | 42 | 1.68 | (1) 60 in., T8 lamp | Electronic | 40 | 1 | 40 | 36 | 0.90 |
| (2) 36 in., T12 ES lamp | Mag-Std | 25 | 2 | 50 | 73 | 1.46 | (2) 60 in., T8 lamp | Electronic | 40 | 2 | 80 | 72 | 0.90 |
| (1) 36 in., T12 HO lamp | Mag-Std | 50 | 1 | 50 | 70 | 1.40 | (3) 60 in., T8 lamp | Electronic | 40 | 3 | 120 | 106 | 0.88 |
| (2) 36 in., T12 HO lamp | Mag-Std | 50 | 2 | 100 | 114 | 1.14 | (4) 60 in., T8 lamp | Electronic | 40 | 4 | 160 | 134 | 0.84 |
| (2) 36 in., T12 lamp | Mag-ES | 30 | 2 | 60 | 74 | 1.23 | (1) 72 in., T12 lamp | Mag-Std | 55 | 1 | 55 | 76 | 1.38 |
| (2) 36 in., T12 ES lamp | Mag-ES | 25 | 2 | 50 | 66 | 1.32 | (2) 72 in., T12 lamp | Mag-Std | 55 | 2 | 110 | 122 | 1.11 |
| (1) 36 in., T12 lamp | Electronic | 30 | 1 | 30 | 31 | 1.03 | (3) 72 in., T12 lamp | Mag-Std | 55 | 3 | 165 | 202 | 1.22 |
| (1) 36 in., T12 ES lamp | Electronic | 25 | 1 | 25 | 26 | 1.04 | (4) 72 in., T12 lamp | Mag-Std | 55 | 4 | 220 | 244 | 1.11 |
| (1) 36 in., T8 lamp | Electronic | 25 | 1 | 25 | 24 | 0.96 | (1) 72 in., T12 HO lamp | Mag-Std | 85 | 1 | 85 | 120 | 1.41 |
| (2) 36 in., T12 lamp | Electronic | 30 | 2 | 60 | 58 | 0.97 | (2) 72 in., T12 HO lamp | Mag-Std | 85 | 2 | 170 | 220 | 1.29 |
| (2) 36 in., T12 ES lamp | Electronic | 25 | 2 | 50 | 50 | 1.00 | (1) 72 in., T12 VHO lamp | Mag-Std | 160 | 1 | 160 | 180 | 1.13 |
| (2) 36 in., T8 lamp | Electronic | 25 | 2 | 50 | 46 | 0.92 | (2) 72 in., T12 VHO lamp | Mag-Std | 160 | 2 | 320 | 330 | 1.03 |
| (2) 36 in., T8 HO lamp | Electronic | 25 | 2 | 50 | 50 | 1.00 | (2) 72 in., T12 lamp | Mag-ES | 55 | 2 | 110 | 122 | 1.11 |
| (2) 36 in., T8 VHO lamp | Electronic | 25 | 2 | 50 | 70 | 1.40 | (4) 72 in., T12 lamp | Mag-ES | 55 | 4 | 220 | 244 | 1.11 |
| (1) 48 in., T12 lamp | Mag-Std | 40 | 1 | 40 | 55 | 1.38 | (2) 72 in., T12 HO lamp | Mag-ES | 85 | 2 | 170 | 194 | 1.14 |
| (2) 48 in., T12 lamp | Mag-Std | 40 | 2 | 80 | 92 | 1.15 | (4) 72 in., T12 HO lamp | Mag-ES | 85 | 4 | 340 | 388 | 1.14 |
| (3) 48 in., T12 lamp | Mag-Std | 40 | 3 | 120 | 140 | 1.17 | (1) 72 in., T12 lamp | Electronic | 55 | 1 | 55 | 68 | 1.24 |
| (4) 48 in., T12 lamp | Mag-Std | 40 | 4 | 160 | 184 | 1.15 | (2) 72 in., T12 lamp | Electronic | 55 | 2 | 110 | 108 | 0.98 |
| (1) 48 in., T12 ES lamp | Mag-Std | 34 | 1 | 34 | 48 | 1.41 | (3) 72 in., T12 lamp | Electronic | 55 | 3 | 165 | 176 | 1.07 |
| (2) 48 in., T12 ES lamp | Mag-Std | 34 | 2 | 68 | 82 | 1.21 | (4) 72 in., T12 lamp | Electronic | 55 | 4 | 220 | 216 | 0.98 |
| (3) 48 in., T12 ES lamp | Mag-Std | 34 | 3 | 102 | 100 | 0.98 | (1) 96 in., T12 ES lamp | Mag-Std | 60 | 1 | 60 | 75 | 1.25 |
| (4) 48 in., T12 ES lamp | Mag-Std | 34 | 4 | 136 | 164 | 1.21 | (2) 96 in., T12 ES lamp | Mag-Std | 60 | 2 | 120 | 128 | 1.07 |
| (1) 48 in., T12 ES lamp | Mag-ES | 34 | 1 | 34 | 43 | 1.26 | (3) 96 in., T12 ES lamp | Mag-Std | 60 | 3 | 180 | 203 | 1.13 |
| (2) 48 in., T12 ES lamp | Mag-ES | 34 | 2 | 68 | 72 | 1.06 | (4) 96 in., T12 ES lamp | Mag-Std | 60 | 4 | 240 | 256 | 1.07 |
| (3) 48 in., T12 ES lamp | Mag-ES | 34 | 3 | 102 | 115 | 1.13 | (1) 96 in., T12 ES HO lamp | Mag-Std | 95 | 1 | 95 | 112 | 1.18 |
| (4) 48 in., T12 ES lamp | Mag-ES | 34 | 4 | 136 | 144 | 1.06 | (2) 96 in., T12 ES HO lamp | Mag-Std | 95 | 2 | 190 | 227 | 1.19 |
| (1) 48 in., T8 lamp | Mag-ES | 32 | 1 | 32 | 35 | 1.09 | (3) 96 in., T12 ES HO lamp | Mag-Std | 95 | 3 | 285 | 380 | 1.33 |
| (2) 48 in., T8 lamp | Mag-ES | 32 | 2 | 64 | 71 | 1.11 | (4) 96 in., T12 ES HO lamp | Mag-Std | 95 | 4 | 380 | 454 | 1.19 |
| (3) 48 in., T8 lamp | Mag-ES | 32 | 3 | 96 | 110 | 1.15 | (1) 96 in., T12 ES VHO lamp | Mag-Std | 185 | 1 | 185 | 205 | 1.11 |
| (4) 48 in., T8 lamp | Mag-ES | 32 | 4 | 128 | 142 | 1.11 | (2) 96 in., T12 ES VHO lamp | Mag-Std | 185 | 2 | 370 | 380 | 1.03 |
| (1) 48 in., T12 ES lamp | Electronic | 34 | 1 | 34 | 32 | 0.94 | (3) 96 in., T12 ES VHO lamp | Mag-Std | 185 | 3 | 555 | 585 | 1.05 |
| (2) 48 in., T12 ES lamp | Electronic | 34 | 2 | 68 | 60 | 0.88 | (4) 96 in., T12 ES VHO lamp | Mag-Std | 185 | 4 | 740 | 760 | 1.03 |
| (3) 48 in., T12 ES lamp | Electronic | 34 | 3 | 102 | 92 | 0.90 | (2) 96 in., T12 ES lamp | Mag-ES | 60 | 2 | 120 | 123 | 1.03 |
| (4) 48 in., T12 ES lamp | Electronic | 34 | 4 | 136 | 120 | 0.88 | (3) 96 in., T12 ES lamp | Mag-ES | 60 | 3 | 180 | 210 | 1.17 |
| (1) 48 in., T8 lamp | Electronic | 32 | 1 | 32 | 32 | 1.00 | (4) 96 in., T12 ES lamp | Mag-ES | 60 | 4 | 240 | 246 | 1.03 |
| (2) 48 in., T8 lamp | Electronic | 32 | 2 | 64 | 60 | 0.94 | (2) 96 in., T12 ES HO lamp | Mag-ES | 95 | 2 | 190 | 207 | 1.09 |
| (3) 48 in., T8 lamp | Electronic | 32 | 3 | 96 | 93 | 0.97 | (4) 96 in., T12 ES HO lamp | Mag-ES | 95 | 4 | 380 | 414 | 1.09 |

## Table 7-15 Typical Nonincandescent Light Fixtures (*Concluded*)

| Description | Ballast | Watts/Lamp | Lamps/Fixture | Lamp Watts | Fixture Watts | Special Allowance Factor | Description | Ballast | Watts/Lamp | Lamps/Fixture | Lamp Watts | Fixture Watts | Special Allowance Factor |
|---|---|---|---|---|---|---|---|---|---|---|---|---|---|
| (1) 96 in., T12 ES lamp | Electronic | 60 | 1 | 60 | 69 | 1.15 | (1) 96 in., T8 HO lamp | Electronic | 59 | 1 | 59 | 68 | 1.15 |
| (2) 96 in., T12 ES lamp | Electronic | 60 | 2 | 120 | 110 | 0.92 | (1) 96 in., T8 VHO lamp | Electronic | 59 | 1 | 59 | 71 | 1.20 |
| (3) 96 in., T12 ES lamp | Electronic | 60 | 3 | 180 | 179 | 0.99 | (2) 96 in., T8 lamp | Electronic | 59 | 2 | 118 | 109 | 0.92 |
| (4) 96 in., T12 ES lamp | Electronic | 60 | 4 | 240 | 220 | 0.92 | (3) 96 in., T8 lamp | Electronic | 59 | 3 | 177 | 167 | 0.94 |
| (1) 96 in., T12 ES HO lamp | Electronic | 95 | 1 | 95 | 80 | 0.84 | (4) 96 in., T8 lamp | Electronic | 59 | 4 | 236 | 219 | 0.93 |
| (2) 96 in., T12 ES HO lamp | Electronic | 95 | 2 | 190 | 173 | 0.91 | (2) 96 in., T8 HO lamp | Electronic | 86 | 2 | 172 | 160 | 0.93 |
| (4) 96 in., T12 ES HO lamp | Electronic | 95 | 4 | 380 | 346 | 0.91 | (4) 96 in., T8 HO lamp | Electronic | 86 | 4 | 344 | 320 | 0.93 |
| (1) 96 in., T8 lamp | Electronic | 59 | 1 | 59 | 58 | 0.98 | | | | | | | |
| **Circular Fluorescent Fixtures** | | | | | | | | | | | | | |
| Circlite, (1) 20 W lamp | Mag-PH | 20 | 1 | 20 | 20 | 1.00 | (2) 8 in. circular lamp | Mag-RS | 22 | 2 | 44 | 52 | 1.18 |
| Circlite, (1) 22 W lamp | Mag-PH | 22 | 1 | 22 | 20 | 0.91 | (1) 12 in. circular lamp | Mag-RS | 32 | 1 | 32 | 31 | 0.97 |
| Circline, (1) 32 W lamp | Mag-PH | 32 | 1 | 32 | 40 | 1.25 | (2) 12 in. circular lamp | Mag-RS | 32 | 2 | 64 | 62 | 0.97 |
| (1) 6 in. circular lamp | Mag-RS | 20 | 1 | 20 | 25 | 1.25 | (1) 16 in. circular lamp | Mag-Std | 40 | 1 | 40 | 35 | 0.88 |
| (1) 8 in. circular lamp | Mag-RS | 22 | 1 | 22 | 26 | 1.18 | | | | | | | |
| **High-Pressure Sodium Fixtures** | | | | | | | | | | | | | |
| (1) 35 W lamp | HID | 35 | 1 | 35 | 46 | 1.31 | (1) 250 W lamp | HID | 250 | 1 | 250 | 295 | 1.18 |
| (1) 50 W lamp | HID | 50 | 1 | 50 | 66 | 1.32 | (1) 310 W lamp | HID | 310 | 1 | 310 | 365 | 1.18 |
| (1) 70 W lamp | HID | 70 | 1 | 70 | 95 | 1.36 | (1) 360 W lamp | HID | 360 | 1 | 360 | 414 | 1.15 |
| (1) 100 W lamp | HID | 100 | 1 | 100 | 138 | 1.38 | (1) 400 W lamp | HID | 400 | 1 | 400 | 465 | 1.16 |
| (1) 150 W lamp | HID | 150 | 1 | 150 | 188 | 1.25 | (1) 1000 W lamp | HID | 1000 | 1 | 1000 | 1100 | 1.10 |
| (1) 200 W lamp | HID | 200 | 1 | 200 | 250 | 1.25 | | | | | | | |
| **Metal Halide Fixtures** | | | | | | | | | | | | | |
| (1) 32 W lamp | HID | 32 | 1 | 32 | 43 | 1.34 | (1) 250 W lamp | HID | 250 | 1 | 250 | 295 | 1.18 |
| (1) 50 W lamp | HID | 50 | 1 | 50 | 72 | 1.44 | (1) 400 W lamp | HID | 400 | 1 | 400 | 458 | 1.15 |
| (1) 70 W lamp | HID | 70 | 1 | 70 | 95 | 1.36 | (2) 400 W lamp | HID | 400 | 2 | 800 | 916 | 1.15 |
| (1) 100 W lamp | HID | 100 | 1 | 100 | 128 | 1.28 | (1) 750 W lamp | HID | 750 | 1 | 750 | 850 | 1.13 |
| (1) 150 W lamp | HID | 150 | 1 | 150 | 190 | 1.27 | (1) 1000 W lamp | HID | 1000 | 1 | 1000 | 1080 | 1.08 |
| (1) 175 W lamp | HID | 175 | 1 | 175 | 215 | 1.23 | (1) 1500 W lamp | HID | 1500 | 1 | 1500 | 1610 | 1.07 |
| **Mercury Vapor Fixtures** | | | | | | | | | | | | | |
| (1) 40 W lamp | HID | 40 | 1 | 40 | 50 | 1.25 | (1) 250 W lamp | HID | 250 | 1 | 250 | 290 | 1.16 |
| (1) 50 W lamp | HID | 50 | 1 | 50 | 74 | 1.48 | (1) 400 W lamp | HID | 400 | 1 | 400 | 455 | 1.14 |
| (1) 75 W lamp | HID | 75 | 1 | 75 | 93 | 1.24 | (2) 400 W lamp | HID | 400 | 2 | 800 | 910 | 1.14 |
| (1) 100 W lamp | HID | 100 | 1 | 100 | 125 | 1.25 | (1) 700 W lamp | HID | 700 | 1 | 700 | 780 | 1.11 |
| (1) 175 W lamp | HID | 175 | 1 | 175 | 205 | 1.17 | (1) 1000 W lamp | HID | 1000 | 1 | 1000 | 1075 | 1.08 |

Abbreviations: Mag = electromagnetic; ES = energy saver; Std = standard; HID = high-intensity discharge; HO = high output; VHO = very high output; PH = preheat; RS = rapid start

$$q_2 = U_f A_f (t_p - t_{fa}) \quad (7\text{-}22)$$

$$q_3 = 1.1 Q(t_p - t_r) \quad (7\text{-}23)$$

$$q_{lp} - q_2 - q_1 - q_3 = 0 \quad (7\text{-}24)$$

$$Q = \frac{q_r + q_1}{1.1(t_r - t_s)} \quad (7\text{-}25)$$

where

$q_1$ = heat gain to space from plenum through ceiling, Btu/h

$q_2$ = heat loss from plenum through floor above, Btu/h

$q_3$ = heat gain "pickup" by return air, Btu/h

$Q$ = return airflow, cfm

$q_{lp}$ = light heat gain to plenum via return air, Btu/h

$q_{lr}$ = light heat gain to space, Btu/h

$q_f$ = heat gain from plenum below, through floor, Btu/h

$q_w$ = heat gain from exterior wall, Btu/h

$q_r$ = space cooling load, Btu/h, including appropriate treatment of $q_{lr}$, $q_f$, and/or $q_w$

$t_p$ = plenum temperature, °F

$t_r$ = space temperature, °F

$t_{fa}$ = space temperature of floor above, °F

$t_s$ = supply air temperature, °F

By substituting Equations (7-21), (7-22), (7-23), and (7-25) into the heat balance equation (7-24), $t_p$ can be found as the resultant return air temperature or plenum temperature. The results, although rigorous and best solved by computer, are important in determining the cooling load, which affects equipment size selection, future energy consumption, and other factors.

Equations (7-21) through (7-25) are simplified to illustrate the heat balance relationship. Heat gain into a return air plenum is not limited to the heat of lights alone. Exterior walls directly exposed to the ceiling space will transfer heat directly

**Table 7-16 Lighting Power Densities Using Space-by-Space Method**
*(Table 2, Chapter 18, 2013 ASHRAE Handbook—Fundamentals)*

| Common Space Types* | LPD, W/ft² | Building-Specific Space Types* | LPD, W/ft² | Building-Specific Space Types* | LPD, W/ft² |
|---|---|---|---|---|---|
| Atrium | | Automotive | | Library | |
|   First 40 ft in height | 0.03 per ft (height) |   Service/repair | 0.67 |   Card file and cataloging | 0.72 |
| | | Bank/office | |   Reading area | 0.93 |
|   Height above 40 ft | 0.02 per ft (height) |   Banking activity area | 1.38 |   Stacks | 1.71 |
| | | Convention center | | Manufacturing | |
| Audience/seating area—permanent | |   Audience seating | 0.82 |   Corridor/transition | 0.41 |
|   For auditorium | 0.79 |   Exhibit space | 1.45 |   Detailed manufacturing | 1.29 |
|   For performing arts theater | 2.43 | Courthouse/police station/penitentiary | |   Equipment room | 0.95 |
|   For motion picture theater | 1.14 |   Courtroom | 1.72 |   Extra high bay (>50 ft floor-to-ceiling height) | 1.05 |
| | |   Confinement cells | 1.10 | | |
| Classroom/lecture/training | 1.24 |   Judges' chambers | 1.17 |   High bay (25 to 50 ft floor-to-ceiling height) | 1.23 |
| Conference/meeting/multipurpose | 1.23 |   Penitentiary audience seating | 0.43 | | |
| Corridor/transition | 0.66 |   Penitentiary classroom | 1.34 |   Low bay (<25 ft floor-to-ceiling height) | 1.19 |
| | |   Penitentiary dining | 1.07 | | |
| Dining area | 0.65 | Dormitory | | Museum | |
|   For bar lounge/leisure dining | 1.31 |   Living quarters | 0.38 |   General exhibition | 1.05 |
|   For family dining | 0.89 | Fire stations | |   Restoration | 1.02 |
| Dressing/fitting room for performing arts theater | 0.40 |   Engine room | 0.56 | Parking garage | |
| | |   Sleeping quarters | 0.25 |   Garage area | 0.19 |
| | | Gymnasium/fitness center | | Post office | |
| Electrical/mechanical | 0.95 |   Fitness area | 0.72 |   Sorting area | 0.94 |
| Food preparation | 0.99 |   Gymnasium audience seating | 0.43 | Religious buildings | |
| | |   Playing area | 1.20 |   Audience seating | 1.53 |
| Laboratory | | Hospital | |   Fellowship hall | 0.64 |
|   For classrooms | 1.28 |   Corridor/transition | 0.89 |   Worship pulpit, choir | 1.53 |
|   For medical/industrial/research | 1.81 |   Emergency | 2.26 | Retail | |
| | |   Exam/treatment | 1.66 |   Dressing/fitting room | 0.87 |
| Lobby | 0.90 |   Laundry/washing | 0.60 |   Mall concourse | 1.10 |
|   For elevator | 0.64 |   Lounge/recreation | 1.07 |   Sales area | 1.68 |
|   For performing arts theater | 2.00 |   Medical supply | 1.27 | Sports arena | |
|   For motion picture theater | 0.52 |   Nursery | 0.88 |   Audience seating | 0.43 |
| | |   Nurses' station | 0.87 |   Court sports arena—class 4 | 0.72 |
| Locker room | 0.75 |   Operating room | 1.89 |   Court sports arena—class 3 | 1.20 |
| Lounge/recreation | 0.73 |   Patient room | 0.62 |   Court sports arena—class 2 | 1.92 |
| | |   Pharmacy | 1.14 |   Court sports arena—class 1 | 3.01 |
| Office | |   Physical therapy | 0.91 |   Ring sports arena | 2.68 |
|   Enclosed | 1.11 |   Radiology/imaging | 1.32 | Transportation | |
|   Open plan | 0.98 |   Recovery | 1.15 |   Air/train/bus—baggage area | 0.76 |
| | | Hotel/highway lodging | |   Airport—concourse | 0.36 |
| Restrooms | 0.98 |   Hotel dining | 0.82 |   Waiting area | 0.54 |
| Sales area | 1.68 |   Hotel guest rooms | 1.11 |   Terminal—ticket counter | 1.08 |
| Stairway | 0.69 |   Hotel lobby | 1.06 | Warehouse | |
| Storage | 0.63 |   Highway lodging dining | 0.88 |   Fine material storage | 0.95 |
| Workshop | 1.59 |   Highway lodging guest rooms | 0.75 |   Medium/bulky material storage | 0.58 |

*Source*: ASHRAE Standard 90.1-2010.    *In cases where both a common space type and a building-specific type are listed, the building-specific space type applies.

to or from the return air. For single-story buildings or the top floor of a multistory building, the roof heat gain or loss enters or leaves the ceiling plenum rather than entering or leaving the conditioned space directly. The supply air quantity calculated by Equation 7-25 is for the conditioned space under consideration only, and it is assumed equal to the return air quantity.

The amount of airflow through a return plenum above a conditioned space may not be limited to that supplied into the space under consideration; it will, however, have no noticeable effect on plenum temperature if the surplus comes from an adjacent plenum operating under similar conditions. Where special conditions exist, heat balance Equations (7-21) through (7-25) must be modified appropriately. Finally, even though the building's thermal storage has some effect, the amount of heat entering the return air is small and may be considered as convective for calculation purposes.

**Power.** Instantaneous heat gain from equipment operated by electric motors within a conditioned space is calculated as follows:

$$q_{em} = 2545(P/E_M)F_{LM}F_{MU} \qquad (7\text{-}26)$$

where

$q_{em}$ = heat equivalent of equipment operation, Btu/h

$P$ = motor horsepower rating

$E_M$ = motor efficiency, as decimal fraction < 1.0

$F_{LM}$ = motor load factor, 1.0 or decimal fraction < 1.0

$F_{UM}$ = motor use factor, 1.0 or decimal fraction < 1.0

# Nonresidential Cooling and Heating Load Calculations

**Table 7-17 Lighting Heat Gain Parameters for Typical Operating Conditions**
*(Table 3, Chapter 18, 2013 ASHRAE Handbook—Fundamentals)*

| Luminaire Category | Space Fraction | Radiative Fraction | Notes |
|---|---|---|---|
| Recessed fluorescent luminaire without lens | 0.64 to 0.74 | 0.48 to 0.68 | • Use middle values in most situations<br>• May use higher space fraction, and lower radiative fraction for luminaire with side-slot returns<br>• May use lower values of both fractions for direct/indirect luminaire<br>• May use higher values of both fractions for ducted returns |
| Recessed fluorescent luminaire with lens | 0.40 to 0.50 | 0.61 to 0.73 | • May adjust values in the same way as for recessed fluorescent luminaire without lens |
| Downlight compact fluorescent luminaire | 0.12 to 0.24 | 0.95 to 1.0 | • Use middle or high values if detailed features are unknown<br>• Use low value for space fraction and high value for radiative fraction if there are large holes in luminaire's reflector |
| Downlight incandescent luminaire | 0.70 to 0.80 | 0.95 to 1.0 | • Use middle values if lamp type is unknown<br>• Use low value for space fraction if standard lamp (i.e. A-lamp) is used<br>• Use high value for space fraction if reflector lamp (i.e. BR-lamp) is used |
| Non-in-ceiling fluorescent luminaire | 1.0 | 0.5 to 0.57 | • Use lower value for radiative fraction for surface-mounted luminaire<br>• Use higher value for radiative fraction for pendant luminaire |

*Source:* Fisher and Chantrasrisalai (2006).

The motor use factor may be applied when motor use is known to be intermittent with significant nonuse during all hours of operation (i.e., overhead door operator, and so forth). For conventional applications, its value is 1.0.

The motor load factor is the fraction of the rated load delivered under the conditions of the cooling load estimate. In Equation 7-26, both the motor and the driven equipment are assumed to be within the conditioned space. If the motor is outside the space or airstream, Equation 7-27 is used:

$$q_{em} = 2545 P F_{LM} F_{UM} \qquad (7\text{-}27)$$

When the motor is inside the conditioned space or airstream but the driven machine is outside, Equation 7-28 is used:

$$q_{em} = 2545 P [(1 - E_M)/E_M] F_{LM} F_{UM} \qquad (7\text{-}28)$$

Equation 7-28 also applies to a fan or pump in the conditioned space that exhausts air or pumps fluid outside that space.

Average efficiencies, and related data representative of typical electric motors, generally derived from the lower efficiencies reported by several manufacturers of open, drip-proof motors, are given in Table 7-18. Unless the manufacturers' technical literature indicates otherwise, the heat gain may be divided equally between radiant and convective components for subsequent cooling load calculations.

Table 7-19 gives minimum efficiencies and related data representative of typical electric motors from ASHRAE Standard 90.1-2010. The actual value should be obtained from the manufacturer. If the motor is underloaded or overloaded, efficiency could vary from the manufacturer's listing.

**Appliances.** In a cooling load estimate, heat gain from all appliances—electrical, gas, or steam—should be considered. Food preparation equipment is among the most common types of heat-producing appliances found in conditioned areas. Appliance surfaces contribute most of the heat to commercial kitchens. When installed under an effective hood, cooling load is independent of the fuel or energy used for similar equipment performing the same operations. Because the heat is primarily radiant energy from the appliance surfaces and cooking utensils, convected and latent heat are negligible.

To establish a heat gain value, actual input values and various factors, efficiencies, or other judgmental modifiers are preferred. Where specific rating data are unavailable (nameplate missing, equipment not yet purchased, and so forth), recommended heat gains tabulated in this chapter may be used. In estimating the appliance load, probabilities of simultaneous use and operation for different appliances located in the same space must be considered.

Radiation contributes up to 32% of the hourly heat input to hooded appliances (for a conservative radiation factor $F_{RA}$ = 0.32). Radiant heat temperature rises can be substantially reduced by shielding the fronts of cooking appliances. These reductions amount to 61% with glass panels and 78% with polished aluminum shielding. A floor-slot air curtain in front of appliances reduces the radiant temperature rise by 15%.

For each meal served, the heat transferred to the dining space is approximately 50 Btu/h, 75% of which is sensible and 25% is latent.

The maximum hourly input can be estimated as 50% of the total nameplate or catalog input $q_i$ ratings because of the diversity of appliance use and the effect of thermostatic controls, giving a usage factor $F_{UA}$ = 0.50. Therefore, the maximum hourly heat gain $q_m$ for generic types of **electric and steam appliance**s installed under a hood can be estimated from the following equations:

$$q_a = q_i F_{UA} F_{RA} \qquad (7\text{-}29)$$

or

$$q_a = 0.16 q_i \qquad (7\text{-}30)$$

Direct fuel-fired cooking appliances require more heat input than electric or steam equipment of the same type and size. In the case of gas fuel, the American Gas Association established an overall figure of approximately 60% more. Where appliances are installed under an effective hood, only radiant heat

Table 7-18  Heat Gain from Typical Electric Motors

| Motor Nameplate or Rated Horsepower | Motor Type | Nominal rpm | Full Load Motor Efficiency, % | A Motor in, Driven Equipment in, Btu/h | B Motor out, Driven Equipment in, Btu/h | C Motor in, Driven Equipment out, Btu/h |
|---|---|---|---|---|---|---|
| 0.05 | Shaded pole | 1500 | 35 | 360 | 130 | 240 |
| 0.08 | Shaded pole | 1500 | 35 | 580 | 200 | 380 |
| 0.125 | Shaded pole | 1500 | 35 | 900 | 320 | 590 |
| 0.16 | Shaded pole | 1500 | 35 | 1160 | 400 | 760 |
| 0.25 | Split phase | 1750 | 54 | 1180 | 640 | 540 |
| 0.33 | Split phase | 1750 | 56 | 1500 | 840 | 660 |
| 0.50 | Split phase | 1750 | 60 | 2120 | 1270 | 850 |
| 0.75 | 3-Phase | 1750 | 72 | 2650 | 1900 | 740 |
| 1 | 3-Phase | 1750 | 75 | 3390 | 2550 | 850 |
| 1.5 | 3-Phase | 1750 | 77 | 4960 | 3820 | 1140 |
| 2 | 3-Phase | 1750 | 79 | 6440 | 5090 | 1350 |
| 3 | 3-Phase | 1750 | 81 | 9430 | 7640 | 1790 |
| 5 | 3-Phase | 1750 | 82 | 15,500 | 12,700 | 2790 |
| 7.5 | 3-Phase | 1750 | 84 | 22,700 | 19,100 | 3640 |
| 10 | 3-Phase | 1750 | 85 | 29,900 | 24,500 | 4490 |
| 15 | 3-Phase | 1750 | 86 | 44,400 | 38,200 | 6210 |
| 20 | 3-Phase | 1750 | 87 | 58,500 | 50,900 | 7610 |
| 25 | 3-Phase | 1750 | 88 | 72,300 | 63,600 | 8680 |
| 30 | 3-Phase | 1750 | 89 | 85,700 | 76,300 | 9440 |
| 40 | 3-Phase | 1750 | 89 | 114,000 | 102,000 | 12,600 |
| 50 | 3-Phase | 1750 | 89 | 143,000 | 127,000 | 15,700 |
| 60 | 3-Phase | 1750 | 89 | 172,000 | 153,000 | 18,900 |
| 75 | 3-Phase | 1750 | 90 | 212,000 | 191,000 | 21,200 |
| 100 | 3-Phase | 1750 | 90 | 283,000 | 255,000 | 28,300 |
| 125 | 3-Phase | 1750 | 90 | 353,000 | 318,000 | 35,300 |
| 150 | 3-Phase | 1750 | 91 | 420,000 | 382,000 | 37,800 |
| 200 | 3-Phase | 1750 | 91 | 569,000 | 509,000 | 50,300 |
| 250 | 3-Phase | 1750 | 91 | 699,000 | 636,000 | 62,900 |

Table 7-19  Minimum Nominal Full-Load Efficiency for 60 HZ NEMA General Purpose Electric Motors (Subtype I) Rated 600 Volts or Less (Random Wound)*
*(Table 4, Chapter 18, 2013 ASHRAE Handbook—Fundamentals)*

Minimum Nominal Full Load Efficiency (%) for Motors Manufactured on or after December 19, 2010

| | Open Drip-Proof Motors | | | Totally Enclosed Fan-Cooled Motors | | |
|---|---|---|---|---|---|---|
| Number of Poles ⇒ | 2 | 4 | 6 | 2 | 4 | 6 |
| Synchronous Speed (RPM) ⇒ | 3600 | 1800 | 1200 | 3600 | 1800 | 1200 |
| **Motor Horsepower** | | | | | | |
| 1 | 77.0 | 85.5 | 82.5 | 77.0 | 85.5 | 82.5 |
| 1.5 | 84.0 | 86.5 | 86.5 | 84.0 | 86.5 | 87.5 |
| 2 | 85.5 | 86.5 | 87.5 | 85.5 | 86.5 | 88.5 |
| 3 | 85.5 | 89.5 | 88.5 | 86.5 | 89.5 | 89.5 |
| 5 | 86.5 | 89.5 | 89.5 | 88.5 | 89.5 | 89.5 |
| 7.5 | 88.5 | 91.0 | 90.2 | 89.5 | 91.7 | 91.0 |
| 10 | 89.5 | 91.7 | 91.7 | 90.2 | 91.7 | 91.0 |
| 15 | 90.2 | 93.0 | 91.7 | 91.0 | 92.4 | 91.7 |
| 20 | 91.0 | 93.0 | 92.4 | 91.0 | 93.0 | 91.7 |
| 25 | 91.7 | 93.6 | 93.0 | 91.7 | 93.6 | 93.0 |
| 30 | 91.7 | 94.1 | 93.6 | 91.7 | 93.6 | 93.0 |
| 40 | 92.4 | 94.1 | 94.1 | 92.4 | 94.1 | 94.1 |
| 50 | 93.0 | 94.5 | 94.1 | 93.0 | 94.5 | 94.1 |
| 60 | 93.6 | 95.0 | 94.5 | 93.6 | 95.0 | 94.5 |
| 75 | 93.6 | 95.0 | 94.5 | 93.6 | 95.4 | 94.5 |
| 100 | 93.6 | 95.4 | 95.0 | 94.1 | 95.4 | 95.0 |
| 125 | 94.1 | 95.4 | 95.0 | 95.0 | 95.4 | 95.0 |
| 150 | 94.1 | 95.8 | 95.4 | 95.0 | 95.8 | 95.8 |
| 200 | 95.0 | 95.8 | 95.4 | 95.4 | 96.2 | 95.8 |
| 250 | 95.0 | 95.8 | 95.4 | 95.8 | 96.2 | 95.8 |
| 300 | 95.4 | 95.8 | 95.4 | 95.8 | 96.2 | 95.8 |
| 350 | 95.4 | 95.8 | 95.4 | 95.8 | 96.2 | 95.8 |
| 400 | 95.8 | 95.8 | 95.8 | 95.8 | 96.2 | 95.8 |
| 450 | 95.8 | 96.2 | 96.2 | 95.8 | 96.2 | 95.8 |
| 500 | 95.8 | 96.2 | 96.2 | 95.8 | 96.2 | 95.8 |

Source: ASHRAE Standard 90.1-2010
*Nominal efficiencies established in accordance with NEMA *Standard* MG1. Design A and Design B are National Electric Manufacturers Association (NEMA) design class designations for fixed-frequency small and medium AC squirrel-cage induction motors.

adds to the cooling load; convected and latent heat from the cooking process and combustion products are exhausted and do not enter the kitchen. To compensate for 60% higher input ratings, Equation 7-31 must be used with **fuel-fired appliances**, since the appliance surface temperatures are the same and the extra heat input combustion products are exhausted to outdoors. This correction is made by introducing a flue loss factor ($F_{FL}$) of 1.60 as follows:

$$q_a = (q_i F_{UA} F_{RA})/F_{FL} \quad (7\text{-}31)$$

or

$$q_a = 0.10 q_i \quad (7\text{-}32)$$

Factors for seven typical electrical and steam appliances are found in Table 7-20.

**Unhooded Equipment.** For all cooking appliances not installed under an exhaust hood or directly vent-connected and located in the conditioned area, the heat gain may be estimated as 50% ($F_U = 0.50$) or the rated hourly input, regardless of the type of energy or fuel used. On average, 34% of the heat may be assumed to be latent and the remaining 66% sensible. Note that cooking appliances ventilated by "ductless" hoods should

Table 7-20  Heat Gain Factors of Typical Appliances under Hoods

| Appliance | Usage Factor $F_U$ | Radiation Factor $F_R$ | Load Factor $F_L = F_U F_R$ Elec/Steam |
|---|---|---|---|
| Griddle | 0.16 | 0.45 | 0.07 |
| Fryer | 0.06 | 0.43 | 0.03 |
| Convection oven | 0.42 | 0.17 | 0.07 |
| Charbroiler | 0.83 | 0.29 | 0.24 |
| Open-top range without oven | 0.34 | 0.46 | 0.16 |
| Hot-top range without oven | 0.79 | 0.47 | 0.37 |
| with oven | 0.59 | 0.48 | 0.28 |
| Steam cooker | 0.13 | 0.30 | 0.04 |

be treated as unhooded appliances from the perspective of estimating heat gain. In other words, all energy consumed by the appliance and all moisture produced by the cooking process is introduced to the kitchen as a sensible or latent cooling load.

**Recommended Heat Gain Values.** As an alternative procedure, Table 7-21 lists recommended rates of heat gain from typical commercial cooking appliances. The data in the "with

# Nonresidential Cooling and Heating Load Calculations

**Table 7-21A  Recommended Rates of Radiant and Convective Heat Gain from Unhooded Electric Appliances During Idle (Ready-to-Cook) Conditions**

*(Table 5A, Chapter 18, 2013 ASHRAE Handbook—Fundamentals)*

| Appliance | Energy Rate, Btu/h Rated | Standby | Sensible Radiant | Sensible Convective | Latent | Total | Usage Factor $F_u$ | Radiation Factor $F_r$ |
|---|---|---|---|---|---|---|---|---|
| Cabinet: hot serving (large), insulated* | 6,800 | 1,200 | 400 | 800 | 0 | 1,200 | 0.18 | 0.33 |
| Cabinet: hot serving (large), uninsulated | 6,800 | 3,500 | 700 | 2,800 | 0 | 3,500 | 0.51 | 0.2 |
| Cabinet: proofing (large)* | 17,400 | 1,400 | 1,200 | 0 | 200 | 1,400 | 0.08 | 0.86 |
| Cabinet: proofing (small-15 shelf) | 14,300 | 3,900 | 0 | 900 | 3,000 | 3,900 | 0.27 | 0 |
| Coffee brewing urn | 13,000 | 1,200 | 200 | 300 | 700 | 1,200 | 0.09 | 0.17 |
| Drawer warmers, 2-drawer (moist holding)* | 4,100 | 500 | 0 | 0 | 200 | 200 | 0.12 | 0 |
| Egg cooker | 10,900 | 700 | 300 | 400 | 0 | 700 | 0.06 | 0.43 |
| Espresso machine* | 8,200 | 1,200 | 400 | 800 | 0 | 1,200 | 0.15 | 0.33 |
| Food warmer: steam table (2-well-type) | 5,100 | 3,500 | 300 | 600 | 2,600 | 3,500 | 0.69 | 0.09 |
| Freezer (small) | 2,700 | 1,100 | 500 | 600 | 0 | 1,100 | 0.41 | 0.45 |
| Hot dog roller* | 3,400 | 2,400 | 900 | 1,500 | 0 | 2,400 | 0.71 | 0.38 |
| Hot plate: single burner, high speed | 3,800 | 3,000 | 900 | 2,100 | 0 | 3,000 | 0.79 | 0.3 |
| Hot-food case (dry holding)* | 31,100 | 2,500 | 900 | 1,600 | 0 | 2,500 | 0.08 | 0.36 |
| Hot-food case (moist holding)* | 31,100 | 3,300 | 900 | 1,800 | 600 | 3,300 | 0.11 | 0.27 |
| Microwave oven: commercial (heavy duty) | 10,900 | 0 | 0 | 0 | 0 | 0 | 0 | 0 |
| Oven: countertop conveyorized bake/finishing* | 20,500 | 12,600 | 2,200 | 10,400 | 0 | 12,600 | 0.61 | 0.17 |
| Panini* | 5,800 | 3,200 | 1,200 | 2,000 | 0 | 3,200 | 0.55 | 0.38 |
| Popcorn popper* | 2,000 | 200 | 100 | 100 | 0 | 200 | 0.1 | 0.5 |
| Rapid-cook oven (quartz-halogen)* | 41,000 | 0 | 0 | 0 | 0 | 0 | 0 | 0 |
| Rapid-cook oven (microwave/convection)* | 24,900 | 4,100 | 1,000 | 3,100 | 0 | 1,000 | 0.16 | 0.24 |
| Reach-in refrigerator* | 4,800 | 1,200 | 300 | 900 | 0 | 1,200 | 0.25 | 0.25 |
| Refrigerated prep table* | 2,000 | 900 | 600 | 300 | 0 | 900 | 0.45 | 0.67 |
| Steamer (bun) | 5,100 | 700 | 600 | 100 | 0 | 700 | 0.14 | 0.86 |
| Toaster: 4-slice pop up (large): cooking | 6,100 | 3,000 | 200 | 1,400 | 1,000 | 2,600 | 0.49 | 0.07 |
| Toaster: contact (vertical) | 11,300 | 5,300 | 2,700 | 2,600 | 0 | 5,300 | 0.47 | 0.51 |
| Toaster: conveyor (large) | 32,800 | 10,300 | 3,000 | 7,300 | 0 | 10,300 | 0.31 | 0.29 |
| Toaster: small conveyor | 5,800 | 3,700 | 400 | 3,300 | 0 | 3,700 | 0.64 | 0.11 |
| Waffle iron | 3,100 | 1,200 | 800 | 400 | 0 | 1,200 | 0.39 | 0.67 |

*Source*: Swierczyna et al. (2008, 2009). Items with an asterisk appear only in Swierczyna (2009).

hood" columns assume installation under a properly designed exhaust hood connected to a mechanical fan exhaust system.

**Hospital and Laboratory Equipment.** Hospital and laboratory equipment items are major sources of heat gain in conditioned spaces. Care must be taken in evaluating the probability and duration of simultaneous usage when many components are concentrated in one area, such as a laboratory, an operating room, etc. Commonly, heat gain from equipment in a laboratory ranges from 15 to 70 Btu/h·ft² or, in laboratories with outdoor exposure, as much as four times the heat gain from all other sources combined.

**Medical Equipment.** It is more difficult to provide generalized heat gain recommendations for medical equipment than for general office equipment because medical equipment is much more varied in type and in application. Some heat gain testing has been done and can be presented, but the equipment included represents only a small sample of the type of equipment that may be encountered.

The data presented for medical equipment in Table 7-22 are relevant for portable and bench-top equipment. Medical equipment is very specific and can vary greatly from application to application. The data are presented to provide guidance in only the most general sense. For large equipment, such as MRI, engineers must obtain heat gain from the manufacturer.

**Laboratory Equipment.** Equipment in laboratories is similar to medical equipment in that it will vary significantly from space to space. Chapter 16 of the 2011 *ASHRAE Handbook—HVAC Applications* discusses heat gain from equipment, which may range from 5 to 25 W/ft² in highly automated laboratories. Table 7-23 lists some values for laboratory equipment, but, as is the case for medical equipment, it is for general guidance only.

## Office Equipment

Computers, printers, copiers, etc., can generate very significant heat gains, sometimes greater than all other gains combined. ASHRAE research project RP-822 developed a method to measure the actual heat gain from equipment in buildings and the radiant/convective percentages (Hosni et al. 1998; Jones et al. 1998). This methodology was then incorporated into ASHRAE research project RP-1055 and applied to a wide range of equipment (Hosni et al. 1999) as a follow-up to independent research by Wilkins and McGaffin (1994) and Wilkins et al. (1991). Komor (1997) found similar results. Analysis of measured data showed that results for office equipment could be generalized, but results from laboratory and hospital equipment proved too diverse. The following general guidelines for office equipment are a result of these studies.

**Table 7-21B  Recommended Rates of Radiant Heat Gain from Hooded Electric Appliances During Idle (Ready-to-Cook) Conditions**
*(Table 5B, Chapter 18, 2013 ASHRAE Handbook—Fundamentals)*

| Appliance | Energy Rate, Btu/h Rated | Energy Rate, Btu/h Standby | Rate of Heat Gain, Btu/h Sensible Radiant | Usage Factor $F_u$ | Radiation Factor $F_r$ |
|---|---|---|---|---|---|
| Broiler: underfired 3 ft | 36,900 | 30,900 | 10,800 | 0.84 | 0.35 |
| Cheesemelter* | 12,300 | 11,900 | 4,600 | 0.97 | 0.39 |
| Fryer: kettle | 99,000 | 1,800 | 500 | 0.02 | 0.28 |
| Fryer: open deep-fat, 1-vat | 47,800 | 2,800 | 1,000 | 0.06 | 0.36 |
| Fryer: pressure | 46,100 | 2,700 | 500 | 0.06 | 0.19 |
| Griddle: double sided 3 ft (clamshell down)* | 72,400 | 6,900 | 1,400 | 0.1 | 0.2 |
| Griddle: double sided 3 ft (clamshell up)* | 72,400 | 11,500 | 3,600 | 0.16 | 0.31 |
| Griddle: flat 3 ft | 58,400 | 11,500 | 4,500 | 0.2 | 0.39 |
| Griddle-small 3 ft* | 30,700 | 6,100 | 2,700 | 0.2 | 0.44 |
| Induction cooktop* | 71,700 | 0 | 0 | 0 | 0 |
| Induction wok* | 11,900 | 0 | 0 | 0 | 0 |
| Oven: combi: combi-mode* | 56,000 | 5,500 | 800 | 0.1 | 0.15 |
| Oven: combi: convection mode | 56,000 | 5,500 | 1,400 | 0.1 | 0.25 |
| Oven: convection full-size | 41,300 | 6,700 | 1,500 | 0.16 | 0.22 |
| Oven: convection half-size* | 18,800 | 3,700 | 500 | 0.2 | 0.14 |
| Pasta cooker* | 75,100 | 8,500 | 0 | 0.11 | 0 |
| Range top: top off/oven on* | 16,600 | 4,000 | 1,000 | 0.24 | 0.25 |
| Range top: 3 elements on/oven off | 51,200 | 15,400 | 6,300 | 0.3 | 0.41 |
| Range top: 6 elements on/oven off | 51,200 | 33,200 | 13,900 | 0.65 | 0.42 |
| Range top: 6 elements on/oven on | 67,800 | 36,400 | 14,500 | 0.54 | 0.4 |
| Range: hot-top | 54,000 | 51,300 | 11,800 | 0.95 | 0.23 |
| Rotisserie* | 37,900 | 13,800 | 4,500 | 0.36 | 0.33 |
| Salamander* | 23,900 | 23,300 | 7,000 | 0.97 | 0.3 |
| Steam kettle: large (60 gal) simmer lid down* | 110,600 | 2,600 | 100 | 0.02 | 0.04 |
| Steam kettle: small (40 gal) simmer lid down* | 73,700 | 1,800 | 300 | 0.02 | 0.17 |
| Steamer: compartment: atmospheric* | 33,400 | 15,300 | 200 | 0.46 | 0.01 |
| Tilting skillet/braising pan | 32,900 | 5,300 | 0 | 0.16 | 0 |

*Source*: Swierczyna et al. (2008, 2009). Items with an asterisk appear only in Swierczyna (2009).

**Nameplate Versus Measured Energy Use.** Nameplate data rarely reflect the actual power consumption of office equipment. Actual power consumption is assumed to equal total (radiant plus convective) heat gain, but its ratio to the nameplate value varies widely. ASHRAE research project RP-1055 (Hosni et al. 1999) found that, for general office equipment with nameplate power consumption of less than 1000 W, the actual ratio of total heat gain to nameplate ranged from 25% to 50%, but when all tested equipment is considered, the range is broader. Generally, if the nameplate value is the only information known and no actual heat gain data are available for similar equipment, it is conservative to use 50% of nameplate as heat gain and more nearly correct if 25% of nameplate is used. Much better results can be obtained, however, by considering heat gain to be predictable based on the type of equipment. However, if the device has a mainly resistive internal electric load (e.g., a space heater), the nameplate rating may be a good estimate of its peak energy dissipation.

**Computers.** Based on tests by Hosni et al. (1999) and Wilkins and McGaffin (1994), nameplate values on computers should be ignored when performing cooling load calculations. Table 7-24 presents typical heat gain values for computers with varying degrees of safety factor.

**Monitors.** Based on monitors tested by Hosni et al. (1999), heat gain for cathode ray tube (CRT) monitors correlates approximately with screen size as

$$q_{mon} = 5S - 20 \qquad (7\text{-}33)$$

where
 $q_{mon}$ = sensible heat gain from monitor, W
 $S$ = nominal screen size, in.
Table 7-24 shows typical values.

Flat-panel monitors have replaced CRT monitors in many workplaces. Power consumption, and thus heat gain, for flat-panel displays are significantly lower than for CRTs. Consult manufacturers' literature for average power consumption data for use in heat gain calculations.

**Laser Printers.** Hosni et al. (1999) found that power consumption, and therefore the heat gain, of laser printers depended largely on the level of throughput for which the printer was designed. Smaller printers tend to be used more intermittently, and larger printers may run continuously for longer periods.

Table 7-25 presents data on laser printers. These data can be applied by taking the value for continuous operation and then applying an appropriate diversity factor. This would likely be most appropriate for larger open office areas. Another approach, which may be appropriate for a single room or small area, is to take the value that most closely matches the expected operation of the printer with no diversity.

**Copiers.** Hosni et al. (1999) also tested five photocopy machines, including desktop and office (freestanding high-

Table 7-21C  Recommended Rates of Radiant Heat Gain from Hooded Gas Appliances During Idle (Ready-to-Cook) Conditions
*(Table 5C, Chapter 18, 2013 ASHRAE Handbook—Fundamentals)*

| Appliance | Energy Rate, Btu/h Rated | Energy Rate, Btu/h Standby | Rate of Heat Gain, Btu/h Sensible Radiant | Usage Factor $F_u$ | Radiation Factor $F_r$ |
|---|---|---|---|---|---|
| Broiler: batch* | 95,000 | 69,200 | 8,100 | 0.73 | 0.12 |
| Broiler: chain (conveyor) | 132,000 | 96,700 | 13,200 | 0.73 | 0.14 |
| Broiler: overfired (upright)* | 100,000 | 87,900 | 2,500 | 0.88 | 0.03 |
| Broiler: underfired 3 ft | 96,000 | 73,900 | 9,000 | 0.77 | 0.12 |
| Fryer: doughnut | 44,000 | 12,400 | 2,900 | 0.28 | 0.23 |
| Fryer: open deep-fat, 1 vat | 80,000 | 4,700 | 1,100 | 0.06 | 0.23 |
| Fryer: pressure | 80,000 | 9,000 | 800 | 0.11 | 0.09 |
| Griddle: double sided 3 ft (clamshell down)* | 108,200 | 8,000 | 1,800 | 0.07 | 0.23 |
| Griddle: double sided 3 ft (clamshell up)* | 108,200 | 14,700 | 4,900 | 0.14 | 0.33 |
| Griddle: flat 3 ft | 90,000 | 20,400 | 3,700 | 0.23 | 0.18 |
| Oven: combi: combi-mode* | 75,700 | 6,000 | 400 | 0.08 | 0.07 |
| Oven: combi: convection mode | 75,700 | 5,800 | 1,000 | 0.08 | 0.17 |
| Oven: convection full-size | 44,000 | 11,900 | 1,000 | 0.27 | 0.08 |
| Oven: conveyor (pizza) | 170,000 | 68,300 | 7,800 | 0.4 | 0.11 |
| Oven: deck | 105,000 | 20,500 | 3,500 | 0.2 | 0.17 |
| Oven: rack mini-rotating* | 56,300 | 4,500 | 1,100 | 0.08 | 0.24 |
| Pasta cooker* | 80,000 | 23,700 | 0 | 0.3 | 0 |
| Range top: top off/oven on* | 25,000 | 7,400 | 2,000 | 0.3 | 0.27 |
| Range top: 3 burners on/oven off | 120,000 | 60,100 | 7,100 | 0.5 | 0.12 |
| Range top: 6 burners on/oven off | 120,000 | 120,800 | 11,500 | 1.01 | 0.1 |
| Range top: 6 burners on/oven on | 145,000 | 122,900 | 13,600 | 0.85 | 0.11 |
| Range: wok* | 99,000 | 87,400 | 5,200 | 0.88 | 0.06 |
| Rethermalizer* | 90,000 | 23,300 | 11,500 | 0.26 | 0.49 |
| Rice cooker* | 35,000 | 500 | 300 | 0.01 | 0.6 |
| Salamander* | 35,000 | 33,300 | 5,300 | 0.95 | 0.16 |
| Steam kettle: large (60 gal) simmer lid down* | 145,000 | 5,400 | 0 | 0.04 | 0 |
| Steam kettle: small (10 gal) simmer lid down* | 52,000 | 3,300 | 300 | 0.06 | 0.09 |
| Steam kettle: small (40 gal) simmer lid down | 100,000 | 4,300 | 0 | 0.04 | 0 |
| Steamer: compartment: atmospheric * | 26,000 | 8,300 | 0 | 0.32 | 0 |
| Tilting skillet/braising pan | 104,000 | 10,400 | 400 | 0.1 | 0.04 |

*Source*: Swierczyna et al. (2008, 2009). Items with an asterisk appear only in Swierczyna (2009).

Table 7-21D  Recommended Rates of Radiant Heat Gain from Hooded Solid Fuel Appliances During Idle (Ready-to-Cook) Conditions
*(Table 5D, Chapter 18, 2013 ASHRAE Handbook—Fundamentals)*

| Appliance | Energy Rate, Btu/h Rated | Rate of Heat Gain, Btu/h Standby | Rate of Heat Gain, Btu/h Sensible | Usage Factor $F_u$ | Radiation Factor $F_r$ |
|---|---|---|---|---|---|
| Broiler: solid fuel: charcoal | 40 lb | 42,000 | 6200 | N/A | 0.15 |
| Broiler: solid fuel: wood (mesquite)* | 40 lb | 49,600 | 7000 | N/A | 0.14 |

*Source*: Swierczyna et al. (2008, 2009). Items with an asterisk appear only in Swierczyna (2009).

volume copiers) models. Larger machines used in production environments were not addressed. Table 7-25 summarizes the results. Desktop copiers rarely operate continuously, but office copiers frequently operate continuously for periods of an hour or more. Large, high-volume photocopiers often include provisions for exhausting air outdoors; if so equipped, the direct-to-space or system makeup air heat gain needs to be included in the load calculation. Also, when the air is dry, humidifiers are often operated near copiers to limit static electricity; if this occurs during cooling mode, their load on HVAC systems should be considered.

**Miscellaneous Office Equipment.** Table 7-26 presents data on miscellaneous office equipment such as vending machines and mailing equipment.

**Diversity.** The ratio of measured peak electrical load at equipment panels to the sum of the maximum electrical load of each individual item of equipment is the usage diversity. A small, one- or two-person office containing equipment listed in Tables 7-24 to 7-26 usually contributes heat gain to the space at the sum of the appropriate listed values. Progressively larger areas with many equipment items always experience some degree of usage diversity resulting from whatever percentage of such equipment is not in operation at any given time.

Wilkins and McGaffin (1994) measured diversity in 23 areas within five different buildings totaling over 275,000 ft². Diversity was found to range between 37 and 78%, with the average (normalized based on area) being 46%. Figure 7-7 illustrates the relationship between nameplate, sum of peaks, and actual electrical load with diversity accounted for, based on the average of the total area tested. Data on actual diversity can be used as a guide, but diversity varies significantly with

**Table 7-21E  Recommended Rates of Radiant and Convective Heat Gain from Warewashing Equipment During Idle (Standby) or Washing Conditions**
*(Table 5E, Chapter 18, 2013 ASHRAE Handbook—Fundamentals)*

| Appliance | Energy Rate, Btu/h Rated | Standby/ Washing | Unhooded Sensible Radiant | Unhooded Sensible Convective | Latent | Total | Hooded Sensible Radiant | Usage Factor $F_u$ | Radiation Factor $F_r$ |
|---|---|---|---|---|---|---|---|---|---|
| Dishwasher (conveyor type, chemical sanitizing) | 46,800 | 5700/43,600 | 0 | 4450 | 13490 | 17940 | 0 | 0.36 | 0 |
| Dishwasher (conveyor type, hot-water sanitizing) standby | 46,800 | 5700/N/A | 0 | 4750 | 16970 | 21720 | 0 | N/A | 0 |
| Dishwasher (door-type, chemical sanitizing) washing | 18,400 | 1200/13,300 | 0 | 1980 | 2790 | 4770 | 0 | 0.26 | 0 |
| Dishwasher (door-type, hot-water sanitizing) washing | 18,400 | 1200/13,300 | 0 | 1980 | 2790 | 4770 | 0 | 0.26 | 0 |
| Dishwasher* (under-counter type, chemical sanitizing) standby | 26,600 | 1200/18,700 | 0 | 2280 | 4170 | 6450 | 0 | 0.35 | 0.00 |
| Dishwasher* (under-counter type, hot-water sanitizing) standby | 26,600 | 1700/19,700 | 800 | 1040 | 3010 | 4850 | 800 | 0.27 | 0.34 |
| Booster heater* | 130,000 | 0 | 500 | 0 | 0 | 0 | 500 | 0 | N/A |

*Note*: Heat load values are prorated for 30% washing and 70% standby.   *Source*: Swierczyna et al. (2008, 2009). Items with an asterisk appear only in Swierczyna (2009).

**Table 7-22  Recommended Heat Gain from Typical Medical Equipment**
*(Table 6, Chapter 18, 2013 ASHRAE Handbook—Fundamentals)*

| Equipment | Nameplate, W | Peak, W | Average, W |
|---|---|---|---|
| Anesthesia system | 250 | 177 | 166 |
| Blanket warmer | 500 | 504 | 221 |
| Blood pressure meter | 180 | 33 | 29 |
| Blood warmer | 360 | 204 | 114 |
| ECG/RESP | 1440 | 54 | 50 |
| Electrosurgery | 1000 | 147 | 109 |
| Endoscope | 1688 | 605 | 596 |
| Harmonical scalpel | 230 | 60 | 59 |
| Hysteroscopic pump | 180 | 35 | 34 |
| Laser sonics | 1200 | 256 | 229 |
| Optical microscope | 330 | 65 | 63 |
| Pulse oximeter | 72 | 21 | 20 |
| Stress treadmill | N/A | 198 | 173 |
| Ultrasound system | 1800 | 1063 | 1050 |
| Vacuum suction | 621 | 337 | 302 |
| X-ray system | 968 | | 82 |
| X-ray system | 1725 | 534 | 480 |
| X-ray system | 2070 | | 18 |

*Source*: Hosni et al. (1999).

**Table 7-23  Recommended Heat Gain from Typical Laboratory Equipment**
*(Table 7, Chapter 18, 20013 ASHRAE Handbook—Fundamentals)*

| Equipment | Nameplate, W | Peak, W | Average, W |
|---|---|---|---|
| Analytical balance | 7 | 7 | 7 |
| Centrifuge | 138 | 89 | 87 |
| Centrifuge | 288 | 136 | 132 |
| Centrifuge | 5500 | 1176 | 730 |
| Electrochemical analyzer | 50 | 45 | 44 |
| Electrochemical analyzer | 100 | 85 | 84 |
| Flame photometer | 180 | 107 | 105 |
| Fluorescent microscope | 150 | 144 | 143 |
| Fluorescent microscope | 200 | 205 | 178 |
| Function generator | 58 | 29 | 29 |
| Incubator | 515 | 461 | 451 |
| Incubator | 600 | 479 | 264 |
| Incubator | 3125 | 1335 | 1222 |
| Orbital shaker | 100 | 16 | 16 |
| Oscilloscope | 72 | 38 | 38 |
| Oscilloscope | 345 | 99 | 97 |
| Rotary evaporator | 75 | 74 | 73 |
| Rotary evaporator | 94 | 29 | 28 |
| Spectronics | 36 | 31 | 31 |
| Spectrophotometer | 575 | 106 | 104 |
| Spectrophotometer | 200 | 122 | 121 |
| Spectrophotometer | N/A | 127 | 125 |
| Spectro fluorometer | 340 | 405 | 395 |
| Thermocycler | 1840 | 965 | 641 |
| Thermocycler | N/A | 233 | 198 |
| Tissue culture | 475 | 132 | 46 |
| Tissue culture | 2346 | 1178 | 1146 |

*Source*: Hosni et al. (1999).

occupancy. The proper diversity factor for an office of mail-order catalog telephone operators is different from that for an office of sales representatives who travel regularly.

ASHRAE research project RP-1093 derived diversity profiles for use in energy calculations (Abushakra et al. 2004; Claridge et al. 2004). Those profiles were derived from available measured data sets for a variety of office buildings, and indicated a range of peak weekday diversity factors for lighting ranging from 70 to 85% and for receptacles (appliance load) between 42 and 89%.

**Heat Gain per Unit Area.** Wilkins and Hosni (2000, 2011) and Wilkins and McGaffin (1994) summarized research on a heat gain per unit area basis. Diversity testing showed that the actual heat gain per unit area, or load factor, ranged from 0.44 to 1.08 W/ft², with an average (normalized based on area) of 0.81 W/ft². Spaces tested were fully occupied and highly automated, comprising 21 unique areas in five buildings, with a computer and monitor at every workstation. Table 7-27 presents a range of load factors with a subjective description of the type of space to which they would apply. The medium load density is likely to be appropriate for most standard office spaces. Medium/heavy or heavy load densities may be encountered

# Nonresidential Cooling and Heating Load Calculations

**Table 7-24  Recommended Heat Gain from Typical Computer Equipment**
*(Table 8, Chapter 18, 2013 ASHRAE Handbook Fundamentals)*

| Equipment | Description | Nameplate Power, W | Average Power, W | Radiant Fraction |
|---|---|---|---|---|
| Desktop computer[a] | Manufacturer A (model A); 2.8 GHz processor, 1 GB RAM | 480 | 73 | 0.10[a] |
| | Manufacturer A (model B); 2.6 GHz processor, 2 GB RAM | 480 | 49 | 0.10[a] |
| | Manufacturer B (model A); 3.0 GHz processor, 2 GB RAM | 690 | 77 | 0.10[a] |
| | Manufacturer B (model B); 3.0 GHz processor, 2 GB RAM | 690 | 48 | 0.10[a] |
| | Manufacturer A (model C); 2.3 GHz processor, 3 GB RAM | 1200 | 97 | 0.10[a] |
| Laptop computer[b] | Manufacturer 1; 2.0 GHz processor, 2 GB RAM, 17 in. screen | 130 | 36 | 0.25[b] |
| | Manufacturer 1; 1.8 GHz processor, 1 GB RAM, 17 in. screen | 90 | 23 | 0.25[b] |
| | Manufacturer 1; 2.0 GHz processor, 2 GB RAM, 14 in. screen | 90 | 31 | 0.25[b] |
| | Manufacturer 2; 2.13 GHz processor, 1 GB RAM, 14 in. screen, tablet PC | 90 | 29 | 0.25[b] |
| | Manufacturer 2; 366 MHz processor, 130 MB RAM (4 in. screen) | 70 | 22 | 0.25[b] |
| | Manufacturer 3; 900 MHz processor, 256 MB RAM (10.5 in. screen) | 50 | 12 | 0.25[b] |
| Flat-panel monitor[c] | Manufacturer X (model A); 30 in. screen | 383 | 90 | 0.40[c] |
| | Manufacturer X (model B); 22 in. screen | 360 | 36 | 0.40[c] |
| | Manufacturer Y (model A); 19 in. screen | 288 | 28 | 0.40[c] |
| | Manufacturer Y (model B); 17 in. screen | 240 | 27 | 0.40[c] |
| | Manufacturer Z (model A); 17 in. screen | 240 | 29 | 0.40[c] |
| | Manufacturer Z (model C); 15 in. screen | 240 | 19 | 0.40[c] |

*Source*: Hosni and Beck (2008).

[a]Power consumption for newer desktop computers in operational mode varies from 50 to 100 W, but a conservative value of about 65 W may be used. Power consumption in sleep mode is negligible. Because of cooling fan, approximately 90% of load is by convection and 10% is by radiation. Actual power consumption is about 10 to 15% of nameplate value.

[b]Power consumption of laptop computers is relatively small: depending on processor speed and screen size, it varies from about 15 to 40 W. Thus, differentiating between radiative and convective parts of the cooling load is unnecessary and the entire load may be classified as convective. Otherwise, a 75/25% split between convective and radiative components may be used. Actual power consumption for laptops is about 25% of nameplate values.

[c]Flat-panel monitors have replaced cathode ray tube (CRT) monitors in many workplaces, providing better resolution and being much lighter. Power consumption depends on size and resolution, and ranges from about 20 W (for 15 in. size) to 90 W (for 30 in.). The most common sizes in workplaces are 19 and 22 in., for which an average 30 W power consumption value may be used. Use 60/40% split between convective and radiative components. In idle mode, monitors have negligible power consumption. Nameplate values should not be used.

**Table 7-25  Recommended Heat Gain from Typical Laser Printers and Copiers**
*(Table 9, Chapter 18, 2013 ASHRAE Handbook Fundamentals)*

| Equipment | Description | Nameplate Power, W | Average Power, W | Radiant Fraction |
|---|---|---|---|---|
| Laser printer, typical desktop, small-office type[a] | Printing speed up to 10 pages per minute | 430 | 137 | 0.30[a] |
| | Printing speed up to 35 pages per minute | 890 | 74 | 0.30[a] |
| | Printing speed up to 19 pages per minute | 508 | 88 | 0.30[a] |
| | Printing speed up to 17 pages per minute | 508 | 98 | 0.30[a] |
| | Printing speed up to 19 pages per minute | 635 | 110 | 0.30[a] |
| | Printing speed up to 24 page per minute | 1344 | 130 | 0.30[a] |
| Multifunction (copy, print, scan)[b] | Small, desktop type | 600 | 30 | d |
| | | 40 | 15 | d |
| | Medium, desktop type | 700 | 135 | d |
| Scanner[b] | Small, desktop type | 19 | 16 | d |
| Copy machine[c] | Large, multiuser, office type | 1750 | 800 (idle 260 W) | d (idle 0.00[c]) |
| | | 1440 | 550 (idle 135 W) | d (idle 0.00[c]) |
| | | 1850 | 1060 (idle 305 W) | d (idle 0.00[c]) |
| Fax machine | Medium | 936 | 90 | d |
| | Small | 40 | 20 | d |
| Plotter | Manufacturer A | 400 | 250 | d |
| | Manufacturer B | 456 | 140 | d |

*Source*: Hosni and Beck (2008).

[a]Various laser printers commercially available and commonly used in personal offices were tested for power consumption in print mode, which varied from 75 to 140 W, depending on model, print capacity, and speed. Average power consumption of 110 W may be used. Split between convection and radiation is approximately 70/30%.

[b]Small multifunction (copy, scan, print) systems use about 15 to 30 W; medium-sized ones use about 135 W. Power consumption in idle mode is negligible. Nameplate values do not represent actual power consumption and should not be used. Small, single-sheet scanners consume less than 20 W and do not contribute significantly to building cooling load.

[c]Power consumption for large copy machines in large offices and copy centers ranges from about 550 to 1100 W in copy mode. Consumption in idle mode varies from about 130 to 300 W. Count idle-mode power consumption as mostly convective in cooling load calculations.

[d]Split between convective and radiant heat gain was not determined for these types of equipment.

but can be considered extremely conservative estimates even for densely populated and highly automated spaces. Table 7-28 indicates applicable diversity factors.

**Radiant/Convective Split.** ASHRAE research project RP-1482 (Hosni and Beck 2008) is examining the radiant/convective split for common office equipment; the most important differentiating feature is whether the equipment had a cooling fan. Footnotes in Tables 7-24 and 7-25 summarizes those results.

### 7.3.5 Ventilation and Infiltration Air

Wind and pressure differences cause outdoor air to infiltrate into the cracks around doors and windows, resulting in

**Table 7-26  Recommended Heat Gain from Miscellaneous Office Equipment**
*(Table 10, Chapter 18, 2013 ASHRAE Handbook—Fundamentals)*

| Equipment | Maximum Input Rating, W | Recommended Rate of Heat Gain, W |
|---|---|---|
| Mail-processing equipment | | |
| Folding machine | 125 | 80 |
| Inserting machine, 3600 to 6800 pieces/h | 600 to 3300 | 390 to 2150 |
| Labeling machine, 1500 to 30,000 pieces/h | 600 to 6600 | 390 to 4300 |
| Postage meter | 230 | 150 |
| Vending machines | | |
| Cigarette | 72 | 72 |
| Cold food/beverage | 1150 to 1920 | 575 to 960 |
| Hot beverage | 1725 | 862 |
| Snack | 240 to 275 | 240 to 275 |
| Other | | |
| Bar code printer | 440 | 370 |
| Cash registers | 60 | 48 |
| Check processing workstation, 12 pockets | 4800 | 2470 |
| Coffee maker, 10 cups | 1500 | 1050 W sens., 1540 Btu/h latent |
| Microfiche reader | 85 | 85 |
| Microfilm reader | 520 | 520 |
| Microfilm reader/printer | 1150 | 1150 |
| Microwave oven, 1 ft$^3$ | 600 | 400 |
| Paper shredder | 250 to 3000 | 200 to 2420 |
| Water cooler, 32 qt/h | 700 | 350 |

**Table 7-27  Recommended Load Factors for Various Types of Offices**
*(Table 11, Chapter 18, 2013 ASHRAE Handbook—Fundamentals)*

| Type of Use | Load Factor, W/ft$^2$ | Description |
|---|---|---|
| 100% Laptop, light | 0.25 | 167 ft$^2$/workstation, all laptop use, 1 printer per 10, speakers, misc. |
| medium | 0.33 | 125 ft$^2$/workstation, all laptop use, 1 printer per 10, speakers, misc. |
| 50% Laptop, light | 0.40 | 167 ft$^2$/workstation, 50% laptop / 50% desktop, 1 printer per 10, speakers, misc. |
| medium | 0.50 | 125 ft$^2$/workstation, 50% laptop / 50% desktop, 1 printer per 10, speakers, misc. |
| 100% Desktop, light | 0.60 | 167 ft$^2$/workstation, all desktop use, 1 printer per 10, speakers, misc. |
| medium | 0.80 | 125 ft$^2$/workstation, all desktop use, 1 printer per 10, speakers, misc. |
| 100% Desktop, two monitors | 1.00 | 125 ft$^2$/workstation, all desktop use, 2 monitors, 1 printer per 10, speakers, misc. |
| 100% Desktop, heavy | 1.50 | 85 ft$^2$/workstation, all desktop use, 2 monitors, 1 printer per 8, speakers, misc. |
| 100% Desktop, full on | 2.00 | 85 ft$^2$/workstation, all desktop use, 2 monitors, 1 printer per 8, speakers, misc., no diversity. |

Source: Wilkins and Hosni (2011).

localized sensible and latent heat gains. Also, some outdoor ventilation air is needed to eliminate any odors. This outdoor air imposes a cooling and dehumidifying load on the cooling coil because heat and/or moisture must be removed from the air. Heat gains due to infiltration and ventilation can be computed using equations in Chapter 5.

These equations are valid for calculating the cooling load due to infiltration of outdoor air and also due to the positive introduction of air for ventilation, provided it is introduced

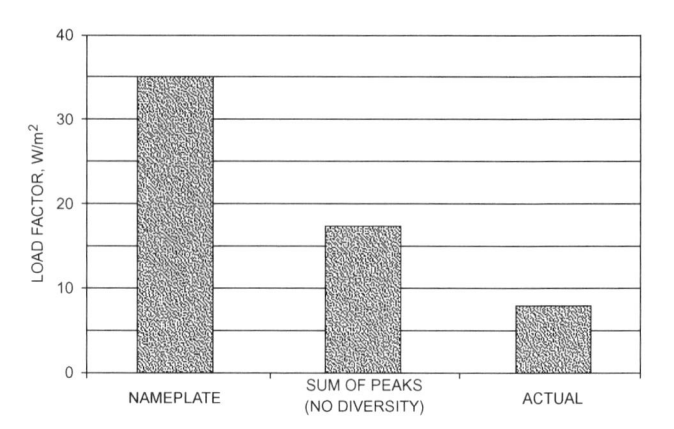

*Fig. 7-7  Office Equipment Load Factor Comparison*
*(Wilkins and McGaffin 1994)*
*(Figure 4, Chapter 18, 2013 ASHRAE Handbook—Fundamentals)*

**Table 7-28  Recommended Diversity Factors for Office Equipment**
*(Table 12, Chapter 18, 2013 ASHRAE Handbook—Fundamentals)*

| Device | Recommended Diversity Factor |
|---|---|
| Desktop computer | 75% |
| LCD monitor | 60% |
| Notebook computer | 75% |

### 7.3.6 Moisture Transfer Through Permeable Building Materials

In the usual comfort air-conditioning application, moisture transfer through walls is often neglected, since the actual rate is small and the corresponding latent heat gain is insignificant. On the other hand, industrial applications frequently call for a low moisture content in a conditioned space. Here, moisture transfer cannot be neglected, as the latent heat gain accompanying this transfer may be of greater magnitude than any other latent heat gain. Under these conditions, proper calculation of the moisture transfer due to both air infiltration and diffusion through building materials is important.

### 7.3.7 Miscellaneous Sources of Heat

Fans that circulate air through HVAC systems add energy to the system by one or all of the following processes:

*Temperature rise in the airstream from fan inefficiency.* Depending on the equipment, fan efficiencies generally range between 50 and 70%, with an average value of 65%. Thus, some 35% of the energy required by the fan appears as instantaneous heat gain to the air being transported.

*Temperature rise in the airstream as a consequence of air static and velocity pressure.* The fan energy that creates pressure to move air spreads throughout the entire air transport system in the process of conversion into sensible heat. Designers commonly assume that the temperature change equivalent of this heat occurs in one point in the system, depending on fan location.

*Temperature rise from heat generated by motor and drive inefficiencies.* The relatively small gains from fan motors and drives are normally disregarded unless the motor and/or drive are physically located within the conditioned airstream. Equations (7-26), (7-27), and (7-28) may be used to estimate heat gains from typical motors. Belt drive losses are often estimated as 3% of the rated motor power.

The location of each fan relative to other elements (primarily the cooling coil), the type of system (single zone, multizone, double-duct, terminal reheat, VAV, etc.), and the type of equipment control (space temperature alone, space temperature and relative humidity, and so forth) must be known before the analysis can be completed. A fan located upstream of the cooling coil (blow-through supply fan, return air fan, outdoor air fan) adds the heat equivalent to its inefficiency to the airstream temperature at that point; the cooling coil sees this as elevated entering dry-bulb temperature. A fan located downstream of the cooling coil raises the dry-bulb temperature of air leaving the cooling coil. This rise can be offset by reducing the cooling coil temperature, or alternatively, by increasing airflow across the cooling coil.

Unless return air duct systems are extensive or subjected to rigorous conditions, only the heat gained or lost by supply duct systems is significant. It is estimated as a percentage of space sensible cooling load (normally about 1%) and applied to the air dry-bulb temperature of the air leaving the coil in the form of an equivalent temperature reduction.

Losses from air leakage out of (or into) ductwork or equipment can be greater than conventional duct heat gain or loss but is normally about the same or less. Outward duct leakage is a direct loss of cooling and/or dehumidifying capacity and must be offset by increased airflow (sometimes reduced supply air temperature) unless it enters the conditioned space directly. Inward duct leakage causes temperature and/or humidity variations, but it is often ignored under ordinary circumstances due to the low temperature and pressure differentials involved.

A well-designed and properly installed duct system should not leak more than 1 to 3% of the total system airflow. HVAC equipment and volume control units connected into a duct system should be delivered from manufacturers with allowable leakage rates not exceeding 1 or 2% of maximum airflow. Where duct systems are specified to be sealed and leak-tested, both low- and high-pressure types can be constructed and required to fall in this range. Latent heat considerations are frequently ignored.

Poorly designed or installed duct systems can have leakage rates of 10 to 30%. Leakage from low-pressure lighting troffer connections lacking proper taping and sealing runs up to 35% or more of the terminal air supply. Such extremes can ruin the validity of any load calculation. As such, they may not affect overall system loads enough to cause problems; they will, however, always adversely impact required supply air quantities for most air-conditioning systems. Also, using uninsulated supply ductwork running through return air plenums results in high thermal leakage, loss of space cooling capability by the supply air, and condensation difficulties during a warm start-up.

**Table 7-29  Convective and Radiant Percentages of Total Sensible Heat Gain**

| Heat Gain Source | Radiant Heat, % | Convective Heat, % |
|---|---|---|
| Transmitted solar, no inside shade | 100 | 0 |
| Window solar, with inside shade | 63 | 37 |
| Absorbed (by fenestration) solar | 63 | 37 |
| Fluorescent lights, suspended, unvented | 67 | 33 |
| Fluorescent lights, recessed, vented to return air | 59 | 41 |
| Fluorescent lights, recessed, vented to return air and supply air | 19 | 81 |
| Incandescent lights | 80 | 20 |
| People | 70 | 30 |
| Conduction, exterior walls | 63 | 37 |
| Conduction, exterior roofs | 84 | 16 |
| Infiltration and ventilation | 0 | 100 |
| Machinery and appliances | 20 to 80 | 80 to 20 |

*Sources*: Pedersen et al. (1998), Hosni et al. (1999).

## 7.4 Description of Radiant Time Series (RTS)

Design cooling loads are based on the assumption of **steady-periodic conditions** (i.e., the design day's weather, occupancy, and heat gain conditions are identical to those for preceding days such that the loads repeat on an identical 24 h cyclical basis). Thus, the heat gain for a particular component at a particular hour is the same as 24 h prior, which is the same as 48 h prior, etc. This assumption is the basis for the RTS derivation from the HB method.

Cooling load calculations must address two time-delay effects inherent in building heat transfer processes: (1) delay of conductive heat gain through opaque massive exterior surfaces (walls, roofs, or floors) and (2) delay of radiative heat gain conversion to cooling loads.

Exterior walls and roofs conduct heat due to temperature differences between outdoor and indoor air. In addition, solar energy on exterior surfaces is absorbed, then transferred by conduction to the building interior. Due to the mass and thermal capacity of the wall or roof construction materials, there is a substantial time delay in heat input at the exterior surface becoming heat gain at the interior surface.

As described earlier in this chapter, most heat sources transfer energy to a room by a combination of convection and radiation. (See Table 7-29). The convection part of heat gain immediately becomes cooling load. The radiation part must first be absorbed by the finishes and mass of the interior room surfaces and becomes cooling load only when it is later transferred by convection from those surfaces to the room air. Thus, radiant heat gains become cooling loads over a delayed period of time.

### Overview of the RTS Method

Figure 7-8 gives an overview of the radiant time series method. In the calculation of solar radiation, transmitted solar heat gain through windows, sol-air temperature, and infiltration, the RTS method is exactly the same as previous simplified methods (TFM and TETD/TA). Important areas that are different include the computation of conductive heat gain, the splitting of all heat gains into radiant and convective portions, and the conversion of radiant heat gains into cooling loads.

The RTS method accounts for both conduction time delay and radiant time delay effects by multiplying hourly heat gains by 24 h time series. The time series multiplication, in effect, distributes heat gains over time. Series coefficients, which are called **radiant time factors** and **conduction time factors**, are derived using the heat balance method. Radiant time factors reflect the percentage of an earlier radiant heat gain that becomes cooling load during the current hour. Likewise, conduction time factors reflect the percentage of an earlier heat gain at the exterior of a wall or roof that becomes heat gain at the inside during the current hour. By definition, each radiant or conduction time series must total 100%.

These series can be used to easily compare the time-delay impact of one construction versus another. This ability to compare choices is of particular benefit in the design process, when all construction details may not have been decided. Comparison can illustrate the magnitude of difference between the choices, allowing the engineer to apply judgment and make more informed assumptions in estimating the load.

### RTS Procedure

The general procedure for calculating cooling load for each load component (lights, people, walls, roofs, windows, appliances, etc.) with RTS is as follows:

1. Calculate 24 h profile of component heat gain for design day (for conduction, first account for conduction time delay by applying conduction time series).
2. Split heat gains into radiant and convective parts (see Table 7-27 for radiant and convective fractions).
3. Apply appropriate radiant time series to radiant part of heat gains to account for time delay in conversion to cooling load.
4. Sum convective part of heat gain and delayed radiant part of heat gain to determine cooling load for each hour for each cooling load component.

After calculating cooling loads for each component for each hour, sum those to determine the total cooling load for each hour and select the hour with the peak load for design of the air-conditioning system. This process should be repeated for multiple design months to determine the month when the peak load occurs, especially with windows on southern exposures (northern exposure in southern latitudes), which can result in higher peak room cooling loads in winter months than in summer.

### Conduction Heat Gain

In the RTS method, conduction through exterior walls and roofs is calculated using conduction time series (CTS). Wall and roof conductive heat input at the exterior is defined by the familiar conduction Equation 7-34 as

$$q_{i,\theta-n} = UA(t_{e,\theta-n} - t_{rc}) \qquad (7\text{-}34)$$

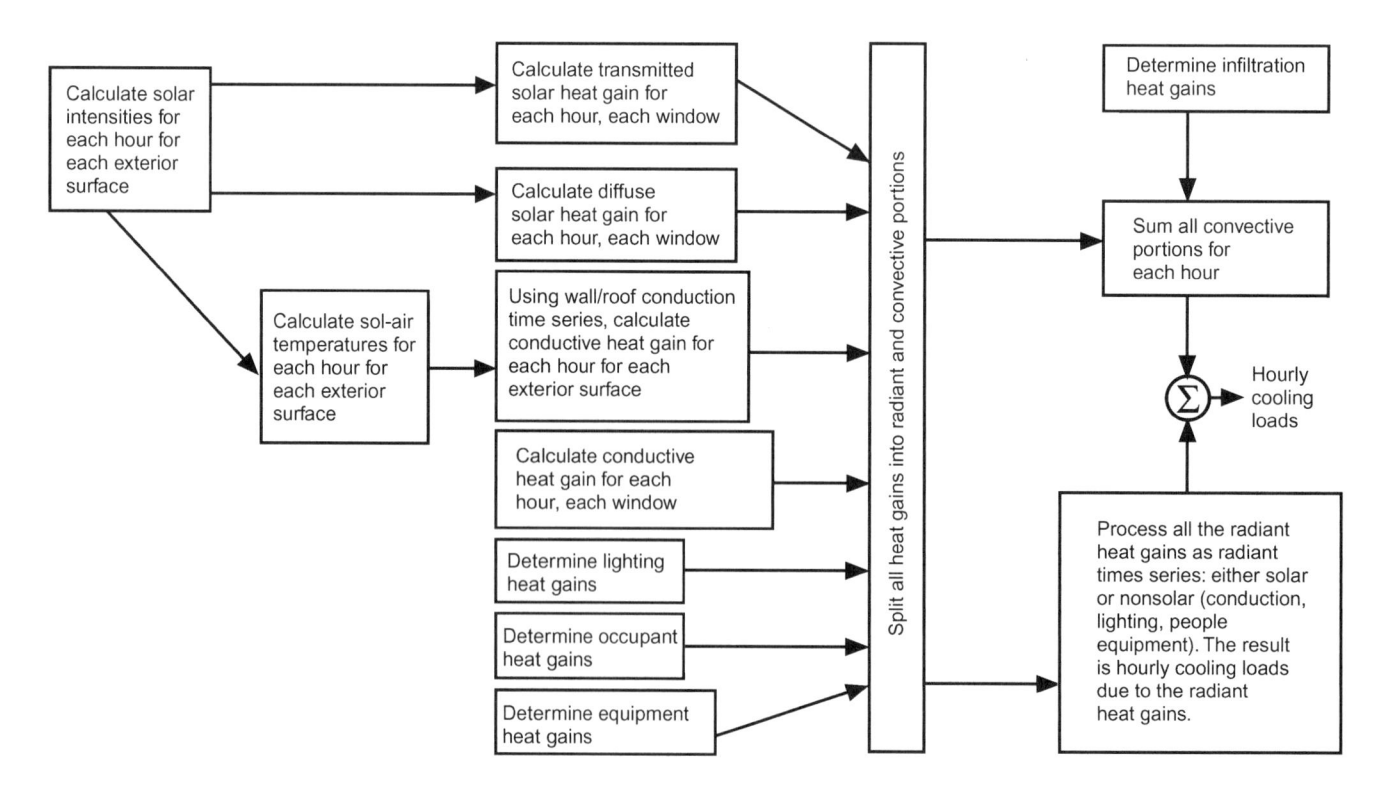

*Fig. 7-8 Overview of Radiant Time Series Method*
*(Figure 8, Chapter 18, 2009 ASHRAE Handbook—Fundamentals)*

where

$q_{i,\theta-n}$ = conductive heat input for the surface $n$ hours ago
$U$ = overall heat transfer coefficient for the surface
$A$ = surface area, ft$^2$
$t_{e,\theta-n}$ = sol-air temperature, °F, $n$ hours ago
$t_{rc}$ = presumed constant room air temperature, °F

Conductive heat gain through walls or roofs can be calculated using conductive heat inputs for the current and past 23 h and conduction time series, as illustrated in Equation 7-35:

$$q_\theta = c_0 q_{i,\theta} + c_1 q_{i,\theta-1} + c_2 q_{i,\theta-2} + c_3 q_{i,\theta-3} + \ldots + c_{23} q_{i,\theta-23} \qquad (7\text{-}35)$$

where

$q_\theta$ = hourly conductive heat gain, Btu/h, for the surface
$q_{i,\theta}$ = heat input for the current hour
$q_{i,\theta-n}$ = heat input $n$ hours ago
$c_0, c_1$, etc. = conduction time factors

Conduction time factors for representative wall and roof types are included in Tables 7-30 and 7-31. Those values were derived by first calculating conduction transfer functions for each example wall and roof construction. The assumption of steady-periodic heat input conditions for design load calculations allowed the conduction transfer functions to be reformulated into periodic response factors as demonstrated by Spitler and Fisher (1999a). The periodic response factors were further simplified by dividing the 24 periodic response factors by the respective overall wall or roof U-factor to form the conduction time series (CTS). The CTS factors can then be used in Equation 7-35 and provide a means for comparison of time delay characteristics between different wall and roof constructions. Construction material data used in the calculations for walls and roofs included in Tables 7-30 and 7-31 are listed in Table 7-32.

Heat gains calculated for walls or roofs using periodic response factors (and thus CTS) are identical to those calculated using conduction transfer functions for the steady periodic conditions assumed in design cooling load calculations.

## 7.5 Cooling Load Calculation Using RTS

The instantaneous cooling load is defined as the rate at which heat energy is convected to the zone air at a given point in time. The computation of cooling load is complicated by the radiant exchange between surfaces, furniture, partitions, and other mass in the zone. Most heat gain sources transfer energy by both convection and radiation. Radiative heat transfer introduces to the process a time dependency that is not easily quantified. Radiation is absorbed by the thermal masses in the zone and then later transferred by convection into the space. This process creates a time lag and dampening effect. The convection portion of heat gains, on the other hand, is assumed to immediately become cooling load in the hour in which that heat gain occurs.

Heat balance procedures calculate the radiant exchange between surfaces based on their surface temperatures and emissivities, but they typically rely on estimated "radiative-convective splits" to determine the contribution of internal loads, including people, lighting, appliances, and equipment, to the radiant exchange. The radiant time series procedure further simplifies the heat balance procedure by also relying on an estimated radiative-convective split of wall and roof conductive heat gain instead of simultaneously solving for the instantaneous convective and radiative heat transfer from each surface, as is done in the heat balance procedure.

Thus, the cooling load for each load component (lights, people, walls, roofs, windows, appliances, etc.) for a particular hour is the sum of the convective portion of the heat gain for that hour plus the time-delayed portion of radiant heat gains for that hour and the previous 23 h. Table 7-29 contains recommendations for splitting each of the heat gain components into convective and radiant portions.

The radiant time series method converts the radiant portion of hourly heat gains to hourly cooling loads using radiant time factors, the coefficients of the radiant time series. Radiant time factors are used to calculate the cooling load for the current hour on the basis of current and past heat gains. The radiant time series for a particular zone gives the time-dependent response of the zone to a single pulse of radiant energy. The series shows the portion of the radiant pulse that is convected to the zone air for each hour. Thus, $r_0$ represents the fraction of the radiant pulse convected to the zone air in the current hour $r_1$ in the previous hour, and so on. The radiant time series thus generated is used to convert the radiant portion of hourly heat gains to hourly cooling loads.

Two different radiant time series are used: **solar**, for directly transmitted solar heat gain (radiant energy assumed to be distributed to the floor and furnishings only), and **nonsolar** for all other types of heat gains (radiant energy assumed to be uniformly distributed on all internal surfaces). Nonsolar RTS apply to radiant heat gains from people, lights, appliances, walls, roofs, and floors. Also, for diffuse solar heat gain and direct solar heat gain from fenestration with inside shading (blinds, drapes, etc.), the nonsolar RTS should be used. Radiation from those sources is assumed to be more uniformly distributed onto all room surfaces.

Representative solar and nonsolar RTS data for light, medium, and heavyweight constructions are provided in Tables 7-33 and 7-34. Those were calculated using the zone characteristics listed in Table 7-35.

**Comparison of RTS with Previous Methods.** The user may question what benefits may be expected now that the TFM, TETD/TA, and CLTD/CLF procedures presented in earlier editions have been superseded (*not* invalidated or discredited). The primary benefit will be improved accuracy, with reduced dependency upon purely subjective input (such as determining a proper time-averaging period for TETD/TA or ascertaining appropriate safety factors to add to the "rounded off" TFM results). As a generic example, the space sensible cooling load for the traditional little ASHRAE store building

### Table 7-30 Wall Conduction Time Series (CTS)
(Table 16, Chapter 18, *2013 ASHRAE Handbook—Fundamentals*)

| | CURTAIN WALLS | | | STUD WALLS | | | | EIFS | | | BRICK WALLS | | | | | | | | | |
|---|---|---|---|---|---|---|---|---|---|---|---|---|---|---|---|---|---|---|---|---|
| **Wall Number =** | 1 | 2 | 3 | 4 | 5 | 6 | 7 | 8 | 9 | 10 | 11 | 12 | 13 | 14 | 15 | 16 | 17 | 18 | 19 | 20 |
| **U-Factor, Btu/h·ft²·°F** | 0.075 | 0.076 | 0.075 | 0.074 | 0.074 | 0.071 | 0.073 | 0.118 | 0.054 | 0.092 | 0.101 | 0.066 | 0.050 | 0.102 | 0.061 | 0.111 | 0.124 | 0.091 | 0.102 | 0.068 |
| **Total R** | 13.3 | 13.2 | 13.3 | 13.6 | 13.6 | 14.0 | 13.8 | 8.5 | 18.6 | 10.8 | 9.9 | 15.1 | 20.1 | 9.8 | 16.3 | 9.0 | 8.1 | 11.0 | 9.8 | 14.6 |
| **Mass, lb/ft²** | 6.3 | 4.3 | 16.4 | 5.2 | 17.3 | 5.2 | 13.7 | 7.5 | 7.8 | 26.8 | 42.9 | 44.0 | 44.2 | 59.6 | 62.3 | 76.2 | 80.2 | 96.2 | 182.8 | 136.3 |
| **Thermal Capacity, Btu/ft²·°F** | 1.5 | 1.0 | 3.3 | 1.2 | 3.6 | 1.6 | 3.0 | 1.8 | 1.9 | 5.9 | 8.7 | 8.7 | 8.7 | 11.7 | 12.4 | 15.7 | 15.3 | 19.0 | 38.4 | 28.4 |
| **Hour** | | | | | | | | | | **Conduction Time Factors, %** | | | | | | | | | | |
| 0 | 18 | 25 | 8 | 19 | 6 | 7 | 5 | 11 | 2 | 1 | 0 | 0 | 0 | 1 | 2 | 2 | 1 | 3 | 4 | 3 |
| 1 | 58 | 57 | 45 | 59 | 42 | 44 | 41 | 50 | 25 | 2 | 5 | 4 | 1 | 1 | 2 | 2 | 1 | 3 | 4 | 3 |
| 2 | 20 | 15 | 32 | 18 | 33 | 32 | 34 | 26 | 31 | 6 | 14 | 13 | 7 | 2 | 2 | 2 | 3 | 3 | 4 | 3 |
| 3 | 4 | 3 | 11 | 3 | 13 | 12 | 13 | 9 | 20 | 9 | 17 | 17 | 12 | 5 | 3 | 4 | 6 | 3 | 4 | 4 |
| 4 | 0 | 0 | 3 | 1 | 4 | 4 | 4 | 3 | 11 | 9 | 15 | 15 | 13 | 8 | 5 | 5 | 7 | 3 | 4 | 4 |
| 5 | 0 | 0 | 1 | 0 | 1 | 1 | 2 | 1 | 5 | 9 | 12 | 12 | 13 | 9 | 6 | 6 | 8 | 4 | 4 | 4 |
| 6 | 0 | 0 | 0 | 0 | 1 | 0 | 1 | 0 | 3 | 8 | 9 | 9 | 11 | 9 | 7 | 6 | 8 | 4 | 4 | 5 |
| 7 | 0 | 0 | 0 | 0 | 0 | 0 | 0 | 0 | 2 | 7 | 7 | 7 | 9 | 9 | 7 | 7 | 8 | 5 | 4 | 5 |
| 8 | 0 | 0 | 0 | 0 | 0 | 0 | 0 | 0 | 1 | 6 | 5 | 5 | 7 | 8 | 7 | 7 | 8 | 5 | 4 | 5 |
| 9 | 0 | 0 | 0 | 0 | 0 | 0 | 0 | 0 | 0 | 6 | 4 | 4 | 6 | 7 | 7 | 6 | 7 | 5 | 4 | 5 |
| 10 | 0 | 0 | 0 | 0 | 0 | 0 | 0 | 0 | 0 | 5 | 3 | 3 | 5 | 7 | 6 | 6 | 6 | 5 | 4 | 5 |
| 11 | 0 | 0 | 0 | 0 | 0 | 0 | 0 | 0 | 0 | 5 | 2 | 2 | 4 | 6 | 6 | 6 | 6 | 5 | 5 | 5 |
| 12 | 0 | 0 | 0 | 0 | 0 | 0 | 0 | 0 | 0 | 4 | 2 | 2 | 3 | 5 | 5 | 5 | 5 | 5 | 5 | 5 |
| 13 | 0 | 0 | 0 | 0 | 0 | 0 | 0 | 0 | 0 | 4 | 1 | 2 | 2 | 4 | 5 | 5 | 4 | 5 | 5 | 5 |
| 14 | 0 | 0 | 0 | 0 | 0 | 0 | 0 | 0 | 0 | 3 | 1 | 2 | 2 | 4 | 5 | 5 | 4 | 5 | 5 | 5 |
| 15 | 0 | 0 | 0 | 0 | 0 | 0 | 0 | 0 | 0 | 3 | 1 | 1 | 1 | 3 | 4 | 4 | 3 | 5 | 4 | 5 |
| 16 | 0 | 0 | 0 | 0 | 0 | 0 | 0 | 0 | 0 | 3 | 1 | 1 | 1 | 3 | 4 | 4 | 3 | 5 | 4 | 4 |
| 17 | 0 | 0 | 0 | 0 | 0 | 0 | 0 | 0 | 0 | 2 | 1 | 1 | 1 | 2 | 3 | 4 | 3 | 4 | 4 | 4 |
| 18 | 0 | 0 | 0 | 0 | 0 | 0 | 0 | 0 | 0 | 2 | 0 | 0 | 1 | 2 | 3 | 3 | 2 | 4 | 4 | 4 |
| 19 | 0 | 0 | 0 | 0 | 0 | 0 | 0 | 0 | 0 | 2 | 0 | 0 | 1 | 2 | 3 | 3 | 2 | 4 | 4 | 4 |
| 20 | 0 | 0 | 0 | 0 | 0 | 0 | 0 | 0 | 0 | 2 | 0 | 0 | 0 | 1 | 3 | 3 | 2 | 4 | 4 | 4 |
| 21 | 0 | 0 | 0 | 0 | 0 | 0 | 0 | 0 | 0 | 1 | 0 | 0 | 0 | 1 | 2 | 2 | 1 | 4 | 4 | 4 |
| 22 | 0 | 0 | 0 | 0 | 0 | 0 | 0 | 0 | 0 | 1 | 0 | 0 | 0 | 1 | 2 | 2 | 1 | 4 | 4 | 3 |
| 23 | 0 | 0 | 0 | 0 | 0 | 0 | 0 | 0 | 0 | 0 | 0 | 0 | 0 | 0 | 1 | 1 | 1 | 3 | 4 | 3 |
| **Total Percentage** | 100 | 100 | 100 | 100 | 100 | 100 | 100 | 100 | 100 | 100 | 100 | 100 | 100 | 100 | 100 | 100 | 100 | 100 | 100 | 100 |
| **Layer ID from outside to inside (see Table 7-30)** | F01 | F01 | F01 | F01 | F01 | F01 | F01 | F01 | F01 | F01 | F01 | F01 | F01 | F01 | F01 | F01 | F01 | F01 | F01 | F01 |
| | F09 | F08 | F10 | F08 | F10 | F11 | F07 | F06 | F06 | F06 | M01 | M01 | M01 | M01 | M01 | M01 | M01 | M01 | M01 | M01 |
| | F04 | F04 | F04 | G03 | G03 | G02 | G03 | I01 | I01 | I01 | F04 | F04 | F04 | F04 | F04 | F04 | F04 | F04 | F04 | F04 |
| | I02 | I02 | I02 | I04 | I04 | I04 | I04 | G03 | G03 | G03 | I01 | G03 | I01 | I01 | M03 | I01 | I01 | I01 | I01 | M15 |
| | F04 | F04 | F04 | G01 | G01 | G04 | G01 | F04 | I04 | M03 | G03 | I04 | G03 | M03 | I04 | M05 | M01 | M13 | M16 | I04 |
| | G01 | G01 | G01 | F02 | F02 | F02 | F02 | G01 | G01 | F04 | F04 | G01 | I04 | F02 | G01 | G01 | F02 | F04 | F04 | G01 |
| | F02 | F02 | F02 | — | — | — | — | F02 | F02 | G01 | G01 | F02 | G01 | — | F02 | F02 | — | G01 | G01 | F02 |
| | — | — | — | — | — | — | — | — | — | F02 | F02 | — | F02 | — | — | — | — | F02 | F02 | — |

**Wall Number Descriptions**

1. Spandrel glass, R-10 insulation board, gyp board
2. Metal wall panel, R-10 insulation board, gyp board
3. 1 in. stone, R-10 insulation board, gyp board
4. Metal wall panel, sheathing, R-11 batt insulation, gyp board
5. 1 in. stone, sheathing, R-11 batt insulation, gyp board
6. Wood siding, sheathing, R-11 batt insulation, 1/2 in. wood
7. 1 in. stucco, sheathing, R-11 batt insulation, gyp board
8. EIFS finish, R-5 insulation board, sheathing, gyp board
9. EIFS finish, R-5 insulation board, sheathing, R-11 batt insulation, gyp board
10. EIFS finish, R-5 insulation board, sheathing, 8 in. LW CMU, gyp board
11. Brick, R-5 insulation board, sheathing, gyp board
12. Brick, sheathing, R-11 batt insulation, gyp board
13. Brick, R-5 insulation board, sheathing, R-11 batt insulation, gyp board
14. Brick, R-5 insulation board, 8 in. LW CMU
15. Brick, 8 in. LW CMU, R-11 batt insulation, gyp board
16. Brick, R-5 insulation board, 8 in. HW CMU, gyp board
17. Brick, R-5 insulation board, brick
18. Brick, R-5 insulation board, 8 in. LW concrete, gyp board
19. Brick, R-5 insulation board, 12 in. HW concrete, gyp board
20. Brick, 8 in. HW concrete, R-11 batt insulation, gyp board

(used for example purposes since the 1940s) was calculated by means of the heat balance procedure and independently calculated by application of the radiant time series procedure, with each set of results plotted as one of the load profile curves of Figure 7-9. Also plotted on this chart are the corresponding curves produced by the TFM and TETD/TA methodologies in the 1997 edition of this chapter. Users may draw their own conclusions from this chart.

As part of the presentation of this method, **RTS Method Load Calculation spreadsheets are available at www.ashrae.org/PrinciplesofHVAC7.**

## 7.6 Heating Load Calculations

Techniques for estimating design heating load for commercial, institutional, and industrial applications are

## Nonresidential Cooling and Heating Load Calculations

**Table 7-30  Wall Conduction Time Series (CTS) *(Concluded)***
(Table 16, Chapter 18, *2013 ASHRAE Handbook—Fundamentals*)

| | CONCRETE BLOCK WALL | | | | | | PRECAST AND CAST-IN-PLACE CONCRETE WALLS | | | | | | | | |
|---|---|---|---|---|---|---|---|---|---|---|---|---|---|---|---|
| Wall Number = | 21 | 22 | 23 | 24 | 25 | 26 | 27 | 28 | 29 | 30 | 31 | 32 | 33 | 34 | 35 |
| U-Factor, Btu/h·ft²·°F | 0.067 | 0.059 | 0.073 | 0.186 | 0.147 | 0.121 | 0.118 | 0.074 | 0.076 | 0.115 | 0.068 | 0.082 | 0.076 | 0.047 | 0.550 |
| Total *R* | 14.8 | 16.9 | 13.7 | 5.4 | 6.8 | 8.2 | 8.4 | 13.6 | 13.1 | 8.7 | 14.7 | 12.2 | 13.1 | 21.4 | 1.8 |
| Mass, lb/ft² | 22.3 | 22.3 | 46.0 | 19.3 | 21.9 | 34.6 | 29.5 | 29.6 | 53.8 | 59.8 | 56.3 | 100.0 | 96.3 | 143.2 | 140.0 |
| Thermal Capacity, Btu/ft²·°F | 4.8 | 4.8 | 10.0 | 4.1 | 4.7 | 7.4 | 6.1 | 6.1 | 10.8 | 12.1 | 11.4 | 21.6 | 20.8 | 30.9 | 30.1 |
| Hour | Conduction Time Factors, % | | | | | | | | | | | | | | |
| 0 | 0 | 1 | 0 | 1 | 0 | 1 | 1 | 0 | 1 | 2 | 1 | 3 | 1 | 2 | 1 |
| 1 | 4 | 1 | 2 | 11 | 3 | 1 | 10 | 8 | 1 | 2 | 2 | 3 | 2 | 2 | 2 |
| 2 | 13 | 5 | 8 | 21 | 12 | 2 | 20 | 18 | 3 | 3 | 3 | 4 | 5 | 3 | 4 |
| 3 | 16 | 9 | 12 | 20 | 16 | 5 | 18 | 18 | 6 | 5 | 6 | 5 | 8 | 3 | 7 |
| 4 | 14 | 11 | 12 | 15 | 15 | 7 | 14 | 14 | 8 | 6 | 7 | 6 | 9 | 5 | 8 |
| 5 | 11 | 10 | 11 | 10 | 12 | 9 | 10 | 11 | 9 | 6 | 8 | 6 | 9 | 5 | 8 |
| 6 | 9 | 9 | 9 | 7 | 10 | 9 | 7 | 8 | 9 | 6 | 8 | 6 | 8 | 6 | 8 |
| 7 | 7 | 8 | 8 | 5 | 8 | 8 | 5 | 6 | 9 | 6 | 7 | 5 | 7 | 6 | 8 |
| 8 | 6 | 7 | 7 | 3 | 6 | 8 | 4 | 4 | 8 | 6 | 7 | 5 | 6 | 6 | 7 |
| 9 | 4 | 6 | 6 | 2 | 4 | 7 | 3 | 3 | 7 | 6 | 6 | 5 | 6 | 6 | 6 |
| 10 | 3 | 5 | 5 | 2 | 3 | 6 | 2 | 2 | 7 | 5 | 6 | 5 | 5 | 6 | 6 |
| 11 | 3 | 4 | 4 | 1 | 3 | 6 | 2 | 2 | 6 | 5 | 5 | 5 | 5 | 5 | 5 |
| 12 | 2 | 4 | 3 | 1 | 2 | 5 | 1 | 2 | 5 | 5 | 5 | 4 | 4 | 5 | 4 |
| 13 | 2 | 3 | 2 | 1 | 2 | 4 | 1 | 1 | 4 | 5 | 4 | 4 | 4 | 5 | 4 |
| 14 | 2 | 3 | 2 | 0 | 1 | 4 | 1 | 1 | 4 | 4 | 4 | 4 | 3 | 4 | 4 |
| 15 | 1 | 3 | 2 | 0 | 1 | 3 | 1 | 1 | 3 | 4 | 3 | 4 | 3 | 4 | 3 |
| 16 | 1 | 2 | 1 | 0 | 1 | 3 | 0 | 1 | 2 | 4 | 3 | 4 | 3 | 4 | 3 |
| 17 | 1 | 2 | 1 | 0 | 1 | 2 | 0 | 0 | 2 | 3 | 3 | 4 | 2 | 4 | 3 |
| 18 | 1 | 2 | 1 | 0 | 0 | 2 | 0 | 0 | 1 | 3 | 2 | 4 | 2 | 4 | 2 |
| 19 | 0 | 1 | 1 | 0 | 0 | 2 | 0 | 0 | 1 | 3 | 2 | 3 | 2 | 3 | 2 |
| 20 | 0 | 1 | 1 | 0 | 0 | 2 | 0 | 0 | 1 | 3 | 2 | 3 | 2 | 3 | 2 |
| 21 | 0 | 1 | 1 | 0 | 0 | 2 | 0 | 0 | 1 | 3 | 2 | 3 | 2 | 3 | 1 |
| 22 | 0 | 1 | 1 | 0 | 0 | 1 | 0 | 0 | 1 | 3 | 2 | 3 | 1 | 3 | 1 |
| 23 | 0 | 1 | 0 | 0 | 0 | 1 | 0 | 0 | 1 | 2 | 2 | 2 | 1 | 3 | 1 |
| Total Percentage | 100 | 100 | 100 | 100 | 100 | 100 | 100 | 100 | 100 | 100 | 100 | 100 | 100 | 100 | 100 |
| Layer ID from outside to inside (see Table 7-30) | F01 M03 I04 G01 F02 — | F01 M08 I04 G01 F02 — | F01 F07 M05 I04 G01 F02 | F01 M08 F02 — — — | F01 M08 F04 G01 F02 — | F01 M09 F04 G01 F02 — | F01 M11 I01 F04 G01 F02 | F01 M11 I04 G01 F02 — | F01 M11 I02 M11 F02 — | F01 F06 I01 M13 G01 F02 | F01 M13 I04 G01 F02 — | F01 F06 I02 M15 G01 F02 | F01 M15 I04 G01 F02 — | F01 M16 I05 G01 F02 — | F01 M16 F02 — — — |
| | Wall Number Descriptions | | | | | | | | | | | | | | |

21. 8 in. LW CMU, R-11 batt insulation, gyp board
22. 8 in. LW CMU with fill insulation, R-11 batt insulation, gyp board
23. 1 in. stucco, 8 in. HW CMU, R-11 batt insulation, gyp board
24. 8 in. LW CMU with fill insulation
25. 8 in. LW CMU with fill insulation, gyp board
26. 12 in. LW CMU with fill insulation, gyp board
27. 4 in. LW concrete, R-5 board insulation, gyp board
28. 4 in. LW concrete, R-11 batt insulation, gyp board
29. 4 in. LW concrete, R-10 board insulation, 4 in. LW concrete
30. EIFS finish, R-5 insulation board, 8 in. LW concrete, gyp board
31. 8 in. LW concrete, R-11 batt insulation, gyp board
32. EIFS finish, R-10 insulation board, 8 in. HW concrete, gyp board
33. 8 in. HW concrete, R-11 batt insulation, gyp board
34. 12 in. HW concrete, R-19 batt insulation, gyp board
35. 12 in. HW concrete

essentially the same as those for estimating design cooling loads, with the following exceptions:
- Credit for solar or internal heat gains is not included.
- Thermal storage effect of building structure is ignored.

This simplified approach is justified because it evaluates worst-case conditions that can reasonably occur during a heating season. Thus, the worst-case load is based on the following:
- Design interior and exterior conditions
- Infiltration and/or ventilation
- No solar effect
- No heat gains from lights, people, and appliances

Typical new commercial and retail spaces have nighttime unoccupied periods at a setback temperature where no ventilation is required, building lights and equipment are off, and heat loss is primarily through conduction. Before being occupied, buildings are warmed to the occupied temperature. During occupied time, building lights, equipment, and people heat gains can offset conductive heat loss, although some perimeter heat may be required, leaving ventilation load as the primary heating load.

Table 7-31 Roof Conduction Time Series (CTS), Layers, U-Factors, Mass and Thermal Capacity
(Table 17, Chapter 18, *2009 ASHRAE Handbook—Fundamentals*)

| | SLOPED FRAME ROOFS | | | | | | WOOD DECK | | METAL DECK ROOFS | | | | | CONCRETE ROOFS | | | | | |
|---|---|---|---|---|---|---|---|---|---|---|---|---|---|---|---|---|---|---|---|
| Roof Number | 1 | 2 | 3 | 4 | 5 | 6 | 7 | 8 | 9 | 10 | 11 | 12 | 13 | 14 | 15 | 16 | 17 | 18 | 19 |
| U-factor, Btu/h·ft²·°F | 0.044 | 0.040 | 0.045 | 0.041 | 0.042 | 0.041 | 0.69 | 0.058 | 0.080 | 0.065 | 0.057 | 0.036 | 0.052 | 0.054 | 0.052 | 0.051 | 0.056 | 0.055 | 0.042 |
| Total R | 22.8 | 25.0 | 22.2 | 24.1 | 23.7 | 24.6 | 14.5 | 17.2 | 12.6 | 15.4 | 17.6 | 27.6 | 19.1 | 18.6 | 19.2 | 19.7 | 18.0 | 18.2 | 23.7 |
| Mass, lb/ft² | 5.5 | 4.3 | 2.9 | 7.1 | 11.4 | 7.1 | 10.0 | 11.5 | 4.9 | 6.3 | 5.1 | 5.6 | 11.8 | 30.6 | 43.9 | 57.2 | 73.9 | 97.2 | 74.2 |
| Thermal Capacity, Btu/ft²·°F | 1.3 | 0.8 | 0.6 | 2.3 | 3.6 | 2.3 | 3.7 | 3.9 | 1.4 | 1.6 | 1.4 | 1.6 | 2.8 | 6.6 | 9.3 | 12.0 | 16.3 | 21.4 | 16.2 |
| Hour | Conduction Time Factors, % | | | | | | | | | | | | | | | | | | |
| 0 | 6 | 10 | 27 | 1 | 1 | 1 | 0 | 1 | 18 | 4 | 8 | 1 | 0 | 1 | 2 | 2 | 2 | 3 | 1 |
| 1 | 45 | 57 | 62 | 17 | 17 | 12 | 7 | 3 | 61 | 41 | 53 | 23 | 10 | 2 | 2 | 2 | 2 | 3 | 2 |
| 2 | 33 | 27 | 10 | 31 | 34 | 25 | 18 | 8 | 18 | 35 | 30 | 38 | 22 | 8 | 3 | 3 | 5 | 3 | 6 |
| 3 | 11 | 5 | 1 | 24 | 25 | 22 | 18 | 10 | 3 | 14 | 7 | 22 | 20 | 11 | 6 | 4 | 6 | 5 | 8 |
| 4 | 3 | 1 | 0 | 14 | 13 | 15 | 15 | 10 | 0 | 4 | 2 | 10 | 14 | 11 | 7 | 5 | 7 | 6 | 8 |
| 5 | 1 | 0 | 0 | 7 | 6 | 10 | 11 | 9 | 0 | 1 | 0 | 4 | 10 | 10 | 8 | 6 | 7 | 6 | 8 |
| 6 | 1 | 0 | 0 | 4 | 3 | 6 | 8 | 8 | 0 | 1 | 0 | 2 | 7 | 9 | 8 | 6 | 6 | 6 | 7 |
| 7 | 0 | 0 | 0 | 2 | 1 | 4 | 6 | 7 | 0 | 0 | 0 | 0 | 5 | 7 | 7 | 6 | 6 | 6 | 7 |
| 8 | 0 | 0 | 0 | 0 | 0 | 2 | 5 | 6 | 0 | 0 | 0 | 0 | 4 | 6 | 7 | 6 | 6 | 6 | 6 |
| 9 | 0 | 0 | 0 | 0 | 0 | 1 | 3 | 5 | 0 | 0 | 0 | 0 | 3 | 5 | 6 | 6 | 5 | 5 | 5 |
| 10 | 0 | 0 | 0 | 0 | 0 | 1 | 3 | 5 | 0 | 0 | 0 | 0 | 2 | 5 | 5 | 6 | 5 | 5 | 5 |
| 11 | 0 | 0 | 0 | 0 | 0 | 1 | 2 | 4 | 0 | 0 | 0 | 0 | 1 | 4 | 5 | 5 | 5 | 5 | 5 |
| 12 | 0 | 0 | 0 | 0 | 0 | 0 | 1 | 4 | 0 | 0 | 0 | 0 | 1 | 3 | 5 | 5 | 4 | 5 | 4 |
| 13 | 0 | 0 | 0 | 0 | 0 | 0 | 1 | 3 | 0 | 0 | 0 | 0 | 1 | 3 | 4 | 5 | 4 | 4 | 4 |
| 14 | 0 | 0 | 0 | 0 | 0 | 0 | 1 | 3 | 0 | 0 | 0 | 0 | 0 | 3 | 4 | 4 | 4 | 4 | 3 |
| 15 | 0 | 0 | 0 | 0 | 0 | 0 | 1 | 3 | 0 | 0 | 0 | 0 | 0 | 2 | 3 | 4 | 4 | 4 | 3 |
| 16 | 0 | 0 | 0 | 0 | 0 | 0 | 0 | 2 | 0 | 0 | 0 | 0 | 0 | 2 | 3 | 4 | 3 | 4 | 3 |
| 17 | 0 | 0 | 0 | 0 | 0 | 0 | 0 | 2 | 0 | 0 | 0 | 0 | 0 | 2 | 3 | 4 | 3 | 4 | 3 |
| 18 | 0 | 0 | 0 | 0 | 0 | 0 | 0 | 2 | 0 | 0 | 0 | 0 | 0 | 1 | 3 | 3 | 3 | 3 | 2 |
| 19 | 0 | 0 | 0 | 0 | 0 | 0 | 0 | 2 | 0 | 0 | 0 | 0 | 0 | 1 | 2 | 3 | 3 | 3 | 2 |
| 20 | 0 | 0 | 0 | 0 | 0 | 0 | 0 | 1 | 0 | 0 | 0 | 0 | 0 | 1 | 2 | 3 | 3 | 3 | 2 |
| 21 | 0 | 0 | 0 | 0 | 0 | 0 | 0 | 1 | 0 | 0 | 0 | 0 | 0 | 1 | 2 | 3 | 3 | 3 | 2 |
| 22 | 0 | 0 | 0 | 0 | 0 | 0 | 0 | 1 | 0 | 0 | 0 | 0 | 0 | 1 | 2 | 3 | 2 | 2 | 2 |
| 23 | 0 | 0 | 0 | 0 | 0 | 0 | 0 | 0 | 0 | 0 | 0 | 0 | 0 | 1 | 1 | 2 | 2 | 2 | 2 |
| | 100 | 100 | 100 | 100 | 100 | 100 | 100 | 100 | 100 | 100 | 100 | 100 | 100 | 100 | 100 | 100 | 100 | 100 | 100 |
| Layer ID from outside to inside (see Table 7-30) | F01 | F01 | F01 | F01 | F01 | F01 | F01 | F01 | F01 | F01 | F01 | F01 | F01 | F01 | F01 | F01 | F01 | F01 | F01 |
| | F08 | F08 | F08 | F12 | F14 | F15 | F13 | F13 | F13 | F13 | F13 | F13 | M17 | F13 | F13 | F13 | F13 | F13 | F13 |
| | G03 | G03 | G03 | G05 | G05 | G05 | G03 | G03 | G03 | G03 | G03 | G03 | F13 | G03 | G03 | G03 | G03 | G03 | M14 |
| | F05 | F05 | F05 | F05 | F05 | F05 | I02 | I02 | I02 | I02 | I03 | I02 | G03 | I03 | I03 | I03 | I03 | I03 | F05 |
| | I05 | I05 | I05 | I05 | I05 | I05 | G06 | G06 | F08 | F08 | F08 | I03 | I03 | M11 | M12 | M13 | M14 | M15 | I05 |
| | G01 | F05 | F03 | F05 | F05 | F05 | F03 | F05 | F03 | F05 | F03 | F08 | F08 | F03 | F03 | F03 | F03 | F03 | F16 |
| | F03 | F16 | — | G01 | G01 | G01 | — | F16 | — | F16 | — | — | F03 | — | — | — | — | — | F03 |
| | — | F03 | — | F03 | F03 | F03 | — | F03 | — | F03 | — | — | — | — | — | — | — | — | — |

**Roof Number Descriptions**

1. Metal roof, R-19 batt insulation, gyp board
2. Metal roof, R-19 batt insulation, suspended acoustical ceiling
3. Metal roof, R-19 batt insulation
4. Asphalt shingles, wood sheathing, R-19 batt insulation, gyp board
5. Slate or tile, wood sheathing, R-19 batt insulation, gyp board
6. Wood shingles, wood sheathing, R-19 batt insulation, gyp board
7. Membrane, sheathing, R-10 insulation board, wood deck
8. Membrane, sheathing, R-10 insulation board, wood deck, suspended acoustical ceiling
9. Membrane, sheathing, R-10 insulation board, metal deck
10. Membrane, sheathing, R-10 insulation board, metal deck, suspended acoustical ceiling
11. Membrane, sheathing, R-15 insulation board, metal deck
12. Membrane, sheathing, R-10 plus R-15 insulation boards, metal deck
13. 2-in. concrete roof ballast, membrane, sheathing, R-15 insulation board, metal deck
14. Membrane, sheathing, R-15 insulation board, 4-in. LW concrete
15. Membrane, sheathing, R-15 insulation board, 6-in. LW concrete
16. Membrane, sheathing, R-15 insulation board, 8-in. LW concrete
17. Membrane, sheathing, R-15 insulation board, 6-in. HW concrete
18. Membrane, sheathing, R-15 insulation board, 8-in. HW concrete
19. Membrane, 6-in HW concrete, R-19 batt insulation, suspended acoustical ceiling

A combined warm-up/safety allowance of 20% to 25% is fairly common but varies depending on the particular climate, building use, and type of construction. Engineering judgment must be applied for the particular project.

## 7.7 Design Loads Calculation Example

To illustrate the cooling and heating design load calculation procedures presented in this chapter, as taken from Chapter 18, 2013 *ASHRAE Handbook—Fundamentals*, an example problem was developed by the ASHRAE Technical Committee responsible for Chapter 18.

This example problem has been developed based on the ASHRAE headquarters building located in Atlanta, Georgia. This example is a two-story office building of approximately 35,000 ft², including a variety of common office functions and occupancies. In addition to demonstrating calculation procedures, a hypothetical design/construction process is discussed to illustrate (1) application of load calculations and (2)

### Table 7-32 Thermal Properties and Code Numbers of Layers Used in Wall and Roof Descriptions for Tables 7-28 and 7-29
(Table 18, Chapter 18, *2013 ASHRAE Handbook—Fundamentals*)

| Layer ID | Description | Thickness, in. | Conductivity, Btu·in/h·ft²·°F | Density, lb/ft³ | Specific Heat, Btu/lb·°F | Resistance, ft²·°F·h/Btu | R | Mass, lb/ft² | Thermal Capacity, Btu/ft²·°F | Notes |
|---|---|---|---|---|---|---|---|---|---|---|
| F01 | Outside surface resistance | — | — | — | — | 0.25 | 0.25 | — | — | 1 |
| F02 | Inside vertical surface resistance | — | — | — | — | 0.68 | 0.68 | — | — | 2 |
| F03 | Inside horizontal surface resistance | — | — | — | — | 0.92 | 0.92 | — | — | 3 |
| F04 | Wall air space resistance | — | — | — | — | 0.87 | 0.87 | — | — | 4 |
| F05 | Ceiling air space resistance | — | — | — | — | 1.00 | 1.00 | — | — | 5 |
| F06 | EIFS finish | 0.375 | 5.00 | 116.0 | 0.20 | — | 0.08 | 3.63 | 0.73 | 6 |
| F07 | 1 in. stucco | 1.000 | 5.00 | 116.0 | 0.20 | — | 0.20 | 9.67 | 1.93 | 6 |
| F08 | Metal surface | 0.030 | 314.00 | 489.0 | 0.12 | — | 0.00 | 1.22 | 0.15 | 7 |
| F09 | Opaque spandrel glass | 0.250 | 6.90 | 158.0 | 0.21 | — | 0.04 | 3.29 | 0.69 | 8 |
| F10 | 1 in. stone | 1.000 | 22.00 | 160.0 | 0.19 | — | 0.05 | 13.33 | 2.53 | 9 |
| F11 | Wood siding | 0.500 | 0.62 | 37.0 | 0.28 | — | 0.81 | 1.54 | 0.43 | 10 |
| F12 | Asphalt shingles | 0.125 | 0.28 | 70.0 | 0.30 | — | 0.44 | 0.73 | 0.22 | |
| F13 | Built-up roofing | 0.375 | 1.13 | 70.0 | 0.35 | — | 0.33 | 2.19 | 0.77 | |
| F14 | Slate or tile | 0.500 | 11.00 | 120.0 | 0.30 | — | 0.05 | 5.00 | 1.50 | |
| F15 | Wood shingles | 0.250 | 0.27 | 37.0 | 0.31 | — | 0.94 | 0.77 | 0.24 | |
| F16 | Acoustic tile | 0.750 | 0.42 | 23.0 | 0.14 | — | 1.79 | 1.44 | 0.20 | 11 |
| F17 | Carpet | 0.500 | 0.41 | 18.0 | 0.33 | — | 1.23 | 0.75 | 0.25 | 12 |
| F18 | Terrazzo | 1.000 | 12.50 | 160.0 | 0.19 | — | 0.08 | 13.33 | 2.53 | 13 |
| G01 | 5/8 in. gyp board | 0.625 | 1.11 | 50.0 | 0.26 | — | 0.56 | 2.60 | 0.68 | |
| G02 | 5/8 in. plywood | 0.625 | 0.80 | 34.0 | 0.29 | — | 0.78 | 1.77 | 0.51 | |
| G03 | 1/2 in. fiberboard sheathing | 0.500 | 0.47 | 25.0 | 0.31 | — | 1.06 | 1.04 | 0.32 | 14 |
| G04 | 1/2 in. wood | 0.500 | 1.06 | 38.0 | 0.39 | — | 0.47 | 1.58 | 0.62 | 15 |
| G05 | 1 in. wood | 1.000 | 1.06 | 38.0 | 0.39 | — | 0.94 | 3.17 | 1.24 | 15 |
| G06 | 2 in. wood | 2.000 | 1.06 | 38.0 | 0.39 | — | 1.89 | 6.33 | 2.47 | 15 |
| G07 | 4 in. wood | 4.000 | 1.06 | 38.0 | 0.39 | — | 3.77 | 12.67 | 4.94 | 15 |
| I01 | R-5, 1 in. insulation board | 1.000 | 0.20 | 2.7 | 0.29 | — | 5.00 | 0.23 | 0.07 | 16 |
| I02 | R-10, 2 in. insulation board | 2.000 | 0.20 | 2.7 | 0.29 | — | 10.00 | 0.45 | 0.13 | 16 |
| I03 | R-15, 3 in. insulation board | 3.000 | 0.20 | 2.7 | 0.29 | — | 15.00 | 0.68 | 0.20 | 16 |
| I04 | R-11, 3-1/2 in. batt insulation | 3.520 | 0.32 | 1.2 | 0.23 | — | 11.00 | 0.35 | 0.08 | 17 |
| I05 | R-19, 6-1/4 in. batt insulation | 6.080 | 0.32 | 1.2 | 0.23 | — | 19.00 | 0.61 | 0.14 | 17 |
| I06 | R-30, 9-1/2 in. batt insulation | 9.600 | 0.32 | 1.2 | 0.23 | — | 30.00 | 0.96 | 0.22 | 17 |
| M01 | 4 in. brick | 4.000 | 6.20 | 120.0 | 0.19 | — | 0.65 | 40.00 | 7.60 | 18 |
| M02 | 6 in. LW concrete3 block | 6.000 | 3.39 | 32.0 | 0.21 | — | 1.77 | 16.00 | 3.36 | 19 |
| M03 | 8 in. LW concrete block | 8.000 | 3.44 | 29.0 | 0.21 | — | 2.33 | 19.33 | 4.06 | 20 |
| M04 | 12 in. LW concrete block | 12.000 | 4.92 | 32.0 | 0.21 | — | 2.44 | 32.00 | 6.72 | 21 |
| M05 | 8 in. concrete block | 8.000 | 7.72 | 50.0 | 0.22 | — | 1.04 | 33.33 | 7.33 | 22 |
| M06 | 12 in. concrete block | 12.000 | 9.72 | 50.0 | 0.22 | — | 1.23 | 50.00 | 11.00 | 23 |
| M07 | 6 in. LW concrete block (filled) | 6.000 | 1.98 | 32.0 | 0.21 | — | 3.03 | 16.00 | 3.36 | 24 |
| M08 | 8 in. LW concrete block (filled) | 8.000 | 1.80 | 29.0 | 0.21 | — | 4.44 | 19.33 | 4.06 | 25 |
| M09 | 12 in. LW concrete block (filled) | 12.000 | 2.04 | 32.0 | 0.21 | — | 5.88 | 32.00 | 6.72 | 26 |
| M10 | 8 in. concrete block (filled) | 8.000 | 5.00 | 50.0 | 0.22 | — | 1.60 | 33.33 | 7.33 | 27 |
| M11 | 4 in. lightweight concrete | 4.000 | 3.70 | 80.0 | 0.20 | — | 1.08 | 26.67 | 5.33 | |
| M12 | 6 in. lightweight concrete | 6.000 | 3.70 | 80.0 | 0.20 | — | 1.62 | 40.00 | 8.00 | |
| M13 | 8 in. lightweight concrete | 8.000 | 3.70 | 80.0 | 0.20 | — | 2.16 | 53.33 | 10.67 | |
| M14 | 6 in. heavyweight concrete | 6.000 | 13.50 | 140.0 | 0.22 | — | 0.44 | 70.00 | 15.05 | |
| M15 | 8 in. heavyweight concrete | 8.000 | 13.50 | 140.0 | 0.22 | — | 0.59 | 93.33 | 20.07 | |
| M16 | 12 in. heavyweight concrete | 12.000 | 13.50 | 140.0 | 0.22 | — | 0.89 | 140.0 | 30.10 | |
| M17 | 2 in. LW concrete roof ballast | 2.000 | 1.30 | 40 | 0.20 | — | 1.54 | 6.7 | 1.33 | 28 |

*Notes*: The following notes give sources for the data in this table.
1. Chapter 26, Table 1 for 7.5 mph wind
2. Chapter 26, Table 1 for still air, horizontal heat flow
3. Chapter 26, Table 1 for still air, downward heat flow
4. Chapter 26, Table 3 for 1.5 in. space, 90°F, horizontal heat flow, 0.82 emittance
5. Chapter 26, Table 3 for 3.5 in. space, 90°F, downward heat flow, 0.82 emittance
6. EIFS finish layers approximated by Chapter 26, Table 4 for 3/8 in. cement plaster, sand aggregate
7. Chapter 33, Table 3 for steel (mild)
8. Chapter 26, Table 4 for architectural glass
9. Chapter 26, Table 4 for marble and granite
10. Chapter 26, Table 4, density assumed same as Southern pine
11. Chapter 26, Table 4 for mineral fiberboard, wet molded, acoustical tile
12. Chapter 26, Table 4 for carpet and rubber pad, density assumed same as fiberboard
13. Chapter 26, Table 4, density assumed same as stone
14. Chapter 26, Table 4 for nail-base sheathing
15. Chapter 26, Table 4 for Southern pine
16. Chapter 26, Table 4 for expanded polystyrene
17. Chapter 26, Table 4 for glass fiber batt, specific heat per glass fiber board
18. Chapter 26, Table 4 for clay fired brick
19. Chapter 26, Table 4, 16 lb block, 8 in. × 16 in. face
20. Chapter 26, Table 4, 19 lb block, 8 in. × 16 in. face
21. Chapter 26, Table 4, 32 lb block, 8 in. × 16 in. face
22. Chapter 26, Table 4, 33 lb normal weight block, 8 in. × 16 in. face
23. Chapter 26, Table 4, 50 lb normal weight block, 8 in. × 16 in. face
24. Chapter 26, Table 4, 16 lb block, vermiculite fill
25. Chapter 26, Table 4, 19 lb block, 8 in. × 16 in. face, vermiculite fill
26. Chapter 26, Table 4, 32 lb block, 8 in. × 16 in. face, vermiculite fill
27. Chapter 26, Table 4, 33 lb normal weight block, 8 in. × 16 in. face, vermiculite fill
28. Chapter 26, Table 4 for 40 lb/ft³ LW concrete

Table 7-33 Representative Nonsolar RTS Values for Light to Heavy Construction

| | Light | | | | | | Medium | | | | | | Heavy | | | | | | Interior Zones | | | | | |
|---|---|---|---|---|---|---|---|---|---|---|---|---|---|---|---|---|---|---|---|---|---|---|---|---|
| | | | | | | | | | | | | | | | | | | | Light | | Medium | | Heavy | |
| | With Carpet | | | No Carpet | | | With Carpet | | | No Carpet | | | With Carpet | | | No Carpet | | | With Carpet | No Carpet | With Carpet | No Carpet | With Carpet | No Carpet |
| % Glass | 10% | 50% | 90% | 10% | 50% | 90% | 10% | 50% | 90% | 10% | 50% | 90% | 10% | 50% | 90% | 10% | 50% | 90% | | | | | | |
| Hour | Radiant Time Factor, % | | | | | | | | | | | | | | | | | | | | | | | |
| 0 | 47 | 50 | 53 | 41 | 43 | 46 | 46 | 49 | 52 | 31 | 33 | 35 | 34 | 38 | 42 | 22 | 25 | 28 | 46 | 40 | 46 | 31 | 33 | 21 |
| 1 | 19 | 18 | 17 | 20 | 19 | 19 | 18 | 17 | 16 | 17 | 16 | 15 | 9 | 9 | 9 | 10 | 9 | 9 | 19 | 20 | 18 | 17 | 9 | 9 |
| 2 | 11 | 10 | 9 | 12 | 11 | 11 | 10 | 9 | 8 | 11 | 10 | 10 | 6 | 6 | 5 | 6 | 6 | 6 | 11 | 12 | 10 | 11 | 6 | 6 |
| 3 | 6 | 6 | 5 | 8 | 7 | 7 | 6 | 5 | 5 | 8 | 7 | 7 | 4 | 4 | 4 | 5 | 5 | 5 | 6 | 8 | 6 | 8 | 5 | 5 |
| 4 | 4 | 4 | 3 | 5 | 5 | 5 | 4 | 3 | 3 | 6 | 5 | 5 | 4 | 4 | 4 | 5 | 5 | 4 | 4 | 5 | 3 | 6 | 4 | 5 |
| 5 | 3 | 3 | 2 | 4 | 3 | 3 | 2 | 2 | 2 | 4 | 4 | 4 | 4 | 3 | 3 | 4 | 4 | 4 | 3 | 4 | 2 | 4 | 4 | 4 |
| 6 | 2 | 2 | 2 | 3 | 3 | 2 | 2 | 2 | 2 | 4 | 3 | 3 | 3 | 3 | 3 | 4 | 4 | 4 | 2 | 3 | 2 | 4 | 3 | 4 |
| 7 | 2 | 1 | 1 | 2 | 2 | 2 | 1 | 1 | 1 | 3 | 3 | 3 | 3 | 3 | 3 | 4 | 4 | 4 | 2 | 2 | 1 | 3 | 3 | 4 |
| 8 | 1 | 1 | 1 | 1 | 1 | 1 | 1 | 1 | 1 | 3 | 2 | 2 | 3 | 3 | 3 | 4 | 3 | 3 | 1 | 1 | 1 | 3 | 3 | 4 |
| 9 | 1 | 1 | 1 | 1 | 1 | 1 | 1 | 1 | 1 | 2 | 2 | 2 | 3 | 3 | 2 | 3 | 3 | 3 | 1 | 1 | 1 | 2 | 3 | 3 |
| 10 | 1 | 1 | 1 | 1 | 1 | 1 | 1 | 1 | 1 | 2 | 2 | 2 | 3 | 2 | 2 | 3 | 3 | 3 | 1 | 1 | 1 | 2 | 3 | 3 |
| 11 | 1 | 1 | 1 | 1 | 1 | 1 | 1 | 1 | 1 | 2 | 2 | 2 | 2 | 2 | 2 | 3 | 3 | 3 | 1 | 1 | 1 | 2 | 2 | 3 |
| 12 | 1 | 1 | 1 | 1 | 1 | 1 | 1 | 1 | 1 | 1 | 1 | 1 | 2 | 2 | 2 | 3 | 3 | 3 | 1 | 1 | 1 | 1 | 2 | 3 |
| 13 | 1 | 1 | 1 | 0 | 1 | 0 | 1 | 1 | 1 | 1 | 1 | 1 | 2 | 2 | 2 | 3 | 3 | 2 | 1 | 1 | 1 | 1 | 2 | 3 |
| 14 | 0 | 0 | 1 | 0 | 1 | 0 | 1 | 1 | 1 | 1 | 1 | 1 | 2 | 2 | 2 | 3 | 2 | 2 | 1 | 0 | 1 | 1 | 2 | 3 |
| 15 | 0 | 0 | 1 | 0 | 0 | 0 | 1 | 1 | 1 | 1 | 1 | 1 | 2 | 2 | 2 | 2 | 2 | 2 | 0 | 0 | 1 | 1 | 2 | 3 |
| 16 | 0 | 0 | 0 | 0 | 0 | 0 | 1 | 1 | 1 | 1 | 1 | 1 | 2 | 2 | 2 | 2 | 2 | 2 | 0 | 0 | 1 | 1 | 2 | 3 |
| 17 | 0 | 0 | 0 | 0 | 0 | 0 | 1 | 1 | 1 | 1 | 1 | 1 | 2 | 2 | 2 | 2 | 2 | 2 | 0 | 0 | 1 | 1 | 2 | 2 |
| 18 | 0 | 0 | 0 | 0 | 0 | 0 | 1 | 1 | 1 | 1 | 1 | 1 | 2 | 2 | 1 | 2 | 2 | 2 | 0 | 0 | 1 | 1 | 2 | 2 |
| 19 | 0 | 0 | 0 | 0 | 0 | 0 | 0 | 1 | 0 | 0 | 1 | 1 | 2 | 2 | 1 | 2 | 2 | 2 | 0 | 0 | 1 | 0 | 2 | 2 |
| 20 | 0 | 0 | 0 | 0 | 0 | 0 | 0 | 0 | 0 | 0 | 1 | 1 | 2 | 1 | 1 | 2 | 2 | 2 | 0 | 0 | 0 | 0 | 2 | 2 |
| 21 | 0 | 0 | 0 | 0 | 0 | 0 | 0 | 0 | 0 | 0 | 1 | 1 | 2 | 1 | 1 | 2 | 2 | 2 | 0 | 0 | 0 | 0 | 2 | 2 |
| 22 | 0 | 0 | 0 | 0 | 0 | 0 | 0 | 0 | 0 | 0 | 1 | 0 | 1 | 1 | 1 | 2 | 2 | 2 | 0 | 0 | 0 | 0 | 1 | 2 |
| 23 | 0 | 0 | 0 | 0 | 0 | 0 | 0 | 0 | 0 | 0 | 0 | 0 | 1 | 1 | 1 | 2 | 2 | 1 | 0 | 0 | 0 | 0 | 1 | 2 |
| | 100 | 100 | 100 | 100 | 100 | 100 | 100 | 100 | 100 | 100 | 100 | 100 | 100 | 100 | 100 | 100 | 100 | 100 | 100 | 100 | 100 | 100 | 100 | 100 |

Table 7-34 Representative Solar RTS Values for Light to Heavy Construction

| | Light | | | | | | Medium | | | | | | Heavy | | | | | |
|---|---|---|---|---|---|---|---|---|---|---|---|---|---|---|---|---|---|---|
| | With Carpet | | | No Carpet | | | With Carpet | | | No Carpet | | | With Carpet | | | No Carpet | | |
| % Glass | 10% | 50% | 90% | 10% | 50% | 90% | 10% | 50% | 90% | 10% | 50% | 90% | 10% | 50% | 90% | 10% | 50% | 90% |
| Hour | Radiant Time Factor, % | | | | | | | | | | | | | | | | | |
| 0 | 53 | 55 | 56 | 44 | 45 | 46 | 52 | 54 | 55 | 28 | 29 | 29 | 47 | 49 | 51 | 26 | 27 | 28 |
| 1 | 17 | 17 | 17 | 19 | 20 | 20 | 16 | 16 | 15 | 15 | 15 | 15 | 11 | 12 | 12 | 12 | 13 | 13 |
| 2 | 9 | 9 | 9 | 11 | 11 | 11 | 8 | 8 | 8 | 10 | 10 | 10 | 6 | 6 | 6 | 7 | 7 | 7 |
| 3 | 5 | 5 | 5 | 7 | 7 | 7 | 5 | 4 | 4 | 7 | 7 | 7 | 4 | 4 | 3 | 5 | 5 | 5 |
| 4 | 3 | 3 | 3 | 5 | 5 | 5 | 3 | 3 | 3 | 6 | 6 | 6 | 3 | 3 | 3 | 4 | 4 | 4 |
| 5 | 2 | 2 | 2 | 3 | 3 | 3 | 2 | 2 | 2 | 5 | 5 | 5 | 2 | 2 | 2 | 4 | 4 | 4 |
| 6 | 2 | 2 | 2 | 3 | 2 | 2 | 2 | 1 | 1 | 4 | 4 | 4 | 2 | 2 | 2 | 3 | 3 | 3 |
| 7 | 1 | 1 | 1 | 2 | 2 | 2 | 1 | 1 | 1 | 4 | 3 | 3 | 2 | 2 | 2 | 3 | 3 | 3 |
| 8 | 1 | 1 | 1 | 1 | 1 | 1 | 1 | 1 | 1 | 3 | 3 | 3 | 2 | 2 | 2 | 3 | 3 | 3 |
| 9 | 1 | 1 | 1 | 1 | 1 | 1 | 1 | 1 | 1 | 3 | 3 | 3 | 2 | 2 | 2 | 3 | 3 | 3 |
| 10 | 1 | 1 | 1 | 1 | 1 | 1 | 1 | 1 | 1 | 2 | 2 | 2 | 2 | 2 | 2 | 3 | 3 | 3 |
| 11 | 1 | 1 | 1 | 1 | 1 | 1 | 1 | 1 | 1 | 2 | 2 | 2 | 2 | 1 | 1 | 3 | 3 | 2 |
| 12 | 1 | 1 | 1 | 1 | 1 | 0 | 1 | 1 | 1 | 2 | 2 | 2 | 1 | 1 | 1 | 2 | 2 | 2 |
| 13 | 1 | 1 | 0 | 1 | 0 | 0 | 1 | 1 | 1 | 2 | 2 | 2 | 2 | 1 | 1 | 2 | 2 | 2 |
| 14 | 1 | 0 | 0 | 0 | 0 | 0 | 1 | 1 | 1 | 1 | 1 | 1 | 2 | 1 | 1 | 2 | 2 | 2 |
| 15 | 1 | 0 | 0 | 0 | 0 | 0 | 1 | 1 | 1 | 1 | 1 | 1 | 1 | 1 | 1 | 2 | 2 | 2 |
| 16 | 0 | 0 | 0 | 0 | 0 | 0 | 1 | 1 | 1 | 1 | 1 | 1 | 1 | 1 | 1 | 2 | 2 | 2 |
| 17 | 0 | 0 | 0 | 0 | 0 | 0 | 1 | 1 | 1 | 1 | 1 | 1 | 1 | 1 | 1 | 2 | 2 | 2 |
| 18 | 0 | 0 | 0 | 0 | 0 | 0 | 1 | 1 | 1 | 1 | 1 | 1 | 1 | 1 | 1 | 2 | 2 | 2 |
| 19 | 0 | 0 | 0 | 0 | 0 | 0 | 0 | 0 | 0 | 1 | 1 | 1 | 1 | 1 | 1 | 2 | 2 | 2 |
| 20 | 0 | 0 | 0 | 0 | 0 | 0 | 0 | 0 | 0 | 1 | 1 | 1 | 1 | 1 | 1 | 2 | 2 | 2 |
| 21 | 0 | 0 | 0 | 0 | 0 | 0 | 0 | 0 | 0 | 0 | 0 | 0 | 1 | 1 | 1 | 2 | 2 | 2 |
| 22 | 0 | 0 | 0 | 0 | 0 | 0 | 0 | 0 | 0 | 0 | 0 | 0 | 1 | 1 | 1 | 2 | 1 | 1 |
| 23 | 0 | 0 | 0 | 0 | 0 | 0 | 0 | 0 | 0 | 0 | 0 | 0 | 1 | 1 | 1 | 2 | 1 | 1 |
| | 100 | 100 | 100 | 100 | 100 | 100 | 100 | 100 | 100 | 100 | 100 | 100 | 100 | 100 | 100 | 100 | 100 | 100 |

**Table 7-35  RTS Representative Zone Construction for Tables 7-31 and 7-32**

| Construction Class | Exterior Wall | Roof/Ceiling | Partitions | Floor | Furnishings |
|---|---|---|---|---|---|
| Light | steel siding, 2 in. insulation, air space, 3/4 in. gyp | 4 in. LW concrete, ceiling air space, acoustic tile | 3/4 in. gyp, air space, 3/4 in. gyp | acoustic tile, ceiling air space, 4 in. LW concrete | 1 in. wood at 50% of floor area |
| Medium | 4 in. face brick, 2 in. insulation, air space, 3/4 in. gyp | 4 in. HW concrete, ceiling air space, acoustic tile | 3/4 in. gyp, air space, 3/4 in. gyp | acoustic tile, ceiling air space, 4 in. HW concrete | 1 in. wood at 50% of floor area |
| Heavy | 4 in. face brick, 8 in. HW concrete air space, 2 in. insulation, 3/4 in. gyp | 8 in. HW concrete, ceiling air space, acoustic tile | 3/4 in. gyp, 8 in. HW concrete block, 3/4 in. gyp | acoustic tile, ceiling air space, 8 in. HW concrete | 1 in. wood at 50% of floor area |

the need to develop reasonable assumptions when specific data is not yet available, as often occurs in everyday design processes.

Table 7-36 provides a summary of RTS load calculation procedures.

### 7.7.1 Single-Room Example

Calculate the peak heating and cooling loads for the office room shown in Figure 7-10, for Atlanta, Georgia. The room is on the second floor of a two-story building and has two vertical exterior exposures, with a flat roof above. Calculate the peak heating and cooling loads for the office room shown in Figure 7-10, for Atlanta, Georgia. The room is on the second floor of a two-story building and has two vertical exterior exposures, with a flat roof above.

**Room Characteristics.**

*Area*: 130 ft$^2$.

*Floor*: Carpeted 5 in. concrete slab on metal deck above a conditioned space.

*Roof*: Flat metal deck topped with rigid closed-cell polyisocyanurate foam core insulation (R = 30), and light-colored membrane roofing. Space above 9 ft suspended acoustical tile ceiling is used as a return air plenum. Assume 30% of the cooling load from the roof is directly absorbed in the return airstream without becoming room load. Use roof U = 0.032 Btu/h·ft$^2$·°F.

*Spandrel wall*: Spandrel bronze-tinted glass, opaque, backed with air space, rigid mineral fiber insulation (R = 5.0), mineral fiber batt insulation (R = 13), and 5/8 in. gypsum wall board. Use spandrel wall U = 0.077 Btu/h·ft$^2$·°F.

*Brick wall*: Light-brown-colored face brick (4 in.), lightweight concrete block (6 in.), rigid continuous insulation (R = 5), mineral fiber batt insulation (R = 13), and gypsum wall board (5/8 in.). Use brick wall U = 0.08 Btu/h·ft$^2$·°F.

*Windows*: Double glazed, 1/4 in. bronze-tinted outdoor pane, 1/2 in. air space and 1/4 in. clear indoor pane with light-colored interior miniblinds. Window normal solar heat gain coefficient (SHGC) = 0.49. Windows are nonoperable and mounted in aluminum frames with thermal breaks having overall combined U = 0.56 Btu/h·ft$^2$·°F (based on Type 5d from Tables 4 and 10 of Chapter 15 of 2013 *ASHRAE Handbook—Fundamentals*). Indoor attenuation coefficients (IACs) for indoor miniblinds are based on light venetian blinds (assumed louver reflectance = 0.8 and louvers positioned at 45° angle) with heat-absorbing double glazing

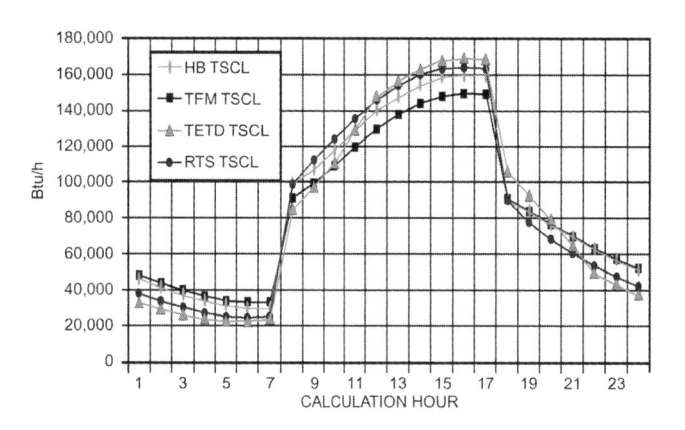

*Fig. 7-9 Load Profile Comparison*
*(Figure 13, Chapter; 29, 2001 ASHRAE Handbook—Fundamentals)*

(Type 5d from Table 13B of Chapter 15 of 2013 *ASHRAE Handbook—Fundamentals*), IAC(0) = 0.74, IAC(60) = 0.65, IAD(diff) = 0.79, and radiant fraction = 0.54. Each window is 6.25 ft 1.91 m wide by 6.4 ft tall for an area per window = 40 ft$^2$.

*South exposure*:

| | |
|---|---|
| Orientation | = 30° east of true south |
| Window area | = 40 ft$^2$ |
| Spandrel wall area | = 60 ft$^2$ |
| Brick wall area | = 60 ft$^2$ |

*West exposure*: 

| | |
|---|---|
| Orientation | = 60° west of south |
| Window area | = 40 ft$^2$ |
| Spandrel wall area | = 60 ft$^2$ |
| Brick wall area | = 40 ft$^2$ |

*Occupancy*: 1 person from 8:00 AM to 5:00 PM.

*Lighting*: One 4-lamp pendant fluorescent 8 ft type. The fixture has four 32 W T-8 lamps plus electronic ballasts (special allowance factor 0.85 per manufacturer's data), for a total of 110 W for the room. Operation is from 7:00 AM to 7:00 PM. Assume 0% of the cooling load from lighting is directly absorbed in the return airstream without becoming room load, per Table 3 in Chapter 18 of 2013 *ASHRAE Handbook—Fundamentals*.

*Equipment*: One computer and a personal printer are used, for which an allowance of 1 W/ft$^2$ is to be accommodated by the cooling system, for a total of 130 W for the room. Operation is from 8:00 AM to 5:00 PM.

## Table 7-36 Summary of RTS Load Calculation Procedures

| Equation | Equation No. in Chapter |
|---|---|

### External Heat Gain

*Sol-Air Temperature*

$$t_e = t_o + \frac{\alpha E_t}{h_o} - \frac{\varepsilon \Delta R}{h_o} \quad (30)$$

where
- $t_e$ = sol-air temperature, °F
- $t_o$ = outdoor air temperature, °F
- $a$ = absorptance of surface for solar radiation
- $E_t$ = total solar radiation incident on surface, Btu/h·ft²
- $h_o$ = coefficient of heat transfer by long-wave radiation and convection at outer surface, Btu/h·ft²·°F
- $\varepsilon$ = hemispherical emittance of surface
- $\Delta R$ = difference between long-wave radiation incident on surface from sky and surroundings and radiation emitted by blackbody at outdoor air temperature, Btu/h·ft²; 20 for horizontal surfaces; 0 for vertical surfaces

*Wall and Roof Transmission*

$$q_\theta = c_0 q_{i,\theta} + c_1 q_{i,\theta-1} + c_2 q_{i,\theta-2} + \cdots + c_{23} q_{i,\theta-23} \quad (32)$$

$$q_{i,\theta-n} = UA(t_{e,\theta-n} - t_{rc}) \quad (31)$$

where
- $q_\theta$ = hourly conductive heat gain for surface, Btu/h
- $q_{i,\theta}$ = heat input for current hour
- $q_{i,\theta-n}$ = conductive heat input for surface n hours ago, Btu/h
- $c_0, c_1$, etc. = conduction time factors
- $U$ = overall heat transfer coefficient for surface, Btu/h·ft²·°F
- $A$ = surface area, ft²
- $t_{e,\theta-n}$ = sol-air temperature $n$ hours ago, °F
- $t_{rc}$ = presumed constant room air temperature, °F

*Fenestration Transmission*

$$q_c = UA(T_{out} - T_{in}) \quad (15)$$

where
- $q$ = fenestration transmission heat gain, Btu/h
- $U$ = overall U-factor, including frame and mounting orientation from Table 4 of Chapter 15, Btu/h·ft²·°F
- $A$ = window area, ft²
- $T_{in}$ = indoor temperature, °F
- $T_{out}$ = outdoor temperature, °F

*Fenestration Solar*

$$q_b = AE_{t,b}\,\text{SHGC}(\theta)\,\text{IAC}(\theta,\Omega) \quad (13)$$

$$q_d = A(E_{t,d} + E_{t,r})\langle\text{SHGC}\rangle_D\,\text{IAC}_D \quad (14)$$

where
- $q_b$ = beam solar heat gain, Btu/h
- $q_d$ = diffuse solar heat gain, Btu/h
- $A$ = window area, ft²
- $E_{t,b}, E_{t,d}$, and $E_{t,r}$ = beam, sky diffuse, and ground-reflected diffuse irradiance, calculated using equations in Chapter 14
- SHGC(θ) = beam solar heat gain coefficient as a function of incident angle θ; may be interpolated between values in Table 10 of Chapter 15
- $\langle\text{SHGC}\rangle_D$ = diffuse solar heat gain coefficient (also referred to as hemispherical SHGC); from Table 10 of Chapter 15
- IAC(θ.Ω) = indoor solar attenuation coefficient for beam solar heat gain coefficient; = 1.0 if no indoor shading device. IAC(θ.Ω) is a function of shade type and, depending on type, may also be a function of beam solar angle of incidence θ and shade geometry
- $\text{IAC}_D$ = indoor solar attenuation coefficient for diffuse solar heat gain coefficient; = 1.0 if not indoor shading device. $\text{IAC}_D$ is a function of shade type and, depending on type, may also be a function of shade geometry

*Partitions, Ceilings, Floors Transmission*

$$q = UA(t_b - t_i) \quad (33)$$

where
- $q$ = heat transfer rate, Btu/h
- $U$ = coefficient of overall heat transfer between adjacent and conditioned space, Btu/h·ft²·°F
- $A$ = area of separating section concerned, ft²
- $t_b$ = average air temperature in adjacent space, °F
- $t_i$ = air temperature in conditioned space, °F

### Internal Heat Gain

*Occupants*

$$q_s = q_{s,\text{per}} N$$
$$q_l = q_{l,\text{per}} N$$

where
- $q_s$ = occupant sensible heat gain, Btu/h
- $q_l$ = occupant latent heat gain, Btu/h
- $q_{s,\text{per}}$ = sensible heat gain per person, Btu/h·person; see Table 1
- $q_{l,\text{per}}$ = latent heat gain per person, Btu/h·person; see Table 1
- $N$ = number of occupants

*Lighting*

$$q_{el} = 3.41 W F_{ul} F_{sa} \quad (1)$$

where
- $q_{el}$ = heat gain, Btu/h
- $W$ = total light wattage, W
- $F_{ul}$ = lighting use factor
- $F_{sa}$ = lighting special allowance factor
- 3.41 = conversion factor

*Electric Motors*

$$q_{em} = 2545(P/E_M) F_{UM} F_{LM} \quad (2)$$

where
- $q_{em}$ = heat equivalent of equipment operation, Btu/h
- $P$ = motor power rating, hp
- $E_M$ = motor efficiency, decimal fraction <1.0
- $F_{UM}$ = motor use factor, 1.0 or decimal fraction <1.0
- $F_{LM}$ = motor load factor, 1.0 or decimal fraction <1.0
- 2545 = conversion factor, Btu/h·hp

*Hooded Cooking Appliances*

$$q_s = q_{\text{input}} F_U F_R$$

where
- $q_s$ = sensible heat gain, Btu/h
- $q_{\text{input}}$ = nameplate or rated energy input, Btu/h
- $F_U$ = usage factor; see Tables 5B, 5C, 5D
- $F_R$ = radiation factor; see Tables 5B, 5C, 5D

# Nonresidential Cooling and Heating Load Calculations

Table 7-36  Summary of RTS Load Calculation Procedures (*Concluded*)

| Equation | Equation No. in Chapter | Equation | Equation No. in Chapter |
|---|---|---|---|
| For other appliances and equipment, find $q_s$ for  Unhooded cooking appliances: Table 5A  Other kitchen equipment: Table 5E  Hospital and laboratory equipment: Tables 6 and 7  Computers, printers, scanners, etc.: Tables 8 and 9  Miscellaneous office equipment: Table 10 | | *Radiant Portion of Sensible Cooling Load*  $$Q_{i,r} = Q_{r,\theta}$$  $Q_{r,\theta} = r_0 q_{r,\theta} + r_1 q_{r,\theta-1} + r_2 q_{r,\theta-2} + r_3 q_{r,\theta-3} + \cdots + r_{23} q_{r,\theta-23}$  where | (34) |

Find $q_l$ for
    Unhooded cooking appliances: Table 5A
    Other kitchen equipment: Table 5E

***Ventilation and Infiltration Air Heat Gain***

$$q_s = 1.10 Q_s \Delta t \quad (10)$$

$$q_l = 60 \times 0.075 \times 1076 Q_s \Delta W = 4840 Q_s \Delta W \quad (11)$$

where
    $q_s$ = sensible heat gain due to infiltration, Btu/h
    $q_l$ = latent heat gain due to infiltration, Btu/h
    $Q_s$ = infiltration airflow at standard air conditions, cfm
    $t_o$ = outdoor air temperature, °F
    $t_i$ = indoor air temperature, °F
    $W_o$ = outdoor air humidity ratio, lb/lb
    $W_i$ = indoor air humidity ratio, lb/lb
    1.10 = air sensible heat factor at standard air conditions, Btu/h·cfm
    4840 = air latent heat factor at standard air conditions, Btu/h·cfm

***Instantaneous Room Cooling Load***

$$Q_s = \Sigma Q_{i,r} + \Sigma Q_{i,c}$$

$$Q_l = \Sigma q_{i,l}$$

where
    $Q_s$ = room sensible cooling load, Btu/h
    $Q_{i,r}$ = radiant portion of sensible cooling load for current hour, resulting from heat gain element $i$, Btu/h
    $Q_{i,c}$ = convective portion of sensible cooling load, resulting from heat gain element $i$, Btu/h
    $Q_l$ = room latent cooling load, Btu/h
    $q_{i,l}$ = latent heat gain for heat gain element $i$, Btu/h

[Right column continuation:]

$Q_{r,\theta}$ = radiant cooling load $Q_r$ for current hour θ, Btu/h
$q_{r,\theta}$ = radiant heat gain for current hour, Btu/h
$q_{r,\theta-n}$ = radiant heat gain n hours ago, Btu/h
$r_0, r_1$, etc. = radiant time factors; see Table 19 for radiant time factors for nonsolar heat gains: wall, roof, partition, ceiling, floor, fenestration transmission heat gains, and occupant, lighting, motor, appliance heat gain. Also used for fenestration diffuse solar heat gain; see Table 20 for radiant time factors for fenestration beam solar heat gain.

$$q_{r,\theta} = q_{i,s} F_r$$

where
    $q_{i,s}$ = sensible heat gain from heat gain element $i$, Btu/h
    $F_r$ = fraction of heat gain that is radiant.

Data Sources:
    Wall transmission: see Table 14
    Roof transmission: see Table 14
    Floor transmission: see Table 14
    Fenestration transmission: see Table 14
    Fenestration solar heat gain: see Table 14, Chapter 18 and Tables 13A to 13G, Chapter 15
    Lighting: see Table 3
    Occupants: see Tables 1 and 14
    Hooded cooking appliances: see Tables 5B, 5C, and 5D
    Unhooded cooking appliances: see Table 5A
    Other appliances and equipment: see Tables 5E, 8, 9, 10, and 14
    Infiltration: see Table 14

*Convective Portion of Sensible Cooling Load*

$$Q_{i,c} = q_{i,c}$$

where $q_{i,c}$ is convective portion of heat gain from heat gain element $i$, Btu/h.

$$q_{i,c} = q_{i,s}(1 - F_r)$$

where
    $q_{i,s}$ = sensible heat gain from heat gain element $i$, Btu/h
    $F_r$ = fraction of heat gain that is radiant; see row for radiant portion for sources of radiant fraction data for individual heat gain elements

---

***Infiltration***: For purposes of this example, assume the building is maintained under positive pressure during peak cooling conditions and therefore has no infiltration. Assume that infiltration during peak heating conditions is equivalent to one air change per hour.

***Weather data***: Per Chapter 14 of 2013 *ASHRAE Handbook—Fundamentals*, for Atlanta, Georgia, latitude = 33.64, longitude = 84.43, elevation = 1027 ft above sea level, 99.6% heating design dry-bulb temperature = 21.5°F. For cooling load calculations, use 5% dry-bulb/coincident wet-bulb monthly design day profile calculated per Chapter 14. See Table 7-37 or temperature profiles used in these examples.

***Indoor design conditions***: 72°F for heating; 75°F with 50% rh for cooling.

**Cooling Loads Using RTS Method.** Traditionally, simplified cooling load calculation methods have estimated the total cooling load at a particular design condition by independently calculating and then summing the load from each component (walls, windows, people, lights, etc). Although the actual heat transfer processes for each component do affect each other, this simplification is appropriate for design load calculations and useful to the designer in understanding the relative contribution of each component to the total cooling load.

Cooling loads are calculated with the RTS method on a component basis similar to previous methods. The following example parts illustrate cooling load calculations for individual components of this single room for a particular hour and month. Equations used are summarized in Table 7-36.

**Part 1. Internal cooling load using radiant time series.** Calculate the cooling load from lighting at 3:00 PM for the previously described office.

**Solution:**

First calculate the 24 h heat gain profile for lighting, then split those heat gains into radiant and convective portions, apply the appropriate RTS to the radiant portion, and sum the convective and radiant cooling load components to determine total cooling load at the designated time. Using Equation 1, the lighting heat gain profile, based on the occupancy schedule indicated is

$q_1 = (110 \text{ W})3.41(0\%) = 0$
$q_2 = (110 \text{ W})3.41(0\%) = 0$
$q_3 = (110 \text{ W})3.41(0\%) = 0$
$q_4 = (110 \text{ W})3.41(0\%) = 0$
$q_5 = (110 \text{ W})3.41(0\%) = 0$
$q_6 = (110 \text{ W})3.41(0\%) = 0$
$q_7 = (110 \text{ W})3.41(100\%) = 375$
$q_8 = (110 \text{ W})3.41(100\%) = 375$
$q_9 = (110 \text{ W})3.41(100\%) = 375$
$q_{10} = (110 \text{ W})3.41(100\%) = 375$
$q_{11} = (110 \text{ W})3.41(100\%) = 375$
$q_{12} = (110 \text{ W})3.41(100\%) = 375$
$q_{13} = (110 \text{ W})3.41(100\%) = 375$
$q_{14} = (110 \text{ W})3.41(100\%) = 375$
$q_{15} = (110 \text{ W})3.41(100\%) = 375$
$q_{16} = (110 \text{ W})3.41(100\%) = 375$
$q_{17} = (110 \text{ W})3.41(100\%) = 375$
$q_{18} = (110 \text{ W})3.41(100\%) = 375$
$q_{19} = (110 \text{ W})3.41(0\%) = 0$
$q_{20} = (110 \text{ W})3.41(0\%) = 0$
$q_{21} = (110 \text{ W})3.41(0\%) = 0$
$q_{22} = (110 \text{ W})3.41(0\%) = 0$
$q_{23} = (110 \text{ W})3.41(0\%) = 0$
$q_{24} = (110 \text{ W})3.41(0\%) = 0$

The convective portion is simply the lighting heat gain for the hour being calculated times the convective fraction for non-in-ceiling fluorescent luminaire (pendant), from Table 3 of Chapter 18, 2013 *ASHRAE Handbook—Fundamentals*:

$$Q_{c,15} = (375)(43\%) = 161.3 \text{ Btu/h}$$

The radiant portion of the cooling load is calculated using lighting heat gains for the current hour and past 23 h, the radiant fraction from Table 3 of Chapter 18, 2013 *ASHRAE Handbook—Fundamentals* (57%), and radiant time series from Table 19, in accordance with Equation 34. From Table 19 of Chapter 18, 2013 *ASHRAE Handbook—Fundamentals*, select the RTS for medium-weight construction, assuming 50% glass and carpeted floors as representative of the described construction. Thus, the radiant cooling load for lighting is

$Q_{r,15} = r_0(0.48)q_{15} + r_1(0.48)q_{14} + r_2(0.48)q_{13} + r_3(0.48)q_{12}$
$\quad + \cdots + r_{23}(0.48)q_{16}$
$= (0.49)(0.57)(375) + (0.17)(0.57)(375)$
$+ (0.09)(0.57)(375) + (0.05)(0.57)(375) + (0.03)(0.57)(375)$

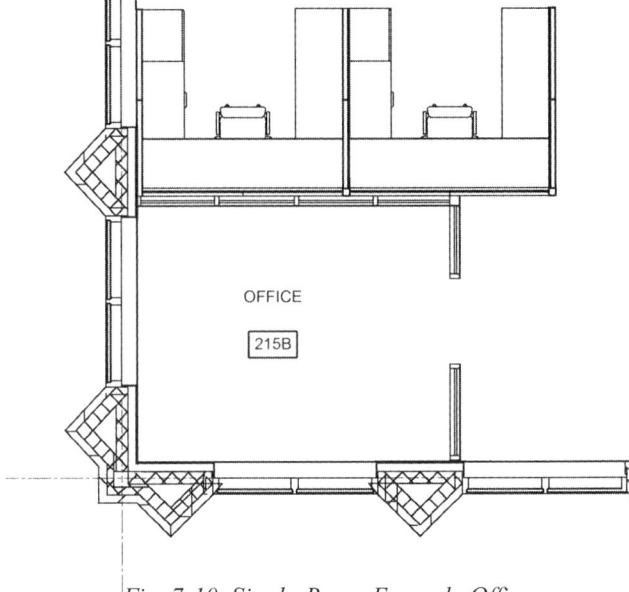

*Fig. 7-10 Single-Room Example Office*
*(Figure 16, Chapter 18, 2013 ASHRAE Handbook—Fundamentals)*

$+ (0.02)(0.57)(375) + (0.02)(0.57)(375) + (0.01)(0.57)(375)$
$+ (0.01)(0.57)(375) + (0.01)(0.57)(0) + (0.01)(0.57)(0)$
$+ (0.01)(0.57)(0) + (0.01)(0.57)(0) + (0.01)(0.57)(0)$
$+ (0.01)(0.5748)(0) + (0.01)(0.57)(0) + (0.01)(0.57)(0)$
$+ (0.01)(0.57)(0)$
$+ (0.01)(0.57)(0) + (0.01)(0.57)(0) + (0.00)(0.57)(0)$
$+ (0.00)(0.57)(375) + (0.00)(0.57)(375)$
$+ (0.00)(0.57)(375) = 190.3 \text{ Btu/h}$

The total lighting cooling load at the designated hour is thus

$$Q_{\text{light}} = Q_{c,15} + Q_{r,15} = 161.3 + 190.3 = 351.6 \text{ Btu/h}$$

See Table 7-38 for the office's lighting usage, heat gain, and cooling load profiles.

**Part 2. Wall cooling load using sol-air temperature, conduction time series and radiant time series.** Calculate the cooling load contribution from the spandrel wall section facing 60° west of south at 3:00 PM local standard time in July for the previously described office.

**Solution:**

Determine the wall cooling load by calculating (1) sol-air temperatures at the exterior surface, (2) heat input based on sol-air temperature, (3) delayed heat gain through the mass of the wall to the interior surface using conduction time series, and (4) delayed space cooling load from heat gain using radiant time series.

First, calculate the sol-air temperature at 3:00 PM local standard time (LST) (4:00 PM daylight saving time) on July 21 for a vertical, dark-colored wall surface, facing 60° west of south, located in Atlanta, Georgia (latitude = 33.64, longitude = 84.43), solar taub = 0.440 and taud = 2.202 from monthly Atlanta weather data for July (Table 1 in Chapter 14, 2013 *ASHRAE Handbook—Fundamentals*). From Table 7-3, the calculated outdoor design temperature for that month and time is 92°F. The ground reflectivity is assumed $\rho_g = 0.2$.

## Nonresidential Cooling and Heating Load Calculations

Table 7-37  Monthly/Hourly Design Temperatures (5% Conditions) for Atlanta, GA, °F

| Hour | January db | January wb | February db | February wb | March db | March wb | April db | April wb | May db | May wb | June db | June wb | July db | July wb | August db | August wb | September db | September wb | October db | October wb | November db | November wb | December db | December wb |
|---|---|---|---|---|---|---|---|---|---|---|---|---|---|---|---|---|---|---|---|---|---|---|---|---|
| 1 | 44.8 | 44.0 | 47.4 | 45.7 | 53.1 | 48.7 | 59.4 | 54.3 | 66.4 | 61.9 | 72.2 | 66.7 | 73.9 | 68.9 | 73.8 | 69.3 | 69.5 | 64.9 | 60.1 | 56.8 | 52.5 | 50.8 | 46.4 | 46.3 |
| 2 | 44.0 | 43.4 | 46.6 | 45.1 | 52.1 | 48.2 | 58.5 | 53.9 | 65.6 | 61.6 | 71.4 | 66.4 | 73.0 | 68.7 | 73.1 | 69.1 | 68.7 | 64.6 | 59.3 | 56.4 | 51.7 | 50.3 | 45.6 | 45.6 |
| 3 | 43.4 | 43.0 | 46.0 | 44.8 | 51.5 | 47.9 | 57.8 | 53.6 | 65.0 | 61.4 | 70.8 | 66.2 | 72.4 | 68.5 | 72.5 | 68.9 | 68.2 | 64.4 | 58.7 | 56.2 | 51.0 | 50.0 | 45.0 | 45.0 |
| 4 | 42.8 | 42.6 | 45.3 | 44.4 | 50.8 | 47.5 | 57.2 | 53.3 | 64.4 | 61.2 | 70.2 | 66.0 | 71.8 | 68.3 | 71.9 | 68.7 | 67.6 | 64.2 | 58.1 | 55.9 | 50.4 | 49.6 | 44.4 | 44.4 |
| 5 | 42.4 | 42.3 | 44.9 | 44.1 | 50.3 | 47.3 | 56.7 | 53.1 | 64.0 | 61.0 | 69.8 | 65.9 | 71.4 | 68.2 | 71.5 | 68.6 | 67.2 | 64.1 | 57.7 | 55.7 | 50.0 | 49.4 | 44.0 | 44.0 |
| 6 | 42.8 | 42.6 | 45.3 | 44.4 | 50.8 | 47.5 | 57.2 | 53.3 | 64.4 | 61.2 | 70.2 | 66.0 | 71.8 | 68.3 | 71.9 | 68.7 | 67.6 | 64.2 | 58.1 | 55.9 | 50.4 | 49.6 | 44.4 | 44.4 |
| 7 | 44.2 | 43.6 | 46.8 | 45.3 | 52.4 | 48.3 | 58.8 | 54.0 | 65.8 | 61.7 | 71.6 | 66.5 | 73.2 | 68.8 | 73.2 | 69.1 | 68.9 | 64.7 | 59.5 | 56.5 | 51.9 | 50.4 | 45.8 | 45.8 |
| 8 | 47.7 | 46.0 | 50.4 | 47.5 | 56.3 | 50.3 | 62.6 | 55.6 | 69.3 | 63.0 | 75.1 | 67.6 | 76.7 | 69.8 | 76.5 | 70.2 | 72.2 | 65.9 | 63.0 | 58.1 | 55.4 | 52.4 | 49.1 | 48.2 |
| 9 | 51.6 | 48.7 | 54.4 | 50.0 | 60.7 | 52.4 | 67.0 | 57.5 | 73.1 | 64.5 | 78.9 | 68.8 | 80.6 | 71.0 | 80.2 | 71.3 | 75.8 | 67.2 | 66.8 | 59.8 | 59.4 | 54.6 | 52.9 | 50.9 |
| 10 | 55.0 | 51.1 | 57.9 | 52.2 | 64.6 | 54.4 | 70.8 | 59.1 | 76.6 | 65.8 | 82.3 | 69.9 | 84.1 | 72.0 | 83.5 | 72.3 | 79.0 | 68.3 | 70.3 | 61.3 | 62.9 | 56.6 | 56.3 | 53.2 |
| 11 | 58.1 | 53.2 | 61.1 | 54.1 | 68.0 | 56.1 | 74.3 | 60.6 | 79.6 | 66.9 | 85.4 | 70.9 | 87.2 | 73.0 | 86.4 | 73.2 | 81.9 | 69.3 | 73.3 | 62.7 | 66.0 | 58.3 | 59.2 | 55.3 |
| 12 | 60.1 | 54.7 | 63.2 | 55.4 | 70.3 | 57.2 | 76.5 | 61.5 | 81.7 | 67.7 | 87.4 | 71.6 | 89.2 | 73.6 | 88.4 | 73.8 | 83.8 | 70.0 | 75.4 | 63.6 | 68.1 | 59.5 | 61.2 | 56.7 |
| 13 | 61.8 | 55.8 | 64.9 | 56.5 | 72.2 | 58.1 | 78.4 | 62.3 | 83.3 | 68.3 | 89.0 | 72.1 | 90.9 | 74.1 | 89.9 | 74.3 | 85.3 | 70.6 | 77.0 | 64.3 | 69.8 | 60.4 | 62.8 | 57.8 |
| 14 | 62.8 | 56.5 | 65.9 | 57.1 | 73.3 | 58.7 | 79.5 | 62.8 | 84.3 | 68.7 | 90.0 | 72.4 | 91.9 | 74.4 | 90.9 | 74.6 | 86.3 | 70.9 | 78.0 | 64.8 | 70.8 | 61.0 | 63.8 | 58.5 |
| 15 | 62.8 | 56.5 | 65.9 | 57.1 | 73.3 | 58.7 | 79.5 | 62.8 | 84.3 | 68.7 | 90.0 | 72.4 | 91.9 | 74.4 | 90.9 | 74.6 | 86.3 | 70.9 | 78.0 | 64.8 | 70.8 | 61.0 | 63.8 | 58.5 |
| 16 | 61.6 | 55.6 | 64.6 | 56.3 | 71.9 | 58.0 | 78.1 | 62.2 | 83.1 | 68.2 | 88.8 | 72.0 | 90.7 | 74.0 | 89.7 | 74.2 | 85.2 | 70.5 | 76.8 | 64.3 | 69.6 | 60.3 | 62.6 | 57.7 |
| 17 | 59.9 | 54.5 | 63.0 | 55.3 | 70.1 | 57.1 | 76.3 | 61.4 | 81.5 | 67.6 | 87.2 | 71.5 | 89.0 | 73.5 | 88.2 | 73.8 | 83.6 | 69.9 | 75.2 | 63.5 | 67.9 | 59.4 | 61.0 | 56.6 |
| 18 | 57.9 | 53.1 | 60.9 | 54.0 | 67.8 | 56.0 | 74.0 | 60.5 | 79.4 | 66.9 | 85.2 | 70.8 | 87.0 | 72.9 | 86.2 | 73.2 | 81.7 | 69.3 | 73.1 | 62.6 | 65.8 | 58.2 | 59.0 | 55.2 |
| 19 | 54.8 | 51.0 | 57.7 | 52.0 | 64.3 | 54.3 | 70.6 | 59.0 | 76.4 | 65.7 | 82.1 | 69.9 | 83.9 | 72.0 | 83.3 | 72.3 | 78.9 | 68.2 | 70.1 | 61.3 | 62.7 | 56.5 | 56.1 | 53.1 |
| 20 | 52.6 | 49.4 | 55.4 | 50.6 | 61.8 | 53.0 | 68.1 | 58.0 | 74.2 | 64.9 | 79.9 | 69.2 | 81.7 | 71.3 | 81.2 | 71.6 | 76.8 | 67.5 | 67.9 | 60.3 | 60.4 | 55.2 | 53.9 | 51.6 |
| 21 | 50.8 | 48.1 | 53.5 | 49.4 | 59.7 | 52.0 | 66.0 | 57.1 | 72.3 | 64.2 | 78.1 | 68.6 | 79.8 | 70.7 | 79.5 | 71.1 | 75.0 | 66.9 | 66.0 | 59.4 | 58.5 | 54.2 | 52.1 | 50.3 |
| 22 | 48.9 | 46.8 | 51.6 | 48.3 | 57.7 | 50.9 | 64.0 | 56.2 | 70.5 | 63.5 | 76.3 | 68.0 | 78.0 | 70.2 | 77.7 | 70.5 | 73.3 | 66.3 | 64.2 | 58.6 | 56.7 | 53.1 | 50.3 | 49.0 |
| 23 | 47.5 | 45.9 | 50.2 | 47.4 | 56.1 | 50.2 | 62.4 | 55.5 | 69.1 | 62.9 | 74.9 | 67.5 | 76.5 | 69.8 | 76.4 | 70.1 | 72.0 | 65.8 | 62.8 | 58.0 | 55.2 | 52.3 | 49.0 | 48.1 |
| 24 | 46.1 | 44.9 | 48.7 | 46.4 | 54.4 | 49.4 | 60.8 | 54.8 | 67.7 | 62.4 | 73.4 | 67.1 | 75.1 | 69.3 | 75.0 | 69.7 | 70.6 | 65.3 | 61.4 | 57.3 | 53.7 | 51.5 | 47.6 | 47.1 |

Table 7-38  Cooling Load Component: Lighting, Btu/h

| Hour | Usage Profile, % | Heat Gain, Btu/h Total | Convective 43% | Radiant 57% | Nonsolar RTS Zone Type 8, % | Radiant Cooling Load | Total Sensible Cooling Load | % Lighting to Return 26% | Room Sensible Cooling Load |
|---|---|---|---|---|---|---|---|---|---|
| 1 | 0 | — | — | — | 49 | 26 | 26 | — | 26 |
| 2 | 0 | — | — | — | 17 | 26 | 26 | — | 26 |
| 3 | 0 | — | — | — | 9 | 24 | 24 | — | 24 |
| 4 | 0 | — | — | — | 5 | 21 | 21 | — | 21 |
| 5 | 0 | — | — | — | 3 | 19 | 19 | — | 19 |
| 6 | 0 | — | — | — | 2 | 17 | 17 | — | 17 |
| 7 | 100 | 375 | 161 | 214 | 2 | 120 | 281 | — | 281 |
| 8 | 100 | 375 | 161 | 214 | 1 | 154 | 315 | — | 315 |
| 9 | 100 | 375 | 161 | 214 | 1 | 171 | 332 | — | 332 |
| 10 | 100 | 375 | 161 | 214 | 1 | 180 | 341 | — | 341 |
| 11 | 100 | 375 | 161 | 214 | 1 | 184 | 345 | — | 345 |
| 12 | 100 | 375 | 161 | 214 | 1 | 186 | 347 | — | 347 |
| 13 | 100 | 375 | 161 | 214 | 1 | 188 | 349 | — | 349 |
| 14 | 100 | 375 | 161 | 214 | 1 | 188 | 349 | — | 349 |
| 15 | 100 | 375 | 161 | 214 | 1 | 190 | 352 | — | 352 |
| 16 | 100 | 375 | 161 | 214 | 1 | 192 | 354 | — | 354 |
| 17 | 100 | 375 | 161 | 214 | 1 | 195 | 356 | — | 356 |
| 18 | 100 | 375 | 161 | 214 | 1 | 197 | 358 | — | 358 |
| 19 | 0 | — | — | — | 1 | 94 | 94 | — | 94 |
| 20 | 0 | — | — | — | 1 | 60 | 60 | — | 60 |
| 21 | 0 | — | — | — | 0 | 43 | 43 | — | 43 |
| 22 | 0 | — | — | — | 0 | 34 | 34 | — | 34 |
| 23 | 0 | — | — | — | 0 | 30 | 30 | — | 30 |
| 24 | 0 | — | — | — | 0 | 28 | 28 | — | 28 |
| Total | | 4,501 | 1,936 | 2,566 | 1 | 2,566 | 4,501 | — | 4,501 |

Sol-air temperature is calculated using Equation 30. For the dark-colored wall, $\alpha/h_o = 0.30$, and for vertical surfaces, $\varepsilon \Delta R/h_o = 0$. The solar irradiance $E_t$ on the wall must be determined using the equations in Chapter 14, 2013 *ASHRAE Handbook—Fundamentals*:

### Solar Angles:
$\psi$ = southwest orientation = +60°
$\Sigma$ = surface tilt from horizontal (where horizontal = 0°) = 90° for vertical wall surface
3:00 PM LST = hour 15

Calculate solar altitude, solar azimuth, surface solar azimuth, and incident angle as follows:

From Table 2 in Chapter 14 of 2013 *ASHRAE Handbook—Fundamentals*, solar position data and constants for July 21 are
ET = –6.4 min
$\delta$ = 20.4°
$E_o$ = 419.8 Btu/h·ft²
Local standard meridian (LSM) for Eastern Time Zone = 75°.

### Apparent solar time AST
$$\begin{aligned} AST &= LST + ET/60 + (LSM - LON)/15 \\ &= 15 + (-6.4/60) + [(75 - 84.43)/15] \\ &= 14.2647 \end{aligned}$$

### Hour angle H, degrees
$$\begin{aligned} H &= 15(AST - 12) \\ &= 15(14.2647 - 12) \\ &= 33.97° \end{aligned}$$

### Solar altitude $\beta$
$$\begin{aligned} \sin \beta &= \cos L \cos \delta \cos H + \sin L \sin \delta \\ &= \cos(33.64)\cos(20.4)\cos(33.97) + \sin(33.64)\sin(20.4) \\ &= 0.841 \\ \beta &= \sin^{-1}(0.841) = 57.2° \end{aligned}$$

### Solar azimuth $\phi$
$$\begin{aligned} \cos \phi &= (\sin \beta \sin L - \sin \delta)/(\cos \beta \cos L) \\ &= [(\sin(57.2)\sin(33.64) - \sin(20.4)]/[\cos(57.2)\cos(33.64)] \\ &= 0.258 \\ \phi &= \cos^{-1}(0.253) = 75.05° \end{aligned}$$

### Surface-solar azimuth $\gamma$
$$\begin{aligned} \gamma &= \phi - \psi \\ &= 75.05 - 60 \\ &= 15.05° \end{aligned}$$

### Incident angle $\theta$
$$\begin{aligned} \cos \theta &= \cos \beta \cos g \sin \Sigma + \sin \beta \cos \Sigma \\ &= \cos(57.2)\cos(15.05)\sin(90) + \sin(57.2)\cos(90) \\ &= 0.523 \\ \theta &= \cos^{-1}(0.523) = 58.45° \end{aligned}$$

### Beam normal irradiance $E_b$
$E_b = E_o \exp(-\tau_b m^{ab})$
$m$ = relative air mass
= $1/[\sin \beta + 0.50572(6.07995 + \beta)^{-1.6364}]$, $\beta$ expressed in degrees
= 1.18905
ab = beam air mass exponent
= $1.454 - 0.406\tau_b - 0.268\tau_d + 0.021\tau_b\tau_d$
= 0.7055705
$E_b$ = $419.8 \exp[-0.556(1.8905^{0.7055705})]$
= 255.3 Btu/h·ft²

### Surface beam irradiance $E_{t,b}$
$$\begin{aligned} E_{t,b} &= E_b \cos \theta \\ &= (255.3)\cos(58.5) \\ &= 133.6 \text{ Btu/h·ft}^2 \end{aligned}$$

### Ratio Y of sky diffuse radiation on vertical surface to sky diffuse radiation on horizontal surface
$$\begin{aligned} Y &= 0.55 + 0.437 \cos \theta + 0.313 \cos^2 \theta \\ &= 0.55 + 0.437 \cos(58.45) + 0.313 \cos^2(58.45) \\ &= 0.8644 \end{aligned}$$

### Diffuse irradiance $E_d$ – Horizontal surfaces
$E_d = E_o \exp(-\tau_d m^{ad})$
ad = diffuse air mass exponent
= $0.507 + 0.205\tau_b - 0.080\tau_d - 0.190\tau_b\tau_d$
= 0.2369528
$E_d = E_o \exp(-\tau_d m^{ad})$
= $419.8 \exp[-2.202(1.8905^{0.2369528})]$
= 42.3 Btu/h·ft²

### Diffuse irradiance $E_d$ – Vertical surfaces
$$\begin{aligned} E_{t,d} &= E_d Y \\ &= (42.3)(0.864) \\ &= 36.6 \text{ Btu/h·ft}^2 \end{aligned}$$

### Ground reflected irradiance $E_{t,r}$
$$\begin{aligned} E_{t,r} &= (E_b \sin \beta + E_d)\rho_g(1 - \cos \Sigma)/2 \\ &= [255.3 \sin(57.2) + 42.3](0.2)[1 - \cos(90)]/2 \\ &= 25.7 \text{ Btu/h·ft}^2 \end{aligned}$$

### Total surface irradiance $E_t$
$$\begin{aligned} E_t &= E_D + E_d + E_r \\ &= 133.6 + 36.6 + 25.7 \\ &= 195.9 \text{ Btu/h·ft}^2 \end{aligned}$$

### Sol-air temperature [from Equation 30]:
$$\begin{aligned} T_e &= t_o + \alpha E_t/h_o - \varepsilon \Delta R/h_o \\ &= 91.9 + (0.30)(195.9) - 0 \\ &= 150.7°F \end{aligned}$$

This procedure is used to calculate the sol-air temperatures for each hour on each surface. Because of the tedious solar angle and intensity calculations, using a simple computer spreadsheet or other computer software can reduce the effort involved. A spreadsheet was used to calculate a 24 h sol-air temperature profile for the data of this example. See Table 7-39A for the solar angle and intensity calculations and Table 7-39B for the sol-air temperatures for this wall surface and orientation.

# Nonresidential Cooling and Heating Load Calculations

### Table 7-39A  Wall Component of Solar Irradiance

| Local Standard Hour | Apparent Solar Time | Hour Angle $H$ | Solar Altitude $\beta$ | Solar Azimuth $\varphi$ | Air Solar Mass $m$ | $E_b$, Direct Normal Btu/h·ft² | Surface Incident Angle $\theta$ | Surface Direct Btu/h·ft² | $E_d$, Diffuse Horizontal, Btu/h·ft² | Ground Diffuse Btu/h·ft² | $Y$ Ratio | Sky Diffuse Btu/h·ft² | Subtotal Diffuse Btu/h·ft² | Total Surface Irradiance Btu/h·ft² |
|---|---|---|---|---|---|---|---|---|---|---|---|---|---|---|
| 1 | 0.26 | −176 | −36 | −175 | 0.0 | 0.0 | 117.4 | 0.0 | 0.0 | 0.0 | 0.4500 | 0.0 | 0.0 | 0.0 |
| 2 | 1.26 | −161 | −33 | −159 | 0.0 | 0.0 | 130.9 | 0.0 | 0.0 | 0.0 | 0.4500 | 0.0 | 0.0 | 0.0 |
| 3 | 2.26 | −146 | −27 | −144 | 0.0 | 0.0 | 144.5 | 0.0 | 0.0 | 0.0 | 0.4500 | 0.0 | 0.0 | 0.0 |
| 4 | 3.26 | −131 | −19 | −132 | 0.0 | 0.0 | 158.1 | 0.0 | 0.0 | 0.0 | 0.4500 | 0.0 | 0.0 | 0.0 |
| 5 | 4.26 | −116 | −9 | −122 | 0.0 | 0.0 | 171.3 | 0.0 | 0.0 | 0.0 | 0.4500 | 0.0 | 0.0 | 0.0 |
| 6 | 5.26 | −101 | 3 | −113 | 16.91455 | 16.5 | 172.5 | 0.0 | 5.7 | 0.6 | 0.4500 | 2.6 | 3.2 | 3.2 |
| 7 | 6.26 | −86 | 14 | −105 | 3.98235 | 130.7 | 159.5 | 0.0 | 19.8 | 5.2 | 0.4500 | 8.9 | 14.1 | 14.1 |
| 8 | 7.26 | −71 | 27 | −98 | 2.22845 | 193.5 | 145.9 | 0.0 | 29.3 | 11.6 | 0.4500 | 13.2 | 24.8 | 24.8 |
| 9 | 8.26 | −56 | 39 | −90 | 1.58641 | 228.3 | 132.3 | 0.0 | 36.0 | 18.0 | 0.4500 | 16.2 | 34.2 | 34.2 |
| 10 | 9.26 | −41 | 51 | −81 | 1.27776 | 248.8 | 118.8 | 0.0 | 40.7 | 23.5 | 0.4500 | 18.3 | 41.8 | 41.8 |
| 11 | 10.26 | −26 | 63 | −67 | 1.11740 | 260.9 | 105.6 | 0.0 | 43.8 | 27.7 | 0.4553 | 19.9 | 47.6 | 47.6 |
| 12 | 11.26 | −11 | 74 | −39 | 1.04214 | 266.9 | 92.6 | 0.0 | 45.4 | 30.1 | 0.5306 | 24.1 | 54.2 | 54.2 |
| 13 | 12.26 | 4 | 76 | 16 | 1.02872 | 268.0 | 80.2 | 45.5 | 45.7 | 30.6 | 0.6332 | 29.0 | 59.6 | 105.1 |
| 14 | 13.26 | 19 | 69 | 57 | 1.07337 | 264.3 | 68.7 | 96.2 | 44.7 | 29.1 | 0.7505 | 33.6 | 62.7 | 158.8 |
| 15 | 14.26 | 34 | 57 | 75 | 1.18905 | 255.3 | 58.45 | 133.6 | 42.3 | 25.7 | 0.8644 | 36.6 | 62.3 | 195.9 |
| 16 | 15.26 | 49 | 45 | 86 | 1.41566 | 239.2 | 50.4 | 152.4 | 38.4 | 20.7 | 0.9555 | 36.7 | 57.4 | 209.9 |
| 17 | 16.26 | 64 | 32 | 94 | 1.86186 | 212.2 | 45.8 | 148.1 | 32.7 | 14.6 | 1.0073 | 33.0 | 47.6 | 195.7 |
| 18 | 17.26 | 79 | 20 | 102 | 2.89735 | 165.3 | 45.5 | 115.8 | 24.7 | 8.1 | 1.0100 | 24.9 | 33.1 | 148.9 |
| 19 | 18.26 | 94 | 8 | 109 | 6.84406 | 76.0 | 49.7 | 49.1 | 13.0 | 2.4 | 0.9631 | 12.5 | 14.9 | 64.0 |
| 20 | 19.26 | 109 | −3 | 117 | 0.0 | 0.0 | 57.5 | 0.0 | 0.0 | 0.0 | 0.8755 | 0.0 | 0.0 | 0.0 |
| 21 | 20.26 | 124 | −14 | 127 | 0.0 | 0.0 | 67.5 | 0.0 | 0.0 | 0.0 | 0.7630 | 0.0 | 0.0 | 0.0 |
| 22 | 21.26 | 139 | −23 | 138 | 0.0 | 0.0 | 79.0 | 0.0 | 0.0 | 0.0 | 0.6452 | 0.0 | 0.0 | 0.0 |
| 23 | 22.26 | 154 | −30 | 151 | 0.0 | 0.0 | 91.3 | 0.0 | 0.0 | 0.0 | 0.5403 | 0.0 | 0.0 | 0.0 |
| 24 | 23.26 | 169 | −35 | 167 | 0.0 | 0.0 | 104.2 | 0.0 | 0.0 | 0.0 | 0.4618 | 0.0 | 0.0 | 0.0 |

### Table 7-39B  Wall Component of Sol-Air Temperatures, Heat Input, Heat Gain, Cooling Load

| Local Standard Hour | Total Surface Irradiance Btu/h·ft² | Outdoor Temp., °F | Sol-Air Temp., °F | Indoor Temp., °F | Heat Input, Btu/h | CTS Type 1, % | Heat Gain, Btu/h Total | Convective 54% | Radiant 46% | Nonsolar RTS Zone Type 8, % | Radiant Cooling Load, Btu/h | Total Cooling Load, Btu/h |
|---|---|---|---|---|---|---|---|---|---|---|---|---|
| 1 | 0.0 | 73.9 | 73.9 | 75 | −5 | 18 | 1 | 1 | 1 | 49 | 16 | 16 |
| 2 | 0.0 | 73.0 | 73.0 | 75 | −9 | 58 | −4 | −2 | −2 | 17 | 12 | 10 |
| 3 | 0.0 | 72.4 | 72.4 | 75 | −12 | 20 | −9 | −5 | −4 | 9 | 10 | 5 |
| 4 | 0.0 | 71.8 | 71.8 | 75 | −15 | 4 | −12 | −6 | −5 | 5 | 8 | 2 |
| 5 | 0.0 | 71.4 | 71.4 | 75 | −17 | 0 | −14 | −8 | −7 | 3 | 7 | −1 |
| 6 | 3.2 | 71.8 | 72.8 | 75 | −10 | 0 | −15 | −8 | −7 | 2 | 6 | −2 |
| 7 | 14.1 | 73.2 | 77.4 | 75 | 11 | 0 | −8 | −4 | −4 | 2 | 7 | 2 |
| 8 | 24.8 | 76.7 | 84.1 | 75 | 42 | 0 | 11 | 6 | 5 | 1 | 11 | 17 |
| 9 | 34.2 | 80.6 | 90.8 | 75 | 73 | 0 | 39 | 21 | 18 | 1 | 18 | 39 |
| 10 | 41.8 | 84.1 | 96.7 | 75 | 100 | 0 | 69 | 37 | 32 | 1 | 27 | 64 |
| 11 | 47.6 | 87.2 | 101.5 | 75 | 122 | 0 | 96 | 52 | 44 | 1 | 36 | 88 |
| 12 | 54.2 | 89.2 | 105.5 | 75 | 141 | 0 | 119 | 64 | 55 | 1 | 43 | 108 |
| 13 | 105.1 | 90.9 | 122.4 | 75 | 219 | 0 | 150 | 81 | 69 | 1 | 53 | 134 |
| 14 | 158.8 | 91.9 | 139.6 | 75 | 298 | 0 | 214 | 115 | 98 | 1 | 71 | 186 |
| 15 | 195.9 | 91.9 | 150.7 | 75 | 350 | 0 | 285 | 154 | 131 | 1 | 93 | 247 |
| 16 | 209.9 | 90.7 | 153.7 | 75 | 363 | 0 | 337 | 182 | 155 | 1 | 114 | 296 |
| 17 | 195.7 | 89.0 | 147.7 | 75 | 336 | 0 | 353 | 191 | 162 | 1 | 127 | 318 |
| 18 | 148.9 | 87.0 | 131.7 | 75 | 262 | 0 | 329 | 177 | 151 | 1 | 128 | 306 |
| 19 | 64.0 | 83.9 | 103.1 | 75 | 130 | 0 | 257 | 139 | 118 | 1 | 115 | 253 |
| 20 | 0.0 | 81.7 | 81.7 | 75 | 31 | 0 | 147 | 79 | 67 | 1 | 86 | 165 |
| 21 | 0.0 | 79.8 | 79.8 | 75 | 22 | 0 | 58 | 32 | 27 | 0 | 56 | 88 |
| 22 | 0.0 | 78.0 | 78.0 | 75 | 14 | 0 | 27 | 14 | 12 | 0 | 38 | 52 |
| 23 | 0.0 | 76.5 | 76.5 | 75 | 7 | 0 | 15 | 8 | 7 | 0 | 27 | 35 |
| 24 | 0.0 | 75.1 | 75.1 | 75 | 0 | 0 | 8 | 4 | 4 | 0 | 20 | 25 |

Conductive heat gain is calculated using Equations 31 and 32. First, calculate the 24 h heat input profile using Equation 31 and the sol-air temperatures for a southwest-facing wall with dark exterior color:

$q_{i,1}$ = (0.077)(60)(73.9 – 75)= –5 Btu/h
$q_{i,2}$ = (0.077)(60)(73 – 75)= –9
$q_{i,3}$ = (0.077)(60)(72.4 – 75)= –12
$q_{i,4}$ = (0.077)(60)(71.8 – 75)= –15
$q_{i,5}$ = (0.077)(60)(71.4 – 75)= –17
$q_{i,6}$ = (0.077)(60)(72.8 – 75)= –10
$q_{i,7}$ = (0.077)(60)(77.4 – 75)= 11
$q_{i,8}$ = (0.077)(60)(84.1 – 75)= 42
$q_{i,9}$ = (0.077)(60)(90.8 – 75)= 73
$q_{i,10}$ = (0.077)(60)(96.7 – 75)= 100
$q_{i,11}$ = (0.077)(60)(101.5 – 75)= 122
$q_{i,12}$ = (0.077)(60)(105.5 – 75)= 141
$q_{i,13}$ = (0.077)(60)(122.4 – 75)= 219
$q_{i,14}$ = (0.077)(60)(139.6 – 75)= 298
$q_{i,15}$ = (0.077)(60)(150.7 – 75)= 350
$q_{i,16}$ = (0.077)(60)(153.7 – 75)= 363
$q_{i,17}$ = (0.077)(60)(147.7 – 75) = 336
$q_{i,18}$ = (0.077)(60)(131.7 – 75)= 262
$q_{i,19}$ = (0.077)(60)(103.1 – 75) = 130
$q_{i,20}$ = (0.077)(60)(81.7 – 75) = 31
$q_{i,21}$ = (0.077)(60)(79.8 – 75) = 22
$q_{i,22}$ = (0.077)(60)(78.0 – 75) = 14
$q_{i,23}$ = (0.077)(60)(76.5 – 75) = 7
$q_{i,24}$ = (0.077)(60)(75.1 – 75) = 0

Next, calculate wall heat gain using conduction time series. The preceding heat input profile is used with conduction time series to calculate the wall heat gain. From Table 16 of Chapter 18, 2013 *ASHRAE Handbook—Fundamentals*, the most similar wall construction is wall number 1. This is a spandrel glass wall that has similar mass and thermal capacity. Using Equation 32, the conduction time factors for wall 1 can be used in conjunction with the 24 h heat input profile to determine the wall heat gain at 3:00 PM LST:

$q_{15} = c_0 q_{i,15} + c_1 q_{i,14} + c_2 q_{i,13} + c_3 q_{i,12} + \cdots + c_{23} q_{i,14}$
= (0.18)(350) + (0.58)(298) + (0.20)(219) + (0.04)(141)
+ (0.00)(122) + (0.00)(100) + (0.00)(73) + (0.00)(42)
+ (0.00)(11) + (0.00)(–10) + (0.00)(–17) + (0.00)(–15)
+ (0.00)(–12) + (0.00)(–9) + (0.00)(–5) + (0.00)(0)
+ (0.00)(7) + (0.00)(14) + (0.00)(22) + (0.00)(31)
+ (0.00)(130) + (0.00)(262) + (0.00)(336) +
(0.00)(363) = 285 Btu/h

Because of the tedious calculations involved, a spreadsheet is used to calculate the remainder of a 24 h heat gain profile indicated in Table 7-39B for the data of this example.

Finally, calculate wall cooling load using radiant time series. Total cooling load for the wall is calculated by summing the convective and radiant portions. The convective portion is simply the wall heat gain for the hour being calculated times the convective fraction for walls from Table 14 of Chapter 18, 2013 *ASHRAE Handbook—Fundamentals* (54%):

$Q_c$ = (285)(0.54) = 154 Btu/h

The radiant portion of the cooling load is calculated using conductive heat gains for the current and past 23 h, the radiant fraction for walls from Table 14 of Chapter 18, 2013 *ASHRAE Handbook—Fundamentals* (46%), and radiant time series from Table 19, in accordance with Equation 34. From Table 19, select the RTS for medium-weight construction, assuming 50% glass and carpeted floors as representative for the described construction. Use the wall heat gains from Table 39B for 24 h design conditions in July. Thus, the radiant cooling load for the wall at 3:00 PM is

$Q_{r,15} = r_0(0.46)q_{i,15} + r_1(0.46)\ q_{i,14} + r_2(0.46)\ q_{i,13} + r_3(0.46)\ q_{i,12}$
$+ \cdots + r_{23}(0.46)\ q_{i,16}$
= (0.49)(0.46)(285) + (0.17)(0.46)(214) +
(0.09)(0.46)(150)
+ (0.05)(0.46)(119) + (0.03)(0.46)(96) +
(0.02)(0.46)(69)
+ (0.02)(0.46)(39) + (0.01)(0.46)(11) +
(0.01)(0.46)(–8)
+ (0.01)(0.46)(–15) + (0.01)(0.46)(–14) +
(0.01)(0.46)(–12)
+ (0.01)(0.46)(–9) + (0.01)(0.46)(–4) +
(0.01)(0.46)(1)
+ (0.01)(0.46)(8) + (0.01)(0.46)(15) +
(0.01)(0.46)(27) + (0.01)(0.46)(58) +
(0.01)(0.46)(147) + (0.00)(0.46)(257)
+ (0.00)(0.46)(329) + (0.00) (0.46)(353) +
(0.00)(0.46)(337) = 93 Btu/h

The total wall cooling load at the designated hour is thus

$Q_{wall} = Q_c + Q_{r15}$ = 154 + 93 = 247 Btu/h

Again, a simple computer spreadsheet or other software is necessary to reduce the effort involved. A spreadsheet was used with the heat gain profile to split the heat gain into convective and radiant portions, apply RTS to the radiant portion, and total the convective and radiant loads to determine a 24 h cooling load profile for this example, with results in Table 39B.

**Part 3. Window cooling load using radiant time series.** Calculate the cooling load contribution, with and without indoor shading (venetian blinds) for the window area facing 60° west of south at 3:00 PM in July for the conference room example.

**Solution:**

First, calculate the 24 h heat gain profile for the window, then split those heat gains into radiant and convective portions, apply the appropriate RTS to the radiant portion, then sum the convective and radiant cooling load components to determine total window cooling load for the time. The window heat gain components are calculated using Equations (13) to (15). From Part 2, at hour 15 LST (3:00 PM):

$E_{t,b}$ = 133.6 Btu/h·ft²
$E_{t,d}$ = 36.6 Btu/h·ft²

# Nonresidential Cooling and Heating Load Calculations

Table 7-40 Window Component of Heat Gain (No Blinds or Overhang)

| Local Std. Hour | Beam Normal, Btu/h·ft² | Surface Incident Angle | Surface Beam, Btu/h·ft² | Beam SHGC | Adjusted Beam IAC | Beam Solar Heat Gain, Btu/h | Diffuse Horiz. $E_d$, Btu/h·ft² | Ground Diffuse, Btu/h·ft² | Y Ratio | Sky Diffuse, Btu/h·ft² | Subtotal Diffuse, Btu/h·ft² | Hemis. SHGC | Diff. Solar Heat Gain, Btu/h | Outside Temp., °F | Conduction Heat Gain, Btu/h | Total Window Heat Gain, Btu/h |
|---|---|---|---|---|---|---|---|---|---|---|---|---|---|---|---|---|
| 1 | 0.0 | 117.4 | 0.0 | 0.000 | 1.000 | 0 | 0.0 | 0.0 | 0.4500 | 0.0 | 0.0 | 0.410 | 0 | 73.9 | −25 | −25 |
| 2 | 0.0 | 130.9 | 0.0 | 0.000 | 1.000 | 0 | 0.0 | 0.0 | 0.4500 | 0.0 | 0.0 | 0.410 | 0 | 73.0 | −45 | −45 |
| 3 | 0.0 | 144.5 | 0.0 | 0.000 | 1.000 | 0 | 0.0 | 0.0 | 0.4500 | 0.0 | 0.0 | 0.410 | 0 | 72.4 | −58 | −58 |
| 4 | 0.0 | 158.1 | 0.0 | 0.000 | 1.000 | 0 | 0.0 | 0.0 | 0.4500 | 0.0 | 0.0 | 0.410 | 0 | 71.8 | −72 | −72 |
| 5 | 0.0 | 171.3 | 0.0 | 0.000 | 1.000 | 0 | 0.0 | 0.0 | 0.4500 | 0.0 | 0.0 | 0.410 | 0 | 71.4 | −81 | −81 |
| 6 | 16.5 | 172.5 | 0.0 | 0.000 | 0.000 | 0 | 5.7 | 0.6 | 0.4500 | 2.6 | 3.2 | 0.410 | 52 | 71.8 | −72 | −19 |
| 7 | 130.7 | 159.5 | 0.0 | 0.000 | 0.000 | 0 | 19.8 | 5.2 | 0.4500 | 8.9 | 14.1 | 0.410 | 231 | 73.2 | −40 | 191 |
| 8 | 193.5 | 145.9 | 0.0 | 0.000 | 0.000 | 0 | 29.3 | 11.6 | 0.4500 | 13.2 | 24.8 | 0.410 | 406 | 76.7 | 38 | 444 |
| 9 | 228.3 | 132.3 | 0.0 | 0.000 | 0.000 | 0 | 36.0 | 18.0 | 0.4500 | 16.2 | 34.2 | 0.410 | 560 | 80.6 | 125 | 686 |
| 10 | 248.8 | 118.8 | 0.0 | 0.000 | 0.000 | 0 | 40.7 | 23.5 | 0.4500 | 18.3 | 41.8 | 0.410 | 686 | 84.1 | 204 | 890 |
| 11 | 260.9 | 105.6 | 0.0 | 0.000 | 0.000 | 0 | 43.8 | 27.7 | 0.4553 | 19.9 | 47.6 | 0.410 | 781 | 87.2 | 273 | 1055 |
| 12 | 266.9 | 92.6 | 0.0 | 0.000 | 0.000 | 0 | 45.4 | 30.1 | 0.5306 | 24.1 | 54.2 | 0.410 | 890 | 89.2 | 318 | 1208 |
| 13 | 268.0 | 80.2 | 45.5 | 0.166 | 1.000 | 302 | 45.7 | 30.6 | 0.6332 | 29.0 | 59.6 | 0.410 | 977 | 90.9 | 356 | 1635 |
| 14 | 264.3 | 68.7 | 96.2 | 0.321 | 1.000 | 1234 | 44.7 | 29.1 | 0.7505 | 33.6 | 62.7 | 0.410 | 1028 | 91.9 | 379 | 2640 |
| 15 | 255.3 | 58.4 | 133.6 | 0.398 | 1.000 | 2126 | 42.3 | 25.7 | 0.8644 | 36.6 | 62.3 | 0.410 | 1021 | 91.9 | 379 | 3526 |
| 16 | 239.2 | 50.4 | 152.4 | 0.438 | 1.000 | 2670 | 38.4 | 20.7 | 0.9555 | 36.7 | 57.4 | 0.410 | 942 | 90.7 | 352 | 3964 |
| 17 | 212.2 | 45.8 | 148.1 | 0.448 | 1.000 | 2656 | 32.7 | 14.6 | 1.0073 | 33.0 | 47.6 | 0.410 | 781 | 89.0 | 314 | 3751 |
| 18 | 165.3 | 45.5 | 115.8 | 0.449 | 1.000 | 2080 | 24.7 | 8.1 | 1.0100 | 24.9 | 33.1 | 0.410 | 542 | 87.0 | 269 | 2892 |
| 19 | 76.0 | 49.7 | 49.1 | 0.441 | 1.000 | 865 | 13.0 | 2.4 | 0.9631 | 12.5 | 14.9 | 0.410 | 244 | 83.9 | 199 | 1309 |
| 20 | 0.0 | 57.5 | 0.0 | 0.403 | 0.000 | 0 | 0.0 | 0.0 | 0.8755 | 0.0 | 0.0 | 0.410 | 0 | 81.7 | 150 | 150 |
| 21 | 0.0 | 67.5 | 0.0 | 0.330 | 0.000 | 0 | 0.0 | 0.0 | 0.7630 | 0.0 | 0.0 | 0.410 | 0 | 79.8 | 108 | 108 |
| 22 | 0.0 | 79.0 | 0.0 | 0.185 | 0.000 | 0 | 0.0 | 0.0 | 0.6452 | 0.0 | 0.0 | 0.410 | 0 | 78.0 | 67 | 67 |
| 23 | 0.0 | 91.3 | 0.0 | 0.000 | 1.000 | 0 | 0.0 | 0.0 | 0.5403 | 0.0 | 0.0 | 0.410 | 0 | 76.5 | 34 | 34 |
| 24 | 0.0 | 104.2 | 0.0 | 0.000 | 1.000 | 0 | 0.0 | 0.0 | 0.4618 | 0.0 | 0.0 | 0.410 | 0 | 75.1 | 2 | 2 |

$E_r$ = 25.7 Btu/h·ft²

θ = 58.45°

From Chapter 15, Table 10, 2013 *ASHRAE Handbook—Fundamentals* for glass type 5d,

$$\text{SHGC}(\theta) = \text{SHGC}(58.45) = 0.3978 \text{ (interpolated)}$$

$$\langle \text{SHGC} \rangle_D = 0.41$$

From Chapter 15, Table 13B, 2013 *ASHRAE Handbook—Fundamentals*, for light-colored blinds (assumed louver reflectance = 0.8 and louvers positioned at 45° angle) on double-glazed, heat-absorbing windows (Type 5d from Table 13B of Chapter 15), IAC(0) = 0.74, IAC(60) = 0.65, IAC(diff) = 0.79, and radiant fraction = 0.54. Without blinds, IAC = 1.0. Therefore, window heat gain components for hour 15, without blinds, are

$$q_{b15} = AE_{t,b}\text{SHGC}(\theta)(\text{IAC}) = (40)(133.6)(0.3978)(1.00)$$
$$= 2126 \text{ Btu/h}$$

$$q_{d15} = A(E_{t,d} + E_r)\langle \text{SHGC} \rangle_D(\text{IAC})$$
$$= (40)(36.6 + 25.7)(0.41)(1.00) = 1021 \text{ Btu/h}$$

$$q_{c15} = UA(t_{out} - t_{in}) = (0.56)(40)(91.9 - 75) = 379 \text{ Btu/h}$$

This procedure is repeated to determine these values for a 24 h heat gain profile, shown in Table 7-30.

Total cooling load for the window is calculated by summing the convective and radiant portions. For windows with indoor shading (blinds, drapes, etc.), the direct beam, diffuse, and conductive heat gains may be summed and treated together in calculating cooling loads. However, in this example, the window does not have indoor shading, and the direct beam solar heat gain should be treated separately from the diffuse and conductive heat gains. The direct beam heat gain, without indoor shading, is treated as 100% radiant, and solar RTS factors from Table 20 are used to convert the beam heat gains to cooling loads. The diffuse and conductive heat gains can be totaled and split into radiant and convective portions according to Table 14, and nonsolar RTS factors from Table 19 are used to convert the radiant portion to cooling load.

The solar beam cooling load is calculated using heat gains for the current hour and past 23 h and radiant time series from Table 20, in accordance with Equation 39. From Table 20, select the solar RTS for medium-weight construction, assuming 50% glass and carpeted floors for this example. Using Table 7-40 values for direct solar heat gain, the radiant cooling load for the window direct beam solar component is

$$Q_{b,15} = r_0 q_{b,15} + r_1 q_{b,14} + r_2 q_{b,13} + r_3 q_{b,12} + \cdots + r_{23} q_{b,16}$$
$$= (0.54)(2126) + (0.16)(1234) + (0.08)(302) +$$
$$(0.04)(0) + (0.03)(0) + (0.02)(0) + (0.01)(0) +$$
$$(0.01)(0) + (0.01)(0) + (0.01)(0) + (0.01)(0) +$$
$$(0.01)(0) + (0.01)(0) + (0.01)(0) + (0.01)(0) +$$
$$(0.01)(0) + (0.01)(0) + (0.01)(0) + (0.01)(0)$$
$$+ (0.00)(0) + (0.00)(865) + (0.00)(2080) +$$
$$(0.00)(2656) + (0.00)(2670) = 1370 \text{ Btu/h}$$

### Table 7-41  Window Component of Cooling Load (No Blinds or Overhang)

| | Unshaded Direct Beam Solar (if AC = 1) | | | | | | Shaded Direct Beam (AC < 1.0) + Diffuse + Conduction | | | | | | | | | |
|---|---|---|---|---|---|---|---|---|---|---|---|---|---|---|---|---|
| Local Standard Hour | Beam Heat Gain, Btu/h | Convective 0%, Btu/h | Radiant 100%, Btu/h | Solar RTS, Zone Type 8, % | Radiant Btu/h | Cooling Load, Btu/h | Beam Heat Gain, Btu/h | Diffuse Heat Gain, Btu/h | Conduction Heat Gain, Btu/h | Total Heat Gain, Btu/h | Convective 54%, Btu/h | Radiant 46%, Btu/h | Non-solar RTS, Zone Type 8 | Radiant Btu/h | Cooling Load, Btu/h | Window Cooling Load, Btu/h |
| 1  | 0    | 0 | 0    | 54 | 119  | 119  | 0    | 0    | −25  | −25  | −13 | −11 | 49% | 59  | 45   | 165  |
| 2  | 0    | 0 | 0    | 16 | 119  | 119  | 0    | 0    | −45  | −45  | −24 | −21 | 17% | 49  | 24   | 144  |
| 3  | 0    | 0 | 0    | 8  | 119  | 119  | 0    | 0    | −58  | −58  | −31 | −27 | 9%  | 41  | 9    | 129  |
| 4  | 0    | 0 | 0    | 4  | 119  | 119  | 0    | 0    | −72  | −72  | −39 | −33 | 5%  | 32  | −6   | 113  |
| 5  | 0    | 0 | 0    | 3  | 119  | 119  | 0    | 0    | −81  | −81  | −44 | −37 | 3%  | 25  | −19  | 100  |
| 6  | 0    | 0 | 0    | 2  | 119  | 119  | 0    | 52   | −72  | −19  | −10 | −9  | 2%  | 32  | 22   | 141  |
| 7  | 0    | 0 | 0    | 1  | 119  | 119  | 0    | 231  | −40  | 191  | 103 | 88  | 2%  | 78  | 181  | 301  |
| 8  | 0    | 0 | 0    | 1  | 116  | 116  | 0    | 406  | 38   | 444  | 240 | 204 | 1%  | 148 | 388  | 504  |
| 9  | 0    | 0 | 0    | 1  | 104  | 104  | 0    | 560  | 125  | 686  | 370 | 315 | 1%  | 225 | 596  | 700  |
| 10 | 0    | 0 | 0    | 1  | 83   | 83   | 0    | 686  | 204  | 890  | 481 | 409 | 1%  | 300 | 780  | 863  |
| 11 | 0    | 0 | 0    | 1  | 56   | 56   | 0    | 781  | 273  | 1055 | 569 | 485 | 1%  | 365 | 935  | 991  |
| 12 | 0    | 0 | 0    | 1  | 29   | 29   | 0    | 890  | 318  | 1208 | 652 | 556 | 1%  | 426 | 1078 | 1108 |
| 13 | 302  | 0 | 302  | 1  | 172  | 172  | 0    | 977  | 356  | 1333 | 720 | 613 | 1%  | 480 | 1200 | 1372 |
| 14 | 1234 | 0 | 1234 | 1  | 715  | 715  | 0    | 1028 | 379  | 1406 | 759 | 647 | 1%  | 521 | 1281 | 1995 |
| 15 | 2126 | 0 | 2126 | 1  | 1370 | 1370 | 0    | 1021 | 379  | 1400 | 756 | 644 | 1%  | 541 | 1297 | 2666 |
| 16 | 2670 | 0 | 2670 | 1  | 1893 | 1893 | 0    | 942  | 352  | 1294 | 699 | 595 | 1%  | 530 | 1229 | 3122 |
| 17 | 2656 | 0 | 2656 | 1  | 2090 | 2090 | 0    | 781  | 314  | 1094 | 591 | 503 | 1%  | 487 | 1078 | 3168 |
| 18 | 2080 | 0 | 2080 | 1  | 1890 | 1890 | 0    | 542  | 269  | 811  | 438 | 373 | 1%  | 411 | 849  | 2739 |
| 19 | 865  | 0 | 865  | 1  | 1211 | 1211 | 0    | 244  | 199  | 444  | 240 | 204 | 1%  | 302 | 542  | 1753 |
| 20 | 0    | 0 | 0    | 0  | 549  | 549  | 0    | 0    | 150  | 150  | 81  | 69  | 1%  | 196 | 277  | 826  |
| 21 | 0    | 0 | 0    | 0  | 322  | 322  | 0    | 0    | 108  | 108  | 58  | 49  | 0%  | 145 | 203  | 525  |
| 22 | 0    | 0 | 0    | 0  | 213  | 213  | 0    | 0    | 67   | 67   | 36  | 31  | 0%  | 112 | 148  | 361  |
| 23 | 0    | 0 | 0    | 0  | 157  | 157  | 0    | 0    | 34   | 34   | 18  | 15  | 0%  | 89  | 107  | 265  |
| 24 | 0    | 0 | 0    | 0  | 128  | 128  | 0    | 0    | 2    | 2    | 1   | 1   | 0%  | 72  | 73   | 201  |

### Table 7-42  Window Component of Cooling Load (With Blinds, No Overhang)

| | Unshaded Direct Beam Solar (if AC = 1) | | | | | | Shaded Direct Beam (AC < 1.0) + Diffuse + Conduction | | | | | | | | | |
|---|---|---|---|---|---|---|---|---|---|---|---|---|---|---|---|---|
| Local Standard Hour | Beam Heat Gain, Btu/h | Convective 0%, Btu/h | Radiant 100%, Btu/h | Solar RTS, Zone Type 8, % | Radiant Btu/h | Cooling Load, Btu/h | Beam Heat Gain, Btu/h | Diffuse Heat Gain, Btu/h | Conduction Heat Gain, Btu/h | Total Heat Gain, Btu/h | Convective 54%, Btu/h | Radiant 46%, Btu/h | Non-solar RTS, Zone Type 8 | Radiant, Btu/h | Cooling Load, Btu/h | Window Cooling Load, Btu/h |
| 1  | 0 | 0 | 0 | 1 | 0 | 0 | 0    | 0   | −25 | −25  | −11  | −13  | 49% | 105  | 94   | 94   |
| 2  | 0 | 0 | 0 | 0 | 0 | 0 | 0    | 0   | −45 | −45  | −21  | −24  | 17% | 90   | 70   | 70   |
| 3  | 0 | 0 | 0 | 0 | 0 | 0 | 0    | 0   | −58 | −58  | −27  | −31  | 9%  | 81   | 54   | 54   |
| 4  | 0 | 0 | 0 | 0 | 0 | 0 | 0    | 0   | −72 | −72  | −33  | −39  | 5%  | 72   | 39   | 39   |
| 5  | 0 | 0 | 0 | 0 | 0 | 0 | 0    | 0   | −81 | −81  | −37  | −44  | 3%  | 63   | 26   | 26   |
| 6  | 0 | 0 | 0 | 0 | 0 | 0 | 0    | 41  | −72 | −30  | −14  | −16  | 2%  | 70   | 56   | 56   |
| 7  | 0 | 0 | 0 | 0 | 0 | 0 | 0    | 183 | −40 | 143  | 66   | 77   | 2%  | 114  | 180  | 180  |
| 8  | 0 | 0 | 0 | 0 | 0 | 0 | 0    | 321 | 38  | 359  | 165  | 194  | 1%  | 183  | 348  | 348  |
| 9  | 0 | 0 | 0 | 0 | 0 | 0 | 0    | 443 | 125 | 568  | 261  | 307  | 1%  | 260  | 522  | 522  |
| 10 | 0 | 0 | 0 | 0 | 0 | 0 | 0    | 542 | 204 | 746  | 343  | 403  | 1%  | 331  | 674  | 674  |
| 11 | 0 | 0 | 0 | 0 | 0 | 0 | 0    | 617 | 273 | 891  | 410  | 481  | 1%  | 391  | 801  | 801  |
| 12 | 0 | 0 | 0 | 0 | 0 | 0 | 0    | 703 | 318 | 1021 | 470  | 551  | 1%  | 443  | 913  | 913  |
| 13 | 0 | 0 | 0 | 0 | 0 | 0 | 196  | 772 | 356 | 1325 | 609  | 715  | 1%  | 540  | 1149 | 1149 |
| 14 | 0 | 0 | 0 | 0 | 0 | 0 | 802  | 812 | 379 | 1992 | 916  | 1076 | 1%  | 751  | 1668 | 1668 |
| 15 | 0 | 0 | 0 | 0 | 0 | 0 | 1388 | 807 | 379 | 2574 | 1184 | 1390 | 1%  | 987  | 2171 | 2171 |
| 16 | 0 | 0 | 0 | 0 | 0 | 0 | 1784 | 744 | 352 | 2880 | 1325 | 1555 | 1%  | 1170 | 2495 | 2495 |
| 17 | 0 | 0 | 0 | 0 | 0 | 0 | 1816 | 617 | 314 | 2747 | 1263 | 1483 | 1%  | 1221 | 2484 | 2484 |
| 18 | 0 | 0 | 0 | 0 | 0 | 0 | 1458 | 428 | 269 | 2156 | 992  | 1164 | 1%  | 1103 | 2094 | 2094 |
| 19 | 0 | 0 | 0 | 0 | 0 | 0 | 624  | 193 | 199 | 1017 | 468  | 549  | 1%  | 774  | 1242 | 1242 |
| 20 | 0 | 0 | 0 | 0 | 0 | 0 | 0    | 0   | 150 | 150  | 69   | 81   | 1%  | 434  | 503  | 503  |
| 21 | 0 | 0 | 0 | 0 | 0 | 0 | 0    | 0   | 108 | 108  | 49   | 58   | 0%  | 290  | 339  | 339  |
| 22 | 0 | 0 | 0 | 0 | 0 | 0 | 0    | 0   | 67  | 67   | 31   | 36   | 0%  | 209  | 240  | 240  |
| 23 | 0 | 0 | 0 | 0 | 0 | 0 | 0    | 0   | 34  | 34   | 15   | 18   | 0%  | 160  | 176  | 176  |
| 24 | 0 | 0 | 0 | 0 | 0 | 0 | 0    | 0   | 2   | 2    | 1    | 1    | 0%  | 128  | 129  | 129  |

# Nonresidential Cooling and Heating Load Calculations

This process is repeated for other hours; results are listed in Table 7-41.

For diffuse and conductive heat gains, the radiant fraction according to Table 14 is 46%. The radiant portion is processed using nonsolar RTS coefficients from Table 19. The results are listed in Tables 30 and 31. For 3:00 PM, the diffuse and conductive cooling load is 1297 Btu/h.

The total window cooling load at the designated hour is thus

$$Q_{window} = Q_b + Q_{diff + cond} = 1370 + 1297 = 2667 \text{ Btu/h}$$

Again, a computer spreadsheet or other software is commonly used to reduce the effort involved in calculations. The spreadsheet illustrated in Table 7-40 is expanded in Table 7-41 to include splitting the heat gain into convective and radiant portions, applying RTS to the radiant portion, and totaling the convective and radiant loads to determine a 24 h cooling load profile for a window without indoor shading.

If the window has an indoor shading device, it is accounted for with the indoor attenuation coefficients (IAC), the radiant fraction, and the RTS type used. If a window has no indoor shading, 100% of the direct beam energy is assumed to be radiant and solar RTS factors are used. However, if an indoor shading device is present, the direct beam is assumed to be interrupted by the shading device, and a portion immediately becomes cooling load by convection. Also, the energy is assumed to be radiated to all surfaces of the room, therefore nonsolar RTS values are used to convert the radiant load into cooling load.

IAC values depend on several factors: (1) type of shading device, (2) position of shading device relative to window, (3) reflectivity of shading device, (4) angular adjustment of shading device, as well as (5) solar position relative to the shading device. These factors are discussed in detail in Chapter 15 of 2013 *ASHRAE Handbook—Fundamentals*. For this example with venetian blinds, the IAC for beam radiation is treated separately from the diffuse solar gain. The direct beam IAC must be adjusted based on the profile angle of the sun. At 3:00 PM in July, the profile angle of the sun relative to the window surface is 58°. Calculated using Equation 39 from Chapter 15, the beam IAC = 0.653. The diffuse IAC is 0.79. Thus, the window heat gains, with light-colored blinds, at 3:00 PM are

$$q_{b15} = AE_D \text{ SHGC}(\theta)(IAC) = (40)(133.6)(0.3978)(0.653) = 1388 \text{ Btu/h}$$

$$q_{d15} = A(E_d + E_r)\langle SHGC \rangle_D (IAC)_D = (40)(36.6 + 25.7)(0.41)(0.79) = 807 \text{ Btu/h}$$

$$q_{c15} = UA(t_{out} - t_{in}) = (0.56)(40)(91.9 - 75) = 379 \text{ Btu/h}$$

Because the same radiant fraction and nonsolar RTS are applied to all parts of the window heat gain when indoor shading is present, those loads can be totaled and the cooling load calculated after splitting the radiant portion for processing with nonsolar RTS. This is illustrated by the spreadsheet results in Table 7-42. The total window cooling load with venetian blinds at 3:00 PM = 2171 Btu/h.

**Part 4. Window cooling load using radiant time series for window with overhang shading.** Calculate the cooling load contribution for the previous example with the addition of a 10 ft overhang shading the window.

**Solution:**

In Chapter 15 of 2013 *ASHRAE Handbook—Fundamentals*, methods are described and examples provided for calculating the area of a window shaded by attached vertical or horizontal projections. For 3:00 PM LST in July, the solar position calculated in previous examples is

$$\text{Solar altitude } \beta = 57.2°$$

$$\text{Solar azimuth } \phi = 75.1°$$

$$\text{Surface-solar azimuth } \gamma = 15.1°$$

From Chapter 15 of 2013 *ASHRAE Handbook—Fundamentals*, Equation 33, profile angle $\Omega$ is calculated by

$$\tan \Omega = \tan \beta / \cos \gamma = \tan(57.2)/\cos(15.1) = 1.6087$$

$$\Omega = 58.1°$$

From Chapter 15 of 2013 *ASHRAE Handbook—Fundamentals*, Equation 40, shadow height $S_H$ is

$$S_H = P_H \tan \Omega = 10(1.6087) = 16.1 \text{ ft}$$

Because the window is 6.4 ft tall, at 3:00 PM the window is completely shaded by the 10 ft deep overhang. Thus, the shaded window heat gain includes only diffuse solar and conduction gains. This is converted to cooling load by separating the radiant portion, applying RTS, and adding the resulting radiant cooling load to the convective portion to determine total cooling load. Those results are in Table 7-43. The total window cooling load = 1098 Btu/h.

**Part 5. Room cooling load total.** Calculate the sensible cooling loads for the previously described office at 3:00 PM in July.

**Solution:**

The steps in the previous example parts are repeated for each of the internal and external loads components, including the southeast facing window, spandrel and brick walls, the southwest facing brick wall, the roof, people, and equipment loads. The results are tabulated in Table 7-44. The total room sensible cooling load for the office is 3674 Btu/h at 3:00 PM in July. When this calculation process is repeated for a 24 h design day for each month, it is found that the peak room sensible cooling load actually occurs in July at hour 14 (2:00 PM solar time) at 3675 Btu/h as indicated in Table 7-45.

Although simple in concept, these steps involved in calculating cooling loads are tedious and repetitive, even using the "simplified" RTS method; practically, they should be performed using a computer spreadsheet or other program. The calculations should be repeated for multiple design conditions (i.e., times of day, other months) to determine the maximum cooling load for mechanical equipment sizing. Example

Table 7-43 Window Component of Cooling Load (With Blinds and Overhang)

| Local Standard Hour | Overhang and Fins Shading |||||  Shaded Direct Beam (AC < 1.0) + Diffuse + Conduction ||||||||
|---|---|---|---|---|---|---|---|---|---|---|---|---|---|---|
| | Surface Solar Azimuth | Profile Angle | Shadow Width, ft | Shadow Height, ft | Direct Sunlit Area, ft² | Beam Heat Gain, Btu/h | Diffuse Heat Gain, Btu/h | Conduction Heat Gain, Btu/h | Total Heat Gain, Btu/h | Convective 54%, Btu/h | Radiant 46%, Btu/h | Non-solar RTS, Zone Type 8 | Radiant, Btu/h | Cooling Load, Btu/h | Window Cooling Load, Btu/h |
| 1 | −235 | 52 | 0.0 | 0.0 | 0.0 | 0 | 0 | −25 | −25 | −13 | −11 | 49% | 55 | 42 | 42 |
| 2 | −219 | 40 | 0.0 | 0.0 | 0.0 | 0 | 0 | −45 | −45 | −24 | −21 | 17% | 43 | 19 | 19 |
| 3 | −204 | 29 | 0.0 | 0.0 | 0.0 | 0 | 0 | −58 | −58 | −31 | −27 | 9% | 36 | 4 | 4 |
| 4 | −192 | 19 | 0.0 | 0.0 | 0.0 | 0 | 0 | −72 | −72 | −39 | −33 | 5% | 28 | −11 | −11 |
| 5 | −182 | 9 | 0.0 | 0.0 | 0.0 | 0 | 0 | −81 | −81 | −44 | −37 | 3% | 20 | −23 | −23 |
| 6 | −173 | −3 | 0.0 | 0.0 | 0.0 | 0 | 41 | −72 | −30 | −16 | −14 | 2% | 26 | 10 | 10 |
| 7 | −165 | −15 | 0.0 | 0.0 | 0.0 | 0 | 183 | −40 | 143 | 77 | 66 | 2% | 64 | 141 | 141 |
| 8 | −158 | −28 | 0.0 | 0.0 | 0.0 | 0 | 321 | 38 | 359 | 194 | 165 | 1% | 122 | 316 | 316 |
| 9 | −150 | −43 | 0.0 | 0.0 | 0.0 | 0 | 443 | 125 | 568 | 307 | 261 | 1% | 189 | 496 | 496 |
| 10 | −141 | −58 | 0.0 | 0.0 | 0.0 | 0 | 542 | 204 | 746 | 403 | 343 | 1% | 253 | 656 | 656 |
| 11 | −127 | −73 | 0.0 | 0.0 | 0.0 | 0 | 617 | 273 | 891 | 481 | 410 | 1% | 310 | 791 | 791 |
| 12 | −99 | −87 | 0.0 | 0.0 | 0.0 | 0 | 703 | 318 | 1021 | 551 | 470 | 1% | 363 | 914 | 914 |
| 13 | −44 | 80 | 0.0 | 6.4 | 0.0 | 0 | 772 | 356 | 1128 | 609 | 519 | 1% | 409 | 1018 | 1018 |
| 14 | −3 | 69 | 0.0 | 6.4 | 0.0 | 0 | 812 | 379 | 1190 | 643 | 548 | 1% | 443 | 1085 | 1085 |
| 15 | 15 | 58 | 0.0 | 6.4 | 0.0 | 0 | 807 | 379 | 1186 | 640 | 545 | 1% | 457 | 1098 | 1098 |
| 16 | 26 | 48 | 0.0 | 6.4 | 0.0 | 0 | 744 | 352 | 1096 | 592 | 504 | 1% | 449 | 1040 | 1040 |
| 17 | 34 | 38 | 0.0 | 6.4 | 0.0 | 0 | 617 | 314 | 930 | 502 | 428 | 1% | 412 | 915 | 915 |
| 18 | 42 | 26 | 0.0 | 4.9 | 18.9 | 344 | 428 | 269 | 1041 | 562 | 479 | 1% | 427 | 990 | 990 |
| 19 | 49 | 12 | 0.0 | 2.2 | 53.0 | 414 | 193 | 199 | 806 | 435 | 371 | 1% | 380 | 816 | 816 |
| 20 | 57 | −6 | 0.0 | 0.0 | 0.0 | 0 | 0 | 150 | 150 | 81 | 69 | 1% | 219 | 300 | 300 |
| 21 | 67 | −32 | 0.0 | 0.0 | 0.0 | 0 | 0 | 108 | 108 | 58 | 49 | 0% | 154 | 212 | 212 |
| 22 | 78 | −64 | 0.0 | 0.0 | 0.0 | 0 | 0 | 67 | 67 | 36 | 31 | 0% | 113 | 150 | 150 |
| 23 | 91 | 87 | 0.0 | 0.0 | 0.0 | 0 | 0 | 34 | 34 | 18 | 15 | 0% | 87 | 106 | 106 |
| 24 | 107 | 67 | 0.0 | 0.0 | 0.0 | 0 | 0 | 2 | 2 | 1 | 1 | 0% | 70 | 71 | 71 |

spreadsheets for computing each cooling load component using conduction and radiant time series have been compiled and are available from ASHRAE. To illustrate the full building example discussed previously, those individual component spreadsheets have been compiled to allow calculation of cooling and heating loads on a room by room basis as well as for a "block" calculation for analysis of overall areas or buildings where detailed room-by-room data are not available.

### 7.7.2 Single-Room Example Peak Heating Load

Although the physics of heat transfer that creates a heating load is identical to that for cooling loads, a number of traditionally used simplifying assumptions facilitate a much simpler calculation procedure. As described in the Heating Load Calculations section, design heating load calculations typically assume a single outdoor temperature, with no heat gain from solar or internal sources, under steady-state conditions. Thus, space heating load is determined by computing the heat transfer rate through building envelope elements ($UA\Delta T$) plus heat required because of outdoor air infiltration.

**Part 6. Room heating load.** Calculate the room heating load for the previous described office, including infiltration airflow at one air change per hour.

**Solution:**

Because solar heat gain is not considered in calculating design heating loads, orientation of similar envelope elements may be ignored and total areas of each wall or window type combined. Thus, the total spandrel wall area = 60 + 60 = 120 ft², total brick wall area = 60 + 40 = 100 ft², and total window area = 40 + 40 = 80 ft². For this example, use the U-factors that were used for cooling load conditions. In some climates, higher prevalent winds in winter should be considered in calculating U-factors (see Chapter 25 of 2013 ASHRAE Handbook—Fundamentals for information on calculating U-factors and surface heat transfer coefficients appropriate for local wind conditions). The 99.6% heating design dry-bulb temperature for Atlanta is 21.5°F and the indoor design temperature is 72°F. The room volume with a 9 ft ceiling = 9 × 130 = 1170 ft³. At one air change per hour, the infiltration airflow = 1 × 1170/60 = 19.5 cfm. Thus, the heating load is

| | | |
|---|---|---|
| Windows: | 0.56 × 80 × (72 − 21.5) = | 2262 Btu/h |
| Spandrel Wall: | 0.077 × 120 × (72 − 21.5) = | 467 |
| Brick Wall: | 0.08 × 100 × (72 − 21.5) = | 404 |
| Roof: | 0.032 × 130 × (72 − 21.5) = | 210 |
| Infiltration: | 19.5 × 1.1 × (72 − 21.5) = | 1083 |
| Total Room Heating Load: | | 4426 Btu/h |

## 7.8 Problems

**7.1** The exterior windows are of double insulating glass with 0.25 in. (6-mm) airspace and have metal sashes. Determine the design U-factor for cooling for the window.

Table 7-44  Single-Room Example Cooling Load (July 3:00 PM) for ASHRAE Example Office Building, Atlanta, GA

Table 7-45  Single-Room Example Peak Cooling Load (Sept. 5:00 PM) for ASHRAE Example Office Building, Atlanta, GA

**7.2** A store in Lafayette, Indiana, is on the northeast corner of an intersection with one street running due north. The bottom of the show windows are 2 ft, 6 in. above the sidewalk; the show windows are 7 ft high. An aluminum awning with a 3 in. rise per horizontal foot is to be hung with the bottom strut at the window header. Both south and west awnings are to have the same dimensions.

(a) What minimum distance should the strut extend from the building to keep the shade line on the windows at 3 PM sun time?

(b) Which face of the building governs the awning dimensions?

(c) Where will the shade line be at 3 PM on the other face of the building?

(d) What is the elevation of the top of the awnings above the sidewalk?

**7.3** Calculate the solar radiation entering through clear glass as shown at right. [Ans 692 Btu/h]

**7.4** Solve the following:

(a) Determine the solar angle of incidence for a vertical wall facing 15° west of south when the sun has an azimuth of 79.2° west of south and an altitude of 75.7°. [Ans: 83.8°]

(b) Find the solar incident angle (for direct solar radiation) for a vertical surface facing southeast at 8:30 AM CST on October 22 at 32° N latitude and 95° W longitude. [Ans: 28.4°]

**7.5** What environmental factors affect the solar intensity reaching the earth's surface?

**7.6** Determine the heat being dissipated by 50 pendant mounted fluorescent luminaires with four 40 W lamps in each luminaire.

**268**　　　　　　　　　　　　　　　　　　　　　　　　　　　　　　　　　　　　　　　　　Principles of HVAC

**Elevation**
Wall orientation:
South facing

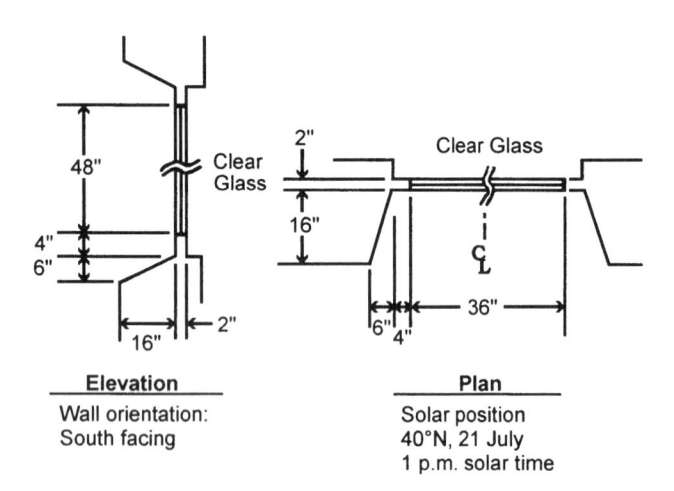

**Plan**
Solar position
40°N, 21 July
1 p.m. solar time

**7.7** How much sensible, latent, and total heat is contributed by 50 customers shopping in a drugstore?

**7.8** A 1 hp motor driving a pump is located in a space to be air conditioned. Determine heat dissipated to the space from the motor and pump. [Ans: 3390 Btu/h]

**7.9** Calculate the heat gain to a room from a 21-lb deep fat fryer (8 kW input rating) if (a) hooded and (b) nonhooded.

**7.10** Calculate the maximum heat gain through the floor for a room directly over a boiler room. The air temperature at the underside of the floor is 100°F, and the room air temperature desired close to the floor is 70°F. The floor is 4 in. concrete with vinyl tile finish.

[Ans: 21 Btu/h·ft$^2$]

**7.11** An air-conditioning unit serves an office having the following areas:

| Description | Size | Occupancy |
|---|---|---|
| General office | 25 ft by 50 ft | 75 ft$^2$ per person |
| Director's room | 25 ft by 25 ft | 16 people |
| Conference room | 10 ft by 25 ft | Plush furnishings |
| 5 private offices | 10 ft by 10 ft | Smoking permitted |

What quantity of outdoor air must be brought into the air-conditioning unit for ventilation?

**7.12** Suppose the fan of the air-conditioning unit in Problem 7.11 supplies 3200 cfm to the ductwork.

(a) How many air changes per hour are being used? (Assume a ceiling height of 9 ft.)
(b) What is the percentage of outdoor air?
(c) Suppose the minimum recommended quantities of total air and outdoor air are used, what will be the percentage of outdoor air?

**7.13** A small parts assembly area in a factory has a working force of 25 men and occupies a space 27.4 m by 9.1 m with a 3 m ceiling. Smoking is not allowed. Determine

(a) Sensible heat load from the occupants
(b) Latent heat load from the occupants
(c) Moisture added from the occupants
(d) The minimum volume of outdoor air for ventilation
(e) Suitable summer design inside dry-bulb temperature

**7.14** The office portion of a multistory commercial building, located at 40° N latitude, is shown in the sketch on the next page. Neglecting any outdoor air load, for August 21 determine the cooling load at 4 PM if it is located at 40° N.

Inside design: 75°F, $W = 0.0102$

Outside design: 95°F, 22°F daily range, $W = 0.0168$

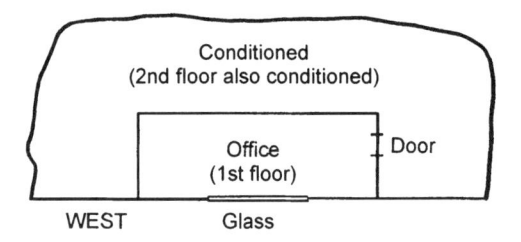

*For Office Portion:*
West wall: Net area = 230 ft$^2$, $U = 0.333$
　　4 in. red face brick, 4 in. low weight concrete block, and 0.75 in. plaster
Window: Area = 90 ft$^2$, $U = 0.59$
　　double pane, regular sheet, insulating glass
People: 4, moderately active office work
Lights: 800 W fluorescent, on continuously

**7.15** Calculate the cooling load for the office shown in the diagram at right.
　Location: Indianapolis, Indiana
　Occupancy: 180 people
　Factory temperature: Assume 5°F below outside ambient
　Equipment: 35 business machines, each 110 V at 4.0 A
　East and south walls: 8 in. concrete block,
　　$U = 0.3$ Btu/h·ft$^2$·°F
　North and west walls: Concrete block,
　　$U = 0.5$ Btu/h·ft$^2$·°F
　Floor transmission: Neglect
　Ceiling: $U = 0.5$ Btu/h·ft$^2$·°F
　Lights: 6 W/ft$^2$
　Windows: Insulating glass, 0.25 in. gray plate outside, regular plate inside
　Direct-expansion packaged air-conditioning unit will be used.
　Locate equipment in factory west of office.
　Calculate the heat gain for this office.

**7.16** An architect provides the following data for a new cafeteria in Tampa, Florida:
　Size: 80 ft wide by 100 ft deep; 30 ft deep area in rear is for kitchen and storage. Wall separating this area from main dining area is made of concrete block.
　Walls: 4 in. concrete block with an outside facing of 1/2 in. cement mortar; 3/8 in. air space filled with fiberglass

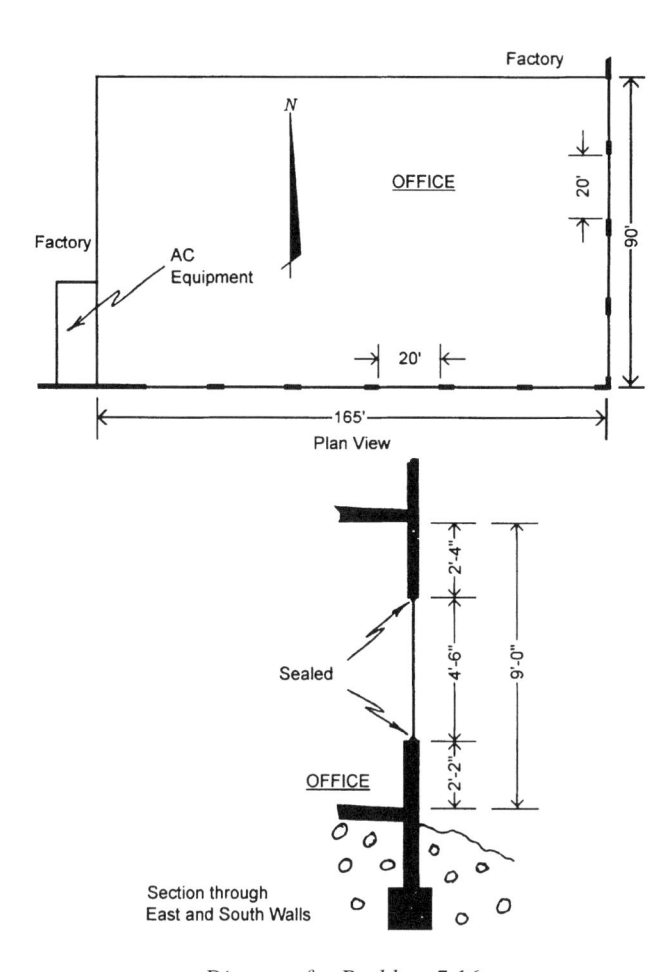

*Diagram for Problem 7.16*

insulation, backed by 1 mil aluminum foil; 3/8 in. gypsum wallboard; two coats light green paint; front wall has a canopy projection of 4 ft (located 12 ft above ground level).

Windows: On front wall: 80% of front wall is composed of 1/4 in. gray plate glass, backed by fiberglass draperies of a medium-colored yarn of a close weave; windows are well sealed. On side walls: No windows. On back walls: No windows.

Roof: Flat, stone aggregation on tar base outside; 3/8 in. roofing board; 1/2 in. roof insulation; metal decking; 2 ft air space between steel beams (for lighting fixtures and ductwork); suspended acoustical ceiling.

Floor: 8 in. concrete slab with 1/4 in. floor tile: 1 in. edge insulation.

Doors: Two sets of double swinging doors on opposite sides of front wall; single bank; 7 ft wide (total of two doors); 7 ft high; peak traffic expected is 300 people per hour. There are two service entrance doors on rear, normally closed, 3 ft by 7 ft, with slight cracks all around.

Occupancy: Maximum expected occupancy is 300 between the hours of 5:00 PM and 7:00 PM.

Internal equipment: Usual cooking, washing, and food service storage trays for a cafeteria; a refrigerated food locker is in the rear.

Lighting: Indirect neon lights on ceiling.

Location: Front faces southwest; located in Tampa, Florida.

Calculate the sensible cooling load (using your best judgment as to the number of people present at off-hours) for 10 AM, 2 PM, and 6 PM. Consider the cafeteria as a single zone and plan to air condition the kitchen also.

**7.17** Size the cooling system by determining the design cooling loads for the following pharmacy building to be built in Tulsa, Oklahoma.

Overall size: 60 ft by 130 ft by 12 ft high [long sides on east and west]

Walls: 8-inch concrete block with normal weight sand and gravel aggregazte with 1-in. stucco on the outside and 3/4-in. cement plaster with sand aggregate on the inside

Doors: Double 2-1/4-in. solid core doors (6 ft by 7 ft) on north and south

Window: One 6 ft by 90 ft double glass with thermal break frames, 1/8-in. thick glass with 3/8-in. air gap, with translucent roller shades, on west side

Roof/Ceiling: 4-in. lightweight concrete with 3/8-in. built-up roofing on the exterior and 3/4-in. cement plaster with sand aggregate on the inside

Carefully state all assumptions.

**7.18** Solve the following:

(a) A 115 ft by 10 ft high wall in Minneapolis, Minnesota, consists of face brick, a 3/4-in. air gap, 8-in. cinder aggregate concrete blocks, 1-in. organic bonded glass fiber insulation, and 4-in. clay tile interior. Determine the design heat loss through the wall in winter in Btu/h.

(b) If the wall of Part (a) is converted to 60% single-glazed glass, what is the winter design heat loss through the total wall in Btu/h?

## 7.9  Bibliography

Abushakra, B., J.S. Haberl, and D.E. Claridge. 2004. Overview of literature on diversity factors and schedules for energy and cooling load calculations (1093-RP). *ASHRAE Transactions* 110(1):164-176.

Alereza, T., and J.P. Breen, III. 1984. Estimates of recommended heat gain due to commercial appliances and equipment. *ASHRAE Transactions* 90(2A):25-58.

Armstrong, P.R., C.E. Hancock, III, and J.E. Seem. 1992a. Commercial building temperature recovery—Part I: Design procedure based on a step response model. *ASHRAE Transactions* 98(1):381-396.

Armstrong, P.R., C.E. Hancock, III, and J.E. Seem. 1992b. Commercial building temperature recovery—Part II: Experiments to verify the step response model. *ASHRAE Transactions* 98(1):397-410.

ASHRAE. 1975. *Procedure for determining heating and cooling loads for computerized energy calculations, algorithms for building heat transfer subroutines.*

ASHRAE. 1979. *Cooling and heating load calculation manual.*

ASHRAE. 2010. Energy standard for building except low-rise residential buildings. ANSI/ASHRAE/IESNA *Standard* 90.1-2010.

ASHRAE. 2010. Ventilation for acceptable indoor air quality. ANSI/ASHRAE *Standard* 62.1-2010.

ASHRAE. 2010. Thermal environmental conditions for human occupancy. ANSI/ASHRAE *Standard* 55-2010.

ASHRAE. 2012. Updating the climatic design conditions in the *ASHRAE Handbook—Fundamentals* (RP-1613). ASHRAE Research Project, *Final Report*.

ASTM. 2008. Practice for estimate of the heat gain or loss and the surface temperatures of insulated flat, cylindrical, and spherical systems by use of computer programs. Standard C680-08. American Society for Testing and Materials, West Conshohocken, PA.

Bauman, F., S. Schiavon, T. Webster, and K.H. Lee. 2010. Cooling load design tool for UFAD systems. *ASHRAE Journal* (September):62-71.http://escholarship.org/uc/item/9d8430v3.

Bauman, F., T. Webster, P. Linden, and F. Buhl. 2007. Energy performance of UFAD systems. CEC-500-2007-050, *Final Report* to CEC PIER Buildings Program. Center for the Built Environment, University of California, Berkeley. http://www.energy.ca.gov/2007publications/CEC -500-2007-050/index.html

Bauman, F.S., and A. Daly. 2013. *Underfloor air distribution (UFAD) design guide*, 2nd ed. ASHRAE.

BLAST Support Office. 1991. *BLAST user reference.* University of Illinois, Urbana–Champaign.

Bliss, R.J.V. 1961. Atmospheric radiation near the surface of the ground. *Solar Energy* 5(3):103.

Buffington, D.E. 1975. Heat gain by conduction through exterior walls and roofs—Transmission matrix method. *ASHRAE Transactions* 81(2):89.

Burch, D.M., B.A. Peavy, and F.J. Powell. 1974. Experimental validation of the NBS load and indoor temperature prediction model. *ASHRAE Transactions* 80(2):291.

Burch, D.M., J.E. Seem, G.N. Walton, and B.A. Licitra. 1992. Dynamic evaluation of thermal bridges in a typical office building. *ASHRAE Transactions* 98:291-304.

Butler, R. 1984. The computation of heat flows through multilayer slabs. *Building and Environment* 19(3):197-206.

Ceylan, H.T., and G.E. Myers. 1985. Application of response-coefficient method to heat-conduction transients. *ASHRAE Transactions* 91:30-39.

Chantrasrisalai, C., D.E. Fisher, I. Iu, and D. Eldridge. 2003. Experimental validation of design cooling load procedures: The heat balance method. *ASHRAE Transactions* 109(2):160-173.

Chiles, D.C., and E.F. Sowell. 1984. A counter-intuitive effect of mass on zone cooling load response. *ASHRAE Transactions* 91(2A):201-208.

Chorpening, B.T. 1997. The sensitivity of cooling load calculations to window solar transmission models. *ASHRAE Transactions* 103(1).

Claridge, D.E., B. Abushakra, J.S. Haberl, and A. Sreshthaputra. 2004. Elec-tricity diversity profiles for energy simulation of office buildings (RP-1093). *ASHRAE Transactions* 110(1):365-377.

Clarke, J.A. 1985. *Energy simulation in building design.* Adam Hilger Ltd., Boston.

Davies, M.G. 1996. A time-domain estimation of wall conduction transfer function coefficients. *ASHRAE Transactions* 102(1):328-208.

Eldridge, D., D.E. Fisher, I. Iu, and C. Chantrasrisalai. 2003. Experimental validation of design cooling load procedures: Facility design (RP-1117). *ASHRAE Transactions* 109(2):151-159.

Falconer, D.R., E.F. Sowell, J.D. Spitler, and B.B. Todorovich. 1993. Electronic tables for the ASHRAE load calculation manual. *ASHRAE Transactions* 99(1):193-200.

Feng, J., S. Schiavon, and F. Bauman. 2012. Comparison of zone cooling load for radiant and air conditioning systems. Proceedings of the International Conference on Building Energy and Environment. Boulder, CO. http://escholarship.org/uc/item/9g24f38j.

Fisher, D.E., and C. Chantrasrisalai. 2006. Lighting heat gain distribution in buildings (RP-1282). ASHRAE Research Project, *Final Report*.

Fisher, D.E., and C.O. Pedersen. 1997. Convective heat transfer in building energy and thermal load calculations. *ASHRAE Transactions* 103(2): 137-148.

Fisher, D.R. 1998. New recommended heat gains for commercial cooking equipment. *ASHRAE Transactions* 104(2):953-960.

Gordon, E.B., D.J. Horton, and F.A. Parvin. 1994. Development and application of a standard test method for the performance of exhaust hoods with commercial cooking appliances. *ASHRAE Transactions* 100(2).

Harris, S.M., and F.C. McQuiston. 1988. A study to categorize walls and roofs on the basis of thermal response. *ASHRAE Transactions* 94(2): 688-714.

Hittle, D.C. 1981. Calculating building heating and cooling loads using the frequency response of multilayered slabs, Ph.D. dissertation, Department of Mechanical and Industrial Engineering, University of Illinois, Urbana-Champaign.

Hittle, D.C. 1999. The effect of beam solar radiation on peak cooling loads. *ASHRAE Transactions* 105(2):510-513.

Hittle, D.C., and C.O. Pedersen. 1981. Calculating building heating loads using the frequency of multi-layered slabs. *ASHRAE Transactions* 87(2):545-568.

Hittle, D.C., and R. Bishop. 1983. An improved root-finding procedure for use in calculating transient heat flow through multilayered slabs. *International Journal of Heat and Mass Transfer* 26:1685-1693.

Hosni, M.H., and B.T. Beck. 2008. Update to measurements of office equipment heat gain data (RP-1482). ASHRAE Research Project, *Progress Report*.

Hosni, M.H., B.W. Jones, and H. Xu. 1999. Experimental results for heat gain and radiant/convective split from equipment in buildings. *ASHRAE Transactions* 105(2):527-539.

Hosni, M.H., B.W. Jones, J.M. Sipes, and Y. Xu. 1998. Total heat gain and the split between radiant and convective heat gain from office and laboratory equipment in buildings. *ASHRAE Transactions* 104(1A):356-365.

Incropera, F.P., and D.P DeWitt. 1990. *Fundamentals of heat and mass transfer*, 3rd ed. Wiley, New York.

Iu, I., and D.E. Fisher. 2004. Application of conduction transfer functions and periodic response factors in cooling load calculation procedures. *ASHRAE Transactions* 110(2):829-841.

Iu, I., C. Chantrasrisalai, D.S. Eldridge, and D.E. Fisher. 2003. experimental validation of design cooling load procedures: The radiant time series method (RP-1117). *ASHRAE Transactions* 109(2):139-150.

Jones, B.W., M.H. Hosni, and J.M. Sipes. 1998. Measurement of radiant heat gain from office equipment using a scanning radiometer. *ASHRAE Transactions* 104(1B):1775-1783.

Karambakkam, B.K., B. Nigusse, and J.D. Spitler. 2005. A one-dimensional approximation for transient multi-dimensional conduction heat transfer in building envelopes. *Proceedings of the 7th Symposium on Building Physics in the Nordic Countries*, The Icelandic Building Research Institute, Reykjavik, vol. 1, pp. 340-347.

Kerrisk, J.F., N.M. Schnurr, J.E. Moore, and B.D. Hunn. 1981. The custom weighting-factor method for thermal load calculations in the DOE-2 computer program. *ASHRAE Transactions* 87(2):569-584.

Kimura and Stephenson. 1968. Theoretical study of cooling loads caused by lights. *ASHRAE Transactions* 74(2):189-197.

Komor, P. 1997. Space cooling demands from office plug loads. *ASHRAE Journal* 39(12):41-44.

Kusuda, T. 1967. *NBSLD, the computer program for heating and cooling loads for buildings*. BSS 69 and NBSIR 74-574. National Bureau of Standards.

Kusuda, T. 1969. Thermal response factors for multilayer structures of various heat conduction systems. *ASHRAE Transactions* 75(1):246.

Latta, J.K., and G.G. Boileau. 1969. Heat losses from house basements. *Canadian Building* 19(10):39.

LBL. 2003. *WINDOW 5.2: A PC program for analyzing window thermal performance for fenestration products*. LBL-44789. Windows and Daylighting Group. Lawrence Berkeley Laboratory, Berkeley.

Liesen, R.J., and C.O. Pedersen. 1997. An evaluation of inside surface heat balance models for cooling load calculations. *ASHRAE Transactions* 103(2):485-502.

Marn, W.L. 1962. Commercial gas kitchen ventilation studies. *Research Bulletin* 90(March). Gas Association Laboratories, Cleveland, OH.

Mast, W.D. 1972. Comparison between measured and calculated hour heating and cooling loads for an instrumented building. ASHRAE *Symposium Bulletin* 72(2).

McBridge, M.F., C.D. Jones, W.D. Mast, and C.F. Sepsey. 1975. Field validation test of the hourly load program developed from the ASHRAE algorithms. *ASHRAE Transactions* 1(1):291.

McClellan, T.M., and C.O. Pedersen. 1997. Investigation of outdoor heat balance models for use in a heat balance cooling load calculation procedure. *ASHRAE Transactions* 103(2):469-484.

McQuiston, F.C., and J.D. Spitler. 1992. *Cooling and heating load calculation manual*, 2nd ed. ASHRAE.

Miller, A. 1971. *Meteorology*, 2nd ed. Charles E. Merrill, Columbus.

Mitalas, G.P. 1968. Calculations of transient heat flow through walls and roofs. *ASHRAE Transactions* 74(2):182-188.

Mitalas, G.P. 1969. An experimental check on the weighting factor method of calculating room cooling load. *ASHRAE Transactions* 75(2):22.

Mitalas, G.P. 1972. Transfer function method of calculating cooling loads, heat extraction rate, and space temperature. *ASHRAE Journal* 14(12):52.

Mitalas, G.P. 1973. Calculating cooling load caused by lights. *ASHRAE Transactions* 75(6):7.

Mitalas, G.P. 1978. Comments on the Z-transfer function method for calculating heat transfer in buildings. *ASHRAE Transactions* 84(1):667-674.

Mitalas, G.P., and D.G. Stephenson. 1967. Room thermal response factors. *ASHRAE Transactions* 73(2): III.2.1.

Mitalas, G.P., and J.G. Arsenault. 1970. Fortran IV program to calculate Z-transfer functions for the calculation of transient heat transfer through walls and roofs. *Use of Computers for Environmental Engineering Related to Buildings*, pp. 633-668. National Bureau of Standards, Gaithersburg, MD.

Mitalas, G.P., and K. Kimura. 1971. A calorimeter to determine cooling load caused by lights. *ASHRAE Transactions* 77(2):65.

Nevins, R.G., H.E. Straub, and H.D. Ball. 1971. Thermal analysis of heat removal troffers. *ASHRAE Transactions* 77(2):58-72.

NFPA. 2012. Health care facilities code. *Standard* 99-2012. National Fire Protection Association, Quincy, MA.

Nigusse, B.A. 2007. *Improvements to the radiant time series method cooling load calculation procedure*. Ph.D. dissertation, Oklahoma State University.

Ouyang, K., and F. Haghighat. 1991. A procedure for calculating thermal response factors of multi-layer walls—State space method. *Building and Environment* 26(2):173-177.

Parker, D.S., J.E.R. McIlvaine, S.F. Barkaszi, D.J. Beal, and M.T. Anello. 2000. *Laboratory testing of the reflectance properties of roofing material*. FSEC-CR670-00. Florida Solar Energy Center, Cocoa.

Peavy, B.A. 1978. A note on response factors and conduction transfer functions. *ASHRAE Transactions* 84(1):688-690.

Peavy, B.A., F.J. Powell, and D.M. Burch. 1975. Dynamic thermal performance of an experimental masonry building. NBS *Building Science Series* 45 (July).

Pedersen, C.O., D.E. Fisher, and R.J. Liesen. 1997. Development of a heat balance procedure for calculating cooling loads. *ASHRAE Transactions* 103(2):459-468.

Pedersen, C.O., D.E. Fisher, J.D. Spitler, and R.J. Liesen. 1998. *Cooling and heating load calculation principles*. ASHRAE.

Rees, S.J., J.D. Spitler, M.G. Davies, and P. Haves. 2000. Qualitative comparison of North American and U.K. cooling load calculation methods. *International Journal of Heating, Ventilating, Air-Conditioning and Refrigerating Research* 6(1):75-99.

Rock, B.A. 2005. A user-friendly model and coefficients for slab-on-grade load and energy calculation. *ASHRAE Transactions* 111(2):122-136.

Rock, B.A., and D.J. Wolfe. 1997. A sensitivity study of floor and ceiling plenum energy model parameters. *ASHRAE Transactions* 103(1):16-30.

Romine, T.B., Jr. 1992. Cooling load calculation: Art or science? *ASHRAE Journal*, 34(1):14.

Rudoy, W. 1979. Don't turn the tables. *ASHRAE Journal* 21(7):62.

Rudoy, W., and F. Duran. 1975. Development of an improved cooling load calculation method. *ASHRAE Transactions* 81(2):19-69.

Schiavon, S., F. Bauman, K.H. Lee, and T. Webster. 2010a. Simplified cal-culation method for design cooling loads in underfloor air distribution (UFAD) systems. *Energy and Buildings* 43(1-2):517-528. http://escholarship.org/uc/item/5w53c7kr.

Schiavon, S., F. Bauman, K.H. Lee, and T. Webster. 2010c. Development of a simplified cooling load design tool for underfloor air distribution systems. *Final Report* to CEC PIER Program, July. http://escholarship.org/uc/item/6278m12z.

Schiavon, S., K.H. Lee, F. Bauman, and T. Webster. 2010b. Influence of raised floor on zone design cooling load in commercial buildings. *Energy and Buildings* 42(5):1182-1191. http://escholarship.org/uc/item/2bv611dt.

Seem, J.E., S.A. Klein, W.A. Beckman, and J.W. Mitchell. 1989. Transfer functions for efficient calculation of multi-dimensional transient heat transfer. *Journal of Heat Transfer* 111:5-12.

Smith, V.A., R.T. Swierczyna, and C.N. Claar. 1995. Application and enhancement of the standard test method for the performance of commercial kitchen ventilation systems. *ASHRAE Transactions* 101(2).

Sowell, E.F. 1988a. Cross-check and modification of the DOE-2 program for calculation of zone weighting factors. *ASHRAE Transactions* 94(2).

Sowell, E.F. 1988b. Load calculations for 200,640 zones. *ASHRAE Transactions* 94(2):716-736.

Sowell, E.F., and D.C. Chiles. 1984a. Characterization of zone dynamic response for CLF/CLTD tables. *ASHRAE Transactions* 91(2A):162-178.

Sowell, E.F., and D.C. Chiles. 1984b. Zone descriptions and response characterization for CLF/CLTD calculations. *ASHRAE Transactions* 91(2A): 179-200.

Spitler, J.D. 1996. *Annotated guide to load calculation models and algorithms.* ASHRAE.

Spitler, J.D., and D.E. Fisher. 1999a. Development of periodic response factors for use with the radiant time series method. *ASHRAE Transactions* 105(2):491-509.

Spitler, J.D., and D.E. Fisher. 1999b. On the relationship between the radiant time series and transfer function methods for design cooling load calculations. *International Journal of Heating, Ventilating, Air-Conditioning and Refrigerating Research* (now *HVAC&R Research*) 5(2):125-138.

Spitler, J.D., and F.C. McQuiston. 1993. Development of a revised cooling and heating calculation manual. *ASHRAE Transactions* 99(1):175-182.

Spitler, J.D., D.E. Fisher, and C.O. Pedersen. 1997. The radiant time series cooling load calculation procedure. *ASHRAE Transactions* 103(2).

Spitler, J.D., F.C. McQuiston, and K.L. Lindsey. 1993. The CLTD/SCL/CLF cooling load calculation method. *ASHRAE Transactions* 99(1): 183-192.

Spitler, J.D., S.J. Rees, and P. Haves. 1998. Quantitive comparison of North American and U.K. cooling load calculation procedures—Part 1: Methodology, Part II: Results. *ASHRAE Transactions* 104(2):36-46, 47-61.

Stephenson, D.G. 1962. Method of determining non-steady-state heat flow through walls and roofs at buildings. *Journal of the Institution of Heating and Ventilating Engineers* 30:5.

Stephenson, D.G., and G.P. Mitalas. 1967. Cooling load calculation by thermal response factor method. *ASHRAE Transactions* 73(2):III.1.1.

Stephenson, D.G., and G.P. Mitalas. 1971. Calculation of heat transfer functions for multi-layer slabs. *ASHRAE Transactions* 77(2):117-126.

Sun, T.-Y. 1968. Computer evaluation of the shadow area on a window cast by the adjacent building. *ASHRAE Journal* (September).

Sun, T.-Y. 1968. Shadow area equations for window overhangs and side-fins and their application in computer calculation. *ASHRAE Transactions* 74(1):I-1.1 to I-1.9.

Swierczyna, R., P. Sobiski, and D. Fisher. 2008. Revised heat gain and capture and containment exhaust rates from typical commercial cooking appliances (RP-1362). ASHRAE Research Project, *Final Report*.

Swierczyna, R., P.A. Sobiski, and D.R. Fisher. 2009 (forthcoming). Revised heat gain rates from typical commercial cooking appliances from RP-1362. *ASHRAE Transactions* 115(2).

Talbert, S.G., L.J. Canigan, and J.A. Eibling. 1973. An experimental study of ventilation requirements of commercial electric kitchens. *ASHRAE Transactions* 79(1):34.

Todorovic, B. 1982. Cooling load from solar radiation through partially shaded windows, taking heat storage effect into account. *ASHRAE Transactions* 88(2):924-937.

Todorovic, B. 1984. Distribution of solar energy following its transmittal through window panes. *ASHRAE Transactions* 90(1B):806-815.

Todorovic, B. 1987. The effect of the changing shade line on the cooling load calculations. In ASHRAE videotape, *Practical applications for cooling load calculations.*

Todorovic, B. 1989. *Heat storage in building structure and its effect on cooling load; Heat and mass transfer in building materials and structure.* Hemisphere Publishing, New York.

Todorovic, B., and D. Curcija. 1984. Calculative procedure for estimating cooling loads influenced by window shading, using negative cooling load method. *ASHRAE Transactions* 2:662.

Todorovic, B., L. Marjanovic, and D. Kovacevic. 1993. Comparison of different calculation procedures for cooling load from solar radiation through a window. *ASHRAE Transactions* 99(2):559-564.

Walton, G. 1983. *Thermal analysis research program reference manual.* National Bureau of Standards.

Webster, T., F. Bauman, F. Buhl, and A. Daly. 2008. Modeling of underfloor air distribution (UFAD) systems. SimBuild 2008, University of California, Berkeley.

Wilkins, C.K. 1998. Electronic equipment heat gains in buildings. *ASHRAE Transactions* 104(1B):1784-1789.

Wilkins, C.K., and M. Hosni. 2011. Plug load design factors. *ASHRAE Journal* 53(5):30-34.

Wilkins, C.K., and M.H. Hosni. 2000. Heat gain from office equipment. *ASHRAE Journal* 42(6):33-44.

Wilkins, C.K., and M.R. Cook. 1999. Cooling loads in laboratories. *ASHRAE Transactions* 105(1):744-749.

Wilkins, C.K., and N. McGaffin. 1994. Measuring computer equipment loads in office buildings. *ASHRAE Journal* 36(8):21-24.

Wilkins, C.K., R. Kosonen, and T. Laine. 1991. An analysis of office equipment load factors. *ASHRAE Journal* 33(9):38-44.

York, D.A., and C.C. Cappiello. 1981. *DOE-2 engineers manual* (Version 2.1A). Lawrence Berkeley Laboratory and Los Alamos National Laboratory.

## SI Tables and Figures

### Table 7-1 SI  Sol-Air Temperatures ($t_e$) for July 21, 40° N Latitude

$$t_e = t_o + \alpha I_t/h_o - \varepsilon \Delta R/h_o$$

| Time | Air Temp. $t_o$, °C | \multicolumn{9}{c}{Light Colored Surface, $\alpha/h_o = 0.026$} | Time | Air Temp. $t_o$, °C | \multicolumn{9}{c}{Dark Colored Surface, $\alpha/h_o = 0.052$} |
|---|---|---|---|---|---|---|---|---|---|---|---|---|---|---|---|---|---|---|---|---|
| | | N | NE | E | SE | S | SW | W | NW | HOR | | | N | NE | E | SE | S | SW | W | NW | HOR |
| 1 | 25.4 | 25.4 | 25.4 | 25.4 | 25.4 | 25.4 | 25.4 | 25.4 | 25.4 | 21.5 | 1 | 25.4 | 25.4 | 25.4 | 25.4 | 25.4 | 25.4 | 25.4 | 25.4 | 25.4 | 21.5 |
| 2 | 24.9 | 24.9 | 24.9 | 24.9 | 24.9 | 24.9 | 24.9 | 24.9 | 24.9 | 21.0 | 2 | 24.9 | 24.9 | 24.9 | 24.9 | 24.9 | 24.9 | 24.9 | 24.9 | 24.9 | 21.0 |
| 3 | 24.4 | 24.4 | 24.4 | 24.4 | 24.4 | 24.4 | 24.4 | 24.4 | 24.4 | 20.5 | 3 | 24.4 | 24.4 | 24.4 | 24.4 | 24.4 | 24.4 | 24.4 | 24.4 | 24.4 | 20.5 |
| 4 | 24.1 | 24.1 | 24.1 | 24.1 | 24.1 | 24.1 | 24.1 | 24.1 | 24.1 | 20.2 | 4 | 24.1 | 24.1 | 24.1 | 24.1 | 24.1 | 24.1 | 24.1 | 24.1 | 24.1 | 20.2 |
| 5 | 24.0 | 24.1 | 24.2 | 24.2 | 24.1 | 24.0 | 24.0 | 24.0 | 24.0 | 20.1 | 5 | 24.0 | 24.2 | 24.4 | 24.3 | 24.1 | 24.0 | 24.0 | 24.0 | 24.0 | 20.2 |
| 6 | 24.2 | 27.2 | 34.5 | 35.5 | 29.8 | 25.1 | 25.1 | 25.1 | 25.1 | 22.9 | 6 | 24.2 | 30.2 | 44.7 | 46.7 | 35.4 | 26.0 | 26.0 | 26.0 | 26.0 | 25.5 |
| 7 | 24.8 | 27.3 | 38.1 | 41.5 | 35.2 | 26.5 | 26.4 | 26.4 | 26.4 | 28.1 | 7 | 24.8 | 29.7 | 51.5 | 58.2 | 45.6 | 28.2 | 28.0 | 28.0 | 28.0 | 35.4 |
| 8 | 25.8 | 28.1 | 38.0 | 43.5 | 38.9 | 28.2 | 28.0 | 28.0 | 28.0 | 33.8 | 8 | 25.8 | 30.5 | 50.1 | 61.2 | 52.1 | 30.7 | 30.1 | 30.1 | 30.1 | 45.8 |
| 9 | 27.2 | 29.9 | 35.9 | 43.1 | 41.2 | 31.5 | 29.8 | 29.8 | 29.8 | 39.2 | 9 | 27.2 | 32.5 | 44.5 | 58.9 | 55.1 | 35.8 | 32.3 | 32.3 | 32.3 | 55.1 |
| 10 | 28.8 | 31.7 | 33.4 | 40.8 | 41.8 | 35.4 | 31.8 | 31.7 | 31.7 | 43.9 | 10 | 28.8 | 34.5 | 38.0 | 52.8 | 54.9 | 42.0 | 34.7 | 34.5 | 34.5 | 62.8 |
| 11 | 30.7 | 33.7 | 34.0 | 37.4 | 41.1 | 39.0 | 34.2 | 33.7 | 33.7 | 47.7 | 11 | 30.7 | 36.8 | 37.2 | 44.0 | 51.5 | 47.4 | 37.7 | 36.8 | 36.8 | 68.5 |
| 12 | 32.5 | 35.6 | 35.6 | 35.9 | 39.1 | 41.4 | 39.1 | 35.9 | 35.6 | 50.1 | 12 | 32.5 | 38.7 | 38.7 | 39.3 | 45.7 | 50.4 | 45.7 | 39.3 | 38.7 | 71.6 |
| 13 | 33.8 | 36.8 | 36.8 | 36.8 | 37.3 | 42.1 | 44.2 | 40.5 | 37.1 | 50.8 | 13 | 33.8 | 39.9 | 39.9 | 39.9 | 40.8 | 50.5 | 54.6 | 47.1 | 40.3 | 71.6 |
| 14 | 34.7 | 37.6 | 37.6 | 37.6 | 37.7 | 41.3 | 47.7 | 46.7 | 39.3 | 49.8 | 14 | 34.7 | 40.4 | 40.4 | 40.4 | 40.6 | 47.9 | 60.8 | 58.7 | 43.9 | 68.7 |
| 15 | 35.0 | 37.7 | 37.6 | 37.6 | 37.6 | 39.3 | 49.0 | 50.9 | 43.7 | 47.0 | 15 | 35.0 | 40.3 | 40.1 | 40.1 | 40.1 | 43.6 | 62.9 | 66.7 | 52.3 | 62.9 |
| 16 | 34.7 | 37.0 | 36.9 | 36.9 | 36.9 | 37.1 | 47.8 | 52.4 | 46.9 | 42.7 | 16 | 34.7 | 39.4 | 39.0 | 39.0 | 39.0 | 39.6 | 61.0 | 70.1 | 59.0 | 54.7 |
| 17 | 33.9 | 36.4 | 35.5 | 35.5 | 35.5 | 35.6 | 44.3 | 50.6 | 47.2 | 37.2 | 17 | 33.9 | 38.8 | 37.1 | 37.1 | 37.1 | 37.3 | 54.7 | 67.3 | 60.6 | 44.5 |
| 18 | 32.7 | 35.7 | 33.6 | 33.6 | 33.6 | 33.6 | 38.3 | 44.0 | 43.0 | 31.4 | 18 | 32.7 | 38.7 | 34.5 | 34.5 | 34.5 | 34.5 | 43.9 | 55.2 | 53.2 | 34.0 |
| 19 | 31.3 | 31.4 | 31.3 | 31.3 | 31.3 | 31.3 | 31.4 | 31.5 | 31.5 | 27.4 | 19 | 31.3 | 31.5 | 31.3 | 31.3 | 31.3 | 31.3 | 31.4 | 31.6 | 31.7 | 27.5 |
| 20 | 29.8 | 29.8 | 29.8 | 29.8 | 29.8 | 29.8 | 29.8 | 29.8 | 29.8 | 25.9 | 20 | 29.8 | 29.8 | 29.8 | 29.8 | 29.8 | 29.8 | 29.8 | 29.8 | 29.8 | 25.9 |
| 21 | 28.6 | 28.6 | 28.6 | 28.6 | 28.6 | 28.6 | 28.6 | 28.6 | 28.6 | 24.7 | 21 | 28.6 | 28.6 | 28.6 | 28.6 | 28.6 | 28.6 | 28.6 | 28.6 | 28.6 | 24.7 |
| 22 | 27.5 | 27.5 | 27.5 | 27.5 | 27.5 | 27.5 | 27.5 | 27.5 | 27.5 | 23.6 | 22 | 27.5 | 27.5 | 27.5 | 27.5 | 27.5 | 27.5 | 27.5 | 27.5 | 27.5 | 23.6 |
| 23 | 26.6 | 26.6 | 26.6 | 26.6 | 26.6 | 26.6 | 26.6 | 26.6 | 26.6 | 22.7 | 23 | 26.6 | 26.6 | 26.6 | 26.6 | 26.6 | 26.6 | 26.6 | 26.6 | 26.6 | 22.7 |
| 24 | 26.0 | 26.0 | 26.0 | 26.0 | 26.0 | 26.0 | 26.0 | 26.0 | 26.0 | 22.1 | 24 | 26.0 | 26.0 | 26.0 | 26.0 | 26.0 | 26.0 | 26.0 | 26.0 | 26.0 | 22.1 |
| Avg. | 29.0 | 30.0 | 32.0 | 33.0 | 32.0 | 31.0 | 32.0 | 33.0 | 32.0 | 32.0 | Avg. | 29.0 | 32.0 | 35.0 | 37.0 | 37.0 | 34.0 | 37.0 | 37.0 | 35.0 | 40.0 |

*Note*: Sol-air temperatures are calculated based on $\varepsilon \Delta R/h_o = -3.9$°C for horizontal surfaces and 0°C for vertical surfaces.

# Nonresidential Cooling and Heating Load Calculations

**Table 7-4 SI** Visible Transmittance ($T_v$), Solar Heat Gain Coefficient (SHGC), Solar Transmittance ($T$), Front Reflectance ($R^f$), Back Reflectance ($R^b$), and Layer Absorptance $A_n^f$ for Glazing and Window Systems

*(Table 10, Chapter 15, 2013 ASHRAE Handbook—Fundamentals)*

| | | | | | \multicolumn{7}{c}{Center-of-Glazing Properties} | \multicolumn{4}{c}{Total Window SHGC at Normal Incidence} | \multicolumn{4}{c}{Total Window $T_v$ at Normal Incidence} |
|---|---|---|---|---|---|---|---|---|---|---|---|---|---|---|---|---|---|---|---|
| | | \multicolumn{2}{c}{Glazing System} | | \multicolumn{7}{c}{Incidence Angles} | \multicolumn{2}{c}{Aluminum} | \multicolumn{2}{c}{Other Frames} | \multicolumn{2}{c}{Aluminum} | \multicolumn{2}{c}{Other Frames} |
| ID | Glass Thick., mm | | | Center Glazing $T_v$ | | Normal 0.00 | 40.00 | 50.00 | 60.00 | 70.00 | 80.00 | Hemis., Diffuse | Operable | Fixed | Operable | Fixed | Operable | Fixed | Operable | Fixed |

### Uncoated Single Glazing

| ID | Thk | Type | $T_v$ | Prop | Normal | 40 | 50 | 60 | 70 | 80 | Hemis | Op-Al | Fx-Al | Op-Ot | Fx-Ot | Op-Al | Fx-Al | Op-Ot | Fx-Ot |
|---|---|---|---|---|---|---|---|---|---|---|---|---|---|---|---|---|---|---|---|
| 1a | 3 | CLR | 0.90 | SHGC | 0.86 | 0.84 | 0.82 | 0.78 | 0.67 | 0.42 | 0.78 | 0.78 | 0.79 | 0.70 | 0.76 | 0.80 | 0.81 | 0.72 | 0.79 |
| | | | | $T$ | 0.83 | 0.82 | 0.80 | 0.75 | 0.64 | 0.39 | 0.75 | | | | | | | | |
| | | | | $R^f$ | 0.08 | 0.08 | 0.10 | 0.14 | 0.25 | 0.51 | 0.14 | | | | | | | | |
| | | | | $R^b$ | 0.08 | 0.08 | 0.10 | 0.14 | 0.25 | 0.51 | 0.14 | | | | | | | | |
| | | | | $A_1^f$ | 0.09 | 0.10 | 0.10 | 0.11 | 0.11 | 0.11 | 0.10 | | | | | | | | |
| 1b | 6 | CLR | 0.88 | SHGC | 0.81 | 0.80 | 0.78 | 0.73 | 0.62 | 0.39 | 0.73 | 0.74 | 0.74 | 0.66 | 0.72 | 0.78 | 0.79 | 0.70 | 0.77 |
| | | | | $T$ | 0.77 | 0.75 | 0.73 | 0.68 | 0.58 | 0.35 | 0.69 | | | | | | | | |
| | | | | $R^f$ | 0.07 | 0.08 | 0.09 | 0.13 | 0.24 | 0.48 | 0.13 | | | | | | | | |
| | | | | $R^b$ | 0.07 | 0.08 | 0.09 | 0.13 | 0.24 | 0.48 | 0.13 | | | | | | | | |
| | | | | $A_1^f$ | 0.16 | 0.17 | 0.18 | 0.19 | 0.19 | 0.17 | 0.17 | | | | | | | | |
| 1c | 3 | BRZ | 0.68 | SHGC | 0.73 | 0.71 | 0.68 | 0.64 | 0.55 | 0.34 | 0.65 | 0.67 | 0.67 | 0.59 | 0.65 | 0.61 | 0.61 | 0.54 | 0.60 |
| | | | | $T$ | 0.65 | 0.62 | 0.59 | 0.55 | 0.46 | 0.27 | 0.56 | | | | | | | | |
| | | | | $R^f$ | 0.06 | 0.07 | 0.08 | 0.12 | 0.22 | 0.45 | 0.12 | | | | | | | | |
| | | | | $R^b$ | 0.06 | 0.07 | 0.08 | 0.12 | 0.22 | 0.45 | 0.12 | | | | | | | | |
| | | | | $A_1^f$ | 0.29 | 0.31 | 0.32 | 0.33 | 0.33 | 0.29 | 0.31 | | | | | | | | |
| 1d | 6 | BRZ | 0.54 | SHGC | 0.62 | 0.59 | 0.57 | 0.53 | 0.45 | 0.29 | 0.54 | 0.57 | 0.57 | 0.50 | 0.55 | 0.48 | 0.49 | 0.43 | 0.48 |
| | | | | $T$ | 0.49 | 0.45 | 0.43 | 0.39 | 0.32 | 0.18 | 0.41 | | | | | | | | |
| | | | | $R^f$ | 0.05 | 0.06 | 0.07 | 0.11 | 0.19 | 0.42 | 0.10 | | | | | | | | |
| | | | | $R^b$ | 0.05 | 0.68 | 0.66 | 0.62 | 0.53 | 0.33 | 0.10 | | | | | | | | |
| | | | | $A_1^f$ | 0.46 | 0.49 | 0.50 | 0.51 | 0.49 | 0.41 | 0.48 | | | | | | | | |
| 1e | 3 | GRN | 0.82 | SHGC | 0.70 | 0.68 | 0.66 | 0.62 | 0.53 | 0.33 | 0.63 | 0.64 | 0.64 | 0.57 | 0.62 | 0.73 | 0.74 | 0.66 | 0.72 |
| | | | | $T$ | 0.61 | 0.58 | 0.56 | 0.52 | 0.43 | 0.25 | 0.53 | | | | | | | | |
| | | | | $R^f$ | 0.06 | 0.07 | 0.08 | 0.12 | 0.21 | 0.45 | 0.11 | | | | | | | | |
| | | | | $R^b$ | 0.06 | 0.07 | 0.08 | 0.12 | 0.21 | 0.45 | 0.11 | | | | | | | | |
| | | | | $A_1^f$ | 0.33 | 0.35 | 0.36 | 0.37 | 0.36 | 0.31 | 0.35 | | | | | | | | |
| 1f | 6 | GRN | 0.76 | SHGC | 0.60 | 0.58 | 0.56 | 0.52 | 0.45 | 0.29 | 0.54 | 0.55 | 0.55 | 0.49 | 0.53 | 0.68 | 0.68 | 0.61 | 0.67 |
| | | | | $T$ | 0.47 | 0.44 | 0.42 | 0.38 | 0.32 | 0.18 | 0.40 | | | | | | | | |
| | | | | $R^f$ | 0.05 | 0.06 | 0.07 | 0.11 | 0.20 | 0.42 | 0.10 | | | | | | | | |
| | | | | $R^b$ | 0.05 | 0.06 | 0.07 | 0.11 | 0.20 | 0.42 | 0.10 | | | | | | | | |
| | | | | $A_1^f$ | 0.47 | 0.50 | 0.51 | 0.51 | 0.49 | 0.40 | 0.49 | | | | | | | | |
| 1g | 3 | GRY | 0.62 | SHGC | 0.70 | 0.68 | 0.66 | 0.61 | 0.53 | 0.33 | 0.63 | 0.64 | 0.64 | 0.57 | 0.62 | 0.55 | 0.56 | 0.50 | 0.55 |
| | | | | $T$ | 0.61 | 0.58 | 0.56 | 0.51 | 0.42 | 0.24 | 0.53 | | | | | | | | |
| | | | | $R^f$ | 0.06 | 0.07 | 0.08 | 0.12 | 0.21 | 0.44 | 0.11 | | | | | | | | |
| | | | | $R^b$ | 0.06 | 0.07 | 0.08 | 0.12 | 0.21 | 0.44 | 0.11 | | | | | | | | |
| | | | | $A_1^f$ | 0.33 | 0.36 | 0.37 | 0.37 | 0.37 | 0.32 | 0.35 | | | | | | | | |
| 1h | 6 | GRY | 0.46 | SHGC | 0.59 | 0.57 | 0.55 | 0.51 | 0.44 | 0.28 | 0.52 | 0.54 | 0.54 | 0.48 | 0.52 | 0.41 | 0.41 | 0.37 | 0.40 |
| | | | | $T$ | 0.46 | 0.42 | 0.40 | 0.36 | 0.29 | 0.16 | 0.38 | | | | | | | | |
| | | | | $R^f$ | 0.05 | 0.06 | 0.07 | 0.10 | 0.19 | 0.41 | 0.10 | | | | | | | | |
| | | | | $R^b$ | 0.05 | 0.06 | 0.07 | 0.10 | 0.19 | 0.41 | 0.10 | | | | | | | | |
| | | | | $A_1^f$ | 0.49 | 0.52 | 0.54 | 0.54 | 0.52 | 0.43 | 0.51 | | | | | | | | |
| 1i | 6 | BLUGRN | 0.75 | SHGC | 0.62 | 0.59 | 0.57 | 0.54 | 0.46 | 0.30 | 0.55 | 0.57 | 0.57 | 0.50 | 0.55 | 0.67 | 0.68 | 0.60 | 0.66 |
| | | | | $T$ | 0.49 | 0.46 | 0.44 | 0.40 | 0.33 | 0.19 | 0.42 | | | | | | | | |
| | | | | $R^f$ | 0.06 | 0.06 | 0.07 | 0.11 | 0.20 | 0.43 | 0.11 | | | | | | | | |
| | | | | $R^b$ | 0.06 | 0.06 | 0.07 | 0.11 | 0.20 | 0.43 | 0.11 | | | | | | | | |
| | | | | $A_1^f$ | 0.45 | 0.48 | 0.49 | 0.49 | 0.47 | 0.38 | 0.48 | | | | | | | | |

### Reflective Single Glazing

| ID | Thk | Type | $T_v$ | Prop | Normal | 40 | 50 | 60 | 70 | 80 | Hemis | Op-Al | Fx-Al | Op-Ot | Fx-Ot | Op-Al | Fx-Al | Op-Ot | Fx-Ot |
|---|---|---|---|---|---|---|---|---|---|---|---|---|---|---|---|---|---|---|---|
| 1j | 6 | SS on CLR 8% | 0.08 | SHGC | 0.19 | 0.19 | 0.19 | 0.18 | 0.16 | 0.10 | 0.18 | 0.18 | 0.18 | 0.16 | 0.17 | 0.07 | 0.07 | 0.06 | 0.07 |
| | | | | $T$ | 0.06 | 0.06 | 0.06 | 0.05 | 0.04 | 0.03 | 0.05 | | | | | | | | |
| | | | | $R^f$ | 0.33 | 0.34 | 0.35 | 0.37 | 0.44 | 0.61 | 0.36 | | | | | | | | |
| | | | | $R^b$ | 0.50 | 0.50 | 0.51 | 0.53 | 0.58 | 0.71 | 0.52 | | | | | | | | |
| | | | | $A_1^f$ | 0.61 | 0.61 | 0.60 | 0.58 | 0.52 | 0.37 | 0.57 | | | | | | | | |
| 1k | 6 | SS on CLR 14% | 0.14 | SHGC | 0.25 | 0.25 | 0.24 | 0.23 | 0.20 | 0.13 | 0.23 | 0.24 | 0.24 | 0.21 | 0.22 | 0.12 | 0.13 | 0.11 | 0.12 |
| | | | | $T$ | 0.11 | 0.10 | 0.10 | 0.09 | 0.07 | 0.04 | 0.09 | | | | | | | | |

**Table 7-4 SI** Visible Transmittance ($T_v$), Solar Heat Gain Coefficient (SHGC), Solar Transmittance ($T$), Front Reflectance ($R^f$), Back Reflectance ($R^b$), and Layer Absorptance $A_n^f$ for Glazing and Window Systems (*Continued*)

*(Table 10, Chapter 15, 2013 ASHRAE Handbook—Fundamentals)*

| | | Glazing System | | | | | Center-of-Glazing Properties | | | | | | | Total Window SHGC at Normal Incidence | | | | Total Window $T_v$ at Normal Incidence | | | |
|---|---|---|---|---|---|---|---|---|---|---|---|---|---|---|---|---|---|---|---|---|---|
| | | | | | | | | | Incidence Angles | | | | | Aluminum | | Other Frames | | Aluminum | | Other Frames | |
| ID | Glass Thick., mm | | Center Glazing $T_v$ | | | Normal 0.00 | 40.00 | 50.00 | 60.00 | 70.00 | 80.00 | Hemis., Diffuse | | Operable | Fixed | Operable | Fixed | Operable | Fixed | Operable | Fixed |
| | | | | | $R^f$ | 0.26 | 0.27 | 0.28 | 0.31 | 0.38 | 0.57 | 0.30 | | | | | | | | | |
| | | | | | $R^b$ | 0.44 | 0.44 | 0.45 | 0.47 | 0.52 | 0.67 | 0.46 | | | | | | | | | |
| | | | | | $A_1^f$ | 0.63 | 0.63 | 0.62 | 0.60 | 0.55 | 0.39 | 0.60 | | | | | | | | | |
| 1l | 6 | SS on CLR 20% | 0.20 | | SHGC | 0.31 | 0.30 | 0.30 | 0.28 | 0.24 | 0.16 | 0.28 | | 0.29 | 0.29 | 0.26 | 0.28 | 0.18 | 0.18 | 0.16 | 0.18 |
| | | | | | $T$ | 0.15 | 0.15 | 0.14 | 0.13 | 0.11 | 0.06 | 0.13 | | | | | | | | | |
| | | | | | $R^f$ | 0.21 | 0.22 | 0.23 | 0.26 | 0.34 | 0.54 | 0.25 | | | | | | | | | |
| | | | | | $R^b$ | 0.38 | 0.38 | 0.39 | 0.41 | 0.48 | 0.64 | 0.41 | | | | | | | | | |
| | | | | | $A_1^f$ | 0.64 | 0.64 | 0.63 | 0.61 | 0.56 | 0.40 | 0.60 | | | | | | | | | |
| 1m | 6 | SS on GRN 14% | 0.12 | | SHGC | 0.25 | 0.25 | 0.24 | 0.23 | 0.21 | 0.14 | 0.23 | | 0.24 | 0.24 | 0.21 | 0.22 | 0.11 | 0.11 | 0.10 | 0.11 |
| | | | | | $T$ | 0.06 | 0.06 | 0.06 | 0.06 | 0.04 | 0.03 | 0.06 | | | | | | | | | |
| | | | | | $R^f$ | 0.14 | 0.14 | 0.16 | 0.19 | 0.27 | 0.49 | 0.18 | | | | | | | | | |
| | | | | | $R^b$ | 0.44 | 0.44 | 0.45 | 0.47 | 0.52 | 0.67 | 0.46 | | | | | | | | | |
| | | | | | $A_1^f$ | 0.80 | 0.80 | 0.78 | 0.76 | 0.68 | 0.48 | 0.75 | | | | | | | | | |
| 1n | 6 | TI on CLR 20% | 0.20 | | SHGC | 0.29 | 0.29 | 0.28 | 0.27 | 0.23 | 0.15 | 0.27 | | 0.27 | 0.27 | 0.24 | 0.26 | 0.18 | 0.18 | 0.16 | 0.18 |
| | | | | | $T$ | 0.14 | 0.13 | 0.13 | 0.12 | 0.09 | 0.06 | 0.12 | | | | | | | | | |
| | | | | | $R^f$ | 0.22 | 0.22 | 0.24 | 0.26 | 0.34 | 0.54 | 0.26 | | | | | | | | | |
| | | | | | $R^b$ | 0.40 | 0.40 | 0.42 | 0.44 | 0.50 | 0.65 | 0.43 | | | | | | | | | |
| | | | | | $A_1^f$ | 0.65 | 0.65 | 0.64 | 0.62 | 0.57 | 0.40 | 0.62 | | | | | | | | | |
| 1o | 6 | TI on CLR 30% | 0.30 | | SHGC | 0.39 | 0.38 | 0.37 | 0.35 | 0.30 | 0.20 | 0.35 | | 0.36 | 0.36 | 0.32 | 0.35 | 0.27 | 0.27 | 0.24 | 0.26 |
| | | | | | $T$ | 0.23 | 0.22 | 0.21 | 0.19 | 0.16 | 0.09 | 0.20 | | | | | | | | | |
| | | | | | $R^f$ | 0.15 | 0.15 | 0.17 | 0.20 | 0.28 | 0.50 | 0.19 | | | | | | | | | |
| | | | | | $R^b$ | 0.32 | 0.33 | 0.34 | 0.36 | 0.43 | 0.60 | 0.36 | | | | | | | | | |
| | | | | | $A_1^f$ | 0.63 | 0.65 | 0.64 | 0.62 | 0.57 | 0.40 | 0.62 | | | | | | | | | |

***Uncoated Double Glazing***

| | | | | | | | | | | | | | | | | | | | | | |
|---|---|---|---|---|---|---|---|---|---|---|---|---|---|---|---|---|---|---|---|---|---|
| 5a | 3 | CLR CLR | 0.81 | | SHGC | 0.76 | 0.74 | 0.71 | 0.64 | 0.50 | 0.26 | 0.66 | | 0.69 | 0.70 | 0.62 | 0.67 | 0.72 | 0.73 | 0.65 | 0.71 |
| | | | | | $T$ | 0.70 | 0.68 | 0.65 | 0.58 | 0.44 | 0.21 | 0.60 | | | | | | | | | |
| | | | | | $R^f$ | 0.13 | 0.14 | 0.16 | 0.23 | 0.36 | 0.61 | 0.21 | | | | | | | | | |
| | | | | | $R^b$ | 0.13 | 0.14 | 0.16 | 0.23 | 0.36 | 0.61 | 0.21 | | | | | | | | | |
| | | | | | $A_1^f$ | 0.10 | 0.11 | 0.11 | 0.12 | 0.13 | 0.13 | 0.11 | | | | | | | | | |
| | | | | | $A_2^f$ | 0.07 | 0.08 | 0.08 | 0.08 | 0.07 | 0.05 | 0.07 | | | | | | | | | |
| 5b | 6 | CLR CLR | 0.78 | | SHGC | 0.70 | 0.67 | 0.64 | 0.58 | 0.45 | 0.23 | 0.60 | | 0.64 | 0.64 | 0.57 | 0.62 | 0.69 | 0.70 | 0.62 | 0.69 |
| | | | | | $T$ | 0.61 | 0.58 | 0.55 | 0.48 | 0.36 | 0.17 | 0.51 | | | | | | | | | |
| | | | | | $R^f$ | 0.11 | 0.12 | 0.15 | 0.20 | 0.33 | 0.57 | 0.18 | | | | | | | | | |
| | | | | | $R^b$ | 0.11 | 0.12 | 0.15 | 0.20 | 0.33 | 0.57 | 0.18 | | | | | | | | | |
| | | | | | $A_1^f$ | 0.17 | 0.18 | 0.19 | 0.20 | 0.21 | 0.20 | 0.19 | | | | | | | | | |
| | | | | | $A_2^f$ | 0.11 | 0.12 | 0.12 | 0.12 | 0.10 | 0.07 | 0.11 | | | | | | | | | |
| 5c | 3 | BRZ CLR | 0.62 | | SHGC | 0.62 | 0.60 | 0.57 | 0.51 | 0.39 | 0.20 | 0.53 | | 0.57 | 0.57 | 0.50 | 0.55 | 0.55 | 0.56 | 0.50 | 0.55 |
| | | | | | $T$ | 0.55 | 0.51 | 0.48 | 0.42 | 0.31 | 0.14 | 0.45 | | | | | | | | | |
| | | | | | $R^f$ | 0.09 | 0.10 | 0.12 | 0.16 | 0.27 | 0.49 | 0.15 | | | | | | | | | |
| | | | | | $R^b$ | 0.12 | 0.13 | 0.15 | 0.21 | 0.35 | 0.59 | 0.19 | | | | | | | | | |
| | | | | | $A_1^f$ | 0.30 | 0.33 | 0.34 | 0.36 | 0.37 | 0.34 | 0.33 | | | | | | | | | |
| | | | | | $A_2^f$ | 0.06 | 0.06 | 0.06 | 0.06 | 0.05 | 0.03 | 0.06 | | | | | | | | | |
| 5d | 6 | BRZ CLR | 0.47 | | SHGC | 0.49 | 0.46 | 0.44 | 0.39 | 0.31 | 0.17 | 0.41 | | 0.45 | 0.45 | 0.40 | 0.43 | 0.42 | 0.42 | 0.38 | 0.41 |
| | | | | | $T$ | 0.38 | 0.35 | 0.32 | 0.27 | 0.20 | 0.08 | 0.30 | | | | | | | | | |
| | | | | | $R^f$ | 0.07 | 0.08 | 0.09 | 0.13 | 0.22 | 0.44 | 0.12 | | | | | | | | | |
| | | | | | $R^b$ | 0.10 | 0.11 | 0.13 | 0.19 | 0.31 | 0.55 | 0.17 | | | | | | | | | |
| | | | | | $A_1^f$ | 0.48 | 0.51 | 0.52 | 0.53 | 0.53 | 0.45 | 0.50 | | | | | | | | | |
| | | | | | $A_2^f$ | 0.07 | 0.07 | 0.07 | 0.07 | 0.06 | 0.04 | 0.07 | | | | | | | | | |
| 5e | 3 | GRN CLR | 0.75 | | SHGC | 0.60 | 0.57 | 0.54 | 0.49 | 0.38 | 0.20 | 0.51 | | 0.55 | 0.55 | 0.49 | 0.53 | 0.67 | 0.68 | 0.60 | 0.66 |
| | | | | | $T$ | 0.52 | 0.49 | 0.46 | 0.40 | 0.30 | 0.13 | 0.43 | | | | | | | | | |
| | | | | | $R^f$ | 0.09 | 0.10 | 0.12 | 0.16 | 0.27 | 0.50 | 0.15 | | | | | | | | | |
| | | | | | $R^b$ | 0.12 | 0.13 | 0.15 | 0.21 | 0.35 | 0.60 | 0.19 | | | | | | | | | |
| | | | | | $A_1^f$ | 0.34 | 0.37 | 0.38 | 0.39 | 0.39 | 0.35 | 0.37 | | | | | | | | | |
| | | | | | $A_2^f$ | 0.05 | 0.05 | 0.05 | 0.04 | 0.04 | 0.03 | 0.04 | | | | | | | | | |

**Table 7-4 SI** Visible Transmittance ($T_v$), Solar Heat Gain Coefficient (SHGC), Solar Transmittance ($T$), Front Reflectance ($R^f$), Back Reflectance ($R^b$), and Layer Absorptance $A_n^f$ for Glazing and Window Systems (*Continued*)

*(Table 10, Chapter 15, 2013 ASHRAE Handbook—Fundamentals)*

| | | | | | | Center-of-Glazing Properties | | | | | | | Total Window SHGC at Normal Incidence | | | | Total Window $T_v$ at Normal Incidence | | | |
|---|---|---|---|---|---|---|---|---|---|---|---|---|---|---|---|---|---|---|---|---|
| | **Glazing System** | | | | | Incidence Angles | | | | | | | Aluminum | | Other Frames | | Aluminum | | Other Frames | |
| ID | Glass Thick., mm | | Center Glazing $T_v$ | | Normal 0.00 | 40.00 | 50.00 | 60.00 | 70.00 | 80.00 | Hemis., Diffuse | | Operable | Fixed | Operable | Fixed | Operable | Fixed | Operable | Fixed |
| 5f | 6 | GRN CLR | 0.68 | SHGC | 0.49 | 0.46 | 0.44 | 0.39 | 0.31 | 0.17 | 0.41 | | 0.45 | 0.45 | 0.40 | 0.43 | 0.61 | 0.61 | 0.54 | 0.60 |
| | | | | $T$ | 0.39 | 0.36 | 0.33 | 0.29 | 0.21 | 0.09 | 0.31 | | | | | | | | | |
| | | | | $R^f$ | 0.08 | 0.08 | 0.10 | 0.14 | 0.23 | 0.45 | 0.13 | | | | | | | | | |
| | | | | $R^b$ | 0.10 | 0.11 | 0.13 | 0.19 | 0.31 | 0.55 | 0.17 | | | | | | | | | |
| | | | | $A_1^f$ | 0.49 | 0.51 | 0.05 | 0.53 | 0.52 | 0.43 | 0.50 | | | | | | | | | |
| | | | | $A_2^f$ | 0.05 | 0.05 | 0.05 | 0.05 | 0.04 | 0.03 | 0.05 | | | | | | | | | |
| 5g | 3 | GRY CLR | 0.56 | SHGC | 0.60 | 0.57 | 0.54 | 0.48 | 0.37 | 0.20 | 0.51 | | 0.55 | 0.55 | 0.49 | 0.53 | 0.50 | 0.50 | 0.45 | 0.49 |
| | | | | $T$ | 0.51 | 0.48 | 0.45 | 0.39 | 0.29 | 0.12 | 0.42 | | | | | | | | | |
| | | | | $R^f$ | 0.09 | 0.09 | 0.11 | 0.16 | 0.26 | 0.48 | 0.14 | | | | | | | | | |
| | | | | $R^b$ | 0.12 | 0.13 | 0.15 | 0.21 | 0.34 | 0.59 | 0.19 | | | | | | | | | |
| | | | | $A_1^f$ | 0.34 | 0.37 | 0.39 | 0.40 | 0.41 | 0.37 | 0.37 | | | | | | | | | |
| | | | | $A_2^f$ | 0.05 | 0.06 | 0.06 | 0.05 | 0.05 | 0.03 | 0.05 | | | | | | | | | |
| 5h | 6 | GRY CLR | 0.41 | SHGC | 0.47 | 0.44 | 0.42 | 0.37 | 0.29 | 0.16 | 0.39 | | 0.43 | 0.43 | 0.38 | 0.42 | 0.36 | 0.37 | 0.33 | 0.36 |
| | | | | $T$ | 0.36 | 0.32 | 0.29 | 0.25 | 0.18 | 0.07 | 0.28 | | | | | | | | | |
| | | | | $R^f$ | 0.07 | 0.07 | 0.08 | 0.12 | 0.21 | 0.43 | 0.12 | | | | | | | | | |
| | | | | $R^b$ | 0.10 | 0.11 | 0.13 | 0.18 | 0.31 | 0.55 | 0.17 | | | | | | | | | |
| | | | | $A_1^f$ | 0.51 | 0.54 | 0.56 | 0.57 | 0.56 | 0.47 | 0.53 | | | | | | | | | |
| | | | | $A_2^f$ | 0.07 | 0.07 | 0.07 | 0.06 | 0.05 | 0.03 | 0.06 | | | | | | | | | |
| 5i | 6 | BLUGRN CLR | 0.67 | SHGC | 0.50 | 0.47 | 0.45 | 0.40 | 0.32 | 0.17 | 0.43 | | 0.46 | 0.46 | 0.41 | 0.44 | 0.60 | 0.60 | 0.54 | 0.59 |
| | | | | $T$ | 0.40 | 0.37 | 0.34 | 0.30 | 0.22 | 0.10 | 0.32 | | | | | | | | | |
| | | | | $R^f$ | 0.08 | 0.08 | 0.10 | 0.14 | 0.24 | 0.46 | 0.13 | | | | | | | | | |
| | | | | $R^b$ | 0.11 | 0.11 | 0.14 | 0.19 | 0.31 | 0.55 | 0.17 | | | | | | | | | |
| | | | | $A_1^f$ | 0.47 | 0.49 | 0.50 | 0.51 | 0.50 | 0.42 | 0.48 | | | | | | | | | |
| | | | | $A_2^f$ | 0.06 | 0.06 | 0.06 | 0.05 | 0.04 | 0.03 | 0.05 | | | | | | | | | |
| 5j | 6 | HI-P GRN CLR | 0.59 | SHGC | 0.39 | 0.37 | 0.35 | 0.31 | 0.25 | 0.14 | 0.33 | | 0.36 | 0.36 | 0.32 | 0.35 | 0.53 | 0.53 | 0.47 | 0.52 |
| | | | | $T$ | 0.28 | 0.26 | 0.24 | 0.20 | 0.15 | 0.06 | 0.22 | | | | | | | | | |
| | | | | $R^f$ | 0.06 | 0.07 | 0.08 | 0.12 | 0.21 | 0.43 | 0.11 | | | | | | | | | |
| | | | | $R^b$ | 0.10 | 0.11 | 0.13 | 0.19 | 0.31 | 0.55 | 0.17 | | | | | | | | | |
| | | | | $A_1^f$ | 0.62 | 0.65 | 0.65 | 0.65 | 0.62 | 0.50 | 0.63 | | | | | | | | | |
| | | | | $A_2^f$ | 0.03 | 0.03 | 0.03 | 0.03 | 0.02 | 0.01 | 0.03 | | | | | | | | | |

*Reflective Double Glazing*

| | | | | | | | | | | | | | | | | | | | | |
|---|---|---|---|---|---|---|---|---|---|---|---|---|---|---|---|---|---|---|---|---|
| 5k | 6 | SS on CLR 8%, CLR | 0.07 | SHGC | 0.13 | 0.12 | 0.12 | 0.11 | 0.10 | 0.06 | 0.11 | | 0.13 | 0.13 | 0.11 | 0.12 | 0.06 | 0.06 | 0.06 | 0.06 |
| | | | | $T$ | 0.05 | 0.05 | 0.04 | 0.04 | 0.03 | 0.01 | 0.04 | | | | | | | | | |
| | | | | $R^f$ | 0.33 | 0.34 | 0.35 | 0.37 | 0.44 | 0.61 | 0.37 | | | | | | | | | |
| | | | | $R^b$ | 0.38 | 0.37 | 0.38 | 0.40 | 0.46 | 0.61 | 0.40 | | | | | | | | | |
| | | | | $A_1^f$ | 0.61 | 0.61 | 0.60 | 0.58 | 0.53 | 0.37 | 0.56 | | | | | | | | | |
| | | | | $A_2^f$ | 0.01 | 0.01 | 0.01 | 0.01 | 0.01 | 0.01 | 0.01 | | | | | | | | | |
| 5l | 6 | SS on CLR 14%, CLR | 0.13 | SHGC | 0.17 | 0.17 | 0.16 | 0.15 | 0.13 | 0.08 | 0.16 | | 0.17 | 0.16 | 0.14 | 0.15 | 0.12 | 0.12 | 0.10 | 0.11 |
| | | | | $T$ | 0.08 | 0.08 | 0.08 | 0.07 | 0.05 | 0.02 | 0.07 | | | | | | | | | |
| | | | | $R^f$ | 0.26 | 0.27 | 0.28 | 0.31 | 0.38 | 0.57 | 0.30 | | | | | | | | | |
| | | | | $R^b$ | 0.34 | 0.33 | 0.34 | 0.37 | 0.44 | 0.60 | 0.36 | | | | | | | | | |
| | | | | $A_1^f$ | 0.63 | 0.64 | 0.64 | 0.63 | 0.61 | 0.56 | 0.60 | | | | | | | | | |
| | | | | $A_2^f$ | 0.02 | 0.02 | 0.02 | 0.02 | 0.02 | 0.02 | 0.02 | | | | | | | | | |
| 5m | 6 | SS on CLR 20%, CLR | 0.18 | SHGC | 0.22 | 0.21 | 0.21 | 0.19 | 0.16 | 0.09 | 0.20 | | 0.21 | 0.21 | 0.18 | 0.20 | 0.16 | 0.16 | 0.14 | 0.16 |
| | | | | $T$ | 0.12 | 0.11 | 0.11 | 0.09 | 0.07 | 0.03 | 0.10 | | | | | | | | | |
| | | | | $R^f$ | 0.21 | 0.22 | 0.23 | 0.26 | 0.34 | 0.54 | 0.25 | | | | | | | | | |
| | | | | $R^b$ | 0.30 | 0.30 | 0.31 | 0.34 | 0.41 | 0.59 | 0.33 | | | | | | | | | |
| | | | | $A_1^f$ | 0.64 | 0.64 | 0.63 | 0.62 | 0.57 | 0.41 | 0.61 | | | | | | | | | |
| | | | | $A_2^f$ | 0.03 | 0.03 | 0.03 | 0.03 | 0.02 | 0.02 | 0.03 | | | | | | | | | |
| 5n | 6 | SS on GRN 14%, CLR | 0.11 | SHGC | 0.16 | 0.16 | 0.15 | 0.14 | 0.12 | 0.08 | 0.14 | | 0.16 | 0.16 | 0.14 | 0.14 | 0.10 | 0.10 | 0.09 | 0.10 |
| | | | | $T$ | 0.05 | 0.05 | 0.05 | 0.04 | 0.03 | 0.01 | 0.04 | | | | | | | | | |
| | | | | $R^f$ | 0.14 | 0.14 | 0.16 | 0.19 | 0.27 | 0.49 | 0.18 | | | | | | | | | |
| | | | | $R^b$ | 0.34 | 0.33 | 0.34 | 0.37 | 0.44 | 0.60 | 0.36 | | | | | | | | | |
| | | | | $A_1^f$ | 0.80 | 0.80 | 0.79 | 0.76 | 0.69 | 0.49 | 0.76 | | | | | | | | | |
| | | | | $A_2^f$ | 0.01 | 0.01 | 0.01 | 0.01 | 0.01 | 0.01 | 0.01 | | | | | | | | | |
| 5o | 6 | TI on CLR 20%, CLR | 0.18 | SHGC | 0.21 | 0.20 | 0.19 | 0.18 | 0.15 | 0.09 | 0.18 | | 0.20 | 0.20 | 0.18 | 0.19 | 0.16 | 0.16 | 0.14 | 0.16 |
| | | | | $T$ | 0.11 | 0.10 | 0.10 | 0.08 | 0.06 | 0.03 | 0.09 | | | | | | | | | |
| | | | | $R^f$ | 0.22 | 0.22 | 0.24 | 0.27 | 0.34 | 0.54 | 0.26 | | | | | | | | | |
| | | | | $R^b$ | 0.32 | 0.31 | 0.32 | 0.35 | 0.42 | 0.59 | 0.35 | | | | | | | | | |

**Table 7-4 SI** Visible Transmittance ($T_v$), Solar Heat Gain Coefficient (SHGC), Solar Transmittance ($T$), Front Reflectance ($R^f$), Back Reflectance ($R^b$), and Layer Absorptance $A_n^f$ for Glazing and Window Systems (*Continued*)

*(Table 10, Chapter 15, 2013 ASHRAE Handbook—Fundamentals)*

|  |  |  |  |  | Center-of-Glazing Properties |  |  |  |  |  |  | Total Window SHGC at Normal Incidence |  |  |  | Total Window $T_v$ at Normal Incidence |  |  |  |
|---|---|---|---|---|---|---|---|---|---|---|---|---|---|---|---|---|---|---|---|
|  | Glazing System |  |  |  | Incidence Angles |  |  |  |  |  |  | Aluminum |  | Other Frames |  | Aluminum |  | Other Frames |  |
| ID | Glass Thick., mm |  | Center Glazing $T_v$ |  | Normal 0.00 | 40.00 | 50.00 | 60.00 | 70.00 | 80.00 | Hemis., Diffuse | Operable | Fixed | Operable | Fixed | Operable | Fixed | Operable | Fixed |
| 5p | 6 | TI on CLR 30%, CLR | 0.27 | $A_1^f$ | 0.65 | 0.66 | 0.65 | 0.63 | 0.58 | 0.41 | 0.62 | | | | | | | | |
|  |  |  |  | $A_2^f$ | 0.02 | 0.02 | 0.02 | 0.02 | 0.02 | 0.01 | 0.02 | | | | | | | | |
|  |  |  |  | SHGC | 0.29 | 0.28 | 0.27 | 0.25 | 0.20 | 0.12 | 0.25 | 0.27 | 0.27 | 0.24 | 0.26 | 0.24 | 0.24 | 0.22 | 0.24 |
|  |  |  |  | $T$ | 0.18 | 0.17 | 0.16 | 0.14 | 0.10 | 0.05 | 0.15 | | | | | | | | |
|  |  |  |  | $R^f$ | 0.15 | 0.15 | 0.17 | 0.20 | 0.29 | 0.51 | 0.19 | | | | | | | | |
|  |  |  |  | $R^b$ | 0.27 | 0.27 | 0.28 | 0.31 | 0.40 | 0.58 | 0.31 | | | | | | | | |
|  |  |  |  | $A_1^f$ | 0.64 | 0.64 | 0.63 | 0.62 | 0.58 | 0.43 | 0.61 | | | | | | | | |
|  |  |  |  | $A_2^f$ | 0.04 | 0.04 | 0.04 | 0.04 | 0.03 | 0.02 | 0.04 | | | | | | | | |

*Low-e Double Glazing, e = 0.2 on surface 2*

| 17a | 3 | LE CLR | 0.76 | SHGC | 0.65 | 0.64 | 0.61 | 0.56 | 0.43 | 0.23 | 0.57 | 0.59 | 0.60 | 0.53 | 0.58 | 0.68 | 0.68 | 0.61 | 0.67 |
|---|---|---|---|---|---|---|---|---|---|---|---|---|---|---|---|---|---|---|---|
|  |  |  |  | $T$ | 0.59 | 0.56 | 0.54 | 0.48 | 0.36 | 0.18 | 0.50 | | | | | | | | |
|  |  |  |  | $R^f$ | 0.15 | 0.16 | 0.18 | 0.24 | 0.37 | 0.61 | 0.22 | | | | | | | | |
|  |  |  |  | $R^b$ | 0.17 | 0.18 | 0.20 | 0.26 | 0.38 | 0.61 | 0.24 | | | | | | | | |
|  |  |  |  | $A_1^f$ | 0.20 | 0.21 | 0.21 | 0.21 | 0.20 | 0.16 | 0.20 | | | | | | | | |
|  |  |  |  | $A_2^f$ | 0.07 | 0.07 | 0.08 | 0.08 | 0.07 | 0.05 | 0.07 | | | | | | | | |
| 17b | 6 | LE CLR | 0.73 | SHGC | 0.60 | 0.59 | 0.57 | 0.51 | 0.40 | 0.21 | 0.53 | 0.55 | 0.55 | 0.49 | 0.53 | 0.65 | 0.66 | 0.58 | 0.64 |
|  |  |  |  | $T$ | 0.51 | 0.48 | 0.46 | 0.41 | 0.30 | 0.14 | 0.43 | | | | | | | | |
|  |  |  |  | $R^f$ | 0.14 | 0.15 | 0.17 | 0.22 | 0.35 | 0.59 | 0.21 | | | | | | | | |
|  |  |  |  | $R^b$ | 0.15 | 0.16 | 0.18 | 0.23 | 0.35 | 0.57 | 0.22 | | | | | | | | |
|  |  |  |  | $A_1^f$ | 0.26 | 0.26 | 0.26 | 0.26 | 0.25 | 0.19 | 0.25 | | | | | | | | |
|  |  |  |  | $A_2^f$ | 0.10 | 0.11 | 0.11 | 0.11 | 0.10 | 0.07 | 0.10 | | | | | | | | |

*Low-e Double Glazing, e = 0.2 on surface 3*

| 17c | 3 | CLR LE | 0.76 | SHGC | 0.70 | 0.68 | 0.65 | 0.59 | 0.46 | 0.24 | 0.61 | 0.64 | 0.64 | 0.57 | 0.62 | 0.68 | 0.68 | 0.1 | 0.67 |
|---|---|---|---|---|---|---|---|---|---|---|---|---|---|---|---|---|---|---|---|
|  |  |  |  | $T$ | 0.59 | 0.56 | 0.54 | 0.48 | 0.36 | 0.18 | 0.50 | | | | | | | | |
|  |  |  |  | $R^f$ | 0.17 | 0.18 | 0.20 | 0.26 | 0.38 | 0.61 | 0.24 | | | | | | | | |
|  |  |  |  | $R^b$ | 0.15 | 0.16 | 0.18 | 0.24 | 0.37 | 0.61 | 0.22 | | | | | | | | |
|  |  |  |  | $A_1^f$ | 0.11 | 0.12 | 0.13 | 0.13 | 0.14 | 0.15 | 0.12 | | | | | | | | |
|  |  |  |  | $A_2^f$ | 0.14 | 0.14 | 0.14 | 0.13 | 0.11 | 0.07 | 0.13 | | | | | | | | |
| 17d | 6 | CLR LE | 0.73 | SHGC | 0.65 | 0.63 | 0.60 | 0.54 | 0.42 | 0.21 | 0.56 | 0.59 | 0.60 | 0.53 | 0.58 | 0.65 | 0.66 | 0.58 | 0.64 |
|  |  |  |  | $T$ | 0.51 | 0.48 | 0.46 | 0.41 | 0.30 | 0.14 | 0.43 | | | | | | | | |
|  |  |  |  | $R^f$ | 0.15 | 0.16 | 0.18 | 0.23 | 0.35 | 0.57 | 0.22 | | | | | | | | |
|  |  |  |  | $R^b$ | 0.14 | 0.15 | 0.17 | 0.22 | 0.35 | 0.59 | 0.21 | | | | | | | | |
|  |  |  |  | $A_1^f$ | 0.17 | 0.19 | 0.20 | 0.21 | 0.22 | 0.22 | 0.19 | | | | | | | | |
|  |  |  |  | $A_2^f$ | 0.17 | 0.17 | 0.17 | 0.15 | 0.13 | 0.07 | 0.16 | | | | | | | | |
| 17e | 3 | BRZ LE | 0.58 | SHGC | 0.57 | 0.54 | 0.51 | 0.46 | 0.35 | 0.18 | 0.48 | 0.52 | 0.52 | 0.46 | 0.51 | 0.52 | 0.52 | 0.46 | 0.51 |
|  |  |  |  | $T$ | 0.46 | 0.43 | 0.41 | 0.36 | 0.26 | 0.12 | 0.38 | | | | | | | | |
|  |  |  |  | $R^f$ | 0.12 | 0.12 | 0.14 | 0.18 | 0.28 | 0.50 | 0.17 | | | | | | | | |
|  |  |  |  | $R^b$ | 0.14 | 0.15 | 0.17 | 0.23 | 0.35 | 0.60 | 0.21 | | | | | | | | |
|  |  |  |  | $A_1^f$ | 0.31 | 0.34 | 0.35 | 0.37 | 0.38 | 0.35 | 0.34 | | | | | | | | |
|  |  |  |  | $A_2^f$ | 0.11 | 0.11 | 0.10 | 0.10 | 0.08 | 0.04 | 0.10 | | | | | | | | |
| 17f | 6 | BRZ LE | 0.45 | SHGC | 0.45 | 0.42 | 0.40 | 0.35 | 0.27 | 0.14 | 0.38 | 0.42 | 0.42 | 0.37 | 0.40 | 0.40 | 0.41 | 0.36 | 0.40 |
|  |  |  |  | $T$ | 0.33 | 0.30 | 0.28 | 0.24 | 0.17 | 0.07 | 0.26 | | | | | | | | |
|  |  |  |  | $R^f$ | 0.09 | 0.09 | 0.10 | 0.14 | 0.23 | 0.44 | 0.13 | | | | | | | | |
|  |  |  |  | $R^b$ | 0.13 | 0.14 | 0.16 | 0.21 | 0.34 | 0.58 | 0.20 | | | | | | | | |
|  |  |  |  | $A_1^f$ | 0.48 | 0.51 | 0.52 | 0.54 | 0.53 | 0.45 | 0.50 | | | | | | | | |
|  |  |  |  | $A_2^f$ | 0.11 | 0.11 | 0.10 | 0.09 | 0.07 | 0.04 | 0.09 | | | | | | | | |
| 17g | 3 | GRN LE | 0.70 | SHGC | 0.55 | 0.52 | 0.50 | 0.44 | 0.34 | 0.17 | 0.46 | 0.50 | 0.51 | 0.45 | 0.49 | 0.62 | 0.63 | 0.56 | 0.62 |
|  |  |  |  | $T$ | 0.44 | 0.41 | 0.38 | 0.33 | 0.24 | 0.11 | 0.36 | | | | | | | | |
|  |  |  |  | $R^f$ | 0.11 | 0.11 | 0.13 | 0.17 | 0.27 | 0.48 | 0.16 | | | | | | | | |
|  |  |  |  | $R^b$ | 0.14 | 0.15 | 0.17 | 0.23 | 0.35 | 0.60 | 0.21 | | | | | | | | |
|  |  |  |  | $A_1^f$ | 0.35 | 0.38 | 0.39 | 0.41 | 0.42 | 0.37 | 0.38 | | | | | | | | |
|  |  |  |  | $A_2^f$ | 0.11 | 0.10 | 0.10 | 0.09 | 0.07 | 0.04 | 0.09 | | | | | | | | |
| 17h | 6 | GRN LE | 0.61 | SHGC | 0.41 | 0.39 | 0.36 | 0.32 | 0.25 | 0.13 | 0.34 | 0.38 | 0.38 | 0.34 | 0.36 | 0.54 | 0.55 | 0.49 | 0.54 |
|  |  |  |  | $T$ | 0.29 | 0.26 | 0.24 | 0.21 | 0.15 | 0.06 | 0.23 | | | | | | | | |
|  |  |  |  | $R^f$ | 0.08 | 0.08 | 0.09 | 0.13 | 0.22 | 0.43 | 0.13 | | | | | | | | |

**Table 7-4 SI** Visible Transmittance ($T_v$), Solar Heat Gain Coefficient (SHGC), Solar Transmittance ($T$), Front Reflectance ($R^f$), Back Reflectance ($R^b$), and Layer Absorptance $A_n^f$ for Glazing and Window Systems (*Continued*)

(Table 10, Chapter 15, 2013 ASHRAE Handbook—Fundamentals)

| | | Glazing System | | | | | Center-of-Glazing Properties | | | | | | Total Window SHGC at Normal Incidence | | | | Total Window $T_v$ at Normal Incidence | | | |
|---|---|---|---|---|---|---|---|---|---|---|---|---|---|---|---|---|---|---|---|---|
| | | | | | | | Incidence Angles | | | | | | Aluminum | | Other Frames | | Aluminum | | Other Frames | |
| ID | Glass Thick., mm | | Center Glazing $T_v$ | | Normal 0.00 | 40.00 | 50.00 | 60.00 | 70.00 | 80.00 | Hemis., Diffuse | | Operable | Fixed | Operable | Fixed | Operable | Fixed | Operable | Fixed |
| | | | | $R^b$ | 0.13 | 0.14 | 0.16 | 0.21 | 0.34 | 0.58 | 0.20 | | | | | | | | | |
| | | | | $A_1^f$ | 0.53 | 0.57 | 0.58 | 0.59 | 0.58 | 0.48 | 0.56 | | | | | | | | | |
| | | | | $A_2^f$ | 0.10 | 0.09 | 0.09 | 0.08 | 0.06 | 0.03 | 0.08 | | | | | | | | | |
| 17i | 3 | GRY LE | 0.53 | SHGC | 0.54 | 0.51 | 0.49 | 0.44 | 0.33 | 0.17 | 0.46 | | 0.50 | 0.50 | 0.44 | 0.48 | 0.47 | 0.48 | 0.42 | 0.47 |
| | | | | $T$ | 0.43 | 0.40 | 0.38 | 0.33 | 0.24 | 0.11 | 0.35 | | | | | | | | | |
| | | | | $R^f$ | 0.11 | 0.11 | 0.13 | 0.17 | 0.27 | 0.48 | 0.16 | | | | | | | | | |
| | | | | $R^b$ | 0.14 | 0.15 | 0.17 | 0.22 | 0.35 | 0.60 | 0.21 | | | | | | | | | |
| | | | | $A_1^f$ | 0.36 | 0.39 | 0.40 | 0.42 | 0.42 | 0.38 | 0.39 | | | | | | | | | |
| | | | | $A_2^f$ | 0.10 | 0.10 | 0.10 | 0.09 | 0.07 | 0.04 | 0.09 | | | | | | | | | |
| 17j | 6 | GRY LE | 0.37 | SHGC | 0.39 | 0.37 | 0.35 | 0.31 | 0.24 | 0.13 | 0.33 | | 0.36 | 0.36 | 0.32 | 0.35 | 0.33 | 0.33 | 0.30 | 0.33 |
| | | | | $T$ | 0.27 | 0.25 | 0.23 | 0.20 | 0.14 | 0.06 | 0.21 | | | | | | | | | |
| | | | | $R^f$ | 0.09 | 0.09 | 0.11 | 0.14 | 0.23 | 0.44 | 0.14 | | | | | | | | | |
| | | | | $R^b$ | 0.13 | 0.14 | 0.16 | 0.22 | 0.34 | 0.58 | 0.20 | | | | | | | | | |
| | | | | $A_1^f$ | 0.55 | 0.58 | 0.59 | 0.59 | 0.58 | 0.48 | 0.56 | | | | | | | | | |
| | | | | $A_2^f$ | 0.09 | 0.09 | 0.08 | 0.07 | 0.06 | 0.03 | 0.08 | | | | | | | | | |
| 17k | 6 | BLUGRN LE | 0.62 | SHGC | 0.45 | 0.42 | 0.40 | 0.35 | 0.27 | 0.14 | 0.37 | | 0.42 | 0.42 | 0.37 | 0.40 | 0.55 | 0.56 | 0.50 | 0.55 |
| | | | | $T$ | 0.32 | 0.29 | 0.27 | 0.23 | 0.17 | 0.07 | 0.26 | | | | | | | | | |
| | | | | $R^f$ | 0.09 | 0.09 | 0.10 | 0.14 | 0.23 | 0.44 | 0.13 | | | | | | | | | |
| | | | | $R^b$ | 0.13 | 0.14 | 0.16 | 0.21 | 0.34 | 0.58 | 0.20 | | | | | | | | | |
| | | | | $A_1^f$ | 0.48 | 0.51 | 0.53 | 0.54 | 0.54 | 0.45 | 0.51 | | | | | | | | | |
| | | | | $A_2^f$ | 0.11 | 0.10 | 0.10 | 0.09 | 0.07 | 0.03 | 0.09 | | | | | | | | | |
| 17l | 6 | HI-P GRN LE | 0.55 | 0.241 | 0.34 | 0.31 | 0.30 | 0.26 | 0.20 | 0.11 | 0.28 | | 0.32 | 0.32 | 0.28 | 0.30 | 0.49 | 0.50 | 0.44 | 0.48 |
| | | | | $T$ | 0.22 | 0.19 | 0.18 | 0.15 | 0.10 | 0.04 | 0.17 | | | | | | | | | |
| | | | | $R^f$ | 0.07 | 0.07 | 0.08 | 0.11 | 0.20 | 0.41 | 0.11 | | | | | | | | | |
| | | | | $R^b$ | 0.13 | 0.14 | 0.16 | 0.21 | 0.33 | 0.58 | 0.20 | | | | | | | | | |
| | | | | $A_1^f$ | 0.64 | 0.67 | 0.68 | 0.68 | 0.66 | 0.53 | 0.65 | | | | | | | | | |
| | | | | $A_2^f$ | 0.08 | 0.07 | 0.06 | 0.06 | 0.04 | 0.02 | 0.06 | | | | | | | | | |

*Low-e Double Glazing, e = 0.1 on surface 2*

| | | | | | | | | | | | | | | | | | | | | |
|---|---|---|---|---|---|---|---|---|---|---|---|---|---|---|---|---|---|---|---|---|
| 21a | 3 | LE CLR | 0.76 | SHGC | 0.65 | 0.64 | 0.62 | 0.56 | 0.43 | 0.23 | 0.57 | | 0.59 | 0.60 | 0.53 | 0.58 | 0.68 | 0.68 | 0.61 | 0.67 |
| | | | | $T$ | 0.59 | 0.56 | 0.54 | 0.48 | 0.36 | 0.18 | 0.50 | | | | | | | | | |
| | | | | $R^f$ | 0.15 | 0.16 | 0.18 | 0.24 | 0.37 | 0.61 | 0.22 | | | | | | | | | |
| | | | | $R^b$ | 0.17 | 0.18 | 0.20 | 0.26 | 0.38 | 0.61 | 0.24 | | | | | | | | | |
| | | | | $A_1^f$ | 0.20 | 0.21 | 0.21 | 0.21 | 0.20 | 0.16 | 0.20 | | | | | | | | | |
| | | | | $A_2^f$ | 0.07 | 0.07 | 0.08 | 0.08 | 0.07 | 0.05 | 0.07 | | | | | | | | | |
| 21b | 6 | LE CLR | 0.72 | SHGC | 0.60 | 0.59 | 0.57 | 0.51 | 0.40 | 0.21 | 0.53 | | 0.55 | 0.55 | 0.49 | 0.53 | 0.64 | 0.65 | 0.58 | 0.63 |
| | | | | $T$ | 0.51 | 0.48 | 0.46 | 0.41 | 0.30 | 0.14 | 0.43 | | | | | | | | | |
| | | | | $R^f$ | 0.14 | 0.15 | 0.17 | 0.22 | 0.35 | 0.59 | 0.21 | | | | | | | | | |
| | | | | $R^b$ | 0.15 | 0.16 | 0.18 | 0.23 | 0.35 | 0.57 | 0.22 | | | | | | | | | |
| | | | | $A_1^f$ | 0.26 | 0.26 | 0.26 | 0.26 | 0.25 | 0.19 | 0.25 | | | | | | | | | |
| | | | | $A_2^f$ | 0.10 | 0.11 | 0.11 | 0.11 | 0.10 | 0.07 | 0.10 | | | | | | | | | |

*Low-e Double Glazing, e = 0.1 on surface 3*

| | | | | | | | | | | | | | | | | | | | | |
|---|---|---|---|---|---|---|---|---|---|---|---|---|---|---|---|---|---|---|---|---|
| 21c | 3 | CLR LE | 0.75 | SHGC | 0.60 | 0.58 | 0.56 | 0.51 | 0.40 | 0.22 | 0.52 | | 0.55 | 0.55 | 0.49 | 0.53 | 0.67 | 0.68 | 0.60 | 0.66 |
| | | | | $T$ | 0.48 | 0.45 | 0.43 | 0.37 | 0.27 | 0.13 | 0.40 | | | | | | | | | |
| | | | | $R^f$ | 0.26 | 0.27 | 0.28 | 0.32 | 0.42 | 0.62 | 0.31 | | | | | | | | | |
| | | | | $R^b$ | 0.24 | 0.24 | 0.26 | 0.29 | 0.38 | 0.58 | 0.28 | | | | | | | | | |
| | | | | $A_1^f$ | 0.12 | 0.13 | 0.14 | 0.14 | 0.15 | 0.15 | 0.13 | | | | | | | | | |
| | | | | $A_2^f$ | 0.14 | 0.15 | 0.15 | 0.16 | 0.16 | 0.10 | 0.15 | | | | | | | | | |
| 21d | 6 | CLR LE | 0.72 | SHGC | 0.56 | 0.55 | 0.52 | 0.48 | 0.38 | 0.20 | 0.49 | | 0.51 | 0.52 | 0.46 | 0.50 | 0.64 | 0.65 | 0.58 | 0.63 |
| | | | | $T$ | 0.42 | 0.40 | 0.37 | 0.32 | 0.24 | 0.11 | 0.35 | | | | | | | | | |
| | | | | $R^f$ | 0.24 | 0.24 | 0.25 | 0.29 | 0.38 | 0.58 | 0.28 | | | | | | | | | |
| | | | | $R^b$ | 0.20 | 0.20 | 0.22 | 0.26 | 0.34 | 0.55 | 0.25 | | | | | | | | | |
| | | | | $A_1^f$ | 0.19 | 0.20 | 0.21 | 0.22 | 0.23 | 0.22 | 0.21 | | | | | | | | | |
| | | | | $A_2^f$ | 0.16 | 0.17 | 0.17 | 0.17 | 0.16 | 0.10 | 0.16 | | | | | | | | | |
| 21e | 3 | BRZ LE | 0.57 | SHGC | 0.48 | 0.46 | 0.44 | 0.40 | 0.31 | 0.17 | 0.42 | | 0.44 | 0.44 | 0.39 | 0.43 | 0.51 | 0.51 | 0.46 | 0.50 |
| | | | | $T$ | 0.37 | 0.34 | 0.32 | 0.27 | 0.20 | 0.08 | 0.30 | | | | | | | | | |

**Table 7-4 SI** Visible Transmittance ($T_v$), Solar Heat Gain Coefficient (SHGC), Solar Transmittance ($T$), Front Reflectance ($R^f$), Back Reflectance ($R^b$), and Layer Absorptance $A_n^f$ for Glazing and Window Systems (*Continued*)

*(Table 10, Chapter 15, 2013 ASHRAE Handbook—Fundamentals)*

| | | | | | | | | | | | | | | | | | | | | |
|---|---|---|---|---|---|---|---|---|---|---|---|---|---|---|---|---|---|---|---|---|
| | | Glazing System | | Center Glazing $T_v$ | | \multicolumn{7}{c}{Center-of-Glazing Properties} | \multicolumn{4}{c}{Total Window SHGC at Normal Incidence} | \multicolumn{4}{c}{Total Window $T_v$ at Normal Incidence} |
| | | | | | | \multicolumn{7}{c}{Incidence Angles} | \multicolumn{2}{c}{Aluminum} | \multicolumn{2}{c}{Other Frames} | \multicolumn{2}{c}{Aluminum} | \multicolumn{2}{c}{Other Frames} |
| ID | Glass Thick., mm | | | | | Normal 0.00 | 40.00 | 50.00 | 60.00 | 70.00 | 80.00 | Hemis., Diffuse | Operable | Fixed | Operable | Fixed | Operable | Fixed | Operable | Fixed |
| 21f | 6 | BRZ LE | | 0.45 | $R^f$ | 0.18 | 0.17 | 0.19 | 0.22 | 0.30 | 0.50 | 0.21 | | | | | | | | |
| | | | | | $R^b$ | 0.23 | 0.23 | 0.25 | 0.29 | 0.37 | 0.57 | 0.28 | | | | | | | | |
| | | | | | $A_1^f$ | 0.34 | 0.37 | 0.38 | 0.39 | 0.39 | 0.35 | 0.37 | | | | | | | | |
| | | | | | $A_2^f$ | 0.11 | 0.12 | 0.12 | 0.12 | 0.11 | 0.07 | 0.11 | | | | | | | | |
| | | | | | SHGC | 0.39 | 0.37 | 0.35 | 0.31 | 0.24 | 0.13 | 0.33 | 0.36 | 0.36 | 0.32 | 0.35 | 0.40 | 0.41 | 0.36 | 0.40 |
| | | | | | T | 0.27 | 0.24 | 0.22 | 0.19 | 0.13 | 0.05 | 0.21 | | | | | | | | |
| | | | | | $R^f$ | 0.12 | 0.12 | 0.13 | 0.16 | 0.24 | 0.44 | 0.16 | | | | | | | | |
| | | | | | $R^b$ | 0.19 | 0.20 | 0.22 | 0.25 | 0.34 | 0.55 | 0.24 | | | | | | | | |
| | | | | | $A_1^f$ | 0.51 | 0.54 | 0.55 | 0.56 | 0.55 | 0.46 | 0.53 | | | | | | | | |
| | | | | | $A_2^f$ | 0.10 | 0.10 | 0.10 | 0.10 | 0.09 | 0.05 | 0.10 | | | | | | | | |
| 21g | 3 | GRN LE | | 0.68 | SHGC | 0.46 | 0.44 | 0.42 | 0.38 | 0.30 | 0.16 | 0.40 | 0.42 | 0.43 | 0.38 | 0.41 | 0.61 | 0.61 | 0.54 | 0.60 |
| | | | | | T | 0.36 | 0.32 | 0.30 | 0.26 | 0.18 | 0.08 | 0.28 | | | | | | | | |
| | | | | | $R^f$ | 0.17 | 0.16 | 0.17 | 0.20 | 0.29 | 0.48 | 0.20 | | | | | | | | |
| | | | | | $R^b$ | 0.23 | 0.23 | 0.25 | 0.29 | 0.37 | 0.57 | 0.27 | | | | | | | | |
| | | | | | $A_1^f$ | 0.38 | 0.41 | 0.42 | 0.43 | 0.43 | 0.38 | 0.40 | | | | | | | | |
| | | | | | $A_2^f$ | 0.10 | 0.11 | 0.11 | 0.11 | 0.10 | 0.06 | 0.10 | | | | | | | | |
| 21h | 6 | GRN LE | | 0.61 | SHGC | 0.36 | 0.33 | 0.31 | 0.28 | 0.22 | 0.12 | 0.30 | 0.34 | 0.34 | 0.30 | 0.32 | 0.54 | 0.55 | 0.49 | 0.54 |
| | | | | | T | 0.24 | 0.21 | 0.19 | 0.16 | 0.11 | 0.05 | 0.18 | | | | | | | | |
| | | | | | $R^f$ | 0.11 | 0.10 | 0.11 | 0.14 | 0.22 | 0.43 | 0.14 | | | | | | | | |
| | | | | | $R^b$ | 0.19 | 0.20 | 0.22 | 0.25 | 0.34 | 0.55 | 0.24 | | | | | | | | |
| | | | | | $A_1^f$ | 0.56 | 0.59 | 0.61 | 0.61 | 0.59 | 0.48 | 0.58 | | | | | | | | |
| | | | | | $A_2^f$ | 0.09 | 0.09 | 0.09 | 0.08 | 0.08 | 0.04 | 0.08 | | | | | | | | |
| 21i | 3 | GRY LE | | 0.52 | SHGC | 0.46 | 0.44 | 0.42 | 0.38 | 0.30 | 0.16 | 0.39 | 0.42 | 0.43 | 0.38 | 0.41 | 0.46 | 0.47 | 0.42 | 0.46 |
| | | | | | T | 0.35 | 0.32 | 0.30 | 0.25 | 0.18 | 0.08 | 0.28 | | | | | | | | |
| | | | | | $R^f$ | 0.16 | 0.16 | 0.17 | 0.20 | 0.28 | 0.48 | 0.20 | | | | | | | | |
| | | | | | $R^b$ | 0.23 | 0.23 | 0.25 | 0.29 | 0.37 | 0.57 | 0.27 | | | | | | | | |
| | | | | | $A_1^f$ | 0.39 | 0.42 | 0.43 | 0.44 | 0.44 | 0.38 | 0.41 | | | | | | | | |
| | | | | | $A_2^f$ | 0.10 | 0.11 | 0.11 | 0.11 | 0.10 | 0.06 | 0.10 | | | | | | | | |
| 21j | 6 | GRY LE | | 0.37 | SHGC | 0.34 | 0.32 | 0.30 | 0.27 | 0.21 | 0.12 | 0.28 | 0.32 | 0.32 | 0.28 | 0.30 | 0.33 | 0.33 | 0.30 | 0.33 |
| | | | | | T | 0.23 | 0.20 | 0.18 | 0.15 | 0.11 | 0.04 | 0.17 | | | | | | | | |
| | | | | | $R^f$ | 0.11 | 0.11 | 0.12 | 0.15 | 0.23 | 0.44 | 0.15 | | | | | | | | |
| | | | | | $R^b$ | 0.20 | 0.20 | 0.22 | 0.25 | 0.34 | 0.55 | 0.24 | | | | | | | | |
| | | | | | $A_1^f$ | 0.58 | 0.60 | 0.61 | 0.61 | 0.59 | 0.48 | 0.59 | | | | | | | | |
| | | | | | $A_2^f$ | 0.08 | 0.08 | 0.08 | 0.08 | 0.07 | 0.04 | 0.08 | | | | | | | | |
| 21k | 6 | BLUGRN LE | | 0.62 | SHGC | 0.39 | 0.37 | 0.34 | 0.31 | 0.24 | 0.13 | 0.33 | 0.36 | 0.36 | 0.32 | 0.35 | 0.55 | 0.56 | 0.50 | 0.55 |
| | | | | | T | 0.28 | 0.25 | 0.23 | 0.20 | 0.14 | 0.06 | 0.22 | | | | | | | | |
| | | | | | $R^f$ | 0.12 | 0.12 | 0.13 | 0.16 | 0.24 | 0.44 | 0.16 | | | | | | | | |
| | | | | | $R^b$ | 0.23 | 0.23 | 0.25 | 0.28 | 0.37 | 0.57 | 0.27 | | | | | | | | |
| | | | | | $A_1^f$ | 0.51 | 0.54 | 0.56 | 0.56 | 0.55 | 0.46 | 0.53 | | | | | | | | |
| | | | | | $A_2^f$ | 0.08 | 0.09 | 0.08 | 0.08 | 0.08 | 0.05 | 0.08 | | | | | | | | |
| 21l | 6 | HI-P GRN W/LE CLR | | 0.57 | SHGC | 0.31 | 0.30 | 0.29 | 0.26 | 0.21 | 0.12 | 0.27 | 0.29 | 0.29 | 0.26 | 0.28 | 0.51 | 0.51 | 0.46 | 0.50 |
| | | | | | T | 0.22 | 0.21 | 0.19 | 0.17 | 0.12 | 0.06 | 0.18 | | | | | | | | |
| | | | | | $R^f$ | 0.07 | 0.07 | 0.09 | 0.13 | 0.22 | 0.46 | 0.12 | | | | | | | | |
| | | | | | $R^b$ | 0.23 | 0.23 | 0.24 | 0.28 | 0.37 | 0.57 | 0.27 | | | | | | | | |
| | | | | | $A_1^f$ | 0.67 | 0.68 | 0.67 | 0.66 | 0.62 | 0.46 | 0.65 | | | | | | | | |
| | | | | | $A_2^f$ | 0.04 | 0.05 | 0.05 | 0.05 | 0.04 | 0.03 | 0.04 | | | | | | | | |

*Low-e Double Glazing, e = 0.05 on surface 2*

| | | | | | | | | | | | | | | | | | | | | |
|---|---|---|---|---|---|---|---|---|---|---|---|---|---|---|---|---|---|---|---|---|
| 25a | 3 | LE CLR | | 0.72 | SHGC | 0.41 | 0.40 | 0.38 | 0.34 | 0.27 | 0.14 | 0.36 | 0.38 | 0.38 | 0.34 | 0.36 | 0.64 | 0.65 | 0.58 | 0.63 |
| | | | | | T | 0.37 | 0.35 | 0.33 | 0.29 | 0.22 | 0.11 | 0.31 | | | | | | | | |
| | | | | | $R^f$ | 0.35 | 0.36 | 0.37 | 0.40 | 0.47 | 0.64 | 0.39 | | | | | | | | |
| | | | | | $R^b$ | 0.39 | 0.39 | 0.40 | 0.43 | 0.50 | 0.66 | 0.42 | | | | | | | | |
| | | | | | $A_1^f$ | 0.24 | 0.26 | 0.26 | 0.27 | 0.28 | 0.23 | 0.26 | | | | | | | | |
| | | | | | $A_2^f$ | 0.04 | 0.04 | 0.04 | 0.04 | 0.03 | 0.03 | 0.04 | | | | | | | | |

**Table 7-4 SI** Visible Transmittance ($T_v$), Solar Heat Gain Coefficient (SHGC), Solar Transmittance ($T$), Front Reflectance ($R^f$), Back Reflectance ($R^b$), and Layer Absorptance $A_n^f$ for Glazing and Window Systems (*Continued*)

*(Table 10, Chapter 15, 2013 ASHRAE Handbook—Fundamentals)*

| | | | | | | | Center-of-Glazing Properties | | | | | | Total Window SHGC at Normal Incidence | | | | Total Window $T_v$ at Normal Incidence | | | |
|---|---|---|---|---|---|---|---|---|---|---|---|---|---|---|---|---|---|---|---|---|
| | | Glazing System | | | | | Incidence Angles | | | | | | Aluminum | | Other Frames | | Aluminum | | Other Frames | |
| ID | Glass Thick., mm | | Center Glazing $T_v$ | | Normal 0.00 | 40.00 | 50.00 | 60.00 | 70.00 | 80.00 | Hemis., Diffuse | | Operable | Fixed | Operable | Fixed | Operable | Fixed | Operable | Fixed |
| 25b | 6 | LE CLR | 0.70 | SHGC | 0.37 | 0.36 | 0.34 | 0.31 | 0.24 | 0.13 | 0.32 | | 0.34 | 0.34 | 0.30 | 0.33 | 0.62 | 0.63 | 0.56 | 0.62 |
| | | | | $T$ | 0.30 | 0.28 | 0.27 | 0.23 | 0.17 | 0.08 | 0.25 | | | | | | | | | |
| | | | | $R^f$ | 0.30 | 0.30 | 0.32 | 0.35 | 0.42 | 0.60 | 0.34 | | | | | | | | | |
| | | | | $R^b$ | 0.35 | 0.35 | 0.35 | 0.38 | 0.44 | 0.60 | 0.37 | | | | | | | | | |
| | | | | $A_1^f$ | 0.34 | 0.35 | 0.35 | 0.36 | 0.35 | 0.28 | 0.34 | | | | | | | | | |
| | | | | $A_2^f$ | 0.06 | 0.07 | 0.07 | 0.06 | 0.06 | 0.04 | 0.06 | | | | | | | | | |
| 25c | 6 | BRZ W/LE CLR | 0.42 | SHGC | 0.26 | 0.25 | 0.24 | 0.22 | 0.18 | 0.10 | 0.23 | | 0.25 | 0.25 | 0.22 | 0.23 | 0.37 | 0.38 | 0.34 | 0.37 |
| | | | | $T$ | 0.18 | 0.17 | 0.16 | 0.14 | 0.10 | 0.05 | 0.15 | | | | | | | | | |
| | | | | $R^f$ | 0.15 | 0.16 | 0.17 | 0.21 | 0.29 | 0.51 | 0.20 | | | | | | | | | |
| | | | | $R^b$ | 0.34 | 0.34 | 0.35 | 0.37 | 0.44 | 0.60 | 0.37 | | | | | | | | | |
| | | | | $A_1^f$ | 0.63 | 0.63 | 0.63 | 0.61 | 0.57 | 0.42 | 0.60 | | | | | | | | | |
| | | | | $A_2^f$ | 0.04 | 0.04 | 0.04 | 0.04 | 0.03 | 0.03 | 0.04 | | | | | | | | | |
| 25d | 6 | GRN W/LE CLR | 0.60 | SHGC | 0.31 | 0.30 | 0.28 | 0.26 | 0.21 | 0.12 | 0.27 | | 0.29 | 0.29 | 0.26 | 0.28 | 0.53 | 0.54 | 0.48 | 0.53 |
| | | | | $T$ | 0.22 | 0.21 | 0.20 | 0.17 | 0.13 | 0.06 | 0.18 | | | | | | | | | |
| | | | | $R^f$ | 0.10 | 0.10 | 0.12 | 0.16 | 0.25 | 0.48 | 0.15 | | | | | | | | | |
| | | | | $R^b$ | 0.35 | 0.34 | 0.35 | 0.37 | 0.44 | 0.60 | 0.37 | | | | | | | | | |
| | | | | $A_1^f$ | 0.64 | 0.64 | 0.64 | 0.63 | 0.59 | 0.43 | 0.62 | | | | | | | | | |
| | | | | $A_2^f$ | 0.05 | 0.05 | 0.05 | 0.05 | 0.04 | 0.03 | 0.05 | | | | | | | | | |
| 25e | 6 | GRY W/LE CLR | 0.35 | SHGC | 0.24 | 0.23 | 0.22 | 0.20 | 0.16 | 0.09 | 0.21 | | 0.23 | 0.23 | 0.20 | 0.21 | 0.31 | 0.32 | 0.28 | 0.31 |
| | | | | $T$ | 0.16 | 0.15 | 0.14 | 0.12 | 0.09 | 0.04 | 0.13 | | | | | | | | | |
| | | | | $R^f$ | 0.12 | 0.13 | 0.15 | 0.18 | 0.26 | 0.49 | 0.17 | | | | | | | | | |
| | | | | $R^b$ | 0.34 | 0.34 | 0.35 | 0.37 | 0.44 | 0.60 | 0.37 | | | | | | | | | |
| | | | | $A_1^f$ | 0.69 | 0.69 | 0.68 | 0.67 | 0.62 | 0.45 | 0.66 | | | | | | | | | |
| | | | | $A_2^f$ | 0.03 | 0.03 | 0.03 | 0.03 | 0.03 | 0.02 | 0.03 | | | | | | | | | |
| 25f | 6 | BLUE W/LE CLR | 0.45 | SHGC | 0.27 | 0.26 | 0.25 | 0.23 | 0.18 | 0.11 | 0.24 | | 0.26 | 0.25 | 0.22 | 0.24 | 0.40 | 0.41 | 0.36 | 0.40 |
| | | | | $T$ | 0.19 | 0.18 | 0.17 | 0.15 | 0.11 | 0.05 | 0.16 | | | | | | | | | |
| | | | | $R^f$ | 0.12 | 0.12 | 0.14 | 0.17 | 0.26 | 0.49 | 0.16 | | | | | | | | | |
| | | | | $R^b$ | 0.34 | 0.34 | 0.35 | 0.37 | 0.44 | 0.60 | 0.37 | | | | | | | | | |
| | | | | $A_1^f$ | 0.66 | 0.66 | 0.65 | 0.64 | 0.60 | 0.44 | 0.63 | | | | | | | | | |
| | | | | $A_2^f$ | 0.04 | 0.04 | 0.04 | 0.04 | 0.04 | 0.03 | 0.04 | | | | | | | | | |
| 25g | 6 | HI-P GRN W/LE CLR | 0.53 | SHGC | 0.27 | 0.26 | 0.25 | 0.23 | 0.18 | 0.11 | 0.23 | | 0.26 | 0.25 | 0.22 | 0.24 | 0.47 | 0.48 | 0.42 | 0.47 |
| | | | | $T$ | 0.18 | 0.17 | 0.16 | 0.14 | 0.10 | 0.05 | 0.15 | | | | | | | | | |
| | | | | $R^f$ | 0.07 | 0.07 | 0.09 | 0.13 | 0.22 | 0.46 | 0.12 | | | | | | | | | |
| | | | | $R^b$ | 0.35 | 0.34 | 0.35 | 0.38 | 0.44 | 0.60 | 0.37 | | | | | | | | | |
| | | | | $A_1^f$ | 0.71 | 0.72 | 0.71 | 0.69 | 0.64 | 0.47 | 0.68 | | | | | | | | | |
| | | | | $A_2^f$ | 0.04 | 0.04 | 0.04 | 0.04 | 0.03 | 0.02 | 0.04 | | | | | | | | | |
| *Triple Glazing* | | | | | | | | | | | | | | | | | | | | |
| 29a | 3 | CLR CLR CLR | 0.74 | SHGC | 0.68 | 0.65 | 0.62 | 0.54 | 0.39 | 0.18 | 0.57 | | 0.62 | 0.62 | 0.55 | 0.60 | 0.66 | 0.67 | 0.59 | 0.65 |
| | | | | $T$ | 0.60 | 0.57 | 0.53 | 0.45 | 0.31 | 0.12 | 0.49 | | | | | | | | | |
| | | | | $R^f$ | 0.17 | 0.18 | 0.21 | 0.28 | 0.42 | 0.65 | 0.25 | | | | | | | | | |
| | | | | $R^b$ | 0.17 | 0.18 | 0.21 | 0.28 | 0.42 | 0.65 | 0.25 | | | | | | | | | |
| | | | | $A_1^f$ | 0.10 | 0.11 | 0.12 | 0.13 | 0.14 | 0.14 | 0.12 | | | | | | | | | |
| | | | | $A_2^f$ | 0.08 | 0.08 | 0.09 | 0.09 | 0.08 | 0.07 | 0.08 | | | | | | | | | |
| | | | | $A_3^f$ | 0.06 | 0.06 | 0.06 | 0.06 | 0.05 | 0.03 | 0.06 | | | | | | | | | |
| 29b | 6 | CLR CLR CLR | 0.70 | SHGC | 0.61 | 0.58 | 0.55 | 0.48 | 0.35 | 0.16 | 0.51 | | 0.56 | 0.56 | 0.50 | 0.54 | 0.62 | 0.63 | 0.56 | 0.62 |
| | | | | $T$ | 0.49 | 0.45 | 0.42 | 0.35 | 0.24 | 0.09 | 0.39 | | | | | | | | | |
| | | | | $R^f$ | 0.14 | 0.15 | 0.18 | 0.24 | 0.37 | 0.59 | 0.22 | | | | | | | | | |
| | | | | $R^b$ | 0.14 | 0.15 | 0.18 | 0.24 | 0.37 | 0.59 | 0.22 | | | | | | | | | |
| | | | | $A_1^f$ | 0.17 | 0.19 | 0.20 | 0.21 | 0.22 | 0.21 | 0.19 | | | | | | | | | |
| | | | | $A_2^f$ | 0.12 | 0.13 | 0.13 | 0.13 | 0.12 | 0.08 | 0.12 | | | | | | | | | |
| | | | | $A_3^f$ | 0.08 | 0.08 | 0.08 | 0.08 | 0.06 | 0.03 | 0.08 | | | | | | | | | |
| 29c | 6 | HI-P GRN CLR CLR | 0.53 | SHGC | 0.32 | 0.29 | 0.27 | 0.24 | 0.18 | 0.10 | 0.26 | | 0.30 | 0.30 | 0.26 | 0.29 | 0.47 | 0.48 | 0.42 | 0.47 |
| | | | | $T$ | 0.20 | 0.17 | 0.15 | 0.12 | 0.07 | 0.02 | 0.15 | | | | | | | | | |
| | | | | $R^f$ | 0.06 | 0.07 | 0.08 | 0.11 | 0.20 | 0.41 | 0.11 | | | | | | | | | |

**Table 7-4 SI** Visible Transmittance ($T_v$), Solar Heat Gain Coefficient (SHGC), Solar Transmittance ($T$), Front Reflectance ($R^f$), Back Reflectance ($R^b$), and Layer Absorptance $A_n^f$ for Glazing and Window Systems (*Continued*)

*(Table 10, Chapter 15, 2013 ASHRAE Handbook—Fundamentals)*

| ID | Glass Thick., mm | Glazing System | Center Glazing $T_v$ | | Normal 0.00 | 40.00 | 50.00 | 60.00 | 70.00 | 80.00 | Hemis., Diffuse | Aluminum Operable | Aluminum Fixed | Other Frames Operable | Other Frames Fixed | Aluminum Operable | Aluminum Fixed | Other Frames Operable | Other Frames Fixed |
|---|---|---|---|---|---|---|---|---|---|---|---|---|---|---|---|---|---|---|---|
| | | | | $R^b$ | 0.13 | 0.14 | 0.16 | 0.22 | 0.35 | 0.57 | 0.20 | | | | | | | | |
| | | | | $A_1^f$ | 0.64 | 0.67 | 0.68 | 0.68 | 0.66 | 0.53 | 0.65 | | | | | | | | |
| | | | | $A_2^f$ | 0.06 | 0.06 | 0.05 | 0.05 | 0.05 | 0.03 | 0.05 | | | | | | | | |
| | | | | $A_3^f$ | 0.04 | 0.04 | 0.04 | 0.03 | 0.02 | 0.01 | 0.04 | | | | | | | | |

*Triple Glazing, e = 0.2 on surface 2*

| ID | Glass Thick., mm | Glazing System | Center Glazing $T_v$ | | Normal | 40.00 | 50.00 | 60.00 | 70.00 | 80.00 | Hemis., Diffuse | Al Op | Al Fix | Oth Op | Oth Fix | Al Op | Al Fix | Oth Op | Oth Fix |
|---|---|---|---|---|---|---|---|---|---|---|---|---|---|---|---|---|---|---|---|
| 32a | 3 | LE CLR CLR | 0.68 | SHGC | 0.60 | 0.58 | 0.55 | 0.48 | 0.35 | 0.17 | 0.51 | 0.55 | 0.55 | 0.49 | 0.53 | 0.61 | 0.61 | 0.54 | 0.60 |
| | | | | T | 0.50 | 0.47 | 0.44 | 0.38 | 0.26 | 0.10 | 0.41 | | | | | | | | |
| | | | | $R^f$ | 0.17 | 0.19 | 0.21 | 0.27 | 0.41 | 0.64 | 0.25 | | | | | | | | |
| | | | | $R^b$ | 0.19 | 0.20 | 0.22 | 0.29 | 0.42 | 0.63 | 0.26 | | | | | | | | |
| | | | | $A_1^f$ | 0.20 | 0.20 | 0.20 | 0.21 | 0.21 | 0.17 | 0.20 | | | | | | | | |
| | | | | $A_2^f$ | 0.08 | 0.08 | 0.08 | 0.09 | 0.08 | 0.07 | 0.08 | | | | | | | | |
| | | | | $A_3^f$ | 0.06 | 0.06 | 0.06 | 0.06 | 0.05 | 0.03 | 0.06 | | | | | | | | |
| 32b | 6 | LE CLR CLR | 0.64 | SHGC | 0.53 | 0.50 | 0.47 | 0.41 | 0.29 | 0.14 | 0.44 | 0.49 | 0.49 | 0.43 | 0.47 | 0.57 | 0.58 | 0.51 | 0.56 |
| | | | | T | 0.39 | 0.36 | 0.33 | 0.27 | 0.17 | 0.06 | 0.30 | | | | | | | | |
| | | | | $R^f$ | 0.14 | 0.15 | 0.17 | 0.21 | 0.31 | 0.53 | 0.20 | | | | | | | | |
| | | | | $R^b$ | 0.16 | 0.16 | 0.19 | 0.24 | 0.36 | 0.57 | 0.22 | | | | | | | | |
| | | | | $A_1^f$ | 0.28 | 0.31 | 0.31 | 0.34 | 0.37 | 0.31 | 0.31 | | | | | | | | |
| | | | | $A_2^f$ | 0.11 | 0.11 | 0.11 | 0.11 | 0.10 | 0.08 | 0.11 | | | | | | | | |
| | | | | $A_3^f$ | 0.08 | 0.08 | 0.08 | 0.07 | 0.05 | 0.03 | 0.07 | | | | | | | | |

*Triple Glazing, e = 0.2 on surface 5*

| ID | Glass Thick., mm | Glazing System | Center Glazing $T_v$ | | Normal | 40.00 | 50.00 | 60.00 | 70.00 | 80.00 | Hemis., Diffuse | Al Op | Al Fix | Oth Op | Oth Fix | Al Op | Al Fix | Oth Op | Oth Fix |
|---|---|---|---|---|---|---|---|---|---|---|---|---|---|---|---|---|---|---|---|
| 32c | 3 | CLR CLR LE | 0.68 | SHGC | 0.62 | 0.60 | 0.57 | 0.49 | 0.36 | 0.16 | 0.52 | 0.57 | 0.57 | 0.50 | 0.55 | 0.61 | 0.61 | 0.54 | 0.60 |
| | | | | T | 0.50 | 0.47 | 0.44 | 0.38 | 0.26 | 0.10 | 0.41 | | | | | | | | |
| | | | | $R^f$ | 0.19 | 0.20 | 0.22 | 0.29 | 0.42 | 0.63 | 0.26 | | | | | | | | |
| | | | | $R^b$ | 0.18 | 0.19 | 0.21 | 0.27 | 0.41 | 0.64 | 0.25 | | | | | | | | |
| | | | | $A_1^f$ | 0.11 | 0.12 | 0.13 | 0.14 | 0.15 | 0.15 | 0.13 | | | | | | | | |
| | | | | $A_2^f$ | 0.09 | 0.10 | 0.10 | 0.10 | 0.10 | 0.08 | 0.10 | | | | | | | | |
| | | | | $A_3^f$ | 0.11 | 0.11 | 0.11 | 0.10 | 0.08 | 0.04 | 0.10 | | | | | | | | |
| 32d | 6 | CLR CLR LE | 0.64 | SHGC | 0.56 | 0.53 | 0.50 | 0.44 | 0.32 | 0.15 | 0.47 | 0.51 | 0.52 | 0.46 | 0.50 | 0.57 | 0.58 | 0.1 | 0.56 |
| | | | | T | 0.39 | 0.36 | 0.33 | 0.27 | 0.17 | 0.06 | 0.30 | | | | | | | | |
| | | | | $R^f$ | 0.16 | 0.16 | 0.19 | 0.24 | 0.36 | 0.57 | 0.22 | | | | | | | | |
| | | | | $R^b$ | 0.14 | 0.15 | 0.17 | 0.21 | 0.31 | 0.53 | 0.20 | | | | | | | | |
| | | | | $A_1^f$ | 0.17 | 0.19 | 0.20 | 0.21 | 0.22 | 0.22 | 0.19 | | | | | | | | |
| | | | | $A_2^f$ | 0.13 | 0.14 | 0.14 | 0.14 | 0.13 | 0.10 | 0.13 | | | | | | | | |
| | | | | $A_3^f$ | 0.15 | 0.16 | 0.15 | 0.14 | 0.12 | 0.05 | 0.14 | | | | | | | | |

*Triple Glazing, e = 0.1 on surface 2 and 5*

| ID | Glass Thick., mm | Glazing System | Center Glazing $T_v$ | | Normal | 40.00 | 50.00 | 60.00 | 70.00 | 80.00 | Hemis., Diffuse | Al Op | Al Fix | Oth Op | Oth Fix | Al Op | Al Fix | Oth Op | Oth Fix |
|---|---|---|---|---|---|---|---|---|---|---|---|---|---|---|---|---|---|---|---|
| 40a | 3 | LE CLR LE | 0.62 | SHGC | 0.41 | 0.39 | 0.37 | 0.32 | 0.24 | 0.12 | 0.34 | 0.38 | 0.38 | 0.34 | 0.36 | 0.55 | 0.56 | 0.50 | 0.55 |
| | | | | T | 0.29 | 0.26 | 0.24 | 0.20 | 0.13 | 0.05 | 0.23 | | | | | | | | |
| | | | | $R^f$ | 0.30 | 0.30 | 0.31 | 0.34 | 0.41 | 0.59 | 0.33 | | | | | | | | |
| | | | | $R^b$ | 0.30 | 0.30 | 0.31 | 0.34 | 0.41 | 0.59 | 0.33 | | | | | | | | |
| | | | | $A_1^f$ | 0.25 | 0.27 | 0.28 | 0.30 | 0.32 | 0.27 | 0.28 | | | | | | | | |
| | | | | $A_2^f$ | 0.07 | 0.08 | 0.08 | 0.08 | 0.07 | 0.06 | 0.07 | | | | | | | | |
| | | | | $A_3^f$ | 0.08 | 0.09 | 0.09 | 0.09 | 0.07 | 0.04 | 0.08 | | | | | | | | |
| 40b | 6 | LE CLR LE | 0.59 | SHGC | 0.36 | 0.34 | 0.32 | 0.28 | 0.21 | 0.10 | 0.30 | 0.34 | 0.34 | 0.30 | 0.32 | 0.53 | 0.53 | 0.47 | 0.52 |
| | | | | T | 0.24 | 0.21 | 0.19 | 0.16 | 0.10 | 0.03 | 0.18 | | | | | | | | |
| | | | | $R^f$ | 0.34 | 0.34 | 0.35 | 0.38 | 0.44 | 0.61 | 0.37 | | | | | | | | |
| | | | | $R^b$ | 0.23 | 0.23 | 0.25 | 0.28 | 0.36 | 0.56 | 0.27 | | | | | | | | |
| | | | | $A_1^f$ | 0.24 | 0.25 | 0.26 | 0.28 | 0.30 | 0.25 | 0.26 | | | | | | | | |
| | | | | $A_2^f$ | 0.10 | 0.11 | 0.11 | 0.11 | 0.10 | 0.07 | 0.10 | | | | | | | | |
| | | | | $A_3^f$ | 0.09 | 0.09 | 0.09 | 0.08 | 0.07 | 0.03 | 0.08 | | | | | | | | |

*Triple Glazing, e = 0.05 on surface 2 and 4*

| ID | Glass Thick., mm | Glazing System | Center Glazing $T_v$ | | Normal | 40.00 | 50.00 | 60.00 | 70.00 | 80.00 | Hemis., Diffuse | Al Op | Al Fix | Oth Op | Oth Fix | Al Op | Al Fix | Oth Op | Oth Fix |
|---|---|---|---|---|---|---|---|---|---|---|---|---|---|---|---|---|---|---|---|
| 49 | 3 | LE LE CLR | 0.58 | SHGC | 0.27 | 0.25 | 0.24 | 0.21 | 0.16 | 0.08 | 0.23 | 0.26 | 0.25 | 0.22 | 0.25 | 0.52 | 0.52 | 0.46 | 0.51 |
| | | | | T | 0.18 | 0.17 | 0.16 | 0.13 | 0.08 | 0.03 | 0.14 | | | | | | | | |
| | | | | $R^f$ | 0.41 | 0.41 | 0.42 | 0.44 | 0.50 | 0.65 | 0.44 | | | | | | | | |
| | | | | $R^b$ | 0.46 | 0.45 | 0.46 | 0.48 | 0.53 | 0.68 | 0.47 | | | | | | | | |

**Table 7-4 SI** Visible Transmittance ($T_v$), Solar Heat Gain Coefficient (SHGC), Solar Transmittance ($T$), Front Reflectance ($R^f$), Back Reflectance ($R^b$), and Layer Absorptance $A_n^f$ for Glazing and Window Systems (*Continued*)

*(Table 10, Chapter 15, 2013 ASHRAE Handbook—Fundamentals)*

| | Glazing System | | | | Center-of-Glazing Properties | | | | | | | Total Window SHGC at Normal Incidence | | | | Total Window $T_v$ at Normal Incidence | | | |
|---|---|---|---|---|---|---|---|---|---|---|---|---|---|---|---|---|---|---|---|
| | | | | | Incidence Angles | | | | | | | Aluminum | | Other Frames | | Aluminum | | Other Frames | |
| ID | Glass Thick., mm | | Center Glazing $T_v$ | | Normal 0.00 | 40.00 | 50.00 | 60.00 | 70.00 | 80.00 | Hemis., Diffuse | Operable | Fixed | Operable | Fixed | Operable | Fixed | Operable | Fixed |
| 50 | 6 | LE LE CLR | 0.55 | $A_1^f$ | 0.27 | 0.28 | 0.28 | 0.29 | 0.30 | 0.24 | 0.28 | | | | | | | | |
| | | | | $A_2^f$ | 0.12 | 0.12 | 0.12 | 0.12 | 0.11 | 0.07 | 0.12 | | | | | | | | |
| | | | | $A_3^f$ | 0.02 | 0.02 | 0.02 | 0.02 | 0.01 | 0.01 | 0.02 | | | | | | | | |
| | | | | SHGC | 0.26 | 0.25 | 0.23 | 0.21 | 0.16 | 0.08 | 0.22 | 0.25 | 0.25 | 0.21 | 0.24 | 0.49 | 0.0 | 0.44 | 0.48 |
| | | | | $T$ | 0.15 | 0.14 | 0.12 | 0.10 | 0.07 | 0.02 | 0.12 | | | | | | | | |
| | | | | $R^f$ | 0.33 | 0.33 | 0.34 | 0.37 | 0.43 | 0.60 | 0.36 | | | | | | | | |
| | | | | $R^b$ | 0.39 | 0.38 | 0.38 | 0.40 | 0.46 | 0.61 | 0.40 | | | | | | | | |
| | | | | $A_1^f$ | 0.34 | 0.36 | 0.36 | 0.37 | 0.36 | 0.28 | 0.35 | | | | | | | | |
| | | | | $A_2^f$ | 0.15 | 0.15 | 0.15 | 0.14 | 0.12 | 0.08 | 0.14 | | | | | | | | |
| | | | | $A_3^f$ | 0.03 | 0.03 | 0.03 | 0.03 | 0.02 | 0.01 | 0.03 | | | | | | | | |

KEY:
CLR = clear, BRZ = bronze, GRN = green, GRY = gray, BLUGRN = blue-green, SS = stainless steel reflective coating, TI = titanium reflective coating
Reflective coating descriptors include percent visible transmittance as *x*%.
HI-P GRN = high-performance green tinted glass, LE = low-emissivity coating

$T_v$ = visible transmittance, $T$ = solar transmittance, SHGC = solar heat gain coefficient, and H. = hemispherical SHGC
ID #s refer to U-factors in Table 4, except for products 49 and 50.

**Table 7-8 SI** Interior Solar Attenuation Coefficients (IAC) for Single or Double Glazings Shaded by Interior Venetian Blinds or Roller Shades

| | Nominal Thickness Each Pane, mm | Glazing Solar Transmittance[b] | | Glazing SHGC | IAC | | | | |
|---|---|---|---|---|---|---|---|---|---|
| | | | | | Venetian Blinds | | Roller Shades | | |
| Glazing System[a] | | Outer Pane | Single or Inner Pane | | Medium | Light | Opaque Dark | Opaque White | Translucent Light |
| ***Single Glazing Systems*** | | | | | | | | | |
| Clear, residential | 3 | | 0.87 to 0.80 | 0.86 | 0.75 | 0.68 | 0.82 | 0.40 | 0.40 |
| Clear, commercial | 6 to 13 | | 0.80 to 0.71 | 0.82 | | | | | |
| Clear, pattern | 3 to 13 | | 0.87 to 0.79 | | | | | | |
| Heat absorbing, pattern | 3 | | | 0.59 | | | | | |
| Tinted | 5, 5.5 | | 0.74, 0.71 | | | | | | |
| Above glazings, automated blinds[e] | | | | 0.86 | 0.64 | 0.59 | | | |
| Above glazings, tightly closed vertical blinds | | | | 0.85 | 0.30 | 0.26 | | | |
| Heat absorbing[f] | 6 | | 0.46 | 0.59 | 0.84 | 0.78 | 0.66 | 0.44 | 0.47 |
| Heat absorbing, pattern | 6 | | | | | | | | |
| Tinted | 3, 6 | | 0.59, 0.45 | | | | | | |
| Heat absorbing or pattern | | | 0.44 to 0.30 | 0.59 | 0.79 | 0.76 | 0.59 | 0.41 | 0.47 |
| Heat absorbing | 10 | | 0.34 | | | | | | |
| Heat absorbing or pattern | | | 0.29 to 0.15 | | | | | | |
| | | | 0.24 | 0.37 | 0.99 | 0.94 | 0.85 | 0.66 | 0.73 |
| Reflective coated glass | | | 0.26 to 0.52 | 0.83 | 0.75 | | | | |
| ***Double Glazing Systems***[g] | | | | | | | | | |
| Clear double, residential | 3 | 0.87 | 0.87 | 0.76 | 0.71 | 0.66 | 0.81 | 0.40 | 0.46 |
| Clear double, commercial | 6 | 0.80 | 0.80 | 0.70 | | | | | |
| Heat absorbing double[f] | 6 | 0.46 | 0.8 | 0.47 | 0.72 | 0.66 | 0.74 | 0.41 | 0.55 |
| Reflective double | | | | 0.17 to 0.35 | 0.90 | 0.86 | | | |
| ***Other Glazings*** (*Approximate*) | | | | | 0.83 | 0.77 | 0.74 | 0.45 | 0.52 |
| *± Range of Variation*[h] | | | | | 0.15 | 0.17 | 0.16 | 0.21 | 0.21 |

[a] Systems listed in the same table block have same IAC.
[b] Values or ranges given for identification of appropriate IAC value; where paired, solar transmittances and thicknesses correspond. SHGC is for unshaded glazing at normal incidence.
[c] Typical thickness for residential glass.
[d] From measurements by Van Dyke and Konen (1982) for 45° open venetian blinds, 35° solar incidence, and 35° profile angle.
[e] Use these values only when operation is automated for exclusion of beam solar (as opposed to daylight maximization). Also applies to tightly closed horizontal blinds.
[f] Refers to gray, bronze, and green tinted heat-absorbing glass (on exterior pane in double glazing)
[g] Applies either to factory-fabricated insulating glazing units or to prime windows plus storm windows.
[h] The listed approximate IAC value may be higher or lower by this amount, due to glazing/shading interactions and variations in the shading properties (e.g., manufacturing tolerances).

### Table 7-9 SI  Between-Glass Solar Attenuation Coefficients (BAC) for Double Glazing with Between-Glass Shading

| Type of Glass | Nominal Thickness, Each Pane | Solar Transmittance[a] Outer Pane | Solar Transmittance[a] Inner Pane | Description of Air Space | Venetian Blinds Light | Venetian Blinds Medium | Louvered Sun Screen |
|---|---|---|---|---|---|---|---|
| Clear out, Clear in | 2.43, 3 mm | 0.87 | 0.87 | Shade in contact with glass or shade separated from glass by air space. | 0.33 | 0.36 | 0.43 |
| Clear out, Clear in | 6 mm | 0.80 | 0.80 | Shade in contact with glass-voids filled with plastic. | — | — | 0.49 |
| Heat-absorbing[b] out, Clear in |  |  |  | Shade in contact with glass or shade separated from glass by air space. | 0.28 | 0.30 | 0.37 |
|  | 6 mm | 0.46 | 0.80 | Shade in contact with glass-voids filled with plastic. | — | — | 0.41 |

[a] Refer to manufacturers' literature for exact values.
[b] Refers to gray, bronze and green tinted heat-absorbing glass.

### Table 7-14 SI  Rates of Heat Gain from Occupants[a,b,c]
*(Table 1, Chapter 18, 2013 ASHRAE Handbook—Fundamentals)*

| Degree of Activity |  | Total Heat, W Adult Male | Total Heat, W Adjusted, M/F[a] | Sensible Heat, W | Latent Heat, W | % Sensible Heat that is Radiant[b] Low $V$ | % Sensible Heat that is Radiant[b] High $V$ |
|---|---|---|---|---|---|---|---|
| Seated at theater | Theater, matinee | 115 | 95 | 65 | 30 |  |  |
| Seated at theater, night | Theater, night | 115 | 105 | 70 | 35 | 60 | 27 |
| Seated, very light work | Offices, hotels, apartments | 130 | 115 | 70 | 45 |  |  |
| Moderately active office work | Offices, hotels, apartments | 140 | 130 | 75 | 55 |  |  |
| Standing, light work; walking | Department store; retail store | 160 | 130 | 75 | 55 | 58 | 38 |
| Walking, standing | Drug store, bank | 160 | 145 | 75 | 70 |  |  |
| Sedentary work | Restaurant[c] | 145 | 160 | 80 | 80 |  |  |
| Light bench work | Factory | 235 | 220 | 80 | 140 |  |  |
| Moderate dancing | Dance hall | 265 | 250 | 90 | 160 | 49 | 35 |
| Walking 4.8 km/h; light machine work | Factory | 295 | 295 | 110 | 185 |  |  |
| Bowling[d] | Bowling alley | 440 | 425 | 170 | 255 |  |  |
| Heavy work | Factory | 440 | 425 | 170 | 255 | 54 | 19 |
| Heavy machine work; lifting | Factory | 470 | 470 | 185 | 285 |  |  |
| Athletics | Gymnasium | 585 | 525 | 210 | 315 |  |  |

Notes:
1. Tabulated values are based on 24°C room dry-bulb temperature. For 27°C room dry bulb, the total heat remains the same, but the sensible heat values should be decreased by approximately 20%, and the latent heat values increased accordingly.
2. Also refer to Table 4, Chapter 8, for additional rates of metabolic heat generation.
3. All values are rounded to nearest 5 W.

[a] Adjusted heat gain is based on normal percentage of men, women, and children for the application listed, with the postulate that the gain from an adult female is 85% of that for an adult male, and that the gain from a child is 75% of that for an adult male.
[b] Values approximated from data in Table 6, Chapter 8, where is air velocity with limits shown in that table.
[c] Adjusted heat gain includes 18 W for food per individual (9 W sensible and 9 W latent).
[d] Figure one person per alley actually bowling, and all others as sitting (117 W) or standing or walking slowly (231 W).

Table 7-16 SI  Heat Gain from Typical Electric Motors

| Motor Nameplate or Rated Horsepower | (kW) | Motor Type | Nominal rpm | Full Load Motor Efficiency, % | A Motor in, Driven Equipment in, Watt | B Motor out, Driven Equipment in, Watt | C Motor in, Driven Equipment out, Watt |
|---|---|---|---|---|---|---|---|
| 0.05 | (0.04) | Shaded pole | 1500 | 35 | 105 | 35 | 70 |
| 0.08 | (0.06) | Shaded pole | 1500 | 35 | 170 | 59 | 110 |
| 0.125 | (0.09) | Shaded pole | 1500 | 35 | 264 | 94 | 173 |
| 0.16 | (0.12) | Shaded pole | 1500 | 35 | 340 | 117 | 223 |
| 0.25 | (0.19) | Split phase | 1750 | 54 | 346 | 188 | 158 |
| 0.33 | (0.25) | Split phase | 1750 | 56 | 439 | 246 | 194 |
| 0.50 | (0.37) | Split phase | 1750 | 60 | 621 | 372 | 249 |
| 0.75 | (0.56) | 3-Phase | 1750 | 72 | 776 | 557 | 217 |
| 1 | (0.75) | 3-Phase | 1750 | 75 | 993 | 747 | 249 |
| 1.5 | (1.1) | 3-Phase | 1750 | 77 | 1453 | 1119 | 334 |
| 2 | (1.5) | 3-Phase | 1750 | 79 | 1887 | 1491 | 396 |
| 3 | (2.2) | 3-Phase | 1750 | 81 | 2763 | 2238 | 525 |
| 5 | (3.7) | 3-Phase | 1750 | 82 | 4541 | 3721 | 817 |
| 7.5 | (5.6) | 3-Phase | 1750 | 84 | 6651 | 5596 | 1066 |
| 10 | (7.5) | 3-Phase | 1750 | 85 | 8760 | 7178 | 1315 |
| 15 | (11.2) | 3-Phase | 1750 | 86 | 13 009 | 11 192 | 1820 |
| 20 | (14.9) | 3-Phase | 1750 | 87 | 17 140 | 14 913 | 2230 |
| 25 | (18.6) | 3-Phase | 1750 | 88 | 21 184 | 18 635 | 2545 |
| 30 | (22.4) | 3-Phase | 1750 | 89 | 25 110 | 22 370 | 2765 |
| 40 | (30) | 3-Phase | 1750 | 89 | 33 401 | 29 885 | 3690 |
| 50 | (37) | 3-Phase | 1750 | 89 | 41 900 | 37 210 | 4600 |
| 60 | (45) | 3-Phase | 1750 | 89 | 50 395 | 44 829 | 5538 |
| 75 | (56) | 3-Phase | 1750 | 90 | 62 115 | 55 962 | 6210 |
| 100 | (75) | 3-Phase | 1750 | 90 | 82 918 | 74 719 | 8290 |
| 125 | (93) | 3-Phase | 1750 | 90 | 103 430 | 93 172 | 10 342 |
| 150 | (110) | 3-Phase | 1750 | 91 | 123 060 | 111 925 | 11 075 |
| 200 | (150) | 3-Phase | 1750 | 91 | 163 785 | 149 135 | 14 738 |
| 250 | (190) | 3-Phase | 1750 | 91 | 204 805 | 186 346 | 18 430 |

**Table 7-21A SI  Recommended Rates of Radiant and Convective Heat Gain from Unhooded Electric Appliances During Idle (Ready-to-Cook) Conditions**

*(Table 5A, Chapter 18, 2013 ASHRAE Handbook—Fundamentals)*

| Appliance | Energy Rate, W Rated | Energy Rate, W Standby | Sensible Radiant | Sensible Convective | Latent | Total | Usage Factor $F_U$ | Radiation Factor $F_R$ |
|---|---|---|---|---|---|---|---|---|
| Cabinet: hot serving (large), insulated* | 1993 | 352 | 117 | 234 | 0 | 352 | 0.18 | 0.33 |
| hot serving (large), uninsulated | 1993 | 1026 | 205 | 821 | 0 | 1026 | 0.51 | 0.20 |
| proofing (large)* | 5099 | 410 | 352 | 0 | 59 | 410 | 0.08 | 0.86 |
| proofing (small 15-shelf) | 4191 | 1143 | 0 | 264 | 879 | 1143 | 0.27 | 0.00 |
| Coffee brewing urn | 3810 | 352 | 59 | 88 | 205 | 352 | 0.08 | 0.17 |
| Drawer warmers, 2-drawer (moist holding)* | 1202 | 147 | 0 | 0 | 59 | 59 | 0.12 | 0.00 |
| Egg cooker | 3194 | 205 | 88 | 117 | 0 | 205 | 0.06 | 0.43 |
| Espresso machine* | 2403 | 352 | 117 | 234 | 0 | 352 | 0.15 | 0.33 |
| Food warmer: steam table (2-well-type) | 1495 | 1026 | 88 | 176 | 762 | 1026 | 0.69 | 0.08 |
| Freezer (small) | 791 | 322 | 147 | 176 | 0 | 322 | 0.41 | 0.45 |
| Hot dog roller* | 996 | 703 | 264 | 440 | 0 | 703 | 0.71 | 0.38 |
| Hot plate: single burner, high speed | 1114 | 879 | 264 | 615 | 0 | 879 | 0.79 | 0.30 |
| Hot-food case (dry holding)* | 9115 | 733 | 264 | 469 | 0 | 733 | 0.08 | 0.36 |
| Hot-food case (moist holding)* | 9115 | 967 | 264 | 528 | 176 | 967 | 0.11 | 0.27 |
| Microwave oven: commercial (heavy duty) | 3194 | 0 | 0 | 0 | 0 | 0 | 0 | 0.00 |
| Oven: countertop conveyorized bake/finishing* | 6008 | 3693 | 645 | 3048 | 0 | 3693 | 0.61 | 0.17 |
| Panini* | 1700 | 938 | 352 | 586 | 0 | 938 | 0.55 | 0.38 |
| Popcorn popper* | 586 | 59 | 29 | 29 | 0 | 59 | 0.1 | 0.50 |
| Rapid-cook oven (quartz-halogen)* | 12 016 | 0 | 0 | 0 | 0 | 0 | 0 | 0.00 |
| Rapid-cook oven (microwave/convection)* | 7297 | 1202 | 293 | 909 | 0 | 293 | 0.16 | 0.24 |
| Reach-in refrigerator* | 1407 | 352 | 88 | 264 | 0 | 352 | 0.25 | 0.25 |
| Refrigerated prep table* | 586 | 264 | 176 | 88 | 0 | 264 | 0.45 | 0.67 |
| Steamer (bun) | 1495 | 205 | 176 | 29 | 0 | 205 | 0.14 | 0.86 |
| Toaster: 4-slice pop up (large): cooking | 1788 | 879 | 59 | 410 | 293 | 762 | 0.49 | 0.07 |
| contact (vertical) | 3312 | 1553 | 791 | 762 | 0 | 1553 | 0.47 | 0.51 |
| conveyor (large) | 9613 | 3019 | 879 | 2139 | 0 | 3019 | 0.31 | 0.29 |
| small conveyor | 1700 | 1084 | 117 | 967 | 0 | 1084 | 0.64 | 0.11 |
| Waffle iron | 909 | 352 | 234 | 117 | 0 | 352 | 0.39 | 0.67 |

*Items with an asterisk appear only in Swierczyna et al. (2009); all others appear in both Swierczyna et al. (2008) and (2009).

**Table 7-21B SI  Recommended Rates of Radiant Heat Gain from Hooded Electric Appliances During Idle (Ready-to-Cook) Conditions**

*(Table 5B, Chapter 18, 2013 ASHRAE Handbook—Fundamentals)*

| Appliance | Energy Rate, W Rated | Energy Rate, W Standby | Sensible Radiant | Usage Factor $F_U$ | Radiation Factor $F_R$ |
|---|---|---|---|---|---|
| Broiler: underfired 900 mm | 10 814 | 9056 | 3165 | 0.84 | 0.35 |
| Cheesemelter* | 3605 | 3488 | 1348 | 0.97 | 0.39 |
| Fryer: kettle | 29 014 | 528 | 147 | 0.02 | 0.28 |
| Fryer: open deep-fat, 1-vat | 14 008 | 821 | 293 | 0.06 | 0.36 |
| Fryer: pressure | 13 511 | 791 | 147 | 0.06 | 0.19 |
| Griddle: double sided 900 mm (clamshell down)* | 21 218 | 2022 | 410 | 0.10 | 0.20 |
| Griddle: double sided 900 mm (clamshell up)* | 21 218 | 3370 | 1055 | 0.16 | 0.31 |
| Griddle: flat 900 mm | 17 115 | 3370 | 1319 | 0.20 | 0.39 |
| Griddle-small 900 mm* | 8997 | 1788 | 791 | 0.20 | 0.44 |
| Induction cooktop* | 21 013 | 0 | 0 | 0.00 | 0.00 |
| Induction wok* | 3488 | 0 | 0 | 0.00 | 0.00 |
| Oven: combi: combi-mode* | 16 411 | 1612 | 234 | 0.10 | 0.15 |
| Oven: combi: convection mode | 16 412 | 1612 | 410 | 0.10 | 0.25 |
| Oven: convection full-size | 12 103 | 1964 | 440 | 0.16 | 0.22 |
| Oven: convection half-size* | 5510 | 1084 | 147 | 0.20 | 0.14 |
| Pasta cooker* | 22 010 | 2491 | 0 | 0.11 | 0.00 |
| Range top: top off/oven on* | 4865 | 1172 | 293 | 0.24 | 0.25 |
| Range top: 3 elements on/oven off | 15 005 | 4513 | 1846 | 0.30 | 0.41 |
| Range top: 6 elements on/oven off | 15 005 | 9730 | 4074 | 0.65 | 0.42 |
| Range top: 6 elements on/oven on | 19 870 | 10 668 | 4250 | 0.54 | 0.40 |
| Range: hot-top | 15 826 | 15 035 | 3458 | 0.95 | 0.23 |
| Rotisserie* | 11 107 | 4044 | 1319 | 0.36 | 0.33 |
| Salamander* | 7004 | 6829 | 2051 | 0.97 | 0.30 |
| Steam kettle: large (225 L), simmer lid down* | 32 414 | 762 | 29 | 0.02 | 0.04 |
| Steam kettle: small (150 L), simmer lid down* | 21 599 | 528 | 88 | 0.02 | 0.17 |
| Steamer: compartment: atmospheric* | 9789 | 4484 | 59 | 0.46 | 0.01 |
| Tilting skillet/braising pan | 9642 | 1553 | 0 | 0.16 | 0.00 |

* Items with an asterisk appear only in Swierczyna et al. (2009); all others appear in both Swierczyna et al. (2008) and (2009).

# Nonresidential Cooling and Heating Load Calculations

**Table 7-21C SI    Recommended Rates of Radiant Heat Gain from Hooded Gas Appliances During Idle (Ready-to-Cook) Conditions**

*(Table 5C, Chapter 18, 2013 ASHRAE Handbook—Fundamentals)*

| Appliance | Energy Rate, W Rated | Energy Rate, W Standby | Rate of Heat Gain, W Sensible Radiant | Usage Factor $F_U$ | Radiation Factor $F_R$ |
|---|---|---|---|---|---|
| Broiler: batch* | 27 842 | 20 280 | 2374 | 0.73 | 0.12 |
| Broiler: chain (conveyor) | 38 685 | 28 340 | 3869 | 0.73 | 0.14 |
| Broiler: overfired (upright)* | 29 307 | 25 761 | 733 | 0.88 | 0.03 |
| Broiler: underfired 900 mm | 28 135 | 21 658 | 2638 | 0.77 | 0.12 |
| Fryer: doughnut | 12 895 | 3634 | 850 | 0.28 | 0.23 |
| Fryer: open deep-fat, 1 vat | 23 446 | 1377 | 322 | 0.06 | 0.23 |
| Fryer: pressure | 23 446 | 2638 | 234 | 0.11 | 0.09 |
| Griddle: double sided 900 mm (clamshell down)* | 31 710 | 2345 | 528 | 0.07 | 0.23 |
| Griddle: double sided 900 mm (clamshell up)* | 31 710 | 4308 | 1436 | 0.14 | 0.33 |
| Griddle: flat 900 mm | 26 376 | 5979 | 1084 | 0.23 | 0.18 |
| Oven: combi: combi-mode* | 22 185 | 1758 | 117 | 0.08 | 0.07 |
| Oven: combi: convection mode | 22 185 | 1700 | 293 | 0.08 | 0.17 |
| Oven: convection full-size | 12 895 | 3488 | 293 | 0.27 | 0.08 |
| Oven: conveyor (pizza) | 49 822 | 20 017 | 2286 | 0.40 | 0.11 |
| Oven: deck | 30 772 | 6008 | 1026 | 0.20 | 0.17 |
| Oven: rack mini-rotating* | 16 500 | 1319 | 322 | 0.08 | 0.24 |
| Pasta cooker* | 23 446 | 6946 | 0 | 0.30 | 0.00 |
| Range top: top off/oven on* | 7327 | 2169 | 586 | 0.30 | 0.27 |
| Range top: 3 burners on/oven off | 35 169 | 17 614 | 2081 | 0.50 | 0.12 |
| Range top: 6 burners on/oven off | 35 169 | 35 403 | 3370 | 1.01 | 0.10 |
| Range top: 6 burners on/oven on | 42 495 | 36 018 | 3986 | 0.85 | 0.11 |
| Range: wok* | 29 014 | 25 614 | 1524 | 0.88 | 0.06 |
| Rethermalizer* | 26 376 | 6829 | 3370 | 0.26 | 0.49 |
| Rice cooker* | 10 257 | 147 | 88 | 0.01 | 0.60 |
| Salamander* | 10 257 | 9759 | 1553 | 0.95 | 0.16 |
| Steam kettle: large (225 L) simmer lid down* | 42 495 | 1583 | 0 | 0.04 | 0.00 |
| Steam kettle: small (38 L) simmer lid down* | 15 240 | 967 | 88 | 0.06 | 0.09 |
| Steam kettle: small (150 L) simmer lid down | 29 307 | 1260 | 0 | 0.04 | 0.00 |
| Steamer: compartment: atmospheric* | 7620 | 2432 | 0 | 0.32 | 0.00 |
| Tilting skillet/braising pan | 30 479 | 3048 | 117 | 0.10 | 0.04 |

*Items with an asterisk appear only in Swierczyna et al. (2009); all others appear in both Swierczyna et al. (2008) and (2009).

**Table 7-21D SI    Recommended Rates of Radiant Heat Gain from Hooded Solid Fuel Appliances During Idle (Ready-to-Cook) Conditions**

*(Table 5D, Chapter 18, 2013 ASHRAE Handbook—Fundamentals)*

| Appliance | Energy Rate, Rated | Rate of Heat Gain, Standby | Rate of Heat Gain, Sensible | Usage Factor $F_U$ | Radiation Factor $F_R$ |
|---|---|---|---|---|---|
| Broiler: solid fuel: charcoal | | | | N/A | 0.15 |
| Broiler: solid fuel: wood (mesquite)* | | | | N/A | 0.14 |

*Items with an asterisk appear only in Swierczyna et al. (2009); all others appear in both Swierczyna et al. (2008) and (2009).

**Table 7-21E SI    Recommended Rates of Radiant and Convective Heat Gain from Warewashing Equipment During Idle (Standby) or Washing Conditions**

*(Table 5E, Chapter 18, 2013 ASHRAE Handbook—Fundamentals)*

| Appliance | Energy Rate, W Rated | Energy Rate, W Standby/Washing | Unhooded Sensible Radiant | Unhooded Sensible Convective | Unhooded Latent | Unhooded Total | Hooded Sensible Radiant | Usage Factor $F_U$ | Radiation Factor $F_R$ |
|---|---|---|---|---|---|---|---|---|---|
| Dishwasher (conveyor type, chemical sanitizing) | 13 716 | 1671/12 778 | 0 | 1304 | 3954 | 5258 | 0 | 0.36 | 0 |
| Dishwasher (conveyor type, hot-water sanitizing) standby | 13 716 | 1671/N/A | 0 | 1392 | 4973 | 6366 | 0 | N/A | 0 |
| Dishwasher (door-type, chemical sanitizing) washing | 5393 | 352/3898 | 0 | 580 | 818 | 1398 | 0 | 0.26 | 0 |
| Dishwasher (door-type, hot-water sanitizing) washing | 5393 | 352/3898 | 0 | 580 | 818 | 1398 | 0 | 0.26 | 0 |
| Dishwasher* (under-counter type, chemical sanitizing) standby | 7796 | 352/5480 | 0 | 668 | 1222 | 1890 | 0 | 0.35 | 0.00 |
| Dishwasher* (under-counter type, hot-water sanitizing) standby | 7796 | 498/5774 | 234 | 305 | 882 | 1421 | 234 | 0.27 | 0.34 |
| Booster heater* | 38 099 | 0 | 147 | 0 | 0 | 0 | 147 | 0 | N/A |

*Items with an asterisk appear only in Swierczyna et al. (2009); all others appear in both Swierczyna et al. (2008) and (2009).

*Note*: Heat load values are prorated for 30% washing and 70% standby.

# Chapter 8

# ENERGY ESTIMATING METHODS

This chapter discusses general techniques, as well as several simplified methods, for estimating the energy consumption of heating and cooling systems, including the variable-base degree-day and the bin concepts. Details of the more complete and sophisticated procedures may be found in Chapter 19, Energy Estimating and Modeling Methods, of the 2013 *ASHRAE Handbook—Fundamentals*. Complementary material on *energy* may be found in Chapter 34, Energy Resources, of the 2013 *ASHRAE Handbook—Fundamentals*.

## 8.1 General Considerations

### 8.1.1 Energy Resources and Sustainability

Because energy used in buildings and facilities comprises a significant amount of the total energy used for all purposes and, thus, affects energy resources, ASHRAE recognizes the "effect of its technology on the environment and natural resources to protect the welfare of posterity" (ASHRAE 2003). Many governmental agencies regulate energy conservation, often through the procedures to obtain building permits. Required efficiency values for building energy use strongly influence selection of HVAC&R systems and equipment. The HVAC&R industry deals with energy forms as they occur on or arrive at a building site. Generally, these forms are fossil fuels (natural gas, oil, and coal) and electricity. Solar and wind energy are also available at most sites, as in low-level geothermal energy for heat pumps. The term *energy source* refers to on-site energy in the form in which it arrives at or occurs in a site (e.g., electricity, gas, oil, coal). *Energy resource* refers to the raw energy that (1) is extracted from the earth, (2) is used to generate the energy sources delivered to a building site (e.g., coal used to generate electricity), or (3) occurs naturally and is available at a site (solar, wind, or geothermal energy).

The energy requirements and fuel consumption of building HVAC systems have a direct impact on the cost of operating a building and an indirect impact on the environment. This chapter takes an introductory look at methods for estimating energy use as a guide in design, for standards/code compliance, and for economic optimization. These energy estimating methods can provide quantitative energy and cost comparisons among design alternatives. A primary objective of building energy analysis is economic—to determine which of the available options has the lowest total (lifecycle) cost. Several building energy codes and standards allow the use of an energy analysis to demonstrate compliance with the energy performance goals in the code. In fact, this use of energy analysis programs may be more prevalent than actual comparative energy studies. The large number of uncontrolled and/or unknown factors related to actual building use and HVAC system control generally preclude the use of such methods for determining absolute energy consumption and should not be used to predict future energy bills.

### 8.1.2 Energy Estimating Techniques

Although the procedures used for estimating energy requirements vary widely in their degree of complexity, they share three elements. These elements are the calculation of (1) space load, (2) secondary equipment load, and (3) primary equipment energy requirements. *Secondary* refers to equipment that distributes the heating, cooling, or ventilating medium to the conditioned spaces; *primary* refers to the central plant equipment that converts fuel or electric energy for heating or cooling.

The **first step** in calculating energy requirements is to determine the **space load**, which is the amount of energy that must be added to or extracted from a space to maintain thermal comfort. The simplest procedures assume that the energy required to maintain comfort is a function of a single parameter—the outdoor dry-bulb temperature. More accurate methods consider solar effects, internal gains, heat storage in the walls and interiors, and the effects of wind on both the building envelope heat transfer and infiltration. The most sophisticated procedures are based on hourly profiles for climatic conditions and operational characteristics for a number of typical days of the year or a full 8760 hours of operation.

The **second step** translates the space load into a load on the **secondary system** equipment. This step must include calculation of all forms of energy required by the secondary system, which may include electrical energy to operate fans and/or pumps, as well as energy in the form of heated or chilled water.

The **third step** calculates the fuel and energy required by the **primary equipment** to meet these loads. It considers efficiencies and part-load characteristics of the equipment. Often, the different forms of energy, such as electricity, natural gas, or oil, must be tracked. In some cases where calculations are done to ensure compliance with codes or standards, these energies must be converted to the energy source or resource consumed, as opposed to that delivered to the building boundary.

Energy calculations often lead to an economic analysis that aims to establish the cost-effectiveness of conservation measures. Thus, a thorough energy analysis provides intermediate data, such as time of energy usage and maximum demand, so that utility charges can be accurately estimated. Although not part of the energy calculations, estimated capital equipment costs should be included in the complete analysis.

The sophistication of the calculation procedure used can often be inferred from the number of separate ambient conditions and/or time increments used in the calculations. A simple procedure may use only one measure, such as annual degree-days, and is appropriate only for simple systems and applications. Such methods are called *single-method measures*. Improved accuracy may be obtained by using more information, such as the number of hours anticipated under particular conditions of operation. These methods, of which the *bin method* is the most well known, are called *simplified multiple-measure methods*. The most elaborate methods perform energy balance calculations at each hour over some period of analysis, typically one year. These *detailed simulation methods* require hourly weather data, as well as hourly estimates of internal loads such as lighting and occupants.

Because systems that consume energy in buildings are nonlinear, dynamic, and very complex, few methods other than **computer modeling** are available for accurately calculating energy consumption. The most accurate methods for calculating building energy consumption are the most costly because of their intense computational requirements and high degree of user expertise. However, the cost of the computer facilities and the software itself are typically a small fraction of the total cost of running a building energy analysis. The major costs are for learning to use the program and for the time involved in using it.

The US Department of Energy maintains an up-to-date listing of building energy software with links to other sites that describe energy modeling and load estimating tools at http://www.energytoolsdirectory.gov. A number of these software programs are available for downloading without cost.

## 8.2 Component Modeling and Loads

### 8.2.1 Loads

After peak loads have been evaluated, equipment with capacity sufficient to offset these loads must be selected. Air supplied to the space must be at the proper conditions to satisfy both the sensible and latent loads. However, peak load occurs but a few times each year, whereas partial load operation exists most of the time. With operation predominately at part load, partial load analysis for energy use and fuel cost is often as important as the sizing procedure.

Calculating instantaneous space load is a key step in any building energy simulation. The **heat balance method** and the **weighing factor method** are two methods used for these calculations. The weighing factor method and the heat balance method use conduction transfer functions (or their equivalents) to calculate transmission heat gain or loss. The principal difference is in the methods used to calculate the subsequent internal heat transfers to the room. Experience with both methods has indicated largely the same results, provided the weighing factors are determined for the specific building under analysis.

### 8.2.2 Secondary System Components

Secondary HVAC systems generally include all elements of the overall building energy system between a central heating and cooling plant and the building zones. The precise definition depends heavily on the building design. A secondary system typically includes air-handling equipment, air distribution systems with the associated ductwork, dampers, fans, and heating, cooling, and humidity conditioning equipment. Secondary systems also include the liquid distribution systems between the central plant and the zone and air-handling equipment, including piping, valves, and pumps.

To the extent that the secondary system consumes energy and transfers energy between the building and central plant, energy analysis can be performed by characterizing the energy consumption of the individual components and the energy transferred among system components. In fact, few of the secondary components consume energy directly, except for fans, pumps, furnaces, direct-expansion air-conditioning package units with gas-fired heaters, and inline heaters. Secondary components are divided into two categories: distribution components and heat and mass transfer components.

The distribution system of an HVAC system affects energy consumption in two ways. First fans and pumps consume electrical energy directly, based on the flow and pressures under which the device operates. Ducts and dampers, or pipes and valves, and the system control strategies affect the flow and pressures at the fan. Second, thermal energy is often transferred to (or from) the fluid due to heat transfer through pipes and ducts and due to the electrical input to fans and pumps. The analysis of system components should, therefore, account for both direct electrical energy consumption and thermal energy transfer.

Strictly speaking, performance calculations of the fan and air distribution systems in a building require a detailed pressure balance on the entire network.

While a detailed analysis of a distribution system requires flow and pressure balancing among the components, nearly all commercially available methods of energy analysis approximate the effect of the interactions with part-load performance curves. This procedure eliminates the need to calculate pressure drop through the distribution system at off-design conditions. The exact shape of the part-load curve depends on the effect of flow control on the pressure and fan efficiency and may be calculated using a detailed analysis.

### 8.2.3 Primary System Components

Primary HVAC systems consume energy and deliver heating and cooling to a building, usually through secondary systems. Primary equipment generally includes chillers, boilers, cooling towers, cogeneration equipment, and plant-level thermal storage equipment. In particular, primary equipment

# Energy Estimating Methods

generally represents the major energy-consuming equipment of a building, so accurate characterization of building energy use relies on accurate modeling of primary equipment energy consumption.

The energy consumption characteristics of primary equipment generally depend on equipment design, load conditions, environmental conditions, and equipment control strategies. For example, chiller performance depends on the basic equipment design features (e.g., heat exchange surfaces, compressor design), the temperatures and flow through the condenser and evaporator, and the methods for controlling the chiller at different loads and operating conditions (e.g., inlet guide vane control on centrifugal chillers to maintain leaving chilled water temperature setpoint). In general, these variables that dictate energy consumption vary constantly and require calculations on an hourly basis.

While many secondary components are readily described by fundamental engineering principles (e.g., heat exchangers, valves), the complex nature of most primary equipment has discouraged the use of first-principle models for energy calculations. Instead, the energy consumption characteristics of primary equipment have traditionally been modeled using simple equations developed by running regression analyses on manufacturers' published design data. Because published data are generally only available for full-load design conditions, additional correction functions are used to correct the full-load data to part-load conditions. The functional form of the regression equations and correction functions takes many forms, including exponentials, Fourier series, and, most of the time, second- or third-order polynomials. The selection of an appropriate functional form depends on the behavior of the equipment. In some cases, energy consumption is calculated using direct interpolation from tables of data. However, this method often requires excessive data input and computer memory.

## 8.3 Overall Modeling Strategies

In developing a simulation model for building energy prediction, two basic issues must be considered—modeling of components or subsystems and the overall modeling strategy. Modeling of components results in sets of equations describing the individual components. Overall modeling strategy refers to the *sequence* and *procedures* used to solve these equations. The accuracy of results and the computer resources required to achieve these results depend on the modeling strategy.

In most building energy programs, the load models are executed for every space for every hour of the simulation period. (Practically all models use one hour as the time-step, which excludes any information on phenomena occurring in a shorter time span.) The load model is followed by running models for every secondary system, one at a time, for every hour of the simulation. Finally, the plant simulation model is executed again for the entire period. Each sequential execution processes the *fixed* output of the preceding step.

This procedure is illustrated in Fig. 8-1. The solid lines represent data passed from one model to the next. The dashed lines represent information, usually provided by the user, about one model to the preceding model.

Because of this loads-systems-plants sequence, certain phenomena cannot be modeled precisely. For example, if the heat balance method for computing loads is used and some component in the system simulation model cannot meet the load, the program can only report the current load. In actuality, the space temperature should readjust until the load matches the equipment capacity, but this cannot be modeled because the loads have been precalculated and fixed. If the weighting factor method is used for loads, this problem is partially overcome because loads are continually readjusted during the system simulation. However, the weighting factor technique is based on linear mathematics, and wide departures of room temperatures from those used during execution of the load program can introduce errors.

A similar problem arises in plant simulation. For example, in an actual building, as the load on the central plant varies the supply, chilled water temperature also varies. This variation, in turn, affects the capacity of the secondary system equipment. In an actual building, when the central plant becomes overloaded, space temperatures should rise to reduce the load. However, in most energy-estimating programs, this condition cannot occur; thus, only the overload condition can be reported. These are some of the penalties associated with decoupling of the load, system, and plant models.

An alternative strategy, in which all calculations are performed at each time step, is conceivable. Here the load, system, and plant equations are solved simultaneously at each time interval. With this strategy, unmet loads and imbalances cannot occur; conditions at the plant are immediately reflected to the secondary system and then to the load model, forcing them to readjust to the instantaneous conditions throughout the building. The results of this modeling strategy are superior to those currently available, although the magnitude and importance of the improvement are uncertain.

The principal disadvantage of the alternative approach, and the reason that it has not been widely used, is that it demands more computing resources.

An economic model, as included in Fig. 8-1, calculates energy costs (and sometimes capital costs) based on the estimated required input energy. Thus, the simulation model calculates energy usage and cost for any given input weather

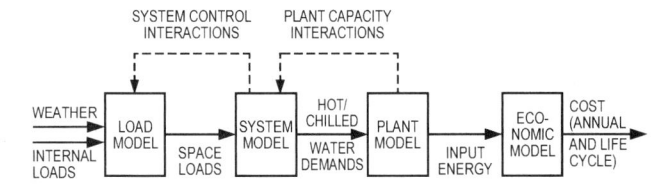

*Fig. 8-1 Overall Modeling Strategy*
(Figure 10, Chapter 19, 2013 ASHRAE Handbook—Fundamentals)

and internal loads. By applying this model (i.e., determining output for given inputs) at each hour (or other suitable interval), the hour-by-hour energy consumption and cost can be determined. Maintaining running sums of these quantities yields monthly or annual energy usage and costs.

These models only compare design alternatives; a large number of uncontrolled and unknown factors usually rules out such models for accurate prediction of utility bills.

Traditionally, most energy analysis programs include a set of preprogrammed models that represent various systems, such as variable air volume, terminal reheat, multizone, etc. In this scheme, the equations for each system are arranged so they can be solved sequentially. If this is not possible, then the smallest number of equations that must be solved simultaneously is solved using an appropriate technique. Furthermore, individual equations may vary from hour to hour in the simulation, depending on controls and operating conditions. For example, a dry coil uses different equations than a wet coil.

The primary disadvantage of this scheme is that it is relatively inflexible—to modify a system, the program source code may have to be modified and recompiled. Alternative strategies have viewed the system as a series of components (e.g. fan, coil, pump, duct, pipe, damper, thermostat) that may be organized in a component library. Users of the program specify the connections between the components. The program then resolves the specification of components and connections into a set of simultaneous equations.

## 8.4 Integration of System Models

Energy calculations for secondary systems involve construction of the complete system from the set of HVAC components. For example, a VAV system is a single-path system that controls zone temperature by modulating the airflow while maintaining a constant supply air temperature. VAV terminal units, located at each zone, adjust the quantity of air reaching each zone depending on its load requirements. Reheat coils may be included to provide required heating for perimeter zones.

This VAV system simulation consists of a central air-handling unit and a VAV terminal unit with reheat coil located at each zone, as shown in Fig. 8-2. The central air-handling unit includes a tan cooling coil, preheat coil, and an outdoor air economizer. The supply air leaving the air-handling unit is controlled to a fixed setpoint. The VAV terminal unit at each zone varies the airflow to meet the cooling load. As the zone cooling load decreases, the VAV terminal unit decreases the zone airflow until the unit reaches its minimum position. If the cooling load continues to decrease, the reheat coil will be activated to meet the zone load. As the supply air volume leaving the unit decreases, the fan power consumption will also be reduced. A variable-speed drive is used to control the supply fan.

The simulation is based on system characteristics and zone design requirements. For each zone, the inputs include the sensible and latent loads, the zone setpoint temperature, and the minimum zone supply air mass flow. System characteristics include the supply air temperature setpoint, the entering water temperature of the reheat, preheat, and cooling coils, the minimum mass flow of outdoor air, and the economizer temperature/enthalpy setpoint for minimum airflow.

The algorithm for performing the calculations for this VAV system is shown in Fig. 8-3. The algorithm directs sequential calculations of system performance. Calculations proceed from the zones forward along the return air path to the cooling coil inlet and back through the supply air path to the cooling oil discharge.

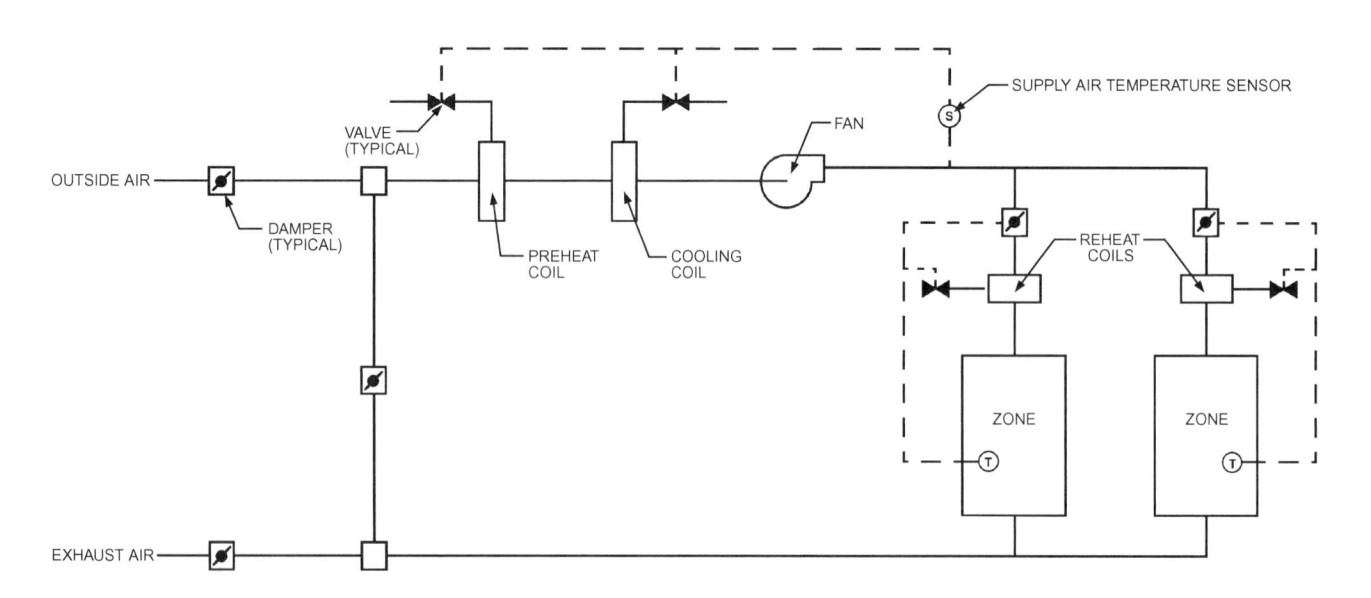

*Fig. 8-2 Schematic of Variable Air Volume System with Reheat*
*(Figure 15, Chapter 19, 2013 ASHRAE Handbook—Fundamentals)*

# Energy Estimating Methods

This basic algorithm for simulation of a VAV system might be used in conjunction with a heat-balance type of load calculation. For a weighting factor approach, it would have to be modified to allow zone temperatures to vary and a consequent readjustment of zone loads. It should also be enhanced to allow for possible limits on reheat temperature and/or cooling coil limits, zone humidity limits, outdoor air control (economizers), and/or heat-recovery devices, zone exhaust, return air fan, heat gain in the return air path because of lights, the presence of baseboard-type heaters, and more realistic control profiles. Most current building energy programs incorporate these and other features as user options, as well as algorithms for other types of systems.

The method chosen to analyze a building's energy use is determined by the purpose for the investigation.

**Forward modeling** begins with a description of the building system or component of interest and defines the building being modeled according to its physical description. For example, the building geometry, geographical location, type of HVAC system, wall insulation, etc., may be defined. The primary benefits of this method are that it is based on sound engineering principles and it has widespread acceptance and use in major public domain simulation codes (e.g., BLAST, DOE-2, and EnergyPlus). Figure 8-4 is a flow chart that illustrates the ordering of the analysis that is typically performed by a building energy simulation program.

**Inverse modeling** is based on the empirical behavior of the building as it relates to one or more driving forces. This approach is referred to as system identification, parameter identification, or inverse modeling. In this modeling approach, a structure or physical configuration of the building or system is assumed first and then important parameters are identified by a statistical analysis.

## 8.5 Degree-Day Methods

Degree-day methods are the simplest methods for energy analysis and are appropriate if the building use and the efficiency of the HVAC equipment are constant. Where efficiency or conditions of use vary with outdoor temperature, the consumption can be calculated for different values of outdoor temperature and multiplied by the corresponding number of hours; this approach is used in various bin methods. When the indoor temperature is allowed to fluctuate or when interior gains vary, models other than simple steady-state models must be used.

Single-measure methods for estimating cooling energy are less established than those for heating, primarily because the indoor-outdoor temperature difference during cooling is

```
BEGIN LOOP  Calculate zone related design requirements
    • Calculate required supply airflow to meet zone load
    • Sum actual zone mass air flow rate
    • Sum zone latent loads
    IF zone equals last zone THEN Exit Loop
END LOOP
    • Calculate system return air temperature from zone temps
    • Assume an initial cooling coil leaving air humidity ratio

BEGIN LOOP Iterate on cooling coil leaving air humidity ratio
    • Calculate return air humidity ratio from latent loads
    • Calculate supply fan power consumption and
      entering fan air temperature
    • Calculate mixed air temperature and humidity
      ratio using an economizer cycle
    IF mixed air temperature is less than design
      supply air temperature THEN
        • Calculate preheat coil load
    ELSE
        • Calculate cooling coil load and leaving air
          humidity ratio
    ENDIF
    IF cooling coil leaving air humidity ratio converged
    THEN Exit Loop
END LOOP

BEGIN LOOP Calculate the zone reheat coil loads
    IF zone supply air temperature is greater than system
      design supply air temperature THEN
        • Calculate reheat coil load
          (Subroutine: COILINV/HCDET)
    ENDIF
    • Sum reheat coil loads for all zones
    IF zone equals last zone THEN Exit Loop
END LOOP
```

*Fig. 8-3 Algorithm for Calculating Performance of VAV with Reheat System*
(Figure 16, Chapter 19, 2013 ASHRAE Handbook—Fundamentals)

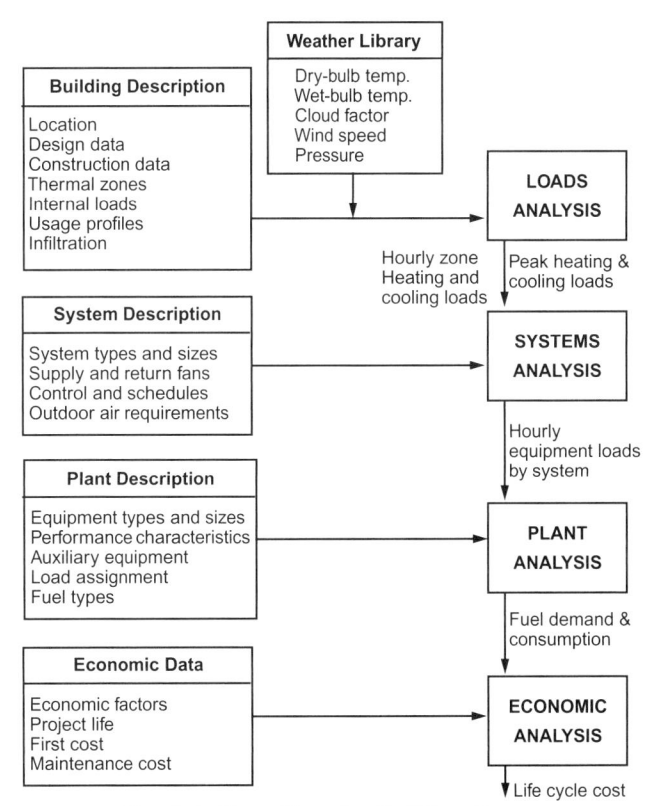

*Fig. 8-4 Flow Chart for Building Energy Simulation Program*
(Ayres and Stamper 1995)
(Figure 1, Chapter 19, 2013 ASHRAE Handbook—Fundamentals)

typically much smaller than under heating conditions. As a result, cooling loads depend more on such factors as solar gain and internal loads (besides temperature) than do heating loads. Since these loads largely depend on specific building features, attempts to correlate cooling energy requirements against a single climate parameter have not been very successful. Nonetheless, a companion to the heating degree-day procedure has emerged, the *cooling degree-day method*, and is widely used.

Even in an age when computers can easily calculate the energy consumption of a building, the concepts of degree-days and balance-point temperature remain valuable tools. The severity of a climate can be characterized concisely in terms of degree-days. Also, the degree-day method and its generalizations can provide a simple estimate of annual loads, which can be accurate if the indoor temperature and internal gains are relatively constant and if the heating or cooling systems are to operate for a complete season. Thus, basic steady-state methods continue to be important.

Virtually all the energy-consumption estimating techniques used prior to the last few years are based on an assumed or measured set of average conditions taken from a relatively small sampling of buildings. These techniques use steady-state models based on monthly average weather and operating conditions and, for the degree-day method in particular, assume that energy use depends only on the difference between indoor and outdoor dry-bulb temperature. Thus, use of the degree-day method is restricted to small structures with envelope-dominated heating and cooling loads. Other methods should be used for larger commercial or industrial buildings, where internal, cooling-only zones are prevalent or where cooling loads are not linearly dependent on the outdoor-to-indoor temperature difference.

Any estimating method produces a more reliable result over a long period of operation than over a short period. Nearly all of the methods in use give reasonable results over a full annual heating season, but estimates for shorter periods, such as a month, may produce inaccurate results. As the period of the estimate shortens, there is more chance that some factor not directly taken into account in the estimating method will deviate from its long-term average value and thus lead to an error in the predicted energy requirement.

## 8.5.1 Balance-Point Temperature and Degree-Days

The *balance-point temperature* is the average outdoor temperature at which the building requires neither heating nor cooling from the HVAC system. The degree-day procedure recognizes that the heating equipment needs to meet only the heating not covered by internal sources such as lights, equipment, occupants, and solar gain. In other words, these sources provide heat down to the balance-point temperature, and below that, energy requirements of the space are proportional to the difference between the balance-point and the outside temperature.

The balance-point temperature for heating $t_{bal}$ of a building is defined as that value of the outdoor temperature $t_o$ at which, for the specified value of the interior temperature $t_i$, the total heat gain $q_{gain}$ is equal to all space heat gains from sun, occupants, lights, and so forth.

The steady-state heating requirements of a building $q_h$ can be expressed as

$$q_h = K(t_i - t_o) - q_{gain} \qquad (8\text{-}1)$$

where
  $K$ = building loss coefficient (design heat loss divided by the design temperature difference)
  $t_i$ = indoor temperature
  $t_o$ = outdoor temperature
  $q_{gain}$ = all space heat gains not attributed to the space-conditioning system

The balance temperature is determined by setting $q_h = 0$ and solving for $t_o = t_{bal}$, so

$$t_{bal} = t_i - q_{gain}/K \qquad (8\text{-}2)$$

Heating is needed only when $t_o$ drops below $t_{bal}$.

With $t_{o,av}$ being the daily average value of the outdoor temperature for a particular day, the heating degree-days $DD_h(t_{bal})$ for that day are obtained as $(t_{bal} - t_{o,av})$ and for any particular period as

$$DD_h(t_{bal}) = (1 \text{ day})\Sigma(t_{bal} - t_{o,av}) \qquad (8\text{-}3)$$

where the summation may extend over a month, the heating season, or the entire year. In connection with degree-days, the balance temperature is also known as the base of the degree-days.

Cooling degree-days can be calculated using an expression analogous to heating degree-days:

$$DD_c(t_{bal}) = (1 \text{ day})\Sigma(t_o - t_{bal}) \qquad (8\text{-}4)$$

While the definition of the balance-point temperature is the same as that for heating, in a given building its numerical value for cooling is generally different from that for heating because $q_{gain}$, $K$, and/or $t_i$ can be different.

## 8.5.2 Seasonal Efficiency

The seasonal efficiency of heating equipment $\eta$ depends on such factors as steady-state efficiency, sizing, cycling effects, and energy conservation devices installed. Sometimes it is much lower than and other times it is comparable to steady-state efficiency. Expressions have been developed from information supplied by the National Institute of Standards and Technology that estimate the seasonal efficiency for a variety of furnaces if information on rated input and output is available. For the case where heat loss from the ducting is neglected, the series of equations is

$$\eta = \eta_{si} CF_{pl} \qquad (8\text{-}5)$$

where $\eta_{ss}$ is the steady-state efficiency (rated output/input).

The term $CF_{pl}$ is a trait of the part-load efficiency of the heating equipment, which may be calculated as follows:

*Gas-forced air furnaces*
With pilot:
$$CF_{pl} = 0.6328 + 0.5738 RLC - 0.3323(RLC)^2$$

With intermittent ignition:
$$CF_{pl} = 0.7791 + 0.1983 RLC - 0.0711(RLC)^2$$

*Oil furnaces without stack damper*
$$CF_{pl} = 0.7092 + 0.6515 RLC - 0.4711(RLC)^2$$

*Resistance electric furnaces*
$$CF_{pl} = 1.0$$

RLC is defined as follows:
$$RLC = \frac{K(t_{bal} - t_{od})}{CHT}$$

where
 $K$ = building loss coefficient
 $t_{od}$ = outside design temperature
 CHT = rated output of equipment

### 8.5.3 Heating Degree-Day Method

The degree-day procedure for estimating heating energy requirements is based on the assumption that, on a long-term average, solar and internal gains offset heat loss when the mean daily outdoor temperature is equal to the balance-point temperature and that fuel consumption is proportional to the difference between the mean daily temperature and the balance-point temperature. In other words, if the balance-point temperature is 60°F, then on a day when the mean temperature is 50°F (10°F below 60°F), 10 times as much fuel is consumed as on days when the mean temperature is 59°F (1°F below 60°F). This basic concept can be represented in an equation stating that energy consumption is directly proportional to the number of degree-days in the estimation period with the heat loss per degree difference being constant.

Estimating the theoretical seasonal energy requirement of a conventional heating system (gas furnace, oil furnace, electric furnace, and so forth) using the degree-day method is simple because the efficiency of the system is assumed to be constant regardless of outdoor temperature. The theoretical heating requirement is calculated as

$$\frac{\text{Design heat loss (Btu/h)} \times 24 \times \text{degree-days}}{\text{Design temperature difference}}$$

The general form of the degree-day equation is

$$E = \frac{q_L(DD)24}{\eta(HV)\Delta t} C_D \qquad (8\text{-}6)$$

where
 $E$ = fuel or energy consumption for the estimate period, units of fuel
 $q_L$ = design heat loss, including infiltration and ventilation, Btu/h
 DD = number of degree-days for the estimate period
 $\Delta t$ = design temperature difference, °F
 $\eta$ = efficiency of heating system [Eq. (8-5)], also designated on an annual basis as the annual fuel utilization efficiency (AFUE)
 HV = heating value of fuel, Btu/unit of fuel
 $C_D$ = 1 if DD are based on $t_{bal}$, or
 $C_D$ = 0.77 if 65°F is arbitrarily assumed as the balance temperature

Typical heating values HV are given in Table 8-1; additional values for $\eta$ are provided in Chapter 19.

Heating degree-days or degree-hours for a balance-point temperature of 65°F have been widely tabulated based on the observation that this has represented average conditions in typical buildings in the **past**. Today, use of the 65°F base will considerably overestimate the energy consumption due to improved building construction as well as increased internal loads in recent years. The errors inherent in the established base of 65°F (18.3°C) may be adjusted by using an empirical correction factor $C_D$. The best single value for $C_D$ is taken from the DOE HSPF methodology, which uses the correction factor 0.77 as the multiplier for the design heat loss per degree to provide a more appropriate building loss coefficient $K$ to account for internal heat gains in energy estimation. However, the best approach is to avoid the arbitrary 65°F base, using the variable-base degree-day approach instead.

Both heating degree days and cooling degree days can be found to several base temperatures in Figure 4-4 for many locations in the US. The CD-ROM of the 2013 *ASHRAE Handbook—Fundamentals* contains this data for many other locations.

**Example 8-1** A residence located in St. Joseph, Missouri, has a design heating load of 84,000 Btu/h. Determine

(a) Gallons of No. 2 fuel oil used for heating season with warm air system

(b) Electrical energy used for heating if electric baseboard units are used

Assume the indoor design temperature is 72°F and outdoor design temperature is 10.2°F.

**Table 8-1 Heating Values**

| Fuel | Heating Value of Fuel |
|---|---|
| Natural gas | 1050 Btu/ft$^3$ (39 MJ/m$^3$) |
| Propane | 90,000 Btu/gal (25 GJ/m$^3$) |
| No. 2 fuel oil | 140,000 Btu/gal (39 GJ/m$^3$) |

**Solution:**

From Equation (8-6) [with $\eta = 1$ and HV = 1],

$$E = \frac{84{,}000 \times 5435 \times 24}{[72-(10.2)]} 0.77$$

$$= 136{,}500{,}000 \text{ Btu/h}$$

(a) No. 2 fuel oil $= \dfrac{136.5 \times 10^6}{140{,}000 \times 0.75}$ (assumed)

$$= 1300 \text{ gallons}$$

(b) Heating electrical energy $= \dfrac{136.5 \times 10^6}{3413} = 40{,}000$ kWh

### 8.5.4 Variable-Base Heating Degree-Day

The variable-base degree-day method (VBDD) is a generalization of the degree-day method. It retains the original degree-day concept but counts degree-days based on the actual balance-point temperature rather than the now outdated 65°F.

The degree-day method, like any steady-state method, is unreliable for estimating the consumption during mild weather. Despite such problems, the degree-day method (using an appropriate base temperature) can give remarkably accurate results for the annual heating energy of single-zone buildings dominated by gains through the walls and roof and/or ventilation.

Table 8-2 provides the degree-days in °F to several bases for various locations in the United States. Annual degree-day tabulations for seven base temperatures in °C at 1195 stations in Canada are in Canadian Climate Normals (Atmospheric Environmental Service 1981).

Heating and cooling degree-day summary data for over 4000 US stations are available online at http://www.5.ncdc.noaa.gov/climatenormals/CLIM81_Sup_02.pdf (NCDC 2002a, 2002b). This publication presents annual heating degree-day normals to the following bases (°F): 65, 60, 57, 55, 50, 45, and 40; and annual cooling degree-day normals to the following bases (°F): 70, 65, 60, 57, 55, 50, and 45.

### 8.5.5 Cooling Degree-Day Method

Cooling degree-days are also available to several base temperatures as provided in Table 8-2.

Cooling energy is predicted in a similar fashion as described for heating earlier:

$$E_C = \frac{q_g(\text{CDD})24}{1000(\text{SEER})\Delta t_d} \qquad (8\text{-}7)$$

where
$E_C$ = energy consumed for cooling, kWh
$q_g$ = design cooling load, Btu/h (kW)
CDD = cooling degree-days
$\Delta t_d$ = cooling design temperature difference, °F (°C)
SEER = seasonal energy efficiency ratio, (Btu/h)/W (W/W)

The variable-base cooling degree-day procedure for estimating cooling energy requirements is similar to the procedure for heating. If no cooling is done by ventilation, the balance temperature $t_{bal}$ is calculated in the same way as the balance temperature for heating, except that summer values for cooling set temperature $t_i$, total heat gains $q_g$, and loss coefficient $K$ must be used. The only significant difference from heating in the calculation procedure is that $q_g$ customarily includes a latent term $q_l$ for cooling.

The degree-day method assumes that $t_{bal}$ is constant, which is not well satisfied in practice. Solar gains are zero at night, and internal gains tend to be highest during the evening. Also, during the intermediate season, heat gains can be eliminated, and the onset of mechanical cooling can be postponed by opening windows or increasing the ventilation. Therefore, cooling degree-hours can be used to better represent the period when equipment is operating than cooling degree-days because degree-days assume uninterrupted equipment operation as long as there is a cooling load.

**Example 8-2** For a residence located in El Paso, Texas, the design cooling load is 10.7 kW (36,500 Btu/h). Determine:

(a) Annual energy requirements for cooling
(b) Cost of this energy if a unit having an SEER of 11.5 is selected and the electric rate is 6.9¢/kWh

**Solution:**

Table 8-2: Cooling degree-days = 2098
Summer design = 98.3°F

(a) Using cooling degree method,

Cooling $= [36{,}500/(98.3 - 78)] \times 2098 \times 24$

$$= 90{,}534{,}000 \text{ Btu}$$

$$E_C = (90{,}534{,}000)/(1000 \times 11.5) = 7872 \text{ kWh}$$

(b) Cost $= 7872$ kWh $\times \$0.069$/kWh $= \$543$

**Example 8-3** A small dental office, 30 ft by 100 ft, located in downtown Chicago, Illinois, has design heating and cooling loads of 240,000 Btu/h and 85,000 Btu/h, respectively. Design conditions of 72°F inside and −3°F outside were used for winter while 78°F inside and 91°F outside were used for summer. The average interior heat gain during the winter has been estimated at 8 kW. Estimate the annual heating and cooling energy requirements and the corresponding energy costs using the national average rates given in the table below, if:

(a) Electric baseboard heating units and a "high-efficiency" air conditioner (SEER = 14) are used
(b) A condensing gas furnace and the same air conditioner are used

| Unit Fuel Prices | |
|---|---|
| Natural gas | $0.62/therm |
| No. 2 fuel oil | $1.06/gallon |
| Electricity | $0.072/kWh |

**Solution:**

The heat loss per °F, $K = 240{,}000/[72 - (-3)] = 3200$

The balance temperature with an internal gain of 8 kW is determined from Eq. 8-2 as

$$t_{bal} = 72 - (8 \times 3413)/3200$$

$$= 72 - 8.5 = 63.5°\text{F}$$

# Energy Estimating Methods

**Table 8-2   Degree-Days to Several Bases (in °F) for Various US Locations**

| State | Heating Base 65 | 60 | 55 | 50 | 45 | Cooling Base 65 | 60 | 55 | 50 | 45 |
|---|---|---|---|---|---|---|---|---|---|---|
| **ALABAMA** | | | | | | | | | | |
| Birmingham | 2844 | 1995 | 1333 | 838 | 488 | 1928 | 2916 | 4073 | 5403 | 6877 |
| Huntsville | 3302 | 2414 | 1670 | 1103 | 686 | 1808 | 2747 | 3828 | 5090 | 6492 |
| Mobile | 1684 | 1062 | 619 | 330 | 148 | 2577 | 3780 | 5162 | 6698 | 8342 |
| Montgomery | 2269 | 1508 | 945 | 547 | 282 | 2238 | 3302 | 4568 | 5991 | 7551 |
| **ALASKA** | | | | | | | | | | |
| Anchorage AP | 10911 | 9122 | 7492 | 6081 | 4896 | 0 | 36 | 224 | 647 | 1279 |
| Annette AP | 7053 | 5315 | 3773 | 2513 | 1543 | 14 | 98 | 386 | 940 | 1803 |
| Barrow AP | 20265 | 18440 | 16615 | 14803 | 13009 | 0 | 0 | 0 | 13 | 44 |
| Barter Island AP | 19994 | 18169 | 16344 | 14528 | 12738 | 0 | 0 | 0 | 9 | 44 |
| Bethel AP | 13203 | 11404 | 9695 | 8140 | 6835 | 0 | 26 | 142 | 411 | 938 |
| Bettles AP | 15925 | 14180 | 12548 | 11060 | 9718 | 17 | 97 | 289 | 626 | 1110 |
| Big Delta AP | 13698 | 11985 | 10410 | 8977 | 7735 | 34 | 145 | 395 | 787 | 1370 |
| Cold Bay AP | 9865 | 8040 | 6230 | 4532 | 3095 | 0 | 0 | 16 | 138 | 533 |
| Fairbanks AP | 14345 | 12661 | 11115 | 9714 | 8451 | 52 | 196 | 467 | 898 | 1468 |
| Gulkana | 13938 | 12162 | 10507 | 8985 | 7648 | 9 | 63 | 228 | 537 | 1027 |
| Homer | 10364 | 8539 | 6745 | 5133 | 3840 | 0 | 0 | 24 | 240 | 777 |
| Juneau AP | 9007 | 7222 | 5557 | 4107 | 2925 | 0 | 39 | 197 | 573 | 1219 |
| King Salmon AP | 11582 | 9773 | 8047 | 6563 | 5304 | 0 | 12 | 112 | 456 | 1023 |
| Kodiak | 8860 | 7049 | 5327 | 3819 | 2593 | 0 | 7 | 117 | 436 | 1032 |
| Kotzebue AP | 16039 | 14237 | 12491 | 10852 | 9337 | 0 | 23 | 102 | 268 | 598 |
| McGrath AP | 14487 | 12736 | 11107 | 9634 | 8348 | 14 | 88 | 284 | 642 | 1184 |
| Nome AP | 14325 | 12503 | 10721 | 9047 | 7528 | 0 | 0 | 46 | 197 | 503 |
| St. Paul Island AP | 11119 | 9294 | 7469 | 5667 | 4021 | 0 | 0 | 0 | 24 | 199 |
| Shemya AP | 9735 | 7910 | 6085 | 4298 | 2693 | 0 | 0 | 0 | 30 | 254 |
| Summit FAA AP | 14368 | 12556 | 10790 | 9146 | 7640 | 0 | 10 | 71 | 253 | 578 |
| Talkeetna | 11708 | 9934 | 8306 | 6848 | 5609 | 6 | 57 | 254 | 620 | 1207 |
| Unalakleet | 14027 | 12238 | 10515 | 8943 | 7565 | 0 | 31 | 138 | 391 | 842 |
| Yakutat AP | 9533 | 7711 | 5942 | 4420 | 3181 | 0 | 0 | 56 | 362 | 947 |
| **ARIZONA** | | | | | | | | | | |
| Flagstaff | 7322 | 5776 | 4421 | 3267 | 2299 | 140 | 416 | 894 | 1562 | 2418 |
| Phoenix | 1552 | 899 | 431 | 165 | 45 | 3508 | 4680 | 6039 | 7596 | 9297 |
| Prescott FAA AP | 4456 | 3303 | 2321 | 1507 | 883 | 882 | 1560 | 2400 | 3414 | 4612 |
| Tucson | 1752 | 1050 | 541 | 229 | 65 | 2814 | 3937 | 5253 | 6765 | 8431 |
| Winslow | 4733 | 3623 | 2683 | 1882 | 1249 | 1203 | 1921 | 2802 | 3828 | 5018 |
| Yuma | 1005 | 507 | 211 | 59 | 8 | 4195 | 5518 | 7045 | 8719 | 10498 |
| **ARKANSAS** | | | | | | | | | | |
| Fort Smith | 3336 | 2442 | 1687 | 1075 | 613 | 2022 | 2949 | 4015 | 5239 | 6595 |
| Little Rock | 3354 | 2442 | 1687 | 1075 | 624 | 1925 | 2843 | 3908 | 5128 | 6496 |
| **CALIFORNIA** | | | | | | | | | | |
| Bakersfield | 2185 | 1367 | 760 | 371 | 147 | 2179 | 3185 | 4400 | 5835 | 7437 |
| Bishop | 4313 | 3179 | 2230 | 1437 | 848 | 1037 | 1728 | 2603 | 3641 | 4875 |
| Blue Canyon | 5704 | 4271 | 3037 | 2015 | 1206 | 302 | 698 | 1283 | 2079 | 3106 |
| Daggett FAA AP | 2203 | 1420 | 824 | 410 | 166 | 2729 | 3765 | 5004 | 6415 | 7996 |
| Eureka | 4679 | 2925 | 1494 | 607 | 194 | 0 | 55 | 460 | 1414 | 2816 |
| Fresno | 2650 | 1724 | 995 | 493 | 205 | 1671 | 2563 | 3667 | 4986 | 6525 |
| Long Beach | 1606 | 772 | 292 | 70 | 8 | 985 | 1982 | 3325 | 4928 | 6696 |
| Los Angeles Int'l. | 1819 | 833 | 295 | 66 | 7 | 615 | 1464 | 2755 | 4348 | 6115 |
| Los Angeles Civic Center | 1245 | 522 | 158 | 26 | 0 | 1185 | 2289 | 3747 | 5442 | 7244 |
| Mount Shasta | 5890 | 4458 | 3215 | 2177 | 1338 | 286 | 680 | 1263 | 2045 | 3035 |
| Oakland | 2909 | 1570 | 714 | 263 | 61 | 128 | 622 | 1598 | 2963 | 4587 |
| Red Bluff | 2688 | 1762 | 1018 | 505 | 208 | 1904 | 2803 | 3895 | 5196 | 6727 |
| Sacramento | 2843 | 1837 | 1043 | 493 | 186 | 1159 | 1971 | 3011 | 4286 | 5812 |
| Sacramento City | 2587 | 1627 | 893 | 406 | 148 | 1291 | 2158 | 3249 | 4584 | 6151 |
| Sandberg | 4427 | 3177 | 2107 | 1250 | 622 | 800 | 1374 | 2123 | 3100 | 4293 |
| San Diego | 1507 | 648 | 213 | 42 | 9 | 722 | 1694 | 3084 | 4746 | 6532 |
| San Francisco | 3042 | 1668 | 769 | 289 | 67 | 108 | 550 | 1496 | 2832 | 4438 |
| San Francisco Fed. Bldg. | 3080 | 1576 | 608 | 169 | 25 | 39 | 368 | 1230 | 2619 | 4298 |
| Santa Maria | 3053 | 1624 | 690 | 229 | 42 | 84 | 484 | 1377 | 2738 | 4380 |
| Stockton | 2806 | 1835 | 1072 | 537 | 219 | 1259 | 2100 | 3167 | 4455 | 5958 |
| **COLORADO** | | | | | | | | | | |
| Alamosa | 8609 | 7029 | 5654 | 4473 | 3457 | 88 | 329 | 780 | 1428 | 2227 |
| Colorado Springs | 6473 | 5131 | 3954 | 2949 | 2089 | 461 | 945 | 1592 | 2417 | 3383 |
| Denver–Stapleton | 6016 | 4723 | 3601 | 2653 | 1852 | 625 | 1159 | 1857 | 2739 | 3759 |
| Denver–City | 5505 | 4246 | 3175 | 2271 | 1533 | 742 | 1312 | 2071 | 2993 | 4074 |
| Eagle AP | 8426 | 6864 | 5505 | 4319 | 3317 | 117 | 385 | 845 | 1487 | 2313 |
| Grand Junction | 5605 | 4441 | 3425 | 2551 | 1814 | 1140 | 1810 | 2619 | 3565 | 4653 |
| Pueblo | 5394 | 4221 | 3220 | 2351 | 1628 | 981 | 1632 | 2456 | 3412 | 4514 |
| **CONNECTICUT** | | | | | | | | | | |
| Bridgeport | 5461 | 4264 | 3216 | 2321 | 1583 | 735 | 1362 | 2140 | 3064 | 4152 |
| Hartford | 6350 | 5085 | 3971 | 3005 | 2173 | 584 | 1143 | 1855 | 2715 | 3706 |
| **DISTRICT OF COLUMBIA** | | | | | | | | | | |
| Washington, D.C.–Dulles AP | 5005 | 3881 | 2898 | 2054 | 1380 | 940 | 1636 | 2474 | 3458 | 4616 |
| Washington, D.C.–Nat'l. AP | 4211 | 3182 | 2293 | 1563 | 984 | 1415 | 2210 | 3152 | 4237 | 5489 |
| **DELAWARE** | | | | | | | | | | |
| Wilmington Ncastle | 4940 | 3824 | 2839 | 2003 | 1330 | 992 | 1697 | 2537 | 3525 | 4675 |
| **FLORIDA** | | | | | | | | | | |
| Apalachicola | 1361 | 792 | 426 | 189 | 67 | 2663 | 3930 | 5377 | 6967 | 8669 |
| Daytona Beach | 897 | 480 | 215 | 71 | 25 | 2919 | 4321 | 5881 | 7563 | 9341 |

Table 8-2 Degree-Days to Several Bases (in °F) for Various US Locations (*Continued*)

| State | Heating Base 65 | 60 | 55 | 50 | 45 | Cooling Base 65 | 60 | 55 | 50 | 45 |
|---|---|---|---|---|---|---|---|---|---|---|
| Fort Myers | 457 | 189 | 56 | 9 | 0 | 3711 | 5265 | 6958 | 8741 | 10553 |
| Jacksonville | 1327 | 788 | 429 | 194 | 70 | 2596 | 3890 | 5349 | 6938 | 8641 |
| Key West | 59 | 7 | 0 | 0 | 0 | 4888 | 6660 | 8474 | 10299 | 12124 |
| Lakeland | 678 | 330 | 128 | 40 | 7 | 3298 | 4774 | 6398 | 8135 | 9927 |
| Miami | 206 | 54 | 8 | 0 | 0 | 4038 | 5715 | 7494 | 9306 | 11131 |
| Orlando | 733 | 370 | 151 | 48 | 9 | 3226 | 4686 | 6291 | 8016 | 9806 |
| Pensacola | 1578 | 991 | 575 | 302 | 135 | 2695 | 3932 | 5341 | 6893 | 8551 |
| Tallahassee | 1563 | 996 | 550 | 279 | 116 | 2563 | 3792 | 5200 | 6755 | 8415 |
| Tampa | 718 | 364 | 151 | 50 | 14 | 3366 | 4836 | 6447 | 8172 | 9963 |
| West Palm Beach | 299 | 88 | 27 | 0 | 0 | 3786 | 5408 | 7159 | 8960 | 10785 |
| **GEORGIA** | | | | | | | | | | |
| Athens | 2975 | 2084 | 1370 | 830 | 462 | 1722 | 2661 | 3767 | 5052 | 6508 |
| Atlanta | 3095 | 2189 | 1461 | 911 | 524 | 1589 | 2511 | 3604 | 4880 | 6316 |
| Augusta | 2547 | 1729 | 1106 | 652 | 348 | 1995 | 2999 | 4204 | 5573 | 7094 |
| Columbus | 2378 | 1600 | 1010 | 593 | 313 | 2143 | 3188 | 4429 | 5831 | 7375 |
| Macon | 2240 | 1492 | 934 | 536 | 271 | 2294 | 3373 | 4643 | 6068 | 7626 |
| Rome | 3342 | 2422 | 1653 | 1068 | 637 | 1615 | 2514 | 3576 | 4816 | 6210 |
| Savannah | 1952 | 1258 | 751 | 413 | 185 | 2317 | 3444 | 4766 | 6248 | 7851 |
| **HAWAII** | | | | | | | | | | |
| Hilo | 0 | 0 | 0 | 0 | 0 | 3066 | 4887 | 6712 | 8537 | 10362 |
| Honolulu | 0 | 0 | 0 | 0 | 0 | 4221 | 6046 | 7871 | 9695 | 11521 |
| Kahului | 0 | 0 | 0 | 0 | 0 | 3732 | 5555 | 7380 | 9205 | 11030 |
| Lihue AP | 0 | 0 | 0 | 0 | 0 | 3719 | 5535 | 7360 | 9185 | 11010 |
| **IDAHO** | | | | | | | | | | |
| Boise | 5833 | 4533 | 3399 | 2434 | 1626 | 714 | 1233 | 1929 | 2793 | 3811 |
| Idaho Falls | 8619 | 7129 | 5800 | 4609 | 3590 | 237 | 573 | 1064 | 1703 | 2511 |
| Lewiston | 5464 | 4158 | 3050 | 2112 | 1366 | 657 | 1186 | 1886 | 2780 | 3856 |
| Pocatello | 7063 | 5687 | 4454 | 3401 | 2504 | 437 | 883 | 1477 | 2252 | 3177 |
| **ILLINOIS** | | | | | | | | | | |
| Cairo | 3833 | 2895 | 2090 | 1447 | 925 | 1806 | 2687 | 3710 | 4893 | 6197 |
| Chicago Midway AP | 6127 | 4952 | 3912 | 2998 | 2219 | 925 | 1575 | 2361 | 3272 | 4317 |
| Chicago O'Hare AP | 6497 | 5245 | 4163 | 3220 | 2404 | 664 | 1243 | 1986 | 2863 | 3872 |
| Decatur | 5344 | 4247 | 3293 | 2461 | 1778 | 1197 | 1925 | 2797 | 3791 | 4932 |
| Moline | 6395 | 5202 | 4170 | 3263 | 2462 | 893 | 1530 | 2324 | 3236 | 4262 |
| Peoria | 6098 | 4930 | 3910 | 3013 | 2239 | 968 | 1631 | 2431 | 3360 | 4412 |
| Rockford | 6845 | 5600 | 4507 | 3555 | 2713 | 714 | 1298 | 2032 | 2899 | 3883 |
| Springfield | 5558 | 4437 | 3468 | 2615 | 1913 | 1116 | 1821 | 2670 | 3654 | 4772 |
| **INDIANA** | | | | | | | | | | |
| Evansville | 4624 | 3578 | 2685 | 1929 | 1327 | 1364 | 2139 | 3064 | 4140 | 5367 |
| Fort Wayne | 6209 | 4992 | 3930 | 2996 | 2193 | 748 | 1358 | 2117 | 3008 | 4030 |
| Indianapolis | 5577 | 4430 | 3431 | 2568 | 1856 | 974 | 1653 | 2478 | 3441 | 4554 |
| South Bend | 6462 | 5213 | 4118 | 3156 | 2333 | 695 | 1271 | 2002 | 2865 | 3867 |
| **IOWA** | | | | | | | | | | |
| Burlington Radio KBUR | 6149 | 4988 | 3970 | 3077 | 2308 | 994 | 1657 | 2466 | 3396 | 4447 |
| Des Moines | 6710 | 5521 | 4470 | 3546 | 2728 | 928 | 1561 | 2335 | 3235 | 4243 |
| Dubuque | 7277 | 5992 | 4871 | 3893 | 3028 | 606 | 1146 | 1850 | 2697 | 3657 |
| Mason City AP | 7901 | 6586 | 5430 | 4425 | 3529 | 580 | 1088 | 1763 | 2581 | 3508 |
| Sioux City | 6953 | 5745 | 4674 | 3738 | 2898 | 932 | 1545 | 2298 | 3182 | 4177 |
| Spencer | 7770 | 6474 | 5329 | 4334 | 3448 | 641 | 1171 | 1857 | 2685 | 3620 |
| Waterloo | 7415 | 6153 | 5040 | 4070 | 3199 | 675 | 1236 | 1950 | 2806 | 3760 |
| **KANSAS** | | | | | | | | | | |
| Concordia | 5623 | 4498 | 3509 | 2646 | 1912 | 1302 | 1998 | 2832 | 3795 | 4886 |
| Dodge City | 5046 | 3963 | 3011 | 2184 | 1512 | 1411 | 2153 | 3022 | 4025 | 5176 |
| Goodland | 6119 | 4891 | 3804 | 2857 | 2041 | 925 | 1515 | 2253 | 3131 | 4139 |
| Russell FAA AP | 5312 | 4220 | 3259 | 2423 | 1735 | 1485 | 2219 | 3081 | 4076 | 5210 |
| Topeka | 5243 | 4152 | 3203 | 2378 | 1700 | 1361 | 2093 | 2974 | 3974 | 5112 |
| Wichita | 4687 | 3654 | 2750 | 1977 | 1346 | 1673 | 2464 | 3386 | 4438 | 5638 |
| **KENTUCKY** | | | | | | | | | | |
| Covington | 5070 | 3965 | 3001 | 2189 | 1527 | 1080 | 1798 | 2654 | 3672 | 4834 |
| Lexington | 4729 | 3652 | 2743 | 1963 | 1350 | 1197 | 1941 | 2858 | 3904 | 5120 |
| Louisville | 4640 | 3584 | 2676 | 1906 | 1303 | 1268 | 2032 | 2942 | 4005 | 5227 |
| **LOUISIANA** | | | | | | | | | | |
| Alexandria | 2200 | 1443 | 880 | 490 | 234 | 2193 | 3260 | 4525 | 5958 | 7531 |
| Baton Rouge | 1670 | 1036 | 582 | 295 | 120 | 2585 | 3775 | 5150 | 6685 | 8340 |
| Lake Charles | 1498 | 908 | 500 | 240 | 91 | 2739 | 3978 | 5391 | 6956 | 8638 |
| New Orleans, Aud. Park | 1343 | 805 | 439 | 202 | 82 | 2876 | 4165 | 5622 | 7215 | 8916 |
| New Orleans, N.O. | 1465 | 893 | 492 | 239 | 96 | 2706 | 3960 | 5383 | 6956 | 8638 |
| Shreveport | 2167 | 1438 | 883 | 490 | 233 | 2538 | 3634 | 4906 | 6335 | 7903 |
| **MAINE** | | | | | | | | | | |
| Bangor | 7950 | 6496 | 5222 | 4103 | 3122 | 268 | 640 | 1194 | 1896 | 2740 |
| Caribou | 9632 | 8044 | 6634 | 5409 | 4319 | 128 | 365 | 784 | 1379 | 2118 |
| Old Town FAA AP | 8648 | 7133 | 5800 | 4628 | 3589 | 209 | 519 | 1016 | 1660 | 2454 |
| Portland | 7498 | 6035 | 4764 | 3658 | 2705 | 252 | 616 | 1169 | 1890 | 2758 |
| **MARYLAND** | | | | | | | | | | |
| Baltimore | 4729 | 3631 | 2682 | 1873 | 1236 | 1108 | 1840 | 2708 | 3728 | 4918 |
| **MASSACHUSETTS** | | | | | | | | | | |
| Blue Hill | 6335 | 5020 | 3885 | 2895 | 2071 | 457 | 968 | 1659 | 2493 | 3498 |
| Boston | 5621 | 4383 | 3313 | 2405 | 1659 | 661 | 1250 | 2000 | 2920 | 4000 |

Table 8-2  Degree-Days to Several Bases (in °F) for Various US Locations (*Continued*)

| State | Heating Base 65 | 60 | 55 | 50 | 45 | Cooling Base 65 | 60 | 55 | 50 | 45 |
|---|---|---|---|---|---|---|---|---|---|---|
| Nantucket AP | 5929 | 4520 | 3323 | 2311 | 1513 | 284 | 708 | 1332 | 2143 | 3170 |
| Worcester | 6848 | 5498 | 4326 | 3296 | 2421 | 387 | 863 | 1514 | 2303 | 3259 |
| **MICHIGAN** | | | | | | | | | | |
| Alpena | 8518 | 6982 | 5635 | 4473 | 3464 | 208 | 497 | 981 | 1642 | 2459 |
| Detroit | 6419 | 5167 | 4072 | 3113 | 2280 | 654 | 1227 | 1961 | 2823 | 3815 |
| Flint | 7041 | 5705 | 4540 | 3529 | 2640 | 438 | 923 | 1586 | 2399 | 3335 |
| Grand Rapids | 6801 | 5514 | 4383 | 3396 | 2524 | 575 | 1108 | 1807 | 2646 | 3598 |
| Houghton Lake | 8347 | 6861 | 5579 | 4455 | 3486 | 250 | 590 | 1132 | 1832 | 2689 |
| Lansing | 6904 | 5608 | 4464 | 3470 | 2595 | 535 | 1059 | 1747 | 2578 | 3528 |
| Marquette | 8351 | 6835 | 5517 | 4379 | 3378 | 216 | 531 | 1031 | 1725 | 2549 |
| Muskegon | 6890 | 5550 | 4390 | 3373 | 2482 | 469 | 953 | 1620 | 2428 | 3360 |
| Sault Ste Marie | 9193 | 7614 | 6215 | 5017 | 3971 | 139 | 386 | 816 | 1443 | 2217 |
| Traverse City AP | 7698 | 6272 | 5035 | 3953 | 3013 | 376 | 773 | 1362 | 2101 | 2989 |
| **MINNESOTA** | | | | | | | | | | |
| Duluth | 9756 | 8185 | 6793 | 5581 | 4540 | 176 | 425 | 864 | 1482 | 2259 |
| Intnl Falls | 10547 | 8995 | 7623 | 6413 | 5348 | 176 | 454 | 908 | 1523 | 2283 |
| Minneapolis–St. Paul | 8159 | 6842 | 5677 | 4668 | 3765 | 585 | 1097 | 1758 | 2575 | 3491 |
| Rochester | 8227 | 6868 | 5682 | 4643 | 3733 | 474 | 943 | 1579 | 2370 | 3280 |
| St. Cloud | 8868 | 7481 | 6255 | 5187 | 4241 | 426 | 862 | 1468 | 2220 | 3098 |
| **MISSISSIPPI** | | | | | | | | | | |
| Jackson | 2300 | 1548 | 988 | 590 | 319 | 2316 | 3394 | 4664 | 6086 | 7639 |
| Meridian | 2388 | 1621 | 1042 | 623 | 339 | 2231 | 3289 | 4538 | 5940 | 7483 |
| **MISSOURI** | | | | | | | | | | |
| Columbia Region | 5078 | 3997 | 3064 | 2259 | 1605 | 1269 | 2009 | 2901 | 3919 | 5089 |
| Kansas City | 5161 | 4089 | 3157 | 2351 | 1694 | 1421 | 2169 | 3061 | 4085 | 5249 |
| St Louis | 4750 | 3701 | 2798 | 2031 | 1419 | 1475 | 2252 | 3174 | 4232 | 5445 |
| St Joseph | 5435 | 4341 | 3378 | 2544 | 1847 | 1334 | 2064 | 2925 | 3911 | 5046 |
| Springfield | 4570 | 3517 | 2611 | 1844 | 1235 | 1382 | 2149 | 3068 | 4126 | 5342 |
| **MONTANA** | | | | | | | | | | |
| Billings | 7265 | 5898 | 4697 | 3641 | 2766 | 498 | 951 | 1581 | 2354 | 3298 |
| Butte | 9719 | 8059 | 6557 | 5225 | 4078 | 58 | 222 | 545 | 1038 | 1718 |
| Cut Bank AP | 9033 | 7474 | 6096 | 4907 | 3886 | 140 | 406 | 856 | 1489 | 2299 |
| Dillon AP | 8354 | 6821 | 5457 | 4255 | 3237 | 199 | 492 | 953 | 1570 | 2382 |
| Glasgow | 8969 | 7572 | 6329 | 5238 | 4302 | 438 | 867 | 1449 | 2185 | 3074 |
| Great Falls | 7652 | 6248 | 5022 | 3965 | 3074 | 339 | 760 | 1365 | 2132 | 3066 |
| Havre | 8687 | 7282 | 6073 | 5005 | 4104 | 395 | 818 | 1432 | 2191 | 3113 |
| Helena | 8190 | 6710 | 5389 | 4247 | 3258 | 256 | 606 | 1105 | 1786 | 2629 |
| Kalispell | 8554 | 6959 | 5542 | 4304 | 3233 | 117 | 348 | 755 | 1342 | 2096 |
| Lewistown FAA AP | 8586 | 7038 | 5676 | 4487 | 3467 | 192 | 468 | 933 | 1567 | 2379 |
| Miles City AP | 7889 | 6562 | 5392 | 4369 | 3479 | 752 | 1252 | 1905 | 2706 | 3641 |
| Missoula | 7931 | 6410 | 5066 | 3884 | 2876 | 188 | 497 | 970 | 1616 | 2428 |
| **NEBRASKA** | | | | | | | | | | |
| Grand Island | 6420 | 5224 | 4166 | 3239 | 2434 | 1036 | 1662 | 2428 | 3326 | 4345 |
| Lincoln AP | 6218 | 5062 | 4040 | 3139 | 2362 | 1148 | 1809 | 2611 | 3536 | 4585 |
| Lincoln | 6012 | 4875 | 3870 | 2993 | 2234 | 1187 | 1865 | 2685 | 3634 | 4701 |
| Norfolk | 6981 | 5745 | 4663 | 3710 | 2863 | 925 | 1520 | 2263 | 3131 | 4118 |
| North Platte | 6743 | 5470 | 4345 | 3354 | 2509 | 802 | 1359 | 2060 | 2898 | 3874 |
| Omaha–Eppley | 6049 | 4907 | 3911 | 3037 | 2290 | 1173 | 1862 | 2691 | 3637 | 4715 |
| Omaha–North | 6601 | 5400 | 4349 | 3427 | 2624 | 949 | 1573 | 2346 | 3249 | 4270 |
| Scottsbluff | 6774 | 5473 | 4304 | 3289 | 2415 | 666 | 1188 | 1845 | 2653 | 3605 |
| Valentine | 7300 | 6006 | 4859 | 3847 | 2956 | 736 | 1267 | 1945 | 2758 | 3692 |
| **NEVADA** | | | | | | | | | | |
| Elko | 7483 | 6027 | 4714 | 3586 | 2625 | 342 | 706 | 1228 | 1910 | 2785 |
| Ely | 7814 | 6327 | 5004 | 3826 | 2829 | 207 | 550 | 1052 | 1694 | 2526 |
| Las Vegas | 2601 | 1770 | 1120 | 625 | 306 | 2946 | 3938 | 5114 | 6443 | 7950 |
| Lovelock FAA | 5990 | 4695 | 3550 | 2579 | 1747 | 684 | 1217 | 1894 | 2743 | 3470 |
| Reno | 6022 | 4612 | 3387 | 2360 | 1534 | 329 | 739 | 1344 | 2150 | 3140 |
| Tonopah | 5900 | 4610 | 3492 | 2532 | 1723 | 631 | 1167 | 1869 | 2739 | 3753 |
| Winnemucca | 6629 | 5241 | 3994 | 2931 | 2015 | 407 | 845 | 1423 | 2185 | 3096 |
| **NEW HAMPSHIRE** | | | | | | | | | | |
| Concord | 7360 | 5967 | 4757 | 3682 | 2762 | 349 | 781 | 1394 | 2150 | 3051 |
| Mount Washington | 13878 | 12053 | 10253 | 8534 | 6960 | 0 | 01 | 25 | 132 | 379 |
| **NEW JERSEY** | | | | | | | | | | |
| Atlantic City | 4946 | 3783 | 2784 | 1941 | 1267 | 864 | 1533 | 2349 | 3339 | 4485 |
| Atlantic City Marina | 4693 | 3534 | 2530 | 1713 | 1076 | 835 | 1503 | 2317 | 3333 | 4517 |
| Newark | 5034 | 3911 | 2920 | 2074 | 1391 | 1024 | 1721 | 2543 | 3533 | 4677 |
| Trenton | 4947 | 3818 | 2832 | 1996 | 1323 | 968 | 1661 | 2493 | 3482 | 4634 |
| **NEW MEXICO** | | | | | | | | | | |
| Albuquerque | 4292 | 3234 | 2330 | 1557 | 963 | 1316 | 2080 | 2996 | 4053 | 5288 |
| Clayton | 5207 | 3999 | 2966 | 2089 | 1374 | 767 | 1380 | 2176 | 3120 | 4231 |
| Roswell | 3697 | 2729 | 1898 | 1226 | 706 | 1560 | 2417 | 3412 | 4566 | 5872 |
| Truth or Consequences | 3392 | 2447 | 1636 | 1007 | 542 | 1558 | 2429 | 3447 | 4647 | 6008 |
| Tucumcari FAA | 4047 | 3015 | 2135 | 1415 | 858 | 1357 | 2148 | 3096 | 4200 | 5467 |
| Zuni FAA | 5515 | 4507 | 3381 | 2437 | 1648 | 473 | 983 | 1685 | 2567 | 3605 |
| **NEW YORK** | | | | | | | | | | |
| Albany | 6888 | 5596 | 4451 | 3457 | 2595 | 574 | 1111 | 1787 | 2619 | 3583 |
| Binghamton | 7285 | 5908 | 4714 | 3677 | 2767 | 369 | 820 | 1452 | 2231 | 3151 |
| Buffalo | 6927 | 5591 | 4429 | 3403 | 2508 | 437 | 928 | 1590 | 2388 | 3319 |
| Massena FAA | 8237 | 6827 | 5596 | 4510 | 3552 | 343 | 759 | 1352 | 2088 | 2958 |

Table 8-2  Degree-Days to Several Bases (in °F) for Various US Locations (*Continued*)

| State | Heating Base 65 | 60 | 55 | 50 | 45 | Cooling Base 65 | 60 | 55 | 50 | 45 |
|---|---|---|---|---|---|---|---|---|---|---|
| New York Central Park | 4848 | 3739 | 2771 | 1958 | 1299 | 1068 | 1784 | 2636 | 3653 | 4814 |
| New York JFK Int'l. AP | 5184 | 4023 | 2994 | 2130 | 1422 | 861 | 1520 | 2321 | 3278 | 4395 |
| New York LaGuardia | 4909 | 3787 | 2806 | 1980 | 1311 | 1048 | 1752 | 2587 | 3589 | 4740 |
| Oswego East | 6792 | 5444 | 4274 | 3243 | 2376 | 435 | 915 | 1570 | 2360 | 3319 |
| Rochester | 6719 | 5417 | 4285 | 3291 | 2434 | 531 | 1062 | 1750 | 2580 | 3549 |
| Syracuse | 6678 | 5379 | 4250 | 3267 | 2429 | 551 | 1081 | 1778 | 2621 | 3607 |
| **NORTH CAROLINA** | | | | | | | | | | |
| Asheville | 4237 | 3129 | 2224 | 1488 | 937 | 872 | 1587 | 2508 | 3595 | 4868 |
| Cape Hatteras | 2731 | 1846 | 1166 | 702 | 380 | 1550 | 2485 | 3635 | 4991 | 6500 |
| Charlotte | 3218 | 2300 | 1552 | 984 | 585 | 1596 | 2503 | 3579 | 4842 | 6263 |
| Greensboro | 3825 | 2811 | 1984 | 1324 | 825 | 1341 | 2158 | 3149 | 4318 | 5640 |
| Raleigh–Durham | 3514 | 2542 | 1744 | 1123 | 670 | 1394 | 2242 | 3273 | 4482 | 5850 |
| Wilmington | 2433 | 1632 | 1028 | 610 | 321 | 1964 | 2995 | 4225 | 5622 | 7162 |
| **NORTH DAKOTA** | | | | | | | | | | |
| Bismarck | 9044 | 7656 | 6425 | 5326 | 4374 | 487 | 928 | 1518 | 2248 | 3116 |
| Fargo | 9271 | 7891 | 6663 | 5573 | 4615 | 473 | 919 | 1515 | 2251 | 3122 |
| Minot FAA | 9407 | 7964 | 6685 | 5564 | 4573 | 370 | 758 | 1299 | 2002 | 2837 |
| Williston | 9161 | 7753 | 6504 | 5387 | 4450 | 422 | 841 | 1415 | 2128 | 3011 |
| **OHIO** | | | | | | | | | | |
| Akron–Canton | 6224 | 4971 | 3883 | 2936 | 2129 | 634 | 1205 | 1943 | 2820 | 3839 |
| Cincinnati Abbe Obs | 4844 | 3763 | 2830 | 2040 | 1412 | 1188 | 1931 | 2819 | 3864 | 5060 |
| Cincinnati AP | 5070 | 3965 | 3001 | 2189 | 1527 | 1080 | 1798 | 2654 | 3672 | 4834 |
| Cleveland | 6154 | 4901 | 3819 | 2876 | 2079 | 613 | 1183 | 1926 | 2807 | 3836 |
| Columbus | 5702 | 4513 | 3480 | 2597 | 1846 | 809 | 1449 | 2244 | 3183 | 4257 |
| Dayton | 5641 | 4483 | 3468 | 2600 | 1866 | 936 | 1603 | 2414 | 3370 | 4460 |
| Mansfield | 5818 | 4618 | 3573 | 2679 | 1917 | 818 | 1445 | 2225 | 3152 | 4219 |
| Toledo Express | 6381 | 5136 | 4049 | 3091 | 2274 | 685 | 1268 | 2001 | 2870 | 3877 |
| Youngstown | 6426 | 5145 | 4032 | 3054 | 2232 | 518 | 1065 | 1774 | 2621 | 3623 |
| **OKLAHOMA** | | | | | | | | | | |
| Oklahoma City | 3695 | 2760 | 1962 | 1326 | 809 | 1876 | 2768 | 3788 | 4980 | 6289 |
| Tulsa | 3680 | 2750 | 1950 | 1306 | 778 | 1949 | 2850 | 3865 | 5052 | 6347 |
| **OREGON** | | | | | | | | | | |
| Astoria | 5295 | 3620 | 2233 | 1215 | 570 | 13 | 159 | 5961 | 1415 | 2598 |
| Burns | 7212 | 5740 | 4436 | 3299 | 2343 | 289 | 649 | 1161 | 1851 | 2724 |
| Eugene | 4739 | 3313 | 2141 | 1226 | 607 | 239 | 638 | 1286 | 2201 | 3417 |
| Meacham | 7863 | 6249 | 4817 | 3556 | 2495 | 103 | 317 | 712 | 1275 | 2034 |
| Medford | 4930 | 3614 | 2496 | 1577 | 882 | 562 | 1077 | 1779 | 2685 | 3813 |
| North Bend AP | 4688 | 2985 | 1642 | 756 | 292 | 0 | 131 | 597 | 1553 | 2913 |
| Pendleton | 5240 | 3968 | 2868 | 1970 | 1264 | 656 | 1211 | 1935 | 2858 | 3982 |
| Portland | 4792 | 3385 | 2234 | 1333 | 708 | 300 | 711 | 1378 | 2309 | 3520 |
| Redmond AP | 6643 | 5106 | 3767 | 2621 | 1680 | 170 | 459 | 943 | 1620 | 2512 |
| Salem | 4852 | 3424 | 2246 | 1317 | 667 | 232 | 620 | 1272 | 2169 | 3355 |
| Sexton Summit | 6430 | 4859 | 3477 | 2311 | 1374 | 137 | 381 | 837 | 1499 | 2386 |
| **PACIFIC** | | | | | | | | | | |
| Guam | 0 | 0 | 0 | 0 | 0 | 5011 | 6836 | 8661 | 10486 | 12311 |
| Johnston AP | 0 | 0 | 0 | 0 | 0 | 5086 | 6911 | 8736 | 10561 | 12386 |
| Koror | 0 | 0 | 0 | 0 | 0 | 6008 | 7833 | 9658 | 11483 | 13308 |
| Kwajalein AP | 0 | 0 | 0 | 0 | 0 | 6164 | 7989 | 9814 | 11639 | 13464 |
| Majuro AP | 0 | 0 | 0 | 0 | 0 | 5904 | 7729 | 9554 | 11379 | 13204 |
| Pago Pago AP | 0 | 0 | 0 | 0 | 0 | 5325 | 7150 | 8975 | 10800 | 12625 |
| Ponape | 0 | 0 | 0 | 0 | 0 | 5652 | 7477 | 9302 | 11127 | 12952 |
| Truk, Moen I, AP | 0 | 0 | 0 | 0 | 0 | 5888 | 7713 | 9536 | 11363 | 13188 |
| Wake | 0 | 0 | 0 | 0 | 0 | 5455 | 7280 | 9105 | 10930 | 12755 |
| Yap AP | 0 | 0 | 0 | 0 | 0 | 5916 | 7741 | 9566 | 11391 | 13216 |
| **PENNSYLVANIA** | | | | | | | | | | |
| Allentown | 5827 | 4618 | 3550 | 2633 | 1843 | 772 | 1392 | 2150 | 3053 | 4088 |
| Bradford AP | 7804 | 6294 | 5006 | 3894 | 2931 | 170 | 482 | 1022 | 1735 | 2596 |
| Erie | 6851 | 5485 | 4304 | 3283 | 2411 | 373 | 832 | 1482 | 2282 | 3235 |
| Harrisburg | 5224 | 4087 | 3097 | 2238 | 1541 | 1025 | 1711 | 2545 | 3511 | 4644 |
| Philadelphia | 4865 | 3753 | 2788 | 1965 | 1312 | 1104 | 1817 | 2671 | 3679 | 4849 |
| Pittsburgh City | 5278 | 4135 | 3138 | 2294 | 1603 | 948 | 1630 | 2456 | 3440 | 4573 |
| Pittsburgh AP | 5930 | 4694 | 3637 | 2720 | 1938 | 647 | 1240 | 2004 | 2914 | 3961 |
| Wilkes-Barre/Scranton | 6277 | 5018 | 3923 | 2972 | 2149 | 608 | 1181 | 1909 | 2778 | 3783 |
| Williamsport | 5981 | 4757 | 3695 | 2764 | 1971 | 698 | 1299 | 2059 | 2952 | 3986 |
| **PUERTO RICO** | | | | | | | | | | |
| San Juan | 0 | 0 | 0 | 0 | 0 | 4982 | 6807 | 8632 | 10457 | 12282 |
| **RHODE ISLAND** | | | | | | | | | | |
| Block Island | 5771 | 4432 | 3289 | 2306 | 1517 | 359 | 844 | 1523 | 2368 | 3409 |
| Providence | 5972 | 4682 | 3565 | 2599 | 1803 | 532 | 1067 | 1774 | 2625 | 3662 |
| **SOUTH CAROLINA** | | | | | | | | | | |
| Charleston AP | 2146 | 1406 | 864 | 496 | 240 | 2078 | 3163 | 4454 | 5903 | 7478 |
| Charleston City | 1904 | 1230 | 741 | 412 | 188 | 2354 | 3502 | 4839 | 6334 | 7937 |
| Columbia | 2598 | 1783 | 1154 | 686 | 374 | 2087 | 3094 | 4292 | 5647 | 7159 |
| Florence | 2566 | 1748 | 1127 | 676 | 374 | 1952 | 2960 | 4171 | 5538 | 7060 |
| Greenville–Spartanburg | 3163 | 2246 | 1493 | 921 | 519 | 1573 | 2477 | 3552 | 4809 | 6229 |
| **SOUTH DAKOTA** | | | | | | | | | | |
| Aberdeen | 8617 | 7267 | 6078 | 5014 | 4087 | 566 | 1046 | 1678 | 2440 | 3337 |
| Huron | 8055 | 6751 | 5600 | 4582 | 3678 | 711 | 1239 | 1912 | 2714 | 3641 |
| Pierre AP | 7677 | 6401 | 5271 | 4273 | 3409 | 858 | 1406 | 2102 | 2928 | 3889 |

# Energy Estimating Methods

Table 8-2  Degree-Days to Several Bases (in °F) for Various US Locations (*Continued*)

| State | Heating Base 65 | 60 | 55 | 50 | 45 | Cooling Base 65 | 60 | 55 | 50 | 45 |
|---|---|---|---|---|---|---|---|---|---|---|
| Rapid City | 7324 | 5982 | 4799 | 3762 | 2868 | 661 | 1148 | 1796 | 2575 | 3511 |
| Sioux Falls | 7838 | 6543 | 5401 | 4394 | 3498 | 719 | 1253 | 1933 | 2746 | 3681 |
| **TENNESSEE** | | | | | | | | | | |
| Bristol | 4306 | 3255 | 2373 | 1646 | 1093 | 1107 | 1880 | 2823 | 3922 | 5197 |
| Chattanooga | 3505 | 2574 | 1785 | 1180 | 737 | 1636 | 2526 | 3566 | 4791 | 6169 |
| Knoxville | 3478 | 2557 | 1775 | 1187 | 744 | 1569 | 2475 | 3518 | 4753 | 6135 |
| Memphis | 3227 | 2352 | 1624 | 1058 | 640 | 2029 | 2984 | 4077 | 5339 | 6744 |
| Nashville | 3696 | 2758 | 1964 | 1338 | 852 | 1694 | 2576 | 3613 | 4812 | 6151 |
| Oak Ridge | 3944 | 2955 | 2119 | 1445 | 933 | 1367 | 2202 | 3187 | 4338 | 5656 |
| **TEXAS** | | | | | | | | | | |
| Abilene | 2610 | 1801 | 1162 | 664 | 342 | 2466 | 3481 | 4670 | 5995 | 7498 |
| Amarillo | 4183 | 3156 | 2278 | 1548 | 976 | 1433 | 2230 | 3177 | 4274 | 5527 |
| Austin | 1737 | 1097 | 620 | 316 | 127 | 2903 | 4095 | 5443 | 6962 | 8600 |
| Brownsville | 650 | 336 | 146 | 54 | 19 | 3874 | 5385 | 7020 | 3753 | 10543 |
| Corpus Christi | 930 | 514 | 243 | 98 | 28 | 3474 | 4880 | 6438 | 8111 | 9872 |
| Dallas | 2290 | 1544 | 949 | 526 | 250 | 2755 | 3835 | 5073 | 6467 | 8016 |
| Del Rio | 1523 | 923 | 494 | 230 | 801 | 3363 | 4596 | 5986 | 7548 | 9222 |
| El Paso | 2678 | 1833 | 1149 | 653 | 326 | 2098 | 3077 | 4229 | 5548 | 7048 |
| Fort Worth | 2382 | 1616 | 1007 | 562 | 274 | 2587 | 3642 | 4862 | 6239 | 7775 |
| Galveston | 1224 | 704 | 369 | 157 | 54 | 3004 | 4312 | 5800 | 7413 | 9139 |
| Houston | 1434 | 864 | 471 | 215 | 81 | 2889 | 4147 | 5576 | 7150 | 8835 |
| Laredo No. 2 | 876 | 481 | 230 | 87 | 32 | 4137 | 5568 | 7143 | 8824 | 10593 |
| Lubbock | 3545 | 2603 | 1807 | 1163 | 666 | 1647 | 2535 | 3559 | 4745 | 6068 |
| Lufkin AP | 1940 | 1253 | 731 | 385 | 163 | 2592 | 3730 | 5033 | 6512 | 8114 |
| Midland | 2621 | 1808 | 1159 | 656 | 333 | 2245 | 3258 | 4434 | 5757 | 7258 |
| Port Arthur | 1518 | 924 | 504 | 238 | 86 | 2798 | 4028 | 5431 | 6990 | 8669 |
| San Angelo | 2240 | 1498 | 918 | 493 | 227 | 2702 | 3789 | 5031 | 6432 | 7993 |
| San Antonio | 1570 | 956 | 518 | 242 | 92 | 2994 | 4206 | 5594 | 7146 | 8818 |
| Victoria | 1227 | 702 | 364 | 150 | 51 | 3140 | 4440 | 5925 | 7537 | 9262 |
| Waco | 2058 | 1357 | 807 | 437 | 195 | 2863 | 3988 | 5271 | 6717 | 8303 |
| Wichita Falls | 2904 | 2061 | 1384 | 832 | 451 | 2611 | 3594 | 4741 | 6015 | 7458 |
| **UTAH** | | | | | | | | | | |
| Blanding | 6163 | 4869 | 3732 | 2757 | 1912 | 600 | 1129 | 1827 | 2670 | 3646 |
| Bryce Canyon AP | 9133 | 7459 | 5949 | 4616 | 3480 | 41 | 193 | 505 | 1005 | 1686 |
| Cedar City AP | 6137 | 4833 | 3690 | 2717 | 1897 | 615 | 1130 | 1813 | 2671 | 3678 |
| Milford | 6412 | 5109 | 3957 | 2969 | 2121 | 688 | 1212 | 1885 | 2721 | 3704 |
| Salt Lake City | 5983 | 4733 | 3633 | 2676 | 1864 | 927 | 1502 | 2221 | 3094 | 4108 |
| Wendover | 5760 | 4558 | 3511 | 2621 | 1870 | 1137 | 1760 | 2538 | 3475 | 4547 |
| **VERMONT** | | | | | | | | | | |
| Burlington | 7876 | 6488 | 5270 | 4190 | 3246 | 396 | 833 | 1440 | 2180 | 3066 |
| **VIRGINIA** | | | | | | | | | | |
| Lynchburg | 4233 | 3172 | 2269 | 1536 | 966 | 1100 | 1861 | 2783 | 3873 | 5128 |
| Norfolk | 3488 | 2516 | 1710 | 1100 | 663 | 1441 | 2284 | 3315 | 4530 | 5918 |
| Richmond | 3939 | 2916 | 2061 | 1388 | 866 | 1353 | 2157 | 3127 | 4276 | 5580 |
| Roanoke | 4307 | 3234 | 2326 | 1580 | 1011 | 1030 | 1778 | 2690 | 3771 | 5029 |
| Wallops Island | 4240 | 3170 | 2268 | 1531 | 978 | 1107 | 1865 | 2788 | 3881 | 5149 |
| **WASHINGTON** | | | | | | | | | | |
| Olympia | 5530 | 3970 | 2653 | 1617 | 854 | 101 | 365 | 880 | 1657 | 2731 |
| Omak | 6858 | 5476 | 4253 | 3230 | 2355 | 522 | 965 | 1573 | 2367 | 3324 |
| Quillayute | 5951 | 4232 | 2750 | 1603 | 813 | 8 | 116 | 458 | 1137 | 2172 |
| Seattle–Tacoma | 5185 | 3657 | 2386 | 1416 | 731 | 129 | 423 | 984 | 1832 | 2981 |
| Seattle (Urban) | 4727 | 3269 | 2091 | 1194 | 602 | 183 | 549 | 1197 | 2127 | 3358 |
| Spokane | 6835 | 5420 | 4173 | 3088 | 2188 | 388 | 797 | 1377 | 2120 | 3040 |
| Stampede Pass | 9400 | 7643 | 6006 | 4532 | 3256 | 16 | 83 | 274 | 623 | 1176 |
| Walla Walla | 4835 | 3616 | 2600 | 1760 | 1126 | 862 | 1471 | 2279 | 3260 | 4457 |
| Yakima | 6009 | 4655 | 3483 | 2502 | 1688 | 479 | 945 | 1604 | 2452 | 3455 |
| **WEST INDIES** | | | | | | | | | | |
| Swan Island | 0 | 0 | 0 | 0 | 0 | 5809 | 7634 | 9459 | 11284 | 13109 |
| **WEST VIRGINIA** | | | | | | | | | | |
| Beckley | 5615 | 4356 | 3279 | 2390 | 1652 | 490 | 1061 | 1809 | 2745 | 3833 |
| Charleston | 4590 | 3500 | 2590 | 1809 | 1216 | 1055 | 1790 | 2699 | 3750 | 4981 |
| Elkins | 5975 | 4659 | 3533 | 2616 | 1834 | 389 | 905 | 1601 | 2508 | 3555 |
| Huntington | 4624 | 3533 | 2624 | 1843 | 1249 | 1098 | 1829 | 2746 | 3790 | 5020 |
| Parkersburg | 4817 | 3720 | 2786 | 1987 | 1363 | 1045 | 1770 | 2657 | 3686 | 4888 |
| **WISCONSIN** | | | | | | | | | | |
| Eau Claire AP | 8388 | 7033 | 5832 | 4786 | 3860 | 459 | 928 | 1554 | 2331 | 3231 |
| Green Bay | 8098 | 6689 | 5473 | 4405 | 3478 | 386 | 805 | 1411 | 2168 | 3066 |
| La Crosse AP | 7417 | 6158 | 5050 | 4088 | 3219 | 695 | 1264 | 1978 | 2841 | 3798 |
| Madison | 7730 | 6373 | 5188 | 4156 | 3250 | 460 | 923 | 1572 | 2361 | 3279 |
| Milwaukee | 7444 | 6080 | 4898 | 3860 | 2946 | 450 | 911 | 1554 | 2342 | 3252 |
| **WYOMING** | | | | | | | | | | |
| Casper | 7555 | 6167 | 4914 | 3813 | 2857 | 458 | 895 | 1468 | 2193 | 3061 |
| Cheyenne | 7255 | 5825 | 4562 | 3452 | 2512 | 327 | 734 | 1288 | 2003 | 2886 |
| Lander | 7869 | 6471 | 5207 | 4080 | 3140 | 383 | 814 | 1376 | 2078 | 2965 |
| Rock Springs AP | 8410 | 6922 | 5592 | 4412 | 3393 | 227 | 563 | 1059 | 1703 | 2515 |
| Sheridan | 7708 | 6298 | 5024 | 3935 | 3000 | 446 | 860 | 1411 | 2147 | 3037 |

From Table 8-2, the degree-days for Chicago, Midway, to this base can be obtained by interpolation to be 5775. For summer, the CDD to base 63.5 are found from Table 8-2 to be 1120.

(a) From Eq. 8-6, the heating energy requirements are

$$E_H = \frac{240{,}000 \times 5775 \times 24}{0.95 \times 3413 \times 75} = 129{,}900 \text{ kWh}$$

(assuming 5% loss from baseboard units through wall)

Heating cost = 129,900 × $0.072 = $9350

From Eq. 8-7, the cooling energy is determined:

$$E_C = \frac{85{,}000 \times 1120 \times 24}{1000 \times 14 \times (91 - 78)} = 12{,}550 \text{ kWh}$$

Cooling cost = 12,550 × $0.072 = $904

(b) From Eq. 8-6, the heating energy requirements are

$$E_H = \frac{240{,}000 \times 5775 \times 24}{0.93 \times 100{,}000 \times 75} = 4770 \text{ therm}$$

(with 1 therm = 100,000 Btu and assuming 93% AFUE from Chapter 19)

Heating cost = 4770 × $0.062 = $2960

## 8.6 Bin Method (Heating and Cooling)

For many applications, the degree-day method should not be used, even with the variable-base method, because the heat loss coefficient $K$, the efficiency $\eta$ of the HVAC system, or the balance-point temperature may not be sufficiently constant. The efficiency of a heat pump, for example, varies strongly with outdoor temperature, or the efficiency of the HVAC equipment may be affected indirectly by $t_o$ when the efficiency varies with the load, which is a common situation for boilers and chillers. Furthermore, in most commercial buildings, the occupancy has a pronounced pattern, which affects heat gain, indoor temperature, and ventilation rate.

In such cases, a steady-state calculation can yield good results for the annual energy consumption if different temperature intervals and time periods are evaluated separately. This approach is known as the *bin method*, because the energy consumption $E_{bin}$ is calculated for several values of the outdoor temperature $t_o$ and multiplied by the number of hours $N_{bin}$ in the temperature interval (bin) centered around that temperature:

$$E_{bin} = N_{bin} K_{tot}(t_{bal} - t_o)/\eta \qquad (8\text{-}8)$$

This equation is evaluated for each bin, and the total energy consumption is the sum of $E_{bin}$ over all bins:

$$E = \Sigma E_{bin} \qquad (8\text{-}9)$$

In the United States, the necessary data are widely available. The bins are usually in 5°F increments and are often collected in three daily 8-hour shifts. Mean coincident wet-bulb temperature data (for each dry-bulb bin) are used to calculate latent cooling loads from infiltration and ventilation. The bin method considers both occupied and unoccupied building conditions and gives credit for internal loads by adjusting the balance point. For many applications, the number of calculations can be minimized. A residential heat pump (heating mode) could be evaluated using only the bins below the balance point without the three-shift breakdown.

The data included in Table 8-3 are annual totals for various cities in the United States. ASHRAE (1995) and USAF (1978) include monthly data and data further separated into time intervals throughout the day.

**Modified Bin Method.** Various refinements, such as the seasonal variation of solar gains, can be included in a bin calculation. If a separate calculation is done for each month, the heat gain could be based on the average solar heat gain of the month. The diurnal variation of solar gains can be accounted for by calculating the average solar gain for each of the hourly time periods of the bin method. If such a detailed calculation of solar gains is not necessary, a linear correlation of monthly average solar heat gains with monthly average outdoor temperature could be assumed. Using bin data for the corresponding periods, the calculation can also account for the operating schedules of commercial buildings.

The modified bin method has the advantage of allowing the use of diversified (part-load) rather than peak-load values to establish the load as a function of outdoor dry-bulb temperature. This method also allows both secondary and primary (plant) HVAC equipment effects to be included in the energy calculation. The modified bin method permits the user to predict more accurately effects such as reheat and heat recovery that can only be guessed at with the degree-day or conventional bin methods.

In the modified bin method, average solar gain profiles, average equipment and lighting use profiles, and cooling load temperature difference (CLTD) values are used to characterize time-dependent diversified loads. The CLTDs approximate the transient effects of building mass. Time dependencies resulting from scheduling are averaged over a selected period, or multiple calculation periods are established. The duration of a calculation period determines the number of bin hours included. Normally, two calculation periods, representing occupied and unoccupied hours, are sufficient. The method can be further refined by making calculations on a monthly, not annual, basis.

**Degree-Day Data from Bin Data.** To calculate degree-days from hourly bin data, the base or balance temperature must first be determined. When $t_{bal}$ is known, the following summation can be used for any time scale, either monthly or annually, or for several periods of a day on either a monthly or annual basis.

$$\text{DD}_h(t_{bal}) = [\Sigma(t_{bal} - t_{bin})N_{bin}]/24$$

where

$t_{bin}$ = temperature at center of bin

$N_{bin}$ = number of hours in bin at $t_{bin}$

Cooling degree-days are calculated analogously from

# Energy Estimating Methods

**Table 8-3  Hourly Weather Occurrences**

| Location | 72 | 67 | 62 | 57 | 52 | 47 | 42 | 37 | 32 | 27 | 22 | 17 | 12 | 7 | 2 | −3 | −8 | −13 | −18 |
|---|---|---|---|---|---|---|---|---|---|---|---|---|---|---|---|---|---|---|---|
| Albany, NY | 588 | 733 | 740 | 708 | 652 | 625 | 647 | 769 | 793 | 574 | 404 | 278 | 184 | 110 | 63 | 32 | 10 | 5 | 4 |
| Albuquerque, NM | 767 | 831 | 719 | 651 | 687 | 734 | 741 | 689 | 552 | 346 | 154 | 66 | 21 | 4 | 1 | 1 | | | |
| Atlanta, GA | 1185 | 926 | 823 | 784 | 735 | 676 | 598 | 468 | 271 | 112 | 44 | 19 | 8 | 2 | | | | | |
| Bakersfield, CA | 831 | 898 | 966 | 977 | 908 | 746 | 541 | 247 | 77 | 7 | | | | | | | | | |
| Birmingham, AL | 1138 | 908 | 805 | 742 | 668 | 614 | 528 | 433 | 292 | 143 | 69 | 17 | 6 | 3 | | | | | |
| Bismarck, ND | 454 | 566 | 614 | 606 | 563 | 520 | 518 | 604 | 653 | 550 | 474 | 371 | 338 | 292 | 278 | 208 | 131 | 77 | 80 |
| Boise, ID | 492 | 575 | 643 | 702 | 786 | 798 | 878 | 829 | 522 | 307 | 148 | 53 | 26 | 14 | 6 | 2 | | | |
| Boston, MA | 676 | 819 | 804 | 781 | 766 | 757 | 828 | 848 | 674 | 429 | 256 | 151 | 74 | 35 | 4 | 9 | 1 | | |
| Buffalo, NY | 646 | 772 | 760 | 700 | 666 | 624 | 647 | 756 | 849 | 602 | 426 | 267 | 170 | 81 | 5 | 24 | 2 | | |
| Burlington VT | 573 | 670 | 703 | 694 | 655 | 603 | 637 | 716 | 752 | 561 | 491 | 336 | 272 | 216 | 135 | 81 | 39 | 17 | 8 |
| Casper, WY | 423 | 532 | 592 | 642 | 606 | 670 | 782 | 831 | 806 | 683 | 495 | 325 | 200 | 116 | 73 | 45 | 30 | 15 | 5 |
| Charleston, SC | 1267 | 1090 | 889 | 787 | 651 | 576 | 434 | 321 | 192 | 79 | 27 | 5 | | | | | | | |
| Charleston, WV | 912 | 949 | 767 | 689 | 661 | 667 | 607 | 633 | 630 | 356 | 252 | 135 | 73 | 22 | 7 | 1 | | | |
| Charlotte, NC | 1115 | 908 | 839 | 752 | 730 | 684 | 634 | 515 | 360 | 166 | 64 | 23 | 5 | 2 | | | | | |
| Chattanooga, TN | 1021 | 895 | 775 | 722 | 713 | 679 | 642 | 553 | 414 | 228 | 113 | 45 | 4 | 4 | 2 | | | | |
| Chicago, IL | 762 | 769 | 653 | 592 | 569 | 543 | 591 | 800 | 822 | 551 | 335 | 196 | 117 | 85 | 59 | 25 | 12 | 3 | |
| Cincinnati, OH | 879 | 843 | 726 | 639 | 611 | 599 | 627 | 698 | 711 | 460 | 249 | 131 | 68 | 44 | 18 | 8 | 2 | | |
| Cleveland, OH | 763 | 831 | 723 | 641 | 638 | 607 | 620 | 754 | 806 | 578 | 355 | 201 | 111 | 47 | 22 | 11 | 2 | | |
| Columbus, OH | 774 | 820 | 720 | 648 | 622 | 603 | 658 | 772 | 730 | 502 | 280 | 169 | 94 | 40 | 20 | 10 | 4 | 1 | |
| Corpus Christi, TX | 1175 | 1041 | 748 | 551 | 444 | 302 | 180 | 83 | 27 | 9 | 3 | | | | | | | | |
| Dallas, TX | 831 | 795 | 693 | 656 | 629 | 576 | 504 | 371 | 231 | 91 | 34 | 17 | 4 | 1 | | | | | |
| Denver, CO | 549 | 684 | 783 | 731 | 678 | 704 | 692 | 717 | 721 | 553 | 359 | 216 | 119 | 78 | 36 | 22 | 6 | 1 | 1 |
| Des Moines, IA | 707 | 751 | 681 | 600 | 585 | 512 | 510 | 627 | 747 | 557 | 405 | 281 | 211 | 152 | 104 | 59 | 23 | 8 | 1 |
| Detroit, MI | 721 | 783 | 695 | 633 | 592 | 566 | 595 | 808 | 884 | 618 | 377 | 248 | 131 | 61 | 17 | 4 | 1 | | |
| El Paso, TX | 933 | 839 | 749 | 760 | 687 | 611 | 494 | 369 | 233 | 34 | 104 | 10 | 2 | | | | | | |
| Ft. Wayne, IN | 728 | 777 | 699 | 608 | 569 | 552 | 601 | 725 | 905 | 596 | 381 | 205 | 124 | 69 | 40 | 19 | 6 | 1 | |
| Fresno, CA | 709 | 803 | 921 | 1006 | 1036 | 952 | 673 | 426 | 168 | 34 | | | | | | | | | |
| Grand Rapids, MI | 634 | 739 | 712 | 647 | 571 | 565 | 554 | 742 | 938 | 690 | 469 | 293 | 172 | 78 | 31 | 10 | 1 | 1 | |
| Great Falls, MT | 407 | 520 | 636 | 754 | 822 | 830 | 832 | 813 | 698 | 533 | 355 | 218 | 167 | 136 | 118 | 101 | 68 | 51 | 62 |
| Harrisburg, PA | 807 | 824 | 737 | 692 | 635 | 659 | 722 | 888 | 749 | 427 | 222 | 125 | 52 | 18 | 4 | 1 | | | |
| Hartford, CT | 617 | 755 | 751 | 752 | 649 | 575 | 683 | 807 | 825 | 552 | 370 | 233 | 153 | 77 | 33 | 11 | 3 | 2 | |
| Houston, TX | 1172 | 980 | 772 | 681 | 570 | 452 | 291 | 141 | 64 | 18 | 4 | 2 | | | | | | | |
| Indianapolis, IN | 821 | 815 | 722 | 585 | 586 | 579 | 605 | 712 | 791 | 551 | 293 | 152 | 97 | 60 | 35 | 13 | 3 | 2 | |
| Jackson, MS | 1168 | 922 | 790 | 677 | 618 | 605 | 484 | 367 | 224 | 103 | 41 | 6 | 2 | 2 | 1 | | | | |
| Jacksonville, FL | 1334 | 975 | 879 | 692 | 530 | 355 | 288 | 154 | 83 | 24 | 2 | | | | | | | | |
| Kansas City, MO | 761 | 723 | 601 | 572 | 553 | 562 | 628 | 591 | 625 | 407 | 265 | 175 | 99 | 51 | 21 | 4 | | | |
| Knoxville, TN | 1056 | 889 | 746 | 675 | 672 | 689 | 648 | 590 | 456 | 217 | 101 | 41 | 21 | 7 | 2 | | | | |
| Las Vegas, NV | 651 | 644 | 699 | 786 | 769 | 716 | 591 | 396 | 194 | 44 | 7 | 1 | | | | | | | |
| Little Rock, AR | 940 | 803 | 725 | 672 | 638 | 669 | 605 | 509 | 363 | 172 | 50 | 23 | 5 | 1 | | | | | |
| Los Angeles, CA | 881 | 1654 | 2193 | 1904 | 1054 | 428 | 107 | 10 | | | | | | | | | | | |
| Louisville, KY | 869 | 758 | 693 | 654 | 619 | 634 | 649 | 703 | 631 | 332 | 169 | 97 | 45 | 25 | 8 | 3 | 1 | | |
| Lubbock, TX | 833 | 829 | 688 | 700 | 642 | 618 | 620 | 546 | 490 | 346 | 180 | 86 | 33 | 7 | 5 | 1 | | | |
| Memphis, TN | 977 | 798 | 715 | 690 | 618 | 633 | 614 | 532 | 374 | 196 | 74 | 25 | 10 | 4 | | | | | |
| Miami, FL | 1705 | 810 | 452 | 277 | 147 | 71 | 26 | 4 | | | | | | | | | | | |
| Milwaukee WI | 597 | 753 | 749 | 634 | 585 | 591 | 611 | 774 | 913 | 659 | 421 | 285 | 176 | 116 | 83 | 47 | 18 | 4 | 3 |
| Minneapolis, MN | 621 | 690 | 695 | 602 | 588 | 482 | 500 | 560 | 632 | 609 | 514 | 383 | 311 | 246 | 186 | 119 | 62 | 31 | 16 |
| Mobile, AL | 1411 | 1038 | 882 | 698 | 609 | 506 | 377 | 214 | 109 | 49 | 7 | 3 | | | | | | | |
| Nashville, TN | 933 | 838 | 738 | 697 | 637 | 619 | 627 | 565 | 463 | 263 | 132 | 67 | 28 | 9 | 3 | 1 | 1 | | |
| New Orleans, LA | 1189 | 987 | 850 | 692 | 621 | 449 | 282 | 128 | 47 | 9 | 2 | | | | | | | | |
| New York, NY | 926 | 877 | 754 | 745 | 722 | 796 | 838 | 858 | 603 | 330 | 188 | 82 | 26 | 10 | 1 | | | | |
| Oklahoma City, OK | 881 | 769 | 717 | 173 | 643 | 645 | 611 | 641 | 570 | 468 | 287 | 77 | 36 | 12 | 3 | 1 | | | |
| Omaha, NB | 726 | 721 | 606 | 558 | 539 | 543 | 543 | 655 | 663 | 511 | 390 | 287 | 189 | 135 | 93 | 40 | 15 | 1 | |
| Philadelphia, PA | 863 | 809 | 735 | 710 | 663 | 701 | 758 | 818 | 654 | 335 | 189 | 100 | 32 | 9 | | | | | |
| Phoenix, AZ | 762 | 776 | 767 | 769 | 659 | 540 | 391 | 182 | 57 | 8 | | | | | | | | | |
| Pittsburg, PA | 722 | 910 | 799 | 678 | 637 | 587 | 631 | 688 | 569 | 774 | 360 | 233 | 159 | 60 | 30 | 7 | 1 | | |
| Portland, ME | 407 | 627 | 780 | 808 | 760 | 748 | 722 | 839 | 820 | 599 | 408 | 293 | 190 | 109 | 60 | 29 | 15 | 5 | 1 |
| Portland, OR | 373 | 581 | 1001 | 1316 | 1274 | 1271 | 1238 | 772 | 343 | 123 | 40 | 10 | 4 | 1 | | | | | |
| Raleigh, NC | 1087 | 937 | 848 | 762 | 707 | 672 | 638 | 527 | 410 | 236 | 103 | 38 | 11 | 1 | | | | | |
| Reno, NV | 418 | 477 | 572 | 690 | 845 | 909 | 890 | 829 | 733 | 530 | 387 | 277 | 101 | 37 | 15 | 4 | 1 | | |
| Richmond, VA | 953 | 850 | 784 | 745 | 690 | 673 | 699 | 632 | 478 | 285 | 138 | 67 | 19 | 2 | 1 | | | | |
| Sacramento, CA | 630 | 773 | 1071 | 1329 | 1298 | 1049 | 701 | 355 | 9 | 8 | | | | | | | | | |
| Salt Lake City, UT | 569 | 615 | 614 | 635 | 682 | 685 | 755 | 831 | 798 | 564 | 328 | 158 | 80 | 41 | 16 | 2 | | | |
| San Antonio, TX | 1086 | 943 | 789 | 669 | 569 | 445 | 387 | 190 | 94 | 31 | 11 | 4 | 1 | 1 | | | | | |
| San Francisco, CA | 285 | 665 | 1264 | 2341 | 2341 | 1153 | 449 | 99 | 10 | | | | | | | | | | |
| Seattle, WA | 258 | 448 | 750 | 1272 | 1462 | 1445 | 1408 | 914 | 427 | 104 | 39 | 20 | 3 | | | | | | |
| Shreveport, LA | 1063 | 886 | 772 | 679 | 619 | 609 | 516 | 361 | 200 | 72 | 23 | 6 | 2 | | | | | | |
| Sioux Falls, SD | 566 | 684 | 669 | 605 | 522 | 498 | 501 | 625 | 712 | 585 | 520 | 448 | 293 | 208 | 152 | 102 | 59 | 43 | 18 |
| St. Louis, MO | 823 | 728 | 646 | 575 | 585 | 578 | 620 | 671 | 650 | 411 | 219 | 134 | 77 | 40 | 15 | 7 | 1 | | |
| Syracuse, NY | 627 | 735 | 723 | 717 | 656 | 641 | 651 | 720 | 830 | 547 | 392 | 282 | 190 | 102 | 55 | 23 | 5 | 2 | 2 |
| Tampa, FL | 1387 | 1187 | 877 | 570 | 345 | 216 | 137 | 48 | 10 | 1 | | | | | | | | | |
| Waco, TX | 909 | 830 | 701 | 622 | 651 | 558 | 501 | 354 | 216 | 84 | 24 | 3 | 1 | | | | | | |
| Washington, D.C. | 950 | 766 | 740 | 673 | 690 | 684 | 790 | 744 | 542 | 254 | 138 | 54 | 17 | 2 | | | | | |
| Wichita, KS | 758 | 709 | 641 | 603 | 589 | 592 | 611 | 584 | 607 | 426 | 273 | 161 | 85 | 45 | 14 | 3 | 1 | | |

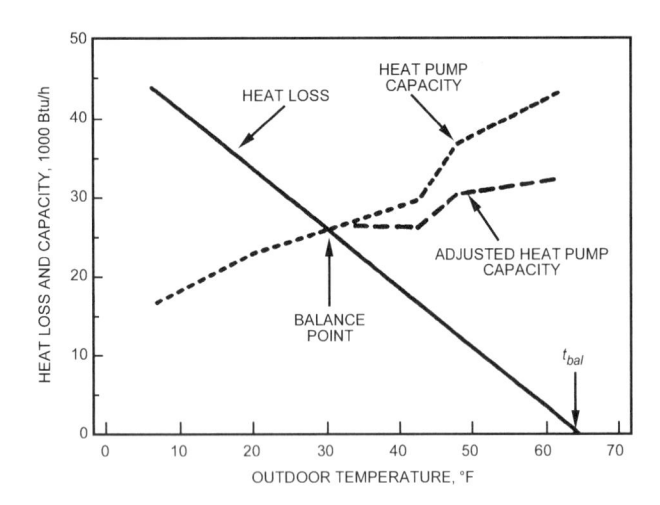

*Fig. 8-5 Heat Pump Capacity and Building Load*
(Figure 15, Chapter 19, *2009 ASHRAE Handbook—Fundamentals*)

$$\mathrm{DD}_c(t_{\mathrm{bal}}) = [\Sigma(t_{\mathrm{bin}} - t_{\mathrm{bal}})N_{\mathrm{bin}}]/24$$

This method generally produces degree-day values slightly higher than published values from NOAA or the National Climatic Data Center. This small but systematic deviation can be suppressed by ignoring degree-days during the swing seasons when totals are less than a minimum, e.g., 50 to 100 Fahrenheit degree-days per month.

Weather data for use with the bin method are available from ASHRAE and in *Engineering Weather Data* (US Air Force Manual 88-29 1978). When time-of-occurrence bin data are not required, the hourly weather occurrence information in Table 8-3 may be used.

The basic data form for making a bin analysis is provided in Table 8-4.

**Example 8-4** Estimate the energy requirements for a residence located in Washington, DC, with a design heat loss of 40,000 Btu/h at 53°F design temperature difference. The inside design temperature is 70°F. Average internal heat gains are estimated to be 4280 Btu/h. Assume a 3 ton heat pump with the characteristics given in Columns E and H of Table 8-5 and in Figure 8-5.

**Solution:** The design heat loss is based on no internal heat generation. The heat pump system energy input is the net heat requirement of the space (i.e., envelope loss minus internal heat generation). The net heat loss per degree and the heating/cooling balance temperature may be computed:

$$HL/\Delta t = 40{,}000/53 = 755 \text{ Btu/°F-h}$$

$$t_{\mathrm{bal}} = 70 - (4280/755) = 64.3\text{°F}$$

Table 8-5 is then computed, resulting in an estimate of total electrical energy consumption for heat pump and supplemental heating of 9578 kWh.

## 8.7 Problems

**8.1** The total design heating load on a residence in New York City is 32.8 kW (112,000 Btu/h) for an indoor temperature of 72°F. The furnace is *off* from June through September. Estimate:

(a) The annual energy requirement for heating
(b) The annual heating cost if electric heat is used with the single rate of 16¢/kWh, $/yr
(c) The maximum savings effected if the thermostat is set back to 65°F between 10 P.M. and 6 A.M., $/yr

**8.2** Determine the cost per 1000 Btu of supplying heat in your territory for (a) oil, (b) gas, (c) direct electric heating, and (d) an air-source heat pump. In calculating gas and oil costs, assume heating plant efficiencies of 80% and 75%, respectively. In calculating heat pump costs, assume a condensing refrigerant temperature of 110°F, an evaporating refrigerant temperature 10°F below the local average winter outdoor air temperature, and an actual COP 70% of that for the reversed Carnot cycle. Assume a compressor mechanical efficiency of 80%, with the compressor located in the airstream being heated.

**8.3** A home is located in Cleveland, Ohio, and has a design heat loss of 112,000 Btu/h at an inside design temperature of 72°F and an outside design temperature of 0°F. The home has an oil-fired furnace. Find the savings in gallons of fuel oil if the owner lowers the temperature in the home to 68°F between 10 P.M. and 6 A.M. every day during January.
[Ans: 13.2 gal (50 L)]

**8.4** The total design heating load on a residence in Kansas City, Missouri, is 32.8 kW (112,000 Btu/h). The furnace is *off* during June through September. Estimate:

(a) Annual energy requirement for heating
(b) Annual heating cost if No. 2 fuel oil is used in a furnace with an efficiency of 80% (assume fuel oil costs 68¢/L)
(c) Maximum savings effected if the thermostat is set back from 22.2 to 18.3°C (72 to 65°F) between 10 PM and 6 AM in $/yr

**8.5** A residence located in Tulsa, Oklahoma, has a design heating load of 20 kW and a design cooling load of 9.4 kW. Determine the following:

(a) Heating energy requirements, kWh
(b) Litres of No. 2 fuel oil per season if used as heating fuel
(c) Litres of natural gas per season if used as heating fuel
(d) kWh of electric energy if used as heating fuel with baseboard units
(e) kWh of electric energy if used for air-conditioning system having $\mathrm{COP}_{\mathrm{seasonal}} = 3.4$
(f) Total airflow rate in L/s if a warm air system is used
(g) Total steam flow in kg/s if a steam system is used

**8.6** Estimate the annual energy costs for heating and cooling a residence located in Cleveland, Ohio, having design loads of 65,000 Btu/h (heating) and 30,000 Btu/h (cooling) based on a 75°F indoor temperature. In winter the thermostat is set back

# Energy Estimating Methods

**Table 8-4  Bin Data Form**

| | Climate | | House | Heat Pump | | | | | | | | Supplemental | |
|---|---|---|---|---|---|---|---|---|---|---|---|---|---|
| A | B | C | D | E | F | G | H | I | J | K | L | M | N |
| Temp. Bin, °F | Temp. Diff. $t_{bal} - t_{bin}$ | Weather Data Bin, hours | Heat Loss Rate, 1000 Btu/h | Heat Pump Integrated Heating Capacity, 1000 Btu/h | Cycling Capacity Adjustment Factor[a] | Adjusted Heat Pump Capacity, 1000 Btu/h[b] | Rated Electric Input, kW | Operating Time Fraction[c] | Heat Pump Supplied Heating, $10^6$ Btu[d] | Seasonal Heat Pump Elec. Consumption, kWh[e] | Space Load, $10^6$ Btu[f] | Supplemental Heating Required, kWh[g] | Total Electric Energy Consumption |
| 62 | | | | | | | | | | | | | |
| 57 | | | | | | | | | | | | | |
| 52 | | | | | | | | | | | | | |
| 47 | | | | | | | | | | | | | |
| 42 | | | | | | | | | | | | | |
| 37 | | | | | | | | | | | | | |
| 32 | | | | | | | | | | | | | |
| 27 | | | | | | | | | | | | | |
| 22 | | | | | | | | | | | | | |
| 17 | | | | | | | | | | | | | |
| 12 | | | | | | | | | | | | | |
| 7 | | | | | | | | | | | | | |
| 2 | | | | | | | | | | | | | |
| –3 | | | | | | | | | | | | | |

**TOTALS:**

a Cycling Capacity Adjustment Factor = $1 - Cd(1 - x)$, where $Cd$ = degradation coefficient (default = 0.25 unless part load factor is known) and $x$ = building heat loss per unit capacity at temperature bin. Cycling capacity = 1 at the balance point and below.

b Col G = Col E × Col F
c Operating Time Factor equals smaller of 1 or Col D/Col G
d Col J = (Col I × Col G × Col C)/1000
e Col K = Col I × Col H × Col C

f Col L = Col C × Col D/1000
g Col M = (Col L – Col J) × $10^6$/3413
h Col N = Col K + Col M

**Table 8-5 Calculation of Annual Heating Energy Consumption for Example 8-4**
*(Table 8, Chapter 19, 2009 ASHRAE Handbook—Fundamentals)*

| Climate | | | House | Heat Pump | | | | | | | Supplemental | | |
|---|---|---|---|---|---|---|---|---|---|---|---|---|---|
| A | B | C | D | E | F | G | H | I | J | K | L | M | N |
| Temp. Bin, °F | Temp. Diff., $t_{bal} - t_{bin}$ | Weather Data Bin, h | Heat Loss Rate, 1000 Btu/h | Heat Pump Integrated Heating Capacity, 1000 Btu/h | Cycling Capacity Adjustment Factor[a] | Adjusted Heat Pump Capacity, 1000 Btu/h[b] | Rated Electric Input, kW | Operating Time Fraction[c] | Heat Pump Supplied Heating, $10^6$ Btu[d] | Seasonal Heat Pump Electric Consumption, KWh[e] | Space Load, $10^6$ Btu[f] | Supplemental Heating Required, kWh[g] | Total Electric Energy Consumption[h] |
| 62 | 2.3 | 740 | 1.8 | 44.3 | 0.760 | 33.7 | 3.77 | 0.05 | 1.30 | 146 | 1.30 | — | 146 |
| 57 | 7.3 | 673 | 5.5 | 41.8 | 0.783 | 32.7 | 3.67 | 0.17 | 3.72 | 417 | 3.72 | — | 417 |
| 52 | 12.3 | 690 | 9.3 | 39.3 | 0.809 | 31.8 | 3.56 | 0.29 | 6.42 | 719 | 6.42 | — | 719 |
| 47 | 17.3 | 684 | 13.1 | 36.8 | 0.839 | 30.9 | 3.46 | 0.42 | 8.95 | 1002 | 8.95 | — | 1002 |
| 42 | 22.3 | 790 | 16.9 | 29.9 | 0.891 | 26.6 | 3.23 | 0.63 | 13.31 | 1614 | 13.31 | — | 1614 |
| 37 | 27.3 | 744 | 20.6 | 28.3 | 0.932 | 26.4 | 3.15 | 0.78 | 15.35 | 1833 | 15.35 | — | 1833 |
| 32 | 32.3 | 542 | 24.4 | 26.6 | 0.979 | 26.0 | 3.07 | 0.94 | 13.22 | 1559 | 13.22 | — | 1559 |
| 27 | 37.3 | 254 | 28.2 | 25.0 | 1.000 | 25.0 | 3.00 | 1.00 | 6.35 | 762 | 7.16 | 236 | 998 |
| 22 | 42.3 | 138 | 31.9 | 23.4 | 1.000 | 23.4 | 2.92 | 1.00 | 3.23 | 403 | 4.41 | 345 | 748 |
| 17 | 47.3 | 54 | 35.7 | 21.8 | 1.000 | 21.8 | 2.84 | 1.00 | 1.18 | 153 | 1.93 | 220 | 373 |
| 12 | 52.3 | 17 | 39.5 | 19.3 | 1.000 | 19.3 | 2.74 | 1.00 | 0.33 | 47 | 0.67 | 101 | 147 |
| 7 | 57.3 | 2 | 43.3 | 16.8 | 1.000 | 16.8 | 2.63 | 1.00 | 0.03 | 5 | 0.09 | 16 | 21 |
| 2 | 62.3 | 0 | 47.0 | 14.3 | 1.000 | — | — | — | — | — | — | — | — |
| | | | | | | | | Totals: | 73.39 | 8660 | 76.52 | 917 | 9578 |

[a] Cycling Capacity Adjustment Factor = $1 - C_d(1 - x)$, where $C_d$ = degradation coefficient (default = 0.25 unless part load factor is known) and $x$ = building heat loss per unit capacity at temperature bin. Cycling capacity = 1 at the balance point and below. The cycling capacity adjustment factor should be 1.0 at all temperature bins if the manufacturer includes cycling effects in the heat pump capacity (Column E) and associated electrical input (Column H).
[b] Column G = Column E × Column F
[c] Operating Time Factor equals smaller of 1 or Column D/Column G
[d] Column J = (Column I × Column G × Column C)/1000
[e] Column K = Column I × Column H × Column C
[f] Column L = Column C × Column D/1000
[g] Column M = (Column L − Column J) × $10^6$/3413
[h] Column N = Column K + Column M

to 60°F for 10 hours each night. The furnace is on from October 1 through May 31. Electric baseboard heat is used. The air conditioner has an SEER of 7.3 (Btu/h)/W. Electricity costs 0.0725 $/kWh year-round.

[Ans: $2414]

**8.7** A residence in St. Joseph, Missouri, has a design heating load of 68,000 Btu/h when design indoor and outdoor temperatures are 75°F and 3°F, respectively. The furnace is *off* from June through September. Determine the fuel and energy requirements for heating in:

(a) Btu/yr
(b) Gallons of No. 2 fuel oil/yr
(c) Cubic feet of natural gas/yr
(d) kWh/yr
(e) Total airflow rate in cfm if a warm air system is used
(f) Total steam flow in lb/h if a steam system is used
(g) Total water flow rate in gpm if a hydronic system is used
(h) Total electric power in kW if electric heating is used

**8.8** For a residence located in New Orleans, Louisiana, the design cooling load is 12 kW (41,000 Btu/h). Determine:

(a) Annual energy requirements for cooling, kWh
(b) Cost of this energy if the electric rate is 6.5¢/kWh

**8.9** An office building located in Springfield, Missouri, has a heat loss of 2,160,000 Btu/h for design condition of 75°F inside and 10°F outside. The heating system is operational between October 1 and April 30. Determine:

(a) Annual energy usage for heating
(b) Estimated fuel cost if No. 2 fuel oil is used having a heating value of 140,000 Btu/gal and costing $2.50/gal

[Ans: $2.8 \times 10^9$ Btu; $71,800]

**8.10** A small football promotion office is being designed for Jacksonville, Florida. The design heating and cooling loads are 61,200 and 55,400 Btu/h, respectively, based on 99.6% and 1% outdoor design dry-bulb temperatures. Balance point has been estimated as 65°F.

(a) Select an appropriate heat pump from the XYZ Corporation models listed on the next page and estimate the energy costs for summer and winter if electricity is 8¢/kWh.
(b) Compare the heating energy cost for the heat pump to that for a condensing gas furnace with natural gas costing $1.20 per therm.

**8.11** A 1980 ft² residence located in Cincinnati, Ohio, has design heating and cooling loads of 74,000 Btu/h and 35,000 Btu/h, respectively. Determine:

(a) Heating energy requirements, Btu
(b) Gallons of No. 2 fuel oil if 75% efficient oil-fired warm air system is used
(c) Therms of natural gas if 88% efficient gas-fired warm air system is used
(d) kWh of electricity if 98% efficient baseboard units are used
(e) Required airflow, cfm, for warm air systems
(f) kWh of electricity if heat pump system (WA-36 specifications follow) including supplementary electric resistance heat is used
(g) kWh of electricity for cooling for air conditioner with SEER of 8.5 using degree-day estimation
(h) Required airflow, cfm, for air conditioning

# Energy Estimating Methods

**HEAT PUMP MODELS—XYZ CORPORATION for Problem 8.10—**
**Performance Data at ARI Standard Conditions**

| | Cooling Capacity | | | | | | | Heating Capacity, 70°F Indoor Air | | | | | | DOE |
|---|---|---|---|---|---|---|---|---|---|---|---|---|---|---|
| | Design Conditions: ARI Rating Temperatures 80°F DB, 67°F WB Indoor, Return Air; 95°F DB Outdoor | | | | | | | Outdoor Air 47°F DB/43°F WB DOE High Temperature | | | Outdoor Air 17°F DB/15°F WB DOE Low Temperature | | | Region IV |
| Model Numbers | ARI Std. Cap. Btu/h | Net Sens. Cooling Cap. | Net Lat. Btu/h | Single Phase SEER | Total W | ARI Noise Rating | Approx. CFM | Btu/h | Power Input W | COP | Btu/h | Power Input W | COP | HSPF Btu/W·h |
| A018 | 18,200 | 13,700 | 4,500 | 11.20 | 1820 | 7.0 | 650 | 18,400 | 1671 | 3.25 | 9,700 | 1255 | 2.25 | 8.15 |
| A024 | 24,200 | 18,000 | 6,200 | 10.80 | 2513 | 7.0 | 850 | 24,800 | 2162 | 3.35 | 14,400 | 1718 | 2.45 | 8.85 |
| A030 | 30,000 | 22,000 | 8,000 | 10.60 | 3151 | 7.2 | 1050 | 31,800 | 2885 | 3.25 | 18,600 | 2263 | 2.40 | 8.65 |
| A036 | 35,800 | 25,500 | 10,300 | 10.50 | 3837 | 7.4 | 1250 | 39,000 | 3590 | 3.15 | 22,000 | 2763 | 2.35 | 8.25 |
| A042 | 42,500 | 31,900 | 10,600 | 11.20 | 4250 | 7.6 | 1450 | 43,500 | 3803 | 3.35 | 24,600 | 2929 | 2.45 | 8.85 |
| A048 | 49,500 | 36,800 | 12,700 | 10.50 | 5269 | 7.6 | 1650 | 51,000 | 4596 | 3.25 | 30,000 | 3578 | 2.45 | 8.80 |
| A060 | 60,000 | 43,700 | 16,300 | 10.50 | 6250 | 7.8 | 2050 | 66,000 | 6050 | 3.20 | 37,000 | 4612 | 2.35 | 8.50 |

### A030 — Heating, Indoor Air Conditions, 70°F DB

| Outdoor Temperature | Btu/h | W | COP | EER |
|---|---|---|---|---|
| −18 | 8900 | 1800 | 1.45 | 4.94 |
| −13 | 9700 | 1830 | 1.55 | 5.30 |
| −8 | 10600 | 1880 | 1.65 | 5.64 |
| −3 | 11800 | 1940 | 1.78 | 6.08 |
| 2 | 13200 | 2010 | 1.92 | 6.57 |
| 7 | 14800 | 2090 | 2.07 | 7.08 |
| 12 | 16600 | 2170 | 2.24 | 7.65 |
| 17 | 18600 | 2260 | 2.41 | 8.23 |
| 22 | 20500 | 2360 | 2.55 | 8.69 |
| 27 | 22700 | 2470 | 2.69 | 9.19 |
| 32 | 24900 | 2570 | 2.84 | 9.69 |
| 37 | 27300 | 2690 | 2.97 | 10.15 |
| 42 | 29600 | 2800 | 3.10 | 10.57 |
| 47 | 31800 | 2890 | 3.22 | 11.00 |
| 52 | 34500 | 3020 | 3.35 | 11.42 |
| 57 | 37000 | 3140 | 3.45 | 11.78 |
| 62 | 39400 | 3250 | 3.55 | 12.12 |
| 67 | 41800 | 3350 | 3.66 | 12.48 |
| 72 | 44100 | 3450 | 3.75 | 12.78 |
| 77 | 46400 | 3550 | 3.83 | 13.07 |
| 82 | 48600 | 3640 | 3.91 | 13.35 |

### A036 — Heating, Indoor Air Conditions, 70°F DB

| Outdoor Temperature | Btu/h | W | COP | EER |
|---|---|---|---|---|
| −18 | 9300 | 2150 | 1.27 | 4.33 |
| −13 | 10300 | 2210 | 1.37 | 4.66 |
| −8 | 11700 | 2270 | 1.51 | 5.15 |
| −3 | 13300 | 2350 | 1.66 | 5.66 |
| 2 | 15100 | 2440 | 1.81 | 6.19 |
| 7 | 17200 | 2550 | 1.98 | 6.75 |
| 12 | 19500 | 2650 | 2.15 | 7.33 |
| 17 | 22000 | 2760 | 2.34 | 7.97 |
| 22 | 24500 | 2910 | 2.47 | 8.42 |
| 27 | 27300 | 3050 | 2.62 | 8.95 |
| 32 | 30100 | 3190 | 2.76 | 9.44 |
| 37 | 33100 | 3330 | 2.91 | 9.94 |
| 42 | 36100 | 3480 | 3.04 | 10.37 |
| 47 | 39000 | 3610 | 3.17 | 10.80 |
| 52 | 42200 | 3770 | 3.28 | 11.19 |
| 57 | 45200 | 3910 | 3.39 | 11.56 |
| 62 | 48300 | 4050 | 3.49 | 11.93 |
| 67 | 51200 | 4190 | 3.58 | 12.22 |
| 72 | 54100 | 4310 | 3.68 | 12.55 |
| 77 | 56900 | 4440 | 3.75 | 12.82 |
| 82 | 59600 | 4550 | 3.84 | 13.10 |

### A048 — Heating, Indoor Air Conditions, 70°F DB

| Outdoor Temperature | Btu/h | W | COP | EER |
|---|---|---|---|---|
| −18 | 10900 | 2780 | 1.15 | 3.92 |
| −13 | 13200 | 2860 | 1.35 | 4.62 |
| −8 | 15700 | 2950 | 1.56 | 5.32 |
| −3 | 18400 | 3060 | 1.76 | 6.01 |
| 2 | 21200 | 3180 | 1.95 | 6.67 |
| 7 | 24200 | 3310 | 2.14 | 7.31 |
| 12 | 27300 | 3450 | 2.32 | 7.91 |
| 17 | 30000 | 3580 | 2.46 | 8.38 |
| 22 | 33800 | 3780 | 2.63 | 8.99 |
| 27 | 37300 | 3930 | 2.78 | 9.49 |
| 32 | 40800 | 4100 | 2.92 | 9.95 |
| 37 | 44400 | 4280 | 3.04 | 10.37 |
| 42 | 48000 | 4460 | 3.15 | 10.76 |
| 47 | 51000 | 4600 | 3.25 | 11.09 |
| 52 | 55400 | 4810 | 3.37 | 11.52 |
| 57 | 59200 | 4990 | 3.48 | 11.86 |
| 62 | 63000 | 5160 | 3.58 | 12.21 |
| 67 | 66700 | 5330 | 3.67 | 12.51 |
| 72 | 70500 | 5490 | 3.76 | 12.84 |
| 77 | 74200 | 5640 | 3.85 | 13.16 |
| 82 | 78000 | 5790 | 3.95 | 13.47 |

### A060 — Heating, Indoor Air Conditions, 70°F DB

| Outdoor Temperature | Btu/h | W | COP | EER |
|---|---|---|---|---|
| −18 | 16800 | 3360 | 1.46 | 5.00 |
| −13 | 18300 | 3470 | 1.55 | 5.27 |
| −8 | 20400 | 3600 | 1.66 | 5.67 |
| −3 | 23000 | 3760 | 1.79 | 6.12 |
| 2 | 26000 | 3940 | 1.93 | 6.60 |
| 7 | 29400 | 4130 | 2.09 | 7.12 |
| 12 | 33200 | 4350 | 2.24 | 7.63 |
| 17 | 37000 | 4610 | 2.35 | 8.03 |
| 22 | 41700 | 4810 | 2.54 | 8.67 |
| 27 | 46300 | 5050 | 2.69 | 9.17 |
| 32 | 51000 | 5300 | 2.82 | 9.62 |
| 37 | 56000 | 5560 | 2.95 | 10.07 |
| 42 | 61000 | 5810 | 3.08 | 10.50 |
| 47 | 66000 | 6050 | 3.20 | 10.91 |
| 52 | 71200 | 6320 | 3.30 | 11.27 |
| 57 | 76300 | 6560 | 3.41 | 11.63 |
| 62 | 81300 | 6790 | 3.51 | 11.97 |
| 67 | 86200 | 7020 | 3.60 | 12.28 |
| 72 | 91000 | 7230 | 3.69 | 12.59 |
| 77 | 95600 | 7420 | 3.78 | 12.88 |
| 82 | 99900 | 7600 | 3.85 | 13.14 |

**Performance Data for Model WA-36 Heat Pump for Problem 8-11**

| Air Temperature, °F | Heat Pump Output, 1000 Btu | Heat Pump Input, kW |
|---|---|---|
| 62 | 44 | 4.5 |
| 57 | 43 | 4.4 |
| 52 | 41 | 4.3 |
| 47 | 39 | 4.1 |
| 42 | 36 | 4.0 |
| 37 | 33 | 3.9 |
| 32 | 30 | 3.7 |
| 27 | 27 | 3.6 |
| 22 | 24 | 3.5 |
| 17 | 22 | 3.3 |
| 12 | 19 | 3.2 |
| 7 | 17 | 3.1 |
| 2 | 15 | 2.9 |
| −3 | 13 | 2.8 |

**8.12** A small commercial building located in Oklahoma City, Oklahoma, has design loads of 245,000 Btu/h, *heating*, and 162,000 Btu/h, *cooling*. Balance point for the building has been estimated at 65°F. Determine:

(a) Annual energy requirements for heating, Btu
(b) Fuel cost using LPG at $2.50/gallon, $
(c) Fuel cost using electric baseboard units with electricity at 6.7¢/kWh, $
(d) Savings if setback to 55°F is effected between 10 P.M. and 6 A.M., Monday through Saturday, and all day Sunday, %
(e) Cooling season energy cost using cooling degree-days if conditioner has SEER of 11.5 and electricity is 7¢/kWh

**8.13** A small commercial building in Indianapolis, Indiana, has design heating and cooling loads of 98,000 Btu/h and 48,000 Btu/h, respectively. Internal heat gains throughout the winter are relatively steady at 4.5 kW. Electricity costs 7.1¢/kWh. Estimate:

(a) Annual heating cost if baseboard electric resistance units are used.
(b) Annual cooling cost with a conventional vapor compression air-cooled unit, using your choice of method.

Select a heat pump system for the building from the XYZ Corporation models. Determine the

(a) Annual heating cost
(b) Annual cooling cost

**8.14** A small (2200 ft$^2$) food mart store located in Charlotte, NC, has design heating and cooling loads of 94,500 Btu/h and 57,400 Btu/h, respectively, based on inside design temperatures of 72°F (winter) and 78°F (summer). The store is open 24 hours a day and has a relatively constant internal load due to lights, food cases, people, etc., of 3.3 W/ft$^2$. Select a suitable heat pump for the XYZ Corporation and estimate its operating energy costs for both summer and winter if the price of electricity is 7.4¢/kWh.

## 8.8 Bibliography

ASHRAE. 1995. *Bin and degree hour weather data for simplified energy calculations.*

ASHRAE. 2003. *ASHRAE Energy Position Document.*

ASHRAE. 2013. *2013 ASHRAE Handbook—Fundamentals*, Chapter 19, Energy Estimating and Modeling Methods, and Chapter 34, Energy Resources.

Grumman, D., ed. 2003. *ASHRAE GreenGuide*. Atlanta: American Society of Heating, Refrigerating and Air-Conditioning Engineers, Inc.

NCDC. 2002. Annual degree-days to selected bases (1971-2000). In *Climatography of the U.S. #81*. National Climatic Data Center, Asheville, NC.

USAF. 1978. Engineering weather data. US Air Force *Manual* 88-29. Superintendent of Documents, Government Printing Office, Washington, D.C. 26402.

# Chapter 9

# DUCT AND PIPE SIZING

This chapter discusses the design of systems for conveying air and water. Chapter 21 of the 2013 *ASHRAE Handbook—Fundamentals* has further details on the design of duct systems. Chapter 20 from the same source gives details on space air diffusion. Chapter 22 of the 2013 *ASHRAE Handbook—Fundamentals* has additional details on pipe sizing, while Chapters 21, 44, and 46 in the 2012 *ASHRAE Handbook—HVAC Systems and Equipment* have further information on fans, pumps, pipes, tubes, and fittings.

## 9.1 Duct Systems

An air-conditioning system must not only condition air, it must also distribute conditioned air throughout the space. Usually the conditioning fluid is distributed from a central equipment location to the individual spaces requiring environment control. For example, a fan and duct system distributes air, and a pump and piping system distributes water. Conditioned air is distributed into the room by air diffusers or grilles.

An objective of duct system design is to provide a system that, within prescribed limits of velocities, noise intensity, and space available for ducts, efficiently transmits the required flow rate of air to each space while maintaining a proper balance between investment and operating cost. When the heating, cooling, or ventilation load is established, the total flow rate of air required can be determined by methods shown in Chapters 17 and 18 of the 2013 *ASHRAE Handbook—Fundamentals*. The size of the duct system governs frictional losses and thereby the size of fan and power required to operate the duct system.

### 9.1.1 Pressure Changes

Resistance to airflow imposed by the supply duct system must be overcome by mechanical energy, which is ordinarily supplied by a fan. Resistance also is imposed by the return-air system, which must also be overcome by the fan. In air-conditioning and ventilating work, the pressure differences are ordinarily so small that the equations for incompressible flow can be applied. Additional simplification is obtained by considering the air to be at the standard density of 0.075 lb/ft³ (1.2 kg/m³).

At any cross section in a duct, the total pressure $p_t$ is the sum of the static pressure $p_s$ and the velocity pressure $p_v$.

$$p_t = p_s + p_v \qquad (9\text{-}1)$$

Pressures are normally expressed in inches of water (Pa). The velocity pressure is then given by

$$p_v = \left(\frac{V}{cf_1}\right)^2 \qquad (9\text{-}2)$$

where $V$ is defined by the equation

$$V = cf_2(Q/A) \qquad (9\text{-}3)$$

where

$p_v$ = velocity pressure, in. of water (Pa)
$V$ = mean velocity of fluid, ft/min (m/s)
$Q$ = airflow rate, ft³/min (L/s)
$A$ = cross-sectional area of duct, in² (mm²)
$cf_1$ = conversion factor, 4005 (1.29)
$cf_2$ = conversion factor, 144 (1000)

The following table relates the air velocity to the velocity pressure for a range of velocities that might be encountered in HVAC duct systems. These values were calculated from Equation (9-2) assuming a standard density of air as 0.075 $\text{lb}_m/\text{ft}^3$.

| $V$ fpm | $P_v$ in. water | $V$ fpm | $P_v$ in. water | $V$ fpm | $P_v$ in. water |
|---|---|---|---|---|---|
| 500 | 0.016 | 1800 | 0.202 | 4400 | 1.21 |
| 600 | 0.022 | 2000 | 0.249 | 4600 | 1.32 |
| 700 | 0.031 | 2200 | 0.302 | 4800 | 1.44 |
| 800 | 0.040 | 2400 | 0.359 | 5000 | 1.56 |
| 900 | 0.050 | 2600 | 0.421 | 5200 | 1.69 |
| 1000 | 0.062 | 2800 | 0.489 | 5400 | 1.82 |
| 1100 | 0.075 | 3000 | 0.561 | 5600 | 1.96 |
| 1200 | 0.090 | 3200 | 0.638 | 5800 | 2.10 |
| 1300 | 0.105 | 3400 | 0.721 | 6000 | 2.24 |
| 1400 | 0.122 | 3600 | 0.808 | 6200 | 2.40 |
| 1500 | 0.140 | 3800 | 0.900 | 6400 | 2.55 |
| 1600 | 0.160 | 4000 | 0.998 | 6600 | 2.72 |
| 1700 | 0.180 | 4200 | 1.10 | 6800 | 2.88 |

If the air is not at this standard density, Equation (9-4) may be used in place of Equation (9-2):

$$p_v = \rho\left(\frac{V}{cf_3}\right)^2 \qquad (9\text{-}4)$$

where

$\rho$ = air density, lb/ft³ (kg/m³)
$V$ = fluid mean velocity, ft/min (m/s)
$p_v$ = velocity pressure, in. of water (Pa)
$cf_3$ = conversion factor, 1097 (1.414)

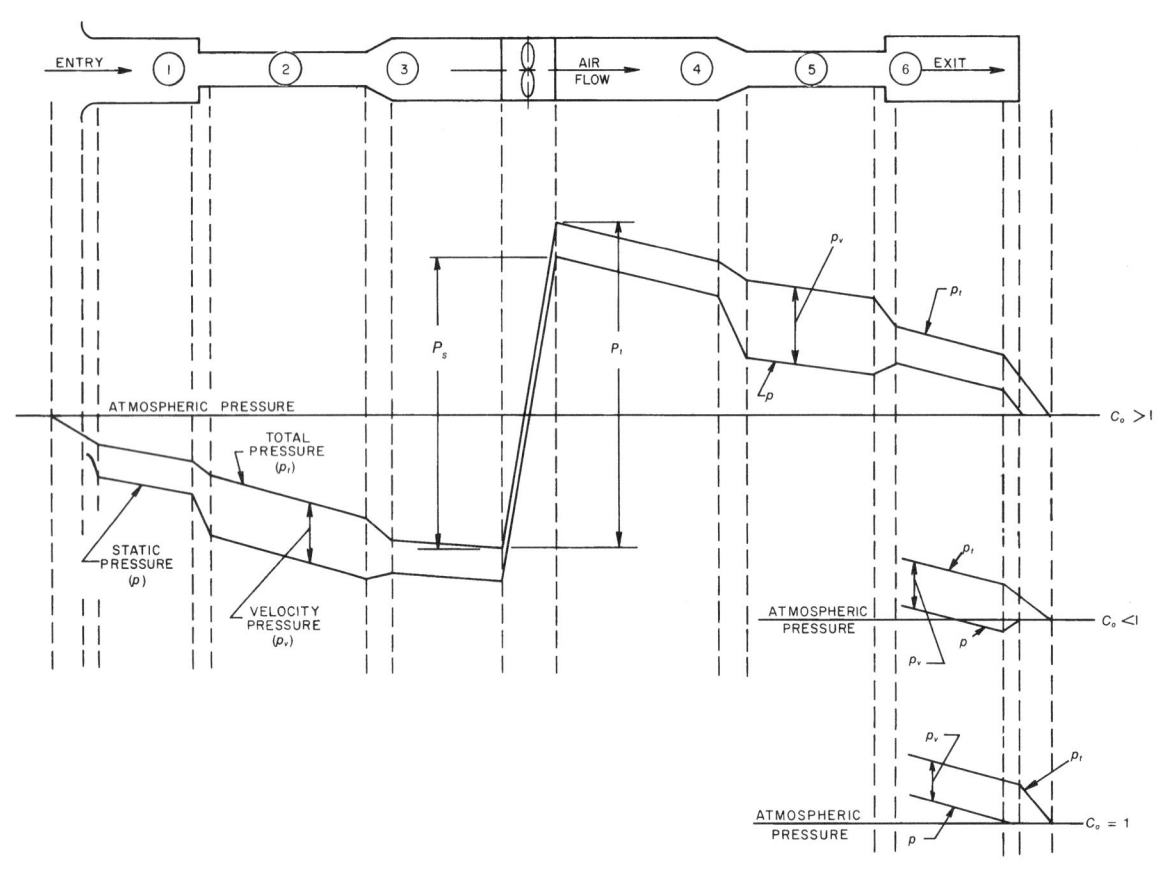

*Fig. 9-1 Pressure Changes During Flow in Ducts*
*(Figure 7, Chapter 21, 2013 ASHRAE Handbook—Fundamentals)*

The total pressure $p_t$ is a measure of the total available energy at a cross section. In any duct system, the total pressure always decreases in the direction of the airflow. Static and velocity pressure are mutually convertible and either increase or decrease in the flow direction.

Total and static pressure changes in a simplified fan/duct system are shown in Figure 9-1. This illustrative system consists of a fan with both supply and return air ductwork. Also shown are the total and static pressure gradients referenced to atmospheric pressure.

For all constant-area sections, such as ducts and elbows, the total and static pressure losses are equal. In the case of ducts, the losses are entirely frictional, while the losses in constant-area fitting are frictional and dynamic.

At diverging sections 3 and 6, the velocity pressure decreases, the absolute value of the total pressure decreases, and the absolute value of the static pressure may increase. The increase in static pressure as shown at these sections is known as **static regain**.

At converging sections 1 and 4, the velocity pressure increases in the direction of airflow, and the absolute value of both the total and static pressure decreases. The static or total pressure loss from upstream to downstream is the difference in total/static pressure between the two sections.

At the exit, total pressure loss depends on the shape of the fitting and the flow characteristics. Exit loss coefficients can be greater than, less than, or equal to, one. For this variation of coefficients, the total and static pressure grade lines are shown in Figure 9-1. Note that when the loss coefficient is less than one, the static pressure upstream of the exit is less than atmospheric pressure (negative). The static pressure upstream of the discharge fitting can be calculated by subtracting the upstream velocity pressure from the upstream total pressure.

The entry loss also depends on the shape of the fitting. The total pressure immediately downstream of the entrance equals the difference between the upstream pressure, which is zero (atmospheric pressure), and the loss through the fitting. The static pressure at the entrance is zero, and immediately downstream, the difference between static pressure is negative, algebraically equal to the total pressure (negative) and the velocity pressure (always positive), or $p_s = p_t - p_v$.

The total system resistance to airflow is noted by $p_t$ in Figure 9-1. The fan inlet and outlet system effect factors due to the interaction of the fan and system are not shown; only system resistances are shown. To obtain the fan static pressure $p_s$ requirement for selecting a fan, knowing the systems' total pressure, use

$$p_s = p_t - p_{v,o} \tag{9-5}$$

where the subscript $o$ refers to the discharge area of the fan.

# Duct and Pipe Sizing

Static pressure is used as the basis for system design; total pressure determines the actual mechanical energy that must be supplied to the system. Total pressure always decreases in the direction of flow. Note, however, in Figure 9-1, the static pressure decreases and then increases in the direction of flow. Moreover, it can even become negative (below atmospheric). Therefore, in dealing with static pressure, distinction must always be made between static pressure **loss** and static pressure **change** as a result of conversion of velocity pressure.

## 9.1.2 Circular Equivalents of Ducts

**Rectangular Ducts.** An air-handling system is usually sized first for round ducts. Then, if rectangular ducts are desired, duct sizes are selected to provide flow rates equivalent to those of the round ducts originally selected.

Rectangular ducts of aspect ratios not exceeding 8:1 usually have the same friction pressure loss for equal lengths and mean velocities of flow as do circular ducts of the same hydraulic diameter. When duct sizes are expressed in terms of hydraulic diameter (4 times area divided by perimeter), and when equations for friction loss in round and rectangular ducts are equated for equal flow rate and equal length, Equation (9-6a), giving the circular equivalent of a rectangular duct, is obtained.

$$d_c = \frac{1.30 ab^{0.625}}{(a+b)^{0.250}} \quad (9\text{-}6a)$$

where

$a$ = length of one side of rectangular duct, in. (mm)
$b$ = length of adjacent side of rectangular duct, in. (mm)
$d_c$ = circular equivalent of a rectangular duct for equal friction and capacity, in. (mm)

The circular equivalents of rectangular ducts for equal friction and flow rate for aspect ratios not greater than 8:1, based on Equation (9-6a), are given in Table 9-1a. Note that the mean velocity in a rectangular duct is less than its circular equivalent. Frictional losses are then obtained from Figure 9-2.

Multiplying or dividing the length of each side of a duct by a constant is the same as multiplying or dividing the equivalent round size by the same constant. Thus, if the circular equivalent of an 80 in. by 24 in. duct is required, it is twice that of a 40 in. by 12 in. duct, or 2 × 23.0 = 46.0 in.

**Flat Oval Ducts.** To convert round ducts to flat oval sizes, use Table 9-1b, which is based on Equation (9-6b) (Heyt and Diaz 1975), the circular equivalent of a flat oval duct for equal airflow, resistance, and length.

$$D_e = \frac{1.55 AR^{0.625}}{P^{0.250}} \quad (9\text{-}6b)$$

where AR is the cross-sectional area of flat oval duct defined as

$$AR = (\pi a^2/4) + a(A - a)$$

and the perimeter $P$ is calculated by

$$P = \pi a + 2(A - a)$$

where

$P$ = perimeter of flat oval duct, in.
$A$ = major axis of flat oval duct, in.
$a$ = minor axis of flat oval duct, in.

Friction losses are then obtained from Figure 9-2.

## 9.1.3 Friction Losses

Pressure drop in a straight duct is caused by surface friction. This friction loss can be calculated by using the Air Friction Charts (Figure 9-2). The charts were built from the basic flow equation for pressure loss in circular ducts:

$$\Delta p_{fr} = f_D \, (cf_4 \, L/D) \, p_v \quad (9\text{-}7)$$

where

$\Delta p_{fr}$ = friction loss in terms of total pressure, in. of water (Pa)
$f_D$ = dimensionless friction factor, which for air-conditioning work depends on Reynolds number and relative roughness of the conduit. Approximate values of $f$ were taken from Moody (1944) where $\varepsilon$ = 0.0003 ft (0.09 mm). It numerically equals the reciprocal of the number of duct diameters required to cause a pressure loss equal to one velocity pressure. (Figure 13, Chapter 3, 2013 *ASHRAE Handbook—Fundamentals*)
$L$ = length of duct, ft (m)
$D$ = inside diameter of duct, in. (mm)
$p_v$ = velocity pressure of mean velocity, in. of water (Pa)
$cf_4$ = conversion factor, 12 (1).

The air friction chart is based on air with a density of 0.075 lb/ft³ (1.2 kg/m³) flowing through average, clean, round, galvanized metal ducts with beaded slip couplings on 48 in. (1220 mm) centers.

Variations in air temperature of the order of ±20°F from 70°F (±11°C from 20°C) have little effect on duct friction. Therefore, Figure 9-2 may be used for all air systems with temperatures from 50 to 90°F (10 to 32°C). However, the values found in the charts must be corrected for systems carrying air at much higher temperatures. To determine the friction loss in such systems, the actual flow rate or velocity existing at the nonstandard conditions must be used. For duct materials other than those indicated, and for significant variations in temperature and barometric pressure/elevation, correction factors should be applied to the Air Friction Chart values. Details concerning these correction factors can be found in Chapter 21 of the 2009 *ASHRAE Handbook—Fundamentals*.

## 9.1.4 Dynamic Losses

Wherever eddying flow is present, a greater loss in total pressure takes place than would occur in steady flow through a similar length of straight duct having a uniform cross section.

## Table 9-1a Equivalent Rectangular Duct Dimension
*(Table 2, Chapter 21 of the 2013 ASHRAE Handbook—Fundamentals)*

| Circular Duct Diameter, in. | \multicolumn{17}{c}{Length One Side of Rectangular Duct (*a*), in.} |
|---|---|---|---|---|---|---|---|---|---|---|---|---|---|---|---|---|---|---|
| | 4 | 5 | 6 | 7 | 8 | 9 | 10 | 12 | 14 | 16 | 18 | 20 | 22 | 24 | 26 | 28 | 30 | 32 | 34 | 36 |
| | \multicolumn{20}{c}{Length Adjacent Side of Rectangular Duct (*b*), in.} |
| 5 | 5 | | | | | | | | | | | | | | | | | | | |
| 5.5 | 6 | 5 | | | | | | | | | | | | | | | | | | |
| 6 | 8 | 6 | | | | | | | | | | | | | | | | | | |
| 6.5 | 9 | 7 | 6 | | | | | | | | | | | | | | | | | |
| 7 | 11 | 8 | 7 | | | | | | | | | | | | | | | | | |
| 7.5 | 13 | 10 | 8 | 7 | | | | | | | | | | | | | | | | |
| 8 | 15 | 11 | 9 | 8 | | | | | | | | | | | | | | | | |
| 8.5 | 17 | 13 | 10 | 9 | | | | | | | | | | | | | | | | |
| 9 | 20 | 15 | 12 | 10 | 8 | | | | | | | | | | | | | | | |
| 9.5 | 22 | 17 | 13 | 11 | 9 | | | | | | | | | | | | | | | |
| 10 | 25 | 19 | 15 | 12 | 10 | 9 | | | | | | | | | | | | | | |
| 10.5 | 29 | 21 | 16 | 14 | 12 | 10 | | | | | | | | | | | | | | |
| 11 | 32 | 23 | 18 | 15 | 13 | 11 | 10 | | | | | | | | | | | | | |
| 11.5 | | 26 | 20 | 17 | 14 | 12 | 11 | | | | | | | | | | | | | |
| 12 | | 29 | 22 | 18 | 15 | 13 | 12 | | | | | | | | | | | | | |
| 12.5 | | 32 | 24 | 20 | 17 | 15 | 13 | | | | | | | | | | | | | |
| 13 | | 35 | 27 | 22 | 18 | 16 | 14 | 12 | | | | | | | | | | | | |
| 13.5 | | 38 | 29 | 24 | 20 | 17 | 15 | 13 | | | | | | | | | | | | |
| 14 | | | 32 | 26 | 22 | 19 | 17 | 14 | | | | | | | | | | | | |
| 14.5 | | | 35 | 28 | 24 | 20 | 18 | 15 | | | | | | | | | | | | |
| 15 | | | 38 | 30 | 25 | 22 | 19 | 16 | 14 | | | | | | | | | | | |
| 16 | | | 45 | 36 | 30 | 25 | 22 | 18 | 15 | | | | | | | | | | | |
| 17 | | | | 41 | 34 | 29 | 25 | 20 | 17 | 16 | | | | | | | | | | |
| 18 | | | | 47 | 39 | 33 | 29 | 23 | 19 | 17 | | | | | | | | | | |
| 19 | | | | 54 | 44 | 38 | 33 | 26 | 22 | 19 | 18 | | | | | | | | | |
| 20 | | | | | 50 | 43 | 37 | 29 | 24 | 21 | 19 | | | | | | | | | |
| 21 | | | | | 57 | 48 | 41 | 33 | 27 | 23 | 20 | | | | | | | | | |
| 22 | | | | | 64 | 54 | 46 | 36 | 30 | 26 | 23 | 20 | | | | | | | | |
| 23 | | | | | | 60 | 51 | 40 | 33 | 28 | 25 | 22 | | | | | | | | |
| 24 | | | | | | 66 | 57 | 44 | 36 | 31 | 27 | 24 | 22 | | | | | | | |
| 25 | | | | | | | 63 | 49 | 40 | 34 | 29 | 26 | 24 | | | | | | | |
| 26 | | | | | | | 69 | 54 | 44 | 37 | 32 | 28 | 26 | 24 | | | | | | |
| 27 | | | | | | | 76 | 59 | 48 | 40 | 35 | 31 | 28 | 25 | | | | | | |
| 28 | | | | | | | | 64 | 52 | 43 | 38 | 33 | 30 | 27 | 26 | | | | | |
| 29 | | | | | | | | 70 | 56 | 47 | 41 | 36 | 32 | 29 | 27 | | | | | |
| 30 | | | | | | | | 76 | 61 | 51 | 44 | 39 | 35 | 31 | 29 | 28 | | | | |
| 31 | | | | | | | | 82 | 66 | 55 | 47 | 41 | 37 | 34 | 31 | 29 | | | | |
| 32 | | | | | | | | 89 | 71 | 59 | 51 | 44 | 40 | 36 | 33 | 31 | | | | |
| 33 | | | | | | | | 96 | 76 | 64 | 54 | 48 | 42 | 38 | 35 | 33 | 30 | | | |
| 34 | | | | | | | | | 82 | 68 | 58 | 51 | 45 | 41 | 37 | 35 | 32 | | | |
| 35 | | | | | | | | | 88 | 73 | 62 | 54 | 48 | 44 | 40 | 37 | 34 | 32 | | |
| 36 | | | | | | | | | 95 | 78 | 67 | 58 | 51 | 46 | 42 | 39 | 36 | 34 | | |
| 37 | | | | | | | | | 101 | 83 | 71 | 62 | 55 | 49 | 45 | 41 | 38 | 36 | 34 | |
| 38 | | | | | | | | | 108 | 89 | 76 | 66 | 58 | 52 | 47 | 44 | 40 | 38 | 36 | |
| 39 | | | | | | | | | | 95 | 80 | 70 | 62 | 55 | 50 | 46 | 43 | 40 | 37 | 36 |
| 40 | | | | | | | | | | 101 | 85 | 74 | 65 | 58 | 53 | 49 | 45 | 42 | 39 | 37 |
| 41 | | | | | | | | | | 107 | 91 | 78 | 69 | 62 | 56 | 51 | 47 | 44 | 41 | 39 |
| 42 | | | | | | | | | | 114 | 96 | 83 | 73 | 65 | 59 | 54 | 50 | 46 | 44 | 41 |
| 43 | | | | | | | | | | 120 | 102 | 88 | 77 | 69 | 62 | 57 | 53 | 49 | 46 | 43 |
| 44 | | | | | | | | | | | 107 | 93 | 81 | 73 | 66 | 60 | 55 | 51 | 48 | 45 |
| 45 | | | | | | | | | | | 113 | 98 | 86 | 76 | 69 | 63 | 58 | 54 | 50 | 47 |
| 46 | | | | | | | | | | | 120 | 103 | 90 | 80 | 72 | 66 | 61 | 56 | 53 | 49 |
| 47 | | | | | | | | | | | 126 | 108 | 95 | 84 | 76 | 69 | 64 | 59 | 55 | 52 |
| 48 | | | | | | | | | | | 133 | 114 | 100 | 89 | 80 | 73 | 67 | 62 | 58 | 54 |
| 49 | | | | | | | | | | | 140 | 120 | 105 | 93 | 84 | 76 | 70 | 65 | 60 | 56 |
| 50 | | | | | | | | | | | 147 | 126 | 110 | 98 | 88 | 80 | 73 | 68 | 63 | 59 |
| 51 | | | | | | | | | | | | 132 | 115 | 102 | 92 | 83 | 76 | 71 | 66 | 61 |
| 52 | | | | | | | | | | | | 139 | 121 | 107 | 96 | 87 | 80 | 74 | 69 | 64 |
| 53 | | | | | | | | | | | | 145 | 127 | 112 | 100 | 91 | 83 | 77 | 71 | 67 |
| 54 | | | | | | | | | | | | 152 | 133 | 117 | 105 | 95 | 87 | 80 | 74 | 70 |
| 55 | | | | | | | | | | | | | 139 | 123 | 110 | 99 | 91 | 84 | 78 | 72 |
| 56 | | | | | | | | | | | | | 145 | 128 | 114 | 104 | 95 | 87 | 81 | 75 |
| 57 | | | | | | | | | | | | | 151 | 134 | 119 | 108 | 98 | 91 | 84 | 78 |
| 58 | | | | | | | | | | | | | 158 | 139 | 124 | 112 | 102 | 94 | 87 | 81 |
| 59 | | | | | | | | | | | | | 165 | 145 | 130 | 117 | 107 | 98 | 91 | 85 |
| 60 | | | | | | | | | | | | | 172 | 151 | 135 | 122 | 111 | 102 | 94 | 88 |

# Duct and Pipe Sizing

**Table 9-1b  Equivalent Flat Oval Duct Dimensions**
*(Table 3, Chapter 21 of the 2013 ASHRAE Handbook—Fundamentals)*

| Circular Duct Diameter, in. | \multicolumn{10}{c}{Minor Axis a, in.} |||||||||| | Circular Duct Diameter, in. | \multicolumn{10}{c}{Minor Axis a, in.} ||||||||||
|---|---|---|---|---|---|---|---|---|---|---|---|---|---|---|---|---|---|---|---|---|---|
|  | 3 | 4 | 5 | 6 | 7 | 8 | 9 | 10 | 11 | 12 | 14 | 16 |  | 8 | 9 | 10 | 11 | 12 | 14 | 16 | 18 | 20 | 22 | 24 |
|  | \multicolumn{10}{c}{Major Axis A, in.} |||||||||| |  | \multicolumn{10}{c}{Major Axis A, in.} ||||||||||
| 5 | 8 | | | | | | | | | | | | 19 | 46 | — | 34 | — | 28 | 23 | 21 | | | | |
| 5.5 | 9 | 7 | | | | | | | | | | | 20 | 50 | — | 38 | — | 31 | 27 | 24 | 21 | | | |
| 6 | 11 | 9 | | | | | | | | | | | 21 | 58 | — | 43 | — | 34 | 28 | 25 | 23 | | | |
| 6.5 | 12 | 10 | 8 | | | | | | | | | | 22 | 65 | — | 48 | — | 37 | 31 | 29 | 26 | | | |
| 7 | 15 | 12 | 10 | 8 | | | | | | | | | 23 | 71 | — | 52 | — | 42 | 34 | 30 | 27 | | | |
| 7.5 | 19 | 13 | — | 9 | | | | | | | | | 24 | 77 | — | 57 | — | 45 | 38 | 33 | 29 | 26 | | |
| 8 | 22 | 15 | 11 | — | | | | | | | | | 25 | | 63 | — | 50 | 41 | 36 | 32 | 29 | | | |
| 8.5 | | 18 | 13 | 11 | 10 | | | | | | | | 26 | | 70 | — | 56 | 45 | 38 | 34 | 31 | | | |
| 9 | | 20 | 14 | 12 | — | 10 | | | | | | | 27 | | 76 | — | 59 | 49 | 41 | 37 | 34 | | | |
| 9.5 | | 21 | 18 | 14 | 12 | — | | | | | | | 28 | | | 65 | 52 | 46 | 40 | 36 | | | | |
| 10 | | | 19 | 15 | 13 | 11 | | | | | | | 29 | | | 72 | 58 | 49 | 43 | 39 | 35 | | | |
| 10.5 | | | 21 | 17 | 15 | 13 | 12 | | | | | | 30 | | | 78 | 61 | 54 | 46 | 40 | 38 | | | |
| 11 | | | | 19 | 16 | 14 | — | 12 | | | | | 31 | | | 81 | 67 | 57 | 49 | 44 | 39 | 37 | | |
| 11.5 | | | | 20 | 18 | 16 | 14 | — | | | | | 32 | | | | 71 | 60 | 53 | 47 | 42 | 40 | | |
| 12 | | | | 23 | 20 | 17 | 15 | 13 | | | | | 33 | | | | 77 | 66 | 56 | 51 | 46 | 41 | | |
| 12.5 | | | | 25 | 21 | — | — | 15 | 14 | | | | 34 | | | | | 69 | 59 | 55 | 47 | 44 | | |
| 13 | | | | 28 | 23 | 19 | 17 | 16 | — | 14 | | | 35 | | | | | 76 | 65 | 58 | 50 | 46 | | |
| 13.5 | | | | 30 | — | 21 | 18 | — | 16 | — | | | 36 | | | | | 79 | 68 | 61 | 53 | 49 | | |
| 14 | | | | 33 | — | 22 | 20 | 18 | 17 | 15 | | | 37 | | | | | | 71 | 64 | 57 | 52 | | |
| 14.5 | | | | 36 | — | 24 | 22 | 19 | — | 17 | | | 38 | | | | | | 78 | 67 | 60 | 55 | | |
| 15 | | | | 39 | — | 27 | 23 | 21 | 19 | 18 | | | 40 | | | | | | | 77 | 69 | 62 | | |
| 16 | | | | 45 | — | 30 | — | 24 | 22 | 20 | 17 | | 42 | | | | | | | | 75 | 68 | | |
| 17 | | | | 52 | — | 35 | — | 27 | 24 | 21 | 19 | | 44 | | | | | | | | 82 | 74 | | |
| 18 | | | | 59 | — | 39 | — | 30 | — | 25 | 22 | 19 | | | | | | | | | | | | |

The amount of this loss in excess of straight duct friction is termed **dynamic loss**.

Dynamic losses result from flow disturbances caused by fittings that change the airflow path's direction and/or area. These fittings include entries, exits, transitions, and junctions. Idelchik et al. (1986) discuss parameters affecting fluid resistance of fittings and presents loss coefficients in three forms: tables, curves, and equations.

The following dimensionless coefficient is used for fluid resistance since this coefficient has the same value in dynamically similar streams (i.e., streams with geometrically similar stretches), equal values of Reynolds number, and equal values of other criteria necessary for dynamic similarity. The fluid resistance coefficient represents the ratio of total pressure loss to velocity pressure at the referenced cross section.

$$C = \frac{\Delta p_j}{\rho(V/1097)^2} = \frac{\Delta p_j}{p_v} \qquad (9\text{-}8 \text{ I-P})$$

$$C = \frac{\Delta p_j}{\rho(V^2/2)} = \frac{\Delta p_j}{p_v} \qquad (9\text{-}8 \text{ SI})$$

where
 $C$ = local loss coefficient, dimensionless
 $\Delta p_j$ = fitting total pressure loss, in. of water (Pa)
 $\rho$ = density, $lb_m/ft^3$ (kg/m$^3$)
 $V$ = velocity, fpm (m/s)
 $p_v$ = velocity pressure, in. of water (Pa)

Dynamic losses occur along a duct length and cannot be separated from frictional losses. For ease of calculation, dynamic losses are assumed to be concentrated at a section (local) and to exclude friction. Frictional losses must be considered only for relatively long fittings. Generally, fitting friction losses are accounted for by measuring duct lengths from the centerline of one fitting to that of the next fitting. For fittings closely coupled (less than six hydraulic diameters apart), the flow pattern entering subsequent fittings differs from the flow pattern used to determine loss coefficients. Adequate data for these situations are unavailable.

For all fittings, except junctions, calculate the total pressure loss $\Delta p_j$ at a section by

$$\Delta p_j = C_o p_{v,o} \qquad (9\text{-}9)$$

where the subscript $o$ is the cross section at which the velocity pressure is referenced. The dynamic loss is based on the actual velocity in the duct, not the velocity in an equivalent noncircular duct. Where necessary (unequal area fittings), convert a loss coefficient from section $o$ to section $i$ by Equation (9-10), where $V$ is the velocity at the respective sections.

$$C_i = \frac{C_o}{(V_i/V_o)^2} \qquad (9\text{-}10)$$

For converging and diverging flow junctions, total pressure losses through the straight (main) section are calculated as

$$\Delta p_j = C_{c,s} p_{v,c} \qquad (9\text{-}10a)$$

For total pressure losses through the branch section,

$$\Delta p_j = C_{c,b} p_{v,c} \qquad (9\text{-}10b)$$

where $P_{v,c}$ is the velocity pressure at the common section $c$, and $C_{c,s}$ and $C_{c,b}$ are losses for the straight (main) and branch flow paths, respectively, each referenced to the velocity pressure at section $c$. To convert junction local loss coefficients referenced to straight and branch velocity pressures, use Equation (9-10c).

$$C_i = \frac{C_{c,i}}{(V_i/V_c)^2} \quad (9\text{-}10c)$$

where

$C_i$ = local loss coefficient referenced to section being calculated (see subscripts), dimensionless

$C_{c,i}$ = straight ($C_{c,s}$) or branch ($C_{c,b}$) local loss coefficient referenced to dynamic pressure at the common section, dimensionless

$V_i$ = velocity at section to which $C_i$ is being referenced

$V_c$ = velocity at common section

*Subscripts*:

$b$ = branch

$s$ = straight (main) section

$c$ = common section

The junction of two parallel streams moving at different velocities is characterized by turbulent mixing of the streams, accompanied by pressure losses. In the course of this mixing, an exchange of momentum takes place between the particles moving at different velocities, finally resulting in the equalization of the velocity distributions in the common stream. The jet with higher velocity loses a part of its kinetic energy by transmitting it to the slower moving jet. The loss in total pressure before and after mixing is always large and positive for the higher velocity jet and increases with an increase in the amount of energy transmitted to the lower velocity jet. Consequently, the local loss coefficient, defined by Equation (9-8), will always be positive. The energy stored in the lower velocity jet increases as a result of mixing. The loss in total pressure and the local loss coefficient can, therefore, also have negative values for the lower velocity jet (Idelchik et al. 1986).

### 9.1.5 Ductwork Sectional Losses

Total pressure loss in a duct section is calculated by combining Equations (9-7) and (9-8) in terms of $\Delta p$, where $\Sigma C$ is the sum of local loss coefficients within the duct section. Each fitting loss coefficient must be referenced to that section's velocity pressure.

$$\Delta p = \left(\frac{12 fL}{D} + \Sigma C\right)\rho(V/1097)^2 \quad (9\text{-}11 \text{ I-P})$$

$$\Delta p = \left(\frac{1000 fL}{D} + \Sigma C\right)\rho(V^2/2) \quad (9\text{-}11 \text{ SI})$$

### 9.1.6 System Analysis

Since $p_t = p_s + p_v$

$$\Delta p_t = p_{t,1} - p_{t,2} \quad (9\text{-}12)$$

where $\Delta p_t$ is the change in total pressure in the direction of airflow between stations 1 and 2. For all main and branch ducts in a system, including both supply and return air ductwork, Equation (9-12) may be used for any fitting/duct or section of ductwork. The path with the greatest resistance to flow, usually the longest and most complicated, is known as the critical path. The minimum total pressure to be developed by the fan is the summation of duct/fitting losses throughout the **critical path** of the system, plus the fan system effect factors. Thus,

$$\Delta p_t = \sum_{i=1}^{n} \Delta p_i + \text{SEF}_s + \text{SEF}_d \quad (9\text{-}13)$$

where

$\Delta p_t$ = fan's total pressure, in. of water (Pa)

$\Delta p_i$ = component total pressure loss, in. of water (Pa)

$n$ = number of ducts/fittings in critical path of system

$\text{SEF}_s$ = system effect factor due to fan inlet conditions, in. of water (Pa)

$\text{SEF}_d$ = system effect due to fan outlet conditions, in. of water (Pa)

For some simple systems, $\text{SEF}_s$ and $\text{SEF}_d$ can be negligible. However, for typical systems, these quantities must be estimated. Also, duct heat transfer must be considered in duct system design. Information concerning these items is given in Chapter 21 of the 2013 *ASHRAE Handbook—Fundamentals*.

### 9.1.7 Duct Design

This discussion refers to central station ducts for commercial and industrial heating, ventilating, and air-conditioning systems. The equal friction and static regain methods given yield the static pressure required to overcome the resistance of the ductwork, including the supply outlets and return intakes. The fan selected for the duct system must produce not only this pressure but also the additional pressure required by the central equipment such as washers or spray chambers, heating or cooling coils, and filters. Pressure losses of these components should be obtained from the manufacturers' catalogs.

Rules that should be followed in the design of ducts are

1. Convey air as directly as possible at the permissible velocities to obtain the desired results with minimum noise and greatest economy of power, material, and space.
2. Avoid sudden changes in air direction or velocity. When sudden changes are necessary at bends, use turning vanes to minimize the pressure loss.
3. Where the greatest air-carrying capacity per unit area of sheet metal is desired, make rectangular ducts as nearly square as possible. Avoid aspect ratios (ratio of width to depth) greater than 8 to 1. Where possible, maintain a ratio of 4 to 1 or less.
4. Ducts should be constructed of smooth material, such as steel or aluminum sheet metal. For ducts made from other materials, allow for the change in roughness.

# Duct and Pipe Sizing

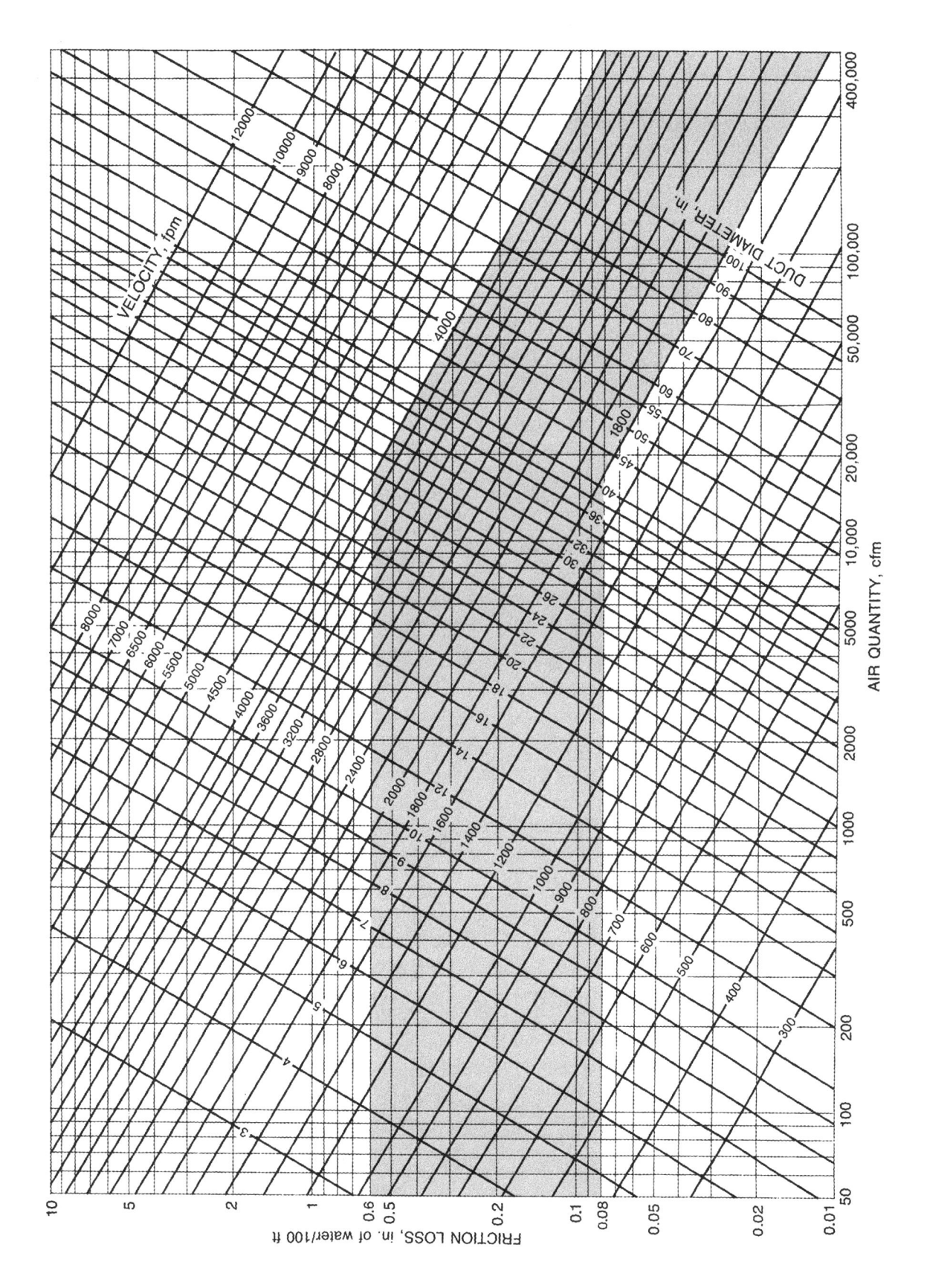

*Fig. 9-2 Friction Chart for Round Duct ($\rho = 0.075\ lb_m/ft^3$ and $\varepsilon = 0.0003\ ft$)*
*(Figure 9, Chapter 21, 2013 ASHRAE Handbook—Fundamentals)*

5. A reasonable estimate of the flow resistances offered by the system can be obtained through the following design procedures. However, in actual installations, resistances may vary considerably from the calculated values because of variation in the smoothness of materials, types of joints used, and the ability to fabricate the system in accordance with the design. Select fans and motors to provide at least a slight factor of safety, and install dampers in each branch outlet for balancing the system.

6. Avoid obstructing ducts with piping, conduits, or structural members. Unavoidable duct obstructions must be streamlined with an easement or a tear-drop, the length of which should be at least three times the thickness of the tear-drop.

### 9.1.8 Design Velocities

It is impossible to give specific rules for selecting duct velocities and duct shapes (rectangular, round, or oval) without considering cost and system constraints. An ideal design has minimum owning and operating costs when all constraints on the design are considered. The velocity and friction loss rate ranges indicated in Figure 9-2 are offered as preliminary design values. In the constant velocity design method, ignore the limits on friction loss rate. Do not use Figure 9-2 indiscriminately, as noise generation throughout a system increases as the velocity increases.

A summary of recommended velocities for HVAC components encountered in built-up systems is presented in Table 9-2. Final component selection should be based on the various chapters in 2012 *ASHRAE Handbook—HVAC Systems and Equipment* or from manufacturers.

### 9.1.9 Design Methods

The transmission of air at high velocities has gained wide acceptance in comfort air-conditioning and ventilation systems. This acceptance is due partly to improved fans and special sound attenuation and control equipment and partly to improved design and installation methods based on a better understanding of the design and installation of high-velocity air-conditioning systems.

The design of high-velocity duct systems involves a compromise between reduction of duct size and the consequent need for higher fan power. While the duct size and air velocities are governed in large part by the space available, the maximum velocities (given later in this section) should not be exceeded without carefully examining all factors.

The following principles and laws apply to all duct systems regardless of the design method used or the numerical values obtained.

1. The measure of the amount of energy required to move air from one location to another is the change (decrease) in the total pressure within the system.
2. The total pressure $p_t$ at any location within a system is a measure of the total mechanical energy at that location. It is the sum of the static pressure and the velocity pressure.

**Table 9-2  Typical Design Velocities for HVAC Components**
*(Table 6, Chapter 21 of the 2013 ASHRAE Handbook—Fundamentals)*

| Component | Face Velocity, fpm |
|---|---|
| Terminal Units | Inlet Velocity, Maximum: 2000 |
|  | Velocity Pressure, Minimum: 0.02 in. water |
| Louvers[a] |  |
| Intake |  |
| 7000 cfm and greater | 400 |
| Less than 7000 cfm | See Figure 17 |
| Exhaust |  |
| 5000 cfm and greater | 500 |
| Less than 5000 cfm | See Figure 17 |
| Filters[b] |  |
| Panel filters |  |
| Viscous impingement | 200 to 800 |
| Dry extended-surface |  |
| Flat (low efficiency) | Duct velocity |
| Pleated media (intermediate efficiency) | Up to 750 |
| HEPA | 250 to 500 |
| Renewable media filters |  |
| Moving-curtain viscous impingement | 500 |
| Heating Coils[c] |  |
| Steam and hot water | 500 to 1000 |
|  | 200 min., 1500 max. |
| Electric |  |
| Open wire | Refer to mfg. data |
| Finned tubular | Refer to mfg. data |
| Dehumidifying Coils[d] | 400 to 500 |
| Air Washers[e] |  |
| Spray type | Refer to mfg. data |
| Cell type | Refer to mfg. data |
| High-velocity spray type | 1200 to 1800 |

[a]Based on assumptions presented in text.
[b]Abstracted from Chapter 29, 2012 *ASHRAE Handbook—HVAC Systems and Equipment*.
[c]Abstracted from Chapter 27, 2012 *ASHRAE Handbook—HVAC Systems and Equipment*.
[d]Abstracted from Chapter 23, 2012 *ASHRAE Handbook—HVAC Systems and Equipment*.
[e]Abstracted from Chapter 41, 2012 *ASHRAE Handbook—HVAC Systems and Equipment*.

3. In any duct system, the total pressure always decreases in the direction of airflow.
4. In any system having two or more branches, the losses in total pressure between the fan and the end of each branch are the same.
5. Static pressure and velocity pressure are mutually convertible and can either increase or decrease in the direction of flow. For example, in a straight run of duct, the static pressure decreases, the velocity pressure remains constant, and the total pressure (their sum) decreases. In a gradually diverging section (area increase), the velocity pressure decreases, the static pressure increases, and the total pressure remains the same (neglecting the small friction loss).

The most common methods of air duct design are (1) equal friction, (2) velocity reduction, and (3) static regain and variations such as total pressure. No single duct design method automatically produces the most economical duct system for all conditions; the system design with the minimum owning and operating cost depends on both the application and ingenuity of the designer.

**Equal Friction Method.** The principle of this method is to size a system's ductwork for a constant pressure loss per unit length of duct. At higher airflow rates, it may be necessary to limit the velocity so as not to generate objectionable noises. For an initial design, the friction loss per unit length of duct for the corresponding recommended velocities is given in Figure 9-2.

Once the system is sized, the total pressure losses for the main and branch sections from junction-to-junction/fan/ terminal may be calculated and the total pressure grade line plotted. To optimize a system, additional designs are necessary to establish the annual system and power cost curves, and to find the minimum point on the total owning and operating cost curve. For system costs, only the incremental differences due to the redesign need to be considered.

After the system has been designed and the total pressure grade line plotted, sections of ductwork may be redesigned to achieve an approximate balance at the junctions without relying entirely on balancing dampers.

**Velocity Reduction Method.** This method consists of selecting the velocity at the fan discharge and designing for progressively lower velocities in the main duct at each junction or branch duct. For the selected velocities and known airflow rates, the various duct diameters may be read directly from Figure 9-2 and the equivalent rectangular sizes obtained from Equation (9-6). The return air ductwork is sized similarly, starting with the highest velocity at the fan suction and decreasing progressively in the direction of the return intakes. With the ducts sized and the fittings known, the total pressure losses can be calculated, the pressure gradients plotted, and the maximum pressure loss or critical path of the system established.

A refinement of this method involves sizing the branch ducts to dissipate the pressure available at the entrance to each. The pressure loss of the ductwork between the fan and first branch take-off is subtracted from the total fan pressure to obtain the available pressure at the first junction. Through trial, a branch velocity is found that results in the branch pressure loss being equal to or less than the pressure available. The procedure is repeated for each branch.

If the fan is specified so that the total pressure available for the system is known, the method consists of finding, through trial, the velocities in the main and branch ducts that will result in a pressure loss equal to or less than the pressure available. The branch ducts are sized as discussed previously.

**Static Regain Method.** In the static regain method, the ducts are sized so that the increase in static pressure (static regain) at each take-off offsets the pressure loss of the succeeding section of ductwork. This method is especially suited for high-velocity installations having long runs with many take-offs or terminal units. Approximately the same static pressure exists at the entrance to each branch, which simplifies outlet or terminal unit selection and system balancing. With the ductwork sized by this method, the system's total pressure losses can be calculated. A disadvantage of this method is that excessively large ducts (low velocities) result at the ends of long duct runs.

The total pressure design method is an adaptation of the static regain method. This method is advantageous since the intermediate system pressures and control of duct sizes and velocities are known.

### 9.1.10 Duct Design Procedures

The general procedure for duct design is as follows:
1. Study the plans of the building and arrange the supply/return outlets to provide proper distribution of air within each space. Adjust calculated actual air quantities for duct heat gains or losses and duct leakage. Adjust the supply, return, and/or exhaust air quantities to meet space pressurization requirements.
2. Select outlet sizes from manufacturers' data. (Refer to Chapter 20, Space Air Diffusion, of the 2013 *ASHRAE Handbook—Fundamentals*.)
3. Sketch the duct system, connect the supply outlets and return intakes with the central station apparatus, and avoid all structural obstructions and equipment. Be aware of system space allocations. Use round duct whenever feasible.
4. Divide the system into sections and number each section. A duct system should be divided at all points where flow, size, or shape changes. Assign fittings to the section toward the supply and return (or exhaust) terminals.
5. Size ducts by the selected design method. Calculate system total pressure loss; then select the fan (refer to Chapter 21 in the 2012 *ASHRAE Handbook—HVAC Systems and Equipment*).
6. Lay out the system in detail. If duct routing and fittings vary significantly from the original design, recalculate the pressure losses. Reselect the fan if necessary.
7. Resize duct sections to approximately balance pressures at each junction.
8. Analyze the design for objectionable noise levels and specify sound attenuators as necessary.

### 9.1.11 Fitting Loss Coefficients

A duct fitting database, which includes more than 220 round, rectangular, and flat oval fittings, is available from ASHRAE (2012) in electronic form with the capability to be linked to duct design programs. The fittings are numbered (coded) as shown in Table 9-3.

For convenience, a selection of dynamic loss coefficients from various sources is given in Table 9-4 for use with the problems in this book. **Use of the 2012 *ASHRAE Duct Fitting Database* is recommended for actual projects.**

### 9.1.12 Automated Duct Design

Duct design calculations have been automated by computers. Automated duct design offers features such as (1) standardization of duct design, (2) stored loss coefficients for fittings, (3) stored duct construction and thermal insulation standards, (4) balancing analysis, (5) noise analysis, (6) duct

Table 9-3 Duct Fitting Codes

| Fitting Function | Geometry | Category | Sequence Number |
|---|---|---|---|
| S: **S**upply | **D**: round (**D**iameter) | 1. Entries | 1,2,3,...*n* |
|  |  | 2. Exits |  |
| E: **E**xhaust/ Return | R: **R**ectangular | 3. Elbows |  |
|  |  | 4. Transitions |  |
|  | F: **F**lat **o**val | 5. Junctions |  |
| C: **C**ommon (supply and return) |  | 6. Obstructions |  |
|  |  | 7. Fan and system interactions |  |
|  |  | 8. Duct-mounted equipment |  |
|  |  | 9. Dampers |  |
|  |  | 10. Hoods |  |

heat gain or loss analysis, (7) material takeoffs, and (8) documentation. For available duct design programs and hardware requirements, refer to Chapter 21 of the 2009 *ASHRAE Handbook—Fundamentals*.

Duct and Pipe Sizing    319

*Table 9-4  Fitting Loss Coefficients*

## ENTRIES

**1-1  Duct Mounted in Wall (Hood, Nonenclosing, Flanged, and Unflanged) (Idelchik et al. 1986, Diagram 3-1)**

**General.** If entry has a screen, use Fitting 6-7 to calculate screen resistance.

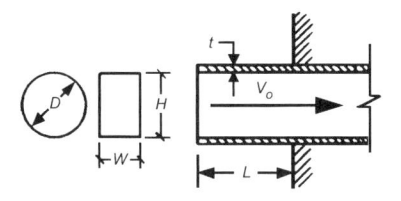

Rectangular: $D = 2HW/(H + W)$

| | $C_o$ | | | | | |
|---|---|---|---|---|---|---|
| | | | $L/D$ | | | |
| $t/D$ | 0 | 0.002 | 0.01 | 0.05 | 0.2 | 0.5 | ≥1.0 |
| ≈0 | 0.50 | 0.57 | 0.68 | 0.80 | 0.92 | 1.0 | 1.0 |
| 0.02 | 0.50 | 0.51 | 0.52 | 0.55 | 0.66 | 0.72 | 0.72 |
| ≥0.05 | 0.50 | 0.50 | 0.50 | 0.50 | 0.50 | 0.50 | 0.50 |

**1-2  Smooth Converging Bellmouth Without End Wall (Idelchik et al. 1986, Diagram 34)**

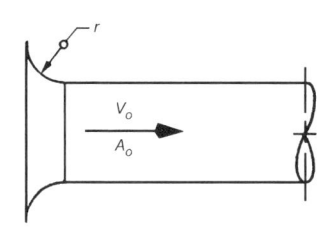

| $r/D$ | 0 | 0.01 | 0.02 | 0.03 | 0.04 | 0.05 |
|---|---|---|---|---|---|---|
| $C_o$ | 1.0 | 0.87 | 0.74 | 0.61 | 0.51 | 0.40 |
| $r/D$ | 0.06 | 0.08 | 0.10 | 0.12 | 0.16 | ≥0.20 |
| $C_o$ | 0.32 | 0.20 | 0.15 | 0.10 | 0.06 | 0.03 |

**1-3  Smooth Converging Bellmouth with End Wall (Idelchik et al. 1986, Diagram 3-4)**

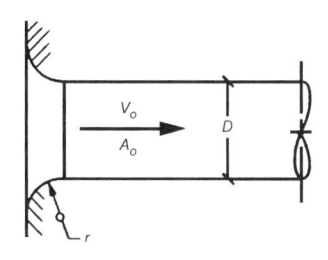

| $r/D$ | 0 | 0.01 | 0.02 | 0.03 | 0.04 | 0.05 |
|---|---|---|---|---|---|---|
| $C_o$ | 0.50 | 0.44 | 0.37 | 0.31 | 0.26 | 0.22 |
| $r/D$ | 0.06 | 0.08 | 0.10 | 0.12 | 0.16 | ≥0.20 |
| $C_o$ | 0.20 | 0.15 | 0.12 | 0.09 | 0.06 | 0.03 |

**1-4  Conical Converging Bellmouth Without End Wall, Round and Rectangular**

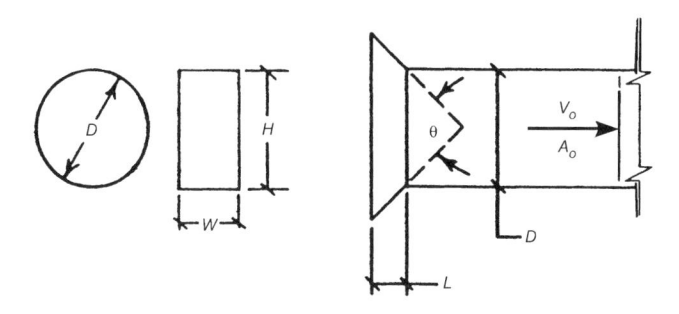

Rectangular: $D = 2HW/(H + W)$

| | $C_o$ | | | | | | | | | |
|---|---|---|---|---|---|---|---|---|---|---|
| | $\theta$, degrees | | | | | | | | | |
| $L/D$ | 0 | 10 | 20 | 30 | 45 | 60 | 90 | 120 | 150 | 180 |
| 0.025 | 1.0 | 0.96 | 0.93 | 0.90 | 0.85 | 0.80 | 0.72 | 0.64 | 0.57 | 0.50 |
| 0.05 | 1.0 | 0.93 | 0.86 | 0.80 | 0.73 | 0.67 | 0.60 | 0.56 | 0.52 | 0.50 |
| 0.10 | 1.0 | 0.80 | 0.67 | 0.55 | 0.46 | 0.41 | 0.41 | 0.43 | 0.46 | 0.50 |
| 0.25 | 1.0 | 0.68 | 0.45 | 0.30 | 0.21 | 0.17 | 0.21 | 0.28 | 0.38 | 0.50 |
| 0.60 | 1.0 | 0.46 | 0.27 | 0.18 | 0.14 | 0.13 | 0.19 | 0.27 | 0.37 | 0.50 |
| 1.0 | 1.0 | 0.32 | 0.20 | 0.14 | 0.11 | 0.10 | 0.16 | 0.24 | 0.35 | 0.50 |

**1-5  Conical Converging Bellmouth with End Wall, Round and Rectangular (Idelchik et al. 1986, Diagram 3-7)**

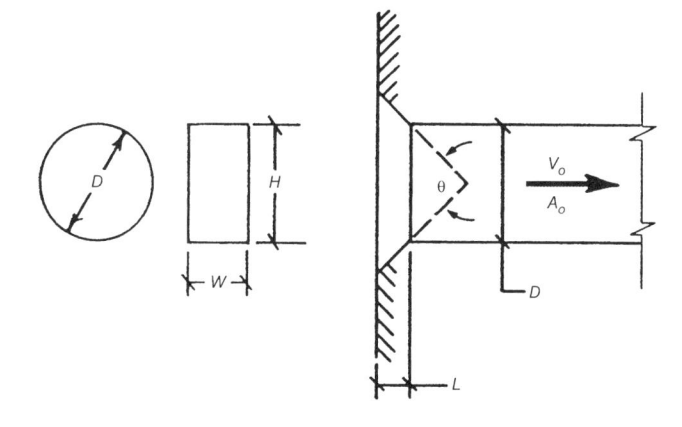

Rectangular: $D = 2HW/(H + W)$

| | $C_o$ | | | | | | | | | |
|---|---|---|---|---|---|---|---|---|---|---|
| | $\theta$, degrees | | | | | | | | | |
| $L/D$ | 0 | 10 | 20 | 30 | 45 | 60 | 90 | 120 | 150 | 180 |
| 0.025 | 0.50 | 0.47 | 0.45 | 0.43 | 0.41 | 0.40 | 0.42 | 0.44 | 0.46 | 0.50 |
| 0.05 | 0.50 | 0.45 | 0.41 | 0.36 | 0.32 | 0.30 | 0.34 | 0.39 | 0.44 | 0.50 |
| 0.075 | 0.50 | 0.42 | 0.35 | 0.30 | 0.25 | 0.23 | 0.28 | 0.35 | 0.43 | 0.50 |
| 0.10 | 0.50 | 0.39 | 0.32 | 0.25 | 0.21 | 0.18 | 0.25 | 0.33 | 0.41 | 0.50 |
| 0.15 | 0.50 | 0.37 | 0.27 | 0.20 | 0.16 | 0.15 | 0.23 | 0.31 | 0.40 | 0.50 |
| 0.60 | 0.50 | 0.27 | 0.18 | 0.13 | 0.11 | 0.12 | 0.20 | 0.30 | 0.40 | 0.50 |

## 1-6  Intake Hood (Idelchik et al. 1986, Diagram 3-18)

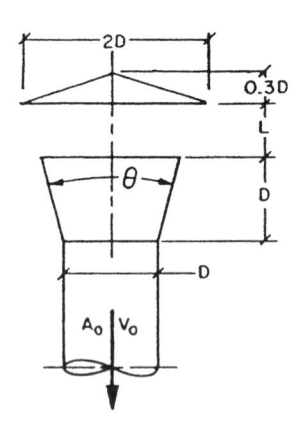

| θ, degrees | $C_o$ L/D |||||||||
|---|---|---|---|---|---|---|---|---|---|
| | 0.1 | 0.2 | 0.3 | 0.4 | 0.5 | 0.6 | 0.8 | 1.0 | 4.0 |
| 0 | 2.63 | 1.83 | 1.53 | 1.39 | 1.31 | 1.19 | 1.08 | 1.06 | 1.0 |
| 15 | 1.32 | 0.77 | 0.60 | 0.48 | 0.41 | 0.30 | 0.28 | 0.25 | 0.25 |

## 1-7  Hood, Tapered, Flanged or Unflanged (Brandt and Steffy 1946)

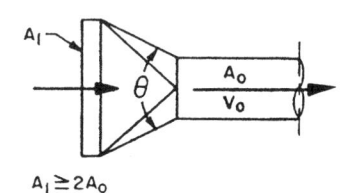

θ is major angle for rectangular hoods

| Hood Shape: Round ||||||||||
|---|---|---|---|---|---|---|---|---|---|
| θ, degrees | 0 | 20 | 40 | 60 | 80 | 100 | 120 | 140 | 160 | 180 |
| $C_o$ | 1.0 | 0.11 | 0.06 | 0.09 | 0.14 | 0.18 | 0.27 | 0.32 | 0.43 | 0.50 |

| Hood Shape: Square or Rectangular ||||||||||
|---|---|---|---|---|---|---|---|---|---|
| θ, degrees | 0 | 20 | 40 | 60 | 80 | 100 | 120 | 140 | 160 | 180 |
| $C_o$ | 1.0 | 0.19 | 0.13 | 0.16 | 0.21 | 0.27 | 0.33 | 0.43 | 0.53 | 0.62 |

## 1-8  Orifice, Sharp-Edged, Inlet Duct (Idelchik et al. 1986; Diagrams 3-12, 3-14, and 4-19)

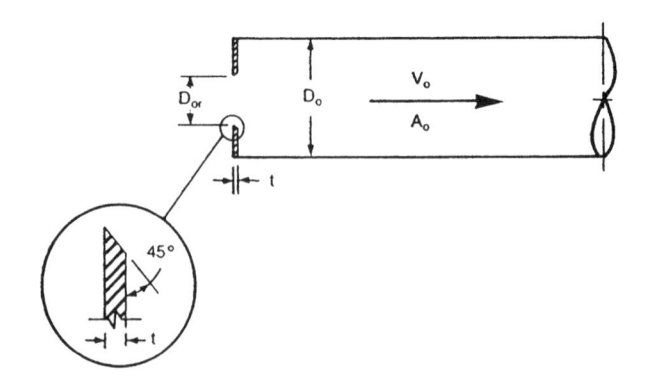

$t/D_{or} \leq 0.015$
$Re = D_o V_o / \nu$
$A_{or}$: orifice area

| | $C_o$ ||||||||
|---|---|---|---|---|---|---|---|---|
| | $A_{or}/A_o$ ||||||||
| $Re \times 10^{-3}$ | 0.2 | 0.3 | 0.4 | 0.5 | 0.6 | 0.7 | 0.8 | 0.9 |
| 4 | 45 | 18 | 7.9 | 3.9 | 2.3 | 1.3 | 0.83 | 0.51 |
| 10 | 49 | 20 | 9.2 | 4.4 | 2.7 | 1.5 | 0.96 | 0.59 |
| 20 | 50 | 21 | 9.3 | 4.9 | 2.9 | 1.6 | 1.1 | 0.65 |
| 100 | 55 | 23 | 11.0 | 5.6 | 3.3 | 1.9 | 1.2 | 0.75 |

## EXITS

**General.** If exit has a screen, use Fitting 6-7 to calculate screen resistance.

### 2-1  Exit, Abrupt, Round and Rectangular (Idelchik et al. 1986, Diagram 11-1)

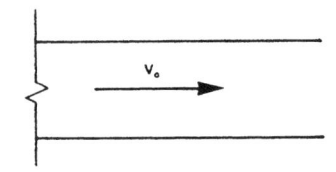

*Uniform Velocity Distribution*

$$C_o = 1.0$$

*Exponential, Sinusoidal, Asymmetrical, and Parabolic Velocity Distribution*

$C_o$ varies from 1.0 to 3.67. For details, consult Idelchik (1986), Diagram 11-1.

### 2-2  Exit, Abrupt, Round and Rectangular, with End (Idelchik et al. 1986, Diagrams 5-2 and 5-4)

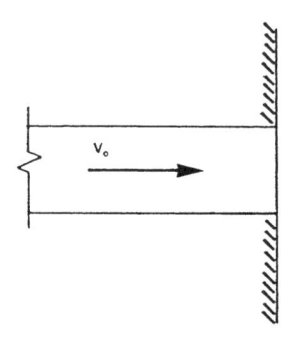

$C_o = 0.88$

### 2-3  Exit, Duct Flush with Wall, Flow along Wall (Idelchik et al. 1986, Diagram 11-2)

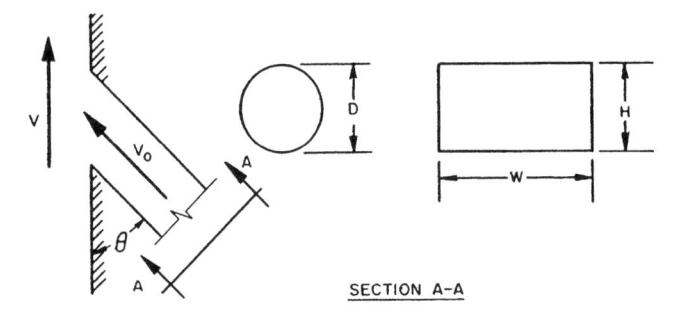

# Duct and Pipe Sizing

**Round**

| θ, degrees | $C_o$ |  |  |  |  |
|---|---|---|---|---|---|
|  | $V/V_o$ |  |  |  |  |
|  | 0 | 0.5 | 1.0 | 1.5 | 2.0 |
| ≤45 | 1.0 | 1.0 | 1.1 | 1.3 | 1.6 |
| 60 | 1.0 | 0.90 | 1.1 | 1.4 | 1.6 |
| 90 | 1.0 | 0.80 | 0.95 | 1.4 | 1.7 |
| 120 | 1.0 | 0.80 | 0.95 | 1.3 | 1.7 |
| 150 | 1.0 | 0.82 | 0.83 | 1.0 | 1.3 |

**Rectangular**

| Aspect Ratio (H/W) | θ, degrees | $C_o$ |  |  |  |  |
|---|---|---|---|---|---|---|
|  |  | $V/V_o$ |  |  |  |  |
|  |  | 0 | 0.5 | 1.0 | 1.5 | 2.0 |
| ≤0.2 | ≤90 | 1.0 | 0.95 | 1.2 | 1.5 | 1.8 |
|  | 120 | 1.0 | 1.1 | 1.1 | 1.4 | 1.9 |
|  | 150 | 1.0 | 0.95 | 0.95 | 1.4 | 1.8 |
| 0.5-2.0 | ≤45 | 1.0 | 1.0 | 1.1 | 1.3 | 1.6 |
|  | 60 | 1.0 | 0.90 | 1.1 | 1.4 | 1.6 |
|  | 90 | 1.0 | 0.80 | 0.95 | 1.4 | 1.7 |
|  | 120 | 1.0 | 0.80 | 0.95 | 1.3 | 1.7 |
|  | 150 | 1.0 | 0.82 | 0.83 | 1.0 | 1.3 |
| ≥5 | 45 | 1.0 | 0.92 | 0.93 | 1.1 | 1.3 |
|  | 60 | 1.0 | 0.87 | 0.87 | 1.0 | 1.3 |
|  | 90 | 1.0 | 0.82 | 0.80 | 0.97 | 1.2 |
|  | 120 | 1.0 | 0.80 | 0.76 | 0.90 | 0.98 |

**2-4 Exit, Round, Diverging (Idelchik et al. 1986, Diagram 11-3)**

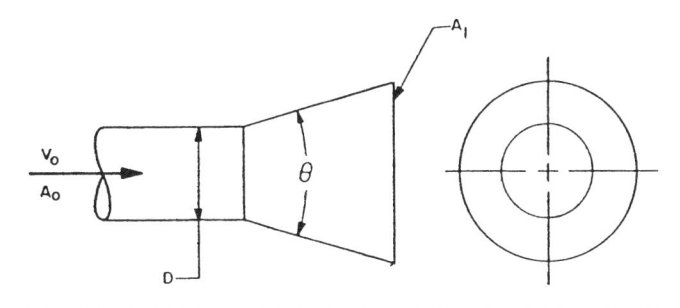

| $A_1/A_o$ | $C_o$ |  |  |  |  |  |  |
|---|---|---|---|---|---|---|---|
|  | θ, degrees |  |  |  |  |  |  |
|  | 8 | 10 | 14 | 20 | 30 | 45 | ≥60 |
| 2 | 0.36 | 0.33 | 0.37 | 0.51 | 0.90 | 1.0 | 1.0 |
| 4 | 0.24 | 0.21 | 0.28 | 0.40 | 0.70 | 0.99 | 1.0 |
| 6 | 0.20 | 0.19 | 0.26 | 0.37 | 0.67 | 0.99 | 1.0 |
| 10 | 0.18 | 0.16 | 0.24 | 0.36 | 0.68 | 0.99 | 1.0 |
| 16 | 0.16 | 0.16 | 0.20 | 0.36 | 0.66 | 0.99 | 1.0 |

**2-5 Exit, Round, with End Wall Transition (Idelchik et al. 1986, Diagram 5-8)**

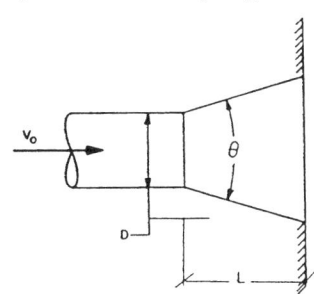

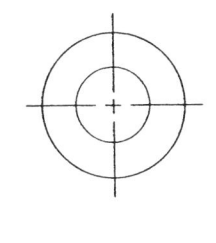

θ = optimum angle

| L/D | 0.5 | 1.0 | 2.0 | 3.0 | 4.0 | 5.0 | 6.0 | 8.0 | 10 | 12 | 14 |
|---|---|---|---|---|---|---|---|---|---|---|---|
| θ, degrees | 34 | 24 | 16 | 13 | 11 | 10 | 9 | 8 | 7 | 6 | 6 |
| $C_o$ | 0.41 | 0.32 | 0.24 | 0.20 | 0.17 | 0.15 | 0.14 | 0.12 | 0.11 | 0.11 | 0.10 |

**2-6 Exit, Rectangular, Two Sides Parallel, Diverging, Symmetrical (Idelchik et al. 1986, Diagram 11-6)**

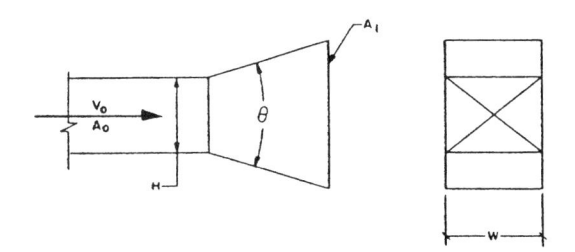

$0.5 \leq H/W \leq 2.0$

| $A_1/A_o$ | $C_o$ |  |  |  |  |  |  |
|---|---|---|---|---|---|---|---|
|  | θ, degrees |  |  |  |  |  |  |
|  | 8 | 10 | 14 | 20 | 30 | 45 | ≥60 |
| 2 | 0.50 | 0.51 | 0.56 | 0.63 | 0.80 | 0.96 | 1.0 |
| 4 | 0.34 | 0.38 | 0.48 | 0.63 | 0.76 | 0.91 | 1.0 |
| 6 | 0.32 | 0.34 | 0.41 | 0.56 | 0.70 | 0.84 | 0.96 |

**2-7 Exit, Rectangular, with Wall, Two Sides Parallel, Symmetrical, Diverging (Idelchik et al. 1986, Diagram 5-10)**

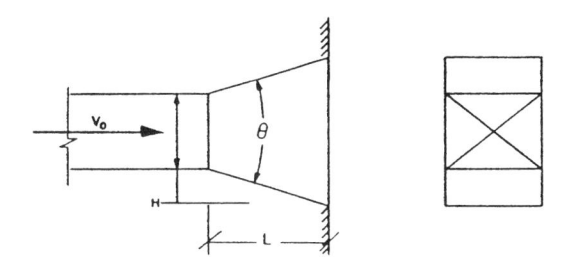

θ = optimum angle

| L/H | 0.5 | 1.0 | 2.0 | 3.0 | 4.0 | 5.0 | 6.0 | 8.0 | 10 | 12 | 14 |
|---|---|---|---|---|---|---|---|---|---|---|---|
| θ, degrees | 50 | 35 | 25 | 21 | 18 | 16 | 15 | 13 | 12 | 11 | 10 |
| $C_o$ | 0.53 | 0.44 | 0.35 | 0.31 | 0.28 | 0.25 | 0.24 | 0.22 | 0.20 | 0.19 | 0.19 |

**2-8 Exit, Rectangular, Pyramidal, Diverging (Idelchik et al. 1986, Diagram 11-5)**

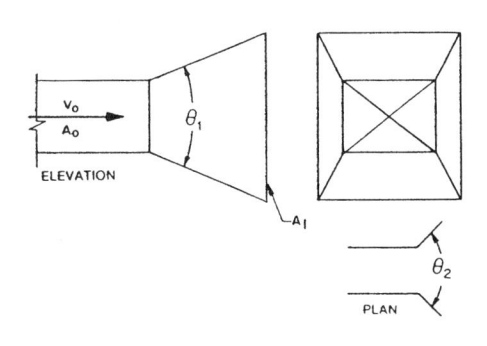

θ is larger of $θ_1$ and $θ_2$

|  | $C_o$ | | | | | | |
|---|---|---|---|---|---|---|---|
| | θ, degrees | | | | | | |
| $A_1/A_o$ | 8 | 10 | 14 | 20 | 30 | 45 | ≥60 |
| 2 | 0.65 | 0.68 | 0.74 | 0.82 | 0.92 | 1.1 | 1.1 |
| 4 | 0.53 | 0.60 | 0.69 | 0.78 | 0.90 | 1.0 | 1.1 |
| 6 | 0.50 | 0.57 | 0.66 | 0.77 | 0.91 | 1.0 | 1.1 |
| 10 | 0.45 | 0.53 | 0.64 | 0.74 | 0.85 | 0.97 | 1.1 |

**2-9 Exit, Rectangular, with Wall, Pyramidal, Diverging (Idelchik et al. 1986, Diagram 5-9)**

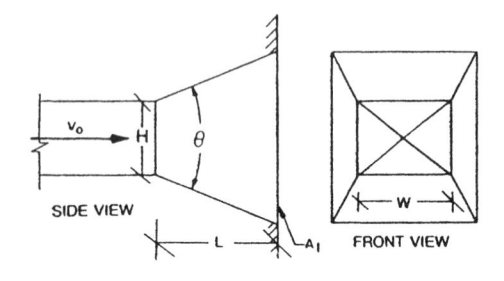

$D = 2HW/(H + W)$
θ = optimum angle

| $L/D$ | 0.5 | 1.0 | 2.0 | 3.0 | 4.0 | 5.0 | 6.0 | 8.0 | 10 | 12 | 14 |
|---|---|---|---|---|---|---|---|---|---|---|---|
| θ, degrees | 26 | 19 | 13 | 11 | 9 | 8 | 7 | 6 | 6 | 5 | 5 |
| $C_o$ | 0.49 | 0.40 | 0.30 | 0.26 | 0.23 | 0.21 | 0.19 | 0.17 | 0.16 | 0.15 | 0.14 |

**2-10 Exit, Discharge to Atmosphere from a 90° Elbow, Rectangular and Round (Note: Elbow Loss Included) (Idelchik et al. 1986, Diagram 11-14)**

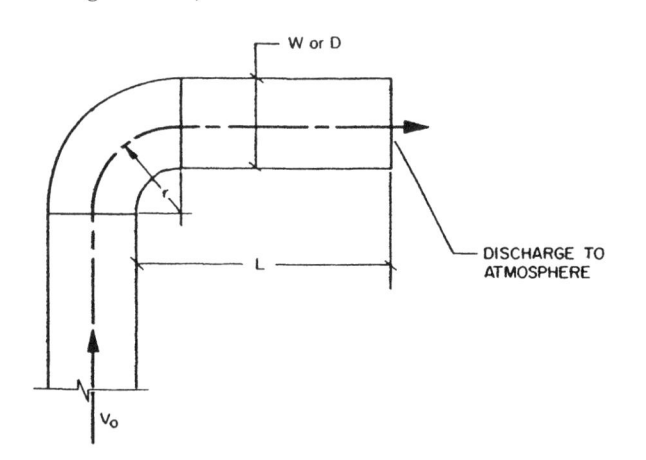

Rectangular

|  | $C_o$ | | | | | | | | | |
|---|---|---|---|---|---|---|---|---|---|---|
| | $L/W$ | | | | | | | | | |
| $r/W$ | 0 | 0.5 | 1.0 | 1.5 | 2.0 | 3.0 | 4.0 | 6.0 | 8.0 | 12.0 |
| 0 | 3.0 | 3.1 | 3.2 | 3.0 | 2.7 | 2.4 | 2.2 | 2.1 | 2.1 | 2.0 |
| 0.75 | 2.2 | 2.2 | 2.1 | 1.8 | 1.7 | 1.6 | 1.6 | 1.5 | 1.5 | 1.5 |
| 1.0 | 1.8 | 1.5 | 1.4 | 1.4 | 1.3 | 1.3 | 1.2 | 1.2 | 1.2 | 1.2 |
| 1.5 | 1.5 | 1.2 | 1.1 | 1.1 | 1.1 | 1.1 | 1.1 | 1.1 | 1.1 | 1.1 |
| 2.5 | 1.2 | 1.1 | 1.1 | 1.0 | 1.0 | 1.0 | 1.0 | 1.0 | 1.0 | 1.0 |

Round ($r/D = 1.0$)

| $L/D$ | 0.9 | 1.3 |
|---|---|---|
| $C_o$ | 1.5 | 1.4 |

**2-11 Exhaust Hood (Idelchik et al. 1986, Diagram 11-16)**

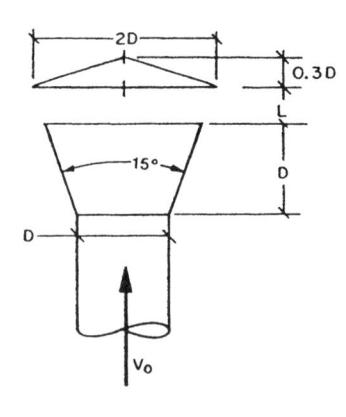

Poor Design—Should Not Be Used (see Chapter 14, Figure 13)

| $L/D$ | 0.1 | 0.2 | 0.25 | 0.3 | 0.35 | 0.4 | 0.5 | 0.6 | 0.8 | 1.0 |
|---|---|---|---|---|---|---|---|---|---|---|
| $C_o$ | 2.6 | 1.2 | 1.0 | 0.80 | 0.70 | 0.65 | 0.60 | 0.60 | 0.60 | 0.60 |

**2-12 Stackhead (Idelchik et al. 1986, Diagram 11-23)**

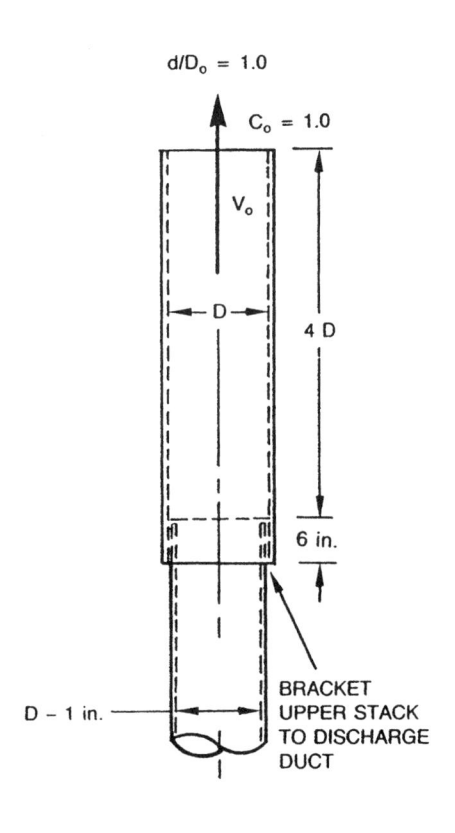

| $d/D$ | 0.3 | 0.4 | 0.5 | 0.6 | 0.7 | 0.8 | 0.9 | 1.0 |
|---|---|---|---|---|---|---|---|---|
| $C_o$ | 130 | 41 | 17 | 8.1 | 4.4 | 2.6 | 1.6 | 1.0 |

# Duct and Pipe Sizing

## ELBOWS

**3-1 Elbow, Smooth Radius (Die Stamped), Round (Locklin 1950, Equation A-10)**

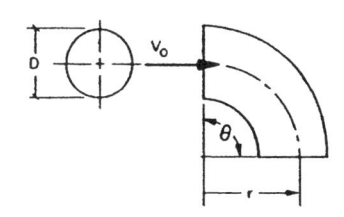

$C_o = K_q C'_o$

**Coefficients for 90° Elbows**

| r/D | 0.5 | 0.75 | 1.0 | 1.5 | 2.0 | 2.5 |
|---|---|---|---|---|---|---|
| $C'_o$ | 0.71 | 0.33 | 0.22 | 0.15 | 0.13 | 0.12 |

**Angle Correction Factors $K_q$ (Idelchik et al. 1986, Diagram 6-1):**

| θ, degrees | 0 | 20 | 30 | 45 | 60 | 75 | 90 | 110 | 130 | 150 | 180 |
|---|---|---|---|---|---|---|---|---|---|---|---|
| $K_\theta$ | 0 | 0.31 | 0.45 | 0.60 | 0.78 | 0.90 | 1.00 | 1.13 | 1.20 | 1.28 | 1.40 |

**3-2 Elbows; 3-, 4-, and 5-Pieces, Round (Locklin 1950, Figure 10)**

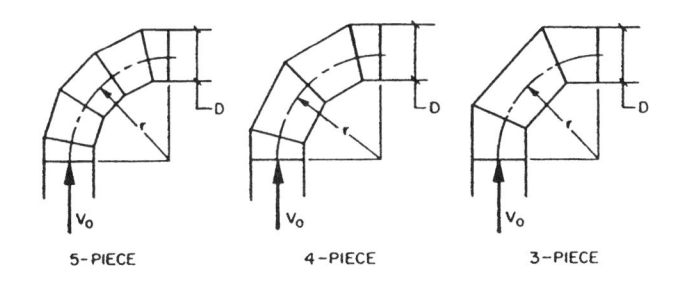

**Coefficients for 90° Elbows ($C'_o$)**

| No. of Pieces | r/D |||| 
|---|---|---|---|---|
|  | 0.75 | 1.0 | 1.5 | 2.0 |
| 5 | 0.46 | 0.33 | 0.24 | 0.19 |
| 4 | 0.50 | 0.37 | 0.27 | 0.24 |
| 3 | 0.54 | 0.42 | 0.34 | 0.33 |

**Angle Correction Factors $K_q$ (Idelchik et al. 1986, Diagram 6-1)**

| θ, degrees | 0 | 20 | 30 | 45 | 60 | 75 | 90 | 110 | 130 | 150 | 180 |
|---|---|---|---|---|---|---|---|---|---|---|---|
| $K_\theta$ | 0 | 0.31 | 0.45 | 0.60 | 0.78 | 0.90 | 1.00 | 1.13 | 1.20 | 1.28 | 1.40 |

**3-3 Elbow, Mitered, Round (Idelchik et al. 1986, Diagram 6-5)**

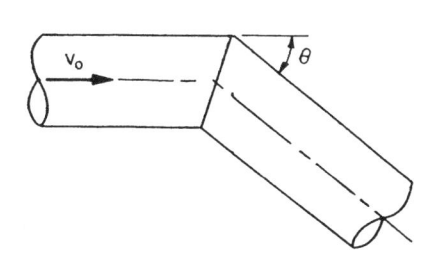

$C_o = K_{Re} C'_o$

| θ, degrees | 20 | 30 | 45 | 60 | 75 | 90 |
|---|---|---|---|---|---|---|
| $C'_o$ | 0.08 | 0.16 | 0.34 | 0.55 | 0.81 | 1.2 |

**Reynolds Number Correction Factors: M/hc**

| $Re \times 10^{-4}$ | 1 | 2 | 3 | 4 | 6 | 8 | 10 | ≥14 |
|---|---|---|---|---|---|---|---|---|
| $K_{Re}$ | 1.40 | 1.26 | 1.19 | 1.14 | 1.09 | 1.06 | 1.04 | 1.0 |

**3-4 Elbows, 30,° Z-Shaped, Round**

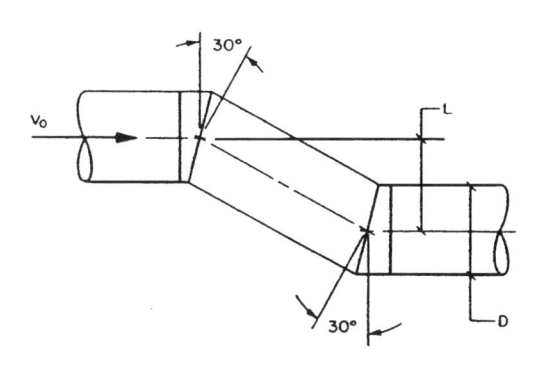

$C_o = K_{Re} C'_o$

| L/D | 0 | 0.5 | 1.0 | 1.5 | 2.0 | 2.5 | 3.0 |
|---|---|---|---|---|---|---|---|
| $C'_o$ | 0 | 0.15 | 0.15 | 0.16 | 0.16 | 0.16 | 0.16 |

**Reynolds Number Correction Factors**

| $Re \times 10^{-4}$ | 1 | 2 | 3 | 4 | 6 | 8 | 10 | ≥14 |
|---|---|---|---|---|---|---|---|---|
| $K_{Re}$ | 1.40 | 1.26 | 1.19 | 1.14 | 1.09 | 1.06 | 1.04 | 1.0 |

**3-5 Elbow, Without Vanes, Rectangular (Idelchik et al. 1986, Diagram 6-1)**

**Smooth Radius**

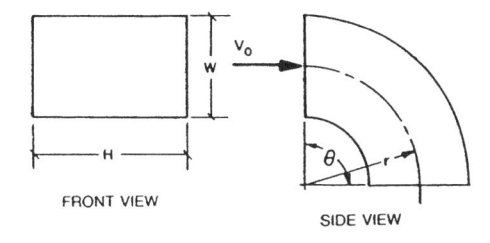

$C_o = K_\theta K_{Re} C'_o$

**90°, Sharp Throat Radius Heel (r/W = 0.5)**

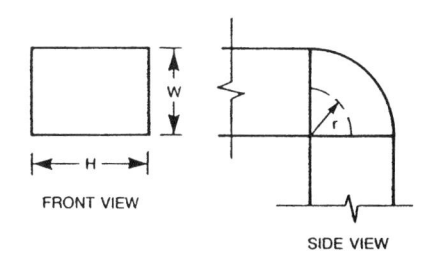

$C_o = K_{Re} C'_o$

**Coefficients for 90° Elbows ($C'_o$)**

| | H/W | | | | | | | | | | |
|---|---|---|---|---|---|---|---|---|---|---|---|
| r/W | 0.25 | 0.5 | 0.75 | 1.0 | 1.5 | 2.0 | 3.0 | 4.0 | 5.0 | 6.0 | 8.0 |
| 0.5 | 1.3 | 1.3 | 1.2 | 1.2 | 1.1 | 1.0 | 1.0 | 1.1 | 1.1 | 1.2 | 1.2 |
| 0.75 | 0.57 | 0.52 | 0.48 | 0.44 | 0.40 | 0.39 | 0.39 | 0.40 | 0.42 | 0.43 | 0.44 |
| 1.0 | 0.27 | 0.25 | 0.23 | 0.21 | 0.19 | 0.18 | 0.18 | 0.19 | 0.20 | 0.21 | 0.21 |
| 1.5 | 0.22 | 0.20 | 0.19 | 0.17 | 0.15 | 0.14 | 0.14 | 0.15 | 0.16 | 0.17 | 0.17 |
| 2.0 | 0.20 | 0.18 | 0.16 | 0.15 | 0.14 | 0.13 | 0.13 | 0.14 | 0.14 | 0.15 | 0.15 |

**Angle Correction Factor**

| θ, degrees | 0 | 20 | 30 | 45 | 60 | 75 | 90 | 110 | 130 | 150 | 180 |
|---|---|---|---|---|---|---|---|---|---|---|---|
| $K_\theta$ | 0 | 0.31 | 0.45 | 0.60 | 0.78 | 0.90 | 1.00 | 1.13 | 1.20 | 1.28 | 1.40 |

**Reynolds Number Correction Factor ($K_{Re}$)**

| | Re × 10⁻⁴ | | | | | | | | |
|---|---|---|---|---|---|---|---|---|---|
| r/W | 1 | 2 | 3 | 4 | 6 | 8 | 10 | 14 | ≥20 |
| 0.5 | 1.40 | 1.26 | 1.19 | 1.14 | 1.09 | 1.06 | 1.04 | 1.0 | 1.0 |
| ≥0.75 | 2.0 | 1.77 | 1.64 | 1.56 | 1.46 | 1.38 | 1.30 | 1.15 | 1.0 |

### 3-6 Elbow, Mitered, Rectangular (Idelchik et al. 1986, Diagram 6-5)

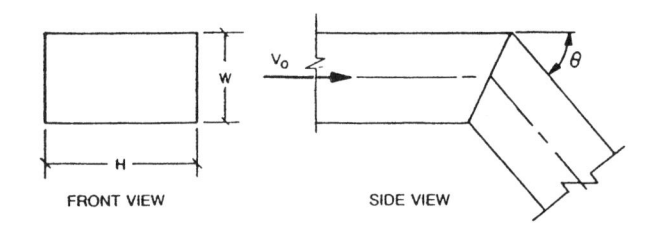

$C_o = K_{Re} C'_o$

| | $C'_o$ | | | | | | | | | | |
|---|---|---|---|---|---|---|---|---|---|---|---|
| θ, degrees | 0.25 | 0.5 | 0.75 | 1.0 | 1.5 | 2.0 | 3.0 | 4.0 | 5.0 | 6.0 | 8.0 |
| 20 | 0.08 | 0.08 | 0.08 | 0.07 | 0.07 | 0.07 | 0.06 | 0.06 | 0.05 | 0.05 | 0.05 |
| 30 | 0.18 | 0.17 | 0.17 | 0.16 | 0.15 | 0.15 | 0.13 | 0.13 | 0.12 | 0.12 | 0.11 |
| 45 | 0.38 | 0.37 | 0.36 | 0.34 | 0.33 | 0.31 | 0.28 | 0.27 | 0.26 | 0.25 | 0.24 |
| 60 | 0.60 | 0.59 | 0.57 | 0.55 | 0.52 | 0.49 | 0.46 | 0.43 | 0.41 | 0.39 | 0.38 |
| 75 | 0.89 | 0.87 | 0.84 | 0.81 | 0.77 | 0.73 | 0.67 | 0.63 | 0.61 | 0.58 | 0.57 |
| 90 | 1.3 | 1.3 | 1.2 | 1.2 | 1.1 | 1.1 | 0.98 | 0.92 | 0.89 | 0.85 | 0.83 |

**Reynolds Number Corrections Factors**

| Re × 10⁻⁴ | 1 | 2 | 3 | 4 | 6 | 8 | 10 | ≥14 |
|---|---|---|---|---|---|---|---|---|
| $K_{Re}$ | 1.40 | 1.26 | 1.19 | 1.14 | 1.09 | 1.06 | 1.04 | 1.0 |

### 3-7 Elbow, Smooth Radius with Splitter Vanes, Rectangular (Locklin 1950, Equation 10; Madison and Parker 1936)

**a. One Splitter Vane**

$C_o = K_\theta C'_o$
$R_1 = R/CR$

where

$R$ = throat radius
$R_1$ = splitter vane radius
CR = CURVE RATIO
 (values from Table 3-7.a)
$K_\theta$ = angle factor
 (see Fitting 3-1 for values)

**Table 3-7.a Coefficients for Elbows with One Splitter Vane:**

| | | | $C'_o$ | | | | | | | | | |
|---|---|---|---|---|---|---|---|---|---|---|---|---|
| | | | H/W | | | | | | | | | |
| R/W | r/W | CR | 0.25 | 0.5 | 1.0 | 1.5 | 2.0 | 3.0 | 4.0 | 5.0 | 6.0 | 7.0 | 8.0 |
| 0.05 | 0.55 | 0.218 | 0.52 | 0.40 | 0.43 | 0.49 | 0.55 | 0.66 | 0.75 | 0.84 | 0.93 | 1.0 | 1.1 |
| 0.10 | 0.60 | 0.302 | 0.36 | 0.27 | 0.25 | 0.24 | 0.30 | 0.35 | 0.39 | 0.42 | 0.46 | 0.49 | 0.52 |
| 0.15 | 0.65 | 0.361 | 0.28 | 0.21 | 0.18 | 0.19 | 0.20 | 0.22 | 0.25 | 0.26 | 028 | 0.30 | 0.32 |
| 0.20 | 0.70 | 0.408 | 0.22 | 0.16 | 0.14 | 0.14 | 0.15 | 0.16 | 0.17 | 0.18 | 0.19 | 0.20 | 0.21 |
| 0.25 | 0.75 | 0.447 | 0.18 | 013 | 0.11 | 0.11 | 0.11 | 0.12 | 0.13 | 0.14 | 0.14 | 0.15 | 0.15 |
| 0.30 | 0.80 | 0.480 | 0.15 | 0.11 | 0.09 | 0.09 | 0.09 | 0.09 | 0.10 | 0.10 | 0.11 | 0.11 | 0.12 |
| 0.35 | 0.85 | 0.509 | 0.13 | 0.09 | 0.08 | 0.07 | 0.07 | 0.08 | 0.08 | 0.08 | 0.08 | 0.09 | 0.09 |
| 0.40 | 0.90 | 0.535 | 0.11 | 0.08 | 0.07 | 0.06 | 0.06 | 0.06 | 0.06 | 0.07 | 0.07 | 0.07 | 0.07 |
| 0.45 | 0.95 | 0.557 | 0.10 | 0.07 | 0.06 | 0.05 | 0.05 | 0.05 | 0.05 | 0.05 | 0.06 | 0.06 | 0.06 |
| 0.50 | 1.00 | 0.577 | 0.09 | 0.06 | 0.05 | 0.05 | 0.04 | 0.04 | 0.04 | 0.05 | 0.05 | 0.05 | 0.05 |

**b. Two Splitter Vanes**

$C_o = K_\theta C'_o$
$R_1 = R/CR$
$R_2 = R_1/CR = R/CR^2$

where

$R$ = throat radius
$R_1$ = splitter vane #1 radius
$R_2$ = splitter vane #2 radius
CR = CURVE RATIO
 (value from Table 3-7.b)
$K_\theta$ = angle factor
 (see Note 3 for values)

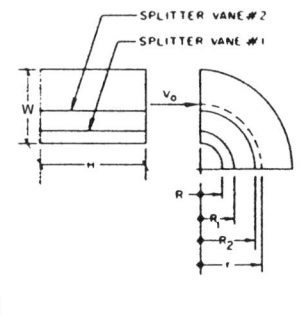

**Table 3-7.b Coefficients for Elbow with Two Splitter Vanes:**

| | | | $C'_o$ | | | | | | | | | |
|---|---|---|---|---|---|---|---|---|---|---|---|---|
| | | | H/W | | | | | | | | | |
| R/W | r/W | CR | 0.25 | 0.5 | 1.0 | 1.5 | 2.0 | 3.0 | 4.0 | 5.0 | 6.0 | 7.0 | 8.0 |
| 0.05 | 0.55 | 0.362 | 0.26 | 0.20 | 0.22 | 0.25 | 0.28 | 0.33 | 0.37 | 0.41 | 0.45 | 0.48 | 0.51 |
| 0.10 | 0.60 | 0.450 | 0.17 | 0.13 | 0.11 | 0.12 | 0.13 | 0.15 | 0.16 | 0.17 | 0.19 | 0.20 | 0.21 |
| 0.15 | 0.65 | 0.507 | 0.12 | 0.09 | 0.08 | 0.08 | 0.08 | 0.09 | 0.10 | 0.10 | 0.11 | 0.11 | 0.11 |
| 0.20 | 0.70 | 0.550 | 0.09 | 0.07 | 0.06 | 0.05 | 0.06 | 0.06 | 0.06 | 0.06 | 0.07 | 0.07 | 0.07 |
| 0.25 | 0.75 | 0.585 | 0.08 | 0.05 | 0.04 | 0.04 | 0.04 | 0.04 | 0.05 | 0.05 | 0.05 | 0.05 | 0.05 |
| 0.30 | 0.80 | 0.613 | 0.06 | 004 | 0.03 | 0.03 | 0.03 | 0.03 | 0.03 | 0.03 | 0.04 | 0.04 | 0.04 |

**c. Three Splitter Vanes**

$C_o = K_\theta C'_o$
$R_1 = R/CR$
$R_2 = R_1/CR = R/CR^2$
$R_3 = R_2/CR = R/CR^3$

where

$R$ = throat radius
$R_1$ = splitter vane #1 radius
$R_2$ = splitter vane #2 radius
$R_3$ = splitter vane #3 radius
CR = curve ratio
 (value from Table 3-7.c)
$K_\theta$ = angle factor
 (see Note 3 for values)

**Table 3-7.c Coefficients for Elbow with Three Splitter Vanes ($C'_o$):**

| | | | H/W | | | | | | | | | |
|---|---|---|---|---|---|---|---|---|---|---|---|---|
| R/W | r/W | CR | 0.25 | 0.5 | 1.0 | 1.5 | 2.0 | 3.0 | 4.0 | 5.0 | 6.0 | 7.0 | 8.0 |
| 0.05 | 0.55 | 0.467 | 0.11 | 0.10 | 0.12 | 0.13 | 0.14 | 0.16 | 0.18 | 0.19 | 0.21 | 0.22 | 0.23 |
| 0.10 | 0.60 | 0.549 | 0.07 | 0.05 | 0.06 | 0.06 | 0.06 | 0.07 | 0.07 | 0.08 | 0.08 | 0.08 | 0.09 |

# Duct and Pipe Sizing

**3-8 Elbow, Mitered, with Single-Thickness Vanes, Rectangular**
(Rozell 1974)

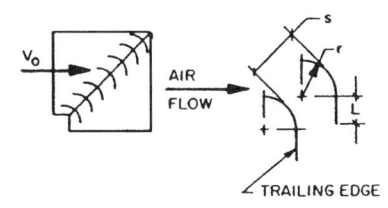

| Design | Dimensions, in. | | | |
|---|---|---|---|---|
| No. | r | s | L | $C_o$ |
| 1[a] | 2.0 | 1.5 | 0.75 | 0.12 |
| 2 | 4.5 | 2.25 | 0 | 0.15 |
| 3 | 4.5 | 3.25 | 1.60 | 0.18 |

[a]When extension of trailing edge is not provided for this vane, losses are approximately unchanged for single elbows, but increase considerably for elbows in series.

**3-9 Elbow, Mitered, with Double-Thickness Vanes, Rectangular**

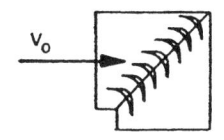

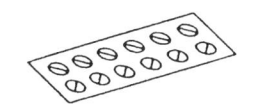

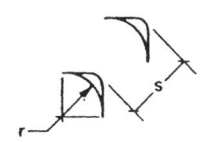

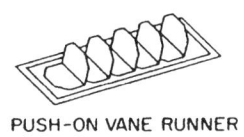

(Rozell 1974)

| Design No. | Dimensions, in. | | Velocity $V_o$, fpm | | | | Remarks |
|---|---|---|---|---|---|---|---|
| | r | s | 1000 | 2000 | 3000 | 4000 | |
| 1 | 2.0 | 1.5 | 0.27 | 0.22 | 0.19 | 0.17 | Embossed Vane Runner |
| 2 | 2.0 | 1.5 | 0.33 | 0.29 | 0.26 | 0.23 | Push-On Vane Runner |
| 3 | 2.0 | 2.13 | 0.38 | 0.31 | 0.27 | 0.24 | Embossed Vane Runner |
| 4 | 4.5 | 3.25 | 0.26 | 0.21 | 0.18 | 0.16 | Embossed Vane Runner |

**3-10 Elbow, Variable Inlet/Outlet Areas, Rectangular**
(Idelchik et al. 1986, Diagram 6-4)

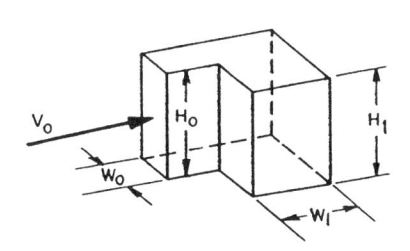

| | $C'_o$ | | | | | |
|---|---|---|---|---|---|---|
| | $W_1/W_o$ | | | | | |
| $H_o/W_o$ | 0.6 | 0.8 | 1.2 | 1.4 | 1.6 | 2.0 |
| 0.25 | 1.8 | 1.4 | 1.1 | 1.1 | 1.1 | 1.1 |
| 1.0 | 1.7 | 1.4 | 1.0 | 095 | 0.90 | 0.84 |
| 4.0 | 1.5 | 1.4 | 0.81 | 0.76 | 0.72 | 0.66 |
| ∞ | 1.5 | 1.0 | 0.69 | 0.63 | 0.60 | 0.55 |

**Reynolds Number Correction Factor**

| $Re \times 10^{-4}$ | 1 | 2 | 3 | 4 | 6 | 8 | 10 | ≥14 |
|---|---|---|---|---|---|---|---|---|
| $K_{Re}$ | 1.40 | 1.26 | 1.19 | 1.14 | 1.09 | 1.06 | 1.04 | 1.0 |

**3-11 Elbows, 90°, Z-Shaped, Rectangular**
(Idelchik et al. 1986, Diagram 6-11)

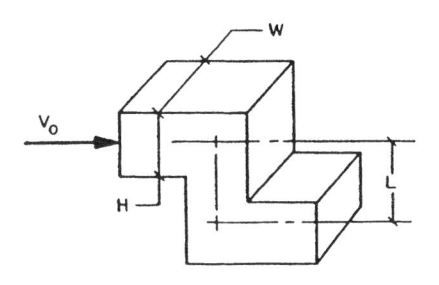

$C = K K_{Re} C'_o$

**Coefficients for W/H = 1.0**

| L/H | 0 | 0.4 | 0.6 | 0.8 | 1.0 | 1.2 | 1.4 | 1.6 | 1.8 | 2.0 |
|---|---|---|---|---|---|---|---|---|---|---|
| $C'_o$ | 0 | 0.62 | 0.90 | 1.6 | 2.6 | 3.6 | 4.0 | 4.2 | 4.2 | 4.2 |
| L/H | 2.4 | 2.8 | 3.2 | 4.0 | 5.0 | 6.0 | 7.0 | 9.0 | 10.0 | ∞ |
| $C'_o$ | 3.7 | 3.3 | 3.2 | 3.1 | 2.9 | 2.9 | 2.8 | 2.7 | 2.5 | 2.3 |

**For W/H Values Other Than 1.0, Apply the Following Factor**

| W/H | 0.25 | 0.50 | 0.75 | 1.0 | 1.5 | 2.0 | 3.0 | 4.0 | 6.0 | 8.0 |
|---|---|---|---|---|---|---|---|---|---|---|
| K | 1.10 | 1.07 | 1.04 | 1.0 | 0.95 | 0.90 | 9.83 | 0.78 | 0.72 | 0.70 |

**Reynolds Number Correction Factor**

| $Re \times 10^{-4}$ | 1 | 2 | 3 | 4 | 6 | 8 | 10 | ≥14 |
|---|---|---|---|---|---|---|---|---|
| $K_{Re}$ | 1.40 | 1.26 | 1.19 | 1.14 | 1.09 | 1.06 | 1.04 | 1.0 |

**3-12 Combined 90° Elbows Lying in Different Planes, Rectangular**

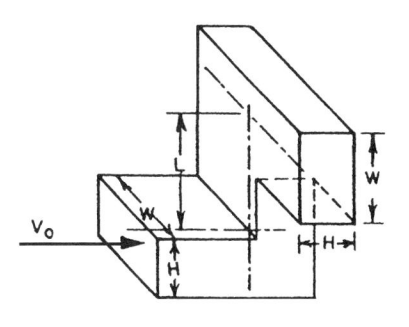

(Idelchik et al. 1986, Diagram 6-11)
$C_o = K K_{Re} C'_o$

**Coefficients for Square Ducts**

| L/W | 0 | 0.4 | 0.6 | 0.8 | 1.0 | 1.2 | 1.4 | 1.6 | 1.8 | 2.0 |
|---|---|---|---|---|---|---|---|---|---|---|
| $C'_o$ | 1.2 | 2.4 | 2.9 | 3.3 | 3.4 | 3.4 | 3.4 | 3.3 | 3.2 | 3.1 |
| L/W | 2.4 | 2.8 | 3.2 | 4.0 | 5.0 | 6.0 | 7.0 | 9.0 | 10.0 | ∞ |
| $C'_o$ | 3.2 | 3.2 | 3.2 | 3.0 | 2.9 | 2.8 | 2.7 | 2.5 | 2.4 | 2.3 |

**Apply the Following Factor for Other Than $H/W = 1.0$**

| H/W | 0.25 | 0.50 | 0.75 | 1.0 | 1.5 | 2.0 | 3.0 | 4.0 | 6.0 | 8.0 |
|---|---|---|---|---|---|---|---|---|---|---|
| K | 1.10 | 1.07 | 1.04 | 1.0 | 0.95 | 0.90 | 0.83 | 0.78 | 0.72 | 0.70 |

**Reynolds Number Correction Factor**

| Re × 10⁻⁴ | 1 | 2 | 3 | 4 | 6 | 8 | 10 | ≥14 |
|---|---|---|---|---|---|---|---|---|
| $K_{Re}$ | 1.40 | 1.26 | 1.19 | 1.14 | 1.09 | 1.06 | 1.04 | 1.0 |

### 3-13 Offset, S-Shaped (Gooseneck), Rectangular and Round (Idelchik et al. 1986, Diagram 6-16)

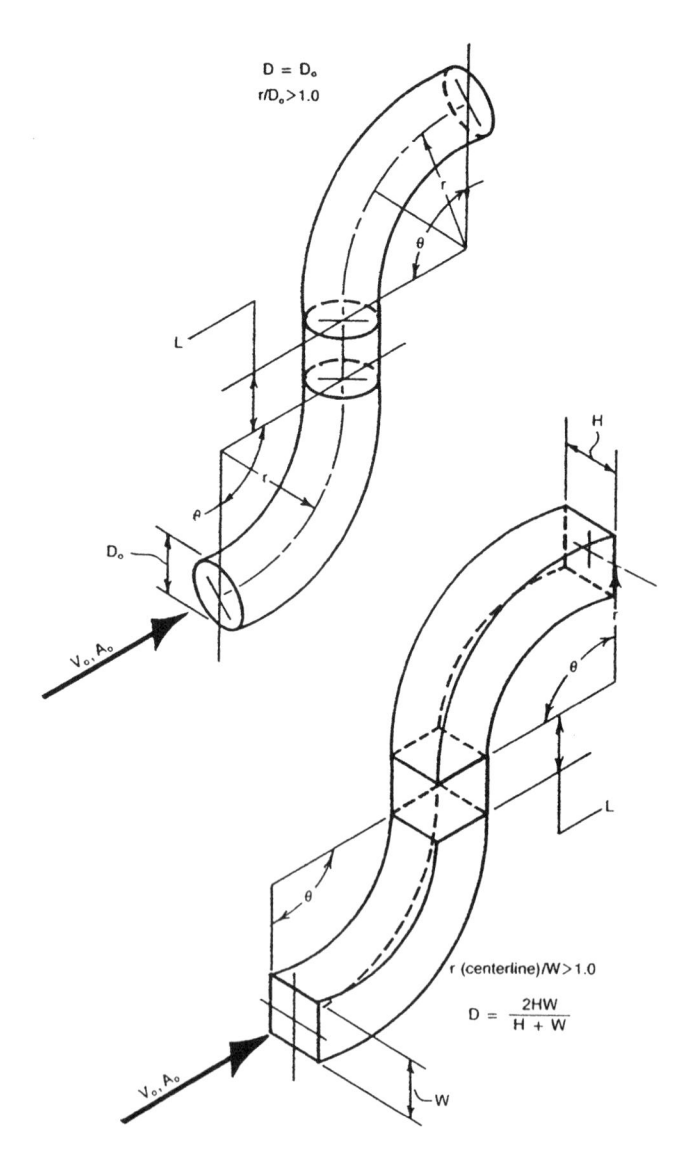

$C_o = KC'_o$
where
 $C_o$ = offset loss coefficient
 $C'_o$ = single elbow loss coefficient (see Fittings 3-1, 3-2, and 3-5)

| θ, degrees | K, L/D = 0 | 1 | 2 | 3 | 4 | 6 | 8 | 10 |
|---|---|---|---|---|---|---|---|---|
| 15 | 0.20 | 0.42 | 0.60 | 0.78 | 0.94 | 1.16 | 1.20 | 1.15 |
| 30 | 0.40 | 0.65 | 0.88 | 1.16 | 1.20 | 1.18 | 1.12 | 1.06 |
| 45 | 0.60 | 1.06 | 1.20 | 1.23 | 1.20 | 1.08 | 1.03 | 1.08 |
| 60 | 1.05 | 1.38 | 1.37 | 1.28 | 1.15 | 1.06 | 1.16 | 1.30 |
| 75 | 1.50 | 1.58 | 1.46 | 1.30 | 1.27 | 1.30 | 1.37 | 1.47 |
| 90 | 1.70 | 1.67 | 1.40 | 1.37 | 1.38 | 1.47 | 1.55 | 1.63 |

| θ, degrees | K, L/D = 12 | 14 | 16 | 18 | 20 | 25 | ≥40 |
|---|---|---|---|---|---|---|---|
| 15 | 1.08 | 1.05 | 1.02 | 1.00 | 1.10 | 1.25 | 2.0 |
| 30 | 1.06 | 1.15 | 1.28 | 1.40 | 1.50 | 1.70 | 2.0 |
| 45 | 1.17 | 1.30 | 1.42 | 1.55 | 1.65 | 1.80 | 2.0 |
| 60 | 1.42 | 1.54 | 1.66 | 1.76 | 1.85 | 1.95 | 2.0 |
| 75 | 1.57 | 1.68 | 1.75 | 1.80 | 1.88 | 1.97 | 2.0 |
| 90 | 1.70 | 1.76 | 1.82 | 1.88 | 1.92 | 1.98 | 2.0 |

### 3-14 Offset, S-Shaped in Two Planes 90° Apart, Rectangular and Round (Idelchik et al. 1986, Diagram 6-16)

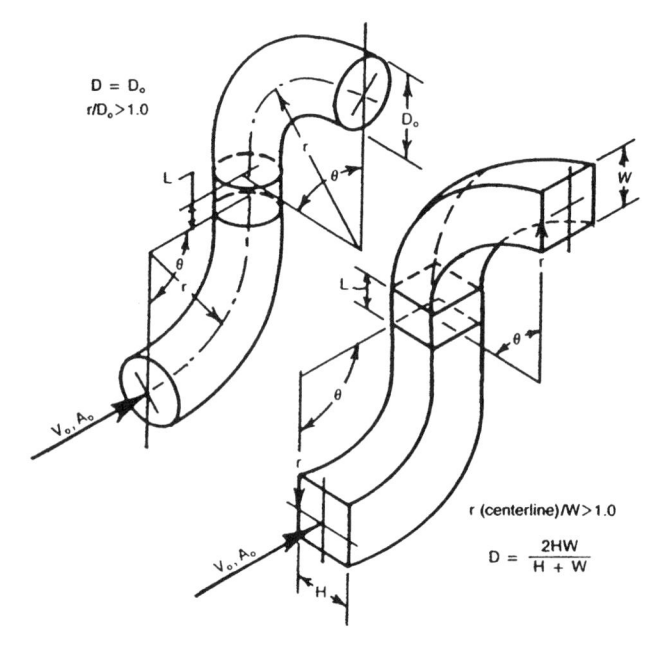

$C_o = KC'_o$
where
 $C_o$ = offset loss coefficient
 $C'_o$ = single elbow loss coefficient (see Fittings 3-1, 3-2, and 3-5)

| θ, degrees | K, L/D = 0 | 1 | 2 | 3 | 4 | 6 | 8 |
|---|---|---|---|---|---|---|---|
| 60 | 2.0 | 1.90 | 1.50 | 1.35 | 1.30 | 1.20 | 1.25 |
| 90 | 2.0 | 1.80 | 1.60 | 1.55 | 1.55 | 1.65 | 1.80 |

| θ, degrees | K, L/D = 10 | 12 | 14 | 20 | 25 | 40 |
|---|---|---|---|---|---|---|
| 60 | 1.50 | 1.63 | 1.73 | 1.85 | 1.95 | 2.0 |
| 90 | 1.90 | 1.93 | 1.98 | 2.0 | 2.0 | 2.0 |

# Duct and Pipe Sizing

## 3-15 Elbows (4), 45°, Smooth Radius, Rectangular, Arranged to Go Around an Obstruction (SMACNA 1981, Table 6-14K)

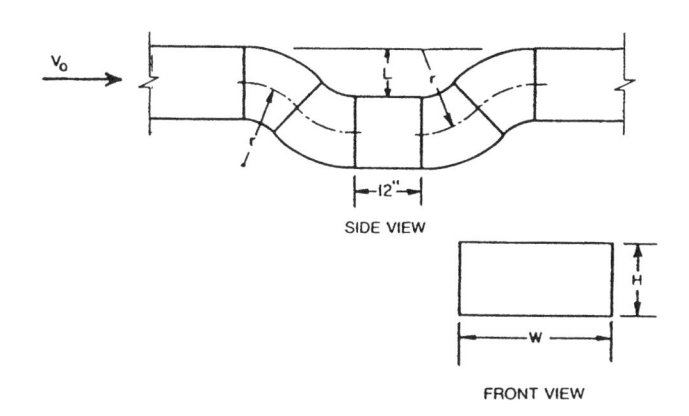

$W/H = 4$, $r/H = 1.5$, $L = 1.5H$

| $V_o$, fpm | 800 | 1200 | 1600 | 2000 | 2400 |
|---|---|---|---|---|---|
| $C_o$ | 0.18 | 0.22 | 0.24 | 0.25 | 0.26 |

## TRANSITIONS

### 4-1 Transition, Round (Idelchik et al. 1986, Diagrams 5-2 and 5-22)

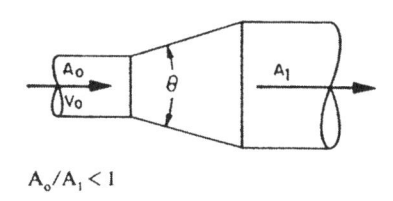

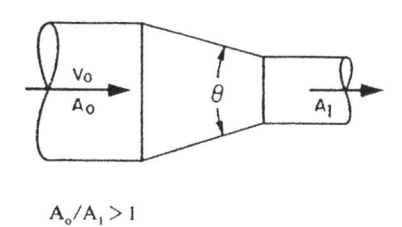

|  | $C_o$ |||||||||| 
|---|---|---|---|---|---|---|---|---|---|---|
|  | θ, degrees ||||||||||
| $A_o/A_1$ | 10 | 15 | 20 | 30 | 45 | 60 | 90 | 120 | 150 | 180 |
| 0.06 | 0.21 | 0.29 | 0.38 | 0.60 | 0.84 | 0.88 | 0.88 | 0.88 | 0.88 | 0.88 |
| 0.1 | 0.21 | 0.28 | 0.38 | 0.59 | 0.76 | 0.80 | 0.83 | 0.84 | 0.83 | 0.83 |
| 0.25 | 0.16 | 0.22 | 0.30 | 0.46 | 0.61 | 0.68 | 0.64 | 0.63 | 0.62 | 0.62 |
| 0.5 | 0.11 | 0.13 | 0.19 | 0.32 | 0.33 | 0.33 | 0.32 | 0.31 | 0.30 | 0.30 |
| 1 | 0 | 0 | 0 | 0 | 0 | 0 | 0 | | | |
| 2 | 0.20 | 0.20 | 0.20 | 0.20 | 0.22 | 0.24 | 0.48 | 0.72 | 0.96 | 1.0 |
| 4 | 0.80 | 0.64 | 0.64 | 0.64 | 0.88 | 1.1 | 2.7 | 4.3 | 5.6 | 6.6 |
| 6 | 1.8 | 1.4 | 1.4 | 1.4 | 2.0 | 2.5 | 6.5 | 10 | 13 | 15 |
| 10 | 5.0 | 5.0 | 5.0 | 5.0 | 6.5 | 8.0 | 19 | 29 | 37 | 43 |

### 4-2 Transition, Rectangular, Two Sides Parallel, Symmetrical (Idelchik et al. 1986, Diagram 5-5)[a]

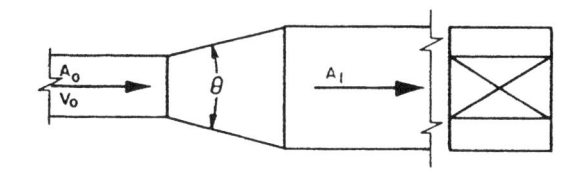

|  | $C_o$ |||||||||| 
|---|---|---|---|---|---|---|---|---|---|---|
|  | θ, degrees ||||||||||
| $A_o/A_1$ | 10 | 15 | 30 | 30 | 45 | 60 | 90 | 120 | 150 | 180 |
| 0.06 | 0.26 | 0.27 | 0.40 | 0.56 | 0.71 | 0.86 | 1.00 | 0.99 | 0.98 | 0.98 |
| 0.1 | 0.24 | 0.26 | 0.36 | 0.53 | 0.69 | 0.82 | 0.93 | 0.93 | 0.92 | 0.91 |
| 0.25 | 0.17 | 0.19 | 0.22 | 0.42 | 0.60 | 0.68 | 0.70 | 0.69 | 0.67 | 0.66 |
| 0.5 | 0.14 | 0.13 | 0.15 | 0.24 | 0.35 | 0.37 | 0.38 | 0.37 | 0.36 | 0.35 |
| 1 | 0 | 0 | 0 | 0 | 0 | 0 | 0 | 0 | 0 | 0 |
| 2 | 0.23 | 0.20 | 0.20 | 0.20 | 0.24 | 0.28 | 0.54 | 0.78 | 1.0 | 1.1 |
| 4 | 0.81 | 0.64 | 0.64 | 0.64 | 0.88 | 1.1 | 2.8 | 4.4 | 5.7 | 6.6 |
| 6 | 1.8 | 1.4 | 1.4 | 1.4 | 2.0 | 2.5 | 6.6 | 10 | 13 | 15 |
| 10 | 5.0 | 5.0 | 5.0 | 5.0 | 6.5 | 8.0 | 19 | 29 | 37 | 43 |

[a]$A_o/A_1 > 1$ is tentative (adapted from Fitting 4-1 data).

### 4-3 Transition, Rectangular, Three Sides Straight[a]

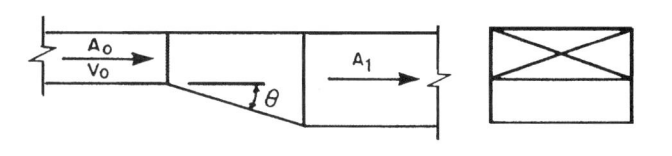

|  | $C_o$ |||||||
|---|---|---|---|---|---|---|---|
|  | θ, degrees |||||||
| $A_o/A_1$ | 10 | 15 | 20 | 30 | 45 | 60 | 90 |
| 0.06 | 0.26 | 0.27 | 0.40 | 0.56 | 0.71 | 0.86 | 1.00 |
| 0.1 | 0.24 | 0.26 | 0.36 | 0.53 | 0.69 | 0.82 | 0.93 |
| 0.25 | 0.17 | 0.19 | 0.22 | 0.42 | 0.60 | 0.68 | 0.70 |
| 0.5 | 0.14 | 0.13 | 0.15 | 0.24 | 0.35 | 0.37 | 0.38 |
| 1 | 0 | 0 | 0 | 0 | 0 | 0 | 0 |
| 2 | 0.23 | 0.20 | 0.20 | 0.20 | 0.24 | 0.28 | 0.54 |
| 4 | 0.81 | 0.64 | 0.64 | 0.64 | 0.88 | 1.1 | 2.8 |
| 6 | 1.8 | 1.4 | 1.4 | 1.4 | 2.0 | 2.5 | 6.6 |
| 10 | 5.0 | 5.0 | 5.0 | 5.0 | 6.5 | 8.0 | 19 |

[a]Tentative (assumed same as Fitting 4-2 data).

### 4-4 Transition, Rectangular, Pyramidal (Idelchik et al. 1986, Diagram 5-4)[a]

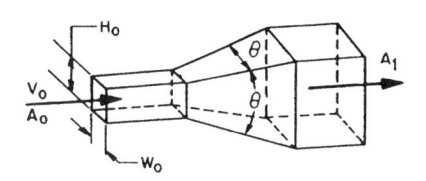

| | $C_o$ | | | | | | | | | |
|---|---|---|---|---|---|---|---|---|---|---|
| | θ, degrees | | | | | | | | | |
| $A_o/A_1$ | 10 | 15 | 20 | 30 | 45 | 60 | 90 | 120 | 150 | 180 |
| 0.06 | 0.26 | 0.30 | 0.44 | 0.54 | 0.53 | 0.65 | 0.77 | 0.88 | 0.95 | 0.98 |
| 0.1 | 0.24 | 0.30 | 0.43 | 0.50 | 0.53 | 0.64 | 0.75 | 0.84 | 0.89 | 0.91 |
| 0.25 | 0.20 | 0.25 | 0.34 | 0.36 | 0.45 | 0.52 | 0.58 | 0.62 | 0.64 | 0.64 |
| 0.5 | 0.14 | 0.15 | 0.20 | 0.21 | 0.25 | 0.30 | 0.33 | 0.33 | 0.33 | 0.32 |
| 1 | 0 | 0 | 0 | 0 | 0 | 0 | 0 | 0 | 0 | 0 |
| 2 | 0.23 | 0.22 | 0.21 | 0.20 | 0.22 | 0.2 | 0.49 | 0.74 | 0.99 | 1.1 |
| 4 | 0.84 | 0.68 | 0.68 | 0.64 | 0.88 | 1.1 | 2.7 | 4.3 | 5.6 | 6.6 |
| 6 | 1.8 | 1.5 | 1.5 | 1.4 | 2.0 | 2.5 | 6.5 | 10 | 13 | 15 |
| 10 | 5.0 | 5.0 | 5.1 | 5.0 | 6.5 | 8.0 | 19 | 29 | 37 | 43 |

[a] $A_o/A_1 >1$ is tentative (adapted from Fitting 4-1 data).

### 4-5 Transition, Round/Rectangular
(Idelchik et al. 1986, Diagram 5-27)

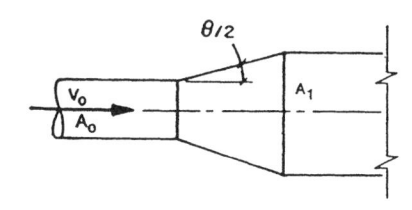

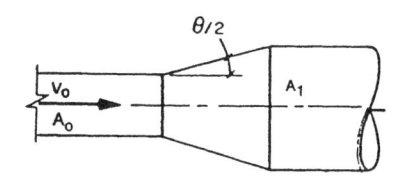

| | $C_o$ | | | | | | | | |
|---|---|---|---|---|---|---|---|---|---|
| | θ, degrees | | | | | | | | |
| $A_1/A_o$ | 8 | 10 | 14 | 20 | 30 | 45 | 60 | 90 | 180 |
| 2 | 0.14 | 0.15 | 0.20 | 0.25 | 0.30 | 0.33 | 0.33 | 0.33 | 0.30 |
| 4 | 0.20 | 0.25 | 0.34 | 0.45 | 0.52 | 0.58 | 0.62 | 0.64 | 0.64 |
| 6 | 0.21 | 0.30 | 0.42 | 0.53 | 0.63 | 0.72 | 0.78 | 0.79 | 0.79 |
| ≥10 | 0.24 | 0.30 | 0.43 | 0.53 | 0.64 | 0.75 | 0.84 | 0.89 | 0.88 |

### 4-6 Transition, Rectangular to Round
(Idelchik et al. 1986, Diagram 5-26)

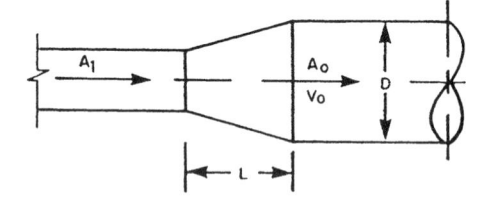

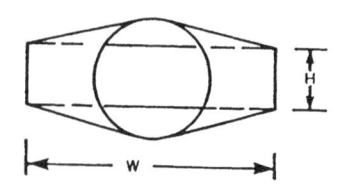

$\text{Re} = DV_o/v$
$B = W/H(A_o/A_1)^2$

$C_o$ for $A_o > A_1$ and $H < D$

| | | B | | | | | | | |
|---|---|---|---|---|---|---|---|---|---|
| $\text{Re} \times 10^{-4}$ | $L/D$ | 0.1 | 0.5 | 1 | 2 | 5 | 10 | 20 | 50 |
| 1 | 1 | 0.46 | 0.48 | 0.50 | 0.55 | 0.70 | 0.94 | 1.4 | 2.9 |
| | 1.5 | 0.46 | 0.47 | 0.49 | 0.53 | 0.64 | 0.82 | 1.2 | 2.3 |
| | 2 | 0.46 | 0.47 | 0.48 | 0.50 | 0.58 | 0.71 | 0.98 | 1.8 |
| | 3 | 0.45 | 0.46 | 0.46 | 0.48 | 0.51 | 0.57 | 0.69 | 1.0 |
| | 4 | 0.45 | 0.46 | 0.46 | 0.46 | 0.48 | 0.51 | 0.56 | 0.73 |
| | 5 | 0.45 | 0.46 | 0.46 | 0.47 | 0.49 | 0.53 | 0.61 | 0.84 |
| 2 | 1 | 0.41 | 0.43 | 0.46 | 0.51 | 0.66 | 0.90 | 1.4 | 2.9 |
| | 1.5 | 0.41 | 0.43 | 0.45 | 0.48 | 0.59 | 0.78 | 1.1 | 2.2 |
| | 2 | 0.41 | 0.42 | 0.44 | 0.46 | 0.54 | 0.67 | 0.93 | 1.7 |
| | 3 | 0.41 | 0.42 | 0.42 | 0.43 | 0.47 | 0.53 | 0.64 | 0.99 |
| | 4 | 0.41 | 0.41 | 0.41 | 0.42 | 0.44 | 0.46 | 0.52 | 0.69 |
| | 5 | 0.41 | 0.41 | 0.42 | 0.43 | 0.45 | 0.49 | 0.57 | 0.80 |
| 5 | 1 | 0.31 | 0.33 | 0.35 | 0.40 | 0.55 | 0.79 | 1.3 | 2.8 |
| | 1.5 | 0.31 | 0.32 | 0.34 | 0.38 | 0.49 | 0.67 | 1.0 | 2.1 |
| | 2 | 0.31 | 0.32 | 0.33 | 0.36 | 0.43 | 0.57 | 0.83 | 1.6 |
| | 3 | 0.30 | 0.31 | 0.31 | 0.33 | 0.36 | 0.42 | 0.54 | 0.88 |
| | 4 | 0.30 | 0.31 | 0.31 | 0.31 | 0.33 | 0.36 | 0.41 | 0.58 |
| | 5 | 0.30 | 0.31 | 0.31 | 0.32 | 0.34 | 0.38 | 0.46 | 0.70 |
| 10 | 1 | 0.19 | 0.21 | 0.23 | 0.28 | 0.43 | 0.68 | 1.2 | 2.6 |
| | 1.5 | 0.19 | 0.20 | 0.22 | 0.26 | 0.37 | 0.55 | 0.92 | 2.0 |
| | 2 | 0.19 | 0.20 | 0.21 | 0.24 | 0.31 | 0.45 | 0.71 | 1.5 |
| | 3 | 0.19 | 0.19 | 0.20 | 0.21 | 0.24 | 0.30 | 0.42 | 0.77 |
| | 4 | 0.18 | 0.19 | 0.19 | 0.19 | 0.21 | 0.24 | 0.29 | 0.46 |
| | 5 | 0.18 | 0.19 | 0.19 | 0.20 | 0.22 | 0.26 | 0.34 | 0.58 |
| 20 | 1 | 0.07 | 0.09 | 0.12 | 0.17 | 0.31 | 0.56 | 1.1 | 2.5 |
| | 1.5 | 0.07 | 0.09 | 0.10 | 0.14 | 0.25 | 0.43 | 0.80 | 0.9 |
| | 2 | 0.07 | 0.08 | 0.09 | 0.12 | 0.20 | 0.33 | 0.59 | 1.4 |
| | 3 | 0.07 | 0.07 | 0.08 | 0.09 | 0.13 | 0.18 | 0.30 | 0.65 |
| | 4 | 0.07 | 0.07 | 0.07 | 0.08 | 0.10 | 0.12 | 0.18 | 0.34 |
| | 5 | 0.07 | 0.07 | 0.08 | 0.08 | 0.11 | 0.15 | 0.22 | 0.46 |
| 50 | 1 | 0.01 | 0.03 | 0.05 | 0.10 | 0.25 | 0.30 | 0.99 | 2.5 |
| | 1.5 | 0.01 | 0.02 | 0.04 | 0.08 | 0.19 | 0.37 | 0.74 | 1.8 |
| | 2 | 0.01 | 0.02 | 0.03 | 0.06 | 0.13 | 0.27 | 0.53 | 1.3 |
| | 3 | 0 | 0.01 | 0.02 | 0.03 | 0.06 | 0.12 | 0.24 | 0.58 |
| | 4 | 0 | 0.01 | 0.01 | 0.01 | 0.03 | 0.06 | 0.11 | 0.28 |
| | 5 | 0 | 0.01 | 0.01 | 0.02 | 0.04 | 0.08 | 0.16 | 0.40 |

$C_o$ for $A_o < A_1$ and $H < D$

| | | B | | | | | |
|---|---|---|---|---|---|---|---|
| $\text{Re} \times 10^{-4}$ | $L/D$ | 0.1 | 1 | 5 | 10 | 20 | 50 |
| 1 | 1 | 0.27 | 0.27 | 0.28 | 0.29 | 0.31 | 0.37 |
| | 3 | 0.27 | 0.27 | 0.28 | 0.29 | 0.30 | 0.33 |
| | 4 | 0.27 | 0.27 | 0.28 | 0.28 | 0.29 | 0.32 |
| | 5 | 0.27 | 0.27 | 0.27 | 0.27 | 0.27 | 0.27 |
| 2 | 1 | 0.25 | 0.25 | 0.26 | 0.27 | 0.29 | 0.35 |
| | 3 | 0.25 | 0.25 | 0.25 | 0.26 | 0.28 | 0.33 |
| | 4 | 0.25 | 0.25 | 0.25 | 0.26 | 0.27 | 0.30 |
| | 5 | 0.25 | 0.25 | 0.25 | 0.25 | 0.25 | 0.25 |
| 5 | 1 | 0.18 | 0.18 | 0.19 | 0.20 | 0.22 | 0.28 |
| | 3 | 0.18 | 0.18 | 0.19 | 0.20 | 0.22 | 0.27 |
| | 4 | 0.18 | 0.18 | 0.19 | 0.19 | 0.20 | 0.23 |
| | 5 | 0.18 | 0.18 | 0.18 | 0.18 | 0.18 | 0.18 |
| 10 | 1 | 0.11 | 0.11 | 0.12 | 0.13 | 0.15 | 0.21 |
| | 3 | 0.11 | 0.11 | 0.12 | 0.13 | 0.14 | 0.19 |
| | 4 | 0.11 | 0.11 | 0.12 | 0.12 | 0.13 | 0.16 |
| | 5 | 0.11 | 0.11 | 0.11 | 0.11 | 0.11 | 0.11 |
| 20 | 1 | 0.04 | 0.04 | 0.05 | 0.06 | 0.08 | 0.14 |
| | 3 | 0.04 | 0.04 | 0.05 | 0.06 | 0.07 | 0.12 |
| | 4 | 0.04 | 0.04 | 0.05 | 0.05 | 0.06 | 0.09 |
| | 5 | 0.04 | 0.04 | 0.04 | 0.04 | 0.04 | 0.04 |
| 50 | 1 | 0 | 0 | 0.01 | 0.02 | 0.04 | 0.10 |
| | 3 | 0 | 0 | 0.01 | 0.02 | 0.04 | 0.09 |
| | 4 | 0 | 0 | 0.01 | 0.01 | 0.02 | 0.05 |
| | 5 | 0 | 0 | 0 | 0 | 0 | 0 |

# Duct and Pipe Sizing

## 4-7 Transition, Rectangular to Round, Stepped, Conical (Idelchik et al. 1986, Diagram 4-9)

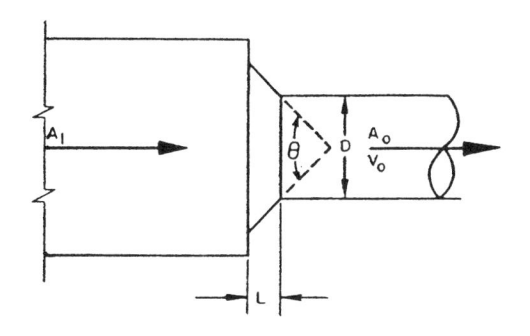

| | | | | | $C_o$ | | | | | |
|---|---|---|---|---|---|---|---|---|---|---|
| | | | | | $\theta$, degrees | | | | | |
| $A_o/A_1$ | $L/H$ | 0 | 10 | 20 | 30 | 45 | 60 | 90 | 120 | 150 | 180 |
| 0.1 | 0.025 | 0.46 | 0.43 | 0.42 | 0.40 | 0.38 | 0.37 | 0.38 | 0.40 | 0.43 | 0.46 |
| | 0.05 | 0.46 | 0.42 | 0.38 | 0.33 | 0.30 | 0.28 | 0.31 | 0.36 | 0.41 | 0.46 |
| | 0.075 | 0.46 | 0.39 | 0.32 | 0.28 | 0.23 | 0.21 | 0.26 | 0.32 | 0.39 | 0.46 |
| | 0.1 | 0.46 | 0.36 | 0.30 | 0.23 | 0.19 | 0.17 | 0.23 | 0.30 | 0.38 | 0.46 |
| | 0.15 | 0.46 | 0.34 | 0.25 | 0.18 | 0.15 | 0.14 | 0.21 | 0.29 | 0.37 | 0.46 |
| | 0.3 | 0.46 | 0.31 | 0.22 | 0.16 | 0.13 | 0.13 | 0.20 | 0.28 | 0.37 | 0.46 |
| | 0.6 | 0.46 | 0.25 | 0.17 | 0.12 | 0.10 | 0.11 | 0.19 | 0.27 | 0.36 | 0.46 |
| 0.25 | 0.025 | 0.40 | 0.38 | 0.36 | 0.35 | 0.33 | 0.32 | 0.33 | 0.35 | 0.37 | 0.40 |
| | 0.05 | 0.40 | 0.36 | 0.33 | 0.29 | 0.26 | 0.24 | 0.27 | 0.31 | 0.35 | 0.40 |
| | 0.075 | 0.40 | 0.34 | 0.28 | 0.24 | 0.20 | 0.19 | 0.23 | 0.28 | 0.34 | 0.40 |
| | 0.1 | 0.40 | 0.31 | 0.26 | 0.20 | 0.17 | 0.14 | 0.20 | 0.26 | 0.33 | 0.40 |
| | 0.15 | 0.40 | 0.30 | 0.22 | 0.16 | 0.13 | 0.12 | 0.18 | 0.25 | 0.32 | 0.40 |
| | 0.3 | 0.40 | 0.27 | 0.19 | 0.14 | 0.11 | 0.11 | 0.18 | 0.25 | 0.32 | 0.40 |
| | 0.6 | 0.40 | 0.22 | 0.14 | 0.10 | 0.09 | 0.10 | 0.16 | 0.24 | 0.32 | 0.40 |
| 0.5 | 0.025 | 0.30 | 0.28 | 0.27 | 0.25 | 0.24 | 0.24 | 0.25 | 0.26 | 0.27 | 0.30 |
| | 0.05 | 0.30 | 0.27 | 0.24 | 0.21 | 0.19 | 0.18 | 0.20 | 0.23 | 0.26 | 0.30 |
| | 0.075 | 0.30 | 0.25 | 0.21 | 0.18 | 0.15 | 0.14 | 0.17 | 0.21 | 0.25 | 0.30 |
| | 0.1 | 0.30 | 0.23 | 0.19 | 0.15 | 0.12 | 0.11 | 0.15 | 0.19 | 0.24 | 0.30 |
| | 0.15 | 0.30 | 0.22 | 0.16 | 0.12 | 0.09 | 0.09 | 0.13 | 0.18 | 0.24 | 0.30 |
| | 0.3 | 0.30 | 0.20 | 0.14 | 0.10 | 0.08 | 0.08 | 0.13 | 0.18 | 0.24 | 0.30 |
| | 0.6 | 0.30 | 0.16 | 0.11 | 0.08 | 0.07 | 0.07 | 0.12 | 0.17 | 0.23 | 0.30 |
| 0.8 | 0.025 | 0.15 | 0.14 | 0.13 | 0.13 | 0.12 | 0.12 | 0.12 | 0.13 | 0.14 | 0.15 |
| | 0.05 | 0.15 | 0.13 | 0.12 | 0.11 | 0.10 | 0.09 | 0.10 | 0.12 | 0.13 | 0.15 |
| | 0.075 | 0.15 | 0.13 | 0.10 | 0.09 | 0.08 | 0.07 | 0.08 | 0.10 | 0.13 | 0.15 |
| | 0.1 | 0.15 | 0.12 | 0.10 | 0.07 | 0.06 | 0.05 | 0.07 | 0.10 | 0.12 | 0.15 |
| | 0.15 | 0.15 | 0.11 | 0.08 | 0.06 | 0.05 | 0.04 | 0.07 | 0.09 | 0.12 | 0.15 |
| | 0.3 | 0.15 | 0.10 | 0.07 | 0.05 | 0.04 | 0.04 | 0.07 | 0.09 | 0.12 | 0.15 |
| | 0.6 | 0.15 | 0.08 | 0.05 | 0.04 | 0.03 | 0.04 | 0.06 | 0.09 | 0.12 | 0.15 |

## JUNCTIONS (Tees, Wyes, Crosses)

### 5-1 Wye, 30°, Converging (Idelchik et al. 1986, Diagram 7-1)

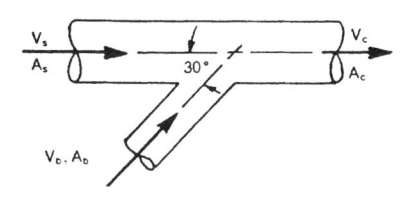

Branch, $C_{c,b}$

| | | | $A_b/A_c$ | | | | |
|---|---|---|---|---|---|---|---|
| $Q_b/Q_c$ | 0.1 | 0.2 | 0.3 | 0.4 | 0.6 | 0.8 | 1.0 |
| 0 | −1.0 | −1.0 | −1.0 | −0.9 | −0.9 | −0.9 | −0.9 |
| 0.1 | 0.21 | −0.46 | −0.57 | −0.51 | 0.53 | −0.54 | −0.54 |
| 0.2 | 3.1 | 0.37 | −0.06 | −0.16 | 0.23 | −0.24 | −0.28 |
| 0.3 | 7.6 | 1.5 | 0.50 | 0.15 | 0.04 | −0.06 | −0.08 |
| 0.4 | 14 | 3.0 | 1.2 | 0.42 | 0.19 | 0.13 | 0.12 |
| 0.5 | 21 | 4.6 | 1.8 | 0.53 | 0.24 | 0.19 | 0.15 |
| 0.6 | 30 | 6.4 | 2.6 | 0.77 | 0.35 | 0.25 | 0.17 |
| 0.7 | 41 | 8.5 | 3.4 | 0.99 | 0.42 | 0.28 | 0.22 |
| 0.8 | 54 | 12 | 4.2 | 1.2 | 0.47 | 0.29 | 0.25 |
| 0.9 | 58 | 14 | 5.3 | 1.4 | 0.49 | 0.29 | 0.22 |
| 1.0 | 84 | 17 | 6.3 | 1.6 | 0.49 | 0.21 | 0.15 |

Main, $C_{c,s}$

| | | | $A_b/A_c$ | | | | |
|---|---|---|---|---|---|---|---|
| $Q_b/Q_c$ | 0.1 | 0.2 | 0.3 | 0.4 | 0.6 | 0.8 | 1.0 |
| 0 | 0 | 0 | 0 | 0 | 0 | 0 | 0 |
| 0.1 | 0.02 | 0.11 | 0.13 | 0.15 | 0.16 | 0.17 | 0.17 |
| 0.2 | −0.33 | 0.01 | 0.13 | 0.19 | 0.24 | 0.27 | 0.29 |
| 0.3 | −1.1 | −0.25 | −0.01 | 0.10 | 0.22 | 0.30 | 0.35 |
| 0.4 | −2.2 | −0.75 | −0.30 | −0.05 | 0.17 | 0.26 | 0.36 |
| 0.5 | −3.6 | −1.4 | −0.70 | −0.35 | 0 | 0.21 | 0.32 |
| 0.6 | −5.4 | −2.4 | −1.3 | −0.70 | −0.20 | 0.06 | 0.25 |
| 0.7 | −7.6 | −3.4 | −2.0 | −1.2 | −0.50 | −0.15 | 0.10 |
| 0.8 | −10 | −4.6 | −2.7 | −1.8 | −0.90 | −0.43 | −0.15 |
| 0.9 | −13 | −6.2 | −3.7 | −2.6 | −1.4 | −0.80 | −0.45 |
| 1.0 | −16 | −7.7 | −4.8 | −3.4 | −1.9 | −1.2 | −0.75 |

### 5-2 Wye, 45° Converging, Round (Idelchik et al. 1986, Diagram 7-2)

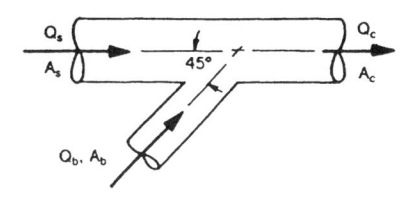

$A_s = A_c$

$C_{c,b}$

| | | | $A_b/A_c$ | | | | |
|---|---|---|---|---|---|---|---|
| $Q_b/Q_c$ | 0.1 | 0.2 | 0 3 | 0.4 | 0.6 | 0.8 | 1.0 |
| 0 | −1.0 | −1.0 | −1.0 | −0.90 | −0.90 | −0.80 | −0.90 |
| 0.1 | 0.24 | −0.45 | −0.56 | −0.50 | −0.52 | −0.53 | −0.53 |
| 0.2 | 3.2 | 0.54 | −0.02 | −0.14 | −0.21 | −0.23 | −0.23 |
| 0.3 | 8.0 | 1.6 | 0.60 | 0.23 | 0.06 | 0 | −0.02 |
| 0.4 | 14 | 3.2 | 1.3 | 0.52 | 0.25 | 0.18 | 0.15 |
| 0.5 | 22 | 5.0 | 2.1 | 0.65 | 0.33 | 0.25 | 0.22 |
| 0.6 | 32 | 7.0 | 3.0 | 0.91 | 0.81 | 0.61 | 0.51 |
| 0.7 | 43 | 9.2 | 3.9 | 1.2 | 0.56 | 0.39 | 0.33 |
| 0.8 | 56 | 12 | 4.9 | 1.5 | 0.66 | 0.39 | 0.36 |
| 0.9 | 71 | 15 | 6.2 | 1.8 | 0.72 | 0.44 | 0.35 |
| 1.0 | 87 | 19 | 7.4 | 2.0 | 0.78 | 0.44 | 0.32 |

$C_{c,s}$

| | | | $A_b/A_c$ | | | | |
|---|---|---|---|---|---|---|---|
| $Q_b/Q_c$ | 0.1 | 0.2 | 0.3 | 0.4 | 0.6 | 0.8 | 1.0 |
| 0 | 0 | 0 | 0 | 0 | 0 | 0 |
| 0.1 | 0.05 | 0.12 | 0.14 | 0.16 | 0.17 | 0.17 | 0.17 |
| 0.2 | −0.20 | 0.17 | 0.22 | 0.27 | 0.27 | 0.29 | 0.31 |
| 0.3 | −0.76 | −0.13 | 0.08 | 0.20 | 0.28 | 0.32 | 0.40 |
| 0.4 | −1.7 | −0.50 | −0.12 | 0.08 | 0.26 | 0.36 | 0.41 |
| 0.5 | −2.8 | −1.0 | −0.49 | −0.13 | 0.16 | 0.30 | 0.40 |
| 0.6 | −4.3 | −1.7 | −0.87 | −0.45 | −0.04 | 0.20 | 0.33 |
| 0.7 | −6.1 | −2.6 | −1.4 | −0.85 | −0.25 | 0.08 | 0.25 |
| 0.8 | −8.1 | 3.6 | −2.1 | −1.3 | −0.55 | −0.17 | 0.06 |
| 0.9 | −10 | 4.8 | −2.8 | −1.9 | −0.88 | −0.40 | −0.18 |
| 1.0 | −13 | −6.1 | −3.7 | −2.6 | −1.4 | −0.77 | −0.42 |

**330**  Principles of HVAC

## 5-3 Tee, Converging, Round (Idelchik et al. 1986, Diagram 7-4)

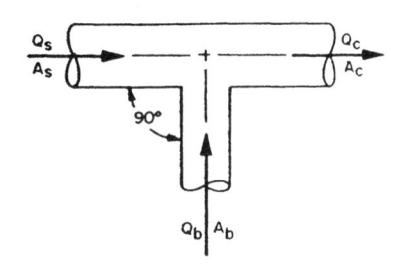

$A_s = A_c$

### Branch, $C_{c,b}$

| $Q_s/Q_c$ | $A_b/A_c$ = 0.1 | 0.2 | 03 | 0.4 | 0.6 | 0.8 | 1.0 |
|---|---|---|---|---|---|---|---|
| 0 | −1.0 | −1.0 | −1.0 | −0.90 | −0.90 | −0.90 | −0.90 |
| 0.1 | 0.40 | −0.37 | −0.51 | −0.46 | −0.50 | −0.51 | −0.52 |
| 0.2 | 3.8 | 0.72 | 0.17 | −0.02 | −0.14 | −0.18 | −0.24 |
| 0.3 | 9.2 | 2.3 | 1.0 | 0.44 | 0.21 | 0.11 | −0.08 |
| 0.4 | 16 | 4.3 | 2.1 | 0.94 | 0.54 | 0.40 | 0.32 |
| 0.5 | 26 | 6.8 | 3.2 | 1.1 | 0.66 | 0.49 | 0.42 |
| 0.6 | 37 | 9.7 | 4.7 | 1.6 | 0.92 | 0.69 | 0.57 |
| 0.7 | 43 | 13 | 6.3 | 2.1 | 1.2 | 0.88 | 0.72 |
| 0.8 | 65 | 17 | 7.9 | 2.7 | 1.5 | 1.1 | 0.86 |
| 0.9 | 82 | 21 | 9.7 | 3.4 | 1.8 | 1.2 | 0.99 |
| 1.0 | 101 | 26 | 12 | 4.0 | 2.1 | 1.4 | 1.1 |

### Main

| $Q_s/Q_c$ | 0 | 0.1 | 0.2 | 0.3 | 0.4 | 0.5 | 0.6 | 0.7 | 0.8 | 0.9 | 1.0 |
|---|---|---|---|---|---|---|---|---|---|---|---|
| $C_{c,s}$ | 0 | 0.16 | 0.27 | 0.38 | 0.46 | 0.53 | 0.57 | 0.59 | 0.60 | 0.59 | 0.55 |

## 5-4 Wye, 30° Converging, Round, Conical Main (Sepsy and Pies 1973)

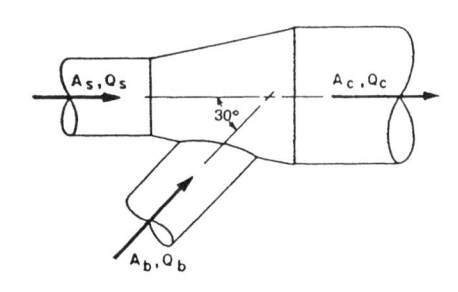

### Branch, $C_{c,b}$

| $A_s/A_c$ | $A_b/A_c$ | $Q_b/Q_s$ = 0.2 | 0.4 | 0.6 | 0.8 | 1.0 | 1.2 | 1.4 | 1.6 | 1.8 | 2.0 |
|---|---|---|---|---|---|---|---|---|---|---|---|
| 0.3 | 0.2 | −2.4 | −0.11 | 1.8 | 3.4 | 4.8 | 6.0 | 7.1 | 8.0 | 8.9 | 9.7 |
|  | 0.3 | −2.8 | −1.3 | 0.14 | 0.72 | 1.4 | 2.0 | 2.4 | 2.8 | 3.2 | 3.5 |
| 0.4 | 0.2 | −1.4 | 0.61 | 2.3 | 3.8 | 5.2 | 6.3 | 7.3 | 8.3 | 9.1 | 9.8 |
|  | 0.3 | −1.8 | −0.54 | 0.42 | 1.2 | 1.8 | 2.3 | 2.7 | 3.1 | 3.4 | 3.7 |
|  | 0.4 | −1.9 | −0.89 | −0.17 | 0.36 | 0.76 | 1.1 | 1.3 | 1.5 | 1.7 | 1.9 |
| 0.5 | 0.2 | −0.82 | 0.97 | 2.6 | 4.0 | 5.3 | 6.4 | 7.4 | 8.3 | 9.1 | 9.9 |
|  | 0.3 | −1.2 | −0.15 | 0.71 | 1.4 | 2.0 | 2.5 | 2.9 | 3.3 | 3.6 | 3.9 |
|  | 0.4 | −1.4 | −0.54 | 0.06 | 0.50 | 0.85 | 1.1 | 1.3 | 1.5 | 1.7 | 1.8 |
|  | 0.5 | −1.4 | −0.66 | −0.15 | 0.21 | 0.48 | 0.68 | 0.84 | 0.97 | 1.1 | 1.2 |
| 0.6 | 0.2 | −0.52 | 1.2 | 2.7 | 4.1 | 5.3 | 6.4 | 7.4 | 8.3 | 9.1 | 9.9 |
|  | 03 | −0.93 | 0.06 | 0.85 | 1.5 | 2.1 | 2.6 | 3.0 | 3.4 | 3.7 | 4.0 |
|  | 0.4 | −1.1 | −0.37 | 0.16 | 0.55 | 0.86 | 1.1 | 1.3 | 1.4 | 1.6 | 1.8 |
|  | 0.5 | −1.1 | −0.49 | −0.06 | 0.25 | 0.48 | 0.66 | 0.79 | 0.90 | 1.0 | 1.1 |
|  | 0.6 | −1.2 | −0.55 | −0.15 | 0.12 | 0.31 | 0.45 | 0.56 | 0.65 | 0.71 | 0.77 |

### Branch, $C_{c,b}$

| $A_s/A_c$ | $A_b/A_c$ | $Q_b/Q_c$ = 0.2 | 0.4 | 0.6 | 0.8 | 1.0 | 1.2 | 1.4 | 1.6 | 1.8 | 2.0 |
|---|---|---|---|---|---|---|---|---|---|---|---|
| 0.8 | 0.2 | −0.27 | 1.3 | 2.7 | 4.0 | 5.2 | 6.3 | 7.3 | 8.2 | 9.0 | 9.7 |
|  | 0.3 | −0.67 | 0.18 | 0.90 | 1.5 | 2.0 | 2.5 | 2.9 | 3.3 | 3.6 | 4.0 |
|  | 0.4 | −0.85 | −0.27 | 0.16 | 0.49 | 0.75 | 0.97 | 1.2 | 1.3 | 1.4 | 1.6 |
|  | 0.5 | −0.90 | −0.40 | −0.07 | 0.18 | 0.36 | 0.50 | 0.61 | 0.70 | 0.78 | 0.84 |
|  | 0.6 | −0.92 | −0.46 | −0.16 | 0.04 | 0.18 | 0.29 | 0.37 | 0.44 | 0.49 | 0.53 |
|  | 0.7 | −0.93 | −0.49 | −0.21 | −0.03 | 0.10 | 0.19 | 0.25 | 0.30 | 0.34 | 0.37 |
|  | 0.8 | −0.93 | −0.50 | −0.24 | −0.07 | 0.05 | 0.13 | 0.19 | 0.23 | 0.27 | 0.29 |
| 1.0 | 0.2 | −0.26 | 1.2 | 2.6 | 3.9 | 5.1 | 6.1 | 7.1 | 8.0 | 8.8 | 9.5 |
|  | 0.3 | −0.65 | 0.12 | 0.79 | 1.4 | 1.9 | 2.4 | 2.8 | 3.1 | 3.5 | 3.8 |
|  | 0.4 | −0.83 | −0.34 | 0.04 | 0.33 | 0.58 | 0.78 | 0.95 | 1.1 | 1.2 | 1.3 |
|  | 0.5 | −0.89 | −0.48 | −0.20 | 0 | 0.15 | 0.27 | 0.37 | 0.45 | 0.51 | 0.57 |
|  | 0.6 | −0.91 | −0.54 | −0.31 | −0.14 | −0.03 | 0.06 | 0.12 | 0.18 | 0.22 | 0.25 |
|  | 0.8 | −0.91 | −0.59 | −0.38 | −0.25 | −0.16 | −0.10 | −0.06 | −0.03 | −0.01 | 0.01 |
|  | 1.0 | −0.93 | −0.60 | −0.40 | −0.28 | −0.20 | −0.14 | −0.11 | −0.08 | −0.07 | −0.06 |

### Main, $C_{c,s}$

| $A_s/A_c$ | $A_b/A_c$ | $Q_b/Q_c$ = 0.2 | 0.4 | 0.6 | 0.8 | 1.0 | 1.2 | 1.4 | 1.6 | 1.8 | 2.0 |
|---|---|---|---|---|---|---|---|---|---|---|---|
| 0.3 | 0.2 | 4.5 | 2.8 | 1.5 | 0.56 | −0.17 | −0.74 | −1.2 | −1.6 | −1.9 | −2.1 |
|  | 0.3 | 4.6 | 3.1 | 2.0 | 1.2 | 0.57 | 0.08 | −0.30 | −0.62 | −0.89 | −1.1 |
| 0.4 | 0.2 | 1.6 | 0.85 | 0.16 | −0.43 | −0.92 | −1.3 | −1.7 | −1.9 | −2.2 | −2.4 |
|  | 0.3 | 1.7 | 1.1 | 0.58 | 0.13 | −0.24 | −0.56 | −0.82 | −1.1 | −1.3 | −1.4 |
|  | 0.4 | 1.8 | 1.3 | 0.80 | 0.42 | 0.11 | −0.15 | −0.37 | −0.55 | −0.72 | −0.86 |
| 0.5 | 0.2 | 0.67 | 0.18 | −0.33 | −0.79 | −1.2 | −1.5 | −1.8 | −2.1 | −2.3 | −2.5 |
|  | 0.3 | 0.75 | 0.42 | 0.07 | −0.25 | −0.54 | −0.80 | −1.0 | −1.2 | −1.4 | −1.5 |
|  | 0.4 | 0.80. | 0.55 | 0.28 | 0.03 | −0.20 | −0.40 | −0.57 | −0.73 | −0.86 | −0.98 |
|  | 0.5 | 0.82 | 0.62 | 0.41 | 0.20 | 0.02 | −0.15 | −0.29 | −0.42 | −0.53 | −0.63 |
| 0.6 | 0.2 | 0.26 | −0.11 | −0.54 | −0.95 | −1.3 | −1.6 | −1.9 | −2.1 | −2.4 | −0.25 |
|  | 0.3 | 0.34 | 0.13 | −0.14 | −0.42 | −0.67 | −0.90 | −0.11 | −0.13 | −0.14 | −0.16 |
|  | 0.4 | 0.39 | 0.25 | 0.06 | −0.14 | −0.33 | −0.51 | −0.66 | −0.80 | −0.93 | −1.0 |
|  | 0.5 | 0.41 | 0.32 | 0.18 | 0.03 | −1.2 | −0.26 | −0.38 | −0.50 | −0.60 | −0.69 |
|  | 0.6 | 0.43 | 0.37 | 0.26 | 0.14 | 0.02 | −0.09 | −0.19 | −0.29 | −0.37 | −0.45 |
| 0.8 | 0.2 | −0.01 | −0.30 | −0.67 | −1.1 | −1.4 | −1.7 | −2.0 | −2.2 | −2.4 | −2.6 |
|  | 0.3 | 0.07 | −0.07 | −0.29 | −0.58 | −0.76 | −0.97 | −1.2 | −1.3 | −1.5 | −1.6 |
|  | 0.4 | 0.11 | 0.05 | −0.09 | −0.26 | −0.42 | −0.58 | −0.72 | −0.85 | −0.97 | −1.1 |
|  | 0.5 | 0.14 | 0.12 | 0.03 | −0.09 | −0.21 | −0.34 | −0.45 | −0.55 | −0.64 | −0.73 |
|  | 0.6 | 0.15 | 0.17 | 0.11 | 0.02 | −0.07 | −0.17 | −0.26 | −0.34 | −0.42 | −0.49 |
|  | 0.7 | 0.17 | 0.21 | 0.17 | 0.11 | 0.03 | −0.05 | −0.12 | −0.19 | −0.26 | −0.32 |
|  | 0.5 | 0.17 | 0.23 | 0.22 | 0.17 | 0.11 | 0.05 | −0.02 | −0.07 | −0.13 | −0.18 |
| 1.0 | 0.2 | −0.05 | −0.33 | −0.70 | −1.1 | −1.4 | −1.7 | −2.0 | −2.2 | −2.4 | −2.6 |
|  | 0.3 | 0.03 | −0.10 | −0.31 | −0.55 | −0.78 | −0.98 | −1.2 | −1.3 | −1.5 | −1.6 |
|  | 0.4 | 0.07 | 0.02 | −0.12 | −0.28 | −0.44 | −0.59 | −0.73 | −0.86 | −0.98 | −1.1 |
|  | 0.5 | 0.09 | 0.09 | 0.01 | −0.11 | −0.23 | −0.35 | −0.46 | −0.56 | −0.65 | −0.74 |
|  | 0.6 | 0.11 | 0.14 | 0.09 | 0 | −0.09 | −0.18 | −0.27 | −0.35 | −0.43 | −0.50 |
|  | 0.8 | 0.13 | 0.20 | 0.19 | 0.15 | 0.09 | 0.03 | −0.03 | −0.08 | −0.14 | −0.19 |
|  | 1.0 | 0.14 | 0.24 | 0.25 | 0.24 | 0.20 | 0.16 | 0.12 | 0.08 | 0.04 | 0 |

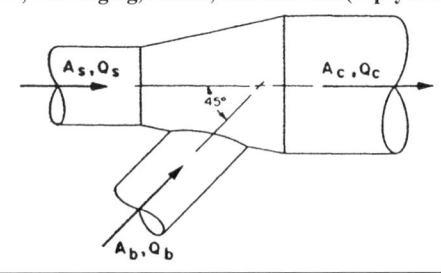

## 5-5 Wye, 45°, Converging, Round, Conical Main (Sepsy and Pies 1973)

### Branch, $C_{c,b}$

| $A_s/A_c$ | $A_b/A_c$ | $Q_b/Q_s$ = 0.2 | 0.4 | 0.6 | 0.8 | 1.0 | 1.2 | 1.4 | 1.6 | 1.8 | 2.0 |
|---|---|---|---|---|---|---|---|---|---|---|---|
| 0.3 | 0.2 | −2.4 | −0.01 | 2.0 | 3.8 | 5.3 | 6.6 | 7.8 | 8.9 | 9.8 | 11 |
|  | 0.3 | −2.8 | −1.2 | 0.12 | 1.1 | 1.9 | 2.6 | 3.2 | 3.7 | 4.2 | 4.6 |
| 0.4 | 0.2 | −1.2 | 0.93 | 2.8 | 4.5 | 5.9 | 7.2 | 8.4 | 9.5 | 10 | 11 |
|  | 0.3 | −1.6 | −0.27 | 0.81 | 1.7 | 2.4 | 3.0 | 3.6 | 4.1 | 4.5 | 4.9 |
|  | 0.4 | −1.8 | −0.72 | 0.07 | 0.66 | 1.1 | 1.5 | 1.8 | 2.1 | 2.3 | 2.5 |
| 0.5 | 0.2 | −0.46 | 1.5 | 3.3 | 4.9 | 6.4 | 7.7 | 8.8 | 9.9 | 11 | 12 |
|  | 0.3 | −0.94 | 0.25 | 1.2 | 2.0 | 2.7 | 3.3 | 3.8 | 4.2 | 4.7 | 5.0 |
|  | 0.4 | −1.1 | −0.24 | 0.42 | 0.92 | 1.3 | 1.6 | 1.9 | 2.1 | 2.3 | 2.5 |
|  | 0.5 | −1.2 | −0.38 | 0.18 | 0.58 | 0.88 | 1.1 | 1.3 | 1.5 | 1.6 | 1.7 |

# Duct and Pipe Sizing

| | | Branch, $C_{c,b}$ | | | | | | | | | |
|---|---|---|---|---|---|---|---|---|---|---|---|
| | | $Q_b/Q_s$ | | | | | | | | | |
| $A_s/A_c$ | $A_b/A_c$ | 0.2 | 0.4 | 0.6 | 0.8 | 1.0 | 12 | 1.4 | 1.6 | 1.8 | 2.0 |
| 0.6 | 0.2 | −0.55 | 1.3 | 3.1 | 4.7 | 6.1 | 7.4 | 8.6 | 9.6 | 11 | 12 |
| | 0.3 | −1.1 | 0 | 0.88 | 1.6 | 2.3 | 2.8 | 3.3 | 3.7 | 4.1 | 4.5 |
| | 0.4 | −1.2 | −0.48 | 0.10 | 0.54 | 0.89 | 1.2 | 1.4 | 1.6 | 1.8 | 2.0 |
| | 0.5 | −1.3 | −0.62 | −0.14 | 0.21 | 0.47 | 0.68 | 0.85 | 0.99 | 1.1 | 1.2 |
| | 0.6 | −1.3 | −0.69 | −0.26 | 0.04 | 0.26 | 0.42 | 0.57 | 0.66 | 0.75 | 0.82 |
| 0.8 | 0.2 | 0.06 | 1.8 | 3.5 | 5.1 | 6.5 | 7.8 | 8.9 | 10 | 11 | 12 |
| | 0.3 | −0.52 | 0.35 | 1.1 | 1.7 | 2.3 | 2.8 | 3.2 | 3.6 | 3.9 | 4.2 |
| | 0.4 | −0.67 | −0.05 | 0.43 | 0.80 | 1.1 | 1.4 | 1.6 | 1.8 | 1.9 | 2.1 |
| | 0.6 | −0.75 | −0.27 | 0.05 | 0.28 | 0.45 | 0.58 | 0.68 | 0.76 | 0.83 | 0.88 |
| | 0.7 | −0.77 | −0.31 | −0.02 | 0.18 | 0.32 | 0.43 | 0.50 | 0.56 | 0.61 | 0.65 |
| | 0.8 | −0.78 | −0.34 | −0.07 | 0.12 | 0.24 | 0.33 | 0.39 | 0.44 | 0.47 | 0.50 |
| 1.0 | 0.2 | 0.40 | 2.1 | 3.7 | 5.2 | 6.6 | 7.8 | 9.0 | 11 | 11 | 12 |
| | 0.3 | −0.21 | 0.54 | 1.2 | 1.8 | 2.3 | 2.7 | 3.1 | 3.7 | 3.7 | 4.0 |
| | 0.4 | −0.33 | 0.21 | 0.62 | 0.96 | 1.2 | 1.5 | 1.7 | 2.0 | 2.0 | 2.1 |
| | 0.5 | −0.38 | 0.05 | 0.37 | 0.60 | 0.79 | 0.93 | 1.1 | 1.2 | 1.2 | 1.3 |
| | 0.6 | −0.41 | −0.02 | 0.23 | 0.42 | 0.55 | 0.66 | 0.73 | 0.80 | 0.85 | 0.89 |
| | 0.8 | −0.44 | −0.10 | 0.11 | 0.24 | 0.33 | 0.39 | 0.43 | 0.46 | 0.47 | 0.48 |
| | 1.0 | −0.46 | −0.14 | 0.05 | 0.16 | 0.23 | 0.27 | 0.29 | 0.30 | 0.30 | 0.29 |

| | | Main, $C_{c,s}$ | | | | | | | | | |
|---|---|---|---|---|---|---|---|---|---|---|---|
| | | $Q_b/Q_s$ | | | | | | | | | |
| $A_s/A_c$ | $A_b/A_c$ | 0.2 | 0.4 | 0.6 | 0.8 | 1.0 | 12 | 1.4 | 1.6 | 1.8 | 2.0 |
| 0.3 | 0.2 | 5.3 | −0.01 | 2.0 | 1.1 | 0.34 | −0.20 | −0.61 | −0.93 | −1.2 | −1.4 |
| | 0.3 | 5.4 | 3.7 | 2.5 | 1.6 | 1.0 | 0.53 | 0.16 | −0.14 | −0.38 | −0.58 |
| 0.4 | 0.2 | 1.9 | 1.1 | 0.46 | −0.07 | −0.49 | −0.83 | −1.1 | −1.3 | −1.5 | −1.7 |
| | 0.3 | 2.0 | 1.4 | 0.81 | 0.42 | 0.08 | −0.20 | −0.43 | −0.62 | −0.78 | −0.92 |
| | 0.4 | 2.0 | 1.5 | 1.0 | 0.68 | 0.39 | 0.16 | −0.04 | −0.21 | −0.35 | −0.47 |
| 0.5 | 0.2 | 0.77 | 0.34 | −0.09 | −0.48 | −0.81 | −1.1 | −1.3 | −1.5 | −1.7 | −1.8 |
| | 0.3 | 0.85 | 0.56 | 0.25 | −0.03 | −0.27 | −0.48 | −0.67 | −0.82 | −0.96 | −1.1 |
| | 0.4 | 0.88 | 0.66 | 0.43 | 0.21 | 0.02 | −0.15 | −0.30 | −0.42 | −0.54 | −0.64 |
| | 0.5 | 0.91 | 0.73 | 0.54 | 0.36 | 0.21 | 0.06 | −0.06 | −0.17 | −0.26 | −0.35 |
| 0.6 | 0.2 | 0.30 | 0 | −0.34 | −0.67 | −0.96 | −1.2 | −1.4 | −1.6 | −1.8 | −1.9 |
| | 0.3 | 0.37 | 0.21 | −0.02 | −0.24 | −0.44 | −0.63 | −0.79 | −0.93 | −1.1 | −1.2 |
| | 0.4 | 0.40 | 0.31 | 0.16 | −0.1 | −0.16 | −0.30 | −0.43 | −0.54 | −0.64 | −0.73 |
| | 0.5 | 0.43 | 0.37 | 0.26 | 0.14 | 0.02 | −0.09 | −0.20 | −0.29 | −0.37 | −0.45 |
| | 0.6 | 0.44 | 0.41 | 0.33 | 0.24 | 0.14 | 0.05 | −0.03 | −0.11 | −0.18 | −0.25 |
| 0.8 | 0.2 | −0.06 | −0.27 | −0.57 | −0.86 | −1.1 | −1.4 | −1.6 | −1.7 | −1.9 | −2.0 |
| | 0.3 | 0 | −0.08 | −0.25 | −0.43 | −0.62 | −0.78 | −0.93 | −1.1 | −1.2 | −1.3 |
| | 0.4 | 0.04 | 0.02 | −0.08 | −0.21 | −0.34 | −0.46 | −0.57 | −0.67 | −0.77 | −0.85 |
| | 0.5 | 0.06 | 0.08 | 0.02 | −0.06 | −0.16 | −0.25 | −0.34 | −0.42 | −0.50 | −0.57 |
| | 0.6 | 0.07 | 0.12 | 0.09 | 0.03 | −0.04 | −0.11 | −0.18 | −0.25 | −0.31 | −0.37 |
| | 0.7 | 0.08 | 0.15 | 0.14 | 0.10 | 0.05 | −0.01 | −0.07 | −0.12 | −0.17 | −0.22 |
| | 0.8 | 0.09 | 0.17 | 0.18 | 0.16 | 0.11 | 0.07 | 0.02 | −0.02 | −0.07 | −0.11 |
| 1.0 | 0.2 | −0.19 | −0.39 | −0.67 | −0.96 | −1.2 | −1.5 | −1.6 | −1.8 | −2.0 | −2.1 |
| | 0.3 | −0.12 | −0.19 | −0.35 | −0.54 | −0.71 | −0.87 | −1.0 | −1.2 | −1.3 | −1.4 |
| | 0.4 | −0.09 | −0.10 | −0.19 | −0.31 | −0.43 | −0.55 | −0.66 | −0.77 | −0.86 | −0.94 |
| | 0.5 | −0.07 | −0.04 | −0.09 | −0.17 | −0.26 | −0.35 | −0.44 | −0.52 | −0.59 | −0.66 |
| | 0.6 | −0.06 | 0 | −0.02 | −0.07 | −0.14 | −0.21 | −0.28 | −0.34 | −0.40 | −0.46 |
| | 0.8 | −0.04 | 0.06 | 0.07 | 0.05 | 0.02 | −0.03 | −0.07 | −0.12 | −0.16 | −0.20 |
| | 1.0 | −0.3 | 0.09 | 0.13 | 0.13 | 0.11 | 0.08 | 0.06 | 0.03 | −0.01 | −0.03 |

## 5-6 Tee, Converging, Rectangular (Idelchik 1986, Diagram 7-11)

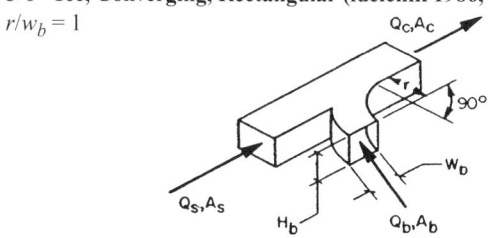

$r/w_b = 1$

| | | Branch, $C_{c,b}$ | | | | | | | | |
|---|---|---|---|---|---|---|---|---|---|---|
| | | $Q_c/Q_c$ | | | | | | | | |
| $A_b/A_s$ | $A_b/A_c$ | 0.1 | 0.2 | ~0.3 | 0.4 | 0.5 | 0.6 | 0.7 | 0.8 | 0.9 |
| 0.33 | 0.25 | −1.2 | −0.40 | 0.40 | 1.6 | 3.0 | 4.8 | 6.8 | 8.9 | 11 |
| 0.5 | 0.5 | −0.50 | −0.20 | 0 | 0.25 | 0.45 | 0.70 | 1.0 | 1.5 | 2.0 |
| 0.67 | 0.5 | −1.0 | −0.60 | −0.20 | 0.10 | 0.30 | 0.60 | 1.0 | 1.5 | 2.0 |
| 1.0 | 0.5 | −2.2 | −1.5 | −0.95 | −0.50 | 0 | 0.40 | 0.80 | 1.3 | 1.9 |
| 1.0 | 1.0 | −0.60 | −0.30 | −0.10 | −0.04 | 0.13 | 0.21 | 0.29 | 0.36 | 0.42 |
| 1.33 | 1.0 | −1.2 | −0.80 | −0.40 | −0.20 | 0 | 0.16 | 0.24 | 0.32 | 0.38 |
| 2.0 | 1.0 | −2.1 | −1.4 | −0.90 | −0.50 | −0.20 | 0 | 0.20 | 0.25 | 0.30 |

| | | Main, $C_{c,s}$ | | | | | | | |
|---|---|---|---|---|---|---|---|---|---|
| | | $Q_b/Q_s$ | | | | | | | |
| $A_b/A_s$ | $A_b/A_c$ | 0.1 | 0.2 | 0.3 | 0.4 | 0.5 | 0.6 | 0.7 | 0.8 | 0.9 |
| 0.33 | 0.25 | 0.30 | 0.30 | 0.20 | −0.10 | −0.45 | −0.92 | −1.5 | −2.0 | −2.6 |
| 0.5 | 0.5 | 0.17 | 0.16 | 0.10 | 0 | −0.08 | −0.18 | −0.27 | −0.37 | −0.46 |
| 0.67 | 0.5 | 0.27 | 0.35 | 0.32 | 0.25 | 0.12 | −0.03 | −0.23 | −0.42 | −0.58 |
| 1.0 | 0.5 | 1.2 | 1.1 | 0.90 | 0.65 | 0.35 | 0 | −0.40 | −0.80 | −1.3 |
| 1.0 | 1.0 | 0.18 | 0.24 | 0.27 | 0.26 | 0.23 | 0.18 | 0.10 | 0 | −0.12 |
| 1.33 | 1.0 | 0.75 | 0.36 | 0.38 | 0.35 | 0.27 | 0.18 | 0.05 | −0.08 | −0.22 |
| 2.0 | 1.0 | 0.80 | 0.87 | 0.80 | 0.68 | 0.55 | 0.40 | 0.25 | 0.08 | −0.10 |

## 5-7 Tee, Converging, Round Tap to Rectangular Main (SMACNA 1981, Table 6-9C)

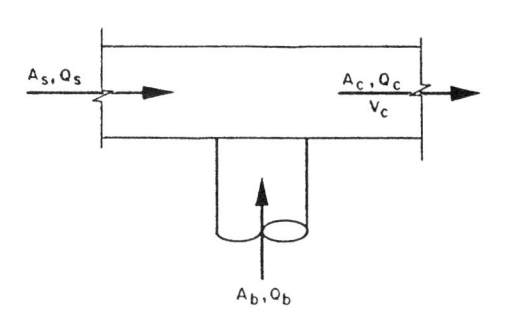

| $A_b/A_s$ | $A_s/A_c$ | $A_b/A_c$ |
|---|---|---|
| 0.5 | 1.0 | 0.5 |

| | Branch, $C_{c,b}$ | | | | | | | | | |
|---|---|---|---|---|---|---|---|---|---|---|
| | $Q_b/Q_c$ | | | | | | | | | |
| $V_c$ (fpm) | 0.1 | 0.2 | 0.3 | 0.4 | 0.5 | 0.6 | 0.7 | 0.8 | 0.9 | 1.0 |
| <1200 | −0.63 | −0.55 | 0.13 | 0.23 | 0.78 | 1.30 | 1.93 | 3.10 | 4.88 | 5.60 |
| >1200 | −0.49 | −0.21 | 0.23 | 0.60 | 1.27 | 2.06 | 2.75 | 3.70 | 4.93 | 5.95 |

For main coefficient ($C_{c,s}$), see Fitting 5-3.

## 5-8 Tee, Converging, Rectangular Main and Tap (SMACNA 1981, Table 6-9D)

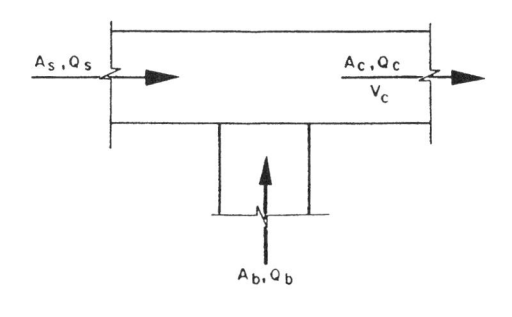

| $A_b/A_s$ | $A_s/A_c$ | $A_b/A_c$ |
|---|---|---|
| 0.5 | 1.0 | 0.5 |

| | Branch, $C_{c,b}$ | | | | | | | | | |
|---|---|---|---|---|---|---|---|---|---|---|
| $V_c$ | $Q_b/Q_c$ | | | | | | | | | |
| (fpm) | 0.1 | 0.2 | 0.3 | 0.4 | 0.5 | 0.6 | 0.7 | 0.8 | 0.9 | 1.0 |
| <1200 | −0.75 | −0.53 | −0.03 | 0.33 | 1.03 | 1.10 | 2.15 | 2.93 | 4.18 | 4.78 |
| >1200. | −0.69 | −0.21 | 0.23 | 0.67 | 1.17 | 1.66 | 2.67 | 3.36 | 3.93 | 5.13 |

For main coefficient ($C_{c,s}$), see Fitting 5-3.

## 5-9 Converging, Rectangular Main and Tap (45° Entry) (SMACNA 1981, Table-9F)

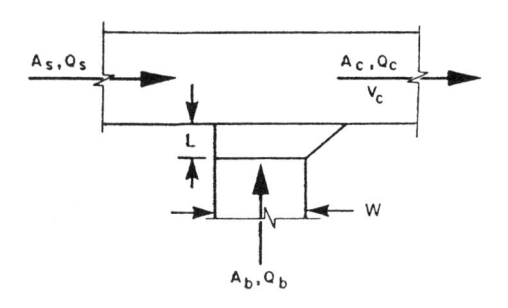

$L = 0.25\,W$, 3 in. min.

| $A_b/A_s$ | $A_s/A_c$ | $A_b/A_c$ |
|---|---|---|
| 0.5 | 1.0 | 0.5 |

| | Branch, $C_{c,b}$ | | | | | | | | | |
|---|---|---|---|---|---|---|---|---|---|---|
| $V_c$ | $Q_b/Q_c$ | | | | | | | | | |
| (fpm) | 0.1 | 0.2 | 0.3 | 0.4 | 0.5 | 0.6 | 0.7 | 0.8 | 0.9 | 1.0 |
| <1200 | –0.83 | –0.68 | –0.30 | 0.28 | 0.55 | 1.03 | 1.50 | 1.93 | 2.50 | 3.03 |
| >1200 | –0.72 | –0.52 | –0.23 | 0.34 | 0.76 | 1.14 | 1.83 | 2.01 | 2.90 | 3.63 |

For main coefficient ($C_{c,s}$), see Fitting 5-3.

## 5-10 Tee, Diverging, Round, Conical Branch (Jones and Sepsy 1969, Figure 12)

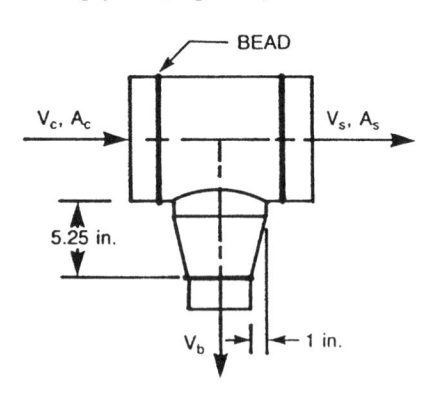

$A_c = A_s$

| | Branch | | | | | | | | | |
|---|---|---|---|---|---|---|---|---|---|---|
| $V_b/V_s$ | 0 | 0.2 | 0.4 | 0.6 | 0.8 | 1.0 | 1.2 | 1.4 | 1.6 | 1.8 | 2.0 |
| $C_{c,b}$ | 1.0 | 0.85 | 0.74 | 0.62 | 0.52 | 0.42 | 0.36 | 0.32 | 0.32 | 0.37 | 0.52 |

For main loss coefficient ($C_{c,s}$), see Fitting 5-23.

## 5-11 Wye, 45°, Diverging, Round, Conical Branch (Jones and Sepsy 1969, Figure 14)

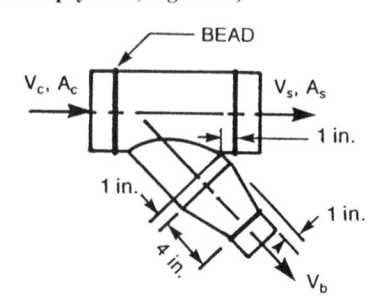

$A_c = A_s$

| | Branch | | | | | | | | | |
|---|---|---|---|---|---|---|---|---|---|---|
| $V_b/V_c$ | 0 | 0.2 | 0.4 | 0.6 | 0.8 | 1.0 | 1.2 | 1.4 | 1.6 | 1.8 | 2.0 |
| $C_{c,b}$ | 1.0 | 0.84 | 0.61 | 0.41 | 0.27 | 0.17 | 0.12 | 0.12 | 0.14 | 0.18 | 0.27 |

For main loss coefficient ($C_{c,s}$), see Fitting 5-23.

## 5-12 Tee, Diverging, Round, with 90° Elbow, Branch 90° to Main (Jones and Sepsy 1969, Figure 17)

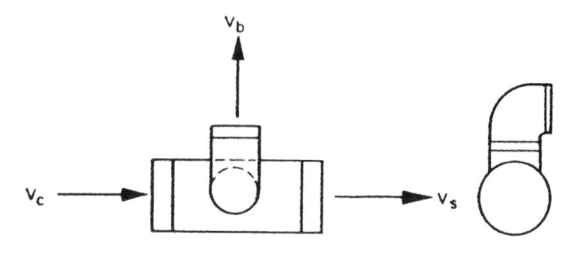

$A_r = A_s$

| | Branch | | | | | | | | | |
|---|---|---|---|---|---|---|---|---|---|---|
| $V_b/V_c$ | 0 | 0.2 | 0.4 | 0.6 | 0.8 | 1.0 | 1.2 | 1.4 | 1.6 | 1.8 | 2.0 |
| $C_{c,b}$ | 1.0 | 1.03 | 1.08 | 1.18 | 1.33 | 1.56 | 1.86 | 2.2 | 2.6 | 3.0 | 3.4 |

For main loss coefficient ($C_{c,s}$), see Fitting 5-23.

## 5-13 Tee, Diverging, Round, with 45° Elbow, Branch 90° to Main (Jones and Sepsy 1969, Figure 18)

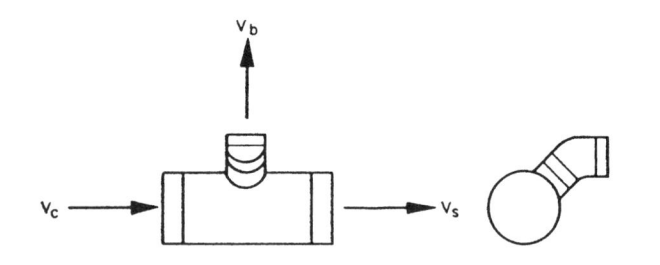

$A_c = A_s$

| | Branch | | | | | | | | | |
|---|---|---|---|---|---|---|---|---|---|---|
| $V_b/V_c$ | 0 | 0.2 | 0.4 | 0.6 | 0.8 | 1.0 | 1.2 | 1.4 | 1.6 | 1.8 | 2.0 |
| $C_{c,b}$ | 1.0 | 1.32 | 1.51 | 1.60 | 1.65 | 1.74 | 1.87 | 2.0 | 2.2 | 2.5 | 2.7 |

For main loss coefficient ($C_{c,s}$), see Fitting 5-23.

## 5-14 Tee, Diverging, Round (Conical Branch), with 45° Elbow, Branch 90° to Main (Jones and Sepsy 1969, Figure 19)

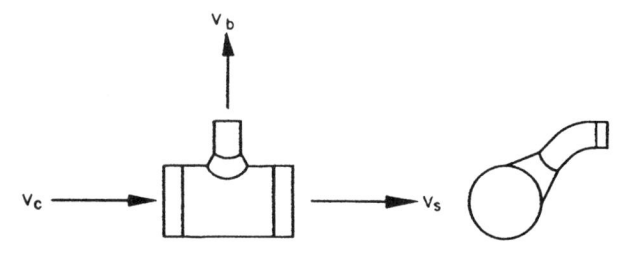

For tee geometry, see Fitting 5-10.

$A_c = A_s$

| | Branch | | | | | | | | | |
|---|---|---|---|---|---|---|---|---|---|---|
| $V_b/V_s$ | 0 | 0.2 | 0.4 | 0.6 | 0.8 | 1.0 | 1.2 | 1.4 | 1.6 | 1.8 | 2.0 |
| $C_{c,b}$ | 1.0 | 0.94 | 0.88 | 0.84 | 0.80 | 0.82 | 0.84 | 0.87 | 0.90 | 0.95 | 1.02 |

For main loss coefficient ($C_{c,s}$), see Fitting 5-23.

# Duct and Pipe Sizing

**5-15 Wye, 45°, Round, with 60° Elbow, Branch 90° to Main (Jones and Sepsy 1969, Figure 3)**

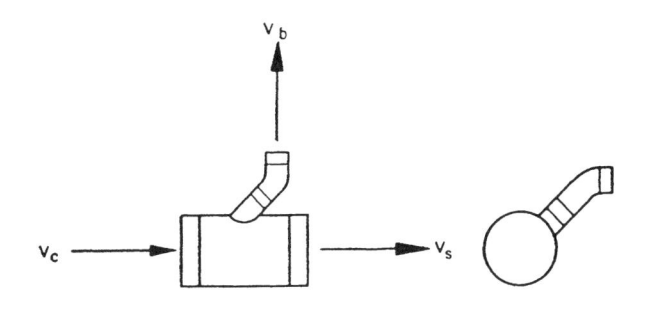

$A_c = A_s$

| | Branch | | | | | | | | | | |
|---|---|---|---|---|---|---|---|---|---|---|---|
| $V_b/V_c$ | 0 | 0.2 | 0.4 | 0.6 | 0.8 | 1.0 | 1.2 | 1.4 | 1.6 | 1.8 | 2.0 |
| $C_{c,b}$ | 1.0 | 0.88 | 0.77 | 0.68 | 0.65 | 0.69 | 0.73 | 0.88 | 1.14 | 1.54 | 2.2 |

For main loss coefficient ($C_{c,s}$), see Fitting 5-23.

**5-16 Wye, 45°, Diverging, Round (Conical Branch), with 60° Elbow, Branch 90° to Main (Jones and Sepsy 1969, Figure 20)**

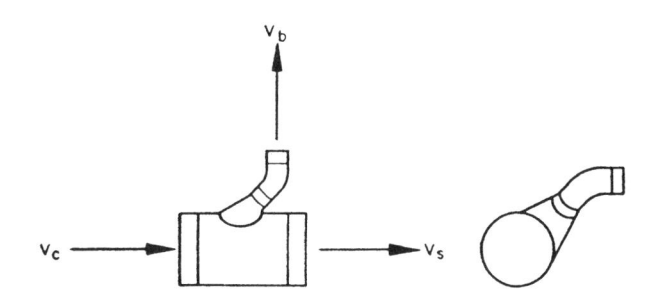

For wye geometry, see Fitting 5-11.

$A_c = A_s$

| | Branch | | | | | | | | | | |
|---|---|---|---|---|---|---|---|---|---|---|---|
| $V_b/V_c$ | 0 | 0.2 | 0.4 | 0.6 | 0.8 | 1.0 | 1.2 | 1.4 | 1.6 | 1.8 | 2.0 |
| $C_{c,b}$ | 1.0 | 0.82 | 0.63 | 0.52 | 0.45 | 0.42 | 0.41 | 0.40 | 0.41 | 0.45 | 0.56 |

For main loss coefficient ($C_{c,s}$), see Fitting 5-23.

**5-17 Wye, 48°, Diverging, Conical Main and Branch, with 45° Elbow,**

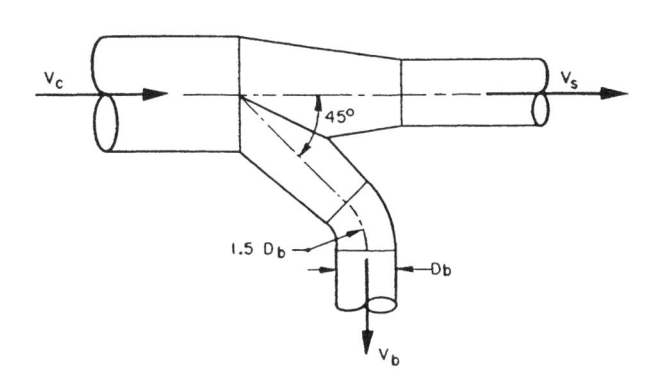

| | Branch | | | | | | | |
|---|---|---|---|---|---|---|---|---|
| $V_b/V_c$ | 0.2 | 0.4 | 0.6 | 0.7 | 0.8 | 0.9 | 1.0 | 1.1 | 1.2 |
| $C_{c,b}$ | 0.76 | 0.60 | 0.52 | 0.50 | 0.51 | 0.52 | 0.56 | 0.61 | 0.68 |
| $V_b/V_c$ | 1.4 | 1.6 | 1.8 | 2.0 | 2.2 | 2.4 | 2.6 | 2.8 | 3.0 |
| $C_{c,b}$ | 0.86 | 1.1 | 1.4 | 1.8 | 2.2 | 2.6 | 3.1 | 3.7 | 4.2 |

| | Main | | | | | | | | |
|---|---|---|---|---|---|---|---|---|---|
| $V_s/V_c$ | 0.2 | 0.4 | 0.6 | 0.8 | 1.0 | 1.2 | 1.4 | 1.6 | 1.8 | 2.0 |
| $C_{c,s}$ | 0.14 | 0.06 | 0.05 | 0.09 | 0.18 | 0.30 | 0.46 | 0.64 | 0.84 | 1.0 |

**5-18 Tee, Diverging, Round, with 60° Elbow, Branch 45° to Main (Jones and Sepsy 1969, Figure 22)**

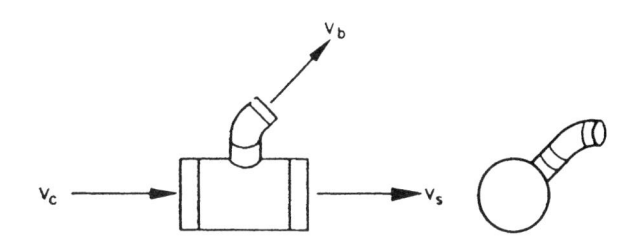

$A_c = A_s$

| | Branch | | | | | | | | | | |
|---|---|---|---|---|---|---|---|---|---|---|---|
| $V_b/V_c$ | 0 | 0.2 | 0.4 | 0.6 | 0.8 | 1.0 | 1.2 | 1.4 | 1.6 | 1.8 | 2.0 |
| $C_{c,b}$ | 1.0 | 1.06 | 1.15 | 1.29 | 1.45 | 1.65 | 1.89 | 2.2 | 2.5 | 2.9 | 3.3 |

For main loss coefficient ($C_{c,s}$), see Fitting 5-23.

**5-19 Tee, Diverging, Round (Conical Branch), with 60° Elbow, Branch 45° to Main (Jones and Sepsy 1969, Figure 23)**

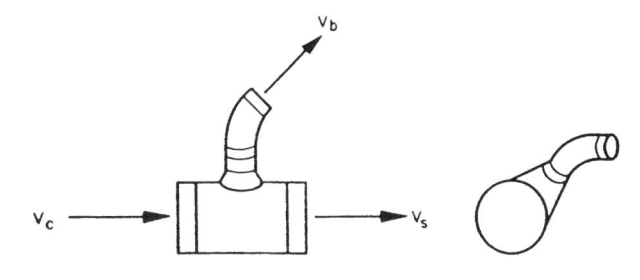

For tee geometry, see Fitting 5-10.

$A_c = A_s$

| | Branch | | | | | | | | | | |
|---|---|---|---|---|---|---|---|---|---|---|---|
| $V_b/V_c$ | 0 | 0.2 | 0.4 | 0.6 | 0.8 | 1.0 | 1.2 | 1.4 | 1.6 | 1.8 | 2.0 |
| $C_{c,b}$ | 1.0 | 0.95 | 0.90 | 0.86 | 0.81 | 0.79 | 0.79 | 0.81 | 0.86 | 0.96 | 1.10 |

For main loss coefficient ($C_{c,s}$), see Fitting 5-23.

**5-20 Wye, 45°, Diverging, Round, with 30° Elbow, Branch 45° to Main (Jones and Sepsy 1969, Figure 2)**

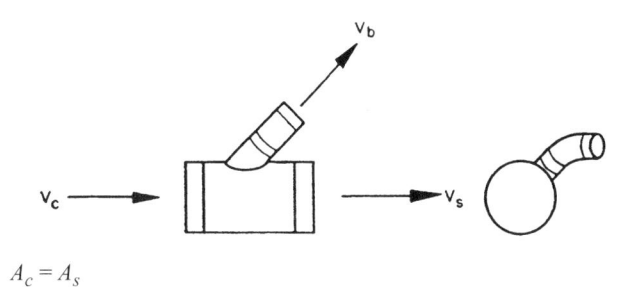

$A_c = A_s$

**334**  **Principles of HVAC**

| Branch | | | | | | | | | | | |
|---|---|---|---|---|---|---|---|---|---|---|---|
| $V_b/V_c$ | 0 | 0.2 | 0.4 | 0.6 | 0.8 | 1.0 | 1.2 | 1.4 | 1.6 | 1.8 | 2.0 |
| $C_{c,b}$ | 1.0 | 0.84 | 0.72 | 0.62 | 0.54 | 0.50 | 0.56 | 0.71 | 0.92 | 1.22 | 1.66 |

For main loss coefficient ($C_{c,s}$), see Fitting 5-23.

### 5-21 Wye, 45°, Diverging, Round (Conical Branch), with 30° Elbow, Branch 45° to Main (Jones and Sepsy 1969, Figure 24)

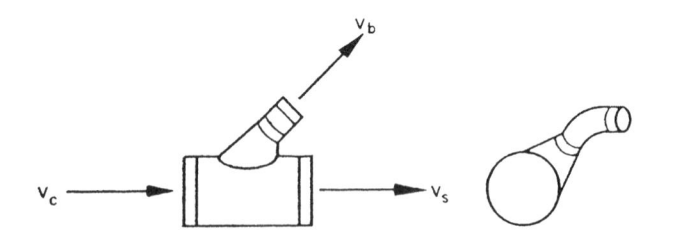

For wye geometry, see Fitting 5-11.

$A_c = A_s$

| Branch | | | | | | | | | | | |
|---|---|---|---|---|---|---|---|---|---|---|---|
| $V_b/V_c$ | 0 | 0.2 | 0.4 | 0.6 | 0.8 | 1.0 | 1.2 | 1.4 | 1.6 | 1.8 | 2.0 |
| $C_{c,b}$ | 1.0 | 0.93 | 0.71 | 0.55 | 0.44 | 0.42 | 0.42 | 0.44 | 0.47 | 0.54 | 0.62 |

For main loss coefficient ($C_{c,s}$), see Fitting 5-23.

### 5-22 Tee, Diverging, Rectangular (Idelchik et al. 1986, Diagram 7-21)

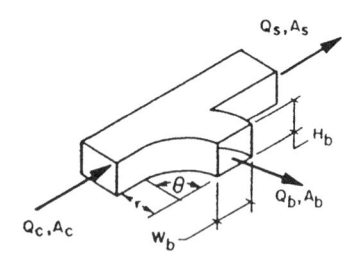

$\theta = 90°$
$r/W_b = 1.0$

**Branch, $C_{c,b}$**

| | | $Q_b/Q_c$ | | | | | | | | |
|---|---|---|---|---|---|---|---|---|---|---|
| $A_b/A_s$ | $A_b/A_f$ | 0.1 | 0.2 | 0.3 | 0.4 | 0.5 | 0.6 | 0.7 | 0.8 | 0.9 |
| 0.25 | 0.25 | 0.55 | 0.50 | 0.60 | 0.85 | 1.2 | 1.8 | 3.1 | 4.4 | 6.0 |
| 0.33 | 0.25 | 0.35 | 0.35 | 0.50 | 0.80 | 1.3 | 2.0 | 2.8 | 3.8 | 5.0 |
| 0.5 | 0.5 | 0.62 | 0.48 | 0.40 | 0.40 | 0.48 | 0.60 | 0.78 | 1.1 | 1.5 |
| 0.67 | 0.5 | 0.52 | 0.40 | 0.32 | 0.30 | 0.34 | 0.44 | 0.62 | 0.92 | 1.4 |
| 1.0 | 0.5 | 0.44 | 0.38 | 0.38 | 0.41 | 0.52 | 0.68 | 0.92 | 1.2 | 1.6 |
| 1.0 | 1.0 | 0.67 | 0.55 | 0.46 | 0.37 | 0.32 | 0.29 | 0.29 | 0.30 | 0.37 |
| 1.33 | 1.0 | 0.70 | 0.60 | 0.51 | 0.42 | 0.34 | 0.28 | 0.26 | 0.26 | 0.29 |
| 2.0 | 1.0 | 0.60 | 0.52 | 0.43 | 0.33 | 0.24 | 0.17 | 0.15 | 0.17 | 0.21 |

**Main, $C_{c,s}$**

| | | $Q_b/Q_c$ | | | | | | | | |
|---|---|---|---|---|---|---|---|---|---|---|
| $A_b/A_s$ | $A_b/A_c$ | 0.1 | 0.2 | 0.3 | 0.4 | 0.5 | 0.6 | 0.7 | 0.8 | 0.9 |
| 0.25 | 0.25 | −0.01 | −0.03 | −0.01 | 0.05 | 0.13 | 0.21 | 0.29 | 0.38 | 0.46 |
| 0.33 | 0.25 | 0.08 | 0 | −0.02 | −0.01 | 0.02 | 0.08 | 0.16 | 0.24 | 0.34 |
| 0.5 | 0.5 | −0.03 | −0.06 | −0.05 | 0 | 0.06 | 0.12 | 0.19 | 0.27 | 0.35 |
| 0.67 | 0.5 | 0.04 | −0.02 | −0.04 | −0.03 | −0.01 | 0.04 | 0.12 | 0.23 | 0.37 |
| 1.0 | 0.5 | 0.72 | 0.48 | 0.28 | 0.13 | 0.05 | 0.04 | 0.09 | 0.18 | 0.30 |
| 1.0 | 1.0 | −0.02 | −0.04 | −0.04 | −0.01 | 0.06 | 0.13 | 0.22 | 0.30 | 0.38 |
| 1.33 | 1.0 | 0.10 | 0 | 0.01 | −0.03 | −0.01 | 0.03 | 0.10 | 0.20 | 0.30 |
| 2.0 | 1.0 | 0.62 | 0.38 | 0.23 | 0.23 | 0.08 | 0.05 | 0.06 | 0.10 | 0.20 |

### 5-23 Wye, Diverging, Rectangular and Round (Idelchik et al. 1986, Diagrams 7-15 and 7-17)

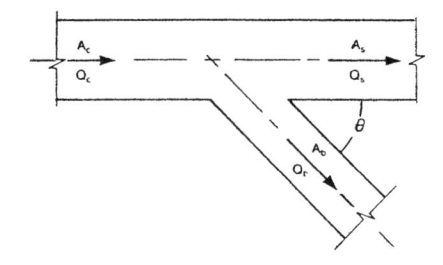

$A_c = A_s$; $H_b = H_c$, where $H$ is height of rectangular duct

$\theta = 30°$

**Branch, $C_{c,b}$**

| | $Q_b/Q_c$ | | | | | | | | |
|---|---|---|---|---|---|---|---|---|---|
| $A_b/A_c$ | 0.1 | 0.2 | 0.3 | 0.4 | 0.5 | 0.6 | 0.7 | 0.8 | 0.9 |
| 0.8 | 0.75 | 0.55 | 0.40 | 0.28 | 0.21 | 0.16 | 0.15 | 0.16 | 0.19 |
| 0.7 | 0.72 | 0.51 | 0.36 | 0.25 | 0.18 | 0.15 | 0.16 | 0.20 | 0.26 |
| 0.6 | 0.69 | 0.46 | 0.31 | 0.21 | 0.17 | 0.16 | 0.20 | 0.28 | 0.39 |
| 0.5 | 0.65 | 0.41 | 0.26 | 0.19 | 0.18 | 0.22 | 0.32 | 0.47 | 0.67 |
| 0.4 | 0.59 | 0.33 | 0.21 | 0.20 | 0.27 | 0.40 | 0.62 | 0.92 | 1.3 |
| 0.3 | 0.55 | 0.28 | 0.24 | 0.38 | 0.76 | 1.3 | 2.0 | 3.0 | 4.1 |
| 0.2 | 0.40 | 0.26 | 0.58 | 1.3 | 2.5 | 4.1 | 6.1 | 8.6 | 11.0 |
| 0.1 | 0.28 | 1.5 | 4.3 | 8.3 | 15.0 | — | — | — | — |

$\theta = 45°$

**Branch, $C_{c,b}$**

| | $Q_b/Q_c$ | | | | | | | | |
|---|---|---|---|---|---|---|---|---|---|
| $A_b/A_c$ | 0.1 | 0.2 | 0.3 | 0.4 | 0.5 | 0.6 | 0.7 | 0.8 | 0.9 |
| 0.8 | 0.78 | 0.62 | 0.49 | 0.40 | 0.34 | 0.31 | 0.32 | 0.35 | 0.40 |
| 0.7 | 0.71 | 0.59 | 0.47 | 0.38 | 0.34 | 0.32 | 0.35 | 0.41 | 0.50 |
| 0.6 | 0.74 | 0.56 | 0.44 | 0.37 | 0.35 | 0.36 | 0.43 | 0.54 | 0.68 |
| 0.5 | 0.71 | 0.52 | 0.41 | 0.38 | 0.40 | 0.45 | 0.59 | 0.78 | 1.0 |
| 0.4 | 0.66 | 0.47 | 0.40 | 0.43 | 0.54 | 0.69 | 0.95 | 1.3 | 1.7 |
| 0.3 | 0.66 | 0.48 | 0.52 | 0.73 | 1.2 | 1.8 | 2.7 | 3.7 | 4.9 |
| 0.2 | 0.56 | 0.56 | 1.0 | 1.8 | 3.2 | 4.9 | 7.1 | 9.6 | 13.0 |
| 0.1 | 0.60 | 2.1 | 5.1 | 9.3 | 16.0 | — | — | — | — |

$\theta = 60°$

**Branch, $C_{c,b}$**

| | $Q_b/Q_c$ | | | | | | | | |
|---|---|---|---|---|---|---|---|---|---|
| $A_b/A_c$ | 0.1 | 0.2 | 0.3 | 0.4 | 0.5 | 0.6 | 0.7 | 0.8 | 0.9 |
| 0.8 | 0.83 | 0.71 | 0.62 | 0.56 | 0.52 | 0.50 | 0.53 | 0.60 | 0.68 |
| 0.7 | 0.82 | 0.69 | 0.61 | 0.56 | 0.54 | 0.54 | 0.60 | 0.70 | 0.82 |
| 0.6 | 0.81 | 0.68 | 0.60 | 0.58 | 0.58 | 0.61 | 0.72 | 0.87 | 1.1 |
| 0.5 | 0.79 | 0.66 | 0.61 | 0.62 | 0.68 | 0.76 | 0.94 | 1.2 | 1.5 |
| 0.4 | 0.76 | 0.65 | 0.65 | 0.74 | 0.89 | 1.1 | 1.4 | 1.8 | 2.3 |
| 0.3 | 0.80 | 0.75 | 0.89 | 1.2 | 1.8 | 2.6 | 3.5 | 4.6 | 6.0 |
| 0.2 | 0.77 | 0.96 | 1.6 | 2.5 | 4.0 | 6.0 | 8.3 | 11.0 | — |
| 0.1 | 1.0 | 2.9 | 6.2 | 10.0 | — | — | — | — | — |

$\theta = 90°$

**Branch, $C_{c,b}$**

| | $Q_b/Q_c$ | | | | | | | | |
|---|---|---|---|---|---|---|---|---|---|
| $A_b/A_c$ | 0.1 | 0.2 | 0.3 | 0.4 | 0.5 | 0.6 | 0.7 | 0.8 | 0.9 |
| 0.8 | 0.95 | 0.92 | 0.92 | 0.93 | 0.94 | 0.95 | 1.1 | 1.2 | 1.4 |
| 0.7 | 0.95 | 0.94 | 0.95 | 0.98 | 1.0 | 1.1 | 1.2 | 1.4 | 1.6 |
| 0.6 | 0.96 | 0.97 | 1.0 | 1.1 | 1.1 | 1.2 | 1.4 | 1.7 | 2.0 |
| 0.5 | 0.97 | 1.0 | 1.1 | 1.2 | 1.4 | 1.5 | 1.8 | 2.1 | 2.5 |
| 0.4 | 0.99 | 1.1 | 1.3 | 1.5 | 1.7 | 2.0 | 2.4 | 3.0 | 3.6 |
| 0.3 | 1.1 | 1.4 | 1.8 | 2.3 | 3.2 | 4.3 | 5.5 | 6.9 | 8.5 |
| 0.2 | 1.3 | 1.9 | 2.9 | 4.1 | 6.2 | 8.5 | 11.0 | — | — |
| 0.1 | 2.1 | 4.8 | 8.9 | 14.0 | — | — | — | — | — |

# Duct and Pipe Sizing

| | Main | | | | | | | | |
|---|---|---|---|---|---|---|---|---|---|
| $V_s/V_c$ | 0 | 0.1 | 0.2 | 0.3 | 0.4 | 0.5 | 0.6 | 0.8 | 1.0 |
| $C_{c,s}$ | 0.40 | 0.32 | 0.26 | 0.20 | 0.14 | 0.10 | 0.06 | 0.02 | 0 |

## 5-24 Diverging Wye, Rectangular
(Idelchik et al. 1986, Diagrams 7-16 and 7-17)

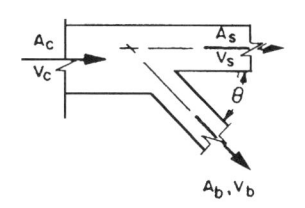

$\theta = 15°$ to $90°$ and $A_c = A_s + A_b$

| | Branch, $C_{c,b}$ | | | | | | | | | | | | |
|---|---|---|---|---|---|---|---|---|---|---|---|---|---|
| $\theta$, | $V_b/V_c$ | | | | | | | | | | | | |
| deg. | 0.1 | 0.2 | 0.3 | 0.4 | 0.5 | 0.6 | 0.8 | 1.0 | 1.2 | 1.4 | 1.6 | 1.8 | 2.0 |
| 15 | 0.81 | 0.65 | 0.51 | 0.38 | 0.28 | 0.20 | 0.11 | 0.06 | 0.14 | 0.30 | 0.51 | 0.76 | 1.0 |
| 30 | 0.84 | 0.69 | 0.56 | 0.44 | 0.34 | 0.26 | 0.19 | 0.15 | 0.15 | 0.30 | 0.51 | 0.76 | 1.0 |
| 45 | 0.87 | 0.74 | 0.63 | 0.54 | 0.45 | 0.38 | 0.29 | 0.24 | 0.23 | 0.30 | 0.51 | 0.76 | 1.0 |
| 60 | 0.90 | 0.82 | 0.79 | 0.66 | 0.59 | 0.53 | 0.43 | 0.36 | 0.33 | 0.39 | 0.51 | 0.76 | 1.0 |
| 90 | 1.0 | 1.0 | 1.0 | 1.0 | 1.0 | 1.0 | 1.0 | 1.0 | 1.0 | 1.0 | 1.0 | 1.0 | 1.0 |

| | Main, $C_{c,s}$ | | | | | |
|---|---|---|---|---|---|---|
| $\theta$, degrees | 15-60 | 90 | | | | |
| | | $A_s/A_c$ | | | | |
| $V_s/V_c$ | 0-1.0 | 0-0.4 | 0.5 | 0.6 | 0.7 | $\geq 0.8$ |
| 0 | 1.0 | 1.0 | 1.0 | 1.0 | 1.0 | 1.0 |
| 0.1 | 0.81 | 0.81 | 0.81 | 0.81 | 0.81 | 0.81 |
| 0.2 | 0.64 | 0.64 | 0.64 | 0.64 | 0.64 | 0.64 |
| 0.3 | 0.50 | 0.50 | 0.52 | 0.52 | 0.50 | 0.50 |
| 0.4 | 0.36 | 0.36 | 0.40 | 0.38 | 0.37 | 0.36 |
| 0.5 | 0.25 | 0.25 | 0.30 | 0.28 | 0.27 | 0.25 |
| 0.6 | 0.16 | 0.16 | 0.23 | 0.20 | 0.18 | 0.16 |
| 0.8 | 0.04 | 0.04 | 0.17 | 0.10 | 0.07 | 0.04 |
| 1.0 | 0 | 0 | 0.20 | 0.10 | 0.05 | 0 |
| 1.2 | 0.07 | 0.07 | 0.36 | 0.21 | 0.14 | 0.07 |
| 1.4 | 0.39 | 0.39 | 0.79 | 0.59 | 0.39 | — |
| 1.6 | 0.90 | 0.90 | 1.4 | 1.2 | — | — |
| 1.8 | 1.8 | 1.8 | 2.4 | — | — | — |
| 2.0 | 3.2 | 3.2 | 4.0 | — | — | — |

## 5-25 Tee, Diverging, Rectangular Main to Round Tap (SMACNA 1981, Table 6-10T)

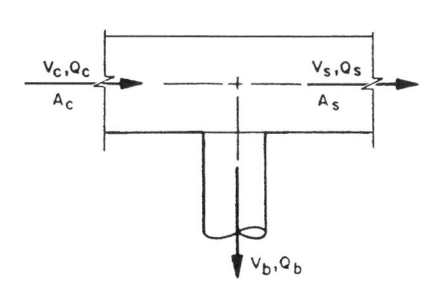

$A_c = A_s$

| | Branch, $C_{c,b}$ | | | | | | | | |
|---|---|---|---|---|---|---|---|---|---|
| | $Q_b/Q_s$ | | | | | | | | |
| $V_b/V_c$ | 0.1 | 0.2 | 0.3 | 0.4 | 0.5 | 0.6 | 0.7 | 0.8 | 0.9 |
| 0.2 | 1.00 | | | | | | | | |
| 0.4 | 1.01 | 1.07 | | | | | | | |
| 0.6 | 1.14 | 1.10 | 1.08 | | | | | | |
| 0.8 | 1.18 | 1.31 | 1.12 | 1.13 | | | | | |
| 1.0 | 1.30 | 1.38 | 1.20 | 1.23 | 1.26 | | | | |
| 1.2 | 1.46 | 1.58 | 1.45 | 1.31 | 1.39 | 1.48 | | | |
| 1.4 | 1.70 | 1.82 | 1.65 | 1.51 | 1.56 | 1.64 | 1.71 | | |
| 1.6 | 1.93 | 2.06 | 2.00 | 1.85 | 1.70 | 1.76 | 1.80 | 1.88 | |
| 1.8 | 2.06 | 2.17 | 2.10 | 2.13 | 2.06 | 1.98 | 1.99 | 2.00 | 2.07 |

For main coefficient ($C_{c,s}$), see Fitting 5-23.

## 5-26 Tee, Diverging, Rectangular Main to Round Tap (Conical)
(Inoue et al. 1980, Korst et al. 1950)

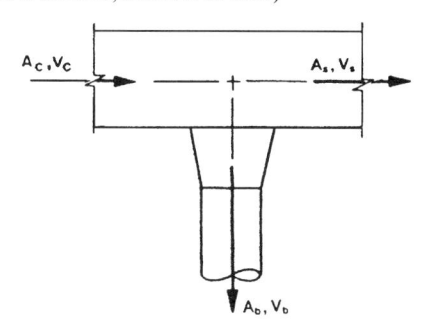

$A_c = A_s$

| | Branch | | | | | |
|---|---|---|---|---|---|---|
| $V_b/V_c$ | 0.40 | 0.50 | 0.75 | 1.0 | 1.3 | 1.5 |
| $C_{c,b}$ | 0.80 | 0.83 | 0.90 | 1.0 | 1.1 | 1.4 |

For main coefficient ($C_{c,s}$), see Fitting 5-23.

## 5-27 Tee, Diverging, Rectangular Main, and Tap (45° Entry)
(SMACNA 1981, Table 6-10N)

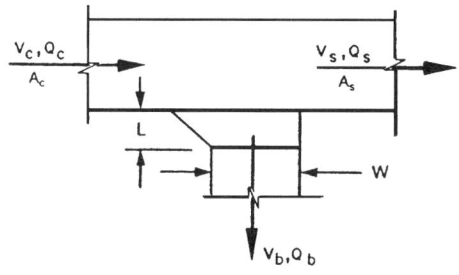

Recommended[a]

$L = 0.25W$, 3 in. min.

$A_c = A_s$

| | Branch, $C_{c,b}$ | | | | | | | | |
|---|---|---|---|---|---|---|---|---|---|
| | $Q_b/Q_s$ | | | | | | | | |
| $V_b/V_c$ | 0.1 | 0.2 | 0.3 | 0.4 | 0.5 | 0.6 | 0.7 | 0.8 | 0.9 |
| 0.2 | 0.91 | | | | | | | | |
| 0.4 | 0.81 | 0.79 | | | | | | | |
| 0.6 | 0.77 | 0.72 | 0.70 | | | | | | |
| 0.8 | 0.78 | 0.73 | 0.69 | 0.66 | | | | | |
| 1.0 | 0.78 | 0.98 | 0.85 | 0.79 | 0.74 | | | | |
| 1.2 | 0.90 | 1.11 | 1.16 | 1.23 | 1.03 | 0.86 | | | |
| 1.4 | 1.19 | 1.22 | 1.26 | 1.29 | 1.54 | 0.25 | 0.92 | | |
| 1.6 | 1.35 | 1.42 | 1.55 | 1.59 | 1.63 | 1.50 | 1.31 | 1.09 | |
| 1.8 | 1.44 | 1.50 | 1.75 | 1.74 | 1.72 | 2.24 | 1.63 | 1.40 | 1.17 |

For main coefficient ($C_{c,s}$), see Fitting 5-23.
[a]For performance study, see SMACNA (1987).

### 5-28 Tee, Diverging, Rectangular Main, and Tap[a] (SMACNA 1981, Table 10Q)

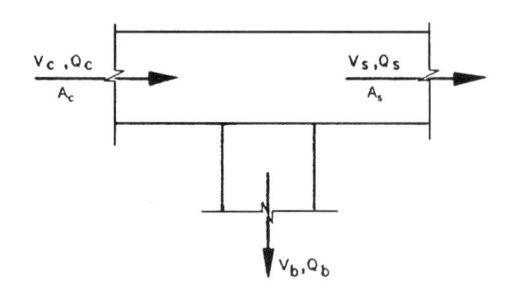

$A_c = A_s$

| | Branch, $C_{c,b}$ | | | | | | | | |
|---|---|---|---|---|---|---|---|---|---|
| | $Q_b/Q_c$ | | | | | | | | |
| $V_b/V_c$ | 0.1 | 0.2 | 0.3 | 0.4 | 0.5 | 0.6 | 0.7 | 0.8 | 0.9 |
| 0.2 | 1.03 | | | | | | | | |
| 0.4 | 1.04 | 1.01 | | | | | | | |
| 0.6 | 1.11 | 1.03 | 1.05 | | | | | | |
| 0.8 | 1.16 | 1.21 | 1.17 | 1.12 | | | | | |
| 1.0 | 1.38 | 1.40 | 1.30 | 1.36 | 1.27 | | | | |
| 1.2 | 1.52 | 1.61 | 1.68 | 1.91 | 1.47 | 1.66 | | | |
| 1.4 | 1.79 | 2.01 | 1.90 | 2.31 | 2.28 | 2.20 | 1.95 | | |
| 1.6 | 2.07 | 2.28 | 2.13 | 2.71 | 2.99 | 2.81 | 2.09 | 2.20 | |
| 1.8 | 2.32 | 2.54 | 2.64 | 3.09 | 3.72 | 2.48 | 2.21 | 2.57 | 2.32 |

For main coefficient ($C_{c,s}$) see Fitting 5-23.
[a]For performance study see SMACNA (1987).

### 5-29 Tee, Diverging, Rectangular Main and Tap (45° Entry), with Damper (SMACNA 1981, Table 6-10P)

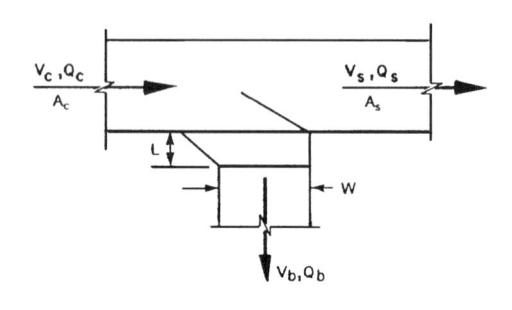

Poor; should not be used.[a]
$L = 0.25W$, 3 in. min
$A_c = A_s$

| | Branch, $C_{c,b}$ | | | | | | | | |
|---|---|---|---|---|---|---|---|---|---|
| | $Q_b/Q_c$ | | | | | | | | |
| $V_b/V_c$ | 0.1 | 0.2 | 0.3 | 0.4 | 0.5 | 0.6 | 0.7 | 0.8 | 0.9 |
| 0.2 | 0.61 | | | | | | | | |
| 0.4 | 0.46 | 0.61 | | | | | | | |
| 0.6 | 0.43 | 0.50 | 0.54 | | | | | | |
| 0.8 | 0.39 | 0.43 | 0.62 | 0.53 | | | | | |
| 1.0 | 0.34 | 0.57 | 0.77 | 0.73 | 0.68 | | | | |
| 1.2 | 0.37 | 0.64 | 0.85 | 0.98 | 1.07 | 0.83 | | | |
| 1.4 | 0.57 | 0.71 | 1.04 | 1.16 | 1.54 | 1.36 | 1.18 | | |
| 1.6 | 0.89 | 1.08 | 1.28 | 1.30 | 1.69 | 2.09 | 1.81 | 1.47 | |
| 1.8 | 1.33 | 1.34 | 2.04 | 1.78 | 1.90 | 2.40 | 2.77 | 2.23 | 1.92 |

For main coefficient ($C_{c,s}$), see Fitting 5-31.
[a]For performance study, see SMACNA (1987).

### 5-30 Tee, Diverging, Rectangular Main and Tap, with Damper (SMACNA 1981, Table 6-10R)

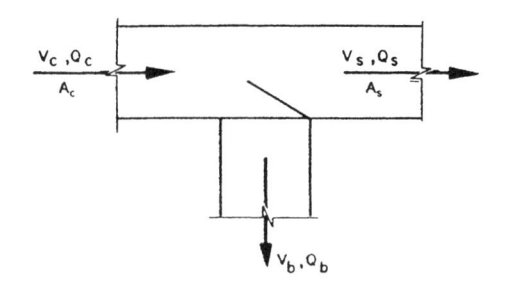

Poor; should not be used.[a]
$A_c = A_s$

| | Branch, $C_{c,b}$ | | | | | | | | |
|---|---|---|---|---|---|---|---|---|---|
| | $Q_b/Q_c$ | | | | | | | | |
| $V_b/V_c$ | 0.1 | 0.2 | 0.3 | 0.4 | 0.5 | 0.6 | 0.7 | 0.8 | 0.9 |
| 0.2 | 0.58 | | | | | | | | |
| 0.4 | 0.67 | 0.64 | | | | | | | |
| 0.6 | 0.78 | 0.76 | 0.75 | | | | | | |
| 0.8 | 0.88 | 0.98 | 0.81 | 1.01 | | | | | |
| 1.0 | 1.12 | 1.05 | 1.08 | 1.18 | 1.29 | | | | |
| 1.2 | 1.49 | 1.48 | 1.40 | 1.51 | 1.70 | 1.91 | | | |
| 1.4 | 2.10 | 2.21 | 2.25 | 2.29 | 2.32 | 2.48 | 2.53 | | |
| 1.6 | 2.72 | 3.30 | 2.84 | 3.09 | 3.30 | 3.19 | 3.29 | 3.16 | |
| 1.8 | 3.42 | 4.58 | 3.65 | 3.92 | 4.20 | 4.15 | 4.14 | 4.10 | 4.05 |

For main coefficient ($C_{c,s}$) see Fitting 5-31.
[a]For performance study see SMACNA (1987).

### 5-31 Tee, Diverging, Rectangular, with Extractor (SMACNA 1981, Table 6-10S)

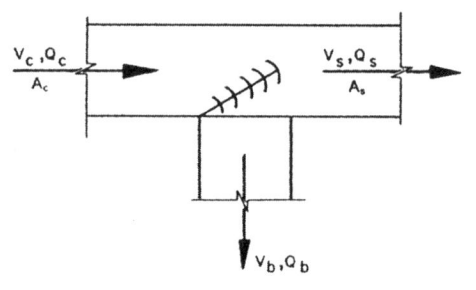

Poor; should not be used.[a]
$A_c = A_s$

| | Branch, $C_{c,b}$ | | | | | | | | |
|---|---|---|---|---|---|---|---|---|---|
| | $Q_b/Q_c$ | | | | | | | | |
| $V_b/V_c$ | 0.1 | 0.2 | 0.3 | 0.4 | 0.5 | 0.6 | 0.7 | 0.8 | 0.9 |
| 0.2 | 0.60 | | | | | | | | |
| 0.4 | 0.62 | 0.69 | | | | | | | |
| 0.6 | 0.74 | 0.80 | 0.82 | | | | | | |
| 0.8 | 0.99 | 1.10 | 0.95 | 0.90 | | | | | |
| 1.0 | 1.48 | 1.12 | 1.41 | 1.24 | 1.21 | | | | |
| 1.2 | 1.91 | 1.33 | 1.43 | 1.52 | 1.55 | 1.64 | | | |
| 1.4 | 2.47 | 1.67 | 1.70 | 2.04 | 1.86 | 1.98 | 2.47 | | |
| 1.6 | 3.17 | 2.40 | 2.33 | 2.53 | 2.31 | 2.51 | 3.13 | 3.25 | |
| 1.8 | 3.85 | 3.37 | 2.89 | 3.23 | 3.09 | 3.03 | 3.30 | 3.74 | 4.11 |

| | Main | | | | | | | | |
|---|---|---|---|---|---|---|---|---|---|
| $V_b/V_c$ | 0.2 | 0.4 | 0.6 | 0.8 | 1.0 | 1.2 | 1.4 | 1.6 | 1.8 |
| $C_{c,s}$ | 0.03 | 0.04 | 0.07 | 0.12 | 0.13 | 0.14 | 0.27 | 0.30 | 0.25 |

[a]For performance study, see SMACNA (1987).

# Duct and Pipe Sizing

## 5-32 Symmetrical Wye, Dovetail, Rectangular
(Idelchik et al. 1986, Diagram 7-24)

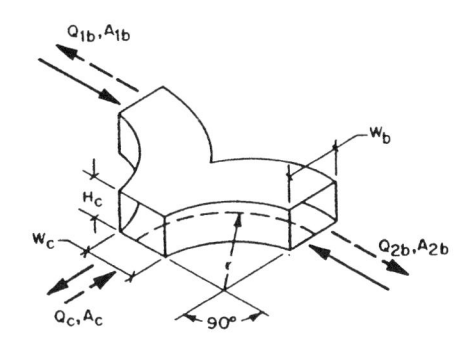

$r/W_c = 1.5$
$Q_{1b}/Q_c = Q_{2b}/Q_c = 0.5$

**Converging**

| $A_{1b}/A_c$ or $A_{2b}/A_c$ | 0.50 | 1.0 |
|---|---|---|
| $C_{c,1b}$ or $C_{c,2b}$ | 0.23 | 0.07 |

**Diverging**

| $A_{1b}/A_c$ or $A_{2b}/A_c$ | 0.50 | 1.0 |
|---|---|---|
| $C_{c,1b}$ or $C_{c,2b}$ | 0.30 | 0.25 |

## 5-33 Wye, Rectangular and Round
(Idelchik et al. 1986, Diagram 7-30)

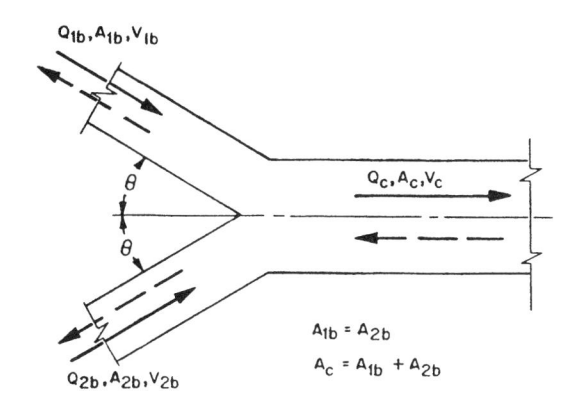

$A_{1b} = A_{2b}$
$A_c = A_{1b} + A_{2b}$

**Converging**    $C_{c,1b}$ or $C_{c,2b}$

| θ, deg. | $Q_{1b}/Q_c$ or $Q_{2b}/Q_c$ |||||||||||
|---|---|---|---|---|---|---|---|---|---|---|---|
| | 0 | 0.1 | 0.2 | 03 | 0.4 | 0.5 | 0.6 | 0.7 | 0.8 | 0.9 | 1.0 |
| 15 | −2.6 | −1.9 | −1.3 | −0.77 | −0.30 | 0.10 | 0.41 | 0.67 | 0.85 | 0.97 | 1.0 |
| 30 | −2.1 | −1.5 | −1.0 | −0.53 | −0.10 | 0.28 | 0.69 | 0.91 | 1.1 | 1.4 | 1.6 |
| 45 | −1.3 | −0.93 | −0.55 | −0.16 | 0.20 | 0.56 | 0.92 | 1.3 | 1.6 | 2.0 | 2.3 |

**Diverging**    $C_{c,1b}$ or $C_{c,2b}$

| θ, deg. | $V_{1b}/V_c$ or $V_{2b}/V_c$ ||||||||||||
|---|---|---|---|---|---|---|---|---|---|---|---|---|
| | 0.1 | 0.2 | 03 | 0.4 | 0.5 | 0.6 | 0.8 | 1.0 | 1.2 | 1.4 | 1.6 | 1.8 | 2.0 |
| 15 | 0.81 | 0.65 | 0.51 | 0.38 | 0.28 | 0.20 | 0.11 | 0.06 | 0.14 | 0.30 | 0.51 | 0.76 | 1.0 |
| 30 | 0.84 | 0.69 | 0.56 | 0.44 | 0.34 | 0.26 | 0.19 | 0.15 | 0.15 | 0.30 | 0.51 | 0.76 | 1.0 |
| 45 | 0.87 | 0.74 | 0.63 | 0.54 | 0.45 | 0.38 | 0.29 | 0.24 | 0.23 | 0.30 | 0.51 | 0.76 | 1.0 |
| 60 | 0.90 | 0.82 | 0.79 | 0.66 | 0.59 | 0.53 | 0.43 | 0.36 | 0.33 | 0.39 | 0.51 | 0.76 | 1.0 |
| 90 | 1.0 | 1.0 | 1.0 | 1.0 | 1.0 | 1.0 | 1.0 | 1.0 | 1.0 | 1.0 | 1.0 | 1.0 | 1.0 |

## 5-34 Wye (Double), 45° Rectangular and Round
(Idelchik et al. 1986, Diagram 7-27)

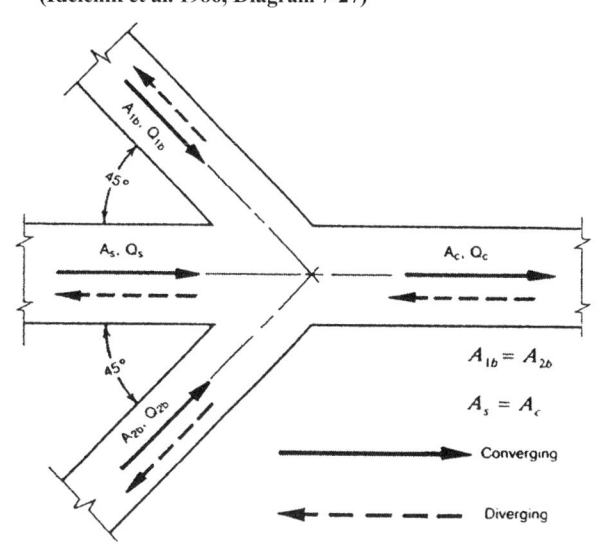

$A_{1b} = A_{2b}$
$A_s = A_c$

**Converging Flow**

Branch, $C_{c,b}$

| | $Q_{1b}/Q_c$ |||||||
|---|---|---|---|---|---|---|---|
| $Q_{2b}/Q_{1b}$ | 0 | 0.1 | 0.2 | 0.3 | 0.4 | 0.5 | 0.6 |
| | $A_{1b}/A_c = 0.2$ |||||||
| 0.5 | −1.0 | −0.36 | 0.59 | 1.8 | 3.2 | 4.9 | 6.8 |
| 1.0 | −1.0 | −0.24 | 0.63 | 1.7 | 2.6 | 3.7 | — |
| 2.0 | −1.0 | −0.19 | 0.21 | 0.04 | — | — | — |
| | $A_{1b}/A_c = 0.4$ |||||||
| 0.5 | −1.0 | −0.48 | −0.02 | 0.58 | 0.92 | 1.3 | 16 |
| 1.0 | −1.0 | −0.36 | 0.17 | 0.55 | 0.72 | 0.78 | — |
| 2.0 | −1.0 | −0.18 | 0.16 | −0.06 | — | — | — |
| | $A_{1b}/A_c = 0.6$ |||||||
| 0.5 | −1.0 | −0.50 | −0.07 | 0.31 | 0.60 | 0.82 | 0.92 |
| 1.0 | −1.0 | −0.37 | 0.12 | 0.55 | 0.60 | 0.52 | — |
| 2.0 | −1.0 | −0.18 | 0.26 | 0.16 | — | — | — |
| | $A_{1b}/A_c = 1.0$ |||||||
| 0.5 | −1.0 | −0.51 | −0.09 | 0.25 | 0.50 | 0.65 | 0.64 |
| 1.0 | −1.0 | −0.37 | 0.13 | 0.46 | 0.61 | 0.54 | — |
| 2.0 | −1.0 | −0.15 | 0.38 | 0.42 | — | — | — |

Main, $C_{c,s}$

| | $Q_s/Q_c$ |||||||||||
|---|---|---|---|---|---|---|---|---|---|---|---|
| $Q_{2b}/Q_{1b}$ | 0 | 0.1 | 0.2 | 0.3 | 0.4 | 0.5 | 0.6 | 0.7 | 0.8 | 0.9 | 1.0 |
| | $A_{1b}/A_c = 0.2$ |||||||||||
| 0.5 & 2.0 | −2.9 | −1.9 | −1.3 | −0.80 | −0.56 | −0.23 | −0.01 | 0.16 | 0.22 | 0.15 | 0 |
| 1.0 | −2.5 | −1.9 | −1.3 | −0.80 | −0.42 | −0.12 | 0.08 | 0.20 | 0.22 | 0.15 | 0 |
| | $A_{1b}/A_c = 0.4$ |||||||||||
| 0.5 & 2.0 | −0.98 | −0.61 | −0.30 | −0.05 | 0.14 | 0.26 | 0.33 | 0.34 | 0.28 | 0.17 | 0 |
| 1.0 | −0.77 | −0.44 | −0.16 | 0.05 | 0.21 | 0.31 | 0.36 | 0.35 | 0.29 | 0.17 | 0 |
| | $A_{1b}/A_{cc} = 0.6$ |||||||||||
| 0.5 & 2.0 | −0.32 | 0.08 | 0.11 | 0.27 | 0.37 | 0.43 | 0.44 | 0.40 | 0.31 | 0.18 | 0 |
| 1.0 | −0.18 | −0.04 | 0.21 | 0.34 | 0.42 | 0.46 | 0.46 | 0.41 | 0.31 | 0.18 | 0 |
| | $A_{1b}/A_c = 1.0$ |||||||||||
| 0.5 & 2.0 | | 0.11 | 0.36 | 0.46 | 0.53 | 0.57 | 0.56 | 0.52 | 0.44 | 0.33 | 0.18 | 0 |
| 1.0 | | 0.29 | 0.42 | 0.51 | 0.57 | 0.58 | 0.58 | 0.54 | 0.45 | 0.33 | 0.18 | 0 |

**Diverging Flow:** Use Fitting 5-23.

## 5-35 Cross, 90°, Rectangular and Round (Idelchik et al. 1986, Diagram 7-29)

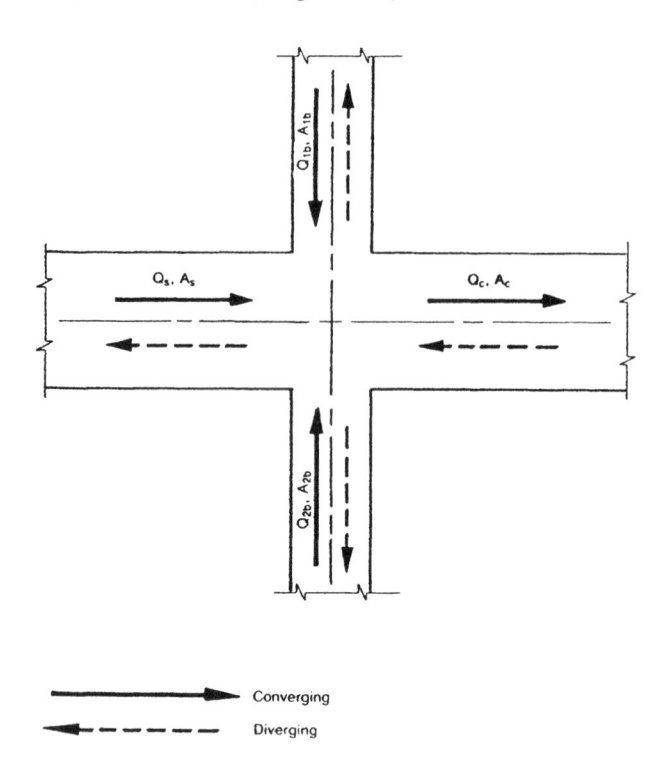

$A_{1b} = A_{2b}$
$A_s = A_c$

### Converging Flow

**Branch, $C_{c,b}$**

| $Q_{2b}/Q_{1b}$ | \multicolumn{7}{c}{$Q_{1b}/Q_c$ or $Q_{2b}/Q_c$} |
|---|---|---|---|---|---|---|---|
| | 0 | 0.1 | 0.2 | 0.3 | 0.4 | 0.5 | 0.6 |
| \multicolumn{8}{c}{$A_{1b}/A_c = 0.2$} |
| 0.5 | −0.85 | −0.10 | 1.1 | 2.7 | 4.8 | 7.3 | 10 |
| 1.0 | −0.85 | −0.05 | 1.4 | 3.1 | 5.1 | 7.4 | — |
| 2.0 | −0.85 | −0.31 | 1.8 | 3.4 | — | — | — |
| \multicolumn{8}{c}{$A_{1b}/A_c = 0.4$} |
| 0.5 | −0.85 | −0.29 | 0.34 | 1.0 | 1.8 | 2.6 | 3.4 |
| 1.0 | −0.85 | −0.14 | 0.60 | 1.3 | 2.1 | 2.7 | — |
| 2.0 | −0.85 | 0.12 | 1.0 | 1.7 | — | — | — |
| \multicolumn{8}{c}{$A_{1b}/A_c = 0.6$} |
| 0.5 | −0.85 | −0.32 | 0.20 | 0.72 | 1.2 | 1.7 | 2.1 |
| 1.0 | −0.85 | −0.18 | 0.46 | 1.0 | 1.5 | 1.9 | — |
| 2.0 | −0.85 | 0.09 | 0.88 | 1.4 | — | — | — |
| \multicolumn{8}{c}{$A_{1b}/A_c = 0.8$} |
| 0.5 | −0.85 | −0.33 | 0.13 | 0.61 | 1.0 | 1.4 | 1.7 |
| 1.0 | −0.85 | −0.18 | 0.41 | 0.91 | 1.3 | 1.5 | — |
| 2.0 | −0.85 | 0.08 | 0.83 | 1.3 | — | — | — |
| \multicolumn{8}{c}{$A_{1b}/A_c = 1.0$} |
| 0.5 | −0.85 | −0.34 | 0.13 | 0.56 | 0.93 | 1.3 | 1.5 |
| 1.0 | −0.85 | −0.19 | 0.39 | 0.86 | 1.2 | 1.4 | — |
| 2.0 | −0.85 | 0.07 | 0.81 | 1.2 | — | — | — |

**Main**

| $Q_s/Q_c$ | 0 | 0.1 | 0.2 | 0.3 | 0.4 | 0.5 |
|---|---|---|---|---|---|---|
| $C_{c,s}$ | 1.2 | 1.2 | 1.2 | 1.1 | 1.1 | 0.96 |
| $Q_s/Q_c$ | 0.6 | 0.7 | 0.8 | 0.9 | 1.0 | |
| $C_{c,s}$ | 0.85 | 0.72 | 0.56 | 0.39 | 0.20 | |

**Diverging Flow:** Use Fitting 5-23.

## OBSTRUCTIONS

### 6-1 Damper, Butterfly, Round (Idelchik et al. 1986, Diagram 9-16; Zolotov 1967)

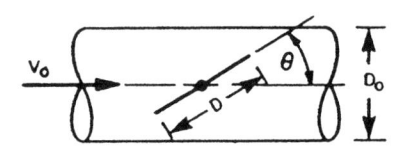

| | \multicolumn{10}{c}{$C_o$} |
|---|---|---|---|---|---|---|---|---|---|---|
| | \multicolumn{10}{c}{θ, degrees} |
| $D/D_o$ | 0 | 10 | 20 | 30 | 40 | 50 | 60 | 70 | 75 | 80 | 85 |
| 0.5 | 019 | 0.27 | 0.37 | 0.49 | 0.61 | 0.74 | 0.86 | 0.96 | 0.99 | 1.0 | 1.0 |
| 0.6 | 0.19 | 0.32 | 0.48 | 0.69 | 0.94 | 1.2 | 1.5 | 1.7 | 1.8 | 1.9 | 1.9 |
| 0.7 | 0.19 | 0.37 | 0.64 | 1.0 | 1.5 | 2.1 | 2.8 | 3.5 | 3.7 | 3.9 | 4.1 |
| 0.8 | 0.19 | 0.45 | 0.87 | 1.6 | 2.6 | 4.1 | 6.1 | 8.4 | 9.4 | 10 | 10 |
| 0.9 | 0.19 | 0.54 | 1.2 | 2.5 | 5.0 | 9.6 | 17 | 30 | 38 | 45 | 50 |
| 1.0 | 0.19 | 0.67 | 1.8 | 4.4 | 11 | 32 | 113 | — | — | — | — |

### 6-2 Damper, Butterfly, Rectangular (Idelchik et al. 1986, Diagram 9-17; Zolotov 1967)

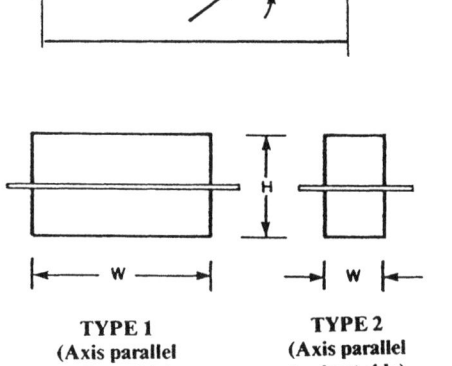

**TYPE 1** (Axis parallel to long side)
**TYPE 2** (Axis parallel to short side)

| | | \multicolumn{9}{c}{$C_o$} |
|---|---|---|---|---|---|---|---|---|---|
| | | \multicolumn{9}{c}{θ, degrees} |
| Type | H/W | 0 | 10 | 20 | 30 | 40 | 50 | 60 | 65 | 70 |
| 1 | <0.25 | 0.04 | 0.30 | 1.1 | 3.0 | 8.0 | 23 | 60 | 100 | 190 |
| 1 | 0.25–1.0 | 0.08 | 0.33 | 1.2 | 3.3 | 9.0 | 26 | 70 | 128 | 210 |
| 2 | >1.0 | 0.13 | 0.35 | 1.3 | 3.6 | 10 | 29 | 80 | 155 | 230 |

# Duct and Pipe Sizing

**6-3 Damper, Gate, Round (Idelchik et al. 1986, Diagram 9-5)**

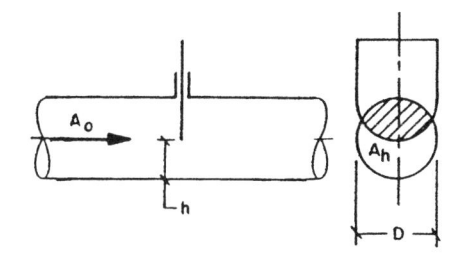

| h/D | 0.2 | 0.3 | 0.4 | 0.5 | 0.6 | 0.7 | 0.8 | 0.9 |
|---|---|---|---|---|---|---|---|---|
| $A_h/A_o$ | 0.25 | 0.38 | 0.50 | 0.61 | 0.71 | 0.81 | 0.90 | 0.96 |
| $C_o$ | 35 | 10 | 4.6 | 2.1 | 0.98 | 0.44 | 0.17 | 0.06 |

**6-4 Damper, Gate, Rectangular (Idelchik et al. 1986, Diagram 9-5)**

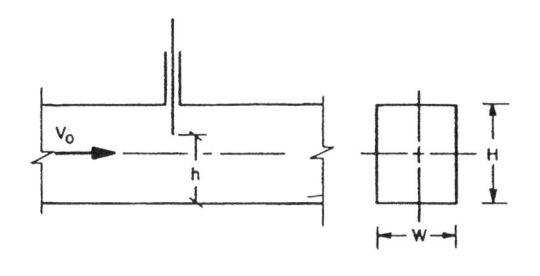

| | h/H | | | | | | |
|---|---|---|---|---|---|---|---|
| H/W | 0.3 | 0.4 | 0.5 | 0.6 | 0.7 | 0.8 | 0.9 |
| 0.5 | 14 | 6.9 | 3.3 | 1.7 | 0.83 | 0.32 | 0.09 |
| 1.0 | 19 | 8.8 | 4.5 | 2.4 | 1.2 | 0.55 | 0.17 |
| 1.5 | 20 | 9.1 | 4.7 | 2.7 | 1.2 | 0.47 | 0.11 |
| 2.0 | 18 | 8.8 | 4.5 | 2.3 | 1.1 | 0.51 | 0.13 |

**6-5 Damper, Rectangular, Parallel Blades (Brown and Fellows 1957)**

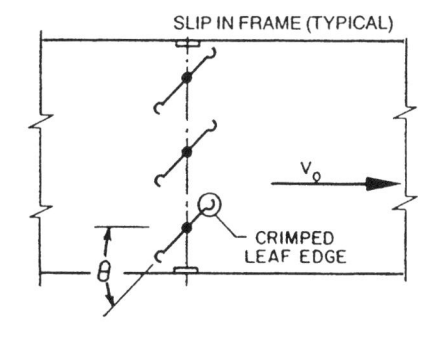

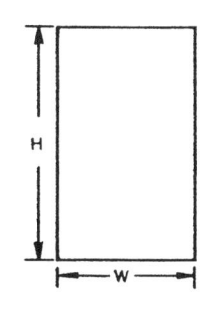

$$L/R = \frac{NW}{2(H+W)}$$

where

$N$ = number of damper blades
$W$ = duct dimension parallel to blade axis, in.
$H$ = duct height, in.
$L$ = sum of damper blade lengths, in.
$R$ = perimeter of duct, in.

| | $C_o$ | | | | | | | |
|---|---|---|---|---|---|---|---|---|
| | $\theta$, degrees | | | | | | | |
| L/R | 0 | 10 | 20 | 30 | 40 | 50 | 60 | 70 |
| 0.3 | 0.52 | 0.79 | 1.4 | 2.3 | 5.0 | 9 | 14 | 32 |
| 0.4 | 0.52 | 0.85 | 1.5 | 2.4 | 5.0 | 9 | 16 | 38 |
| 0.5 | 0.52 | 0.92 | 1.5 | 2.4 | 5.0 | 9 | 18 | 45 |
| 0.6 | 0.52 | 0.92 | 1.5 | 2.4 | 5.4 | 9 | 21 | 45 |
| 0.8 | 0.52 | 0.92 | 1.5 | 2.5 | 5.4 | 9 | 22 | 55 |
| 1.0 | 0.52 | 1.0 | 1.6 | 2.6 | 5.4 | 10 | 24 | 65 |
| 1.5 | 0.52 | 1.0 | 1.6 | 2.7 | 5.4 | 10 | 28 | 102 |

**6-6 Damper, Rectangular, Opposed Blades (Brown and Fellows 1957)**

$$L/R = \frac{NW}{2(H+W)}$$

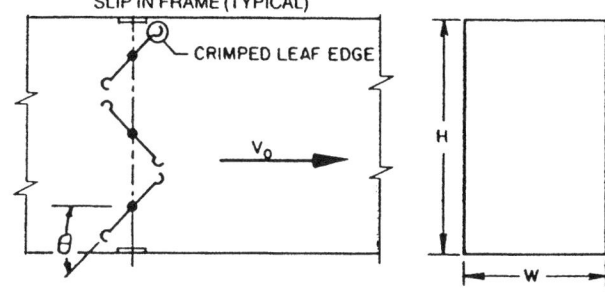

See Fitting 6-5 for definition of terms.

| | $C_o$ | | | | | | | |
|---|---|---|---|---|---|---|---|---|
| | $\theta$, degrees | | | | | | | |
| L/R | 0 | 10 | 20 | 30 | 40 | 50 | 60 | 70 |
| 0.3 | 0.52 | 0.85 | 2.1 | 4.1 | 9 | 21 | 73 | 284 |
| 0.4 | 0.52 | 0.92 | 2.2 | 5.0 | 11 | 28 | 100 | 332 |
| 0.5 | 0.52 | 1.0 | 2.3 | 5.4 | 13 | 33 | 122 | 377 |
| 0.6 | 0.52 | 1.0 | 2.3 | 6.0 | 14 | 38 | 148 | 411 |
| 0.8 | 0.52 | 1.1 | 2.4 | 6.6 | 18 | 54 | 188 | 495 |
| 1.0 | 0.52 | 1.2 | 2.7 | 7.3 | 21 | 65 | 245 | 547 |
| 1.5 | 0.52 | 1.4 | 3.2 | 9.0 | 28 | 107 | 361 | 677 |

**6-7 Obstruction, Screen, Round and Rectangular (Idelchik et al. 1986, Diagram 8-6)**

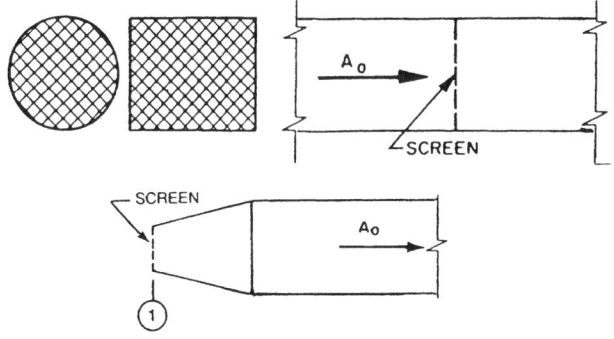

$n$ = free area ratio of screen
$A_o$ = area of duct
$A_1$ = cross-sectional area of duct or fitting where screen is located

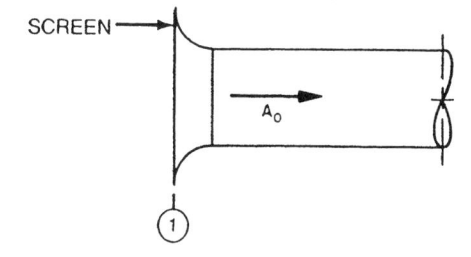

| | $C_o$ | | | | | | | |
|---|---|---|---|---|---|---|---|---|
| | n | | | | | | | |
| $A_1/A_o$ | 0.3 | 0.4 | 0.5 | 0.6 | 0.7 | 0.8 | 0.9 | 1.0 |
| 0.2 | 155 | 75 | 42 | 24 | 15 | 8.0 | 3.5 | 0 |
| 0.3 | 69 | 33 | 19 | 11 | 6.4 | 3.6 | 1.6 | 0 |
| 0.4 | 39 | 19 | 10 | 6.1 | 3.6 | 2.0 | 0.88 | 0 |
| 0.6 | 17 | 8.3 | 4.7 | 2.7 | 1.6 | 0.89 | 0.39 | 0 |
| 0.8 | 9.7 | 4.7 | 2.7 | 1.5 | 0.91 | 0.50 | 0.22 | 0 |
| 1.0 | 6.2 | 3.0 | 1.7 | 0.97 | 0.58 | 0.32 | 0.14 | 0 |
| 1.2 | 4.3 | 2.1 | 1.2 | 0.67 | 0.40 | 0.22 | 0.10 | 0 |
| 1.4 | 3.2 | 1.5 | 0.87 | 0.49 | 0.30 | 0.16 | 0.07 | 0 |
| 1.6 | 2.4 | 1.2 | 0.66 | 0.38 | 0.23 | 0.12 | 0.05 | 0 |
| 2.0 | 1.6 | 0.75 | 0.43 | 0.24 | 0.15 | 0.08 | 0.04 | 0 |
| 2.5 | 0.99 | 0.48 | 0.27 | 0.16 | 0.09 | 0.05 | 0.02 | 0 |
| 3.0 | 0.69 | 0.33 | 0.19 | 0.11 | 0.06 | 0.04 | 0.02 | 0 |
| 4.0 | 0.39 | 0.19 | 0.11 | 0.06 | 0.04 | 0.02 | 0.01 | 0 |
| 6.0 | 0.17 | 0.08 | 0.05 | 0.03 | 0.02 | 0.01 | 0 | 0 |

**6-8 Obstruction, Perforated Plate, Thick, Round, and Rectangular (Idelchik et al. 1986, Diagram 8-6)**

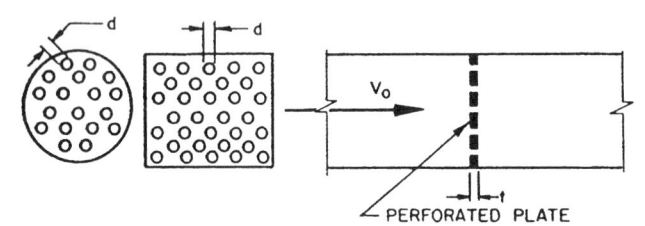

$t/d \geq 0.015$
$A_{or} = \pi d^2/4$
$n = \Sigma A_{or}/A_o$
where
  $A_o$ = area of duct
  $A_{or}$ = orifice area
  $d$ = diameter of perforated hole
  $n$ = free area ratio of plate dimensionless
  $t$ = plate thickness

| | $C_o$ | | | | | | | | |
|---|---|---|---|---|---|---|---|---|---|
| | n | | | | | | | | |
| $t/d$ | 0.20 | 0.25 | 0.30 | 0.40 | 0.50 | 0.60 | 0.70 | 0.80 | 0.90 |
| 0.015 | 52 | 30 | 18 | 8.3 | 4.0 | 2.0 | 0.97 | 0.42 | 0.13 |
| 0.2 | 48 | 28 | 17 | 7.7 | 3.8 | 1.9 | 0.91 | 0.40 | 0.13 |
| 0.4 | 46 | 27 | 17 | 7.4 | 3.6 | 1.8 | 0.88 | 0.39 | 0.13 |
| 0.6 | 42 | 24 | 15 | 6.6 | 3.2 | 1.6 | 0.80 | 0.36 | 0.13 |

**6-9 Obstruction, Smooth Cylinder in Round and Rectangular Ducts (Idelchik et al. 1986, Diagram 10-1)**

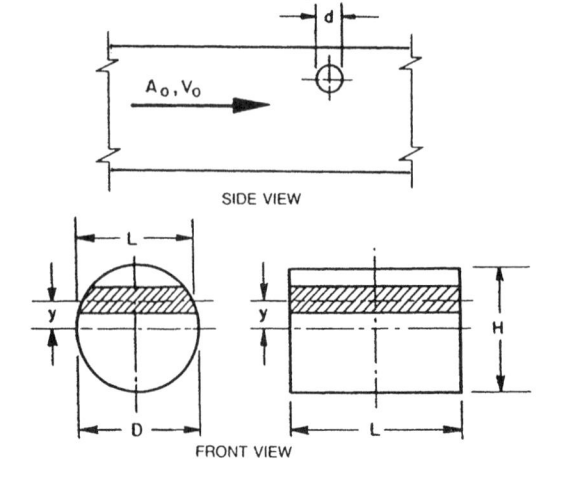

$S_m/A_o < 0.3$
$S_m = dL$
$Re' = dV_o/v$
$C_o = KC'_o$

| | $C'_o$ | | | |
|---|---|---|---|---|
| | $S_m/A_o$ | | | |
| Re' | 0.05 | 0.10 | 0.15 | 0.20 |
| 0.1 | 3.9 | 8.4 | 14 | 19 |
| 0.5 | 1.5 | 3.2 | 5.2 | 7.1 |
| 1 | 0.66 | 1.4 | 2.3 | 3.2 |
| 5 | 0.30 | 0.64 | 1.1 | 1.4 |
| 10 | 0.17 | 0.38 | 0.62 | 0.84 |
| 50 | 0.11 | 0.24 | 0.38 | 0.52 |
| 100 | 0.10 | 0.21 | 0.35 | 0.47 |
| 500 to 200,000 | 0.07 | 0.15 | 0.24 | 0.33 |
| $3 \times 10^5$ | 0.07 | 0.16 | 0.26 | 0.35 |
| $4 \times 10^5$ | 0.05 | 0.11 | 0.19 | 0.25 |
| $5 \times 10^5$ | 0.04 | 0.09 | 0.14 | 0.19 |
| $6 \times 10^5$ to $10^6$ | 0.02 | 0.05 | 0.07 | 0.10 |

For obstruction offset from the centerline, use the following factors:

| $y/D$ or $y/H$ | 0 | 0.05 | 0.10 | 0.15 | 0.20 | 0.25 | 0.30 | 0.35 | 0.40 |
|---|---|---|---|---|---|---|---|---|---|
| K | 1.0 | 0.97 | 0.93 | 0.89 | 0.84 | 0.79 | 0.74 | 0.67 | 0.58 |

**6-10 Round Duct, Depressed to Avoid an Obstruction (SMACNA 1981, Table 6-14I)**

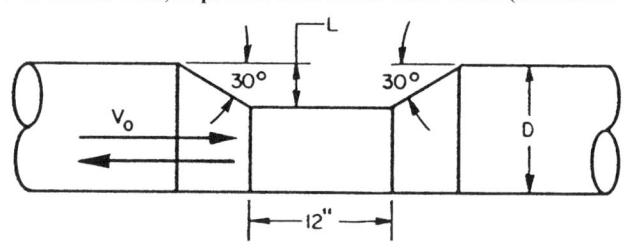

$L/D = 0.33$
$C_o = 0.24$

**6-11 Rectangular Duct, Depressed to Avoid an Obstruction (SMACNA 1981, Table 6-14J)**

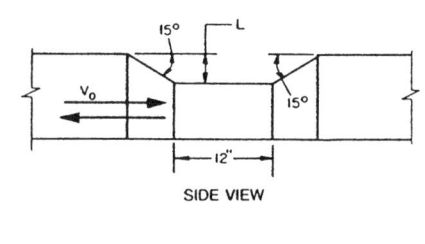

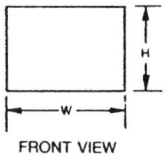

| | $C_o$ | | | |
|---|---|---|---|---|
| | $L/H$ | | | |
| W/H | 0.125 | 0.15 | 0.25 | 030 |
| 1.0 | 0.26 | 0.30 | 0.33 | 0.35 |
| 4.0 | 0.10 | 0.14 | 0.22 | 0.30 |

# Duct and Pipe Sizing

## FAN-SYSTEM CONNECTIONS

### 7-1 Fans Discharging into a Plenum (AMCA 1973, Figure 19)

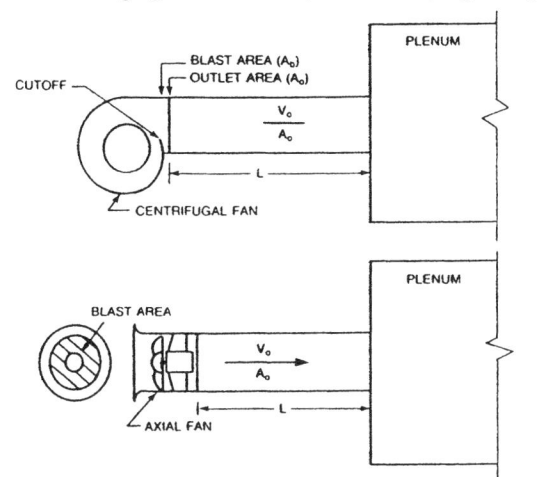

Calculate effective duct length.

$$(V_o > 2500 \text{ fpm: }) L_e = V\sqrt{A_o}/10{,}600$$

$$V_o \leq 2500 \text{ fpm: } L_e = \sqrt{A_o}/4.3$$

where

$V_o$ = duct velocity, fpm
$L_e$ = effective duct length, ft
$A_o$ = duct area, in$^2$

| | $C_o$ | | | | |
|---|---|---|---|---|---|
| | | | $L/L_e$ | | |
| $A_b/A_o$ | 0 | 0.12 | 0.25 | 0.5 | ≥1.0 |
| 0.4 | 2.0 | 1.0 | 0.40 | 0.18 | 0 |
| 0.5 | 2.0 | 1.0 | 0.40 | 0.18 | 0 |
| 0.6 | 1.0 | 0.66 | 0.33 | 0.14 | 0 |
| 0.7 | 0.8 | 0.40 | 0.14 | 0 | 0 |
| 0.8 | 0.47 | 0.22 | 0.10 | 0 | 0 |
| 0.9 | 0.22 | 0.14 | 0 | 0 | 0 |
| 1.0 | 0 | 0 | 0 | 0 | 0 |

### 7-2 Single Width Single Inlet (SWSI) Fan with an Outlet Duct Elbow (AMCA 1973, Figure 22)

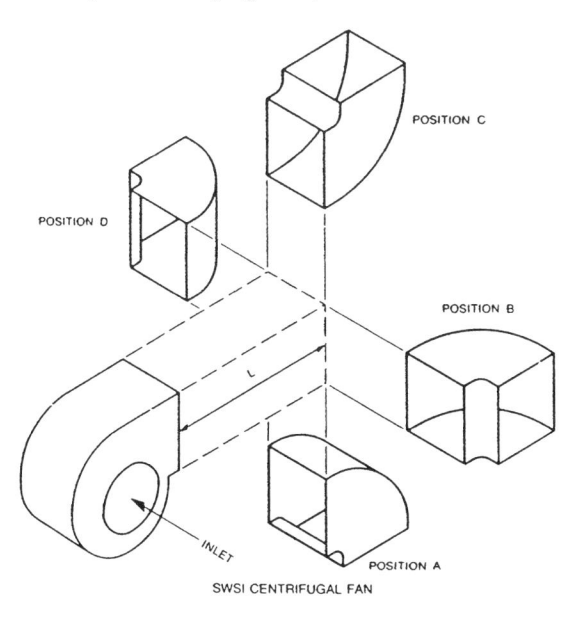

$A_b$ = centrifugal fan blast area (see Fitting 7-1)
$A_o$ = duct/outlet area (see Fitting 7-1)

To calculate effective duct length $L_e$, see Fitting 7-1.

| | | $C_o$ | | | | |
|---|---|---|---|---|---|---|
| | Outlet Elbow | | | $L/L_e$ | | |
| $A_b/A_o$ | Position | 0 | 0.12 | 0.25 | 0.5 | ≥1.0 |
| 0.4 | A | 3.2 | 2.7 | 1.8 | 0.84 | 0 |
| | B | 4.0 | 3.3 | 2.2 | 1.0 | 0 |
| | C | 5.8 | 4.8 | 3.2 | 1.5 | 0 |
| | D | 5.8 | 4.8 | 3.2 | 1.5 | 0 |
| 0.5 | A | 2.3 | 1.9 | 1.3 | 0.60 | 0 |
| | B | 2.8 | 2.4 | 1.6 | 0.72 | 0 |
| | C | 4.0 | 3.3 | 2.2 | 1.0 | 0 |
| | D | 4.0 | 3.3 | 2.2 | 1.0 | 0 |
| 0.6 | A | 1.6 | 1.3 | 0.88 | 0.40 | 0 |
| | B | 2.0 | 1.7 | 1.1 | 0.52 | 0 |
| | C | 2.9 | 2.4 | 1.6 | 0.76 | 0 |
| | D | 2.9 | 2.4 | 1.6 | 0.76 | 0 |
| 0.7 | A | 1.1 | 0.88 | 0.60 | 0.28 | 0 |
| | B | 1.3 | 1.1 | 0.72 | 0.36 | 0 |
| | C | 2.0 | 1.6 | 1.1 | 0.52 | 0 |
| | D | 2.0 | 1.6 | 1.1 | 0.52 | 0 |
| 0.8 | A | 0.76 | 0.64 | 0.44 | 0.20 | 0 |
| | B | 0.96 | 0.80 | 0.52 | 0.24 | 0 |
| | C | 1.4 | 1.2 | 0.76 | 0.36 | 0 |
| | D | 1.4 | 1.2 | 0.76 | 0.36 | 0 |
| 0.9 | A | 0.60 | 0.48 | 0.32 | 0.16 | 0 |
| | B | 0.76 | 0.64 | 0.44 | 0.20 | 0 |
| | C | 1.1 | 0.92 | 0.64 | 0.28 | 0 |
| | D | 1.1 | 0.92 | 0.64 | 0.28 | 0 |
| 1.0 | A | 0.56 | 0.48 | 0.32 | 0.16 | 0 |
| | B | 0.68 | 0.56 | 0.36 | 0.16 | 0 |
| | C | 1.0 | 0.84 | 0.56 | 0.26 | 0 |
| | D | 1.0 | 0.84 | 0.56 | 0.16 | 0 |

### 7-3 Double Width Double Inlet (DWDI) Fan with an Outlet Duct Elbow (AMCA 1973, Figure 22)

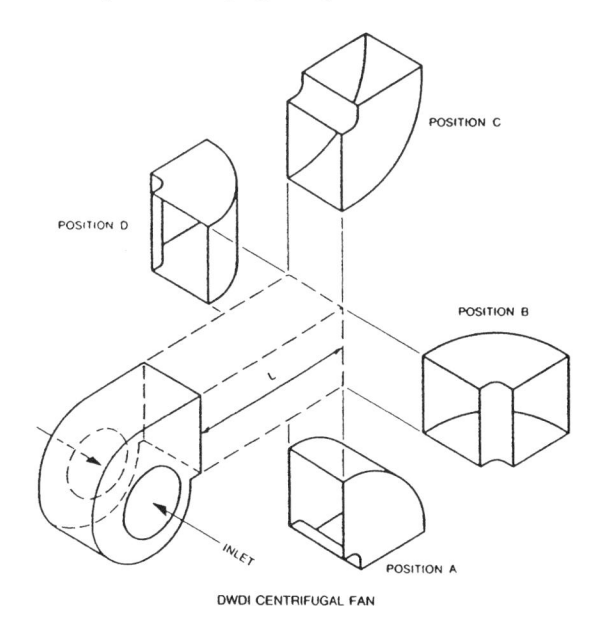

$A_b$ = centrifugal fan blast area (see Fitting 7-1)
$A_o$ = duct/outlet area (see Fitting 7-1)

To calculate effective duct length $L_e$, see Fitting 7-1.

342                                                                                                                    Principles of HVAC

| $A_b/A_o$ | Outlet Elbow Position | $C_o$ $L/L_e$ 0 | 0.12 | 0.25 | 0.5 | ≥1.0 |
|---|---|---|---|---|---|---|
| 0.4 | A | 3.2 | 2.7 | 1.8 | 0.84 | 0 |
|  | B | 5.0 | 4.2 | 2.8 | 1.3 | 0 |
|  | C | 5.8 | 4.8 | 3.2 | 1.5 | 0 |
|  | D | 4.9 | 4.1 | 2.7 | 1.3 | 0 |
| 0.5 | A | 2.3 | 1.9 | 1.3 | 0.60 | 0 |
|  | B | 3.6 | 3.0 | 2.0 | 0.90 | 0 |
|  | C | 4.0 | 3.3 | 2.2 | 1.0 | 0 |
|  | D | 3.4 | 2.8 | 1.9 | 0.88 | 0 |
| 0.6 | A | 1.6 | 1.3 | 0.88 | 0.40 | 0 |
|  | B | 2.5 | 2.1 | 1.4 | 0.65 | 0 |
|  | C | 2.9 | 2.4 | 1.6 | 0.76 | 0 |
|  | D | 2.5 | 2.1 | 1.4 | 0.65 | 0 |
| 0.7 | A | 1.1 | 0.88 | 0.60 | 0.28 | 0 |
|  | B | 1.7 | 1.4 | 0.90 | 0.45 | 0 |
|  | C | 2.0 | 1.6 | 1.1 | 0.52 | 0 |
|  | D | 1.7 | 1.4 | 0.92 | 0.44 | 0 |
| 0.8 | A | 0.76 | 0.64 | 0.44 | 0.20 | 0 |
|  | B | 1.2 | 1.0 | 0.65 | 0.30 | 0 |
|  | C | 1.4 | 1.2 | 0.76 | 0.36 | 0 |
|  | D | 1.2 | 0.99 | 0.65 | 0.31 | 0 |
| 0.9 | A | 0.60 | 0.48 | 0.32 | 0.16 | 0 |
|  | B | 0.95 | 0.80 | 0.55 | 0.25 | 0 |
|  | C | 1.1 | 0.92 | 0.64 | 0.28 | 0 |
|  | D | 0.95 | 0.78 | 0.54 | 0.24 | 0 |
| 1.0 | A | 0.56 | 0.48 | 0.32 | 0.16 | 0 |
|  | B | 0.85 | 0.70 | 0.45 | 0.20 | 0 |
|  | C | 1.0 | 0.84 | 0.56 | 0.28 | 0 |
|  | D | 0.85 | 0.71 | 0.48 | 0.24 | 0 |

**7-4  Nonuniform Elbow into a Fan Inlet Induced by a 90° Round Smooth Radius Elbow Without Vanes (AMCA 1973, Fig. 27)**

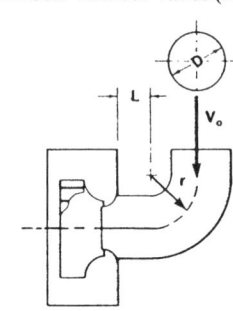

| $r/D$ | $C_o$ $L/D$ 0 | 2.0 | ≥5.0 |
|---|---|---|---|
| 0.75 | 1.4 | 0.80 | 0.40 |
| 1.0 | 1.2 | 0.66 | 0.33 |
| 1.5 | 1.1 | 0.60 | 0.33 |
| 2.0 | 1.0 | 0.53 | 0.33 |
| 3.0 | 0.66 | 0.40 | 0.22 |

**7-5  Nonuniform Elbow into a Fan Inlet Induced by 90° Mitered and Multipiece Elbows Without Vanes (AMCA 1973, Fig. 29)**

**Mitered**

| $L/D$ | 0 | 2.0 | ≥5.0 |
|---|---|---|---|
| $C_o$ | 3.2 | 2.0 | 1.0 |

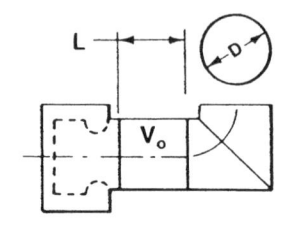

**Three-Piece**

| $r/D$ | $C_o$ $L/D$ 0 | 2.0 | ≥5.0 |
|---|---|---|---|
| 0.50 | 2.5 | 1.6 | 0.80 |
| 0.75 | 1.6 | 1.0 | 0.47 |
| 1.0 | 1.2 | 0.66 | 0.33 |
| 1.5 | 1.1 | 0.60 | 0.33 |
| 2.0 | 1.0 | 0.53 | 0.33 |
| 3.0 | 0.80 | 0.47 | 0.26 |

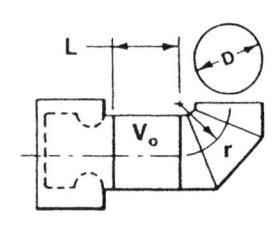

**Four-Piece or more**

| $r/D$ | $C_o$ $L/D$ 0 | 2.0 | ≥5.0 |
|---|---|---|---|
| 0.50 | 1.8 | 1.0 | 0.53 |
| 0.75 | 1.4 | 0.80 | 0.40 |
| 1.0 | 1.2 | 0.66 | 0.33 |
| 1.5 | 1.1 | 0.60 | 0.33 |
| 2.0 | 1.0 | 0.53 | 0.33 |
| 3.0 | 0.66 | 0.40 | 0.22 |

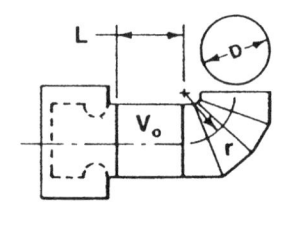

**7-6  Nonuniform Elbow into a Fan Inlet Induced by a 90° Square Smooth Radius Elbow (AMCA 1973, Figure 35A)**

**Square Elbow with an Inlet Transition,[a] No Vanes**

| $r/D$ | $C_o$ $L/D$ 0 | 2.5 | ≥6.0 |
|---|---|---|---|
| 0.50 | 2.5 | 1.6 | 0.80 |
| 0.75 | 2.0 | 1.2 | 0.66 |
| 1.0 | 1.2 | 0.66 | 0.33 |
| 1.5 | 1.0 | 0.57 | 0.30 |
| 2.0 | 0.8 | 0.47 | 0.26 |

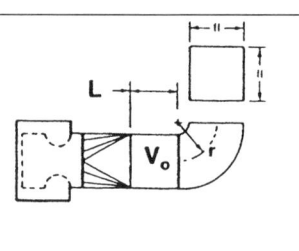

**Square Elbow with an Inlet Transition,[a] Full Radius Vanes Equally Spaced**

| $r/D$ | $C_o$ $L/D$ 0 | 2.5 | ≥6.0 |
|---|---|---|---|
| 0.50 | 0.80 | 0.47 | 0.26 |
| 1.0 | 0.53 | 0.33 | 0.18 |
| 1.5 | 0.40 | 0.28 | 0.16 |
| 2.0 | 1.26 | 0.22 | 0.14 |

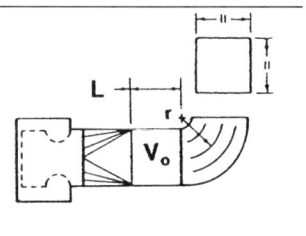

**Square Elbow with an Inlet Transition,[a] Short Vanes per Fitting 3.8**

| $r/D$ | $C_o$ $L/D$ 0 | 2.5 | ≥6.0 |
|---|---|---|---|
| 0.50 | 0.80 | 0.47 | 0.26 |
| 1.0 | 0.53 | 0.33 | 0.18 |
| 1.5 | 0.40 | 0.28 | 0.16 |
| 2.0 | 0.26 | 0.22 | 0.14 |

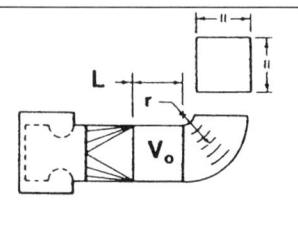

[a]The inside area of the square duct ($H \times H$) is equal to the inside area circumscribed by the fan inlet collar. The maximum angle of any converging element of the transition is 15° and, for a diverging element, 7.5°.

# Duct and Pipe Sizing

## 7-7 Fans Located in Plenums and Cabinet Enclosures (AMCA 1973, Figure 35A)

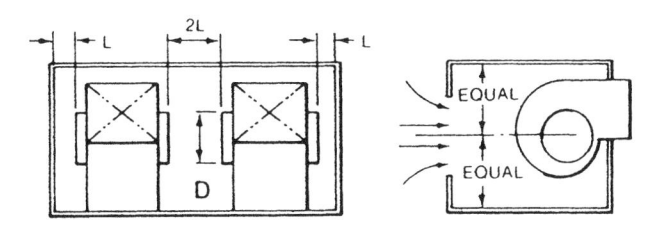

| L | $C_o$ |
|---|---|
| 0.75 D | 0.22 |
| 0.5 D | 0.40 |
| 0.4 D | 0.53 |
| 0.3 D | 0.80 |
| 0.2 D | 1.2 |

## 7-8 Fan Without an Outlet Diffuser (AMCA 1973, Figure 19)

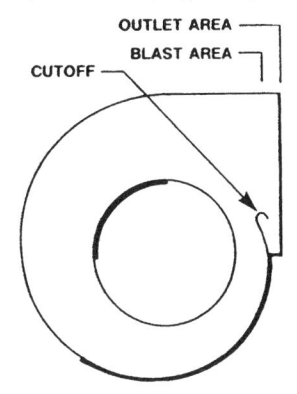

Poor; should not be used.

| $A_b/A_o$ | 0.4 | 0.5 | 0.6 | 0.7 | 0.8 | 0.9 | 1.0 |
|---|---|---|---|---|---|---|---|
| $C_o$ | 2.0 | 2.0 | 1.0 | 0.80 | 0.47 | 0.22 | 0 |

## 7-9 Plane Asymmetric Diffuser at Fan Outlet Without Ductwork (Idelchik et al. 1986, Diagram 11-11)

| θ, degrees | $C_o$ $A_1/A_o$ ||||||
|---|---|---|---|---|---|---|
| | 1.5 | 2.0 | 2.5 | 3.0 | 3.5 | 4.0 |
| 10 | 0.51 | 0.34 | 0.25 | 0.21 | 0.18 | 0.17 |
| 15 | 0.54 | 0.36 | 0.27 | 0.24 | 0.22 | 0.20 |
| 20 | 0.55 | 0.38 | 0.31 | 0.27 | 0.25 | 0.24 |
| 25 | 0.59 | 0.43 | 0.37 | 0.35 | 0.33 | 0.33 |
| 30 | 0.63 | 0.50 | 0.46 | 0.44 | 0.43 | 0.42 |
| 35 | 0.65 | 0.56 | 0.53 | 0.52 | 0.51 | 0.50 |

If diffuser has a screen, use Fitting 6-7 to calculate screen resistance.

## 7-10 Pyramidal Diffuser at Fan Outlet Without Ductwork (Idelchik et al. 1986, Diagram 11-11)

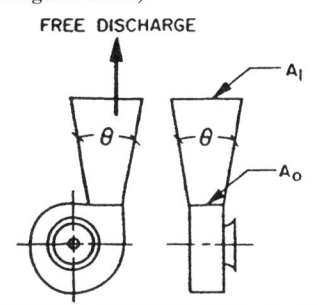

| θ, degrees | $C_o$ $A_1/A_o$ ||||||
|---|---|---|---|---|---|---|
| | 1.5 | 2.0 | 2.5 | 3.0 | 3.5 | 4.0 |
| 10 | 0.54 | 0.42 | 0.37 | 0.34 | 0.32 | 0.31 |
| 15 | 0.67 | 0.58 | 0.53 | 0.51 | 0.50 | 0.51 |
| 20 | 0.75 | 0.67 | 0.65 | 0.64 | 0.64 | 0.65 |
| 25 | 0.80 | 0.74 | 0.72 | 0.70 | 0.70 | 0.72 |
| 30 | 0.85 | 0.78 | 0.76 | 0.75 | 0.75 | 0.76 |

If diffuser has a screen, use Fitting 6-7 to calculate screen resistance.

## 7-11 Plane Symmetric Diffuser at Fan Outlet with Ductwork (Idelchik et al. 1986, Diagram 5-12)

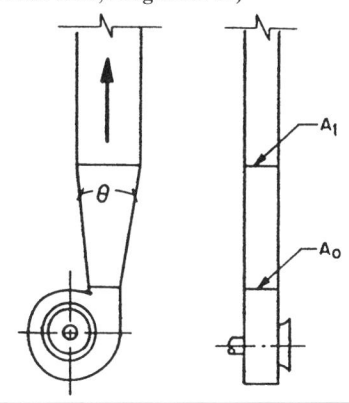

| θ, degrees | $C_o$ $A_1/A_o$ ||||||
|---|---|---|---|---|---|---|
| | 1.5 | 2.0 | 2.5 | 3.0 | 3.5 | 4.0 |
| 10 | 0.05 | 0.07 | 0.09 | 0.10 | 0.11 | 0.11 |
| 15 | 0.06 | 0.09 | 0.11 | 0.13 | 0.13 | 0.14 |
| 20 | 0.07 | 0.10 | 0.13 | 0.15 | 0.16 | 0.16 |
| 25 | 0.08 | 0.13 | 0.16 | 0.19 | 0.21 | 0.23 |
| 30 | 0.16 | 0.29 | 0.39 | 0.32 | 0.34 | 0.35 |
| 35 | 0.24 | 0.34 | 0.39 | 0.44 | 0.48 | 0.50 |

## 7-12 Plane Asymmetric Diffuser at Fan Outlet with Ductwork (Idelchik et al. 1986, Diagram 5-13)

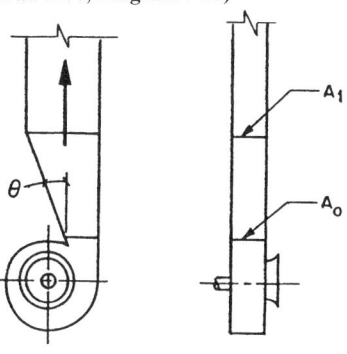

| θ, degrees | $C_o$ |||||| 
|---|---|---|---|---|---|---|
| | $A_1/A_o$ |||||| 
| | 1.5 | 2.8 | 2.5 | 3.0 | 3.5 | 4.0 |
| 10 | 0.08 | 0.09 | 0.10 | 0.10 | 0.11 | 0.11 |
| 15 | 0.10 | 0.11 | 0.12 | 0.13 | 0.14 | 0.15 |
| 20 | 0.12 | 0.14 | 0.15 | 0.16 | 0.17 | 0.18 |
| 25 | 0.15 | 0.18 | 0.21 | 0.23 | 0.25 | 0.26 |
| 30 | 0.18 | 0.25 | 0.30 | 0.33 | 0.35 | 0.35 |
| 35 | 0.21 | 0.31 | 0.38 | 0.41 | 0.43 | 0.44 |

**7-13 Plane Asymmetric Diffuser at Fan Outlet with Ductwork**
(Idelchik et al. 1986, Diagram 5-14)

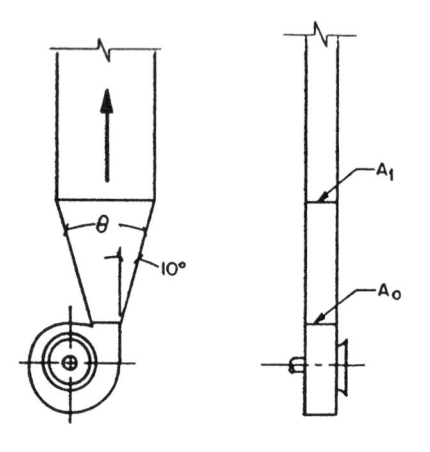

| θ, degrees | $C_o$ |||||| 
|---|---|---|---|---|---|---|
| | $A_1/A_o$ |||||| 
| | 1.5 | 2.8 | 2.5 | 3.0 | 3.5 | 4.0 |
| 10 | 0.05 | 0.08 | 0.11 | 0.13 | 0.13 | 0.14 |
| 15 | 0.06 | 0.10 | 0.12 | 0.14 | 0.15 | 0.15 |
| 20 | 0.07 | 0.11 | 0.14 | 0.15 | 0.16 | 0.16 |
| 25 | 0.09 | 0.14 | 0.18 | 0.20 | 0.21 | 0.22 |
| 30 | 0.13 | 0.18 | 0.23 | 0.26 | 0.28 | 0.29 |
| 35 | 0.15 | 0.23 | 0.28 | 0.33 | 0.35 | 0.36 |

**7-14 Plane Asymmetric Diffuser at Fan Outlet with Ductwork**
(Idelchik et al. 1986, Diagram 5-15)

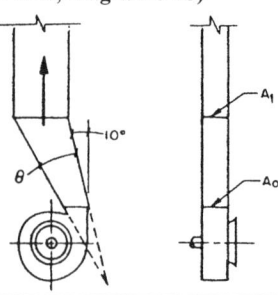

| θ, degrees | $C_o$ |||||| 
|---|---|---|---|---|---|---|
| | $A_1/A_o$ |||||| 
| | 1.5 | 2.8 | 2.5 | 3.0 | 3.5 | 4.0 |
| 10 | 0.11 | 0.13 | 0.14 | 0.14 | 0.14 | 0.14 |
| 15 | 0.13 | 0.15 | 0.16 | 0.17 | 0.18 | 0.18 |
| 20 | 0.19 | 0.22 | 0.24 | 0.26 | 0.28 | 0.30 |
| 25 | 0.29 | 0.32 | 0.35 | 0.37 | 0.39 | 0.40 |
| 30 | 0.36 | 0.42 | 0.46 | 0.49 | 0.51 | 0.51 |
| 35 | 0.44 | 0.54 | 0.61 | 0.64 | 0.66 | 0.66 |

**7-15 Pyramidal Diffuser at Fan Outlet with Ductwork**
(Idelchik et al. 1986, Diagram 5-16)

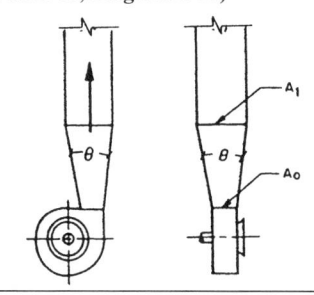

| θ, degrees | $C_o$ |||||| 
|---|---|---|---|---|---|---|
| | $A_1/A_o$ |||||| 
| | 1.5 | 2.8 | 2.5 | 3.0 | 3.5 | 4.0 |
| 10 | 0.10 | 0.18 | 0.21 | 0.23 | 0.24 | 0.25 |
| 15 | 0.23 | 0.33 | 0.38 | 0.40 | 0.42 | 0.44 |
| 20 | 0.31 | 0.43 | 0.48 | 0.53 | 0.56 | 0.58 |
| 25 | 0.36 | 0.49 | 0.55 | 0.58 | 0.62 | 0.64 |
| 30 | 0.42 | 0.53 | 0.59 | 0.64 | 0.67 | 0.69 |

# Duct and Pipe Sizing 345

**Example 9.1.** The supply duct system for the 100% outside makeup air system of a clean room as shown in the following sketch is to use rectangular, galvanized ducting throughout. An air velocity of 500 fpm is to be used. The following data are known about the components: (a) elbows are mitered, design 2 type (3-8), (b) the flow split is accomplished with a symmetrical wye (5-32), (c) the loss coefficient for the combination of transition section and outlet diffuser $C_c$ is 1.86, (d) the HEPA filter is to be 99.9% efficient, and (e) the coil is a 4-row Series 56 with 12 fins per inch. Data for the filter and coil can be obtained from the given data.

Size each duct and determine the total pressure drop for the supply duct.

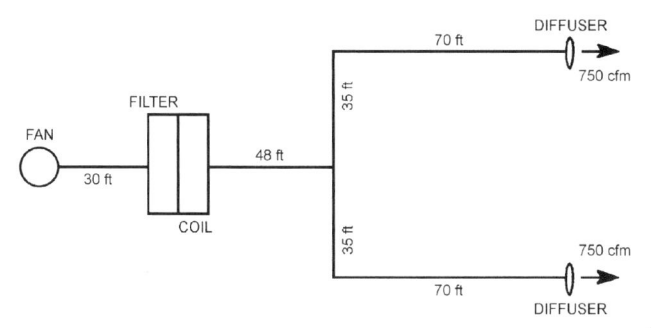

**Pressure Losses, Inches of Water**

| Filter Pressure Losses | Velocity, fpm | |
|---|---|---|
| | 250 (Clean/Dirty)* | 500 (Clean/Dirty)* |
| Panel 2 in. pleated 30% efficiency | 0.08 / 0.90 | 0.28 / 0.90 |
| Panel 4 in. pleated 30% efficiency | 0.07 / 0.90 | 0.27 / 0.90 |
| Bag 22 in. deep 60-65% efficiency | 0.12 / 1.00 | 0.30 / 1.00 |
| Bag 22 in. deep 80-85% efficiency | 0.28 / 1.00 | .045 / 1.00 |
| Bag 22 in. deep 90-95% efficiency | 0.50 / 1.50 | 0.70 / 1.50 |
| Cartridge 12 in. deep 60-65% efficiency | 0.15 / 1.50 | 0.29 / 1.50 |
| Cartridge 12 in. deep 80-85% efficiency | 0.27 / 1.50 | 0.50 / 1.50 |
| Cartridge 12 in. deep 90-95% efficiency | 0.34 / 1.50 | 0.68 / 1.50 |
| HEPA 12 in. deep, 99.97% DOP** | 0.60 / 2.00 | 1.20 / 2.00 |

\* Unit air flow performance should be selected halfway between initial (clean) and final (dirty) filter pressure loss, or as specified.
\*\* HEPA's shown are high flow capacity; rated for 500 fpm face velocity.

**Pressure Loss Per Row, Inches of Water**

| Fin Type | Fins per inch | Face Velocity, fpm | | | | | |
|---|---|---|---|---|---|---|---|
| | | 200 | 300 | 400 | 500 | 600 | 700 |
| Series 56 (5/8 in. tubes) | 8 | 0.050 | 0.092 | 0.140 | 0.196 | 0.257 | 0.324 |
| | 10 | 0.063 | 0.112 | 0.170 | 0.234 | 0.304 | 0.379 |
| | 12 | 0.075 | 0.133 | 0.199 | 0.271 | 0.360 | 0.435 |

**Solution:**

$$P_v = \left(\frac{500}{4005}\right)^2 = 0.0156 \text{ in. wg.}$$

Main Duct: 78 ft long, 1500 cfm, 500 fpm

$$A = \frac{1500}{500} = 3 \text{ ft}^2 = \frac{\pi D_e^2}{4}$$

$D_e = 1.95$ ft = 23.5 in.
equivalent rectangular = 30 × 16 to 42 × 12
Select 32 × 16 (lower aspect ratio and easier to match with coil face)
From Fig. 9.2, $\Delta p/100$ ft = 0.013 in. w.g.

$$\Delta p_M = 0.017 \times \frac{78}{100} = 0.013 \text{ in. w.g.}$$

Branches: 105 ft, 750 cfm, 500 fpm

$$750/500 = 1.5 \text{ ft}^2$$

$D_e = 1.38$ ft = 16.6 in.
equivalent rectangular = 16 × 16 to 34 × 8
From Fig. 9.2 $\Delta p/100$ ft = 0.022

$$\Delta p_B = 0.022 \times \frac{105}{100} = 0.023 \text{ in. wg.}$$

Elbows: (type 3-8-2) $C_o = 0.15$ $\Delta p_e = C_o p_v$

$$\Delta p_e = 0.15 (0.0156 \text{ in. w.g.}) = 0.0023 \text{ in. w.g.}$$

Wye: (type 5-32) $C_o = 0.30$

$$\Delta p_Y = 0.30 (0.0156) = 0.0047 \text{ in. w.g.}$$

Diffusers: $C_o = 1.86$

$$\Delta p_D = 1.86 (0.0156) = 0.029 \text{ in. w.g.}$$

$$\begin{aligned}\Delta p_{\text{TOTAL}} &= \Delta p_M + \Delta p_F + \Delta p_C + \Delta p_Y \\ &\quad + \Delta p_B + \Delta p_e + \Delta p_D \\ &= 0.010 + 1.60(\text{avg}) + (4 \times 0.271) + 0.0047 \\ &\quad + 0.023 + 0.0023 + 0.029 \\ &= 2.75 \text{ in. w.g.}\end{aligned}$$

**Example 9.2.** Find the total pressure loss in the straight-through section of a 90° cylindrical tee. The velocity in the main upstream section is 2000 fpm and in the main downstream is 1500 fpm. The velocity in the tee branch is 1060 fpm. Also, calculate the total pressure loss between the straight through section and the branch. The value of $A_b/A_c$ is 0.6.

**Solution:**

(a) For the straight-through section

$$V_c = 2000 \text{ fpm } V_s = 1500 \text{ fpm } V_b = 1060 \text{ fpm}$$

From Equation (9-2),

$$p_{vc} = (2000/4005)^2 = 0.25 \text{ in. of water}$$

For Fitting 5-23, Table 9-4, with $\theta = 90°$, and $V_s/V_c = 0.75$:

$$c_{c,s} = 0.03$$

By Equation (9-10a),

$$\Delta p_t = C_{c,s} P_{vc}$$
$$\Delta p_t = 0.03 (0.25) = 0.0075 \text{ in. of water (negligible)}$$

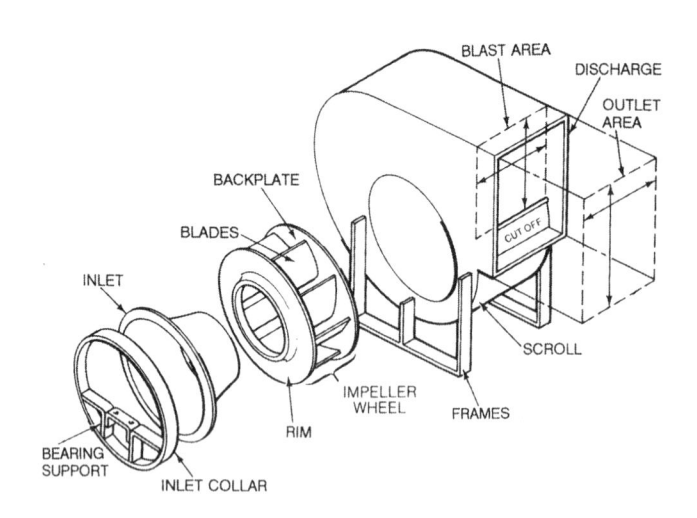

*Fig. 9-3 Centrifugal Fan Components*
*(Figure 1, Chapter 21, 2012 ASHRAE Handbook—HVAC Systems and Equipment)*

(b) For the branch section for θ = 90° and $V_b/V_c = 0.53$ and $A_b/A_c = 0.06$, for Fitting 5-23, Table 9-4

$$c_{c,b} = 1.01$$

By Equation (9-10b),

$$\Delta p_1 = C_{c,b}$$

$$\Delta p_t = 1.01\,(0.25) = 0.25 \text{ in. w.g.}$$

## 9.2 Fans

A fan is an air pump, a machine that creates a pressure difference and causes airflow. The impeller imparts to the air static and kinetic energy, which varies in proportion, depending on the type of fan.

Fans are classified as centrifugal fans or axial flow fans, according to the direction of airflow through the impeller. The general configuration of a centrifugal fan is shown in Figure 9-3. A similar description of an axial flow fan is shown in Figure 9-4. In addition to these two types, there are several subdivisions of each of the general types. A comparison of typical characteristics of some fan types is shown in Table 9-5.

### 9.2.1 Principles of Operation

Centrifugal fan impellers produce pressure from two related sources: (1) the centrifugal force created by rotating the air column enclosed between the blades and (2) the kinetic energy imparted to the air by its velocity leaving the impeller. This velocity is a combination of rotative velocity of the impeller and air speed relative to the impeller. When the blades are inclined forward, these two velocities are cumulative, when backward, oppositional. The forward-curved fans depend less on centrifugal force for pressure and more on velocity pressure conversion in the scroll. Conversely, fans with backward-curved blades build up more pressure by centrifugal force and less by velocity conversion

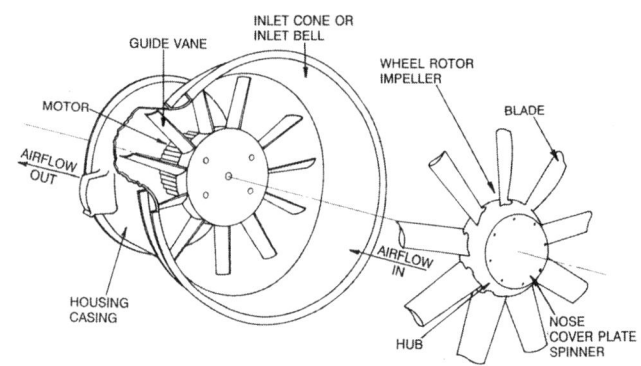

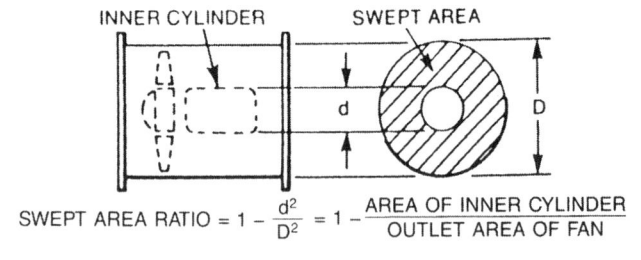

$$\text{SWEPT AREA RATIO} = 1 - \frac{d^2}{D^2} = 1 - \frac{\text{AREA OF INNER CYLINDER}}{\text{OUTLET AREA OF FAN}}$$

Note: The swept area ratio in axial fans is equivalent to the blast area ratio in centrifugal fans.

*Fig. 9-4 Axial Fan Components*
*(Figure 2, Chapter 21, 2012 ASHRAE Handbook—HVAC Systems and Equipment)*

in the scroll. However, since the buildup of pressure by centrifugal force is a more efficient form of energy transfer than the conversion of velocity, backward-curved blade fans generally are more efficient than forward-curved blade fans.

Axial flow fans produce all of their static pressure from the change in velocity of the air passing through the impeller. These fans are divided into three subtypes. Propeller fans, customarily used at free delivery, are of relatively simple construction. The impellers usually have a small hub-to-tip ratio and are mounted in an orifice plate or inlet ring. Tubeaxial fans are mounted in a cylindrical tube. They have reduced running clearance and operate at higher tip speeds, which give the tubeaxial fan a higher static pressure capability than the propeller fan. Vaneaxial fans are essentially tubeaxial fans with guide vanes that give improved pressure characteristics and efficiency.

### 9.2.2 Definitions

**Volume flow rate**, handled by the fan, is the number of cubic feet of air per minute expressed at fan inlet conditions.

**Fan total pressure rise** is the fan total pressure at outlet minus the fan total pressure at inlet, in. of water.

**Fan velocity pressure** is the pressure corresponding to the average velocity determined from the volume flow rate and fan outlet area, in. of water.

**Fan static pressure rise** is the fan total pressure rise diminished by the fan velocity pressure. The fan inlet velocity

# Duct and Pipe Sizing

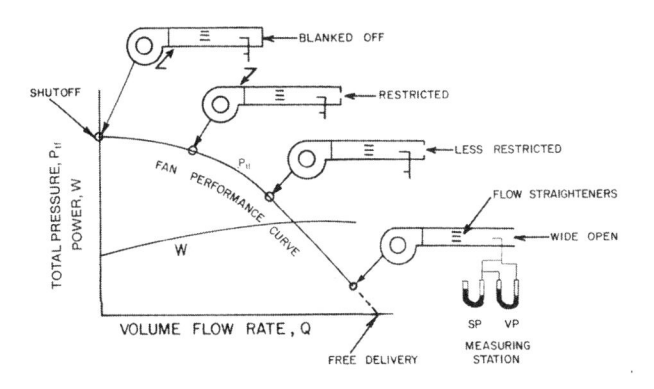

*Fig. 9-5 Method of Obtaining Fan Performance Curve*
*(Figure 3, Chapter 21, 2012 ASHRAE Handbook—HVAC Systems and Equipment)*

head is assumed equal to zero for fan rating purposes, in. of water.

**Power output** of a fan is expressed as horsepower and is based on the fan volume flow rate and the fan total pressure.

**Power input** of a fan is expressed as horsepower and is measured as power delivered to the fan shaft.

**Mechanical efficiency** (or fan total efficiency) of a fan is the ratio of power output to power input.

**Static efficiency** of a fan is the mechanical efficiency multiplied by the ratio of fan static pressure to fan total pressure.

**Point of rating** may be any point on the fan performance curve. For each case, the particular point on the curve must be specifically defined.

## 9.2.3 Fan Testing

The pilot tube duct traverse can be used to explain the procedure by which a constant speed fan performance curve can be obtained (Figure 9-5). Fans are tested in accordance with *ASHRAE Standard* 51 and various AMCA standards.

The fan is tested from **shutoff** conditions to nearly **free delivery** conditions. At shutoff, the duct is completely blanked off; at free delivery, the outlet resistance is reduced to zero. Between these two conditions, various flow restrictions are placed on the end of the duct to simulate various conditions on the fan. Sufficient points are obtained to define the curve between the shutoff point and free delivery conditions.

Fans designed for duct systems are tested with a length of duct between the fan and the measuring station. This length of duct smoothes the flow of the fan and provides stable, uniform flow conditions at the plane of measurement. The measured pressures are corrected back to fan outlet conditions. Fans designed for use without ducts are tested without ductwork.

Not all sizes are tested for rating. Test information may be used to calculate the performance of larger fans that are geometrically similar, but it should not be extrapolated to smaller fans. For the performance of one fan to be determined from the known performance of another, the two fans must be dynamically similar. Strict dynamic similarity

**Table 9-5 Fan Laws**
*(Table 2, Chapter 21, 2012 ASHRAE Handbook—HVAC Systems and Equipment)*

| Law No. | Dependent Variables | | | Independent Variables |
|---|---|---|---|---|
| 1a | $Q_1$ | = | $Q_2$ × | $(D_1/D_2)^3 (N_1/N_2)$ |
| 1b | $p_1$ | = | $p_2$ × | $(D_1/D_2)^2 (N_1/N_2)^2 \rho_1/\rho_2$ |
| 1c | $W_1$ | = | $W_2$ × | $(D_1/D_2)^5 (N_1/N_2)^3 \rho_1/\rho_2$ |
| 2a | $Q_1$ | = | $Q_2$ × | $(D_1/D_2)^2 (p_1/p_2)^{1/2} (\rho_2/\rho_1)^{1/2}$ |
| 2b | $N_1$ | = | $N_2$ × | $(D_2/D_1) (p_1/p_2)^{1/2} (\rho_2/\rho_1)^{1/2}$ |
| 2c | $W_1$ | = | $W_2$ × | $(D_1/D_2)^2 (p_1/p_2)^{3/2} (\rho_2/\rho_1)^{1/2}$ |
| 3a | $N_1$ | = | $N_2$ × | $(D_2/D_1)^3 (Q_1/Q_2)$ |
| 3b | $p_1$ | = | $p_2$ × | $(D_2/D_1)^4 (Q_1/Q_2)^2 \rho_1/\rho_2$ |
| 3c | $W_1$ | = | $W_2$ × | $(D_2/D_1)^4 (Q_1/Q_2)^3 \rho_1/\rho_2$ |

*Notes:*
1. Subscript 1 denotes the variable for the fan under consideration. Subscript 2 denotes the variable for the tested fan.
2. For all fans laws $(\eta_t)_1 = (\eta_t)_2$ and (Point of rating)$_1$ = (Point of rating)$_2$.
3. $p$ equals either $p_{tf}$ or $p_{sf}$.

requires that the important nondimensional parameters vary in only insignificant ways. These nondimensional parameters include those that affect the aerodynamic characteristics, such as Mach number, Reynolds number, surface roughness, and gap size. (For more specific information, the manufacturer's application manual or engineering data should be consulted.)

## 9.2.4 Fan Laws

Fan laws relate the performance variables for any geometrically similar series of fans (Table 9-5). The variables involved are fan size $D$, rotational speed $N$, gas density $\rho$, volume flow rate $Q$, pressure $p_t$ or $p_s$, power $H$ (either air or shaft), and mechanical efficiency $\eta_1$.

Fan laws mathematically express the fact that when two fans are both members of a geometrically similar series, their performance curves are homologous. At the same point of rating (i.e., at the same relative point on the fan performance curve), efficiencies are equal. Point of rating is sometimes expressed as a stated percent of free delivery airflow. Another method of describing point of rating is the static pressure-velocity pressure ratio $p_s/p_v$.

Unless otherwise identified, fan performance data are based on a standard air density of 0.075 lb/ft³ (1.2 kg/m³). With constant size and speed, the power and pressure varies directly as the ratio of gas density to standard air density.

The application of the fan laws for a change in fan speed $N$ to a specific size fan is illustrated in Figure 9-6. The computed P curve is derived from the base curve. For example, point E ($N_1 = 650$) is computed from point D ($N_2 = 600$) as follows:

At D, $Q_2 = 6000$ and $p_{tf2} = 1.13$

Using Fan Law 1a at Point E

$$Q_1 = 6000 (650/600) = 6500$$

Using Fan Law 1b

$$p_{t1} = 1.13 (650/600)^2 = 1.33$$

### Table 9-6 Types of Fans
*(Table 1, Chapter 21, 2012 ASHRAE Handbook—HVAC Systems and Equipment)*

| | Type | Impeller Design | Housing Design |
|---|---|---|---|
| **Centrifugal Fans** | Airfoil | Blades of airfoil contour curved away from direction of rotation. Deep blades allow efficient expansion within blade passages. Air leaves impeller at velocity less than tip speed. For given duty, has highest speed of centrifugal fan designs. | Scroll design for efficient conversion of velocity pressure to static pressure. Maximum efficiency requires close clearance and alignment between wheel and inlet. |
| | Backward-Inclined Backward-Curved | Single-thickness blades curved or inclined away from direction of rotation. Efficient for same reasons as airfoil fan. | Uses same housing configuration as airfoil design. |
| | Radial (R) Radial Tip (Rt) | Higher pressure characteristics than airfoil, backward-curved, and backward-inclined fans. Curve may have a break to left of peak pressure and fan should not be operated in this area. Power rises continually to free delivery. | Scroll similar to and often identical to other centrifugal fan designs. Fit between wheel and inlet not as critical as for airfoil and backward-inclined fans. |
| | Forward-Curved | Flatter pressure curve and lower efficiency than the airfoil, backward-curved, and backward-inclined. Do not rate fan in the pressure curve dip to the left of peak static pressure. Power rises continually toward free delivery. | Scroll similar to and often identical to other centrifugal fan designs. Fit between wheel and inlet not as critical as for airfoil and backward-inclined fans. |
| | Plenum/Plug | Plenum and plug fans typically use airfoil, backward inclined, or backward curved impellers in a single inlet configuration. Relative benefits of each impeller are the same as those described for scroll housed fans. | Plenum and plug fans are unique in that they operate with no housing. The equivalent of a housing, or plenum chamber (dashed line), depends on the application. The components of the drive system for the plug fan are located outside the airstream. |
| **Axial Fans** | Propeller | Low efficiency. Limited to low-pressure applications. Usually low-cost impellers have two or more blades of single thickness attached to relatively small hub. Primary energy transfer by velocity pressure. | Simple circular ring, orifice plate, or venturi. Optimum design is close to blade tips and forms smooth airfoil into wheel. |
| | Tubeaxial | Somewhat more efficient and capable of developing more useful static pressure than propeller fan. Usually has 4 to 8 blades with airfoil or single-thickness cross section. Hub is usually less than half the fan tip diameter. | Cylindrical tube with close clearance to blade tips. |
| | Vaneaxial | Good blade design gives medium- to high-pressure capability at good efficiency. Most efficient have airfoil blades. Blades may have fixed, adjustable, or controllable pitch. Hub is usually greater than half fan tip diameter. | Cylindrical tube with close clearance to blade tips. Guide vanes upstream or downstream from impeller increase pressure capability and efficiency. |

# Duct and Pipe Sizing

## Table 9-6 Types of Fans (Continued)

| Performance Curves* | Performance Characteristics | Applications |
|---|---|---|
| | Highest efficiency of all centrifugal fan designs and peak efficiencies occur at 50 to 60% of wide-open volume.<br>Fan has a non-overloading characteristic, which means power reaches maximum near peak efficiency and becomes lower, or self-limiting, toward free delivery. | General heating, ventilating, and air-conditioning applications.<br>Usually only applied to large systems, which may be low-, medium-, or high-pressure applications.<br>Applied to large, clean-air industrial operations for significant energy savings. |
| | Similar to airfoil fan, except peak efficiency slightly lower.<br>Curved blades are slightly more efficient than straight blades. | Same heating, ventilating, and air-conditioning applications as airfoil fan.<br>Used in some industrial applications where environment may corrode or erode airfoil blade. |
| | Higher pressure characteristics than airfoil and backward-curved fans.<br>Pressure may drop suddenly at left of peak pressure, but this usually causes no problems.<br>Power rises continually to free delivery, which is an overloading characteristic.<br>Curved blades are slightly more efficient than straight blades. | Primarily for materials handling in industrial plants. Also for some high-pressure industrial requirements.<br>Rugged wheel is simple to repair in the field. Wheel sometimes coated with special material.<br>Not common for HVAC applications. |
| | Pressure curve less steep than that of backward-curved fans. Curve dips to left of peak pressure.<br>Highest efficiency occurs at 40 to 50% of wide-open volume.<br>Operate fan to right of peak pressure.<br>Power rises continually to free delivery which is an overloading characteristic. | Primarily for low-pressure HVAC applications, such as residential furnaces, central station units, and packaged air conditioners. |
| | Plenum and plug fans are similar to comparable housed airfoil/backward-curved fans but are generally less efficient because of inefficient conversion of kinetic energy in discharge air stream.<br>They are more susceptible to performance degradation caused by poor installation. | Plenum and plug fans are used in a variety of HVAC applications such as air handlers, especially where direct-drive arrangements are desirable.<br>Other advantages of these fans are discharge configuration flexibility and potential for smaller-footprint units. |
| | High flow rate, but very low pressure capabilities.<br>Maximum efficiency reached near free delivery.<br>Discharge pattern circular and airstream swirls. | For low-pressure, high-volume air-moving applications, such as air circulation in a space or ventilation through a wall without ductwork.<br>Used for makeup air applications. |
| | High flow rate, medium pressure capabilities.<br>Pressure curve dips to left of peak pressure. Avoid operating fan in this region.<br>Discharge pattern circular and airstream rotates or swirls. | Low- and medium-pressure ducted HVAC applications where air distribution downstream is not critical.<br>Used in some industrial applications, such as drying ovens, paint spray booths, and fume exhausts. |
| | High-pressure characteristics with medium-volume flow capabilities.<br>Pressure curve dips to left of peak pressure. Avoid operating fan in this region.<br>Guide vanes correct circular motion imparted by impeller and improve pressure characteristics and efficiency of fan. | General HVAC systems in low-, medium-, and high-pressure applications where straight-through flow and compact installation are required.<br>Has good downstream air distribution.<br>Used in industrial applications in place of tubeaxial fans.<br>More compact than centrifugal fans for same duty. |

**Table 9-6 Types of Fans (*Concluded*)**

| Type | | Impeller Design | Housing Design |
|---|---|---|---|
| Mixed-Flow | Mixed-Flow | Combination of axial and centrifugal characteristics. Ideally suited in applications in which the air has to flow in or out axially. Higher pressure characteristic than axial fans. | The majority of mixed-flow fans are in a tubular housing and include outlet turning vanes. Can operate without housing or in a pipe and duct. |
| Cross-flow | Cross-flow (Tangential) | Impeller with forward-curved blades. During rotation the flow of air passes through part of the rotor blades into the rotor. This creates an area of turbulence which, working with the guide system, deflects the airflow through another section of the rotor into the discharge duct of the fan casing. Lowest efficiency of any type of fan. | Special designed housing for 90° or straight through airflow. |
| | Tubular Centri-Fugal | Performance similar to backward-curved fan except capacity and pressure are lower. Lower efficiency than backward-curved fan. Performance curve may have a dip to the left of peak pressure. | Cylindrical tube similar to vaneaxial fan, except clearance to wheel is not as close. Air discharges radially from wheel and turns 90° to flow through guide vanes. |
| Other Designs | Power Roof Ventilators — Centrifugal | Low-pressure exhaust systems such as general factory, kitchen, warehouse, and some commercial installations. Provides positive exhaust ventilation, which is an advantage over gravity-type exhaust units. Centrifugal units are slightly quieter than axial units. | Normal housing not used, because air discharges from impeller in full circle. Usually does not include configuration to recover velocity pressure component. |
| | Power Roof Ventilators — Axial | Low-pressure exhaust systems such as general factory, kitchen, warehouse, and some commercial installations. Provides positive exhaust ventilation, which is an advantage over gravity-type exhaust units. Hood protects fan from weather and acts as safety guard. | Essentially a propeller fan mounted in a supporting structure. Air discharges from annular space at bottom of weather hood. |

# Duct and Pipe Sizing

**Table 9-6  Types of Fans (*Concluded*)**

| Performance Curves* | Performance Characteristics | Applications |
|---|---|---|
|  | Characteristic pressure curve between axial fans and centrifugal fans. Higher pressure than axial fans and higher volume flow than centrifugal fans. | Similar HVAC applications to centrifugal fans or in applications where an axial fan cannot generate sufficient pressure rise. |
|  | Similar to forward-curved fans. Power rises continually to free delivery, which is an overloading characteristic. Unlike all other fans, performance curves include motor characteristics. Lowest efficiency of any fan type. | Low-pressure HVAC systems such as fan heaters, fireplace inserts, electronic cooling, and air curtains. |
|  | Performance similar to backward-curved fan, except capacity and pressure are lower. Lower efficiency than backward-curved fan because air turns 90°. Performance curve of some designs is similar to axial flow fan and dips to left of peak pressure. | Primarily for low-pressure, return air systems in HVAC applications. Has straight-through flow. |
|  | Usually operated without ductwork; therefore, operates at very low pressure and high volume. | Centrifugal units are somewhat quieter than axial flow units. Low-pressure exhaust systems, such as general factory, kitchen, warehouse, and some commercial installations. Low first cost and low operating cost give an advantage over gravity-flow exhaust systems. |
|  | Usually operated without ductwork; therefore, operates at very low pressure and high volume. | Low-pressure exhaust systems, such as general factory, kitchen, warehouse, and some commercial installations. Low first cost and low operating cost give an advantage over gravity-flow exhaust systems. |

*These performance curves reflect general characteristics of various fans as commonly applied. They are not intended to provide complete selection criteria, because other parameters, such as diameter and speed, are not defined.

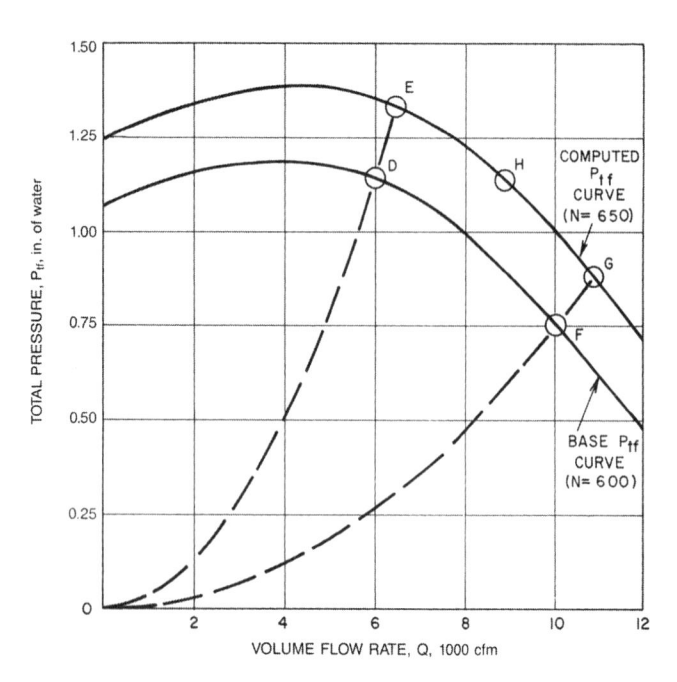

*Fig. 9-6 Example Application of Fan Laws*
*(Figure 4, Chapter 21, 2012 ASHRAE Handbook—HVAC Systems and Equipment)*

Therefore, the completed $P_{tf1}$, $N = 650$ curve may be generated by computing additional points from data on the base curve, such as point G from point F. If equivalent points of rating are joined, as illustrated by the dotted lines in Figure 9-6, they will lie on parabolas.

Each point on the base $P_{tf}$ curve determines only one point on the computed curve. For example, there is no way to calculate point H from either point D or point F. It is totally unrelated to either one of these points. Point H is, however, related to some point between these two points on the base $P_{tf}$ curve, and only that point can be used to locate point H. Furthermore, there is no way that point D can be used to calculate the position of point F on the base $P_{tf}$ curve. The entire base curve must be defined by test.

### 9.2.5 Duct System Characteristics

A simplified duct system with three 90° elbows is shown in Figure 9-7. These elbows represent the resistance offered by the ductwork, heat exchangers, cabinets, dampers, grilles, and other system components. A given rate of airflow through a given system requires a definite total pressure in the system. The resulting total pressure varies as the volume flow rate squared.

The following Equation (9-14) is true for turbulent airflow systems. Heating, ventilating, and air-conditioning systems generally follow this law closely, and no serious error is introduced by its use.

$$\frac{\Delta p_2}{\Delta p_1} = \left(\frac{Q_1}{Q_2}\right)^2 \qquad (9\text{-}14)$$

The discussion in this chapter is limited to turbulent flow, which is the flow regime in which most fans operate. In

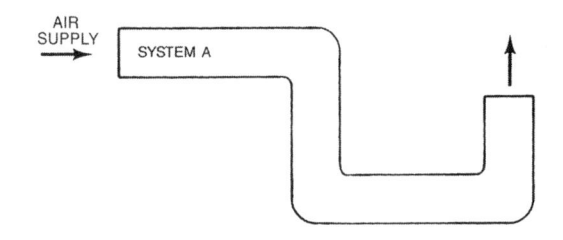

*Fig. 9-7 Simplified Duct System with Resistance to Flow Represented by Three 90° Elbows*
*(Figure 9, Chapter 21, 2012 ASHRAE Handbook—HVAC Systems and Equipment)*

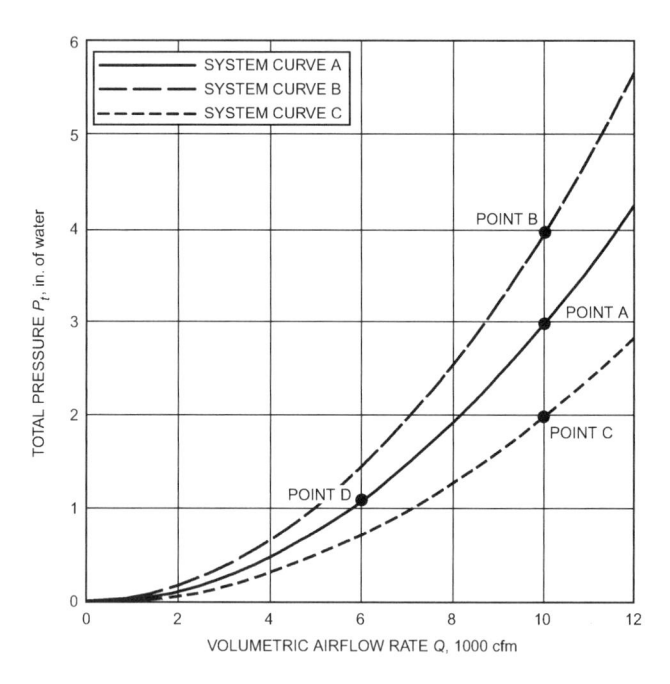

*Fig. 9-8 Example System Total Pressure Loss ($\Delta p$) Curves*
*(Figure 10, Chapter 21, 2012 ASHRAE Handbook—HVAC Systems and Equipment)*

some systems, particularly constant or variable volume air conditioning, the air-handling devices and associated controls may produce system resistance curves that deviate widely from Equation (9-14), even though each element of the system may be described by this equation.

Note that Equation (9-14) permits plotting a turbulent flow system's pressure loss ($\Delta p$) curve from one known operating condition (see Figure 9-5). The fixed system must operate at some point on this system curve as the volume flow rate changes. As an example, at point A, curve A, Figure 9-8, when the flow rate through a duct system such as that shown in Figure 9-7 is 10,000 cfm, the total pressure drop is 3 in. of water. If these values are substituted in Equation (9-14) for $\Delta p_1$ and $Q_1$, other points of the system's $\Delta p$ curve can be determined.

For 6000 cfm

$\Delta p_2 = 3$ in. of water $(6000/10{,}000)^2 = 1.08$ in. of water

# Duct and Pipe Sizing

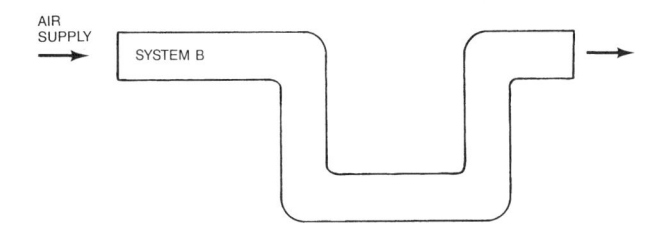

*Fig. 9-9 Resistance Added to System of Figure 9-7*
*(Figure 11, Chapter 21, 2012 ASHRAE Handbook—HVAC Systems and Equipment)*

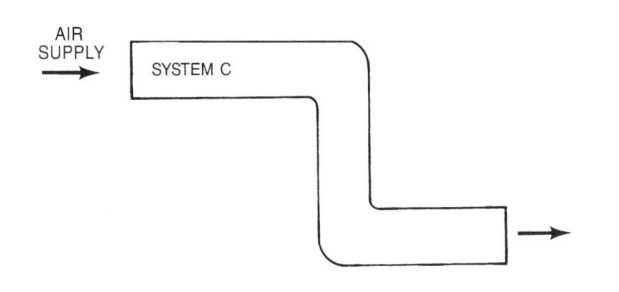

*Fig. 9-10 Resistance Removed from that of Figure 9-7*
*(Figure 12, Chapter 21, 2012 ASHRAE Handbook—HVAC Systems and Equipment)*

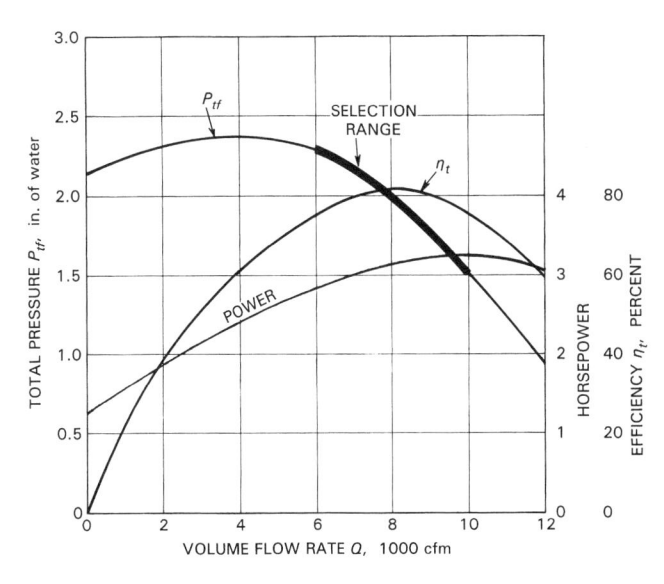

*Fig. 9-11 Typical Manufacturer's Fan Performance Curve (Curve shows performance of a fixed fan size running at a fixed speed.)*
*(Figure 13, Chapter 20, 2008 ASHRAE Handbook—HVAC Systems and Equipment)*

If a change is made in the system so that the total pressure at design flow rate is increased, the system will no longer operate on the previous $\Delta p$ curve; a new curve will be defined.

For example, in Figure 9-9, an additional elbow has been added to the schematic duct system shown in Figure 9-7, which increases the total pressure of the system. If the total pressure at 10,000 cfm is increased by 1.0 in. of water, the system total pressure drop at this point will now be 4.0 in. of water, as shown by point B in Figure 9-8. Following the procedure outlined above, a series of $\Delta p$ points may be computed and a new curve plotted ($\Delta p$ curve B in Figure 9-8).

If the system in Figure 9-7 is now changed by removing one of the schematic elbows, the resulting system total pressure drops below the total pressure resistance (Figure 9-10). The new $\Delta p$ curve is shown in Curve C of Figure 9-8. For curve C, a total pressure reduction of 1.0 in. has been assumed when 10,000 cfm flows through the system; thus, the point of operation is at 2.0 in. of water as shown by point C. These three $\Delta P$ curves all follow the relationship expressed in Equation (9-14). Note also that these curves result from changes within the system itself and do not change the fan performance. During the design phase, such system total pressure changes may be due to studies of alternative duct routing, studies of differences in duct sizes, allowance for future duct extensions, or the effect of the design safety factor being applied to the system.

In an actual operating system, these three $\Delta p$ curves could represent three system characteristic lines that result from three different positions of a throttling control damper. Curve C is the most open position, and curve B is the most closed position of the three positions illustrated. A control damper actually forms a continuous series of these $\Delta p$ curves as it moves from a wide open position to a completely closed position and covers a much wider range of operation than illustrated here. Such curves could also represent the clogging of turbulent flow filters in a system.

### 9.2.6 Fan Selection

After the system pressure loss curve of the air distribution system has been defined, a fan can be selected to meet the system's requirements. Fan manufacturers present performance data in either the tabular (multi-rating table) or the graphic (curve) as in Figure 9-11, which shows the performance of a fan of a fixed size running at fixed speed. Figure 9-11b is an example of a manufacturer's fan table.

Multi-rating tables usually provide only performance data within the recommended (safe) operating range. The optimum selection area or peak efficiency point is identified in various ways by the different manufacturers. Performance data in the usual fan tables are based on arbitrary increments of flow rate and pressure. In these tables, adjacent data, either horizontally or vertically, represent different points of operation (i.e., different points of rating) on the fan performance curve. These points of rating depend solely on the fan's characteristics; they cannot be obtained one from the other by use of the fan laws. However, these points are usually close together, so intermediate points may be interpolated arithmetically with accuracy adequate for fan selection.

The selection of a fan for a particular air distribution system requires the fan pressure characteristics to fit the system pressure characteristics. Thus, the total system must be evaluated and the flow requirements and resistances existing at the fan inlet and outlet must be known. The direct effect that certain types of installations have on fan performance must also be considered. Performance pressure changes, known as system effect factors, are a direct effect that must be added to the system resistance before fan selection.

Fan speed and power requirements are then calculated using one of the methods available from fan manufacturers. These may consist of the multi-rating tables as mentioned or of single-speed or multi-speed performance curves or graphs.

The point of operation selected must be at a desirable point on the fan curve, so that maximum efficiency and resistance to stall and pulsation can be attained (Figure 9-12). On systems where more than one point of operation is encountered during operation, the range of performance must be evaluated to determine how the selected fan will react within this complete range. This is particularly true in variable volume systems, where the fan not only experiences a change in performance, but the entire system does not follow the relationship defined in Equation (9-14). In these cases, actual losses in the system at performance extremes must be evaluated.

## 9.3 Air-Diffusing Equipment

Supply air outlets and diffusing equipment introduce air into a conditioned space to obtain a desired indoor atmospheric environment. Return and exhaust air are removed from the space through return and exhaust inlets. Various types of diffusing equipment are available as standard manufactured products. Refer to Chapter 20 of the 2013 *ASHRAE Handbook—Fundamentals* and Chapter 20 of the 2012 *ASHRAE Handbook—HVAC Systems and Equipment* for additional details concerning space air distribution.

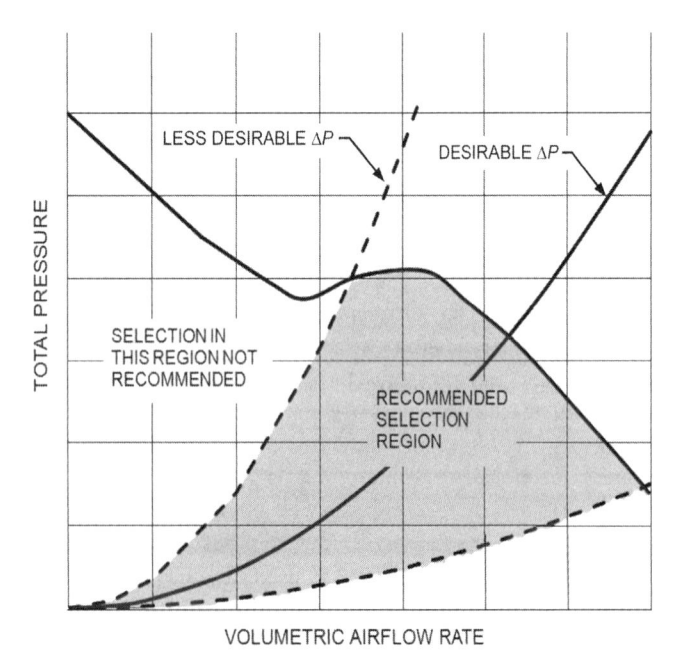

*Fig. 9-12 Desirable Combination of $p_{tf}$ and $\Delta p$ Curves*
*(Figure 14, Chapter 21, 2012 ASHRAE Handbook—HVAC Systems and Equipment)*

Air diffusion in warm air heating, ventilating, and air-conditioning systems should create the proper combination of temperature, humidity, and air motion in the occupied zone of the conditioned room [floor to 6 ft (1.8 m) above floor level]. Standard limits have been established as **acceptable, effective draft temperature**. This term comprises air temperature, air motion, relative humidity, and their physiological effect on the human body. Any variation from accepted standards of one of these elements results in discomfort to the occupants. Such discomfort may arise due to excessive room air temperature variations (horizontally, vertically, or both), excessive air motion (draft), failure to deliver or distribute the air according to the load requirements at the different locations, or overly rapid fluctuation of room temperature.

To define the difference $\theta$ in effective draft temperature between any point in the occupied zone and the control condition, investigators have used the following equation:

$$\theta = (t_x - t_c) - a(V_x - b) \quad (9\text{-}15)$$

where

$t_x$ = local airstream dry-bulb temperature, °F
$V_x$ = local airstream velocity, ft/min
$t_c$ = average room dry-bulb temperature, °F
$b$ = 30
$a$ = 0.07

Equation (9-15) accounts for the feeling of coolness produced by air motion. It also shows that a 1.8°F temperature change is equal to a 25 fpm velocity change. In summer, the local airstream temperature $t_x$ is below the control temperature. Hence, both temperature and velocity terms are negative when the velocity $V_x$ is greater than 30 fpm, and both of them add to the feeling of coolness. If, in winter, $t_x$ is above the control temperature, any air velocity above 30 fpm subtracts from the feeling of warmth produced by $t_x$. Therefore, it is usually possible to have zero difference in effective temperature between location "x" and the control point in winter, but not in summer.

Conditioned air is normally supplied to air outlets at velocities much higher than would be acceptable in the occupied zone. The conditioned air temperature may be above, below, or equal to the temperature of the air in the occupied zone. Proper air distribution, therefore, calls for (1) entrainment of room air by the primary airstream outside the zone of occupancy so that air motion and temperature differences are reduced to acceptable limits before the air enters the occupied zone and (2) counteraction of the natural convection and radiation effects within the room.

### 9.3.1 Supply Air Outlets

The correct types of outlets, properly sized and properly located, control the air pattern within the space to obtain proper air motion and temperature equalization in the occupied zone. Four types of supply outlets are commonly available: (1) grille outlets, (2) slot diffuser outlets, (3) ceiling diffuser outlets, and (4) perforated ceiling panels. These outlets have different con-

# Duct and Pipe Sizing

## SIZE 273

Wheel diameter: 27″  
Fan outlet area: 4.19 ft²  
Maximum BHP = $2.87\left(\dfrac{RPM}{1000}\right)^3$

| CFM | OV | ½″ SP RPM | BHP | 1″ SP RPM | BHP | 1½″ SP RPM | BHP | 2″ SP RPM | BHP | 2½″ SP RPM | BHP | 3″ SP RPM | BHP | 4″ SP RPM | BHP | 5″ SP RPM | BHP | 6″ SP RPM | BHP | 7″ SP RPM | BHP | 8″ SP RPM | BHP |
|---|---|---|---|---|---|---|---|---|---|---|---|---|---|---|---|---|---|---|---|---|---|---|---|
| 3351 | 800 | 487 | 0.33 | 643 | 0.68 | 783 | 1.11 | 904 | 1.57 | 1010 | 2.07 | 1108 | 2.61 | 1280 | 3.74 | 1430 | 4.96 | 1568 | 6.29 | 1692 | 7.68 | 1809 | 9.15 |
| 3770 | 900 | 511 | 0.38 | 652 | 0.75 | 784 | 1.19 | 904 | 1.68 | 1010 | 2.21 | 1107 | 2.76 | 1279 | 3.95 | 1432 | 5.25 | 1566 | 6.61 | 1694 | 8.07 | 1808 | 9.56 |
| 4198 | 1000 | 539 | 0.44 | 665 | 0.82 | 788 | 1.27 | 904 | 1.78 | 1011 | 2.34 | 1106 | 2.92 | 1279 | 4.17 | 1429 | 5.51 | 1567 | 6.93 | 1691 | 8.41 | 1809 | 10.0 |
| 4608 | 1100 | 569 | 0.51 | 683 | 0.91 | 797 | 1.37 | 908 | 1.91 | 1011 | 2.47 | 1107 | 3.08 | 1278 | 4.38 | 1430 | 5.77 | 1568 | 7.27 | 1691 | 8.79 | 1808 | 10.4 |
| 5027 | 1200 | 601 | 0.58 | 705 | 1.01 | 810 | 1.48 | 915 | 2.02 | 1015 | 2.62 | 1108 | 3.24 | 1278 | 4.61 | 1429 | 6.04 | 1567 | 7.58 | 1691 | 9.17 | 1810 | 10.8 |
| 5446 | 1300 | 634 | 0.67 | 730 | 1.11 | 828 | 1.61 | 925 | 2.16 | 1020 | 2.77 | 1110 | 3.41 | 1279 | 4.82 | 1429 | 6.31 | 1556 | 7.91 | 1691 | 9.56 | 1808 | 11.2 |
| 5865 | 1400 | 668 | 0.77 | 758 | 1.23 | 847 | 1.74 | 939 | 2.31 | 1027 | 2.92 | 1115 | 3.59 | 1279 | 5.02 | 1429 | 6.58 | 1566 | 8.22 | 1692 | 9.94 | 1809 | 11.7 |
| 6284 | 1500 | 704 | 0.88 | 787 | 1.36 | 871 | 1.88 | 955 | 2.46 | 1040 | 3.11 | 1122 | 3.77 | 1282 | 5.26 | 1431 | 6.86 | 1565 | 8.53 | 1692 | 10.3 | 1809 | 12.1 |
| 6703 | 1600 | 739 | 1.01 | 818 | 1.51 | 896 | 2.05 | 975 | 2.64 | 1054 | 3.28 | 1133 | 3.98 | 1288 | 5.51 | 1431 | 7.13 | 1566 | 8.87 | 1691 | 10.7 | 1808 | 12.5 |
| 7541 | 1800 | 813 | 1.28 | 884 | 1.84 | 953 | 2.42 | 1022 | 3.05 | 1092 | 3.72 | 1163 | 4.44 | 1303 | 6.01 | 1440 | 7.71 | 1568 | 9.51 | 1692 | 11.4 | 1806 | 13.4 |
| 8379 | 2000 | 887 | 1.62 | 952 | 2.22 | 1015 | 2.86 | 1077 | 3.52 | 1140 | 4.23 | 1203 | 4.97 | 1329 | 6.58 | 1455 | 8.34 | 1578 | 10.2 | 1696 | 12.2 | 1809 | 14.3 |
| 9217 | 2200 | 962 | 2.01 | 1024 | 2.69 | 1081 | 3.36 | 1138 | 4.07 | 1194 | 4.81 | 1252 | 5.61 | 1367 | 7.27 | 1482 | 9.08 | 1595 | 11.0 | 1708 | 13.0 | 1816 | 15.2 |
| 10055 | 2400 | 1039 | 2.49 | 1096 | 3.21 | 1149 | 3.94 | 1202 | 4.71 | 1253 | 5.47 | 1306 | 6.31 | 1411 | 8.05 | 1515 | 9.88 | 1622 | 11.8 | 1726 | 13.9 | 1829 | 16.2 |
| 10893 | 2600 | 1115 | 3.01 | 1170 | 3.81 | 1220 | 4.59 | 1268 | 5.41 | 1317 | 6.23 | 1364 | 7.08 | 1460 | 8.87 | 1558 | 10.8 | 1654 | 12.8 | 1753 | 15.0 | 1850 | 17.3 |
| 11731 | 2800 | 1192 | 3.62 | 1244 | 4.49 | 1293 | 5.35 | 1337 | 6.18 | 1382 | 7.07 | 1427 | 7.96 | 1517 | 9.87 | 1606 | 11.8 | 1696 | 13.9 | 1787 | 16.1 | 1878 | 18.4 |
| 12569 | 3000 | 1270 | 4.32 | 1319 | 5.25 | 1364 | 6.14 | 1408 | 7.07 | 1451 | 8.01 | 1492 | 8.95 | 1576 | 10.9 | 1658 | 12.9 | 1741 | 15.0 | 1826 | 17.3 | 1912 | 19.8 |
| 13407 | 3200 | 1349 | 5.13 | 1395 | 6.11 | 1438 | 7.06 | 1480 | 8.05 | 1520 | 9.03 | 1559 | 10.0 | 1637 | 12.0 | 1714 | 14.1 | 1793 | 16.4 | 1870 | 18.7 | 1952 | 21.2 |
| 14245 | 3400 | 1472 | 6.02 | 1472 | 7.07 | 1513 | 8.09 | 1553 | 9.12 | 1591 | 10.1 | 1628 | 11.2 | 1703 | 13.3 | 1774 | 15.5 | 1850 | 17.9 | 1923 | 20.3 | 1997 | 22.7 |

## SIZE 303

Wheel diameter: 30″  
Fan outlet area: 5.17 ft²  
Maximum BHP = $4.83\left(\dfrac{RPM}{1000}\right)^3$

| CFM | OV | ½″ SP RPM | BHP | 1″ SP RPM | BHP | 1½″ SP RPM | BHP | 2″ SP RPM | BHP | 2½″ SP RPM | BHP | 3″ SP RPM | BHP | 4″ SP RPM | BHP | 5″ SP RPM | BHP | 6″ SP RPM | BHP | 7″ SP RPM | BHP | 8″ SP RPM | BHP |
|---|---|---|---|---|---|---|---|---|---|---|---|---|---|---|---|---|---|---|---|---|---|---|---|
| 4135 | 800 | 438 | 0.41 | 579 | 0.84 | 704 | 1.36 | 813 | 1.94 | 910 | 2.56 | 997 | 3.21 | 1151 | 4.61 | 1286 | 6.12 | 1410 | 7.76 | 1523 | 9.47 | 1628 | 11.2 |
| 4652 | 900 | 460 | 0.47 | 586 | 0.92 | 705 | 1.46 | 813 | 2.07 | 910 | 2.72 | 996 | 3.41 | 1150 | 4.87 | 1288 | 6.47 | 1409 | 8.14 | 1524 | 9.96 | 1629 | 11.8 |
| 5169 | 1000 | 484 | 0.54 | 598 | 1.01 | 710 | 1.58 | 814 | 2.21 | 909 | 2.88 | 996 | 3.61 | 1150 | 5.14 | 1288 | 6.81 | 1410 | 8.55 | 1524 | 10.4 | 1628 | 12.3 |
| 5686 | 1100 | 512 | 0.62 | 614 | 1.11 | 717 | 1.69 | 817 | 2.34 | 910 | 3.05 | 996 | 3.81 | 1150 | 5.41 | 1286 | 7.12 | 1411 | 8.96 | 1524 | 10.8 | 1630 | 12.9 |
| 6203 | 1200 | 541 | 0.72 | 634 | 1.23 | 730 | 1.83 | 823 | 2.51 | 913 | 3.23 | 996 | 4.01 | 1150 | 5.67 | 1286 | 7.45 | 1409 | 9.35 | 1524 | 11.3 | 1628 | 13.4 |
| 6720 | 1300 | 571 | 0.83 | 657 | 1.37 | 744 | 1.98 | 832 | 2.66 | 917 | 3.41 | 998 | 4.21 | 1151 | 5.94 | 1286 | 7.79 | 1409 | 9.74 | 1522 | 11.7 | 1629 | 13.9 |
| 7237 | 1400 | 601 | 0.95 | 682 | 1.51 | 762 | 2.14 | 844 | 2.85 | 926 | 3.62 | 1003 | 4.43 | 1152 | 6.22 | 1286 | 8.12 | 1409 | 10.1 | 1522 | 12.2 | 1628 | 14.4 |
| 7754 | 1500 | 633 | 1.08 | 709 | 1.68 | 783 | 2.32 | 860 | 3.05 | 935 | 3.83 | 1011 | 4.67 | 1154 | 6.48 | 1287 | 8.46 | 1409 | 10.5 | 1522 | 12.7 | 1628 | 15.0 |
| 8271 | 1600 | 655 | 1.23 | 736 | 1.85 | 806 | 2.53 | 877 | 3.26 | 949 | 4.06 | 1020 | 4.92 | 1158 | 6.78 | 1288 | 8.79 | 1409 | 10.9 | 1522 | 13.1 | 1626 | 15.5 |
| 9305 | 1800 | 731 | 1.58 | 795 | 2.27 | 857 | 2.98 | 920 | 3.77 | 983 | 4.59 | 1046 | 5.48 | 1172 | 7.41 | 1295 | 9.51 | 1413 | 11.7 | 1522 | 14.1 | 1628 | 16.6 |
| 10339 | 2000 | 798 | 1.99 | 857 | 2.74 | 913 | 3.53 | 970 | 4.37 | 1025 | 5.22 | 1083 | 6.15 | 1197 | 8.14 | 1311 | 10.3 | 1420 | 12.6 | 1528 | 15.1 | 1628 | 17.6 |
| 11373 | 2200 | 866 | 2.48 | 922 | 3.32 | 972 | 4.14 | 1023 | 5.02 | 1074 | 5.93 | 1126 | 6.91 | 1230 | 8.96 | 1333 | 11.2 | 1437 | 13.6 | 1537 | 16.1 | 1636 | 18.8 |
| 12407 | 2400 | 935 | 3.07 | 986 | 3.96 | 1034 | 4.86 | 1081 | 5.79 | 1129 | 6.79 | 1175 | 7.78 | 1270 | 9.92 | 1364 | 12.2 | 1459 | 14.6 | 1553 | 17.2 | 1646 | 20.1 |
| 13441 | 2600 | 1003 | 3.72 | 1053 | 4.71 | 1097 | 5.67 | 1141 | 6.65 | 1184 | 7.68 | 1227 | 8.72 | 1313 | 10.9 | 1402 | 13.3 | 1488 | 15.8 | 1577 | 18.5 | 1665 | 21.3 |
| 14475 | 2800 | 1072 | 4.46 | 1119 | 5.53 | 1163 | 6.59 | 1204 | 7.66 | 1244 | 8.71 | 1283 | 9.82 | 1365 | 12.1 | 1445 | 14.6 | 1526 | 17.1 | 1608 | 19.9 | 1689 | 22.8 |
| 15509 | 3000 | 1142 | 5.33 | 1187 | 6.47 | 1227 | 7.57 | 1267 | 8.72 | 1305 | 9.88 | 1343 | 11.0 | 1417 | 13.4 | 1491 | 15.9 | 1568 | 18.7 | 1642 | 21.4 | 1720 | 24.4 |
| 16543 | 3200 | 1213 | 6.32 | 1255 | 7.53 | 1294 | 8.71 | 1331 | 9.92 | 1368 | 11.1 | 1402 | 12.3 | 1473 | 14.8 | 1544 | 17.5 | 1613 | 20.2 | 1685 | 23.2 | 1756 | 26.1 |
| 17577 | 3400 | 1284 | 7.42 | 1324 | 8.72 | 1361 | 9.98 | 1397 | 11.2 | 1431 | 12.5 | 1465 | 13.8 | 1532 | 16.4 | 1598 | 19.2 | 1664 | 22.0 | 1730 | 25.0 | 1797 | 28.1 |

## SIZE 363

Wheel diameter: 37½″  
Fan outlet area: 7.66 ft²  
Maximum BHP = $12.8\left(\dfrac{RPM}{1000}\right)^3$

| CFM | OV | ½″ SP RPM | BHP | 1″ SP RPM | BHP | 1½″ SP RPM | BHP | 2″ SP RPM | BHP | 2½″ SP RPM | BHP | 3″ SP RPM | BHP | 4″ SP RPM | BHP | 5″ SP RPM | BHP | 6″ SP RPM | BHP | 7″ SP RPM | BHP | 8″ SP RPM | BHP |
|---|---|---|---|---|---|---|---|---|---|---|---|---|---|---|---|---|---|---|---|---|---|---|---|
| 6127 | 800 | 360 | 0.58 | 475 | 1.21 | 578 | 1.97 | 669 | 2.81 | 750 | 3.69 | 823 | 4.64 | 953 | 6.68 | 1065 | 8.84 | 1166 | 11.1 | 1260 | 13.6 | 1348 | 16.1 |
| 6893 | 900 | 377 | 0.67 | 481 | 1.33 | 580 | 2.11 | 668 | 2.99 | 784 | 3.93 | 821 | 4.93 | 952 | 7.08 | 1063 | 9.33 | 1166 | 11.7 | 1261 | 14.3 | 1348 | 17.0 |
| 7659 | 1000 | 397 | 0.78 | 491 | 1.46 | 582 | 2.26 | 669 | 3.18 | 747 | 4.16 | 820 | 5.21 | 950 | 7.46 | 1064 | 9.86 | 1166 | 12.3 | 1260 | 15.0 | 1347 | 17.8 |
| 8425 | 1100 | 419 | 0.91 | 504 | 1.61 | 589 | 2.44 | 670 | 3.37 | 748 | 4.41 | 819 | 5.49 | 948 | 7.82 | 1061 | 10.3 | 1165 | 12.9 | 1260 | 15.7 | 1347 | 18.6 |
| 9191 | 1200 | 442 | 1.03 | 520 | 1.77 | 598 | 2.63 | 675 | 3.59 | 749 | 4.54 | 819 | 5.77 | 946 | 8.21 | 1060 | 10.7 | 1164 | 13.5 | 1258 | 16.4 | 1346 | 19.4 |
| 9957 | 1300 | 466 | 1.19 | 539 | 1.97 | 611 | 2.85 | 682 | 3.82 | 752 | 4.89 | 820 | 6.06 | 945 | 8.55 | 1058 | 11.2 | 1162 | 14.1 | 1257 | 17.1 | 1345 | 20.2 |
| 10723 | 1400 | 491 | 1.36 | 559 | 2.18 | 626 | 3.09 | 693 | 4.09 | 758 | 5.18 | 823 | 6.36 | 946 | 8.96 | 1058 | 11.7 | 1160 | 14.6 | 1255 | 17.7 | 1344 | 20.9 |
| 11489 | 1500 | 517 | 1.55 | 580 | 2.41 | 643 | 3.35 | 705 | 4.37 | 767 | 5.49 | 829 | 6.69 | 947 | 9.34 | 1057 | 12.1 | 1158 | 15.1 | 1253 | 18.4 | 1340 | 21.6 |
| 12255 | 1600 | 543 | 1.76 | 603 | 2.67 | 662 | 3.64 | 719 | 4.69 | 778 | 5.82 | 836 | 7.05 | 950 | 9.73 | 1058 | 12.6 | 1157 | 15.7 | 1252 | 19.0 | 1339 | 22.4 |
| 13787 | 1800 | 596 | 2.23 | 651 | 3.26 | 703 | 4.31 | 754 | 5.41 | 806 | 6.61 | 859 | 7.88 | 962 | 10.6 | 1062 | 13.6 | 1158 | 16.9 | 1251 | 20.3 | 1336 | 23.9 |
| 15319 | 2000 | 650 | 2.81 | 701 | 3.93 | 748 | 5.08 | 795 | 6.27 | 842 | 7.52 | 888 | 8.84 | 981 | 11.6 | 1075 | 14.8 | 1166 | 18.1 | 1253 | 21.7 | 1338 | 25.4 |
| 16851 | 2200 | 704 | 3.47 | 752 | 4.71 | 796 | 5.96 | 838 | 7.21 | 881 | 8.55 | 923 | 9.92 | 1009 | 12.8 | 1094 | 16.0 | 1178 | 19.4 | 1261 | 23.1 | 1342 | 27.0 |
| 18383 | 2400 | 759 | 4.24 | 804 | 5.61 | 845 | 6.93 | 885 | 8.31 | 924 | 9.71 | 963 | 11.1 | 1041 | 14.2 | 1119 | 17.5 | 1197 | 21.0 | 1274 | 24.7 | 1351 | 28.7 |
| 19915 | 2600 | 815 | 5.13 | 857 | 6.61 | 869 | 8.08 | 934 | 9.54 | 971 | 11.0 | 1007 | 12.6 | 1078 | 15.7 | 1150 | 19.1 | 1221 | 22.7 | 1294 | 26.6 | 1364 | 30.5 |
| 21447 | 2800 | 871 | 6.15 | 911 | 7.75 | 948 | 9.33 | 984 | 10.9 | 1017 | 12.4 | 1051 | 14.1 | 1119 | 17.4 | 1184 | 20.9 | 1252 | 24.7 | 1319 | 28.6 | 1386 | 32.7 |
| 22979 | 3000 | 926 | 7.27 | 966 | 9.05 | 1000 | 10.7 | 1035 | 12.4 | 1066 | 14.0 | 1098 | 15.7 | 1162 | 19.3 | 1223 | 22.9 | 1286 | 26.8 | 1348 | 30.8 | 1410 | 35.0 |
| 24511 | 3200 | 984 | 8.61 | 1020 | 10.4 | 1053 | 12.2 | 1086 | 14.0 | 1117 | 15.8 | 1147 | 17.6 | 1206 | 21.3 | 1266 | 25.2 | 1324 | 29.2 | 1381 | 33.2 | 1440 | 37.5 |
| 26043 | 3400 | 1040 | 10.0 | 1075 | 12.0 | 1108 | 14.0 | 1138 | 15.8 | 1168 | 17.7 | 1197 | 19.7 | 1253 | 23.6 | 1309 | 27.6 | 1363 | 31.6 | 1418 | 35.9 | | |

BHP shown does not include belt drive losses.  
Performance shown is for General Purpose Fans with outlet ducts and with or without inlet ducts.

*Fig. 9-11b Manufacturer's Fan Performance Table*

struction features and physical configurations and differ in their performance characteristics (air pattern).

### 9.3.2 Outlet Selection and Location Procedure

The following procedure is generally used in selecting outlet locations and types of outlets:

1. Determine the amount of air to be supplied to each room. Refer to Chapter 3 to determine air quantities for heating and cooling. Determine the amount of outdoor air to be introduced according to ventilation requirements of appropriate codes.
2. Select the type and quantity of outlets for each room, considering such factors as air quantity required, distance available for throw or radius of diffusion, structural characteristics, and architectural concepts. Table 9-7 is based on experience and typical ratings of various types of outlets. It may be used as a guide to the types of outlets applicable to various room air loadings. Special conditions such as ceiling heights greater than 8 to 12 ft (2.4 to 3.6 m), exposed duct mounting, and so forth, as well as product modifications and unusual conditions of room occupancy can all modify the information in this table. Manufacturers' rating

*Table 9-7  Guide to Use of Various Outlets*

| Type of Outlet | Air Loading of Floor Space, cfm/ft² [L/(s·m²)] | Approximate Maximum Air Changes per hour for 10 ft (3 m) Ceiling |
|---|---|---|
| Grille | 0.6 to 1.2 (3 to 6) | 7 |
| Slot | 0.8 to 2.0 (4 to 10) | 12 |
| Perforated panel | 0.9 to 3.0 (5 to 15) | 18 |
| Ceiling diffuser | 0.9 to 5.0 (5 to 25) | 30 |
| Perforated ceiling | 1.0 to 10 (5 to 50) | 60 |

data should be consulted to determine the suitability of the specific outlets.

3. Locate outlets to distribute the air as uniformly as possible throughout the room. Outlets may be sized and located to distribute air in various portions of the room in proportion to the heat gain or loss in those areas.
4. Select proper outlet size from manufacturers' ratings according to the air quantities, discharge velocities, distribution patterns, and sound levels. Note manufacturers' recommendations with regard to use, method of installation, minimum velocities, maximum temperature differentials, and any air distribution characteristics that may limit the performance of the outlet. Give special attention to obstructions to the normal air-distribution pattern.

### 9.3.3 Noise Control

Sound at an outlet is composed of the sound generation of the outlet (a function of the discharge velocity) and the transmission of systemic noise (a function of the size of the outlet). Higher-frequency sounds are caused by excessive outlet velocity, but they may also be the result of sounds generated in the duct by the moving airstream. Lower-pitched sounds are generally caused by mechanical equipment noise transmitted through the duct system and outlet. The cause of high-frequency sounds can be pinpointed as outlet or system sounds by removing the outlet during operation. A reduction in sound level indicates that the outlet is causing noise. If the sound level remains essentially unchanged, then the system is at fault.

**Example 9.3.** A 12 in. by 18 in. high sidewall grille with an 11.25 in. by 17.25 in. core area (80% free area) has been selected. Calculate the throw to 50 fpm, 100 fpm, and 150 fpm if the airflow is 600 cfm.

**Note:** Chapter 20 of the 2013 *ASHRAE Handbook—Fundamentals* is required.

**Solution:**

From Table 1, Chapter 20, 2013 *ASHRAE Handbook—Fundamentals*, the centerline velocity constant $K = 5.0$.

Use Equation 5, Chapter 20 of the 2013 *ASHRAE Handbook—Fundamentals*, to calculate the maximum throw for an outlet.

$$X = \frac{KQ}{V_x \sqrt{A_o}}$$

$$X = \frac{5 \times 600}{V_x \sqrt{11.25 \times 17.25 / 144}} = \frac{2920}{V_x}$$

Solving for 50 fpm throw, $X = 2920/50 = 58$ ft.

But according to Figure 3, Chapter 20, 2013 *ASHRAE Handbook—Fundamentals*, 50 fpm is in Zone 4, which is typically 20% less than calculated, or

$$X = 58 \times 0.80 = 46 \text{ ft}$$

Solving for 100 fpm throw, $X = 2920/100 = 29$ ft
Solving for 150 fpm throw, $X = 2920/150 = 19$ ft

## 9.4  Pipe, Tube, and Fittings

### 9.4.1 Selection and Application

Listed in Tables 9-8 and 9-9 are the common sizes and dimensions for pipes and tubing used in the HVAC&R industry. Regulatory codes and voluntary standards of such organizations as the American Society of Mechanical Engineers (ASME) and American Society for Testing and Materials (ASTM) should be considered when selecting and applying these components.

### 9.4.2 Materials

Table 9-10 is a guide to materials used in heating and air conditioning. While steel, iron, and copper materials are most commonly used, iron and steel alloys, copper alloys, nickel and nickel alloys, and nonmetallic pipe are finding increasing applications.

# Duct and Pipe Sizing

### Table 9-8  Steel Pipe Data
*(Table 2, Chapter 46, 2012 ASHRAE Handbook—HVAC Systems and Equipment)*

| Nominal Size, in. | Pipe OD, in. | Schedule Number or Weight[a] | Wall Thickness $t$, in. | Inside Diameter $d$, in. | Surface Area Outside, ft²/ft | Surface Area Inside, ft²/ft | Cross Section Metal Area, in² | Cross Section Flow Area, in² | Weight Pipe, lb/ft | Weight Water, lb/ft | Mfr. Process | Joint Type[b] | Working Pressure[c] ASTM A53 B to 400°F, psig |
|---|---|---|---|---|---|---|---|---|---|---|---|---|---|
| 1/4 | 0.540 | 40 ST | 0.088 | 0.364 | 0.141 | 0.095 | 0.125 | 0.104 | 0.424 | 0.045 | CW | T | 188 |
|  |  | 80 XS | 0.119 | 0.302 | 0.141 | 0.079 | 0.157 | 0.072 | 0.535 | 0.031 | CW | T | 871 |
| 3/8 | 0.675 | 40 ST | 0.091 | 0.493 | 0.177 | 0.129 | 0.167 | 0.191 | 0.567 | 0.083 | CW | T | 203 |
|  |  | 80 XS | 0.126 | 0.423 | 0.177 | 0.111 | 0.217 | 0.141 | 0.738 | 0.061 | CW | T | 820 |
| 1/2 | 0.840 | 40 ST | 0.109 | 0.622 | 0.220 | 0.163 | 0.250 | 0.304 | 0.850 | 0.131 | CW | T | 214 |
|  |  | 80 XS | 0.147 | 0.546 | 0.220 | 0.143 | 0.320 | 0.234 | 1.087 | 0.101 | CW | T | 753 |
| 3/4 | 1.050 | 40 ST | 0.113 | 0.824 | 0.275 | 0.216 | 0.333 | 0.533 | 1.13 | 0.231 | CW | T | 217 |
|  |  | 80 XS | 0.154 | 0.742 | 0.275 | 0.194 | 0.433 | 0.432 | 1.47 | 0.187 | CW | T | 681 |
| 1 | 1.315 | 40 ST | 0.133 | 1.049 | 0.344 | 0.275 | 0.494 | 0.864 | 1.68 | 0.374 | CW | T | 226 |
|  |  | 80 XS | 0.179 | 0.957 | 0.344 | 0.251 | 0.639 | 0.719 | 2.17 | 0.311 | CW | T | 642 |
| 1-1/4 | 1.660 | 40 ST | 0.140 | 1.380 | 0.435 | 0.361 | 0.669 | 1.50 | 2.27 | 0.647 | CW | T | 229 |
|  |  | 80 XS | 0.191 | 1.278 | 0.435 | 0.335 | 0.881 | 1.28 | 2.99 | 0.555 | CW | T | 594 |
| 1-1/2 | 1.900 | 40 ST | 0.145 | 1.610 | 0.497 | 0.421 | 0.799 | 2.04 | 2.72 | 0.881 | CW | T | 231 |
|  |  | 80 XS | 0.200 | 1.500 | 0.497 | 0.393 | 1.068 | 1.77 | 3.63 | 0.765 | CW | T | 576 |
| 2 | 2.375 | 40 ST | 0.154 | 2.067 | 0.622 | 0.541 | 1.07 | 3.36 | 3.65 | 1.45 | CW | T | 230 |
|  |  | 80 XS | 0.218 | 1.939 | 0.622 | 0.508 | 1.48 | 2.95 | 5.02 | 1.28 | CW | T | 551 |
| 2-1/2 | 2.875 | 40 ST | 0.203 | 2.469 | 0.753 | 0.646 | 1.70 | 4.79 | 5.79 | 2.07 | CW | W | 533 |
|  |  | 80 XS | 0.276 | 2.323 | 0.753 | 0.608 | 2.25 | 4.24 | 7.66 | 1.83 | CW | W | 835 |
| 3 | 3.500 | 40 ST | 0.216 | 3.068 | 0.916 | 0.803 | 2.23 | 7.39 | 7.57 | 3.20 | CW | W | 482 |
|  |  | 80 XS | 0.300 | 2.900 | 0.916 | 0.759 | 3.02 | 6.60 | 10.25 | 2.86 | CW | W | 767 |
| 4 | 4.500 | 40 ST | 0.237 | 4.026 | 1.178 | 1.054 | 3.17 | 12.73 | 10.78 | 5.51 | CW | W | 430 |
|  |  | 80 XS | 0.337 | 3.826 | 1.178 | 1.002 | 4.41 | 11.50 | 14.97 | 4.98 | CW | W | 695 |
| 6 | 6.625 | 40 ST | 0.280 | 6.065 | 1.734 | 1.588 | 5.58 | 28.89 | 18.96 | 12.50 | ERW | W | 696 |
|  |  | 80 XS | 0.432 | 5.761 | 1.734 | 1.508 | 8.40 | 26.07 | 28.55 | 11.28 | ERW | W | 1209 |
| 8 | 8.625 | 30 | 0.277 | 8.071 | 2.258 | 2.113 | 7.26 | 51.16 | 24.68 | 22.14 | ERW | W | 526 |
|  |  | 40 ST | 0.322 | 7.981 | 2.258 | 2.089 | 8.40 | 50.03 | 28.53 | 21.65 | ERW | W | 643 |
|  |  | 80 XS | 0.500 | 7.625 | 2.258 | 1.996 | 12.76 | 45.66 | 43.35 | 19.76 | ERW | W | 1106 |
| 10 | 10.75 | 30 | 0.307 | 10.136 | 2.814 | 2.654 | 10.07 | 80.69 | 34.21 | 34.92 | ERW | W | 485 |
|  |  | 40 ST | 0.365 | 10.020 | 2.814 | 2.623 | 11.91 | 78.85 | 40.45 | 34.12 | ERW | W | 606 |
|  |  | XS | 0.500 | 9.750 | 2.814 | 2.552 | 16.10 | 74.66 | 54.69 | 32.31 | ERW | W | 887 |
|  |  | 80 | 0.593 | 9.564 | 2.814 | 2.504 | 18.92 | 71.84 | 64.28 | 31.09 | ERW | W | 1081 |
| 12 | 12.75 | 30 | 0.330 | 12.090 | 3.338 | 3.165 | 12.88 | 114.8 | 43.74 | 49.68 | ERW | W | 449 |
|  |  | ST | 0.375 | 12.000 | 3.338 | 3.141 | 14.58 | 113.1 | 49.52 | 48.94 | ERW | W | 528 |
|  |  | 40 | 0.406 | 11.938 | 3.338 | 3.125 | 15.74 | 111.9 | 53.48 | 48.44 | ERW | W | 583 |
|  |  | XS | 0.500 | 11.750 | 3.338 | 3.076 | 19.24 | 108.4 | 65.37 | 46.92 | ERW | W | 748 |
|  |  | 80 | 0.687 | 11.376 | 3.338 | 2.978 | 26.03 | 101.6 | 88.44 | 43.98 | ERW | W | 1076 |
| 14 | 14.00 | 30 ST | 0.375 | 13.250 | 3.665 | 3.469 | 16.05 | 137.9 | 54.53 | 59.67 | ERW | W | 481 |
|  |  | 40 | 0.437 | 13.126 | 3.665 | 3.436 | 18.62 | 135.3 | 63.25 | 58.56 | ERW | W | 580 |
|  |  | XS | 0.500 | 13.000 | 3.665 | 3.403 | 21.21 | 132.7 | 72.04 | 57.44 | ERW | W | 681 |
|  |  | 80 | 0.750 | 12.500 | 3.665 | 3.272 | 31.22 | 122.7 | 106.05 | 53.11 | ERW | W | 1081 |
| 16 | 16.00 | 30 ST | 0.375 | 15.250 | 4.189 | 3.992 | 18.41 | 182.6 | 62.53 | 79.04 | ERW | W | 421 |
|  |  | 40 XS | 0.500 | 15.000 | 4.189 | 3.927 | 24.35 | 176.7 | 82.71 | 76.47 | ERW | W | 596 |
| 18 | 18.00 | ST | 0.375 | 17.250 | 4.712 | 4.516 | 20.76 | 233.7 | 70.54 | 101.13 | ERW | W | 374 |
|  |  | 30 | 0.437 | 17.126 | 4.712 | 4.483 | 24.11 | 230.3 | 81.91 | 99.68 | ERW | W | 451 |
|  |  | XS | 0.500 | 17.000 | 4.712 | 4.450 | 27.49 | 227.0 | 93.38 | 98.22 | ERW | W | 530 |
|  |  | 40 | 0.562 | 16.876 | 4.712 | 4.418 | 30.79 | 223.7 | 104.59 | 96.80 | ERW | W | 607 |
| 20 | 20.00 | 20 ST | 0.375 | 19.250 | 5.236 | 5.039 | 23.12 | 291.0 | 78.54 | 125.94 | ERW | W | 337 |
|  |  | 30 XS | 0.500 | 19.000 | 5.236 | 4.974 | 30.63 | 283.5 | 104.05 | 122.69 | ERW | W | 477 |
|  |  | 40 | 0.593 | 18.814 | 5.236 | 4.925 | 36.15 | 278.0 | 122.82 | 120.30 | ERW | W | 581 |

[a] Numbers are schedule numbers per ASME *Standard* B36.10M; ST = Standard Weight; XS = Extra Strong.

[b] T = Thread; W = Weld

[c] Working pressures were calculated per ASME B31.9 using furnace butt-weld (continuous weld, CW) pipe through 4 in. and electric resistance weld (ERW) thereafter. The allowance A has been taken as

(1) 12.5% of *t* for mill tolerance on pipe wall thickness, *plus*
(2) An arbitrary corrosion allowance of 0.025 in. for pipe sizes through NPS 2 and 0.065 in. from NPS 2½ through 20, *plus*
(3) A thread cutting allowance for sizes through NPS 2.

Because the pipe wall thickness of threaded standard pipe is so small after deducting the allowance A, the mechanical strength of the pipe is impaired. It is good practice to limit standard weight threaded pipe pressure to 90 psig for steam and 125 psig for water.

## Table 9-9 Copper Tube Data
*(Table 3, Chapter 46, 2012 ASHRAE Handbook—HVAC Systems and Equipment)*

| Nominal Diameter, in. | Type | Wall Thickness $t$, in. | Outside $D$, in. | Inside $d$, in. | Outside, ft²/ft | Inside, ft²/ft | Metal Area, in² | Flow Area, in² | Tube, lb/ft | Water, lb/ft | Annealed, psig | Drawn, psig |
|---|---|---|---|---|---|---|---|---|---|---|---|---|
| 1/4 | K | 0.035 | 0.375 | 0.305 | 0.098 | 0.080 | 0.037 | 0.073 | 0.145 | 0.032 | 851 | 1596 |
|  | L | 0.030 | 0.375 | 0.315 | 0.098 | 0.082 | 0.033 | 0.078 | 0.126 | 0.034 | 730 | 1368 |
| 3/8 | K | 0.049 | 0.500 | 0.402 | 0.131 | 0.105 | 0.069 | 0.127 | 0.269 | 0.055 | 894 | 1676 |
|  | L | 0.035 | 0.500 | 0.430 | 0.131 | 0.113 | 0.051 | 0.145 | 0.198 | 0.063 | 638 | 1197 |
|  | M | 0.025 | 0.500 | 0.450 | 0.131 | 0.118 | 0.037 | 0.159 | 0.145 | 0.069 | 456 | 855 |
| 1/2 | K | 0.049 | 0.625 | 0.527 | 0.164 | 0.138 | 0.089 | 0.218 | 0.344 | 0.094 | 715 | 1341 |
|  | L | 0.040 | 0.625 | 0.545 | 0.164 | 0.143 | 0.074 | 0.233 | 0.285 | 0.101 | 584 | 1094 |
|  | M | 0.028 | 0.625 | 0.569 | 0.164 | 0.149 | 0.053 | 0.254 | 0.203 | 0.110 | 409 | 766 |
| 5/8 | K | 0.049 | 0.750 | 0.652 | 0.196 | 0.171 | 0.108 | 0.334 | 0.418 | 0.144 | 596 | 1117 |
|  | L | 0.042 | 0.750 | 0.666 | 0.196 | 0.174 | 0.093 | 0.348 | 0.362 | 0.151 | 511 | 958 |
| 3/4 | K | 0.065 | 0.875 | 0.745 | 0.229 | 0.195 | 0.165 | 0.436 | 0.641 | 0.189 | 677 | 1270 |
|  | L | 0.045 | 0.875 | 0.785 | 0.229 | 0.206 | 0.117 | 0.484 | 0.455 | 0.209 | 469 | 879 |
|  | M | 0.032 | 0.875 | 0.811 | 0.229 | 0.212 | 0.085 | 0.517 | 0.328 | 0.224 | 334 | 625 |
| 1 | K | 0.065 | 1.125 | 0.995 | 0.295 | 0.260 | 0.216 | 0.778 | 0.839 | 0.336 | 527 | 988 |
|  | L | 0.050 | 1.125 | 1.025 | 0.295 | 0.268 | 0.169 | 0.825 | 0.654 | 0.357 | 405 | 760 |
|  | M | 0.035 | 1.125 | 1.055 | 0.295 | 0.276 | 0.120 | 0.874 | 0.464 | 0.378 | 284 | 532 |
| 1-1/4 | K | 0.065 | 1.375 | 1.245 | 0.360 | 0.326 | 0.268 | 1.217 | 1.037 | 0.527 | 431 | 808 |
|  | L | 0.055 | 1.375 | 1.265 | 0.360 | 0.331 | 0.228 | 1.257 | 0.884 | 0.544 | 365 | 684 |
|  | M | 0.042 | 1.375 | 1.291 | 0.360 | 0.338 | 0.176 | 1.309 | 0.682 | 0.566 | 279 | 522 |
|  | DWV | 0.040 | 1.375 | 1.295 | 0.360 | 0.339 | 0.168 | 1.317 | 0.650 | 0.570 | 265 | 497 |
| 1-1/2 | K | 0.072 | 1.625 | 1.481 | 0.425 | 0.388 | 0.351 | 1.723 | 1.361 | 0.745 | 404 | 758 |
|  | L | 0.060 | 1.625 | 1.505 | 0.425 | 0.394 | 0.295 | 1.779 | 1.143 | 0.770 | 337 | 631 |
|  | M | 0.049 | 1.625 | 1.527 | 0.425 | 0.400 | 0.243 | 1.831 | 0.940 | 0.792 | 275 | 516 |
|  | DWV | 0.042 | 1.625 | 1.541 | 0.425 | 0.403 | 0.209 | 1.865 | 0.809 | 0.807 | 236 | 442 |
| 2 | K | 0.083 | 2.125 | 1.959 | 0.556 | 0.513 | 0.532 | 3.014 | 2.063 | 1.304 | 356 | 668 |
|  | L | 0.070 | 2.125 | 1.985 | 0.556 | 0.520 | 0.452 | 3.095 | 1.751 | 1.339 | 300 | 573 |
|  | M | 0.058 | 2.125 | 2.009 | 0.556 | 0.526 | 0.377 | 3.170 | 1.459 | 1.372 | 249 | 467 |
|  | DWV | 0.042 | 2.125 | 2.041 | 0.556 | 0.534 | 0.275 | 3.272 | 1.065 | 1.416 | 180 | 338 |
| 2-1/2 | K | 0.095 | 2.625 | 2.435 | 0.687 | 0.637 | 0.755 | 4.657 | 2.926 | 2.015 | 330 | 619 |
|  | L | 0.080 | 2.625 | 2.465 | 0.687 | 0.645 | 0.640 | 4.772 | 2.479 | 2.065 | 278 | 521 |
|  | M | 0.065 | 2.625 | 2.495 | 0.687 | 0.653 | 0.523 | 4.889 | 2.026 | 2.116 | 226 | 423 |
| 3 | K | 0.109 | 3.125 | 2.907 | 0.818 | 0.761 | 1.033 | 6.637 | 4.002 | 2.872 | 318 | 596 |
|  | L | 0.090 | 3.125 | 2.945 | 0.818 | 0.771 | 0.858 | 6.812 | 3.325 | 2.947 | 263 | 492 |
|  | M | 0.072 | 3.125 | 2.981 | 0.818 | 0.780 | 0.691 | 6.979 | 2.676 | 3.020 | 210 | 394 |
|  | DWV | 0.045 | 3.125 | 3.035 | 0.818 | 0.795 | 0.435 | 7.234 | 1.687 | 3.130 | 131 | 246 |
| 3-1/2 | K | 0.120 | 3.625 | 3.385 | 0.949 | 0.886 | 1.321 | 8.999 | 5.120 | 3.894 | 302 | 566 |
|  | L | 0.100 | 3.625 | 3.425 | 0.949 | 0.897 | 1.107 | 9.213 | 4.291 | 3.987 | 252 | 472 |
|  | M | 0.083 | 3.625 | 3.459 | 0.949 | 0.906 | 0.924 | 9.397 | 3.579 | 4.066 | 209 | 392 |
| 4 | K | 0.134 | 4.125 | 3.857 | 1.080 | 1.010 | 1.680 | 11.684 | 6.510 | 5.056 | 296 | 555 |
|  | L | 0.110 | 4.125 | 3.905 | 1.080 | 1.022 | 1.387 | 11.977 | 5.377 | 5.182 | 243 | 456 |
|  | M | 0.095 | 4.125 | 3.935 | 1.080 | 1.030 | 1.203 | 12.161 | 4.661 | 5.262 | 210 | 394 |
|  | DWV | 0.058 | 4.125 | 4.009 | 1.080 | 1.050 | 0.741 | 12.623 | 2.872 | 5.462 | 128 | 240 |
| 5 | K | 0.160 | 5.125 | 4.805 | 1.342 | 1.258 | 2.496 | 18.133 | 9.671 | 7.846 | 285 | 534 |
|  | L | 0.125 | 5.125 | 4.875 | 1.342 | 1.276 | 1.963 | 18.665 | 7.609 | 8.077 | 222 | 417 |
|  | M | 0.109 | 5.125 | 4.907 | 1.342 | 1.285 | 1.718 | 18.911 | 6.656 | 8.183 | 194 | 364 |
|  | DWV | 0.072 | 5.125 | 4.981 | 1.342 | 1.304 | 1.143 | 19.486 | 4.429 | 8.432 | 128 | 240 |
| 6 | K | 0.192 | 6.125 | 5.741 | 1.603 | 1.503 | 3.579 | 25.886 | 13.867 | 11.201 | 286 | 536 |
|  | L | 0.140 | 6.125 | 5.845 | 1.603 | 1.530 | 2.632 | 26.832 | 10.200 | 11.610 | 208 | 391 |
|  | M | 0.122 | 6.125 | 5.881 | 1.603 | 1.540 | 2.301 | 27.164 | 8.916 | 11.754 | 182 | 341 |
|  | DWV | 0.083 | 6.125 | 5.959 | 1.603 | 1.560 | 1.575 | 27.889 | 6.105 | 12.068 | 124 | 232 |
| 8 | K | 0.271 | 8.125 | 7.583 | 2.127 | 1.985 | 6.687 | 45.162 | 25.911 | 19.542 | 304 | 570 |
|  | L | 0.200 | 8.125 | 7.725 | 2.127 | 2.022 | 4.979 | 46.869 | 19.295 | 20.280 | 224 | 421 |
|  | M | 0.170 | 8.125 | 7.785 | 2.127 | 2.038 | 4.249 | 47.600 | 16.463 | 20.597 | 191 | 358 |
|  | DWV | 0.109 | 8.125 | 7.907 | 2.127 | 2.070 | 2.745 | 49.104 | 10.637 | 21.247 | 122 | 229 |
| 10 | K | 0.338 | 10.125 | 9.449 | 2.651 | 2.474 | 10.392 | 70.123 | 40.271 | 30.342 | 304 | 571 |
|  | L | 0.250 | 10.125 | 9.625 | 2.651 | 2.520 | 7.756 | 72.760 | 30.054 | 31.483 | 225 | 422 |
|  | M | 0.212 | 10.125 | 9.701 | 2.651 | 2.540 | 6.602 | 73.913 | 25.584 | 31.982 | 191 | 358 |
| 12 | K | 0.405 | 12.125 | 11.315 | 3.174 | 2.962 | 14.912 | 100.554 | 57.784 | 43.510 | 305 | 571 |
|  | L | 0.280 | 12.125 | 11.565 | 3.174 | 3.028 | 10.419 | 105.046 | 40.375 | 45.454 | 211 | 395 |
|  | M | 0.254 | 12.125 | 11.617 | 3.174 | 3.041 | 9.473 | 105.993 | 36.706 | 45.863 | 191 | 358 |

Working Pressure[a,b,c] ASTM B88 to 250°F

[a] When using soldered or brazed fittings, the joint determines the limiting pressure.
[b] Working pressures were calculated using ASME *Standard* B31.9 allowable stresses. A 5% mill tolerance has been used on the wall thickness. Higher tube ratings can be calculated using the allowable stress for lower temperatures.
[c] If soldered or brazed fittings are used on hard drawn tubing, use the annealed ratings. Full-tube allowable pressures can be used with suitably rated flare or compression-type fittings.

# Duct and Pipe Sizing

**Table 9-10  Application of Pipe, Fitting and Valves for Heating and Air-Conditioning**
*(Table 5, Chapter 46, 2012 ASHRAE Handbook—HVAC Systems and Equipment)*

| Application | Pipe Material | Weight | Joint Type | Fitting Class | Fitting Material | System Temperature, °F | Maximum Pressure at Temperature[a], psig |
|---|---|---|---|---|---|---|---|
| **Recirculating Water** 2 in. and smaller | Steel (CW) | Standard | Thread | 125 | Cast iron | 250 | 125 |
| | Copper, hard | Type L | Braze or silver solder[b] | | Wrought copper | 250 | 150 |
| | PVC | Sch 80 | Solvent | Sch 80 | PVC | 75 | |
| | CPVC | Sch 80 | Solvent | Sch 80 | CPVC | 150 | |
| | PB | SDR-11 | Heat fusion | | PB | 160 | |
| | | | Insert crimp | | Metal | 160 | |
| 2.5 to 12 in. | A53 B ERW Steel | Standard | Weld | Standard | Wrought steel | 250 | 400 |
| | | | Flange | 150 | Wrought steel | 250 | 250 |
| | | | Flange | 125 | Cast iron | 250 | 175 |
| | | | Flange | 250 | Cast iron | 250 | 400 |
| | | | Groove | | MI or ductile iron | 230 | 300 |
| | PB | SDR-11 | Heat fusion | | PB | 160 | |
| **Steam and Condensate** 2 in. and smaller | Steel (CW) | Standard[c] | Thread | 125 | Cast iron | | 90 |
| | | | Thread | 150 | Malleable iron | | 90 |
| | A53 B ERW Steel | Standard[c] | Thread | 125 | Cast iron | | 100 |
| | | | Thread | 150 | Malleable iron | | 125 |
| | A53 B ERW Steel | XS | Thread | 250 | Cast iron | | 200 |
| | | | Thread | 300 | Malleable iron | | 250 |
| 2.5 to 12 in. | Steel | Standard | Weld | Standard | Wrought steel | | 250 |
| | | | Flange | 150 | Wrought steel | | 200 |
| | | | Flange | 125 | Cast iron | | 100 |
| | A53 B ERW Steel | XS | Weld | XS | Wrought steel | | 700 |
| | | | Flange | 300 | Wrought steel | | 500 |
| | | | Flange | 250 | Cast iron | | 200 |
| **Refrigerant** | Copper, hard | Type L or K | Braze | | Wrought copper | | |
| | A53 B SML Steel | Standard | Weld | | Wrought steel | | |
| **Underground Water** Through 12 in. | Copper, hard | Type K | Braze or silver solder[b] | | Wrought copper | 75 | 350 |
| Through 6 in. | Ductile iron | Class 50 | MJ | MJ | Cast iron | 75 | 250 |
| | PB | SDR 9, 11 | Heat fusion | | PB | 75 | |
| | | SDR 7, 11.5 | Insert crimp | | Metal | 75 | |
| **Potable Water, Inside Building** | Copper, hard | Type L | Braze or silver solder[b] | | Wrought copper | 75 | 350 |
| | Steel, galvanized | Standard | Thread | 125 | Galv. cast iron | 75 | 125 |
| | | | | 150 | Galv. mall. iron | 75 | 125 |
| | PB | SDR-11 | Heat fusion | | PB | 75 | |
| | | | Insert crimp | | Metal | 75 | |

[a] Maximum allowable working pressures have been derated in this table. Higher system pressures can be used for lower temperatures and smaller pipe sizes. Pipe, fittings, joints, and valves must all be considered.

[b] Lead- and antimony-based solders should not be used for potable water systems. Brazing and silver solders should be employed.

[c] Extra strong pipe is recommended for all threaded condensate piping to allow for corrosion.

Codes, dimensional standards, and material specifications cover service requirements, which consider such effects as corrosion, scale, thermal or mechanical fatigue, and metallurgical instability at high temperatures. Copper and red brass pipe have always been used in heating, ventilating, refrigeration, and water supply installations because of their corrosion resistance. However, copper and brass are not compatible with ammonia and should not be used in ammonia refrigeration systems.

Plastic pipe has become popular in heating and air-conditioning installations because of its flexibility in handling, lower labor costs, and resistance to corrosive fluids. Pressure and temperature are the basic limitations to be considered. Basically, two kinds of plastics are used:

1. **Thermoplastic**, which melts under heat, hardens when cooled, and can be melted and reworked again and again.
2. **Thermoset**, which hardens and fuses under heat and pressure into a permanent shape and cannot be remelted.

Some thermoplastics remain flexible and have moderate strength for water service where temperature and pressures are moderate. Thermoplastics may be joined by solvent welding and hot-gas welding. Both pipe and fittings are available in IPS sizes.

Epoxy or polyester resins, usually reinforced with glass fibers, are the major ingredient of thermosetting piping. They are available in commercial pipe sizes, joined by either standard screwed or socket-type fittings. Commercial polyester pipe is suitable for temperatures up to 250°F (120°C).

### 9.4.3 Pipe Fittings

Metal pipe is joined by welding one length to another or by using fittings. Pipe fittings are made in such forms as elbows, tees, couplings, reducers, and unions; they may be screwed, flanged, or welded. The type of fitting used is determined by the pressure of the fluid being carried or by the intended use of the pipeline. The materials generally used are steel, cast iron, malleable iron (heat treated cast iron), copper, brass, stainless steel, alloy steel, or bronze. The material used also depends upon the pressure and fluid characteristics.

Fittings are designated and sized in accordance with ANSI specifications and are identified by their nominal pipe sizes. In the case of reducing tees, crosses, and Y-branches (laterals), the size of the largest run opening is given first, followed by the size of the opening at the opposite end of the run. Where the fitting is a tee or Y-branch (lateral), the size of the outlet is given last. Where the fitting is a cross, the largest side outlet opening is the third dimension given followed by the opening opposite. Where an external thread is wanted, the word male follows the size of that opening; female indicates an internal thread.

### 9.4.4 Pipe Sizing

Specific procedures for sizing various fluid flow systems using piping or tubing are given in Chapter 22 of the 2013 *ASHRAE Handbook—Fundamentals* and are required for solution.

**Example 9.4.** Select the design pipe sizes for the two-pipe, forced circulation hot water system shown in the figure below. A 20°F drop in the water temperature has been selected and the water leaves the boiler at 190°F. The numbers in parentheses are the measured linear runs of pipe. The dashed line is the return line pipe. The piping in the system is to be Type L copper tube.

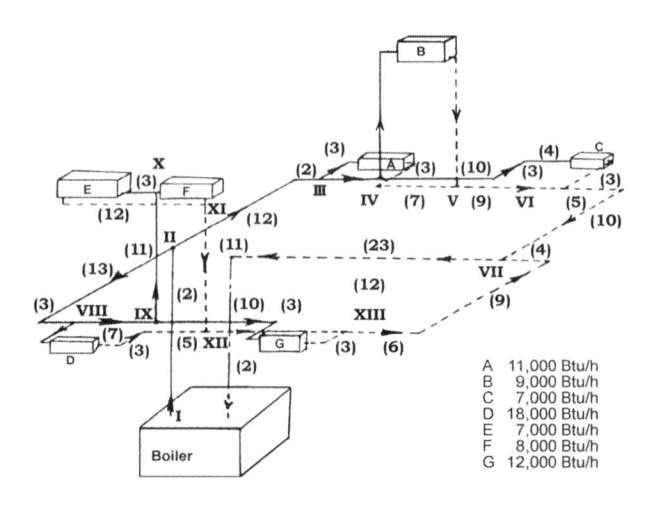

The pressure loss through each convector is given as:

A 1600 milli-inches  E 1425 milli-inches
B 1500 milli-inches  F 1450 milli-inches
C 1450 milli-inches  G 1550 milli-inches
D 1600 milli-inches

**Solution:**

First, calculate the gallons per minute circulated through each circuit of the system. For the ABC circuit, the total heat delivered is

$$11,000 + 9000 + 7000 = 27,000 \text{ Btu/h}$$

The relation between heat flow and flow rate is

$$q = 60 W c_p (t_{in} - t_{out}), \text{ or}$$

$$q = 490 Q (t_{in} - t_{out})$$

where

$q$ = output of the convectors, Btu/h
$Q$ = flow rate, gpm
$(t_{in} - t_{out})$ = temperature drop of the water across convector

For this circuit

$$Q = 27,000/(490 \times 20) = 2.76 \text{ gpm}$$

For circuit DEFG

$$q = 45,000 \text{ Btu/h}$$

and

$$Q = 45,000/(490 \times 20) = 4.59 \text{ gpm}$$

Next, select a design friction loss for the system. Assume a design value of 2.5 ft/100 ft (0.3 in./ft). Now the pipe can be sized for each branch in the system, using Figure 5, Chapter 22 of the 2013 *ASHRAE Handbook—Fundamentals*, on the supply side:

for I-II      at 7.34 gpm (72,000 Btu/h), use 1-1/4 in.
II-III        at 2.75 gpm (27,000 Btu/h), use 3/4 in.
III-IV        at 1.63 gpm (16,000 Btu/h), use 5/8 in.
Convector A   at 1.12 gpm (11,000 Btu/h), use 5/8 in.
Convector B   at 0.92 gpm (0,000 Btu/h), use 1/2 in.
Convector C   at 0.71 gpm (7,000 Btu/h), use 1/2 in.
II-VIII       at 4.59 gpm (45,000 Btu/h), use 1 in.
VIII-IX       at 2.75 gpm (27,000 Btu/h), use 3/4 in.
IX-X          at 1.53 gpm (15,000 Btu/h), use 5/8 in.
Convector D   at 1.84 gpm (18,000 Btu/h), use 3/4 in.
Convector E   at 0.71 gpm (7,000 Btu/h), use 1/2 in.
Convector F   at 0.82 gpm (8,000 Btu/h), use 1/2 in.
Convector G   at 1.22 gpm (12,000 Btu/h), use 5/8 in.

and on the return side

for V-VI      at 2.04 gpm (20,000 Btu/h), use 3/4 in.
VI-VII        at 2.75 gpm (27,000 Btu/h), use 3/4 in.
XI-XII        at 1.53 gpm (15,000 Btu/h), use 5/8 in.
XII-XIII      at 3.34 gpm (33,000 Btu/h), use 1 in.
XIII-VII      at 4.59 gpm (45,000 Btu/h), use 1 in.
VII-I         at 7.34 gpm (72,000 Btu/h), use 1-1/4 in.

Next, find the longest run in the system. Apparently, this is the circuit from I-II-IX-X-Convector E-XI-XII-I. The other possible longest run could be one that includes convector F or one that includes convector B. Both should be checked. Assume here that the convector E run is the longest. The pressure or head loss is found from Figure 5, Chapter 22 of the 2013 *ASHRAE Handbook—Fundamentals*. For example, for 7.34 gpm in a 1-1/4 in. pipe, the friction loss is 1.46 ft/100 ft.

For section I-II the loss is 2 × 1.46 = 2.92 ft/100 ft. The friction losses in the other straight sections are found in a similar manner:

| | |
|---|---|
| I-II | 2.92 ft/100 ft |
| II-VIII | 24.66 ft/100 ft |
| VII-IX | 17.50 ft/100 ft |
| IX-X | 22.92 ft/100 ft |
| X-Convector | 4.37 ft/100 ft |
| Convector-XI | 17.50 ft/100 ft |
| XI-XII | 23.83 ft/100 ft |
| XII-XIII | 12.0 ft/100 ft |
| XIII-VII | 15.0 ft/100 ft |
| VII-I | 33.5 ft/100 ft |
| Total = | 174.2 ft/100 ft |

The friction losses for the fittings are as follows:

*Elbow* (using Figure 5 and Table 10, Chapter 22 of the 2013 *ASHRAE Handbook—Fundamentals*)

| | |
|---|---|
| 2.5 × 1.66 | = 4.17 ft/100 ft |
| 2 × 1.2 × 1.46 | = 3.50 ft/100 ft |
| 2 × 2.5 × 1.66 | = 8.3 ft/100 ft |
| 1.0 × 3.3 × 1.46 | = 4.83 ft/100 ft |
| Total | = 20.8 ft/100 ft |

*Tees* (using Figure 5, Figure 7, and Table 10, Chapter 22 of the 2013 *ASHRAE Handbook—Fundamentals*) at

| | |
|---|---|
| II | 8.3 ft/100 ft |
| IX | 8.17 ft/100 ft |
| X | 3.50 ft/100 ft |
| XI | 8.83 ft/100 ft |
| XII | 4.83 ft/100 ft |
| XIII | 8.0 ft/100 ft |
| VII | 10.4 ft/100 ft |
| Total | = 57.5 ft/100 ft |

Convector loss = 1425 milli-inch (1.425 in. of water per foot of pipe or 11.87 ft/100 ft)

Total fitting losses

$$= 20.8 + 57.5 + 11.87 = 90.2 \text{ ft}/100 \text{ ft}$$

The total friction loss for the longest run is then

$$174.2 + 90.2 = 264.4 \text{ ft}/100 \text{ ft}$$

Select a boiler and pump combination that will supply 72,000 Btu/h and circulate 7.62 gpm with a developed head of 264.4 ft/100 ft (friction loss).

**Table 9-11  Affinity Laws for Pumps**

| Impeller Diameter | Speed | Density | To Correct For | Multiply By |
|---|---|---|---|---|
| Constant | Variable | Constant | Flow | $\left(\dfrac{\text{New Speed}}{\text{Old Speed}}\right)$ |
| | | | Head | $\left(\dfrac{\text{New Speed}}{\text{Old Speed}}\right)^2$ |
| | | | Power | $\left(\dfrac{\text{New Speed}}{\text{Old Speed}}\right)^3$ |
| Variable | Constant | Constant | Flow | $\left(\dfrac{\text{New Speed}}{\text{Old Speed}}\right)$ |
| | | | Head | $\left(\dfrac{\text{New Speed}}{\text{Old Speed}}\right)^2$ |
| | | | Power | $\left(\dfrac{\text{New Speed}}{\text{Old Speed}}\right)^3$ |
| Constant | Constant | Variable | Power | $\left(\dfrac{\text{New Density}}{\text{Old Density}}\right)$ |

*Notes*:
(a) Some friction loss will occur through the boiler, which must be added to the total friction loss of the system.
(b) Balancing cocks will need to be placed in the various circuits to balance the flow to each convector. Include the friction loss for these balancing cocks (valves) in the calculation.
(c) Air vent valves, as well as drain valves, should also be put into the system. Also include the friction loss due to these fittings.

## 9.5 Pumps

Pumps can be classified into three broad categories: (1) reciprocating, (2) rotary, and (3) centrifugal. Both reciprocating and rotary pumps are **positive displacement** pumps. They discharge a fixed amount of liquid for a given speed. The primary moving element of a reciprocating pump is a piston or plunger, while for a rotary pump, it is a rotor. Rotary pumps include gear pumps, vane pumps, lobe pumps, screw pumps, and cam action pumps.

Centrifugal pumps can be classified by the style of impeller—single or double suction, closed or open, radial, Francis, axial—as well as the casing—volute, diffuser, and concentric.

In pumps with radial or Francis impellers, centrifugal force develops the pressure energy and is a function of the impeller peripheral velocity. The liquid enters at the eye of the impeller and energy is added to the liquid by the impeller. The casing collects the liquid as it leaves the impeller and guides it out the discharge of the pump.

In the heating, ventilating, and air-conditioning industry, the most frequently used pump has a radial or Francis enclosed impeller and a volute casing and is single stage.

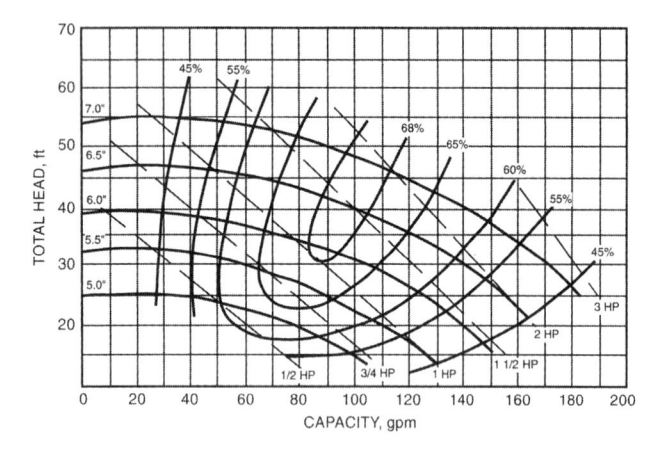

*Fig. 9-13 Typical Pump Performance Curves*
*(Figure 11, Chapter 44, 2012 ASHRAE Handbook—HVAC Systems and Equipment)*

### 9.5.1 Pump Laws

Approximate centrifugal pump performance comparisons of flow, head, and power, depending on impeller diameter, speed, and liquid density are given in Table 9-11.

### 9.5.2 Pump Performance Curves

Performance of a pump is commonly shown graphically, as in Figure 9-13, which relates the flow (gpm and L/s), the pressure produced (ft and Pa), the power required (bhp and kW), the hydraulic efficiency in percent, the shaft speed (rpm and r/s), and the net positive suction head (ft and Pa, absolute) required for normal operation of the pump with various impeller diameters. Pump curves present the average results obtained from testing several pumps of the same design under standardized test conditions. Manufacturers should be consulted for pump applications that differ considerably from ordinary practice.

The maximum efficiency of a pump is not always the most important feature in making a selection. If the pump is to operate at its peak efficiency, system resistance must also be considered. Systems with no corrosion and systems protected against corrosion should not be designed from pipe friction loss tables containing high corrosion allowances. Such tables show excessive pressure losses, which require selecting a larger pump than necessary. Pumps for systems using liquids for heating and cooling should be hydraulically and mechanically designed for quiet operation, durability, simple service, minimum maintenance, and minimum suction requirements rather than for minimum cost or size.

### 9.5.3 Hydronic System Characteristics

Hydronic systems in HVAC are all of the loop type (open or closed); the water is circulated through the system and returned to the pumps. No appreciable amount of water is lost from the system, except in the cooling tower, where evaporative cooling occurs.

Hydronic systems are either full flow or throttling flow. Full flow systems are usually found on residential or small commercial systems where pump motors are small and the energy waste caused by constant flow is not appreciable. Larger systems use flow control valves. These valves control flow in the system in accordance with the heating or cooling load imposed on the system. Hydronic systems can also operate continuously or intermittently. Most hot water or chilled water systems are in continuous operation as long as a heating or cooling load exists on the system. Condensate and boiler feed pumps are often of the intermittent type, starting and stopping as the water level changes in condensate tanks or boilers.

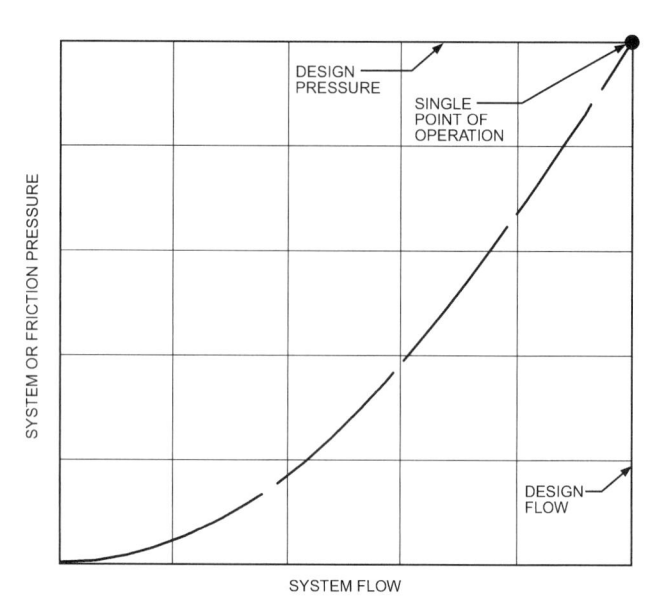

*Fig. 9-14 Typical System Curve*
*(Figure 15, Chapter 44, 2012 ASHRAE Handbook—HVAC Systems and Equipment)*

Pumps move water through hydronic systems, overcoming the friction caused by liquid flow through equipment, piping, fittings, and valves. The system head curve graphically describes the flow-head relationship for a hydronic system from minimum to maximum flow (Figure 9-14). The total head consists of the independent head and the system friction head. The independent head is unaffected by the total flow in the system. Typical **independent heads** are: (1) static rise to the top of a cooling tower in condenser water systems; (2) boiler pressure in condensate and boiler feed systems; and (3) control valve and heating or cooling coil friction in hot and chilled water systems.

The head loss of a control valve and its coil is independent of total flow in a system because these head losses can occur at any time, regardless of the total flow on the system; that is, a particular coil may require its maximum flow and, therefore, maximum head loss for the coil and its valve, even though the total flow in the entire system is at a minimum with minimum system friction head.

The other component, **system friction head**, depends on total system flow, increasing or decreasing with total system flow.

# Duct and Pipe Sizing

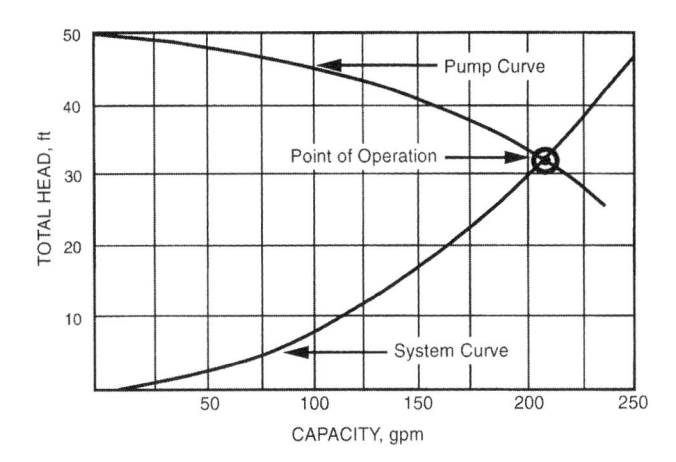

*Fig. 9-15 System and Pump Curves*
*(Figure 17, Chapter 44, 2012 ASHRAE Handbook—HVAC Systems and Equipment)*

Plotting the system head against the system flow produces a curve as shown in Figure 9-14. If no independent head exists in a hydronic system, the system friction curve becomes the system curve as shown in Figure 9-15. Domestic and small commercial heating and cooling systems without control valves are typical systems without independent head. In actual operation, since flow is not throttled, flow occurs at only one point in the system: at the intersection of the pump head-capacity curve and the system curve (Figure 9-15).

Additional data concerning pump design, selection, and operation can be found in Chapter 44 of the 2012 *ASHRAE Handbook—HVAC Systems and Equipment*.

## 9.6  Problems

**9.1** The air velocity in a human occupied zone should not exceed (a) 10 to 25 ft/min, (b) 25 to 40 ft/min, (c) 40 to 64 ft/min, (d) none of the above.

**9.2** You are to select the type of outlets for a home to be constructed in Houston, Texas. Discuss your selection of outlets and locations for each of the following combinations: (a) Group A or Group C, (b) Group B or Group E.

**9.3** What velocity of air is necessary at a location in a room such that most people will feel neither cool nor warm? Assume that the local temperature is equal to the control temperature of 24.4°C. [Ans: 0.152 m/s]

**9.4** Solve the following problems.
(a) Find the airflow through a 12 in. by 24 in. (305 mm by 610 mm) duct if the static pressure is measured at 0.5 in. of water (125 Pa) and total pressure is measured at 0.54 in. of water (135 Pa)
(b) The pressure difference available to a 60 ft (18.3 m) length of circular duct is 0.2 in. of water (50 Pa). The duct has an ID of 12 in. (305 mm). What rate of airflow is expected?

**9.5** For a residential air-conditioning system, one branch duct must supply 207 cfm to one of the rooms. The branch duct has a run of 16 ft. (a) Determine the branch duct size and the pressure drop from the main duct to the room, and (b) specify the supply and return grille sizes for the room.

**9.6** How large a duct is required to carry 20,000 cfm (9400 L/s) of air if the velocity is not to exceed 1600 fpm (8.1 m/s)? [Ans: 48 in. (1220 mm)]

**9.7** Given the duct system shown below, plot $p_v$, $\Delta p_s$, and $\Delta p_t$ for the flow through the system.

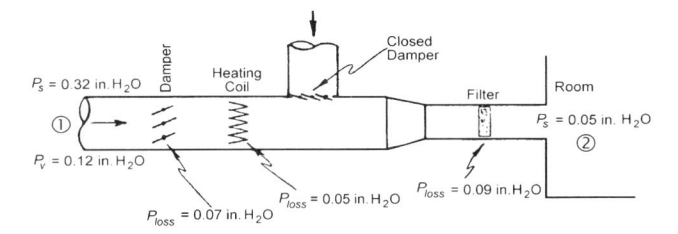

**9.8** Solve the following.
(a) For *Problem 9.7*, find the frictional pressure loss between points (1) and (2).
(b) How can the static pressure be increased in a duct system as the air moves away from the fan?

**9.9** Determine the dynamic loss of total pressure that occurs in an abrupt expansion from a 1 ft² (0.093 m²) duct to a 2 ft² (0.186 m²) duct carrying 1000 cfm (470 L/s) of air. [Ans: 0.02 in. of water (5 Pa)]

**9.10** Determine the friction loss when circulating 20,000 cfm (9430 L/s) of air at 75°F (23.9°C) through 150 ft (45.7 m) of 36 in. (0.914 m) diameter galvanized steel duct.

**9.11** Find the equivalent rectangular duct for equal friction and capacity for the duct in: *Problem 9.6*, one side is 26 in.

**9.12** Find the pressure loss between points A and D for the 12 in. by 12 in. duct shown below. Air at standard conditions is being supplied at the rate of 2000 cfm in galvanized duct of average construction. Elbows No. 1 and No. 2 have center line radii of 13 in. and 24 in., respectively. [Ans: 0.33 in. wg]

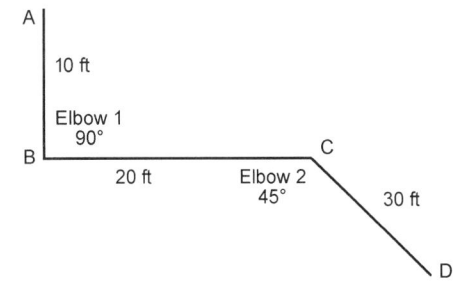

**9.13** Analyze the air-handling system shown in the following diagram. Determine if a damper is needed in either section (d) or (b), and if so, in what section. There is a damper located in section (u) so that the proper static pressure can be maintained in section (u). If a damper is needed, what is the pressure loss across the damper?

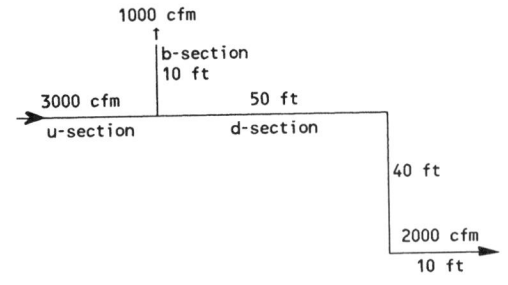

$\rho = 0.075$ lb/ft$^3$    $f = 0.02$
$D = 12$ in.    $(R/D)_{elbows} = 1$

**9.14** A 1 ft high by 3 ft wide main duct carries 2000 cfm of air to a branch where 1500 cfm continues in the 1 ft by 2 ft straight-through section and 500 cfm goes into the branch. Find the actual static pressure regain and the total pressure loss in the straight-through section if the static regain coefficient is 0.80. If the branch takeoff is a 45° cylindrical Y, find the static pressure loss in this section.

**9.15** The supply ductwork for an office space is shown in the following diagram. Size the ductwork by the equal-friction method and calculate the pressure drop. Assume a maximum duct depth of 12 in. and that all duct take-offs are straight rectangular takeoffs. [Ans: $\Delta P = 0.211$ in. wg]

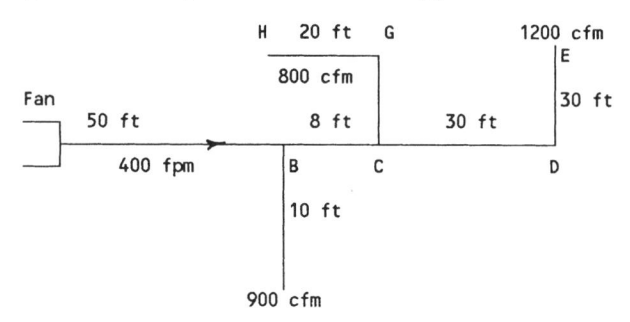

**9.16** The following duct system contains circular, galvanized duct. The velocity in the ducts is to be 2000 fpm, and each outlet is to handle 2000 cfm. Each outlet grille has a pressure loss of 0.12 in. of water. Estimate the required pressure increase of the fan. [Ans: 0.97 in. of water]

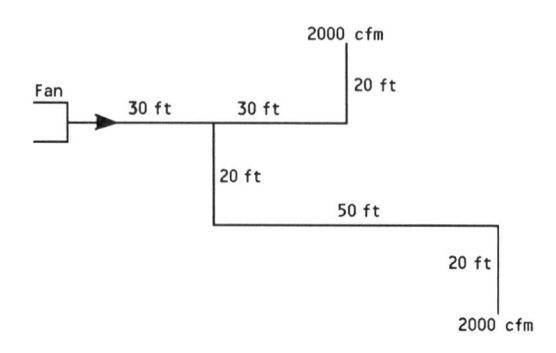

**9.17** Select a fan for the following system. The radius ratio of the elbows is 1.0 and the elbows are three-piece. The pipe is circular. Calculate the frictional pressure loss for the system and the total capacity required by the fan.

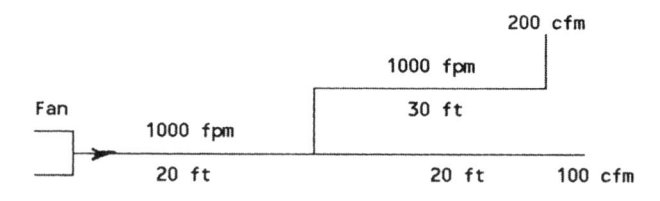

**9.18** (a) Estimate the total pressure loss between points (1) and (2) and between (1) and (3) in the following take-off.

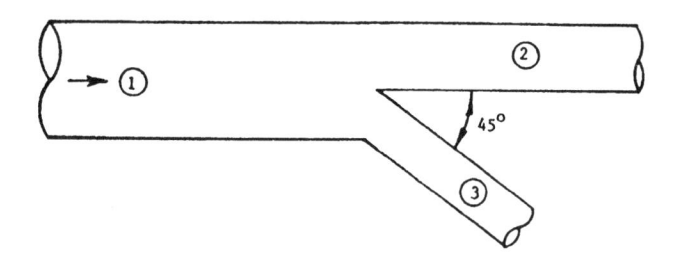

when:
$V_1 = 8.12$ m/s    $Q_1 = 1510$ L/s
$V_2 = 6.1$ m/s    $Q_2 = 1227$ L/s
$V_3 = 3.05$ m/s    $Q_3 = 283$ L/s

The duct is rectangular, of commercial fabrication, and has mastic tape joints. [Ans: 0.0064 in. wg]

(b) Estimate the static pressure at (3) if the static pressure at (1) is 1.0 in. of water. [Ans: 1.05 in. wg]

**9.19** Solve the following problems:
(a) What is the expected approximate frictional pressure from (1) to (2) in the following length of duct. Assume round ducts, clean sheet metal, and air at standard temperature and pressure.

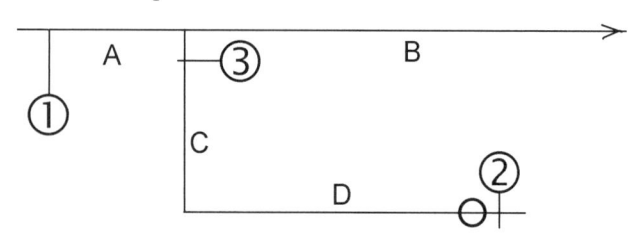

Elbow radius = 36 in.
Grille loss = 0.1 in w.g. at 600 fpm

| Duct | cfm | Vel., fpm | length |
|---|---|---|---|
| A | 2000 | 1000 | 40 ft |
| B | 1000 | 600 | – |
| C | 1000 | 600 | 30 ft |
| D | 1000 | 600 | 30 ft |

# Duct and Pipe Sizing

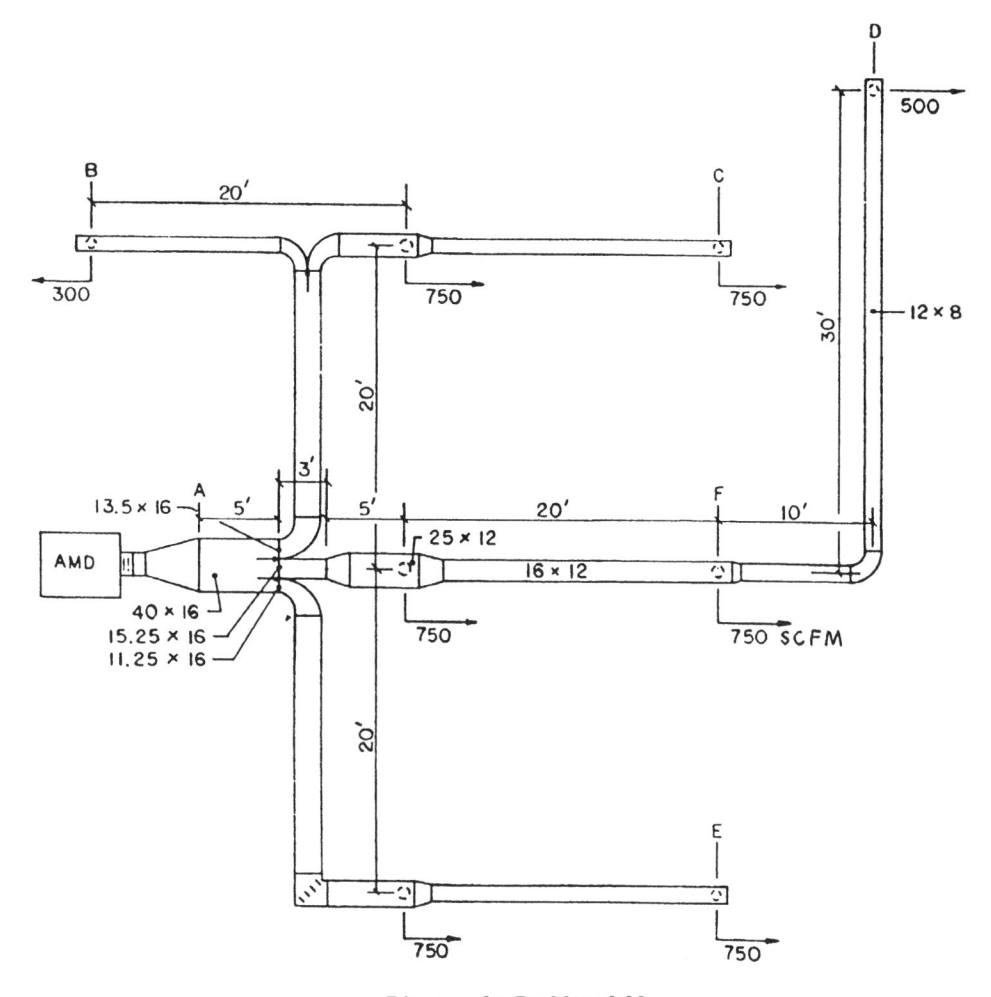

**Diagram for Problem 9.20**

(b) If the static pressure at (3) is 0.350 in. of water, what friction drop will be required of a damper at 68°F?

(c) What size duct would be required (for ducts C and D) if the damper is eliminated? What is the velocity in the line? Assume a temperature of 68°F. Assume that the static pressure at (3) is still 0.35 in. of water and that the $R/D$ of the elbow is 2. Also assume that the grille loss is linear with velocity.

**9.20** The pressure (energy) loss between A-B, A-C, A-D, and A-E must be equal if a proper air balance is to be achieved. The static pressure required in the duct at points B, C, D, and E to produce the proper flow from the air-diffusing terminal units (ceiling diffusers) is assumed to be uniform and to be 0.10 in. of water. Ignoring interference losses due to terminal unit take-off,

1. Calculate the total pressure loss between A and D.
2. Size the ducts between A and B.
3. Size the ducts between A and C.
4. Size the ducts between A and E.
5. What would the static pressure be at F?

*Note*: In practical application, minor static pressure imbalances up to 0.05 in. of water can be absorbed by adjustment of dampers installed in air-diffusing terminal units and further minor adjustments of diverting dampers, in branch fittings. For greater imbalances, duct sizing must be modified and/or butterfly dampers installed in the branch ducts.

**9.21** Determine the equivalent feet of pipe for a 2 in. (50 mm) open gate valve and a 2 in. (50 mm) open globe valve at a flow velocity of 5 fps (1.5 m/s). [Ans: 2.95 ft, 70.8 ft]

**9.22** A convector unit is rated at 1.8 gpm (0.113 L/s) and has a 3.4 ft (10 kPa) pressure loss at rated flow. Estimate the pressure loss with a flow of 2.3 gpm through the convector. [Ans: 6 ft (1.8 m)]

**9.23** Size the system shown in *Example 9.5* for a 10°F temperature drop.

**9.24** Size the system shown in *Example 9.5* for 30°F temperature drop.

**9.25** Size the system shown for iron pipe. The water leaves the boiler at 200°F and has a 20°F temperature drop. The convectors have a loss given by the equation: Loss (milli-inches) = 0.3 (Btu/h output). Note: 1 milli-inch = 0.001 in. of water.

Assume a 3 ft rise is needed to get to the convectors and then a 3 ft drop to return to the boiler. What head must be developed by the pump and what flow rate (gpm) is required?

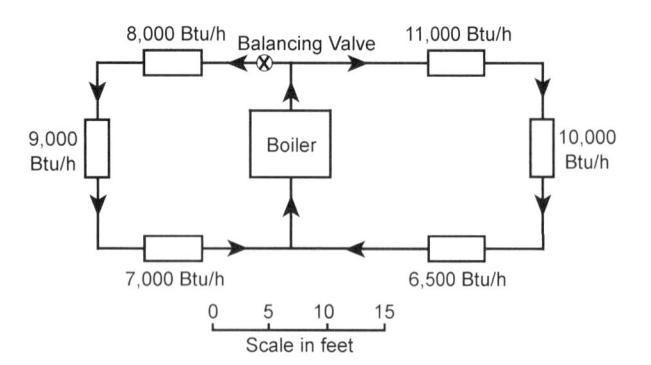

**Diagram for Problem 9.25** (not labeled, shown above): Loop with boiler in center, balancing valve at top, convectors labeled 8,000 Btu/h, 11,000 Btu/h, 10,000 Btu/h, 6,500 Btu/h, 7,000 Btu/h, 9,000 Btu/h. Scale in feet: 0, 5, 10, 15.

**9.26** Rework *Problem 9.25* using Type L copper tubing.

**9.27** A steam system requires 15,000 lb/h of steam at an initial pressure of 120 psig. The design pressure drop is to be 6 psi per 100 ft. Determine the size of schedule 40 pipe required and the velocity in the steam pipe. [Ans: 3.5 in.]

**9.28** Determine the pipe sizes for the refrigeration systems shown in the following figure.

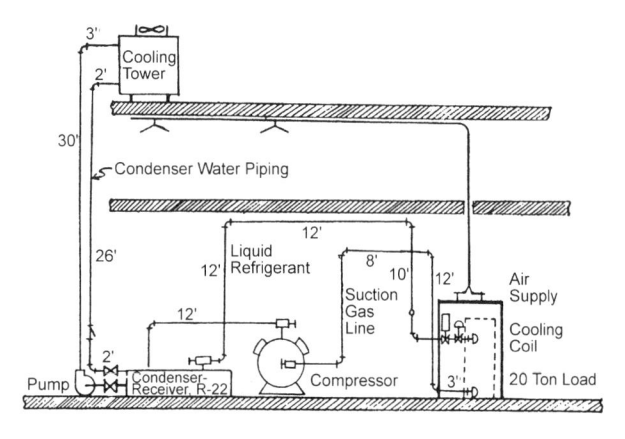

**Diagram for Problem 9.28**

(a) Refrigerant lines using R-22
(b) Condenser water lines

**9.29** A gas appliance has an input rating of 80,000 Btu/h and is operated with natural gas (specific gravity = 0.60) having a heating value of 1050 Btu/ft$^3$. What size of supply pipe is necessary when the equivalent length is 70 ft and a pressure loss of 0.3 in. of water is allowable?

**9.30** A fan operating at 1200 rpm has been delivering 6500 cfm against a static head of 3.25 in. of water and a total head of 5.25 in. of water. The air temperature is 130°F, gage temperature, 90°F, and the input power is 6.1 kW. During a time of power difficulty, the operator notices that the static head is now 2.36 in. of water. There has been no change in the system. Find:

(a) New capacity in cfm
(b) New power input in kW
(c) Original efficiency of the fan (%)
[Ans: 5570 cfm, 3.82 kw, 65.5%]

**9.31** A certain damper design introduces a head loss of 0.5 velocity heads when wide open. A damper of this design is to be installed in a 12 in. by 30 in. duct that handles 3000 cfm. The pressure drop in the undampered system is 1.5 in. of water. If the pressure drop through the damper when wide open is to be 5% of the total system resistance, how much cross-sectional area in the duct should the damper occupy?

**9.32** What effect on the following parameters does a variation in air density have for a fan operating in a system:
(a) flow rate
(b) developed head
(c) horsepower

**9.33** A fan delivers 1500 cfm (708 L/s) of dry air at 65°F (18.3°C) against a static pressure of 0.20 in. of water (50 Pa) and requires 0.10 bhp. Find the volume circulated, the static pressure, and the bhp required to deliver the same weight of air when the air temperature is increased to 165°F (73.9°C). (Note: Atmospheric pressure is constant.) [Ans: 0.168 in. wg, 0.084 hp]

**9.34** Should fans be placed before or after air heaters? Why?

**9.35** A 40 in. by 24 in. rectangular duct conveying 12,000 cfm of standard air divides into 3 branches (see figure). Branch A carries 6000 cfm for 100 ft, B carries 4000 cfm for 150 ft, and C carries 2000 for 35 ft.

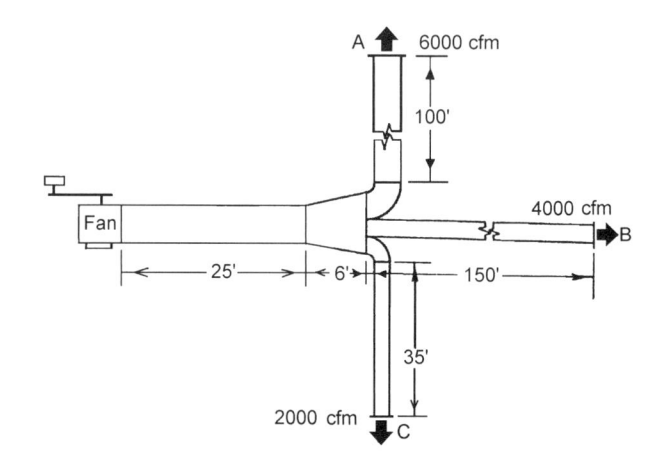

**Diagram for Problem 9.35**

(a) Size each branch for equal total friction of 0.15 in. of water. Do not exceed upper velocity limit of 2000 fpm.
(b) What is the total friction loss if the same quantity of air, 12,000 cfm of air at 150°F and 14.0 psia, is passed through the same system as part (a)?
(c) For a fan selected for part (a), at what percentage of the speed in part (a) must the fan run to satisfy part (b)?

**9.36** A centrifugal fan operating at 2400 rpm delivers 20,000 cfm of air through a 32 in. diameter duct against a static pressure of 4.8

in. of water. The air is 40°F. The barometer is 29.0 in. Hg. Determine the horsepower input if the efficiency is 70%. If the fan size, gas density, and duct system remain the same, calculate the horsepower required if operated at 3200 rpm. [Ans: 60 hp]

**9.37** Compute the efficiency of Fan 303 (Fig. 9-11b) when delivering 15,500 cfm at 4 in. static pressure (SP).

**9.38** Develop and explain the following relations for fan performance:
(a) HP = CFM × $\Delta P/6350\eta_f$
(b) kWH = HP $(0.746)$ hours/$\eta_m$
(c) $\Delta t_f = \Delta P(0.371)/\eta_f$
(d) HP ~ CFM$^3$

**9.39** A water pump develops a total head of 200 ft. The pump efficiency is 80% and the motor efficiency is 87.5%. If the power rate is 1.5 cents per kilowatt-hour, what is the power cost for pumping 1000 gallons? [Ans: 1.35 cents in 1 h]

**9.40** For a certain system it is required to select a pump that will deliver 2400 gpm (150 L/s) at a total head of 360 ft (110 m), and a pump shaft speed of 2400 rpm. What type of pump would you suggest? [Ans: centrifugal]

**9.41** A pump delivers 1400 gpm of water. The inlet pipe is 4 in. nominal and the outlet pipe is 2 in. nominal standard pipe. The water temperature is 40°F. The surface of the inlet supply is 40 ft higher than the pump centerline. The discharge gage, which is 22 ft above the pump centerline, reads 180 psi. If the pump and motor combined efficiency is 60%, calculate the necessary input to the motor in kilowatts. [Ans: 290 kW]

**9.42** A pump is required to force 9250 lb/h (4200 kg/h) of water at 165°F (74°C) through a heating system against a total resistance of 82,300 milli-inches of water (20.5 kPa). If the mechanical efficiency of the pump is 65%, find the required horsepower input.

**9.43** How many horsepower are required to pump 66 gpm (4.16 L/s) against 60 ft (18.3 m) of head assuming 75% efficiency?

**9.44** Solve the following problems:

(a) A certain system is found to have losses due to frictional effects according to the equation $H = 0.001$ (gpm)$^2$ where $H$ is in ft of water. The system is handling water at 160°F. For a design capacity of 300 gpm, what is the head developed by the pump and the bhp if the pump efficiency is 80%?

(b) What would be the theoretical maximum length of suction in order to prevent cavitation if the level of the supply tank is below the centerline of the pump? Assume atmospheric pressure to be 14.7 psi.

(c) If a capacity of 400 gpm is desired, what would be the speed ratio $n_2/n_1$ for the same pump, density of fluid, and system?

(d) Should a backward- or forward-curved blade pump be chosen? Would you make arrangements for a priming system for the pump? [Ans: 90 ft, 8.4 hp, 23.6 ft, 1.33]

## 9.7 References

Abushakra, B., I.S. Walker, and M.H. Sherman. 2002. A study of pressure losses in residential air distribution systems. Proceedings of the ACEEE Summer Study 2002, American Council for an Energy Efficient Economy, Washington, D.C. LBNL Report 49700. Lawrence Berkeley National Laboratory, CA.

Abushakra, B., I.S. Walker, and M.H. Sherman. 2004. Compression effects on pressure loss in flexible HVAC ducts. International Journal of HVAC&R Research (now HVAC&R Research) 10(3):275-289.

ACCA. 2009. Manual D-Residential duct systems. Air Conditioning Contractors of America, Washington, DC.

ACGIH. 2010. Industrial ventilation: A manual of recommended practice for design, 27th ed. American Conference of Governmental Industrial Hygienists, Lansing, MI.

AHRI. 2008. Procedure for estimating occupied space sound levels in the application of air terminals and air outlets. Standard 885-2008 with Addendum 1. Air-Conditioning, Heating, and Refrigeration Institute, Arlington, VA.

AIVC. 1999. Improving ductwork-A time for tighter air distribution systems. F.R. Carrié, J. Andersson, and P. Wouters, eds. The Air Infiltration and Ventilation Centre, Coventry, UK. Available at http://www.aivc.org/frameset/frameset.html?./Publications/guides/tp1999_4.htm~mainFrame.

AMCA. 2007. Laboratory methods for testing fans for certified aerodynamic performance rating. ANSI/AMCA Standard 210-07. Also ANSI/ASHRAE/AMCA Standard 51-07.

AMCA. 2011a. Fans and systems. AMCA Publication 201-02 (R2011). Air Movement and Control Association International, Arlington Heights, IL.

AMCA. 2011b. Field performance measurement of fan systems. AMCA Publication 203-90 (R2011). Air Movement and Control Association International, Arlington Heights, IL.

AMCA. 2012. Certified ratings program-Product rating manual for air control. AMCA Publication 511-10 (Rev. 8/12). Air Movement and Control Association International, Arlington Heights, IL.

AMCA. 2012. Laboratory method of testing louvers for rating. ANSI/ AMCA Standard 500-L-12. Air Movement and Control Association International, Arlington Heights, IL.

AMCA. 2012. Laboratory methods of testing dampers for rating. ANSI/AMCA Standard 500-D-12. Air Movement and Control Association International, Arlington Heights, IL.

ASHRAE. 2006. Energy conservation in existing buildings. ANSI/ASHRAE/IESNA Standard 100-2006.

ASHRAE. 2007. Energy-efficient design of low-rise residential buildings. ANSI/ASHRAE Standard 90.2-2007.

ASHRAE. 2008a. Method of testing HVAC air ducts and fittings. ANSI/ASHRAE/SMACNA Standard 126.

ASHRAE. 2008b. Methods of testing air terminal units. ANSI/ASHRAE Standard 130.

ASHRAE. 2010. Energy standard for buildings except low-rise residential buildings. ANSI/ASHRAE/IESNA Standard 90.1-2010.

ASHRAE. 2012. ASHRAE duct fitting database, v. 6.00.01. Also available as an iOS app.

Behls, H.F. 1971. Computerized calculation of duct friction. Building Science Series 39, p. 363. National Institute of Standards and Technology, Gaithersburg, MD.

Brown, R.B. 1973. Experimental determinations of fan system effect factors. In Fans and systems, ASHRAE Symposium Bulletin LO-73-1, Louisville, KY (June).

CEN. 2004. Ventilation for buildings-Ductwork-Measurement of ductwork surface area. European Standard EN 14239-2004. European Committee for Standardization, Brussels.

Clarke, M.S., J.T. Barnhart, F.J. Bubsey, and E. Neitzel. 1978. The effects of system connections on fan performance. ASHRAE Transactions 84(2): 227-263.

Colebrook, C.F. 1938-1939. Turbulent flow in pipes, with particular reference to the transition region between the smooth and rough pipe laws. Journal of the Institution of Civil Engineers 11:133.

Culp, C.H. 2011. HVAC flexible duct pressure loss measurements. ASHRAE Research Project RP-1333, Final Report.

Farquhar, H.F. 1973. System effect values for fans. In Fans and systems, ASHRAE Symposium Bulletin LO-73-1, Louisville, KY (June).

Griggs, E.I., and F. Khodabakhsh-Sharifabad. 1992. Flow characteristics in rectangular ducts. ASHRAE Transactions 98(1):116-127.

Griggs, E.I., W.B. Swim, and G.H. Henderson. 1987. Resistance to flow of round galvanized ducts. ASHRAE Transactions 93(1):3-16.

Heyt, J.W., and M.J. Diaz. 1975. Pressure drop in flat-oval spiral air duct. ASHRAE Transactions 81(2):221-232.

Huebscher, R.G. 1948. Friction equivalents for round, square and rectangular ducts. ASHVE Transactions 54:101-118.

Hutchinson, F.W. 1953. Friction losses in round aluminum ducts. ASHVE Transactions 59:127-138.

Idelchik, I.E., M.O. Steinberg, G.R. Malyavskaya, and O.G. Martynenko. 1994. Handbook of hydraulic resistance, 3rd ed. CRC Press/Begell House, Boca Raton.

Jones, C.D. 1979. Friction factor and roughness of United Sheet Metal Company spiral duct. United Sheet Metal, Division of United McGill Corp., Westerville, OH (August). Based on data in Friction loss tests, United Sheet Metal Company Spiral Duct, Ohio State University Engineering Experiment Station, File T-1011, September 1958.

Klote, J.H., J.A. Milke, P.G. Tumbull, A. Kashef, and M.J. Ferreira. 2012. Handbook of smoke control engineering. ASHRAE.

Kulkarni, D., S. Khaire, and S. Idem. 2009. Pressure loss of corrugated spiral duct. ASHRAE Transactions 115(1).

Madison, R.D., and W.R. Elliot. 1946. Friction charts for gases including correction for temperature, viscosity and pipe roughness. ASHVE Journal (October).

McGill. 1988. Round vs. rectangular duct. Engineering Report147, United McGill Corp. (contact McGill Airflow Technical Service Department), Westerville, OH.

McGill. 1995. Flat oval vs. rectangular duct. Engineering Report 150, United McGill Corp. (contact McGill Airflow Technical Service Department), Westerville, OH.

Meyer, M.L. 1973. A new concept: The fan system effect factor. In Fans and Systems, ASHRAE Symposium Bulletin LO-73-1, Louisville, KY (June).

Moody, L.F. 1944. Friction factors for pipe flow. ASME Transactions 66:671.

NFPA. 2008. Fire protection handbook. National Fire Protection Association, 20th ed. Quincy, MA.

NFPA. 2012. Installation of air-conditioning and ventilating systems. ANSI/NFPA Standard 90A. National Fire Protection Association, Quincy, MA.

Osborne, W.C. 1966. Fans. Pergamon, London.

Schaffer, M.E. 2005. A practical guide to noise and vibration control for HVAC systems, 2nd ed. ASHRAE.

Swim, W.B. 1978. Flow losses in rectangular ducts lined with fiberglass. ASHRAE Transactions 84(2):216.

Swim, W.B. 1982. Friction factor and roughness for airflow in plastic pipe. ASHRAE Transactions 88(1):269.

Tsal, R.J., and M.S. Adler. 1987. Evaluation of numerical methods for ductwork and pipeline optimization. ASHRAE Transactions 93(1):17-34.

Tsal, R.J., H.F. Behls, and R. Mangel. 1988. T-method duct design, Part I: Optimization theory; Part II: Calculation procedure and economic analysis. ASHRAE Transactions 94(2):90-111.

Tsal, R.J., H.F. Behls, and R. Mangel. 1990. T-method duct design, Part III: Simulation. ASHRAE Transactions 96(2).

UL. 2012. Fire dampers. UL Standard 555, 7th ed. Underwriters Laboratories, Northbrook, IL.

UL. 2012. Smoke dampers. UL Standard 555S, 4th ed. Underwriters Laboratories, Northbrook, IL.

UL. Published annually. Building materials directory. Underwriters Laboratories, Northbrook, IL.

UL. Published annually. Fire resistance directory. Underwriters Laboratories, Northbrook, IL.

Wright, D.K., Jr. 1945. A new friction chart for round ducts. ASHVE Transactions 51:303-316.

# SI Figures and Tables

### Table 9-1a SI  Equivalent Rectangular Duct Dimension
*(Table 2, Chapter 21, 2013 ASHRAE Handbook—Fundamentals, SI version)*

| Lgth Adj.[b] | \multicolumn{19}{c}{Length of One Side of Rectangular Duct (*a*), mm} |
|---|---|---|---|---|---|---|---|---|---|---|---|---|---|---|---|---|---|---|---|
|  | 100 | 125 | 150 | 175 | 200 | 225 | 250 | 275 | 300 | 350 | 400 | 450 | 500 | 550 | 600 | 650 | 700 | 750 | 800 | 900 |
|  | \multicolumn{19}{c}{Circular Duct Diameter, mm} |
| 100 | 109 | | | | | | | | | | | | | | | | | | | |
| 125 | 122 | 137 | | | | | | | | | | | | | | | | | | |
| 150 | 133 | 150 | 164 | | | | | | | | | | | | | | | | | |
| 175 | 143 | 161 | 177 | 191 | | | | | | | | | | | | | | | | |
| 200 | 152 | 172 | 189 | 204 | 219 | | | | | | | | | | | | | | | |
| 225 | 161 | 181 | 200 | 216 | 232 | 246 | | | | | | | | | | | | | | |
| 250 | 169 | 190 | 210 | 228 | 244 | 259 | 273 | | | | | | | | | | | | | |
| 275 | 176 | 199 | 220 | 238 | 256 | 272 | 287 | 301 | | | | | | | | | | | | |
| 300 | 183 | 207 | 229 | 248 | 266 | 283 | 299 | 314 | 328 | | | | | | | | | | | |
| 350 | 195 | 222 | 245 | 267 | 286 | 305 | 322 | 339 | 354 | 383 | | | | | | | | | | |
| 400 | 207 | 235 | 260 | 283 | 305 | 325 | 343 | 361 | 378 | 409 | 437 | | | | | | | | | |
| 450 | 217 | 247 | 274 | 299 | 321 | 343 | 363 | 382 | 400 | 433 | 464 | 492 | | | | | | | | |
| 500 | 227 | 258 | 287 | 313 | 337 | 360 | 381 | 401 | 420 | 455 | 488 | 518 | 547 | | | | | | | |
| 550 | 236 | 269 | 299 | 326 | 352 | 375 | 398 | 419 | 439 | 477 | 511 | 543 | 573 | 601 | | | | | | |
| 600 | 245 | 279 | 310 | 339 | 365 | 390 | 414 | 436 | 457 | 496 | 533 | 567 | 598 | 628 | 656 | | | | | |
| 650 | 253 | 289 | 321 | 351 | 378 | 404 | 429 | 452 | 474 | 515 | 553 | 589 | 622 | 653 | 683 | 711 | | | | |
| 700 | 261 | 298 | 331 | 362 | 391 | 418 | 443 | 467 | 490 | 533 | 573 | 610 | 644 | 677 | 708 | 737 | 765 | | | |
| 750 | 268 | 306 | 341 | 373 | 402 | 430 | 457 | 482 | 506 | 550 | 592 | 630 | 666 | 700 | 732 | 763 | 792 | 820 | | |
| 800 | 275 | 314 | 350 | 383 | 414 | 442 | 470 | 496 | 520 | 567 | 609 | 649 | 687 | 722 | 755 | 787 | 818 | 847 | 875 | |
| 900 | 289 | 330 | 367 | 402 | 435 | 465 | 494 | 522 | 548 | 597 | 643 | 686 | 726 | 763 | 799 | 833 | 866 | 897 | 927 | 984 |
| 1000 | 301 | 344 | 384 | 420 | 454 | 486 | 517 | 546 | 574 | 626 | 674 | 719 | 762 | 802 | 840 | 876 | 911 | 944 | 976 | 1037 |
| 1100 | 313 | 358 | 399 | 437 | 473 | 506 | 538 | 569 | 598 | 652 | 703 | 751 | 795 | 838 | 878 | 916 | 953 | 988 | 1022 | 1086 |
| 1200 | 324 | 370 | 413 | 453 | 490 | 525 | 558 | 590 | 620 | 677 | 731 | 780 | 827 | 872 | 914 | 954 | 993 | 1030 | 1066 | 1133 |
| 1300 | 334 | 382 | 426 | 468 | 506 | 543 | 577 | 610 | 642 | 701 | 757 | 808 | 857 | 904 | 948 | 990 | 1031 | 1069 | 1107 | 1177 |
| 1400 | 344 | 394 | 439 | 482 | 522 | 559 | 595 | 629 | 662 | 724 | 781 | 835 | 886 | 934 | 980 | 1024 | 1066 | 1107 | 1146 | 1220 |
| 1500 | 353 | 404 | 452 | 495 | 536 | 575 | 612 | 648 | 681 | 745 | 805 | 860 | 913 | 963 | 1011 | 1057 | 1100 | 1143 | 1183 | 1260 |
| 1600 | 362 | 415 | 463 | 508 | 551 | 591 | 629 | 665 | 700 | 766 | 827 | 885 | 939 | 991 | 1041 | 1088 | 1133 | 1177 | 1219 | 1298 |
| 1700 | 371 | 425 | 475 | 521 | 564 | 605 | 644 | 682 | 718 | 785 | 849 | 908 | 964 | 1018 | 1069 | 1118 | 1164 | 1209 | 1253 | 1335 |
| 1800 | 379 | 434 | 485 | 533 | 577 | 619 | 660 | 698 | 735 | 804 | 869 | 930 | 988 | 1043 | 1096 | 1146 | 1195 | 1241 | 1286 | 1371 |
| 1900 | 387 | 444 | 496 | 544 | 590 | 633 | 674 | 713 | 751 | 823 | 889 | 952 | 1012 | 1068 | 1122 | 1174 | 1224 | 1271 | 1318 | 1405 |
| 2000 | 395 | 453 | 506 | 555 | 602 | 646 | 688 | 728 | 767 | 840 | 908 | 973 | 1034 | 1092 | 1147 | 1200 | 1252 | 1301 | 1348 | 1438 |
| 2100 | 402 | 461 | 516 | 566 | 614 | 659 | 702 | 743 | 782 | 857 | 927 | 993 | 1055 | 1115 | 1172 | 1226 | 1279 | 1329 | 1378 | 1470 |
| 2200 | 410 | 470 | 525 | 577 | 625 | 671 | 715 | 757 | 797 | 874 | 945 | 1013 | 1076 | 1137 | 1195 | 1251 | 1305 | 1356 | 1406 | 1501 |
| 2300 | 417 | 478 | 534 | 587 | 636 | 683 | 728 | 771 | 812 | 890 | 963 | 1031 | 1097 | 1159 | 1218 | 1275 | 1330 | 1383 | 1434 | 1532 |
| 2400 | 424 | 486 | 543 | 597 | 647 | 695 | 740 | 784 | 826 | 905 | 980 | 1050 | 1116 | 1180 | 1241 | 1299 | 1355 | 1409 | 1461 | 1561 |
| 2500 | 430 | 494 | 552 | 606 | 658 | 706 | 753 | 797 | 840 | 920 | 996 | 1068 | 1136 | 1200 | 1262 | 1322 | 1379 | 1434 | 1488 | 1589 |
| 2600 | 437 | 501 | 560 | 616 | 668 | 717 | 764 | 810 | 853 | 935 | 1012 | 1085 | 1154 | 1220 | 1283 | 1344 | 1402 | 1459 | 1513 | 1617 |
| 2700 | 443 | 509 | 569 | 625 | 678 | 728 | 776 | 822 | 866 | 950 | 1028 | 1102 | 1173 | 1240 | 1304 | 1366 | 1425 | 1483 | 1538 | 1644 |
| 2800 | 450 | 516 | 577 | 634 | 688 | 738 | 787 | 834 | 879 | 964 | 1043 | 1119 | 1190 | 1259 | 1324 | 1387 | 1447 | 1506 | 1562 | 1670 |
| 2900 | 456 | 523 | 585 | 643 | 697 | 749 | 798 | 845 | 891 | 977 | 1058 | 1135 | 1208 | 1277 | 1344 | 1408 | 1469 | 1529 | 1586 | 1696 |

| Lgth Adj.[b] | \multicolumn{19}{c}{Length of One Side of Rectangular Duct (*a*), mm} |
|---|---|---|---|---|---|---|---|---|---|---|---|---|---|---|---|---|---|---|---|
|  | 1000 | 1100 | 1200 | 1300 | 1400 | 1500 | 1600 | 1700 | 1800 | 1900 | 2000 | 2100 | 2200 | 2300 | 2400 | 2500 | 2600 | 2700 | 2800 | 2900 |
|  | \multicolumn{19}{c}{Circular Duct Diameter, mm} |
| 1000 | 1093 | | | | | | | | | | | | | | | | | | | |
| 1100 | 1146 | 1202 | | | | | | | | | | | | | | | | | | |
| 1200 | 1196 | 1256 | 1312 | | | | | | | | | | | | | | | | | |
| 1300 | 1244 | 1306 | 1365 | 1421 | | | | | | | | | | | | | | | | |
| 1400 | 1289 | 1354 | 1416 | 1475 | 1530 | | | | | | | | | | | | | | | |
| 1500 | 1332 | 1400 | 1464 | 1526 | 1584 | 1640 | | | | | | | | | | | | | | |
| 1600 | 1373 | 1444 | 1511 | 1574 | 1635 | 1693 | 1749 | | | | | | | | | | | | | |
| 1700 | 1413 | 1486 | 1555 | 1621 | 1684 | 1745 | 1803 | 1858 | | | | | | | | | | | | |
| 1800 | 1451 | 1527 | 1598 | 1667 | 1732 | 1794 | 1854 | 1912 | 1968 | | | | | | | | | | | |
| 1900 | 1488 | 1566 | 1640 | 1710 | 1778 | 1842 | 1904 | 1964 | 2021 | 2077 | | | | | | | | | | |
| 2000 | 1523 | 1604 | 1680 | 1753 | 1822 | 1889 | 1952 | 2014 | 2073 | 2131 | 2186 | | | | | | | | | |
| 2100 | 1558 | 1640 | 1719 | 1793 | 1865 | 1933 | 1999 | 2063 | 2124 | 2183 | 2240 | 2296 | | | | | | | | |
| 2200 | 1591 | 1676 | 1756 | 1833 | 1906 | 1977 | 2044 | 2110 | 2173 | 2233 | 2292 | 2350 | 2405 | | | | | | | |
| 2300 | 1623 | 1710 | 1793 | 1871 | 1947 | 2019 | 2088 | 2155 | 2220 | 2283 | 2343 | 2402 | 2459 | 2514 | | | | | | |
| 2400 | 1655 | 1744 | 1828 | 1909 | 1986 | 2060 | 2131 | 2200 | 2266 | 2330 | 2393 | 2453 | 2511 | 2568 | 2624 | | | | | |
| 2500 | 1685 | 1776 | 1862 | 1945 | 2024 | 2100 | 2173 | 2243 | 2311 | 2377 | 2441 | 2502 | 2562 | 2621 | 2678 | 2733 | | | | |
| 2600 | 1715 | 1808 | 1896 | 1980 | 2061 | 2139 | 2213 | 2285 | 2355 | 2422 | 2487 | 2551 | 2612 | 2672 | 2730 | 2787 | 2842 | | | |
| 2700 | 1744 | 1839 | 1929 | 2015 | 2097 | 2177 | 2253 | 2327 | 2398 | 2466 | 2533 | 2598 | 2661 | 2722 | 2782 | 2840 | 2896 | 2952 | | |
| 2800 | 1772 | 1869 | 1961 | 2048 | 2133 | 2214 | 2292 | 2367 | 2439 | 2510 | 2578 | 2644 | 2708 | 2771 | 2832 | 2891 | 2949 | 3006 | 3061 | |
| 2900 | 1800 | 1898 | 1992 | 2081 | 2167 | 2250 | 2329 | 2406 | 2480 | 2552 | 2621 | 2689 | 2755 | 2819 | 2881 | 2941 | 3001 | 3058 | 3115 | 3170 |

[a] Table based on $D_e = 1.30(ab)^{0.625}/(a+b)^{0.25}$.  [b] Length of adjacent side of rectangular duct (*b*), mm.

Table 9-1b SI  Equivalent Flat Oval Duct Dimensions
*(Table 3, Chapter 21, 2013 ASHRAE Handbook—Fundamentals, SI version)*

| Circular Duct Diameter, mm | \multicolumn{17}{c}{Minor Axis $a$, mm} |
|---|---|---|---|---|---|---|---|---|---|---|---|---|---|---|---|---|---|
| | 70 | 100 | 125 | 150 | 175 | 200 | 250 | 275 | 300 | 325 | 350 | 375 | 400 | 450 | 500 | 550 | 600 |
| | \multicolumn{17}{c}{Major Axis $A$, mm} |
| 125 | 205 | | | | | | | | | | | | | | | | |
| 140 | 265 | 180 | | | | | | | | | | | | | | | |
| 160 | 360 | 235 | 190 | | | | | | | | | | | | | | |
| 180 | 475 | 300 | 235 | 200 | | | | | | | | | | | | | |
| 200 | | 380 | 290 | 245 | 215 | | | | | | | | | | | | |
| 224 | | 490 | 375 | 305 | — | 240 | | | | | | | | | | | |
| 250 | | | 475 | 385 | 325 | 290 | | | | | | | | | | | |
| 280 | | | | 485 | 410 | 360 | — | 285 | | | | | | | | | |
| 315 | | | | 635 | 525 | — | — | 345 | 325 | | | | | | | | |
| 355 | | | | 840 | — | 580 | 460 | 425 | 395 | 375 | | | | | | | |
| 400 | | | | 1115 | — | 760 | — | 530 | 490 | 460 | 435 | | | | | | |
| 450 | | | | 1490 | — | 995 | — | 675 | — | 570 | 535 | 505 | | | | | |
| 500 | | | | | | 1275 | — | 845 | — | 700 | 655 | 615 | 580 | | | | |
| 560 | | | | | | 1680 | — | 1085 | — | 890 | 820 | 765 | 720 | | | | |
| 630 | | | | | | | | 1425 | — | 1150 | 1050 | 970 | 905 | 810 | | | |
| 710 | | | | | | | | | | 1505 | 1370 | 1260 | 1165 | 1025 | | | |
| 800 | | | | | | | | | | | 1800 | 1645 | 1515 | 1315 | 1170 | 1065 | |
| 900 | | | | | | | | | | | | 2165 | 1985 | 1705 | 1500 | 1350 | |
| 1000 | | | | | | | | | | | | | | 2170 | 1895 | 1690 | |
| 1120 | | | | | | | | | | | | | | | 2455 | 2170 | 1950 |
| 1250 | | | | | | | | | | | | | | | | 2795 | 2495 |

# Duct and Pipe Sizing

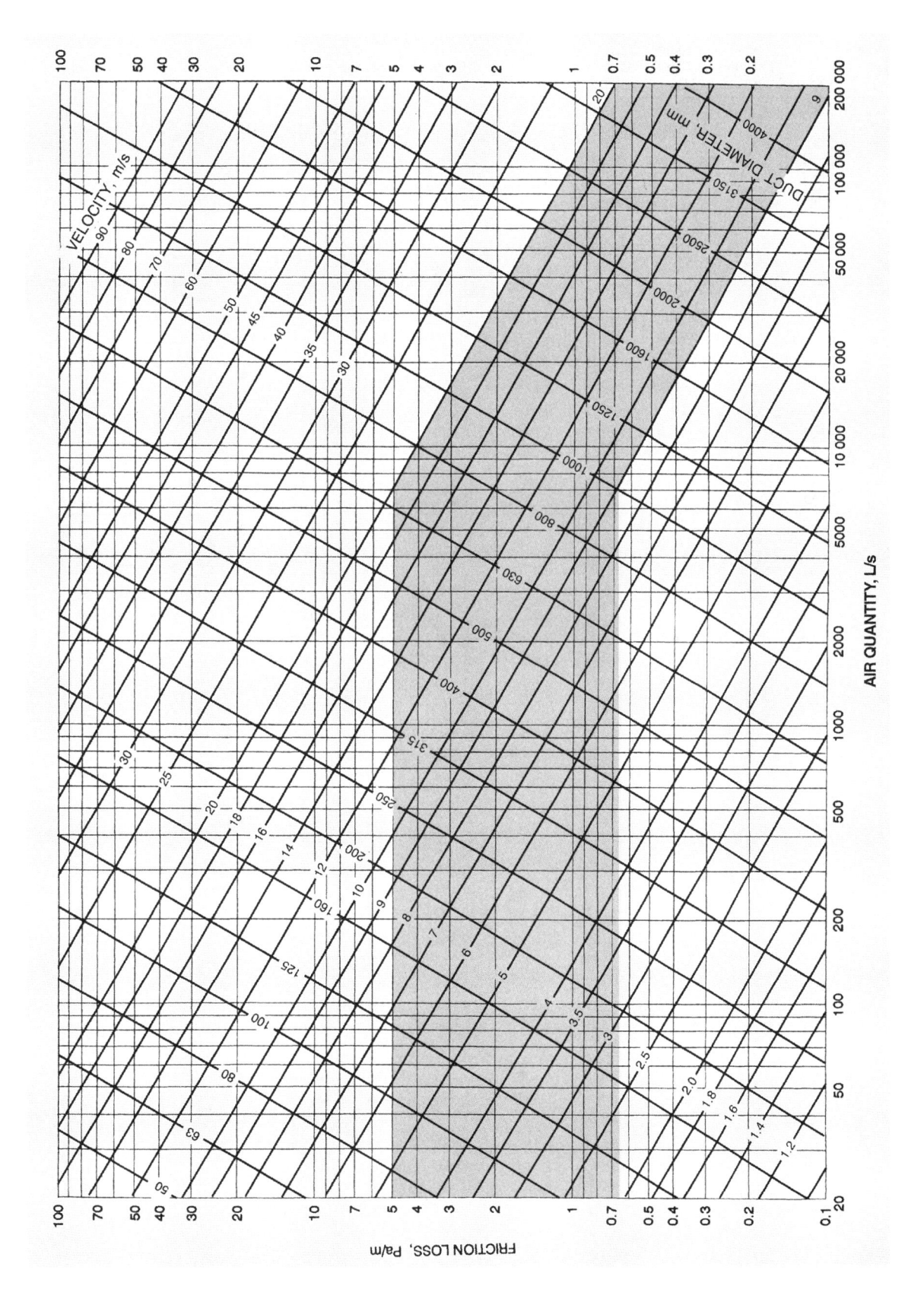

*Fig. 9-2 SI Friction of Air in Straight Ducts*
(Figure 9, Chapter 21, 2013 ASHRAE Handbook—Fundamentals)

## Table 9-8 SI  Steel Pipe Data
*(Table 2, Chapter 46, 2012 ASHRAE Handbook—HVAC Systems and Equipment, SI version)*

| U.S. Nominal Size, in. | Nominal Size, mm | Schedule[a] | Wall Thickness $t$, mm | Inside Diameter $d$, mm | Surface Area Outside, $m^2/m$ | Surface Area Inside, $m^2/m$ | Cross Section Metal Area, $mm^2$ | Cross Section Flow Area, $mm^2$ | Mass Pipe, kg/m | Mass Water, kg/m | Mfr. Process | Joint Type[b] | Working Pressure[c] ASTM A53 B to 200°C, kPa (gage) |
|---|---|---|---|---|---|---|---|---|---|---|---|---|---|
| 1/4 | 8 | 40 ST | 2.24 | 9.25 | 0.043 | 0.029 | 80.6 | 67.1 | 0.631 | 0.067 | CW | T | 1296 |
|  |  | 80 XS | 3.02 | 7.67 | 0.043 | 0.024 | 101.5 | 46.2 | 0.796 | 0.046 | CW | T | 6006 |
| 3/8 | 10 | 40 ST | 2.31 | 12.52 | 0.054 | 0.039 | 107.7 | 123.2 | 0.844 | 0.123 | CW | T | 1400 |
|  |  | 80 XS | 3.20 | 10.74 | 0.054 | 0.034 | 140.2 | 90.7 | 1.098 | 0.091 | CW | T | 5654 |
| 1/2 | 15 | 40 ST | 2.77 | 15.80 | 0.067 | 0.050 | 161.5 | 196.0 | 1.265 | 0.196 | CW | T | 1476 |
|  |  | 80 XS | 3.73 | 13.87 | 0.067 | 0.044 | 206.5 | 151.1 | 1.618 | 0.151 | CW | T | 5192 |
| 3/4 | 20 | 40 ST | 2.87 | 20.93 | 0.084 | 0.066 | 214.6 | 344.0 | 1.68 | 0.344 | CW | T | 1496 |
|  |  | 80 XS | 3.91 | 18.85 | 0.084 | 0.059 | 279.7 | 279.0 | 2.19 | 0.279 | CW | T | 4695 |
| 1 | 25 | 40 ST | 3.38 | 26.64 | 0.105 | 0.084 | 318.6 | 557.6 | 2.50 | 0.558 | CW | T | 1558 |
|  |  | 80 XS | 4.55 | 24.31 | 0.105 | 0.076 | 412.1 | 464.1 | 3.23 | 0.464 | CW | T | 4427 |
| 1 1/4 | 32 | 40 ST | 3.56 | 35.05 | 0.132 | 0.110 | 431.3 | 965.0 | 3.38 | 0.965 | CW | T | 1579 |
|  |  | 80 XS | 4.85 | 32.46 | 0.132 | 0.102 | 568.7 | 827.6 | 4.45 | 0.828 | CW | T | 4096 |
| 1 1/2 | 40 | 40 ST | 3.68 | 40.89 | 0.152 | 0.128 | 515.5 | 1 313 | 4.05 | 1.313 | CW | T | 1593 |
|  |  | 80 XS | 5.08 | 38.10 | 0.152 | 0.120 | 689.0 | 1 140 | 5.40 | 1.140 | CW | T | 3972 |
| 2 | 50 | 40 ST | 3.91 | 52.50 | 0.190 | 0.165 | 690.3 | 2 165 | 5.43 | 2.165 | CW | T | 1586 |
|  |  | 80 XS | 5.54 | 49.25 | 0.190 | 0.155 | 953 | 1 905 | 7.47 | 1.905 | CW | T | 3799 |
| 2 1/2 | 65 | 40 ST | 5.16 | 62.71 | 0.229 | 0.197 | 1 099 | 3 089 | 8.62 | 3.089 | CW | W | 3675 |
|  |  | 80 XS | 7.01 | 59.00 | 0.229 | 0.185 | 1 454 | 2 734 | 11.40 | 2.734 | CW | W | 5757 |
| 3 | 80 | 40 ST | 5.49 | 77.93 | 0.279 | 0.245 | 1 438 | 4 769 | 11.27 | 4.769 | CW | W | 3323 |
|  |  | 80 XS | 7.62 | 73.66 | 0.279 | 0.231 | 1 946 | 4 261 | 15.25 | 4.261 | CW | W | 5288 |
| 4 | 100 | 40 ST | 6.02 | 102.26 | 0.359 | 0.321 | 2 048 | 8 213 | 16.04 | 8.213 | CW | W | 2965 |
|  |  | 80 XS | 8.56 | 97.18 | 0.359 | 0.305 | 2 844 | 7 417 | 22.28 | 7.417 | CW | W | 4792 |
| 6 | 150 | 40 ST | 7.11 | 154.05 | 0.529 | 0.484 | 3 601 | 18 639 | 28.22 | 18.64 | ERW | W | 4799 |
|  |  | 80 XS | 10.97 | 146.33 | 0.529 | 0.460 | 5 423 | 16 817 | 42.49 | 16.82 | ERW | W | 8336 |
| 8 | 200 | 30 | 7.04 | 205.0 | 0.688 | 0.644 | 4 687 | 33 000 | 36.73 | 33.01 | ERW | W | 3627 |
|  |  | 40 ST | 8.18 | 202.7 | 0.688 | 0.637 | 5 419 | 32 280 | 42.46 | 32.28 | ERW | W | 4433 |
|  |  | 80 XS | 12.70 | 193.7 | 0.688 | 0.608 | 8 234 | 29 460 | 64.51 | 29.46 | ERW | W | 7626 |
| 10 | 250 | 30 | 7.80 | 257.5 | 0.858 | 0.809 | 6 498 | 52 060 | 50.91 | 52.06 | ERW | W | 3344 |
|  |  | 40 ST | 9.27 | 254.5 | 0.858 | 0.800 | 7 683 | 50 870 | 60.20 | 50.87 | ERW | W | 4178 |
|  |  | XS | 12.70 | 247.7 | 0.858 | 0.778 | 10 388 | 48 170 | 81.39 | 48.17 | ERW | W | 6116 |
|  |  | 80 | 15.06 | 242.9 | 0.858 | 0.763 | 12 208 | 46 350 | 95.66 | 46.35 | ERW | W | 7453 |
| 12 | 300 | 30 | 8.38 | 307.1 | 1.017 | 0.965 | 8 307 | 74 060 | 65.09 | 74.06 | ERW | W | 3096 |
|  |  | ST | 9.53 | 304.8 | 1.017 | 0.958 | 9 406 | 72 970 | 73.70 | 72.97 | ERW | W | 3641 |
|  |  | 40 | 10.31 | 303.2 | 1.017 | 0.953 | 10 158 | 72 190 | 79.59 | 72.21 | ERW | W | 4020 |
|  |  | XS | 12.70 | 298.5 | 1.017 | 0.938 | 12 414 | 69 940 | 97.28 | 69.96 | ERW | W | 5157 |
|  |  | 80 | 17.45 | 289.0 | 1.017 | 0.908 | 16 797 | 65 550 | 131.62 | 65.57 | ERW | W | 7419 |
| 14 | 350 | 30 ST | 9.53 | 336.6 | 1.117 | 1.057 | 10 356 | 88 970 | 81.15 | 88.96 | ERW | W | 3316 |
|  |  | 40 | 11.10 | 333.4 | 1.117 | 1.047 | 12 013 | 87 290 | 94.13 | 87.30 | ERW | W | 3999 |
|  |  | XS | 12.70 | 330.2 | 1.117 | 1.037 | 13 681 | 85 610 | 107.21 | 85.63 | ERW | W | 4695 |
|  |  | 80 | 19.05 | 317.5 | 1.117 | 0.997 | 20 142 | 79 160 | 157.82 | 79.17 | ERW | W | 7453 |
| 16 | 400 | 30 ST | 9.53 | 387.4 | 1.277 | 1.217 | 11 876 | 117 800 | 93.06 | 117.8 | ERW | W | 2903 |
|  |  | 40 XS | 12.70 | 381.0 | 1.277 | 1.197 | 15 708 | 114 000 | 123.09 | 114.0 | ERW | W | 4109 |
| 18 | 450 | ST | 9.53 | 438.2 | 1.436 | 1.376 | 13 396 | 150 800 | 104.98 | 150.8 | ERW | W | 2579 |
|  |  | 30 | 11.10 | 435.0 | 1.436 | 1.367 | 15 556 | 148 600 | 121.90 | 148.6 | ERW | W | 3110 |
|  |  | XS | 12.70 | 431.8 | 1.436 | 1.357 | 17 735 | 146 450 | 138.97 | 146.4 | ERW | W | 3654 |
|  |  | 40 | 14.27 | 428.7 | 1.436 | 1.347 | 19 863 | 144 300 | 155.65 | 144.3 | ERW | W | 4185 |
| 20 | 500 | 20 ST | 9.53 | 489.0 | 1.596 | 1.536 | 14 916 | 187 700 | 116.88 | 187.4 | ERW | W | 2324 |
|  |  | 30 XS | 12.70 | 482.6 | 1.596 | 1.516 | 19 762 | 182 900 | 154.85 | 182.9 | ERW | W | 3289 |
|  |  | 40 | 15.06 | 477.9 | 1.596 | 1.501 | 23 325 | 179 400 | 182.78 | 179.4 | ERW | W | 4006 |

[a] Numbers are schedule numbers per ASME *Standard* B36.10M; ST = Standard; XS = Extra Strong.
[b] T = Thread; W = Weld
[c] Working pressures were calculated per ASME B31.9 using furnace butt-weld (continuous weld, CW) pipe through 100 mm and electric resistance weld (ERW) thereafter. The allowance A has been taken as
(1) 12.5% of $t$ for mill tolerance on pipe wall thickness, *plus*
(2) An arbitrary corrosion allowance of 0.64 mm for pipe sizes through NPS 2 and 1.65 mm from NPS 2 1/2 through 20, *plus*
(3) A thread cutting allowance for sizes through NPS 2.

Because the pipe wall thickness of threaded standard pipe is so small after deducting allowance A, the mechanical strength of the pipe is impaired. It is good practice to limit standard threaded pipe pressure to 620 kPa (gage) for steam and 860 kPa (gage) for water.

## Table 9-9 SI  Copper Tube Data

*(Table 2, Chapter 46, 2012 ASHRAE Handbook—HVAC Systems and Equipment, SI version)*

| U.S. Nominal Size, in. | Type | Wall Thickness $t$, mm | Outside $D$, mm | Inside $d$, mm | Outside, m²/m | Inside, m²/m | Metal Area, mm² | Flow Area, mm² | Tube, kg/m | Water, kg/m | Annealed MPa | Drawn MPa |
|---|---|---|---|---|---|---|---|---|---|---|---|---|
| 1/4 | K | 0.89 | 9.53 | 7.75 | 0.030 | 0.0244 | 24 | 47 | 0.216 | 0.047 | 5.868 | 11.004 |
|  | L | 0.76 | 9.53 | 8.00 | 0.030 | 0.0250 | 21 | 50 | 0.188 | 0.050 | 5.033 | 9.432 |
| 3/8 | K | 1.24 | 12.70 | 10.21 | 0.040 | 0.0320 | 45 | 82 | 0.400 | 0.082 | 6.164 | 11.556 |
|  | L | 0.89 | 12.70 | 10.92 | 0.040 | 0.0344 | 33 | 94 | 0.295 | 0.094 | 4.399 | 8.253 |
|  | M | 0.64 | 12.70 | 11.43 | 0.040 | 0.0360 | 24 | 103 | 0.216 | 0.103 | 3.144 | 5.895 |
| 1/2 | K | 1.24 | 15.88 | 13.39 | 0.050 | 0.0421 | 57 | 141 | 0.512 | 0.141 | 4.930 | 9.246 |
|  | L | 1.02 | 15.88 | 13.84 | 0.050 | 0.0436 | 48 | 151 | 0.424 | 0.151 | 4.027 | 7.543 |
|  | M | 0.71 | 15.88 | 14.45 | 0.050 | 0.0454 | 34 | 164 | 0.302 | 0.164 | 2.820 | 5.282 |
| 5/8 | K | 1.24 | 19.05 | 16.56 | 0.060 | 0.0521 | 70 | 215 | 0.622 | 0.215 | 4.109 | 7.702 |
|  | L | 1.07 | 19.05 | 16.92 | 0.060 | 0.0530 | 60 | 225 | 0.539 | 0.225 | 3.523 | 6.605 |
| 3/4 | K | 1.65 | 22.23 | 18.92 | 0.070 | 0.0594 | 106 | 281 | 0.954 | 0.281 | 4.668 | 8.757 |
|  | L | 1.14 | 22.23 | 19.94 | 0.070 | 0.0628 | 75 | 312 | 0.677 | 0.312 | 3.234 | 6.061 |
|  | M | 0.81 | 22.23 | 20.60 | 0.070 | 0.0646 | 55 | 333 | 0.488 | 0.333 | 2.303 | 4.309 |
| 1 | K | 1.65 | 28.58 | 25.27 | 0.090 | 0.0792 | 139 | 502 | 1.249 | 0.502 | 3.634 | 6.812 |
|  | L | 1.27 | 28.58 | 26.04 | 0.090 | 0.0817 | 109 | 532 | 0.973 | 0.532 | 2.792 | 5.240 |
|  | M | 0.89 | 28.58 | 26.80 | 0.090 | 0.0841 | 77 | 564 | 0.691 | 0.564 | 1.958 | 3.668 |
| 1 1/4 | K | 1.65 | 34.93 | 31.62 | 0.110 | 0.0994 | 173 | 785 | 1.543 | 0.785 | 2.972 | 5.571 |
|  | L | 1.40 | 34.93 | 32.13 | 0.110 | 0.1009 | 147 | 811 | 1.316 | 0.811 | 2.517 | 4.716 |
|  | M | 1.07 | 34.93 | 32.79 | 0.110 | 0.1030 | 114 | 845 | 1.015 | 0.845 | 1.924 | 3.599 |
|  | DWV | 1.02 | 34.93 | 32.89 | 0.110 | 0.1033 | 108 | 850 | 0.967 | 0.850 | 1.827 | 3.427 |
| 1 1/2 | K | 1.83 | 41.28 | 37.62 | 0.130 | 0.1183 | 226 | 1 111 | 2.025 | 1.111 | 2.786 | 5.226 |
|  | L | 1.52 | 41.28 | 38.23 | 0.130 | 0.1201 | 190 | 1 148 | 1.701 | 1.148 | 2.324 | 4.351 |
|  | M | 1.24 | 41.28 | 38.79 | 0.130 | 0.1219 | 157 | 1 181 | 1.399 | 1.182 | 1.896 | 3.558 |
|  | DWV | 1.07 | 41.28 | 39.14 | 0.130 | 0.1228 | 135 | 1 203 | 1.204 | 1.203 | 1.627 | 3.048 |
| 2 | K | 2.11 | 53.98 | 49.76 | 0.170 | 0.1564 | 343 | 1 945 | 3.070 | 1.945 | 2.455 | 4.606 |
|  | L | 1.78 | 53.98 | 50.42 | 0.170 | 0.1585 | 292 | 1 997 | 2.606 | 1.997 | 2.069 | 3.951 |
|  | M | 1.47 | 53.98 | 51.03 | 0.170 | 0.1603 | 243 | 2 045 | 2.171 | 2.045 | 1.717 | 3.220 |
|  | DWV | 1.07 | 53.98 | 51.84 | 0.170 | 0.1628 | 177 | 2 111 | 1.585 | 2.111 | 1.241 | 2.331 |
| 2 1/2 | K | 2.41 | 66.68 | 61.85 | 0.209 | 0.1942 | 487 | 3 004 | 4.35 | 3.004 | 2.275 | 4.268 |
|  | L | 2.03 | 66.68 | 62.61 | 0.209 | 0.1966 | 413 | 3 079 | 3.69 | 3.079 | 1.917 | 3.592 |
|  | M | 1.65 | 66.68 | 63.37 | 0.209 | 0.1990 | 337 | 3 154 | 3.02 | 3.154 | 1.558 | 2.917 |
| 3 | K | 2.77 | 79.38 | 73.84 | 0.249 | 0.2320 | 666 | 4 282 | 5.96 | 4.282 | 2.193 | 4.109 |
|  | L | 2.29 | 79.38 | 74.80 | 0.249 | 0.2350 | 554 | 4 395 | 4.95 | 4.395 | 1.813 | 3.392 |
|  | M | 1.83 | 79.38 | 75.72 | 0.249 | 0.2378 | 446 | 4 503 | 3.98 | 4.503 | 1.448 | 2.717 |
|  | DWV | 1.14 | 79.38 | 77.09 | 0.249 | 0.2423 | 281 | 4 667 | 2.51 | 4.667 | 0.903 | 1.696 |
| 3 1/2 | K | 3.05 | 92.08 | 85.98 | 0.289 | 0.2701 | 852 | 5 806 | 7.62 | 5.806 | 2.082 | 3.903 |
|  | L | 2.54 | 92.08 | 87.00 | 0.289 | 0.2733 | 714 | 5 944 | 6.39 | 5.944 | 1.738 | 3.254 |
|  | M | 2.11 | 92.08 | 87.86 | 0.289 | 0.2761 | 596 | 6 063 | 5.33 | 6.063 | 1.441 | 2.703 |
| 4 | K | 3.40 | 104.78 | 97.97 | 0.329 | 0.3078 | 1084 | 7 538 | 9.69 | 7.538 | 2.041 | 3.827 |
|  | L | 2.79 | 104.78 | 99.19 | 0.329 | 0.3115 | 895 | 7 727 | 8.00 | 7.727 | 1.675 | 3.144 |
|  | M | 2.41 | 104.78 | 99.95 | 0.329 | 0.3139 | 776 | 7 846 | 6.94 | 7.846 | 1.448 | 2.717 |
|  | DWV | 1.47 | 104.78 | 101.83 | 0.329 | 0.3200 | 478 | 8 144 | 4.27 | 8.144 | 0.883 | 1.655 |
| 5 | K | 4.06 | 130.18 | 122.05 | 0.409 | 0.3834 | 1610 | 11 699 | 14.39 | 11.70 | 1.965 | 3.682 |
|  | L | 3.18 | 130.18 | 123.83 | 0.409 | 0.3889 | 1266 | 12 042 | 11.32 | 12.04 | 1.531 | 2.875 |
|  | M | 2.77 | 130.18 | 124.64 | 0.409 | 0.3917 | 1108 | 12 201 | 9.91 | 12.20 | 1.338 | 2.510 |
|  | DWV | 1.83 | 130.18 | 126.52 | 0.409 | 0.3975 | 737 | 12 572 | 6.59 | 12.57 | 0.883 | 1.655 |
| 6 | K | 4.88 | 155.58 | 145.82 | 0.489 | 0.4581 | 2309 | 16 701 | 20.64 | 16.70 | 1.972 | 3.696 |
|  | L | 3.56 | 155.58 | 148.46 | 0.489 | 0.4663 | 1698 | 17 311 | 15.18 | 17.31 | 1.434 | 2.696 |
|  | M | 3.10 | 155.58 | 149.38 | 0.489 | 0.4694 | 1484 | 17 525 | 13.27 | 17.53 | 1.255 | 2.351 |
|  | DWV | 2.11 | 155.58 | 151.36 | 0.489 | 0.4755 | 1016 | 17 993 | 9.09 | 17.99 | 0.855 | 1.600 |
| 8 | K | 6.88 | 206.38 | 192.61 | 0.648 | 0.6050 | 4314 | 29 137 | 38.56 | 29.14 | 2.096 | 3.930 |
|  | L | 5.08 | 206.38 | 196.22 | 0.648 | 0.6163 | 3212 | 30 238 | 28.71 | 30.24 | 1.544 | 2.903 |
|  | M | 4.32 | 206.38 | 197.74 | 0.648 | 0.6212 | 2741 | 30 710 | 24.50 | 30.71 | 1.317 | 2.468 |
|  | DWV | 2.77 | 206.38 | 200.84 | 0.648 | 0.6309 | 1771 | 31 680 | 15.83 | 31.62 | 0.841 | 1.579 |
| 10 | K | 8.59 | 257.18 | 240.00 | 0.808 | 0.7541 | 6705 | 45 241 | 59.93 | 45.15 | 2.096 | 3.937 |
|  | L | 6.35 | 257.18 | 244.48 | 0.808 | 0.7681 | 5004 | 46 942 | 44.73 | 46.94 | 1.551 | 2.910 |
|  | M | 5.38 | 257.18 | 246.41 | 0.808 | 0.7742 | 4259 | 47 686 | 38.07 | 47.69 | 1.317 | 2.468 |
| 12 | K | 10.29 | 307.98 | 287.40 | 0.968 | 0.9028 | 9621 | 64 873 | 85.99 | 64.87 | 2.103 | 3.937 |
|  | L | 7.11 | 307.98 | 293.75 | 0.968 | 0.9229 | 6722 | 67 771 | 60.09 | 67.77 | 1.455 | 2.724 |
|  | M | 6.45 | 307.98 | 295.07 | 0.968 | 0.9269 | 6112 | 68 382 | 54.63 | 68.38 | 1.317 | 2.468 |

Working Pressure[a,b,c] ASTM B88 to 120°C

[a] When using soldered or brazed fittings, the joint determines the limiting pressure.
[b] Working pressures were calculated using ASME *Standard* B31.9 allowable stresses. A 5% mill tolerance has been used on the wall thickness. Higher tube ratings can be calculated using the allowable stress for lower temperatures.
[c] If soldered or brazed fittings are used on hard drawn tubing, use the annealed ratings. Full-tube allowable pressures can be used with suitably rated flare or compression-type fittings.

### Table 9-10 SI  Application of Pipe, Fittings, and Valves for Heating and Air Conditioning
*(Table 2, Chapter 46, 2012 ASHRAE Handbook—HVAC Systems and Equipment, SI version)*

| Application | Pipe Material | Type | Joint Type | Fitting Class | Fitting Material | System Temperature, °C | Maximum Pressure at Temperature[a], kPa (gage) |
|---|---|---|---|---|---|---|---|
| **Recirculating Water** | Steel (CW) | Standard | Thread | 125 | Cast iron | 120 | 860 |
| 50 mm and smaller | Copper, hard | Type L | Braze or silver solder[b] | | Wrought copper | 120 | 1030 |
| | PVC | Sch 80 | Solvent | Sch 80 | PVC | 24 | |
| | CPVC | Sch 80 | Solvent | Sch 80 | CPVC | 65 | |
| | PB | SDR-11 | Heat fusion | | PB | 70 | |
| | | | Insert crimp | | Metal | 70 | |
| 65 to 300 mm | A53 B ERW Steel | Standard | Weld | Standard | Wrought steel | 120 | 2760 |
| | | | Flange | 150 | Wrought steel | 120 | 1720 |
| | | | Flange | 125 | Cast iron | 120 | 1200 |
| | | | Flange | 250 | Cast iron | 120 | 2760 |
| | | | Groove | | MI or ductile iron | 110 | 2070 |
| | PB | SDR-11 | Heat fusion | | PB | 70 | |
| **Steam and Condensate** | Steel (CW) | Standard[c] | Thread | 125 | Cast iron | | 620 |
| 50 mm and smaller | | | Thread | 150 | Malleable iron | | 620 |
| | A53 B ERW Steel | Standard[c] | Thread | 125 | Cast iron | | 690 |
| | | | Thread | 150 | Malleable iron | | 860 |
| | A53 B ERW Steel | XS | Thread | 250 | Cast iron | | 1380 |
| | | | Thread | 300 | Malleable iron | | 1720 |
| 65 to 300 mm | Steel | Standard | Weld | Standard | Wrought steel | | 1720 |
| | | | Flange | 150 | Wrought steel | | 1380 |
| | | | Flange | 125 | Cast iron | | 690 |
| | A53 B ERW Steel | XS | Weld | XS | Wrought steel | | 4830 |
| | | | Flange | 300 | Wrought steel | | 3450 |
| | | | Flange | 250 | Cast iron | | 1380 |
| **Refrigerant** | Copper, hard | Type L or K | Braze | | Wrought copper | | |
| | A53 B SML Steel | Standard | Weld | | Wrought steel | | |
| **Underground Water** | | | | | | | |
| Through 300 mm | Copper, hard | Type K | Braze or silver solder[b] | | Wrought copper | 24 | 2410 |
| Through 150 mm | Ductile iron | Class 50 | MJ | MJ | Cast iron | 24 | 1720 |
| | PB | SDR 9, 11 | Heat fusion | | PB | 24 | |
| | | SDR 7, 11.5 | Insert crimp | | Metal | 24 | |
| **Potable Water,** | Copper, hard | Type L | Braze or silver solder[b] | | Wrought copper | 24 | 2410 |
| **Inside Building** | Steel, galvanized | Standard | Thread | 125 | Galv. cast iron | 24 | 860 |
| | | | | 150 | Galv. mall. iron | 24 | 860 |
| | PB | SDR-11 | Heat fusion | | PB | 24 | |
| | | | Insert crimp | | Metal | 24 | |

[a] Maximum allowable working pressures have been derated in this table. Higher system pressures can be used for lower temperatures and smaller pipe sizes. Pipe, fittings, joints, and valves must all be considered.

[b] Lead- and antimony-based solders should not be used for potable water systems. Brazing and silver solders should be used.

[c] Extra strong pipe is recommended for all threaded condensate piping to allow for corrosion.

# Duct and Pipe Sizing

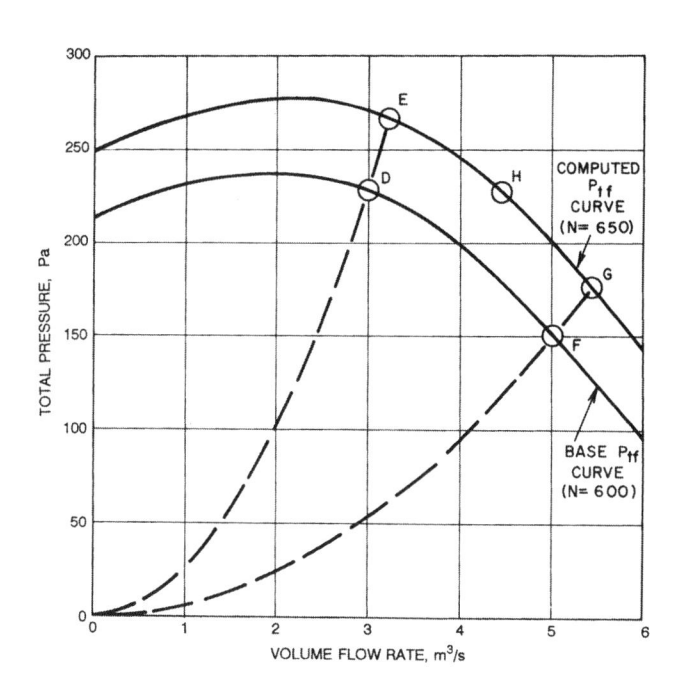

*Fig. 9-6 SI Application of Fan Laws*
(Figure 4, Chapter 21, 2012 ASHRAE Handbook—HVAC Systems and Equipment, SI version)

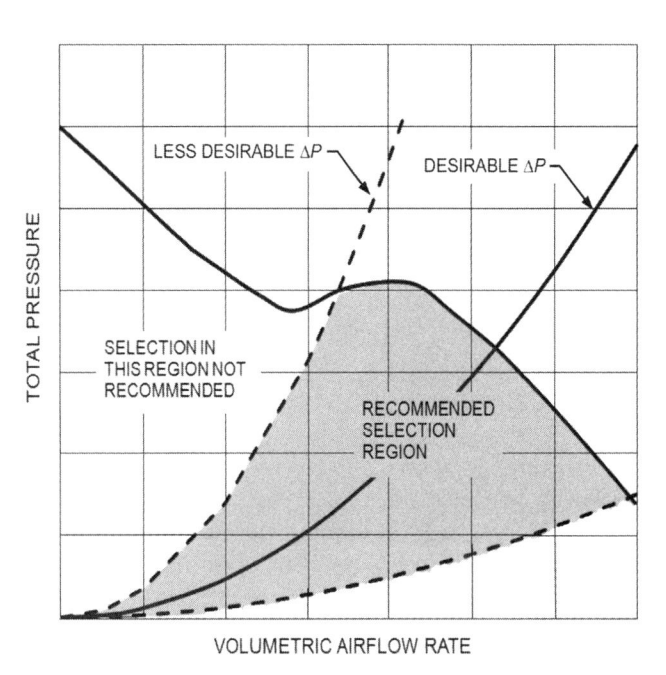

*Fig. 9-12 SI Desirable Combination of $p_{tf}$ and $\Delta p$ Curves*
(Figure 14, Chapter 21, 2012 ASHRAE Handbook—HVAC Systems and Equipment, SI version)

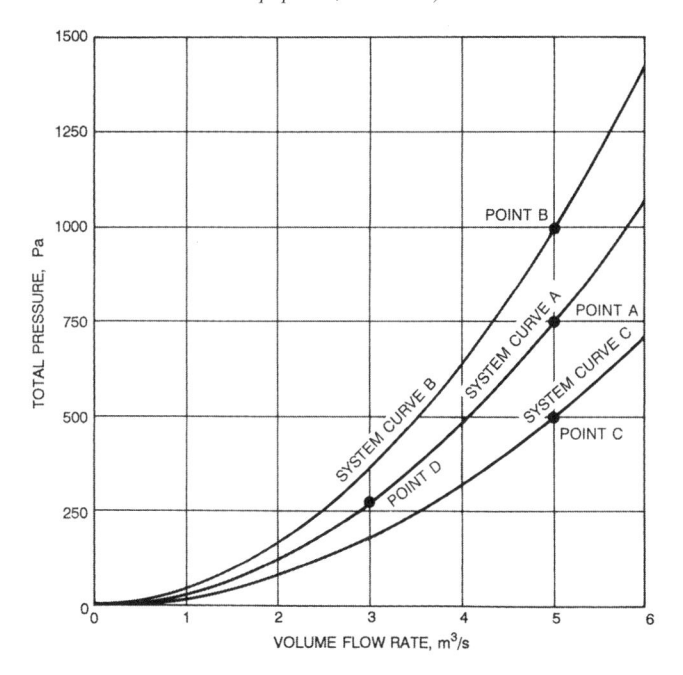

*Fig. 9-8 SI System Total Pressure Loss ($\Delta p$) Curves*
(Figure 10, Chapter 21, 2012 ASHRAE Handbook—HVAC Systems and Equipment, SI version)

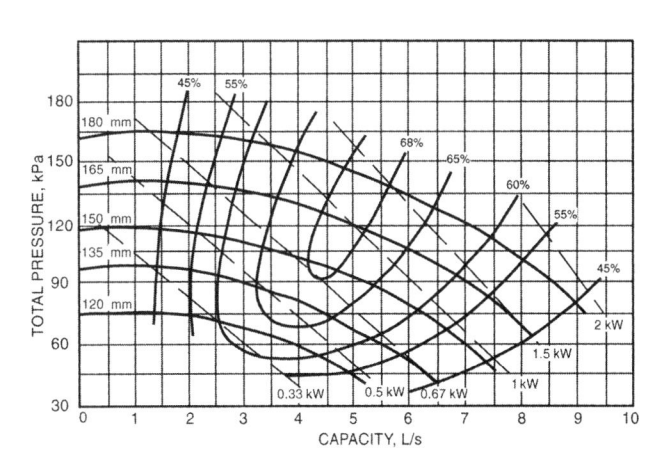

*Fig. 9-13 SI Pump Performance Curves*
(Figure 11, Chapter 44, 2012 ASHRAE Handbook—HVAC Systems and Equipment, SI version)

# Chapter 10

# ECONOMIC ANALYSES AND LIFE-CYCLE COSTS

This chapter presents the fundamentals for doing a simple engineering economic analysis for heating and air-conditioning systems and refrigeration installations. This information is reproduced from Chapter 37 of the 2011 *ASHRAE Handbook—HVAC Applications*.

## 10.1 Introduction

Owning and operating cost information for the HVAC system should be part of the investment plan of a facility. This information can be used for preparing annual budgets, managing assets, and selecting design options. Table 10-1 shows a representative form that summarizes these costs.

A properly engineered system must also be economical, but this is difficult to assess because of the complexities surrounding effective money management and the inherent difficulty of predicting future operating and maintenance expenses. Complex tax structures and the time value of money can affect the final engineering decision. This does not imply use of either the cheapest or the most expensive system; instead, it demands intelligent analysis of financial objectives and the owner's requirements.

Certain tangible and intangible costs or benefits must also be considered when assessing owning and operating costs. Local codes may require highly skilled or certified operators for specific types of equipment. This could be a significant cost over the life of the system. Similarly, intangible items such as aesthetics, acoustics, comfort, safety, security, flexibility, and environmental impact may vary by location and be important to a particular building or facility.

## 10.2 Owning Costs

The following elements must be established to calculate annual owning costs: (1) initial cost, (2) analysis or study period, (3) interest or discount rate, and (4) other periodic costs such as insurance, property taxes, refurbishment, or disposal fees. Once established, these elements are coupled with operating costs to develop an economic analysis, which may be a simple payback evaluation or an in-depth analysis such as outlined in the section on Economic Analysis Techniques.

### 10.2.1 Initial Cost

Major decisions affecting annual owning and operating costs for the life of the building must generally be made before completing contract drawings and specifications. To achieve the best performance and economics, alternative methods of solving the engineering problems peculiar to each project should be compared in the early stages of design. Oversimplified estimates can lead to substantial errors in evaluating the system.

The evaluation should lead to a thorough understanding of installation costs and accessory requirements for the system(s) under consideration. Detailed lists of materials, controls, space and structural requirements, services, installation labor, and so forth can be prepared to increase accuracy in preliminary cost estimates. A reasonable estimate of capital cost of components may be derived from cost records of recent installations of comparable design or from quotations submitted by manufacturers and contractors, or by consulting commercially available cost-estimating guides and software. Table 10-2 shows a representative checklist for initial costs.

### 10.2.2 Analysis Period

The time frame over which an economic analysis is performed greatly affects the results. The analysis period is usually determined by specific objectives, such as length of planned ownership or loan repayment period. However, as the length of time in the analysis period increases, there is a diminishing effect on net present-value calculations. The chosen analysis period is often unrelated to the equipment depreciation period or service life, although these factors may be important in the analysis.

## 10.3 Service Life

For many years, the Owning and Operating Costs chapter of *ASHRAE Handbook—HVAC Applications* included estimates of service lives for various HVAC system components, based on a survey conducted in 1976 under ASHRAE research project RP-186 (Akalin 1978). These estimates have been useful to a generation of practitioners, but changes in technology, materials, manufacturing techniques, and maintenance practices now call into question the continued validity of the original estimates. Consequently, ASHRAE research project TRP-1237 developed an Internet-based data collection tool

### Table 10-1 Owning and Operating Cost Data and Summary

**OWNING COSTS**

I. Initial Cost of System _____

II. Periodic Costs
   A. Income taxes _____
   B. Property taxes _____
   C. Insurance _____
   D. Rent _____
   E. Other periodic costs _____
   **Total Periodic Costs** _____

III. Replacement Cost _____
IV. Salvage Value _____

**Total Owning Costs** _____

**OPERATING COSTS**

V. Annual Utility, Fuel, Water, etc., Costs
   A. Utilities
     1. Electricity _____
     2. Natural gas _____
     3. Water/sewer _____
     4. Purchased steam _____
     5. Purchased hot/chilled water _____
   B. Fuels
     1. Propane _____
     2. Fuel oil _____
     3. Diesel _____
     4. Coal _____
   C. On-site generation of electricity _____
   D. Other utility, fuel, water, etc., costs _____
   *Total* _____

VI. Annual Maintenance Allowances/Costs
   A. In-house labor _____
   B. Contracted maintenance service _____
   C. In-house materials _____
   D. Other maintenance allowances/costs _____
     (e.g., water treatment)
   *Total* _____

VII. Annual Administration Costs _____

**Total Annual Operating Costs** _____

**TOTAL ANNUAL OWNING AND OPERATING COSTS** _____

### Table 10-2 Initial Cost Checklist

**Energy and Fuel Service Costs**
Fuel service, storage, handling, piping, and distribution costs
Electrical service entrance and distribution equipment costs
Total energy plant

**Heat-Producing Equipment**
Boilers and furnaces
Steam-water converters
Heat pumps or resistance heaters
Makeup air heaters
Heat-producing equipment auxiliaries

**Refrigeration Equipment**
Compressors, chillers, or absorption units
Cooling towers, condensers, well water supplies
Refrigeration equipment auxiliaries

**Heat Distribution Equipment**
Pumps, reducing valves, piping, piping insulation, etc.
Terminal units or devices

**Cooling Distribution Equipment**
Pumps, piping, piping insulation, condensate drains, etc.
Terminal units, mixing boxes, diffusers, grilles, etc.

**Air Treatment and Distribution Equipment**
Air heaters, humidifiers, dehumidifiers, filters, etc.
Fans, ducts, duct insulation, dampers, etc.
Exhaust and return systems
Heat recovery systems

**System and Controls Automation**
Terminal or zone controls
System program control
Alarms and indicator system
Energy management system

**Building Construction and Alteration**
Mechanical and electric space
Chimneys and flues
Building insulation
Solar radiation controls
Acoustical and vibration treatment
Distribution shafts, machinery foundations, furring

### Table 10-3 Median Service Life

| Equipment Type | Median Service Life, Years | Total No. of Units | No. of Units Replaced |
|---|---|---|---|
| DX air distribution equipment | >24 | 1907 | 284 |
| Chillers, centrifugal | >25 | 234 | 34 |
| Cooling towers, metal | >22 | 170 | 24 |
| Boilers, hot-water, steel gas-fired | >22 | 117 | 24 |
| Controls, pneumatic | >18 | 101 | 25 |
| electronic | >7 | 68 | 6 |
| Potable hot-water heaters, electric | >21 | 304 | 36 |

and database on HVAC equipment service life and maintenance costs, to allow equipment owning and operating cost data to be continually updated and current. The database was seeded with information gathered from a sample of 163 commercial office buildings located in major metropolitan areas across the United States. Abramson et al. (2005) provide details on the distribution of building size, age, and other characteristics. Table 10-3 presents estimates of median service life for various HVAC components in this sample.

Median service life in Table 10-3 is based on analysis of survival curves, which take into account the units still in service and the units replaced at each age (Hiller 2000). Conditional and total survival rates are calculated for each age, and the percent survival over time is plotted. Units still in service are included up to the point where the age is equal to their current age at the time of the study. After that point, these units are censored (removed from the population). Median service life in this table indicates the highest age at which the survival rate remains at or above 50% while the sample size is 30 or more.

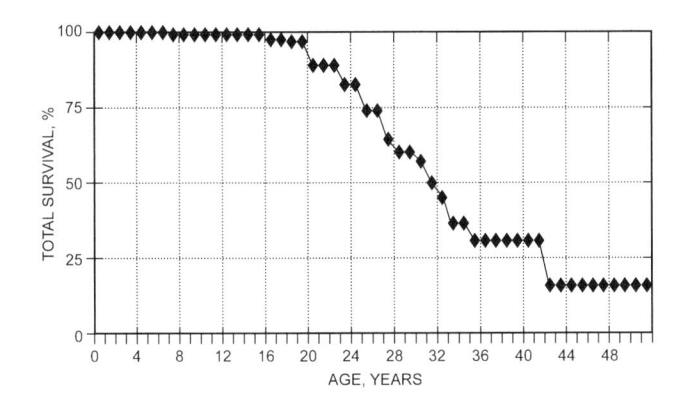

*Fig. 10-1 Survival Curve for Centrifugal Chillers*
*[Based on data in Abramson et al. (2005)]*

There is no hard-and-fast rule about the number of units needed in a sample before it is considered statistically large enough to be representative, but usually the number should be larger than 25 to 30 (Lovvorn and Hiller 2002). This rule-of-thumb is used because each unit removal represents greater than a 3% change in survival rate as the sample size drops below 30, and that percentage increases rapidly as the sample size gets even smaller.

The database initially developed and seeded under research project TRP-1237 (Abramson et al. 2005) is now available online, providing engineers with equipment service life and annual maintenance costs for a variety of building types and HVAC systems. The database can be accessed at www.ashrae.org/database.

As of the end of 2009 this database contained more than 300 building types, with service life data on more than 38,000 pieces of equipment.

The database allows users to access up-to-date information to determine a range of statistical values for equipment owning and operating costs. Users are encouraged to contribute their own service life and maintenance cost data, further expanding the utility of this tool. Over time, this input will provide sufficient service life and maintenance cost data to allow comparative analysis of many different HVAC systems types in a broad variety of applications. Data can be entered by logging into the database and registering, which is free. With this, ASHRAE is providing the necessary methods and information to assist in using life-cycle analysis techniques to help select the most appropriate HVAC system for a specific application. This system of collecting data also greatly reduces the time between data collection and when users can access the information.

Figure 10-1 presents the survival curve for centrifugal chillers, based on data in Abramson et al. (2005). The point at which survival rate drops to 50% based on all data in the survey is 31 years. However, because the sample size drops below the statistically relevant number of 30 units at 25 years, the median service life of centrifugal chillers can only be stated with confidence as >25 years.

Table 10-4 compares the estimates of median service life in Abramson et al. (2005) with those developed with those in Akalin (1978). Most differences are on the order of one to five years.

Estimated service life of new equipment or components of systems not listed in Table 10-3 or 10-4 may be obtained from manufacturers, associations, consortia, or governmental agencies. Because of the proprietary nature of information from some of these sources, the variety of criteria used in compiling the data, and the diverse objectives in disseminating them, extreme care is necessary in comparing service life from different sources. Designs, materials, and components of equipment listed in Tables 10-3 and 10-4 have changed over time and may have altered the estimated service lives of those equipment categories. Therefore, establishing equivalent comparisons of service life is important.

As noted, service life is a function of the time when equipment is replaced. Replacement may be for any reason, including, but not limited to, failure, general obsolescence, reduced reliability, excessive maintenance cost, and changed system requirements (e.g., building characteristics, energy prices, environmental considerations). Service lives shown in the tables are based on the age of the equipment when it was replaced, regardless of the reason it was replaced.

Locations in potentially corrosive environments and unique maintenance variables affect service life. Examples include the following:

- **Coastal and marine environments**, especially in tropical locations, are characterized by abundant sodium chloride (salt) that is carried by sea spray, mist, or fog.
- Many owners require equipment specifications stating that HVAC equipment located along coastal waters will have corrosion-resistant materials or coatings. Design criteria for systems installed under these conditions should be carefully considered.
- **Industrial** applications provide many challenges to the HVAC designer. It is very important to know if emissions from the industrial plant contain products of combustion from coal, fuel oils, or releases of sulfur oxides ($SO_2$, $SO_3$) and nitrogen oxides ($NO_x$) into the atmosphere. These gases typically accumulate and return to the ground in the form of acid rain or dew.
- Not only is it important to know the products being emitted from the industrial plant being designed, but also the adjacent upwind or downwind facilities. HVAC system design for a plant located downwind from a paper mill requires extraordinary corrosion protection or recognition of a reduced service life of the HVAC equipment.
- **Urban** areas generally have high levels of automotive emissions as well as abundant combustion by-products. Both of these contain elevated sulfur oxide and nitrogen oxide concentrations.
- **Maintenance** factors also affect life expectancy. The HVAC designer should temper the service life expectancy of equipment with a maintenance factor. To achieve the estimated service life values in Table 10-3, HVAC equipment must be maintained properly, including good filter-changing practices and good

Table 10-4 Comparison of Service Life Estimates

| Equipment Item | Abramson et al. (2005) | Akalin (1978) | Equipment Item | Abramson et al. (2005) | Akalin (1978) | Equipment Item | Abramson et al. (2005) | Akalin (1978) |
|---|---|---|---|---|---|---|---|---|
| **Air Conditioners** | | | **Air Terminals** | | | **Condensers** | | |
| Window unit | N/A* | 10 | Diffusers, grilles, and registers | N/A* | 27 | Air-cooled | N/A | 20 |
| Residential single or split package | N/A* | 15 | Induction and fan-coil units | N/A* | 20 | Evaporative | N/A* | 20 |
| Commercial through-the-wall | N/A* | 15 | VAV and double-duct boxes | N/A* | 20 | **Insulation** | | |
| Water-cooled package | >24 | 15 | **Air washers** | N/A* | 17 | Molded | N/A* | 20 |
| **Heat pumps** | | | **Ductwork** | N/A* | 30 | Blanket | N/A* | 24 |
| Residential air-to-air | N/A* | 15[b] | **Dampers** | N/A* | 20 | **Pumps** | | |
| Commercial air-to-air | N/A* | 15 | **Fans** | N/A* | | Base-mounted | N/A* | 20 |
| Commercial water-to-air | >24 | 19 | Centrifugal | N/A* | 25 | Pipe-mounted | N/A* | 10 |
| **Roof-top air conditioners** | | | Axial | N/A* | 20 | Sump and well | N/A* | 10 |
| Single-zone | N/A* | 15 | Propeller | N/A* | 15 | Condensate | N/A* | 15 |
| Multizone | N/A* | 15 | Ventilating roof-mounted | N/A* | 20 | **Reciprocating engines** | N/A* | 20 |
| **Boilers, Hot-Water (Steam)** | | | **Coils** | | | **Steam turbines** | N/A* | 30 |
| Steel water-tube | >22 | 24 (30) | DX, water, or steam | N/A* | 20 | **Electric motors** | N/A* | 18 |
| Steel fire-tube | | 25 (25) | Electric | N/A* | 15 | **Motor starters** | N/A* | 17 |
| Cast iron | N/A* | 35 (30) | **Heat Exchangers** | | | **Electric transformers** | N/A* | 30 |
| Electric | N/A* | 15 | Shell-and-tube | N/A* | 24 | **Controls** | | |
| **Burners** | N/A* | 21 | **Reciprocating compressors** | N/A* | 20 | Pneumatic | N/A* | 20 |
| **Furnaces** | | | **Packaged Chillers** | | | Electric | N/A* | 16 |
| Gas- or oil-fired | N/A* | 18 | Reciprocating | N/A* | 20 | Electronic | N/A* | 15 |
| **Unit heaters** | | | Centrifugal | >25 | 23 | **Valve actuators** | | |
| Gas or electric | N/A* | 13 | Absorption | N/A* | 23 | Hydraulic | N/A* | 15 |
| Hot-water or steam | N/A* | 20 | **Cooling Towers** | | | Pneumatic | N/A* | 20 |
| **Radiant heaters** | | | Galvanized metal | >22 | 20 | Self-contained | | 10 |
| Electric | N/A* | 10 | Wood | N/A* | 20 | | | |
| Hot-water or steam | N/A* | 25 | Ceramic | N/A* | 34 | | | |

*N/A: Not enough data yet in Abramson et al. (2005). Note that data from Akalin (1978) for these categories may be outdated and not statistically relevant. Use these data with caution until enough updated data are accumulated in Abramson et al.

maintenance procedures. For example, chilled-water coils with more than four rows and close fin spacing are virtually impossible to clean even using extraordinary methods; they are often replaced with multiple coils in series, with a maximum of four rows and lighter fin spacing.

## 10.4 Depreciation

Depreciation periods are usually set by federal, state, or local tax laws, which change periodically. Applicable tax laws should be consulted for information on depreciation.

## 10.5 Interest or Discount Rate

Most major economic analyses consider the opportunity cost of borrowing money, inflation, and the time value of money. **Opportunity cost** of money reflects the earnings that investing (or lending) the money can produce. **Inflation** (price escalation) decreases the purchasing or investing power (value) of future money because it can buy less in the future. **Time value** of money reflects the fact that money received today is more useful than the same amount received a year from now, even with zero inflation, because the money is available earlier for reinvestment.

The cost or value of money must also be considered. When borrowing money, a percentage fee or interest rate must normally be paid. However, the interest rate may not necessarily be the correct cost of money to use in an economic analysis. Another factor, called the **discount rate**, is more commonly used to reflect the true cost of money [see Fuller and Petersen (1996) for detailed discussions]. Discount rates used for analyses vary depending on individual investment, profit, and other opportunities. Interest rates, in contrast, tend to be more centrally fixed by lending institutions.

To minimize the confusion caused by the vague definition and variable nature of discount rates, the U.S. government has specified particular discount rates to be used in economic analyses relating to federal expenditures. These discount rates are updated annually (Rushing et al. 2010) but may not be appropriate for private-sector economic analyses.

## 10.6 Periodic Costs

Regularly or periodically recurring costs include insurance, property taxes, income taxes, rent, refurbishment expenses, disposal fees (e.g., refrigerant recycling costs), occasional major repair costs, and decommissioning expenses.

**Insurance.** Insurance reimburses a property owner for a financial loss so that equipment can be repaired or replaced. Insurance often indemnifies the owner from liability, as well.

# Economic Analyses and Life-Cycle Costs

Financial recovery may include replacing income, rents, or profits lost because of property damage or machinery failure.

Some of the principal factors that influence the total annual insurance premium are building size, construction materials, amount and size of mechanical equipment, geographic location, and policy deductibles. Some regulations set minimum required insurance coverage and premiums that may be charged for various forms of insurable property.

**Property Taxes.** Property taxes differ widely and may be collected by one or more agencies, such as state, county, or local governments or special assessment districts. Furthermore, property taxes may apply to both real (land, buildings) and personal (everything else) property. Property taxes are most often calculated as a percentage of assessed value, but are also determined in other ways, such as fixed fees, license fees, registration fees, etc. Moreover, definitions of assessed value vary widely in different geographic areas. Tax experts should be consulted for applicable practices in a given area.

**Income Taxes.** Taxes are generally imposed in proportion to net income, after allowance for expenses, depreciation, and numerous other factors. Special tax treatment is often granted to encourage certain investments. Income tax professionals can provide up-to-date information on income tax treatments.

**Other Periodic Costs.** Examples of other costs include changes in regulations that require unscheduled equipment refurbishment to eliminate use of hazardous substances, and disposal costs for such substances.

**Replacement Costs and Salvage Value.** Replacement costs and salvage value should be evaluated when calculating owning cost. Replacement cost is the cost to remove existing equipment and install new equipment. Salvage value is the value of equipment or its components for recycling or other uses. Equipment's salvage value may be negative when removal, disposal, or decommissioning costs are considered.

## 10.7 Operating Costs

Operating costs are those incurred by the actual operation of the system. They include costs of fuel and electricity, wages, supplies, water, material, and maintenance parts and services. Energy is a large part of total operating costs. Chapter 19 of the 2013 *ASHRAE Handbook—Fundamentals* outlines how fuel and electrical requirements are estimated. Because most energy management activities are dictated by economics, the design engineer must understand the utility rates that apply. Electric rates are usually more complex than gas or water rates. In addition to general commercial or institutional electric rates, special rates may exist such as time of day, interruptible service, on-peak/off-peak, summer/winter, and peak demand. Electric rate schedules vary widely in North America. The design engineer should work with local utility companies to identify the most favorable rates and to understand how to qualify for them.

### 10.7.1 Electrical Energy

The total cost of electricity is determined by a rate schedule and is usually a combination of several components: consumption (kilowatt-hours), demand (kilowatts) fuel adjustment charges, special allowances or other adjustments, and applicable taxes. Of these, consumption and demand are the major cost components and the ones the owner or facility manager may be able to affect.

**Electricity Consumption Charges.** Most electric rates have step-rate schedules for consumption, and the cost of the last unit consumed may be substantially different from that of the first. The last unit is usually cheaper than the first because the fixed costs to the utility may already have been recovered from earlier consumption costs. Because of this, the energy analysis cannot use average costs to accurately predict savings from implementation of energy conservation measures. Average costs will overstate the savings possible between alternative equipment or systems; instead, marginal (or incremental) costs must be used.

To reflect time-varying operating costs or to encourage peak shifting, electric utilities may charge different rates for consumption according to the time of use and season, with higher costs occurring during the peak period of use.

**Fuel Adjustment Charge.** Because of substantial variations in fuel prices, electric utilities may apply a fuel adjustment charge to recover costs. This adjustment may not be reflected in the rate schedule. The fuel adjustment is usually a charge per unit of consumption and may be positive or negative, depending on how much of the actual fuel cost is recovered in the energy consumption rate. The charge may vary monthly or seasonally.

**Allowances or Adjustments.** Special discounts or rates may be available for customers who can receive power at higher voltages or for those who own transformers or similar equipment. Special rates or riders may be available for specific interruptible loads such as domestic water heaters.

Certain facility electrical systems may produce a low power factor [i.e., ratio of real (active) kilowatt power to apparent (reactive) kVA power], which means that the utility must supply more current on an intermittent basis, thus increasing their costs. These costs may be passed on as an adjustment to the utility bill if the power factor is below a level established by the utility.

When calculating power bills, utilities should be asked to provide detailed cost estimates for various consumption levels. The final calculation should include any applicable special rates, allowances, taxes, and fuel adjustment charges.

**Demand Charges.** Electric rates may also have demand charges based on the customer's peak kilowatt demand. Whereas consumption charges typically cover the utility's operating costs, demand charges typically cover the owning costs.

Demand charges may be formulated in a variety of ways:
- *Straight charge*. Cost per kilowatt per month, charged for the peak demand of the month.

**Table 10-5 Electricity Data Consumption and Demand for Atlanta Example Building, 2003 to 2004**

| | Billing Days | Consumption, kWh | Actual Demand, kW | Billing Demand, kW | Total Cost, US$ |
|---|---|---|---|---|---|
| Jan. 2003 | 29 | 57,120 | 178 | 185 | 4,118 |
| Feb. 2003 | 31 | 61,920 | 145 | 185 | 4,251 |
| Mar. 2003 | 29 | 60,060 | 140 | 185 | 4,199 |
| Apr. 2003 | 29 | 62,640 | 154 | 185 | 4,271 |
| May. 2003 | 33 | 73,440 | 161 | 185 | 4,569 |
| Jun. 2003 | 26 | 53,100 | 171 | 185 | 4,007 |
| Jul. 2003 | 32 | 67,320 | 180 | 185 | 4,400 |
| Aug. 2003 | 30 | 66,000 | 170 | 185 | 4,364 |
| Sep. 2003 | 32 | 63,960 | 149 | 171 | 4,127 |
| Oct. 2003 | 30 | 55,260 | 122 | 171 | 3,865 |
| Nov. 2003 | 27 | 46,020 | 140 | 171 | 3,613 |
| Dec. 2003 | 34 | 61,260 | 141 | 171 | 4,028 |
| **Total 2003** | 362 | 670,980 | | | 49,812 |
| Jan. 2004 | 31 | 59,040 | 145 | 171 | 3,967 |
| Feb. 2004 | 29 | 54,240 | 159 | 171 | 3,837 |
| Mar. 2004 | 20 | 37,080 | 122 | 171 | 2,584 |
| Apr. 2004 | 12 | 22,140 | 133 | 171 | 1,547 |
| May. 2004 | 34 | 64,260 | 148 | 171 | 4,110 |
| Jun. 2004 | 29 | 63,720 | 148 | 171 | 4,321 |
| Jul. 2004 | 30 | 69,120 | 169 | 169 | 4,458 |
| Aug. 2004 | 32 | 73,800 | 170 | 170 | 4,605 |
| Sep. 2004 | 29 | 64,500 | 166 | 166 | 4,281 |
| Oct. 2004 | 30 | 60,060 | 152 | 161 | 3,866 |
| Nov. 2004 | 32 | 65,760 | 128 | 161 | 4,018 |
| Dec. 2004 | 31 | 51,960 | 132 | 161 | 3,646 |
| **Total 2004** | 339 | 685,680 | | | 45,240 |

- *Excess charge.* Cost per kilowatt above a base demand, (e.g., 50 kW) which may be established each month.
- *Maximum demand (ratchet).* Cost per kilowatt for maximum annual demand, which may be reset only once a year. This established demand may either benefit or penalize the owner.
- *Combination demand.* Cost per hour of operation of demand. In addition to a basic demand charge, utilities may include further demand charges as demand-related consumption charges.

The actual demand represents the peak energy use averaged over a specific period, usually 15, 30, or 60 min. Accordingly, high electrical loads of only a few minutes' duration may never be recorded at the full instantaneous value. Alternatively, peak demand is recorded as the average of several consecutive short periods (i.e., 5 min out of each hour).

The particular method of demand metering and billing is important when load shedding or shifting devices are considered. The portion of the total bill attributed to demand may vary greatly, from 0% to as high as 70%.

- *Real-time* or *time-of-day rates.* Cost of electricity at time of use. An increasing number of utilities offer these rates. End users who can shift operations or install electric load-shifting equipment, such as thermal storage, can take advantage of such rates. Because these rates usually reflect a utility's overall load profile and possibly the availability

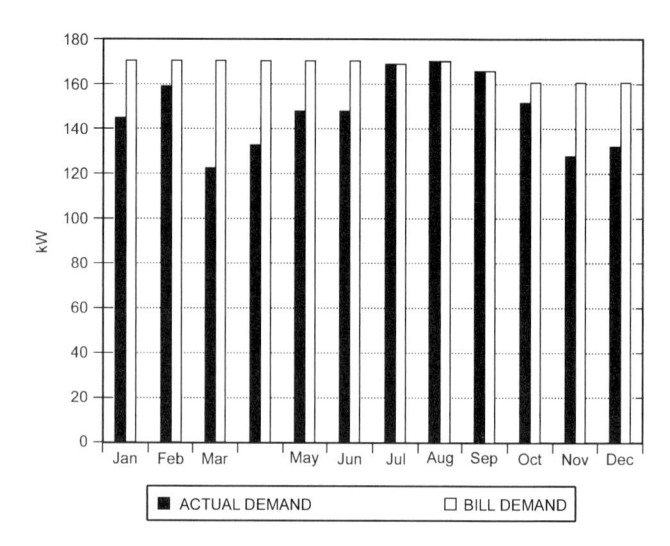

*Fig. 10-2 Bill Demand and Actual Demand for Atlanta Example Building, 2004*

of specific generating resources, contact with the supplying utility is essential to determine whether these rates are a reasonable option for a specific application.

**Understanding Electric Rates.** To illustrate a typical commercial electric rate with a ratchet, electricity consumption and demand data for an example building are presented in Table 10-5.

The example building in Table 10-5 is on a ratcheted rate, and bill demand is determined as a percentage of actual demand in the summer. How the ratchet operates is illustrated in Figure 10-2.

Table 10-5 shows that the actual demand in the first six months of 2004 had no effect on the billing demand, and therefore no effect on the dollar amount of the bill. The same is true for the last three months of the year. Because of the ratchet, the billing demand in the first half of 2004 was set the previous summer. Likewise, billing demand for the last half of 2004 and first half of 2005 was set by the peak actual demand of 180 kW in July 2003. This tells the facility manager to pay attention to demand in the summer months (June to September) and that demand is not a factor in the winter (October to May) months for this particular rate. (Note that Atlanta's climate is hot and humid; in other climates, winter electric demand is an important determinant of costs.) Consumption must be monitored all year long.

Understanding the electric rates is key when evaluating the economics of energy conservation projects. Some projects save electrical demand but not consumption; others save mostly consumption but have little effect on demand. Electric rates must be correctly applied for economic analyses to be accurate. Chapter 56 in the 2011 *ASHRAE Handbook—Applications* contains a thorough discussion of various electric rates.

### 10.7.2 Natural Gas

**Rates.** Conventional natural gas rates are usually a combination of two main components: (1) utility rate or base

# Economic Analyses and Life-Cycle Costs

charges for gas consumption and (2) purchased gas adjustment (PGA) charges.

Although gas is usually metered by volume, it is often sold by energy content. The utility rate is the amount the local distribution company charges per unit of energy to deliver the gas to a particular location. This rate may be graduated in steps; the first 100 units of gas consumed may not be the same price as the last 100 units. The PGA is an adjustment for the cost of the gas per unit of energy to the local utility. It is similar to the electric fuel adjustment charge. The total cost per unit is then the sum of the appropriate utility rate and the PGA, plus taxes and other adjustments.

**Interruptible Gas Rates and Contract/Transport Gas.** Large industrial plants usually have the ability to burn alternative fuels and can qualify for special interruptible gas rates. During peak periods of severe cold weather, these customers' supply may be curtailed by the gas utility, and they may have to switch to propane, fuel oil, or some other back-up fuel. The utility rate and PGA are usually considerably cheaper for these interruptible customers than they are for firm-rate (non-interruptible) customers.

Deregulation of the natural gas industry allows end users to negotiate for gas supplies on the open market. The customer actually contracts with a gas producer or broker and pays for the gas at the source. Transport fees must be negotiated with the pipeline companies carrying the gas to the customer's local gas utility. This can be a very complicated administrative process and is usually economically feasible only for large gas users. Some local utilities have special rates for delivering contract gas volumes through their system; others simply charge a standard utility fee (PGA is not applied because the customer has already negotiated with the supplier for the cost of the fuel itself).

When calculating natural gas bills, be sure to determine which utility rate and PGA and/or contract gas price is appropriate for the particular interruptible or firm-rate customer. As with electric bills, the final calculation should include any taxes, prompt payment discounts, or other applicable adjustments.

## 10.7.3 Other Fossil Fuels

Propane, fuel oil, and diesel are examples of other fossil fuels in widespread use. Calculating the cost of these fuels is usually much simpler than calculating typical utility rates.

The cost of the fuel itself is usually a simple charge per unit volume or per unit mass. The customer is free to negotiate for the best price. However, trucking or delivery fees must also be included in final calculations. Some customers may have their own transport trucks, but most seek the best delivered price. If storage tanks are not customer-owned, rental fees must be considered. Periodic replacement of diesel-type fuels may be necessary because of storage or shelf-life limitations and must also be considered. The final fuel cost calculation should include any of these costs that are applicable, as well as appropriate taxes.

It is usually difficult, however, to relate usage of stored fossil fuels (e.g., fuel oil) with their operating costs. This is because propane or fuel oil is bought in bulk and stored until needed, and normally not metered or measured as it is consumed, whereas natural gas and electricity are utilities and are billed for as they are used.

**Energy Source Choices.** In planning for a new facility, the designer may undertake energy master planning. One component of **energy master planning** is choice of fuels. Typical necessary decisions include, for example, whether the building should be heated by electricity or natural gas, how service hot water should be produced, whether a hybrid heating plant (i.e., a combination of both electric and gas boilers) should be considered, and whether emergency generators should be fueled by diesel or natural gas.

Engineers should consider histories or forecasts of price volatility when selecting energy sources. In addition to national trending, local energy price trends from energy suppliers can be informative. These evaluations are particularly important where relative operating costs parity exists between various fuel options, or where selecting more efficient equipment may help mitigate utility price concerns.

Many sources of historic and projected energy costs are available for reference. In addition to federal projections, utility and energy supplier annual reports and accompanying financial data may provide insight into future energy costs. Indicators such as constrained or declining energy supply or production may be key factors in projecting future energy pricing trends. Pricing patterns that suggest unusual levels of energy price volatility should be carefully analyzed and tested at extreme predicted price levels to assess potential effects on system operating costs.

## 10.7.4 Water and Sewer Costs

Water and sewer costs should not be overlooked in economic analyses. Fortunately, these rates are usually very simple and straightforward: commonly, a charge per hundred cubic feet (CCF) for water and a different charge per CCF for sewer. Because water consumption is metered and sewage is not, most rates use the water consumption quantity to compute the sewer charge. If an owner uses water that is not returned to the sewer, there may be an opportunity to receive a credit or refund. Owners frequently use irrigation meters for watering grounds when the water authority has a special irrigation rate with no sewer charge. Another opportunity that is sometimes overlooked is to separately meter makeup water for cooling towers. This can be done with an irrigation meter if the costs of setting the meter can be justified; alternatively, it may be done by installing an in-line water meter for the cooling tower, in which case the owner reports the usage annually and applies for a credit or refund.

Because of rising costs of water and sewer, water recycling and reclamation is becoming more cost effective. For example, it may now be cost effective in some circumstances to capture cooling coil condensate and pump it to a cooling tower for makeup water.

Table 10-6  Comparison of Maintenance Costs Between Studies

| Survey | Cost per ft², as Reported Mean | Median | Consumer Price Index | Cost per ft², 2004 Dollars Mean | Median |
|---|---|---|---|---|---|
| Dohrmann and Alereza (1983) | $0.32 | $0.24 | 99.6 | $0.61 | $0.46 |
| Abramson et al. (2005) | $0.47 | $0.44 | 188.9 | $0.47 | $0.44 |

### 10.7.5 Maintenance Costs

The quality of maintenance and maintenance supervision can be a major factor in overall life-cycle cost of a mechanical system. The maintenance cost of mechanical systems varies widely depending upon configuration, equipment locations, accessibility, system complexity, service duty, geography, and system reliability requirements. Maintenance costs can be difficult to predict, because each system or facility is unique.

Dohrmann and Alereza (1986) obtained maintenance costs and HVAC system information from 342 buildings located in 35 states in the United States. In 1983 U.S. dollars, data collected showed a mean HVAC system maintenance cost of $0.32/ft² per year, with a median cost of $0.24/ft² per year. Building age has a statistically significant but minor effect on HVAC maintenance costs. Analysis also indicated that building size is not statistically significant in explaining cost variation. The type of maintenance program or service agency that building management chooses can also have a significant effect on total HVAC maintenance costs. Although extensive or thorough routine and preventive maintenance programs cost more to administer, they usually extend equipment life; improve reliability; and reduce system downtime, energy costs, and overall life-cycle costs.

Some maintenance cost data are available, both in the public domain and from proprietary sources used by various commercial service providers. These sources may include equipment manufacturers, independent service providers, insurers, government agencies (e.g., the U.S. General Services Administration), and industry-related organizations [e.g., the Building Owners and Managers Association (BOMA)] and service industry publications. More traditional, widely used products and components are likely to have statistically reliable records. However, design changes or modifications necessitated by industry changes, such as alternative refrigerants, may make historical data less relevant.

Newer HVAC products, components, system configurations, control systems and protocols, and upgraded or revised system applications present an additional challenge. Care is required when using data not drawn from broad experience or field reports. In many cases, maintenance information is proprietary or was sponsored by a particular entity or group. Particular care should be taken when using such data. It is the user's responsibility to obtain these data and to determine their appropriateness and suitability for the application being considered.

ASHRAE research project TRP-1237 (Abramson et al. 2005) developed a standardized Internet-based data collection tool and database on HVAC equipment service life and maintenance costs. The database was seeded with data on 163 buildings from around the country. Maintenance cost data were gathered for total HVAC system maintenance costs from 100 facilities. In 2004 dollars, the mean HVAC maintenance cost from these data was $0.47/ft², and the median cost was $0.44/ft². Table 10-6 compares these figures with estimates reported by Dohrmann and Alereza (1983), both in terms of contemporary dollars, and in 2004 dollars, and shows that the cost per square foot varies widely between studies.

**Estimating Maintenance Costs.** Total HVAC maintenance cost for new and existing buildings with various types of equipment may be estimated several ways, using several resources. Equipment maintenance requirements can be obtained from the equipment manufacturers for large or custom pieces of equipment. Estimating in-house labor requirements can be difficult; BOMA (2003) provides guidance on this topic. Many independent mechanical service companies provide preventative maintenance contracts. These firms typically have proprietary estimating programs developed through their experience, and often provide generalized maintenance costs to engineers and owners upon request, without obligation.

When evaluating various HVAC systems during design or retrofit, the absolute magnitude of maintenance costs may not be as important as the relative costs. Whichever estimating method or resource is selected, it should be used consistently throughout any evaluation. Mixing information from different resources in an evaluation may provide erroneous results.

Applying simple costs per unit of building floor area for maintenance is highly discouraged. Maintenance costs can be generalized by system types. When projecting maintenance costs for different HVAC systems, the major system components need to be identified with a required level of maintenance. The potential long-term costs of environmental issues on maintenance costs should also be considered.

**Factors Affecting Maintenance Costs.** Maintenance costs are primarily a measure of labor activity. System design, layout, and configuration can significantly affect the amount of time and effort required for maintenance and, therefore, the maintenance cost. Factors to consider when evaluating maintenance costs include the following:

- **Quantity and type of equipment.** Each piece of equipment requires a core amount of maintenance and time, regardless of its size or capacity. A greater number of similar pieces of equipment are generally more expensive to maintain than larger but fewer units. For example, one manufacturer suggests the annual maintenance for centrifugal chillers is 24 h for a nominal 1000 ton chiller and 16 h for a nominal 500 ton chiller. Therefore, the total maintenance labor for a 1000 ton chiller plant with two 500 ton chillers would be 32 h, or 1/3 more than a single 1000 ton chiller.

- **Equipment location and access.** The ability to maintain equipment in a repeatable and cost-effective manner is significantly affected by the equipment's location and accessibility. Equipment that is difficult to access increases the amount of time required to maintain it, and therefore increases maintenance cost. Equipment maintenance requiring erection of ladders and scaffolding or hydraulic lifts increases maintenance costs while likely reducing the quantity and quality of maintenance performed. Equipment location may also dictate an unusual working condition that could require more service personnel than normal. For example, maintenance performed in a confined space (per OSHA definitions) requires an additional person to be present, for safety reasons.
- **System run time.** The number of hours of operation for a HVAC system affects maintenance costs. Many maintenance tasks are dictated by equipment run time. The greater the run time, the more often these tasks need to be performed.
- **Critical systems.** High-reliability systems require more maintenance to ensure uninterrupted system operation. Critical system maintenance is also usually performed with stringent shutdown and failsafe procedures that tend to increase the amount of time required to service equipment. An office building system can be turned off for a short time with little effect on occupants, allowing maintenance almost any time. Shutdown of a hospital operating room or pharmaceutical manufacturing HVAC system, on the other hand, must be coordinated closely with the operation of the facility to eliminate risk to patients or product. Maintenance on critical systems may sometimes incur labor premiums because of unusual shutdown requirements.
- **System complexity.** More complex systems tend to involve more equipment and sophisticated controls. Highly sophisticated systems may require highly skilled service personnel, who tend to be more costly.
- **Service environment.** HVAC systems subjected to harsh operating conditions (e.g., coastal and marine environments) or environments like industrial operations may require more frequent and/or additional maintenance.
- **Local conditions.** The physical location of the facility may require additional maintenance. Equipment in dusty or dirty areas or exposed to seasonal conditions (e.g., high pollen, leaves) may require more frequent or more difficult cleaning of equipment and filters. Additional maintenance tasks may be needed.
- **Geographical location.** Maintenance costs for remote locations must consider the cost of getting to and from the locations. Labor costs for the number of anticipated trips and their duration for either in-house or outsourced service personnel to travel to and from the site must be added to the maintenance cost to properly estimate the total maintenance cost.
- **Equipment age.** The effect of age on equipment repair costs varies significantly by type of HVAC equipment. Technologies in equipment design and application have changed significantly, affecting maintenance costs.
- **Available infrastructure.** Maintenance costs are affected by the availability of an infrastructure that can maintain equipment, components, and systems. Available infrastructure varies on a national, regional, and local basis and is an important consideration in the HVAC system selection process.

## 10.8 Economic Analysis Techniques

Analysis of overall owning and operating costs and comparisons of alternatives require an understanding of the cost of lost opportunities, inflation, and the time value of money. This process of economic analysis of alternatives falls into two general categories: simple payback analysis and detailed economic analyses (life-cycle cost analyses).

A simple payback analysis reveals options that have short versus long paybacks. Often, however, alternatives are similar and have similar paybacks. For a more accurate comparison, a more comprehensive economic analysis is warranted. Many times it is appropriate to have both a simple payback analysis and a detailed economic analysis. The simple payback analysis shows which options should not be considered further, and the detailed economic analysis determines which of the viable options are the strongest. The strongest options can be accepted or further analyzed if they include competing alternatives.

### 10.8.1 Simple Payback

In the simple payback technique, a projection of the revenue stream, cost savings, and other factors is estimated and compared to the initial capital outlay. This simple technique ignores the cost of borrowing money (interest) and lost opportunity costs. It also ignores inflation and the time value of money.

**Example 10-1.** Equipment item 1 costs $10,000 and will save $2000 per year in operating costs; equipment item 2 costs $12,000 and saves $3000 per year. Which item has the best simple payback?

Item 1 $10,000($2000/yr) = 5-year simple payback
Item 2 $12,000/($3000/yr) = 4-year simple payback

Because analysis of equipment for the duration of its realistic life can produce a very different result, the simple payback technique should be used with caution.

### 10.8.2 More Sophisticated Economic Analysis Methods

Economic analysis should consider details of both positive and negative costs over the analysis period, such as varying inflation rates, capital and interest costs, salvage costs, replacement costs, interest deductions, depreciation allowances, taxes, tax credits, mortgage payments, and all other costs associated with a particular system. See the section on Symbols at the end of this chapter for definitions of variables.

**Present-Value (Present Worth) Analysis.** All sophisticated economic analysis methods use the basic principles of present value analysis to account for the time value of money.

The total present value (present worth) for any analysis is determined by summing the present worths of all individual items under consideration, both future single-payment items and series of equal future payments. The scenario with the highest present value is the preferred alternative.

*Single-Payment Present-Value Analysis.* The cost or value of money is a function of the available interest rate and inflation rate. The future value $F$ of a present sum of money $P$ over $n$ periods with compound interest rate $i$ per period is

$$F = P(1 + i)^n \quad (10\text{-}1)$$

Conversely, the present value or present worth $P$ of a future sum of money $F$ is given by

$$P = F/(1 + i)^n \quad (10\text{-}2)$$

or

$$P = F \times \text{PWF}(i,n)_{sgl} \quad (10\text{-}3)$$

where the single-payment present-worth factor $\text{PWF}(i,n)_{sgl}$ is defined as

$$\text{PWF}(i,n)_{sgl} = 1/(1 + i)^n \quad (10\text{-}4)$$

**Example 10-2.** Calculate the value in 10 years at 10% per year interest of a system presently valued at $10,000

$$F = P(1 + i)^n = \$10{,}000\,(1 + 0.1)^{10} = \$25{,}937.42$$

**Example 10-3.** Using the present-worth factor for 10% per year interest and an analysis period of 10 years, calculate the present value of a future sum of money valued at $10,000. (Stated another way, determine what sum of money must be invested today at 10% per year interest to yield $10,000 10 years from now.)

$$P = F \times \text{PWF}(i,n)_{sgl}$$

$$P = \$10{,}000 \times 1/(1 + 0.1)^{10}$$

$$= \$3855.43$$

*Series of Equal Payments.* The present-worth factor for a series of future equal payments (e.g., operating costs) is given by

$$\text{PWF}(i,n)_{ser} = \frac{(1 + i)^n - 1}{i(1 + i)^n} \quad (10\text{-}5)$$

The present value $P$ of those future equal payments (PMT) is then the product of the present-worth factor and the payment [i.e., $P = \text{PWF}(i,n)_{ser} \times \text{PMT}$].

The number of future equal payments to repay a present value of money is determined by the capital recovery factor (CRF), which is the reciprocal of the present-worth factor for a series of equal payments:

$$\text{CRF} = \text{PMT}/P \quad (10\text{-}6)$$

$$\text{CRF}(i,n)_r = \frac{i(1 + i)^n}{(1 + i)^n - 1} = \frac{i}{1 - (1 + i)^{-n}} \quad (10\text{-}7)$$

The CRF is often used to describe periodic uniform mortgage or loan payments.

Note that when payment periods other than annual are to be studied, the interest rate must be expressed per appropriate period. For example, if monthly payments or return on investment are being analyzed, then interest must be expressed per month, not per year, and $n$ must be expressed in months.

**Example 10-4.** Determine the present value of an annual operating cost of $1000 per year over 10 years, assuming 10% per year interest rate.

$$\text{PWF}(i,n)_{ser} = [(1 + 0.1)^{10} - 1]/[0.1(1 + 0.1)^{10}] = 6.14$$

$$P = \$1000(6.14) = \$6140$$

**Example 10-5.** Determine the uniform monthly mortgage payments for a loan of $100,000 to be repaid over 30 years at 10% per year interest. Because the payment period is monthly, the payback duration is 30(12) = 360 monthly periods, and the interest rate per period is 0.1/12 = 0.00833 per month.

$$\text{CRF}(i,n) = 0.00833(1 + 0.00833)^{360}/[(1 + 0.00833)^{360} - 1]$$

$$= 0.008773$$

$$\text{PMT} = P(\text{CRF})$$

$$= \$100{,}000(0.008773)$$

$$= \$877.30 \text{ per month}$$

**Improved Payback Analysis.** This somewhat more sophisticated payback approach is similar to the simple payback method, except that the cost of money (interest rate, discount rate, etc.) is considered. Solving Equation (7) for $n$ yields the following:

$$n = \frac{\ln[\text{CRF}/(\text{CRF} - i)]}{\ln(1 + i)} \quad (10\text{-}8)$$

Given known investment amounts and earnings, CRFs can be calculated for the alternative investments. Subsequently, the number of periods until payback has been achieved can be calculated using Equation (10-8).

**Example 10-6.** Compare the years to payback of the same items described in Example 10-1 if the value of money is 10% per year.

Item 1
- cost = $10,000
- savings = $2000/year
- CRF = $2000/$10,000 = 0.2
- $n$ = ln[0.2/(0.2 − 0.1)]/ln(1 + 0.1) = 7.3 years

Item 2
- cost = $12,000
- savings = $3000/year
- CRF = $3000/$12,000 = 0.25
- $n$ = ln[0.25/(0.25 − 0.1)]/ln(1 + 0.1) = 5.4 years

If years to payback is the sole criteria for comparison, Item 2 is preferable because the investment is repaid in a shorter period of time.

**Accounting for Inflation.** Different economic goods may inflate at different rates. Inflation reflects the rise in the real cost of a commodity over time and is separate from the time

# Economic Analyses and Life-Cycle Costs

value of money. Inflation must often be accounted for in an economic evaluation. One way to account for inflation is to substitute effective interest rates that account for inflation into the equations given in this chapter.

The effective interest rate $i'$, sometimes called the real rate, accounts for inflation rate $j$ and interest rate $i$ or discount rate $i_d$; it can be expressed as follows (Kreider and Kreith 1982):

$$i' = \frac{1+i}{1+j} - 1 = \frac{i-j}{1+j} \qquad (10\text{-}9)$$

Different effective interest rates can be applied to individual components of cost. Projections for future fuel and energy prices are available in the *Annual Supplement to NIST Handbook 135* (Rushing et al. 2010).

**Example 10-7.** Determine the present worth $P$ of an annual operating cost of $1000 over 10 years, given a discount rate of 10% per year and an inflation rate of 5% per year.

$$i' = (0.1 - 0.05)/(1 + 0.05) = 0.0476$$

$$\text{PWF}(i',n)_{\text{ser}} = \frac{(1 + 0.0476)^{10} - 1}{0.0476(1 + 0.0476)^{10}} = 7.813$$

$$P = \$1000(7.813) = \$7813$$

The following are three common methods of present-value analysis that include life-cycle cost factors (life of equipment, analysis period, discount rate, energy escalation rates, maintenance cost, etc., as shown in Table 10-1). These comparison techniques rely on the same assumptions and economic analysis theories but display the results in different forms. They also use the same definition of each term. All can be displayed as a single calculation or as a cash flow table using a series of calculations for each year of the analysis period.

**Internal Rate of Return.** The internal rate of return (IRR) method calculates a return on investment over the defined analysis period. The annual savings and costs are not discounted, and a cash flow is established for each year of the analysis period, to be used with an initial cost (or value of the loan). Annual recurring and special (nonannual) savings and costs can be used. The cash flow is then discounted until a calculated discount rate is found that yields a net present value of zero. This method assumes savings are reinvested at the same calculated rate of return; therefore, the calculated rates of return can be overstated compared to the actual rates of return.

Another version of this is the **modified** or **adjusted internal rate of return (MIRR or AIRR)**. In this version, reinvested savings are assumed to have a given rate of return on investment, and the financed moneys a given interest rate. The cash flow is then discounted until a calculated discount rate is found that yields a net present value of zero. This method gives a more realistic indication of expected return on investment, but the difference between alternatives can be small.

The most straightforward method of calculating the AIRR requires that the SIR for a project (relative to its base case) be calculated first. Then the AIRR can be computed easily using the following equation:

$$\text{AIRR} = (1 + r)(\text{SIR})^{1/N} - 1 \qquad (10\text{-}10)$$

where $r$ is the reinvestment rate and $N$ is the number of years in the study period. Using the SIR of 12.6 from Equation (10-10) and a reinvestment rate of 3% [the minimum acceptable rate of return (MARR)], the AIRR is found as follows:

$$\text{AIRR}_{\text{A:BC}} = (1 + 0.03)(12.6)^{1/20} - 1 = 0.1691$$

Because an AIRR of 16.9% for the alternative is greater than the MARR, which in this example is the FEMP discount rate of 3%, the project alternative is considered to be cost effective in this application.

**Life-Cycle Costs.** This method of analysis compares the cumulative total of implementation, operating, and maintenance costs. The total costs are discounted over the life of the system or over the loan repayment period. The costs and investments are both discounted and displayed as a total combined life-cycle cost at the end of the analysis period. The options are compared to determine which has the lowest total cost over the anticipated project life.

**Example 10-9.** A municipality is evaluating two different methods of providing chilled water for cooling a government office building: purchasing chilled water from a central chilled-water utility service in the area, or installing a conventional chiller plant. Because the municipality is not a tax-paying entity, the evaluation does not need to consider taxes, allowing for either a current or constant dollar analysis.

The first-year price of the chilled-water utility service contract is $65,250 per year, and is expected to increase at a rate of 2.5% per year.

The chiller and cooling tower would cost $220,000, with an expected life of 20 years. A major overhaul ($90,000) of the chiller is expected to occur in year ten. Annual costs for preventative maintenance ($1400), labor ($10,000), water ($2000) and chemical treatments ($1800) are all expected to keep pace with inflation, which is estimated to average 3% annually over the study period. The annual electric cost ($18,750) is expected to increase at a rate of 5% per year. The municipality uses a discount rate of 8% to evaluate financial decisions.

Which option has the lowest life-cycle cost?

**Solution:**

Table 10-7 compares the two alternatives. For the values provided, alternative 1 has a 20-year life cycle cost of $769,283 and alternative 2 has a 20-year life cycle cost of $717,100. If LCC is the only basis for the decision, alternative 2 is preferable because it has the lower life-cycle cost.

**Computer Analysis.** Many computer programs are available that incorporate economic analysis methods. These range from simple calculations developed for popular spreadsheet applications to more comprehensive, menu-driven computer programs. Commonly used examples of the latter include Building Life-Cycle Cost (BLCC) and PC-ECONPACK.

BLCC was developed by the National Institute of Standards and Technology (NIST) for the U.S. Department of Energy (DOE). The program follows criteria established by the Federal Energy Management Program (FEMP) and the Office of Management and Budget (OMB). It is intended for

**Table 10-7  Two Alternative LCC Examples for Example 10-9**

**Alternative 1: Purchase Chilled Water from Utility**

| | \multicolumn{11}{c}{Year} |
|---|---|---|---|---|---|---|---|---|---|---|---|
| | 0 | 1 | 2 | 3 | 4 | 5 | 6 | 7 | 8 | 9 | 10 |
| First costs | — | — | — | — | — | — | — | — | — | — | — |
| Chilled-water costs | | $65,250 | $66,881 | $68,553 | $70,267 | $72,024 | $73,824 | $75,670 | $77,562 | $79,501 | $81,488 |
| Replacement costs | | — | — | — | — | — | — | — | — | — | — |
| Maintenance costs | | — | — | — | — | — | — | — | — | — | — |
| Net annual cash flow | | 65,250 | 66,881 | 68,553 | 70,267 | 72,024 | 73,824 | 75,670 | 77,501 | 79,501 | 81,488 |
| Present value of cash flow | | 60,417 | 57,340 | 54,420 | 51,648 | 49,018 | 46,522 | 44,153 | 41,904 | 39,770 | 37,745 |

| | \multicolumn{10}{c}{Year} |
|---|---|---|---|---|---|---|---|---|---|---|
| | 11 | 12 | 13 | 14 | 15 | 16 | 17 | 18 | 19 | 20 |
| Financing annual payments | — | — | — | — | — | — | — | — | — | — |
| Chilled-water costs | $83,526 | $85,614 | $87,754 | $89,948 | $92,197 | $94,501 | $96,864 | $99,286 | $101,768 | $104,312 |
| Replacement costs | — | — | — | — | — | — | — | — | — | — |
| Maintenance costs | — | — | — | — | — | — | — | — | — | — |
| Net annual cash flow | 83,526 | 85,614 | 87,754 | 89,948 | 92,197 | 94,501 | 96,864 | 99,286 | 101,768 | 104,312 |
| Present value of cash flow | 35,823 | 33,998 | 32,267 | 30,624 | 29,064 | 27,584 | 26,179 | 24,846 | 23,581 | 22,380 |

20-year life-cycle cost   $769,823

**Alternative 2: Install Chiller and Tower**

| | \multicolumn{11}{c}{Year} |
|---|---|---|---|---|---|---|---|---|---|---|---|
| | 0 | 1 | 2 | 3 | 4 | 5 | 6 | 7 | 8 | 9 | 10 |
| First costs | $220,000 | — | — | — | — | — | — | — | — | — | — |
| Energy costs | | $18,750 | $19,688 | $20,672 | $21,705 | $22,791 | $23,930 | $25,127 | $26,383 | $27,702 | $29,087 |
| Replacement costs | | — | — | — | — | — | — | — | — | — | 90,000 |
| Maintenance costs | | 15,200 | 15,656 | 16,126 | 16,609 | 17,108 | 17,621 | 18,150 | 18,694 | 19,255 | 19,833 |
| Net annual cash flow | 220,000 | 33,950 | 35,344 | 36,798 | 38,315 | 39,898 | 41,551 | 43,276 | 45,077 | 46,957 | 138,920 |
| Present value of cash flow | 220,000 | 31,435 | 30,301 | 29,211 | 28,163 | 27,154 | 26,184 | 25,251 | 24,354 | 23,490 | 64,347 |

| | \multicolumn{10}{c}{Year} |
|---|---|---|---|---|---|---|---|---|---|---|
| | 11 | 12 | 13 | 14 | 15 | 16 | 17 | 18 | 19 | 20 |
| Financing annual payments | — | — | — | — | — | — | — | — | — | — |
| Energy costs | $30,542 | $32,069 | $33,672 | $35,356 | $37,124 | $38,980 | $40,929 | $42,975 | $45,124 | $47,380 |
| Replacement costs | — | — | — | — | — | — | — | — | — | — |
| Maintenance costs | 20,428 | 21,040 | 21,672 | 22,322 | 22,991 | 23,681 | 24,392 | 25,123 | 25,877 | 26,653 |
| Net annual cash flow | 50,969 | 53,109 | 55,344 | 57,678 | 60,115 | 62,661 | 65,320 | 68,099 | 71,001 | 74,034 |
| Present value of cash flow | 21,860 | 21,090 | 20,350 | 19,637 | 18,951 | 18,290 | 17,654 | 17,042 | 16,452 | 15,884 |

20-year life-cycle cost   $717,100

---

evaluation of energy conservation investments in nonmilitary government buildings; however, it is also appropriate for similar evaluations of commercial facilities.

PC-ECONPACK, developed by the U.S. Army Corps of Engineers for use by the DOD, uses economic criteria established by the OMB. The program performs standardized life-cycle cost calculations such as net present value, equivalent uniform annual cost, SIR, and discounted payback period.

## 10.9  Reference Equations

Table 10-8 lists commonly used discount formulas as addressed by NIST. Refer to NIST *Handbook* 135 (Fuller and Petersen 1996) for detailed discussions.

## 10.10  Problems

**10.1** If $1000 is invested at 8% interest, determine the value of this money in 10 years.

**10.2** Find the present worth of money that will have a value of $35,000 in 3 years with an interest rate of 9%.

**10.3** $1000 is invested at the end of each year for 10 years. Interest is 11%. Find the amount accumulated. [$16,722]

**10.4** If $100,000 is invested at 8% interest, find the yearly withdrawal that will use up the money in 20 years.

**10.5** The cost of a new heat pump system is $3000 with an expected lifetime of 20 years. Neglect energy and maintenance costs. Find the annual cost if the salvage value is $0 and the interest rate is 8%.

Economic Analyses and Life-Cycle Costs 389

Table 10-8  Commonly Used Discount Formulas

| Name | Algebraic Form[a,b] | Name | Algebraic Form[a,b] |
|---|---|---|---|
| Single compound-amount (SCA) equation | $F = P \cdot [(1+d)^n]$ | Uniform compound-amount (UCA) equation | $F = A \cdot \left[\dfrac{(1+d)^n - 1}{d}\right]$ |
| Single present-value (SPV) equation | $P = F \cdot \left[\dfrac{1}{(1+d)^n}\right]$ | Uniform present-value (UPV) equation | $P = A \cdot \left[\dfrac{(1+d)^n - 1}{d(1+d)^n}\right]$ |
| Uniform sinking-fund (USF) equation | $A = F \cdot \left[\dfrac{d}{(1+d)^n - 1}\right]$ | Modified uniform present-value (UPV*) equation | $P = A_0 \cdot \left(\dfrac{1+e}{d-e}\right) \cdot \left[1 - \left(\dfrac{1+e}{1+d}\right)^n\right]$ |
| Uniform capital recovery (UCR) equation | $A = P \cdot \left[\dfrac{d(1+d)^n}{(1+d)^n - 1}\right]$ | | |

where

$A$ = end-of-period payment (or receipt) in a uniform series of payments (or receipts) over $n$ periods at $d$ interest or discount rate
$A_0$ = initial value of a periodic payment (receipt) evaluated at beginning of study period

$A_t = A_0(1+e)^t$, where $t = 1, \ldots, n$
$d$ = interest or discount rate
$e$ = price escalation rate per period

Source: NIST Handbook 135 (Fuller and Petersen 1996).

[a] Note that the USF, UCR, UCA, and UPV equations yield undefined answers when $d = 0$. The correct algebraic forms for this special case would be as follows: USF formula, $A = F/N$; UCR formula, $A = P/N$; UCA formula, $F = An$. The UPV* equation also yields an undefined answer when $e = d$. In this case, $P = A_0 \cdot n$.

[b] The terms by which known values are multiplied are formulas for the factors found in discount factor tables. Using acronyms to represent the factor formulas, the discounting equations can also be written as $F = P \times \text{SCA}$, $P = F \times \text{SPV}$, $A = F \times \text{USF}$, $A = P \times \text{UCR}$, $F = \text{UCA}$, $P = A \times \text{UPV}$, and $P = A_0 \times \text{UPV*}$.

**10.6** A new heating system has a cost of $15,000 and a salvage value of $5000, independent of age. The new system saves $1400 per year in fuel cost. Calculate the breakeven point if $i = 9\%$. Neglect maintenance costs.

**10.7** A new high-efficiency cooling system costs $60,000 and saves $7500 in energy costs each year. The system has a salvage value of $10,000 in 20 years. Compute the rate of return. Neglect maintenance costs. [11.25%]

**10.8** The costs of two small heat pump units A and B are $1000 and $1200 and the annual operating costs are $110 and $100, respectively. The interest rate is 8% and the amortization is selected as 20 years. Compare the systems on the basis of present worth.

**10.9** Compare the units in Problem 10.8 on the basis of uniform annual costs.

**10.10** An installation is going to require a 500 ton chiller. An annual energy analysis for this office building application shows that the required ton-hours over the year will be 2,100,000. The economic data are given below.

| | Chiller A | Chiller B |
|---|---|---|
| Average chiller efficiency | 0.73 kW/ton | 0.63 kW/ton |
| Initial cost | $221,500 | $240,500 |
| Installation cost | $19,000 | $19,000 |
| Electricity cost | 6¢/kWh | 5.9¢/kWh |
| Maintenance costs | $9,500 | $10,000 |
| Estimated life | 20 years | 20 years |

Perform a simple payback analysis for this option. [1.4 yrs]

## 10.11  Symbols

| | | |
|---|---|---|
| AIRR | = | modified or adjusted internal rate of return (MIRR or AIRR) |
| $c$ | = | cooling system adjustment factor |
| $C$ | = | total annual building HVAC maintenance cost |
| $C_e$ | = | annual operating cost for energy |
| $C_{s,\text{assess}}$ | = | assessed system value |
| $C_{s,\text{init}}$ | = | initial system cost |
| $C_{s,\text{salv}}$ | = | system salvage value at end of study period |
| $C_y$ | = | uniform annualized mechanical system owning, operating, and maintenance costs |
| CRF | = | capital recovery factor |
| CRF($i,n$) | = | capital recovery factor for interest rate $i$ and analysis period $n$ |
| CRF($i',n$) | = | capital recovery factory for interest rate $i'$ for items other than fuel and analysis period $n$ |
| CRF($i'',n$) | = | capital recovery factor for fuel interest rate $i''$ and analysis period $n$ |
| CRF($i_m,n$) | = | capital recovery factor for loan or mortgage rate $i_m$ and analysis period $n$ |
| $d$ | = | distribution system adjustment factor |
| $D_k$ | = | depreciation during period $k$ |
| $D_{k,\text{SL}}$ | = | depreciation during period $k$ from straight-line depreciation method |
| $D_{k,\text{SD}}$ | = | depreciation during period $k$ from sum-of-digits depreciation method |
| $F$ | = | future value of sum of money |
| $h$ | = | heating system adjustment factor |
| $i$ | = | compound interest rate per period |
| $i_d$ | = | discount rate per period |
| $i_m$ | = | market mortgage rate |

$i'$ = effective interest rate for all but fuel
$i''$ = effective interest rate for fuel
$I$ = insurance cost per period
ITC = investment tax credit
$j$ = inflation rate per period
$j_e$ = fuel inflation rate per period
$k$ = end of period(s) during which replacement(s), repair(s), depreciation, or interest are calculated
$M$ = maintenance cost per period
$n$ = number of periods under analysis
$P$ = present value of a sum of money
$P_k$ = outstanding principle on loan at end of period $k$
PMT = future equal payments
PWF = present worth factor
PWF$(i_d,k)$ = present worth factor for discount rate $i_d$ at end of period $k$
PWF$(i',k)$ = present worth factor for effective interest rate $i'$ at end of period $k$
PWF$(i,n)_{sgl}$ = single payment present worth factor
PWF$(i,n)_{ser}$ = present worth factor for a series of future equal payments
$R_k$ = net replacement, repair, or disposal costs at end of period $k$
SIR = savings-to-investment ratio
$T_{inc}$ = net income tax rate
$T_{prop}$ = property tax rate
$T_{salv}$ = tax rate applicable to salvage value of system

## 10.12 References

Abramson, B., D. HermaZn, and L. Wong. 2005. Interactive Web-based owning and operating cost database (TRP-1237). ASHRAE Research Project, *Final Report*.

Akalin, M.T. 1978. Equipment life and maintenance cost survey (RP-186). *ASHRAE Transactions* 84(2):94-106.

BOMA. 2003. *Preventive maintenance and building operation efficiency.* Building Owners and Managers Association, Washington, D.C.

DOE. 2007. The international performances measurement and verification protocol (IPMVP). *Publication* No. DOE/EE-0157. U.S. Department of Energy. http://www.ipmvp.org/.

Dohrmann, D.R. and T. Alereza. 1986. Analysis of survey data on HVAC maintenance costs (RP-382). *ASHRAE Transactions* 92(2A):550-565.

Fuller, S.K. and S.R. Petersen. 1996. Life-cycle costing manual for the federal energy management program, 1995 edition. NIST *Handbook* 135. National Institute of Standards and Technology, Gaithersburg, MD. http://fire.nist.gov/bfrl pubs/build96/art121.html.

Hiller, C.C. 2000. Determining equipment service life. *ASHRAE Journal* 42(8):48-54.

Kreider, J. and F. Kreith. 1982. *Solar heating and cooling: Active and passive design.* McGraw-Hill, New York.

Lovvorn, N.C. and C.C. Hiller. 2002. Heat pump life revisited. *ASHRAE Transactions* 108(2):107-112.

NIST. Building life-cycle cost computer program (BLCC 5.2-04). Available from the U.S. Department of Energy Efficiency and Renewable Energy Federal Energy Management Program, Washington, D.C. http://www. eere.energy.gov/femp/ information/download_blcc.cfm#blcc5.

OMB. 1992. Guidelines and discount rates for benefit-cost analysis of federal programs. *Circular* A-94. Office of Management and Budget, Washington, D.C. Available at http://www.whitehouse.gov/OMB/circulars/a094/a094.html.

Rushing, A.S., Kneifel, J.D., and B.C. Lippiatt. 2010. Energy price indices and discount factors for life-cycle cost analysis—2010. *Annual Supplement to NIST Handbook* 135 and *NBS Special Publication* 709. NISTIR 85-3273-25. National Institute of Standards and Technology, Gaithersburg, MD. http://www1.eere.energy.gov/femp/pdfs/ashb10.pdf

## 10.13 Bibliography

ASHRAE. 1999. HVAC maintenance costs (RP-929). ASHRAE Research Project, *Final Report*.

ASTM. 2004. Standard terminology of building economics. *Standard* E833-04. American Society for Testing and Materials, International, West Conshohocken, PA.

Easton Consultants. 1986. Survey of residential heat pump service life and maintenance Issues. AGA S-77126. American Gas Association, Arlington, VA.

Haberl, J. 1993. Economic calculations for ASHRAE *Handbook*. ESL-TR-93/04-07. Energy Systems Laboratory, Texas A&M University, College Station. http://repository.tamu.edu/bitstream/handle /1969.1/2113/ESL-TR-93-04-07.pdf?sequence=1.

Kurtz, M. 1984. *Handbook of engineering economics: Guide for engineers, technicians, scientists, and managers.* McGraw-Hill, New York.

Lovvorn, N.C. and C.C. Hiller. 1985. A study of heat pump service life. *ASHRAE Transactions* 91(2B):573-588.

Quirin, D.G. 1967. *The capital expenditure decision.* Richard D. Win, Inc., Homewood, IL.

U.S. Department of Commerce, Bureau of Economic Analysis. (Monthly) *Survey of current business.* U.S. Department of Commerce Bureau of Economic Analysis, Washington, D.C. http://www.bea.gov/scb/index. htm.

USA-CERL. 1998. Life cycle cost in design computer program (WinLCCID 98). Available from Building Systems Laboratory, University of Illinois, Urbana-Champaign.

U.S. Department of Labor. 2005. *Annual percent changes from 1913 to present.* Bureau of Labor Statistics. Available from http://www.bls.gov/cpi.

USACE. PC-ECONPACK computer program (ECONPACK 4.0.2). U.S. Army Corps of Engineers, Huntsville, AL. http://www.hnd.usace.army. mil/paxspt/econ/download .aspx.

# Chapter 11

# AIR-CONDITIONING SYSTEM CONCEPTS

An air-conditioning system maintains desired environmental conditions within a space. In almost every application there are a myriad of options available to the designer to satisfy the basic goal. It is in the selection and combination of these options that the engineer must consider all the parameters relating to the project.

Air-conditioning systems are categorized by how they control temperature and humidity in the conditioned area. They are also segregated to accomplish specific purposes by special equipment arrangement. This chapter considers elements that constitute the system. Chapter 12 describes those systems that are used to solve the psychrometric problem and control the environmental conditions of the occupied space, and Chapter 13 discusses the design of hot water and chilled water systems. Chapters 14 through 19 describe the various components of the air-conditioning system.

## 11.1 System Objectives and Categories

The two fundamental objectives of the air-conditioning system are to control the quality of the air in the conditioned space and the thermodynamic properties of the air. Additionally, in conjunction with other aspects of the space and system design, they become the major element in providing for the thermal comfort in the space.

### 11.1.1 Air Quality

Air quality is usually obtained through a process of ventilation combined with filtration. Ventilation consists of the removal of contaminants from the indoor air by replacing the contaminated air with uncontaminated air from the outdoors. This replacement is generally achieved through a process of dilution or, in some cases, displacement. The amount of outdoor air required is determined by the rate and type of contaminant generation, generally as prescribed by ASHRAE Standard 62.

In some systems, the outdoor air is mixed with the return air from the space at the inlet to the air-conditioning system and the mixture, then, conditioned to the thermodynamic state necessary to control the space temperature and humidity. However, this method need not be employed, and often it is preferable to provide total thermodynamic conditioning of the outdoor ventilation air separate and apart from the space conditioning unit or directly to the space via a separate ventilating supply air system. The separate ventilating system, properly applied, is preferable when using variable air volume systems because of the assurance of adequate ventilation at all conditions of load and improved humidity control in humid climates with most temperature control systems.

Another benefit of the separate ventilation air-conditioning unit is that it can be used in conjunction with in-room devices designed to control the room temperature (such as fan-coil units, radiant panels, or induction coil devices). These separate ventilation air-conditioning systems are commonly called dedicated outdoor air systems (DOAS).

### 11.1.2 Thermodynamic Conditioning

The thermodynamic conditioning of the air is the process of simultaneously cooling and dehumidifying the air or heating (and sometimes humidifying) it to maintain comfort conditions within the space. In most systems designed for human comfort, the system control responds to dry-bulb temperature, with the dehumidification being provided as a noncontrolled by-product of the cooling process.

The fundamental equation describing a sensible cooling or heating process is

$$q_s = 1.1 Q \Delta t \qquad (11\text{-}1)$$

where

$q_s$ = heating or sensible cooling load or capacity, Btu/h
$Q$ = flow quantity of supply air, ft$^3$/min or cfm
1.1 = constant for standard air, Btu·min/h·°F·ft$^3$
$\Delta t$ = difference between room temp. and supply air temp.

For design conditions, i.e., the airflow and temperature differential required at the design system load (maximum system capacity), the sensible design load is calculated and Equation (11-1) is usually solved for $Q$ with the designer selecting a $\Delta t$. For cooling, the $\Delta t$ is equal to $t_r - t_s$, where $t_s$ is the supply air temperature, usually selected as the dew-point temperature necessary to provide the latent cooling requirement at design conditions.

Then, where the myriad of different types of air-conditioning systems arise from are the different methods that are employed to change the capacity as the load reduces. Considering Equation (11-1), if the load $q_s$ reduces (or changes), the only methods by which the equation can be kept in balance is by reducing either the $Q$ or the $\Delta t$ or some combination of the two. Thermodynamically, four different methods are used to achieve these changes:

- Heat-Cool-Off
- Dual Stream
- Reheat
- Variable Air Volume

### 11.1.3 Heat-Cool-Off Systems

A heat-cool-off system is any system that responds to the need for changes in capacity by varying the temperature of the air supplied by the unit, usually while maintaining a constant airflow. There are many configurations of heat-cool-off systems, both two-position and modulating, and although it is not common, they can be both variable flow rate $Q$ or variable $\Delta t$ but are usually the latter. The essential feature that defines a heat-cool-off system is that it is limited to a *single* control zone. Examples of heat-cool-off systems are single-zone air-handling units, residential heating/cooling units, and room fan-coil units.

### 11.1.4 Dual-Stream Systems

Dual-stream systems are applicable to cases in which multiple control zones are to be served from a single air-conditioning unit. They are constant flow ($Q$), variable $\Delta t$ systems. They achieve the variation in supply air temperature (from full cooling load to full heating load) by mixing a high-temperature stream with a low-temperature stream. The two most common dual-stream systems are double-duct and multi-zone systems. The induction system is another special configuration of a dual-stream system.

### 11.1.5 Reheat Systems

Reheat systems also apply to cases in which multiple control zones are served from a single air-conditioning unit. However, single-zone units that function as heat-cool-off *can* be thermodynamically configured as reheat systems when applied to controlled dehumidifying requirements. This system is a constant flow, variable $\Delta t$ system. In the multiple zoned reheat system (often called **terminal reheat systems**) the central air-handling system generally conditions the air to a fixed or controlled dew-point temperature and the cool air is reheated to the temperature required to satisfy the space sensible load for each control zone.)

### 11.1.6 Variable Air Volume (VAV) Systems

These systems, like the dual-stream and terminal reheat systems, are usually applied in cases in which multiple control zones are served from a single air-conditioning unit. The main difference between variable air volume systems and other multiple-zone systems is that the VAV system varies the flow of the supply air rather than the temperature. Two characteristics of the VAV system become immediately evident:

- The VAV system can either heat or cool at any given time, but it cannot do both at the same time. In order to do both, it must be combined with one of the other systems (i.e., VAV-reheat or VAV-dual stream).
- Conceptually, as the load reduces the air flow rate reduces and as the air flow rate reduces the fan power reduces, ideally as the cube of the flow.

Since, in most applications, the supply fan systems consume more energy annually than the refrigeration systems, the use of VAV systems offer significant potential benefits in energy conservation.

Most systems employ a combination of two or more of the above, such as dual stream/reheat, VAV/dual stream, VAV/reheat, heat-cool-off/reheat, etc. The following discussions will subdivide the system types into air systems and water systems, consistent with the types discussed in the 2012 *ASHRAE Handbook—HVAC Systems and Equipment*. And, although the physical description and common name of the systems more closely describes their physical configuration, the designer is encouraged to always relate the system for purposes of modeling, diagnostics, or analyses to its generic thermodynamic type as described above.

## 11.2 System Selection and Design

A "system" is that device or assembly of devices that provides the environmental conditions that the engineer determines are required in the space. It could range from a simple window air-conditioning unit and portable space heater to a fully integrated environmental control system designed to maintain all of the ideal comfort conditions in numerous control zones of a major building complex.

The most important and significant contribution that the design engineer provides in the process of designing an HVAC system is in the "selection" of the system. That selection affects virtually all aspects of the building for the life of the building. The ingredients that go into the system selection include not only the rigid and quantifiable engineering aspects but the economic, psychological, and social aspects involved in constructing, running, operating, and using the building.

This chapter and those that follow relating to system selections are intended to prepare the student for this fascinating task.

## 11.3 Design Parameters

The design engineer is responsible for considering various systems and selecting the one that will best satisfy all of the design parameters. It is imperative that the designer and the owner collaborate on identifying the goals of the design. Some of the parameters that should be considered are

- Load dynamics
- Performance requirements
- Availability of equipment
- Capacity
- Spatial requirements
- First cost
- Energy consumption
- Operating cost
- Simplicity
- Reliability
- Flexibility (short range and long range)

- Operations requirements
- Serviceability
- Maintainability
- Availability of service
- Availability of replacement components
- Environmental requirements of space
- Environmental requirements of the community

The degree of success of the design of any system is directly related to the accuracy with which the designer (1) identifies the design parameters and (2) achieves them.

Because these factors are interrelated, the owner and system designer must consider how each factor affects the others. The relative impact of these parameters differs with different owners and often changes from one project to another for the same owner.

## 11.4 Performance Requirements

In addition to goals for providing the desired environment for human comfort, the designer must be aware of and account for other goals the owner may require. These goals may include

- Supporting a process, such as the operation of computer equipment
- Promoting an aseptic environment
- Increasing sales
- Increasing net rental income
- Increasing the salability of a property

Typical concerns of owners include first cost compared to operating cost, the extent and frequency of maintenance and whether that maintenance requires entering the occupied space, the expected frequency of failure of a system, the impact of a failure, and the time required to correct the failure. Each of these concerns has a different priority, depending on the owner's goals.

The owner can only make appropriate value judgments if the designer provides complete information on the advantages and disadvantages (i.e., the impact) of each option. Just as the owner does not usually know the relative advantages and disadvantages of different systems, the designer rarely knows all the owner's financial and functional goals. Hence, it is important to involve the owner in the selection of a system.

## 11.5 Focusing on System Options

Following the establishment of the design parameters, including the performance requirements, there are four fundamental features and constraints of the system that will inevitably assist the designer in focusing on the type of system. They are (1) the cooling and heating loads, (2) the zoning requirements, (3) the need for heating and ventilation, and (4) the architectural constraints.

### 11.5.1 Cooling Load

Establishing the cooling and heating loads often narrows the choice to systems that will satisfy all of the part-load requirements, fit within the available space, and are compatible with the building architecture. Chapter 7 covers determination of the magnitude and characteristics of the cooling and heating loads and their variation with time and operating conditions. By establishing the capacity requirement, the equipment size can be estimated. Then, the choice may be narrowed to those systems that work well on projects within a certain size range.

### 11.5.2 Zoning Requirements

Loads vary over time due to changes in the occupancy, weather, activities, and solar exposure. Each space with a different load dynamic requires a different control zone to maintain reasonably constant thermal conditions. Some areas with special requirements in dry-bulb or wet-bulb control points may need individual control or individual systems, independent of the rest of the building. Variations in indoor conditions that are acceptable in one space may be unacceptable in other areas of the same building. The extent of zoning, the degree of control required in each zone, and the space required for individual controlled spaces also narrow the system choices.

No matter how efficiently a particular system operates or how economical it may be to install, it cannot be considered if it (1) cannot maintain the desired interior environment within an acceptable tolerance under all conditions and occupant activities and (2) does not physically fit into the building without being objectionable and non-maintainable.

### 11.5.3 Heating and Ventilation

Cooling and humidity control are often the basis of sizing air-conditioning components and subsystems, but the system may also provide other functions, such as heating and ventilation. For example, if the system provides large quantities of outdoor air for ventilation or to replace air exhausted from the building, only systems that transport large air volumes need to be considered. In this situation, the ventilation system requires a large air-handling and duct distribution system, which may eliminate some systems.

Effective delivery of heat to an area may be an equally important factor in system selection. A distribution system that offers high efficiency and comfort for cooling may be a poor choice for heating. This performance compromise may be small for one application in one climate, but it may be unacceptable in another that has more demanding heating requirements.

### 11.5.4 Architectural Constraints

Air-conditioning systems and the associated distribution systems often occupy a significant amount of space. Major components may also require special support from the structure. The size and appearance of terminal devices (whether they are diffusers, fan-coil units, or radiant panels) have an effect on the architectural design because they are visible from the occupied space.

Other factors that limit the selection of a system include (1) acceptable noise levels, (2) the space available to house equipment and its location relative to the occupied space, (3) the space available for distribution pipes and ducts, and (4) the acceptability of components obtruding into the occupied space, both physically and visually.

## 11.6 Narrowing the Choice

Chapter 12 covers the types of systems categorized by all-air systems, air-and-water systems, all-water systems, and unitary refrigerant-based systems. Each section includes an evaluation component, which briefly summarizes the advantages and disadvantages of various systems. Comparing the features against the list of design parameters and their relative importance usually allows identification of two or three systems that best satisfy the project criteria. In making subjective choices, it is helpful to keep notes on all systems considered and the reasons for eliminating those that are unacceptable.

In most cases, two or three systems must be selected: a **secondary system** (or distribution system), which delivers heating or cooling to the occupied space, and a **primary system**, which converts energy from fuel or electricity, and in some cases, an intermediate system, which conveys energy between the primary system and the secondary system. Chapter 13 discusses the most common type of intermediate systems—water systems. The systems are, to a great extent, independent, so that several secondary systems may work with different primary and intermediate systems. In some cases, however, only one secondary system may be suitable for a particular primary system.

Once subjective analysis has identified two or three systems—and sometimes only one choice may remain—detailed quantitative evaluations of each system must be made. All systems considered should provide satisfactory performance to meet the design parameter goals. The owner or designated decision maker then needs specific data on each system to make an informed choice. Chapter 8 outlines methods for estimating annual energy costs, and Chapter 10 describes economic analyses and life-cycle costing, which can be used to compare the overall economics of systems.

**Example 11-1.** A dual-stream cooling system is a double-duct system with 55°F saturated air in the cold duct and 95°F db/60°F dewpoint air in the hot duct. The air in the cold duct is conditioned by cooling a mixed air stream (i.e., return air and outdoor air from 80°F db/66.8°F wb to 55°F saturated. The air in the hot duct is conditioned by heating the same mixed air stream to 95°F db with no addition or removal of water vapor. The room or space conditions are 75°F db/50% rh.

(a) A given zone has a sensible design cooling load of 12,960 Btu/h. What airflow (in cfm) must be supplied?

**Solution:**

From Equation (11-1)

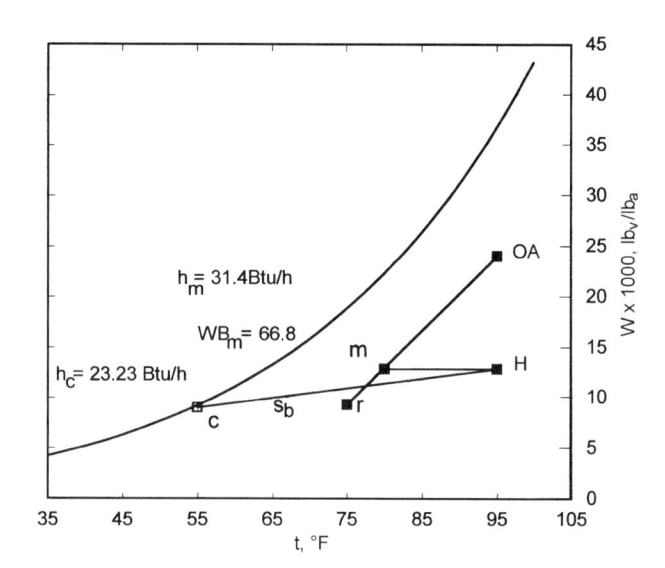

$$q_s = 1.1 Q \Delta t$$

$$Q = \frac{q}{1.1(t_r - t_s)} = \frac{12960}{1.1(75-55)}$$

$$Q = 590 \text{ cfm}$$

(b) The load reduces, and when the space sensible load is 50% of design or 6480 Btu/h, what is the condition of the supply air (dry-bulb temperature and wet-bulb temperature)?

**Solution:**

Dry-bulb temperature

From Equation (11-1)

$$t_{sdb} = t_r - \frac{q}{1.1 Q} = 75 - \frac{6480}{1.1(600)}$$

$$t_s = 65°F \text{ dry bulb}$$

Wet-bulb temperature

The wet-bulb temperature is determined where the mixing line between $H$ (air at "m" heated to 95°F) and C on the psychrometric chart shown in the figure cross the 65°F db line. Thus

$$t_{swb} = 59.7°F \text{ wet bulb}$$

(c) At the half-load condition, approximately what airflow from the cold duct and the hot duct will be required?

**Solution:**

From the following equation for mixed air

$$t_m = t_c + \frac{m_H}{m_C + m_H}(t_H - t_C)$$

Solve for

$$\frac{Q_H}{Q_C} = \frac{m_H}{m_C + m_H} = \frac{t_m - t_C}{t_H - t_C} = \frac{65-55}{95-55} = 0.25$$

Then $Q_H = 0.25 Q = 0.25(590) = 148$ cfm

$Q_C = 590 - 148 = 442$ cfm

(d) At the half sensible load condition, approximately what is the load on the cooling coil (in Btu/h) related to this zone?

**Solution:**
From the following energy equation

$$q_C = \dot{m}(h_m - h_C) = 60\rho_a Q(h_m - h_C)$$
$$q_C = 60(0.075)Q(31.4 - 23.23) = 4.5(442)(8.17)$$
$$= 16{,}250 \text{ Btu/h}$$

(e) At the half sensible load condition, approximately what is the load on the heating coil related to this zone?

**Solution:**
From Equation (11-1)

$$q_H = 1.1(148)(95 - 65) = 4880 \text{ Btu/h}$$

(f) At full-load conditions, approximately what is the load on the cooling coil related to this zone? The heating coil?

**Solution:**

$$q_C = 4.5Q(h_m - h_C)$$
$$= 4.5(590)(31.4 - 23.23)$$
$$= 22690 \text{ Btu/h}$$
$$q_H = 0$$

## 11.7 Energy Considerations of Air Systems

Those systems that move energy from place to place within a building, such as the air distribution systems and liquid fluid circulation systems, are called the energy transport systems. In first selecting and then designing these systems, the power and energy required are seen to be two of the basic selection parameters.

The impact of the power is twofold. First, if the power is less, the first cost will be less because the machinery is smaller. Second, if the power is less, the energy consumption will be less (other things being equal) because the energy consumed is simply the product of the power and the operating hours.

### 11.7.1 Air System Power and Energy

Equation (11-1), the heat capacity equation, is the first consideration in the design of energy effective systems. It becomes evident that if the designer can do anything to reduce the load, which is a power term (Btu/h or tons of cooling), the system and machinery will be smaller, thus will cost less and use less energy when operated.

The fan power equation is

$$P = \frac{Q \Delta p_t}{6350 \eta_f} \quad (11\text{-}2)$$

where

$P$ = fan power, hp
$Q$ = air circulation rate, ft³/min
$\Delta p_t$ = fan total pressure rise, in. w.g.
$\eta_f$ = fan efficiency, decimal

The equation shows that reducing the flow rate or the system pressure or increasing the fan efficiency will reduce the power. The flow rate can be reduced by (1) reducing the load [Equation (11-1)], (2) by making a careful load analysis that does not include excessive or hidden uncertainty or "safety factors," or (3) by increasing the temperature difference between the supply and room air. [i.e., $(t_r - t_s)$].

The fan pressure equals the pressure loss requirement of the conditioner and the distribution system, a term that can be controlled by the system designer. Friction losses in the distribution system can be expressed by the Darcy-Weisbach equation,

$$\Delta h_f = Cf \frac{L}{D} \frac{\rho V^2}{2g_c} \quad (11\text{-}3)$$

where

$\Delta h_f$ = friction loss, in. w.g.
$f$ = friction factor, dimensionless
$L$ = length of duct, ft
$D$ = diameter of duct, ft
$V$ = velocity of air in duct, ft/s
$g_c$ = units conversion constant, $\text{lb}_m \cdot \text{ft/s}^2 \cdot \text{lb}_f$
$C$ = unit conversion factor, 0.1923, $\frac{\text{in. w.g. ft}^2}{\text{lb}_f}$
$\rho$ = density of air, $\text{lb}_m/\text{ft}^3$

The methods, then, to reduce the fan pressure requirement (fluid head loss) are:

- Limit the length of duct runs to the minimum possible
- Increase the diameter (or equivalent diameter)
- Reduce the velocity
- Reduce the roughness of interior surfaces, which reduces the friction factor

Additionally, the duct fittings, in most systems, create a significant amount of the pressure losses (see Chapter 9). All fittings, in both the supply and return ductwork, should be designed for minimum pressure losses, which are a function of the fitting construction and the velocity head $V^2/2g$. (Also see Chapter 21, 2013 *ASHRAE Handbook—Fundamentals*.) To ensure continued operation with low pressure drops, any device such as a damper, coil, turning vanes, etc., which could result in a blockage to air flow, should be provided with an inspection and access port.

An additional consideration regarding the pressure is the temperature rise across the fan. The temperature rise is expressed by the equation:

$$\Delta t_f = \frac{0.371 \Delta p_t}{\eta_f} \quad (11\text{-}4)$$

where

$\Delta t_f$ = temperature rise across fan, °F
$\Delta p_t$ = total pressure rise across fan, in. w.g.
$\eta_f$ = fan efficiency, decimal

Equation (11-4) shows that if the fan efficiency is 74%, the temperature rise in °F would equal one-half the pressure rise in in. w.g. For example, a 74% efficient fan producing 4 in. w.g. pressure would raise the air temperature 2°F. In most systems, this temperature rise becomes part of the cooling load, thus requiring the use of yet more energy.

Regarding fan efficiencies, fans should always be selected at the maximum efficiency point on their curves, and it is highly recommended that a designer always use a fan curve when selecting a fan so that the anticipated range of operation can be analyzed.

The fan energy equation (11-5) is the power equation multiplied by the hours of operation and with the appropriate terms for motor efficiency and conversion of horsepower to kilowatts.

$$q_{fan} = \frac{Q \Delta p_t \theta}{8512 \eta_f \eta_m} \quad (11\text{-}5)$$

where

$q_{fan}$ = fan energy, kWh
$Q$ = air circulation rate, ft³/min
$\Delta p_t$ = total fan pressure, in. w.g.
$\theta$ = time of operation, hours
$\eta_f$ = fan efficiency
$\eta_m$ = motor efficiency

The only variables that have been incorporated in this equation that were not included in the power equation (11-2) are the hours of operation and the motor efficiency. In selecting and designing a system, accommodation should always be made to maintain an unoccupied building under controlled conditions while shutting down the major energy consuming devices or operating them at a low energy consumption idle mode.

Regarding motor efficiencies, the use of high-efficiency motors is always recommended for fan or pump drives.

Another form of the fan energy equation for *a fixed or given system* (i.e., the system curve is constant) is:

$$P_{fan} = \frac{Q^3}{6350 \eta_f C_s^2} \quad (11\text{-}6)$$

$$q_{fan} = \frac{Q^3 \theta}{8512 \eta_f \eta_m C_s^2} \quad (11\text{-}7)$$

where $C_s$ is the system constant in ft³/[min (in w.g.)$^{0.5}$].

This equation shows that, if the air delivery rate can be reduced for a given system, the energy consumption is reduced as the cube of the flow rate (e.g., 20% flow reduction equates to 49% energy reduction, 50% flow reduction equates to 87.5% energy reduction, etc.). This characteristic explains one of the major benefits of using a variable flow (VAV) system instead of a variable $\Delta t$ system.

## 11.7.2 Water System Power and Energy

The relationships of power and energy consumption in water systems are similar to those in air systems, except for the change in the constants and dimensions. The fundamental power and energy equations for pumps in chilled and heating water systems are:

$$P_p = \frac{Q \Delta h}{3960 \eta_p} \quad (11\text{-}8)$$

$$q_p = \frac{Q \Delta h \theta}{5308 \eta_p \eta_m} \quad (11\text{-}9)$$

where

$P_p$ = pump horsepower
$q_p$ = pump energy, kWh
$Q$ = water circulating rate, gallons/minute (gpm)
$\Delta h$ = pump head, feet of water
$\theta$ = time of operation, hours
$\eta_p$ = pump efficiency, decimal
$\eta_m$ = motor efficiency, decimal

The principles of energy effective design discussed above for air systems apply equally to water systems.

**Example 11-2.** An air system is designed to handle 30,000 cfm at a total pressure of 6 in. w.g. If the ideal fan horsepower can be calculated by the use of Equation (11-2) with the fan efficiency term set at 100% efficiency,

(a) What is the ideal fan horsepower?

$$P = \frac{29630(6)}{6350} = 28.0 \text{ hp}$$

(b) If the fan were 70% efficient, what would be the power requirement?

$$P = \frac{28.0}{0.70} = 40.0 \text{ hp}$$

(c) If the pressure required was reduced to 4 in. w.g. by modifying the distribution system, what power would be required?

$$P = \frac{29630(4)}{6350(0.70)} = 26.7 \text{ hp}$$

(d) If after installing the system, it was decided the capacity could be reduced 20% by reducing the fan speed, what would the power requirement be?

$$P = P_c \left( \frac{\text{New flow rate}}{\text{Original flow rate}} \right)^3$$

$$= 26.7(0.8)^3 = 13.7 \text{ hp}$$

## 11.8 Basic Central Air-Conditioning and Distribution System

The basic secondary system is an all-air, single-zone, air-conditioning system. It may be designed to supply a constant air volume or a variable air volume and for low-,

# Air-Conditioning System Concepts

medium-, or high-velocity air distribution. Normally, the equipment is located outside the conditioned area, in a basement, penthouse, service area, or outdoors on the roof. It can, however, be installed within the area if conditions permit. The equipment can be adjacent to the primary heating and refrigeration equipment or at considerable distance from it by circulating refrigerant, chilled water, hot water, or steam for energy transfer.

## 11.8.1 Applications

Some central system applications are (1) spaces with uniform loads, (2) small spaces requiring precision control, (3) multiple systems for large areas, (4) systems for complete environmental control, and (5) primary sources of conditioned air for other subsystems.

## 11.8.2 Spaces with Uniform Loads

Spaces with uniform loads are generally those with relatively large open areas and small external loads, such as theaters, auditoriums, department stores, and the public spaces of many buildings. Adjustment for minor variations in the air-conditioning load in parts of the space can be made by supplying more or less air in the original design and balance of the system.

In office buildings, the interior areas generally meet these criteria as long as local areas of relatively intense and variable heat sources, such as computer rooms and conference rooms, are treated separately. In these applications, dwarf partitions allow wider diffusion of the conditioned air and equalization of temperatures. These areas usually require year-round cooling, and any isolated spaces with limited occupancy may require special evaluation, as discussed in Chapter 3 of the 2011 *ASHRAE Handbook—HVAC Applications*.

In most single-room commercial applications, temperature variations up to 4°F at the outside walls are usually considered acceptable for tenancy requirements. However, these variations should be carefully determined and limited during design. If people sit or work near the outside walls or if they are isolated by partitions, supplementary heating equipment may be required at the walls, depending on the outdoor design temperature in winter and the thermal characteristics of the wall. If the wall surface temperature is calculated to be more than 10°F below the room temperature, some special consideration must be given to the placing of heat at the perimeter.

## 11.8.3 Spaces Requiring Precise Control

These spaces are usually isolated rooms within a larger building (such as a computer room, auditorium, etc.) and have stringent requirements for such things as cleanliness, humidity, temperature control, and/or air distribution. Central system components should be selected and assembled to meet the exact requirements of the area.

## 11.8.4 Multiple Systems for Large Areas

In large buildings such as hangars, factories, large stores, office buildings, and hospitals, practical considerations sometimes require installation of multiple central single-zone systems. The size of the individual system is determined by an evaluation of the relative design parameters.

## 11.8.5 Systems for Environmental Control

ASHRAE Standard 62 specifies ventilation rates for acceptable indoor air quality (IAQ). All-air systems often provide the necessary air supply to dilute the contaminants in the air in controlled spaces. These applications consist of combinations of supply, return, and exhaust systems that circulate the diluting air through the space. The designer must consider the terminal systems used because establishing adequate dilution volumes is closely related to design criteria, occupancy type, air delivery, and scavenging methods. Indoor air quality is an essential consideration in the design of all systems.

Cleanliness of the air supply also relates directly to the level of environmental control desired. Suitable air filtration should be incorporated in the central system upstream from the air-moving and temperature control equipment. Some applications, such as hospitals, require downstream filtration as well.

Chapter 29 of the 2012 *ASHRAE Handbook—HVAC Systems and Equipment* has information on air cleaners.

## 11.8.6 Primary Sources for Other Systems

Systems for separate control of multiple zones are described in Chapter 12. These secondary systems may move a constant or variable supply of conditioned air for ventilation and temperature control and handle some or all of the air-conditioning load. Use of a secondary system may reduce the amount of conditioned air required to be delivered by the central system, thereby reducing the size of and the space required for ductwork. Ductwork size can be reduced further either by moving air at high velocities or by delivering air at reduced temperatures. However, high-velocity system design must consider the resultant high pressure, sound levels, and power and energy requirements. System design for low-temperature air delivery must consider the minimum required ventilation rates and insulation and condensation problems associated with lower temperatures as well as the additional refrigeration power.

## 11.8.7 Central System Performance

Figure 11-1 shows a typical draw-through central system that supplies conditioned air to a single zone or to multiple zones. A blow-through configuration may also be used if space or other conditions dictate or require. The quantity and quality of supplied air are fixed by space requirements and determined as indicated in Chapters 6 and 7. Air gains and loses heat by contacting heat transfer surfaces and by mixing with air of another condition. Some of this mixing is intentional, as at the outdoor air intake; for others, mixing is the result of the physical characteristics of a particular component, as when untreated air passes without contacting the fins of a coil (bypass factor).

All treated and untreated air must be thoroughly mixed for maximum performance of heat transfer surfaces and for uniform temperatures in the airstream. Stratified, parallel paths of treated and untreated air must be avoided, particularly in the

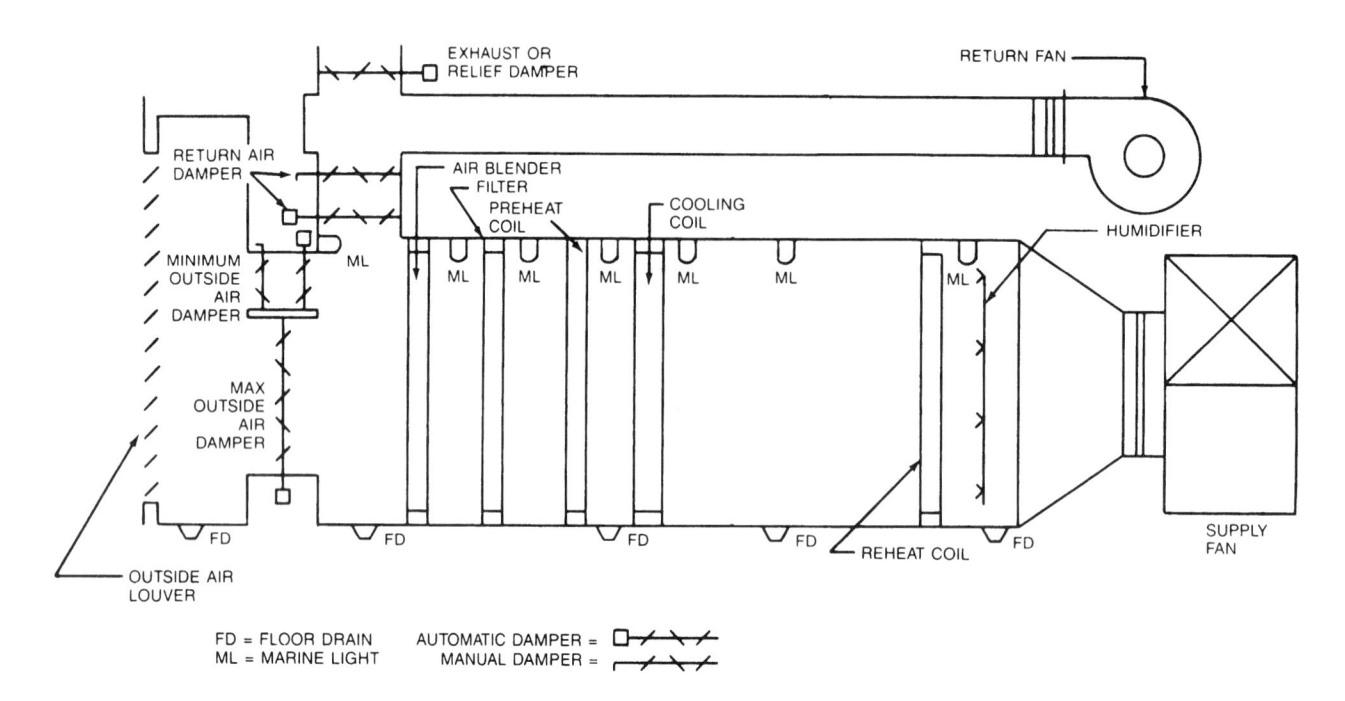

*Fig. 11-1 Equipment Arrangement for Central System Draw-Through Unit*

vertical plane of systems using double-inlet or multiple-wheel fans. Because these fans do not completely mix the air, different temperatures can occur in branches coming from opposite sides of the supply duct.

## 11.9 Smoke Management

Air-conditioning systems are often used for smoke control during fires. Controlled air flow can provide smoke-free areas for occupant evacuation and fire fighter access. Space pressurization attempts to create a low-pressure area at the smoke source, surrounding it with high-pressure spaces. For more information, see Chapter 53 of the 2011 *ASHRAE Handbook—HVAC Applications*. The ASHRAE publication *Principles of Smoke Management Systems* (Klote and Milke 2002) has detailed information.

## 11.10 Components

### 11.10.1 Air-Conditioning Units

To determine a system's air-handling requirement, the designer must consider the function and physical characteristics of the space to be conditioned and the air volume and thermal exchange capacities required. Then, the various components may be selected and arranged.

Further, the designer should consider economics in component selection. Both initial cost and operating costs affect design decisions. The designer should not arbitrarily design for a 500 fpm face velocity, which has been common for selection of cooling coils and other components. Filter and coil selection at 300 to 400 fpm, with resultant lower pressure loss, could produce a substantial payback in constant volume systems.

Figure 11-1 shows a general arrangement of the components for a single-zone, all-air central system suitable for year-round air conditioning. With this arrangement, close control of temperature and humidity are possible. All these components are seldom used simultaneously in a comfort application. Although Figure 11-1 indicates a built-up system, most of the components are available from many manufacturers completely assembled or in subassembled sections that can be bolted together in the field. These units are called **air-handling units**.

### 11.10.2 Return and Relief Air Fans

A return air fan is optional but is essential for the proper operation of a so-called air economizer or free cooling systems unless the excess intake air can be relieved directly from the space. It provides a positive return and exhaust from the conditioned area, preventing overpressurization of the space when mixing dampers provide cooling with outdoor air at times when the outdoor air temperature is at or lower than the required supply air temperature.

The return air fan prevents excess ambient pressure in the conditioned space(s) and reduces the static pressure the supply fan has to work against.

The supply fan(s) must be carefully matched with the return fans. The return air fan should handle a smaller amount of air to account for the building ventilation air requirements and to ensure a slight positive pressure in the conditioned space.

In many situations, a relief (or exhaust) air fan may be used instead of a return fan. A relief air fan relieves ventilation air introduced during air economizer operation and operates only when this control cycle is in effect.

# Air-Conditioning System Concepts

When a relief air fan is used, the supply fan must be designed for the total supply and return static pressure in the system and to operate without the relief air fan during the non-economizer mode of operation. During the economizer mode of operation for a constant volume system, the relief fan must be controlled to exhaust at a rate that tracks the quantity of outdoor air introduced, to ensure a slight positive pressure in the conditioned space, as with the return air fan system.

When a mixing chamber and economizer system are utilized with a variable air volume system, special precautions must be taken in the control of the return air or relief air fans to ensure that adequate ventilation air is provided at all times. These systems become so complex that the use of an outdoor air-return air mixing chamber is not recommended for air-handling units serving VAV systems. For these systems, ventilation provided by a dedicated outdoor air system (DOAS) is preferable (see Chapter 12 for more discussion).

## 11.10.3 Automatic Dampers

Opposed blade dampers for the outdoor, return, and relief airstreams provides the highest degree of control. Section 11.10.7 on mixing plenums covers the conditions that dictate the use of parallel blade dampers.

Pressure relationships between various sections must be considered to ensure that automatic dampers are properly sized for wide open and modulating pressure drops.

## 11.10.4 Relief Openings

Relief openings in large buildings should be constructed similarly to outdoor air intakes, but they may require motorized or self-acting backdraft dampers to prevent high wind pressure or stack action from causing the airflow to reverse when the automatic dampers are open. The pressure loss through relief openings should be 0.10 in. w.g. or less unless they are provided with a relief fan. Low-leakage dampers, such as those for outdoor intakes, should always be used.

Relief dampers sized for the same air velocity as the maximum outdoor air dampers facilitate control when an air economizer cycle is used. The relief air opening should be located so that the exhaust air does not short-circuit to the outdoor air intake.

## 11.10.5 Return Air Dampers

The negative pressure in the outdoor air intake plenum is a function of the resistance or static pressure loss through the outdoor air louvers, damper, and duct. The positive pressure in the relief air plenum is, likewise, a function of the static pressure loss through the exhaust or relief damper, the exhaust duct between the plenum and outside, and the relief louver. The pressure drop through the return air damper must accommodate the pressure difference between the positive-pressure relief air plenum and the negative-pressure outdoor air intake plenum. Proper sizing of this damper facilitates both air balancing and mixing.

## 11.10.6 Outdoor Air Intakes

Resistance through outdoor intakes varies widely, depending on construction. Frequently, architectural considerations dictate the type and style of louver. The designer should ensure that the louvers selected offer minimum pressure loss, preferable not more than 0.10 in. w.g. High-efficiency, low-pressure louvers that effectively limit carryover of rain or snow are available. Flashing installed at the outside wall and weep holes or a floor drain will carry away rain and melted snow entering the intake. Cold regions may require a snow baffle to direct fine snow particles to a low-velocity area below the dampers. Outdoor dampers should be low-leakage types with special gasketed edges and special end treatment. **When mixing chambers are employed, separate damper sections with separate damper operators are strongly recommended for the minimum outdoor air needed for ventilation.** The maximum outdoor air needed for economizer cycles is then drawn through the outdoor air economizer damper.

## 11.10.7 Mixing Plenums

To achieve effective mixing, many designers prefer to mix in the ductwork some distance upstream of the unit inlet. If a mixing plenum is employed, careful consideration must be given to the objective of total mixing with no stratification under **all** conditions of operation.

If the equipment is alongside outdoor louvers in a wall, the minimum outdoor air damper should be located as close as possible to the return damper connection. An outdoor air damper sized for 1500 fpm usually provides acceptable control. The pressure difference between the relief plenum and outdoor intake plenum must be taken through the return damper section. A higher velocity through the return air damper—high enough to cause this loss at its full open position—enhances air balance and provides for better mixing. To create maximum turbulence and mixing, return air dampers should be set so that any deflection of air is toward the outdoor air. Parallel blade dampers may aid mixing.

When using a plenum or chamber, mixing dampers should be placed across the full width of the unit, even though the location of the return duct makes it more convenient to return air through the side. When return dampers are placed at one side, return air passes through one side of a double-inlet fan, and cold outdoor air passes through the other. If the air return must enter the side, some form of air blender or mixing device is necessary.

Although opposed blade dampers offer better control, properly proportioned parallel blade dampers are more effective for mixing airstreams of different temperatures. If parallel blades are used, each damper should be mounted so that its partially opened blades direct the airstreams toward the other damper for maximum mixing.

Baffles that direct the two airstreams to impinge on each other at right angles and in multiple jets create the turbulence required to mix the air thoroughly. In some instances, unit heaters or propeller fans have been used for mixing, regardless of the final type and configuration of dampers. Otherwise, the preheat coil will waste heat, or the cooling coil may freeze. Low-leakage outdoor air dampers minimize leakage during shutdown.

Coil freezing can be a serious problem with chilled water, heating water, or steam coils. Full flow circulation of water during freezing weather, or even reduced flow with a small recirculating pump, discourages coil freezing. Further, it can provide a source of off-season chilled water in air-and-water systems. Antifreeze solutions or complete coil draining also prevent coil freezing. However, because it is difficult, if not impossible, to drain most cooling coils completely, caution should be exercised if this option is considered (see Chapter 13).

### 11.10.8 Filter Section

Control of air cleanliness depends heavily on the filter. Unless the filter is regularly maintained, system resistance is increased and airflow diminishes. Accessibility is a primary consideration in filter selection and location. In a built-up system, there should be a minimum of 3 ft between the upstream face of the filter bank and any obstruction. Other requirements for filters can be found in Chapters 29 and 30 of the 2012 *ASHRAE Handbook—HVAC Systems and Equipment*, and ASHRAE Standard 52.1.

Good mixing of outdoor and return air is also necessary for good filter performance. A poorly placed outdoor air duct or a poor duct connection to the mixing plenum can cause uneven loading of the filter and poor distribution of air through the coil section.

### 11.10.9 Preheat Coil

The preheat coil should have relatively wide fin spacing, be accessible for easy cleaning, and be protected by filters. If the preheat coil is located in the minimum outdoor airstream rather than in the mixed airstream as shown in Figure 11-1, it should not heat the air to an exit temperature above 35 to 45°F; preferably, it should become inoperative at outdoor temperatures of 45°F. Inner distributing tube or integral face and bypass coils are preferable with steam, and full steam pressure should be applied at entering air temperatures below 40°F. Hot water preheat coils should have a constant flow circulating pump and should be piped for parallel heat flow so that the coldest air will contact the warmest coil surface first.

### 11.10.10 Cooling Coil

In this section, sensible and latent heat are removed from the air. In many finned coils, some air passes through without contacting (i.e., being thermally affected by) the fins or tubes. The amount of this bypass can vary from 30% for a four-row coil at 700 fpm to less than 2% for an eight-row coil at 300 fpm.

The dew point of the air mixture leaving a four-row coil might satisfy a comfort installation with 25% or less outdoor air, a small internal latent load, and sensible temperature control only. For close control of room conditions for precision work, a deeper coil is required.

A central station unit that is the primary source of conditioned ventilation air for other subsystems, such as in an air-and-water system or for a VAV system, does not need to supply as much air to a space. In this case, the primary air furnishes outdoor air for ventilation and handles space dehumidification and some sensible cooling. This application normally requires deeper, cleanable coils with flat fins and sometimes utilizes sprays. (See DOAS units in Chapter 12.) To prevent water carryover and allow proper cleaning and dew-point temperature control, fin spacing should be 0.125 in. minimum with flat fins and a minimum depth of six rows.

### 11.10.11 Reheat Coil Section

Reheat is discouraged, unless it is required to satisfy humidity control requirements. For energy conservation purposes, consideration may be given to using some form of recovered energy for a reheat source.

Heating coils located in the reheat position, as shown in Figure 11-1, in addition to being used for temperature control purposes, are frequently used for warm-up.

Hot water heating coils provide the highest degree of control. Oversized coils, particularly steam, can stratify the airflow; thus, where cost-effective, inner distributing coils are preferable for steam applications. Electric coils may also be used.

### 11.10.12 Humidifiers

For comfort installations not requiring close control, moisture can be added to the air by mechanical atomizers or point-of-use electric or ultrasonic humidifiers. Proper location of this equipment prevents stratification of moist air in the system.

Steam grid humidifiers with dew-point control often are used for humidity control. In this application, the heat of evaporation should be replaced by heating the recirculated water rather than by increasing the size of the preheat coil. It is not possible, of course, to add moisture to saturated air, even with a steam grid humidifier. Air in a laboratory or other application that requires close humidity control must be reheated after leaving a dehumidifier coil before moisture can be added. This requires reconsideration of air discharge temperatures and quantities.

The capacity of the humidifying equipment should not exceed the expected peak load by more than 10%. If the humidity is controlled from a sensor in the room or the return air, a limiting humidistat and fan interlock may be needed in the supply duct. This prevents condensation and mold or mildew growth within the ductwork when temperature controls call for cooler air. Many humidifiers add some sensible heat that should be accounted for in the psychrometric analysis.

It is quite difficult to prevent a steam humidifier from supersaturating the airstream, which, of course, can result in moisture in the air-handling unit, ductwork, or the space. This moisture can then become a source of mold or mildew. A preferred method of adding moisture to the air is with an evaporating matt or a sprayed coil in the outdoor airstream, downstream of a preheat coil (see Chapter 2).

### 11.10.13 Supply Air Fan

Axial flow, centrifugal, or plug fans may be chosen as supply air fans for straight-through flow applications. In factory-fabricated units, more than one centrifugal fan may be tied to the same shaft. If headroom permits, a single-inlet fan should be chosen when air enters at right angles to the flow of air through the equipment. This permits a more gradual transition from the fan to the duct and increases the static regain in the velocity pressure conversion.

To minimize inlet losses, the distance between the casing walls and the fan inlet should be at least equal to the diameter of the fan wheel. With a single-inlet fan, the length of the transition section should be at least half the width or height of the casing, whichever is longer.

If fans blow through the equipment, the air distribution through the downstream components needs analyzing, and baffles should be used to ensure uniform air distribution. See Chapter 21 of the 2012 *ASHRAE Handbook—HVAC Systems and Equipment*.

## 11.11 Air Distribution

### 11.11.1 Ductwork

Ductwork should deliver conditioned air to an area as directly, quietly, and economically as possible. Structural features of the building generally require some compromise and often limit depth. Chapter 9 describes ductwork design in detail and gives several methods of sizing duct systems.

It is imperative that the designer coordinate duct design with the structure. In commercially developed projects, it is common that great effort is made to reduce floor-to-floor dimensions. The resultant decrease in the available interstitial space left for ductwork is a major design challenge.

### 11.11.2 Room Terminals

In some instances, such as in single-zone, all-air systems, the air may enter from the supply air ductwork directly into the conditioned space through a grille or diffuser.

In multiple zoned air systems, an intermediate device controls air temperature and/or volume. Various devices are available, including (1) an air-water induction terminal, which includes a coil or coils in the induced airstream to condition the return air before it mixes with the primary air and enters the space; (2) an all-air induction terminal, which controls the volume of primary air, induces ceiling plenum air, and distributes the mixture through low-velocity ductwork to the space; (3) a fan-powered mixing box, which uses a fan to accomplish the mixing rather than depending on the induction principle; (4) a VAV box, which varies the amount of air delivered with no induction; (5) a VAV reheat terminal; (6) a variable volume, constant velocity diffuser; (7) a double-duct mixing box; and (8) a terminal reheat device.

### 11.11.3 Insulation

In all new construction (except low-rise residential buildings), air-handling duct and plenums installed as part of an HVAC air distribution system should be thermally insulated in accordance with ASHRAE Standard 90.1. See Chapter 21 of the 2013 *ASHRAE Handbook—Fundamentals* for further discussion and calculation methodology.

### 11.11.4 Ceiling Plenums

The space between a hung ceiling and the floor slab above is used frequently as a return plenum to reduce sheet metal work and remove heat from the plenum. Local and national codes should be consulted before using this approach in new design, because most codes prohibit combustible material in a return air ceiling plenum.

Entrance lobby ceilings with lay-in panels do not work well as return plenums where negative pressures from high-rise elevators or stack effects of high-rise buildings may occur. If the plenum leaks to the low-pressure area, the tiles may lift and drop out when the outside door is opened and closed. Return plenums directly below a roof deck have substantially higher return air heat gain or losses than a ducted return, which has the advantage of reducing the heat gain to or loss from the space.

### 11.11.5 Controls

Controls should be automatic and *simple* for best operating and maintenance effectiveness. Operations should follow a natural sequence—depending on the space need, one controlling thermostat closes a normally open heating valve, opens the outdoor air mixing dampers, or opens the cooling valve. In some applications, an enthalpy controller, which compares the heat content of outdoor air to that of return air, then opens the outdoor air damper when enthalpy of the outdoors is less than return air or space enthalpy and thus reduces the refrigeration load. On other systems, a dry-bulb control saves the cost of the enthalpy control and approaches these savings when an optimum changeover temperature, near the design dew point, is established.

A minimum outdoor air damper with separate motor, selected for a velocity of 1500 fpm, is preferred to one large outdoor air damper with minimum stops. A separate damper simplifies air balancing.

For control system fundamentals see Chapter 7, 2013 *ASHRAE Handbook—Fundamentals*, and Chapter 47, 2011 *ASHRAE Handbook—HVAC Applications*.

### 11.11.6 Vibration Isolation

Vibration and sound isolation equipment is required for most central system fan installations. Standard mountings of fiberglass, ribbed rubber, neoprene mounts, and springs are available for both fans and prefabricated units. If the fan manufacturer supplies the vibration isolators, it is common practice to supply neoprene pads for speeds above 1200 rpm, rubber-in-shear isolators (0.5 in. deflection) for speeds between 700 and 1200 rpm, and springs with 1 in. deflection for speeds below 700 rpm.

Special consideration is required in the isolation of rotating machinery such as fans and pumps when they are equipped with variable-speed drives because of the relationship between frequency (i.e., rotational speed) and the effectiveness of vibration isolation devices.

The transmissibility of vibration isolators is defined as the ratio of the transmitted force to the impressed force, or

$$TR = \frac{\text{Transmitted force}}{\text{Impressed force}} \qquad (11\text{-}10)$$

Thus, the smaller the transmissibility, the more effective is the isolator. It is common practice to select vibration isolators with a transmissibility between 0.10 and 0.05.

The relationship of the transmissibility to the system frequency is

$$TR = \frac{1}{(f/f_n)^2 - 1} \qquad (11\text{-}11)$$

where

$TR$ = transmissibility
$f$ = frequency of the rotating mass, Hz
$f_n$ = natural frequency of isolator system, Hz

Then, the frequency of the rotating mass in Hz is

$$f = N/60 \qquad (11\text{-}12)$$

where

$N$ = rotational speed, rpm
60 = seconds per minute

and

$$f_n = \frac{1}{2\pi}\sqrt{\frac{g}{y}} \qquad (11\text{-}13)$$

where

$g$ = gravitational constant, 386 in./s²
$y$ = static deflection of isolator, in.

The solution to Equation (11-11) for rotating fan speeds of 0 to 1000 rpm and static deflections of 0 to 3 in. is plotted in Figure 11-2 for transmissibility of 0.05, 0.10, 1, and ∞. The graph demonstrates that for a given static deflection as the speed is decreased, the transmissibility increases to the point where the isolator loses all of its effectiveness when $f/f_n = \sqrt{2}$, below which the isolator becomes an amplifier. Thus, extra care should be taken when selecting and designing isolators for a variable-speed device. One method of ensuring effective vibration control is to design the isolator at the minimum speed at which the system will operate rather than at the design or maximum speed.

Ductwork connections should be made with fireproof fiber cloth sleeves having considerable slack, but without offset between the fan outlet and rigid duct. Misalignment between the duct and fan outlet can cause turbulence, generate noise, and reduce system efficiency. Electrical and piping connections to vibration-isolated equipment should be made with flexible conduit and flexible pipe connections. Special considerations are required in seismic zones.

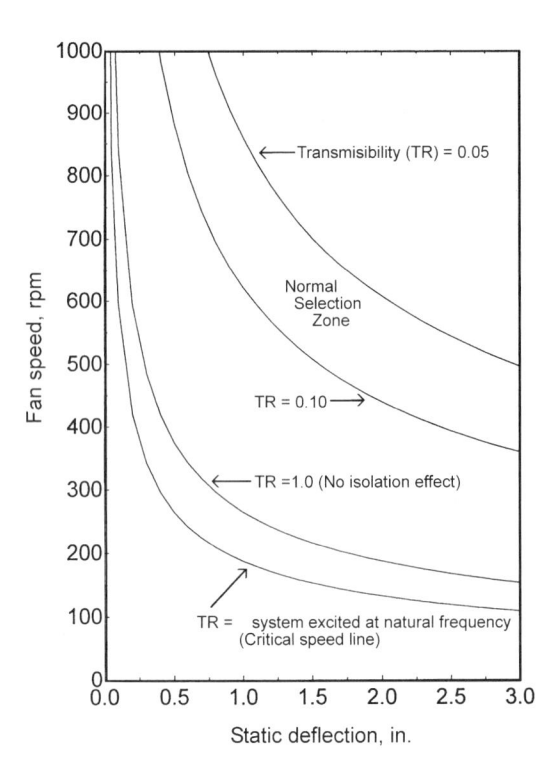

*Fig. 11-2 Static Deflection of Isolator Versus Fan Speed*

Equipment noise transmitted through the ductwork can be reduced by sound-absorbing units, acoustical lining, and other means of attenuation. Sound transmitted through the return and relief ducts should not be overlooked. Acoustical lining sufficient to adequately attenuate any objectionable system noise or locally generated noise should be considered. Chapter 48 of the 2011 *ASHRAE Handbook—HVAC Applications*, Chapter 8 of the 2013 *ASHRAE Handbook—Fundamentals*, and ASHRAE Standard 68 have further information on sound and vibration control.

The designer must account for seismic restraint requirements for the seismic zone in which the particular project is located.

**Example 11-3.** A fan is selected to be operated at 720 rpm at design conditions and is supplied with a spring vibration isolator with 1 in. static deflection.

(a) What will be the transmissibility of the fan with the selected isolator mount?

**Solution:**

$$f = N/60 = 720/60 = 12 \text{ Hz}$$

$$f_n = \frac{1}{2\pi}\sqrt{\frac{g}{y}} = \frac{1}{2\pi}\sqrt{\frac{386}{1}} = 3.13 \text{ Hz}$$

# Air-Conditioning System Concepts

$$TR = \frac{1}{(f/f_n)^2 - 1} = \frac{1}{(12/3.13)^2 - 1} = 0.07$$

(b) If the fan is driven by a variable-speed drive controlled to reduce the speed to 37% of the design value, what will the transmissibility be at the minimum speed?

$$N_b = 0.37(720) = 266.4 \text{ rpm}$$

$$f = 266.4/60 = 4.44 \text{ Hz}$$

$$TR = \frac{1}{(4.44/3.13)^2 - 1} = 1$$

The isolator would provide no isolation.

## 11.12 Space Heating

Although steam is an acceptable medium for central system preheat or reheat coils, low-temperature hot water provides a much more uniform means of perimeter and other heating devices located within the space. Individual automatic control of each terminal provides the ideal space comfort. A control system that varies the water temperature inversely with the change in outdoor air temperature provides water temperatures that produce acceptable results without individual room or space control in limited applications. To produce the best results, the most satisfactory ratio of indoor air temperature to that outdoors can be set after the installation is completed and actual operating conditions are ascertained.

Multiple perimeter spaces on one exposure served by a central system may be heated by supplying warm air from the central system. Areas that have heat gain from lights and occupants and no heat loss require cooling in winter as well as in summer, as do perimeter areas in which solar gains in winter combined with internal gains can exceed transmission losses.

In systems with mixing chambers, little or no heating of the return and outdoor air is required when the space is occupied. Local codes dictate the amount of outdoor air required, which is generally based on the requirements of ASHRAE Standard 62. For example, with return air at 75°F and outdoor air at 0°F, the temperature of a 25% outdoor/75% return air mixture would be 56°F, which is close to the temperature of the air supplied to cool such a space in summer. In this instance, a preheat coil installed in the minimum outdoor airstream to warm the outdoor air can produce overheating, unless it is sized as previously recommended. Assuming good mixing, a preheat coil located in the mixed airstream prevents this problem. The outdoor air damper should be kept closed until room temperatures are reached during warm-up. A return air thermostat can terminate the warm-up period.

When a central air-handling unit supplies both perimeter and interior spaces, the supply air must be cool to handle the interior zones. Additional control is needed to heat the perimeter spaces properly. Reheating the air is the simplest solution, but it is not acceptable by some energy codes. An acceptable solution is to vary the volume of air to the perimeter and combine it with a terminal heating coil or a separate perimeter heating system, either baseboard, overhead air heating, or a fan-powered mixing box with supplemental heat. The perimeter heating should be individually controlled and integrated with the cooling control.

## 11.13 Primary Systems

The type of central heating and cooling equipment used for air-conditioning systems in large buildings depends chiefly on economic factors, once the total required capacity has been determined. Component choice depends on such factors as the type of fuel available, environmental protection required, structural support, and available space.

Rising energy costs have fostered many designs to recover the internal heat from lights, people, and equipment to reduce the size of the heating plant. Chapters 9 and 26 in the 2012 *ASHRAE Handbook—HVAC Systems and Equipment* describe several heat recovery devices. Also, see Chapter 16 of this book.

The search for energy savings has extended to cogeneration or total energy systems in which on-site power generation has been added to the heating and air-conditioning project. The economics of this function is determined by gas and electric rate differentials, investment cost of the plant, and by the ratio and coincidence of electric to heat demands for the project. In these systems, reject heat from the prime movers can be put into the heating system and the cooling equipment, either to drive the turbines of centrifugal compressors or to serve absorption chillers. Chapter 16 of this book and Chapter 7 of the 2012 *ASHRAE Handbook—HVAC Systems and Equipment* present further details on cogeneration or total energy systems.

Among the largest installations of central mechanical equipment are the central cooling and heating plants serving groups of large buildings. These plants provide higher diversity and sometimes operate more efficiently and with lower maintenance and labor costs than individual plants.

The economics of these systems requires extensive analysis. Boilers, gas and steam turbine-driven centrifugals, and absorption chillers may be installed in combination in one plant. In large buildings with core areas that require cooling while perimeter areas require heating, one of several heat reclaim systems could heat the perimeter to save energy. Chapter 7 of the 2012 *ASHRAE Handbook—HVAC Systems and Equipment* gives details of these combinations, and Chapters 11, 12, and 15 of the 2012 *ASHRAE Handbook—HVAC Systems and Equipment* give design details of central plants. Also see Chapter 13.

Most large buildings, however, have their own central heating and cooling plant in which the choice of equipment depends on the following:

- Required capacity and type of usage
- Costs and types of energy available
- Location of the equipment room
- Type of air distribution system(s)
- Owning and operating costs

Many electric utilities impose severe penalties for peak summertime power use or, alternatively, offer incentives for off-peak use. This policy has renewed interest in both water and ice thermal storage systems. The storage capacity installed for summertime load leveling may also be available for use in the winter, making heat reclaim and storage a more viable option. With ice-storage, the low-temperature ice water can provide colder air than that available from a conventional system. Use of high water temperature rise and lower temperature air results in lower pumping and fan energy and, in some instances, offsets the energy penalty due to the lower refrigeration temperature required to make ice.

### 11.13.1 Heating Equipment

Steam and hot water boilers for heating are manufactured for high or low pressure and use gas, oil, or electricity and, sometimes, coal or waste material for fuel. Low-pressure boilers are rated for a working pressure of 15 psig for steam and 160 psig for water, with a maximum temperature limitation of 250°F. Package boilers, with all components and controls assembled as a unit, are available. Electrode or resistance-type electric boilers are available for either hot water or steam generation. Boilers and furnaces designed for higher efficiencies have the ability to condense the water vapor in the combustion chamber, but the boilers must operate with water temperatures much lower than noncondensing boilers, usually limited to about 125°F. For further information, see Chapter 19.

Where steam or hot water is supplied from a central plant, as on some university campuses and in downtown areas of large cities, the service entrance to the building must conform to utility standards. The utility should be contacted at the beginning of the project to determine availability, cost, and the specific requirements of the service.

### 11.13.2 Fuels

Chapter 19 gives fuel types, properties, and proper combustion factors and includes information for the design, selection, and operation of fuel selection and automatic fuel-burning equipment.

### 11.13.3 Refrigeration Equipment

The major types of refrigeration equipment used in large systems are reciprocating compressors, helical rotary compressors, screw rotary compressors, centrifugal compressors, and absorption chillers. See Chapter 18 for further discussion of refrigeration equipment, including the general size ranges of available equipment.

Reciprocating, helical or screw rotary, and centrifugal compressors are usually driven by electric motors; however, they can, and sometimes are, driven by natural gas or diesel engines and gas and steam turbines. The compressors may be purchased as part of a refrigeration chiller consisting of compressor, drive, chiller-evaporator, condenser, and necessary safety and operating controls.

Reciprocating and helical or screw rotary compressor units are frequently used in packaged as well as in field-assembled systems, with air-cooled or evaporative condensers arranged for remote installation. Centrifugal compressors are used usually only in packaged chillers.

Absorption chillers are water cooled. Most use a lithium bromide/water or water/ammonia cycle and are generally available in the following four configurations: (1) direct fired, (2) indirect fired by low-pressure steam or hot water, (3) indirect fired by high-pressure steam or hot water, and (4) fired by hot exhaust gas. Small direct-fired chillers are single-effect machines with capacities of 3.5 to 25 tons. Larger indirect-fired, single-effect and double-effect chillers in the 100 to 2000 ton capacity range are available.

Low-pressure steam at 12 psig or hot water heats the generator of single-effect absorption chillers with capacities from 50 to 1600 tons. Double-effect machines use higher pressure steam up to 150 psig or hot water at an equivalent temperature (365°F). Absorption chillers of this type are available from 350 to 2000 tons.

The absorption chiller is sometimes combined with steam turbine-driven centrifugal compressors in large installations. Steam from the noncondensing turbine is piped to the generator of the absorption machine. When a centrifugal chiller is driven by a gas turbine or an engine, an absorption machine generator may be fed with steam or hot water from the jacket and/or an exhaust gas heat exchanger.

### 11.13.4 Cooling Towers

Water is usually cooled by the atmosphere for use in water-cooled condensers. Either natural draft or mechanical draft cooling towers or spray ponds are used for the cooling. Of these, the mechanical draft tower, which may be of the forced draft, induced draft, or ejector type, is used for most designs because it does not depend on the wind. Air-conditioning systems use towers ranging from small package units of 5 to 500 tons or field-erected towers with multiple cells in virtually unlimited sizes.

The tower must be winterized if required for operation at ambient outdoor dry-bulb temperatures below 35°F. Winterizing includes the capability of bypassing water directly into the tower return line (either automatically or manually, depending on the installation) and of heating the tower pan water to a temperature above freezing. (See Chapter 14 in the 2012 *ASHRAE Handbook—HVAC Systems and Equipment*.)

Heat may be added by steam or hot water coils or electric resistance heaters in the tower basin. Also, it is often necessary to provide an electric heating cable on the condenser water and makeup water pipes and to insulate these heat-traced sections to prevent the pipes from freezing. Special controls are required when it is necessary to operate the cooling tower at or near freezing conditions.

Where the cooling tower will not operate in freezing weather, provisions for draining the tower and piping are necessary. Draining is the most effective way to prevent the tower and piping from freezing.

Careful attention must also be given to water treatment to keep maintenance required in the refrigeration machine absorbers and/or condensers to a minimum.

Cooling towers may also be used as a cooling source for the building in low-temperature seasons by filtering and directly circulating the condenser water through the chilled water circuit, by cooling the chilled water with the cooling tower water in a separate heat exchanger, or by using the heat exchangers in the refrigeration equipment to produce thermal cooling. Towers are usually selected in multiples so that they may be run at reduced capacity and shut down for maintenance in cool weather. Chapter 18 includes further design and application details.

### 11.13.5 Air-Cooled Condensers

Air-cooled condensers pass outdoor air over a dry coil to condense the refrigerant. This results in a higher condensing temperature and, thus, a larger power input at peak condition; however, over the year this peak time may be relatively short. The air-cooled condenser is popular in small reciprocating and helical or screw rotary systems because of its low first cost and lower cost maintenance requirements.

Recent emphasis on energy efficiency and water conservation have led to the concept of hybrid air/water-cooled refrigeration plants that use water-cooled chillers for peak loads in warm-weather months, and smaller air-cooled units for cold-weather operation.

### 11.13.6 Evaporative Condensers

Evaporative condensers pass air over refrigerant condensing coils sprayed with water, thus taking advantage of adiabatic saturation to lower the condensing temperature. As with the cooling tower, freeze prevention and close control of water treatment are required for successful operation. The lower power consumption of the refrigeration system and the much smaller footprint from the use of the evaporative versus the air-cooled condenser are gained at the expense of the cost of the water used and increased maintenance costs.

### 11.13.7 Pumps

Pumps used in heating and air-conditioning systems are usually centrifugal pumps. Pump configurations include horizontal split case with a double-suction impeller, or end suction pumps, either close-coupled or flexibly connected.

Major applications for pumps in the equipment room are as follows:

- Chilled water
- Heating water
- Condenser water
- Steam condensate pumps
- Boiler feed pumps
- Fuel oil

When the pumps handle hot liquids or have high inlet pressure drops, the required net positive suction head (NPSH) must not exceed the NPSH available at the pump. It is common practice to provide spare pumps or spare critical components to maintain system continuity in case of a pump failure. See Chapter 13 and Chapters 14 and 44 of the 2012 *ASHRAE Handbook—HVAC Systems and Equipment*, for more design information on pumps and pumping systems. Chapters 9 in this book and 13 also discuss pump selection.

### 11.13.8 Piping

Air-conditioning piping systems can be divided into two parts: the piping in the main equipment room and the piping required to deliver heat or chilled water to the air-handling equipment and terminal devices throughout the building. The air-handling system piping follows procedures detailed in Chapters 9 and 13.

The major piping in the main equipment room includes fuel lines, refrigerant piping, and steam and water connections. Chapter 11 of the 2012 *ASHRAE Handbook—HVAC Systems and Equipment* includes more information on piping for steam systems, and Chapter 13 of the same volume has information on heating and chilled and dual-temperature water systems, as does Chapter 14. Chapter 22 in the 2013 *ASHRAE Handbook—Fundamentals* presents data on sizing steam, hydronic, fuel oil, natural gas, and service water piping.

### 11.13.9 Instrumentation

All equipment must have adequate gages, thermometers, flow-meters, balancing devices, and dampers for effective system testing, balancing, monitoring, commissioning, and operations. In addition, capped thermometer wells, gage cocks, plugged duct openings, volume dampers, and access/inspection, maintenance doors or ports should be installed at strategic points for inspection service and system balancing. Chapter 38 of the 2011 *ASHRAE Handbook—HVAC Applications* indicates the locations and types of fittings required.

A central control console or personal computer to monitor the many system points should be considered for any large, air-conditioning system. A computer terminal permits a single operator or building manager to monitor and perform functions at any point in the building to increase occupant comfort and to free maintenance staff for other duties. Chapter 47 of the 2011 *ASHRAE Handbook—HVAC Applications* describes these systems in detail.

## 11.14 Space Requirements

In the initial phases of building design, the engineer seldom has sufficient information to finalize the designs. Therefore, most experienced engineers have evaluation criteria to estimate the building space needed.

The air-conditioning system selected, the building configuration, and other variables govern the space required for the mechanical system. The final design is usually a compromise between what the engineer recommends and what the architect can accommodate. Where designers cannot negotiate a space large enough to suit the installation, the judgment of the owners should be called upon.

Although few buildings are identical in design and concept, some basic criteria are applicable to most buildings and help allocate space that approximates the final requirements.

These space requirements are often expressed as a percentage of the total building floor area.

### 11.14.1 Mechanical, Electrical, and Plumbing Facilities

The total mechanical and electrical space requirements range from 4 to 9% of the gross building area, the majority of buildings falling within the 6 to 9% range.

Most of the facilities should be centrally located to minimize long duct, pipe, and conduit runs and sizes; simplify shaft layouts; and centralize maintenance and operation. A central location also reduces pump and fan motor power, which may reduce building operating costs. But for many reasons, it is often impossible to centrally locate all the mechanical, electrical, and plumbing facilities within the building. In any case, the equipment should be located to minimize space requirements, centralize maintenance and operation, and simplify the electrical system.

Equipment rooms generally require clear ceiling height ranging from 12 to 20 ft, depending on equipment sizes and the complexity of ductwork, piping, and conduit.

The main electrical transformer and switchgear rooms should be located as close to the incoming electrical service as practical. If there is an emergency generator, it should be located considering (1) proximity to emergency electrical loads and sources of combustion and cooling air, (2) ease of properly venting exhaust gases to the outdoors, and (3) provisions for noise control.

The main plumbing equipment room usually contains gas and domestic water meters, the domestic hot water system, the fire protection system, and various other elements such as compressed air, special gases, and vacuum, ejector, and sump pump systems. Some water and gas utilities require a remote outdoor meter location.

The heating and air-conditioning equipment room houses the boiler or pressure-reducing station, or both; the refrigeration machines, including the chilled water and condensing water pumps; converters for furnishing hot or cold water for air conditioning; control air compressors; steam condensate pumps; and other miscellaneous equipment. Local codes and ASHRAE *Standard 15* should be consulted for special equipment room requirements.

In high-rise buildings, it is often economical to locate the refrigeration plant at the top or intermediate floors, or in a roof-level penthouse. The electrical service and structural costs will rise, but these may be offset by reducing costs for condenser and chilled water piping, energy consumption, and equipment because of the lower operating pressure. The boiler plant may also be placed at roof level, which eliminates the need for a chimney through the building.

Gas fuel may be more desirable because oil storage and pumping present added design and operating problems, and the cost of oil leak detection and prevention may be substantial. Heat recovery systems, in conjunction with the refrigeration equipment, can appreciably reduce the heating plant size in buildings with large core areas.

Additional space may be needed for a telephone terminal room, a pneumatic tube equipment room, an incinerator, or other equipment.

Many buildings, especially larger ones, need cooling towers, which often present problems. If the cooling tower is on the ground, it should be at least 100 ft away from the building for two reasons: (1) to reduce tower noise in the building and (2) to keep cool season discharge air from fogging the building's windows. Towers must also be located so as to prevent any possibility of tower discharge being mixed into the ventilation air intakes, to prevent the possibility of *Legionella* entering the building. Towers should be kept the same distance from parking lots to avoid staining car finishes with water treatment chemicals. When the tower is on the roof, its vibration and noise must be isolated from the building. Some towers are less noisy than others, and some have attenuation housings to reduce noise levels. These options should be explored before selecting a tower.

The bottom of roof-mounted towers, especially larger ones, must be set on a steel frame 4 to 5 ft above the roof to allow room for piping and proper tower and roof maintenance. Pumps below the tower should be designed for adequate net positive suction head, but they must be installed to prevent the draining of the piping on shutdown.

### 11.14.2 Fan Rooms and Rooftop Units

Fan rooms should preferably be located as close as possible to the space to be conditioned to reduce both installation cost, fan power, and fan energy.

The number of fan rooms depends largely on the geometry of the buildings and use schedules of the spaces. Buildings with large floor areas often have multiple fan rooms on each floor. Many high-rise buildings, however, may have one fan room serving 10 to 20 floors—one serving the lower floors, one serving the middle of the building, and one at the roof level serving the top portion of the building.

Life safety is a very important factor in fan room location. Chapter 53 of the 2011 *ASHRAE Handbook—HVAC Applications* discusses principles of fire spread and smoke management.

It is very common, particularly on smaller and low-rise buildings, to locate air-handling units or packaged HVAC equipment on the roof instead of in a fan room or mechanical equipment room. Many units are designed for rooftop installation and are furnished with a roof curb for mounting. However, it is the responsibility of the system designer to select machinery that will withstand the severity of the climate and that can be easily and conveniently accessed for normal maintenance and emergency service and to provide wearing surfaces on the roof for the maintenance and service personnel, tools, and replacement components.

### 11.14.3 Interior Shafts

In tall buildings, interior shaft space is necessary to accommodate return and exhaust air; interior supply air and hot, chilled, and condenser water piping; steam and return piping;

# Air-Conditioning System Concepts

electrical closets; telephone closets; plumbing piping; and, possibly, pneumatic tubes and conveyer systems.

The shafts must be clear of stairs, elevators, and structural beams on at least two sides to allow maximum headroom when the pipes and ducts come out above the ceiling. In general, duct shafts having an aspect ratio of 2:1 to 4:1 are easier to develop than large square shafts. The rectangular shape also makes it easier to go from the equipment in the fan rooms to the shafts.

The size, number, and location of shafts is important in multi-story buildings. Vertical duct distribution systems with little horizontal branch ductwork are desirable because they are usually less costly; easier to balance; create less conflict with pipes, beams, and lights; and enable the architect to design lower floor-to-floor heights.

The number of shafts is a function of building geometry, but, in larger buildings, it is usually more economical in cost and space to have several smaller shafts rather than one large shaft. Separate supply and exhaust duct shafts may be desired to reduce the number of duct crossovers. **From 10% to 15% additional shaft space should be allowed for future expansion and modifications.** The additional space also reduces the initial installation cost.

## 11.14.4 Equipment Access

Properly designed mechanical equipment rooms must allow for the movement of large, heavy equipment in, out, and within the building. Equipment replacement and maintenance can be very costly if access is not planned properly. Many designers (and some machinery codes) require a minimum of 36 to 48 in. clear aisles between all machines and equipment.

Because systems vary greatly, it is difficult to estimate space requirements for refrigeration and boiler rooms without making block layouts of the system selected. Block layouts allow the engineer to develop the most efficient arrangement of the equipment, with adequate access and serviceability. Block layouts can also be used in preliminary discussions with the owner and architect. Only then can the engineer obtain verification of the estimates and provide a workable and economical design.

**Example 11-4.** A new law office building is being designed for St. Louis, Missouri, to use the HVAC system shown in the sketch below. Minimum outdoor air of 15 cfm/person for meeting the ventilation requirements of the anticipated 125 occupants will be maintained throughout the year. During summer operation, the cooling coil supplies air to the conditioned space at 55°F, 90% relative humidity. The space summer design loads and designs conditions are

323,000 Btu/h sensible (gain);
89,000 Btu/h latent (gain)
Outdoor: 94°F db, 75°F wb; Indoor: 78°F

The 55% efficient fan must produce a 3 in. w.g. pressure increase to overcome the friction in the duct system. Determine:

(a) Size of cooling coil, ft$^2$ of face area
(b) Rating of chiller unit, Btu/h
(c) Fan motor size, hp
(d) Outdoor air damper face area, ft$^2$

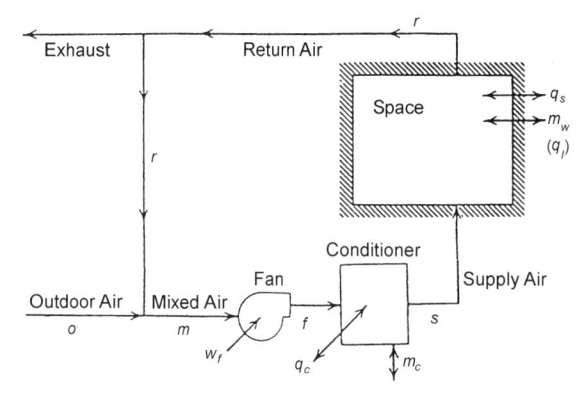

**Solution:**

$$m_w = 89000/1100 = 80.9 \text{ lb/h}$$

$$m_a = \frac{323000}{0.244(78-55)} = 57555 \text{ lb/h}$$

$$m_o = \frac{125 \times 15 \times 60}{13.33} = 8440 \text{ lb/h}$$

$$m_r = 57555 - 8440 = 49115 \text{ lb/h}$$

$$W_r = W_s + \frac{m_w}{m_s} = 0.0083 + \frac{80.9}{57555} = 0.00971$$

$$W_m = \frac{8440(0.0144) + 49115(0.00971)}{57555} = 0.0104 \text{ lb/lb}$$

$$h_m = \frac{8440(38.5) + 49115(29.4)}{57555} = 30.7 \text{ Btu/lb}$$

$$t_m = 80.4°F; \phi = 48\%$$

$$h_f = h_m + w_f = 30.7 + 0.49 = 31.2 \text{ Btu/lb}$$

$$W_f = 0.0104 \text{ lb/lb}$$

$$Q = \frac{57555(13.33)}{60} = 12800 \text{ cfm at std. conditions}$$

**Coil Size** (select 400 fpm face velocity)

$$A_F = 12800/400 = 32 \text{ ft}^2$$

**Chiller Size**

$$m_a[h_f - h_s - (W_f - W_s)h_c] + Q_c = 0$$

$$57555[31.2 - 22.3 - (0.0104 - 0.0083)23] + Q_c = 0$$

$$Q_c = -509,000 \text{ Btu/h}$$

$$Q_{\text{chiller}} \approx Q_{\text{coil}} = 509,000 \text{ Btu/h}$$

**Fan Motor Size**

$$P_f = \frac{57555(0.49)}{2545} = 11 \text{ hp}$$

**Outdoor Air Damper Size**

$$OA = 125 \times 15 = 1875 \text{ cfm}$$

If the recommended face velocity is 1500 fpm:

$$A_D = 1875/1500 = 1.25 \text{ ft}^2$$

## 11.15 Problems

**11.1** A room is to be cooled to a temperature of 75°F and a relative humidity of 50%. If there is negligible latent load within the space, what is the highest temperature at which the conditioned air can be supplied ($t_s$)? Why?

**11.2** A room has a total space cooling load of 20 tons and a sensible heat ratio of 0.90. If the conditioned air is to be supplied at 20°F less than the room temperature, how much air must be circulated?

**11.3** What are the four generic types of air systems expressed by thermodynamic methods?

**11.4** What are the 18 fundamental parameters that must be addressed in the selection and design of an HVAC system?

**11.5** Before designing a system, the cooling and heating load for each room in a building must be calculated? Why?

**11.6** If outdoor air at 95°F dry bulb and 78°F wet bulb is cooled to 75°F dry bulb without any dehumidification, what will the relative humidity be?

**11.7** In the air-handling unit of Figure 11-1, under design conditions, the outdoor air temperature is 95°F dry bulb and 78°F wet bulb and the space temperature is 75°F and 50% rh. The supply fan handles 60,000 cfm of air at 55°F saturated (entering the fan).

If the minimum outdoor air dampers are sized for 6000 cfm of ventilation air, what is the state point (dry-bulb and wet-bulb temperatures) of the mixed air?

**11.8** A constant-flow air-handling system is designed to circulate 60,000 cfm of air at a total fan pressure rise of 6 in. w.g. The system is designed to operate continuously. The fan efficiency is 70% and the motor efficiency is 90%.
(a) How much power (hp) is required to drive the fan?
(b) What will be the annual fan energy consumption?

**11.9** If, in the above problem, the sensible space load were reduced by 25% by using a more energy effective building envelope and improved lighting system, and this change were accommodated by reducing the air flow rate at the same fan efficiencies, what would be the reduction in annual fan energy?

**11.10** It is desired to transfer a given quantity of heat energy from one location to another location in a building. Two methods being considered are either by an air system operating at 4 in. w.g. total pressure or by a water system with a pump head of 40 ft. Calculate the ratio of fan power required for an air system to pump power required for a water system with the following system variables:

| | |
|---|---|
| Fan efficiency | 70% |
| Pump efficiency | 80% |
| Air $\Delta t$ | 20°F |
| Water $\Delta t$ | 40°F |

**11.11** A fan with a variable-speed drive is selected to operate at 900 rpm, and it is installed on a spring isolator mount with 1 in. static deflection. Determine:
(a) Transmissibility of the isolator
(b) Minimum speed that the unit can be operated at before the transmissibility is 0.50

**11.12** Specify typical temperatures for the following:
(a) Air leaving a gas-fired warm air furnace
(b) Air leaving a heat pump condenser
(c) Air leaving the cooling coil of a residential air conditioner
(d) Air leaving the cooling coil of a commercial air conditioner
(e) Hot water entering the convectors (radiators) of a hydronic system
(f) Hot water returning to the boiler from the convectors

**11.13** An air-conditioned room has a sensible cooling load of 200,000 Btu/h, a latent cooling load of 50,000 Btu/h, an occupancy of 20 people, and is maintained at 76°F dry bulb and 64°F wet bulb. Twenty-five percent of the air entering the room is vented through cracks and hoods. Outdoor air is assumed to be at design conditions of 95°F dry bulb and 76°F wet bulb. Conditioned air leaves the apparatus and enters the room at 60°F dry bulb.

Use the following letters to designate state points:

    **A** Outside design conditions
    **B** Inside design conditions
    **C** Air entering apparatus (mixed air temperature)
    **D** Air entering room (supply air temperature)
(a) Complete the following table:

| Point | Dry Bulb | Wet Bulb | h | W |
|---|---|---|---|---|
| A | | | | |
| B | | | | |
| C | | | | |
| D | | | | |

(b) Calculate the room SHR.
(c) What air quantity must enter the room?
(d) What is the apparatus load in tons?
(e) What is the load of the outdoor air? In lb per hour? In cfm?
(f) Does the room load plus the outdoor air load equal the coil load?

**11.14** A space has a sensible heat loss of 60,000 Btu/h and a latent loss of 20,000 Btu/h. The space is to be maintained at 70°F and 40% rh. The air that passes through the conditioner is 90% recirculated and 10% outdoor air at 40°F and 20% rh. The conditioner consists of an adiabatic saturator and a heating coil. Estimate the temperature and humidity ratio of the air entering the conditioned space. What is the flow rate in cfm? How much heat is added by the coil to the air in Btu/h? How much water is added to the air by the adiabatic saturator (lb/h)?

**11.15** Air at 800 ft³/min leaves a residential air conditioner at 65°F with 40% rh. The return air from the rooms has average dry- and wet-bulb temperatures of 75°F and 65°F, respectively. Determine:

(a) Size of the unit in tons (12,000 Btu/h = 1 ton)
(b) Rate of dehumidification
 [Ans: 2.58 tons, 20.16 lb/h (9.1 kW, 2.54 g/s)]

**11.16** In an air-conditioning unit 6000 cfm at 80°F dry bulb, 60% rh, and standard atmospheric pressure, enter the unit. The leaving condition of the air is 57°F dry bulb and 90% rh. Calculate:

(a) Cooling capacity of the air-conditioning unit, tons
(b) Rate of water removal from the unit, lb/h
(c) Sensible heat load on the conditioner, Btu/h
(d) Latent heat load on the conditioner, Btu/h
(e) Dew point of the air leaving the conditioner, °F

**11.17** A space in an industrial building has a winter sensible heat loss of 200,000 Btu/h and a negligible latent heat load (latent losses to outside are made up by latent gains within the space). The space is to be maintained at 75°F and 50% rh. Due to the nature of the process, 100% outdoor air is required for ventilation. The outdoor air conditions can be taken as saturated air at 20°F. The amount of ventilation air required is 7000 cfm and the air is to be preheated, humidified with an adiabatic saturator, and then reheated. The temperature out of the adiabatic saturator is to be maintained at 60°F dry bulb. Calculate the following:

(a) Temperature of the air entering the preheater
(b) Temperature of the air entering the space to be heated
(c) Heat supplied to preheat coil, Btu/h
(d) Heat supplied to reheat coil, Btu/h
(e) Quantity of makeup water added to adiabatic saturator, gpm
(f) Temperature of the spray water
(g) Show the processes and label points on the psychrometric diagram

**11.18** In winter, a meeting room with a large window is to be maintained at comfort conditions. The inside glass temperature on the design day is 40°F. Condensation on the window is highly undesirable. The room is to accommodate 18 adult males [250 Btu/h (sensible) and 200 Btu/h (latent) per person]. The heat loss through the walls, ceiling, and floor is 33,600 Btu/h. There are 640 watts of lights in the room.

(a) Determine the sensible heat loss or gain.
(b) Specify the desired interior dry-bulb temperature and relative humidity.
(c) If the heating system provides air at 95°F, determine the required airflow (cfm) and the maximum relative humidity permissible in the incoming air.
 [Ans: 26,916 Btu/h; 75°F, 28%; 1246 cfm, 14%]

**11.19** A zone in a building has a sensible load of 20.5 kW (70,000 Btu/h) and a latent load of 8.8 kW (30,000 Btu/h). The zone is to be maintained at 25°C (77°F) and 50% rh.

(a) Calculate the conditions ($t$ and $W$) of the entering air to the zone if the air leaves the coil saturated.
(b) What flow rate is required in order to maintain the space temperatures?
(c) If a mixture of 50% return air and 50% outdoor air at 31.6°C (97°F) and 60% rh enters the air conditioner, what is the refrigeration load?

**11.20** Sketch (with line diagrams) and list the advantages, disadvantages, and typical uses of the following systems:
(a) Fan-coil units
(b) Terminal reheat system
(c) Multizone system
(d) Double-duct system
(e) Variable-volume system
(f) Induction system

**11.21** A general office building in St. Louis, Missouri, has a winter sensible space heating load of 1,150,000 Btu/h for design conditions of 75°F and –5°F. The heating system operates with 25% outdoor air mixed with return air.
(a) Schematically draw the flow diagram and label, including temperatures and flow rates at each location.
(b) Specify the necessary furnace size.

**11.22** For the building of Problem 11.23, determine:
(a) Annual energy requirements for heating, Btu
(b) Annual fuel cost using No. 2 fuel oil at $1.60/gal

**11.23** To provide comfort conditions for a general office building, 38 ft by 80 ft by 8 ft, an air-treating unit consisting of cooling coil, heating coil, and humidifier is provided for this space with the flow diagram as shown. Indoor design conditions are: summer, 78°F/60% rh; winter, 72°F/25% rh.

Ninety people are normally employed doing light work while seated. The building is in Kansas City, Missouri. Fan operation is constant all year long. Ventilation rate is 15 cfm/person.

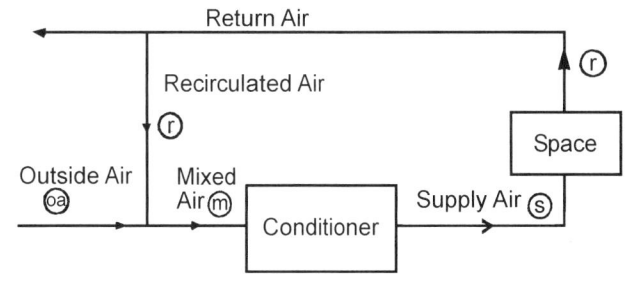

**Winter:** Sensible space heat loss is 189,000 Btu/h at design conditions, latent load is negligible. Maximum supply air temperature is 155°F.

**Summer:** Sensible space heat gain is 101,200 Btu/h at design conditions. Latent load is due entirely to the occupancy. The minimum supply air temperature from the cooling coil is 58°F.

(a) Determine the fan size (scfm) needed to provide sufficient air
(b) Size the heating unit needed, Btu/h
(c) Size the cooling coil needed, Btu/h
(d) Size the humidifier, gal/h

**11.24** A view of the air-conditioning system for a building in Denver, Colorado (elevation = 5000 ft; barometric pressure = 12.23 psi), is given. Outdoor air at the rate of 2500 cfm is required for ventilation. Other conditions at summer design are

*Space Loads*
   Sensible = 410,000 Btu/h
   Latent = 220,000 Btu/h

*Outdoor Air:* 91°F, 30% rh

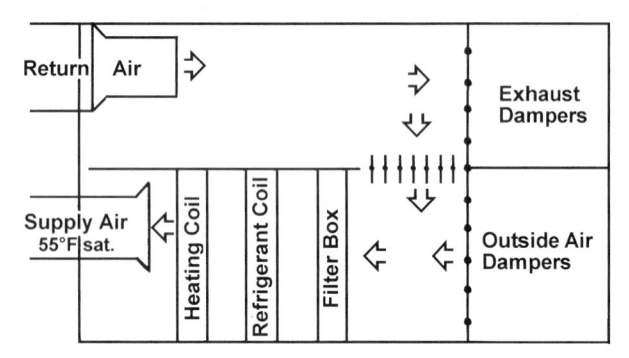

For an indoor design temperature of 78°F, determine

1. Supply airflow, lb/h
2. Supply airflow, cfm
3. Relative humidity at return, %
4. Size of cooling unit, Btu/h
5. Latent component of (4)
6. Sensible component of (4)
7. Sensible cooling load due to outdoor air, Btu/h

## 11.16  Bibliography

ASHRAE. 2012. Chapter 1, *ASHRAE Handbook—HVAC Systems and Equipment.*

Klote, J.H. and J.A. Milke. 2002. *Principles of Smoke Management.* ASHRAE.

Wheeler, A.E. 1997. Air Handling Unit Design for Energy Conservation. *ASHRAE Journal* 39(June).

# Chapter 12

# SYSTEM CONFIGURATIONS

In this chapter the various types of HVAC systems that provide cooling, dehumidifying, heating, and humidifying are described. The way the systems are identified with regard to the fluid used for carrying the energy and/or moisture is presented. Details about these systems are given in Chapters 1 through 6 in the 2012 *Handbook—HVAC Systems and Equipment*.

## 12.1 Introduction

It has been common practice for many years to classify systems as all-air systems, air and water systems, and all-water systems. However, the current publication of the 2012 *ASHRAE Handbook—HVAC Systems and Equipment* has subdivided systems into two categories: (1) decentralized cooling and heating (Chapter 2), and (2) central cooling and heating (Chapter 3). As high-performance systems evolve, though, there are a growing number of hybrid system options that combine central system technology with decentralized technology.

**Decentralized Systems.** In the 2012 *ASHRAE Handbook—HVAC Systems and Equipment*, Chapter 2, the decentralized systems are those that are generally located in or near the space to be conditioned and convert an available energy form to useful cooling, heating, or humidity control within that device. Most of these systems have historically been referred to as unitary systems. That chapter gives the following as examples of decentralized HVAC systems:

- Window air conditioners
- Through-the-wall room HVAC units
- Air-cooled heat pump systems
- Water-cooled heat pump systems
- Multiple-unit variable refrigerant flow systems
- Residential and light commercial split systems
- Self-contained (floor-by-floor) systems
- Outdoor package systems
- Packaged, special-procedure units (e.g., for computer rooms)

The outside package systems would include the very common host of rooftop units, generally located just above the space that they condition, and the multiple-unit systems would include multiple-evaporator minisplit systems, which have become quite popular.

**Central Cooling and Heating.** According to Chapter 3, Central Cooling and Heating, of the 2012 *ASHRAE Handbook—HVAC Systems and Equipment*, central cooling and/or heating plants generate cooling and/or heating in one location for distribution to multiple locations in one building or an entire campus or neighborhood, and represent approximately 25% of HVAC system applications. Central cooling and heating systems are used in almost all classes of buildings, but particularly in large buildings and complexes or where there is a high density of energy use. They are especially suited to applications where maximizing equipment service life and using energy and operational workforce efficiently are important.

The following facility types are good candidates for central cooling and/or heating systems:

- Campus environments with distribution to several buildings
- High-rise facilities
- Large office buildings
- Large public assembly facilities, entertainment complexes, stadiums, arenas, and convention and exhibition centers
- Urban centers (e.g., city centers/districts)
- Shopping malls
- Large condominiums, hotels, and apartment complexes
- Educational facilities
- Hospitals and other health care facilities
- Industrial facilities (e.g., pharmaceutical, manufacturing)
- Large museums and similar institutions
- Locations where waste heat is readily available (result of power generation or industrial processes)

The following are advantages and disadvantages of central cooling and heating systems:

**Advantages**

- Primary cooling and heating can be provided at all times, independent of the operation mode of equipment and systems outside the central plant.
- Using larger but fewer pieces of equipment generally reduces the facility's overall operation and maintenance cost. It also allows wider operating ranges and more flexible operating sequences.
- A centralized location minimizes restrictions on servicing accessibility.
- Energy-efficient design strategies, energy recovery, thermal storage, and energy management can be simpler and more cost-effective to implement.
- Multiple energy sources can be applied to the central plant, providing flexibility and leverage when purchasing fuel.
- Standardizing equipment can be beneficial for redundancy and stocking replacement parts. However, strategically

selecting different-sized equipment for a central plant can provide better part-load capability and efficiency.
- Standby capabilities (for firm capacity/redundancy) and back-up fuel sources can easily be added to equipment and plant when planned in advance.
- Equipment operation can be staged to match load profile and taken offline for maintenance.
- District cooling and heating can be provided.
- A central plant and its distribution can be economically expanded to accommodate future growth (e.g., adding new buildings to the service group).
- Load diversity can substantially reduce the total equipment capacity requirement.
- Submetering secondary distribution can allow individual billing of cooling and heating users outside the central plant.
- Major vibration and noise-producing equipment can be grouped away from occupied spaces, making acoustic and vibration controls simpler. Acoustical treatment can be applied in a single location instead of many separate locations.
- Issues such as cooling tower plume and plant emissions are centralized, allowing a more economic solution.

**Disadvantages**

- Equipment may not be readily available, resulting in long lead-time for production and delivery.
- Equipment may be more complicated than decentralized equipment, and thus require a more knowledgeable equipment operator.
- A central location within or adjacent to the building is needed.
- Additional equipment room height may be needed.
- Depending on the fuel source, large underground or surface storage tanks may be required on site. If coal is used, space for storage bunker(s) will be needed.
- Access may be needed for large deliveries of fuel (oil or coal).
- Heating plants require a chimney and possibly emission permits, monitoring, and treatments.
- Multiple equipment manufacturers are required when combining primary and ancillary equipment.
- System control logic may be complex.
- First costs can be higher.
- Special permitting may be required.
- Safety requirements are increased.
- A large pipe distribution system may be necessary (which may actually be an advantage for some applications).

## 12.2 Selecting the System

Willis Carrier defined air conditioning as follows:

Air conditioning is the control of the humidity of the air by adding or removing moisture from the air, the control of the temperature of the air by heating or cooling the air, the control of the purity of the air by filtering or washing the air, and the control of air motion and ventilation.

Thus, in selecting a system to air-condition a building, the engineer must ensure that, for each space or room in the building, each of the five objectives (i.e., humidity, temperature, purity, air motion, and ventilation) are accomplished within the range necessary to achieve human comfort or satisfy an industrial parameter range.

Many times, it could be that, to achieve best control over all five properties and still satisfy all of the other design parameters (see section 11.3), more than a single system is required. As an example, control of ventilation, humidity, and air purity may best be provided with a dedicated outdoor air system, whereas control of the temperature and air motion may best be achieved with an in-space unit such as a fan-coil unit or unitary heat pump.

## 12.3 Multiple-Zone Control Systems

This discussion of central station air-handling systems serving multiple zones follows Chapter 4, "Air Handling and Distribution," of the 2012 *ASHRAE Handbook—HVAC Systems and Equipment*.

### 12.3.1 Constant Volume, Variable $\Delta t$

While maintaining constant airflow, constant-volume systems change the supply air temperature in response to the space load (Figure 12-1).

**Single-Zone Systems.** The simplest all-air system is a supply unit serving a single zone. The unit can be installed either in or remote from the space it serves, and may operate with or without distribution ductwork. Ideally, this system responds completely to the space needs, and well-designed control systems maintain temperature and humidity closely and efficiently. Single-zone systems often involve short ductwork with low pressure drop and thus low fan energy, and can be shut down when not required without affecting operation of adjacent areas, offering further energy savings. A return or relief fan may be needed, depending on system capacity and whether 100% outdoor air is used for cooling as part of an economizer

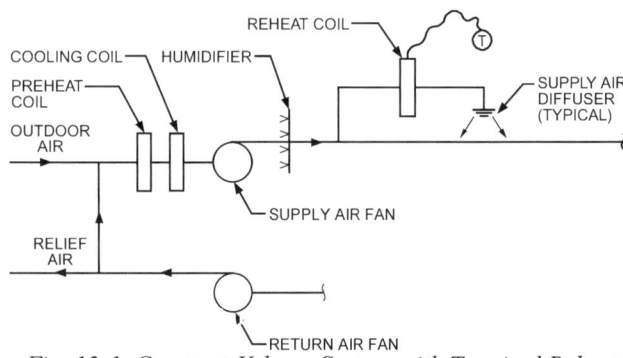

*Fig. 12-1 Constant-Volume System with Terminal Reheat*
*(Figure 9, Chapter 4, 2012 ASHRAE Handbook—*
*HVAC Systems and Equipment)*

# System Configurations

cycle. Relief fans can be eliminated if overpressurization can be relieved by other means, such as gravity dampers.

**Multiple-Zone Terminal Reheat Systems.** Multiple-zone reheat is a modification of the single-zone system (Figure 12-1). It provides (1) zone or space control for areas of unequal loading, (2) simultaneous heating or cooling of perimeter areas with different exposures, and (3) close control for temperature, humidity, and space pressure in process or comfort applications. As the word *reheat* implies, heat is added as a secondary simultaneous process to either preconditioned (cooled, humidified, etc.) primary air or recirculated room air. Relatively small low-pressure systems place reheat coils in the ductwork at each zone. More complex designs include high-pressure primary distribution ducts to reduce their size and cost, and pressure reduction devices to maintain a constant volume for each reheat zone.

The system uses conditioned air from a central unit, generally at a fixed cold-air temperature that is low enough to meet the maximum cooling load. Thus, all supply air is always cooled the maximum amount, regardless of the current load. Heat is added to the airstream in each zone to avoid overcooling that zone, for every zone except the zone experiencing peak cooling demand. The result is very high energy use, and therefore use of this system is restricted by ASHRAE *Standard* 90.1. However, the supply air temperature from the unit can be varied, with proper control, to reduce the amount of reheat required and associated energy consumption. Care must be taken to avoid high internal humidity when the temperature of air leaving the cooling coil is allowed to rise during cooling.

In cold weather, when a reheat system heats a space with an exterior exposure, the reheat coil must not only replace the heat lost from the space, but also must offset the cooling of the supply air (enough cooling to meet the peak load for the space), further increasing energy consumption. If a constant-volume system is oversized, reheat energy becomes excessive.

In commercial applications, use of a constant-volume reheat system is generally discouraged in favor of variable-volume or other systems. Constant-volume reheat systems may continue to be applied in hospitals, laboratories, and other critical applications where variable airflow may be detrimental to proper pressure relationships (e.g., for infection control).

**Dual-Duct Systems.** A dual-duct system conditions all the air in a central apparatus and distributes it to conditioned spaces through two ducts, one carrying cold air and the other carrying warm air. In each conditioned zone, air valve terminals mix warm and cold air in proper proportion to satisfy the space temperature and pressure control (Figure 12-2). Dual-duct systems may be designed as constant volume or variable air volume; a dual-duct, constant-volume system uses more energy than a single-duct VAV system. As with other VAV systems, certain primary-air configurations can cause high relative humidity in the space during the cooling season.

Dual-duct, constant-volume systems using a single supply fan were common through the mid-1980s, and were used frequently as an alternative to constant-volume reheat systems.

Today, dual-fan, dual-duct are preferred over the former, based on energy performance. There are two types of dual-duct, single-fan application: with reheat, and without.

**Single Fan With Reheat.** There are two major differences between this and a conventional terminal reheat system: (1) reheat is applied at a central point in the fan unit hot deck instead of at individual zones (Figure 12-2), and (2) only part of the supply air is cooled by the cooling coil (except at peak cooling demand); the rest of the supply is heated by the hot-deck coil during most hours of operation. This uses less heating and cooling energy than the terminal reheat system where all the air is cooled to full cooling capacity for more operating hours, and then all of it is reheated as required to match the space load. Fan energy is constant because airflow is constant.

**Single Fan Without Reheat.** This system has no heating coil in the fan unit hot deck and simply pushes a mixture of outside and recirculated air through the hot deck. A problem occurs during periods of high outside humidity and low internal heat load, causing the space humidity to rise rapidly unless reheat is added. This system has limited use in most modern buildings because they are not capable of maintaining comfort conditions in many climatic conditions. A single-fan, no-reheat dual-duct system does not use any extra energy for reheat, but fan energy is constant regardless of space load.

## 12.3.2 Variable Volume (VAV), Constant or Variable Δ*t*

A VAV system (Figure 12-3) controls temperature in a space by varying the quantity of supply air rather than varying the supply air temperature. A VAV terminal unit at the zone varies the quantity of supply air to the space. The supply air temperature is held relatively constant. Although supply air temperature can be moderately reset depending on the season, it must always be low enough to meet the cooling load in the most demanding zone and to maintain appropriate humidity. VAV systems can be applied to interior or perimeter zones, with common or separate fans, with common or separate air temperature control, and with or without auxiliary heating devices. The greatest energy saving associated with VAV occurs at the perimeter zones, where variations in solar load and outside temperature allow the supply air quantity to be reduced.

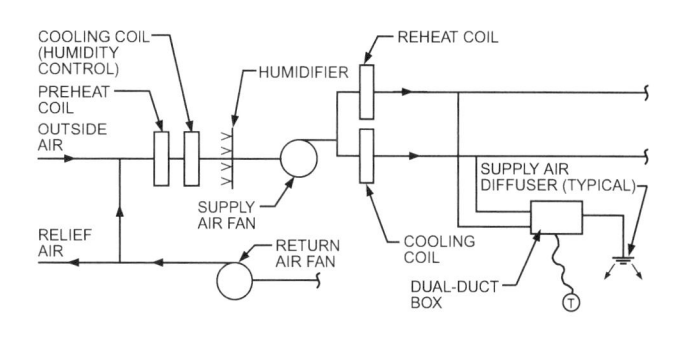

*Fig. 12-2 Single-Fan, Dual-Duct System*
*(Figure 11, Chapter 4, 2012 ASHRAE Handbook—HVAC Systems and Equipment)*

Humidity control is a potential problem with VAV systems. If humidity is critical, as in certain laboratories, process work, etc., constant-volume airflow may be required.

Other measures must also maintain enough air circulation through the room to achieve acceptable ventilation and air movement. The human body is more sensitive to elevated air temperatures when there is little air movement. Minimum air circulation can be maintained during reduced load by (1) raising the supply air temperature of the entire system, which could increase space humidity, or supplying reheat on a zone-by-zone basis; (2) providing auxiliary heat in each room independent of the air system; (3) using individual-zone recirculation and blending varying amounts of supply and room air or supply and ceiling plenum air with fan-powered VAV terminal units, or, if design permits, at the air-handling unit; (4) recirculating air with a VAV induction unit; or (5) providing a dedicated recirculation fan to increase airflow.

VAV reheat can ensure close room space pressure control with the supply terminal functioning in sync with associated room exhaust. A typical application might be a fume hood VAV exhaust with constant open sash velocity (e.g., 85 or 100 fpm) or occupied/ unoccupied room hood exhaust (e.g., 100 fpm at sash in occupied periods and 60 fpm in unoccupied periods).

**Dual-Conduit.** This method is an extension of the single-duct VAV system: one supply duct offsets exterior transmission cooling or heating loads by its terminal unit with or without auxiliary heat, and the other supply air path provides cooling throughout the year. The first airstream (primary air) operates as a constant-volume system, and the air temperature is varied to offset transmission only (i.e., it is warm in winter and cool in summer). Often, however, the primary-air fan is limited to operating only during peak heating and cooling periods to further reduce energy use. When calculating this system's heating requirements, the cooling effect of secondary air must be included, even though the secondary system operates at minimum flow. The other airstream, or secondary air, is cool year-round and varies in volume to match the load from solar heating, lights, power, and occupants. It serves both perimeter and interior spaces.

**Variable Diffuser.** The discharge aperture of this diffuser is reduced to keep discharge velocity relatively constant while reducing conditioned supply airflow. Under these conditions, the induction effect of the diffuser is kept high, cold air mixes in the space, and the room air distribution pattern is more nearly maintained at reduced loads. These devices are of two basic types: one has a flexible bladder that expands to reduce the aperture, and the other has a diffuser plate that moves. Both devices are pressure-dependent, which must be considered in duct-distribution system design. They are either powered by the system or pneumatically or electrically driven.

Since the single-duct VAV system cannot provide heat, it is often combined with some configuration of reheat or dual-stream system, devices that are available in many configurations, including the following (see Figure 12-3).

**VAV Reheat.** This simple VAV system integrates heating at the terminal unit. It is applied to systems requiring full heating and cooling flexibility in interior and exterior zones. The terminal units are set to maintain a predetermined minimum throttling ratio, which is established as the lowest air quantity necessary to (1) offset the heating load, (2) limit the maximum humidity, (3) provide reasonable air movement within the space, and (4) provide required ventilation air. Note, (2) and (4) do not apply if a separate ventilation system is used. (See section 12.4.)

Variable-air-volume with reheat permits airflow to be reduced as the first step in control; heat is then initiated as the second step. Compared to constant-volume reheat, this procedure reduces energy consumption because the amount of primary air to be cooled and secondary air to be heated is reduced in additon to the reduction in fan energy.

A feature can be provided to isolate the availability of reheat during the summer, except in situations where low airflow should be avoided or where an increase in humidity causes discomfort (e.g., in conference rooms when the lights are turned off).

**VAV Induction.** The VAV induction system uses a terminal unit to reduce cooling capacity by simultaneously reducing primary air and inducing room or ceiling air to maintain a

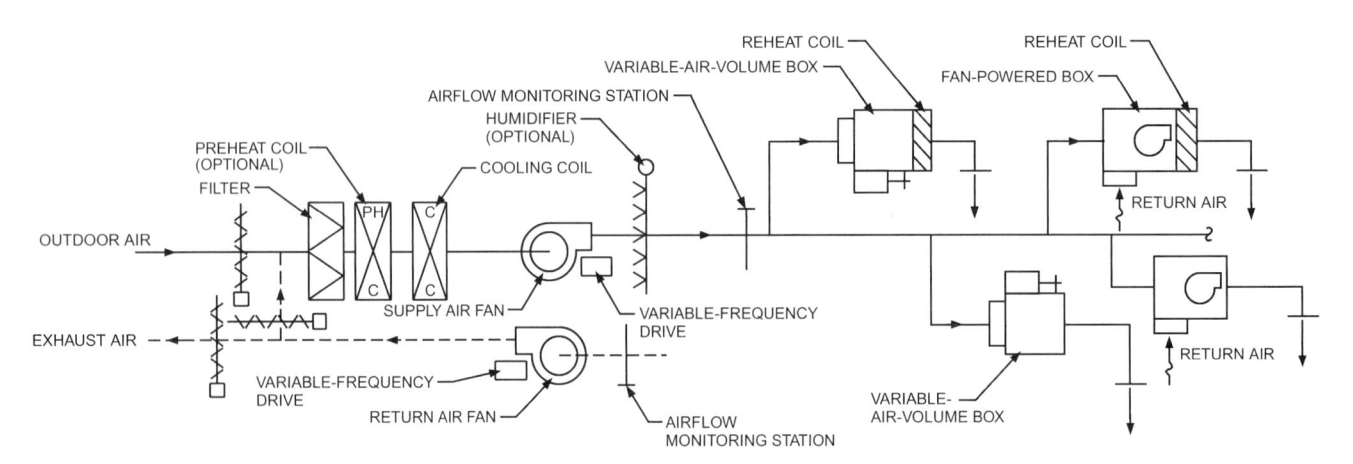

*Fig. 12-3 Variable-Air-Volume System with Reheat and Induction and Fan-Powered Devices*
*(Figure 10, Chapter 4, 2012 ASHRAE Handbook—HVAC Systems and Equipment)*

# System Configurations

relatively constant circulating air volume to the room. The primary air quantity decreases with load, retaining the savings of reduced fan power, while quantity of the the air supplied to the space is kept relatively constant to avoid the effect of low-velocity "dumping" or low air movement.

The terminal device is usually located in the ceiling cavity to recover heat from lights. This allows the induction box to be used without reheat coils in internal spaces. In cold climates, provisions must be made for morning warm-up and night heating. Also, interior spaces with a roof load must have heat supplied either separately in the ceiling cavity or at the terminal.

**Fan-Powered.** Fan-powered systems are available in either parallel or series airflow. In parallel flow units, the fan is located outside the primary airstream to allow intermittent fan operation. In series units, the fan is located within the primary airstream and runs continuously when the zone is occupied. Fan-powered systems, both series and parallel, are often selected because they maintain higher air circulation through a room at low loads while still retaining some of the energy advantages of VAV systems.

As the cold primary air valve modulates from maximum to minimum (or closed), the unit recirculates more plenum air. In a perimeter zone, a hot water heating coil, electric heater, baseboard heater, or remote radiant heater can be sequenced with the cooling to offset external heat losses. Between heating and cooling operations, a dead band, in which the fan recirculates ceiling air only, is provided. This operation permits heat from lights to be used for space heating for improved energy efficiency. During unoccupied periods in cold climates, the main supply air-handling unit remains off and individual fan-powered heating zone terminals are cycled to maintain required space temperature, thereby reducing energy consumption.

Both parallel and series systems use the heat from lights in the ceiling plenum, and both may be provided with filters.

**Parallel Arrangement—Intermittent Fan.** In this device, primary air is modulated in response to cooling demand and energizes an integral fan at a predetermined reduced primary flow to deliver ceiling air to offset heating demand. These devices are primarily used in perimeter zones where auxiliary hot water or electric heating is required. The induction fan operating range normally overlaps the range of the primary air valve. A back-draft damper on the terminal fan prevents conditioned air from backflowing into the return air plenum when the terminal fan is off.

**Series Arrangement—Constant Fan.** A constant-volume (series) fan-powered box mixes primary air with air from the ceiling space using a continuously operating fan; this provides a relatively constant volume to the space. These devices are used for interior or perimeter zones and are supplied with or without an auxiliary heating coil. They can be used to mix primary and return air to raise the temperature of the air supplied to the space such as with low-temperature air systems.

## 12.4 Ventilation and Dedicated Outdoor Air Systems (DOAS)

The major source of water vapor in building spaces in warm, humid climates is the outdoor air. Since the control of both ventilation and humidity in those climates is a major requirement for acceptable indoor air quality and human comfort, the use of a dedicated unit to introduce and condition the outdoor ventilation air source is gaining favor over the use of mixing chambers, which depend on the temperature control psychrometric system to provide both the temperature and humidity controls.

Separating the control of the ventilation air from the room temperature control requires the use of a 100% outdoor air unit, sized to provide only the required ventilation air and designed to filter contaminants from the outdoor air, then provide the humidity control for the entire building space served by the unit and the temperature control for the ventilation air only. A typical unit of this type is shown diagrammatically in Figure 12-4.

The amount of air supplied by the ventilation air conditioning (VAC) unit must be no less than that required by either ASHRAE Standard 62 or that necessary to make up for all of the building exhaust plus some additional for building pressurization, whichever is greater.

The outdoor air enters the unit through intake louvers and a two-position intake damper (open-close), then through a filter section designed to remove undesirable particulate and/or chemical impurities from the outdoor air. The heating coil is designed to heat the air from winter outdoor temperatures up to a desired supply air temperature. This coil is not necessary in climates that do not experience winter temperatures below 70°F. The other feature of this coil is that it must be designed to prevent freezing of the heating fluid in below-freezing climates. With water coils, this is best accomplished with variable-temperature constant-flow design, and with steam coils it is best achieved with some form of face and bypass control, most desirably integrated face-and-bypass design. Needless to say, low-temperature freeze protection is always required, which should have its sensor on the leaving face of the coil and, on a signal of approaching freezing temperature leaving the

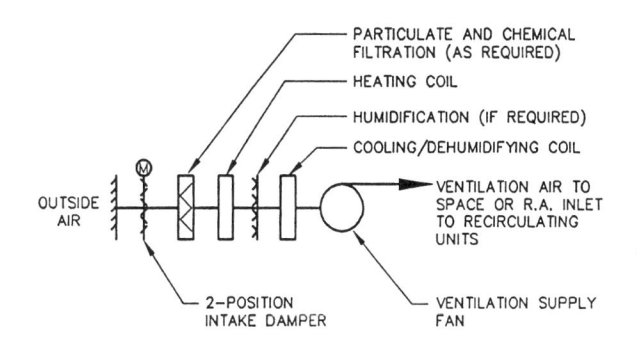

*Fig. 12-4 Dedicated Outdoor Air System (DOAS) Unit*

coil, should close the intake dampers, shut down the fan, and signal an alarm.

The cooling-dehumidifying coil can be either a chilled water coil or a refrigerant direct expansion coil. It should be sized for 400 fpm maximum velocity, have wide fin spacing (8 fins per inch maximum) to prevent condensate bridging and carryover and should have flat plate fins to allow ready inspection of fouling and related cleaning. Since all of the building dehumidification is achieved at this point, the coil should be close to 100% efficient, which will usually require a deep coil with a minimum of eight rows. The leaving dew-point temperature should be equal to the dew point required in the space less whatever is required to absorb the latent space load.

Winter humidification is generally not recommended or required for human comfort. If, however, it is required for process or safety purposes (such as in hospital operating rooms, museums, or rare book libraries), the humidifier should be placed between the heating coil and the cooling coil.

In systems wherein critical humidity control is required on both the humidifying and dehumidifying cycles, this can be achieved quite effectively by using a sprayed coil and simply sequencing the cooling coil valve with the heating coil valve and controlling with a discharge air controller. (The dry-bulb temperature will be equal to the dew-point temperature since the air leaving a sprayed coil is saturated.) Certain precautions must be taken in designing sprayed-coil units: (1) the pan and coil must be readily accessible for cleaning, and (2) high-temperature shutdown and alarm controls must be provided to protect against microbial contamination.

The discharge air from the VAC unit can be supplied directly into the space through a separate ventilation distribution system or can be supplied into the return air side of the recirculating air-handling system(s).

When a VAC unit is used, the space temperature and air circulation control can be provided either with a constant-volume or variable-volume recirculating air unit or with a unit or system located within the space.

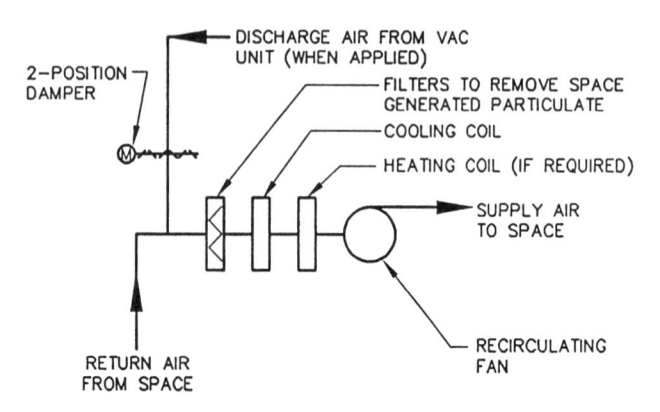

*Fig. 12-5 Recirculating Unit*

## 12.5 All-Air System with DOAS Unit

Under these conditions, with the ventilation requirements and the humidity control handled by the DOAS unit, the air-handling unit becomes a simple sensible cooling and heating unit, as shown in Figure 12-5. It can be configured to handle any kind of terminal control, and the supply air temperature can be increased or decreased as desired without reheat since it is not necessary that the cooling coil provide any humidity control.

## 12.6 Air-and-Water Systems with DOAS Unit

If control of the space temperature is achieved by heating and cooling devices located within the space, such as fan-coil units or radiant panels, the ventilation air can be introduced directly into the space. Generally this air can be introduced at the dew-point temperature when the system is in a dehumidifying mode. If the load dynamics dictate the need for a higher supply air temperature, it is necessary to continue to cool the air to the design dew point and then reheat it to the supply temperature required.

Many earlier systems concepts combined large air circulation systems with in-space heating-cooling devices in some integrated fashion. These systems were generally categorized as air-water systems. The most common type of air-water system was the so-called induction system in which the air was introduced through high-velocity nozzles, inducing a stream of room air across a cold or warm coil to provide the ultimate control of room temperature. A later configuration of the induction system is the chilled beam, in which the induction heating and cooling coils are above the ceiling. (See Chapter 20 of the 2012 *ASHRAE Handbook—HVAC Systems and Equipment*.)

Following the publication of ASHRAE Standard 90-1975, the improvement in the thermal quality of building envelopes changed the dynamics of system requirements. The benefits of the air-and-water systems gave way to the all-air systems. But with the improved envelope designs came another issue, the need for a reliable and predictable supply of ventilation air, which was addressed by ASHRAE Standard 62. The result of these two events has led to another generation of the concept of air-water systems. These are the systems that utilize a dedicated outdoor air system (DOAS; see section 12.5) to provide the ventilation air and humidity control (and often the ambient air motion) and a heating-cooling device located within the space, such as a fan-coil unit, radiant panels, or a chilled-beam device to provide the space temperature control.

## 12.7 In-Space Temperature Control Systems

In-space temperature control systems generally utilize either hot or chilled water for the space conditioning or devices that generate the heating or cooling directly from

# System Configurations

electricity, such as resistance heating or unitary refrigeration and/or heat pumps. Such devices could include, but are not limited to, the following:

Heating only:
  Baseboard radiation
  Radiators and convectors
  Finned tube radiation
  Unit heaters (cabinet or "propeller")
  Radiant panels

Cooling only:
  Fan-coil units (two-pipe)
  Radiant panels
  Mini-split evaporators
  Variable refrigerant volume terminals
  Package terminal air conditioners (PTAC units)
  Water-cooled package units
  Chilled-beam induction units

Heating and Cooling:
  Fan-coil units (two-pipe and four-pipe)
  Radiant panels (two-pipe and four-pipe)
  Mini-split evaporators with electric heat
  PTAC heat pump
  Water-source heat pump
  Hot/chilled-beam induction units

A complete description of each of these devices is included in the equipment chapters that follow or in the appropriate chapter(s) of the 2012 *ASHRAE Handbook—HVAC Systems and Equipment*. Those devices that require the support of an infrastructure system are discussed below. Suffice it to say that none of these in-space temperature control systems has the capability to control humidity (except as a by-product) or provide controlled quantities of ventilation air.

## 12.7.1 Fan-Coil Units

Basic elements of fan-coil units are a finned-tube coil, filter, and fan section (Figure 12-6). The fan recirculates air continuously from the space through the coil, which contains either hot or chilled water. The unit may contain an additional electric resistance, steam, or hot water heating coil.

A cleanable or replaceable 35% efficiency filter, located upstream of the fan, prevents clogging the coil with dirt or lint entrained in the recirculated air. It also protects the motor and fan and reduces the level of airborne contaminants within the conditioned space. The fan-coil unit is equipped with an insulated drain pan. The fan and motor assembly should be arranged for quick removal for servicing and cleaning the coil. Most manufacturers furnish units with cooling performance that is ARI certified. The prototypes of the units should be tested and labeled by Underwriters' Laboratories (UL), or Engineering Testing Laboratories (ETL), as required by some codes.

Fan-coil units are for recirculation heating and cooling only. If ventilation is provided to or through a fan-coil unit, it must be provided by a DOAS unit (see section 12.5).

Room fan-coil units are generally available in nominal sizes of 200, 300, 400, 600, 800, and 1200 cfm, usually with multispeed fan motors. Ventilation should always be provided through a DOAS that engages each room or space.

**Basic Components.** Room fan-coil units are available in many configurations. Figure 12-7 shows several vertical units. Low vertical units are available for use under windows with low sills; however, in some cases, the low silhouette is achieved by compromising such features as filter area, serviceability, and cabinet style.

Floor-to-ceiling, chase-enclosed units are available in which the water and condensate drain risers are part of the factory-furnished unit. Supply and return air systems must be isolated from each other to prevent air and sound interchange between rooms.

Horizontal overhead units may be fitted with ductwork on the discharge to supply several outlets. A single unit may serve several rooms (e.g., in an apartment unit where individual room control is not essential and a common air return is feasible). High static pressure units with larger fan motors handle the higher pressure drops of units with ductwork.

Central ventilation air from the DOAS unit may be connected to the inlet plenums of the units or introduced directly into the space. If this is done, provisions should be made to ensure that this air is pretreated and held at a temperature to not cause occupant discomfort when the fan-coil unit is off. Coil selection must be based on the temperature of the entering mixture of primary and recirculated air, and the air leaving the coil must satisfy the room sensible cooling and heating requirements. Horizontal models conserve floor space and usually cost less, but when located overhead in furred ceilings, they create problems such as condensate collection and disposal, mixing of return air from other rooms, leakage of pans causing damage to ceilings, difficulty of access for maintenance and service, and related IAQ concerns.

Vertical models give better results in climates of extremely cold temperatures, since heating is enhanced by under-window or exterior wall locations. Vertical units can be operated

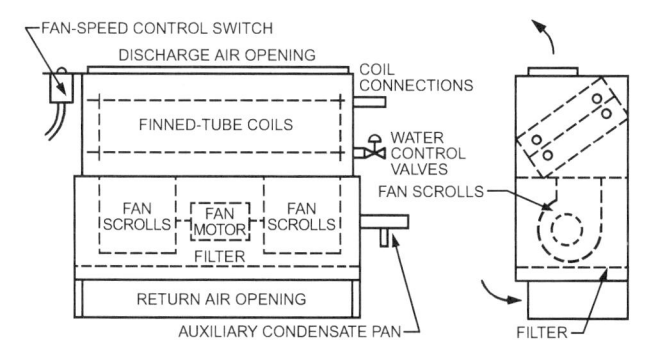

*Fig. 12-6 Typical Fan-Coil Unit*
(Figure 1, Chapter 5, 2012 ASHRAE Handbook—HVAC Systems and Equipment)

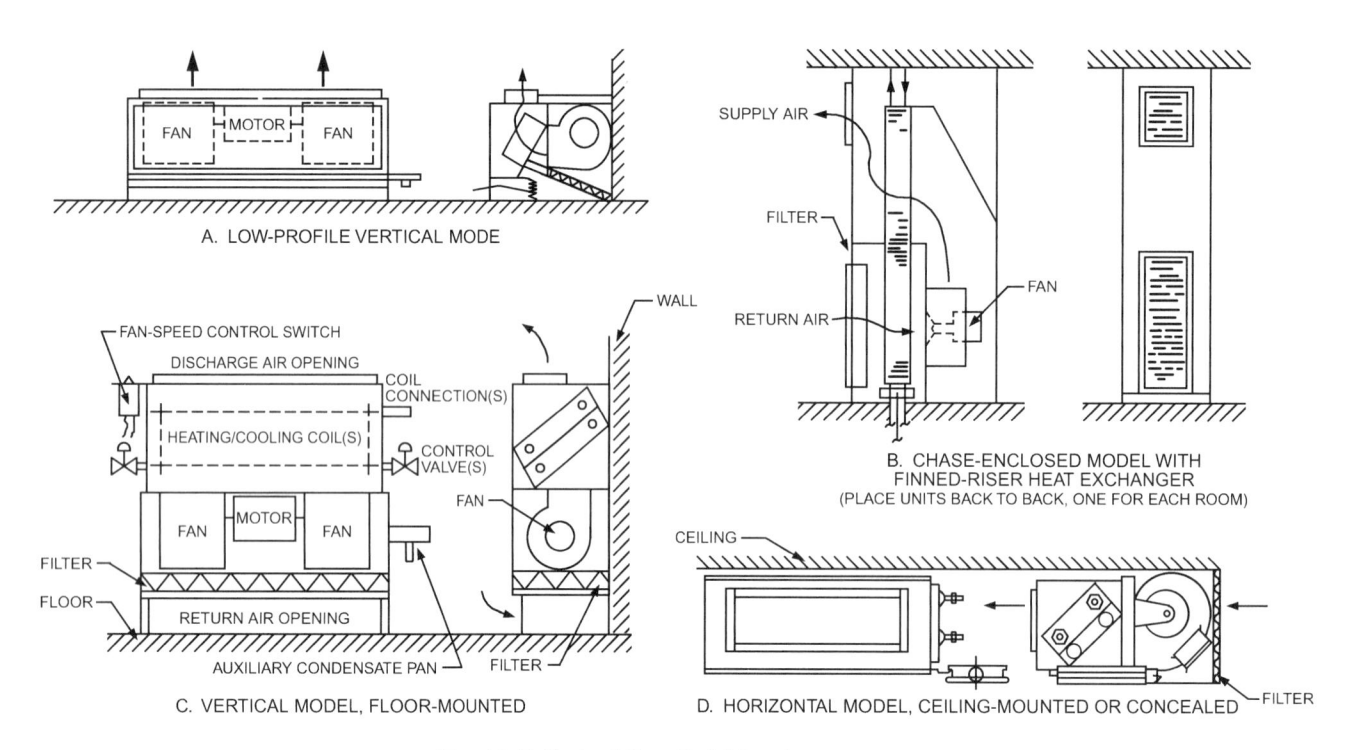

*Fig. 12-7 Typical Fan-Coil Unit Arrangements*
(Figure 1, Chapter 5, 2008 *ASHRAE Handbook—HVAC Systems and Equipment*)

as convectors with the fans turned off during winter night operation.

**Selection.** Some designers size fan-coil units for nominal cooling at the medium-speed setting when a three-speed control switch is provided. This method ensures quieter operation within the space and adds a safety factor in that capacity can be increased by operating at high speed. Sound power ratings are available from some manufacturers, and, as with any in-the-room unit, sound is a very important design parameter.

Only the sensible space heating and cooling loads need to be handled by the terminal fan-coil units when outdoor air is pretreated by a dedicated outdoor air system. If the ventilation air from the DOAS is supplied at or below dew-point temperature in design cooling weather, the sensible cooling capacity thereby provided can be deducted from the needed capacity of the fan-coil unit.

**Wiring.** Fan-coil conditioner fans are driven by small motors generally of the shaded pole or capacitor type, with internal overload protection. Operating wattage of even the largest sizes rarely exceeds 300 W at the high-speed setting. Running current rarely exceeds 2.5 A.

Almost all motors on units sold in the United States are wired for 120 V, single-phase, 60 Hz current, and they provide multiple (usually three) fan speeds and an off position or continual speed variation. Other voltages and power characteristics may be encountered, depending on the location, and should be investigated before selecting the fan motor characteristics. Many manufacturers are providing variable speed electronically commutated motors (ECM), which tend to be extremely quiet, and have many control system benefits.

In planning the wiring circuit, local and national electrical codes must be followed. Wiring methods generally provide separate electrical circuits for fan-coil units and do not connect them into the lighting circuit.

Separate electrical circuits connected to a central panel allow the building control system to turn off unit fans from a central point whenever is desired.

**Piping.** Even when outdoor air is pretreated, a condensate removal system should be installed on the terminal units. This precaution ensures that moisture condensed from air from an unexpected open window that bypasses the ventilation system is carried away. Drain pans should be an integral feature of all units. Condensate drain lines should be oversized to avoid clogging with dirt and other materials, and provision should be made for periodic cleaning of the condensate drain system. Condensation may occur on the outside of the drain piping, which requires that these pipes be insulated.

**Capacity Control.** Fan-coil unit capacity can be controlled by coil water flow, fan speed, or a combination of these. The units can be thermostatically controlled by either return air or wall thermostats.

Room thermostats are preferred where fan speed control is used. Return air thermostats do not give a reliable index of room temperature when the fan is off.

**Maintenance.** Room fan-coil units are equipped with either cleanable or disposable filters that should be cleaned or replaced when dirty. Good filter maintenance improves sanitation and provides full airflow, ensuring full capacity. The frequency of cleaning varies with the application. Applications in apartments, hotels, and hospitals usually

# System Configurations

require more frequent filter service because of lint. The condensate drain pan and drain system must be cleaned or flushed periodically to prevent overflow and microbiological buildup.

## 12.7.2 Water Distribution

Chilled and hot water must run to the fan-coil units. The piping arrangement determines the quality of performance, ease of operation, and initial cost of the system.

**Two-Pipe Changeover.** This method has low initial cost and supplies either chilled water or hot water through the same piping system (see Chapter 13). The fan-coil unit has a single coil, and room temperature controls reverse their action, depending on whether hot or cold water is available at the unit coil.

This system works well in warm weather when all rooms need cooling and in cold weather when all rooms need heat. The two-pipe system **does not have the simultaneous heating or cooling capability** that is required for most facilities during intermediate seasons when some rooms need cooling and others need heat. This problem can be especially troublesome if a single piping zone supplies the entire building. This difficulty may be partly overcome by dividing the piping into zones based on solar exposure. Then each zone may be operated to heat or cool, independent of the others. However, one room may still require cooling while another room on the same solar exposure requires heating—particularly if the building is partially shaded by an adjacent building.

Another difficulty of the two-pipe changeover system is the need for frequent changeover from heating to cooling, which complicates the operation and increases energy consumption to the extent that it may become impractical. For example, two-pipe changeover system hydraulics must consider the water expansion (and relief) that occurs during the cycling from cooling to heating.

The designer should consider the disadvantages of the two-pipe system carefully; many installations of this type waste energy and have been unsatisfactory in climates where frequent changeover is required and where interior loads require cooling simultaneously as exterior spaces require heat. Furthermore, most building occupants demand the ability to select either heating or cooling at any time as the thermal conditions *or* their personal metabolism dictates.

Any two-pipe system must be carefully analyzed. In any case, *they are not recommended for commercial buildings*.

**Two-pipe changeover with partial electric resistance heat.** This arrangement provides simultaneous heating and cooling in intermediate seasons by using a small electric resistance heater in the fan-coil unit. The unit can handle heating requirements in mild weather, typically down to 40°F, while continuing to circulate chilled water to handle any cooling requirements. When the outdoor temperature drops sufficiently to require heating in excess of the electric heater capacity, the water system must be changed over to hot water.

**Four-Pipe Distribution.** The four-pipe system with separate heating and coolng coils provides the best fan-coil system performance. It provides (1) all-season availability of heating and cooling at each unit, (2) no summer/winter changeover requirement, (3) simpler operation, and (4) use of any heating fuel, heat recovery, or solar heat. In addition, it can be controlled to maintain a "dead band" between heating and cooling so that there is no possibility of simultaneous heating and cooling with the same unit.

**Central Equipment.** Central equipment size is based on the block load of the entire building at the time of building peak load, not on the sum of individual fan-coil unit peak loads. Cooling load should include appropriate diversity factors for lighting and occupant loads. Heating load is based on maintaining the unoccupied building at design temperature, plus an additional allowance for pickup capacity if the building temperature is set back at night.

If water supply temperature or quantities are to be reset at times other than at peak load, the adjusted settings must be adequate for the most heavily loaded space in the building. An analysis of individual room load variations is required.

If the side exposed to the sun or interior zone loads require chilled water in cold weather, the use of condenser water with a water-to-water interchanger may be considered. Varying refrigeration loads requires the water chiller to operate satisfactorily under all conditions.

Ventilation requires a dedicated outdoor air unit complete with heating and cooling coils, filters, and fans to handle the ventilation load. An additional advantage of the DOAS unit is that, if it is sized for the internal latent load, the terminal cooling coils remain dry.

**Example 12-1.** A two-zone building in St. Louis, Missouri, has the hourly heating and cooling loads given in the following tables. Assuming that the January day is the coldest for the year, and the August day is the warmest for the year, size the basic components and sketch the equipment arrangement for each of the following types of systems:

1. Separate 4-pipe fan-coil units using chilled water supplied at 45°F for cooling and hot water supplied at 190°F for heating.
2. Variable volume with reheat (turndown to 50% of design airflow) with cooling coil discharge at 58°F all year long.
3. Double-duct or multizone with design cold and hot deck temperatures of 58°F and 130°F, respectively (winter), and 58°F and 85°F, respectively (summer).

Space Design Conditions
   Summer:     78°F db
   Winter:      72°F db, 30% rh

Outdoor Design Conditions
   Summer:     94°F/75°F wb
   Winter:      6°F

Ventilation (Outside) Air Requirements
   Zone 1:     550 cfm
   Zone 2:     400 cfm

Design Pressure Drop for Duct System
   2.8 in. w.g. for Systems 2, 3
   1.0 in. w.g. for System 1

|  | JANUARY | | | | | AUGUST | | | |
|--|--|--|--|--|--|--|--|--|--|
| | Sensible Load | | Latent Load | | | Sensible Load | | Latent Load | |
| Hour | Zone 1 Btu/Hr | Zone 2 Btu/Hr | Zone 1 Btu/Hr | Zone 2 Btu/Hr | Hour | Zone 1 Btu/Hr | Zone 2 Btu/Hr | Zone 1 Btu/Hr | Zone 2 Btu/Hr |
| 1 | -32985. | -32485. | 1500. | 1800. | 1 | 1027. | 527. | 1500. | 1800. |
| 2 | -33568. | -33068. | 1500. | 1800. | 2 | 1820. | 1320. | 1500. | 1800. |
| 3 | -33918. | -33418. | 1500. | 1800. | 3 | 2426. | 1926. | 1500. | 1800. |
| 4 | -34034. | -33534. | 1500. | 1800. | 4 | 2659. | 2159. | 1500. | 1800. |
| 5 | -34337. | -33837. | 1500. | 1800. | 5 | 3195. | 2695. | 1500. | 1800. |
| 6 | -34477. | -33977. | 1500. | 1800. | 6 | 416. | 7621. | 1500. | 1800. |
| 7 | -34360. | -33860. | 1500. | 1800. | 7 | 3263. | 22148. | 1500. | 1800. |
| 8 | -26830. | -23173. | 1500. | 1800. | 8 | 10136. | 29670. | 1500. | 1800. |
| 9 | -13237. | -13728. | 1500. | 1800. | 9 | 18660. | 31112. | 1500. | 1800. |
| 10 | -5232. | -14699. | 1500. | 1800. | 10 | 31591. | 34088. | 1500. | 1800. |
| 11 | 2636. | -18162. | 1500. | 1800. | 11 | 38869. | 27952. | 1500. | 1800. |
| 12 | 7092. | -21509. | 1500. | 1800. | 12 | 42982. | 23170. | 1500. | 1800. |
| 13 | 11608. | -20380. | 1500. | 1800. | 13 | 48478. | 23717. | 1500. | 1800. |
| 14 | 16073. | -20729. | 1500. | 1800. | 14 | 54945. | 22592. | 1500. | 1611. |
| 15 | 13690. | -21782. | 1500. | 1800. | 15 | 45420. | 17379. | 1500. | 1471. |
| 16 | 192. | -23353. | 1500. | 1800. | 16 | 41390. | 16062. | 1500. | 1564. |
| 17 | -27088. | -26588. | 1500. | 1800. | 17 | 36070. | 13277. | 1500. | 1800. |
| 18 | -28183. | -27683. | 1500. | 1800. | 18 | 21612. | 9042. | 1500. | 1800. |
| 19 | -29115. | -28615. | 1500. | 1800. | 19 | 6175. | 6675. | 1500. | 1800. |
| 20 | -30188. | -29688. | 1500. | 1800. | 20 | 4264. | 4764. | 1500. | 1800. |
| 21 | -30980. | -30480. | 1500. | 1800. | 21 | 2632. | 3132. | 1500. | 1800. |
| 22 | -31587. | -31087. | 1500. | 1800. | 22 | 1187. | 1687. | 1500. | 1800. |
| 23 | -32309. | -31809. | 1500. | 1800. | 23 | 185. | 685. | 1500. | 1800. |
| 24 | -32752. | -32252. | 1500. | 1800. | 24 | 515. | 15. | 1500. | 1800. |

**Solution:**

St. Louis

Winter: $t_o = 6°F$, 100% rh → $h_o = 2.61$, $W_o = 0.0011$
$t_i = 72°F$, 30% → $h_i = 22.86$, $W_i = 0.0051$

Summer: $t_o = 94°F/75°F$ → $h_o = 38.85$, $W_o = 0.0148$
$t_i = 78°F$

Cold Deck: 58°F (if $\phi = 100\%$); $h_s = 25.36$, $w_s = 0.0105$

**System No. 1: Four-Pipe Fan-Coil Units**

($\Delta p = 1.0$ in. w.g.)

**Zone 1**

$$Q_1 = \frac{54945}{1.10(78-58)} = 2498 \text{ cfm} \qquad Q_o = 550 \text{ cfm}$$

$$m_1 = \frac{2498(60)}{13.33} = 11,200 \text{ lb/h}$$

**Summer:**

$$W_r = 0.0105 + \frac{1500/1100}{2498 \times 60/13.33} = 0.0106$$

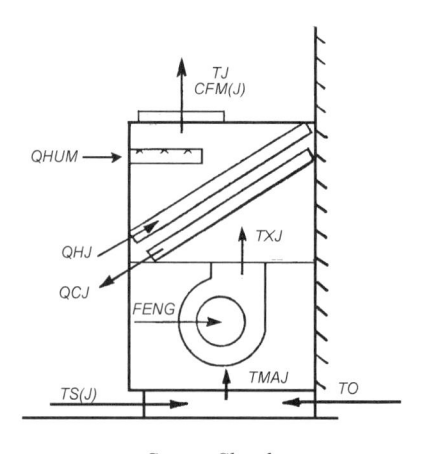

*System Sketch*

$t_r = 78 \quad h_r = 30.34$

$$h_m = \frac{1948(30.34) + 550(38.85)}{2498} = 32.21$$

$$W_m = \frac{1948(0.0106) + 550(0.0148)}{2498} = 0.0115$$

# System Configurations

$$W_{\text{fan1}} = \frac{Q \Delta p}{6350} = \frac{2498(1)}{6350} = 0.4 \text{ hp}$$

(2500 cfm, 1 in. w.g.)

$$h_f = 32.21 + \frac{0.4(2545)}{11,200} = 32.3$$

$$q_{cc} = 11,200[32.3 - 25.36 - (0.0115 - 0.0105)26]$$
$$= 77,400 \text{ Btu/h}$$

**Winter:**

$$t_s = 72 + \frac{34477}{1.10(2498)} = 84.5°\text{F}$$

$$W_s = 0.0051 - \frac{1500/1100}{11,200} = 0.00498$$

$$W_m = \frac{1948(0.0051) + 550(0.0011)}{2498} = 0.00422$$

$$t_m = \frac{1948(72) + 550(6)}{2498} = 57.5°\text{F}$$

$$t_f = 57.5 + \frac{0.4(2545)}{11,200(0.244)} = 57.9°\text{F}$$

$$q_{hc} = 2498(1.1)(84.5 - 57.9) = 73,000 \text{ Btu/h heaters}$$

$$m_w = 11,200(0.00498 - 0.00422) = 8.5 \text{ lb/h humidifiers}$$

**Zone 2**

$$Q_2 = \frac{34088}{1.10(78-58)} = 1549 \text{ cfm} \quad Q_0 = 400 \text{ cfm}$$

$$m_2 = \frac{1549(60)}{13.33} = 6970 \text{ lb/h}$$

**Summer:**

$$W_r = 0.015 + \frac{(1800/1100)}{6970} = 0.01073, \quad h_r = 30.5$$

$$W_m = \frac{1139(0.01073) + 400(0.0148)}{1549} = 0.0117$$

$$h_m = 32.45$$

$$W_{\text{fan2}} = \frac{1549 \times 1}{6350} = 0.244 \text{ hp}, \quad h_f = 32.53$$

$$q_{cc} = \frac{1549 \times 60}{13.33}[32.53 - 25.36 - (0.0117 - 0.0105)26]$$
$$= 49,800 \text{ Btu/h}$$

**Winter:**

$$t_o = 72 + \frac{33977}{1.10(1549)} = 91.9°\text{F}$$

$$W_s = 0.0051 - \frac{1800/1100}{6970} = 0.00486$$

$$W_m = \frac{1149(0.0051) + 400(0.0011)}{1549} = 0.00407$$

$$t_m = \frac{1149(72) + 400(6)}{1549} = 55.0°\text{F}$$

$$t_f = 55.0 + \frac{0.24(2545)}{6970(0.244)} = 55.4°\text{F}$$

$$q_{Hc} = 1549(1.1)(91.9 - 55.4) = 52,200 \text{ Btu/h}$$

$$m_w = 6970(0.00486 - 0.00407) = 5.5 \text{ lb/h}$$

## System No. 2: VAV with Reheat

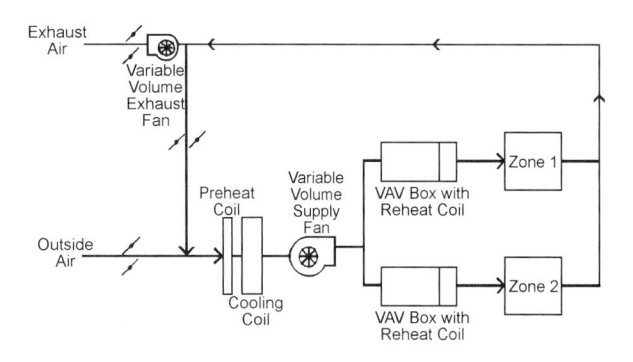

*System Sketch*

Design Peak Cooling = $77,537 = 1.10 Q_T(78-58)$

$Q_T = 3524$ cfm  $m = 15,860$ lb/h

Space 1 airflow: $54,945 = 1.10 Q(78-58)$
 $Q_1 = 2498$ cfm (max), 1249 cfm (min)

Space 2 airflow: $34,088 = 1.10 Q(78-58)$
 $Q_2 = 1549$ cfm (max), 775 cfm (min)

$$W_{\text{fan}} = \frac{3524(2.8)(62.4)(60)}{12(778)(2545)} = 1.6 \text{ hp}$$

$$\Delta h_f = \frac{1.6(2545)}{15,860} = 0.26 \quad \Delta t_f = \frac{\Delta h_f}{c_p} = 1.1°\text{F}$$

Fan: 3500 scfm; $\Delta_P = 2.8$ in. w.g.; 1.6 hp motor

$$W_r = W_s + \frac{m_{w1} + m_{w2}}{m_a}$$

$$= 0.0105 + \frac{3111}{15,860(1100)} = 0.01068$$

$t_r = 78°\text{F} \quad h_r = 30.42$

$$h_m = \frac{2574(30.42) + 950(38.85)}{3524} = 32.69 \quad h_f = 32.95$$

$$W_m = \frac{2574(0.01068) + 950(0.0148)}{3524} = 0.0118 = W_f$$

$$q_{cc} = 15,860[32.95 - 25.36 - (0.0118 - 0.0105)26]$$
$$= 119,800 \text{ Btu/h}$$

**Winter:**

#1 $q_1 = 34477 = 1.10(1249)(t_1 - 72) \quad t_1 = 97°\text{F}$
#2 $q_2 = 33977 = 1.10(775)(t_2 - 72) \quad t_2 = 112°\text{F}$

$q_{rh1} = 1.10(1249)(97 - 58) = 53,600$ Btu/h
$q_{rh2} = 1.10(775)(112 - 58) = 46,000$ Btu/h

$$W_m = \frac{950(0.0011) + 1074(0.0051)}{(950 + 1074)} = 0.00322$$

$$W_s = 0.0051 - \frac{3300/1100}{15,860/2} = 0.00472$$

$$m_w = \frac{15860}{2}(0.00472 - 0.00322) = 11.9 \text{ lb/h}$$

**System No. 3: Multizone**

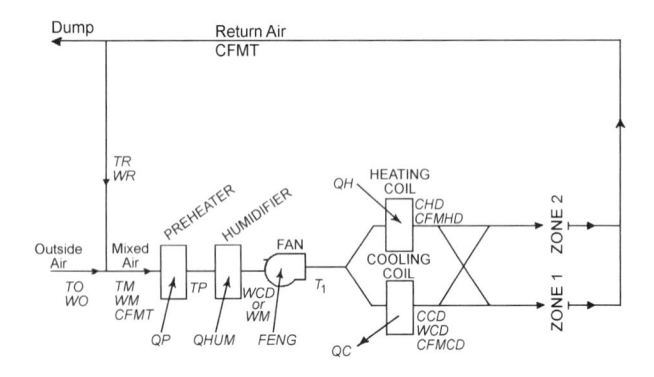

*System Sketch*

$$Q_{1,s\text{MAX}} = \frac{54,945}{1.10(78-58)} = 2498 \text{ cfm}$$

$$Q_{1,w\text{MAX}} = \frac{34,477}{1.10(130-72)} = 540 \text{ cfm (use 2498 cfm)}$$

$$Q_{2,s\text{MAX}} = \frac{34,088}{1.10(78-58)} = 1549 \text{ cfm}$$

$$Q_{2,w\text{MAX}} = \frac{33,977}{1.10(130-72)} = 533 \text{ cfm (use 1549 cfm)}$$

$$m = 4047 \times 60/13.33 = 18,200$$

Total cfm = 2498 + 1549 = 4047

Fan

$$W_f = \frac{4047(2.8)}{6350} = 1.8 \text{ hp}, \quad 4047 \text{ cfm}, \quad 2.8 \text{ in. w.g.}$$

$$\Delta h_f = \frac{1.8(2545)}{18,200} = 0.25 \quad \Delta t = 0.25/0.244 \approx 1°\text{F}$$

**Summer:**

$$W_r = 0.0105 + \frac{3111/1100}{18,200} = 0.0106 \quad t_r = 78°\text{F}$$

$$h_r = 30.34$$

$$h_m = \frac{3097(30.34) + 950(38.85)}{4047} = 32.34$$

$$W_m = \frac{3097(0.0106) + 950(0.0148)}{4047} = 0.01158$$

$$h_f = 32.34 + 0.25 = 32.59$$

$$1.10 Q_{c2}(78-58) = 22,592 + 1.10(1549 - Q_{c2})(85-78)$$

$$Q_{c2} = 1162 \text{ cfm}; \quad Q_{cc,\text{max}} = 2498 + 1162 = 3660 \text{ cfm}$$

$$m_{cc} = \frac{3660(60)}{13.33} = 16,474 \text{ lb/h}$$

$$q_{cc} = 16,474[32.59 - 25.36 - (0.01158 + 0.0105)26]$$
$$= 118,640 \text{ Btu/h}$$

**Winter:**

$$m_w = \frac{950(60)}{13.33}(0.005 - 0.0011) - \frac{3300}{1100}$$
$$= 13.67 \text{ lb/h}$$

#1 $1.10 Q_{H1}(130-72) = 34,477 + 1.10(2498 - Q_{H1})(72-58)$
$Q_{H1} = 921$ cfm

#2 $1.10 Q_{H2}(130-72) = 33,977 + 1.10(1549 - Q_{H2})(72-58)$
$Q_{H2} = 730$ cfm

$$Q_H = 921 + 730 = 1651 \text{ cfm}$$

$$t_m = \frac{3097(72) + 950(6)}{4047} = 54.5°\text{F} \quad t_f = 55.5°\text{F}$$

$$q_{Hc} = 1.10(1651)(130 - 55.5) = 135,300 \text{ Btu/h}$$

The humidity quantities are different in the three systems (14.0, 11.9, and 13.7 lb/h). The following could be reasons for this. Sometimes the humidity ratio $W$ was rounded, and when differences are taken errors can occur. In the VAV system, all the air goes through the cooling coil and is dehumidified, while in the multi-zone, some of the air goes through the cooling coil, but not all. Also, in the calculations, the correct value for the specific volume (v) at each point was not always used for convenience reasons. For mixing of the air streams, the volume flow rate was used for convenience rather than the mass flow rate. Also, with a temperature control only thermostat being used, the relative humidity in some spaces for some systems may float rather than stay at the design value. The procedures used in the example are typical in the industry and with the various assumptions in the equation development, it is not unusual to see these kinds of small differences. HVAC is not an exact science.

## 12.8 Problems

**12.1** From an energy consumption perspective, list the four fundamental psychrometric system types from least consumption to most consumption.

**12.2** In a VAV system with series fan powered terminals, why must all of the terminal fans be running prior to turning on the system fan?

**12.3** What is the advantage of a parallel fan-powered terminal over a series fan-powered terminal?

**12.4** What is the purpose of using a fan-powered terminal in a variable-air-volume system?

**12.5**

(a) Why do some VAV systems also use dual-duct or reheat features?

# System Configurations

(b) In your own words, describe the operating sequence of the zone or terminal control of

(1) A VAV system
(2) A VAV reheat system
(3) A dual-duct VAV system

**12.6** What is the primary advantage of a dedicated outdoor air system (DOAS)?

**12.7** Why is a high-pressure primary system fan required with an induction system?

**12.8** Are fan-coil units with direct connections to the outdoors recommended as an acceptable method for providing ventilation air? Why?

**12.9** Size the basic components and sketch the equipment arrangement if the HVAC system now under consideration for the building of Example 12-1 is a triple-deck multizone (hot, cold, and neutral decks).

**12.10** Size the basic components and sketch the equipment arrangement if the HVAC system now under consideration for the building of Example 12-1 is a variable volume, dual fan, dual duct.

**12.11** A small single-zone classroom building is being designed for Knoxville, Tennessee, to use the HVAC system shown in the sketch. Minimum outdoor air for meeting the ventilation requirements of the anticipated 550 occupants will be maintained throughout the year. Fan speed will be changed between summer and winter. The duct system will be designed so that at summer air flow rate the pressure drop does not exceed 3.75 in. w.g. At winter design conditions, the air is heated to 130°F at which temperature it is supplied to the conditioned space. The winter conditioning unit includes both a heating coil and a humidifier supplied with city water at 60°F. The humidistat in the return airstream maintains the design relative humidity of 30% in winter. During summer operation, the cooling coil supplied air to the conditioned space at 58°F. The space design loads are

Summer: 423,000 Btu/h sensible (gain)
139,000 Btu/h latent (gain)
Winter: 645,000 Btu/h sensible (loss)
negligible latent

Size the following system components:

(a) Cooling coil, Btu/h and ft² of face area
(b) Chiller unit, Btu/h
(c) Heating coil, Btu/h and ft² of face area
(d) Boiler, Btu/h
(e) Humidifier, gph

Select an appropriate air handler from the following data.

**12.12** A double-duct system is to be used for air conditioning of a two-zone building. At winter design outdoor temperature

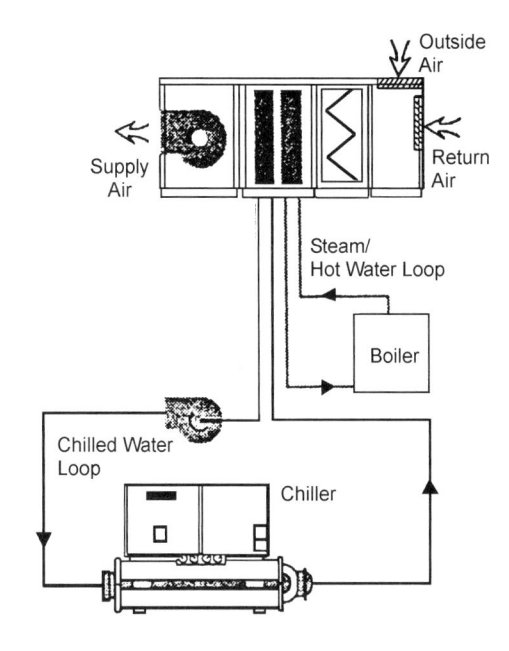

| | Unit Physical Data (Approximate) | | |
|---|---|---|---|
| Unit Size | Design cfm | Unit Coil Face Area*, ft² | Max Unit Wt., lb |
| 3 | 1,660 | 2.34 – 3.32 | |
| 6 | 2,930 | 4.31 – 5.86 | |
| 8 | 3,770 | 5.49 – 7.54 | |
| 10 | 4,820 | 7.01 – 9.64 | |
| 12 | 6,150 | 9.46 – 12.3 | ≤ 3,600 |
| 14 | 7,110 | 10.2 – 14.2 | |
| 17 | 8,400 | 12.3 – 16.8 | |
| 21 | 10,390 | 15.0 – 20.8 | |
| 25 | 12,190 | 17.8 – 24.4 | |
| 30 | 14,505 | 21.2 – 29.0 | |
| 35 | 17,050 | 26.72 – 34.10 | |
| 40 | 19,650 | 30.78 – 39.30 | ≤ 4,500 |
| 50 | 24,715 | 34.22 – 49.43 | |
| 66 | 32,815 | 48.13 – 65.63 | ≤ 6,000 |
| 80 | 39,375 | 56.88 – 78.75 | (all modules) |
| 100 | 50,180 | 73.44 – 100.4 | |

* Actual face area varies with unit coil type.

of 0°F, exterior SPACE 1 has a design sensible heat *loss* of 112,000 Btu/h while interior SPACE 2 has a net sensible heat *gain* of 23,500 Btu/h. At summer design outdoor conditions of 95°F db and 75°F wb, SPACE 1 has a design sensible heat gain of 67,000 Btu/h while SPACE 2 experiences a design sensible heat gain of 49,000 Btu/h. Interior design temperatures of both spaces is 75°F, all year long. Duct pressure drop is 3.1 in. water. Outdoor air requirement is 1400 cfm.

Calculate the size of

(a) Fan (scfm, pressure, motor horsepower)
(b) Heating coil (Btu/h).

**12.13** To maintain necessary close control of humidity and temperature required for a computer room, the reheat air conditioning system shown in the sketch is used. Space loads for the computer room include a heat load of 85,000 Btu/h and a moisture load of 42 lb/h. The return air conditions from the space must be exactly 50% relative humidity and 78°F. After mixing of the outside ventilation air with return air, the mixed air is at 80°F dry bulb with a relative humidity of 0.0114 lb/lb. The air is then cooled to saturation at 50°F by the cooling coil. There is a 2°F temperature rise across the fan. Air flow is controlled by a humidistat in the return air duct. The thermostat controls the temperature leaving the reheater.

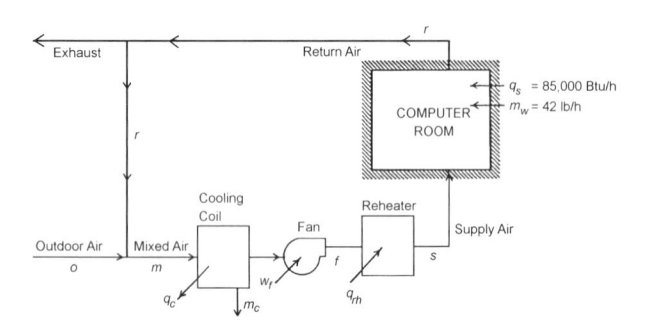

*Sketch for Problem 12.13*

Size the reheater (kW) and the cooling coil (Btu/h). From a manufacturer's catalog, select an appropriate electric resistance reheater coil. From a manufacturer's catalog, select an appropriate chilled water cooling coil.

**12.14** A small commercial building located in St. Louis, Missouri is to be conditioned using a variable-air-volume (VAV) system with reheat, as shown in the following sketch. At this stage of the process, preliminary sizing of the central cooling unit, of the reheaters, and of the fan (scfm) is to take place. There are four zones (separately thermostated spaces) in the building. Supply air from the cooling coil is maintained at 55°F during the summer and 58°F during the winter. Relative humidity off the coil is approximately 90% in both cases. Minimum outdoor air of 4000 scfm is maintained at all times (just don't ask how). The VAV boxes are not to be cut back beyond 50% of rated flow. The design conditions and calculated design load for each zone are as follow:

Zone 1
  Winter inside temperature = 72°F
  Winter design heat loss = –55,000 Btu/h (a loss)
  Summer inside temperature = 78°F
  Summer design heat gains = 124,000 Btu/h (sensible) and 31,000 Btu/h (latent)
Zone 2 (an interior space)
  Winter inside temperature = 78°F
  Winter design heat loss = 40,000 Btu/h (a gain)
  Summer inside temperature = 78°F
  Summer design heat gains = 220,000 Btu/h (sensible) and 71,000 Btu/h (latent)
Zone 3 (an interior space)
  Winter inside temperature = 78°F
  Winter design heat loss = 115,000 Btu/h (a gain)
  Summer inside temperature = 78°F
  Summer design heat gains = 140,000 Btu/h (sensible) and 42,000 Btu/h (latent)

Zone 4
  Winter inside temperature = 72°F
  Winter design heat loss = –180,000 Btu/h (a loss)
  Summer inside temperature = 78°F
  Summer design heat gains = 210,000 Btu/h (sensible) and 52,500 Btu/h (latent)

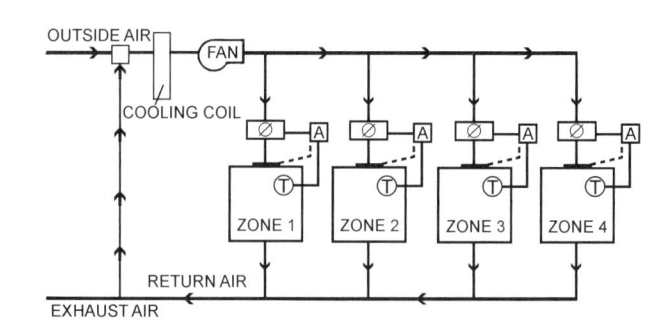

*Sketch for Problem 12.14*

**12.15** A commercial three-zone office building is being designed for St. Louis, Missouri where summer outdoor design conditions are 94°F db and 75°F wb and winter outdoor design conditions are 3°F and 100% rh. Each zone is to contain 10,000 ft$^2$ of floor space. A blow-through multizone unit will be used with cold deck temperature maintained at 58°F all year long and with hot deck temperature varying from a maximum of 130°F at winter design to 85°F during the summer. The amount of outdoor air is to equal the recommended 20 cfm per person. Design occupancy is to be 10 people per 1000 ft$^2$ of floor area. The duct system will be designed so that the pressure drop does not exceed 2.0 in. w.g. Fan efficiency is estimated at 65%. In winter, the control humidistat in the common return air duct is set at 30% rh. Due to the building orientation and internal zoning, all spaces will experience their peak loads at the same time. The space design loads at indoor design temperatures of 78°F summer and 72°F winter are

Summer
Zone 1: 116,000 Btu/h sensible, 43,000 Btu/h latent (gains)
Zone 2: 290,000 Btu/h Sensible, 59,000 Btu/h Latent (gains)
Zone 3: 190,000 Btu/h sensible, 39,000 Btu/h latent (gains)

Winter
Zone 1: –215,000 Btu/h sensible (loss), negligible latent
Zone 2: 110,000 Btu/h sensible (gain), negligible latent
Zone 3: –171,000 Btu/h sensible (loss), negligible latent

Conduct the preliminary sizing of the fan (scfm and horsepower), cooling coil (scfm and Btu/h), heating coil (scfm and Btu/h), and humidifier (gal/h). Provide a completely labeled sketch of the system.

**12.16** An air-conditioning unit takes in 2000 cfm of outdoor air at 95°F dry bulb and 76°F wet bulb, and 6000 cfm of return air at 78°F dry bulb and 50% rh. The conditioned air leaves the chilled water coil at 52°F dry bulb and 90% rh.

(a) What is the refrigeration load on the chiller in tons?

(b) Assume the conditioned air were reheated to 58.5°F dry bulb with electric heaters. What would be the operating cost of these heaters at 2.5 cents per kWh?

**12.17** In Problem 12.16, assume 2000 cfm of return air bypasses the chilled water coil and is used for reheat.

(a) How does the final condition of the air compare with the reheated air in part (b) of 12.16? [Ans: 58°F, $w = 0.0081$]

(b) Comment on the ability of the leaving air to absorb latent load in the conditioned space. [Ans: Less than in Example 12-1]

**12.18** For the building and reheat system shown below, determine:

(a) Fan rating, scfm
(b) Return air relative humidity at summer design conditions, %
(c) Size cooling coil, Btu/h
(d) Size reheat coils, Btu/h and scfm for each

*Winter*: Outside 6°F, $w = 0.001$; indoor 72°F, no humidity control.

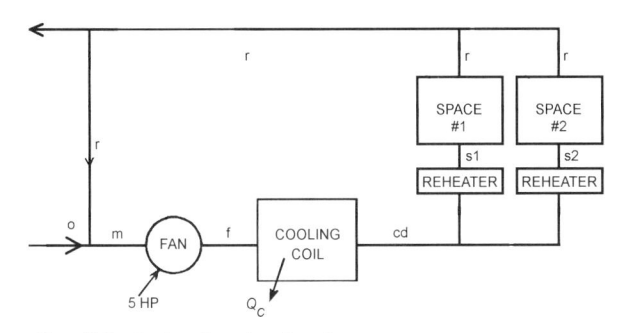

Sensible design heating loads
  Space 1: 162,000 Btu/h
  Space 2: 143,000 Btu/h

*Summer*: Outdoor 95°F dry bulb, 78°F wet bulb; indoor 78°F.

Sensible design cooling loads
  Space 1: 64,500 Btu/h
  Space 2: 55,000 Btu/h
Latent design loads (moisture produced)
  Space 1: 38 lb/h
  Space 2: 26 lb/h

*Year-round*: 10% by mass outdoor air required for ventilation.
  Conditions of cooling coil: 58°F, 90% rh.

**12.19** A basic reheat system has been retrofitted with an improved control system. For the operating conditions shown in the sketch below and with all thermostats set at 78°F, for what cooling coil discharge temperature $T$ should the logic system of the controller be calling if there is no humidity override?

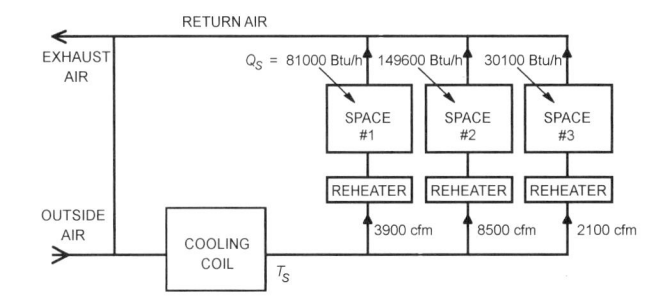

## 12.9 Bibliography

ASHRAE. 2008. *2008 ASHRAE Handbook—HVAC Systems and Equipment*.

ASHRAE. 2009. *Dedicated Outdoor Air Systems*.

Delp, W.W., H.J. Sauer, Jr., R.H. Howell, and B. Subbarao. 1993. Control of outside air and building pressurization in VAV systems. *ASHRAE Transactions* 99(1).

Sauer, H.J., Jr. and R.H. Howell. 1992. Estimating the indoor air quality and energy performance of VAV systems. *ASHRAE Journal* 34(July).

# Chapter 13

# HYDRONIC HEATING AND COOLING SYSTEM DESIGN

This chapter provides information useful for the design of hydronic heating and cooling systems. It provides information on the classification of systems, system descriptions, and design procedures. Additional information can be obtained from the 2012 *ASHRAE Handbook—HVAC Systems and Equipment* and the 2011 *ASHRAE Handbook—HVAC Applications*.

## 13.1 Introduction

Water systems that convey heat to or from a conditioned space or process with hot or chilled water are called hydronic systems. The water flows through piping that connects a boiler, water heater, or chiller to suitable terminal heat transfer units, usually located near or at the space or process.

Water systems can be classified by (1) operating temperature, (2) flow generation, (3) pressurization, (4) piping arrangement, and (5) pumping arrangement.

Classified by flow generation, hydronic heating systems may be (1) gravity systems, which use the difference in density between the supply and return water columns of a circuit or system to circulate water, or (2) forced systems, in which a pump, usually driven by an electric motor, maintains the flow. Gravity systems are only used for heating and are seldom used today.

Water systems can be either once-through or recirculating systems. This chapter describes forced recirculating systems.

### 13.1.1 Principles

The design of effective and economical water systems is affected by complex relationships between the various system components. The design water temperature, flow rate, piping layout, pump selection, terminal unit selection, and control method are all interrelated. The size and complexity of the system determines the importance of these relationships to the total system operating success. In the United States, present hydronic heating system design practice originated in residential heating applications, where a temperature drop ($\Delta t$) of 20°F was used to determine flow rate. Besides producing satisfactory operation and economy in small systems, this $\Delta t$ enabled simple calculations because 1 gpm conveys approximately 10,000 Btu/h. However, almost universal use of hydronic systems for both heating and cooling of large buildings and building complexes has rendered this simplified approach obsolete.

### 13.1.2 Temperature Classifications

Water systems can be classified by operating temperature as follows.

**Low-temperature water (LTW) system.** This hydronic heating system operates within the pressure and temperature limits of the ASME Boiler and Pressure Vessel Code for low-pressure boilers. The maximum allowable working pressure for low-pressure boilers is 160 psig, with a maximum temperature limitation of 250°F. The usual maximum working pressure for boilers for LTW systems is 30 psig, although boilers specifically designed, tested, and stamped for higher pressures are frequently used. Steam-to-water or water-to-water heat exchangers are also used for heating low-temperature water. Low-temperature water systems are used in buildings ranging from small, single dwellings to very large and complex structures.

### 13.1.3 Condensing Systems

Condensing systems are a special class of low-temperature water systems in which the water is heated in a boiler that is designed to condense the water vapor contained in the flue gases. The condensing boilers are fueled with natural gas (methane), from which the vapor condenses at approximately 130°F. In order to remove the latent heat from the combustion products, the heating water must enter the boiler below that temperature and be of adequate flow to remain below that temperature while absorbing the latent heat in that section of the heat exchanger. (See section 19.5.)

**Medium-temperature water (MTW) and high-temperature water (HTW) systems.** MTW systems operate at temperatures between 250 and 350°F, with pressures not exceeding 160 psig. The usual design supply temperature is approximately 250 to 325°F, with a usual pressure rating of 150 psig for boilers and equipment.

**HTW systems** operate at temperatures over 350°F and usually at pressures of about 300 psig. The maximum design supply water temperature is usually about 400°F, with a pressure rating for boilers and equipment of about 300 psig. The pressure-temperature rating of each component must be checked against the system's design characteristics. The use of MTW and HTW systems is usually limited to large campus or district-type distribution systems.

**Chilled water (CW) system.** This hydronic cooling system normally operates with a design supply water temperature of 40 to 55°F, usually 44 or 45°F, and at a pressure of up to 120 psig. Antifreeze or brine solutions may be used for applications (usually process or low-dew-point applications) that require temperatures below 40°F or for coil freeze protection. Direct well water systems can use supply temperatures of 60°F or higher.

**Dual-temperature water (DTW) system.** This hydronic combination heating and cooling system circulates hot and/or chilled water through common piping and terminal heat transfer apparatus. These systems operate within the pressure and temperature limits of LTW systems, with usual winter design supply water temperatures of about 100 to 150°F and summer supply water temperatures of 40 to 45°F.

Terminal heat transfer units include convectors, cast-iron radiators, baseboard and commercial finned-tube units, fan-coil units, chilled beams, radiant panels, snow-melting panels, and air-handling unit coils. A large storage tank may be included in the system to store energy to use when such heat input devices as the boiler or a solar energy collector are not supplying adequate energy.

## 13.2 Closed Water Systems

Because most hot and chilled water systems are closed, this chapter addresses only closed systems. The fundamental difference between a closed and an open water system is the interface of the water with a compressible gas (such as air) or an elastic surface (such as a diaphragm). A closed water system is defined as one with no more than one point of interface with a compressible gas or surface. This definition is fundamental to understanding the hydraulic dynamics of these systems. Earlier literature referred to a system with an open or vented expansion tank as an "open" system, but such a system is actually a closed system; the atmospheric interface of the tank simply establishes the system pressure at that point.

An open system, on the other hand, has more than one such interface. For example, a cooling tower system has at least two points of interface: the tower basin and the discharge pipe or nozzles entering the tower. One of the major differences in hydraulics between open and closed systems is that certain hydraulic characteristics of open systems cannot occur in closed systems. For example, in contrast to the hydraulics of an open system, in a closed system (1) flow cannot be motivated by static head differences, (2) pumps do not provide static lift, and (3) the entire piping system is always filled with water.

### 13.2.1 Basic System

Figure 13-1 shows the fundamental components of a closed hydronic system. Actual systems generally have additional components such as valves, vents, regulators, etc., but they are not essential to the basic principles underlying the system.

These fundamental components are

- Loads
- Source
- Expansion chamber
- Pump
- Distribution system

Theoretically, a hydronic system could operate with only these five components.

The components are subdivided into two groups: thermal and hydraulic. The thermal components consist of the load, the source, and the expansion chamber. The hydraulic components consist of the distribution system, the pump, and the expansion chamber. The expansion chamber is the only component that serves both a thermal and a hydraulic function.

### 13.2.2 Thermal Components

**Loads.** The load is the point where heat flows out of or into the system to or from the space or process; it is the independent variable to which the remainder of the system must respond. Outward heat flow characterizes a heating system, and inward heat flow characterizes a cooling system. The quantity of heating or cooling is calculated by one of the following means.

**Sensible heating or cooling.** The rate of heat entering or leaving an airstream is expressed as follows:

$$q = 60 Q_a \rho_a c_p \Delta t \qquad (13\text{-}1)$$

where

$q$ = heat transfer rate to or from air, Btu/h
$Q_a$ = airflow rate, cfm
$\rho_a$ = density of air, lb/ft$^3$
$c_p$ = specific heat of air, Btu/lb·°F
$\Delta t$ = temperature increase or decrease of air, °F

For standard air with a density of 0.075 lb/ft$^3$ and a specific heat of 0.244 Btu/lb·°F, Equation (13-1) becomes

$$q = 1.1 Q_a \Delta t \qquad (13\text{-}2)$$

The heat exchanger or coil must then transfer this heat from or to the water. The rate of sensible heat transfer to or from the heated or cooled medium in a specific heat exchanger is a function of the heat transfer surface area, the mean temperature difference between the water and the medium, and the overall heat transfer coefficient, which itself is a function of the fluid velocities, properties of the medium, geometry of the heat transfer surfaces, and other factors. The rate of heat transfer may be expressed by

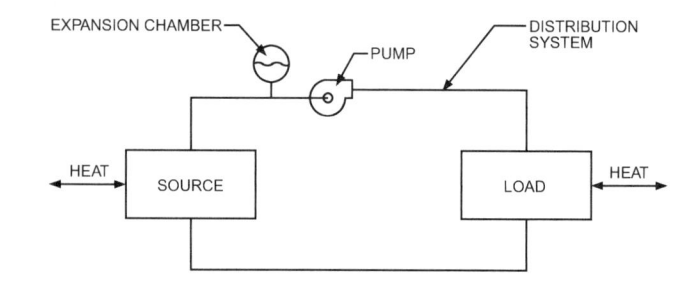

*Fig. 13-1 Hydronic System—Fundamental Components*
(Figure 1, Chapter 13, 2012 ASHRAE Handbook—HVAC Systems and Equipment)

# Hydronic Heating and Cooling System Design

$$q = UA(\text{LMTD}) \quad (13\text{-}3)$$

where

$q$ = heat transfer rate through heat exchanger, Btu/h
$U$ = overall coefficient of heat transfer, Btu/h·ft$^2$·°F
$A$ = heat transfer surface area, ft$^2$
LMTD = logarithmic mean temperature difference, heated or cooled medium to water, °F

**Cooling and dehumidification.** The rate of heat removal from the cooled medium when both sensible cooling and dehumidification are present is expressed by

$$q_t = w \Delta h \quad (13\text{-}4)$$

where

$q_t$ = total heat transfer rate from cooled medium, Btu/h
$w$ = mass flow rate of cooled medium, lb/h
$\Delta h$ = enthalpy difference between entering and leaving conditions of cooled medium, Btu/lb

Expressed for an air-cooling coil, this equation becomes

$$q_t = 60 Q_a \rho_a \Delta h \quad (13\text{-}5)$$

which, for standard air with a density of 0.075 lb/ft$^3$, reduces to

$$q_t = 4.5 Q_a \Delta h \quad (13\text{-}6)$$

**Heat transferred to or from water.** The rate of heat transfer to or from the water is a function of the flow rate, the specific heat, and the temperature rise or drop of the water as it passes through the heat exchanger. The heat transferred to or from the water is expressed by

$$q_w = m c_p \Delta t \quad (13\text{-}7)$$

where

$q_w$ = heat transfer rate to or from water, Btu/h
$m$ = mass flow rate of water, lb/h
$c_p$ = specific heat of water, Btu/lb·°F
$\Delta t$ = water temperature increase or decrease across unit, °F

With water systems, it is common to express the flow rate as volumetric flow, in which case Equation (13-7) becomes

$$q_w = 8.02 \rho_w c_p Q_w \Delta t \quad (13\text{-}8)$$

where

$Q_w$ = water flow rate, gpm
$\rho_w$ = density of water, lb/ft$^3$

For standard conditions in which the density is 62.4 lb/ft$^3$ and the specific heat is 1 Btu/lb·°F, Equation (13-8) becomes

$$q_w = 500 Q_w \Delta t \quad (13\text{-}9)$$

Equation (13-8) or (13-9) can be used to express the heat transfer across a single load or source device, or any quantity of such devices connected across a piping system. In the design or diagnosis of a system, the load side may be balanced with the source side using these equations.

**Heat carrying capacity of piping.** Equations (13-8) and (13-9) are also used to express the heat carrying capacity of the piping or distribution system or any portion thereof. When the existing temperature differential $\Delta t$, sometimes called the temperature range, is identified for any flow rate $Q_w$ through the piping, $q_w$ is called the heat carrying capacity.

Load systems can be any system in which heat is conveyed to or from the water for heating or cooling the space or process. Most load systems are basically a water-to-air finned-coil heat exchanger or a water-to-water exchanger. The specific configuration is usually used to describe the load device. The most common configurations include the following:

Heating load devices
Preheat coils in central units
Heating coils in central units
Zone or central unit reheat coils
Finned-tube radiation
Baseboard radiation
Convectors
Unit heaters
Fan-coil units
Induction unit and chilled beam coils
Water-to-water heat exchangers
Radiant heating panels
Snow-melting panels
Cooling load devices
Coils in central units
Fan-coil units
Induction unit and chilled beam coils
Radiant cooling panels
Water-to-water heat exchangers

**Source.** The source is the point where heat is added to (heating) or removed from (cooling) the system. Ideally, the amount of energy entering or leaving the source equals the amount entering or leaving through the load system(s). Under steady-state conditions, the load energy and source energy are equal and opposite. Also, when properly measured or calculated, temperature differentials and flow rates across the source and loads are all equal. Equations (13-8) and (13-9) are used to express the source capacities as well as the load capacities.

Any device that can be used to heat or cool water under controlled conditions can be used as a source device. The most common source devices for heating and cooling systems are the following:

Heating source devices
    Hot water generator or boiler
    Steam-to-water heat exchanger
    Water-to-water heat exchanger
    Solar heating panels
    Heat recovery or salvage heat device, (e.g., water jacket of an internal combustion engine)
    Exhaust gas heat exchanger

Incinerator heat exchanger
Heat pump condenser
Air-to-water heat exchanger (heat recovery coil)

Cooling source devices
Vapor compression chiller
Thermal absorption chiller
Heat pump evaporator
Air-to-water heat exchanger (heat recovery coil)
Water-to-water heat exchanger

The two primary considerations in selecting a source device are the design capacity and the part-load capability, sometimes called the turndown ratio. The turndown ratio, expressed in percent of design capacity, is

$$\text{Turndown ratio} = 100 \frac{\text{Minimum capacity}}{\text{Design capacity}} \quad (13\text{-}10)$$

The reciprocal of the turndown ratio is sometimes used (for example, a turndown ratio of 25% may also be expressed as a turndown ratio of 4:1).

The turndown ratio has a significant effect on performance; not considering the source system's part-load capability has been responsible for many systems that either do not function properly or do so at the expense of excess energy consumption. The turndown ratio has a significant impact on the ultimate equipment and/or system design selection.

**System Temperatures.** Design temperatures and temperature ranges are selected by considering the performance requirements and the economics of the components. For a cooling system that must maintain 50% rh at 75°F, the dew-point temperature is 55°F, which sets the maximum return water temperature at about 55°F (60°F maximum); on the other hand, the lowest reasonable temperature for refrigeration, considering the freezing point, energy consumption, and economics, is about 40°F. This temperature spread then sets constraints for a chilled water system.

For a heating system, the maximum hot water temperature is normally established by the ASME low-pressure code as 250°F, and with space temperature requirements of little over 75°F, the actual operating supply temperatures and the temperature ranges are set by the design of the load devices. The most economical systems related to distribution and pumping favor the use of the maximum possible temperature range ($\Delta t$). However, for better boiler fuel efficiency, the condensing boiler temperature limits may prevail. For systems with condensing boilers, the maximum temperature is around 130°F.

**Expansion Chamber.** The expansion chamber (also called an expansion or compression tank) serves both a thermal function and a hydraulic function. In its thermal function the tank provides a space into which the noncompressible liquid can expand or from which it can contract as the liquid undergoes volumetric changes with changes in temperature. To allow for this expansion or contraction, the expansion tank provides an interface point between the system fluid and a compressible gas. Note: By definition, a closed system can have only one such interface; thus, a system designed to function as a closed system should have only one expansion chamber.

Expansion tanks are of three basic configurations: (1) a closed tank, which contains a captured volume of compressed air and water, with an air water interface (sometimes called a plain steel tank); (2) an open tank (i.e., a tank open to the atmosphere); and (3) a diaphragm tank, in which a flexible membrane is inserted between the air and the water (another configuration of a diaphragm tank is the bladder tank).

In the plain steel tank and the open tank, gases can enter the water through the interface and can adversely affect performance. Thus, current design practice normally uses diaphragm or bladder tanks.

Sizing the tank is a primary thermal concern when designing the system. However, prior to sizing the tank, the control or elimination of air must be considered. The amount of air that will be absorbed and can be held in solution with the water is expressed by Henry's equation (Pompei 1981):

$$x = p/H \quad (13\text{-}11)$$

where
$x$ = solubility of air in water (% by volume)
$p$ = absolute pressure
$H$ = Henry's constant

Henry's constant, however, is constant only for a given temperature (Figure 13-2). Combining the data of Figure 13-2 (Himmelblau 1960) with Equation (13-11) results in the solubility diagram of Figure 13-3. With that diagram, the solubility can be determined if the temperature and pressure are known.

If the water is not saturated with air, it will absorb air at the air/water interface until the point of saturation has been reached. Once absorbed, the air moves through the water

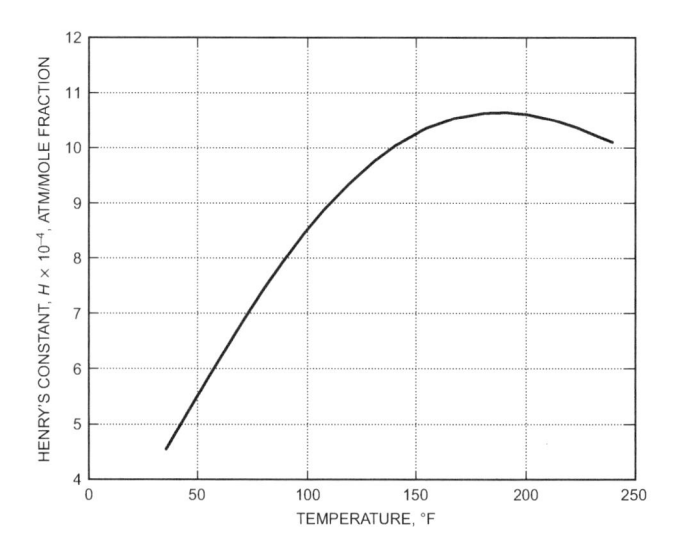

*Fig. 13-2  Henry's Constant Versus
Temperature for Air and Water*
*(Coad 1980a)*
*(Figure 2, Chapter 13, 2012 ASHRAE Handbook—
HVAC Systems and Equipment)*

# Hydronic Heating and Cooling System Design

either by mass migration or by molecular diffusion until the water is uniformly saturated.

If the air/water solution changes to a state that reduces solubility, the excess air will be released as a gas. For example, if the air/water interface is at a high pressure, the water will absorb air to its limit of solubility at that point; if at another point the pressure is less, some of the dissolved air will be released. In the design of systems with open or plain steel expansion tanks, the tank is commonly used as the major air control or release point in the system.

The following equations are used to size the three common configurations of expansion tanks (Coad 1980a):

For closed tanks with air/water interface,

$$V_t = V_s \frac{[(v_2/v_1) - 1] - 3\alpha\Delta t}{(P_a/P_1) - (P_a - P_2)} \quad (13\text{-}12)$$

For open tanks with air/water interface,

$$V_t = 2\{V_s[(v_2/v_1) - 1] - 3\alpha\Delta t\} \quad (13\text{-}13)$$

For diaphragm tanks,

$$V_t = V_s \frac{[(v_2/v_1) - 1] - 3\alpha\Delta t}{1 - (P_1/P_2)} \quad (13\text{-}14)$$

where

$V_t$ = volume of expansion tank, gal
$V_s$ = volume of water in system, gal
$t_1$ = lower temperature, °F
$t_2$ = higher temperature, °F
$P_a$ = atmospheric pressure, psia
$P_1$ = pressure at lower temperature, psia
$P2$ = pressure at higher temperature, psia
$v_1$ = specific volume of water at lower temperature, ft$^3$/lb
$v_2$ = specific volume of water at higher temperature, ft$^3$/lb
$\alpha$ = linear coefficient of thermal expansion, in/in °F
   = $6.5 \times 10^{-6}$ in./in. °F for steel
   = $9.5 \times 10^{-6}$ in./in. °F for copper
$\Delta t$ = $(t_2 - t_1)$, °F

As an example, the lower temperature for a heating system is usually normal ambient temperature at fill conditions (e.g., 50°F) and the higher temperature is the operating supply water temperature for the system. For a chilled water system, the lower temperature is usually the design chilled water supply temperature, and the higher temperature is ambient temperature (e.g., 95°F). However, in very large central systems that remain at operating temperatures the $\Delta t$ is quite small since it is the average system water temperature, which remains almost constant except for central variance. For a dual-temperature hot/chilled system, the lower temperature is the chilled water design supply temperature, and the higher temperature is the heating water design supply temperature.

For specific volume and saturation pressure of water at various temperatures, see Table 3 in Chapter 1 of the 2013 *ASHRAE Handbook—Fundamentals*, or any other comprehensive steam table.

At the tank connection point, the pressure in closed tank systems increases as the water temperature increases. Pressures at the expansion tank are generally set by the following parameters:

- The pressure at the lower temperature is usually selected to hold a positive pressure at the highest point in the system (usually about 10 psig).
- The pressure at the higher temperature is normally set by the maximum pressure allowable at the location of the safety relief valve(s) without opening them.

Other considerations are to ensure that (1) the pressure at no point in the system will ever drop below the saturation pressure at the operating system temperature and (2) all pumps have sufficient net positive suction head (NPSH) available to prevent cavitation.

**Example 13-1.** Size an expansion tank for a water heating system that will operate at 180 to 220°F. The minimum pressure at the tank is 10 psig (24.7 psia) and the maximum pressure is 25 psig (39.7 psia). (Atmospheric pressure is 14.7 psia.) The volume of water is 3000 gal. The piping is steel.

1. Calculate the required size for a closed tank with an air/water interface.

**Solution:** For lower temperature $t_1$, use 40°F.

From Table 3 in Chapter 1 of the 2013 *ASHRAE Handbook—Fundamentals*,

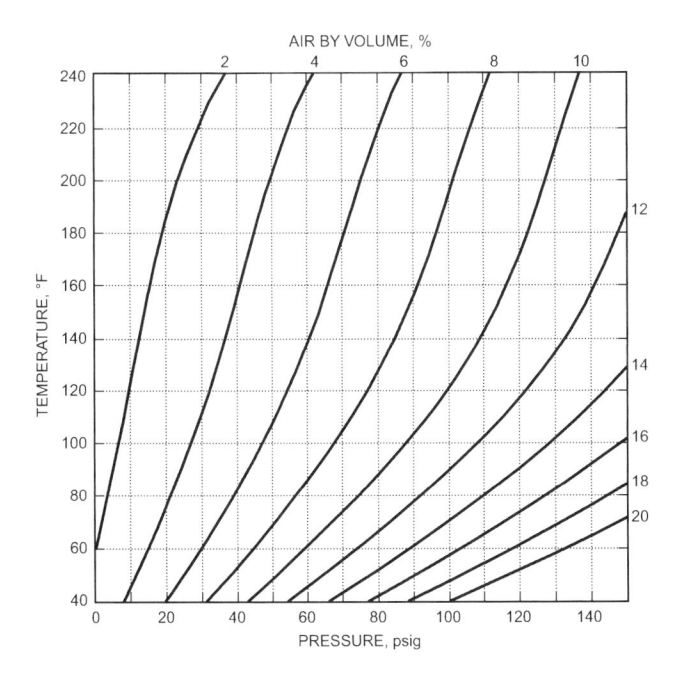

*Fig. 13-3 Solubility Versus Temperature and Pressure for Air/Water Solutions*
*(Figure 3, Chapter 13, 2012 ASHRAE Handbook—HVAC Systems and Equipment)*

$v_1$ (at 40°F) = 0.01602 ft³/lb and $v_2$ (at 220°F) = 0.01677 ft³/lb

Using Equation (13-12),

$$V_t = 3000 \frac{[(0.01677/0.01602) - 1] - 3(6.5 \times 10^{-6})(220 - 40)}{(14.7/24.7) - (14.7/39.7)}$$

$$= 578 \text{ gal}$$

If a diaphragm tank were to be used in lieu of the plain steel tank, what tank size would be required?

**Solution:** Using Equation (13-14),

$$V_t = 3000 \frac{[(0.01677/0.01602) - 1] - 3(6.5 \times 10^{-6})(220 - 40)}{1 - (24.7/39.7)}$$

$$= 119 \text{ gal}$$

### 13.2.3 Hydraulic Components

**Distribution System.** The distribution system is the piping system connecting the various other components of the system. The primary considerations in designing this system are (1) sizing the piping to handle the heating or cooling capacity required, (2) arranging the piping to ensure flow in the quantities required at design conditions and at all other loads, and (3) assuring that the air is purged from the system so the water can flow freely.

The flow requirement of the pipe is determined by Equation (13-8) or (13-9). After $\Delta t$ is established based on the thermal requirements, either of these equations (as applicable) can be used to determine the flow rate. First-cost economics and energy consumption make it advisable to design for the greatest practical $\Delta t$ because the flow rate is inversely proportional to $\Delta t$; that is, if $\Delta t$ doubles, the flow rate is reduced by half.

The three related variables in sizing the pipe are flow rate, pipe size, and pressure drop. The primary consideration in selecting a design pressure drop is the relationship between the economics of first cost and energy costs.

Once the distribution system is designed, the pressure loss at design flow is calculated by the methods discussed in Chapter 9 in this book or in Chapter 22 of the 2013 *ASHRAE Handbook—Fundamentals*. The relationship between flow rate and pressure loss can be expressed by

$$Q = C_v \sqrt{\Delta p} \tag{13-15}$$

where

$Q$ = system flow rate, gpm

$\Delta p$ = pressure drop in system, psi

$C_v$ = system constant (sometimes called valve coefficient, which is discussed in Chapter 46 in 2012 *ASHRAE Handbook—HVAC Systems and Equipment*)

Equation (13-15) may be modified as follows:

$$Q = C_s \sqrt{\Delta h} \tag{13-16}$$

where

$\Delta h$ = system head loss, ft of fluid [$\Delta h = \Delta p/\rho$]

$C_s$ = system constant
[$C_s = 0.67 C_v$ for water with a density = 62.4 lb/ft³]

Equations (13-15) and (13-16) are the system constant form of the Darcy-Weisbach equation. If the flow rate and head loss are known for a system, Equation (13-16) may be used to calculate the system constant $C_s$. From this calculation, the pressure loss can be determined at any other flow rate. Equation (13-16) can be graphed as a system curve (Figure 13-4).

The system curve changes if anything occurs that changes the flow/pressure drop characteristics. Examples of this are a strainer that starts to block or a control valve closing, either of which increases the head loss at any given flow rate, thus changing the system curve in a direction from curve A to curve B in Figure 13-4.

**Pump or Pumping System.** Centrifugal pumps are the type most commonly used in hydronic systems (see Chapter 44 in 2012 *ASHRAE Handbook—HVAC Systems and Equipment* or Chapter 9 in this book). Circulating pumps used in water systems can vary in size from small in-line circulators delivering 5 gpm at 6 or 7 ft head to base-mounted or vertical pumps handling hundreds or thousands of gallons per minute, with pressures limited only by the characteristics of the system. Pump operating characteristics must be carefully matched to system operating requirements.

**Pump Curves and Water Temperature.** Performance characteristics of centrifugal pumps are described by pump curves, which plot flow versus head or pressure, as well as by efficiency and power information. The point at which a pump operates is the point at which the pump curve intersects the system curve (Figure 13-5).

A complete piping system follows the same water flow/pressure drop relationships as any component of the system [see Equation (13-16)]. Thus, the pressure required for

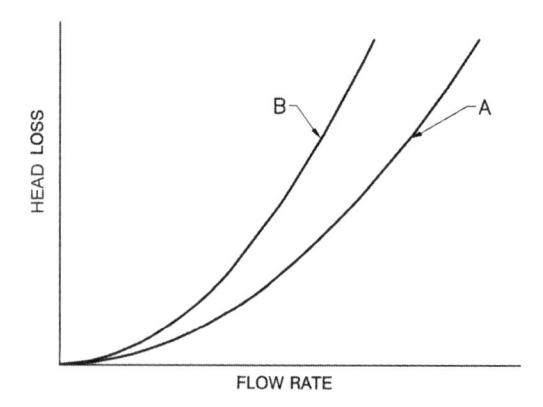

*Fig. 13-4 Typical System Curves for Closed System*

any proposed flow rate through the system may be determined and a system curve constructed. A pump may be selected by using the calculated system pressure at the design flow rate as the base point value.

Figure 13-6 illustrates how a shift of the system curve to the right affects system flow rate. This shift can be caused by incorrectly calculating the system pressure drop by using arbitrary safety factors or overstated pressure drop charts. Variable system flow caused by control valve operation or improperly balanced systems (subcircuits having substantially lower pressure drops than the longest circuit) can also cause a shift to the right.

Pumps for closed-loop piping systems should have a flat pressure characteristic and should operate slightly to the left of the peak efficiency point on their curves. This characteristic permits the system curve to shift to the right without causing undesirable pump operation, overloading, or reduction in available pressure across circuits with large pressure drops.

Many dual-temperature systems are designed so that the chillers are bypassed during the winter months. The chiller pressure drop, which may be quite high, is thus eliminated from the system pressure drop, and the pump operating point shift to the right may be quite large. For such systems, system curve analysis should be used to check winter operating points.

Operating points may be highly variable, depending on (1) load conditions, (2) the types of control valves used, and (3) the piping circuitry and heat transfer elements. In general, the best selection will be:

- For design flow rates calculated using pressure drop charts that illustrate actual closed-loop hydronic system piping pressure drops
- To the left of the maximum efficiency point of the pump curve to allow shifts to the right caused by system circuit unbalance, direct-return circuitry applications, and modulating three-way valve applications
- A pump with a flat curve to compensate for unbalanced circuitry and to provide a minimum pressure differential increase across two-way control valves

**Parallel Pumping.** When pumps are applied in parallel, each pump operates at the same head, and provides its share of the system flow at that pressure (Figure 13-7). Generally, pumps of equal size are used, and the parallel pump curve is established by doubling the flow of the single pump curve (with identical pumps).

Plotting a system curve across the parallel pump curve shows the operating points for both single and parallel pump operation (Figure 13-7). Note that single-pump operation does not yield 50% flow. The system curve crosses the single pump curve considerably to the right of its operating point when both pumps are running. This leads to two important concerns: (1) the pumps must be powered to prevent overloading during single-pump operation, and (2) a single pump can provide standby service of up to 80% of design flow; the actual amount depends on the specific pump curve and system curve.

**Series Pumping.** When pumps are operated in series, each pump operates at the same flow rate and provides its share of the total pressure at that flow. A system curve plotted across the series pump curve shows the operating points for both single and series pump operation (Figure 13-8). Note that the single pump can provide up to 80% flow for standby and at a lower power requirement.

Series pump installations are often used in two-pipe heating and cooling systems so that both pumps operate during the cooling season to provide maximum flow and head, while only a single pump operates during the heating season. Note that both parallel and series pump applications require that the actual pump operating points be used to accurately determine the pumping point. Adding artificial safety factor head, using improper pressure drop charts, or incorrectly calculating pressure drops may lead to an unwise selection.

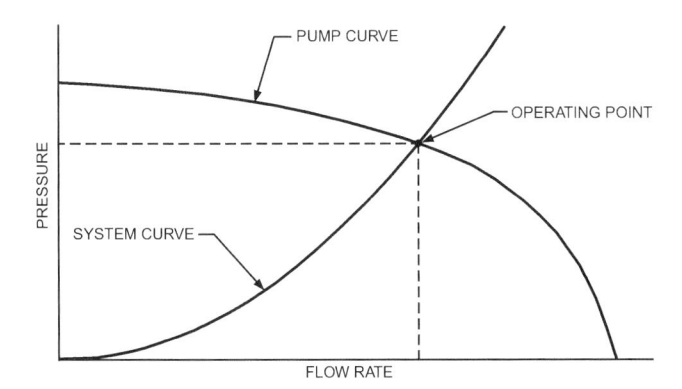

*Fig. 13-5  Pump Curve and System Curve*
*(Figure 5, Chapter 13, 2012 ASHRAE Handbook—HVAC Systems and Equipment)*

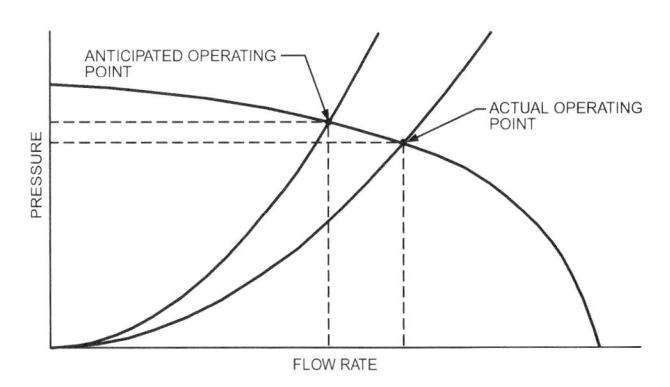

*Fig. 13-6  Shift of System Curve due to Circuit Unbalance*
*(Figure 6, Chapter 13, 2008 ASHRAE Handbook—HVAC Systems and Equipment)*

**Multiple-Pump Systems.** Care must be taken in designing systems with multiple pumps to ensure that if pumps ever operate in either parallel or series, such operation is fully understood and considered by the designer. Pumps performing unexpectedly in series or parallel have been the cause of performance problems in hydronic systems. Typical problems resulting from pumps functioning in parallel and series when not anticipated by the designer are the following.

Parallel. With pumps of unequal pressures, one pump may create a pressure across the other pump in excess of its cutoff pressure, causing flow through the second pump to diminish significantly or to cease. This phenomenon can cause flow problems or pump damage.

Series. With pumps of different flow capacities, the pump of greater capacity may overflow the pump of lesser capacity, which could cause damaging cavitation in the smaller pump and could actually cause a pressure drop rather than a pressure rise across that pump. In other circumstances, unexpected series operation can cause excessively high or low pressures that can damage system components.

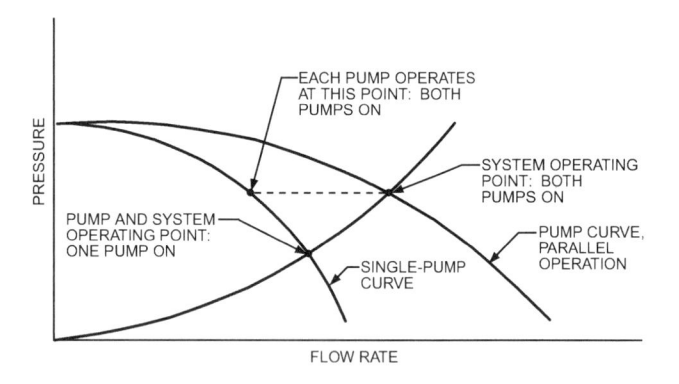

*Fig. 13-7 Operating Conditions for Parallel Pump Installation*
(Figure 8, Chapter 13, 2012 ASHRAE Handbook—
HVAC Systems and Equipment)

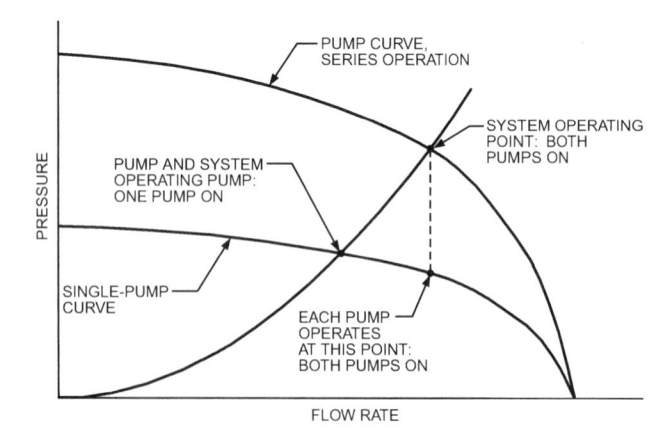

*Fig. 13-8 Operating Conditions for Series Pump Installation*
(Figure 9, Chapter 13, 2012 ASHRAE Handbook—
HVAC Systems and Equipment)

**Standby Pump Provision.** If total flow standby capacity is required, a properly valved standby pump of equal capacity is installed to operate when the normal pump is inoperable. A single standby may be provided for several similarly sized pumps. Parallel or series pump installation can provide up to 80% standby, as stated above, which is often sufficient.

**Compound Pumping.** In larger systems, compound pumping, also known as primary-secondary pumping, is often employed to provide system advantages that would not be available with a single pumping system. The concept of compound pumping is illustrated in Figure 13-9.

In Figure 13-9, Pump No. 1 can be referred to as the source or primary pump and Pump No. 2 as the distribution or secondary pump. The short section of pipe between A and B is called the common pipe because it is common to both the source and distribution circuits. Other terms used for the common pipe are the decoupling line and the neutral bridge. In the design of compound systems, the common pipe should be kept as short and as large in diameter as practical to minimize the pressure loss between those two points. Care must be taken, however, to ensure adequate length in the common pipe to prevent recirculation from entry or exit turbulence. There should never be a valve or check valve in the common pipe. If these conditions are met and the pressure loss in the common pipe can be assumed to be zero, then neither pump will affect the other. Then, except for the system static pressure at any given point, the circuits can be designed and analyzed and will function dynamically independently of one another.

In Figure 13-9, if Pump No. 1 has the same flow capacity in its circuit as Pump No. 2 has in its circuit, all of the flow entering Point A from Pump No. 1 will leave in the branch supplying Pump No. 2, and no water will flow in the common pipe. Under this condition, the water entering the load will be at the same temperature as that leaving the source.

If the flow capacity of Pump No. 1 exceeds that of Pump No. 2, some water will flow downward in the common pipe. Under this condition, Tee A is a diverting tee, and Tee B becomes a mixing tee. Again, the temperature of the fluid entering the load is the same as that leaving the source. However, because of the mixing taking place at Point B, the tem-

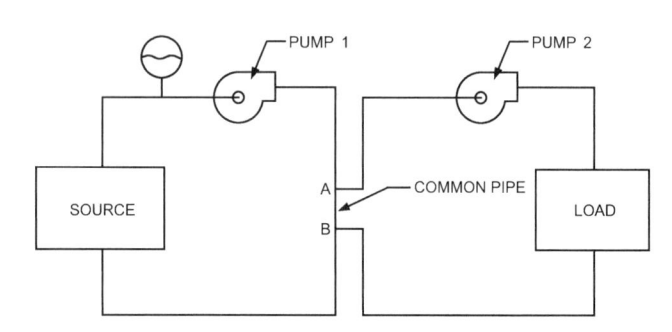

*Fig. 13-9 Compound Pumpimg (Primary-Secondary Pumping)*
(Figure 10, Chapter 13, 2012 ASHRAE Handbook—
HVAC Systems and Equipment)

# Hydronic Heating and Cooling System Design

perature of the water returning to the source is between the source supply temperature and the load return temperature.

On the other hand, if the flow capacity of Pump No. 1 is less than that of Pump No. 2, then Point A becomes a mixing point because some water must recirculate upward in the common pipe from Point B. Under this condition, the temperature of the water entering the load is between the supply water temperature from the source and the return water temperature from the load.

For example, if Pump No. 1 circulates 25 gpm of water leaving the source at 200°F, and Pump No. 2 circulates 50 gpm of water leaving the load at 100°F, then the water temperature entering the load is

$$t_{load} = 200 - (25/50)(200 - 100) = 150°F$$

The following are some advantages of compound circuits:

1. They enable the designer to achieve different water temperatures and temperature ranges in different elements of the system.
2. They decouple the circuits hydraulically, thereby making the control, operation, and analysis of large systems much less complex. Hydraulic decoupling also prevents unwanted series or parallel operation.
3. Circuits can be designed for different flow characteristics. For example, a chilled water load system can be designed with two-way valves for better control and energy conservation while the source system operates at constant flow to protect the water in the evaporator from freezing.

**Expansion Chamber.** As a hydraulic device, the expansion tank serves as the reference pressure point in the system, analogous to a ground in an electrical system (Lockhart and Carlson 1953). Where the tank connects to the piping, the pressure equals the pressure of the air in the tank plus or minus any fluid pressure due to the elevation difference between the tank liquid surface and the pipe (Figure 13-10).

As previously stated, a closed system should have only one expansion chamber. The presence of more than one chamber or of excessive amounts of undissolved air in a piping system can cause the closed system to behave in unexpected (but understandable) ways, causing extensive damage from shock waves or water hammer.

With a single chamber on a system, assuming isothermal conditions for the air, the air pressure can change only as a result of displacement by the water. The only thing that can cause the water to move into or out of the tank (assuming no water is being added to or removed from the system) is expansion or contraction of the water in the system. Thus, in sizing the tank, thermal expansion is related to the pressure extremes of the air in the tank [Equations (13-12), (13-13), and (13-14)].

The point of connection of the tank should be based on the pressure requirements of the system, remembering that the pressure at the tank connection will not change as the pump is turned on or off. For example, consider a system containing an expansion tank at 30 psig and a pump with a pump head of 23.1 ft (10 psig). Figure 13-11 shows alternative locations for connecting the expansion tank; in either case, with the pump off, the pressure will be 30 psig on both the pump suction and discharge. With the tank on the pump suction side, when the pump is turned on, the pressure increases on the discharge side by an amount equal to the pump pressure (Figure 13-11A). With the tank connected on the discharge side of the pump, the pressure decreases on the suction side by the same amount (Figure 13-11B).

Other considerations relating to the tank connection include the following:

- A tank open to the atmosphere must be located above the highest point in the system.
- A tank with an air/water interface is generally used with an air control system that continually revents the air into the tank. For this reason, it should be connected at a point where air can best be released.
- Within reason, the lower the pressure in a tank, the smaller is the tank [see Equations (13-12) and (13-14)]. Thus, in a vertical system, the higher the tank is placed, the smaller it can be.

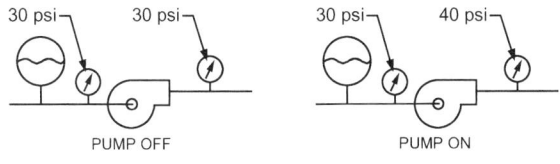

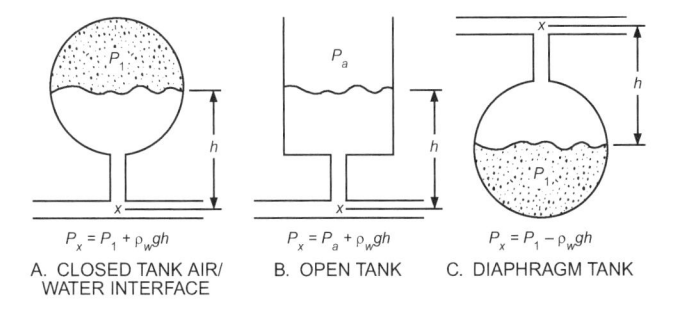

$P_x = P_1 + \rho_w gh$
A. CLOSED TANK AIR/WATER INTERFACE

$P_x = P_a + \rho_w gh$
B. OPEN TANK

$P_x = P_1 - \rho_w gh$
C. DIAPHRAGM TANK

*Fig. 13-10 Tank Pressure Related to "System" Pressure*
(Figure 15, Chapter 13, 2012 ASHRAE Handbook—HVAC Systems and Equipment)

A. TANK ON PUMP SUCTION SIDE

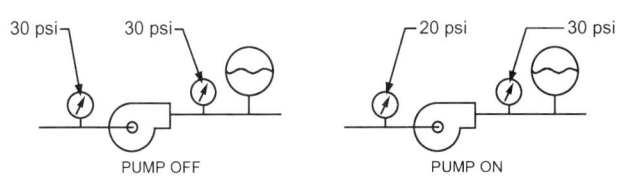

B. TANK ON PUMP DISCHARGE SIDE

*Fig. 13-11 Effect of Expansion Tank Location with Respect to Pump Pressure*
(Figure 16, Chapter 13, 2012 ASHRAE Handbook—HVAC Systems and Equipment)

## 13.3 Design Considerations

### 13.3.1 Piping Circuits

Hydronic systems are designed with many different configurations of piping circuits. In addition to simple preference by the design engineer, the method of arranging the circuiting can be dictated by such factors as the shape or configuration of the building, the economics of installation, energy economics, the nature of the load, part-load capabilities or requirements, and others.

Each piping system is a network; the more extensive the network, the more complicated it is to understand, analyze, or control. Thus, a major design objective is to maintain the highest degree of simplicity.

Load distribution circuits are of four general types:

- Full series
- Diverting series
- Parallel direct return
- Parallel reverse return

**Series Circuit.** A simple series circuit is shown in Figure 13-12. Series loads generally have the advantage of both lower piping costs and higher temperature drops that result in smaller pipe sizes and lower energy consumption. A disadvantage is that the different circuits cannot be controlled separately. Simple series circuits are generally limited to residential and small commercial standing radiation heating systems. Figure 13-13 shows a typical layout of such a system with two zones for residential or small commercial heating.

**Diverting Series.** The simplest diverting series circuit diverts some of the flow from the main piping circuit through a special diverting tee to a load device (usually standing radiation) that has a low pressure drop. This system is generally limited to heating systems in residential or small commercial applications.

Figure 13-14 illustrates a typical one-pipe diverting tee circuit. For each terminal unit, a supply and a return tee are installed on the main. One of the two tees is a special diverting tee that creates a pressure drop in the main flow to divert part of the flow to the unit. One (return) diverting tee is usually sufficient for upfeed (units above the main) systems. Two special fittings (supply and return tees) are usually required to overcome thermal pressure in downfeed units. Special tees are proprietary; therefore, manufacturer's literature must be consulted for flow rates and pressure drop data on these devices. Unit selection can be only approximate without these data.

One-pipe diverting series circuits allow manual or automatic control of flow to individual heating units. On-off rather than flow modulation control is preferred because of the relatively low pressure drop allowable through the control valve in the diverted flow circuit. This system is likely to cost more than the series loop because extra branch pipe and fittings, including special tees, are required. Each unit usually requires a manual air vent because of the low water velocity through the unit. The length and load imposed on a one-pipe circuit are usually small because of these limitations.

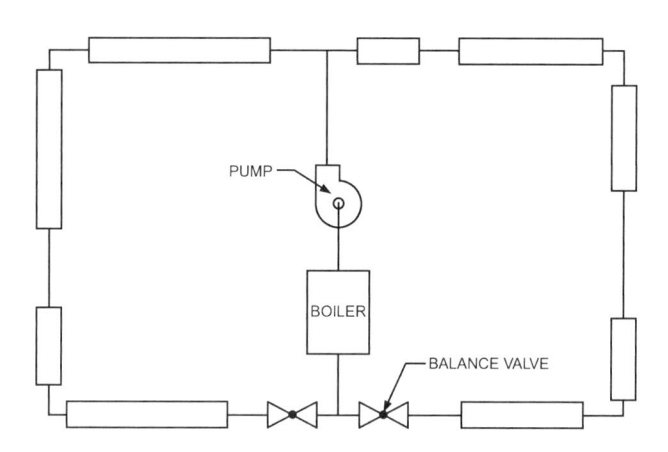

*Fig. 13-13   Series Loop System*
*(Figure 18, Chapter 13, 2012 ASHRAE Handbook—HVAC Systems and Equipment)*

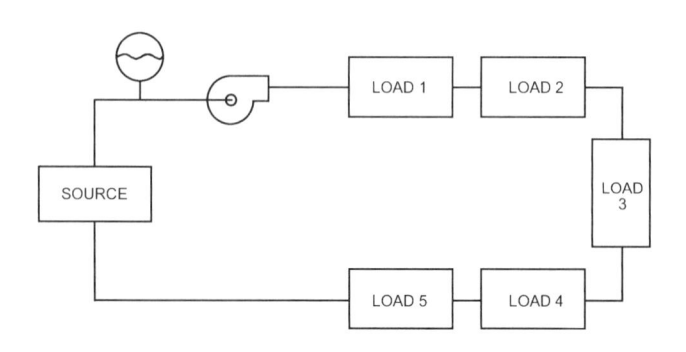

*Fig. 13-12   Flow Diagram of Simple Series Circuit*
*(Figure 17, Chapter 13, 2012 ASHRAE Handbook—HVAC Systems and Equipment)*

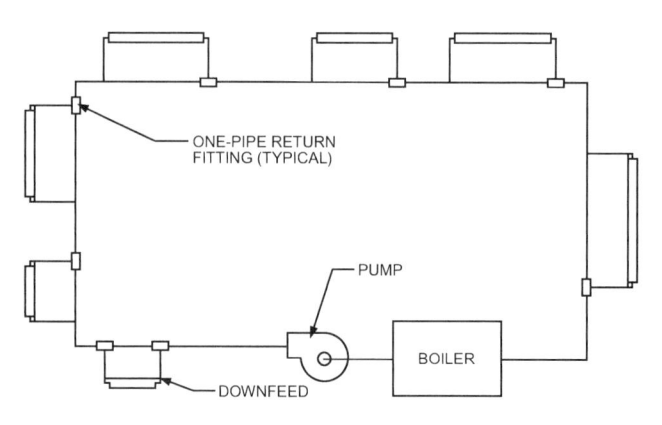

*Fig. 13-14   One-Pipe Diverting Tee System*
*(Figure 19, Chapter 13, 2012 ASHRAE Handbook—HVAC Systems and Equipment)*

# Hydronic Heating and Cooling System Design

Because only a fraction of the main flow is diverted in a one-pipe circuit, the flow rate and pressure drop are less variable. When two or more one-pipe circuits are connected to the same two-pipe mains, the circuit flow will need to be mechanically balanced. After balancing, sufficient flow must be maintained in each one-pipe circuit to ensure adequate flow diversion to the loads.

When coupled with compound pumping systems, series circuits can be applied to multiple control zones on larger commercial or institutional systems (Figure 13-15). Note that in the series circuit with compound pumping, the load pumps need not be equal in capacity to the system pump. If, for example, load pump LP-1 circulates less flow ($Q_{LP1}$) than system pump SP-1 ($Q_{SP1}$), the temperature difference across Load 1 would be greater than the circuit temperature difference between A and B (i.e., water would flow in the common pipe from A to B). If, on the other hand, the load pump LP-2 is equal in flow capacity to the system pump SP-1, the temperature differentials across Load 2 and across the system from C to D would be equal and no water would flow in the common pipe. If $Q_{LP3}$ exceeds $Q_{SP1}$, mixing occurs at Point E and, in a heating system, the temperature of the water entering pump LP-3 would be lower than that available from the system leaving load connection D.

Thus, a series circuit using compound or load pumps offers many design options. Each of the loads shown in Figure 13-15 could also be a complete piping circuit or network.

**Parallel Piping.** These networks are the most commonly used in hydronic systems because they allow the same temperature water to be available to all loads. The two types of parallel networks are direct return and reverse return (Figure 13-16).

In the direct-return system, the length of supply and return piping through the subcircuits is unequal, which may cause unbalanced flow rates and require careful balancing to provide each subcircuit with design flow. Ideally, the reverse-return system provides nearly equal total lengths for all terminal circuits.

Direct-return piping has been successfully applied where the designer has guarded against major flow unbalance by:

1. Providing for pressure drops in the subcircuits or terminals that are significant percentages of the total, usually establishing pressure drops for close subcircuits at higher values than those for the far subcircuits by use of balancing values.
2. Minimizing distribution piping pressure drop (in the limit, if the distribution piping loss is zero and the loads are of equal flow resistance, the system is inherently balanced).
3. Including balancing devices and some means of measuring flow at each terminal or branch circuit
4. Using control valves with a high head loss at the terminals

### 13.3.2 Capacity Control of Load System

The two alternatives for controlling the capacity of hydronic systems are on-off control and variable-flow or modulating control. The on-off option is generally limited to smaller systems or components (e.g., residential or small commercial) and individual components of larger systems. In smaller systems where the entire building is a single zone, control is accomplished by cycling the source device (the boiler or chiller) on and off. Usually a space thermostat allows the chiller or boiler to run, then a water temperature thermostat (aquastat) controls the capacity of the chiller(s) or boiler(s) as a function of supply or return water temperature. The pump can be either cycled with the load device (usually the case in a residential heating system) or left running (usually done in commercial hot or chilled water systems).

In these single-zone applications, the piping design requires no special consideration for control. Where multiple zones of control are required, the various load devices are controlled first, then the source system capacity is controlled to follow the capacity requirement of the loads.

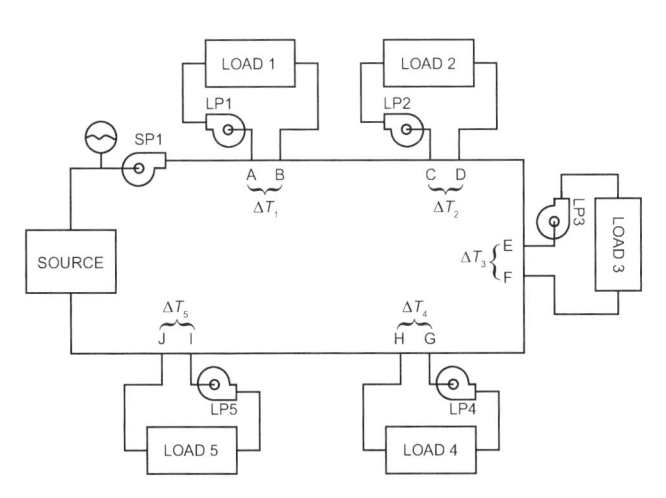

*Fig. 13-15  Series Circuit with Load Pumps*
*(Figure 20, Chapter 13, 2012 ASHRAE Handbook—HVAC Systems and Equipment)*

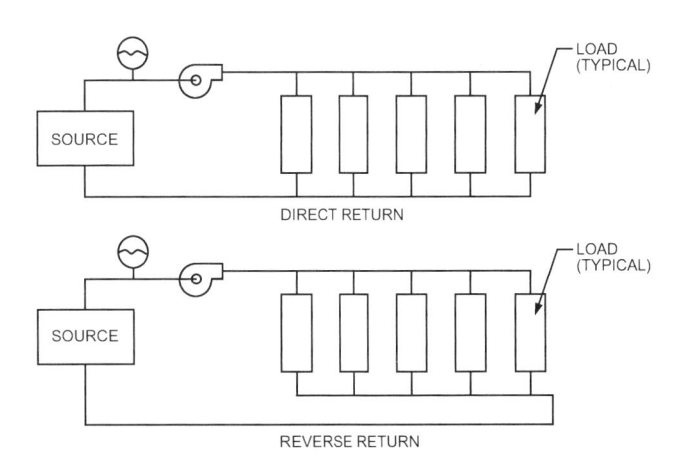

*Fig. 13-16  Direct- and Reverse-Return Two-Pipe Systems*
*(Figure 21, Chapter 13, 2012 ASHRAE Handbook—HVAC Systems and Equipment)*

**438**  Principles of HVAC

Control valves are commonly used to control loads. These valves control the capacity of each load by varying the amount of water flow through the load device when load pumps are not used. Control valves for hydronic systems are straight-through (two-way) valves and three-way valves (Figure 13-17). The effect of either valve is to vary the amount of water flowing through the load device.

With a two-way valve (Figure 13-17A), as the valve strokes from full-open to full-closed, the quantity of water flowing through the load gradually decreases from design flow to no flow. With a three-way mixing valve (Figure 13-17B) in one position, the valve is open from Port A to AB, with Port B closed off. In that position, all the flow is through the load. As the valve moves from the A-AB position to the B-AB position, some of the water bypasses the load by flowing through the bypass line, thus decreasing flow through the load. At the end of the stroke, Port A is closed, and all of the fluid will flow from B to AB with no flow through the load. Thus, the three-way mixing valve has the same effect on the load as the two-way valve—as the load reduces, the quantity of water flowing through the load decreases.

The effect on load control with the three-way diverting valve (Figure 13-17C) is the same as with the mixing valve in a closed system—the flow is either directed through the load or through the bypass in proportion to the load. Because of the dynamics of valve operation, diverting valves are more complex in design and are thus more expensive than mixing valves; because they accomplish the same function as the simpler mixing valve, they are seldom used in closed hydronic systems.

In terms of load control, a two-way valve and a three-way valve perform identical functions—they both vary the flow through the load as the load changes. The fundamental difference between the two-way valve and the three-way valve is that as the source or distribution system sees the load, the two-way valve provides a variable flow load response and the three-way valve provides a constant flow load response.

According to Equation (13-9), the load $q$ is proportional to the product of $Q$ and $\Delta t$. Ideally, as the load changes, $Q$ changes, while $\Delta t$ remains fixed. However, as the system sees it, as the load changes with the two-way valve, $Q$ varies and $\Delta t$ is fixed; whereas with a three-way valve, $\Delta t$ varies and $Q$ is fixed. This principle is illustrated in Figure 13-18. An understanding of this concept is fundamental to the design or analysis of hydronic systems.

The flow characteristics of two-way and three-way valve ports are described in Chapter 47, "Design and Application of Controls," of the 2011 *ASHRAE Handbook—HVAC Applications* and must be understood. The equal percentage characteristic is recommended for proportional control of the load flow for two-way and three-way valves; the bypass flow port of three-way valves should have the linear characteristic to maintain a uniform flow during part-load operation.

### 13.3.3 Sizing Control Valves

For stable control, the pressure drop in the control valve at the full-open position should be no less than one-half the pressure drop in the branch. For example, in Figure 13-18 the pressure drop at full-open position for the two-way valve should equal one-half the pressure drop from A to B, and for the three-way valve, the full-open pressure drop should be half that from C to D. The pressure drop in the bypass balancing valve in the three-way valve circuit should be set to equal that in the coil (load).

Control valves should be sized on the basis of the valve coefficient $C_V$. For more information, see the section on control valve sizing in Chapter 47 in the 2012 *ASHRAE Handbook—HVAC Systems and Equipment*.

If a system is to be designed with multiple zones of control such that load response is to be by constant flow through the load and variable $\Delta t$, control cannot be achieved by valve control alone; a load pump is required.

Several control arrangements of load pump and control valve configurations are shown in Figure 13-19. Note that in all three configurations the common pipe has no restriction or check valve. In all configurations there is no difference in control as seen by the load. However, the basic differences in control are:

1. With the two-way modulation valve configuration (Figure 13-19A), the distribution system sees a variable flow and a constant $\Delta t$, whereas with both three-way configurations, the distribution system sees a constant flow and a variable $\Delta t$.

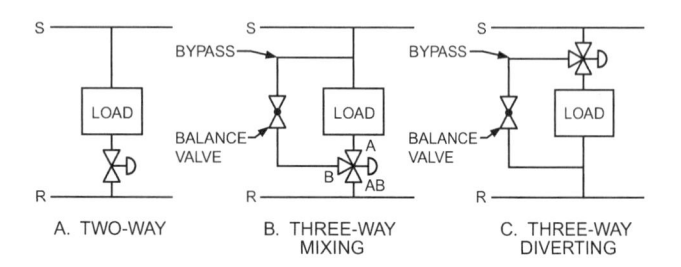

*Fig. 13-17  Load Control Valves*
(Figure 22, Chapter 13, 2012 ASHRAE Handbook—
HVAC Systems and Equipment)

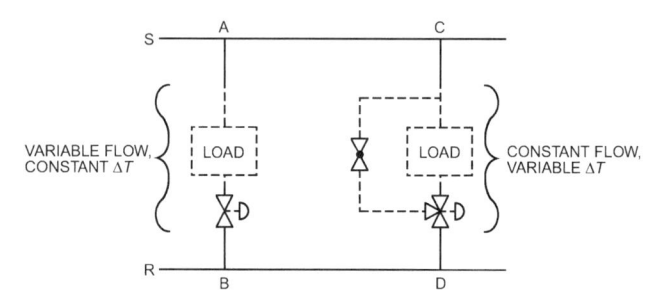

*Fig. 13-18  System Flow with Two-Way and
Three-Way Valves*
(Figure 23, Chapter 13, 2012 ASHRAE Handbook—
HVAC Systems and Equipment)

# Hydronic Heating and Cooling System Design

2. Configuration B differs from C in that the pressure required through the three-way valve in Figure 13-19B is provided by the load pump, while in Figure 13-19C it is provided by the distribution pump(s).

## 13.3.4 Low-Temperature Heating Systems

These systems are used for heating spaces or processes directly, as with standing radiation and process heat exchangers, or indirectly, through air-handling unit coils for preheating, for reheating, or in hot water unit heaters. These systems have generally been designed with supply water temperatures from 180 to 240°F and temperature drops from 20 to 100°F. With increasing use of condensing boilers, systems with temperature ranges between 110°F and 130°F are becoming more commonly used.

In the United States, hot water heating systems were historically designed for a 200°F supply water temperature and a 20°F temperature drop. This practice evolved from earlier gravity system designs and provides convenient design relationships for heat transfer coefficients related to coil tubing and finned-tube radiation and for calculations (1 gallon per minute conveys approximately 10,000 Btu/h at 20°F $\Delta t$). Because many terminal devices still require these flow rates, it is important to recognize this relationship in selecting devices and designing systems.

However, the greater the temperature range (and related lower flow rate) that can be applied, the less costly the system is to install and operate. A lower flow rate requires smaller and less expensive piping, less secondary building space, and smaller pumps. Also, smaller pumps require less electrical energy.

**Nonresidential Heating Systems.** Possible approaches to enhancing the economics of large heating systems include (1) higher supply temperatures, (2) compound pumping, and (3) terminal equipment designed for smaller flow rates. The three techniques may be used either singly or in combination.

Using higher supply water temperatures achieves higher temperature drops and smaller flow rates. Terminal units with a reduced heating surface can be used. These smaller terminals are not necessarily less expensive, however, because their required operating temperatures and pressures may increase manufacturing costs and the problems of pressurization, corrosion, expansion, and control. System components may not increase in cost uniformly with temperature but rather in steps conforming to the three major temperature classifications. Within each classification, the most economical design uses the highest temperature in that classification.

Primary-secondary or compound pumping reduces the size and cost of the distribution system and also may use larger flows and lower temperatures in the terminal or secondary circuits. A primary pump circulates water in the primary distribution system while one or more secondary pumps circulate the terminal circuits. The connection between primary and secondary circuits provides complete hydraulic isolation of both circuits and permits a controlled interchange of water between the two. Thus, a high supply water temperature can be used in the primary circuit at a low flow rate and high temperature drop, while a lower temperature and conventional temperature drop can be used in the secondary circuit(s).

For example, a system could be designed with primary-secondary pumping in which the supply temperature from the boiler was 240°F, the supply temperature in the secondary was 200°F, and the return temperature was 180°F. This design results in a conventional 20°F $\Delta t$ in the secondary zones but permits the primary circuit to be sized on the basis of a 60°F drop. This primary-secondary pumping arrangement is most advantageous with terminal units such as convectors and finned radiation, which are generally unsuited for small flow rate design.

A fourth technique is to put certain loads in series utilizing a combination of control valves and compound pumping (Figure 13-20). In the system illustrated, the capacity of the boiler or heat exchanger is $2 \times 10^6$ Btu/h, and each of the four loads is $0.5 \times 10^6$ Btu/h. Under design conditions, the system is designated for an 80°F water temperature drop, and the loads each provide 20°F of the total $\Delta t$. The loads in these systems, as well as the smaller or simpler systems in residential or commercial applications, can be connected in a direct-return or a reverse-return piping system. The different features of each load are as follows:

1. The domestic hot water heat exchanger has a two-way valve and is thus arranged for variable flow (while the main distribution circuit provides constant flow for the boiler circuit).
2. The finned-tube radiation circuit is a 20°F $\Delta t$ circuit with the design entering water temperature reduced to and controlled at 200°F.
3. The reheat coil circuit takes a 100°F temperature drop for a very low flow rate.
4. The preheat coil circuit provides constant flow through the coil to keep it from freezing.

When loads such as water-to-air heating coils in low-temperature water systems are valve controlled (flow varies), they have a heating characteristic of flow versus capacity as

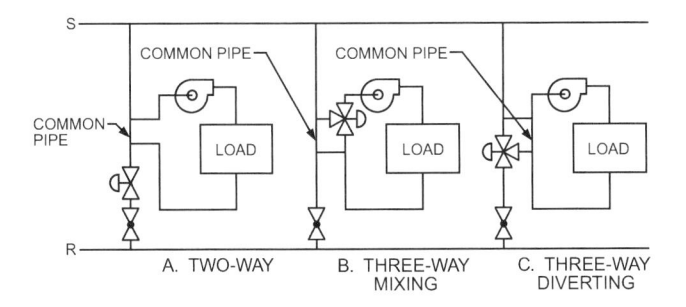

*Fig. 13-19 Load Pumps with Valve Control*
*(Figure 29, Chapter 13, 2012 ASHRAE Handbook—HVAC Systems and Equipment)*

shown in Figure 13-21 for 20°F and 60°F temperature drops, respectively. For a 20°F Δ*t* coil, 50% flow provides approximately 90% capacity; valve control will tend to be unstable. For this reason, proportional temperature control is required, and equal percentage characteristic two-way valves should be selected such that 10% flow is achieved with 50% valve lift. This combination of the valve characteristic and the heat transfer characteristic of the coil makes the control linear with respect to the control signal. This type of control can be obtained only with equal percentage two-way valves and can be further enhanced if piped with a secondary pump arrangement as shown in Figure 13-19A. See Chapter 47 of the 2011 *ASHRAE Handbook—HVAC Applications* for further information on automatic controls.

### 13.3.5 Chilled-Water Systems

Designers have less latitude in selecting supply water temperatures for cooling applications because there is only a narrow range of water temperatures low enough to provide adequate dehumidification and high enough to avoid chiller freeze-up. Circulated water quantities can be reduced by selecting proper air quantities and heat transfer surface at the terminals. Terminals suited for a 12°F rise rather than an 8°F rise reduce circulated water quantity and pump power by one-third and increase chiller efficiency.

A proposed system should be evaluated for the desired balance between installation cost, operating cost, and energy efficiency. Table 13-1 shows the effect of coil circuiting and chilled water temperature on water flow and temperature rise. The coil rows, fin spacing, air-side performance, and cost are identical for all selections. Morabito (1960) showed how such changes in coil circuiting affect the overall system. Considering the investment cost of piping and insulation versus the operating cost of refrigeration and pumping motors, higher temperature rises (e.g., 16 to 24°F temperature rise at about 1.0 to 1.5 gpm per ton of cooling) can be applied on chilled-water systems with long distribution piping runs; larger flow rates should be used only where reasonable in close-coupled systems.

For the most economical design, the minimum flow rate to each terminal heat exchanger is calculated. For example, if one terminal can be designed for an 18°F rise, another for 14°F, and others for 12°F, the highest rise to each terminal should be used, rather than designing the system for an overall temperature rise based on the smallest capability.

The control system selected also influences the design water flow. For systems using multiple terminal units, diversity factors can be applied to flow quantities before sizing pump and piping mains if exposure or use prevents the unit design loads from occurring simultaneously and if two-way valves are used for water flow control. If air-side control (e.g., face-and-bypass or fan cycling) or three-way valves on the

**Table 13-1 Chilled-Water Coil Performance**
*(Table 1, Chapter 13, 2012 ASHRAE Handbook—HVAC Systems and Equipment)*

| Coil Circuiting | Chilled-Water Inlet Temp., °F | Coil Pressure Drop, psi | Chilled-Water Flow gpm/ton | Chilled-Water Temp. Rise, °F |
|---|---|---|---|---|
| Full[a] | 45 | 1.0 | 2.2 | 10.9 |
| Half[b] | 45 | 5.5 | 1.7 | 14.9 |
| Full[a] | 40 | 0.5 | 1.4 | 17.1 |
| Half[b] | 40 | 2.5 | 1.1 | 21.8 |

Note: Table is based on cooling air from 81°F dry bulb, 67°F wet bulb to 58°F dry bulb, 56°F wet bulb.
a Full circuiting (also called single circuit). Water at the inlet temperature flows simultaneously through all tubes in a plane transverse to airflow; it then flows simultaneously through all tubes, in unison, in successive planes (i.e., rows) of the coil.
b Half circuiting. Tube connections are arranged so there are half as many circuits as there are tubes in each plane (row) thereby using higher water velocities through the tubes. This circuiting is used with small water quantities.

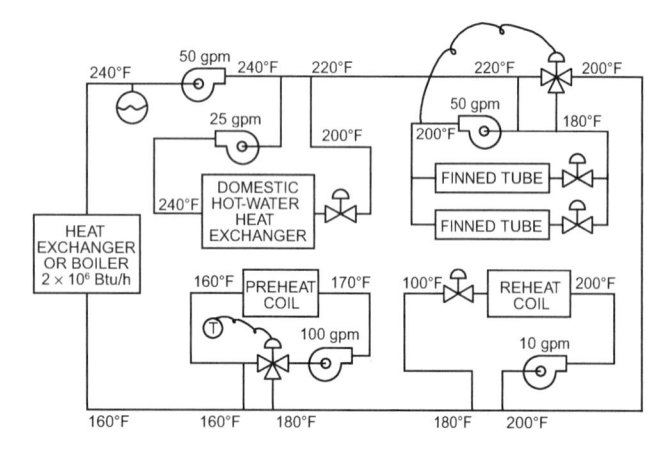

*Fig. 13-20 Example of Series-Connected Loading*
*(Figure 31, Chapter 13, 2012 ASHRAE Handbook—HVAC Systems and Equipment)*

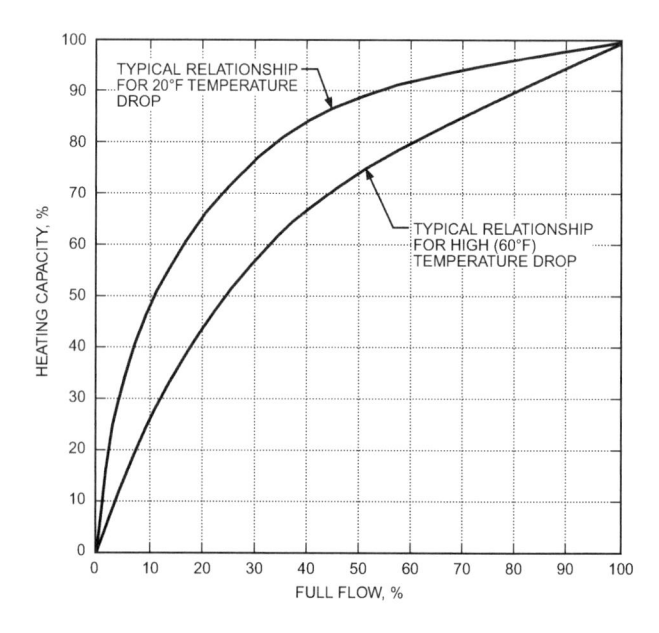

*Fig. 13-21 Heat Emission Versus Flow Characteristic of Typical Hot Water Heating Coil*
*(Figure 32, Chapter 13, 2012 ASHRAE Handbook—HVAC Systems and Equipment)*

# Hydronic Heating and Cooling System Design

water side are used, diversity should not be a consideration in pump and piping design, although it should be considered in the chiller selection.

A primary consideration with chilled-water system design is the control of the source systems at reduced loads. The constraints on the temperature parameters are (1) a water freezing temperature of 32°F, (2) economics of the refrigeration system in generating chilled water, and (3) the dew-point temperature of the air at nominal indoor comfort conditions (55°F dew point at 75°F and 50% rh). These parameters have led to the common practice of designing for a supply chilled-water temperature of 44 to 45°F and a return water temperature between 55 and 64°F.

Historically, most chilled-water systems have used three-way control valves to achieve constant water flow through the chillers. However, as systems have become larger, as designers have turned to multiple chillers for reliability and controllability, and as energy efficiency has become an increasing concern, the use of two-way valves and source pumps for the chillers has greatly increased.

A typical configuration of a small chilled-water system using two parallel chillers and loads with three-way valves is illustrated in Figure 13-22. Note that the flow is essentially constant. A simple energy balance [Equation (13-9)] dictates that with a constant flow rate, at one-half of design load, the water temperature differential drops to one-half of design. At this load, if one of the chillers is turned off, the return water circulating through the off chiller mixes with the supply water. This mixing raises the temperature of the supply chilled water and can cause a loss of control if the designer does not consider this operating mode.

A typical configuration of a large chilled-water system with multiple chillers and loads and compound piping is shown in Figure 13-23. This system provides variable flow, essentially constant supply temperature chilled water, multiple chillers, more stable two-way control valves, and the advantage of adding chilled-water storage with little additional complexity.

One design issue illustrated in Figure 13-23 is the placement of the common pipe for the chillers. With the common pipe as shown on the left side of the chillers, the chillers will unload from left to right. With the common pipe in the alternate location shown, the chillers will unload equally in proportion to their capacity (i.e., equal percentage).

### 13.3.6 Dual-Temperature Systems

Dual-temperature systems are used when the same load devices and distribution systems are used for both heating and cooling (e.g., fan-coil units and central station air-handling unit coils). In the design of dual-temperature systems, the cooling cycle design usually dictates the requirements of the load heat exchangers and distribution systems. Basically, dual-temperature systems are of three different configurations, each requiring different design techniques:

1. Two-pipe systems
2. Four-pipe common load systems
3. Four-pipe independent load systems

**Two-Pipe Systems.** In a two-pipe system, the load devices and the distribution system circulate chilled water

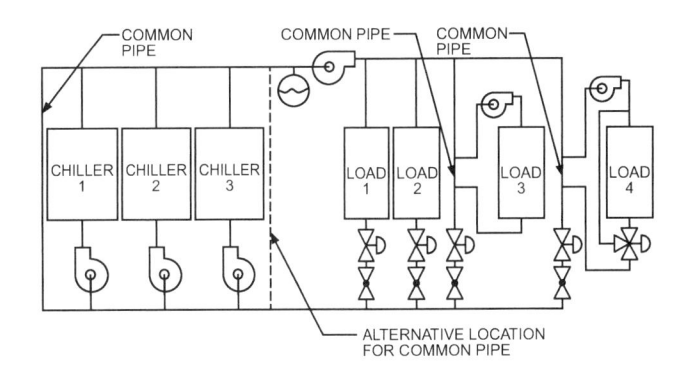

*Fig. 13-23  Variable-Flow Chilled-Water System*
(Figure 36, Chapter 13, 2012 ASHRAE Handbook—
HVAC Systems and Equipment)

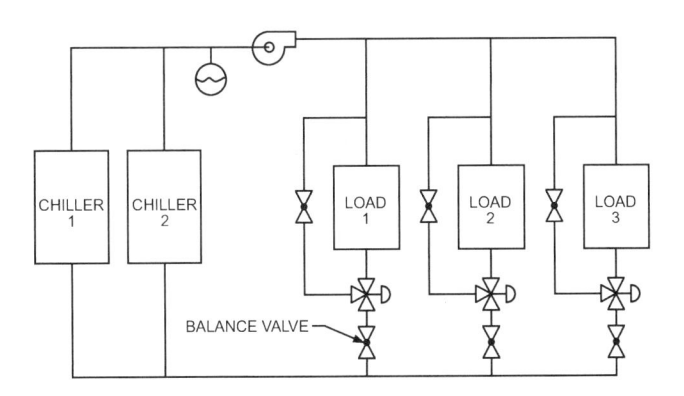

*Fig. 13-22  Constant-Flow Chilled-Water System*
(Figure 35, Chapter 13, 2012 ASHRAE Handbook—
HVAC Systems and Equipment)

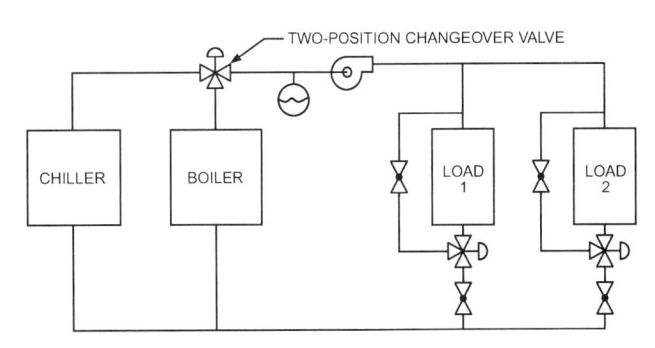

*Fig. 13-24  Simplified Diagram of Two-Pipe System*
(Figure 37, Chapter 13, 2012 ASHRAE Handbook—
HVAC Systems and Equipment)

when cooling is required and hot water when heating is required (Figure 13-24). Design considerations for these systems include the following:

- Loads must all require cooling or heating coincidentally; that is, if cooling is required for some loads and heating for other loads at a given time, this type of system should not be used.
- When designing the system, the flow and temperature requirements for both the cooling and the heating media must be calculated first. The load and distribution system should be designed for the more stringent, and the water temperatures and temperature differential should be dictated by the other mode.
- The changeover procedure should be designed such that the chiller evaporator is not exposed to damaging high water temperatures and the boiler is not subjected to damaging low water temperatures. To accommodate these limiting requirements, the changeover of a system from one mode to the other requires considerable time. If rapid load swings are anticipated, a two-pipe system should not be selected, even though it is the least costly of the three options.

**Four-Pipe Common Load Systems.** In the four-pipe common load system, load devices are used for both heating and cooling, as in the two-pipe system. The four-pipe common load system differs from the two-pipe system in that both heating and cooling are available to each load device, and the changeover from one mode to the other takes place at each individual load device, or grouping of load devices, rather than at the source. Thus, some of the load systems can be in the cooling mode while others are in the heating mode. Figure 13-25 is a flow diagram of a four-pipe common load system, with multiple loads and a single boiler and chiller.

Although many of these systems have been installed, many have not performed successfully due to problems in implementing the design concepts.

One problem that must be addressed is the expansion tank connection(s). Many four-pipe systems were designed with two expansion tanks—one for the cooling circuit and one for the heating circuit. However, with multiple loads, these circuits become hydraulically interconnected, thus creating a system with two expansion chambers. The preferred method of handling the expansion tank connection sets the point of reference pressure equal in both circuits (Figure 13-25).

Another potential problem is the mixing of hot and chilled water. At each load connection, two three-way valves are required: a mixing valve on the inlet and a diverting valve on the outlet. These valves operate in unison in just two positions—opening either Port B to AB or Port A to AB. If, for example, the valve on the outlet does not seat tightly and Load 1 is indexed to cooling and Load 2 is indexed to heating, return heating water from Load 2 will flow into the chilled-water circuit, and return chilled water from Load 1 will flow into the heating-water circuit. The probability of this occurring increases as the number of loads increases because the number of control valves increases.

Another disadvantage of this system is that the loads have no individual capacity control as far as the water system is concerned. That is, each valve must be positioned to either full heating or full cooling with no control in between.

Because of these disadvantages, four-pipe common load systems should be limited to those applications in which there cannot be independent load circuits, such as radiant ceiling panels, chilled beams, or induction unit coils.

**Four-Pipe Independent Load Systems.** The four-pipe independent load system is preferred for those hydronic applications in which some of the loads are in the heating mode while others are in the cooling mode. Control is simpler and more reliable than for the common load systems, and in many applications, the four-pipe independent load system is more reliable and less costly to install. Also, the flow through the individual loads can be modulated, providing both the control capability for variable capacity and the opportunity for variable flow in either or both circuits.

A simplified example of a four-pipe independent load system with two loads, one boiler, and two chillers is shown in Figure 13-26. Note that both hydronic circuits are essentially

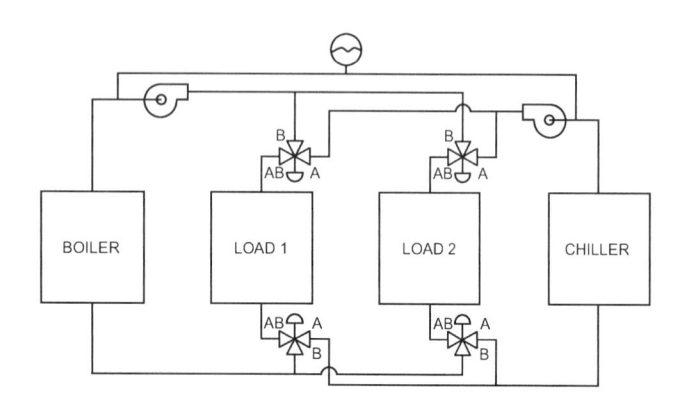

*Fig. 13-25  Four-Pipe Common Load System*
(Figure 38, Chapter 13, 2012 ASHRAE Handbook—HVAC Systems and Equipment)

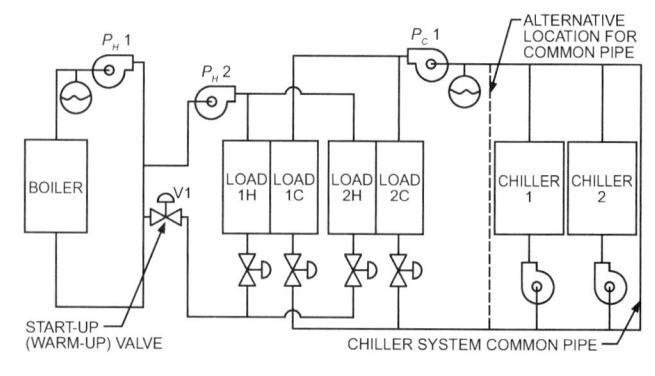

*Fig. 13-26  Four-Pipe Independent Load System*
(Figure 39, Chapter 13, 2012 ASHRAE Handbook—HVAC Systems and Equipment)

independent, so that each can be designed with disregard for the other system. Although both circuits in the figure are shown as variable-flow distribution systems, they could be constant flow (three-way valves) or one variable flow and one constant flow. Generally, the control modulates the two load valves in sequence with a dead band at the control midpoint.

This type of system offers additional flexibility when some selective loads are arranged for heating only or cooling only, such as unit heaters or preheat coils. Then, central station systems can be designed for humidity control with reheat through configuration at the coil locations and with proper control sequences.

### 13.3.7 Other Design Considerations

**Makeup and Fill Water Systems.** Generally, a hydronic system is filled with water through a valved connection to a domestic water source, with a service valve, a backflow preventer, and a pressure gage. (The domestic water source pressure must exceed the system fill pressure.)

Because the expansion chamber is the reference pressure point in the system, the water makeup point is usually located at or near the expansion chamber.

Many designers prefer to install automatic makeup valves, which consist of a pressure-regulating valve in the makeup line. However, the quantity of water being made up must be monitored to identify leakage, which causes scaling, oxygen corrosion, gaseous air, and related problems in the system.

**Safety Relief Valves.** Safety relief valves should be installed at any point at which pressures can be expected to exceed the safe limits of the system components. Causes of excessive pressures include:

- Overpressurization from fill system
- Pressure increases due to thermal expansion
- Surges caused by momentum changes (shock or water hammer)

Overpressurization from the fill system could occur due to an accident in filling the system or due to the failure of an automatic fill regulator. To prevent this, a safety relief valve is usually installed at the fill location. Figure 13-27 shows a typical piping configuration for a system with a plain steel or air/water interface expansion tank. Note that no valves are installed between the hydronic system piping and the safety relief valve. This is a mandatory design requirement if the valve in this location is also to serve as a protection against pressure increases due to thermal expansion.

As previously stated, the expansion chamber is installed in a hydronic system to allow for the volumetric changes that accompany water temperature changes. However, if any part of the system is configured such that it can be isolated from the expansion tank and its temperature can increase while it is isolated, then overpressure relief should be provided.

The relationship between pressure change due to temperature change and the temperature change in a piping system is expressed by the following equation:

$$\Delta p = \frac{(\beta - 3\alpha)\Delta t}{(5/4)(D/E\Delta r) + \gamma} \quad (13\text{-}17)$$

where
- $\Delta p$ = pressure increase, psi
- $\beta$ = volumetric coefficient of thermal expansion of water, 1/°F
- $\alpha$ = linear coefficient of thermal expansion for piping material, 1/°F
- $\Delta t$ = water temperature increase, °F
- $D$ = pipe diameter, in.
- $E$ = modulus of elasticity of piping material, psi
- $\gamma$ = volumetric compressibility of water, in.$^2$/lb
- $\Delta r$ = thickness of pipe wall, in.

Figure 13-28 shows a solution to Equation (13-17) demonstrating the pressure increase caused by any given temperature increase for 1 in. and 10 in. steel piping. If the temperature in a chilled water system with piping spanning sizes between 1 and 10 in. were to increase by 15°F, the pressure would increase between 340 and 420 psi, depending on the average pipe size in the system.

Safety relief should be provided to protect boilers, heat exchangers, cooling coils, chillers, and the entire system when the expansion tank is isolated for air charging or other service. As a minimum, the ASME Boiler Code requires that a dedicated safety relief valve be installed on each boiler and that isolating or service valves be provided on the supply and return connections to each boiler.

Potential forces caused by shock waves or water hammer should also be considered in design. Chapter 22 of the 2013 *ASHRAE Handbook—Fundamentals* discusses the causes of shock forces and the methodology for calculating the magnitude of these forces.

**Air Elimination.** If air and other gases are not eliminated from the flow circuit, they may cause binding in the terminal heat transfer elements, corrosion, noise, reduced pumping capacity or flow in a circuit, and loss of hydraulic stability. A

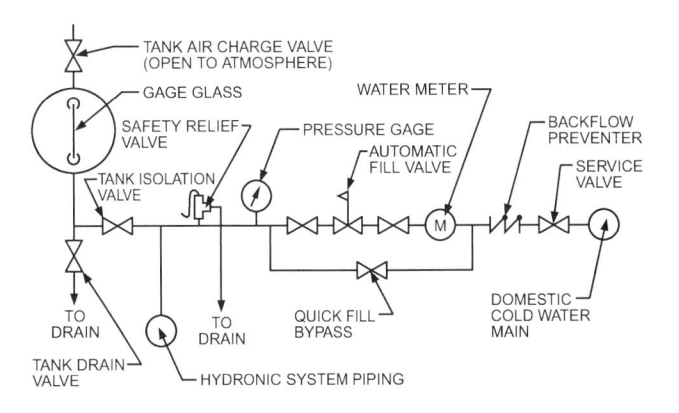

*Fig. 13-27 Typical Makeup Water and Expansion Tank Piping Configuration for Plain Steel Expansion Tank*
*(Figure 40, Chapter 13, 2012 ASHRAE Handbook—HVAC Systems and Equipment)*

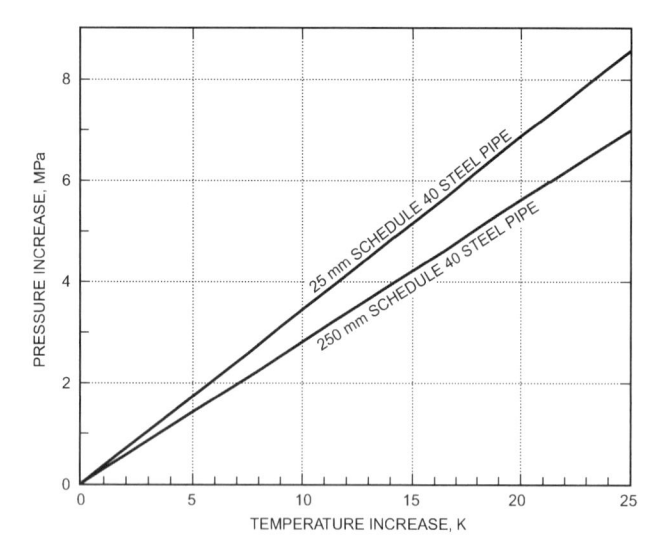

*Fig. 13-28 Pressure Increase Resulting from Thermal Expansion as Function of Temperature Increase*
*(Figure 41, Chapter 13, 2012 ASHRAE Handbook—HVAC Systems and Equipment)*

closed tank without a diaphragm can be installed at the point of the lowest solubility of air in water (see Figure 13-3). When a diaphragm tank is used, air in the system can be removed by an air separator and air elimination valve installed at the point of lowest solubility. Manual vents should be installed at high points to remove all trapped air during initial operation and to ensure that the system is tight. Shutoff valves should be installed on any automatic air removal device to permit servicing without draining the system.

**Drain and Shutoff.** All low points should have drains. Separate shutoff and draining of individual equipment and circuits should be possible so that the entire system does not have to be drained to service a particular item. Whenever a device or section of the system is isolated, and the water in that section or device could increase in temperature following isolation, overpressure safety relief protection as discussed above must be provided.

**Balance Fittings.** Balance fittings or valves and a means of measuring flow quantity should be applied as needed to permit balancing of individual terminals and subcircuits.

**Pitch.** Piping need not pitch but can run level, providing that flow velocities exceeding 1.5 fps are maintained or a diaphragm tank is used.

**Strainers.** Strainers should be used where necessary to protect system elements. Strainers in the pump suction must be checked regularly to prevent cavitation. Large separating chambers can serve as main air venting points. Automatic control valves or other devices operating with small clearances require protection from pipe scale, gravel, and welding slag, which may readily pass through the pump and its protective separator. Individual fine mesh strainers may therefore be required ahead of each such device.

**Thermometers.** Thermometers or thermometer wells for temperature sensing and control calibration should be installed to assist the system operator in routine operation and troubleshooting. Permanent temperature sensors or thermometers, with the correct scale range and separate sockets, should be used at all points where temperature readings are regularly needed. Thermometer wells should be installed where readings will be needed only during start-up and infrequent trouble-shooting. If a central monitoring system is provided, a calibration well should be installed adjacent to each sensing point in insulated piping systems.

**Flexible Connectors and Expansion Compensation.** Flexible connectors are sometimes installed at pumps and machinery to reduce pipe stress. See Chapter 48 of the 2011 *ASHRAE Handbook—HVAC Applications* for vibration isolation information. Expansion, flexibility, and hanger and support information is in Chapter 46 of the 2012 *ASHRAE Handbook—HVAC Systems and Equipment Handbook*.

**Gage Cocks.** Gage cocks or quick-disconnect test ports should be installed at points requiring pressure readings. Gages permanently installed in the system will deteriorate because of vibration and pulsation and will, therefore, be unreliable. It is good practice to install gage cocks and provide the operator with several quality gages for diagnostic purposes.

**Insulation.** Insulation should be applied to minimize pipe thermal loss and to prevent condensation during chilled water operation (see Chapter 23 of the 2013 *ASHRAE Handbook—Fundamentals*). On chilled-water systems, special rigid metal sleeves or shields should be installed at all hanger and support points, and all valves should be provided with extended bonnets to allow for the full insulation thickness without interference with the valve operation.

**Condensate Drains.** Condensate drains from dehumidifying coils should be trapped and piped to an open-sight plumbing drain. Traps should be deep enough to overcome the air pressure differential between drain inlet and room, which ordinarily will not exceed 2 in. of water. Pipe should be noncorrosive and insulated to prevent moisture condensation. Depending on the quantity and temperature of condensate, plumbing drain lines may require insulation to prevent sweating.

**Common Pipe.** In compound (primary-secondary) pumping systems, the common pipe is used to dynamically decouple the two pumping circuits. Ideally, there is no pressure drop in this section of piping; however, in actual systems, it is recommended that this section of piping be a minimum of 10 diameters in length to reduce the likelihood of unwanted mixing resulting from velocity (kinetic) energy or turbulence.

## 13.4 Design Procedures

### 13.4.1 Preliminary Equipment Layout

**Flows in Mains and Laterals.** Regardless of the method used to determine the flow through each item of terminal equipment, the desired result should be listed in terms of mass flow on the preliminary plans or in a schedule of flow rates for the piping system. (In the design of small systems and chilled-water systems, the determination may be made in terms of volumetric flow).

In an equipment schedule or on the plans, starting from the most remote terminal and working toward the pump, progressively list the cumulative flow in each of the mains and branch circuits in the distribution system.

**Preliminary Pipe Sizing.** For each portion of the piping circuit, select a tentative pipe size from the unified flow chart in Chapter 22 of the 2013 *ASHRAE Handbook—Fundamentals*, using a value of pipe friction loss ranging from 0.75 to 4 ft per 100 ft (approximately 0.1 to 0.5 in./ft).

Residential piping size is often based on pump preselection using pipe sizing tables, which are available from the Hydronics Institute or from manufacturers.

**Preliminary Pressure Drop.** Using the preliminary pipe sizing indicated above, determine the pressure drop through each portion of the piping. The total pressure drop in the longest or highest head loss circuit determines the maximum pressure drop through the piping, including the terminals and control valves, that must be made available by the pump.

**Preliminary Pump Selection.** The preliminary selection should be based on the pump's ability to fulfill the determined capacity requirements. It should be selected at a point left of center on the pump curve and should not overload the motor. Because pressure drop in a flow system varies as the square of the flow rate, the flow variation between the nearest size of stock pump and an exact point selection will be relatively minor.

## 13.4.2 Final Pipe Sizing and Pressure Drop Determination

**Final Piping Layout.** Examine the overall piping layout to determine whether pipe sizes in some areas need to be readjusted. Several principal circuits should have approximately equal pressure drops so that excessive pressures are not needed to serve a small portion of the building.

Consider both the initial costs of the pump and piping system and the pump's power and energy requirement when determining final system friction loss. Lower heads and larger piping are more energy-efficient and are generally more economical when longer amortization periods are considered, especially in larger systems. However, in small systems such as in residences, it may be most economical to select the pump first and design the piping system to meet the available pressure. In all cases, adjust the piping system design and pump selection until the optimum design is found.

**Final Pressure Drop.** When the final piping layout has been established, determine the friction loss for each section of the piping system from the pressure drop charts (Chapter 22 of the 2013 *ASHRAE Handbook—Fundamentals*) for the mass flow rate in each portion of the piping system.

After calculating the friction loss at design flow for all sections of the piping system and all fittings, terminal units, and control valves, sum them for several of the longest piping circuits to determine the pressure against which the pump must operate at design flow.

**Final Pump Selection.** After completing the final pressure drop calculations, select the pump by plotting a system curve and pump curve and selecting the pump or pump assembly that operates closest to the calculated design point.

### 13.4.3 Freeze Prevention

All circulating water systems require precautions to prevent freezing, particularly in makeup air applications in cold climates (1) where coils are exposed to outdoor air at below-freezing temperatures, (2) where undrained chilled water coils are in the winter airstream, or (3) where piping passes through unheated spaces. Freezing will not occur as long as flow is maintained and the water is at least warm. Unfortunately, during extremely cold weather or in the event of a power failure, water flow and temperature cannot be guaranteed. Additionally, continuous pumping can be energy-intensive and cause system wear. The following are precautions to avoid flow stoppage or damage from freezing:

1. Select all load devices (such as preheat coils) that are subjected to outdoor air temperatures for constant flow, variable $\Delta t$ control.
2. Position the coil valves of all cooling coils with valve control that are dormant in winter months to the full-open position at those times.
3. If intermittent pump operation is used as an economy measure, use an automatic override to operate both chilled water and heating water pumps in below-freezing weather.
4. Select pump starters that automatically restart after power failure (i.e., maintain contact control).
5. Select non-overloading pumps.
6. Instruct operating personnel never to shut down pumps in subfreezing weather.
7. Do not use aquastats, which can stop a pump, in boiler circuits.
8. Avoid sluggish circulation, which may cause air binding or dirt deposit. Properly balance and clean systems. Provide proper air control or means to eliminate air.
9. Install low-temperature-detection thermostats that have phase-change capillaries wound in a serpentine pattern across the leaving face of the upstream coil.

In fan equipment handling outdoor air, take precautions to avoid stratification of air entering the coil. The best methods for proper mixing of indoor and outdoor air are the following:

1. Select dampers for pressure drops adequate to provide stable control of mixing, preferably with dampers installed several equivalent diameters upstream of the air-handling unit.
2. Design intake and approach duct systems to promote natural mixing.
3. Select heating coils with circuiting to allow parallel flow of air and water.

Freeze-up may still occur with any of these precautions. If an antifreeze solution is not used, water should circulate at all times. Valve-controlled elements should have low-limit thermostats, and sensing elements should be located to ensure accurate air temperature readings. Primary-secondary pumping of coils with three-way valve injection (as in Figure 13-19) is advantageous.

### 13.4.4 Antifreeze Solutions

In systems in danger of freeze-up, water solutions of ethylene glycol and propylene glycol are commonly used. Freeze protection may be needed (1) in snow-melting applications (see Chapter 51 of the 2011 *ASHRAE Handbook—HVAC Applications*); (2) in systems subjected to 100% outdoor air, where the methods outlined above may not provide absolute antifreeze protection; (3) in isolated parts or zones of a heating system where intermittent operation or long runs of exposed piping increase the danger of freezing; and (4) in process cooling applications requiring temperatures below 40°F. Although using ethylene glycol or propylene glycol is comparatively expensive and tends to create corrosion problems unless suitable inhibitors are used, it may be the only practical solution in many cases.

Solutions of triethylene glycol, as well as certain other heat transfer fluids, may also be used. However, ethylene glycol and propylene glycol are the most common substances used in hydronic systems because they are less costly and provide the most effective heat transfer.

**Heat Transfer and Flow.** Chapter 31 of the 2013 *ASHRAE Handbook—Fundamentals* presents density, specific heat, thermal conductivity, and viscosity of various aqueous solutions of ethylene glycol and propylene glycol.

System heat transfer rate is affected by relative density and specific heat according to the following equation:

$$q_w = 500 \, Q(\rho/\rho_w) c_p \, \Delta t \tag{13-18}$$

where

$q_w$ = total heat transfer rate, Btu/h
$Q$ = flow rate, gpm
$\rho$ = fluid density, lb/ft$^3$
$\rho_w$ = density of water at 60°F, lb/ft$^3$
$c_p$ = specific heat of fluid, Btu/lb·°F
$\Delta t$ = temperature increase or decrease, °F

**Effect on Heat Source or Chiller.** Generally, ethylene glycol solutions should not be used directly in a boiler because of the danger of chemical corrosion caused by glycol breakdown on direct heating surfaces. However, properly inhibited glycol solutions can be used in low-temperature water systems directly in the heating boiler if proper operation can be ensured. Automobile antifreeze solutions are not recommended because the silicate inhibitor can cause fouling, pump seal wear, fluid gelation, and reduced heat transfer. The area or zone requiring the antifreeze protection can be isolated with a separate heat exchanger or converter. Glycol solutions are used directly in water chillers in many cases.

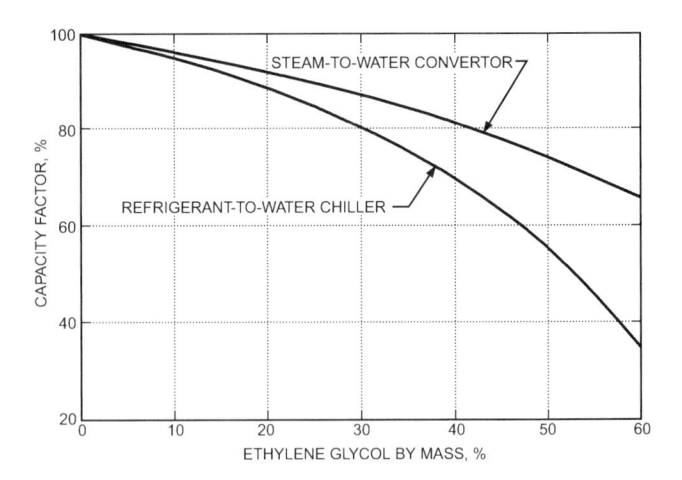

*Fig. 13-29 Example of Effect of Aqueous Ethylene Glycol Solutions on Heat Exchanger Output*
(Figure 42, Chapter 13, 2012 ASHRAE Handbook—HVAC Systems and Equipment)

Glycol solutions affect the output of a heat exchanger by changing the film coefficient of the surface contacting the solution. This change in film coefficient is caused primarily by viscosity changes. Figure 13-29 illustrates typical changes in output for two types of heat exchangers, Curve A for a steam-to-liquid converter and Curve B for a refrigerant-to-liquid chiller. The curves are plotted for one set of operating conditions only and reflect the change in ethylene glycol concentration as the only variable. Propylene glycol has a similar effect on heat exchanger output.

Because many other variables, such as liquid velocity, steam or refrigerant loading, temperature difference, and unit construction, affect the overall coefficient of a heat exchanger, designers should consult manufacturers' ratings when selecting such equipment. The curves indicate only the relative magnitude of these output changes.

**Effect on Terminal Units.** Because the effect of glycol on the capacity of terminal units may vary widely with temperature, the manufacturer's rating data should be consulted when selecting heating or cooling units in glycol systems.

**Effect on Pump Performance.** Centrifugal pump characteristics are affected to some degree by glycol solutions because of viscosity changes. Figure 13-30 shows these effects on pump capacity, head, and efficiency. Figures in Chapter 31 of the 2013 *ASHRAE Handbook—Fundamentals* plot the viscosity of aqueous ethylene glycol and propylene glycol. Centrifugal pump performance is normally cataloged for water at 60 to 80°F. Hence, absolute viscosity effects below 1.1 centipoise can safely be ignored as far as pump performance is concerned. In intermittently operated systems, such as snow-melting applications, viscosity effects at start-up may decrease flow enough to slow pickup.

**Effect on Piping Pressure Loss.** The friction loss in piping also varies with viscosity changes. Figure 13-31 gives cor-

# Hydronic Heating and Cooling System Design

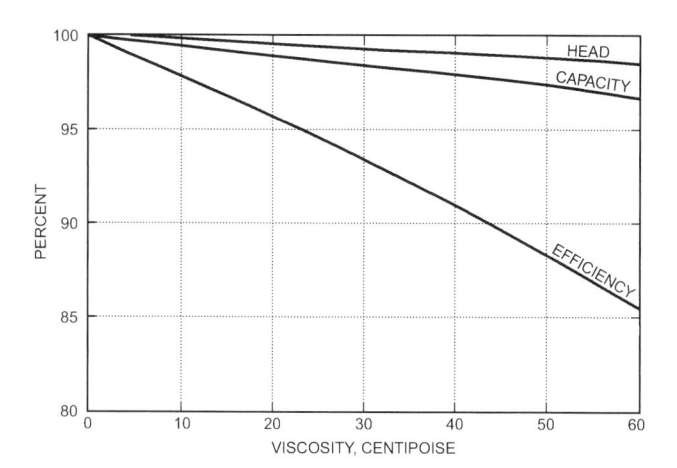

*Fig. 13-30 Effect of Viscosity on Pump Characteristics*
(Figure 43, Chapter 13, 2012 ASHRAE Handbook—HVAC Systems and Equipment)

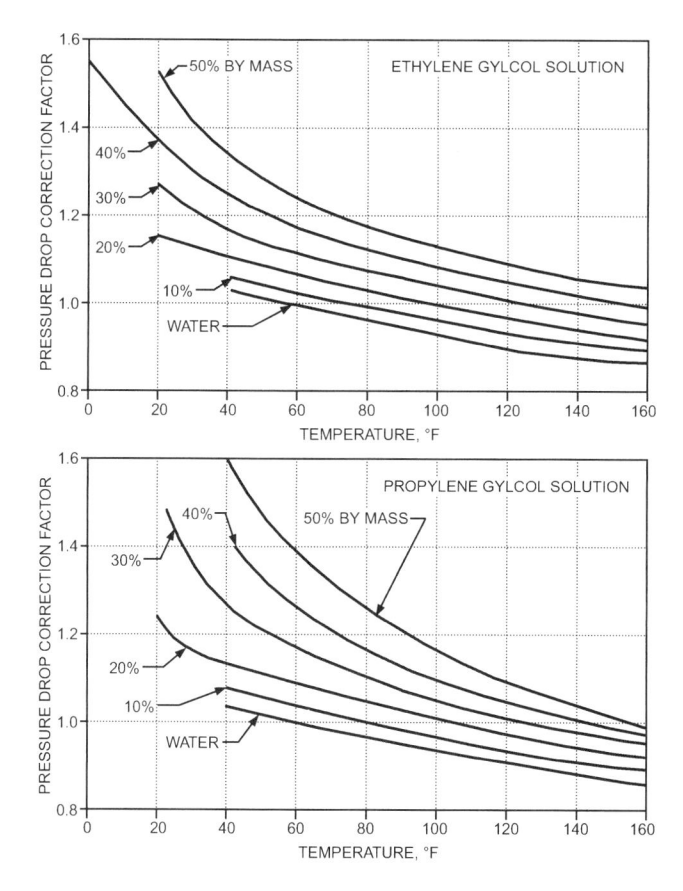

*Fig. 13-31 Pressure Drop Correction for Glycol Solutions*
(Figure 44, Chapter 13, 2012 ASHRAE Handbook—HVAC Systems and Equipment)

rection factors for various ethylene glycol and propylene glycol solutions. These factors are applied to the calculated pressure loss for water. No correction is needed for ethylene glycol and propylene glycol solutions above 160°F.

**Installation and Maintenance.** Because glycol solutions are comparatively expensive, the smallest possible concentrations to produce the desired antifreeze properties should be used. The total water content of the system should be calculated carefully to determine the required amount of glycol (Craig et al. 1993). The solution can be mixed outside the system in drums or barrels and then pumped in. Air vents should be watched during filling to prevent loss of solution. The system and the cold water supply should not be permanently connected, so automatic fill valves are usually not used.

Ethylene glycol and propylene glycol must include an inhibitor to help prevent corrosion. Solutions should be checked regularly using a suitable refractometer to determine glycol concentration. The following precautions regarding the use of inhibited glycol solutions should be taken to extend their service life and to preserve equipment:

1. Before injecting the glycol solution, thoroughly clean and flush the system.
2. Use waters that are soft and low in chloride and sulfate ions to prepare the solution whenever possible.
3. Limit the maximum operating temperature to 250°F in a closed hydronic system. In a heat exchanger, limit glycol film temperatures to 300 to 350°F (steam pressures 120 psi or less) to prevent deterioration of the solution.
4. Check the concentration of inhibitor regularly, following procedures recommended by the glycol manufacturer.

## 13.5 Problems

**13.1** What is the maximum temperature at which a heating water system can be operated if the boiler (hot water generator) is rated as low pressure by the ASME Boiler and Pressure Vessel Code?

**13.2** Sketch the fundamental components for a chilled water system with a single load and source and a capacity of 100 tons of cooling.
(a) What is the water circulation rate (gpm) required if the temperature range of the water is 12°F.
(b) If the head loss in the system is 60 feet, and the pump is 80% efficient, what is the pump horsepower? Motor size?
(c) If the motor is 90% efficient and it operates for one-third of the total hours in the year, what is the annual energy consumption of the pump?

**13.3** Calculate the size of the expansion tank for a hot water heating system of 1,200,000 Btu/h heating capacity if the tank is a closed tank with an air/water interface and the following system parameters are known:

| | |
|---|---|
| Supply water temperature | 210°F |
| Ambient temperature | 60°F |
| Fill pressure (at tank) | 30 psig |
| Maximum operating pressure (at tank) | 35 psig |
| System water volume | 6,000 gallons |
| Steel piping system material | |

**13.4** What size diaphragm tank would be required for the above system?

**13.5** In a given chilled water system, the pump head required at 640 gpm is 80 ft.
(a) What is the system constant, $C_s$?
(b) Plot the system curve from 0 to 800 gpm.

**13.6** In a chilled water system, the pump is located in a basement equipment room with the expansion tank connected to the pump suction. The pump is the lowest point in the system and the highest point is a pipe in the penthouse, which is 115 feet above the pump. The dynamic head losses in the system are:

| | |
|---|---|
| Piping and fittings | 30 ft |
| Chiller | 20 ft |
| Control valve | 10 ft |
| Cooling coil | 10 ft |

When the system is filled (at 95°F ambient temperature) it is desired to have a pressure of 10 psig at the highest point in the system, which will reduce to 5 psig when the water temperature reduces to 45°F.
(a) What operating pressures ($p_1$, $p_2$) should the expansion tank be designed for?
(b) What pump head is required?
(c) With the pump off and a cold (45°F) system, what is the pressure at the pump suction? The pump discharge?
(d) With the pump on and a cold (45°F) system, what is the pressure at the pump suction? The pump discharge?

**13.7** In your own words, explain the difference between a three-way control valve and a two-way control valve as they affect the hydraulics of the system.

**13.8** A control valve is to be sized for a cooling coil with a capacity of 30 tons of cooling. The water temperature entering the coil is at 44°F with a 12°F $\Delta t$. It is determined that the valve should have a pressure drop of 5 psi. What is the required $C_v$ of the valve?

**13.9** A section of 1in. steel pipe in a chilled water system at 50 psig is in a pipe chase and is located between two service valves. With a cold system (45°F) the section is isolated by closing off the two service valves. If the chase is at a temperature of 95°F and the pipe reaches thermal equilibrium with the chase, what will the final pressure in the pipe be?

## 13.6  Bibliography

ASHRAE. 2007. *2007 ASHRAE Handbook—HVAC Applications*.

ASHRAE. 2008. *2008 ASHRAE Handbook—HVAC Systems and Equipment*.

ASHRAE. 2009. *2009 ASHRAE Handbook—Fundamentals*.

ASME. 1995. Boiler and Pressure Vessel Codes. American Society of Mechanical Engineers, New York.

Carlson, G.F. 1981a. The design influence of air on hydronic systems. *ASHRAE Transactions* 87(1):1293-1300.

Coad, W.J. 1980a. Expansion tanks. *Heating/Piping/Air Conditioning* (May).

Coad, W.J. 1980b. Air in hydronic systems. *Heating/Piping/Air Conditioning* (July).

Coad, W.J. 1985. Variable flow in hydronic systems for improved stability, simplicity, and energy economics. *ASHRAE Transactions* 91(1B):224-237.

Coad, W.J. 1985. Variable flow in hydronic systems for improved stability, simplicity, and energy economics. *ASHRAE Transactions* 91(1B):224-237.

Craig, N.C., B.W. Jones, and D.L. Fenton. 1993. Glycol concentration requirements for freeze burst protection. *ASHRAE Transactions* 99(2):200-209.

Himmelblau, D.M. 1960. Solubilities of inert gases in water. *Journal of Chemical and Engineering Data* 5(1).

Hull, R.E. 1981. Effect of air on hydraulic performance of the HVAC system. *ASHRAE Transactions* 87(1):1301-1325.

Lockhart, H.A. and G.E. Carlson. 1953. Compression tank selection for hot water heating systems. *ASHVE Journal* 25(4):132-139. Also in *ASHVE Transactions* 59:55-76.

Morabito, B.R. 1960. How higher cooling coil differentials affect system economics. *ASHRAE Journal* 2(8):60.

Pierce, J.D. 1963. Application of fin tube radiation to modern hot water heating systems. *ASHRAE Journal* 5(2):72.

Pompei, E. 1981. Air in hydronic systems: How Henry's law tells us what happens. *ASHRAE Transactions* 87(1):1326-1342.

Stewart, W.E. and C.L. Dona. 1987. Water flow rate limitations. *ASHRAE Transactions* 93(2):811-825.

# Chapter 14

# UNITARY AND ROOM AIR CONDITIONERS

This chapter discusses the availability of the various types of unitary units and room air conditioners. Additional details on this equipment can be found in Chapters 49 and 50 in the 2012 *ASHRAE Handbook—HVAC Systems and Equipment*.

## 14.1 Unitary Air Conditioners

Unitary air-conditioning equipment is an assembly of factory-matched refrigerant cycle devices for inclusion as components in field-designed air-conditioning systems. Some of the many types of unitary air conditioners available include the following characteristics:

**Arrangement:** Single package or split system (i.e., an indoor evaporator and blower and a separate, usually outdoor, compressor and condenser unit).
**Heat rejection:** Air cooled, water cooled, evaporative condenser.
**Unit exterior:** Decorative for in-space applications, functional for equipment room and ducts, weatherproofed for outdoors.
**Placement:** Floor standing, wall mounted, ceiling suspended, roof mounted.
**Indoor air:** Vertical upflow or downflow, horizontal flow, 90° and 180° turns, with fan, or for use with forced-air furnaces.
**Locations:** *Indoor*—exposed with plenums or furred in ductwork concealed in closets, attic, crawlspaces, basements, garages, utility rooms, or equipment rooms. Wall, window, or transom mounted.
*Outdoor*—rooftop, wall mounted, or on ground
**Heat:** May be combined with electric heat, gas heat, hot water, or steam coil.

Unitary air conditioners, in contrast to room air conditioners, include fans capable of operating with ductwork, although some units may be applied with supply air plenums. Heat pumps are also offered in many of the same types and capacities as unitary air conditioners. Packaged reciprocating and centrifugal water chillers are considered to be unitary air conditioners, particularly when applied with unitary chilled-water blower coil units.

Single-package air conditioners are depicted in Figures 14-1 through 14-4. Split systems and condensing units with coils and with blower coil units are shown in Figures 14-5 through 14-7.

The many combinations of coil configurations, evaporator temperatures, air-handling arrangements, refrigerating capacities, and variations thereof that are available in central systems are seldom possible with unitary systems. Consequently, a higher level of design ingenuity and performance the many smaller interlocked and independent systems have is required to develop acceptable system performance from unitary equipment.

Unitary equipment tends to serve zoned systems, with each zone served by its own unit. The room conditioner or packaged terminal air conditioner (PTAC) carries this concept to relatively small rooms.

For large single spaces where central systems are at their best advantage, multiple central systems are often advantageous because as load sources move within the larger space, more flexibility than one central system.

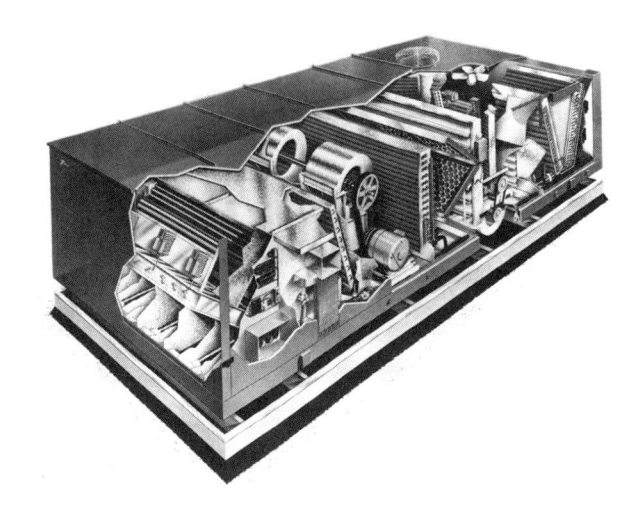

*Fig. 14-1 Typical Rooftop Air-Cooled Single-Package Air Conditioner*
(Figure 1, Chapter 49, 2012 ASHRAE Handbook—HVAC Systems and Equipment)

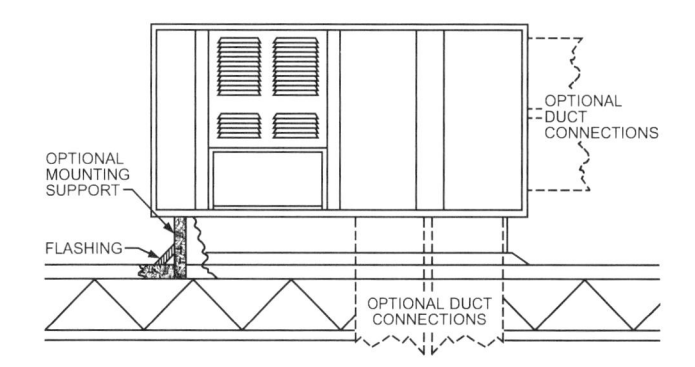

*Fig. 14-2 Rooftop Installation of a Single-Package Unit*
(Figure 5, Chapter 49, 2012 ASHRAE Handbook—HVAC Systems and Equipment)

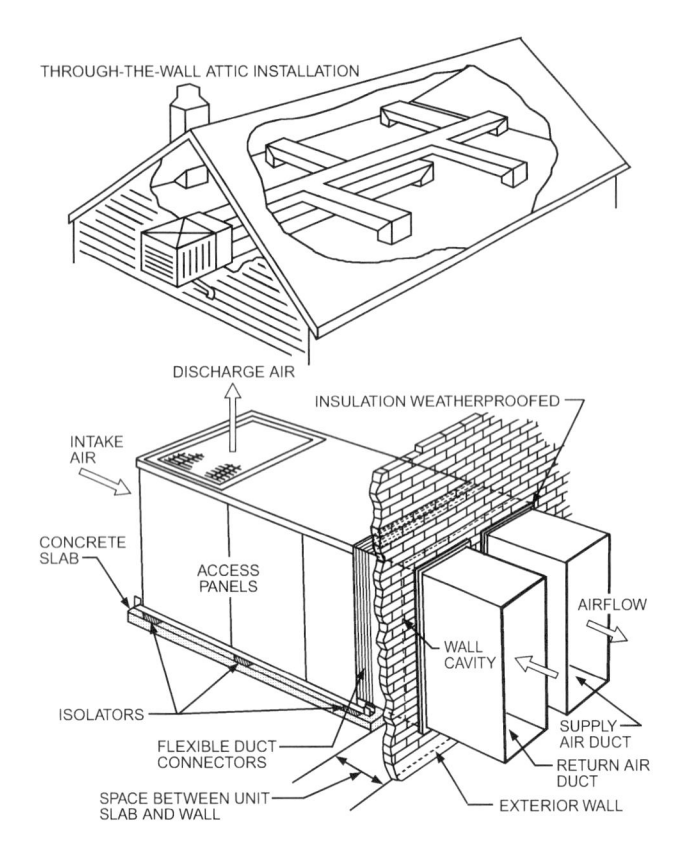

*Fig. 14-3 Typical Through-the-Wall Air-Cooled Single-Package Unit*
(Figure 7, Chapter 49, 2012 ASHRAE Handbook—HVAC Systems and Equipment)

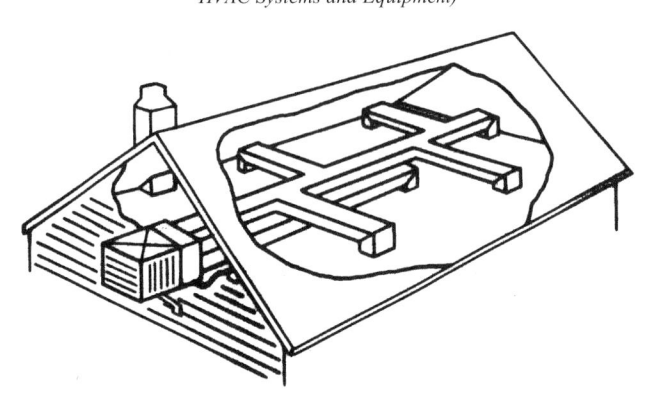

*Fig. 14-4 Through-the-Wall Installation of a Single-Package Unit*
(Figure 7, Chapter 49, 2012 ASHRAE Handbook—HVAC Systems and Equipment)

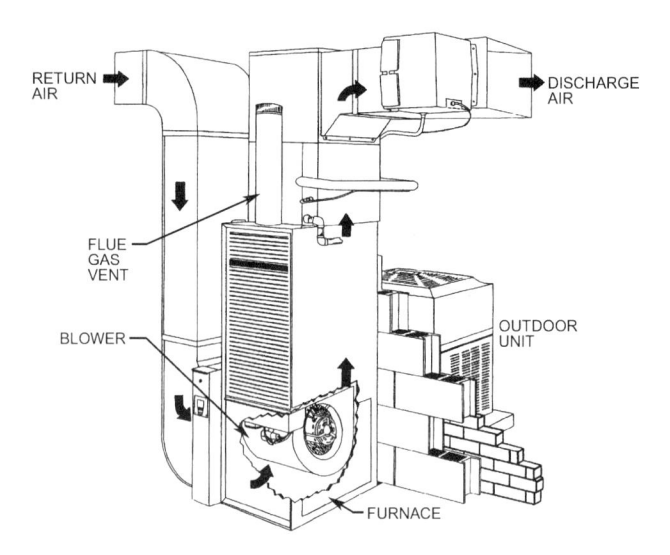

*Fig. 14-5 Residential Installation of Split-System Air-Cooled Condensing Unit with Indoor Coil and Upflow Furnace*
(Figure 8, Chapter 49, 2012 ASHRAE Handbook—HVAC Systems and Equipment)

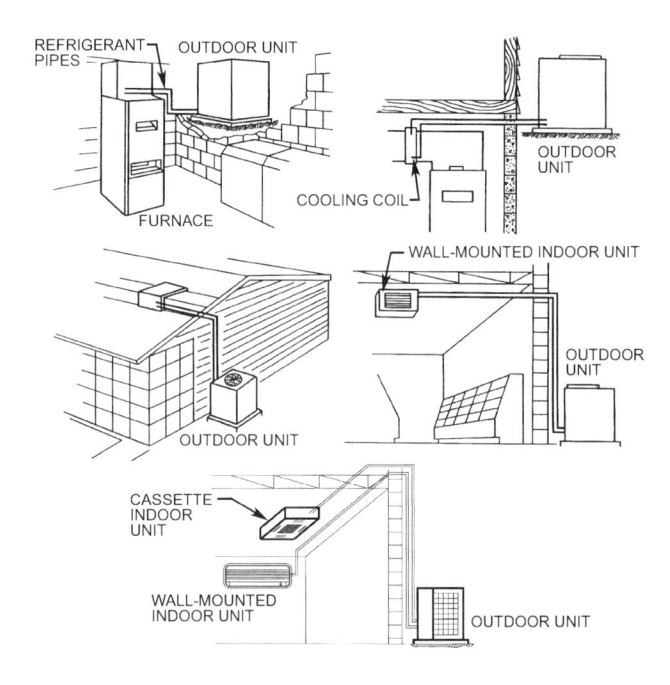

*Fig. 14-6 Outdoor Installations of Split-System Air-Cooled Condensing Units with Coil and Upflow Furnace or with Indoor Blower Coils*
(Figure 9, Chapter 49, 2012 ASHRAE Handbook—HVAC Systems and Equipment)

However, rooms with less than 0.5 ton (2 kW) or more than 25 ton (100 kW) cooling loads are seldom conditioned by their own single unit in multiple-unit systems. Multiple-unit systems may provide the following advantages over central system alternatives:

- Simple and inexpensive individual room control
- Individual air distribution for each room, usually with convenient and simple adjustment
- Heating and cooling capability at all times, independent of the mode of operation of other spaces in the building
- Consistent performance assured by manufacturer-matched components
- Generally have published certified ratings and performance data
- Single source of accountability because manufacturer assembles components
- Manufacturer instructions and multiple-unit nature simplify and systematize installation through repetition of tasks

# Unitary and Room Air Conditioners

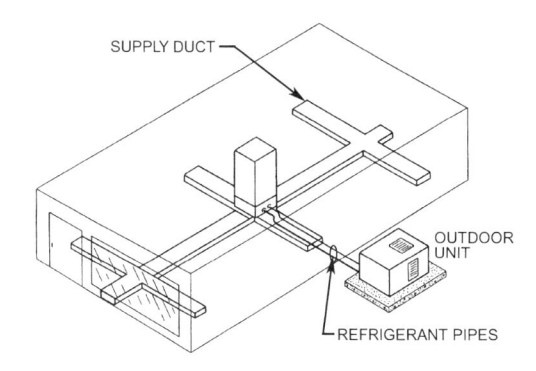

*Fig. 14-7 Outdoor Installation of Split-System Air-Cooled Condensing Unit with Indoor Coil and Downflow Furnace*
(Figure 10, Chapter 49, 2012 ASHRAE Handbook—HVAC Systems and Equipment)

- Only one terminal zone or conditioner is affected in the event of equipment malfunction
- Often saves some space
- Usually quick availability and installation are possible
- Often lower initial cost
- Responsibility for performance of complete package(s) rests with one manufacturer and its agents who provide information on application, installation, maintenance, and service
- Equipment serving spaces that become vacant can be turned off locally or from a central point without affecting occupied spaces

Multiple-unit systems may have the following disadvantages:

- Limited performance options are available because airflow and cooling coil and condenser sizing are fixed.
- Not generally suited for effective humidity control, except when using special purpose equipment such as packaged units for a computer room. Poor humidity control can result in mold and mildew growth within the space.
- Energy use may be greater than for central systems if efficiency of the unitary equipment is less than that of the combined central system components.
- Winter cooling by outdoor air economizers is not always available.
- Air distribution control may be limited.
- Operating sound levels can be high.
- Ventilation capabilities are limited by equipment design.
- Engineered ventilation and humidity control must be provided by a supplementary system, usually a dedicated outdoor air system (DOAS).
- Overall appearance can be unappealing.
- Air filtration options are limited.
- Maintenance may be difficult because of the many pieces of equipment and their locations.

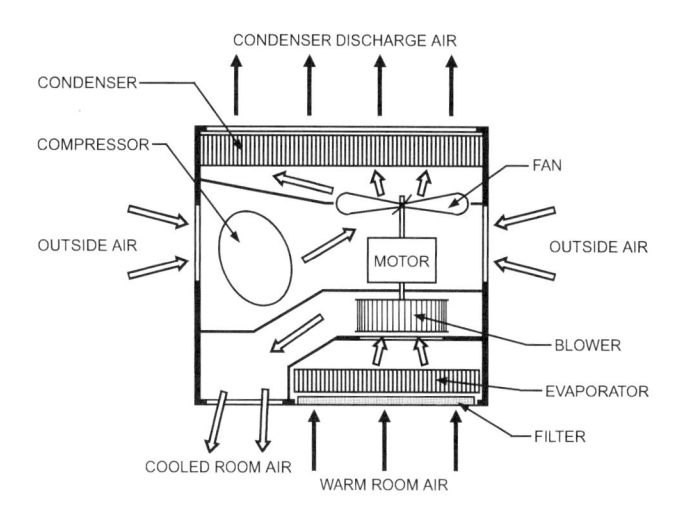

*Fig. 14-8 Schematic View of Window Air Conditioner*
(Figure 1, Chapter 50, 2012 ASHRAE Handbook—HVAC Systems and Equipment)

## 14.2 Combined Unitary and Dedicated Outdoor Air Systems

Combining some type of unitary system with a dedicated outdoor air system (DOAS) can provide very good comfort conditions at a reasonable cost and low energy consumption (see Chapter 12). The DOAS is designed to provide the ventilation, humidity control, and a high level of filtration of the outdoor air, and control of room temperature and air motion is assigned to the unitary room unit. In some cases, with a well-designed supply of DOAS air, that unit can also provide the room air motion, or at least supplement it so the fan of the unitary system can be cycled with the need for cooling or heating.

## 14.3 Window Air Conditioners

A window air conditioner is an encased assembly designed as a unit primarily for mounting in a window. These units are designed for comfort cooling and provide delivery of conditioned air to the room either without ducts or with very short ducts up to a maximum of about 48 in. (1.2 m).

A window air conditioner cools, dehumidifies, filters or cleans, and circulates room air. Ventilation may also be provided by introducing outdoor air into the room and/or by exhausting room air to the outside. Some conditioners provide heating by reverse cycle (heat pump) operation or by electric resistance elements.

A typical window air conditioner is shown diagrammatically in Figure 14-8. Warm room air passes over the cooling coil, giving up its sensible and latent heat. The conditioned air is then circulated in the room by a fan or blower.

The cooling and heating capacities of window air conditioners are always measured and stated in terms of Btu/h (W). A wide range of capacities is available [from approximately 4000 to 36,000 Btu/h (1.2 to 10.5 kW)].

The design of a window air conditioner is usually based on one or more of the following criteria, any one of which automatically limits the freedom of the designer in overall system design:

- Lowest initial cost
- Lowest operating cost (highest efficiency)
- Low sound level
- Physical chassis size
- An unusual chassis shape (minimal depth, height, etc.)
- An amperage limitation (7.5 A, 12 A, etc.)
- Weight

The basic design is a carefully selected group of components consisting of an evaporator, a condenser, a compressor, one or more fan motors, blower wheels for evaporator and condenser airflow, and an expansion device, usually consisting of one or more capillary tubes.

## 14.4 Through-the-Wall Conditioner System

A through-the-wall system is an air-cooled room air conditioner designed for mounting through the wall and normally capable of providing both heating and cooling. Design and manufacturing specifications range from appliance grade to heavy-duty commercial grade. The latter is called *packaged terminal air conditioner* (PTAC) and is defined as such by ARI in the "Packaged Terminal Air Conditioners" subsection of *Air Conditioning Heat Transfer Products*; all others are covered by AHAM *Standard* CN-l.

**System Concept and Description**. The through-the-wall concept incorporates a complete air-cooled refrigeration and air-handling system in an individual package, using space normally occupied by the building wall for equipment, with the remainder projecting inside the room.

Each packaged terminal air conditioner has a self-contained, direct-expansion cooling system, heating coil (electric, hot water, or steam), and packaged controls. Two general configurations used consist of (1) wall box, heat section, room cabinet, outdoor louver, and cooling chassis (Figure 14-9) and (2) a combination wall sleeve and room cabinet, combination heating and cooling chassis, and outdoor louver (Figure 14-10).

The exterior louver is installed flush (or nearly so) with the outside wall of the building and receives a variety of architectural treatments, such as emphasizing the existing louvers and using them as aesthetic highlights, hiding them completely with solid or pierced wall coverings, or designing the building so that the louvers blend into the overall exterior.

**Advantages**. The initial cost of the through-the-wall system in multiroom applications is considerably less than central systems adapted to simultaneously heat or cool each room under control of the room occupants. The cost differ-

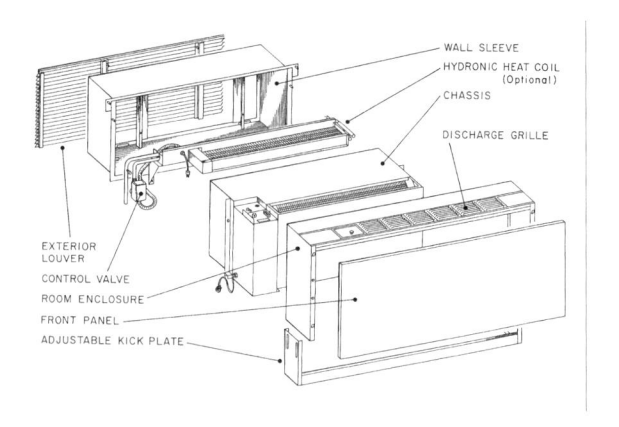

*Fig. 14-9 Packaged Terminal Air Conditioner with Separate Heat Section and Cooling Chassis*

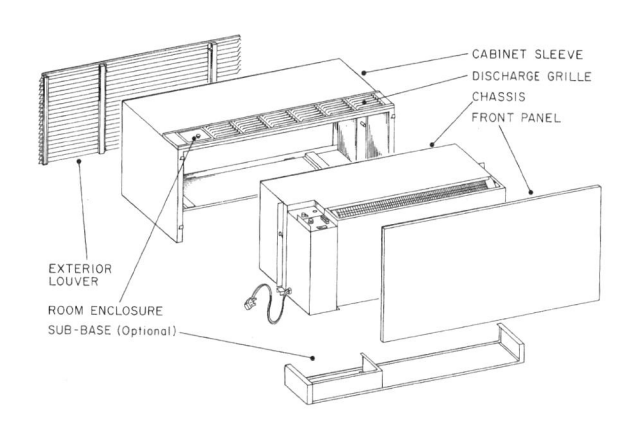

*Fig. 14-10 Packaged Terminal Air Conditioner with Combination Heating and Cooling Chassis*

ences may run as high as 20% to 50%, depending on the design and components.

Because a through-the-wall system has no system auxiliaries, the energy consumption may be lower than for central systems. For this reason, economical comparisons between central systems and through-the-wall conditioners should include all system components, including fans, pumps, and heat rejection devices. Keep in mind that fan and motor efficiencies will be significantly lower for the through-the-wall air conditioner.

A through-the-wall system requires less space because it requires no ductwork or equipment rooms. This space savings can range from 5% to 15% of the total building space when the savings are considered on a floor-by-floor basis.

**Limitations**. The through-the-wall system is limited to multizone systems and generally cannot be used economically in large spaces requiring more than three units per zone. However, where the packaged terminal air-conditioner system is coordinated with a well-designed core system, it can be economically used for large office areas while allowing for maximum flexibility in moving partitions. Through-the-wall units are more likely to be limited by their ability to throw air across the room.

Current products do not include individual conditioner humidifier systems. Humidification and controlled dehumidification can be achieved with a through-the-wall system but must be done through a separate dedicated outdoor air system. The through-the-wall system should not be used where there are high sensible load requirements coming from a concentration of heat-producing equipment, such as is used in computer server rooms or radio and television stations.

In commercial or public buildings through-the-wall or PTAC units should not be used for ventilation. Ventilation should be provided with a supplementary DOAS.

**Applications.** Through-the-wall systems are often applied in multiple zone applications. The system lends itself to both low- and high-rise buildings. The system is most generally applied in

- Office buildings
- Motels and hotels
- Apartments and dormitories
- Schools and other educational buildings
- Nursing homes

This system is also applicable for renovating existing buildings because all or part of the existing heating system can be used. This equipment could cause less disruption and construction-forced sacrifice of rentable space than alternative systems. The result is an automatically controlled heating system and a self-contained cooling system, except for the companion DOAS. The major disadvantage of most PTAC systems is the relatively high sound level in the occupied space.

## 14.5 Typical Performance

Specially constructed equipment cannot be justified for small commercial and residential applications. Furthermore, these applications generally have a higher sensible heat factor (SHF), so dehumidification is not as critical as in large commercial buildings. Therefore, the equipment is manufactured to operate at or near one set of conditions.

For example, typical residential and light commercial cooling equipment operates with a coil SHF of 0.75 to 0.8 with the air entering the coil at about 80°F (27°C) dry-bulb and 67°F (19°C) wet-bulb temperature. This equipment usually has a capacity of less than 10 tons (35 kW). When the peak cooling load and latent heat requirements are appropriate, this less expensive type of equipment can be considered.

Selected equipment should be within the range of 95 to 115% of the peak cooling load. The air quantity is specified by the manufacturer for each unit and is about 400 cfm/ton (50 L/s per kW). The total air quantity is then divided among the various rooms according to the cooling load of each room. Typical performance data for residential and light commercial cooling equipment are listed in Table 14-1.

**Example 14.1** An air-conditioning unit selected for a residence has a rated total cooling capacity of 36,000 Btu/h and a sensible cooling capacity of 27,000 Btu/h. The

**Table 14-1** Typical Residential or Light Commercial Cooling Coil Performance Data (Split System)

| Capacity | | Airflow | | Coil Pressure Loss | |
|---|---|---|---|---|---|
| Btu/h | kW | cfm | m³/s | in. of water | Pa |
| 18,000 | 5.3 | 600 | 0.28 | 0.18 | 45 |
| 24,000 | 7.0 | 800 | 0.38 | 0.30 | 75 |
| 30,000 | 8.8 | 1050 | 0.50 | 0.13 | 32 |
| 41,000 | 10.3 | 1270 | 0.60 | 0.20 | 50 |
| 48,000 | 14.1 | 1750 | 0.83 | 0.25 | 62 |
| 59,000 | 17.3 | 2140 | 1.01 | 0.30 | 75 |

manufacturer also lists a SEER of 13 for this unit. The unit is expected to operate for 1900 hours during each cooling season.

a) Determine the latent cooling capacity.
b) What is the sensible heat ratio for the unit?
c) What is the expected energy use per cooling season?

**Solution:**

a) $Q_{latent} = Q_{Total} - Q_{sensible}$
$Q_{latent} = 36,000 - 27,000 = 9000$ Btu/h

b) Sensible Heat Ratio (SHR) = $\dfrac{Q_{sensible}}{Q_{Total}}$

$SHR = \dfrac{27,000}{36,000} = 0.75$

c) $\dfrac{36,000 \text{ Btu/h} \times 1900 \text{ h}}{13 \text{ Btu/h} \cdot \text{W} \times 1000 \text{ W/kW}} = 5262$ kWh

## 14.6 Minisplits, Multisplits, and Variable-Refrigerant-Flow (VRF) Systems

A minisplit is a packaged air-conditioning (cooling) system that is supplied as two components. The indoor component includes and evaporator coil, a blower or fan, and an expansion device. It is generally finished for installing in the finished space, and is of small capacity (0.5 to 1 ton [1.75 to 3.5 kW]). The outdoor unit is matched in capacity to the indoor unit, and contains the compressor, condenser, and control package.

A multisplit system uses indoor evaporator-blower units similar to the minisplit, but will connect several of the indoor units to one outdoor condensing (compressor/condenser) unit, with various capacity ranges up to about 8 tons (28 kW).

A variable-refrigerant-flow (VRF) system typically consists of a condensing section housing compressor(s) and condenser heat exchanger interconnected by a single set of refrigerant piping to multiple indoor direct-expansion (DX) evaporator fan-coil units. Thirty or more DX fan coil units can be connected to a single condensing section, depending on system design, and with capacity ranging from 0.5 to 8 tons (1.75 to 28 kW).

The DX fan coils are constant air volume, but use variable refrigerant flow through an electronic expansion valve. The electronic expansion valve reacts to several temperature-sensing devices such as return air, inlet and outlet refrigerant temperatures, or suction pressure. The electronic expansion valve modulates to maintain the desired set point.

### 14.6.1 Application

VRF systems are most commonly air-to-air, but are also available in a water-source (water-to-refrigerant) configuration. They can be configured for simultaneous heating and cooling operation, i.e., operating on a heat pump cycle with liquid, suction, and hot gas lines to each unit that contains the changeover valve assembly with some indoor fan coil units operating in heating and some in cooling, depending on requirements of each building zone.

Indoor units are typically direct-expansion evaporators using individual electronic expansion devices and dedicated microprocessor controls for individual control. Each indoor unit can be controlled by individual thermostat. The outdoor unit may connect several indoor evaporator or heat pump units with capacities 130% or more than the outdoor condensing unit capacity.

### 14.6.2 Categories

VRF equipment is divided into three general categories: residential, light commercial, and applied. Residential equipment is single-phase unitary equipment with a cooling capacity of 65,000 Btu/h (19 kW) or less. Light commercial equipment is generally three-phase, with cooling capacity greater than 65,000 Btu/h, (19 kW) and is designed for small businesses and commercial properties. Applied equipment has cooling capacity higher than 135,000 Btu/h (40 kW) and is designed for large commercial buildings.

### 14.6.3 Refrigerant Circuit and Components

VRF heat pump systems use a two-pipe (liquid and suction gas) system; simultaneous heat and cool systems use the same system, as well as a hot gas line and flow device that determines the proper routing of refrigerant gas to a particular indoor unit.

VRF systems use a sophisticated refrigerant circuit that monitors mass flow, oil flow, and balance to ensure optimum performance. This is accomplished in unison with variable-speed compressors and condenser fan motors. Both of these components adjust their frequency in reaction to changing mass flow conditions and refrigerant operating pressures and temperatures. A dedicated microprocessor continuously monitors and controls these key components to ensure proper refrigerant is delivered to each indoor unit in cooling or heating.

### 14.6.4 Heating and Defrost Operation

In heating mode, VRF systems typically must defrost like any mechanical heat pump, using reverse cycle valves to temporarily operate the outdoor coil in cooling mode. Oil return and balance with the refrigerant circuit is managed by the microprocessor to ensure that any oil entrained in the low side of the system is brought back to the high side by increasing the refrigerant velocity using a high-frequency operation performed automatically based on hours of operation.

## 14.7 Water-Source Heat Pumps

A water-source heat pump (WSHP) is a single-package reverse-cycle heat pump that uses water as the heat source for heating and as the heat sink for cooling. The water supply may be a recirculating closed loop, a well, a lake, or a stream. Water for closed-loop heat pumps is usually circulated at 2 to 3 gpm per ton (0.04 to 0.05 L/s per kW) of cooling capacity. **A groundwater heat pump (GWHP)** can operate with considerably less water flow. The main components of a WSHP refrigeration system are a compressor, refrigerant-to-water heat exchanger, refrigerant-to-air heat exchanger, refrigerant expansion devices, and refrigerant-reversing valve.

Designs of packaged WSHPs range from horizontal units located primarily above the ceiling or on the roof, to vertical units usually located in basements or equipment rooms, to

*Horizontal*

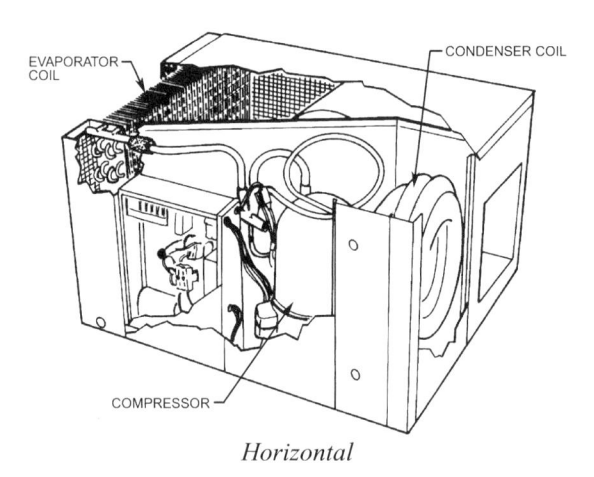

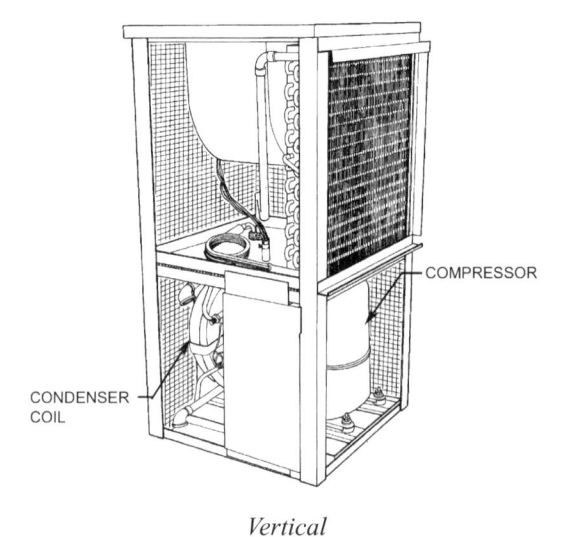

*Vertical*

*Fig. 14-11 Typical Arrangements of Water-Source Heat Pump*
*(Figures 14 and 15, Chapter 49, 2012 ASHRAE Handbook—HVAC Systems and Equipment)*

console units located in the conditioned space. Figure 14-11 illustrates typical designs.

**Systems.** WHSPs are used in a variety of systems, such as:

- Water-loop heat pump systems
- Groundwater heat pump systems
- Closed-loop surface-water heat pump systems
- Surface-water heat pump systems
- Ground-coupled heat pump systems

A **water-loop heat pump (WLHP)** uses a circulating water loop as the heat source and heat sink. When loop water temperature exceeds a certain level during cooling, a cooling tower dissipates heat from the water loop into the atmosphere. When loop water temperature drops below a prescribed level during heating, heat is added to the circulating loop water, usually with a boiler. In multiple-unit installations, some heat pumps may operate in cooling mode while others operate in heating, and controls are needed to keep loop water temperature within the prescribed limits. In commercial applications water-loop heat pumps should be used in conjunction with a dedicated outdoor air system, which provides the ventilation and humidity control.

A **groundwater heat pump (GWHP)**, sometimes referred to as a **geothermal heat pump**, should more accurately be referred to as a ground-coupled heat pump. When installing water pipes or refrigerant piping to serve as a heat exchanger with the ground above the water table, on the warm cycle (building cooling), the earth tends to shrink away from the warm pipes, forming an air space between the pipes and the earth. The ground-coupled heat pump utilizes a deep water well (usually 200 to 400 ft deep). Into this well is inserted a pipe loop: a supply and return line with a U-bend at the bottom. The piping is then encased in a heat transfer grout through which it transfers heat to or from the ground as the season requires.

Installing this type of system requires detailed knowledge of the climate; site; soil temperature, moisture content, and thermal characteristics; and performance, design, and installation of water-to-earth heat exchangers.

**Entering Water Temperatures.** These various water sources provide a wide range of entering water temperatures to WSHPs. Entering water temperatures vary not only by water source but also by climate and time of year. Because of the wide range of entering water or brine temperatures encountered, it is not feasible to design a universal packaged product that can handle the full range of possibilities effectively. Therefore, WSHPs are rated for performance at a number of standard rating conditions.

## 14.8 Problems

**14.1** An air-cooled packaged air conditioning unit with a hot water heating coil is to be used to condition a small office suite in a high-rise office building. The unit has a total cooling capacity of three tons of refrigeration, and the power requirement to the compressor is 1 kW per ton of cooling.

How many cfm of air must be brought into the condenser from an ambient outdoor temperature of 95°F db and 78°F wb if the condensing temperature is to be 115°F with a 10°F approach to the leaving air temperature?

**14.2** If the ductwork supplying the air to and from the condenser section in Problem 14.1 were sized for a velocity of 800 ft/min, what would be the cross-sectional area of the ductwork?

a. From the outdoors to the condenser?

b. From the condenser back to the outdoors?

**14.3** In passing through the condenser coil, the air would be heated a a constant humidity ratio. Air at 95°F db and 78°F wb ($w$ = 117.49 gr/lb) heated to 115°F db has a final specific volume ($v$) of 14.85 ft³/lb.

**14.4** If the packaged air-conditioning unit of Problem 14.1 were provided with a water-cooled condenser instead of an air-cooled unit, and 1) the water was supplied at 85°F, 2) the leaving water temperature was 95°F, and 3) the condensing temperature was 105°F, what would be

a. The Carnot COP between 40°F suction temperature and the 90°F condensing temperature?

b. The Carnot COP between the 40°F suction temperature and the 105°F condensing temperature of the Problem 14.1 air-cooled unit?

**14.5** Assuming that the actual power requirement for the cooling cycles of Problems 14.1 and 14.3 were proportioned in the same relationship as the Carnot COPs of Problem 14.3, what would be the kW per ton for the water-cooled unit of Problem 14.3?

**14.6** How many gallons per minute of water would be required for the water-cooled unit of Problem 14.3?

## 14.1 Bibliography

ASHRAE. 2012. Chapter 2, Decentralized Cooling and Heating. 2012 *ASHRAE Handbook—HVAC Systems and Equipment*.

ASHRAE. 2012. Chapter 49, Unitary Air Conditioners and Unitary Heat Pumps. *2012 ASHRAE Handbook—HVAC Systems and Equipment*.

ASHRAE. 2012. Chapter 50, Room Air Conditioners, Packaged Terminal Air Conditioners, and Dehumidifiers. *2012 ASHRAE Handbook—HVAC Systems and Equipment*.

# Chapter 15

# PANEL HEATING AND COOLING SYSTEMS

This chapter discusses the principles and equipment available for panel heating and cooling systems. Additional details can be found in Chapter 6, "Panel Heating and Cooling," in the 2012 *ASHRAE Handbook—HVAC Systems and Equipment*.

## 15.1 General

Radiant panel systems combine temperature control of room surfaces with central air conditioning. Radiant surfaces may be located in the floor, walls, or ceiling, and the temperature is maintained by circulating water or air or by electric resistance. Where heating-cooling panel systems are used in commercial and institutional applications, they must be supplemented with a dedicated outdoor air system (DOAS) that must provide the ventilation, filtration, air motion, and all of the space humidity control. On the cooling cycle, the system must ensure that the room dew-point temperature is always above the lowest panel or chilled-water supply pipe temperature. A controlled-temperature surface is called a **radiant panel** if 50% or more of the heat transfer is by radiation to other surfaces.

Residential heating-only applications usually consist of pipe coils embedded in wood or masonry floors or plaster ceilings. This construction serves well where loads are relatively stable and where solar effects are minimized by building design. However, in buildings with large glass areas and rapid load changes, the slow response, lag, and override effect of concrete or masonry panels is unsatisfactory. Lightweight metal panel ceiling systems quickly respond to load changes and are used for cooling as well as heating in commercial and institutional applications.

Warm air and electric heating elements are used where local factors influence such use. In the warm air system, air is supplied to a cavity behind, under, or encapsulated in the panel surface. The air may leave the cavity through a normal diffuser and flow into the room. These systems are used as floor radiant panels in schools and in floors subject to extreme cold, such as over an overhang. Electric heating elements embedded in the floor or ceiling construction and unitized electric ceiling panels are used in residences, apartments, and various applications for local spot heating. Two factors to consider when using electric radiant panels are local electric codes and the relative difference between electric and fossil fuel heating costs.

The radiant panel is often located in the ceiling of a room. A ceiling is used because it sees all other surfaces and objects in the room; it is not subject to unpredictable coverings, as are floors; for heating, higher surface temperatures can be used; it is of smaller mass and therefore has quicker response to load changes; radiant cooling can be incorporated; and, in the case of the metal ceiling system, the piping is accessible for servicing.

Ceiling panel systems commonly used are an outgrowth of the perforated metal, suspended, acoustical ceiling. These radiant ceiling systems are usually designed into buildings where the features of the suspended acoustical ceiling can be combined with panel heating and cooling. The panels can be designed as small units to fit the building module and provide extensive flexibility for zoning and control, or, for maximum economy, the panels can be arranged as large continuous areas. Two types of metal ceiling systems are available. One type consists of lightweight aluminum panels, usually 12 in. by 24 in. (305 mm by 610 mm), that are attached in the field to 0.5 in. (15 mm) galvanized pipe coils. The second type consists of a copper coil metallurgically bonded to the aluminum face sheet forming a modular panel. Modular panels are available in sizes up to approximately 36 in. by 60 in. (910 mm by 1520 mm) and are held in position by various ceiling suspension systems.

The arrangement of components in radiant panel systems is similar to other air-water systems. Room temperature conditions are primarily maintained by a combination of direct transfer of radiant energy, and by convective heating and cooling. The room heating and cooling loads are calculated in the conventional manner. Manufacturers generally rate their equipment in the form of total performance, which can be applied directly to the calculated room load for heating and to the room sensible load for cooling.

These are the principal advantages of panel heating and cooling systems:

1. If they are properly designed; because of the low airflow quantities these systems can be very energy efficient.
2. Panel systems do not require any mechanical heat exchange equipment at the outside walls, thus simplifying the wall, floor, and structural systems.
3. All pumps, fans, filters, and so forth, are centrally located, thereby centralizing maintenance and operation.
4. Cooling or heating may be obtained during any season, without central zoning or seasonal changeover, when four-pipe systems are used.
5. Supply air quantities usually do not exceed those required for ventilation and dehumidification.
6. No mechanical equipment requiring maintenance or repair is placed within the occupied space, except possibly the control valves.

7. Draperies and curtains can be installed at the outside wall without interfering with heating and cooling systems.
8. The modular panel provides flexibility to meet changes in partitioning.
9. No space is required within the air-conditioned room for the mechanical equipment. This feature is especially valuable when compared to other conditioning methods for applications in existing buildings, hospital patient rooms, and other applications where space is at a premium and where maximum cleanliness is essential.
10. A common central air system for ventilation and dehumidification can serve both the interior and perimeter zones.
11. Wet surface cooling coils are eliminated from the occupied space, thus reducing the potential for septic contamination.

Other essential factors when considering the use of panel systems are as follows:

1. Evaluate early to plan an optimum physical arrangement of the building to take full advantage of the panel system.
2. Select recessed lighting fixtures, air diffusers, hung ceiling, and other ceiling devices to provide the maximum ceiling area possible for use as radiant panels.
3. The air-side design must maintain the room dew-point temperature below the lowest temperature of panel surface at all times to eliminate any possibility of condensation on the panels. The systems must be interlocked to shut down the chilled water to the panels if the dehumidifying system fails.
4. As with any hydronic system, design the piping system to avoid noises from entrained air, high velocity or high pressure drop devices, or from pump and pipe vibrations.
5. Anticipate thermal expansion of the ceiling and other devices in or adjacent to the ceiling.

## 15.2 Types

The most common forms of panels applied in panel heating and cooling systems are

- Metal ceiling panels
- Embedded piping in ceilings, walls, or floors (heat only)
- Air-heated floors
- Electrically heated ceilings or floors

**Metal Ceiling Panels.** Metal ceiling panels are often integrated into a system that both heats and cools. In such a system, a source of dehumidified ventilation air is required. This system must provide all of the ventilation air and all of the humidity control, as well as pressurize the building to avoid any significant infiltration. In such a system, various amounts of forced air are supplied year-round. (See section 12.4.)

A metal ceiling panel system using copper tubing metallurgically bonded to an aluminum panel is shown in Figure 15-1. This panel can be mounted into various ceiling suspension systems.

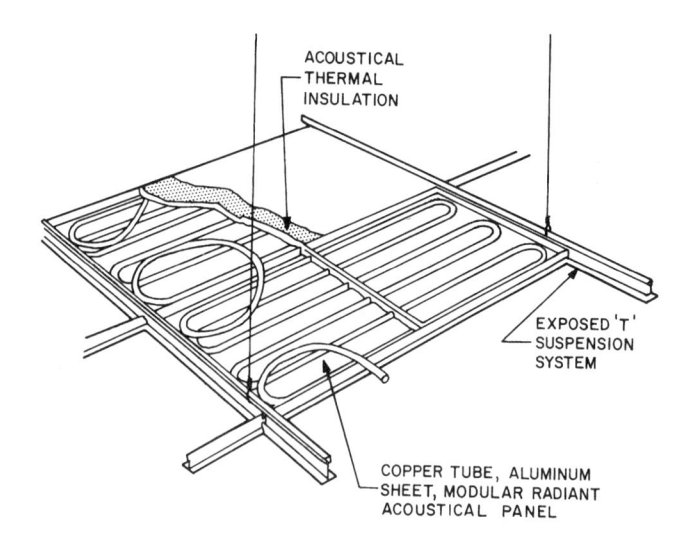

*Fig. 15-1 Metal Ceiling Panels Metallurgically Bonded to Copper Tubing*
(Figure 14, Chapter 6, 2012 ASHRAE Handbook—HVAC Systems and Equipment)

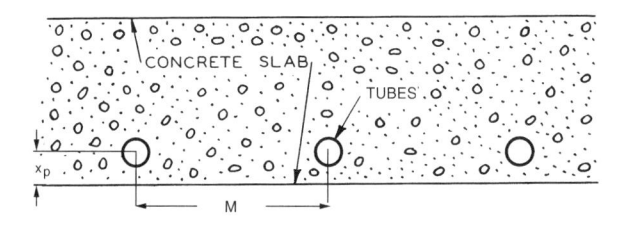

*Fig. 15-2 Coils in Structural Concrete Slab*
(Figure 17, Chapter 6, 2012 ASHRAE Handbook—HVAC Systems and Equipment)

Two-pipe and four-pipe distribution systems have been used successfully with metal ceiling panels. Common design practice calls for a 20°F (11°C) drop for heating across a given grid and a 5°F (3°C) rise for cooling, but higher temperature differentials may apply in some cases.

Some ceiling installations require that active grids cover only a part of the room, and consequently, compatible matching standard acoustical panels are normally used for the remaining ceiling area.

**Embedded Piping in Ceilings, Walls, and Floors.** When piping is embedded in ceilings, the construction used is generally one of the following:

1. Pipe or tube is embedded in the lower portion of a concrete slab, generally within an inch of its lower surface. If plaster is to be applied to the concrete, the piping may be placed directly on the wood forms. If the slab is to be used without plaster finish, then the piping should be installed not less than 0.75 in. (19 mm) above the undersurface of the slab. This method of construction is shown in Figure 15-2. The minimum coverage must be in compliance with the local building code requirements.

2. Pipe or tube is embedded in a metal lath and plaster ceiling. If the lath is suspended to form a hung ceiling, both the lath

# Panel Heating and Cooling Systems

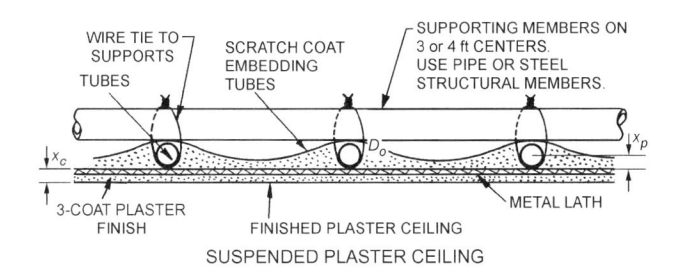

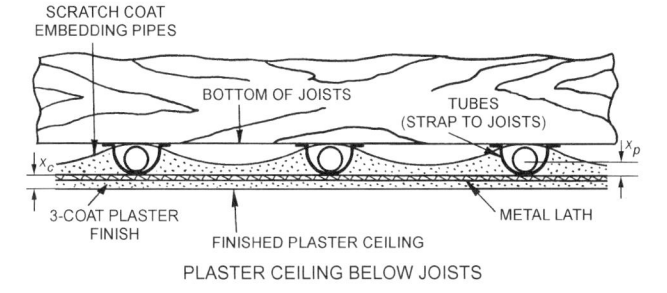

*Fig. 15-3 Coils in Plaster Above Lath*
*(Figure 18, Chapter 6, 2012 ASHRAE Handbook—*
*HVAC Systems and Equipment)*

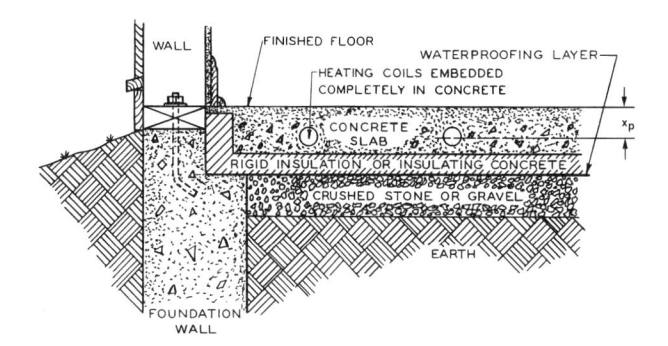

*Fig. 15-4 Coils in Floor Slab on Grade*
*(Figure 20, Chapter 6, 2012 ASHRAE Handbook—*
*HVAC Systems and Equipment)*

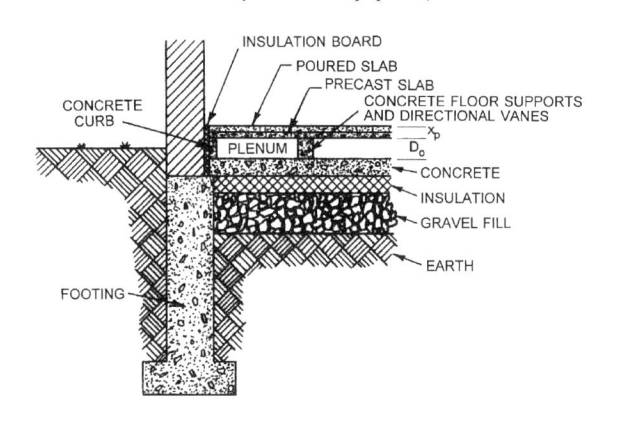

*Fig. 15-5 Warm Air Floor Panel Construction*
*(Figure 27, Chapter 6, 2012 ASHRAE Handbook—*
*HVAC Systems and Equipment)*

and the heating coils are securely wired to the supporting members in such a way that the lath is below, but in good contact with, the coils. Plaster is then applied to the metal lath, with care being taken to embed the coil.

3. Copper tube of the smaller diameters or cross-linked polyethylene (PEX) tubing is attached to the underside of a wire lath or gypsum lath. Plaster is then applied to the lath to embed the tube (Figure 15-3).

4. Other forms of ceiling construction are composition board, wood paneling, etc., with warm water piping, tube, or channels built into the panel sections.

Coils are usually of the sinuous type, although some header or grid-type coils have been used in ceilings. Coils may be of plastic (PEX), ferrous, or nonferrous pipe or tube, with coil pipes spaced from 4.5 to 9 in. (115 to 230 mm) on centers, depending on the required output, pipe or tube size, and other factors.

Although not so universally used as ceiling panels, wall panels may be constructed by any of the methods described for ceilings.

The construction for piping embedded in floors depends on whether (1) the floor is laid on grade or (2) the floor is above grade.

**On-Grade Floor.** Plastic (PEX), ferrous, and nonferrous pipe and tube are used in floor slabs which rest on grade. The coils are constructed as either sinuous, continuous pipe coils or arranged as heater coils with the pipes spaced from 6 to 18 in. (150 to 460 mm) on centers. The coils are generally installed with 1.5 to 4 in. (40 to 100 mm) of cover above the coils. Insulation should be used to reduce the perimeter and reverse side losses. Illustrated in Figure 15-4 is the application of pipe coils in slabs resting on grade. Coils should be embedded completely and should not rest on an interface. Any supports used for positioning the heating coils should be nonabsorbent and inorganic.

**Above-Grade Floor.** Where the coils are embedded in structural load-supporting slabs above grade, construction codes may affect their position. Otherwise, the coil piping is installed in the same manner as described for slabs resting on grade. Except the pipes should be installed in the wearing (finish) concrete rather than in the structural concrete.

**Air-Heated Floors.** Several methods have been devised to warm interior room surfaces by circulating heated air through passages in the floor. In some cases, the heated air is recirculated in a closed system. In others, all or part of the air is passed through the room on its way back to the furnace or air-handling unit to provide supplementary heating and ventilation (Figure 15-5).

**Electrically Heated Ceilings.** Several types of electric resistance units are available for heating interior room surfaces. These include (1) electric heating cables that may be embedded in concrete or plaster or laminated in drywall ceiling construction; (2) prefabricated electric heating panels to be attached to room surfaces; and (3) electrically heated fabrics or other materials for application to, or incorporation into, finished room surfaces.

**Ceiling Cables.** The details of ceiling cable installation for plastered and drywall construction is shown in Figure 15-6.

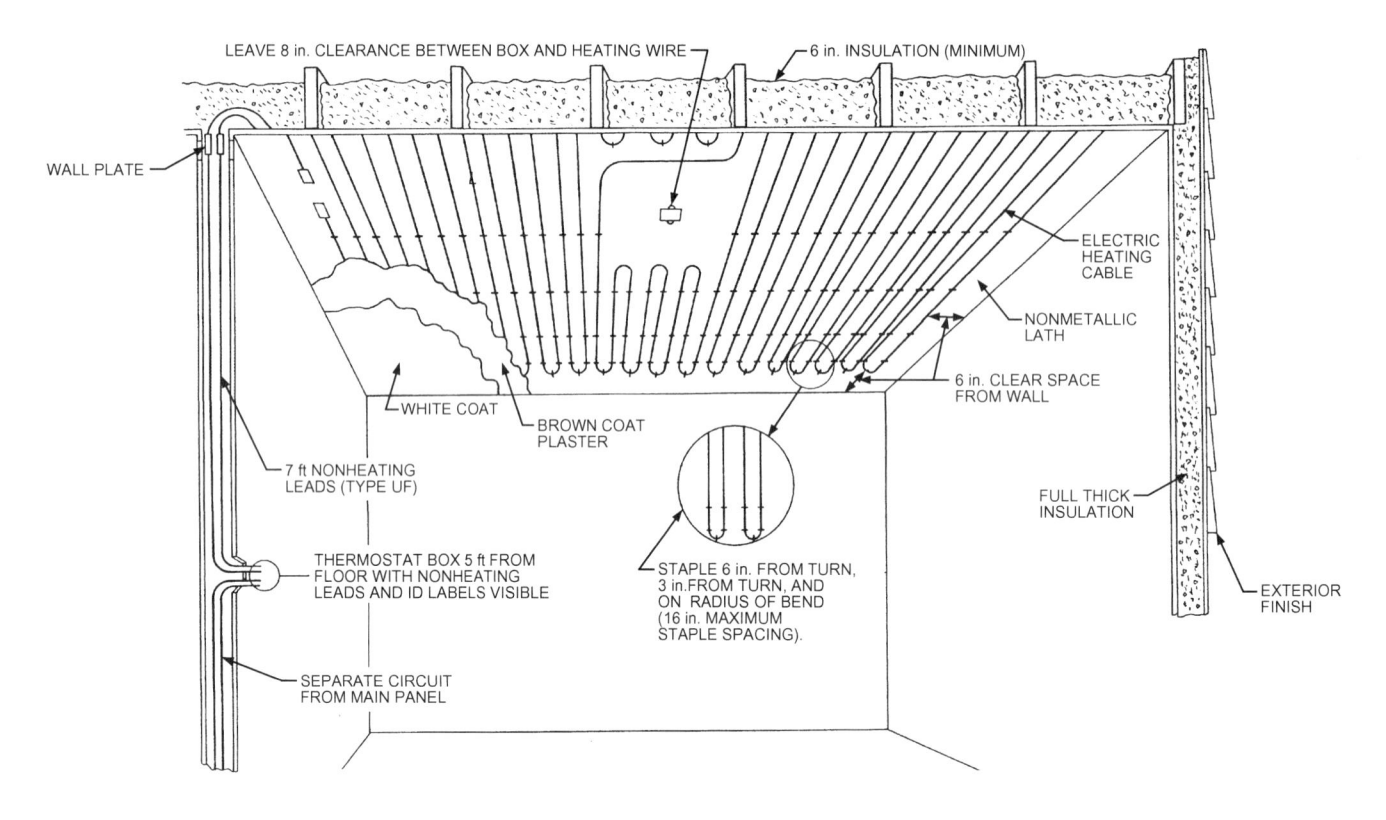

*Fig. 15-6 Electric Heating Panel for Wet Plastered Ceiling*
*(Figure 25, Chapter 6, 2012 ASHRAE Handbook—HVAC Systems and Equipment)*

**Electric Heating Panels.** A variety of prefabricated electric heating panels are used for either supplemental or full room heating. These panels are available in sizes from 2 ft by 4 ft to 6 ft by 12 ft (0.6 m by 1.2 m to 1.8 m by 3.6 m). They are constructed from a variety of materials such as gypsum board, glass, steel, and vinyl. Different panels have rated inputs varying from 10 to 95 W/ft$^2$ (108 to 1023 W/m$^2$) for 120, 208, and 240 V service. Maximum operating temperatures vary from about 100 to about 300°F (38 to 49°C) depending on watt density.

Panel heating elements may be embedded conductors, laminated conductive coatings, or printed circuits. Nonheating leads are connected and furnished as part of the panel.

Some panels may be cut to fit available space; others must be installed as received. Panels may be either flush or surface mounted. In some cases, they are finished as part of the ceiling. Rigid panels that are about 1 in. (25 mm) thick and weigh about 25 lb (11 kg) each are available to fit standard 2 ft by 4 ft (0.6 m by 1.2 m) modular tee-bar ceilings.

Cable embedded in walls, similar to ceiling construction, is occasionally found in Europe. Because of possible damage due to nails driven for hanging pictures or because of building alteration, most codes prohibit such panels in the United States.

Some of the prefabricated panels described in the preceding section are also used for wall panel heating.

Electric heating cable assemblies, such as those used for ceiling panels, are sometimes used for concrete floor heating systems.

## 15.3 Design Steps

Panel design requires specification of the following: panel area, size and location of the heating elements in the panel, insulation on the reverse side and edge of the panel, required input to panel, and temperature of the heating elements. Specific procedures are given in Chapter 6 of the 2012 *ASHRAE Handbook—HVAC Systems and Equipment*. The procedure is summarized as follows:

1. Calculate heat loss for each room.
2. Determine the available area for panels in each room.
3. Calculate the required unit panel output.
4. Determine the required panel surface temperature.
5. Select the means of heating the panel and the size and location of the heating elements.
6. Select insulation for the reverse side and edge of panel.
7. Determine panel heat loss and required input to the panel.
8. Determine the other temperatures that are required or developed.
9. Design the system for heating the panels in accordance with conventional practice and manufacturers recommendations.

In the steps outlined for design, the effect of each assumption or choice on comfort should be considered carefully. The following general rules should be followed:

1. Place panels near cold areas where heat losses occur.
2. Do not use high-temperature ceiling panels in very low ceilings.

# Panel Heating and Cooling Systems

3. Keep floor panels temperatures at or below 85°F (30°C).

**Example 15-1** The living room in a home is occupied by adults in light clothing and engaged in sedentary activity. The room has a net outside wall area of 275 ft$^2$ with a surface temperature of 54°F, 45 ft$^2$ of glass with a surface temperature of 20°F; 540 ft$^2$ of ceiling with a surface temperature of 60°F; 670 ft$^2$ of partitions with a surface temperature of 70°F; and 540 ft$^2$ of floor with a surface temperature of 70°F. If the air movement is 20 fpm, determine the air temperature necessary for comfort.

**Solution:**

Mean radiant temperature = MRT =

$$\text{MRT} = \frac{275(54) + 45(20) + 540(60) + 670(70) + 540(70)}{275 + 45 + 540 + 670 + 540}$$

MRT = 64.1°F

for sedentary activity with light clothing at 20 fpm from Figure 4-3:

$$t_{\text{dry bulb}} = 90°F \text{ for comfort}$$

## 15.4 Problems

**15.1** A room has a net outside wall area of 300 ft$^2$ that has a surface temperature of 55°F; 50 ft$^2$ of glass with a surface temperature of 30°F; 560 ft$^2$ of ceiling with a surface temperature of 70°F; and 560 ft$^2$ with a surface temperature of 70°F. Estimate the average unheated surface temperature or the area-weighted mean radiant temperature. [Ans: 65.6°F]

**15.2** For the room in Problem 15.1, estimate the following:

(a) radiant output for a 100 ft$^2$ heating panel with a panel surface temperature of 120°F

(b) natural convection output for the ceiling panel when the air temperature is 70°F

**15.3** A room has 1500 ft$^2$ of surface area and 320 ft$^2$ is to be heated. The average unheated surface temperature in the room is 67°F. The air temperature in the room is 75°F. The room is occupied by adults in light clothing at a sedentary activity. Determine the surface temperature of the heated panel necessary to produce comfort if the air velocity is 20 fpm. [Ans: 131°F]

**15.4** For Problem 15.3, determine the total heat transferred by the ceiling heating panel.

## 15.5 Bibliography

ASHRAE. 2012. Chapter 6, Panel Heating and Cooling, 2012 *ASHRAE Handbook—HVAC Systems and Equipment*.

# Chapter 16

# HEAT PUMP, COGENERATION, AND HEAT RECOVERY SYSTEMS

This chapter discusses applied heat pump systems, heat recovery systems, and cogeneration systems. Specific details on these subjects can be found in Chapters 7 and 9 of the 2012 *ASHRAE Handbook—HVAC Systems and Equipment*.

## 16.1 General

As described in Chapter 2, a heat pump extracts heat from a source and transfers it to a sink at a higher temperature. According to this definition, all pieces of refrigeration equipment, including air conditioners and chillers with refrigeration cycles, are heat pumps. In engineering, however, the term *heat pump* is generally reserved for equipment that heats for beneficial purposes, rather than that which removes heat for cooling only. Dual-mode heat pumps alternately provide heating, cooling, or both simultaneously. Heat reclaim heat pumps provide heating only or simultaneous heating and cooling. An applied heat pump requires field engineering for the specific application, in contrast to the use of a manufacturer-designed unitary product. Applied heat pumps include built-up heat pumps (field- or custom-assembled from components) and industrial process heat pumps. Most current heat pumps use a vapor compression (modified Rankine) cycle or absorption cycle. Any of the other refrigeration cycles discussed in Chapter 2 of the 2013 *ASHRAE Handbook—Fundamentals* are also suitable. Although most heat pump compressors are powered by electric motors, use is also made of engine and turbine drives that can add engine coolant or "waste" heat to the heat generated. Applied heat pumps are most commonly used for heating and cooling buildings, but they are occasionally used for domestic and service water heating, pool heating, and industrial process heating.

Applied heat pumps with capacities from 24,000 to 150,000,000 Btu/h (7 to 45,000 kW) operate in many facilities. Some machines are capable of output water temperatures up to 220°F and steam pressures up to 60 psig [415 kPa (gage)].

Compressors in large systems vary from one or more reciprocating, scroll, or screw types to single- or multistaged centrifugal types. A single or central system is often used, but in some instances, multiple or unitary heat pump systems are used (Chapter 14) to facilitate zoning. Heat sources include the ground, well water, surface water, gray water, solar energy, the air, internal building heat, and a hydronic water circuit that is heated or cooled. Compression can be single-stage or multistage. Frequently, heating and cooling are supplied simultaneously to separate zones.

Decentralized systems with water loop heat pumps are common, using multiple water-source heat pumps connected to a common circulating water loop. They can also include ground coupling, heat rejectors (cooling towers and dry coolers), supplementary heaters (boilers and steam heat exchangers), loop reclaim heat pumps, solar collection devices, and thermal storage.

Community and district heating and cooling systems can utilize both centralized and distributed heat pump systems.

## 16.2 Types of Heat Pumps

Heat pumps are classified by (1) heat source and sink, (2) heating and cooling distribution fluid, (3) thermodynamic cycle, (4) building structure, (5) size and configuration, and (6) limitation of the source and sink. Table 16-1 shows the more common types of closed vapor-compression cycle heat pumps for heating and cooling service.

**Air-to-Air Heat Pumps.** This type of heat pump is quite common and is particularly suitable for factory-built unitary heat pumps. It is widely used in residential and commercial applications (see Chapter 14). The first diagram in Table 16-1 is a typical refrigeration circuit.

In other air-to-air heat pump systems, air circuits can be interchanged by motor-driven or manually operated dampers to obtain either heated or cooled air for the conditioned space. In that system, one heat exchanger coil is always the evaporator, and the other is always the condenser. Conditioned air passes over the evaporator during the cooling cycle, and outdoor air passes over the condenser. Damper positioning causes the change from cooling to heating.

**Water-to-Air Heat Pumps.** These heat pumps rely on water as the heat source and sink and use air to transmit heat to or from the conditioned space. (See the second diagram in Table 16-1.) They include the following:

- *Groundwater heat pumps*, which use groundwater from wells as a heat source and/or sink. They can either circulate source water directly to the heat pump or use an intermediate fluid in a closed loop, similar to the ground-coupled heat pump.

- *Surface water heat pumps*, which use surface water from a lake, pond, or stream as a heat source or sink. As with

**Table 16-1 Common Types of Heat Pumps**
*(Figure 5, Chapter 9, 2012 ASHRAE Handbook—HVAC Systems and Equipment)*

| Heat Source and Sink | Distribution Fluid | Thermal Cycle | Diagram |
|---|---|---|---|
| Air | Air | Refrigerant changeover | |
| Water | Air | Refrigerant changeover | |
| Water | Water | Water changeover | |
| Ground-coupled (or Closed-loop ground-source) | Air | Refrigerant changeover | |
| Ground-source, Direct-expansion | Air | Refrigerant changeover | |

ground-coupled and groundwater heat pumps, these systems can either circulate source water directly to the heat pump or use an intermediate fluid in a closed loop.

- *Internal-source heat pumps*, which use high internal cooling load generated in buildings either directly or with storage. These include water-loop heat pumps and variable refrigerant flow heat pump systems.
- *Solar-assisted heat pumps*, which rely on low-temperature solar energy as the heat source. Solar heat pumps may resemble water-to-air, or other types, depending on the form of solar heat collector and the type of heating and cooling distribution system.
- *Wastewater-source heat pumps*, which use sanitary waste heat or laundry waste heat as a heat source. Waste fluid can be introduced directly into the heat pump evaporator after waste filtration, or it can be taken from a storage tank, depending on the application. An intermediate loop may also be used for heat transfer between the evaporator and the waste heat source.

**Water-to-Water Heat Pumps.** These heat pumps use water as the heat source and sink for cooling and heating. Heating/cooling changeover can be done in the refrigerant circuit, but it is often more convenient to perform the switching in the water circuits, as shown in the third diagram of Table 16-1. Although the diagram shows direct admittance of the water source to the evaporator, in some cases, it may be necessary to apply the water source indirectly through a heat exchanger (or double-wall evaporator) to avoid contaminating the closed chilled-water system, which is normally treated. Another configuration employs a closed-circuit condenser water system that is a water chiller of which the condenser water is a hydronic heating simultaneously with the chilled-water serving as a chilled water circuit.

**Ground-Coupled Heat Pumps.** These use the ground as a heat source and sink. A heat pump may have a refrigerant-to-water heat exchanger or may be direct-expansion (DX). Both types are shown in Table 16-1. In systems with refrigerant-to-water heat exchangers, a water or antifreeze solution is pumped through horizontal, vertical, or coiled pipes embedded in the ground. Direct-expansion ground-coupled heat pumps use refrigerant in direct-expansion, flooded, or recirculation evaporator circuits for the ground pipe coils.

A common configuration of ground-coupled heat pump employs a deep well (usually 6 in. diameter and several hundred feet deep). Into the well is inserted a supply and return water pipe loop of high-pressure plastic pipe with a U bend at the bottom, which serves as a heat exchanger to either reject heat to or obtain heat from the ground. After inserting the pipe, the well is filled with a heat transfer grout that holds the pipe in place and protects the water table from contamination by any undesirable surface materials. Depending upon the depth of the well, the depth of the water table, the heat transfer characteristics of the well construction, and the soil, the capacity of each well is usually between 3 and 5 tons (10 and 18 kW) of heat rejection capacity. They are often used singly for residential applications, and in multiple "fields" of wells, spaced 20 to 40 ft (6 to 12 m) apart, for commercial and institutional installations. These systems are sometimes called geothermal heat pumps.

Soil type, moisture content, composition, density, and uniformity close to the surrounding field areas affect the success of this method of heat exchange of any ground-coupled heat exchange. With some piping materials, the material of construction for the pipe and the corrosiveness of the local soil and underground water may affect the heat transfer and service life. In a variation of this cycle, all or part of the heat from the evaporator plus the heat of compression are transferred to a water-cooled condenser. This condenser heat is then available for uses such as heating air or domestic hot water.

Additional heat pump types include the following:

**Air-to-Water Heat Pumps Without Changeover.** These are commonly called *heat pump water heaters*.

**Refrigerant-to-Water Heat Pumps.** These condense a refrigerant by the cascade principle. Cascading pumps the heat to a higher temperature, where it is rejected to water or another liquid. This type of heat pump can also serve as a condensing unit to cool almost any fluid or process. More than one heat source can be used to offset those times when insufficient heat is available from the primary source.

## 16.3 Heat Sources and Sinks

Table 16-2 shows the principal media used as heat sources and sinks. Selecting a heat source and sink for an application is primarily influenced by geographic location, climate, initial cost, availability, and type of structure. Table 16-2 presents various factors to be considered for each medium.

**Air.** Outdoor air is a universal heat source and sink medium for heat pumps and is widely used in residential and light commercial systems. Extended-surface, forced-convection heat transfer coils transfer heat between the air and refrigerant. Typically, the surface area of outdoor coils is considerably larger than that of indoor coils. The volume of outdoor air handled is also greater than the volume of indoor air handled. During heating, the temperature of the evaporating refrigerant is generally 10°F to 20°F (5 to 10 °C) less than the outdoor air temperature.

When selecting or designing an air-source heat pump, two factors in particular must be considered: (1) the local outdoor air temperature and (2) frost formation.

As the outdoor temperature decreases, the heating capacity of an air-source heat pump decreases. This makes equipment selection for a given outdoor heating design temperature more critical for an air-source heat pump than for a fuel-fired system. Equipment must be sized for as low a balance point as is practical for heating, often requiring much more compressor capacity for heating than for cooling. Many heat pumps utilize auxillary heating in cold climates and only utilize the heat pump to the outdoor air limits of the air-conditioning compressor.

### Table 16-2 Heat Pump Sources and Sinks
*(Table 1, Chapter 9, 2012 ASHRAE Handbook—HVAC Systems and Equipment)*

| | | Suitability | | Availability | | Cost | | Temperature | | Common Practice | |
|---|---|---|---|---|---|---|---|---|---|---|---|
| Medium | Examples | Heat Source | Heat Sink | Location Relative to Need | Coincidence with Need | Installed | Operation and Maintenance | Level | Variation | Use | Limitations |
| **AIR** | | | | | | | | | | | |
| Outdoor | Ambient air | Good, but efficiency and capacity in heating mode decrease with decreasing outdoor air temperature | Good, but efficiency and capacity in cooling mode decrease with increasing outdoor air temperature | Universal | Continuous | Low | Moderate | Variable | Generally extreme | Most common, many standard products | Defrosting and supplemental heat usually required |
| Exhaust | Building ventilation | Excellent | Fair | Excellent if planned for in building design | Excellent | Low to moderate | Low unless exhaust is laden with dirt or grease | Excellent | Very low | Excellent as energy-conservation measure | Insufficient for typical loads |
| **WATER** | | | | | | | | | | | |
| Well[*] | Groundwater well may also provide a potable water source | Excellent | Excellent | Poor to excellent, practical depth varies by location | Continuous | Low if existing well used or shallow wells suitable; can be high otherwise | Low, but periodic maintenance required | Generally excellent, varies by location | Extremely stable | Common | Water disposal and required permits may limit; may require double-wall exchangers; may foul or scale |
| Surface | Lakes, rivers, oceans | Excellent for large water bodies or high flow rates | Excellent for large water bodies or high flow rates | Limited, depends on proximity | Usually continuous | Depends on proximity and water quality | Depends on proximity and water quality | Usually satisfactory | Depends on source | Available, particularly for fresh water | Often regulated or prohibited; may clog, foul, or scale |
| Tap (city) | Municipal water supply | Excellent | Excellent | Excellent | Continuous | Low | Low energy cost, but water use and disposal may be costly | Excellent | Usually very low | Use is decreasing due to regulations | Use or disposal may be regulated or prohibited; may corrode or scale |
| Condensing | Cooling towers, refrigeration systems | Excellent | Poor to good | Varies | Varies with cooling loads | Usually low | Moderate | Favorable as heat source | Depends on source | Available | Suitable only if heating need is coincident with heat rejection |
| Closed loops | Building water-loop heat pump systems | Good, loop may need supplemental heat | Favorable, may need loop heat rejection | Excellent if designed as such | As needed | Low | Low to moderate | As designed | As designed | Very common | Most suitable for medium or large buildings |
| Waste | Raw or treated sewage, gray water | Fair to excellent | Fair, varies with source | Varies | Varies, may be adequate | Depends on proximity, high for raw sewage | Varies, may be high for raw sewage | Excellent | Usually low | Uncommon, practical only in large systems | Usually regulated; may clog, foul, scale, or corrode |
| **GROUND**[a] | | | | | | | | | | | |
| Ground-coupled | Buried or submerged fluid loops | Good if ground is moist, otherwise poor | Fair to good if ground is moist, otherwise poor | Depends on soil suitability | Continuous | High to moderate | Low | Usually good | Low, particularly for vertical systems | Rapidly increasing | High initial costs for ground loop |
| Direct-expansion | Refrigerant circulated in ground coil | Varies with soil conditions | Varies with soil conditions | Varies with soil conditions | Continuous | High | High | Varies by design | Generally low | Extremely limited | Leak repair very expensive; requires large refrigerant quantities |
| **SOLAR ENERGY** | | | | | | | | | | | |
| Direct or heated water | Solar collectors and panels | Fair | Poor, usually unacceptable | Universal | Highly intermittent, night use requires storage | Extremely high | Moderate to high | Varies | Extreme | Very limited | Supplemental source or storage required |
| **INDUSTRIAL PROCESS** | | | | | | | | | | | |
| Process heat or exhaust | Distillation, molding, refining, washing, drying | Fair to excellent | Varies, often impractical | Varies | Varies | Varies | Generally low | Varies | Varies | Varies | May be costly unless heat need is near rejected source |

[a] Groundwater-source heat pumps are also considered ground-source heat pump systems.

When the surface temperature of an outdoor air coil is 32°F (0°C) or less, with a corresponding outdoor air dry-bulb temperature 4°F to 10°F (2°C to 5°C) higher, frost may form on the coil surface. If allowed to accumulate, frost inhibits heat transfer; therefore, the outdoor coil must be defrosted periodically. The number of defrosting operations is influenced by the climate, air-coil design, and the hours of operation. Experience shows that, generally, little defrosting is required when outdoor air conditions are below 17°F and 60% rh. However, under very humid conditions, when small suspended water droplets are present in the air, the rate of frost deposit may be about three times as great as predicted from psychrometric theory and the heat pump may require defrosting after as little as 20 minutes of operation. The loss of available heating capacity caused by frosting should be considered when sizing an air-source heat pump utilizing outdoor air.

Following commercial refrigeration practice, early designs of air-source heat pumps had relatively wide fin spacing of 4 to 5 fins/in., (5 to 6 mm apart) based on the theory that this would minimize defrosting frequency. However, experience has shown that effective hot-gas defrosting allows much closer fin spacing and reduces the system's size and bulk. In current practice, fin spacings of 10 to 20 fins/in. (1.5 to 2.5 mm apart) are sometimes used.

In many institutional and commercial buildings, some air must be continuously exhausted year-round. This exhaust air can be used as a heat source for some configurations of a heat recovery system.

High humidity caused by indoor swimming pools causes condensation on ceiling structural members, walls, windows, and floors and causes discomfort to spectators. Traditionally, outdoor air and dehumidification coils with reheat from a boiler that also heats the pool water are used. This is ideal for air-to-air and air-to-water heat pumps because energy costs can be reduced. Suitable materials must be chosen so that heat pump components are resistant to corrosion from chlorine and high humidity.

**Water.** Water can be a satisfactory heat source, subject to the considerations listed in Table 16-2. City water is seldom used because of cost and municipal restrictions. Groundwater (well water) is particularly attractive as a heat source because of its relatively high and nearly constant temperature. Water temperature depends on source depth and climate but in the United States generally ranges from 40°F (4.5°C) in northern areas to 70°F (21°C) in southern areas. Frequently, sufficient water is available from wells. In some locations, under strict regulations, water can be reinjected into the aquifer. This use is nonconsumptive and, with proper design, only the water temperature changes. Water quality should be analyzed, and the possibility of scale formation and corrosion should be considered. In some instances, it may be necessary to separate the well fluid from the equipment with an additional heat exchanger. Special consideration must also be given to filtering and settling ponds for specific fluids. Other considerations are the costs of drilling, piping, pumping, and a means for disposal of used water. Information on well water availability, temperature, and chemical and physical analysis is available from US Geological Survey offices in most locations.

Heat exchangers may also be submerged in open ponds, lakes, or streams. When surface or stream water is used as a source, the temperature drop across the evaporator in winter may need to be limited to prevent freeze-up.

In industrial applications, waste process water (e.g., spent warm water in laundries, plant effluent, and warm condenser water) may be a heat source for heat pump operation.

Sewage, which often has temperatures higher than that of surface or groundwater, may be an acceptable heat source. Secondary effluent (treated sewage) is usually preferred, but untreated sewage may be used successfully with proper heat exchanger design.

Use of water during cooling follows the conventional practice for water-cooled condensers.

Water-to-refrigerant heat exchangers are generally direct-expansion or flooded water coolers, usually shell-and-coil or shell-and-tube. Brazed-plate heat exchangers may also be used. In large applied heat pumps, the water is usually reversed instead of the refrigerant.

**Ground.** The ground is used extensively as a heat source and sink, with heat transfer through buried coils. Soil composition, which varies widely from wet clay to sandy soil, has a predominant effect on thermal properties and expected overall performance. The heat transfer process in soil depends on transient heat flow. Thermal diffusivity is a dominant factor and is difficult to determine without local soil data. Thermal diffusivity is the ratio of thermal conductivity to the product of density and specific heat. The soil's moisture content influences its thermal conductivity.

There are three primary types of ground-source heat pumps: (1) groundwater, which is discussed in the previous section; (2) direct-expansion, in which the ground-to-refrigerant heat exchanger is buried underground; and (3) ground-coupled (also called closed-loop ground-source and geothermal), in which a secondary loop (sometimes with a brine) connects the ground-to-water and water-to-refrigerant heat exchangers (see Table 16-1).

Ground loops can be placed either horizontally or vertically. A horizontal system consists of single or multiple serpentine heat exchanger pipes buried 3 to 6 ft (1 to 2 m) apart in a horizontal plane at a depth 3 to 6 ft (1 to 2 m) below grade. Pipes may be buried deeper, but excavation costs and temperature must be considered. Horizontal systems can also use coiled loops referred to as *slinky coils*. A vertical system uses a concentric tube or U-tube heat exchanger. The design of ground-coupled heat exchangers is covered in Chapter 34 of the 2011 *ASHRAE Handbook—HVAC Applications*.

**Solar Energy.** Solar energy may be used either as the primary heat source or in combination with other sources. Air, surface water, shallow groundwater, and shallow ground-source systems all use solar energy indirectly. The principal advantage of using solar energy directly is that, when available, it provides heat at a higher temperature than the indirect sources, increasing the heating coefficient of performance. Compared to solar heating without a heat pump, the collector efficiency and capacity are increased because a lower collector temperature is required.

Research and development of solar-source heat pumps has been concerned with two basic types of systems: direct and indirect. The direct system places refrigerant evaporator tubes in a solar collector, usually a flat-plate type. Research shows that a collector without glass cover plates can also extract heat from the outdoor air. The same surface may then serve as a condenser using outdoor air as a heat sink for cooling

An indirect system circulates either water or air through the solar collector. When air is used, the collector may be controlled in such a way that (1) the collector can serve as an outdoor air preheater, (2) the outdoor air loop can be closed so that all source heat is derived from the sun, or (3) the collector can be disconnected from the outdoor air serving as the source or sink.

## 16.4 Cogeneration

**Cogeneration** designates on-site electrical generating systems that salvage byproduct or waste heat from the generating process. The magnitude, duration, and coincidence of electrical and thermal loads must be analyzed, and prime movers and waste heat recovery systems must be selected to determine system feasibility and design. The basic components of the cogeneration plant are (1) prime mover, (2) generator, (3) waste heat recovery systems, (4) control systems, and (5) connections to building mechanical and electrical services.

The normal prime movers are reciprocating internal combustion engines, combustion gas turbines, expansion turbines, and steam boiler-turbine combinations. These units convert the heat in the fuel (liquid, solid, gaseous, or nuclear) into rotating shaft energy.

Use of the prime mover heat determines overall system efficiency and is one of the critical factors of economic feasibility. Two kinds of energy are available from the prime mover: (1) mechanical energy from the shaft and (2) heat energy remaining after the fuel or steam has acted on the shaft.

Shaft loads (generators, centrifugal chillers, compressors, process equipment) require a given amount of rotating mechanical energy. Once the prime mover has been selected to provide the required shaft output, it has a fixed relationship to system efficiency that is dependent upon the prime mover fuel versus load and the load versus heat balance curves.

Steam turbine drives can be arranged to extract steam at intermediate turbine stages. The waste heat value of a steam turbine is the enthalpy of the steam at the point it is extracted from the turbine or at the turbine's exhaust outlet. This steam, reduced in pressure and temperature by the amount of shaft

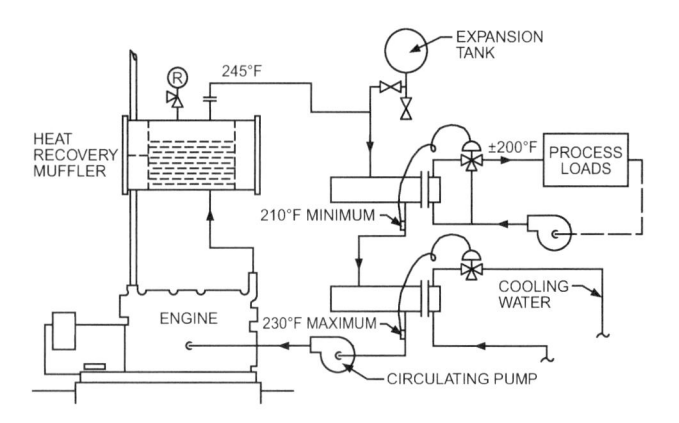

*Fig. 16-1 Typical Reciprocating Engine Heat Recovery System*
(Figure 47, Chapter 7, 2012 ASHRAE Handbook—HVAC Systems and Equipment)

work, can be fed to heat exchange equipment, absorption chillers, and steam turbine-driven centrifugal chillers.

The gas turbine cycle has a thermal efficiency of approximately 20%, with the remainder of the fuel energy exhausted or radiated. A minimum exhaust temperature of approximately 300°F (150°C) is required to prevent condensation. The recoverable heat per unit of power is greater for a gas turbine than for a reciprocating engine because the power is less per unit of fuel input. Because gas turbine exhaust contains a large percentage of excess air, afterburners or boost burners may be installed in the exhaust to create a supplementary boiler system. This system can provide additional steam or level the steam production during reduced turbine loads.

In all reciprocating internal combustion engines except small air-cooled units, heat can be reclaimed from the lubricating system, jacket cooling water system, and the exhaust.

Coolant fluids and lubricating oil are generally circulated to remove excess heat conducted into the power train during combustion and heat from friction. Some engines are constructed to convert cooling water to steam within the engine.

The approximate distribution of input fuel energy is as follows:

| | |
|---|---|
| Useful work | 33% |
| Friction and radiation | 7% |
| Rejected in jacket water | 30% |
| Rejected in exhaust | 30% |

Of course, neither all of the exhaust heat nor all of the jacket heat can be recovered. Good design could result in the usable portion of the jacket heat and the exhaust heat being about 70% of that shown, or 21% of the input for each.

Depending on engine design, these amounts vary. However, they do indicate that overall cycle thermal efficiency can be greatly improved by waste heat recovery systems if there is a beneficial use for the recovered heat.

**Example 16-1.** A 200 kW internal combustion engine power unit produces 2570 lb/h (0.32 kg/s) of exhaust gas at 950°F

(510°C). The exhaust gas mixture has a specific heat of 0.252 Btu/lb·°F (1.06 kJ/(kg·K)]. Energy in the exhaust gas is to be used in a waste heat boiler to produce dry saturated steam at 280°F (138°C) from water supplied at 60°F (16°C). The exhaust gas is cooled during the process from 950°F to 400°F (510°C to 204°C). Determine the quantity of steam that can be produced, lb/h.

**Solution:**

From Table 2-1 (Chapter 2 of this book), Thermodynamic Properties of Water:

$h_g$ [at 280°F (138°C)] = 1173.94 Btu/lb (2730.88 kJ/kg)

$h_f$ [at 60°F (16°C)] = 28.07 Btu/lb (67.16 kJ/kg)

Equating the heat transfer rate from the exhaust gas to that for the water/steam, yields

$$m_g c_p (t_{g,in} - t_{g,out}) = m_s (h_g - h_f)$$

$$2570(0.252)(950 - 400) = m_s(1173.4 - 28.07)$$

$$m_s = 311 \text{ lb/h } (0.039 \text{ kg/s})$$

## 16.5 Heat Recovery Terminology and Concepts

The following definitions serve as an introduction to heat recovery systems.

**Balanced heat recovery.** Occurs when internal heat gain equals recovered heat and no external heat is introduced to the conditioned space. Maintaining balance may require raising the temperature of recovered heat.

**Break-even temperature.** The outdoor temperature at which total heat losses from conditioned spaces equal internally generated heat gains.

**Changeover temperature.** The outdoor temperature the designer selects as the point of changeover from cooling to heating by the HVAC system.

**External heat.** Heat generated from sources outside the conditioned area. This heat from gas, oil, steam, electricity, or solar sources supplements internal heat and internal process heat sources. Recovered internal heat can reduce the demand for external heat.

**Internal heat.** The total passive heat generated within the conditioned space. It includes heat generated by lighting, computers, business machines, occupants, and mechanical and electrical equipment such as fans, pumps, compressors, and transformers.

**Internal process heat.** Heat from industrial activities and sources such as wastewater, boiler flue gas, coolants, exhaust air, and some waste materials. This heat is normally wasted unless equipment is included to extract it for further use.

**Pinch technology.** An energy analysis tool that uses vector analysis to evaluate all heating and cooling utilities in a process. Composite curves created by adding the vectors allow identification of a "pinch" point, which is the best thermal location for a heat pump.

**Recovered (or reclaimed) heat.** Comes from internal heat sources. It is used for space heating, domestic or service water heating, air reheat in air conditioning, process heating in industrial applications, or other similar purposes. Recovered heat may be stored for later use.

**Stored heat.** Heat from external or recovered heat sources that is held in reserve for later use.

**Usable temperature.** The temperature or range of temperatures at which heat energy can be absorbed, rejected, or stored for use within the system.

**Waste heat.** Heat rejected from the building (or process) because its temperature is too low for economical recovery or direct use or storage capacity is not available.

### 16.5.1 Definition of Balanced Heat Recovery Systems

In an ideal heat recovery system, all components work year-round to recover all of the internal heat before adding external heat. Any excess heat is either stored or rejected. Such an idealized goal is identified as a balanced heat recovery system.

When the outdoor temperature drops significantly, or when the building is shut down (e.g., on nights and weekends), internal heat gain may be insufficient to meet the space heating requirements. Then, a balanced system provides heat from storage or an external source. When internal heat is again generated, the external heat is automatically reduced to maintain proper temperature in the space. There is a time delay before equilibrium is reached. The size of the equipment and the external heat source can be reduced in a balanced system that includes storage. Regardless of the system, a heat balance analysis establishes the merits of balanced heat recovery at various outdoor temperatures.

Outdoor air less than 55°F to 65°F (13 to 18°C) may be used to cool building spaces with an air economizer cycle. When considering this method of cooling, the space required by ducts, air shafts, and fans, as well as the increased filtering requirements to remove contaminants and the hazard of possible freeze-up of dampers and coils must be weighted against alternatives such as using deep row coils with antifreeze fluids and efficient heat exchange. Innovative use of heat pump principles may give considerable energy savings and more satisfactory human comfort than an air economizer. In any case, hot and cold air should not be mixed (if avoidable) to control zone temperatures because it wastes energy.

Many buildings, especially those with computers or large interior areas, generate more heat than can be used for most of the year. Operating cost is minimized when the system changes over from net heating to net cooling at the break-even outdoor temperature at which the building heat loss equals the internal heat load. If heat is unnecessarily rejected or added to the space, the changeover temperature varies from the natural break-even temperature, and operating costs increase. Heating costs can be reduced or eliminated if excess heat is stored for later distribution. The concept of ideal heat balance in an overall building project or a single space requires that one of the following takes place on demand:

- Heat must be removed.
- Heat must be added.
- Heat recovered must exactly balance the heat required, in which case heat should be neither added nor removed.

In small air-conditioning projects serving only one space, either cooling or heating satisfies the thermostat demand. If humidity control is not required, operation is simple. Assuming both heating and cooling are available, automatic controls will respond to the thermostat to supply either. A system should not heat and cool the same space simultaneously.

Multiroom buildings commonly require heating in some rooms and cooling in others. Optimum design considers the building as a whole and transfers excess internal heat from one area to another, as required, without introducing external heat that would require waste heat disposal at the same time. The heat balance concept is violated when this occurs.

Humidity control must also be considered. Any system should add or remove only enough heat to maintain the desired temperature and control the humidity. Large percentages of outdoor air with high wet-bulb temperatures, as well as certain types of humidity control, may require reheat, which could upset the desirable balance. Usually, humidity control can be obtained without upsetting the balance. When reheat is unavoidable, internally transferred heat from heat recovery should always be used to the extent it is available before using an external heat source such as a boiler. However, the effect of the added reheat must be analyzed because it affects the heat balance and may have to be treated as a variable internal load.

When a building requires heat and the refrigeration plant is not in use, dehumidification is not usually required and the outdoor air is dry enough to compensate for any internal moisture gains. This should be carefully reviewed for each design.

**Heat Balance Studies.** The following examples illustrate situations that can occur in nonrecovery and unbalanced heat recovery situation. Figure 16-2 shows the major components of a building that comprise the total air-conditioning load. Values above the zero line are cooling loads, and values below the zero line are heating loads. On an individual basis, the ventilation and conduction loads cross the zero line, which indicated that these loads can be a heating or a cooling load, depending on outdoor temperature. Solar and internal loads are always a cooling load and are, therefore, above the zero line.

Figure 16-3 combines all of the loads shown in Figure 16-2. The graph is obtained by plotting the conduction load of a building at various outdoor temperatures and then adding or subtracting the other loads at each temperature. The project load lines, with and without solar effect, cross the zero line at 16°F (–9°C) and 30°F (–1°C), respectively. These are the outdoor temperatures for the plotted conditions when the naturally created internal load exactly balances the loss.

As plotted, this heat balance diagram includes only the building loads with no allowance for additional external heat from a boiler or other source. If external heat is necessary because of system design, the diagram should include the additional heat.

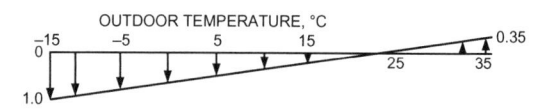

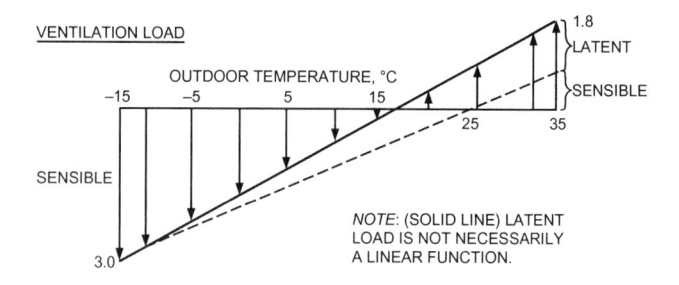

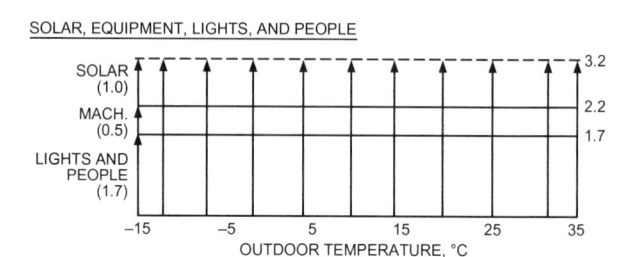

*Fig. 16-2 Major Load Components*
(Figure 31, Chapter 9, 2012 ASHRAE Handbook—HVAC Systems and Equipment)

Figure 16-4 illustrates what happens when heat recovery is not used. It is assumed that with a temperature of 70°F (21°C), heat from an external source is added to balance conduction through the building's skin in increasing amounts down to the minimum outdoor temperature winter design condition. Figure 16-4 also adds the heat required for the outdoor air intake. The outdoor air, comprising part or all of the supply air, must be heated from outdoor temperature to room temperature. Only the temperature range above the room temperature is effective for heating to balance the perimeter conduction loss.

These loads are plotted at the minimum outdoor winter design temperature, resulting in a new line passing through points A, D, and E. This line crosses the zero line at –35°F (–37°C), which becomes the artificially created break-even temperature rather than 30°F (–1°C), when not allowing for solar effect. When the sun shines, the added solar heat at the minimum design temperature would further drop the –35°F (–37°C) break-even temperature. Such a design adds more heat than the overall project requires and does not use balanced heat recovery to use the available internal heat. This problem is most evident during mild weather on systems not designed to take full advantage of internally generated heat year-round.

The following are two examples of situations that can be shown in a heat balance study:

1. As the outdoor air wet-bulb temperature drops, the total heat of the air falls. If a mixture of outdoor and recirculated

# Heat Pump, Cogeneration, and Heat Recovery Systems

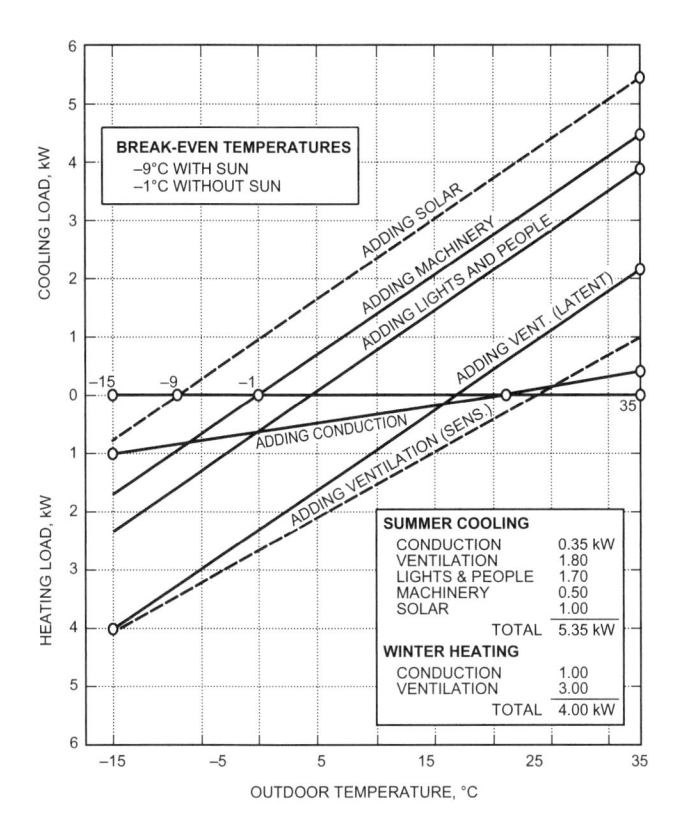

*Fig. 16-3 Composite Plot of Loads in Fig. 16-2*
(Figure 32, Chapter 9, 2012 ASHRAE Handbook—
HVAC Systems and Equipment)

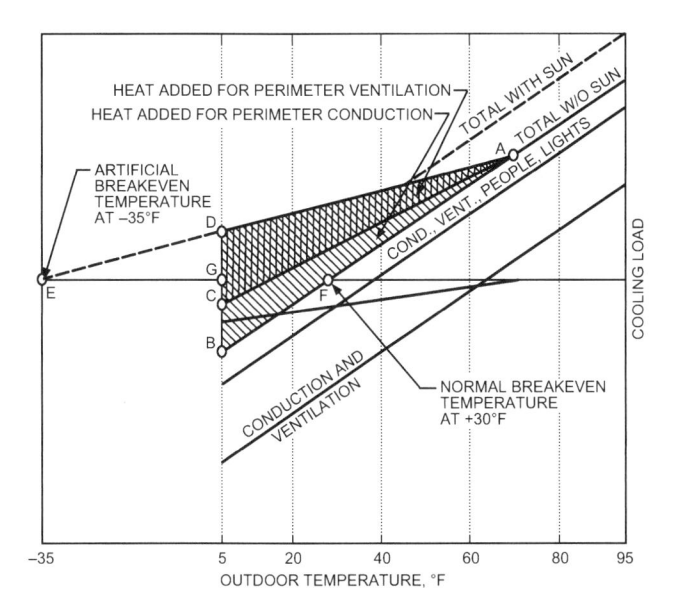

*Fig. 16-4 Non-Heat Recovery System*
(Figure 33, Chapter 9, 2012 ASHRAE Handbook—
HVAC Systems and Equipment)

air is cooled to 55°F (13°C) in summer and the same dry-bulb temperature is supplied by an economizer cycle for interior space cooling in winter, there will be an entirely different result. As the outdoor wet-bulb temperature drops below 55°F (13°C), each unit volume of air introduced does more cooling. To make matters more difficult, this increased cooling is latent cooling, which requires adding latent heat to prevent too low a relative humidity, yet this air is intended to cool. The extent of this added external heat for free cooling is shown to be very large when plotted on a heat balance analysis at 0°F (–18°C) outside temperature.

Figure 16-4 is typical for many current non-heat-recovery systems. There may be a need for cooling, even at the minimum design temperature, but the need to add external heat for humidification can be eliminated by using available internal heat. When this asset is thrown away and external heat is added, operation is less efficient.

Some systems recover heat from exhaust air to heat the incoming air. When a system operates below its natural break-even temperature $t_{be}$, such as 30°F (–1°C) or 16°F (–9°C) (shown in Figure 16-3), the heat recovered from exhaust air is useful and beneficial. This assumes that only the available internal heat is used and that no supplementary heat is added at or above $t_{be}$. Above $t_{be}$, the internal heat is sufficient and any recovered heat would become excessive heat to be removed by more outdoor air or refrigeration.

If heat is added to a central system to create an artificial $t_{be}$ of –35°F (–37°C) as in Figure 16-4, any recovered heat above –35°F (–37°C) requires an equivalent amount of heat removal elsewhere. If the project were in an area with a minimum design temperature of 0°F (–18°C), heat recovery from exhaust air could be a liability at all times for the conditions stipulated in Figure 16-3. This does not mean that the value of heat recovered from exhaust air should be forgotten. The emphasis should be on recovering heat from exhaust air rather than on adding external heat.

2. A heat balance shows that insulation, double glazing, and so forth can be extremely valuable in some situations. However, these practices may be undesirable in certain regions during the heating season, when excess heat must usually be removed from large buildings. For instance, for minimum winter design temperatures of approximately 35°F to 40°F (1.7°C to 4.5°C), it is improbable that the interior core of a large office building will ever reach its break-even temperature. The temperature lag for shutdown periods, such as nights and weekends, at minimum design conditions could never economically justify the added cost of double-pane windows. Therefore, double-pane windows merely require the amount of heat saved to be removed elsewhere. However, in cold climates the double-pane windows may be necessary to provide comfort.

## 16.6 Heat Recovery Systems

Figures 16-5 through 16-9 show several possible heat recovery/simultaneous heating-cooling systems. Figure 16-5 illustrates one method of using water as the heat source or sink and as the heating and cooling medium. The compressor, evaporator, condenser, refrigerant piping, and accessories are

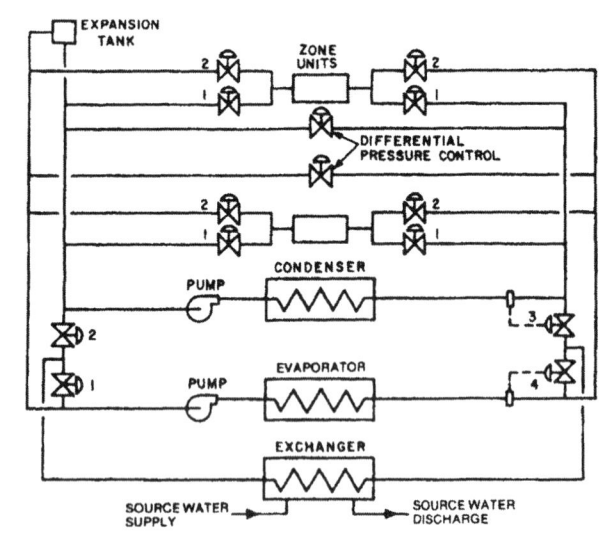

*Fig. 16-5 Water-to-Water Heat Pump Cycle*

essentially standard and are available as a factory-packaged water-to-water heat pumps (see Chapters 14 and 15).

The cycle is flexible, and the heating or cooling medium is instantly available at all times. Heating can be provided exclusively to the zone conditioners by closing valves 2 and 3 and opening valves 1 and 4. With the valves in these positions, the water is divided into two separate circuits. The warm water circuit consists of the condenser (where the heat is supplied by the high-temperature refrigerant), valve 1, zone conditioners, and a circulating pump. The cold water circuit consists of the evaporator (where heat is taken from the water by the low-temperature refrigerant), valve 4, a heat exchanger (where heat is taken from the source water), valve 1, and a circulating pump. The refrigerating compressor operates to maintain the desired leaving water temperature from the condenser.

Similarly, cooling can be exclusively obtained in the cycle of Figure 16-5 by opening valves 2 and 3 and closing valves 1 and 4. With this arrangement, the cold water circuit consists of the evaporator (where heat is removed from the water by the low-temperature refrigerant), valve 2, zone conditioners, and a circulating pump. The warm water circuit consists of the condenser (which receives the heat from the refrigerant), valve 3, a heat exchanger (where heat is rejected to the source water), valve 2, and a circulating pump. The refrigerating compressor operates to maintain the desired water temperature leaving the evaporator.

During the intermediate season, simultaneous heating and cooling can be provided by the cycle shown in Figure 16-5. Valves 3 and 4 are modulated when valves 1 and 2 are open. Valve 3 is adjusted to maintain 85 to 140°F (29 to 60°C) water in the condenser circuit and valve 4 to maintain 45 to 50°F (7 to 10°C) in the evaporator circuit. The excess heating or cooling effect is discharged to the exchanger, which passes it on to the source water.

The source or sink water, if of suitable quality, can be supplied directly to the condenser and evaporator instead of using an exchanger (Figure 16-5). This eliminates one heat transfer surface and its performance penalty.

A water loop heat pump (WLHP) cycle that combines load transfer characteristics with water-to-air heat pump units is illustrated in Figure 16-6. Each module or space has one or more water-to-air heat pumps. The units in both the building core and perimeter areas are connected hydronically with a common two-pipe system. Each unit cools conventionally, supplying air to the individual module and rejecting the heat removed to the two-pipe system through its integral condenser. The total heat gathered by the two-pipe system is expelled through a common heat rejection device. This device often includes a closed circuit evaporative cooler with an integral spray pump. If and when some of the modules, particularly on the northern side, require heat, the individual units switch (by means of four-way refrigerant valves) into the heating cycle. The units derive their heat source from the two-pipe water loop, basically obtaining heat from a relatively high temperature source, that is, the condenser water of the other units. When only heating is required, all units are in the heating cycle and, consequently, an external heat input source is needed to provide heating capability. The heat of compression contrib-

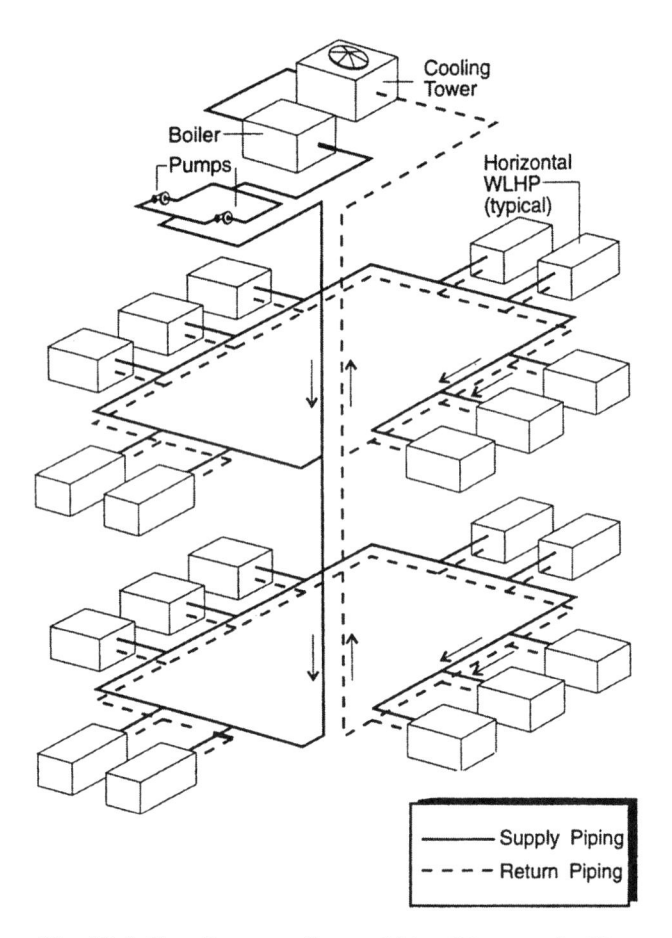

*Fig. 16-6 Heat Recovery System Using Water-to-Air Heat Pump in Closed Loop*
(Figure 27, Chapter 9, 2012 ASHRAE Handbook—HVAC Systems and Equipment)

# Heat Pump, Cogeneration, and Heat Recovery Systems

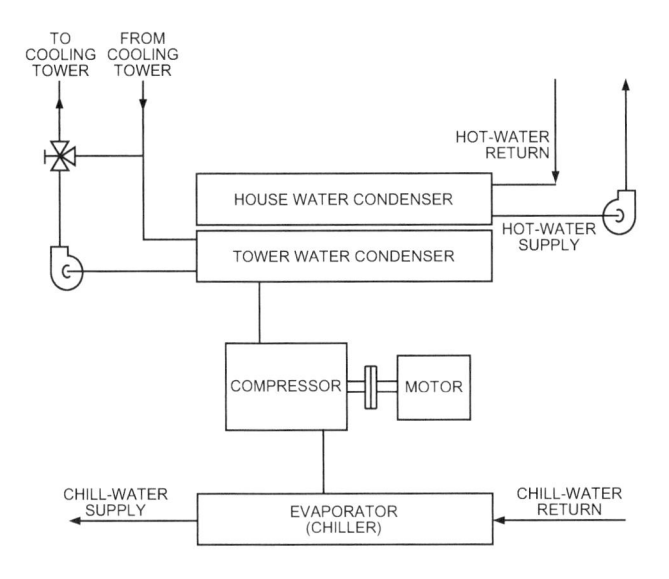

*Fig. 16-7 Heat Transfer Heat Pump with Double-Bundle Condenser*
(Figure 23, Chapter 9, 2012 ASHRAE Handbook—HVAC Systems and Equipment)

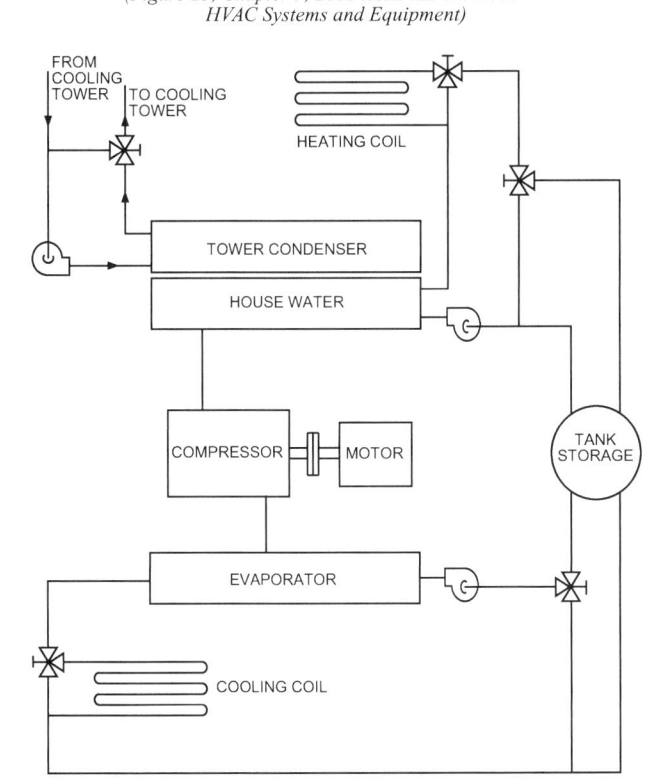

*Fig. 16-8 Heat Transfer System with Storage Tank*
(Figure 24, Chapter 9, 2012 ASHRAE Handbook—HVAC Systems and Equipment)

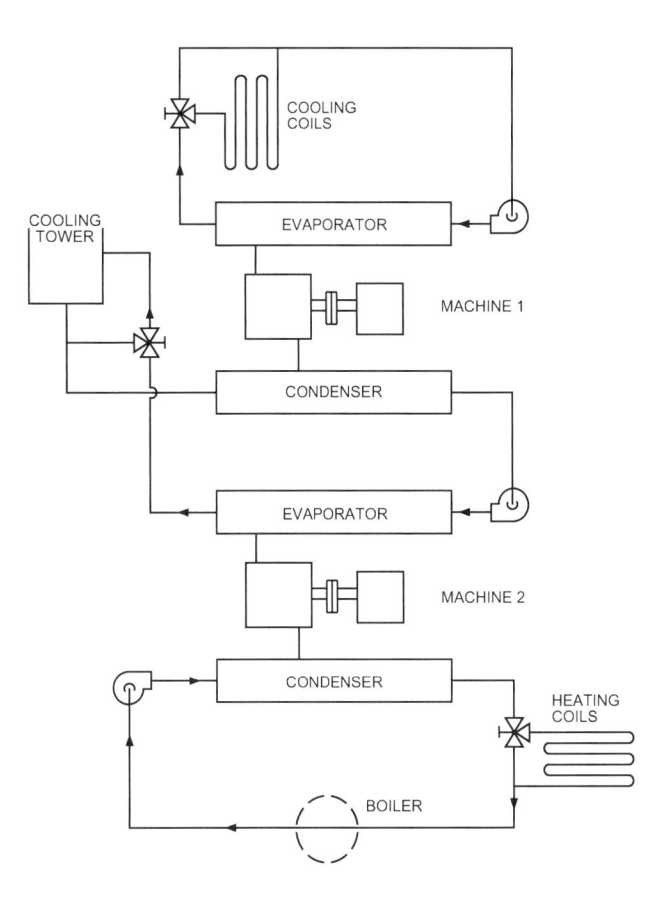

*Fig. 16-9 Multistage (Cascade) Heat Transfer System*
(Figure 25, Chapter 9, 2012 ASHRAE Handbook—HVAC Systems and Equipment)

In many large buildings, internal heat gains require year-round chiller operation. This internal heat is often discharged through a cooling tower. Prudent design may dictate cascade systems with chillers in parallel or series. Manufacturers can assist with custom components to meet a wide range of load and temperature requirements. The double-bundle condenser working with a reciprocating or centrifugal compressor is most often used in this application. Figure 16-7 shows the basic configuration of this system, which makes heat available in the range of 100 to 130°F (38 to 54°C). The warm water is supplied as a secondary function of the heat pump and represents recovered heat.

Figure 16-8 shows a similar cycle, except that a storage tank has been added, enabling the system to store heat during occupied hours by raising the temperature of the water in the tank. During unoccupied hours, water from the tank is gradually fed to the evaporator providing load for the compressor and condenser that heats the building during off hours.

Figure 16-9 is another transfer system capable of generating 130 to 140°F (54 to 60°C) or warmer water whenever there is a cooling load by cascading two compressors hydronically. In this configuration, one chiller can be considered as a chiller only and the second unit as a heating-only heat pump.

utes to the heat source. The water loop is usually 60 to 90°F (15 to 32°C) and, therefore, seldom requires piping insulation.

A water-to-water heat pump can be added in the closed water loop before the heat rejection device for further heat reclaim. This heat pump reuses the heat and can provide domestic hot water or elevate water temperatures in a storage tank to be bled back into the loop.

## 16.7 Problems

**16.1** A heat pump is used in place of a furnace for heating a house. In winter, when the outdoor air temperature is 15°F, the heat loss from the house is 100,000 Btu/h if the inside is maintained at 70°F. (a) Determine the minimum electric power (Carnot) required to operate the heat pump. (b) Determine the actual electric power to operate the heat pump with a heating COP of 3.

**16.2** An air-source heat pump is to be used for both air conditioning and heating of a residence, maintaining the interior at 80°F in summer with an outdoor air temperature of 95°F and a cooling load of 36,000 Btu/h. As a heat pump, it is to maintain 70°F in winter with an outdoor air temperature of 2°F and a heating load of 52,000 Btu/h.

Select a heat pump from the table in Problem 8.13, sized for cooling. What size resistance heater is required at the winter design condition?

**16.3** A 100,000 ft$^2$ building design has a design electrical load of 5 W/ft$^2$. A reciprocating natural gas engine cogeneration plant is to serve the building. The engine-generator is sized for the electrical load, with salvaged heat being used for heating and for driving a single-effect absorption chiller. The design heating load is 3,000,000 Btu/h. The design cooling load is 250 tons; the absorber requires 20,000 Btu/ton·h input.

Calculate hourly design operating costs for heating and cooling. Any shortfall in heating from recovered heat must be made up by a boiler. Any shortfall in cooling by the absorber with recovered heat must be made up by the boiler as input to the absorber.

Compare design operating costs with hourly design operating costs using conventional equipment (purchased electricity for the building and for cooling with an electric chiller at 1.0 kW/ton, purchased gas for a boiler for heating). Use $1.00 per therm, boiler efficiency of 80% for fuel cost, $0.10/kWh for purchased electricity cost.

## 16.8 Bibliography

Anantapantula, V.S. and H.J. Sauer, Jr. 1994. Heat Recovery and the economizer for HVAC systems. *ASHRAE Journal* (November):48-53.

ASHRAE. 2004. Chapter 8, Applied Heat Pump and Heat Recovery Systems. *2004 ASHRAE Handbook—HVAC Systems and Equipment.*

ASHRAE. 1996. *Practical Guide to Cool Storage Projects.*

ASHRAE. 1995. *Commercial/Institutional Ground-Source Heat Pump Engineering Manual.*

Dorgan, C. E. and J. S. Elleson. 1993. *Design Guide for Cool Thermal Storage*, ASHRAE.

Orlando, J. 1996. *Cogeneration Design Guide*, ASHRAE.

Sauer, H. J., Jr. and R. H. Howell. 1988. "Design Guidelines for Use of an Economizer with Heat Recovery," *ASHRAE Transactions* 94(2).

## SI Figures

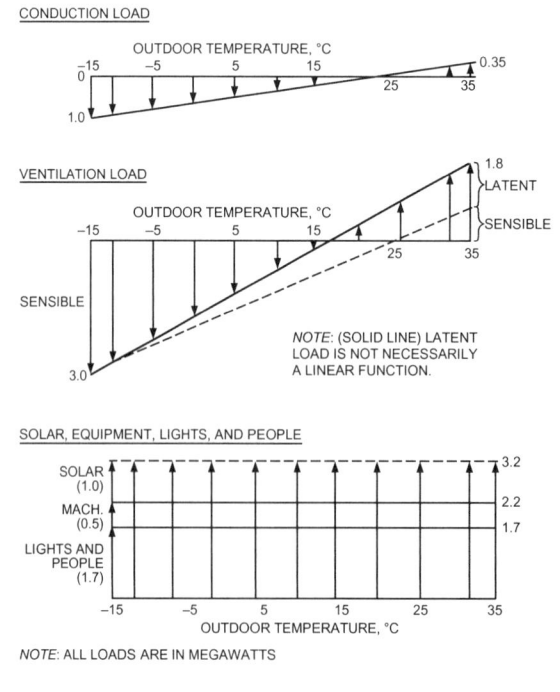

*Fig. 16-2 SI Major Load Components*
(Figure 31, Chapter 12, 2012 ASHRAE Handbook—
HVAC Systems and Equipment)

# Heat Pump, Cogeneration, and Heat Recovery Systems

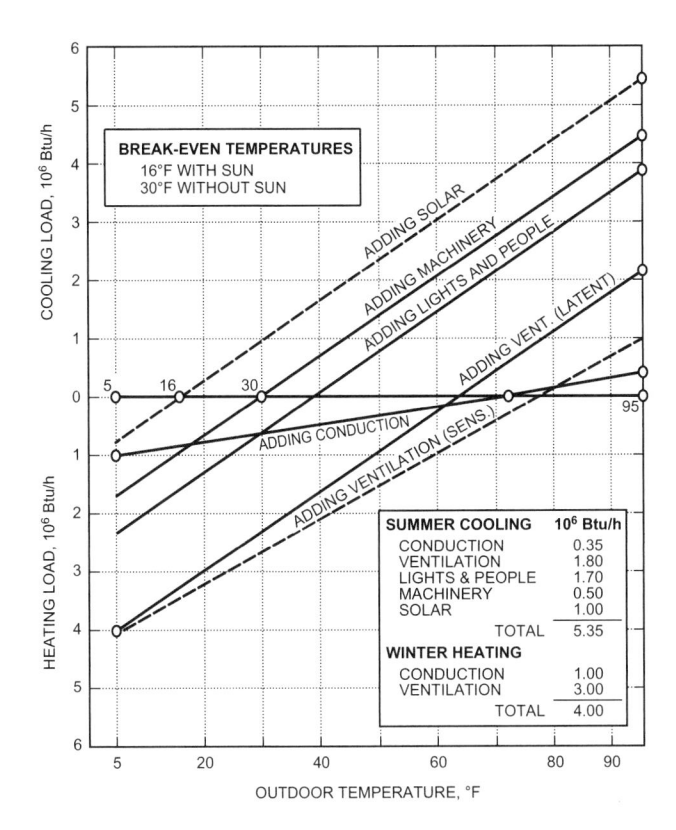

*Fig. 16-3 SI  Composite Plot of Loads in Fig. 16-2*
(Figure 32, Chapter 9, 2012 ASHRAE Handbook—
HVAC Systems and Equipment)

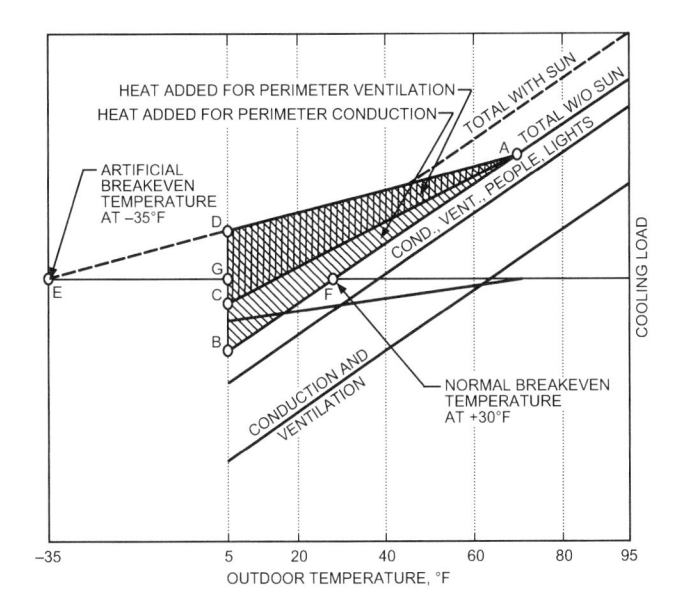

*Fig. 16-4 SI  Nonheat Recovery-System*
(Figure 33, Chapter 9, 2012 ASHRAE Handbook—
HVAC Systems and Equipment)

# Chapter 17

# AIR-PROCESSING EQUIPMENT

Cooling, heating, humidifying, dehumidifying, and cleaning of air are some of the processes for which mechanical equipment is needed in HVAC systems. These kinds of air-processing devices are examined in this chapter, along with air-to-air energy recovery equipment. For additional information on these topics, the reader is referred to Chapters 21 through 29, and 41 of the *2012 ASHRAE Handbook—HVAC Systems and Equipment*. Chapters 3 and 4 of the *2013 ASHRAE Handbook—Fundamentals* provide the heat transfer background for much of this equipment.

## 17.1 Air-Handling Equipment

Air-handling units consist of the equipment that filter, heat, humidify, cool, dehumidify, and move the air. The three major types of air handlers are

- Factory-fabricated units
- Built-up (field-erected) units
- Customized units

Each category contains both single-zone and multizone units.

Factory-fabricated units usually contain the fan and cooling and/or heating coils and filter assembled in a cabinet, usually of sheet metal. Mixing boxes are also included as desired. Standard arrangements and options available with factory-assembled units include

- Draw-through or blow-through
- Horizontal or vertical assembly
- Chilled water or refrigerant cooling coils
- Hot water, steam, or electric heating coils
- Forward-curved, backward-curved, airfoil, axial, or plug fans
- Flat or "V" bank filter sections
- Mixing boxes with damper assemblies

Factory-fabricated units are usually less expensive than field-erected built-up systems and the delivery time is usually shorter than for the purchase and assembly of a built-up unit.

Rooftop units may be considered a special type of factory-fabricated unit, being similar to such indoor units, except that they are located on the roof, saving valuable space that would otherwise be needed for the mechanical equipment room. Packaged rooftop units are sometimes completely self-contained including the condensing unit and usually either a gas-fired or electric heating section. On the negative side, rooftop units generally require higher maintenance and have a shorter service life than units protected in mechanical equipment rooms.

Built-up air-handling units consist of separate casings enclosing fans, coils, filters, mixing boxes, and plenums. Built-up unit casings may be factory made or fabricated by a local sheet metal contractor. Dimensions of built-up units can often be varied for the floor area available, being limited only by the coil and fan selections. There is almost no limit to the capacity of built-up units.

Customized air-handling units are normally selected to provide either a higher level of quality or special component configurations for a special application.

## 17.2 Cooling Coils

Ceiling coils are used to cool an airstream under forced convection. Such equipment may consist of a single coil section or several individual coil sections built into banks. Coils are also used extensively as components in central station air-handling units, room terminals, and in factory-assembled, self-contained air conditioners.

Chilled water, brines, or volatile refrigerants are the usual cooling media. Cooling coils that use relatively high temperature water usually do not dehumidify the air; however, most coil equipment is designed to remove sensible heat and dehumidify simultaneously. The coil assembly should include a means to protect the coil from dirt accumulation and to keep dust and foreign matter out of the conditioned space.

### 17.2.1 Coil Construction and Arrangement

Coils include bare tubes or pipe, and those with extended or finned surfaces. The design of coils with extended surfaces on the air side considers the materials, fin size and spacing, ratio of extended surface area to that of the tube area, tube nesting center dimensions, staggered or in-line tube arrangement, and use of turbulators.

Staggered tubes increase the total heat transfer over that of the in-line arrangement, and turbulators may also be used to enhance total heat transfer efficiency. The surface arrangement has a great effect on the air-film heat transfer resistance and associated air-side pressure drop. Several arrangements are illustrated in Figure 17-1.

In fin or extended surface coils, the external surface of the tubes is **primary**, and the fin surface is **secondary**. The primary surface consists of rows of round tubes or pipes, which may be staggered or placed in line with respect to the airflow. Some tubes are flattened or have nonround internal passageways. The inside surface of the tubes is usually smooth and plain, but some designs have internal fins or

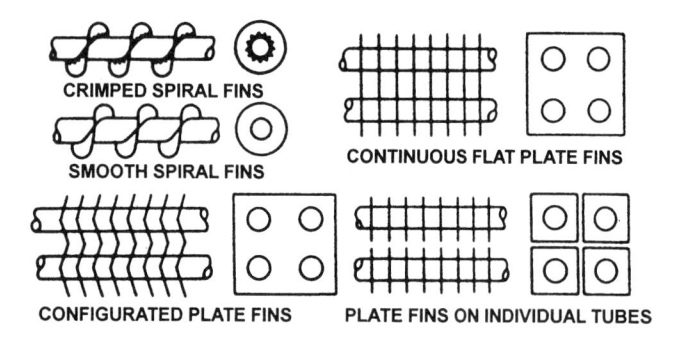

*Fig. 17-1 Types of Fin-Coil Arrangements*

turbulence promoters (either fabricated or extruded) to provide turbulence and additional inside surface area for enhancing performance. Individual tubes are generally interconnected by return bends, sometimes together with hairpin tubes to form the required serpentine arrangement of multipass tube circuits. Some flattened tubes are folded into continuous serpentine circuits with fins metallurgically bonded between adjacent tube passes. Numerous fin arrangements are used; the most common are smooth spiral, crimped spiral, flat plate, and configured plate. A good thermal bond between the tube and the fin must be maintained permanently to ensure low resistance to heat transfer from fin to tube.

In some coils, fins are wound under pressure onto the tubes in order to upset the metal slightly at the fin root. They are then coated with solder while the fin and tube are still revolving to ensure a uniform solder coating. In other types of coils, the spiral fin may be knurled into a shallow groove on the exterior of the tube. The tube may be expanded after the fins are assembled, or the tube-hole flanges of a flat or configurated fin may be made to override those in the preceding fin and so compress them on the tube. Some construction techniques even form the fin from the material of the tube itself.

Cooling coils for water or for volatile refrigerants frequently have aluminum fins and copper tubes, although copper fins on copper tubes are also used and the combination of aluminum fins on aluminum tubes is finding use. Adhesive bonding is sometimes used in making header connections, return bends, and fin-tube joints, particularly for aluminum-to-aluminum joints. Many types of lightweight, extended-surface cooling coils are made for both heating and cooling. Tube outside diameters are commonly 1/4, 3/16, 3/8, 5/8, 3/4, and 1 in. (6.4, 9.5 12.7, 15.9, 19.1, and 25 mm), and fins are spaced from 3 to 14 per inch (1.8 to 8.5 mm apart). The tube spacing varies from about 5/8 to 2 1/2 in. (16 to 64 mm) on centers, depending on the width of individual fins and other performance considerations. Fin spacing should be chosen for the duty to be performed, with special attention paid to air friction, prevention of water carryover, possible lint accumulation, and, especially at lower temperatures, frost accumulation.

## 17.2.2 Water Coils

Water coil performance depends on eliminating air from the water circuit and properly distributing the water. Unless

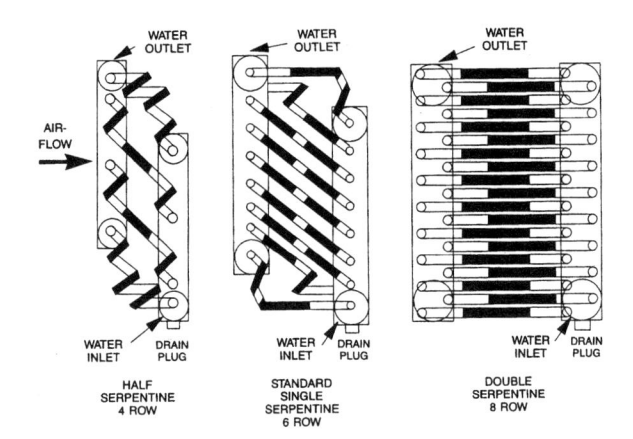

*Fig. 17-2 Typical Water Circuit Arrangements*

the system is vented, air may accumulate in the coil circuits, which reduces thermal performance and may cause noise or vibration. Air vent connections are usually provided on the coil water headers. Depending on performance requirements, the water velocity inside tubes usually ranges from approximately 1 to 8 ft/s (0.3 to 2.4 m/s), and the design water pressure drop across coils varies from about 5 to 50 ft of water (15 to 150 kPa).

A variety of circuit arrangements, and combinations thereof, for varying the number of parallel water flow passes within the tube core are usually available. Some typical arrangements are shown in Figure 17-2.

## 17.2.3 Direct-Expansion Refrigerant Coils

Direct-expansion coils present more complex problems of cooling fluid distribution than water or brine coils. The coil should be effectively and uniformly cooled throughout, and the compressor must be protected from entrained, unevaporated refrigerant. Direct-expansion coils are used on both flooded and dry-expansion refrigeration systems.

The **flooded system** is mainly used for low-temperature applications where a small temperature difference between the air and refrigerant is advantageous. However, a relatively large volume of refrigerant is required, together with extra components such as a surge tank and interconnecting piping. Other applications of direct-expansion coils generally use dry expansion. For **dry-expansion systems**, the most commonly used refrigerant liquid metering devices are the capillary tube and the thermostatic expansion valve.

The **capillary tube system** is applied on evaporator coils in factory-assembled, self-contained air conditioners up to approximately 10 tons (35 kW) capacity and is used extensively on the smaller capacity models such as window or room units. In this system, the bore and length of a capillary tube are sized so that at full load, just enough refrigerant liquid is metered from the condenser to the evaporator coil to be completely evaporated.

The **thermostatic expansion valve system** is common for all dry-expansion coil applications, particularly for field-assembled coil sections in central air-handling units, as well as for the larger factory-assembled hermetic air conditioners.

# Air-Processing Equipment

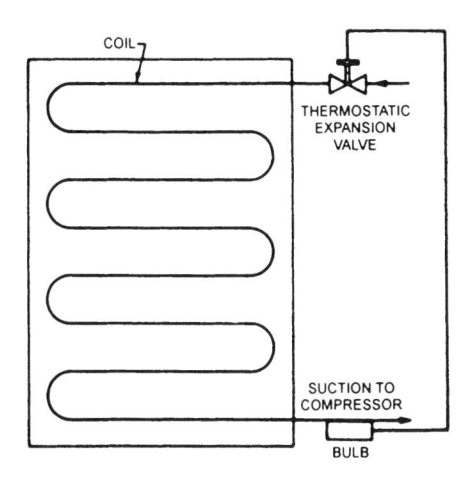

*Fig. 17-3 Dry Expansion Coil with Thermostatic Expansion Valve*

A schematic typical of a coil and thermostatic expansion valve assembly is shown in Figure 17-3. The thermostatic expansion valve automatically regulates the rate of refrigerant liquid flow to the coil in direct proportion to the rate of evaporation of refrigerant liquid in the coil, thereby maintaining the superheat at the coil suction outlet within the usual predetermined limits of 6 to 10°F (3 to 6°C).

## 17.2.4 Coil Selection

When selecting a coil, the following factors must be considered:

- Job requirements: cooling, dehumidifying, and the capacity required to maintain balance with other system components, for example, compressor equipment in the case of direct-expansion coils.
- Temperature of entering air: dry bulb only if there is no dehumidification; dry and wet bulb if moisture is to be removed.
- Temperature of entering, chilled water or evaporating pressure of refrigerant and type of refrigerant.
- Temperature of leaving air dry bulb and wet bulb or dewpoint.
- Available cooling media and its operating temperature and quantity.
- Space and dimensional limitations.
- Air quantity limitations.
- Allowable frictional resistances in cooling media piping system (including coils).
- Allowable frictional resistances in air circuit (including coils).
- Characteristics of individual coil designs.
- Individual installation requirements, such as the type of automatic control to be used or the presence of a corrosive atmosphere.
- Coil air face velocity.
- To prevent water carryover from dehumidifying coils, face velocities should not exceed 500 fpm (2.54 m/s), and fin density should be no greater than 8 fins per inch (fins 3 mm apart).

Coil ratings and selection procedures are usually presented in one of two ways:

**Basic Data Method.** Coil performance parameters are published in the form of tables or charts from which the coil row depth is calculated after determining the required coil sensible and total heat capacities and other design variables from the job conditions. The initial selection generally indicates a nonintegral row depth requirement. It is frequently necessary to recheck and reselect the coil to match more closely the required air and cooling fluid conditions with the integral row depth actually installed, particularly for dehumidifying coils. The method is generally used in selecting coils or coil banks for field assembly, since there is a vast number of size and row depth combinations available.

**Unit Rating Method.** Performance for specific combinations of coil face area and row depth are presented in tables or charts. This method provides a direct selection of specific coils to match the required capacity under the job conditions. This method is frequently used in selecting coils for central station air-handling units and also in determining performance for factory-assembled, self-contained air conditioners.

With either method, various combinations of coil face area, row depth, air velocity, and air quantity may be chosen to do the same job. Coil selection requires understanding each case, and selection should be based on an economic analysis of the plant as a whole.

Most coil manufacturers and some commercial software providers have programs to assist in the selection and diagnosis of cooling coils.

Table 17-1 provides an illustration of coil performance data as may be found in a typical cooling coil catalog.

Coils that operate wet, particularly those with enhanced (configurated) fins tend to build up dirt on the fin and tube surfaces so they must be capable of being, first, visually inspected for fouling and, second, cleaned. Both of these are difficult without sufficient space on each side of the coil. Fin spacing closer than 1/8 of an inch or enhanced fins are not recommended.

Improper selection of the cooling coil is the most common cause of performance failures in air-conditioning systems.

## 17.2.5 Application Range

Based on information in AHRI Standard 410, dry surface (sensible cooling) coils and dehumidifying coils (which both cool and dehumidify), particularly for field-assembled coil banks, factory-assembled coil banks, or factory-assembled central station air conditioners using different combinations of coils, are usually rated within the following limits:

| | |
|---|---|
| Entering air temperature | 65 to 100°F (18 to 38°C) |
| Entering air wet bulb | 60 to 85°F (15 to 30°C) |
| Air face velocity | 300 to 800 fpm (1.5 to 4.0 m/s) (sometimes as low as 200 and as high as 1500 fpm) |
| Refrigerant saturated temperature | 30 to 55°F (−1 to 13°C) at coil suction outlet [refrigerant vapor superheat at coil suction outlet is 6°F (3.3°C) or higher |

## Table 17-1 Typical Cooling Coil Catalog Data

*[Table: 44EWT Cooling Coil Catalog Data, EDB/EWB=80/67, showing values for WTR=8 and WTR=10, across air velocities 400–700 FPM, with rows organized by FPS (2, 4, 6, 8), ROW (2, 4, 6, 8), and FPF (80, 100, 120, 140, 160). Columns give MBH, LDB, LWB.]*

NOTE:
1. MBh = MBh/ft² Coil Face Area
   WTR = Water Temperature Rise (F)
   fps = Water Velocity (ft/sec)
   LDB = Leaving Dry Bulb (F)
   LWB = Leaving Wet Bulb (F)
   EWT = Entering Water Temperature
2. When using turbulators, make selection based on double the actual water velocity.

*Note:* 1. EWT = entering water temperature, °F  
  MBH = 1000 Btu/h per square foot of coil area  
  EDB = entering dry bulb, °F  
  EWB = entering wet bulb, °F  
  FPS = water velocity, feet per second  
  fpm = air velocity, feet per minute  
  WTR = water temperature rise, °F  
  LDB = leaving dry bulb, °F  
  LWB = leaving wet bulb, °F  
  FPF = fins per foot  
2. When using turbulators, make selection based on double the actual water velocity.

# Air-Processing Equipment

Entering liquid — 35 to 65°F (2 to 18°C) temperature

Water flow rate — 1.2 to 6 gpm per ton (0.02 to 0.1 L/s per kW) [equivalent to a water temperature rise of from 4 to 20°F (2 to 11°C)]

Water velocity — 1 to 8 fps (0.3 to 2.4 m/s)

## 17.2.6 Determining Refrigeration Load

The following method of calculating refrigeration load shows a division of the true sensible and latent heat loss of the air which is accurate within the limitations of the data. This division does not correspond to load determination obtained from approximate factors or constants.

The total refrigeration load $q_t$ of a cooling and dehumidifying coil (or air washer) per unit mass dry air is indicated in Figure 17-4 and consists of the following components:

1. The sensible heat $q_s$ removed from the dry air and moisture in cooling from entering temperature $t_1$ to leaving temperature $t_2$.
2. The latent heat $q_l$ removed to condense the moisture from $W_1$ to $W_2$.
3. The heat leaving the system as liquid condensate.

Items 1, 2 and 3 may be related by

$$q_t = q_s + q_l - q_w \qquad (17\text{-}1)$$

If only the total heat value is desired, it may be computed by

$$q_t = (h_1 - h_2) - (W_1 - W_2)h_{f4} \qquad (17\text{-}2)$$

where

$h_1$ and $h_2$ = enthalpy of moist air at points 1 and 2, respectively

$W_1$ and $W_2$ = humidity ratio at points 1 and 2, respectively

$h_{f4}$ = enthalpy of saturated liquid at final temperature

$W_1 - W_2$ = mass of water vapor condensed per unit mass of air

If a breakdown into latent and sensible heat components is desired, the following relations may be used. The latent heat may be found from

$$q_l = (h_1 - h_3) - (W_1 - W_2)h_{f2} \qquad (17\text{-}3)$$

where

$h_{g1}$ = enthalpy of saturated water vapor at temperature $t_1$

$h_{g3}$ = enthalpy defined by temperature $T_1$, and humidity ratio $W_2$

The sensible heat may be shown to be

$$q_s = (h_3 - h_2) + (W_1 + W_2)(h_{f2} - h_{f4}) \qquad (17\text{-}4)$$

The final condensate temperature $t_4$ leaving the system is subject to substantial variations, depending on the method of coil installation, as affected by such factors as coil face orientation, airflow direction and air duct insulation. In practice, $t_4$ is frequently the same as the leaving wet-bulb temperature. Within the normal air-conditioning range, precise values of $t_4$ are not necessary since energy in the condensate removed from the air usually represents about 0.5 to 1.5% of the total refrigeration load. Values needed to calculate moist air properties can be found on the ASHRAE Psychrometric Chart (Chart 1) and Tables 2 and 3 of Chapter 1 of the 2013 *ASHRAE Handbook—Fundamentals*.

**Example 17-1** Air enters a coil at 90°F dry bulb, 75°F wet bulb; it leaves at 61°F dry bulb, 58°F wet bulb; leaving liquid water is at 54°F.

**Solution:**

From the ASHRAE Psychrometric Chart, find the following:

$h_1$ = 38.37 Btu/lb$_m$ dry air

$h_2$ = 25.06 Btu/lb$_m$ dry air

$h_3$ = 32.14 Btu/lb$_m$ dry air

$W_1$ = 106.63 grains water vapor/lb$_m$ dry air

$W_2$ = 67.04 grains water vapor/lb$_m$ dry air

From Table 3, Chapter 1, of the 2013 *ASHRAE Handbook—Fundamentals*, find the following:

$h_{f4}$ = 22.07 Btu/lb$_m$ liquid water

From Equation (17-2), the total heat is:

$$q_t = (h_1 - h_2) - (W_1 - W_2)h_{f4}$$

$$= (38.37 - 25.06) - \frac{(106.63 - 67.04)}{7000}(22.07)$$

$$= 13.19 \text{ Btu/lb}_m \text{ dry air}$$

From Equation (17-3), the latent heat is

$$q_l = (h_1 - h_3) - (W_1 - W_2)h_{f2}$$

$$= (38.37 - 32.14) - \frac{(106.63 - 67.04)}{7000}29.08$$

$$= 6.07 \text{ Btu/lb}_m \text{ dry air}$$

From Equation (17-4), the sensible heat is

$$q_s = (h_3 - h_2) + (W_1 - W_2)(h_{f2} - h_{f4})$$

$$(32.14 - 25.06) + \frac{(106.63 - 67.04)}{7000}(29.08 - 22.07)$$

$$7.12 \text{ Btu}_m \text{ dry air}$$

The heat content of the leaving condensate is

$$q_w = (W_1 - W_2)h_{f4}$$

$$\frac{(106.63 - 67.04)}{7000}22.07$$

$$0.12 \text{ ( Btu/lb}_m \text{ dry air)}$$

## 17.3 Heating Coils

Generally, extended-surface coils are used for heating air with steam or hot water.

### 17.3.1 Steam Coils

For proper performance of steam heating coils, condensate and air or other noncondensables must be rapidly eliminated and the steam must be uniformly distributed to the individual tubes. Noncondensable gases (such as air or carbon dioxide) that remain in the coil cause chemical corrosion and result in early coil failure.

Steam is distributed uniformly by methods such as:

- Individual orifices in the tubes
- Distributing plates in the steam headers
- Perforated, small-diameter, inner steam-distributing tubes extending to the larger tubes of the primary surface

Coils with perforated inner tubes are constructed with different arrangements such as:

- Supply and return on one end, with the incoming steam used to heat the leaving condensate
- Supply and return on opposite ends
- Supply and return on one end and a supply on the opposite end

Properly designed and selected steam distribution coils distribute steam throughout the full length of all primary tubes, even when the leaving air temperature is controlled by modulating the steam supply through a steam-metering valve. Thus, more uniform leaving air temperatures are produced over the entire length and face of the coil than from a single-tube coil.

Piping, controls, and installation must be designed to protect the coils from freezing due to incomplete draining of condensate. When the entering air temperature is 32°F (0°C) or below, the steam supply to the coil should not be modulated. A series of coils in the airstream, with each coil sized to be on or completely off in a specific sequence depending on the entering air temperature, has less likelihood for a freeze-up. Bypass dampers are also common. When less than full load conditions occur, air is bypassed around the steam coil with full steam being kept on the coil. In this system, the high-velocity jets of low-temperature air must not impinge on the coil when the face dampers are in a partially closed position.

### 17.3.2 Water Coils

In order to produce the desired heating capacity, hot water comfort heating systems usually require no more than one or two rows of tubes in the direction of airflow. Various circuits are used to produce the most efficient capacity without excessive water pressure drop through the coil. The relative directions of fluid flows influence the performance of heat transfer surfaces. In air-heating coils with only one row of tubes, the air flows at right angles relative to the heating medium. In coils with more than one row of tubes in the direction of the airflow, the heating medium in the tubes may be circuited in various ways.

Although crossflow is common in steam-heating coils, parallel flow and counterflow arrangements are common in water coils. Counterflow is the preferred arrangement to obtain the highest possible mean temperature difference.

A single-tube serpentine circuit on small-size booster heaters requires small water quantities up to a maximum of approximately 4 or 5 gpm (0.25 or 0.32 L/s). With this arrangement, a single tube handles the entire water quantity, provided the tube is circuited such that it makes a number of passes across the airstream.

Commonly selected water circuits are illustrated in Figure 17-5. When entering air temperatures are below freezing (an antifreeze brine is sometimes used), piping the coil for parallel flow rather than counterflow should be considered, placing the highest water temperature on the entering air side. Coils piped for counterflow have water enter the coil in the tube row on the exit air side of the coil. Coils piped for parallel flow have water enter the tube row on the entering air side of the coil.

If air temperatures near or below freezing are anticipated, full design water flow rate should be ensured whenever the temperature approaches freezing. These coils are usually provided with a coil pump that provides constant flow through the pump with varying water temperature for reduced load control.

### 17.3.3 Electric Heating Coils

An electric heating coil consists of a length of resistance wire (commonly nickel/chromium) to which a voltage is applied. The resistance wire may be bare or sheathed. The

*Psychrometric Performance of a Cooling and Dehumidifying Coil*

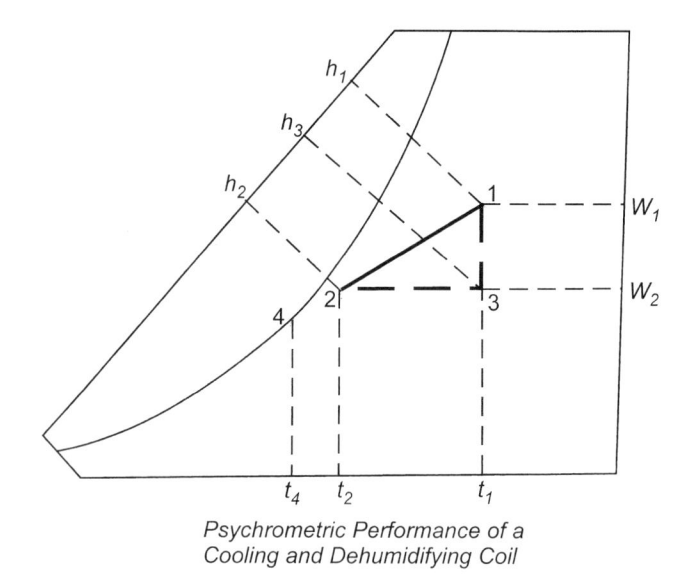

Fig. 17-4 *Psychrometric Performance of Cooling and Dehumidifying Coil*

# Air-Processing Equipment

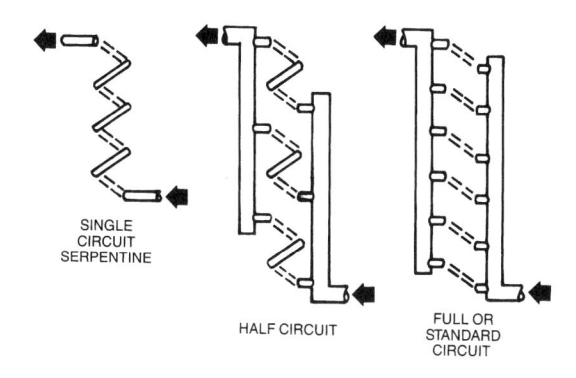

*Fig. 17-5 Water Circuit Arrangements for Heating Coils*

sheathed coil is a resistance wire encased by an electrically insulating layer such as magnesium oxide, which is encased in a finned steel tube. The sheathed coils are more expensive, have a higher air-side pressure drop, and require more space. The outer surface temperature of sheathed coils is lower, the coils are mechanically stronger, and electrical contact with body or housing is prevented.

## 17.3.4 Coil Ratings

Steam and hot water coils are usually rated within these limits, which may be exceeded for special applications:

| | |
|---|---|
| Air face velocity | Between 200 and 1500 fpm (1 to 8 m/s), based on air at standard density of 0.075 lb/ft$^3$ (1.2 kg/m$^3$) |
| Entering air temperature | –20 to 100°F (–30 to 38°C) for steam coils; 0 to 100°F (–20 to 38°C) for hot water coils |
| Steam pressures | From 2 to 250 psia (14 to 1700 kPa) at the coil steam supply connection (pressure drop through the steam control valve must be considered) |
| Hot water | Between 120 and 250°F (50 and 120°C) temperature |
| Water velocities | From 0.5 to 8 fps (0.2 to 2.5 m/s) |

Individual installations vary widely, but the following values can be used as a guide. The most common air face velocities used are between 500 and 1000 fpm (2.5 and 5 m/s). Delivered air temperatures vary from about 72°F (22°C) for ventilation only to about 150°F (66°C) for complete heating. Steam pressures vary from 2 to 15 psig [15 to 100 kPa (gage)], with 5 psig [35 kPa (gage)] being the most common. A minimum steam pressure of 5 psig [35 kPa (gage)] is recommended for systems with entering air temperatures below freezing. Water temperatures for comfort heating are commonly between 120 and 200°F (50 and 93°C), with water velocities between 4 and 6 fps (1.2 and 1.8 m/s). Water quantity is usually based on about 20 to 40°F (10 to 20°C) temperature drop through the coil. Air resistance is usually limited to 0.4 to 0.6 in. (100 to 150 Pa) of water for commercial buildings and to about 1 in. (250 Pa) for industrial buildings. High-temperature water systems have water temperatures commonly between 300 and 400°F (150 and 200°C), with up to 100°F (55°C) drops through the coil.

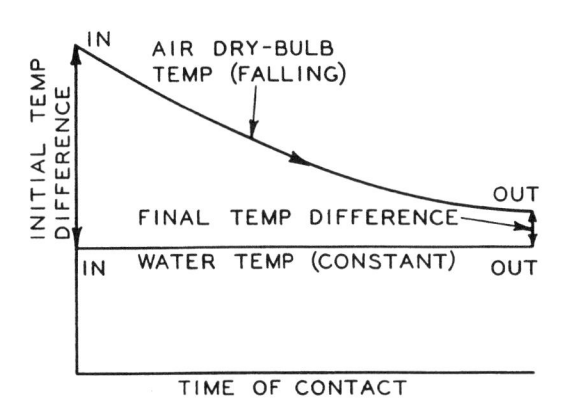

*Fig. 17-6 Interaction of Air and Water in Evaporative Air Coolers*

## 17.4 Evaporative Air-Cooling Equipment

In its broadest sense, the evaporative cooling principle applies to all equipment that exchanges sensible heat for latent heat. Cooling towers, evaporative condensers, vacuum cooling apparatus, air washers, spray coil dehumidifiers, and packaged evaporative coolers cool air by evaporation. This equipment falls into two general categories: (1) apparatus for air cooling and (2) apparatus for heat rejection.

Evaporative air cooling evaporates water into an airstream. Illustrated in Figure 17-6 are thermodynamic changes that take place between air and water that are in direct contact in the moving airstream. The continuously recirculated water has reached an equilibrium temperature that equals the entering air wet-bulb temperature. The heat and mass transfer process between the air and water lowers the air dry-bulb temperature and increases the humidity ratio at constant wet-bulb temperature.

The extent to which the leaving air temperature approaches the thermodynamic wet-bulb temperature of the entering air, or the extent to which complete saturation is approached, is conveniently expressed as cooling or saturation efficiency and is defined as

$$e_c = (t_1 - t_2)/(t_1 - t') \qquad (17\text{-}5)$$

where

$e_c$ = cooling or saturation efficiency
$t_1$ = dry-bulb temperature of the entering air
$t_2$ = dry-bulb temperature of the leaving air
$t'$ = thermodynamic wet-bulb temperature of entering air

If warm or cold unrecirculated water is used, the air can be heated and humidified or cooled and dehumidified.

Evaporative air-cooling equipment can be placed in two general classes, direct and indirect. In the direct system, air is cooled by direct contact with the water, either from the wetted

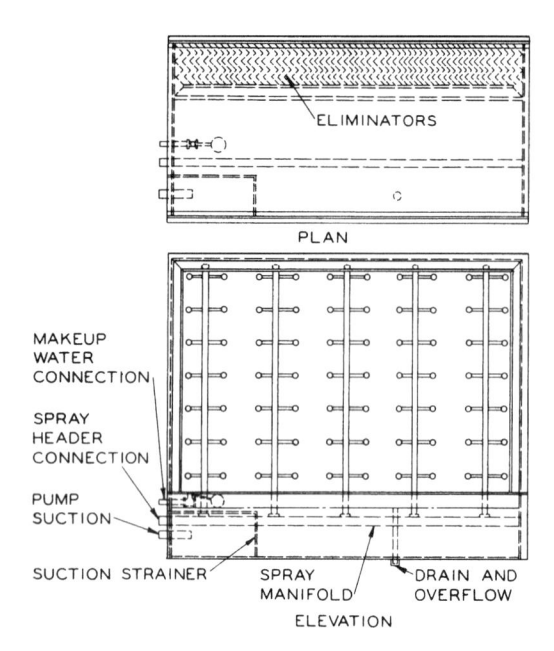

*Fig. 17-7 Single-Bank Air Washer*

surface of an extended-surface material (as in packaged air coolers) or with a series of sprays (as in an air washer). In the indirect system, air is cooled in a heat exchanger by a secondary stream of air and water that has been evaporatively cooled, such as by a cooling tower and cooling coil.

By applying recirculating, regenerating principles, temperatures below the initial wet bulb may be produced. However, the cost of these more complex devices has restricted their use.

## 17.5   Air Washers

A spray-type air washer consists of a chamber or casing containing a spray nozzle system, a tank for collecting the spray water as it falls, and an eliminator section at the discharge to remove the entrained drops of water from the air. A pump recirculates water at a rate in excess of the evaporation rate. Heat and mass transfer between the air and the water cools the water.

Construction features of conventional spray-type air washers are shown in Figure 17-7. Requirements of air washer operation are:

- Uniform distribution of air across the spray chamber
- Air velocities from 300 to 700 fpm (1.5 to 3.5 m/s) in the washer chamber
- Adequate spray water broken up into fine droplets at pressures from 20 to 40 psig [140 to 280 kPa (gage)]
- Good spray distribution across the airstream
- Sufficient length of travel through the spray and wetted surfaces
- Elimination of free moisture from the outlet air

Spray water requirements for spray-type air washers used for washing or evaporative cooling vary from 4 gpm per 1000 cfm with a single bank to 10 gpm per 1000 cfm for double banks (0.5 to 1.3 L/s per m³/s). Pumping heads usually range from 55 to 150 ft (160 to 450 kPa) of water, depending on such factors as spray pressure, height of apparatus, and pressure losses in pipe and strainers.

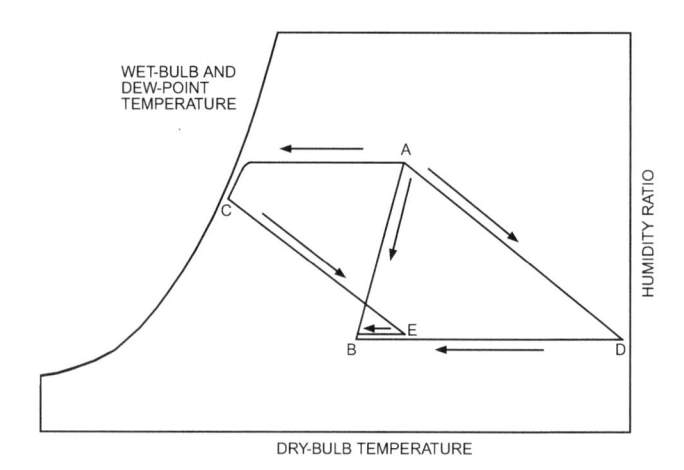

*Fig. 17-8  Methods of Dehumidification*
(Figure 1, Chapter 24, 2012 ASHRAE Handbook—HVAC Systems and Equipment)

## 17.6   Dehumidification

Dehumidification is the reduction of the water content of air, gases, or other fluids. Dehumidification is normally limited to equipment operating at essentially atmospheric pressures, built to standards similar to other types of air-handling equipment.

Drying of gases has become an increasingly important operation. Some commercial applications include the following:

- Lowering the relative humidity to facilitate manufacturing and handling of hygroscopic materials
- Air conditioning for comfort (in combination with cooling, under certain design conditions, such as high moisture load in comparison to sensible cooling load)
- Providing protective atmospheres for the heat treatment of metals
- Maintaining controlled humidity conditions in warehouses and caves for storage
- Preserving ships and other surplus equipment, which would otherwisedeteriorate
- Condensation and corrosion control
- Drying air for wind tunnels
- Drying natural gas
- Drying instrument air and plant air
- Drying of process and industrial gases
- Dehydration of liquids

### 17.6.1  Methods of Dehumidification

Dehumidification can be accomplished by compression, refrigeration, liquid sorption, solid sorption, or combinations of these systems.

# Air-Processing Equipment

Three methods by which sorbent materials or sorbent equipment dehumidify air are illustrated in the skeleton psychrometric chart shown in Figure 17-8. Air at point A can be dehumidified and cooled to point B by a liquid sorption system with intercooling directly. Alternatively, it may be dehumidified in a solid sorption unit by precooling and dehumidifying from point A to point C, desiccating from point C to point E, and finally cooling to point B. Air could also be dehumidified with solid sorption equipment by desiccating from point A to point D and then by refrigeration from point D to point B.

**Compression Dehumidification.** Compressing a gas to be dehumidified reduces its absolute moisture content, but it generally produces a saturated condition at the elevated pressure. This method is uneconomical but is of value for pressure systems since part of the moisture is removed by compression of the gas; the remaining moisture may be removed by cooling alone, with sorption, or both, depending on the final dew point required. Expansion of high-pressure gas also lowers the dew point.

**Refrigeration Dehumidification.** Refrigerating gas below its dew point is the most common method of dehumidification. This method is advantageous when the gas (1) is comparatively warm, (2) has a high moisture content, and (3) requires an outlet dew point above 40°F (5°C). Frequently, refrigeration is used in combination with sorption dehumidifiers to obtain an extremely low dew point that is difficult to achieve with refrigeration alone.

**Direct Sorption Dehumidification.** Sorbent materials used in dehumidification equipment may either be liquids or solids. The performance of the different sorption dehumidification machines is, to some extent, a function of the sorbent used. Sorbents either retain water on their surface (adsorption) or chemically combine with water (absorption). In regenerative equipment, the water must be removed from the sorbent to regenerate it for further use.

Nonregenerative equipment uses hygroscopic salts such as calcium chloride, urea, or sodium chloride. In regenerative systems, the sorbent is usually a form of silica, alumina gel, activated alumina, molecular sieves, lithium chloride salt, lithium chloride solution, or glycol solution.

In liquid sorption dehumidification systems, the gas passes through sprays of a liquid sorbent such as lithium chloride or a glycol solution. The sorbent in active state has a vapor pressure below that of the gas to be dehumidified and absorbs moisture from the gas stream. The sorbent solution during the process of absorption becomes diluted with moisture, which during regeneration is given up to an outdoor airstream in which the solution is heated. A partial bleed-off of the solution is used for continuous reconcentration of the sorbent in a closed circuit between the spraying or contactor unit and the regenerator unit.

In solid sorption, the gas stream passes through or over granular beds of fixed desiccant structures. A number of commercially available desiccants may be used, depending on such factors as the character of the gas to be dried, inlet temperature, moisture levels, and required final dew point. Outdoor air is passed through beds or layers of the sorbent, which in its active state has a vapor pressure below that of the gas to be dehumidified and absorbs moisture from the gas stream. After becoming saturated with moisture, the desiccant is periodically reactivated to give up previously absorbed moisture to an outdoor air or gas stream.

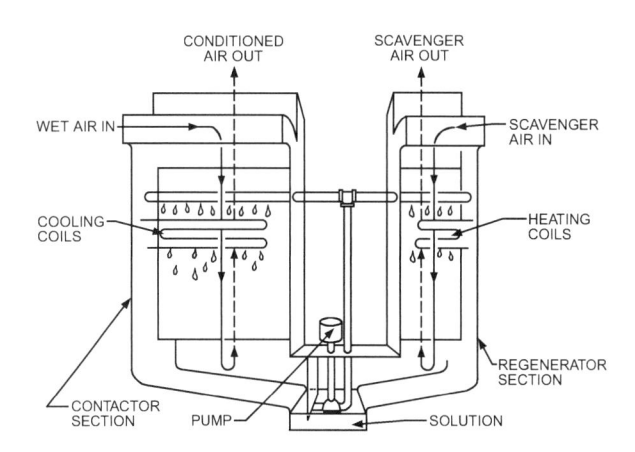

*Fig. 17-9 Flow Diagram for Liquid-Absorbent Dehumidifier*
(Figure 2, Chapter 24 2012 ASHRAE Handbook—HVAC Systems and Equipment)

## 17.6.2 Liquid Absorption Equipment

The flow diagram for a typical liquid absorption system with extended-surface contactor coils is shown in Figure 17-9. For dehumidifying operation, the strong absorbent solution is pumped from the sump of the unit and sprayed over the contactor coils.

Air to be conditioned passes over the contactor coils and comes in contact with the hygroscopic solution. Airflow can be either parallel with or counter to the sprayed solution flow, depending on the space and application requirements.

The degree of dehumidification depends on the concentration, temperature, and characteristics of the hygroscopic solution. Moisture is absorbed from the air by the solution due to the vapor pressure difference between the air and the liquid absorbent. The moisture content of the outlet air can be precisely maintained by varying the coolant flow in the coil to control the absorbent solution contact temperature. The absorbent solution is maintained at the proper concentration by continuous regeneration.

The heat generated in absorbing moisture from the air consists of the latent heat of condensation of water vapor and the heat of solution, or the heat of mixing, of the water and the absorbent. The heat of mixing varies with the liquid absorbent used and with the concentration and temperature of the absorbent. The total heat removal required by the conditioner coil consists of the heat of absorption, sensible heat removed from the air, and the residual heat load added by the regeneration process.

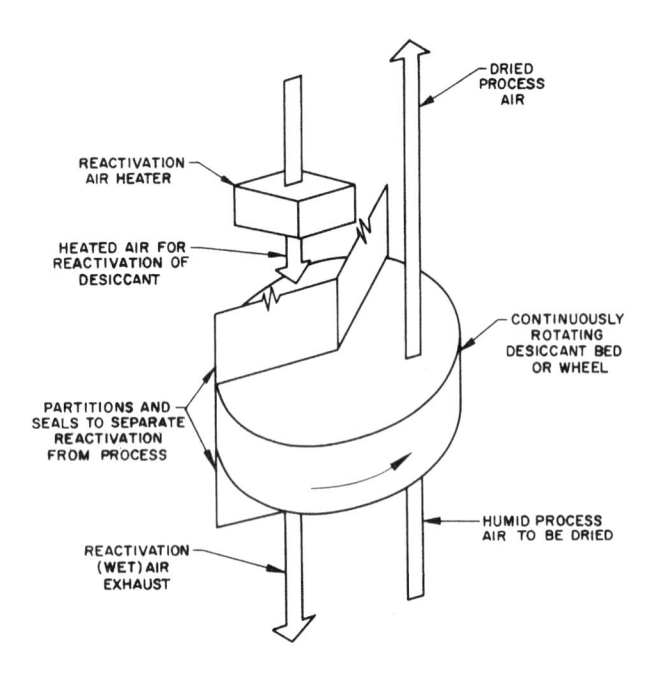

*Fig. 17-10 Rotary Dehumidification Unit*

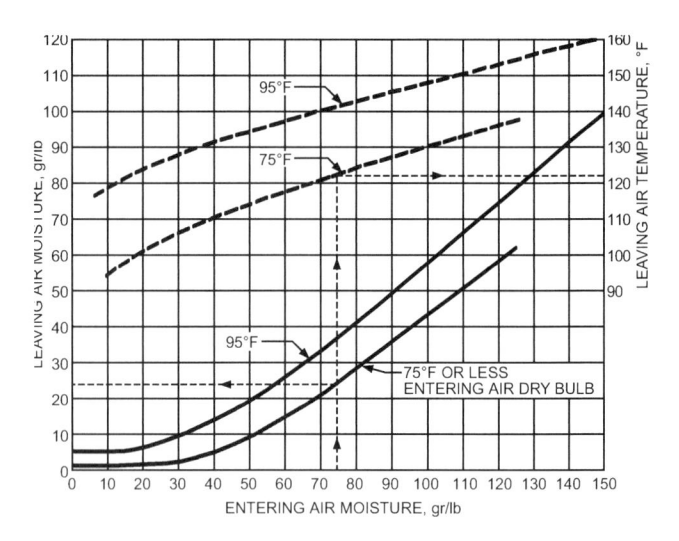

*Fig. 17-11 Performance Data for Rotary Solid Absorption Dehumidifier*
*(Figure 14, Chapter 24, 2012 ASHRAE Handbook—HVAC Systems and Equipment)*

### 17.6.3 Solid Sorption

Dehumidification by use of a solid desiccant, such as silica gel, molecular sieves, activated alumina, or hygroscopic salts, may be performed under either static or dynamic conditions. In the static method, the air or gas is not forced through the desiccant. Instead, the air immediately surrounding the desiccant dries, and through convection and vapor diffusion, water vapor from surrounding areas reaches the desiccant where it is absorbed. In dynamic dehumidification, the air or gas being dried flows through the desiccant bed or structure. In the air-conditioning industry, which is primarily concerned with operation at atmospheric pressure, an air-moving device such as a fan forces the gas through the desiccant bed and a heater or other means periodically reactivates the desiccant.

### 17.6.4 Solid Absorption Equipment

The arrangement of major components of a typical solid absorption system is shown in Figure 17-10. To achieve dehumidification, moist air is passed through a desiccant structure, where the water vapor in the air is absorbed by the desiccant. The systems employ a desiccant structure that may be a disc, drum, or wheel, filled or saturated with an absorbent, such as lithium chloride. The desiccant structure rotates slowly through a heated stream of air where the desiccant is desorbed or reactivated. By continuous rotation, freshly reactivated portions of the desiccant are always available for drying.

The amount of drying depends on the temperature and absolute humidity of the air and the useful concentration of desiccant. The useful concentration of the desiccant is affected by:

- Quantity of desiccant in relation to the mass flow of air and water vapor
- Energy (amount of heat and temperature) for reactivation
- Frequency of reactivation (speed of rotation)

Performance typical of a solid absorption dehumidifier using a fixed desiccant structure is shown in Figure 17-11.

### 17.7 Humidification

When selecting humidification equipment, both the environmental conditions of the occupancy or process and the characteristics of the building enclosure must be considered. These may not always be compatible and a compromise solution may be necessary.

Air washers and evaporative coolers may be used as humidifiers but are usually selected to provide some additional function, such as air cooling or air cleaning, as discussed in the section on air washers.

**Residential Humidifiers for Central Air Systems.** This type of unit depends on airflow in the heating system for evaporation and distribution. General principles of operation and description of equipment are as follows:

**Pan Type.** The humidification rate varies with temperature, humidity, and velocity of the air in the system (Figure 17-12).

**Wetted Elements.** These units use an open textured, wetted media, through or over which air is circulated. The evaporating surface may take the form of a fixed pad, wetted by sprays or by water flowing through by gravity from a header at the top, or the pad may be a paddle wheel, drum, or belt rotating through a water reservoir. Figures 17-13 and 17-14 are examples of this type of humidifier.

**Residential Humidifiers for Nonducted Applications.** Many portable or room humidifiers are available for use in

# Air-Processing Equipment

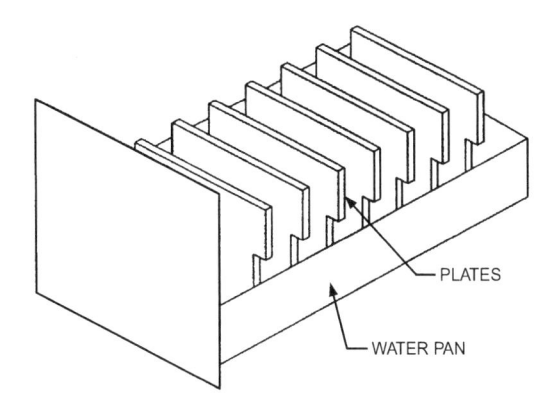

*Fig. 17-12 Pan-Type Humidifier*

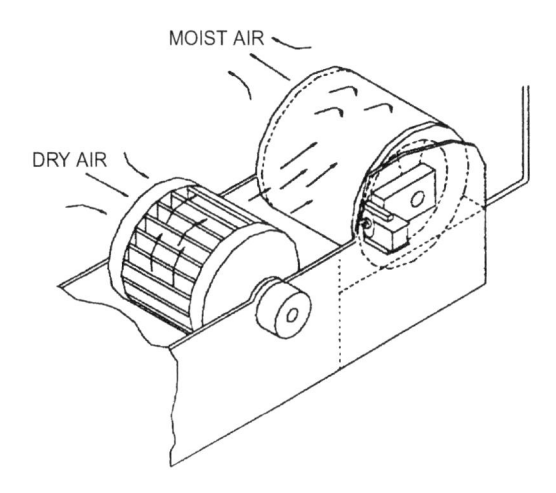

*Fig. 17-13 Wetted-Drum Humidifier*

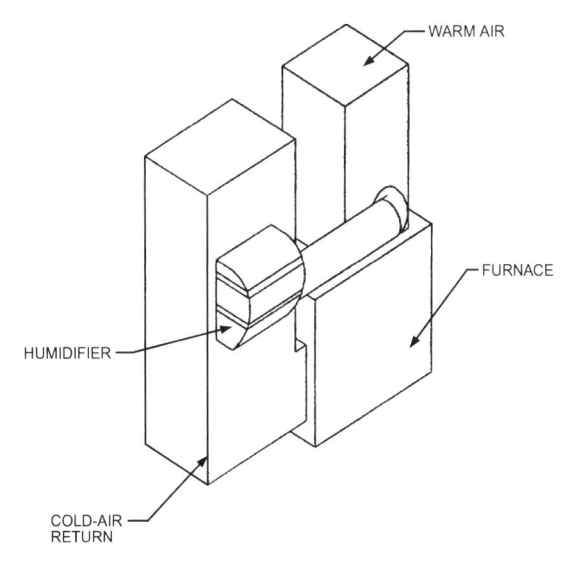

*Fig. 17-14 Bypass Wetted-Element Humidifier*

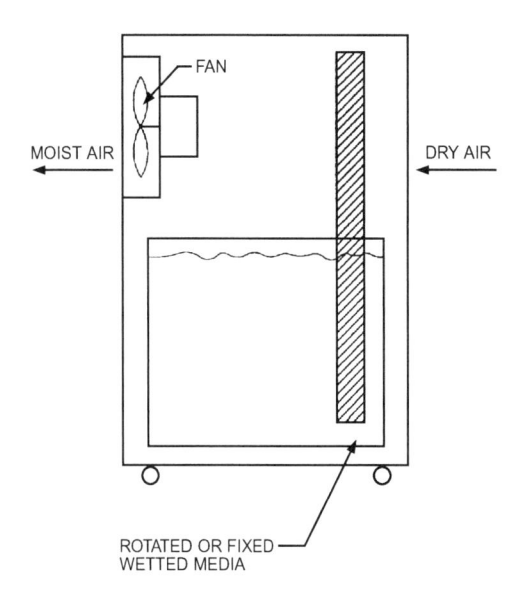

*Fig. 17-15 Portable Humidifier*

residences and apartments heated by nonducted systems, such as hydronic or electric, or where the occupant is prevented from making a permanent installation. An example of this type of humidifier is shown in Figure 17-15.

**Industrial/Commercial Humidifiers for Central Air Systems.** Humidification equipment commonly used in central air-handling systems incorporates a heated water pan, direct steam injection, or atomizers. Specific types are shown in Figure 17-16 and discussed briefly in the following paragraphs.

**Heated Pan.** These units offer a broad range of capacity and may be heated by an electric element, steam, or hot water coil (see Figure 17-16A). Some units are designed to be attached directly to the underside of a system duct. Others are provided with a fan or a steam hose, which allows them to be installed remote from the ductwork.

**Steam.** Direct steam injection humidifiers cover a wide range of designs and capacities. Since water vapor is steam at low pressure and temperature, the whole process can be simplified by introducing steam directly into the air to be humidified. This method is essentially an isothermal process because the temperature of the air remains constant as the moisture is added in vapor form. The steam control valve may be modulating or two-position in response to a humidity controller. The steam may be from an external source, as in the enclosed grid, cup, or jacketed dry steam humidifiers, or produced within the humidifier, as in the self-contained type.

- *Enclosed steam grid humidifiers* (Figure 17-16B) should be used on low steam pressures to prevent splashing of condensate in the duct.

- A *cup* or *pot-type humidifier* (Figure 17-16C) is usually attached under a system duct. Steam is attached tangentially to the inner periphery of the cup by one or more steam inlets, depending on the capability of the unit. The steam supply line should have a suitable steam trap.

- A *jacketed steam humidifier* uses an integral steam valve with a steam-jacketed duct-traversing dispersing tube and condensate separator to prevent condensate from being introduced into the airstream (Figure 17-16D). An inverted bucket-type steam trap is required to drain the separating chamber.

The aforementioned humidifiers inject steam directly from the boiler into the space or duct system. Some boiler treatment chemicals can be discharged, which can affect indoor air quality.

- A *self-contained steam humidifier* converts tap water to steam by electrical energy using either the electrode boiler principle or resistance heating. This steam is injected into the duct system through a dispersion manifold (Figure 17-16E), or the humidifier may be freestanding for nonducted applications.

**Atomizing humidifiers** with optional filter eliminator (Figure 17-16F). Centrifugal atomizers use a high-speed disc that slings water through a fine comb to create a fine mist that is introduced directly into the air where it is evaporated. The ability of the air to absorb the moisture depends on temperature, air velocity, and moisture content. Where mineral fallout from hard water is a problem, optional filter eliminators may be added to remove mineral dust from humidified air, or water demineralizers may be installed. Additional atomizing methods use nozzles; one uses water pressure and the other uses both air and water, as shown in Figure 17-16G. Mixing air and water streams at combined pressures atomizes water into a fine mist, which is evaporated in the room or air duct.

**Wetted-element humidifier.** Wetted-element humidifiers have a wetted media, sometimes in modular configurations, through or over which air is circulated to evaporate water. This unit depends on airflow for evaporation; the rate varies with temperature, humidity, and velocity of the air.

## 17.8 Sprayed Coil Humidifiers/Dehumidifiers

A special adaptation of an air washer can be used very effectively to simultaneously control both the temperature and humidity of an outdoor airstream as in a dedicated outdoor air system (DOAS).

The cooling coil system provided with a deep cooling coil (usually 8 to 12 rows with flat [nonenhanced] fins spaced not less than 1/8 in. apart and with the entire coil assembly and housing made of a noncorrosive material such as stainless steel.) (Figure 17-17A)

The coil is located in a stainless steel section of the air-handling unit over a drain pan with an air spray and recirculating pump. The air spray sprays water from the pump into the airstream against the cooling coil. The water tends to approach the average temperature of the coil and the leaving airstream dew-point temperature. The coil/spray assembly essentially has a coil efficiency of 100% (by pass factor of 0), and the coil-leaving air is saturated at the temperature of the water.

Since the system also functions as an air washer, it very effectively removes particulate (and some chemical) contamination from the air.

Another benefit is that it serves as both a humidifier and a dehumidifier as necessary, depending upon the dew-point temperature or humidity ratio of the leaving air stream compared to the entering air stream Figure 17-17B is a skeleton psychometric chart illustrating the performance as a humidifier and as a dehumidifier. It is controlled by a single discharge temperature sensor set at the desired dew point temperature and operating the heating coil and chilled water coil in sequence. For freeze protection, the heating coil is usually equipped with a constant flow pump or an integrated face and bypass damper.

The sprayed coil DOAS is very effective in maintaining a fixed dew point temperature for critical applications such as art museums, archived storage, rare book libraries, critical humidity manufacturing processes, surgical suites, etc.

Since the sprayed coil and water system functions as an air washer, it requires scheduled blow down and cleaning to remove the particulate collected from the pan and the coil surfaces. Also, to prevent bacterial growth, the safety controls should shut the system down if the water temperature exceeds 58°F.

## 17.9 Air Cleaners

In buildings, dust content is usually less than 0.2 mg/m$^3$ (0.01 × 10$^{-6}$ lb$_m$/ft$^3$) of air. This concentration is in contrast to exhaust gases from processes and flue gases where dust concentration typically ranges from 200 to 40 000 mg/m$^3$ (10 × 10$^{-6}$ to 2000 × 10$^{-6}$ lb$_m$/ft$^3$).

With certain exceptions, the air cleaners described in this section are not applicable to the cleaning of exhaust gases. Atmospheric dust is a complex mixture of smokes, mists, fumes, dry granular particles, and fibers (such particles, when suspended in a gas, are called aerosols). A sample of atmospheric dust gathered at any point generally contains materials common to that locality, together with other components that originated at a distance but have been transported by air currents or diffusion.

These components vary with the geography of the locality in question, the season of the year, the direction and strength of the wind, and the proximity of dust sources. A sample of atmospheric dust usually contains soot and smoke, silica, clay, decayed animal and vegetable matter, organic materials in the form of lint and plant fibers, and metallic fragments. It may also contain living organisms, such as mold spores, bacteria, and plant pollens, which may cause diseases or allergic responses. Chapter 11 of the 2013 *ASHRAE Handbook—Fundamentals* contains further information on atmospheric contaminants.

The particles in the atmosphere can range in size from less than 0.01 mm to dimensions of lint, leaves, and insects.

# Air-Processing Equipment

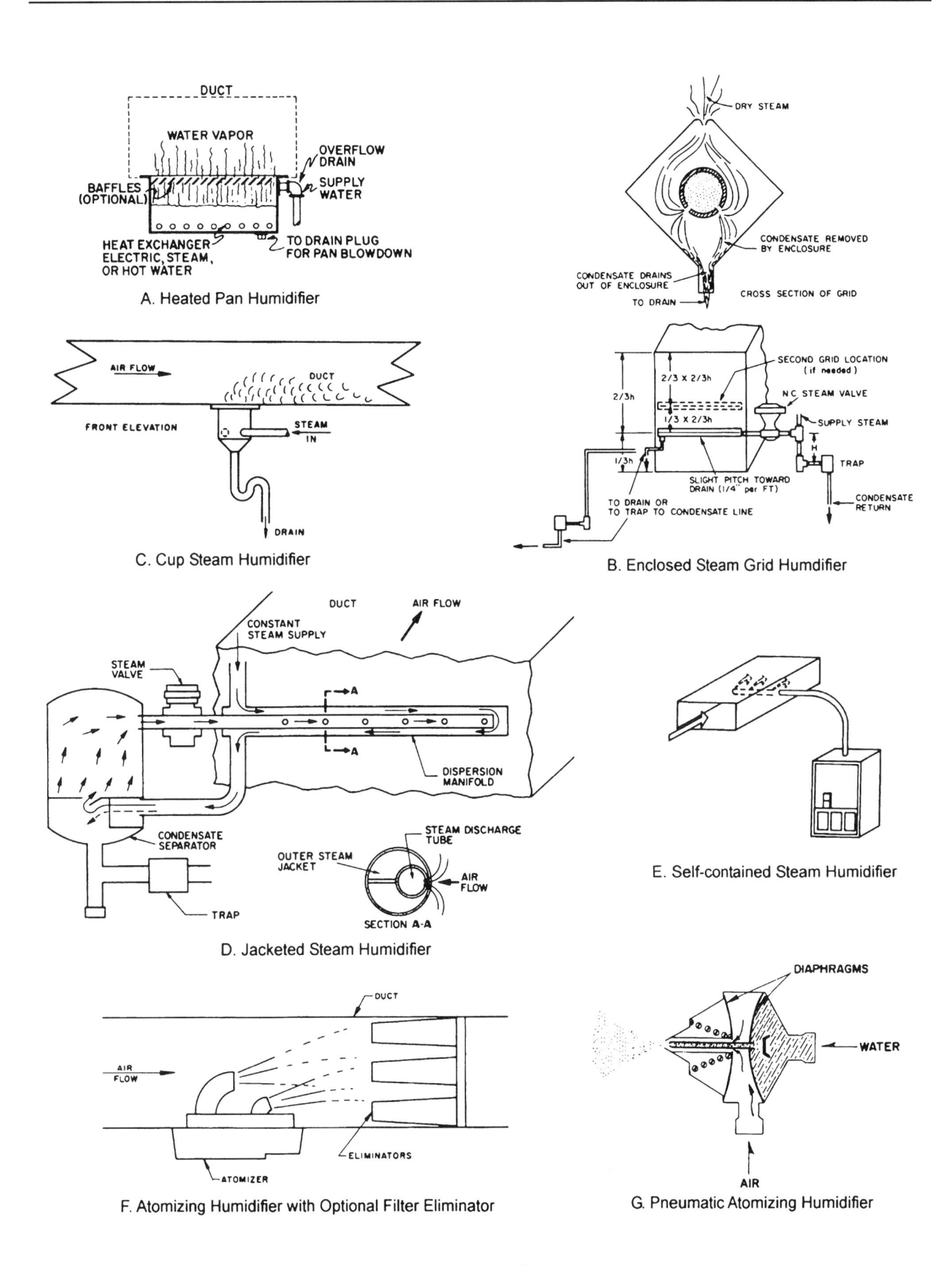

*Fig. 17-16 Commercial Humidifiers*

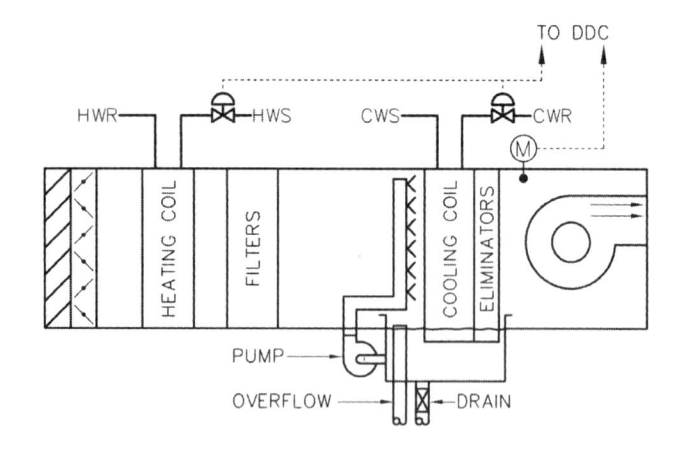

*Fig. 17-17A Sprayed Coil DOAS*

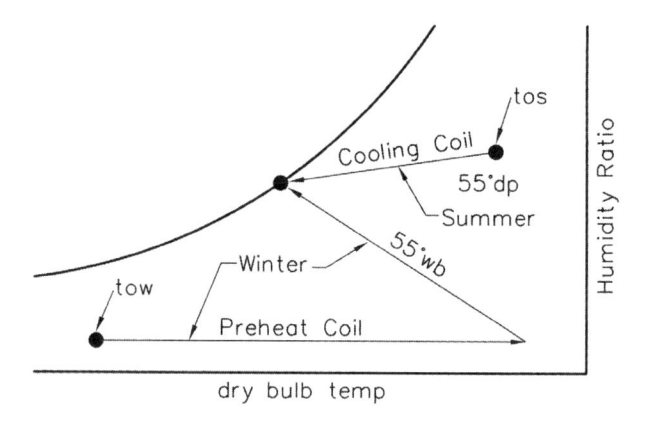

*Fig. 17-17B Psychrometric Process for Sprayed Coil DOAS*

Almost all conceivable shapes and sizes are represented. This wide variety makes it impossible to design one cleaner that is best for all applications. For example, in industrial ventilation, only the coarser dust particles may need to be removed from the airstream for cleanliness of the structure and protection of mechanical equipment. In other applications, surface discoloration must be prevented. Unfortunately, the smaller components of atmospheric dust are the worst offenders in smudging and discoloring building interiors. Electronic air cleaners or high-efficiency dry filters are required for small particle removal. In clean room applications or when radioactive or other dangerous particles are present, extremely high-efficiency mechanical filters should be used.

The most important characteristics of aerosols affecting the performance of an air filter include particle size and shape, density, and concentration. The most important of these is particle size. Data on the sizes and characteristics of airborne particulate matter and the wide range of particle size encountered are given in Figure 17-18. Cleaning efficiency is also affected by the velocity of the airstream. The degree of air cleanliness required is a major factor influencing filter design and selection. Removal of particles becomes progressively more difficult as particle size decreases.

Cost considerations (both in initial investment and maintenance), space requirements, and airflow resistance, in addition to wide-ranging criteria as to degree of air cleanliness, have resulted in a wide variety of commercial air cleaners.

To evaluate filters and air cleaners properly for a particular application, the following factors should be considered:

- Degree of air cleanliness required
- Disposal of dust after it is removed from the air
- Amount and type of dust in the air filtered
- Operating resistance to airflow (pressure drop)
- Space available for filtration equipment
- Cost of maintaining or replacing filters
- Initial cost of the system

Figure 17-19 shows typical locations at filters and lists some applications of filters classified according to their efficiencies and type.

### 17.9.1 Types of Air Cleaners

Air cleaners fall into the following three categories:

1. *Fibrous media unit filters* in which accumulating dust load causes pressure drop to increase up to some maximum permissible value. During this period, efficiency normally increases; however, at high dust loads, dust may adhere poorly to filter fibers, and efficiency drops. Filters in such condition should be replaced or reconditioned, as should filters that have reached their terminal (maximum permissible) pressure drop. This category includes both viscous impingement and dry air filters.

2. *Renewable media filters* in which fresh media is introduced into the airstream as needed to maintain essentially constant resistance. This also maintains essentially constant efficiency.

3. *Electronic air cleaners*, which have essentially constant pressure drop and efficiency unless their precipitating elements become severely dust-loaded.

### 17.9.2 Fibrous Media Unit Filters

**Viscous Impingement Filters.** These flat panel filters consist of coarse fibers and have a high porosity. The filter media are coated with a viscous substance, such as oil, which acts as an adhesive for particles that impinge on the fibers. Design air velocity through the media is usually in the range of 250 to 700 fpm (1 to 4 m/s). These filters are characterized by low pressure drop, low cost, and good efficiency on lint, but low efficiency on normal atmospheric dust. They are commonly 1/2 to 4 in. (13 to 100 mm) thick; 1 to 2 in. (25 to 50 mm) nominal thickness is most popular. The thicker configurations have a high dust-holding capacity. Unit panels are available in standard and special sizes up to about 24 in. by 24 in. (610 mm by

# Air-Processing Equipment

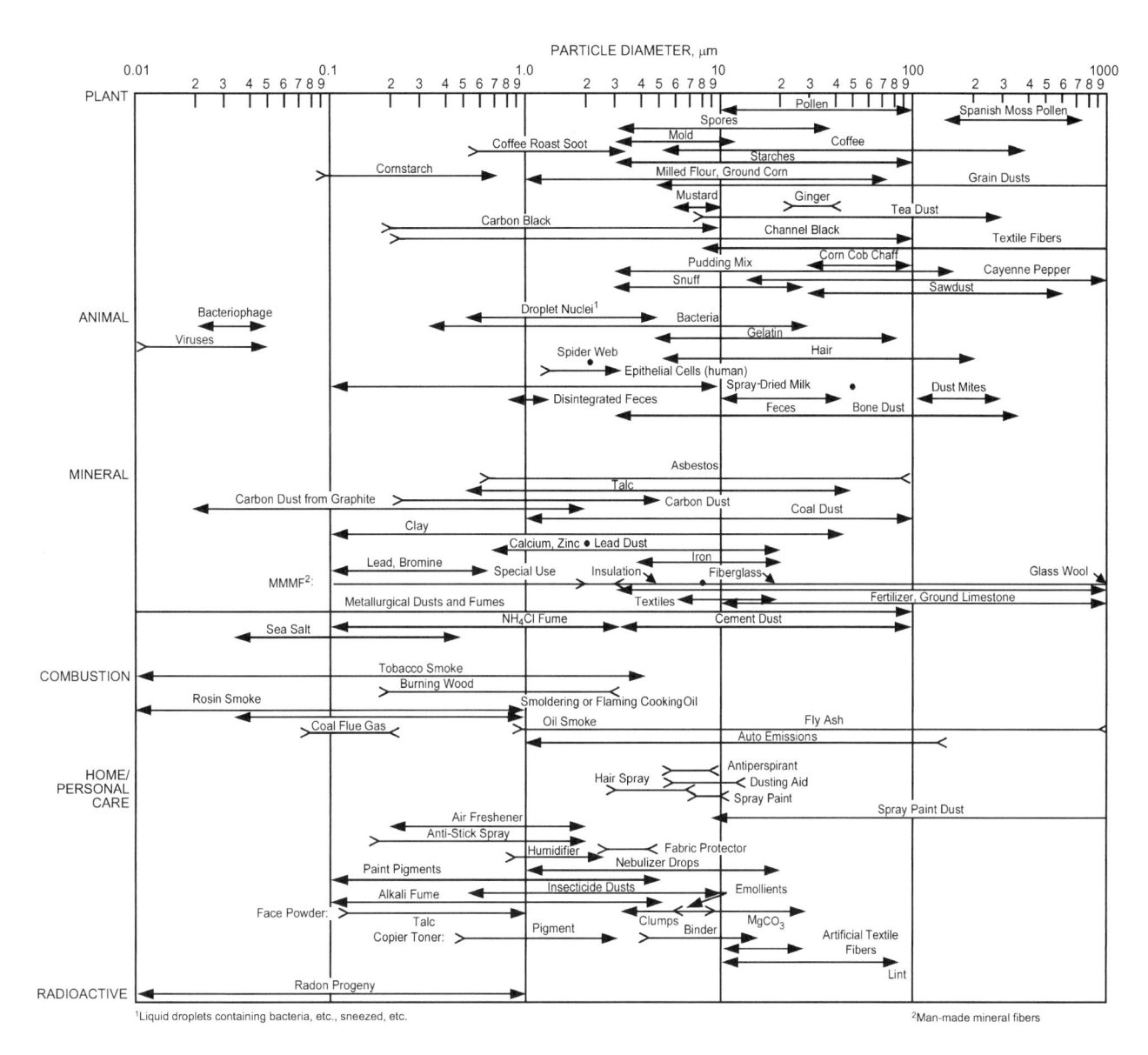

*Fig. 17-18 Sizes of Indoor Particles*
(Owen et al. 1992)
(Figure 3, Chapter 11, 2013 ASHRAE Handbook—Fundamentals)

610 mm). This type of filter is often used as a prefilter to higher efficiency filters.

Many different materials have been used as the filtering medium, including coarse (15 to 60 μm diameter) glass fibers, animal hair, vegetable fibers, synthetic fibers, metallic wools, expanded metals and foils, crimped screens, random-matted wire, and synthetic open-cell foams.

Although viscous impingement filters usually operate in the range of 300 to 600 fpm (1.5 to 3 m/s), they may operate at higher velocities. The limiting factor other than increased flow resistance is the danger of blowing off agglomerates of collected dust and the viscous coating on the filter.

The rate of filter loading depends on the type and concentration of the dirt being handled and on the operating cycle of the system. Manometers or draft gages are often installed to measure the pressure drop across the filter bank and thereby indicate when the filter needs servicing. The final allowable pressure drop may vary from one type of filter to another. The decline in filter efficiency that occurs when all the viscous coating has been absorbed by the collected dust, rather than the increased resistance due to dust load, may be the limiting factor in operating life.

**Dry Air Filters.** The media used in dry air filters are random fiber mats or blankets of varying thicknesses, fiber sizes, and densities. Media of bonded glass fiber, cellulose fibers, wool felt, synthetics, and other materials have been used commercially. The medium in filters of this class is frequently supported by a wire frame in the form of pockets or V-shaped pleats. In other designs, the media may be self-supporting because of inherent rigidity or because airflow inflates it into extended form. Pleating of the media provides a high ratio of media area to face area, thus allowing reason-

492                                                                                                                    Principles of HVAC

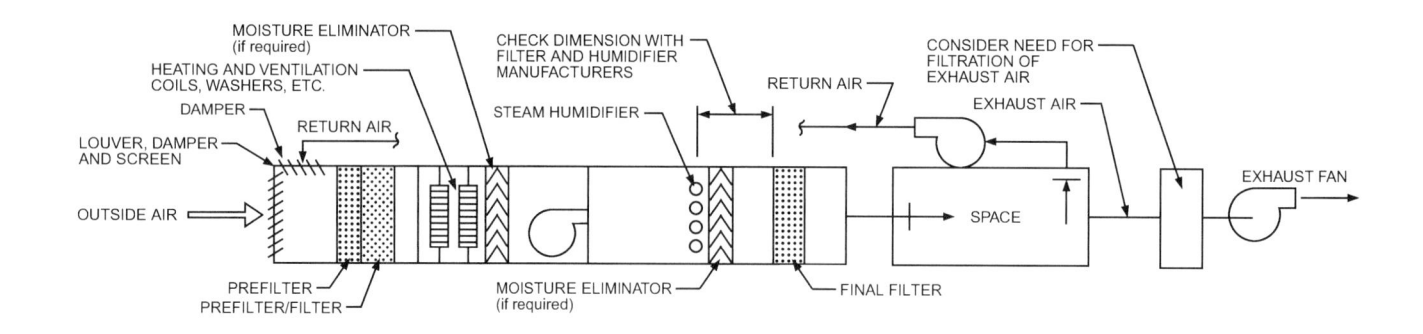

| Standard 52.2 MERV | Intended Standard 52.1 Value | Arrestance Value | Example Range of Contaminants Controlled | Example Applications | Sample Air Cleaner Type(s) |
|---|---|---|---|---|---|
| **HEPA filters** | | | | | |
| MERV 20 | N/A | | 0.12 to 0.5 μm particles: virus (unattached), carbon dust, sea salt, radon progeny, combustion smoke | Cleanroom, pharmaceutical manufacturing and exhaust, radioactive material handling and exhaust, orthopedic and organ transplant surgery, carcinogenic materials, welding fumes | SULPA >99.999% 0.1 to 0.2 μm IEST type F (ceiling panel) |
| MERV 19 | | | | | ULPA >99.999% 0.3 μm IEST type D (ceiling panel) |
| MERV 18 | | | | | HEPA > 99.99% 0.3 μm IEST type C (ceiling or up to 12 in.deep) |
| MERV 17 | | | | | HEPA > 99.97% 0.3 μm IEST type A (box style 6 to 12 in. deep) |
| **E-1 Range** | | | | | |
| MERV 16 | Intended to replace 70 to 98% dust-spot efficiency filters | >99% | 0.3 to 1.0 μm size range: bacteria, smoke (ETS), paint pigments, face powder, some virus, droplet nuclei, insecticide dusts, soldering fumes | Day surgery, general surgery, hospital general ventilation, turbo equipment, compressors, welding/soldering air cleaners, prefilters to HEPAs, LEED for existing (EB) and new (NC) commercial buildings, smoking lounges | Box-style wet-laid or lofted fiberglass, box-style synthetic media, minipleated synthetic or fiberglass paper, depths from 4 to 12 in., Pocket filters of fiberglass or synthetic media 12 to 36 in. |
| MERV 15 | | >99% | | | |
| MERV 14 | | >98% | | | |
| MERV 13 | | >97% | | | |
| **E-2 Range** | | | | | |
| MERV 12 | Intended to replace 50 to 80% dust-spot efficiency filters | >97% | 1.0 to 3.0 μm size range: milled flour, lead dust, combustion soot, *Legionella*, coal dust, some bacteria, process grinding dust | Food processing facilities, air separation plants, commercial buildings, better residential, industrial air cleaning, prefiltration to higher-efficiency filters, schools, gymnasiums | Box-style wet-laid or lofted fiberglass, box-style synthetic media, minipleated synthetic or fiberglass paper, depths from 2 to 12 in. Pocket filters either rigid or flexible in synthetic or fiberglass, depths from 12 to 36 in. |
| MERV 11 | | >95% | | | |
| MERV 10 | | >95% | | | |
| MERV 9 | | >90% | | | |
| **E-3 Range** | | | | | |
| MERV 8 | Intended to replace 20 to 60% dust-spot efficiency filters | >90% | 3.0 to 10 μm size range: pollens, earth-origin dust, mold spores, cement dust, powdered milk, snuff, hair spray mist | General HVAC filtration, industrial equipment filtration, commercial property, schools, prefilter to high-efficiency filters, paint booth intakes, electrical/phone equipment protection | Wide range of pleated media, ring panels, cubes, pockets in synthetic or fiberglass, disposable panels, depths from 1 to 24 in. |
| MERV 7 | | >90% | | | |
| MERV 6 | | >85% | | | |
| MERV 5 | | >85% | | | |
| MERV 4 | <20% | >70% | Arrestance method | Protection from blowing large particle dirt and debris, industrial environment ventilation air | Inertial separators |
| MERV 3 | <20% | >70% | | | |
| MERV 2 | <20% | >65% | | | |
| MERV 1 | <20% | <65% | | | |

*Note*: MERV for non-HEPA/ULPA filters also includes test airflow rate, but it is not shown here because it is of no significance for the purposes of this table.
N/A = not applicable.

*Fig. 17-19 Typical Filter Locations and Cross Reference and Application Guidelines*
*(Figure 3 and Table 2, Chapter 29, 2012 ASHRAE Handbook—HVAC Systems and Equipment)*

able pressure drop despite the density and fineness of the media.

The efficiency of dry air filters is usually higher than that of viscous impingement filters, and the variety of media available makes it possible to furnish almost any degree of cleaning efficiency desired. Most dry filter media and filter configurations also have higher dust-holding capacities than viscous impingement filters. Coarse prefilters placed ahead of high-efficiency dry filters may be economically justified by the even longer life they give the main filters.

**Electronic Air Cleaners.** These ionizing filters are efficient, low pressure drop devices for removing fine dust and smoke particles. Collector plates are often coated with a special oil as an adhesive. Cleaning is generally accomplished by washing the cells in place with hot water from a water hose or by a fixed or moving nozzle system.

Electrical forces drive most particles to the collecting surface but cannot hold them there. In fact, after a particle touches the collecting surface, the electrical force reverses and tends to pull it off, and the dust is held only by intermolecular adhesion forces. It is, therefore, very important with the washed type of electronic air cleaner to ensure that either the dust is naturally adherent or that the plates are always covered with adhesive.

Electronic air cleaners, however, are often used without any adhesive treatment on the plates. Under such conditions, the precipitator forms agglomerates that eventually blow off the plates. They must be followed downstream by a secondary filter or storage section. The dry agglomerates produced in the precipitator are allowed to blow off and be caught by the downstream filter. An automatic replaceable media filter for catching these agglomerates gives a combination with a high degree of cleaning efficiency and also the convenient maintenance associated with an automatic filter.

## 17.10 Air-to-Air Energy Recovery Equipment

Energy can be recovered from exhaust air, as well as transferred from one location to another. This can be done using rotary devices, heat pipe heat exchangers, coil heat recovery loops, twin tower heat recovery loops, fixed plate exchangers, and thermosiphon heat exchangers.

Table 17-2 provides comparative data for common types of air-to-air energy recovery devices. The following sections describe the construction, operation, and unique features of the various devices.

### 17.10.1 Performance Rating of Equipment

Performance of air-to-air heat exchangers is usually expressed in terms of their effectiveness in transferring (1) sensible energy (dry-bulb temperature), (2) latent energy (humidity ratio), or (3) total energy (enthalpy). The effectiveness $\varepsilon$ of a heat exchanger is defined as follows:

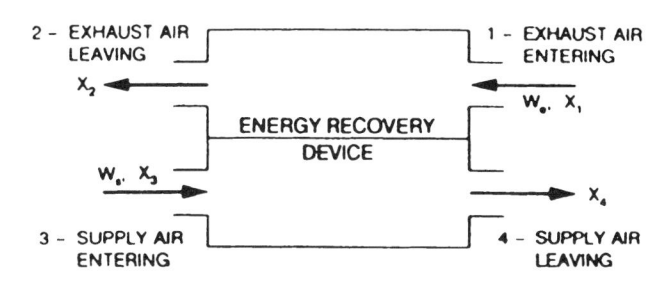

*Fig. 17-20 Nomenclature for Effectiveness Evaluation*

$$\varepsilon = \frac{\text{Actual transfer for given device}}{\text{Maximum possible transfer between airstreams}}$$

Referring to Figure 17-20,

$$= \frac{W_e(X_1 - X_2)}{W_{min}(X_1 - X_3)} = \frac{W_s(X_4 - X_3)}{W_{min}(X_1 - X_3)} \quad (17\text{-}6)$$

where

$\varepsilon$ = sensible, latent, or total heat effectiveness
$X$ = dry-bulb temperature, humidity ratio, or enthalpy at locations indicated in Figure 17-20

For the latent and total heat effectiveness,

$W_s$ = mass flow rate of supply
$W_e$ = mass flow rate of exhaust

For the sensible heat effectiveness,

$W_s$ = (specific heat)(mass flow rate) for the supply
$W_e$ = (specific heat)(mass flow rate) for the exhaust
$W_{min}$ = smaller of $W_s$ and $W_e$

The leaving supply air condition is then:

$$X_2 = X_1 - \varepsilon(X_1 - X_3)(W_{min}/W_s) \quad (17\text{-}7)$$

The leaving exhaust air condition is:

$$X_4 = X_3 + \varepsilon(X_1 - X_3)(W_{min}/W_e) \quad (17\text{-}8)$$

The effectiveness of a particular air-to-air energy recovery device is a function of several variables, including the supply and exhaust mass flow rates and the energy transfer characteristics of the device. Because of this combination, performance data must be established for each device.

### 17.10.2 Economics of Air-to-Air Energy Recovery

In analyzing the advisability of any air-to-air energy recovery application, one must usually consider both first and operating costs.

*First Cost*

- Energy recovery device
- Installing energy recovery device
- Additional cost of building space to accommodate apparatus
- Additional ductwork to accommodate device or cost of piping for liquid energy transfer

**Table 17-2  Comparison of Air-to-Air Energy Recovery Devices**
*(Table 3, Chapter 26, 2012 ASHRAE Handbook—HVAC Systems and Equipment)*

|  | Fixed Plate | Membrane Plate | Energy Wheel | Heat Wheel | Heat Pipe | Runaround Coil Loop | Thermosiphon | Twin Towers |
|---|---|---|---|---|---|---|---|---|
| Airflow arrangements | Counterflow Cross-flow | Counterflow Cross-flow | Counterflow Parallel flow | Counterflow | Counterflow Parallel flow | — | Counterflow Parallel flow | — |
| Equipment size range, cfm | 50 and up | 50 and up | 50 to 74,000 and up | 50 to 74,000 and up | 100 and up | 100 and up | 100 and up | — |
| Typical sensible effectiveness ($m_s = m_e$), % | 50 to 80 | 50 to 75 | 50 to 85 | 50 to 85 | 45 to 65 | 55 to 65 | 40 to 60 | 40 to 60 |
| Typical latent effectiveness,* % | — | 50 to 72 | 50 to 85 | 0 | — | — | — | — |
| Total effectiveness,* % | — | 50 to 73 | 50 to 85 | — | — | — | — | — |
| Face velocity, fpm | 200 to 1000 | 200 to 600 | 500 to 1000 | 400 to 1000 | 400 to 800 | 300 to 600 | 400 to 800 | 300 to 450 |
| Pressure drop, in. of water | 0.4 to 4 | 0.4 to 2 | 0.4 to 1.2 | 0.4 to 1.2 | 0.6 to 2 | 0.6 to 2 | 0.6 to 2 | 0.7 to 1.2 |
| EATR, % | 0 to 5 | 0 to 5 | 0.5 to 10 | 0.5 to 10 | 0 to 1 | 0 | 0 | 0 |
| OACF | 0.97 to 1.06 | 0.97 to 1.06 | 0.99 to 1.1 | 1 to 1.2 | 0.99 to 1.01 | 1.0 | 1.0 | 1.0 |
| Temperature range, °F | –75 to 1470 | 15 to 120 | –65 to 1470 | –65 to 1470 | –40 to 105 | –50 to 930 | –40 to 105 | –40 to 115 |
| Typical mode of purchase | Exchanger only Exchanger in case Exchanger and blowers Complete system | Exchanger only Exchanger in case Exchanger and external blowers Complete system | Exchanger only Exchanger in case Exchanger and blowers Complete system | Exchanger only Exchanger in case Exchanger and blowers Complete system | Exchanger only Exchanger in case Exchanger and blowers Complete system | Coil only Complete system | Exchanger only Exchanger in case | Complete system |
| Advantages | No moving parts Low pressure drop Easily cleaned | No moving parts Low pressure drop Low air leakage | Moisture or mass transfer Compact large sizes Low pressure drop Available on all ventilation system platforms | Compact large sizes Low pressure drop Easily cleaned | No moving parts except tilt Fan location not critical Allowable pressure differential up to 2 psi | Exhaust airstream can be separated from supply air Fan location not critical | No moving parts Exhaust airstream can be separated from supply air Fan location not critical | Latent transfer from remote airstreams Efficient microbiological cleaning of both supply and exhaust airstreams |
| Limitations | Large size at higher flow rates | Few suppliers Long-term maintenance and performance unknown | Supply air may require some further cooling or heating Some EATR without purge | Some EATR without purge | Effectiveness limited by pressure drop and cost Few suppliers | Predicting performance requires accurate simulation model | Effectiveness may be limited by pressure drop and cost Few suppliers | Few suppliers Maintenance and performance unknown |
| Heat rate control (HRC) methods | Bypass dampers and ducting | Bypass dampers and ducting | Bypass dampers and wheel speed control | Bypass dampers and wheel speed control | Tilt angle down to 10% of maximum heat rate | Bypass valve or pump speed control | Control valve over full range | Control valve or pump speed control over full range |

*Rated effectiveness values are for balanced flow conditions. Effectiveness values increase slightly if flow rates of either or both airstreams are higher than flow rates at which testing is done.

EATR = Exhaust Air Transfer Ratio
OACF = Outdoor Air Correction Factor

- Larger fans and/or motors to overcome air pressure loss of energy recovery device
- Additional air filtration required (if any)
- Capacity controls
- Any auxiliary heaters required for frost control
- Savings of boiler or heating equipment due to reduced design load
- Savings of heating coils and associated piping and pumps due to reduced design capacity
- Savings of chiller or cooling plant due to reduced design load
- Savings of cooling coils and associated piping and pumps due to reduced design capacity
- Savings of electric power requirement

*Operating Cost*

- Maintaining the energy recovery device
- Operating fans to overcome additional static pressure
- Maintaining additional filtration (if any)

# Air-Processing Equipment

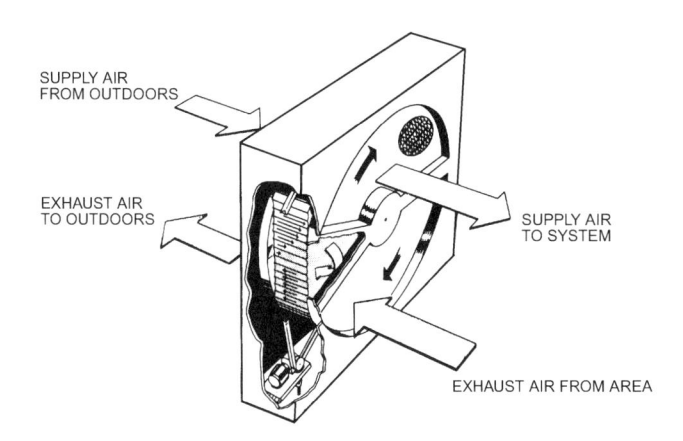

*Fig. 17-21 Rotary Energy Wheel*
(Figure 7, Chapter 26, 2012 ASHRAE Handbook—HVAC Systems and Equipment)

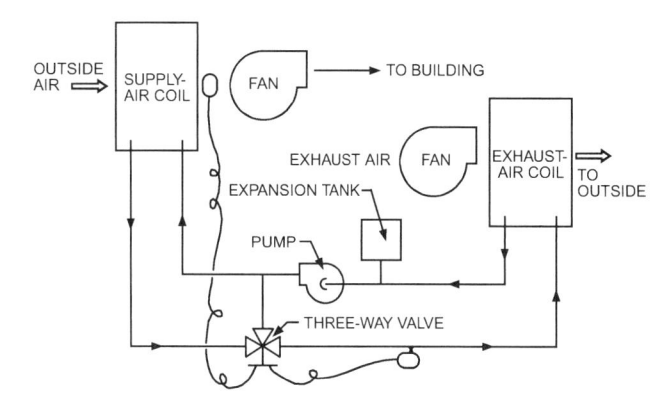

*Fig. 17-22 Arrangement of Coil Energy Recovery Loop*
(Figure 14, Chapter 26, 2012 ASHRAE Handbook—HVAC Systems and Equipment)

- Operating energy recovery device drive, pumps, controls, and defrost heaters
- Savings of annual heating energy, based on weather data for the system location
- Savings of annual cooling energy, based on weather data for the system location

## 17.10.3 Rotary Air-to-Air Energy Exchangers

A rotary air-to-air energy exchanger, often called a *heat wheel*, is a revolving cylinder filled with an air-permeable medium with a large internal surface area for contact with the air passing through it. Adjacent supply and exhaust airstreams each flow through half the exchanger in a counterflow pattern (Figure 17-21). Media material may be selected to recover either sensible heat only or total heat (sensible heat plus latent heat). With total heat recovery, the unit is called an *energy wheel*.

Sensible heat is recovered (transferred) as the medium picks up and stores heat from the hot airstream and gives it up to the cold one. Latent heat is transferred as the medium (1) condenses moisture from the airstream having the higher humidity ratio (by means of absorption for liquid desiccants and adsorption for solid desiccants) with a simultaneous release of heat and (2) then releases the moisture through evaporation (and heat pickup) into the airstream with the lower humidity ratio. Thus, more moist air is dried while drier air is humidified. In total heat transfer, both sensible and latent heat recovery processes take place simultaneously.

Choice of structural material for the casing, rotor structure, and medium of a rotary energy exchanger is influenced by the contaminants, dew point, and temperature of the exhaust air, as well as by the properties of the supply air. Aluminum and steel are the usual casing and rotor materials for normal comfort ventilating systems. Exchanger media are fabricated from metal, mineral, or man-made materials and classified as providing either random flow or directionally oriented flow through their structures.

The performance of a rotary energy exchanger is defined by the exchanger's effectiveness and the media pressure drops. Face velocities for most energy recovery applications range from 500 to 800 fpm (2.5 to 4.0 m/s). Low face velocities give lower pressure drop, higher effectiveness, and lower operating costs, but require larger size units with higher capital costs and more installation space.

Typical pressure drops for various types of media at 500 fpm (2.5 m/s) vary from 0.4 to 0.7 in. (100 to 170 kPa) of water. (Consult the manufacturer's catalog in each case for actual data.) Average effectiveness values for sensible and total heat exchangers lie in the 70 to 85% range for equal supply and exhaust air mass flow rates and usual exchanger face velocities.

Rotary energy wheels are available in single units to 68,000 cfm (32 m$^3$/s) capacity; they are usually not larger than 14 ft (4.25 m) in diameter due to difficulty in shipping, erecting, and fitting into buildings. Multiple units may be used to provide greater single-system capacities. Units are available for temperatures from −70 to 1500°F (−60 to 800°C). When installed horizontally (vertical airflow), units greater than 8 ft (2.4 m) in diameter may require special structural considerations due to their size and weight.

## 17.10.4 Coil Energy Recovery Loops

The coil energy recovery or "runaround loop" system uses extended-surface, finned-tube water coils placed in the supply and exhaust airstreams of a building or process. The coils are connected in a closed loop via counterflow piping and an intermediate heat transfer fluid of water (typically) or a freeze-protected solution pumped through the coils.

This system allows energy to transfer from the warmer to the cooler airstream. In a typical comfort-to-comfort application, the system is seasonally reversible—the supply air preheats when outdoor air is cooler than the exhaust air and precools when the outdoor air is warmer. This system operates generally for sensible heat recovery only.

As with other air-to-air energy recovery equipment, measures must be taken to prevent potential freezing of exhaust air condensate. A dual-purpose, three-way temperature control valve is used to prevent exhaust coil frosting. The valve is controlled to maintain an entering solution temperature to the exhaust coil of not less than 30°F (−1°C). This is accomplished

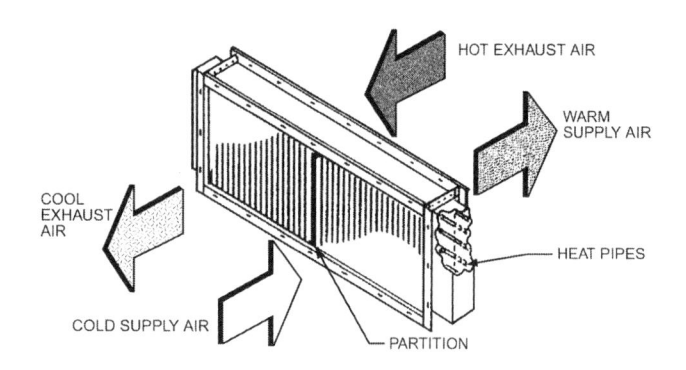

*Fig. 17-23 Heat Pipe Assembly*
(Figure 16, Chapter 26, 2012 ASHRAE Handbook—HVAC Systems and Equipment)

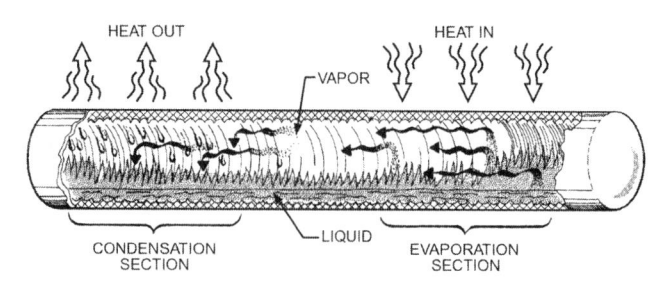

*Fig. 17-24 Heat Pipe*
(Figure 17, Chapter 26, 2012 ASHRAE Handbook—HVAC Systems and Equipment)

by bypassing some of the warmer solution from the supply air coil. The valve can also ensure that a prescribed leaving air temperature from the supply air coil is not exceeded for those applications where the energy recovered must be limited.

This system affords a high degree of flexibility, which makes it well-suited for renovation and industrial applications. The system accommodates remotely located supply and exhaust ducts. It also allows simultaneous energy recovery from multiple supply and exhausts. A basic arrangement of the coil energy recovery loop is depicted in Figure 17-22.

### 17.10.5 Heat Pipe Heat Exchangers

A heat pipe heat exchanger has no moving parts; it is a passive energy recovery device. Although it appears similar to a standard steam or chilled water coil, it differs in two major aspects. As shown in Figure 17-23, each tube is an individual heat pipe, that operates independently and acts as a superconductor of heat. Secondly, the heat pipe heat exchanger is divided into two airflow paths. Hot air passes through one side of the exchanger and cold air through the other side in the opposite direction, that is, in a counterflow arrangement. Sensible energy from the hot air is transferred by the heat pipes to the other side of the exchanger, where it is captured by the cold air, thereby warming it. Although the heat pipes span the width of the unit, a sealed partition separates the two airstreams, preventing any cross-contamination between them.

Figure 17-24 is a schematic of a heat pipe. A heat pipe is a tube that is fabricated with a capillary wick structure, filled with a refrigerant or suitable change-of-phase heat transfer fluid, and permanently sealed.

Basically, the heat pipe operates on a condensation/evaporation cycle that is continuous as long as there is a temperature difference to drive the process. Energy transfer within a heat pipe is accomplished with a very small temperature drop; a heat pipe is essentially an isothermal device.

Thermal energy applied to either end of the pipe causes the fluid at the end to vaporize. The vapor then travels to the other end of the pipe where the removal of thermal energy causes the vapor to condense into liquid again, thus giving up the latent heat of condensation. The condensed liquid then flows back to the evaporator section (i.e., the hot end) to be reused, completing the cycle.

Heat pipes have a finite heat transfer capacity affected by such factors as wick design, tube diameter, working fluid, and tube orientation relative to horizontal. Design face velocities for heat pipe heat exchangers range from 400 to 800 fpm (2.0 to 4.1 m/s), with 450 to 550 fpm (2.3 to 2.8 m/s) most common. Design face velocities are generally established on allowable pressure drop rather than recovery performance.

Pressure drops at 60% effectiveness range from 0.4 to 0.7 in. of water at (100 to 170 Pa) 400 fpm (2.0 m/s) up to 1.5 to 2.0 in. of water (370 to 500 Pa) at 800 fpm (4.1 m/s). Recovery performance, or effectiveness, decreases with increasing velocity, but the effect is not as pronounced as with pressure drop.

Available fin designs include continuous corrugated plate fins, continuous flat plate fins, and spiral fins. These fin designs and tube spacing cause the variation in pressure drop, noted previously, at a given face velocity.

Figure 17-25 presents a typical effectiveness curve for various face velocities and rows of tubes. As the number of rows increases, effectiveness increases at a decreasing rate. For example, doubling the rows of tubes in a 60% effective heat exchanger increases the effectiveness to 75%.

### 17.10.6 Fixed Plate Exchangers

The plate heat exchanger is a static device that has no leakage between airstreams. Since it uses no secondary heat transfer medium, such as water or refrigerant, its temperature range is the broadest of all air-to-air energy recovery equipment.

A fixed surface plate exchanger can be classified as (1) a **pure-plate heat exchanger**, consisting of only a primary heat transfer surface, or (2) a **plate-fin heat exchanger**, which is made up of alternate layers of separate plates and interconnecting fins. The pure-plate exchanger is usually a counterflow design, whereas the basic plate-fin exchanger is a crossflow design with combinations sometimes arranged to approach a counterflow unit. Counterflow provides the greatest temperature difference for maximum heat transfer, but crossflow can sometimes give more convenient air connections.

Fixed surface plate exchangers have no moving parts. Alternate layers of plates, separated and sealed (referred to as

# Air-Processing Equipment

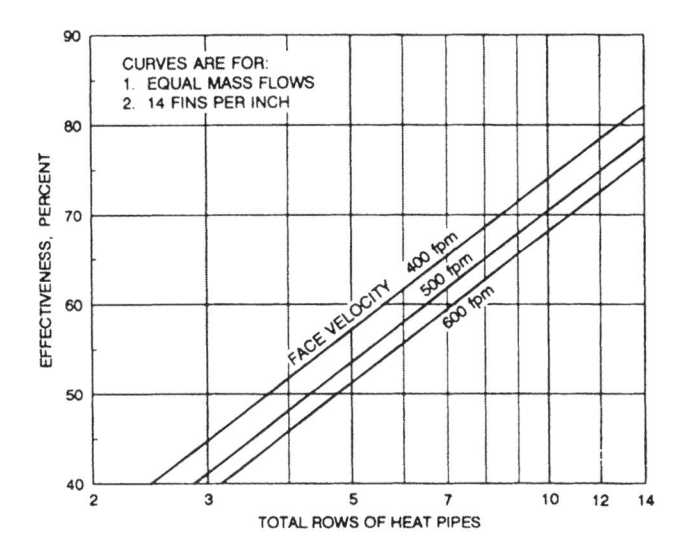

*Fig. 17-25 Heat Pipe Exchanger Effectiveness*
*(Figure 18, Chapter 26, 2012 ASHRAE Handbook—HVAC Systems and Equipment)*

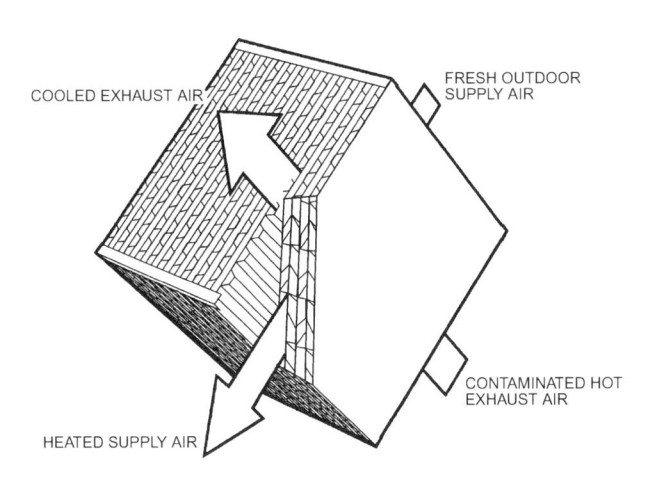

*Fig. 17-26 Fixed Plate Heat Exchanger*
*(Figure 4, Chapter 25, 2008 ASHRAE Handbook—HVAC Systems and Equipment)*

the heat exchanger core), form the exhaust and supply airstream passages (Figure 17-26). Plate spacings range from 0.1 to 0.5 in. (2.5 to 13 mm), depending on the design and application. Heat transfers directly from the warm airstreams through the separating plates into the cool airstreams.

Normally, both latent heat of condensation, from moisture condensed as the temperature of the warm (exhaust) airstream drops below its dew point, and sensible heat are conducted through the separating plates into the cool (supply) airstream. Thus, latent energy but not actual moisture may be transferred. Recovering upward of 80% of the available waste exhaust heat is not uncommon.

Fixed-plate heat exchangers can be made from permeable microporous membranes designed to maximize moisture and energy transfer between airstreams while minimizing air transfer. Suitable permeable microporous membranes for this emerging technology include cellulose, polymers, and other synthetic materials such as hydrophilic electrolyte. Hydrophilic electrolytes are made from sulphonation chemistry techniques and contain charged ions that attract polar water molecules; adsorption and desorption of water occur in vapor state.

Plate exchangers are of many proprietary designs, with weights, sizes, and flow patterns depending on the manufacturer (Figure 17-27). Most manufacturers of plate exchangers offer the equipment in modular design.

### 17.10.7 Thermosiphon Heat Exchangers

Thermosiphon heat exchangers use the natural gravity circulation of a boiling and condensing intermediate fluid to transfer energy between exhaust and supply airstreams. They may be classified as sealed tube thermosiphons and coil loop thermosiphons. These two types are illustrated in Figures 17-28 and 17-29.

The sealed tube type is similar to a heat pipe and is often given that name. The only distinction made between the two is that heat pipes are usually considered to use, if not solely rely on, capillary forces to cause the intermediate liquid to flow from the cold to the hot end of the tubes, whereas the thermosiphon tubes rely only on gravity. The coil loop type is similar in appearance to the coil energy recovery loop discussed previously. The most obvious difference is the absence of a circulating pump in the thermosiphon loop and the need for evaporator and condenser coils rather than single-phase liquid coils.

**Example 17-2** Determine the leaving conditions for an energy wheel exchanger with the following conditions:

*Cooling*

1. Design conditions: outdoor air = 94°F (34.4°C) dry bulb, 77°F (25°C) wet bulb
2. Design conditions: space = 75°F (23.9°C) dry bulb, 62.5°F (16.9°C) wet bulb

*Heating*

1. Design conditions: outdoor air = 0°F (−17.8°) dry bulb, −3°F (−19.4°C) wet bulb
2. Design conditions: space = 75°F (23.9°C) dry bulb, 59.5°F (15.3°C) wet bulb

*Cooling and Heating*

Exchanger effectiveness = 80% on sensible heat and 80% on latent heat. Equal mass flow rates.

**Solution:**

Dry-bulb temperature and humidity ratio are calculated from Equation (17-8).

*Summer dry-bulb temperature*

$$X_2 = X_1 - \varepsilon(X_1 - X_2) = 94 - 0.8(94 - 75)$$
$$= 78.8°F (26.0°C)$$

*Summer humidity ratio*

$$X_2 = X_1 - \varepsilon(X_1 - X_2) = 0.0162 - 0.8(0.0162 - 0.0092)$$

**498**  Principles of HVAC

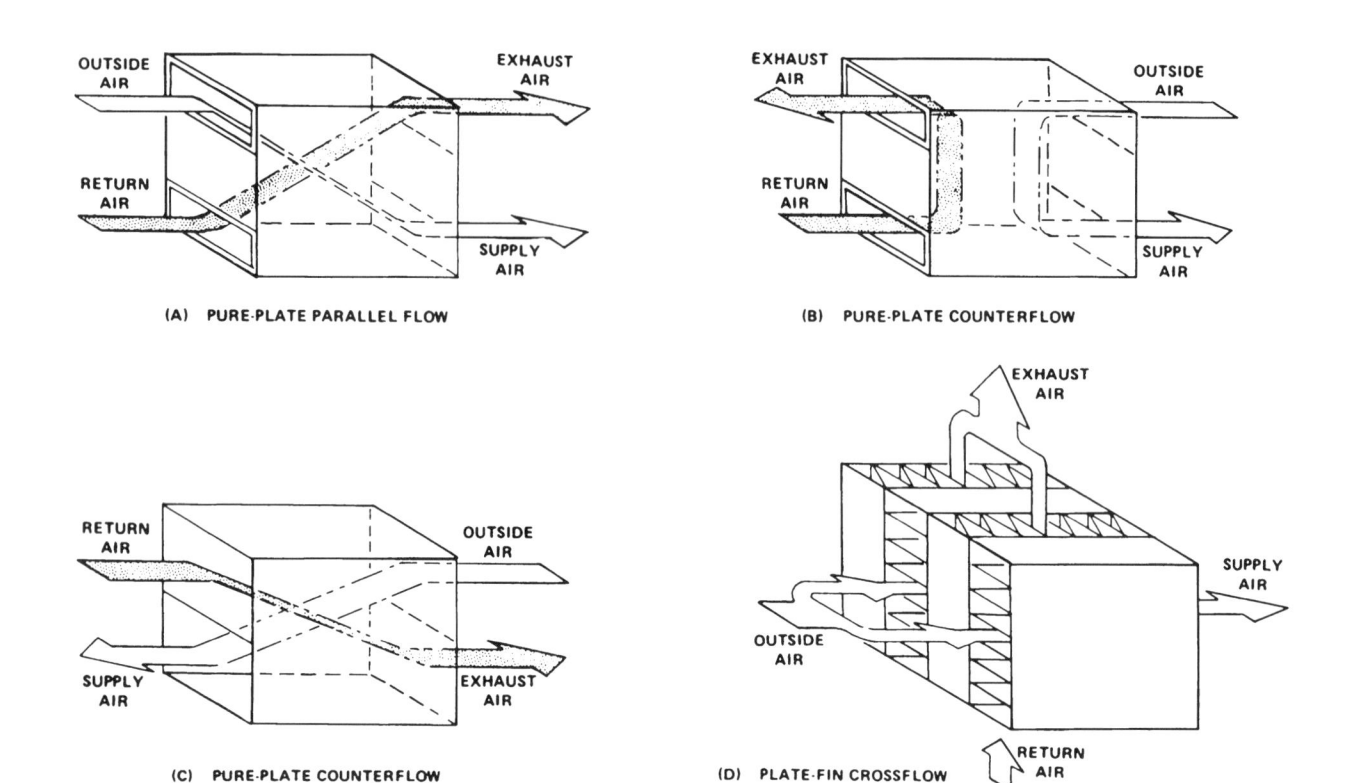

*Fig. 17-27 Pure-Plate and Plate-Fin Models*

= 0.0106 lb/lb (0.0106 kg/kg)

*Winter dry-bulb temperature*

$$X_2 = X_1 - \varepsilon(X_1 - X_2) = 0 - 0.8(0 - 75)$$
$$= 60°F \text{ dry bulb } (15.5°C)$$

*Winter humidity ratio*

$$X_2 = X_1 - \varepsilon(X_1 - X_2) = 0.0001 - 0.8(0.0001 - 0.0074)$$
$$= 0.0059 \text{ lb/lb } (0.0059 \text{ kg/kg})$$

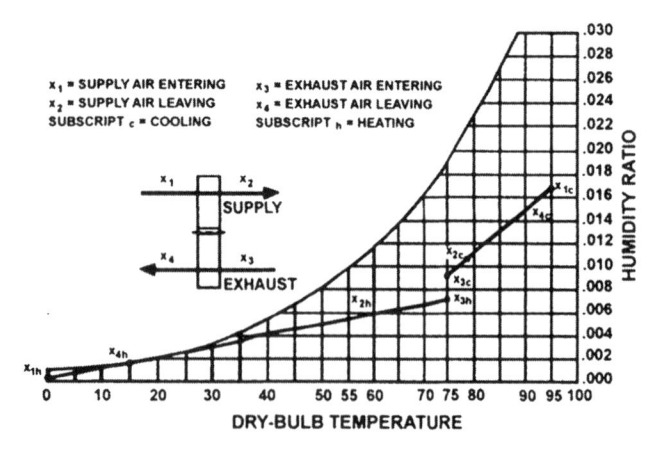

**Conditions for Example 17-2**

**Example 17-3** Determine the leaving conditions for an energy wheel exchanger with the following comfort conditions:

*Cooling*

1. Design conditions. Outdoor air = 95°F (35°C) dry bulb, 78°F (25.6°C) wet bulb, and 41.3 Btu/lb (96.1 kJ/kg).
2. Design conditions. Space = 75°F (23.9°C) dry bulb, 62.5°F (17°C) wet bulb, and 28.3 Btu/lb (65.8 kJ/kg) of dry air.

*Heating*

1. Design conditions. Outdoor air = 20°F (–6.7°C) dry bulb and 5.0 Btu/lb (11.6 kJ/kg) of dry air.
2. Design conditions. Space = 75°F (23.9°C) dry bulb and 22.4 Btu/lb (52.1 kJ/kg) of dry air.

*Cooling and Heating*

Exchanger effectiveness = 80% on sensible heat and 65% on total heat. Equal mass flow rates.

**Solution:**

Dry-bulb temperature and enthalpy are calculated from Equation (17-8).

*Summer dry-bulb temperature*

$$X_2 = X_1 - \varepsilon(X_1 - X_3) = 95 - 0.8(95 - 75)$$
$$= 79°F (26.1°C)$$

*Summer enthalpy*

$$X_2 = X_1 - \varepsilon(X_1 - X_3) = 41.3 - 0.65(41.3 - 28.3)$$

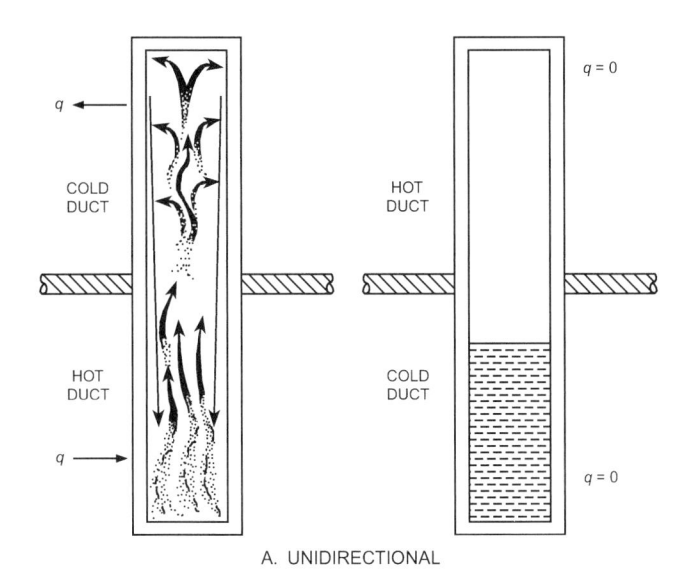

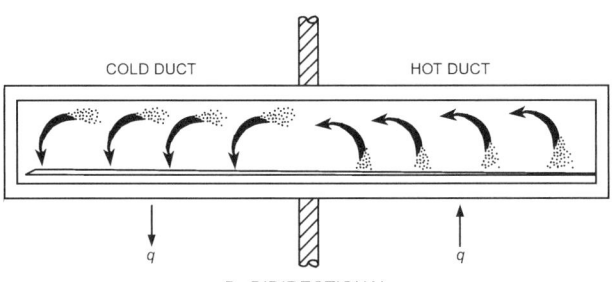

# Air-Processing Equipment

*Fig. 17-28 Sealed-Tube Thermosiphons*

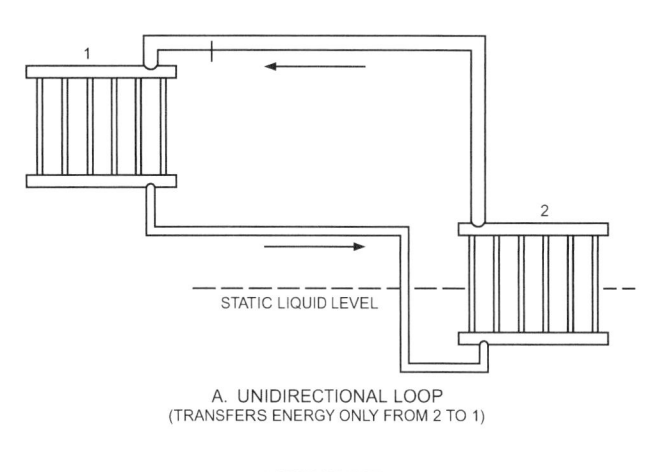

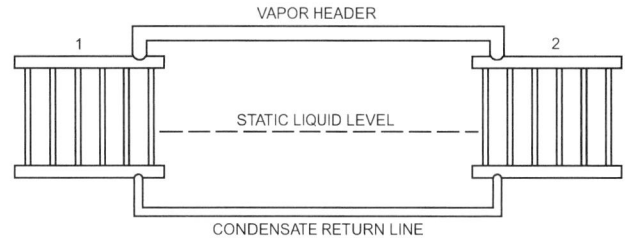

*Fig. 17-29 Coil-Type Thermosiphon Loops*

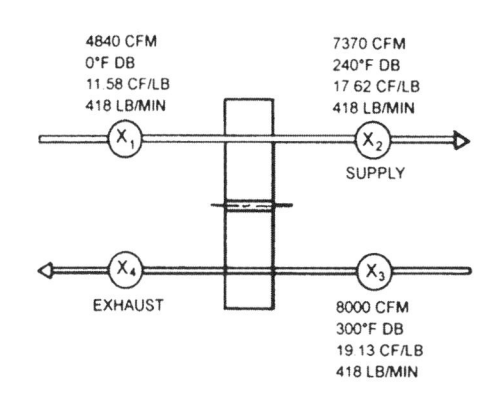

**Schematic and Conditions for Example 17-4**

= 32.9 Btu/lb (76.5 kJ/kg)

*Winter dry-bulb temperature*

$$X_2 = X_1 - \varepsilon(X_1 - X_3) = 20 - 0.8(20 - 75)$$
$$= 64°F (17.8°C)$$

*Heating enthalpy*

$$X_2 = X_1 - \varepsilon(X_1 - X_3) = 0.5 - 0.65(5.0 - 22.4)$$
$$= 16.3 \text{ Btu/lb} (37.9 \text{ kJ/kg})$$

**Example 17-4** Calculate recovered temperatures and volumes (measured at point 2), as well as energy savings, for a heat wheel exchanger handling 8000 cfm (3.8 m³/s) of process exhaust at 300°F (149°C) measured at point 3, considered dry air. Assume an equal mass of makeup air at winter design conditions of 0°F (−17.8°C). Exchanger effectiveness is 80% at equal mass flows. Neglect cross leakage and purge volume.

**Solution:**

1. For conditions at 300°F (149°C)

   8000 ft³/min / 19.13 ft³/lb = 418 lb/min (189.3 kg/min)

2. For conditions of 0°F (−17.8°C)

   $$v_1 = 11.58 \text{ ft}^3/\text{lb}$$

   418 lb/min × 11.58 ft³/lb = 4840 cfm at Point 1 (2.28 m³/s)

3. Temperature at Point 2 from Equation (17-8):

   $$X_2 = X_1 - \varepsilon(X_1 - X_3) = 0 - 0.8(0 - 300)$$
   $$= 240°F \text{ dry bulb } (115.5°C)$$

4. Specific volume at 240°F dry bulb (115.5°C)

   $$v_2 = 17.62 \text{ ft}^3/\text{lb}$$

   418 lb/min × 17.62 ft³/lb = 7370 cfm (3.5 m³/s)

5. Savings at winter design conditions
   [specific heat = 0.242 Btu/(lb·°F)]

   418 lb/min × 60 min/h × 0.242 Btu/(lb·°F) × (240 − 0)
   = 1,460,000 Btu/h (428 kW)

**Example 17-5** Determine the leaving conditions for a heat pipe exchanger with the following comfort conditions:

*Cooling*

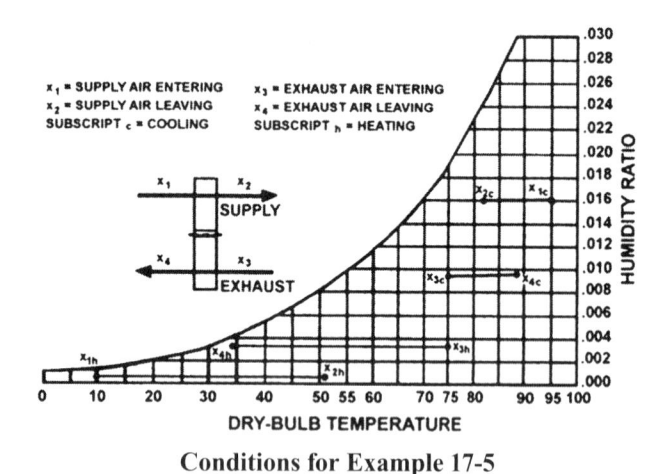

**Conditions for Example 17-5**

1. Design conditions, outdoor air = 95°F dry bulb (35°C)
2. Design conditions, space = 75°F dry bulb (23.9°C)

*Heating*

1. Design conditions, outdoors = 10°F dry bulb (–12.2°C)
2. Design conditions, space = 75°F dry bulb (23.9°C)

*Cooling and Heating*

Exchanger effectiveness = 65% on sensible heat at equal mass flows.

**Solution:**

Dry-bulb temperature is calculated from Equation (17-8).

*Cooling dry-bulb temperature*

$$X_2 = X_1 - \varepsilon(X_1 - X_3) = 95 - 0.65(95 - 75)$$
$$= 82°F (27.8°C)$$

*Heating dry-bulb temperature*

$$X_2 = X_1 - \varepsilon(X_1 - X_3) = 10 - 0.65(10 - 75)$$
$$= 52.3°F (11.3°C)$$

## 17.11 Economizers

An economizer uses outdoor air to reduce the refrigeration required to provide cooling for the building when the outdoor air dry-bulb temperature is low enough to provide for the sensible cooling needs or reduce the refrigeration requirements to do so. Either the air-side or the water-side economizer is an attractive option for reducing energy costs with self-contained HVAC systems. Although climate often dictates which economizer type is selected, either one can provide advantages.

The air-side economizer includes an outdoor air damper, relief damper, return air damper, filters, actuator, and linkage. Economizer controls are usually a factory-installed option. The air-side economizer takes advantage of cool outdoor air to either assist mechanical cooling or, if the outdoor air is sufficiently cool, provide total system cooling. However, if the building has significant simultaneous heating and cooling requirements, the interaction of the economizer with an installed heat recovery system must be thoroughly analyzed.

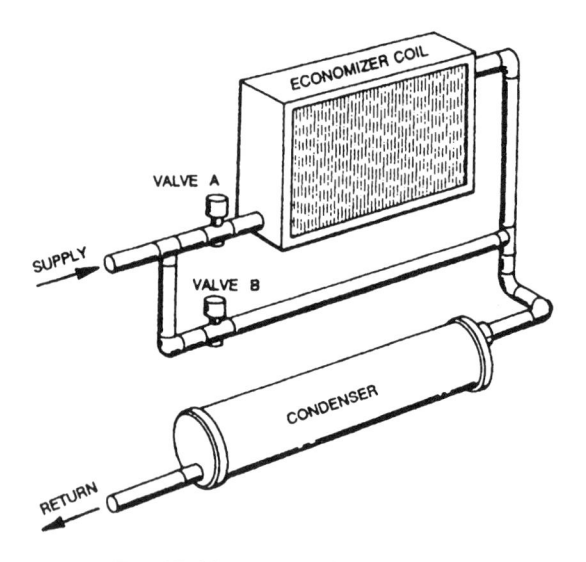

Self-contained units usually do not include return air fans. It is necessary to include a variable-volume relief fan unit when air-side economizers are employed. The relief fan volume is generally controlled with discharge dampers in response to building space pressure. The relief fan is off and discharge dampers are closed when the air-side economizer is inactive. However, the installed cost of an air-side economizer is generally higher than that for a water-side economizer.

Typically, in an air-side economizer, an enthalpy sensor or dry-bulb temperature probe energizes the unit to bring in outdoor air as the first stage of cooling. An outdoor air damper modulates the flow to meet a design temperature, and when outdoor air can no longer provide enough cooling, the refrigeration system is energized.

The water-side economizer consists of a water coil located upstream of the main cooling coil. All economizer control valves, piping between the economizer coil and the condenser, and economizer control wiring can be factory installed. The water-side economizer takes advantage of low cooling tower water temperature (approaching ambient wet-bulb temperature) to either precool the entering air, assist mechanical cooling, or, if the cooling water is cold enough, to provide total system cooling. If the economizer is unable to maintain the supply air setpoint for VAV units or zone set point for constant-volume units, factory-mounted controls integrate economizer and compressor operation to meet cooling requirements.

Cooling water flow rate is controlled by two valves (Figure 17-30), one at the economizer coil inlet (A) and one in the bypass loop to the condenser (B). Two control methods are common—constant water flow and variable water flow.

*Fig. 17-30 Water-Side Economizer*

## 17.12 Problems

**17.1** Air enters a coil at 95°F dry-bulb and 78°F wet-bulb temperature and leaves at 62°F dry-bulb and 60°F wet-bulb temperature. The condensate is assumed to be at a temperature of 56°F. Find the total, latent, and sensible cooling loads on the coil with air at 14.7 psia.

**17.2** Air enters a direct expansion coil at 85°F (29.4°C) dry bulb and 70°F (21.1°C) wet bulb and leaves at 62°F (16.7°C) dry bulb and 90% rh.

(a) How much sensible heat and how much latent heat is removed from the air by the coil?

(b) How much condensate drains off the coil?

**17.3** Air enters a direct-expansion coil at 90°F (32.2°C) dry bulb 60% rh, and leaves the coil at 60°F (15.6°C) dry bulb, 95% rh. Find:

(a) heat removed from air

(b) moisture condensed from air

(c) SHR for the condition line

[Ans: (a) 19.2 Btu/lb (44.6 kJ/kg), (b) 0.008 lb/lb (0.008 kg/kg), (c) 0.45]

**17.4** Water flowing at 60 lb/min and at 51°F is chilled in an evaporator to 40°F. The heat transfer area is 20 ft² and the heat exchanger has an overall heat transfer coefficient of 60 Btu/h·ft²·°F. The direct-expansion evaporator uses R-12 and operates at 35°F. Find the evaporator effectiveness.

**17.5** Outdoor air at 35°F and 70% rh is supplied to an air-conditioning apparatus. Recirculated air is returned from the plant at 69°F dry bulb and 40% rh; 8100 cfm of outdoor air mixes with 18,900 cfm of recirculated air. The mixture is heated by a steam coil and humidified by a pan humidifier to final conditions of 115°F dry bulb and 20% rh.

(a) What steam flow, in pounds per hour, should be supplied to the heating coil?

(b) Estimate the steam consumption of the humidifier.

**17.6** Outdoor air (8000 cfm) at 10°F dry bulb and 50% rh enters the central apparatus of a split heating system. It is tempered to 55°F dry bulb. Then, it flows through a spray humidifier where the leaving sump water is maintained at 50°F. The spray humidifier has a performance factor of 0.80. After leaving the humidifier, the air flows through a steam heating coil and is heated to 70°F dry bulb.

(a) What is the final relative humidity and humidity ratio of the air as it leaves the heating coil? [Ans: 40%]

(b) Assume steam at 2 psig and 90% quality is supplied to the tempering coil, the sump water heat exchanger, and the heating coil. How many pounds of steam per hour should be supplied to each? [Ans.: 170.7 lb]

**17.7** The heat exchanger for the spray water in Problem 17.6 is out of service for maintenance. The split heating system is operating as specified except that the sump water is recirculated. Assume makeup water to the sump is 37°F and saturating effectiveness is equal to the performance factor.

(a) What is the final relative humidity and humidity ratio of the air leaving the heating coil?

(b) What is the steam rate (lb/h) for the tempering coils and for the heating coil?

**17.8** Air at 105°F dry bulb (40.6°C) and 75% rh passes through a chilled water spray. Air leaves the spray chamber at 45°F dry bulb (7.2°C) saturated. How many grains of moisture per pound of entering air are condensed?

**17.9** A building space is to be maintained at 70°F and 35% rh when outdoor design temperature is 10°F. Design heat losses from the space are 250,000 Btu/h, sensible, and 45,000 Btu/h, latent. Ventilation requires that 1500 cfm of outdoor air be used. Supply air is to be at 120°F. Determine:

(a) the amount of supply air required, lb/h and cfm

(b) the capacity of the heating coil, Btu/h, if:

1) the humidifier is a spray washer using recirculated spray water with makeup water provided at 60°F

2) the humidifier is a steam humidifier using dry, saturated steam at 17.2 psia

(c) the capacity of the humidifier, lb/h.

The conditioning equipment and nomenclature are shown in the following sketch.

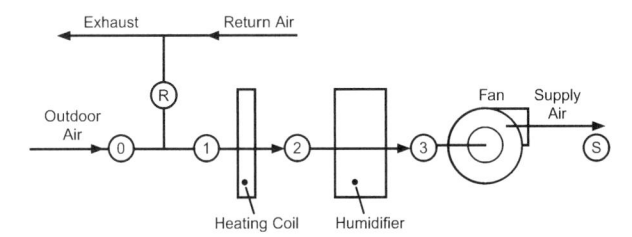

**17.10** A spray-type air washer is to be used for humidification as well as cleaning of 9000 scfm of air. Inlet conditions to the washer are 75°F db and 48 F wb. Desired humidity ratio at outlet is 0.005 lb_w/lb_a. Determine (a) the necessary humidification efficiency of the washer, %, and (b) the makeup water requirements (humidifying capacity) of the unit, $lb_w/h$.

**17.11** A heat pipe air-to-air energy recovery device is being considered for a system requiring 9000 scfm of outdoor air. Initially, a separate preheater was planned for bringing the outdoor air from its −2°F design ambient outdoor temperature to 40°F. Determine (a) the rating (Btu/h) and (b) the sensible effectiveness (%) to specify for the heat pipe unit if it is to eliminate the need for the air preheater. [Ans: 415,800 Btu/h, 57%]

**17.12** The HVAC system for a hospital operating room, which requires 100% outdoor air, is shown in the following figure and includes an air-to-air heat pipe energy recovery unit having a sensible effectiveness of 73%. The air leaving the cooling coil is maintained at 58°F, 90% rh, all year long. During winter operation, air leaves the heater at 130°F. Fan speed is changed between summer and winter operation. Design duct system pressure drop (summer) is 3.25 in. water.

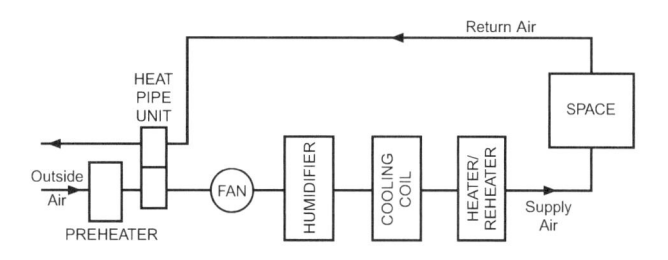

1. At winter design conditions (indoor: 72°F and 30% rh; outdoor: 5°F and 100% rh) the space load is 235,000 Btu/h (sensible) with negligible latent load. Determine (a) the necessary size of heating unit (Btu/h) both with and without the energy recovery unit and (b) the humidifier size (gal/day). Neglect fan effects.

2. At summer design conditions (indoor: 78°F; outdoor: 95°F db/76°F wb), the space cooling loads are 146,000 Btu/h (sensible) and 79,000 Btu/h (latent). Determine (a) fan size (hp and scfm), (b) sensible coil load, Btu/h, (c) latent coil load, Btu/h, and (d) necessary size of cooling unit, Btu/h, both with and without the energy recovery unit. Include fan effects.

## 17.13 Bibliography

ASHRAE. 1991. Method of testing air-to-air heat exchangers. *ASHRAE Standard 84-1991*.

ASHRAE. 2008. *2008 ASHRAE Handbook—HVAC Systems and Equipment*.

Gudac, G.J., M.A. Mueller, J.J. Bosch, R.H. Howell, and H.J. Sauer, Jr. 1981. "Effectiveness and Pressure Drop Characteristics of Various Types of Air-to-Air Energy Recovery Systems," *ASHRAE Transactions* 87(1).

Sauer, H. J., Jr. and R. H. Howell. 1981. "Promise and Potential of Air-to-Air Energy Recovery Systems," *ASHRAE Transactions* 87(1).

# Air-Processing Equipment

# SI Table

### Table 17-2 SI  Comparison of Air-to-Air Energy Recovery Devices
*(Table 3, Chapter 26, 2012 ASHRAE Handbook—HVAC Systems and Equipment, SI Version)*

|  | Fixed Plate | Membrane Plate | Energy Wheel | Heat Wheel | Heat Pipe | Runaround Coil Loop | Thermosiphon | Twin Towers |
|---|---|---|---|---|---|---|---|---|
| Airflow arrangements | Counterflow Cross-flow | Counterflow Cross-flow | Counterflow Parallel flow | Counterflow | Counterflow Parallel flow | — | Counterflow Parallel flow | — |
| Equipment size range, cfm | 50 and up | 50 and up | 50 to 74,000 and up | 50 to 74,000 and up | 100 and up | 100 and up | 100 and up | — |
| Typical sensible effectiveness ($m_s = m_e$), % | 50 to 80 | 50 to 75 | 50 to 85 | 50 to 85 | 45 to 65 | 45 to 65 | 40 to 60 | 40 to 60 |
| Typical latent effectiveness,* % | — | 50 to 72 | 50 to 85 | 0 | — | — | — | — |
| Total effectiveness,* % | — | 50 to 73 | 50 to 85 | — | — | — | — | — |
| Face velocity, fpm | 200 to 1000 | 200 to 600 | 500 to 1000 | 400 to 1000 | 400 to 800 | 300 to 600 | 400 to 800 | 300 to 450 |
| Pressure drop, in. of water | 0.4 to 4 | 0.4 to 2 | 0.4 to 1.2 | 0.4 to 1.2 | 0.6 to 2 | 0.6 to 2 | 0.6 to 2 | 0.7 to 1.2 |
| EATR, % | 0 to 2 | 0 to 5 | 0.5 to 10 | 0.5 to 10 | 0 to 1 | 0 | 0 | 0 |
| OACF | 0.97 to 1.06 | 0.97 to 1.06 | 0.99 to 1.1 | 1 to 1.2 | 0.99 to 1.01 | 1.0 | 1.0 | 1.0 |
| Temperature range, °F | −75 to 1470 | −40 to 120 | −65 to 1470 | −65 to 1470 | −40 to 104 | −50 to 930 | −40 to 104 | −40 to 115 |
| Typical mode of purchase | Exchanger only Exchanger in case Exchanger and blowers Complete system | Exchanger only Exchanger in case Exchanger and blowers Complete system | Exchanger only Exchanger in case Exchanger and blowers Complete system | Exchanger only Exchanger in case Exchanger and blowers Complete system | Exchanger only Exchanger in case Exchanger and blowers Complete system | Coil only Complete system | Exchanger only Exchanger in case | Complete system |
| Advantages | No moving parts Low pressure drop Easily cleaned | No moving parts Low pressure drop Low air leakage Moisture/mass transfer | Moisture/mass transfer Compact large sizes Low pressure drop Available on all ventilation system platforms | Compact large sizes Low pressure drop Easily cleaned | No moving parts except tilt Fan location not critical Allowable pressure differential up to 2 psi | Exhaust airstream can be separated from supply air Fan location not critical | No moving parts Exhaust airstream can be separated from supply air Fan location not critical | Latent transfer from remote airstreams Efficient microbiological cleaning of both supply and exhaust airstreams |
| Limitations | Large size at higher flow rates | Few suppliers Long-term maintenance and performance unknown | Supply air may require some further cooling or heating Some EATR without purge | Some EATR with purge | Effectiveness limited by pressure drop and cost Few suppliers | Predicting performance requires accurate simulation model | Effectiveness may be limited by pressure drop and cost Few suppliers | Few suppliers Maintenance and performance unknown |
| Heat rate control (HRC) methods | Bypass dampers and ducting | Bypass dampers and ducting | Bypass dampers and wheel speed control | Bypass dampers and wheel speed control | Tilt angle down to 10% of maximum heat rate | Bypass valve or pump speed control | Control valve over full range | Control valve or pump speed control over full range |

*Rated effectiveness values are for balanced flow conditions. Effectiveness values increase slightly if flow rates of either or both airstreams are higher than flow rates at which testing is done.

EATR = exhaust air transfer ratio
OACF = outdoor air correction factor

# Chapter 18

# REFRIGERATION EQUIPMENT

This chapter provides a relatively brief treatment of the systems and components used for providing the cooling requirements of building HVAC systems. Primary topics are vapor compression refrigeration, absorption refrigeration, and cooling towers. Additional information can be obtained from the 2010 *ASHRAE Handbook—Refrigeration* and the 2012 *ASHRAE Handbook—HVAC Systems and Equipment*.

## 18.1 Mechanical Vapor Compression

The basic components of the mechanical vapor compression cycle are the compressor, condenser, expansion device, and evaporator (Figure 18-1). The basic principles of the vapor compression cycle are detailed in Chapter 2 of this book. Additional information is also provided in Chapter 2 of the 2013 *ASHRAE Handbook—Fundamentals*.

### 18.1.1 Compressors

The compressor is one of the essential parts of the compression refrigeration system and serves both to provide the necessary increase in pressure of the refrigerant vapor and as a refrigerant pump to circulate the refrigerant through the system in a continuous cycle.

There are two basic types of compressors: positive displacement and dynamic. Positive-displacement compressors increase the pressure of refrigerant vapor by reducing the volume of the compressor chamber through work applied to the compressor's mechanism. This class of compressor includes reciprocating, rolling piston, rotary vane, single screw, double screw, trochoidal, and scroll. Dynamic compressors increase the pressure of refrigerant vapor by a continuous transfer of angular momentum from the rotating member to the vapor followed by the conversion of this momentum into a pressure rise. Centrifugal compressors function based on these principles.

Compressor performance is the result of design constraints involving physical limitations of the refrigerant, compressor, and motor, while attempting to provide the following:

- Greatest trouble-free life expectancy
- Most refrigeration effect for the least power input
- Lowest applied cost
- Wide range of operating conditions
- Acceptable vibration and sound level

Two useful measures of compressor performance are capacity (which is related to compressor volume displacement) and efficiency. Compressor refrigerating capacity is the rate of heat removal by the refrigerant pumped by the compressor in a refrigerating system at the evaporator. Capacity equals the product of the mass flow rate of refrigerant pumped by the compressor and the difference in specific enthalpies of the refrigerant when it leaves the evaporator and when it enters the evaporator.

**Reciprocating Compressors.** Most reciprocating compressors are single acting, using pistons driven directly through a pin and connecting rod from the crankshaft. Double-acting compressors are not extensively used.

The halocarbon compressor is the most widely used and is manufactured in three designs: (1) open, (2) semihermetic or bolted hermetic, and (3) welded-shell hermetic. Ammonia compressors are manufactured only in the open design, in which the driveshaft extends through a seal in the crankcase for an external drive.

In hermetic compressors, the motor and compressor are contained within the same pressure vessel; the motor shaft is integral with the compressor crankshaft, and the motor is in contact with the refrigerant. A hermetic compressor is shown in Figure 18-2. A semi-, bolted, accessible, or serviceable hermetic compressor is bolted together and may be repaired in the field. The motor compressor in a welded shell (sealed) hermetic compressor is mounted inside a steel shell, which in turn is sealed by welding. Table 18-1 shows combinations of common design features and Table 18-2 gives typical performance values for halocarbon refrigerant compressors.

Capacity data are given in Figure 18-3, which is a typical set of curves for a four-cylinder semi-hermetic compressor,

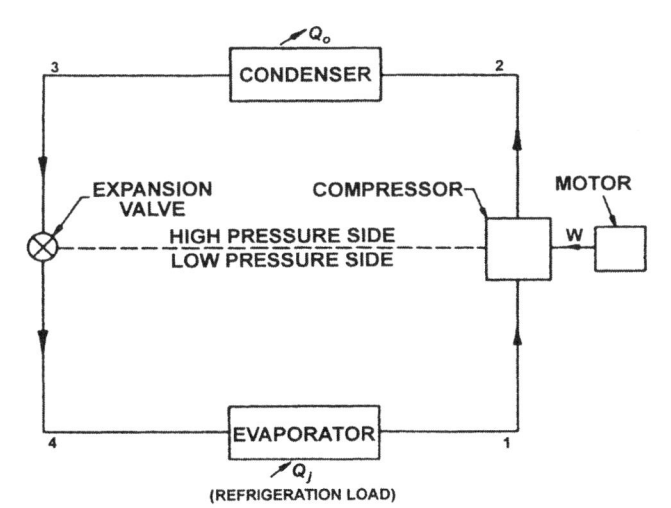

*Fig. 18-1 Simplified Equipment Diagram for the Basic Vapor-Compression Cycle*

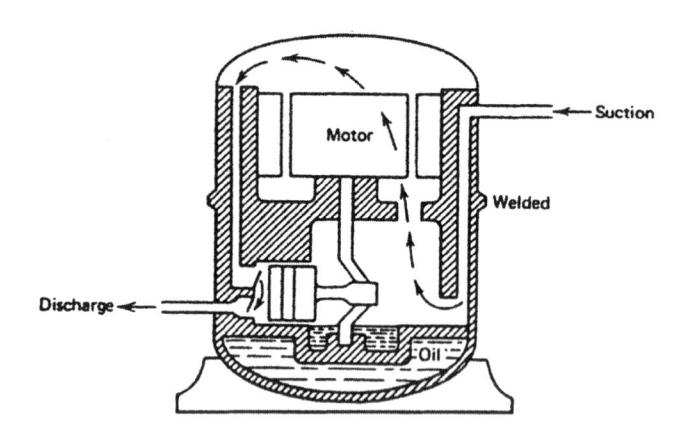

*Fig. 18-2 Hermetic Compressor*

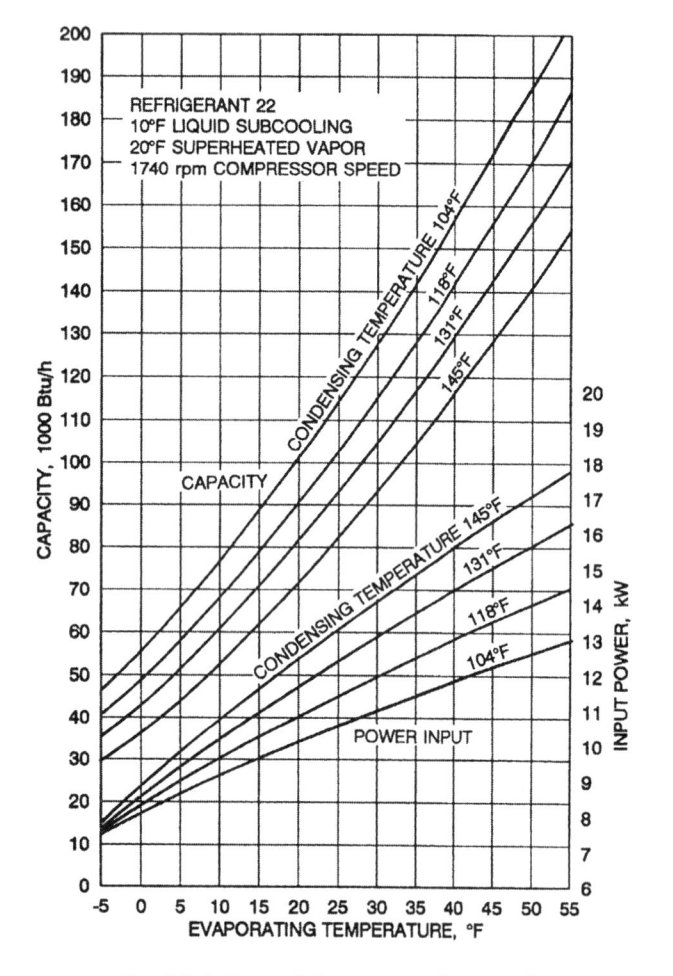

*Fig. 18-3 Typical Capacity and Power for Reciprocating Compressor*
(Figure 8, Chapter 38, 2012 ASHRAE Handbook—HVAC Systems and Equipment)

2 3/8 in. (60 mm) bore, 1 3/4 in. (44 mm) stroke, 1740 rpm, operating with R-22. A set of power curves for the same compressor is also shown.

Reciprocating compressors are most commonly used for systems in the range of 0.5 to 100 tons (2 to 350 kW) and larger. They are used in unitary heat pumps and, in most cases, are either fully or accessibly hermetic.

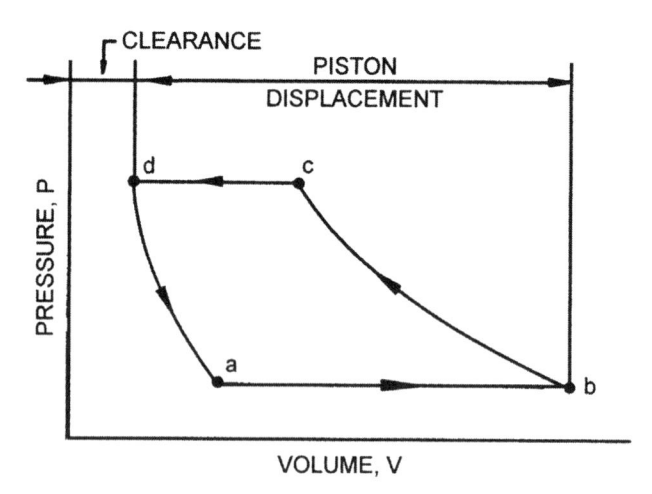

*Fig. 18-4 Cycle for Idealized Piston Compressor*

One of the important thermodynamic considerations for this compressor is the effect of the clearance volume (i.e., the volume occupied by the refrigerant within the compressor that is not displaced by the moving member). The effect is illustrated, in the case of the piston-type compressor, by considering the clearance volume between the piston and the cylinder head when the piston is at top dead-center position. The clearance gas remaining in this space after the compressed gas is discharged from the cylinder reexpands as the piston moves downward, preventing a fresh charge into the cylinder until the pressure falls to the inlet (suction) pressure (see Figure 18-4). As a consequence, the volume (and mass) of refrigerant entering the cylinder is less than the volume swept by the piston. This effect is quantitatively expressed by the volumetric efficiency $e_v$ as

$$\frac{e_v}{100} = \frac{m_a}{m_i}$$

where

$m_a$ = actual mass of new gas entering the compressor
$m_i$ = theoretical mass, equal to piston displacement divided by specific volume of refrigerant vapor at suction conditions

The volumetric efficiency due only to reexpansion of the clearance volume gas can be calculated as follows:

$$\frac{e_v}{100} = 1 + C\left(\frac{v_s}{v_d}\right)$$

where

$C$ = clearance ratio = $(V_b - V_a)/(V_b - V_d)$
$v_s$ = specific volume of refrigerant at suction conditions
$v_d$ = specific volume of refrigerant at discharge conditions

$V_a$, $V_b$, and $V_d$ = the volumes at the locations given in Figure 18-4

# Refrigeration Equipment

Table 18-1  Typical Design Features of Reciprocating Compressors

| Item | Halo-, Fluoro-, or Hydrocarbon Open | Semi-hermetic | Welded Hermetic | Ammonia Open | Item | Halo-, Fluoro-, or Hydrocarbon Open | Semi-hermetic | Welded Hermetic | Ammonia Open |
|---|---|---|---|---|---|---|---|---|---|
| 1. Number of cylinders: one to | 16 | 12 | 6 | 16 | 10. Bearings | | | | |
| | | | | | a. Sleeve, antifriction | X | X | X | X |
| 2. Power range | 0.17 hp 125 W | 0.50.35 to 150 hp 110 kW | 0.170.12 to 25 hp 20 kW | 10 hp 7.5 kW | b. Tapered roller | X | | | X |
| | and up | | | and up | 11. Capacity control, if provided: manual or automatic | | | | |
| 3. Cylinder arrangement | | | | | | | | | |
| a. Vertical, V or W, radial | X | X | | | a. Suction valve lifting | X | X | X | X |
| b. Radial, horizontal opposed | | | X | | b. Bypass-cylinder heads to suction | X | X | X | X |
| c. Horizontal, vertical V or W | | X | | X | c. Closing inlet | X | X | | X |
| 4. Drive | | | | | d. Adjustable clearance | X | X | | X |
| a. Electric motor | | X | X | | e. Variable-speed | X | X | X | X |
| | | | | | 12. Materials | | | | |
| b. Direct drive, V belt chain, gear, by electric motor or engine | X | | | X | Motor insulations and rubber materials must be compatible with refrigerant and lubricant mixtures; otherwise, no restrictions | | X | X | |
| 5. Lubrication: splash or force feed, flooded | X | X | X | X | No copper or brass | | | | X |
| 6. Suction and discharge valves: ring plate or ring or reed flexing | X | X | X | X | 13. Lubricant return | | | | |
| | | | | | a. Crankcase separated from suction manifolds, oil return check valves, equalizers, spinners, foam breakers | X | X | | X |
| 7. Suction and discharge valve arrangement | | | | | | | | | |
| a. Suction and discharge valves in head | X | X | X | X | b. Crankcase common with suction manifold | | | X | |
| b. Uniflow: suction valves in top of piston, suction gas entering through cylinder walls; discharge valves in head | X | | | X | 14. Synchronous fixed speeds, rpm | 250 to 3600 | 1500 to 3600 | 1500 to 3600 | 250 to 1500 |
| | | | | | 15. Pistons | | | | |
| 8. Cylinder cooling | | | | | a. Aluminum or cast iron | X | X | X | X |
| a. Suction-gas-cooled | X | X | X | X | b. Ringless | X | X | X | X |
| b. Water jacket cylinder wall, head, or cylinder wall and head | X | | | X | c. Compression and oil-control rings | X | X | X | X |
| | | | | | 16. Connecting rod | | | | |
| c. Air-cooled | X | X | X | X | Split rod with removable cap or solid eccentric strap | X | X | X | X |
| d. Refrigerant-cooled heads | X | | | X | 17. Mounting | | | | |
| 9. Cylinder head | | | | | Internal spring mount | | X | X | |
| a. Spring-loaded | X | X | X | X | External spring mount | | X | X | |
| b. Bolted | X | X | X | X | Rigidly mounted on base | X | X | | X |

The actual volumetric efficiency is affected by other factors such as cylinder wall heating due to friction and pressure drops through the inlet and discharge valves and is best obtained by actual laboratory measurements of the amount of refrigerant compressed and delivered by the compressor. The difference between actual and predicted volumetric efficiency, considering only clearance volume effects, is illustrated in Figure 18-5.

**Rotary Compressors.** Rotary compressors operate with a circular, or rotary, motion instead of reciprocating motion. Their positive-displacement compression process is nonreversing and either continuous or cyclical, depending on the type of mechanism. Most are direct drive machines.

The rolling piston rotary compressor is shown in Figure 18-6; the rotary vane type is shown in Figure 18-7. These two machines are similar in size, weight, thermodynamic

**Table 18-2 Typical Performance Values**

| | Operating Conditions and Refrigerants | | | |
|---|---|---|---|---|
| | R-404a | R-134a | R-22 | R-22 |
| Compressor Size and Type | Evap. Temp. = −40°F<br>Cond. Temp. = 105°F<br>Suction Gas = 65°F<br>Subcooling = 0°F | Evap. Temp. = 0°F<br>Cond. Temp. = 110°F<br>Suction Gas = 65°F<br>Subcooling = 0°F | Evap. Temp. = 40°F<br>Cond. Temp. = 105°F<br>Suction Gas = 55°F<br>Subcooling = 0°F | Evap. Temp. = 45°F<br>Cond. Temp. = 130°F<br>Suction Gas = 65°F<br>Subcooling = 0°F |
| Large, over 25 hp<br>  Open<br>  Hermetic | 0.21 tons/hp<br>3.15 Btu/h per W | 0.40 tons/hp<br>6.00 Btu/h per W | 1.05 tons/hp<br>14.2 Btu/h per W | 1.07 tons/hp<br>10.4 Btu/h per W |
| Medium, 5 to 25 hp<br>  Open<br>  Hermetic | 0.19 tons/hp<br>2.89 Btu/h per W | 0.37 tons/hp<br>5.60 Btu/h per W | 1.00 ton/hp<br>14.0 Btu/h per W | 1.00 tons/hp<br>10.2 Btu/h per W |
| Small, under 5 hp<br>  Open<br>  Hermetic | —<br>— | —<br>3.80 Btu/h per W | —<br>13.8 Btu/h per W | —<br>10.0 Btu/h per W |

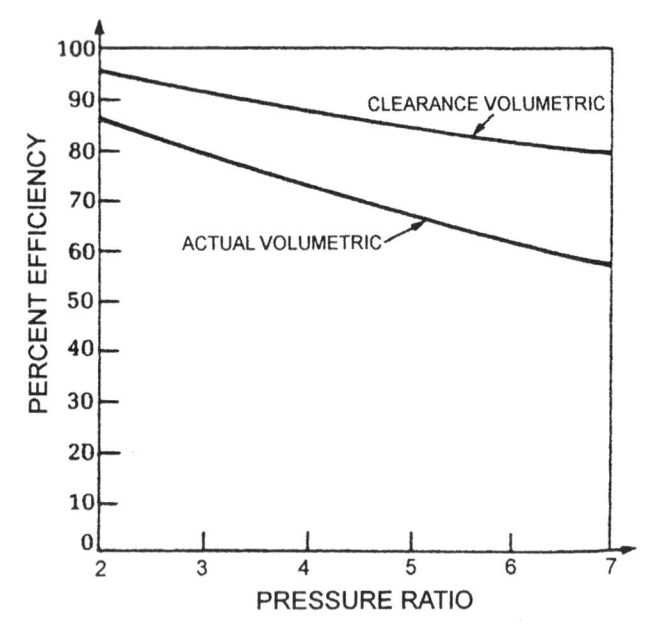

*Fig. 18-5 Volumetric Efficiency*

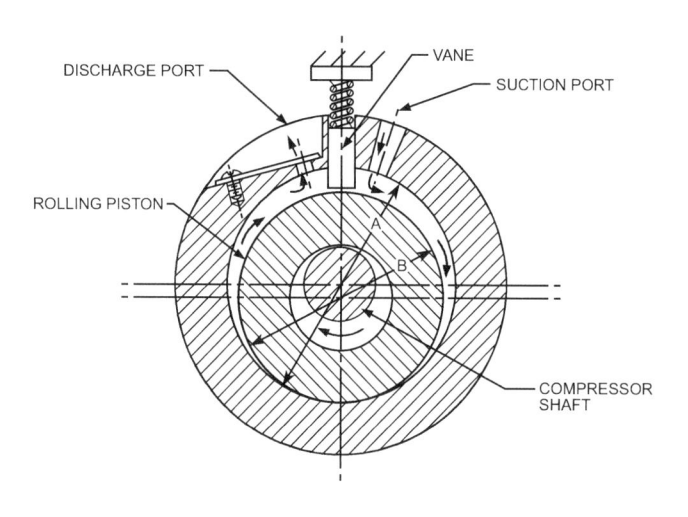

*Fig. 18-6 Rolling Piston Compressor*
(Figure 10, Chapter 38, 2012 ASHRAE Handbook—
HVAC Systems and Equipment)

performance, field of applications, range of capacities, durability, and sound level.

Internal leakage is controlled in rotary compressors through hydrodynamic sealing; thus, precision fits and optimum clearance are design requirements. The hydrodynamic sealing depends on clearances, surface speed, oil viscosity, and surface finish of the parts. Smoother finishes and closer clearances are used with low-viscosity oil in small machines. Larger machines have greater clearances and usually use a higher-viscosity oil.

Rotary compressor performance is characterized by high volumetric efficiency due to the small clearance volume and by correspondingly low reexpansion loss.

The **rolling piston compressor** uses a roller mounted on an eccentric shaft. A single vane or blade positioned in the nonrotating cylindrical housing reciprocates as the eccentrically moving roller turns. Rolling piston compressors are used in household refrigerators and air-conditioning units in sizes up to about 3 hp (2 kW).

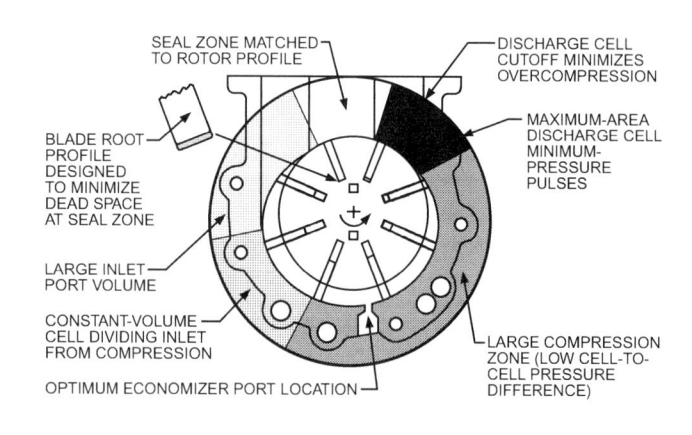

*Fig. 18-7 Rotary Vane Compressor*
(Figure 13, Chapter 38, 2012 ASHRAE Handbook—
HVAC Systems and Equipment)

# Refrigeration Equipment

Displacement for this compressor can be calculated from the following equation:

$$V_d = \frac{\pi H (A^2 - B^2)}{4}$$

where

$V_d$ = displacement
$H$ = cylinder block height
$A$ = cylinder diameter
$B$ = roller diameter

Suction gas is directly piped into the suction port of the compressor, and the compressed gas is discharged into the compressor housing shell. This high-side shell design is used because of the simplicity of its lubrication system and the absence of oiling and compressor cooling problems. Compressor performance is also improved because this arrangement minimizes heat transfer to the suction gas and reduces gas leakage areas.

The performance typical of rolling piston compressors is illustrated in Figure 18-8.

**The rotating vane compressor** has a rotor concentric with the shaft, with vanes in the rotor; this assembly is off center with respect to the cylindrical housing. An oval shaped bore produces a double lobe or a two-cylinder compressor. Rotary vane compressors have a low weight-to-displacement ratio, which, in combination with their compact size, makes them suitable for transport applications. Small compressors in the 3 to 50 hp (2 to 40 kW) range are single staged, for a saturated suction temperature range of –40 to 45°F (–40 to 7°C) at saturated condensing temperatures of up to 140°F (60°C). By employing a second stage, low-temperature applications down to –60°F (–50°C) are possible. Currently, R-22, R-502, and R-717 refrigerants are used.

**Screw Compressors.** The helical rotary compressor, or the screw compressor, belongs to the class of positive-displacement compressors. Screw compressors currently in production for refrigeration and air-conditioning applications comprise two distinct types: single screw and twin screw. Both are conventionally used in the fluid injection mode where sufficient fluid cools and seals the compressor. Single-screw compressors have the capability to operate at pressure ratios above 20:1 single stage. The capacity range currently available is from 20 to 1300 tons (70 to 4600 kW).

The single-screw compressor consists of a single cylindrical main rotor that works with a pair of gaterotors. Both the main rotor and gaterotors can vary widely in terms of form and mutual geometry. Figure 18-9 shows the design normally encountered in refrigeration.

The main rotor has six helical grooves, with a cylindrical periphery and a globoid (or hourglass shape) root profile. The two identical gaterotors each have 11 teeth and are located on opposite sides of the main rotor. The casing enclosing the main rotor has two slots, which allow the teeth of the gaterotors to pass through them. Two diametrically

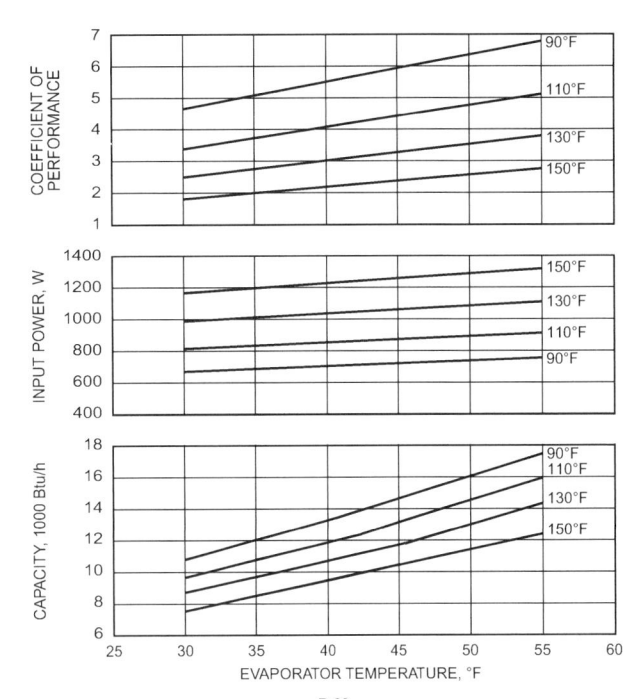

*Fig. 18-8 Typical Rolling Piston Compressor Performance*
*(Figure 11, Chapter 38, 2012 ASHRAE Handbook—HVAC Systems and Equipment)*

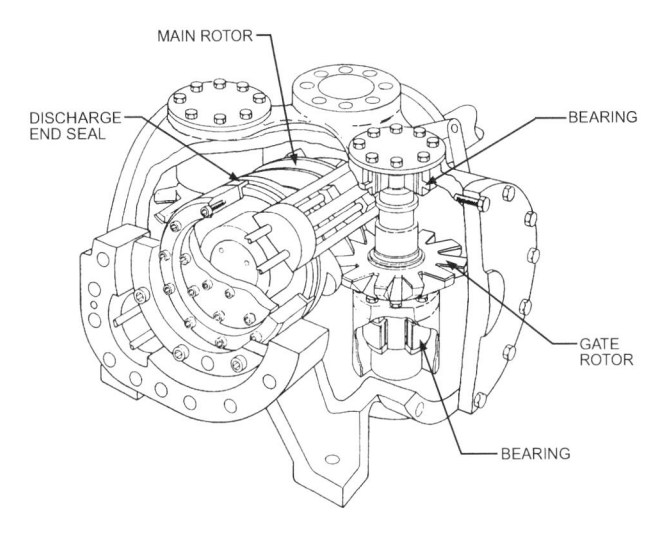

*Fig. 18-9 Section of Single-Screw Compressor*
*(Figure 14, Chapter 38, 2012 ASHRAE Handbook—HVAC Systems and Equipment)*

opposed discharge ports use a common discharge manifold located in the casting. The compressor is driven through the main rotor shaft, and the gaterotors follow by direct meshing action at 6:11 ratio of the main rotor speed. The geometry of the single-screw compressor is such that 100% of the gas compression power is transferred directly from the main rotor to the gas. No power (other than small frictional losses) is transferred across the meshing points to the gaterotors.

Compression is obtained by direct volume reduction with pure rotary motion as illustrated in Figure 18-10.

The four basic continuous phases of the working cycle are as follows:

**Suction.** As a lobe of the male rotor begins to unmesh from an interlobe space in the female rotor, a void is created and gas is drawn in through the inlet port. As the rotors continue to turn, the interlobe space increases in size and gas flows continuously into the compressor. Prior to the point at which the interlobe space leaves the inlet port, the entire length of the interlobe space is completely filled with gas.

**Transfer.** As rotation continues, the trapped gas pocket in the interlobe space is moved circumferentially around the compressor housing at constant suction pressure.

**Compression.** Further rotation starts meshing of another male lobe with the female interlobe space on the suction end and progressively squeezes (compresses) the gas in the direction of the discharge port. Thus, the occupied volume of the trapped gas within the interlobe space is decreased and the gas pressure consequently increased.

**Discharge.** At a point determined by the design built-in volume ratio, the discharge port is uncovered and the compressed gas is discharged by further meshing of the lobe and interlobe space.

During the remeshing period of compression and discharge, a fresh charge is drawn through the inlet on the opposite side of the meshing point. With four male lobes rotating at 3600 rpm, four interlobe volumes are filled and discharged per revolution, providing 14,400 discharges per minute or

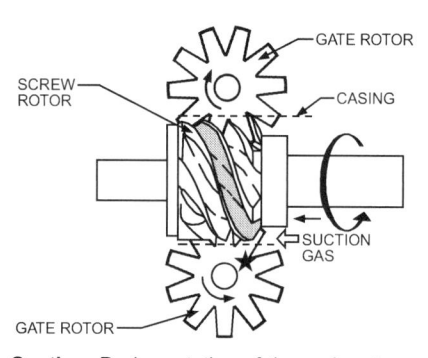

**Suction.** During rotation of the main rotor, a typical groove in open communication with the suction chamber gradually fills with suction gas. The tooth of the gaterotor in mesh with the groove acts as an aspirating piston.

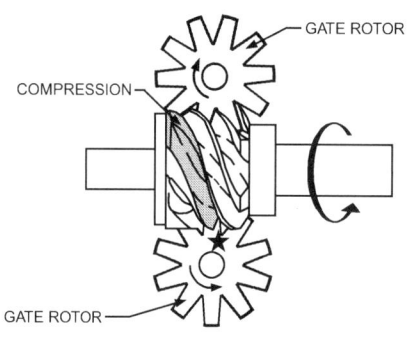

**Compression.** As the main rotor turns, the groove engages a tooth on the gaterotor and is covered simultaneously by the cylindrical main rotor casing. The gas is trapped in the space formed by the three sides of the groove, the casing, and the gaterotor tooth. As rotation continues, the groove volume decreases and compression occurs.

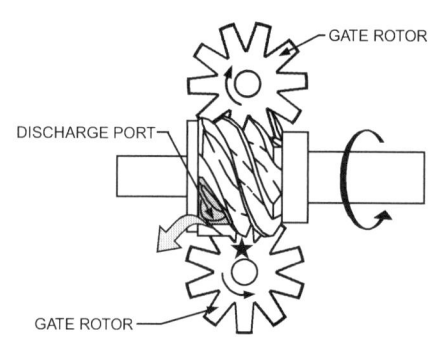

**Discharge.** At the geometrically fixed point where the leading edge of the groove and the edge of the discharge port coincide, compression ceases, and the gas discharges into the delivery line until the groove volume has been reduced to zero.

*Fig. 18-10 Sequence of Compression Process in Single-Screw Compressor*
*(Figure 15, Chapter 38, 2012 ASHRAE Handbook—HVAC Systems and Equipment)*

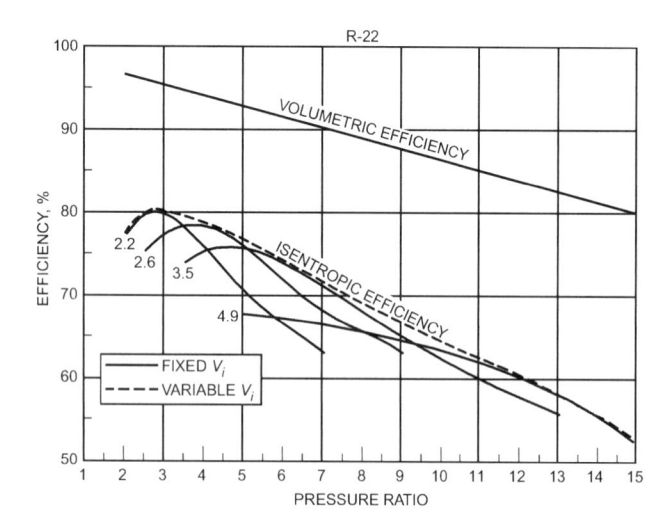

*Fig. 18-11 Typical Screw Compressor Performance with R-22*
*(Figure 25, Chapter 38, 2012 ASHRAE Handbook—HVAC Systems and Equipment)*

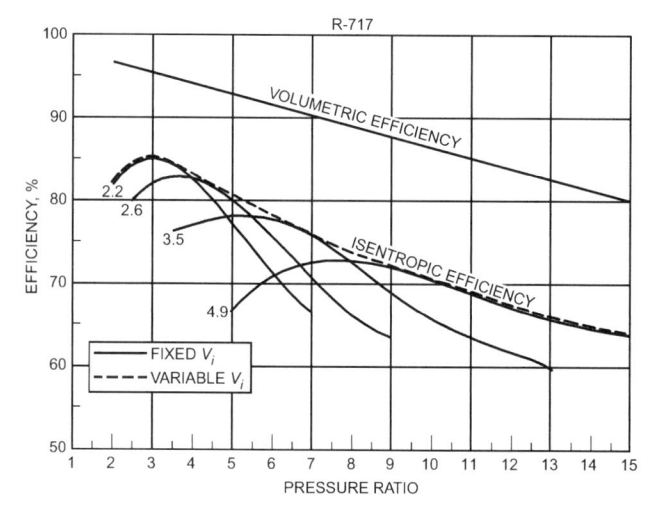

*Fig. 18-12 Typical Screw Compressor Performance with R-717 (Ammonia)*
*(Figure 26, Chapter 38, 2012 ASHRAE Handbook—HVAC Systems and Equipment)*

240 per second. Since the intake and discharge cycles overlap effectively, a smooth, continuous flow of gas results.

Figures 18-11 and 18-12 show typical efficiencies of all single-screw compressor designs. High isentropic and volumetric efficiencies are the result of internal compression, the absence of suction and discharge valves and their losses, and extremely small clearance volumes. The curves show the importance of selecting the correct volume ratio in fixed volume ratio compressors.

**Twin screw** is a common designation for double helical rotary screw compressors. A twin-screw compressor consists of two mating helically grooved rotors—male (lobes) and female (flutes or gullies) in a stationary housing with inlet and outlet gas ports (Figure 18-13).

While operating, some twin-screw compressors adjust the volume ratio of the compressor to the most efficient ratio for whatever system pressures are encountered. The comparative efficiencies of fixed and variable volume ratio screw compressors are shown in Figure 18-14 for full-load operation on ammonia and R-22 refrigerants. The greater the change in either suction or condensing pressure a given system experiences, the more benefits are possible with a variable volume ratio. Efficiency improvements as high as 30% are possible, depending on the application, refrigerant, and system operating range. Hermetic screw compressors are commercially available through 400 tons (1.4 MW) of refrigeration using R-22.

**Scroll compressors** are rotary motion, positive-displacement machines that compress with two interfitting, spiral-shaped scroll members. They are currently used in residential and commercial air-conditioning and heat pump applications as well as in automotive air-conditioning systems. Capacities range from 10,000 to 170,000 Btu/h (3 to 50 kW). To function effectively, the scroll compressor requires close tolerance machining of the scroll members, which has become possible only recently due to current advances in manufacturing technology. This positive-displacement, rotary motion compressor includes performance features, such as high efficiency and low noise.

Scroll members are typically a geometrically matched pair, assembled 180° out of phase. Each scroll member is open on one end of the vane and bound by a base plate on the other. The two scrolls are fitted to form pockets between their respective base plates and various lines of contact between their vane walls. One scroll is held fixed, while the other moves in an orbital path with respect to the first. The flanks of the scrolls remain in contact, although the contact locations move progressively inward. Relative rotation between the pair is prevented by a coupling. An alternative approach creates relative orbital motion via two scrolls synchronously rotating about noncoincident axes.

Compression is accomplished by sealing suction gas in pockets of a given volume at the outer periphery of the scrolls and progressively reducing the size of these pockets as the scroll relative motion moves them inward toward the discharge port. Figure 18-15 shows the sequence of suction, compression, and discharge phases.

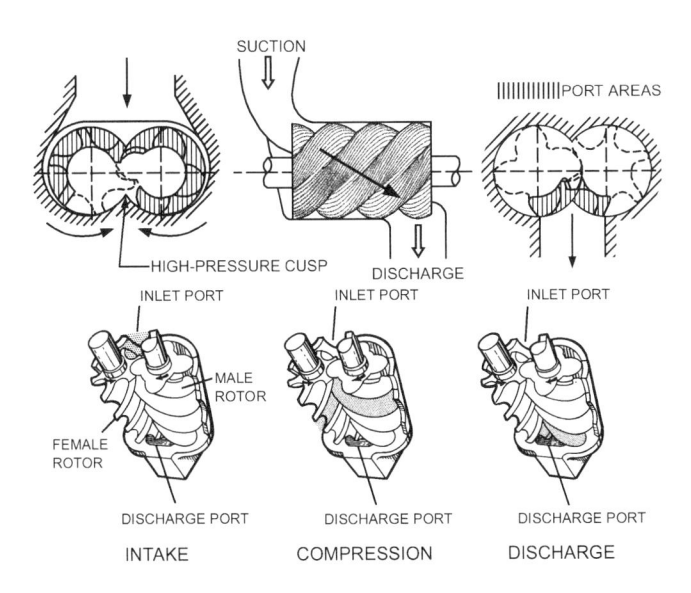

*Fig. 18-13 Twin-Screw Compressor*
*(Adapted from Figures 29 and 30, Chapter 38, 2012 ASHRAE Handbook—HVAC Systems and Equipment)*

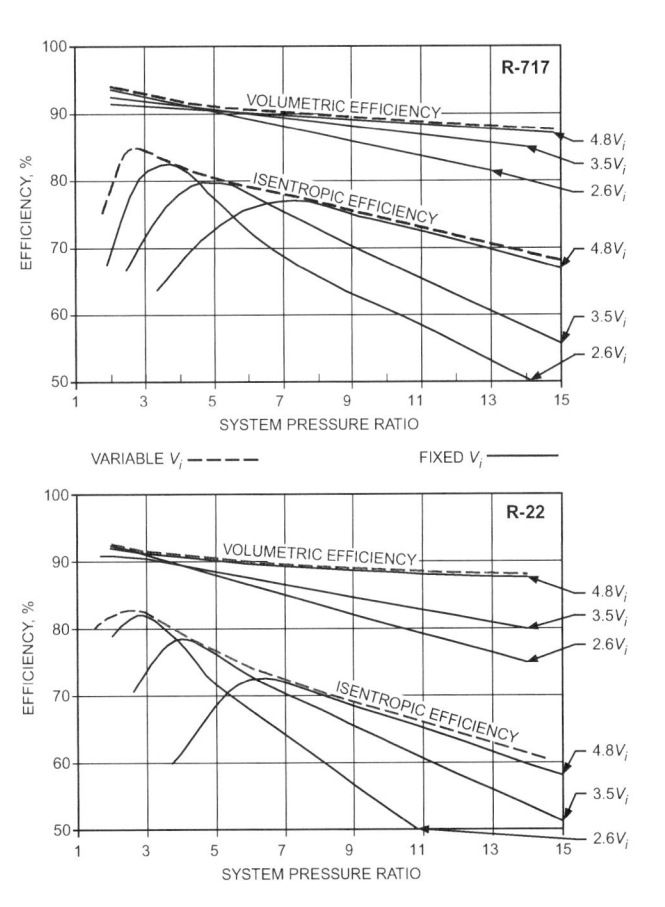

*Fig. 18-14 Twin-Screw Compressor Efficiency Curves*
*(Figure 34, Chapter 38, 2012 ASHRAE Handbook—HVAC Systems and Equipment)*

As the outermost pockets are sealed off (Figure 18-15a), the trapped gas is at suction pressure and has just entered the compression process. At stages (b) through (f), orbiting motion moves the gas toward the center of the scroll pair, and pressure rises as pocket volumes are reduced. At stage (g), the gas reaches the central discharge port and begins to exit the scrolls. Stages (a) through (h) show that two distinct compression paths operate simultaneously in a scroll set. The discharge process is nearly continuous, since new pockets reach the discharge stage very shortly after the previous discharge pockets have been evacuated.

Both high-side and low-side shells are available. In the former, the entire compressor is at discharge pressure, except for the outer areas of the scroll set. Suction gas is introduced into the suction port of the scrolls through piping, which keeps it discrete from the rest of the compressor. Discharge gas is directed into the compressor shell, which acts as a plenum. In the low-side type, most of the shell is at suction pressure, and the discharge gas exiting the scrolls is routed outside the shell, sometimes through a discrete or integral plenum.

Scroll technology offers an advantage in performance for a number of reasons. Large suction and discharge ports reduce pressure losses incurred in the suction and discharge processes. Physical separation of these processes also reduces heat transfer to the suction gas. The absence of valves and reexpansion volumes and the continuous flow process results in high volumetric efficiency over a wide range of operation conditions. Figure 18-16 illustrates this effect.

The built-in volume ratio can be designed for lowest over- or undercompression at typical demand conditions (2.5 to 3.5 pressure ratio for air conditioning). Isentropic efficiency in the range of 70% is possible at such pressure ratios, and it remains quite close to the efficiency of other compressor types at high pressure ratios. Scroll compressors offer a flatter capacity versus outdoor ambient curve than reciprocating products, which means that they can more closely approach indoor requirements at high demand conditions. As a result, the heat pump mode requires less supplemental heating; the cooling mode is more comfortable because cycling is less as demand decreases. Scroll compressors available in the United States are typically specified as producing ARI operating efficiencies (COP) in the range of 3.10 to 3.34.

**Trochoidal compressors** are small, rotary, positive-displacement compressors that can run at high speed up to 9000 rpm. They are manufactured in various configurations. Trochoidal curvatures can be produced by the rolling motion of one circle outside or inside the circumference of a basic circle, producing either epitrochoids or hypotrochoids, respectively. Both types of trochoids can be used either as a cylinder or piston form, so that four types of trochoidal machines can be designed (Figure 18-17).

In each case, the counterpart of the trochoid member always has one apex more than the trochoid itself. In the case of a trochoidal cylinder, the apexes of the piston show slipping along the inner cylinder surface; for trochoidal piston design, the piston shows a gear-like motion. As seen in Figure 18-17, a built-in theoretical pressure ratio disqualifies many configurations as valid concepts for refrigeration compressor design. Because of additional valve ports, clearances, etc., and the resulting decrease in the built-in maximum theoretical pres-

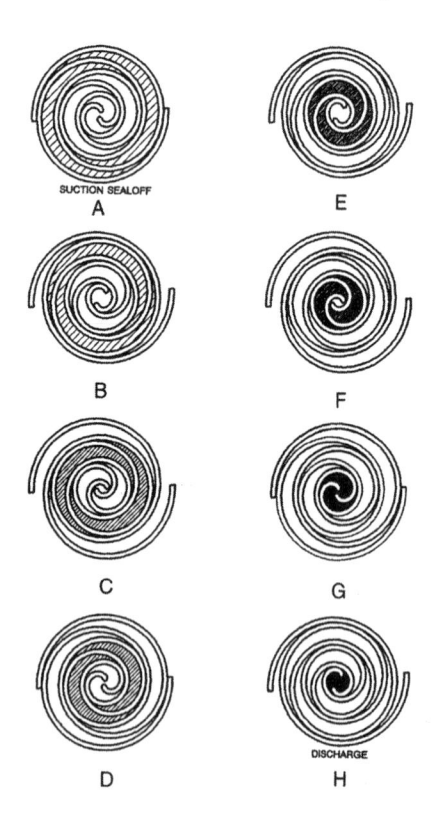

*Fig. 18-15 Scroll Compression Process*
*(Figure 39, Chapter 38, 2012 ASHRAE Handbook—HVAC Systems and Equipment)*

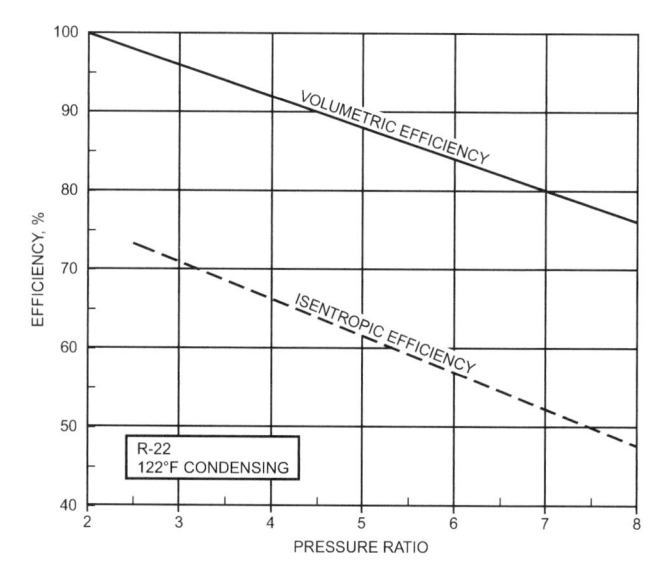

*Fig. 18-16 Volumetric and Isentropic Efficiency versus Pressure Ratio for Scroll Compressor*
*(Figure 41, Chapter 38, 2012 ASHRAE Handbook—HVAC Systems and Equipment)*

# Refrigeration Equipment

## EPITROCHOIDS AS CYLINDER

$i = 1:2$, $\varepsilon = 140$, $\phi_{max} = 19.5°$

$i = 2:3$, $\varepsilon = 15.5$, $\phi_{max} = 30°$

$i = 3:4$, $\varepsilon = 7.5$, $\phi_{max} = 56.4°$

$i = 4:5$, $\varepsilon = 6.0$, $\phi_{max} = 56.4°$

## EPITROCHOIDS AS PISTON

$i = 1:2$, $\varepsilon > 100$, $\phi_{max} = 19.5°$

$i = 2:3$, $\varepsilon > 100$, $\phi_{max} = 30°$

$i = 3:4$, $\varepsilon > 100$, $\phi_{max} = 41.8°$

$i = 4:5$, $\varepsilon > 100$, $\phi_{max} = 56.4°$

## HYPOTROCHOIDS AS CYLINDER

$i = 1:2$, $\varepsilon = 0$, $\phi_{max} = 9.6°$

$i = 2:3$, $\varepsilon = 2.7$, $\phi_{max} = 19.5°$

$i = 3:4$, $\varepsilon = 5$, $\phi_{max} = 30°$

$i = 4:5$, $\varepsilon = 10.4$, $\phi_{max} = 41.6°$

## HYPOTROCHOIDS AS PISTON

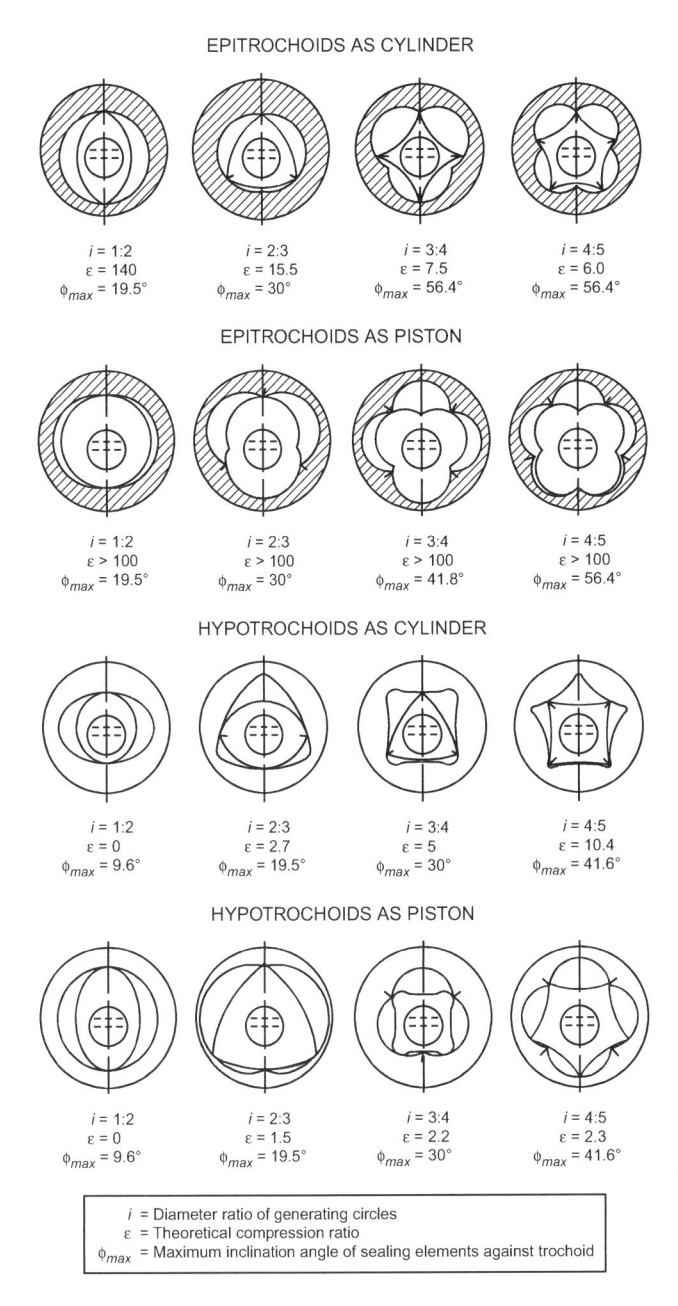

$i = 1:2$, $\varepsilon = 0$, $\phi_{max} = 9.6°$

$i = 2:3$, $\varepsilon = 1.5$, $\phi_{max} = 19.5°$

$i = 3:4$, $\varepsilon = 2.2$, $\phi_{max} = 30°$

$i = 4:5$, $\varepsilon = 2.3$, $\phi_{max} = 41.6°$

$i$ = Diameter ratio of generating circles
$\varepsilon$ = Theoretical compression ratio
$\phi_{max}$ = Maximum inclination angle of sealing elements against trochoid

*Fig. 18-17 Possible Versions of Epitrochoidal and Hypotrochoidal Machines*
*(Figure 44, Chapter 38, 2012 ASHRAE Handbook—HVAC Systems and Equipment)*

sure ratio, only the first two types with epitrochoidal cylinders, and all candidates with epitrochoidal pistons, can be used for compressor technology. The latter, however, require sealing elements on the cylinder as well as on the side plates, which does not allow the design of a closed sealing borderline.

In the past, trochoidal machines were designed much like those of today. However, like other positive-displacement rotary concepts that could not tolerate oil injection, early trochoidal equipment failed because of sealing problems. The invention of a closed sealing border by Wankel changed this. Today, the Wankel trochoidal compressor with a three-sided epitorchoidal piston (motor) and two-envelope cylinder (casing) is built in capacities of up to 2 tons. The sequence of operation of a Wankel rotary compressor is illustrated in Figure 18-18.

**Centrifugal compressors**, or turbocompressors, are characterized by a continuous exchange of angular momentum between a rotating mechanical element and a steadily flowing fluid. Because their flows are continuous, turbomachines have greater volumetric capacities, size-for-size, than do positive-displacement devices. For effective momentum exchange, their rotative speeds must be higher, but little vibration or wear results because of the steadiness of the motion and the absence of contacting parts.

In centrifugal compressors, the suction flow enters the rotating element, or impeller, in the axial direction and is discharged radially at a higher velocity. This dynamic head is then converted to static head, or pressure, through a diffusion process, which generally begins within the impeller and ends in a radial diffuser and scroll outboard of the impeller.

Centrifugal compressors are used in a variety of refrigeration and air-conditioning installations, but primarily in packaged water chillers. Suction flow rates range between 60 and 30,000 cfm (0.03 and 14 m$^3$/s), with rotational speeds between 1800 and 90,000 rpm. However, the high angular velocity associated with a low volumetric flow establishes a minimum practical capacity for most centrifugal applications. The upper capacity limit is determined by physical size, a 30,000 cfm (14 m$^3$/s) compressor being about 6 or 7 ft (2 m) in diameter.

A centrifugal compressor can be single stage, having only one impeller, or it can be multistage, having two or more impellers mounted in the same casing as shown in

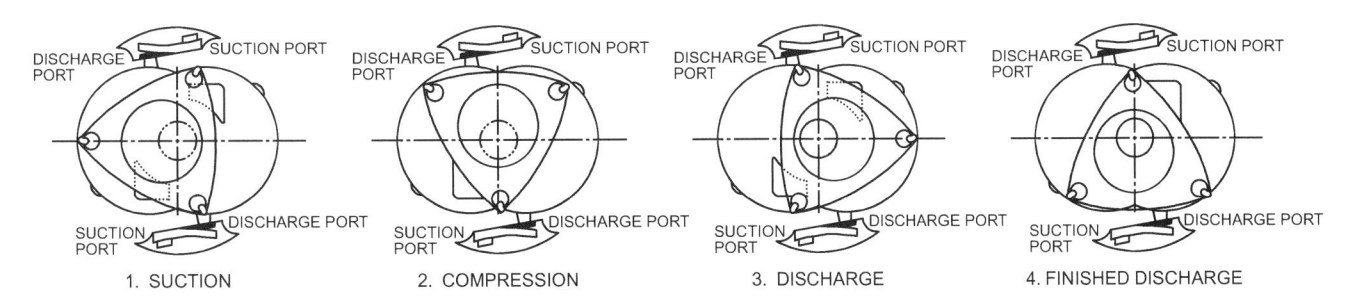

1. SUCTION
2. COMPRESSION
3. DISCHARGE
4. FINISHED DISCHARGE

*Fig. 18-18 Sequence of Operation of Wankel Rotary Compressor*
*(Figure 46, Chapter 38, 2012 ASHRAE Handbook—Systems and Equipment)*

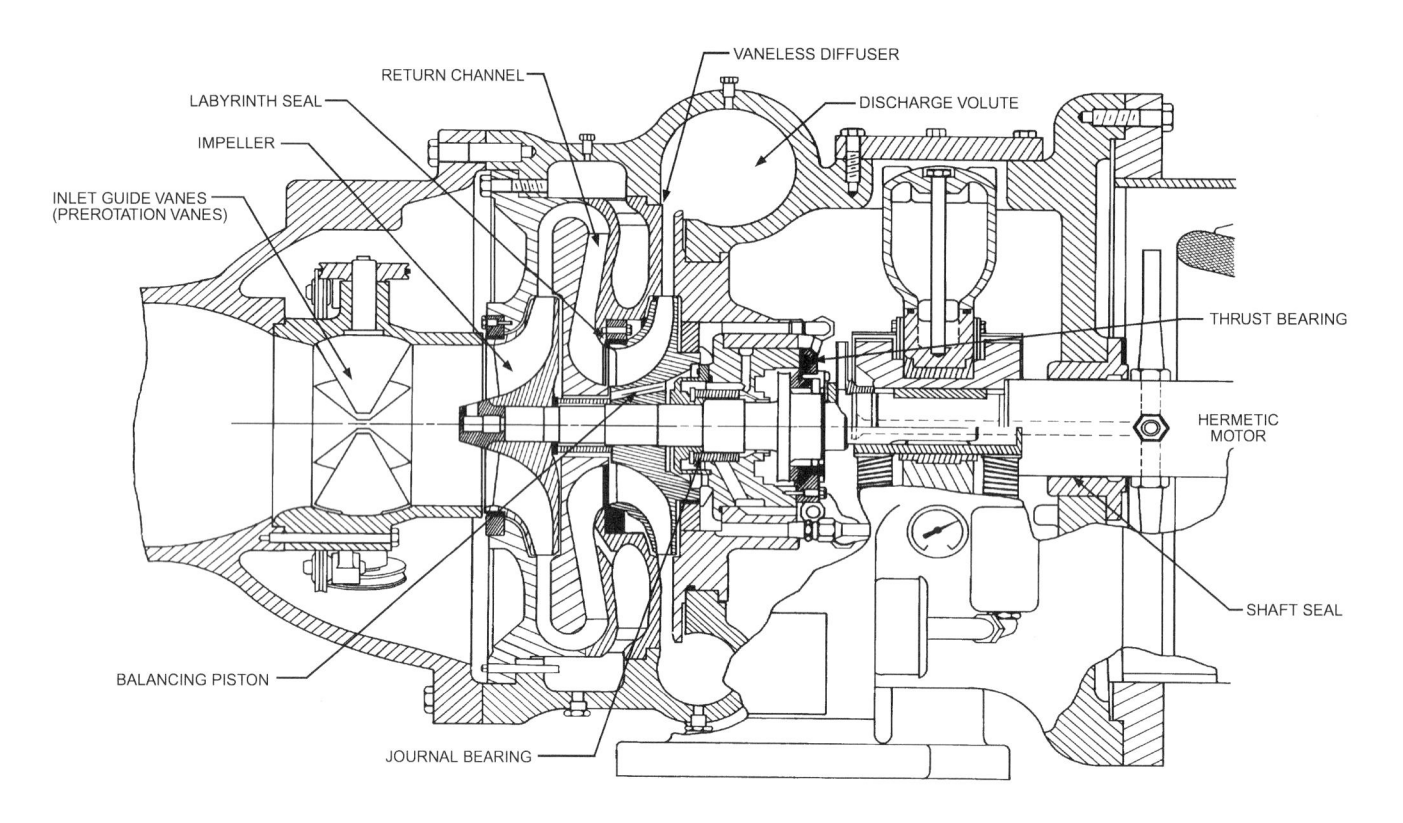

*Fig. 18-19 Centrifugal Refrigeration Compressor*
*(Figure 47, Chapter 38, 2012 ASHRAE Handbook—
HVAC Systems and Equipment)*

Figure 18-19. For process refrigeration applications, a compressor can have as many as ten stages.

The suction gas generally passes through a set of adjustable inlet guide vanes or an external suction damper before entering the impeller. The vanes (or suction damper) are used for capacity control.

Suction temperatures are usually between 50 and −150°F (10 and −100°C), with suction pressures between 2 and 100 psia (14 and 700 kPa) and discharge pressures up to 300 psia (2100 kPa). Pressure ratios range between 2 and 30. Almost any refrigerant can be used.

The momentum exchange, or energy transfer, between a centrifugal impeller and a flowing refrigerant is expressed by the following equation:

$$W_i = u_i c_u / g$$

where

$W_i$ = impeller work input per unit mass of refrigerant, ft·lb$_f$/lb$_m$

$u_i$ = impeller blade tip speed, ft/s

$c_u$ = tangential component of refrigerant velocity leaving impeller blades, ft/s

$g$ = gravitational constant, 32.17 lb$_m$·ft/lb$_f$s$^2$

These velocities are shown in Figure 18-20, where refrigerant flows out from between impeller blades with relative velocity $b$ and absolute velocity $c$. The relative angle $\beta$ is a few

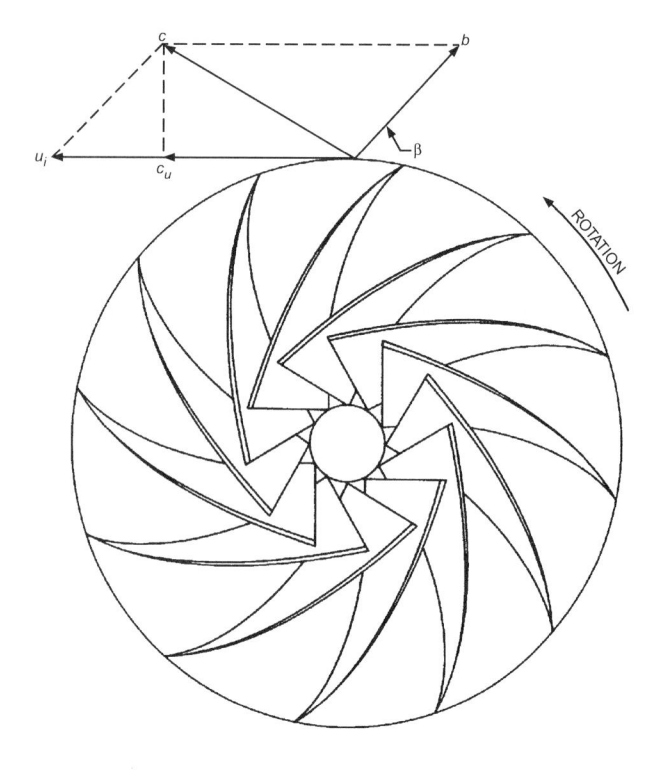

*Fig. 18-20 Impeller Exit Velocity Diagram*
*(Figure 51, Chapter 38, 2012 ASHRAE Handbook—
HVAC Systems and Equipment)*

# Refrigeration Equipment

degrees smaller than the blade angle because of a phenomenon known as slip. This equation assumes that the refrigerant enters the impeller without any tangential velocity component or swirl. This is generally the case at design flow conditions.

At least at low refrigerant flow rates, the tip speed of the impeller and the tangential velocity of the refrigerant are nearly identical. Thus, for a mass flow rate $m$, the ideal power can be estimated by

$$P = mc_u^2 = mu_i^2$$

Another expression for the ideal power input comes from the first law of thermodynamics:

$$P = m\Delta h_i$$

where $\Delta h_i$ is the isentropic change in enthalpy across the compressor.

Equating the two expressions for power yields an order-of-magnitude estimate of the tip speed:

$$u_i^2 = \Delta h_i(g) \quad \text{ft/s}$$

**Example 18-1** Estimate the impeller tip speed needed to compress R-717 (ammonia) from saturated vapor at 20°F to a pressure corresponding to a condensing temperature of 100°F.

**Solution:**

From Figure 18, Chapter 30, 2013 *ASHRAE Handbook—Fundamentals*:

$$\Delta h_i = 718 - 617 = 101 \text{ Btu/lb}$$

The tip speed is

$$u_i = [(32.2)(101)(778)]^{1/2} = 1591 \text{ ft/s}$$

*Note:* $g = 32.2$ ft/s$^2$ and 778 ft-lb/Btu is a conversion factor.

Some of the work done by the impeller increases the refrigerant pressure, while the remainder only increases its kinetic energy. The ratio of pressure-producing work to total work is known as the impeller reaction. Since this varies from about 0.4 to 0.7, an appreciable amount of kinetic energy leaves the impeller with magnitude $c^2/2g$. To convert this kinetic energy into additional pressure, a diffuser is located after the impeller. Radial vaneless diffusers are most common, but vaned, scroll, and conical diffusers are also used. In a multistage compressor, the flow leaving the first diffuser is guided to the inlet of the second impeller and so on through the machine. The total compression work input is the sum of the individual stage inputs provided that the mass flow rate is constant throughout the compressor:

$$W = \Delta W_i$$

## 18.1.2 Condensers

The condenser removes (from the refrigerant gas) the heat of compression and the heat absorbed by the refrigerant in the evaporator. The refrigerant is thereby converted back into the liquid phase at the condenser pressure and is available for reexpansion into the evaporator. The common forms of condensers may be classified on the basis of the cooling medium

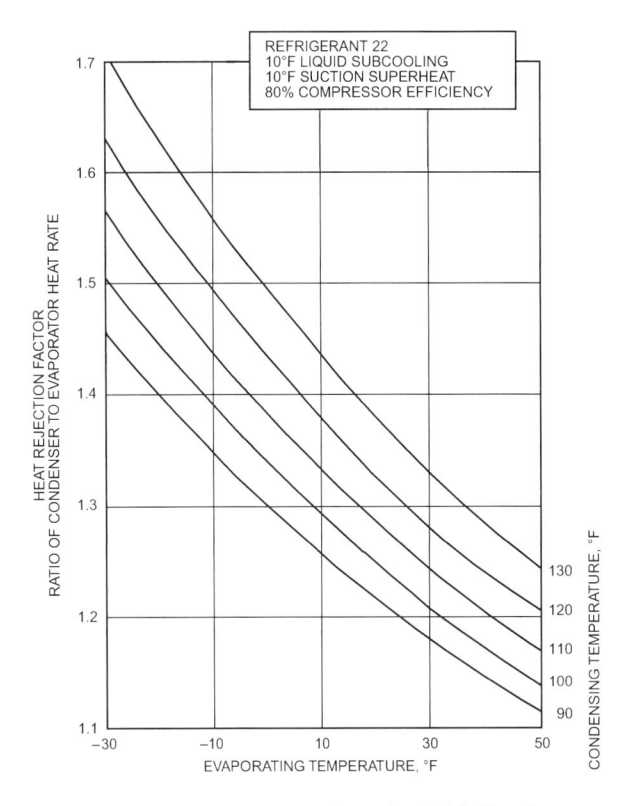

*Fig. 18-21 Heat Rejection Rate for R-22 Condenser*
(Figure 1, Chapter 39, 2012 ASHRAE Handbook—
HVAC Systems and Equipment)

as (1) water-cooled, (2) air-cooled, and (3) evaporative (air and water) cooled.

Water-cooled condensers consist of the following types:

- Shell-and-tube (vertical)
- Shell-and-tube (horizontal)
- Shell-and-coil (horizontal and vertical)
- Double pipe
- Atmospheric

The selection of a water-cooled condenser depends on the cooling load, the refrigerant used, the source and temperature of the available cooling water, the amount of water that can be circulated, the condenser location, the required operating pressures, and the maintainability.

The heat rejection rate of the condenser for each unit of refrigeration produced in the evaporator may be estimated from Figure 18-21. Similar plots can be prepared for other refrigerants from tables of thermodynamic properties. In practice, the heat removed is 5 to 10% higher than the theoretical values because of losses during compression.

An accurate determination of the heat rejection requirement $q_o$ can usually be made from known values of evaporator load $q_i$ and the heat equivalent of the actual power required $q_w$ for compression:

$$q_o = q_i + q_w \quad \text{Btu/h} \tag{18-1}$$

*Note:* $q_w$ is reduced by any independent heat rejection processes such as oil cooling and motor cooling.

The volumetric flow rate $Q$ of condensing water required may be found from the following equation:

$$Q = \frac{q_o}{\rho c_p (t_2 - t_1)} \quad \text{ft}^3/\text{h} \quad (18\text{-}2)$$

where
- $q_o$ = heat rejection rate, Btu/h
- $\rho$ = density of water, lb/ft$^3$
- $t_1$ = temperature of water entering condenser, °F
- $t_2$ = temperature of water leaving condenser, °F
- $c_p$ = specific heat of water, Btu/lb·°F

The heat rejection rate may also be determined as:

$$q_o = UA\Delta t_m \quad \text{Btu/h} \quad (18\text{-}3)$$

where
- $U$ = overall heat transfer coefficient, Btu/h·ft$^2$·°F
- $A$ = surface area associated with $U$, ft$^2$
- $\Delta t_m$ = mean temperature difference, °F

The computation of overall heat transfer in a water-cooled condenser with water inside the tubes may be made from calculated or test-derived heat transfer coefficients of the water and refrigerant sides, from physical measurements of the condenser tubes, and from a fouling factor on the water side, by using

$$U_o = \frac{1}{(S_r/h_w) + S_R r_{fw} + (X/k)(A_o/A_m) + 1/h_r \phi_w} \quad (18\text{-}4)$$

where
- $U_o$ = overall heat transfer coefficient, based on the external surface and the log mean temperature difference, between the external and internal fluids, Btu/h·ft$^2$·°F [W/(m$^2$·K)]
- $S_R$ = ratio of external to internal surface area
- $h_w$ = internal or water side film coefficient, Btu/h·ft$^2$·°F [W/(m$^2$·K)]
- $r_{fw}$ = fouling resistance on water side, ft$^2$·°F·h/Btu (m$^2$·K/W)
- $X$ = thickness of tube wall, ft (m)
- $k$ = thermal conductivity of tube material, Btu/h·ft·°F [W/(m·K)]
- $A_o/A_m$ = ratio of external surface to mean heat transfer area of metal wall
- $h_r$ = external, or refrigerant side coefficient, Btu/h·ft$^2$·°F [W/(m$^2$·K)]
- $\phi_w$ = weighted fin efficiency (100% for bare tubes)

Values of the water-side coefficient may be calculated from equations in Chapter 4 of the 2013 *ASHRAE Handbook—Fundamentals*. For turbulent flow, at Reynolds numbers exceeding 10,000 in horizontal tubes and using average water temperatures, the general equation is

$$\frac{h_w D}{k} = 0.023 \left(\frac{DG}{\mu}\right)^{0.8} \left(\frac{c_p \mu}{k}\right)^{0.4} \quad (18\text{-}5)$$

where
- $D$ = inside tube diameter, ft (m)
- $k$ = thermal conductivity of water, Btu/h·ft·°F [W/(m·K)]
- $G$ = mass velocity of water, lb/s·ft$^2$ (kg/s·m$^2$)
- $\mu$ = viscosity of water, lb/ft·h (mPa·s)
- $c_p$ = specific heat of water at constant pressure, Btu/lb·°F [(kJ/kg·K)]

*Note*: The constant (0.023) in this equation applies only to tubes with plain inside diameters.

Because of its strong influence on the value of $h_w$, water velocity should be maintained as high as permitted by water pressure drop considerations. Maximum velocities with clean water of 6 to 10 ft/s are commonly used. A minimum velocity of 3 ft/s is considered good practice when the water quality is such that noticeable fouling or corrosion could result. With clean water, the velocity may be lower if dictated by conservation or low supply temperature considerations.

Factors that influence the value of $h_r$ are

- Type of refrigerant being condensed
- Geometry of condensing surface (plain tube, outside diameter, finned tube, fin spacing, height, and cross-section profile)
- Condensing temperature
- Condensing rate, in terms of mass velocity or rate of heat transferred
- Arrangement of tubes in bundle
- Vapor distribution and flow rate
- Condensate drainage

Values of the refrigerant side coefficients may be estimated from correlations shown in Chapters 4 or 5 of the 2013 *ASHRAE Handbook—Fundamentals*.

**Example 18-2** Estimate the volumetric flow rate of condensing water required for the condenser of an R-22 water chilling unit assumed to be operating at a condensing temperature of 100°F, and evaporating temperature of 40°F, an entering condensing water temperature of 86°F, a leaving condensing water temperature of 95°F, and a refrigeration load of 100 tons.

**Solution:**
From Figure 18-21, the heat rejection factor is found to be 1.17.

$q_o = 100 \times 1.17 = 117$ tons = 1,404,000 Btu/h

$\rho = 62.2$ lb/ft$^3$ at 90.5°F

$c_p = 1$ Btu/lb·°F

From Equation (18-2),

$Q = 1{,}404{,}000/[62.2 \times 1(95 - 86)] = 2500$ ft$^3$/h = 310 gpm

A typical horizontal closed shell-and-tube ammonia condenser is shown in Figure 18-22.

**Air-Cooled Condensers.** The heat transfer process in an air-cooled condenser has three main phases: (1) desuperheating, (2) condensing, and (3) subcooling. The changes of state of R-134a passing through the condenser coil and the corresponding temperature change of the cooling air as it passes through the coil are shown in Figure 18-23. Desuperheating, condensing, and subcooling zones vary 5 to 10%, depending on the entering gas temperature and the leaving

# Refrigeration Equipment

liquid temperature, but Figure 18-23 is typical for most of the commonly used refrigerants.

Condensing occurs in approximately 85% of the condenser area at a substantially constant temperature. The drop in condensing temperature is due to the friction loss through the condenser coil.

Coils in air-cooled condensers are commonly constructed of copper, aluminum, or steel tubes, ranging from 1/4 to 3/4 in (8 to 20 mm). diameter. Copper is easy to use in manufacturing and requires no protection against corrosion. Aluminum requires exact manufacturing methods, and special protection must be provided if aluminum to copper joints are made. Steel tubing is used, but weather protection must be provided.

Fins are used to improve the air-side heat transfer. Fins are usually made of aluminum, but copper and steel are also used. The most common forms are plate fins making a coil bank, plate fins individually fastened to the tube, or a fin spirally wound onto the tube. Other forms such as plain tube-fin extrusions or tube extrusions with accordion type fins are also used. The number of fins per inch varies from 4 to 30 (0.8 to 6.4 mm fin spacing). The most common range is 8 to 18 (1.4 to 3.22 mm spacing).

**Evaporative Condensers.** As with water-cooled and air-cooled condensers, evaporative condensers reject heat from a condensing vapor into the environment. In an evaporative condenser, hot high-pressure vapor from the compressor discharge circulates through a condensing coil that is continually wetted on the outside by a recirculating water system. As seen in Figure 18-24, air is simultaneously directed over the coil, causing a small portion of the recircu-

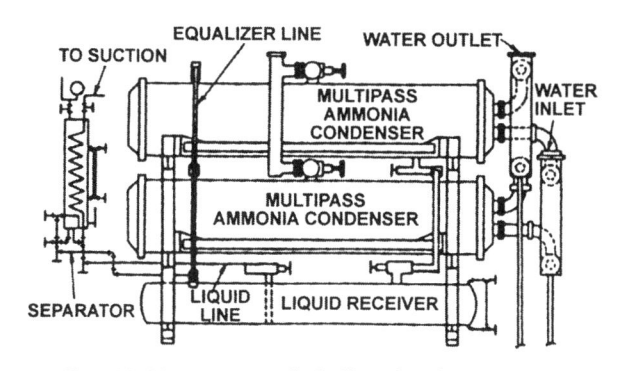

*Fig. 18-22 Horizontal Shell-and-Tube Ammonia Condenser and Receiver*

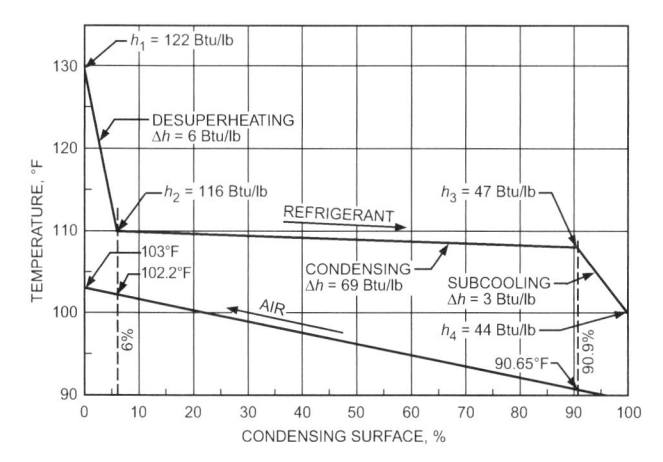

*Fig. 18-23 Temperature and Enthalpy Changes in Air-Cooled Condenser with R-134a*
(Figure 6, Chapter 39, 2012 ASHRAE Handbook—
HVAC Systems and Equipment)

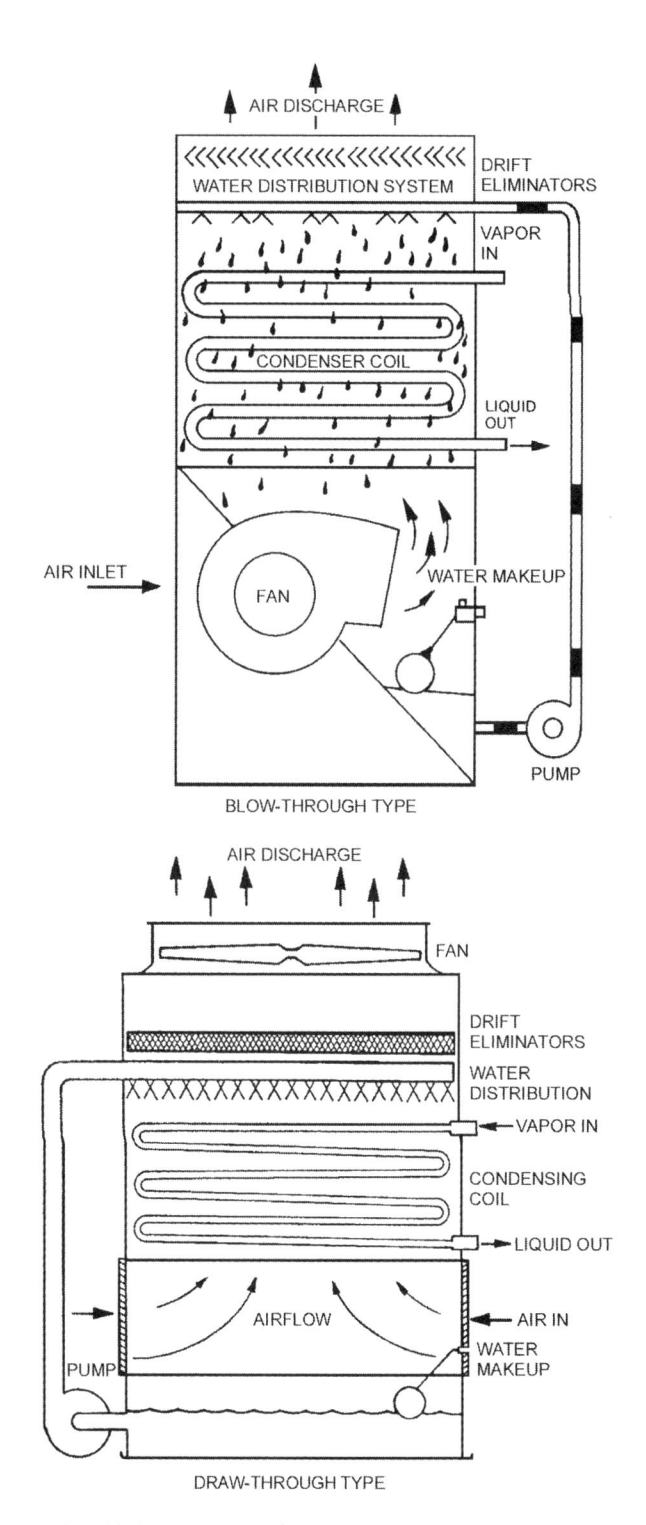

*Fig. 18-24 Functional View of Evaporative Condenser*
(Adapted from Figure 10, Chapter 39, 2012 ASHRAE Handbook—
HVAC Systems and Equipment)

lated water to evaporate. This evaporation removes heat from the coil, thus cooling and condensing the vapor.

Evaporative condensers reduce the water pumping and chemical treatment requirements associated with cooling tower/refrigerant condenser systems. In comparison with an air-cooled condenser, an evaporative condenser requires less coil surface and airflow to reject the same heat, or alternatively, greater operating efficiencies can be achieved by operating at a lower condensing temperature.

The evaporative condenser can operate at a lower condensing temperature than an air-cooled condenser because the air-cooled condenser is limited by the ambient dry-bulb temperature. In the evaporative condenser, heat rejection is limited by the ambient wet-bulb temperature, which is normally 14 to 24°F (8 to 13°C) lower than the ambient dry bulb. The evaporative condenser also provides lower condensing temperatures than the cooling tower/water-cooled condenser because the heat transfer/mass transfer steps are reduced from two (between the refrigerant and the cooling water and between the water and ambient air) to one step (refrigerant directly to ambient wet bulb). While both the water-cooled condenser/cooling tower combination and the evaporative condenser use evaporative heat rejection, the former has added a second step of nonevaporative heat transfer from the condensing refrigerant to the circulating water, requiring more surface area. Evaporative condensers are, therefore, the most compact for a given capacity.

### 18.1.3 Refrigerant Expansion and Control Devices

Any refrigeration system requires that the flow of refrigerant be controlled. Valves are used to start, stop, direct, and modulate the flow of refrigerant to satisfy load requirements. To ensure satisfactory performance, valves should be adequately protected from foreign material, excessive moisture, and corrosion. Such protection is accomplished by installing properly sized strainers and driers.

**Thermostatic Expansion Valves.** The thermostatic expansion valve controls the flow rate of liquid refrigerant entering the evaporator in response to the superheat of the refrigerant gas leaving the evaporator. It keeps the entire evaporator active, without permitting unevaporated refrigerant liquid to be returned through the suction line to the compressor. The thermostatic expansion valve does so by controlling the mass flow rate of refrigerant entering the evaporator so that it equals the rate at which the refrigerant can be completely vaporized in the evaporator by heat absorption. Since the thermostatic expansion valve is operated by the superheated refrigerant gas leaving the evaporator and is responsive to changes in superheat of this gas, a portion of the evaporator must be devoted to superheating the refrigerant gas.

Unlike the constant pressure expansion valve, the thermostatic expansion valve is not limited to constant load applications. It is used to control refrigerant flow to all types of direct-expansion evaporators in air-conditioning, commercial, low-temperature, and ultra-low-temperature refrigeration systems.

A schematic cross section of the thermostatic expansion valve, with the principal components identified, is shown in Figure 18-25. Three forces are shown that govern thermostatic expansion valve operation:

$p_1$ = vapor pressure of the thermostatic element (a function of the bulb temperature), which is applied to the top of the diaphragm and acts to open valve

$p_2$ = evaporator pressure, which is applied underneath the diaphragm through the equalizer passage, and acts in a closing direction

$p_3$ = pressure equivalent of the superheat spring force, which is applied underneath the diaphragm, and is also a closing force

At any constant operating condition, these forces are balanced and $p_1 = p_2 + p_3$.

An additional force is that arising from the unbalanced pressure across the valve port. It can affect thermostatic expansion valve operation to a degree. For the configuration shown in Figure 18-26, the force due to port unbalance is the product of the pressure drop across the port and the difference in area of the port and the stem, and it would be an opening force. In other designs, depending on the direction of flow through the valve, the port unbalance might result in a closing force.

The principal effect of port unbalance is on valve control stability. As with any modulating control, if the ratio of power element area to port area is kept large, the unbalanced port effect is minor. Large capacity valves are made with

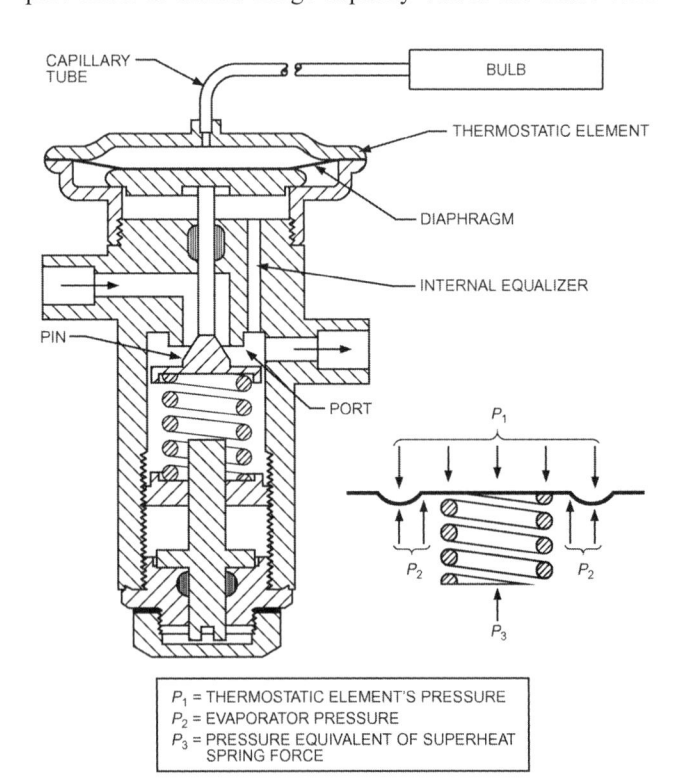

*Fig. 18-25 Typical Thermostatic Expansion Valve*
*(Figure 10, Chapter 11, 2010 ASHRAE Handbook—Refrigeration)*

# Refrigeration Equipment

double-ported, or semibalanced, construction to minimize the effect of unbalanced pressure.

An evaporator using R-22 and operating at a saturation temperature of 40°F (4.4°C) and a pressure of 68.5 psig (472 kPa) is shown in Figure 18-26. Liquid refrigerant enters the expansion valve, is reduced in pressure and temperature at the valve port, and enters the evaporator at point A as a mixture of saturated liquid and vapor. As flow continues through the evaporator, more and more of the boiling refrigerant is evaporated. The refrigerant temperature remains at 40°F (4.4°C) until the liquid portion is completely evaporated by the absorption of heat at point B. From this point, additional heat absorption increases the temperature and superheats the refrigerant gas, while the pressure remains constant at 68.5 psig (472 kPa), until at point C (the outlet of the evaporator), the refrigerant gas temperature is 50°F (10°C). At this point, the superheat is 10°F (from 40 to 50°F) [5.6°C (from 4.4 to 10°C)].

An increase in the heat load on the evaporator increases the temperature of the refrigerant gas leaving the evaporator. The bulb of the thermostatic expansion valve senses this increase; the thermostatic charge pressure ($p_1$) increases and causes the valve to open wider. The increased flow rate results in a higher evaporator pressure ($p_2$) and a balanced control point is established again. Conversely, a decrease in the heat load on the evaporator decreases the temperature of the refrigerant gas leaving the evaporator and causes the thermostatic expansion valve pin to move in a closing direction.

External pressure equalizing thermostatic expansion valves are also used. A pressure line is connected between the valve and the suction side of the evaporator. This connection compensates for the frictional pressure loss in the evaporator. A common technique for this type of valve installation is illustrated in Figure 18-27.

**Constant Pressure Expansion Valves.** The constant pressure expansion valve is operated by the evaporator or valve outlet pressure to regulate the mass flow rate of liquid refrigerant entering the evaporator and thereby maintain this pressure at a constant value.

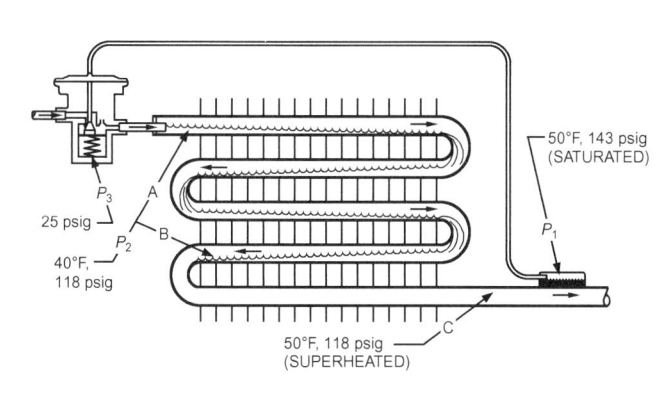

Figure 18-28 shows a schematic cross section of a constant pressure expansion valve. The valve has an adjustable spring that exerts its force on top of the diaphragm in an opening direction and a spring beneath the diaphragm that exerts its force in a closing direction. Evaporator pressure admitted beneath the diaphragm, through either the internal or external equalizer passage, combines with the closing spring to counterbalance the opening spring pressure.

With the valve set and feeding refrigerant at a given pressure, a small increase in the evaporator pressure forces the diaphragm upward and causes the valve pin to move in a closing direction, thereby restricting refrigerant flow and limiting evaporator pressure. When the evaporator pressure, because of a decrease in load, drops below the valve setting, the top spring pressure moves the valve pin in an opening direction, thereby increasing the refrigerant flow in an effort to raise the evaporator pressure to the balanced valve setting. This valve controls the evaporation of the liquid refrigerant in the evaporator at a constant temperature.

**Electric Expansion Valves.** Application of an electric expansion valve requires a valve, controller, and control sensors. The control sensors may include pressure transducers,

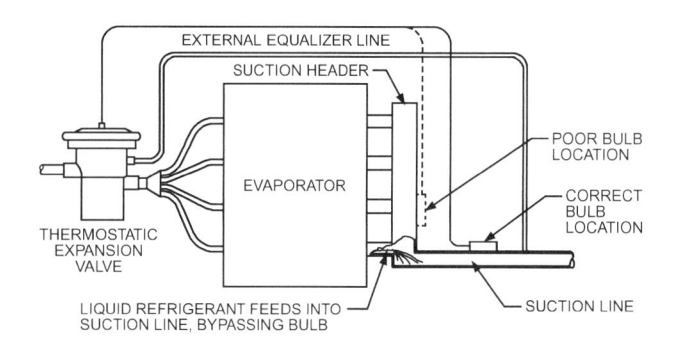

*Fig. 18-27 Bulb Location for Thermostatic Expansion Valve*
*(Figure 16, Chapter 11, 2010 ASHRAE Handbook—Refrigeration)*

*Fig. 18-26 Thermostatic Expansion Valve Controlling Flow of Liquid R-22 Entering Evaporator (Assuming R-22 Charge in Bulb)*
*(Figure 12, Chapter 11, 2010 ASHRAE Handbook—Refrigeration)*

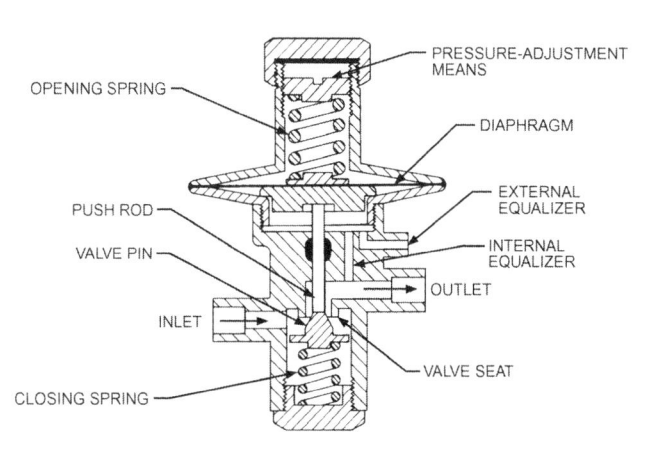

Valve is used with either internal or external equalizer, but not with both.

*Fig. 18-28 Constant Pressure Expansion Valve*
*(Figure 25, Chapter 11, 2010 ASHRAE Handbook—Refrigeration)*

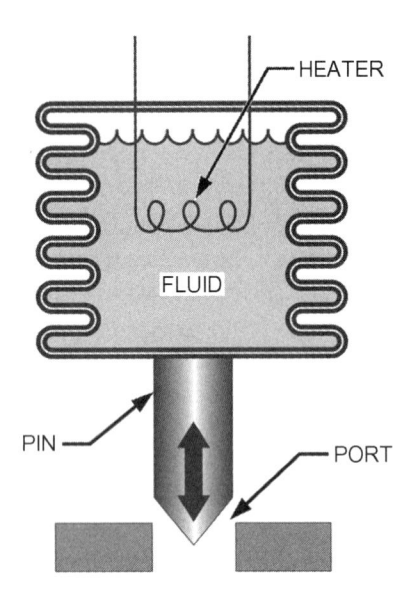

*Fig. 18-29 Fluid-Filled Heat-Motor Valve*
*(Figure 20, Chapter 11, 2010 ASHRAE Handbook—Refrigeration)*

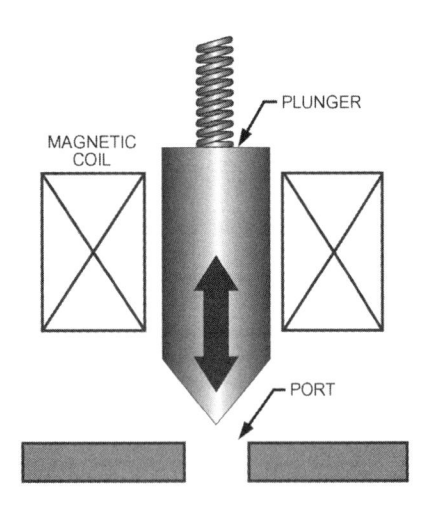

*Fig. 18-30 Magnetically Modulated Valve*
*(Figure 21, Chapter 11, 2010 ASHRAE Handbook—Refrigeration)*

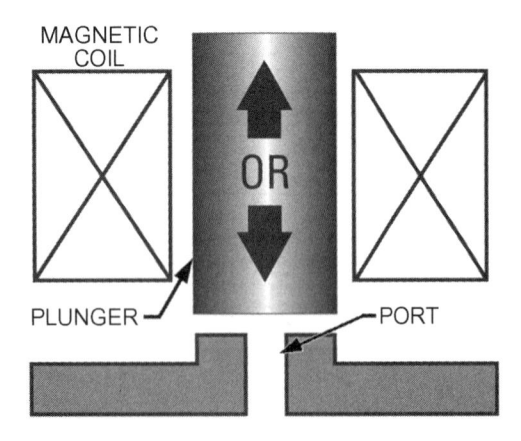

*Fig. 18-31 Pulse-Width-Modulated Valve*
*(Figure 22, Chapter 11, 2010 ASHRAE Handbook—Refrigeration)*

thermistors, resistance temperature devices (RTDs), or other pressure and temperature sensors. See Chapter 36 in the 2013 *ASHRAE Handbook—Fundamentals* for a discussion of instrumentation. Specific types should be discussed with the electric valve and electronic controller manufacturers to ensure compatibility of all components.

Electric valves typically have four basic types of actuation:
- Heat-motor operated
- Magnetically modulated
- Pulse-width-modulated (on/off type)
- Step-motor-driven

**Heat-motor valves** may be one of two types. In one type, one or more bimetallic elements are heated electrically, causing them to deflect. The bimetallic elements are linked mechanically to a valve pin or poppet; as the bimetallic element deflects, the valve pin or poppet follows the element movement. In the second type, a volatile fluid is contained within an electrically heated chamber so that the regulated temperature (and pressure) is controlled by electrical power input to the heater. The regulated pressure acts on a diaphragm or bellows, which is balanced against atmospheric air pressure or refrigerant pressure. The diaphragm is linked to a pin or poppet, as shown in Figure 18-29.

A **magnetically modulated** (analog) valve functions by modulation of an electromagnet; a solenoid armature compresses a spring progressively as a function of magnetic force (Figure 18-30). The modulating armature may be connected to a valve pin or poppet directly or may be used as the pilot element to operate a much larger valve. When the modulating armature operates a pin or poppet directly, the valve may be of a pressure-balanced port design so that pressure differential has little or no influence on valve opening.

The **pulse-width-modulated valve** is an on/off solenoid valve with special features that allow it to function as an expansion valve through a life of millions of cycles (Figure18-31). Although the valve is either fully opened or closed, it operates as a variable metering device by rapidly pulsing the valve open and closed. For example, if 50% flow is needed, the valve will be open 50% of the time and closed 50% of the time. The duration of each opening, or pulse, is regulated by the electronics.

A **step motor** is a multiphase motor designed to rotate in discrete fractions of a revolution, based on the number of signals or "steps" sent by the controller. The controller tracks the number of steps and can offer fine control of the valve position with a high level of repeatability. Step motors are used in instrument drives, plotters, and other applications where accurate positioning is required. When used to drive expansion valves, a lead screw changes the rotary motion of the rotor to a linear motion suitable for moving a valve pin or poppet (Figure 18-32A). The lead screw may be driven directly from the rotor, or a reduction gearbox may be placed between the motor and lead screw. The motor may be hermetically sealed within the refrigerant environment, or the rotor may be enclosed in a thin-walled, nonmagnetic, pressuretight metal tube, similar to those used in solenoid valves, which is surrounded by the stator such that the rotor is in the refrigerant environment and the stator is outside the refrigerant environment. In some designs, the motor and gearbox can operate outside the refrigerant system with an appropriate stem seal (Figure 18-32B).

Electric expansion valves may be controlled by either digital or analog electronic circuits. Electronic control gives additional flexibility over traditional mechanical valves to consider control schemes that would otherwise be impossible, including stopped or full flow when required.

The electric expansion valve, with properly designed electronic controllers and sensors, offers a refrigerant flow control means that is not refrigerant specific, has a very wide load range, can often be set remotely, and can respond to a variety of input parameters.

**Evaporator Pressure Regulators.** The evaporator pressure regulator (back pressure regulator) regulates the evaporator pressure (pressure entering the regulator) at a constant value. It is used in the evaporator outlet or suction line to prevent frosting on the coil or to keep the leaving air temperature from lowering under light load conditions. These pressure regulators are commonly used on multiple evaporators served by a single compressor or when different suction pressures are required by multiple evaporator coils.

As illustrated in Figure 18-33, the inlet pressure acts on the bottom of the seat disk and is opposed by the adjusting spring. The outlet pressure acts on the underside of the bellows and the top of the seating disk, and, since the effective areas of the bellows and the port are equal, the two forces cancel and the valve is responsive to inlet pressure only. When the evaporator pressure rises above the force exerted by the spring, the valve moves in the opening direction. When the evaporator pressure drops below the force exerted by the spring, the valve moves in the closing direction. In actual operation, the valve assumes a throttling position to balance system load.

**Capillary Tubes.** Every refrigerating unit requires a pressure-reducing device to meter the refrigerant flow to the low side in accordance with the system demands. The capillary tube is popular for smaller unitary hermetic equipment, such as household refrigerators and freezers, dehumidifiers, and room air conditioners. It is also used in larger units such as unitary air conditioners in sizes up to 10 tons (35 kW) capacity. The capillary operates on the principle that liquid passes through it more readily than does gas. It consists of a small diameter line that connects the outlet of the condenser to the inlet of the evaporator. It is sometimes soldered to the outer surface of the suction line for heat exchange purposes.

Assume that a condenser-to-evaporator capillary has been sized to permit the desired flow of refrigerant with a liquid seal at its inlet. If a system unbalance occurs so that some gas (uncondensed refrigerant) enters the capillary, this gas tends to considerably reduce the mass flow of refrigerant with little or no change in the system pressures. If the opposite type of unbalance occurs, liquid refrigerant backs up in the condenser. This condition tends to cause subcooling and increases the mass flow of refrigerant. Thus, a capillary properly sized for the application tends to automatically compensate for load and system variations and gives acceptable performance over a wide range of operating conditions.

A refrigerating system is operating at the **condition of capacity balance** when the resistance of the capillary is sufficient to maintain a liquid seal at its entrance without excess liquid accumulating in the high side of the system (Figure 18-34). Only one such capacity balance point exists for any given compressor discharge pressure.

### 18.1.4 Evaporators for Liquid Chillers

A liquid cooler (hereafter called a cooler) is a component of a refrigeration system in which the refrigerant is evaporated to produce a cooling effect on a fluid (usually water or brine). Various types of water and brine coolers, as well as refrigerant flow control, capacity range, and refrigerants commonly used, are listed in Table 18-3.

In the **direct-expansion cooler**, the refrigerant is expanded into the inside of the tubes and vaporizes completely before leaving. The fluid being cooled is circulated

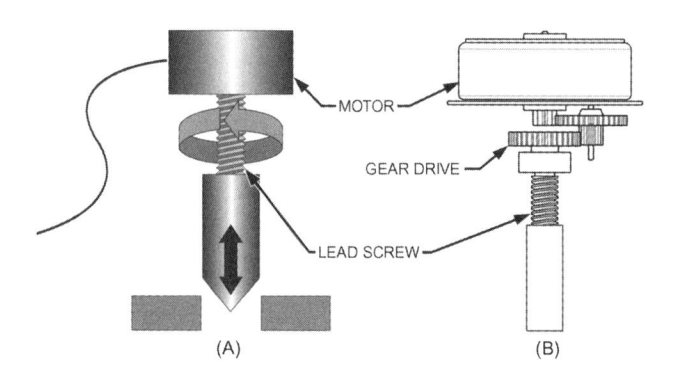

*Fig. 18-32 Step Motor with (A) Lead Screw and (B) Gear Drive with Stem Seal*
*(Figure 23, Chapter 11, 2010 ASHRAE Handbook—Refrigeration)*

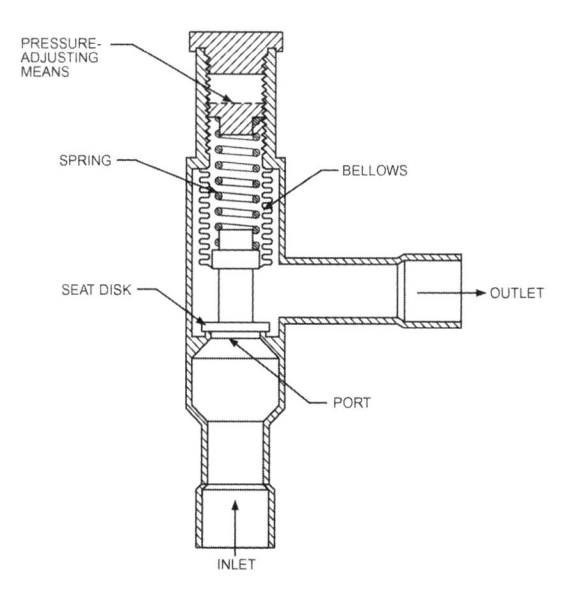

*Fig. 18-33 Direct-Acting Evaporator Pressure Regulator*
*(Figure 25, Chapter 11, 2010 ASHRAE Handbook—Refrigeration)*

on the outside of the tube surface within an enclosing shell. These coolers are usually used with positive-displacement compressors, such as reciprocating, rotary, or rotary screw compressors, to cool water or brine. Shell-and-tube is the most common arrangement, although tube-in-tube and brazed plate cooler are also available.

Figure 18-35 shows a typical shell-and-tube cooler. A series of baffles channels the fluid throughout the shell side. The baffles increase the velocity of the fluid, thereby increasing its heat transfer coefficient. The velocity of the fluid flowing perpendicular to the tubes should be at least 2 ft/s (0.6 m/s) to clean the tubes and less than 10 ft/s (3 m/s) to prevent erosion.

Distribution is critical in direct-expansion coolers. If some tubes are fed more refrigerant than others, they tend to bleed liquid refrigerant into the suction line. Since most direct-expansion coolers are controlled to a given suction super-heat, the remaining tubes must produce a higher superheat to evaporate the liquid bleeding through. This unbalance causes poor heat transfer. Uniform distribution is often achieved by a spray distributor. Most direct-expansion coolers are designed for horizontal mounting.

In a **flooded cooler**, the refrigerant vaporizes on the outside of tubes, which are submerged in liquid refrigerant within a closed shell. The fluid flows through the tubes as shown in Figure 18-36. Flooded coolers are usually used with rotary screw or centrifugal compressors to cool water or brine.

Refrigerant liquid/vapor mixture usually feeds into the bottom of the shell through a distributor that distributes the refrigerant vapor equally under the tubes. The relatively warm fluid in the tubes heats the refrigerant liquid surrounding the tubes, causing it to boil. As bubbles rise up through

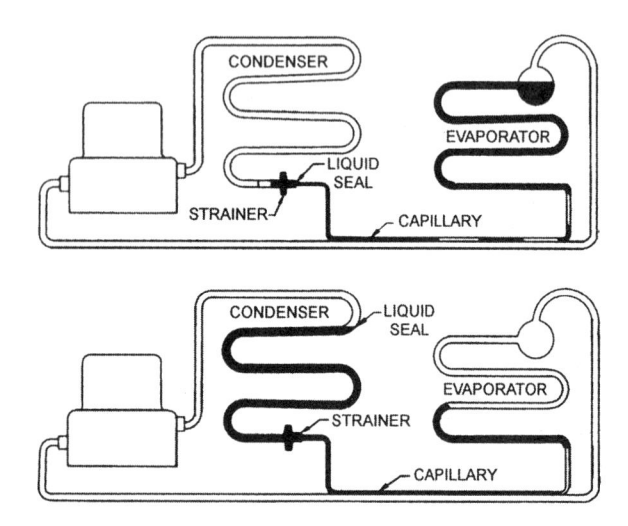

*Notes:*
1. Capillary selected for capacity balance conditions. Liquid seal at capillary inlet but no excess liquid in condenser. Compressor discharge and suction pressure normal. Evaporator properly charged.
2. Too much capillary resistance—liquid refrigerant backs up in condenser and causes evaporator to be undercharged. Compressor discharge pressure may be abnormally high. Suction pressure below normal. Bottom of condenser subcooled.

*Fig. 18-34 Effect of Capillary Tube Selection on Refrigerant Distribution*
*(Figure 47, Chapter 11, 2010 ASHRAE Handbook—Refrigeration)*

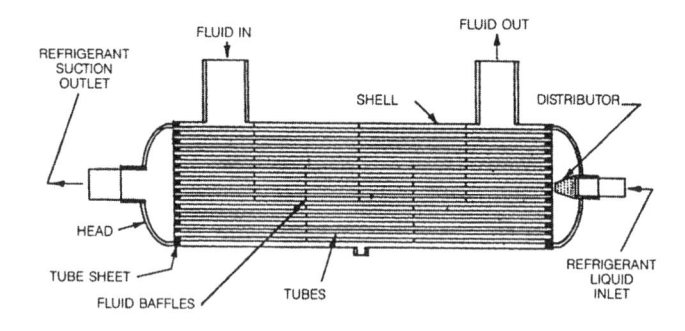

*Fig. 18-35 Direct-Expansion Shell-and-Tube Cooler*
*(Figure 1, Chapter 42, 2012 ASHRAE Handbook—HVAC Systems and Equipment)*

**Table 18-3 Types of Coolers**
*(Table 1, Chapter 42, 2012 ASHRAE Handbook—HVAC Systems and Equipment)*

| Type of Cooler | Subtype | Usual Refrigerant Feed Device | Usual Capacity Range, tons kW | Commonly Used Refrigerants |
|---|---|---|---|---|
| Direct-expansion | Shell-and-tube | Thermal expansion valve | 2 to 1000 7 to 3500 | 12, 22, 134a, 404A, 407C, 410A, 500, 502, 507A, 717 |
| | | Electronic modulation valve | 2 to 1000 7 to 3500 | |
| | Tube-in-tube | Thermal expansion valve | 5 to 25 18 to 90 | 12, 22, 134a, 717 |
| | Brazed-plate | Thermal expansion valve | 0.6 to 200 2 to 700 | 12, 22, 134a, 404A 407C, 410A, 500, 502, 507A, 508B, 717, 744 |
| | Semiwelded plate | Thermal expansion valve | 50 to 1990 175 to 7000 | 12, 22, 134a, 500, 502, 507A, 717, 744 |
| Flooded | Shell-and-tube | Low-pressure float | 25 to 2000 90 to 7000 | 11, 12, 22, 113, 114 |
| | | High-pressure float | 25 to 6000 90 to 21 100 | 123, 134a, 500, 502, 507A, 717 |
| | | Fixed orifice(s) | 25 to 6000 90 to 21 100 | |
| | | Weir | 25 to 6000 90 to 21 100 | |
| | Spray shell-and-tube | Low-pressure float | 50 to 10,000 180 to 35 000 | 11, 12, 13B1, 22 |
| | | High-pressure float | 50 to 10,000 180 to 35 000 | 113, 114, 123, 134a |
| | Brazed-plate | Low-pressure float | 0.6 to 200 2 to 700 | 12, 22, 134a, 500, 502, 507A, 717, 744 |
| | Semiwelded plate | Low-pressure float | 50 to 1990 175 to 7000 | 12, 22, 134a, 500, 502, 507A, 717, 744 |
| Baudelot | Flooded | Low-pressure float | 10 to 100 35 to 350 | 22, 717 |
| | Direct-expansion | Thermal expansion valve | 5 to 25 18 to 90 | 12, 22, 134a, 717 |
| Shell-and-coil | — | Thermal expansion valve | 2 to 10 7 to 35 | 12, 22, 134a, 717 |

# Refrigeration Equipment

the space between tubes, the liquid surrounding the tubes becomes increasingly bubbly (or foamy, if much oil is present). The refrigerant vapor must be separated from the mist generated by the boiling refrigerant. The simplest separation method is provided by a dropout area between the top row of tubes and the suction connections. If this dropout area is insufficient, a coalescing filter may be required between the tubes and connectors.

The size of tubes, number of tubes, and number of passes should be determined to maintain the fluid velocity typically between 3 and 10 ft/s (1 and 3 m/s). In some cases, the minimum velocity may be determined by a lower Reynolds number limit to ensure turbulent flow. Flooded shell-and-tube coolers are generally unsuitable for other than horizontal orientation.

A **spray cooler** is similar to a flooded shell-and-tube cooler except that the refrigerant liquid is recirculated through spray nozzles located above the top tubes. None of the tubes is submerged in liquid.

A **shell-and-coil cooler** is a tank containing the fluid to be cooled with a simple coiled tube used to cool the fluid. This type of cooler has the advantage of cold fluid storage to offset peak loads. In some models, the tank can be opened for cleaning. Most applications are at low capacities (e.g., for bakeries, for photographic laboratories, and to cool drinking water).

The coiled tube containing the refrigerant can be either inside the tank (Figure 18-37) or attached to the outside of the tank in a way that permits heat transfer.

The rate at which heat is transferred in the evaporator is given by the following equation:

$$q = U_o A_o \Delta t_m \quad \text{Btu/h} \quad (18\text{-}6)$$

where

$q$ = heat transfer rate, Btu/h (W)
$U_o$ = overall heat transfer coefficient based on outside surface, Btu/h·ft²·°F [W/(m²·K)]
$A_o$ = outside surface area, ft² (m²)
$\Delta t_m$ = logarithmic mean temperature difference, °F (°C)

Details on determining these quantities are given in Chapter 4 of the 2013 *ASHRAE Handbook—Fundamentals*. Listed in Table 18-4 are approximate minimum and maximum values for $U_o$.

### 18.1.5 Refrigerants

The choice of a refrigerant for a particular application often depends on properties not directly related to its ability to remove heat. Such properties are flammability, toxicity, density, viscosity, availability, and environmental acceptability. As a rule, the selection of a refrigerant is a compromise between conflicting desirable properties. For example, the pressure in the evaporator should be as high as possible, and at the same time, a low condensing pressure is desirable.

Tables 18-5 and 18-6 provide the ASHRAE standard designation of refrigerant and refrigerant blend data and safety classifications given in ANSI/ASHRAE Standard 34. Table 18-7 lists the basic physical properties of these refrigerants. Table 18-8 gives the comparative refrigerant performance per unit (ton) of refrigeration.

A discussion of the properties of various refrigerants, as well as their relative performance characteristics, is presented in Chapter 29 of the 2013 *ASHRAE Handbook—Fundamentals*. Complete thermodynamic and thermophysical properties for the refrigerants may be found in Chapter 30 of the 2013 *ASHRAE Handbook—Fundamentals*.

## 18.2 Absorption Air-Conditioning and Refrigeration Equipment

Absorption refrigeration cycles are heat-operated cycles in which a secondary fluid, the **absorbent**, is used to absorb the **primary fluid**, a gaseous refrigerant, which has been vaporized in the evaporator. The basic absorption cycle is shown in Figure 18-38.

Chapter 2 of the 2013 *ASHRAE Handbook—Fundamentals* discusses operating principles and thermodynamics of the basic absorption cycle and other information on the thermodynamics of workable absorbent-refrigerant combinations. A complete thermodynamic analysis of the absorption cycle is complex. However, a detailed analysis is not necessary to understanding the operating principles of the cycle.

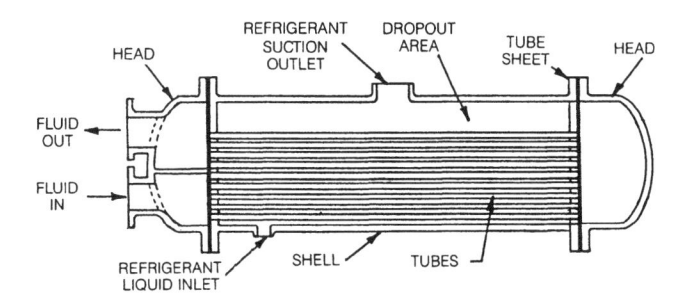

*Fig. 18-36 Flooded Shell-and-Tube Cooler*
(Figure 2, Chapter 42, 2012 ASHRAE Handbook—
HVAC Systems and Equipment)

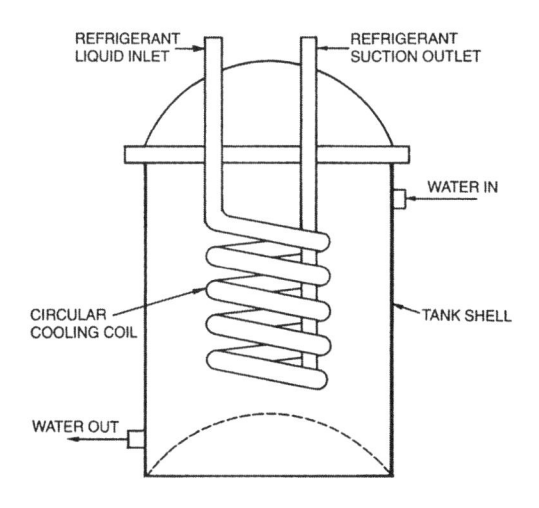

*Fig. 18-37 Shell-and-Coil Cooler*
(Figure 5, Chapter 42, 2012 ASHRAE Handbook—
HVAC Systems and Equipment)

Table 18-4  Overall Heat Transfer Coefficients for Liquid Coolers

| Type of Evaporator | Overall $U$, Btu/h·ft$^2$·°F (W/(m$^2$·K)) Minimum | Overall $U$, Btu/h·ft$^2$·°F (W/(m$^2$·K)) Maximum | Surface Side Basis for $U$ |
|---|---|---|---|
| Flooded shell-and-plain-tube (water to Refrigerants 12, 22, and 717) | 130 (740) | 190 (1080) | Refrigerant |
| Flooded shell-and-finned-tube (water to Refrigerants 12, 22, or 500) | 90 (510) | 170 (970) | Refrigerant |
| Flooded shell-and-plain-tube (brine to Refrigerant 717) | 45 (260) | 100 (570) | Refrigerant |
| Flooded shell-and-plain-tube (brine to Refrigerants 12, 22, or 502) | 30 (170) | 90 (510) | Refrigerant |
| Direct-expansion, shell-and-plain-tube (water to Refrigerants 12, 22, and 717) (Refrigerant in Tubes) | 80 (450) | 220 (1250) | Liquid |
| Direct-expansion, shell-and-internal-finned-tubes (water to Refrigerants 12 or 22) (Refrigerant in Tubes) | 160 (910) | 250 (1420) | Liquid |
| Direct-expansion, shell-and-plain-tube (brine to Refrigerants 12, 22, 717, or 502) (Refrigerant in Tubes) | 60 (340) | 140 (790) | Liquid |
| Direct-expansion, shell-and-internal-finned-tubes (nonsalt brines to Refrigerants 12, 22, or 502) | 60 (340) | 170 (970) | Liquid |
| Shell-and-plain-tube coil (water in shell) (Refrigerant 12, 22, or 717 in coil) | 10 (57) | 25 (140) | Liquid |
| Baudelot cooler, flooded (Refrigerant 12 or 22 to water) | 100 (570) | 200 (1130) | Liquid |
| Baudelot cooler, direct expansion (Refrigerant 717 to water) | 60 (340) | 150 (850) | Liquid |
| Baudelot cooler, direct expansion (Refrigerant 12 or 22 to water) | 60 (340) | 120 (680) | Liquid |
| Double-pipe cooler (Refrigerant 717 to water) | 50 (280) | 150 (850) | Liquid |
| Double-pipe cooler (Refrigerant 717 to water) | 50 (280) | 125 (710) | Liquid |
| Tank-and-agitator, coil type water cooler (flooded, Refrigerant 717) | 80 (450) | 125 (710) | Liquid |
| Tank-and-agitator, coil type water cooler (flooded, Refrigerant 12, 22, or 500) | 60 (340) | 100 (570) | Liquid |
| Tank, ammonia (Refrigerant 717) to brine cooling, coils between cans in ice | 15 (85) | 40 (230) | Liquid |
| Tank-and-agitator, coil type water cooler (flooded, Refrigerant 717) | 80 (450) | 110 (620) | Liquid |

Table 18-5  Refrigerant Data and Safety Classifications

| Refrigerant Number | Chemical Name[a,b] | Chemical Formula[a] | Molecular Mass[a] | Normal Boiling Point,[a] °F°C | Safety Group |
|---|---|---|---|---|---|
| **Methane Series** | | | | | |
| 11 | Trichlorofluoromethane | $CCl_3F$ | 137.4 | 75 24 | A1 |
| 12 | Dichlorodifluoromethane | $CCl_2F_2$ | 120.9 | −22 −30 | A1 |
| 12B1 | Bromochlorodifluoromethane | $CBrClF_2$ | 165.4 | 25 −4 | |
| 13 | Chlorotrifluoromethane | $CClF_3$ | 104.5 | −115 −81 | A1 |
| 13B1 | Bromotrifluoromethane | $CBrF_3$ | 148.9 | −72 −58 | A1 |
| 14 | Tetrafluoromethane (carbon tetrafluoride) | $CF_4$ | 88.0 | −198 −128 | A1 |
| 21 | Dichlorofluoromethane | $CHCl_2F$ | 102.9 | 48 9 | B1 |
| 22 | Chlorodifluoromethane | $CHClF_2$ | 86.5 | −41 | A1 |
| 23 | Trifluoromethane | $CHF_3$ | 70.0 | −116 −82 | A1 |
| 30 | Dichloromethane (methylene chloride) | $CH_2Cl_2$ | 84.9 | 104 40 | B2 |
| 31 | Chlorofluoromethane | $CH_2ClF$ | 68.5 | 16 −9 | |
| 32 | Difluoromethane (methylene fluoride) | $CH_2F_2$ | 52.0 | −62 −52 | A2L |
| 40 | Chloromethane (methyl chloride) | $CH_3Cl$ | 50.4 | −12 −24 | B2 |
| 41 | Fluoromethane (methyl fluoride) | $CH_3F$ | 34.0 | −109 −78 | |
| 50 | Methane | $CH_4$ | 16.0 | −259 −161 | A3 |
| **Ethane Series** | | | | | |
| 113 | 1,1,2-trichloro-1,2,2-trifluoroethane | $CCl_2FCClF_2$ | 187.4 | 118 48 | A1 |
| 114 | 1,2-dichloro-1,1,2,2-tetrafluoroethane | $CClF_2CClF_2$ | 170.9 | 38 4 | A1 |
| 115 | Chloropentafluoroethane | $CClF_2CF_3$ | 154.5 | −38 −39 | A1 |
| 116 | Hexafluoroethane | $CF_3CF_3$ | 138.0 | −109 −78 | A1 |
| 123 | 2,2-dichloro-1,1,1-trifluoroethane | $CHCl_2CF_3$ | 153.0 | 81 27 | B1 |
| 124 | 2-chloro-1,1,1,2-tetrafluoroethane | $CHClFCF_3$ | 136.5 | 10 −12 | A1 |
| 125 | Pentafluoroethane | $CHF_2CF_3$ | 120.0 | −55 −48 | A1 |
| 134a | 1,1,1,2-tetrafluoroethane | $CH_2FCF_3$ | 102.0 | −15 −26 | A1 |
| 141b | 1,1-dichloro-1-fluoroethane | $CH_3CCl_2F$ | 117.0 | 90 32 | |
| 142b | 1-chloro-1,1-difluoroethane | $CH_3CClF_2$ | 100.5 | 14 −10 | A2 |
| 143a | 1,1,1-trifluoroethane | $CH_3CF_3$ | 84.0 | −53 −47 | A2L |
| 152a | 1,1-difluoroethane | $CH_3CHF_2$ | 66.0 | −11 −24 | A2 |
| 170 | Ethane | $CH_3CH_3$ | 30.0 | −128 −89 | A3 |

# Refrigeration Equipment

**Table 18-5  Refrigerant Data and Safety Classifications (Continued)**

| Refrigerant Number | Chemical Name[a,b] | Chemical Formula[a] | Molecular Mass[a] | Normal Boiling Point,[a] °F°C | Safety Group |
|---|---|---|---|---|---|
| **Ethers** | | | | | |
| E170 | Dimethyl ether | $CH_3OCH_3$ | 46.0 | –13–25 | A3 |
| **Propane Series** | | | | | |
| 218 | Octafluoropropane | $CF_3CF_2CF_3$ | 188.0 | –35–37 | A1 |
| 227ea | 1,1,1,2,3,3,3-heptafluoropropane | $CF_3CHFCF_3$ | 170.0 | 3–16 | A1 |
| 236fa | 1,1,1,3,3,3-hexafluoropropane | $CF_3CH_2CF_3$ | 152.0 | 29–1 | A1 |
| 245fa | 1,1,1,3,3-pentafluoropropane | $CF_3CH_2CHF_2$ | 134.0 | 5915 | B1 |
| 290 | Propane | $CH_3CH_2CH_3$ | 44.0 | –44–42 | A3 |
| **Cyclic Organic Compounds** (see Table 2 for blends) | | | | | |
| C318 | Octafluorocyclobutane | $-(CF_2)_4-$ | 200.0 | 21–6 | A1 |
| **Miscellaneous Organic Compounds** | | | | | |
| **Hydrocarbons** | | | | | |
| 600 | Butane | $CH_3CH_2CH_2CH_3$ | 58.1 | 310 | A3 |
| 600a | 2-methylpropane (isobutane) | $CH(CH_3)_2CH_3$ | 58.1 | 11–12 | A3 |
| 601 | Pentane | $CH_3(CH_2)_3CH_3$ | 72.15 | 9736.1 | A3 |
| 601a | 2-methylbutane (isopentane) | $(CH_3)_2CHCH_2CH_3$ | 72.15 | 8227.8 | A3 |
| **Oxygen Compounds** | | | | | |
| 610 | Ethyl ether | $CH_3CH_2OCH_2CH_3$ | 74.1 | 9435 | |
| 611 | Methyl formate | $HCOOCH_3$ | 60.0 | 8932 | B2 |
| **Sulfur Compounds** | | | | | |
| 620 | (Reserved for future assignment) | | | | |
| **Nitrogen Compounds** | | | | | |
| 630 | Methanamine (methyl amine) | $CH_3NH_2$ | 31.1 | 20–7 | |
| 631 | Ethanamine (ethyl amine) | $CH_3CH_2(NH_2)$ | 45.1 | 6217 | |
| **Inorganic Compounds** | | | | | |
| 702 | Hydrogen | $H_2$ | 2.0 | –423–253 | A3 |
| 704 | Helium | He | 4.0 | –452–269 | A1 |
| 717 | Ammonia | $NH_3$ | 17.0 | –28–33 | B2L |
| 718 | Water | $H_2O$ | 18.0 | 212100 | A1 |
| 720 | Neon | Ne | 20.2 | –411–246 | A1 |
| 728 | Nitrogen | $N_2$ | 28.1 | –320–196 | A1 |
| 732 | Oxygen | $O_2$ | 32.0 | –297–183 | |
| 740 | Argon | Ar | 39.9 | –303–186 | A1 |
| 744 | Carbon dioxide | $CO_2$ | 44.0 | –109–78[c] | A1 |
| 744A | Nitrous oxide | $N_2O$ | 44.0 | –129–90 | |
| 764 | Sulfur dioxide | $SO_2$ | 64.1 | 14–10 | B1 |
| **Unsaturated Organic Compounds** | | | | | |
| 1150 | Ethene (ethylene) | $CH_2=CH_2$ | 28.1 | –155–104 | A3 |
| 1234yf | 2,3,3,3-tetrafluoro-1-propene | $CF_3CF=CH_2$ | 114.0 | –20.9–29.4 | A2L |
| 1234ze(E) | Trans-1,3,3,3-tetrafluoro-1-propene | $CF_3CH=CHF$ | 114.0 | –2.2–19.0 | A2L |
| 1270 | Propene (propylene) | $CH_3CH=CH_2$ | 42.1 | –54–48 | A3 |

*Source*: ANSI/ASHRAE *Standard* 34-2010.
[a]Chemical name, chemical formula, molecular mass, and normal boiling point are not part of this standard.
[b]Preferred chemical name is followed by the popular name in parentheses.
[c]Sublimes.

**Table 18-6  Data and Safety Classifications for Refrigerant Blends**
*(Table 2, Chapter 29, 2013 ASHRAE Handbook—Fundamentals)*

| Refrigerant Number | Composition (Mass %) | Composition Tolerances | Molecular Mass[a] | Normal Bubble Point, °F | Normal Dew Point, °F | Safety Group |
|---|---|---|---|---|---|---|
| **Zeotropes** | | | | | | |
| 400 | R-12/114 (must be specified) | | | | | A1 |
| 401A | R-22/152a/124 (53.0/13.0/34.0) | (±2.0 /+0.5,–1.5/±1.0) | 94.4 | –29.9 | –19.8 | A1 |
| 401B | R-22/152a/124 (61.0/11.0/28.0) | (±2/+0.5,–1.5/±1.0) | 92.8 | –32.3 | –23.4 | A1 |
| 401C | R-22/152a/124 (33.0/15.0/52.0) | (±2/+0.5,1.5/±1.0) | 101 | –22.9 | –10.8 | A1 |
| 402A | R-125/290/22 (60.0/2.0/38.0) | (±2.0/+0.1,–1.0/±2.0) | 101.6 | –56.6 | –52.6 | A1 |
| 402B | R-125/290/22 (38.0/2.0/60.0) | (±2/+0.1,–1/±2) | 94.7 | –53.0 | –48.8 | A1 |

## Table 18-6 Data and Safety Classifications for Refrigerant Blends (Continued)
*(Table 2, Chapter 29, 2013 ASHRAE Handbook—Fundamentals)*

| Refrigerant Number | Composition (Mass %) | Composition Tolerances | Molecular Mass[a] | Normal Bubble Point, °F | Normal Dew Point, °F | Safety Group |
|---|---|---|---|---|---|---|
| 403A | R-290/22/218 (5.0/75.0/20.0) | (+0.2,−2/±2/±2) | 92 | −47.2 | −44.1 | A1 |
| 403B | R-290/22/218 (5.0/56.0/39.0) | (+0.2,−2/±2/±2) | 103.3 | −46.8 | −44.1 | A1 |
| 404A | R-125/143a/134a (44.0/52.0/4.0) | (±2/±1/±2) | 97.6 | −51.9 | −50.4 | A1 |
| 405A | R-22/152a/142b/C318 (45.0/7.0/5.5/42.5) | (±2/±1/±1 /±2) sum of R-152a and R-142b = (+0.0, −2.0) | 111.9 | −27.2 | −12.1 | |
| 406A | R-22/600a/142b (55.0/4.0/41.0) | (±2/±1/±1) | 89.9 | −26.9 | −10.3 | A2 |
| 407A | R-32/125/134a (20.0/40.0/40.0) | (±2/±2/±2) | 90.1 | −49.4 | −37.7 | A1 |
| 407B | R-32/125/134a (10.0/70.0/20.0) | (±2/±2/±2) | 102.9 | −52.2 | −44.3 | A1 |
| 407C | R-32/125/134a (23.0/25.0/52.0) | (±2/±2/±2) | 86.2 | −46.8 | −34.1 | A1 |
| 407D | R-32/125/134a (15.0/15.0/70.0) | (±2/±2/±2) | 91 | −38.9 | −26.9 | A1 |
| 407E | R-32/125/134a (25.0/15.0/60.0) | (±2,±2,±2) | 83.8 | −45.0 | −32.1 | A1 |
| 407F | R-32/125/134a (30.0/30.0/40.0) | (±2,±2,±2) | 82.1 | −51.0 | −39.5 | A1 |
| 408A | R-125/143a/22 (7.0/46.0/47.0) | (±2/±1/±2) | 87 | −49.9 | −49.0 | A1 |
| 409A | R-22/124/142b (60.0/25.0/15.0) | (±2/±2/±1) | 97.4 | −31.7 | −17.5 | A1 |
| 409B | R-22/124/142b (65.0/25.0/10.0) | (±2/±2/±1) | 96.7 | −33.7 | −21.5 | A1 |
| 410A | R-32/125 (50.0/50.0) | (+0.5,−1.5/+1.5,−0.5) | 72.6 | −60.9 | −60.7 | A1 |
| 410B | R-32/125 (45.0/55.0) | (±1/±1) | 75.6 | −60.7 | −60.5 | A1 |
| 411A | R-1270/22/152a (1.5/87.5/11.0) | (+0,−1/+2,−0/+0,−1) | 82.4 | −39.5 | −35.0 | A2 |
| 411B | R-1270/22/152a (3.0/94.0/3.0) | (+0,−1/+2,−0/+0,−1) | 83.1 | −42.9 | −42.3 | A2 |
| 412A | R-22/218/142b (70.0/5.0/25.0) | (±2/±2/±1) | 92.2 | −33.5 | −19.8 | A2 |
| 413A | R-218/134a/600a (9.0/88.0/3.0) | (±1/±2/±0,−1) | 104 | −20.7 | −17.7 | A2 |
| 414A | R-22/124/600a/142b (51.0/28.5/4.0/16.5) | (±2/±2/±0.5/+0.5,−1) | 96.9 | −29.2 | −14.4 | A1 |
| 414B | R-22/124/600a/142b (50.0/39.0/1.5/9.5) | (±2/±2/±0.5/+0.5,−1) | 101.6 | −29.9 | −15.0 | A1 |
| 415A | R-22/152a (82.0/18.0) | (±1/±1) | 81.9 | −35.5 | −30.5 | A2 |
| 415B | R-22/152a (25.0/75.0) | (±1/±1) | 70.2 | −17.8 | −15.2 | A2 |
| 416A | R-134a/124/600 (59.0/39.5/1.5) | (+0.5,−1/+1,−0.5/+1,−0.2) | 111.9 | −10.1 | −7.2 | A1 |
| 417A | R-125/134a/600 (46.6/50.0/3.4) | (±1.1/±1/+0.1,0.4) | 106.7 | −36.4 | −27.2 | A1 |
| 417B | R-125/134a/600 (79.0/18.3/2.7) | (±1/±1/+0.1,−0.5) | 113.1 | −48.8 | −42.7 | A1 |
| 418A | R-290/22/152a (1.5/96.0/2.5) | (±0.5/±1/±0.5) | 84.6 | −42.2 | −40.2 | A2 |
| 419A | R-125/134a/E170 (77.0/19.0/4.0) | (±1/±1/±1) | 109.3 | −44.7 | −32.8 | A2 |
| 420A | R-134a/142b (88.0/12.0) | (±1,−0/+0,−1) | 101.8 | −13.0 | −11.6 | A1 |
| 421A | R-125/134a (58.0/42.0) | (±1/±1) | 111.8 | −41.5 | −31.9 | A1 |
| 421B | R-125/134a (85.0/15.0) | (±1/±1) | 116.9 | −50.2 | −44.6 | A1 |
| 422A | R-125/134a/600a (85.1/11.5/3.4) | (±1/±1/+0.1,−0.4) | 113.6 | −51.7 | −47.4 | A1 |
| 422B | R-125/134a/600a (55.0/42.0/3.0) | (±1/±1/+0.1,−0.5) | 108.5 | −40.9 | −32.2 | A1 |
| 422C | R-125/134a/600a (82.0/15.0/3.0) | (±1/±1/+0.1,−0.5) | 116.3 | −49.5 | −44.2 | A1 |
| 422D | R-125/134a/600a (65.1/31.5/3.4) | (+0.9,−1.1/±1/+0.1,−0.4) | 109.9 | −45.8 | −37.1 | A1 |
| 423A | R-134a/227ea (52.5/47.5) | (±1/±1) | 126 | −11.6 | −10.3 | A1 |
| 424A | R-125/134a/600a/600/601a (50.5/47.0/0.9/1.0/0.6) | (±1/±1/+0.1,−0.2/+0.1,−0.2/+0.1,−0.2) | 108.4 | −38.4 | −27.9 | A1 |
| 425A | R-32/134a/227ea (18.5/69.5/12.0) | (±0.5/±0.5/±0.5) | 90.3 | −36.6 | −24.3 | A1 |
| 426A[a] | R-125/134a/600a/601a (5.1/93.0/1.3/0.6) | (±1/±1/+0.1,−0.2/+0.1,−0.2) | 101.6 | −19.3 | −16.1 | A1 |
| 427A[a] | R-32/125/143a/134a (15.0/25.0/10.0/50.0) | (±2/±2/±2/±2) | 90.4 | −45.4 | −33.3 | A1 |
| 428A[a] | R-125/143a/290/600a (77.5/20.0/0.6/1.9) | (±1/±1/+0.1,−0.2/+0.1,−0.2) | 107.5 | −54.9 | −53.5 | A1 |
| 429A | R-E170/152a/600a (60.0/10.0/30.0) | (±1/±1/±1) | 50.8 | −14.8 | −14.1 | A3 |
| 430A | R-152a/600a (76.0/24.0) | (±1/±1) | 64 | −17.7 | −17.3 | A3 |
| 431A | R-290/152a (71.0/29.0) | (±1/±1) | 48.8 | −45.6 | −45.6 | A3 |
| 432A | R-1270/E170 (80.0/20.0) | (±1/±1) | 42.8 | −51.9 | −50.1 | A3 |
| 433A | R-1270/290 (30.0/70.0) | (±1/±1) | 43.5 | −48.3 | −47.6 | A3 |
| 433B | R-1270/290 (5.0/95.0) | (±1/±1) | 44 | −44.9 | −44.5 | A3 |
| 433C | R-1270/290 (25.0/75.0) | (±1/±1) | 43.6 | −47.7 | −47.0 | A3 |
| 434A | R-125/143a/134a/600a (63.2/18.0/16.0/2.8) | (±1/±1/±1/+0.1,−0.2) | 105.7 | −49.0 | −44.1 | A1 |
| 435A | R-E170/152a (80.0/20.0) | (±1/±1) | 49.04 | −15.0 | −14.6 | A3 |
| 436A | R-290/600a (56.0/44.0) | (±1/±1) | 49.33 | −29.7 | −16.2 | A3 |
| 436B | R-290/600a (52.0/48.0) | (±1/±1) | 49.87 | −28.1 | −13.0 | A3 |
| 437A | R-125/134a/600/601 (19.5/78.5/1.4/0.6) | (+0.5,−1.8/+1.5,−0.7/+0.1,−0.2/+0.1/−0.2) | 103.7 | −27.2 | −20.6 | A1 |
| 438A | R-32/125/134a/600/601a (8.5/45.0/44.2/1.7/0.6) | (+0.5,−1.5/+1.5/±1.5/+0.1,−0.2/+0.1/−0.2) | 99.1 | −45.4 | −33.5 | A1 |
| 439A | R-32/125/600a (50.0/47.0/3.0) | (±1/±1) | 71.2 | −61.6 | −61.2 | A2 |
| 440A | R-290/134a/152a (0.6/1.6/97.8) | (±0.1/±0.6/±0.5) | 66.2 | −13.9 | −11.7 | A2 |
| 441A | R-170/290/600a/600 (3.1/54.8/6.0/36.1) | (±0.3/±2/±0.6/±2) | 48.2 | −43.4 | −4.7 | A3 |
| 442A | R-32/125/134a/152a/227ea (31.0/31.0/30.0/3.0/5.0) | (±1.0/±1.0/±1.0/+0.5/±1.0) | 81.77 | −51.7 | −39.8 | A1 |

# Refrigeration Equipment

**Table 18-7 Physical Properties of Selected Refrigerants[a]**
*(Table 5, Chapter 29, 2013 ASHRAE Handbook—Fundamentals)*

| Number | Refrigerant Chemical Name or Composition (% by Mass) | Chemical Formula | Molecular Mass | Boiling Pt.[f] (NBP) at 14.696 psia, °F 101.325 kPa, °C | Freezing Point, °F °C | Critical Temperature, °F °C | Critical Pressure, psi kPa | Critical Density, lb/ft³ kg/m³ | Refractive Index of Liquid[b,c] |
|---|---|---|---|---|---|---|---|---|---|
| 728 | Nitrogen | N$_2$ | 28.013 | | | | | | 1.205 (83 K) |
| | | | | –320.44 –195.8 | –346 –210.0 | –232.528 –146.96 | 492.53 3395.8 | 19.56 313.3 | 589.3 nm |
| 729 | Air | — | 28.959 | –317.65 –194.25 | — | –221.062 –140.59 | 549.63 789.6 | 20.97 335.94 | — |
| 740 | Argon | Ar | 39.948 | | | | | | 1.233 (84 K) |
| | | | | –302.53 –185.85 | –308.812 –189.34 | –188.428 –122.46 | 705.34 863.0 | 33.44 535.6 | 589.3 nm |
| 732 | Oxygen | O$_2$ | 31.999 | | | | | | 1.221 (92 K) |
| | | | | –297.328 –182.96 | –361.822 –218.79 | –181.426 –118.57 | 731.45 043.0 | 27.23 436.14 | 589.3 nm |
| 50 | Methane | CH$_4$ | 16.043 | –258.664 –161.48 | –296.428 –182.46 | –116.6548 –82.586 | 667.14 599.2 | 10.15 162.66 | — |
| 14 | Tetrafluoromethane | CF$_4$ | 88.005 | –198.49 –128.05 | –298.498 –183.61 | –50.152 –45.64 | 543.93 750.0 | 39.06 625.66 | — |
| 170 | Ethane | C$_2$H$_6$ | 30.07 | –127.4764 –88.581 | –297.01 –182.8 | 89.92 432.72 | 706.64 872.2 | 12.87 206.18 | — |
| 508A | R-23/116 (39/61) | — | 100.1 | –125.73 –87.60 | — | 50.34 610.192 | 529.53 650.8 | 35.43 567.58 | — |
| 508B | R-23/116 (46/54) | — | 95.394 | –125.68 –87.6 | — | 52.17 011.205 | 547.03 771.6 | 35.49 568.45 | — |
| 23 | Trifluoromethane | CHF$_3$ | 70.014 | –115.6324 –82.018 | –247.234 –155.13 | 79.057 426.143 | 700.84 832 | 32.87 526.5 | — |
| 13 | Chlorotrifluoromethane | CClF$_3$ | 104.46 | –114.664 –81.48 | –294.07 –181.15 | 83.93 28.85 | 562.63 879 | 36.39 582.88 | 1.146 (25)[2] |
| 744 | Carbon dioxide | CO$_2$ | 44.01 | –109.12 –78.4[d] | –69.8044 –56.558[e] | 87.760 430.978 | 1070.07 377.3 | 29.19 467.6 | 1.195 (15) |
| 504 | R-32/115 (48.2/51.8) | — | 79.249 | –72.23 –57.906 | — | 143.85 62.138 | 642.34 428.8 | 31.51 504.68 | — |
| 32 | Difluoromethane | CH$_2$F$_2$ | 52.024 | –60.9718 –51.651 | –214.258 –136.81 | 172.58 978.105 | 838.65 782.0 | 26.47 424 | — |
| 410A | R-32/125 (50/50) | — | 72.585 | –60.5974 –51.446 | — | 160.444 471.358 | 711.14 902.6 | 28.69 459.53 | — |
| 125 | Pentafluoroethane | C$_2$HF$_5$ | 120.02 | –54.562 –48.09 | –149.134 –100.63 | 150.841 466.023 | 524.73 617.7 | 35.81 573.58 | — |
| 1270 | Propylene | C$_3$H$_6$ | 42.08 | –53.716 –47.62 | –301.35 –185.2 | 195.91 91.061 | 660.64 554.8 | 14.36 230.03 | 1.3640 (–50)[1] |
| 143a | Trifluoroethane | CH$_3$CF$_3$ | 84.041 | –53.0338 –47.241 | –169.258 –111.81 | 162.872 672.707 | 545.53 761.0 | 26.91 431.0 | — |
| 507A | R-125/143a (50/50) | — | 98.859 | –52.1338 –46.741 | — | 159.110 670.617 | 537.43 705 | 30.64 490.77 | — |
| 404A | R-125/143a/134a (44/52/4) | — | 97.604 | –51.1996 –46.222 | — | 161.682 872.046 | 540.83 728.9 | 30.37 486.53 | — |
| 502 | R-22/115 (48.8/51.2) | — | 111.63 | –49.3132 –45.174 | — | 178.71 80.507 | 582.64 016.8 | 35.50 568.70 | — |
| 407C | R-32/125/134a (23/25/52) | — | 86.204 | –46.5286 –43.627 | — | 186.861 286.034 | 671.54 629.8 | 30.23 484.23 | — |
| 290 | Propane | C$_3$H$_8$ | 44.096 | –43.805 –42.11 | –305.72 –187.62 | 206.1396.74 | 616.58 4251.2 | 13.76 220.4 | 1.3397 (–42) |
| 22 | Chlorodifluoromethane | CHClF$_2$ | 86.468 | –41.458 –40.81 | –251.356 –157.42 | 205.06 196.145 | 723.74 990.0 | 32.70 523.84 | 1.234 (25)[2] |
| 115 | Chloropentafluoroethane | CClF$_2$CF$_3$ | 154.47 | –38.65 –39.25 | –146.92 –99.39 | 175.91 79.95 | 453.83 129.0 | 38.38 614.8 | 1.221 (25)[2] |
| 500 | R-12/152a (73.8/26.2) | — | 99.303 | –28.4854 –33.603 | — | 215.762 102.09 | 604.64 168.6 | 30.91 495.1 | — |
| 717 | Ammonia | NH$_3$ | 17.03 | –27.9886 –33.327 | –107.779 –77.655 | 270.05 132.25 | 1643.71 1 333.0 | 14.05 225.0[d] | 1.325 (16.5) |
| 12 | Dichlorodifluoromethane | CCl$_2$F$_2$ | 120.91 | –21.5536 –29.752 | –250.69 –157.05 | 233.546 111.97 | 599.94 136.1 | 35.27 565.0 | 1.288 (25)[2] |
| R-1234yf | 2,3,3,3-tetrafluoroprop-1-ene | CF$_3$CF=CH$_2$ | 114.04 | –21.01 –29.45 | | 202.4 694.7 | 490.55 3382.2 | 29.66 8475.55 | |
| 134a | Tetrafluoroethane | CF$_3$CH$_2$F | 102.03 | –14.9332 –26.074 | –153.94 –103.3 | 213.908 101.06 | 588.84 059.3 | 31.96 511.9 | — |
| 152a | Difluoroethane | CHF$_2$CH$_3$ | 66.051 | –11.2414 –24.023 | –181.462 –118.59 | 235.868 113.26 | 655.14 516.8 | 22.97 368 | — |
| R-1234ze (E) | Trans-1,3,3,3-tetrafluoropropene | CF$_3$CH=CHF | 114.04 | –2.11 –18.95 | | 228.87 109.37 | 527.29 3636.3 | 30.54 2489.24 | |
| 124 | Chlorotetrafluoroethane | CHClFCF$_3$ | 136.48 | 10.4666 –11.963 | –326.47 –199.15 | 252.104 122.28 | 525.73 624.3 | 34.96 560.0 | — |
| 600a | Isobutane | C$_4$H$_{10}$ | 58.122 | 10.852 –11.75 | –254.96 –159.42 | 274.39 134.66 | 526.343 629.05 | 14.08 225.5 | 1.3514 (–25)[1] |
| 142b | Chlorodifluoroethane | CClF$_2$CH$_3$ | 100.5 | 15.53 –9.15 | –202.774 –130.43 | 278.798 137.11 | 590.34 055.0 | 27.84 466.0 | — |
| C318 | Octafluorocyclobutane | C$_4$F$_8$ | 200.03 | 21.245 –5.975 | –39.64 –39.8 | 239.414 115.23 | 402.82 777.5 | 38.70 619.97 | — |
| 600 | Butane | C$_4$H$_{10}$ | 58.122 | 31.118 –0.49 | –216.86 –138.27 | 305.564 151.98 | 550.63 796.0 | 14.23 227.94 | 1.3562 (–15)[1] |
| 114 | Dichlorotetrafluoroethane | CClF$_2$CClF$_2$ | 170.92 | 38.454 83.586 | –134.54 –92.5 | 294.224 145.68 | 472.43 257.0 | 36.21 579.97 | 1.294 (25) |
| 11 | Trichlorofluoromethane | CCl$_3$F | 137.37 | 74.674 423.708 | –166.846 –110.47 | 388.328 197.96 | 639.34 407.6 | 34.59 554.0 | 1.362 (25)[2] |
| 123 | Dichlorotrifluoroethane | CHCl$_2$CF$_3$ | 152.93 | 82.08 27.823 | –160.87 –107.15 | 362.624 183.68 | 531.13 661.8 | 34.34 550.0 | — |
| 141b | Dichlorofluoroethane | CCl$_2$FCH$_3$ | 116.95 | 89.69 32.05 | –154.25 –103.5 | 399.83 204.4 | 610.94 212.0 | 28.63 458.6 | — |

527

**Table 18-7  Physical Properties of Selected Refrigerants[a]  (Continued)**
*(Table 5, Chapter 29, 2013 ASHRAE Handbook—Fundamentals)*

| | Refrigerant | | | Boiling Pt.[f] (NBP) | | | | | |
|---|---|---|---|---|---|---|---|---|---|
| Number | Chemical Name or Composition (% by Mass) | Chemical Formula | Molecular Mass | at 14.696 psia, °F 101.325 kPa, °C | Freezing Point, °F °C | Critical Temperature, °F °C | Critical Pressure, psi kPa | Critical Density, lb/ft³ kg/m³ | Refractive Index of Liquid[b,c] |
| 113 | Trichlorotrifluoro-ethane | CCl₂FClF₂ | 187.38 | 117.65 347.585 | –33.19 –36.22 | 417.30 8214.06 | 492.03 392.2 | 34.96 560.0 | 1.357 (25)[2] |
| 718[3] | Water | H₂O | 18.015 | 211.95 3299.974 | 32.018 0.01 | 705.11 373.95 | 3200.12 2 064.0 | 20.103 22.0 | — |

*Notes*:
[a] Data from NIST (2010) REFPROP v. 9.0.
[b] Temperature of measurement (°C, unless kelvin is noted) shown in parentheses. Data from CRC (1987), unless otherwise noted.
[c] For the sodium D line.
[d] Sublimes.
[e] At 76.4 psi 527 kPa.
[f] Bubble point used for blends

*References*:
[1] Kirk and Othmer (1956).
[2] *Bulletin* B-32A (DuPont).
[3] *Handbook of Chemistry* (1967).

The absorption cycle and the mechanical compression cycle have in common the evaporation and condensation of a refrigerant liquid; these processes occur at two pressure levels within the unit. The two cycles differ in that the absorption cycle uses a pump and a heat-operated generator to produce the pressure differential, whereas the mechanical compression cycle uses a compressor; the absorption cycle substitutes physiochemical processes for the purely mechanical processes of the compression cycle. Both cycles require energy for operation: heat in the absorption cycle, mechanical energy in the compression cycle.

Of the many combinations that have been tried, only the lithium bromide-water and the ammonia-water cycles remain in common use for air-conditioning. In addition, ammonia-water absorption equipment has been used in large industrial applications requiring low temperatures for process work.

Figure 18-40 is a typical schematic diagram of machines available in the form of indirect-fired liquid chillers in capacities of 50 to 1500 tons (180 to 5300 kW).

**Generators** (concentrators) are tube bundles submerged in the solution, heated by steam or hot liquids.

**Condensers** are tube bundles located in the vapor space over the generator and shielded from carryover of salt by eliminators. Cooling water to the condenser first passes through the absorber.

**Absorbers** are tube bundles over which strong absorbent is sprayed. Refrigerant vapor is condensed into the absorbent, releasing heat to the cooling water passing through.

**Evaporators** (coolers) are tube bundles over which the refrigerant water is sprayed and evaporated. The liquid to be cooled passes inside the tubes.

**Solution heat exchangers** are of all steel shell-and-tube construction.

**Solution and evaporator pumps** are generally electric-motor-driven centrifugal pumps of hermetic design that use the cycle fluids for cooling and lubrication.

**Purgers** are used to remove noncondensable gases. Noncondensable gases present in small quantities can raise the total pressure in the absorber sufficiently to significantly change the evaporator pressure. Small pressure increases cause appreciable change in the refrigerant evaporating temperature.

**Expansion devices** commonly used in absorption machines are usually an orifice or fixed restriction, which controls the flow of refrigerant liquid between the condenser and the evaporator.

Lithium bromide-water cycle absorption machines meet load variations and maintain chilled water temperature control by varying the rate of reconcentration of the absorbent solution. At any given constant load, the chilled water temperature is maintained by a temperature difference between refrigerant and chilled water. In turn, the refrigerant temperature is maintained by the absorber being supplied with a flow rate and concentration of solution, and by the absorber cooling water temperature.

Load changes are reflected by corresponding changes in chilled water temperature. A load reduction, for example, results in less temperature difference needed in the evaporator and a reduced requirement for solution flow or concentration. The resultant chilled water temperature drop is met by adjusting the rate of reconcentration to match the reduced requirements of the absorber.

The coefficient of performance (COP) of a lithium bromide-water cycle absorption machine operating at 45° leaving chilled water temperature, 85° entering condenser water temperature and 12 psig steam pressure is typically in the range of 0.65 to 0.70. Whenever chilled water temperatures are above the nominal, or condensing water temperatures are below the nominal, a COP as high as 0.70 can be reached. Reversing the temperature conditions cited reduces the COP to below 0.60. A coefficient of performance of 0.68 corresponds approximately to a steam rate of 18 lb/h per ton of refrigeration (1.45 kW/kW).

Absorption machines can be made with a two-stage generator. Such a unit may be called **dual effect**. Figure 18-40 is a schematic diagram of a nominally single-shell design with a two-stage generator. The first-effect generator receives the external heat, which boils refrigerant from the weak absorbent. This hot refrigerant vapor then goes to a second generator and supplies heat for further refrigerant vaporization from the absorbent of intermediate concentration, which flows from the first generator and is cooled by passing through a first-stage heat economizer. Other than the generator, all components of the single-stage lithium bro-

# Refrigeration Equipment

### Table 18-8  Comparative Refrigerant Performance per Ton of Refrigeration

| Refrigerant Number | Chemical Name or Composition (% by mass) | Evaporator Pressure, psia | Condenser Pressure, psia | Compression Ratio | Net Refrigerating Effect, Btu/lb | Refrigerant Circulated, lb/min | Liquid Circulated, gal/min | Specific Volume of Suction Gas, ft³/lb | Compressor Displacement, ft³/min | Power Consumption, hp | Coefficient of Performance | Compressor Discharge Temp., °F |
|---|---|---|---|---|---|---|---|---|---|---|---|---|
| *Evaporator −25°F/Condenser 86°F* | | | | | | | | | | | | |
| 744 | Carbon dioxide | 195.7 | 1046.2 | 5.35 | 56.8 | 3.52 | 0.711 | 0.457 | 1.61 | 2.779 | 1.698 | 196.3 |
| 170 | Ethane | 146.8 | 675.1 | 4.6 | 66.0 | 3.03 | 1.314 | 0.878 | 2.66 | 2.805 | 1.681 | 136.2 |
| 1270 | Propylene | 28.8 | 189.3 | 6.57 | 115.7 | 1.73 | 0.416 | 3.63 | 6.28 | 1.637 | 2.88 | 120.3 |
| 507A | R-125/143a (50/50) | 28.8 | 211.7 | 7.34 | 43.5 | 4.60 | 0.54 | 1.52 | 6.98 | 1.833 | 2.573 | 100.6 |
| 404A | R-125/143a/134a (44/52/4) | 27.6 | 206.1 | 7.46 | 45.1 | 4.44 | 0.521 | 1.61 | 7.13 | 1.817 | 2.595 | 102.1 |
| 502 | R-22/115 (48.8/51.2) | 26.5 | 189.2 | 7.14 | 42.1 | 4.76 | 0.48 | 1.48 | 7.06 | 1.722 | 2.739 | 106.3 |
| 22 | Chlorodifluoromethane | 22.1 | 172.9 | 7.81 | 66.8 | 3.00 | 0.307 | 2.32 | 6.95 | 1.589 | 2.967 | 149.8 |
| 717 | Ammonia | 16.0 | 169.3 | 10.61 | 463.9 | 0.43 | 0.087 | 16.7 | 7.19 | 1.569 | 3.007 | 285.6 |
| *Evaporator 20°F/Condenser 86°F* | | | | | | | | | | | | |
| 744 | Carbon dioxide | 421.9 | 1046.2 | 2.48 | 55.7 | 3.59 | 0.726 | 0.203 | 0.73 | 1.342 | 3.514 | 142.3 |
| 170 | Ethane | 293.6 | 675.1 | 2.3 | 70.1 | 2.85 | 1.238 | 0.421 | 1.20 | 1.314 | 3.588 | 115.8 |
| 32 | Difluoromethane | 94.7 | 279.6 | 2.95 | 111.2 | 1.80 | 0.229 | 0.902 | 1.62 | 0.797 | 5.924 | 139.4 |
| 410A | R-32/125 (50/50) | 93.2 | 273.6 | 2.94 | 73.5 | 2.72 | 0.316 | 0.651 | 1.77 | 0.815 | 5.78 | 115.8 |
| 507A | R-125/143a (50/50) | 72.9 | 211.7 | 2.9 | 49.4 | 4.05 | 0.476 | 0.616 | 2.50 | 0.848 | 5.564 | 93.5 |
| 404A | R-125/143a/134a (44/52/4) | 70.5 | 206.1 | 2.92 | 51.1 | 3.92 | 0.46 | 0.649 | 2.54 | 0.842 | 5.598 | 94.3 |
| 1270 | Propylene | 69.1 | 189.3 | 2.74 | 126.6 | 1.58 | 0.381 | 1.58 | 2.50 | 0.79 | 5.975 | 102.8 |
| 502 | R-22/115 (48.8/51.2) | 66.3 | 189.2 | 2.86 | 47.1 | 4.25 | 0.429 | 0.619 | 2.63 | 0.813 | 5.799 | 95.8 |
| 22 | Chlorodifluoromethane | 57.8 | 172.9 | 2.99 | 71.3 | 2.80 | 0.287 | 0.935 | 2.62 | 0.772 | 6.105 | 118.0 |
| 407C | R-32/125/134a (23/25/52) | 57.5 | 183.7 | 3.19 | 71.9 | 2.78 | 0.296 | 0.942 | 2.62 | 0.795 | 5.93 | 111.0 |
| 290 | Propane | 55.8 | 156.5 | 2.8 | 124.1 | 1.61 | 0.399 | 1.89 | 3.05 | 0.787 | 5.987 | 94.8 |
| 717 | Ammonia | 48.2 | 169.3 | 3.51 | 478.5 | 0.42 | 0.084 | 5.91 | 2.47 | 0.754 | 6.254 | 179.8 |
| 1234yf | 2,3,3,3-tetrafluoropropene* | 36.3 | 113.6 | 3.13 | 51.8 | 3.86 | 0.43 | 1.15 | 4.44 | 0.809 | 5.835 | 86.0 |
| 134a | Tetrafluoroethane | 33.1 | 111.7 | 3.37 | 65.8 | 3.04 | 0.307 | 1.41 | 4.28 | 0.778 | 6.063 | 94.7 |
| 1234ze(E) | Trans-1,3,3,3-tetrafluoropropene* | 24.4 | 83.9 | 3.44 | 60.0 | 3.33 | 0.349 | 1.74 | 5.81 | 0.782 | 6.03 | 86.0 |
| 600a | Isobutane* | 17.9 | 58.7 | 3.29 | 119.5 | 1.67 | 0.368 | 4.78 | 7.99 | 0.764 | 6.171 | 86.0 |
| *Evaporator 45°F/Condenser 86°F* | | | | | | | | | | | | |
| 32 | Difluoromethane | 147.7 | 279.6 | 1.89 | 112.2 | 1.78 | 0.223 | 0.577 | 1.03 | 0.445 | 10.602 | 116.4 |
| 410A | R-32/125 (50/50) | 145.0 | 273.6 | 1.89 | 75.2 | 2.66 | 0.308 | 0.416 | 1.11 | 0.455 | 10.379 | 103.7 |
| 502 | R-22/115 (48.8/51.2) | 102.0 | 189.2 | 1.85 | 49.6 | 4.03 | 0.407 | 0.404 | 1.63 | 0.451 | 10.474 | 91.8 |
| 407C | R-32/125/134a (23/25/52) | 92.8 | 183.7 | 1.98 | 74.7 | 2.68 | 0.284 | 0.588 | 1.57 | 0.443 | 10.655 | 102.7 |
| 22 | Chlorodifluoromethane | 90.8 | 172.9 | 1.9 | 73.5 | 2.72 | 0.279 | 0.604 | 1.64 | 0.433 | 10.885 | 104.5 |
| 290 | Propane | 85.3 | 156.5 | 1.84 | 130.7 | 1.53 | 0.379 | 1.26 | 1.92 | 0.439 | 10.743 | 90.7 |
| 717 | Ammonia | 81.0 | 169.3 | 2.09 | 484.9 | 0.41 | 0.083 | 3.61 | 1.49 | 0.421 | 11.186 | 137.4 |
| 500 | R-12/152a (73.8/26.2) | 66.5 | 127.6 | 1.92 | 64.7 | 3.09 | 0.331 | 0.725 | 2.24 | 0.432 | 10.925 | 94.2 |
| 1234yf | 2,3,3,3-tetrafluoropropene* | 58.1 | 113.6 | 1.96 | 55.5 | 3.61 | 0.402 | 0.726 | 2.62 | 0.444 | 10.623 | 86.0 |
| 12 | Dichlorodifluoromethane | 56.3 | 107.9 | 1.92 | 54.6 | 3.67 | 0.34 | 0.719 | 2.64 | 0.429 | 11.004 | 91.6 |
| 134a | Tetrafluoroethane | 54.7 | 111.7 | 2.04 | 69.2 | 2.89 | 0.292 | 0.868 | 2.51 | 0.433 | 10.903 | 90.6 |
| 1234ze(E) | Trans-1,3,3,3-tetrafluoropropene* | 40.6 | 83.9 | 2.06 | 64.1 | 3.12 | 0.327 | 1.07 | 3.34 | 0.433 | 10.899 | 86.0 |
| 600a | Isobutane* | 29.2 | 58.7 | 2.01 | 127.4 | 1.57 | 0.345 | 3.01 | 4.72 | 0.425 | 11.084 | 86.0 |
| 600 | Butane* | 19.5 | 41.1 | 2.11 | 140.5 | 1.42 | 0.301 | 4.57 | 6.50 | 0.42 | 11.226 | 86.0 |
| 123 | Dichlorotrifluoroethane | 6.5 | 15.9 | 2.44 | 66.9 | 2.99 | 0.246 | 5.3 | 15.85 | 0.414 | 11.397 | 86.0 |
| 113 | Trichlorotrifluoroethane* | 3.1 | 7.9 | 2.57 | 59.2 | 3.38 | 0.26 | 9.41 | 31.81 | 0.413 | 11.409 | 86.0 |

*Superheat required
*Source*: Data from NIST CYCLE_D 4.0, zero subcool, zero superheat unless noted, no line losses, 100% efficiencies, average temperatures.

mide-water absorption units are common to the two-stage units. The advantage of the dual-effect unit is higher performance, with steam rates approximately two-thirds those of single-stage machines. Heat source temperature for the dual-effect unit is over 120°F (67°C) higher than for the single-effect unit, requiring higher steam pressures.

## 18.3  Cooling Towers

A cooling tower, through a combination of mass and energy transfer, cools water by exposing it as an extended surface to the atmosphere. Water to be cooled is distributed in the tower by spray nozzles, splash bars, or film-type fill, which exposes a very large water surface area to atmospheric air. The airflow may be caused by mechanical means, by convection currents due to variation in density, or by natural wind currents. The airflow is either crossflow or counterflow. *Crossflow* describes air flowing horizontally in the filled portion of the tower or normal to the water flow, whereas *counterflow* implies the airstream rises vertically or countercurrent to a falling stream of water.

**Counterflow mechanical-draft towers** are principally found in air-conditioning applications. The main advantage of counterflow is its adaptability to restrictive space limitations. Factory-assembled towers often use centrifugal blowers in forced-draft configurations. The field-erected designs are usually induced-draft units with axial flow fans.

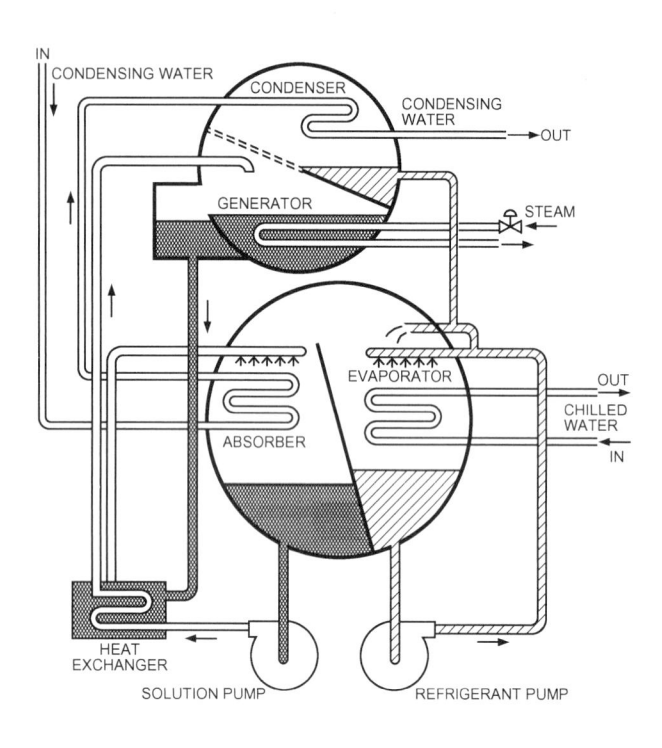

*Fig. 18-38 Two-Shell Lithium Bromide Cycle Water Chiller*
*(Figure 2, Chapter 18, 2010 ASHRAE Handbook—Refrigeration)*

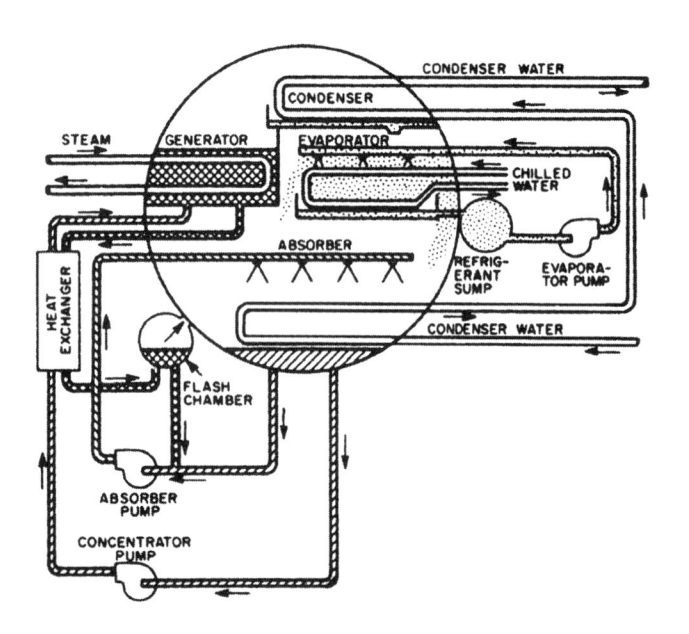

*Fig. 18-39 Diagram of One-Shell Lithium Bromide Cycle Water Chiller*

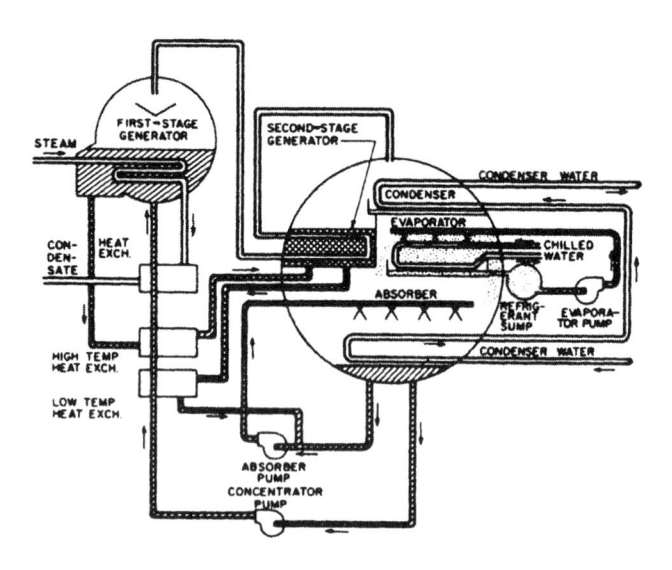

*Fig. 18-40 Diagram of One-Shell Lithium Bromide Cycle Water Chiller with Two-Stage Generator*

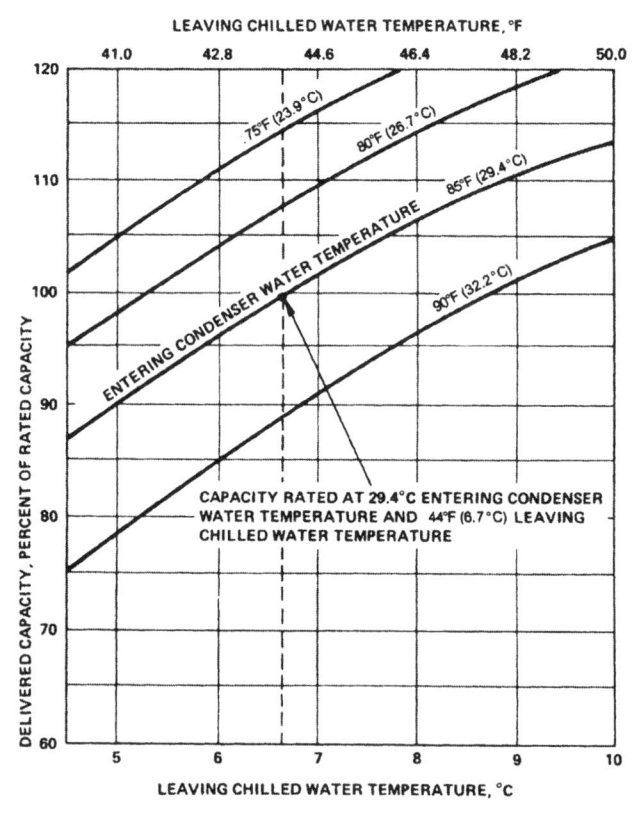

*Fig. 18-41 Performance Characteristics of Lithium Bromide Cycle Water Chiller*

# Refrigeration Equipment

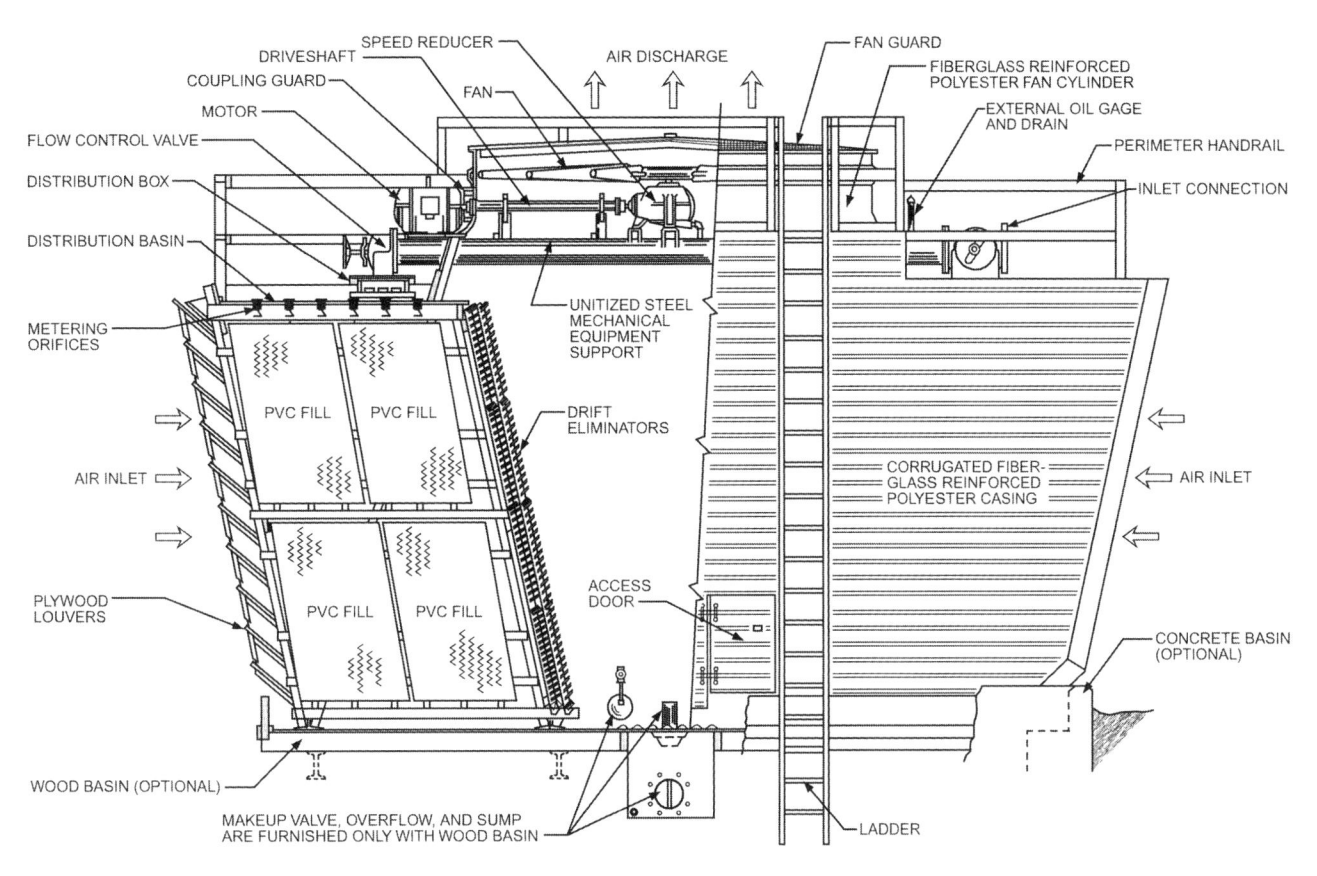

*Fig. 18-42 Crossflow Induced-Draft Tower*
(Figure 13, Chapter 40, 2012 ASHRAE Handbook—HVAC Systems and Equipment)

**Crossflow towers** are widely used in air-conditioning, process, and industrial applications. Crossflow towers have: (1) low air-side pressure drop in relation to high transfer surface areas and (2) the inherent capability to obtain uniform distributional characteristics of both the air and water streams.

The thermal capability of any cooling tower may be defined by the following parameters:

- Entering and leaving water temperatures
- Entering air wet-bulb temperature
- Water flow rate

The variations in tower performance associated with changes in these parameters are discussed in Chapter 40 of the 2012 *ASHRAE Handbook—HVAC Systems and Equipment*.

The thermal capability of cooling towers for air-conditioning applications is usually stated in terms of nominal refrigeration tonnage based on heat dissipation of 15,000 Btu/h (1.25 kW/kW) per ton and a water circulation rate of 3 gpm per ton (0.054 L/s per kW) cooled from 95 to 85°F (35 to 39.4°C) at 78°F (25.6°C) wet-bulb temperature. For industrial applications, nominal tonnage ratings are not used and the performance capability of the cooling tower is usually stated in terms of flow rate at specified operating conditions (entering and leaving water temperature and entering air wet-bulb temperature).

Fans in the mechanical-draft tower provide a positive and constant airflow. Since performance does not depend on the wind, mechanical-draft towers may be designed for exacting conditions. The fans may operate to provide forced or induced draft, depending on their location at the inlet or outlet of the tower. The tower may be crossflow (Figure 18-42) or counterflow (Figure 18-43). The addition of the fan makes it possible to design wider towers that are more compact than the tall, narrow atmospheric towers.

Factory-assembled cooling towers in both crossflow and counterflow designs are available. Water distribution is by gravity or low-pressure flume with crossflow design, and spray nozzles are used on counterflow units.

A major consideration in the selection of cooling towers is the power requirement per ton of refrigeration, since the tower is a parasitic energy burden.

## 18.3.1 Spray Ponds

Heat dissipates from the surface of a body of water by evaporation, radiation, and convection. A spray pond divides the water into small droplets, greatly extending the water surface and bringing it into contact with the air. Heat transfer is largely due to evaporative cooling. Temperature control, large space requirements, limited ability to approach the wet-bulb temperature, and winter operational difficulties have generally ruled out the spray pond in favor of more compact and more controllable mechanical-draft or hyperbolic towers.

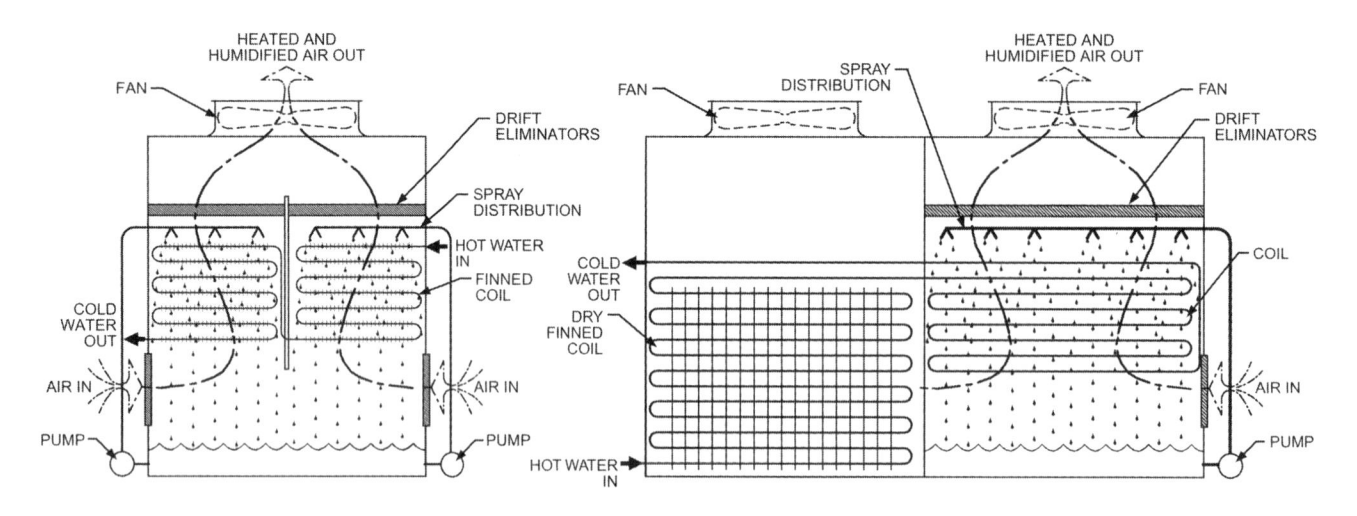

*Fig. 18-43 Counterflow Forced-Draft Cooling Tower*
(Figure 16, Chapter 40, 2012 ASHRAE Handbook—HVAC Systems and Equipment)

## 18.4 Problems

**18.1** A condenser used in a refrigeration system has a capacity of 10 tons at a 40°F evaporating temperature. When 20 gpm of cooling water enters at 75°F, the condensing temperature is 90°F. The manufacturer claims a U-factor of 95 Btu/h·ft²·°F, with a heat transfer area of 83 ft². Are these claims reasonable? Why?

**18.2** Given a compressor using R-22 condensing at 80°F (26.7°C) and evaporating at 20°F (–6.7°C), find the enthalpy of the refrigerant when it enters the

(a) compressor

(b) condenser

(c) evaporator

Find the power required for the compressor.

[Ans: (a) 106.4, (b) 116.5, (c) 33.1 Btu/lb, and 0.65 HP/ton]

**18.3** What is the maximum theoretical COP of a refrigeration device operating between 0°F and 75°F (–17.8°C and 23.9°C). Why is this theoretical limit difficult to obtain? [Ans: 6.14]

**18.4** A reference book on refrigeration indicates that a compressor using R-22 requires a displacement of 40.59 cfm per ton for evaporation at –100°F and condensing at –30°F. Is this correct? Substantiate your answer with calculations based on knowledge of R-22 for these conditions. Also, verify the mass flow rate in lb per min.

**18.5** An R-134a refrigerating system develops 10 tons of refrigeration when operating at 100°F condensing and +10°F evaporating, with no liquid subcooling or vapor superheating. Determine the volume of the refrigerant leaving the expansion valve in cubic feet per minute.

**18.6** An expansion device has a mass flow rate for R-134a given by

$$m = 60 + 0.25\, \Delta p$$

where $m$ = flow rate in lb/min and $\Delta p$ = pressure drop across the valve in psi.

For an evaporator temperature of 0°F and a condenser temperature of 100°F, estimate the piston displacement required for a compressor if $C = 0.04$ and the polytropic compression coefficient $n = 1.1$ for the compression process. [Ans: 236 cfm]

**18.7** A liquid-to-suction heat exchanger is installed in an R-134a system to cool liquid that comes from the condenser with vapor that flows from the evaporator. The evaporator generates 10 tons (35.17 kW) of refrigeration at 30°F (–1.1°C). Liquid leaves the condenser saturated at 100°F (37.8°C), vapor leaves the evaporator saturated, and vapor leaves the heat exchanger at a temperature of 50°F (10°C). What is the flow rate of the refrigerant?

**18.8** An eight-cylinder ammonia compressor is designed to operate at 800 rpm and deliver 30 tons of refrigeration. The evaporator is to operate at 10°F with a condensing temperature of 100°F. The vapor enters the compressor at 30°F. The ammonia leaves the condenser as saturated liquid. If the average piston speed is to be 600 ft/min and the actual volumetric efficiency at this condition is 83%, find the bore of the compressor.

**18.9** A condenser is to be selected for a system that generates 30 tons (105.5 kW) of refrigeration at 10°F (–12.2°C). The condenser is to operate at 110°F (43.3°C) and is cooled with 90 gpm (5.68 L/s) of water at 85°F (29.4°C). If the expected *U-factor* of the condenser is 130 Btu/h·ft²·°F [738 W/(m²·K)], calculate the condensing area required.

**18.10** A cooling tower cools water by passing it through a stream of air. If 1000 cfm of air at 95°F dry bulb and 78°F wet bulb enters the tower and leaves saturated at 84°F, to what temperature can this air cool water that enters at 110°F with a flow of 80 lb/min? What is the makeup water rate?

## 18.5 Bibliography

ASHRAE. 2012. *2012 ASHRAE Handbook—HVAC Systems and Equipment.*

ASHRAE. 2013. *2013 ASHRAE Handbook—Fundamentals.*

Sauer, H.J., Jr. and Howell, R.H. 1983. *Heat Pump Systems.* Wiley Interscience, New York.

Stoecker, W. and Jones, J. 1987. *Refrigeration and Air Conditioning*, 3rd ed. McGraw-Hill, New York.

## SI Tables

### Table 18-3 SI  Types of Coolers

| Type of Cooler | Subtype | Usual Refrigerant Feed Device | Usual Capacity Range, kW | Commonly Used Refrigerants |
|---|---|---|---|---|
| Direct-expansion | Shell-and-tube | Thermal expansion valve | 7 to 1800 | 12, 22, 134a, 404A, 407C, 410A, 500, 502, 507A, 717 |
|  |  | Electronic modulation valve | 7 to 1800 |  |
|  | Tube-in-tube | Thermal expansion valve | 18 to 90 | 12, 22, 134a, 717 |
|  | Brazed-plate | Thermal expansion valve | 2 to 700 | 12, 22, 134a, 404A 407C, 410A, 500, 502, 507A, 508B, 717, 744 |
|  | Semiwelded plate | Thermal expansion valve | 175 to 7000 | 12, 22, 134a, 500, 502, 507A, 717, 744 |
| Flooded | Shell-and-tube | Low-pressure float | 90 to 7000 | 11, 12, 22, 113, 114 |
|  |  | High-pressure float | 90 to 21 100 | 123, 134a, 500, 502, 507A, 717 |
|  |  | Fixed orifice(s) | 90 to 21 100 |  |
|  |  | Weir | 90 to 21 100 |  |
|  | Spray shell-and-tube | Low-pressure float | 180 to 35 000 | 11, 12, 13B1, 22 |
|  |  | High-pressure float | 180 to 35 000 | 113, 114, 123, 134a |
|  | Brazed-plate | Low-pressure float | 2 to 700 | 12, 22, 134a, 500, 502, 507A, 717, 744 |
|  | Semiwelded plate | Low-pressure float | 175 to 7000 | 12, 22, 134a, 500, 502, 507A, 717, 744 |
| Baudelot | Flooded | Low-pressure float | 35 to 350 | 22, 717 |
|  | Direct-expansion | Thermal expansion valve | 18 to 90 | 12, 22, 134a, 717 |
| Shell-and-coil | — | Thermal expansion valve | 7 to 35 | 12, 22, 134a, 717 |

### Table 18-6 SI  Physical Properties of Selected Refrigerants[a]
(Table 3, chapter 19, *2005 ASHRAE Handbook—Fundamentals* SI)

| No. | Chemical Name or Composition (% by Mass) | Chemical Formula | Molecular Mass | Boiling Pt. (NBP) at 101.325 kPa, °C | Freezing Point, °C | Critical Temperature, °C | Critical Pressure, kPa | Critical Density, kg/m$^3$ | Refractive Index of Liquid[b,c] |
|---|---|---|---|---|---|---|---|---|---|
| 728 | Nitrogen | N$_2$ | 28.013 | −195.8 | −210.0 | −146.96 | 3395.8 | 313.3 | 1.205 (83 K) 589.3 nm |
| 729 | Air | — | 28.959 | −194.25 | — | −140.59 | 3789.6 | 335.94 | — |
| 740 | Argon | Ar | 39.948 | −185.85 | −189.34 | −122.46 | 4863.0 | 535.6 | 1.233 (84 K) 589.3 nm |
| 732 | Oxygen | O$_2$ | 31.999 | −182.96 | −218.79 | −118.57 | 5043.0 | 436.14 | 1.221 (92 K) 589.3 nm |
| 50 | Methane | CH$_4$ | 16.043 | −161.48 | −182.46 | −82.586 | 4599.2 | 162.66 | — |
| 14 | Tetrafluoromethane | CF$_4$ | 88.005 | −128.05 | −183.61 | −45.64 | 3750.0 | 625.66 | — |
| 170 | Ethane | C$_2$H$_6$ | 30.07 | −88.598 | −182.8 | 32.18 | 4871.8 | 206.58 | — |
| 503 | R-23/13 (40.1/59.9) | — | 87.247 | −87.76 | — | 18.417 | 4280.5 | 565.68 | — |
| 508A[4] | R-23/116 (39/61) | — | 100.1 | −87.377 | — | 10.844 | 3668.2 | 570.62 | — |
| 508B[4] | R-23/116 (46/54) | — | 95.394 | −87.344 | — | 11.827 | 3789 | 572.13 | — |
| 23 | Trifluoromethane | CHF$_3$ | 70.014 | −82.018 | −155.13 | 26.143 | 4832 | 526.5 | — |
| 13 | Chlorotrifluoromethane | CClF$_3$ | 104.46 | −81.48 | −181.15 | 28.85 | 3879 | 582.88 | 1.146 (25)[2] |
| 744 | Carbon dioxide | CO$_2$ | 44.01 | −78.4[d] | −56.558[e] | 30.978 | 7377.3 | 467.6 | 1.195 (15) |
| 504 | R-32/115 (48.2/51.8) | — | 79.249 | −57.695 | — | 61.084 | 433.7 | 504.62 | — |
| 32 | Difluoromethane | CH$_2$F$_2$ | 52.024 | −51.651 | −136.81 | 78.105 | 5782.0 | 424 | — |
| 410A | R-32/125 (50/50) | — | 72.585 | −51.443 | — | 71.358 | 4902.6 | 459.53 | — |
| 125 | Pentafluoroethane | C$_2$HF$_5$ | 120.02 | −48.09 | −100.63 | 66.023 | 3617.7 | 573.58 | — |
| 1270 | Propylene | C$_3$H$_6$ | 42.08 | −47.69 | −185.2 | 92.42 | 4664.6 | 223.39 | 1.3640 (−50)[1] |
| 143a | Trifluoroethane | CH$_3$CF$_3$ | 84.041 | −47.241 | −111.81 | 72.707 | 3761.0 | 431.0 | — |
| 507A | R-125/143a (50/50) | — | 98.859 | −46.741 | — | 70.617 | 3705 | 490.77 | — |
| 404A | R-125/143a/134a (44/52/4) | — | 97.604 | −46.222 | — | 72.046 | 3728.9 | 486.53 | — |
| 502 | R-22/115 (48.8/51.2) | — | 111.63 | −45.174 | — | 80.153 | 3917.6 | 566.03 | — |
| 407C | R-32/125/134a (23/25/52) | — | 86.204 | −43.627 | — | 86.034 | 4629.8 | 484.23 | — |
| 290 | Propane | C$_3$H$_8$ | 44.096 | −42.09 | −187.67 | 96.675 | 4247.1 | 218.5 | 1.3397 (−42) |
| 22 | Chlorodifluoromethane | CHClF$_2$ | 86.468 | −40.81 | −157.42 | 96.145 | 4990.0 | 523.84 | 1.234 (25)[2] |
| 115 | Chloropentafluoroethane | CClF$_2$CF$_3$ | 154.47 | −38.94 | −99.39 | 79.95 | 3120.0 | 613.1 | 1.221 (25)[2] |
| 500 | R-12/152a (73.8/26.2) | — | 99.303 | −33.603 | — | 102.09 | 4168.6 | 495.1 | — |
| 717 | Ammonia | NH$_3$ | 17.03 | −33.327 | −77.655 | 132.25 | 11333.0 | 225.0[d] | 1.325 (16.5) |
| 12 | Dichlorodifluoromethane | CCl$_2$F$_2$ | 120.91 | −29.752 | −157.05 | 111.97 | 4136.1 | 565.0 | 1.288 (25)[2] |
| 134a | Tetrafluoroethane | CF$_3$CH$_2$F | 102.03 | −26.074 | −103.3 | 101.06 | 4059.3 | 511.9 | — |
| 152a | Difluoroethane | CHF$_2$CH$_3$ | 66.051 | −24.023 | −118.59 | 113.26 | 4516.8 | 368 | — |
| 124 | Chlorotetrafluoroethane | CHClFCF$_3$ | 136.48 | −11.963 | −199.15 | 122.28 | 3624.3 | 560.0 | — |
| 600a | Isobutane | C$_4$H$_{10}$ | 58.122 | −11.67 | −159.59 | 134.67 | 3640.0 | 224.35 | 1.3514 −25)[1] |
| 142b | Chlorodifluoroethane | CClF$_2$CH$_3$ | 100.5 | −9.15 | −130.43 | 137.11 | 4070.0 | 446.0 | — |
| C318 | Octafluorocyclobutane | C$_4$F$_8$ | 200.03 | −5.975 | −39.8 | 115.23 | 2777.5 | 619.97 | |
| 600 | Butane | C$_4$H$_{10}$ | 58.122 | −0.55 | −138.28 | 151.98 | 3796.0 | 227.84 | 1.3562 (−15)[1] |
| 114 | Dichlorotetrafluoroethane | CClF$_2$CClF$_2$ | 170.92 | 3.586 | −94.15 | 145.68 | 3257.0 | 579.97 | 1.294 (25) |
| 11 | Trichlorofluoromethane | CCl$_3$F | 137.37 | 23.708 | −110.47 | 197.96 | 4407.6 | 554.0 | 1.362 (25)[2] |
| 123 | Dichlorotrifluoroethane | CHCl$_2$CF$_3$ | 152.93 | 27.823 | −107.15 | 183.68 | 3661.8 | 550.0 | — |
| 141b | Dichlorotrifluoroethane | CCl$_2$FCH$_3$ | 116.95 | 32.05 | −103.3 | 206.81 | 4460.0 | 460.0 | — |
| 113 | Trichlorotrifluoroethane | CCl$_2$FCClF$_2$ | 187.38 | 47.585 | −36.22 | 214.06 | 3392.2 | 560.0 | 1.357 (25)[2] |
| 718[3] | Water | H$_2$O | 18.015 | 99.974 | 0.01 | 373.95 | 22 064.0 | 322.0 | — |

Note:
[a] Data from *ASHRAE Thermodynamic Properties of Refrigerants* (Stewart et al. 1986) or from Lemmon et al. (2002), unless otherwise noted.
[b] Temperature of measurement (°C, unless kelvin is noted) shown in parentheses. Data from *CRC Handbook of Chemistry and Physics* (CRC 1987), unless otherwise noted.
[c] For the sodium D line.
[d] Sublimes.
[e] At 527 kPa.

References:
[1] Kirk and Othmer (1956).
[2] *Bulletin B-32A* (DuPont).
[3] *Handbook of Chemistry* (1967).
[4] *NIST Standard Reference Database 23*, v.7.

Table 18-7 SI  Comparative Refrigerant Performance per Kilowatt of Refrigeration
(Table 7, chapter 19, *2005 ASHRAE Handbook—Fundamentals* SI)

| No. | Refrigerant Chemical Name or Composition (% by mass) | Evaporator Pressure, MPa | Condenser Pressure, MPa | Compression Ratio | Net Refrigerating Effect, kJ/kg | Refrigerant Circulated, g/s | Liquid Circulated, L/s | Specific Volume of Suction Gas, m³/kg | Compressor Displacement, L/s | Power Consumption, kW | Coefficient of Performance | Compressor Discharge Temp., K |
|---|---|---|---|---|---|---|---|---|---|---|---|---|
| 170 | Ethane | 1.608 | 4.639 | 2.88 | 161.71 | 6.10 | 0.0219 | 0.0338 | 0.206 | 0.365 | 2.70 | 323 |
| 744 | Carbon dioxide | 2.254 | 7.18 | 3.19 | 133.23 | 3.88 | 0.0064 | 0.0168 | 0.065 | 0.192 | 2.69 | 343 |
| 1270 | Propylene | 0.358 | 1.304 | 3.64 | 286.17 | 3.46 | 0.0070 | 0.1299 | 0.449 | 0.220 | 4.50 | 315 |
| 290 | Propane | 0.286 | 1.075 | 3.76 | 277.90 | 3.53 | 0.0073 | 0.1562 | 0.551 | 0.218 | 4.50 | 309 |
| 502 | R-22/115 (48.8/51.2) | 0.343 | 1.312 | 3.83 | 105.95 | 9.43 | 0.0079 | 0.0508 | 0.479 | 0.228 | 4.38 | 311 |
| 507A | R-125/143a (50/50) | 0.379 | 1.459 | 3.85 | 110.14 | 9.07 | 0.0089 | 0.0508 | 0.461 | 0.239 | 4.18 | 308 |
| 404A | R-125/143a/134a (44/52/4) | 0.365 | 1.42 | 3.89 | 114.15 | 8.75 | 0.0086 | 0.0537 | 0.470 | 0.237 | 4.21 | 309 |
| 410A | R-32/125 (50/50) | 0.478 | 1.872 | 3.92 | 167.89 | 5.84 | 0.0056 | 0.0545 | 0.318 | 0.222 | 4.41 | 324 |
| 125 | Pentafluoroethane | 0.403 | 1.561 | 3.87 | 85.30 | 11.41 | 0.0098 | 0.0394 | 0.449 | 0.244 | 3.99 | 304 |
| 22 | Chlorodifluoro-methane | 0.295 | 1.187 | 4.02 | 162.67 | 6.13 | 0.0052 | 0.0779 | 0.478 | 0.214 | 4.66 | 326 |
| 12 | Dichlorodifluoro-methane | 0.181 | 0.741 | 4.09 | 117.02 | 8.49 | 0.0066 | 0.0923 | 0.784 | 0.212 | 4.70 | 311 |
| 500 | R-12/152a (73.8/26.2) | 0.214 | 0.876 | 4.09 | 139.68 | 7.08 | 0.0063 | 0.0939 | 0.665 | 0.212 | 4.66 | 314 |
| 407C | R-32/125/134a (23/25/52) | 0.288 | 1.26 | 4.38 | 163.27 | 6.11 | 0.0054 | 0.0805 | 0.492 | 0.222 | 4.50 | 321 |
| 600a | Isobutane* | 0.088 | 0.403 | 4.58 | 263.91 | 3.76 | 0.0069 | 0.4073 | 1.533 | 0.215 | 4.62 | 303 |
| 134a | Tetrafluoroethane | 0.163 | 0.767 | 4.71 | 148.03 | 6.71 | 0.0056 | 0.1214 | 0.814 | 0.216 | 4.60 | 310 |
| 124 | Chlorotetrafluoro-ethane* | 0.088 | 0.443 | 5.03 | 117.83 | 8.41 | 0.0063 | 0.1711 | 1.439 | 0.214 | 4.62 | 303 |
| 717 | Ammonia | 0.235 | 1.162 | 4.94 | 1103.14 | 0.90 | 0.0015 | 0.5117 | 0.463 | 0.210 | 4.76 | 372 |
| 600 | Butane* | 0.056 | 0.283 | 5.05 | 292.24 | 3.53 | 0.0062 | 0.6446 | 2.274 | 0.218 | 4.74 | 303 |
| 11 | Trichlorofluoro-methane | 0.02 | 0.125 | 6.25 | 155.95 | 6.36 | 0.0043 | 0.7689 | 4.891 | 0.197 | 5.02 | 316 |
| 123 | Dichlorotrifluoro-ethane | 0.016 | 0.109 | 6.81 | 142.28 | 7.02 | 0.0048 | 0.8914 | 6.259 | 0.204 | 4.90 | 306 |
| 113 | Trichlorotrifluoro-ethane* | 0.007 | 0.054 | 7.71 | 122.58 | 7.84 | 0.0051 | 1.6818 | 13.187 | 0.200 | 4.81 | 303 |

*Superheat required.

# Chapter 19

# HEATING EQUIPMENT

In this chapter the principles of combustion and data concerning various types of fuels common to HVAC systems are discussed. The information in this chapter has been extracted from Chapter 28 in the 2013 *ASHRAE Handbook—Fundamentals* and Chapters 28 and 31 through 37 in the 2012 *ASHRAE Handbook—HVAC Systems and Equipment*.

## 19.1 Fuels and Combustion

### 19.1.1 Principles of Combustion

**Combustion** is the chemical process in which an oxidant reacts rapidly with a fuel to liberate stored energy as thermal energy, generally in the form of high-temperature gases. The oxidant is usually oxygen in the air.

Conventional hydrocarbon fuels contain primarily hydrogen and carbon, in either elemental form or in various compounds. Their complete combustion produces mainly carbon dioxide and water; however, small quantities of carbon monoxide and partially reacted flue gas constituents may form. Most conventional fuels also contain small amounts of sulfur, oxidized to $SO_2$ or $SO_3$ during combustion, and noncombustible substances (mineral matter or ash, water, and inert gases). Flue gas is the product of complete or incomplete combustion, including excess air (if present) but not dilution air.

The rate of fuel combustion depends on (1) the chemical reaction rate of combustible fuel constituents with oxygen, (2) the rate at which oxygen is supplied to fuel (mixing of air and fuel), and (3) the temperature in the combustion region. The reaction rate is fixed by fuel selection. Increasing the mixing rate or temperature increases the combustion rate. In complete combustion of hydrocarbon fuels, all hydrogen and carbon in the fuel are oxidized to $H_2O$ and $CO_2$.

For complete combustion, excess oxygen or excess air must generally be supplied beyond the amount theoretically required to oxidize the fuel; this is usually expressed as a percentage of the air required to completely oxidize the fuel. Incomplete combustion occurs when a fuel element is not completely oxidized in the combustion process. For example, a hydrocarbon may not completely oxidize to carbon dioxide and water, but it may form partially oxidized compounds such as carbon monoxide, aldehydes, and ketones. Conditions that promote incomplete combustion include

- Insufficient air and fuel mixing, causing local fuel-rich and fuel-lean zones.
- Insufficient air supply to the flame, providing less than the required quantity of oxygen.
- Insufficient reactant residence time in the flame, preventing completion of combustion reactions.
- Flame impingement on a cold surface, quenching combustion reactions.
- Too-low flame temperature, slowing combustion reactions.

Incomplete combustion uses fuel inefficiently, contributes to air pollution, and can be hazardous because of carbon monoxide production.

Combustion of oxygen with the combustible elements and compounds in fuels occurs according to fixed chemical principles, including:

- Chemical reaction equations.
- Law of matter conservation. The mass of each element in the reaction products must equal the mass of that element in the reactants.
- Law of combining mass. Chemical compounds are formed by elements combining in fixed mass relationships.
- Chemical reaction rates.

Oxygen for combustion is normally obtained from air, which is a physical mixture of nitrogen, oxygen, small amounts of water vapor, carbon dioxide, and inert gases. For practical combustion calculations, dry air consists of 20.95% oxygen and 79.05% inert gases, primarily nitrogen, by volume, or 23.15% oxygen and 76.85% inert gases by mass. For calculation purposes, nitrogen is assumed to pass through the combustion process unchanged (although small quantities of nitrogen oxides are known to form).

The quantity of heat generated by complete combustion of a unit of specific fuel is constant and is called the **heating value**, **heat of combustion**, or **caloric value** of that fuel. The heating value of a fuel can be determined by measuring the heat evolved during combustion of a known quantity of the fuel in a calorimeter, or it can be estimated from chemical analysis of the fuel and the heating value of the several chemical elements in the fuel.

Higher heating value, gross heating value, or total heating value is determined when water vapor in fuel combustion products is condensed and the latent heat of vaporization is included in the fuel's heating value. Conversely, lower heating value or net heating value is obtained when latent heat of vaporization is not included. When the heating value of a fuel is specified without designating higher or lower, it generally means the higher heating value in the United States. (lower heating value is mainly used for internal combustion engine fuels.)

Table 19-1  Heating Values of Selected Fuels

| Substance | Molecular Symbol | Higher Heating Values,[a] Btu/lb | Lower Heating Values,[a] Btu/lb | Specific Volume,[b] ft³/lb | Higher Heating Values,[c] MJ/kg | Lower Heating Values,[c] MJ/kg | Density,[d] kg/m³ |
|---|---|---|---|---|---|---|---|
| Carbon (to CO) | C | 3950 | 3950 | | 9.188 | 9.188 | |
| Carbon (to $CO_2$) | C | 14,093 | 14,093 | | 32.788 | 32.780 | |
| Carbon Monoxide | CO | 4347 | 4347 | 13.5 | 10.111 | 10.111 | 1.187 |
| Hydrogen | $H_2$ | 61,095 | 51,623 | 188.0 | 142.107 | 118.680 | 0.085 |
| Methane | $CH_4$ | 23,875 | 21,495 | 23.6 | 55.533 | 49.997 | 0.679 |
| Ethane | $C_2H_6$ | 22,323 | 20,418 | 12.5 | 51.922 | 47.492 | 1.28 |
| Propane | $C_3H_8$ | 21,669 | 19,937 | 8.36 | 50.402 | 46.373 | 1.92 |
| Butane | $C_4H_{10}$ | 21,321 | 19,678 | 6.32 | 49.593 | 45.771 | 2.53 |
| Ethylene | $C_2H_4$ | 21,636 | 20,275 | | 50.325 | 47.160 | |
| Propylene | $C_3H_6$ | 21,048 | 19,687 | 9.01 | 48.958 | 45.792 | 1.78 |
| Acetylene | $C_2H_2$ | 21,502 | 20,769 | 14.3 | 50.028 | 48.309 | 1.120 |
| Sulfur (to $SO_2$) | S | 3980 | 3980 | | 9.257 | 9.257 | |
| Sulfur (to $SO_3$) | S | 5940 | 5940 | | 13.816 | 13.816 | 1.456 |
| Hydrogen Sulfide | $H_2S$ | 7097 | 6537 | 11.0 | 16.508 | 15.205 | |

a All values corrected to 60 °F, 30 in. Hg dry. For gases saturated with water vapor at 60°F, deduct 1.74% of the value to adjust for gas volume displaced by water vapor.
b At 32 °F and 29.92 in. Hg
c All values corrected to 16°C, 101.4 kPa dry. For gases saturated with water vapor at 16°F, deduct 1.74% of the value to adjust for gas volume displaced by water vapor.
d At 32 °F and 29.92 in. Hg

Heating values are usually expressed in Btu/ft³ (kJ/m³) for gaseous fuels, Btu/gal (kJ/m³) for liquid fuels, and Btu/lb (kJ/kg) for solid fuels. Heating values are always given in relation to a certain reference temperature and pressure, usually 60, 68, or 77°F and 14.696 psia (15, 20, or 25°C and 103.325 kPa), depending on the particular industry practice. Heating values of several substances and common fuels are listed in Table 19-1.

When combustion is incomplete, not all fuel is completely oxidized, and the heat released is less than the heating value of the fuel. Therefore, the quantity of heat produced per unit of fuel consumed decreases, implying lower combustion efficiency.

Not all heat released during combustion can be used effectively. The greatest heat loss is in the form of increased temperature (thermal energy) of hot exhaust gases above the temperature of incoming air and fuel. Other heat losses include radiative and convective heat transfer from outer walls of combustion equipment to the environment.

### 19.1.2 Fuels

Generally, hydrocarbon fuels are classified according to physical state (gaseous, liquid, or solid). Different types of combustion equipment are usually needed to burn fuels in different physical states. Gaseous fuels can be burned in premix or diffusion burners that take advantage of the gaseous state. Liquid fuel burners must include a means of atomizing or vaporizing fuel into small droplets or to a vapor for burning and must provide adequate mixing of fuel and air. Solid fuel combustion equipment must (1) heat fuel to vaporize sufficient volatiles to initiate and sustain combustion, (2) provide residence time to complete combustion, and (3) provide space for ash containment.

Principal uses of fuel include space heating and cooling of residential, commercial, industrial, and institutional buildings; service water heating; steam generation; and refrigeration. Fuels for these applications are natural and liquefied petroleum gases, fuel oils, diesel and gas turbine fuels (for total energy applications), and coal.

**Gaseous Fuels.** Heating and cooling applications are presently limited to natural and liquefied petroleum gases. Natural gas is a nearly odorless and colorless gas that accumulates in the upper parts of oil and gas wells. Raw natural gas is a mixture of methane (55 to 98%), higher hydrocarbons (primarily ethane), and noncombustible gases. Some constituents, principally water vapor, hydrogen sulfide, helium, liquefied petroleum gases, and gasoline, are removed prior to distribution.

Heating values of natural gases vary from 900 to 1200 Btu/ft³ (34 to 45 MJ/m³); the usual range is 1000 to 1050 Btu/ft³ (37 to 39 MJ/m³) at sea level. Three liquefied petroleum gases—butane, propane, and a mixture of the two—are commercially available.

Commercial propane consists primarily of propane but generally contains about 5 to 10% propylene. It has a heating value of about 21,560 Btu/lb (50.15 MJ/kg) or about 2500 Btu/ft³ (93 MJ/m³) of gas. At atmospheric pressure, commercial propane has a boiling point of about −40°F (−40°C). The low boiling point of propane makes it usable during winter in cold climates such as the northern United States. It is available in cylinders, bottles, tank trucks, or tank cars.

Commercial butane consists primarily of butane but may contain up to 5% butylene. It has a heating value of about 21,180 Btu/lb (49.26 MJ/kg) or 3200 Btu/ft³ (119 MJ/m³).

Commercial propane-butane mixtures with varying ratios of propane and butane are available. Their properties generally fall between those of the unmixed fuels. Propane-air and butane-air mixtures are used in place of natural gas in small communities and by natural gas companies at peak loads. Typical heating values are listed in Table 19-2.

**Liquid Fuels.** Liquid fuels, with few exceptions, are mixtures of hydrocarbons refined from crude petroleum. Fuel oils for heating are broadly classified as distillate fuel oils

**Table 19-2  Heating Values of Gaseous Fuels**

| Gas | Btu/ft$^3$ | MJ/m$^3$ | Specific Gravity Air = 1.0 |
|---|---|---|---|
| Natural | 1030 | 38.4 | 0.60 |
| Propane | 2500 | 93.1 | 1.53 |
| Butane | 3175 | 118.3 | 2.00 |

**Table 19-3  Typical API Gravity, Density, and Heating Value of Standard Grades of Fuel Oil**

| Grade No. | API Gravity | Density, lb/gal$^a$ | Heating Value, Btu/gal$^b$ |
|---|---|---|---|
| 1 | 38 to 45 | 6.950 to 6.675 | 137,000 to 132,900 |
| 2 | 30 to 38 | 7.296 to 6.960 | 141,800 to 137,000 |
| 4 | 20 to 28 | 7.787 to 7.396 | 148,100 to 143,100 |
| 5L | 17 to 22 | 7.940 to 7.686 | 150,000 to 146,800 |
| 5H | 14 to 18 | 8.080 to 7.890 | 152,000 to 149,400 |
| 6 | 8 to 15 | 8.448 to 8.053 | 155,900 to 151,300 |

$^a$1 lb/gal = 120 kg/m$^3$
$^b$1 Btu/gal = 279 kJ/m$^3$

(lighter oils) or residual fuel oils (heavier oils). ASTM has established specifications for fuel oil properties that subdivide the oils into various grades. Grades No. 1 and 2 are distillate fuel oils. Grades 4,5 (Light), 5 (Heavy), and 6 are residual fuel oils. Specifications for the grades are based on required characteristics of fuel oils for use in different types of burners. Typical gravity and heating values of standard grades of fuel oil are shown in Table 19-3.

**Solid Fuels.** Solid fuels include coal, coke, wood, and waste products of industrial and agricultural operations. Of these, only coal is widely used. Chemically, coal consists of carbon, hydrogen, oxygen, nitrogen, sulfur, and a mineral residue, ash. Heating values may be reported on an as-received, dry, dry and mineral-matter-free, or moist and mineral-matter-free basis. Higher heating values of coals are frequently reported with their proximate analysis. When more specific data are lacking, the higher heating value of higher quality coals can be calculated by the Dulong Formula:

Higher Heating Value (Btu/lb)
$$= 14{,}544C + 62{,}028[H - (O/8)] + 4050S \quad (19\text{-}1a)$$

or  Higher Heating Value (MJ/kg)
$$= 33.829C + 144.28[H - (O/8)] + 9.42S \quad (19\text{-}1b)$$

where C, H, O, and S are the mass fractions of carbon, hydrogen, oxygen, and sulfur in the coal. Typical values for coal are listed in Table 19-4.

### 19.1.3 Combustion Calculations

Calculations of the quantity of air required for combustion and quantity of flue gas products generated during combustion are frequently needed for sizing system components and as input to efficiency calculations. Other calculations, such as values for excess air and theoretical CO$_2$, are useful in estimating combustion system performance. Analysis of the flue gas products is often done using an Orsat analysis. This is a measure in percent by volume of the CO$_2$, CO, and O$_2$ in the dry products. The quantity of water vapor in the

**Table 19-4  Heating Value of Coal**

| | Heating Value As Received | |
|---|---|---|
| Rank | Btu/lb | MJ/kg |
| Anthracite | 12,700 | 29.5 |
| Semianthracite | 13,600 | 31.6 |
| Low-volatile bituminous | 14,350 | 33.4 |
| Medium-volatile bituminous | 14,000 | 32.6 |
| High-volatile bituminous A | 13,800 | 32.1 |
| High-volatile bituminous B | 12,500 | 29.1 |
| High-volatile bituminous C | 11,000 | 25.6 |
| Subbituminous B | 9000 | 20.9 |
| Subbituminous C | 8500 | 19.8 |
| Lignite | 6900 | 16.0 |

products is taken to be zero for this calculation. It is then assumed that the % N$_2$ in the products is 100% minus the % CO$_2$, % CO, and % O$_2$. Details on combustion calculations are given in Chapter 28 of the 2009 *ASHRAE Handbook—Fundamentals*.

**Example 19-1.** Carbon burns with 160% theoretical air. Combustion goes to completion. Determine: (a) the air/fuel ratio by mass and (b) the Orsat gas analysis and dew point of the products.

*Solution*:
*Theoretical*:
$$C + O_2 + 3.76N_2 \rightarrow CO_2 + 3.76N_2$$

*Actual*:
$$C + 1.6O_2 + 1.6(3.76)N_2 \rightarrow CO_2 + 6.02N_2 + 0.6O_2$$

(a) A/F = [1.6(32) + 1.6(3.76)28]/12 = 18.3 lb$_{air}$/lb$_{fuel}$

(b) 
| | | | |
|---|---|---|---|
| CO$_2$ | 1.0 moles/7.62 | = | 13.1% |
| O$_2$ | 0.6 moles/7.62 | = | 7.9% ORSAT |
| N$_2$ | 6.02 moles/7.6 | = | 79.0% |
| | 7.62 moles | = | 100.0% |

No dew point because there is no water vapor in the products.

*Note*: Assuming air is 79% N$_2$, 21% O$_2$ by volume:

$$1 \text{ mol } O_2 + 3.76 \text{ mol } N_2 = 4.76 \text{ mol air}$$

**Example 19-2.** The flue gas analysis of a hydrocarbon fuel on a percent by volume dry basis shows CO$_2$ = 12.4%, O$_2$ = 3.2%, CO = 0.1%, H$_2$ = 0.2%, and N$_2$ = 84.1%. Determine the air/fuel ratio by volume.

*Solution*:

$$C_xH_y + (84.1/3.76)O_2 + 84.1 N_2 \rightarrow 12.4 CO_2 + 3.2 O_2$$
$$+ 0.1 CO + 0.2 H_2 + 84.1 N_2 + z H_2O$$

| | |
|---|---|
| x = 12.4 + 0.1 = 12.5 | Carbon balance |
| 22.4 = 12.4 + 3.2 + 0.05 + z/2 | O$_2$ balance |
| z = 13.5 | |
| y = 0.4 + 27 = 27.4 | Hydrogen balance |

$$C_{12.5}H_{27.4} + 22.4\,O_2 + 84.1\,N_2 \rightarrow 12.4\,CO_2 + 3.2\,O_2$$
$$+ 0.1\,CO + 0.2\,H_2 + 84.1\,N_2 + 13.5\,H_2O$$

A/F = (22.4 moles $O_2$ + 84.1 moles $N_2$)/mole fuel

A/F = 106.5/1 by volume

## 19.2 Burners

When heating is required within a controlled environment, it is normally done using hot water, warm air, steam, radiant heat, or electric heat. Water, air, and steam can be heated using gas, oil, coal, or electric energy. In some instances, energy for heating can be provided by heat recovery techniques or solar energy.

### 19.2.1 Domestic Gas Burning Equipment

A gas burner is used to convey gas or a mixture of gas and air to the combustion zone. Burners are of the atmospheric injection, luminous flame, or power burner types.

Domestic gas burners may be classified as either those types designed for central heating plants or those designed for unit application. Gas-designed units and conversion burners are available for the several kinds of central systems and for other applications where the units are installed in the heated space. Central heating appliances include warm air furnaces and steam or hot water boilers.

Warm air furnaces are of two types, gravity and forced air. Forced warm-air furnaces use a motor-driven blower to circulate air over the heat exchanger and through the ducts. A draft hood is attached to the outlet of the furnace and replaces the barometric damper.

Gas-designed boilers for hot water or steam heating are available in cast-iron, steel, and nonferrous metals. Burners are located beneath the sections; the flue gases pass upward between the sections to the flue collector. Some boilers are designed to provide domestic hot water, using tankless or instantaneous heaters.

Some gas furnaces and boilers are available with sealed combustion chambers. These units have no draft hood and are called **direct vent appliances**. Combustion air is piped, usually through a side wall from outdoors, directly to the combustion chamber.

The air intake pipe terminates in the same location as the flue gas vent, sometimes as concentric pipes, with an appropriate terminal covering or cowl. No chimney or vertical flue is needed with such units.

Floor furnaces are used in mild climates for auxiliary heating, heating of single rooms, and, in some cases, heating of several rooms. Floor furnaces are not usually considered central heating systems. However, some are furnished with a circulating fan and as many as eight takeoff ducts. This type of floor furnace is a central heating system. In such a system, piping is below the floor as in other central warm air systems.

Space heaters are generally used for heating a single room or limited area. They differ from central heating equipment in the extent of distribution systems used. Both natural convection and forced-circulation systems distribute the warm air; however, no ductwork distributes heat beyond the immediate area occupied by the heater. Domestic, gas-fired space heaters include floor furnaces (without ductwork), vented wall furnaces, vented and unvented room heaters, radiant heaters, baseboard-sealed combustion system wall furnaces, wall heaters, and unit heaters. Some small room heaters are operated without vents, but they must be used with caution. Both manual and automatic controls are used. An automatic pilot or ignition system must be a part of any control system.

Under controlled conditions, an oil burner combines fuel oil and air for combustion. Fuel oil may be either **atomized** or **vaporized** for the combustion process. Air for combustion is supplied by natural or mechanical draft. Ignition is generally accomplished by an electric spark, gas pilot flame, oil pilot flame, or combination of these. Burners of different types operate with luminous or nonluminous flame. Operation may be continuous, modulating, or intermittent with high-low flame.

Residential oil burners are ordinarily used in the range of 0.5 to 3.5 gph (0.5 to 3.7 mL/s) fuel consumption rate. However, burners up to 7.0 gph (7.2 mL/s) sometimes fall in the residential classification because of similarities in controls and standards. Burner capacity of 7.0 gph (7.35 mL/s) and above is classified as commercial-industrial. Generally, No. 2 fuel oil is used, although burners in the residential size range can also operate on No. 1 fuel oil. In addition to boilers and furnaces for space heating, burners in the 0.5 to 1.0 gph (0.5 to 1.0 mL/s) size range are also used for separate tank type residential hot water heaters, infrared heaters, space heaters, and other commercial equipment.

Most burners manufactured (over 95%) are high-pressure, atomizing gun burners. While other types of burners are still in operation, only a few of these types are currently in production.

The high-pressure, atomizing gun burner illustrated in Figure 19-1 supplies oil to the atomizing nozzle at 100 to 300 psi (700 to 2100 kPa). A blower supplies air for combustion, and a damper or other device regulates the air supply at the burner. Ignition is usually accomplished by a high voltage electric spark, which may be intermittent, sometimes called either **constant ignition** ("on" when the burner motor is on) or **interrupted** (on only to start combustion). Typically, these burners fire into a combustion chamber in which negative draft is maintained.

Present high-pressure atomizing gun burner design includes retention heads and residential burner motors operating at 3450 rpm instead of 1725 rpm. The retention head assists combustion by providing better air-oil mixing, turbulence, and shear. Using the higher-speed 3450 rpm motors (often combined with a retention head design) results in a more compact burner having equal capacity to one operating at 1725 rpm, and one that has a wide tolerance for varying combustion chamber and draft conditions.

Vaporizing burners are designed for use with No. 1 fuel oil. Fuel is ignited by manual pilot or electric spark. The combustion process usually provides enough heat for oil vaporization.

Oil-designated boiler-burner units for hydronic (hot water or steam) heating are constructed of cast iron, steel, and nonferrous metals. The oil burner is usually located at, or near,

# Heating Equipment

the base of the boiler; the flue gases pass upward and around the heat transfer sections and then out the chimney connector. Most oil-powered boiler burner units provide domestic hot water by incorporating a water heating coil (tankless heater) within the boiler.

Warm air furnaces with oil burner units are normally of forced-air design. A blower circulates air over the heat exchanger and through the duct. Upflow, counterflow, horizontal, or downflow furnaces are available. Most forced warm air furnaces make provisions for installing direct expansion air conditioning directly in the unit.

Additional details concerning commercial and industrial gas and/or oil burning equipment, as well as solid fuel burning equipment, can be found in Chapter 31 of the 2012 *ASHRAE Handbook—HVAC Systems and Equipment*. Details about various types of steam and hot water boilers are in Chapter 32 of the above reference.

## 19.3 Residential Furnaces

### 19.3.1 Equipment Variations

Residential furnaces provide heated air through a system of ductwork into the space being heated. They are of two basic types:

**Fuel-Burning Furnaces.** Combustion takes place within a combustion chamber. Circulating air passes over the outside surfaces of the heat exchanger so that it does not contact the fuel or the products of combustion. The products of combustion are passed to the outside atmosphere through a vent.

**Electric Furnaces.** A resistance-type heating element heats the circulating air either directly or through a metal sheath enclosing the resistance element.

Residential furnaces may be further categorized by:

- Fuel type
- Mounting arrangement
- Airflow direction
- Combustion system
- Installation location

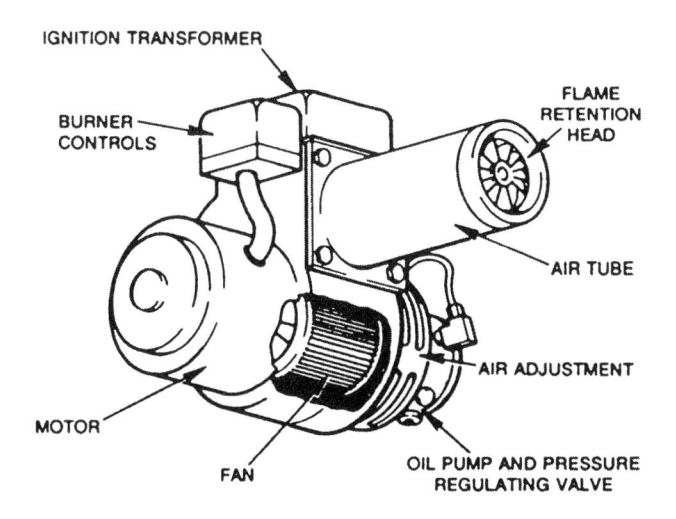

*Fig. 19-1 High-Pressure Atomizing Oil Burner*

A forced air furnace is a self-enclosed appliance used to heat air that is circulated through the enclosure and discharged directly into the space being heated or air that is conveyed through ducts to the space to be heated. An induced draft gas-fired unit is illustrated in Figure 19-2.

In fuel-burning furnaces, combustion occurs within a metal-walled heat exchanger. Air passes over the outside surfaces of the heat exchanger and the heat is transferred through the heat exchanger walls; the circulating air does not come in contact with the fuel or the products of combustion. The products of combustion are conveyed to the outside atmosphere through a flue or vent.

In an electric furnace, the heating element, heated by its electric resistance, heats the circulating air either directly or through a metal sheath enclosing the resistance element.

Residential forced air furnaces usually have a capacity under 250,000 Btu/h (74 kW) output. Residential furnaces are often used in commercial installations. The reverse is usually not true since more than one residential furnace would usually be used in an extremely large home rather than a furnace intended for commercial use.

The furnace may be designed and manufactured to combine components in a variety of ways. The relative positions of the components in the different types of furnaces are described as follows:

1. In a **horizontal furnace**, the blower is located beside the heat exchanger (Figure 19-3). The air enters one end and travels horizontally through the blower, over the heat

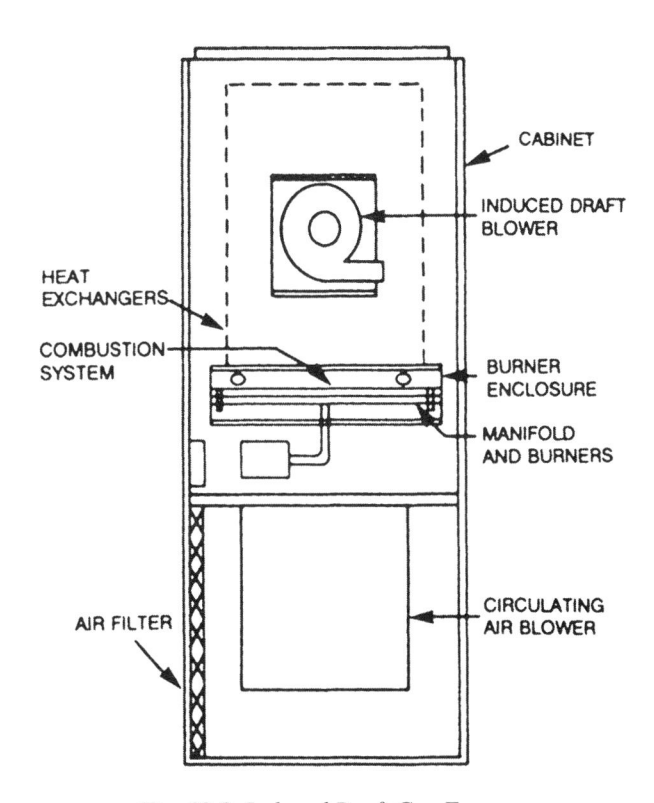

*Fig. 19-2 Induced Draft Gas Furnace*
(Figure 1, Chapter 33, 2012 ASHRAE Handbook—HVAC Systems and Equipment)

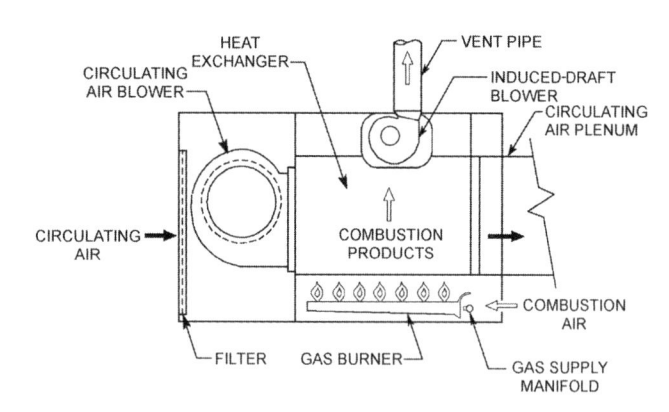

*Fig. 19-3 Horizontal Forced-Warm-Air Furnace*
(Figure 4, Chapter 33, 2012 ASHRAE Handbook—
HVAC Systems and Equipment)

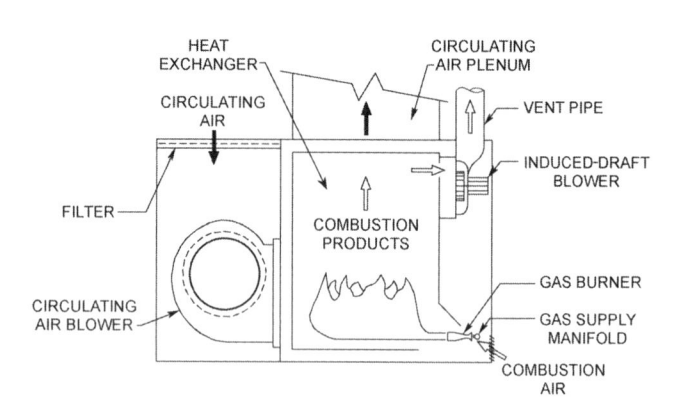

*Fig. 19-5 Basement Forced-Warm-Air Furnace*
(Figure 5, Chapter 33, 2012 ASHRAE Handbook—
HVAC Systems and Equipment)

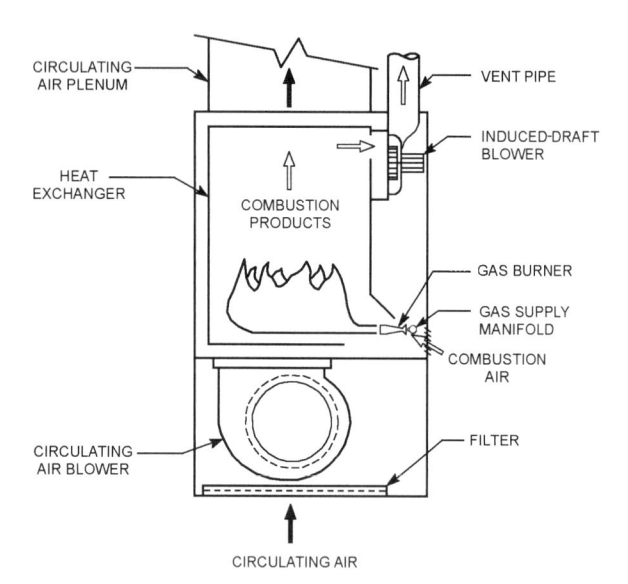

*Fig. 19-4 Upflow Forced-Warm-Air Furnace*
(Figure 2, Chapter 33, 2012 ASHRAE Handbook—
HVAC Systems and Equipment)

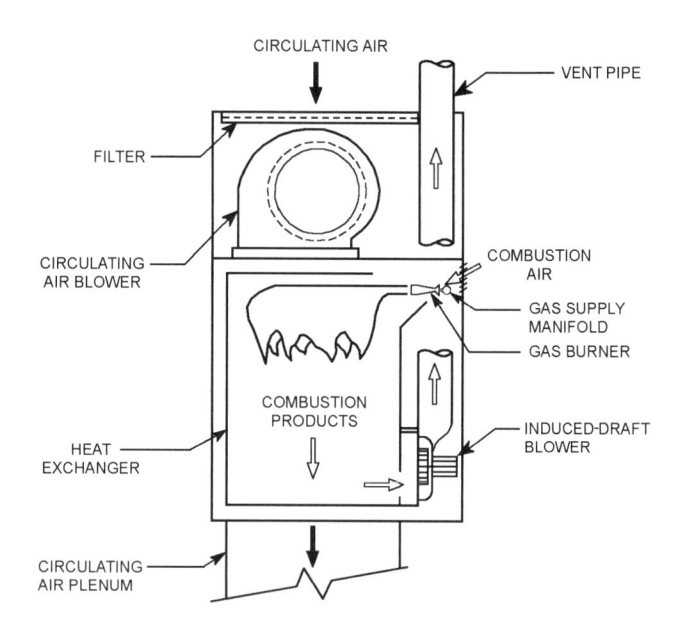

*Fig. 19-6 Downflow (Counterflow) Warm-Air Furnace*
(Figure 3, Chapter 33, 2012 ASHRAE Handbook—
HVAC Systems and Equipment)

exchanger, and discharges out of the opposite end. Such furnaces are used for locations with limited head room (e.g., attics and crawlspaces). These units are often designed to allow for component rearrangement so that airflow may be from left to right or from right to left.

2. The blower in an **upflow furnace** is beneath the heat exchanger and discharges vertically upward (Figure 19-4). Air enters through the bottom or the side of the blower compartment and leaves at the top. Such furnaces may be used in utility rooms on the first floor of homes without basements or in basements with the return air ducted down to the blower compartment entrance.

3. The **basement furnace**, which is a variation of the upflow furnace, requires less head room (Figure 19-5). The blower is located at the bottom beside the heat exchanger. Air enters the top of the cabinet, flows down through the blower, discharges over the heat exchanger, and leaves vertically at the top.

4. The blower in the **downflow** or **counterflow furnace** is located above the heat exchanger and discharges downward (Figure 19-6). Air enters at the top and discharges vertically at the bottom. This style of furnace is used in a house without a basement that has a perimeter heating system.

Further discussion about furnaces and space heaters can be found in the 2012 *ASHRAE Handbook—HVAC Systems and Equipment*.

### 19.3.2 Capacity Selection

Heating capacity depends on several variables that may operate alone or in combination.

**Design heating requirement of building.** The heat loss of the structure can be calculated by using the procedures in Chapters 6 or 7.

**Additional heating required if furnace is operating on night setback cycle.** During the morning recovery period,

# Heating Equipment

additional capacity is required to bring the conditioned space temperature up to the desired level.

**Internal loads.** Normally, the heat gain from internal loads is neglected when selecting a furnace, but if the internal loads are constant, they should be used to reduce the required capacity of the furnace. This is particularly applicable in nonresidential applications.

**Energy required for humidification.** Humidification energy depends on the desired level of relative humidity and the rate at which moisture must be supplied to maintain the specified level. Net moisture requirements must take into account internal gains as the result of people, equipment, and appliances, and losses through migration in exterior surfaces, plus air infiltration.

**Influence of off-peak storage devices.** A storage device, when used in conjunction with a furnace, decreases the required capacity of the furnace. The storage device can supply the additional capacity required during the morning recovery of a night setback cycle or reduce the daily peak loads to assist in load shedding.

**Influence of backup systems.** A furnace can exist as a backup to a solar system, a heat recovery system, or a structure requiring multiple units for uninterrupted service. Oversizing results in higher initial costs, possible increased operating costs, and decreased comfort control. Undersizing produces unacceptable comfort control near the design conditions.

**Capacity to accommodate air conditioning.** Even if air conditioning is not initially planned, the cabinet should be large enough to accept a cooling coil that satisfies the design cooling load. The blower and motor should have sufficient capacity to provide increased airflow rates typically required in air-conditioning applications.

## 19.3.3 Performance Criteria

Some typical efficiencies encountered in performance criteria of a furnace are (1) steady-state efficiency, (2) use efficiency, and (3) annual fuel use efficiency.

These efficiencies are generally used by the furnace industry in the following manner:

**Steady-state efficiency** is the efficiency of a furnace when it is operated under equilibrium conditions. It is calculated by measuring the energy input, subtracting the losses for exhaust gases, and then dividing by the fuel input (cabinet loss not included), i.e.,

$$SS\,(\%) = 100(\text{Fuel input} - \text{Flue loss} - \text{Condensate loss})/\text{Fuel input}$$

For furnaces tested under the isolated combustion system (ICS) method, cabinet heat loss must also be deducted from the energy input.

$$SS\,(\%)(ICS) = 100(\text{Fuel input} - \text{Flue loss} - \text{Condensate loss} - \text{Jacket loss})/\text{Fuel input}$$

**Utilization efficiency** is obtained from an empirical equation developed by Kelly et al. (1978) by starting with 100% efficiency and deducting losses for exhausted latent and sensible heat, cyclic effects, infiltration, and pilot-burner effect.

**Table 19-5 Typical Values of Efficiency**
*(Table 1, Chapter 33, 2012 ASHRAE Handbook—HVAC Systems and Equipment)*

| Type of Gas Furnace | AFUE, % Indoor | ICS[a] |
|---|---|---|
| 1. Natural-draft with standing pilot | 64.5 | 63.9[b] |
| 2. Natural-draft with intermittent ignition | 69.0 | 68.5[b] |
| 3. Natural-draft with intermittent ignition and auto vent damper | 78.0 | 68.5[b] |
| 4. Fan-assisted combustion with standing pilot or intermittent ignition | 80.0 | 78.0 |
| 5. Same as 4, except with improved heat transfer | 82.0 | 80.0 |
| 6. Direct vent, natural-draft with standing pilot, preheat | 66.0 | 64.5[b] |
| 7. Direct vent, fan-assisted combustion, and intermittent ignition | 80.0 | 78.0 |
| 8. Fan-assisted combustion (induced-draft) | 80.0 | 78.0 |
| 9. Condensing | 90.0 | 88.0 |
| **Type of Oil Furnace** | **Indoor** | **ICS[a]** |
| 1. Standard—pre-1992 | 71.0 | 69.0[b] |
| 2. Standard—post-1992 | 80.0 | 78.0 |
| 3. Same as 2, with improved heat transfer | 81.0 | 79.0 |
| 4. Same as 3, with automatic vent damper | 82.0 | 80.0 |
| 5. Condensing | 91.0 | 89.0 |

[a] Isolated combustion system (estimate).
[b] Pre-1992 design (see text).

**Annual fuel utilization efficiency** (AFUE) is the same as **utilization efficiency**, except that losses from a standing pilot during nonheating season time are deducted. The equation for AFUE can also be found in Kelly et al. (1978) or ASHRAE Standard 103.

Federal law requires manufacturers of furnaces to use AFUE to rate efficiency. Typical values of AFUE for various types of furnaces are listed in Table 19-5.

## 19.4 Commercial Furnaces

### 19.4.1 Equipment Variations

The basic difference between residential and commercial furnaces is the size and heating capacity of the equipment. The heating capacity of a commercial furnace may range from 150,000 Btu/h (44 kW) to over 2,000,000 Btu/h (600 kW). Generally, furnaces with output capacities less than 320,000 Btu/h (93 kW) are classified as light commercial, and those above 320,000 Btu/h (93 kW) are large commercial equipment. In addition to the difference in capacity, commercial equipment is constructed from material having increased structural strength and more sophisticated control systems.

Light commercial heating equipment comes in many flow arrangements and design variations. Some are identical to residential equipment, while others are unique to commercial applications. Some commercial units function as a part of a ducted system; others operate as unducted space heaters.

**Ducted Equipment.** Upflow gas-fired commercial furnaces are available up to 300,000 Btu/h (88 kW) and supply enough airflow to handle up to 10 tons (35 kW) of air conditioning. These furnaces may develop high static pressure, have belt-driven blowers, and frequently consist of two standard upflow furnaces tied together in a side-by-side arrangement.

Horizontal gas-fired duct furnaces are also available for built-in light commercial systems. This furnace is not equipped with its own blower but is designed for uniform airflow across the entire furnace.

Duct furnaces are normally certified for operation either upstream or downstream of an air-conditioner cooling coil. Electric duct furnaces are available in a great range of sizes and are suitable for operation in upflow, downflow, or horizontal positions. These units are also used to supply auxiliary heat with the indoor section of a split-heat pump.

The most common commercial furnace is the combination package unit, which is sometimes called a combination rooftop unit. They are available as air-conditioning units with liquified petroleum gas (LPG) and natural gas furnaces, electric resistance heaters, or heat pumps. Combination oil-heat-electric-cool units are not commonly available. Combination units come in a full range of sizes covering air-conditioning ratings from 5 to 50 tons (18 to 176 kW) with matched furnaces supplying heat-to-cool ratios of approximately 1.5:1.

Combination units of 15 tons (53 kW) and under are available as single-zone units. The entire unit must be in either the heating or cooling mode. All air delivered by the unit is at the same temperature. Large combination units in the 15 to 50 ton (50 to 180 kW) range are available as single-zoned units, as are small units; however, they are also available as multi-zone units. A multizone unit supplies conditioned air at several different zones of a building in response to individual thermostats controlling those zones. These units are capable of supplying heating to one or more zones at the same time cooling is being supplied to other zones.

Large combination units are normally available only in a curbed configuration; i.e., the units are mounted on a rooftop over a curbed opening in the roof. The supply and return air enter through the bottom of the unit. Smaller units may be available for either curbed or uncurbed mounting. In either case, the unit is always connected to ductwork within the building to distribute the conditioned air.

**Unducted Heaters**. Three types of commercial heating equipment used as unducted space heaters are unit heaters, infrared heaters, and floor furnaces.

**Unit heaters** are available from about 25,000 to 320,000 Btu/h (7 to 94 kW). Normally they are mounted from ceiling hangers and blow air across the heat exchanger into the heated space. Natural gas, LPG, and electric heat units are available.

**Infrared heaters** are mounted from ceiling hangers and transmit heat downward by radiation. Low- and medium-intensity infrared heaters are compact, self-contained, direct-heating devices. They are used in hangars, factories, warehouses, foundries, greenhouses, and gymnasiums and for areas such as loading docks, racetrack stands, under marquees, outdoor restaurants, and around swimming pools.

Low-temperature radiant heaters are used in office buildings and other commercial buildings.

These heaters can be used with variable-air-volume systems. Infrared heating units may be electric, gas fired, or oil fired. They consist of an infrared source or generator operating in a temperature range of 350 to 5000°F (180 to 2800°C), with the specific temperature determined by the energy source, size, and application. Reflectors can be used to distribute radiation in specific patterns. Common configurations for gas-fired and electric infrared heaters are shown in Figures 19-7 and 19-8, respectively.

Radiant heaters transfer energy directly to solid objects. As floor and objects are warmed by the infrared energy, they, in turn, reradiate heat to solid objects, as well as transfer heat to the air by convection. Dry-bulb temperatures can be maintained slightly less than the mean radiant temperature. With convective heat, the converse is true. Since human comfort is determined by the average of mean radiant and dry-bulb temperatures, dry-bulb temperature may be lowered when heating with radiation. Heat loss to ventilating air and by transmission is correspondingly lower, as is energy consumption. Because buildings heated by infrared require less heating fuel for a given application, equipment that handles only 80 to 85% of the design heat loss calculated by ASHRAE methodology is typically installed.

**Floor furnaces** are used as large area unducted heaters. They are available in capacities ranging from 200,000 to 2,000,000 Btu/h (60 to 590 kW). Floor furnaces direct heated air through nozzles for task heating or use air circulators to heat large industrial spaces.

### 19.4.2 System Design and Equipment Selection

The procedure for design and selection of a commercial furnace is similar to that for a residential furnace. First, the design capacity of the heating system must be determined, considering structure heat loss, recovery load, internal load, humidification, off-peak storage, waste heat recovery, and backup capacity. Since most commercial buildings use setback periods during weekends, evenings, or other long periods of inactivity, the recovery load is important, as are internal loads and waste heat recovery.

Commercial sizing criteria are essentially the same as for residential furnaces. The furnace should be oversized 30% above the total load if setback is anticipated. Since combination units must be sized accurately for the cooling load, it is possible that the smallest gas-fired capacity available is larger than the 30% oversize value. This is especially true for the warmer climates of the United States.

Commercial units have about the same efficiency as residential units. Two-stage gas valves are frequently used with commercial furnaces, but the efficiency of a two-stage system may be lower than that of a single-stage system. At a reduced firing rate, the excess combustion airflow through the burners increases, which decreases the steady-state operating efficiency of the furnace. Multistage furnaces with

# Heating Equipment 545

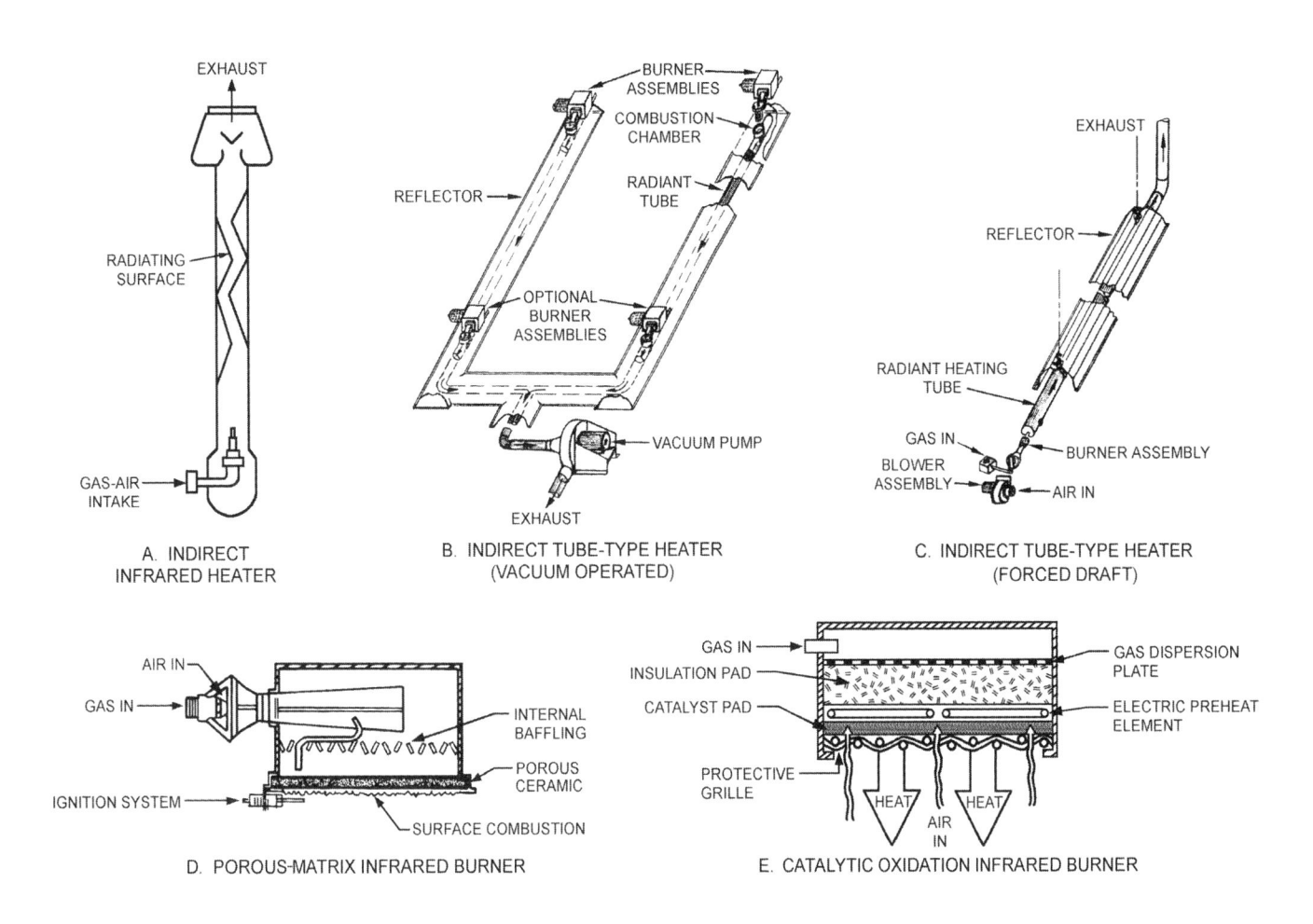

*Fig. 19-7 Gas-Fired Infrared Heaters*
*(Figure 1, Chapter 16, 2012 ASHRAE Handbook—HVAC Systems and Equipment)*

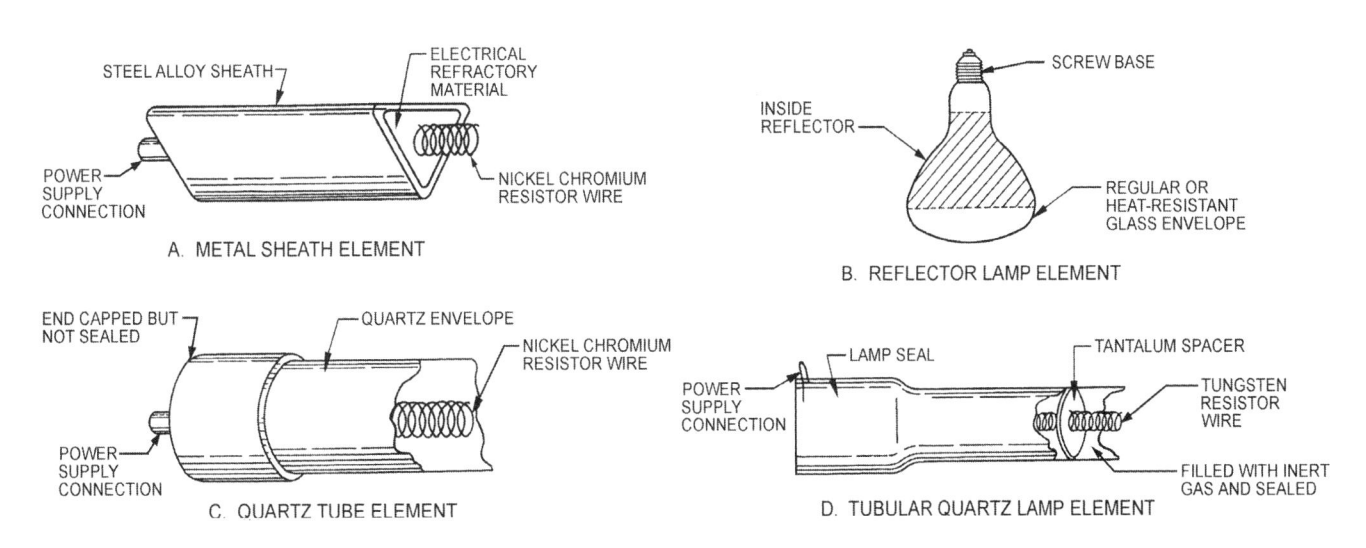

*Fig. 19-8 Electric Infrared Heaters*
*(Figure 2, Chapter 16, 2012 ASHRAE Handbook—HVAC Systems and Equipment)*

multistage thermostats and controls provide more uniform distribution of heat within the building.

The useful life of commercial heating and cooling equipment is about 20 years.

## 19.5 Boilers

A boiler is a pressure vessel designed to transfer heat produced by combustion to a fluid. Heat is normally produced by combustion, but electrical resistance elements or electrodes acting directly on the fluid are also used. In boilers of interest to ASHRAE, the fluid is usually water, in the form of liquid or steam. If the fluid being heated is air, the heat-exchange device is called a furnace. The firebox, or combustion space, of some boilers is also called the furnace.

Excluding special and unusual fluids, materials, and methods, a boiler is a cast-iron, steel, or copper pressure vessel heat exchanger, designed with and for fuel-burning devices and other equipment (1) to burn fossil fuels (or use electric current) and (2) to transfer the released heat to water (in water boilers) or to water and steam (in steam boilers). Boiler heating surface is the area of fluid-backed surface exposed to the products of combustion, or the fire-side surface. Various codes and standards define allowable heat transfer rates in terms of heating surface. Boiler design provides for connections to a piping system, which delivers heated fluid to the point of use and returns the cooled fluid to the boiler.

### 19.5.1 Boiler Classifications

Boilers may be grouped into classes based on such criteria as working pressure and temperature, fuel used, shape and size, use (such as heating or process), and steam or water.

**Working Pressure/Temperature**. With few exceptions, all boilers are constructed to meet the ASME *Boiler and Pressure Vessel Code*.

**Low-pressure boilers** are constructed for maximum working pressures of 15 psi (103 kPa) steam and up to 160 psi (1100 kPa) hot water. Hot water boilers are limited to 250°F (120°C) operating temperature.

**Medium- and high-pressure boilers** are designed to operate above 15 psi (103 kPa) steam or above 160 psi (1100 kPa) water or 250°F (120°C) water boilers.

Steam boilers are available in standard sizes of up to 100,000 lb steam/h (60,000 to 100,000,000 Btu/h or 17 to 30,000 kW), many of which are used for space heating in both new and existing systems. On larger installations, they may also provide steam for auxiliary uses, such as hot-water heat exchangers, absorption cooling, laundry, and sterilizers. In addition, many steam boilers provide steam at various temperatures and pressures for a wide variety of industrial processes.

Water boilers are available in standard sizes of up to 100,000,000 Btu/h (30 MW) from 50,000 Btu/h (15 kW), many of which are in the low-pressure class and are used for space heating in both new and existing systems. Some water boilers may be equipped with either internal or external heat exchangers to supply domestic (service) hot water.

Every steam or water boiler is rated at the maximum working pressure determined by the ASME *Boiler Code Section* (or other code) under which it is constructed and tested. When installed, it must also be equipped with safety controls and pressure relief devices mandated by such code provisions.

**Fuel Used**. Boilers may be designed to burn coal, wood, various grades of fuel oil, various types of fuel gas, or to operate as electric boilers. A boiler designed for one specific fuel type may not be convertible to another type of fuel. Some boiler designs can be adapted to burn coal, oil, or gas. Several designs allow firing with oil or gas by burner conversion or by using a dual-fuel burner.

**Construction Materials**. Most boilers, other than special or unusual models, are made of cast iron or steel. Some small boilers are made of copper or copper-clad steel.

Cast iron boilers are constructed of individually cast sections, assembled into blocks (assemblies) or sections. Push or screw nipples, gaskets, or an external header join the sections pressure-tight and provide passages for the water, steam, and products of combustion. The number of sections assembled determines boiler size and energy rating. Sections may be vertical or horizontal, the vertical design being the most common. The boiler may be dry-base (the firebox is beneath the fluid-backed sections), wet-leg (the firebox top and sides are enclosed by fluid-backed sections), or wet-base (the firebox is surrounded by fluid-backed sections, except for necessary openings).

Steel boilers are fabricated into one assembly of a given size and rating, usually by welding. The heat-exchange surface past the firebox usually is an assembly of vertical, horizontal, or slanted tubes. The tubes may be firetube (flue gas inside, heated fluid outside) or watertube (fluid inside, hot gas outside). The tubes may be in one or more passes. Dry-base, wet-leg, or wet-base design may be used. Most small steel heating boilers are of dry-base, vertical firetube design. Larger boilers usually have horizontal or slanted tubes; both firetube and watertube designs are used. A popular design for medium and large steel boilers is the Scotch, or Scotch Marine, which is characterized by a central fluid-backed cylindrical firebox, surrounded by firetubes in one or more passes, all within the outer shell.

Cast iron boilers range in size from 35,000 to 13,000,000 Btu/h (10 to 3770 kW) gross output. Steel boilers range in size from 50,000 Btu/h (15 kW) to the largest boilers made.

**Condensing or Noncondensing Boilers**. Because higher boiler efficiencies can be achieved with lower water temperatures, condensing boilers purposely allow the flue gas water vapor in the boiler to condense and drain. Illustrated in Figure 19-9 is a typical relationship of overall boiler efficiency to return water temperature. The dew point of 130°F (55°C) shown varies with the percent of hydrogen in the fuel and the $CO_2$ (or excess air) in the flue gas.

Low return water temperatures and condensing boilers are particularly important because they are so efficient at part-

# Heating Equipment

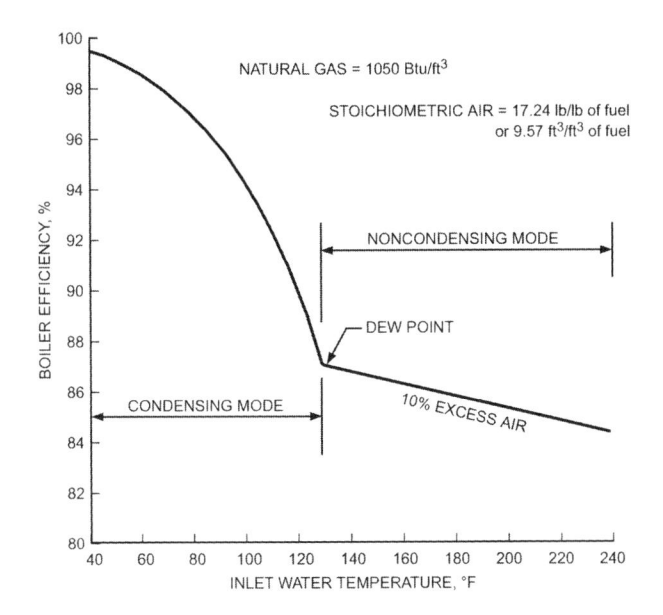

*Fig. 19-9 Boiler Efficiency*
*(Figure 6, Chapter 32, 2012 ASHRAE Handbook—HVAC Systems and Equipment)*

load operation when high water temperatures are not required. For example, a hot water heating system that operates under light load conditions at 80°F (27°C) return water temperature has a potential overall boiler efficiency of 97% when operated with natural gas of the specifications applicable for Figure 19-9.

## 19.5.2 Selection Parameters

Boiler selection should be based on a competent review of the following parameters:

**All Boilers**
- ASME Code Section, under which the boiler is constructed and tested
- Net boiler output capacity, Btu/h (kW)
- Total heat transfer surface, ft² (m²)
- Water content, lb (kg)
- Auxiliary power requirements, kWh
- Internal water-flow patterns
- Cleaning provisions for all heat transfer surfaces
- Operational efficiency
- Space requirements and piping arrangement
- Water treatment requirements
- Operating personnel requirements
- Maintenance requirements
- Regulatory emission limitations

**Fuel-Fired Boilers**
- Combustion space (furnace volume), ft² (m²)
- Internal flow patterns of combustion products
- Combustion air and venting requirements
- Fuel availability

**Steam Boilers**
- Steam space, ft³ (m³)
- Steam quality

### Electric Boilers

Electric boilers are in a separate class. Since no combustion occurs, no boiler heating surface and no flue openings are necessary. Heating surface is the surface of the electric elements or electrodes immersed in the boiler water. The design of electric boilers is largely determined by the shape and heat release of the electric heating elements used.

## 19.5.3 Efficiency: Input and Output Ratings

Efficiency of fuel-burning boilers is defined by combustion efficiency and overall efficiency. Overall efficiency of electric boilers is in the 92 to 96% range.

**Combustion efficiency** is input minus stack (chimney) loss, divided by input, and ranges from 75 to 86% for most noncondensing, mechanically fired boilers. Condensing boilers operate in the range of 88% to over 95% efficiency.

**Overall efficiency** is gross output divided by input. Gross output is measured in the steam or water leaving the boiler and depends on individual installation characteristics. Overall efficiency is lower than combustion efficiency by the percentage of heat lost from the outside surface of the boiler (this loss is usually termed **radiation loss**). Overall efficiency can be determined only by testing under fixed conditions. Approximate combustion efficiency can be determined under any operating condition by measuring operating flue-gas temperature and percentage $CO_2$ or $O_2$ and by consulting a chart or table for the fuel being used.

Heating boilers are usually rated according to standards developed by (1) The Hydronics Institute [formerly the Institute of Boiler and Radiator Manufacturers (IBR) and The Steel Boiler Institute (SBI)], (2) The American Gas Association (AGA), and (3) The American Boiler Manufacturers Association (ABMA).

## 19.5.4 Boiler Sizing

Boiler sizing is the selection of boiler output capacity to meet connected load. The boiler gross output is the rate of heat delivered by the boiler to the system under continuous firing at rated input. Net rating (IBR rating) is gross output minus a fixed percentage to allow for an estimated average piping heat loss plus an added load for initially heating the water in a system (sometimes called pickup).

Piping loss is variable. If all piping is run within the space defined as load, loss is zero. If piping runs through unheated spaces, heat loss from the piping may be higher than accounted for by the fixed net rating factor. Pickup is also variable. Pickup factor may be unnecessary when actual connected load is less than design load. On the design's coldest day, extra system output (boiler and radiation) is needed to pick up the load from a shutdown or low night setback. If night setback is not used, or if no extended shutdown occurs, no pickup load exists. Standby system capacity for pickup, if needed, can be in the form of excess capacity in base-load boilers or in a standby boiler.

### 19.5.5 Control of Boiler Input and Output

Boiler controls regulate the rate of fuel input (on-off, step-firing, or modulating) in response to a control signal representing load change, so that average boiler output equals load within some accepted control tolerance. Boiler controls include safety controls that shut off fuel flow when unsafe conditions develop.

Steam boilers are operated by boiler-mounted, pressure-actuated controls, which vary the input of fuel to the boiler. Common examples of controls are on-off, high-low-off, and modulating. Modulating controls continuously vary the fuel input from 100% down to a selected minimum point. The ratio of maximum to minimum is called the **turndown ratio**. The minimum input is usually between 5% and 25%, i.e., 20:1 to 4:1, and depends on the size and type of fuel-burning apparatus.

### 19.5.6 Hydronic Systems

Hot water heating systems are frequently called **hydronic systems**. Water systems can be classified by (1) temperature, (2) flow generation, (3) pressurization, (4) piping arrangement, and (5) pumping arrangement. Water systems are either once-through or recirculating systems.

The two types of hot water heating systems classified by flow generation are (1) the **gravity system**, which uses the difference in density between the supply and return water columns of a circuit or system to circulate water, and (2) the **forced system**, in which a pump, usually driven by an electric motor, maintains the flow.

**Low-temperature hot water systems** (LTHW) are the most widely used heating systems for residential, commercial, and institutional systems where loads consist primarily of space heating and domestic water heating and do not exceed 5,000,000 Btu/h (1.5 MW) total. The maximum allowable working pressure for low-pressure heating boilers is 160 psia (1100 kPa), with a maximum temperature limitation of 250°F (120°C). The usual maximum working pressure for boilers for LTHW systems is 30 psia (200 kPa), although boilers specifically designed, tested, and stamped for higher pressures may frequently be used with working pressures to 160 psia (1100 kPa). Steam-to-water or water-to-water heat exchangers are also often used.

**Medium-temperature hot water systems** (MTHW) (350°F [175°C] or less) are most commonly used for space heating in large commercial and institutional buildings or in industrial applications with process loads, and where total loads range from $5 \times 10^6$ to $20 \times 10^9$ Btu/h (1.5 to 6 MW). In a medium-temperature system, the usual design supply temperature is above 250°F (120°C) and below 350°F (175°C) with a usual pressure rating for boilers and equipment of 150 psia (1030 kPa).

**High-temperature hot water systems** (HTHW) are generally limited to campus-type district heating installations and to applications requiring process heating temperatures of 350°F (175°C) or higher.

The design of water systems is affected by complex relationships between the various system components. The design water temperature, flow rate, piping layout, pump selection, terminal unit selection, and control method are all interrelated. The size and complexity of the system determines the importance of these relationships in affecting the total operating success. Present hot water heating system design practice originated in residential heating applications where a 20°F (11°C) temperature drop (TD) was used to determine flow rate. Besides producing satisfactory operation and economy in small systems, this TD enabled simple calculations because 1 gpm conveys 10,000 Btu/h (1 L/s conveys 41 kW with a 10°C TD).

Elements of a high-temperature water system are illustrated in Figure 19-10. Requirements for such a system include a high limit control, a safety relief valve, and other safety controls and devices on the boiler, and a boiler efficiency with an accepted test rating.

## 19.6 Terminal Units

Radiators, convectors, baseboard, and finned tube units are commonly used to distribute heat to the space provided by the steam or LTHW systems. They supply heat through a combination of radiation and convection. In general, these units are placed at the points of greatest heat loss of the space such as under windows, along exposed walls, and at door openings.

A **radiator** is generally considered a sectional cast-iron unit of column, large tube, or small tube type (Fig. 19-11).

A **convector** is a heat-distributing unit that operates with gravity-circulated air. The heating element contains two or more tubes with headers at both ends. The heating element is surrounded by an enclosure with an air inlet below and an air outlet above (Fig. 19-11).

**Baseboard** and **baseboard radiation heat distributing units** are installed at the base of walls in place of the conventional baseboard. They may be made of cast iron, with a substantial portion of the front face directly exposed to the room, or with a finned-tube element in a sheet metal enclosure (Fig. 19-11).

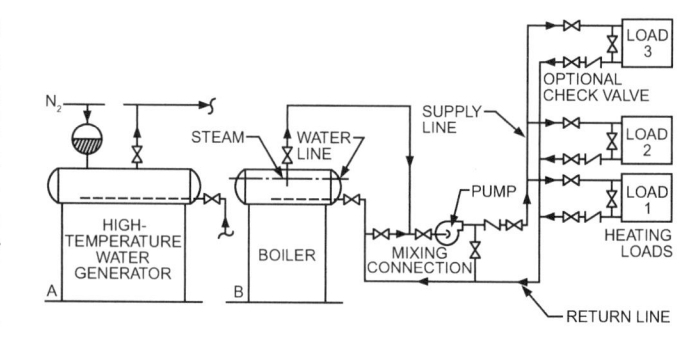

*Fig. 19-10 Elements of High-Temperature Water System*
*(Figure 2, Chapter 15, 2012 ASHRAE Handbook—HVAC Systems and Equipment)*

# Heating Equipment

**Finned tube** or **fin-tube** refers to heat-distributing units fabricated from metallic tubing with metallic fins bonded to the tube. They operate with gravity-circulated room air and may be installed bare, with an expanded metal grille, a cover, or an enclosure having top, front, or inclined outlets.

## 19.6.1 Radiators

The small-tube type radiators, with a length of only 1 3/4 in. (44 mm) per section, occupy less space than the older column and large-tube units and are particularly suited to installation in recesses. Shown in Figure 19-12 are the types of units now being manufactured.

## 19.6.2 Convectors

Convectors are made in a wide variety of depths, sizes, and lengths in both cabinet and enclosure style. The heating elements are fabricated in ferrous and nonferrous metals. The air enters the enclosure below the heating element, is heated as it passes through the element, and leaves the enclosure through the outlet grille located above the heating element. Factory-assembled units composed of a heating element and enclosure are widely used. These may be free standing, wall hung, or recessed and may have outlet grilles and arched inlets or inlet grilles (Figure 19-11).

## 19.6.3 Baseboard Units

Baseboard heat-distributing units are divided into two types: (1) radiant-convector and (2) finned tube.

A **radiant-convector baseboard heating unit** is made of cast iron or steel. The units have air openings at the top and bottom to permit circulation of room air over the wall side of the unit. The wall side of the unit has an extended surface to provide increased heat output. A large portion of the heat is transferred by convection (Fig. 19-11).

A **finned tube baseboard heating unit** has a finned-tube heating element that is concealed by a long, low, sheet-metal enclosure or cover. Most of the heat is transferred to the room by convection. Optimum comfort for room occupants is obtained when units are installed along as much of the exposed wall as possible.

The baseboard unit has the following advantages: (1) it is normally placed along the cold walls and under areas where the greatest heat loss occurs, (2) it is inconspicuous, (3) it offers minimum interference with furniture placement, and (4) it distributes heat near the floor. This last characteristic reduces the floor-to-ceiling temperature gradient to about 2 to 4°F (1 to 2°C) and helps produce uniform temperatures throughout the room. It also makes baseboard heat-distributing units adaptable to homes without basements, where cold floors are prevalent.

## 19.7 Electric Heating

For many applications, the compactness, simplicity, responsiveness, accuracy of control, safety, and cleanliness of electric heating may outweigh its disadvantages. Electric space heating is often used where minimum initial cost is the dominating factor.

### 19.7.1 Electric Heating Elements

Electric heating elements usually are composed of metal-alloy wire or ribbons, nonmetallic carbon compounds in rod or other shapes, or printed circuits. Heating elements may have exposed resistor coils mounted on insulators, metallic resistors embedded within refractory insulation encased in a protective metal sheath, or a printed circuit on glass sheets or vitrified panels. Fins or extended surfaces may be used to

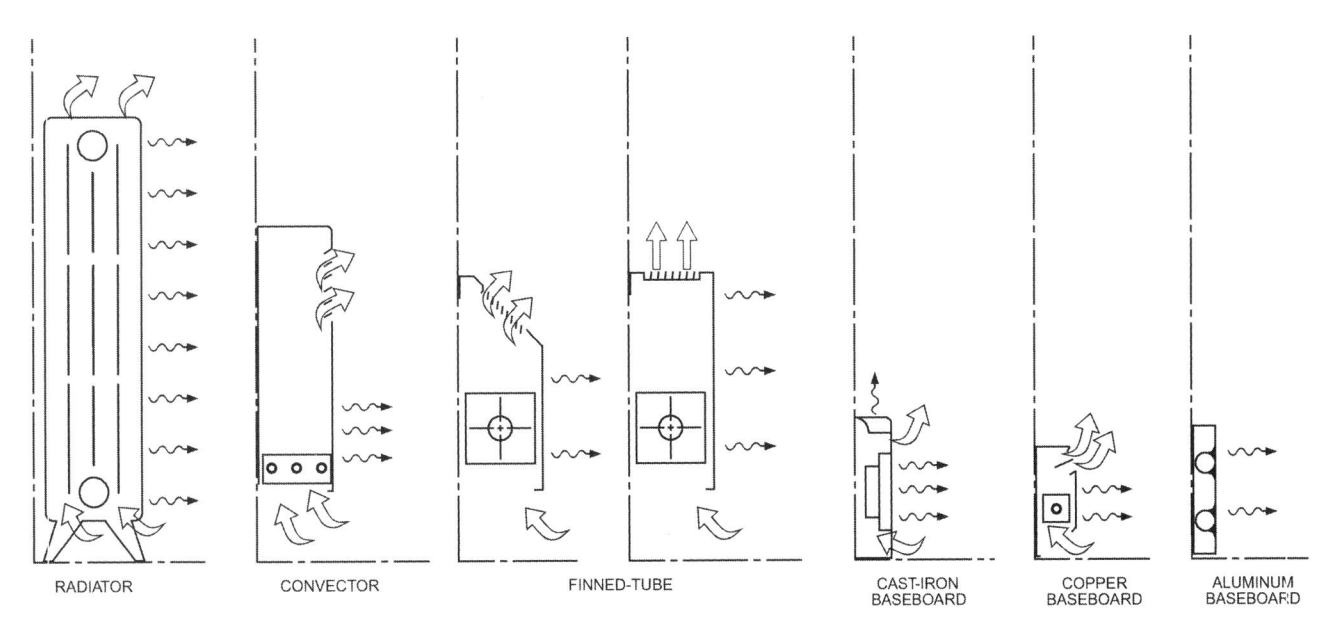

*Fig. 19-11 Terminal Units*
*(Figure 1, Chapter 36, 2012 ASHRAE Handbook—HVAC Systems and Equipment)*

increase heat dissipation. Elements are made in many forms. Metal or oxide conductive films on glass and ceramics have been used, usually in panel form. Tubular elements may be immersed in liquids, used bare, formed into coils, or cast into metal.

Cloth fabrics, incorporating flexible resistor wires, are used for low-temperature purposes such as heating pads, sheets, blankets, aviators' clothing, window draperies, and some radiant panel heating installations. Paper or fabric incorporating a surface resistor on the back is applied for ceiling panel heating installations.

### 19.7.2 Electric Space Heating Systems

Types of electric heating equipment and complete heating systems in use are listed in Table 19-6. The sequence shown is for convenient reference only and does not indicate the relative performance or extent of use. Installations of all types listed are in successful operation, but operating characteristics depend largely on proper application.

### 19.7.3 Calculating Capacities

The procedure outlined in Chapters 17 and 18 of the 2013 *ASHRAE Handbook—Fundamentals* for calculating heating load of residential and nonresidential buildings may be used for electric systems. The NEMA *Manual for Electric Comfort Conditioning*, published by the National Electrical Manufacturers Association, gives heat loss factors and describes methods for calculating load directly in kilowatts.

The most economical electric heating systems from an operating standpoint are decentralized, with the thermostat provided on each unit or for each room. This permits each room to compensate for heat contributed by auxiliary sources such as sunshine, lighting, and appliances. Such an arrangement also gives a better diversity of the power demand because the electric loads from all units of an installation do not coincide. Manual switches allow the heat to be turned off or allow for reduced temperatures in rooms that

**Table 19-6  Types of Electric Space-Heating Equipment**

**Decentralized Systems**
**Natural Convection Units**
  Floor drop-in units
  Wall insert and surface-mounted heaters
  Baseboard convectors
  Hydronic baseboard convectors with immersion elements
**Forced Air units**
  Unit ventilators
  Unit heaters
  Wall insert heaters
  Baseboard heaters
  Floor drop-in heaters
**Radiant Units (High Intensity)**
  Radiant wall, insert, or surface-mounted (open ribbon or wire element)
  Metal-sheathed element with focusing reflector
  Quartz tube element with focusing reflector
  Quartz lamp with focusing reflector
  Heat lamps
  Valance (cove) heaters
**Radiant Panel Systems (Low Intensity)**
  Radiant ceiling with embedded conductors
  Prefabricated panels
  Radiant floor with embedded conductors
  Radiant-convector panel heaters

**Centralized Systems**
**Heated Water Systems**
  Electric boiler
  Electric boiler with hydronic off-peak storage
  Heat pumps
  Integrated heat recovery systems
**Steam Systems**
  Electric boiler with immersion element or electrode
**Heated-Air Systems**
  Duct heaters
  Electric furnaces
  Heat pumps
  Integrated heat recovery systems
  Unit ventilators
  Self-contained heating and cooling units
  Storage units (ceramic, water)

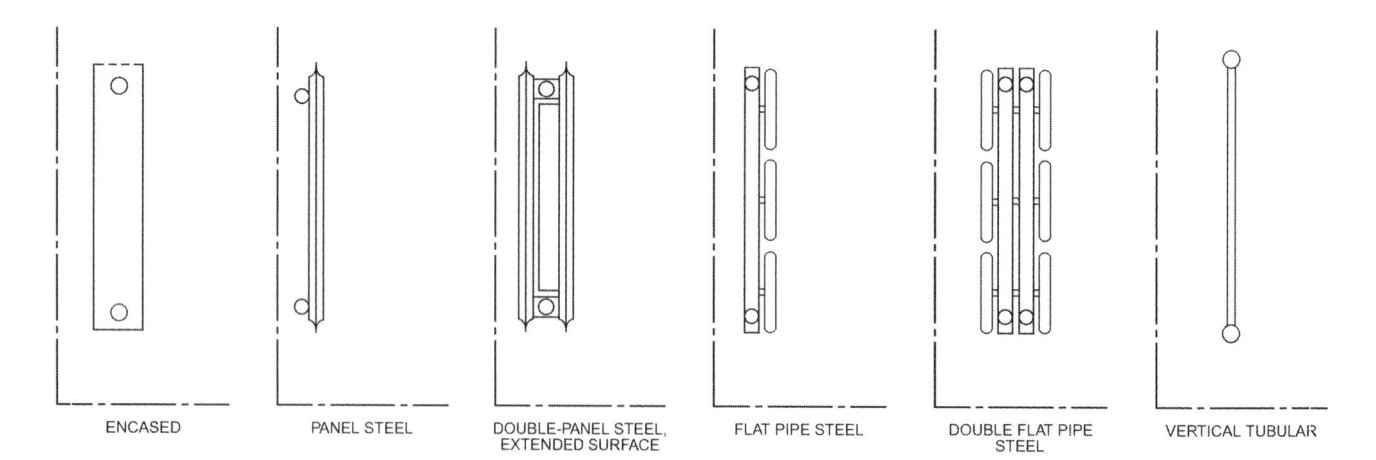

*Fig. 19-12 Typical Radiators*
*(Figure 2, Chapter 36, 2012 ASHRAE Handbook—HVAC Systems and Equipment)*

# Heating Equipment

are not in use. When manual switches are used, adequate capacity for warmup should be provided.

Residences and other buildings designed for electric heat should be well insulated. Depending on the climate, storm or multiple-glazed windows and storm doors are generally recommended to reduce heat loss and further increase comfort. Weather stripping doors and windows effectively reduces air infiltration.

## 19.8 Problems

*In all problems, assume air is 79% $N_2$ and 21% $O_2$.*

**19.1** Set up the necessary combustion equations and determine the mass of air required to burn 0.45 kg (1 lb) of pure carbon to equal masses of CO and $CO_2$.

**19.2** The gravimetric analysis of a gaseous mixture is: $CO_2$ = 32%, $O_2$ = 54.5%, and $N_2$ = 11.5%. The mixture is at a pressure of 20.7 kPa (3 psia). Determine (a) the volumetric analysis and (b) the partial pressure of each component.

**19.3** A liquid petroleum fuel, $C_2H_6OH$, is burned in a space heater at atmospheric pressure.

(a) For combustion with 20% excess air, determine the air/fuel ratio by mass, the mass of water formed by combustion per pound of fuel, and the dew point of the combustion products. [Ans: 11.45, 1.34, 133.9°F]

(b) For combustion with 80% theoretical air, determine the dry analysis of the exhaust gases in percentage by volume. [Ans: 5.94% $CO_2$, 11.04% CO, 83.02% $N_2$]

**19.4** Find the air/fuel ratio by mass when benzene ($C_6H_6$) burns with theoretical air and determine the dew point at atmospheric pressure of the combustion products if an air/fuel ratio of 20:1 by mass is used.

**19.5** A diesel engine uses 30 $lb_m$ of fuel per hour (3.8 g/s) when the brake output is 75 hp. If the heating value of the fuel is 19,600 Btu/lb (45,600 kJ/kg), what is the brake thermal efficiency of the engine?

**19.6** Methane ($CH_4$) is burned with air at atmospheric pressure. The Orsat analysis of the flue gas gives: $CO_2$ = 10.00%, $O_2$ = 2.41%, CO = 0.52%, and $N_2$ = 87.07%. Balance the combustion equation and determine the air-fuel ratio, the percent theoretical air, and the percent excess air. [Ans: 10.48 (vol.), 18.89 (mass), 110.1%, 10.1%]

**19.7** Fuel oil composed of $C_{16}H_{32}$ is burned with the chemically correct air-fuel ratio. Find:

(a) Moisture formed per kg of fuel; moisture formed per lb of fuel
(b) Partial pressure of the water vapor, kPa; water vapor, psia
(c) Percentage of $CO_2$ in the stack gases on an Orsat basis
(d) Volume of exhaust gases per unit mass of oil, if the gas is at 260°C (500°F) and 102 kPa (14.8 psia).

**19.8** Determine the composition of a hydrocarbon fuel if the Orsat analysis gives: $CO_2$ = 8.0%, CO = 1.0%, $O_2$ = 8.7%, and $N_2$ = 82.3%.

**19.9** Determine the air/fuel ratio by mass when a liquid fuel of 16% hydrogen and 84% carbon by mass is burned with 15% excess air. [Ans: 17.49]

**19.10** Compute the compositions of the flue gases on a percent by volume on dry basis (same as Orsat) resulting from the combustion of $C_8H_{18}$ with 85% theoretical air.

**19.11** A liquid petroleum fuel having a hydrogen to carbon ratio of 0.169 by mass is burned in a heater with an air/fuel ratio of 17 by mass. Determine (a) the volumetric analysis on both wet and dry bases of the exhaust gases and (b) the dew point of the exhaust gas.

**19.12** Compare the heating value for semianthracite coal as given in Table 8, Chapter 28, 2009 *ASHRAE Handbook—Fundamentals* with the value predicted using the Dulong Formula. [Ans: 1.24% difference]

**19.13** Natural gas with a volumetric composition of 93.32% methane, 4.17% ethane, 0.69% propane, 0.19% butane, 0.05% pentane, 0.98% carbon dioxide, and 0.61% nitrogen burns with 30% excess air. Calculate the volume of dry air at 60°F, 30 in. Hg (15.6°C, 101.5 kPa) used to burn 1000 ft$^3$ (28.3 mL) of gas at 68°F and 29.92 in. Hg (20°C and 101.4 kPa) and find the dew point of the combustion products.

**19.14** The proximate analysis of a coal is: moisture = 4.33%, volatile matter = 40.21%, fixed carbon = 45.07%, and ash = 10.39%. The heating value was determined as 29 000 kJ/kg (12,490 Btu/lb). Find the ASTM rank of the coal.

**19.15** A fuel oil shows an API gravity of 36. Calculate the specific gravity at 60°F and the pounds per gallon of fuel. Estimate the ASTM grade. [Ans: 0.845, 7.05, No. 2]

**19.16** A representative No. 4 fuel oil has a gravity of 25 degrees API and the following composition: carbon = 87.4%, hydrogen = 10.7%, sulfur = 1.2%, nitrogen = 0.2, moisture = 0, and solids = 0.5%.

(a) Estimate its higher heating value.
(b) Compute the mass of air required to burn, theoretically, 1 gallon of the fuel.

**19.17** The following data were taken from a test on an oil-fired furnace:

Fuel rate = 20 gal oil/h
Specific gravity of fuel oil = 0.89% by mass
Hydrogen in fuel = 14.7%
Temperature of fuel for combustion = 80°F
Temperature of entering combustion air = 80°F
Relative humidity of entering air = 45%
Temperature of flue gases leaving furnace = 550°F

(a) Calculate the heat loss in water vapor in products formed by combustion.
(b) Calculate the heat loss in water vapor in the combustion air.
[Ans: 1672.5 Btu/lb (3888 kJ/kg), 29.4 Btu/lb (68.3 kJ/kg)]

**19.18** An office building requires 2901 MJ (2.75 × 10$^9$ Btu) of heat for the winter season. Compute the seasonal heating costs, if the following fuel is used:

(a) Bituminous coal: 31,380 kJ/kg (13,500 Btu/lb), $70.00 per ton
(b) No. 2 fuel oil: 38,500 kJ/L (38,000 Btu/gal), $2.75 per gallon.

Assume that the conversion efficiency is 75% for the oil and 61% for the coal.

**19.19** Saturated air at 41°F dry bulb (5°C) enters a furnace; it leaves the furnace at 110°F dry bulb (43.3°C) and 0.00543 lb$_v$/lb$_a$ (0.00543 kg/kg) and circulates through a factory. Air leaves the factory at 65°F dry bulb (18.3°C) and 63°F wet bulb (17.2°C).

(a) What is the sensible and latent heat change for the air passing through the factory?
(b) State whether the air gains or loses sensible and latent heat during each process. [Ans: $q_s$ = –10.8 Btu/lb (–25.2 kJ/kg), $q_l$ = +7 Btu/lb (16.3 kJ/kg)]

**19.20** A plant is maintained at 70°F dry bulb, 60% rh, and has a low-pressure steam heating system. A makeup air system is being added to the plant and it has been decided that the input air should be 10,127 cfm. Outside design conditions are –1°F dry bulb, 50% rh. The plant is 250 ft by 560 ft and normally has 325 people working per shift.

(a) What are the total steam requirements for the heating coil and the humidifier?
(b) What capacity should the humidifier have in pounds of water per hour?

**19.21** A residence with a design heating load of 26 kW (89,000 Btu/h) is to use an oil-fired warm air system with forced circulation. Return air to the furnace is at 22.2°C (72°F). Specify the following:

**This problem requires catalog data.**
(a) Supply air temperature
(b) Airflow rate
(c) Make and catalog number of suitable furnace

**19.22** A residence with a design heating load of 16 kW (55,000 Btu/h) is to use a forced circulation hot-water baseboard radiator system. The baseboard units house copper tubing with aluminum fins and operate with the inlet air temperature at 18.3°C (65°F). Specify the following:

**This problem requires catalog data.**
(a) Hot water inlet temperature and outlet temperature
(b) Total water flow rate
(c) Total length of radiator panel for house
(d) Location of panels
(e) Make and catalog number of suitable hot water heater

**19.23** For the residence of Problem 19-22, electric baseboard units replace the hot water system. Specify:

**This problem requires catalog data.**

(a) Total rating of electric system, kW
(b) Total length of baseboard units

**19.24** A large classroom has a winter design heat loss of 19.9 kW (68,000 Btu/h) with installed forced circulation hot-water baseboard radiators. The baseboard units house copper tubing with aluminum fins and operate with the inlet air temperature at 18.3°C (65°F). Specify the following:

**This problem requires catalog data.**
(a) Hot water inlet temperature and outlet temperature
(b) Water flow rate
(c) Length of radiator panel

**19.25** A large classroom has a winter design load of 26 kW (89,000 Btu/h). A forced circulation warm air system is to be used with return air at 23.3°C (74°F). Specify

(a) Supply air temperature
(b) Airflow rate
[Ans: $t_s$ ≈ 135°F, 1330 cfm]

**19.26** For a heat loss from the space to be conditioned of $Q$, write the expression for determining

(a) Amount of air L/s (cfm) that must be supplied if a hot air system is used
(b) Amount of hot water L/s (gpm) that must be supplied if a hydronic system is used
(c) Amount of steam kg/h (lb/h) that must be supplied if a steam heating system is used
(d) Size, in watts, of electric heaters required if electric heat is used

**19.27** List the steps taken when designing a forced-circulation hot water heating system.

**19.28** Compute the increase in length of 28.3 m (93 ft) of steel steam pipe when the average steam temperature is 113°C (235°F) and the air is 21°C (70°F). The pipe was installed during a period when the temperature was 15.6°C (60°F). [Ans: 1.27 in.]

**19.29** The total mass of steel in the boiler and piping of a school's heating system is 9080 kg (20,000 lb). The piping and boiler also contain 6810 kg (15,000 lb) of water. After a weekend shutdown, the temperature of the system is 10°C (50°F). The operating temperature is 93°C (200°F).

(a) Assuming the system should be warmed up in one hour, determine the required furnace size.
[Ans: 764 kW (2,610,000 Btu/h)]
(b) For a furnace size of 146 kW (500,000 Btu/h) output, when should the furnace be started to be up to the operating temperature of 93°C (200°F) by 7:30 AM Monday morning? [Ans: 2:17 AM Sunday]

## 19.9 Bibliography

ASHRAE. 2013. Chapter 28, *2013 ASHRAE Handbook—Fundamentals*.

ASHRAE. 2012. Chapters 28 and 31 through 37, *2012 ASHRAE Handbook—HVAC Systems and Equipment*.

Kelly, G.E., J.G. Chi, and M. Kuklewicz. 1978. *Recommended testing and calculation procedures for determining the seasonal performance of residential central furnaces and boilers*. Available from National Technical Information Service, Springfield, VA (Order No. PB289484).

Strehlow, R.A. 1984. *Combustion Fundamentals*. McGraw-Hill, New York.

NFPA. 1992. Chimneys, Fireplaces, Vents and Solid Fuel-Burning Appliances. *Standard* 211-92. National Fire Protection Association, Quincy, MA.

# Chapter 20

# HEAT EXCHANGE EQUIPMENT

Heat transfer plays a vital role in heating, refrigerating, and air-conditioning as can be seen by the many sections in previous chapters dealing with equipment whose main function is the exchange of heat, thus providing either heating or cooling. The *ASHRAE Handbook* series has a number of chapters devoted to heat transfer and heat transfer applications: in the *2013 Fundamentals*, Chapter 4, Heat Transfer, and Chapter 5, Two-Phase Flow; and 2012 *Systems and Equipment*, Chapter 23, Air-Cooling and Dehumidifying Coils, Chapter 27, Air-Heating Coils, and Chapter 48, Heat Exchangers. This chapter briefly reviews the fundamentals of *applied* heat transfer and illustrates the basic approach to heat exchanger design and analysis.

## 20.1 Modes of Heat Transfer

*Heat transfer* or *heat* (the "transfer" is redundant) can be defined as the transfer of energy from one region or one body to another due to a ***temperature difference***. Heat transfer is as universal as gravity since differences in temperature exist all over the universe. Unlike gravity, however, heat transfer is governed not by a single relationship but by a combination of various independent laws of physics. Heat transfer is generally divided into three distinct modes: *conduction*, *convection*, and *radiation*. Strictly speaking, only conduction and radiation are both separate and purely heat transfer processes, since convection also involves mass transfer and includes conduction.

### 20.1.1 Conduction

*Conduction* is the mode of heat transfer whereby energy is transported between parts of an opaque, stationary medium or between two media in direct physical contact. In gases, conduction is due to the elastic collision of molecules; in liquids and electrically non-conducting solids, it is due primarily to longitudinal oscillations of the lattice structure. In metals, thermal conduction takes place in the same manner as electrical conduction; that is, with the movement of the free electrons.

The theory of heat transfer by conduction was first proposed by Jean B. Fourier in a noted work, published in 1822 in Paris, titled *Theorie analytique de la chaleur*. Fourier's law gives the heat transfer rate past any plane by the following:

$$q = -kA\frac{\partial T}{\partial n} \quad (20\text{-}1)$$

where $k$ is the thermal conductivity of the material, $A$ is the area normal to the flow of heat, $T$ is the temperature, and $n$ is the distance in the direction of heat flow. The partial derivative $\partial T/\partial n$ is the temperature gradient in the direction of the heat flow. The minus sign indicates that heat flows of its own accord only in the direction of decreasing temperature (from hot to cold), in accordance with the Second Law of Thermodynamics. The thermal conductivity $k$ is the specific property of matter that indicates a material's ability to transfer heat, expressed as energy transferred per unit time per unit area per unit temperature gradient. Table 20-1 provides a few order of magnitude values of thermal conductivity.

If heat flows in more than one direction, or there are temperature variations in more than one direction, Equation 20-1 cannot be directly integrated but is only a start in the development of the three-dimensional *general conduction equation*. However, for the following two simple, but important, cases to HVAC applications shown in Figure 20-1, direct integration is possible with these results:

**Case 1.** The slab or plane wall under steady-state conditions:

$$q = kA\frac{(T_1 - T_2)}{L} = \frac{T_1 - T_2}{L/(kA)} \quad (20\text{-}2)$$

where $L$ is the wall thickness, $T_1$ is the temperature at $x = 0$, and $T_2$ is the temperature at $x = L$. The quantity $L/(kA)$ can be considered the "thermal resistance" to the flow of heat.

**Case 2.** The hollow cylinder (tubes, pipes, etc.) under steady-state conditions:

$$q = \frac{2\pi kL(T_1 - T_2)}{\ln(r_2/r_1)} = \frac{T_1 - T_2}{[\ln(r_2/r_1)/(2\pi kL)]} \quad (20\text{-}3)$$

where $L$ is the length of the cylinder, $T_1$ is the temperature at the inner radius $r_1$, and $T_2$ is temperature at the outer radius $r_2$. Here, the quantity $\ln(r_2/r_1)/(2\pi kL)$ is considered the "thermal resistance."

### 20.1.2 Convection

*Convection* is the mode of heat transfer whereby energy is transported by the combined action of conduction, energy storage, and mixing motion. Convection occurs between a solid surface and a moving fluid. When the fluid movement is produced by other than the heat transfer process itself (such as by a fan or pump), the convection is termed *forced*

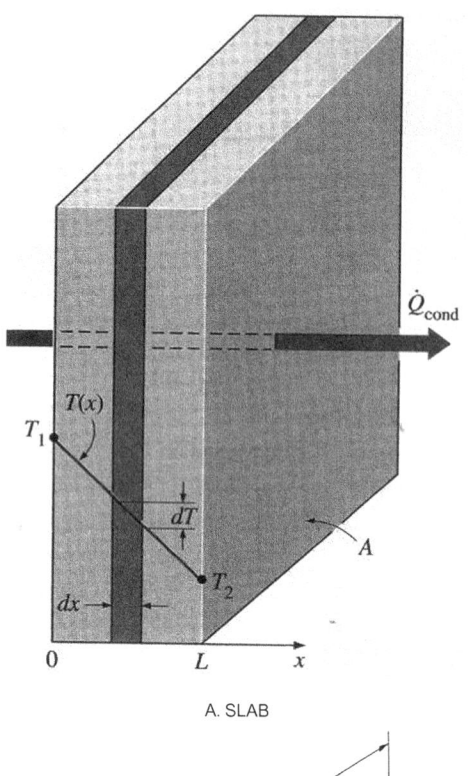

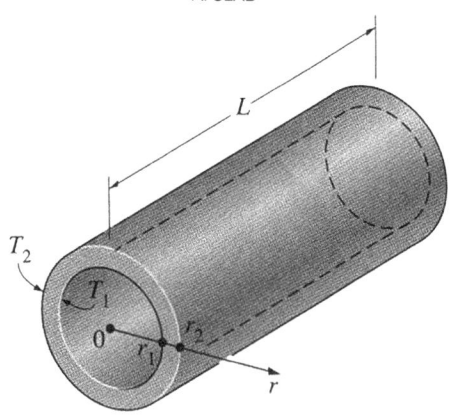

Table 20-1  Order of Magnitude of Thermal Conductivity at Room Temperature

| Material | k, W/(m·K) | K, Btu/h·ft·°F |
|---|---|---|
| Copper | 400 | 220 |
| Aluminum | 200 | 110 |
| Mild Steel | 64 | 40 |
| Stainless Steel | 15 | 10 |
| Concrete | 1.4 | 0.8 |
| Glass | 1.1 | 0.6 |
| Water | 0.6 | 0.4 |
| Wood | 0.10 | 0.06 |
| Polyvinyl chloride (PVC) | 0.10 | 0.06 |
| Fiberglass (medium density) | 0.04 | 0.02 |
| Air | 0.04 | 0.02 |

where $h$ is the convective heat transfer coefficient (also called the "film coefficient"), $A$ is the surface area in contact with the fluid, $T_s$ ($T_w$ sometimes used) is the surface or wall temperature, and $T_f$ ($T_\infty$ often used) is the fluid temperature outside the boundary layer. For convection, the quantity $1/hA$ is taken as the thermal resistance. Unfortunately, Equation (20-4) is actually no more nor less than the defining equation for the convection coefficient, h, and should be written:

$$h = q/[A(T_s - T_f)]$$

The equations for the governing laws for convection actually consist of five partial differential equations, namely:

- Conservation of Mass Equation
- Conservation of Energy Equation
- Momentum Equations (three, one per direction).

The $h$ value could be obtained from solving these equations for $q$, which is then substituted in Eq. (20-4). Due to the mathematical complexity of the problem, most available values of the convection coefficient have been determined experimentally with empirical correlations provided for future reference and use in similar situations.

In almost every case, the fluid properties found in the correlations for predicting h depend significantly on temperature. In the case of density (and kinematic viscosity), there is also a pressure effect for gases. The temperature dependence means that there may be a significant variation in the quantities through the region of fluid near the surface (through the *boundary layer*). The accuracy of the predictive correlations depends upon the temperature(s) used for evaluating these thermodynamic and thermophysical properties. For convection occurring on an exterior surface, the *film temperature*, which is the average of the surface temperature and the undisturbed fluid temperature, is normally required. For internal flow, the *bulk mean fluid temperature* (also called the *average bulk temperature*), which is generally the average of the mean fluid inlet temperature and the mean fluid outlet temperature, is used as the temperature at which to evaluate the fluid properties. The bulk temperature is also called the "mixing cup" temperature as it represents an energy weighted average temperature. For viscous fluids, the

*Fig. 20-1 Fourier's Law Applied to Two Simple Cases: (a) Slab and (b) Hollow Cylinder*

convection. When the only motion is due to the heat transfer and the fact that warmer fluids are less dense and will naturally tend to rise, the convection is termed *free* or *natural* convection. A combination of free and forced convection may occur and is termed *mixed* convection.

The quite complex phenomenon of convection was analyzed by Sir Isaac Newton in 1701 resulting in the often called "Newton's Law of Cooling,"

$$q = hA(T_s - T_f) = \frac{(T_s - T_f)}{(1/hA)} \quad (20\text{-}4)$$

# Heat Exchange Equipment

Table 20-2  Approximate Ranges
of Convective Heat Transfer Coefficients

| Flow and Fluid | $h_c$, W/(m²·K) | $h$, Btu/h·ft²·°F |
|---|---|---|
| Free convection, air | 10 | 2 |
| Free convection, water | 50 | 10 |
| Forced convection, air | 100 | 20 |
| Forced convection, water | 500–10,000 | 100 |
| Condensing vapor | 5000–50,000 | 1000–10,000 |
| Boiling liquid | 1000–100,000 | 200–20,000 |

correlation may also include the viscosity evaluated at the surface temperature.

Many engineering applications involve convection transport in noncircular tubes. At least to a first approximation, many of the circular tube results may be applied by using an effective diameter, also termed the hydraulic diameter, as the characteristic length and is defined as

$$D_e = D_h = 4A_c/P$$

where $A_c$ and $P$ are the flow cross-sectional area and the wetted perimeter, respectively. It is this diameter that is used in the calculation of $Re_D$ and the $Nu_D$.

Table 20-2 provides the typical range of convective heat transfer coefficients for several common processes.

## 20.1.3 Radiation

*Radiation* is the mode of heat transfer wherein energy is emitted by one surface (converted from internal energy), transmitted as electromagnetic waves, and then absorbed by a receiving surface. All bodies emit radiant heat continually, with the intensity depending upon the temperature and the nature of the surface. Radiant heat is emitted by a body in the form of finite patches, or quanta, of energy. Their motion in space is similar to the propagation of light and is approximated as traveling in a straight line (slight curvature is neglected). There are many types of electromagnetic radiation, including radio waves, x-rays, gamma rays, as well as light and thermal radiation. The thermal radiation region is from about 0.1 to 100 microns, whereas the visible light portion of the spectrum is very narrow, extending from about 0.40 to 0.65 microns.

Conduction and convection heat transfer rates are driven primarily by temperature gradients and somewhat by temperature due to temperature-dependent properties; however, radiative heat transfer rates are driven by the fourth power of the absolute temperature and increase rapidly with temperature. Unlike conduction and convection, no medium is required to transmit electromagnetic energy and thermal radiation is assumed to pass undiminished through a vacuum and transparent gases. Although the rate of emission of energy is independent of the surroundings, the net heat transfer rate by radiation depends on the temperatures and spatial relationships of all surfaces involved.

The starting point for analyzing radiation heat transfer is with the answers to the three questions:

1. How much radiation is emitted (sent out) by any body?
2. Where does it go?
3. What happens when it gets there?

Depending upon the application, there are actually three answers to Question 1:

1a. How much radiation is emitted at a particular wavelength?
1b. How much radiation is emitted over all wavelengths?
1c. How much radiation is emitted between any two wavelengths?

**Radiation Emitted.** The rate at which thermal radiation is emitted by an ideal surface (perfect emitter) is dependent on its absolute temperature and the wavelength. Such a surface is also a perfect absorber (i.e., it absorbs all incident radiant energy) and is called a ***blackbody***. Planck in 1901 showed that the spectral distribution of energy radiated by a blackbody at an absolute temperature T is given by:

$$E_{b\lambda} = \frac{C_1 \lambda^{-5}}{e^{C_2/\lambda T} - 1} \qquad (20\text{-}5)$$

where

$\lambda$ = wavelength, μm
$T$ = temperature, K
$C_1$ = 3.743 × 10⁸ W·μ/m² (1.187 × 10⁸ Btu·μm⁴/h·ft²)
$C_2$ = 1.4387 × 10⁴ μm/K (2.5896 × 10⁴ μm·°R)

The symbol $E_{b\lambda}$ is used to denote the emitted flux per wavelength (monochromatic emissive power) and is defined as the energy emitted per unit surface area at wavelength lambda per unit wavelength interval around lambda. Equation (20-5), called Planck's distribution law, or just ***Planck's law***, is the basic equation for thermal radiation.

The total radiation emitted over all wavelengths for a blackbody may be obtained by the direct integration of Planck's law:

$$E_b = \int_0^\infty E_{b\lambda}\, d\lambda = \sigma T^4 \qquad (20\text{-}6)$$

The constant $\sigma$ has the value of 5.67 × 10⁻⁸ W/m²·K⁴ (0.1714 × 10⁻⁸ Btu/h·ft²·°R⁴). This expression was deduced by Stefan in 1879 from experimental data. Boltzmann, in 1884, using classical thermodynamics, derived the expression and placed it on firm theoretical ground. The equation is called the ***Stefan-Boltzmann law***.

The radiation emitted between any two wavelengths can also be obtain by integrating Planck's law with the two wavelengths as upper and lower limits of the integral. However, the integration is not easy but fortunately has been accomplished and recorded as ***blackbody radiation functions*** in form(s) readily used, such as provided in Table 20-3.

The blackbody radiation function, $f_{0\text{-}\lambda}$, represents the fraction of radiation emitted from a blackbody at temperature $T$ in the wavelength range from $\lambda = 0$ to $\lambda$. The values of $f$ are

Table 20-3  Blackbody Radiation Function, $f_{0-\lambda}$

| $\lambda T$, μm·K | $f_\lambda$ | $\lambda T$, μm·K | $f_\lambda$ |
|---|---|---|---|
| 200 | 0.000000 | 6200 | 0.754140 |
| 400 | 0.000000 | 6400 | 0.769234 |
| 600 | 0.000000 | 6600 | 0.783199 |
| 800 | 0.000016 | 6800 | 0.796129 |
| 1000 | 0.000321 | 7000 | 0.808109 |
| 1200 | 0.002134 | 7200 | 0.819217 |
| 1400 | 0.007790 | 7400 | 0.829527 |
| 1600 | 0.019718 | 7600 | 0.839102 |
| 1800 | 0.039341 | 7800 | 0.848005 |
| 2000 | 0.066728 | 8000 | 0.856288 |
| 2200 | 0.100888 | 8500 | 0.874608 |
| 2400 | 0.140256 | 9000 | 0.890029 |
| 2600 | 0.183120 | 9500 | 0.903085 |
| 2800 | 0.227897 | 10,000 | 0.914199 |
| 3000 | 0.273232 | 10,500 | 0.923710 |
| 3200 | 0.318102 | 11,000 | 0.931890 |
| 3400 | 0.361735 | 11,500 | 0.939959 |
| 3600 | 0.403607 | 12,000 | 0.945098 |
| 3800 | 0.443382 | 13,000 | 0.955139 |
| 4000 | 0.480877 | 14,000 | 0.962898 |
| 4200 | 0.516014 | 15,000 | 0.969981 |
| 4400 | 0.548796 | 16,000 | 0.973814 |
| 4600 | 0.579280 | 18,000 | 0.980860 |
| 4800 | 0.607559 | 20,000 | 0.985602 |
| 5000 | 0.633747 | 25,000 | 0.992215 |
| 5200 | 0.658970 | 30,000 | 0.995340 |
| 5400 | 0.680360 | 40,000 | 0.997967 |
| 5600 | 0.701046 | 50,000 | 0.998953 |
| 5800 | 0.720158 | 75,000 | 0.999713 |
| 6000 | 0.737818 | 100,000 | 0.999905 |

listed in Table 20-3 as a function of $\lambda T$, where $\lambda$ is in μm and $T$ is in K. The fraction of radiant energy emitted by a blackbody at temperature $T$ over a finite wavelength band from $\lambda_1$ to $\lambda$ is determined from

$$f_{\lambda_1 - \lambda_2}(T) = f_{\lambda_2}(T) - f_{\lambda_1}(T) \qquad (20\text{-}7)$$

**Radiation Between Any Two Surfaces.** Since thermal radiation is taken to travel in straight lines, the determination of "how much goes where" becomes a matter of geometry. Hence the factor used to quantitatively describe the fraction of the radiation leaving a surface and going to another surface is called the **geometric factor, angle factor, configuration factor,** or **view factor**. *View factor* will be used here.

The view factor $F_{ij}$ by definition is the *fraction* of the *total radiation* leaving surface $i$ that *directly* falls upon surface $j$. Figure 20-2 shows the geometry for determining the view factor between two surfaces, 1 and 2. The resulting equation for the view factor is

$$F_{12} = F_{A_1 \to A_2} = \frac{1}{A_1}\int_{A_2}\int_{A_1}\frac{\cos\theta_1 \cos\theta_2}{\pi r^2}dA_1 dA_2 \qquad (20\text{-}8)$$

where $dA_1$ and $dA_2$ are elemental areas of the two surfaces, $r$ is the distance between $dA_1$ and $dA_2$, and $\theta_1$ and $\theta_2$ are the angles between the respective normals to $dA_1$ and $dA_2$ and the connecting line $r$. The solution of this equation in closed form is difficult, if not impossible, for all geometries.

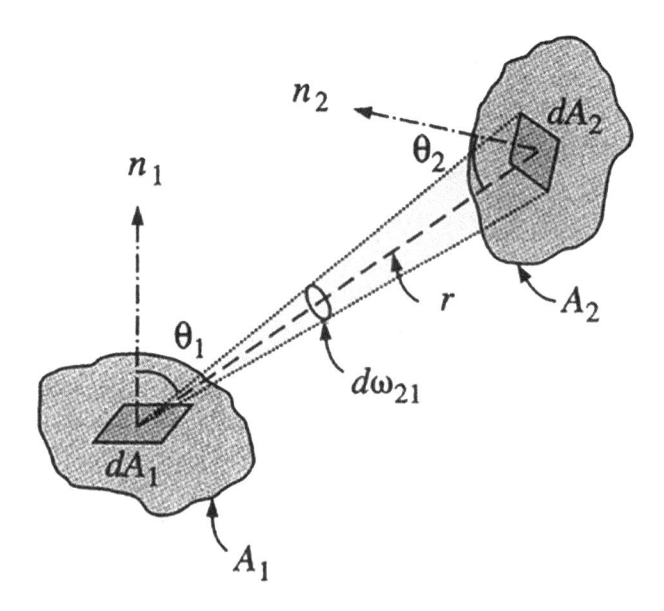

*Fig. 20-2 View Factor Nomenclature*

Numerical, graphical, and mechanical techniques have all provided alternative methods, and numerical values of the view factor for many geometries encountered in engineering may be found in the literature. It must be emphasized that the expression for the view factor is based on the assumption that the directional distribution of radiation leaving a surface is diffuse and uniformly distributed.

Two special properties play a very important role in obtaining numerical values for the complete set of view factors between the surfaces exchanging radiation. If there are $n$ surfaces forming an enclosure, then

$$\sum_{j=1}^{n} F_{ij} = 1 \qquad (20\text{-}9)$$

The other is the reciprocal relationship: $F_{ij}A_i = F_{ji}A_j$. In both cases, it is important to note that $F_{jj}$ is not necessarily 0 since a concave surface may irradiate ("see") itself.

Numerical values of view factors for three common geometries are provided in Table 20-4 and Figure 20-3.

**Radiation Falling on Surface.** When radiant energy falls on a surface, portions may be absorbed in, reflected from, or transmitted through the material as shown in Figure 20-4.

Therefore, based on conservation of energy,

$$\alpha + \rho + \tau = 1 \qquad (20\text{-}10)$$

where

$\alpha$ = fraction of incident radiation absorbed (absorptivity or absorptance)

$\rho$ = fraction of incident radiation reflected (reflectivity or reflectance)

Table 20-4  View Factors for Three-Dimensional Geometries

| Geometry | Relation |
|---|---|
| Aligned parallel rectangles | $\bar{X} = X/L, \quad \bar{Y} = Y/L$ $F_{ij} = \frac{2}{\pi \bar{X} \bar{Y}} \Big\{ \ln \Big[ \frac{(1+\bar{X}^2)(1+\bar{Y}^2)}{1+\bar{X}^2+\bar{Y}^2} \Big]^{1/2}$ $+ \bar{X}(1+\bar{Y}^2)^{1/2} \tan \frac{\bar{X}}{(1+\bar{Y}^2)^{1/2}}$ $+ \bar{Y}(1+\bar{X}^2)^{1/2} \tan^{-1} \frac{\bar{Y}}{(1+\bar{X}^2)^{1/2}}$ $- \bar{X} \tan^{-1} \bar{X} - \bar{Y} \tan^{-1} \bar{Y} \Big\}$ |
| Coaxial parallel disks | $R_i = R_i/L, \quad R_j = r_j/L$ $S = 1 + \frac{1+R_j^2}{R_i^2}$ $F_{ij} = \frac{1}{2} \Big\{ S - [S^2 - 4(r_j/r_i)^2]^{1/2} \Big\}$ |
| Perpendicular rectangles with a common edge | $H = Z/X, \quad W = (Y/X)$ $F_{ij} = \frac{1}{\pi W} \Big( W \tan^{-1} \frac{1}{W} + H \tan^{-1} \frac{1}{H}$ $- (H^2+W^2)^{1/2} \tan^{-1} \frac{1}{(H^2+W^2)^{1/2}}$ $+ \frac{1}{4} \ln \Big\{ \frac{(1+W^2)(1+H^2)}{1+W^2+H^2}$ $\times \Big[ \frac{W^2(1+W^2+H^2)}{(1+W^2)(W^2+H^2)} \Big]^{W^2}$ $\times \Big[ \frac{H^2(1+H^2+W^2)}{(1+H^2)(H^2+W^2)} \Big]^{H^2} \Big\} \Big)$ |

Table 20-5  Emittance Values of Common Materials

| Material and Surface Condition | Total Hemispherical Emittance | Solar Absorptance |
|---|---|---|
| Aluminum | | |
|   Foil | 0.05 | 0.15 |
|   Alloy, as received | 0.04 | 0.37 |
|   Weather alloy | 0.20 | 0.54 |
| Asphalt (Roofing/Pavement) | 0.88 | |
| Brick | 0.90 | 0.63 |
| Concrete, rough | 0.91 | 0.60 |
| Copper | | |
|   Electroplated | 0.03 | 0.47 |
|   Oxidized plate | 0.76 | |
| Frost, rime | 0.99 | |
| Glass (smooth) | 0.91 | |
| Gravel | 0.30 | |
| Ice (smooth) | 0.97 | |
| Iron | | |
|   Wrought, polished | 0.29 | |
|   Wrought, dull | 0.91 | |
| Marble | | |
|   Polished | 0.89 | |
|   Smooth | 0.56 | |
| Paints | | |
|   Black | | |
|     Flat | 0.97 | 0.98 |
|     Gloss | 0.90 | |
|   White | | |
|     Acrylic resin | 0.90 | 0.26 |
|     Gloss | 0.85 | |
| Skin | 0.95 | |
| Soil | 0.94 | |
| Snow (fresh) | 0.82 | 0.13 |
| Stainless Steel | | |
|   Polished | 0.60 | 0.37 |
|   Dull | 0.2 | |
| Vegetation | 0.94 | |
| Water | 0.90 | 0.98 |
| Wood (smooth) | 0.84 | |

$\tau$ = fraction of incident radiation transmitted (transmissivity or transmittance)

For the many materials encountered in HVAC practice (other than fenestrations) that are opaque in the infrared region, $\tau = 0$, and thus $\rho = 1 - \alpha$.

**Actual Surfaces (Nonblack Bodies).** Materials and surfaces of engineering interest show marked divergences from the Stephan-Boltzmann and Planck laws. Actual surfaces emit and absorb less readily and are called nonblack. The emittance (or emissivity) $\varepsilon$ of the actual surface is defined as the ratio of the radiation emitted by the surface to the radiation emitted by a blackbody at the same temperature. Emissivity is a function of the material, the condition of its surface, and its temperature. In general, the emissivity of a surface may vary with wavelengths. To overcome this complexity, gray surface behavior ($\varepsilon$ = constant over all wavelengths) actually do approximate this condition, at least in some regions of the spectrum. However, one must be especially careful at high temperature. The emissive power of a non-black surface, at temperature $T$, is given by

$$E = \varepsilon E_b = \varepsilon \sigma T^4 \qquad (20\text{-}11)$$

where $\varepsilon$ is the total hemispherical emittance (or emissivity) and is a strong function of the condition and temperature of the actual surface. Table 20-5 provides approximate emittance values of some common materials and surface finishes at room and solar temperatures. In general, both $\varepsilon$ and $\alpha$ of a surface depend on the temperature and the wavelength of the radiation. **Kirchhoff's law** of radiation states that the emittance and the absorptance of a surface at a given temperature and wavelength are equal. In many practical applications, the surface temperature and the temperature of the source of incident radiation (major exception – solar radiation) are of the same order of magnitude, and the average absorptance of a surface is taken to be equal to its average emittance ($\alpha = \varepsilon$).

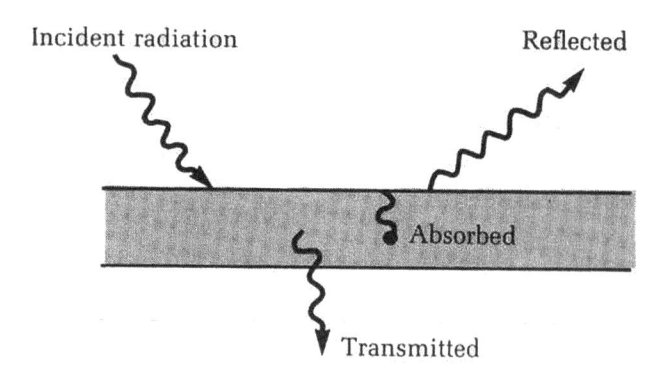

*Fig. 20-3 Radiation Incident on Surface*

### 20.1.4 Net Radiant Energy Loss from a Surface

When radiation heat transfer is involved in the energy balance on a surface, the net heat gain (or loss) by radiation from the surface is the quantity of interest. The general problem of determining the radiation exchange in an enclosure consisting of *n* surfaces, which may see one another by no others, requires the solution of n linear algebraic equations to account for the possibly infinite number of reflections of radiation from the participating surfaces. The current method of determining this quantity is called the ***radiosity method***. It begins with the definitions of the two terms, radiosity and irradiation, and is applicable subject to the following conditions:

- Each surface is opaque, gray, isothermal, and uniformly irradiated.
- The emission and the reflections from each surface are diffuse.

### 20.1.5 Radiosity

The *radiosity J* is the total radiation leaving a surface, per unit area, per unit time. It includes both the emitted and the reflected amounts and can be expressed for surface *i* as

$$J_i = E_i + \rho G_i = \varepsilon_i \sigma T_i^4 + (1-\varepsilon_i) G_i \quad (20\text{-}12)$$

where $G_i$ is the radiation falling upon surface *i*.

### 20.1.6 Irradiation

The *irradiation G* is the total radiation falling on a surface, per unit area, per unit time. The radiation incident on the *i*th surface is

$$A_i G_i = J_1 A_1 F_{1i} + J_2 A_2 F_{2i} + J_3 A_3 F_{3i} + \cdots$$

or

$$G_i = \sum_{j=1}^{n} J_j F_{ij}$$

(using the reciprocal rule).

The radiosity of surface *i* is $J_i = \varepsilon_i E_{bi} + (1-G_i)$; substituting for $G_i$ gives

$$J_i = \varepsilon_i E_{bi} + (1-\varepsilon_i)\sum_{k}^{n} J_j F_{ik}; \quad (20\text{-}13)$$

$$i = 1, 2, \ldots, n$$

a system of *n* linear equations in the *n* unknowns $J_i$. Upon solving the simultaneous equations and obtaining the values for the *J*s, the net radiant heat loss from each surface is obtained from

(20-14)

$$q_i = [J_i - G_i]A_i$$

$$q_i = A_i\left(J_i - \sum_{j=1}^{n} F_{ij} J_j\right) = \frac{E_{bi} - J_i}{(1-\varepsilon_i)/E_i A_i}$$

For the special case when only two surfaces are involved, the net loss can be written as

$$q_i = \frac{E_{b1} - E_{b2}}{\dfrac{1-\varepsilon_1}{\varepsilon_1 A_1} + \dfrac{1}{F_{1\text{-}2} A_1} + \dfrac{1-\varepsilon_2}{\varepsilon_2 A_2}}$$

$$= \frac{\sigma(T_1^4 - T_2^4)}{\dfrac{1-\varepsilon_1}{\varepsilon_1 A_1} + \dfrac{1}{F_{1\text{-}2} A_1} + \dfrac{1-\varepsilon_2}{\varepsilon_2 A_2}} \quad (20\text{-}15)$$

### 20.1.7 Radiation and Combined Heat Transfer Coefficients

A special case that occurs frequently involves radiation exchange between a small surface at $T_s$ and a much larger, isothermal surface that completely surrounds the smaller one. The *surroundings* could, for example, be the walls of a room whose temperature $T_{sur}$ differs from that of an enclosed surface ($T_{sur} \neq T_s$). For such a condition, the *net* rate of radiation heat transfer *from* the surface, per unit area of the surface, is

$$q = \varepsilon A \sigma(T_s^4 - T_{sur}^4) \quad (20\text{-}16)$$

The $T^4$ dependence of radiant heat transfer complicates engineering calculations. When $T_1$ and $T_2$ are not too different, it is convenient to linearize Eq. (20-16) by factoring the term $(\sigma T_1^4 - \sigma T_2^4)$ to obtain

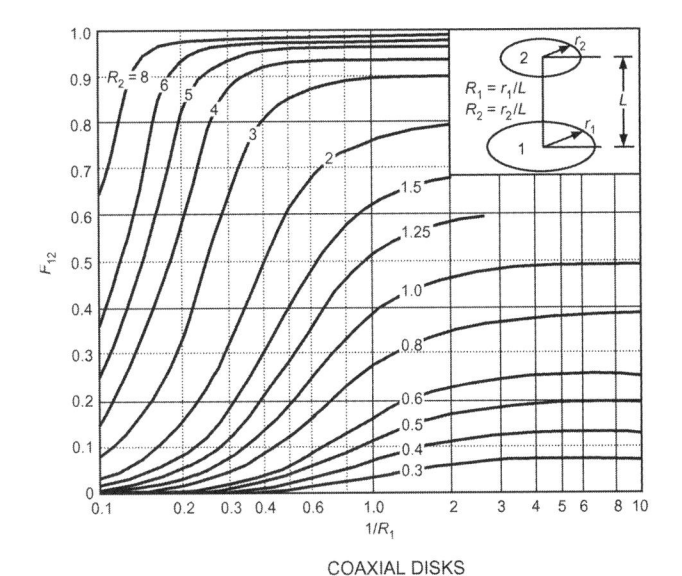

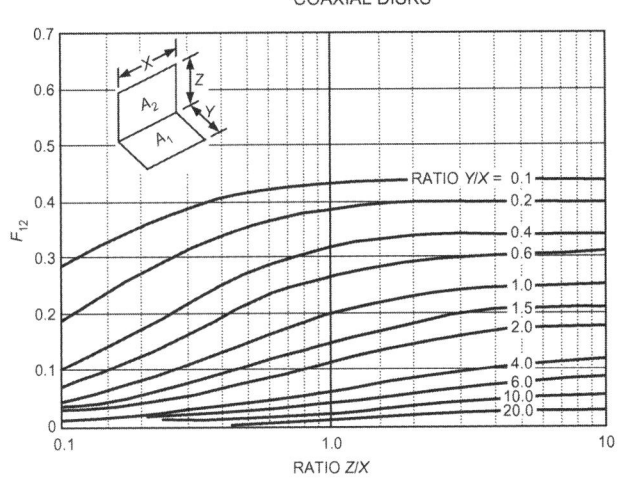

# Heat Exchange Equipment

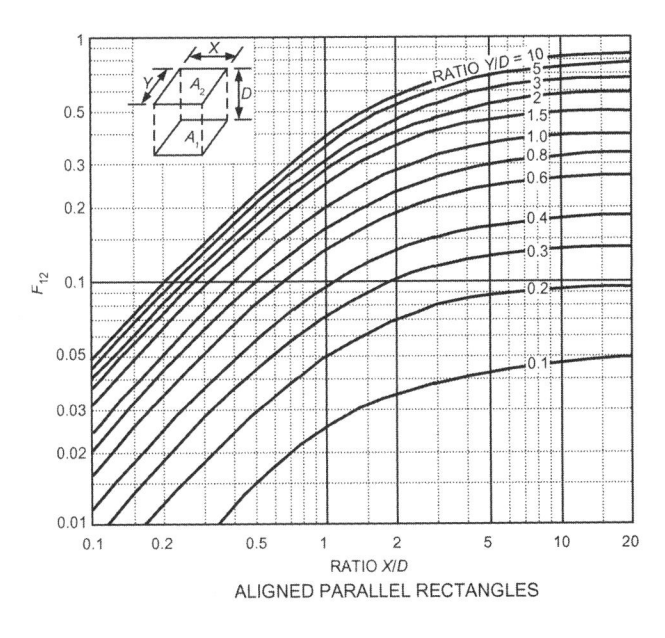

*Fig. 20-4 View Factor Graphs for Common Geometries*

$$q_{12} = \varepsilon_1 A_1 (T_1^2 + T_2^2)(T_1 + T_2)(T_1 - T_2) \quad (20\text{-}17)$$
$$\cong \varepsilon_1 A_1 \sigma (4 T_m^3)(T_1 - T_2)$$

for $T_1 \cong T_2$, where $T_m$ is the mean of $T_1$ and $T_2$. This result can be written more concisely as

$$q \cong A_1 h_r (T_1 - T_2)$$

where $h_r = 4\varepsilon_1 \sigma T_m^3$ is called the **radiation heat transfer coefficient**, in Btu/h·ft²·°F (W/m²·K).

Heat transfer from surfaces is usually a combination of convection and radiation. It is assumed that these modes are additive, and therefore a combined surface coefficient can be used to estimate the heat flow to/from a surface:

$$h_o = h_c + h_r$$

where

$h_o$ = overall surface coefficient, Btu/h·ft²·°F (W/m²·K)
$h_c$ = convection coefficient, Btu/h·ft²·°F (W/m²·K)
$h_r$ = radiation coefficient, Btu/h·ft²·°F (W/m²·K)

Assuming the radiant environment is equal to the temperature of the ambient air, the heat loss/gain at the surface can be calculated as

$$q = h_o A (T_{\text{surf}} - T_{\text{amb}}) \quad (20\text{-}18)$$

Radiation is usually significant relative to conduction or natural convection, but negligible relative to forced convection. Thus, radiation in forced-convection applications is usually disregarded, especially when the surfaces involved have low emissivities and low to moderate temperatures.

## 20.2 Heat Exchangers

A heat exchanger is a device that permits the transfer of heat from a warm fluid to a cooler fluid through an intermediate surface without mixing of the two fluids. The correct sizing and selection of heat exchangers is probably the most important single factor in designing an efficient and economical building HVAC&R system. Whether the heat exchanger is selected as an off-the-shelf item or designed especially for the application, the following factors are normally considered:

- Thermal performance
- Cost
- Pressure drop
- Space requirements
- Serviceability.

The main types of heat exchangers found in HVAC&R systems are the: finned-tube (coil), shell-and-tube, and plate. Sketches of each are given in Figure 20-5.

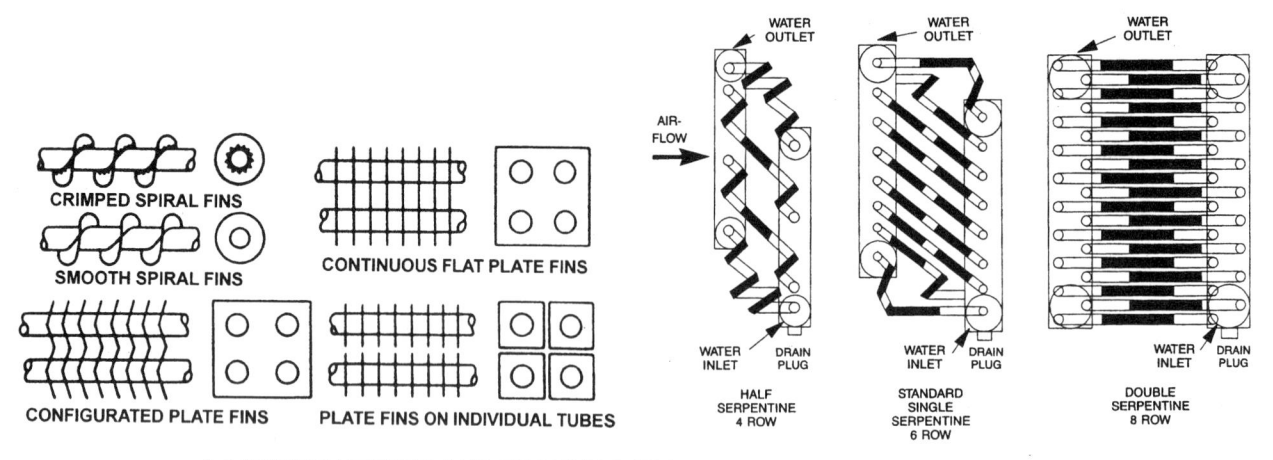

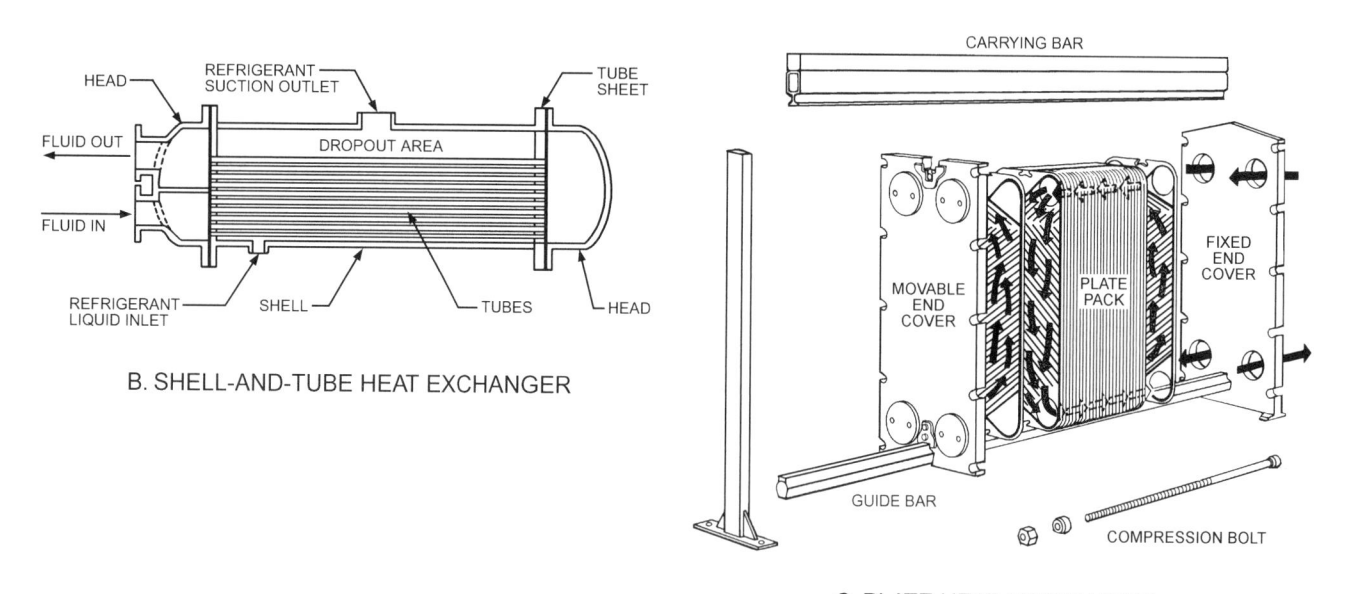

*Fig. 20-5 Sketches of the Three Types of Heat Exchangers: (A) Finned-Tube, (B) Shell-and-Tube, and (C) Plate (Part A: Figure 1, Chapter 23; Part B: Figure 2, Chapter 42; and Part C: Figure 14, Chapter 48 in 2012 ASHRAE Handbook—HVAC Systems and Equipment)*

## 20.2.1 Plate Fin (Extended-Surface) Coils

Most coils in HVAC systems consist of tubes with fins attached to their outer surface. Air flows over the outside of the tubes and refrigerant, steam, or water flows inside the tubes. The purpose of the fins is to increase the surface area on the air side where the convection coefficient is usually much lower than on the refrigerant, steam, or water side.

The *face area* of the coil is the cross sectional area of the air stream at the entrance of the coil and is obtained from the *length* (sometimes called the *width*) of the coil multiplied by the *height* of the coil. The *face velocity* of the air is the volume flow rate of the air divided by the face area. The *surface area* of the coil is the heat transfer surface area in contact with the air. The *number of rows* of tubes and the *depth* of the coil are measured in the direction of the airflow.

In comparison to bare tube coils of the same capacity, finned coils are much more compact, less weight, and usually less expensive. The secondary surface area of a finned coil may be 10 to 40 times that of the bare tubes. The primary surface is that of the tubes or pipes. The secondary surface (fins) consists of thin metal plates or a spiral ribbon uniformly spaced or wound along the length of the primary surface and in intimate contact with it. The bond between fin and tube is a significant parameter in the thermal performance of the coil. The bonding is usually accomplished by expanding the tubes (often copper) into the tube holes in the plate fins (often sheets of aluminum). The tube holes are often punched with a formed fin collar which both provides contact area as well as a means of spacing the fins uniformly along the length of the tubes. Figure 20.5(a) illustrates several finned-tube heat exchangers (coils).

More information of cooling coils can be found in Section 17.2 of this book while more on heating coils appears in Section 17.3.

### 20.2.2 Shell-and-Tube Heat Exchangers

The most common type of heat exchanger in industrial applications is probably the shell-and-tube heat exchanger. This type also finds extensive use in HVAC&R applications involving water cooled units and water and brine chillers. Figure 20.5(b) provides a sketch of the shell-and-tube heat exchanger.

Shell-and-tube heat exchangers can handle from a single tube to a large number of tubes packed in a shell with their axes parallel to that of the shell. Heat transfer takes place as one fluid flows inside the tubes while the other fluid moves outside the tubes through the shell. Baffles are usually placed in the shell to force the shell-side fluid to maintain uniform and good contact with the outside of the tubes. Shell-and-tube heat exchangers are generally classified according to the number of tube and shell passes involved.

### 20.2.3 Plate heat exchangers

The plate-and-frame (or just plate) type of heat exchanger continues to find additional use in HVAC applications, including the water side economizer. Plate heat exchangers consist of metal plate pairs arranged to provide separate flow paths (channels) for the two fluids. Heat transfer occurs across the plate walls. The hot and cold fluids flow in alternate passages, and thus each cold fluid stream is surrounded by two hot fluid streams, resulting in very effective heat transfer. Figure 20.5(c) provides a pictorial view of the plate exchanger and illustrates the flow paths.

The exchangers have multiple channels in series that are mounted on a frame and clamped or welded together. The rectangular plates have an opening or port at each corner. When assembled, the plates are sealed such that the ports provide manifolds to distribute the fluids through the separate flow paths. The clamped type of plate heat exchanger can be easily enlarged to meet higher heat transfer rates by simply mounting more plates. Plate exchangers are particularly well suited for liquid-to-liquid heat exchange applications, but also find use as condensers and evaporators.

## 20.3   Basic Heat Exchanger Design Equation

By applying the fundamentals of heat transfer to heat exchangers whose purpose it is to transfer heat from one fluid to another, the overall heat flow from one fluid across a barrier to a second fluid is often expressed as:

$$q = UA \, \Delta T \qquad (20\text{-}19)$$

where
- $q$ = rate of heat transfer, Btu/h or kW
- $A$ = surface area of material separating the two fluids, ft$^2$ or m$^2$
- $U$ = overall coefficient of heat transfer, Btu/h·ft$^2$·°F or W/m$^2$·K
- $\Delta T$ = *mean* temperature difference between the hot and cold fluids, °F or K

Rearranging this equation yields the basic design equation for a heat exchanger as

$$A = \frac{q}{U \, \Delta T_m} \qquad (20\text{-}20)$$

where $A$ is the total heat transfer area required in the exchanger. Thus, it will now be necessary to first calculate (estimate) $q$, $U$, and $\Delta T$, as discussed in the following sections.

## 20.4   Estimation of Heat Load

The usual first step in designing a heat exchanger is to use the First Law of Thermodynamics to make an energy balance for (a) estimating the heat load (duty) of the heat exchanger, and probably (b) the required flow rate of one of the fluids. Since the heat exchanger is normally assumed to be overall adiabatic (only energy exchange takes place within the heat exchanger, from the hot fluid to the cold fluid), the heat load is calculated in the general case from

$$q = m_h(h_{h,\,in} - h_{h,\,out}) = m_c(h_{c,\,out} - h_{c,\,in}) \qquad (20\text{-}21)$$

where $m_h$ and $m_c$ are the mass flow rates of the hot and cold fluids and $h_{h,in}$, $h_{h,out}$, $h_{c,in}$, and $h_{c,out}$ are the respective enthalpies. When there is no change in phase, the enthalpy change can be replaced as follows:

$$\begin{aligned} h_{out} - h_{in} &= c_p(T_{out} - T_{in}) \\ \text{or} \quad h_{in} - h_{out} &= c_p(T_{in} - T_{out}) \end{aligned} \qquad (20\text{-}22)$$

where $c_p$ is the specific heat of the particular fluid.

Upon specifying the function of the particular heat exchanger (e.g., cool a known amount of hot fluid from one temperature to another), Eq. (20-20) can be used to calculate the required heat transfer rate from the hot fluid (also the rate to the cold fluid) as well as to determine either the amount of the other fluid (if both inlet and outlet temperatures are known) or determine its outlet temperature (if its flow rate and inlet temperature are known).

## 20.5   Mean Temperature Difference

The temperature of the fluids flowing through a heat exchanger generally varies from location to location as heat is transferred from the hotter to the colder fluid. There is no single temperature difference serving as the driving force for the heat transfer and thus the rate of heat transfer with this varying temperature difference must be obtained by integrating

$$dq = U\, dA\, \Delta T$$

over the heat transfer area $A$ along the length of the heat exchanger. The result, for either concurrent (parallel) or countercurrent (counterflow) flow conditions yields

$$q = UA(\Delta T_a - \Delta T_b)/\ln(\Delta T_a/\Delta T_b)$$

where $a$ and $b$ refer to the two ends of the heat exchanger.

The concept of an appropriate mean temperature difference, a single temperature difference which results in the same heat flow value, is useful and widely used in engineering practice:

$$q = UA\, \Delta T_{mean}$$

For parallel or counterflow conditions, this mean temperature difference is therefore

$$\Delta T_{mean} = (\Delta T_a - \Delta T_b)/\ln(\Delta T_a/\Delta T_b) \quad (20\text{-}23)$$

and is called the log mean temperature difference or LMTD.

For stream conditions other than the ideal counterflow (such as the common cross flow), a correction factor $F$ is applied to the LMTD obtained as if the flow had been pure counterflow. Examples of these corrections factors are provided in Figure 20-6 (Bowman et al. 1940).

The resulting equation for the heat transfer rate becomes

$$q = UAF\, \Delta T_{m,cf} \quad (20\text{-}24)$$

For cases where at least one fluid temperature remains constant (e.g., evaporation or condensation), the correction factor is unity regardless of the flow pattern.

## 20.6 Estimation of the Overall Heat Transfer Coefficient $U$

The governing equations for the design of heat exchangers are as follows:

$$Q = U_o A_o F\, \Delta T_{m,cf} = U_i A_i F\, \Delta T_{m,cf} \quad (20\text{-}24\text{a})$$

where
- $Q$ = amount of heat transfer for the coil to do
- $U$ = overall heat transfer coefficient
- $F$ = temperature difference correction
- $\Delta T$ = log mean temperature difference
- $A_i, A_o$ = internal or outside areas

$$U = 1/\Sigma R \quad (20\text{-}24\text{b})$$

where the $R$s are resistances to the flow defined by

- $A/H_i A_i$ = internal convective resistance
- $AR_{fi}/A_i$ = internal fouling in the pipes

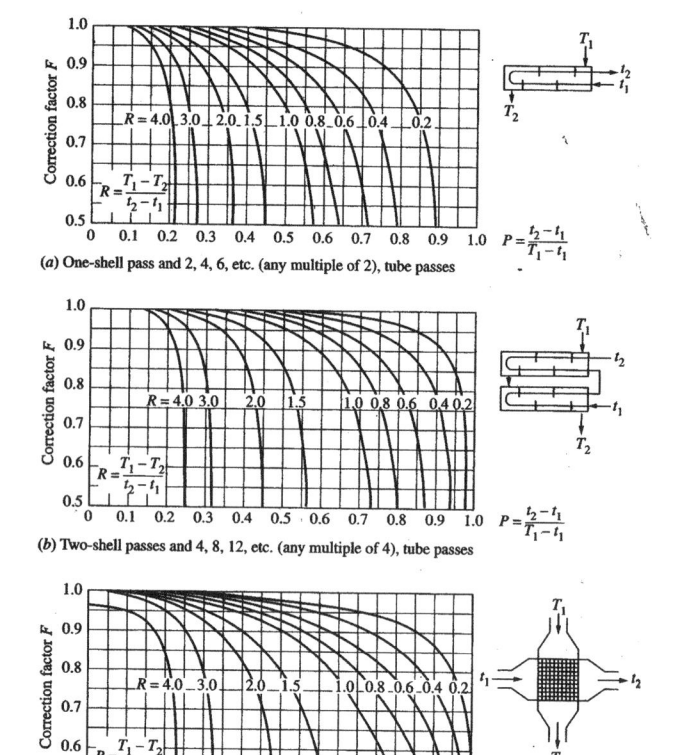

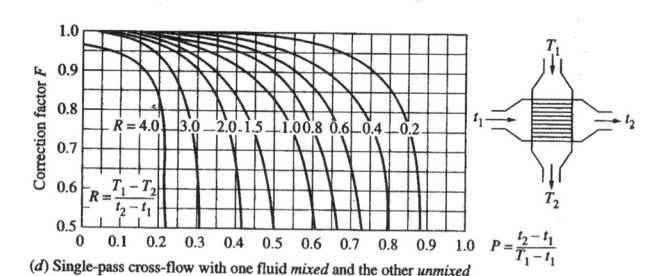

*Fig. 20-6 LMTD Correction Factors for Several Flow Configurations*
(Bowman et al. 1940)

- $A \ln(d_o/d_i)/(2\Pi kl)$ = conductive resistance through wall of piping
- $AR_{fo}/A_o$ = outside fouling resistance
- $AR_c/A_o$ = contact resistance between tube and fin
- $A/[h_o(A_{unfin} + \phi A_{fin})]$ = outside convective resistance

$$\Delta T_{m,cf} = (\Delta T_a - \Delta T_b)/\ln(\Delta T_a - \Delta T_b) \quad (20\text{-}24\text{c})$$

These equations become

$$Q = \frac{F\, \Delta T_{m,cf}}{\dfrac{1}{h_i A_i} + \dfrac{R_{fi}}{A_i} + \dfrac{\ln(D_o/D_i)}{2\pi kL} + \dfrac{R_{fo}}{A_o} + \dfrac{R_c}{A_o} + \dfrac{1}{h_o(A_u + \phi A_f)}}$$

$$(20\text{-}25)$$

# Heat Exchange Equipment

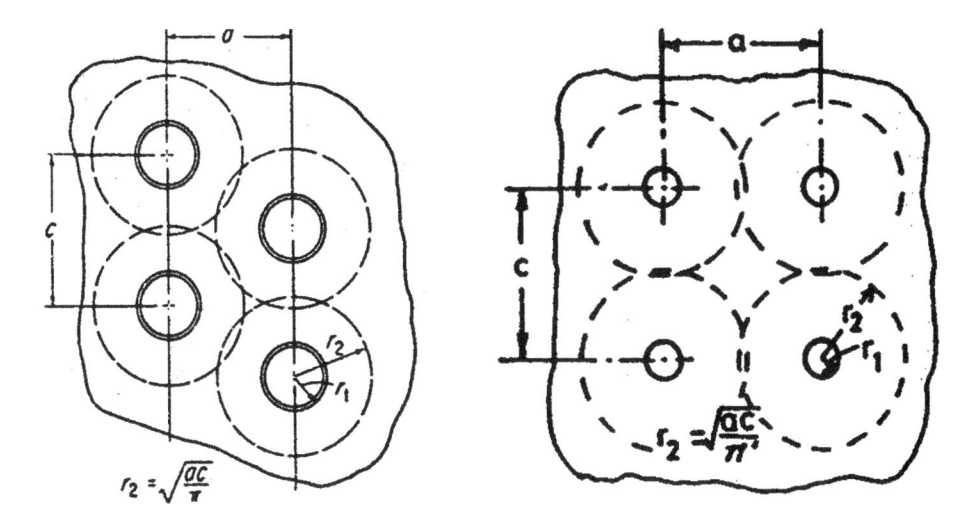

*Fig. 20-7 Approximate Method for Obtaining Efficiency of Common Flat Plate Fin*

A complicated heat transfer phenomenon is considerably simplified by the assumption of boundary layers or films between the barrier wall and the fluids that offer resistance to heat flow. This mechanism is represented by the following equation, which assumes a constant overall heat transfer coefficient in the entire heat exchanger. This is not unreasonable when the temperature change in each fluid is small and therefore there is little change of physical properties between the inlet and the outlet.

$$U = \frac{1}{\frac{1}{h_o} + \frac{1}{R_{fo}} + \frac{d_w}{k_w}\left(\frac{A_o}{A_{mean}}\right) + \frac{1}{R_{fi}}\left(\frac{A_o}{A_i}\right) + \frac{1}{h_i}\left(\frac{A_o}{A_i}\right)} \quad (20\text{-}25a)$$

where

$U$ = overall heat transfer coefficient, Btu/h·ft²·°F (W/m²·K)

$h_o$ = film coefficient of fluid outside tube, Btu/h·ft²·°F (W/m²·K)

$h_i$ = film coefficient of fluid inside tube, Btu/h·ft²·°F (W/m²·K)

$R_{fo}$ = fouling coefficient outside of tube, Btu/h·ft²·°F (W/m²·K)

$R_{fi}$ = fouling coefficient inside of tube, Btu/h·ft²·°F (W/m²·K)

$d_w$ = thickness of tube wall, ft (m)

$k_w$ = thermal conductivity of tube, Btu·ft/h·ft²·°F (W/m·K)

$A_o/A_i$ = ratio of outside to inside tube surface

$A_{mean}$ = average tube area per unit length, ft²/ft (m²/m)

$A_o$ = outside tube area per unit length, ft²/ft (m²/m)

The accuracy of this relationship is limited by the reliability of the correlations for calculating the individual film coefficients, and by the arbitrary selection of fouling coefficients.

## 20.7 Extended Surfaces, Fin Efficiency, and Fin-Tube Contact Resistance

When the tube is finned on the air side to enhance heat transfer, the total heat transfer surface on the finned side becomes

$$A_s = A_{total} = A_{fin} + A_{unfinned}$$

where $A_{fin}$ is the surface area of the fins and $A_{unfinned}$ is the area of the unfinned portion of the tube surface. However, when determining the heat transfer rate using the overall coefficient $U$, the appropriate total surface area to use is the effective area which means de-rating the area of the fins due to the internal thermal resistance within the fins. Thus, the appropriate surface area is

$$A = A_{unfinned} + \phi A_{fin} \quad (20\text{-}26)$$

The rectangular-plate fin of uniform thickness is commonly used in finned coils for heating or cooling air. It is not possible to obtain an exact mathematical solution for the efficiency of such a fin. It can be shown that an adequate approximation is to assume that the fin area served by each tube is equivalent in performance to a flat circular-plate fin of equal area. Figure 20-7 shows the method for determining the equivalent outer radius for this method. The corresponding efficiency for the flat plate fin can then be obtained from Figure 20-8.

As detailed in Chapter 4 of the 2013 *ASHRAE Handbook—Fundamentals*, the approximate fin efficiencies can also be calculated as provided in Figure 20-9.

The most common means of bonding the fins to the tubes on common heating and cooling coils is by mechanical expansion. Results of Sheffield et al. (1985) have shown that,

a. rectangular tube array

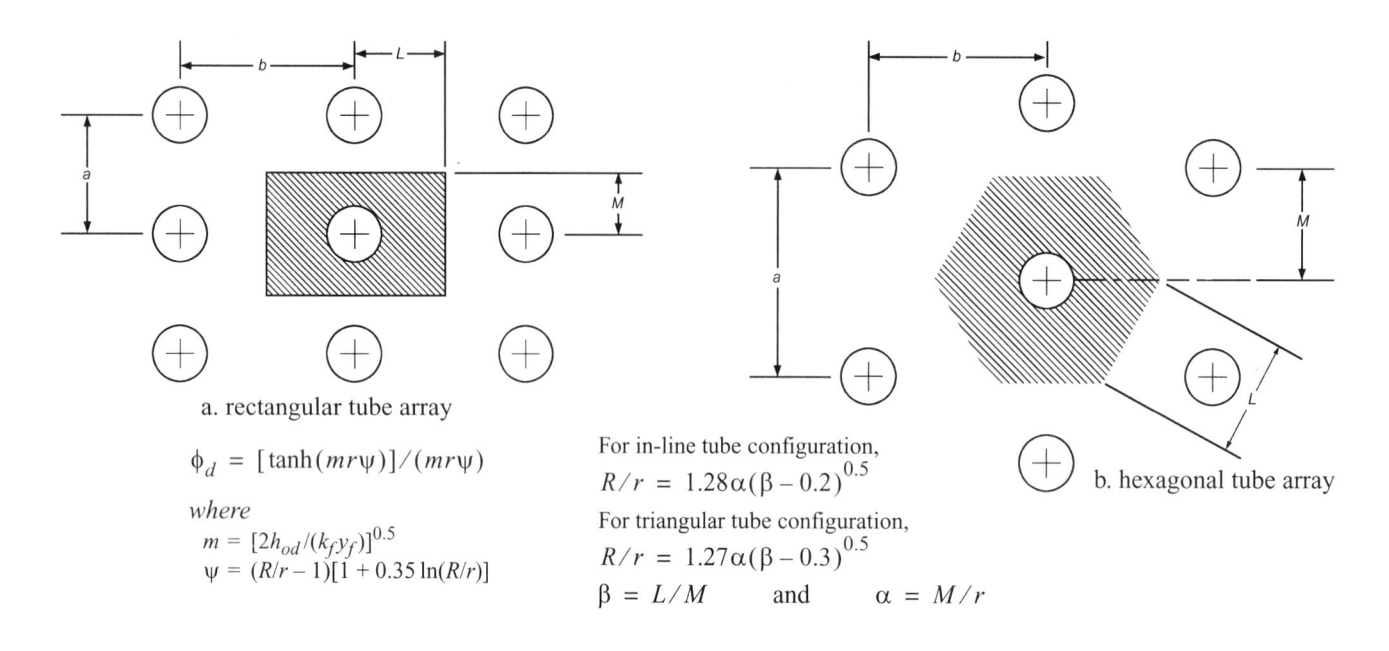

$$\phi_d = [\tanh(mr\psi)]/(mr\psi)$$

where
$m = [2h_{od}/(k_f y_f)]^{0.5}$
$\psi = (R/r - 1)[1 + 0.35 \ln(R/r)]$

For in-line tube configuration,
$R/r = 1.28\alpha(\beta - 0.2)^{0.5}$

For triangular tube configuration,
$R/r = 1.27\alpha(\beta - 0.3)^{0.5}$

$\beta = L/M$ and $\alpha = M/r$

b. hexagonal tube array

*Fig. 20-8 Approximate Equations for Plate Fin Efficiency*

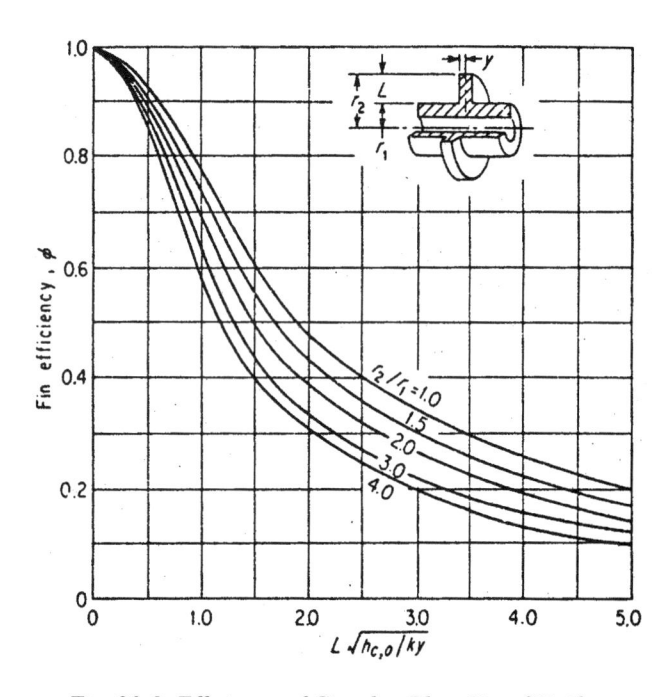

*Fig. 20-9 Efficiency of Circular-Plate Fin of Uniform Thickness*
(Source: Gardner 1945)

for properly expanded tubes, there is a relatively narrow range of values of thermal resistance due to the fact that the tubes and fins are only in contact over a relative small area due to surface asperities, non-roundness, and other factors. As reported by Sheffield et al., a reasonable value for the conductance between mechanically expanded copper tubes and aluminum fins is 3750 Btu/h·ft²·°F (21 293 W/m²·K).

**Table 20-6  Fouling Factors for Water, h·ft²·°F/Btu (m²·K/W)**

| | Water Velocity, ft/s (m/s) | |
|---|---|---|
| | 3 (1) and less | Over 3 (1) |
| Type of Water | IP (SI) | IP (SI) |
| Cooling tower and spray pond | | |
|   Treated makeup | 0.001 (0.00018) | 0.001 (0.00018) |
|   Untreated | 0.003 (0.0005) | 0.003 (0.0005) |
| River water (average) | 0.002 (0.00035) | 0.001 (0.00018) |
| Hard | 0.003 (0.0005) | 0.002 (0.00035) |
| Distilled or closed cycle condensate | 0.001 (0.00018) | 0.001 (0.00018) |

The thermal contact resistance (TCR) is the inverse, or 0.000267 h·ft²·°F/Btu (0.000047 m²·K/W).

## 20.8 Fouling Factors

After a period of operation, the heat transfer surfaces of a heat exchanger may become coated with various deposits from the fluids or may become corroded as a result of interaction between the fluids and the surface material. This coating represents an additional resistance to the flow of heat and results in decreased heat transfer performance. The effect is accounted for by a fouling factor, or fouling resistance, $R_f$, which is then to be added to the other resistances in the thermal path between the two fluids.

The most common type of fouling is the precipitation of solids from the fluid onto the heat transfer surface. Other types of fouling include chemical fouling (e.g., corrosion) and biological fouling from algae growth.

# Heat Exchange Equipment

Table 20-7  Fouling Factors for Various Fluids

| Type of Fluid | Fouling Factor h·ft²·F/Btu | m²·K/W |
|---|---|---|
| **Gases and vapors** | | |
| Steam (non-oil bearing) | 0.0005 | 0.00009 |
| Refrigerant vapors (oil bearing) | 0.002 | 0.00035 |
| Refrigerant vapors (pure) | 0 | 0 |
| Compressed air | 0.002 | 0.00035 |
| Industrial organic heat transfer media | 0.001 | 0.00018 |
| **Liquids** | | |
| Refrigerant liquids | 0.001 | 0.00018 |
| Industrial organic heat transfer media | 0.001 | 0.00018 |

The fouling factor depends upon the tube material, the nature of the fluid, and the fluid velocity. Fluid velocities less than about 3 ft/s (0.9 m/s) tend toward excess fouling. A few example values of fouling factors are provided in Table 20-6 and Table 20-7. Considerable uncertainty exists in these values, and they should be used cautiously. More comprehensive tables are available from TEMA (Tubular Exchanger Manufacturers Association) (tema.org).

There is little published data on the rate of fouling for heat exchangers in typical air conditioning and refrigeration service. For many years, the basic reference has been the TEMA Standard. The air conditioning industry has for decades commonly used an assumed fouling level of 0.0005 h·ft²·°F/Btu (0.00009 m²·K/W) in both condensers and coolers. Occasionally, where a condenser was to use river water, engineers would specify as much as 0.0020 h·ft²·°F/Btu (000035 m²·K/W) fouling. Based on more recent studies by the Air-Conditioning, Heating, and Refrigeration Institute (AHRI, formerly ARI), it appears reasonable to specify a fouling factor of 0.00025 h·ft²·°F/Btu (0.000044 m²·K/W) for

- closed-loop liquid chillers
- condensers served by well-maintained cooling towers

## 20.9 Convective Heat Transfer Coefficients $h_i$ and $h_o$

The determination of accurate values for the inside and outside convective heat transfer coefficients is critical to the accurate evaluation of heat exchanger performance. Unfortunately, even today, the available correlations for predicting these coefficients often leave much to be desired, particularly if a phase change is occurring.

The correlations presented in the following subsections are included herein primarily as (a) examples, and (b) to provide sample working relations for use with both the example heat exchanger design problems and the homework problems at the end of the chapter. The reader is referred to the current technical literature in heat transfer (e.g., *International Journal of Heat and Mass Transfer, Journal of Heat Transfer, International Journal of HVAC & Refrigeration Research*) for improved correlations.

### 20.9.1 Single-Phase Internal Flow in Tubes

The simplified relation of McAdams (1940) is widely used for turbulent single-phase flow in tubes and pipes"

$$\text{Nu} = 0.023(\text{Re})^{0.8}(\text{Pr})^{1/3} \quad (20\text{-}27)$$

The relation is relatively simple, but gives maximum errors of ±25% in the range of $0.67 < \text{Pr} < 100$. A more accurate correlation, which is also applicable for rough ducts, has been developed by Petukhov and coworkers at the Moscow Institute for High Temperature:

$$\text{Nu} = (\text{Re}\,\text{Pr}/X)(f/8)(\mu_b/\mu_w)^n \quad (20\text{-}28)$$

where

$$X = 1.07 + 12.7(\text{Pr}^{2/3} - 1)(f/8)^{1/2}$$

and

$$n = \begin{cases} 0.11 & \text{heating } (T_w > T_b) \\ 0.25 & \text{cooling } (T_w < T_b) \\ 0 & \text{gases} \end{cases}$$

This correlation is applicable for fully developed turbulent flow in the range

$$10^4 < \text{Re} < 5 \times 10^6$$
$$2 < \text{Pr} < 140 \quad \text{with 5 to 6\% error}$$
$$0.5 < \text{Pr} < 2000 \quad \text{with 10\% error}$$
$$0.08 < \mu_w/\mu_b < 40$$

All properties, except $\mu_w$, are evaluated at the bulk temperature. For smooth tubes and pipes, the friction factor is evaluated by

$$f = (1.82 \log \text{Re} - 1.64)^{-2} \quad (20\text{-}29)$$

### 20.9.2 Forced Convection Boiling in Tubes

Correlations for forced convection have been developed for boiling refrigerants in horizontal tubes. All are restricted to test conditions for particular refrigerants, and one should be careful in applying them to conditions outside the test range.

Bo Pierre introduced the load factor, $K_f = J \Delta x\, h_{fg}/L$, which effectively combines the Boiling and Martinelli numbers. In the load factor expression, $J$ is joules equivalent of heat (778 ft·lb$_f$/Btu [1 J/J]) and $\Delta x$ is the change in quality that occurred during the evaporation process. Bo Pierre correlated R-12 and R-22 for a wide range of operating conditions with separate correlations for complete and incomplete evaporation. These correlations for the

Nusselt number (Nu) with two-phase (*tp*) flow are as follows:

$$\text{Nu}_{tp} = 0.009(\text{Re}^2 K_f)^{0.5}$$
$$\text{for } 10^9 < \text{Re}^2 K_f < 0.7 \times 10^{12} \quad (20\text{-}30)$$

and exit vapor quality < 90% (incomplete evaporation)

$$\text{Nu}_{tp} = 0.0082(\text{Re}^2 K_f)^{0.5}$$
$$\text{for } 10^9 < \text{Re}^2 K_f < 0.7 \times 10^{12} \quad (20\text{-}31)$$

and up to 11°F (6.1°C) superheat (complete evaporation)

### 20.9.3 Forced Convection Condensation in Tubes

Condensers used for refrigeration and air-conditioning systems often involve vapor condensation inside horizontal tubes. Unfortunately, conditions within the tube are complicated and depend strongly on the velocity of the vapor flowing through the tube.

If this velocity is small, the condensate flow is from the upper portion of the tube to the bottom, from whence it flows in a longitudinal direction with the vapor. For low vapor velocities such that

$$\text{Re}_{v,i} = \left(\frac{\rho_v m_{m,v} D}{\mu_v}\right)_i < 35{,}000$$

where *i* refers to the tube inlet. An expression of the form

$$\bar{h}_D = 0.555 \left[\frac{g\rho_l(\rho_l - \rho_v)k_l^3 h'_{fg}}{\mu_l(T_{sat} - T_s)D}\right]^{1/4} \quad (20\text{-}32)$$

is recommended where, for this case, the modified latent heat is

$$h'_{fg} \equiv h_{fg} + \frac{3}{8}c_{p,l}(T_{sat} - T_s) \quad (20\text{-}33)$$

At higher vapor velocities, the two-phase flow regime becomes annular, and the following correlation is preferred.

$$hD/k_l = 0.026 \text{Pr}_l^{1/3}[\text{Re}_l + \text{Re}_v(\rho_l/\rho_s)^{1/2}]^{0.8} \quad (20\text{-}34)$$

where

$$\text{Re}_l = (4M_l/\pi D \mu_l) \qquad \text{Re}_v = (4M_v/\pi D \mu_v)$$

Here, $M_l$ and $M_v$ are, respectively, the mass flow rates of liquid and vapor. This expression is valid for Re > 20,000.

### 20.9.4 Condensation on Horizontal Tubes

One of the earliest investigations into laminar film condensation on horizontal tubes was carried out by Nusselt (1916). By applying a force and energy balance to the condensate film, Nusselt arrived at the following equation for condensation from a single horizontal tube:

$$h = 0.729[g\rho_l(\rho_l - \rho_v)k_l^3 h_{fg}/D_o \mu_l(T_{sat} - T_w)]^{1/4} \quad (20\text{-}35)$$

For a vertical tier of *N* horizontal tubes, the average convection coefficient (over the *N* tubes) may be expressed as

$$h = 0.729 \left[\frac{g\rho_l(\rho_l - \rho_v)k_l^3 h_{fg}}{N\mu_l(T_{sat} - T_s)D}\right]$$

That is, $h_N = hN^{-1/4}$, where *h* is the heat transfer for the first (upper) tube. Such an arrangement is often used in condenser design. The reduction in *h* with increasing *N* is due to an increase in the film thickness for each successive tube.

### 20.9.5 Boiling from Horizontal Tubes

The correlations for heat transfer under fully developed nucleate boiling conditions have been divided into two main groups: those based upon direct curve fitting of experimental data banks, called strictly empirical, and those based upon a physical model, but ultimately curve fitted by experimental results, called semi-empirical. The procedure is based on a straightforward reasoning, according to which, nucleate boiling heat transfer correlations, even those of the second group, can be reduced to a product of powers of the transport properties. These properties can be written in terms of reduced primary thermodynamic properties, such as pressure and temperature $p_r$ and $T_r$, as in the Law of Corresponding States. Thus, in principle, all the heat transfer correlations could be reduced to a product of powers of $p_r$ and $T_r$, presenting a single and common form depending on numerical coefficient and exponents that can be obtained by fitting experimental data.

This correlation uses molecular weight, reduced pressure and surface roughness as the correlation parameters and can be written as

$$h = 95 \cdot q^{0.67} M^{-0.5} \left(\frac{p}{p_c}\right)^{0.12 - 0.21 \log_{10} R_p}$$
$$\times \left(-\log_{10}\frac{p}{p_c}\right)^{-0.55} \quad (20\text{-}36)$$

where *M* is the molecular weight and $R_p$ is the roughness of the surface.

It is generally assumed that commercial-finish copper tubes have a surface roughness of 0.4 μm.

**Rohsenow's Correlation.** The first and still most widely used correlation for heat transfer in nucleate pool boiling was proposed by Rohsenow (1973) using experimental data on pool boiling from many different fluids as a guide, Rohsenow obtained

# Heat Exchange Equipment

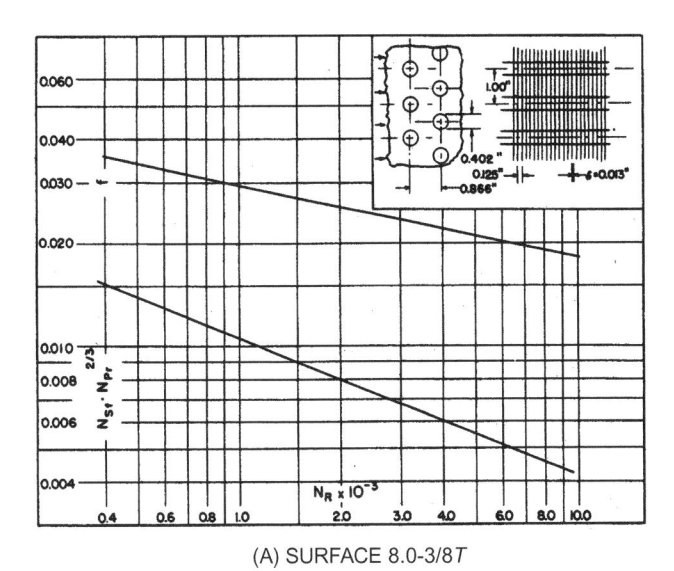

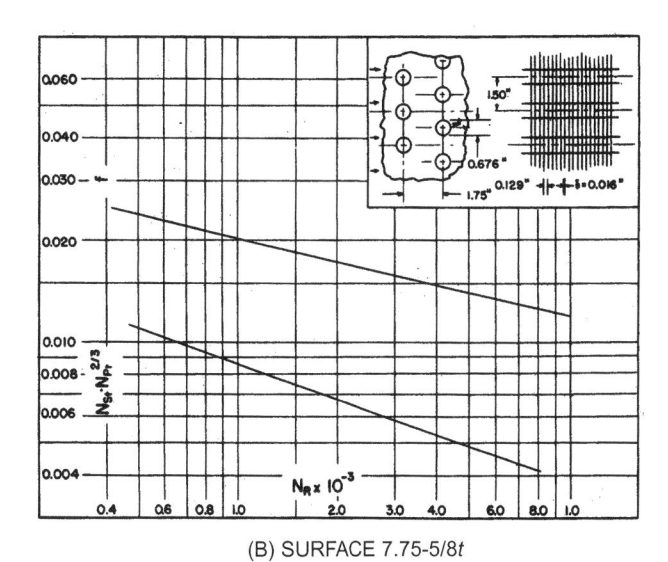

*Fig. 20-10 Dimensional Data for Two Finned-Tube Surfaces*

(A) SURFACE 8.0-3/8T

(B) SURFACE 7.75-5/8t

$$h = \mu_l h_{fg}\left[\frac{g(\rho_l - \rho_v)}{\sigma}\right]^{1/2}\left(\frac{c_{p,l}}{C_{sf} h_{fg} \Pr_l^n}\right)^3 \Delta T_x^2 \quad (20\text{-}37)$$

where

- $\mu_l$ = viscosity of the liquid, lb/ft²·s (kg/m·s)
- $h_{fg}$ = enthalpy of vaporization, Btu/lb (J/kg)
- $g$ = gravitational acceleration, ft²/s (m²/s)
- $\rho_l$ = density of liquid, lb/ft³ (kg/m³)
- $\rho_v$ = density of vapor, lb/ft³ (kg/m³)
- $\sigma$ = surface tension of liquid-vapor interface, lb$_f$/ft (N/m)
- $c_{pl}$ = specific heat of the liquid, Btu/lb·°F (J/kg·°C)
- $\Delta T_x = \Delta T_x = T_s - T_{sat}$
- $T_s$ = surface temperature, °F (°C)
- $T_{sat}$ = saturation temperature of the fluid, °F (°C)
- $C_{sf}$ = experimental constant that depends on surface-fluid combination
- $\Pr_l$ = Prandtl number of the liquid
- $n$ = experimental constant that depends on the fluid = 1.0 for water, 1.7 for other fluids

The Rohsenow method correlates data for all types of nucleate-boiling processes, including pool boiling of saturated or subcooled liquids. Unfortunately, Rohsenow's Correlation can be used **only if** the $C_{sf}$ value is known and the common values are given in Table 20-8. Fortunately, values for several common refrigerations have recently been reported, as shown in Table 20-9.

## 20.9.6 Airflow Across Finned Tubes

The outside heat transfer coefficient is a very crucial parameter which has to be estimated accurately in the design and performance simulation of finned-tube heat exchangers. It is often the controlling factor in the estimation of the overall heat transfer coefficient for the exchanger when air is the external heat transfer fluid. Most correlations have been obtained experimentally and are (a) valid only for specific surfaces and (b) proprietary for most configured fin surfaces.

**Table 20-8  Values of $C_{sf}$ for Various Fluid-Surface Combinations**
*[Incropera (2007)]*

| Fluid-Surface Combination | $C_{sf}$ | n |
|---|---|---|
| Water-copper | | |
| Scored | 0.0068 | 1.0 |
| Polished | 0.0130 | 1.0 |
| Water-stainless steel | | |
| Chemically etched | 0.0130 | 1.0 |
| Mechanically polished | 0.0130 | 1.0 |
| Ground and polished | 0.0060 | 1.0 |
| Water-brass | 0.0060 | 1.0 |
| Water-nickel | 0.006 | 1.0 |
| Water-platinum | 0.0130 | 1.0 |
| n-Pentane-copper | | |
| Polished | 0.0154 | 1.7 |
| Lapped | 0.0049 | 1.7 |
| Benzene-chromium | 0.0101 | 1.7 |
| Ethyl alcohol-chromium | 0.0027 | 1.7 |

*Source*: Incropera (2007).

Table 20-12 provides dimensional data for two plate fin-and-tube arrangements consisting of aluminum fins bonded to copper tubes (Schedule 18). Figure 20-10 gives a schematic of each surface and presents the corresponding heat transfer correlation for the external surface. The friction factor for the outside surface is also shown. The principal dimensionless groups governing these correlations are the Stanton, Prandtl, and Reynolds numbers:

$$\text{St} = h/Gc_p \qquad \Pr = c_p \mu/k \qquad \text{Re} = GD_h/\mu$$

where $G$ is the mass velocity defined as

$$G = m/A_{\min}$$

where $m$ = total mass flow rate of the fluid and $A_{\min}$ = minimum free-flow cross-sectional area in the coil. The hydraulic diameter ($D_h = 4r_h$) is specified on each figure.

569

Table 20-9  $C_{sf}$s for Refrigerants

| Refrigerant | Value of $C_{sf}$ |
|---|---|
| R-12 | 0.008339 |
| R-22 | 0.007947 |
| R-123 | 0.006706 |
| R-134a | 0.006232 |
| R-407C | 0.007269 |
| R-410A | 0.008294 |

## 20.10  Calculation of Heat Exchanger Surface Area and Overall Size

Design methods for shell-and-tube and finned-tube heat exchangers are outlined in Tables 20-10 and 20-11, respectively.

## 20.11  Fluids and Their Thermophysical Properties

Proper evaluation of the necessary thermodynamic and thermophysical properties of the working fluids in heat exchangers is most important. Properties need to be evaluated at the correct temperature (and pressure, if gas or vapor) which may be the average bulk fluid temperature (average between in and out), the saturation temperature, the "film" temperature (average between fluid and wall), and/or the surface temperature. Unfortunately, there is no single "correct" temperature to use but varies primarily with the correlation selected for predicting the convective heat transfer coefficient.

Unfortunately, very limited property data is included here due to space limitations (see Table 2.2 for R-134a table). However, Chapters 29, 30, and 33 of the 2013 *ASHRAE Handbook—Fundamentals* are an excellent source of such data. In addition, most heat transfer textbooks include some tables, and there are rather extensive reference handbooks either on heat transfer or thermophysical properties. And, of course, the Web is another valuable source of such data.

## 20.12  Example Finned-Tube Heat Exchanger Design

*Task:* An aluminum tube with $k$ = 1290 Btu·in/h·ft$^2$·°F, ID = 1.8 in., and OD = 2 in. has circular aluminum fins $\delta$ = 0.04 in. thick with an outer diameter of $D_{fin}$ = 3.9 in. There are $N'$ = 76 fins per foot of tube length. Steam condenses inside the tube at $t_i$ = 392°F with a large heat transfer coefficient on the inner tube surface. Air at $t_\infty$ = 77°F is heated by the steam. The heat transfer coefficient outside the tube is 7 Btu/h·ft$^2$·°F. Find the rate of heat transfer per foot of tube length.

*Solution:* From Figure 20-8 efficiency curve, the efficiency of these circular fins is

$$L = (D_{fin} - OD)/2 = (3.9 - 2)/2 = 0.95 \text{ in.}$$
$$f_2/f_1 = \frac{3.9/2}{2/2} = 1.95 \text{ in.}$$
$$L\sqrt{\frac{h}{ky}} = 0.95 \text{ in.} \sqrt{\frac{7 \text{ Btu/h·ft}^2\text{·°F}}{(1290 \text{ Btu·in/h·ft}^2\text{·°F})(0.02 \text{ in.})}}$$
$$\phi = 0.89$$

The fin area for $L$ = 1 ft is

$$A_s = N'L \times 2\pi(D_{fin}^2 - OD^2)/4 = 1338 \text{ in}^2 = 9.29 \text{ ft}^2$$

The unfinned area for $L$ = 1 ft is

$$A_p = \pi \times OD \times L(1 - N'\delta)$$
$$= \pi(2/12) \text{ ft} \times 1 \text{ ft}(1 - 76 \times 0.04/12)$$
$$= 0.39 \text{ ft}^2$$

and the total area $A = A_s + A_p$ = 9.68 ft$^2$. Surface efficiency is

$$\phi_s = \frac{\phi A_f + A_s}{A} = 0.894$$

and resistance of the finned surface is

$$R_s = \frac{1}{\phi_s hA} = 0.0165 \text{ h·°F/Btu}$$

Tube wall resistance is

$$R_{wall} = \frac{\ln(OD/ID)}{2\pi L k_{tube}} = \frac{\ln(2/1.8)}{2\pi(1 \text{ ft})(1290/12) \text{ Btu·in/h·ft}^2\text{·°F}}$$
$$= 1.56 \times 10^{-4} \text{ h·°F/Btu}$$

The rate of heat transfer is then

$$q = \frac{t_i - t_\infty}{R_s + R_{wall}} = 18{,}912 \text{ Btu/h}$$

## 20.13  Problems

**20.1** A hot-water coil is to be sized (designed) to heat 5000 cfm of air from 70°F to 130°F with the coil face velocity based on duct space selected as 500 fpm. The inlet and outlet water temperatures are 190°F and 170°F, respectively. Determine: coil height, width, number of rows, depth.

**20.2** A supplementary cooling coil is being added to the building's HVAC system to provide additional cooling for a computer room. The cooling coil will do sensible cooling only, taking 1500 scfm of preconditioned air from 70°F db and 60°F wb to 55°F db. A 50/50 antifreeze (ethylene glycol) solution at 50°F will be supplied to the coil. Prepare a preliminary design for the coil using Surface 8.0-3/8T with copper tubes having a final wall thickness of 0.036 in.

**20.3** Design a steam coil to heat 8500 scfm of outdoor air from 0°F to 45°F with a face velocity of 600 fpm. Low-pressure saturated steam at 5 psig is used. Surface 7.75-5/8T is to be examined first.

# Heat Exchange Equipment

Table 20-10  Design Procedure for Shell-and-Tube Heat Exchanger

1. FIND Q OF THE CONDENSER
2. CHOOSE TUBE SIZE
3. CALCULATE MEAN TEMPERATURE
4. CALCULATE REYNOLDS NUMBER
5. USING   Q=M*CP*DELT CALCULATE MASS FLOW RATE FOR WATER
6. USING M=RHO*AREA*VELOCITY*N TUBES CALCULATE NUMBER OF TUBES
7. CALCULATE  LOG MEAN TEMPERATURE DIFFERENCE USING COUNTERFLOW EXAMPLE
8. USING   Q=U*AREA*TEMP LOG MEAN  CALCULATE VALUE OF   U*L
9. CALCULATE CONVECTIVE COEFFICIENT (H) FOR INSIDE TUBES
10. CALCULATE CONVECTIVE COEFFICIENT (H) FOR OUTSIDE TUBES
11. ASSUME A SURFACE TEMPERATURE FOR OUTER SURFACE OF TUBE, FIND PROPERTIES AT THE MEAN TEMPERATURE
12. LOOK UP FOULING FACTORS FOR WATER AND REFRIGERANT
13. ASSUME COPPER TUBING FOR CONDUCTIVITY (K)
14. CALCULATE NUSSELT NUMBER AND OVERALL RESISTANCE (U)
15. VERIFY ASSUMED TEMPERATURE
16. REPEAT PROCESS 10-15 IF NECESSARY
17. WHEN FINAL  H AND U ARE FOUND THEN L MAY BE FOUND FROM EQUATION IN LINE 8

Table 20-11  Design Procedure for Finned-Tube Heat Exchanger

1. DO CALCULATIONS WITH BOTH SURFACES
2. ASSUME TUBE OF GAGE 16 FOR BOTH GIVING DIFFERENCE OF .13 INCHES BETWEEN Do AND Di
3. USING M=RHO*AREA*VELOCITY*N TUBES CALCULATE NUMBER OF TUBES
4. USING   HEIGHT=N(TUBES)*SPACING    CALCULATE HEIGHT
5. USING  M=RHO*AREA*VELOCITY OF FACE     CALCULATE AREA OF FACE
6. USING A(FACE)=HEIGHT*WIDTH
7. CALCULATE REYNOLDS NUMBER INSIDE TUBES
8. USING GIVEN VALUE FOR DENSITY OF AIR USE TM=70 F
9. CALCULATE NUSSELT NUMBER, FIND  H
10. CALCULATE REYNOLDS NUMBER FOR AIR USING HYDRAULIC DIAMETER
11. USING FIGURE FOR SURFACE FIND    $StPr^{2/3}$, FIND H FROM WITHIN St NUMBER
12. FIND CONTACT RESISTANCE ASSUMING Al FINS AND Cu TUBES
13. ASSUME FINS ARE ANNULAR (CIRCULAR) AND HAVE A RADIUS EQUAL TO THE SPACING BETWEEN THE TUBES
14. CALCULATE  ALPHA AND BETA FOR ANNULAR FIN SHAPE
15. CALCULATE  $K_{an}$ FOR ANNULAR FINS, FIND ETA
16. CALCULATE  U  VALUE

**20.4** Design both the heating coil and the shell-and-tube water heat exchanger for the heating system shown in the sketch provided.

**20.5** Design the evaporator/condenser for a cascade low-temperature refrigeration system using R-410 in the high-temperature loop and R-22 in the low-temperature loop. The shell-and-tube heat exchanger will use standard size copper tubes with a steel pipe as a shell. R-22 at the rate of 0.130 kg/s is to be condensed from saturated vapor to saturated liquid at a pressure of 0.91 MPa as it flows through the tubes. R-410A surrounds the tubes and evaporates under pool boiling conditions at a pressure of 1.1 MPa. The exterior of the heat exchanger shell is to be well insulated. Space limits the length of the exchanger to 2 m.

**20.6** A shell-and-tube heat exchanger is to cool 1 L/s of water from 15°C to 5°C using R-22 evaporating at 50 kPa on the outside of the tubes. Tubes are to be of copper with a 1.41 cm ID and 1.59 cm OD. Maximum water velocity in the tubes is to be 2 m/s. Design the heat exchanger including specification of its duty (thermal rating) in kW, the design U-factor, the number of tubes per pass, and the length of the exchanger. *Boiling performance may be obtained from the ASHRAE data provided in the figure provided.*

Table 20-12 Dimensional Data for Two Finned-Tube Surfaces

| (a) Surface 8.0-3/8T | (b) Surface 7.75-5/8T |
|---|---|
| Tube outside diameter = 0.402 in. | Tube outside diameter = 0.676 in. |
| Fin pitch = 8 per in. | Fin pitch = 7.75 per in. |
| Fin thickness = 0.013 in. | Fin thickness = 0.016 in. |
| Hydraulic diameter = 0.001192 ft | Hydraulic diameter = 0.0114 ft |
| Free-flow area/face area = 0.534 | Free-flow area/face area = 0.481 |
| Fin area/total external area = 0.913 | Fin area/total external area = 0.913 |
| Total external area/inside tube area = 12.5 | Total external area/inside tube area = 20.5 |
| Total external area/outside tube area = 10.3 | Total external area/outside tube area = 17.6 |
| Total outside area/face area = 12.9 | Total outside area/face area = 24.6 |
| Heat transfer area/total volume = 179 ft$^2$/ft$^3$ | Heat transfer area/total volume = 169 ft$^2$/ft$^3$ |

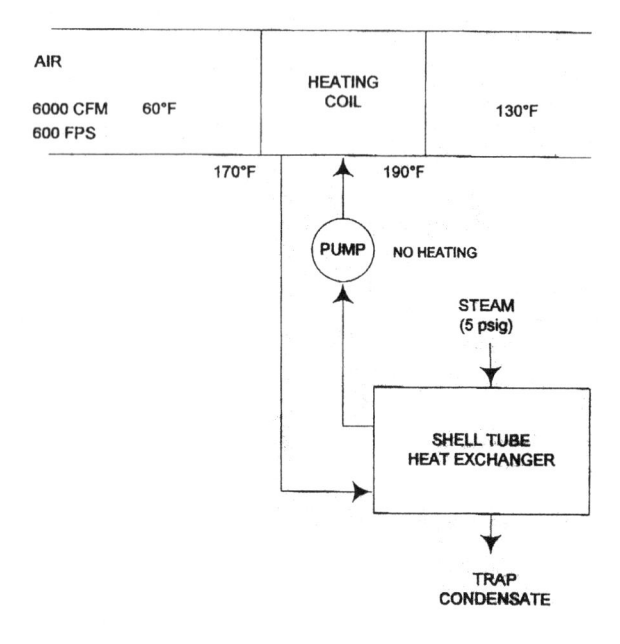

*Diagram for Problem 20.4*

*Diagram for Problem 20.5*

*Diagram for Problem 20.6*

## 20.14 Bibliography

ASHRAE. 2013. *2013 ASHRAE Handbook—Fundamentals.*

ASHRAE. 2012. *2012 ASHRAE Handbook—HVAC Systems and Equipment.*

Ackers, W.W., H.A. Dean, and O.K. Crosser. 1961. Condensing heat transfer within horizontal tubes. Chemical Engr. Pro. Symposium Ser. Vol. 55, No. 29.

R.A. Bowman, A.C. Mueller, and W.M. Nagel. 1940. Mean Temperature Difference in Design. *ASME Transactions* 62.

Gardner, K.A. 1945. *ASME Transactions* 67:625.

Incropera, F.P., D.P. DeWitt, T.L. Bergman, and S.A. Lavine. Fundamentals of Heat and Mass Transfer, 6th ed., John Wiley and Sons, New York..

Kays, W.M. and A.L. London. 1964. *Compact Heat Exchangers*, 2nd ed. McGraw-Hill, New York.

McAdams, W.H. 1940. Review and Summary of Developments in Heat Transfer by Conduction and Convection. *AICHE Transactions* 36.

Nusselt, W. 1916. Die Oberflachenkondensation des Wasserdampfes. *Zeitung Verein Deutscher Ingenieure* 60:541-569.

Pierre, B.S. 1955. S. F. Review, A. B. Svenska, Vol. 2, No. 1, Flakabriken, Stockholm, Sweden.

Rohsenow, W.M., and J.P. Hartnett, eds. 1973. *Handbook of Heat Transfer*. McGraw-Hill, New York.

Sheffield, J.W., M. AbuEbid, and H.J. Sauer, Jr. 1985. Finned Tube Contact Conductance—Empirical Correlation of Contact Conductance (RP-295). *ASHRAE Transactions* 91(2):100-117.

TEMA. 2007. *TEMA Standard,* 9th Ed. http://www.tema.org/highlig8.html

# Appendix A
# SI FOR HVAC&R

This guide conforms to ANSI SI 10-1997, *Standard for Use of the International System of Units (SI): The Modern Metric System*. See ANSI SI 10 for more information and a complete list of conversion factors with more significant digits.

## SI PRACTICE

## 1 General

**1.1** The International System of Units (SI) consists of seven base units, listed in Table 1, and numerous derived units, which are combinations of base units (Table 2).

**Table 1  SI Base Units**

| Quantity | Name | Symbol |
|---|---|---|
| length | metre | m |
| mass | kilogram | kg |
| time | second | s |
| electric current | ampere | A |
| thermodynamic temperature | kelvin | K |
| amount of substance | mole | mol |
| luminous intensity | candela | cd |

## 2 Units

**2.1** In SI each physical quantity has only one unit. The base and derived units may be modified by prefixes as indicated in Section 4. All derived units are defined by simple formulas using the base units. The basic simplicity of the system can only be kept by adhering to the approved units.

**2.2 Angle.** The unit of plane angle is the radian. The degree and its decimal fractions may be used, but the minute and second should not be used.

**2.3 Area.** The unit of area is the square metre. Large areas are expressed in hectares (ha) or square kilometres (km$^2$). The hectare is restricted to land or sea areas and equals 10 000 m$^2$.

**2.4 Energy.** The unit of energy, work, and quantity of heat is the joule (J). The kilowatthour (kWh) is presently permitted as an alternative in electrical applications but should not be introduced in new applications.

$$1 \text{ kilowatthour (kWh)} = 3.6 \text{ megajoules (MJ)}$$

The unit of power and heat flow rate is the watt (W).

$$1 \text{ watt (W)} = 1 \text{ joule per second (J/s)}$$

**2.5 Force.** The unit of force is the newton (N). The newton is also used in derived units that include force.

*Examples*:  pressure or stress = N/m$^2$ = Pa (pascal)

work = N·m = J (joule)

power = N·m/s = W (watt)

**Table 2  Some SI Derived Units**

| Quantity | Expression in Other SI Units | Name | Symbol |
|---|---|---|---|
| acceleration | | | |
|   angular | rad/s$^2$ | | |
|   linear | m/s$^2$ | | |
| angle | | | |
|   plane | dimensionless | radian | rad |
|   solid | dimensionless | steradian | sr |
| area | m$^2$ | | |
| Celsius temperature | K | degree Celsius | °C |
| conductivity, thermal | W/(m·K) | | |
| density | | | |
|   heat flux | W/m$^2$ | | |
|   mass | kg/m$^3$ | | |
| energy, enthalpy | | | |
|   work, heat | N·m | joule | J |
|   specific | J/kg | | |
| entropy | | | |
|   heat capacity | J/K | | |
|   specific | J/(kg·K) | | |
| flow, mass | kg/s | | |
| flow, volume | m$^3$/s | | |
| force | kg·m/s$^2$ | newton | N |
| frequency | | | |
|   periodic | 1/s | hertz | Hz |
|   rotating | rev/s | | |
| inductance | Wb/A | henry | H |
| magnetic flux | V·s | weber | |
| moment of a force | N·m | | |
| potential, electric | W/A | volt | V |
| power, radiant flux | J/s | watt | W |
| pressure, stress | N/m$^2$ | pascal | Pa |
| resistance, electric | V/a | ohm | Ω |
| velocity | | | |
|   angular | rad/s | | |
|   linear | m/s | | |
| viscosity | | | |
|   dynamic (absolute)(m) | Pa·s | | |
|   kinematic (n) | m$^2$/s | | |
| volume | m$^3$ | | |
| volume, specific | m$^3$/kg | | |

**2.6 Length.** The unit of length is the metre. The millimetre is used on architectural or construction drawings and mechanical or shop drawings. The symbol *mm* does not need to be placed after each dimension; a note, "All dimensions in mm" is sufficient.

The centimetre is used only for cloth, clothing sizes, and anatomical measurements.

The metre is used for topographical and plot plans. It is always written with a decimal and three figures following the decimal, i.e., 38.560.

**2.7 Mass.** The unit of mass is the kilogram (kg). The unit of mass is the only unit whose name, for historical reasons, contains a prefix. Names of multiples of the unit mass are formed by attaching prefixes to the word gram. The megagram, Mg (1000 kg, metric ton or tonne, t), is the appropriate unit for describing large masses. Do not use the term *weight* when *mass* is intended.

**2.8 Pressure.** The unit of stress or pressure, force per unit area, is the newton per square metre. This unit is called the *pascal* (Pa). SI has no equivalent symbol for psig or psia. If a misinterpretation is likely, spell out Pa (absolute) or Pa (gage).

**2.9 Volume.** The unit of volume is the cubic metre. Smaller units are the litre, L (m³/1000); millilitre, mL; and microlitre, μL. No prefix other than m or μ is used with litre.

**2.10 Temperature.** The unit of thermodynamic (absolute) temperature is the Kelvin. Celsius temperature is measured in degrees Celsius. Temperature intervals may be measured in kelvins or degrees Celsius and are the same in either scale. Thermodynamic temperature is related to Celsius temperature as follows:

$$t_c = T - T_0$$

where

$t_c$ = Celsius temperature, °C

$T$ = thermodynamic temperature, kelvins (K)

$T_0$ = 273.15 K by definition

**2.11 Time.** The unit of time is the second, which should be used in technical calculations. However, where time relates to life customs or calendar cycles, the minute, hour, day, and other calendar units may be necessary.

*Exception*: Revolutions per minute may be used, but revolutions per second is preferred.

## 3 Symbols

**3.1** The correct use of symbols is important because an incorrect symbol may change the meaning of a quantity. Some SI symbols are listed in Table 3.

**3.2** SI has no abbreviations—only symbols. Therefore, no periods follow a symbol except at the end of a sentence.

*Examples*: SI, *not* S.I.; s, *not* sec; A, *not* amp

**3.3** Symbols appear in lowercase unless the unit name has been taken from a proper name. In this case the first letter of the symbol is capitalized.

*Examples*: m, metre; W, watt; Pa, pascal

*Exception*: L, litre

**3.4** Symbols and prefixes are printed in upright (roman) type regardless of the type style in surrounding text.

*Example*: . . . *a distance of* 56 km *between* . . .

**3.5** Unit symbols are the same whether singular or plural.

*Examples*: 1 kg, 14 kg; 1 mm, 25 mm

**Table 3   SI Symbols**

| Symbol | Name | Quantity | Formula |
|---|---|---|---|
| A | ampere | electric current | base unit |
| a | atto | prefix | $10^{-18}$ |
| Bq | becquerel | activity (of a radio nuclide) | 1/s |
| C | coulomb | quantity of electricity | A·s |
| °C | degree Celsius | temperature | °C = K |
| c | centi | prefix | $10^{-2}$ |
| cd | candela | luminous intensity | base unit |
| d | deci | prefix | $10^{-1}$ |
| da | deka | prefix | $10^{1}$ |
| E | exa | prefix | $10^{18}$ |
| F | farad | electric capacitance | C/V |
| f | femto | prefix | $10^{-15}$ |
| G | giga | prefix | $10^{9}$ |
| Gy | gray | absorbed dose | J/kg |
| g | gram | mass | kg/1000 |
| H | henry | inductance | Wb/A |
| Hz | hertz | frequency | 1/s |
| h | hecto | prefix | $10^{2}$ |
| ha | hectare | area | 10 000 m² |
| J | joule | energy, work, heat | N·m |
| K | kelvin | temperature | base unit |
| k | kilo | prefix | $10^{3}$ |
| kg | kilogram | mass | base unit |
| L | litre | volume | m³/1000 |
| lm | lumen | luminous flux | cd·sr |
| lx | lux | illuminance | lm/m² |
| M | mega | prefix | $10^{6}$ |
| m | metre | length | base unit |
| m | milli | prefix | $10^{-3}$ |
| mol | mole | amount of substance | base unit |
| μ | micro | prefix | $10^{-6}$ |
| N | newton | force | kg·m/s² |
| n | nano | prefix | $10^{-9}$ |
| Ω | ohm | electric resistance | V/A |
| P | peta | prefix | $10^{15}$ |
| Pa | pascal | pressure, stress | N/m² |
| p | pico | prefix | $10^{-12}$ |
| rad | radian | plane angle | dimensionless |
| S | siemens | electric conductance | A/V |
| Sv | sievert | dose equivalent | J/kg |
| s | second | time | base unit |
| sr | steradian | solid angle | dimensionless |
| T | tera | prefix | $10^{12}$ |
| T | tesla | magnetic flux density | Wb/m² |
| t | tonne, metric ton | mass | 1000 kg; Mg |
| V | volt | electric potential | W/A |
| W | watt | power, radiant flux | J/s |
| Wb | weber | magnetic flux | V·s |

**3.6** Leave a space between the value and the symbol.

*Examples*: 55 mm, *not* 55mm; 100 W, *not* 100W

*Exception*: No space is left between the numerical value and symbol for degree Celsius and degree of plane angle.

Note: Symbol for degree Celsius is °C; for coulomb, C.

*Examples*: 20°C, not 20 °C or 20° C; 45°, not 45 °

**3.7** Do not mix symbols and names in the same expression.

*Examples*: m/s or metres per second,

*not* metres/second; *not* metres/s

J/kg or joules per kilogram,

*not* joules/kilogram; *not* joules/kg

**3.8** Symbol for product—use the raised dot ( · )

*Examples*: N·m; mPa·s; W/(m²·K)

**3.9** Symbol for quotient—use one of the following forms:

*Examples*: m/s or $\frac{m}{s}$ or use negative exponent

Note: Use only one solidus ( / ) per expression.

**3.10** Place modifying terms such as electrical, alternating current, etc., parenthetically after the symbol with a space in between.

*Examples*: MW (e); *not* MWe; *not* MW(e)

V (ac); *not* Vac; *not* V(ac)

kPa (gage); *not* kPa(gage); *not* KPa gage

## 4 Prefixes

**4.1** Most prefixes indicate orders of magnitude in steps of 1000. Prefixes provide a convenient way to express large and small numbers and to eliminate nonsignificant digits and leading zeros in decimal fractions. Some prefixes are listed in Table 4.

*Examples*: 126 000 watts is the same as 126 kilowatts

0.045 metre is the same as 45 millimetres

65 000 metres is the same as 65 kilometres

**4.2** To realize the full benefit of the prefixes when expressing a quantity by numerical value, choose a prefix so that the number lies between 0.1 and 1000. For simplicity, give preference to prefixes representing 1000 raised to an integral power (i.e., μm, mm, km).

*Exceptions*:

1. In expressing area and volume, the prefixes hecto, deka, deci, and centi are sometimes used; for example, cubic decimetre (L), square hectometre (hectare), cubic centimetre.

2. Tables of values of the same quantity.

3. Comparison of values.

4. For certain quantities in particular applications. For example, the millimetre is used for linear dimensions in engineering drawings even when the values lie far outside the range of 0.1 mm to 1000 mm; the centimetre is usually used for body measurements and clothing sizes.

**4.3 Compound units.** A compound unit is a derived unit expressed with two or more units. The prefix is attached to a unit in the numerator.

*Examples*: V/m *not* mV/mm

mN·m *not* N·mm (torque)

MJ/kg not kJ/g

**Table 4    SI Prefixes**

| Prefix | Pronunciation | Symbol | Represents | |
|---|---|---|---|---|
| exa | ex′a (a as in about) | E | $10^{18}$ | |
| peta | pet′a (e as in pet, a as in about) | P | $10^{15}$ | |
| tera | as in *terra* firma | T | $10^{12}$ | |
| giga | jig′ (i as in jig, a as in about) | G | $10^{9}$ | |
| mega | as in *mega*phone | M | $10^{6}$ | |
| kilo | kill′oh | k | $10^{3}$ | = 1000 |
| hecto | heck′ toe | h* | $10^{2}$ | = 100 |
| deka | deck′a (a as in about) | da* | $10^{1}$ | = 10 |
| deci | as in *deci*mal | d* | $10^{-1}$ | = 0.1 |
| centi | as in *centi*pede | c* | $10^{-2}$ | = 0.01 |
| milli | as in *mili*tary | m | $10^{-3}$ | = 0.001 |
| micro | as in *micro*phone | μ | $10^{-6}$ | |
| nano | nan′oh (an as in ant) | n | $10^{-9}$ | |
| pico | peek′oh | p | $10^{-12}$ | |

*See paragraph 4.2 regarding use of this prefix.

**4.4** Compound prefixes formed by a combination of two or more prefixes are not used. Use only one prefix.

*Examples*: 2 nm *not* 2 mμm

6 m³ *not* 6 kL

6 MPa *not* 6 kkPa

**4.5 Exponential Powers.** An exponent attached to a symbol containing a prefix indicates that the multiple (of the unit with its prefix) is raised to the power of 10 expressed by the exponent.

*Examples*: $1 \text{ mm}^3 = (10^{-3} \text{ m})^3 = 10^{-9} \text{ m}^3$

$1 \text{ ns}^{-1} = (10^{-9} \text{ s})^{-1} = 10^9 \text{ s}^{-1}$

$1 \text{ mm}^2/\text{s} = (10^{-3} \text{ m})^2/\text{s} = 10^{-6} \text{ m}^2/\text{s}$

## 5 Numbers

**5.1 Large Numbers.** International practice separates the digits of large numbers into groups of three, counting from the decimal to the left and to the right, and inserts a space to separate the groups. In numbers of four digits, the space is not necessary except for uniformity in tables.

*Examples*: 2.345 678; 73 846; 635 041; 600.000;

0.113 501; 7 258

**5.2 Small Numbers.** When writing numbers less than one, always put a zero before the decimal marker.

*Example*: 0.046

**5.3 Decimal Marker.** The recommended decimal marker is a dot on the line (period). (In some countries, a comma is used as the decimal marker.)

**5.4 Billion.** Because billion means a thousand million in the United States and a million million in most other countries, avoid using the term in technical writing.

**5.5 Roman Numerals.** Do not use M to indicate thousands (MBtu for a thousand Btu), nor MM to indicate millions, nor C to indicate hundreds because they conflict with SI prefixes.

**Table 5  SI Units for HVAC&R Catalogs**

| Quantity | Unit | Quantity | Unit | Quantity | Unit |
|---|---|---|---|---|---|
| **Boilers** | | **Diffusers and Grilles** | | **Pumps** | |
| Heat output | kW | Air volume flow rate | m$^3$/s, L/s | Mass flow rate | kg/s |
| Heat input | kW | Airflow pressure loss | Pa | Volume flow rate | L/s |
| Heat release | kW/m$^2$ | Velocity | m/s | Power input (to drive) | kW |
| Steam generation rate | kg/s | | | Developed pressure | kPa |
| Fuel firing rate: | | **Fans** | | Operating pressure | kPa |
|   solid | kg/s | Air volume flow rate | m$^3$/s, L/s | Rotational frequency | rev/s (rpm)* |
|   gaseous | L/s | Power input (to drive) | kW | | |
|   liquid | kg/s, L/s | Fan static pressure | Pa | **Space Heating Apparatus** | |
| Volume flow rate (combust. products) | m$^3$/s, L/s | Fan total pressure | Pa | Heat output | kW |
| Power input (to drives) | kW | Rotational frequency | rev/s (rpm)* | Airflow volume flow rate | m$^3$/s, L/s |
| Operating pressure | kPa | Outlet velocity | m/s | Power input (to drive) | kW |
| Hydraulic resistance | kPa | | | Primary medium mass flow rate | kg/s |
| Draft conditions | Pa | **Air Filters** | | Hydraulic resistance | kPa |
| | | Air volume flow rate | m$^3$/s, L/s | Operating pressure | kPa |
| **Coil, Cooling and Heating** | | Static pressure loss | Pa | Airflow static pressure loss | Pa |
| Heat exchange rate | kW | Face area | m$^2$ | | |
| Primary medium: | | | | **Vessels** | |
|   mass flow rate | kg/s | **Fuels** | | Operating pressure | kPa |
|   hydraulic resistance | kPa | Heating value: | | Volumetric capacity | m$^3$, L |
| Air volume flow rate | m$^3$/s, L/s |   solid | MJ/kg | | |
| Airflow static pressure loss | Pa |   gaseous | MJ/m$^3$ | **Air Washers** | |
| Face area | m$^2$ |   liquid | MJ/kg | Volume flow rate: | |
| Fin spacing, center to center | mm | | |   air | m$^3$/s, L/s |
| | | **Heat Exchangers** | |   water | m$^3$/s, L/s |
| **Controls and Instruments** | | Heat output | kW | Mass flow rate, water | kg/s |
| Flow rate: | | Mass flow rate | kg/s | Power input (to drive) | kW |
|   mass | kg/s | Hydraulic resistance | kPa | Airflow static pressure loss | Pa |
|   volume | m$^3$/s, L/s, mL/s | Operating pressure | kPa | Hydraulic resistance | kPa |
| Operating pressure | kPa | Flow velocity | m/s | | |
| Hydraulic resistance | kPa | Heat exchange surface | m$^2$ | **Water Chillers** | |
| Rotational frequency | rev/s (rpm)* | Fouling factor | m$^2$/W | Cooling capacity | kW |
| | | | | Mass flow rate, water | kg/s |
| **Cooling Towers** | | **Induction Terminals** | | Power input (to drive) | kW |
| Heat extraction rate | kW | Heating or cooling output | kW | Refrigerant pressure | kPa |
| Volume flow rate: | | Primary air volume flow rate | m$^3$/s, L/s | Hydraulic resistance | kPa |
|   air | m$^3$/s, L/s | Primary air static pressure loss | Pa | | |
|   water | m$^3$/s, L/s | Secondary water mass flow rate | kg/s | *Acceptable | |
| Power input (to drive) | kW | Secondary water hydraulic resistance | kPa | | |

## 6  Words

**6.1**  The units in the international system of units are called SI units—not Metric Units and not SI Metric Units.

(Inch-Pound units are called I-P units—not conventional units, not U.S. customary units, not English units, and not Imperial units.)

**6.2**  Treat all spelled out names as nouns. Therefore, do not capitalize the first letter of a unit except at the beginning of a sentence or in capitalized material such as a title.

*Examples*: watt; pascal; ampere; volt; newton; kelvin

*Exception*: Always capitalize the first letter of Celsius.

**6.3**  Do not begin a sentence with a unit symbol—either rearrange the words or write the unit name in full.

**6.4**  Use plurals for spelled out words when required by the rules of grammar.

*Examples*: metre — metres; henry — henries; kilogram — kilograms; kelvin — kelvins

*Irregular*: hertz — hertz; lux — lux; siemens — siemens

**Table 6  Typical Densities (kg/m$^3$ at 20°C)**

| Gases (101.325 kPa) | | Liquids | | Solids | |
|---|---|---|---|---|---|
| butane | 2.412 | mercury | 13 550 | lead | 11 300 |
| propane | 1.829 | sulphuric acid | 1 830 | copper | 8 900 |
| oxygen | 1.330 | refrigerant 12 | 1 329 | steel | 7 830 |
| air, dry | 1.204 | glycerine | 1 264 | cast iron | 7 200 |
| carbon dioxide | 1.970 | battery electr. | 1 260 | aluminum | 2 700 |
| air, 50% rh | 1.191 | refrigerant 22 | 1 213 | glass | 2 500 |
| acetylene | 1.173 | water | 998 | concrete | 2 300 |
| nitrogen | 1.164 | mineral oil | 900 | brick | 1 920 |
| natural gas | 0.719 | kerosene | 820 | hardwood | 750 |
| helium | 0.166 | ethyl alcohol | 791 | softwood | 540 |
| hydrogen | 0.083 | gasoline | 730 | fiberglass board | 80 |
| | | propane | 580 | polystyrene | 20 |

**6.5**  Do not put a space or hyphen between the prefix and unit name.

*Examples*: kilometre not kilo metre or kilo-metre;

milliwatt not milli watt or milli-watt

**6.6**  When a prefix ends with a vowel and the unit name begins with a vowel, retain and pronounce both vowels.

*Example*: kiloampere

*Exceptions*: hectare; kilohm; megohm

**6.7** When compound units are formed by multiplication, leave a space between units that are multiplied.

*Examples*: newton metre, *not* newton-metre; volt ampere, *not* volt-ampere

**6.8** Use the modifier squared or cubed after the unit name.

*Example*: metre per second squared

*Exception*: For area or volume, place the modifier before the units. *Example*: square millimetre; cubic metre

**6.9** When compound units are formed by division, use the word *per*, not a solidus ( / ).

*Examples*: metre per second, *not* metre/second; watt per square metre, *not* watt/square metre

## TEMPERATURE CONVERSION
*(exact)*

$t_C = (t_F - 32)/1.8$  $\quad t_F = 1.8\, t_C + 32$

$t_C = T - 273.15$  $\quad t_F = T_R - 459.67$

$T = T_R/1.8$  $\quad T_R = 1.8T$

$T = t_C + 273.15$  $\quad T_R = t_F + 459.67$

where

$t_C$ = Celsius temperature, °C
$T$ = thermodynamic (absolute) temperature, kelvins (K)
$t_F$ = Fahrenheit temperature, °F
$T_R$ = thermodynamic (absolute) temperature, degrees Rankine (°R)

and  °C = K = 1.8°F°F = °R = °C/1.8

## PHYSICAL PROPERTIES

### Atmospheric Pressure

Standard pressure = 101.325 kPa, exact value by definition (approximately 29.921 in. Hg at 32°F; 760 mm Hg at 0°C; 14.696 psi at 32°F).

### Gravity

Standard acceleration = 9.806 65 m/s², exact value by definition (approximately 32.1740 ft/s²).

### Standard Air

Dry air at 101.325 kPa and 20°C (density ≈ 1.204 kg/m³)
Specific heat (constant pressure), $c_p$ = 1.006 kJ/(kg·K)

### Heating of Air

| | | |
|---|---|---|
| Sensible heat | $q_s = 1.2\, Q\Delta t$ | |
| Latent heat | $q_l = 3.0\, Q\Delta w$ | |
| Total heat | $q_t = 1.2\, Q\Delta h$ | |

where

$\Delta t$ = temperature difference, K or °C
$\Delta w$ = moisture content difference, g/kg (dry air)
$\Delta h$ = enthalpy difference, kJ/kg (dry air)
$Q$ = volume flow rate, m³/s (standard air)
$q_s, q_l, q_t$ = heat flow, kW

### Water

Heat of vaporization
  at 101.325 kPa and 100°C = 2257 kJ/kg

Heat of fusion at 0°C = 334 kJ/kg

# CONVERSION FACTORS

When making conversions, remember that a converted value is no more precise than the original value. Round off the final value to the same number of significant figures as those in the original value.

CAUTION: The conversion values are rounded to three or four significant figures, which is sufficiently accurate for most applications. See ANSI SI 10 for additional conversions with more significant figures.

| Multiply | By | To Obtain |
|---|---|---|
| acre | 0.4047 | ha |
| atmosphere, standard | *101.325 | kPa |
| bar | *100 | kPa |
| barrel (42 US gal, petroleum) | 159 | L |
| Btu, (International Table) | 1.055 | kJ |
| Btu/ft$^2$ | 11.36 | kJ/m$^2$ |
| Btu·ft/h·ft$^2$·°F | 1.731 | W/(m·K) |
| Btu·in/h·ft$^2$·°F (thermal conductivity, $k$) | 0.1442 | W/(m·K) |
| Btu/h | 0.2931 | W |
| Btu/h·ft | 0.9615 | W/m |
| Btu/h·ft$^2$ | 3.155 | W/m$^2$ |
| Btu/h·ft$^2$·°F (heat transfer coefficient, $U$) | 5.678 | W/(m$^2$·K) |
| Btu/lb | *2.326 | kJ/kg |
| Btu/lb·°F (specific heat, $c_p$) | 4.184 | kJ/(kg·K) |
| bushel | 0.03524 | m$^3$ |
| calorie, (thermochemical) | *4.184 | J |
| calorie, nutrition (kilocalorie) | *4.184 | kJ |
| candle, candlepower | *1.0 | cd |
| centipoise, dynamic vicosity, µ | *1.00 | mPa·s |
| centistokes, kinematic viscosity, ν | *1.00 | mm$^2$/s |
| clo | 0.155 | m$^2$·K/W |
| dyne/cm$^2$ | *0.100 | Pa |
| EDR hot water (150 Btu/h) | 44.0 | W |
| EDR steam (240 Btu/h) | 70.3 | W |
| fuel cost comparison at 100% eff. | | |
| cents per gallon (no. 2 fuel oil) | 0.0677 | $/GJ |
| cents per gallon (no. 6 fuel oil) | 0.0632 | $/GJ |
| cents per gallon (propane) | 0.113 | $/GJ |
| cent per kWh | 2.78 | $/GJ |
| cents per therm | 0.0948 | $/GJ |
| ft | *0.3048 | m |
| ft | *304.8 | mm |
| ft/min, fpm | *0.00508 | m/s |
| ft/s, fps | *0.3048 | m/s |
| ft of water | 2.99 | kPa |
| ft of water per 100 ft of pipe | 0.0981 | kPa/m |
| ft$^2$ | 0.09290 | m$^2$ |
| ft$^2$·h·°F/Btu (thermal resistance, $R$) | 0.176 | m$^2$·K/W |
| ft$^2$/s, kinematic viscosity, ν | 92 900 | mm$^2$/s |
| ft$^3$ | 28.32 | L |
| ft$^3$ | 0.02832 | m$^3$ |
| ft$^3$/h, cfh | 7.866 | mL/s |
| ft$^3$/min, cfm | 0.4719 | L/s |
| ft$^3$/s, cfs | 28.32 | L/s |
| footcandle | 10.76 | lx |
| ft·lb$_f$ (torque or moment) | 1.36 | N·m |
| ft·lb$_f$ (work) | 1.36 | J |
| ft·lb$_f$/lb (specific energy) | 2.99 | J/kg |
| ft·lb$_f$/min (power) | 0.0226 | W |
| gallon, US (*231 in$^3$) | 3.785 | L |
| gph | 1.05 | mL/s |
| gpm | 0.0631 | L/s |
| gpm/ft$^2$ | 0.6791 | L/(s·m$^2$) |
| gpm/ton refrigeration | 0.0179 | mL/J |
| grain (1/7000 lb) | 0.0648 | g |
| gr/gal | 17.1 | g/m$^3$ |
| horsepower (boiler)(33,470 Btu/h) | 9.81 | kW |
| horsepower (550 ft·lb$_f$/s) | 0.746 | kW |
| inch | *25.4 | mm |
| inch of mercury (60°F) | 3.377 | kPa |
| inch of water (60°F) | 248.8 | Pa |
| **To Obtain** | **By** | **Divide** |

| Multiply | By | To Obtain |
|---|---|---|
| in/100 ft (thermal expansion) | 0.833 | mm/m |
| in·lb$_f$ (torque or moment) | 113 | mN·m |
| in$^2$ | 645 | mm$^2$ |
| in$^3$ (volume) | 16.4 | mL |
| in$^3$/min (SCIM) | 0.273 | mL/s |
| in$^3$ (section modulus) | 16 400 | mm$^3$ |
| in$^4$ (section moment) | 416 200 | mm$^4$ |
| km/h | 0.278 | m/s |
| kWh | *3.60 | MJ |
| kW/1000 cfm | 2.12 | kJ/m$^3$ |
| kilopond (kg force) | 9.81 | N |
| kip (1000 lb$_f$) | 4.45 | kN |
| kip/in$^2$ (ksi) | 6.895 | MPa |
| litre | *0.001 | m$^3$ |
| MBtuh (1000 Btu/h) | 0.2931 | kW |
| met | 58.15 | W/m$^2$ |
| micron (µm) of mercury (60°F) | 133 | mPa |
| mil (0.001 in.) | *25.4 | mm |
| mile | 1.61 | km |
| mile, nautical | 1.85 | km |
| mph | 1.61 | km/h |
| mph | 0.447 | m/s |
| millibar | *0.100 | kPa |
| mm of mercury (60°F) | 0.133 | kPa |
| mm of water (60°F) | 9.80 | Pa |
| ounce (mass, avoirdupois) | 28.35 | g |
| ounce (force of thrust) | 0.278 | N |
| ounce (liquid, US) | 29.6 | mL |
| ounce (avoirdupois) per gallon | 7.49 | kg/m$^3$ |
| perm (permeance) | 57.45 | ng/(s·m$^2$·Pa) |
| perm inch (permeability) | 1.46 | ng/(s·m·Pa) |
| pint (liquid, US) | 473 | mL |
| pound | | |
| lb (mass) | 0.4536 | kg |
| lb (mass) | 453.6 | g |
| lb$_f$ (force or thrust) | 4.45 | N |
| lb/ft (uniform load) | 1.49 | kg/m |
| lb$_m$/(ft·h) (dynamic viscosity, µ) | 0.413 | mPa·s |
| lb$_m$/(ft·s) (dynamic viscosity, µ) | 1490 | mPa·s |
| lb$_f$·s/ft$^2$ (dynamic viscosity, µ) | 47 880 | mPa·s |
| lb/min | 0.00756 | kg/s |
| lb/h | 0.126 | g/s |
| lb/h (steam at 212°F)(970 Btu/h) | 0.284 | kW |
| lb$_f$/ft$^2$ | 47.9 | Pa |
| lb/ft$^2$ | 4.88 | kg/m$^2$ |
| lb/ft$^3$ (density, ρ) | 16.0 | kg/m$^3$ |
| lb/gallon | 120 | kg/m$^3$ |
| ppm (by mass) | *1.00 | mg/kg |
| psi | 6.895 | kPa |
| quad (10$^{15}$ Btu) | 1.06 | EJ |
| quart (liquid, US) | 0.946 | L |
| revolutions per minute (rpm) | *1/60 | Hz |
| square (100 ft$^2$) | 9.29 | m$^2$ |
| tablespoon (approx.) | 15 | mL |
| teaspoon (approx.) | 5 | mL |
| therm (100,000 Btu) | 105.5 | MJ |
| ton, short (2000 lb) | 0.907 | Mg; t (tonne) |
| ton, refrigeration (12,000 Btu/h) | 3.517 | kW |
| torr (1 mm Hg at 0°C) | 133 | Pa |
| watt per square foot | 10.8 | W/m$^2$ |
| yd | *0.9144 | m |
| yd$^2$ | 0.836 | m$^2$ |
| yd$^3$ | 0.7646 | m$^3$ |
| **To Obtain** | **By** | **Divide** |

Note: In this list the kelvin (K) expresses temperature intervals. The degree Celsius symbol (°C) is often used for this purpose as well.

*Conversion factor is exact.

# Appendix B

# SYSTEMS DESIGN PROBLEMS

## 1 Combination Water Chillers

(Centrifugal and Absorption Machines in Series)

*Given*:

1. The 1000 ton turbine-driven centrifugal compressor in the figure below is supplied with steam at 30,000 lb/h at 125 psig. The turbine exhaust pressure is 15 psig. The temperature rise through the condenser is 10°F.
2. A 1500 ton lithium bromide water chiller uses exhaust steam at 12 psig from the steam turbine. It has a 4-pass evaporator with a pressure drop of 15 ft. The leaving temperature for the 4800 gpm condenser water is 105.3°F.
3. Water velocity for the condenser and chilled-water piping is limited to 10 ft/s.
4. Chilled-water supply temperature is 40°; chilled-water return temperature is 52°F.
5. Cooling tower design data: 95°F dry bulb, 76°F wet bulb, 9°F approach.

*Required*:

1. Calculate the overall steam rate in pounds per hour per ton for the refrigeration plant.
2. Calculate the chilled-water flow rate in gpm.
3. What is the temperature of water off the tower? What is the temperature of the water entering the cooling tower?
4. In the evaporator of the centrifugal compressor, what is the pressure drop and water velocity in the tubes?
5. In the condenser of the centrifugal compressor, what is the pressure drop and water velocity in the tubes?
6. Using Schedule 40 pipe, what size pipe would you use for the condenser water piping to each machine (a,b), the cooling tower (c), and the chilled-water piping (d)?

## 2 Absorption Chiller Selection

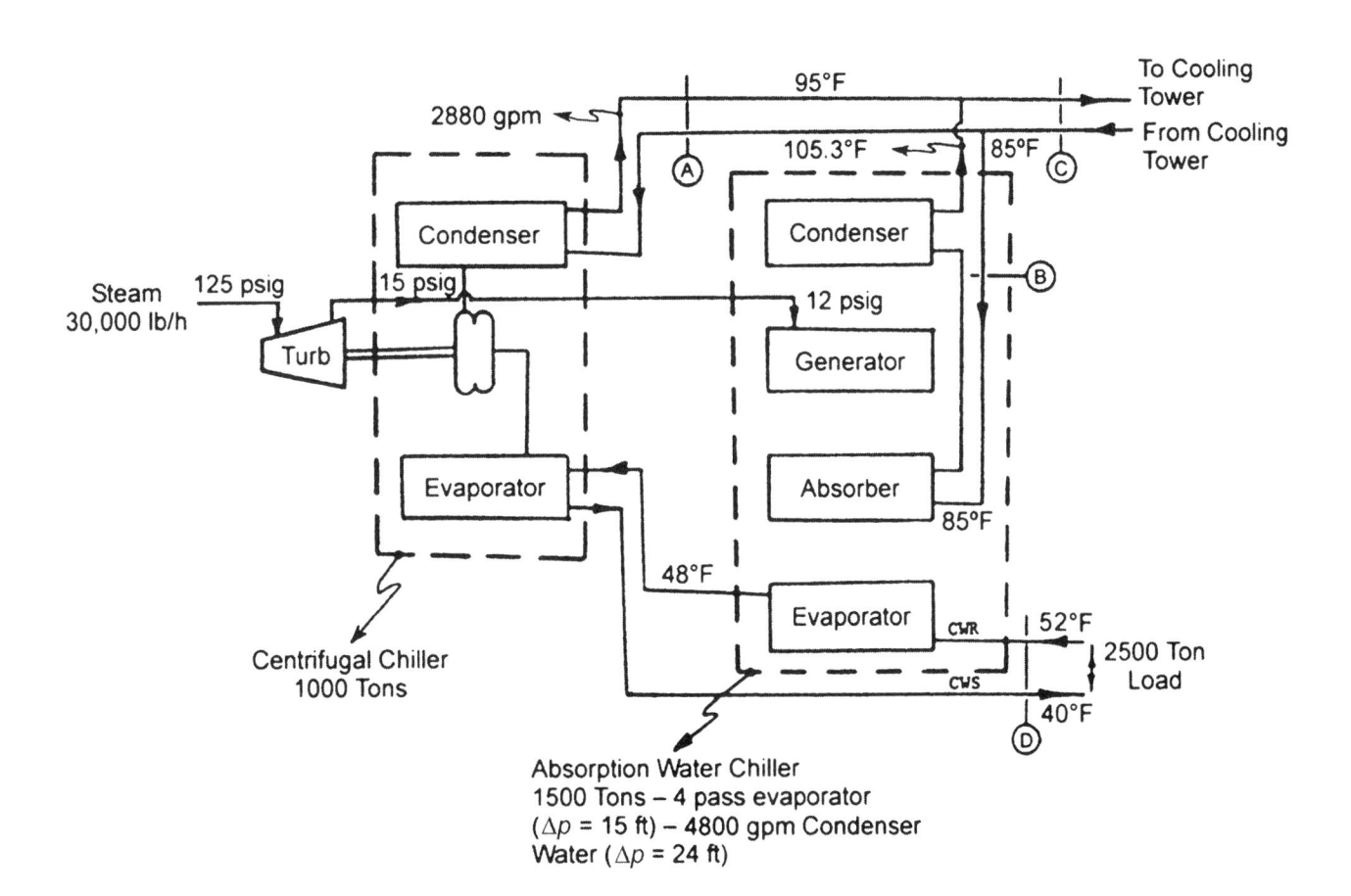

A small college is to be built in the Santa Fe, New Mexico, area (elevation 7000 ft). You have the assignment to design the mechanical systems for this project. You decide to recommend a central plant for both heating (steam) and cooling (chilled water). Since the available fuel is relatively inexpensive, you decide to use absorption refrigeration to keep the electrical demand as low as possible and to make use of the steam boilers that would otherwise be idle in the summertime.

Your preliminary analysis indicates that the first four buildings to be built will have the following characteristics:

| Building | Area, ft$^2$ | Estimated ft$^2$/ton | Estimated Total Tons |
|---|---|---|---|
| A | 75,000 | 400 | 190 |
| B | 50,000 | 325 | 150 |
| C | 65,000 | 350 | 185 |
| D | 25,000 | 300 | 85 |

In addition, you decide to install sufficient additional capacity to handle 100,000 ft$^2$ of building (you do not know exactly what type of building it will be, so average and estimate). Also estimate a 5% loss in capacity in the piping and distribution system.

Steam is available at 30 psi and can be reduced to any pressure you desire. You decide to have 42°F to 58°F or a 16°F temperature difference (TD0 as your chilled-water design temperatures. Condensing water is available from the cooling tower at 80°F. The maximum allowable pressure drop through the chiller is 40 ft for the condenser-absorber and 20 ft for the evaporator. You want at least a 0.001 fouling factor on the condenser and a 0.0005 factor on the evaporator.

A. Select and specify an absorption chiller to handle the determined capacity. Indicate water and steam flow rates and unit pump motor horsepower requirements.

B. What is the rate of refrigerant flow at maximum load?

C. What is the theoretical pump horsepower required for the chilled water and condensing water due to the pressure drop through the unit?

D. What is the total hourly purchased energy requirement for this chiller (electrical + thermal)? How does this compare with an equivalent motor-driven centrifugal machine? Which one is more economical to operate (assume electricity costs five times as much as steam)?

E. What is the cooling tower requirement for the absorption machine compared with a centrifugal machine? How might this affect your answer to question D above?

## 3 Owning and Operating Costs

Management has decided to move a subsidiary of your company to Indianapolis, Indiana. This will necessitate a new office building. The chief engineer was asked to estimate the owning and operating cost for the refrigeration and summer air-conditioning services in the building. The chief just called you in this morning and now it is your estimate. Here is all the information available.

*Building*: Five-story, 160 ft by 300 ft gross with 90% of floor area air conditioned

*Refrigeration*: reciprocating compressors, R-134a, 95°F condensing temperature, 40°F evaporator temperature

*Power rate*: $0.07 per kWh. Includes both demand and energy charge.

*Water rate*: $0.90 per 1000 gal

*Water rate for condenser*: 3.0 gpm per ton for full load operating hour

*Operating hours for auxiliaries*: 5 1/2 days per week

*Power requirements:*

  Fan, air: 0.4 bhp per 1000 cfm
  Fan, cooling tower: 0.05 bhp per ton
  Pump, chilled water: 1.10 bhp per ton
  Pump, cooling tower: 0.09 bhp per ton

*Annual operating labor and maintenance*: 4 1/2% of first cost (average for life of equipment).

*Interest*: 7%

*Taxes*: $2.00 per $100 of assessed valuation.

*Assessed valuation*: 25% of first cost.

## 4 Animal Rooms

Facilities are to house laboratory animals at control temperature, humidity, air motion, odor, and bacterial count. Design conditions vary widely depending on whether the animals are subjected to test environments or simply quartered. A general range of design temperatures and relateve humidity is tabulated in Table 1.

Recommended tolerances are ±2°F dry bulb and ±5% rh at the point of control. Low temperature gradients are desirable. The maximum spread vertically and horizontally within the cage zones should be limited to within ±1.5°F of the control point. Air motion limits in the cage zone are 35 to 50 fpm for general applications and 25 to 35 fpm in quarters for mice. See the diagram on the following page.

The approximate amount of heat released by laboratory animals at rest and during normal activity is shown in Table 2.

Conformance to temperature and humidity requirements requires control on a room basis and preferably on a module basis, because cage loading, occupancy distribution, and animal heat release are variable. A module constitutes two rows of cages with a working aisle separation. Temperature and velocity gradient control requires low supply air to room air temperature differential, overhead high induction diffusion, uniform horizontal and vertical air distribution, and low return outlets.

Odor control within animal rooms requires 100% outdoor air for odor removal from recirculated air. Unidirectional single pass airflow through the room is an aid for lowering odor levels. Air rates range from 10 to 20 air changes per hour depending on the animal occupancy and density.

# Systems Design Problems

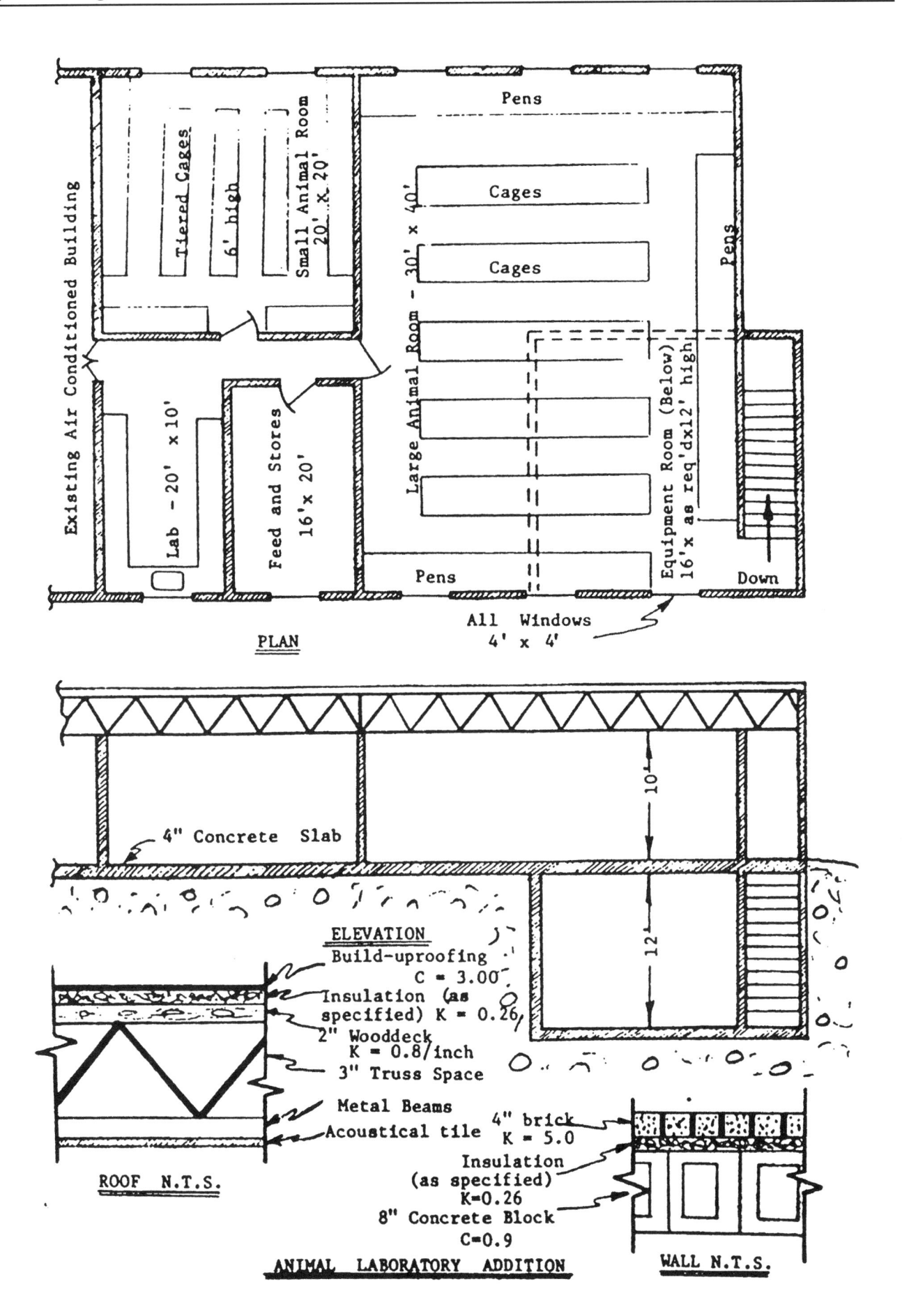

## PLAN

- Existing Air Conditioned Building
- Tiered Cages 6' high
- Small Animal Room 20' x 20'
- Pens
- Cages
- Large Animal Room – 30' x 40'
- Pens
- Lab – 20' x 10'
- Feed and Stores 16' x 20'
- Equipment Room (Below) 16' x as req'd x 12' high
- Pens
- Down
- All Windows 4' x 4'

## ELEVATION

- 4" Concrete Slab
- 10'
- 12'

## ROOF N.T.S.

- Build-up roofing C = 3.00
- Insulation (as specified) K = 0.26
- 2" Wooddeck K = 0.8/inch
- 3" Truss Space
- Metal Beams
- Acoustical tile

## WALL N.T.S.

- 4" brick K = 5.0
- Insulation (as specified) K = 0.26
- 8" Concrete Block C = 0.9

### ANIMAL LABORATORY ADDITION

**Table 1  Animal Room Temperature**

| Animal | Temp., °F | Relative Humidity, % |
|---|---|---|
| Mice | 73 to 77 | 45 to 50 |
| Rats | 73 to 77 | 45 to 50 |
| Guinea Pigs | 72 to 74 | 45 to 50 |
| Rabbits | 70 to 72 | 45 to 50 |
| Cats | 75 to 77 | 45 to 50 |
| Dogs | 70 to 72 | 45 to 50 |
| Monkeys | 76 to 79 | 75 |

**Table 2  Heat Generated by Laboratory Animals**

| | | Heat Generation, Btu/h per Animal | | |
|---|---|---|---|---|
| | | Response (Basal) | Normally Active (est.) | | |
| Animal | Mass, g | Total Heat | Sensible | Latent | Total |
| Mouse | 21 | 0.6 | 3.3 | 1.7 | 5.0 |
| Hamster | 118 | 1.65 | 10 | 3 | 13 |
| Pigeon | 275 | 4.63 | 15 | 3 | 18 |
| Rat | 300 | 4.46 | 22 | 11 | 33 |
| Guinea Pig | 410 | 5.80 | 32 | 15 | 47 |
| Chicken | 2,100 | 18.8 | 64 | 12 | 76 |
| Rabbit | 2,600 | 19.3 | 61 | 18 | 79 |
| Cat | 3,000 | 25.1 | 75 | 25 | 100 |
| Monkey | 4,200 | 34.5 | 92 | 46 | 138 |
| Dog | 16,000 | 87.5 | 250 | 120 | 370 |
| Goat | 36,000 | 137.0 | 410 | 130 | 540 |
| Sheep | 45,000 | 192.0 | 560 | 190 | 750 |
| Pig | 250,000 | 718.0 | 2100 | 700 | 2800 |

**Table 3  Average Odor-Free Requirements**

| Animal | Mass, g | Gross Space ft$^3$/animal | Odor-Free Air cfm/animal |
|---|---|---|---|
| Mice | 21 | 1.0 | 0.10 |
| Rats | 200 | 3.5 | 0.75 |
| Guinea pigs | 410 | 6.0 | 1.5 |
| Chickens | 2,100 | 8.0 | 2.0 |
| Rabbits | 2,600 | 10.0 | 2.0 |
| Cats | 3,000 | 35.0 | 8.0 |
| Monkeys | 3,200 | 100.0 | 20.0 |
| Dogs | 14,000 | 150.0 | 50.0 |

Odor-free air rates for various animal occupancies are listed in Table 3. Control of odor dissemination to adjoining spaces requires that the animal room be maintained under negative pressure or that air locks be employed. Bacterial control require high-efficiency filtration or germicidal treatment.

Conditions in animal rooms must be continuously maintained, which requires year-round availability of refrigeration and, in some cases, dual air-conditioning facilities and emergency power for motor drives and control instrument energy.

An air-conditioning system is to be designed for an animal room for housing laboratory animals. The area is a separate wing of a research facility and is to have its own completely independent system to provide year-round temperature and humidity control. The large animal room will have a maximum of 80 dogs or their equivalent and the small animal room will contain mice, rats, rabbits and a few monkeys, with an equivalent heat release of 200 rabbits.

Insulation and window construction shall be selected as follows:

| Design Temperature | Insulation | Windows (all sealed) |
|---|---|---|
| +30°F and above | None | Single pane |
| +10°F to +30°F | 1 in. | Single pane |
| −20°F to +10°F | 2 in. | Two pane |
| below −20°F | 3 in. | Two pane |

$U$ for single pane = 1.13 Btu/h·ft$^2$·°F
$U$ for double pane = 0.45 Btu/h·ft$^2$·°F

*Lights:*

In animal rooms: 1 W/ft$^2$
In feed room: 1 W/ft$^2$ (fluorescent)
In laboratory: 4 W/ft$^2$

*For laboratory*: 3 bunsen burners, one 1100 W sterilizer

*For feed room*: 1 hp electric food grinder

Perform the following:

(a) Calculate the heating loads for each room.
(b) Calculate the cooling loads for each room.
(c) 1. Calculate the air quantities (cfm) for each room.
   2. Make a schematic diagram of the apparatus required for air conditioning. (Heat air with hot water from hot water boiler).
(d) Calculate the areas of the cooling coil required and the face area (frontal area).
(e) 1. Make a single line duct layout.
   2. Size the ducts.
   3. Select the grilles, diffusers, and other components.
(f) 1. Select filter pressure loss.
   2. Calculate pressure losses in other parts of the air distributing system, coils, and outdoor louvers.
   3. Calculate the total pressure loss.
   4. Select fan, fans or units.
(g) 1. Select a refrigeration water chilling unit or units.
   2. Select a hot water boiler.
   3. Select pump or pumps for chilled water circulation.
(h) 1. Draw a plan of the building on 22 in. × 17 in. paper (1/2 in. = 1 ft 0 in.).
   2. Draw the mechanical system designed on the plan. (Make sure mechanical system outline is heavier than building outline.)
   3. Draw a plan and elevation of equipment room showing main items of equipment and ducts (1/2 in. = 1 ft, 0 in.).

## 5  Greenhouse

An environmental control system is to be developed for a greenhouse located in the Lafayette, Indiana, area. This greenhouse is a rigid frame structure, 40 ft wide and 100 ft long, with a roof slope of 6/12 and a sidewall height of 6 ft. The covering material is a single layer of polyethylene plastic.

# Systems Design Problems

Determine the following, listing your sources of data and assumptions used:

1. What is the maximum heat requirement for the house in Btu/h if the inside temperature is not to fall below 60°F?
2. On a clear winter day, January 21, with an outside noon temperature of 30°F and 65% rh, how much heat (Btu/h) or ventilation (cfm) is required to maintain an inside temperature somewhere between 60 and 70°F at 12 noon solar time? (Consider sensible heat only.)
3. Reconsider Part 2 assuming that half of the solar radiation load is used to evaporate water from the plants, soil, and floor and manifests itself as a latent heat load. A further restriction of a maximum 50% inside rh is added to prevent condensation on the inside of the plastic covering. Under these conditions, what is the amount of heat (Btu/h) and/or ventilation (cfm) required to maintain an inside temperature of 75°F?
4. On a clear summer day, July 21, with an outside noon temperature of 85° and 50% rh, what is the minimum ventilation rate (cfm) required to prevent the inside temperature from exceeding 90°F? (Consider sensible heat only.)
5. A pad evaporative cooler is to be considered for cooling the greenhouse on a clear day, August 21, with a noon temperature of 90°F and 40% rh. The entering air, after passing through the pad, has 90% rh. Using the ventilation rate from Part 4, what is the exhaust air temperature? How many square feet of pad are required if the velocity through the pad is restricted to 150 fpm?

## 6 Drying Room

You are making a feasibility study for a 90,000 ft$^3$ drying room built inside a factory. An air conditioner is located outside the drying room. Air enters the conditioner from the plant at 82°F dry bulb, 50% rh, and is distributed to the drying room at 82°F dry bulb, 10% rh. The air leaves the drying room at 112°F dry bulb, 90% rh and is exhausted from the factory through an exhaust system. Product passes through the drying room on a belt conveyor entering at 3600 lb/h at 128°F dry bulb, 40% moisture content.

Assume 10 air changes per hour through the drying room. Assume no heat or moisture flow through the drying room structure.

(a) What is the refrigeration load on the conditioning apparatus in tons?
(b) What is the moisture content of the product leaving the drying room?

You report the results to the product engineer who wants to know if you can dry the product to 20% moisture. The engineer says the 112°F leaving air is maximum; otherwise; the product will be too hot and will have to be cooled after it leaves the drying room to prevent checking.

You study the conveyor and feel it will be possible to modify the air distribution arrangement in the drying room. This will give more intimate contact between air and product and perhaps the same air volume flow rate can be used.

(c) What will be the humidity ratio of the air leaving the drying room for a 20% moisture content of the product? (You retain the 10 air changes per hour and the 112°F dry-bulb leaving air.)

## 7 Air Washer

A chiller supplies water to a washer. The water leaves the chiller at 44°F and is returned at 54°F. Design conditions in the building are 78°F dry bulb and 64.5°F wet bulb. A mixture of outdoor and recirculated air enters the washer at 88°F dry bulb and 72°F wet bulb. The total air circulated is 33,000 cfm. The washer has a performance factor of 0.85. The washer is a two-bank single-stage design.

1. What is the refrigeration load on the chiller?
2. What water quantity should it be handling in gpm?
3. What room SHR will these operating conditions exactly satisfty?
4. How many nozzles would you expect to find in the washer assuming 2 gpm per nozzle?
5. What should be the cross-sectional area of the washer assuming a velocity of 500 fpm?
6. Suppose the design engineer made a mistake in estimating the head on the chilled water circulating pump. The operating head was lower than the selection point for the pump. From the characteristic pump curve you estimate, the pump is now delivering 30% more water than in Part 2. Assuming the same performance factor, load, and air flow, what is your answer to Part 3? Discuss the effect of the conditions in the building.

## 8 Two-Story Building

As both architect and chief mechanical engineer of your own consulting engineering firm, you are responsible for the design of the building and for the sizing and selection of the major components of the HVAC system for the building. A report on your design, analysis, and recommendations for the HVAC system must be prepared for the sponsor and his staff. Sizing and selection are to cover only the heating, cooling, and humidifying equipment with cost estimates. Estimates of annual HVAC energy use and cost are to be included.

Simple tables of design conditions, building data, results, and recommended equipment should be included in the body of report. A cover letter is mandatory. Completely labeled sketches and diagrams should be included as appropriate. Detailed calculations and/or computer printouts are to be included as appended material.

The 24,000 ft$^2$, two-floor office building is to be located in Louisville, Kentucky. There are to be six, separately thermostated zones, three on each floor. The east and north ends are the be combined into one zone, the west and south into a second, and a center portion into the third. The zones are each approximately the same size. The east and north zones are to contain a minimum of 70% glass in the exterior walls while the exterior walls of the west and south zones are to contain between 20 and 40% glass.

Since the owner will be picking up the utility bills, he or she is somewhat energy conscious. However, the owner is also very concerned about the first cost of the building and does not plan extreme departures from common building practices and materials. A rooftop installation is planned to conserve interior space.

As the engineer and architect hired by the owner-to-be of the building, you are to design the building, specifying the layout, wall and ceiling descriptions, types of glass, doors, etc. Neglect details such as closets, room dividers, halls, etc. Items to be determined include:

Design heating load for each space.
Design cooling load for each space.
Projected energy requirements for heating and for cooling.
Sizing of major components (heating unit, cooling unit, fan, humidifier).
Sizing and layout of ducts.
Specific recommendations for equipment and fuels.
Estimated initial cost of HVAC equipment and annual operating cost.

## 9  Motel

As both architect and chief mechanical engineer of your own consulting engineering firm, you are responsible for the design of the building and for the sizing and selection of the major components of the HVAC system for the building. A report on your design, analysis, and recommendations for the HVAC system must be prepared for the sponsor and his staff.

Sizing and selection are to cover only the heating, cooling, and humidifying equipment with cost estimates. Estimates of annual HVAC energy use and cost are to be included.

Simple tables of design conditions, building data, results and recommended equipment should be included in the body of the report. A cover letter is *mandatory*. Completely labeled sketches and diagrams should be included as appropriate. Detailed calculations are to be included as appended material.

The 24-unit (plus office) single building motel is to be located in St. Louis, Missouri, in the Lambert Airport area. Each unit is to be 12 ft by 24 ft with the office twice the size of a regular room. Two-thirds of the units are to be nonsmoking. Each unit will be conditioned with its own packaged terminal air conditioner (PTAC) or packaged terminal heat pump (PTHP).

Since the owner will be picking up the utility bills, he or she is somewhat energy conscious. However, the owner is also very concerned about the first cost of the building and does not plan extreme departures from common building practices and materials.

As the engineer and architect hired by the owner-to-be of the building, you are to design the building, specifying the layout, wall and ceiling descriptions, types of glass, doors, etc. Items to be determined include:

Design heating load for each space.
Design cooling load for each space.
Projected energy requirements for heating and for cooling.
Sizing of major components (heating unit, cooling unit, fan, humidifier).
Specific recommendations for equipment and fuels.
Estimated initial cost of HVAC equipment and annual operating cost.

## 10  Building Renovation

As chief mechanical engineer of a consulting engineering firm, you are responsible for the design, sizing, and selection of the major components of the HVAC system for the building. A report on your design, analysis, and recommendations for the HVAC system must be prepared for the sponsor and his staff.

Sizing and selection are to cover the heating, cooling, and humidifying equipment with cost estimates. Estimates of annual HVAC energy use, *both before and after* the modifications, are to be included.

Simple tables of design conditions, building data, results, and recommended equipment should be included in the body of report. A cover letter is *mandatory*.

Completely labeled sketches and diagrams should be included as appropriate.

Detailed calculations and/or computer printouts are to be included as appended material.

The project concerns the complete renovation of an office building located in St. Louis, Missouri, and shown in the following sketch.

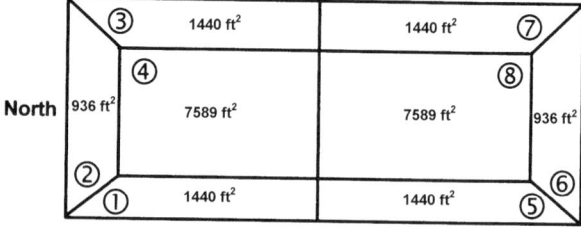

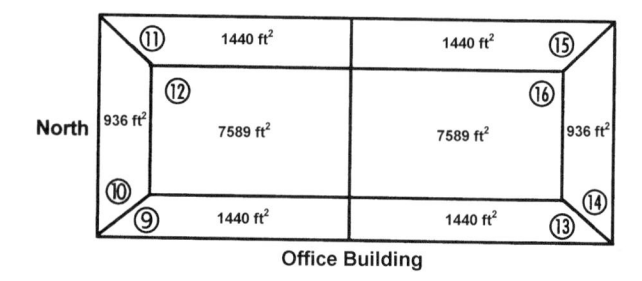

*Sketch of Project 10 Building.*

# Systems Design Problems

The building is divided into 16 separately thermostated zones. Physical description and building operation (base case) data are

Building roof area: 22,810 ft$^2$
Building floor area: 45,620 ft$^2$
Ceiling height: 8.5 ft
Building exterior wall area: 9,460 ft$^2$ (net)
Building glass area: 7,536 ft$^2$
Building thermal mass: M (medium)
Uniform internal load density: 2.9 W/ft$^2$
Occupancy: 408 people (uniformly distributed)
Original U-factors:

Roof—0.25 Btu/h·ft$^2$·°F
Walls—0.20 Btu/h·ft$^2$·°F
Glass—1.0 Btu/h·ft$^2$·°F
Shading coefficient for glass: 0.6

Originally built in the late 1960s with an all-electric reheat HVAC system, the renovated building will replace the inefficient reheat system with all-electric packaged terminal air conditioners (PTACs) in each space. In the exterior zones, the PTACs will include provisions for ventilation air. For the interior zones, the PTACs will operate without outdoor air provisions and a separate rooftop makeup air system will be used for ventilation requirements. When renovated, all walls and ceilings will include an additional 2 inches of glass fiber, organic bonded rigid insulation, The windows will be upgraded to double pane, 1/4 in. air gap, aluminum frame with thermal break (nonoperable). The internal load density is estimated to have increased over the years from the original 2.9 W/ft$^2$ to 6.0 W/ft$^2$.

Include the following items:

Design heating load for each space.
Design cooling load for each space.
Projected annual energy requirements for heating and for cooling.
Sizing and selection of major components.
Sizing and layout of ducts for interior zones.
Estimated initial cost of HVAC equipment and annual operating cost.
Potential problem areas with this type of equipment.

## 11  Building with Neutral Deck Multizone

As the chief mechanical engineer of your own consulting engineering firm, you are responsible for preliminary design of the HVAC system for the building shown in the figure at right. Selection of all major components is to be included. A report on your analysis and recommendations for the HVAC system must be prepared for the sponsor and his staff.

Sizing and selection will cover all heat exchangers (coils), fans, refrigeration units, boilers and/or other heating equipment, humidifiers, cooling towers, heat reclaim and/or air-to-air energy recovery equipment, and pumps. Piping, ducting, and related fittings need not be sized nor selected at this time.

Simple tables of design conditions, building data, results and recommended equipment should be included in the body of the report. A cover letter is mandatory. Completely labeled sketches and diagrams should be included as appropriate. Detailed calculations and appropriate manufacturers' catalog data are to be included as appended material.

Building Location: Atlanta, GA

The ventilation requirements are to be in accordance with ASHRAE Standard 62. Anticipate occupancy rate is 10 persons per 1000 ft$^2$ of floor space. Design pressure drop for the ducting system is 3.25 in. of water. The design loads are given in the following table.

**Design Loads (Btu)**

| Zone | Heating (sensible) | Cooling (sensible) | Cooling (latent) |
|---|---|---|---|
| 1 | −95,000 | 164,000 | 47,000 |
| 2 | +33,000 | 157,000 | 14,000 |
| 3 | −98,000 | 199,000 | 40,000 |
| 4 | −276,000 | 567,000 | 72,000 |

The winter latent load is negligible.

Type of HVAC System:  Multizone with neutral deck

Primary Systems:  R-22 condensing unit and DX coil, multiple gas-fired boilers for steam coil

Auxiliary Equipment:  Spray washer, heat pipe air-to-air energy recovery system, air-side economizer

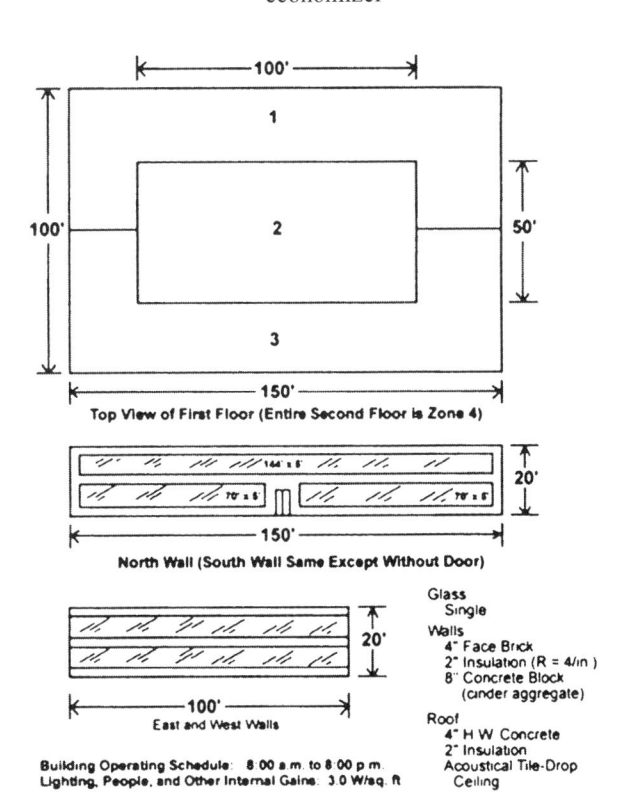

*Figure for Project 11.*